THE ESSENTIAL DICTIONARY OF SCIENCE

THE ESSENTIAL DICTIONARY OF SCIENCE

General Editor:
John O. E. Clark

BARNES & NOBLE
NEW YORK

Helicon Publishing is a division of Research Machines.

Text design by Lundquist Design, New York

2004 Barnes & Noble Books

ISBN 0-7607-4616-8

MP 9 8 7 6 5 4 3 2

Acknowledgements

General Editor:
JOHN O. E. CLARK

Editors
JOHN J O'CONNOR
DAVID APPLIN
DR. KEITH HUTTON

Managing Editor
ELENA SOFTLEY

Preface

As scientific knowledge develops and grows, the number of scientific terms has grown exponentially. To remain current and useful, a modern dictionary of science cannot be an unchanging repository of facts; it must incorporate new terms and the latest discoveries in scientific theory and research.

The Essential Dictionary of Science is designed to be an authoritative research tool; useful for the serious student of science, yet also accessible and appealing to the general reader. This new dictionary offers clear, concise explanations of over 8,000 entries, including the latest terms and concepts. Crafted specifically to enhance the everyday reader's understanding of science, this book defines terms from the realms of physics, chemistry, biology, astronomy, mathematics, and the earth sciences. The entries are thoroughly cross-referenced and include over 250 diagrams and other illustrations; 1,000 biographies of important scientists; and over 1,000 weblinks to the most informative internet sites.

In addition, *The Essential Dictionary of Science* contains a thematically-arranged 65-page appendix that includes information ranging from the planets of the solar system and the periodic tables to key dates in the history of science and a roster of Nobel Prize winners.

A Note on Format:

Headwords are in blue bold type, and the alphabetization is letter-by-letter. Subhead words (terms that are defined within the headword definitions) are also in bold, but in black type.
Cross-references (words that are defined elsewhere) are indicated by a † dagger immediately preceding the entry being cross-referenced. A cross reference is offered when another entry contains material that is relevant.

THE ESSENTIAL DICTIONARY OF SCIENCE

A

A in physics, symbol for †ampere, a unit of electrical current.

aa in earth science, type of lava that has a rough surface, which may be jagged or spiny.

abacus ancient calculating device made up of a frame of parallel wires on which beads are strung. The method of calculating with a handful of stones on a "flat surface" (Latin *abacus*) was familiar to the Greeks and Romans, and used by earlier peoples, possibly even in ancient Babylon; it survives in the more sophisticated bead-frame form of the Russian *schoty* and the Japanese *soroban*. The abacus has been superseded by the electronic calculator.

The wires of a bead-frame abacus define place value (for example, in the decimal number system each successive wire, counting from right to left, would stand for ones, tens, hundreds, thousands, and so on) and beads are slid to the top of each wire in order to represent the digits of a particular number. On a simple decimal abacus, for example, the number 8,493 would be entered by sliding three beads on the first wire (three ones), nine beads on the second wire (nine tens), four beads on the third wire (four hundreds), and eight beads on the fourth wire (eight thousands).

Vintage Calculators http://www.vintagecalculators.com/index.html

This Web site features articles and photographs on many calculators and other mathematical machines. There are sections devoted to the development of the different types of mechanical and electronic calculators, and information on collecting them.

Abbe, Ernst (1840–1905) German physicist who, working with Carl Zeiss, greatly improved the design and quality of optical instruments, particularly the compound microscope. This enabled researchers to observe micro-organisms and internal cellular structures for the first time.

Abbe, Ernst http://www-history.mcs.st-andrews.ac.uk/Mathematicians/Abbe.html

Biography and portrait of the 19th-century German physicist. The history of Abbe's work is also described and several references are quoted which contain either biographical details or examples of Abbe's mathematical works.

Abd-er-Rahman ibn-Umar alternative transcription of Al †Sufi.

abdomen in vertebrates, the part of the body below the †thorax, containing the digestive organs; in insects and other arthropods, it is the hind part of the body. In mammals, the abdomen is separated from the thorax by the †diaphragm, a sheet of muscular tissue; in arthropods, commonly by a narrow constriction. In mammals, the female reproductive organs are in the abdomen. In insects and spiders, it is characterized by the absence of limbs.

Abel, John Jacob (1857–1938) US biochemist, discoverer of †adrenaline. He studied the chemical composition of body tissues, and this led, in 1898, to the discovery of adrenaline, the first hormone to be identified, which Abel called epinephrine. He later became the first to isolate †amino acids from blood.

Abel, Niels Henrik (1802–1829) Norwegian mathematician. He demonstrated that the general quintic equation $ax^5 + bx^4 + cx^3 + dx^2 + ex + f = 0$ could not be solved algebraically. Subsequent work covered elliptic functions, integral equations, infinite series, and the binomial theorem.

Abel, Niels

http://www.shu.edu/html/teaching/math/reals/history/abel.html

Interesting account of the life of the pioneer in the field of elliptic functions. His desperate craving for knowledge, the chance to travel, and academic recognition are chronicled together with his early death from tuberculosis. There is also an assessment of his contribution to modern mathematics.

Abelian functions in mathematics, functions of the form $f(x,y)dx$, where y is an algebraic function of x, and $f(x,y)$ is an algebraic function of x and y.

aberration in biology, an abnormal structure or deviation from the type.

aberration of starlight apparent displacement of a star from its true position, due to the combined effects of the speed of light and the speed of the Earth in orbit around the

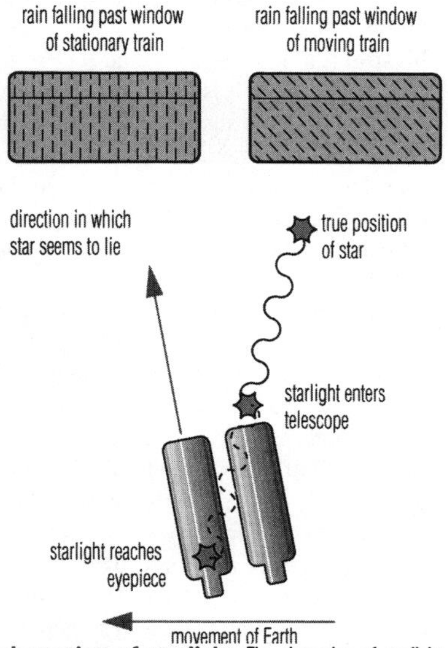

aberration of starlight The aberration of starlight is an optical illusion caused by the motion of the Earth. Rain falling appears vertical when seen from the window of a stationary train; when seen from the window of a moving train, the rain appears to follow a sloping path. In the same way, light from a star—falling—down a telescope seems to follow a sloping path because the Earth is moving. This causes an apparent displacement, or aberration, in the position of the star.

Sun (about 30 kps/18.5 mps).

During a year the apparent position of a star describes a curve around its true position. This curve is an ellipse except when the star lies on the †ecliptic (when it is a line along which the star appears to move backwards and forwards) or when it is in the pole of the ecliptic (when it is a circle). As aberration depends upon the ratio between the velocity of light and the velocity of the Earth, the determination of this "constant of aberration" provides a method of calculating an approximate value for one of these velocities if the other is known.

aberration, optical any of a number of defects that impair the image in an optical instrument. Aberration occurs because of minute variations in lenses and mirrors, and because different parts of the light †spectrum are reflected or refracted by varying amounts.

In **chromatic aberration** the image is surrounded by colored fringes, because light of different wavelengths is brought to different focal points by a lens. In **spherical aberration** the image is blurred because different parts of a spherical lens or mirror have different focal lengths. In **astigmatism** the image appears elliptical or cross-shaped because of an irregularity in the curvature of the lens. In **coma** the images appear progressively elongated towards the edge of the field of view. Elaborate computer programs are now used to design lenses in which the aberrations are minimized.

aberration, optical The main defects,or aberrations, of optical systems. Chromatic aberration, or colored fringes around images, arises because light of different colors is focused at different points by a lens, causing a blurred image. Spherical aberration arises because light that passes through the center of the lens is focused at a different point from light passing through the edge of the lens. Astigmatism arises if a lens has different curvatures in the vertical and horizontal directions. Coma arises because light passing directly through a lens is focused at a different point to light entering the lens from an angle. Helicon

abiotic factor non-living variable within the ecosystem, affecting the life of organisms. Examples include temperature, light, and water. Abiotic factors can be harmful to the environment, as when sulfur dioxide emissions from power stations produce acid rain.

ablation the loss of snow and ice from a †glacier by melting and evaporation. It is the opposite of †accumulation.

Ablation is most significant near the snout (foot or end) of a glacier, since temperatures tend to be higher at lower altitudes. The rate of ablation also varies according to the time of year, being greatest during the summer. If total ablation exceeds total accumulation for a particular glacier, then the glacier will retreat.

abrasion in earth science, type of erosion in which rock fragments scrape and grind away a surface. It is also known as †corrasion. The rock fragments may be carried by rivers, wind, ice, or the sea. †Striations, or grooves, on rock surfaces are common abrasions, caused by the scratching of rock by debris carried in glacier ice.

abrasive (Latin "to scratch away") substance used for cutting and polishing or for removing small amounts of the surface of hard materials. There are two types: natural and artificial abrasives, and their hardness is measured using the †Mohs scale. Natural abrasives include quartz, sandstone, pumice, diamond, emery, and corundum; artificial abrasives include rouge, whiting, and carborundum.

abscissa in †coordinate geometry, the *x*-coordinate of a point – that is, the horizontal distance of that point from the vertical or *y*-axis. For example, a point with the coordinates (4, 3) has an abscissa of 4. The *y*-coordinate of a point is known as the †ordinate.

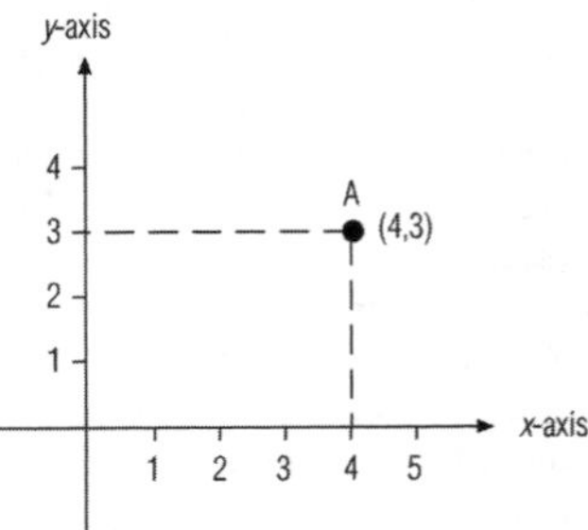

abscissa The horizontal distance from the y-axis in coordinate geometry. In this diagram the abscissa is 4. Helicon

abscissin or **abscissic acid**, plant hormone found in all higher plants. It is involved in the process of †abscission and also inhibits stem elongation, germination of seeds, and the sprouting of buds.

abscission in botany, the controlled separation of part of a plant from the main plant body – most commonly, the falling of leaves or the dropping of fruit controlled by †abscissin. In †deciduous plants the leaves are shed before the winter or dry season, whereas †evergreen plants drop their leaves continually throughout the year. Fruitdrop, the abscission of fruit while still immature, is a naturally occurring process.

Abscission occurs after the formation of an abscission zone at the point of separation. Within this, a thin layer of cells, the abscission layer, becomes weakened and breaks down through the conversion of pectic acid to pectin. Consequently the leaf, fruit, or other part can easily be dislodged by wind or rain. The process is thought to be controlled by the amount of †auxin present. Fruitdrop is particularly common in fruit trees such as apples, and orchards are often sprayed with artificial auxin as a preventive measure.

absolute (of a value) in computing, real and unchanging. For example, an **absolute address** is a location in memory and an **absolute cell reference** is a single fixed cell in a spreadsheet display. The opposite of absolute is relative.

absolute age in earth science, actual age of a fossil, rock, or stratum, normally expressed in years. It is usually measured by †radiometric dating or †dendrochronology (study of fossilized tree rings).

absolute humidity in physics, amount of water vapor in air expressed as the mass of vapor (in kilograms) per cubic meter.

absolute magnitude in astronomy, measure of the intrinsic brightness of a celestial body in contrast to its apparent brightness or †magnitude as seen from Earth.

For a non-self-luminous body, such as an asteroid, the absolute magnitude is the magnitude it would appear to have if it were one astronomical unit (149.6 million km/92.6 million mi) from both the Sun and the Earth with the phase angle zero. For a self-luminous body, such as a star or galaxy, the absolute magnitude is the magnitude it would appear to have if it were at a distance of 10 parsecs or 32.616 light years.

absolute temperature temperature given by an absolute scale that is independent of the properties of thermometric substances. Absolute thermodynamic temperature, proposed by William Kelvin, is defined according to the principles of thermodynamics alone, but for experimental work is closely approximated by the International Temperature Scale of 1990 (ITS-90). It is usually expressed in units of degrees Celsius (or centigrade) from †absolute zero, and freezing point is then 273.15 K.

absolute value or **modulus**, in mathematics, the value, or magnitude, of a number irrespective of its sign. The absolute value of a number n is written $|n|$ (or sometimes as mod n), and is defined as the positive square root of n^2. For example, the numbers –5 and 5 have the same absolute value:

$|5| = |-5| = 5$

For a †complex number, the absolute value is its distance to the origin when it is plotted on an †Argand diagram, and can be calculated (without plotting) by applying †Pythagoras' theorem. By definition, the absolute value of any complex number $a + ib$ (where a and b are real numbers and i is the square root of –1) is given by the expression:

$|a + ib| = \sqrt{(a^2 + b^2)}$

absolute weight weight of a body considered apart from all modifying influences such as the atmosphere. To determine its absolute weight, the body must, therefore, be weighed in a vacuum or allowance must be made for buoyancy.

absolute zero lowest temperature theoretically possible according to kinetic theory, zero kelvin (0 K), equivalent to –273.15°C/–459.67°F, at which molecules are in their lowest energy state. Although the third law of †thermodynamics indicates the impossibility of reaching absolute zero, in practice temperatures of less than a billionth of a degree above absolute zero have been reached. Near absolute zero, the physical properties of some materials change substantially; for example, some metals lose their electrical resistance and become superconducting.

absolutism in physics, the theories that space and time are absolutes. These theories were held by Isaac Newton and were defended on his behalf by Samuel Clarke in his controversy with Gottfried Leibniz who was in many ways a precursor of Albert Einstein and his theory of relativity.

absorptance formerly **absorbtivity**; symbol a, in physics, measure of the ability of a body to absorb radiation. It is reached by determining luminous flux or radiant flux absorbed by a material, divided by the flux incident upon it.

absorption in physics, taking up of matter or energy of one substance by another, such as a liquid by a solid (ink by blotting paper) or a gas by a liquid (ammonia by water). In physics, absorption is the phenomenon by which a substance retains the energy of radiation of particular wavelengths; for example, a piece of blue glass absorbs all visible light except the wavelengths in the blue part of the spectrum; it also refers to the partial loss of energy resulting from light and other electromagnetic waves passing through a medium. In nuclear physics, absorption is the capture by elements, such as boron, of neutrons produced by fission in a reactor.

absorption lines in astronomy, dark lines in the spectrum of a hot object due to the presence of absorbing material along the line of sight. Absorption lines are caused by atoms absorbing light from the source at sharply defined wavelengths. Numerous absorption lines in the spectrum of the Sun (†Fraunhofer lines) allow astronomers to study the composition of the Sun's outer layers. Absorption lines in the spectra of stars give clues to the composition of interstellar gas and dust.

absorption of light phenomenon whereby light energy striking a medium may be partially or totally lost by not being transmitted, reflected, or scattered. For example, red wine appears red because it transmits the red part of the spectrum but absorbs the yellow, green, and blue parts.

The absorbed light energy is transformed to heat but the amount of heat produced is normally too small to be noticed. Thus light absorption is responsible for many, though not all, of the colors seen in everyday objects.

absorption spectroscopy or **absorptiometry**, in analytical chemistry, a technique for determining the identity or amount present of a chemical substance by measuring the amount of electromagnetic radiation the substance absorbs at specific wavelengths; see †spectroscopy.

absorption trough in astronomy, range of †wavelengths (around 21 cm/8 in) at which atomic hydrogen absorbs (or emits) radiation; this is a concept used in the attempt to detect †intergalactic matter.

abstract algebra in mathematics, generalization of †algebra; the word "abstract" merely draws attention to the level of generality.

abstract group in mathematics, generalization of a group; the word "abstract" merely indicates that a group is to be considered with reference not to a specific example but to its more general properties (which are shared therefore by other examples).

abundant number in mathematics, †natural number that is less than the sum of its divisors (factors).

abyssal fauna one of the three divisions of marine fauna, the others being littoral or shallow-water, and pelagic or surface fauna. Abyssal fauna exist in the greatest depths of the ocean, from about 2,000–6,000 m/6,500–19,500 ft.

The cold temperature, absence of light, and pressure of water cause various modifications in the abyssal fauna. Many are blind, while others see by means of the phosphorescent glow emitted from their own bodies and those of other fish. Their organs of touch are often highly developed. In color they are very vivid; scarlet, violet, orange and purple being the predominant colors.

As no plants can grow in the abyssal depths because of lack of light, the fauna are carnivorous, catching in their wide jaws the falling debris of the organisms that exist above them.

abyssal zone dark ocean region 2,000–6,000 m/6,500–19,500 ft deep; temperature 4°C/39°F. Three-quarters of the area of the deep-ocean floor lies in the abyssal zone, which is too far from the surface for photosynthesis to take place. Some fish and crustaceans living there are blind or have their own light sources. The region above is the bathyal zone; the region below, the hadal zone.

abzyme in biotechnology, an artificially created antibody that can be used like an enzyme to accelerate reactions.

AC in physics, abbreviation for †alternating current.

acceleration rate of change of the velocity of a moving body. For example, an object falling towards the ground covers more distance in each successive time interval. Therefore, its velocity is changing with time and the object is accelerating. It is usually measured in meters per second per second (m s^{-2}) or feet per second per second (ft s^{-2}). Acceleration = change in velocity/time taken. Because velocity is a †vector quantity (possessing both magnitude and direction), a body traveling at constant speed may be said to be accelerating if its direction of motion changes. According to Newton's second law of motion, a body will accelerate only if it is acted upon by an unbalanced, or resultant, †force. †Acceleration of free fall is the acceleration of a body falling freely under the influence of the Earth's gravitational field; it varies slightly at different latitudes and altitudes. The value adopted internationally for gravitational acceleration is 9.806 m s^{-2}/32.174 ft s^{-2}.

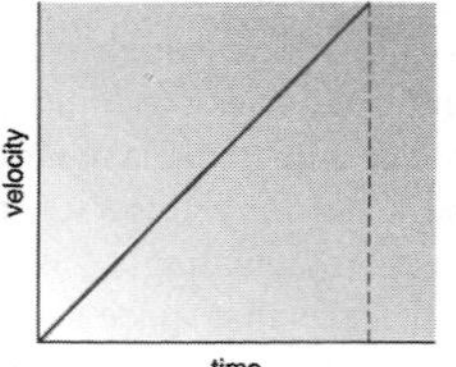

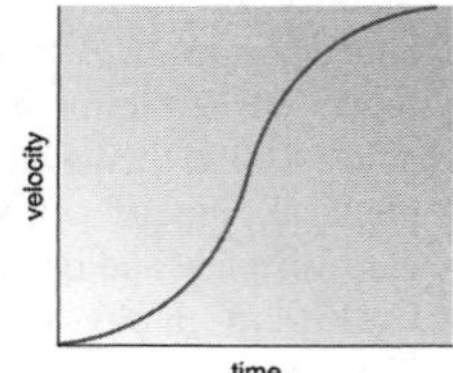

acceleration Acceleration can be depicted graphically by mapping velocity against time. Where acceleration is constant the graph is a straight line. Helicon

Acceleration and Forces http://library.thinkquest.org/16600/beginner/acceleration.shtml
Acceleration is defined in a multitude of ways on this text-only page. An example is given of different concepts of acceleration and of common pitfalls in its understanding.

acceleration of free fall or **acceleration due to gravity**; symbol *g*, in physics, constant acceleration of an object moving freely under gravity near the Earth (ignoring friction or air resistance). The standard value at the Earth's surface is 9.806 m s^{-2}/32.174 ft s^{-2}; it decreases with altitude.

acceleration, secular in astronomy, continuous and non-periodic change in orbital velocity of one body around another, or the axial rotation period of a body.

An example is the axial rotation of the Earth. This is gradually slowing down owing to the gravitational effects of the Moon and the resulting production of tides, which have a frictional effect on the Earth. However, the angular †momentum of the Earth-Moon system is maintained, because the momentum lost by the Earth is passed to the Moon. This results in an increase in the Moon's orbital period and a consequential moving away from the Earth. The overall effect is that the Earth's axial rotation period is increasing by about 15 millionths of a second per year, and the Moon is receding from the Earth at about 4 cm/1.5 in per year.

acceleration, uniform in physics, acceleration in which the velocity of a body changes by equal amounts in successive time intervals. Uniform acceleration is represented by a straight line on a †speed–time graph.

accelerator in physics, a device to bring charged particles (such as protons and electrons) up to high speeds and energies, at which they can be of use in industry, medicine, and pure physics. At low energies, accelerated particles can be used to produce the image on a television screen and (by means of a †cathode-ray tube) generate X-rays, destroy tumor cells, or kill bacteria. When high-energy particles collide with other particles, the fragments formed reveal the nature of the fundamental forces.

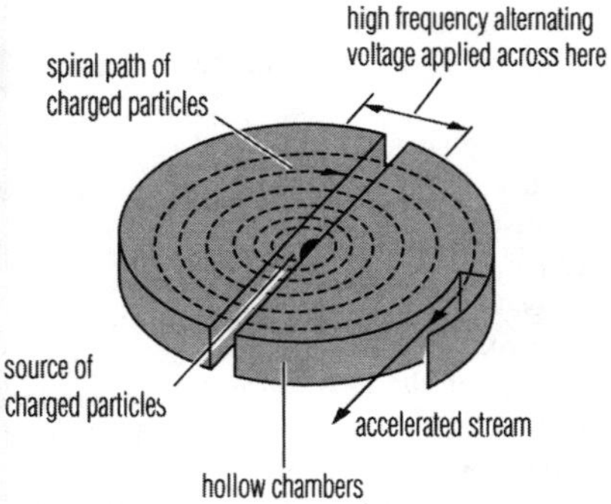

accelerator The cyclotron, an early accelerator, consisted of two D-shaped hollow chambers enclosed in a vacuum. An alternating voltage was applied across the gap between the hollows. Charged particles spiralled outward from the centre, picking up energy and accelerating each time they passed through the gap. Helicon

The first accelerators used high voltages (produced by †Van de Graaff generators) to generate a strong, unvarying electric field. Charged particles were accelerated as they passed through the electric field. However, because the voltage produced by a generator is limited, these accelerators were replaced by machines where the particles passed through regions of alternating electric fields, receiving a succession of small pushes to accelerate them. The first of these accelerators was the **linear accelerator** or **linac**. The linac consists of a line of metal tubes, called drift tubes, through which the particles travel. The particles are accelerated by electric fields in the gaps between the drift tubes.

Another way of making repeated use of an electric field is to bend the path of a particle into a circle so that it passes repeatedly through the same electric field. The first accelerator to use this idea was the **cyclotron** pioneered in the early 1930s by US physicist Ernest Lawrence. One of the world's most powerful accelerators is the 2 km/1.25 mi diameter machine at †Fermilab near Batavia, Illinois, USA. This machine, the Tevatron, accelerates protons and antiprotons and then collides them at energies up to a thousand billion electron volts (or 1 TeV, hence the name of the machine). The largest accelerator is the †Large Electron Positron Collider at †CERN near Geneva, operational 1989–2000, which has a circumference of 27 km/16.8 mi around which electrons and positrons are accelerated before being allowed to collide. The world's longest linac is also a colliding beam machine: the Stanford Linear Collider, in California, in which electrons and positrons are accelerated along a straight track, 3.2 km/2 mi long, and then steered to a head-on collision with other particles, such as protons and neutrons. Such experiments have been instrumental in revealing that protons and neutrons are made up of smaller elementary particles called †quarks.

accelerator nerve in biology, nerve that conducts impulses to the heart. On stimulation of the cardiac sympathetic nerves, the rate and strength of the heartbeat increases.

accelerometer apparatus, either mechanical or electromechanical, for measuring †acceleration or deceleration – that is, the rate of increase or decrease in the †velocity of a moving object.

The mechanical types have a spring-supported mass with a damper system, with indication of acceleration on a scale on which a light beam is reflected from a mirror on the mass. The electromechanical types use (1) a slide wire, (2) a strain gauge, (3) variable inductance, or (4) a piezoelectric or similar device that produces electrically measurable effects of acceleration.

Accelerometers are used to measure the efficiency of the braking systems on road and rail vehicles; those used in aircraft and spacecraft can determine accelerations in several directions simultaneously. There are also accelerometers for detecting vibrations in machinery.

accent in mathematics, symbol used to express feet and inches, for example 2'6" = 2 ft 6 in, and minutes and seconds as subdivisions of an angular †degree, for example 60' = 60 minutes, 30" = 30 seconds.

access time in computing, the time taken by a computer, after an instruction has been given, to read from or write to †memory.

acclimation or **acclimatization**, physiological changes induced in an organism by exposure to new environmental conditions. When humans move to higher altitudes, for example, the number of red blood cells rises to increase the oxygen-carrying capacity of the blood in order to compensate for the lower levels of oxygen in the air.

In evolutionary terms, the ability to acclimate is an important adaptation as it allows the organism to cope with the environmental changes occurring during its lifetime.

accommodation in biology, the ability of the †eye to focus on near or far objects by changing the shape of the lens.

For an object to be viewed clearly its image must be precisely focused on the retina, the light-sensitive layer of cells at the rear of the eye. Close objects can be seen when the lens takes up a more spherical shape, far objects when the lens is flattened. These changes in shape are caused by the movement of ligaments attached to a ring of ciliary muscles lying beneath the iris.

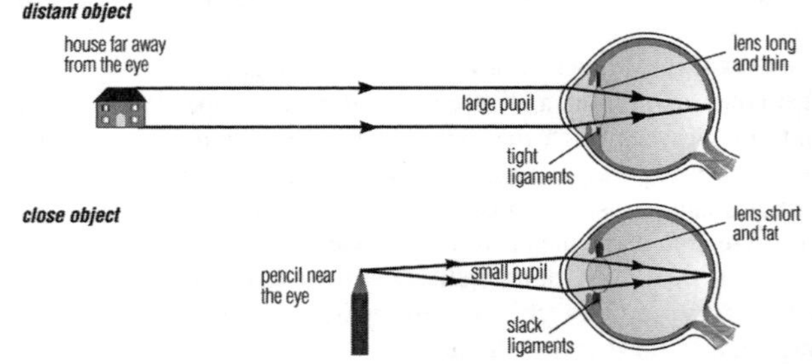

accommodation The mechanism by which the shape of the lens in the eye is changed so that clear images of objects, whether distant or near, can be focused on the retina. Helicon

From about the age of 40, the lens in the human eye becomes less flexible, causing the defect of vision known as **presbyopia** or lack of accommodation. People with this defect need different spectacles for reading and distance vision.

accretion in astrophysics, process by which an object gathers up surrounding material by gravitational attraction, thus simultaneously increasing in mass and releasing gravitational energy. Accretion onto compact objects, such as †white dwarfs, †neutron stars, and †black holes, can release large amounts of gravitational energy and is believed to be the power source for active galaxies. Accreted material falling towards a star may form a swirling disc of material known as an †accretion disc that can be a source of X-rays. Earth, and other planetary bodies in the Solar System, are thought to have formed by accretion of nebular material and †planetesimals.

accretion disc in astronomy, flattened ring of gas and dust orbiting an object in space, such as a star or †black hole. The orbiting material is accreted (gathered in) from a neighboring object such as another star. Giant accretion discs are thought to exist at the centers of some galaxies and †quasars.

If the central object of the accretion disc has a strong gravitational field, as with a neutron star or a black hole, gas falling onto the accretion disc releases energy, which heats the gas to extreme temperatures and emits short-wavelength radiation, notably X-rays.

accumulation in earth science, the addition of snow and ice to a †glacier. It is the opposite of †ablation. Snow is added through snowfall and avalanches, and is gradually compressed to form ice. Although accumulation occurs at all parts of a glacier, it is most significant at higher altitudes near the glacier's start where temperatures are lower.

accumulator in computing, a special register, or memory location, in the †arithmetic and logic unit of the computer processor. It is used to hold the result of a calculation temporarily or to store data that is being transferred.

accumulator in electricity, a storage †battery – that is, a group of rechargeable secondary cells. A familiar example is the lead–acid car battery.

An ordinary 12-volt car battery consists of six lead–acid cells that are continually recharged when the motor is running by the car's alternator or dynamo. It has electrodes of lead and lead oxide in an electrolyte of sulfuric acid. Another common type of accumulator is the "nife" or NiFe cell, which has electrodes of nickel and iron in a potassium hydroxide electrolyte.

accuracy in mathematics, a measure of the precision of a †number. The degree of accuracy depends on how many figures or †decimal places are used in †rounding off the number. For example, the result of a calculation or measurement (such as 13.429314) might be rounded off to three decimal places (13.429), to two decimal places (13.43), to one decimal place (13.4), or to the nearest whole number (13). The first answer is more accurate than the second, the second more accurate than the third, and so on.

Accuracy also refers to a range of †errors. For example, an accuracy of ± 5% means that a value may lie between 95% and 105% of a given answer.

acesulfame-K non-carbohydrate sweetener that is up to 300 times as sweet as sugar. It is used in soft drinks and desserts.

acetaldehyde common name for †ethanal.

acetamide common name for †ethanamide.

acetate common name for †ethanoate.

acetic acid common name for †ethanoic acid.

acetic anhydride common name for †ethanoic anhydride.

acetic ester common name for †ethyl acetate (ethyl ethanoate).

acetone common name for †propanone.

acetonitrile formerly **methyl cyanide**; CH_3CN, clear liquid, miscible with water, which, unless freshly purified (when it has a faint, pleasant smell), retains the mouse like odor of ethanamide, from which it is prepared by dehydration, using phosphorus pentoxide as the dehydrating reagent. Its reactions are typical of volatile, reactive nitriles (containing the monovalent group –CN).

acetyl organic group that would result from the elimination of hydroxyl from ethanoic acid; it therefore corresponds to the formula CH_3CO. It is not stable in the free state, but is looked upon as the radical of such compounds as acetyl chloride (CH_3COCl).

acetylacetone $CH_3COCH_2COCH_3$, colorless pungent liquid, boiling point 194°C/381.2°F. This diketone is unusual in that it exists to the extent of about 85% as the †enol form, the enol form being considerably stabilized by hydrogen bonding between the enol OH proton and the carbonyl oxygen. Acetylacetone is an important reagent for the preparation of †chelate compounds with a wide range of metals.

acetylation or **ethanoylation**, in chemistry, process for the introduction of an †acetyl group into an organic molecule that includes an †amide, †hydroxyl, or †thiol group via the displacement of a hydrogen atom from that grouping.

acetylcholine **ACh**, chemical that serves as a †neurotransmitter, communicating nerve impulses between the cells of the nervous system. It is largely associated with the transmission of impulses across the †synapse (junction) between nerve and muscle cells, causing the muscles to contract.

ACh is produced in the synaptic knob (a swelling at the end of a nerve cell) and stored in vesicles until a nerve impulse triggers its discharge across the synapse. When the ACh reaches the membrane of the receiving cell it binds with a specific site and brings about depolarization – a reversal of the electric charge on either side of the membrane – causing a fresh impulse (in nerve cells) or a contraction (in muscle cells). Its action is short-lived because it is quickly destroyed by the enzyme cholinesterase.

Anticholinergic drugs have a number of uses in medicine to block the action of ACh, thereby disrupting the passage of nerve impulses and relaxing certain muscles, for example in premedication before surgery.

acetyl coenzyme A or **acetyl CoA**, compound active in processes of metabolism. It is a heat-stable coenzyme with an acetyl group ($-COCH_3$) attached by a sulfur linkage. This linkage is a high-energy bond and the acetyl group can easily be donated to other compounds. Acetyl groups donated in this way play an important part in glucose breakdown as well as in fatty acid and steroid synthesis. It is involved in the †Krebs cycle, the cyclical pathway involved in the intracellular metabolism of foodstuffs.

acetylene common name for †ethyne.

acetylsalicylic acid chemical name for the painkilling drug †aspirin.

achene dry, one-seeded †fruit that develops from a single †ovary and does not split open to disperse the seed. Achenes commonly occur in groups – for example, the fruiting heads of buttercup *Ranunculus* and clematis. The outer surface may be smooth, spiny, ribbed, or tuberculate, depending on the species.

Achernar or **Alpha Eridani**, brightest star in the constellation Eridanus, and the ninth-brightest star in the sky. It is a hot, luminous, blue star with a true luminosity 250 times that of the Sun. It is 144 light years away from the Sun.

Achilles tendon tendon at the back of the ankle attaching the calf muscles to the heel bone. It is one of the largest tendons in the human body, and can resist great tensional strain, but is sometimes ruptured by contraction of the muscles in sudden extension of the foot.

Ancient surgeons regarded wounds in this tendon as fatal, probably because of the Greek legend of Achilles, which relates how the mother of the hero Achilles dipped him when an infant into the River Styx, so that he became invulnerable except for the heel by which she held him.

achiral in chemistry, describing a molecule that is not optically active (see †optical activity) because its symmetry allows it to be superimposed on its mirror image.

achondrite stony †meteorite representing differentiated planetary material. Because differentiation is an igneous process, these are igneous rocks or breccias (broken fragments) of igneous rocks. They comprise about 15% of all meteorites and lack the †chondrules (silicate spherules) found in †chondrites.

achromatic lens combination of lenses made from materials of different refractive indexes, constructed in such a way as to minimize chromatic aberration (which in a single lens causes colored fringes around images because the lens diffracts the different wavelengths in white light to slightly different extents).

acid in chemistry, compound that releases hydrogen ions (H^+ or protons) in the presence of an ionizing solvent (usually water). Acids react with †bases to form salts, and they act as solvents. Strong acids are corrosive; dilute acids have a sour or sharp taste, although in some organic acids this may be partially masked by other flavor characteristics. The strength of an acid is measured by its hydrogen-ion concentration, indicated by the †pH value. All acids have a pH below 7.0.

Acids can be classified as monobasic, dibasic, tribasic, and so forth, according to their basicity (the number of hydrogen atoms available to react with a base) and degree of ionization (how many of the available hydrogen atoms dissociate in water). Dilute sulfuric acid is classified as a strong (highly ionized), dibasic acid.

Inorganic acids include boric, carbonic, hydrochloric, hydrofluoric, nitric, phosphoric, and sulfuric. Organic acids include ethanoic (acetic), benzoic, citric, methanoic (formic), lactic, oxalic, and salicylic, as well as complex substances such as †nucleic acids and †amino acids.

Sulfuric, nitric, and hydrochloric acids are sometimes referred to as the mineral acids. Most naturally occurring acids are found as organic compounds, such as the fatty acids R–COOH and sulfonic acids $R–SO_3H$, where R is an organic group.

Acid and Base pH Tutorial http://www.chem.ubc.ca/courseware/pH/launch.html

Aimed at university students with no knowledge of this branch of chemistry, this well-designed tutorial takes you through the basics of acid–base chemistry, including interactive exercises to test your knowledge.

Acids http://www.purchon.com/chemistry/acids.htm

Offers a clear explanation of the nature and functions of acids. A general definition is followed by more detailed descriptions of some common acids.

acid amide any organic compound that may be regarded as being derived from ammonia by the substitution of acid or acyl groups for atoms of hydrogen. They are described as primary, secondary, or tertiary, according to the number of atoms of hydrogen displaced. Thus the general formula for a primary amide is $RCONH_2$. The main acid amides are †ethanamide and methanamide.

acidic oxide oxide of a †non-metal. Acidic oxides are covalent compounds. Those that dissolve in water, such as sulfur dioxide, give acidic solutions.

$SO_2 + H_2O \longleftrightarrow H_2SO_{3(aq)} \longleftrightarrow H^+_{(aq)} + HSO_3^-{}_{(aq)}$

All acidic oxides react with alkalis to form salts.

$CO_2 + NaOH \rightarrow NaHCO_3$

acid rain acidic precipitation thought to be caused mainly by the release into the atmosphere of sulfur dioxide (SO_2) and oxides of nitrogen (NO_x), which dissolve in pure rainwater making it acidic. Sulfur dioxide is formed by the burning of fossil fuels, such as coal, that contain high quantities of sulfur; nitrogen oxides are produced by various industrial activities and are present in car exhaust fumes.

Acidity is measured on the †pH scale, where the value of 0 represents liquids and solids that are completely acidic and 14 represents those that are highly alkaline. Distilled water is neutral and has a pH of 7. Normal rain has a value of 5.6. It is slightly acidic due to the presence of carbonic acid formed by the mixture of CO_2 and rainwater. Acid rain has values of 5.6 or less on the pH scale.

Acid deposition occurs not only as **wet precipitation** (mist, snow, or rain), but also comes out of the atmosphere as dry particles (**dry deposition**) or is absorbed directly by lakes, plants, and masonry as gases. Acidic gases can travel over 500 km/310 mi a day, so acid rain can be considered an example of transboundary (international) pollution.

Acid rain is linked with damage to and the death of forests and lake organisms in Scandinavia, Europe, and eastern North America. It is increasingly common in countries such as China and India that are industrializing rapidly. It also results in damage to buildings and statues. According to the UK Department of the Environment figures, emissions of sulfur dioxide from power stations would have to be decreased by 81% in order to stop the damage.

acid rock igneous rock that contains more than 60% by weight silicon dioxide, SiO_2, such as a granite or rhyolite. Along with the terms **basic rock** and **ultrabasic rock** it is part of an outdated classification system based on the erroneous belief that silicon in rocks is in the form of silicic acid. Geologists today are more likely to use the descriptive term †felsic rock or report the amount of SiO_2 in percentage by weight.

acid salt chemical compound formed by the partial neutralization of a dibasic or tribasic †acid (one that contains two or three replaceable hydrogen atoms). Although a salt, it still contains replaceable hydrogen, so it may undergo the typical reactions of an acid. Examples are sodium hydrogen sulfate ($NaHSO_4$) and acid phosphates.

aclinic line magnetic equator, an imaginary line near the Equator, where a compass needle balances horizontally, the attraction of the north and south magnetic poles being equal.

aclinic line The magnetic equator, or the line at which the attraction of both magnetic poles is equal. Along the aclinic line, a compass needle swinging vertically will settle in a horizontal position. Helicon

acoustic coupler device that enables computer data to be transmitted and received through a normal telephone handset; the handset rests on the coupler to make the connection. A small speaker within the device is used to convert the computer's digital output data into sound signals, which are then picked up by the handset and transmitted through the telephone system. At the receiving telephone, a second acoustic coupler or modem converts the sound signals back into digital data for input into a computer.

Unlike a †modem, an acoustic coupler does not require direct connection to the telephone system. However, interference from background noise means that the quality of transmission is poorer than with a modem, and more errors are likely to arise.

acoustic ohm *c.*g.s. unit of acoustic impedance (the ratio of the sound pressure on a surface to the sound flux through the surface). It is analogous to the ohm as the unit of electrical †impedance.

acoustics in general, the experimental and theoretical science of sound and its transmission; in particular, that branch of the science that has to do with the phenomena of sound in a particular space such as a room or theatre. In architecture, the sound-reflecting character of an internal space.

Acoustic engineering is concerned with the technical control of sound, and involves architecture and construction, studying control of vibration, soundproofing, and the elimination of noise. It also includes all forms of sound recording and reinforcement, the hearing and perception of sounds, and hearing aids.

Acoustic Illusions

http://cips02.physik.uni-bonn.de/~scheller/acoustic-illusions/main.html

Dedicated to the Shepard effect: a scale that gives the listener the impression of an endlessly rising melody, when in fact the pitch of the tones does not rise. Sound files illustrate this, and accompany an account of an experiment in which the Shepard effect was applied to JS Bach's Das Musikalische Opfer/Musical Offering.

acquired character feature of the body that develops during the lifetime of an individual, usually as a result of repeated use or disuse, such as the enlarged muscles of a weightlifter.

French naturalist Jean Baptiste †Lamarck's theory of evolution assumed that acquired characters were passed from parent to offspring. Modern evolutionary theory does not recognize the inheritance of acquired characters because there is no reliable scientific evidence that it occurs, and because no mechanism is known whereby bodily changes can influence the genetic material. The belief that this does not occur is known as †central dogma.

acre traditional English land measure equal to 4,840 square yards (4,047 sq m/0.405 ha). Originally meaning a field, it was the area that a yoke of oxen could plough in a day.

As early as Edward I's reign, the acre was standardized by statute for official use, although local variations in Ireland, Scotland, and some English counties continued. It may be subdivided into 160 square rods (one square rod equaling 25.29 sq m/30.25 sq yd).

acre-foot unit sometimes used to measure large volumes of water, such as the capacity of a reservoir (equal to its area in acres multiplied by its average depth in feet). One acre-foot equals 1,233.5 cu m/43,560 cu ft or the amount of water covering one acre to a depth of one foot.

acridine $C_{13}H_9N$, heterocyclic organic compound that occurs in coal tar. It is crystalline, melting at 108°C/226.4°F. Acridine is extracted by dilute acids but can also be obtained synthetically. It is used to make dyes and drugs.

acrolein common name for †propenal.

Acrux or **Alpha Crucis**, brightest star in the constellation of †Crux, marking one of the four points of the †Southern Cross, and the 13th-brightest star in the night sky. It is a double star comprising two blue-white stars, and is 360 light years away from the Sun. Together with nearby †Gacrux, it points towards the south celestial pole.

acrylic acid common name for propenoic acid.

acrylonitrile or **vinyl cyanide**; CH_2:CHCN, colorless liquid compound used in the production of acrylic fibers and synthetic rubbers. It polymerizes by a free radical mechanism in the presence of an appropriate catalyst. Commercially, the polymer is woven, and the yarn is known as Orlon.

Acrylonitrile is made by reacting propene, ammonia, and air with a catalyst at 450°C/842°F, a process developed in the 1960s from research into ways of utilizing a worldwide surplus of propene derived from ethene manufacture.

ACTH abbreviation of **adrenocorticotrophic hormone**, †hormone secreted by the anterior lobe of the †pituitary gland. It controls the production of corticosteroid hormones by the †adrenal gland and is commonly produced as a response to stress.

actin in zoology, protein that occurs in muscles. See †myosin.

actinic radiation in physics, historical term for radiation, such as †ultraviolet radiation, that can cause a chemical reaction (as in photography).

actinide any of a series of 15 radioactive metallic chemical elements with atomic numbers 89 (actinium) to 103 (lawrencium). Elements 89 to 95 occur in nature; the rest of the series are synthesized elements only. Actinides are grouped together because of their chemical similarities (for example, they are all bivalent), the properties differing only slightly with atomic number. The series is set out in a band in the †periodic table of the elements, as are the †lanthanides.

actinium chemical symbol Ac, (Greek *aktis* "ray") white, radioactive, metallic element, the first of the actinide series, atomic number 89, relative atomic mass 227; it is a weak emitter of high-energy alpha particles.

Actinium occurs with uranium and radium in †pitchblende and other ores, and can be synthesized by bombarding radium with neutrons. The longest-lived isotope, Ac-227, has a half-life of 21.8 years (all the other isotopes have very short half-lives). Chemically, it is exclusively trivalent, resembling in its reactions the lanthanides and the other actinides. Actinium was discovered in 1899 by the French chemist André Debierne (1874–1949).

actinometer (Greek "ray measure") instrument devised by English astronomer John †Herschel in 1825 to measure the amount of heat received from the Sun by a given surface in a given time.

The term is now sometimes used to describe any instrument having this purpose, for example a †pyrometer.

action potential in biology, a change in the †potential difference (voltage) across the membrane of a nerve cell when an impulse passes along it. A change in potential (from about –60 to +45 millivolts) accompanies the passage of sodium and potassium ions across the membrane.

activation analysis in analytical chemistry, a technique used to reveal the presence and amount of minute impurities in a substance or element. A sample of a material that may contain traces of a certain element is irradiated with †neutrons, as in a reactor. Gamma rays emitted by the material's radioisotopes have unique energies and relative intensities, similar to the spectral lines from a luminous gas. Measurements and interpretation of the gamma-ray spectrum, using data from standard samples for comparison, provide information on the amounts of impurities present.

activation energy in chemistry, the minimum energy required in order to start a chemical reaction. Some elements and compounds will react together merely by bringing them into contact (spontaneous reaction). For others it is necessary to supply energy (heat, radiation, or electrical charge) in order to start the reaction, even if there is ultimately a net output of energy. This initial energy is the activation energy. The point at which the reaction begins is known as the energy barrier. When the energy barrier is reached, the chemical bonds in the reactants are broken, enabling them to proceed from reactants to products.

In some reactions, such as the †combustion of †fuels, the activation energy required for the chemical reaction to take place is very small, resulting in a rapid reaction. Other chemical reactions, such as the †rusting of iron (a type of †oxidation) have a very large energy barrier and take place slowly.

A chemical equation only describes the †**energy of reaction**; the activation energy is not shown. The total chemical energy involved can be represented in an **energy level diagram**; this also shows whether a reaction is †exothermic (giving off energy) or †endothermic (absorbing energy).

active galaxy in astronomy, type of galaxy that emits vast quantities of energy from a small region at its center, the active galactic nucleus (AGN). Active galaxies are subdivided into †radio galaxies, †Seyfert galaxies, †BL Lacertae objects, and †quasars.

Active galaxies are thought to contain black holes with a mass 10^8 times that of the Sun, drawing stars and interstellar gas towards them in a process of †accretion. The gravitational energy released by the infalling material is the power source for the AGN. Some of the energy may appear as a pair of opposed jets emerging from the nucleus. The orientation of the jets to the line of sight and their interaction with surrounding material determines the type of active galaxy that is seen by observers. See also †starburst galaxy.

active region in astronomy, area on the surface of the Sun where strong magnetic fields cause disturbances such as †prominences, †solar flares, and †sunspots.

active transport movement of molecules or ions across a cell membrane using †energy provided by †respiration. Examples of substances that can be actively transported across membranes are sodium ions and glucose.

Energy is needed because the movement occurs against a concentration gradient, with substances being moved from an area of low concentration to an area where there is a higher concentration. Active transport is therefore quite different from †diffusion, which requires no input of energy. In diffusion the movement is in the opposite direction – from an area of high concentration to an area where the concentration is low. An example of diffusion is the movement of oxygen into the blood vessels of the lungs.

activity in physics, number of particles emitted in one second by a radioactive source. The term is used to describe the radioactivity or the potential danger of that source. The unit of activity is the becquerel (Bq), named after the French physicist Henri Becquerel, who discovered radioactivity.

activity series in chemistry, alternative name for †reactivity series.

acute angle angle between 0° and 90°; that is, an amount of turn that is less than a quarter of a circle.

acylation in chemistry, process for the introduction of an †acyl group into an organic molecule. It is usually achieved by the substitution of an active hydrogen, such as in a †hydroxyl group, by reaction with a †carboxylic acid.

acyl group in chemistry, organic group of general formula —CO—R, where R is an †alkyl group. Acyl groups are derived from †carboxylic acids by the removal of the †hydroxyl group.

Ada high-level computer-programming language, developed and owned by the US Department of Defense, designed for use in situations in which a computer directly controls a process or machine, such as a military aircraft. The language took more than five years to specify, and became commercially available only in the late 1980s. It is named after English mathematician Ada Augusta †Byron.

Adams, John Couch (1819–1892) English astronomer. He mathematically deduced the existence of the planet Neptune in 1845 from the effects of its gravitational pull on the motion of Uranus, although it was not found until 1846 by J G †Galle. Adams also studied the Moon's motion, the Leonid meteors, and terrestrial magnetism.

Adams, John Couch http://www-history.mcs.st-andrews.ac.uk/Mathematicians/Adams.html

Biography and portrait of the famous 19th-century English astronomer. Details of Adam's work leading to the discovery of Neptune and other work on orbits and gravitation are also available at this site. It also features reference sources in the form of books and journals that contain Adams's work and biographical details.

Adams, Walter Sydney (1876–1956) US astronomer. He developed the use of spectroscopy in the study of stars and planets, and found that luminosity and the relative intensities of particular spectral lines could distinguish giant stars from dwarf stars. Spectra could also be used to study the physical properties, motions, and distances of stars.

adaptation (Latin *adaptare* "to fit to") in biology, any feature in the structure or function of an organism that allows it to survive and reproduce more effectively in its environment. Much adaptation is inherited (see †inheritance) and is the result of many thousands of years of †evolution. It is thought to occur as a result of random variation in the genetic make-up of organisms coupled with †natural selection. Species become extinct when they are no longer adapted to their environment.

Adaptations to cope with changing seasons can be quite different from one organism to another. Some plants die back to survive the winter and have specially protected structures such as tight buds on a twig or a bulb underground. These structures may store food so that growth can start rapidly in spring. A carrot, for example, stores sugars for this reason. Some animals, such as swallows, may migrate, while other animals, such as the dormouse, may hibernate, allowing the body temperature and heart rate to drop in the winter.

Adaptations also occur as a result of an animal evolving into an efficient predator. A predator may have forward facing eyes, acute vision and sense of smell, and have claws, talons, or a beak for killing. The prey also adapt as a result of evolution. Prey may have eyes on the side of their heads for a wide field of view, acute hearing and sense of smell, be easily startled, be nocturnal, and be camouflaged.

In evolution, adaptation is the result of †natural selection. Inherited variation usually occurs in a species. If the environment changes in some way, such as the winters becoming more severe, those individuals possessing genes that allow them to survive and reproduce more successfully pass their genes to their offspring more than those that do not. Some genes therefore become more common in the species, leading to an evolutionary change that enables the species to be better adapted to the changed environment.

Desert Plant Survival http://www.desertusa.com/du_plantsurv.html

Attractive and readable page on the mechanisms plants use to survive in the hostile environment of the desert. Part of a larger site on desert survival, this page describes the main features of the xerophytes (plants, such as cacti, that have altered their physical structure for the sake of survival) and the phreatophytes, which adapt to their arid habitat by growing roots long enough to reach the water table.

adaptive optics in optics, system that can be adjusted to overcome distortions caused by the Earth's atmosphere or gravity; for example, an astronomical telescope.

adaptive radiation in evolution, the formation of several species, with †adaptations to different ways of life, from a single ancestral type. Adaptive radiation is likely to occur whenever members of a species migrate to a new habitat with unoccupied ecological niches. It is thought that the lack of competition in such niches allows sections of the migrant population to develop new adaptations, and eventually to become new species.

The colonization of newly formed volcanic islands has led to the development of many unique species. The 13 species of Darwin's finch on the Galapagos Islands, for example, are probably descended from a single species from the South American mainland. The parent stock evolved into different species that now occupy a range of diverse niches.

ADC in electronics, abbreviation for †analog-to-digital converter.

adder electronic circuit in a computer or calculator that carries out the process of adding two binary numbers. A separate adder is needed for each pair of binary †bits to be added. Such circuits are essential components of a computer's †arithmetic and logic unit (ALU).

adding machine device for adding (and usually subtracting, multiplying, and dividing) numbers, operated mechanically or electromechanically; now largely superseded by electronic calculators.

addition in arithmetic, the operation of combining two numbers to form a sum; thus, $7 + 4 = 11$. It is one of the four basic operations of arithmetic (the others are subtraction, multiplication, and division).

addition polymerization †polymerization reaction in which a single monomer gives rise to a single polymer, with no other reaction products.

Addition polymerization occurs in †alkenes, hydrocarbons containing double bonds. The alkenes, such as †ethene, are the monomers, the small starting molecules. In addition polymerization, ethene undergoes an addition reaction with itself. As one molecule joins to a second, a long molecular chain is built up. The long molecular chain is called a polymer. In the case of ethene, †polythene is formed. It is made up of repeating units of the monomer.

addition reaction chemical reaction in which the atoms of an element or compound react with a double bond or triple bond in an organic compound by opening up one of the bonds and becoming attached to it, for example:

$CH_2 = CH_2 + HCl \rightarrow CH_3CH_2Cl$

Another example is the addition of hydrogen atoms to †unsaturated compounds in vegetable oils to produce margarine. Addition reactions are used to make polymers from †alkenes.

additive process in physics, result of mixing colored lights to form new colors, as when combinations of red, green, and blue are mixed to form all the colors in a television picture or transparency film. See also †color.

Adelard of Bath (lived 12th century) English philosopher and student of mathematics and science. He translated Euclid's *Stoicheia/Elements* from Arabic into Latin and wrote *De Eodem et Diverso/On Identity and Difference* (1113–33), which establishes him as the founder of the doctrine of indifference, according to which the reality of universals is reconciled with that of individuals. This attempt at reconciliation distinguished him from other Platonists.

adenine in biochemistry, one of four base molecules that form part of the †base pairs in the DNA molecule, where it is always paired with the purine †thymine.

Adenine is also part of the genetic code of RNA, where it pairs with the pyrimidine uracil.

adenoids masses of lymphoid tissue, similar to †tonsils, located in the upper part of the throat, behind the nose. They are part of a child's natural defenses against the entry of germs but usually shrink and disappear by the age of ten.

Adenoids may swell and grow, particularly if infected, and block the breathing passages. If they become repeatedly infected, they may be removed surgically (**adenoidectomy**), usually along with the tonsils.

adenosine triphosphate energy-rich compound present in cells. See †ATP.

adenovirus any of a group of DNA-containing viruses that cause respiratory infections and conjunctivitis.

ADH abbreviation for antidiuretic hormone, in biology, part of the system maintaining a correct salt/water balance in vertebrates.

Its release is stimulated by the †hypothalamus in the brain, which constantly receives information about salt concentration from receptors situated in the neck. In conditions of water shortage increased ADH secretion from the brain will cause more efficient conservation of water in the kidney, so that fluid is retained by the body. When an animal is able to take in plenty of water, decreased ADH secretion will cause the urine to become dilute and plentiful. The system allows the body to compensate for a varying fluid intake and maintain a correct balance.

Adhara or **Epsilon Canis Majoris**, star 600 light years from Earth. US astronomers concluded in 1995 that it plays a greater part in ionizing hydrogen in our region of the Galaxy than all the other 3 million stars lying closer to our local cloud.

Most of space contains unionized hydrogen in tiny amounts but Adhara is connected to our local cloud by a tunnel of almost hydrogen-free space. This means its ionizing radiation reaches our local cloud without obstacle.

The surface temperature of Adhara is 21,000 K (20,700°C/37,292°F), almost four times that of the Sun, and it is the brightest source of extreme ultraviolet radiation apart from the Sun.

adhesive substance that sticks two surfaces together. Natural adhesives (glues) include gelatin in its crude industrial form (made from bones, hide fragments, and fish offal) and vegetable gums. Synthetic adhesives include thermoplastic and thermosetting resins, which are often stronger than the substances they join; mixtures of epoxy resin and hardener that set by chemical reaction; and elastomeric (stretching) adhesives for flexible joints. Superglues are fast-setting adhesives used in very small quantities.

adiabatic in biology and physics, describing a process that occurs without loss or gain of heat, especially the expansion or contraction of a gas in which a change takes place in the pressure or volume, although no heat is allowed to enter or leave. Adiabatic processes can be both non-reversible and approximately reversible.

adipic acid common name for †hexanedioic acid.

adipocere wax like substance produced by the exposure of fleshy tissue to moisture with the exclusion of air, as in the earth or under water. Human bodies in moist burial places often undergo this change.

adipose tissue type of †connective tissue of vertebrates that serves as an energy reserve, and also pads some organs. It is commonly called fat tissue, and consists of large spherical cells filled with fat. In mammals, major layers are in the inner layer of skin and around the kidneys and heart.

Fatty acids are transported to and from it via the blood system. An excessive amount of adipose tissue is developed in the course of some disorders, especially obesity.

adjacent angles pair of angles meeting at a common vertex (corner) and sharing a common arm. Two adjacent angles lying on the same side of a straight line add up to 180° and are said to be supplementary.

adjacent side in a †right-angled triangle, the side that is next to a given angle but is not the hypotenuse (the side opposite the right angle). The third side is the **opposite side** to the given angle.

Adler, Alfred (1870–1937) Austrian psychologist. He saw the "will to power" as more influential in accounting for human behavior than the sexual drive. A dispute over this theory led to the dissolution of his ten-year collaboration with psychiatry's founder Sigmund †Freud.

The concepts of inferiority complex and overcompensation originated with Adler.

admittance symbol *Y*, in physics, response of an electric circuit to an alternating current. It is equal to the reciprocal of †impedance and is measured in siemens.

adolescence in the human life cycle, the period between the beginning of †puberty and adulthood.

ADP abbreviation for adenosine diphosphate, the chemical product formed in cells when †ATP breaks down to release energy.

adrenal gland or **suprarenal gland**, triangular endocrine gland situated on top of the †kidney. The adrenals are soft and yellow, and consist of two parts: the cortex and medulla. The **cortex** (outer part) secretes various steroid hormones and other hormones that control salt and water metabolism and regulate the use of carbohydrates, proteins, and fats. The **medulla** (inner part) secretes the hormones adrenaline and noradrenaline which, during times of stress, cause the heart to beat faster and harder, increase blood flow to the heart and muscle cells, and dilate airways in the lungs, thereby delivering more oxygen to cells throughout the body and in general preparing the body for "fight or flight."

adrenaline or **epinephrine**, hormone secreted by the medulla of the †adrenal glands. Adrenaline is synthesized from a closely related substance, noradrenaline, and the two hormones are released into the bloodstream in situations of fear or stress.

Adrenaline's action on the †liver raises blood-sugar levels by stimulating glucose production and its action on adipose tissue raises blood fatty-acid levels; it also increases the heart rate, increases blood flow to muscles, reduces blood flow to the skin with the production of sweat, widens the smaller breathing tubes (bronchioles) in the lungs, and dilates the pupils of the eyes.

adrenocorticotrophic hormone hormone secreted by the anterior lobe of the †pituitary gland; see †ACTH.

Adrian, Edgar Douglas (1889–1977) 1st Baron Adrian, English physiologist who was awarded a Nobel Prize for Physiology or Medicine in 1932 for his work with Charles †Sherrington in the field of nerve impulses and the function of the nerve cell. Adrian was also one of the first to study the electrical activity of the brain. He was made a baron in 1955.

adsorption taking up of a gas or liquid at the surface of another substance, most commonly a solid (for example, activated charcoal adsorbs gases). It involves molecular attraction at the surface, and should be distinguished from †absorption (in which a uniform solution results from a gas or liquid being incorporated into the bulk structure of a liquid or solid).

advanced gas-cooled reactor AGR, type of †nuclear reactor favored in Western Europe. The AGR uses a fuel of enriched uranium dioxide in stainless-steel cladding and a moderator of graphite. Carbon dioxide gas is pumped through the reactor core to extract the heat produced by the †fission of the uranium. The heat is transferred to water in a steam generator, and the steam drives a turbogenerator to produce electricity.

advection fog †fog formed by warm moist air meeting a cold ocean current or flowing over a cold land surface. It is common in coastal areas where warm air from the sea meets the colder land surface, particularly at night or during the winter.

adventitious in botany, arising in an abnormal position, as in roots developing on the stem of a cutting or buds developing on roots.

aeolian deposits in geology, sediments carried, formed, eroded, or deposited by the wind. Such sediments include desert sands and dunes as well as deposits of windblown silt, called loess, carried long distances from deserts and from stream sediments derived from the melting of glaciers.

aeon or **eon**, in astronomical terms, 1,000 million years.

aerated water water that has had air (oxygen) blown through it. Such water supports aquatic life and prevents the growth of putrefying bacteria. Polluted waterways may be restored by artificial aeration.

aerenchyma plant tissue with numerous air-filled spaces between the cells. It occurs in the stems and roots of many aquatic plants where it aids buoyancy and facilitates transport of oxygen around the plant.

aerial or **antenna**, in radio and television broadcasting, a conducting device that radiates or receives electromagnetic waves. The design of an aerial depends principally on the wavelength of the signal. Long waves (hundreds of meters in wavelength) may employ long wire aerials; short waves (several centimeters in wavelength) may employ rods and dipoles; microwaves may also use dipoles – often with reflectors arranged like a toast rack – or highly directional parabolic dish aerials. Because microwaves travel in straight lines, requiring line-of-sight communication, microwave aerials are usually located at the tops of tall masts or towers.

aerial oxidation in chemistry, a reaction in which air is used to oxidize another substance, as in the contact process for the manufacture of sulfuric acid:

$$2SO_2 + O_2 \longleftrightarrow 2SO_3$$

and in the †souring of wine, where alcohol (ethanol) is oxidized by atmospheric oxygen to produce vinegar (ethanoic, or acetic, acid).

aerobic in biology, describing those organisms that require †oxygen in order to survive. Aerobic organisms include all plants and animals and many micro-organisms. They use oxygen (usually dissolved in water) to release the energy contained in food molecules such as glucose in a process called aerobic respiration. Oxygen is used to break down carbohydrates into †carbon dioxide and †water, releasing †energy, which is used to drive many processes within the cells.

Aerobic reactions occur inside every cell and lead to the formation of energy-rich †ATP, which is used by the cell for driving its metabolic processes.

Most aerobic organisms die in the absence of oxygen, but certain organisms and cells, such as those found in muscle tissue, can function for short periods anaerobically (without oxygen). Some micro-organisms – certain bacteria, yeasts, and internal parasites – can live without oxygen and these are said to be †anaerobic.

aerobic respiration in plant and animal cells, a complex process of chemical reactions in which oxygen is used to break †glucose into †carbon dioxide and †water. Energy is released in this process. The starting and finishing points of this process would be the same if glucose was burned. During burning the energy that is released is all in the form of heat. In aerobic respiration, however, the energy is released in a controlled way and less is released as heat. Most of the released energy is used to drive various processes in the cell, such as growth or movement.

If insufficient oxygen reaches human muscles in severe exercise, such as a sprint, the muscles can carry on respiring using anaerobic respiration for a short time.

aerodynamics branch of fluid physics that studies the forces exerted by air or other gases in motion. Examples include the airflow around bodies moving at speed through the atmosphere (such as land vehicles, bullets, rockets, and aircraft), the behavior of gas in engines and furnaces, air conditioning of buildings, the deposition of snow, the operation of air-cushion vehicles (hovercraft), wind loads on buildings and bridges, bird and insect flight, musical wind instruments, and meteorology. For maximum efficiency, the aim is usually to design the shape of an object to produce a streamlined flow, with a minimum of turbulence in the moving air. The behavior of aerosols or the pollution of the atmosphere by foreign particles are other aspects of aerodynamics.

aerogel light, transparent, highly porous material composed of more than 90% air. Such materials are formed from silica, metal oxides, and organic chemicals, and are produced by drying gels – networks of linked molecules suspended in a liquid – so that air fills the spaces previously occupied by the liquid. They are excellent heat insulators and have unusual optical, electrical, and acoustic properties.

Aerogels were first produced by US scientist Samuel Kristler in the early 1930s by drying silica gels at high temperatures and pressures.

aerogenerator wind-powered electricity generator. These range from large models used in arrays on wind farms (see †wind turbine) to battery chargers used on yachts.

aeronautics science of travel through the Earth's atmosphere, including aerodynamics, aircraft structures, jet and rocket propulsion, and aerial navigation. It is distinguished from †astronautics, which is the science of travel through space.

In **subsonic aeronautics** (below the speed of sound), aerodynamic forces increase at the rate of the square of the speed.

Trans-sonic aeronautics covers the speed range from just below to just above the speed of sound and is crucial to aircraft design. Ordinary sound waves move at about 1,225 kph/760 mph at sea level, and air in front of an aircraft moving slower than this is "warned" by the waves so that it can move aside. However, as the flying speed approaches that of the sound waves, the warning is too late for the air to escape, and the aircraft pushes the air aside, creating shock waves, which absorb much power and create design problems. On the ground the shock waves give rise to a †sonic boom. It was once thought that the speed of sound was a speed limit to aircraft, and the term sound barrier came into use.

Supersonic aeronautics is concerned with speeds above that of sound and in one sense may be considered a much older study than aeronautics itself, since the study of the flight of bullets (†ballistics) was undertaken soon after the introduction of firearms. **Hypersonics** is the study of airflows and forces at speeds above five times that of sound (Mach 5); for example, for guided missiles, space rockets, and advanced concepts such as HOTOL (horizontal takeoff and landing). For all flight speeds, streamlining is necessary to reduce the effects of air resistance.

Astronavigation (navigation by reference to the stars) is used in aircraft as well as in ships and is a part of aeronautics.

Beginner's Guide to Aeronautics http://www.grc.nasa.gov/WWW/K-12/airplane/

NASA's introduction to aeronautics. Tutorials, practical experiments (build your own wind tunnel!), and information about aerodynamics, rockets, and space flight. An index of aerodynamics makes browsing the documents very easy.

aerosol particles of liquid or solid suspended in a gas. Fog is a common natural example. Aerosol cans contain material packed under pressure with a device for releasing it as a fine spray. Most aerosols used †chlorofluorocarbons (CFCs) as propellants until these were found to cause destruction of the †ozone layer in the stratosphere.

The international community has agreed to phase out the use of CFCs, but most so-called "ozone-friendly" aerosols also use ozone-depleting chemicals, although they are not as destructive as CFCs. Some of the products sprayed, such as pesticides, can be directly toxic to humans.

aestivation in zoology, a state of inactivity and reduced metabolic activity, similar to †hibernation, that occurs during the dry season in species such as lungfish and snails. In botany, the term is used to describe the way in which flower petals and sepals are folded in the buds. It is an important feature in †plant classification.

aether alternative form of †ether, the hypothetical medium once believed to permeate all of space.

afferent nerve nerve that carries impulses towards the central nervous system (that is, the spinal cord or brain).

affine geometry geometry that preserves parallelism and the ratios between intervals on any line segment.

affinity in chemistry, the force of attraction (see †bond) between atoms that helps to keep them in combination in a molecule. The term is also applied to attraction between molecules, such as those of biochemical significance (for example, between †enzymes and substrate molecules). This is the basis for affinity †chromatography, by which biologically important compounds are separated.

The atoms of a given element may have a greater affinity for the atoms of one element than for another (for example, hydrogen has a great affinity for chlorine, with which it easily and rapidly combines to form hydrogen chloride, but has little or no affinity for argon).

afforestation planting of trees in areas that have not previously held forests. (**Reafforestation** is the planting of trees in deforested areas.) Trees may be planted (1) to provide timber and wood pulp; (2) to provide firewood in countries where this is an energy source; (3) to bind soil together and prevent soil erosion; and (4) to act as windbreaks.

Afforestation is a controversial issue because while many ancient woodlands of mixed trees are being lost, the new plantations consist almost exclusively of conifers. It is claimed that such plantations acidify the soil and conflict with the interests of †biodiversity (they replace more ancient and ecologically valuable species and do not sustain wildlife).

afterbirth in mammals, the placenta, umbilical cord, and ruptured membranes that become detached from the uterus and expelled soon after birth.

afterimage persistence of an image on the retina of the eye after the object producing it has been removed. This leads to persistence of vision, a necessary phenomenon for the illusion of continuous movement in films and television. The term is also used for the persistence of sensations other than vision.

after-ripening process undergone by the seeds of some plants before germination can occur. The length of the after-ripening period in different species may vary from a few weeks to many months.

It helps seeds to germinate at a time when conditions are most favorable for growth. In some cases the embryo is not fully mature at the time of dispersal and must develop further before germination can take place. Other seeds do not germinate even when the embryo is mature, probably owing to growth inhibitors within the seed that must be leached out or broken down before germination can begin.

agar jellylike carbohydrate, obtained from seaweeds. It is used mainly in microbiological experiments as a culture medium for growing bacteria and other micro-organisms. The agar is resistant to breakdown by micro-organisms, remaining a solid jelly throughout the course of the experiment.

Agassiz, (Jean) Louis Rodolphe (1807–1873) Swiss-born US paleontologist and geologist who developed the idea of the ice age. He established his name through his work on the classification of fossil fishes. Unlike Charles Darwin, he did not believe that individual species themselves changed, but that new species were created from time to time.

agate cryptocrystalline (with crystals too small to be seen with an optical microscope) silica, SiO_2, composed of cloudy and banded †chalcedony, sometimes mixed with †opal, which forms in rock cavities.

Agate stones, being hard, are also used to burnish and polish gold applied to glass and ceramics and as clean vessels for grinding rock samples in laboratories.

agglutination in biology, the clumping together of †antigens, such as red blood cells or bacteria, to form larger, visible clumps, under the influence of †antibodies. Because each antigen clumps only in response to its particular antibody, agglutination provides a way of determining blood groups and the identity of unknown bacteria.

aggression in biology, behavior used to intimidate or injure another organism (of the same or of a different species), usually for the purposes of gaining territory, a mate, or food. Aggression often involves an escalating series of threats aimed at intimidating an opponent without having to engage in potentially dangerous physical contact. Aggressive signals include roaring by red deer, snarling by dogs, the fluffing-up of feathers by birds, and the raising of fins by some species of fish.

Agnatha one of the five classes of fish, represented by the hagfish and lamprey. They are primitive craniate vertebrates, which have no true jaw and a sucker like mouth.

Agnesi, Maria Gaetana (1718–1799) Italian mathematician. She produced a 1,000-page textbook, *Istituzioni analitiche ad uso della gioventu italiana/Foundations of Analysis for the Use of Italian Youth* (1748), which provided an integrated treatment of algebra, analytical geometry, and calculus. She is acknowledged as the first woman mathematician.

Agnesi, Maria Gaetana http://www-history.mcs.st-andrews.ac.uk/Mathematicians/Agnesi.html

Biography and portrait of the first recognized woman mathematician. Agnesi's most famous curve, known as the witch of Agnesi, is shown on this site. Thirteen literature references containing further details of her works and biography are also listed.

agonist in biology, a †muscle that contracts and causes a movement. Contraction of an agonist is complemented by relaxation of its †antagonist. For example, the biceps (in the front of the upper arm) bends the elbow while the triceps (lying behind the biceps) straightens the arm.

AGR abbreviation for †**advanced gas-cooled reactor**, a type of nuclear reactor.

agriculture (Latin *ager* "field," *colere* "to cultivate") the practice of farming, including the cultivation of the soil (for raising crops) and the raising of domesticated animals. The units for managing agricultural production vary from smallholdings and individually owned farms to corporate-run farms and collective farms run by entire communities or by the government.

Crops are cultivated for human or animal food, or as industrial crops such as cotton and sisal. For successful production, the land must be prepared (ploughed, cultivated, harrowed, and rolled), seed must be planted and the growing plants nurtured. This may involve †fertilizers, †irrigation, pest control by chemicals, and monitoring of acidity or nutrients. When the crop has grown, it must be harvested and, depending on the crop, processed in a variety of ways before it is stored or sold. Greenhouses allow cultivation of plants in cold climates. †Hydroponics allows commercial cultivation of crops using nutrient-enriched water instead of soil. Special methods, such as terracing, may be adopted to allow cultivation in steep regions and to retain topsoil in mountainous areas with heavy rainfall.

Animals are raised for wool, milk, leather, dung (as fuel), or meat. They may be semi domesticated, such as reindeer, or fully domesticated but nomadic (where naturally growing or cultivated food supplies are sparse), or kept on a farm. Animal farming involves rearing, feeding, breeding, gathering the produce (eggs, milk, or wool), slaughtering, and further processing such as tanning.

Sustainable Agriculture Network http://www.sare.org/

US-government-funded organization dedicated to the global exchange of information on sustainable farming systems. This well-organized site highlights the issues through interesting profiles of individual farming families who have moved away from conventional cultivation techniques. Findings of research projects are presented and there are links to a number of US sustainable agriculture sites.

agrochemical artificially produced chemical used in modern, intensive agricultural systems. Agrochemicals include nitrate and phosphate fertilizers, pesticides, some animal-feed additives, and pharmaceuticals. Many are responsible for pollution and almost all are avoided by organic farmers.

agroforestry agricultural system that involves planting and maintaining trees on land where crops and animals are also raised. The trees may be used for timber, fuel, food, or animal fodder. Trees may be interplanted with other crops in order to provide shade, reduce soil erosion, and increase moisture retention in the soil. Leaves dropped by the trees also improve the fertility of the soil.

agronomy study of crops and soils, a branch of agricultural science. Agronomy includes such topics as selective breeding (of plants and animals), irrigation, pest control, and soil analysis and modification.

AI abbreviation for †artificial intelligence and artificial insemination.

AIDS acronym for **acquired immune deficiency syndrome**, most serious of all the sexually transmitted diseases (STDs). It is caused by the †retrovirus human immunodeficiency virus (†HIV), and is transmitted in body fluids, such as blood, salvia, semen, and vaginal secretions. AIDS is the world's most deadly infectious disease, and the fourth leading global cause of death.

Aiken, Howard Hathaway (1900–1973) US mathematician and computer pioneer. In 1939, in conjunction with engineers from IBM, he started work on the design of an automatic calculator using standard business-machine components. In 1944 the team completed one of the first computers, the Automatic Sequence Controlled Calculator (known as the Harvard Mark I), a programmable computer controlled by punched paper tape and using punched cards.

air the mixture of gases making up the Earth's †atmosphere.

airglow faint and variable light in the Earth's atmosphere produced by chemical reactions (the recombination of ionized particles) in the ionosphere.

air mass large body of air with particular characteristics of temperature and humidity. An air mass forms when air rests over an area long enough to pick up the conditions of that area. When an air mass moves to another area it affects the †weather of that area, but its own characteristics become modified in the process. For example, an air mass formed over the Sahara will be hot and dry, becoming cooler as it moves northwards. Air masses that meet form **fronts**.

There are four types of air mass. **Tropical continental** (Tc) air masses form over warm land and **Tropical maritime** (Tm) masses form over warm seas. Air masses that form over cold land are called **Polar continental** (Pc) and those forming over cold seas are called **Polar** or **Arctic maritime** (Pm or Am).

The weather of the UK is affected by a number of air masses which, having different characteristics, bring different weather conditions. For example, an Arctic air mass brings cold conditions whereas a Tropical air mass brings hot conditions.

air passages in biology, the nose, pharynx, larynx, trachea, and bronchi. When a breath is taken, air passes through high narrow passages on each side of the nose where it is warmed and moistened and particles of dust are removed. Food and air passages meet and cross in the pharynx. The larynx lies in front of the lower part of the pharynx and it is the organ where the voice is produced using the vocal cords. The air passes the glottis (the opening between the vocal cords) and enters the trachea. The trachea leads into the chest and divides above the heart into two bronchi. The bronchi carry the air to the lungs and they subdivide to form a succession of fine tubes and, eventually, a network of capillaries that allow the exchange of gases between the inspired air and the blood.

air pollution contamination of the atmosphere caused by the discharge, accidental or deliberate, of a wide range of toxic airborne substances. Often the amount of the released substance is relatively high in a certain locality, so the harmful effects become more noticeable. The cost of preventing any discharge of pollutants into the air is prohibitive, so attempts are more usually made to reduce the amount of discharge gradually and to disperse it as quickly as possible by using a very tall chimney, or by intermittent release.

One major air pollutant is sulfur dioxide (SO_2), produced from the burning of †fossil fuels. It dissolves in atmospheric moisture to produce sulfurous acid (†acid rain):

$SO_2 + H_2O = H_2SO_3$

Air Pollution – U.S. Environmental Protection Agency http://www.epa.gov/air

EPA's website discusses air pollution issues.

air pressure in earth science, pressure of the atmosphere at any place on Earth; see †atmospheric pressure.

air pump device used to pump air from one vessel to another, or to evacuate a vessel altogether to produce a †vacuum.

air sac in birds, a thin-walled extension of the lungs. There are nine of these and they extend into the abdomen and bones, effectively increasing lung capacity. The sacs subdivide into further air spaces, which partially replace the marrow in many of the bird's bones. The air space in these bones assists flight by making them lighter. In mammals, it is another name for the alveoli in the lungs, and in some insects, for widening of the trachea.

Airy, George Biddell (1801–1892) English astronomer. He installed a transit telescope at the Royal Observatory at Greenwich, England, and accurately measured †Greenwich Mean Time by the stars as they crossed the meridian. Knighted in 1872.

Airy, George Biddell http://www-history.mcs.st-andrews.ac.uk/Mathematicians/Airy.html

Biography and portrait of the famous English astronomer. Airy's collaborative work with John Couch Adams that led to the discovery of Neptune, and their work on orbits and gravitation are described here, as is Airy's work on the nephroid. Comments are also made on Airy's work in engineering, and literature references to Airy's works are listed.

Airy disc in astronomy, small disc of light that forms the image of a point source such as a star, as produced by a telescope lens or mirror and resulting from the spreading of the light by †diffraction. It was named after the English astronomer George †Airy.

alabaster naturally occurring fine-grained white or light-colored translucent form of †gypsum, often streaked or mottled. A soft material, it is easily carved, but seldom used for outdoor sculpture.

albedo fraction of the incoming light reflected by a body such as a planet. A body with a high albedo (near 1) is very bright, while a body with a low albedo (near 0) is dark. The Moon has an average albedo of 0.12, Venus 0.76, and Earth 0.37.

Albirea or **Beta Cygnii**, in astronomy, double star in †Cygnus, the second brightest star in the constellation.

albumin any of a group of sulfur-containing †proteins. The best known is in the form of egg white (albumen); others occur in milk, and as a major component of serum. Many vegetables and fluids also contain albumins. They are soluble in water and dilute salt solutions, and are coagulated by heat.

The presence of serum albumin in the urine, termed albuminuria or proteinuria, may be indicative of kidney or heart disease.

alcohol in chemistry, any member of a group of organic chemical compounds characterized by the presence of one or more aliphatic OH (hydroxyl) groups in the molecule, and which form †esters with acids. The main uses of alcohols are as solvents for gums, resins, lacquers, and varnishes; in the making of dyes; for essential oils in perfumery; and for medical substances in pharmacy. The alcohol produced naturally in the †fermentation process and consumed as part of alcoholic beverages is called †ethanol. When consumed the effects of alcohol include poisoning at high concentrations, and changes in the functioning of human nerve cells.

Alcohols may be liquids or solids, according to the size and complexity of the molecule. A **monohydric alcohol** contains only one †hydroxyl group in each molecule. The five simplest alcohols form a series in which the number of carbon and hydrogen atoms in the molecule increases progressively, each one having an extra CH_2 (methylene) group: methanol or wood spirit (methyl alcohol, CH_3OH); ethanol (ethyl alcohol, C_2H_5OH); propanol (propyl alcohol, C_3H_7OH); butanol (butyl alcohol, C_4H_9OH); and pentanol (amyl alcohol, $C_5H_{11}OH$). The lower alcohols are liquids that mix with water; the higher alcohols, such as pentanol, are oily liquids immiscible with water; and the highest are waxy solids – for example, hexadecanol (cetyl alcohol, $C_{16}H_{33}OH$) and melissyl alcohol ($C_{30}H_{61}OH$), which occur in sperm-whale oil and beeswax, respectively. Alcohols containing the CH_2OH group are primary; those containing CHOH are secondary; while those containing COH are tertiary.

alcoholic solution solution produced when a solute is dissolved in ethanol.

alcoholometry determination of the proportion of alcohol in a liquid, using a specially calibrated hydrometer.

Sikes's hydrometer is used for determining the strength of spirits, and Atkins's has a series of scales for any "wort," or beer in the making.

Alcor in astronomy, white star in the constellation †Ursa Major that forms an optical double with †Mizar (which is in turn also a true †double star).

Alcyone or **Eta Tauri**, brightest member of the Pleiades cluster, a third-magnitude blue star. It is one of the cluster of stars in the zodiacal constellation of Taurus.

Aldebaran or **Alpha Tauri**, brightest star in the constellation Taurus and the 14th-brightest star in the night sky; it marks the eye of the "bull." Aldebaran is a red giant 65 light years away from the Sun, shining with a true luminosity of about 100 times that of the Sun.

aldehyde any of a group of organic chemical compounds prepared by oxidation of primary alcohols, so that the OH (hydroxyl) group loses its hydrogen to give an oxygen joined by a double bond to a carbon atom (the aldehyde group, with the formula CHO).

Alder, Kurt (1902–1958) German organic chemist who with Otto †Diels was awarded the Nobel Prize for Chemistry in 1950 for the discovery and development of the diene synthesis in 1928, a fundamental process that has become known as the Diels–Alder reaction. It is used in organic chemistry to synthesize cyclic (ring) compounds, including many that can be made into plastics and others – which normally occur only in small quantities in plants and other natural sources – that are the starting materials for various drugs and dyes.

aldol or **3-hydroxybutanal**; $CH_3CH(OH)CH_2CHO$, in chemistry, colorless, viscous liquid, boiling point 83°C/181°F. It is prepared by treatment of †ethanal with a dilute acid or base.

aldol condensation in chemistry, base-catalyzed reaction between two †aldehyde molecules resulting in the elimination of a water molecule and the formation of b-hydroxyaldehydes. The reaction is named after aldol, the simplest b-hydroxyaldehyde.

aldose in chemistry, member of a group of †monosaccharides (simple sugars) that have a potential †aldehyde group. The sugars in this group are named according the number of carbon atoms in the molecule; for example, aldopentose has five carbon atoms.

Alembert, Jean Le Rond d' (1717–1783) French mathematician, encyclopedist, and theoretical physicist. In association with Denis Diderot, he helped plan the great *Encyclopédie* , for which he also wrote the "Discours préliminaire" (1751). He framed several theorems and principles – notably **d'Alembert's principle** – in dynamics and celestial mechanics, and devised the theory of partial differential equations.

D'Alembert, Jean Le Rond http://www-history.mcs.st-andrews.ac.uk/history/Mathematicians/D'Alembert.html

Biography and portrait of the French physicist and mathematician famous for his works on the use of differential equations in physics. In addition to a brief history of d'Alembert's work there are links to lengthy descriptions of his fields of expertise, and also a short publications list for use as a reference.

alexandrite rare gemstone variety of the mineral chrysoberyl (beryllium aluminum oxide, $BeAl_2O_4$), which is green in daylight but appears red in artificial light.

Alferov, Zhores Ivanovich (1930–) Russian physicist who with German physicist Herbert †Kroemer and US electrical engineer Jack S †Kilby shared the Nobel Prize for Physics in 2000 for their contributions to the field of information and communication technology. Alferov played a leading role in the development of †semiconductor heterostructures used in high-speed electronics applications and †optoelectronics (the branch of electronics concerned with the development of devices that respond to both electrons and photons).

Alfvén, Hannes Olof Gösta (1908–1995) Swedish astrophysicist. He shared the Nobel Prize for Physics in 1970 for his fundamental contributions to †plasma physics, particularly in the field of magnetohydrodynamics (MHD) – the study of plasmas in magnetic fields.

algebra branch of mathematics in which the general properties of numbers are studied by using symbols, usually letters, to represent variables and unknown quantities. For example, the algebraic statement:

$(x + y)^2 = x^2 + 2xy + y^2$

is true for all values of x and y. For instance, the †substitution $x = 7$ and $y = 3$ gives:

$(7 + 3)^2 = 7^2 + 2(7 \times 3) + 3^2 = 100$

An algebraic †expression that has one or more variables (denoted by letters) is a †polynomial equation. A polynomial equation has the form:

$f(x) = a_n x^n + a_{n-1} x^{n-1} + \ldots + a_2 x^2 + a_1 x + a_0$

where a_n, a_{n-1}, …, a_0 are all constants, n is a positive integer, and a_n x 0. Examples of polynomials are:

$f(x) = 3x^4 + 2x^2 + 1$

or

$f(x) = x^5 - 18x + 71$

or

$f(x) = 2x + 3$

Algebra is used in many areas of mathematics – for example, †arithmetic progressions, or number sequences, and Boolean algebra (the latter is used in working out the logic for computers).

In ordinary algebra the same operations are carried on as in arithmetic, but, as the symbols are capable of a more generalized and extended meaning than the figures used in arithmetic, it facilitates calculation where the numerical values are not known, or are inconveniently large or small, or where it is desirable to keep them in an analyzed form.

For example, the following table shows the cost of gas for heating:

therms used	cost (£)
10	20.00
20	25.00
30	30.00
40	35.00
50	40.00

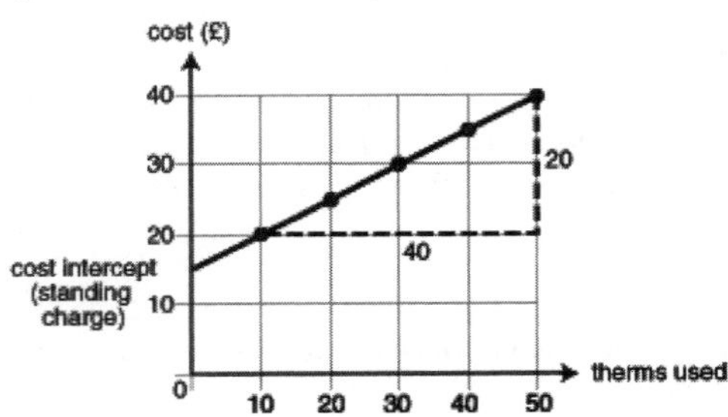

There is a connecting rule between the cost and the number of therms used.

Gradient = change in cost/change in therms:

= 40 – 20/50 – 10 = 20/40 = £0.5 per therm

Cost intercept = £15 (the intercept is the standing charge).

Since this is a straight line †graph, a †linear equation connecting the cost and therms used can be created:

cost = 0.5 therms + 15 or c = 0.5 t + 15

A straight line graph can be represented by the general formula:

$y = mx + c$

where c is the y-intercept, m is the gradient, and (x,y) are the points on the line.

order of calculation The †simplification of an algebraic †equation or expression must be completed in a set order. The procedure follows the rules of †BODMAS – any elements in **b**rackets should always be calculated first, followed by power **o**f (or index), **d**ivision, **m**ultiplication, **a**ddition, and **s**ubtraction. For example, to solve the equation:

$3(2x - x - 1) = 2(x + 3 + 4)$

collect the like terms and work out the brackets:

$3(x - 1) = 2(x + 7)$

multiply out the brackets:

$3x - 3 = 2x + 14$

collect the xs on the left-hand side of the equation:

$3x - 3 - 2x = 14$

then solve for x:

$x - 3 = 14$ $x = 14 + 3$ $x = 17$

Inequations or †inequalities may be solved using similar rules. When multiplying or dividing by a negative value, however, the direction of the inequality must be reversed, for example: $-x > 5$ is equivalent to $x < -5$.

quadratic equation A †quadratic equation is a polynomial equation of second degree (that is, an equation containing as its highest power the square of a variable, such as x^2). The general †formula of such equations is:

$ax^2 + bx + c = 0$

in which the †coefficients a, b, and c are real numbers, and only the coefficient a cannot equal 0. Some quadratic equations can be solved by factorization (see †factor (algebra)), or the values of x can be found by using the formula for the general solution.

$x = [-b + \sqrt{(b^2 - 4ac)}]/2a$ or

$x = [-b - \sqrt{(b^2 - 4ac)}]/2a$

Depending on the value of the discriminant $b^2 - 4ac$, a quadratic equation has two real, two equal, or two complex roots (solutions).

When $b^2 - 4ac > 0$, there are two distinct real roots.

When $b^2 - 4ac = 0$, there are two equal real roots.

When $b^2 - 4ac < 0$, there are two distinct complex roots.

simultaneous equations If there are two or more algebraic equations that contain two or more unknown quantities that may have a unique solution, they can be solved simultaneously as †simultaneous equations. For example, in the case of two linear equations with two unknown variables, such as:

(i) $3y + x = 6$ and

(ii) $3y - 2x = 6$

the solution will be those unique values of x and y that are valid for both equations. Linear simultaneous equations can be solved by using algebraic manipulation to eliminate one of the variables. For example, subtracting equation (ii) from equation (i) gives:

$3y - 3y + x + 2x = 6 - 6$

So $x = 0$, and substituting this value into (ii) gives:

$3y = 6$

So $y = 2$.

Another method is to rearrange (i) to give:

$x = 6 - 3y$

Substituting this into (ii) gives:

$3y - 2(6 - 3y) = 6$

Multiplying out the brackets gives:

$3y - 12 + 6y = 6$

So $9y = 18$, and $y = 2$.

"Algebra" was originally the name given to the study of equations. In the 9th century, the Arab mathematician Muhammad ibn-Musa al-†Khwarizmi used the term *al-jabr* for the process of adding equal quantities to both sides of an equation. When his treatise was later translated into Latin, *al-jabr* became "algebra" and the word was adopted as the name for the whole subject.

Algebra Help http://www.algebrahelp.com/

Great resource containing algebra lessons, beginning with the basics, then moving on to the more complex issue of simplifying equations. There are numerous worksheets that demonstrate the workings as well as giving the answers. In addition, there are some excellent online calculators that will solve equations, complete the square, and simplify or factorize equations for you.

Fundamental Theorem of Algebra
http://www-history.mcs.st-andrews.ac.uk/~history/HistTopics/Fund_theorem_of_algebra.html
Web site run by St Andrews University, Scotland, chronicling the history of the theorem of algebra. The site also provides biographical details on many of the main characters involved in its development.

algebraic curve geometrical curve that can be precisely described by an (algebraic) equation.

algebraic fraction in mathematics, fraction in which letters are used to represent numbers – for example, $^{a}/_{b}$, $^{xy}/_{z^2}$, and $^{1}/_{(x+y)}$. Like numerical fractions, algebraic fractions may be simplified or factorized. Two equivalent algebraic fractions can be cross-multiplied; for example, if
$a/b = c/d$
then $ad = bc$
(In the same way, the two equivalent numerical fractions $^2/_3$ and $^4/_6$ can be cross-multiplied to give cross-products that are both 12.)
factorization Algebraic fractions can be simplified by factorization, that is by taking out those factors that are common to both the numerator and denominator. For example, the algebraic fraction:
$(x^2 - 2x - 8)(x^2 + 5x + 4)/(x^2 - 16)(x + 1)$
can be simplified as follows:
$(x^2 - 2x - 8)$ factorizes as $(x - 4)(x + 2)$
$(x^2 + 5x + 4)$ factorizes as $(x + 4)(x + 1)$
$(x^2 - 16)$ factorizes as $(x - 4)(x + 4)$
so the $(x - 4)$, $(x + 4)$, and $(x + 1)$ from the denominator cancel out those in the numerator, leaving only $(x + 2)$.
addition and subtraction As with numerical fractions, to add or subtract algebraic fractions a common denominator must be found. The easiest way to do this is to multiply the denominators together. For example:
$3/x - 4/x + 2 = 3(x + 2) - 4x/x(x + 2)$
which can be simplified to:
$3x + 6 - 4x/x(x + 2) = 6 - x/x^2 + 2x$

algebraic numbers in mathematics, numbers that satisfy a polynomial equation with rational coefficients: for example, Ö2 solves $x^2 - 2 = 0$. †Real numbers that are not algebraic are called †transcendental numbers. Although there is an infinity of algebraic numbers, there is in fact a "larger" infinity of transcendental numbers.

algebraic topology in mathematics, study of surfaces and similar but more general objects in higher dimensions, using algebraic techniques. It is based upon †homology.

alginate in chemistry, salt of alginic acid, $(C_6H_8O_6)_n$, obtained from brown seaweeds and used in textiles, paper, food products, and pharmaceuticals.

ALGOL contraction of algorithmic language, in computing, an early high-level programming language, developed in the 1950s and 1960s for scientific applications. A general-purpose language, ALGOL is best suited to mathematical work and has an algebraic style. Although no longer in common use, it has greatly influenced more recent languages, such as Ada and Pascal.

Algol or **Beta Persei**, †eclipsing binary, a pair of orbiting stars in the constellation Perseus, one of which eclipses the other every 69 hours, causing its brightness to drop by two-thirds.
The brightness changes were first explained in 1782 by English amateur astronomer John Goodricke. He pointed out that the changes between magnitudes 2.2 and 3.5 repeated themselves exactly after an interval of 2.867 days and supposed this to be due to two stars orbiting round and eclipsing each other.

algorithm procedure or series of steps that can be used to solve a problem.
In computer science, it describes the logical sequence of operations to be performed by a program. A †flow chart is a visual representation of an algorithm.
The word derives from the name of 9th-century Arab mathematician Muhammad ibn-Musa al-†Khwarizmi .

Alhazen, Ibn al-Haytham (*c.* 965–1038) also known as **Abu Ali al-Hassan ibn al-Haytham**, Arabian scientist. He was the author of the *Kitab al-Manazir/Book of Optics*, translated into Latin as *Opticae thesaurus* (1572). For centuries it remained the most comprehensive and authoritative treatment of optics in both East and West.

alicyclic in chemistry, describing a cyclic carbon compound that does not have the conjugated double-bond structure characteristic of †aromatic ring compounds. The name is derived from aliphatic-cyclic. †Cyclohexane is an example of this class of compounds.

alimentary canal tube through which food passes in animals – it extends from the mouth to the anus and forms a large part of the digestive system. In human adults, it is about 9 m/30 ft long, consisting of the mouth cavity, pharynx, esophagus, stomach, and the small and large intestines. It is also known as the gut. It is a complex organ, specifically adapted for †digestion and the absorption of food. Enzymes from the wall of the canal and from other associated organs, such as the pancreas, speed up the digestive process.

The muscles in the wall of the alimentary canal contract, mixing food with the enzymes and slowly pushing it along in the direction of the anus in a process known as †peristalsis. Dietary †fiber encourages this movement. The constant stream of enzymes from the canal wall and from the pancreas assists the breakdown of food molecules into smaller, soluble nutrient molecules, which are absorbed through the canal wall into the bloodstream and carried to individual cells. The wall of the canal receives an excellent supply of blood and is folded so as to increase its surface area. These two adaptations ensure efficient absorption of nutrient molecules.

Alinda in astronomy, asteroid (member of the †Amor group) that was discovered in 1918 when it approached close to the Earth. It is 4 km/2.5 mi in diameter.

aliphatic compound any organic chemical compound in which the carbon atoms are joined in straight chains, as in hexane (C_6H_{14}), or in branched chains, as in 2-methylpentane ($CH_3CH(CH_3)CH_2CH_2CH_3$). Aliphatic compounds have bonding electrons localized within the vicinity of the bonded atoms. †Cyclic compounds that do not have delocalized electrons are also aliphatic, as in the alicyclic compound cyclohexane (C_6H_{12}) or the heterocyclic piperidine ($C_5H_{11}N$). Compare †aromatic compound.

alizarin or **1,2-dihydroxy-anthraquinone**; $C_6H_4(CO)_2C_6H_2(OH)_2$, derivative from anthraquinone. It is now prepared synthetically from anthracene. Alizarin crystallizes in dark red prisms and sublimes in orange-colored needles, melting at 290°C/554°F. It is almost insoluble in water, but dissolves in alcohol. It yields with metallic oxides magnificently colored insoluble compounds called "lakes," to which it owes its great value for dyeing purposes. Ferric oxide with alizarin gives a violet-black compound, chromium oxide a claret, calcium oxide a blue, and aluminum and tin give different shades of red.

alk resin obtained from the turpentine tree *Pistacia terebinthus*, which grows chiefly in the Mediterranean region. A yellow to green liquid, Chian or Chio turpentine, is distilled from it.

alkahest (or alcahest) hypothetical universal dissolvent of the alchemists, first mentioned by †Paracelsus. It was believed to be the means by which the quintessence of any substance could be extracted and confined within the philosopher's stone.

alkali in chemistry, a †base that is soluble in water. Alkalis neutralize acids, and solutions of alkalis are soapy to the touch. The strength of an alkali is measured by its hydrogen-ion concentration, indicated by the †pH value. They may be divided into strong and weak alkalis: a strong alkali (for example, potassium hydroxide, KOH) ionizes completely when dissolved in water, whereas a weak alkali (for example, ammonium hydroxide, NH_4OH) exists in a partially ionized state in solution. All alkalis have a pH above 7.0.

The hydroxides of metals are alkalis. Those of sodium and potassium are corrosive; both were historically derived from the ashes of plants.

The four main alkalis are sodium hydroxide (caustic soda, NaOH); potassium hydroxide (caustic potash, KOH); calcium hydroxide (slaked lime or limewater, $Ca(OH)_2$); and aqueous ammonia ($NH_{3(aq)}$). Their solutions all contain the hydroxide ion OH^-, which gives them a characteristic set of properties.

alkali metal any of a group of six metallic elements with similar chemical properties: †lithium, †sodium, †potassium, †rubidium, †cesium, and †francium. They form a linked group (Group 1) in the †periodic table of the elements. They each have a valency of one and have very low densities (lithium, sodium, and potassium float on water); in general they are reactive, soft, low-melting-point metals. Because of their reactivity they are only found as compounds in nature.

alkaline-earth metal any of a group of six metallic elements with similar bonding properties: beryllium, magnesium, calcium, strontium, barium, and radium. They form a linked group in the †periodic table of the elements. They are strongly basic, bivalent (have a valency of two), and occur in nature only in compounds.

alkaloid any of a number of physiologically active and frequently poisonous substances contained in some plants. They are usually organic bases and contain nitrogen. They form salts with acids and, when soluble, give alkaline solutions.

Substances in this group are included by custom rather than by scientific rules. Examples include morphine, cocaine, quinine, caffeine, strychnine, nicotine, and atropine.

In 1992, epibatidine, a chemical extracted from the skin of an Ecuadorian frog, was identified as a member of an entirely new class of alkaloid. It is an organochlorine compound, which is rarely found in animals, and a powerful painkiller, about 200 times as effective as morphine.

alkane member of a group of †hydrocarbons having the general formula C_nH_{2n+2}, commonly known as **paraffins**. As they contain only single †covalent bonds, alkanes are said to be saturated. Lighter alkanes, such as methane, ethane, propane, and butane, are colorless gases; heavier ones are liquids or solids. In nature they are found in natural gas and petroleum.

Properties and Reactions of Alkanes
http://mychemistrypage.future.easyspace.com/Organic/Alkanes_and_Alkenes/Properties_of_Alkanes.htm
Part of a larger comprehensive site on physical, inorganic, and organic chemistry, this page details the properties of the chemical group alkanes and includes useful graphs and tables.

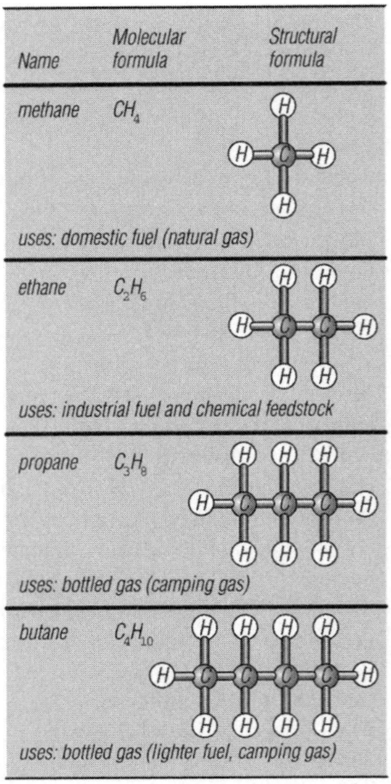

Name	Molecular formula	Structural formula
methane	CH_4	
uses: domestic fuel (natural gas)		
ethane	C_2H_6	
uses: industrial fuel and chemical feedstock		
propane	C_3H_8	
uses: bottled gas (camping gas)		
butane	C_4H_{10}	
uses: bottled gas (lighter fuel, camping gas)		

alkane The lighter alkanes methane, ethane, propane, and butane, showing the aliphatic chains, where a hydrogen atom bonds to a carbon atom at all available sites. Helicon

alkene member of the group of †hydrocarbons having the general formula C_nH_{2n}, formerly known as **olefins**. Alkenes are unsaturated compounds, characterized by one or more double bonds between adjacent carbon atoms. Lighter alkenes, such as †ethene and propene, are gases, obtained from the †cracking of oil fractions. Alkenes react by addition, and many useful compounds, such as †polythene and bromoethane, are made from them.

alkyl any organic radical of the formula C_nH_{2n+1}; the chief members are methyl, ethyl, propyl, butyl, and amyl. These radicals are not stable in the free state but are found combined in a large number of types of organic compounds such as alcohols, esters, aldehydes, ketones, and halides.

alkylation in chemistry, acid-catalyzed reaction that is used to introduce †alkyl groups into an organic molecule. The reaction is especially important to the petrochemical industry for the conversion of alkene gases, such as †propene, into heavier hydrocarbon components of gasoline.

alkyne member of the group of †hydrocarbons with the general formula C_nH_{2n-2}, formerly known as the **acetylenes**. They are unsaturated compounds, characterized by one or more triple bonds between adjacent carbon atoms. Lighter alkynes, such as ethyne, are gases; heavier ones are liquids or solids.

ethene propene butene

— single bond
= double bond
H hydrogen
C carbon

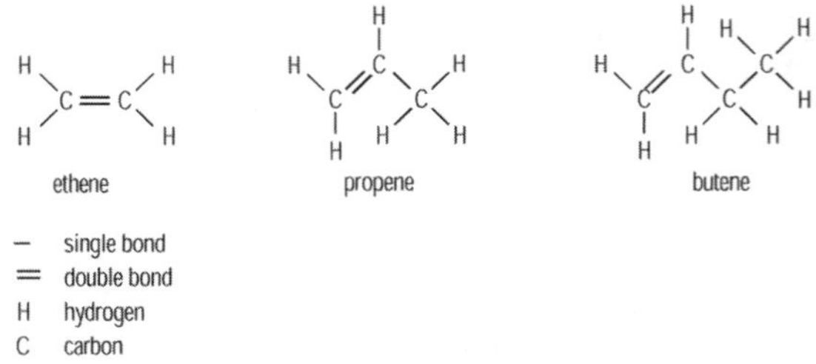

alkene The alkenes ethene (C_2H_4), propene ($CH_3CH{=}CH_2$), and butene (C_4H_8). Alkenes all have the general formula C_nH_{2n}. Helicon

allantois bladder in the embryo of reptiles, birds, and mammals, that grows outside the embryo into the wall of the yolk sac of reptiles and birds and under the †chorion of mammals. In mammals, blood vessels in the allantois carry blood to the †placenta; in reptiles and birds the blood vessels permit respiration. As they develop, the vessels become the umbilical vein and arteries.

allele one of two or more alternative forms of a †gene at a given position (locus) on a †chromosome, caused by a difference in the sequence of †DNA. This is best explained with examples. A gene that controls eye color in humans may have two alternative forms – an allele that can produce blue eyes, and an allele that produces brown eyes. In a plant that occurs in tall and short forms, there may be an allele that tends to produce tall plants and an alternative allele that produces short plants.

The individual genes that form a pair of alleles are located at exactly the same point along a chromosome. Organisms with two sets of chromosomes (diploids), such as animals and plants, have chromosomes that are found as matching pairs in the †nucleus of each cell. This means that there will always be two genes for a characteristic in a cell. If the same allele is present twice, the organism is said to be †homozygous for this characteristic. If, however, one chromosome contains one allele and the other chromosome a contrasting allele, the organism is said to be †heterozygous.

In a heterozygous organism the appearance of the organism (phenotype) may be determined by one allele and not the other. The allele that determines the phenotype is said to be dominantly expressed; it shows †dominance over other alleles. The expression of the other allele is described as being recessive.

allene any of a class of dienes (hydrocarbons with two double bonds) with adjacent double bonds. The simplest example is $CH_2=C=CH_2$, allene itself. Because of the stereochemistry of the double bonds, the terminal hydrogen atoms lie in planes mutually at right angles. Allenes behave mainly as typical unsaturated compounds, but are less stable than dienes with nonadjacent double bonds.

allergy special sensitivity of the body that makes it react with an exaggerated response of the natural immune defense mechanism to the introduction of an otherwise harmless foreign substance (**allergen**).

allogamy or **cross-fertilization**, (Greek *allos* "other," *gamos* "marriage") transfer of the pollen of one flower to the †pistil of another in the fertilization of flowering plants.

allometry in biology, a regular relationship between a given feature (for example, the size of an organ) and the size of the body as a whole, when this relationship is not a simple proportion of body size. Thus, an organ may increase in size proportionately faster, or slower, than body size does. For example, a human baby's head is much larger in relation to its body than is an adult's.

allopatric in biology, describing or relating to groups within a species population whose gene flow has been interrupted by a climatic, geographical, or geological barrier. Before separation, the species could interbreed but once divided, the sub-groups diverge genetically as each is subject to slightly different selective pressures compared with its neighbor. As a result of the separation the different subgroups evolve over time into new species (†speciation).

allosteric effect regulatory effect on an enzyme that takes place at a site distinct from that enzyme's catalytic site. For example, in a chain of enzymes the end product may act on an enzyme in the chain to regulate its own production.

Allosteric effects occur because effector molecules are able to bring about conformational changes within the enzyme. This may lead to disruption of the active site, the inability of the substrate molecule (the molecule undergoing change) to bind, or the inability of the products of the reaction to be released.

allotropy property whereby an element can exist in two or more forms (allotropes), each possessing different physical properties but the same state of matter (gas, liquid, or solid). The allotropes of carbon are diamond, fullerene, and graphite. Sulfur has several allotropes (flowers of sulfur, plastic, rhombic, and monoclinic). These solids have different crystal structures, as do the white and gray forms of tin and the black, red, and white forms of phosphorus.

Oxygen exists as two gaseous allotropes: one used by organisms for respiration (O_2), and the other a poisonous pollutant, ozone (O_3).

alloy metal blended with some other metallic or non-metallic substance to give it special qualities, such as resistance to corrosion, greater hardness, or tensile strength. The atoms in a †metal are held together by the †metallic bond. In a pure metal the atoms are all the same size and can slip over each other if a force is applied. In an alloy, the presence of different sized atoms prevents such dislocations from weakening the metal. Useful alloys include bronze, brass, cupronickel, duralumin, German silver, gunmetal, pewter, solder, steel, and stainless steel.

Among the oldest alloys is bronze (mainly an alloy of copper and tin), the widespread use of which ushered in the Bronze Age. Complex alloys are now common; for example, in dentistry, where a cheaper alternative to gold is made of chromium, cobalt, molybdenum, and titanium. Among the most recent alloys are superplastics: alloys that can stretch to double their length at specific temperatures, permitting, for example, their injection into moulds as easily as plastic.

alluvial deposit layer of broken rocky matter, or sediment, formed from material that has been carried in suspension by a river or stream and dropped as the velocity of the current decreases. River plains and deltas are made entirely of alluvial deposits, but smaller pockets can be found in the beds of upland torrents.

Alluvial deposits can consist of a whole range of particle sizes, from boulders down through cobbles, pebbles, gravel, sand, silt, and clay. The raw materials are the rocks and soils of upland areas that are loosened by erosion and washed away by mountain streams. Much of the world's richest farmland lies on alluvial deposits. These deposits can also provide an economic source of minerals. River currents produce a sorting action, with particles of heavy material deposited first while lighter materials are washed downstream. Hence heavy minerals such as gold and tin, present in the original rocks in small amounts, can be concentrated and deposited on stream beds in commercial quantities. Such deposits are called "placer ores."

alluvial fan roughly triangular sedimentary formation found at the base of slopes. An alluvial fan results when a sediment-laden stream or river rapidly deposits its load of gravel and silt as its speed is reduced on entering a plain.

The surface of such a fan slopes outward in a wide arc from an apex at the mouth of the steep valley. A small stream carrying a load of coarse particles builds a shorter, steeper fan than a large stream carrying a load of fine particles. Over time, the fan tends to become destroyed piecemeal by the continuing headward and downward erosion leveling the slope.

alluvium sediments laid down by streams and rivers. The most common constituents are clay, silt, and gravel. The loose, unconsolidated material forms features such as river terraces, †flood plains, and †deltas.

Alluvium is deposited along the river channel where the water's velocity is too low to transport the river's load – for example, on the inside bend of a †meander.

allyl in chemistry, the propenyl group, an unsaturated organic radical corresponding to the formula CH_2:$CHCH_2$–.

The chief compounds are allyl alcohol (propenol), with the general properties of a primary alcohol; allyl iodide, a colorless liquid with an odor of garlic; allyl bromide, a heavy liquid obtained by treating allyl alcohol with phosphorus tribromide; allyl sulfide, or "oil of garlic," obtained by macerating garlic; and allyl isothiocyanate, or "oil of mustard," occurring in black mustard seeds.

allyl-metal complex coordination compound of a metal and the organic allyl (CH_2:$CHCH_2$–) group. Allyl-metal complexes are very reactive to oxygen, inflaming in air. The allyl groups are easily replaced by other ligands such as the carbonyl group, CO.

There are two principal types: the sigma complex, in which the allyl group is directly and singly bonded to the metal, as in any of the organometallic compounds, for example in a manganese complex, CH_2:$CHCH_2$–$Mn(CO)_5$; and the pi-complex, where delocalization of the pi-electrons in the allyl system gives an allyl ligand which replaces two monodentate ligands on the metal (for example $CH_2CHCH_2Mn(CO)_4$).

almucantar small circle on the †celestial sphere parallel to the horizon; a circle of altitude.

Alpha Centauri brightest star in the constellation Centaurus; see †Rigil Kent.

alpha chain particular secondary structure of the polypeptide chain (of a protein) brought about by hydrogen bonding between adjacent †peptide units. Alpha chains occur, for example, in the †hemoglobin molecule.

alpha decay spontaneous alteration of the nucleus of a radioactive atom, which transmutes the atom from one atomic number to another through the emission of a helium nucleus (known as an †alpha particle). As a result, the atomic number decreases by two and the atomic weight decreases by four. See also †radioactivity.

alphanumeric data data made up of any of the letters of the alphabet and any digit from 0 to 9. The classification of data according to the type or types of character contained enables computer validation systems to check the accuracy of data: a computer can be programmed to reject entries that contain the wrong type of character. For example, a person's name would be rejected if it contained any numeric data, and a bank-account number would be rejected if it contained any alphabetic data. A car's registration number, by comparison, would be expected to contain alphanumeric data but no punctuation marks.

alpha particle or **alpha ray**, positively charged (2+), high-energy particle emitted from the nucleus of a radioactive atom. It is one of the products of the spontaneous disintegration of radioactive elements (see †radioactivity) such as radium and thorium, and is identical to the nucleus of a helium atom (^{4}He) – that is, it consists of two protons and two neutrons. The process of emission, **alpha decay**, transforms one element into another, decreasing the atomic number by two and the atomic mass by four. Plutonium-239 (^{239}Pu) is an example of a material that emits alpha particles.

Because of their large mass, alpha particles have a short range of only a few centimeters in air. They have a low penetrating power and can be stopped by a sheet of paper or aluminum. They have a strongly ionizing effect (see †ionizing radiation) on the molecules that they strike, and are therefore capable of damaging living cells. Alpha particles traveling in a vacuum are deflected slightly by magnetic and electric fields.

alpha ray alternative name for †alpha particle.

Alpher, Ralph Asher (1921–) US scientist who carried out the first quantitative work on nucleosynthesis and in 1948 was the first to predict the existence of primordial †background radiation, which is now regarded as one of the major pieces of evidence for the validity of the †Big Bang model of the universe.

al-Sufi Persian astronomer; see †Sufi, al-.

Altair or **Alpha Aquilae**, brightest star in the constellation Aquila and the 13th-brightest star in the night sky. It is a white star about 16 light years away from the Sun and forms the Summer Triangle with the stars Deneb (in the constellation Cygnus) and Vega (in Lyra).

altazimuth astronomical instrument designed for observing the altitude and †azimuth of a celestial object. It is essentially a large precision †theodolite that is an instrument for the measurement of horizontal and vertical angles.

Alter, David (1807–1881) US inventor and physicist. He put forward in 1854 the idea that each element has a characteristic spectrum, and that spectroscopic analysis of a substance can therefore be used to identify the elements present.

alternate angles pair of angles that lie on opposite sides and at opposite ends of a transversal (a line that cuts two or more lines in the same plane). The alternate angles formed by a transversal of two parallel lines are equal.

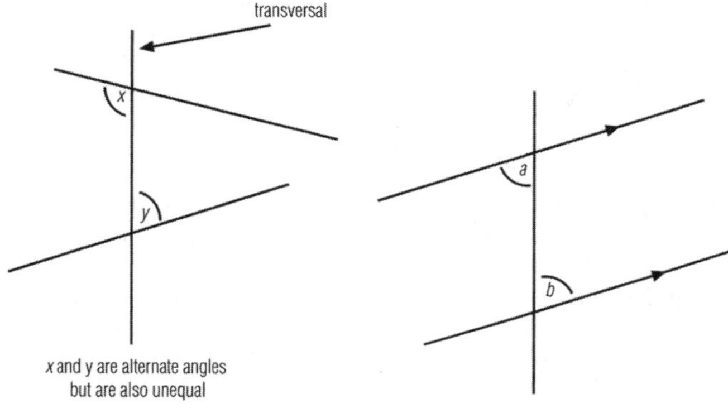

alternate angles Where a transversal (a line that cuts two or more lines in the same plane) cuts through a pair of parallel lines, the alternate angles a and b are equal. Helicon

alternating current (AC), electric current that flows for an interval of time in one direction and then in the opposite direction; that is, a current that flows in alternately reversed directions through or around a circuit. Electric energy is usually generated as alternating current in a power station, and alternating currents may be used for both power and lighting.

If half an amp moves in one direction past a point in half a second and in the next half a second half an amp reverses and moves past the same point in the opposite direction, then in total one coulomb of electrons or one amp has passed the point in one second. The current flow is 1 amp AC.

The advantage of alternating current over direct current (DC), as from a battery, is that its voltage can be raised or lowered economically by a transformer; high voltage for generation and transmission, and low voltage for safe utilization. Railways, factories, and domestic appliances, for example, use alternating current.

alternation of generations typical life cycle of terrestrial plants and some seaweeds, in which there are two distinct forms occurring alternately: **diploid** (having two sets of chromosomes) and **haploid** (one set of chromosomes). The diploid generation produces haploid spores by †meiosis, and is called the sporophyte, while the haploid generation produces gametes (sex cells), and is called the gametophyte. The gametes fuse to form a diploid †zygote, which develops into a new sporophyte; thus the sporophyte and gametophyte alternate.

alternative energy see †energy, alternative.

alternator electricity †generator that produces an alternating current.

altimetry process of measuring altitude, or elevation. Satellite altimetry involves using an instrument – commonly a laser – to measure the distance between the satellite and the ground.

altitude or **elevation,** in astronomy, angular distance of an object above the horizon, ranging from 0° on the horizon to 90° at the zenith. Together with the †azimuth, it forms a system of horizontal coordinates for specifying the positions of celestial bodies.

altitude in geometry, the perpendicular distance from a †vertex (corner) of a figure, such as a triangle, to the base (the side opposite the vertex).

altitude measurement of height, usually given in meters (or feet) above sea level.

altruism in biology, helping another individual of the same species to reproduce more effectively, as a direct result of which the altruist may leave fewer offspring itself. Female honey bees (workers) behave altruistically by rearing sisters in order to help their mother, the queen bee, reproduce, and forgo any possibility of reproducing themselves.

ALU abbreviation for †arithmetic and logic unit.

alula in zoology, group of small feathers on the outer front edge of the wing of a bird. Each alula acts as a slot through which air rushes, keeping the airflow over each wing free of turbulence. As it climbs, a bird spreads the alula, preventing stalling.

alum any double sulfate of a monovalent metal or radical (such as sodium, potassium, or ammonium) and a trivalent metal (such as aluminum, chromium, or iron). The commonest alum is the double sulfate of potassium and aluminum, $K_2Al_2(SO_4)_4.24H_2O$, a white crystalline powder that is readily soluble in water. It is used in curing animal skins. Other alums are used in papermaking and to fix dyes in the textile industry.

alumina or **corundum**; Al_2O_3, oxide of aluminum, widely distributed in clays, slates, and shales. It is formed by the decomposition of the feldspars in granite and used as an abrasive. Typically it is a white powder, soluble in most strong acids or caustic alkalis but not in water. Impure alumina is called "emery." Rubies, sapphires, and topaz are corundum gemstones. It is the chief component of †bauxite.

aluminate in chemistry, any salt of aluminic acid, H_3AlO_3, a weak acid that is a tautomeric (see †tautomerism) form of †aluminum hydroxide $Al(OH)_3$.

aluminum chemical symbol Al, lightweight, silver-white, ductile and malleable, metallic element, atomic number 13, relative atomic mass 26.9815, melting point 658°C/1,216°F. It is the third most abundant element (and the most abundant metal) in the Earth's crust, of which it makes up about 8.1% by mass. It is nonmagnetic, an excellent conductor of electricity, and oxidizes easily, the layer of oxide on its surface making it highly resistant to tarnish.

pure aluminum Aluminum is a reactive element which forms stable compounds, so a great deal of energy is needed in order to separate aluminum from its ores, and the pure metal was not readily obtainable until the middle of the 19th century. Commercially, it is prepared by the electrolysis of alumina (aluminum oxide), which is obtained from the ore †bauxite. In its pure state aluminum is a weak metal, but when combined with elements such as copper, silicon, or magnesium it forms alloys of great strength.

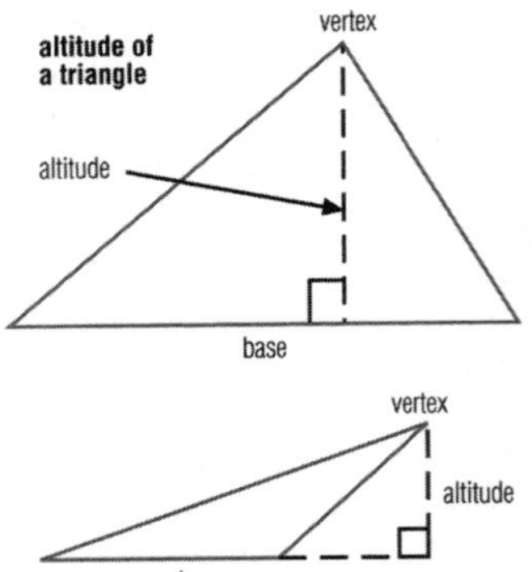

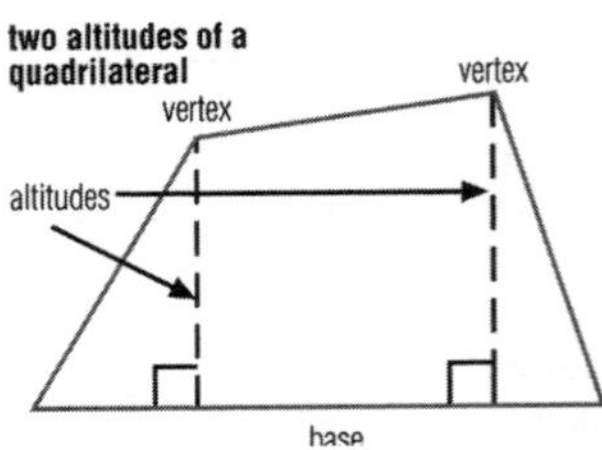

altitude The altitude of a figure is the perpendicular distance from a vertex (corner) to the base (the side opposite the vertex). Helicon

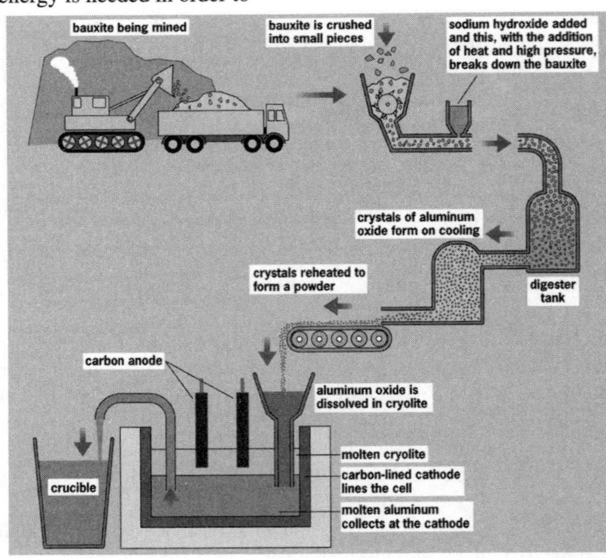

aluminum The extraction of aluminum is a two-stage process. In the first stage bauxite is crushed to extract aluminum oxide (alumina). In the second stage pure aluminum is extracted from the alumina by electrolysis. Helicon

aluminum chloride $AlCl_3$, white solid made by direct combination of aluminum and chlorine.

$$2Al + 3Cl_2 \rightarrow 2AlCl_3$$

The anhydrous form is a typical covalent compound.

aluminum hydroxide or **alumina cream**; $Al(OH)_3$, gelatinous precipitate formed when a small amount of alkali solution is added to a solution of an aluminum salt.

$$Al_{(aq)} + 3OH_{(aq)} \rightarrow Al(OH)_{3(s)}$$

It is an †amphoteric compound as it readily reacts with both acids and alkalis.

aluminum ore raw material from which aluminum is extracted. The main ore is †bauxite, a mixture of minerals, found in economic quantities in Australia, Guinea, the West Indies, and several other countries.

aluminum oxide or **alumina**; Al_2O_3, white solid formed by heating aluminum hydroxide. It is an †amphoteric oxide, since it reacts readily with both acids and alkalis, and it is used as a refractory (furnace lining) and in column †chromatography.

Alvarez, Luis Walter (1911–1988) US physicist. He led the research team that discovered the X_0 subatomic particle in 1959. He also made many other breakthroughs in fundamental physics, †accelerators, and radar. He worked on the US atom bomb for two years, at Chicago and at Los Alamos, New Mexico, during World War II. He was awarded the Nobel Prize for Physics in 1968 for his work in elementary-particle physics, and discovery of resonance states, using the hydrogen bubble chamber and data analysis.

alveolus plural **alveoli**, one of the many thousands of tiny air sacs in the †lungs. By the process of †diffusion †oxygen is transported from air in the alveoli into the blood, while †carbon dioxide diffuses out of the blood and into the air in the alveoli. This process is called gas exchange. To aid this process, the alveoli have very thin walls, which are folded in order to increase their surface area and allow rapid gas exchange, and are surrounded by blood capillaries. The air in the alveoli is replenished as a result of †breathing or ventilation. Smoking and exposure to other pollutants can cause the loss of the folds in the alveoli, which causes difficulties in breathing and the condition called emphysema.

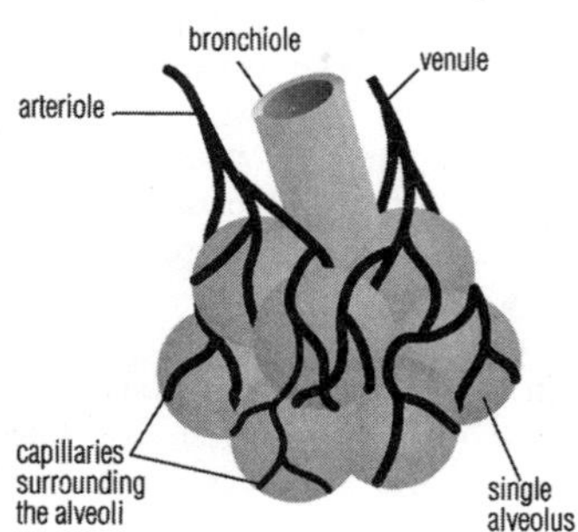

alveolus The tiny air sacs, called alveoli, in the lungs are covered with a network of blood capillaries, allowing oxygen to enter the blood. The alveoli of an adult have a total surface area of 70 sq m/750 sq ft. Helicon

AM in physics, abbreviation for †amplitude modulation.

amalgam any alloy of mercury with other metals. Most metals will form amalgams, except iron and platinum. Amalgam is used in dentistry for filling teeth, and usually contains copper, silver, and zinc as the main alloying ingredients. This amalgam is pliable when first mixed and then sets hard, but the mercury leaches out and may cause a type of heavy-metal poisoning.

Amalgamation, the process of forming an amalgam, is a technique sometimes used to extract gold and silver from their ores. The ores are ground to a fine sand and brought into contact with mercury, which dissolves the gold and silver particles. The amalgam is then heated to distil off the mercury, leaving a residue of silver and gold. The mercury is recovered and reused.

Amalgamation to extract gold from its ore has been in use since Roman times.

Amalthea in astronomy, small moon of Jupiter that orbits close to the surface once every 12 hours. It is ellipsoidal in shape with the longest dimension (270 km/168 mi) oriented towards Jupiter.

amatol explosive consisting of ammonium nitrate and TNT (trinitrotoluene) in almost any proportions.

americium chemical symbol Am, radioactive metallic element of the †actinide series, atomic number 95, relative atomic mass 243.13; it was first synthesized in 1944. It occurs in nature in minute quantities in †pitchblende and other uranium ores, where it is produced from the decay of neutron-bombarded plutonium, and is the element with the highest atomic number that occurs in nature. It is synthesized in quantity only in nuclear reactors by the bombardment of plutonium with neutrons. Its longest-lived isotope is Am-243, with a half-life of 7,650 years.

The element was named by US nuclear chemist Glenn †Seaborg, one of the team who first synthesized it. Ten isotopes are known.

Ames, Adelbert (1880–1955) US scientist who studied optics and the psychology of visual perception. He concluded that much of what a person sees depends on what he or she expects to see, based (consciously or unconsciously) on previous experience.

amethyst variety of †quartz, SiO_2, colored violet by the presence of small quantities of impurities such as manganese or iron; used as a semi-precious stone. Amethysts are found chiefly in the Ural Mountains, India, the USA, Uruguay, and Brazil.

amicable number in mathematics, one of a pair of numbers, *m* and *n*, if *m* is the sum of the proper numbers that can be divided into *n*, that is *n*'s divisors, and vice versa.

amide any organic chemical derived from a fatty acid by the replacement of the hydroxyl group (—OH) by an amino group (—NH_2).

One of the simplest amides is ethanamide (acetamide, CH_3CONH_2), which has a strong odor.

amine any of a class of organic chemical compounds in which one or more of the hydrogen atoms of ammonia (NH_3) have been replaced by other groups of atoms.

Methyl amines have unpleasant ammonia odors and occur in decomposing fish. They are all gases at ordinary temperature.

Aromatic amine compounds include aniline, which is used in dyeing.

amino acid water-soluble organic †molecule, mainly composed of carbon, oxygen, hydrogen, and nitrogen, containing both a basic amino group (NH_2) and an acidic carboxyl (COOH) group. They are small molecules able to pass through membranes. When two or more amino acids are joined together, they are known as †peptides; †proteins are made up of peptide chains folded or twisted in characteristic shapes.

Many different proteins are found in the cells of living organisms, but they are all made up of the same 20 amino acids, joined together in varying combinations (although other types of amino acid do occur infrequently in nature). Eight of these, the **essential amino acids**, cannot be synthesized by humans and must be obtained from the diet. Children need a further two amino acids that are not essential for adults. Other animals also need some preformed amino acids in their diet, but green plants can manufacture all the amino acids they need from simpler molecules, relying on energy from the Sun and minerals (including nitrates) from the soil.

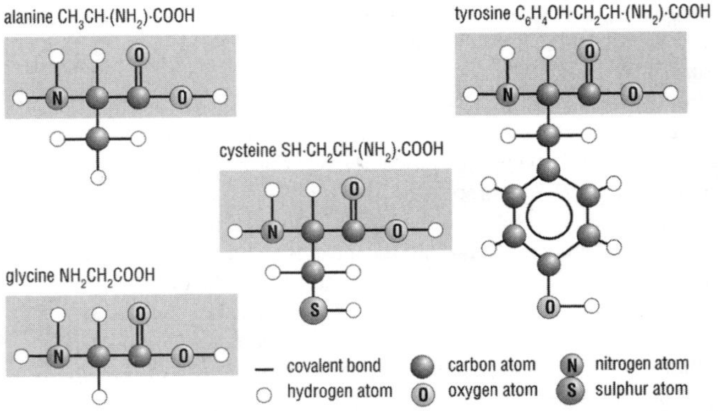

alanine $CH_3CH{\cdot}(NH_2){\cdot}COOH$

tyrosine $C_6H_4OH{\cdot}CH_2CH{\cdot}(NH_2){\cdot}COOH$

cysteine $SH{\cdot}CH_2CH{\cdot}(NH_2){\cdot}COOH$

glycine NH_2CH_2COOH

— covalent bond ○ hydrogen atom ● carbon atom (O) oxygen atom (N) nitrogen atom (S) sulphur atom

amino acid Amino acids are natural organic compounds that make up proteins and can thus be considered the basic molecules of life. There are 20 different common amino acids. They consist mainly of carbon, oxygen, hydrogen, and nitrogen. Each amino acid has a common core structure (consisting of two carbon atoms, two oxygen atoms, a nitrogen atom, and four hydrogen atoms) to which is attached a variable group, known as the R group. In glycine, the R group is a single hydrogen atom; in alanine, the R group consists of a carbon and three hydrogen atoms (methyl group). Helicon

Amino Acids http://www.chemie.fu-berlin.de/chemistry/bio/amino-acids_en.html

Small but interesting site giving the names and chemical structures of all the common amino acids. The information is available in both English and German.

Molecular Expressions: The Amino Acid Collection http://micro.magnet.fsu.edu/aminoacids/index.html

Fascinating collection of images showing what all the known amino acids look like when photographed through a microscope. There is also a detailed article about the different amino acids.

amino-acid racemization used in archaeological dating, a rotation process converting L-isomer molecular structures in the †amino acids of living tissue to D-isomers after death. These two structures, one the mirror image of the other, are distinguishable by their chemical behavior. The concentration ratio of L-isomers to D-isomers in material to be dated may be calibrated against bone samples that have been radiocarbon dated, but the method is still experimental.

The rate at which racemization proceeds depends upon the amino acid involved and the material in which it survives. In fossil bone, isoleucine is the amino acid examined for dating samples over a million years old, and aspartic acid for samples thought to be 0.1–0.3 million years old. In wood, proline and hydroxyproline are examined for these time periods.

amino group in chemistry, group consisting of one nitrogen and two hydrogen atoms, NH_2. It is an essential component of all †amino acids, being joined directly to a carbon atom in the molecule.

ammeter instrument that measures electric current (flow of charge per unit time), usually in †amperes, through a conductor. It should not to be confused with a †voltmeter, which measures potential difference between two points in a circuit. The ammeter is placed in series (see †series circuit) with the component through which current is to be measured, and is constructed with a low internal resistance in order to prevent the reduction of that current as it flows through the instrument itself. A common type is the †moving-coil meter, which measures direct current (DC), but can, in the presence of a rectifier (a device which converts alternating current to direct current), measure alternating current (AC) also. Hot-wire, moving-iron, and dynamometer ammeters can be used for both DC and AC.

ammonia NH_3, colorless pungent-smelling gas, lighter than air and very soluble in water. It is made on an industrial scale by the †Haber (or Haber–Bosch) process, and used mainly to produce nitrogenous fertilizers, nitric acid, and some explosives.

In aquatic organisms and some insects, nitrogenous waste (from the breakdown of amino acids) is excreted in the form of ammonia, rather than as urea in mammals.

ammoniacal solution in chemistry, a solution produced by dissolving a solute in aqueous ammonia.

ammonium bicarbonate or **ammonium hydrogen carbonate**; NH_4HCO_3, in chemistry, white crystalline powder. It can be formed by the decomposition of ammonium carbonate and is used as a component in †baking powder.

ammonium carbamate NH_2–CO–ONH_3, in chemistry, ammonium salt of carbamic acid (NH_2–CO–OH). It can be formed by a reaction between carbon dioxide and ammonia in dry conditions. It is present in most commercially produced ammonium carbonate.

ammonium carbonate $(NH_4)_2CO_3$, white, crystalline solid that readily sublimes at room temperature into its constituent gases: ammonia, carbon dioxide, and water. It was formerly known as sal volatile acid and used in †smelling salts.

ammonium chloride or **sal ammoniac**; NH_4Cl, volatile salt that forms white crystals around volcanic craters. It is prepared synthetically for use in "dry-cell" batteries, fertilizers, and dyes.

ammonium nitrate NH_4NO_3, colorless, crystalline solid, prepared by †neutralization of nitric acid with ammonia; the salt is crystallized from the solution. It sublimes on heating.

amnion innermost of three membranes that enclose the embryo within the egg (reptiles and birds) or within the uterus (mammals). It contains the amniotic fluid that helps to cushion the embryo.

amoeba plural **amoebae**, one of the simplest living animals, consisting of a single cell and belonging to the †protozoa group. The body consists of colorless protoplasm. Its activities are controlled by the nucleus, and it feeds by flowing round and engulfing organic debris. It reproduces by †binary fission. Some species of amoeba are harmful parasites.

Amor in astronomy, asteroid that orbits inside the orbit of Mars and can make close approaches to the Earth. It was discovered in 1932 and is 1 km/0.6 mi in diameter.

amorphous in physics, describing any solid that is not crystalline, for example glass. Many apparently amorphous substances are actually made up of many tiny crystals.

amount of substance symbol *n*, in physics, measure of the number of elementary entities (such as atoms, ions, or molecules) in a substance, for which the SI unit is the †mole.

amp in physics, abbreviation for †**ampere**, a unit of electrical current.

ampere symbol A, SI unit of electrical current. Electrical current (a flow of negative charge) is measured in a similar way to water current, in terms of an amount per unit time; one ampere (amp) represents a flow of one coulomb per second, which is about 6.28×10^{18} †electrons per second.

When a charge (Q) of one coulomb is flowing at a given point in a material in a time (t) of one second, a current (I) of one ampere is flowing. This can be represented as a formula: $Q = I \times t$. The instrument used to measure the flow of current through a conductor is called an †ammeter.

The ampere is defined as the current that produces a specific magnetic force (2×10^{-7} newtons per meter) between two long, straight, parallel conductors placed 1 m/3.3 ft apart in a vacuum. It is named after the French scientist André †Ampère.

Ampère, André-Marie (1775–1836) French physicist and mathematician who made many discoveries in electromagnetism and electrodynamics. He followed up the work of Hans †Oersted on the interaction between magnets and electric currents, developing a rule for determining the direction of the magnetic field associated with an electric current. The unit of electric current, the ampere, is named after him.

Ampère, André http://www-history.mcs.st-andrews.ac.uk/Mathematicians/Ampere.html

Biography and portrait of the famous French physicist and mathematician. This Web site also contains a brief history of Ampère's work and a list of 24 publications that may be used as a reference source.

Ampère's rule rule developed by French physicist André-Marie Ampère connecting the direction of an electric current and its associated magnetic currents. It states that around a wire carrying a current towards the observer, the magnetic field curls in the anticlockwise direction. This assumes the conventional direction of current flow (from the positive to the negative terminal of a battery).

amphetamine or **speed**, powerful synthetic stimulant. Benzedrine was the earliest amphetamine marketed, used as a "pep pill" in World War II to help soldiers to overcome fatigue, and until the 1970s amphetamines were prescribed by doctors as an appetite suppressant for weight loss; as an antidepressant, to induce euphoria; and as a stimulant, to increase alertness.

Indications for its use today are very restricted because of severe side effects, including addiction. It is a sulfate or phosphate form of $C_9H_{13}N$.

amphibian (Greek "double life") member of the vertebrate class Amphibia, which generally spend their larval (tadpole) stage in fresh water, transferring to land at maturity (after †metamorphosis) and generally returning to water to breed. Like fish and reptiles, they continue to grow throughout life, and cannot maintain a temperature greatly differing from that of their environment. The class contains about 4,500 known species, 4,000 of which are frogs and toads, 390 salamanders, and 160 caecilians (wormlike in appearance).

According to analysis of statistics from over 900 amphibian populations taken 1950–97, by Canadian and Swiss researchers, and published in 2000, amphibian populations have been declining at a rate of about 4% per year. US biologists concluded that the marked worldwide decline in amphibian populations is unlikely to have a single cause. Possible causes have been cited as increased ultraviolet radiation exposure, pollution, global climate change, and infection.

amphibole any one of a large group of rock-forming silicate minerals with an internal structure based on double chains of silicon and oxygen, and with a general formula $X_2Y_5Si_8O_{22}(OH)_2$; closely related to †pyroxene. Amphiboles form orthorhombic, monoclinic, and triclinic †crystals.

Amphiboles occur in a wide range of igneous and metamorphic rocks. Common examples are †hornblende (X = Ca, Y = Mg, Fe, Al) and tremolite (X = Ca, Y = Mg).

amphoteric of a chemical compound, being able to behave either as an †acid or as a †base depending on their environment. For example, the metals aluminum and zinc, and the oxides and hydroxides, act as bases in acidic solutions and as acids in alkaline solutions.

Amino acids and proteins are also amphoteric, as they contain both a basic (amino, $—NH_2$) and an acidic (carboxyl, —COOH) group.

amplifier electronic device that increases the strength of a signal, such as a radio signal. The ratio of output signal strength to input signal strength is called the gain of the amplifier. As well as achieving high gain, an amplifier should be free from distortion and able to operate over a range of frequencies. Practical amplifiers are usually complex circuits, although simple amplifiers can be built from single transistors or valves.

amplitude or **argument**, in mathematics, the angle in an †Argand diagram between the line that represents the complex number and the real (positive horizontal) axis. If the complex number is written in the form *r*(cos q + i sin q), where *r* is radius and i = Ö(−1), the amplitude is the angle q (theta).

The amplitude is also the peak value of an oscillation.

amplitude in physics, maximum displacement of an oscillation from the equilibrium position. For a †transverse wave motion, as in electromagnetic waves, it is the height of a crest (or the depth of a trough). For a †longitudinal wave, such as a sound wave, amplitude is the maximum distance a particle is pushed (due to compression) or pulled (due to rarefaction) from its resting position. A quiet sound has a lower amplitude and a loud sound has a higher amplitude. For a louder sound, more sound energy enters the ear every second. Amplitude is generally denoted by *a*.

AM (amplitude modulation) radio broadcasting is a method of transmitting audio-frequency signals that would otherwise travel not very far in space on their own and would have to be transmitted one at a time. The amplitude of the low-frequency audio signal is made to modulate (vary slightly) the amplitude of a continuously transmitted radio **carrier wave** of a higher frequency. In this way the **modulating signal** is imprinted on the carrier wave.

amplitude modulation AM, method by which radio waves are altered for the transmission of broadcasting signals. AM waves are constant in frequency, but the †amplitude of the transmitting wave varies in accordance with the signal being broadcast.

ampulla in the inner †ear, a slight swelling at the end of each semicircular canal, able to sense the motion of the head.

The sense of balance depends largely on sensitive hairs within the ampullae responding to movements of fluid within the canal.

amygdala almond-shaped region of the †brain adjacent to the hippocampus, that links the cortex, responsible for conscious thought, with the regions controlling emotions. It is involved in interpreting fear-provoking information and linking it to fear responses. For example, where the amygdala is damaged, patients are unable to recognize fearful expressions.

Emotionally charged events are more easily recalled than neutral events, and in 1996 US researchers demonstrated a link between the amygdala and the memorizing of emotionally loaded images.

The amygdala is involved in assessment of character in unfamiliar people by using knowledge of past experiences to determine whether a person is, for example, trustworthy or friendly. Those people with damaged amygdalas were found, in a 1998 study, to be unable to make these judgments.

amygdalin $C_{20}H_{27}O_{11}N$, glucoside found in bitter almonds, cherry kernels, and other vegetable products.

amyl alcohol former name for †pentanol.

amylase †enzyme that breaks down starch into a complex sugar that can be used in the body. It occurs widely in both plants and animals. In humans it is found in saliva and in the pancreatic digestive juices that drain into the †alimentary canal.

Starch is an important food for humans and is found in plants such as potatoes and in wheat grains. It is made of large and insoluble molecules, but under the action of amylase produces sugars that are soluble

– these can be absorbed by the body.

Human amylase has an optimum pH of 7.2–7.4. Like most enzymes amylase is denatured by temperatures above 60°C.

anabatic wind warm wind that blows uphill in steep-sided valleys in the early morning. As the sides of a valley warm up in the morning the air above is also warmed and rises up the valley to give a gentle breeze. By contrast, a †katabatic wind is cool and blows down a valley at night.

anabolic steroid any †hormone of the †steroid group that stimulates muscular tissue growth. Its use in medicine is limited to the treatment of some anemias and breast cancers; it may help to break up blood clots. Side effects include aggressive behavior, masculinization in women, and, in children, reduced height.

anabolism process of building up body tissue, promoted by the influence of certain hormones. It is the constructive side of †metabolism, as opposed to †catabolism.

anabranch (Greek *ana* "again") stream that branches from a main river, then reunites with it. For example, the Great Anabranch in New South Wales, Australia, leaves the Darling near Menindee, and joins the Murray below the Darling–Murray confluence.

anaerobic not requiring oxygen for the release of energy from food molecules such as glucose. An organism is described as anaerobic if it does not require †oxygen in order to survive. Instead, anaerobic organisms use anaerobic respiration to obtain energy from food. Most anaerobic organisms are microorganisms such as bacteria, yeasts, and internal parasites that live in places where there is never much oxygen, such as in the mud at the bottom of a lake or pond, or in the alimentary canal. Anaerobic organisms release much less of the available energy from their food than do †aerobic organisms.

Obligate anaerobes, such as certain primitive bacteria, cannot function in the presence of oxygen, but **facultative anaerobes**, such as the fermenting yeasts and most bacteria, can function with or without oxygen.

analgesic agent for relieving pain. †Opiates alter the perception or appreciation of pain and are effective in controlling "deep" visceral (internal) pain. Non-opiates, such as †aspirin, paracetamol, and NSAIDs (nonsteroidal anti-inflammatory drugs), relieve musculoskeletal pain and reduce inflammation in soft tissues.

Pain is felt when electrical stimuli travel along a nerve pathway, from peripheral nerve fibers to the brain via the spinal cord. An anesthetic agent acts either by preventing stimuli from being sent (local), or by removing awareness of them (general). Analgesic drugs act on both.

Temporary or permanent analgesia may be achieved by injection of an anesthetic agent into, or the severing of, a nerve. Implanted devices enable patients to deliver controlled electrical stimulation to block pain impulses. Production of the body's natural opiates, †endorphins, can be manipulated by techniques such as relaxation and biofeedback. However, for the severe pain of, for example, terminal cancer, opiate analgesics are required.

US researchers found in 1996 that some painkillers were more effective and provided longer-lasting relief for women than men.

analog or analogue (of a quantity or device) changing continuously; by contrast, a †digital quantity or device varies in series of distinct steps. For example, an analog clock measures time by means of a continuous movement of hands around a dial, whereas a digital clock measures time with a numerical display that changes in a series of discrete steps.

Most computers are digital devices. Therefore, any signals and data from an analog device must be passed through a suitable †analog-to-digital converter before they can be received and processed by computer. Similarly, output signals from digital computers must be passed through a †digital-to-analog converter before they can be received by an analog device.

analog computer computing device that performs calculations through the interaction of continuously varying physical quantities, such as voltages (as distinct from the more common †digital computer, which works with discrete quantities). An analog computer is said to operate in real time (corresponding to time in the real world), and can therefore be used to monitor and control other events as they happen.

Although common in engineering since the 1920s, analog computers are not general-purpose computers, but specialize in solving †differential calculus and similar mathematical problems. The earliest analog computing device is thought to be the flat, or planispheric, astrolabe, which originated in about the 8th century.

analog signal in electronics, current or voltage that conveys or stores information, and varies continuously in the same way as the information it represents (compare †digital signal). Analog signals are prone to interference and distortion.

The bumps in the grooves of a vinyl record form a mechanical analog of the sound information stored, which is then converted into an electrical analog signal by the record player's pick-up device.

analog-to-digital converter **ADC**, electronic circuit that converts an analog signal into a digital one. Such a circuit is needed to convert the signal from an analog device into a digital signal for input into a computer. For example, many †sensors designed to measure physical quantities, such as temperature and pressure, produce an analog signal in the form of voltage and this must be passed through an ADC before a computer can process the information. A †digital-to-analog converter performs the opposite process.

analogous in biology, term describing a structure that has a similar function to a structure in another organism, but not a similar evolutionary path. For example, the wings of bees and of birds have the same purpose – to give powered flight – but have different origins. Compare †homologous.

analogy in mathematics and logic, a form of argument or process of reasoning from one case to another parallel case. Arguments from analogy generally have the following form: if some event or thing has the properties *a* and *b*, and if another event or thing has the properties *b* and *c*, then the former event or thing has the property *c*, too. Arguments from analogy are not always sound and can mislead. False analogies arise when the cases are insufficiently similar to support the reasoning. For example, a whale lives in water and resembles a fish, but we cannot conclude from this that it is a fish. When arguments from analogy are compressed, they are called metaphors.

analysis in chemistry, the determination of the composition of substances; see †analytical chemistry.

analysis branch of mathematics concerned with limiting processes on axiomatic number systems; †calculus of variations and infinitesimal calculus is now called analysis.

analyst or **business analyst** or **systems analyst**, job classification for computer personnel. They are responsible for looking at procedures within organizations and determining the different types of data used, how it flows around the organization, and what processes use that data. The resulting analysis document enables computer programmers to design and write software to computerize the procedures.

analytical chemistry branch of chemistry that deals with the determination of the chemical composition of substances. **Qualitative analysis** determines the identities of the substances in a given sample; **quantitative analysis** determines how much of a particular substance is present.

Simple qualitative techniques exploit the specific, easily observable properties of elements or compounds – for example, the flame test makes use of the different flame colors produced by metal cations when their compounds are held in a hot flame. More sophisticated methods, such as those of †spectroscopy, are required where substances are present in very low concentrations or where several substances have similar properties.

Most quantitative analyses involve initial stages in which the substance to be measured is extracted from the test sample, and purified. The final analytical stages (or "finishes") may involve measurement of the substance's mass (gravimetry) or volume (volumetry, titrimetry), or a number of techniques initially developed for qualitative analysis, such as fluorescence and absorption spectroscopy, chromatography, electrophoresis, and polarography. Many modern methods enable quantification by means of a detecting device that is integrated into the extraction procedure (as in gas–liquid chromatography).

Analytical Chemistry Basics http://www.chem.vt.edu/chem-ed/ac-basic.html

Detailed online course, designed for those at undergraduate level, which provides the user with an introduction to some of the fundamental concepts and methods of analytical chemistry. Some of the sections included are gravimetric analysis, titration, and spectroscopy.

analytical engine programmable computing device designed by English mathematician Charles †Babbage in 1833.

It was based on the †difference engine but was intended to automate the whole process of calculation. It introduced many of the concepts of the digital computer but, because of limitations in manufacturing processes, was never built.

Among the concepts introduced were input and output, an arithmetic unit, memory, sequential operation, and the ability to make decisions based on data. It would have required at least 50,000 moving parts. The design was largely forgotten until some of Babbage's writings were rediscovered in 1937.

analytical geometry another name for †coordinate geometry.

anastigmatic in optics, describing a lens that is constructed so as to minimize †astigmatism. Its curvature is different in two axes at right angles.

anatomy study of the structure of the body and its component parts, especially the †human body, as distinguished from physiology, which is the study of bodily functions.

Herophilus of Chalcedon (*c.* 330–*c.* 260 BC) is regarded as the founder of anatomy. In the 2nd century AD, the Graeco-Roman physician Galen produced an account of anatomy that was the only source of anatomical knowledge until *On the Working of the Human Body* (1543) by Belgian physician Andreas

Vesalius. In 1628, English physician William Harvey published his demonstration of the circulation of the blood. With the invention of the microscope, Italian physiologist Marcello Malpighi and Dutch microscopist Anton van Leeuwenhoek were able to found the study of histology. In 1747, Albinus, with the help of the artist Wandelaar, produced the most exact account of the bones and muscles, and in 1757–65 Swiss biologist Albrecht von Haller gave the most complete and exact description of the organs that had yet appeared. Among the anatomical writers of the early 19th century are the surgeon Charles Bell, Jonas Quain, and Henry Gray. Radiographic anatomy (using X-rays; see †radiography) has been one of the triumphs of the 20th century, which has also been marked by immense activity in embryological investigation.

Human Anatomy Online http://www.innerbody.com/htm/body.html

Interactive guide to the human body. Choose from ten body "systems" to learn more about them. For each system there is a large diagram with "hotspots," each of which can be clicked on to open up a description in the right-hand frame.

Anaximander (*c.* 610–*c.* 546 BC) Greek astronomer and philosopher. He claimed that the Earth was a cylinder three times wider than it is deep, motionless at the center of the universe, and that the celestial bodies were fire seen through holes in the hollow rims of wheels encircling the Earth. According to Anaximander, the first animals came into being from moisture and the first humans grew inside fish, emerging once fully developed.

andalusite aluminum silicate, Al_2SiO_5, a white to pinkish mineral crystallizing as square- or rhombus-based prisms. It is common in metamorphic rocks formed from clay sediments under low-pressure conditions. Andalusite, kyanite, and sillimanite are all polymorphs of Al_2SiO_5.

Anderson, Carl David (1905–1991) US physicist. He shared the Nobel Prize for Physics in 1936 for his discovery in 1932 of the positive electron (positron). His discovery of another particle, the muon, in 1937 launched elementary-particle physics.

Anderson, Philip Warren (1923–) US physicist. He shared the Nobel Prize for Physics in 1977 with his senior colleague John †Van Vleck for his theoretical work on the behavior of electrons in magnetic, noncrystalline solids.

andesite volcanic igneous rock, intermediate in silica content between rhyolite and basalt. It is characterized by a large quantity of feldspar †minerals, giving it a light color. Andesite erupts from volcanoes at destructive plate margins (where one plate of the Earth's surface moves beneath another; see †plate tectonics), including the Andes, from which it gets its name.

AND gate in electronics, a type of †logic gate.

androecium male part of a flower, comprising a number of †stamens.

androgen general name for any male sex hormone, of which †testosterone is the most important.

They are all †steroids and are principally involved in the production of male †secondary sexual characteristics (such as beard growth).

Andromeda major constellation of the northern hemisphere, visible in autumn. Its main feature is the †Andromeda galaxy. The star Alpha Andromedae forms one corner of the Square of Pegasus. It is named after the princess of Greek mythology.

Andromeda galaxy galaxy 2.2 million light years away from Earth in the constellation Andromeda, and the most distant object visible to the naked eye. It is the largest member of the †Local Group of galaxies.

Like the Milky Way, it is a spiral orbited by several companion galaxies but contains about twice as many stars as the Milky Way. It is about 200,000 light years across.

AND rule rule used for finding the combined probability of two or more independent events both occurring. If two events E_1 and E_2 are independent (have no effect on each other) and the probabilities of their taking place are p_1 and p_2, respectively, then the combined probability p that both E_1 and E_2 will happen is given by:

$p = p_1 \times p_2$

For example, if a blue die and a red die are thrown together, the probability of a blue six is $^1/_6$, and the probability of a red six is $^1/_6$. Therefore, the probability of both a red six and a blue six being thrown is $^1/_6 \times {}^1/_6 = {}^1/_{36}$.

By contrast, the **OR rule** is used for finding the probability of either one event or another taking place.

anechoic chamber room designed to be of high sound absorbency. All surfaces inside the chamber are covered by sound-absorbent materials such as rubber. The walls are often covered with inward-facing pyramids of rubber, to minimize reflections back into the interior of the chamber. It is used for experiments in †acoustics and for testing audio equipment.

anemometer device for measuring wind speed and liquid flow. The most basic form, the **cup-type**

anemometer, consists of cups at the ends of arms, which rotate when the wind blows. The speed of rotation indicates the wind speed.

Vane-type anemometers have vanes, like a small windmill or propeller, which rotate when the wind blows. **Pressure-tube anemometers** use the pressure generated by the wind to indicate speed. The wind blowing into or across a tube develops a pressure, proportional to the wind speed, which is measured by a manometer or pressure gauge. **Hot-wire anemometers** work on the principle that the rate at which heat is transferred from a hot wire to the surrounding air is a measure of the air speed. Wind speed is determined by measuring either the electric current required to maintain a hot wire at a constant temperature, or the variation of resistance while a constant current is maintained.

anemophily type of †pollination in which the pollen is carried on the wind. Anemophilous flowers are usually unscented, have either very reduced petals and sepals or lack them altogether, and do not produce nectar. In some species they are borne in †catkins. Male and female reproductive structures are commonly found in separate flowers. The male flowers have numerous exposed stamens, often on long filaments; the female flowers have long, often branched, feathery stigmas.

aneroid barometer kind of †barometer.

anesthetic drug that produces loss of sensation or consciousness; the resulting state is **anesthesia**, in which the patient is insensitive to stimuli. Anesthesia may also happen as a result of nerve disorder.

Ever since the first successful operation in 1846, on a patient rendered unconscious by ether, advances have been aimed at increasing safety and control. Sedatives may be given before the anesthetic to make the process easier. The level and duration of unconsciousness are managed precisely. Where general anesthesia may be inappropriate (for example, in childbirth, for a small procedure, or in the elderly), many other techniques are available. A topical substance may be applied to the skin or tissue surface; a local agent may be injected into the tissues under the skin in the area to be treated; or a regional block of sensation may be achieved by injection into a nerve. Spinal anesthetic, such as epidural, is injected into the tissues surrounding the spinal cord, producing loss of feeling in the lower part of the body.

Virtual Anesthesia Textbook http://www.virtual-anaesthesia-textbook.com/

Huge source of information on all aspects of anesthesia. Though primarily designed for anesthetists, a non-medical reader wishing to learn about the history and practice of modern anesthetics would find a lot of interest on this massive site. A search engine facilitates navigation. There are also links to anesthesia sites around the world.

angiosperm flowering plant in which the seeds are enclosed within an ovary, which ripens into a fruit. Angiosperms are divided into †monocotyledons (single seed leaf in the embryo) and †dicotyledons (two seed leaves in the embryo). They include the majority of flowers, herbs, grasses, and trees except conifers.

There are over 250,000 different species of angiosperm, found in a wide range of habitats. Like †gymnosperms, they are seed plants, but differ in that ovules and seeds are protected within the carpel. Fertilization occurs by male gametes passing into the ovary from a pollen tube. After fertilization the ovule develops into the seed while the ovary wall develops into the fruit.

angle in mathematics, the amount of turn or rotation; it may be defined by a pair of rays (half-lines) that share a common endpoint (†vertex) but do not lie on the same line. Angles are measured in †degrees (°) or †radians (rads or ^{c}) – a complete turn or circle being 360° or 2p rads.

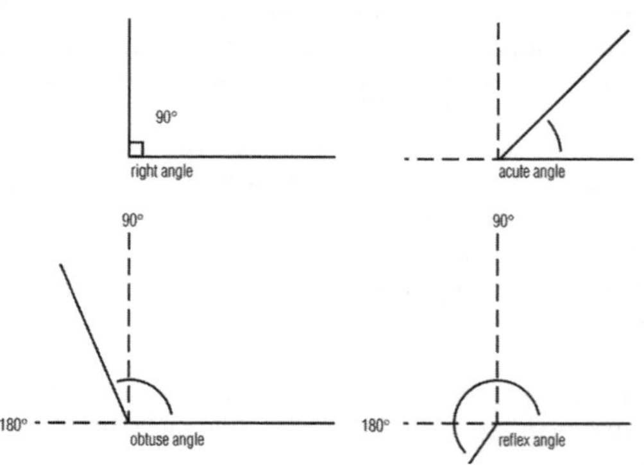

angle The four types of angle, as classified by their degree measures. No angle is classified as having a measure of 180°, as by definition such an "angle" is actually a straight line. Helicon

All angles around a point on a straight line add up to 180°. All angles around a point add up to 360°.

Angles are classified generally by their degree measures: †**acute angles** are less than 90°; **right angles** are exactly 90° (a quarter turn) and are created by two †perpendicular lines crossing; †**obtuse angles** are greater than 90° but less than 180° (a straight line); †**reflex angles** are greater than 180° but less than 360°; **supplementary angles** add up to 180°.

angles in triangles A †triangle has three interior angles that together add up to 180°. In an equilateral triangle these angles are equal (60°). The exterior angles of a triangle (those produced if one side is extended beyond the triangle) are equal to the sum of the opposite internal angles.

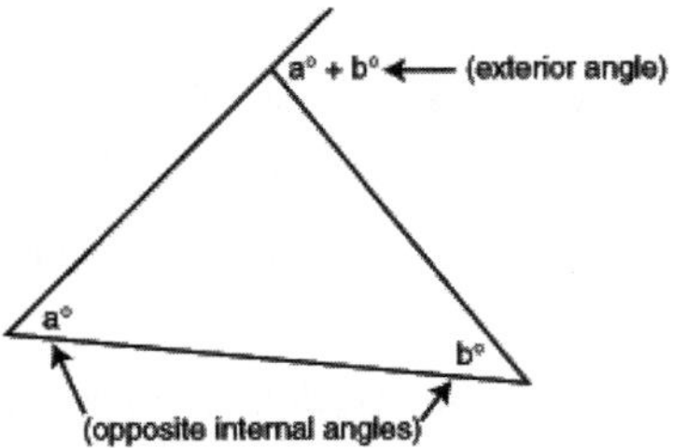

Unknown angles in a right-angled triangle can be worked out using †trigonometry.

angles in polygons Regular †polygons have three types of angle:

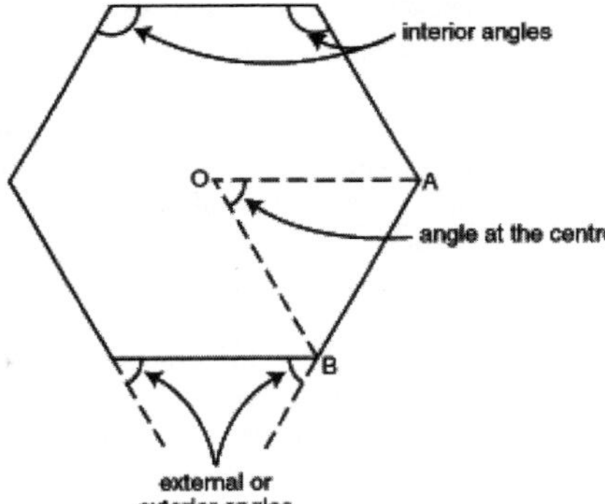

The angle at the center is equal to the exterior angle and it is found by dividing 360° by the number of sides in the polygon.

For example, the angle at the center of a hexagon is 60°:

$360 \div 6 = 60°$

angles at parallel lines Angles created by a line crossing two †parallel lines are known as †opposite angles, †corresponding angles, or †alternate angles.

angles in circles Angles within a †circle can be calculated using †circle theorems.

angle of declination angle at a particular point on the Earth's surface between the direction of the true or geographic North Pole and the magnetic north pole. The angle of declination has varied over time because of the slow drift in the position of the magnetic north pole.

angle of dip or **angle of inclination**, angle at a particular point on the Earth's surface between the direction of the Earth's magnetic field and the horizontal; see †magnetic dip.

angle of incidence angle between a ray of light striking a surface (incident ray) and the normal to that surface. For a mirror it is equal to the †angle of reflection.

angle of parallelism in non-Euclidean geometry involving the application of classical geometrical principles to non-flat surfaces, the angle at which a line perpendicular to one of two parallel lines meets the other – which may also be 90°, or may be less.

angle of reflection angle between a ray of light reflected from a mirror and the normal (perpendicular) to that mirror. The law of reflection states that the angle of reflection is equal to the †angle of incidence.

angle of refraction angle between a refracted ray of light and the normal to the surface at which †refraction occurred. When a ray passes from air into a denser medium such as glass, it is bent towards the normal so that the angle of refraction is less than the †angle of incidence.

angstrom symbol Å, unit of length equal to 10^{-10} meters or one-ten-millionth of a millimeter, used for atomic measurements and the wavelengths of electromagnetic radiation. It is named after the Swedish scientist Anders †Ångström.

Ångström, Anders Jonas (1814–1874) Swedish astrophysicist who worked in spectroscopy and solar physics. In 1861 he identified the presence of hydrogen in the Sun. His outstanding *Recherches sur le spectre solaire* (1868) presented an atlas of the solar spectrum with measurements of 1,000 spectral lines expressed in units of one-ten-millionth of a millimeter, the unit which later became the angstrom.

angular acceleration in physics, acceleration of a spinning object about its rotational axis, analogous to linear acceleration.

angular frequency symbol w, in physics, for a spinning object, the number of complete rotations per unit time, for which the unit is the †hertz.

angular magnification symbol *M*, in physics, for a lens or optical instrument, the angle subtended at the eye by an image divided by the angle subtended there by the actual object.

angular momentum in physics, a type of †momentum.

angular velocity symbol w, for a rotating or revolving object, rate of change of angular displacement, usually measured in degrees or radians per second.

anhydride chemical compound obtained by the removal of water from another compound; usually a dehydrated acid. For example, sulfur(VI) oxide (sulfur trioxide, SO_3) is the anhydride of sulfuric acid (H_2SO_4).

anhydrite naturally occurring anhydrous calcium sulfate ($CaSO_4$). It is used commercially for the manufacture of plaster of Paris and builders' plaster.

anhydrous of a chemical compound, containing no water. If the water of crystallization is removed from blue crystals of copper(II) sulfate, a white powder (anhydrous copper sulfate) results. Liquids from which all traces of water have been removed are also described as being anhydrous.

aniline or **phenylamine**; $C_6H_5NH_2$, (Portuguese *anil* "indigo") one of the simplest aromatic chemicals (a substance related to benzene, with its carbon atoms joined in a ring). When pure, it is a colorless oily liquid; it has a characteristic odor, and turns brown on contact with air. It occurs in coal tar, and is used in the rubber industry and to make drugs and dyes. It is highly poisonous.

Aniline was discovered in 1826, and was originally prepared by the dry distillation of indigo, hence its name.

animal or **metazoan**, (Latin *anima* "breath," "life") member of the †kingdom Animalia, one of the major categories of living things, the science of which is **zoology**. Animals are all multicellular †heterotrophs (they obtain their energy from organic substances produced by other organisms); they have eukaryotic cells (the genetic material is contained within a distinct nucleus), which are bounded by a thin cell membrane rather than the thick cell wall of plants. Most animals are capable of moving around for at least part of their life cycle.

In the past, it was common to include the single-celled †protozoa with the animals, but these are now classified as protists, together with single-celled plants. Thus all animals are multicellular. The oldest land animals known date back 440 million years. Their remains were found 1990 in a sandstone deposit near Ludlow, Shropshire, UK, and included fragments of two centipedes a few centimeters long and a primitive spider measuring about one millimeter.

Electronic Zoo http://netvet.wustl.edu/e-zoo.htm

Vast collection of links to animal Web sites and images (there is even a section devoted to fictional animals, from Animaniacs to Wile E Coyote), plus NetVet – animal welfare information and links to veterinary sites.

Virtual Library: Zoos http://www.mindspring.com/~zoonet/www_virtual_lib/zoos.html

Comprehensive collection of links to zoo sites, with many links to related Web sites, such as specific animal pages, images, and sounds.

animal behavior scientific study of the behavior of animals, either by comparative psychologists (with an interest mainly in the psychological processes involved in the control of behavior) or by ethologists (with an interest in the biological context and relevance of behavior; see †ethology).

animal cell †cell in an animal, which, like all cells, has a †cell membrane, †cytoplasm, and a †nucleus, but in which the surface consists of the cell surface membrane only – it does not have a cell wall. The cell surface membrane keeps the cell together by being strong, even though it is very thin and flexible. The membrane also controls what enters and leaves the cell. For example, nutrients need to be able to enter and waste materials to leave. The cell contains cytoplasm, in which is located a nucleus. This controls the activities of the cell.

An animal is made up of many thousands of cells, which are used as building blocks. However, as with a house, different types of building blocks are used in different places. Groups of similar cells form a †tissue. An example of this is the tissue forming the brain. One of the specialized cells found in this tissue is the neuron or nerve cell. The brain is an †organ made up of a number of different tissues.

The blood is a tissue consisting mainly of another type of specialized cell – the red blood cell. Other specialized cells include the †epithelium and sperm cell.

animism in anthropology, the belief that everything, whether animate or inanimate, possesses a soul or spirit. It is a fundamental system of belief in certain religions, particularly those of some pre-industrial societies. Linked with this is the worship of natural objects such as stones and trees, thought to harbor spirits (naturism); fetishism; and ancestor worship.

In psychology and physiology, animism is the view of human personality that attributes human life and behavior to a force distinct from matter. In developmental psychology, an animistic stage in the early thought and speech of the child has been described, notably by Swiss psychologist Jean Piaget.

In philosophy, the view that in all things consciousness or something mind-like exists.

In religious theory, the conception of a spiritual reality behind the material one: for example, beliefs in the soul as a shadowy duplicate of the body capable of independent activity, both in life and death.

anion †ion carrying a negative charge. An anion is formed from an †atom by the gain of electrons, a process known as †**ionic bonding**. †Non-metallic elements form anions. During †electrolysis, anions in the electrolyte move towards the anode (positive electrode).

An electrolyte, such as the salt zinc chloride ($ZnCl_2$), is dissociated in aqueous solution or in the molten state into doubly charged Zn^{2+} †cations and singly-charged Cl^- anions. During electrolysis, the zinc cations flow to the cathode (to become discharged and liberate zinc metal) and the chloride anions flow to the anode (to become discharged and form chlorine gas).

anisotropic or **aelotropic**, describing a substance that has different physical properties in different directions. Some crystals, for example, have different refractive indices in different directions.

annelid any segmented worm of the phylum Annelida. Annelids include earthworms, leeches, and marine worms such as lugworms.

They have a distinct head and soft body, which is divided into a number of similar segments shut off from one another internally by membranous partitions, but there are no jointed appendages. Annelids are noted for their ability to regenerate missing parts of their bodies.

annihilation in nuclear physics, a process in which a particle and its "mirror image" particle called an antiparticle collide and disappear, with the creation of a burst of energy. The energy created is equivalent to the mass of the colliding particles in accordance with the †mass–energy equation. For example, an electron and a positron annihilate to produce a burst of high-energy X-rays. Not all particle–antiparticle interactions result in annihilation; the exception concerns the group called †mesons, which belong to the class of particles that are composed of †quarks and their antiquarks. See †antimatter.

Anning, Mary (1799–1847) English fossil collector. In 1811 she discovered in a Dorset cliff the fossil skeleton of an ichthyosaur, now in the Natural History Museum, London. She also discovered the first plesiosaur in 1821 and *Dimorphodon*, the first pterodactyl, in 1828.

annual plant plant that completes its life cycle within one year, during which time it germinates, grows to maturity, bears flowers, produces seed, and then dies.

annual rings or **growth rings**, concentric rings visible on the wood of a cut tree trunk or other woody stem. Each ring represents a period of growth when new †xylem is laid down to replace tissue being converted into wood (secondary xylem). The wood formed from xylem produced in the spring and early summer has larger and more numerous vessels than the wood formed from xylem produced in autumn when growth is slowing down. The result is a clear boundary between the pale spring wood and the denser, darker autumn wood. Annual rings may be used to estimate the age of the plant (see †dendrochronology), although occasionally more than one growth ring is produced in a given year.

annual variation in earth science, regular change in the strength and direction of the Earth's magnetic field, which reaches a maximum during years of high †sunspot activity. It must be allowed for by navigators using a compass.

annular eclipse solar †eclipse in which the Moon does not completely obscure the Sun and a thin ring of sunlight remains visible. Annular eclipses occur when the Moon is at its farthest point from the Earth.

annulus (Latin "ring") in geometry, the plane area between two concentric circles, making a flat ring.

anode positive electrode of an electrolytic cell, towards which negative particles (anions), usually in solution, are attracted. See †electrolysis.

In a thermionic valve, cathode ray tube, or similar device, electrons are drawn to the anode after being emitted from the †cathode.

anodizing process that increases the resistance to †corrosion of a metal, such as aluminum, by building up a protective oxide layer on the surface. The natural corrosion resistance of aluminum is provided by a thin film of aluminum oxide; anodizing increases the thickness of this film and thus the corrosion protection.

It is so called because the metal becomes the †anode in an electrolytic bath containing a solution of, for example, sulfuric or chromic acid as the †electrolyte. During †electrolysis oxygen is produced at the anode, where it combines with the metal to form an oxide film.

anomalous dispersion in physics, behavior of a transparent medium whose refractive index decreases with decreasing wavelength of light (normally refractive index increases with decreasing wavelength). It occurs mainly with ultraviolet light, which is strongly absorbed by the medium.

anomalous expansion of water expansion of water as it is cooled from 4°C to 0°C. This behavior is unusual because most substances contract when they are cooled. It means that water has a greater density at 4°C than at 0°C. Hence ice floats on water, and the water at the bottom of a pool in winter is warmer than at the surface. As a result lakes and ponds freeze slowly from the surface downwards, usually remaining liquid near the bottom, where aquatic life is more likely to survive.

anorthosite in earth science, coarse-grained igneous rock rich in plagioclase †feldspar and formed well below the surface. It is found on Earth and in the highland regions of the Moon.

anoxia or **hypoxia**, in biology, deprivation of oxygen, a condition that rapidly leads to collapse or death, unless immediately reversed.

antacid any substance that neutralizes stomach acid, such as sodium hydrogencarbonate (sodium bicarbonate) or magnesium hydroxide ("milk of magnesia"). Antacids are weak †bases, swallowed as solids or emulsions. They may be taken between meals to relieve symptoms of hyperacidity, such as pain, bloating, nausea, and "heartburn." Excessive or prolonged need for antacids should be investigated medically.

antagonist in biology, a †muscle that relaxes in response to the contraction of its agonist muscle. See †antagonistic muscles.

antagonistic muscles in the body, a pair of †muscles working together to allow coordinated movement of the skeletal joints. Muscles can only exert a force and do work by contracting. An example is the antagonistic pair of muscles used in bending and straightening the arm. To bend the arm requires one set of muscles – the biceps – to contract, while another set – the †triceps – relaxes. If the arm is to be straightened, the reverse happens. The individual components of antagonistic pairs can be classified into extensors (muscles that straighten a limb) and flexors (muscles that bend a limb).

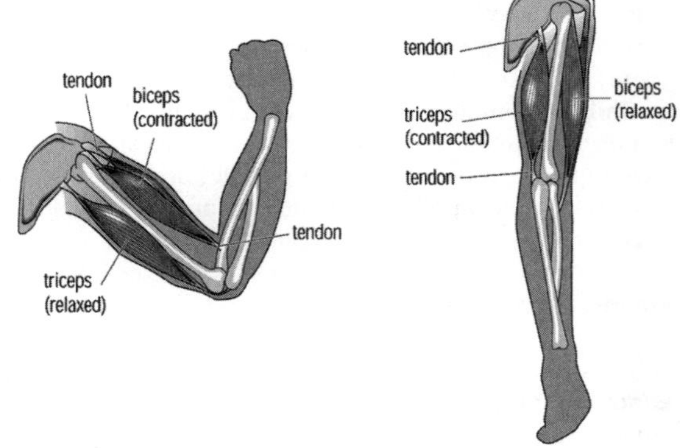

antagonistic muscle Even simple movements such as bending and straightening the arm require muscle pairs to contract and relax synchronously. Helicon

Antarctic Circle imaginary line that encircles the South Pole at latitude 66° 32' S. The line encompasses the continent of Antarctica and the Antarctic Ocean.

The region south of this line experiences at least one night in the southern summer during which the Sun never sets, and at least one day in the southern winter during which the Sun never rises.

Antares or **Alpha Scorpii**, brightest star in the constellation Scorpius and the 15th-brightest star in the night sky. It is a red supergiant several hundred times larger than the Sun and perhaps 10,000 times as luminous. It lies about 420 light years away from the Sun, and varies in brightness.

antenna in radio and television, another name for †aerial.

antenna in zoology, an appendage ("feeler") on the head. Insects, centipedes, and millipedes each have one pair of antennae but there are two pairs in crustaceans, such as shrimps. In insects, the antennae are involved with the senses of smell and touch; they are frequently complex structures with large surface areas that increase the ability to detect scents.

anterior in biology, the front of an organism, usually the part that goes forward first when the animal is moving. In higher organisms the anterior end of the nervous system, over the course of evolution, has developed into a brain with associated receptor organs able to detect stimuli including light and chemicals.

anther in a flower, the terminal part of a stamen in which the †pollen grains are produced. It is usually borne on a slender stalk or filament, and has two lobes, each containing two chambers, or pollen sacs, within which the pollen is formed.

antheridium organ producing the male gametes, †antherozoids, in algae, bryophytes (mosses and liverworts), and pteridophytes (ferns, club mosses, and horsetails). It may be either single-celled, as in most algae, or multicellular, as in bryophytes and pteridophytes.

antherozoid motile (or independently moving) male gamete produced by algae, bryophytes (mosses and liverworts), pteridophytes (ferns, club mosses, and horsetails), and some gymnosperms (notably the cycads). Antherozoids are formed in an antheridium and, after being released, swim by means of one or more †flagella, to the female gametes. Higher plants have nonmotile male gametes contained within †pollen grains.

anthracene white, glistening, crystalline, tricyclic, aromatic hydrocarbon with a faint blue fluorescence when pure. Its melting point is about 216°C/421°F and its boiling point 351°C/664°F. It occurs in the high-boiling-point fractions of coal tar, where it was discovered in 1832 by the French chemists Auguste Laurent (1808–1853) and Jean Dumas (1800–1884).

anthracite (from Greek *anthrax* "coal") hard, dense, shiny variety of †coal, containing over 90% carbon and a low percentage of ash and impurities, which causes it to burn without flame, smoke, or smell. Because of its purity, anthracite gives off relatively little sulfur dioxide when burnt.

Anthracite gives intense heat, but is slow-burning and slow to light; it is therefore unsuitable for use in open fires. Its characteristic composition is thought to be due to the action of bacteria in disintegrating the coal-forming material when it was laid down during the †Carboniferous period.

Among the chief sources of anthracite coal are Pennsylvania in the USA; South Wales, UK; the Donbas, Ukraine and Russia; and Shanxi province, China.

anthrax disease of livestock, occasionally transmitted to humans, usually via infected hides and fleeces. It may also be used as a weapon in biological warfare. It develops as black skin pustules or severe pneumonia. Treatment is possible with antibiotics, and vaccination is effective.

anthropic principle idea that "the universe is the way it is because if it were different we would not be here to observe it." The principle arises from the observation that if the laws of science were even slightly different, it would have been impossible for intelligent life to evolve. For example, if the strengths of the fundamental forces were only slightly different, stars would have been unable to burn hydrogen and produce the chemical elements that make up our bodies.

antibiotic drug that kills or inhibits the growth of bacteria and fungi.

The earliest antibiotics, the †penicillins, came into use from 1941 and were quickly joined by chloramphenicol, the cephalosporins, erythromycins, tetracyclines, and aminoglycosides. A range of broad-spectrum antibiotics, the 4-quinolones, was developed in 1989, of which ciprofloxacin was the first. Each class and individual antibiotic acts in a different way and may be effective against either a broad spectrum or a specific type of disease-causing agent. Use of antibiotics has become more selective as side effects, such as toxicity, allergy, and resistance, have become better understood. Bacteria have the ability to develop resistance following repeated or subclinical (insufficient) doses, so more advanced antibiotics and synthetic antimicrobials are continually required to overcome them.

Antibiotics: How Do Antibiotics Work?

http://www.pharmacology.about.com/health/pharmacology/library/weekly/aa970514.htm

Introduction to the use and importance of antibiotics, in easy-to-understand language. The site includes a full list of all antibacterial drugs, a discussion of the growing problem of bacterial resistance, and links to other sites providing additional information on these topics.

antibody protein molecule produced in the blood by †lymphocytes in response to the presence of foreign or invading substances (†antigens); such substances include the proteins carried on the surface of infecting micro-organisms. Antibody production is only one aspect of †immunity in vertebrates.

Each antibody acts against only one kind of antigen, and combines with it to form a "complex." This action may render antigens harmless, or it may destroy micro-organisms by setting off chemical changes that cause them to self-destruct.

In other cases, the formation of a complex will cause antigens to form clumps that can then be detected and engulfed by white blood cells, such as †macrophages and †phagocytes.

Each bacterial or viral infection will bring about the manufacture of a specific antibody, which will then fight the disease. Many diseases can only be contracted once because antibodies remain in the blood after the infection has passed, preventing any further invasion. Vaccination boosts a person's resistance by causing the production of antibodies specific to particular infections.

Large quantities of specific antibodies can now be obtained by the monoclonal technique (see †monoclonal antibody).

Antibodies were discovered in 1890 by the German physician Emil von Behring and the Japanese bacteriologist Shibasaburo †Kitasato.

Antibody Resource Page http://www.antibodyresource.com/

Fascinating access site for all you could ever want to know about antibodies and some things you'd probably rather not know! This site contains several educational resources (aimed at university level), but also contains images, animations, and descriptions of research into many different types of antibodies.

anticlastic curvature in physics, saddle-shaped curvature on the upper surface of a horizontal beam that is loaded (and thus bent downward) at each of its ends.

anticline in geology, rock layers or beds folded to form a convex arch (seldom preserved intact) in which older rocks comprise the core. Where relative ages of the rock layers, or stratigraphic ages, are not known, convex upward folded rocks are referred to as antiforms.

The fold of an anticline may be undulating or steeply curved. A step like bend in otherwise gently dipping or horizontal beds is a **monocline**. The opposite of an anticline is a †syncline.

anticlockwise direction of rotation, opposite of the way the hands of a clock turn.

anticoagulant substance that inhibits the formation of blood clots. Common anticoagulants are heparin, produced by the liver and some white blood cells, and derivatives of coumarin, such as warfarin. Anticoagulants are used medically in the prevention and treatment of thrombosis and heart attacks. Anticoagulant substances are also produced by blood-feeding animals, such as mosquitoes, leeches, and vampire bats, to keep the victim's blood flowing.

Most anticoagulants prevent the production of thrombin, an enzyme that induces the formation from blood plasma of fibrinogen, to which blood platelets adhere and form clots.

anticyclone area of high atmospheric pressure caused by descending air, which becomes warm and dry. Winds radiate from a calm center, taking a clockwise direction in the northern hemisphere and an anticlockwise direction in the southern hemisphere. Anticyclones are characterized by clear weather and the absence of rain and violent winds. In summer they bring hot, sunny days and in winter they bring fine, frosty spells, although fog and low cloud are not uncommon in the UK. Blocking anticyclones, which prevent the normal air circulation of an area, can cause summer droughts and severe winters.

antidiuretic hormone pituitary hormone that prevents excessive fluid loss. See †ADH.

antiferromagnetic material material with a very low magnetic †susceptibility that increases with temperature up to a certain temperature, called the †Néel temperature. Above the Néel temperature, the material is only weakly attracted to a strong magnet.

antifreeze substance added to a water-cooling system (for example, that of a car) to prevent it freezing in cold weather.

antigen any substance that causes the production of †antibodies by the body's immune system. Common antigens include the proteins carried on the surface of bacteria, viruses, and pollen grains. The proteins of incompatible blood groups or tissues also act as antigens, which has to be taken into account in medical procedures such as blood transfusions and organ transplants.

antigravity in astronomy, hypothetical force of repulsion between objects because of their mass, as postulated to explain such phenomena as the repulsion of light by Kerr (spinning) †black holes.

antiknock substance added to †gasoline to reduce knock in car engines, caused by premature combusting. It is a mixture of dibromoethane and tetraethyl lead. Its use in leaded gasoline has resulted in atmospheric pollution by lead compounds.

antilogarithm or **antilog**, inverse of †logarithm, or the number whose logarithm to a given base is a given number. If $y = \log_a x$, then $x = \text{antilog}_a\, y$.

antimatter in physics, form of matter in which most of the attributes (such as electrical charge, magnetic moment, and spin) of †elementary particles are reversed. These †antiparticles can be created in particle accelerators, such as those at †CERN in Geneva, Switzerland, and at †Fermilab in the USA. In 1996 physicists at CERN created the first atoms of antimatter: nine atoms of antihydrogen survived for 40 nanoseconds (40 billionths of a second).

antimony chemical symbol Sb, silver-white, brittle, semi metallic element (a metalloid), atomic number 51, relative atomic mass 121.75. Its chemical symbol comes from Latin *stibium*. It occurs chiefly as the ore stibnite, and is used to make alloys harder; it is also used in photosensitive substances in color photography, optical electronics, fireproofing, pigments, and medicine. It was employed by the ancient Egyptians in a mixture to protect the eyes from flies.

antinode in physics, the position in a †standing wave pattern at which the amplitude of vibration is greatest (compare †node). The standing wave of a stretched string vibrating in the fundamental mode has one antinode at its midpoint. A vibrating air column in a pipe has an antinode at the pipe's open end and at the place where the vibration is produced.

anti-oxidant any substance that prevents deterioration of fats, oils, paints, plastics, and rubbers by oxidation. When used as food additives, anti-oxidants prevent fats and oils from becoming rancid when exposed to air, and thus extend their shelf life.

Vegetable oils contain natural anti-oxidants, such as vitamin E, which prevent spoilage, but anti-oxidants are nevertheless added to most oils. They are not always listed on food labels because if a food manufacturer buys an oil to make a food product, and the oil has anti-oxidant already added, it may not list it on the label of the product.

antiparticle in nuclear physics, a particle corresponding in mass and properties to a given †elementary particle but with the opposite electrical charge, magnetic properties, or coupling to other fundamental forces. For example, an electron carries a negative charge whereas its antiparticle, the positron, carries a positive one. When a particle and its antiparticle collide, they destroy each other, in the process called "annihilation," their total energy being converted to lighter particles and/or photons. A substance consisting entirely of antiparticles is known as †antimatter.

antipodal in mathematics, describing two points on a sphere where the line joining them passes through the center of the sphere. For example, the North and South Poles are antipodal.

antipodes (Greek "opposite feet") places at opposite points on the globe.

antipsychiatry school of thought derived from the ideas of US psychoanalyst Thomas Szasz, R D †Laing, and others, who have argued strenuously against the usefulness both of the concept of mental illness and of psychiatric treatment in general.

Szasz has argued, for example, that in most cases of incapacity due to psychological causes it is inappropriate to talk about illness. Not only can psychiatrists differ widely in their diagnoses, but it is also often difficult to determine whether someone is actually mentally ill. Furthermore, labeling someone as mentally ill can have unfortunate consequences owing to the prejudice that persists in society. Adherents of antipsychiatry see society as the cause of mental illness, application of the label "mentally ill" as an attempt to cover up this fact, and treatment of the so-called mentally ill as a postponement of society's obligation to deal with the root of the problem.

antiseptic any substance that kills or inhibits the growth of micro-organisms. The use of antiseptics was pioneered by Joseph Lister. He used carbolic acid (†phenol), which is a weak antiseptic; antiseptics such as TCP are derived from this.

Lister's work was extended by surgeons such as William Cheyne, professor of surgery at Kings College Hospital, London, and author of *Antiseptic Surgery* (1882).

antivirus software computer software that detects †viruses and/or cleans viruses from an infected computer system. Antivirus programs can be designed to catch all viruses entering a system, or to catch only those viruses delivered by a particular method, such as e-mail. In order for the software to remain effective it must be updated regularly with new signature files to enable the software to recognize the latest viruses; this can usually be done over the Internet.

Antlia or **Airpump**, small, faint constellation in the southern hemisphere, identified with an air pump.

anus or **anal canal**, opening at the end of the †alimentary canal which allows undigested foods and other waste materials to pass out of the body in the form of feces. In humans the term is also used to describe the last 4 cm/1.5 in of the alimentary canal. It is normally kept closed by a ring of muscle called a sphincter. A common medical condition in humans associated with the anus is hemorrhoids (piles).

The anus is found in all types of multicellular animal except the coelenterates (sponges) and the platyhelminths (flatworms), which have a mouth only.

aorta the body's main †artery, arising from the left ventricle of the heart in birds and mammals. Carrying freshly oxygenated blood, it arches over the top of the heart and descends through the trunk, finally splitting in the lower abdomen to form the two iliac arteries. Arteries branching off the arch of the aorta carry blood to the upper body. Loss of elasticity in the aorta provides evidence of atherosclerosis, which may lead to heart disease.

In fish a ventral aorta carries deoxygenated blood from the heart to the †gills, and the dorsal aorta carries oxygenated blood from the gills to other parts of the body.

apastron point at which an object traveling in an elliptical orbit around a star is at its farthest from the star. The term is usually applied to the position of the minor component of a †binary star in relation to the primary. Its opposite is the †periastron.

apatite common calcium phosphate mineral, $Ca_5(PO_4)_3(F,OH,Cl)$. Apatite has a hexagonal structure and occurs widely in igneous rocks, such as pegmatite, and in contact metamorphic rocks, such as marbles. It is used in the manufacture of fertilizer and as a source of phosphorus. Carbonate hydroxylapatite, $Ca_5(PO_4,CO_3)_3(OH)_2$, is the chief constituent of tooth enamel and, together with other related phosphate minerals, is the inorganic constituent of bone. Apatite ranks 5 on the †Mohs scale of hardness.

aperture size of the opening admitting light into an optical instrument such as a telescope or camera.

aperture synthesis in astronomy, technique used in †radio astronomy in which several small radio dishes are linked together over an area that can be many kilometers in diameter. This is done to simulate the performance of one very large †radio telescope.

apetalous plant plant in which the petals are suppressed.

apex the highest point of a triangle, cone, or pyramid – that is, the vertex (corner) opposite a given base.

aphelion point at which an object, traveling in an elliptical orbit around the Sun, is at its farthest from the Sun. This is a solar-orbit apoapsis. The Earth is at its aphelion on 5 July.

Apianus, Petrus (1495–1552) born Peter Bienewitz, German astronomer, cartographer, and instrument maker. His *Cosmographia* (1524), translated into all the major European languages, was the most popular text of its time on astronomy and geography. It included the suggestion that lunar distances might be used to determine longitude.

aplanatic lens lens that is free from spherical aberration and coma. See †aberration, optical.

apochromatic lens lens constructed of three or more different types of glass so that it corrects chromatic aberration at three different wavelengths.

apogee point at which a spacecraft, or other object, traveling in an elliptical orbit around the Earth is at its farthest from the Earth. This is an Earth-orbit apoapsis.

Apollo asteroid member of a group of †asteroids whose orbits cross that of the Earth. They are named after the first of their kind, Apollo, discovered in 1932 by German astronomer Karl Reinmuth and then lost until 1973. Apollo asteroids are so small and faint that they are difficult to see except when close to the Earth (Apollo is about 2 km/1.2 mi across).

Apollonius of Perga (*c.* 262–*c.* 190 BC) Greek mathematician, called "the Great Geometer." In his work *Konica/The Conics* he showed that a plane intersecting a cone will generate an ellipse, a parabola, or a hyperbola, depending on the angle of intersection. In astronomy, he used a system of circles called epicycles and deferents to explain the motion of the planets; this system, as refined by Ptolemy, was used until the Renaissance.

Apollonius of Perga http://www-history.mcs.st-andrews.ac.uk/history/Mathematicians/Apollonius.html
Biography of the Greek philosopher and mathematician Apollonius of Perga. Excerpts from an early 18th-century translation of Apollonius' greatest work, Conics, are included on this Web site, as are links to other important works.

Apollonius' problem problem set by Apollonius of Perga, to describe a circle touching three other given circles.

apoptosis or **cell suicide**, self-destruction of a cell. All cells contain genes that cause them to self-destruct if damaged, diseased, or as part of the regulation of cell numbers during the organism's normal development. Many cancer cells have mutations in genes controlling apoptosis, so understanding apoptosis may lead to new cancer treatments where malfunctioning cells can be instructed to destroy themselves.

During apoptosis, a cell first produces the enzymes needed for self-destruction before shrinking to a characteristic spherical shape with balloon-like bumps on its outer surface. The enzymes break down its contents into small fragments that are easily digestible by surrounding cells.

aposematic coloration in biology, the technical name for warning coloration markings that make a dangerous, poisonous, or foul-tasting animal particularly conspicuous and recognizable to a predator. Examples include the yellow and black stripes of bees and wasps, and the bright red or yellow colors of many poisonous frogs and snakes. See also †mimicry.

apparent depth depth that a transparent material such as water or glass appears to have when viewed from above. This is less than its real depth because of the †refraction that takes place when light passes into a less dense medium. The ratio of the real depth to the apparent depth of a transparent material is equal to its †refractive index.

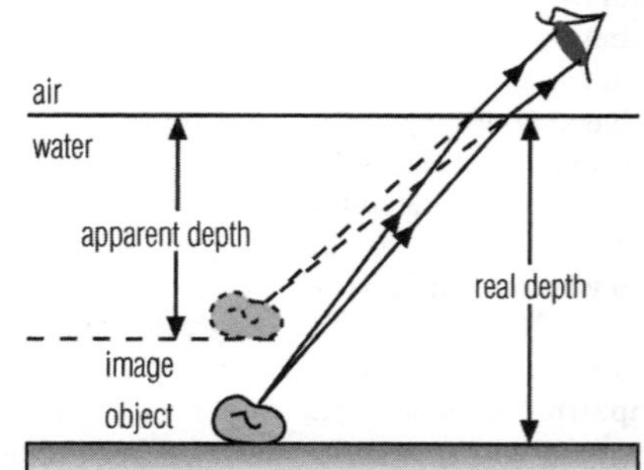

apparent depth The apparent depth of an object in water when viewed from above. This is less than the object's real depth because of refractive effects: the light wave bends as it passes from the denser water into the less dense air. Helicon

appendix short, blind-ended tube attached to the †cecum. It has no known function in humans, but in herbivores it may be large, containing millions of bacteria that secrete enzymes to digest grass (as no vertebrate can secrete enzymes that will digest cellulose, the main constituent of plant cell walls).

The appendix secretes antibodies into the gut and so plays a role in the body's immune system, though it is not a vital one, as patients survive perfectly well following removal of the appendix.

applet in computing, mini-software application. Examples of applets include Microsoft WordPad, the simple word processor in Microsoft Windows or the single-purpose applications that appear on the World Wide Web, written in †Java. These include animations such as a moving ticker tape of stock prices.

Today Java is the most common programming language used to create applets for the World Wide Web, though early enthusiasm for its potential has died down. Development of a number of high-profile Web-applet based applications has been stopped because of problems with performance and compatibility.

Appleton, Edward Victor (1892–1965) British physicist. He worked at Cambridge under Ernest †Rutherford from 1920. He proved the existence of the Kennelly-Heaviside layer (now called the †E-layer) in the atmosphere, and the †Appleton layer beyond it, and was involved in the initial work on the atom bomb. He was awarded the Nobel Prize for Physics in 1947 for his work on the physics of the upper atmosphere. He was made KCB in 1941 and GBE in 1946.

Appleton layer or **F-layer**, band containing ionized gases in the Earth's upper atmosphere, at a height of 150–1,000 km/94–625 mi, above the †E-layer (formerly the Kennelly-Heaviside layer). It acts as a dependable reflector of radio signals as it is not affected by atmospheric conditions, although its ionic composition varies with the sunspot cycle.

The Appleton layer has the highest concentration of free electrons and ions of the atmospheric layers. It is named after the English physicist Edward †Appleton.

application in computing, program or job designed for the benefit of the end user. Examples of **general-purpose** application programs include †word processors, †desktop publishing programs, †databases, †spreadsheet packages, and graphics programs (see †CAD and †CAM). **Application-specific** programs include payroll and stock control systems. Applications may also be **custom designed** to solve a specific problem, not catered for in other types of application.

The term is used to distinguish such programs from those that control the computer (†systems programs) or assist the programmer, such as a †compiler.

application in mathematics, a curved line that connects a series of points (or "nodes") in the smoothest possible way. The shape of the curve is governed by a series of complex mathematical formulae.

In computing, a program designed for the end user.

approximation rough estimate of a given value. For example, for †pi (which has a value of 3.1415926 correct to seven decimal places), 3 is an approximation to the nearest whole number.

appulse in astronomy, apparent close approach of any two heavenly objects, as viewed from Earth, usually because they happen to align from the observer's point of view; their true distance apart may remain very large.

apsis plural **apsides**, either of two points in an orbit, one being at maximum and the other at minimum distance from the controlling central body. The †apogee and †perigee of the Moon are apsides, as are the †aphelion and †perihelion of a planet.

Apus in astronomy, faint minor constellation in the southern hemisphere near Crux (the Southern Cross); its name means bird of paradise.

aqua fortis former name for †nitric acid.

aquamarine blue variety of the mineral †beryl. A semi-precious gemstone, it is used in jewelry.

aqua regia (Latin "royal water") mixture of three parts concentrated hydrochloric acid and one part concentrated nitric acid, which dissolves "noble" metals such as gold and platinum.

aquarium tank or similar container used for the study and display of living aquatic plants and animals. The same name is used for institutions that exhibit aquatic life. These have been common since Roman times, but the first modern public aquarium was opened in Regent's Park, London, in 1853. A recent development is the oceanarium or seaquarium, a large display of marine life forms.

FINS – Fish Information Service http://www.actwin.com/fish/index.php

Huge source of practical information on all aspects of keeping fish in aquariums. The needs of all manner of tropical and temperate fish, marine, and freshwater fish are covered in detail. With the help of FINS you can learn whatever you want about ichthyology.

Aquarius zodiacal constellation a little south of the celestial equator near Pegasus. Aquarius is represented as a man pouring water from a jar. The Sun passes through Aquarius from late February to early March. In astrology, the dates for Aquarius, the 11th sign of the zodiac, are between about 20 January and 18 February (see †precession).

aquatic living in water. All life on Earth originated in the early oceans, because the aquatic environment has several advantages for organisms. Dehydration is almost impossible, temperatures usually remain stable, and the density of water provides physical support.

Life forms that cannot exist out of water, amphibians that take to the water on occasions, animals that are also perfectly at home on land, and insects that spend a stage of their life cycle in water can all be described as aquatic. Aquatic plants are known as †hydrophytes.

aqueous humour watery fluid found in the chamber between the cornea and lens of the vertebrate eye. Similar to blood serum in composition, it is constantly renewed.

aqueous solution solution in which the solvent is water.

aquifer a body of rock through which appreciable amounts of water can flow. The rock of an aquifer must be porous and permeable (full of interconnected holes) so that it can conduct water. Aquifers are an important source of fresh water, for example for drinking and irrigation, in many arid areas of the world, and are exploited by the use of †artesian wells.

An aquifer may be underlain, overlain, or sandwiched between less permeable layers, called **aquicludes** or **aquitards**, which impede water movement. Sandstones and porous limestones make the best aquifers.

Edwards Aquifer Home Page http://www.edwardsaquifer.net/

Guide to the Edwards Aquifer (a rock formation containing water) in Texas – one of the most prolific artesian aquifers in the world.

Aquila constellation on the celestial equator (see †celestial sphere). Its brightest star is first-magnitude †Altair, flanked by the stars Beta and Gamma Aquilae. It is represented by an eagle.

Nova Aquilae, which appeared June 1918, shone for a few days nearly as brightly as Sirius.

Ara in astronomy, faint minor constellation in the southern hemisphere between †Triangulum Australe and †Scorpius; its name means altar.

arabin or **arabinic acid** or **gummic acid**, constituent of gum arabic, which is made up of the potassium and calcium salts of arabin. Arabinic acid is a complex polyurenic substance containing galacto-uronic acid, galactose, arabinose, and rhamnose as structural units.

arachnid or **arachnoid**, type of arthropod of the class Arachnida, including spiders, scorpions, ticks, and mites. They differ from insects in possessing only two main body regions, the cephalothorax and the abdomen, and in having eight legs.

Arachnida class of the phylum Arthropoda that includes spiders, scorpions, ticks, and mites; see †arachnid.

Arago, (Dominique) François (Jean) (1786–1853) French physicist and astronomer who made major contributions to the early study of electromagnetism. In 1820 he found that iron enclosed in a wire coil could be magnetized by the passage of an electric current. In 1824, he was the first to observe the ability of a rotating copper disc to deflect a magnetic needle, the phenomenon of magnetic rotation.

aragonite white, yellowish or gray mineral, calcium carbonate, $CaCO_3$, a denser, harder polymorph (substance having the same chemical composition but different structure) of calcite.

Secreted by corals, mollusks and green algae, it is an important constituent of shallow marine muds and pearl.

arboretum collection of trees. An arboretum may contain a wide variety of species or just closely related species or varieties – for example, different types of pine tree.

arc in geometry, a section of a curved line or circle. A circle has three types of arc: a **semicircle**, which is exactly half of the circle; **minor arcs**, which are less than the semicircle; and **major arcs**, which are greater than the semicircle.

An arc of a circle is measured in degrees, according to the angle formed by joining its two ends to the center of that circle. A semicircle is therefore 180°, whereas a minor arc will always be less than 180° (acute or obtuse) and a major arc will always be greater than 180° but less than 360° (reflex).

arch in geomorphology (the study of landforms), any natural bridge-like land feature formed by †erosion. Most sea arches are formed from the wave erosion of a †headland where the backs of two caves have met and broken through. The roof of the arch eventually collapses to leave part of the headland isolated in the sea as a †stack. A natural bridge is formed on land by wind or water erosion and spans a valley or ravine.

In some cases, as at Stair Hole, Dorset, England, an arch is formed when the sea has battered a hole through a cliff of hard rock (such as limestone) and has removed soft rock (such as clay) behind.

Archaea group of micro-organisms that are without a nucleus and have a single chromosome. Most taxonomists now classify these bacteria in their own kingdom, separate from other bacteria. All are strict anaerobes, that is, they are killed by oxygen. This is thought to be a primitive condition and to indicate that Archaea are related to the earliest life forms, which appeared about 4 billion years ago, when there was little oxygen in the Earth's atmosphere. They are found in undersea vents, hot springs, the Dead Sea, and salt pans, and have even adapted to refuse tips.

Archaea was originally classified as bacterial, but in 1996 when the genome of *Methanococcus jannaschii* (an archaeaon that lives in undersea vents at temperatures around 100°C/212°F) was sequenced, US geneticists found that 56% of its genes were unlike those of any other organism, making Archaea unique.

In 1994 US biologists detected archaeaons in the Antarctic (where they make up 30% of the single-celled marine biomass), Arctic, Mediterranean, and Baltic Sea.

Archaean or **Archaeozoic**, widely used term for the earliest era of geological time; the first part of the Precambrian **Eon**, spanning the interval from the formation of Earth to about 2,500 million years ago.

archegonium (Greek *arche* "origin," *gonos* "offspring") female sex organ found in bryophytes (mosses and liverworts), pteridophytes (ferns, club mosses, and horsetails), and some †gymnosperms. It is a multicellular, flask-shaped structure consisting of two parts: the swollen base or venter containing the egg cell, and the long, narrow neck. When the egg cell is mature, the cells of the neck dissolve, allowing the passage of the male gametes, or †antherozoids.

Archimedes (*c.* 287–212 BC) Greek mathematician and philosopher who made major discoveries in geometry, hydrostatics, and mechanics, and established the sciences of statics and hydrostatics. He formulated a law of fluid displacement (Archimedes' principle), and is credited with the invention of the Archimedes screw, a cylindrical device for raising water. His method of finding mathematical proof to substantiate experiment and observation became the method of modern science in the High Renaissance.

Archimedes Home Page http://www.mcs.drexel.edu/~crorres/Archimedes/contents.html

Extensive material on the ancient Greek philosopher and mathematician Archimedes. The site includes sections on his life and travels, as well as extracts from ancient sources about his inventions and some famous problems associated with him.

Archimedes' principle in physics, the principle that the weight of the liquid displaced by a floating body is equal to the weight of the body. The principle is often stated in the form: "an object totally or partially submerged in a fluid displaces a volume of fluid that weighs the same as the apparent loss in weight of the object (which, in turn, equals the upwards force, or upthrust, experienced by that object)." It was discovered by the Greek mathematician Archimedes.

archipelago group of islands, or an area of sea containing a group of islands. The islands of an archipelago are usually volcanic in origin, and they sometimes represent the tops of peaks in areas around continental margins flooded by the sea.

Volcanic islands are formed either when a hot spot within the Earth's mantle produces a chain of volcanoes on the surface, such as the Hawaiian Archipelago, or at a destructive plate margin (see †plate tectonics) where the subduction of one plate beneath another produces an arc-shaped island group called an "island arc," such as the Aleutian Archipelago. Novaya Zemlya in the Arctic Ocean, the northern extension of the Ural Mountains, resulted from continental flooding.

architecture in computing, overall design of a computer system, encompassing both hardware and software. The architecture of a particular system includes the specifications of individual components and the ways they interact. Because the operating system defines how these elements interact with each other and with application software, it is also included in the term.

Computer Architecture Page http://www.cs.wisc.edu/~arch/www/

Directory of computer architecture resources on the Internet. This simple Web site has links to computer manufacturers, research groups, software tools, and other resources.

archive collection of computer files. The term is commonly used to refer to the files created by †data compression programs, such as the popular PKZIP, which contain one or more files. On the Internet it is also used to refer to a large store of files from which visitors can select the ones they want.

arc minute symbol ', unit for measuring small angles, used in geometry, surveying, map-making, and astronomy. An arc minute is one-sixtieth of a degree and is divided into 60 arc seconds (symbol "). Small distances in the sky, as between two close stars or the apparent width of a planet's disc, are expressed in minutes and seconds of arc.

Arctic Circle imaginary line that encircles the North Pole at latitude 66° 30' north. Within this line there is at least one day in the summer during which the Sun never sets, and at least one day in the winter during which the Sun never rises.

Arctic Circle http://arcticcircle.uconn.edu/

Well-written site with information about all aspects of life in the Arctic. There are sections on history, natural resources, the rights of indigenous peoples, and issues of environmental concern.

Arcturus or **Alpha Boötis**, brightest star in the constellation Boötes and the fourth-brightest star in the night sky. Arcturus is a red giant about 28 times larger than the Sun and 70 times more luminous, 36 light years away from the Sun.

are metric unit of area, equal to 100 square meters (119.6 sq yd); 100 ares make one †hectare.

area size of a surface. It is measured in square units, usually square centimeters (cm^2), square meters (m^2), or square kilometers (km^2). **Surface area** is the area of the outer surface of a solid. †Integration may be used to determine the area of shapes enclosed by curves.

The areas of †two-dimensional (2D) shapes with straight edges are determined using the area of a rectangle. For example, the area of a parallelogram is found by taking a triangle from one end and moving it to the other, forming a rectangle.

arête German **grat**; North American **combe-ridge**, sharp narrow ridge separating two †glacial troughs (U-shaped valleys), or †corries. They are formed by intense †freeze–thaw weathering on the sides of mountains. The typical U-shaped cross-sections of glacial troughs give arêtes very steep sides. Arêtes are common in glaciated mountain regions such as the Rockies, the Himalayas, and the Alps. There are also several in the UK, for example Striding Edge and Swirral Edge in the English Lake District.

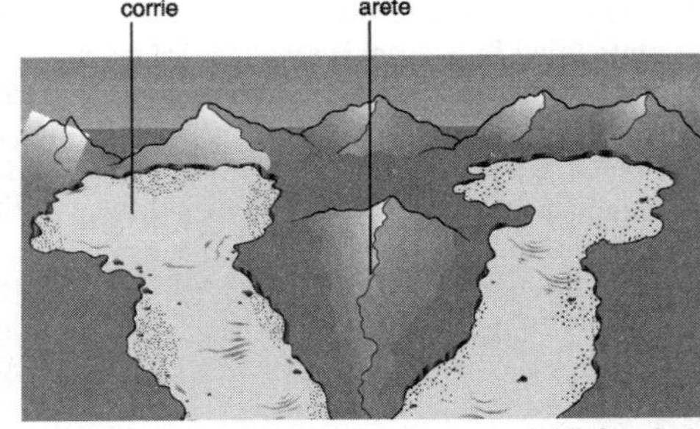

arête An arête is a steep-sided ridge between two glacially-formed valleys, or corries. If three corries all erode towards the same point, a triangular arête may be formed. Helicon

Argand diagram in mathematics, a method for representing complex numbers by Cartesian coordinates (x, y). Along the x-axis (horizontal axis) are plotted the real numbers, and along the y-axis (vertical axis) the nonreal, or †imaginary, numbers.

Argelander, Friedrich Wilhelm August (1799–1875) Prussian astronomer. He made a catalogue of 324,198 stars in the northern hemisphere, the *Bonner Durchmusterung/Bonn Survey* (1859–62). His *Uranometrica nova* (1843) introduced the "estimation by steps" method for determining stellar magnitudes with the naked eye.

Argo very large, bright constellation of the southern hemisphere, represented as a ship. It was described by Greek Astronomer †Ptolemy in the second century AD, but now, to avoid confusion, is split into four: †Carina, "the Keel"; †Puppis, "the Stern"; †Vela, "the Sails"; and †Pyxis, "the Compass." The last was formed from stars in Malus, "the Mast," another former subdivision of Argo that has not survived as a separate constellation.

argol crude product deposited on the bottom of a cask during alcoholic fermentation. It is used for the preparation of "cream of tartar" and tartaric acid.

The argol is partially purified by recrystallization from hot water, then dissolved in water, and boiled with chalk, when calcium tartrate, $CaC_4H_4O_6$, is precipitated. After washing, the tartrate is treated with sulfuric acid, which sets free the tartaric acid.

argon chemical symbol Ar, (Greek *argos* "idle") colorless, odorless, non-metallic, gaseous element, atomic number 18, relative atomic mass 39.948. It is grouped with the †noble gases (rare gases) in Group 0 of the †periodic table of the elements. It was long believed not to react with other substances, but observations now indicate that it can be made to combine with boron fluoride to form compounds. It constitutes almost 1% of the Earth's atmosphere, and was discovered in 1894 by British chemists John Rayleigh and William Ramsay after all oxygen and nitrogen had been removed chemically from a sample of air. It is used in electric discharge tubes and argon lasers.

Discovery of Argon – A Case Study in Scientific Method http://webserver.lemoyne.edu/faculty/giunta/acspaper.html

Account of the background to Ramsay and Raleigh's discovery of argon. This is part of a course designed to show the influence of scientific method on the progress of science. The reasons for the failure of earlier chemists to isolate the inert gas are also set out here.

argument in astronomy, angular distance of an orbiting spacecraft or other body from the object being orbited. For example, the **argument of periapsis** is the angle between the periapsis (closest point of the orbiting body to the object it orbits) and the ascending node (point where the orbit intersects the plane of the †ecliptic, moving from south to north of the ecliptic). The **argument of perihelion** is the angle between the ascending node and the perihelion (closest point between the Sun and an orbiting body).

argument in mathematics, a specific value of the independent variable of a †function of x. It is also another name for †amplitude.

arid region in earth science, a region that is very dry and has little vegetation. Aridity depends on temperature, rainfall, and evaporation, and so is difficult to quantify, but an arid area is usually defined as one that receives less than 250 mm/10 in of rainfall each year. (By comparison, New York City receives 1,120 mm/44 in per year.) There are arid regions in North Africa, Pakistan, Australia, the USA, and elsewhere. Very arid regions are †deserts.

Ariel in astronomy, innermost of the five major moons (or satellites) of the planet Uranus. It was officially discovered in 1851 by English astronomer William Lassell, but was probably seen by Russian astronomer Otto †Struve four years earlier.

Ariel is thought to consist primarily of water ice. It has an estimated diameter of 1,160 km/721 mi and orbits Uranus at an average distance of 191,020 km/118,432 mi with a period (the time it takes to circle the planet) of 2.52 days. Images from the Voyager 2 probe, taken in January 1986, show craters and channels that may be caused by volcanic activity.

Aries zodiacal constellation in the northern hemisphere between Pisces and Taurus, near Auriga, represented as the legendary ram whose golden fleece was sought by Jason and the Argonauts.

Its most distinctive feature is a curve of three stars of decreasing brightness. The brightest of these is Hamal or Alpha Arietis, 65 light years from Earth.

The Sun passes through Aries from late April to mid-May. In astrology, the dates for Aries, the first sign of the zodiac, are between about 21 March and 19 April (see †precession). The spring †equinox once lay in Aries, but has now moved into Pisces through the effect of the Earth's precession (wobble).

aril accessory seed cover other than a †fruit; it may be fleshy and sometimes brightly colored, woody, or hairy. In flowering plants (†angiosperms) it is often derived from the stalk that originally attached the ovule to the ovary wall. Examples of arils include the bright-red, fleshy layer surrounding the yew seed (yews are †gymnosperms so they lack true fruits), and the network of hard filaments that partially covers the nutmeg seed and yields the spice known as mace.

Another aril, the horny outgrowth found towards one end of the seed of the castor-oil plant *Ricinus communis*, is called a caruncle. It is formed from the integuments (protective layers enclosing the ovule) and develops after fertilization.

Aristarchus of Samos (*c.* 320–*c.* 250 BC) Greek astronomer. The first to argue that the Earth moves around the Sun, he was ridiculed for his beliefs. He was also the first astronomer to estimate (quite inaccurately) the sizes of the Sun and Moon and their distances from the Earth.

Aristotle's lantern jaw apparatus of the sea urchins, consisting of five hard, pointed teeth in sockets formed by five ossicles. They are used to scrape algae and seaweeds from the rocks for food.

arithmetic branch of mathematics concerned with the study of numbers and their properties. The fundamental operations of arithmetic are addition, subtraction, multiplication, and division. Raising to powers (for example, squaring or cubing a number), the extraction of roots (for example, square roots), percentages, fractions, and ratios are developed from these operations.

Forms of simple arithmetic existed in prehistoric times. In China, Egypt, Babylon, and early civilizations generally, arithmetic was used for commercial purposes, records of taxation, and astronomy. During the Dark Ages in Europe, knowledge of arithmetic was preserved in India and later among the Arabs. European mathematics revived with the development of trade and overseas exploration. Hindu-Arabic numerals replaced Roman numerals, allowing calculations to be made on paper, instead of by the †abacus.

The essential feature of this number system was the introduction of zero, which allows us to have a **place–value** system. The decimal numeral system employs ten numerals (0,1,2,3,4,5,6,7,8,9) and is said to operate in "base ten." In a base-ten number, each position has a value ten times that of the position to its immediate right; for example, in the number 23 the numeral 3 represents three units (ones), and the numeral 2 represents two tens. The Babylonians, however, used a complex base-sixty system, residues of which are found today in the number of minutes in each hour and in angular measurement (6 x 60 degrees). The Mayas used a base-twenty system.

There have been many inventions and developments to make the manipulation of the arithmetic processes easier, such as the invention of †logarithms by Scottish mathematician John †Napier in 1614 and of the slide rule in the period 1620–30. Since then, many forms of ready reckoners, mechanical and electronic calculators, and computers have been invented.

Modern computers fundamentally operate in base two, using only two numerals (0,1), known as a **binary** system. In binary, each position has a value twice as great as the position to its immediate right, so that for example binary 111 (or 111_2) is equal to 7 in the decimal system, and binary 1111 (or 1111_2) is equal to 15. Because the main operations of subtraction, multiplication, and division can be reduced mathematically to addition, digital computers carry out calculations by adding, usually in binary numbers in which the numerals 0 and 1 can be represented by off and on pulses of electric current.

Modular or modulo arithmetic, sometimes known as residue arithmetic or clock arithmetic, can take only a specific number of digits, whatever the value. For example, in modulo 4 (mod 4) the only values any number can take are 0, 1, 2, or 3. In this system, 7 is written as 3 mod 4, and 35 is also 3 mod 4. Notice 3 is the residue, or remainder, when 7 or 35 is divided by 4. This form of arithmetic is often illustrated on a circle. It deals with events recurring in regular cycles, and is used in describing the functioning of gasoline engines, electrical generators, and so on. For example, in the mod 12, the answer to a question as to what time it will be in five hours if it is now ten o'clock can be expressed 10 + 5 = 3.

arithmetic and logic unit ALU, in a computer, the part of the †central processing unit (CPU) that performs the basic arithmetic and logic operations on data.

arithmetic mean average of a set of *n* numbers, obtained by adding the numbers and dividing by *n*. For example, the arithmetic mean of the set of 5 numbers 1, 3, 6, 8, and 12 is (1 + 3 + 6 + 8 + 12) ÷ 5 = 30 ÷ 5 = 6.

The term †average is often used to refer only to the arithmetic mean, even though the mean is in fact only one form of average (the others include †median and †mode). In addition it is useful to know the †range or spread of the data.

arithmetic progression or **arithmetic sequence**, sequence of numbers or terms that have a common difference between any one term and the next in the sequence. For example, 2, 7, 12, 17, 22, 27, ... is an arithmetic sequence with a common difference of 5.

The *n*th term in any arithmetic progression can be found using the formula:

$n\text{th term} = a + (n - 1)d$

where *a* is the first term and *d* is the common difference.

An **arithmetic series** is the sum of the terms in an arithmetic sequence. The sum *S* of *n* terms is given by:

$S = n/2[2a + (n - 1)d]$

arkose in earth science, type of †sandstone composed of grains of quartz and feldspar, derived from crystalline rocks such as gneiss or granite. It may be gray, pink, or red, often colored by iron oxide impurities.

arm fore or upper limb in humans from the shoulder to the wrist. The humerus, or bone of the upper arm, fits at the top into the scapula, or shoulder blade. At the lower end it articulates with the radius and ulna, the bones of the forearm. These are articulated with each other and with the wrist bones.

The deltoid is the large muscle that forms the shoulder cap and serves to raise the arm from the side; it runs from the shoulder blade to the middle of the humerus. The chief muscles engaged in lowering the arm again are the *latismus dorsi*, which runs forwards and upwards from the back and side of the chest wall to be attached to the front of the neck of the humerus; the *coraco-brachialis*, attached to the middle of the front of the humerus; and the *pectoralis major*, which forms the front fold of the armpit.

The arm is flexed by the biceps, running from shoulder to elbow joint in front, and is straightened by the triceps, running from shoulder to elbow at the back. The muscles of the forearm consist of the flexors and extensors of the wrist and fingers.

armature in a motor or generator, the wire-wound coil that carries the current and rotates in a magnetic field. (In alternating-current machines, the armature is sometimes stationary.) The pole piece of a permanent magnet or electromagnet and the moving, iron part of a †solenoid, especially if the latter acts as a switch, may also be referred to as armatures.

armillary sphere earliest known astronomical device, in use from the 3rd century BC. It showed the Earth at the center of the universe, surrounded by a number of movable metal rings representing the Sun, Moon, and planets. The armillary sphere was originally used to observe the heavens and later for teaching navigators about the arrangements and movements of the heavenly bodies.

aromatic compound organic chemical compound in which some of the bonding electrons are delocalized (shared among several atoms within the molecule and not localized in the vicinity of the atoms involved in bonding). The commonest aromatic compounds have ring structures, the atoms compris-

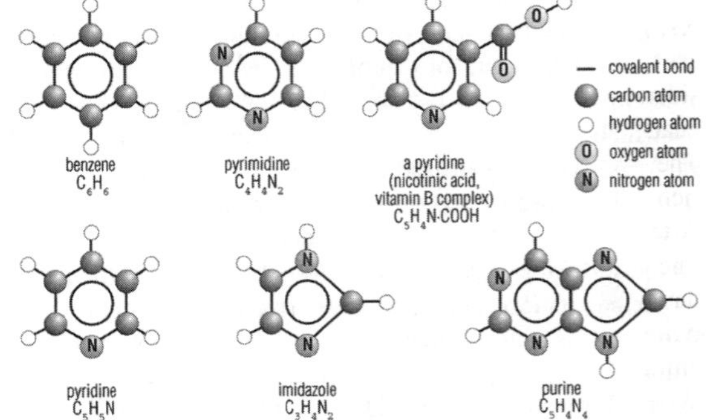

aromatic compound Compounds whose molecules contain the benzene ring, or variations of it, are called aromatic. The term was originally used to distinguish sweet-smelling compounds from others. Helicon

ing the ring being either all carbon or mostly carbon with one or more different atoms (usually nitrogen, sulfur, or oxygen). Typical examples are benzene (C_6H_6) and pyridine (C_5H_5N).

Arp, Halton Christian (1927–) US astronomer. Arp has worked particularly on the identification of galaxies. He published *Atlas of Peculiar Galaxies* (1965). He also carried out the first photometric work on the Magellanic Clouds – the nearest extragalactic system.

array collection of numbers (or letters representing numbers) arranged in rows and columns. A †matrix is an array shown inside a pair of brackets; it indicates that the array should be treated as a single entity.

Arrhenius, Svante August (1859–1927) Swedish scientist, the founder of physical chemistry. He was awarded the Nobel Prize for Chemistry in 1903 for his study of electrolysis. In 1905 he predicted global warming as a result of carbon dioxide emission from burning fossil fuels.

"Dissociation of Substances Dissolved in Water" by Svante Arrhenius

http://dbhs.wvusd.k12.ca.us/Chem-History/Arrhenius-dissociation.html

Extract from the above paper that appeared in Zeitschrift fur physikalische Chemie, I, 631, in 1887. In the paper Arrhenius discusses the dissociation of certain substances in water, an observation that led to deductions on electrolysis and his Nobel Prize in 1903.

Arrhenius theory of dissociation in chemistry, description of the behavior of aqueous solutions of †acids and †bases in terms of dissociation of ions, proposed by Swedish physical chemist Svante August †Arrhenius. He suggested that in aqueous solutions, acids dissociate to liberate hydrogen ions and, similarly, bases dissociate to form hydroxyl ions. This description explains why acids and alkaline solutions conduct electricity.

arsenate in chemistry, member of a group of compounds formed by the reaction of †arsenic oxide with a base. Compounds of this type are used in a wide range of commercial applications from printing chemicals to insecticides.

arsenic chemical symbol As, brittle, grayish-white, semi metallic element (a metalloid), atomic number 33, relative atomic mass 74.92. It occurs in many ores and occasionally in its elemental state, and is widely distributed, being present in minute quantities in the soil, the sea, and the human body. In larger quantities, it is poisonous. The chief source of arsenic compounds is as a by-product from metallurgical processes. It is used in making semiconductors, alloys, and solders.

arsenic poisoning As it is a cumulative poison, its presence in food and drugs is very dangerous. The symptoms of arsenic poisoning are vomiting, diarrhea, tingling and possibly numbness in the limbs, and collapse. It featured in some drugs, including Salvarsan, the first specific treatment for syphilis. Its name derives from the Latin *arsenicum*. The maximum safe level for arsenic in drinking water, as recommended by the World Health Organization, is 10 micrograms per liter.

arsenic oxide in chemistry, highly poisonous white powder found in two forms, As_2O_3 and As_2O_5. The more common form, As_2O_3, is also known as white arsenic. Arsenic oxide is used as a general-purpose rodent poison and herbicide.

arsine arsenic hydride, AsH_3, prepared by the action of dilute sulfuric acid on sodium or zinc arsenides. Arsine is a colorless, highly poisonous gas with an unpleasant odor, and is a very powerful reducing agent. It burns in air with a clear blue flame, giving water vapor and a black deposit of arsenic.

arsphenamine organic compound of arsenic discovered by Paul Ehrlich in 1909, used to treat syphilis. It was given the proprietary name †Salvarsan. Neoarsphenamine (neosalvarsan) followed arsphenamine and was used until 1945, when penicillin superseded it.

arteriole in biology, blood vessel that branches from an artery, which branches further to form capillaries. The diameter of an arteriole is intermediate between that of an †artery and a †capillary.

artery blood vessel that carries †blood from the †heart to any part of the body. It is built to withstand considerable pressure, having thick walls that contain †muscle and elastic fibers. As blood pulses out of the heart, arteries expand to allow for the increase in pressure – this elasticity helps the blood to flow evenly. The †pulse or pressure wave generated can be felt at the wrist. Not all arteries carry oxygenated (oxygen-rich) blood – the pulmonary arteries convey deoxygenated (oxygen-poor) blood from the heart to the lungs. Arteries are flexible, elastic tubes, consisting of three layers. The middle layer is muscular and its rhythmic contraction aids the pumping of blood around the body. In middle and old age, artery walls become damaged. Some of this damage is caused by the build-up of fatty deposits (see †fats). This reduces elasticity and narrows the space (or bore) through which the blood can flow. This condition is known as hardening of the arteries or atherosclerosis and can lead to high blood pressure, loss of circulation, and death. When it affects the arteries supplying blood to the muscle of the heart wall it is called heart disease. People with heart disease run an increased risk of having a heart attack. In a heart attack blood vessels supplying the heart muscles are blocked by a blood clot and this can result in sudden death. Research indicates that for some people a typical Western diet, high in fat from animals, increases the chances of heart disease developing.

artesian well well that is supplied with water rising naturally from an underground water-saturated rock layer (†aquifer). The water rises from the aquifer under its own pressure. Such a well may be drilled into an aquifer that is confined by impermeable rocks both above and below. If the water table (the top of the region of water saturation) in that aquifer is above the level of the well head, hydrostatic pressure will force the water to the surface.

Artesian wells are often overexploited because their water is fresh and easily available, and they eventually become unreliable. There is also some concern that pollutants such as pesticides or nitrates can seep into the aquifers.

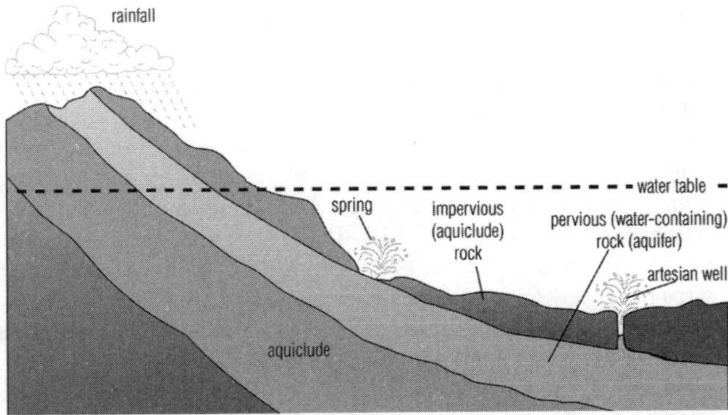

artesian well In an artesian well, water rises from an underground water-containing rock layer under its own pressure. Rain falls at one end of the water-bearing layer, or aquifer, and percolates through the layer. The layer fills with water up to the level of the water table. Water will flow from a well under its own pressure if the well head is below the level of the water table. Helicon

arthropod member of the phylum Arthropoda; an invertebrate animal with jointed legs and a segmented body with a horny or chitinous casing (exoskeleton), which is shed periodically and replaced as the animal grows. Included are arachnids such as spiders and mites, as well as crustaceans, millipedes, centipedes, and insects.

artificial intelligence AI, branch of science concerned with creating computer programs that can perform actions comparable with those of an intelligent human. AI research covers such areas as planning (for robot behavior), language understanding, pattern recognition, and knowledge representation.

The possibility of artificial intelligence was first proposed by the English mathematician Alan †Turing in 1950. Early AI programs, developed in the 1960s, attempted simulations of human intelligence or were aimed at general problem-solving techniques. By the mid-1990s, scientists were concluding that AI was more difficult to create than they had imagined. It is now thought that intelligent behavior depends as much on the knowledge a system possesses as on its reasoning power. Present emphasis is on †knowledge-based systems, such as †expert systems, while research projects focus on †neural networks, which attempt to mimic the structure of the human brain.

On the †Internet, small bits of software that automate common routines or attempt to predict human likes or behavior based on past experience are called intelligent agents or bots.

artificial radioactivity natural and spontaneous radioactivity arising from radioactive isotopes or elements that are formed when elements are bombarded with subatomic particles – protons, neutrons, or electrons – or small nuclei.

Artin, Emil (1898–1962) Austrian mathematician. Artin, who made important contributions to the development of class field theory and the theory of hypercomplex numbers, was one of the creators of modern algebra.

aryl any organic †radical derived from an †aromatic compound; for example, the phenyl radical (—C_6H_5) is derived from benzene (C_6H_6).

asbestos any of several related minerals of fibrous structure that offer great heat resistance because of their nonflammability and poor conductivity. Commercial asbestos is generally either made from serpentine ("white" asbestos) or from sodium iron silicate ("blue" asbestos). The fibers are woven together or bound by an inert material. Over time the fibers can work loose and, because they are small enough to float freely in the air or be inhaled, asbestos usage is now strictly controlled; exposure to its dust can cause cancer.

ASCII acronym for American Standard Code for Information Interchange, in computing, coding system in which numbers are assigned to letters, digits, and punctuation symbols. Although computers work in code based on the †binary number system, ASCII numbers are usually quoted as decimal or †hexadecimal numbers. For example, the decimal number 45 (binary 0101101) represents a hyphen, and 65 (binary 1000001) a capital A. The first 32 codes are used for control functions, such as carriage return and backspace.

Strictly speaking, ASCII is a 7-bit binary code, allowing 128 different characters to be represented, but an eighth bit is often used to provide †parity or to allow for extra characters. The system is widely used for the storage of text and for the transmission of data between computers.

ASCII Chart and Other Resources http://www.jimprice.com/jim-asc.htm

Features a brief introduction to American Standard Code for Information Interchange (ASCII), as well as an ASCII chart, a series of related questions and answers, and numerous other ASCII resources.

Ascomycetes large class of fungi (15,000 species in 1,950 genera) whose reproductive spores (ascospores) are contained in a sac-like structure known as the ascus (plural asci). It is possible to arrange the Ascomycetes in a series of groups of increasing complexity from yeast like forms having no ascocarps (the ascus-bearing structure) to those having an elaborate macroscopic ascocarp. The largest group of ascomycete fungi are the †lichens. Ascomycetes contains the Discomycetes, where the ascocarp is an apothecium, cup-shaped, and the Pyrenomycetes where the ascocarp is a perithecium, flask-shaped or rounded.

Ascomycota http://tolweb.org/tree?group=Ascomycota&contgroup=Fungi

Good source of information on this large group of fungi. A colorful home page shows photos of baker's yeast, penicillium, and other well-known ascomycetes. Among the interesting and well-presented facts about these fungi is that they have more than one reproductive option open to them. The text is supported with diagrams and photos. Further sources of information from professional plant pathologists are indicated.

ascorbic acid or **vitamin C**; $C_6H_8O_6$, relatively simple organic acid found in citrus fruits and vegetables. It is soluble in water and destroyed by prolonged boiling, so soaking or overcooking of vegetables reduces their vitamin C content. Lack of ascorbic acid results in scurvy.

In the human body, ascorbic acid is necessary for the correct synthesis of †collagen. Lack of vitamin C causes skin sores or ulcers, tooth and gum problems, and burst capillaries (scurvy symptoms) owing to an abnormal type of collagen replacing the normal type in these tissues.

The Australian billygoat plum, *Terminalia ferdiandiana*, is the richest natural source of vitamin C, containing 100 times the concentration found in oranges.

asexual reproduction reproduction that does not involve the manufacture and fusion of sex cells (gametes) from two parents. Asexual reproduction has advantages in that there is no need to search for a mate; every asexual organism can reproduce on its own. Asexual reproduction can therefore lead to a rapid population build-up. However, every new organism produced by asexual reproduction is genetically identical to the parent – a †clone.

In evolutionary terms, the disadvantage of asexual reproduction arises from the fact that only identical individuals (clones) are produced – there is no variation. In agriculture and horticulture, where standardized production is needed, this is useful. Taking cuttings of a good †variety of fruit tree is an example of artificial asexual reproduction. However, in the wild, an asexual population that cannot adapt to a changing environ-

asexual reproduction Examples of asexual reproduction. Asexual reproduction is the simplest form of reproduction, occurring in many plants and simple animals. Strawberry plants can reproduce by sending out runners; onion plants form bulbs; and potato plants form tubers. Amoebas divide into two (binary fission) and hydra form new hydra by budding. The offspring are always genetically identical to the parent. Helicon

ment or evolve defenses against a new disease is at risk of extinction. Many asexually reproducing organisms are therefore capable of reproducing sexually as well.

Asexual reproduction is very common in micro-organisms. But there are also many plants that use it naturally. The blackberry or bramble spreads by allowing its stems to root where they touch the ground. However, the blackberry also reproduces sexually using its flowers.

ashen light in astronomy, faint glow occasionally reported in the dark hemisphere of †Venus when the planet is in a crescent phase. The glow probably originates from the high temperatures at the surface and in the lower atmosphere of Venus.

aspartame noncarbohydrate sweetener used in foods under the trade name Nutrasweet. It is about 200 times as sweet as sugar and, unlike saccharine, has no aftertaste.

The aspartame molecule consists of two amino acids (aspartic acid and phenylalanine) linked by a methylene ($-CH_2-$) group. It breaks down slowly at room temperature and rapidly at higher temperatures. It is not suitable for people who suffer from phenylketonuria.

aspect in earth sciences the direction in which a slope faces. In the northern hemisphere a slope with a southerly aspect receives more sunshine than other slopes and is therefore better suited for growing crops that require many hours of sunshine in order to ripen successfully. Vineyards in northern Europe are usually situated on south-facing slopes.

asphalt mineral mixture containing semisolid brown or black †bitumen, used in the construction industry. Asphalt is mixed with rock chips to form paving material, and the purer varieties are used for insulating material and for waterproofing masonry. It can be produced artificially by the distillation of †petroleum.

The availability of recycled colored glass led in 1988 to the invention of **glassphalt**, asphalt that is 15% crushed glass. It is used to pave roads in New York.

Considerable natural deposits of asphalt occur around the Dead Sea and in the Philippines, Cuba, Venezuela, and Trinidad. Bituminous limestone occurs at Neufchâtel, France.

aspirin acetylsalicylic acid, a popular pain-relieving drug (†analgesic) developed in the late 19th century as a household remedy for aches and pains. It relieves pain and reduces inflammation and fever. It is the world's most widely used drug.

Aspirin was first refined from salicylic acid by German chemist Felix Hoffman, and marketed in 1899. Although salicylic acid occurs naturally in bark of willow *Salix alba* (and has been used for pain relief since 1763) the acetyl derivative is less bitter and less likely to cause vomiting.

Aspirin – Molecule of the Month http://www.bris.ac.uk/Depts/Chemistry/MOTM/aspirin/aspirin.htm

Interactive presentation on aspirin with information on its properties and uses and a rotatable image of its molecule from the Department of Chemistry at Bristol University, UK, for those with Java-enabled browsers.

assay in chemistry, the determination of the quantity of a given substance present in a sample. Usually it refers to determining the purity of precious metals.

The assay may be carried out by "wet" methods, when the sample is wholly or partially dissolved in some reagent (often an acid), or by "dry" or "fire" methods, in which the compounds present in the sample are combined with other substances.

assembler in computing, program that translates a program written in an assembly language into a complete †machine code program that can be executed by a computer. Each instruction in the assembly language is translated into only one machine-code instruction.

assembly language low-level computer-programming language closely related to a computer's internal codes. It consists chiefly of a set of short sequences of letters (mnemonics), which are translated, by a program called an assembler, into †machine code for the computer's †central processing unit (CPU) to follow directly. In assembly language, for example, "JMP" means "jump" and "LDA" means "load accumulator." Assembly code is used by programmers who need to write very fast or efficient programs.

Because they are much easier to use, high-level languages are normally used in preference to assembly languages. An assembly language may still be used in some cases, however, particularly when no suitable high-level language exists or where a very efficient machine-code program is required.

Assembler Zone http://www.programmersheaven.com/zone5/index.htm

Great programming resource. The Assembler Zone has information, code, tools, and source code in a variety of languages, including assembler. It is packed with news and information, it has useful downloads, discussion forums, and it features programming books recommended by users of the Web site.

assimilation in animals, the process by which absorbed food molecules, circulating in the blood, pass into the cells and are used for growth, tissue repair, and other metabolic activities. The actual destiny of each food molecule depends not only on its type, but also on the body requirements at that time.

associative operation in mathematics, an operation in which the outcome is independent of the grouping of the numbers or symbols concerned. For example, multiplication is associative, as 4 x (3 x 2) = (4 x 3) x 2 = 24; however, division is not, as 12 ÷ (4 ÷ 2) = 6, but (12 ÷ 4) ÷ 2 = 1.5. Compare †commutative operation and †distributive operation.

assortative mating in †population genetics, selective mating in a population between individuals that are genetically related or have similar characteristics. If sufficiently consistent, assortative mating can theoretically result in the evolution of new species without geographical isolation (see †speciation).

astatine chemical symbol At, (Greek *astatos* "unstable") non-metallic, radioactive element, atomic number 85, relative atomic mass 210. It is a member of the †halogen group, and is found at the bottom of Group 7 of the †periodic table of the elements. It is very rare in nature. Astatine is highly unstable, with at least 19 isotopes; the longest lived has a †half-life of about eight hours.

asterism group of stars, not necessarily a complete constellation, forming an easily recognizable figure, such as the †Plough (part of †Ursa Major), the †Pleiades, or †Orion's belt.

asteroid any of many thousands of small bodies, made of rock and minerals, which orbit the Sun. Most lie in a region called the **asteroid belt** between the orbits of Mars and Jupiter, and are thought to be fragments left over from the formation of the Solar System. About 100,000 asteroids may exist, but their total mass is only a few hundredths of the mass of the Moon. These rocky fragments range in size from 1 km/0.6 mi to 900 km/560 mi in diameter.

Asteroid Introduction http://www.solarviews.com/eng/asteroid.htm

What is the difference between an asteroid and a meteorite? Find out at this site, which contains a table of statistics about asteroids, a chronology of asteroid exploration, and images of selected asteroids.

Asteroids and Comets http://nssdc.gsfc.nasa.gov/planetary/planets/asteroidpage.html

Documents spacecraft encounters with asteroids and comets, and outlines future missions. The site contains a wealth of useful background information, an image gallery, and links to many important asteroid resources on the Internet.

Doomsday Asteroid http://www.pbs.org/wgbh/nova/spacewatch/

Companion to the US Public Broadcasting Service television program Nova, this page provides information on comets: what they are made of, where they come from, and why they are important to us. It includes images and information about the Hale-Bopp and Hyakutake comets, including a number of images from the Hubble Space Telescope. You can also view images of Venus. There is a recipe for making your own comet, courtesy of the Jet Propulsion Laboratory, and a list of resources for further research.

asthenosphere layer within Earth's †mantle lying beneath the †lithosphere, typically beginning at a depth of approximately 100 km/63 mi and extending to depths of approximately 260 km/160 mi. Sometimes referred to as the "weak sphere," it is characterized by being weaker and more elastic than the surrounding mantle.

The asthenosphere's elastic behavior and low viscosity allow the overlying, more rigid plates of lithosphere to move laterally in a process known as †plate tectonics. Its elasticity and viscosity also allow overlying crust and mantle to move vertically in response to gravity to achieve isostatic equilibrium (see †isostasy).

astigmatism aberration occurring in the lens of the eye. It results when the curvature of the lens differs in two perpendicular planes, so that rays in one plane may be in focus while rays in the other are not. With astigmatic eyesight, the vertical and horizontal cannot be in focus at the same time; correction is by the use of a cylindrical lens that reduces the overall focal length of one plane so that both planes are seen in sharp focus.

astrolabe ancient navigational instrument, forerunner of the sextant. Astrolabes usually consisted of a flat disc with a sighting rod that could be pivoted to point at the Sun or bright stars. From the altitude of the Sun or star above the horizon, the local time could be estimated.

Astrolabe – An Instrument with a Past and a Future http://www.astrolabes.org/

Good guide to this ancient astronomical computer. The long history of the instrument, and its importance in the history of astrology and astronomy, is explained with the help of good photographs. A section on the parts of the astrolabe explains how the instrument works and its uses.

astrometry measurement of the precise positions of stars, planets, and other bodies in space. Such information is needed for practical purposes including accurate timekeeping, surveying and navigation, and calculating orbits and measuring distances in space. Astrometry is not concerned with the surface features or the physical nature of the body under study.

Before telescopes, astronomical observations were simple astrometry. Precise astrometry has shown that stars are not fixed in position but have a †proper motion caused as they and the Sun orbit the Milky Way galaxy. The nearest stars also show †parallax (apparent change in position), from which their distances can be calculated. Above the distorting effects of the atmosphere, satellites such as Hipparcos, launched by the European Space Agency in 1989, can make even more precise measurements than ground telescopes, so refining the distance scale of space.

astronautics science of space travel. It was formerly subsumed under †aeronautics.

Both subjects include **hypersonics**, the study of airflows and forces at speeds above five times that of sound (Mach 5), as experienced by space rockets and guided missiles.

Astronautix http://www.rocketry.com/mwade/spaceflt.htm

Enormous online encyclopedia of world spacecraft, equipment, missions, and people. The site has a very detailed chronology of space flight and an exhaustive reference list. Fascinating in-depth articles describe Russian space and aeronautic projects, with lots of rare and unusual drawings and photographs.

Basics of Space Flight http://www.jpl.nasa.gov/basics/

Illustrated tutorials covering essential aspects of space flight, including the space environment, flight projects, and operations. The site has a useful glossary of technical terms and abbreviations.

astronomical color index difference in a star's brightness when measured on two selected †wavelengths, in order to determine the star's temperature. Cooler stars emit more light at longer wavelengths (and so appear redder than hot stars). Modern methods involve †photoelectric filtering and the †UBV photometry system.

astronomical unit AU, unit equal to the mean distance of the Earth from the Sun: 149.6 million km/92.96 million mi. It is used to describe planetary distances. Light travels this distance in approximately 8.3 minutes.

astronomy science of the celestial bodies: the Sun, the Moon, and the planets; the stars and galaxies; and all other objects in the universe. It is concerned with their positions, motions, distances, and physical conditions and with their origins and evolution. Astronomy thus divides into fields such as astrophysics, celestial mechanics, and †cosmology. See also †gamma-ray astronomy, †infrared astronomy, †radio astronomy, †ultraviolet astronomy, and †X-ray astronomy.

Greek astronomers Astronomy is perhaps the oldest recorded science; there are observational records from ancient Babylonia, China, Egypt, and Mexico. The first true astronomers, however, were the Greeks, who deduced the Earth to be a sphere and attempted to measure its size. Ancient Greek astronomers included †Thales and †Pythagoras. †Eratosthenes of Cyrene measured the size of the Earth with considerable accuracy. Star catalogues were drawn up, the most celebrated being that of Hipparchus. The *Almagest*, by †Ptolemy of Alexandria, summarized Greek astronomy and survived in its Arabic translation. The Greeks still regarded the Earth as the center of the universe, although this was doubted by some philosophers, notably †Aristarchus of Samos, who maintained that the Earth moves around the Sun.

Ptolemy, the last famous astronomer of the Greek school, died in about AD 180, and little progress was made for some centuries.

Arab revival The Arabs revived the science, developing the astrolabe and producing good star catalogues. Unfortunately, a general belief in the pseudoscience of astrology continued until the end of the Middle Ages (and has been revived from time to time).

the Sun at the center The dawn of a new era came in 1543, when a Polish canon, †Copernicus, published a work entitled *De revolutionibus orbium coelestium/On the Revolutions of the Heavenly Spheres*, in which he demonstrated that the Sun, not the Earth, is the center of our planetary system. (Copernicus was wrong in many respects – for instance, he still believed that all celestial orbits must be perfectly circular.) Tycho †Brahe, a Dane, increased the accuracy of observations by means of improved instruments allied to his own personal skill, and his observations were used by German mathematician Johannes †Kepler to prove the validity of the Copernican system. Considerable opposition existed, however, for removing the Earth from its central position in the universe; the Catholic Church was openly hostile to the idea, and, ironically, Brahe never accepted the idea that the Earth could move around the Sun. Yet before the end of the 17th century, the theoretical work of Isaac †Newton had established celestial mechanics.

Galileo and the telescope The refracting telescope was invented about 1608, by Hans †Lippershey in Holland, and was first applied to astronomy by Italian scientist †Galileo in the winter of 1609–10. Immediately, Galileo made a series of spectacular discoveries. He found the four largest satellites of Jupiter, which gave strong support to the Copernican theory; he saw the craters of the Moon, the phases of Venus, and the myriad faint stars of our †Galaxy, the Milky Way.

Galileo's most powerful telescope magnified only 30 times, but it was not long before larger telescopes were built and official observatories were established.

Galileo's telescope was a refractor; that is to say, it collected its light by means of a glass lens or object glass. Difficulties with his design led Newton, in 1671, to construct a reflector, in which the light is collected by means of a curved mirror.

further discoveries In the 17th and 18th centuries astronomers were mostly concerned with positional measurements. Uranus was discovered in 1781 by William †Herschel, and this was soon followed by the discovery of the first four asteroids, Ceres in 1801, Pallas in 1802, Juno in 1804, and Vesta in 1807. In 1846 Neptune was located by Johann †Galle, following calculations by British astronomer John Couch †Adams and French astronomer Urbain Jean Joseph †Leverrier. Also significant was the first measurement of the distance of a star, when in 1838 the German astronomer Friedrich †Bessel measured the †parallax of the star 61 Cygni, and calculated that it lies at a distance of about 6 light years (about half the modern value).

Astronomical spectroscopy was developed, first by Fraunhofer in Germany and then by people such as Pietro Angelo †Secchi and William Huggins, while Gustav †Kirchhoff successfully interpreted the spectra of the Sun and stars. By the 1860s good photographs of the Moon had been obtained, and by the end of the century photographic methods had started to play a leading role in research.

galaxies William Herschel investigated the shape of our Galaxy during the latter part of the 18th century and concluded that its stars are arranged roughly in the form of a double-convex lens. Basically Herschel was correct, although he placed our Sun near the center of the system; in fact, it is well out towards the edge, and lies 25,000 light years from the galactic nucleus. Herschel also studied the luminous "clouds" or nebulae, and made the tentative suggestion that those nebulae capable of resolution into stars might be separate galaxies, far outside our own Galaxy.

It was not until 1923 that US astronomer Edwin Hubble, using the 2.5 m/100 in reflector at the Mount Wilson Observatory, was able to verify this suggestion. It is now known that the "spiral nebulae" are galaxies in their own right, and that they lie at immense distances. The most distant galaxy visible to the naked eye, the Great Spiral in †Andromeda, is 2.2 million light years away; the most remote galaxy so far measured lies over 10 billion light years away. It was also found that galaxies tended to form groups, and that the groups were apparently receding from each other at speeds proportional to their distances.

a growing universe This concept of an expanding and evolving universe at first rested largely on †Hubble's law, relating the distance of objects to the amount their spectra shift towards red – the †red shift. Subsequent evidence derived from objects studied in other parts of the †electromagnetic spectrum, at radio and X-ray wavelengths, has provided confirmation. †Radio astronomy established its place in probing the structure of the universe by demonstrating in 1954 that an optically visible distant galaxy was identical with a powerful radio source known as Cygnus A. Later analysis of the comparative number, strength, and distance of radio sources suggested that in the distant past these, including the †quasars discovered in 1963, had been much more powerful and numerous than today. This fact suggested that the universe has been evolving from an origin, and is not of infinite age as expected under a †steady-state theory.

The discovery in 1965 of microwave background radiation was evidence for the enormous temperature of the giant explosion, or Big Bang, that brought the universe into existence.

further exploration Although the practical limit in size and efficiency of optical telescopes has apparently been reached, the siting of these and other types of telescope at new observatories in the previously neglected southern hemisphere has opened fresh areas of the sky to search. Australia has been in the forefront of these developments. The most remarkable recent extension of the powers of astronomy to explore the universe is in the use of rockets, satellites, space stations, and space probes. Even the range and accuracy of the conventional telescope may be greatly improved free from the Earth's atmosphere. When the USA launched the Hubble Space Telescope into permanent orbit in 1990, it was the most powerful optical telescope yet constructed, with a 2.4 m/94.5 in mirror. It detects celestial phenomena seven times more distant (up to 14 billion light years) than any Earth-based telescope.

See also †black hole and †infrared radiation.

Astronomy Course for Students Using the Internet http://darkskyinstitute.org/astronomy.html

Well-presented and interesting introduction to astronomy including sections on "Finding your way around the sky," "Stars, nebulae, and star clusters," and "Messages from the cosmos."

Bad Astronomy http://www.badastronomy.com/

Pages designed and maintained by Phil Plait, an astronomer and teacher, which are dedicated to exposing misinformation and myths about astronomy. As well as keeping you up-to-date on astronomical events, this site also has a section called "Bite-sized astronomy" which contains an essay explaining one concept about the science, which is changed on a weekly basis.

Practical Guide to Astronomy http://www.hotliquidmagma.com/space/index.html

Well-illustrated guide to astronomy that contains explanations of many aspects of the subject, including the "Big Bang" theories of

English physicist Stephen Hawking, a list of early astronomers and their key discoveries, and an in-depth look at all the main elements of our Solar System.

Spacewatch http://www.space.com/spacewatch/

Updated daily with a brief of what is currently visible in the night sky and what to look out for, the site has many useful features including star charts and graphics showing the present alignment of the planets. Pages devoted to space weather give details of expected solar, auroral, and meteor activity.

astrophotography use of photography in astronomical research. The first successful photograph of a celestial object was the daguerreotype plate of the Moon taken by English Scientist John W Draper in March 1840. The first photograph of a star, Vega, was taken by US astronomer William Bond in 1850. Modern-day astrophotography uses techniques such as †charge-coupled devices (CCDs), which are often carried aboard a spacecraft.

astrophysics study of the physical nature of stars, galaxies, and the universe. It began with the development of spectroscopy in the 19th century, which allowed astronomers to analyze the composition of stars from their light. Astrophysicists view the universe as a vast natural laboratory in which they can study matter under conditions of temperature, pressure, and density that are unattainable on Earth.

asymmetric atom in chemistry, atom joined to three or more other atoms in such a configuration that it cannot be superimposed on its mirror image. This type of atom forms the basis for †optical activity of a molecule.

asymptote in †coordinate geometry, a straight line that a curve approaches progressively more closely but never reaches. The x and y axes are asymptotes to the graph of xy = constant (a rectangular †hyperbola).

If a point on a curve approaches a straight line such that its distance from the straight line is d, then the line is an asymptote to the curve if limit d tends to zero as the point moves towards infinity. Among †conic sections (curves obtained by the intersection of a plane and a double cone), a hyperbola has two asymptotes, which in the case of a rectangular hyperbola are at right angles to each other.

atactic polymer in chemistry, long-chain hydrocarbon polymer in which any attached groups have no specific orientation with regard to the main carbon chain (see also †isotactic polymer and †syntactic polymer).

atavism (Latin *atavus* "ancestor") in genetics, the reappearance of a characteristic not apparent in the immediately preceding generations; in psychology, the manifestation of primitive forms of behavior.

Aten in astronomy, asteroid that orbits in less than a year within the orbit of Venus. It was discovered in 1976 and is 0.8 km/0.5 mi in diameter.

atmosphere mixture of gases surrounding a planet. Planetary atmospheres are prevented from escaping by the pull of gravity. On Earth, atmospheric pressure decreases with altitude. In its lowest layer, the atmosphere consists of nitrogen (78%) and oxygen (21%), both in molecular form (two atoms bonded together) and argon (1%). Small quantities of other gases are important to the chemistry and physics of the Earth's atmosphere, including water, carbon dioxide, and traces of other noble gases (rare gases), as well as ozone. The atmosphere plays a major part in the various cycles of nature (the †water cycle, the †carbon cycle, and the †nitrogen cycle). It is the principal industrial source of nitrogen, oxygen, and argon, which are obtained by the †fractional distillation of liquid air.

The Earth's atmosphere is divided into four regions of atmosphere classified by temperature.

troposphere This is the lowest level of the atmosphere (altitudes from 0 to 10 km/6 mi) and it is heated to an average temperature of 15°C/59°F by the Earth, which in turn is warmed by infrared and visible radiation from the Sun. Warm air cools as it rises in the troposphere and this rising of warm air causes rain and most other weather phenomena. The temperature at the top of the troposphere is approximately –60°C/–76°F.

stratosphere Temperature increases with altitude in this next layer (from 10 km/6 mi to 50 km/31 mi), from –60°C/–76°F to near 0°C/32°F.

mesosphere Temperature again decreases with altitude through the mesosphere (50 km/31 mi to 80 km/50 mi), from 0°C/32°F to below –100°C/–148°F.

thermosphere In the highest layer (80 km/50mi to about 700 km/450 mi), temperature rises with altitude to extreme values of thousands of degrees. The meaning of these extreme temperatures can be misleading. High thermosphere temperatures represent little heat because they are defined by motions among so few atoms and molecules spaced widely apart from one another.

atmosphere symbol atm; or **standard atmosphere**, in physics, a unit of pressure equal to 760 mmHg, 1013.25 millibars, or 1.01325×10^5 pascals, or newtons per square meter. The actual pressure exerted by the atmosphere fluctuates around this value, which is assumed to be standard at sea level and 0°C/32°F, and is used when dealing with very high pressures.

atmospheric circulation large-scale movement of air within the lower †atmosphere. Warm air at the †Equator rises, creating a zone of low pressure. This air moves towards the poles, losing energy and becoming cooler and denser, until it sinks back to the surface at around 30° latitude, creating an area of high pressure. At the surface, air moves from this high pressure zone back towards the low pressure zone at the Equator, completing a circulatory movement.

atmospheric pollution contamination of the atmosphere with the harmful by-products of human activity; see †air pollution.

atmospheric pressure pressure at any point on the Earth's surface that is due to the weight of the column of air above it; it therefore decreases as altitude increases, because there is less air above. Particles in the air exert a force (pressure) against surfaces; when large numbers of particles press against a surface, the overall effect is known as air pressure. At sea level the average pressure is 101 kilopascals (1,013 millibars, or 760 mm Hg, or 14.7 lb per sq in, or 1 atmosphere). Changes in atmospheric pressure, measured with a barometer, are used in weather forecasting. Areas of relatively high pressure are called †anticyclones; areas of low pressure are called †depressions.

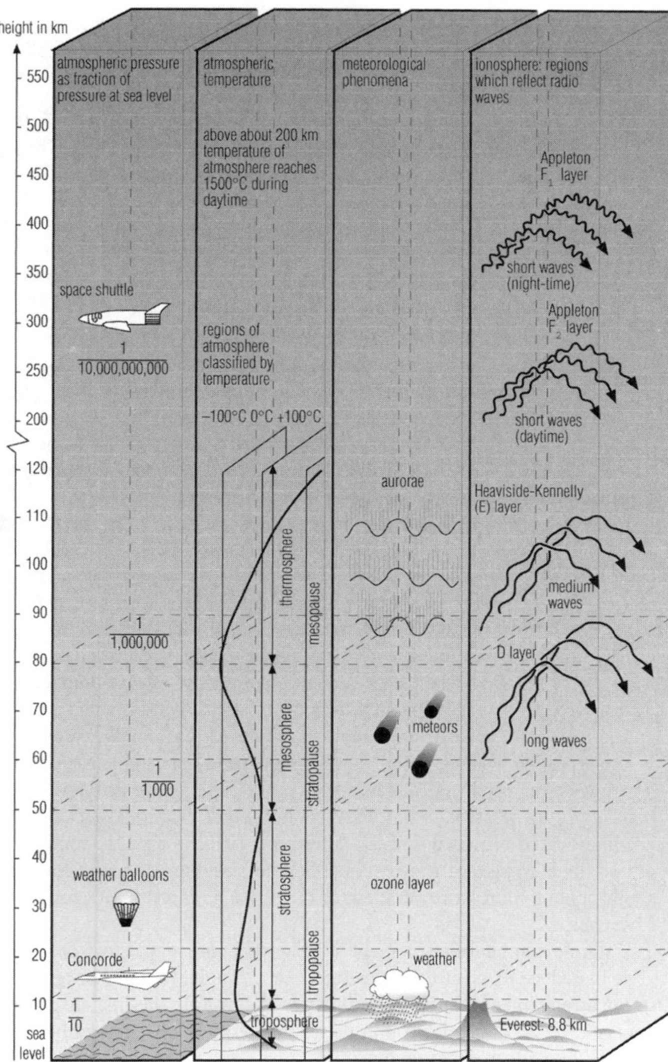

atmosphere All but 1% of the Earth's atmosphere lies in a layer 30 km/19 mi above the ground. At a height of 5,500 m/18,000 ft, air pressure is half that at sea level. The temperature of the atmosphere varies greatly with height; this produces a series of layers, called the troposphere, stratosphere, mesosphere, and thermosphere. Helicon

For every square meter of a surface, a force of 10 tonnes of air pressure is exerted. This air pressure does not crush objects because they exert an equal amount of force to balance the air pressure. At higher altitudes the air is thinner and the air pressure is lower. Here, water boils at a temperature less than that at sea level. In space there are no air particles and so no air pressure is exerted on an astronaut's body. Astronauts wear spacesuits that supply an air pressure against their bodies.

A barometer is an instrument used to measure air pressure. If a glass tube is filled with mercury and turned upside down with its end in a bowl of mercury, then the height of the column of mercury is held by the air particles pressing on the mercury in the bowl. This measures the air pressure. A standard measurement for atmospheric pressure at sea level is a column of mercury 760 mm/30 in high. At higher altitudes the height of the mercury column would be less, as the air pressure is lower.

atoll continuous or broken circle of †coral reef and low coral islands surrounding a lagoon.

atom (Greek *atomos* "undivided") smallest unit of matter that can take part in a chemical reaction, and which cannot be broken down chemically into anything simpler. An atom is made up of protons and neutrons in a central nucleus (except for hydrogen, which has a single proton in its nucleus) surrounded by electrons (see †atomic structure). The atoms of the various elements differ in atomic number, relative atomic mass, and chemical behavior.

Atoms are much too small to be seen by even the most powerful optical microscope (the largest, cesium, has a diameter of 0.0000005 mm/0.00000002 in), and they are in constant motion. However, modern electron microscopes, such as the †scanning tunneling microscope (STM) and the †atomic force microscope (AFM), can produce images of individual atoms and molecules.

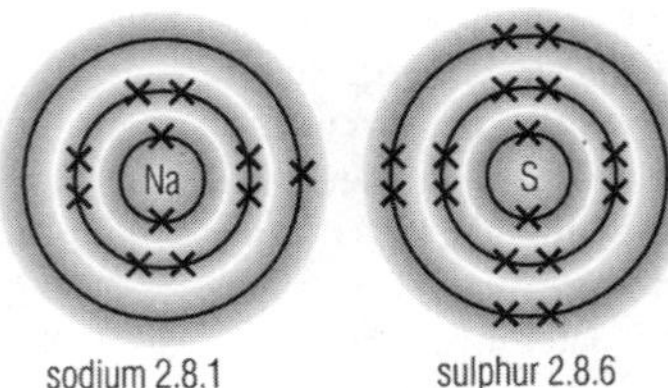

atom, electronic structure The arrangement of electrons in a sodium atom and a sulfur atom. The number of electrons in a neutral atom gives that atom its atomic number: sodium has an atomic number of 11 and sulfur has an atomic number of 16. Helicon

Atoms and Nuclei http://www.triumf.ca/safety/rpt/rpt_1/node1.html

Covering all aspects of atomic structure, this site contains separate pages on "Atomic structure," "Atomic mass," "Atomic number," "Chemical symbols," "Isotopes," and "Symbols for isotopes." A good resource for learning the basics.

Common Sense Science http://www.commonsensescience.org/

Alternative science site that explores new, "common sense" models for elementary particles, the atom, and the nucleus.

Theory of Atoms in Molecules http://www.chemistry.mcmaster.ca/faculty/bader/aim/

Canadian research paper about atoms that assumes some level of knowledge about the subject. The introductory topics include sections on "What is an atom?" and "What is a bond?" and will be of interest to the GCSE student or general reader.

atom, electronic structure of arrangement of electrons around the nucleus of an atom, in distinct energy levels, also called atomic †orbitals, or shells. These shells can be regarded as a series of concentric spheres, each of which can contain a certain maximum number of electrons; the noble gases (rare gases) have an arrangement in which every shell contains this number (see †noble gas structure). The energy levels are usually numbered beginning with the shell nearest to the nucleus. The outermost shell is known as the †valency shell as it contains the valence electrons.

The lowest energy level, or innermost shell, can contain no more than two electrons. Outer shells are considered to be stable when they contain eight electrons but additional electrons can sometimes be accommodated provided that the outermost shell has a stable configuration. Electrons in unfilled shells are available to take part in chemical bonding, giving rise to the concept of valency. In ions, the electron shells contain more or fewer electrons than are required for a neutral atom, resulting in negative or positive charges.

The atomic number of an element indicates the number of electrons in a neutral atom. From this it is possible to deduce its electronic structure. For example, sodium has atomic number 11 ($Z = 11$) and its electronic arrangement (configuration) is two electrons in the first energy level, eight electrons in the second energy level and one electron in the third energy level – generally written as 2.8.1. Similarly for sulfur ($Z = 16$), the electron arrangement is 2.8.6. The electronic structure dictates whether two elements will combine by ionic or covalent bonding (see †bond) or not at all.

atomic clock timekeeping device regulated by various periodic processes occurring in atoms and molecules, such as atomic vibration or the frequency of absorbed or emitted radiation.

The first atomic clock was the **ammonia clock**, invented at the US National Bureau of Standards in 1948. It was regulated by measuring the speed at which the nitrogen atom in an ammonia molecule vibrated back and forth. The rate of molecular vibration is not affected by temperature, pressure, or other external influences, and can be used to regulate an electronic clock.

A more accurate atomic clock is the **cesium clock**. Because of its internal structure, a cesium atom produces or absorbs radiation of a very precise frequency (9,192,631,770 Hz) that varies by less than one part in 10 billion. This frequency has been used to define the second, and is the basis of atomic clocks used in international timekeeping.

Hydrogen maser clocks, based on the radiation from hydrogen atoms, are the most accurate. The hydrogen maser clock at the US Naval Research Laboratory, Washington, DC, is estimated to lose one second in 1,700,000 years. Cooled hydrogen maser clocks could theoretically be accurate to within one second in 300 million years.

Atomic clocks are so accurate that minute adjustments must be made periodically to the length of the year to keep the calendar exactly synchronized with the Earth's rotation, which has a tendency to slow down. There have been 17 adjustments made since 1972 adding a total of 20 seconds to the calendar. In 1997 the northern hemisphere's summer was longer than usual – by one second. An extra second was added to the world's time at precisely 23 hours, 59 minutes, and 60 seconds on 30 June 1997. The adjustment was called for by the International Earth Rotation Service in Paris, which monitors the difference between Earth time and atomic time.

atomic energy another name for †nuclear energy.

atomic force microscope **AFM**, microscope developed in the late 1980s that produces a magnified image using a diamond probe, with a tip so fine that it may consist of a single atom, dragged over the surface of a specimen to "feel" the contours of the surface. In effect, the tip acts like the stylus of a record player, reading the surface. The tiny up-and-down movements of the probe are converted to an image of the surface by computer and displayed on a screen. The AFM is useful for examination of biological specimens since, unlike the †scanning tunneling microscope, the specimen does not have to be electrically conducting.

atomic heat product of †relative atomic mass and †specific heat capacity. It is approximately constant for many solid elements and equal to 6 calories per gram atom per degree (25.2 joules per mole per kelvin).

atomicity number of atoms of an †element that combine together to form a molecule. A molecule of oxygen (O_2) has atomicity 2; sulfur (S_8) has atomicity 8.

atomic mass see †relative atomic mass.

atomic mass unit or **dalton**; symbol u, unit of mass that is used to measure the relative mass of atoms and molecules. It is equal to one-twelfth of the mass of a carbon-12 atom, which is approximately the mass of a proton or 1.66×10^{-27} kg. The †relative atomic mass of an atom has no units; thus oxygen-16 has an atomic mass of 16 daltons but a relative atomic mass of 16.

atomic number or **proton number**; symbol *Z*, number of protons in the nucleus of an atom. It is equal to the positive charge on the nucleus. In a neutral atom, it is also equal to the number of electrons surrounding the nucleus. The chemical elements are arranged in the †periodic table of the elements according to their atomic number. †Nuclear notation is used to label an atom according to the composition of its nucleus.

atomic orbital region in space occupied by an electron associated with the nucleus of an atom. Atomic orbitals have various shapes, depending on the energy level of the electron and the degree of †hybridization. Atomic orbitals overlap to form molecular orbitals, or chemical bonds between atoms.

atomic physics former name for †nuclear physics.

atomic radiation energy given out by disintegrating atoms during †radioactive decay, whether natural or synthesized. The energy may be in the form of fast-moving particles, known as †alpha particles and †beta particles, or in the form of high-energy electromagnetic waves known as †gamma radiation. Overlong exposure to atomic radiation can lead to radiation sickness.

Radiation biology studies the effect of radiation on living organisms. Exposure to atomic radiation is linked to chromosomal damage, cancer, and, in laboratory animals at least, hereditary disease.

atomic radius effective radius of an atom. Atomic radii vary periodically with atomic number, being largest for the alkali metals and smallest for the rare gases.

atomic size or **atomic radius**, size of an atom expressed as the radius in †angstroms or other units of length.

The sodium atom has an atomic radius of 1.57 angstroms (1.57×10^{-8} cm). For metals, the size of the atom is always greater than the size of its ion. For non-metals the reverse is true.

atomic structure internal structure of an †atom.

the nucleus The core of the atom is the **nucleus**, a dense body only one ten-thousandth the diameter of the atom itself. The simplest nucleus, that of hydrogen, comprises a single stable positively charged particle, the **proton**. Nuclei of other ele-

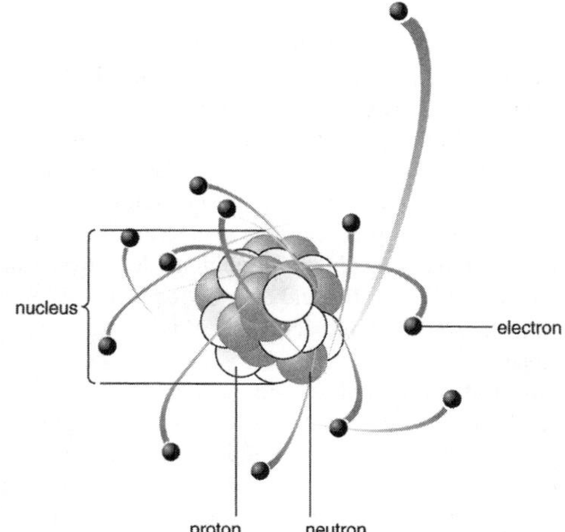

atomic structure The structure of a sodium atom. The nucleus is composed of 11 protons and 12 neutrons. Eleven electrons orbit the nucleus in 3 orbits: 2 in the inner orbit, 8 in the middle, and 1 in the outer. Helicon

ments contain more protons and additional particles, called **neutrons**, of about the same mass as the proton but with no electrical charge. Each element has its own characteristic nucleus with a unique number of protons, the atomic number. The number of neutrons may vary. Where atoms of a single element have different numbers of neutrons, they are called †isotopes. Although some isotopes tend to be unstable and exhibit †radioactivity, all those of a single element have identical chemical properties.

electrons The nucleus is surrounded by a number of moving **electrons**, each of which has a negative charge equal to the positive charge on a proton, but which has a mass of only $^1/_{1,836}$ times as much. In a neutral atom, the nucleus is surrounded by the same number of electrons as it contains protons. According to †quantum theory, the position of an electron is uncertain; it may be found at any point. However, it is more likely to be found in some places than others. The region of space in which an electron is most likely to be found is called an atomic †orbital. The chemical properties of an element are determined by the ease with which its atoms can gain or lose electrons.

attraction and repulsion According to the theory of fundamental †forces, atoms are held together by the electrical forces of attraction between each negative electron and the positive protons within the nucleus. The latter repel one another with enormous forces; a nucleus holds together only because an even stronger force, called the **strong nuclear force**, attracts the protons and neutrons to one another. The strong force acts over a very short range – the protons and neutrons must be in virtual contact with one another. If, therefore, a fragment of a complex nucleus, containing some protons, becomes only slightly loosened from the main group of neutrons and protons, the natural repulsion between the protons will cause this fragment to fly apart from the rest of the nucleus at high speed. It is by such fragmentation of atomic nuclei (nuclear †fission) that nuclear energy is released.

atomic time time as given by †atomic clocks, which are regulated by natural resonance frequencies of particular atoms, and display a continuous count of seconds.

In 1967 a new definition of the second was adopted in the SI system of units: the duration of 9,192,631,770 periods of the radiation corresponding to the transition between two hyperfine levels of the ground state of the cesium-133 atom. The International Atomic Time Scale is based on clock data from a number of countries; it is a continuous scale in days, hours, minutes, and seconds from the origin on 1 January 1958, when the Atomic Time Scale was made 0 h 0 min 0 sec when Greenwich Mean Time was at 0 h 0 min 0 sec.

atomic volume volume of one gram-atom of an element.

ATP abbreviation for **adenosine triphosphate**, nucleotide molecule found in all cells. It can yield large amounts of energy, and is used to drive the thousands of biological processes needed to sustain life, growth, movement, and reproduction. Green plants use light energy to manufacture ATP as part of the process of †photosynthesis. In animals, ATP is formed by the breakdown of glucose molecules, usually obtained from the carbohydrate component of a diet, in a series of reactions termed †respiration. It is the driving force behind muscle contraction and the synthesis of complex molecules needed by individual cells.

atrium either of the two upper chambers of the heart. The left atrium receives freshly oxygenated blood from the lungs via the pulmonary vein; the right atrium receives deoxygenated blood from the †vena cava. Atrium walls are thin and stretch easily to allow blood into the heart. On contraction, the atria force blood into the thick-walled ventricles, which then give a second, more powerful beat.

atropine alkaloid derived from belladonna, a plant with toxic properties. It acts as an anticholinergic, inhibiting the passage of certain nerve impulses. It is used in premedication, to reduce bronchial and gastric secretions. It is also administered as a mild antispasmodic drug, and to dilate the pupil of the eye.

attachment way of incorporating a file into an e-mail message for transmission. Within a single system, such as a corporate local area network (LAN) or a commercial online service, binary files can be sent intact. Over the Internet, attached files must be encoded into †ASCII characters and then decoded by the receiver.

attenuation in biology and medicine, the reduction in the virulence of a pathogenic micro-organism by culturing it in unfavorable conditions, by drying or heating it, or by subjecting it to chemical treatment. Attenuated viruses, for example, are used in vaccines.

attractor in mathematics, subset that all nearby solution curves or trajectories lead to in a dynamical system. A point that is in the attractor will stay there.

attrition in earth science, the process by which particles of rock are rounded and gradually reduced in size by hitting one another as they are transported by rivers, wind, or the sea.

The rounding of particles is a good indication of how far they have been transported. This is particularly true for particles carried by rivers, which become more rounded and smaller as the distance downstream increases. Thus the load of a river is often large and jagged in its upper course, and smaller and rounder in its lower course.

audiofrequency in physics, sound frequency within the range of human hearing (between 20 Hz and about 20 kHz). Microphones and loudspeakers employ audiofrequency electrical signals.

auditory canal tube leading from the outer †ear opening to the eardrum. It is found only in animals whose eardrums are located inside the skull, principally mammals and birds.

Audubon, John James (1785–1851) US naturalist and artist. In 1827, after extensive travels and observations of birds, he published the first part of his *Birds of North America*, with a remarkable series of color plates. Later he produced a similar work on North American quadrupeds.

Auger effect in physics, ejection of an electron from an atom without the simultaneous production of gamma rays or X-rays. Instead of being emitted, a photon undergoes an internal conversion into an electron within the atom.

Auriga constellation of the northern hemisphere, represented as a charioteer. Its brightest star is the first-magnitude †Capella, about 42 light years from the Sun; Epsilon Aurigae is an †eclipsing binary star with a period of 27 years, the longest of its kind (last eclipse in 1983).

aurora colored light in the night sky near the Earth's magnetic poles, called **aurora borealis** ("northern lights") in the northern hemisphere and **aurora australis** ("southern lights") in the southern hemisphere. Although auroras are usually restricted to the polar skies, fluctuations in the †solar wind occasionally cause them to be visible at lower latitudes. An aurora is usually in the form of a luminous arch with its apex towards the magnetic pole, followed by arcs, bands, rays, curtains, and coronae, usually green but often showing shades of blue and red, and sometimes yellow or white. Auroras are caused at heights of over 100 km/60 mi by a fast stream of charged particles from solar flares and low-density "holes" in the Sun's corona. These are guided by the Earth's magnetic field towards the north and south magnetic poles, where they enter the upper atmosphere and bombard the gases in the atmosphere, causing them to emit visible light.

Auroras: Paintings in the Sky http://www.exploratorium.edu/learning_studio/auroras/

Take a tour of the auroras. This page provides an in-depth explanation for the aurora phenomenon, includes a number of images taken from Earth and space, and sheds light on where they are best viewed. You can view a video clip of the aurora borealis and listen to audio clips of David Stern of NASA.

autoimmunity in medicine, condition in which the body's immune responses are mobilized not against "foreign" matter, such as invading germs, but against the body itself. Diseases considered to be of autoimmune origin include myasthenia gravis, rheumatoid arthritis, and lupus erythematosus.

In autoimmune diseases T-lymphocytes reproduce to excess to home in on a target (properly a foreign disease-causing molecule); however, molecules of the body's own tissue that resemble the target may also be attacked, for example insulin-producing cells, resulting in insulin-dependent diabetes; if certain joint membrane cells are attacked, then rheumatoid arthritis may result; and if myelin, the basic protein of the nervous system, then multiple sclerosis results. In 1990 in Israel a T-cell vaccine was produced that arrests the excessive reproduction of T-lymphocytes attacking healthy target tissues.

autolysis in biology, the destruction of a †cell after its death by the action of its own †enzymes, which break down its structural molecules.

automorphic function in mathematics, function that in relation to a †group of †transformations has a value on the transformed point identical with the value on the original point.

autonomic nervous system in mammals, the part of the nervous system that controls those functions not controlled voluntarily, including the heart rate, activity of the intestines, and the production of sweat. There are two divisions of the autonomic nervous system. The **sympathetic** system responds to stress, when it speeds the heart rate, increases blood pressure, and generally prepares the body for action. The **parasympathetic** system is more important when the body is at rest, since it slows the heart rate, decreases blood pressure, and stimulates the digestive system.

At all times, both types of autonomic nerves carry signals that bring about adjustments in visceral organs. The actual rate of heartbeat is the net outcome of opposing signals. Today, it is known that the word "autonomic" is misleading – the reflexes managed by this system are actually integrated by commands from the brain and spinal cord (the central nervous system).

autophagy process in which a cell synthesizes substances and then metabolizes and absorbs them for its own sustenance.

autopolyploid in genetics, organism that has more than the normal two sets (†diploid number) of chromosomes. The sets all originate from a single species.

autoradiography in biology, a technique for following the movement of molecules within an organism, especially a plant, by labeling with a radioactive isotope that can be traced on photographs. It is commonly used to study †photosynthesis, where the pathway of radioactive carbon dioxide can be traced as it moves through the various chemical stages.

autosome any †chromosome in the cell other than a sex chromosome. Autosomes are of the same number and kind in both males and females of a given species.

autotroph any living organism that synthesizes organic substances from inorganic molecules by using light or chemical energy. Autotrophs are the primary producers in all food chains since the materials they synthesize and store are the energy sources of all other organisms. All green plants and many planktonic organisms are autotrophs, using sunlight to convert carbon dioxide and water into sugars by †photosynthesis.

The total †biomass of autotrophs is far greater than that of animals, reflecting the dependence of animals on plants, and the ultimate dependence of all life on energy from the Sun – green plants convert light energy into a form of chemical energy (food) that animals can exploit. Some bacteria use the chemical energy of sulfur compounds to synthesize organic substances. It is estimated that 10% of the energy in autotrophs can pass into the next stage of the †food chain, the rest being lost as heat or indigestible matter. See also †heterotroph.

autumnal equinox see †equinox.

auxin plant †hormone that regulates stem and root growth in plants. Auxins influence many aspects of plant growth and development, including cell enlargement, inhibition of development of axillary buds, †tropisms, and the initiation of roots. Auxin affects cell division mainly at the tip, because it is here that cell division in a stem or root mainly occurs. Just behind the tip the cells grow in size under the influence of auxins, causing the stem or root to grow longer. Auxin therefore affects the amount of elongation here too.

Synthetic auxins are used in hormone rooting powders to encourage cuttings to root. They are also used in some weed killers, where high auxin concentrations cause such rapid growth that the plants die. Other uses include the prevention of premature fruit drop in orchards.

The most common naturally occurring auxin is known as indoleacetic acid, or IAA. It is produced in the shoot apex and transported to other parts of the plant.

avalanche (from French *avaler* "to swallow") fall or flow of a mass of snow and ice down a steep slope under the force of gravity. Avalanches occur because of the unstable nature of snow masses in mountain areas.

Changes of temperature, sudden sound, or earth-borne vibrations may trigger an avalanche, particularly on slopes of more than 35°. The snow compacts into ice as it moves, and rocks may be carried along, adding to the damage caused.

Avalanches leave slide tracks, long gouges down the mountainside that can be up to 1 km/0.6 mi long and 100 m/330 ft wide. These slides have a similar beneficial effect on †biodiversity as do forest fires, clearing the land of snow and mature mountain forest and so enabling plants and shrubs that cannot grow in shade, to recolonize and creating wildlife corridors.

Avalanche Awareness http://nsidc.org/snow/avalanche/index.html

Description of avalanches, what causes them, and how to minimize dangers if caught in one. There is advice on how to determine the stability of a snowpack, what to do if caught out, and how to locate people trapped under snow. Nobody skiing off piste should set off without reading this.

average in statistics, a term used inexactly to indicate the typical member of a set of †data. It usually refers to the †arithmetic mean. The term is also used to refer to the middle member of the set when it is sorted in ascending or descending order (the †median), and the most commonly occurring item of data (the †mode), as in "the average family."

In addition, it is useful to know the †range or spread of data.

Using and Handling Data http://www.mathsisfun.com/data.html

Description of the three main types of average: mean, median, and mode. Each section quickly explains what the term means and illustrates how to find it by using worked examples. There are links to other GCSE exam topics on the same site.

Avery, Oswald Theodore (1877–1955) Canadian-born US bacteriologist. His work on transformation in bacteria established in 1944 that †DNA (deoxyribonucleic acid) is responsible for the transmission of heritable characteristics. He also proved that polysaccharides play an important part in immunity.

Avogadro, Amedeo (1776–1856) Conte di Quaregna, Italian physicist, one of the founders of physical chemistry, who proposed †Avogadro's hypothesis on gases in 1811. His work enabled scientists to calculate †Avogadro's number, and still has relevance for atomic studies.

Avogadro's hypothesis in chemistry, the law stating that equal volumes of all gases, when at the same temperature and pressure, have the same numbers of molecules. One †mole of any gas contains 6.023 x 10^{23} particles and occupies 24 dm^3 at room temperature and pressure. The type of gas does not make any difference. The law was first put forward by Italian chemist Amedeo Avogadro.

Avogadro's Hypothesis http://www.carlton.paschools.pa.sk.ca/chemical/molemass/avogadro.htm

Scientific biography of the ground-breaking 18th-century scientist Amadeo Avogadro, who hypothesized that that "equal volumes of gases, at the same temperature and pressure, contain equal numbers of molecules." The page opens with a discussion of the nature of scientific thought, moves on to describe Avogadro's discoveries and ends by explaining the number named in honor of Avogadro, who died in obscurity, his distinction between atoms and molecules as yet unrecognized.

Avogadro's number or **Avogadro's constant**, number of carbon atoms in 12 g of the carbon-12 isotope (6.022045 x 10^{23}). It is named after Italian chemist Amedeo Avogadro. The †relative atomic mass of any element, expressed in grams, contains this number of atoms and is called a †mole. For example, one mole of any substance contains 6.022 x 10^{23} particles. One mole of carbon has a mass of 12 g.

avoirdupois system of units of mass based on the pound (0.45 kg), which consists of 16 ounces (each of 16 drams) or 7,000 grains (each equal to 65 mg).

Axelrod, Julius (1912–) US neuropharmacologist who was awarded a Nobel Prize for Physiology or Medicine in 1970 for his work with the biophysicists Bernard †Katz and Ulf von Euler, on neurotransmitters (the chemical messengers of the brain).

axil upper angle between a leaf (or bract) and the stem from which it grows. Organs developing in the axil, such as shoots and buds, are termed axillary, or lateral.

axiom in mathematics, a statement that is assumed to be true and upon which theorems are proved by using logical deduction; for example, two straight lines cannot enclose a space. The Greek mathematician Euclid used a series of axioms that he considered could not be demonstrated in terms of simpler concepts to prove his geometrical theorems.

axis plural **axes**, in geometry, one of the reference lines by which a point on a graph may be located. The horizontal axis is usually referred to as the *x*-axis, and the vertical axis as the *y*-axis. The term is also used to refer to the imaginary line about which an object may be said to be symmetrical (**axis of symmetry**) – for example, the diagonal of a square – or the line about which an object may revolve (**axis of rotation**).

axon long threadlike extension of a †nerve cell that conducts electrochemical impulses away from the cell body towards other nerve cells, or towards an effector organ such as a muscle. Axons terminate in †synapses, junctions with other nerve cells, muscles, or glands.

axoplasm in a nerve fiber, the †cytoplasm of an †axon, containing the †neurofibrils.

azeotropic mixture liquid mixture whose composition does not change upon or during boiling. The boiling point of such a mixture remains constant. Examples include ethanol and water.

azimuth in astronomy, angular distance of an object eastwards along the horizon, measured from due north, between the astronomical †meridian (the vertical circle passing through the center of the sky and the north and south points on the horizon) and the vertical circle containing the celestial body whose position is to be measured.

azobenzene C_6H_6–N = N–C_6H_6, in chemistry, orange-red crystalline solid, melting point 68°C/154°F. It is usually prepared by the partial reduction of †nitrobenzene.

azo dye synthetic dye containing the azo group of two nitrogen atoms (N—N) connecting aromatic ring compounds. Azo dyes are usually red, brown, or yellow, and make up about half the dyes produced. They are manufactured from aromatic †amines.

azulene $C_{10}H_8$, in chemistry, double-ring aromatic hydrocarbon. It is a blue-violet crystalline solid, melting point 99°C/210°F. It readily undergoes a substitution reaction at 270°C/518°F to form †naphthalene.

B

Baade, (Wilhelm Heinrich) Walter (1893–1960) German-born US astronomer who made observations that doubled the distance, scale, and age of the universe. He discovered that stars are in two distinct populations according to their age, known as Population I (the younger) and Population II (the older). Later, he found that †Cepheid variable stars of Population I are brighter than had been supposed and that distances calculated from them were wrong.

Babbage, Charles (1792–1871) English mathematician who devised a precursor of the computer. He designed an †analytical engine, a general-purpose mechanical computing device for performing different calculations according to a program input on punched cards (an idea borrowed from the Jacquard loom). This device was never built, but it embodied many of the principles on which digital computers are based.

Babbage, Charles http://ei.cs.vt.edu/~history/Babbage.html

Extended biography of the visionary mathematician, industrialist, and misanthrope. This very full and entertaining account of his life is supported by a number of pictures of Babbage, his plan of the difference engine, and the actual difference engine built by the British Science Museum in 1991. There is a full bibliography.

Babcock, Harold Delos (1882–1968) US astronomer and physicist. He measured the Sun's general magnetic field in 1948 and studied the relationship between sunspots and local magnetic fields. He also did important work in spectroscopy.

bacillus genus of rod-shaped †bacteria that occur everywhere in the soil and air. Some are responsible for diseases such as †anthrax, or for causing food spoilage.

backcross cross between an offspring and one of its parents, or an individual that is genetically identical to one of its parents. It is a breeding technique used to determine the genetic makeup of an individual organism.

back emf in physics, †electromotive force (emf) that opposes the normal flow of current in a circuit or circuit component, usually caused by electromagnetic induction.

A back emf is generated in the armature of an electric motor if the coils are mechanically turned. Polarization in an electrolytic cell, caused by a layer of hydrogen bubbles on the cathode, also causes a back emf.

background radiation radiation that is always present in the environment. By far the greater proportion (87%) of it is emitted from natural sources. †Alpha particles, †beta particles, and †gamma radiation are emitted by the traces of radioactive minerals that occur naturally in the environment and even in the human body (for example, by breathing in 14C). Radioactive gases such as †radon and thoron are found in soil and may seep upwards into buildings. Radiation from space (†cosmic radiation) also contributes to the background level.

The **background count** is the count registered on a †Geiger counter when no other radioactive source is nearby.

back-up in computing, a copy file that is transferred to another medium, usually a disk or tape. The purpose of this is to have available a copy of a file that can be restored in case of a fault in the system or the file itself. Back-up files are also created by many applications (with the extension .BAC or .BAK); a version is therefore available of the original file before it was modified by the current application.

Basic Computing http://www.computerbits.com/archive/1998/0400/crla9804.html

Article explaining the "mystery of backing up" files and hard drives. It explains why backing up is such an important safety measure, and looks at the procedures which should be carried out to ensure data is secure.

backwash the retreat of a wave that has broken on a †beach. When a wave breaks, water rushes up the beach as †swash and is then drawn back towards the sea as backwash.

Bacon, Francis (1561–1626) 1st Baron Verulam and Viscount St Albans, English philosopher, politician, and writer, a founder of modern scientific research. His works include *Essays* (1597, revised and augmented 1612 and 1625), characterized by pith and brevity; *The Advancement of Learning* (1605), a seminal work discussing scientific method; *Novum Organum* (1620), in which he redefined the task of natural science, seeing it as a means of empirical discovery and a method of increasing human power over nature; and *The New Atlantis* (1626), describing a utopian state in which scientific knowledge is systematically sought and exploited. He was briefly Lord Chancellor in 1618 but lost his post through corruption.

bacteria singular **bacterium**, microscopic single-celled organisms lacking a membrane-bound nucleus. Bacteria, like fungi and †viruses, are micro-organisms – organisms that are so small they can only be seen using a microscope. They are organisms that are simpler than the cells of animals, plants, and fungi in that they lack a nucleus. Bacteria are widespread, being present in soil, air, and water, and as parasites on and in other living things. In fact, they occur anywhere life can exist. Some parasitic bacteria cause disease by producing toxins, but others are harmless and can even benefit their hosts. Bacteria usually reproduce by †binary fission (dividing into two equal parts), and, on average, this occurs every 20 minutes. Only 4,000 species of bacteria are known (1998), although bacteriologists believe that around 3 million species may actually exist. Certain types of bacteria are vital in many food and industrial processes, while others play an essential role in the †nitrogen cycle, which maintains soil fertility. They can be the first organisms of a food chain, by acting as decomposers of dead plant and animal remains. This helps to recycle nutrients.

Bacteria can be grown on the surface of agar jelly in dishes in the laboratory and have been studied in detail. This has led to the development of antibiotics, chemicals that kill or inhibit other micro-organisms such as bacteria. Sometimes the chemical is not harmful to humans and so can be used to treat disease. Some antibiotics work against a range of bacteria and in many situations, such as penicillins. Others are quite specific. For example, neomycin sulfate is especially active against bacteria that cause infections of the middle ear. Unfortunately, there are increasing numbers of bacteria that are resistant to antibiotics. New antibiotics need to be discovered so that antibiotic-resistant bacteria can be destroyed, but this is only occurring very slowly.

What the Heck is an E. coli? http://people.ku.edu/~jbrown/ecoli.html

Explains the basics behind this bacterium, which can cause food poisoning, including information on the dangerous strain of the bacterium and how it developed. It contains guidelines to reduce the risk of infection. There are a number of links throughout the article to sites expanding on issues raised.

bacteriology study of †bacteria.

bacteriophage virus that attacks †bacteria, commonly called a phage. Such viruses are now useful vectors in genetic engineering for introducing modified DNA.

badlands barren landscape cut by erosion into a maze of ravines, pinnacles, gullies, and sharp-edged ridges. Areas in South Dakota and Nebraska, USA, are examples.

Baekeland, Leo Hendrik (1863–1944) Belgian-born US chemist. He invented †Bakelite, the first commercial plastic, made from formaldehyde (methanal) and †phenol. He also made a photographic paper, Velox, which could be developed in artificial light.

Baer, Karl Ernst Ritter von (1792–1876) Estonian embryologist who discovered the mammalian ovum in 1827. He made a significant contribution to the systematic study of the development of animals, and showed that an embryo develops from simple to complex, from a homogeneous to a heterogeneous stage.

Baeyer, Johann Friedrich Wilhelm Adolf von (1835–1917) German organic chemist who synthesized the dye indigo in 1880. He discovered barbituric acid in 1863, later to become the parent substance of a major class of hypnotic drugs. In 1888 he carried out the first synthesis of a terpene. He was awarded the Nobel Prize for Chemistry in 1905 for his work in organic dyes and hydroaromatic compounds.

Baily, Francis (1774–1844) British astronomer. He described the light effect called †Baily's beads, observable during a total eclipse of the Sun, in 1836.

Baily's beads bright spots of sunlight seen around the edge of the Moon for a few seconds immediately before and after a total †eclipse of the Sun, caused by sunlight shining between mountains at the Moon's edge. Sometimes one bead is much brighter than the others, producing the so-called **diamond ring** effect. The effect was described in 1836 by the English astronomer Francis Baily, a wealthy stockbroker who retired in 1825 to devote himself to astronomy.

Bainbridge, Kenneth Tompkins (1904–1996) US physicist who was director of the first atomic-bomb test at Alamogordo, New Mexico, in 1945. He also made important innovations in the †mass spectrometer.

Baird, John Logie (1888–1946) Scottish electrical engineer who pioneered television. In 1925 he gave the first public demonstration of television, transmitting an image of a recognizable human face. The following year, he gave the world's first demonstration of true television before an audience of about 50 scientists at the Royal Institution, London. By 1928 Baird had succeeded in demonstrating color television.

Bakelite first synthetic †plastic, created by Belgian-born US chemist Leo †Baekeland in 1909. Bakelite is hard, tough, and heatproof, and is used as an electrical insulator. It is made by the reaction of phenol with methanol (formaldehyde), producing a powdery resin that sets solid when heated. Objects are made by subjecting the resin to compression molding (simultaneous heat and pressure in a mould).

It is one of the thermosetting plastics, which do not remelt when heated, and is often used for electrical fittings.

baking powder mixture of †bicarbonate of soda, an acidic compound, and a nonreactive filler (usually starch or calcium sulfate), used in baking as a raising agent. It gives a light open texture to cakes and scones, and is used as a substitute for yeast in making soda bread.

Several different acidic compounds (for example, tartaric acid, cream of tartar, sodium or calcium acid phosphates, and glucono-delta-lactone) may be used, any of which will react with the sodium hydrogencarbonate, in the presence of water and heat, to release the carbon dioxide that causes the cake mix or dough to rise.

balance apparatus for weighing or measuring mass. The various types include the **beam balance**, consisting of a centrally pivoted lever with pans hanging from each end, and the **spring balance**, in which the object to be weighed stretches (or compresses) a vertical coil spring fitted with a pointer that indicates the weight on a scale. Kitchen and bathroom scales are balances.

balance of nature in ecology, the idea that there is an inherent equilibrium in most †ecosystems, with plants and animals interacting so as to produce a stable, continuing system of life on Earth. The activities of human beings can, and frequently do, disrupt the balance of nature.

Organisms in the ecosystem are adapted to each other – for example, waste products produced by one species are used by another and resources used by some are replenished by others; the oxygen needed by animals is produced by plants while the waste product of animal respiration, carbon dioxide, is used by plants as a raw material in photosynthesis. The †nitrogen cycle, the †water cycle, and the control of animal populations by natural predators are other examples.

ball-and-socket joint joint between bones that allows considerable movement in three dimensions, for instance the joint between the pelvis and the femur (hip joint). To facilitate movement, such joints are rimmed with cartilage and lubricated by synovial fluid. The bones are kept in place by ligaments and moved by muscles.

ballistic galvanometer in physics, moving-coil instrument that measures a sudden pulse of current in terms of the total charge that passes through it. Mechanical damping of the coil lengthens its natural oscillation period so that a rapid discharge is proportional to the coil's maximum deflection.

ballistics study of the motion and impact of projectiles such as bullets, bombs, and missiles. For projectiles from a gun, relevant exterior factors include temperature, barometric pressure, and wind strength; and for nuclear missiles these extend to such factors as the speed at which the Earth turns.

Balmer, Johann Jakob (1825–1898) Swiss physicist and mathematician who developed a formula in 1884 that gave the wavelengths of the light emitted by the hydrogen atom (the hydrogen spectrum). This simple formula played a central role in the development of spectral and atomic theory.

Balmer series visible †spectrum of hydrogen, consisting of a series of distinct †spectral lines with wavelengths in the visible region.

Bamberga in astronomy, one of the largest asteroids and one that reflects sunlight only poorly. It was discovered in 1893 and is 252 km/157 mi in diameter.

bandwidth range of †frequencies over which the capability of a receiver or other electric device does not differ from its peak by a given amount.

Banks, Joseph (1743–1820) English naturalist and explorer. In the position of naturalist, he accompanied Captain James Cook on an expedition from 1768 to 1771 to the southern hemisphere in the *Endeavour*, and brought back 3,600 plants, a large proportion of which had not been classified previously. The *Banksia* genus of shrubs is named after him.

Banting, Frederick Grant (1891–1941) Canadian physician who was awarded a Nobel Prize for Physiology or Medicine in 1923 for his discovery, in 1921, of a technique for isolating the hormone insulin. Banting and his colleague Charles †Best tied off the ducts of the †pancreas to determine the function of the cells known as the †islets of Langerhans and thus made possible the treatment of diabetes. John J R †Macleod, Banting's mentor, shared the prize, and Banting divided his prize with Best.

Banting, Sir Frederick Grant http://www.nobel.se/medicine/laureates/1923/banting-bio.html

Profile of the life and career of the Nobel laureate. It traces his childhood, early interest in divinity, developing interest in diabetes, and his death in a plane crash. There is also a picture of Banting.

bar in earth sciences, deposit of sand or silt formed in a river channel, or a long ridge of sand or pebbles running parallel to a coastline (see †coastal erosion). Coastal bars can extend across estuaries to form **bay bars** and are formed in one of two ways. †Longshore drift can transport material across a bay and deposit it, thereby closing off the bay. Alternatively, an offshore bar (formed where waves touch the seabed and disturb the sediments, causing a small ridge to be formed) may be pushed towards the land as the sea level rises. These bars are greatly affected by the beach cycle. The high tides and high waves of winter erode the beach and deposit the sand as offshore bars. These are known as barrier beaches in the USA.

bar in physics, unit of pressure equal to 10^5 pascals or 10^6 dynes/cm^2, approximately 750 mmHg or 0.987 atm. Its diminutive, the **millibar** (one-thousandth of a bar), is commonly used by meteorologists.

barchan in earth science, type of crescent-shaped sand †dune that has its open end facing downwind, characteristic of places where there is a poor supply of sand.

bar chart in statistics, a way of displaying data, using horizontal or vertical bars. The heights or lengths of the bars are proportional to the quantities they represent.

bar code pattern of bars and spaces that can be read by a computer. Bar codes are widely used in retailing, industrial distribution, and libraries. The code is read by a scanning device; the computer determines the code from the widths of the bars and spaces.

The technique was patented in 1949 but became popular only in 1973, when the food industry in North America adopted the Universal Product Code system.

BarCode 1: A Web Of Information About Bar Code http://www.adams1.com/pub/russadam/barcode1.html
Everything you could want to know about bar codes. Among the many attractions are a comprehensive bar code "Frequently Asked Questions" list, software directory, user forum, and a searchable database of bar code Web sites.

Bardeen, John (1908–1991) US physicist. He was awarded the Nobel Prize for Physics in 1956, with Walter †Brattain and William †Shockley, for the development of the †transistor in 1948 and he became the first double winner of the Nobel Prize for Physics in 1972 (with Leon †Cooper and Robert †Schrieffer) for his work on †superconductivity.

Bardeen–Cooper–Schrieffer theory or **BCS theory**, theory that accounts for the zero electrical resistance of some superconducting metals by invoking an attractive interaction between electrons binding them into "Cooper pairs."

barium chemical symbol Ba, (Greek *barytes* "heavy") soft, silver-white, metallic element, atomic number 56, relative atomic mass 137.33. It is one of the alkaline-earth metals, found in nature as barium carbonate and barium sulfate. As the sulfate it is used in medicine: taken as a suspension (a "barium meal"), its movement along the gut is followed using X-rays. The barium sulfate, which is opaque to X-rays, shows the shape of the gut, revealing any abnormalities of the alimentary canal. Barium is also used in alloys, pigments, and safety matches and, with strontium, forms the emissive surface in cathode-ray tubes. It was first discovered in baryte or heavy spar.

barium sulfate $BaSO_4$, in chemistry, compound found in nature as the mineral barytes or heavy spar. It is insoluble in water and commonly used in the production of paints and pigments. It is also administered medicinally as a contrast medium to allow X-ray examination of the intestines.

barium titanate $BaTiO_3$, in chemistry, important †piezoelectric crystalline ceramic used in a wide range of solid-state applications such as transducer and ultrasonic device manufacture. It is prepared by a reaction at high temperature between barium carbonate and titanium dioxide.

bark protective outer layer on the stems and roots of woody plants, composed mainly of dead cells. To allow for expansion of the stem, the bark is continually added to from within, and the outer surface often becomes cracked or is shed as scales. Trees deposit a variety of chemicals in their bark, including poisons. Many of these chemical substances have economic value because they can be used in the manufacture of drugs. Quinine, derived from the bark of the *Cinchona* tree, is used to fight malarial infections; curare, an anesthetic used in medicine, comes from the *Strychnus toxifera* tree in the Amazonian rainforest.

Bark technically includes all the tissues external to the vascular †cambium (the †phloem, cortex, and periderm), and its thickness may vary from 2.5 mm/0.1 in to 30 cm/12 in or more, as in the giant redwood *Sequoia* where it forms a thick, spongy layer.

Barkhausen, Heinrich Georg (1881–1956) German physicist who did early work in electronics and in 1919 discovered the †Barkhausen effect. In the 1920s he constructed microwave components and UHF oscillators that contributed to the development of German military radar in World War II.

Barkhausen effect in physics, when a piece of ferromagnetic material is placed in an increasing magnetizing field, its magnetic domains suddenly change size or direction. As a result, magnetization of the material proceeds in a series of steps rather than continuously.

Barkla, Charles Glover (1877–1944) English physicist who studied the phenomenon of secondary radiation (the effect whereby a substance subjected to X-rays re-emits secondary X-radiation) and X-ray scattering. He was awarded the Nobel Prize for Physics in 1917 for his discovery that X-ray emissions are a form of transverse electromagnetic radiation, like visible light, and monochromatic.

barn in physics, effective cross-section of an atom or nucleus when it acts as a "target" for incident particles (such as neutrons). It is an area equal to 10^{-28} square meters.

Barnard, Edward Emerson (1857–1923) US observational astronomer. He discovered the fifth satellite of Jupiter in 1892 and the swiftly moving 10th-magnitude Barnard's star in 1916. He was the first to realize that the apparent voids in the Milky Way are in fact dark nebulae of dust and gas.

Barnard, Edward Emerson http://astro.uchicago.edu/yerkes/virtualmuseum/Barnardfull.html

Extensive, illustrated biography of the US astronomer, based at the Yerkes Observatory Museum, USA. This interesting and lively account of the life of the pioneer astronomer is supported by a large number of pictures of Barnard, the telescopes he used, and the comet named after him.

Barnard's star star, 6 light years away from the Sun, in the constellation Ophiuchus. It is the second-closest star to the Sun, after Alpha Centauri, a triple star, the closest component of which, Proxima Centauri, is 4.2 light years away from the Sun. It is a faint red dwarf of 10th magnitude, visible only through a telescope. It is named after the US astronomer Edward E Barnard (1857–1923), who discovered in 1916 that it has the fastest proper motion of any star, 10.3 arc seconds per year.

Some observations suggest that Barnard's star may be accompanied by planets.

barograph device for recording variations in atmospheric pressure. A pen, governed by the movements of an aneroid †barometer, makes a continuous line on a paper strip on a cylinder that rotates over a day or week to create a **barogram**, or permanent record of variations in atmospheric pressure.

barometer instrument that measures atmospheric pressure as an indication of weather. Most often used are the mercury barometer and the aneroid barometer.

In a mercury barometer a column of mercury in a glass tube, roughly 0.75 m/2.5 ft high (closed at one end, curved upwards at the other), is balanced by the pressure of the atmosphere on the open end; any change in the height of the column reflects a change in pressure. In an aneroid barometer, a shallow cylindrical metal box containing a partial vacuum expands or contracts in response to changes in pressure.

baroreceptor in biology, a specialized nerve ending that is sensitive to pressure. There are baroreceptors in various regions of the heart and circulatory system (carotid sinus, aortic arch, atria, pulmonary veins, and left ventricle). Increased pressure in these structures stimulates the baroreceptors, which relay information to the †medulla providing an important mechanism in the control of blood pressure.

barrage in geography, a structure built across a river or estuary in order to manage the water. The Thames barrier is an example of a flood-control barrage. A barrage may be used to regulate the water supply by controlling floods or storing water for †irrigation, or to generate power by, for example, harnessing tidal energy in estuaries.

barrel unit of liquid capacity, the value of which depends on the liquid being measured. It is used for petroleum, a barrel of which contains 159 litres/35 imperial gallons; a barrel of alcohol contains 189 litres/41.5 imperial gallons.

barrier island long island of sand, lying offshore and parallel to the coast.

Some are over 100 km/60 mi in length. Most barrier islands are derived from marine sands piled up by shallow longshore currents that sweep sand parallel to the seashore. Others are derived from former spits, connected to land and built up by drifted sand, that were later severed from the mainland.

Often several islands lie in a continuous row offshore. Coney Island and Jones Beach near New York City are well-known examples, as is Padre Island, Texas. The Frisian Islands are barrier islands along the coast of the Netherlands.

barrier reef †coral reef that lies offshore, separated from the mainland by a shallow lagoon.

Barringer Crater or **Arizona Meteor Crater** or **Coon Butte**, impact crater near Winslow in Arizona, caused by the impact of a 50-m/165-ft iron †meteorite about 50,000 years ago. It is 1.2 km/0.7 mi in diameter and 200 m/660 ft deep, and the walls are raised 50–60 m/165–198 ft above the surrounding desert.

It is named after the US mining engineer Daniel Barringer who proposed in 1902 that it was an impact crater rather than a volcanic feature, an idea confirmed in the 1960s by US astronomer Eugene Shoemaker, who studied the crater for his doctoral research at the California Institute of Technology.

Barrow, Isaac (1630–1677) English mathematician, theologian, and classicist. His *Lectiones geometricae* (1670) contains the essence of the theory of †calculus, which was later expanded by Isaac †Newton and Gottfried †Leibniz.

Barrow, Isaac http://www-groups.dcs.st-andrews.ac.uk/history/Mathematicians/Barrow.html

Part of an archive containing the biographies of the world's greatest mathematicians, this site is devoted to the life and contributions of Isaac Barrow. In addition to biographical information, you will find a list of references about Barrow and links to other essays in the archive that reference him. The text of this essay includes hypertext links to the essays of those mathematicians and thinkers who influenced Barrow. You will also find an image of him, which you can click on to enlarge, and a map of his birthplace.

Bartlett, Neil (1932–) British-born US chemist. In 1962 he prepared the first compound of one of the †noble gases (rare gases), which were previously thought to be incapable of reacting with anything.

Barton, Derek Harold Richard (1918–1998) English organic chemist. Barton investigated the stereochemistry of natural compounds. He showed that their biological activity often depends on the shapes of their molecules and the positions and orientations of key functional groups. He shared the Nobel Prize for Chemistry in 1969 for his creation of the branch of organic chemistry known as conformational analysis.

barycentric calculus †coordinate geometry calculations using a coordinate system devised by August Möbius in which numerical coefficients are assigned to points on a plane, giving the position of a general point by reference to four or more noncoplanar points. Described in this way, the general point thus represents a center of gravity for a distribution of mass at the four (or more) points proportional to the assigned numbers.

baryon in nuclear physics, a heavy subatomic particle made up of three indivisible elementary particles called quarks. The baryons form a subclass of the †hadrons and comprise the nucleons (protons and neutrons) and hyperons.

baryte barium sulfate, $BaSO_4$, the most common mineral of barium. It is white or light-colored, and has a comparatively high density (specific gravity 4.6); the latter property makes it useful in the production of high-density drilling muds (muds used to cool and lubricate drilling equipment). Baryte occurs mainly in ore veins, where it is often found with calcite and with lead and zinc minerals. It crystallizes in the orthorhombic system and can form tabular crystals or radiating fibrous masses.

basal ganglia several masses of gray matter located deep in the †cerebrum, including the caudate nucleus, the globus pallidus and putamen. They are involved with the subconscious regulation of movement. See also †brain.

basal metabolic rate **BMR**, minimum amount of energy needed by the body to maintain life. It is measured when the subject is awake but resting, and includes the energy required to keep the heart beating, sustain breathing, repair tissues, and keep the brain and nerves functioning. The rate varies depending on the height, weight, age, and activity of the person. Measuring the subject's consumption of oxygen gives an accurate value for BMR, because oxygen is needed to release energy from food.

A cruder measure of BMR estimates the amount of heat given off, some heat being released when food is used up. BMR varies from one species to another, and from males to females. In humans, it is highest in children and declines with age. Disease, including mental illness, can make it rise or fall. Hormones from the †thyroid gland control the BMR.

basalt commonest volcanic †igneous rock in the Solar System. Basalt is an †extrusive rock, created by the outpouring of volcanic magma. The magma cools quickly, allowing only small crystals to form. Much of the surfaces of the terrestrial planets Mercury, Venus, Earth, and Mars, as well as the Moon, are composed of basalt. Earth's ocean floor is virtually entirely made of basalt. Basalt is mafic, that is, it contains relatively little †silica: about 50% by weight. It is usually dark gray but can also be green, brown, or black. Its essential constituent minerals are calcium-rich †feldspar, and calcium- and magnesium-rich †pyroxene.

The groundmass may be glassy or finely crystalline, sometimes with large †crystals embedded. Basaltic lava tends to be runny and flows for great distances before solidifying. Successive eruptions of basalt have formed the great plateaus of Colorado and the Deccan plateau region of southwest India. In some places, such as Fingal's Cave in the Inner Hebrides of Scotland and the Giant's Causeway in Antrim, Northern Ireland, shrinkage during the solidification of the molten lava caused the formation of hexagonal columns.

The dark-colored lowland maria regions of the Moon are underlain by basalt. Lunar †mare basalts have higher concentrations of titanium and zirconium and lower concentrations of volatile elements like potassium and sodium relative to terrestrial basalts. Martian basalts are characterized by low ratios of iron to manganese relative to terrestrial basalts, as judged from some Martian meteorites (shergottites, a class of the **SNC** meteorites) and spacecraft analyses of rocks and soils on the Martian surface.

base in chemistry, a substance that accepts hydrogen ions, or protons. A base reacts with an †acid, neutralizing it to form a †salt: acid + base → salt + water. Metal oxides and metal hydroxides are bases; examples include copper oxide and sodium hydroxide. Bases can contain negative ions such as the hydroxide ion (OH^-), which is the strongest base, or be molecules such as ammonia (NH_3). Ammonia is a weak base, as only some of its molecules accept protons.

$OH^- + H^+_{(aq)} \rightarrow H_2O_{(l)}$

$NH_3 + H_2O \rightarrow NH_4^+ + OH^-$

Bases that dissolve in water are called †alkalis.

Inorganic bases are usually oxides or hydroxides of metals, which react with dilute acids to form a salt and water. Many carbonates also react with dilute acids, additionally giving off carbon dioxide.

Definition of Acids and Bases and the Role of Water

http://chemed.chem.purdue.edu/genchem/topicreview/bp/ch11/acidbaseframe.html

This well-organized page contains a table of various definitions of acids and bases. Part of a much larger introductory chemistry site aimed at university students, it shows how scientific knowledge about acids has expanded from the days of Boyle.

base in mathematics, the number of different single-digit symbols used in a particular number system. In our usual (decimal) counting system of numbers (with symbols 0, 1, 2, 3, 4, 5, 6, 7, 8, 9) the base is 10. In the †binary number system, which has only the symbols 1 and 0, the base is two. A base is also a number that, when raised to a particular power (that is, when multiplied by itself a particular number of times as in $10^2 = 10 \times 10 = 100$), has a †logarithm equal to the power. For example, the logarithm of 100 to the base ten is 2.

In geometry, the term is used to denote the line or area on which a polygon or solid stands.

baseflow or **groundwater flow**, movement of water from land to river through rock. It is the slowest form of such water movement, and accounts for the constant flow of water in rivers during times of low rainfall, and makes up the river's base line on a †hydrograph.

base level level, or altitude, at which a river reaches the sea or a lake. The river erodes down to this level. If base level falls (due to uplift or a drop in sea level), †rejuvenation takes place.

base pair in biochemistry, the linkage of two base (purine or pyrimidine) molecules that join the complementary strands of †DNA. Adenine forms a base pair with thymine (or uracil in RNA) and cytosine pairs with guanine in a double-stranded nucleic acid molecule.

One base lies on one strand of the DNA double helix and one on the other, so that the base pairs link the two strands like the rungs of a ladder. In DNA, there are four bases: adenine and guanine (purines) and cytosine and thymine (pyrimidines). Adenine always pairs with thymine and cytosine with guanine.

BASIC acronym for Beginner's All-purpose Symbolic Instruction Code, high-level computer-programming language, developed in 1964, originally designed to take advantage of multi-user systems (which can be used by many people at the same time). The language is relatively easy to learn and is popular among microcomputer users.

Most versions make use of an interpreter, which translates BASIC into †machine code and allows programs to be entered and run with no intermediate translation. Some more recent versions of BASIC allow a †compiler to be used for this process.

BASIC Archives http://www.fys.ruu.nl/~bergmann/basic.html

Detailed guide to this computer-programming language. This site includes an informative "Frequently Asked Questions," giving tips on such topics as "how to print graphics with BASIC," "how to display a BMP or GIF file," and "how to make an EXE file." There is also a brief history of the development of BASIC and a number of links to related sites.

basicity number of replaceable hydrogen atoms in an acid. Nitric acid (HNO_3) is monobasic, sulfuric acid (H_2SO_4) is dibasic, and phosphoric acid (H_3PO_4) is tribasic.

basic rock igneous rock with relatively low silica contents of 45–52% by weight, such as gabbro and basalt. Along with the terms **acid rock** and **ultrabasic rock** it is part of an outdated classification system based on the erroneous belief that silicon in rocks is in the form of silicic acid. Geologists today are more likely to use the descriptive term †mafic rock or report the amount of SiO_2 in percentage by weight.

basidiocarp spore-bearing body, or "fruiting body," of all basidiomycete fungi (see †fungus), except the rusts and smuts. A well known example is the edible mushroom *Agaricus brunnescens*. Other types include globular basidiocarps (puffballs) or flat ones that project from tree trunks (brackets). They are made up of a mass of tightly packed, intermeshed †hyphae.

The tips of these hyphae develop into the reproductive cells, or **basidia**, that form a fertile layer known as the hymenium, or **gills**, of the basidiocarp. Four spores are budded off from the surface of each basidium.

Basidiomycetes large class of fungi, with perhaps 12,000 species, which includes mushrooms, toadstools, bracket fungi, puffballs, smuts, and rusts. The †thallus, the body of the plant, is a mass of branching threads. The reproductive spores are borne on a club-shaped organ called a basidium (plural basidia), and are called basidiospores. There are usually four on each basidium. The basidia are grouped into conspicuous fruiting bodies, from which they are dispersed by the wind. Mushrooms and brackets are fruiting bodies.

Basov, Nikolai Gennadievich (1922–2001) Soviet physicist who, with his compatriot Aleksandr †Prokhorov, developed the microwave amplifier called a †maser. They were both awarded the Nobel

Prize for Physics in 1964 for their work on quantum electronics leading to the construction of oscillators and amplifiers based on the maser-laser principle. They shared the award with Charles †Townes of the USA.

Bassi, Agostino (1773–1856) Italian microbiologist who was the first to demonstrate that microscopic organisms could cause certain infectious diseases. His work preceded that of the Louis †Pasteur, who formulated the theory that germs might cause some diseases, in 1868.

batch file in computing, file that runs a group (batch) of commands. The most commonly used batch file is the †DOS start-up file AUTOEXEC.BAT.

batch processing in computing, system for processing data with little or no operator intervention. Batches of data are prepared in advance to be processed during regular "runs" (for example, each night). This allows efficient use of the computer and is well suited to applications of a repetitive nature, such as a company payroll, or the production of utility bills.

In †**interactive computing**, by contrast, data and instructions are entered while the processing program is running.

Bates, H(enry) W(alter) (1825–1892) English naturalist and explorer. He spent 11 years collecting animals and plants in South America and identified 8,000 new species of insects. He made a special study of †camouflage in animals, and his observation of insect imitation of species that are unpleasant to predators is known as "Batesian mimicry."

Bateson, William (1861–1926) English geneticist. Bateson was one of the founders of the science of †genetics (a term he introduced), and a leading proponent of Austrian biologist Gregor †Mendel's work on heredity. Bateson also made contributions to embryology and to the theory of †evolution.

batholith large, irregular, deep-seated mass of intrusive †igneous rock, usually granite, with an exposed surface of more than 100 sq km/40 sq mi. The mass forms by the intrusion or upswelling of magma (molten rock) through the surrounding rock. Batholiths form the core of some large mountain ranges like the Sierra Nevada of western North America.

According to †plate tectonic theory, magma rises in subduction zones along continental margins where one plate sinks beneath another. The solidified magma becomes the central axis of a rising mountain range, resulting in the deformation (folding and overthrusting) of rocks on either side. Gravity measurements indicate that the downward extent or thickness of many batholiths is some 10–15 km/6–9 mi.

bathyal zone upper part of the ocean, which lies on the continental shelf at a depth of between 200 m/650 ft and 2,000 m/6,500 ft.

Bathyal zones (both temperate and tropical) have greater biodiversity than coral reefs, according to a 1995 study by the Natural History Museum in London. Maximum biodiversity occurs between 1,000 m/3,280 ft and 3,000 m/9,800 ft.

battery any energy-storage device allowing release of electricity on demand. It is made up of one or more electrical †cells. Electricity is produced by a chemical reaction in the cells. There are two types of battery: primary-cell batteries, which are disposable; and secondary-cell batteries, or †accumulators, which are rechargeable. Primary-cell batteries are an extremely uneconomical form of energy, since they produce only 2% of the power used in their manufacture. It is dangerous to try to recharge a primary-cell battery.

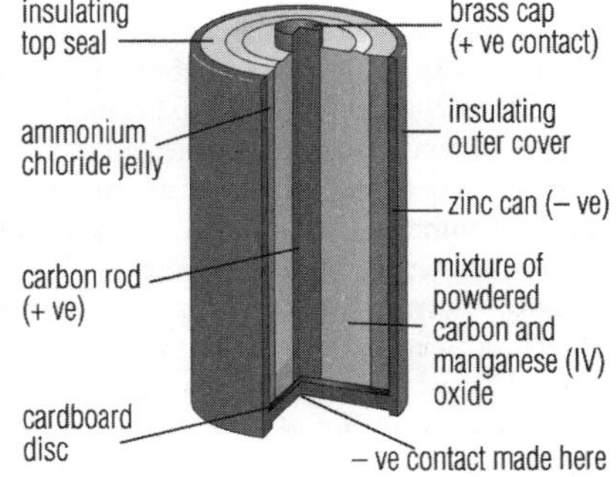

battery The common dry cell relies on chemical changes occurring between the electrodes – the central carbon rod and the outer zinc casing – and the ammonium chloride electrolyte to produce electricity. The mixture of carbon and manganese is used to increase the life of the cell. Helicon

primary cell The **dry cell** is the most common type of primary-cell battery, based on the †Leclanché cell. Dry cells are used in batteries to power, for example, torches, electronic toys, and compact stereo systems. Dry cells consist of a zinc case used as a negative electrode, and a carbon rod as a positive electrode suspended in the center of the case and immersed in a paste of manganese dioxide and ammonium chloride acting as an electrolyte.

The cell depends on the difference in electronegativity between the zinc and carbon to produce electricity. Carbon strongly holds onto its electrons; it is more electronegative. Zinc weakly holds onto its electrons; it is less electronegative. Zinc dissolves in the ammonium chloride and loses two electrons for each atom of zinc. The electrons move towards the carbon electrode:

$Zn \rightarrow Zn^{2+} + 2e^-$

A small amount of charge is produced and zinc becomes negatively charged and carbon becomes positively charged. When a connection is made externally between the positive carbon terminal and the negative zinc terminal, electrons flow to the carbon from the zinc. The charge is neutralized on both electrodes and more of the zinc metal dissolves in the electrolyte to produce more electrons. A dry cell provides a current until the zinc electrode is completely used up.

secondary cell A storage battery of secondary cells gives large amounts of power for a short time and can be recharged. The lead–acid car battery is a secondary-cell battery. The electrolyte is sulfuric acid (†battery acid), the positive electrode is lead peroxide, and the negative electrode is lead. A typical lead–acid battery consists of six lead–acid cells in a case. Each cell produces 2 volts, so the whole battery produces a total of 12 volts.

Hydrogen cells and sodium–sulfur batteries were developed in 1996 to allow cars to run entirely on battery power for up to 160 km/100 mi.

The introduction of rechargeable nickel–cadmium batteries has revolutionized portable electronic news gathering (sound recording, video) and information processing (computing). These batteries offer a stable, short-term source of power free of noise and other electrical hazards.

battery acid †sulfuric acid of approximately 70% concentration used in lead–acid cells (as found in car batteries).

The chemical reaction within the battery that is responsible for generating electricity also causes a change in the acid's composition. This can be detected as a change in its specific gravity: in a fully charged battery the acid's specific gravity is 1.270–1.290; in a half-charged battery it is 1.190–1.210; in a flat battery it is 1.110–1.130.

baud in engineering, a unit of electrical signaling speed equal to one pulse per second, measuring the rate at which signals are sent between electronic devices such as telegraphs and computers.

Bauds were used as a measure to identify the speed of †modems until the early 1990s because at the lower modem speeds available then the baud rate generally equaled the rate of transmission measured in †bps (bits per second). At higher speeds, this is not the case, and modem speeds now are generally quoted in Kbps.

Baudot code five-bit code developed in France by engineer Emil Baudot (1845–1903) in the 1870s. It is still in use for telex.

Baumé scale in physics, liquid relative density (specific gravity) scale on which the density of water is 0° and the density of a 10% salt solution is 10° (at 12.5°C). It is sometimes used to calibrate hydrometers.

bauxite principal ore of †aluminum, consisting of a mixture of hydrated aluminum oxides and hydroxides, generally contaminated with compounds of iron, which give it a red color. It is formed by the †chemical weathering of rocks in tropical climates. Chief producers of bauxite are Australia, Guinea, Jamaica, Russia, Kazakhstan, Suriname, and Brazil.

To extract aluminum from bauxite, high temperatures (about 800°C/1,470°F) are needed to make the ore molten. Electric currents are then passed through the molten ore. The process is economical only if cheap electricity is readily available, usually from a hydroelectric plant.

Bayes, Thomas (1702–1761) English mathematician whose investigations into probability led to what is now known as †Bayes' theorem.

Bayes, Thomas http://www-groups.dcs.st-andrews.ac.uk/history/Mathematicians/Bayes.html

Part of an archive containing the biographies of the world's greatest mathematicians, this site is devoted to the life and contributions of Thomas Bayes. In addition to biographical information, you will find a list of references about Bayes and links to other essays in the archive that reference him. The text of this essay includes hypertext links to the essays of those mathematicians and thinkers who influenced Bayes. You will also find an image of him, which you can click on to enlarge, and a map of his birthplace.

Bayesian statistics form of statistics that uses the knowledge of prior probability together with the probability of actual data to determine posterior probabilities, using Bayes' theorem.

Bayes' theorem in statistics, a theorem relating the †probability of particular events taking place to the probability that events conditional upon them have occurred.

For example, the probability of picking an ace at random out of a pack of cards is $^4/_{52}$. If two cards are picked out, the probability of the second card being an ace is conditional on the first card: if the first card is an ace the probability of drawing a second ace will be $^3/_{51}$; if not it will be $^4/_{51}$. Bayes' theorem gives the probability that given that the second card is an ace, the first card is also.

Bayliss, William Maddock (1860–1924) English physiologist who discovered the digestive hormone secretin, the first hormone to be found, with Ernest †Starling in 1902. During World War I, Bayliss introduced the use of saline (salt water) injections to help the injured recover from shock.

bayou (corruption of French *boyau* "gut") in the southern USA, an †oxbow lake or marshy offshoot of a river.

Bayous may be formed, as in the lower Mississippi, by a river flowing in wide curves or meanders in flat country, and then cutting a straight course across them in times of flood, leaving loops of isolated water behind.

B cell or **B lymphocyte**, type of lymphocyte (white blood cell) that develops in the bone marrow and aids the immune system by producing antibodies. Each B cell produces just one type of †antibody, specific to a single †antigen. Lymphocytes are related to †T cells.

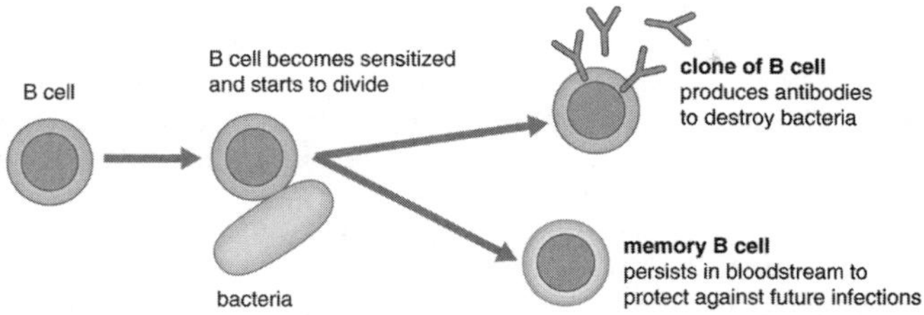

B cell B cells, a type of lymphocyte (white blood cell), have an important role in the body's immune system. When a B cell encounters an invading bacterium it starts to divide, forming two different types of cell. One type is a clone of itself that begins to produce antibodies to fight the infection; the other is a memory cell that will persist in the bloodstream, ready to produce antibodies should re-infection occur. Helicon

beach strip of land bordering the sea, normally consisting of boulders and pebbles on exposed coasts, or †sand on sheltered coasts. Beaches lie between the high- and low-water marks (high and low tides). A †berm, a ridge of sand and pebbles, may be found at the farthest point that the water reaches, generally at high tide.

The material of the beach consists of a rocky debris eroded from exposed rocks and headlands by the processes of †coastal erosion, or material carried in by rivers. The material is transported to the beach, and along the beach, by †longshore drift.

When the energy of the waves decreases, more sand is deposited than is transported, building depositional features such as †spits, †bars, and **tombolos**.

Beadle, George Wells (1903–1989) US biologist who was awarded the Nobel Prize for Physiology or Medicine in 1958 with Edward L †Tatum and Joshua †Lederberg for work in biochemical genetics, forming the "one-gene–one-enzyme" hypothesis (a single gene codes for a single kind of enzyme).

beam balance instrument for measuring mass (or weight). A simple form consists of a beam pivoted at its midpoint with a pan hanging at each end. The mass to be measured, in one pan, is compared with a variety of standard masses placed in the other. When the beam is balanced, the masses' turning effects or moments under gravity, and hence the masses themselves, are equal.

Bear, Great and Little common names (and translations of the Latin) for the constellations †Ursa Major and †Ursa Minor respectively.

bearing direction of a fixed point, or the path of a moving object, from a point of observation. Bearings are angles measured in degrees (°) from the north line in a clockwise direction. A bearing must always have three figures. For instance, north is 000°, northeast is 045°, south is 180°, and southwest is 225°.

True north differs slightly from magnetic north (the direction in which a †compass needle points), hence northeast may be denoted as 045M or 045T, depending on whether the reference line is magnetic (M) or true (T) north. True north also differs slightly from grid north since it is impossible to show a sphere on a flat map.

beat frequency in musical acoustics, fluctuation produced when two notes of nearly equal pitch or †frequency are heard together. Beats result from the †interference between the sound waves of the notes. The frequency of the beats equals the difference in frequency of the notes.

Beaufort, Francis (1774–1857) British admiral, hydrographer to the Royal Navy from 1829; the Beaufort scale and the Beaufort Sea in the Arctic Ocean are named after him. Made Knight Commander of the Bath (KCB) in 1848.

Beaufort scale system of recording wind velocity, devised by Francis Beaufort in 1806. It is a numerical scale ranging from 0 to 17, calm being indicated by 0 and a hurricane by 12; 13–17 indicate degrees of hurricane force.

In 1874 the scale received international recognition; it was modified in 1926. Measurements are made at 10 m/33 ft above ground level.

becquerel symbol Bq, SI unit of †radioactivity, equal to one radioactive disintegration (change in the nucleus of an atom when a particle or ray is given off) per second.

The becquerel is much smaller than the previous standard unit, the †curie (3.7 x 10^{10} Bq). It is named after French physicist Henri †Becquerel.

Becquerel, (Antoine) Henri (1852–1908) French physicist. He was awarded the Nobel Prize for Physics in 1903 for his discovery of penetrating radiation coming from uranium salts, the first indication of spontaneous †radioactivity. He shared the award with Marie and Pierre †Curie.

Biography of A H Becquerel http://www.nobel.se/physics/laureates/1903/becquerel-bio.html

Presentation of the life and discoveries of A H Becquerel, a distinguished scientist who was awarded the Nobel Prize for Physics jointly with Marie and Pierre Curie. This essay mainly focuses on Becquerel's contribution to the understanding of spontaneous radioactivity.

Becquerel, Antoine César (1788–1878) French physicist who pioneered the study of electrochemistry. As the grandfather of Henri †Becquerel, he was the first of several generations of distinguished scientists in the Becquerel family.

bed in geology, a single †sedimentary rock unit with a distinct set of physical characteristics or contained fossils, readily distinguishable from those of beds above and below. Well-defined partings called **bedding planes** separate successive beds or strata.

The depth of a bed can vary from a fraction of a centimeter to several meters, and can extend over any area. The term is also used to indicate the floor beneath a body of water (lake bed), and a layer formed by a fall of particles (ash bed). Beds of Portland and Purbeck limestone are clearly shown at Stair Hole, next to Lulworth Cove, Dorset, England.

bedload material rolled or bounced (by †saltation) along a river bed. The particles carried as bedload are much larger than those carried in suspension in the water. During a flood many heavy boulders may be moved in this way. Such boulders can be seen lying on the river bed during times of normal flow.

Bednorz, Johannes Georg (1950–) German physicist who, with Alexander †Müller, was awarded the Nobel Prize for Physics in 1987 for his discovery of high-temperature †superconductivity in ceramic materials. The discovery of these materials contributed towards the use of superconductors in computers, magnetic levitation trains, and the more efficient generation and distribution of electricity.

Beer's law relationship that states that the amount of light absorbed by a substance is dependent on both the thickness and the molecular concentration of the material it passes through.

beet sugar in chemistry, natural form of sucrose that is extracted from the sugar beet plant, chemically identical to that extracted from sugar cane.

behaviorism school of psychology originating in the USA, of which the leading exponent was John B †Watson.

Behaviorists maintain that all human activity can ultimately be explained in terms of conditioned reactions or reflexes and habits formed in consequence. Leading behaviorists include Ivan †Pavlov and B F †Skinner.

Beijerinck, Martinus Willem (1851–1931) Dutch bacteriologist. Beijerinck published in 1898 his finding that an agent smaller than bacteria could cause diseases. He called this a †virus.

Békésy, Georg von (1899–1972) Hungarian-born US scientist who was awarded a Nobel Prize for Physiology or Medicine in 1961 for resolving the long-standing controversy on how the inner ear functions by discovering the mechanism of stimulation within the cochlea.

bel unit of sound measurement equal to ten †decibels. It is named after Scottish-born US scientist and inventor Alexander Graham †Bell.

Bell, Alexander Graham (1847–1922) Scottish-born US scientist and inventor. He was the first person ever to transmit speech from one point to another by electrical means. This invention – the telephone – was made in 1876, when Bell transmitted speech from Paris, Ontario, to Brantford, Ontario (a distance of 13 km/8 mi). Later Bell experimented with a type of phonograph and, in aeronautics, invented the tricycle undercarriage.

Alexander Graham Bell's Path to the Telephone http://jefferson.village.virginia.edu/albell/homepage.html

Impressive attempt to reconstruct the path towards the invention of the telephone. The presentation aims to communicate the thoughts of Alexander Graham Bell, placing him within the context of other relevant inventions and related ideas. This is achieved mainly through a wealth of graphics retrieved from Bell's experimental notebooks, patents, and other papers accompanied by detailed analysis. An especially worthwhile site for those who are interested in the thought processes behind invention in general, as well as in this particular one.

Bell, John (1928–1990) Northern Irish physicist who in 1964 discovered a paradoxical aspect of †quantum theory: two particles that were once connected are always afterwards interconnected even if they become widely separated. As well as investigating fundamental problems in theoretical physics, Bell contributed to the design of particle accelerators.

Bellatrix or **Gamma Orionis**, second magnitude star that marks the western shoulder of †Orion.

Bell Burnell, (Susan) Jocelyn (1943–) Northern Irish astronomer. In 1967 she discovered the first

†pulsar (rapidly flashing star) with British radio astronomer Antony †Hewish and colleagues at the Mullard Radio Astronomy Observatory, Cambridge, England.

Little Green Men, White Dwarfs, or Pulsars? http://www.bigear.org/vol1no1/burnell.htm

Entertaining first-hand history of the discovery of pulsars. This is the text of an after-dinner speech given by Bell Burnell. It gives an intimate account of the excitement and challenges of gamma and X-ray astronomy. There is also a brief biography of the pioneering astronomer.

Beltrami, Eugenio (1835–1899) Italian mathematician who pioneered modern non-Euclidean geometry. His work ranged over almost the whole field of pure and applied mathematics, but especially theories of surfaces and space of constant curvature.

belt transect in ecology, technique used to estimate the abundance and ground cover of species (plants and sedentary animals). A quadrant is used to record the numbers of species at intervals along the line of a transect between two fixed points.

ben (Scottish Gaelic *beinn* "summit") mountain peak, first element in the names of numerous Scottish mountains such as Ben Lomond, Ben More, and Ben Nevis. The related Brittanic form of *ban* (plural *bannau*) is found in some Welsh place names, including Bannau Brycheiniog (the Brecon Beacons).

Benacerraf, Baruj (1920–) Venezuelan-born US immunologist who was awarded a Nobel Prize for Physiology or Medicine in 1980, together with Jean †Dausset and George D †Snell, for their discovery of immune response genes and the major histocompatibility complex (MHC), the gene complex involved in the immune response. They studied a gene region within one chromosome, which influences the immune system in various ways. The MHC is of great medical and biological significance because of the part it plays in the rejection of transplants made between incompatible individuals.

Benedict solution in chemistry, alkaline aqueous solution of sodium carbonate, copper sulfate, and sodium citrate used as a test for the detection of reducing agents, in particular sugars such as glucose (see †reduction). In the presence of reducing agents, a red precipitate of copper oxide is produced.

Benguela current cold ocean current in the South Atlantic Ocean, moving northwards along the west coast of southern Africa and merging with the south equatorial current at a latitude of 15° S. Its rich plankton supports large, commercially exploited fish populations.

Benioff zone seismically active zone inclined from a deep sea trench. Earthquakes along Benioff zones apparently define a lithospheric plate that descends into the mantle beneath another, overlying plate. The zone is named after Hugo Benioff, a US seismologist who first described this feature.

Bennet bright comet, discovered by South African astronomer J C Bennett in 1969, which became a spectacular sight in the spring of 1970. It passed within 81 million km/50 million mi of the Sun on 20 March and within 104 million km/65 million mi of Earth six days later. At its brightest it reached magnitude zero and had a tail spanning 10°.

benthic describing the environment on the sea floor supporting bottom-dwelling plants (such as seaweeds) and animals (including corals, anemones, sponges, and shellfish).

bentonite soft porous rock consisting mainly of clay minerals, such as montmorillonite, and resembling †fuller's earth, which swells when wet. It is formed by the chemical alteration of glassy volcanic material, such as tuff. Bentonite is used in papermaking, molding sands, drilling muds for oil wells, and as a decolorant in food processing.

benzaldehyde C_6H_5CHO, colorless liquid with the characteristic odor of almonds. It is used as a solvent and in the making of perfumes and dyes. It occurs in certain leaves, such as the cherry, laurel, and peach, and in a combined form in certain nuts and kernels. It can be extracted from such natural sources, but is usually made from †toluene.

benzene C_6H_6, clear liquid hydrocarbon of characteristic odor, occurring in coal tar. It is used as a solvent and in the synthesis of many chemicals.

The benzene molecule consists of a ring of six carbon atoms, all of which are in a single plane, and it is one of the simplest †cyclic compounds. Benzene is the simplest of a class of compounds collectively known as **aromatic compounds**. Some are considered carcinogenic (cancer-inducing).

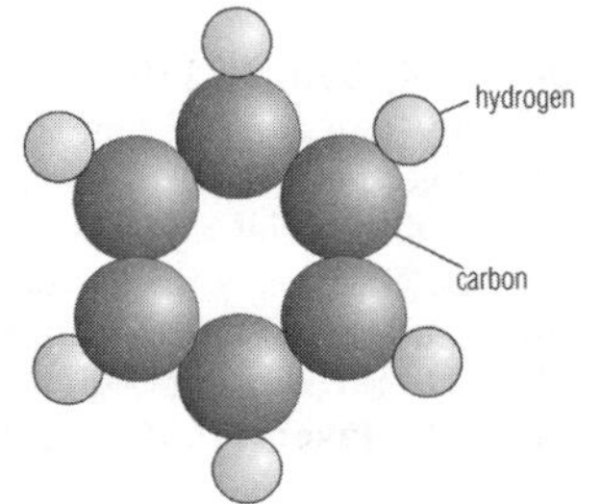

benzene The molecule of benzene consists of six carbon atoms arranged in a ring, with six hydrogen atoms attached. The benzene ring structure is found in many naturally occurring organic compounds. Helicon

benzoic acid C_6H_5COOH, white crystalline solid, sparingly soluble in water, which is used as a preservative for certain foods and as an antiseptic. It is obtained chemically by the direct oxidation of benzaldehyde and occurs in certain natural resins, some essential oils, and as hippuric acid.

benzpyrene one of a number of organic compounds associated with a particular polycyclic ring structure. Benzpyrenes are present in coal tar at low levels and are considered carcinogenic (cancer-inducing). Traces of benzpyrenes are present in wood smoke, and this has given rise to some concern about the safety of naturally smoked foods.

Berg, Paul (1926–) US molecular biologist who, in 1972, spliced and combined into a single hybrid the †DNA from an animal tumor virus (SV40) and the DNA from a bacterial virus, using gene-splicing techniques developed by others. He shared the Nobel Prize for Chemistry in 1980 for his work on the biochemistry of nucleic acids, especially recombinant DNA.

Berger, Hans (1873–1941) German psychiatrist and philosopher of science. He first described the human †electroencephalogram (EEG) in 1929. The differential patterns of cortical electrical activity he observed in alert and relaxed subjects led him to attempt the application of EEG to the study of psychophysical relationships and of conscious processes in general. He saw EEG as a key to the mind–body problem, a problem with which he was preoccupied for much of his life.

Bergius, Friedrich Karl Rudolf (1884–1949) German chemist who invented processes for converting coal into oil and wood into sugar. He shared the Nobel Prize for Chemistry in 1931 with Carl †Bosch for his part in inventing and developing high-pressure industrial methods.

bergschrund deep crevasse that may be found at the head of a †glacier. It is formed as a glacier pulls away from the headwall of the †corrie, or hollow, in which it accumulated.

beriberi nutritional disorder occurring mostly in the tropics and resulting from a deficiency of vitamin B_1 (†thiamine). The disease takes two forms: in one edema (waterlogging of the tissues) occurs; in the other there is severe emaciation. There is nerve degeneration in both forms and many victims succumb to heart failure.

berkelium chemical symbol Bk, synthesized, radioactive, metallic element of the actinide series, atomic number 97, relative atomic mass 247. It was first produced in 1949 by US nuclear chemist Glenn †Seaborg and his team, at the University of California at Berkeley, California, after which it is named.

berm on a beach, a ridge of sand or pebbles running parallel to the water's edge, formed by the action of the waves on beach material. Sand and pebbles are deposited at the farthest extent of swash (advance of water) to form a berm. Berms can also be formed well up a beach following a storm, when they are known as **storm berms**.

The UN coalition forces in the Gulf War adopted the term to mean a temporary defensive barrier formed by bulldozing the desert sand into a ridge.

Bernard, Claude (1813–1878) French physiologist and founder of experimental medicine. Bernard first demonstrated that digestion is not restricted to the stomach, but takes place throughout the small intestine. He discovered the digestive input of the pancreas, several functions of the liver, and the vasomotor nerves that dilate and contract the blood vessels and thus regulate body temperature. This led him to the concept of the *milieu intérieur* ("internal environment") whose stability is essential to good health.

Berners-Lee, Timothy (1955–) English computer scientist and inventor of the †World Wide Web (WWW) in 1990, which he developed while working as a consultant at †CERN (the European Organization for Nuclear Research, Geneva, Switzerland) in the 1980s. He has fought to keep the Web free and accessible to everyone and remains in research, currently serving as director of the W3 Consortium, an industry group that seeks to promote standards and coordinate developments in the Web. *Time* magazine put him on the list of the 100 People of the Century in 1999.

Bernoulli, Daniel (1700–1782) Swiss mathematical physicist. He made important contributions to trigonometry and differential equations (†differentiation). In hydrodynamics he proposed Bernoulli's principle, an early formulation of the idea of conservation of energy.

Bernoulli, Daniel http://www-history.mcs.st-andrews.ac.uk/history/Mathematicians/Bernoulli_Daniel.html

Biography of the Swiss physicist famous for his work on fluid flows. Along with a picture of Bernoulli, there is a family tree detailing the achievements of the remarkable Bernoulli family of scientists and mathematicians. The site also includes links to many of Bernoulli's colleagues and contemporaries.

Bernoulli, Jakob (1654–1705) Swiss mathematician who with his brother Johann pioneered German mathematician Gottfried †Leibniz's calculus. Jakob used calculus to study the forms of many curves arising in practical situations, and studied mathematical probability; **Bernoulli numbers** are named after him.

Bernoulli, Johann (1667–1748) Swiss mathematician who with his brother Jakob Bernoulli pioneered German mathematician Gottfried †Leibniz's calculus. He was the father of Daniel Bernoulli. Johann also contributed to many areas of applied mathematics, including the problem of a particle moving in a gravitational field. He found the equation of the †catenary in 1690 and developed exponential †calculus in 1691.

Bernoulli, Johann http://www.shu.edu/html/teaching/math/reals/history/bernoull.html

Page devoted to the life and ideas of mathematician Johann Bernoulli. In addition to biographical information, you will find links to the equations known as "Bernoulli's inequality," which includes an explanation. You will also find a brief bibliography at the bottom of the page.

Bernoulli numbers or **B numbers**, sequence of †rational numbers that may be represented by the symbolic form $B_n = (B+1)^n$, corresponding to the sequence:

$B_0 = 1$ $B_1 = -\frac{1}{2}$ $B_2 = \frac{1}{6}$

$B_3 = 0$ $B_4 = -\frac{1}{30}$ $B_5 = 0$

$B_6 = \frac{1}{42}$ $B_7 = 0$ $B_8 = -\frac{1}{30}$

and so on. B numbers of odd order (except B_1) are zero; B numbers of even order alternate in sign. They are named after Swiss mathematician Jakob Bernoulli and were first discussed in his work on probability *Ars Conjectandi*.

Bernoulli's principle law stating that the pressure of a fluid varies inversely with speed, an increase in speed producing a decrease in pressure (such as a drop in hydraulic pressure as the fluid speeds up flowing through a constriction in a pipe) and vice versa. The principle also explains the pressure differences on each surface of an aerofoil, which gives lift to the wing of an aircraft. The principle is named after Swiss mathematician and physicist Daniel Bernoulli.

Bernoulli's theorem sum of the †pressure, †potential energy, and †kinetic energy of a fluid flowing along a tube is constant, provided the flow is steady, incompressible, and non-viscous.

Bernstein, Jeremy (1929–) US science writer. A mathematical physicist, he has written many articles and books on various topics of pure and applied science for the nonspecialist reader. He has also sought to give a mathematical analysis and description of the behavior of elementary particles.

berry fleshy, many-seeded †fruit that does not split open to release the seeds. The outer layer of tissue, the exocarp, forms an outer skin that is often brightly colored to attract birds to eat the fruit and thus disperse the seeds. Examples of berries are the tomato and the grape.

A **pepo** is a type of berry that has developed a hard exterior, such as the cucumber fruit. Another is the **hesperidium**, which has a thick, leathery outer layer, such as that found in citrus fruits, and fluid-containing vesicles within, which form the segments.

Berry, Clifford (1918–1963) US scientist and co-developer of the ABC (Atanasoff–Berry Calculator) in 1942. Berry, a graduate student at the University of Iowa, combined with John Atanasoff to build one of the first electronic binary calculating machines. Later he gained an international reputation in the field of mass spectrometry.

beryl in full **beryllium aluminum silicate**, $3BeO.Al_2O_3.6SiO_2$, mineral that forms crystals chiefly in granite. It is the chief ore of beryllium. Two of its gem forms are aquamarine (light-blue crystals) and emerald (dark-green crystals).

beryllium chemical symbol Be, hard, lightweight, silver-white, metallic element, atomic number 4, relative atomic mass 9.012. It is one of the †alkaline-earth metals, with chemical properties similar to those of magnesium. In nature it is found only in combination with other elements and occurs mainly as beryl ($3BeO.Al_2O_3.6SiO_2$). It is used to make sturdy, light alloys and to control the speed of neutrons in nuclear reactors. Beryllium oxide was discovered in 1798 by French chemist Louis-Nicolas Vauquelin (1763–1829), but the element was not isolated until 1828, by Friedrich Wöhler and Antoine-Alexandre-Brutus Bussy independently.

In 1992 large amounts of beryllium were unexpectedly discovered in six old stars in the Milky Way.

Berzelius, Jöns Jakob (1779–1848) Swedish chemist. He accurately determined more than 2,000 relative atomic and molecular masses. In 1813–14, he devised the system of chemical symbols and formulae now in use and proposed oxygen as a reference standard for atomic masses. His discoveries include the elements cerium in 1804, selenium in 1817, and thorium in 1828; he was the first to prepare silicon in its amorphous form and to isolate zirconium. The words "isomerism," "allotropy," and "protein" were coined by him.

Bessel, Friedrich Wilhelm (1784–1846) German astronomer and mathematician. He was the first person to find the approximate distance to a star by direct methods when he measured the †parallax (annual displacement) of the star 61 Cygni in 1838. In mathematics, he introduced the series of functions now known as Bessel functions.

Bessel, Friedrich Wilhelm http://www-history.mcs.st-andrews.ac.uk/history/Mathematicians/Bessel.html
Extensive biography of the German astronomer and mathematician. The Web site contains a clear description of the work of Bessel, and also describes his working relationships with his contemporaries. Also included are several literature references for further reading on the mathematician and a portrait of Bessel.

Bessemer process first cheap method of making †steel, invented by Henry Bessemer in England in 1856. It has since been superseded by more efficient steel-making processes, such as the basic–oxygen process. In the Bessemer process compressed air is blown into the bottom of a converter, a furnace shaped like a cement mixer, containing molten pig iron. The excess carbon in the iron burns out, other impurities form a slag, and the furnace is emptied by tilting.

Best, Charles H(erbert) (1899–1978) Canadian physiologist. He was one of the team of Canadian scientists including Frederick †Banting whose research resulted in 1922 in the discovery of insulin as a treatment for diabetes.

beta chain particular secondary structure of the polypeptide chain (of a protein) that makes it adopt the shape of a pleated sheet.

beta decay disintegration of the nucleus of an atom to produce a beta particle, or high-speed electron, and an electron antineutrino. During beta decay, a neutron in the nucleus changes into a proton, thereby increasing the atomic number by one while the mass number stays the same. The mass lost in the change is converted into kinetic (movement) energy of the beta particle. Beta decay is caused by the weak nuclear force, one of the fundamental †forces of nature operating inside the nucleus.

beta index mathematical measurement of the connectivity of a transport network. If the network is represented as a simplified topological map, made up of nodes (junctions or places) and edges (links), the beta index may be calculated by dividing the number of nodes by the number of edges. If the number of nodes is n and the number of edges is e, then the beta index b is given by the formula:

$b = n/e$

The higher the index number, the better connected the network is. If b is greater than 1, then a complete circuit exists.

beta particle or **beta ray**, electron ejected with great velocity from a radioactive atom that is undergoing spontaneous disintegration. Beta particles are created in the nucleus on disintegration, beta decay, when a neutron converts to a proton (the atomic number increases by one while the atomic mass stays the same) by emitting an electron. The mass lost in the change is converted into †kinetic energy of the beta particle. Strontium-90 (^{90}Sr) is an example of a material that emits beta particles.

Beta particles are more penetrating than †alpha particles, but less so than †gamma radiation; they can travel several meters in air, but are stopped by 2–3 mm/0.08–0.1 in of aluminum. They are less strongly ionizing than alpha particles. Owing to their low mass, beta particles, like cathode rays, are easily deflected by magnetic and electric fields. Beta decay is caused by the weak nuclear force.

beta ray alternative name for †beta particle.

Betelgeuse or **Alpha Orionis**, red supergiant star in the constellation of †Orion. It is the tenth-brightest star in the night sky, although its brightness varies. It is 1,100 million km/700 million mi across, about 800 times larger than the Sun, roughly the same size as the orbit of Mars. It is over 10,000 times as luminous as the Sun, and lies 310 light years from the Sun. Light takes 60 minutes to travel across the giant star. It was the first star whose angular diameter was measured with the Mount Wilson †interferometer in 1920.

Bethe, Hans Albrecht (1906–) German-born US physicist who in 1938 worked out the details of how nuclear mechanisms power the Sun and other stars. He also worked on the first atom bomb but later became a peace campaigner. He was awarded the Nobel Prize for Physics in 1967 for his theory of nuclear reactions, and discoveries concerning production of energy in stars.

Bethe–Weizsäcker cycle see †proton–proton cycle.

Bettelheim, Bruno (1903–1990) Austrian-born US child psychologist. At the University of Chicago he founded a treatment center for emotionally disturbed children based on the principle of a supportive home environment. Among his books are *Love Is Not Enough* (1950) and *The Uses of Enchantment: The Meaning and Importance of Fairy Tales* (1976).

Betti numbers numbers characterizing the connectivity of a variety. They are named after Italian mathematician Enrico Betti.

bevatron in physics, US term for a proton †synchrotron, for example the 6-GeV machine at the University of California.

Bhabha scattering scattering process involving †electrons and †positrons, first determined by Indian physicist Homi Bhabha in 1935.

bhp abbreviation for **brake horsepower**.

bicarbonate common name for †hydrogencarbonate.

bicarbonate indicator pH indicator sensitive enough to show a color change as the concentration of the gas carbon dioxide increases. The indicator is used in photosynthesis and respiration experiments to find out whether carbon dioxide is being liberated. The initial red color changes to yellow as the pH becomes more acidic.

Carbon dioxide, even in the concentrations found in exhaled air, will dissolve in the indicator to form a weak solution of carbonic acid, which will lower the pH and therefore give the characteristic color change.

bicarbonate of soda or **baking soda**; technical name **sodium hydrogencarbonate**, $NaHCO_3$, white crystalline solid that neutralizes acids and is used in medicine to treat acid indigestion. It is also used in baking powders and effervescent drinks.

biconcave in optics, describing a diverging lens that is concave (curved inwards) on both surfaces.

biconvex in optics, describing a converging lens that is convex (bulging outwards) on both surfaces.

bicuspid valve or **mitral valve**, in the left side of the †heart, a flap of tissue that prevents blood flowing back into the atrium when the ventricle contracts.

Biela, Wilhelm (1782–1856) Baron von Biela, Austrian army officer and amateur astronomer. On 27 February 1826 he discovered a faint comet that has become known by his name, although it had been seen before, in 1772 and 1805.

biennial plant plant that completes its life cycle in two years. During the first year it grows vegetatively and the surplus food produced is stored in its †perennating organ, usually the root. In the following year these food reserves are used for the production of leaves, flowers, and seeds, after which the plant dies. Many root vegetables are biennials, including the carrot *Daucus carota* and parsnip *Pastinaca sativa*. Some garden plants that are grown as biennials are actually perennials, for example, the wallflower *Cheiranthus cheiri*.

Biffen, Rowland Harry (1874–1949) English botanist and geneticist who produced several wheat hybrids that are resistant to disease. In the development of better hybrids of wheat, he applied Mendel's laws (see †Mendelism) and demonstrated for the first time that resistance to yellow rust disease of wheat was inherited.

Big Bang in astronomy, the hypothetical "explosive" event that marked the origin of the universe as we know it. At the time of the Big Bang, the entire universe was squeezed into a hot, superdense state. The Big Bang explosion threw this compact material outwards, producing the expanding universe seen today (see †red shift). The cause of the Big Bang is unknown; observations of the current rate of expansion of the universe suggest that it took place about 10–20 billion years ago. The Big Bang theory began modern †cosmology.

According to a modified version of the Big Bang, called the **inflationary theory**, the universe underwent a rapid period of expansion shortly after the Big Bang, which accounts for its current large size and uniform nature. The inflationary theory is supported by the most recent observations of the †cosmic background radiation.

Scientists have calculated that 10^{-36} seconds (equivalent to one million-million-million-million-million-millionth of a second) before the Big Bang, the universe was the size of a pea, and the temperature was 10 billion million million million°C/18 billion million million million°F. One second after the Big Bang, the temperature was about 10 billion°C/18 billion°F.

According to theory, one-tenth of a second after the Big Bang, the temperature and pressure had decreased by many millions of degrees, thus allowing the formation of subatomic particles. After 10 seconds, neutrons had combined with protons to form nuclei of deuterium (an isotope of hydrogen). The nuclei of deuterium then joined together to form helium nuclei. As the universe continued to expand for the next 300,000 years, the temperature cooled to 10,000°C/18,000°F. Under these conditions helium nuclei were able to join with electrons to form helium atoms. Also, hydrogen nuclei joined to form lithium nuclei and thence lithium atoms. After millions of years, at lower temperature and pressure, the force of gravity was able to attract particles together. After millions more years, the universe formed into clumped matter joined together to form galaxies, stars, planets, and moons.

The first detailed images of the universe as it existed 300,000 years after the Big Bang were released by the US National Aeronautics and Space Administration (NASA) in April 2000. The images were created by mapping cosmic background radiation.

Big Crunch in cosmology, possible fate of the universe in which it ultimately collapses to a point following the halting and reversal of the present expansion. See also †Big Bang and †critical density.

Big Dipper North American name for the Plough, the seven brightest and most prominent stars in the constellation †Ursa Major.

bight coastal indentation, crescent-shaped or gently curving, such as the Bight of Biafra in West Africa and the Great Australian Bight.

bile brownish alkaline fluid produced by the †liver. Bile is stored in the gall bladder and is intermittently released into the small intestine (the duodenum), which is part of the †gut, in order to help †digestion. Bile contains chemicals that †emulsify fats. In other words it acts to disperse fat globules into tiny droplets, which speeds up their digestion.

Bile consists of bile salts, bile pigments, cholesterol, and lecithin. The **bile salts** assist in the breakdown and absorption of fats, while **bile pigments** are the breakdown products of old red blood cells, which are passed into the gut to be eliminated with the feces.

billion cardinal number represented by a 1 followed by nine zeros (1,000,000,000 or 10^9), equivalent to a thousand million.

bimetallic strip strip made from two metals each having a different coefficient of †thermal expansion; it therefore bends when subjected to a change in temperature. Such strips are used widely for temperature measurement and control, for instance in the domestic thermostat.

bimodal in statistics, having two distinct peaks of †frequency distribution.

bimolecular reaction in chemistry, any chemical reaction that involves only two reaction molecules or species, for example $H_2 + I_2 \longleftrightarrow 2HI$.

binary compound in chemistry, any compound that is composed of only two elements, for example NaCl.

binary fission in biology, a form of †asexual reproduction, whereby a single-celled organism, such as a bacterium or amoeba, divides into two smaller "daughter" cells.

binary number system system of numbers to †base two, using combinations of the digits 1 and 0. Codes based on binary numbers are used to represent instructions and data in all modern digital computers, the values of the binary digits (contracted to "bits") being stored or transmitted as, for example, open/closed switches, magnetized/unmagnetized disks and tapes, and high/low voltages in circuits.

The value of any position in a binary number increases by powers of 2 (doubles) with each move from right to left (1, 2, 4, 8, 16, and so on). For example, 1011 in the binary number system represents (1 x 8) + (0 x 4) + (1 x 2) + (1 x 1), which adds up to 11 in the decimal system.

The value of any position in a normal decimal, or base-10, number increases by powers of 10 with each move from right to left (1, 10, 100, 1,000, 10,000, and so on). For example, the decimal number 2,567 stands for (2 x 1,000) + (5 x 100) + (6 x 10) + (7 x 1).

binary operation in mathematics, method of combining two elements of a set to get an element of the same set. Examples are addition, subtraction, or multiplication of integers.

binary star pair of stars moving in orbit around their common center of mass. Observations show that most stars are binary, or even multiple – for example, the nearest star system to the Sun, †Rigil Kent (Alpha Centauri).

One of the stars in the binary system Epsilon Aurigae may be the largest star known. Its diameter is 2,800 times that of the Sun. If it were in the position of the Sun, it would engulf Mercury, Venus, Earth, Mars, Jupiter, and Saturn. A †spectroscopic binary is a binary in which two stars are so close together that they cannot be seen separately, but their separate light spectra can be distinguished by a spectroscope.

Another type is the †eclipsing binary, a double star in which the two stars periodically pass in front of each other as seen from Earth. When one star crosses in front of the other, the total light received on Earth from the two stars declines. The first eclipsing binary to be noticed was Algol, in 1670, by Italian astronomer Germniano Montanari.

binding energy in physics, the amount of energy needed to break the nucleus of an atom into the neutrons and protons of which it is made.

Binet, Alfred (1857–1911) French psychologist who introduced the first †intelligence tests in 1905. They were standardized so that the last of a set of graded tests the child could successfully complete gave the level described as "mental age." If the test was passed by most children over 12, for instance, but failed by those younger, it was said to show a mental age of 12. Binet published these in collaboration with Théodore Simon.

Classics in the History of Psychology – "New Methods for the Diagnosis of the Intellectual Level of Subnormals" http://psychclassics.yorku.ca/Binet/binet1.htm

Complete text of French psychologist Alfred Binet's 1905 essay. Also included here is an introduction to the text, as well as a commentary on it.

Binnig, Gerd (1947–) German physicist who was awarded the Nobel Prize for Physics in 1986 for his involvement in the invention of the scanning tunneling electron microscope (STM), an ultra-powerful microscope capable of imaging individual atoms. He shared the award with Ernst †Ruska and Heinrich †Rohrer.

binomial in mathematics, an expression consisting of two terms, such as $a + b$ or $a - b$.

binomial system of nomenclature in biology, the system in which all organisms are identified by a two-part Latinized name. Devised by the biologist †Linnaeus, it is also known as the Linnaean system. The first name is capitalized and identifies the †genus; the second identifies the †species within that genus, for example the bear genus *Ursus* includes *Ursus arctos*, the grizzly bear, and *Ursus maritimus*, the polar bear.

binomial theorem formula whereby any power of a binomial quantity may be found without performing the progressive multiplications.

It was discovered by Isaac †Newton and first published in 1676.

biochemical oxygen demand BOD, amount of dissolved oxygen taken up by micro-organisms in a sample of water. Since these micro-organisms live by decomposing organic matter, and the amount of oxygen used is proportional to their number and metabolic rate, BOD can be used as a measure of the extent to which the water is polluted with organic compounds.

biochemistry science concerned with the chemistry of living organisms: the structure and reactions of proteins (such as enzymes), nucleic acids, carbohydrates, and lipids.

Its study has led to an increased understanding of life processes, such as those by which organisms synthesize essential chemicals from food materials, store and generate energy, and pass on their characteristics through their genetic material. A great deal of medical research is concerned with the ways in which these processes are disrupted. Biochemistry also has applications in agriculture and in the food industry (for instance, in the use of enzymes).

biodegradable capable of being broken down by living organisms, principally bacteria and fungi. In biodegradable substances, such as food and sewage, the natural processes of decay lead to compaction and liquefaction, and to the release of nutrients that are then recycled by the ecosystem.

This process can have some disadvantageous side effects, such as the release of methane, an explosive greenhouse gas. However, the technology now exists for waste tips to collect methane in underground pipes, drawing it off and using it as a cheap source of energy. Nonbiodegradable substances, such as glass, heavy metals, and most types of plastic, present serious problems of disposal.

biodiversity contraction of biological diversity, measure of the variety of the Earth's animal, plant, and microbial species, of genetic differences within species, and of the †ecosystems that support those species. High biodiversity means there are lots of different species in an area. The maintenance of biodiversity is important for ecological stability and as a resource for research into, for example, new drugs and crops.

Estimates of the number of species vary widely because many species-rich ecosystems, such as tropical forests, contain unexplored and unstudied habitats. Among small organisms in particular many are unknown. For example, it is thought that less than 1% of the world's bacterial species have been identified.

The most significant threat to biodiversity comes from the destruction of rainforests and other habitats. It is estimated that 7% of the Earth's surface hosts 50–75% of the world's biological diversity. Costa Rica, for example, has an area less than 10% of the size of France but possesses three times as many vertebrate species.

Biodiversity: Measuring the Variety of Nature http://www.nhm.ac.uk/science/projects/worldmap/

Exhibition on biodiversity and conservation from the Natural History Museum in London, with sections on "Biodiversity value," "Rarity and endemism," and "Conservation priority and gap analysis."

"Biodiversity Loss Threatens New Treatments" http://www.bmj.com/cgi/content/full/316/7140/1261/l

Part of a collection of articles maintained by the British Medical Journal, this page features the full text of a news article originally published on 25 April 1998 that reports on how the loss of biodiversity will adversely affect the future of human health. You can also search the collections under which this article appears, search other articles written by the author of this article, and submit or read responses to this article.

bioengineering application of engineering to biology and medicine. Common applications include the design and use of artificial limbs, joints, and organs, including hip joints and heart valves.

biofeedback in biology, modification or control of a biological system by its results or effects. For example, a change in the position or †trophic level of one species affects all levels above it.

Many biological systems are controlled by negative feedback. When enough of the hormone thyroxine has been released into the blood, the hormone adjusts its own level by "switching off" the gland that produces it. In ecology, as the numbers in a species rise, the food supply available to each individual is reduced. This acts to reduce the population to a sustainable level.

biofeedback in medicine, the use of electrophysiological monitoring devices to "feed back" information about internal processes and thus facilitate conscious control. Developed in the USA in the 1960s, independently by neurophysiologist Barbara Brown and neuropsychiatrist Joseph Kamiya, the technique is effective in alleviating hypertension and preventing associated organic and physiological dysfunctions.

biofouling build-up of barnacles, mussels, seaweed, and other organisms on underwater surfaces, such as ships' hulls. Marine industries worldwide spend at least £1.4 billion controlling biofouling by scraping affected surfaces and painting with antifouling paint.

Antifoulants can contribute to marine pollution as toxic components leach from the paint into surrounding water, for example †tributyl tin. Research is underway to develop natural antifoulants using compounds produced by immobile marine plants and creatures, themselves at risk from biofouling.

biofuel any solid, liquid, or gaseous fuel produced from organic (once living) matter, either directly from plants or indirectly from industrial, commercial, domestic, or agricultural wastes. There are three main methods for the development of biofuels: the burning of dry organic wastes (such as household refuse, industrial and agricultural wastes, straw, wood, and peat); the fermentation of wet wastes (such as animal dung) in the absence of oxygen to produce biogas (containing up to 60% methane), or the fermentation of sugar cane or corn to produce alcohol and esters; and energy forestry (producing fast-growing wood for fuel).

Fermentation produces two main types of biofuels: alcohols and esters. These could theoretically be used in place of fossil fuels but, because major alterations to engines would be required, biofuels are usually mixed with fossil fuels. The EU allows 5% ethanol, derived from wheat, beet, potatoes, or maize, to be added to fossil fuels. A quarter of Brazil's transportation fuel in 1994 was ethanol.

biogenesis biological term coined in 1870 by English scientist Thomas Henry Huxley to express the hypothesis that living matter always arises out of other similar forms of living matter. It superseded the opposite idea of †spontaneous generation or abiogenesis (that is, that living things may arise out of non-living matter).

biogeochemistry emerging branch of †geochemistry involving the study of how chemical elements and their isotopes move between living organisms and geological materials. For example, the analysis of carbon in bone gives biogeochemists information on how the animal lived, its diet, and its environment.

biogeography study of how and why plants and animals are distributed around the world, in the past as well as in the present; more specifically, a theory describing the geographical distribution of †species developed by Robert MacArthur and US zoologist Edward O †Wilson. The theory argues that, for many species, ecological specializations mean that suitable habitats are patchy in their occurrence. For example, for a dragonfly, ponds in which to breed are separated by large tracts of land, and for edelweiss adapted to alpine peaks the deep valleys between cannot be colonized.

bioinformatics in biology, area of study that aims to integrate the large amount of data produced by genomic studies with cell biology in order to develop understanding of the genetic basis of cellular processes. Developments in information technology are important for the storage and analysis of data. Powerful computers are used to help match the base sequences of the genome with the proteins for which they are the code.

biological clock regular internal rhythm of activity, produced by unknown mechanisms, and not dependent on external time signals. Such clocks are known to exist in almost all animals, and also in many plants, fungi, and unicellular organisms; the first biological clock gene in plants was isolated in 1995 by a US team of researchers. In higher organisms, there appears to be a series of clocks of graded importance. For example, although body temperature and activity cycles in human beings are normally "set" to 24 hours, the two cycles may vary independently, showing that two clock mechanisms are involved.

biological control control of pests such as insects and fungi through biological means, rather than the use of chemicals. This can include breeding resistant crop strains; inducing sterility in the pest; infecting the pest species with disease organisms; or introducing the pest's natural predator. Biological control tends to be naturally self-regulating, but as ecosystems are so complex, it is difficult to predict all the consequences of introducing a biological controlling agent.

Ladybirds are sometimes used to control aphids because both adults and larvae feed on them. In 1998 French researchers patented a method of selective breeding to produce hardy, flightless ladybirds for use in biological control, as captive populations are far more effective than mobile ones.

The introduction of the cane toad to Australia 50 years ago to eradicate a beetle that was destroying sugar beet provides an example of the unpredictability of biological control. Since the cane toad is poisonous it has few Australian predators and it is now a pest, spreading throughout eastern and northern Australia at a rate of 35 km/22 mi a year.

biological weathering or **organic weathering**, form of †weathering caused by the activities of living organisms – for example, the growth of roots or the burrowing of animals. Tree roots are probably the most significant agents of biological weathering as they are capable of prising apart rocks by growing into cracks and joints. Plants also give off organic acids that help to break down rocks chemically (see †chemical weathering).

biology (Greek *bios* "life," *logos* "discourse") science of life. Biology includes all the life sciences – for example, anatomy and physiology (the study of the structure of living things), cytology (the study of cells), zoology (the study of animals), botany (the study of plants), ecology (the study of habitats and the interaction of living species), animal behavior, embryology, and taxonomy (classification), and plant breeding. Increasingly biologists have concentrated on molecular structures: biochemistry, biophysics, and genetics (the study of inheritance and variation).

Biological research has come a long way towards understanding the nature of life, and during the late 1990s our knowledge was further extended as the international †Human Genome Project mapped the entire genetic code contained in the 23 pairs of human chromosomes.

Instructional Multimedia – Introductory Biology
http://www.biology.ualberta.ca/facilities/multimedia/index.php?Page=284

With a strong graphical emphasis, this Web site introduces the basic concepts of biology. It is useful as a topic refresher, or as an alternative explanation to a difficult topic. The Web site relies heavily upon Flash technology, which is available via a download link.

MIT Biology Hypertextbook http://web.mit.edu/esgbio/www/

Hypertext biology book, with topics on all the main aspects of biology. Pictures illustrate the text, and each page of the book features easy navigation controls.

Quia: Biology http://www.quia.com/dir/bio/

Twenty biology activities are included on this site, which can be used to highlight the topics that require extra revision. Each activity, on average, contains three types of test: concentration, flashcards, and a word search, although some activities may feature other tests. A list of terms for each topic is also available.

biology, applied practical application of biological knowledge. Applied biology is directed towards the manipulation and control of living organisms and the environment in the light of biological knowledge, since many of the world's most urgent problems, such as overpopulation and disease, food production, environmental pollution, and waste disposal, are born of or fostered by biological ignorance. It covers a vast spectrum of subjects including forestry and fisheries, pest and weed control, pollution control, human and animal nutrition, microbiology, and immunology.

bioluminescence production of light by living organisms. It is a feature of many deep-sea fishes, crustaceans, and other marine animals. On land, bioluminescence is seen in some nocturnal insects such as glow-worms and fireflies, and in certain bacteria and fungi. Light is usually produced by the oxidation of luciferin, a reaction catalyzed by the †enzyme luciferase. This reaction is unique, being the only known biological oxidation that does not produce heat. Animal luminescence is involved in communication, camouflage, or the luring of prey, but its function in some organisms is unclear.

biomass total mass of living organisms present in a given area. It may be used to describe the mass of a particular species (such as earthworm biomass), for a general category (such as herbivore biomass – animals that eat plants), or for everything in a †habitat. Estimates also exist for the entire global plant biomass. Biomass can be the mass of the organisms as they are – wet biomass – or the mass of the organisms after they have been dried to remove all the water – dry biomass. Measurements of biomass can be used to study interactions between organisms, the stability of those interactions, and variations in population numbers. †Growth results in an increase in biomass, so biomass is a good measure of the extent to which organisms thrive in particular habitats. For a plant, biomass increase occurs as a result of the process of †photosynthesis. For a herbivore, biomass increase depends on the availability of plant food. Studying biomass in a habitat is a useful way to see how food is passed from organism to organism along †food chains and through †food webs.

Some two-thirds of the world's population cooks and heats water by burning biomass, usually wood. Plant biomass can be a renewable source of energy as replacement supplies can be grown relatively quickly. Fossil fuels, however, originally formed from biomass, accumulate so slowly that they cannot be considered renewable. The burning of biomass (defined either as natural areas of the ecosystem or as forest, grasslands, and fuel woods) produces 3.5 million tonnes of carbon in the form of carbon dioxide each year, accounting for up to 40% of the world's annual carbon dioxide production.

Plant biomass can be changed into liquid or gaseous fuels to generate electricity or heat, or to fuel internal combustion engines. Fuel from biomass is burned in a reactor to generate heat energy, which is then converted into mechanical energy to turn †turbine blades in a generator to produce electricity. Biomass power stations of 80 megawatts can produce electricity to power approximately 42,000 homes.

Biomass http://www.nrel.gov/clean_energy/bioenergy.html

Information on biomass from the US Department of Energy. There is a clear explanation of the chemical composition of biomass and development of technologies to transform it into usable fuel sources.

Willow Biomass for Energy http://www.esf.edu/pubprog/brochure/willow/willow.htm

Description of the environment friendly use of willow biomass as an energy source. The site goes on to discuss the possibilities of wood energy plantations, which could be producing "green" energy in the near future.

biome broad natural assemblage of plants and animals shaped by common patterns of vegetation and climate. Examples include the †tundra biome, the †rainforest biome, and the †desert biome.

Biomes http://www.furman.edu/~worthen/biomes/map.htm

Introductory page including a world map of the different types of biomes present, and providing links to pages containing more information on deciduous forest, alpine, desert, and rainforest biomes. Each sub-page has lots of detailed information on which animals live there, what the climate is like, and the adaptations plants have made. This is a good source of information for geography case studies.

Major Biomes of the World http://www.runet.edu/~swoodwar/CLASSES/GEOG235/biomes/main.html

Site providing information on the distribution and nature of the world's major biomes. It looks at the structure, growth patterns, and taxonomy of the vegetation, in the context of soil types and climate variations. The site contains supporting materials for students including an online glossary and study hints.

Unexplored Spaces http://magma.nationalgeographic.com/2000/biodiversity/biomes/index.cfm

Visually stunning National Geographic page exploring some of the world's little-known biomes, and showing how much they have to teach scientists about the way in which life is organized. Taking a cobble-bed off the coast of New England, USA, as its starting point, this wide-ranging article branches off into the ocean floor, outer space and other unsuspected sources of knowledge about the processes of life.

biomechanics application of mechanical engineering principles and techniques in the field of medicine and surgery, studying natural structures to improve those produced by humans. For example, mother-of-pearl is structurally superior to glass fiber, and deer antlers have outstanding durability because they are composed of microscopic fibers. Such natural structures may form the basis of "high-tech" composites. Biomechanics has been responsible for many advances in orthopedics, anesthesia, and intensive care. Biomechanical assessment of the requirements for replacement of joints, including evaluation of the stresses and strains between parts, and their reliability, has allowed development of implants with very low friction and long life.

biophysics application of physical laws to the properties of living organisms. Examples include using the principles of †mechanics to calculate the strength of bones and muscles, and †thermodynamics to study plant and animal energetics.

bioreactor sealed vessel in which microbial reactions can take place. The simplest bioreactors involve the slow decay of vegetable or animal waste, with the emission of methane that can be used as fuel. Laboratory bioreactors control pH, acidity, and oxygen content and are used in advanced biotechnological operations, such as the production of antibiotics by genetically engineered bacteria.

biorhythm rhythmic change, mediated by †hormones, in the physical state and activity patterns of certain plants and animals that have seasonal activities. Examples include winter hibernation, spring flowering or breeding, and periodic migration. The hormonal changes themselves are often a response to changes in day length (†photoperiodism); they signal the time of year to the animal or plant. Other biorhythms are innate and continue even if external stimuli such as day length are removed. These include a 24-hour or †circadian rhythm, a 28-day or circalunar rhythm (corresponding to the phases of the Moon), and even a year-long rhythm in some organisms.

Such innate biorhythms are linked to an internal or †biological clock, whose mechanism is still poorly understood.

Often both types of rhythm operate; thus many birds have a circalunar rhythm that prepares them for the breeding season, and a photoperiodic response. There is also a nonscientific and unproven theory that human activity is governed by three biorhythms: the **intellectual** (33 days), the **emotional** (28 days), and the **physical** (23 days). Certain days in each cycle are regarded as "critical," even more so if one such day coincides with that of another cycle.

BIOS acronym for Basic Input/Output System, part of a computer's operating system that handles the basic input and output operation for standard computer †hardware. For example, the BIOS reads the keystrokes from the keyboard, puts information on the display, and sends information to the printer. The small computer programs within the BIOS that carry out these tasks are called device drivers.

biosensor device based on microelectronic circuits that can directly measure medically significant variables for the purpose of diagnosis or monitoring treatment. One such device measures the blood sugar level of diabetics using a single drop of blood, and shows the result on a liquid crystal display within a few seconds.

biosphere narrow zone that supports life on our planet. It is limited to the waters of the Earth, a fraction of its crust, and the lower regions of the atmosphere. The biosphere is made up of all the Earth's †ecosystems. It is affected by external forces such as the Sun's rays, which provide energy, the gravitational effects of the Sun and Moon, and cosmic radiations.

UNESCO Man And The Biosphere Program http://www.unesco.org/mab/

Stylish UNESCO pages containing information on biosphere reserves. Pages answer questions about reserves, explaining their role and importance. Many other aspects of UNESCO's Man and Biosphere Program are also presented on the Web Site.

biosynthesis synthesis of organic chemicals from simple inorganic ones by living cells – for example, the conversion of carbon dioxide and water to glucose by plants during †photosynthesis.

Other biosynthetic reactions produce cell constituents including proteins and fats.

Biot, Jean Baptiste (1774–1862) French physicist who studied the polarization of light. In 1804 he made a balloon ascent to a height of 5 km/3 mi, in an early investigation of the Earth's atmosphere.

Biot, Jean Baptiste http://www-groups.dcs.st-andrews.ac.uk/history/Mathematicians/Biot.html

Part of an archive containing the biographies of the world's greatest mathematicians, this site is devoted to the life and contributions of physicist Jean Baptiste Biot.

biotechnology industrial use of living organisms. Examples of its uses include fermentation, †genetic engineering (gene technology), and the manipulation of reproduction. The brewing and baking industries have long relied on the yeast micro-organism for †fermentation purposes, while the dairy industry employs a range of bacteria and fungi to convert milk into cheeses and yogurts. †Enzymes, whether extracted from cells or produced artificially, are central to most biotechnological applications. Recent advances include genetic engineering, in which single-celled organisms with modified †DNA are used to produce insulin and other drugs.

In 1993, 70% of biotechnology companies were concentrating on human health developments, while only 10% were concerned with applications for food and agriculture. There are many medical and industrial applications of the use of micro-organisms, such as drug production. One important area is the production of antibiotics such as penicillin.

It is thought that biotechnology may be helpful in reducing world food shortages. Micro-organisms grow very quickly in suitable conditions and they often take substances that humans cannot eat and use them to produce foods that we can eat.

Access Excellence http://www.accessexcellence.org/

US-based site for biology teachers sponsored by a biotechnology company, this site has plenty to interest the casual browser as well – particularly its "What's new?" section, with weekly science reports and interviews with scientists making the news, and "About biotech," an in-depth look at the field of biotechnology.

Chemistry and Industry Magazine http://www.chemind.org/CI/index.jsp

Latest developments in the fast-changing world of biotechnology. This online edition of Chemistry and Industry has well-written and readily understandable articles on biotechnology, chemistry, and the pharmaceutical industry. There is also a searchable archive of articles dating back to 1995.

biotic factor organic variable affecting an ecosystem – for example, the changing population of elephants and its effect on the African savannah.

biotic potential the total theoretical reproductive capacity of an individual organism or species under ideal environmental conditions. The biotic potential of many small organisms such as bacteria, annual plants, and small mammals is very high but rarely reached, as other elements of the ecosystem such as predators and nutrient availability keep the population growth in check.

biotin or **vitamin H**, vitamin of the B complex, found in many different kinds of food; egg yolk, liver, legumes, and yeast contain large amounts. Biotin is essential to the metabolism of fats. Its absence from the diet may lead to dermatitis.

biotite dark mica, $K(Mg, Fe)_3Al\ Si_3O_{10}(OH, F)_2$, a common silicate mineral. It is brown to black with shiny surfaces, and like all micas, it splits into very thin flakes along its one perfect cleavage. Biotite is a mineral found in igneous rocks, such as granites, and metamorphic rocks such as schists and gneisses.

biquaternions type of †quaternions devised by English mathematician William Clifford to use specifically in association with linear algebra to represent motions in three-dimensional non-Euclidean space.

Bird, Adrian Peter (1947–) English molecular biologist who demonstrated that the process of DNA methylation provides a switching mechanism by which a cell can control the activity of specific genes in its nucleus. This process is thought to be particularly important for †cell differentiation (the process by which cells become increasingly more specialized).

birefringence another name for †double refraction.

Birkland, Kristian Olaf (1867–1917) Norwegian physicist who is best known for devising (with Norwegian industrial chemist Samuel Eyde) an industrial process for fixing nitrogen. Using inexpensive hydroelectricity, they passed nitrogen (in air) through a large electric spark, so converting the nitrogen into nitrogen monoxide – which could then be made into nitric acid for fertilizers and explosives.

birth act of producing live young from within the body of female animals. Both viviparous and ovoviviparous animals give birth to young. In viviparous animals, embryos obtain nourishment from the mother via a †placenta or other means.

In ovoviviparous animals, fertilized eggs develop and hatch in the oviduct of the mother and gain little or no nourishment from maternal tissues. See also †pregnancy.

bise cold dry northerly wind experienced in southern France and Switzerland.

bisect to divide a line or angle into two equal parts.

bisector a line that bisects an angle or another line (known as a **perpendicular bisector** when it bisects at right angles).

Bishop, (John) Michael (1936–) US virologist and molecular biologist who was awarded a Nobel Prize in Physiology or Medicine in 1989 with Harold †Varmus for their discovery of oncogenes (cancer-causing genes), which, when over-activated or damaged, trigger a normal cell to divide in an uncontrolled fashion.

bismuth chemical symbol Bi, hard, brittle, pinkish-white, metallic element, atomic number 83, relative atomic mass 208.98. It has the highest atomic number of all the stable elements (the elements from atomic number 84 up are radioactive). Bismuth occurs in ores and occasionally as a free metal (†native metal). It is a poor conductor of heat and electricity, and is used in alloys of low melting point and in medical compounds to soothe gastric ulcers. The name comes from the Latin *besemutum*, from the earlier German *Wismut*.

bistable circuit or **flip-flop**, simple electronic circuit that remains in one of two stable states until it receives a pulse (logic 1 signal) through one of its inputs, upon which it switches, or "flips," over to the other state. Because it is a two-state device, it can be used to store binary digits and is widely used in the †integrated circuits used to build computers.

bisulphate another term for †hydrogen sulfate.

bit contraction of binary digit, in computing, a single binary digit, either 0 or 1. A bit is the smallest unit of data stored in a computer; all other data must be coded into a pattern of individual bits. A †byte represents sufficient computer memory to store a single character of data, and usually contains eight bits. For example, in the †ASCII code system used by most microcomputers the capital letter A would be stored in a single byte of memory as the bit pattern 01000001.

The maximum number of bits that a computer can normally process at once is called a **word**. Microcomputers are often described according to how many bits of information they can handle at once. For instance, the first microprocessor, the Intel 4004 (launched in 1971), was a 4-bit device. In the 1970s several different 8-bit computers, many based on the Zilog Z80 or Rockwell 6502 processors, came into common use. In 1981, the IBM Personal Computer (PC) was introduced, using the Intel 8088 processor, which combined a 16-bit processor with an 8-bit data bus. Business micros of the later 1980s began to use 32-bit processors such as the Intel 80386 and Motorola 68030. Machines based on the first 64-bit microprocessor appeared in 1993.

The higher the number of bits a computer can process simultaneously, the more powerful the computer is said to be. However, other factors influence the overall speed of a computer system, such as the †clock rate of the processor and the amount of †RAM available. Tasks that require a high processing speed include sorting a database or doing long, complex calculations in spreadsheets. A system running slowly with a †graphical user interface may benefit more from the addition of extra RAM than from a faster processor.

bit map in computing, pattern of †bits used to describe the organization of data. Bit maps are used to store typefaces or graphic images (bit-mapped or †raster graphics), with 1 representing black (or a color) and 0 white.

Bit maps may be used to store a typeface or font, but a separate set of bit maps is required for each type size. A vector font, by contrast, can be held as one set of data and scaled as required. Bit-mapped graphics are not recommended for images that require scaling (compare †vector graphics – those stored in the form of geometric formulae).

bit-mapped font font held in computer memory as sets of bit maps.

bits per second **bps**, commonly used measure of the speed of transmission of a †modem. In 2000 the fastest modems were rated at 56,600 bps. Modem speeds should conform to standards, known as V numbers, laid down by the Comité Consultatif International Téléphonique et Télégraphique (CCITT) so that modems from different manufacturers can connect to each other. Many modems transfer data much faster than their nominal speeds via techniques such as †data compression.

bitumen involatile, tarry material, containing a mixture of †hydrocarbons (mainly †alkanes), that is the residue from the †fractional distillation of crude oil (unrefined †petroleum). Sometimes the term is restricted to a soft kind of pitch resembling asphalt.

Naturally occurring solid bitumen may have arisen as a residue from the evaporation of petroleum. If evaporation took place from a pool or lake of petroleum, the residue might form a pitch or asphalt lake, such as Pitch Lake in Trinidad. Bitumen was used in ancient times as a mortar, and by the Egyptians for embalming.

bivalent in biology, a name given to the pair of homologous chromosomes during reduction division (†meiosis). In chemistry, the term is sometimes used to describe an element or group with a †valency of two, although the term "divalent" is more common.

bivalve marine or freshwater mollusk whose body is enclosed between two shells hinged together by a ligament on the dorsal side of the body.

The shell is closed by strong "adductor" muscles. Ventrally, a retractile "foot" can be put out to assist movement in mud or sand. Two large platelike gills are used for breathing and also, with the †cilia present on them, make a mechanism for collecting the small particles of food on which bivalves depend. The bivalves form one of the five classes of mollusks, the Lamellibranchiata, otherwise known as Bivalvia or Pelycypoda, containing about 8,000 species.

Bjerknes, Vilhelm Firman Koren (1862–1951) Norwegian scientist whose theory of polar fronts formed the basis of all modern weather forecasting and meteorological studies. He also developed hydrodynamic models of the oceans and the atmosphere and showed how weather prediction could be carried out on a statistical basis, dependent on the use of mathematical models.

Black, Davidson (1884–1934) Canadian anatomist. In 1927, when professor of anatomy at the Union Medical College, Peking (Beijing), he unearthed the remains of Peking man, an example of one of our human ancestors.

Black, Joseph (1728–1799) Scottish physicist and chemist. In 1754 he discovered carbon dioxide (which he called "fixed air"). By his investigations in 1761 of latent heat and specific heat, he laid the foundation for the work of his pupil James Watt.

black body in physics, a hypothetical object that completely absorbs all electromagnetic radiation striking it. It is also a perfect emitter of thermal radiation.

Although a black body is hypothetical, a practical approximation can be made by using a small hole in the wall of a constant-temperature enclosure. The radiation emitted by a black body is of all wavelengths, but with maximum radiation at a particular wavelength that depends on the body's temperature. As the temperature increases, the wavelength of maximum intensity becomes shorter (see †Wien's displacement law). The total energy emitted at all wavelengths is proportional to the fourth power of the absolute temperature (see †Stefan–Boltzmann law). Attempts to explain these facts failed until the development of †quantum theory in 1900.

black-body radiation radiation of a thermal nature emitted by a theoretically perfect emitter of radiation (a †black body) at a certain temperature.

black earth exceedingly fertile soil that covers a belt of land in northeastern North America, Europe, and Asia.

In Europe and Asia it extends from Bohemia through Hungary, Romania, southern Russia, and Siberia, as far as Manchuria, having been deposited when the great inland ice sheets melted at the close of the last †ice age. In North America, it extends from the Great Lakes east through New York State, having been deposited when the last glaciers melted and retreated from the terminal moraine.

Blackett, Patrick Maynard Stuart (1897–1974) Baron Blackett, English physicist. He was awarded the Nobel Prize for Physics in 1948 for work in cosmic radiation and his perfection of the Wilson cloud chamber, an apparatus for tracking ionized particles, with which he confirmed the existence of †positrons.

Blackett, Patrick Maynard Stuart http://www.nobel.se/physics/laureates/1948/blackett-bio.html

Portrait of the physicist and Nobel laureate. It traces the development of his theoretical interests. There is a picture of Blackett and the text is hyperlinked to lead you to pages on related people and places.

black hole object in space whose gravity is so great that nothing can escape from it, not even light. It is thought to form when a massive star shrinks at the end of its life. A black hole sucks in more matter, including other stars, from the space around it. Matter that falls into a black hole is squeezed to infinite density at the center of the hole. Black holes can be detected because gas falling towards them becomes so hot that it emits X-rays.

Black holes containing the mass of millions of stars are thought to lie at the centers of †quasars. Satellites have detected X-rays from a number of objects that may be black holes, but only a small number of likely black holes have been identified in our Galaxy.

Black Holes and Beyond http://archive.ncsa.uiuc.edu/Cyberia/NumRel/BlackHoles.html

Well-written guide to black holes. There is a good balance of text and images, with clear explanatory diagrams. Video and audio clips of scientists explain some of the current theories of the evolution and behavior of black holes.

Blackman, Frederick Frost (1866–1947) English botanist after whom the Blackman reactions of †photosynthesis are named. Leading a successful research group, he worked initially upon respiration in plants and showed that the exchange of CO_2 between the leaves and the air occurred via the stomata (the pores in the epidermis of a plant; see †stoma). His later work continued to apply physiochemical concepts to biology.

bladder hollow elastic-walled organ, which stores the urine, produced in the kidneys. It is present in the †urinary systems of some fishes, most amphibians, some reptiles, and all mammals. Urine enters the bladder through two ureters, one leading from each kidney, and leaves it through the urethra.

Blakeslee, Albert Francis (1874–1954) US botanist who was the first to artificially produce †polyploid plants (plants having three or more sets of chromosomes per cell in cases where the normal number is two sets). He developed the technique using the chemical colchicine to increase the number of chromosomes.

blast freezing industrial method of freezing substances such as foods by blowing very cold air over them. See †deep freezing.

blast furnace smelting furnace used to extract metals from their ores, chiefly pig iron from iron ore. The temperature is raised by the injection of an air blast.

In the extraction of iron the ingredients of the furnace are iron ore (iron(III) oxide), coke (carbon), and limestone (calcium carbonate). The coke is the fuel and provides the agent (carbon monoxide) for the †reduction of the iron ore.

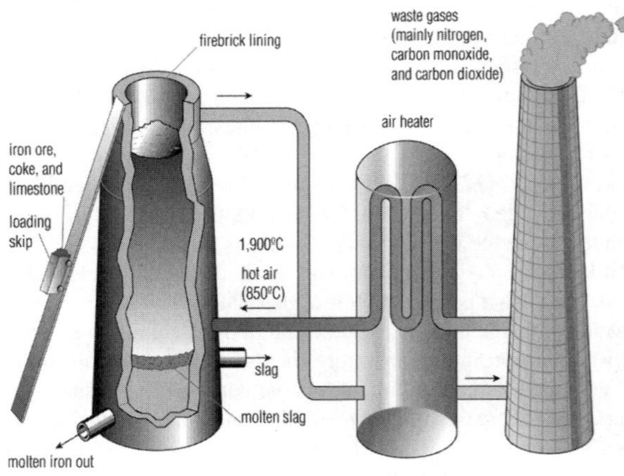

blast furnace The blast furnace is used to extract iron from a mixture of iron ore, coke, and limestone. The less dense impurities float above the molten iron and are tapped off as slag. The molten iron sinks to the bottom of the furnace and is tapped off into moulds referred to as pigs. The iron extracted this way is also known as pig iron. Helicon

blastocyst in mammals, the hollow ball of cells which is an early stage in the development of the †embryo, roughly equivalent to the †blastula of other animal groups.

blastoderm sheet of cells that grows on the surface of a fertilized ovum. In mammals it forms a disc of cells that eventually develops into the embryo between the amniotic cavity and the yolk sac. The †endoderm, †mesoderm, and †ectoderm also develop from the blastoderm.

blastomere in biology, a cell formed in the first stages of embryonic development, after the splitting of the fertilized ovum, but before the formation of the †blastula or blastocyst.

blastula early stage in the development of a fertilized egg, when the egg changes from a solid mass of cells (the morula) to a hollow ball of cells (the blastula), containing a fluid-filled cavity (the blastocoel). See also †embryology.

blazar in astronomy, type of galaxy that has a very active galactic nucleus, with rapid variations in optical brightness and strong radio emissions. The term is formed from a combination of †BL Lacertae object and quasar.

bleaching decolorization of colored materials. The two main types of bleaching agent are **oxidizing bleaches**, which bring about the †oxidation of pigments and include the ultraviolet rays in sunshine, hydrogen peroxide, and chlorine in household bleaches; and **reducing bleaches**, which bring about †reduction and include sulfur dioxide.

Bleach is used in industry to lighten or whiten fabrics, yarns, or fibers. Bleaching processes have been known from antiquity, mainly those acting through sunlight. Both natural and synthetic pigments usually possess highly complex molecules, the color property often being due to only a part of the molecule. Bleaches usually attack only that small part, yielding another substance similar in chemical structure but colorless.

Bleuler, Eugen (1857–1939) Swiss psychiatrist who introduced the concept of schizophrenia in 1911. He was professor of psychiatry at Zürich 1888–1927, where he studied the psychological aspects of mental disorder, using both the experimental methods of Wilhelm Wundt and the psychoanalytical procedures and theories of Sigmund Freud. Later he became critical of psychoanalytic theory.

blight any of a number of plant diseases caused mainly by parasitic species of †fungus, which produce a whitish appearance on leaf and stem surfaces; for example, **potato blight** *Phytophthora infestans*. General damage caused by aphids or pollution is sometimes known as blight.

In 1998 a new virulent strain of *P. infestans*, US-8, was decimating potato and tomato crops throughout the USA and eastern Canada, proving to be resistant to previously effective fungicides.

blind spot area where the optic nerve and blood vessels pass through the retina of the †eye. No visual image can be formed as there are no light-sensitive cells in this part of the retina. Thus the organism is blind to objects that fall in this part of the visual field.

BL Lacertae object starlike object that forms the center of a distant galaxy, with a prodigious energy output. BL Lac objects, as they are called, seem to be related to †quasars and are thought to be the brilliant nuclei of elliptical galaxies. They are so named because the first to be discovered lies in the small constellation Lacerta.

Blobel, Günter (1936–) German cell-biologist who was awarded the Nobel Prize for Physiology or Medicine in 1999 for the discovery that intrinsic signals govern the transport and localization of proteins in the cell.

Bloch, Felix (1905–1983) Swiss-born US physicist who shared the Nobel Prize for Physics in 1952 with US physicist Edward Mills †Purcell for the invention in 1946 of the analytical technique of nuclear magnetic resonance (NMR) †spectroscopy.

block in computing, a group of records treated as a complete unit for transfer to or from backing storage. For example, many disk drives transfer data in 512-byte blocks.

Bloembergen, Nicolaas (1920–) Dutch-born US physicist who was awarded the Nobel Prize for Physics in 1981 for his work on laser †spectroscopy. He used the laser as a tool to excite or break chemical bonds during the study of chemical reactions. He shared the Nobel prize with Arthur †Schawlow and Kai †Siegbahn.

blood fluid pumped by the †heart, that circulates in the †arteries, †veins, and †capillaries of vertebrate animals forming the †bloodstream. The term also refers to the corresponding fluid in those invertebrates that possess a closed †circulatory system. Blood carries nutrients and oxygen to each body cell and removes waste products such as carbon dioxide. It is also important in the immune response and, in many animals, in the distribution of heat throughout the body.

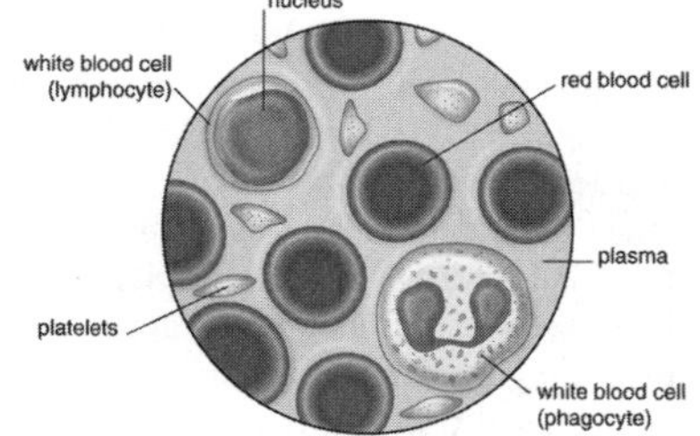

blood Composition of blood. Human blood contains red blood cells, white blood cells (phagocytes and lymphocytes), and platelets, suspended in plasma. Helicon

The adult human body contains about 5.5 l/10 pt of blood (about 5% of the body weight). It is composed of a fluid called †plasma, in which are suspended microscopic cells of three main varieties:

†Red blood cells (erythrocytes) form nearly half the volume of the blood, with about 6 million red blood cells in every milliliter of an adult's blood. They transport †oxygen around the body. Oxygen is absorbed into the millions of blood capillaries surrounding the tiny air sacs of the †lungs and is carried in the blood by †hemoglobin, a red protein within the red blood cells.

†White blood cells (leucocytes) are of various kinds. Some (phagocytes) ingest invading bacteria and so protect the body from disease; these also help to repair injured tissues. Others (lymphocytes) produce

antibodies, which help provide immunity by binding to disease-causing †bacteria and destroying them. Yet others can kill cells infected by †viruses.

Blood †platelets (thrombocytes) assist in the clotting of blood.

Blood cells constantly wear out and die and are replaced from the bone marrow. Red blood cells die at the rate of 200 billion per day but the body produces new cells at an average rate of 9,000 million per hour.

Many different substances are transported by the blood around the body. Following †digestion small food molecules are absorbed into the blood from the †gut. The blood transports these nutrients (for example †glucose) to the cells of the body. Waste products produced by the cells, such as carbon dioxide, are carried by the blood to the lungs to be exhaled. †Hormones (such as estrogen) are transported by the blood from glands to body cells in order to regulate various processes.

Lifeblood http://sln2.fi.edu/biosci/blood/blood.html

Introduction to blood, with graphics and links to other pages where you will find out everything you might ever think to ask about blood. Keywords and concepts are highlighted and provide links to more detailed pages of information.

blood clotting complex series of events (known as the blood clotting cascade) resulting from a series of enzymatic reactions in the blood that prevents excessive bleeding after injury. The result is the formation of a meshwork of protein fibers (fibrin) and trapped blood cells over the cut blood vessels.

When platelets (cell fragments) in the bloodstream come into contact with a damaged blood vessel, they and the vessel wall itself release the enzyme **thrombokinase**, which brings about the conversion of the inactive enzyme **prothrombin** into the active **thrombin**. Thrombin in turn catalyses the conversion of the soluble protein **fibrinogen**, present in blood plasma, to the insoluble **fibrin**. This fibrous protein forms a net over the wound that traps red blood cells and seals the wound; the resulting jellylike clot hardens on exposure to air to form a scab. Calcium, vitamin K, and a variety of enzymes called factors are also necessary for efficient blood clotting. Hemophilia is one of several diseases in which the clotting mechanism is impaired.

blood group any of the types into which blood is classified according to the presence or otherwise of certain †antigens on the surface of its red cells. Red blood cells of one individual may carry molecules on their surface that act as antigens in another individual whose red blood cells lack these molecules. The two main antigens are designated A and B. These give rise to four blood groups: having A only (A), having B only (B), having both (AB), and having neither (O). Each of these groups may or may not contain the †rhesus factor. Correct typing of blood groups is vital in transfusion, since incompatible types of donor and recipient blood will result in coagulation, with possible death of the recipient.

The ABO system was first described by Austrian scientist Karl †Landsteiner in 1902. Subsequent research revealed at least 14 main types of blood group systems, 11 of which are involved with induced †antibody production. Blood typing is also of importance in forensic medicine, cases of disputed paternity, and in anthropological studies.

blood pressure pressure, or tension, of the blood against the inner walls of blood vessels, especially the arteries, due to the muscular pumping activity of the heart. Abnormally high blood pressure (hypertension) may be associated with various conditions or arise with no obvious cause; abnormally low blood pressure (hypotension) occurs in shock and after excessive fluid or blood loss from any cause.

In mammals, the left ventricle of the †heart pumps blood into the arterial system. This pumping is assisted by waves of muscular contraction by the arteries themselves, but resisted by the elasticity of the inner and outer walls of the same arteries. Pressure is greatest when the heart ventricle contracts (**systole**) and lowest when the ventricle relaxes (**diastole**), and pressure is solely maintained by the elasticity of the arteries. Blood pressure is measured in millimeters of mercury (the height of a column on the measuring instrument, a sphygmomanometer). Normal human blood pressure varies with age, but in a young healthy adult it is around 120/80mm Hg; the first number represents the systolic pressure and the second the diastolic. Large deviations from this reading usually indicate ill health.

bloodstream in animals, the means of carrying substances around the body. Oxygen needed by the cells of the body for †respiration is absorbed into the blood at the lungs by †diffusion and is transported to every part of the body by the blood stream. The oxygen is carried by specialized cells – the †red blood cells. Inside each cell there are lots of hemoglobin molecules, which can combine with oxygen and release it when it is needed. Cells that respire produce carbon dioxide. The blood stream carries carbon dioxide from cells to the lungs where it is lost from the body by diffusion. The bloodstream also has an important role in carrying food around the body to cells that need it. This food is absorbed into the blood following †digestion in the †gut.

blood transfusion see †transfusion.

blood vessel tube that conducts blood either away from or towards the heart in multicellular animals. The principal types are †arteries, which conduct blood away from the heart, †veins, which conduct blood towards the heart, and †capillaries, which conduct blood from arteries to veins. Arteries always carry oxygenated blood and veins deoxygenated blood, with the exception of the **pulmonary artery**, which carries deoxygenated blood from the heart to the lungs, and the **pulmonary vein**, which carries oxygenated blood from the lungs to the heart.

bloom whitish powdery or waxlike coating over the surface of certain fruits that easily rubs off when handled. It often contains †yeasts that live on the sugars in the fruit.

The term bloom is also used to describe a rapid increase in number of certain species of algae found in lakes, ponds, and oceans.

Algal blooms may be natural but are often the result of nitrate pollution, in which artificial fertilizers, applied to surrounding fields, leach out into the waterways. This type of bloom can lead to the death of almost every other organism in the water; because light cannot penetrate the algal growth, the plants beneath can no longer photosynthesize and therefore do not release oxygen into the water. Only those organisms that are adapted to very low levels of oxygen survive.

blooming in optics, method of reducing back-reflections from the surface of a lens by coating it with a thin layer (of, for example, magnesium fluoride), in which destructive interference occurs. Coated lenses appear purple because they continue to reflect red and blue light but prevent the reflection of yellow light.

blow-out hollow or depression of bare sand in an area of †dunes on which vegetation grows. Blow-outs are common in coastal dune complexes and are formed by wind erosion, which can be triggered by the destruction of small areas of vegetation by people or animals; lack of sand supply from a beach; and localized dryness.

blubber thick layer of †fat under the skin of marine mammals, which provides an energy store and an effective insulating layer, preventing the loss of body heat to the surrounding water. Blubber has been used (when boiled down) in engineering, food processing, cosmetics, and printing, but all of these products can now be produced synthetically.

blue dwarf, blue giant high-temperature stars (as opposed to red stars). Blue giants are generally on or near the main sequence of the †Hertzsprung-Russell diagram; blue dwarfs represent the very dense, but very small, near-final form of what was once a †red giant.

blue-green algae or **cyanobacteria**, single-celled, primitive organisms that resemble bacteria in their internal cell organization, sometimes joined together in colonies or filaments. Blue-green algae are among the oldest known living organisms and, with bacteria, belong to the kingdom Monera; remains have been found in rocks up to 3.5 billion years old. They are widely distributed in aquatic habitats, on the damp surfaces of rocks and trees, and in the soil.

Blue-green algae and bacteria are prokaryotic organisms. Some can fix nitrogen and thus are necessary to the nitrogen cycle, while others follow a symbiotic existence – for example, living in association with fungi to form lichens. Fresh water can become polluted by nitrates and phosphates from fertilizers and detergents. This eutrophication, or overenrichment, of the water causes multiplication of the algae in the form of algae blooms. The algae multiply and cover the water's surface, remaining harmless until they give off toxins as they decay. These toxins kill fish and other wildlife and can be harmful to domestic animals, cattle, and people.

blue shift in astronomy, manifestation of the †Doppler effect in which an object appears bluer when it is moving towards the observer or the observer is moving towards it (blue light is of a higher frequency than other colors in the spectrum). The blue shift is the opposite of the †red shift.

bluff alternative name for a †river cliff.

BMR abbreviation for †**basal metabolic rate**.

B number abbreviation for †Bernoulli number.

Bode, Johann Elert (1747–1826) German astronomer and mathematician. He contributed greatly to the popularization of astronomy. He published the first atlas of all stars visible to the naked eye, *Uranographia* (1801), and popularized †Bode's law.

Bode's law or **Titius-Bode law**, numerical sequence that gives the approximate distances, in astronomical units (distance between Earth and Sun = one astronomical unit), of the planets from the Sun by adding 4 to each term of the series 0, 3, 6, 12, 24, ... and then dividing by 10. Bode's law predicted the existence of a planet between Mars and Jupiter, which led to the discovery of the asteroid belt.

The "law" breaks down for †Neptune and †Pluto. The relationship was first noted in 1766 by the German mathematician Johann Titius, and developed by German astronomer Johann Bode.

BODMAS mnemonic for **brackets, of, division, multiplication, addition, subtraction**, order in which an arithmetical expression should be calculated. See †algebra.

Bodmer, Walter Fred (1936–) German-born English geneticist who was a pioneer of research into the genetics of the HLA (human leucocyte antigen) histocompatability system, which helps the immune system to distinguish between the body's own cells and foreign, potentially harmful cells, like bacteria, that need to be destroyed and cleared from the body.

bog type of wetland where decomposition is slowed down and dead plant matter accumulates as †peat. Bogs develop under conditions of low temperature, high acidity, low nutrient supply, stagnant water, and oxygen deficiency. Typical bog plants are sphagnum moss, rushes, and cotton grass; insectivorous plants such as sundews and bladderworts are common in bogs (insect prey make up for the lack of nutrients).

Bohm, David Joseph (1917–1992) US-born British physicist who specialized in quantum mechanics but also worked on plasmas, metals, and liquid helium. In 1959 he and his student Yakir Aharanov discovered the Aharanov–Bohm effect, showing that the motions of charged particles can be affected by magnetic fields even if they never enter the regions to which those fields are confined.

Bohr, Aage Niels (1922–) Danish physicist who produced a new model of the nucleus of the atom in 1952, known as the collective model. He shared the Nobel Prize for Physics in 1975 for his discovery of the connection between collective motion and particle motion in atomic nuclei, and the development of the theory of nuclear structure. He was the son of physicist Niels †Bohr.

Bohr, Niels Henrik David (1885–1962) Danish physicist who was awarded the Nobel Prize for Physics in 1922 for his discovery of the structure of atoms and the radiation emanating from them. He pioneered †quantum theory by showing that the nuclei of atoms are surrounded by shells of electrons, each assigned particular sets of quantum numbers according to their orbits. He explained the structure and behavior of the nucleus, as well as the process of nuclear †fission. Bohr also proposed the doctrine of **complementarity**, the theory that a fundamental particle is neither a wave nor a particle, because these are complementary modes of description.

Niels Bohr Institute History http://www.nbi.dk/nbi-history.html

Short survey of the development of the Niels Bohr Institute, so intimately connected with Bohr's life work as a physicist. There are sections on the Institute today, a history from 1929 to 1965, and a picture gallery.

bohrium chemical symbol Bh, synthesized, radioactive element of the †transactinide series, atomic number 107, relative atomic mass 262. It was first synthesized by the Joint Institute for Nuclear Research in Dubna, Russia, in 1976; in 1981 the Laboratory for Heavy-Ion Research in Darmstadt, Germany, confirmed its existence. It was named in 1997 after Danish physicist Niels †Bohr.

The first chemical study of bohrium was published in 2000, after experiments by Swiss researchers produced six atoms of bohrium-267 (half-life 17 seconds). It behaved like a typical group VII element, with chemical similarities to technetium and rhenium.

Bohr magneton symbol m_B, in physics, unit of atomic magnetic moment (that is, the moment of one spin of an electron), equal to 9.27×10^{-24} JT^{-1}.

Bohr model model of the atom conceived by Danish physicist Niels †Bohr in 1913. It assumes that the following rules govern the behavior of electrons: (1) electrons revolve in orbits of specific radius around the nucleus without emitting radiation; (2) within each orbit, each electron has a fixed amount of energy; electrons in orbits farther away from the nucleus have greater energies; (3) an electron may "jump" from one orbit of high energy to another of lower energy causing the energy difference to be emitted as a †photon of electromagnetic radiation such as light; (4) an electron may absorb a photon of radiation and jump from a lower-energy orbit to a higher-energy one. The Bohr model has been superseded by wave mechanics (see †quantum theory).

boiling process of changing a liquid into its vapor, by heating it at the maximum possible temperature for that liquid (see †boiling point) at atmospheric pressure.

boiling point for any given liquid, the temperature at which the application of heat raises the temperature of the liquid no further, but converts it into vapor.

The boiling point of water under normal pressure is 100°C/212°F. The lower the pressure, the lower the boiling point and vice versa.

Bok, Bart J(an) (1906–1983) Dutch-born US astrophysicist who discovered small, circular dark spots in nebulae (now **Bok's globules**). His work broadened our understanding of the formation of stars.

Boksenberg, Alexander (1936–) English astronomer and physicist who devised a light-detecting system that can be attached to telescopes, vastly improving their optical powers. His image photon-counting system (IPCS) revolutionized observational astronomy, enabling him and others to study distant quasars.

Bok's globule small, circular dark spot in a †nebula, with a mass comparable to that of the Sun. Bok's globules are possibly gas clouds in the process of condensing into stars. They were first discovered by Dutch astrophysicist Bart Bok.

bolometer sensitive †thermometer that measures the energy of radiation by registering the change in electrical resistance of a fine wire when it is exposed to heat or light.

The US astronomer Samuel Langley devised it in 1880 for measuring radiation from stars.

bolometric magnitude in astronomy, measure of the brightness of a star over all wavelengths. Bolometric magnitude is related to the total radiation output of the star. See †magnitude.

bolson basin without an outlet, found in desert regions. Bolsons often contain temporary lakes, called playa lakes, and become filled with sediment from inflowing intermittent streams.

Boltzmann, Ludwig Eduard (1844–1906) Austrian physicist who studied the kinetic theory of gases, which explains the properties of gases by reference to the motion of their constituent atoms and molecules. He established the branch of physics now known as statistical mechanics.

Boltzmann, Ludwig http://www-history.mcs.st-andrews.ac.uk/history/Mathematicians/Boltzmann.html

Biography of the physicist and mathematician. The site contains a description of the work of Boltzmann, and also describes his working relationships with his contemporaries, which due to the radical nature of his work were often strained.

Boltzmann constant symbol *k*, in physics, the constant that relates the kinetic energy (energy of motion) of a gas atom or molecule to temperature. Its value is 1.38066×10^{-23} joules per kelvin. It is equal to the gas constant *R*, divided by †Avogadro's number.

bolus mouthful of chewed food mixed with saliva, ready for swallowing. Most vertebrates swallow food immediately, but grazing mammals chew their food a great deal, allowing a mechanical and chemical breakdown to begin.

Bolyai, János (1802–1860) Hungarian mathematician, one of the founders of non-Euclidean geometry. He was the son of Wolfgang Bolyai. He was one of the first to see Euclidean geometry as only one case, and that others were possible. In a paper written in 1823 he described a geometry in which several lines can pass through the point P without intersecting the line L.

Bolzano, Bernardus Placidus Johann Nepomuk (1781–1848) Czech mathematician and philosopher. Bolzano formulated a theory of real †functions and introduced the nondifferentiable **Bolzano function**. He was also able to prove the existence and define the properties of infinite sets.

bomb calorimeter in physics, type of calorimeter (instrument used in the determination of quantities of heat change) used to measure the energy released during the complete combustion of a fuel or food. It consists of an insulated sealed container in which a mixture of the fuel or food and oxygen is ignited (usually electrically) and the resultant rise in temperature measured. This enables the calorific value of the fuel or food to be calculated (usually expressed in joules per kilogram).

bond in chemistry, the result of the forces of attraction that hold together atoms in an element or compound. The principal types of bonding are †ionic, †covalent, †metallic, and †intermolecular (such as hydrogen bonding).

The type of bond formed depends on the elements concerned and their electronic structure. In an ionic or electrovalent bond, common in inorganic compounds, the combining atoms gain or lose electrons to become ions; for example, sodium (Na) loses an electron to form a sodium ion (Na^+) while chlorine (Cl) gains an electron to form a chloride ion (Cl^-) in the ionic bond of sodium chloride (NaCl).

In a covalent bond, the atomic orbitals of two atoms overlap to form a molecular orbital containing two electrons, which are thus effectively shared between the two atoms. Covalent bonds are common in organic compounds, such as the four carbon–hydrogen bonds in methane (CH_4). In a dative covalent or coordinate bond, one of the combining atoms supplies both of the valence electrons in the bond.

A metallic bond joins metals in a crystal lattice, the atoms occupy lattice positions as positive ions, and valence electrons are shared between all the ions in an "electron gas."

In a hydrogen bond, a hydrogen atom joined to an electronegative atom, such as nitrogen or oxygen, becomes partially positively charged, and is weakly attracted to another electronegative atom on a neighboring molecule.

Chemical Bonding Information http://users.senet.com.au/~rowanb/chem/chembond.htm

A concise and clearly-laid-out introduction to chemical bonding, including metallic, ionic, and covalent bonding.

bond energy in chemistry, amount of energy that is needed to break a chemical bond between atoms, usually measured in joules. The energy required to break a specific chemical bond is identical to that generated by the formation of the same bond.

Bondi, Hermann (1919–) Austrian-born British cosmologist. In 1948 he joined with English astronomer and cosmologist Fred †Hoyle and Austrian-born US astronomer Thomas †Gold in developing the steady-state theory of cosmology, which suggested that matter is continuously created in the universe. He was knighted in 1973.

bone hard connective tissue comprising the †skeleton of most vertebrate animals. Bone is composed of a network of collagen fibers impregnated with mineral salts (largely calcium phosphate and calcium carbonate), a combination that gives it great density and strength, comparable in some cases with that of reinforced concrete. Enclosed within this solid matrix are bone cells, blood vessels, and nerves. The interior of the long bones of the limbs consists of a spongy matrix filled with a soft marrow that produces blood cells.

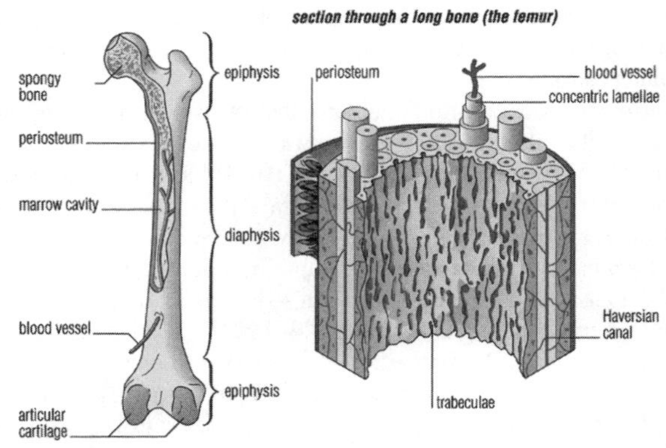

bone The upper end of the thighbone or femur is made up of spongy bone, which has a fine lacework structure designed to transmit the weight of the body. The shaft of the femur consists of hard compact bone designed to resist bending. Fine channels carrying blood vessels, nerves, and lymphatics interweave even the densest bone. Helicon

There are two types of bone: those that develop by replacing †cartilage and those that form directly from connective tissue. The latter, which include the bones of the cranium, are usually platelike in shape and form in the skin of the developing embryo. Humans have 206 distinct bones in the skeleton (see †human body), of which the smallest are the three ossicles in the middle ear. However, a different total is sometimes given, because of a number of fused pairs of bones in the skull that may be counted as one or two bones.

Bones of the Body http://www.meddean.luc.edu/lumen/MedEd/GrossAnatomy /learnem/bones/main_bone.htm
Large resource on most of the bones in the human body. The page is divided into two frames: the left-hand frame lists all of the bones covered, ordered under the headings of "Skull," "Vertebrae," "Thorax," "Upper extremity," "Pelvis," and "Lower extremity." Clicking on a specific bone under one of the headings brings up information in the right-hand frame.

bone marrow substance found inside the cavity of bones. In early life it produces red blood cells but later on lipids (fat) accumulate and its color changes from red to yellow.

Bone marrow may be transplanted in the treatment of some diseases, such as leukemia, using immunosuppressive drugs in the recipient to prevent rejection. Transplants to adult monkeys from early aborted monkey fetuses have successfully bypassed rejection.

Boole, George (1815–1864) English mathematician. His work *The Mathematical Analysis of Logic* (1847) established the basis of modern mathematical logic, and his **Boolean algebra** can be used in designing computers.

Boole, George http://www-history.mcs.st-andrews.ac.uk/history/Mathematicians/Boole.html
Extensive biography of the mathematician. The site contains a clear description of his working relationship with his contemporaries, and also includes the title page of his famous book Investigation of the Laws of Thought. Several literature references for further reading on the mathematician are also listed, and the Web site also features a portrait of Boole.

Boolean algebra set of algebraic rules, named after English mathematician George Boole, in which TRUE and FALSE are equated to 0 and 1. Boolean algebra includes a series of operators (AND, OR, NOT, NAND (NOT AND), NOR, and XOR (exclusive OR)), which can be used to manipulate TRUE and FALSE values (see †truth table). It is the basis of computer logic because the truth values can be directly associated with †bits.

These rules are used in searching databases either locally or across the †Internet via services like AltaVista to limit the number of hits to those which most closely match a user's requirements. A search instruction such as "tennis NOT table" would retrieve articles about tennis and reject those about ping-pong.

boot or **bootstrap**, in computing, the process of starting up a computer. Most computers have a small, built-in boot program that starts automatically when the computer is switched on – its only task is to load a slightly larger program, usually from a hard disk, which in turn loads the main †operating system.

boot disk or **emergency disk**, disk containing the necessary files to †boot a computer without needing to access its hard disk. Boot disks are vital in recovering from virus attacks, when it is not known which files on a computer's hard disk may be infected; in recovering from a system crash which has corrupted existing files; or in correcting mistakes introduced into files necessary for starting up the computer by newly installed software programs.

Boötes constellation of the northern hemisphere represented by a herdsman driving a bear (†Ursa Major) around the pole. Its brightest star is †Arcturus (or Alpha Boötis), which is 36 light years from Earth. The herdsman is assisted by the neighboring †Canes Venatici, "the Hunting Dogs."

borane in chemistry, any hydride of †boron. The simplest member of this group of compounds is borane, BH_3, an unstable compound that readily dimerizes to form diborane B_2H_6.

borax hydrous sodium borate, $Na_2B_4O_7.10H_2O$, found as soft, whitish crystals or encrustations on the shores of hot springs and in the dry beds of salt lakes in arid regions, where it occurs with other borates, halite, and †gypsum. It is used in bleaches and washing powders.

A large industrial source is Borax Lake, California. Borax is also used in glazing pottery, in soldering, as a mild antiseptic, and as a metallurgical flux.

Borda, Jean Charles de (1733–1799) French mathematician and physicist. He introduced new methods and instruments connected with the sciences of navigation, astronomy, and geodesy. He was one of the scientists who framed the metric system adopted in France under the Republican government.

Bordeaux mixture solution made up of equal quantities of copper(II) sulfate and lime in water, used in horticulture and in the wine industry as a †fungicide.

Bordet, Jules Jean Baptiste Vincent (1870–1961) Belgian bacteriologist and immunologist who was awarded a Nobel Prize for Physiology or Medicine in 1919 for his work on immunity. He researched the role of blood serum in the human immune response, and was the first to isolate the whooping-cough bacillus, in 1906.

bore surge of tidal water up an estuary or a river, caused by the funneling of the rising tide by a narrowing river mouth. A very high tide, possibly fanned by wind, may build up when it is held back by a river current in the river mouth. The result is a broken wave, a meter or a few feet high, which rushes upstream.

Famous bores are found in the rivers Severn (England), Seine (France), Hooghly (India), and Chang Jiang (China), where bores of over 4 m/13 ft have been reported.

Borel, Emile Félix-Edouard-Justin (1871–1956) French mathematician. He carried out pioneering work in several areas of mathematics, including game theory, probability, and calculus, and rationalized the theory of functions of real variables.

Borel, Emile Félix Edouard Justin http://www-groups.dcs.st-andrews.ac.uk/history/Mathematicians/Borel.html

Part of an archive containing the biographies of the world's greatest mathematicians, this site is devoted to the life and contributions of Emile Borel. In addition to biographical information, you will find a list of references about Borel and links to other essays in the archive that reference him. The text of this essay includes hypertext links to the essays of those mathematicians and thinkers who influenced Borel. You will also find an image of him, which you can click on to enlarge, and a map of his birthplace.

Borelli, Giovanni Alfonso (1608–1679) Italian scientist who explored the links between physics and medicine and showed how mechanical principles could be applied to animal †physiology. This approach, known as iatrophysics, has proved basic to understanding how the mammalian body works.

boric acid or **boracic acid**; $B(OH)_3$, acid formed by the combination of hydrogen and oxygen with nonmetallic boron. It is a weak antiseptic and is used in the manufacture of glass and enamels. It is also an efficient insecticide against ants and cockroaches.

Born, Max (1882–1970) German-born British physicist. He was awarded the Nobel Prize for Physics in 1954 for fundamental work on the †quantum theory, especially his 1926 discovery that the wave function of an electron is linked to the probability that the electron is to be found at any point.

Born, Max http://www-history.mcs.st-andrews.ac.uk/history/Mathematicians/Born.html

Biography of the German-born British physicist. The Web site details the work of Born, and his relationships with his contemporaries and colleagues. Also included are several literature references for further reading on the physicist, and a portrait photograph of Born.

boron chemical symbol B, non-metallic element, atomic number 5, relative atomic mass 10.811. In nature it is found only in compounds, as with sodium and oxygen in borax. It exists in two allotropic forms (see †allotropy): brown amorphous powder and very hard, brilliant crystals. Its compounds are used in the preparation of boric acid, water softeners, soaps, enamels, glass, and pottery glazes. In alloys it is used to harden steel. Because it absorbs slow neutrons, it is used to make boron carbide control rods for nuclear reactors. It is a necessary trace element in the human diet. The element was named by Humphry Davy, who isolated it in 1808, from *bor*ax + -on, as in carb*on*.

borosilicate in chemistry, member of a group of compounds that are generally prepared as glasses by the fusion of boron oxide and silicon dioxide to which a metal oxide is usually added. The most famous borosilicate glasses are marketed under the trade name Pyrex.

Bosch, Carl (1874–1940) German metallurgist and chemist. He developed the †Haber process from a small-scale technique for the production of ammonia into an industrial high-pressure process that made use of water gas as a source of hydrogen. He shared the Nobel Prize for Chemistry in 1931 with Friedrich †Bergius for his part in inventing and developing high-pressure industrial methods.

Bosch process process for the production of hydrogen developed by German industrial chemist Carl †Bosch. It involves the catalytic reduction of steam with carbon monoxide at high temperature to produce hydrogen. This process was essential to the initial development and economic viability of the †Haber process.

Boscovich, Ruggero Giuseppe (1711–1787) Croatian-born Italian mathematician and astronomer. An early supporter of Newton, he developed a theory, popular in the 19th century, of the atom as a single point with surrounding fields of repulsive and attractive forces.

Bose, Satyendra Nath (1894–1974) Indian physicist who, with †Einstein, formulated the Bose-Einstein statistics, a description of the behavior of elementary particles that have integral (whole-number) spin. He was professor of physics at the University of Calcutta 1945–58. Particles with integral spin are called †bosons after him.

Bose-Einstein condensate BEC, hypothesis put forward in 1924 by German-born US physicist Albert Einstein and Indian physicist Satyendra †Bose, suggesting that when a dense gas is cooled to a little over absolute zero it will condense and its atoms will lose their individuality and act as an organized whole. The first Bose-Einstein condensate was produced in June 1995 by US physicists cooling rubidium atoms to 10 billionths of a degree above absolute zero. The condensate existed for about a minute before becoming rubidium ice.

Bose-Einstein statistics treatment of particles which have integral †spin in †statistical mechanics, enabling properties of systems made up of such particles to be calculated.

boson in physics, an elementary particle whose spin can only take values that are whole numbers or zero. Bosons may be classified as †gauge bosons (carriers of the four fundamental forces) or †mesons. All elementary particles are either bosons or †fermions.

Unlike fermions, more than one boson in a system (such as an atom) can possess the same energy state. That is, they do not obey the Pauli †exclusion principle. When developed mathematically, this statement is known as the Bose-Einstein law, after its discoverers Indian physicist Satyendra Bose and Albert Einstein.

Boss, Lewis (1846–1912) US astronomer. At first a government surveyor, he became dissatisfied with the inaccuracy of the positions of standard stars used. Much of his life was devoted to forming an accurate catalogue of star positions; this work was completed by his son, **Benjamin Boss**, resulting in the *General Catalogue of 33,342 Stars*, published in Washington, DC, in 1937. He was director of Dudley Observatory, Albany, New York, 1876–1912.

botany (Greek *botane* "herb") study of living and fossil †plants, including form, function, interaction with the environment, and classification.

Botany is subdivided into a number of specialized studies, such as the identification and classification of plants (taxonomy), their external formation (plant morphology), their internal arrangement (plant anatomy), their microscopic examination (plant histology), their functioning and life history (plant physiology), and their distribution over the Earth's surface in relation to their surroundings (plant ecology). Palaeobotany concerns the study of fossil plants, while economic botany deals with the utility of plants. †Horticulture, †agriculture, and †forestry are branches of botany.

GardenWeb Glossary of Botanical Terms http://glossary.gardenweb.com/glossary/

Searchable, text-only glossary of well over 2,000 botanical terms. There is a brief explanation of how the search engine works and a selection of criteria with which to refine your search, including "terms" or "definitions," as well as a basic and/or Boolean search.

Bothe, Walther Wilhelm Georg (1891–1957) German physicist who showed in 1929 that the cosmic rays bombarding the Earth are composed not of photons but of more massive particles. He was awarded the Nobel Prize for Physics in 1954 for developing the coincidence method of detecting the emission of electrons.

Boucher de Crèvecoeur de Perthes, Jacques (1788–1868) French geologist whose discovery of paleolithic hand axes in 1837 challenged the accepted view of human history dating only from 4004 BC, as proclaimed by the calculations of Bishop James Ussher.

Bouguer anomaly anomaly in the local gravitational force that is due to the density of rocks rather than local topography, elevation, or latitude. A positive anomaly, for instance, is generally indicative of denser and therefore more massive rocks at or below the surface. A negative anomaly indicates less massive materials. Calculations of Bouguer anomalies are used for mineral prospecting and for understanding the structure beneath the Earth's surface. The Bouguer anomaly is named after its discoverer, the French mathematician Pierre Bouguer, who first observed it in 1735.

boulder clay another name for †till, a type of glacial deposit.

bound see †upper bound and †lower bound.

boundary line around the edge of an area; a perimeter. The boundary of a circle is known as the **circumference**. The boundary that marks the limit of land may be indicated by a post, ditch, hedge, march of stones, road, or river, or it may be indicated by reference to a plan, or to possession of tenants, or by actual measurement.

boundary value in applied mathematics, a natural phenomenon in a given region may be described by functions that satisfy certain differential equations in the interior of the region and take specific values on the boundary of the region. The latter are referred to as boundary values.

Bourbaki, Nicolas Pseudonym adopted by a group of mathematicians, most of them French, who, collectively and anonymously, published a definitive survey of mathematics 1939–67. The group, which at any one time contained about 20 members, was centered at the Ecole Normale Supérieure in Paris. The group's founder was André Weil.

Bourdon gauge instrument for measuring pressure, patented by French watchmaker Eugène Bourdon in 1849. The gauge contains a C-shaped tube, closed at one end. When the pressure inside the tube increases, the tube uncurls slightly causing a small movement at its closed end. A system of levers and gears magnifies this movement and turns a pointer, which indicates the pressure on a circular scale. Bourdon gauges are often fitted to cylinders of compressed gas used in industry and hospitals.

bovine somatotropin BST, hormone that increases an injected cow's milk yield by 10–40%. It is a protein naturally occurring in milk and breaks down within the human digestive tract into harmless amino acids. However, doubts have arisen as to whether such a degree of protein addition could in the long term be guaranteed harmless either to cattle or to humans.

Although no evidence of adverse side effects to consumers has been found, BST was banned in Europe in 1993 until the year 2000. In the USA genetically engineered BST has been in use since February 1994; in Vermont a law requiring milk containing BST to be labeled as such was passed in September 1995. In January 1999, Canada announced a ban on BST. The reasons given for the ban included a 25% increase in the likelihood of mastitis in cows given BST; an 18% increase in infertility; and double the risk of becoming lame.

Bowden, Frank Philip (1903–1968) Australian physicist and chemist who worked mainly in Britain. He studied friction, lubricants, and surface erosion. For example, he realized that the thin layer of water between a ski or an ice skate and the snow or ice is produced not by pressure due to the weight on them but by friction-induced heat caused by irregularities in the sliding surfaces.

Bowen, Ira Sprague (1898–1973) US astrophysicist. He studied the spectra of planetary nebulae and showed that strong green lines in such spectra are due to ionized oxygen and nitrogen under extreme conditions not found on Earth.

Bowen, Norman Levi (1887–1956) Canadian geologist whose work helped found modern petrology (the study of rocks). He demonstrated the principles governing the formation of magma by partial melting, and the fractional crystallization of magma.

Bowman's capsule in the vertebrate kidney, a cup-shaped structure enclosing the glomerulus, which is the initial site of filtration of the blood leading to urine formation.

There are approximately a million of these capsules in a human kidney, each containing a tight knot of capillaries and leading into a kidney tubule or nephron where unwanted fluid and waste molecules are filtered from the blood to be excreted in the form of urine.

Boyle, Robert (1627–1691) Irish chemist and physicist who published *The Sceptical Chymist* (1661), a groundbreaking book that became the definitive text among scientists for decades to follow. He formulated †Boyle's law in 1662. He was a pioneer in the use of experiment and scientific method.

Boyle, Robert http://www-groups.dcs.st-andrews.ac.uk/history/Mathematicians/Boyle.html

Part of an archive containing the biographies of the world's greatest mathematicians, this site is devoted to the life and contributions of chemist and physicist Robert Boyle. In addition to biographical information, you will find a list of references about Boyle and links to other essays in the archive that reference him. The text of this essay includes hypertext links to the essays of those mathematicians and thinkers who influenced Boyle. You will also find an image of him, which you can click on to enlarge, and a map of his birthplace.

Boyle's law law stating that the volume of a given mass of gas at a constant temperature is inversely proportional to its pressure. For example, if the pressure on a gas doubles, its volume will be reduced by a half, and vice versa. The law was discovered in 1662 by Irish physicist and chemist Robert Boyle. See also †gas laws.

If a gas is compressed in a cylinder the volume of the gas decreases. The number of particles of gas in the cylinder remains the same. The particles get closer together, collide with each other more frequently, and the pressure of the gas increases due to the force of the particles colliding. If the pressure is P and the volume is V, then $P = 1/V$. Therefore, as the volume decreases the pressure increases.

Boyle's law can be investigated by using an apparatus consisting of a foot pump attached to a pressure gauge. This is attached to a glass tube (with a scale) containing an oil and trapped air. As more air is pumped into the apparatus from the foot pump and into the oil reservoir, the oil in the glass tube is forced in and the pressure on the trapped air gets higher. A series of pressure and volume readings are taken. It is found that multiplying pressure by volume for each of the readings produces the same result. Therefore, $P \times V$ = constant. The results of P and $1/V$ plotted as a graph give a straight line. Readings from the experiment show that:

$P_1 V_1 = P_2 V_2$

where P_1 and V_1 are the initial pressure and volume of a gas, and P_2 and V_2 are its final pressure and volume.

Boyle's Law http://www.math.montana.edu/frankw/ccp/before-calculus/function/boyle/body.htm

Excellent tutorial on Boyle's law. The Web site outlines Boyle's law, describes a practical investigation, uses a Java applet simulation, then tests the user's understanding.

Boys, Charles Vernon (1855–1944) English inventor and physicist. He designed several scientific instruments, including a very sensitive torsion balance used in 1895 to determine Isaac †Newton's gravitational constant and the mean density of the Earth.

Boys, Sir Charles Vernon http://www-groups.dcs.st-andrews.ac.uk/history/Mathematicians/Boys.html

Part of an archive containing the biographies of the world's greatest mathematicians, this site is devoted to the life and contributions of physicist Vernon Boys. In addition to biographical information, you will find a list of references about Boys and links to other essays in the archive that reference him. The text of this essay includes hypertext links to the essays of those mathematicians and thinkers who influenced Boys.

bps abbreviation for **bits per second**, measure used in specifying data transmission rates.

Bq in physics, symbol for †becquerel, the SI unit of radioactivity (equal to the average number of disintegrations per second in a given time).

brachiopod or **lamp shell**, any member of the phylum Brachiopoda, marine invertebrates with two shells, resembling but totally unrelated to bivalves.

There are about 300 living species; they were much more numerous in past geological ages. They are suspension feeders, ingesting minute food particles from water. A single internal organ, the lophophore, handles feeding, aspiration, and excretion.

brachistochrone in mathematics, curve along which a particle will slide (without friction) from one point to a lower point in the shortest time. Swiss mathematician Jean Bernoulli proposed this as a problem in 1696 and English physicist Isaac †Newton, German mathematician Gottfried †Leibniz, and Jakob †Bernoulli all showed that the answer is a †cycloid.

brackets pairs of signs that show which part of a calculation should be worked out first.

For example, 4(7+3) indicates that 4 is to be multiplied by the result obtained from adding 7 and 3. The mnemonic †BODMAS can help one to remember the order in which an arithmetical expression should be calculated. Brackets may be nested, for example, 4(20–(7+3)), in which case the expression 7+3 within the innermost pair of brackets is evaluated first, the result subtracted from 20 and that result multiplied by 4.

Brackets are also used to contain the argument of a function. For example, $f(x)$.

bract leaflike structure in whose †axil a flower or inflorescence develops. Bracts are generally green and smaller than the true leaves. However, in some plants they may be brightly colored and conspicuous, taking over the role of attracting pollinating insects to the flowers, whose own petals are small; examples include poinsettia *Euphorbia pulcherrima* and bougainvillea.

A whorl of bracts surrounding an †inflorescence is termed an **involucre**. A **bracteole** is a leaflike organ that arises on an individual flower stalk, between the true bract and the †calyx.

Bradley, James (1693–1762) English astronomer. In 1728 he discovered the †aberration of starlight. From the amount of aberration in star positions, he was able to calculate the speed of light. In 1748 he announced the discovery of †nutation (variation in the Earth's axial tilt).

Brady's reagent in chemistry, solution of 2,4-dinitrophenylhydrazine in methanol used as a test to detect the presence of †aldehydes and †ketones. The solution yields a range of yellow to red precipitates when an aldehyde or ketone is added to it.

Bragg, (William) Lawrence (1890–1971) Australian-born British physicist. He shared with his father William †Bragg the Nobel Prize for Physics in 1915 for their research work on X-rays and crystals.

Bragg, William Henry (1862–1942) English physicist. He shared with his son Lawrence †Bragg the Nobel Prize for Physics in 1915 for their research work on X-rays and crystals.

Bragg's law in physics, law that states that the maximum intensity of X-rays diffracted (see †X-ray diffraction) through a crystal occurs when the sine of the complement of the angle of incidence of the X-rays onto the crystal satisfies the relation:

$n\lambda = 2d \sin \theta$

where λ is the wavelength of the radiation, d is the lattice spacing, and n is an integer. The equation was determined by Australian-born British physicist Lawrence Bragg.

Brahe, Tycho (1546–1601) Danish astronomer. His accurate observations of the planets enabled German astronomer and mathematician Johannes †Kepler to prove that planets orbit the Sun in ellipses. Brahe's discovery and report of the 1572 supernova brought him recognition, and his observations of the comet of 1577 proved that it moved in an orbit among the planets, thus disproving Aristotle's view that comets were in the Earth's atmosphere.

Brahe, Tycho http://es.rice.edu/ES/humsoc/Galileo/People/tycho_brahe.html

Full account of the life and work of the Danish astronomer. There are pictures of Brahe, instruments he used, and the observatory he founded. There is an explanation of the Tychonic world view whereby the Earth was held to be the center of the universe while planets revolved around the Sun.

Brahmagupta (598–670) Indian astronomer and mathematician important for his introduction of zero and negative numbers into arithmetic. He also made contributions to geometry and to the solution of †quadratic equations.

braiding the subdivision of a river into several channels caused by deposition of sediment as islets in the channel. Braided channels are common in meltwater streams.

braid theory part of the study of †nodes in three-dimensional space, first devised by Austrian mathematician Emil Artin.

brain in higher animals, a mass of interconnected †nerve cells forming the anterior part of the †central nervous system, whose activities it coordinates and controls. In †vertebrates, the brain is contained by the skull. It is composed of three main regions. At the base of the †brainstem, the **medulla oblongata** contains centers for the control of respiration, heartbeat rate and strength, blood pressure, and thermoregulatory control (the control of body temperature). Overlying this is the **cerebellum**, which is concerned with coordinating complex muscular processes such as maintaining posture and moving limbs, and the control of balance. The **cerebrum** (cerebral hemispheres) are paired outgrowths of the front end of the forebrain, in early vertebrates mainly concerned with the senses, but in higher vertebrates greatly developed and involved in the integration of all sensory input and motor output, and in thought, emotions, memory, and behavior. Sensory information arrives in the cerebrum in the form of nerve impulses that come from receptors – these may be found in sense organs, such as cones (light sensitive cells) in the retina of the eye, which send impulses to the brain along the optic nerve. The cerebrum processes the information received and can cause impulses to be sent out to the body to produce a response, such as moving towards an object that has been seen. Because a decision is made whether to make this kind of movement it is said to be voluntary.

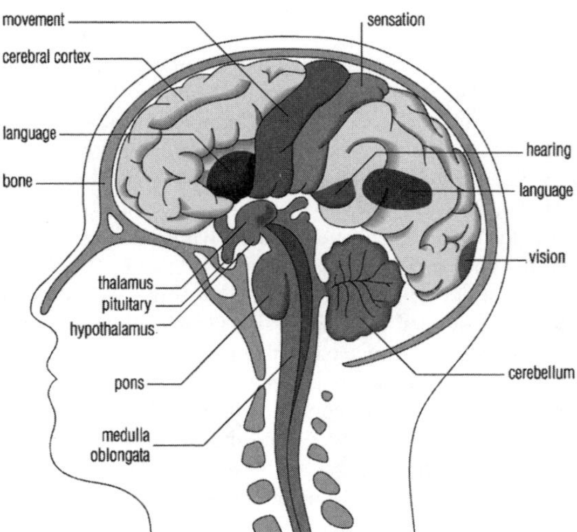

brain The structure of the human brain. At the back of the skull lies the cerebellum, which coordinates reflex actions that control muscular activity. The medulla controls respiration, heartbeat, and blood pressure. The hypothalamus is concerned with instinctive drives and emotions. The thalamus relays signals to and from various parts of the brain. The pituitary gland controls the body's hormones. Distinct areas of the large convoluted cerebral hemispheres that fill most of the skull are linked to sensations, such as hearing and sight, and voluntary activities, such as movement. Helicon

In vertebrates, many of the nerve fibers from the two sides of the body cross over as they enter the brain, so that the left cerebral hemisphere is associated with the right side of the body and vice versa. In humans, a certain asymmetry develops in the two halves of the cerebrum. In right-handed people, the left hemisphere seems to play a greater role in controlling verbal and some mathematical skills, whereas the right hemisphere is more involved in spatial perception. In general, however, skills and abilities are not closely localized. In the brain, nerve impulses are passed across †synapses by neurotransmitters, in the same way as in other parts of the nervous system.

In mammals the cerebrum is the largest part of the brain, carrying the **cerebral cortex**. This consists of a thick surface layer of cell bodies (gray matter), below which fiber tracts (white matter) connect various parts of the cortex to each other and to other points in the central nervous system. As cerebral complexity grows, the surface of the brain becomes convoluted into deep folds. In higher mammals, there are large unassigned areas of the brain that seem to be connected with intelligence, personality, and higher mental faculties. Language is controlled in two special regions usually in the left side of the brain: **Broca's area** governs the ability to talk, and **Wernicke's area** is responsible for the comprehension of spoken and written words. In 1990 scientists at Johns Hopkins University, Baltimore, succeeded in culturing human brain cells.

BodyQuest: The Nervous System http://library.thinkquest.org/10348/find/content/nervous.html

Overview of the way in which the brain sends, receives, and processes the nerve impulses that control everything we do. The page, part of a much larger site on the human body, discusses the roles played by the central, peripheral, and autonomic nervous systems, and provides experiments to answer such questions as "How do we balance?" and "Why do we have two eyes?"

Virtual Body http://www.ehc.com/vbody.asp

Authoritative and interactive anatomical animations, complete with voice-overs, guide you around the whole body, with sections on the brain, digestive system, heart, and skeleton. The Flash plug-in is required to view this site but this can be downloaded simply if it is required.

brainstem region where the top of the spinal cord merges with the undersurface of the brain, consisting largely of the medulla oblongata and midbrain.

The oldest part of the brain in evolutionary terms, the brainstem is the body's life-support center, containing regulatory mechanisms for vital functions such as breathing, heart rate, and blood pressure. It is also involved in controlling the level of consciousness by acting as a relay station for nerve connections to and from the higher centers of the brain.

In many countries, death of the brainstem is now formally recognized as death of the person as a whole. Such cases are the principal donors of organs for transplantation. So-called "beating-heart donors" can be maintained for a limited period by life-support equipment.

Brand, Stewart (1938–) US founder of the *Whole Earth Catalog* (1968) and the WELL (1984). He founded, edited, and published the *Whole Earth Catalog* (1968–85) as well as the *CoEvolution Quarterly* (now known as the *Whole Earth Review*) (1973–84). He is the author of *The Media Lab* (1987), an account of research work carried out at the MIT Media Lab.

Branly, Edouard Eugène Désiré (1844–1940) French physicist and inventor. In 1890 he demonstrated the possibility of detecting radio waves; the apparatus he devised (the coherer) was soon used in the invention of wireless telegraphy and radio.

brass metal †alloy of copper and zinc, with not more than 5% or 6% of other metals. The zinc content ranges from 20% to 45%, and the color of brass varies accordingly from coppery to whitish yellow. Brasses are characterized by the ease with which they may be shaped and machined; they are strong and ductile, resist many forms of corrosion, and are used for electrical fittings, ammunition cases, screws, household fittings, and ornaments.

Brasses are usually classed into those that can be worked cold (up to 25% zinc) and those that are better worked hot (about 40% zinc).

Brattain, Walter Houser (1902–1987) US physicist. He was awarded the Nobel Prize for Physics in 1956 jointly with William †Shockley and John †Bardeen for their work on the development of the transistor, which replaced the comparatively costly and clumsy vacuum tube in electronics.

Braun, Karl Ferdinand (1850–1918) German physicist who was awarded the Nobel Prize for Physics in 1909 for his work on the development of wireless telegraphy. He shared the award with Guglielmo †Marconi. Braun also discovered crystal rectifiers (used in early radios), and invented the oscilloscope in 1895.

breadth thickness, another name for width. The area of a rectangle is given by the formula: area = length times breadth.

breast one of a pair of organs on the chest of the adult human female, also known as a †mammary gland.

Each of the two breasts contains milk-producing cells and a network of tubes or ducts that lead to openings in the nipple.

Milk-producing cells in the breast do not become active until a woman has given birth to a baby. Breast milk is made from substances extracted from the mother's blood as it passes through the breasts, and contains all the nourishment a baby needs. Breast-fed newborns develop fewer infections than bottle-fed babies because of the antibodies and white blood cells contained in breast milk. These are particularly abundant in the †colostrum produced in the first few days of breast-feeding.

breathing or **ventilation**, movement of air into and out of the air passages of an animal, brought about by muscle contraction. It is a form of †gas exchange. Breathing is sometimes referred to as external respiration, for true respiration is a cellular (internal) process. In a mammal, breathing involves the action of the muscles of the diaphragm and the intercostal muscles (between the ribs). When a mammal breathes in, the diaphragm muscles contract, which lowers the diaphragm. The external intercostal muscles contract, which raises the ribs. Lowering the diaphragm and raising the ribs increases the volume of the †thorax. This lowers the pressure of the air inside the lungs in the thorax. The pressure is now lower than atmospheric pressure so air flows into the air passages and the lungs inflate.

Lungs are specialized for gas exchange but are not themselves muscular, consisting of spongy material. In order for oxygen to be passed to the blood and carbon dioxide removed, air is drawn into the lungs (inhaled) by the contraction of the diaphragm and intercostal muscles; relaxation of these muscles enables air to be breathed out (exhaled). The rate of breathing is controlled by the brain. High levels of activity lead to a greater demand for oxygen and an increased rate of breathing.

The Respiratory System http://www.emc.maricopa.edu/faculty/farabee/BIOBK/BioBookRESPSYS.html

Explanation of the respiratory systems of different organisms, from single-celled organisms, flatworms, and jellyfish to larger animals, including humans. Includes a hyperlinked illustrated glossary of key terms.

breathing rate number of times a minute the lungs inhale and exhale. The rate increases during exercise because the muscles require an increased supply of oxygen and nutrients. At the same time very active muscles produce a greater volume of carbon dioxide, a waste gas that must be removed by the lungs via the blood.

The regulation of the breathing rate is under both voluntary and involuntary control, although a person can only forcibly stop breathing for a limited time. The regulatory system includes the use of chemoreceptors, which can detect levels of carbon dioxide in the blood. High concentrations of carbon dioxide, occurring for example during exercise, stimulate a fast breathing rate.

breccia coarse-grained †sedimentary rock, made up of broken fragments (clasts) of pre-existing rocks held together in a fine-grained matrix. It is similar to †conglomerate but the fragments in breccia are jagged in shape.

Bredig, Georg (1868–1944) German physical chemist who devised a method of preparing colloidal solutions in 1898. He studied the catalytic action of colloidal platinum and the "poisoning" of catalysts by impurities.

breed recognizable group of domestic animals, within a species, with distinctive characteristics that have been produced by artificial selection.

breeder reactor or **fast breeder**, alternative name for †fast reactor, a type of nuclear reactor.

breeding in biology, the crossing and selection of animals and plants to change the characteristics of an existing †breed or †cultivar (variety), or to produce a new one. Selective breeding is breeding in which humans choose the parent plants or parent animals.

Ever since humans first carried out agriculture, they have been selecting which farm animals they allow to breed and which seeds to save from their crops to sow the following year. Plant and animal species show inherited variation. This is a result of the processes involved in †sexual reproduction. †Mutation also produces variation that is inherited. Selecting the parents used for breeding makes some variants more common, even though they may have been very rare originally. It can also bring together in one organism characteristics originally found in different organisms of the species.

If selective breeding is carried out over many generations this results in the production of varieties of plants and breeds of animals (see †variety). For example "Cox's," "Golden Delicious," and "Bramley" are different apple varieties. Cattle may be bred for increased meat or milk yield, sheep for thicker or finer wool, and horses for speed or stamina. Plants, such as wheat or corn, may be bred for disease resistance, heavier and more rapid cropping, and hardiness. Selective breeding of animals to produce new varieties helped Charles †Darwin understand the effects of selection and understand its natural equivalent – †natural selection, which can result in evolutionary change.

breeding in nuclear physics, a process in a reactor in which more fissionable material is produced than is consumed in running the reactor.

For example, plutonium-239 can be made from the relatively plentiful (but nonfissile) uranium-238, or uranium-233 can be produced from thorium. The Pu-239 or U-233 can then be used to fuel other reactors.

Bremsstrahlung †electromagnetic radiation produced by the rapid deceleration of charged particles such as electrons, as occurs in the collision between electrons and nuclei.

Brenner, Sydney (1927–) South African scientist and one of the pioneers of genetic engineering. Brenner discovered messenger †RNA (a link between †DNA and the †ribosomes in which proteins are synthesized) in 1960. He received the Lasker Award for Special Achievement in Medical Science in 2000.

Breuer, Josef (1842–1925) Viennese physician, one of the pioneers of psychoanalysis. He applied it successfully to cases of hysteria, and collaborated with †Freud on *Studien über Hysterie/Studies in Hysteria* (1895).

brewster symbol B, unit for measuring the reaction of optical materials to stress, defined in terms of the slowing down of light passing through the material when it is stretched or compressed.

Brewster, David (1781–1868) Scottish physicist who made discoveries about the diffraction and polarization of light, and invented the kaleidoscope in 1816. He was knighted in 1831.

Brewster's law in physics, law that states that the†refractive index of a medium is given by the tangent of the angle at which maximum polarization occurs. It was determined by Scottish physicist David Brewster.

Bridgman, Percy Williams (1882–1961) US physicist. His research into machinery producing high pressure led in 1955 to the creation of synthetic diamonds by General Electric. He was awarded the Nobel Prize for Physics in 1946 for his development of high-pressure physics.

Briggs, Henry (1561–1630) English mathematician, with John †Napier one of the founders of calculation by †logarithms. Briggs's tables remain the basis of those used to this day.

brine common name for a solution of †sodium chloride (NaCl) in water. Brines are used extensively in the food-manufacturing industry for canning vegetables, pickling vegetables (sauerkraut manufacture), and curing meat. Industrially, brine is the source from which †chlorine, caustic soda (†sodium hydroxide), and sodium carbonate are made.

The chlor-alkali industry is based on the †electrolysis of brine, which produces chlorine, hydrogen, and sodium hydroxide. In the Solvay process, sodium carbonate is produced from sodium chloride.

Brinkley, John (1763–1835) English astronomer. After a brilliant career at Cambridge and a short period of working with Nevil †Maskelyne at Greenwich, he was appointed professor of astronomy at Dublin University and shortly afterwards became the first Astronomer Royal for Ireland. At the Dunsink Observatory he investigated the possibility of observing stellar parallaxes with the meridian circle.

brittle material material that breaks suddenly under stress at a point just beyond its elastic limit (see †elasticity). Brittle materials may also break suddenly when given a sharp knock. Pottery, glass, and cast iron are examples of brittle materials. Compare †ductile material.

broadband in computing, a type of data transmission in which a single circuit can carry several channels at once, used for example in cable television. Broadband networking is one way of supplying much greater Internet bandwidth over the existing telephone system.

broad-leaved tree another name for a tree belonging to the †angiosperms, such as ash, beech, oak, maple, or birch. The leaves are generally broad and flat, in contrast to the needlelike leaves of most conifers. See also †deciduous tree.

Brockhouse, Bertram Neville (1918–) Canadian physicist who, with Clifford †Shull, developed neutron diffraction techniques used for studying the structure and properties of matter. Brockhouse designed ingenious instruments with which he recorded the energy of neutrons scattered from various materials. He shared the Nobel Prize for Physics in 1994 for the development of the technique known as "neutron scattering," which led to advances in semiconductor technology.

Broglie, (Louis César Victor) Maurice de (1875–1960) 6th duc de Broglie, French physicist. He worked on X-rays and gamma rays, and helped to establish the Einsteinian description of light in terms of photons. He was the brother of Louis de Broglie.

Broglie, Louis Victor Pierre Raymond de (1892–1987) 7th duc de Broglie, French theoretical physicist. He established that all subatomic particles can be described either by particle equations or by wave equations, thus laying the foundations of wave mechanics. He was awarded the Nobel Prize for Physics in 1929 for his discovery of the wavelike nature of electrons. Succeeded as Duke in 1960.

De Broglie, Louis http://www-history.mcs.st-andrews.ac.uk/history/Mathematicians/Broglie.html

Biographical details and a photograph of Louis de Broglie, the famous French physicist and mathematician. There are also links to de Broglie's most famous work on quantum mechanics and to many of his contemporaries.

bromide salt of the halide series containing the Br^- ion, which is formed when a bromine atom gains an electron.

The term "bromide" is sometimes used to describe an organic compound containing a bromine atom, even though it is not ionic. Modern naming uses the term "bromo-" in such cases. For example, the compound C_2H_5Br is now called bromoethane; its traditional name, still used sometimes, is ethyl bromide.

bromine chemical symbol Br, (Greek *bromos* "stench") dark, reddish-brown, non-metallic element, a volatile liquid at room temperature, atomic number 35, relative atomic mass 79.904. It is a member of the †halogen group, has an unpleasant odor, and is very irritating to mucous membranes. Its salts are known as bromides.

Bromine was formerly extracted from salt beds but is now mostly obtained from sea water, where it occurs in small quantities. Its compounds are used in photography and in the chemical and pharmaceutical industries.

bronchiole small-bore air tube found in the vertebrate lung responsible for delivering air to the main respiratory surfaces. Bronchioles lead off from the larger †bronchus and branch extensively before terminating in the many thousand alveoli that form the bulk of lung tissue.

bronchus one of a pair of large tubes (bronchi) branching off from the windpipe (trachea) and passing into the vertebrate lung. Apart from their size, bronchi differ from the bronchioles in possessing cartilaginous rings, which give rigidity and prevent collapse during breathing movements.

Numerous glands in the wall of the bronchus secrete a slimy mucus, which traps dust and other particles; the mucus is constantly being propelled upwards to the mouth by thousands of tiny hairs or cilia. The bronchus is adversely effected by several respiratory diseases and by smoking, which damages the cilia and therefore the lung-cleansing mechanism.

Brongniart, Alexandre (1770–1847) French naturalist and geologist who first used fossils to date strata of rock and was the first scientist to arrange the geological formations of the Tertiary period in chronological order. He also wrote *Essay on the Classification of Reptiles* (1800), splitting reptiles into four orders: Batrachia (now a separate class, Amphibia), Chelonia, Ophidia, and Sauria.

Bronowski, Jacob (1908–1974) Polish-born British scientist, broadcaster, and writer, who enthusiastically popularized scientific knowledge in several books and in the 13-part television documentary *The Ascent of Man*, issued as a book in 1973.

Brønsted, Johannes Nicolaus (1879–1947) Danish physical chemist whose work in solution chemistry, particularly electrolytes, resulted in a new theory of acids and bases, the theory of proton donors and proton acceptors, published in 1923.

Brønsted on Acids and Bases http://dbhs.wvusd.k12.ca.us/Chem-History/Bronsted-Article.html

Transcript of an article by Brønsted on the properties of acids and bases. The transcript as featured here is not complete, but does cover the essential points in which Brønsted defines the acid and the base. Several examples of his ideal reactions are shown, as well as a brief summary of the properties of the substances.

Broom, Robert (1866–1951) British-born South African paleontologist and physician who proved that *Australopithecus* was bipedal and established its significance as an early hominid.

Brown, Ernest William (1866–1938) English mathematician and astronomer. He studied the effect of gravity on the motions of the planets and smaller members of the Solar System, and published extremely accurate tables of the Moon's movements.

Brown, Ernest William http://www-groups.dcs.st-andrews.ac.uk/history/Mathematicians/Brown.html

Devoted to the life and contributions of Ernest Brown.

Brown, Herbert Charles (1912–) US inorganic chemist who is noted for his research on boron compounds. He manufactured sodium borohydride, a reducing agent (a substance that reduces another substance, see †reduction), and developed a simple technique to synthesize diborane. He also created a new class of compounds, the organoboranes, by reacting diborane with alkenes (unsaturated hydrocarbons containing one or more double bonds). He was awarded the Nobel Prize for Chemistry in 1979 for his use of boron and phosphorus compounds, in organic syntheses.

Brown, Robert (1773–1858) Scottish botanist who in 1827 discovered †Brownian motion. As a botanist, his more lasting work was in the field of plant morphology. He was the first to establish the real basis for the distinction between gymnosperms (conifers) and angiosperms (flowering plants).

Brown, Robert Hanbury (1916–2002) Indian-born British radio astronomer who participated in the early development of radio-astronomy techniques and later in designing a radio interferometer that permits considerably greater resolution in the results provided by radio telescopes.

brown dwarf in astronomy, object less massive than a star but denser than a planet. Brown dwarfs do not have enough mass to ignite nuclear reactions at their centers, but shine by heat released during their contraction from a gas cloud. Groups of brown dwarfs have been discovered, and some astronomers believe that vast numbers of them exist throughout the Galaxy.

The first brown dwarf to be positively identified was Gliese 229B (GI229B), in the constellation Lepus, by US astronomers using images from the Hubble Space Telescope. It is about 50 times as massive as Jupiter but emits only 1% of the radiation of the smallest known star and has a surface temperature of 650°C/1,200°F.

Brownian movement continuous random motion of particles in a fluid medium (gas or liquid) as they are subjected to impact from the molecules of the medium. The phenomenon was explained by German physicist Albert Einstein in 1905 but was observed as long ago as 1827 by the Scottish botanist Robert Brown. Brown was looking at pollen grains in water under a microscope when he noticed the pollen grains were in constant, haphazard motion. The motion of these particles was due to the impact of moving water molecules. It provides evidence for the †kinetic theory of matter.

In order for the irregular motion to be observed, the particles in the medium must be sufficiently small relative to the bombarding molecules for the impact of the bombarding molecules to have an effect. A tennis ball in air, for instance, would not show Brownian motion because the impacts of the moving air molecules on one side of the tennis ball would be balanced by impacts of the molecules on the other side. In other words, the †resultant force of the impacts would be too small.

Brownian movement The irregular movement of dust particles becomes visible in the air when the particles are caught in a ray of sunlight. The tiny dust particles move randomly as they are buffeted by gas molecules in the atmosphere, which are too small to be seen with the naked eye. Helicon

Brownian Motion http://dbhs.wvusd.k12.ca.us/Chem-History/Brown-1829.html
Transcript of "Remarks on Active Molecules" by Robert Brown from Additional Remarks on Active Molecules (1829). The text describes Robert Brown's observations of the random motion of particles, which in turn led to the placement of the subject on millions of school exam papers.

brown ring test in analytical chemistry, a test for the presence of †nitrates.

To an aqueous solution containing the test substance is added iron(II) sulfate. Concentrated sulfuric acid is then carefully poured down the inside wall of the test tube so that it forms a distinct layer at the bottom. The formation of a brown color at the boundary between the two layers indicates the presence of nitrate.

browser in computing, any program that allows the user to search for and view data. Browsers are usually limited to a particular type of data, so, for example, a graphics browser will display graphics files stored in many different file formats. Browsers usually do not permit the user to edit data, but are sometimes able to convert data from one file format to another. Web browsers allow access to the †World Wide Web. Netscape Navigator and Microsoft's Internet Explorer were the leading Web browsers in 2001. They act as a †graphical user interface to information available on the Internet – reading †HTML documents and displaying them as graphical documents which may include images, video, sound, and †hypertext links to other documents.

The first widespread browser for personal computers (PCs) was the text-based program Lynx, which is still used via †gateways from text-based online systems such as Delphi and CIX. Browsers using graphical user interfaces became widely available from 1993 with the release of Mosaic.

Some Web sites contain special pages that might contain animation, 3D scenes, or a particular kind of audio file. To display these kinds of effects, the basic browser technology needs to be enhanced using plug-in technology, such as the Macromedia Flash plug-in which can display sophisticated animations and multimedia effects.

bryophyte member of the Bryophyta, a division of the plant kingdom containing three classes: the Hepaticae (liverwort), Musci (†moss), and Anthocerotae (hornwort). Bryophytes are generally small, low-growing, terrestrial plants with no vascular (water-conducting) system as in higher plants. Their life cycle shows a marked †alternation of generations. Bryophytes chiefly occur in damp habitats and require water for the dispersal of the male gametes (†antherozoids).

In bryophytes, the †sporophyte, consisting only of a spore-bearing capsule on a slender stalk, is wholly or partially dependent on the †gametophyte for water and nutrients. In some liverworts the plant body is a simple †thallus, but in the majority of bryophytes it is differentiated into stem, leaves, and †rhizoids.

BST abbreviation for British Summer Time and †bovine somatotropin.

Btu symbol for British thermal unit: originally it was defined as the heat required to raise the temperature of one pound of water by one degree Fahrenheit. Now defined as equal to 1,055.6 joules.

bubble chamber in physics, a device for observing the nature and movement of atomic particles, and their interaction with radiation. It is a vessel filled with a superheated liquid through which ionizing particles move and collide. The paths of these particles are shown by strings of bubbles, which can be photographed and studied. By using a pressurized liquid medium instead of a gas, it overcomes drawbacks inherent in the earlier †cloud chamber. It was invented by US physicist Donald Glaser in 1952. See †particle detector.

Buchan, Alexander (1829–1907) Scottish meteorologist who studied fluctuations in temperature and weather conditions in the British Isles, and postulated a seasonal forecast. His works include an *Introductory Textbook of Meteorology* (1871), and reports on "Atmospheric Circulation" and "Oceanic Circulation" following the Challenger Expedition (1872–76), an oceanographic survey around the world.

Buchner, Eduard (1860–1917) German chemist who was awarded the Nobel Prize for Chemistry in 1907 for his biochemical research and discovery of the process of cell-free fermentation. In 1897 he observed that fermentation could be produced mechanically, by cell-free extracts. Buchner argued that it was not the whole yeast cell that produced fermentation, but only the presence of the enzyme he named zymase.

Buchner funnel in chemistry, perforated funnel, usually made from porcelain, which holds a filter paper. The funnel is used for the rapid filtration of solids and precipitates from solutions. It was developed and named after German organic chemist Eduard †Buchner.

Buckland, William (1784–1856) English geologist and paleontologist, a pioneer of British geology. He contributed to the descriptive and historical stratigraphy of the British Isles, inferring from the vertical succession of the strata a stage-by-stage temporal development of the Earth's crust.

buckminsterfullerene form of carbon, made up of molecules (buckyballs) consisting of 60 carbon atoms arranged in 12 pentagons and 20 hexagons to form a perfect sphere. It was named after the US architect and engineer Richard Buckminster Fuller because of its structural similarity to the geodesic dome that he designed. See †fullerene.

buckyballs popular name for molecules of †buckminsterfullerene.

bud undeveloped shoot usually enclosed by protective scales; inside is a very short stem and numerous undeveloped leaves, or flower parts, or both. Terminal buds are found at the tips of shoots, while axillary buds develop in the †axils of the leaves, often remaining dormant unless the terminal bud is removed or damaged. Adventitious buds may be produced anywhere on the plant, their formation sometimes stimulated by an injury, such as that caused by pruning.

budding type of †asexual reproduction in which an outgrowth develops from a cell to form a new individual. Most yeasts reproduce in this way.

In a suitable environment, yeasts grow rapidly, forming long chains of cells as the buds themselves produce further buds before being separated from the parent. Simple invertebrates, such as hydra, can also reproduce by budding.

In horticulture, the term is used for a technique of plant propagation whereby a bud (or scion) and a sliver of bark from one plant are transferred to an incision made in the bark of another plant (the stock). This method of †grafting is often used for roses.

buffer in chemistry, mixture of compounds chosen to maintain a steady †pH. The commonest buffers consist of a mixture of a weak organic acid and one of its salts or a mixture of acid salts of phosphoric acid. The addition of either an acid or a base causes a shift in the †chemical equilibrium, thus keeping the pH constant.

buffer in computing, a part of the †memory used to store data temporarily while it is waiting to be used. For example, a program might store data in a printer buffer until the printer is ready to print it.

Buffon, Georges-Louis Leclerc (1707–1788) Comte de Buffon, French naturalist and author of the 18th century's most significant work of natural history, the 44-volume *Histoire naturelle génerale et particulière* (1749–67). In *The Epochs of Nature*, one of the volumes, he questioned biblical chronology for the first time, and raised the Earth's age from the traditional figure of 6,000 years to the seemingly colossal estimate of 75,000 years.

bug in computing, an †error in a program. It can be an error in the logical structure of a program or a syntax error, such as a spelling mistake. Some bugs cause a program to fail immediately; others remain dormant, causing problems only when a particular combination of events occurs. The process of finding and removing errors from a program is called **debugging**.

bulb underground bud with fleshy leaves containing a reserve food supply and with roots growing from its base. Bulbs function in vegetative reproduction and are characteristic of many monocotyledonous plants such as the daffodil, snowdrop, and onion. Bulbs are grown on a commercial scale in temperate countries, such as England and the Netherlands.

bulbil small bulb that develops above ground from a bud. Bulbils may be formed on the stem from axillary buds, as in members of the saxifrage family, or in the place of flowers, as seen in many species of onion *Allium*. They drop off the parent plant and develop into new individuals, providing a means of †vegetative reproduction and dispersal.

Bullard, Edward Crisp (1907–1980) English geophysicist who, with US geologist Maurice Ewing, founded the discipline of marine geophysics. He pioneered the application of the seismic method to study the sea floor. He also studied continental drift before the theory became generally accepted. He was knighted 1953.

bulletin board in computing, a center for the electronic storage of messages, usually accessed over the telephone network via a †modem but also sometimes accessed via Telnet across the Internet. Bulletin board systems (often abbreviated to BBSs) are usually dedicated to specific interest groups, and may carry public and private messages, notices, and programs.

buna rubber in chemistry, synthetic rubber made by the co-polymerization of †butadiene with a number of compounds, notably styrene (Buna-S) and acrylonitrile (Buna-s). It was originally developed in Germany before and during World War I.

Bunsen, Robert Wilhelm (1811–1899) German chemist credited with the invention of the **Bunsen burner**. His name is also given to the carbon–zinc electric cell, which he invented in 1841 for use in arc lamps. In 1860 he discovered two new elements, cesium and rubidium.

Bunsen burner gas burner used in laboratories, consisting of a vertical metal tube through which a fine jet of fuel gas is directed. Air is drawn in through airholes near the base of the tube and the mixture is ignited and burns at the tube's upper opening.

The invention of the burner is attributed to German chemist Robert von Bunsen in 1855, but English chemist and physicist Michael Faraday is known to have produced a similar device at an earlier date. A later refinement was the metal collar that can be turned to close or partially close the airholes, thereby regulating the amount of air sucked in and hence the heat of the burner's flame.

buoyancy lifting effect of a fluid on a body wholly or partly immersed in it. This was studied by †Archimedes in the 3rd century BC.

bur or **burr**, in botany, a type of "false fruit" or †pseudocarp, surrounded by numerous hooks; for instance, that of burdock *Arctium*, where the hooks are formed from bracts surrounding the flowerhead. Burs catch in the feathers or fur of passing animals, and thus may be dispersed over considerable distances.

Burali–Forte's paradox paradox stating that to every collection of ordinal numbers there corresponds an ordinal number greater than any element of the collection. In particular it would follow that the collection of all ordinal numbers is itself an ordinal number. This contradiction demonstrated the need for a rigorous exposition of set theory in which not all collections may be accepted as valid subjects of discourse. There is thus no such thing as a "set of all sets," nor a "set of all ordinals" (indicating that the †foundations of mathematics cannot be expressed in purely logical terms). One therefore distinguishes between sets, which may be manipulated, and †classes, which may not (except in the simplest of circumstances).

Burbidge, (Eleanor) Margaret (1919–) born Margaret Peachey, English astrophysicist who, with her husband Geoffrey Burbidge, discovered processes by which elements are formed in the nuclei of stars. Together they published *Quasi-Stellar Objects* (1967), based on her research. Later, they suggested that quasars and galaxies are linked in some way.

Burbidge, Geoffrey (1925–) English astrophysicist who, with his wife Margaret †Burbidge, made discoveries relating to nucleosynthesis – the creation of elements in space – and studied quasars and galaxies.

burette in chemistry, a piece of apparatus, used in †titration, for the controlled delivery of measured variable quantities of a liquid.

It consists of a long, narrow, calibrated glass tube, with a tap at the bottom, leading to a narrow-bore exit.

Burnell, Jocelyn Bell Northern Irish astronomer. See †Bell Burnell.

burning common name for †combustion.

bursa synovial sac or cleft containing fluid, situated between two moving surfaces of the body. It reduces friction. Bursae may be permanent or temporary and occur in a place where friction has caused irritation of the tissues. Bursae occur between the covering skin and bony projections, for example at the point of the elbow and the knee-cap, but more often occur between †tendons and the surfaces they cross such as †bone, †ligaments, and other tendons. Most bursae are near joints and may communicate with the joint cavity.

burster in astronomy, celestial source of intense rapid pulses of gamma rays or X-rays, generally located in multiple-star systems. The effect is thought to result from the build-up of accreted gas on the surface of a neutron star or in a globular cluster, from which energy is suddenly emitted.

Burt, Cyril Lodowic (1883–1971) English psychologist. A specialist in child and mental development, he argued in *The Young Delinquent* (1925) the importance of social and environmental factors in delinquency. After his death it was claimed that he had falsified experimental results in an attempt to prove his theory that intelligence is largely inherited.

bus in computing, the electrical pathway through which a computer processor communicates with some of its parts and/or peripherals. Physically, a bus is a set of parallel tracks that can carry digital signals; it may take the form of copper tracks laid down on the computer's †printed circuit boards (PCBs), or of an external cable or connection.

A computer typically has three internal buses laid down on its main circuit board: a **data bus**, which carries data between the components of the computer; an **address bus**, which selects the route to be followed by any particular data item traveling along the data bus; and a **control bus**, which is used to decide whether data is written to or read from the data bus. An external **expansion bus** is used for linking the computer processor to peripheral devices, such as modems and printers.

bushel dry or liquid measure equal to eight gallons or four pecks (36.37 l/2,219.36 cu in) in the UK; some US states have different standards according to the goods measured.

butadiene or **buta-1,3-diene**; CH_2:CHCH:CH_2, inflammable gas derived from petroleum, used in making synthetic rubber and resins.

butane C_4H_{10}, †alkane (saturated hydrocarbon) derived from †natural gas and as a product of the †fractional distillation of crude oil (unrefined †petroleum). Liquefied under pressure, it is used as a †fuel for industrial and domestic purposes (for example, in portable cookers).

butanol or **butyl alcohol**, in chemistry, any of four liquid alcohol isomers of general formula $C_4H_{10}O$. The four isomers are butan-1-ol, butan-2-ol, methyl propan-1-ol, and 2,2-dimethyl ethanol. Their main uses are as solvents for resins and lacquers and in the production of butyl esters, which are important in the manufacture of perfumes and flavoring agents.

Butenandt, Adolf Friedrich Johann (1903–1995) German biochemist who was awarded the Nobel Prize for Chemistry in 1939 for his work on sex hormones. He isolated the first sex hormones (oestrone, androsterone, and progesterone), and determined their structure. He shared the Nobel prize with Swiss chemist Leopold †Ruzicka, who synthesized androsterone.

butene C_4H_8, fourth member of the †alkene series of hydrocarbons. It is an unsaturated compound, containing one double bond.

butte steep-sided, flat-topped hill, formed in horizontally layered sedimentary rocks, largely in arid areas. A large butte with a pronounced tablelike profile is a †mesa.

Buttes and mesas are characteristic of semi-arid areas where remnants of resistant rock layers protect softer rock underneath, as in the plateau regions of Colorado, Utah, and Arizona, USA.

butyl rubber in chemistry, synthetic rubber formed by the co-polymerization of isobutene, or isobutylene, with isoprene. It is usually vulcanized (hardened) and used in the manufacture of tire inner tubes and outdoor waterproof sheeting.

Buys-Ballot, Christoph (1817–1890) Dutch meteorologist. He invented a system of weather signals, which was a great aid to international uniformity in meteorological observations. His observations have been formulated in a general law of storms. His works include *Changements periodiques de la temperature/Periodic Temperature Changes* (1847) and, in English, *Suggestions on a Uniform System of Meteorological Observations* (1872–73).

bycatch or **bykill**, in commercial fishing, that part of the catch that is unwanted. Bycatch constitutes approximately 25% of global catch, and consists of a variety of marine life, including fish too small to sell or otherwise without commercial value, seals, dolphins, sharks, turtles, and even seabirds.

Bycatch can sometimes exceed target catch; for example, trawling for shrimp can result in discarded animals outnumbering shrimp by 125–830%. An estimated 55,000 sea turtles die in shrimp trawls annually in US waters. It is not only large nets that cause bycatch: more than 40,000 albatrosses drown each year in the southern hemisphere as a result of becoming hooked by fishing lines when they eat the squid bait set for bluefin tuna.

by-product substance formed incidentally during the production process of some other substance; for example, slag is a by-product of the production of iron in a †blast furnace. The by-product is incidental to the main product. For industrial processes to be economical, by-products must be recycled or used in other ways as far as possible; in this example, slag is used for making roads. Another common example occurs in the petrochemical industry, where crude oil is refined principally to produce petroleum, with a number of by-products being produced, including creosote and paraffin.

Often, a poisonous by-product is removed by transforming it into another substance, which although less harmful is often still inconvenient. For example, the sulfur dioxide produced as a by-product of electricity generation can be removed from the smoke stack using †flue-gas desulfurization. This process produces large amounts of gypsum, some of which can be used in the building industry.

Byron, (Augusta) Ada, Countess of Lovelace (1815–1852) English mathematician, a pioneer in writing programs for Charles †Babbage's analytical engine. In 1983 a new, high-level computer language, Ada, was named after her.

She was the daughter of the poet Lord Byron.

Byron, Ada http://www.agnesscott.edu/lriddle/women/love.htm

Interesting biography of Ada Byron, the woman credited as the first computer programmer. This page also has links to a range of other online resources about this scientist.

byte sufficient computer memory to store a single character of data, such as a letter of the alphabet. The character is stored in the byte of memory as a pattern of †bits (binary digits), using a code such as †ASCII. A byte usually contains eight bits – for example, the capital letter F can be stored as the bit pattern 01000110.

A single byte can specify 256 values, such as the decimal numbers from 0 to 255; in the case of a single-byte †pixel (picture element), it can specify 256 different colors. Three bytes (24 bits) can specify 16,777,216 values. Computer memory size is measured in **kilobytes** (1,024 bytes) or **megabytes** (1,024 kilobytes).

C

°C symbol for degrees †Celsius, sometimes called centigrade.

C in physics, symbol for †coulomb, the SI unit of electrical charge.

C in computing, a high-level, general-purpose programming language popular on minicomputers and microcomputers. Developed in the early 1970s from an earlier language called BCPL, C was first used as the language of the operating system †Unix, though it has since become widespread beyond Unix. It is useful for writing fast and efficient systems programs, such as operating systems (which control the operations of the computer).

C++ in computing, a high-level programming language used in object-oriented applications. It is derived from the language C.

Neil's C++ Stuff http://www.neilstuff.com/

Features a number of tutorials on the programming language C++, covering topics such as "Loops," "Classes," and "Arrays." The information is presented clearly, and there is also a glossary and list of frequently asked questions should you get stuck.

cable unit of length, used on ships, originally the length of a ship's anchor cable or 120 fathoms (219 m/720 ft), but now taken as one-tenth of a †nautical mile (185.3 m/608 ft).

cache memory in computing, a reserved area of the †immediate access memory used to increase the running speed of a computer program.

The cache memory may be constructed from †SRAM, which is faster but more expensive than the normal DRAM. Most programs access the same instructions or data repeatedly. If these frequently used instructions and data are stored in a fast-access SRAM memory cache, the program will run more quickly. In other cases, the memory cache is normal DRAM, but is used to store frequently used instructions and data that would normally be accessed from backing storage. Access to DRAM is faster than access to backing storage so, again, the program runs more quickly. This type of cache memory is often called a **disk cache**.

CAD acronym for **computer-aided design**, use of computers in creating and editing design drawings. CAD also allows such things as automatic testing of designs and multiple or animated three-dimensional views of designs. CAD systems are widely used in architecture, electronics, and engineering, for example in the motor-vehicle industry, where cars designed with the assistance of computers are now commonplace. With a CAD system, picture components are accurately positioned using grid lines. Pictures can be resized, rotated, or mirrored without loss of quality or proportion.

Powerful CAD software allows designers to build virtual designs and manipulate them on screen. For example, an architect can use CAD combined with virtual reality and other graphics programs to "walk" people through a virtual model of a planned building. A related development is †CAM (computer-aided manufacturing).

cadmium chemical symbol Cd, soft, silver-white, ductile, and malleable metallic element, atomic number 48, relative atomic mass 112.40. Cadmium occurs in nature as a sulfide or carbonate in zinc ores. It is a toxic metal that, because of industrial dumping, has become an environmental pollutant. It is used in batteries, electroplating, and as a constituent of alloys used for bearings with low coefficients of friction; it is also a constituent of an alloy with a very low melting point.

Cadmium is also used in the control rods of nuclear reactors, because of its high absorption of neutrons. It was named in 1817 by the German chemist Friedrich Strohmeyer (1776–1835) after the Greek mythological character Cadmus.

Caelum inconspicuous constellation of the southern hemisphere, represented as a sculptor's chisel; discovered in 1752, it was originally named "Caela Sculptoris" by French astronomer Nicolas †Lacaille.

caffeine †alkaloid organic substance found in tea, coffee, and kola nuts; it stimulates the heart and central nervous system. When isolated, it is a bitter crystalline compound, $C_8H_{10}N_4O_2$. Too much caffeine (more than six average cups of tea or coffee a day) can be detrimental to health.

Frequently Asked Questions About Caffeine http://www.cs.unb.ca/~alopez-o/caffaq.html

Source of information on caffeine and coffee. The chemistry of caffeine and its effects are clearly presented. For devotees of good coffee there is advice on various kinds of coffee and how to make the perfect brew. Studies on the side effects of excessive caffeine consumption are also summarized.

Cailletet, Louis Paul (1832–1913) French physicist and inventor. In 1877–78 he was the first to liquefy oxygen, hydrogen, nitrogen, and air. He did it by cooling them below their critical temperatures, first compressing the gas, then cooling it, then allowing it to expand to cool it still further.

Cairns, Hugh John Forster (1922–) English virologist whose research has focused on cancer and influenza. In 1959 he succeeded in carrying out genetic mapping of an animal virus for the first time.

cal symbol for †calorie.

CAL acronym for **computer-assisted learning**, use of computers in education and training: the computer displays instructional material to a student and asks questions about the information given; the student's answers determine the sequence of the lessons.

calamine $ZnCO_3$, zinc carbonate, an ore of zinc. The term also refers to a pink powder made of a mixture of zinc oxide and iron(II) oxide used in lotions and ointments as an astringent for treating, for example, sunburn, eczema, measles rash, and insect bites and stings.

In the USA the term refers to zinc silicate $Zn_4Si_2O_7(OH)_2.H_2O$

calcination †oxidation of substances by roasting in air.

calcite colorless, white, or light-colored common rock-forming mineral, calcium carbonate, $CaCO_3$. It is the main constituent of †limestone and marble and forms many types of invertebrate shell.

Calcite often forms †stalactites and stalagmites in caves and is also found deposited in veins through many rocks because of the ease with which it is dissolved and transported by groundwater; †oolite is a rock consisting of spheroidal calcite grains. It rates 3 on the †Mohs scale of hardness. Large crystals up to 1 m/3 ft have been found in Oklahoma and Missouri, USA. †Iceland spar is a transparent form of calcite used in the optical industry; as limestone it is used in the building industry.

calcium chemical symbol Ca, (Latin *calcis* "lime") soft, silvery-white metallic element, atomic number 20, relative atomic mass 40.08. It is one of the †alkaline-earth metals. It is the fifth most abundant element (the third most abundant metal) in the Earth's crust. It is found mainly as its carbonate $CaCO_3$, which occurs in a fairly pure condition as chalk and limestone (see †calcite). Calcium is an essential component of bones, teeth, shells, milk, and leaves, and it forms 1.5% of the human body by mass.

Calcium ions in animal cells are involved in regulating muscle contraction, blood clotting, hormone secretion, digestion, and glycogen metabolism in the liver. It is acquired mainly from milk and cheese, and its uptake is facilitated by vitamin D. Calcium deficiency leads to chronic muscle spasms (tetany); an excess of calcium may lead to the formation of stones (calculi) in the kidney or gall bladder.

The element was discovered and named by the English chemist Humphry Davy in 1808. Its compounds include slaked lime (calcium hydroxide, $Ca(OH)_2$); plaster of Paris (calcium sulfate, $CaSO_4.\frac{1}{2}H_2O$); calcium phosphate ($Ca_3(PO_4)_2$), the main constituent of animal bones; calcium hypochlorite ($CaOCl_2$), a bleaching agent; calcium nitrate ($Ca(NO_3)_2.4H_2O$), a nitrogenous fertilizer; calcium carbide (CaC_2), which reacts with water to give ethyne (acetylene); calcium cyanamide ($CaCN_2$), the basis of many pharmaceuticals, fertilizers, and plastics, including melamine; calcium cyanide ($Ca(CN)_2$), used in the extraction of gold and silver and in electroplating; and others used in baking powders and fillers for paints.

calcium carbonate $CaCO_3$, white solid, found in nature as limestone, marble, and chalk. It is a valuable resource, used in the making of iron, steel, cement, glass, slaked lime, bleaching powder, sodium carbonate and bicarbonate, and many other industrially useful substances.

calcium chloride $CaCl_2$, deliquescent white solid, melting point 772°C/1,422°F. It is produced as a by-product of the †Solvay process in the manufacture of sodium carbonate and is used as a drying agent.

calcium fluoride CaF_2, colorless solid, melting point 1,360°C/2,480°F. It occurs in nature as the mineral fluorspar.

calcium hydrogencarbonate $Ca(HCO_3)_2$, substance found in †hard water, formed when rainwater passes over limestone rock:

$$CaCO_{3\,(s)} + CO_{2\,(g)} + H_2O_{(l)} \rightarrow Ca(HCO_3)_{2\,(aq)}$$

When this water is boiled it reforms calcium carbonate, removing the hardness; this type of hardness is therefore known as temporary hardness.

calcium hydrogenphosphate $Ca(H_2PO_4)_2$, substance made by heating calcium phosphate with 70% sulfuric acid. It is more soluble in water than calcium phosphate, and is used as a fertilizer.

calcium hydroxide or **slaked lime**; $Ca(OH)_2$, white solid, slightly soluble in water. A solution of calcium hydroxide is called †limewater and is used in the laboratory to test for the presence of carbon dioxide. It is manufactured industrially by adding water to calcium oxide (quicklime) in a strongly †exothermic reaction:

$$CaO + H_2O \rightarrow Ca(OH)_2$$

It is used to reduce soil acidity and as a cheap alkali in many industrial processes.

calcium nitrate $Ca(NO_3)_2$, white crystalline compound that is very soluble in water. The solid decomposes on heating to form oxygen, brown nitrogen(IV) oxide gas, and the white solid calcium oxide:

$$2Ca(NO_3)_2 \rightarrow 2CaO + 4NO_2 + O_2$$

calcium oxide or **quicklime**; CaO, white solid compound, formed by heating †calcium carbonate:

$CaCO_3 \rightarrow CaO + CO_2$

When water is added it forms calcium hydroxide (slaked lime) in an †exothermic reaction:

$CaO + H_2O \rightarrow Ca(OH)_2$

It is a typical basic oxide, turning litmus blue.

calcium phosphate or **calcium orthophosphate**; $Ca_3(PO_4)_2$, white solid, the main constituent of animal bones. It occurs naturally as the mineral apatite and in rock phosphate, and is used in the preparation of phosphate fertilizers.

calcium sulfate $CaSO_4$, white solid compound, found in nature as gypsum and anhydrite. It dissolves slightly in water to form †hard water; this hardness is not removed by boiling, and hence is sometimes called permanent hardness.

calcium superphosphate common name for †calcium hydrogenphosphate.

calculus (Latin "pebble") branch of mathematics which uses the concept of a derivative (see †differentiation) to analyze the way in which the values of a †function vary. Calculus is probably the most widely used part of mathematics. Many real-life problems are analyzed by expressing one quantity as a function of another – position of a moving object as a function of time, temperature of an object as a function of distance from a heat source, force on an object as a function of distance from the source of the force, and so on – and calculus is concerned with such functions.

There are several branches of calculus. Differential and integral calculus, both dealing with small quantities, which during manipulation are made smaller and smaller, compose the **infinitesimal calculus**. **Differential equations** relate to the derivatives of a set of variables and may include the variables. Many give the mathematical models for physical phenomena such as †simple harmonic motion. Differential equations are solved generally by †integration, depending on their degree. If no analytical processes are available, integration can be performed numerically. Other branches of calculus include calculus of variations and calculus of errors.

Rise of Calculus http://www-history.mcs.st-andrews.ac.uk/history/HistTopics/The_rise_of_calculus.html

History of the development of calculus, from the time of the Greeks to the works of Cauchy in the 19th century. The site includes links to the biographical details of the mathematicians responsible for the rise of calculus, and also a brief list of reference publications for further information. Also included are several mathematical demonstrations on the use of calculus.

calculus of variations method of calculation for solving problems in which one of the unknowns cannot be expressed as a number or a finite set of numbers, but is representable as a curve, a function or a system of functions. (A classic problem in the subject is to show that a circle, among all curves of fixed length, encloses the maximum area.)

caldera in geology, a very large basin-shaped †crater. Calderas are found at the tops of volcanoes, where the original peak has collapsed into an empty chamber beneath. The basin, many times larger than the original volcanic vent, may be flooded, producing a crater lake, or the flat floor may contain a number of small volcanic cones, produced by volcanic activity after the collapse.

Typical calderas are Kilauea, Hawaii; Crater Lake, Oregon, USA; and the summit of Olympus Mons, on Mars. Some calderas are wrongly referred to as craters, such as Ngorongoro, Tanzania.

calendar division of the †year into months, weeks, and days and the method of ordering the years. From year one, an assumed date of the birth of Jesus, dates are calculated backwards (BC "before Christ" or BCE "before common era") and forwards (AD, Latin *anno Domini* "in the year of the Lord," or CE "common era"). The **lunar month** (period between one new moon and the next) naturally averages 29.5 days, but the Western calendar uses for convenience a **calendar month** with a complete number of days, 30 or 31 (February has 28). For adjustments, since there are slightly fewer than six extra hours a year left over, they are added to February as a 29th day every fourth year (**leap year**), century years being excepted unless they are divisible by 400. For example, 1896 was a leap year; 1900 was not.

The **month names** in most European languages were probably derived as follows: January from Janus, Roman god; February from *Februar*, Roman festival of purification; March from Mars, Roman god; April from Latin *aperire*, "to open"; May from Maia, Roman goddess; June from Juno, Roman goddess; July from Julius Caesar, Roman general; August from Augustus, Roman emperor; September, October, November, December (originally the seventh to tenth months) from the Latin words meaning seventh, eighth, ninth, and tenth, respectively.

The **days of the week** are Monday named after the Moon; Tuesday from Tiu or Tyr, Anglo-Saxon and Norse god; Wednesday from Woden or Odin, Norse god; Thursday from Thor, Norse god; Friday from Freya, Norse goddess; Saturday from Saturn, Roman god; and Sunday named after the Sun.

All early calendars except the ancient Egyptian were lunar. The word calendar comes from the Latin *Kalendae* or *calendae*, the first day of each month on which, in ancient Rome, solemn proclamation was made of the appearance of the new moon.

The **Western** or **Gregorian calendar** derives from the **Julian calendar** instituted by Julius Caesar in 46 BC. It was adjusted in 1582 by Pope Gregory XIII, who eliminated the accumulated error caused by a faulty calculation of the length of a year and avoided its recurrence by restricting century leap years to those divisible by 400. Other states only gradually changed from Old Style to New Style; Britain and its colonies adopted the Gregorian calendar in 1752, when the error amounted to 11 days, and 3 September 1752 became 14 September (at the same time the beginning of the year was put back from 25 March to 1 January). Russia did not adopt it until the October Revolution of 1917, so that the event (then 25 October) is currently celebrated on 7 November.

The **Jewish calendar** is a complex combination of lunar and solar cycles, varied by considerations of religious observance. A year may have 12 or 13 months, each of which normally alternates between 29 and 30 days; the New Year (Rosh Hashanah) falls between 5 September and 5 October. The calendar dates from the hypothetical creation of the world (taken as 7 October 3761 BC).

The **Chinese calendar** is lunar, with a cycle of 60 years. Both the traditional and, from 1911, the Western calendar are in use in China.

The **Muslim calendar**, also lunar, has 12 months of alternately 30 and 29 days, and a year of 354 days. This results in the calendar rotating around the seasons in a 30-year cycle. The era is counted as beginning on the day Muhammad fled from Mecca in AD 622.

Gregorian Calendar http://es.rice.edu/ES/humsoc/Galileo/Things/gregorian_calendar.html

Very detailed essay that explains the history of calculating and quantifying time, from the Egyptian calendar to the Gregorian calendar.

caliber internal diameter or a bore or pipe.

caliche in earth science, layer of salts or calcium carbonate from percolating groundwater that rises to the surface and evaporates. It is found in deserts on Earth and may have been observed on Mars.

California current cold ocean †current in the eastern Pacific Ocean flowing southwards down the west coast of North America. It is part of the North Pacific †gyre (a vast, circular movement of ocean water).

californium chemical symbol Cf, synthesized, radioactive, metallic element of the actinide series, atomic number 98, relative atomic mass 251. It is produced in very small quantities and used in nuclear reactors as a neutron source. The longest-lived isotope, Cf-251, has a half-life of 800 years. It is named after the state of California, where it was first synthesized in 1950 by US nuclear chemist Glenn Seaborg and his team at the University of California at Berkeley.

calima (Spanish "haze") dust cloud in Europe, coming from the Sahara Desert, which sometimes causes heatwaves and eye irritation.

Callisto second-largest moon of the planet Jupiter, 4,800 km/3,000 mi in diameter, orbiting every 16.7 days at a distance of 1.9 million km/1.2 million mi. Its surface is covered with large craters.

The US space probe Galileo detected oxygen on Callisto in 1997, suggesting that its surface is made of water ice. In 1998 Galileo's magnetometer measured electrical currents near Callisto's surface that are consistent with the existence of a buried salty ocean beneath the icy crust.

callus in botany, a tissue that forms at a damaged plant surface. Composed of large, thin-walled †parenchyma cells, it grows over and around the wound, eventually covering the exposed area.

In animals, a callus is a thickened pad of skin, formed where there is repeated rubbing against a hard surface. In humans, calluses often develop on the hands and feet of those involved in heavy manual work.

Calmette, (Léon Charles) Albert (1863–1933) French bacteriologist. A student of †Pasteur, he developed, with Camille †Guérin, the BCG vaccine against tuberculosis in 1921.

calomel technical name **mercury(I) chloride**; Hg_2Cl_2, white, heavy powder formerly used as a laxative, now used as a pesticide and fungicide.

caloric theory theory that †heat consists of a fluid called "caloric" that flows from hotter to colder bodies. It was abandoned by the mid-18th century.

calorie *c*.g.s. unit of heat, now replaced by the †joule (one calorie is approximately 4.2 joules). It is the heat required to raise the temperature of one gram of water by 1°C. In dietetics, the Calorie or kilocalorie is equal to 1,000 calories.

The kilocalorie measures the energy value of food in terms of its heat output: 28 g/1 oz of protein yields 120 kilocalories, of carbohydrate 110, of fat 270, and of alcohol 200.

calorific value amount of heat generated by a given mass of fuel when it is completely burned. It is measured in joules per kilogram. Calorific values are measured experimentally with a bomb calorimeter.

calorimeter instrument used in physics to measure various thermal properties, such as heat capacity or the heat produced by fuel. A simple calorimeter consists of a heavy copper vessel that is polished (to reduce heat losses by radiation) and covered with insulating material (to reduce losses by convection and conduction).

In a typical experiment, such as to measure the heat capacity of a piece of metal, the calorimeter is filled with water, whose temperature rise is measured using a thermometer when a known mass of the heated metal is immersed in it. Chemists use a bomb calorimeter to measure the heat produced by burning a fuel completely in oxygen.

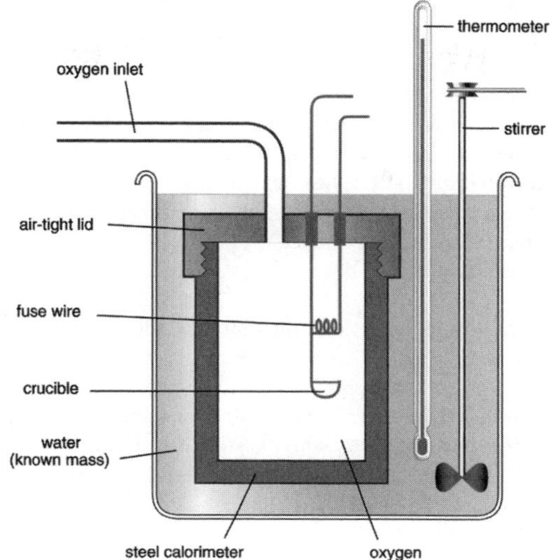

A bomb calorimeter is used to measure the energy that is given out when one mole of a substance is completely burned in oxygen. A known amount of the substance is placed in the crucible, oxygen is added at 20 atmospheres pressure, and then the substance is ignited using the heated wire. As the calorimeter heats, it heats the water. The rise in water temperature is measured and from this the heat generated by the burning substance is calculated. Helicon

calorimetry measurement of the heat-related constants of materials, for example, thermal capacity and †latent heat of vaporization.

Caloris Basin vast impact feature on the surface of †Mercury. It consists of a series of concentric circular mountain ranges up to 1,300 km/807 mi in diameter and 2 km/1.2 mi high. It is believed to have been caused by the impact of a giant asteroid early in the life of the Solar System and was discovered by the Mariner 10 spacecraft in 1974.

Calvin, Melvin (1911–1997) US chemist who was awarded the Nobel Prize for Chemistry in 1961 for his study of the assimilation of carbon dioxide by plants. Using radioactive carbon-14 as a tracer, he determined the biochemical processes of †photosynthesis, in which green plants use †chlorophyll to convert carbon dioxide and water into sugar and oxygen.

Calypso in astronomy, one of Saturn's moons, discovered in 1980. It is tiny, being 30 km/19 mi on its longer dimension, and has an irregular shape. Calypso shares its orbit with another tiny moon, †Telesto, and with the larger moon †Tethys.

calyptra in mosses and liverworts, a layer of cells that encloses and protects the young †sporophyte (spore capsule), forming a sheathlike hood around the capsule. The term is also used to describe the root cap, a layer of †parenchyma cells covering the end of a root that gives protection to the root tip as it grows through the soil. This is constantly being worn away and replaced by new cells from a special †meristem, the calyptrogen.

calyx collective term for the †sepals of a flower, forming the outermost whorl of the †perianth. It surrounds the other flower parts and protects them while in bud. In some flowers, for example, the campions *Silene*, the sepals are fused along their sides, forming a tubular calyx.

CAM acronym for **computer-aided manufacturing**, use of computers to control production processes; in particular, the control of machine tools and robots in factories. In some factories, the whole design and production system is automated by linking †CAD (computer-aided design) to CAM.

Linking flexible CAD/CAM manufacturing to computer-based sales and distribution methods, creating a computer integrated manufacturing system, makes it possible to produce semi-customized goods cheaply and in large numbers. Computers are commonly used to control industrial processes in many industries, from petrochemicals and steel, to electronic engineering and food processing.

cambium in botany, a layer of actively dividing cells (lateral †meristem), found within stems and roots, that gives rise to †secondary growth in perennial plants, causing an increase in girth. There are two main types of cambium: **vascular cambium**, which gives rise to secondary †xylem and †phloem tissues, and **cork cambium** (or phellogen), which gives rise to secondary cortex and cork tissues (see †bark).

Cambrian period period of geological time roughly 570–510 million years ago; the first period of the Paleozoic Era. All invertebrate animal life appeared, and marine algae were widespread. The **Cambrian Explosion** 530–520 million years ago saw the first appearance in the fossil record of modern animal phyla; the earliest fossils with hard shells, such as trilobites, date from this period.

The name comes from Cambria, the medieval Latin name for Wales, where Cambrian rocks are typically exposed and were first described.

Cambrian Period: Life Goes for a Spin
http://www.sciam.com/article.cfm?articleID=000D1CA5-AB38-1C76-9B81809EC588EF21&pageNumber=1&catID=4
Part of a larger site maintained by Scientific American, this page reports on the research of Joseph Kirschvink of the California Institute of Technology which suggests that the so-called "Cambrian Explosion" resulted from a sudden shifting of the Earth's crust. The text includes hyperlinks to further information, and there is also a list of related links, including one to figures, diagrams, and information from Kirschvink's research paper.

Cambridge Catalogues results of five intensive radio-astronomical surveys (1C, 2C, 3C, 4C, and 5C) under the direction of English astronomers Martin Ryle and Antony Hewish, during the 1950s, 1960s, and 1970s, at Cambridge.

Camelopardalis faint constellation of the north polar region, with a long, straggling shape, represented as a giraffe.

camouflage colors or structures that allow an animal to blend with its surroundings to avoid detection by other animals. Camouflage can take the form of matching the background color, of countershading (darker on top, lighter below, to counteract natural shadows), or of irregular patterns that break up the outline of the animal's body. More elaborate camouflage involves closely resembling a feature of the natural environment, as with the stick insect; this is closely akin to †mimicry. Camouflage is also important as a military technique, disguising either equipment, troops, or a position in order to conceal them from an enemy.

camphor $C_{10}H_{16}O$, volatile, aromatic †ketone substance obtained from the camphor tree *Cinnamomum camphora*. It is distilled from chips of the wood, and is used in insect repellents and medicinal inhalants and liniments, and in the manufacture of celluloid.

The camphor tree, a member of the family Lauraceae, is native to China, Taiwan, and Japan.

canal rays streams of positively charged †ions produced from an †anode in a discharge tube, in which gas is subjected to an electric discharge.

Canaries current cold ocean current in the North Atlantic Ocean flowing southwest from Spain along the northwest coast of Africa. It meets the northern equatorial current at a latitude of 20° N.

cancel in mathematics, to simplify a fraction or ratio by dividing both numerator and denominator by the same number (which must be a †common factor of both of them). For example, $5x/25$ cancels to $x/5$ when divided top and bottom by five.

cancer group of diseases characterized by abnormal proliferation of cells. Cancer (malignant) cells are usually degenerate, capable only of reproducing themselves (tumor formation). Malignant cells tend to spread from their site of origin by traveling through the bloodstream or lymphatic system. Cancer kills about 6 million people a year worldwide.

causes There are more than 100 types of cancer. Some, such as lung or bowel cancer, are common; others are rare. The likely causes remain unexplained. Triggering agents (†carcinogens) include chemicals such as those found in cigarette smoke, other forms of smoke, asbestos dust, vehicle exhaust fumes, and many industrial chemicals. Some viruses can also trigger the cancerous growth of cells (see †oncogene), as can X-rays and radioactivity. Dietary factors are important in some cancers; for example, lack of fiber in the diet may predispose people to bowel cancer and a diet high in animal fats and low in fresh vegetables and fruit can increase the risk of breast cancer. Psychological stress may also increase the risk of cancer, more so if the person concerned is not able to control the source of the stress.

cancer genes In some families there is a genetic tendency towards a particular type of cancer. In 1993 researchers isolated the first gene that predisposes individuals to cancer. About 1 in 200 people in the West carry the gene. If the gene mutates, those with the altered gene have a 70% chance of developing colon cancer, and female carriers have a 50% chance of developing cancer of the uterus.

In 1994 a gene that triggers breast cancer was identified. *BRCA1* was found to be responsible for almost half the cases of inherited breast cancer, and most cases of ovarian cancer. In 1995 a link between *BRCA1* and non-inherited breast cancer was discovered. Women with the gene have an 85% chance of developing breast or ovarian cancer during their lifetime. A second breast cancer gene *BRCA2* was identified later in 1995.

The commonest cancer in young men is testicular cancer, the incidence of which has been rising by 3% a year since 1974 (1998). In February 2000 British researchers announced the discovery of a mutant gene that increases incidence of testicular cancer 50-fold. The gene *TGCT1* is inherited from the mother and if a successful screening system is devised 95% of testicular cancer cases will be able to be predicted.

Cancer faintest of the zodiacal constellations (its brightest stars are fourth magnitude). It lies in the northern hemisphere between †Leo and †Gemini, and is represented as a crab. The Sun passes through the constellation during late July and early August. In astrology, the dates for Cancer are between about

22 June and 22 July (see †precession).

Cancer's most distinctive feature is the open star cluster †Praesepe, popularly known as the Beehive, visible to the naked eye as a nebulous patch.

Cancer Genome Project CGP, British project set up in 1999 with the aim of finding variations in the base sequence of human genes that predispose people to develop cancer. This will allow identification of the underlying genetic basis of cancer and understanding of the mechanisms by which cancer develops, and thus to help pinpoint how treatment can be most effective. The project is based at the Sanger Center, Cambridge, England.

candela symbol cd, SI unit of luminous intensity, which replaced the former units of candle and standard candle. It measures the brightness of a light itself rather than the amount of light falling on an object, which is called **illuminance** and measured in †lux.

One candela is defined as the luminous intensity in a given direction of a source that emits monochromatic radiation of frequency 540 x 10^{-12} Hz and whose radiant energy in that direction is 1/683 watt per steradian.

candle cylinder of wax (such as tallow or paraffin wax) with a central wick of string. A flame applied to the end of the wick melts the wax, thereby producing a luminous flame. The wick is treated with a substance such as alum so that it carbonizes but does not rapidly burn out.

Candles and oil lamps were an early form of artificial lighting. Accurately made candles – which burned at a steady rate – were calibrated along their lengths and used as a type of clock. The candle was also the name of a unit of luminous intensity, replaced in 1940 by the †candela (cd), equal to $^1/_{60}$ of the luminance of 1 sq cm of a black body radiator at a temperature of 2,042 K (the temperature of solidification of platinum).

candle-power former scientific measure of luminous intensity, defined as the brightness of a standard candle (made in a specified way) burning at a certain rate. It was replaced by the †candela.

Candolle, Augustin Pyrame de (1778–1841) Swiss botanist who coined the term "taxonomy" to mean the classification of plants on the basis of their gross anatomy in his book *Théorie élémentaire de la botanique* (1813). He posited that plant relationships can be determined by the symmetry of their sexual organs and introduced the concept of †homologous parts, the idea that an organ or structure possessed by different plants may indicate a common ancestor.

cane sugar in chemistry, natural form of sucrose that is extracted from sugar cane, chemically identical to that extracted from the sugar beet plant.

Canes Venatici constellation of the northern hemisphere near †Ursa Major, identified with the hunting dogs of †Boötes, the herder. Its stars are faint, and it contains the Whirlpool galaxy (M51), the first spiral galaxy to be recognized.

It contains many objects of telescopic interest, including the relatively bright †globular cluster M3. The brightest star, a third magnitude double, is called †Cor Caroli or Alpha Canum Venaticorum.

canine in mammalian carnivores, any of the long, often pointed teeth found at the front of the mouth between the incisors and premolars. Canine teeth are used for catching prey, for killing, and for tearing flesh. They are absent in herbivores such as rabbits and sheep, and are much reduced in humans.

Canis Major brilliant constellation of the southern hemisphere, represented (with Canis Minor) as one of the two dogs following at the heel of †Orion. Its main star, †Sirius, is the brightest star in the night sky. Epsilon Canis Majoris is also of the first magnitude, and there are three second magnitude stars.

Canis Minor small constellation along the celestial equator (see †celestial sphere), represented as the smaller of the two dogs of †Orion (the other dog being †Canis Major). Its brightest star is the first magnitude †Procyon.

Procyon and Beta Canis Minoris form what the Arabs called "the Short Cubit," in contrast to "the Long Cubit" formed by †Castor and †Pollux (Alpha and Beta Geminorum).

Cannizaro reaction in chemistry, reaction involving two †aldehydes in an alkaline solution, in which one aldehyde is reduced to become an alcohol and the other is oxidized to produce an acid. The reaction is named after the Italian chemist Stanislao †Cannizzaro.

Cannizzaro, Stanislao (1826–1910) Italian chemist who revived interest in the work of †Avogadro that had, in 1811, revealed the difference between †atoms and †molecules, and so established atomic and molecular weights as the basis of chemical calculations.

Cannon, Annie Jump (1863–1941) US astronomer. She carried out revolutionary work on the classification of stars by examining their spectra. Her system, still used today, has spectra arranged according to temperature into categories labeled O, B, A, F, G, K, M, R, N, and S. O-type stars are the hottest, with surface temperatures ranging from 25,000 to 50,000 K (from 24,700°C/44,450°F to

49,700°C/89,500°F).

Canopus or **Alpha Carinae**, second-brightest star in the night sky (after Sirius), lying in the southern constellation †Carina. It is a yellow-white supergiant about 100 light years from the Sun, and thousands of times more luminous than the Sun.

canopy continuous layer of dense treetop foliage in woodland or forest. The leaves filter the sunlight, reaching the lower layers of foliage, so the nearer to the ground plants grow, the less sunlight they receive. The canopy provides a †habitat for diverse specially adapted species, for example those found in the †rainforest canopy include various monkey species, bats, numerous insects, and birds.

Cantor, Georg Ferdinand Ludwig Philipp (1845–1918) German mathematician who followed his work on number theory and trigonometry by considering the foundations of mathematics. He defined real numbers and produced a treatment of irrational numbers using a series of transfinite numbers. Cantor's set theory has been used in the development of topology and real function theory.

canyon (Spanish *cañon* "tube") deep, narrow valley or gorge running through mountains. Canyons are formed by stream down-cutting, usually in arid areas, where the rate of down-cutting is greater than the rate of weathering, and where the stream or river receives water from outside the area.

There are many canyons in the western USA and in Mexico, for example the Grand Canyon of the Colorado River in Arizona, the canyon in Yellowstone National Park, and the Black Canyon in Colorado.

capacitance, electrical property of a capacitor that determines how much charge can be stored in it for a given potential difference between its terminals. It is equal to the ratio of the electrical charge stored to the potential difference. The SI unit of capacitance is the †farad, but most capacitors have much smaller capacitances, and the microfarad (a millionth of a farad) is the commonly used practical unit.

capacitor or **condenser**, device for storing electric charge, used in electronic circuits; it consists of two or more metal plates separated by an insulating layer called a dielectric (see †capacitance).

Its **capacitance** is the ratio of the charge stored on either plate to the potential difference between the plates.

capacity alternative term for †volume, generally used to refer to the amount of liquid or gas that may be held in a container. Units of capacity include liter and milliliter (metric); pint and gallon.

Capella or **Alpha Aurigae**, brightest star in the constellation †Auriga and the sixth-brightest star in the night sky. It is a visual and spectroscopic binary that consists of a pair of yellow-giant stars 42 light years from the Sun, orbiting each other every 104 days.

Its Latin name means the "the Little Nanny Goat": its kids are the three adjacent stars Epsilon, Eta, and Zeta Aurigae.

capillarity spontaneous movement of liquids up or down narrow tubes, or capillaries. The movement is due to unbalanced molecular attraction at the boundary between the liquid and the tube. If liquid molecules near the boundary are more strongly attracted to molecules in the material of the tube than to other nearby liquid molecules, the liquid will rise in the tube. If liquid molecules are less attracted to the material of the tube than to other liquid molecules, the liquid will fall.

capillary in biology, narrowest blood vessel in vertebrates measuring 0.008–0.02 mm in diameter, barely wider than a †red blood cell. Capillaries are distributed as **beds**, complex networks connecting †arteries and †veins. The function of capillaries is to exchange materials with their surroundings. For this reason, capillary walls are extremely thin, consisting of a single layer of cells through which nutrients, dissolved gases, and waste products can easily pass. This makes the capillaries the main area of exchange between the fluid (†lymph) bathing body tissues and the blood. They provide a large surface area in order to maximize †diffusion.

All body cells lie close to capillaries so they can receive the food and oxygen they require from the capillaries. Networks of capillaries are especially extensive in certain parts of the body. For example, in the †lungs they surround the air sacs (alveoli), taking up †oxygen and releasing †carbon dioxide. They are also found in the lining of the †gut – in little fingerlike projections called villi – where they absorb the products of the †digestion of food.

capillary in physics, a very narrow, thick-walled tube, usually made of glass, such as in a thermometer. Properties of fluids, such as surface tension and viscosity, can be studied using capillary tubes.

capillary condensation in chemistry, mechanism by which vapors can condense inside certain solids due to the relationship between the vapor pressure of the gas and the size of the capillary walls it is wetting. The ability of silica gel to absorb water is explained by this mechanism.

capitulum in botany, a flattened or rounded head (inflorescence) of numerous, small, stalkless flowers. The capitulum is surrounded by a circlet of petal-like bracts and has the appearance of a large, single flower.

Capricornus zodiacal constellation in the southern hemisphere next to †Sagittarius. It is represented as a sea-goat, and its brightest stars are third magnitude. The Sun passes through it from late January to mid-February. In astrology, the dates for Capricornus (popularly known as Capricorn) are between about 22 December and 19 January (see †precession).

capsule in botany, a dry, usually many-seeded fruit formed from an ovary composed of two or more fused †carpels, which splits open to release the seeds. The same term is used for the spore-containing structure of mosses and liverworts; this is borne at the top of a long stalk or seta.

Capsules burst open (dehisce) in various ways, including lengthwise, by a transverse lid – for example, scarlet pimpernel *Anagallis arvensis* – or by a number of pores, either towards the top of the capsule, as in the poppy *Papaver*, or near the base, as in certain species of bellflower *Campanula*.

capture in physics, of the combination of one particle and another, such as an electron and a positive ion (to form a neutral atom) or of a neutron and an atomic nucleus (with the emission of a gamma-ray photon).

captured rotation or **synchronous rotation**, in astronomy, circumstance in which one body in orbit around another, such as the moon of a planet, rotates on its axis in the same time as it takes to complete one orbit. As a result, the orbiting body keeps one face permanently turned towards the body about which it is orbiting. An example is the rotation of our own †Moon, which arises because of the tidal effects of the Earth over a long period of time.

caramel complex mixture of substances produced by heating sugars, without charring, until they turn brown. Caramel is used as coloring and flavoring in foods. Its production in the manufacture of sugar confection gives rise to a toffeelike sweet of the same name.

The intricate chemical reactions involved in the production of caramel (caramelization) are not fully understood, but are known to result in the formation of a number of compounds. Two compounds in particular (acetylformoin and 4-hydroxy-2,5-dimethyl-3-furanone) are thought to contribute to caramel's characteristic flavor.

Commercially, the caramelization process is speeded up by the addition of selected †amino acids.

carapace protective covering of many animals, particularly the arched bony plate characteristic of the order Chelonia (tortoises, terrapins, and turtles), and the shield that protects the fore parts of crustaceans, such as crabs.

carat (Arabic *quirrat* "seed") unit for measuring the mass of precious stones; it is equal to 0.2 g/0.00705 oz, and is part of the troy system of weights. It is also the unit of purity in gold (US "karat"). Pure gold is 24-carat; 22-carat (the purest used in jewelry) is 22 parts gold and two parts alloy (to give greater strength); 18-carat is 75% gold.

Originally, one carat was the weight of a carob seed.

carbanion in chemistry, highly reactive, negatively charged intermediate species produced by the breakage of certain C–C, C–H, C–metal, and C–halogen bonds. Carbanions are formed during †Grignard reactions discovered by French chemist Victor †Grignard.

carbene in chemistry, highly reactive molecule containing a divalent carbon atom, produced as an intermediate species during certain organic chemical reactions.

carbenium ion or **carbonium ion**, in chemistry, positively charged species containing a trivalent carbon atom, R_3C^+, where R is an †alkyl group. It is thought to be an intermediate species in many organic reactions. Although usually short-lived, some carbenium ions are stable and can be isolated.

carbide compound of carbon and one other chemical element, usually a metal, silicon, or boron.

Calcium carbide (CaC_2) can be used as the starting material for many basic organic chemical syntheses, by the addition of water and generation of ethyne (acetylene). Some metallic carbides are used in engineering because of their extreme hardness and strength. Tungsten carbide is an essential ingredient of carbide tools and high-speed tools. The "carbide process" was used during World War II to make organic chemicals from coal rather than from oil.

carbocyclic in chemistry, generic term for any organic compound that is made up of one or more closed rings of carbon atoms, for example cyclohexane and naphthalene.

carbohydrate chemical compounds composed of carbon, hydrogen, and oxygen, with the basic formula $C_m(H_2O)_n$, and related compounds with the same basic structure but modified †functional groups. They are important to living organisms and, as sugar and starch, are an important part of a balanced human diet, providing energy for life processes including growth and movement. Excess carbohydrate intake can be converted into fat and stored in the body.

The simplest carbohydrates are sugars (**monosaccharides**, such as glucose and fructose, and **disaccharides**, such as sucrose), which are soluble compounds, some with a sweet taste. When these basic sugar units are joined together in long chains or branching structures they form **polysaccharides**, such as starch and glycogen.

The simple sugar called †glucose is used in living organisms in †respiration to release energy that can be used for life processes. The sugar we use to sweeten foods is a complex sugar called sucrose. Starch is used by plants as a way of storing energy-rich food, and animals that eat plants take advantage of this. It is a common part of the human diet because it is major constituent of wheat, rice, potatoes, and corn. Humans use a carbohydrate called glycogen as an energy store. This is also made from glucose. It is stored in the †liver and †muscles and can be broken down to supply body cells with glucose. The hormone †insulin regulates this process, causing glucose to be taken in the bloodstream to the liver to be converted to glycogen.

carbolic acid common name for the aromatic compound †phenol.

carbon chemical symbol C, (Latin *carbo, carbonaris* "coal") non-metallic element, atomic number 6, relative atomic mass 12.011. It occurs on its own as diamond, graphite, and as fullerenes (the allotropes), as compounds in carbonaceous rocks such as chalk and limestone, as carbon dioxide in the atmosphere, as hydrocarbons in petroleum, coal, and natural gas, and as a constituent of all organic substances.

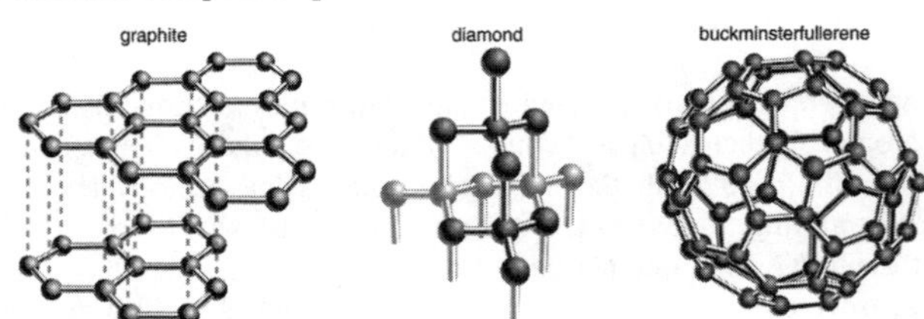

carbon Carbon has three allotropes: diamond, graphite, and the fullerenes. Diamond is strong because each carbon atom is linked to four other carbon atoms. Graphite is made up of layers that slide across one another (giving graphite its qualities as a lubricator); each layer is a giant molecule. In the fullerenes, the carbon atoms form spherical cages. Buckminsterfullerene (shown here) has 60 atoms. Other fullerenes, with 28, 32, 50, 70, and 76 carbon atoms, have also been identified. Helicon

In its amorphous form, it is familiar as coal, charcoal, and soot. The atoms of carbon can link with one another in rings or chains, giving rise to innumerable complex compounds. Of the inorganic carbon compounds, the chief ones are **carbon dioxide** (CO_2), a colorless gas formed when carbon is burned in an adequate supply of air; and **carbon monoxide** (CO), formed when carbon is oxidized in a limited supply of air. **Carbon disulphide** (CS_2) is a dense liquid with a sweetish odor when pure. Another group of compounds is the **carbon halides**, including †carbon tetrachloride (tetrachloromethane, CCl_4).

When added to steel, carbon forms a wide range of alloys with useful properties. In pure form, it is used as a moderator in nuclear reactors; as colloidal graphite it is a good lubricant and, when deposited on a surface in a vacuum, reduces photoelectric and secondary emission of electrons. Carbon is used as a fuel in the form of coal or coke. The radioactive isotope carbon-14 (half-life 5,730 years) is used as a tracer in biological research and in †radiocarbon dating. Analysis of interstellar dust has led to the discovery of discrete carbon molecules, each containing 60 carbon atoms. The C_{60} molecules have been named †buckminsterfullerenes because of their structural similarity to the geodesic domes designed by US architect and engineer Richard Buckminster Fuller.

Exploring the Material World: Three Classroom Teaching Modules
http://www.lbl.gov/MicroWorlds/module_index.html
Attractive presentation on materials, including an interactive tour of current research in the materials sciences. Choose a module to learn how graphite and diamonds can both be forms of the same element, about Kevlar (the material so strong it can stop bullets), and about the delicate chemical balance of marshes and wetlands.

carbonaceous chondrite in astronomy, rare type of meteorite containing up to 5% carbon compounds and chondrules, small round silicaceous inclusions, which give it an overall composition similar to that of the Sun (and believed to have formed at about the same time).

carbonate CO_3^{2-}, ion formed when carbon dioxide dissolves in water; any salt formed by this ion and another chemical element, usually a metal.

Carbon dioxide (CO_2) dissolves sparingly in water (for example, when rain falls through the air) to form carbonic acid (H_2CO_3), which unites with various basic substances to form carbonates. Calcium carbonate ($CaCO_3$) (chalk, limestone, and marble) is one of the most abundant carbonates known, being a constituent of mollusk shells and the hard outer skeletons of crustaceans.

Carbonates give off carbon dioxide when heated or treated with dilute acids.

$CaCO_{3(s)} \rightarrow CaO_{(s)} + CO_{2(g)}$

$CaCo_{3(s)} + 2HCl_{(aq)} \rightarrow CaCl_{2(aq)} + H_2O_{(l)} + CO_{2(g)}$

The latter reaction is used as the laboratory test for the presence of the ion, as it gives an immediate effervescence, with the gas turning limewater (a solution of calcium hydroxide, $Ca(OH)_2$) milky. See †calcium carbonate.

carbonated water water in which carbon dioxide is dissolved under pressure. It forms the basis of many fizzy soft drinks such as soda water and lemonade.

carbonation in earth science, a form of †chemical weathering caused by rainwater that has absorbed carbon dioxide from the atmosphere and formed a weak carbonic acid. The slightly acidic rainwater is then capable of dissolving certain minerals in rocks. Water can also pick up acids when it passes through soil. This water – enriched with organic acid – is also capable of dissolving rock. †Limestone is particularly vulnerable to this form of weathering.

carbon cycle sequence by which †carbon circulates and is recycled through the natural world. Carbon is usually found in a carbon compound of one sort or another and so the carbon cycle is really about the cycling of carbon compounds. Carbon dioxide is released into the atmosphere by most living organisms as a result of †respiration. The CO_2 is taken up and converted into high-energy chemicals – †glucose and other †carbohydrates – during †photosynthesis by plants; the oxygen component is released back into the atmosphere. Some glucose is used by the plant and some is converted into other carbon compounds, making new tissues. However, some of these compounds can be transferred to other organisms. An animal may eat the plant and that animal may be eaten and so on down the food chain. Carbon is also released through the †decomposition of dead plant and animal matter, and the burning of †fossil fuels such as †coal and †oil, which produce carbon dioxide that is released into the atmosphere. The oceans absorb 25–40% of all carbon dioxide released into the atmosphere.

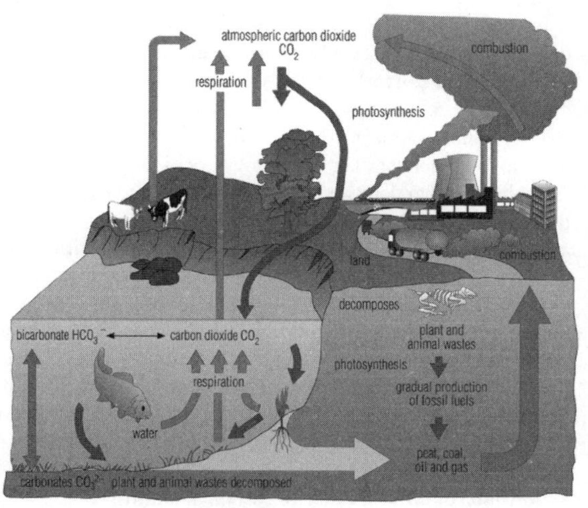

carbon cycle The carbon cycle is necessary for the continuation of life. Since there is only a limited amount of carbon in the Earth and its atmosphere, carbon must be continuously recycled if life is to continue. Other chemicals necessary for life – nitrogen, sulfur, and phosphorus, for example – also circulate in natural cycles. Helicon

The carbon cycle is in danger of being disrupted by the increased burning of fossil fuels, and the destruction of large areas of tropical forests. The rising levels of carbon dioxide in the atmosphere are probably increasing the temperature on Earth (enhanced †greenhouse effect). It is thought that by limiting the production of carbon dioxide through human activities we can slow the rate at which temperatures on Earth will rise.

Carbon Cycle http://www.windows.ucar.edu/tour/link=/earth/Water/co2_cycle.html

Part of a larger site "Windows to the Universe," this site contains three levels of explanation of the carbon cycle. The site assumes you are an "Intermediate" student but you can change this to "Beginner" or "Advanced."

Chemical Carousel: A Trip Around the Carbon Cycle http://www.thinkquest.org/library/lib/site_sum_outside.html

Captain Carbon invites you to explore the scientific process of the carbon cycle and explains the role of carbon in everyday life.

carbon cycle in astrophysics, sequence of nuclear fusion reactions in which carbon atoms act as a catalyst to convert four hydrogen atoms into one helium atom with the release of energy. The carbon cycle is the dominant energy source for ordinary stars of mass greater than about 1.5 times the mass of the Sun. Nitrogen and oxygen are also involved in the sequence so it is sometimes known as the **carbon-nitrogen-oxygen cycle** or **CNO cycle**.

carbon dioxide CO_2, colorless, odorless gas, slightly soluble in †water, and denser than air. It is formed by the complete oxidation of carbon. Carbon dioxide is produced by living things during the processes of †respiration and the †decomposition of organic matter, and it is used up during †photosynthesis. It therefore plays a vital role in the †carbon cycle.

Solid carbon dioxide is called dry ice, as it changes directly from a solid to a gas (sublimes) on warming. It is used as a coolant in its solid form and in the chemical industry.

Its increasing quantity in the atmosphere contributes to the †greenhouse effect and †global warming.

Britain has 1% of the world's population, yet it produces 3% of CO_2 emissions; the USA has 5% of the world's population and produces 25% of CO_2 emissions. Annual releases of carbon dioxide reached 23 billion tonnes in 1997. According to a 1997 estimate by the World Energy Council, carbon dioxide emissions rose by 7.8% between 1986 and 1996.

carbon fiber fine, black, silky, continuous filament of pure carbon produced by heating organic fibers, such as cellulose, in an inert atmosphere, and used for reinforcing plastics, epoxy, and polyester resins. The resulting composite is very stiff and, weight for weight, has four times the strength of high-tensile steel. It is used in the aerospace industry, cars, and electrical and sports equipment.

carbonic acid H_2CO_3, weak, dibasic acid formed by dissolving carbon dioxide in water:

$H_2O + CO_2 \longleftrightarrow H_2CO_3$

It forms two series of salts: †carbonates and †hydrogencarbonates. Fizzy drinks are made by dissolving carbon dioxide in water under pressure; soda water is a solution of carbonic acid.

Carboniferous period period of geological time roughly 362.5 to 290 million years ago, the fifth period of the Paleozoic Era. In the USA it is divided into two periods: the Mississippian (lower) and the Pennsylvanian (upper).

Typical of the lower-Carboniferous rocks are shallow-water †limestones, while upper-Carboniferous rocks have †delta deposits with †coal (hence the name). Amphibians were abundant, and reptiles evolved during this period.

carbonization in chemistry, destructive decomposition of carbon-based substances by heating in the absence of air. The processes by which coal is converted to coke and wood is converted to charcoal are examples of carbonization.

carbon monoxide CO, colorless, odorless gas formed when carbon is oxidized in a limited supply of air. It is a poisonous constituent of car exhaust fumes, forming a stable compound with hemoglobin in the blood, thus preventing the hemoglobin from transporting oxygen to the body tissues.

In industry, carbon monoxide is used as a reducing agent in metallurgical processes – for example, in the extraction of iron in †blast furnaces – and is a constituent of cheap fuels such as water gas. It burns in air with a luminous blue flame to form carbon dioxide.

carbon–nitrogen cycle use of carbon and nitrogen as intermediates in the †nuclear fusion process of the Sun. Cooler stars undergo the †proton–proton cycle.

carbon sequestration disposal of carbon dioxide waste in solid or liquid form. From 1993 energy conglomerates such as Shell, Exxon, and British Coal have been researching ways to reduce their carbon dioxide emissions by developing efficient technologies to trap the gas and store it securely – for example, by burying it or dumping it in the oceans. See also †greenhouse effect.

carbon tetrachloride former name for the chlorinated organic compound †tetrachloromethane.

carbonylation in chemistry, reaction involving carbon monoxide and organic or organometallic compounds to produce products that contain at least one †carbonyl group, $>C = O$.

carbonyl group in chemistry, group consisting of one carbon atom double-bonded to an oxygen atom, $>C = O$. It is the key group in all †aldehydes and †ketones.

Carborundum trademark for a very hard, black abrasive, consisting of silicon carbide (SiC), an artificial compound of carbon and silicon. It is harder than †corundum but not as hard as †diamond.

It was first produced in 1891 by US chemist Edward Acheson (1856–1931).

carboxyl group –COOH, in organic chemistry, the acidic functional group that determines the properties of fatty acids (carboxylic acids) and amino acids.

carboxylic acid organic acid containing the carboxyl group (–COOH) attached to another group (R), which can be hydrogen (giving methanoic acid, HCOOH) or a larger group (up to 24 carbon atoms). When R is a straight-chain alkyl group (such as CH_3 or CH_3CH_2), the acid is known as a †fatty acid.

Examples of carboxylic acids include ethanoic (acetic) acid, found in vinegar, malic acid, found in rhubarb, and citric acid, contained in oranges and lemons.

carburation any process involving chemical combination with carbon, especially the mixing or charging of a gas, such as air, with volatile compounds of carbon (gasoline, kerosene, or fuel oil) in order to increase potential heat energy during combustion. Carburation applies to combustion in the cylinders of reciprocating gasoline engines of the types used in aircraft, road vehicles, or marine vessels. The device by which the liquid fuel is atomized and mixed with air is called a **carburetor**.

carcinogen any agent that increases the chance of a cell becoming cancerous (see †cancer), including various chemical compounds, some viruses, X-rays, and other forms of ionizing radiation. The term is often used more narrowly to mean chemical carcinogens only.

carcinoma malignant †tumor arising from the skin, the glandular tissues, or the mucous membranes that line the gut and lungs.

cardiac pertaining to the †heart.

cardinal number in mathematics, one of the series of numbers 0, 1, 2, 3, 4, ... Cardinal numbers relate to quantity, whereas ordinal numbers (first, second, third, fourth, ...) relate to order.

cardioid heart-shaped curve traced out by a point on the circumference of a circle, resulting from the circle rolling around the edge of another circle of the same diameter.

The polar equation of the cardioid is of the form:

$r = a(1 + \cos q)$ where the polar coordinates of the point are (r, q) and a is a constant.

cardioid The cardioid is the curve formed when one circle rolls around the edge of another circle of the same size. It is named after its heart shape. Helicon

Carina constellation of the southern hemisphere, represented as a ship's keel. Its brightest star is †Canopus, the second brightest in the night sky; it also contains Eta Carinae, a massive and highly luminous star embedded in a gas cloud, perhaps 8,000 light years away from the Sun.

Carina was formerly regarded as part of †Argo, and is situated in one of the brightest parts of the †Milky Way.

Carlsson, Arvid (1923–) Swedish pharmacologist who with US neurobiologists Paul †Greengard and Eric R †Kandel shared the Nobel Prize for Physiology or Medicine in 2000 for their discoveries concerning signal transduction in the nervous system. Carlsson discovered that †dopamine is an important transmitter in the brain, especially in the †basal ganglia, a part of the brain responsible for the control of movement.

carnassial tooth one of a powerful scissorlike pair of molars, found in all mammalian carnivores except seals. Carnassials are formed from an upper premolar and lower molar, and are shaped to produce a sharp cutting surface. Carnivores such as dogs transfer meat to the back of the mouth, where the carnassials slice up the food ready for swallowing.

carnelian semi-precious gemstone variety of †chalcedony consisting of quartz (silica) with iron impurities, which give it a translucent red color. It is found mainly in Brazil, India, and Japan.

carnivore organism that eats other animals. In zoology, a mammal of the order Carnivora.

Carnivores have the greatest range of body size of any mammalian order, from the 100 g/3.5 oz weasel to the 800 kg/1,764 lb polar bear.

The characteristics of the Carnivora are sharp teeth, small incisors, a well-developed brain, a simple stomach, reduced or absent cecum, and incomplete or absent clavicles (collarbones); there are never less than four toes on each foot; the scaphoid and lunar bones are fused in the hand; and the claws are generally sharp and powerful.

Carnot, (Nicolas Léonard) Sadi (1796–1832) French scientist and military engineer who founded the science of thermodynamics. His pioneering work was *Reflexions sur la puissance motrice du feu/On the Motive Power of Fire*, which considered the changes that would take place in an idealized, frictionless steam engine.

Carnot cycle series of changes in the physical condition of a gas in a reversible heat engine, necessarily in the following order: (1) isothermal expansion (without change of temperature), (2) adiabatic expansion (without change of heat content), (3) isothermal compression, and (4) adiabatic compression.

The principles derived from a study of this cycle are important in the fundamentals of heat and †thermodynamics.

carnotite potassium uranium vanadate, $K_2(UO_2)_2(VO_4)_2.3H_2O$, a radioactive ore of vanadium and uranium with traces of radium.

A yellow powdery mineral, it is mined chiefly in the Colorado Plateau, USA; Radium Hill, Australia; and Shaba, the Democratic Republic of Congo (formerly Zaire).

Carnot's theorem theorem in †thermodynamics stating that the efficiency of any (reversible) heat engine depends only on the temperature range through which the machine operates. It was determined by French physicist Sadi Carnot.

carotene naturally occurring pigment of the †carotenoid group. Carotenes produce the orange, yellow, and red colors of carrots, tomatoes, oranges, and crustaceans.

carotenoid any of a group of yellow, orange, red, or brown pigments found in many living organisms, particularly in the †chloroplasts of plants. There are two main types, the **carotenes** and the **xanthophylls**.

Both types are long-chain lipids (†fats).

Some carotenoids act as accessory pigments in †photosynthesis, and in certain algae they are the principal light-absorbing pigments functioning more efficiently than †chlorophyll in low-intensity light. Carotenoids can also occur in organs such as petals, roots, and fruits, giving them their characteristic color, as in the yellow and orange petals of wallflowers *Cheiranthus*. They are also responsible for the autumn colors of leaves, persisting longer than the green chlorophyll, which masks them during the summer.

Carothers, Wallace Hume (1896–1937) US chemist who carried out research into polymerization. He discovered that some polymers were fiber-forming, and in 1931 he produced neoprene, one of the first synthetic rubbers, and in 1935, †nylon.

carotid artery one of a pair of major blood vessels, one on each side of the neck, supplying blood to the head.

carpel female reproductive unit in flowering plants (†angiosperms). It usually comprises an †ovary containing one or more ovules, the stalk or style, and a †stigma at its top which receives the pollen. A flower may have one or more carpels, and they may be separate or fused together. Collectively the carpels of a flower are known as the †gynoecium.

carrier wave wave of †electromagnetic radiation (a radio wave) of constant frequency and †amplitude used in †radio communication. Modulation of the wave allows information to be carried by it.

Carrington, Richard Christopher (1826–1875) English astronomer. By studying sunspots, he established the Sun's axis and rotation. He was the first to record the observation of a solar flare in 1859.

carrying capacity in ecology, the maximum number of animals of a given species that a particular †habitat can support. If the carrying capacity of an ecosystem is exceeded by overpopulation, there will be insufficient resources and one or more species will decline until an equilibrium, or †balance of nature, is restored. Similarly, if the number of species in an environment is less than the carrying capacity, the population will tend to increase until it balances the available resources. Human interference frequently causes disruption to the carrying capacity of an area, for instance by the establishment of too many grazing animals on grassland, the over-culling of a species, or the introduction of a non-indigenous species into an area.

Carson, Rachel Louise (1907–1964) US biologist, writer, and conservationist. Her book *Silent Spring* (1962), attacking the indiscriminate use of pesticides, inspired the creation of the modern environmental movement.

Rachel Carson Homestead http://www.rachelcarson.org/

Information on the life and legacy of the pioneering ecologist from the trust preserving Carson's childhood home. There is a biography of Carson, details of books by and about Carson, and full details of the work of the conservation organizations continuing her work.

Cartesian coordinates in †coordinate geometry, components used to define the position of a point by its perpendicular distance from a set of two or more axes, or reference lines. For a two-dimensional area defined by two axes at right angles (a horizontal *x*-axis and a vertical *y*-axis), the coordinates of a point are given by its perpendicular distances from the *y*-axis and *x*-axis, written in the form (x,y). For example, a point P that lies three units from the *y*-axis and four units from the *x*-axis has Cartesian coordinates (3,4) (see †abscissa and †ordinate).

The Cartesian coordinate system can be extended to any finite number of dimensions (axes), and is used thus in theoretical mathematics. Coordinates can be negative numbers, or a positive and a negative; for example (–4, –7), where the point would be to the left of and below zero on the axes. In three-dimensional coordinate geometry, points are located with reference to a third, *z*-axis, mutually at right angles to the *x* and *y* axes.

Cartesian coordinates are named after the French mathematician, René Descartes. The system is useful in creating technical drawings of machines or buildings, and in computer-aided design (†CAD).

cartilage flexible bluish-white †connective tissue made up of the protein collagen. In cartilaginous fish it forms the skeleton; in other vertebrates it forms the greater part of the embryonic skeleton, and is replaced by †bone in the course of development, except in areas of wear such as bone endings, and the discs between the backbones. It also forms structural tissue in the larynx, nose, and external ear of mammals.

Cartilage does not heal itself, so where injury is severe the joint may need to be replaced surgically. In a 1994 trial, Swedish doctors repaired damaged knee joints by implanting cells cultured from the patient's own cartilage. In 1999 US chemists created an artificial liquid cartilage for use in repairing torn tissue. The cartilage is injected into a wound or damaged joint and will harden with exposure to ultraviolet light. Clinical trials commenced in 2000.

cartography art and practice of drawing †maps, originally with pens and drawing boards, but now mostly with computer-aided drafting programs.

Mathematics of Cartography http://math.rice.edu/~lanius/pres/map/

History of maps and details of the mathematics behind mapmaking. There are math-related mapping problems to solve and a list of cartographical links.

Cartwheel galaxy in astronomy, galaxy that appears to have been formed when two spiral galaxies collided. It is 170,000 light years across and located about 500 million light years distant. It is a rich source of supernovae.

caryopsis dry, one-seeded †fruit in which the wall of the seed becomes fused to the carpel wall during its development. It is a type of †achene, and therefore develops from one ovary and does not split open to release the seed. Caryopses are typical of members of the grass family (Gramineae), including the cereals.

casein main protein of milk, from which it can be separated by the action of acid, the enzyme rennin, or bacteria (souring); it is also the main protein in cheese. Casein is used as a protein supplement in the treatment of malnutrition. It is used commercially in cosmetics, glues, and as a sizing for coating paper.

casparation strip in botany, layer of thickening (mainly †suberin) in the cell walls of the endodermis that encircles the vascular tissue in the roots of seed plants (gymnosperms and angiosperms), separating it from the outer cortex.

Cassegrain (*c.* 1650–1700) French inventor of the system of mirrors used within many modern reflecting telescopes and sometimes used in large refraction telescopes.

Cassegrain telescope or **Cassegrain reflector**, type of reflecting †telescope in which light collected by a concave primary mirror is reflected onto a convex secondary mirror, which in turn directs it back through a hole in the primary mirror to a focus behind it. As a result, the telescope tube can be kept short, allowing equipment for analyzing and recording starlight to be mounted behind the main mirror. All modern large astronomical telescopes are of the Cassegrain type.

It is named after the 17th-century French astronomer †Cassegrain who first devised it as an improvement to the simpler †Newtonian telescope.

Cassini (lived 17th–18th century) Italian family of astronomers and geographers, of whom four generations had charge of the Paris Observatory. **Giovanni Domenico Cassini** (1625–1712) was director from 1669. **Jacques Cassini** (1677–1756), the son of Giovanni Domenico, succeeded his father in 1712. The other members of the family were **César Cassini** (1714–1784), the son of Jacques; and **Jacques Dominique, comte de Cassini** (1748–1845), the son of César. The last was also a mathematician and geographer, who took an active part in the division of France into *départements*, and completed his father's map of the country. Owing to difficulties with the authorities he gave up astronomy in 1794 and lived in retirement.

Cassini, Giovanni Domenico (Jean Dominique) (1625–1712) Italian-born French astronomer. He discovered four moons of Saturn and the gap in the rings of Saturn now called the **Cassini division**.

Cassini, Giovanni Domenico http://www-groups.dcs.st-andrews.ac.uk/history/Mathematicians/Cassini.html

Devoted to the life and contributions of astronomer Giovanni Cassini, this site provides biographical information and a list of references about Cassini, as well as links to essays that reference him.

Cassini division prominent gap in the rings of †Saturn. It was discovered by the Italian-born French astronomer Giovanni †Cassini in 1675. Some 5,000 km/3,100 mi wide, the Cassini division separates the faint outer A-ring from the brighter inner B-ring. It is now known that the division is caused by the gravitational influence of Mimas, one of the moons of Saturn.

Cassiopeia prominent constellation of the northern hemisphere, named after the mother of Andromeda (who was the wife of Perseus). It has a distinctive W-shape, and contains one of the most powerful radio sources in the sky, Cassiopeia A. This is the remains of a †supernova (star explosion) that occurred *c.* AD 1702, too far away to be seen from Earth.

It was in Cassiopeia that Danish astronomer Tycho †Brahe observed a new star in 1572, probably a supernova, since it was visible in daylight and outshone †Venus for ten days.

cassiterite or **tinstone**, mineral consisting of reddish-brown to black stannic oxide (SnO_2), usually found in granite rocks. It is the chief ore of tin. When fresh it has a bright ("adamantine") luster. It was formerly extensively mined in Cornwall, England; today Malaysia is the world's main supplier. Other sources of cassiterite are Africa, Indonesia, and South America.

cast iron cheap but invaluable constructional material, most commonly used for car engine blocks. Cast iron is partly refined pig (crude) †iron, which is very fluid when molten and highly suitable for shaping by casting; it contains too many impurities (for example, carbon) to be readily shaped in any other way. Solid cast iron is heavy and can absorb great shock but is very brittle.

Castor or **Alpha Geminorum**, second-brightest star in the constellation †Gemini and the 23rd brightest star in the night sky. Along with the brighter †Pollux, it forms a prominent pair at the eastern end of Gemini, representing the head of the twins.

Second-magnitude Castor is 45 light years from Earth, and is one of the finest †binary stars in the sky for small telescopes. The two main components orbit each other over a period of 467 years. A third, much fainter, star orbits the main pair over a period probably exceeding 10,000 years. Each of the three visible components is a spectroscopic binary, making Castor a sextuple star system.

catabolism in biology, the destructive part of †metabolism where living tissue is changed into energy and waste products.

It is the opposite of †anabolism. It occurs continuously in the body, but is accelerated during many disease processes, such as fever, and in starvation.

cataclastic rock metamorphic rock, such as a breccia, containing angular fragments of preexisting rock produced by the grinding and crushing action (cataclasis) of †faults.

catadioptric lens in optics, image-forming lens constructed using one or more mirrors as well as lenses. The light path is effectively "folded" back on itself within the lens, allowing a long focal length to be accommodated in a relative short physical length.

catalysis in chemistry, any process that uses a †catalyst to alter the rate of a chemical reaction.

catalyst substance that alters the speed of, or makes possible, a chemical or biochemical reaction but remains unchanged at the end of the reaction. †Enzymes are natural biochemical catalysts. In practice most catalysts are used to speed up reactions.

The catalysts used in the chemical industry are often †transition metals or their compounds. According to the †collision theory, particles must collide before they can react, and the colliding particles must contain enough energy to cause bonds to break. Catalysts increase the †rate of reaction by lowering the amount of energy needed for a successful collision. Therefore, more collisions will be successful.

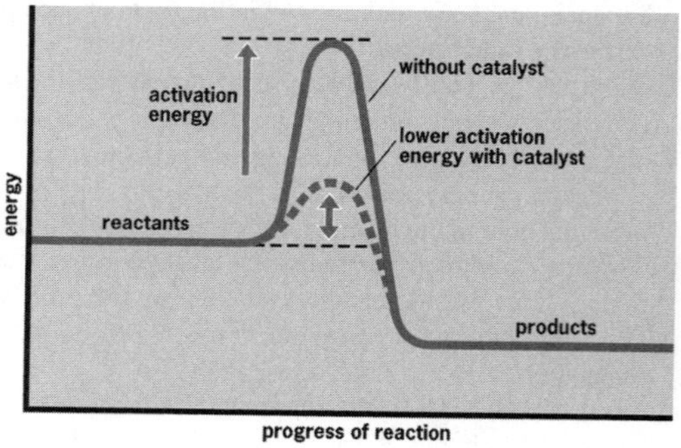

catalyst A graph showing how a reaction speeds up with the addition of a catalyst. Helicon

Catalysts http://www.purchon.com/chemistry/catalyst.htm

Clear explanation of catalysts and what they do. Designed to help GCSE chemists with their revision, this page is part of a larger site of scientific revision aids. The clearly written text includes definitions of important terms such as absorption and intermediate compounds.

catalytic converter device fitted to the exhaust system of a motor vehicle in order to reduce toxic emissions from the engine. It converts the harmful exhaust products that cause †air pollution to relatively harmless ones.

It does this by passing them over a mixture of catalysts coated on a metal or ceramic honeycomb (a structure that increases the surface area and therefore the amount of active catalyst with which the exhaust gases will come into contact). **Oxidation catalysts** (small amounts of precious palladium and platinum metals) convert hydrocarbons (unburnt fuel) and carbon monoxide into carbon dioxide and water, but do not affect nitrogen oxide emissions. **Three-way catalysts** (platinum and rhodium metals) also convert nitrogen oxide gases into nitrogen and oxygen.

catalytic cracking in chemistry, process for breaking down crude oil into lighter †petroleum fractions by passing it over or through a †catalyst. Fluidized bed catalytic cracking is the most widely used modern method for obtaining gasoline from crude oil.

catastrophe theory mathematical theory developed by René Thom in 1972, in which he showed that the growth of an organism proceeds by a series of gradual changes that are triggered by, and in turn trigger, large-scale changes or "catastrophic" jumps. It also has applications in engineering – for example, the gradual strain on the structure of a bridge that can eventually result in a sudden collapse – and has been extended to economic and psychological events.

catastrophism theory that the geological features of the Earth were formed by a series of sudden, violent "catastrophes" beyond the ordinary workings of nature. The theory was largely the work of Georges †Cuvier. It was later replaced by the concepts of †uniformitarianism and †evolution.

catchment area in earth sciences, the area from which water is collected by a river and its tributaries. In the social sciences the term may be used to denote the area from which people travel to obtain a particular service or product, such as the area from which a school draws its pupils.

catecholamine chemical that functions as a †neurotransmitter or a †hormone. Dopamine, adrenaline (epinephrine), and noradrenaline (norepinephrine) are catecholamines.

catenary curve taken up by a flexible cable suspended between two points, under gravity; for example, the curve of overhead suspension cables that hold the conductor wire of an electric railway or tramway.

cathode in chemistry, the negative electrode of an electrolytic cell, towards which positive particles (cations), usually in solution, are attracted. See †electrolysis. A cathode is given its negative charge by connecting it to the negative side of an external electrical supply.

cathode in electronics, the part of an electronic device in which electrons are generated. In a thermionic valve, electrons are produced by the heating effect of an applied current; in a photocell, they are produced by the interaction of light and a semiconducting material. The cathode is kept at a negative potential relative to the device's other electrodes (anodes) in order to ensure that the liberated electrons stream away from the cathode and towards the anodes.

cathode ray stream of fast-moving electrons that travel from a cathode (negative electrode) towards an anode (positive electrode) in a vacuum tube. They carry a negative charge and can be deflected by electric and magnetic fields. Cathode rays focused into fine beams of fast electrons are used in cathode-ray tubes, the electrons' †kinetic energy being converted into light energy as they collide with the tube's fluorescent screen.

cathode-ray oscilloscope **CRO**, instrument used to measure electrical potentials or voltages that vary over time and to display the waveforms of electrical oscillations or signals. Readings are displayed graphically on the screen of a †cathode-ray tube.

The CRO is used as a †voltmeter with any voltage change shown on screen by an up (positive) or down (negative) movement of a bright dot. This dot is produced by a beam of electrons hitting the phosphor layer on the inside of the screen. Over time the dot traces a graph across the screen showing voltage change against time. If voltage remains constant the graph consists of a horizontal line.

cathode-ray tube **CRT**, vacuum tube in which a beam of electrons is produced and focused onto a fluorescent screen. The electrons' kinetic energy is converted into light energy as they collide with the screen. It is an essential component of television receivers, computer visual display units, and †oscilloscopes.

The screen of the CRT is coated on the inside with phosphor, which emits light when struck by an electron beam. The tube itself is glass and coated inside with a black graphite conducting paint, which is connected to one of three anodes. A heated filament heats a metal-oxide coated cathode that emits electrons that pass through a positively charged anode that is held at several thousand volts and accelerates the electrons to a high speed beam. The electrons accumulate on the phosphor of the screen and repel each other back to the conducting graphite paint completing a circuit.

cation †ion carrying a positive charge. During †electrolysis, cations in the electrolyte move to the cathode (negative electrode). Cations are formed from †atoms by loss of electrons during †ionic bonding. †Metals form cations.

catkin in flowering plants (†angiosperms), a pendulous inflorescence, bearing numerous small, usually unisexual flowers. The tiny flowers are stalkless and the petals and sepals are usually absent or much reduced in size. Many types of trees bear catkins, including willows, poplars, and birches. Most plants with catkins are wind-pollinated, so the male catkins produce large quantities of pollen. Some †gymnosperms also have catkin-like structures that produce pollen, for example, the swamp cypress *Taxodium*.

CAT scan or **CT scan**; acronym for **computerized axial tomography scan**, sophisticated method of X-ray imaging. Quick and noninvasive, CAT scanning is used in medicine as an aid to diagnosis, helping to pinpoint problem areas without the need for exploratory surgery. It is also used in archaeology to investigate mummies.

The CAT scanner passes a narrow fan of X-rays through successive "slices" of the suspect body part. These slices are picked up by crystal detectors in a scintillator and converted electronically into cross-sectional images displayed on a viewing screen. Gradually, using views taken from various angles, a three-dimensional picture of the organ or tissue can be built up and irregularities analyzed.

Cauchy, Augustin-Louis (1789–1857) French mathematician who employed rigorous methods of analysis. His prolific output included work on complex functions, determinants, and probability, and on the convergence of infinite series. In calculus, he refined the concepts of the limit and the definite integral.

Cauchy, Augustin Louis http://www-history.mcs.st-andrews.ac.uk/history/Mathematicians/Cauchy.html

Extensive biography of the mathematician. The site contains a description of his working relationship with his contemporaries, and also includes explanations of several aspects of his work. Several literature references for further reading on the mathematician are also listed, and the Web site also features a portrait of Cauchy.

cauda tail, or tail-like appendage; part of the *cauda equina*, a bundle of nerves at the lower of the spinal cord in vertebrates.

caustic describing the harmful action of a chemical on organic tissues, especially that of a base such as sodium hydroxide or potassium hydroxide. Caustic compounds cause chemical burns and have a corrosive effect on organic materials.

caustic curve curve formed by the points of intersection of rays of light reflected or refracted from a curved surface.

caustic soda former name for †sodium hydroxide (NaOH).

Cavalieri, Bonaventura (*c.* 1598–1647) Italian mathematician. He specialized in geometry. At an early age he entered the Society of Jesus, and on its recommendation was appointed professor of mathematics at Bologna University in 1629. In his *Geometria indivisibilibus continuorum nova quadam ratione promota/A certain method for the development of a new geometry for continuous indivisibles* (1635), he expounded his celebrated theory of indivisibles. According to his conception, all space may be regarded as made up of an infinite number of parts, which represent the limit of decomposition that the mind can imagine anything to undergo. This theory had a great influence on the development of the use of calculus for calculating areas. Cavalieri was born in Milan, Italy.

cave passage or tunnel – or series of tunnels – formed underground by water or by waves on a coast. Caves of the former type commonly occur in areas underlain by limestone, such as Kentucky, USA, and many Balkan regions, where the rocks are soluble in water. A **pothole** is a vertical hole in rock caused by water descending a crack; it is thus open to the sky.

Coastal caves are formed where rocks with lines of weakness, like †basalt at tide level, are exposed to severe wave action. The erosion (†corrasion and †corrosion) of the rock layers is increased by subsidence, and the hollow in the cliff face grows still larger because of air compression in the chamber (†hydraulic action). Where the roof of a cave has fallen in, the vent up to the land surface is called a †blowhole. If this grows, finally destroying the cave form, the outside pillars of the cave are known as stacks or columns. The Old Man of Hoy (137 m/449 ft high), in the Orkney Islands, is a fine example of a stack.

Most inland caves are found in †karst (limestone) regions, because †limestone is soluble when exposed to acid water. As the water makes its way along the main joints, fissures, and bedding planes, they are constantly enlarged into potential cave passages, which ultimately join to form a complex network. †Stalactites and stalagmites form due to water that is rich in calcium carbonate dripping from the roof of the cave. The collapse of the roof of a cave produces features such as **natural arches** and **steep-sided gorges**.

Virtual Cave http://library.thinkquest.org/2974/

Browse the mineral wonders unique to the cave environment – from bell canopies and bottlebrushes to splattermites and stalactites.

Cavendish, Henry (1731–1810) English physicist and chemist. He discovered hydrogen (which he called "inflammable air") in 1766, and determined the compositions of water and of nitric acid. The †Cavendish experiment (1798) enabled him to discover the mass and density of the Earth.

Cavendish experiment measurement of the gravitational attraction between large and small lead spheres, which enabled English physicist and chemist Henry Cavendish to calculate in 1798 a mean value for the mass and density of Earth, using Isaac Newton's law of universal gravitation.

cavitation in hydraulics, †erosion of rocks caused by the forcing of air into cracks. Cavitation results from the pounding of waves on the coast and the swirling of turbulent river currents, and exerts great pressure, eventually causing rocks to break apart.

The process is particularly common at waterfalls, where the turbulent falling water contains many air bubbles, which burst and send shock waves into the rocks of the river bed and banks. In addition, as water is forced into cracks in the rock, air within the crack is compressed and literally explodes, helping to break down the rock.

cavitation in earth science, formation of cavities containing a partial vacuum in fluids at high velocities, produced by propellers or other machine parts in hydraulic engines, in accordance with †Bernoulli's principle. When these cavities collapse, pitting, vibration, and noise can occur in the metal parts in contact with the fluids.

Cayley, Arthur (1821–1895) English mathematician who developed matrix algebra, used by Werner †Heisenberg in his elucidation of quantum mechanics. He also developed the study of *n*-dimensional geometry, introducing the concept of the "absolute," and formulated the theory of algebraic invariants.

Cayley, Arthur http://www-groups.dcs.st-andrews.ac.uk/history/Mathematicians/Cayley.html

Devoted to the life and contributions of Arthur Cayley.

cc symbol for †**cubic centimeter**; abbreviation for **carbon copy**.

CCD abbreviation for †charge-coupled device.

cd in physics, symbol for †candela, the SI unit of luminous intensity.

CD-ROM acronym for Compact-Disc Read-Only Memory, computer storage device developed from the technology of the audio compact disc. It consists of a plastic-coated metal disk, on which binary digital information is etched in the form of microscopic pits. This can then be read optically by passing a laser beam over the disk. CD-ROMs typically hold over 600 †megabytes of data, and are used in distributing large amounts of text, graphics, audio, and video, such as encyclopedias, catalogues, technical manuals, and games.

Standard CD-ROMs cannot have information written onto them by computer, but must be manufactured from a master, although recordable CDs, called CD-R disks, have been developed for use as computer disks. A compact disc, CD-RW, which can be overwritten repeatedly by a computer has also been developed. The compact disc, with its enormous storage capability, may eventually replace the magnetic disk as the most common form of backing store for computers.

The technology is being developed rapidly: a standard CD-ROM spins at between 240 and 1,170 rpm, but faster disks have been introduced that speed up data retrieval to many times the standard speed. Research is being conducted into high-density CDs capable of storing many †gigabytes of data, made possible by using multiple layers on the surface of the disk, and by using double-sided disks. The first commercial examples of this research include DVD players and DVD-ROM computer disks launched in 1997.

PhotoCD, developed by Kodak and released in 1992, transfers ordinary still photographs onto CD-ROM disks.

cecum in the †digestive system of animals, a blind-ending tube branching off from the first part of the large intestine, terminating in the appendix. It has no function in humans but is used for the digestion of cellulose by some grass-eating mammals.

The rabbit cecum and appendix contains millions of bacteria that produce cellulase, the enzyme necessary for the breakdown of cellulose to glucose. In order to be able to absorb nutrients released by the breakdown of cellulose, rabbits pass food twice down the intestine. They egest soft pellets that are then re-eaten. This is known as coprophagy.

celestial equator in astronomy, imaginary line around the †celestial sphere where it would be cut into two equal hemispheres by the extension of the plane of the Earth's equator. It is the reference plane from which †declination (coordinate corresponding to Earth's latitude) and †right ascension (coordinate corresponding to longitude) is measured.

celestial mechanics branch of astronomy that deals with the calculation of the orbits of celestial bodies, their gravitational attractions (such as those that produce the Earth's tides), and also the orbits of artificial satellites and space probes. It is based on the laws of motion and gravity laid down by 17th-century English physicist and mathematician Isaac †Newton.

Celestial Mechanics

http://scienceworld.wolfram.com/physics/topics/CelestialMechanics.html

A to Z directory of terms and theorems in celestial mechanics, part of Eric Weisstein's Treasure Trove of Physics. Many of the entries are detailed, with equations, diagrams, and references.

celestial pole in astronomy, two points at opposite ends of the axis about which the †celestial sphere appears to rotate and the zenith, uppermost point of the celestial horizon, immediately above for observers at the Earth's poles. The Earth's axis, if extended, would reach the celestial poles. The north celestial pole lies close to the Pole Star (Polaris).

Celestial Police group of astronomers in Germany between 1800 and 1815, who set out to discover a supposed missing planet thought to be orbiting the Sun between Mars and Jupiter, a region now known to be occupied by types of †asteroid. Although they did not discover the first asteroid (found in 1801), they discovered the second, Pallas (1802), third, Juno (1804), and fourth, Vesta (1807).

celestial sphere imaginary sphere surrounding the Earth, on which the celestial bodies seem to lie. The positions of bodies such as stars, planets, and galaxies are specified by their coordinates on the celestial sphere. The equivalents of latitude and longitude on the celestial sphere are called †declination and †right ascension (which is measured in hours from 0 to 24). The **celestial poles** lie directly above the Earth's poles, and the **celestial equator** lies over the Earth's Equator. The celestial sphere appears to rotate once around the Earth each day, actually a result of the rotation of the Earth on its axis.

celestine or **celestite**, mineral consisting of strontium sulfate, $SrSO_4$, occurring as white or light blue crystals. Celestine occurs in cavity linings associated with calcite, dolomite, or fluorite.

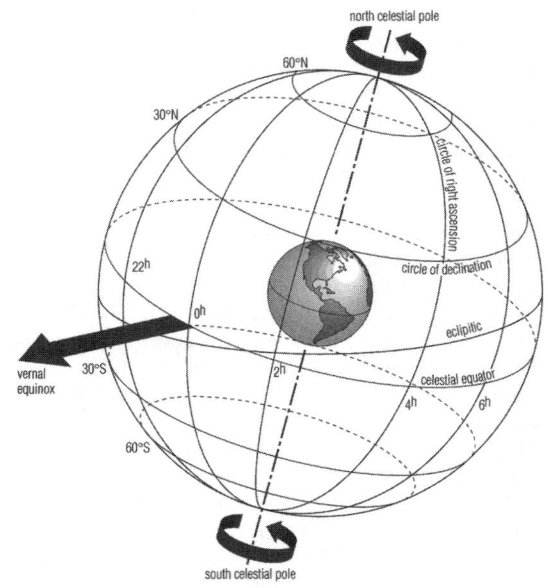

celestial sphere The main features of the celestial sphere. Declination runs from 0° at the celestial equator to 90° at the celestial poles. Right ascension is measured in hours eastwards from the vernal equinox, one hour corresponding to 15° of longitude. Helicon

cell in biology, the basic unit of a living organism. It is the smallest unit capable of independent existence. In organisms, other than the smallest ones, the body of the organism is made up of several cells or many cells. A single cell, therefore, is the smallest unit that shows characteristic features of life, such as †reproduction, †growth, †respiration, response to environmental stimuli, and the ability to take in †mineral salts. †Viruses are particles that are not cells. A virus can only reproduce by "taking over" a cell from another organism. This cell often dies as a result of making many new virus particles.

Some organisms are composed of only one cell, including many bacteria and some fungi, such as yeast. Single-cell organisms are termed **unicellular**, while plants and animals that contain many cells are termed **multicellular** organisms. Organisms such as human beings consist of billions of cells. In organisms made of many cells, groups of cells are specialized to carry out specific functions and are organized into tissues and organs.

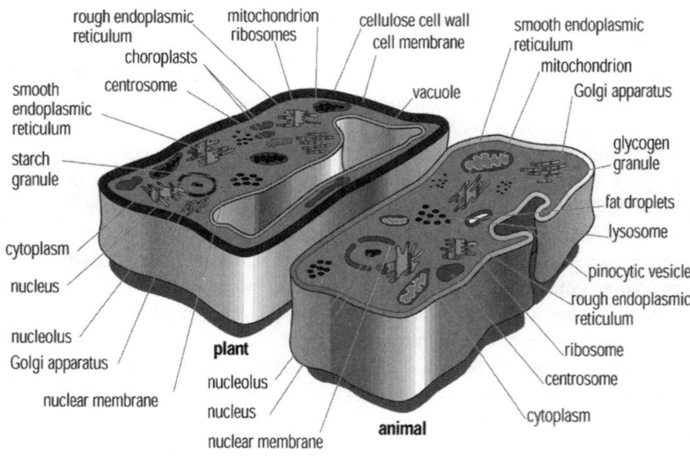

cell structure Typical plant and animal cell. Plant and animal cells share many structures, such as ribosomes, mitochondria, and chromosomes, but they also have notable differences: plant cells have chloroplasts, a large vacuole, and a cellulose cell wall. Animal cells do not have a rigid cell wall but have an outside cell membrane only. Helicon

Cells always have a †cell membrane around them and †cytoplasm inside, and normally a †nucleus. There are differences between plant cells and animal cells; for example, plant cells have a †cell wall made of cellulose outside their cell membrane. This helps to explain why plants look so different from animals. In plants and animals, cells divide by †mitosis or by †meiosis when the organism carries out †sexual reproduction. In both forms of †cell division, the chemical carrying inherited information, †DNA, has to be copied before division. Rarely, mistakes occur, causing †mutations, but it is normally done accurately.

Copying the DNA results in the duplication of structures called †chromosomes in the nucleus. In cell division, the duplicated chromosomes are separated from each other into daughter cells. New cells produced by mitosis are needed to replace cells that die and some cells live only a short time, such as white blood cells that live for only a few days. Within the human body, about 3 billion cells die every minute. Many cells may be lost in normal activities – for example, human skin cells are constantly being worn off and have to be replaced by new cells produced by mitosis. The new cells needed during growth are also produced by mitosis.

In respiration a cell uses chemicals that it has taken in, as nutrients, and breaks them down to release †energy that the cell can use for its life processes.

Talksaver Cell Biology http://www.talksaver.com/cell_biology/01_nutshell.htm

Excellent and well-written guide to the basics of cellular biology. The site contains chapters on DNA and the storage of information within a cell, the way in which cells acquire and process energy, and the applications of cellular biology to all branches of medical and life science.

cell differentiation in developing embryos, the process by which cells acquire their specialization, such as heart cells, muscle cells, skin cells, and brain cells. The seven-day-old human pre-embryo consists of thousands of individual cells, each of which is destined to assist in the formation of individual organs in the body.

Research has shown that the eventual function of a cell, in for example, a chicken embryo, is determined by the cell's position. The embryo can be mapped into areas corresponding with the spinal cord, the wings, the legs, and many other tissues. If the embryo is relatively young, a cell transplanted from one area to another will develop according to its new position. As the embryo develops the cells lose their flexibility and become unable to change their destiny.

cell division process by which a cell divides. †Cells are the basic units of life and they carry out basic functions that are characteristic of living organisms, such as growth and reproduction. Both growth and reproduction usually involve cell division. In plants and animals reproduction can be either †sexual reproduction or †asexual reproduction. Both involve cell division. In sexual reproduction the cell division that is used is †meiosis. In asexual reproduction it is †mitosis that is used. In both forms of cell division, the chemical carrying inherited information, †DNA, has to be copied before division. Rarely, mistakes occur causing †mutations, but it is normally done accurately. Copying the DNA results in the duplication of structures called †chromosomes in the nucleus. In cell division, the duplicated chromosomes are separated from each other into daughter cells.

In sexual reproduction male and female gametes combine. Gametes are produced by meiosis cell division. Usually the male and female gametes are produced by two different individuals. If so, sexual reproduction combines inherited information from the two parents. Most animals and plants reproduce sexually, though many plants also reproduce asexually.

The male and female sex organs of a plant are usually found in the flower. Many flowers contain both male and female organs. The male gametes of a plant are inside the pollen grains, and the female gametes are inside the ovules.

In mammals, the male gametes are called †sperm and are made in the testes. The female gametes are eggs (see †ovum) and are made in the ovaries.

cell, electrical or **voltaic cell** or **galvanic cell**, device in which chemical energy is converted into electrical energy; the popular name is †"battery," but this strictly refers to a collection of cells in one unit. The reactive chemicals of a **primary cell** cannot be replenished, whereas **secondary cells** – such as storage batteries – are rechargeable: their chemical reactions can be reversed and the original condition restored by applying an electric current. It is dangerous to attempt to recharge a primary cell.

cell, electrolytic device to which electrical energy is applied in order to bring about a chemical reaction; see †electrolysis.

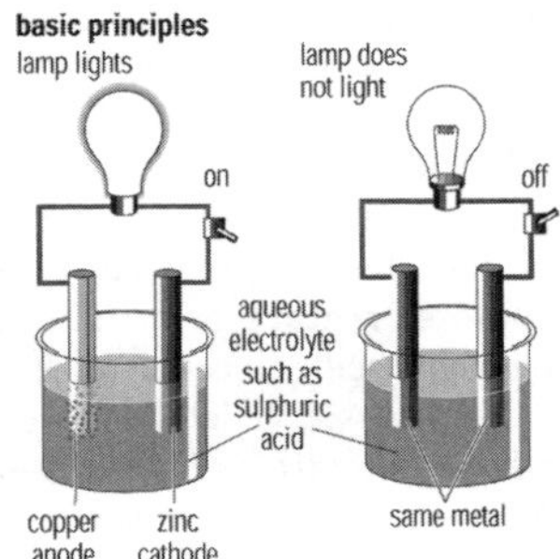

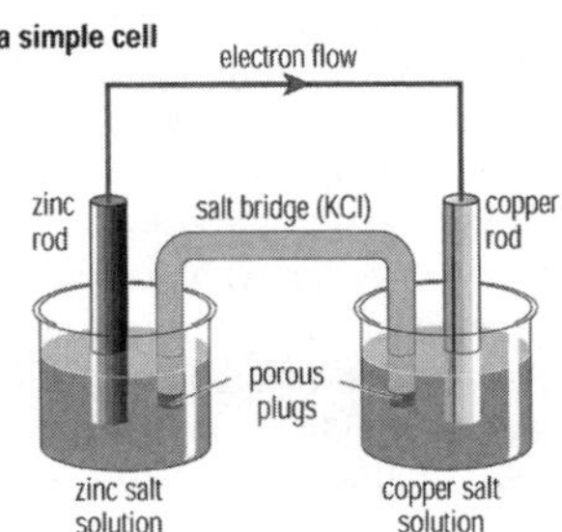

cell, electrical When electrical energy is produced from chemical energy using two metals acting as electrodes in a aqueous solution, it is sometimes known as a galvanic cell or voltaic cell. Here the two metals copper (+) and zinc (–) are immersed in dilute sulfuric acid, which acts as an electrolyte. If a light bulb is connected between the two, an electric current will flow with bubbles of gas being deposited on the electrodes in a process known as polarization. Helicon

cell membrane or **plasma membrane**, thin layer of protein and fat surrounding cells that keeps the cells together and controls substances passing between the cytoplasm and the intercellular space. The cell membrane is semipermeable, allowing some substances to pass through and some not. Generally, small molecules such as water, glucose, and amino acids can penetrate the membrane, while large molecules, such as starch, cannot. Substances often cross the membrane by †diffusion, a spontaneous passage of molecules. Water movement across the membrane is a special case of diffusion known as †osmosis.

Membranes also play a part in †active transport, hormonal response (see †hormones), and cell metabolism.

Membrane Transport Mechanisms http://web.mit.edu/esgbio/www/cb/membranes/transport.html

Concise description of cell diffusion, the means by which materials travel in and out of a cell. With the help of diagrams, the page explains the related concepts of facilitated diffusion and active transport.

cell sap dilute fluid found in the large central vacuole of many plant cells. It is made up of water, amino acids, glucose, and salts. The sap has many functions, including storage of useful materials, and provides mechanical support for non-woody plants.

cell theory theory that regards all living things as being composed of cells and that their replication and growth result from cell division. The theory was proposed by German botanist Matthias Schleiden and German physiologist Theodor Schwann in 1838–39.

celluloid transparent or translucent, highly flammable, plastic material (a †thermoplastic) made from cellulose nitrate and camphor. It was once used for toilet articles, novelties, and photographic film, but has now been replaced by the nonflammable substance †cellulose acetate.

cellulose complex †carbohydrate composed of long chains of glucose units, joined by chemical bonds called glycosidic links. It is the principal constituent of the cell wall of higher plants, and a vital ingredient in the diet of many †herbivores. Molecules of cellulose are organized into long, unbranched microfibrils that give support to the cell wall. No mammal produces the enzyme cellulase, necessary for digesting cellulose; mammals such as rabbits and cows are only able to digest grass because the bacteria present in their gut can manufacture it.

Cellulose is the most abundant substance found in the plant kingdom. It has numerous uses in industry: in rope-making; as a source of textiles (linen, cotton, viscose, and acetate) and plastics (cellophane and celluloid); in the manufacture of nondrip paint; and in such foods as whipped dessert toppings.

Japanese chemists produced the first synthetic cellulose in 1996 and the gene for the plant enzyme that makes cellulose was identified by Australian biologists in 1998.

cellulose acetate or **cellulose ethanoate**, chemical (an †ester) made by the action of acetic acid (ethanoic acid) on cellulose. It is used in making transparent film, especially photographic film; unlike its predecessor, celluloid, it is not flammable.

cellulose nitrate or **nitrocellulose**, series of esters of cellulose with up to three nitrate (NO_3) groups per monosaccharide unit. It is made by the action of concentrated nitric acid on cellulose (for example, cotton waste) in the presence of concentrated sulfuric acid. Fully nitrated cellulose (gun cotton) is explosive, but esters with fewer nitrate groups were once used in making lacquers, rayon, and plastics, such as colored and photographic film, until replaced by the nonflammable cellulose acetate. †Celluloid is based on cellulose nitrate.

cell wall tough outer surface of the cell in plants. It is constructed from a mesh of †cellulose and is very strong and only very slightly elastic so that it protects the cell and holds it in shape. Most living cells are turgid (swollen with water; see †turgor). Water is absorbed by osmosis causing the cell to expand and develop an internal hydrostatic pressure (wall pressure) that acts against the cellulose wall. The result of this turgor pressure is to give the cell, and therefore the plant, rigidity. Plants, or sections of plants, that are not woody are particularly reliant on this form of support.

The cellulose in cell walls plays a vital role in global nutrition. No vertebrate is able to produce cellulase, the enzyme necessary for the breakdown of cellulose into sugar. However, most mammalian herbivores rely on cellulose, using secretions from micro-organisms living in the gut to break it down. Humans cannot digest the cellulose of the cell walls; they possess neither the correct gut micro-organisms nor the necessary grinding teeth. However, cellulose still forms a necessary part of the human diet as †forming (roughage). Because it is not broken down it acts as a signal that the gut needs to contract to move the contents on.

The cell walls of †bacteria and fungi are not made of cellulose, but are just as strong. Some antibiotics kill bacteria by weakening the cell wall. Penicillins work in this way.

Celsius scale of temperature, previously called centigrade, in which the range from freezing to boiling of water is divided into 100 degrees, freezing point being 0 degrees and boiling point 100 degrees.

The degree centigrade (°C) was officially renamed Celsius in 1948 to avoid confusion with the angular measure known as the centigrade (one hundredth of a grade). The Celsius scale is named after the Swedish astronomer Anders †Celsius, who devised it in 1742 but in reverse (freezing point was 100°; boiling point 0°).

Origin of the Celsius Temperature Scale http://www.santesson.com/engtemp.html

Brief but comprehensive description of how Swedish astronomer Anders Celsius devised his temperature scale. The Web site is available in three languages, English, Swedish, and Dutch, and the Swedish cover of Celsius's paper on his temperature scale is also shown here.

Celsius, Anders (1701–1744) Swedish astronomer, physicist, and mathematician who introduced the Celsius scale of temperature. His other scientific works include a paper on accurately determining the shape and size of the Earth, some of the first attempts to gauge the magnitude of the stars in the constellation Aries, and a study of the falling water level of the Baltic Sea.

Celsius, Anders (1701–1744) http://www.astro.uu.se/history/Celsius_eng.html

Biography of Swedish astronomer, physicist, and mathematician Anders Celsius. Famous for his creation of the Celsius temperature scale, the astronomer was also the first to realize that the aurora phenomenon was magnetic in nature. A portrait of Celsius is also available from this page.

cement any bonding agent used to unite particles in a single mass or to cause one surface to adhere to another. **Portland cement** is a powder which when mixed with water and sand or gravel turns into mortar or concrete.

In geology, cement refers to a chemically precipitated material such as carbonate that occupies the interstices of clastic rocks.

The term "cement" covers a variety of materials, such as fluxes and pastes, and also bituminous products obtained from tar. In 1824 English bricklayer Joseph Aspdin (1779–1855) created and patented the first Portland cement, so named because its color in the hardened state resembled that of Portland stone, a limestone used in building.

Cement is made by heating limestone (calcium carbonate) with clay (which contains a variety of silicates along with aluminum). This produces a gray powdery mixture of calcium and aluminum silicates. On addition of water, a complex series of reactions occurs and calcium hydroxide is produced. Cement sets by losing water.

Cenozoic Era or **Cenozoic**, era of geological time that began 65 million years ago and continues to the present day. It is divided into the Tertiary and Quaternary periods. The Cenozoic marks the emergence of mammals as a dominant group, and the rearrangement of continental masses towards their present positions.

centaur in astronomy, class of asteroid whose orbit crosses that of at least one gas giant planet and whose orbital period is not in stable synchrony with that gas giant, meaning they will eventually crash into the planet or evolve in some other way. The name originated because the first of the class to be discovered was named †Chiron, after the wisest of the centaurs from Greek mythology, and by convention all new discoveries in this class are named after centaurs. Most of the centaurs are icy bodies from the †Kuiper belt, which will become comets if their orbits are perturbed to take them nearer to the Sun.

Centaurus large, bright constellation of the southern hemisphere, represented as a centaur. Its brightest star, †Rigil Kent, is a triple star, and contains the closest star to the Sun, Proxima Centauri, which is only 4.2 light years away from the Sun, and 0.1 light years closer than its companions, Alpha Centauri A and B. Omega Centauri, which is just visible to the naked eye as a hazy patch, is the largest and brightest †globular cluster of stars in the sky, 16,000 light years away from the Sun.

Alpha and Beta Centauri are both of the first magnitude and, like Alpha and Beta Ursae Majoris, are known as "the Pointers," as a line joining them leads to †Crux.

Centaurus A in astronomy, galaxy that appears to have been formed when an elliptical galaxy and a spiral galaxy collided. It is rapidly expanding and shows explosive activity, being visible at optical and radio wavelengths; it is also a strong X-ray source. It lies about 15 million light years distant and is surrounded by clouds of expanding dust and gas from the explosions.

centigrade former name for the †Celsius temperature scale.

centimeter symbol cm, unit of length equal to one-hundredth of a meter.

central dogma in genetics and evolution, the fundamental belief that †genes can affect the nature of the physical body, but that changes in the body (†acquired character, for example, through use or accident) cannot be translated into changes in the genes.

central nervous system **CNS**, brain and spinal cord, as distinct from other components of the †nervous system. The CNS integrates all nervous function.

In invertebrates it consists of a paired ventral nerve cord with concentrations of nerve-cell bodies, known as †ganglia in each segment, and a small brain in the head. Some simple invertebrates, such as sponges and jellyfishes, have no CNS but a simple network of nerve cells called a **nerve net**.

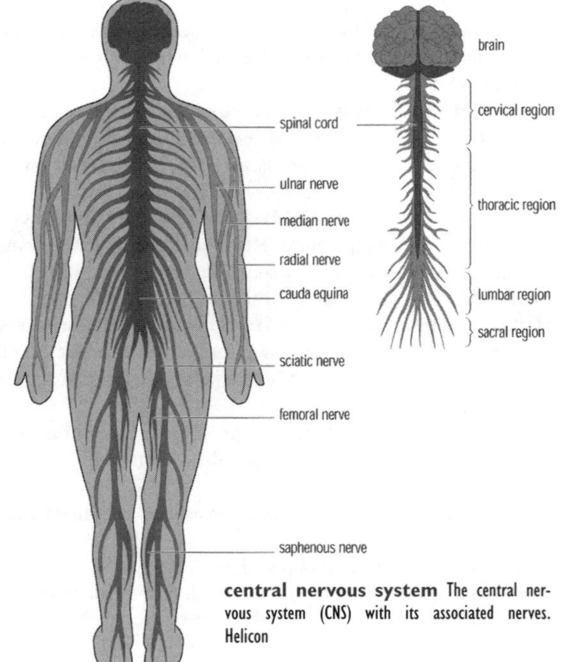

central nervous system The central nervous system (CNS) with its associated nerves. Helicon

central processing unit CPU, main component of a computer, the part that executes individual program instructions and controls the operation of other parts. It is sometimes called the central processor or, when contained on a single integrated circuit, a microprocessor.

The CPU has three main components: the **arithmetic and logic unit** (ALU), where all calculations and logical operations are carried out; a **control unit**, which decodes, synchronizes, and executes program instructions; and the **immediate access memory**, which stores the data and programs on which the computer is currently working. All these components contain †registers, which are memory locations reserved for specific purposes.

Illustrated Guide to CPUs
http://www.karbosguide.com/hardware/module3a1.htm
Series of modules on central processing units (CPUs). The site introduces the design and function of the CPU since it is the key to understanding data processing methodology. It begins by asking "What is a CPU?" and then explores the development of the CPU from the first through to the seventh generation.

PC Technology Guide http://www.pctechguide.com/
Complete guide to the personal computer (PC). It contains comprehensive information on the component parts of a PC, what they are, and how they work. It covers topics such as "Components," "Storage," "Multimedia," and "Input/Output." The topics are covered in detail, yet are written in way that is easy to understand.

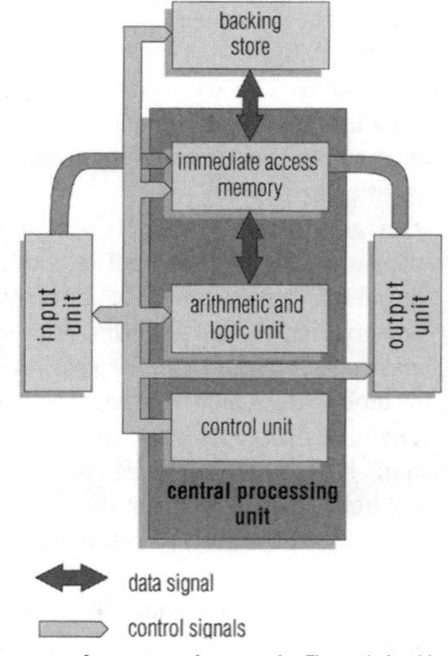

central processing unit The relationship between the three main areas of a computer's central processing unit. The arithmetic and logic unit (ALU) does the arithmetic, using the registers to store intermediate results, supervised by the control unit. Input and output circuits connect the ALU to external memory, input, and output devices. Helicon

center of curvature in optics, the center of the sphere of which the curved surface of a lens or mirror forms part.

center of gravity the point in an object about which its weight is evenly balanced. In a uniform gravitational field, this is the same as the †center of mass.

center of mass point in or near an object at which the whole mass of the object may be considered to be concentrated. A symmetrical homogeneous object such as a sphere or cube has its center of mass at its geometrical center; a hollow object (such as a cup) may have its center of mass in space inside the hollow.

For an object to be in stable equilibrium, a vertical line down through its center of mass must run within the boundaries of its base; if tilted until this line falls outside the base, the object becomes unstable and topples over.

centrifugal force in physics, apparent force arising for an observer moving with a rotating system. For an object of mass m moving with a velocity v in a circle of radius r, the centrifugal force F equals mv^2/r (outwards).

centrifuge apparatus that rotates containers at high speeds, creating centrifugal forces. One use is for separating mixtures of substances of different densities.

A laboratory centrifuge is used to separate small amounts of suspension. Test tubes of suspension are spun around very fast so that the solid gets flung to the bottom. The mixtures are usually spun horizontally in balanced containers ("buckets"), and the rotation sets up centrifugal forces, causing their components to separate according to their densities. A common example is the separation of the lighter plasma from the heavier blood corpuscles in certain blood tests. The **ultracentrifuge** is a very high-speed centrifuge, used in biochemistry for separating †colloids and organic substances; it may operate at several million revolutions per minute. The centrifuges used in the industrial separation of cream from milk, and yeast from fermented wort (infused malt), operate by having mixtures pumped through a continually rotating chamber, the components being tapped off at different points. Large centrifuges are used for physiological research – for example, in astronaut training where bodily responses to gravitational forces many times the normal level are tested.

centriole structure found in the †cells of animals that plays a role in the processes of †meiosis and †mitosis (cell division).

centripetal force force that acts radially inwards on an object moving in a curved path. For example, with a weight whirled in a circle at the end of a length of string, the centripetal force is the tension in the string. For an object of mass m moving with a velocity v in a circle of radius r, the centripetal force F equals mv^2/r (inwards). The reaction to this force is the †centrifugal force.

centroid in mathematics, point on a geometrical figure whose coordinates are the averages of the coordinates of the points making up the figure. For a triangle, the centroid is the common meeting point of the medians (lines joining the vertices to the mid-points of the opposite sides). If a figure is made into a solid of uniform density, the centroid is the center of mass of the solid.

centromere part of the †chromosome where there are no †genes. Under the microscope, it usually appears as a constriction in the strand of the chromosome, and is the point at which the spindle fibers are attached during †meiosis and †mitosis (cell division).

centrosome cell body that contains the †centrioles. During cell division the centrosomes organize the microtubules to form the spindle that divides the chromosomes into daughter cells. Centrosomes were first described in 1887, independently by German biologist Theodor Boveri (1862–1915) and Belgian biologist Edouard van Beneden.

cephalopod any predatory marine mollusk of the class Cephalopoda, with the mouth and head surrounded by tentacles. Cephalopods are the most intelligent, the fastest-moving, and the largest of all animals without backbones, and there are remarkable luminescent forms that swim or drift at great depths. They have the most highly developed nervous and sensory systems of all invertebrates, the eye in some closely paralleling that found in vertebrates. Examples include squid, octopus, and cuttlefish. Shells are rudimentary or absent in most cephalopods.

Typically, they move by swimming with the mantle (fold of outer skin) aided by the arms, but can squirt water out of the siphon (funnel) to propel themselves backwards by jet propulsion. Squid, for example, can escape predators at speeds of 11 kph/7mph. Cephalopods grow very rapidly and may be mature in a year. The female common octopus lays 150,000 eggs after copulation, and stays to brood them for as long as six weeks. After they hatch the female dies, and, although reproductive habits of many cephalopods are not known, it is thought that dying after spawning may be typical.

Cepheid variable yellow supergiant star that varies regularly in brightness every few days or weeks as a result of pulsations. The time that a Cepheid variable takes to pulsate is directly related to its average brightness; the longer the pulsation period, the brighter the star.

This relationship, the **period luminosity law** (discovered by US astronomer Henrietta †Leavitt), allows astronomers to use Cepheid variables as "standard candles" to measure distances within our Galaxy and to nearby galaxies. They are named after their prototype, Delta Cephei, whose light variations were observed in 1784 by English astronomer John Goodricke.

Cepheus constellation of the north polar region, named after King Cepheus of Greek mythology, husband of Cassiopeia and father of Andromeda. It contains the Garnet Star (Mu Cephei), a red supergiant of variable brightness that is one of the reddest-colored stars known, and Delta Cephei, prototype of the †Cepheid variables, which are important both as distance indicators and for the information they give about stellar evolution.

cerebellum part of the brain of †vertebrate animals that controls muscle tone, movement, balance, and coordination. It is relatively small in lower animals such as newts and lizards, but large in birds since flight demands precise coordination. The human cerebellum is also well developed, because of the need for balance when walking or running, and for finely coordinated hand movements.

cerebral pertaining to the brain, especially the part known as the †cerebrum, concerned with higher brain functions.

cerebral cortex or **gray matter**, fissured outer layer of the †cerebrum in the brain. Some 3 mm thick in the adult, it is the most sophisticated part of the brain, responsible for all higher functions and for initiating voluntary movement. Anatomists divide it into four lobes, named after the skull plates beneath which they lie: frontal, parietal, temporal, and occipital.

cerebral hemisphere one of the two halves of the †cerebrum.

cerebrum part of the vertebrate †brain, formed from the two paired cerebral hemispheres, separated by a central fissure. In birds and mammals it is the largest and most developed part of the brain. It is covered with an infolded layer of gray matter, the cerebral cortex, which integrates brain functions. The cerebrum coordinates all voluntary activity.

Ceres largest asteroid, 940 km/584 mi in diameter, and the first to be discovered (by Italian astronomer Giuseppe Piazzi in 1801). Ceres orbits the Sun every 4.6 years at an average distance of 414 million km/257 million mi. Its mass is about 0.014 of that of the Earth's Moon.

cerium chemical symbol Ce, malleable and ductile, gray, metallic element, atomic number 58, relative atomic mass 140.12. It is the most abundant member of the lanthanide series, and is used in alloys, electronic components, nuclear fuels, and lighter flints. It was discovered in 1804 by the Swedish chemists Jöns †Berzelius and Wilhelm Hisinger (1766–1852), and, independently, by German chemist Martin †Klaproth. The element was named after the then recently discovered asteroid Ceres.

CERN particle physics research organization founded in 1954 as a cooperative enterprise among European governments. It has laboratories at Meyrin, near Geneva, Switzerland. It houses the world's largest particle †accelerator, the †Large Electron Positron Collider (LEP), operational 1989–2000, with which notable advances were made in †particle physics.

CERN is the acronym of its original name, **Conseil Européen pour la Recherche Nucléaire**, but has undergone several name changes since then. Its official English name is now European Laboratory for Particle Physics. In 1965 the original laboratory was doubled in size by extension across the border from Switzerland into France. In 1994 the 19 member nations of CERN approved the construction of the Large Hadron Collider. It is expected to cost £1.25 million and to be fully functional in 2005. Permission to begin construction of the Large Hadron Collider (LHC) was granted in 1998. The LHC is being built on the French–Swiss border. The Swiss government gave the go-ahead in early 1998 and the French government approved construction in August 1998.

European Laboratory for Particle Physics http://www.cern.ch/

Information about CERN, the world-class physics laboratory in Geneva. As well as presenting committees, groups, and associations hosted by the Laboratory, this official site offers important scientific material and visual evidence on the current activities and projects. Visitors will also find postings about colloquia, schools, meetings, and other services offered there. A special section is devoted to the history of the World Wide Web and the pioneering contribution of CERN in its conception and expansion.

cervix (Latin "neck") abbreviation for **cervix uteri**, the neck of the womb; see †uterus.

cesium chemical symbol Cs, (Latin *caesius* "bluish-gray") soft, silvery-white, ductile metallic element, atomic number 55, relative atomic mass 132.905. It is one of the †alkali metals that form Group 1 of the periodic table of the elements. The alkali metals increase in reactivity down the group, and cesium, with only the short-lived radioactive francium below it, is the most reactive of them all. In air it ignites spontaneously, and it reacts violently with water. It is the most electropositive of all the elements. It is used in the manufacture of photocells.

The rate of vibration of cesium atoms has been used as the standard of measuring time. Its radioactive isotope Cs-137 (half-life 30.17 years) is a product of fission in nuclear explosions and in nuclear reactors; it is one of the most dangerous waste products of the nuclear industry, being a highly radioactive biological analog of potassium.

cesium clock type of †atomic clock, in which the frequency of transition between the energy levels of two types of cesium-133 atoms in a magnetic field is used to stabilize an oscillator, which produces the radiation that splits the atoms into the two levels. The frequency is 9,192,631,770 Hz and is now used in the formal definition of a second.

Cetus (Latin "whale") large constellation on the celestial equator (see †celestial sphere), represented as a sea monster or a whale. Cetus contains the long-period variable star †Mira, and Tau Ceti, one of the nearest stars, which is visible with the naked eye.

It is named after the sea monster sent to devour Andromeda. Mira is sometimes the most conspicuous object in the constellation, but it is more usually invisible to the naked eye.

CFC abbreviation for †chlorofluorocarbon.

c.g.s. system system of units based on the centimeter, gram, and second, as units of length, mass, and time, respectively. It has been replaced for scientific work by the †SI units to avoid inconsistencies in definition of the thermal calorie and electrical quantities.

Chadwick, James (1891–1974) English physicist. He was awarded the Nobel Prize for Physics in 1935 for his discovery in 1932 of the particle in the nucleus of an atom that became known as the **neutron** because it has no electric charge. He was knighted in 1945.

Chain, Ernst Boris (1906–1979) German-born British biochemist who was awarded a Nobel Prize for Physiology or Medicine in 1945, together with Alexander †Fleming and Howard †Florey (Fleming for his discovery of the bactericidal effect of penicillin, and Chain and Florey for their isolation of penicillin and its development as an antibiotic drug). Chain also discovered penicillinase, an enzyme that destroys penicillin. Chain was knighted in 1969.

chain reaction in chemistry, a succession of reactions, usually involving †free radicals, where the products of one stage are the reactants of the next. A chain reaction is characterized by the continual generation of reactive substances.

A chain reaction comprises three separate stages: **initiation** – the initial generation of reactive species; **propagation** – reactions that involve reactive species and generate similar or different reactive species; and **termination** – reactions that involve the reactive species but produce only stable, nonreactive substances. Chain reactions may occur slowly (for example, the oxidation of edible oils) or accelerate as the number of reactive species increases, ultimately resulting in explosion.

chain reaction in nuclear physics, a fission reaction that is maintained because neutrons released by the splitting of some atomic nuclei themselves go on to split others, releasing even more neutrons. Such a reaction can be controlled (as in a nuclear reactor) by using moderators to absorb excess neutrons. Uncontrolled, a chain reaction produces a nuclear explosion (as in an atom bomb).

chain rule in mathematics, rule for differentiating the composition of two functions. If $y = f(u)$ and $u = g(x)$ then:

$dy/dy = dy/du \times du/dx$

chalaza glutinous mass of transparent albumen supporting the yolk inside birds' eggs. The chalaza is formed as the egg slowly passes down the oviduct, when it also acquires its coiled structure.

chalcedony form of the mineral quartz, SiO_2, in which the crystals are so fine-grained that they are impossible to distinguish with a microscope (cryptocrystalline). Agate, onyx, and carnelian are †gem varieties of chalcedony.

chalcogen or **oxygen group element**, in chemistry, alternative name for any of the five members of group VIa of the periodic table. These five elements are oxygen, sulfur, selenium, tellurium, and polonium.

chalcopyrite copper iron sulfide mineral, $CuFeS_2$, the most common ore of copper. It is brassy yellow in color and may have an iridescent surface tarnish. It occurs in many different types of mineral vein, in rocks ranging from basalt to limestone.

chalk soft, fine-grained, whitish sedimentary rock composed of calcium carbonate, $CaCO_3$, extensively quarried for use in cement, lime, and mortar, and in the manufacture of cosmetics and toothpaste. **Blackboard chalk** in fact consists of †gypsum (calcium sulfate, $CaSO_4.2H_2O$).

Chalk was once thought to derive from the remains of microscopic animals or foraminifera. In 1953, however, it was seen under the electron microscope to be composed chiefly of †coccolithophores, unicellular lime-secreting algae, and hence primarily of plant origin. It is formed from deposits of deep-sea sediments called oozes.

Challis, James (1803–1882) English astronomer. He failed to take the advice of John Couch †Adams on where to search for the planet †Neptune, leaving its discovery to French and German astronomers.

Chamaeleon faint constellation of the south polar region, represented as a chameleon.

Chamberlain, Owen (1920–) US physicist whose graduate studies were interrupted by work on the Manhattan Project at Los Alamos. He shared the Nobel Prize for Physics in 1959 for his work with Italian physicist Emilio †Segrè after World War II, which led to the discovery of the existence of the antiproton.

Chamberlin, Thomas Chrowder (1843–1928) US geophysicist who asserted that the †Earth was far older than then believed. He developed the planetesimal hypothesis for the origin of the Earth and other planetary bodies – that they had been formed gradually by accretion of particles.

chance likelihood, or †probability, of an event taking place, expressed as a fraction or percentage. For example, the chance that a tossed coin will land heads up is 50%.

As a science, it originated when the Chevalier de Méré consulted Blaise †Pascal about how to reduce his gambling losses. In 1664, in correspondence with another mathematician, Pierre de †Fermat, Pascal worked out the foundations of the theory of chance. This underlies the science of statistics.

Chandrasekhar, Subrahmanyan (1910–1995) Indian-born US astrophysicist who was awarded the Nobel Prize for Physics in 1983 for his theoretical studies of the physical processes in connection with the structure and evolution of stars. The †Chandrasekhar limit is the maximum mass of a †white dwarf before it turns into a †neutron star.

Chandrasekhar, Subrahmanyan http://www-groups.dcs.st-andrews.ac.uk/ history/Mathematicians/Chandrasekhar.html
Devoted to the life and contributions of Indian-born US astronomer Subrahmanyan Chandrasekhar, this site provides biographical information, a list of references about Chandrasekhar, and links to essays that reference him. The text of this essay includes hypertext links to the essays of those mathematicians and thinkers who influenced Chandrasekhar. You will also find an image of him, which you can click on to enlarge, and a map of his birthplace, Lahore (now in Pakistan).

Chandrasekhar limit or **Chandrasekhar mass**, in astrophysics, maximum possible mass of a †white dwarf star. The limit depends slightly on the composition of the star but is equivalent to 1.4 times the mass of the Sun. A white dwarf heavier than the Chandrasekhar limit would collapse under its own weight to form a †neutron star or a †black hole. The limit is named after the Indian-born US astrophysicist Subrahmanyan †Chandrasekhar, who developed the theory of white dwarfs in the 1930s.

Chandreasekah–Schönberg limit mass above which the helium core of a star begins to contract (eventually to collapse altogether) The limit is now reckoned as 10–15% of the star's total mass.

change of state in science, change in the physical state (solid, liquid, or gas) of a material. For instance, melting, boiling, and evaporation and their opposites, solidification and condensation, are changes of state. The former set of changes are brought about by heating or decreased pressure; the latter by cooling or increased pressure. These changes involve the absorption or release of heat energy, called †latent heat, even though the temperature of the material does not change during the transition between states. See also †states of matter. Changes of state can be explained by the †kinetic theory of matter. In the unusual change of state called **sublimation**, a solid changes directly to a gas without passing through the liquid state. For example, solid carbon dioxide (dry ice) sublimes to carbon dioxide gas.

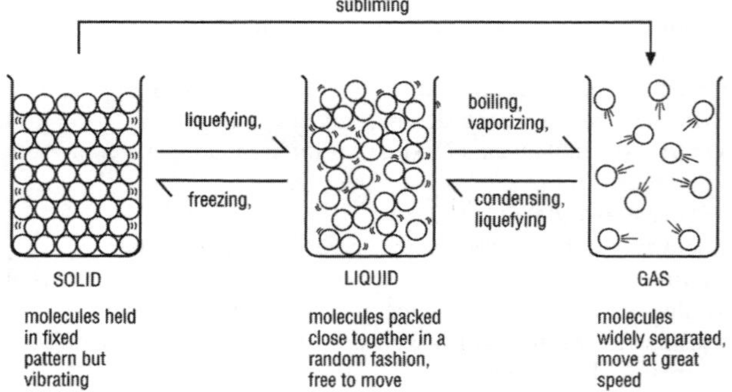

change of state The state (solid, liquid, or gas) of a substance is not fixed but varies with changes in temperature and pressure. Helicon

change of variable or **integration by substitution**, in mathematics, transformation of an integral by making a substitution for the variable being integrated with respect to.

channel efficiency measure of the ability of a river channel to move water and sediment. Channel efficiency can be measured by calculating the channel's †hydraulic radius (cross-sectional area/length of bed and bank). The most efficient channels are generally semicircular in cross-section, and it is this shape that water engineers try to create when altering a river channel to reduce the risk of †flooding.

chaos theory or **chaology** or **complexity theory**, branch of mathematics that attempts to describe irregular, unpredictable systems – that is, systems whose behavior is difficult to predict because there are so many variable or unknown factors. Weather is an example of a chaotic system.

Chaos theory, which attempts to predict the *probable* behavior of such systems, based on a rapid calculation of the impact of as wide a range of elements as possible, emerged in the 1970s with the development of sophisticated computers. First developed for use in meteorology, it has also been used in such fields as economics.

Chaos Club http://www.chaosclub.com/
Introduction to chaos theory, for those with an underdeveloped mathematical background! The site includes a selection of games to illustrate the theory.

Dance of Chance http://polymer.bu.edu/museum/
Examination of fractal patterns in nature, including those of bacterial growth, erosion, metal deposition, and termite trails. Are they dominated by chance or are other factors involved?

chaparral thick scrub country of the southwestern USA. Thorny bushes have replaced what was largely evergreen oak trees.

characteristic in mathematics, the integral (whole-number) part of a †logarithm. The fractional part is the †mantissa.

For example, in base ten, $10^0 = 1$, $10^1 = 10$, $10^2 = 100$, and so on; the powers to which 10 is raised are the characteristics. To determine the power to which 10 must be raised to obtain a number between 10 and 100, say 20 (2 x 10, or log 2 + log 10), the logarithm for 2 is found (0.3010), and the characteristic 1 (log 10) added to make 1.3010.

character set in computing, the complete set of symbols that can be used in a program or recognized by a computer. It may include letters, digits, spaces, punctuation marks, and special symbols.

charcoal black, porous form of †carbon, produced by heating wood or other organic materials in the absence of air. It is used as a fuel in the smelting of metals such as copper and zinc, and by artists for making black line drawings. **Activated charcoal** has been powdered and dried so that it presents a much increased surface area for adsorption; it is used for filtering and purifying liquids and gases – for example, in drinking-water filters and gas masks.

Charcoal was traditionally produced by burning dried wood in a kiln, a process lasting several days. The kiln was either a simple hole in the ground, or an earth-covered mound. Today kilns are of brick or iron, both of which allow the waste gases to be collected and used. Charcoal had many uses in earlier centuries. Because of the high temperature at which it burns (1,100°C/2,012°F), it was used in furnaces and blast furnaces before the development of †coke. It was also used in an industrial process for obtaining ethanoic acid (acetic acid), in producing wood tar and †wood pitch, and (when produced from alder or willow trees) as a component of gunpowder.

Charcot, Jean-Martin (1825–1893) French neurologist who studied hysteria, sclerosis, locomotor ataxia, and senile diseases. Among his pupils was the founder of psychiatry, Sigmund †Freud.

Chardonnet, (Louis-Marie) Hilaire Bernigaud (1839–1924) Comte de Chardonnet, French chemist who developed artificial silk in 1883, the first artificial forming. He also worked on †cellulose nitrate.

Chargaff, Erwin (1905– 2002) Czech-born US biochemist, best known for his work on the base composition of deoxyribonucleic acid (†DNA). In 1950 he demonstrated a simple mathematical relationship between the proportions of nitrogenous bases in DNA. He worked in many fields of biochemistry, publishing on diverse topics such as how the body metabolizes fats and how blood coagulates.

charge see †electric charge.

charge conservation feature of †quantum mechanics, in which reactions between †elementary particles occur in such a way that there is no change in the total charge of the system after the event has occurred.

charge-coupled device CCD, device for forming images electronically, using a layer of silicon that releases electrons when struck by incoming light. The electrons are stored in †pixels and read off into a computer at the end of the exposure. CCDs are used in digital cameras, and have now almost entirely replaced photographic film for applications such as astrophotography, where extreme sensitivity to light is paramount.

charged particle beam high-energy beam of electrons or protons. Such beams are being developed as weapons.

Charles, Jacques Alexandre César (1746–1823) French physicist who studied gases and made the first ascent in a hydrogen-filled balloon in 1783. His work on the expansion of gases led to the formulation of †Charles's law.

Charles's law law stating that the volume of a given mass of gas at constant pressure is directly proportional to its absolute temperature (temperature in kelvin). It was discovered by French physicist Jacques †Charles in 1787, and independently by French chemist Joseph Gay-Lussac in 1802.

The gas increases by 1/273 (0.003663) of its volume at 0°C for each °C rise of temperature. This means that the coefficient of expansion of all gases is the same. The law is only approximately true.

charm in physics, a property possessed by one type of †quark (very small particles found inside protons and neutrons), called the charm quark. The effects of charm are only seen in experiments with particle †accelerators. See †elementary particles.

Chase, Mary Agnes (1869–1963) born Mary Agnes Meara, US botanist and suffragist who made outstanding contributions to the study of grasses. During the course of several research expeditions she collected many plants previously unknown to science, and her work provided much information about naturally occurring cereals and other food crops.

check digit in computing, a digit attached to an important code number as a validation check.

checking procedure in mathematics, a way of making quick checks on the answers to calculations.

There are several methods, such as †rounding the calculation to make an estimate. The most appropriate to use will depend on the type of calculation being done – algebraic equations may be checked by introducing the solution to see whether the equation works.

Common checking procedures include:

reworking: the calculation can be re-done in full

rounding and estimation: the numbers in the calculation can be rounded and an estimate produced

use of inverses: †inverses can be used. For example:

59 x 17 = 1,003 and 1,003 ÷ 17 = 59

48 – 19 = 29 and 29 + 19 = 48

placing the solution in an equation: in algebra, when solving an equation, the solution can be put back into the equation to make sure that it works. For example:

$2x + 6 = 28$, giving $2x = 22$, therefore $x = 11$; so, 2 x 11 + 6 = 28

reasoning: common sense will often help to decide whether an answer is sensible or not. For example, the order of magnitude of an answer can be checked to ensure that it is correct for the question.

checksum in computing, a control total of specific items of data. A checksum is used as a check that data have been input or transmitted correctly. It is used in communications and in, for example, accounts programs.

chelate chemical compound whose molecules consist of one or more metal atoms or charged ions joined to chains of organic residues by coordinate (or dative covalent) chemical †bonds.

The parent organic compound is known as a **chelating agent** – for example, EDTA (ethylene-diaminetetra-acetic acid), used in chemical analysis. Chelates are used in analytical chemistry, in agriculture and horticulture as carriers of essential trace metals, in water softening, and in the treatment of thalassaemia by removing excess iron, which may build up to toxic levels in the body. Metalloproteins (natural chelates) may influence the performance of enzymes or provide a mechanism for the storage of iron in the spleen and plasma of the human body.

chemical in general, any substance of constant composition that has been extracted or purified from its natural state or synthesized by artificial means.

chemical change change that occurs when two or more substances (reactants) interact with each other, resulting in the production of different substances (products) with different chemical compositions. A simple example of chemical change is the burning of carbon in oxygen to produce carbon dioxide (†combustion). Other types of chemical change include †decomposition, †oxidation, and †reduction.

chemical composition in chemistry, general term for the number, type, and distribution of elements and groups that make up a compound.

chemical element alternative name for †element.

chemical equation method of indicating the reactants and products of a chemical reaction by using chemical symbols and formulae. A chemical equation gives two basic pieces of information: (1) the reactants (on the left-hand side) and products (right-hand side); and (2) the reacting proportions (stoichiometry) – that is, how many units of each reactant and product are involved. The equation must balance; that is, the total number of atoms of a particular element on the left-hand side must be the same as the number of atoms of that element on the right-hand side.

$Na_2CO_3 + 2HCl \rightarrow 2NaCl + CO_2 + H_2O$

reactants ➔ *products*

This equation states that one molecule of sodium carbonate combines with two molecules of hydrochloric acid to form two molecules of sodium chloride, one of carbon dioxide, and one of water. Double arrows indicate that the reaction is reversible – in the formation of ammonia from hydrogen and nitrogen, the direction of the reaction depends on the temperature and pressure of the reactants.

$3H_2 + N_2 \longleftrightarrow 2NH_3$

Introduction to Chemical Equations http://www.wissensdrang.com/aufceam.htm

Essay that introduces the concept of chemical equations to the reader. The text is accompanied by explanatory diagrams.

chemical equilibrium condition in which the products of a chemical reaction are formed at the same rate at which they decompose back into the reactants, so that the concentration of each reactant and product remains constant. It is a †reversible reaction; the reaction can happen in both directions.

For example, in the †Haber process, nitrogen and hydrogen combine to form ammonia, but as the ammonia is formed, it breaks down again into nitrogen and hydrogen.

$N_2 + 3H_2 \longleftrightarrow 2NH_3$

At equilibrium, the forward and back reactions occur at the same rate. As fast as ammonia is being formed it is broken down into hydrogen and nitrogen, therefore, the amounts of nitrogen, hydrogen, and ammonia remain constant at equilibrium.

The amounts of reactant and product present at equilibrium are defined by the **equilibrium constant** for that reaction and specific temperature.

chemical family collection of elements that have very similar chemical and physical properties. In the †periodic table of the elements such collections are to be found in the vertical columns (groups). The groups that contain the most markedly similar elements are group I, the †alkali metals; group II, the †alkaline-earth metals; group VII, the †halogens; and group 0, the noble or †rare gases.

chemical oxygen demand COD, measure of water and effluent quality, expressed as the amount of oxygen (in parts per million) required to oxidize the reducing substances present.

Under controlled conditions of time and temperature, a chemical oxidizing agent (potassium permanganate or dichromate) is added to the sample of water or effluent under consideration, and the amount needed to oxidize the reducing materials present is measured. From this the chemically equivalent amount of oxygen can be calculated. Since the reducing substances typically include remains of living organisms, COD may be regarded as reflecting the extent to which the sample is polluted. Compare †biological oxygen demand.

chemical symbol see †symbol, chemical.

chemical weathering form of †weathering brought about by chemical attack on rocks, usually in the presence of water. Chemical weathering involves the breakdown of the original minerals within a rock to produce new minerals (such as †clay minerals, †bauxite, and †calcite). The breakdown of rocks occurs because of chemical reactions between the minerals in the rocks and substances in the environment, such as water, oxygen, and weakly acidic rainwater. Some chemicals are dissolved and carried away from the weathering source, while others are brought in.

Material worn away from rocks by weathering, either as fragments of rock or dissolved material, may be transported and deposited as sediments, which eventually become compacted to form †sedimentary rocks. Thus chemical weathering plays a part in the rock cycle.

Chemical processes involved in weathering include †carbonation (breakdown by weakly acidic rainwater), †hydrolysis (breakdown by water), †hydration (breakdown by the absorption of water), and †oxidation (breakdown by the oxygen in air and water).

The reaction of carbon dioxide gas in the atmosphere with †silicate minerals in rocks to produce carbonate minerals is called the "Urey reaction" after the chemist who proposed it, Harold †Urey. The Urey reaction is an important link between Earth's climate and the geology of the planet. It has been proposed that chemical weathering of large mountain ranges like the Himalayas can remove carbon dioxide from the atmosphere by the Urey reaction (or other more complicated reactions like it), leading to a cooler climate as the †greenhouse effects of the lost carbon dioxide are diminished.

chemiluminescence the emission of light from a substance as a result of a chemical reaction (rather than raising its temperature). See †luminescence.

chemisorption attachment, by chemical means, of a single layer of molecules, atoms, or ions of gas to the surface of a solid or, less frequently, a liquid. It is the basis of catalysis (see †catalyst) and is of great industrial importance.

chemistry branch of science concerned with the study of the structure and composition of the different kinds of matter, the changes that matter may undergo, and the phenomena that occur in the course of these changes.

Organic chemistry is the branch of chemistry that deals with carbon compounds. **Inorganic chemistry** deals with the description, properties, reactions, and preparation of all the elements and their compounds, with the exception of carbon compounds. **Physical chemistry** is concerned with the quantitative explanation of chemical phenomena and reactions, and the measurement of data required for such explanations. This branch studies in particular the movement of molecules and the effects of temperature and pressure, often with regard to gases and liquids.

molecules, atoms, and elements All matter can exist in three states: gas, liquid, or solid. It is composed of minute particles termed **molecules**, which are constantly moving, and may be further divided into †atoms.

Molecules that contain atoms of one kind only are known as **elements**; those that contain atoms of different kinds are called **compounds**.

compounds and mixtures Chemical compounds are produced by a chemical action that alters the arrangement of the atoms in the reacting molecules. Heat, light, vibration, catalytic action, radiation, or pressure, as well as moisture (for ionization), may be necessary to produce a chemical change. Examination and possible breakdown of compounds to determine their components is **analysis**, and the building up of compounds from their components is **synthesis**. When substances are brought together without changing their molecular structures they are said to be **mixtures**.

formulas and equations Symbols are used to denote the elements. The symbol is usually the first letter or letters of the English or Latin name of the element – for example, C for carbon; Ca for calcium; Fe for iron (*ferrum*). These symbols represent one atom of the element; molecules containing more than one atom of an element are denoted by a subscript figure – for example, water is H_2O. In some substances a group of atoms acts as a single entity, and these are enclosed in parentheses in the symbol – for example $(NH_4)_2SO_4$ denotes ammonium sulfate. The symbolic representation of a molecule is known as a **formula**. A figure placed before a formula represents the number of molecules of a substance taking part in, or being produced by, a chemical reaction – for example, $2H_2O$ indicates two molecules of water. Chemical reactions are expressed by means of **equations** as in:

$NaCl + H_2SO_4 \rightarrow NaHSO_4 + HCl$

This equation states the fact that sodium chloride (NaCl) on being treated with sulfuric acid (H_2SO_4) is converted into sodium bisulphate (sodium hydrogen sulfate, $NaHSO_4$) and hydrogen chloride (HCl). See also †chemical equation.

metals, non-metals, and the periodic system Elements are divided into **metals**, which have luster and conduct heat and electricity, and **non-metals**, which usually lack these properties. The **periodic system**, developed by John Newlands in 1863 and established by Dmitri †Mendeleyev in 1869, classified elements according to their relative atomic masses. Those elements that resemble each other in general properties were found to bear a relation to one another by weight, and these were placed in groups or families. Certain anomalies in this system were later removed by classifying the elements according to their atomic numbers. The latter is equivalent to the positive charge on the nucleus of the atom.

ChemTutor http://www.chemtutor.com/

Beginning with the fundamentals, and graduating to "the parts of primary chemistry that have been... the hardest for students to grasp," this site is a useful resource for any chemistry student.

CHEMystery http://www.thinkquest.org/library/lib/site_sum_outside.html?tname=3659&url=3659/

Detailed interactive textbook that includes subjects such as "Atomic structure and bonding," "Equilibrium," and "Thermodynamics." Use the left-hand menu to navigate to a relevant chapter.

Eric Weisstein's World of Chemistry http://scienceworld.wolfram.com/chemistry/

Searchable database of facts relating to chemistry. The author has included an A–Z glossary of terms, history, biographies, and concepts of chemistry and chemical engineering.

General Chemistry Online http://antoine.frostburg.edu/chem/senese/101/index.shtml

This sophisticated Web site is run by the chemistry department of a US university, but its clear text and comprehensive coverage of general chemistry topics make it worth a visit for keen GCSE-level chemists. Searchable databases of common compounds and important chemical terms are two of the most powerful features of a very comprehensive resource.

chemosynthesis method of making †protoplasm (contents of a cell) using the energy from chemical reactions, in contrast to the use of light energy employed for the same purpose in †photosynthesis. The process is used by certain bacteria, which can synthesize organic compounds from carbon dioxide and water using the energy from special methods of †respiration.

Nitrifying bacteria are a group of chemosynthetic organisms that change free nitrogen into a form that can be taken up by plants; nitrobacteria, for example, oxidize nitrites to nitrates. This is a vital part of the †nitrogen cycle. As chemosynthetic bacteria can survive without light energy, they can live in dark and inhospitable regions, including the hydrothermal vents of the Pacific Ocean. Around these vents, where temperatures reach up to 350°C/662°F, the chemosynthetic bacteria are the basis of a food web supporting fishes and other marine life.

chemotaxis in biology, the property that certain cells have of attracting or repelling other cells. For example, white blood cells are attracted to the site of infection by the release of substances during certain types of immune response.

chemotherapy any medical treatment with chemicals. It usually refers to treatment of cancer with cytotoxic and other drugs. The term was coined by the German bacteriologist Paul Ehrlich for the use of synthetic chemicals against infectious diseases.

chemotropism movement by part of a plant in response to a chemical stimulus. The response by the plant is termed "positive" if the growth is towards the stimulus or "negative" if the growth is away from the stimulus.

Fertilization of flowers by pollen is achieved because the ovary releases chemicals that produce a positive chemotropic response from the developing pollen tube.

Cherenkov, Pavel Alexeevich (1904–1990) Soviet physicist. He was awarded the Nobel Prize for

Physics in 1958 for his discovery in 1934 of †Cherenkov radiation; this occurs as a bluish light when charged atomic particles pass through water or other media at a speed in excess of that of light. He shared the award with his colleagues Ilya †Frank and Igor †Tamm for work resulting in a cosmic-ray counter.

Cherenkov detector apparatus through which it is possible to observe the existence and velocity of high-speed particles, important in experimental nuclear physics and in the study of †cosmic radiation. It was originally built to investigate the Cherenkov radiation effect, in which charged particles travel through a medium at a speed greater than that of light in that medium.

Cherenkov radiation type of electromagnetic radiation emitted by charged particles entering a transparent medium at a speed greater than the speed of light in the medium. It appears as a bluish light. Cherenkov radiation can be detected from high-energy cosmic rays entering the Earth's atmosphere. It is named after Pavel Alexseevich †Cherenkov, the Russian physicist who first observed it.

childbirth the expulsion of a baby from its mother's body following †pregnancy. In a broader sense, it is the period of time involving labor and delivery of the baby.

chimera or **chimaera**, in biology, an organism composed of tissues that are genetically different. Chimeras can develop naturally if a †mutation occurs in a cell of a developing embryo, but are more commonly produced artificially by implanting cells from one organism into the embryo of another.

china clay commercial name for †kaolin.

chinook (American Indian "snow-eater") warm dry wind that blows downhill on the east side of the Rocky Mountains of North America. It often occurs in winter and spring when it produces a rapid thaw, and so is important to the agriculture of the area.

chip or **silicon chip**, another name for an †integrated circuit, a complete electronic circuit on a slice of silicon (or other semiconductor) crystal only a few millimeters square.

chiral center in chemistry, that part of an asymmetric molecule that is responsible for †optical activity. A molecule that cannot be superimposed on its mirror image and has the ability to rotate the plane of polarization of polarized light is said to be optically active. In organic compounds, the chiral center is usually a carbon atom that has four different groups attached to it. A large molecule can have more than one chiral center.

chirality in chemistry, property of a molecule that indicates that it cannot be superimposed on its mirror image. Chiral molecules have the same type, number, and ratio of atoms, but differ in the orientation of these atoms within the molecular structure. The term implies "handedness"; the relationship of the molecule with its mirror image is that between a right and left hand (see also †chiral center). The two forms of a chiral compound may have different chemical properties, so it is important to be able to distinguish between them.

Chiron unusual object orbiting between Saturn and Uranus, discovered in 1977 by US astronomer Charles Kowal. Initially classified as an asteroid, it is now believed to be a giant cometary nucleus at least 200 km/120 mi across, composed of ice with a dark crust of carbon dust. It has a 51-year orbit and a coma (cloud of gas and dust) caused by evaporation from its surface, resembling that of a comet. It is classified as a †centaur.

chi-squared function X^2 function, function that in †probability theory provides a test for deviation from a †null hypothesis. It is usually represented as being made up of:

(observed frequency of result – expected frequency of result)^2/expected frequency of result

in which the top line indicates the (squared) deviation from the expected.

chitin complex long-chain compound, or †polymer; a nitrogenous derivative of glucose. Chitin is widely found in invertebrates. It forms the †exoskeleton of insects and other arthropods. It combines with protein to form a covering that can be hard and tough, as in beetles, or soft and flexible, as in caterpillars and other insect larvae. It is insoluble in water and resistant to acids, alkalis, and many organic solvents. In crustaceans such as crabs, it is impregnated with calcium carbonate for extra strength.

Chitin also occurs in some †protozoans and coelenterates (such as certain jellyfishes), in the jaws of annelid worms, and as the cell-wall polymer of fungi. Its uses include coating apples (still fresh after six months), coating seeds, and dressing wounds. In 1993 chemists at North Carolina State University found that it can be used to filter pollutants from industrial waste water.

Chladni, Ernst Florens Friedrich (1756–1827) German physicist, a pioneer in the field of acoustics. He developed an experimental technique whereby sand is vibrated on a metal plate and settles into regular and symmetric patterns (†Chladni's figures), indicating the nodes of the vibration's wave pattern.

Chladni figures visual manifestations of sound waves resulting from the behavior of small particles on plates vibrated in sympathy to a particular sound. They were first recognized by German physicist Ernst Chladni.

chlamydia viruslike bacteria that live parasitically in animal cells, and cause disease in humans and birds. Chlamydiae are thought to be descendants of bacteria that have lost certain metabolic processes. In humans, a strain of chlamydia causes trachoma, a disease found mainly in the tropics (a leading cause of blindness); venereally transmitted chlamydiae cause genital and urinary infections.

A possible link was found between infection with chlamydia and Alzheimer's disease by US researchers in 1998. Chlamydia was found in the brains of 89% of Alzheimer's sufferers within the study, but only 5% of the non-Alzheimer's control group.

A Finnish study published in January 2001 indicated that a woman's risk of developing cervical cancer is 2.5 times higher if she has been previously infected with sexually transmitted chlamydia. Some strains of chlamydia appear to increase the risk by a factor of seven.

chloral or **trichloroethanal**; CCl_3CHO, oily, colorless liquid with a characteristic pungent smell, produced by the action of chlorine on ethanol. It is soluble in water and its compound chloral hydrate is a powerful sleep-inducing agent.

chlorate any salt derived from an acid containing both chlorine and oxygen and possessing the negative ion ClO, ClO_2, ClO_3, or ClO_4. Common chlorates are those of sodium, potassium, and barium. Certain chlorates are used in weedkillers.

chlordane organochlorine †pesticide, used especially against ants and termites.

chlorenchyma in botany, †parenchyma cells that contain chloroplasts. Spongy mesophyll cells in the leaf are an example of chlorenchyma.

chloride Cl^-, negative ion formed when hydrogen chloride dissolves in water, and any salt containing this ion, commonly formed by the action of hydrochloric acid (HCl) on various metals or by direct combination of a metal and chlorine. Sodium chloride (NaCl) is common table salt.

chlorinated solvent any liquid organic compound that contains chlorine atoms, often two or more. These compounds are very effective solvents for fats and greases, but many have toxic properties.

They include trichloromethane (chloroform, $CHCl_3$), tetrachloromethane (carbon tetrachloride, CCl_4), and trichloroethene ($CH_2ClCHCl_2$).

Breath of Fresh Chlorine

http://www.sciam.com/article.cfm?articleID=000E12E6-8481-1C76-9B81809EC588EF21&pageNumber=1&catID=4

Part of a larger site maintained by Scientific American, this page reports on a strain of bacteria that may reduce the level of hazardous waste in the world by removing the chlorine molecules found in noxious chlorinated solvents that degrade our environment. Find out how this bacterium powers itself and whether it has a future in the waste management industry. The text includes hypertext links to further information, and there is also a list of related links at the bottom of the page.

chlorination treatment of water with chlorine in order to disinfect it; also, any chemical reaction in which a chlorine atom is introduced into a chemical compound.

chlorine chemical symbol Cl, (Greek *chloros* "green") greenish-yellow, gaseous, non-metallic element with a pungent odor, atomic number 17, relative atomic mass 35.453. It is a member of the †halogen group and is widely distributed, in combination with the †alkali metals, as chlorides.

Chlorine was discovered in 1774 by the German chemist Karl Scheele, but English chemist Humphry Davy first proved it to be an element in 1810 and named it after its color. In nature it is always found in the combined form, as in hydrochloric acid, produced in the mammalian stomach for digestion. Chlorine is obtained commercially by the electrolysis of concentrated brine and is an important bleaching agent and germicide, used for sterilizing both drinking water and swimming pools. As an oxidizing agent it finds many applications in organic chemistry. The pure gas (Cl_2) is a poison and was used in gas warfare in World War I, where its release seared the membranes of the nose, throat, and lungs, causing pneumonia. Chlorine is a component of chlorofluorocarbons (CFCs) and is partially responsible for the depletion of the †ozone layer; it is released from the CFC molecule by the action of ultraviolet radiation in the upper atmosphere, making it available to react with and destroy the ozone. The concentration of chlorine in the atmosphere in 1997 reached just over 3 parts per billion. It was expected to reach its peak in 1999 and then start falling rapidly due to international action to curb ozone-destroying chemicals.

Chlorine http://c3.org/

Designed to promote better understanding of the science of chlorine chemistry, examining how it contributes to an enhancement of our standard of living and quality of life.

On a Combination of Oxymuriatic Gas and Oxygene Gas http://dbhs.wvusd.k12.ca.us/Chem-History/Davy-Chlorine-1811.html

Transcript of Humphry Davy's submission to the Royal Society naming chlorine and labeling it as an element. The text is taken from Philosophical Transactions of the Royal Society vol. 101, pp 155–162 (1811). Davy had discovered the other half (sodium) of common salt three years earlier.

chlorobenzene C_6H_5Cl, colorless, aromatic liquid, boiling point 132°C/270°F. It is prepared by reacting benzene with chlorine in the presence of an iron or aluminum catalyst. The compound smells of almonds and is used as an industrial solvent and in the synthesis of other organic chemicals, notably phenol and DDT.

chloroethane or **ethyl chloride**; CH_3CH_2Cl, colorless, flammable liquid, boiling point 12.5°C/54.5°F. It is prepared by chlorination of ethane in the presence of a catalyst. It is widely used as a local anesthetic.

chlorofluorocarbon **CFC**, a class of **synthetic chemicals** that are odorless, nontoxic, nonflammable, and chemically inert. The first CFC was synthesized in 1892, but no use was found for it until the 1920s. Their stability and apparently harmless properties made CFCs popular as propellants in †aerosol cans, as refrigerants in refrigerators and air conditioners, as degreasing agents, and in the manufacture of foam packaging. They are now known to be partly responsible for the destruction of the †ozone layer. In June 1990 representatives of 93 nations, including the UK and the USA, agreed to phase out production of CFCs and various other ozone-depleting chemicals by the end of the 20th century.

When CFCs are released into the atmosphere, they drift up slowly into the stratosphere, where, under the influence of ultraviolet radiation from the Sun, they react with ozone (O_3) to form free chlorine (Cl) atoms and molecular oxygen (O_2), thereby destroying the ozone layer that protects Earth's surface from the Sun's harmful ultraviolet rays. The chlorine liberated during ozone breakdown can react with still more ozone, making the CFCs particularly dangerous to the environment. CFCs can remain in the atmosphere for more than 100 years. Replacements for CFCs are being developed, and research into safe methods for destroying existing CFCs is being carried out.

Problem of ChloroFluoroCarbons http://pooh.chem.wm.edu/chemWWW/courses/chem105/projects/group2/page5.html
Graphical presentation of the effect of CFCs on the ozone layer. The information can be readily understood by a general reader wishing to learn more about the chemistry of ozone depletion and why more ultraviolet radiation is reaching the surface of the earth.

chloroform or **trichloromethane**; $CHCl_3$, clear, colorless, toxic, carcinogenic liquid with a characteristic pungent, sickly sweet smell and taste, formerly used as an anesthetic (now superseded by less harmful substances). It is used as a solvent and in the synthesis of organic chemical compounds.

chloromethane or **methyl chloride**; CH_3Cl, colorless, flammable, poisonous gas, boiling point 24°C/75°F. It is usually prepared by a catalytic reaction between hydrogen chloride and methanol, and is mainly used in the production of silicones.

chlorophyll group of pigments including chlorophyll a and chlorophyll b, the green pigments present in †chloroplasts in most plants; it is responsible for the absorption of †light energy during †photosynthesis. The pigment absorbs the red and blue-violet parts of sunlight but reflects the green, thus giving plants their characteristic color. Other chlorophylls include chlorophyll c (in brown algae) and chlorophyll d (found in red algae).

Chlorophyll is found within chloroplasts, present in large numbers in leaves. Cyanobacteria (blue-green algae) and other photosynthetic bacteria also have chlorophyll, though of a slightly different type. Chlorophyll is similar in structure to †hemoglobin, but with magnesium instead of iron as the reactive part of the molecule. Because magnesium is contained in chlorophyll it is considered an essential plant †mineral salt. Magnesium can often be a part of a mixture of †minerals used as a fertilizer.

chloroplast structure (†organelle) within a plant cell containing the green pigment †chlorophyll. Chloroplasts occur in most cells of green plants that are exposed to light, often in large numbers. Typically, they are shaped like a flattened disc, with a double membrane enclosing the stroma, a gel-like matrix. Within the stroma are stacks of fluid-containing cavities, or vesicles, where †photosynthesis occurs, creating †glucose from carbon dioxide and water to be used in the plant's life processes. Sunlight is absorbed by chlorophyll, providing energy that is transferred to the glucose. The glucose may be converted to starch and stored. Starch can then be converted back to glucose to provide energy for the plant at a later stage.

It is thought that the chloroplasts were originally free-living cyanobacteria (blue-green algae), which invaded larger, non-photosynthetic cells and developed a symbiotic relationship with them. Like †mitochondria, they contain a small amount of DNA and divide by fission. Chloroplasts are a type of †plastid.

chlorosis abnormal condition of green plants in which the stems and leaves turn pale green or yellow. The yellowing is due to a reduction in the levels of the green chlorophyll pigments. It may be caused by a deficiency in essential elements (such as magnesium, iron, or manganese), a lack of light, genetic factors, or viral infection.

choke coil in physics, a coil employed to limit or suppress alternating current without stopping direct current, particularly the type used as a "starter" in the circuit of fluorescent lighting.

cholecalciferol or **vitamin D**, fat-soluble chemical important in the uptake of calcium and phosphorous for bones. It is found in liver, fish oils, and margarine. It can be produced in the skin, provided that the skin is adequately exposed to sunlight. Lack of vitamin D leads to rickets and other bone diseases.

cholesterol white, crystalline †sterol found throughout the body, especially in fats, blood, nerve tissue, and bile; it is also provided in the diet by foods such as eggs, meat, and butter. A high level of cholesterol in the blood is thought to contribute to atherosclerosis (hardening of the arteries).

Cholesterol is an integral part of all cell membranes and the starting point for steroid hormones, including the sex hormones. It is broken down by the liver into bile salts, which are involved in fat absorption in the digestive system, and it is an essential component of **lipoproteins**, which transport fats and fatty acids in the blood. **Low-density lipoprotein cholesterol** (LDL-cholesterol), when present in excess, can enter the tissues and become deposited on the surface of the arteries, causing atherosclerosis. **High-density lipoprotein cholesterol** (HDL-cholesterol) acts as a scavenger, transporting fat and cholesterol from the tissues to the liver to be broken down. The composition of HDL-cholesterol can vary and some forms may not be as effective as others. Blood cholesterol levels can be altered by reducing the amount of alcohol and fat in the diet and by substituting some of the saturated fat for polyunsaturated fat, which gives a reduction in LDL-cholesterol. A 1999 US study of children with high levels of cholesterol found no evidence that controlling cholesterol levels through diet is harmful. Another 1999 US study suggested that cholesterol-lowering drugs are as beneficial for older men and women as they are for the middle-aged. HDL-cholesterol can be increased by exercise.

chondrite type of †meteorite characterized by †chondrules (small spherules), typically about 1 mm/0.04 inches in diameter, made up of silicate material.

chondrule in astronomy, small, round mass of the †silicate minerals olivine and orthopyroxene. Chondrites (stony †meteorites) are characterized by the presence of chondrules.

Chondrules are thought to be mineral grains that condensed from hot gas in the early Solar System, most of which were later incorporated into larger bodies from which the †planets formed.

chord in geometry, a straight line joining any two points on a curve. The chord that passes through the center of a circle (its longest chord) is the diameter. The longest and shortest chords of an ellipse (a regular oval) are called the major and minor axes, respectively.

Chordata phylum of animals, members of which are called †chordates.

chordate animal belonging to the phylum Chordata, which includes vertebrates, sea squirts, amphioxi, and others. All these animals, at some stage of their lives, have a supporting rod of tissue (notochord or backbone) running down their bodies.

Chordates are divided into three major groups: †tunicates, cephalochordates, and craniates (including all vertebrates).

chorion outermost of the three membranes enclosing the embryo of reptiles, birds, and mammals; the †amnion is the innermost membrane.

choroid layer found at the rear of the †eye beyond the retina. By absorbing light that has already passed through the retina, it stops back-reflection and so prevents blurred vision.

chromate in chemistry, any substance that contains an oxygen-containing chromium species, in particular any salts of chromium oxide, CrO_3. Chromates are strongly colored, yellow, orange, and red compounds and are used in tanning and as pigments.

chromatic aberration optical effect commonly found in simple lens instruments in which colored fringes are seen around an image as a result of the †wavelength-dependence of the †refractive index of glass.

chromatin nucleoprotein found in †chromosomes and thought to be the molecular substance of heredity. It is readily stained by basic dyes and is therefore easily identified and studied under the microscope.

chromatography (Greek *chromos* "color") technique for separating or analyzing a mixture of gases, liquids, or dissolved substances. This is brought about by means of two immiscible substances, one of which (**the mobile phase**) transports the sample mixture through the other (**the stationary phase**). The mobile phase may be a gas or a liquid; the stationary phase may be a liquid or a solid, and may be in a column, on paper, or in a thin layer on a glass or plastic support. The components of the mixture are adsorbed or impeded by the stationary phase to different extents and therefore become separated. The technique is used for both qualitative and quantitative analyses in biology and chemistry.

In **paper chromatography**, the mixture separates because the components have differing solubilities in the solvent flowing through the paper and in the chemically bound water of the paper.

In **thin-layer chromatography**, a wafer-thin layer of adsorbent medium on a glass plate replaces the filter paper. The mixture separates because of the differing solubilities of the components in the solvent flowing up the solid layer, and their differing tendencies to stick to the solid (adsorption). The same principles apply in **column chromatography**.

In **gas–liquid chromatography**, a gaseous mixture is passed into a long, coiled tube (enclosed in an oven) filled with an inert powder coated in a liquid. A carrier gas flows through the tube. As the mixture proceeds along the tube it separates as the components dissolve in the liquid to differing extents or stay as a gas. A detector locates the different components as they emerge from the tube. The technique is very powerful, allowing tiny quantities of substances (fractions of parts per million) to be separated and analyzed.

Preparative chromatography is carried out on a large scale for the purification and collection of one or more of a mixture's constituents; for example, in the recovery of protein from abattoir wastes.

Analytical chromatography is carried out on far smaller quantities, often as little as one microgram (one-millionth of a gram), in order to identify and quantify the component parts of a mixture. It is used to determine the identities and amounts of amino acids in a protein, and the alcohol content of blood and urine samples. The technique was first used in the separation of colored mixtures into their component pigments.

Chromatography

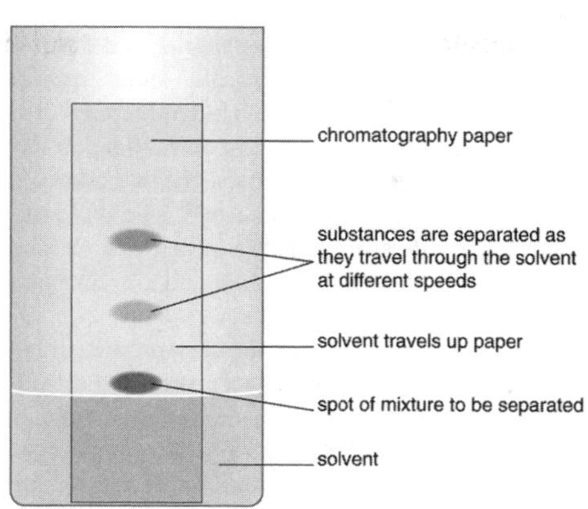

chromatography Paper chromatography utilizes the fact that different substances dissolve at different rates. As the solvent travels up the paper it dissolves the mixture, the components of which travel at different speeds and so become separated. Helicon

chromatophore in zoology, pigment-containing cell in the skin. It may change shape or distribution of pigment in response to nervous or hormonal stimulation causing color change, as in the chameleon, for example.

chromite $FeCr_2O_4$, iron chromium oxide, the main chromium ore. It is one of the †spinel group of minerals, and crystallizes in dark-colored octahedra of the cubic system. Chromite is usually found in association with ultrabasic and basic rocks; in Cyprus, for example, it occurs with †serpentine, and in South Africa it forms continuous layers in a layered †intrusion.

chromium chemical symbol Cr, (Greek *chromos* "color") hard, brittle, gray-white, metallic element, atomic number 24, relative atomic mass 51.996. It takes a high polish, has a high melting point, and is very resistant to corrosion. It is used in chromium electroplating, in the manufacture of stainless steel and other alloys, and as a catalyst. Its compounds are used for tanning leather and for †alums. In human nutrition it is a vital trace element. In nature, it occurs chiefly as chrome iron ore or chromite ($FeCr_2O_4$). Kazakhstan, Zimbabwe, and Brazil are sources.

The element was named in 1797 by the French chemist Louis Vauquelin after its brightly colored compounds.

chromium ore essentially the mineral †chromite, $FeCr_2O_4$, from which chromium is extracted. South Africa and Zimbabwe are major producers.

chromogranin protein released by the †adrenal gland, along with the hormone adrenaline, during times of stress. There are three types, chromagranin A, B, and C. The function of chromagranins is poorly understood, but they do have an antibacterial affect that could boost the immune system when it is suppressed during times of stress. Chromagranin A also stimulates †insulin release and relaxes blood vessels – functions similarly reduced by adrenaline during stress.

chromophore in chemistry, group of atoms that is responsible for making a compound colored. Examples of such groups are –C = C–, –N = N–, and –C = O.

chromosome structures in a cell †nucleus that carry the many thousands of †genes, in sequence, that determine the characteristics of an organism. There are 46 chromosomes in a normal human cell. Each chromosome normally consists of one very long strand (or molecule) of DNA, coiled and folded to produce a compact structure. The exception is just before cell division when each chromosome contains two strands of DNA, a result of the copying of each molecule of DNA. The point on a chromosome where a particular gene occurs is known as its locus. Most higher organisms have two copies of each chromosome, together known as a **homologous pair** (they are †diploid) but some have only one (they are †haploid). See also †mitosis and †meiosis.

In a working cell chromosomes exist in a less dense form called chromatin and cannot be seen using a laboratory light microscope. However, during cell division they get shorter and fatter and so become visible. When pictures are seen of chromosomes, they have usually been taken in this state during mitosis. In such pictures of human chromosomes it can be seen that most chromosomes in a cell have a matching chromosome of exactly the same size. These are known as homologous pairs. However, one pair is not matched. These are the sex chromosomes – a short, male determining one called the †Y chromosome and the †X chromosome. Males have an X and a Y chromosome and females have two Xs.

In a sex cell (gamete) that has been produced by meiosis, the number of chromosomes is halved. Only one chromosome from each pair of homologous chromosomes is found in a gamete.

Cytogenetics Gallery http://www.pathology.washington.edu/Cytogallery/

Online gallery of chromosomes from the Pathology Department of the University of Washington, Seattle. This in-depth look at the chromosome contains a concise definition of the subject as well as a collection of images used to show how scientists pin-point chromosomal abnormalities.

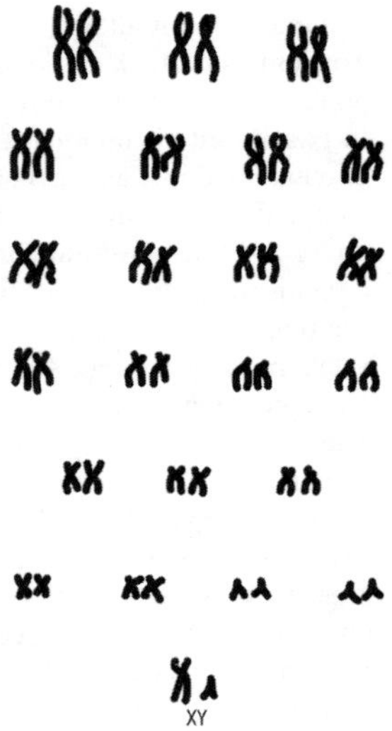

XY

chromosome The 23 pairs of chromosomes of a normal human male. Helicon

chromosome map description of the position of †genes along a †chromosome.

chromosphere (Greek *chromos* "color," *sphaira* "sphere") layer of mostly hydrogen gas about 10,000 km/6,000 mi deep above the visible surface of the Sun (the photosphere). It appears pinkish red during †eclipses of the Sun.

chronobiology area of study concerned with the relationship between biochemical, physiological, and behavioral activities and their synchrony with daily, seasonal, and annual changes. These rhythmic interactions occur as a result of †biorhythms and the †biological clock, the self-sustained oscillations of cellular activities.

chronometer instrument for measuring time precisely, originally used at sea. It is designed to remain accurate through all conditions of temperature and pressure. The first accurate marine chronometer, capable of an accuracy of half a minute a year, was made in 1761 by John Harrison in England.

chrysotile mineral in the †serpentine group, $Mg_3Si_2O_5(OH)_4$. A soft, fibrous, silky mineral, the primary source of asbestos.

Church, Alonzo (1903–1995) US mathematician. In 1936 he published the first precise definition of a calculable function, and so contributed enormously to the systematic development of the theory of †algorithms. The solving of algorithmic problems involves the construction of an algorithm capable of solving a given set with respect to some other set, and if such an algorithm cannot be constructed, it signifies that the problem is unsolvable. From English mathematician Alan †Turing's thesis, Church proved that there were no algorithms for a class of quite elementary arithmetical questions.

chyme general term for the stomach contents. Chyme resembles a thick creamy fluid and is made up of partly digested food, hydrochloric acid, and a range of enzymes.

The muscular activity of the stomach churns this fluid constantly, continuing the mechanical processes initiated by the mouth. By the time the chyme leaves the stomach for the duodenum, it is a smooth liquid ready for further digestion and absorption by the small intestine.

cilia singular **cilium**, small hairlike organs on the surface of some cells, particularly the cells lining the upper respiratory tract. Their wavelike movements waft particles of dust and debris towards the exterior. Some single-celled organisms move by means of cilia. In multicellular animals, they keep lubricated surfaces clear of debris. They also move food in the digestive tracts of some invertebrates.

ciliary muscle ring of muscle surrounding and controlling the lens inside the vertebrate eye, used in †accommodation (focusing). Suspensory ligaments, resembling spokes of a wheel, connect the lens to the ciliary muscle and pull the lens into a flatter shape when the muscle relaxes. When the muscle is relaxed the lens has its longest †focal length and focuses rays from distant objects. On contraction, the lens returns to its normal spherical state and therefore has a shorter focal length and focuses images of near objects.

ciliate aquatic protozoan of the phylum Ciliophora, characterized by possessing †cilia (hairlike organs) at some stage in its life cycle.

Most ciliates are asymmetrical organisms whose shape is definite (unlike the amoebae). There are free-swimming forms such as *Paramecium*, sessile (fixed) forms, such as *Vorticella*, and some may be found in colonies.

cinnabar mercuric sulfide mineral, HgS, the only commercially useful ore of mercury. It is deposited in veins and impregnations near recent volcanic rocks and hot springs. The mineral itself is used as a red pigment, commonly known as **vermilion**. Cinnabar is found in the USA (California), Spain (Almadén), Peru, Italy, and Slovenia.

cinnamic acid or **3-phenylpropenoic acid**; $C_6H_5CH = CHCOOH$, in chemistry, colorless crystalline solid, melting point 133°C/271°F. It can be prepared by heating benzaldehyde with ethanoic anhydride and sodium ethanoate. It is used in the preparation of perfumes, artificial flavorings, and cosmetics.

cipher mathematical term for †zero.

circadian rhythm metabolic rhythm found in most organisms, which generally coincides with the 24-hour day. Its most obvious manifestation is the regular cycle of sleeping and waking, but body temperature and the concentration of †hormones that influence mood and behavior also vary over the day. In humans, alteration of habits (such as rapid air travel round the world) may result in the circadian rhythm being out of phase with actual activity patterns, causing malaise until it has had time to adjust.

In mammals the circadian rhythm is controlled by the suprachiasmatic nucleus in the †hypothalamus. US researchers discovered a second circadian control mechanism in 1996; they found that cells within the retina also produced the hormone †melatonin. In 1997 US geneticists identified a gene, *clock*, in chromosome 5 in mice, that regulated the circadian rhythm.

Circinus small constellation of the southern hemisphere, represented as a pair of compasses.

Circinus X-1 in astronomy, strong X-ray source in the constellation Circinus, which flares every 16.6 days. It probably consists of a binary combination of a black hole and an ordinary star that emits X-rays as material passes from the star into the black hole. There is some evidence to suggest that one of the components is a neutron star, not a black hole.

circle perfectly round shape, the path of a point that moves so as to keep a constant distance from a fixed point (the **center**). A circle has a †**radius** (the distance from any point on the circle to the center), a †**circumference** (the boundary of the circle, part of which is called an †**arc**), †**diameters** (straight lines crossing the circle through the center), †**chords** (lines joining two points on the circumference), †**tangents** (lines that touch the circumference at one point only), †**sectors** (regions inside the circle between two radii), and †**segments** (regions between a chord and the circumference).

The ratio of the distance all around the circle (the circumference) to the diameter is an †irrational number called p (**pi**), roughly equal to 3.1416.

A circle of radius *r* and diameter *d* has a circumference $C = pd$, or $C = 2pr$, and an area $A = pr^2$. The area of a circle can be shown by dividing it into very thin sectors and reassembling them to make an approximate rectangle. The proof of $A = pr^2$ can be done only by using †integral calculus.

Angles within a circle can be calculated using the †circle theorems.

StudyWeb http://library.thinkquest.org/11962/data/algindex.HTML

Basic properties of circles. This site aims to cover issues such as the center, radius, and the equation of the circle. There are exercises and examples included, along with a link to math-related content on the rest of the site.

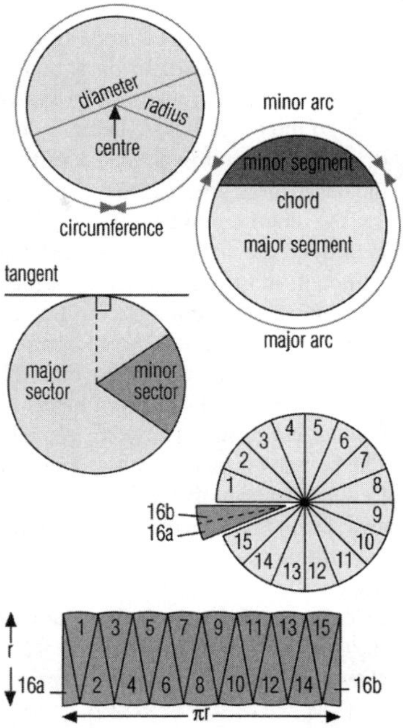

circle Technical terms used in the geometry of the circle; the area of a circle can be seen to equal pr^2 by dividing the circle into segments that form a rectangle. Helicon

circle theorem in geometry, the relationship between †lines and †angles within a †circle. These can be used in further geometrical †proofs.

circuit in physics or electrical engineering, an arrangement of electrical components connected by a conducting material through which a current can flow. There are two basic circuits, series and parallel. In a †series circuit, the components are connected end to end so that the current flows through all components one after the other. In a †parallel circuit, components are connected side by side so that part of the current passes through each component. A circuit diagram shows in graphical form how components are connected together, using standard symbols for the components. If the circuit is unbroken, it is a closed circuit and current flows. If the circuit is broken, it becomes an open circuit and no current flows.

Circuit http://library.thinkquest.org/10796/ch14/ch14.htm?tqskip=1

Site with helpful graphics and in-depth case studies of series and parallel circuits. There are also interactive questions included to test the user's knowledge, with answers available online at the press of a button.

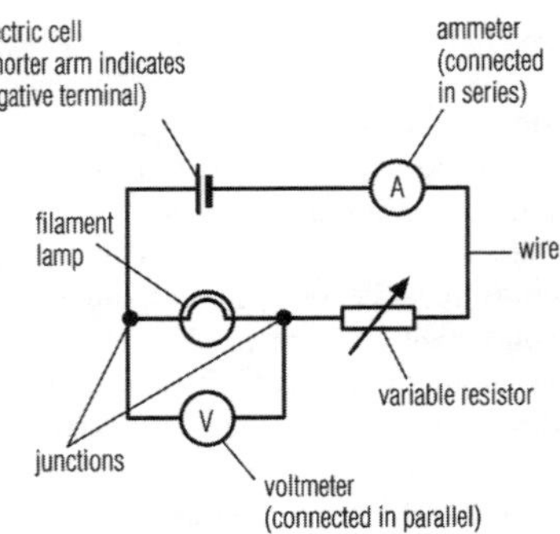

circuit diagram A circuit diagram shows in graphical form how the components of an electric circuit are connected together. Each component is represented by an internationally recognized symbol, and the connecting wires are shown by straight lines. A dot indicates where wires join. Helicon

circular function in mathematics, one of the standard trigonometric functions sine, cosine, tangent, or their reciprocals. They are so called because the functions sine and cosine may be used to parameterize the points of a circle. In the same way, the hyperbolic functions can be used to parameterize a hyperbola.

circulatory system system of vessels in an animal's body that transports essential substances (†blood or other circulatory fluid) to and from the different parts of the body. It was first discovered and described by English physician William †Harvey. All animals except for the most simple – such as sponges, jellyfish, sea anemones, and corals – have some type of circulatory system. Some invertebrates (animals without a backbone), such as insects, spiders, and most shellfish, have an "open" circulatory system which consists of a simple network of tubes and hollow spaces. Other invertebrates have pumplike structures that send blood through a system of blood vessels. All vertebrates (animals with a backbone), including humans, have a "closed" circulatory system which principally consists of a pumping organ – the †heart – and a network of blood vessels.

Fish have a single circulatory system in which blood passes once around the body before returning to a two-chambered heart. In birds and mammals, there is a †double circulatory system – the lung or pulmonary circuit and the body or systemic circuit. Blood is first pumped from the heart to the †lungs and back to the heart, before being pumped to the remainder of the body and back. The heart is therefore a double pump and is divided into two halves. In all vertebrates, blood flows in one direction. Valves in the heart, large arteries, and veins prevent backflow, and the muscular walls of the arteries assist in pushing the blood around the body. A network of tiny †capillaries carries the blood from arteries to veins. It is through the walls of capillaries that materials are transported to and from the blood.

Although most animals have a heart or hearts to pump the blood, in some small invertebrates normal body movements circulate the fluid. In the **open system**, found in snails and other mollusks, the blood (more correctly called †haemolymph) passes from the arteries into a body cavity (haemocoel), and from here is gradually returned by other blood vessels to the heart, via the gills. Insects and other arthropods have an open system with a heart. In the **closed system** of earthworms, blood flows directly from the main artery to the main vein, via smaller lateral vessels in each body segment.

The human circulatory system performs a number of functions: it supplies the cells of the body with the food and oxygen they need to survive (see †nutrition); it carries carbon dioxide and other waste products away from the cells; it helps to regulate the temperature of the body; and protects the body from disease. In addition, the system transports hormones, which help to regulate the activities of various parts of the body.

Circulatory System: Online Learning

http://library.thinkquest.org/25896/?tqskip=1

Wide-ranging guide to the circulatory system. This attractive Web site contains pages on blood, arteries, veins, and capillaries, the heart, and diseases of the circulatory system. With accessible text and clear, attractive diagrams, this is an excellent resource.

circumcircle in mathematics, circle that touches all vertices of a triangle or any other polygon.

circumference in geometry, the curved line that encloses a curved plane figure, for example a †circle or an ellipse. Its length varies according to the nature of the curve, and may be ascertained by the appropriate formula. The circumference of a circle is pd or 2pr, where d is the diameter of the circle, r is its radius, and p is the constant pi, approximately equal to 3.1416.

circumpolar in astronomy, term used to describe celestial objects that remain above the horizon at all times and do not set as seen from a given location. The amount of sky that is circumpolar depends on the observer's latitude on Earth. At the Earth's poles, all the visible sky is circumpolar, but at the Earth's Equator none of it is circumpolar.

circumpolar star any star that appears to circle around the Earth's North and/or South Pole without either rising or setting, as a result of the motion of the Earth. The radius of the area around the North Pole in which such stars occur is equal to the latitude of the place of observation. Thus at the pole all visible stars are circumpolar, while at the Equator none are. From London, England, for example, all stars within 52° of the North Pole are circumpolar.

circumscribe in geometry, to surround a figure with a circle which passes through all the vertices of the figure. Any triangle may be circumscribed and so may any regular polygon. Only certain quadrilaterals may be circumscribed (their opposite angles must add up to 180°).

cirque French name for a †corrie, a steep-sided armchair-shaped hollow in a mountainside.

cis- prefix used in †stereochemistry to distinguish an †isomer that has two substituents or groupings on the same side of the main axis or plane of the molecule. The isomer with the two on opposite sides is denoted by the prefix **trans-**.

cistron in genetics, the segment of †DNA that is required to synthesize a complete polypeptide chain. It is the molecular equivalent of a †gene.

citric acid $HOOCCH_2C(OH)(COOH)CH_2COOH$, organic acid widely distributed in the plant kingdom; it is found in high concentrations in citrus fruits and has a sharp, sour taste. At one time it was commercially prepared from concentrated lemon juice, but now the main source is the fermentation of sugar with certain moulds.

citric acid cycle in biochemistry, another term for †Krebs cycle.

citrulline an †amino acid not commonly found in proteins. It was first isolated from the water melon *Citrullus vulgaris* and may also be isolated after digestion of the †enzyme caseinogen by †trypsin. Citrulline is involved in the process of †urea excretion.

cladistics method of biological classification that uses a formal step-by-step procedure for objectively assessing the extent to which organisms share particular characteristics, and for assigning them to taxonomic groups called **clades**. Clades comprise all the species descended from a known or inferred common ancestor plus the ancestor itself, and may be large – consisting of a hierarchy of other clades.

cladode in botany, a flattened stem that is leaflike in appearance and function. It is an adaptation to dry conditions because a stem contains fewer †stomata than a leaf, and water loss is thus minimized. The true leaves in such plants are usually reduced to spines or small scales. Examples of plants with cladodes are butcher's-broom *Ruscus aculeatus*, asparagus, and certain cacti. Cladodes may bear flowers or fruit on their surface, and this distinguishes them from leaves.

cladogram in biology, branched diagram used in †cladistics to

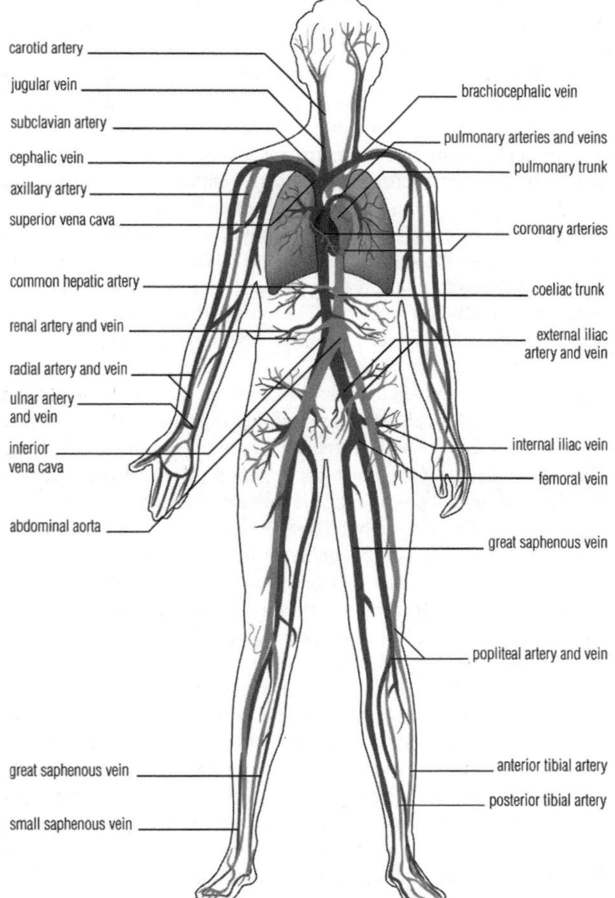

circulatory system Blood flows through 96,500 km/60,000 mi of arteries and veins, supplying oxygen and nutrients to organs and limbs. Oxygen-poor blood (blue) circulates from the heart to the lungs where oxygen is absorbed. Oxygen-rich blood (red) flows back to the heart and is then pumped round the body through the aorta, the largest artery, to smaller arteries and capillaries. Here oxygen and nutrients are exchanged with carbon dioxide and waste products and the blood returns to the heart via the veins. Waste products are filtered by the liver, spleen, and kidneys, and nutrients are absorbed from the stomach and small intestine. Helicon

represent the relationships between taxa based on the order in which species split off from one another during evolution.

Clairaut, Alexis Claude (1713–1765) French mathematician who studied celestial mechanics. His *Théorie de la lune/Theory of the Moon* (1752) explains the lunar apogee (at which the Moon is furthest from the Earth in its elliptical orbit), which had been omitted by Isaac Newton. He also calculated the path of Halley's Comet, and wrote several papers on the orbit of the Moon. He evolved a theorem connecting the gravity on the surface of a rotating ellipsoid with the compression, and the centrifugal force at the equator.

Claisen condensation in chemistry, specific condensation reaction between an †ester and either another ester or a †ketone in the presence of sodium hydroxide to produce a combined molecule with the loss of water.

Clark, Wilfrid Edward Le Gros (1895–1971) English anatomist and surgeon. His research made a major contribution to the understanding of the structural anatomy of the †brain. By emphasizing the importance of relating structure to function, he had a profound influence on the teaching of anatomy.

Clark cell in physics, standard electrical cell with a mercury sulfate-coated mercury cathode and a zinc anode in a zinc sulfate electrolyte. It produces an emf of 1.434 volts (at 15°C/59°F).

Clarke, Cyril Astley (1907–2000) English physician and amateur lepidopterist. He made many contributions to medical and scientific journals particularly on the prevention of rhesus hemolytic disease of the newborn and on the evolution of mimicry in swallowtail butterflies.

class in biological classification, a subdivision of †phylum and forms a group of related †orders. For example, all mammals belong to the class Mammalia and all birds to the class Aves. Among plants, all class names end in "idae" (such as Asteridae) and among fungi in "mycetes"; there are no equivalent conventions among animals. Related classes are grouped together in a phylum.

class in mathematics another name for a †set.

class field theory theory involving the mathematical structure known as a †field, dealing specifically with those that extend a given field in a special kind of way.

classification in biology, the arrangement of organisms into a hierarchy of groups on the basis of their similarities. The basic grouping is a †species, several of which may constitute a †genus, which in turn are grouped into families, and so on up through orders, classes, phyla (in plants, sometimes called divisions), and finally to kingdoms. The system that is used is one that reflects the evolutionary origin of the organisms. In other words, organisms belonging one group are thought to have evolved from a common ancestor at some time in the past.

All organisms are given a name formed from the species name and the genus name. This name is therefore called a binomial. The binomial name of humans is *Homo sapiens*. *Homo* is the genus name and *sapiens* the species name. Humans belong to the animal kingdom.

The oldest method of classification, called **phenetic classification**, aims to classify organisms on the basis of as many as possible of their observable characteristics: their morphology, anatomy, physiology, and so on. Greek philosopher Theophrastus adopted this method in the 4th century BC, when he classified plants into trees, shrubs, undershrubs, and herbs.

Awareness of evolutionary theory, however, led to the development of **phylogenetic classification**, which aims to classify organisms in a way that mirrors their evolutionary (see †evolution) and genetic relationships (see †genetics). Species are grouped according to shared characteristics believed to be derived from common ancestors (care being taken to exclude shared characteristics known to be due to †convergent evolution – such as the wings of bats and birds). In practice, most present-day systems of classification compromise between the phenetic and the phylogenetic approaches.

Center for Biodiversity: Study of Plants and Animals http://www.inhs.uiuc.edu/cbd/main/generalinfo.html

If you have ever wondered about the mysteries of astacology, this site will provide the answers, as well as introducing other, less obscure, branches of taxonomic biology such as herpetology, mammology, and botany. Linked to the Illinois Natural History Survey's vast collection of photographs, this is an excellent resource for anyone wanting to explore the vast diversity of life on Earth.

Classification of Living Things http://anthro.palomar.edu/animal/default.htm

Well-written introduction to the principles of Linnaean classification. This Web site explains the general principles of taxonomy, with lavish use of photography, diagrams, and hypertext links. It follows this with a number of tables illustrating taxonomic relationships, such as the five kingdoms of life. A table of human classification shows how humans fit into the scheme.

Linnaeus, Carolus http://www.ucmp.berkeley.edu/history/linnaeus.html

Profile of the life and legacy of the Swedish "father of taxonomy," Carl Linnaeus. A biography traces how his childhood interest in plants led to him becoming the greatest observational biologist of his day. There is also a discussion as to whether he could be considered an evolutionist.

classify in mathematics, to put into separate classes, or †sets, which may be uniquely defined.

class interval in statistics, the range of each class of data, used when arranging large amounts of †raw data into †grouped data. To obtain an idea of the distribution, the data are broken down into convenient classes (commonly 6–16), which must be mutually exclusive and are usually equal in width to enable †histograms to be drawn. The class boundaries should clearly define the range of each class. When dealing with †discrete data, suitable intervals would be, for example, 0–2, 3–5, 6–8, and so on. When dealing with †continuous data, suitable intervals might be $170 \leq X < 180$, $180 \leq X < 190$, $190 \leq X < 200$, and so on.

clastic rock in earth science, type of sedimentary rock consisting mainly of broken fragments of a parent rock deposited by some transport mechanism. Examples include †conglomerate, †sandstone, and †shale.

clathrate compound formed when the small molecules of one substance fill in the holes in the structural lattice of another, solid, substance – for example, sulfur dioxide molecules in ice crystals. Clathrates are therefore intermediate between mixtures and true compounds (which are held together by †ionic or covalent chemical bonds).

clathration in chemistry, a method of removing water from an aqueous solution, and therefore increasing the solution's concentration, by trapping it in a matrix with rare gases such as freons.

Claude, Georges (1870–1960) French industrial chemist, responsible for inventing neon signs. In 1896 he discovered that acetylene, normally explosive, could be safely transported when dissolved in acetone. He later demonstrated that neon gas could be used to provide a bright red light in signs. These were displayed publicly for the first time at the Paris Motor Show in 1910.

Claude process in chemistry, process for producing industrial quantities of liquefied air, developed by French chemist Georges †Claude. Cooling is achieved by expanding the gas using a series of pistons, which carry out work, further cooling the gas and helping the liquefaction process.

clausius in engineering, a unit of †entropy (the loss of energy as heat in any physical process). It is defined as the ratio of energy to temperature above absolute zero.

Clausius, Rudolf Julius Emanuel (1822–1888) German physicist, one of the founders of the science of †thermodynamics. In 1850 he enunciated its second law: heat cannot pass from a colder to a hotter body.

Clausius–Clapeyron equation relationship between the pressure, temperature, and latent heat in a change of state, for example, liquid to gas. It was determined by German physicist Rudolf Clausius and French mathematician Emile Clapeyron (1799–1864).

clavicle (Latin *clavis* "key") the collar bone of many vertebrates. In humans it is vulnerable to fracture, since falls involving a sudden force on the arm may result in very high stresses passing into the chest region by way of the clavicle and other bones. It is connected at one end with the sternum (breastbone), and at the other end with the shoulder-blade, together with which it forms the arm socket. The wishbone of a chicken is composed of its two fused clavicles.

claw hard, hooked, pointed outgrowth of the digits of mammals, birds, and most reptiles. Claws are composed of the protein keratin, and grow continuously from a bundle of cells in the lower skin layer. Hooves and nails are modified structures with the same origin as claws.

clay very fine-grained †sedimentary deposit that has undergone a greater or lesser degree of consolidation. When moistened it is plastic, and it hardens on heating, which renders it impermeable. It may be white, gray, red, yellow, blue, or black, depending on its composition. Clay minerals consist largely of hydrous silicates of aluminum and magnesium together with iron, potassium, sodium, and organic substances. The crystals of clay minerals have a layered structure, capable of holding water, and are responsible for its plastic properties. According to international classification, in mechanical analysis of soil, clay has a grain size of less than 0.002 mm/0.00008 in.

clay mineral one of a group of hydrous silicate minerals that form most of the fine-grained particles in clays. Clay minerals are normally formed by weathering or alteration of other silicate minerals. Virtually all have sheet silicate structures similar to the †micas. They exhibit the following useful properties: loss of water on heating; swelling and shrinking in different conditions; cation exchange with other media; and plasticity when wet. Examples are kaolinite, illite, and montmorillonite.

Kaolinite $Al_2Si_2O_5(OH)_4$ is a common white clay mineral derived from alteration of aluminum silicates, especially feldspars. Illite contains the same constituents as kaolinite, plus potassium, and is the main mineral of clay sediments, mudstones, and shales; it is a weathering product of feldspars and other silicates. Montmorillonite contains the constituents of kaolinite plus sodium and magnesium; along with related magnesium- and iron-bearing clay minerals, it is derived from alteration and weathering of mafic igneous rocks. Kaolinite (the mineral name for kaolin or china clay) is economically important in the ceramic and paper industries. Illite, along with other clay minerals, may also be used in ceramics. Montmorillonite is the chief constituent of fuller's earth, and is also used in drilling muds (muds used to cool and lubricate drilling equipment). Vermiculite (similar to montmorillonite) will expand on heating to produce a material used in insulation.

cleanliness unit unit for measuring air pollution: the number of particles greater than 0.5 micrometers in diameter per cubic foot of air. A more usual measure is the weight of contaminants per cubic meter of air.

cleavage in geology and mineralogy, the tendency of a rock or mineral to split along defined, parallel planes related to its internal structure; the clean splitting of slate is an example. It is a useful distinguishing feature in rock and mineral identification. Cleavage occurs as a result of realignment of component minerals during deformation or metamorphism. It takes place where bonding between atoms is weakest, and cleavages may be perfect, good, or poor, depending on the bond strengths; a given rock or mineral may possess one, two, three, or more orientations along which it will cleave.

Some minerals have no cleavage, for example, quartz will fracture to give curved surfaces similar to those of broken glass. Some other minerals, such as apatite, have very poor cleavage that is sometimes known as a parting. Micas have one perfect cleavage and therefore split easily into very thin flakes. Pyroxenes have two good cleavages and break (less perfectly) into long prisms. Galena has three perfect cleavages parallel to the cube edges, and readily breaks into smaller and smaller cubes. Baryte has one perfect cleavage plus good cleavages in other orientations.

cleistogamy production of flowers that never fully open and that are automatically self-fertilized. Cleistogamous flowers are often formed late in the year, after the production of normal flowers, or during a period of cold weather, as seen in several species of violet *Viola*.

Clemmensen reduction in chemistry, process for the selective reduction of †aldehydes and †ketones, in the presence of hydrochloric acid and zinc amalgam, to form their corresponding hydrocarbons.

client in †client–server architecture, software that enables a user to access a store of data or programs on a server. On the Internet, client software is the software that users need to run on home computers in order to be able to use services such as the World Wide Web. Examples of client programs on a personal computer are Web browser clients, such as Netscape Navigator, Microsoft Internet Explorer, or Opera. The Web browser client communicates with a Web server to send and retrieve data.

client–server architecture in computing, a system in which the mechanics of looking after data are separated from the programs that use the data. For example, the "server" might be a central database, typically located on a large computer that is reserved for this purpose. The "client" would be an ordinary program that requests data from the server as needed.

Most Internet services are examples of client–server applications, including the World Wide Web, FTP, Telnet, and Gopher.

climate combination of weather conditions at a particular place over a period of time – usually a minimum of 30 years. A climate classification encompasses the averages, extremes, and frequencies of all meteorological elements such as temperature, atmospheric pressure, precipitation, wind, humidity, and sunshine, together with the factors that influence them.

The primary factors that influence the climate of an area are: latitude (as a result of the Earth's rotation and orbit); ocean currents; large-scale movements of wind belts and air masses over the Earth's surface; temperature differences between land and sea surfaces; topography; continent positions; and vegetation. In the long term, changes in the Earth's orbit and the angle of its axis inclination also affect climate. Climatologists are especially concerned with the influences of human activity on climate change, among the most important of which, at both local and global levels, are those currently linked with †ozone depleters, the †greenhouse effect, and †global warming.

World Environmental Changes Landmarks http://www.bbc.co.uk/education/landmarks/

Well-designed and attractive page from the British Broadcasting Corporation's (BBC) Education site looking at the effect the increasing human population has on the environment and climate. The site is broken down into five different environment types (wetlands, drylands, cities, the sea, and forests) with information, case studies from around the world, and activities for each. The high-tech version also includes animations.

climate change change in the †climate of an area or of the whole world over an appreciable period of time. That is, a single winter that is colder than average does not indicate climate change. It is the change in average weather conditions from one period of time (30–50 years) to the next. Climate fluctuations are natural phenomena, but there is increasing evidence to suggest that human industrial activity affects the global climate, particularly in terms of †global warming.

Causes of Climate Change http://www.geog.ouc.bc.ca/physgeog/contents/7y.html

From a much larger site "Fundamentals of Physical Geography" set up by a Canadian University and covering many aspects of the area of study. The site is clear and well organized and contains maps and diagrams.

Global Climate Change Information Program (GCCIP) http://www.doc.mmu.ac.uk/aric/arichome.html

Established in October 1991, the GCCIP provides an information link between scientists (both natural and social), politicians, economists, and the general public, on the subjects of climate change and air quality. Their Web site includes a number of essays on related issues.

climate model computer simulation, based on physical and mathematical data, of a climate system, usually the global (rather than local) climate. It is used by researchers to study such topics as the possible long-term disruptive effects of the greenhouse gases, or of variations in the amount of radiation given off by the Sun.

climatology study of climate, its global variations and causes.

Climatologists record mean daily, monthly, and annual temperatures and monthly and annual rainfall totals, as well as maximum and minimum values. Other data collected relate to pressure, humidity, sunshine, cold cover, and the frequency of days of frost, snow, hail, thunderstorms, and gales. The main facts are summarized in tables and climatological atlases published by nearly all the national meteorological services of the world. Climatologists also study climates of the past (paleoclimates) by gathering information from such things as tree rings, deep sea sediments, and ice cores – all of which record various climate factors as they form.

climax community group of plants and animals that is best able to exploit the environment in which it exists. It is brought about by †succession (a change in the species present) and represents the point at which succession ceases to occur.

climax vegetation the plants in a †climax community.

climograph diagram that shows both the average monthly temperature and †precipitation of a place.

cline in biology, gradual change in any measurable characteristic of a population (or species).

clinical psychology branch of psychology dealing with the understanding and treatment of health problems, particularly mental disorders. The main problems dealt with include anxiety, phobias, depression, obsessions, sexual and marital problems, drug and alcohol dependence, childhood behavioral problems, psychoses (such as schizophrenia), mental disability, and brain disease (such as dementia) and damage. Other areas of work include forensic psychology (concerned with criminal behavior) and health psychology.

Assessment procedures assess intelligence and cognition (for example, in detecting the effects of brain damage) by using psychometric tests. **Behavioral approaches** are methods of treatment that apply learning theories to clinical problems. **Behavior therapy** helps people change unwanted behaviors (such as phobias, obsessions, sexual problems) and to develop new skills (such as improving social interactions). **Behavior modification** relies on operant conditioning, making selective use of rewards (such as praise) to change behavior. This is helpful for children, the mentally disabled, and for patients in institutions, such as mental hospitals. **Cognitive therapy** is an approach to treating emotional problems, such as anxiety and depression, by teaching people how to deal with negative thoughts and attitudes. **Counseling**, developed by Carl Rogers, is widely used to help people solve their own problems. **Psychoanalysis**, as developed by Sigmund Freud and Carl Jung, is little used by clinical psychologists today. It emphasizes childhood conflicts as a source of adult problems.

clint one of a number of flat-topped limestone blocks that make up a †limestone pavement. Clints are separated from each other by enlarged joints called grykes.

clitoris (Greek *kleitoris* "little hill") in anatomy, part of the female reproductive system. The glans of the clitoris is visible externally. It connects to a pyramid-shaped pad of erectile tissue. Attached to this are two "arms" that extend backwards into the body towards the anus and are approximately 9 cm/3.5 inches in length. Between these arms are the clitoral bulbs, lying one on each side of the vaginal cavity.

During sexual intercourse the clitoris swells and stimulation of it results in orgasm. In its swollen state it presses on the urethra and closes it, preventing harmful bacteria from entering the bladder.

Anatomically, the structure of the clitoris has been largely overlooked, until the work of an Australian urologist revealed it to be much larger than previously thought, in 1998.

clo unit of thermal insulation of clothing. Standard clothes have an insulation of about 1 clo; the warmest clothing is about 4 clo per 2.5 cm/1 in of thickness. See also †tog.

cloaca the common posterior chamber of most vertebrates into which the digestive, urinary, and reproductive tracts all enter; a cloaca is found in most reptiles, birds, and amphibians; many fishes; and, to a reduced degree, marsupial mammals. Placental mammals, however, have a separate digestive opening (the anus) and urinogenital opening. The cloaca forms a chamber in which products can be stored before being voided from the body via a muscular opening, the cloacal aperture.

clock rate frequency of a computer's internal electronic clock. Every computer contains an electronic clock, which produces a sequence of regular electrical pulses used by the control unit to synchronize the components of the computer and regulate the †fetch–execute cycle by which program instructions are processed.

A fixed number of time pulses is required in order to execute each particular instruction. The speed at which a computer can process instructions therefore depends on the clock rate: increasing the clock rate will decrease the time required to complete each particular instruction.

Clock rates are measured in **megahertz** (MHz), or millions of pulses a second. Microcomputers commonly have a clock rate of 8–50 MHz.

clockwise the direction in which the hands of a traditional clock turn.

clone exact replica – in genetics, any one of a group of genetically identical cells or organisms. An identical †twin is a clone; so too are bacteria living in the same colony. "Clone" also describes genetically engineered replicas of DNA sequences.

British scientists confirmed in February 1997 that they had cloned an adult sheep from a single cell to produce a lamb with the same genes as its mother. A cell was taken from the udder of the mother sheep, and its DNA combined with an unfertilized egg that had had its DNA removed. The fused cells were grown in a laboratory and then implanted into the uterus of a surrogate mother sheep. The resulting lamb, Dolly, came from an animal that was six years old. However, in 1999, Dolly was revealed not to be an exact clone – research showed her mitochondria to have come mainly from the egg cell rather than the udder cell. This was the first time cloning had been achieved using cells other then reproductive cells. The cloning breakthrough has ethical implications, as the same principle could be used with human cells and eggs. The news was met with international calls to prevent the cloning of humans. The UK, Spain, Germany, Canada, and Denmark already have laws against cloning humans, as do some individual states in the USA (legislators introduced bills to ban human cloning and associated research nationally in March 1997). France and Portugal also have very restrictive laws on cloning.

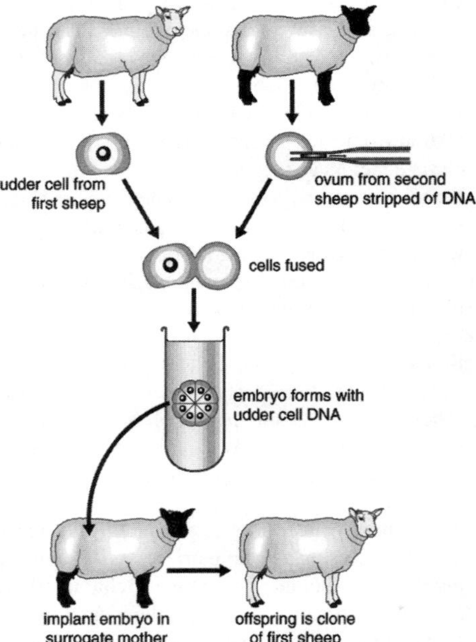

clone The cloning of Dolly the sheep by the Roslin Institute in Edinburgh was a genetic milestone. It was the first successful clone produced using genetic material from an adult (udder) cell rather than from a gamete (egg or sperm). The DNA from the udder cell was fused with an ovum stripped of its own DNA. The fused cells divided in vitro to form an embryo that was then implanted into a surrogate mother. The resulting lamb was a clone of the ewe that had provided the udder cell. Helicon

Dr Frankenstein, I presume? http://www.salon.com/feb97/news/news2970224.html

Interview with the man who made the first cloned mammal using adult cells, Dolly. Embryologist Dr Ian Wilmut speaks to Andrew Ross about his worries and his future projects, about the distinction between science fiction and human cloning, and about what could go wrong with researching this delicate area.

Hello Dolly http://whyfiles.org/034clone/

Imaginative and humorous look at the issue of cloning, presented by the Why Files, an online magazine on the science behind the news. Genes, chromosomes, DNA, ethical considerations, and the "lucky lamb" itself are all covered here. The magazine also features a series of interviews with specialists on genetics, bioethics, and other fields related to the sheep-cloning experiment.

I Can Do That http://www.eurekascience.com/ICanDoThat/

Eureka! Science offers the "I Can Do That!" Web pages as "a painless way to find fun facts about DNA, RNA, cells, protein and cloning." Interactive tests and printable worksheets accompany every section.

closed in mathematics, descriptive of a set of data for which an operation (such as addition or multiplication) done on any members of the set gives a result that is also a member of the set.

For example, the set of even numbers is closed with respect to addition, because two even numbers added to each other always give another even number.

closed curve curve of which the end point coincides with the initial point, for example, a circle or an ellipse.

cloud water vapor condensed into minute water particles that float in masses in the atmosphere. Clouds, like fogs or mists, which occur at lower levels, are formed by the cooling of air containing water vapor, which generally condenses around tiny dust particles.

Clouds are classified according to the height at which they occur, and by their shape. **Cirrus** and **cirrostratus** clouds occur at around 10 km/33,000 ft. The former, sometimes called mares'-tails, consist of minute specks of ice and appear as feathery white wisps, while cirrostratus clouds stretch across the sky as a thin white sheet. Three types of cloud are found at 3–7 km/10,000–23,000 ft: cirrocumulus, altocumulus, and altostratus. **Cirrocumulus** clouds occur in small or large rounded tufts, sometimes arranged in the pattern called mackerel sky. **Altocumulus** clouds are similar, but larger, white clouds, also arranged in lines. **Altostratus** clouds are like heavy cirrostratus clouds and may stretch across the sky as a gray sheet. **Stratocumulus** clouds are generally lower, occurring at 2–6 km/6,500–20,000 ft. They are dull gray clouds that give rise to a leaden sky that may not yield rain. Two types of clouds, **cumulus** and **cumulonimbus**, are placed in a special category because they are produced by daily ascending air currents, which take moisture into the cooler regions of the atmosphere. Cumulus clouds have a flat base generally at 1.4 km/4,500 ft where condensation begins, while the upper part is dome-shaped and extends to about 1.8 km/6,000 ft. Cumulonimbus clouds have their base at much the same level, but extend much higher, often up to over 6 km/20,000 ft. Short heavy showers and sometimes thunder may accompany them. **Stratus** clouds, occurring below 1–2.5 km/3,000–8,000 ft, have the appearance of sheets parallel to the horizon and are like high-level fogs.

In addition to their essential role in the water cycle, clouds are important in the regulation of radiation in the Earth's atmosphere. They reflect short-wave radiation from the Sun, and absorb and re-emit long-wave radiation from the Earth's surface.

Clouds and Precipitation http://ww2010.atmos.uiuc.edu/(Gh)/guides/mtr/cld/home.rxml

Illustrated guide to how clouds form and to the various different types. The site contains plenty of images and a glossary of key terms in addition to further explanations of the various types of precipitation.

cloud chamber apparatus, now obsolete, for tracking ionized particles. It consists of a vessel fitted with a piston and filled with air or other gas, saturated with water vapor. When the volume of the vessel is suddenly expanded by moving the piston outwards, the vapor cools and a cloud of tiny droplets forms on any nuclei, dust, or ions present. As fast-moving ionizing particles collide with the air or gas molecules, they show as visible tracks.

Much information about interactions between such particles and radiations has been obtained from photographs of these tracks. The system was improved upon by the use of liquid hydrogen or helium instead of air or gas (see †particle detector) and by the †spark chamber. The cloud chamber was devised in 1897 by Charles Thomson Rees Wilson (1869–1959) at Cambridge University.

clusec unit for measuring the power of a vacuum pump.

cluster group of stars or of †galaxies, usually with some recognizably systematic configuration. It appears that both types of cluster are a structural feature of the universe, which form over the passage of time.

cm symbol for †**centimeter**.

CNO cycle in astrophysics, carbon-nitrogen-oxygen cycle, alternative name for †carbon cycle.

coacervate collection of particles in an emulsion that can be reversed into droplets of liquid before they flocculate.

coal black or blackish mineral substance formed from the compaction of ancient plant matter in tropical swamp conditions. It is used as a fuel and in the chemical industry. Coal is classified according to the proportion of carbon it contains. The main types are †anthracite (shiny, with about 90% carbon), **bituminous coal** (shiny and dull patches, about 75% carbon), and **lignite** (woody, grading into peat, about 50% carbon). Coal can be burned to produce heat energy, for example in power stations to produce electricity. Coal burning is one of the main causes of †acid rain, which damages buildings and can be detrimental to aquatic and plant life.

coal gas fuel gas comprising mostly hydrogen and methane and carbon monoxide produced by processing coal.

coal mining extraction of coal from the Earth's crust. Coal mines may be opencast, adit, or deepcast. The least expensive is opencast but this may result in scars on the landscape.

Coalsack dark nebular cloud of dust and gas in the southern-hemisphere constellation of †Crux (the

Southern Cross), visible as a dark patch silhouetted against the stars of the Milky Way.

coal tar black oily material resulting from the destructive distillation of bituminous coal.

The †fractional distillation of coal tar yields light oil, middle oil, heavy oil, and anthracene oil; the residue is called pitch. On further distillation a large number of substances are obtained, about 200 of which have been isolated. They are used as dyes and in medicines.

coastal erosion the erosion of the land by the constant battering of the sea, primarily by the processes of hydraulic action, corrasion, attrition, and corrosion. †Hydraulic action occurs when the force of the waves compresses air pockets in coastal rocks and cliffs. The air expands explosively, breaking the rocks apart. It is also the force of the water on the cliff. During severe gales this can be as high as 6 tonnes/cm^3 – the force of a bulldozer. Rocks and pebbles flung by waves against the cliff face wear it away by the process of †corrasion, or abrasion as it is also known. Chalk and limestone coasts are often broken down by †solution (also called †corrosion). †Attrition is the process by which the eroded rock particles themselves are worn down, becoming smaller and more rounded.

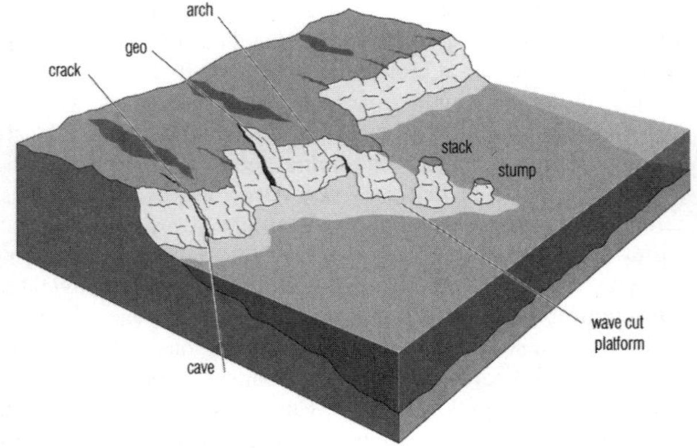

coastal erosion Typical features of coastal erosion: from the initial cracks in less resistant rock through to arches, stacks, and stumps that can occur as erosion progresses. Helicon

Frost shattering (or †freeze–thaw), caused by the expansion of frozen water in cracks, and †biological weathering, caused by the burrowing of rock-boring mollusks and plants, also lead to the breakdown of rock.

Where resistant rocks form †headlands, the sea erodes the coast in successive stages. First it exploits weaknesses such as faults and cracks to form caves. Then it gradually wears away the interior of the caves and enlarges them. In some cases the roofs may be broken through to form blowholes. In other cases the caves at either side of a headland may unite to form a natural arch. When the roof of the arch collapses, a †stack is formed. This may be worn down further to produce a †stump.

Coasts in Crisis http://pubs.usgs.gov/circular/c1075/

Full text of a US Geological Survey publication. This detailed site uses intelligent text and photographs to show a coastline under siege. It describes the various types of coastline, methods of shoring them up, and the human and meteorological forces attacking the coast. This is an excellent resource for anyone studying coastal erosion, deposition, and protection.

Virtual Fieldwork http://www.users.globalnet.co.uk/~drayner/indexvf.htm

Field studies are the best way to learn about geographical concepts; take the next best thing – an Internet field trip – by selecting from this page. The choices include coastal erosion in Kent, UK, and coastal sand dunes and management in north Wales, UK. Users are able to study digital photographs of the area, read the accompanying notes, then take a short test to find out how much they have learned.

coastal protection measures taken to prevent †coastal erosion. Many stretches of coastline are so severely affected by erosion that beaches are swept away, threatening the livelihood of seaside resorts, and buildings become unsafe.

To reduce erosion, several different forms of coastal protection are used. Structures such as sea walls attempt to prevent waves reaching the cliffs by deflecting them back to sea. Such structures are expensive and of limited success. Adding sediment (beach nourishment) to make a beach wider causes waves to break early so that they have less power when they reach the cliffs. Wooden or concrete barriers called groynes may also be constructed at right angles to the beach in order to block the movement of sand along the beach (†longshore drift). However, this has the effect of "starving" beaches downshore: "protection" of one area usually means destruction of another.

Rock armour refers to large blocks of stone dumped on a beach or at the base of a cliff to reduce erosion. **Hard engineering** refers to constructed/built devices, while **soft engineering** refers to natural features such as salt marshes or sand dunes, which may help to protect against erosion.

Coastal protection may also refer to the process of simply leaving the coast to the elements but removing the harmful factor of human population and development.

coaxial cable electric cable that consists of a solid or stranded central conductor insulated from and surrounded by a solid or braided conducting tube or sheath. It can transmit the high-frequency signals used in television, telephone, and other telecommunications transmissions.

cobalt chemical symbol Co, (German *Kobalt* "evil spirit") hard, lustrous, gray, metallic element, atomic number 27, relative atomic mass 58.933. It is found in various ores and occasionally as a free metal, sometimes in metallic meteorite fragments. It is used in the preparation of magnetic, wear-resistant, and high-strength alloys; its compounds are used in inks, paints, and varnishes.

The isotope Co-60 is radioactive (half-life 5.3 years) and is produced in large amounts for use as a source of gamma rays in industrial radiography, research, and cancer therapy. Cobalt was named in 1730 by Swedish chemist Georg Brandt (1694–1768); the name derives from the fact that miners considered its ore malevolent because it interfered with copper production.

cobalt-60 radioactive (half-life 5.3 years) isotope produced by neutron radiation of cobalt in heavy-water reactors, used in large amounts for gamma rays in cancer therapy, industrial radiography, and research, substituting for the much more costly radium.

cobalt chloride $CoCl_2$, compound that exists in two forms: the hydrated salt ($CoCl_2.6H_2O$), which is pink, and the anhydrous salt, which is blue. The anhydrous form is used as an indicator because it turns pink if water is present. When the hydrated salt is gently heated the blue anhydrous salt is reformed.

cobalt ore cobalt is extracted from a number of minerals, the main ones being **smaltite**, $(CoNi)As_3$; **linnaeite**, Co_3S_4; **cobaltite**, CoAsS; and **glaucodot**, (CoFe)AsS.

All commercial cobalt is obtained as a by-product of other metals, usually associated with other ores, such as copper. The Democratic Republic of Congo (formerly Zaire) is the largest producer of cobalt, and it is obtained there as a by-product of the copper industry. Other producers include Canada and Morocco. Cobalt is also found in the manganese nodules that occur on the ocean floor, and was successfully refined in 1988 from the Pacific Ocean nodules, although this process has yet to prove economic.

COBOL acronym for **common business-oriented language**, high-level computer-programming language, designed in the late 1950s for commercial data-processing problems; it has become the major language in this field. COBOL features powerful facilities for file handling and business arithmetic. Program instructions written in this language make extensive use of words and look very much like English sentences. This makes COBOL one of the easiest languages to learn and understand.

coccolithophorid microscopic, planktonic marine alga, which secretes a calcite shell. The shells (coccoliths) of coccolithophores are a major component of deep sea ooze. Coccolithophores were particularly abundant during the late †Cretaceous period and their remains form the northern European chalk deposits, such as the white cliffs of Dover.

coccus plural **cocci**, member of a group of globular bacteria, some of which are harmful to humans. The cocci contain the subgroups **streptococci**, where the bacteria associate in straight chains, and **staphylococci**, where the bacteria associate in branched chains.

cochlea part of the inner †ear. It is equipped with approximately 10,000 hair cells, which move in response to sound waves and thus stimulate nerve cells to send messages to the brain. In this way they turn vibrations of the air into electrical signals.

Cockcroft, John Douglas (1897–1967) British physicist. He was awarded the Nobel Prize for Physics in 1951 for his work in 1932 when, working with the Irish physicist Ernest †Walton, with whom he shared the award, they succeeded in splitting the nucleus of an atom for the first time. He was knighted in 1948, and awarded the Order of Merit in 1957.

Cockerell, Christopher Sydney (1910–1999) English engineer who invented the hovercraft in the 1950s.

COD abbreviation for †**chemical oxygen demand**, a measure of water and effluent quality.

code expression of an †algorithm in a †programming language. The term is also used as a verb, to describe the act of programming.

codominance in genetics, the failure of a pair of alleles, controlling a particular characteristic, to show the classic recessive-dominant relationship. Instead, aspects of both alleles may show in the phenotype. For example, the human blood group AB shows the phenotype effect of both A and B codominant genes. The snapdragon shows codominance in respect to color. Two alleles, one for red petals and the other for white, will produce a pink color if the alleles occur together as a heterozygous form.

codon or **coding triplet**, in genetics, a triplet of bases (see †base pair) in a molecule of DNA or RNA that directs the placement of a particular amino acid during the process of protein (polypeptide) synthesis. There are 64 codons in the †genetic code.

coefficient number part in front of an algebraic term, signifying multiplication. For example, in the expression $4x^2 + 2xy - x$, the coefficient of x^2 is 4 (because $4x^2$ means $4 \times x^2$), that of xy is 2, and that of x is –1 (because $-1 \times x = -x$).

In general algebraic expressions, coefficients are represented by letters that may stand for numbers; for example, in the equation $ax^2 + bx + c = 0$, a, b, and c are coefficients, which can take any number.

coefficient of expansion quantity that describes the amount of expansion undergone by a material for a degree rise in temperature. It is expressed as the increase in length per unit length, per degree centigrade. For example, the metal with the lowest coefficient of expansion is Invar, at 2.3×10^{-6} m of expansion for every meter in length, per degree Celsius rise in temperature.

coefficient of relationship probability that any two individuals share a given gene by virtue of being descended from a common ancestor. In sexual reproduction of †diploid species, an individual shares half its genes with each parent, with its offspring, and (on average) with each sibling; but only a quarter (on average) with its grandchildren or its siblings' offspring; an eighth with its great-grandchildren, and so on.

In certain species of insects (for example honey bees), females have only one set of chromosomes (inherited from the mother), so that sisters are identical in genetic make-up; this produces a different set of coefficients. These coefficients are important in calculations of †inclusive fitness.

coelenterate any freshwater or marine organism of the phylum Coelenterata, having a body wall composed of two layers of cells. They also possess stinging cells. Examples are jellyfish, hydra, and coral.

coelom in all but the simplest animals, the fluid-filled cavity that separates the body wall from the gut and associated organs, and allows the gut muscles to contract independently of the rest of the body.

coenzyme small organic nonprotein compound that attaches to an †enzyme and is necessary for its correct functioning. Tightly bound coenzymes are known as prosthetic groups; more loosely bound ones are called cofactors. The coenzyme itself is not usually changed during a reaction. If it is, it is usually converted rapidly back to its original form. Well-known coenzymes include NAD, ATP, and coenzyme A.

coercive force in physics, intensity of magnetizing field needed to reduce a magnetic material's flux density to zero. The maximum coercive force, required for a magnetically saturated material, is termed the coercivity.

coevolution evolution of those structures and behaviors within a species that can best be understood in relation to another species. For example, some insects and flowering plants have evolved together: insects have produced mouthparts suitable for collecting pollen or drinking nectar, and plants have developed chemicals and flowers that will attract insects to them. Parasites often evolve and speciate with their hosts.

Coevolution occurs because both groups of organisms, over millions of years, benefit from a continuing association, and will evolve structures and behaviors that maintain this association.

cofactor in mathematics, of an element in the ith row and jth column of a †determinant, the quantity obtained by deleting the ith row and jth column and multiplying the determinant of the matrix that is left by $(-1)^{i+j}$. It is used as part of the process of evaluating a determinant.

cognition in psychology, a general term covering the functions involved in synthesizing information – for example, perception (seeing, hearing, and so on), attention, memory, and reasoning.

cognitive dissonance state of psychological tension occurring when a choice has to be made between two equally attractive or equally unpleasant alternatives. The dissonance is greater the closer the alternatives are in attractiveness or unpleasantness.

Dissonance usually remains after the decision has been made and this motivates efforts to achieve a state of equilibrium, or consonance. The concept, first described by US psychologist Leon Festinger (1919–), has been one of the most influential in †social psychology and has led to much experimental research. Studies have focused on conditions that enhance or minimize dissonance and on how it can be resolved; for example, changes in the cognition or awareness of the decision situation, changes in attitude following compliance to a request to perform (typically, a role-playing task not normally voluntarily undertaken), and the effect of incentives on attitude changes following such tasks.

cognitive psychology study of information-processing functions in humans and animals, covering their role in learning, memory, reasoning, and language development. Cognitive psychologists use a number of experimental techniques, including laboratory-based research with normal and brain-damaged subjects, as well as computer and mathematical models to test and validate theories.

The study of cognition was largely neglected by psychologists for the early part of the 20th century after the demise of †introspection as a method of investigation and the rise of †behaviorism. However, several influential theorists, such as US psychologist Edward Chase Tolman (1886–1959), continued to argue that in order to comprehend fully the determinants of behavior, cognitive processes must be studied and understood, and in 1957 Noam Chomsky's examination of behaviorist approaches to language acquisition appeared. With the rise of telecommunications technology and digital computing, such theorists as English psychologist Donald Broadbent (1926–) developed information-processing models of the brain, later elaborated, for example, by German-born US psychologist Ulrich Neisser (1928–). More recently, the limitations of these approaches, for example, in elaborating the role of emotion and motivation in cognitive processes, have become the focus of attention.

cognitive therapy or **cognitive behavior therapy**, treatment for emotional disorders such as depression and anxiety states. It encourages the patient to challenge the distorted and unhelpful thinking that is characteristic of depression, for example. The treatment may include behavior therapy.

Cohen, Stanley (1922–) US biochemist who was awarded a Nobel Prize for Physiology or Medicine in 1986 jointly with Rita †Levi-Montalcini for their work to isolate and characterize **growth factors**, small proteins that regulate the growth of specific types of cells, such as nerve and epidermal cells.

coherence in physics, property of two or more waves of a beam of light or other electromagnetic radiation having the same frequency and the same †phase, or a constant phase difference.

cohomology or **cohomology theory**, algebraic study, using group theory, of geometric objects with specific reference to the operation of finding a boundary. Cohomology theory represents a modification of †homology in which it is possible both to add and to multiply †classes.

coil, electric length of conducting wire that has been formed into a long series of loops. Used to form, for example, transformers, solenoids, and electromagnets.

coke clean, light fuel produced, along with town gas, when coal is strongly heated in an airtight oven. Coke contains 90% carbon and makes a useful domestic and industrial fuel (used, for example, in the iron and steel industries).

The process was patented in England in 1622, but it was only in 1709 that Abraham Darby devised a commercial method of production.

cold-blooded of animals, dependent on the surrounding temperature; see †poikilothermy.

cold dark matter in cosmology, unseen material consisting of slow-moving particles believed by some cosmologists to make up the bulk of the matter in the universe. See †dark matter.

cold fusion in nuclear physics, the fusion of atomic nuclei at room temperature. If cold fusion were possible it would provide a limitless, cheap, and pollution-free source of energy, and it has therefore been the subject of research around the world.

In 1989 Martin Fleischmann and Stanley Pons of the University of Utah, USA, claimed that they had achieved cold fusion in the laboratory, but their results could not be substantiated. The University of Utah announced in 1998 that they would allow the cold fusion patent to elapse, given that the work of Pons and Fleischmann has never been reproduced.

coleoptile the protective sheath that surrounds the young shoot tip of a grass during its passage through the soil to the surface. Although of relatively simple structure, most coleoptiles are very sensitive to light, ensuring that seedlings grow upwards.

collagen protein that is the main constituent of †connective tissue. Collagen is present in skin, cartilage, tendons, and ligaments. Bones are made up of collagen, with the mineral calcium phosphate providing increased rigidity.

Collagen is made by fibroblast cells that are found in connective tissue. The collagen molecule consists of three protein chains, each about 1,000 amino acids in length, which are entwined in a triple helix. There are 15 known types of collagen that can be divided into fibrillar and nonfibrillar. **Fibrillar collagens** are found in cartilage, tendons, skin, and bones and the fibrillar structure gives these great strength. **Nonfibrillar collagens** are found separating the basement membranes, the epithelial layers that line tube and in the cornea. They form networks and sheets.

collagen therapy Collagen is used cosmetically to enhance lips and smooth out wrinkles. The collagen, which is extracted mainly from cow hides, is injected into the skin. For wrinkle treatment it is injected into the dermis over six sessions, which have to be repeated after several months as the implants degrade.

collective unconscious in psychology, a shared pool of memories, ideas, modes of thought, and so on, which, according to the Swiss psychiatrist Carl Jung, comes from the life experience of one's ancestors, indeed from the entire human race. It coexists with the personal †unconscious, which contains the material of individual experience, and may be regarded as an immense depository of ancient wisdom.

Primal experiences are represented in the collective unconscious by archetypes, symbolic pictures, or personifications that appear in dreams and are the common element in myths, fairy tales, and the literature of the world's religions. Examples include the serpent, the sphinx, the Great Mother, the anima (representing the nature of woman), and the mandala (representing balanced wholeness, human or divine).

collenchyma plant tissue composed of relatively elongated cells with thickened cell walls, in particular at the corners where adjacent cells meet.

It is a supporting and strengthening tissue found in nonwoody plants, mainly in the stems and leaves.

colligative property property that depends only on the concentration of particles in a solution and not on the nature of the particles. Such properties include osmotic pressure (see †osmosis), elevation of †boiling point, depression of †freezing point, and lowering of †vapor pressure.

collimator optical device for producing a nondivergent beam of light, or any device for limiting the size and angle of spread of a beam of radiation or particles. The term is also used for a small telescope attached to a larger optical instrument to fix its line of sight.

collinear in mathematics, lying on the same straight line.

collision theory theory that explains how chemical reactions take place and why rates of reaction alter. For a reaction to occur the reactant particles must collide. Only a certain fraction of the total collisions cause chemical change; these are called **successful collisions**. The successful collisions have sufficient energy (activation energy) at the moment of impact to break the existing bonds and form new bonds, resulting in the products of the reaction. Increasing the concentration of the reactants and raising the temperature bring about more collisions and therefore more successful collisions, increasing the rate of reaction.

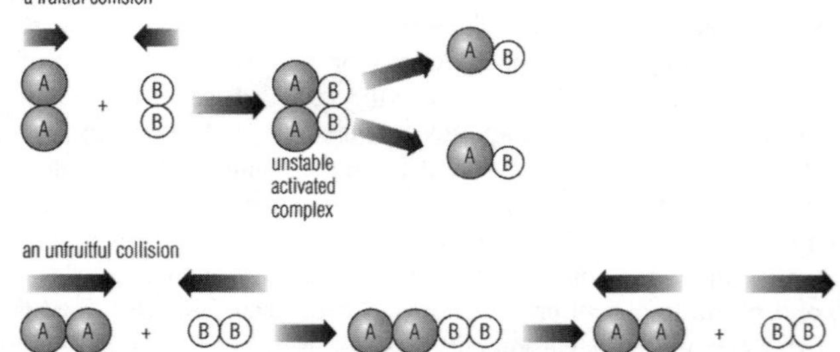

collision theory Collision theory explains how chemical reactions occur and why rates of reaction differ. For a reaction to occur, particles must collide. If the collision causes a chemical change it is referred to as a fruitful collision. Helicon

When a †catalyst undergoes collision with the reactant molecules, less energy is required for the chemical change to take place, and hence more collisions have sufficient energy for reaction to occur. The reaction rate therefore increases.

collision zone in plate tectonics, a region in which two pieces of continental crust collide after an intervening ocean has disappeared. It may be an area of mountain building and volcanic and seismic activity.

colloid substance composed of extremely small particles of one material (the dispersed phase) evenly and stably distributed in another material (the continuous phase). The size of the dispersed particles (1–1,000 nanometers across) is less than that of particles in suspension but greater than that of molecules in true solution. Colloids involving gases include **aerosols** (dispersions of liquid or solid particles in a gas, as in fog or smoke) and **foams** (dispersions of gases in liquids).

Those involving liquids include **emulsions** (in which both the dispersed and the continuous phases are liquids) and **sols** (solid particles dispersed in a liquid). Sols in which both phases contribute to a molecular three-dimensional network have a jellylike form and are known as **gels**; gelatin, starch "solution," and silica gel are common examples.

Steel is a solid colloid.

Milk is a natural emulsion of liquid fat in a watery liquid; synthetic emulsions such as some paints and cosmetic lotions have chemical emulsifying agents to stabilize the colloid and stop the two phases from separating out. Colloids were first studied thoroughly by the British chemist Thomas Graham, who defined them as substances that will not diffuse through a semipermeable membrane (as opposed to what he termed crystalloids, solutions of inorganic salts, which will diffuse through).

Colombo, Matteo Realdo (*c.* 1516–1559) Italian anatomist who was the first European to describe pulmonary circulation, the process of blood circulating from the heart to the lungs and back, and oxygenation of the blood.

colon in anatomy, the main part of the large intestine, between the cecum and rectum. Water and mineral salts are absorbed from undigested food in the colon, and the residue passes as feces towards the rectum.

colonization in ecology, the spread of species into a new habitat, such as a freshly cleared field, a new motorway verge, or a recently flooded valley. The first species to move in are called **pioneers**, and may establish conditions that allow other animals and plants to move in (for example, by improving the condition of the soil or by providing shade). Over time a range of species arrives and the habitat matures; early colonizers will probably be replaced, so that the variety of animal and plant life present changes.

This is known as †succession.

colorimeter in physics, instrument that analyses, compares, or reproduces colors in terms of a mixture of each of the three primary colors.

colostrum clear yellowish fluid produced from the breast during the first few days after giving birth. It contains antibodies from the mother, white blood cells (particularly macrophages), and fibronectin, a protein that increases their antimicrobial activity, all of which serve to protect the newborn from infection.

color in physics, quality or wavelength of light emitted or reflected from an object. Visible white light consists of electromagnetic radiation of various wavelengths, and if a beam is refracted through a prism, it can be spread out into the visible spectrum (that can be detected by the human eye), in which the various colors correspond to different wavelengths. From long to short wavelengths (from about 700 to 400 nanometers) the colors are red, orange, yellow, green, blue, indigo, and violet.

The color of grass is green because grass absorbs all the colors from the spectrum and only transmits or reflects the wavelength corresponding to green. A sheet of white paper reflects all the colors of the spectrum from its surface; black objects absorb all the colors of the spectrum.

All colors can be obtained from mixing proportions of red, green, and blue light. These are known as primary colors. Different color filters can also produce light of different colors. For example, a red filter only transmits red light, the remaining colors of the spectrum being absorbed by the filter.

Mixing red, green, and blue light in the correct proportions produces white light. When these colors are mixed in different proportions, secondary colors are formed, such as cyan, magenta, and yellow. For example, blue + red = magenta, red + green = yellow, and green + blue = cyan. Yellow light is reflected from the surfaces of some flowers as the petals absorb blue light. Red and green light are reflected back, and these mix to give the sensation of yellow.

Color Matters http://www.colormatters.com/entercolormatters.html

Examination of the interaction between color and the disciplines of physiology, psychology, philosophy, architecture, and optics.

color in elementary particle physics, a property of †quarks analogous to electric charge but having three states, denoted by red, green, and blue. The color of a quark is changed when it emits or absorbs a †gluon. The term has nothing to do with color in its usual sense. See †quantum chromodynamics.

color index in astronomy, measure of the color of a star made by comparing its brightness through different colored filters. It is defined as the difference between the measurements of the †magnitude of the star given by two standard photometric filters. Color index is directly related to the surface temperature of a star and its spectral classification.

color vision the ability of the eye to recognize different frequencies in the visible spectrum as colors. In most vertebrates, including humans, color vision is due to the presence on the †retina of three types of light-sensitive cone cell, each of which responds to a different primary color (red, green, or blue).

Color vision is one of the ways in which the brain can acquire knowledge of the unchanging characteristics of objects. Perceived colors are functions of the state of the brain, as well as of physical features of objects. They remain more or less stable, and objects remain recognizable, in spite of the continuously changing illumination in which they are seen, a phenomenon known as **color constancy**.

columbium symbol Cb, former name for the chemical element †niobium. The name is still used occasionally in metallurgy.

column vertical list of numbers or terms, especially in matrices.

coma in astronomy, hazy cloud of gas and dust that surrounds the nucleus of a †comet.

coma in optics, one of the geometrical aberrations of a lens, whereby skew rays from an object make a comet-shaped spot on the image plane instead of meeting at a point.

Coma Berenices constellation of the northern hemisphere, represented as Queen Berenice's hair.

combe or **coombe**, steep-sided valley found on the scarp slope of a chalk †escarpment. The inclusion of "combe" in a placename usually indicates that the underlying rock is chalk.

combination in mathematics, a selection of a number of objects from some larger number of objects when no account is taken of order within any one arrangement. For example, 123, 213, and 312 are regarded as the same combination of three digits from 1234. Combinatorial analysis is used in the study of †probability.

The number of ways of selecting r objects from a group of n is given by the formula:

$n!/[r!(n - r)!]$ (see †factorial). This is usually denoted by ${}^nC_r$.

combine in †probability theory, to work out the chances of two or more events occurring at the same time.

combustion burning, defined in chemical terms as the rapid combination of a substance with oxygen, accompanied by the evolution of heat and usually light. A slow-burning candle flame and the explosion of a mixture of gasoline vapor and air are extreme examples of combustion. Combustion is an †exothermic reaction as heat energy is given out.

comet small, icy body orbiting the Sun, usually on a highly elliptical path that takes it beyond the planet Pluto. A comet consists of a central nucleus a few kilometers across and is made mostly of ice mixed with gas and dust. As a comet approaches the Sun its nucleus heats up, releasing gas and dust, which form a coma (comet head) up to 100,000 km/60,000 mi wide. Gas and dust stream away from the coma to form one or more tails, which may extend for millions of kilometers. Some comets, such as Halley's comet, stay within Pluto's orbit for most of the time. Comets are normally visible during sunset or sunrise.

Comets are of many different types, characterized by their orbits, their composition (the ratio of ice to dust, and the amount of frozen volatiles other than water ice, such as methane and carbon monoxide), and their size. Most comets approach the Sun on a hyperbolic orbit and are seen only once; others (the periodic comets) return regularly in elliptical orbits. Famous examples of the periodic comets are †Halley's comet, which has a period of 76 years and is one of the largest comets known, with a nucleus about 15 x 7 x 7 km/9 x 4 x 4 mi across, and comet Encke, which has one of the shortest periods at only 3.3 years. Current thinking is that the nonperiodic comets, and those with very long periods, mostly originate in the †Oort cloud, which lies far beyond the orbit of Pluto and which may contain billions of protocomets, only a few of which are gravitationally perturbed into the inner Solar System each decade. The orbits of the periodic comets suggest a different source, and this is the †Kuiper belt, a zone just beyond the orbit of Neptune. This, too, contains a huge number of bodies, most of them too small to be detected by current techniques, although the count is rising steadily. A dozen or more comets are discovered every year, some by amateur astronomers.

Asteroid and Comet Impact Hazards http://impact.arc.nasa.gov/index.html

Overview of, and the latest news on, asteroid and comet impact hazards from NASA's Ames Space Science Division, with the last Spaceguard Survey Report and a list of future near-Earth objects (NEOs).

Comet Observation Home Page http://encke.jpl.nasa.gov/

Links to information on comet observation, including definitions and explanations of technical terms, recent news and observations, photographs, and links to the NASA and Jet Propulsion Laboratory home pages.

comfort index estimate of how tolerable conditions are for humans in hot climates. It is calculated as the temperature in degrees Fahrenheit plus a quarter of the relative †humidity, expressed as a percentage. If the sum is less than 95, conditions are tolerable for those unacclimatized to the tropics.

command language in computing, a set of commands and the rules governing their use, by which users control a program. For example, an †operating system may have commands such as SAVE and DELETE, or a payroll program may have commands for adding and amending staff records.

commensalism in biology, a relationship between two †species whereby one (the commensal) benefits from the association, whereas the other neither benefits nor suffers. For example, certain species of millipede and silverfish inhabit the nests of army ants and live by scavenging on the refuse of their hosts, but without affecting the ants.

common denominator denominator that is a common multiple of, and hence exactly divisible by, all the denominators of a set of fractions, and which therefore enables their sums or differences to be found. For example, $^2/_3$ and $^3/_4$ can both be converted to equivalent fractions of denominator 12, $^2/_3$ being equal to $^8/_{12}$ and $^3/_4$ to $^9/_{12}$. Hence their sum is $^{17}/_{12}$ and their difference is $^1/_{12}$. The **lowest common denominator** (lcd) is the smallest common multiple of the denominators of a given set of fractions.

common difference the difference between any number and the next in an †arithmetic progression. For example, in the set 1, 4, 7, 10, ... , the common difference is 3.

common factor number that will divide two or more others without leaving a remainder. For example, the †factors of 8 are 1, 2, 4, and 8; the factors of 12 are 1, 2, 3, 4, 6 and 12; and the factors of 16 are 1, 2, 4, 8 and 16. The numbers 2 and 4 are factors of all three numbers (8, 12, and 16) and are known as the common factors; 4 is known as the **highest common factor**. A †prime number can only have factors of one and itself.

common gateway interface CGI, on the World Wide Web, a facility for adding scripts to handle user input. It allows a Web server to communicate with other programs running on the same server in order to process data input by visitors to the Web site. CGI scripts "parse" the input data, identifying each element and feeding it to the correct program for action, normally a †search engine or e-mail program. The results are then fed back to the user in the form of search results or sent by e-mail. The CGI also describes how CGI applications should present their output, so that it can be displayed properly by the Web browser. CGI scripts are written in a variety of languages, one of the most common being †Perl.

CGI for the Total Non-Programmer http://www.webteacher.com/cgi/index.html
Guide to programming Common Gateway Interfaces (CGIs) with Perl on a Unix platform. The guide is intended for beginners with no experience of CGI, Perl, or Unix. It begins by explaining what CGIs are, and goes on to guide you through the process of writing your own CGI.

common logarithm another name for a †logarithm to the base ten.

communication in biology, the signaling of information by one organism to another, usually with the intention of altering the recipient's behavior. Signals used in communication may be **visual** (such as the human smile or the display of colorful plumage in birds), **auditory** (for example, the whines or barks of a dog), **olfactory** (such as the odors released by the scent glands of a deer), **electrical** (as in the pulses emitted by electric fish), or **tactile** (for example, the nuzzling of male and female elephants).

communications in computing, see †data communications.

communications satellite relay station in space for sending telephone, television, telex, and other messages around the world. Messages are sent to and from the satellites via ground stations. Most communications satellites are in †geostationary orbit, appearing to hang fixed over one point on the Earth's surface.

The first satellite to carry TV signals across the Atlantic Ocean was Telstar, which was launched into low Earth orbit by the USA on 10 July 1962. The world is now linked by the Intelsat system of communications satellites. Other satellites are used by individual countries for internal communications, or for business or military use. A new generation of satellites, called **direct-broadcast satellites**, are powerful enough to transmit directly to small domestic aerials. The power for such satellites is produced by solar cells. The total energy requirement of a satellite is small; a typical communications satellite needs about 2 kW of power, the same as an electric heater.

community in ecology, an assemblage (group) of plants, animals, and other organisms living within a defined area. Communities are usually named by reference to a dominant feature, such as characteristic plant species (for example, a beech-wood community), or a prominent physical feature (for example, a freshwater-pond community).

commutative operation in mathematics, an operation that is independent of the order of the numbers or symbols concerned. For example, addition is commutative: the result of adding $4 + 2$ is the same as that of adding $2 + 4$; subtraction is not as $4 - 2 = 2$, but $2 - 4 = -2$. Compare †associative operation and †distributive operation.

commutator device in a DC (direct-current) electric motor that reverses the current flowing in the armature coils as the armature rotates.

A DC generator, or †dynamo, uses a commutator to convert the AC (alternating current) generated in the armature coils into DC. A commutator consists of opposite pairs of conductors insulated from one another, and contact to an external circuit is provided by carbon or metal brushes.

compact spaces or **bicompact spaces**, special kinds of topological space exhibiting the property that, internally, every family of open sets whose union is the whole space necessarily contains a finite subfamily whose union is already the whole space. An alternative definition, first formulated with regard to a special class of such spaces, requires every sequence of points to have a converging subsequence. Russian mathematicians initially used the word **bicompact** (not now used in the West) to distinguish between the specialized and the general definitions.

companion star either one of a †binary star system (although usually the less massive), sometimes only detectable by †spectroscopy.

comparative anatomy study of the similarity and differences in the anatomy of different groups of animals. It helps to reveal how animals are related to each other and how they have changed through evolution.

Structures are **homologous** if they have arisen from the same ancestral structure through evolution, but perform either similar or different functions in modern animals. Examples of homologous structures are the wings of birds, the human arm, and the forelimb of whales.

Analogous structures have developed from different ancestral structures, but perform similar functions, such as the wings of insects and those of birds.

comparative psychology branch of psychology concerned with differences in the behavior of various animal species; also the study of animal psychology in general. The most important area of research has been that of learning, covering topics such as †conditioning, †behaviorism, and the effects of reward and punishment on performance.

The effects of various drugs on psychological processes and behavior has also been an important area of study, as has maternal behavior and interactions between mothers and offspring, particularly in mammals, together with the insights gained in our understanding of infant development. A number of experimental techniques are used, including research under laboratory conditions and field studies where the behavior of animals is observed under natural conditions.

compass any instrument for finding direction. The most commonly used is a †magnetic compass, consisting of a thin piece of magnetic material with the north-seeking pole indicated, free to rotate on a pivot and mounted on a compass card on which the points of the compass are marked. When the compass is properly adjusted and used, the north-seeking pole will point to the magnetic north, from which true north can be found from tables of magnetic corrections.

Compasses not dependent on the magnet are gyrocompasses, dependent on the †gyroscope, and radiocompasses, dependent on the use of radio. These are unaffected by the presence of iron and by magnetic anomalies of the Earth's magnetic field, and are widely used in ships and aircraft.

compensation point in biology, the point at which there is just enough light for a plant to survive. At this point all the food produced by †photosynthesis is used up by †respiration. For aquatic plants, the compensation point is the depth of water at which there is just enough light to sustain life (deeper water = less light = less photosynthesis).

competition in ecology, the interaction between two or more organisms, or groups of organisms, that use a common resource in short supply. There can be competition between members of the same species and competition between members of different species. Competition invariably results in a reduction in the numbers of one or both competitors, and in †evolution contributes both to the decline of certain species and to the evolution of †adaptations.

The resources in short supply for which organisms compete may be obvious things, such as †mineral salts for animals and plants, or light for plants. However, there are less obvious resources. For example, competition for suitable nesting sites is important in some species of birds. Competition results in a reduction in breeding success for one or other organism(s). Because of this it is one of the most important aspects of †natural selection, which may result in evolutionary change if the †environment is changing. Competition also results in the distribution of organisms we see in †habitats. It is believed that organisms tend to occur where the pressures of competition are not as great as in other areas. In agriculture cultivation methods are designed to reduce competition. For example, a crop of wheat is sown at a density that minimizes competition within the same species. The plants are grown far enough apart to reduce competition between the roots of neighboring wheat plants for soil †mineral nutrients. The spraying of the ground to kill weeds reduces competition between the wheat and weed plants. Some weeds would grow taller than the wheat and deprive it of light.

competitive exclusion principle in evolution, principle of natural selection whereby similar species are forced to specialize ever more minutely so as not to overlap with each other in a particular niche; if they do not specialize adequately, they die and become extinct.

compiler computer program that translates programs written in a †high-level language into machine code (the form in which they can be run by the computer). The compiler translates each high-level instruction into several machine-code instructions – in a process called **compilation** – and produces a complete independent program that can be run by the computer as often as required, without the original source program being present.

Different compilers are needed for different high-level languages and for different computers. In contrast to using an interpreter, using a compiler adds slightly to the time needed to develop a new program because the machine-code program must be recompiled after each change or correction. Once compiled, however, the machine-code program will run much faster than an interpreted program.

complement protein substance in blood serum which reacts with almost all antibody–antigen systems, lysing (breaking down) the antigen from the antibody and disappearing in the process.

complement in mathematics, the set of the elements within the universal set that are not contained in the designated set. For example, if the universal set is the set of all positive whole numbers and the designated set S is the set of all even numbers, then the complement of S (denoted S') is the set of all odd numbers.

complementary angles two angles that add up to 90°.

complementary colors in physics, any two colors (of light) that mix to produce white. If the colors of the spectrum are arranged around the rim of a wheel, complementary colors (such as blue and orange) lie opposite each other on a diameter of the wheel.

complementary number in number theory, the number obtained by subtracting a number from its base. For example, the complement of 7 in numbers to base 10 is 3. Complementary numbers are necessary in computing, as the only mathematical operation of which digital computers (including pocket calculators) are directly capable is addition. Two numbers can be subtracted by adding one number to the complement of the other; two numbers can be divided by using successive subtraction (which, using complements, becomes successive addition); and multiplication can be performed by using successive addition.

complementation in genetics, the interaction that can occur between two different mutant alleles of a gene in a †diploid organism, to make up for each other's deficiencies and allow the organism to function normally.

completing the square method of converting a quadratic expression such as $x^2 + px + q$ into a perfect square by adding $p^2/4$ and subtracting q in order to solve the quadratic equation $x^2 + px + q = 0$. The steps are:

$$x^2 + px = -q$$

$$x^2 + px + \left(\frac{p}{2}\right)^2 = \left(\frac{p}{2}\right)^2 - q$$

$$\left(x + \frac{p}{2}\right)^2 = \frac{p^2}{4} - q$$

$$x + \frac{p}{2} = \pm\sqrt{\left(\frac{p^2}{4} - q\right)}$$

$$x = -\frac{p}{2} \pm \sqrt{\left(\frac{p^2}{4} - q\right)}$$

complex in psychology, a group of ideas and feelings that have become repressed because they are distasteful to the person in whose mind they arose, but are still active in the depths of the person's unconscious mind, continuing to affect his or her life and actions, even though he or she is no longer fully aware of their existence. Typical examples include the †Oedipus complex and the †inferiority complex.

complex conjugate in mathematics, of a complex number $z = a - ib$, $z = a + ib$ is the conjugate complex number. It is useful for calculating the multiplicative inverse of z since the product $zz = |z|^2$ is a real number.

complex number in mathematics, a number written in the form $a + ib$, where a and b are †real numbers and i is the square root of –1 (that is, $i^2 = -1$); i used to be known as the "imaginary" part of the complex number. Some equations in algebra, such as those of the form

$x^2 + 5 = 0$

cannot be solved without recourse to complex numbers, because the real numbers do not include square roots of negative numbers.

The sum of two or more complex numbers is obtained by adding separately their real and imaginary parts, for example:

$(a + bi) + (c + di) = (a + c) + (b + d)i$

Complex numbers can be represented graphically on an †Argand diagram, which uses rectangular †Cartesian coordinates in which the x-axis represents the real part of the number and the y-axis the imaginary part. Thus the number $z = a + ib$ is plotted as the point (a, b). Complex numbers have applications in various areas of science, such as the theory of alternating currents in electricity.

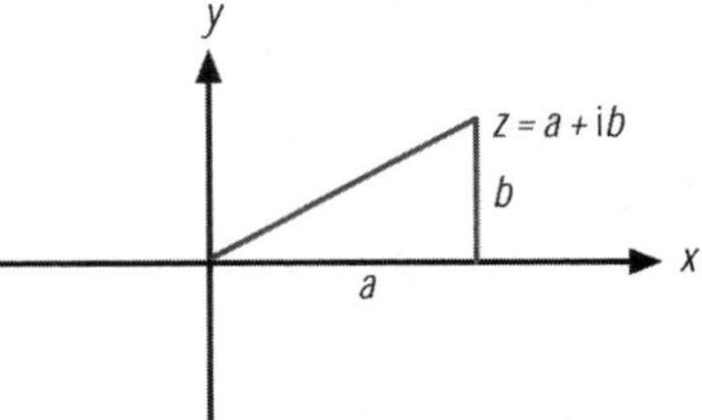

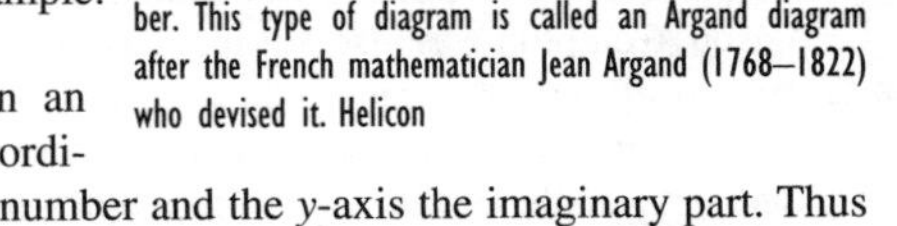
complex number A complex number can be represented graphically as a line whose end-point coordinates equal the real and imaginary parts of the complex number. This type of diagram is called an Argand diagram after the French mathematician Jean Argand (1768–1822) who devised it. Helicon

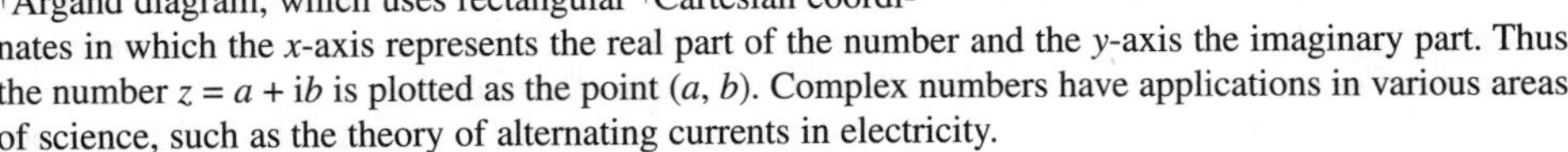

complex number astrophysics basis of †twistor theory.

complex variable †variable representing a †complex number.

component in mathematics, one of the vectors produced when a single vector is resolved into two or more parts. The vector sum of the components gives the original vector.

component in physics, one of two or more vectors, normally at right angles to each other, that add together to produce the same effect as a single resultant vector. Any †vector quantity, such as force, velocity, or electric field, can be resolved into components chosen for ease of calculation. For example, the weight of a body resting on a slope can be resolved into two force components (see †resolution of forces); one normal to the slope and the other parallel to the slope.

composite function in mathematics, a function made up of two or more other functions carried out in sequence, usually denoted by *, as in the relation $(f * g)\, x = f\,[g(x)]$.

Usually, composition is not commutative: $(f * g)$ is not necessarily the same as $(g * f)$.

composite volcano steep-sided conical †volcano formed above a †subduction zone at a destructive †plate margin. It is made up of alternate layers of ash and lava. The †magma (molten rock) associated with composite volcanoes is very thick and often clogs up the vent. This can cause a tremendous build-up of pressure, which, once released, causes a very violent eruption. Examples of composite volcanoes are Mount St Helens in the USA, and Stromboli and Vesuvius in Italy.

Composite volcanoes are usually found in association with island arcs and coastal mountain chains. The magma is mostly derived from plate material, and is rich in silica. This makes a very viscous lava, such as andesite, which solidifies rapidly to form a high, steep-sided volcanic mountain. This magma often clogs the volcanic vent. After the eruption, the crater may collapse to form a caldera.

compost organic material decomposed by bacteria under controlled conditions to make a nutrient-rich natural fertilizer for use in gardening or farming. A well-made compost heap reaches a high temperature during the composting process, killing most weed seeds that might be present.

Compost http://www.ext.vt.edu/pubs/envirohort/426-703/426-703.html

Informative guide to compost from the Virginia Cooperative Extension. It first describes exactly what compost is and what its benefits are. Then it gives detailed information about making your own compost.

compound chemical substance made up of two or more †elements bonded together, so that they cannot be separated by physical means. Compounds are held together by ionic or covalent bonds.

The name of a compound may give a clue to its composition. If the name ends in -ide (with the notable exceptions of hydroxides and ammonium chloride) it usually contain two elements. For example calcium oxide is a compound of calcium and oxygen.

If the name ends in -ate or -ite the compound contains oxygen; compounds ending in -ate have a greater proportion of oxygen than those ending in -ite. For example, sodium sulfate (Na_2SO_4) contains more oxygen than does sodium sulfite (Na_2SO_3).

If the name starts with per- the compound contains extra oxygen. For example, hydrogen peroxide H_2O_2 contains one more oxygen than hydrogen oxide (water) H_2O.

The prefix thio- indicates that the compound contains an atom of sulfur in place of oxygen. For example, sodium thiosulphate ($Na_2S_2O_3$) contains one more sulfur and one less oxygen than the more common sodium sulfate (Na_2SO_4).

The proportions of the different elements in a compound are shown by the chemical formula of that compound. For example, a molecule of sodium sulfate, represented by the formula Na_2SO_4 contains two atoms of sodium, one of sulfur, and four of oxygen.

compound lens in physics, lens that consists of more than one element – which may or may not be in contact – arranged so as to minimize aberrations.

compression in computing, see †data compression.

Compton, Arthur H(olly) (1892–1962) US physicist who in 1923 found that X-rays scattered by such light elements as carbon increased in wavelength. He concluded from this unexpected result that the X-rays were displaying both wavelike and particle-like properties, a phenomenon later named the **Compton effect**. He was awarded the Nobel Prize for Physics in 1927 for his study of the transfer of energy from electromagnetic radiation to a particle. He shared the award with Scottish physicist Charles †Wilson. Compton was also a principal contributor to the development of the atomic bomb.

Compton effect in physics, the increase in wavelength (loss of energy) of a photon by its collision with a free electron (**Compton scattering**). The Compton effect was first demonstrated with X-rays and provided early evidence that electromagnetic waves consisted of particles – photons – which carried both energy and momentum. It is named after US physicist Arthur †Compton.

computer programmable electronic device that processes data and performs calculations and other symbol-manipulation tasks. There are three types: the †digital computer, which manipulates information coded as binary numbers (see †binary number system); the †analog computer, which works with continuously varying quantities; and the **hybrid computer**, which has characteristics of both analog and digital computers. In common usage, when someone refers to a "computer," they tend to mean a digital computer.

There are four types of digital computer, corresponding roughly to their size and intended use. **Microcomputers** are the smallest and most common, used in small businesses, at home, and in schools. They are usually single-user machines. **Minicomputers** (or mid-range computers) are found in medium-sized businesses and university departments. They may support from around 10 to 200 users at once. **Mainframes** (or enterprise servers), which can often service several hundred users simultaneously, are found in large organizations, such as national companies and government departments. **Supercomputers** are mostly used for highly complex scientific tasks, such as analyzing the results of nuclear physics experiments and weather forecasting.

Computer Museum http://www.computerhistory.org/

Well-designed interactive museum, examining the history, development, and future of computer technology. As well as plenty of illustrations and detailed explanations, it is possible to change your route through the museum by indicating whether you are a kid, student, adult, or educator.

computer-aided design use of computers to create and modify design drawings; see †CAD.

computer-aided manufacturing use of computers to regulate production processes in industry; see †CAM.

computer-assisted learning use of computers in education and training; see †CAL.

computer generation any of the five broad groups into which computers may be classified: **first generation**, the earliest computers, developed in the 1940s and 1950s, made from valves and wire circuits; **second generation** from the early 1960s, based on transistors and printed circuits; **third generation** from the late 1960s, using integrated circuits and often sold as families of computers, such as the IBM 360 series; **fourth generation** using †microprocessors, large-scale integration (LSI), and sophisticated programming languages, still in use in the 1990s; and **fifth generation** based on parallel processing and very large-scale integration, currently under development.

computer graphics use of computers to display and manipulate information in pictorial form. Input may be achieved by scanning an image, by drawing with a mouse or stylus on a graphics tablet, or by drawing directly on the screen with a light pen.

The output may be as simple as a pie chart, or as complex as an animated sequence in a science fiction film, or a seemingly three-dimensional engineering blueprint. The drawing is stored in the computer as raster graphics or vector graphics.

Vector graphics are stored in the computer memory by using geometric formulae. They can be transformed (enlarged, rotated, stretched, and so on) without loss of picture resolution. It is also possible to select and transform any of the components of a vector-graphics display because each is separately defined in the computer memory. In these respects vector graphics are superior to raster graphics. They are typically used for drawing applications, allowing the user to create and modify technical diagrams such as designs for houses or cars.

Raster graphics are stored in the computer memory by using a map to record data (such as color and intensity) for every †pixel that makes up the image. When transformed (enlarged, rotated, stretched, and so on), raster graphics become ragged and suffer loss of picture resolution, unlike vector graphics. They are typically used for painting applications, which allow the user to create artwork on a computer screen much as if they were painting on paper or canvas.

Computer graphics are increasingly used in computer-aided design (†CAD), and to generate models and simulations in engineering, meteorology, medicine and surgery, and other fields of science.

Computer Graphics World http://cgw.pennnet.com/home.cfm

Packed with articles, reviews, and up-to-date information. The home page of Computer Graphics World magazine carries comprehensive coverage of the latest graphics news and events. This easily navigated Web site features an impressive and well-documented image gallery.

Exploratories Project http://www.cs.brown.edu/exploratory/

Brown University project that introduces visitors to research into Web-based education. The Exploratories use two- and three-dimensional explorable worlds to offer learning experiences not attainable with conventional educational methods. Aimed at teachers and researchers, the site offers guided tours of subjects such as computer graphics, and links to sites of educational interest.

computer language see †programming language.

computer program coded instructions for a computer; see †program.

computer simulation representation of a real-life situation in a computer program. For example, the program might simulate the flow of customers arriving at a bank. The user can alter variables, such as the number of cashiers on duty, and see the effect.

More complex simulations can model the behavior of chemical reactions or even nuclear explosions. The behavior of solids and liquids at high temperatures can be simulated using †quantum simulation. Computers also control the actions of machines – for example, a †flight simulator models the behavior of real aircraft and allows training to take place in safety. Computer simulations are very useful when it is too dangerous, time consuming, or simply impossible to carry out a real experiment or test.

Society For Modeling and Simulation International http://www.scs.org/

Detailing the structure and activities of the Society for Computer Simulation. There is information about the society's journals, forthcoming conferences, and membership. The Web site links to governmental and research facilities with an interest in computer simulation.

computer terminal device whereby the operator communicates with the computer; see †terminal.

concave of a surface, curving inwards, or away from the eye. For example, a bowl appears concave when viewed from above. In geometry, a concave polygon is one that has an interior angle greater than 180°. Concave is the opposite of †convex.

concave lens lens that possesses at least one surface that curves inwards. It is a diverging lens, spreading out those light rays that have been refracted through it. A concave lens is thinner at its center than at its edges, and is used to correct short-sightedness (myopia).

After light rays have passed through the lens, they appear to come from a point called the principal focus. The distance between the principle focus and the lens is the focal length. A more curved lens will have a smaller focal length and will be a more powerful lens. The image formed by a concave lens is virtual, upright, and smaller than the object, and it cannot be projected onto a screen. The lens formula is used to work out the position and nature of an image formed by a lens: $1/u + 1/v = 1/f$, where u and v are the distances of the object and image from the lens, respectively, and f is the focal length of the lens.

concave mirror curved mirror that reflects light from its inner surface, the curve being inward. It may be either circular or parabolic in section. A concave mirror converges parallel light rays inward to the point of †principal focus. The image formed by a concave mirror is real (reduced and inverted) if the object is not too close to the mirror. A real image is formed at the point of convergence. If the object is close to the mirror then the image formed will be virtual, enlarged, and upright, as the rays of light cannot converge to a point.

Only a parabolic concave mirror has a true, single-point †principal focus for parallel rays. For this reason, parabolic mirrors are used as reflectors to focus light in telescopes, or to focus microwaves in satellite communication systems. The reflector behind a spot lamp or car headlamp is parabolic.

concentration in chemistry, the amount of a substance (†solute) present in a specified amount of a solution. Either amount may be specified as a mass or a volume (liquids only). Common units used are †moles per cubic decimeter, grams per cubic decimeter, grams per 100 cubic centimeters, and grams per 100 grams.

The term also refers to the process of increasing the concentration of a solution by removing some of the substance (†solvent) in which the solute is dissolved. In a **concentrated solution**, the solute is present in large quantities. Concentrated brine is around 30% sodium chloride in water; concentrated caustic soda (caustic liquor) is around 40% sodium hydroxide; and concentrated sulfuric acid is 98% acid.

concentration gradient change in the concentration of a substance from one area to another. Particles, such as sugar molecules, in a fluid move over time so that they become evenly distributed throughout the fluid. In particular, they move from an area of high concentration to an area of low concentration; that is, they diffuse along the concentration gradient (see †diffusion).

This explains why oxygen in the lungs will diffuse into the blood supply. The oxygen molecules are more concentrated in the lungs than they are in the capillaries surrounding the †alveoli (air sacs). As it diffuses along the concentration gradient, oxygen will tend to pass into the blood. Gas exchange therefore depends on the maintenance of a concentration gradient, so that oxygen will continue to diffuse across the respiratory surface.

concentric circles two or more circles that share the same center.

conceptacle flask-shaped cavity found in the swollen tips of certain brown seaweeds, notably the wracks *Fucus*. The gametes are formed within conceptacles and released into the water via a small pore known as an ostiole.

conchoid curve an †algebraic curve represented by an equation of the general form:

$$x^2y^2 = (x - a)^2(c^2 - x^2)$$

conchology branch of zoology that studies mollusks with reference to their shells.

concurrent lines two or more lines passing through a single point; for example, the diameters of a circle are all concurrent at the center of the circle.

condensation in organic chemistry, a reaction in which two organic compounds combine to form a larger molecule, accompanied by the removal of a smaller molecule (usually water). This is also known as an addition–elimination reaction. Polyamides (such as nylon) and polyesters (such as Terylene) are made by †condensation polymerization.

condensation conversion of a vapor to a liquid. This is frequently achieved by letting the vapor come into contact with a cold surface. It is the process by which water vapor turns into fine water droplets to form a †cloud.

Condensation in the atmosphere occurs when the air becomes completely saturated and is unable to hold any more water vapor. As air rises it cools and contracts – the cooler it becomes the less water it can hold.

Rain is frequently associated with warm weather fronts because the air rises and cools, allowing the water vapor to condense as rain. The temperature at which the air becomes saturated is known as the †**dew point**. Water vapor will not condense in air if there are not enough condensation nuclei (particles of dust, smoke, or salt) for the droplets to form on. It is then said to be supersaturated. Condensation is an important part of the †water cycle.

condensation number in physics, the ratio of the number of molecules condensing on a surface to the total number of molecules touching that surface.

condensation polymerization †polymerization reaction in which one or more monomers, with more than one reactive functional group, combine to form a polymer with the elimination of water or another small molecule.

condensation reaction in chemistry, reaction between molecules during which a simple molecule, such as water, is eliminated. For example two monosaccharide molecules can combine by the elimination of water between two hydroxyl (—OH) groups to form a disaccharide.

condenser laboratory apparatus used to condense vapors back to liquid so that the liquid can be recovered. It is used in †distillation and in reactions where the liquid mixture can be kept boiling without the loss of solvent.

condenser in electronic circuits, a former name for a †capacitor.

conditioned stimulus in psychology, an originally neutral stimulus applied in conditioning experiments that evokes a trained or conditioned response. In Pavlov's classical experiments with dogs, the neutral stimulus (the sound of a bell) originally evoked no salivation reflex; but after being presented for a time with an †unconditioned stimulus (food), it became the conditioned stimulus that evoked the conditioned response.

conditioning in psychology, two major principles of behavior modification.

In **classical conditioning**, described by Russian psychologist Ivan Pavlov, a new stimulus can evoke an automatic response by being repeatedly associated with a stimulus that naturally provokes that response. For example, the sound of a bell repeatedly associated with food will eventually trigger salivation, even if sounded without food being presented. In **operant conditioning**, described by US psychologists Edward Lee Thorndike (1874–1949) and B F Skinner, the frequency of a voluntary response can be increased by following it with a reinforcer or reward.

conduction, electrical flow of charged particles through a material giving rise to electric current. Conduction in metals involves the flow of negatively charged free †electrons. Conduction in gases and some liquids involves the flow of †ions that carry positive charges in one direction and negative charges in the other. Conduction in a †semiconductor such as silicon involves the flow of electrons and positive holes.

conduction, heat flow of heat energy (see †energy transfer) through a material without the movement of any part of the material itself (compare †conduction, electrical). Heat energy is present in all materials in the form of the †kinetic energy of their constituent vibrating particles, and may be conducted from one particle to the next in the form of this vibration.

Different materials conduct heat at different rates. This rate is called the thermal conductivity. A good conductor of heat, such as steel, will have a high thermal conductivity and poor conductor of heat, such as air, will have a low thermal conductivity. In general, non-metals, such as wood or glass, are poor conductors of heat.

In the construction industry, the thermal conductivities of the materials to be installed need to be known. A low thermal conductivity indicates that a material is a poor conductor of heat and therefore a good insulator. For example, foam is used to insulate lofts, the air trapped in the foam making it a good insulator. Bricks used in outer walls have tiny air spaces, again allowing air to be trapped and making the bricks good insulators. Insulation can also be achieved with double-glazed windows, with the space between the two glass panes containing air or a vacuum.

conductivity, thermal ($W\ m^{-1}\ K^{-1}$) measure of how well a material conducts heat. A good conductor, such as a metal, has a high conductivity; a poor conductor, called an insulator, has a low conductivity. The measurement of a material's heat-conducting properties is given as its †U-value.

conductor any material that conducts heat or electricity (as opposed to an insulator, or nonconductor). A good conductor has a high electrical or heat conductivity, and is generally a substance rich in loosely-held free electrons, such as a metal. Copper and aluminum are good conductors. A poor conductor (such as the non-metals glass, porcelain, and rubber) has few free electrons and resists the flow of electricity or heat. †Carbon is exceptional in being non-metallic and yet (in some of its forms) a relatively good conductor of heat and electricity. Substances such as †silicon and †germanium, with intermediate conductivities that are improved by heat, light, or impurities, are known as †semiconductors.

Liquids (including water) can also be electrical conductors. Electricity (current) can flow by the movement of charged ions through a solution or a molten salt (electrolyte). This process is called electrolysis.

cone in botany, the reproductive structure of the conifers and cycads; also known as a †strobilus. It consists of a central axis surrounded by numerous, overlapping, scalelike, modified leaves (sporophylls) that bear the reproductive organs. Usually there are separate male and female cones, the former bearing pollen sacs containing pollen grains, and the larger female cones bearing the ovules that contain the ova or egg cells. The pollen is carried from male to female cones by the wind (†anemophily). The seeds develop within the female cone and are released as the scales open in dry atmospheric conditions, which favor seed dispersal.

In some groups (for example, the pines) the cones take two or even three years to reach maturity. The cones of junipers have fleshy cone scales that fuse to form a berrylike structure. One group of †angiosperms, the alders, also bear conelike structures; these are the woody remains of the short female catkins, and they contain the alder †fruits.

cone in zoology, type of light-sensitive cell found in the retina of the †eye.

cone in geometry, a pyramid with a circular base. If the point (vertex) is directly above the center of the circle, it is known as a **right circular cone**. The †volume (V) of this cone is given by the formula $V = 1/3\pi r^2 h$ where h is the perpendicular height and r is the base radius.

A right circular cone is generated by rotating an isosceles triangle about its line of symmetry. The distance from the edge of the base of a cone to the vertex is called the slant height. In a right circular cone of slant height l, the curved surface area is prl, and the area of the base is pr^2. Therefore the total surface area A = prl + pr^2 = p$r(l + r)$.

For example, to find the volume and surface area of a cone with a perpendicular height of 4 cm and radius of 3 cm:

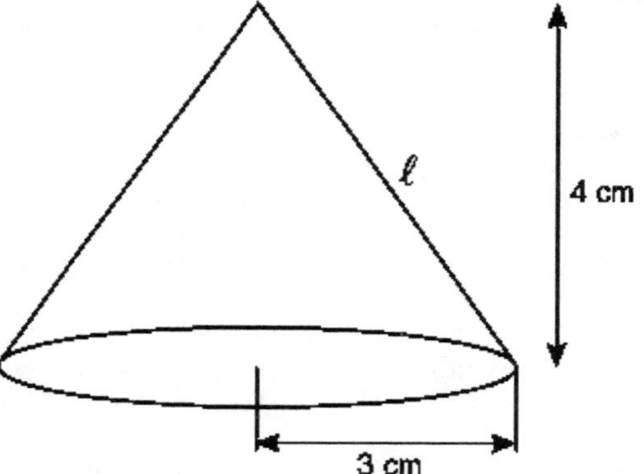

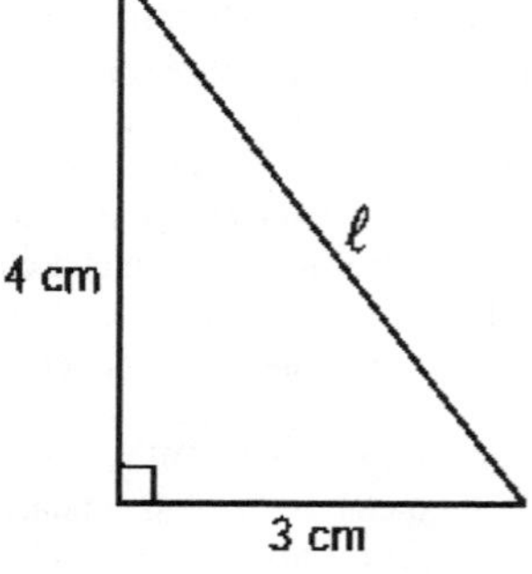

Volume = ¹/₃pr^2h, so

$V = \frac{1}{3}$p x 3^2 x 4

$V = 37.7\text{ cm}^3$ (to 1 decimal place)

Surface area = p$r(l + r)$. First the slant length (l) is calculated:

Using †Pythagoras' theorem,

$l^2 = h^2 + r^2$, so

$l^2 = 4^2 + 3^2 = 16 + 9 = 25$, so

$l = 5$ cm

The surface area can now be calculated:

$A = \pi r(l + r)$, so

$A = \pi \times 3(3 + 5) = 3$p x $8 = 24\pi$, so

$A = 75.4\text{ cm}^2$ (to 1 decimal place)

configuration in chemistry, the arrangement of atoms in a molecule or of electrons in atomic orbitals.

configuration in computing, the way in which a system, whether it be †hardware and/or †software, is set up. A minimum configuration is often referred to for a particular application, and this will usually include a specification of processor, disk and memory size, and peripherals required.

confluence point at which two rivers join, for example the River Thames and River Cherwell at Oxford, England, or the White Nile and Blue Nile at Omdurman, Sudan.

conformation in chemistry, possible spatial arrangements of atoms in a molecule that result from rotation of its constituent groups around a single bond. This term is usually applied only to stable equilibrium configurations of molecules.

conglomerate in geology, a type of †sedimentary rock composed of rounded fragments ranging in size from pebbles to boulders, cemented together in a finer sand or clay material.

congruence in geometry, two sets are congruent if either can be transformed by translations and rotations into the other.

congruent in geometry, having the same shape and size (and area), as applied to two-dimensional or solid figures. With plane congruent figures, one figure will fit on top of the other exactly, though this may first require †rotation, †translation, or †reflection of one of the figures.

conical having the shape of a †cone.

conics study initiated by Apollonius of Perga of how a cone can be "cut" so as to produce circles, ellipses, parabolas, and hyperbolas; he stated "a conic section is the locus of a point that moves so that the ratio of its distance *f* from a fixed point, to its distance *d* from a straight line, is constant." Whether the constant *c* is greater than, equal to, or less than one determines the type of a curve the section represents.

conic section curve obtained when a conical surface is intersected by a plane. If the intersecting plane cuts both extensions of the cone, it yields a †hyperbola; if it is parallel to the side of the cone, it produces a †parabola. Other intersecting planes produce †circles or †ellipses.

The Greek mathematician Apollonius of Perga wrote eight books with the title *Conic Sections*, which superseded previous work on the subject by Aristarchus and Euclid.

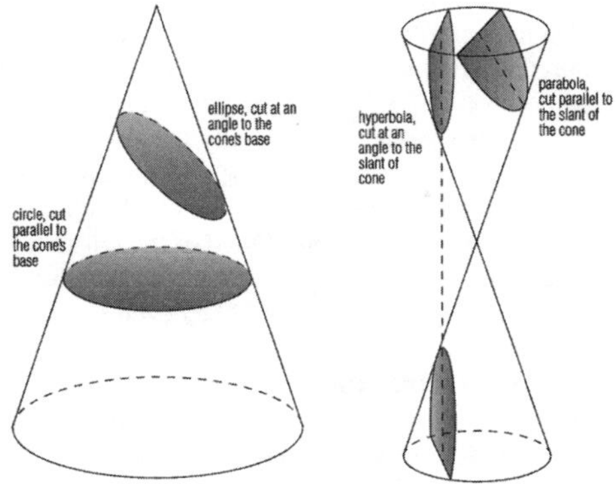

conic section The four types of curve that may be obtained by cutting a single or double right-circular cone with a plane (two-dimensional surface). Helicon

conidium plural **conidia**, asexual spore formed by some fungi at the tip of a specialized †hypha or conidiophore. The conidiophores grow erect, and cells from their ends round off and separate into conidia, often forming long chains. Conidia easily become detached and are dispersed by air movements.

conjugate in mathematics, a term indicating that two elements are connected in some way; for example, $(a + ib)$ and $(a - ib)$ are conjugate complex numbers.

conjugate angles two angles that add up to 360°.

conjugation in biology, temporary union of two single cells (or hyphae in fungi) with at least one of them receiving genetic material from the other: the bacterial equivalent of sexual reproduction. A fragment of the †DNA from one bacterium is passed along a thin tube, the pilus, into another bacterium.

conjugation in organic chemistry, the alternation of double (or triple) and single carbon–carbon bonds in a molecule – for example, in penta-1,3-diene, $H_2C = CH - CH = CH - CH_3$. Conjugation imparts additional stability as the double bonds are less reactive than isolated double bonds.

conjunction in astronomy, alignment of two celestial bodies as seen from Earth. A †superior planet (or other object) is in conjunction when it lies behind the Sun. An †inferior planet (or other object) comes to **inferior conjunction** when it passes between the Earth and the Sun; it is at **superior conjunction** when it passes behind the Sun.

Planetary conjunction takes place when a planet is closely aligned with another celestial object, such as the Moon, a star, or another planet.

Because the orbital planes of the inferior planets are tilted with respect to that of the Earth, they usually pass either above or below the Sun at inferior conjunction. If they line up exactly, a †transit will occur.

conjunctiva membrane covering the front of the vertebrate †eye. It is continuous with the epidermis of the eyelids, and lies on the surface of the cornea.

connective tissue in animals, tissue made up of a noncellular substance, the †extracellular matrix, in which some cells are embedded. Skin, bones, tendons, cartilage, and adipose tissue (fat) are the main connective tissues. There are also small amounts of connective tissue in organs such as the brain and liver, where they maintain shape and structure.

connectivity in mathematics, measure of how well connected places are in a transport network. A simple measure is the †beta index.

consciousness the state of being aware of oneself and one's surroundings, without hindrance from sleep, illness, drugs, or hypnotism. This awareness is not purely of external events or phenomena, but also of one's own feelings, beliefs, and mental events.

Such introspective self-awareness, as opposed to merely responding to external stimuli, is generally taken to be a prerequisite for consciousness. This sidesteps the question of animal consciousness, which is largely believed to be very different or even nonexistent.

conservation in the life sciences, action taken to protect and preserve the natural world, usually from pollution, overexploitation, and other harmful features of human activity. The late 1980s saw a great increase in public concern for the environment, with membership of conservation groups, such as Friends of the Earth, Greenpeace, and the US Sierra Club, rising sharply and making the green movement an increasingly-powerful political force. Globally the most important issues include the depletion of atmospheric ozone by the action of †chlorofluorocarbons (CFCs), the build-up of carbon dioxide in the atmosphere (thought to contribute to the †greenhouse effect), and †deforestation.

Conservation may be necessary to prevent an endangered species from dying out in an area or even becoming extinct. But conservation of particular †habitats may be as important, if not more important. Habitat loss is believed to be the main cause of the great reduction of biodiversity and the rate of extinction occurring on Earth. There is concern about loss of species and †biodiversity, because living organisms contribute to human health, wealth, and happiness in several ways. Humans often enjoy being in natural environments, especially those who spend much of their lives in towns and cities. Human cultures may be dependent on the natural environment to sustain them and maintain a stable society. This is particularly true of societies in the developing world. The spread of desert into arid areas around the Sahara has probably contributed to the unstable societies of some areas there. There is also an economic argument for conservation. It is believed that many undiscovered useful chemicals may exist within organisms on Earth that could be developed into important drugs – but when a plant or animal becomes extinct, the chemicals it contains are also lost.

Earthwatch Institute http://www.earthwatch.org/

Earthwatch Institute promotes conservation through scientific field research, creating partnerships between scientists, educators, and the public. Members have the opportunity to work with distinguished researchers in a range of different fields; the site includes details of how to join, as well as news of current expeditions and research progress.

conservation of energy principle that states that in a chemical reaction, the total amount of energy in the system remains unchanged. Energy can be transferred from one form into another but cannot be created or destroyed.

In a chemical reaction, for each component there may be changes in energy due to change of physical state, changes in the nature of chemical bonds, and either an input or output of energy. However, there is no net gain or loss of energy. This is true outside of chemical reactions; for example the chemical energy of a battery is transferred into the electrical energy in a circuit, which can then be transferred into the heat and light energy of a bulb, but the total amount of energy is constant.

conservation of mass in chemistry, the principle that states that in a chemical reaction the sum of all the masses of the substances involved in the reaction (reactants) is equal to the sum of all of the masses of the substances produced by the reaction (products) – that is, no matter is gained or lost.

Mass Conservation http://www.iun.edu/~cpanhd/C101webnotes/matter-and-energy/masscons.html

Insightful page on the conservation of mass within chemical reactions. An extremely short history behind the principles precedes a detailed explanation involving the formation of water, and an example of how to calculate the mass of oxygen that has reacted with a metal to form rust.

conservation of momentum in mechanics, a law that states that total †momentum is conserved (remains constant) in all collisions, providing no external resultant force acts on the colliding bodies. The principle may be expressed as an equation used to solve numerical problems: total momentum before collision = total momentum after collision.

constant in mathematics, a fixed quantity or one that does not change its value in relation to †variables. For example, in the algebraic expression $y^2 = 5x - 3$, the numbers 3 and 5 are constants. In physics, certain quantities are regarded as universal constants, such as the speed of light in a vacuum.

constantan or **eureka**, high-resistance alloy of approximately 40% nickel and 60% copper with a very low coefficient of †thermal expansion (measure of expansion on heating). It is used in electrical resistors.

constant composition, law of in chemistry, the law that states that the proportions of the amounts of the elements in a pure compound are always the same and are independent of the method by which the compound was produced.

constant of precession see †precession of the equinoxes.

constellation one of the 88 areas into which the sky is divided for the purposes of identifying and naming celestial objects. The first constellations were simple, arbitrary patterns of stars in which early civilizations visualized gods, sacred beasts, and mythical heroes.

The constellations in use today are derived from a list of 48 known to the ancient Greeks, who inherited some from the Babylonians. The current list of 88 constellations was adopted by the International Astronomical Union, astronomy's governing body, in 1930.

AJ's Cosmic Thing http://www.cpu-zeto.com.pl/hobby/astronomia/CosmicThing/intro.html

Set any time, date, longitude, latitude, and position in the setting window, and this impressive plotter applet will display stationary or moving full-sky images of objects in the sky. Clicking on plotted objects reveals information about the object. There is a technical page for those wishing additional information.

Stars and Constellations http://www.astro.wisc.edu/~dolan/constellations/constellations.html

Hugely informative site on stars and constellations. It includes star charts of all major stars and constellations, details of the origins of the various names, photographs of galaxies and the Milky Way, and details on what stars can be seen at any given time.

construction in geometry, an addition to a figure, which is drawn to help solve a problem or produce a proof. The term is also applied to the accurate drawing of shapes.

constructive margin in †plate tectonics, a boundary between two lithospheric plates, along which new crust is being created. The term usually refers to †divergent margins, where two oceanic plates are moving away from each other. As they diverge, magma (molten rock) wells up to fill the open space, and solidifies, forming new oceanic crust. Similar processes occur where a continent plate is beginning to split apart.

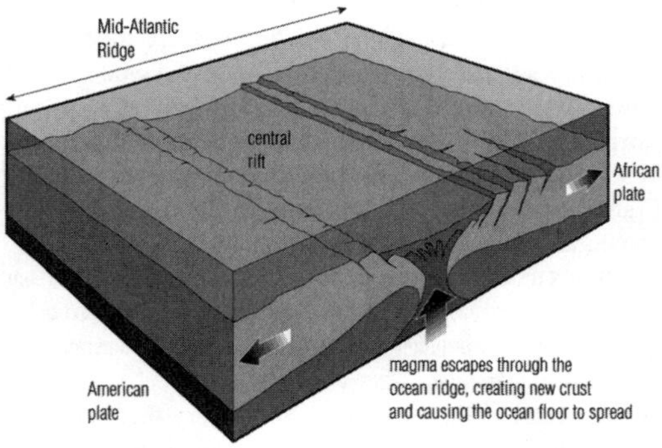

constructive margin A rift in the crustal plates where new material is being formed causes the plates to be pushed apart. This usually occurs as a result of volcanic action. Helicon

contact angle in physics, angle between a liquid and a solid surface, measured inside the liquid. If the liquid wets the surface (such as water and glass) the contact angle is less than 90°; if it does not wet the glass (such as mercury and glass) it is more than 90°.

contact force force or push produced when two objects are pressed together and their surface atoms try to keep them apart. Contact forces always come in pairs – for example, the downward force exerted on a floor by the sole of a person's foot is matched by an equal upward force exerted by the floor on that sole.

contact process main industrial method of manufacturing the chemical †sulfuric acid. Sulfur dioxide is produced by burning sulfur: $S + O_2 \rightarrow SO_2$. Sulfur dioxide and air are passed over a hot (450°C) †catalyst of vanadium (V) oxide: $2SO_2 + O_2 \leftrightarrow 2SO_3$. This reaction is †reversible and so reaches †chemical equilibrium; the conditions are chosen to give the best yield in a short time. The sulfur trioxide produced is absorbed in concentrated sulfuric acid to make fuming sulfuric acid, or oleum ($H_2S_2O_7$). The oleum is then diluted with water to give concentrated sulfuric acid (98%): $H_2S_2O_7 + H_2O \rightarrow 2H_2SO_4$. Unreacted gases are recycled.

container habitat in ecology, small self-contained ecosystem, such as a water pool accumulating in a hole in a tree. Some ecologists believe that much can be learned about larger ecosystems through studying the dynamics of container habitats, which can contain numerous leaf-litter feeders and their predators.

contaminated land land that is considered to pose a health risk to humans because of pollution; usually land that has been the site of industrial activity.

continent any one of the seven large land masses of the Earth, as distinct from the oceans. They are Asia, Africa, North America, South America, Europe, Australia, and Antarctica. Continents are constantly moving and evolving (see †plate tectonics). A continent does not end at the coastline; its boundary is the edge of the shallow continental shelf, which may extend several hundred kilometers out to sea.

Continental crust, as opposed to the crust that underlies the deep oceans, is composed of a wide variety of igneous, sedimentary, and metamorphic rocks. The rocks vary in age from recent (currently forming) to almost 4,000 million years old. Unlike the ocean crust, the continents are not only high standing, but extend to depths as great at 70 km/45 mi under high mountain ranges. Continents, as high, dry masses of rock, are present on Earth because of the density contrast between them and the rock that underlies the oceans. Continental crust is both thick and light, whereas ocean crust is thin and dense. If the crust were the same thickness and density everywhere, the entire Earth would be covered in water.

At the center of each continental mass lies a shield or †craton, a deformed mass of old †metamorphic rocks dating from Precambrian times. The shield is thick, compact, and solid (the Canadian Shield is an example), and is usually worn flat. Around the shield is a concentric pattern of fold mountains, with older ranges, such as the Rockies, closest to the shield, and younger ranges, such as the coastal ranges of North America, farther away. This general concentric pattern is modified when two continental masses have drifted together and they become welded with a great mountain range along the join, the way Europe and northern Asia are joined along the Urals. If a continent is torn apart, the new continental edges have no mountains; for instance, South America has mountains (the Andes) along its western flank, but none along the east where it tore away from Africa 200 million years ago.

continental climate type of climate typical of a large, mid-latitude land mass. This type of climate is to be found in the center of continents between approximately 40° and 60° north of the Equator. The two main areas are the North American prairies and the Russian steppes (both grasslands). Areas with a continental climate are a long way from the oceans, which affects their climate in two ways: they have very low annual rainfall because little atmospheric moisture is available; and their temperature range is very large over a year, because the temperature-moderating effect of the sea has been lost. Maximum mean monthly summer temperatures for continental climates are around 20°C/68°F. During the winter there are several months when the temperature remains below freezing point. Precipitation decreases rapidly towards the east in Russia as distance from the sea increases, whereas in North America totals are lowest to the west in the rain shadow (area sheltered from rain by hills) of the Rockies. In both areas, annual rainfall averages 500 mm/20 in, and there is a threat of drought. The ground is snow-covered for several months between October and April.

Continental Climate http://www.doc.mmu.ac.uk/aric/eae/Climate/Older/Continental_Climate.html

Clear guide to climate and climatic types. The hyperlinked, text-only introduction leads to fuller descriptions and images of the various climate zones including tropical, savannah, desert, temperate, and polar. The influences on climate are also explained with numerous links to other useful areas of information.

continental drift in geology, the theory that, about 250–200 million years ago, the Earth consisted of a single large continent (†Pangaea), which subsequently broke apart to form the continents known today. The theory was first proposed in 1912 by German meteorologist Alfred Wegener, but such vast continental movements could not be satisfactorily explained or even accepted by geologists until the 1960s. The theory of continental drift gave way to the theory of †plate tectonics. Whereas Wegener proposed that continents pushed their way through underlying mantle and ocean floor, plate tectonics states that continents are just part of larger lithospheric plates (which include ocean crust as well) that move laterally over the Earth's surface.

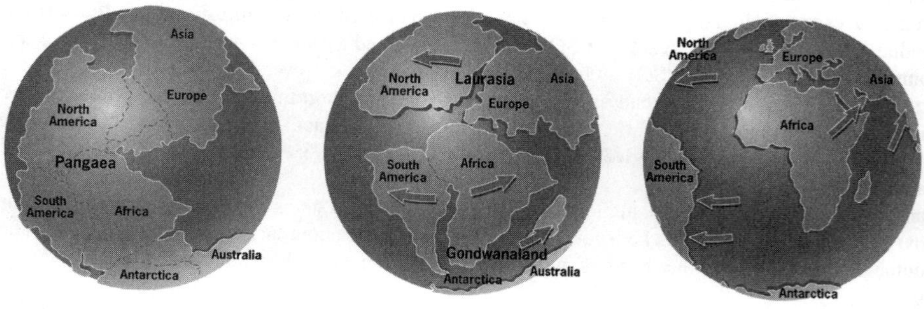

continental drift The changing positions of the Earth's continents. Millions of years ago, there was a single large continent, Pangaea. This split 200 million years ago: the continents had started to move apart, to form Gondwanaland in the south and Laurasia in the north. By 50 million years ago the continents were almost in their present positions. Helicon

continental rise portion of the ocean floor rising gently from the abyssal plain toward the steeper continental slope. The continental rise is a depositional feature formed from sediments transported down the slope mainly by turbidity currents. Much of the continental rise consists of coalescing submarine alluvial fans bordering the continental slope.

continental shelf submerged edge of a continent, a gently sloping plain that extends into the ocean. It typically has a gradient of less than 1°. When the angle of the sea bed increases to 1°–5° (usually several hundred kilometers away from land), it becomes known as the †continental slope.

continental slope sloping, submarine portion of a continent. It extends downward from the edge of the continental shelf. In some places, such as south of the Aleutian Islands of Alaska, continental slopes extend directly to the ocean deeps or abyssal plain. In others, such as the east coast of North America, they grade into the gentler continental rises that in turn grade into the abyssal plains.

contingency table table listing information classified as variable according to two or more independent attributes. Such tables are used commonly in commerce, notably by insurance companies.

continued fraction development of any †real number in the form of a sequence of integers from which approximations to the number may be calculated successively; for example:

$$k_1 + \cfrac{1}{k_1 + \cfrac{1}{k_2 + \cfrac{1}{k_3 + \cfrac{1}{k_4}}}}$$

and so on. The sequence can be finite or infinite. The development will be finite in the case of a rational number, and the calculation will then terminate on reaching the rational number. In the case of an irrational number, a termination will be reached only as the †limit of the sequence of values calculated.

continuity in mathematics, property of functions of a real variable that have an absence of "breaks." A function f is said to be continuous at a point a if:

$$_{x \to a} f(x) = f(a).$$

continuous data in mathematics, †data that can take any of an infinite number of values between whole numbers and so may not be measured completely accurately. This type of data contrasts with †discrete data, in which the variable can only take one of a finite set of values. For example, the sizes of apples on a tree form continuous data, whereas the numbers of apples form discrete data.

continuous function or **continuity**, in mathematics, representation of continuous motion, uniform variation of †transformations. More precisely, a function f is said to be continuous at an argument value x if the function value for arguments close to x can be held down to a value as near to $f(x)$ is required by keeping the argument close enough to x.

continuous spectrum in physics, †spectrum with a continuous range of radiation (either absorbed or emitted), as produced by a solid heated to incandescence.

The spectra of stars may include black lines corresponding to the absorption of radiation by atoms in the star's atmosphere.

continuous variation or **quantitative variation**, slight difference of an individual characteristic, such as height, across a sample of the population. Although there are very tall and very short humans, there are also many people with an intermediate height. The same applies to weight. Continuous variation can be due to the influence of many, rather than a single, genes, or from the influence of individuals' environments.

continuum in mathematics, a †set that is infinite and everywhere continuous, such as the set of points on a line.

contour on a map, a line drawn to join points of equal height. Contours are drawn at regular height intervals; for example, every 10 m. The closer together the lines are, the steeper the slope. Contour patterns can be used to interpret the relief of an area and to identify land forms.

contraceptive any drug, device, or technique that prevents pregnancy. The contraceptive pill (the Pill) contains female hormones that interfere with egg production or the first stage of pregnancy. The "morning-after" pill can be taken up to 72 hours after unprotected intercourse. Barrier contraceptives include condoms (sheaths), femidoms (a female condom), and diaphragms, also called caps or Dutch caps; they prevent the sperm entering the cervix (neck of the womb).

Intrauterine devices, also known as IUDs or coils, cause a slight inflammation of the lining of the womb; this prevents the fertilized egg from becoming implanted.

Other contraceptive methods include †sterilization (women) and vasectomy (men); these are usually nonreversible. "Natural" methods include withdrawal of the penis before ejaculation (coitus interruptus), and avoidance of intercourse at the time of ovulation (rhythm method). These methods are not as reliable as other procedures and are normally only used on religious grounds. A new development is a sponge impregnated with spermicide that is inserted into the vagina. The use of any contraceptive (birth control) is part of family planning. The effectiveness of a contraceptive method is often given as a percentage. To say that a method has 95% effectiveness means that, on average, out of 100 healthy couples using that method for a year, 95 will not conceive.

In 1998, approximately 55% of women of childbearing age worldwide used contraception, according to the UN Population Fund.

Short Visit to the Museum of Contraception http://www.salon.com/07/features/contra.html

Bizarre history of contraception seen through the eyes of a journalist visiting the museum. Many of the "devices" are odd to say the least, for example one of the exhibits in the museum is based around the adage "If a woman were to take a weasel testicle and bones and strap it to her thigh, she would not fall pregnant." The article does not make it clear if any of the methods described have been tested.

contractile root in botany, a thickened root at the base of a corm, bulb, or other organ that helps position it at an appropriate level in the ground. Contractile roots are found, for example, on the corms of plants of the genus *Crocus*. After they have become anchored in the soil, the upper portion contracts, pulling the plant deeper into the ground.

contractile vacuole tiny organelle found in many single-celled freshwater organisms. It slowly fills with water, and then contracts, expelling the water from the cell.

Freshwater protozoa such as *Amoeba* absorb water by the process of †osmosis, and this excess must be eliminated. The rate of vacuole contraction slows as the external salinity is increased, because the osmotic effect weakens; marine protozoa do not have a contractile vacuole.

contradiction in mathematics, assertion that a proposition is both true and false. If an argument leads to such a contradiction, one of the assumptions that was made in the argument must have been false. This is called **proof by contradiction**.

control in biology, the process by which a tissue, an organism, a population, or an ecosystem maintains itself in a balanced, stable state. Blood sugar must be kept at a stable level if the brain is to function properly, and this steady state is maintained by an interaction between the liver, the hormone insulin, and a detector system in the pancreas.

In the ecosystem, the activities of the human race are endangering the balancing mechanisms associated with the atmosphere in general and the †greenhouse effect in particular.

control character in computing, any character produced by depressing the control key (Ctrl) on a keyboard at the same time as another (usually alphabetical) key. The control characters form the first 32 †ASCII characters and most have specific meanings according to the operating system used. They are also used in combination to provide formatting control in many word processors, although the user may not enter them explicitly.

control experiment essential part of a scientifically valid experiment, designed to show that the factor being tested is actually responsible for the effect observed. In the control experiment all factors, apart from the one under test, are exactly the same as in the test experiments, and all the same measurements are carried out. In drug trials, a placebo (a harmless substance) is given alongside the substance being tested in order to compare effects.

convection transfer of heat energy that involves the movement of a fluid (gas or liquid). Fluid in contact with the source of heat expands and tends to rise within the bulk of the fluid. Cooler fluid sinks to take its place, setting up a convection current. This is the principle of natural convection in many domestic hot-water systems and space heaters.

Hot-air balloons use convection currents in order to rise into the air. The air inside the balloon is heated. As the hot air rises, so does the balloon. For the balloon to descend, the air in the balloon is cooled or allowed to escape.

Gravity and air currents allow gliders to fly. The glider falls by the force of gravity acting on it. It gains height in the air from rising, warm currents of air known as thermals. These currents are formed by air being heated by the ground; the heated air becomes less dense and rises.

convection in astronomy, process in the Sun (and possibly other stars) perhaps caused by †solar rotation, that produces the immensely powerful electrical and magnetic fields associated with †sunspots.

convectional rainfall rainfall associated with hot climates, resulting from the uprising of convection currents of warm air. Air that has been warmed by the extreme heating of the ground surface rises to great heights and is abruptly cooled. The water vapor carried by the air condenses and rain falls heavily. Convectional rainfall is often associated with a †thunderstorm.

convection current current caused by the expansion of a liquid, solid, or gas as its temperature rises.

The expanded material, being less dense, rises, while colder, denser material sinks. Material of neutral buoyancy moves laterally. Convection currents arise in the atmosphere above warm land masses or seas, giving rise to †sea breezes and land breezes, respectively. In some heating systems, convection currents are used to carry hot water upwards in pipes.

Convection currents in the hot, solid rock of the Earth's mantle help to drive the movement of the rigid plates making up the Earth's surface (see †plate tectonics).

conventional current direction in which an electric current is considered to flow in a circuit. By convention, the direction is that in which positive-charge carriers would flow – from the positive terminal of a cell to its negative terminal. It is opposite in direction to the flow of electrons. In circuit diagrams, the arrows shown on symbols for components such as diodes and transistors point in the direction of conventional current flow.

In the 19th century, with the early applications of electricity such as starter motors and light bulbs, there was not an exact understanding of the nature of current flow. It was assumed at the time that electricity flowed from the positive terminal to the negative.

convergence in mathematics, the property of a series of numbers in which the difference between consecutive terms gradually decreases. The sum of a converging series approaches a limit as the number of terms tends to †infinity.

convergent evolution or **convergence**, in biology, the independent evolution of similar structures in species (or other taxonomic groups) that are not closely related, as a result of living in a similar way. Thus, birds and bees have wings, not because they are descended from a common winged ancestor, but because their respective ancestors independently evolved flight.

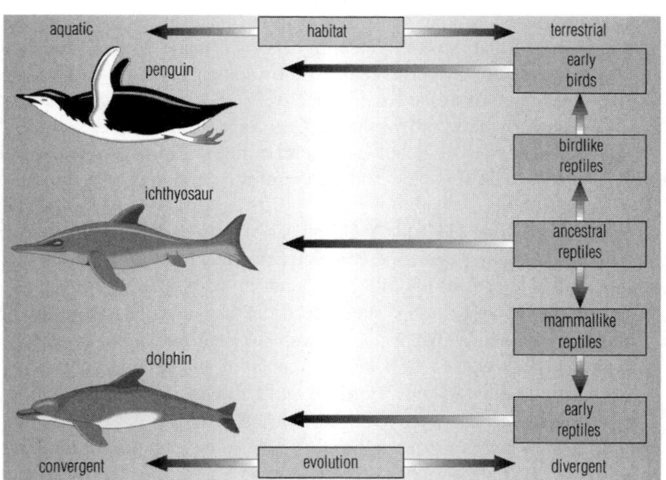

convergent evolution Convergent evolution produced the superficially similar streamlined bodies of the dolphin and penguin. Despite their very different evolutionary paths – one as a mammal, the other as a bird – both have evolved and adapted to the aquatic environment they now inhabit. Helicon

convergent margin or **convergent boundary**, in †plate tectonics, the boundary or active zone between two lithospheric plates that are moving towards one another. Convergent margins are characterized by †folds, reverse faulting, destructive high-magnitude earthquakes, and in some cases volcanic activity.

There are three types of convergent margins: ocean– ocean, ocean–continent, and continent–continent (referring to the types of plates converging). Both ocean–ocean and ocean–continent convergences result in †subduction zones. Examples include the Lesser Antillies (ocean–ocean) and the Andes (ocean–continent). Most ocean–ocean and ocean–continent convergence occurs around the edge of the Pacific Ocean. Because continental crust is too buoyant to sink into the mantle, continent–continent convergences cause large-scale folding, resulting in the formation of mountain ranges such as the Himalayas.

converging lens lens that converges or brings to a focus those light rays that have been refracted by it. It is a †convex lens, with at least one surface that curves outwards, and is thicker towards the center than at the edge. Converging lenses are used to form real images in many †optical instruments, such as cameras and projectors. A converging lens that forms a virtual, magnified image may be used as a †magnifying glass or to correct long-sightedness (†hypermetropia).

converging mirror in physics, concave mirror that reflects a parallel beam of light as a converging beam, which may come to a focus. When such a mirror is used as a reflector, a source of light placed at the focus is reflected away from the mirror as a parallel beam, as employed in headlights and searchlights.

converse in mathematics, the reversed order of a conditional statement; the converse of the statement "if a, then b" is "if b, then a." The converse does not always hold true; for example, the converse of "if $x =$

3, then $x^2 = 9$" is "if $x^2 = 9$, then $x = 3$," which is not true, as x could also be –3.

conversion graph graph for changing values from one unit to another, for example from †Celsius to †Fahrenheit.

conversion table a table for changing values from one unit to another, for example from imperial to metric.

convex of a surface, curving outwards, or towards the eye. For example, the outer surface of a ball appears convex. In geometry, the term is used to describe any polygon possessing no interior angle greater than 180°. Convex is the opposite of †concave.

convex lens lens that possesses at least one surface that curves outwards. It causes light to deviate inward, bringing the rays of light to a focus, and is thus called a †converging lens. A convex lens is thicker at its center than at its edges, and is used to correct long-sightedness (hypermetropism).

The distance between the lens and the point at which all the rays of light converge (the focal point) is called the focal length. The center of the lens is known as the optical center.

At close distance a convex lens forms a magnified virtual image of an object. At further distances the image formed is real, and it can be focused and projected onto a screen behind the lens. The real image formed is upside down, and it can be smaller than, the same size as, or larger than the object depending upon how far the object is from the lens.

Ray diagrams are used to show the position (distance from the lens), size (enlarged or diminished), and nature (real, virtual, inverted, or upright) of an image formed. The lens formula is used to work out the position and nature of an image formed by a lens: $1/u + 1/v = 1/f$, where u and v are the distances of the object and image from the lens, respectively, and f is the focal length of the lens.

convex mirror curved mirror that reflects light from its outer surface, the curve being outward. Rays of light are caused to diverge outward on reflection. It forms a reduced, upright, virtual image. Convex mirrors give a wide field of view and are therefore particularly suitable for car wing mirrors and surveillance purposes in shops.

Cooper, Leon Niels (1930–) US physicist who in 1955 began work on the phenomenon of †superconductivity. He proposed that at low temperatures electrons would be bound in pairs (since known as **Cooper pairs**) and in this state electrical resistance to their flow through solids would disappear. He was awarded the Nobel Prize for Physics in 1972 for his work on the theory of superconductivity. He shared the award with John †Bardeen and J Robert †Schrieffer.

Cooper pairs see †Bardeen–Cooper–Schrieffer theory.

coordinate in geometry, a number that defines the position of a point relative to a point or axis (reference line). †Cartesian coordinates define a point by its perpendicular distances from two or more axes drawn through a fixed point mutually at right angles to each other. †Polar coordinates define a point in a †plane by its distance from a fixed point and direction from a fixed line.

When working on the two-dimensional plane the coordinates can be plotted on a †graph.

The coordinates of the points A, B, C, and D are: A (3, 1), B (–3, 2),

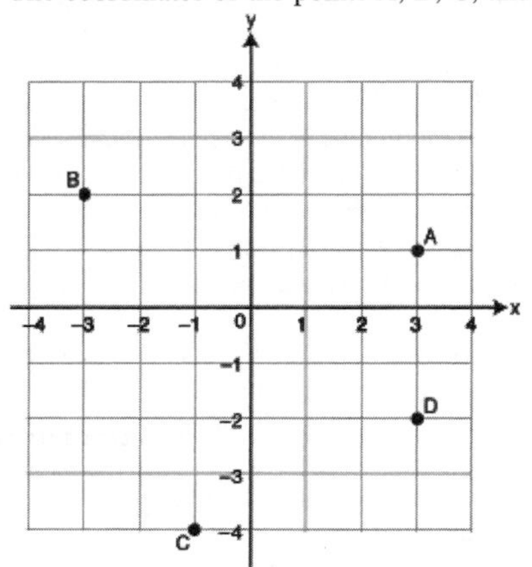

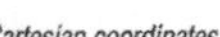
Cartesian coordinates

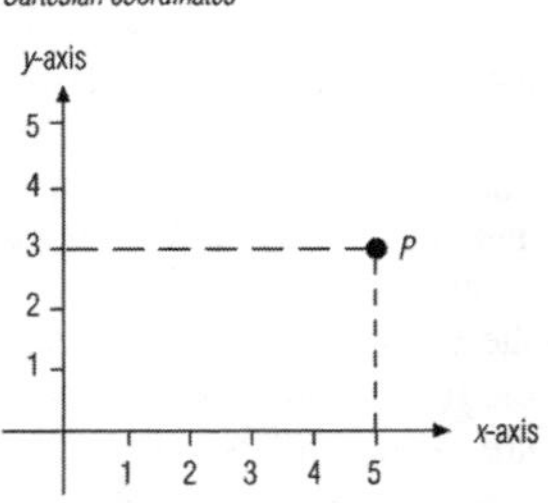

the Cartesian coordinates of *P* are (5,3)

Polar coordinates

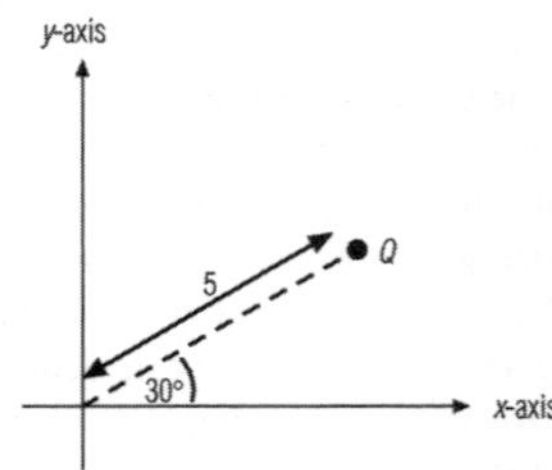

the Polar coordinates of *Q* are (5,30°)

coordinate Coordinates are numbers that define the position of points in a plane or in space. In the Cartesian coordinate system, a point in a plane is charted based upon its location along intersecting horizontal and vertical axes. In the polar coordinate system, a point in a plane is defined by its distance from a fixed point and direction from a fixed line. Helicon

C (–1, –4), D (3, –2). Coordinates can, therefore, be both positive numbers, both negative numbers, or a positive and a negative number. The *x*- and *y*-axes intersect to create four †quadrants.

Once plotted, coordinates can be joined up to obtain †straight-line graphs from linear equations or curves from †quadratic equations and †cubic equations. The †intercept and †gradient can be calculated.

†Distance–time graphs and †conversion graphs are both obtained by plotting coordinates.

coordinate geometry or **analytical geometry**, system of geometry in which points, lines, shapes, and surfaces are represented by algebraic expressions. In plane (two-dimensional) coordinate geometry, the plane is usually defined by two axes at right angles to each other, the horizontal *x*-axis and the vertical *y*-axis, meeting at O, the origin. A point on the plane can be represented by a pair of †Cartesian coordinates, which define its position in terms of its distance along the *x*-axis and along the *y*-axis from O. These distances are, respectively, the *x* and *y* coordinates of the point.

Lines are represented as equations; for example, $y = 2x + 1$ gives a straight line, and $y = 3x^2 + 2x$ gives a †parabola (a curve). The graphs of varying equations can be drawn by plotting the coordinates of points that satisfy their equations, and joining up the points. One of the advantages of coordinate geometry is that geometrical solutions can be obtained without drawing but by manipulating algebraic expressions. For example, the coordinates of the point of intersection of two straight lines can be determined by finding the unique values of *x* and *y* that satisfy both of the equations for the lines, that is, by solving them as a pair of †simultaneous equations. The curves studied in simple coordinate geometry are the †conic sections (circle, ellipse, parabola, and hyperbola), each of which has a characteristic equation.

coordination compound any of a diverse group of complex chemical compounds characterized by a structure in which several †ligands surround – and are covalently bonded to – a central metal atom. Such compounds may be electrically neutral, or positive or negative ions. Similarly the central metal atom may be neutral, anionic or, rarely, cationic, but it is always one that is able to accept an electron pair(s) to form a coordinate bond(s). The total number of bonds between the central atom and the ligands is the coordination number, which, in general, ranges from two to twelve; four and six are the most common.

coordination number see †coordination compound.

copepod †crustacean of the subclass Copepoda, mainly microscopic and found in plankton.

Copernican model of the universe model of the universe with the Sun at its center (†heliocentric) that replaced the Earth centered (geocentric) †Ptolemaic model, and was thus a considerable improvement. The model, however, still involved †epicycles and the †spheres.

Copernicus, Nicolaus (1473–1543) Polish **Mikolaj Kopernik**, Polish astronomer who believed that the Sun, not the Earth, is at the center of the Solar System, thus defying the Christian church doctrine of the time. For 30 years, he worked on the hypothesis that the rotation and the orbital motion of the Earth are responsible for the apparent movement of the heavenly bodies. His great work *De Revolutionibus Orbium Coelestium/On the Revolutions of the Heavenly Spheres* was the important first step to the more accurate picture of the Solar System built up by Tycho †Brahe, †Kepler, †Galileo, and later astronomers.

Copernican System http://es.rice.edu/ES/humsoc/Galileo/Things/copernican_system.html

Good introduction to Copernicus's work. There are several portraits and pages from early manuscripts showing his concept of the Solar System. Embedded in the text are links to a number of other important astronomers.

coplanar in geometry, describing lines or points that all lie in the same plane.

co-polymer in chemistry, any †polymer that is synthesized by the polymerization of two different types of †monomers, for example butadiene co-polymerized with styrene to produce the synthetic rubber Buna-S.

copper chemical symbol Cu, red-brown, very malleable and ductile, metallic element, atomic number 29, relative atomic mass 63.546. Its symbol comes from the Latin *cuprum*. It is one of the †transition metals in the †periodic table. Copper is used for its durability, pliability, high thermal and electrical conductivity, and resistance to corrosion. It is used in electrical wires and cables, and water pipes and tanks.

Copper was the first metal used systematically for tools by humans; when mined and worked into utensils it formed the technological basis for the Copper Age in prehistory. When alloyed with tin it

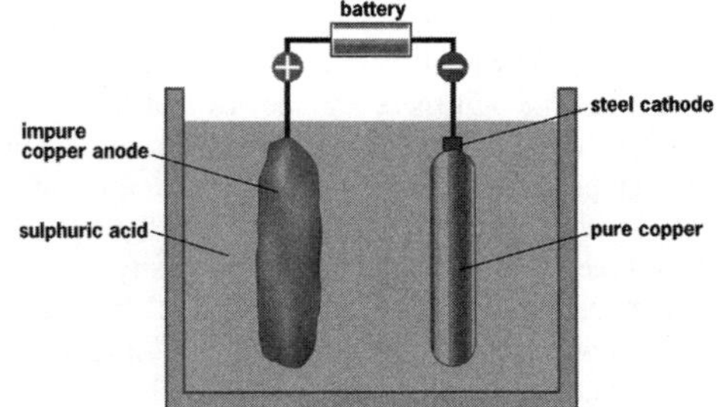

copper Copper purification by electrolysis. Pure copper dissolves from the impure copper anode, which decreases in size. The copper is deposited at the cathode, which increases in size. When the purification process is complete the cathode is melted down and cast into bars to be sold to manufacturers, such as producers of electrical wiring. Helicon forms bronze, which is stronger than pure copper and may hold a sharp edge; the systematic production and use of this †alloy was the basis for the prehistoric Bronze Age. Brass, another hard copper alloy, includes zinc. The element's name comes from the Latin *cyprium* ("Cyprus metal"), due to Cyprus being a major source of copper.

copper(II) carbonate $CuCO_3$, green solid that readily decomposes to form black copper(II) oxide on heating:

$CuCO_3 \rightarrow CuO + CO_2$

It dissolves in dilute acids to give blue solutions with effervescence caused by the giving off of carbon dioxide.

$CuCO_3 + H_2SO_4 \rightarrow CuSO_4 + CO_2 + H_2O$

copper ore any mineral from which copper is extracted, including native copper, Cu; chalcocite, Cu_2S; chalcopyrite, $CuFeS_2$; bornite, Cu_5FeS_4; azurite, $Cu_3(CO_3)_2(OH)_2$; malachite, $Cu_2CO_3(OH)_2$; and chrysocolla, $CuSiO_3.2H_2O$.

Native copper and the copper sulfides are usually found in veins associated with igneous intrusions. Chrysocolla and the carbonates are products of the weathering of copper-bearing rocks. Copper was one of the first metals to be worked, because it occurred in native form and needed little refining. Today the main producers are the USA, Russia, Kazakhstan, Georgia, Uzbekistan, Armenia, Zambia, Chile, Peru, Canada, and the Democratic Republic of Congo (formerly Zaire).

copper(II) oxide CuO, black solid that is readily reduced to copper by carbon, carbon monoxide, or hydrogen if heated with any of these:

$CuO + C \rightarrow Cu + CO$

$CuO + CO \rightarrow Cu + CO_2$

$CuO + H_2 \rightarrow Cu + H_2O$

It is usually made in the laboratory by heating copper(II) carbonate, nitrate, or hydroxide:

$2Cu(NO_3)_2 \rightarrow 2CuO + 4NO_2 + O_2$

Copper(II) oxide is a typical basic oxide, dissolving readily in most dilute acids.

copper(II) sulfate $CuSO_4$, substance usually found as a blue, crystalline, hydrated salt $CuSO_4.5H_2O$ (also called blue vitriol). It is made from the action of dilute sulfuric acid on copper(II) oxide, hydroxide, or carbonate.

$CuO + H_2SO_4 + 4H_2O \rightarrow CuSO_4.5H_2O$

When the hydrated salt is heated gently it loses its water of crystallization and the blue crystals turn to a white powder. The reverse reaction is used as a chemical test for water.

$CuSO_4.5H_2O \leftrightarrow CuSO_4 + 5H_2O$

copulation act of mating in animals with internal †fertilization. Male mammals have a †penis or other organ that is used to introduce spermatozoa into the reproductive tract of the female. Most birds transfer sperm by pressing their cloacas (the openings of their reproductive tracts) together.

coral marine invertebrate of the class Anthozoa in the phylum Cnidaria, which also includes sea anemones and jellyfish. It has a skeleton of lime (calcium carbonate) extracted from the surrounding water. Corals exist in warm seas, at moderate depths with sufficient light. Some coral is valued for decoration or jewelry, for example, Mediterranean red coral *Corallum rubrum*.

Corals live in a symbiotic relationship with microscopic algae (zooxanthellae), which are incorporated into the soft tissue. The algae obtain carbon dioxide from the coral polyps, and the polyps receive nutrients from the algae. Corals also have a relationship to the fish that rest or take refuge within their branches, and which excrete nutrients that make the corals grow faster. The majority of corals form large colonies although there are species that live singly. Their accumulated skeletons make up large coral reefs and atolls. The Great Barrier Reef, to the northeast of Australia, is about 1,600 km/1,000 mi long, has a total area of 20,000 sq km/7,700 sq mi, and adds 50 million tonnes of calcium to the reef each year. The world's reefs cover an estimated 620,000 sq km/240,000 sq mi.

Coral reefs provide a habitat for a diversity of living organisms. In 1997 some 93,000 species were identified. One third of the world's marine fishes live in reefs. The world's first global survey of coral reefs, carried out in 1997, found around 95% of reefs had experienced some damage from overfishing, pollution, dynamiting, poisoning, and the dragging of ships' anchors. A 1998 research showed that nearly two-thirds of the world's coral reefs were at risk, including 80% of the reefs in the Philippines and Indonesia, 66% of those in the Caribbean, and over 50% of those in the Indian Ocean, the Red Sea, and the Gulf of Arabia.

diseases Since the 1990s, coral reefs have been destroyed by previously unknown diseases. The **white plague** attacked 17 species of coral in the Florida Keys, USA, in 1995. The **rapid wasting disease**, discovered in 1997, affects coral reefs from Mexico to Trinidad. In the Caribbean, the fungus *Aspergillus* attacks sea fans, a soft coral. It was estimated in 1997 that around 90% of the coral around the Galapagos islands had been destroyed as a result of "bleaching," a whitening of coral reefs which occurs when the colored algae evacuate the coral. This happens either because the corals produce toxins that are harmful to the algae or because they do not produce sufficient nutrients. Without the algae, the coral crumbles and dies away. Bleaching is widespread all over the Caribbean and the Indo-Pacific. A report published in November 1998 by Reef Check, an international organization, claimed that coral reefs are dying at a record rate throughout the world, and reefs that had existed for hundreds of years had suddenly died in 1998. Worldwide warming has caused bleaching of coral reefs in 50 or more countries in that year.

Corals and Coral Reefs http://www.seaworld.org/infobooks/Coral/home.html

Extensive educational resource about coral, covering everything from classification and anatomy to diet and reproduction. There is also a section on the importance of coral and the measures being taken to conserve it.

Cor Caroli brightest star in †Canes Venatici.

cord unit for measuring the volume of wood cut for fuel. One cord equals 128 cubic feet (3.456 cubic meters), or a stack 8 feet (2.4 m) long, 4 feet (1.2 m) wide, and 4 feet high.

Cordaitales order of Paleozoic gymnosperm plants with straplike leaves, ancestors of the modern true conifers. They were abundant as large trees in Coal Measure and late Carboniferous forests. *Cordaites* is the best-known genus; it grew in mangrove swamps.

cordierite silicate mineral, $(Mg,Fe)_2Al_4Si_5O_{18}$, blue to purplish in color. It is characteristic of metamorphic rocks formed from clay sediments under conditions of low pressure but moderate temperature; it is the mineral that forms the spots in spotted slate and spotted hornfels.

cordillera group of mountain ranges and their valleys, all running in a specific direction, formed by the continued convergence of two tectonic plates (see †plate tectonics) along a line. The term is applied especially to the principal mountain belt of a continent. The Andes of South America are an example.

core in earth science, the innermost part of the Earth. It is divided into an outer core, which begins at a depth of 2,900 km/1,800 mi, and an inner core, which begins at a depth of 4,980 km/3,100 mi. Both parts are thought to consist of iron-nickel alloy. The outer core is liquid and the inner core is solid.

The fact that †seismic shear waves disappear at the mantle–outer core boundary indicates that the outer core is molten, since shear waves cannot travel through fluid. Scientists infer the iron-nickel rich composition of the core from the Earth's density and its †moment of inertia, and the composition of iron meteorites, which are thought to be pieces of cores of small planets. The temperature of the core, as estimated from the melting point of iron at high pressure, is thought to be at least 4,000°C/7,232°F, but remains controversial. The Earth's magnetic field is believed to be the result of the movement of liquid metal in the outer core.

Corey, Elias James (1928–) US organic chemist who was awarded the Nobel Prize for Chemistry in 1990 for the development of **retrosynthetic analysis**, a method of synthesizing complex substances. A prolific worker, Corey has synthesized more than 100 substances for the first time, including terpenes (a family of hydrocarbons found in plant oils) and ginkgolide B (an extract from the ginkgo tree used to control asthma).

Coriolis, Gustave Gaspard de (1792–1843) French physicist. In 1835 he discovered the **Coriolis effect**, which governs the movements of winds in the atmosphere and currents in the ocean. Investigating the movements of moving parts in machines and other systems relative to the fixed parts, Coriolis explained how the rotation of the Earth causes objects moving freely over the surface to follow a curved path relative to the surface. Coriolis was also the first to derive formulas expressing †kinetic energy and mechanical work.

Coriolis effect effect of the Earth's rotation on the atmosphere, oceans, and theoretically all objects moving over the Earth's surface. In the northern hemisphere it causes moving objects and currents to be deflected to the right; in the southern hemisphere it causes deflection to the left. The effect is named after its discoverer, French mathematician Gaspard de Coriolis (1792–1843).

Coriolis Effect http://www.physics.ohio-state.edu/~dvandom/Edu/newcor.html

Subtitled "a (fairly) simple explanation," this site contains a description of the principles involved, and is aimed at non-physicists.

corm short, swollen, underground plant stem, surrounded by protective scale leaves, as seen in the genus *Crocus*. It stores food, provides a means of †vegetative reproduction, and acts as a †perennating organ. During the year, the corm gradually withers as the food reserves are used for the production of leafy, flowering shoots formed from axillary buds. Several new corms are formed at the base of these shoots, above the old corm.

Cormack, Allan MacLeod (1924–1998) South-African born US physicist who was awarded a Nobel

Prize for Physiology or Medicine in 1979 with Godfrey †Hounsfield for their development of the X-ray diagnostic technique computer axial tomography (†CAT scan).

cornea transparent front section of the vertebrate †eye. The cornea is curved and behaves as a fixed lens, so that light entering the eye is partly focused before it reaches the lens.

There are no blood vessels in the cornea and it relies on the fluid in the front chamber of the eye for nourishment. Further protection for the eye is provided by the †conjunctiva. In humans, diseased or opaque parts may be replaced by grafts of corneal tissue from a donor.

Cornforth, John Warcup (1917–) Australian chemist. Using †radioisotopes as markers, he found out how cholesterol is manufactured in the living cell and how enzymes synthesize chemicals that are mirror images of each other (optical †isomers).

He shared the Nobel Prize for Chemistry in 1975 with Swiss chemist Vladimir †Prelog for his work in the stereochemistry of enzyme-catalyzed reactions. He was knighted in 1977.

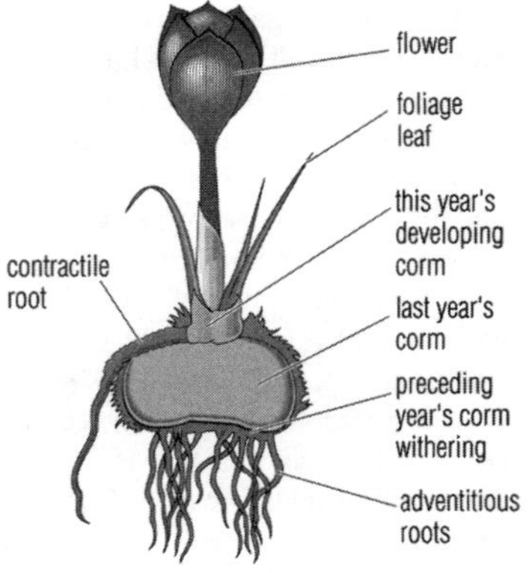

corm Corms, found in plants such as the gladiolus and crocus, are underground storage organs. They provide the food for growth during adverse conditions such as cold or drought. Helicon

cornified layer the upper layer of the skin where the cells have died, their cytoplasm being replaced by keratin, a fibrous protein also found in nails and hair. Cornification gives the skin its protective waterproof quality.

corolla collective name for the petals of a flower. In some plants the petal margins are partly or completely fused to form a **corolla tube**, for example in bindweed *Convolvulus arvensis*.

corona faint halo of hot (about 2,000,000°C/3,600,000°F) and tenuous gas around the Sun, which boils from the surface. It is visible at solar †eclipses or through a **coronagraph**, an instrument that blocks light from the Sun's brilliant disc. Gas flows away from the corona to form the †solar wind. NASA's High-Energy Solar Spectroscopic Imager mission, launched in 2001, was to study the evolution of energy in the corona.

Corona Australis or **Southern Crown**, small constellation of the southern hemisphere, located near the constellation †Sagittarius. It is similar in size and shape to †Corona Borealis but is not as bright.

Corona Borealis or **Northern Crown**, small but easily recognizable constellation of the northern hemisphere, between †Hercules and †Boötes, traditionally identified with the jeweled crown of Ariadne that was cast into the sky by Bacchus in Greek mythology. Its brightest star is Alphecca (or Gemma), which is 78 light years from Earth.

It contains several variable stars. R Coronae Borealis is normally fairly constant in brightness but fades at irregular intervals and stays faint for a variable length of time. T Coronae Borealis is normally faint, but very occasionally blazes up and for a few days may be visible to the naked eye. It is a recurrent †nova.

coronagraph device for studying the solar †corona at any time of the day. It was first invented by French astronomer Bernard Lyot.

corpus allatum in insects, endocrine (hormone producing) tissue. It is often intimately associated with another endocrine tissue, the †corpus cadiacum. The corpus allatum secretes juvenile hormone (JH), which in the juvenile stages of the insect life cycle maintains the juvenile characteristics of larvae (and nymphs). In female adults JH controls the development of egg follicles.

corpus cardiacum in insects, endocrine (hormone producing) tissue. Neurosecretory hormones (synthesized in the brain) pass to the corpus cardiacum for release. Its connection with neurosecretory cells and intimate association with the †corpus allatum form a tissue complex that is structurally and functionally analogous to the †hypothalamus and †pituitary complex of vertebrates.

corpuscular theory hypothesis about the nature of light championed by Isaac Newton, who postulated that it consists of a stream of particles or corpuscles. The theory was superseded at the beginning of the 19th century by English physicist Thomas Young's wave theory. †Quantum theory and wave mechanics embody both concepts.

corpus luteum glandular tissue formed in the mammalian †ovary after ovulation from the Graafian follicle, a group of cells associated with bringing the egg to maturity. It secretes the hormone progesterone in anticipation of pregnancy.

After the release of an egg the follicle enlarges under the action of luteinizing hormone, released from the pituitary. The corpus luteum secretes the hormone progesterone, which maintains the uterus wall ready for pregnancy. If pregnancy occurs, the corpus luteum continues to secrete progesterone until the fourth month; if pregnancy does not occur the corpus luteum breaks down.

corrasion or **abrasion**, grinding away of solid rock surfaces by particles carried by water, ice, and wind. It is generally held to be the most significant form of †erosion in rivers. As the eroding particles are carried along they are eroded themselves (becoming rounder and smaller) due to the process of †attrition.

correlation degree of relationship between two sets of information. If one set of data increases at the same time as the other, the relationship is said to be **positive** or direct. If one set of data increases as the other decreases, the relationship is **negative** or inverse. If there is no relationship between the two sets of data the relationship is said to be **zero linear correlation**. Correlation can be shown by plotting a †line of best fit on a †scatter diagram. The steeper the line drawn, whether positive or negative, the stronger the correlation.

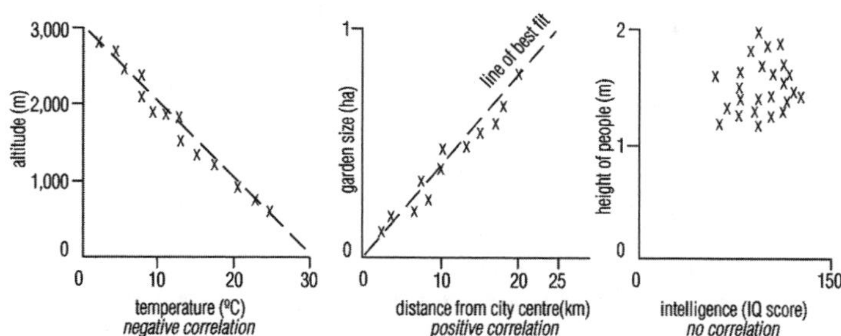

correlation Scattergraphs showing different kinds of correlation. In this way, a causal relationship between two variables may be proved or disproved, provided there are no hidden factors. Helicon

In statistics, such relations are measured by the calculation of †coefficients of correlation. These generally measure correlation on a scale with 1 indicating perfect positive correlation, 0 no correlation at all, and –1 perfect negative correlation. Correlation coefficients for assumed linear relations include the Pearson product moment correlation coefficient (known simply as the correlation coefficient), Kendall's tau correlation coefficient, or Spearman's rho correlation coefficient, which is used in nonparametric statistics (where the data are measured on ordinal rather than interval scales). A high correlation does not always indicate dependence between two variables; it may be that there is a third (unstated) variable upon which both depend.

Correns, Carl Franz Joseph Erich (1864–1933) German botanist and geneticist who is credited with rediscovering †Mendel's laws (Hugo †De Vries is similarly credited).

correspondence in mathematics, the relation between two sets where an operation on the members of one set maps some or all of them onto one or more members of the other. For example, if *A* is the set of members of a family and *B* is the set of months in the year, *A* and *B* are in correspondence if the operation is: "...has a birthday in the month of...."

corresponding angles a pair of equal angles lying on the same side of a transversal (a line that cuts through two or more lines in the same plane), and making an interior and exterior angle with the intersected lines.

corridor, wildlife route linking areas of similar habitat, or between sanctuaries. For example there is a corridor linking the Masai Mara reserve in Kenya and the Serengeti in Tanzania. On a smaller scale, disused railways provide corridors into urban areas for foxes.

Corridors are anything from a few meters to several kilometers in width. The Grider Creek corridor connecting two areas of forest wilderness between the extreme north of California and southern Oregon is 5.5 km/3.5 mi wide and 26 km/17 mi long. The wider the corridor is the greater its conservation value as it will provide additional habitat rather than just a travel route. Corridors have been criticized as pathways for diseases and new predators.

corrie Welsh **cwm**; French, North American **cirque**, Scottish term for a steep-sided armchair-shaped hollow in the mountainside of a glaciated area. The weight and movement of the ice has ground out the bottom and worn back the sides. A corrie is open at the front, and its sides and back are formed of †arêtes. There may be a small lake in the bottom, called a tarn.

A corrie is formed as follows: (1) snow accumulates in a hillside hollow (enlarging the hollow by †nivation), and turns to ice; (2) the hollow is deepened by †abrasion and †plucking; (3) the ice in the corrie moves under the influence of gravity, deepening the hollow still further; (4) since the ice is at the foot of the hollow and moves more slowly, a rock lip (a ridge of solid rock) forms; (5) when the ice melts, a lake or tarn may be formed in the corrie. The steep back wall may be severely weathered by †freeze–thaw weathering, providing material for further abrasion.

corrosion in earth science, an alternative name for †solution, the process by which water dissolves rocks such as limestone.

corrosion eating away and eventual destruction of metals and alloys by chemical attack. The rusting of ordinary iron and steel is the most common form of corrosion. Rusting takes place in moist air, when the iron combines with oxygen and water to form a brown-orange deposit of †rust (hydrated iron oxide). The rate of corrosion is increased where the atmosphere is polluted with sulfur dioxide. Salty road and air conditions accelerate the rusting of car bodies.

Corrosion is largely an electrochemical process, and acidic and salty conditions favor the establishment of electrolytic cells on the metal, which cause it to be eaten away. Other examples of corrosion include the green deposit that forms on copper and bronze, called verdigris, a basic copper carbonate. The tarnish on silver is a corrosion product, a film of silver sulfide.

cortex in biology, the outer part of a structure such as the brain, kidney, or adrenal gland. In botany the cortex includes nonspecialized cells lying just beneath the surface cells of the root and stem.

corticosteroid any of several steroid hormones secreted by the cortex of the †adrenal glands; also synthetic forms with similar properties. Corticosteroids have anti-inflammatory and †immunosuppressive effects and may be used to treat a number of conditions, including rheumatoid arthritis, severe allergies, asthma, some skin diseases, and some cancers. Side effects can be serious, and therapy must be withdrawn very gradually.

The two main groups of corticosteroids include **glucocorticoids** (cortisone, hydrocortisone, prednisone, and dexamethasone), which are essential to carbohydrate, fat, and protein metabolism, and to the body's response to stress; and **mineralocorticoids** (aldosterone, fluorocortisone), which control the balance of water and salt in the body.

corticotrophin-releasing hormone **CRH**, hormone produced by the †hypothalamus that stimulates the adrenal glands to produce the steroid cortisol, essential for normal †metabolism. CRH is also produced by the †placenta and a surge in CRH may trigger the beginning of labor.

cortisol hormone produced by the †adrenal glands. It plays a role in helping the body combat stress and is at its highest level in the blood at dawn.

corundum Al_2O_3, native aluminum oxide and the hardest naturally occurring mineral known, apart from diamond (corundum rates 9 on the Mohs scale of hardness); lack of †cleavage also increases its durability. Its crystals are barrel-shaped prisms of the trigonal system. Varieties of gem-quality corundum are **ruby** (red) and **sapphire** (any color other than red, usually blue). Poorer-quality and synthetic corundum is used in industry, for example as an †abrasive.

Corundum forms in silica-poor igneous and metamorphic rocks. It is a constituent of emery, which is metamorphosed bauxite.

Corvus small constellation of the southern hemisphere, represented as a crow. Its four brightest stars are easily recognizable. It is usually seen as being perched on †Hydra.

cosecant in trigonometry, a †function of an angle in a right-angled triangle found by dividing the length of the hypotenuse (the longest side) by the length of the side opposite the angle. Thus the cosecant of an angle *A*, usually shortened to cosec *A*, is always greater than (or equal to) 1. It is the reciprocal of the sine of the angle, that is, cosec $A = 1/\sin A$.

cosine **cos**, in trigonometry, a †function of an angle in a right-angled †triangle found by dividing the length of the side adjacent to the angle by the length of the hypotenuse (the longest side). This function can be used to find either angles or sides in a right-angled triangle.

cosine rule in trigonometry, a rule that relates the sides and angles of triangles. The rule has the formula:

$$a^2 = b^2 + c^2 - 2bc \cos A$$

where *a*, *b*, and *c* are the lengths of the sides of the triangle, and *A* is the angle opposite *a*.

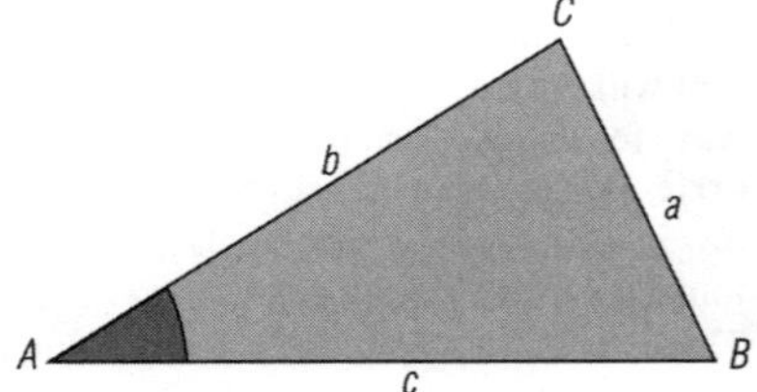

cosine rule The cosine rule is a rule of trigonometry that relates the sides and angles of triangles. It can be used to find an unknown length or angle in a triangle. Helicon

cosmic background radiation or **3° radiation**, electromagnetic radiation left over from the original formation of the universe in the †Big Bang between 10 and 20 billion years ago. It corresponds to an overall background temperature of 2.73 K (–270.4°C/–454.7°F), or 3°C above absolute zero. In 1992 the US Cosmic Background Explorer satellite detected slight "ripples" in the strength of cosmic background radiation that are believed to mark the first stage in the formation of galaxies.

Cosmic background radiation was first detected in 1965 by US physicists Arno Penzias and Robert Wilson, who in 1978 shared the Nobel Prize for Physics for their discovery. On 30 June 2001, NASA launched the Microwave Anisotropy Probe, which will measure and map the temperature of cosmic background radiation over the entire sky.

Cosmic Microwave Background Radiation

http://map.gsfc.nasa.gov/m_uni/uni_101Flucts.html

Mainly text-based page that explains the occurrence of background radiation and provides reasons why the study of it is important. Other links lead to relevant topics, such as the expansion rate and age of the universe.

cosmic censorship in astronomy, theory that the hidden interior within all †event horizons is the same and is always, necessarily, hidden.

cosmic radiation streams of high-energy particles and electromagnetic radiation from outer space, consisting of electrons, protons, alpha particles, light nuclei, and gamma rays, which collide with atomic nuclei in the Earth's atmosphere and produce secondary nuclear particles (chiefly †mesons, such as pions and muons) that shower the Earth. Space shuttles carry dosimeter instruments to measure the levels of cosmic radiation.

Cosmic radiation of lower energy seems to be galactic in origin, while that of high energy is of extragalactic origin. The galactic particles may come from †supernova explosions or †pulsars. At higher energies, other sources must exist, possibly the giant jets of gas that are emitted from some galaxies.

cosmic year time the Sun takes to "orbit" in †galactic rotation: about 225 million years.

cosmid fragment of †DNA, for example from the human genome, inserted into a bacterial cell. The bacterium replicates the fragment along with its own DNA. In this way the fragments are copied for a gene library. Cosmids are characteristically 40,000 base pairs in length. The most commonly used bacterium is *Escherichia coli.* A †yeast artificial chromosome (YAC) works in the same way.

cosmogony (Greek *cosmos* "universe" and *gonia* "creation") study of the origin and evolution of cosmic objects, especially the Solar System.

cosmological principle in astronomy, hypothesis that the expansion of the †universe is perceived to be the same by any observer at any point within it; that is, that the universe is not expanding from any center but all galaxies are moving away from one another.

cosmology branch of astronomy that deals with the structure and evolution of the universe as an ordered whole. Cosmologists construct "model universes" mathematically and compare their large-scale properties with those of the observed universe.

Modern cosmology began in the 1920s with the discovery that the universe is expanding, which suggested that it began in an explosion, the †Big Bang. An alternative – now discarded – view, the †steady-state theory, claimed that the universe has no origin, but is expanding because new matter is being continually created.

Brief History of Cosmology

http://www-history.mcs.st-andrews.ac.uk/~history/HistTopics/Cosmology.html

Based at St Andrews University, Scotland, a site chronicling the history of cosmology from the time of the Babylonians to the Hubble Space Telescope. It includes links to the biographies of the key historical figures responsible for the advancement of the subject and a brief list of references for further reading.

Unsolved Mysteries http://www.pbs.org/wnet/hawking/mysteries/html/myst.html

Part of a larger site on cosmology provided by PBS Online, this page features seven articles that address the most difficult questions regarding the mysteries of our universe. Some of the articles include "Where does matter come from?", "Is time travel possible?", "Where is the missing matter?", "An inhabited universe?", and "Is there a theory for everything?" You can follow the link to the main site to explore other aspects of the cosmos.

cotangent in trigonometry, a †function of an angle in a right-angled triangle found by dividing the length of the side adjacent to the angle by the length of the side opposite it. It is usually written as cotan, or cot and it is the reciprocal of the tangent of the angle, so that cot A = 1/tan A, where A is the angle in question.

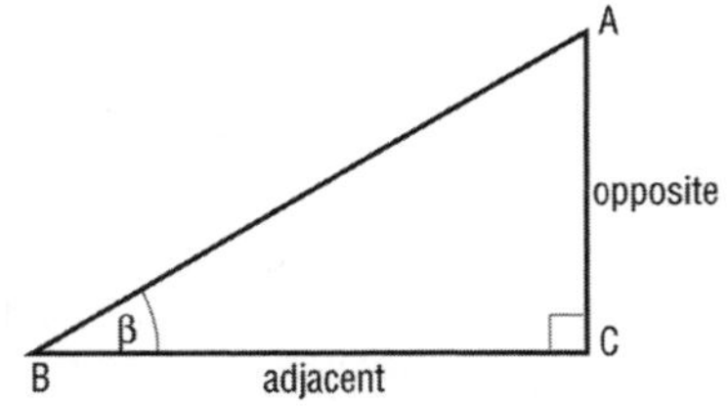

cotangent The cotangent of angle b is equal to the ratio of the length of the adjacent side to the length of the opposite side. Helicon

Cotes, Roger (1682–1716) English mathematician. In 1713 he prepared the second edition of Isaac Newton's *Principia* for publication. His own writings were published posthumously in 1722 under the title *Harmonia mensurarum*.

cotyledon structure in the embryo of a seed plant that may form a "leaf" after germination and is commonly known as a seed leaf. The number of cotyledons present in an embryo is an important character in the classification of flowering plants (†angiosperms).

Monocotyledons (such as grasses, palms, and lilies) have a single cotyledon, whereas dicotyledons (the majority of plant species) have two. In seeds that also contain †endosperm (nutritive tissue), the cotyledons are thin, but where they are the primary food-storing tissue, as in peas and beans, they may be quite large. After germination the cotyledons either remain below ground (hypogeal) or, more commonly, spread out above soil level (epigeal) and become the first green leaves. In gymnosperms there may be up to a dozen cotyledons within each seed.

Coudé telescope in astronomy, type of reflecting telescope in which light reflected from the concave main mirror is reflected back by a small convex mirror onto an angled plain mirror, which directs the light into the eyepiece.

Coué, Emile (1857–1926) French psychological healer, the pioneer of autosuggestion. He coined the slogan "Every day, in every way, I am getting better and better." Couéism reached the height of its popularity in the 1920s.

coulomb symbol C, SI unit of electrical charge. One coulomb is the amount of charge transferred by a current of one †ampere in one second. The unit is named after French scientist Charles Augustin de †Coulomb.

Coulomb, Charles Augustin de (1736–1806) French scientist, inventor of the †torsion balance for measuring the force of electric and magnetic attraction. The coulomb was named after him. In the fields of structural engineering and friction, Coulomb greatly influenced and helped to develop engineering in the 19th century.

Coulomb field field of force surrounding an electric charge. Its intensity can be deduced from †Coulomb's law.

Coulomb's law in physics, law that states that the†force between two charged bodies varies directly as the product of the two charges, and inversely as the square of the distance between them. It was determined by French physicist Charles Coulomb.

Coulson, Charles Alfred (1910–1974) English theoretical chemist. He developed a molecular orbital theory which is an extension of atomic †quantum theory and deals with "allowed" states of electrons in association with two or more atomic nuclei, treating a molecule as a whole. He was thus able to explain properly phenomena such as the structure of †benzene and other conjugated systems, and invoked what he called partial †valency to account for the bonding in such compounds as diborane. He developed many mathematical techniques for solving chemical and physical problems.

countable in mathematics, set that is countable if its elements can be put into †one-one correspondence with the elements of the natural numbers N.

counter example method for disproving a statement by finding an example that is not true. For example, the statement: if $x < 4$ then $x^2 < 16$, can be proved or disproved by inserting –5 as the value of x. When $x = -5$, $x^2 = 25$, therefore the statement has been disproved by finding a counter example. Only one counter example is needed to disprove a statement.

count rate in physics, number of particles emitted per unit time by a radioactive source. It is measured by a counter, such as a †Geiger counter, or †ratemeter.

country rock or **host rock**, in earth science, older rock that surrounds an igneous intrusion or mineral veins.

couple in mechanics, a pair of forces acting on an object that are equal in magnitude and opposite in direction, but do not act along the same straight line. The two forces produce a turning effect or moment that tends to rotate the object; however, no single resultant (unbalanced) force is produced and so the object is not moved from one position to another.

The moment of a couple is the product of the magnitude of either of the two forces and the perpendicular distance between those forces. If the magnitude of the force is F newtons and the distance is d meters then the moment, in newton-meters, is given by:

moment = Fd

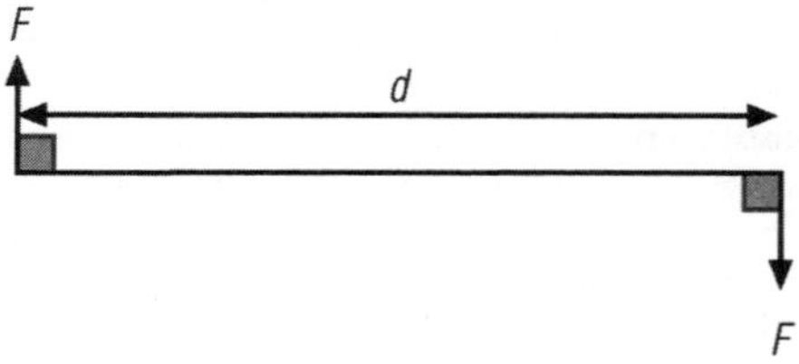

couple Two equal but opposite forces (F) will produce a turning effect on a rigid body, provided that they do not act through the same straight line. The turning effect, or moment, is equal to the magnitude of one of the turning forces multiplied by the perpendicular distance (d) between those two forces. Helicon

courtship behavior exhibited by animals as a prelude to mating. The behavior patterns vary considerably from one species to another, but are often ritualized forms of behavior not obviously related to courtship or mating (for example, courtship feeding in birds).

Courtship ensures that copulation occurs with a member of the opposite sex of the right species. It also synchronizes the partners' readiness to mate and allows each partner to assess the suitability of the other.

Cousteau, Jacques-Yves (1910–1997) French oceanographer. He pioneered the invention of the aqualung in 1943, as well as techniques in underwater filming. In 1951 he began the first of many research voyages in the ship *Calypso*. His film and television documentaries and books established him as a household name.

covalence in chemistry, a form of valence in which two atoms unite by sharing electrons in pairs, so that each atom provides half the shared electrons (see also †covalent bond).

covalent bond chemical †bond produced when two atoms share one or more pairs of electrons (usually each atom contributes an electron). The bond is often represented by a single line drawn between the two atoms. Covalently bonded substances include hydrogen (H_2), water (H_2O), and most organic substances.

Electron Sharing and Covalent Bonds

http://www.chem.ox.ac.uk/vrchemistry/electronsandbonds/intro1.htm

Introduction to electron sharing and covalent bonds. This page outlines what a covalent chemical bond is, and how electrons are involved. Click on the forward arrow under the picture to see an animation of a covalent bond forming, and being broken apart.

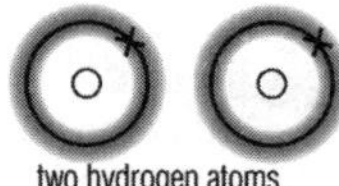

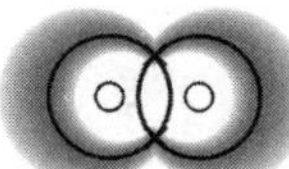

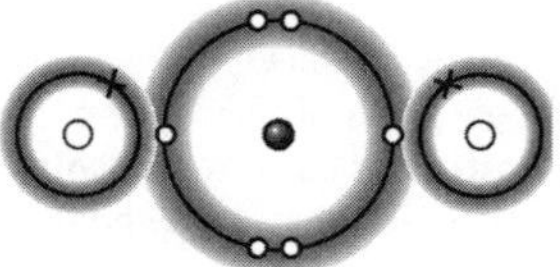

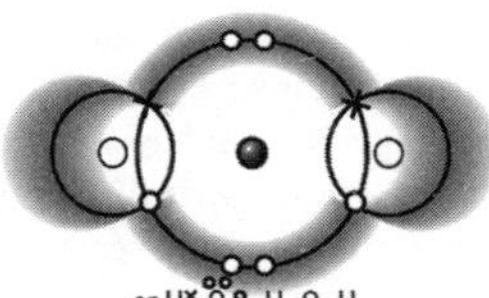

covalent bond The formation of a covalent bond between two hydrogen atoms to form a hydrogen molecule (H2), and between two hydrogen atoms and an oxygen atom to form a molecule of water (H2O). The sharing means that each atom has a more stable arrangement of electrons (its outer electron shells are full). Helicon

CPU in computing, abbreviation for †central processing unit.

CP violation in physics, breaking of a fundamental quantum theory conservation rule by some unstable particles undergoing decay into other particles. C represents charge conjugation that relates particles to †antiparticles and P stands for †parity.

Crab Nebula cloud of gas 6,000 light years from Earth, in the constellation †Taurus. It is the remains of a star that, according to Chinese records, exploded as a †supernova observed as a brilliant point of light on 4 July 1054. At its center is a †pulsar that flashes 30 times a second. It was named by Lord Rosse after its crablike shape.

cracking chemical reaction in which a large †alkane molecule is broken down by heat into a smaller alkane and a small †alkene molecule. The reaction is carried out at a high temperature (600°C/100°F or higher) and often in the presence of a catalyst. Cracking is a commonly used process in the †petrochemical industry.

It is the main method of preparation of alkenes and is also used to manufacture gasoline from the higher-boiling-point †fractions that are obtained by the †fractional distillation of crude oil (unrefined †petroleum).

crag in previously glaciated areas, a large lump of rock that a glacier has been unable to wear away. As the glacier passed up and over the crag, weaker rock on the far side was largely protected from erosion and formed a tapering ridge, or **tail**, of debris.

An example of a crag-and-tail feature is found in Edinburgh in Scotland; Edinburgh Castle was built on the crag (Castle Rock), which dominates the city beneath.

Cram, Donald James (1919– 2001) US chemist who shared the Nobel Prize for Chemistry in 1987 with Jean-Marie †Lehn and Charles J †Pedersen for their work on †molecules with highly selective structure-specific interactions. The work has importance in the synthesis of organic compounds, analytic chemistry, and biochemistry.

Cramer's paradox in mathematics, paradox that although two different cubic curves intersect at nine points, part of the definition of a single cubic curve is that it is itself determined by nine points. It was observed by Swiss mathematician Gabriel Cramer.

Cramer's rule in mathematics, method of solving a simultaneous system of linear equation by using †determinants. It is named after Swiss mathematician Gabriel Cramer.

cranium the dome-shaped area of the vertebrate skull that protects the brain. It consists of eight bony plates fused together by sutures (immovable joints). Fossil remains of the human cranium have aided the development of theories concerning human evolution.

The cranium has been studied as a possible indicator of intelligence or even of personality. The Victorian argument that a large cranium implies a large brain, which in turn implies a more profound intelligence, has been rejected.

Crater small constellation of the southern hemisphere, represented as a cup; it is associated in mythology with Hydra and †Corvus.

crater bowl-shaped depression in the ground, usually round and with steep sides. Craters are formed by explosive events such as the eruption of a volcano or the impact of a meteorite.

The Moon has more than 300,000 craters over 1 km/0.6 mi in diameter, mostly formed by asteroid and meteorite bombardment; similar craters on Earth have mostly been worn away by erosion. Craters are found on all of the other rocky bodies in the Solar System.

Craters produced by impact or by volcanic activity have distinctive shapes, enabling geologists to distinguish likely methods of crater formation on planets in the Solar System. Unlike volcanic craters, impact craters have raised rims and central peaks and are circular, unless the meteorite has an extremely low angle of incidence or the crater has been affected by some later process.

craton or **continental shield**, The relatively stable core of a continent that is not currently affected by tectonics along plate boundaries. Cratons generally consist of highly deformed †metamorphic rock that formed during ancient orogenic explosions.

Cratons exist in the hearts of all the continents, a typical example being the Canadian Shield.

creationism theory concerned with the origins of matter and life, claiming, as does the Bible in Genesis, that the world and humanity were created by a supernatural Creator, not more than 6,000 years ago. It was developed in response to Darwin's theory of †evolution; it is not recognized by most scientists as having a factual basis.

After a trial 1981–82, a US judge ruled as unconstitutional an attempt in Arkansas schools to enforce equal treatment of creationism and evolutionary theory. From 1996 in Alabama, all biology textbooks had to contain a statement that evolution is a controversial theory and not a proven fact. In 1999 the Kansas Board of Education added creationism to the school science curriculum, and removed evolution. However, evolution was restored to the curriculum in 2001.

Center for Scientific Creation http://www.creationscience.com/

Dedicated to researching the case for creation, rather than evolution, as the origin of living species. Most of the evidence is gathered in a book that is heavily promoted on this page. However, the entire book is available online.

creosote black, oily liquid derived from coal tar, used as a wood preservative. Medicinal creosote, which is transparent and oily, is derived from wood tar.

Crêpe Ring rather transparent inner ring (Ring C) of the †Saturn ring system. Its diameter measures about 149,300 km/92,800 mi.

crescent curved shape of the Moon when it appears less than half illuminated. It also refers to any object or symbol resembling the crescent Moon. Often associated with Islam, it was first used by the Turks on their standards after the capture of Constantinople in 1453, and appears on the flags of many Muslim countries. The **Red Crescent** is the Muslim equivalent of the Red Cross.

Cretaceous period of geological time approximately 143–65 million years ago. It is the last period of the Mesozoic era, during which angiosperm (seed-bearing) plants evolved, and dinosaurs reached a peak. The end of the Cretaceous period is marked by a mass extinction of many lifeforms, most notably the dinosaurs. The north European chalk, which forms the white cliffs of Dover, was deposited during the latter half of the Cretaceous, hence the name Cretaceous, which comes from the Latin *creta*, "chalk."

crevasse deep crack in the surface of a glacier; it can reach several meters in depth. Crevasses often occur where a glacier flows over the break of a slope, because the upper layers of ice are unable to stretch and cracks result. Crevasses may also form at the edges of glaciers owing to friction with the bedrock.

Crick, Francis Harry Compton (1916–) English molecular biologist who was awarded a Nobel Prize for Physiology or Medicine in 1962, together with Maurice †Wilkins and James †Watson, for the discovery of the double-helical structure of †DNA and of the significance of this structure in the replication and transfer of genetic information.

Crick, Francis Harry Compton http://kroeber.anthro.mankato.msus.edu/information/biography/abcde/crick_francis.html

Profile of the life and achievements of the pioneer molecular biologist. It traces his upbringing and education and how he brought his knowledge of X-ray diffraction to his work with James Watson in unraveling the structure of DNA. There is a photo, a listing of Crick's major books and articles, and a bibliography.

Crinoidea class of echinoderms containing about 600 living species and more than 2,000 fossil forms; the extinct crinoids are commonly called stone lilies, and the existing species sea lilies and feather stars. Crinoids retain many primitive features, notably in the use of the hydraulic water-vascular system for powering the capture and transportation of food rather than for locomotion. Uniquely amongst living echinoderms, the mouth is on the upper surface. Food particles are collected by the waving arms, and directed to the mouth along food grooves lined with cilia. Most of the living species are non-sessile (free-living), although a stalked stage may occur in development.

classification Crinoidea is in phylum Echinodermata.

crith unit of mass used for weighing gases. One crith is the mass of one liter of hydrogen gas (H_2) at standard temperature and pressure.

critical angle in optics, for a ray of light passing from a denser to a less dense medium (such as from glass to air), the smallest angle of incidence at which the emergent ray grazes the surface of the denser medium – at an angle of refraction of 90°.

When the angle of incidence is less than the critical angle, the ray passes out (is refracted) into the less dense medium; when the angle of incidence is greater than the critical angle, the ray is reflected back into the denser medium (see †total internal reflection).

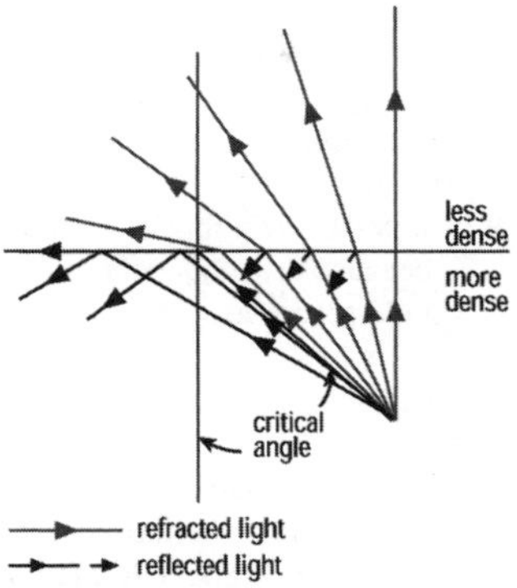

critical angle The critical angle is the angle at which light from within a transparent medium just grazes the surface of the medium. In the diagram, the red beam is at the critical angle. Blue beams escape from the medium, at least partially. Green beams are totally reflected from the surface. Helicon

critical density in cosmology, the minimum average density that the universe must have in order for it to stop expanding at some point in the future.

The precise value depends on †Hubble's constant and so is not fixed, but it is approximately between 10^{-29} and 2 x 10^{-29}g/cm^3, equivalent to a few hydrogen atoms per cubic meter. The density parameter (symbol W) is the ratio of the actual density to the critical density. If $W < 1$, the universe is open and will expand forever. If $W > 1$, the universe is closed and the expansion will eventually halt, to be followed by a contraction (†Big Crunch). Current estimates from visible matter in the universe indicate that W is about 0.01, well below critical density, but unseen dark matter may be sufficient to raise W to somewhere between 0.1 and 2.

critical mass in nuclear physics, the minimum mass of fissile material that can undergo a continuous †chain reaction. Below this mass, too many †neutrons escape from the surface for a chain reaction to carry on; above the critical mass, the reaction may accelerate into a nuclear explosion.

critical path analysis procedure used in the management of complex projects to minimize the amount of time taken. The analysis shows which subprojects can run in parallel with each other, and which have to be completed before other subprojects can follow on.

By identifying the time required for each separate subproject and the relationship between the subprojects, it is possible to produce a planning schedule showing when each subproject should be started and finished in order to complete the whole project most efficiently. Complex projects may involve hundreds of subprojects, and computer †applications packages for critical path analysis are widely used to help reduce the time and effort involved in their analysis.

critical point or **stationary point**, in mathematics, of a real-valued function, a point where the derivative is zero. Such a point is either a maximum, a minimum, or a horizontal point of inflection (point on a curve where the tangent line is horizontal).

critical reaction in a nuclear reactor, a self-sustaining chain reaction in which the number of neutrons being released by the fission of uranium-235 nuclei and the number of neutrons being absorbed by uranium-238 nuclei and by control rods are balanced. If balance is not achieved the reaction will either slow down and cease to generate enough power, or will build up and go out of control, as in a nuclear explosion. Control rods are used to adjust the rate of reaction and maintain balance.

critical temperature in physics, temperature above which a particular gas cannot be converted into a liquid by pressure alone. It is also the temperature at which a magnetic material loses its magnetism (the Curie temperature or point).

CRO abbreviation for †**cathode-ray oscilloscope**.

Cronin, James Watson (1931–) US physicist. He shared the Nobel Prize for Physics in 1980 with Val †Fitch for their work in †particle physics. They showed for the first time that left–right asymmetry is not always preserved when some particles are changed in state from matter to antimatter.

Crookes, William (1832–1919) English scientist whose many chemical and physical discoveries include the metallic element thallium (1861), the radiometer (1875), and the Crookes high-vacuum tube used in X-ray techniques. He was knighted in 1897.

Crookes's radiometer instrument consisting of an evacuated glass dome in which sits over a freely rotating system of vanes, whose opposite sides are white and black respectively. The rotation of the vanes when put near a source of heat is a demonstration of the †kinetic theory of gases. It was devised by English physicist William Crookes.

crop in birds, the thin-walled enlargement of the digestive tract between the esophagus and stomach. It is an effective storage organ especially in seed-eating birds; a pigeon's crop can hold about 500 cereal grains. Digestion begins in the crop, by the moisturizing of food. A crop also occurs in insects and annelid worms.

crop rotation system of regularly changing the crops grown on a piece of land. The crops are grown in a particular order to utilize and add to the nutrients in the soil and to prevent the build-up of insect and fungal pests. Including a legume crop, such as peas or beans, in the rotation helps build up nitrate in the soil, because the roots contain bacteria capable of fixing nitrogen from the air.

A simple seven-year rotation, for example, might include a three-year ley (grassland) followed by two years of wheat and then two years of barley, before returning the land to temporary grass once more. In this way, the cereal crops can take advantage of the build-up of soil fertility that occurs during the period under grass. In the 18th century a four-field rotation was widely adopted; over four years a field might be planted with autumn-sown cereal, followed by a root crop, then spring cereal, and finally a leguminous crop. Innovative farmers such as Charles "Turnip" Townshend improved cultivation techniques.

Cross, Charles Frederick (1855–1935) English chemist who in 1892, with fellow English chemist Edward Bevan, discovered the †viscose process for making artificial silk. They treated cellulose from wood pulp with caustic soda and carbon disulphide. The resulting viscous solution was squirted through spinnerets into dilute sulfuric acid to produce rayon fibers. The solution could also be extruded through a slit to make sheets of cellophane.

crossing over in biology, a process that occurs during †meiosis. While homologous chromosomes are lying alongside each other in pairs, each partner may twist around the other and exchange corresponding chromosomal segments. It is a form of genetic †recombination, which increases variation and thus allows offspring to vary from their parents.

cross-linking in chemistry, lateral linking between two or more long-chain molecules in a †polymer. Cross-linking gives the polymer a higher melting point and makes it harder. Examples of cross-linked polymers include Bakelite and vulcanized rubber.

cross ratio in projective geometry, a ratio expressing a relationship between two other ratios determined by four points on a given line, namely the ratio:

CA/CB:DA/DB

For the purposes of projection, the "point at infinity" may be any one of A, B, C, or D; if the given line is then projected onto another, the cross ratio of the projected points will remain the same – hence the importance of the cross ratio.

cross-section the surface formed when a solid is cut through by a plane perpendicular to its axis.

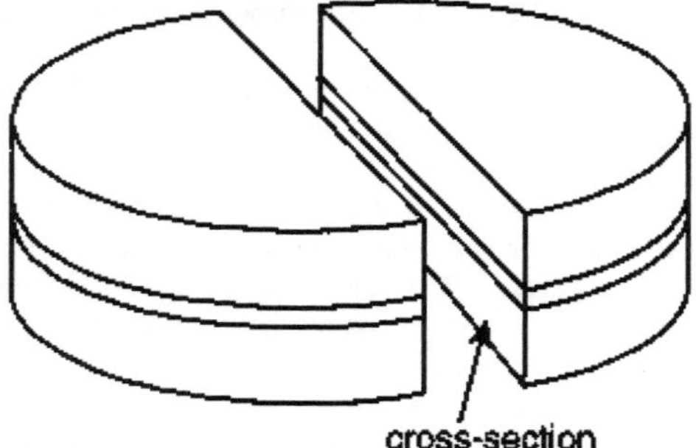

crown ether in chemistry, member of a class of cyclic, planar polyethers usually of the general formula $(-OCH_2CH_2-)_n$ (see †ether). Crown ethers have the ability to strongly bind †alkali metal ions in their center to form metal complexes. They are used to dissolve inorganic salts in solvents in which they were previously insoluble, such as non-polar organic solvents.

CRT abbreviation for †cathode-ray tube.

crude oil unrefined form of †petroleum.

crust rocky outer layer of the Earth, consisting of two distinct parts – the oceanic crust and the continental crust. The **oceanic** crust is on average about 10 km/6 mi thick and consists mostly of basaltic rock overlain by muddy sediments. By contrast, the **continental** crust is largely of granitic composition and has a more complex structure. Because it is continually recycled back into the mantle by the process of subduction, the oceanic crust is in no place older than about 200 million years. However, parts of the continental crust are over 3.5 billion years old.

Beneath a layer of surface sediment, the oceanic crust is made up of a layer of †basalt, followed by a layer of gabbro. The continental crust varies in thickness from about 40 km/25 mi to 70 km/45 mi, being deepest beneath mountain ranges, and thinnest above continental rift valleys. Whereas the oceanic crust is composed almost exclusively of basaltic igneous rocks and sediments, the continental crust is made of a wide variety of †sedimentary rocks, †igneous rocks, and †metamorphic rocks.

crustacean one of the class of arthropods that includes crabs, lobsters, shrimps, woodlice, and barnacles. The external skeleton is made of protein and chitin hardened with lime. Each segment bears a pair of appendages that may be modified as sensory feelers (antennae), as mouthparts, or as swimming, walking, or grasping structures.

Crutzen, Paul (1933–) Dutch meteorologist who shared the Nobel Prize for Chemistry in 1995 with Mexican chemist Mario †Molina and US chemist F Sherwood †Rowland for their work in atmospheric chemistry, particularly concerning the formation and decomposition of ozone. They explained the chemical reactions that are destroying the ozone layer.

Crux constellation of the southern hemisphere, popularly known as the Southern Cross, the smallest of the 88 constellations but one of the brightest, and one of the best known as it is represented on the flags of Australia and New Zealand. Its brightest stars are Alpha Crucis (or Acrux), a †double star about 360 light years from the Sun, and Beta Crucis (or Mimosa).

Near Beta Crucis lies a glittering star cluster known as the Jewel Box, named by John †Herschel. The constellation also contains the Coalsack, a dark nebula silhouetted against the bright starry background of the Milky Way.

cryogenics science of very low temperatures (approaching †absolute zero), including the production of very low temperatures and the exploitation of special properties associated with them, such as the disappearance of electrical resistance (†superconductivity).

Low temperatures can be produced by the Joule–Thomson effect (cooling a gas by making it do work as it expands). Gases such as oxygen, hydrogen, and helium may be liquefied in this way, and temperatures of 0.3 K can be reached. Further cooling requires magnetic methods; a magnetic material, in contact with the substance to be cooled and with liquid helium, is magnetized by a strong magnetic field. The heat generated by the process is carried away by the helium. When the material is then demagnetized, its temperature falls; temperatures of around 10^{-3} K have been achieved in this way. Much lower temperatures, of a few billionths of a kelvin above absolute zero, have been obtained by trapping gas atoms in an "optical molasses" of crossed laser beams, which slow them down. At temperatures near absolute zero, materials can display unusual properties. Some metals, such as mercury and lead, exhibit superconductivity. Liquid helium loses its viscosity and becomes a "superfluid" when cooled to below 2 K; in this state it flows up the sides of its container. Cryogenics has several practical applications. **Cryotherapy** is a process used in eye surgery, in which a freezing probe is briefly applied to repair a break in the retina. Electronic components called Josephson junctions, which could be used in very fast computers, need low temperatures to function. Magnetic levitation (maglev) systems must be maintained at low temperatures. Food can be frozen for years, and freezing eggs, sperm, and pre-embryos is now routine.

Cryogenics http://www.cryogenicsociety.org/

Information site of the Cryogenic Society of America (CSA), a non-profit technical society serving all those interested in any phase of cryogenics. This site provides online access to publications, supported by details of meetings and events.

cryolite rare granular crystalline mineral (sodium aluminum fluoride), Na_3AlF_6, used in the electrolytic reduction of †bauxite to aluminum. It is chiefly found in Greenland.

In the extraction of aluminum from the ore bauxite (impure aluminum oxide), the alumina (pure aluminum oxide) has to be molten for electrolysis to take place. Alumina has a very high melting point (2,047°C/13,717°F), which would require a lot of energy to keep it as a liquid, so the alumina is dissolved in molten cryolite instead. Far less energy is required and a temperature of 950°C/1,750°F can be used.

cryptand in chemistry, member of a class of polycyclic compounds similar to †crown ethers with the exception that donor atoms such as nitrogen and sulfur are located at bridgehead positions to facilitate the binding of cationic species. Essentially, a more efficient three-dimensional form of crown ether.

cryptate in chemistry, complex formed when a cationic species, such as an alkali metal ion, is bound in the central cavity of a †cryptand.

cryptography science of creating and reading codes, for example, those produced by the German Enigma machine used in World War II, those used in the secure transmission of credit card details over the Internet, and those used to ensure the privacy of e-mail messages. Unencoded text (known as plaintext) is converted to an unreadable form (known as cyphertext) by the process of encryption. The recipient must then decrypt the message before it can be read. The breaking of such codes is known as cryptanalysis. No encryption method is completely unbreakable, but cryptanalysis of a strongly encrypted message can be so time-consuming and complex as to be almost impossible.

The growth of the Internet and online commerce has brought an increasing demand for good cryptography. Most Internet cryptography systems involve the use of digital "keys" to encrypt and decrypt messages. In symmetric-key cryptography, both sender and recipient use the same key. An example of this is the Data Encryption Standard (DES), used by the US government. In public-key cryptography, each party has both a freely-available public key (used by anyone to encrypt messages) and a private key (used to decrypt received messages). The program Pretty Good Privacy (PGP) is a popular and effective (and free for non-commercial use) implementation of public-key cryptography. Encrypted messages may be further protected by concealing them within large graphics, audio, or video files – a technique known as steganography.

Codes and Ciphers in the Second World War http://www.codesandciphers.org.uk/

Independent guide to the history, science, and computing behind the work of Allied cryptanalysts at Bletchley Park, England, during World War II. This site offers a detailed technical and historical account of the work of these early computer scientists, including Alan Turing, and their efforts to crack German communications codes. Work here using automated machines to decode complex information was later to lead to the development of modern computer science.

Cryptography http://www.counterpane.com/pitfalls.html

Introduction to the science of encrypting messages. The author introduces various common encrypting systems and explains their relative weaknesses. The site also includes links to other pages that deal more comprehensively with cryptography in the computer age.

cryptosporidium waterborne parasite that causes disease in humans and other animals. It has been found in drinking water in the UK and USA, causing diarrhea, abdominal cramps, vomiting, and fever, and can be fatal in people with damaged immune systems, such as AIDS sufferers or those with leukemia. Just 30 cryptosporidia are enough to cause prolonged diarrhea.

Conventional filtration and chlorine disinfection are ineffective at removing the parasite. Slow sand filtration is the best method of removal, but the existing systems were dismantled in the 1970s because of their slowness.

crystal regular-shaped solid that reflects light. Examples include diamonds, grains of salt, and sugar. Particles forming a crystal are packed in an exact and ordered pattern. When this pattern is repeated many millions of times, the crystal is formed. Such an arrangement of particles, that is regular and repeating, is called a †giant molecular structure.

In †ionic compounds, such as sodium chloride (NaCl), the ions are arranged in a giant ionic lattice, with alternate positive and negative ions in a three-dimensional arrangement. The natural shape of the crystal is the same as the arrangement of ions in the lattice. In sodium chloride the ions form a cubic lattice. Hence sodium chloride crystals are cubic. In diamond there is a giant atomic structure made up of carbon atoms †covalently bonded to each other in a regular, repeating arrangement throughout the whole of the structure.

In metals the atoms are also packed tightly together in a regular pattern. This gives metals like copper a crystalline structure. †Metal crystals are called grains.

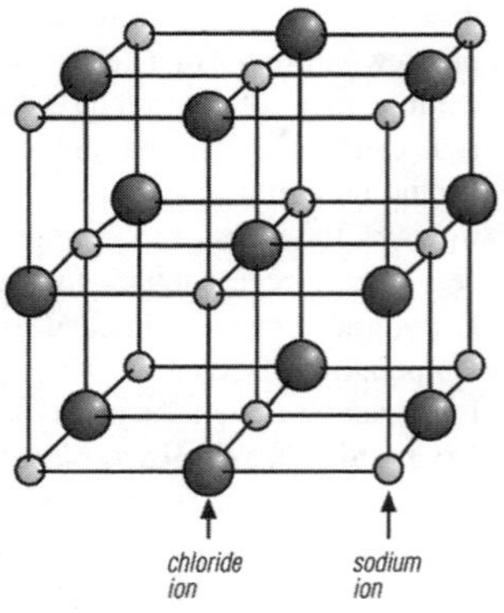

A mineral can often be identified by the shape of its crystals and the system of crystallization determined. For example, extrusive †igneous rock such as basalt contains very small crystals compared with an intrusive igneous rock such as granite. A single crystal can vary in size from a submicroscopic particle to a structure some 30 m/100 ft in length. Crystals fall into seven crystal systems or groups, classified on the basis of the relationship of three or four imaginary axes that intersect at the center of any perfect, undistorted crystal.

crystal The sodium chloride, or common salt, crystal is a regular cubic array of charged atoms (ions) – positive sodium ions and negative chloride ions. Repetition of this structure builds up into cubic salt crystals. Helicon

crystal lattice regular system of points in space (for example, the corners of a cube) about which atoms, molecules, or ions in solids vibrate.

crystallization formation of crystals from a liquid, gas, or solution.

crystallography scientific study of crystals. In 1912 it was found that the shape and size of the repeating atomic patterns (unit cells) in a crystal could be determined by passing X-rays through a sample. This method, known as †X-ray diffraction, opened up an entirely new way of "seeing" atoms. It has been found that many substances have a unit cell that exhibits all the symmetry of the whole crystal; in table salt (sodium chloride, NaCl), for instance, the unit cell is an exact cube.

Many materials were not even suspected of being crystals until they were examined by X-ray crystallography. It has been shown that purified biomolecules, such as proteins and DNA, can form crystals, and such compounds may now be studied by this method. Other applications include the study of metals and their alloys, and of rocks and soils.

Crystallography and Mineralogy http://www.iumsc.indiana.edu/crystmin.html

Understand the shapes and symmetries of crystallography, with these interactive drawings of cubic, tetrahedral, octahedral, and dodecahedral solids (just drag your mouse over the figures to rotate them).

crystalloid substance that, when dissolved in a solvent, can pass through a †semipermeable membrane (as opposed to a †colloid, which cannot).

crystal oscillator in physics, electronic †oscillator based on a piezoelectric crystal such as quartz or Rochelle salt (potassium sodium tartrate tetrahydrate, $KNaC_4H_4O_6.4H_2O$) with a natural frequency of oscillation of from 1 kHz to a few MHz. A voltage between two plates attached to opposite faces of the crystal brings about the oscillation.

crystal system any of the seven crystal systems defined by symmetry, into which all known crystalline substances crystallize. The elements of symmetry used for this purpose are: (1) planes of **mirror symmetry**, across which a mirror image is seen, and (2) axes of **rotational symmetry**, about which, in a 360° rotation of the crystal, equivalent faces are seen two, three, four, or six times. To be assigned to a particular crystal system, a mineral must possess a certain minimum symmetry, but it may also possess additional symmetry elements. Since crystal symmetry is related to internal structure, a given mineral will always crystallize in the same system, although the crystals may not always grow into precisely the same shape. In cases where two minerals have the same chemical composition but different internal structures (for example graphite and diamond, or quartz and cristobalite), they will generally have different crystal systems.

CSH abbreviation for †**corticotrophin-releasing hormone**.

CT-scan another version of †CAT scan or CAT scanner.

CT scanner medical device used to obtain detailed X-ray pictures of the inside of a patient's body. See †CAT scan.

cu abbreviation for **cubic** (measure).

cube in geometry, a solid shape whose faces are all †squares. It has 6 equal-area faces and 12 equal-length edges.

If the length of one edge is l, the †volume (V) of the cube is given by:

$V = l^3$

Its surface †area (A) is calculated by finding the area of one square:

$l \times l = l^2$;

and multiplying it by 6:

$6 \times l^2$

$A = 6l^2$

cube in arithmetic, to multiply a number by itself and then by itself again. For example, 5 cubed = 5^3 = 5 x 5 x 5 = 125. Alternatively, the †cube root of 125 is 5. The term also refers to a number formed by cubing; for example, 1, 8, 27, 64 are the first four cubes.

cube root number that, multiplied by itself, and then by the product, produces the †cube. For example, 3 x 3 x 3 = 27, 3 being the cube root of 27, which is the cube of 3.

cubic centimeter (or meter) the metric measure of volume, corresponding to the volume of a cube whose edges are all 1 cm (or 1 meter) in length.

cubic curve geometrical curve in three-dimensional space. It may be parameterized after a change of variables as $(x, y, z) = (at^3, bt^2, ct)$ and may thus be said to be determined by nine points. See also †Cramer's paradox.

cubic decimeter symbol dm^3, metric measure of volume corresponding to the volume of a cube whose edges are all 1 dm (10 cm) long; it is equivalent to a capacity of one liter.

cubic equation any equation in which the largest power of x is x^3. For example, $x^3 + 3x^2y + 4y^2 = 0$ is a cubic equation.

cubic measure measure of volume, indicated either by the word "cubic" followed by a linear measure, as in "cubic foot," or the word "cubed" after a linear measure, as in "meter cubed."

cubit earliest known unit of length, which originated between 2800 and 2300 BC. It is approximately 50.5 cm/20.6 in long, which is about the length of the human forearm measured from the tip of the middle finger to the elbow.

cuboid six-sided three-dimensional prism whose faces are all †rectangles. A brick is a cuboid.

cuesta alternative name for †escarpment.

cultivar variety of a plant developed by horticultural or agricultural techniques. The term derives from "**culti**vated **var**iety."

culture in biology, the growing of living cells and tissues in laboratory conditions.

cumulative frequency in statistics, the total frequency of a given value up to and including a certain point in a set of data. It is calculated by adding together the frequencies to give a running total and used to draw the cumulative frequency curve, the ogive.

To plot a cumulative frequency diagram, the cumulative frequency is always plotted along the vertical axis and data are always plotted at the top range of the interval: for example, frequency for weekly salary ranges £100–150 and £151–200 would be plotted at 150 and 200.

cuprite **copper(I) oxide**; Cu_2O, ore found in crystalline form or in earthy masses. It is red to black in color, and is often called ruby copper.

cupronickel copper alloy (75% copper and 25% nickel), used in hardware products and for coinage.

curie symbol Ci, former unit of radioactivity, equal to 3.7×10^{10} †becquerels. One gram of radium has a radioactivity of about one curie. It was named after French physicist Pierre Curie.

Curie, Marie (1867–1934) born Maria Sklodowska, Polish scientist who, with husband Pierre Curie, discovered in 1898 two new radioactive elements in pitchblende ores: polonium and radium. They isolated the pure elements in 1902. Both scientists refused to take out a patent on their discovery and were jointly awarded the Nobel Prize for Physics in 1903, with Henri †Becquerel, for their research on radiation phenomena. Marie Curie was also awarded the Nobel Prize for Chemistry in 1911 for the discovery of radium and polonium, and the isolation and study of radium.

Essay about Marie and Pierre Curie http://www.nobel.se/physics/articles/curie/

Extended essay on the life and deeds of Pierre and Marie Curie. The site spans the entirety of the couple's astonishing career and their contribution to the promotion of the understanding of radioactivity. It includes sections on Pierre's and Marie's joint research on radiation phenomena as well as Marie's own work after Pierre's untimely death in 1906: her discovery of the elements radium and polonium, her second Nobel prize, this time in chemistry, her difficulties with the press, and her family life.

Curie temperature the temperature above which a magnetic material cannot be strongly magnetized. Above the Curie temperature, the energy of the atoms is too great for them to join together to form the small areas of magnetized material, or †domains, which combine to produce the strength of the overall magnetization.

curium chemical symbol Cm, synthesized, radioactive, metallic element of the *actinide* series, atomic number 96, relative atomic mass 247. It is produced by bombarding plutonium or americium with neutrons. Its longest-lived isotope has a half-life of 1.7×10^7 years.

current in earth science, flow of a body of water or air, or of heat, moving in a definite direction. Ocean currents are fast-flowing bodies of seawater moved by the wind or by variations in water density between two areas. They are partly responsible for transferring heat from the Equator to the poles and thereby evening out the global heat imbalance. There are three basic types of ocean current: **drift currents** are broad and slow-moving; **stream currents** are narrow and swift-moving; and **upwelling currents** bring cold, nutrient-rich water from the ocean bottom.

Stream currents include the †Gulf Stream and the †Japan (or Kuroshio) Current. Upwelling currents, such as the Gulf of Guinea Current and the Peru (Humboldt) Current, provide food for plankton, which in turn supports fish and sea birds. At approximate five-to-eight-year intervals, the Peru Current that runs from the Antarctic up the west coast of South America, turns warm, with heavy rain and rough seas, and has disastrous results (as in 1982–83) for Peruvian wildlife and for the anchovy industry. The phenomenon is called **El Niño** (Spanish "the Child") because it occurs towards Christmas.

current, alternating see †alternating current.

current density symbol *j*, in physics, for any conductor, the electric current flowing per unit area of its cross-section, expressed in ampere meter^{-2} (Am^{-2}).

current, direct see †direct current.

current, electric see †electric current.

cursor on a computer screen, the symbol that indicates the current entry position (where the next character will appear).

curvature in mathematics, of a curve, the rate of change of the angle of a †tangent to the curve at a given point. It is the †reciprocal of the radius of the circle of curvature, which is the circle that most closely approximates a curve close to the given point.

curve in geometry, the †locus of a point moving according to specified conditions. The circle is the locus of all points equidistant from a given point (the center). Other common geometrical curves are the †ellipse, †parabola, and †hyperbola, which are also produced when a cone is cut by a plane at different angles.

Many curves have been invented for the solution of special problems in geometry and mechanics – for example, the cissoid (the inverse of a parabola) and the †cycloid.

cusp point where two branches of a curve meet and the tangents to each branch coincide.

cuticle the horny noncellular surface layer of many invertebrates such as insects; in botany, the waxy surface layer on those parts of plants that are exposed to the air, continuous except for †stomata and †lenticels. All types are secreted by the cells of the †epidermis. A cuticle reduces water loss and, in arthropods, acts as an †exoskeleton.

cutting technique of vegetative propagation involving taking a section of root, stem, or leaf and treating it so that it develops into a new plant.

Cuvier, Georges (Léopold Chrêtien Frédéric Dagobert) (1769–1832) Baron Cuvier, French comparative anatomist, the founder of paleontology. In 1799 he showed that some species have become extinct by reconstructing extinct giant animals that he believed were destroyed in a series of giant deluges. These ideas are expressed in *Recherches sur les ossiments fossiles de quadrupèdes/Researches on the Fossil Bones of Quadrupeds* (1812) and *Discours sur les révolutions de la surface du globe/Discourse on the Revolutions of the Surface of the Globe* (1825).

Georges Cuvier – Discourse on the Revolutionary Upheavals on the Surface of the Earth

http://www.mala.bc.ca/~johnstoi/cuvier.htm

Complete original text and English translation of Cuvier's celebrated Discours sur les révolutions de la surface du globe. This public domain version has translator's comments and explanatory footnotes to help illustrate Cuvier's argument.

cwt symbol for †hundredweight, a unit of weight equal to 112 pounds (50.802 kg); 100 lb (45.36 kg) in the USA.

cyanamide process process used in the manufacture of calcium cyanamide ($CaCN_2$), a colorless crystalline powder used as a fertilizer under the tradename Nitrolime. Calcium carbide is reacted with nitrogen in an electric furnace.

$CaC_2 + N_2 \rightarrow CaCN_2 + C$

The calcium cyanamide reacts with water in the soil to form the ammonium ion and calcium carbonate. The ammonium is then oxidized to nitrate, which is taken up by plants. Calcium cyanamide can also be converted commercially into ammonia.

cyanide CN^-, ion derived from hydrogen cyanide (HCN), and any salt containing this ion (produced when hydrogen cyanide is neutralized by alkalis), such as potassium cyanide (KCN). The principal cyanides are potassium, sodium, calcium, mercury, gold, and copper. Most cyanides are poisons.

Organic compounds containing a CN group are sometimes called cyanides but are more properly known as nitrites.

cyanobacteria singular **cyanobacterium**, alternative name for †blue-green algae. These organisms are actually not algae but bacteria. The ancestors of modern cyanobacteria generated the oxygen that caused a transformation some 2 billion years ago of the Earth's atmosphere.

cyanocobalamin chemical name for vitamin B_{12}, which is normally produced by micro-organisms in the gut. The richest sources are liver, fish, and eggs. It is essential to the replacement of cells, the maintenance of the myelin sheath that insulates nerve fibers, and the efficient use of folic acid, another vitamin in the B complex. Deficiency can result in pernicious anemia (defective production of red blood cells), and possible degeneration of the nervous system.

cyanoferrate member of a class of chemical compounds formed by the addition of alkali cyanides to solutions of iron salts. This process produces species such as hexacyanoferrate(II), $[Fe(CN)_6]^{4-}$, and hexacyanoferrate (III), $[Fe(CN)_6]^{3-}$. Many cyanoferrates are strongly colored and are used in the manufacture of synthetic inks and dyes. They were formerly known as ferrocyanides and ferricyanides.

cybernetics study of how communication and control mechanisms in machines can be made to imitate those of human beings.

cyclamate derivative of cyclohexysulphamic acid, formerly used as an artificial sweetener, 30 times sweeter than sugar. It was first synthesized in 1937.

Its use in foods was banned in the USA and the UK from 1970, when studies showed that massive doses caused cancer in rats.

cycle in physics, a sequence of changes that moves a system away from, and then back to, its original state. An example is a vibration that moves a particle first in one direction and then in the opposite direction, with the particle returning to its original position at the end of the vibration.

cyclene in chemistry, any cyclic hydrocarbon compound that has at least one double bond in its molecular structure, for example cyclopentadiene C_5H_6.

cyclic AMP **cAMP**, cyclic ^{3,5}-adenosine monophosphate, a major communicator molecule in living cells. It is formed by the action of the enzyme adenyl cyclase on †ATP and has many physiological functions. It plays a part in increasing the level of blood †glucose. †Adrenaline, which increases the level of blood sugar, does so by producing cAMP. Cyclic AMP is also involved in the synthesis of flagellar proteins in †bacteria, and in hormone release, urine regulation, and sensory and nerve processes. In some organisms, cAMP seems to be active in the breakdown of †lipids for the production of energy, when glucose is unavailable.

cyclic compound any of a group of organic chemicals that have rings of atoms in their molecules, giving them a closed-chain structure.

cyclic group in mathematics, group whose elements are all powers of a given element (a generator).

cyclic patterns patterns in which simple ideas are repeated to form more complex designs. Some mathematical †functions show cyclic patterns; for example, mapping round a circle.

cyclic polygon in geometry, a polygon in which each vertex (corner) lies on the circumference of a circle.

cyclic quadrilateral four-sided figure with all four of its vertices lying on the circumference of a circle. The opposite angles of cyclic quadrilaterals add up to 180°, and are therefore said to be supplementary; each external angle (formed by extending a side of the quadrilateral) is equal to the opposite interior angle.

Cycliophora invertebrate phylum containing only one known species *Symbion pandora*. The phylum was discovered in 1995. *S. pandora* are minute (347 mm) bottle-shaped invertebrates living amongst the mouthparts of lobsters. They have threadlike cilia for gathering food and are attached by an adhesive disc. They have a reproductive cycle consisting of both asexual and sexual forms. The cycliophoran's nearest known relatives are bryozoans.

cyclodextrin ring-shaped †glucose molecule chain created in 1993 at Osaka University, Japan. Cyclodextrins are commonly used in food additives, and can also be used as capsules to deliver drugs, as cutters to separate ions and molecules, and as catalysts for chemical reactions.

They generally consist of 6–8 glucose molecules linked together in a ring, leaving a central hole of 0.45–0.8 nanometers, which can hold a small molecule such as benzene. They can be joined together to form tubes even smaller than DNA, the length and width of which can be controlled. They could hypothetically be used in the production of large scale integrated computer systems.

cyclohexane C_6H_{12}, colorless liquid, boiling point 81°C/178°F. It is usually prepared by the catalytic reduction of benzene with hydrogen. It is used as a general solvent for fats and waxes and as a paint remover, and is also important in the manufacture of nylon.

cyclohexanol or **hexalin**; $C_6H_{12}O$, oily colorless liquid, boiling point 160°C/320°F. It is commercially manufactured from phenol using a catalyzed reaction at high temperature and pressure. Derivatives of cyclohexanol are used for the production of plasticizers, celluloid, and detergents.

cyclohexanone or **pimelic ketone**; $C_6H_{10}O$, colorless liquid, strongly smelling of peppermint, boiling point 154–156°C/309–313°F. It can be prepared by passing gaseous cyclohexanol over a heated copper catalyst. It is mainly used as a solvent for lacquers.

cycloid in geometry, a curve resembling a series of arches traced out by a point on the circumference of a circle that rolls along a straight line. Its applications include the study of the motion of wheeled vehicles along roads and tracks.

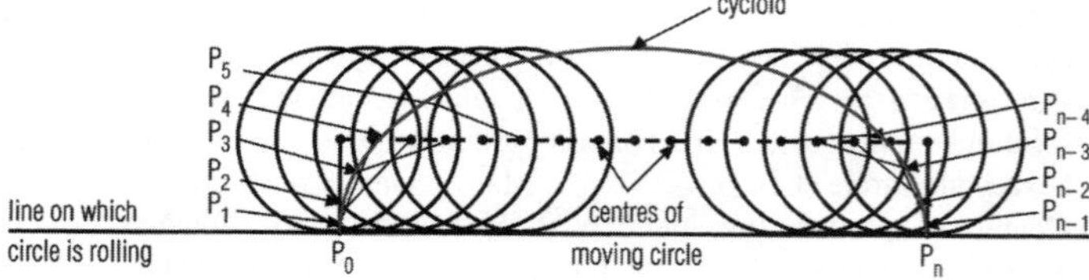

cycloid The cycloid is the curve traced out by a point on a circle as it rolls along a straight line. The teeth of gears are often cut with faces that are arcs of cycloids so that there is rolling contact when the gears are in use. Helicon

Cycloid http://www-

groups.dcs.st-andrews.ac.uk/history/Curves/Cycloid.html
Introduction to cycloids and links to associated curves. If you have a Java-enabled browser you can experiment interactively with this curve and its associates. There are also links to mathematicians who have studied the cycloid, including Galileo, who named the curve, and Cusa, who first studied it.

cyclone alternative name for a †depression, an area of low atmospheric pressure with winds blowing in a anticlockwise direction in the northern hemisphere and in a clockwise direction in the southern hemisphere. A severe cyclone that forms in the tropics is called a tropical cyclone or †hurricane.

cyclotron circular type of particle †accelerator.

Cygnus large prominent constellation of the northern hemisphere, represented as a swan. Its brightest star is first-magnitude Alpha Cygni or †Deneb.
Beta Cygni (Albireo) is a yellow and blue †double star, visible through small telescopes. The constellation contains the North America nebula (named after its shape), the Veil nebula (the remains of a †supernova that exploded about 50,000 years ago), Cygnus A (apparently a double galaxy, a powerful radio source, and the first radio star to be discovered), and the X-ray source Cygnus X-1, thought to mark the position of a black hole. The area is rich in high-luminosity objects, nebulae, and clouds of obscuring matter. Deneb marks the tail of the swan, which is depicted as flying along the Milky Way. Some of the brighter stars form the Northern Cross, the upright being defined by Alpha, Gamma, Eta, and Beta, and the crosspiece by Delta, Gamma, and Epsilon Cygni.

Cygnus-A source in the constellation Cygnus of strong radio emission, possibly caused by the collisions of interstellar dust within two colliding galaxies. It was identified optically in 1952 as a tiny area of magnitude 17.9.

Cygnus X-1 in astronomy, strong X-ray source in the constellation Cygnus and probably the site of the first black hole to be found. It is likely that it consists of a binary combination of a black hole and a blue supergiant star with an orbital period of 5.6 days that emits X-rays as material passes from the star into the black hole. It was discovered in 1966.

cylinder in geometry, a prism with a circular †cross-section. In everyday use, the term applies to a **right cylinder**, in which the curved surface is perpendicular to the base.
As it is a prism, the volume (V) of a cylinder is area of cross-section (pr^2) x height (h), or
$V = pr^2h$
where r is the radius of the base and h is the height of the cylinder.
Its total surface area (A) = the curved surface area + the area of both circular ends. To find the curved surface area, the hollow cylinder area should be imagined stretched flat to form a rectangle:
Curved surface area = circumference x h, or $2\pi r$ x h
Area both circular ends = 2 x πr^2

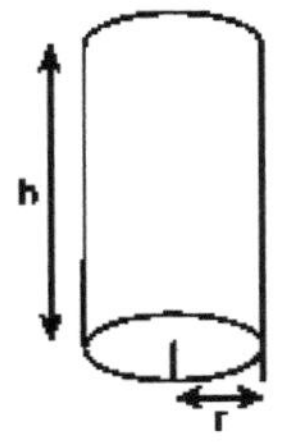

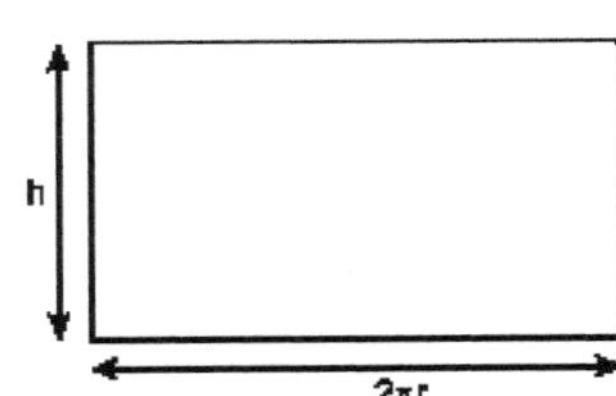

So $A = 2\pi r^2 + 2\pi rh$
and by factorizing:
$A = 2\pi r(h + r)$

cylindrical having the shape of a †cylinder. For example, tins of food, toilet rolls, pipes, and poles are all cylindrical objects.

cytochrome protein responsible for part of the process of †respiration by which food molecules are broken down in †aerobic organisms. Cytochromes are part of the electron transport chain, which uses energized electrons to reduce molecular oxygen (O_2) to oxygen ions (O^{2-}). These combine with hydrogen ions (H^+) to form water (H_2O), the end product of aerobic respiration. As electrons are passed from one cytochrome to another, energy is released and used to make †ATP.

cytokinin †plant hormone that stimulates cell division. Cytokinins affect several different aspects of plant growth and development, but only if †auxin is also present. They may delay the process of senescence, or aging, break the dormancy of certain seeds and buds, and induce flowering.

cytology study of the structure of †cells and their functions. Major advances have been made possible in this field by the development of †electron microscopes.

cytoplasm part of plant and animal cells outside the †nucleus (and outside the large vacuole of plant cells). Strictly speaking, this includes all the †organelles (mitochondria, chloroplasts, and so on) and is the area in which most cell activities take place. However, cytoplasm is often used to refer to the jelly-like matter in which the organelles are embedded (correctly termed the cytosol). Most of the activities in the cytoplasm are chemical reactions (metabolism), for example, protein synthesis.

In many cells, the cytoplasm is made up of two parts: the **ectoplasm** (or plasmagel), a dense gelatinous outer layer concerned with cell movement, and the **endoplasm** (or plasmasol), a more fluid inner part where most of the organelles are found.

cytosine in biochemistry, one of the nucleotide bases that attach to a pentose sugar (ribose or deoxyribose) forming a nucleotide. Nucleic acids (RNA, DNA) are formed from the condensation of nucleotides forming long chains. Cytosine is a pyrimidine and always bonds with the purine †guanine.

cytoskeleton in a living cell, a matrix of protein filaments and tubules that occurs within the cytosol (the liquid part of the cytoplasm). It gives the cell a definite shape, transports vital substances around the cell, and may also be involved in cell movement.

D

DAC abbreviation for †digital-to-analog converter.

Dacron see †Terylene.

d'Alembert French mathematician, encyclopedist, and theoretical physicist; see Jean le Rond d'†Alembert.

Dalton, John (1766–1844) English chemist who proposed the theory of atoms, which he considered to be the smallest parts of matter. He produced the first list of relative atomic masses in "Absorption of Gases" in 1805 and put forward the law of partial pressures of gases (**Dalton's law**).

daminozide trade name **Alar**, chemical formerly used by fruit growers to make apples redder and crisper. In 1989 a report published in the USA found the consumption of daminozide to be linked with cancer, and the US Environment Protection Agency (EPA) called for an end to its use. The makers have now withdrawn it worldwide.

damping in physics, the use of some type of resistance to reduce the amplitude of an oscillation or vibration, such as the swinging needle of a balance or measuring instrument. With **critical damping**, the oscillation ceases in the shortest possible time.

Daniell cell primary cell that uses a †cathode of zinc, immersed in sulfuric acid contained in a porous pot that itself stands in a container of copper sulfate, in which the copper †anode stands. An †electromotive force of about 1.1 volts is produced by this cell.

Danielli, James Frederic (1911–1984) British cell biologist who hypothesized that the molecular structure of the cell membrane was a sandwich of two layers of proteins. In 1943 Danielli and Hugh Davson published their seminal theory on transport of substances across cell membranes in *The Permeability of Natural Membranes*. Their work provided a framework for future physiologists and cell biologists working on the role of the membrane in different cellular activities.

daraf symbol F^{-1}, in physics, unit of elastance (the reciprocal of capacitance, which is expressed in farads). Daraf is farad backwards.

Darboux, Jean Gaston (1842–1917) French mathematician. He contributed immensely to the differential geometry of his time, and to the theory of surfaces. In defining the Riemann integral in 1879, he derived the **Darboux sums** and used the **Darboux integrals**.

Darboux, Jean Gaston http://www-groups.dcs.st-andrews.ac.uk/history/Mathematicians/Darboux.html
Profile of the life and work of the French pioneer of differential geometry. Darboux's achievements are set within the context of the work of other leading mathematicians of his day. His skills as a teacher are also highlighted.

darcy symbol D, *c*.g.s. unit of permeability, used mainly in geology to describe the permeability of rock (for example, to oil, gas, or water).

dark cloud in astronomy, cloud of cold dust and gas seen in silhouette against background stars or an †HII region (region of hot ionized hydrogen).

dark matter matter that, according to certain modern theories of †cosmology, makes up 90–99% of the mass of the universe but so far remains undetected. Dark matter, if shown to exist, would account for many currently unexplained gravitational effects in the movement of galaxies.

Theories of the composition of dark matter include unknown atomic particles (cold dark matter) or fast-moving neutrinos (hot dark matter) or a combination of both.

In 1993 astronomers identified part of the dark matter in the form of stray planets and †brown dwarfs, and, possibly, stars that have failed to ignite. These objects are known as massive astrophysical compact halo objects and may make up approximately half of the dark matter in the Milky Way's halo.

dark reaction series of reactions in †photosynthesis that do not require light. During the dark reaction, carbon dioxide is incorporated into three-carbon sugar phosphate molecules; this reaction is dependant on the †light reaction which does require light.

Dart, Raymond Arthur (1893–1988) Australian-born South African paleontologist and anthropologist who in 1924 discovered the first fossil remains of the australopithecenes, early hominids, near Taungs in Botswana.

Darwin, Charles Robert (1809–1882) English naturalist who developed the modern theory of †evolution and proposed, with Welsh naturalist Alfred Russel †Wallace, the principle of †natural selection. After research in South America and the Galapagos Islands as naturalist on HMS *Beagle* (1831–36), Darwin published *On the Origin of Species by Means of Natural Selection or the Preservation of Favoured Races in the Struggle for Life* (1859). This book explained the evolutionary process through the principles of natural selection and aroused widespread argument and debate among scientists and religious leaders because it disagreed with the literal interpretation of the Book of Genesis in the Bible.

Darwin and Evolution http://www.victorianweb.org/victorian/science/darwin/darwin5.html
Clear introduction to the evolutionary theories of Charles Darwin. The page, part of the Victorian Web site, shows how Darwin came to his ground-breaking scientific conclusions, and shows Darwinism in the context of the work of other scientists such as Linnaeus and Malthus.

Darwin, Charles http://www.literature.org/authors/darwin-charles/
Complete text of Darwin's seminal works On the Origin of Species.

Darwin, Erasmus (1731–1802) British poet, physician, and naturalist; he was the grandfather of Charles Darwin. He anticipated aspects of evolutionary theory, but tended to the French naturalist J B de †Lamarck's interpretation.

Darwin, Francis (1848–1925) English botanist, the third son of Charles Darwin. He was associated with his father's work until the latter's death, and edited his father's biography. In 1884 he was appointed lecturer and reader in botany at the University of Cambridge.

Darwinism theory that the mechanism of biological evolution proceeds by natural selection of adaptive variations. It follows the thinking of 19th-century English naturalist Charles †Darwin.

data singular **datum**, facts, figures, and symbols, especially as stored in computers. The term is often used to mean raw, unprocessed facts, as distinct from information, to which a meaning or interpretation has been applied.

†**Continuous data** is data that can take any of an infinite number of values between whole numbers and so may not be measured completely accurately. This type of data contrasts with †**discrete data**, in which the variable can only take one of a finite set of values. For example, the sizes of apples on a tree form continuous data, whereas the numbers of apples form discrete data.

Box Plots and Box and Whiskers Plots
http://www.math.uncc.edu/~droyster/courses/spring00/maed3103/Box_Plots.htm
Comprehensive coverage of how to create and use box plots. A variety of sample data is given, along with worked examples of how to calculate plots from data. There are also some questions to attempt, using the data provided.

database in computing, a structured collection of data, which may be manipulated to select and sort desired items of information. For example, an accounting system might be built around a database containing details of customers and suppliers. In larger computers, the database makes data available to the various programs that need it, without the need for those programs to be aware of how the data are stored. The term is also sometimes used for simple record-keeping systems, such as mailing lists, in which there are facilities for searching, sorting, and producing records. Examples of database software include Oracle, Sybase, and Microsoft Access.

There are four main types (or "models") of database: relational, object-oriented, hierarchical, and network, of which relational is the most widely used. Object-oriented databases have become more popular for certain types of application, and hybrids like object-relational are also available. In a **relational database** data are viewed as a collection of linked tables. A **free-text database** is one that holds the unstructured text of articles or books in a form that permits rapid searching. A telephone directory stored as a database might allow all the people whose names start with the letter B to be selected by one program, and all those living in Chicago by another.

data capture collecting information for computer processing and analysis. Examples of **automated data capture** include using a †sensor that continuously monitors physical conditions such as temperature, or scanning bar codes to produce detailed receipts at a shop checkout (point-of-sale terminal). **Manual data capture** methods include reading electricity meters, or filling in a form or questionnaire.

data communications sending and receiving data via any communications medium, such as a telephone line. The term usually implies that the data are digital (such as computer data) rather than analog (such as voice messages). However, in the ISDN (Integrated Services Digital Network) system, all data – including voices and video images – are transmitted digitally.

See also †telecommunications.

ISDN Tutorial http://www.ralphb.net/ISDN/
Guide to Integrated Services Digital Network (ISDN), a system that allows data to be transmitted across the world. The guide caters to both beginners and advanced users – it begins with a definition of ISDN and a description of its benefits, and moves on to the various interfaces and protocols. There are also numerous links to related Web sites and resources.

data compression in computing, techniques for reducing the amount of storage needed for a given amount of data. They include word tokenization (in which frequently used words are stored as shorter codes), variable bit lengths (in which common characters are represented by fewer †bits than less common ones), and run-length encoding (in which a repeated value is stored once along with a count).

In **lossless compression** the original file is retrieved unchanged after decompression. Some types of data (sound and pictures) can be stored by **lossy compression** where some detail is lost during compression, but the loss is not noticeable. Lossy compression allows a greater level of compression. The most popular compression program is PKZIP, widely available as shareware.

data processing **DP**; or **electronic data processing (EDP)**, use of computers for performing clerical tasks such as stock control, payroll, and dealing with orders. DP systems are typically †batch systems, running on mainframe computers.

data processing cycle For data to be processed the following cycle of operations must be undergone: data are collected, then input into a computer where they are processed to produce the output. As the output may also be the input of a subsequent process the process is deemed cyclical. While being processed data may undergo other operations such as storage and validation.

data protection safeguarding of information about individuals stored in files and on computers, to protect privacy.

dating in geology, the process of determining the age of minerals, rocks, fossils, and geological formations. There are two types of dating: relative and absolute. **Relative dating** involves determining the relative ages of materials, which is determining the chronological order of formation of particular rocks, fossils, or formations, by means of careful field work. **Absolute dating** is the process of determining the absolute age (that is the age in years) of a mineral, rock, or fossil. Absolute dating is accomplished using methods such as †radiometric dating (measuring the abundances of particular isotopes in a mineral), fission track dating, and even counting annual layers of sediment.

Dausset, Jean Baptiste Gadriel Joachim (1916–) French immunologist who was awarded a Nobel Prize for Physiology or Medicine in 1980 for his work on genetically determined structures on the cell surface that regulate immunological reactions. He was an early pioneer in the study of the human major histocompatability complex (MHC), the system of genes and their corresponding antigens on the surface of cells that enables the body to recognize its own cells.

da Vinci Italian painter, sculptor, architect, engineer, and scientist; see †Leonardo da Vinci.

Davis, William Morris (1850–1934) US physical geographer who analyzed landforms. In the 1890s he developed the organizing concept of a regular cycle of erosion, a theory that dominated geomorphology and physical geography for half a century.

Davisson, Clinton Joseph (1881–1958) US physicist who in 1927 made the first experimental observation of the wave nature of electrons, for which he was awarded the Nobel Prize for Physics in 1937. He shared the award with George †Thomson who carried through the same research independently.

Davy, Humphry (1778–1829) English chemist. He discovered, by electrolysis, the metallic elements †sodium and †potassium in 1807, and †calcium, †boron, †magnesium, †strontium, and †barium in 1808. In addition, he established that †chlorine is an element and proposed that hydrogen is present in all acids. He invented the safety lamp for use in mines where †methane was present, enabling miners to work in previously unsafe conditions. He was knighted for his work in 1812 and made baronet in 1818.

Davy Discovers Sodium and Potassium http://dbhs.wvusd.k12.ca.us/Chem-History/Davy-Na&K-1808.html

Davy's paper to the Royal Society in 1808 titled On Some New Phenomena of Chemical Changes Produced by Electricity, Particularly the Decomposition of Fixed Alkalies, and the Exhibition of the New Substances which Constitute their Bases: and on the General Nature of Alkaline Bodies. Quite apart from it being the longest title ever, it seems astounding that only 190 years ago people did not know that common salt was sodium chloride, this paper marking the leap in knowledge which created a century of chemical discoveries.

Davy, Humphry http://www.rigb.org/heritage/faradaypage.html

Biography of Humphry Davy, from the Royal Institution of Great Britain. The page charts the chemist's controversial career, from his beginnings as an apothecary's assistant in Cornwall, through his early scandalous experiments with nitrous oxide, to the electrochemical theories that made him one of the most famous chemists in Europe.

Dawkins, (Clinton) Richard (1941–) English zoologist, born in Kenya, whose book *The Selfish Gene* (1976) popularized the theories of sociobiology (social behavior in humans and animals in the context of evolution). In *The Blind Watchmaker* (1986) he explained the modern theory of evolution.

Dawkins, Richard http://www.world-of-dawkins.com/default.asp

Biographical information about Richard Dawkins, plus quotes, interviews, papers, articles, and excerpts from his books. This is an unofficial site gathering together a whole host of regularly updated information from a wide variety of sources.

day time taken for the Earth to rotate once on its axis. That part of the Earth's surface at any one time facing the Sun experiences day. As the Earth rotates, these parts move to face away from the Sun, receiving no direct sunlight, and experience night.

The **solar day** is the time that the Earth takes to rotate once relative to the Sun. It is divided into 24 hours, and is the basis of our civil day. The **sidereal day** is the time that the Earth takes to rotate once relative to the stars. It is 3 minutes 56 seconds shorter than the solar day, because the Sun's position against the background of stars as seen from Earth changes as the Earth orbits it.

dBASE family of microcomputer programs used for manipulating large quantities of data; also, a related †fourth-generation language. The first version, dBASE II, was published by Ashton-Tate in 1981; it has since become the basis for a recognized standard for database applications, known as xBase.

DC in physics, abbreviation for **direct current** (electricity).

DDT abbreviation for dichloro-diphenyl-trichloroethane; $(ClC_6H_5)_2CHC(HCl_3)$, insecticide discovered in 1939 by Swiss chemist Paul Müller. It is useful in the control of insects that spread malaria, but resistant strains develop. DDT is highly toxic and persists in the environment and in living tissue. Despite this and its subsequent danger to wildlife, it has evaded a worldwide ban because it remains one of the most effective ways of controlling malaria. China and India were the biggest DDT users in 1999.

deactivation in chemistry, process by which the chemical activity of a substance is reduced or completely eliminated. An example of a deactivation process is the reduction in efficiency of an iron catalyst in the presence of oxygen and water during the industrial production of ammonia.

deamination removal of the amino group ($—NH_2$) from an unwanted †amino acid. This is the nitrogen-containing part, and it is converted into ammonia, uric acid, or urea (depending on the type of animal) to be excreted in the urine.

In vertebrates, deamination occurs in the †liver.

death cessation of all life functions, so that the molecules and structures associated with living things become disorganized and indistinguishable from similar molecules found in nonliving things. In medicine, a person is pronounced dead when the brain ceases to control the vital functions, even if breathing and heartbeat are maintained artificially.

medical definition Death used to be pronounced with the permanent cessation of heartbeat, but the advent of life-support equipment has made this point sometimes difficult to determine. For removal of vital organs in transplant surgery, the World Health Organization in 1968 set out that a potential donor should exhibit no brain–body connection, muscular activity, blood pressure, or ability to breathe spontaneously.

religious belief In religious belief, death may be seen as the prelude to rebirth (as in Hinduism and Buddhism); under Islam and Christianity, there is the concept of a day of judgment and consignment to heaven or hell; Judaism concentrates not on an afterlife but on survival through descendants who honor tradition.

death rate number of deaths per 1,000 of the population of an area over the period of a year. Death rate is a factor in demographic transition.

Death rate is linked to a number of social and economic factors such as standard of living, diet, and access to clean water and medical services. The death rate is therefore lower in wealthier countries; for example, in the USA it is 9/1,000; in Nigeria 18/1,000.

De Bary, Heinrich Anton (1831–1888) German botanist and founder of mycology (the study of fungi). He demonstrated that fungi were not spontaneously generated as many scientists believed, and showed that rusts and smuts were the cause and not the result of disease in cereals. He provided evidence that the spores from the potato blight fungus could cause potato blight in previously healthy plants and was the first to use the term †symbiosis in 1879 to describe the close and mutually beneficial relationship between some plant species.

De Beer, Gavin Rylands (1899–1972) English zoologist. He made important contributions to the theory of †embryology and †evolution. He disproved the germ-layer theory and developed the concept of paedomorphism (the retention of juvenile characteristics of ancestors in mature adults).

de Bono, Edward (Francis Charles Publius) (1933–) Maltese-born British medical doctor and psychologist whose concept of lateral thinking, first expounded in *The Use of Lateral Thinking* (1967), involves thinking round a problem rather than tackling it head on. De Bono has worked with business executives from major corporations and advised governments around the world. He was the founder of the International Creative Forum and was awarded the Pioneer Prize in the field of thinking at the International Conference on Thinking at the Massachusetts Institute of Technology (MIT), Massachusetts, in 1994.

de Broglie French physicists; see †Broglie, Louis, and †Broglie, Maurice.

de Broglie hypothesis cornerstone of quantum physics which relates the wave nature of systems to their particle-like characteristics. For a particle with velocity v and mass m, one can associate a de Broglie wave of wavelength l given by the de Broglie equation:

$l = h/mv$

where h is †Planck's constant. It was determined by French physicist Louis de Broglie.

debugging finding and removing errors, or †bugs, from a computer program or system.

debye unit of electric †dipole moment (of a molecule), equal to 3.335 x 10^{-30} coulomb meter.

Debye, Peter Joseph Willem (1884–1966) Dutch-born US physicist. A pioneer of X-ray powder crystallography, he also worked on polar molecules, dipole moments, molecular structure, and polymers. The Debye–Hückel theory, developed with German chemist Erich †Hückel, concerns the ordering of ions in solution. He was awarded the Nobel Prize for Chemistry in 1936 for his work in molecular structures by investigation of dipole moments and the diffraction of X-rays and electrons in gases.

Debye and Strong Electrolytes http://dbhs.wvusd.k12.ca.us/Chem-History/Debye-Strong-Electrolyte.html

Transcript of Peter Debye's and Erich Hüchel's paper On The Theory Of Electrolytes: I. Freezing Point Depression and Related Phenomena. The paper as presented on this Web site is not complete, but it does cover all the essential points of the paper.

Debye–Hückel theory in chemistry, theory explaining the deviations from the ideal case of a conducting liquid in terms of attraction and repulsion between the charged species (ions) in the solution. The theory was developed by Dutch-born US physicist Peter †Debye and German chemist Erich †Hückel.

decagon in geometry, a ten-sided †polygon.

decalin $C_{10}H_{18}$, colorless liquid existing as two stereoisomers (substances that differ only in the way that the atoms of their molecules are oriented in space), *trans* and *cis*, with boiling points of 185°C/365°F and 198°C/388°F respectively. Decalin is prepared by the catalytic hydrogenation of naphthalene at high temperature and pressure and used mainly as a solvent.

decantation in chemistry, removal of the fluid above a settled suspension (the supernatant fluid), leaving the solid in its original container. The term can also be applied to the removal of the upper layer in a mixture of two †immiscible liquids (liquids that do not mix irreversibly).

decay, radioactive see †radioactive decay.

decibel symbol dB, unit of measure used originally to compare sound intensities and subsequently electrical or electronic power outputs; now also used to compare voltages. A whisper has a sound intensity of 20 dB; 140 dB (a jet aircraft taking off nearby) is the threshold of pain.

An increase of 10 dB is equivalent to a tenfold increase in intensity or power. The decibel scale is used for audibility measurements, as one decibel, representing an increase of about 25%, is about the smallest change the human ear can detect.

deciduous describing trees and shrubs that shed their leaves at the end of the growing season or during a dry season to reduce †transpiration (the loss of water by evaporation).

Most deciduous trees belong to the †angiosperms, plants in which the seeds are enclosed within an ovary, and the term "deciduous tree" is sometimes used to mean "angiosperm tree," despite the fact that many angiosperms are evergreen, especially in the tropics, and a few †gymnosperms, plants in which the seeds are exposed, are deciduous (for example, larches). The term **broad-leaved** is now preferred to "deciduous" for this reason.

decimal fraction in mathematics, †fraction in which the denominator is any higher power of 10. Thus $^3/_{10}$, $^{51}/_{100}$, and $^{23}/_{1,000}$ are decimal fractions and are normally expressed as 0.3, 0.51, and 0.023. The use of †decimals greatly simplifies addition and multiplication of fractions, though not all fractions can be expressed exactly as decimal fractions.

The regular use of the decimal point appears to have been introduced in about 1585, but the occasional use of decimal fractions can be traced back as far as the 12th century.

decimal number system or **denary number system**, most commonly used number system, to the base ten. Decimal numbers do not necessarily contain a decimal point; 563, 5.63, and –563 are all decimal numbers. Other systems are mainly used in computing and include the †binary number system, †octal number system, and †hexadecimal number system.

The decimals 0.3, 0.51, and 0.023 can be expressed as the †decimal fractions $^3/_{10}$, $^{51}/_{100}$, and $^{23}/_{1,000}$. They are all †terminating decimals. These †fractions can equally be expressed as the †percentages 30%, 51%, and 2.3%.

Decimal numbers may be thought of as written under column headings based on the number 10. For example:

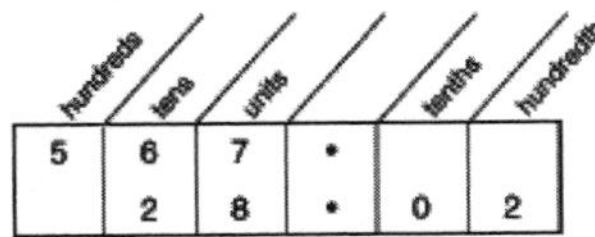

Using the table, 567 stands for 5 hundreds, 6 tens, and 7 units; 28.02 stands for 2 tens, 8 units, and 2

hundredths.

567 has no numbers after the †decimal point, that is 0 **decimal places**. 28.02 has 2 numbers after the decimal point, which is 2 decimal places.

Large decimal numbers may also be expressed in †floating-point notation.

addition and subtraction When adding or subtracting decimals it is important to keep the decimal points underneath each other. For example, to work out 13.56 + 4.08 + 9:

tens	units		tenths	hundredths	
1	3	•	5	6	
	4	•	0	8	+
	9	•	0	0	
2	6	•	6	4	

Using the addition table, the answer is 26.64.

Decimals, Whole Numbers, and Exponents http://www.mathleague.com/help/decwholeexp/decwholeexp.htm

Plain but well laid out introduction to decimals. There are brief lessons covering areas such as adding and multiplying decimals, and square roots, with plenty of examples. The information would be useful as a resource to dip into to gain an overview rather than to read in its entirety.

decimal point dot dividing a decimal number's whole part from its fractional part (the digits to the left of the point are unit digits). It is usually printed on the line but hand written above the line, for example 3·5. Some European countries use a comma to denote the decimal point, for example 3,56.

decision tree in mathematics, alternative name for a †tree diagram.

declination in astronomy, coordinate on the †celestial sphere that corresponds to latitude on the Earth's surface. Declination runs from 0° at the celestial equator to 90° at the north and south celestial poles.

decoder in computing, an electronic circuit used to select one of several possible data pathways. Decoders are, for example, used to direct data to individual memory locations within a computer's immediate access memory.

decomposer in biology, any organism that breaks down dead matter. Decomposers play a vital role in the †ecosystem by freeing important chemical substances, such as nitrogen compounds, locked up in dead organisms or excrement. They feed on some of the released organic matter, but leave the rest to filter back into the soil as dissolved nutrients, or pass in gas form into the atmosphere, for example as nitrogen and carbon dioxide.

The principal decomposers are bacteria and fungi, but earthworms and many other invertebrates are often included in this group. The †nitrogen cycle relies on the actions of decomposers.

decomposition chemical change in which one substance is broken down into two or more simpler substances. In biology, decomposition is the result of the action of decomposer organisms, such as bacteria and fungi. The decomposer organisms obtain food from dead organisms, such as carbon compounds, which are energy-rich. These organisms have an important role in the cycling of carbon compounds as part of the †carbon cycle. The †respiration of the organisms releases †carbon dioxide back into the atmosphere. Other organisms feed on the decomposers and they are part of the decomposer food chain. The decomposition of dead plants and animals allows chemicals to be washed out of the decaying remains into the soil. Many of these are important nutrients that plants can use.

In chemistry, there are three main types of decompositions – thermal, electrolytic, and catalyzed.

Thermal decomposition occurs as a result of heating. For example, copper(II) carbonate decomposes on heating to give copper oxide and carbon dioxide:

$CuCO_3 \rightarrow CuO + CO_2$

Electrolytic decomposition may result when an electrical current is passed through a compound in the molten state or in aqueous solution. For example, molten sodium chloride breaks down into sodium and chlorine:

$2NaCl \rightarrow 2Na + Cl_2$

Catalyzed decomposition describes the process by which decomposition is aided by the presence of a catalyst. For example, hydrogen peroxide decomposes more quickly with the use of manganese(IV) oxide:

$2H_2O_2 \rightarrow 2H_2O + O_2$

decontamination factor in radiological protection, a measure of the effectiveness of a decontamination process. It is the ratio of the original contamination to the remaining radiation after decontamination: 1,000 and above is excellent; 10 and below is poor.

decrepitation in crystallography, unusual feature that accompanies the thermal decomposition of some crystals, such as lead(II) nitrate. When these are heated, they spit and crackle and may jump out of the test tube before they decompose.

Dedekind, (Julius Wilhelm) Richard (1831–1916) German mathematician who made contributions to number theory. In 1872 he introduced the **Dedekind cut** (which divides a line of infinite length representing all †real numbers) to define †irrational numbers in terms of pairs of sequences of rational numbers.

Dedekind's cuts mathematical device by which †irrational numbers can be referred to by means of sets of fractions (rational numbers). It was defined by German mathematician Julius Dedekind.

deep freezing method of preserving food by lowering its temperature to –18°/0°F or below. It stops almost all spoilage processes, although there may be some residual enzyme activity in uncooked vegetables, which is why these are blanched (dipped in hot water to destroy the enzymes) before freezing. Micro-organisms cannot grow or divide while frozen, but most remain alive and can resume activity once defrosted.

Commercial freezing is usually done by one of the following methods: blast, the circulation of air at –40°C/–40°F; contact, in which a refrigerant is circulated through hollow shelves; immersion, for example, fruit in a solution of sugar and glycerol; or cryogenic means, for example, by liquid nitrogen spray. Accelerated freeze-drying (AFD) involves rapid freezing followed by heat drying in a vacuum, for example, prawns for later rehydration. The product does not have to be stored in frozen conditions.

deep-sea trench another term for †ocean trench.

definite integral in mathematics, difference between the values of an †indefinite integral evaluated at two different points. It is written $\int_a^b f(x)dx$ and represents the area between the axis and the graph of the curve (counting areas below the axis as negative).

definition in mathematics, statement that describes a mathematical expression. If the definition is complete all the properties of the expression can be deduced from it.

deflection of light gravitational effect that bends a ray of light. Such an effect was predicted within the general theory of †relativity, although previously considered impossible.

deforestation destruction of forest for timber, fuel, charcoal burning, and clearing for agriculture and extractive industries, such as mining, without planting new trees to replace those lost (reforestation) or working on a cycle that allows the natural forest to regenerate.

Deforestation causes fertile soil to be blown away or washed into rivers, leading to †soil erosion, drought, flooding, and loss of wildlife, and affecting the †biodiversity (biological variety) of ecosystems. It may also increase the carbon dioxide content of the atmosphere and intensify the †greenhouse effect, because there are fewer trees absorbing carbon dioxide from the air for photosynthesis.

Many people are concerned about the rate of deforestation as great damage is being done to the †habitats of plants and animals. Deforestation ultimately leads to famine, and is thought to be partially responsible for the flooding of lowland areas – as for example in Bangladesh – because trees help to slow down water movement.

EO Library: Tropical Deforestation Fact Sheet http://earthobservatory.nasa.gov/Library/Deforestation/
NASA Earth Observatory pages looking at the deforestation of tropical rainforests. Articles examine the causes of deforestation, rates of clearance, and the effects it has. The site also has links to relevant remote-sensing data.

defragmentation program or **disk optimizer**, in computing, a program that rearranges data on a disk so that files are not scattered in many small sections.

degaussing removal of magnetism from a device. The term is usually used in reference to color monitors and other display devices that use a †cathode-ray tube (CRT). External magnetic forces, such as the Earth's natural magnetism or a magnet placed close to the monitor, can cause distorted images and colors. Degaussing works by re-aligning the magnetic fields inside the CRT to compensate for the external magnetism.

In World War II, ships were degaussed to prevent them from detonating magnetic mines. This involved encircling each ship with a conductor, through which an electric current was maintained.

degenerate code in genetics, the triplet code is said to be degenerate because more than one nucleotide triplet (†codon) codes for the same amino acid.

degeneration in biology, a change in the structure or chemical composition of a tissue or organ that interferes with its normal functioning. Examples of degeneration include fatty degeneration, fibroid degeneration (cirrhosis), and calcareous degeneration, all of which are part of natural changes that occur in old age.

The causes of degeneration are often unknown. Heredity often has a role in the degeneration of organs;

for example, fibroid changes in the kidney can be seen in successive generations. Defective nutrition and continued stress on particular organs can cause degenerative changes. Alcoholism can result in cirrhosis of the liver and tuberculosis causes degeneration of the lungs.

de Gennes, Pierre-Gilles (1932–) French physicist who was awarded the Nobel Prize for Physics in 1991 for his work on disordered systems including polymers and liquid crystals, and for the development of mathematical methods for studying the behavior of molecules in a liquid on the verge of solidifying. He showed how mathematical models, developed for studying simpler systems, are applicable to such complicated systems.

degradation in geography, the lowering and flattening of land through †erosion, especially by †rivers. Degradation is also used to describe a reduction in the quality of usefulness of environmental resources such as vegetation or soil.

degree symbol °, in mathematics, a unit of measurement of an angle or arc. A circle or complete rotation is divided into 360°. A degree may be subdivided into 60 minutes (symbol '), and each minute may be subdivided in turn into 60 seconds (symbol ").

Temperature is also measured in degrees, which are divided on a decimal scale. See also †Celsius, and †Fahrenheit.

A degree of latitude is the length along a meridian such that the difference between its north and south ends subtend an angle of 1° at the center of the Earth. A degree of longitude is the length between two meridians making an angle of 1° at the center of the Earth.

degree of freedom in chemistry, in relation to the †phase rule, the number of variables that must be made constant in order to define the state of a chemical system, for example temperature, pressure, and concentration.

Dehmelt, Hans G (1923–) German-born US physicist who was awarded the Nobel Prize for Physics in 1989 for his contribution to the development and application of the ion trap technique, used to store single atoms long enough to make accurate spectroscopic measurement on them. He shared the award with Norman Ramsey and Wolfgang †Paul.

dehydration in chemistry, the removal of water from a substance to give a product with a new chemical formula; it is not the same as drying.

There are two types of dehydration. For substances such as hydrated copper sulfate ($CuSO_4.5H_2O$) that contain †water of crystallization, dehydration means removing this water to leave the anhydrous substance. This may be achieved by heating, and is reversible.

Some substances, such as ethanol, contain the elements of water (hydrogen and oxygen) joined in a different form. **Dehydrating agents** such as concentrated sulfuric acid will remove these elements in the ratio 2:1.

dehydrogenation in chemistry, any process by which the amount of hydrogen in a molecule is decreased to produce the formation of unsaturated bonds, for example: butane → butene + hydrogen ($CH_3CH_2CH_2CH_3 \rightarrow CH_3CH_2CH = CH_2 + H_2$).

Deimos one of the two moons of the planet Mars. It is irregularly shaped, 15 x 12 x 11 km/9 x 7.5 x 7 mi, orbits at a height of 24,000 km/15,000 mi every 1.26 days, and is not as heavily cratered as Mars's other moon, Phobos. Deimos was discovered in 1877 by US astronomer Asaph Hall, and is thought to be an asteroid captured by Mars's gravity.

NASA's Viking 2 probe visited Deimos in 1977, flying within 29 km/18 mi of the Moon's surface and taking detailed photographs of its pitted and boulder-strewn landscape.

Delambre, Jean-Baptiste Joseph (1749–1822) French mathematician and astronomer. With Pierre Méchain (1744–1804), he was appointed by the French government to measure the arc of the meridian between Barcelona, Spain, and Dunkirk, France, in order to establish the value of the meter.

de la Rue, Warren (1815–1889) British astronomer and instrument maker. He was a pioneer in the field of celestial photography; besides inventing the first photoheliographic telescope, he took the first photograph of a solar eclipse in 1860 and used it to prove that the prominences observed during an eclipse are of solar rather than lunar origin.

Delbrück, Max (1906–1981) German-born US biologist who was awarded the Nobel Prize for Physiology or Medicine in 1969 with Salvador †Luria and Alfred †Hershey for their work on the replication mechanism and genetic structure of viruses. He pioneered techniques in molecular biology, studying genetic changes occurring when viruses invade bacteria.

deliquescence phenomenon of a substance absorbing so much moisture from the air that it ultimately dissolves in it to form a solution.

Deliquescent substances make very good drying agents and are used in the bottom chambers of †desiccators. Calcium chloride ($CaCl_2$) is one of the commonest.

Delphinus small but fairly conspicuous constellation on the celestial equator (see †celestial sphere), represented as a dolphin.

delta river sediments deposited when a river flows into a standing body of water with no strong currents, such as a lake, lagoon, sea, or ocean. A delta is the result of fluvial and marine processes. Deposition is enhanced when water is saline because salty water causes small clay particles to adhere together. Other factors influencing deposition include the type of sediment, local geology, sea-level changes, plant growth, and human impact. Some examples of large deltas are those of the Mississippi, Ganges and Brahmaputra, Rhône, Po, Danube, and Nile rivers. The shape of the Nile delta is like the Greek letter *delta* or D, and gave rise to the name. The **arcuate (arc-shaped) delta** of the Nile, found in areas where †longshore drift keeps the seaward edge of the delta trimmed and relatively smooth, is only one form of delta. Others include **birdfoot (projecting) deltas**, where the river brings down enormous amounts of fine silt, as at the Mississippi delta; and **tidal (linear) deltas**, like that of the Mekong, China, where most of the material is swept to one side by sea currents. **Cuspate deltas** are pointed, and are shaped by regular opposing, gentle water movement, as seen at the Ebro and Tiber deltas in Italy.

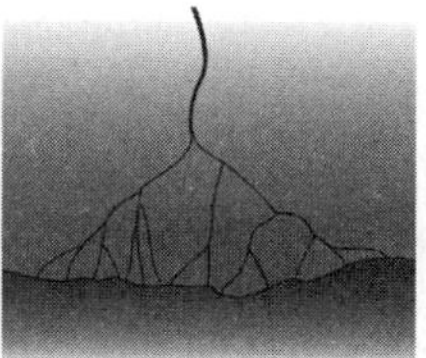

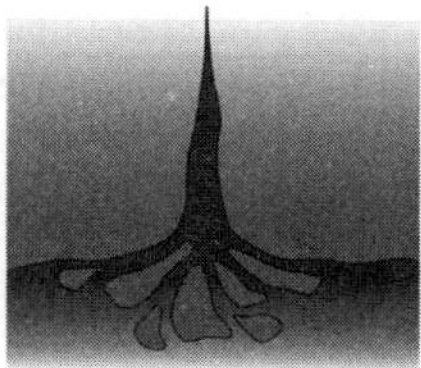

delta An arcuate delta and a bird's foot delta. The Mississippi delta is an example of a bird's foot delta and the Ganges delta is an example of an arcuate delta. Helicon

delta symbols d, D, as a capital, D, in mathematics, the term means "difference"; D*x* represents the difference between consecutive *x* values according to context. German mathematician Leopold Kronecker's delta is a symbol used in the evaluations of †determinants (in matrix theory), to the effect that d(*i*, *j*) = 1 if *i* = *j*, otherwise it equals zero. (It thus measures whether *i* and *j* are different.) The d used by Gottfried Leibniz in his notation for differential calculus, as in d*y*/d*x*, was based on an intended association with the delta.

As a lower-case (small) letter, another delta (d) commonly represents a partial †derivative in †partial differentiation.

demagnetization removal of magnetic properties from a material by hitting it, heating it, or placing it in a coil carrying an alternating current that is gradually reduced to zero. All methods result in the ordered domains of the magnet becoming disordered and randomly oriented.

demodulation in radio, the technique of separating a transmitted audio frequency signal from its modulated radio carrier wave. At the transmitter, the audio frequency signal (representing speech or music, for example) may be made to modulate the amplitude (AM broadcasting) or frequency (FM broadcasting) of a continuously transmitted radio-frequency carrier wave. At the receiver, the signal from the aerial is demodulated to extract the required speech or sound component. In early radio systems, this process was called detection. See †modulation.

de Moivre, Abraham French mathematician; see †Moivre, Abraham de.

de Moivre's equation statement that for integers n $(\cos z + i \sin z)^n = \cos nz + i \sin z$ where i = Ö–1. The equation was determined by French mathematician Abraham de †Moivre.

De Morgan, Augustus (1806–1871) Indian-born British mathematician. He initiated and developed a theory of relations in logic. He devised a symbolism that could express such notions as the contradictory, the converse, and the transitivity of a relation, as well as the union of two relations.

De Morgan, Augustus http://www-history.mcs.st-andrews.ac.uk/history/Mathematicians/De_Morgan.html
Biography and portrait of the British 19th-century mathematician Augustus de Morgan. De Morgan's works are also described. In particular, his four color theorem is examined in some depth.

denaturation in biology, irreversible changes occurring in the structure of proteins such as enzymes, usually caused by changes in pH or temperature, by radiation, or by chemical treatments. An example is the heating of egg albumen resulting in solid egg white.

The enzymes associated with digestion and metabolism become inactive under abnormal conditions. Heat will damage their complex structure so that the usual interactions between enzyme and substrate can no longer occur.

dendrite part of a †nerve cell or neuron. The dendrites are slender filaments projecting from the cell body. They receive incoming messages from many other nerve cells and pass them on to the cell body. If the combined effect of these messages is strong enough, the cell body will send an electrical impulse along the axon (the threadlike extension of a nerve cell). The tip of the axon passes its message to the dendrites of other nerve cells.

dendrochronology or **tree-ring dating**, analysis of the †annual rings of trees to date past events by determining the age of timber. Since annual rings are formed by variations in the water-conducting cells produced by the plant during different seasons of the year, they also provide a means of establishing past climatic conditions in a given area.

Samples of wood are obtained by driving a narrow metal tube into a tree to remove a core extending from the bark to the center. Samples taken from timbers at an archaeological site can be compared with a master core on file for that region or by taking cores from old living trees; the year when they were felled can be determined by locating the point where the rings of the two samples correspond and counting back from the present.

Moisture levels will affect growth, the annual rings being thin in dry years, thick in moist ones, although in Europe ring growth is most affected by temperature change and insect defoliation.

In North America, studies are now extremely extensive, covering many wood types, including sequoia, juniper, and sagebrush. Sequences of tree rings extending back over 8,000 years have been obtained for the southwest USA and northern Mexico by using cores from the bristle-cone pine *Pinus aristata* of the White Mountains, California, which can live for over 4,000 years in that region. The dryness of the southwest USA has preserved wood in its archaeological sites. In wet temperate regions, wood is usually absorbed by soil acidity so that this dating technique cannot be used.

Deneb or **Alpha Cygni**, brightest star in the constellation †Cygnus, and the 20th-brightest star in the night sky. It is one of the greatest supergiant stars known, with a true luminosity of about 60,000 times that of the Sun. Deneb is 1,800 light years from the Sun.

denier unit used in measuring the fineness of yarns, equal to the mass in grams of 9,000 meters of yarn. Thus 9,000 meters of 15 denier nylon, used in nylon stockings, weighs 15 g/0.5 oz, and in this case the thickness of thread would be 0.00425 mm/0.0017 in. The term is derived from the French silk industry; the *denier* was an old French silver coin.

denitrification process occurring naturally in soil, where bacteria break down †nitrates to give nitrogen gas, which returns to the atmosphere.

denominator in mathematics, the bottom number of a fraction, so called because it names the family of the fraction. The top number, or numerator, specifies how many unit fractions are to be taken.

density measure of the compactness of a substance; it is equal to its mass per unit volume and is measured, for example, in kg per cubic meter or lb per cubic foot. Density is a †scalar quantity. The average density D of a mass m occupying a volume V is given by the formula: $D = m/V$. †Relative density is the ratio of the density of a substance to that of water at 4°C/39.2°F.

In photography, density refers to the degree of opacity of a negative; in electricity, current density is the amount of current passing through a cross-sectional area of a conductor (usually given in amperes per square in or per square cm).

Density, Melting Points, and Boiling Points http://www.ncpublicschools.org/curriculum/science/Chemistry/density.htm
Quick reference guide to the density, melting points, and boiling points of the most common chemical compounds.

density wave in astrophysics, concept proposed to account for the existence of spiral arms in †galaxies. In the density wave theory, stars in a spiral galaxy move in elliptical orbits in such a way that they crowd together in waves of temporarily enhanced density that appear as spiral arms. The idea was first proposed by Swedish astronomer Bertil †Lindblad in the 1920s and developed by US astronomers Chia Lin and Frank Shu in the 1960s.

dental formula way of showing the number of teeth in an animal's mouth. The dental formula consists of eight numbers separated by a line into two rows. The four above the line represent the teeth on one side of the upper jaw, starting at the front. If this reads 2 1 2 3 (as for humans) it means two incisors, one canine, two premolars, and three molars (see †tooth). The numbers below the line represent the lower jaw. The total number of teeth can be calculated by adding up all the numbers and multiplying by two.

dentition type and number of teeth in a species. Different kinds of teeth have different functions; a grass-eating animal will have large molars for grinding its food, whereas a meat-eater will need powerful canines for catching and killing its prey. The teeth that are less useful to an animal's lifestyle may be reduced in size or missing altogether. An animal's dentition is represented diagramatically by a †dental formula.

Young children have **deciduous dentition**, popularly known as "milk teeth," the first ones erupting at about six months of age. **Mixed dentition** is present from the ages of about six (when the first milk teeth are shed) to about 12. **Permanent dentition** (up to 32 teeth) is usually complete by the mid-teens, although the third molars (wisdom teeth) may not appear until around the age of 21.

denudation natural loss of soil and rock debris, blown away by wind or washed away by running water, laying bare the rock below. Over millions of years, denudation causes a general lowering of the landscape.

deoxyribonucleic acid full name of †DNA.

depolarizer oxidizing agent used in dry-cell batteries that converts hydrogen released at the negative electrode into water. This prevents the build-up of gas, which would otherwise impair the efficiency of the battery. †Manganese(IV) oxide is often used for this purpose.

depopulation the decline of the population of a given area, usually caused by people moving to other areas for economic reasons, rather than an increase in the †death rate or decrease in birth rate.

depression or **cyclone** or **low**, in meteorology, a region of relatively low atmospheric pressure. In mid-latitudes a depression forms as warm, moist air from the tropics mixes with cold, dry polar air, producing warm and cold boundaries (†fronts) and unstable weather – low cloud and drizzle, showers, or fierce storms. The warm air, being less dense, rises above the cold air to produce the area of low pressure on the ground. Air spirals in towards the center of the depression in an anticlockwise direction in the northern hemisphere, clockwise in the southern hemisphere, generating winds up to gale force. Depressions tend to travel eastwards and can remain active for several days.

depressor nerve nerve that, when stimulated induces reflex vasodilation and thus slows the heartbeat (resulting in a fall in blood pressure).

derivative or **differential coefficient**, in mathematics, the limit of the gradient of a chord linking two points on a curve as the distance between the points tends to zero; for a function of a single variable, $y = f(x)$, it is denoted by $f'(x)$, $Df(x)$, or dy/dx, and is equal to the gradient of the curve.

desalination removal of salt, usually from sea water, to produce fresh water for irrigation or drinking. Distillation has usually been the method adopted, but in the 1970s a cheaper process, using certain polymer materials that filter the molecules of salt from the water by reverse osmosis, was developed.

Desalination plants have been built along the shores of Middle Eastern countries where fresh water is in short supply.

Descartes, René (1596–1650) French philosopher and mathematician. He believed that commonly accepted knowledge was doubtful because of the subjective nature of the senses, and attempted to rebuild human knowledge using as his foundation the dictum *cogito ergo sum* ("I think, therefore I am"). He also believed that the entire material universe could be explained in terms of mathematical physics, and founded coordinate geometry as a way of defining and manipulating geometrical shapes by means of algebraic expressions. †Cartesian coordinates, the means by which points are represented in this system, are named after him. Descartes also established the science of optics, and helped to shape contemporary theories of astronomy and animal behavior.

Descartes, René http://www.newadvent.org/cathen/04744b.htm

Extensive treatment of the life and philosophical, scientific, and mathematical achievements of Renatus Cartesius. The restless travels of the young savant are described before a detailed discussion of his contribution to learning.

Discourse on Method http://www.4literature.net/Rene_Descartes/Discourse_on_the_Method/

Complete text of Descartes' work, which includes the phrase "cogito ergo sum" ("I think, therefore I am").

descriptive geometry branch of mathematics in which three-dimensional objects are represented as two-dimensional (plane) figures, using any of many types of projection.

desert arid area with sparse vegetation (or, in rare cases, almost no vegetation). Soils are poor, and many deserts include areas of shifting sands. Deserts can be either hot or cold. Almost 33% of the Earth's land surface is desert, and this proportion is increasing. Arid land is defined as receiving less than 250 mm/9.75 in rain per year.

The **tropical desert** belts of latitudes from 5° to 30° are caused by the descent of air that is heated over the warm land and therefore has lost its moisture. Other natural desert types are the **continental deserts**, such as the Gobi, that are too far from the sea to receive any moisture; **rain-shadow deserts**, such as California's Death Valley, that lie in the lee of mountain ranges, where the ascending air drops its rain only on the windward slopes; and **coastal deserts**, such as the Namib, where cold ocean currents cause local dry air masses to descend. Desert surfaces are usually rocky or gravelly, with only a small proportion being covered with sand (about 3%). Deserts can be created by changes in climate, or by the human-aided process of desertification.

Characteristics common to all deserts include irregular rainfall of less than 250 mm/9.75 in per year, very high evaporation rates of often 20 times the annual precipitation, and low relative humidity and cloud cover. Temperatures are more variable; tropical deserts have a big diurnal temperature range and very high daytime temperatures (58°C/136.4°F has been recorded at Azizia in Libya), whereas mid-latitude deserts have a wide annual range and much lower winter temperatures (in the Mongolian desert the mean temperature is below freezing point for half the year).

Desert soils are infertile, lacking in †humus and generally gray or red in color. The few plants capable of surviving such conditions are widely spaced, scrubby and often thorny. Long-rooted plants (phreatophytes) such as the date palm and musquite commonly grow along dry stream channels. Salt-loving plants (†halophytes) such as saltbushes grow in areas of highly saline soils and near the edges of †playas (dry saline lakes). †Xerophytes are drought-resistant and survive by remaining leafless during the dry season or by reducing water losses with small waxy leaves. They frequently have shallow and widely branching root systems and store water during the wet season (for example, succulents and cacti with pulpy stems).

desertification spread of deserts by changes in climate, or by human-aided processes. Desertification can sometimes be reversed by special planting (marram grass, trees) and by the use of water-absorbent plastic grains, which, added to the soil, enable crops to be grown. About 30% of land worldwide is affected by desertification (1998), including 1 million hectares in Africa and 1.4 million hectares in Asia. The most rapid desertification is in developed countries such as the USA, Australia, and Spain.

Natural causes of desertification include decreased rainfall, increased temperatures, lowering of the †water table, and †soil erosion.

The human-aided processes leading to desertification include overgrazing, destruction of forest belts, and exhaustion of the soil by intensive cultivation without restoration of fertility – all of which may be prompted by the pressures of an expanding population or by concentration in land ownership. About 135 million people are directly affected by desertification, mainly in Africa, the Indian subcontinent, and South America. The Sahel region in Africa is one example.

desiccator airtight vessel, traditionally made of glass, in which materials may be stored either to dry them or to prevent them, once dried, from reabsorbing moisture.

The base of the desiccator is a chamber in which is placed a substance with a strong affinity for water (such as calcium chloride or silica gel), which removes water vapor from the desiccator atmosphere and from substances placed in it.

de Sitter, Willem Dutch astronomer; see †Sitter, Willem de.

desktop publishing **DTP**, use of microcomputers for small-scale typesetting and page makeup. DTP systems are capable of producing camera-ready pages (pages ready for photographing and printing), made up of text and graphics, with text set in different typefaces and sizes. The page can be previewed on the screen before final printing on a laser printer.

A DTP program is able to import text and graphics from other packages; run text as columns, over pages, and around artwork and other insertions; enable a wide range of fonts; and allow accurate positioning of all elements required to make a page.

DTP Domain http://www.dtpdomain.com/

Useful collection of desktop publishing resources on the Internet, which lists product Web sites for books and software, as well as pages offering tutorials, fonts, clipart, and images.

Internet Resources on Desktop Publishing http://desktoppub.about.com/?once=true&

A comprehensive site covering everything you need to know about desktop publishing. Includes news, software tips, and reviews of a number of DTP packages, as well as a number of links to sites on related topics, including free clip art and fonts.

Desmarest, Nicolas (1725–1815) French naturalist. He became a champion of volcanist geology, countering the widely held belief that all rocks were †sedimentary. He wrote extensively for the *Encyclopédie*. Studying the large basalt deposits of central France, Desmarest traced their origin to ancient volcanic activity in the Auvergne region. In 1768 he produced a detailed study of the geology and eruptive history of the volcanoes responsible. However, he did not believe that all rocks had †igneous origin, and emphasized the critical role of water in the shaping of the Earth's history.

Désormes, Charles Bernard (1777–1862) French physicist and chemist. He determined the ratio of the specific heats of gases in 1819. He did this and almost all his scientific work in collaboration with his son-in-law Nicolas Clément (1779–1841). Clément and Desormes correctly determined the composition of carbon disulphide (CS_2) and carbon monoxide (CO) in 1801–02. In 1806 they elucidated all the chemical reactions that take place during the production of sulfuric acid by the lead chamber method, as used in industrial chemistry. In 1813 they made a study of iodine and its compounds.

detergent surface-active cleansing agent. The common detergents are made from †fats (hydrocarbons) and sulfuric acid, and their long-chain molecules have a type of structure similar to that of †soap molecules: a salt group at one end attached to a long hydrocarbon "tail." They have the advantage over soap in that they do not produce scum by forming insoluble salts with the calcium and magnesium ions present in hard water.

To remove dirt, which is generally attached to materials by means of oil or grease, the hydrocarbon "tails" (soluble in oil or grease) penetrate the oil or grease drops, while the "heads" (soluble in water but insoluble in grease) remain in the water and, being salts, become ionized. Consequently the oil drops become negatively charged and tend to repel one another; thus they remain in suspension and are washed away with the dirt.

Detergents were first developed from coal tar in Germany during World War I, and synthetic organic detergents were increasingly used after World War II.

Domestic powder detergents for use in hot water have alkyl benzene as their main base, and may also include bleaches and fluorescers as whiteners, perborates to free stain-removing oxygen, and water softeners. Environment-friendly detergents contain no phosphates or bleaches. Liquid detergents for washing dishes are based on epoxyethane (ethylene oxide). Cold-water detergents consist of a mixture of various alcohols, plus an ingredient for breaking down the surface tension of the water, so enabling the liquid to penetrate fibers and remove the dirt. When these surface-active agents (surfactants) escape the normal processing of sewage, they cause troublesome foam in rivers; phosphates in some detergents can also cause the excessive enrichment (†eutrophication) of rivers and lakes.

determinant in mathematics, an array of elements written as a square, and denoted by two vertical lines enclosing the array. For a 2 x 2 matrix, the determinant is given by the difference between the products of the diagonal terms. Determinants are used to solve sets of †simultaneous equations by matrix methods.

When applied to transformational geometry, the determinant of a 2 x 2 matrix signifies the ratio of the area of the transformed shape to the original and its sign (plus or minus) denotes whether the image is direct (the same way round) or indirect (a mirror image).

For example

$$\text{if matrix } A = \begin{bmatrix} a & b \\ c & d \end{bmatrix} \text{ then}$$

$$\text{determinant } |A| = ad - bc$$

determinates "known" values, as opposed to indeterminates, "unknown" values.

detritus in biology, the organic debris produced during the †decomposition of animals and plants.

deuterium or **heavy hydrogen**, naturally occurring heavy isotope of hydrogen, mass number 2 (one proton and one neutron), discovered by US chemist Harold Urey in 1932. It is sometimes given the symbol D. In nature, about one in every 6,500 hydrogen atoms is deuterium. Combined with oxygen, it produces "heavy water" (D_2O), used in the nuclear industry.

deuteron nucleus of an atom of deuterium (heavy hydrogen). It consists of one proton and one neutron, and is used in the bombardment of chemical elements to synthesize other elements.

De Vaucouleurs, Gerard Henri (1918–1995) French-born US astronomer. He carried out important research into extragalactic †nebulae. In 1956 he suggested that there is a pattern in the location of nebulae, clusters of stars formerly thought to be randomly scattered. The material he published in 1956 seemed to indicate that a local supercluster of nebulae exists, which includes our own †Milky Way stellar system. He suggested a model in which the great Virgo cluster might be "a dominant congregation not too far from its central region." As evidence for its existence, De Vaucouleurs pointed out the similarity in position and extent of a broad maximum of cosmic radio noise.

development in biology, the process whereby a living thing transforms itself from a single cell into a vastly complicated multicellular organism, with structures, such as limbs, and functions, such as respiration, all able to work correctly in relation to each other. Most of the details of this process remain unknown, although some of the central features are becoming understood.

Apart from the sex cells (†gametes), each cell within an organism contains exactly the same genetic code. Whether a cell develops into a liver cell or a brain cell depends therefore not on which †genes it contains, but on which genes are allowed to be expressed. The development of forms and patterns within an organism, and the production of different, highly specialized cells, is a problem of control, with genes being turned on and off according to the stage of development reached by the organism.

developmental psychology study of development of cognition and behavior from birth to adulthood.

device driver in computing, small piece of software required to tell the operating system how to interact with a particular input or output device or peripheral.

Much work has been done to standardize devices and their interfaces to eliminate the need for individual device drivers. Peripherals such as CD-ROM drives, for example, work with a single standard device driver (in Microsoft Windows, MSCDEX.EXE). Other devices, such as modems and printers, still need an individual driver tailored to work with that specific model.

devil wind minor form of †tornado, usually occurring in fine weather; formed from rising thermals of warm air (as is a †cyclone). A fire creates a similar updraft.

A **fire devil** or **firestorm** may occur in oil-refinery fires, or in the firebombings of cities, for example Dresden, Germany, in World War II.

Devonian period period of geological time 408–360 million years ago, the fourth period of the Paleozoic era. Many desert sandstones from North America and Europe date from this time. The first land plants flourished in the Devonian period, corals were abundant in the seas, amphibians evolved from air-breathing fish, and insects developed on land.

The name comes from the county of Devon in southwest England, where Devonian rocks were first studied.

dew precipitation in the form of moisture that collects on plants and on the ground. It forms after the temperature of the ground has fallen below the †dew point of the air in contact with it. As the temperature falls during the night, the air and its water vapor become chilled, and condensation takes place on the cooled surfaces.

dew point temperature at which the air becomes saturated with water vapor. At temperatures below the dew point, the water vapor condenses out of the air as droplets. If the droplets are large they become deposited on plants and the ground as dew; if small they remain in suspension in the air and form mist or fog.

dewpond drinking pond for farm animals on arid hilltops such as chalk downs. In the UK, dewponds were excavated in the 19th century and lined with mud and clay. It is uncertain where the water comes from but it may be partly rain, partly sea mist, and only a small part dew.

diabase alternative name for †dolerite (a form of basalt that contains very little silica), especially dolerite that has metamorphosed.

diabetes disease that can be caused by reduced production of the †hormone †insulin, or a reduced response of the liver, muscle, and fat cells to insulin. This affects the body's ability to use and regulate sugars effectively. Diabetes mellitus is a disorder of the islets of Langerhans in the †pancreas that prevents the production of insulin. Treatment is by strict dietary control and oral or injected insulin, depending on the type of diabetes.

Insulin increases †glucose uptake from the blood into most body cells, except the brain and liver. In the liver, its effect is to reduce the production of glucose. Insulin works with other hormones to keep the concentration of glucose in the blood stable. A person with diabetes does not have this close regulation and may find that his or her glucose levels rise dangerously after a meal. Normally urine does not contain glucose, but when the blood glucose levels rise in a diabetic person, it appears in the urine. A person with diabetes may also be likely to have glucose levels that are too low at times.

There are two forms of diabetes: type 1, or insulin-dependent diabetes, which usually begins in childhood (early onset) and is an autoimmune condition; and type 2, or non-insulin-dependent diabetes, which occurs in later life (late onset). Without treatment, the patient may go blind, suffer from ulcers, lapse into diabetic coma, and die. Early-onset diabetes tends to be more severe than that developing in later years. There is also a strong link between late-onset diabetes and being seriously overweight (obese). Over 80% of sufferers are obese, though the exact link is not clear.

"Recent Advances: Diabetes" http://www.bmj.com/cgi/content/full/316/7139/1221

Part of a collection of articles maintained by the British Medical Journal, this page features the full text of an 18 April 1998 clinical review regarding the alarming rise in the number of patients with diabetes around the world. The text includes diagrams, and there is a list of references. In addition to the clinical review, you will find a list of related articles from the extensive PubMed collection.

diacaustic †caustic curve formed by refraction.

diagenesis in geology, the physical, chemical, and biological processes by which a sediment becomes a †sedimentary rock. The main processes involved include compaction of the grains, and the cementing of the grains together by the growth of new minerals deposited by percolating groundwater. As a whole, diagenesis is actually a poorly understood process.

diagmagnetism form of magnetism induced in one substance by the magnetic field of another. Its basic cause lies in the shift of the orbital motion of the †electrons of a substance resulting from the external magnetic field of the other substance. It occurs in all materials.

diagonal matrix in mathematics, square †matrix whose only non-zero elements lie on the main diagonal.

diamagnetic material material that is weakly repelled by a magnet. All substances are diamagnetic but the behavior is often masked by stronger forms of magnetism such as †paramagnetic or †ferromagnetic behavior.

Diamagnetism is caused by changes in the orbits of electrons in the substance induced by the applied field. Diamagnetic materials have a small negative magnetic †susceptibility.

diameter straight line joining two points on the circumference of a circle that passes through the center of that circle. It divides a circle into two equal halves.

diamond generally colorless, transparent mineral, an †allotrope of carbon. It is regarded as a precious gemstone, and is the hardest substance known (10 on the †Mohs scale). Industrial diamonds, which may be natural or synthetic, are used for cutting, grinding, and polishing.

Diamond crystallizes in the cubic system as octahedral crystals, some with curved faces and striations. The high refractive index of 2.42 and the high dispersion of light, or "fire," account for the spectral displays seen in polished diamonds.

history Diamonds were known before 3000 BC and until their discovery in Brazil in 1725, India was the principal source of supply. Present sources are Australia, the Democratic Republic of Congo (formerly Zaire), Botswana, Russia (Yakut), South Africa, Namibia, and Angola; the first two produce large volumes of industrial diamonds. Today, about 80% of the world's rough gem diamonds are sold through the De Beers Central Selling Organization in London.

sources Diamonds may be found as alluvial diamonds on or close to the Earth's surface in riverbeds or dried watercourses; on the sea bottom (off southwest Africa); or, more commonly, in diamond-bearing volcanic pipes composed of "blue ground," †kimberlite or lamproite, where the original matrix has penetrated the Earth's crust from great depths. They are sorted from the residue of crushed ground by X-ray and other recovery methods.

varieties There are four chief varieties of diamond: well-crystallized transparent stones, colorless or only slightly tinted, valued as gems; **boart**, poorly crystallized or inferior diamonds; **balas**, an industrial variety, extremely hard and tough; and **carbonado**, or industrial diamond, also called black diamond or carbon, which is opaque, black or gray, and very tough. Industrial diamonds are also produced synthetically from graphite. Some synthetic diamonds conduct heat 50% more efficiently than natural diamonds and are five times greater in strength. This is a great advantage in their use to disperse heat in electronic and telecommunication devices and in the production of laser components.

practical uses Because diamonds act as perfectly transparent windows and do not absorb infrared radiation, they were used aboard NASA space probes to Venus in 1978. The tungsten carbide tools used in steel mills are cut with industrial diamond tools.

cutting Rough diamonds are often dull or greasy before being polished; around 50% are considered "cuttable" (all or part of the diamond may be set into jewelry). Gem diamonds are valued by weight (†carat), cut (highlighting the stone's optical properties), color, and clarity (on a scale from internally flawless to having a large inclusion clearly visible to the naked eye). They are sawn and polished using a mixture of oil and diamond powder. The two most popular cuts are the brilliant, for thicker stones, and the marquise, for shallower ones.

Noted rough diamonds include the Cullinan, or Star of Africa (3,106 carats, over 500 g/17.5 oz before cutting, South Africa, 1905); Excelsior (995.2 carats, South Africa, 1893); and Star of Sierra Leone (968.9 carats, Yengema, 1972).

experiments Using a device known as a †diamond-anvil cell, a moderate force applied to the small tips of two opposing diamonds can be used to attain extreme pressures of millions of atmospheres or more, allowing scientists to subject small amounts of material to conditions that exist deep within planet interiors.

In 1999 US scientists turned a diamond into a metal using a very powerful laser to compress it.

diamond-anvil cell device composed of two opposing cone-shaped diamonds that when squeezed together by a lever-arm exert extreme pressures. The high pressures result from applying force to the small areas of the opposing diamond faces (see †pressure). The device is used to determine the properties of materials at pressures corresponding to those of planetary interiors. One discovery made with the diamond-anvil cell is $MgSiO_3$-perovskite, thought to be the predominant mineral of Earth's lower †mantle.

diamorphine technical term for heroin.

diapause period of suspended development that occurs in some species of insects and other invertebrates, characterized by greatly reduced metabolism. Periods of diapause are often timed to coincide with the winter months, and improve the animal's chances of surviving adverse conditions.

diaphragm in zoology, thin membrane found in the mammalian ear onto which †sound waves impinge, being converted into electrical impulses through movement of fluid in the cochlea, which is connected to the auditory nerve.

diaphragm in telecommunications, thin membrane found in telephone receivers and microphones onto which †sound waves impinge, being converted into electrical impulses by a device that takes its input from the diaphragm movements.

diaphragm in mammals, a thin muscular sheet separating the thorax from the abdomen. It is attached by way of the ribs at either side and the breastbone and backbone, and a central tendon. Arching upwards against the heart and lungs, the diaphragm is important in the mechanics of breathing. It contracts at each inhalation, moving downwards to increase the volume of the chest cavity, and relaxes at exhalation.

diapirism geological process in which a particularly light rock, such as rock salt, punches upwards through the heavier layers above. The resulting structure is called a salt dome, and oil is often trapped in the curled-up rocks at each side.

diastereoisomer in chemistry, an optically active compound that has two or more forms that have the same elemental composition and atomic weight but differ in the spatial arrangement of their molecular structures (see also †enantiomers). The molecular structures of the different isomers cannot be superimposed upon their mirror images. The physical and chemical properties of diastereoisomers are usually different.

diastole in biology, the relaxation of a hollow organ. In particular, the term is used to indicate the resting period between beats of the heart when blood is flowing into it.

diastolic pressure in medicine, measurement due to the pressure of blood against the arterial wall during diastole (relaxation of the heart). It is the lowest †blood pressure during the cardiac cycle. The average diastolic pressure in healthy young adults is about 80 mmHg. The variation of diastolic pressure due to changes in body position and mood is greater than that of systolic pressure. Diastolic pressure is also a more accurate predictor of hypertension (high blood pressure).

diatomic molecule molecule composed of two atoms joined together. In the case of an element such as oxygen (O_2), the atoms are identical.

diazomethane CH_2N_2, highly reactive, poisonous yellow gas that is also explosive. It is usually prepared as a solution in ether by decomposition of nitroso-methyl-urethane with potassium hydroxide and mainly used as a reagent, to introduce methyl groups into organic molecules.

dicarboxylic acid organic compound that contains two carboxyl groups, for example succinic acid.

dichloro-diphenyl-trichloroethane full name of the insecticide †DDT.

dichotomous key in biology, a practical method used to identify species. The key is written so that identification is done in steps. At each step, features of the organism are used to identify which one of two routes through the rest of the key is appropriate to the organism being identified. Dichotomous keys assume a good knowledge of the subject under investigation.

A dichotomous key is best illustrated with an example. Here is the route through a key that was used to identify field maple and sycamore trees: (1) If it is a tree then go to (2). If it is not a tree then go to (5). (2) If leaves are in opposite pairs then go to (3). If leaves are not in opposite pairs then go to (5). (3) If leaves are lobed and arranged like fingers on a hand then go to (4). If leaves are not so arranged then go to (5). (4) If leaves have three lobes then it is *Acer campestre* (field maple). If leaves have five lobes then it is *Acer pseudoplatanus* (sycamore). (5) It is not a field maple or sycamore.

Dicke, Robert Henry (1916– 1997) US physicist. In 1964 he proposed a version of the †Big Bang theory known as the "hot Big Bang": he suggested that the present expansion of the universe had been preceded by a collapse in which high temperatures had been generated. When Arno †Penzias and Robert †Wilson announced they had detected an unexpected and relatively high level of radiation at a wavelength of 7 cm/2.8 in, with a temperature of about 3.5 K (–270°C/–453°F), Dicke proposed that this was cosmic black-body radiation from the hot Big Bang.

dicotyledon major subdivision of the †angiosperms, containing the great majority of flowering plants. Dicotyledons are characterized by the presence of two seed leaves, or †cotyledons, in the embryo, which is usually surrounded by the †endosperm. They generally have broad leaves with netlike veins.

dielectric insulator or nonconductor of electricity, such as rubber, glass, and paraffin wax. An electric field in a dielectric material gives rise to no net flow of electricity. However, the applied field causes electrons within the material to be displaced, creating an electric charge on the surface of the material. This reduces the field strength within the material by a factor known as the dielectric constant (or relative permittivity) of the material. Dielectrics are used in capacitors, to reduce dangerously strong electric fields, and have optical applications.

dielectric constant capacitance of a capacitor (condenser) with a certain dielectric divided by the capacitance of the same capacitor with a vacuum (or, in practice, air) as a dielectric.

Diels, Otto Paul Hermann (1876–1954) German chemist. He and his former assistant, Kurt †Alder, were awarded the Nobel Prize for Chemistry in 1950 for their research into the synthesis of organic compounds.

Diels–Alder reaction in chemistry, a reaction for the synthesis of cyclic organic molecules that involves the addition of an alkene or alkyne, usually containing an activating electronegative group, for example $-CN$ or $-NO_2$, across a compound that has a conjugated †diene (double-bond) component, for example buta-1,3-diene. It was discovered by German chemists Otto †Diels and Kurt †Alder.

diene or **dialkene** or **diolefin**, organic chemical compound that has two carbon-carbon double bonds, for example buta-1,3-diene, $CH_2 = C - C = CH_2$. Dienes have a major use as starting reagents for the †Diels–Alder reaction.

diesel engine †internal-combustion engine that burns a lightweight fuel oil. The diesel engine operates by compressing air until it becomes sufficiently hot to ignite the fuel. It is a piston-in-cylinder engine, like the gasoline engine, but only air (rather than an air-and-fuel mixture) is taken into the cylinder on the first piston stroke (down). The piston moves up and compresses the air until it is at a very high temperature. The fuel oil is then injected into the hot air, where it burns, driving the piston down on its power stroke. For this reason the engine is called a compression-ignition engine.

Diesel engines have sometimes been marketed as "cleaner" than gasoline engines because they do not need lead additives and produce fewer gaseous pollutants. However, they do produce high levels of the tiny black carbon particles called particulates, which are believed to be carcinogenic and may exacerbate or even cause asthma.

The principle of the diesel engine was first explained in England by Herbert Akroyd (1864–1937) in 1890, and was applied practically by Rudolf Diesel in Germany in 1892.

diesel oil lightweight fuel oil used in diesel engines. Like gasoline, it is a petroleum product. When used in vehicle engines, it is also known as **derv** (diesel-engine road vehicle).

diet range of foods eaten by an animal each day; it is also a particular selection of food, or the total amount and choice of food for a specific person or group of people. Most animals require seven kinds of food in their diet: proteins, carbohydrates, fats, vitamins, minerals, water, and roughage. A diet that contains all of these things in the correct amounts and proportions is termed a balanced diet. The amounts and proportions required vary with different animals, according to their size, age, and lifestyle. The †digestive systems of animals have evolved to meet particular needs; they have also adapted to cope with the foods available in the surroundings in which they live. The necessity of finding and processing an appropriate diet is a very basic drive in animal evolution. **Dietetics** is the science of feeding individuals or groups; a dietician is a specialist in this science.

Dietary requirements may vary over the lifespan of an animal, according to whether it is growing, reproducing, highly active, or approaching death. For instance, increased carbohydrate for additional energy, or increased minerals, may be necessary during periods of growth.

An adequate diet for humans is one that supplies the body's daily nutritional needs (see †nutrition), and provides sufficient energy to meet individual levels of activity. The average daily requirement for men is 2,500 calories, but this will vary with age, occupation, and weight; in general, women need fewer calories than men. The energy requirements of active children increase steadily with age, reaching a peak in the late teens. At present, about 450 million people in the world – mainly living in famine or poverty-stricken areas, especially in countries of the developing world – subsist on fewer than 1,500 calories per day. The average daily intake in developed countries is 3,300 calories.

difference in mathematics, the result obtained when subtracting one number from another. Also, those elements of one †set that are not elements of another.

difference engine mechanical calculating machine designed (and partly built in 1822) by the English mathematician Charles †Babbage to produce reliable tables of life expectancy. A precursor of the †analytical engine, it was to calculate mathematical functions by solving the differences between values given to †variables within equations. Babbage designed the calculator so that once the initial values for the variables were set it would produce the next few thousand values without error.

difference equation equation that relates the value of a function at time t to its values at a specified number of past times, from among $t-1$, $t-2$, $t-3$, ...

differences the set of numbers obtained from a sequence by subtracting each element from its successor. For example, in the sequence 1, 4, 9, 16, 25 ..., the differences are 3, 5, 7, 9 ... . The patterns of differences are used to analyze sequences.

differential arrangement of gears in the final drive of a vehicle's transmission system that allows the driving wheels to turn at different speeds when cornering. The differential consists of sets of bevel gears and pinions within a cage attached to the crown wheel. When cornering, the bevel pinions rotate to allow the outer wheel to turn faster than the inner.

differential calculus branch of †calculus involving applications such as the determination of maximum and minimum points and rates of change.

differential drive see †differential gear.

differential equation equation involving †derivatives. In a linear differential equation, the unknown function and its derivatives never appear in a power other than one. Partial differential equations involve unknown functions of several variables, and partial derivatives do therefore appear.

differential gear †gear that allows two shafts to rotate at different rates. Such a system is used in cars to allow the wheels on the outside edge of a corner to rotate relative to those on the inside edge.

differential geometry investigation of geometrical surfaces using †differential calculus.

differential rotation of a stellar cluster or galaxy, the different rates of "orbiting" of stars. Those nearer the center move faster than those at the edge. It is also used to describe the axial rotation of equatorial latitudes faster than polar latitudes of a single body (such as the Sun or a gaseous planet).

differentiation in embryology, the process by which cells become increasingly different and specialized, giving rise to more complex structures that have particular functions in the adult organism. For instance, embryonic cells may develop into nerve, muscle, or bone cells.

differentiation in mathematics, a procedure for determining the †derivative or gradient of the tangent to a curve $f(x)$ at any point x.

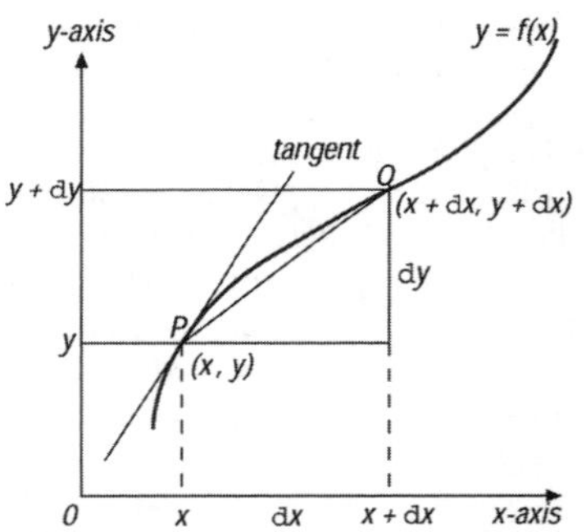

differentiation A mathematical procedure for determining the gradient, or slope, of the tangent to any curve f(x) at any point x. Part of a branch of mathematics called differential calculus, it is used to solve problems involving continuously varying quantities (such as the change in velocity or altitude of a rocket), to find the rates at which these variations occur and to obtain maximum and minimum values for thequantities. Helicon

diffraction the spreading out of waves when they pass through a small gap or around a small object, resulting in some change in the direction of the waves. In order for this effect to be observed, the size of the object or gap must be comparable to or smaller than the †wavelength of the waves. Diffraction occurs with all forms of progressive waves – electromagnetic, sound, and water waves – and explains such phenomena as the ability of long-wave radio waves to bend around hills more easily than short-wave radio waves.

The wavelength of visible light ranges from about 400 to 700 nanometers, several orders of magnitude smaller than radio waves. The gap through which light travels must be extremely small for diffraction to be observed. The slight spreading of a light beam through a narrow slit causes the different wavelengths of light to interfere with each other, producing a pattern of light and dark bands. A **diffraction grating** is a plate of glass or metal ruled with close (some diffraction gratings have from 2,000 to 7,000 lines per inch), equidistant, parallel lines used for separating a wave train such as a beam of incident light into its component frequencies. White light passing through a grating will be separated into its constituent colors. Red light is diffracted more as it has a longer wavelength; blue light is diffracted less as it has a shorter wavelength.

The wavelength of sound is between 0.5 m/1.6 ft and 2.0 m/6.6 ft. When sound waves travel through doorways or between buildings they are diffracted significantly, so that the sound is heard round corners.

diffraction grating polished metallic surface (usually a metallic mirror on a block of glass or quartz) or plastic surface on which has been ruled a great number (in thousands) of thin, parallel lines with small gaps between them, used to split light to produce a †spectrum.

A beam of light traveling through the narrow gaps of a diffraction grating will spread out and be split into the constituent wavelengths of the light. This is known as diffraction. White light passing through a diffraction grating will be split into the spectrum of colors (from red to blue) of which the light is made up. The light is split because each of its constituent wavelengths (colors) is diffracted by a different amount. Red light, having a longer wavelength, is diffracted more than blue light, having a shorter wavelength.

Some diffraction gratings can have from 2,000 to 7,000 lines per inch.

Diffraction Grating http://hyperphysics.phy-astr.gsu.edu/hbase/phyopt/grating.html

"HyperPhysics is an exploration environment for concepts in physics." A really great resource, the HyperPhysics' pages give in-depth treatment of principles with excellent diagrams, worked examples, and interactive tools.

diffusion net spontaneous and random movement of molecules or particles in a fluid (gas or liquid) from a region in which they are at a high concentration to a region of lower concentration, until a uniform concentration is achieved throughout. The difference in concentration between two such regions is called the **concentration gradient**. No mechanical mixing or stirring is involved. For instance, a drop of ink added to water will diffuse down the concentration gradient until evenly mixed.

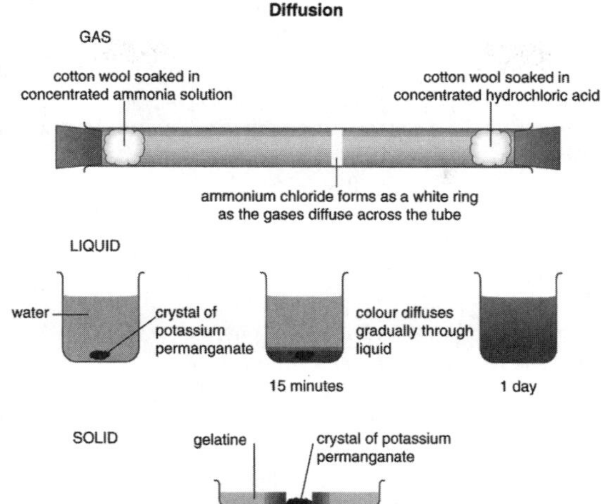

diffusion Diffusion can occur in gases, liquids, and solids.Substances diffuse at different speeds, fastest in gases and slowest in solids. In the gas example, the ring of ammonium chloride forms nearest to the hydrochloric acid end because ammonia diffuses faster than hydrochloric acid. Helicon

Diffusion occurs because particles in a liquid or gas are free to move according to the †kinetic theory of matter. The molecules move randomly, but there is more chance that they will move out of the ink drop than into it, so the molecules diffuse until their color becomes evenly distributed throughout. Diffusion occurs more rapidly across a higher concentration gradient and at higher temperatures.

Diffusion is quite different from the movement of molecules when a fluid is flowing. In this case movement is not random; all molecules are moving together and in the same direction.

In biological systems, diffusion plays an essential role in the transport, over short distances (for example across †cell membranes), of molecules such as nutrients, respiratory gases (†carbon dioxide and †oxygen), and neurotransmitters. It provides the means by which small molecules pass into and out of individual cells and micro-organisms, such as an amoeba, that possess no circulatory system. Plant and animal organs whose function depends on diffusion – such as the lung – have a large surface area. Diffusion of water across a semi-permeable membrane is termed †osmosis. Some important processes that involve diffusion are: the uptake of the products of †digestion from the †gut, gas exchange in the lungs, and gas exchange in the leaves of plants.

One application of diffusion is the separation of isotopes, particularly those of uranium. When uranium hexafluoride diffuses through a porous plate, the ratio of the 235 and 238 isotopes is changed slightly. With sufficient number of passes, the separation is nearly complete. There are large plants in the USA and UK for obtaining enriched fuel for fast nuclear reactors and the fissile uranium-235, originally required for the first atom bombs. Another application is the diffusion pump, used extensively in vacuum work, in which the gas to be evacuated diffuses into a chamber from which it is carried away by the vapor of a suitable medium, usually oil or mercury.

digestion process by which food eaten by an animal is broken down mechanically, and chemically by †enzymes, mostly in the †stomach and †intestines, to make the nutrients available for absorption and cell metabolism. In digestion large molecules of food are broken into smaller, soluble molecules, which are absorbed through the wall of the †gut into the bloodstream and carried to individual cells. The first stage of this may involve just the mixing of the food with water and the crushing and chopping of pieces of food by teeth or the mixing of food as it is squeezed along the gut. The second stage is the breakdown of large molecules by enzymes. The uptake of digested foods is mainly by absorption.

The digestion of fats starts with the action of †bile. This emulsifies lumps of fat into thousands of tiny droplets, allowing the enzyme lipase to break the fats down into glycerol and fatty acids. Lipase is released from the pancreas and digests fat in the small †intestine. Fats are not soluble in water, but glycerol and fatty acids are, and they are easily absorbed into the body.

Not all †carbohydrates can be digested, but one important carbohydrate that is digestible is starch. The digestion of starch begins in the mouth. Saliva contains the enzyme amylase, which breaks starch into the complex sugar maltose. Later on, in the small intestine, amylase is released from the pancreas and finishes off the digestion.

†Protein is partly digested in the †stomach, where the enzyme pepsin breaks large protein molecules into

smaller ones. In the small intestine other enzymes break the small proteins into amino acids. Enzymes in different parts of the gut work best at different levels of acidity, so the body makes the conditions in the stomach, for example, very acid, but in the small intestine, slightly alkaline.

Digestion http://www.geocities.com/jw31645/digest.html

Clear notes and a description of the digestive system, presented in a plain text-only format. The user can also try out the worksheet at the end of the page.

digestive system in the body, all the organs and tissues involved in the digestion of food. In animals, these consist of the mouth, stomach, intestines, and their associated glands. The process of digestion breaks down the food by physical and chemical means into the different elements that are needed by the body for energy and tissue building and repair. Digestion begins in the mouth and continues in the †stomach; from there most nutrients enter the small intestine from where they pass through the intestinal wall into the bloodstream; what remains is stored and concentrated into feces in the large intestine. Birds have two additional digestive organs – the †crop and †gizzard. In smaller, simpler animals such as jellyfish, the digestive system is simply a cavity (coelenteron or enteric cavity) with a "mouth" into which food is taken; the digestible portion is dissolved and absorbed in this cavity, and the remains are ejected back through the mouth.

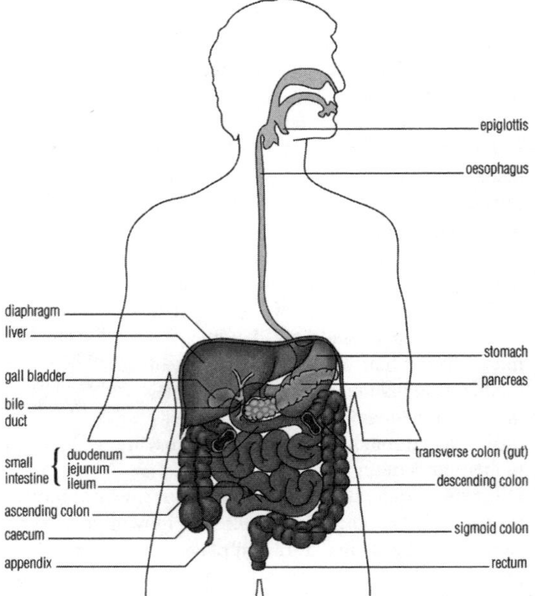

digestive system The human digestive system. When food is swallowed, it is moved down the oesophagus by the action of muscles (peristalsis) into the stomach. Digestion starts in the mouth and continues in the stomach as the food is mixed with enzymes and strong acid. After several hours, the food passes to the small intestine. Here more enzymes are added and digestion is completed. After all nutrients have been absorbed, the indigestible parts pass into the large intestine and thence to the rectum. The liver has many functions, such as storing minerals and vitamins and making bile, which is stored in the gall bladder until needed for the digestion of fats. The pancreas supplies enzymes. The appendix appears to have no function in human beings. Helicon

The digestive system of humans consists primarily of the †alimentary canal, a tube which starts at the mouth, continues with the pharynx, esophagus (or gullet), stomach, large and small intestines, and rectum, and ends at the anus. The food moves through this canal by †peristalsis whereby waves of involuntary muscular contraction and relaxation produced by the muscles in the wall of the gut cause the food to be ground and mixed with various digestive juices. Most of these juices contain digestive enzymes, chemicals that speed up reactions involved in the breakdown of food. Other digestive juices empty into the alimentary canal from the salivary glands, gall bladder, and pancreas, which are also part of the digestive system.

The fats, proteins, and carbohydrates (starches and sugars) in foods contain very complex molecules that are broken down (see †diet; †nutrition) for absorption into the bloodstream: starches and complex sugars are converted to simple sugars; fats are converted to fatty acids and glycerol; and proteins are converted to amino acids and peptides. Foods such as vitamins, minerals, and water do not need to undergo digestion prior to absorption into the bloodstream. The small intestine, which is the main site of digestion and absorption, is subdivided into the duodenum, jejunum, and ileum. Covering the surface of its mucous membrane lining are a large number of small prominences called villi which increase the surface for absorption and allow the digested nutrients to diffuse into small blood-vessels lying immediately under the epithelium.

Your Digestive System and How it Works http://www.niddk.nih.gov/health/digest/pubs/digesyst/newdiges.htm

Health information on the digestive system, detailed enough to interest biology students. Part of a much larger site from the US National Institute of Diabetes and Digestive and Kidney Diseases, it explains what happens as food travels through the body, and discusses the hormones and nerves that regulate the digestive process.

digit any of the numbers from 0 to 9 in the decimal system. Different bases have different ranges of digits. For example, the †hexadecimal system has digits 0 to 9 and A to F, whereas the binary system has two digits (or †bits), 0 and 1.

digital in electronics and computing, a term meaning "coded as numbers." A digital system uses two-state, either on/off or high/low voltage pulses, to encode, receive, and transmit information. A **digital display** shows discrete values as numbers (as opposed to an analog signal, such as the continuous sweep of a pointer on a dial).

Digital electronics is the technology that underlies digital techniques. Low-power, miniature, integrated circuits (chips) provide the means for the coding, storage, transmission, processing, and reconstruction of information of all kinds.

digital computer computing device that operates on a two-state system, using symbols that are internally coded as binary numbers (numbers made up of combinations of the digits 0 and 1); see †computer.

digital data transmission in computing, a way of sending data by converting all signals (whether pictures, sounds, or words) into numeric (normally binary) codes before transmission, then reconverting them on receipt. This virtually eliminates any distortion or degradation of the signal during transmission, storage, or processing.

digital root a digit formed by adding the digits of a number. If this leads to a total of 10 or more, the resultant digits are added. For example, finding the digital root of 365 involves the following computations: 3 + 6 + 5 = 14, followed by 1 + 4 = 5. The digital root of 365 is therefore 5.

digital-to-analog converter electronic circuit that converts a digital signal into an †analog (continuously varying) signal. Such a circuit is used to convert the digital output from a computer into the analog voltage required to produce sound from a conventional loudspeaker.

digital versatile disk full name for †DVD, disk format for storing digital information.

DVD Demystified http://www.dvddemystified.com/dvdfaq.html

Official FAQ site, claiming to be the most reliable source of DVD (digital versatile disk) information on the Web. Updated at least once a month, the site provides in-depth and readable answers to almost any question regarding all DVD formats.

digitizer in computing, a device that converts an analog video signal into a digital format so that video images can be input, stored, displayed, and manipulated by a computer. The term is also sometimes used to refer to a †graphics tablet.

dihybrid cross in genetics, breeding two different traits at the same time, for example seed shape and flower color. Mendel invented the dihybrid cross to determine if different traits of pea plants were inherited independently.

dihybrid inheritance in genetics, a pattern of inheritance observed when two characteristics are studied in succeeding generations. The first experiments of this type, as well as in †monohybrid inheritance, were carried out by Austrian biologist Gregor †Mendel using pea plants.

diketone or **dialkanone**, organic chemical compound that contains two keto groups, >C = O. An example of this class of compound is benzoquinone, $C_6H_4O_2$.

dilute in chemistry, a solution that has a low concentration of solute molecules.

dilution process of reducing the concentration of a solution by the addition of a solvent.

The extent of a dilution normally indicates the final volume of solution required. A fivefold dilution would mean the addition of sufficient solvent to make the final volume five times the original.

dimension in science, any directly measurable physical quantity such as mass (M), length (L), and time (T), and the derived units obtainable by multiplication or division from such quantities. For example, acceleration (the rate of change of velocity) has dimensions (LT^{-2}), and is expressed in such units as km s^{-2}. A quantity that is a ratio, such as relative density or humidity, is dimensionless.

In geometry, the dimensions of a figure are the number of measures needed to specify its size. A point is considered to have zero dimension, a line to have one dimension, a plane figure to have two, and a solid body to have three.

dimer in chemistry, a compound formed by the combination of two identical monomer molecules. The process by which this occurs is called dimerization. An example of this is the dimerization of borane, BH_3 to form diborane, B_2H_6.

dimethyl sulphoxide DMSO; $(CH_3)_2SO$, colorless liquid used as an industrial solvent and an antifreeze. It is obtained as a by-product of the processing of wood to paper.

dimorphism property of a chemical substance that allows it to crystallize in two different forms.

dingbat non-alphanumeric character, such as a star, bullet, or arrow. Dingbats have been combined into PostScript and TrueType fonts for use with word processors and graphics programs.

dinitrogen oxide alternative name for †nitrous oxide, or "laughing gas," one of the nitrogen oxides.

diode combination of a cold anode and a heated cathode, or the semiconductor equivalent, which incorporates a *p–n* junction; see †semiconductor diode. Either device allows the passage of direct current in one direction only, and so is commonly used in a †rectifier to convert alternating current (AC) to direct current (DC).

Semiconductor diodes are made from silicon-based materials primarily used in electronic circuits to allow an electrical current to flow in one direction only. Silicon on its own is not useful and it is therefore doped (mixed) with other materials in order to achieve the desired physical and chemical properties. The impurity atoms occupy various positions in the silicon crystal. Two types of junctions are produced by doping: *p*-type and *n*-type.

***p*-type junction** Silicon has four valence (free) electrons in its outer shell. It is doped with boron atoms, with only three valence electrons in their outer shells. The silicon atoms will share their valence electrons with the valence electrons of the boron atoms. This results in a matrix of silicon atoms doped with boron atoms, which are short of one electron in order to form a fourth chemical bond with the silicon atoms. This is called a *p*-type junction (*p* refers to the boron atom being positive as it is short of one electron).

The *p*-type hole can be occupied by an electron moving into it from a neighboring atom. The *p*-type hole then moves on to a new position in the matrix. As the *p*-type hole moves from one position to another, it is carrying a current with electrons also moving.

***n*-type junction** Phosphorous has five valence electrons, so if phosphorous atoms are added to wafers of silicon they can donate extra electrons. This is called an *n*-type junction (*n* refers to phosphorous being negative due to an extra electron).

***p–n* junction** In a diode, wafers of *p*-type and *n*-type are fused together by heating the wafers and the doping substances in a furnace. An *n*-type silicon wafer is placed on top of a *p*-type silicon wafer to form a junction where the two types come into contact. Electrons flow through the silicon semiconductor in one direction only.

types of diode There are many different types of semiconductor diodes such as light-emitting diodes (LEDs), made from gallium arsenide phosphide and emitting electromagnetic radiation, and light-dependent diodes (photodiodes), allowing a flow of current in the presence of light.

dioecious describing plants with male and female flowers borne on separate individuals of the same species. Dioecism occurs, for example, in the willows *Salix*. It is a way of avoiding self-fertilization.

Diophantine equations algebraic equations involving one or more unknowns (indeterminates) with †integers (whole numbers) as coefficients, to which one or more solutions are sought, also in integers. The classic form is:

$ax + by = c$

Part of the significance of this is that even if not enough information is given to derive a single solution, enough is given to reduce the answer to a definite type. Diophantus, who lived in the third century AD, thus began the investigations into †number theory that still continue.

Diophantus (lived AD 250) Greek mathematician in Alexandria whose *Arithmetica* is one of the first known works on problem solving by algebra, in which both words and symbols are used.

Diophantus of Alexandria http://www-groups.dcs.st-andrews.ac.uk/history/Mathematicians/Diophantus.html

Devoted to the life and contributions of Diophantus of Alexandria. In addition to biographical information, you will find a list of references about Diophantus and links to essays that reference him. The text of this essay includes hypertext links to the essays of those mathematicians and thinkers who influenced Diophantus. You will also find a map of his birthplace.

dioptre optical unit in which the power of a †lens is expressed as the reciprocal of its focal length in meters. The usual convention is that convergent lenses are positive and divergent lenses negative. Short-sighted people need lenses of power about –0.7 dioptre; a typical value for long sight is about +1.5 dioptre.

diorite igneous rock intermediate in composition between mafic (consisting primarily of dark-colored minerals) and felsic (consisting primarily of light-colored minerals) – the coarse-grained plutonic equivalent of †andesite. Constituent minerals include †feldspar and †amphibole or pyroxene with only minor amounts of †quartz.

dioxin any of a family of over 200 organic chemicals, all of which are heterocyclic hydrocarbons (see †cyclic compounds). The term is commonly applied, however, to only one member of the family, 2,3,7,8-tetrachlorodibenzo-*p*-dioxin (2,3,7,8-TCDD), a highly toxic chemical that occurs, for example, as an impurity in the defoliant Agent Orange, used in the Vietnam War, and sometimes in the weedkiller 2,4,5-T. It has been associated with chloracne (a disfiguring skin complaint), birth defects, miscarriages, and cancer.

Disasters involving accidental release of large amounts of dioxin into the environment have occurred at Seveso, Italy, and Times Beach, Missouri, USA. Small amounts of dioxins are released by the burning of a wide range of chlorinated materials (treated wood, exhaust fumes from fuels treated with chlorinated additives, and plastics) and as a side effect of some techniques of paper-making. Dioxin may also be produced as a by-product in the manufacture of the bactericide †hexachlorophene.

The possibility of food becoming contaminated by dioxins in the environment has led the European Union (EU) to decrease significantly the allowed levels of dioxin emissions from incinerators.

dip in earth science, angle at which a structural surface, such as a fault or a bedding plane, is inclined from horizontal. Measured at right angles to the †strike of that surface, it is used together with strike to describe the orientation of geological features in the field. Rocks that are dipping have usually been affected by †folding.

diphenylamine $C_{12}H_{11}N$, (Ph_2NH), colorless crystalline solid with a melting point of 54°C/129°F. It is usually prepared by heating aniline with aniline hydrochloride and commonly used as a stabilizer for cellulose nitrate-based explosives.

diphosphate chemical compound containing two phosphate groups, as in adenosine diphosphate (ADP).

diploblastic in biology, having a body wall composed of two layers. The outer layer is the **ectoderm**, the inner layer is the **endoderm**. This pattern of development is shown by †coelenterates.

diploid having paired †chromosomes in each cell. In sexually reproducing species, one set is derived from each parent, the †gametes, or sex cells, of each parent being †haploid (having only one set of chromosomes) due to †meiosis (reduction cell division).

dip, magnetic angle at a particular point on the Earth's surface between the direction of the Earth's magnetic field and the horizontal. It is measured using a **dip circle**, which has a magnetized needle suspended so that it can turn freely in the vertical plane of the magnetic field. In the northern hemisphere the needle dips below the horizontal, pointing along the line of the magnetic field towards its north pole. At the magnetic north and south poles, the needle dips vertically and the angle of dip is 90°. See also †angle of declination.

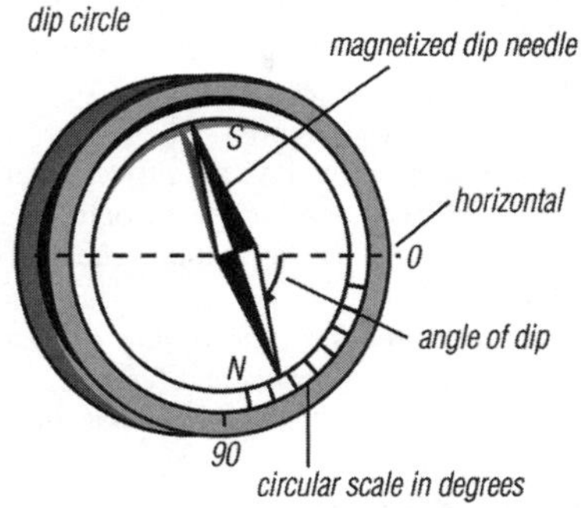

dip A dip circle is used to measure the angle between the direction of the Earth's magnetic field and the horizontal at any point on the Earth's surface. Helicon

dipole uneven distribution of magnetic or electrical characteristics within a molecule or substance so that it behaves as though it possesses two equal but opposite poles or charges, a finite distance apart.

The uneven distribution of electrons within a molecule composed of atoms of different †electronegativities may result in an apparent concentration of electrons towards one end of the molecule and a deficiency towards the other, so that it forms a dipole consisting of apparently separated but equal positive and negative charges. The product of one charge and the distance between them is the **dipole moment**. A bar magnet has a magnetic dipole and behaves as though its magnetism were concentrated in separate north and south magnetic poles.

dipole, magnetic see †magnetic dipole.

Dirac, Paul Adrien Maurice (1902–1984) English physicist who worked out a version of quantum mechanics consistent with special †relativity. The existence of antiparticles, such as the positron (positive electron), was one of its predictions. He shared the Nobel Prize for Physics in 1933 (with Austrian physicist Erwin †Schrödinger) for his work on the development of quantum mechanics.

direct current **DC**, electric current where the electrons (negative charge) flow in one direction, and that does not reverse its flow as †alternating current does. The electricity produced by a battery is direct current. Electromagnets and electric trains use direct current.

If in one second one coulomb of electrons passes a given point, then the current flow is 1 amp DC.

A cathode ray oscilloscope (CRO) is used to display the waveforms that show the pattern of how voltage and current vary over a period of time. The waveforms for DC are straight lines, as the voltage and current do not vary over a period of time. If a resistor is connected in a circuit and measurements of voltage across the resistor and current flowing through it are taken over a period of time, then there will be no change in the voltage and current over this period. A plot of voltage or current against time produces a straight line.

directed number in mathematics, number with a positive (+) or negative (–) sign attached, for example +5 or –5. On a graph, a positive sign shows a movement to the right or upwards; a negative sign indicates movement downwards or to the left.

directory in computing, a list of file names, together with information that enables a computer to retrieve those files from backing storage. The computer operating system will usually store and update a directory on the backing storage to which it refers. So, for example, on each †disk used by a computer a directory file will be created listing the disk's contents.

The term is also used to refer to the area on a disk where files are stored; the main area, the **root** directory, is at the top-most level, and may contain several separate **sub-directories**.

direct proportion in mathematics, term used to describe quantities that increase in the same †ratio. For example, if 5 kg of potatoes cost £1.35, the cost of 12 kg of the potatoes will be in direct proportion to the quantities bought. The ratio of the quantities = 12:5, so the ratio of the prices = 12:5. The cost of 12 kg, therefore, is 12 ÷ 5 x £1.35 = £3.24.

The symbol α is read as "is proportional to." For example, the statement "price is proportional to quantity" could be written as $p \alpha q$, where p stands for price and q stands for quantity.

$a \alpha b$ means $a = kb$ where k is a positive number describing the relationship between a and b. If one pair of values for a and b is known, k can be found.

For example, y is proportional to x and when x is 20, y is 30. Since $y \alpha x$, then

$y = kx$ and $30 = k \times 20$; therefore, $k = 30 \div 20 = 1.5$

The relationship between x and y is, therefore, $y = 1.5x$.

Dirichlet, (Peter Gustav) Lejeune (1805–1859) German mathematician whose work in applying analytical techniques to mathematical theory resulted in the fundamental development of the theory of numbers. He was also a physicist interested in dynamics.

disaccharide sugar made up of two monosaccharides or simple sugars. Sucrose, $C_{12}H_{22}O_{11}$, or table sugar, is a disaccharide.

disc in astronomy, flat, roughly circular region of a spiral or a lenticular (lens-shaped) †galaxy containing stars, †nebulae, and dust clouds orbiting about the nucleus. Discs contain predominantly young stars and regions of star formation. The disc of our own Galaxy is seen from Earth as the band of the †Milky Way.

discharge in a river, the volume of water passing a certain point per unit of time. It is usually expressed in cubic meters per second (cumecs). The discharge of a particular river channel may be calculated by multiplying the channel's cross-sectional area (in square meters) by the velocity of the water (in meters per second).

discharge, electric release of a stored †electric charge.

discharge tube device containing (usually two) electrodes and a vacuum or gas at low pressure; a (high) voltage applied to the electrodes causes an electric discharge to take place between them. A gas-filled tube (such as a neon tube) may emit visible light and other forms of radiation.

discontinuity in mathematics, point at which a function fails to be continuous. If the function is plotted as a graph there will be a break in the graph at such a point.

discontinuous variation in biology, variation that falls into two or more non-overlapping classes, for example Mendel's variations on pea color and shape.

discrete data in mathematics, †data that can take only whole-number or fractional values, that is, distinct values. The opposite is †continuous data, which can take all in-between values. Examples of discrete data include †frequency and †population data. However, measurements of time and other dimensions can give rise to continuous data.

discriminants special functions of the coefficients of an equation, used to find roots of a polynomial equation.

disease condition that disturbs or impairs the normal state of an organism. Diseases can occur in all living things, and normally affect the functioning of cells, tissues, organs, or systems. Diseases are usually characterized by specific symptoms and signs, and can be mild and short-lasting – such as the common cold – or severe enough to decimate a whole species – such as Dutch elm disease. Diseases can be classified as infectious or noninfectious. Infectious diseases are caused by micro-organisms, such as bacteria and viruses, invading the body; they can be spread across a species, or transmitted between one or more species. All other diseases can be grouped together as noninfectious diseases. These can have many causes: they may be inherited (congenital diseases); they may be caused by the ingestion or absorption of harmful substances, such as toxins; they can result from poor nutrition or hygiene; or they may arise from injury or aging. The causes of some diseases are still unknown.

disinfectant agent that kills, or prevents the growth of, bacteria and other micro-organisms. Chemical disinfectants include carbolic acid (phenol, used by Joseph Lister in surgery in the 1870s), ethanol, methanol, chlorine, and iodine.

disjoint set see †universal set.

disk in computing, a common medium for storing large volumes of data (an alternative is †magnetic tape). A **magnetic disk** is rotated at high speed in a disk-drive unit as a read/write (playback or record) head passes over its surfaces to record or read the magnetic variations that encode the data. **Optical disks**, such as †CD-ROM (compact-disc read-only memory) and †WORM (write once, read many times), are also used to store computer data. Data are recorded on the disk surface as etched microscopic pits and are read by a laser-scanning device.

Optical disks have an enormous capacity – ranging from 650 megabytes for CD-ROM to 2.6 gigabytes for magneto-optical drives.

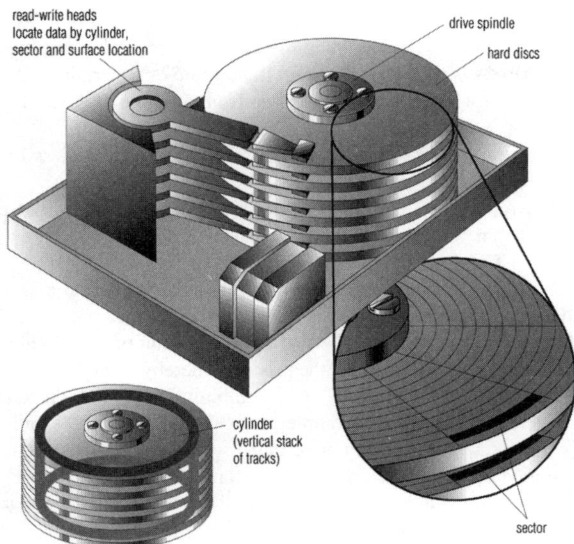

disk A hard disk. Data is stored in sectors within cylinders and is read by a head which passes over the spinning surface of each disk. Helicon

Hard Disk Drives

http://www.pcguide.com/ref/hdd/index-i.htm

Chapter of the Internet PC Guide on hard disk drives. The pages look at the principles, configuration, and design of hard disk drives, and document their history and development.

disk formatting in computing, preparing a blank magnetic disk in order that data can be stored on it. Data are recorded on a disk's surface on circular tracks, each of which is divided into a number of sectors. In formatting a disk, the computer's operating system adds control information such as track and sector numbers, which enables the data stored to be accessed correctly by the disk-drive unit. Some floppy disks, called **hard-sectored disks**, are sold already formatted. However, because different makes of computer use different disk formats, disks are also sold unformatted, or **soft-sectored**, and computers are provided with the necessary †utility program to format these disks correctly before they are used.

dispersal in biology, phase of reproduction during which gametes, eggs, seeds, or offspring move away from the parents into other areas. The result is that overcrowding is avoided and parents do not find themselves in competition with their own offspring. The mechanisms are various, including a reliance on wind or water currents and, in the case of animals, locomotion. The ability of a species to spread widely through an area and to colonize new habitats has survival value in evolution.

dispersion in chemistry, the distribution of the microscopic particles of a †colloid. In colloidal sulfur the dispersion is the tiny particles of sulfur in the aqueous system.

dispersion in physics, a particular property of †refraction in which the angle and velocity of waves passing through a dispersive medium depends upon their frequency. When visible white light passes through a prism it is split into a spectrum (see †electromagnetic waves). This occurs because each component frequency of light, which corresponds to a color, is refracted by a slightly different angle, and so the light is split into its component frequencies (colors). A rainbow is formed when sunlight is dispersed by raindrops.

A white light beam splits up into the colors of which it consists (red, orange, yellow, green, blue, indigo, and violet). Red light has the lowest frequency (longest wavelength) and is the least refracted; violet light has the highest frequency (shortest wavelength) and is the most refracted. The relationship between wavelength and frequency is given by the formula: speed of light = frequency x wavelength (or wavelength = speed of light/frequency).

dispersion in statistics, the extent to which data are spread around a central point (typically the †arithmetic mean).

displacement in psychoanalysis, the transference of an emotion from the original idea with which it is associated to other ideas. It is usually thought to be indicative of †repression in that the emotional content of an unacceptable idea may be expressed without the idea itself becoming conscious.

Sigmund Freud's original discussion of displacement focused on its occurrence in dreams, where strong emotions are often expressed in material that the subject usually finds of indifferent interest. As a defense mechanism, it is also associated with a number of afflictions, including phobias and schizophrenia.

displacement activity in animal behavior, an action that is performed out of its normal context, while the animal is in a state of stress, frustration, or uncertainty. Birds, for example, often peck at grass when uncertain whether to attack or flee from an opponent; similarly, humans scratch their heads when nervous.

displacement reaction chemical reaction in which a less reactive element is replaced in a compound by a more reactive one.

For example, the addition of powdered zinc to a solution of copper(II) sulfate displaces copper metal, which can be detected by its characteristic color:

$Zn_{(s)} + CuSO_{4(aq)} \rightarrow ZnSO_{4(aq)} + Cu_{(s)}$

The copper is taken out of the solution and is deposited as a solid (s).

displacement vector vector that describes how an object has been moved from one position to another.

disproportionation or dismutation splitting of a molecule into two or more simpler molecules.

dissociation in chemistry, the process whereby a single compound splits into two or more smaller products, which may be capable of recombining to form the reactant.

Where dissociation is incomplete (not all the compound's molecules dissociate), a †chemical equilibrium exists between the compound and its dissociation products. The extent of incomplete dissociation is defined by a numerical value (dissociation constant).

distance modulus in astronomy, method of finding the distance to an object in the universe, such as a star or †galaxy, using the difference between the actual and the apparent brightness of the object. The actual brightness is deduced from the object's type and size. The apparent brightness is obtained by direct observation.

distance ratio in a machine, the distance moved by the input force, or effort, divided by the distance moved by the output force, or load. The ratio indicates the movement magnification achieved, and is equivalent to the machine's velocity ratio.

distance–time graph graph used to describe the motion of a body by illustrating the relationship between the distance that it travels and the time taken. Plotting distance (on the vertical axis) against time (on the horizontal axis) produces a graph, the gradient of which at any point is the body's speed at that point (called the instantaneous speed). If the gradient is constant (the graph is a straight line), the body has uniform or constant speed; if the gradient varies (the graph is curved), then so does the speed, and the body may be said to be accelerating or decelerating. The shape of the graph allows the distance traveled, the speed at any point, and the average speed to be worked out.

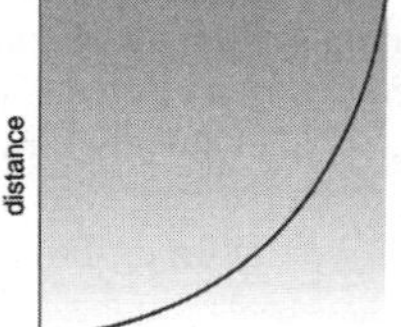

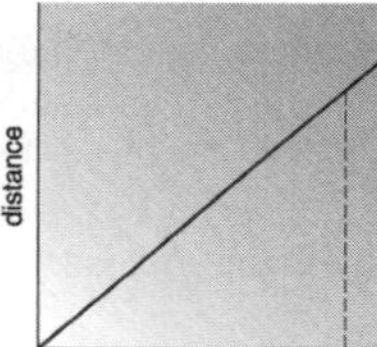

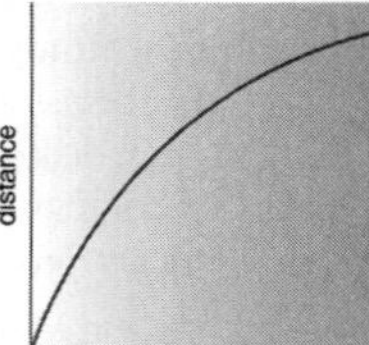

distance-time graph In a distance-time graph distance travelled by an object is plotted against the time it has taken for the object to travel that distance. If the line plotted is straight then the object is travelling at a uniform rate. The line curving upward shows acceleration; a curve downwards plots deceleration. Helicon

distillation technique used to purify liquids or to separate mixtures of liquids possessing different boiling points. **Simple distillation** is used in the purification of liquids or the separation of substances in solution from their solvents – for example, in the production of pure water from a salt solution or the recovery of sodium chloride (table salt) from sea water.

The solution is boiled and the vapors of the solvent rise into a separate piece of apparatus (the condenser) where they are cooled and condensed. The liquid produced (the distillate) is the pure solvent; the substances (the solutes), now in solid form, remain in the distillation vessel.

Mixtures of liquids, such as crude oil (unrefined †petroleum) or aqueous ethanol, are separated by †**fractional distillation**, or fractionation, in a †fractionating column. When the mixture is boiled, the vapors of its most volatile component (the compo-

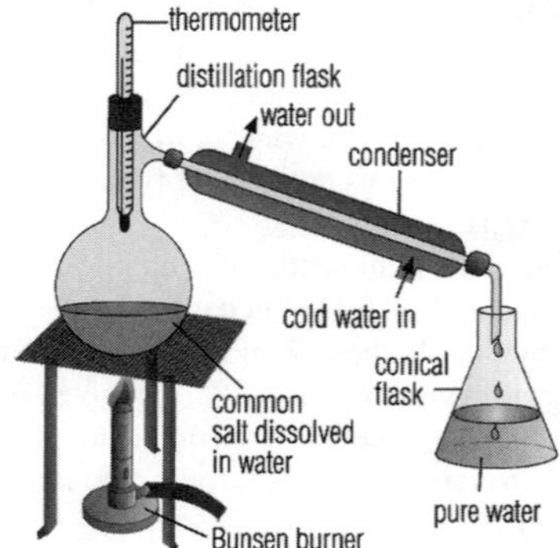

distillation Laboratory apparatus for simple distillation. Other forms of distillation include steam distillation, in which steam is passed into the mixture being distilled, and vacuum distillation, in which air is removed from above the mixture to be distilled. Helicon

nent with the lowest boiling point) rise into the vertical column where they condense to liquid form. As they descend they are reheated to boiling point by the hotter rising vapors of the next component to reach boiling point. This boiling–condensing process occurs repeatedly inside the column. As the column is ascended, progressive enrichment by the lower-boiling-point components occurs; there is thus a temperature gradient inside the column. The vapors of the more volatile components reach the top of the column and enter the condenser for collection before those of the less volatile components. In the fractional distillation of petroleum, groups of compounds possessing similar relative molecular masses and boiling points (the fractions) are tapped off at different points on the column.

The earliest-known reference to the process is to the distillation of wine in the 12th century by Adelard of Bath. The chemical retort used for distillation was invented by Muslims, and was first seen in the West in about 1570.

Introduction to Distillation http://lorien.ncl.ac.uk/ming/distil/distil0.htm

Deals exclusively with distillation, the process of separating different liquids from each other by the application and removal of heat. The page begins with a short paragraph, which defines distillation, and then continues to the processes involved. A variety of links is at the bottom of the page, each of which deals with one specific step of the process.

distributary river that has branched away from a main river. Distributaries are most commonly found on a †delta, where the very gentle gradient and large amounts of silt deposited encourage channels to split.

Channels are said to be **braided** if they branch away from a river and rejoin it at a later point.

distribution in statistics, the pattern of †frequency for a set of †data. This pattern can be displayed as a †frequency diagram.

distributive law in mathematics, law expressing the principle operative in the equation:

$a(b + c) = ab + ac$

distributive operation in mathematics, an operation, such as multiplication, that bears a relationship to another operation, such as addition, such that $a \times (b + c) = (a \times b) + (a \times c)$. For example, 3 x (2 + 4) = (3 x 2) + (3 x 4) = 18. Multiplication may be said to be distributive over addition. Addition is not, however, distributive over multiplication because 3 + (2 x 4) x (3 + 2) x (3 + 4).

divergent in mathematics, describing a sequence or series, where there is no †limit for it to approach; it is the opposite of convergent.

divergent margin in †plate tectonics, the boundary, or active zone between two lithospheric plates that are moving apart. Divergent margins are characterized by extensive normal faulting, low-magnitude earthquakes, and, in most cases, volcanic activity.

The most common type of divergent margin is the **oceanic spreading center**, where two oceanic plates move apart. As they diverge, magma (molten rock) wells up to fill the open space, and solidifies, forming new oceanic crust. These boundaries are characterized by long, segmented volcanic ridges known as †mid-ocean ridges. Mid-ocean ridges, which include the †Mid-Atlantic Ridge and the East Pacific Rise, are found in all the world's oceans.

The second type of divergent margin occurs on the continents, in regions of incipient continental breakup. As the continent stretches, the crust breaks (faults) and thins. The most famous example is the Great Rift Valley in East Africa. Like their oceanic counterparts, continental rifts are characterized by normal faulting and volcanism. If rifting continues, the continent is broken in two, oceanic crust forms between the new continents, and the margin becomes a mid-ocean ridge.

dividend in mathematics, any number that is to be divided by another number. For example in the computation 20 ÷ 4 = 5, 20 is the dividend.

division one of the four basic operations of arithmetic, usually written in the form $a \div b$ or a/b. It is the inverse of †multiplication.

For example, 252 divided by 14 is the same as how many 14s there are in 252; 252 ÷ 14 = 18. The answer can be checked by multiplying 14 by 18; 14 x 18 = 252.

divisor in mathematics, any number that is to be divided into another number. For example in the computation 100 ÷ 25 = 4, 25 is the divisor.

DNA abbreviation for deoxyribonucleic acid, molecular basis of heredity. It is a complex giant molecule that contains, in chemically coded form, the information needed for a cell to make †proteins. In other words it determines the order in which amino acids are joined to make a specific protein in a cell. DNA is a ladderlike double-stranded †nucleic acid, which forms the basis of genetic inheritance in all organisms, except for a few viruses that have only †RNA. DNA is organized into †chromosomes and, in organisms other than bacteria, it is found only in the cell nucleus.

DNA is a ladderlike molecule, which means that it is made up of two halves (the ladder sides), formed of chains of †nucleotide subunits. Each nucleotide contains a deoxyribose sugar, a phosphate, and a base. A set of three bases – known as a **codon** – acts as a blueprint for the manufacture of a particular †amino acid, the subunit of a protein molecule. The two halves are joined together by the bases – a purine (adenine or guanine) or pyrimidine (cytosine or thymine) – forming pairs (the rungs). The bases form into two specific †base pairs: adenine with thymine and guanine with cytosine. The sequence of base pairs along the DNA acts as a code carrying information about the sequence of amino acids in proteins. Three base pairs in sequence (triplet) name an amino acid and the next three name the next amino acid that needs to be joined and so on, to make a specific protein. The specific way in which the pairs form means that the base sequence is preserved from generation to generation. Hereditary information is stored as a specific sequence of bases.

It is important that inherited information is passed on correctly. In the process of DNA replication, which takes place before any cell divides, the two halves of DNA separate and new halves are made. Because of specific base pairing, the inherited information is copied exactly. Despite this, a mistake sometimes occurs and the sequences of bases is altered. This changes the sequence of amino acids in a protein. This is †mutation. Ionizing radiation increases the risk of mutation. In plants and animals DNA is organized into chromosomes and is found in the †nucleus of cells. DNA in bacteria is organized differently. Bacteria have one large circular DNA molecule carrying most of their inherited information. Some bacteria also have small circular molecules of DNA, known as plasmids. These may be used in †genetic engineering to transfer genes from one organism to another.

Molecular Expressions: The DNA Collection http://micro.magnet.fsu.edu/micro/gallery/dna/dna.html

Spectacular gallery of DNA photographic representations in the laboratory as well as in vivo. This site also has links to several other sites offering photographs through a microscope of various substances, including computer chips and various pharmaceutical substances.

DNA fingerprinting or **DNA profiling**, another name for †genetic fingerprinting.

Dobzhansky, Theodosius (1900–1975) adopted name of Feodosy Grigorevich Dobrzhansky, Ukrainian-born US geneticist who established evolutionary genetics as an independent discipline. He showed that genetic variability between individuals of the same species is very high and that this diversity is vital to the process of evolution.

Dobzhansky, Theodosius http://kroeber.anthro.mankato.msus.edu/information/biography/abcde/dobzhansky_theodosius.html

Profile of the pioneering geneticist. It traces his childhood interest in insects, the frustration of his ambitions in the Soviet Union, and his subsequent research in the USA. There are photos of Dobzhansky and a bibliography.

document reader in computing, an input device that reads marks or characters, usually on pre-prepared forms and documents. Such devices are used to capture data by †optical mark recognition (OMR), †optical character recognition (OCR), and mark sensing.

dodecahedron regular solid with 12 pentagonal faces and 12 vertices. It is one of the five regular †polyhedra, or Platonic solids.

Dog Star brightest star in the night sky; see †Sirius.

doldrums area of low atmospheric pressure along the Equator, in the †intertropical convergence zone where the northeast and southeast trade winds converge. The doldrums are characterized by calm or very light winds, during which there may be sudden squalls and stormy weather. For this reason the areas are avoided as far as possible by sailing ships.

dolerite igneous rock formed below the Earth's surface, a form of basalt, containing relatively little silica (mafic in composition).

Dolerite is a medium-grained (hypabyssal) basalt and forms in shallow intrusions, such as dykes, which cut across the rock strata, and sills, which push between beds of sedimentary rock. When exposed at the surface, dolerite weathers into spherical lumps.

Dollfus, Audouin Charles (1924–) French physicist and astronomer. His preferred method of research is to use polarization of light. He was the first to detect a faint atmosphere round the planet Mercury, in 1950, and established that the Moon does not have one. In pursuit of his detailed investigations into Mars, Dollfus made the first French ascent in a stratospheric balloon.

dolomite in mineralogy, white mineral with a rhombohedral structure, calcium magnesium carbonate ($CaMg(CO_3)_2$). Dolomites are common in geological successions of all ages and are often formed when †limestone is changed by the replacement of the mineral calcite with the mineral dolomite.

dolomite in sedimentology, type of limestone rock where the calcite content is replaced by the mineral †dolomite. Dolomite rock may be white, gray, brown, or reddish in color, commonly crystalline. It is used as a building material. The region of the Alps known as the Dolomites is a fine example of dolomite formation.

Domagk, Gerhard (1895–1964) German pathologist who was awarded a Nobel Prize for Physiology or Medicine in 1939 for his discovery of the first antibacterial †sulphonamide drug. In 1932 he found that a coal-tar dye called Prontosil red contains chemicals with powerful antibacterial properties. Sulphanilamide became the first of the sulphonamide drugs, used – before antibiotics were discovered – to treat a wide range of conditions, including pneumonia and septic wounds.

domain in physics, small area in a magnetic material that behaves like a tiny magnet. The magnetism of the material is due to the movement of electrons in the atoms of the domain. In an unmagnetized sample of material, the domains point in random directions, or form closed loops, so that there is no overall magnetization of the sample. In a magnetized sample, the domains are aligned so that their magnetic effects combine to produce a strong overall magnetism.

domain in mathematics, base set of numbers on which a †function works, mapping it on to a second set (the †range).

dome a geologic feature that is the reverse of a basin. It consists of anticlinally folded rocks that dip in all directions from a central high point, resembling an inverted but usually irregular cup.

Such structural domes are the result of pressure acting upward from below to produce an uplifted portion of the crust. Domes are often formed by the upwelling of plastic materials such as salt or magma. The salt domes along the North American Gulf Coast were produced by upwelling ancient sea salt deposits, while the Black Hills of South Dakota are the result of structural domes pushed up by intruding igneous masses.

dominance in genetics, the masking of one †allele (an alternative form of a gene) by another allele. For example, if a †heterozygous person has one allele for blue eyes and one for brown eyes, his or her eye color will be brown. The allele for blue eyes is described as †recessive and the allele for brown eyes as dominant.

For every characteristic of a plant or animal that is inherited, there are always two genes present in the cells that determine the characteristic. By "characteristic" is meant "height" or "eye color" or "ability to make a particular enzyme." If the two genes are identical (homozygous state) the characteristic seen in the organism is determined by either one of the two genes. However, one gene may be different from the other (heterozygous state). If so, the two genes are alleles – contrasting genes for a characteristic. In this case it is possible that one of them determines the characteristic seen and the other does not. The characteristic seen in this case is said to be dominant. The other allele, which does not contribute to the appearance of the organism in this case, will only be expressed and contribute to appearance when no dominant gene is present. This characteristic is said to be recessive.

Donati, Giovanni Battista (1826–1873) Italian astronomer. He discovered six comets 1854–64, and made important contributions to the early development of stellar †spectroscopy and applied spectroscopic methods to the understanding of the nature of comets. He also studied cosmic meteorology. During the 1850s Donati was an enthusiastic comet-seeker, and the most dramatic of his discoveries was named after him. **Donati's comet**, first sighted in 1858, had, in addition to its major tail, two narrow extra tails.

dopamine neurotransmitter, hydroxytyramine $C_8H_{11}NO_2$, an intermediate in the formation of adrenaline. There are special nerve cells (neurons) in the brain that use dopamine for the transmission of nervous impulses. One such area of dopamine neurons lies in the basal ganglia, a region that controls movement. Patients suffering from the tremors of Parkinson's disease show nerve degeneration in this region. Another dopamine area lies in the limbic system, a region closely involved with emotional responses. It has been found that schizophrenic patients respond well to drugs that limit dopamine excess in this area.

Doppler, Christian Johann (1803–1853) Austrian physicist who in 1842 described the †Doppler effect and derived the observed frequency mathematically in Doppler's principle.

Doppler, Christian Andreas http://www-groups.dcs.st-andrews.ac.uk/history/Mathematicians/Doppler.html

Part of an archive containing the biographies of the world's greatest mathematicians, this site is devoted to the life and contributions of physicist Christian Doppler.

Doppler effect change in the observed frequency (or wavelength) of waves due to relative motion between the wave source and the observer. The Doppler effect is responsible for the perceived change in pitch of a siren as it approaches and then recedes, and for the †red shift of light from distant galaxies. It is named after the Austrian physicist Christian Doppler.

Doppler Effect http://archive.ncsa.uiuc.edu/Cyberia/Bima/doppler.html

Discussion of the Doppler effect, including an explanation of how astronomers use it to calculate the speed with which celestial bodies are moving in relation to one another.

Dorado constellation of the southern hemisphere, represented as a goldfish or a swordfish. It is easy to locate, since the Large †Magellanic Cloud marks its southern border. Its brightest star is Alpha Doradus, just under 200 light years from Earth.

One of the most conspicuous objects in the Large Magellanic Cloud is the "Great Looped Nebula" that surrounds 30 Doradus.

dormancy in botany, a phase of reduced physiological activity exhibited by certain buds, seeds, and spores. Dormancy can help a plant to survive unfavorable conditions, as in annual plants that pass the cold winter season as dormant seeds, and plants that form dormant buds.

For various reasons many seeds exhibit a period of dormancy even when conditions are favorable for growth. Sometimes this dormancy can be broken by artificial methods, such as penetrating the seed coat to facilitate the uptake of water (chitting) or exposing the seed to light. Other seeds require a period of †after-ripening.

dorsal in vertebrates, the surface of the animal closest to the backbone. For most vertebrates and invertebrates this is the upper surface, or the surface furthest from the ground. For bipedal primates such as humans, where the dorsal surface faces backwards, then the word is "back."

Not all animals can be said to have a dorsal surface, just as many animals cannot be said to have a front; for example, jellyfish, anemones, and sponges do not have a dorsal surface.

DOS acronym for Disk Operating System, computer †operating system specifically designed for use with disk storage; also used as an alternative name for a particular operating system, †MS-DOS.

DOS Command Index http://www.easydos.com/dosindex.html

Large and detailed index of commands used in the operating system DOS.

dot matrix printer computer printer that produces each character individually by printing a pattern, or matrix, of very small dots. The printing head consists of a vertical line or block of either 9 or 24 printing pins. As the printing head is moved from side to side across the paper, the pins are pushed forwards selectively to strike an inked ribbon and build up the dot pattern for each character on the paper beneath.

A dot matrix printer is more flexible than a daisywheel printer because it can print graphics and text in many different typefaces. It is cheaper to buy and maintain than a laser printer or ink-jet printer, and, because its pins physically strike the paper, is capable of producing carbon copies. However, it is noisy in operation and cannot produce the high-quality printing associated with the nonimpact printers.

double bond two covalent bonds between adjacent atoms, as in the †alkenes (–C = C–) and †ketones (–C = O).

double circulation in mammals, the characteristic †blood circulatory system. The †heart is divided into two halves, each half working as a separate pump. One side of the heart pumps blood to the lungs and back to the heart. The other side of the heart pumps blood to all other parts of the body and back to the heart.

Deoxygenated blood – that is, blood with a reduced content of †oxygen – is pumped to the lungs. Blood returning from the lungs is oxygenated – it is carrying as much oxygen as it is capable of carrying. Oxygenated blood is then pumped by the other half of the heart to all other parts of the body where oxygen is used in †respiration in the body tissues. The deoxygenated blood is returned to the heart.

double decomposition reaction between two chemical substances (usually †salts in solution) that results in the exchange of a constituent from each compound to create two different compounds.

For example, if silver nitrate solution is added to a solution of sodium chloride, there is an exchange of ions yielding sodium nitrate and silver chloride (which is precipitated).

double refraction or **birefringence**, in a crystal, when an unpolarized ray of light entering the crystal is split into two †polarized rays, one of which does obey †Snell's law of refraction, and the other does not. Calcite is such a crystal.

double salt in chemistry, compound formed by crystallization, in definite molar ratios (see †mole), from solutions of two or more single salts. Double salts have distinctive physical properties as a solid but show a mixed behavior related to original individual salts when in solution. †Alums are members of this class of compound.

double star pair of stars that appear close together. Many stars that appear single to the naked eye appear double when viewed through a telescope. Some double stars attract each other due to gravity, and orbit each other, forming a genuine †binary star, but other double stars are at different distances from Earth, and lie in the same line of sight only by chance. Through a telescope both types look the same.

Double stars of the second kind (which are of little astronomical interest) are referred to as "optical pairs," those of the first kind as "physical pairs" or, more usually, "visual binaries." They are the principal source from which is derived our knowledge of stellar masses.

double theta functions †elliptic functions in the form of †theta functions of higher degree.

doubling the cube ancient Greek problem in geometrical construction, to derive a cube of exactly twice the volume of another cube of given measurement, using ruler and compass only.

downhole measurement in geology, any one of a number of experiments performed by instruments lowered down a borehole. Such instruments may be detectors that study vibrations passing through the rock from a generator, or may measure the electrical resistivity of the rock or the natural radiation.

Geophysicists measured heat flow at 10,000 boreholes worldwide 1986–1994 to gather information on climate change. Surface air temperature affects the temperature of underlying rock. Heat travels so slowly that the first 500 m/1,640 ft of the Earth's crust provides a record of the ground temperature for the last thousand years.

Draco large but faint constellation represented as a dragon coiled around the north celestial pole. Due to †precession (Earth's axial wobble), the star Alpha Draconis (Thuban) was the pole star 4,700 years ago. This star seems to have faded, for it is no longer the brightest star in the constellation as it was at the beginning of the 17th century. Gamma Draconis is more than a magnitude brighter. It was extensively observed in the early 18th century by English astronomer James †Bradley, who from its apparent changes in position discovered the †aberration of starlight and †nutation.

drainage basin or **catchment area**, area of land drained by a river system (a river and its tributaries). It includes the surface run-off in the †water cycle, as well as the †water table. Drainage basins are separated by watersheds. A drainage basin is an example of an **open system** because it is open to inputs from outside, such as precipitation, and is responsible for outputs out of the system, such as output of water into the sea and †evaporation of water into the atmosphere.

Draper, Henry (1837–1882) US astronomer. He used a mass spectrograph of his own devising to obtain high-quality spectra of celestial objects. His work is commemorated by the *Henry Draper Catalogue* of stellar spectral types.

Draper Catalogue in astronomy, catalogue of nearly 300,000 stars completed by Harvard Observatory as a memorial to the US astronomer Henry †Draper.

dream series of events or images perceived through the mind during sleep. Their function is unknown, but Sigmund †Freud saw them as wish fulfillment (nightmares being failed dreams prompted by fears of "repressed" impulses). Dreams occur in periods of rapid eye movement (REM) by the sleeper, when the cortex of the brain is approximately as active as in waking hours. Dreams occupy about a fifth of sleeping time.

If a high level of acetylcholine (chemical responsible for transmission of nerve impulses) is present, dreams occur too early in sleep, causing wakefulness, confusion, and depression, which suggests that a form of memory search is involved. Prevention of dreaming, by taking sleeping pills, for example, has similar unpleasant results. For the purposes of (allegedly) foretelling the future, dreams fell into disrepute in the scientific atmosphere of the 18th century.

dreikanter in earth science, eroded stone with three edges, formed by the action of wind-blown sand.

Dreyer, John Louis Emil (1852–1926) Danish astronomer. Resident in Ireland from 1874, he compiled three catalogues which together described more than 13,000 †nebulae and star clusters; these achieved international recognition as standard reference material. He also wrote a biography of Danish astronomer Tycho †Brahe in 1890.

drug any of a range of substances, natural or synthetic, administered to humans and animals as therapeutic agents: to diagnose, prevent, or treat disease, or to assist recovery from injury. Traditionally many drugs were obtained from plants or animals; some minerals also had medicinal value. Today, increasing numbers of drugs are synthesized in the laboratory. It is useful to categorize drugs used by humans into three groups: over-the-counter and prescription drugs (medicines) such as paracetamol and penicillin; recreational drugs such as alcohol and nicotine; and the drugs which are taken illegally if they are not prescribed, such as amphetamines, cannabis, and cocaine.

Drugs are administered in various ways, including orally, by injection, as a lotion or ointment, as a pessary, by inhalation, or by transdermal patch.

One of the first scientific investigations of the effect of a drug on the human body was of digitalis. Digitalis is a mixture of drugs from the foxglove (*Digitalis purpurea*). Its effect is to increase the force of contraction of heart muscle. It is still used to help treat a number of heart problems, such as heart failure.

Drugs prescribed by doctors may have unwanted side effects. Side effects are normally slight, but overdoses and mixing with other drugs can make the side effects worse. These effects can be dangerous. Paracetamol reduces pain, but in large doses it is poisonous and can kill.

Some drugs can become addictive. The most commonly used addictive drug is nicotine. Regular smokers find that the addiction causes it to be very difficult or impossible to give up smoking even though they try. Even more dangerous than nicotine are some mind-affecting addictive drugs. The illegal drugs cocaine and crack are addictive and users find that they have to take more of the drug to get the same effect once they are addicted. Large doses of these drugs can kill.

Internet Drug Index http://www.rxlist.com/

Online list of prescription and non-prescription drugs, with a search engine to help find specific types or products and access to information about indications and side effects too.

Molecular Expressions: The Pharmaceutical Collection http://micro.magnet.fsu.edu/pharmaceuticals/index.html

Fascinating collection of images showing what over 100 drugs look like when recrystallized and photographed through a microscope.

drumlin long streamlined hill created in formerly glaciated areas. Debris (†till) is transported by the glacial icesheet and molded to form an egg-shaped mound, 8–60 m/25–200 ft in height and 0.5–1 km/0.3–0.6 mi in length. Drumlins commonly occur in groups on the floor of †glacial troughs, producing what is called a "basket-of-eggs" landscape.

Drumlins are important indicators of the direction of ice flow, as their blunt ends point upstream, and their gentler slopes trail off downstream.

drupe fleshy †fruit containing one or more seeds that are surrounded by a hard, protective layer – for example cherry, almond, and plum. The wall of the fruit (†pericarp) is differentiated into the outer skin (exocarp), the fleshy layer of tissues (mesocarp), and the hard layer surrounding the seed (endocarp).

The coconut is a drupe, but here the pericarp becomes dry and fibrous at maturity. Blackberries are an aggregate fruit composed of a cluster of small drupes.

druse in earth science, incrustation of mineral crystals that form within a cavity in a rock.

dry cell source of †electromotive force that does not contain a liquid. Usually taken to imply the †Leclanché type of metal and paste cell.

dry ice solid carbon dioxide (CO_2), used as a refrigerant. At temperatures above –79°C/–110.2°F, it sublimes (turns into vapor without passing through a liquid stage) to gaseous carbon dioxide.

DTP abbreviation for †desktop publishing.

duality principle that a law or theorem remains valid if one particular element within that law is exchanged for another equally pertinent element. In projective geometry, a statement of two-dimensional proposition remains valid if the word "point" is exchanged for "line" (and vice versa); a three-dimensional proposition likewise if "point" is exchanged for "plane."

dubnium chemical symbol Db, synthesized, radioactive, metallic element of the †transactinide series, atomic number 105, relative atomic mass 261. Six isotopes have been synthesized, each with very short (fractions of a second) half-lives. Two institutions claim to have been the first to produce it: the Joint Institute for Nuclear Research in Dubna, Russia, in 1967; and the University of California at Berkeley, USA, who disputed the Soviet claim in 1970.

Du Bois-Reymond, Emil Heinrich (1818–1896) German physiologist. He showed the existence of electrical currents in nerves, correctly arguing that it would be possible to transmit nerve impulses chemically. His experimental techniques proved the basis for almost all future work in electrophysiology, and his work with Hermann †Helmholtz and Carl †Ludwig helped establish a link between animal physiology and chemical and physical laws.

ductile describing a material that can sustain large deformations beyond the limits of its †elasticity, without fracture. Metals are very ductile, and may be pulled out into wires, or hammered or rolled into thin sheets without breaking.

ductless gland alternative name for an †endocrine gland.

Dulbecco, Renato (1914–) Italian biologist who was awarded a Nobel Prize for Physiology or Medicine in 1975 with David Baltimore and Howard †Temin for their work on the interactions between tumor-inducing viruses and the genetic material of the cell. Their work was based on the cancer-causing properties of the genes of papovaviruses, circular DNA viruses that integrate into the host cell's DNA, producing a cell clone in which the virus is maintained.

Dulong, Pierre Louis (1785–1838) French chemist and physicist. In 1819 he formulated, together with physicist Alexis †Petit, the †Dulong and Petit law, which states that, for many elements solid at room temperature, the product of †relative atomic mass and †specific heat capacity is approximately constant. He also discovered the explosive nitrogen trichloride in 1811.

Dulong and Petit law in chemistry, law stating that the product of atomic weight and specific heat of a solid element is a constant. The law is not absolute and was later modified to include only metallic elements. It was first observed by French chemist Pierre-Louis †Dulong and French physicist Alexis-Thérèse †Petit in 1819.

Dumas, Jean Baptiste André (1800–1884) French chemist. He made contributions to organic analysis and synthesis, and to the determination of atomic weights (relative atomic masses) through the measurement of vapor densities. In 1833 Dumas worked out an absolute method for the estimation of the amount of nitrogen in an organic compound, which still forms the basis of modern methods of analysis. He went on to correct the atomic masses of 30 elements – half the total number known at that time – referring to the hydrogen value as 1.

dune mound or ridge of wind-drifted sand, common on coasts and in deserts. Loose sand is blown and bounced along by the wind, up the windward side of a dune (†saltation). The sand particles then fall to rest on the lee side, while more are blown up from the windward side. In this way a dune moves gradually downwind. Dunes occur in areas where there is a large supply of sand, strong winds, low rainfall, and some vegetation or obstructions to trap the sand.

In sandy deserts, the typical crescent-shaped dune is called a **barchan**. **Seif dunes** are longitudinal and lie parallel to the wind direction, and **star-shaped dunes** are formed by irregular winds.

dung waste matter excreted by living animals. Dung may also serve as a marker through the addition of scents from the anal glands, whether for determining territorial boundaries or as an indication of status within a group.

Some animals, such as rabbits, may reingest dung immediately after excretion and continue digesting it, a process known as **refection**. In addition to being broken down by bacteria, animal dung provides food for many invertebrates, especially beetles and flies, and provides a habitat for certain species of fungi and plants such as stinging nettles.

duodecimal system system of arithmetic notation using 12 as a base, at one time considered superior to the decimal number system in that 12 has more factors (2, 3, 4, 6) than 10 (2, 5).

It is now superseded by the universally accepted decimal system.

duodenum in vertebrates, a short length of †alimentary canal found between the stomach and the small intestine. Its role is in digesting carbohydrates, fats, and proteins. The smaller molecules formed are then absorbed, either by the duodenum or the ileum.

Entry of food to the duodenum is controlled by the pyloric sphincter, a muscular ring at the outlet of the stomach. Once food has passed into the duodenum it is mixed with bile released from the gall bladder and with a range of enzymes secreted from the pancreas, a digestive gland near the top of the intestine. The bile neutralizes the acidity of the gastric juices passing out of the stomach and aids fat digestion.

dust bowl area in the Great Plains region of North America (Texas to Kansas) that suffered extensive wind erosion as the result of drought and poor farming practice in once-fertile soil. Much of the topsoil was blown away in the droughts of the 1930s and the 1980s.

Similar dust bowls are being formed in many areas today, noticeably across Africa, because of over-cropping and overgrazing.

Dust Bowl http://www.usd.edu/anth/epa/dust.html

Historical information on the US mid-West dust bowl of the 1930s. The site features period photographs and an MPG video clip of a dust storm taken from original film footage (please beware the lengthy download time for this clip).

dust devil small dust storm caused by intense local heating of the ground in desert areas, particularly the Sahara. The air swirls upwards, carrying fine particles of dust with it.

dust storm weather feature caused when large quantities of fine particles are raised into the atmosphere, reducing visibility to less than 1,000 m/3,280 ft. They happen most often in very dry places. Kuwait International Airport, for example, records an average of 27 dust storms per year.

Du Toit, Alexander Logie (1878–1948) South African geologist. His work was to form one of the foundations for the synthesis of †continental drift theory and †plate tectonics that created the geological revolution of the 1960s.

The theory of continental drift put forward by German geophysicist Alfred †Wegener inspired Du Toit's book *A Geological Comparison of South America and South Africa* (1927), in which he suggested that they had probably once been joined. In *Our Wandering Continents* (1937), he maintained that the southern continents had, in earlier times, formed the supercontinent of Gondwanaland, which was distinct from the northern supercontinent of Laurasia.

Dutrochet, (Rene Joachim) Henri (1776–1847) French physiologist who outlined the process of †osmosis (the passive diffusion of water from high concentration to low concentration through a semi-permeable membrane, such as a cell wall) in plants and described various important parts of the plant respiratory mechanism. He was also the first to recognize the role of the pigment chlorophyll in the conversion by plants of carbon dioxide to oxygen (†photosynthesis) and to identify stomata (pores) on the surface of leaves, later recognized as important in the exchange of gases between the plant and its surroundings.

Duve, Christian René de (1917–) English-born Belgian biochemist who was awarded a Nobel Prize for Physiology or Medicine in 1974 with Albert Claude and George †Palade for their work in determining the structural and functional organization of the cell, and in particular the structure and function of †lysosomes (cell organelles containing enzymes that can break down the cell).

Du Vigneaud, Vincent (1901–1978) US biochemist who was awarded the Nobel Prize for Chemistry in 1955 for his investigations into biochemically important sulfur compounds and the first synthesis of a polypeptide hormone, †oxytocin. Du Vigneaud also isolated vitamin H (biotin) and determined its structure.

DVD **digital versatile disk** or **digital video disk**, disk format for storing digital information. DVDs can hold 14 times the data stored on compact discs (CDs). Pre-recorded DVDs have a storage capacity of 4.7 gigabytes and can hold a full-length feature film. As with CDs, information is etched in the form of microscopic pits onto a plastic disk (though the pits are half the size), which is then coated with aluminum. DVDs may have two pitted surfaces per side whereas CDs have only one. The data is read optically using a laser as the disk rotates. A double-layer disk can hold four hours of video.

The Japanese company TDK produced the rewriteable DVD-RAM, capable of holding 2.6 gigabytes, in 1996. The DVD-RW, the rival rewriteable format to DVD-RAM, was developed by Sony and Philips in 1998, and can hold 4.7 gigabytes. A DVD drive now comes as standard on many new PCs, and DVD-RAM has become the standard of choice in the rewriteable market. DVD-RAM drives can store up to 6 gigabytes of data (3 gigabytes on each side). DVD players for home entertainment are now common.

dye substance that, applied in solution to fabrics, stains with a permanent color. Different types of dye are needed for different types of fibers. **Direct dyes** combine with cellulose-based fabrics like cotton, linen, and rayon, to color the fibers. **Indirect dyes** require the presence of another substance (a mordant), with which the fabric must first be treated, to ensure that the dye will remain "fast" during washing. **Vat dyes** are colorless soluble substances that on exposure to air yield an insoluble colored compound that is resistant to water.

Naturally occurring dyes include indigo and madder (alizarin), which are extracted from plants; logwood, produced from wood; and cochineal, a red dye made from crushed insects. Industrial dyes are usually synthetic and are derived from coal or petroleum. English chemist William †Perkin developed the first synthetic dye, mauve, in 1856 and by the early 20th century a wide range of synthetic dyes was available.

dyke in earth science, a sheet of †igneous rock created by the intrusion of magma (molten rock) across layers of pre-existing rock. (By contrast, a sill is intruded *between* layers of rock.) It may form a ridge when exposed on the surface if it is more resistant than the rock into which it intruded. A dyke is also a human-made embankment built along a coastline (for example, in the Netherlands) to prevent the flooding of lowland coastal regions.

Dynamic HTML in computing, the fourth version of hypertext markup language (†HTML), the language used to create Web pages. It is called Dynamic HTML because it enables dynamic effects to be incorporated in pages without the delays involved in downloading Java †applets and without referring back to the server.

Welcome to DHTML School http://www.w3schools.com/dhtml/

Dedicated to Dynamic HTML, an advanced programming language for the Internet, this site offers a Dynamic HTML tutorial, a selection of articles, a number of demonstrations, and sections on Cascading Style Sheets (CSS).

dynamics or **kinetics**, in mechanics, the mathematical and physical study of the behavior of bodies under the action of forces that produce changes of motion in them.

dynamite explosive consisting of a mixture of nitroglycerine and diatomaceous earth (diatomite, an absorbent, chalklike material). It was first devised by Alfred Nobel.

dynamo in physics, a simple †generator or machine for transforming mechanical energy into electrical energy. A dynamo in basic form consists of a powerful field magnet between the poles of which a suitable conductor, usually in the form of a coil (armature), is rotated. The magnetic lines of force are cut by the rotating wire coil, which induces a current to flow through the wire. The mechanical energy of rotation is thus converted into an electric current in the armature.

Present-day dynamos work on the principles described by English physicist Michael Faraday in 1830, that an †electromotive force is developed in a conductor when it is moved in a magnetic field. The dynamo that powers the lights on a bicycle is an example of an alternator, that is, it produces alternating current (AC).

dynamometer instrument for measuring the †power generated by a device.

dynamo theory theory of the origin of the magnetic fields of the Earth and other planets having magnetic fields in which the rotation of the planet as a whole sets up currents within the planet capable of producing a weak magnetic field.

dyne symbol dyn, *c*.g.s. unit of force. 10^5 dynes make one newton. The dyne is defined as the force that will accelerate a mass of one gram by one centimeter per second per second.

Dyson, Frank Watson (1868–1939) English astronomer. He was especially interested in stellar motion and time determination. He was one of a number of astronomers who confirmed the observations of Jacobus †Kapteyn on the proper motions of stars, which indicated that the stars in our Galaxy seemed to be moving in two great streams. These results were later realized to be the first evidence for the rotation of our Galaxy. He initiated the public broadcasting of time signals by the British Broadcasting Corporation over the radio in 1924. He was knighted in 1915 and made a KCB in 1926.

dysprosium chemical symbol Dy, (Greek *dusprositos* "difficult to get near") silver-white, metallic element of the †lanthanide series, atomic number 66, relative atomic mass 162.50. It is among the most magnetic of all known substances and has a great capacity to absorb neutrons.

It was discovered in 1886 by French chemist Paul Lecoq de Boisbaudran (1838–1912).

E

ear organ of hearing in animals. It responds to the vibrations that constitute sound, which are translated into nerve signals and passed to the brain. A mammal's ear consists of three parts: outer ear, middle ear, and inner ear. The **outer ear** is a funnel that collects sound, directing it down a tube to the **eardrum** (tympanic membrane), which separates the outer and **middle ear**. Sounds vibrate this membrane, the mechanical movement of which is transferred to a smaller membrane leading to the **inner ear** by three small bones, the auditory ossicles. Vibrations of the inner ear membrane move fluid contained in the spiral-shaped cochlea, which vibrates hair cells that stimulate the auditory nerve connected to the brain. There are approximately 30,000 sensory hair cells (**stereocilia**). Exposure to loud noise and the process of aging damages the stereocilia, resulting in hearing loss. Three fluid-filled canals of the inner ear detect changes of position; this mechanism, with other sensory inputs, is responsible for the sense of balance.

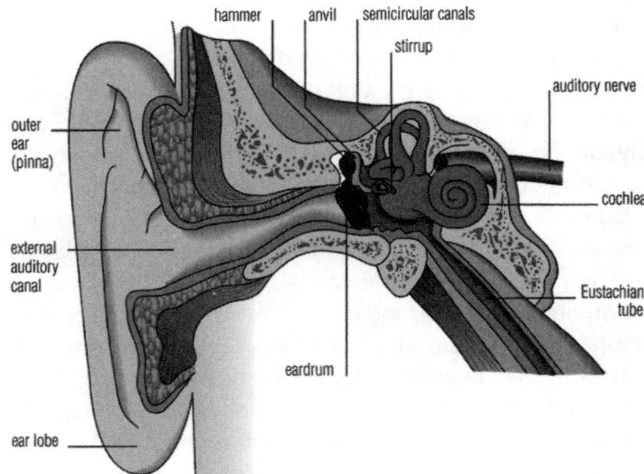

ear The structure of the ear. The three bones of the middle ear – hammer, anvil, and stirrup – vibrate in unison and magnify sounds about 20 times. The spiral-shaped cochlea is the organ of hearing. As sound waves pass down the spiral tube, they vibrate fine hairs lining the tube, which activate the auditory nerve connected to the brain. The semicircular canals are the organs of balance, detecting movements of the head. Helicon

When a loud †noise occurs, muscles behind the eardrum contract automatically, suppressing the noise to enhance perception of sound and prevent injury.

Locating a Mouse by its Sound http://www.hhmi.org/senses/c/c210.htm

As part of a much larger site called "Seeing, Hearing, and Smelling the World," here is a page examining the way we can locate objects through sound alone. It is divided into four sections called "A brain map of auditory space," "The value of having two ears," "Bat sounds and human speech," and "On the trail of a deafness gene." This site makes good use of images and animations to help with the explanations, so it is best viewed with an up-to-date browser.

Range of Human Hearing http://www.sfu.ca/sca/Manuals/ZAAPf/r/range.html

All about sound and hearing. The first page is a brief outline of the upper limits of human hearing in the domains of frequency and intensity. The user can then follow the links in the text for pages containing further definitions of terms, such as sound pressure and waves.

Earth third planet from the Sun. It is almost spherical, flattened slightly at the poles, and is composed of five concentric layers: inner †core, outer core, †mantle, †crust, and atmosphere. About 70% of the surface (including the north and south polar ice caps) is covered with water. The Earth is surrounded by a life-supporting atmosphere and is the only planet on which life is known to exist.

mean distance from the Sun: 149,500,000 km/92,860,000 mi

equatorial diameter: 12,755 km/7,920 mi

circumference: 40,070 km/24,900 mi

rotation period: 23 hours 56 minutes 4.1 seconds

year: (complete orbit, or sidereal period) 365 days 5 hours 48 minutes 46 seconds. The Earth's average speed around the Sun is 30 kps/18.5 mps. The plane of its orbit is inclined to its equatorial plane at an angle of 23.5°; this is the reason for the changing seasons

atmosphere: nitrogen 78.09%; oxygen 20.95%; argon 0.93%; carbon dioxide 0.03%; and less than 0.0001% neon, helium, krypton, hydrogen, xenon, ozone, and radon

surface: land surface 150,000,000 sq km/57,500,000 sq mi (greatest height above sea level 8,872 m/29,118 ft Mount Everest); water surface 361,000,000 sq km/139,400,000 sq mi (greatest depth 11,034 m/36,201 ft †Mariana Trench in the Pacific). The interior is thought to be an inner core about 2,600 km/1,600 mi in diameter, of solid iron and nickel; an outer core about 2,250 km/1,400 mi thick, of molten iron and nickel; and a mantle of mostly solid rock about 2,900 km/1,800 mi thick. The crust and the uppermost layer of the mantle form about twelve major moving plates, some of which carry the continents. The plates are in constant, slow motion, called tectonic drift

satellite: the †Moon

age: 4.6 billion years. The Earth was formed with the rest of the †Solar System by consolidation of interstellar dust. Life began 3.5–4 billion years ago.

Earth Introduction http://www.solarviews.com/solar/eng/earth.htm

Everything you ever wanted to know about planet Earth. This site contains a table of statistics, photographs taken from space, radar-generated images of the planet, animations of the Earth rotating, and more.

earth electrical connection between an appliance and the ground. In the event of a fault in an electrical appliance, for example, involving connection between the live part of the circuit and the outer casing, the current flows to earth, causing no harm to the user.

In most domestic installations, earthing is achieved by a connection to a metal water-supply pipe buried in the ground.

earthquake abrupt motion of the Earth's surface. Earthquakes are caused by the sudden release in rocks of strain accumulated over time as a result of †plate tectonics. The study of earthquakes is called †seismology. Most earthquakes occur along †faults (fractures or breaks) and †Benioff zones. As two plates move past each other they can become jammed. When sufficient strain has accumulated, the rock breaks, releasing a series of elastic waves (†seismic waves) as the plates spring free. The force of earthquakes (magnitude) is measured on the †Richter scale, and their effect (intensity) on the †Mercalli scale. The point at which an earthquake originates is the **focus** or **hypocenter**; the point on the Earth's surface directly above this is the †**epicenter**.

The Alaskan (USA) earthquake of 27 March 1964 ranks as one of the greatest ever recorded, measuring 8.3–8.8 on the Richter scale. The 1906 San Francisco earthquake is among the most famous in history. Its magnitude was 8.3 on the Richter scale. The deadliest, most destructive earthquake in historical times is thought to have been in China in 1556. In 1987 a Californian earthquake was successfully predicted by measurement of underground pressure waves; prediction attempts have also involved the study of such phenomena as the change in gases issuing from the †crust, the level of water in wells, slight deformation of the rock surface, a sequence of minor tremors, and the behavior of animals. The possibility of earthquake prevention is remote. However, rock slippage might be slowed at movement points, or promoted at stoppage points, by the extraction or injection of large quantities of water underground, since water serves as a lubricant. This would ease overall pressure. Human activity can create earthquakes. Mining, water extraction, and oil extraction can cause subsidence that helps generate earthquakes, while the building of large dams and underground nuclear testing have also been linked with earthquakes.

Before and After the Great Earthquake and Fire: Early Films of San Francisco, 1897–1916 http://lcweb2.loc.gov/ammem/papr/sfhome.html

Collection of 26 early films depicting San Francisco before and after the 1906 disaster, including a 1915 travelogue that shows scenes of the rebuilt city and the Panama Pacific Exposition, and a 1916 propaganda film. Should you not wish to download the entire film, each title contains sample still-frames. There is also background information about the earthquake and fire, and a selective bibliography.

Earthquakes and Plate Tectonics http://wwwneic.cr.usgs.gov/neis/plate_tectonics/rift_man.html

US Geological Survey National Earthquake Information Center site, explaining the relationship between plate tectonics and earthquakes.

earth science scientific study of the planet Earth as a whole. The mining and extraction of minerals and gems, the prediction of weather and earthquakes, the pollution of the atmosphere, and the forces that shape the physical world all fall within its scope of study. The emergence of the discipline reflects scientists' concern that an understanding of the global aspects of the Earth's structure and its past will hold the key to how humans affect its future, ensuring that its resources are used in a sustainable way. It is a synthesis of several traditional subjects such as †geology, †meteorology, †oceanography, †geophysics, †geochemistry, and †paleontology.

Earth Sciences http://dir.yahoo.com/science/earth_sciences/

Yahoo! directory of earth science, which provides links to hundreds of earth science sites, organized by discipline, and increasing in number every day.

earthshine in astronomy, faintly lit face of the Moon, illuminated by sunlight reflected from clouds and the surface of the Earth.

east one of the four cardinal points of the compass, indicating that part of the horizon where the Sun rises; when facing north, east is to the right.

Ebbinghaus, Hermann (1850–1909) German experimental psychologist. Influenced by Gustav †Fechner's *Elements of Psychophysics* (1860), he applied quantitative principles to the study of higher mental processes, in particular to human memory.

eccentricity in geometry, a property of a †conic section (circle, ellipse, parabola, or hyperbola). It is the distance of any point on the curve from a fixed point (the focus) divided by the distance of that point from a fixed line (the directrix). A circle has an eccentricity of zero; for an ellipse it is less than one; for a parabola it is equal to one; and for a hyperbola it is greater than one.

Eccles, John Carew (1903–1997) Australian physiologist who was awarded a Nobel Prize for Physiology or Medicine in 1963 with Alan †Hodgkin and Andrew †Huxley for their work on ionic mechanisms involved in the communication or inhibition of impulses across neuron (nerve cell) membranes. In some of his later works, he argued that the mind has an existence independent of the brain.

ecdysis periodic shedding of the †exoskeleton by insects and other arthropods to allow growth. Prior to shedding, a new soft and expandable layer is laid down underneath the existing one. The old layer then splits, the animal moves free of it, and the new layer expands and hardens.

ECG abbreviation for †electrocardiogram.

echinoderm marine invertebrate of the phylum Echinodermata ("spiny-skinned"), characterized by a five-radial symmetry. Echinoderms have a water-vascular system which transports substances around the body. They include starfishes (or sea stars), brittle-stars, sea lilies, sea urchins, and sea cucumbers. The skeleton is external, made of a series of limy plates. Echinoderms generally move by using tube-feet, small water-filled sacs that can be protruded or pulled back to the body.

Echinodermata phylum of invertebrate animals, members of which are called †echinoderms.

echo repetition of a sound wave, or of a †radar or †sonar signal, by reflection from a hard surface such as a wall or building. By accurately measuring the time taken for an echo to return to the transmitter, and by knowing the speed of a radar signal (the speed of light) or a sonar signal (the speed of sound in water), it is possible to calculate the range of the object causing the echo (†echolocation).

The speed of sound can be measured using an echo. The time taken for the sound to be reflected back over a known distance is recorded. The speed of sound can then be calculated using the formula: speed = distance/time.

For an echo to occur, the wavelength of the incident sound must be smaller than the size of the object. Sound obeys the same rules of reflection as for light. If the wavelength of the sound is the same as or greater than the size of the object, then diffraction takes place.

echolocation or **biosonar**, method used by certain animals, notably bats, whales, and dolphins, to detect the positions of objects by using sound. The animal emits a stream of high-pitched sounds, generally at ultrasonic frequencies (beyond the range of human hearing), and listens for the returning echoes reflected off objects to determine their exact location.

The location of an object can be established by the time difference between the emitted sound and its differential return as an echo to the two ears. Echolocation is of particular value under conditions when normal vision is poor (at night in the case of bats, in murky water for dolphins). A few species of bird can also echolocate, including cave-dwelling birds such as some species of swiftlets and the South American oil bird.

The frequency range of bats' echolocation calls is 20–215 kHz. Many species produce a specific and identifiable pattern of sound. Bats vary in the way they use echolocation: some emit pure sounds lasting up to 150 milliseconds, while others use a series of shorter "chirps." Sounds may be emitted through the mouth or nostrils depending on species.

Echolocation was first described in the 1930s, though it was postulated by Italian biologist Lazzaro Spallanzani (1729–1799).

Eckert, J(ohn) Presper (1919–1995) US electronics engineer and mathematician who collaborated with John †Mauchly on the development of the early ENIAC (1946) and UNIVAC 1 (1951) computers.

Eckert, John Presper http://www-groups.dcs.st-andrews.ac.uk/history/Mathematicians/Eckert_John.html

Part of an archive containing the biographies of the world's greatest mathematicians, this site is devoted to the life and contributions of John Eckert. In addition to biographical information, you will find a list of references about Eckert and links to other essays in the archive that reference him. The text of this essay includes hypertext links to the essays of those mathematicians and thinkers who influenced Eckert. You will also find an image of him, which you can click on to enlarge, and a map of his birthplace.

eclipse passage of one astronomical body through the shadow of another. The term is usually used for solar and lunar eclipses. A **solar eclipse** occurs when the Moon is between the Earth and the Sun (which can happen only at new Moon), the Moon blocking the Sun's rays and casting a shadow on the Earth's surface. A **lunar eclipse** occurs when the Earth is between the Moon and the Sun (which can happen only at full Moon), the Earth blocking the Sun's rays and casting a shadow on the Moon's surface.

During a total solar eclipse the Moon appears to cover the Sun's disc completely and day turns into night. This is known as the umbra. A total solar eclipse can last up to 7.5 minutes, and the Sun's †corona can be seen. Between two and five solar eclipses occur each year but each is visible only from a specific area. During a partial solar eclipse sunlight reaches the Earth from around the edge of the Moon. This is known as the pre-umbra. Lunar eclipses can also be partial or total. Total lunar eclipses last for up to 100 minutes; the maximum number each year is three.

Lunar Eclipse Computer
http://aa.usno.navy.mil/data/docs/LunarEclipse.html
Part of a larger site on astronomical data maintained by the US Naval Observatory, this site provides data on recent and upcoming lunar eclipses for any location around the world. The data available includes the local time of each eclipse "event"; the altitude and azimuth of the Moon at each of the events; plus the time of moonrise immediately preceding, and the time of moonset immediately following, the eclipse. Sections of Frequently Asked Questions and research information are also included.

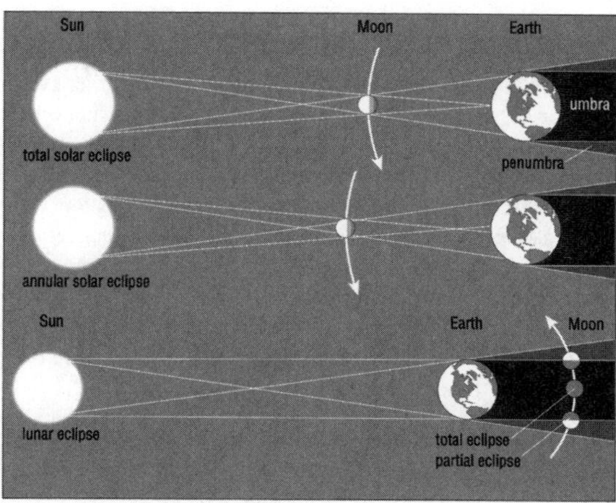

eclipse The two types of eclipse: lunar and solar. A lunar eclipse occurs when the Moon passes through the shadow of the Earth. A solar eclipse occurs when the Moon passes between the Sun and the Earth, blocking out the Sun's light. During a total solar eclipse, when the Moon completely covers the Sun, the Moon's shadow sweeps across the Earth's surface from west to east at a speed of 3,200 kph/2,000 mph. Helicon

Solar Eclipse http://www.eclipse.org.uk/default_hi.htm
Pages dedicated to the eclipse that was visible over the UK on 11 August 1999. This site contains detailed information on the science of eclipses, and includes images of the event.

eclipsing binary binary (double) star in which the two stars periodically pass in front of each other as seen from Earth.

When one star crosses in front of the other the total light received on Earth from the two stars declines. The first eclipsing binary to be noticed was †Algol.

ecliptic path, against the background of stars, that the Sun appears to follow each year as it is orbited by the Earth. It can be thought of as the plane of the Earth's orbit projected onto the †celestial sphere.

The ecliptic is tilted at about 23.5° with respect to the celestial equator, a result of the tilt of the Earth's axis relative to the plane of its orbit around the Sun.

ecliptic coordinates in astronomy, system for measuring the position of astronomical objects on the †celestial sphere with reference to the plane of the Earth's orbit, the †ecliptic.

Ecliptic latitude (symbol b) is measured in degrees from the ecliptic (b = 0°) to the north (b = 90°) and south (b = – 90°) ecliptic poles.

Ecliptic longitude (symbol l) is measured in degrees eastward along the ecliptic (l = 0° to 360°) from a fixed point known as the first point of †Aries or the vernal †equinox. Ecliptic coordinates are often used to measure the positions of the Sun and planets with respect to the Earth.

Ecliptic latitude and longitude are sometimes known as celestial latitude and longitude. The ecliptic longitude of the Sun (solar longitude) is a convenient measure of the position of the Earth in its orbit.

E. coli abbreviation for †*Escherichia coli.*

ecology (Greek *oikos* "house") study of the relationship among organisms and the environments in which they live, including all living and nonliving components. The chief environmental factors governing the distribution of plants and animals are temperature, humidity, soil, light intensity, day length, food supply, and interaction with other organisms. The term ecology was coined by the biologist Ernst Haeckel in 1866.

Ecology may be concerned with individual organisms (for example, behavioral ecology, feeding strategies), with populations (for example, population dynamics), or with entire communities (for example, competition between species for access to resources in an ecosystem, or predator–prey relationships). Applied ecology is concerned with the management and conservation of habitats and the consequences and control of pollution.

ecosystem in ecology, a unit consisting of living organisms and the environment that they live in. A simple example of an ecosystem is a pond. The pond ecosystem includes all the pond plants and animals and also the water and other substances that make up the pond itself. Individual organisms interact with each other and with their environment in a variety of relationships, such as two organisms in competition, predator and prey, or as a food source for other organisms in a †food chain. These relationships are usually complex and finely balanced, and in natural ecosystems should be self-sustaining. However, major changes to an ecosystem, such as climate change, overpopulation, or the removal of a species, may threaten the system's sustainability and result in its eventual destruction. For instance, the removal of a major carnivore predator can result in the destruction of an ecosystem through overgrazing by herbivores. Ecosystems can be large, such as the global ecosystem (the ecosphere), or small, such as the pools that collect water in the branch of a tree, and they can contain smaller systems.

Introduction to the Ecosystem Concept http://www.geog.ouc.bc.ca/physgeog/contents/9j.html

Clear and well-organized hyperlinked explanation of ecosystems, supported by a diagram showing the components of an ecosystem and their inter-relatedness. Part of a much larger site "Fundamentals of Physical Geography" set up by a Canadian University, the page covers many aspects of the area of study.

ectoderm outer layer of cells in an embryo and all the tissues that it gives rise to.

ectoparasite †parasite that lives on the outer surface of its host.

ectoplasm outer layer of a cell's †cytoplasm.

ectotherm "cold-blooded" animal (see †poikilothermy), such as a lizard, that relies on external warmth (ultimately from the Sun) to raise its body temperature so that it can become active. To cool the body, ectotherms seek out a cooler environment.

Eddington, Arthur Stanley (1882–1944) English astrophysicist who studied the motions, equilibrium, luminosity, and atomic structure of the stars. In 1919 his observation of stars during a solar eclipse confirmed German-born US physicist Albert †Einstein's prediction that light is bent when passing near the Sun, in accordance with the general theory of relativity. In *The Expanding Universe* (1933) Eddington expressed the theory that in the spherical universe the outer galaxies, or spiral nebulae, are receding from one another.

Eddington, Arthur Stanley http://www-groups.dcs.st-andrews.ac.uk/history/Mathematicians/Eddington.html

Devoted to the life and contributions of astrophysicist Arthur Eddington. In addition to biographical information, you will find a list of references about Eddington and links to essays that reference him. The text of this essay includes hypertext links to the essays of those mathematicians and thinkers who influenced Eddington. You will also find an image of him, which you can click on to enlarge, and a map of his birthplace.

eddy current electric current induced, in accordance with †Faraday's laws of electromagnetic induction, in a conductor located in a changing magnetic field. Eddy currents can cause much wasted energy in the cores of transformers and other electrical machines.

edge in topology, link on a topological map. Where edges meet and intersect, †nodes are formed.

Edison, Thomas Alva (1847–1931) US scientist and inventor, whose work in the fields of communications and electrical power greatly influenced the world in which we live. With more than 1,000 patents, Edison produced his most important inventions in Menlo Park, New Jersey 1876–87, including the phonograph. He obtained a US patent for the electric light bulb in 1879. He also constructed a system of electric power distribution for consumers, the telephone transmitter, and the megaphone.

Edison, Thomas http://www.si.edu/lemelson/edison/default.asp

Illustrated biography of the US inventor, plus a chronology and bibliography. It includes sections on such topics as his childhood and the electric light.

Edison effect electrical †conduction between a negatively charged filament, and a positively charged †electrode kept together, though separated, in a vacuum chamber.

Edlén, Bengt (1906–1993) Swedish astrophysicist. He resolved the identification of certain lines in spectra of the solar †corona that had misled scientists for the previous 70 years.

educational psychology the work of psychologists primarily in schools, including the assessment of children with achievement problems and advising on problem behavior in the classroom.

EEG abbreviation for †electroencephalogram.

EEPROM acronym for Electrically Erasable Programmable Read-Only Memory, computer memory that can record data and retain it indefinitely. The data can be erased with an electrical charge and new data recorded.

Some EEPROM must be removed from the computer and erased and reprogrammed using a special device. Other EEPROM, called **flash memory**, can be erased and reprogrammed without removal from the computer.

efferent nerve nerve that conducts impulses away from the central nervous system (brain and spinal cord). Most efferent nerves are motor nerves and run to effector organs.

effervescence in chemistry, the generation of large quantities of gas bubbles in a liquid, usually during a chemical reaction.

efficiency in physics, a general term indicating the degree to which a process or device can convert energy from one form to another without loss, or how effectively energy is used, and wasted energy, such as heat and sound, minimized. It is normally expressed as a fraction or a percentage, where 100% indicates conversion with no loss. The efficiency of a machine, for example, is the ratio of the energy output to the energy input; in practice it is always less than 100% because of frictional heat losses.

For example, 75% of the electrical energy of an electric light bulb is converted into heat and only 25% is converted into light. Therefore, an electric light bulb is not an efficient energy converter. Certain electrical machines with no moving parts, such as transformers, can approach 100% efficiency.

efflorescence loss of water or crystallization of crystals exposed to air, resulting in a dry powdery surface.

effluent liquid discharge of waste from an industrial process, usually into rivers or the sea. Effluent is often toxic but is difficult to control and hard to trace.

In some cases, as at Minamata, Japan, where 43 people died of mercury poisoning, effluent can be lethal but usually its toxic effects remain unclear, because it quickly dilutes in the aquatic ecosystem.

egestion removal of undigested food or feces from the gut. In most animals egestion takes place via the anus, although the invertebrate flatworms must use the mouth because their gut has no exit. Egestion refers solely to indigestible matter which is never absorbed into the cells – it should not be confused with †excretion of the waste products of metabolism.

egg in animals, the ovum, or female †gamete (reproductive cell).

After fertilization by a sperm cell, it begins to divide to form an embryo. Eggs may be deposited by the female (†ovipary) or they may develop within her body (†vivipary and †ovovivipary). In the oviparous reptiles and birds, the egg is protected by a shell, and well supplied with nutrients in the form of yolk.

Eggen, Olin Jenck (1919– 1998) US astronomer. His work has included studies of high-velocity stars, †red giants (using narrow- and broadband photometry), and subluminous stars. During the mid-1970s, Eggen completed a study – based on ultraviolet photometry and every available apparent motion – of all red giants brighter than apparent magnitude 5. As a result he was able to classify these stars, categorizing them as very young discs, young discs, and old discs. A few remained unclassifiable (haloes).

egg tooth or **beak tooth**, in birds, the spike on the beak of a chick that develops just before it hatches and helps it to free itself from the shell. It is usually on the upper side of the beak, though in some species it is on the tip.

ego (Latin "I") in psychology, the processes concerned with the self and a person's conception of himself or herself, encompassing values and attitudes. In Freudian psychology, the term refers specifically to the element of the human mind that represents the conscious processes concerned with reality, in conflict with the †id (the instinctual element) and the †superego (the ethically aware element).

Ehrenfels, (Maria) Christian (Julius Leopold Karl) von (1859–1933) Austrian philosopher and psychologist. In his paper *Über Gestalt Qualitäten/On Gestalt Qualities* (1891), he introduced the notion of gestalt to explain observations of wholeness and object-constancy in perception.

Ehrlich, Paul (1854–1915) German bacteriologist and immunologist who was awarded a Nobel Prize for Physiology or Medicine in 1908 with Ilya Mechnikov for their work on immunity. He produced the first cure for syphilis, developing the arsenic compounds, in particular †Salvarsan, that were used in the treatment of syphilis before the discovery of antibiotics.

Eigen, Manfred (1927–) German chemist who worked on extremely rapid chemical reactions (those taking less than 1 millisecond). He was awarded the Nobel Prize for Chemistry in 1967 for his development from 1954 of the technique by which very short bursts of energy could be applied to solutions, disrupting their equilibrium and enabling him to investigate momentary reactions, such as the formation and dissociation of water.

eigenvalue in mathematics, for a matrix A, the number l is said to be an eigenvalue of the matrix if there is a nonzero †vector x such that $Ax = lx$. Eigenvalues are used to derive a change of base to simplify the matrix to one that has entries only on its diagonal. More generally, the number l is an eigenvalue of a linear transformation T if there is a nonzero vector x so that $T(x) = lx$.

eightfold-way in quantum mechanics, scheme for the classification of †elementary particles into families, grouped according to common properties as expressed by various †quantum numbers.

Eijkman, Christiaan (1858–1930) Dutch bacteriologist who was awarded a Nobel Prize for Physiology or Medicine in 1929 for his discovery of a cure for beriberi, a vitamin-deficiency disease. He pioneered the recognition of vitamins as essential to health and identified vitamin B_1 deficiency as the cause of beriberi. He shared the prize with Frederick Gowland †Hopkins, who received the award for his discovery of trace substances, now known as vitamins, that stimulate growth.

Einstein, Albert (1879–1955) German-born US physicist whose theories of †relativity revolutionized our understanding of matter, space, and time. Einstein established that light may have a particle nature. He was awarded the Nobel Prize for Physics in 1921 for his work on theoretical physics, especially the **photoelectric law**. He also investigated Brownian motion, confirming the existence of atoms. His last conception of the basic laws governing the universe was outlined in his †unified field theory, made public in 1953.

Einstein: Still Right After All These Years http://whyfiles.org/052einstein/

Easy-to-understand explanation of some of Einstein's theories. The explanations are grouped under several headings, such as: "The speed of light," "Black holes," and "Gravity."

Life and Theories of Albert Einstein http://www.pbs.org/wgbh/nova/einstein/index.html

Heavily illustrated site on the life and theories of Einstein. There is an illustrated biographical chart, including a summary of his major achievements and their importance to science. The theory of relativity gets an understandably more in-depth coverage, along with photos and illustrations. The pages on his theories on light and time include illustrated explanations and an interactive test. There is also a "Time-traveler" game demonstrating these theories.

einsteinium chemical symbol Es, synthesized, radioactive, metallic element of the actinide series, atomic number 99, relative atomic mass 254.09.

It was produced by the first thermonuclear explosion, in 1952, and discovered in fallout debris in the form of the isotope Es-253 (half-life 20 days). Its longest-lived isotope, Es-254, with a half-life of 276 days, allowed the element to be studied at length. It is now synthesized by bombarding lower-numbered †transuranic elements in particle accelerators. It was first identified by US chemist Albert Ghiorso and his team who named it in 1955 after Albert Einstein, in honor of his theoretical studies of mass and energy.

ejecta in astronomy, any material thrown out of a †crater due to volcanic eruption or the impact of a †meteorite or other object. Ejecta from impact craters on the †Moon often form long bright streaks known as rays, which in some cases can be traced for thousands of kilometers across the lunar surface.

Ekman spiral effect in oceanography, theoretical description of a consequence of the †Coriolis effect on ocean currents, whereby currents flow at an angle to the winds that drive them. It derives its name from the Swedish oceanographer Vagn Ekman (1874–1954).

In the northern hemisphere, surface currents are deflected to the right of the wind direction. The surface current then drives the subsurface layer at an angle to its original deflection. Consequent subsurface layers are similarly affected, so that the effect decreases with increasing depth. The result is that most water is transported at about right-angles to the wind direction. Directions are reversed in the southern hemisphere.

elastic collision in physics, a collision between two or more bodies in which the total †kinetic energy of the bodies is conserved (remains constant); none is converted into any other form of energy. The molecules of a gas may be considered to collide elastically, but large objects may not because some of their kinetic energy will be converted on collision to heat and sound, or used to deform the object.

elasticity in physics, the ability of a solid to recover its shape once deforming forces are removed. An elastic material obeys †Hooke's law, which states that its deformation is proportional to the applied stress up to a certain point, called the **elastic limit**; beyond this point additional stresses will deform it permanently. Elastic materials include metals and rubber; however, all materials have some degree of elasticity.

elastomer any material with rubbery properties that stretches easily and then quickly returns to its original length when released.

Natural and synthetic rubbers and such materials as polychloroprene and butadiene copolymers are elastomers. The convoluted molecular chains making up these materials are uncoiled by a stretching force, but return to their original position when released because there are relatively few crosslinks between the chains.

E-layer formerly **Kennelly-Heaviside layer**, lower regions (90–120 km/56–75 mi) of the †ionosphere, which reflect radio waves, allowing their reception around the surface of the Earth. The E-layer approaches the Earth by day and recedes from it at night.

electrical conduction see †conduction, electrical.

electrical energy form of energy carried by an electric current. It may be converted into other forms of energy such as heat, light, and motion.

Electrical appliances, such as washing machines, vacuum cleaners, radios, and televisions, use electricity in order to do work. Electric motors convert electrical energy into mechanical energy.

electric bell bell that makes use of electromagnetism. A direct current (from a battery) flows through a wire-wound coil on an iron core (an electromagnet). The electromagnet attracts an iron †armature, and a clapper or hammer fixed to the armature strikes the bell. The armature acts as a switch, whose movement causes contact with an adjustable contact point to be broken, so breaking the circuit and switching off the electromagnet. A spring rapidly returns the armature to the contact point, once again closing the circuit, and the process is repeated. The armature oscillates back and forth, causing the clapper or hammer to strike the bell repeatedly.

electric cell device in which chemical energy is converted into electrical energy; see †cell, electrical.

electric charge property of some bodies that causes them to exert forces on each other. Two bodies both with positive or both with negative charges repel each other, whereas bodies with opposite or "unlike" charges attract each other. †Electrons possess a negative charge, and †protons an equal positive charge. The †SI unit of electric charge is the coulomb (symbol C).

A body can be charged by friction, induction, or chemical change, and the charge shows itself as an accumulation of electrons (negative charge) or a loss of electrons (positive charge) on an atom or body. Atoms generally have zero net charge but can gain electrons to become negative ions or lose them to become positive ions. So-called †static electricity, seen in such phenomena as the charging of nylon shirts when they are pulled on or off, or in brushing hair, is in fact the gain or loss of electrons from the surface atoms. A flow of charge (such as electrons through a copper wire) constitutes an electric current; the flow of current is measured in amperes (symbol A).

Static Electricity http://www.sciencemadesimple.com/static.html

Excellent site, explaining the theory simply, but thoroughly. It also includes several easy-to-do experiments to demonstrate the theory behind static electricity.

electric current flow of electrically charged particles through a conducting circuit due to the presence of a †potential difference. The current at any point in a circuit is the amount of charge flowing per second; its SI unit is the ampere (coulomb per second).

Current carries electrical energy from a power supply, such as a battery of electrical cells, to the components of the circuit, where it is converted into other forms of energy, such as heat, light, or motion. It may be either †direct current or †alternating current.

heating effect When current flows in a component possessing resistance, electrical energy is converted into heat energy. If the resistance of the component is R ohms and the current through it is I amperes, then the heat energy W (in joules) generated in a time t seconds is given by the formula: $W = I^2Rt$.

magnetic effect A magnetic field is created around all conductors that carry a current. When a current-bearing conductor is made into a coil it forms an †electromagnet with a †magnetic field that is similar to that of a bar magnet, but which disappears as soon as the current is switched off. The strength of the magnetic field is directly proportional to the current in the conductor – a property that allows a small electromagnet to be used to produce a pattern of magnetism on recording tape that accurately represents the sound or data stored. The direction of the field created around a conducting wire may be predicted by using †Maxwell's screw rule.

motor effect A conductor carrying current in a magnetic field experiences a force, and is impelled to move in a direction perpendicular to both the direction of the current and the direction of the magnetic field. The direction of motion may be predicted by Fleming's left-hand rule (see †Fleming's rules). The magnitude of the force experienced depends on the length of the conductor and on the strengths of the current and the magnetic field, and is greatest when the conductor is at right angles to the field. A conductor wound into a coil that can rotate between the poles of a magnet forms the basis of an †electric motor.

electric field in physics, a region in which a particle possessing electric charge experiences a force owing to the presence of another electric charge. The strength of an electric field, E, is measured in volts per meter (V m^{-1}). It is a type of †electromagnetic field.

An electric field is formed between two metal plates that are parallel to each other and connected to a voltage supply. An electron beam moving through these plates will be deflected slightly towards the positive plate.

electricity all phenomena caused by †electric charge. There are two types of electricity: static and current. Electric charge is caused by an excess or deficit of electrons in a substance, and an electric current is the movement of charge through a material. Materials having equal numbers of positive and negative charges are termed neutral, as the charges balance out. Substances may be electrical conductors, such as metals, which allow the passage of electricity through them readily, or insulators, such as rubber, which are extremely poor conductors. Substances with relatively poor conductivities that increase with a rise in temperature or when light falls on the material are known as †semiconductors. Electric currents also flow through the nerves of organisms. For example, the optic nerve in humans carries electric signals from the eye to the brain. Electricity cannot be seen, but the effects it produces can be clearly seen; for example, a flash of lightning, or the small sparks produced by rubbing a nylon garment.

Electricity is essential to modern society. Electrical devices are used in the home, office, and industry; this is called †mains electricity, and is measured in †kilowatt hours. Electrical energy is used to power devices used for communications such as fax machines, telephones, computers, and satellites.

Electricity and Magnetism http://library.thinkquest.org/12632/magnetism/

Clearly presented explanation of "electric charge," "electromagnetism" and other aspects relating to this area of study.

Verbund: Electricity Glossary http://www.verbund.at/en/extra/help/glossar/index.htm#TopOfPage

Glossary containing definitions of a number of terms to do with electricity. The entries are divided into three sections, "Basic terms," "Units of measurement," and "Engineering," and are explained in an easy-to-understand way.

What Is Electricity? http://www.amasci.com/miscon/whatis.html

Humorous article aiming to answer the question "What is electricity?" It outlines the difficulties in defining electricity, with the use of clear examples. It then describes several ways that the term electricity is used, such as to mean electrical energy, electric field, or electric charge. Many of the key words in the text are linked to useful glossary definitions. The "Applications" section demonstrates real-life applications of physics, such as the workings of the eye. There is also a "Hall of fame" where those who answer the most quiz questions correctly can have their names posted.

electric motor machine that converts electrical energy into mechanical energy. There are various types, including direct-current and induction motors, most of which produce rotary motion. A linear induction motor produces linear (in a straight line) rather than rotary motion. Electric motors and generators have the same components; they differ in the way they are used.

A simple **direct-current motor** consists of a horseshoe-shaped permanent †magnet with a wire-wound coil (†armature) mounted so that it can rotate between the poles of the magnet. The ends of the wire are connected to a †commutator. This reverses the current (from a battery) fed to the coil on each half turn, which rotates because of the mechanical force exerted on a conductor carrying a current in a magnetic field.

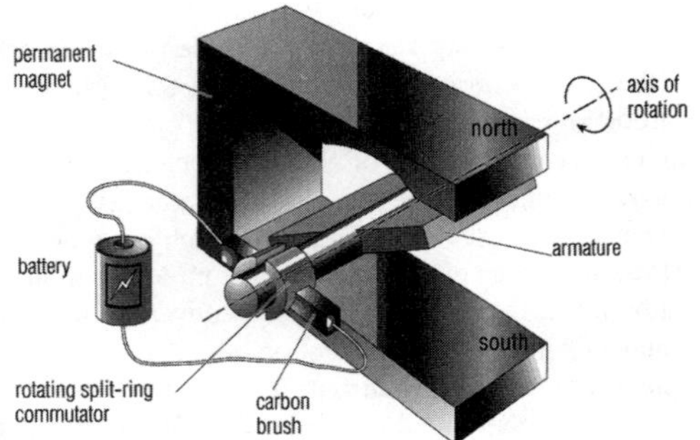

electric motor In an electric motor, magnetic fields generated by electric currents push against each other, causing a shaft (holding the armature) to rotate. Helicon

electric power rate at which an electrical machine uses electrical †energy or converts it into other forms of energy – for example, light, heat, or mechanical energy. Usually measured in watts (equivalent to joules per second), it is equal to the product of the voltage and the current flowing.

In a closed circuit the potential difference (voltage) causes electrons to flow towards the positive potential. Work is done in moving the electrons, and the rate at which this work is done is called the electrical power. If the voltage is V volts and the current is I amps, then the power, P watts, is given by $P = V \times I$.

An electric lamp that passes a current of 0.4 amperes at 250 volts uses 100 watts of electrical power and converts it into light and heat – in ordinary terms it is a 100-watt lamp. An electric motor that requires 6 amperes at the same voltage consumes 1,500 watts (1.5 kilowatts), equivalent to delivering about 2 horsepower of mechanical power.

electrocardiogram ECG, graphic recording of the electrical activity of the heart, as detected by electrodes placed on the skin. Electrocardiography is used in the diagnosis of heart disease.

electrochemical series or **electromotive series**, list of chemical elements arranged in descending order of the ease with which they can lose electrons to form cations (positive ions). An element can be displaced (†displacement reaction) from a compound by any element above it in the series.

electrochemistry branch of science that studies chemical reactions involving electricity. The use of electricity to produce chemical effects, †electrolysis, is employed in many industrial processes, such as electroplating, the manufacture of chlorine, and the extraction of aluminum. The use of chemical reactions to produce electricity is the basis of electrical †cells, such as the dry cell and the †Leclanché cell.

Since all chemical reactions involve changes to the electronic structure of atoms, all reactions are now recognized as electrochemical in nature. Oxidation, for example, was once defined as a process in which oxygen was combined with a substance, or hydrogen was removed from a compound; it is now defined as a process in which electrons are lost.

Electrochemistry is also the basis of new methods of destroying toxic organic pollutants. For example, electrochemical cells that operate with supercritical water (a type of supercritical †fluid) have been developed to combust organic waste materials.

electrode any terminal by which an electric current passes in or out of a conducting substance; for example, the anode or †cathode in an electrolytic cell. The anode is the positive electrode and the cathode is the negative electrode.

The terminals that emit and collect the flow of electrons in thermionic †valves (electron tubes) are also called electrodes: for example, cathodes, plates (anodes), and grids.

electrode potential electric potential between an element and its ions in solution.

electrodynamics branch of physics dealing with electric charges, electric currents, and associated forces. †Quantum electrodynamics (QED) studies the interaction between charged particles and their emission and absorption of electromagnetic radiation. This subject combines quantum theory and relativity theory, making accurate predictions about subatomic processes involving charged particles such as electrons and protons.

electroencephalogram **EEG**, graphic record of the electrical discharges of the brain, as detected by electrodes placed on the scalp. The pattern of electrical activity revealed by electroencephalography is helpful in the diagnosis of some brain disorders, in particular epilepsy.

electrolysis in chemistry, the production of chemical changes by passing an electric current through a solution or molten salt (the electrolyte), resulting in the migration of ions to the electrodes: positive ions (†cations) to the negative electrode (†cathode) and negative ions (†anions) to the positive electrode (†anode).

During electrolysis, the ions react with the electrode, either receiving or giving up electrons. The resultant atoms may be liberated as a gas, or deposited as a solid on the electrode, in amounts that are proportional to the amount of current passed, as discovered by English chemist Michael Faraday. For instance, when acidified water is electrolyzed, hydrogen ions (H^+) at the cathode receive electrons to form hydrogen gas; hydroxide ions (OH^-) at the anode give up electrons to form oxygen gas and water.

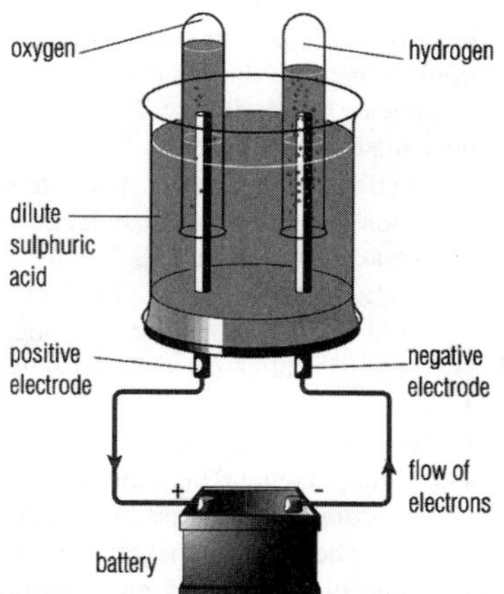

electrolysis Passing an electric current through acidified water (such as diluted sulfuric acid) breaks down the water into its constituent elements – hydrogen and oxygen. Helicon

One application of electrolysis is **electroplating**, in which a solution of a salt, such as silver nitrate ($AgNO_3$), is used and the object to be plated acts as the negative electrode, thus attracting silver ions (Ag^+). Electrolysis is used in many industrial processes, such as coating metals for vehicles and ships, refining †bauxite into aluminum, and the chlor-alkali industry, in which †brine (sodium chloride solution) is electrolyzed to produce †chlorine, hydrogen, and †sodium hydroxide (caustic soda); it also forms the basis of a number of electrochemical analytical techniques, such as polarography.

Electrical Decomposition by Michael Faraday http://dbhs.wvusd.k12.ca.us/Chem-History/Faraday-electrochem.html
Transcript of Faraday's paper in Philosophical Transactions of the Royal Society, 1834, in which Faraday describes for the first time the phenomena of electrolysis. The lengthy text is complete with simple diagrams of Faraday's own equipment.

electrolyte solution or molten substance in which an electric current is made to flow by the movement and discharge of ions in accordance with Faraday's laws of †electrolysis.

The term "electrolyte" is frequently used to denote a substance that, when dissolved in a specified solvent, usually water, dissociates into †ions to produce an electrically conducting medium.

In medicine the term is often used for the ion itself (sodium or potassium, for example). Electrolyte balance may be severely disrupted in illness or injury.

electrolytic cell device in which an externally applied voltage brings about a chemical reaction; see †electrolysis.

electromagnet coil of wire wound around a soft iron core that acts as a magnet when an electric current flows through the wire. Electromagnets have many uses: in switches, electric bells, †solenoids, and metal-lifting cranes.

The strength of the electromagnet can be increased by increasing the current through the wire, changing the material of the core, or by increasing the number of turns in the wire coil. At the north pole of the electromagnet current flows anticlockwise; at the south pole flow is clockwise.

electromagnetic field in physics, region in which a particle with an †electric charge experiences a force. If it does so only when moving, it is in a pure **magnetic field**; if it does so when stationary, it is in an **electric field**. Both can be present simultaneously. For example, a light wave consists of an electric field and a magnetic field traveling simultaneously at right angles to each other.

electromagnetic force one of the four fundamental †forces of nature, the other three being the gravitational force (gravity), the weak nuclear force, and the strong nuclear force. The particle that is the carrier for the electromagnetic force is the †photon.

Sketches of a History of Classical Electromagnetism http://history.hyperjeff.net/electromagnetism.html
Annotated chronology of progress in electromagnetics. Key personalities are linked to biographical information. The site has a brief reading list, and links to relevant tutorials and sites with time lines.

electromagnetic induction in electronics, the production of an †electromotive force (emf) in a circuit by a change of magnetic flux through the circuit or by relative motion of the circuit and the magnetic flux. As a magnet is moved in and out of a coil of wire in a closed circuit an †induced current will be produced. All dynamos and generators produce electricity using this effect. When magnetic tape is driven past the playback head (a small coil) of a tape recorder, the moving magnetic field induces an emf in the head, which is then amplified to reproduce the recorded sounds.

Electromagnetic induction takes place when the magnetic field around a conductor changes. If the magnetic field is made to change quickly, the size of the current induced is larger. A galvanometer can be used to measure the direction of the current. As a magnet is pushed into a coil, the needle on the galvanometer moves in one direction. As the magnet is removed from the coil, the needle moves in the opposite direction.

If the change of magnetic flux is due to a variation in the current flowing in the same circuit, the phenomenon is known as self-induction; if it is due to a change of current flowing in another circuit it is known as mutual induction.

Lenz's law The direction of an electromagnetically-induced current (generated by moving a magnet near a wire or by moving a wire in a magnetic field) will be such as to oppose the motion producing it. This law is named after the German physicist Heinrich Friedrich Lenz (1804–1865), who announced it in 1833.

Faraday's laws English scientist Michael Faraday proposed three laws of electromagnetic induction: (1) a changing magnetic field induces an electromagnetic force in a conductor; (2) the electromagnetic force is proportional to the rate of change of the field; and (3) the direction of the induced electromagnetic force depends on the orientation of the field.

electromagnetic induction, Faraday's law of in physics, law stating that the induced †electromotive force is equal to the rate of decrease of †magnetic flux. The law was determined by English physicist Michael Faraday.

electromagnetic interaction interaction between two charged particles (for example an †electron and a †proton) which appears as a †force (attractive if the two charges are different in sign). In quantum theory, the electromagnetic interaction is carried between particles by †photons.

electromagnetic pollution the electric and magnetic fields set up by high-tension power cables, local electric sub-stations, and domestic items such as electric blankets. There have been claims that these electromagnetic fields are linked to increased levels of cancer, especially leukemia, and to headaches, nausea, dizziness, and depression.

Although the issue has failed to receive official recognition in the UK, physicists there linked electromagnetic pollution with †radon gas as a cause of cancer in 1996. The electric fields were found to attract the radioactive decay products of radon and cause them to vibrate, making them more likely to adhere to skin and mucous membranes. However, in December 1999 British researchers announced that the world's largest study into the safety of electromagnetic fields had not found a link between electromagnetic fields and childhood cancers.

electromagnetic radiation transfer of energy in the form of †electromagnetic waves.

electromagnetic spectrum complete range, over all wavelengths and frequencies, of †electromagnetic waves. These include (in order of decreasing wavelength) radio and television waves, microwaves, infrared radiation, visible light, ultraviolet light, X-rays, and gamma radiation.

The color of sunlight is made up of a whole range of colors. A glass prism can be used to split white light into separate colors that are sensitive to the human eye, ranging from red (longer wavelength) to violet (shorter wavelength). The human eye cannot detect electromagnetic radiation outside this range. Some animals, such as bees, are able to detect ultraviolet light.

Electromagnetic Spectrum http://www.darvill.clara.net/emag/index.htm

Written and designed by a physics teacher for teaching this GCSE topic. The information is supported by diagrams and images, and there is also a quiz. The topics covered are radio waves, microwaves, infrared, visible light, ultraviolet, X-rays, and gamma rays. This is a well-designed and useful site with printable worksheets.

electromagnetic system of units former system of absolute electromagnetic units (emu) based on the †c.g.s. system and having, as its primary electrical unit, the unit magnetic pole. It was replaced by †SI units.

electromagnetic waves oscillating electric and magnetic fields traveling together through space at a speed of nearly 300,000 kps/186,000 mps. Visible light is composed of electromagnetic waves. The **electromagnetic spectrum** is a family of waves that includes radio waves, infrared radiation, visible light, ultraviolet radiation, X-rays, and gamma rays. All electromagnetic waves are †transverse waves. They can be reflected, refracted, diffracted, and polarized. Radio and television waves lie at the **long wavelength–low frequency** end of the spectrum, with wavelengths longer than 10^{-4} m. Infrared radiation has wavelengths between 10^{-4} m and 7 x 10^{-7} m. Visible light has yet shorter wavelengths from 7 x 10^{-7} m to 4 x 10^{-7} m. Ultraviolet radiation is near the **short wavelength–high frequency** end of the spectrum, with wavelengths between 4 x 10^{-7} m and 10^{-8} m. X-rays have wavelengths from 10^{-8} m to 10^{-12} m. Gamma radiation has the shortest wavelengths (less than 10^{-10} m). The different wavelengths and frequencies lend specific properties to electromagnetic waves. While visible light is diffracted by a diffraction grating, X-rays can only be diffracted by crystals. Radio waves are refracted by the atmosphere; visible light is refracted by glass or water.

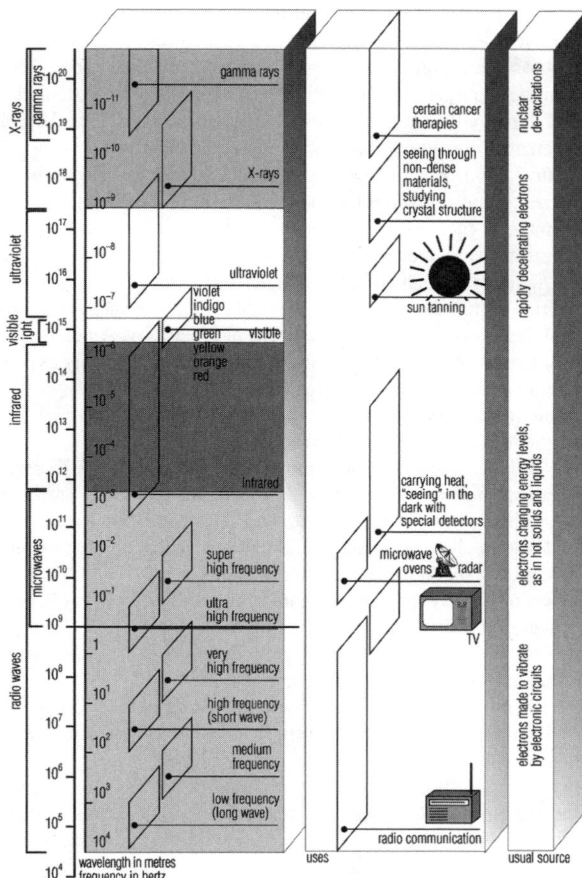

electromagnetic waves Radio waves have the lowest frequency. Infrared radiation, visible light, ultraviolet radiation, X-rays, and gamma rays have progressively higher frequencies. Helicon

Electromagnetic Waves

http://www.colorado.edu/physics/2000/waves_particles/index.html

Great explanation of the electromagnetic spectrum. The pages are very well presented in a conversational style with large, clear text. The site makes good use of Java applets to illustrate the principles behind electromagnetism. Subsequent pages develop the subject further, examining topics such as electric force and electric force fields.

electromagnetism in physics, the properties and interactions of †magnetism and †electric currents. A current flowing through a conductor produces a magnetic field around the conductor; a varying magnetic field interacting with a conductor produces a current in the conductor. This is known as the electromagnetic effect.

electromotive force emf, loosely, the voltage produced by an electric battery or generator in an electrical circuit or, more precisely, the energy supplied by a source of electric power in driving a unit charge around the circuit. The unit is the †volt.

A difference in charge between two points in a material can be created by an external energy source such as a battery. This causes electrons to move so that there is an excess of electrons at one point and a deficiency of electrons at a second point. This difference in charge is stored as electrical potential energy known as emf. It is the emf that causes a current to flow through a circuit.

electron negatively-charged particle with negligible mass. Electrons form the outer portion of all atoms, orbiting the nucleus in groupings called shells. The first shell can hold up to two electrons; the second and third shells can hold up to eight electrons each. The electron arrangement of an element is called its **electronic configuration**; for example, the electronic configuration of the sodium atom is $Na_{(2,8,1)}$. In a neutral atom the number of electrons is equal to the number of protons in the nucleus. This electron structure is responsible for the chemical properties of the atom (see †atomic structure). Electrons are a member of the class of elementary particles known as †leptons.

Electrons carry a charge of 1.602177 x 10^{-19} coulombs and have a mass of 9.109 x 10^{-31} kg, which is $^1/_{1836}$ times the mass of a †proton. Energy can be used to remove electrons from their outer orbits in conducting materials and they can be made to flow through these materials under a potential difference. This flow is known as an electric current or electricity. A beam of electrons will undergo †diffraction (scattering) and produce interference patterns in the same way as †electromagnetic waves such as light; hence they may be regarded as waves as well as particles.

Look Inside the Atom

http://www.aip.org/history/electron/jjhome.htm

Part of the American Institute of Physics site, this page examines J J Thomson's 1897 experiments that led to the discovery of a fundamental building block of matter, the electron. The site includes images and quotes and a section on the legacy of his discovery. Also included is a section on suggested readings and related links.

electron in chemistry, compound that can accept an electron and is reduced in doing so; it can therefore take part in oxidation–reduction reactions.

electron affinity in chemistry, the amount of energy that is released when an electron is added to a neutral atom to form a negatively charged species.

electronegativity ease with which an atom can attract electrons to itself. Electronegative elements attract electrons, so forming negative ions.

US chemist Linus Pauling devised an electronegativity scale to indicate the relative power of attraction of elements for electrons. Fluorine, the most non-metallic element, has a value of 4.0 on this scale; oxygen, the next most non-metallic, has a value of 3.5.

In a covalent bond between two atoms of different electronegativities, the bonding electrons will be located close to the more electronegative atom, creating a †dipole.

electron gun series of †electrodes, including a cathode for producing an electron beam. It plays an essential role in many electronic devices, including †cathode-ray tubes (television tubes) and †electron microscopes.

electronic configuration or **electronic structure**, in chemistry, the arrangement of electrons in the energy shells around an atomic nucleus. See †atom, electronic structure of

electronic mail see †e-mail.

electronic publishing distribution of information using computer-based media such as multimedia and †hypertext in the creation of electronic "books." Critical technologies in the development of electronic publishing were †CD-ROM, with its massive yet compact storage capabilities, and the advent of computer networking with its ability to deliver information instantaneously anywhere in the world.

Some publishers have purchased electronic publishing equipment to enable them to print books on demand, avoiding the need to keep stocks of books which might sell only a small number of copies over several years, and would otherwise go out of print. In addition, global hotel chains were purchasing similar equipment to print international business newspapers on demand via the Internet.

electron microprobe instrument used to determine the relative and absolute abundances of elements at a particular point within an material (such as a mineral).

The electron microprobe shoots electrons at the sample and then measures the electrons that bounce back from the sample (backscatter electrons), the electrons emitted from the sample, and X-rays that are emitted from the sample. The number of electons that bounce back is directly related to elemental mass, and each element emits a characteristic set of X-rays of different energies. The electron microprobe allows for much more detailed chemical analyses of materials than is possible by conventional chemical techniques.

electron microscope instrument that produces a magnified image by using a beam of †electrons instead of light rays, as in an optical †microscope. An **electron lens** is an arrangement of electromagnetic coils that control and focus the beam. Electrons are not visible to the eye, so instead of an eyepiece there is a fluorescent screen or a photographic plate on which the electrons form an image. The wavelength of the electron beam is much shorter than that of light, so much greater magnification and resolution (ability to distinguish detail) can be achieved. The development of the electron microscope has made possible the observation of very minute organisms, viruses, and even large molecules.

A †transmission electron microscope passes the electron beam through a very thin slice of a specimen. A †scanning electron microscope looks at the exterior of a specimen. A †scanning transmission electron microscope (STEM) can produce a magnification of 90 million times. See also †atomic force microscope.

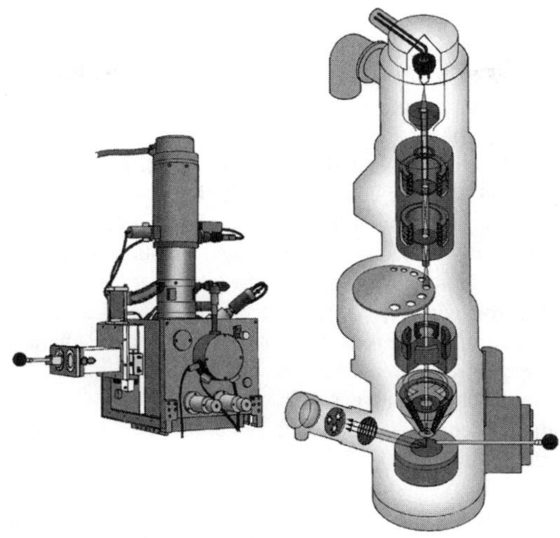

electron microscope The scanning electron microscope. Electrons from the electron gun are focused to a fine point on the specimen surface by the lens systems. The beam is moved across the specimen by the scan coils. Secondary electrons are emitted by the specimen surface and pass through the detector, which produces an electrical signal. The signal is passed to an electronic console, and produces an image on a screen. Helicon

electron pair or **lone pair**, in chemistry, two electrons sharing the same †orbital, but with opposing spins. This type of orbital is usually associated with a covalent or coordinate bond.

electron probe analysis in chemistry, a nondestructive method of elemental analysis. A beam of electrons is focused onto a point on the surface of a sample causing the elements present to emit characteristic X-ray spectra. The spectra can be analyzed both quantitatively and qualitatively.

An example of the use of electron probe analysis would be in determining the percentage of niobium (Nb) impurity in a crystal of potassium titanyl phosphate, $KTiOPO_4$, without damaging the sample.

electrons, delocalized electrons that are not associated with individual atoms or identifiable chemical bonds, but are shared collectively by all the constituent atoms or ions of some chemical substances (such as metals, graphite, and †aromatic compounds).

A metallic solid consists of a three-dimensional arrangement of metal ions through which the delocalized electrons are free to travel. Aromatic compounds are characterized by the sharing of delocalized electrons by several atoms within the molecule.

electrons, localized pair of electrons in a †covalent bond that are located in the vicinity of the nuclei of the two contributing atoms. Such electrons cannot move beyond this area.

electron spin resonance in archaeology, a nondestructive dating method applicable to teeth, bone, heat-treated flint, ceramics, sediments, and stalagmitic concretions. It enables electrons, displaced by natural radiation and then trapped in the structure, to be measured; their number indicates the age of the specimen.

electron transport chain arrangement of substances within the living cell that takes part in energy production. Electron transport is the means by which reduced flavoproteins (FP) and the coenzyme NAD are oxidized in steps, such that energy from certain steps can be used for the manufacture of †ATP by oxidative phosphorylation.

electron volt symbol eV, unit for measuring the energy of a charged particle (†ion or †electron) in terms of the energy of motion an electron would gain from a potential difference of one volt. Because it is so small, more usual units are mega-(million) and giga- (billion) electron volts (MeV and GeV).

electrophilic describing a reagent that readily accepts electrons during a chemical reaction. Such reagents, therefore, typically react at centers of high electron density.

electrophoresis movement of electrically charged solute particles in a †colloid towards the oppositely charged electrode when a pair of electrodes is immersed in the colloidal solution and connected to an external source of direct-current electricity. It is used as an analytical technique similar to †chromatography.

electrophorus means of inducing a charge in one body into another. Usually consists of a metal plate with an insulating handle, the plate being capable of carrying †electrostatic charges.

electroplating deposition of metals upon metallic surfaces by †electrolysis for decorative and/or protective purposes. It is used in the preparation of printing plates, "master" audio discs, and in many other processes.

A current is passed through a bath containing a solution of a salt of the plating metal, the object to be plated being the cathode (negative terminal); the †anode (positive terminal) is either an inert substance or the plating metal. Among the metals most commonly used for plating are zinc, nickel, chromium, cadmium, copper, silver, and gold.

In **electropolishing**, the object to be polished is made the anode in an electrolytic solution and by carefully controlling conditions the high spots on the surface are dissolved away, leaving a high-quality stain-free surface. This technique is useful in polishing irregular stainless-steel articles.

electroporation in biotechnology, technique of introducing foreign †DNA into pollen with strong bursts of electricity which increase the permeability of cell membranes, used in genetic engineering.

electropositivity in chemistry, a measure of the ability of elements (mainly metals) to donate electrons to form positive ions. The greater the metallic character, the more electropositive the element.

electrorheological fluid another name for †smart fluid, a liquid suspension that gels when an electric field is applied across it.

electroscope apparatus for detecting †electric charge. The simple gold-leaf electroscope consists of a vertical conducting (metal) rod ending in a pair of rectangular pieces of gold foil, mounted inside and insulated from an earthed metal case or glass jar. An electric charge applied to the end of the metal rod makes the gold leaves diverge, because they each receive a similar charge (positive or negative) via the rod and so repel each other. The polarity of the charge can be found by bringing up another charge of known polarity and applying it to the metal rod. A like charge has no effect on the gold leaves since they remain repelled. An opposite charge neutralizes the charge on the leaves and causes them to collapse.

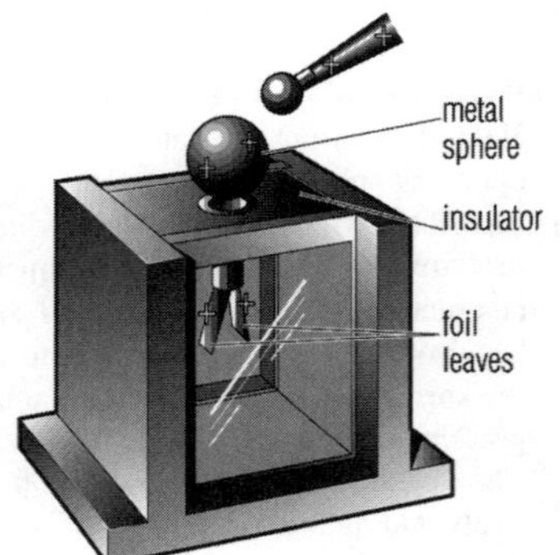

electroscope The electroscope is a simple means of detecting electric charge. The metal foil leaves diverge when a charge is applied to the metal sphere. Helicon

electrostatics study of stationary electric charges and their fields (not currents). See †static electricity.

Charged materials behave differently if they are brought near to each other. Materials of like charge, that is both positive or both negative, will move away from each other (repelled by electrostatic force). Materials of opposite charge, that is one positive and one negative, brought close together will move towards each other (attracted by electrostatic force).

An †electroscope is an instrument that detects charges on objects that have been charged by static electricity.

electrovalent bond another name for an †ionic bond, a chemical bond in which the combining atoms lose or gain electrons to form ions.

element substance that cannot be split chemically into simpler substances. The atoms of a particular element all have the same number of protons in their nuclei (their proton or †atomic number). Elements are classified in the †periodic table of the elements. Of the known elements, 92 are known to occur naturally on Earth (those with atomic numbers 1–92). Those elements with atomic numbers above 96 do not occur in nature and must be synthesized in particle accelerators. Of the elements, 81 are stable; all the others, which include atomic numbers 43, 61, and from 84 up, are radioactive.

Elements are classified as metals, nonmetals, or metalloids (weakly metallic elements) depending on a combination of their physical and chemical properties; about 75% are metallic. Some elements occur abundantly (oxygen, aluminum); others occur moderately or rarely (chromium, neon); some, in particular the radioactive ones, are found in minute (neptunium, plutonium) or very minute (astatine, technetium) amounts. Symbols (devised by Swedish chemist Jöns †Berzelius) are used to denote the elements; the symbol is usually the first letter or letters of the English or Latin name (for example, C for carbon, Ca for calcium, Fe for iron, from the Latin *ferrum*). The symbol represents one atom of the element. Two or more elements bonded together form a **compound** from which they cannot be separated by physical means. Compounds are held together by ionic or covalent bonds. The number of atoms of an element that combine to form a molecule is it **atomicity**. A molecule of oxygen (O_2) has atomicity 2; sulfur (S_8) has atomicity 8.

According to current theories, hydrogen and helium were produced in the †Big Bang at the beginning of the universe. Of the other elements, those up to atomic number 26 (iron) are made by nuclear fusion within the stars. The heavier elements, such as lead and uranium, are produced when an old star explodes; as its center collapses, the gravitational energy squashes nuclei together to make new elements.

Compounds, Elements, Mixtures http://library.thinkquest.org/10429/low/matter/matter.htm

Part of the ChemWeb Online site providing an introduction to chemistry, this page provides a definition of elements, compounds, and mixtures. There are also sections explaining atoms, ions, and molecules, as well as a searchable glossary of chemistry terms.

Web Elements http://www.webelements.com/

Periodic table on the Web, with 12 different categories of information available for each element – from its physical and chemical characteristics to its electronic configuration.

What is an Element? http://www.bbc.co.uk/worldservice/sci_tech/features/elementary/index.shtml

Simple information on elements and atomic structure, part of the BBC World Service elementary science Web site. A good starting point for revision, this page defines key concepts such as atomic number.

element in mathematics, a member of a †set.

elemental analysis in chemistry, any method of analysis that quantitatively determines which elements are present in a compound.

elementary particle in physics, a subatomic particle that is not known to be made up of smaller particles, and so can be considered one of the fundamental units of matter. There are three groups of elementary particles: quarks, leptons, and gauge bosons.

Quarks, of which there are 12 types (up, down, charm, strange, top, and bottom, plus the antiparticles of each), combine in groups of three to produce heavy particles called baryons, and in groups of two to produce intermediate-mass particles called mesons. They and their composite particles are influenced by the strong nuclear force. **Leptons** are particles that do not interact via the strong nuclear force. Again, there are 12 types: the electron, muon, tau; their neutrinos, the electron neutrino, muon neutrino, and tau neutrino; and the antiparticles of each. These particles are influenced by the weak nuclear force, as well as by gravitation and electromagnetism. **Gauge bosons** carry forces between other particles. There are four types: gluon, photon, intermediate vector bosons (W^+, W^-, and Z), and graviton. The gluon carries the strong nuclear force, the photon the electromagnetic force, W^+, W^-, and Z the weak nuclear force, and the graviton, as yet unobserved, the force of gravity (see †forces, fundamental).

Elementary Particles http://www.neutron.anl.gov/hyper-physics/Particles.html

Detailed explanation of elementary particles, "the most basic physical constituents of the universe," with links to other sources of related information.

Virtual Tour of Fermilab Exhibit http://www.fnal.gov/pub/about/tour/index.html

Online guided tour of the particle physics laboratory Fermilab in Illinois, USA. It includes an explanation of the principles of particle physics, a guide to particle accelerators, and an insight into the experiments currently being conducted at the laboratory.

Particle Adventure http://particleadventure.org/particleadventure/index.html

Comprehensive site about particles. It covers many topics including what the world is made of, what holds it together, and decay and annihilations. The subject matter is difficult in places, but the introductions to each topic may be of use to GCSE pupils, and the site is of interest to all those who wish to stretch themselves.

Particle Theory http://www.le.ac.uk/se/centres/sci/selfstudy/atom1.html

Description and explanation of particle theory by Elaine Wilson, of Homerton College, Cambridge. Clear and simple diagrams accompany the concise and insightful text. Also mentioned are isotopes.

elements, the four earth, air, fire, and water. The Greek philosopher Empedocles believed that these four elements made up the fundamental components of all matter and that they were destroyed and renewed through the action of love and discord.

This belief was shared by Aristotle who also claimed that the elements were mutable and contained specific qualities: cold and dry for earth, hot and wet for air, hot and dry for fire, and cold and wet for water. The transformation of the elements formed the basis of medieval alchemy, and the belief that base metals could be turned into gold. The theory of the elements prevailed until the 17th century when Robert †Boyle redefined an element as a substance "simple or unmixed, not made of other bodies" and proposed the existence of a greater number than four.

elevation, angle of an upward angle made with the horizontal.

elevation of boiling point raising of the boiling point of a liquid above that of the pure solvent, caused by a substance being dissolved in it. The phenomenon is observed when salt is added to boiling water; the water ceases to boil because its boiling point has been elevated.

How much the boiling point is raised depends on the number of molecules of substance dissolved. For a single solvent, such as pure water, all substances in the same molecular concentration produce the same elevation of boiling point. The elevation e produced by the presence of a solute of molar concentration C is given by the equation $e = KC$, where K is a constant (called the ebullioscopic constant) for the solvent concerned.

eliminate in mathematics, to remove a variable from a set of equations by substituting for it in terms of the other variables.

ellipse curve joining all points (loci) around two fixed points (foci) such that the sum of the distances from those points is always constant. The diameter passing through the foci is the major axis, and the diameter bisecting this at right angles is the minor axis. An ellipse is one of a series of curves known as †conic sections. A slice across a cone that is not made parallel to, and does not pass through, the base will produce an ellipse.

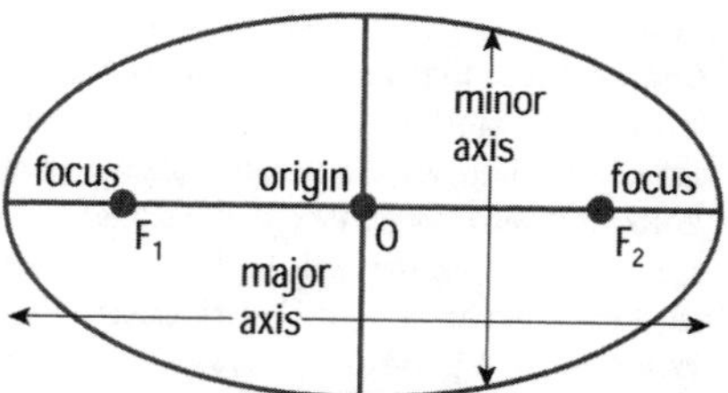

ellipse Technical terms used to describe an ellipse; for all points on the ellipse, the sum of the distances from the two foci, F_1 and F_2, is the same. Helicon

Ellipse

http://www-groups.dcs.st-andrews.ac.uk/history/Curves/Ellipse.html

Introduction to ellipses and links to associated curves. If you have a Java-enabled browser you can experiment interactively with this curve and its associates. There are also links to mathematicians who have studied the ellipse, to the ellipse's particular attributes, and to related Web sites.

elliptical galaxy in astronomy, one of the main classes of †galaxy in the Hubble classification, characterized by a featureless elliptical profile. Unlike spiral galaxies, elliptical galaxies have very little gas or dust and no stars have recently formed within them. They range greatly in size from giant ellipticals, which are often found at the centers of clusters of galaxies and may be strong radio sources, to tiny dwarf ellipticals, containing about a million stars, which are the most common galaxies of any type. More than 60% of known galaxies are elliptical.

elliptic function integral of the general form ò*f*(*x*,Ö*R*)d*x*, where *f* is any rational function of *x* and *R* is a quartic polynomial corresponding to: $a_0x^4 + a_1x^3 + a_2x^2 + a_3x + a_4$ with no multiple roots. French mathematician Adrien Legendre proved that any elliptic integral can be reduced to the sum of an elementary function and of scalar multiples of three special functions. †Abelian functions and †theta functions are both extensions. Elliptic functions are used in the integration of the square root of a cubic or a †quartic, and are thus important to many mathematical operations.

elliptic geometry system of †non-Euclidean geometry developed as the initial form of †Riemann geometry, and regarding all geometrical operations as carried out in "curved" space, for example, as though on the surface of an ellipsoid or sphere. A "straight line" is thus defined (then) as the shortest curve (geodesic) on the curved surface joining two points.

ellipticity in astronomy, of the shape of a planet or †galaxy, the amount of distortion by which it departs from a perfect sphere. The overall ellipticity of the Earth is given as $^1/_{299}$. One class of galaxy is defined in terms of ellipticity, subdivided E0 to E7, according to degree.

elliptic modular functions functions defined in the upper half of an Argand plane that are automorphic relative to a group of modular transformations, that is, transformations *T* such as:

$T(z) = az + b/cz + d$

Elnath or **Beta Tauri**, second brightest star in the constellation Taurus and the 25th brightest in the heavens. It is a blue-white giant star 130 light years distant, with an apparent magnitude of 1.65.

El Niño (Spanish "the child") marked warming of the east Pacific Ocean that occurs when a warm current of water moves from the western Pacific, temporarily replacing the cold Peru Current along the west coast of South America. This results in a reduction in marine plankton, the main food source in the ocean, and fish numbers decline. The atmospheric circulation in the region is also seriously disturbed, and may result in unusual climatic events, for example floods in Peru, and drought in Australia. El Niño events occur at irregular intervals of between two and seven years.

El Niño is believed to be caused by the failure of trade winds and, consequently, of the ocean currents normally driven by these winds. Warm surface waters then flow in from the east. The phenomenon can disrupt the climate of the area disastrously, and has played a part in causing famine in Indonesia, drought and bush fires in the Galapagos Islands, rainstorms in California and South America, and the destruction of Peru's anchovy harvest and wildlife in 1982–83. El Niño contributed to algal blooms in Australia's drought-stricken rivers and an unprecedented number of typhoons in Japan in 1991. It is also thought to have caused the 1997 drought in Australia and contributed to certain ecological disasters such as bush fires in Indonesia.

El Niño usually lasts for about 18 months, but the 1990 occurrence lasted until June 1995; US climatologists estimated this duration to be the longest in 2,000 years. The last prolonged El Niño of 1939–41 caused extensive drought and famine in Bengal. It is understood that there might be a link between El

Niño and †global warming.

El Niño http://www.msc-smc.ec.gc.ca/elnino/what_is/index_e.cfm

Description and brief history of El Niño. This site provides examples of how El Niño can affect different parts of the world, and illustrates how it occurs with animated diagrams. La Niña, the presence of a cooler-than-normal body of water, is also explained.

El Niño Theme Page http://www.pmel.noaa.gov/tao/elnino/nino-home.html

Wealth of scientific information about El Niño (a "disruption of the ocean–atmosphere system in the tropical Pacific") with animated views of the monthly sea changes brought about by it, El Niño-related climate predictions, and forecasts from meteorological centers around the world. It also offers an illuminating "Frequently asked questions" section with basic and more advanced questions as well as an interesting historical overview of the phenomenon starting from 1550.

elongation in astronomy, angular distance between the Sun and a planet or other Solar System object. This angle is 0° at †conjunction, 90° at †quadrature, and 180° at †opposition.

Elsasser, Walter Maurice (1904–1991) German-born US geophysicist. He pioneered analysis of the Earth's former †magnetic fields, which are frozen in rocks. His research in the 1940s yielded the dynamo model of the Earth's magnetic field. The field is explained in terms of the activity of electric currents flowing in the Earth's fluid metallic outer core. The theory premises that these currents are magnified through mechanical motions, rather as currents are sustained in power-station generators.

elution in chemistry, washing of an adsorbed substance from the adsorbing material; it is used, for example, in the separation processes of chromatography and electrophoresis.

elytra horny wing cases characteristic of beetles. The elytra are adapted from the beetles' forewings (only the hindwings are used for flight). They fold over the back, generally meeting in the middle in a straight line, and serve to protect the hindwings and the soft posterior parts of the body.

Elytra are also to be found in earwigs.

In Hemiptera (true bugs) the forewings are hardened over half their length and are known as **hemelytra**. In Orthoptera (crickets, grasshoppers, and locusts) the forewings are greatly thickened and are known as **tegmina**.

e-mail contraction of electronic mail, messages sent electronically from computer to computer via network connections such as Ethernet or the †Internet, or via telephone lines to a host system. Messages once sent are stored on the network or by the host system until the recipient picks them up. As well as text, messages may contain enclosed text files, artwork, or multimedia clips.

Subscribers to an e-mail system type messages in ordinary letter form on a word processor, or microcomputer, and "drop" the letters into a central computer's memory bank by means of a computer/telephone connector (a †modem). The recipient "collects" the letter by calling up the central computer and feeding a unique password into the system. Due to the high speed of delivery, e-mail is cheaper than an equivalent telephone call or fax.

E-mail: An Introduction http://www.webfoot.com/advice/email.top.html

Beginner's guide to e-mail. There is a good introduction followed by sections on why e-mail is different from paper-based communication. The site also includes information on layout, expressing gestures and intonations, and an appendix of e-mail jargon.

embryo early developmental stage of an animal or a plant following fertilization of an ovum (egg cell), or activation of an ovum by †parthenogenesis. In humans, the term embryo describes the fertilized egg during its first seven weeks of existence; from the eighth week onwards it is referred to as a fetus.

In animals the embryo exists either within an egg (where it is nourished by food contained in the yolk), or in mammals, in the †uterus of

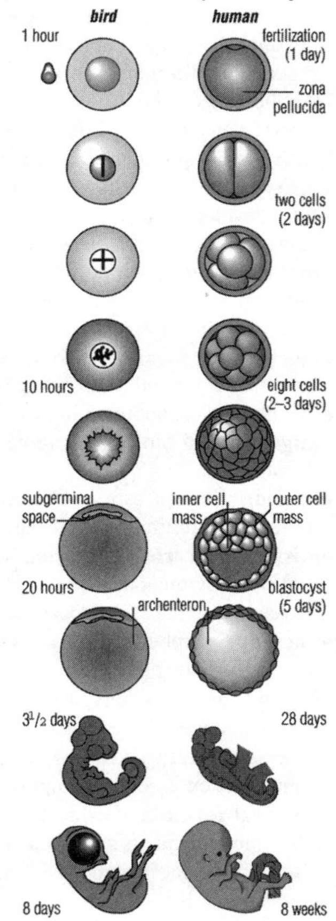

embryo The development of a bird and a human embryo. In the human, division of the fertilized egg, or ovum, begins within hours of conception. Within a week, a hollow, fluid-containing ball – a blastocyte – with a mass of cells at one end has developed. After the third week, the embryo has changed from a mass of cells into a recognizable shape. At four weeks, the embryo is 3 mm/0.1 in long, with a large bulge for the heart and small pits for the ears. At six weeks, the embryo is 1.5 cm/0.6 in long with a pulsating heart and ear flaps. By the eighth week, the embryo (now technically a fetus) is 2.5 cm/1 in long and recognizably human, with eyelids and small fingers and toes. Helicon

the mother. In mammals (except marsupials) the embryo is fed through the †placenta. The plant embryo is found within the seed in higher plants. It sometimes consists of only a few cells, but usually includes a root, a shoot (or primary bud), and one or two †cotyledons, which nourish the growing seedling.

Odyssey of Life http://www.pbs.org/wgbh/nova/odyssey/

Companion to the US Public Broadcasting Service (PBS) television program Nova, this page examines the formation of embryos. It includes time-lapse video sequences of growing embryos. Two leading spokespersons of the evolution/creation debate state their opposing viewpoints regarding the age-old question "How did we get here?" Their debate is carried out online through a series of letters in which they reply to the other's position and expand on their own. There is also an interview with the photographer of the program in which he discusses the tricky techniques and technology required to film living embryos inside the womb.

embryology study of the changes undergone by an organism from its conception as a fertilized ovum (egg) to its emergence into the world at hatching or birth. It is mainly concerned with the changes in cell organization in the embryo and the way in which these lead to the structures and organs of the adult (the process of †differentiation).

Applications of embryology include embryo transplants, both commercial (for example, in building up a prize dairy-cow herd quickly at low cost) and in obstetric medicine (as a method for helping couples with fertility problems to have children).

embryo sac large cell within the ovule of flowering plants that represents the female †gametophyte when fully developed. It typically contains eight nuclei. Fertilization occurs when one of these nuclei, the egg nucleus, fuses with a male †gamete.

emerald clear, green gemstone variety of the mineral †beryl. It occurs naturally in Colombia, the Ural Mountains in Russia, Zimbabwe, and Australia. The green color is caused by the presence of the element chromium in the beryl.

emery black to grayish form of impure †corundum that also contains the minerals magnetite and hematite. It is used as an †abrasive.

emf in physics, abbreviation for †electromotive force.

emigration in ecology, movement of individuals away from a population, contributing to a reduction in its numbers.

emission line in astronomy, bright line in the spectrum of a luminous object caused by †atoms or molecules emitting light at sharply defined †wavelengths.

emission spectroscopy in analytical chemistry, a technique for determining the identity or amount present of a chemical substance by measuring the amount of electromagnetic radiation it emits at specific wavelengths; see †spectroscopy.

emissivity symbol e, in physics, power radiated from a surface compared with that radiated at the same temperature by a perfect black body (which has an emissivity of 1).

emotion in psychology, a powerful feeling; a complex state of body and mind involving, in its bodily aspect, changes in the viscera (main internal organs) and in facial expression and posture, and in its mental aspect, heightened perception, excitement and, sometimes, disturbance of thought and judgment. The urge to action is felt and impulsive behavior may result.

empirical formula chemical formula of a substance in which only the relative proportions of each of its constituent elements are given. The empirical formula does not necessarily reflect a substance's †molecular formula nor its structure. The empirical formula of benzene, for example, is CH, whereas its molecular formula is C_6H_6.

empty set symbol ø, in mathematics, †set with no elements.

emulsify in biology, to disperse †fat as millions of microscopic globules suspended in water. Emulsification is an important stage prior to †digestion of fat in the human gut.

Fat is an important food, but it has to be digested to produce molecules small enough to be absorbed into the body. The enzyme that does this is soluble in water, but the fat itself is not soluble in water. To bring the fat molecules near to the dissolved enzyme, the fat has to be dispersed into millions of tiny globules. The enzyme works on the surfaces of these globules and fat is digested relatively quickly. Chemicals in †bile, produced by the liver, coat the fat and emulsify it. These chemicals act rather like a detergent with one part of the molecule dissolving in the fat and the other dissolving in water.

emulsion stable dispersion of a liquid in another liquid – for example, oil and water in some cosmetic lotions.

enamel in biology, outer layer of the crown (the part of the †tooth projecting above the gum line) that protects the softer dentine within. It is the hardest substance in the body and is formed from crystals of calcium carbonate phosphate, bound together by the protein keratin. The scales of some fish are also covered with enamel.

enantiomer or **mirror image isomer**, in chemistry, †isomer with a center consisting of four different connected atoms (a †chiral center) and whose molecular structures are mirror images of each other.

encephalin naturally occurring chemical produced by nerve cells in the brain that has the same effect as morphine or other derivatives of opium, acting as a natural painkiller. Unlike morphine, encephalins are quickly degraded by the body, so there is no build-up of tolerance to them, and hence no addiction. Encephalins are a variety of †peptides, as are †endorphins, which have similar effects.

Encke, Johann Franz (1791–1865) German astronomer. His work on star charts during the 1840s contributed to the discovery of the planet †Neptune in 1846. He also worked out the path of the †comet that bears his name. The new star charts were compiled from both old and new observations. They were completed in 1859 but were soon improved upon by those of Friedrich †Argelander.

Encke division narrow gap in the rings of †Saturn reported in 1838 by the German astronomer Johann Encke. Only 300 km/186 mi wide, its existence was not generally accepted for another 50 years until it was confirmed by US astronomer James †Keeler.

Encke's Comet comet with the shortest known orbital period, 3.3 years. It is named after German astronomer Johann Encke, who calculated its orbit in 1819 from earlier sightings.

The comet was first seen in 1786 by the French astronomer Pierre Méchain. It is the parent body of the Taurid meteor shower and a fragment of it may have hit the Earth in the †Tunguska Event in 1908.

In 1913 it became the first comet to be observed throughout its entire orbit when it was photographed near its †aphelion (the point in its orbit furthest from the Sun) by astronomers at Mount Wilson Observatory in California, USA.

encryption encoding a message so that it can only be read by the intended recipient. The mathematical calculation of encryption formulae is called †cryptography.

endangered species plant or animal species whose numbers are so few that it is at risk of becoming extinct. Officially designated endangered species are listed by the World Conservation Union (or IUCN). Endangered species are not a new phenomenon; extinction is an integral part of evolution. The replacement of one species by another usually involves the eradication of the less successful form, and ensures the continuance and diversification of life in all forms. However, extinctions induced by humans are thought to be destructive, causing evolutionary dead-ends that do not allow for succession by a more fit species. The great majority of recent extinctions have been directly or indirectly induced by humans; most often by the loss, modification, or pollution of the organism's habitat, but also by hunting for "sport" or for commercial purposes.

According to a 1995 report to Congress by the US Fish and Wildlife Service, although seven of the 893 species listed as endangered under the US Endangered Species Act 1968–93 have become extinct, 40% are no longer declining in number. In February 1996 a private conservation group, Nature Conservancy, reported around 20,000 native US plant and animal species to be rare or imperiled.

According to the Red Data List of endangered species, published in 1996 by the World Conservation Union, 25% of all mammal species (including 46% of primates, 36% of insectivores, and 33% of pigs and antelopes), and 11% of all bird species are threatened with extinction.

Endangered Species http://endangeredspecie.com/

Claiming to offer "The rarest info around," this site is packed with photos and information about endangered species and extinction rates. Sections include "Laws that protect animals," "Endangered species fact sheets," and "Selected species survival plans." There is also a separate children's site.

Endangered Species Home Page http://endangered.fws.gov/index.html

US Fish and Wildlife site with full details of its ongoing conservation programs. The site also contains a "Kids' Corner," a gallery of images, and a link to the latest edition of its Endangered Species Bulletin.

Enders, John Franklin (1897–1985) US virologist who was awarded a Nobel Prize for Physiology or Medicine in 1954 with Thomas Weller and Frederick Robbins for their cultivation of the polio virus in the laboratory. This led to the creation of effective vaccines against polio and measles.

endo- in chemistry, prefix denoting the orientation of an atom or group with respect to the molecule of which it is part. The term denotes that the atom is facing inwards. It is most widely used to describe the configuration of †cyclic compounds.

endocrine gland gland that secretes hormones into the bloodstream to regulate body processes. Endocrine glands are most highly developed in vertebrates, but are also found in other animals, notably insects. In humans the main endocrine glands are the pituitary, thyroid, parathyroid, adrenal, pancreas, ovary, and testis.

Endocrine Disrupters http://whyfiles.org/045env_hormone/index.html

Part of the Why Files project, published by the National Institute for Science Education (NISE) and funded by the National Science

Foundation, this page investigates the harmful effects that synthetic chemicals are having on the endocrine system. The highly readable text, laid out over ten pages, includes information about the endocrine system, how "endocrine disrupters" are adversely effecting the system, and what is being done to solve the problem. Numerous images and diagrams enhance the text throughout. You will also find a comprehensive bibliography of sources for further research.

endocrinology medical specialty devoted to the diagnosis and treatment of hormone disorders.

endocytosis ingestion of material by a cell, including †phagocytosis and †pinocytosis. **Phagocytosis** is the engulfment and ingestion by a white blood cell of bacteria or other foreign particles; **pinocytosis** involves the absorption and ingestion by a cell of surrounding fluid by the folding-in of the cell membrane to form a vesicle which (eventually) releases some of its contents into the cell's †cytoplasm.

endoderm innermost of the three germ layers of an embryo.

endolymph fluid found in the inner †ear, filling the central passage of the cochlea as well as the semicircular canals.

Sound waves traveling into the ear pass eventually through the three small bones of the middle ear and set up vibrations in the endolymph. These vibrations are detected by receptors in the cochlea, which send nerve impulses to the hearing centers of the brain.

endoparasite †parasite that lives inside the body of its host.

endoplasm inner, liquid part of a cell's †cytoplasm.

endoplasmic reticulum **ER**, a membranous system of tubes, channels, and flattened sacs that form compartments within †eukaryotic cells. It stores and transports proteins within cells and also carries various enzymes needed for the synthesis of †fats. The †ribosomes, or the organelles that carry out protein synthesis, are sometimes attached to parts of the ER.

Under the electron microscope, ER looks like a series of channels and vesicles, but it is in fact a large, sealed, baglike structure crumpled and folded into a convoluted mass. The interior of the "bag," the ER lumen, stores various proteins needed elsewhere in the cell, then organizes them into transport vesicles formed by a small piece of ER membrane budding from the main membrane.

endorphin natural substance (a polypeptide) that modifies the action of nerve cells. Endorphins are produced by the pituitary gland and hypothalamus of vertebrates. They lower the perception of pain by reducing the transmission of signals between nerve cells.

Endorphins not only regulate pain and hunger, but are also involved in the release of sex hormones from the pituitary gland. Opiates act in a similar way to endorphins, but are not rapidly degraded by the body, as natural endorphins are, and thus have a long-lasting effect on pain perception and mood. Endorphin release is stimulated by exercise.

endoskeleton internal supporting structure of vertebrates, made up of cartilage or bone. It provides support, and acts as a system of levers to which muscles are attached to provide movement. Certain parts of the skeleton (the skull and ribs) give protection to vital body organs.

Sponges are supported by a network of rigid, or semirigid, spiky structures called spicules; a bath sponge is the proteinaceous skeleton of a sponge.

endosperm nutritive tissue in the seeds of most flowering plants. It surrounds the embryo and is produced by an unusual process that parallels the †fertilization of the ovum by a male gamete. A second male gamete from the pollen grain fuses with two female nuclei within the †embryo sac. Thus endosperm cells are triploid (having three sets of chromosomes); they contain food reserves such as starch, fat, and protein that are utilized by the developing seedling.

In "non-endospermic" seeds, absorption of these food molecules by the embryo is completed early, so that the endosperm has disappeared by the time of germination.

endotherm "warm-blooded," or homeothermic, animal. Endotherms have internal mechanisms for regulating their body temperatures to levels different from the environmental temperature. See †homeothermy.

endothermic reaction chemical reaction that requires an input of energy in the form of heat for it to proceed; the energy is absorbed from the surroundings by the reactants. A sign that this is happening is if the container holding the reactants feels cold and the temperature of the reactants falls. The energy absorbed is represented by the symbol +DH.

The dissolving of sodium chloride in water and the process of photosynthesis are both endothermic changes. In photosynthesis the energy absorbed is in the form of light energy. In an endothermic reaction the energy needed to break the chemical bonds in the reactants is greater than the energy released when bonds are formed in the products. See †energy of reaction.

endotoxin in biology, heat stable complex of protein and lipopolysaccharide that is produced following the death of certain bacteria. Endotoxins are typically produced by the Gram negative bacteria and can cause fever. They can also cause shock by rendering the walls of the blood vessels permeable so that fluid leaks into the tissues and blood pressure falls sharply.

endplate in anatomy, mass of motor nerve endings that penetrate a muscle fiber.

end-point point during a †titration when the two reagents involved are at exact equivalence, that is, when all of each of the reagents has reacted and there is no excess of either.

end user user of a computer program; in particular, someone who uses a program to perform a task (such as word processing or playing a computer game), rather than someone who writes programs (a programmer).

energy capacity for doing †work. This work may be as simple as reading a book, using a computer, or driving a car. Without energy no activity is possible. Energy can exist in many different forms. For example, potential energy (PE) is energy deriving from position; thus a stretched spring has elastic PE, and an object raised to a height above the Earth's surface, or the water in an elevated reservoir, has gravitational PE. Moving bodies possess kinetic energy (KE). Energy can be converted from one form to another, but the total quantity in a system stays the same (in accordance with the †conservation of energy principle). Energy cannot be created or destroyed. For example, as an apple falls it loses gravitational PE but gains KE. Although energy is never lost, after a number of conversions it tends to finish up as the kinetic energy of random motion of molecules (of the air, for example) at relatively low temperatures. This is "degraded" energy that is difficult to convert back to other forms.

energy sources There are two main sources of energy: the Sun, the ultimate source; and decay of radioactive elements in the Earth. Plants use the Sun's energy and convert it into food and oxygen. The remains of plants and animals that lived millions of years ago have been converted into †fossil fuels such as coal, oil, and natural gas.

energy types and transfer For energy to be useful it has to be converted into a form that can do work. Energy occurs in many forms – for example, potential (stored energy), kinetic (movement), chemical, heat, light, electrical, sound, and nuclear energy.

A flat battery in a torch will not light the torch. If the battery is fully charged, it contains enough chemical energy to illuminate the torch bulb. When one body A does work on another body B, A transfers energy to B. The energy transferred is equal to the work done by A on B. Energy is therefore measured in †joules. The rate of doing work or consuming energy is called power and is measured in †watts (joules per second). Energy can be converted from any form into another. A ball resting on a slope possesses †potential energy that is gradually changed into †kinetic energy of rotation and translation as the ball rolls down. As a pendulum swings, energy is constantly being changed from a potential form at the highest points of the swing to kinetic energy at the lowest point. At positions in between these two extremes, the system possesses both kinetic and potential energy in varying proportions. A weightlifter changes chemical energy from the muscles into potential energy of the weight when the weight is lifted. If the weightlifter releases the weight, the potential energy is converted to kinetic energy as it falls, and this in turn is converted to heat energy and sound energy as it hits the floor. A lump of coal and a tank of petrol, together with the oxygen needed for their combustion, have chemical energy. Other sorts of energy include electrical and nuclear energy, light, and sound. However, all of these types are ultimately classifiable as either kinetic or potential energy.

resources So-called energy resources are stores of convertible energy. **Non-renewable resources** include the fossil fuels (coal, oil, and gas) and nuclear-fission "fuels" – for example, uranium-235. The term "fuel" is used for any material from which energy can be obtained. We use up fuel reserves such as coal and oil, and convert the energy they contain into other, useful forms. The chemical energy released by burning fuels can be used to do work. **Renewable resources**, such as wind, tidal, and geothermal power, have so far been less exploited. Hydroelectric projects are well established, and wind turbines and tidal systems are being developed.

energy conservation and efficiency All forms of energy are interconvertible by appropriate processes. Energy is transferred from one form to another, but the sum total of the energy after the conversion is always the same as the initial energy. This is the principle of conservation of energy. This principle can be illustrated by the use of energy flow diagrams, called Sankey diagrams, which show the energy transformations that take place. When a gasoline engine is used to power a car, about 75% of the energy from the fuel is wasted. The total energy input equals the total energy output, but a lot of energy is wasted as heat so that the engine is only about 25% efficient. The combustion of the gasoline–air mixture produces heat energy as well as kinetic energy. All forms of energy tend to be transformed into heat and cannot then readily be converted into other, useful forms of energy.

heat transfer A difference in temperature between two objects in thermal contact leads to the transfer of energy as †heat. Heat is energy transferred due to a temperature difference. Heat is transferred by the movement of particles (that possess kinetic energy) by conduction, convection, and radiation. †Conduction involves the movement of heat through a solid material by the movement of free electrons. For example, thermal energy is lost from a house by conduction through the walls and windows. †Convection involves the transfer of energy by the movement of fluid particles. All objects radiate heat in the form of radiation of electromagnetic waves. Hotter objects emit more energy than cooler objects. Methods of reducing energy transfer as heat through the use of †insulation are important because the world's fuel reserves are limited and heating homes costs a lot of money in fuel bills. Heat transfer from the home can be reduced by a variety of methods, such as loft insulation, cavity wall insulation, and double glazing. The efficiencies of insulating materials in the building industry are compared by measuring their heat-conducting properties, represented by a †U-value. A low U-value indicates a good insulating material.

$E = mc^2$ It is now recognized that mass can be converted into energy under certain conditions, according to Einstein's theory of relativity. This conversion of mass into energy is the basis of atomic power. †Einstein's special theory of †relativity (1905) correlates any gain, E, in energy with a gain, m, in mass, by the equation, $E = mc^2$, in which c is the speed of light. The conversion of mass into energy in accordance with this equation applies universally, although it is only in nuclear reactions that the percentage change in mass is large enough to detect.

EnergyEd http://www.ergon.com.au/energyed/

Light-hearted and educational introduction to electricity and renewable energy with the opportunity to take a virtual tour around a power station.

Heat http://www.brainpop.com/science/energy/heat/index.weml/

Highly graphical Shockwave-based site (this requires a plug-in, but the link is provided on the site). It deals with several topics relating to heat, and contains animation and videos, plus quizzes to test the user's knowledge. In addition, links to other sections of the science site are provided.

Potential and Kinetic Energy http://www.st-agnes.org/~lstinson/webpages/kinpot.htm

Vivid, colorful site detailing the definitions and equations of potential and kinetic energy. Following on from the definitions are examples of each, the last section dealing with how these relate to work and power. There are links to more detailed information on the different types of energy.

energy in biology, the basis for conducting living processes. Much of life involves †energy transfer. Energy is transferred from the surroundings of an organism into its body, and is also transferred within an organism's body. Energy is used by organisms to do things, such as growing or moving. When they do these things energy is transferred from one substance to another or from one place to another.

energy, alternative energy from sources that are renewable and ecologically safe, as opposed to sources that are non-renewable with toxic by-products, such as coal, oil, or gas (fossil fuels), and uranium (for nuclear power). The most important alternative energy source is flowing water, harnessed as †hydroelectric power. Other sources include the oceans' tides and waves (see †wave power), †wind power (harnessed by windmills and wind turbines), the Sun (†solar energy), and the heat trapped in the Earth's crust (†geothermal energy) (see also †cold fusion).

Alternative and Renewable Energy Sources http://www.life.ca/home/index.html

Explains varied energy sources through real-life examples. Energy types covered are wind, solar, and wood. Each example contains detailed information on how the alternative source of energy helps save money and coal, every year. There are links to other pages on the site, such as eco-friendly housing.

Ask an Energy Expert http://www.eere.energy.gov/askanenergyexpert/

The Energy Efficiency and Renewable Energy Clearinghouse (EREC) invites you to submit a question on "energy-efficient and renewable energy technologies."

Center for Alternative Technology http://www.cat.org.uk/

Dedicated to the Center For Alternative Technology in Wales, this includes a virtual tour of the site, pictures, and maps. It also includes a quiz, a question forum, and an extensive list of educational resources.

energy conservation methods of reducing †energy use through insulation, increasing energy efficiency, and changes in patterns of use. Profligate energy use by industrialized countries contributes greatly to air pollution and the †greenhouse effect when it draws on non-renewable energy sources.

It has been calculated that increasing energy efficiency alone could reduce carbon dioxide emissions in several high-income countries by 1–2% a year. The average annual decrease in energy consumption in relation to gross national product 1973–87 was 1.2% in France, 2% in the UK, 2.1% in the USA, and 2.8% in Japan.

Energy Saving http://www.g-k-b.fsnet.co.uk/energy_saving.htm

Well-designed page explaining numerous simple ways to make homes more energy efficient, covering areas such as clothes washing, heating, lighting, and hot water. Further links are provided to other home-orientated Web sites.

energy, electrical see †electrical energy.
energy, kinetic see †kinetic energy.
energy, law of conservation of fundamental principle of physics stating that energy can neither be created nor destroyed, but only changed from one form to another in a closed system.
energy level permitted energy that an electron can have in any particular atom. Energy levels can be calculated using †quantum theory. The permitted energy levels depend mainly on the distance of the electron from the nucleus. See †orbital, atomic.
energy, magnetic see †magnetic energy.
energy of reaction or **enthalpy of reaction** or **heat of reaction**, energy released or absorbed during a chemical reaction, part of the †energy transfer that takes place. In a chemical equation it may be represented by the symbol DH. In a chemical reaction, the energy stored in the reacting molecules is rarely the same as that stored in the product molecules. Depending on which is the greater, energy is either released (an †exothermic reaction) or absorbed (an †endothermic reaction) from the surroundings. The amount of energy released or absorbed by the quantities of substances represented by the chemical equation is the energy of reaction. The principle that the total amount of energy in a given chemical reaction stays the same is known as †conservation of energy.

Examples of chemical reactions that release large amounts of heat energy include the †combustion of †fuels and explosives. For example, when a mixture of hydrogen gas and oxygen gas in a glass jar is ignited by a spark, rapid and large amounts of heat energy are released shattering the glass jar in the process. The spark supplies the †activation energy required for bond breaking to occur in the reactants, enabling the reaction to begin.

$2H_2 + O_2 \rightarrow 2H_2O$ + 115.6 kJ of energy

Some oxidation reactions take place very slowly such as the †rusting of iron. The heat of the chemical reaction is immediately dissipated, and the activation energy required for rusting to take place is very large. The chemical equation describing the rusting of iron is:

$4Fe + 3O_2 \rightarrow 2Fe_2O_3$ + 380 kJ of energy

energy, potential see †potential energy.
energy transfer movement of energy from one place to another or from one substance to another, or the conversion of energy from one form to another. For example, in a car engine, the chemical energy of the fuel is converted to kinetic (motion), heat, and sound energy.

chemical energy transfer In a chemical reaction, energy is transferred between reactants and products, and between the reactants or products and their surroundings. The energy released or absorbed during the chemical reaction is the †**energy of reaction**, or energy of formation. During a chemical reaction the amount of chemical energy stored is equal to the amount of chemical energy released; this principle is known as the law of †conservation of energy. It means that energy cannot be created or destroyed, but can change from one form to another such as chemical energy changing into heat energy.

the living world Life processes of all kinds involve energy transfer; there can be no life without it. Energy may be transferred between an organism and its surroundings, for example when heat is lost or gained. Energy is also transferred inside an organism's body. Energy is used by organisms to do things such as growing and moving.

Nearly all life on Earth depends on energy transferred from the Sun. Plants carry out †photosynthesis in which the green chemical †chlorophyll absorbs sunlight, the energy of sunlight being transferred to the plant. The plant uses this energy to make glucose, which is used as a fuel during †respiration and is also used for growth. When animals eat plants, energy is transferred from the plants to the animals. Chemicals in the plant are absorbed by the animal, and these can then be used as fuel to provide energy for the animal. The same energy transfer occurs when animals eat other animals. So energy transfer is a key part of a †food chain or a food web.

engine device for converting stored energy into useful work or movement. Most engines use a fuel as their energy store. The fuel is burnt to produce heat energy – hence the name "heat engine" – which is then converted into movement. Heat engines can be classified according to the fuel they use (gasoline engine or †diesel engine), or according to whether the fuel is burnt inside (†internal combustion engine) or outside (†steam engine) the engine, or according to whether they produce a reciprocating or a rotary motion (†turbine or Wankel engine).

enlargement mathematical †transformation that changes the size of a shape by multiplying all its proportions by the same scale factor. An enlargement may be either smaller or larger.

For example:

A'B'C'D' is an enlargement of ABCD. The scale factor of the enlargement is 2, because all the lengths of the shape have been doubled and the distances of points A', B', C', and D' from the center of enlargement have also been doubled.

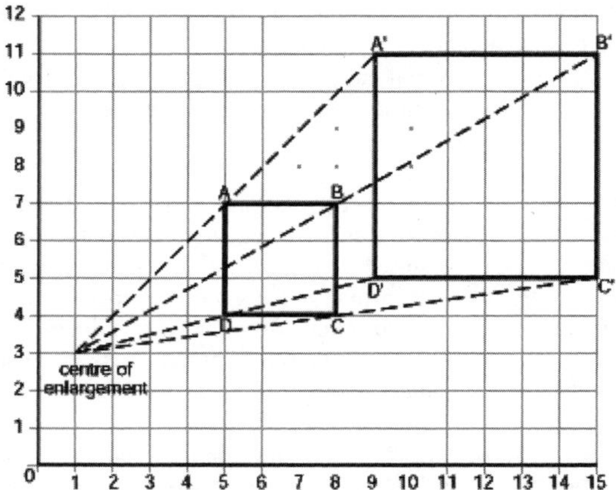

A scale factor greater than 1 will give a new shape that is larger than the original and a scale factor less than 1 will give a new shape that is smaller than the original, but they are both called enlargements.

To find the center of enlargement, the corresponding points of the enlargement and the original shape should be joined, and the lines extended until they all meet. For example:

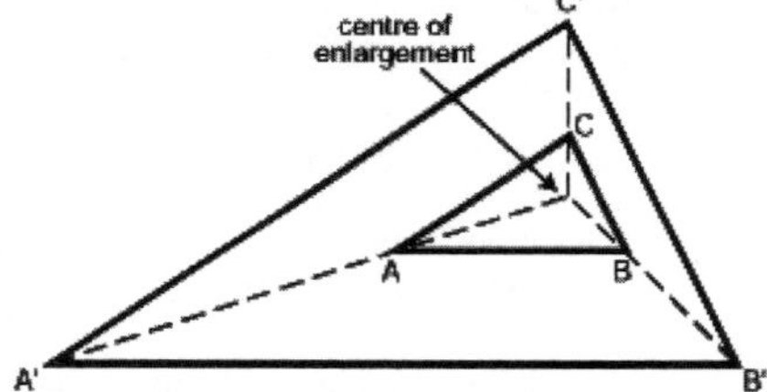

The center of enlargement is at O and the scale factor is 3.

enol in chemistry, unsaturated alcohol form of a compound that is subject to ketone–alcohol †tautomerism, for example 2,4-pentanedione.

enthalpy alternative term for †energy of reaction, the heat energy associated with a chemical change.

entomology study of †insects.

entrainment in earth science, picking up and carrying of one substance, such as rock particles, by another, such as water, air, or ice. The term is also used in †meteorology, when one body of air is incorporated into another, for instance, when clear air is entrained into a cloud.

entropy in †thermodynamics, a parameter representing the state of disorder of a system at the atomic, ionic, or molecular level; the greater the disorder, the higher the entropy. Thus the fast-moving disordered molecules of water vapor have higher entropy than those of more ordered liquid water, which in turn have more entropy than the molecules in solid crystalline ice.

In a closed system undergoing change, entropy is a measure of the amount of energy unavailable for useful work. At †absolute zero (–273.15°C/–459.67°F/0 K), when all molecular motion ceases and order is assumed to be complete, entropy is zero.

E number code number for additives that have been approved for use by the European Commission (EC). The E written before the number stands for European. E numbers do not have to be displayed on lists of ingredients, and the manufacturer may choose to list additives by their name instead. E numbers cover all categories of additives apart from flavorings.

Additives, other than flavorings, that are not approved by the European Commission, but are still used in Britain, are represented by a code number without an E.

envelope in geometry, a curve that touches all the members of a family of lines or curves. For example, a family of three equal circles all touching each other and forming a triangular pattern (like a clover leaf) has two envelopes: a small circle that fits in the space in the middle, and a large circle that encompasses all three circles.

environment in ecology, the sum of conditions affecting a particular organism, including physical surroundings, climate, and influences of other living organisms. Areas affected by †environmental issues include the †biosphere and †habitat. In biology, the environment includes everything outside body cells and fluid surrounding the cells. This means that materials enclosed by part of the body surface that is "folded in" are, in fact, part of the environment and not part of the organism. So the air spaces in human lungs and the contents of the stomach are all part of the environment and not the organism, using these terms correctly. Ecology is the study of the way organisms and their environment interact with each other. Important processes in biology involve the transfer of material between an organism and its environment in exchanges of gases and food, for example during †nutrition, †photosynthesis, or †respiration. In common usage, "the environment" often means the total global environment, without reference to any particular organism. In genetics, it is the external influences that affect an organism's development, and thus its †phenotype.

Environmental Education Network http://envirolink.netforchange.com/

Collaborative effort to place environmental information on the Internet. It is full of multimedia items, and divided into sections for teachers and students. These are further subdivided into sections such as geology, archaeology, and exploration/travel.

Exploring the Environment http://www.cotf.edu/ete/main.html

Learning modules on environment-related earth science topics. Issues covered include biodiversity, land use, global warming, and water pollution. Information pages are accompanied by interactive learning activities.

Jason Project http://www.jasonproject.org/

Engaging presentation of the Jason Project, a non-profit organization founded by Dr Robert Ballard after his discovery of the wreck of RMS Titanic, in order to allow students and teachers from all over the world to take part in global explorations using advanced interactive telecommunications. Details are posted for schools wishing to join the network.

environmental impact assessment **EIA**, in the UK, a process by which the potential environmental impacts of human activities, such as the construction of a power station, dam, or major housing development, are evaluated. The results of an EIA are published and discussed by different levels of government, non-governmental organizations, and the general public before a decision is made on whether or not the project can proceed.

Some developments, notably those relating to national defense, are exempt from EIA. Increasingly studies include the impact not only on the physical environment, but also the socio-economic environment, such as the labor market and housing supply.

environmental impact statement **EIS**, statement predicting the consequences a project or development will have upon the environment. It is intended to help manage environmental change by rating the environmental effects of development against economic factors.

environmental issues matters relating to the damaging effects of human activity on the biosphere, their causes, and the search for possible solutions. The political movement that supports protection of the environment is the green movement. Since the Industrial Revolution, the demands made by both the industrialized and developing nations on the Earth's natural resources are increasingly affecting the balance of the Earth's resources. Over a period of time, some of these resources are renewable – trees can be replanted, soil nutrients can be replenished – but many resources, such as minerals and fossil fuels (coal, oil, and natural gas), are †non-renewable and in danger of eventual exhaustion. In addition, humans are creating many other problems that may endanger not only their own survival, but also that of other species. For instance, †deforestation and †air pollution are not only damaging and radically altering many natural environments, they are also affecting the Earth's climate by adding to the †greenhouse effect and †global warming, while †water pollution is seriously affecting aquatic life, including fish populations, as well as human health.

Environmental pollution is normally taken to mean harm done to the natural environment by human activity. In fact, some environmental pollution can have natural sources, for example volcanic activity, which can cause major air pollution or water pollution and destroy flora and fauna. In terms of environmental issues, however, environmental pollution relates to human actions, especially in connection with energy resources. The demands of the industrialized nations for energy to power machines, provide light, heat, and so on are constantly increasing. The most versatile form of energy is electricity, which can be produced from a wide variety of other energy sources, such as the fossil fuels and nuclear power (produced from uranium). These are all non-renewable resources and, in addition, their extraction, transportation, utilization, and waste products all give rise to pollutants of one form or another. The effects of these pollutants can have consequences not only for the local environment, but also at a global level.

Environment 2000 and Beyond http://www.environment-agency.gov.uk/yourenv/129596/?version=1

Britain's Environment Agency's page on the environment, with access to data collected and studied by the Agency on the topics of "Bathing water quality," "River habitats," and "River gauging stations." There is also more general information on the current "Stresses and strains" on the environment.

USGS Earthshots: Satellite Images of Environmental Change
http://edcwww.cr.usgs.gov/earthshots/slow/tableofcontents
See how Earth's environment is changing by accessing this fascinating collection of satellite pictures from the US Geological Survey. Start by clicking on a highlighted area on a world map to go to a series of detailed comparative images, or select from a huge collection of articles on subjects such as geology, water and wildlife.

environment–heredity controversy
see †nature–nurture controversy.

environmental lapse rate rate at which temperature changes with altitude. Usually, because the atmosphere is heated from below, temperatures decrease as altitude increases at an average rate of –6.4°/11.5°F per 1,000 m/3,280 ft. The rate varies from place to place and according to the time of day. Under certain conditions, the lapse rate can be reversed, so that temperatures increase with altitude. This can be caused by large-scale atmospheric movements, for example where a warm stable air mass overlies a cool, less stable air mass. This is common in †trade wind areas of the tropics. Smaller scale temperature inversions develop at night where surface temperatures fall rapidly, cooling the air near the surface. It can also be caused by cold air flowing down a slope into valleys.

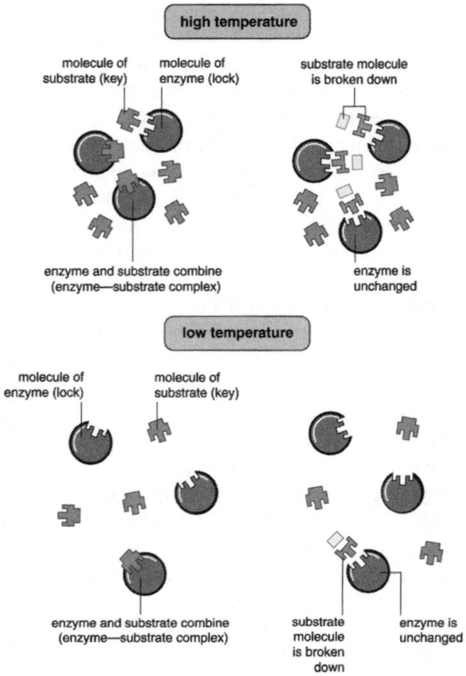

enzyme Enzymes are catalysts that can help break larger molecules into smaller molecules while remaining unchanged themselves. Like a key for a lock, each enzyme is specific to one molecule. Most also function best within a narrow temperature and pH range. As the temperature rises enzymes catalyze more molecules but beyond a certain temperature most enzymes become denatured. Helicon

enzyme biological †catalyst produced in cells, and capable of speeding up the chemical reactions necessary for life. They are large, complex †proteins, usually soluble, and are highly specific, each chemical reaction requiring its own particular enzyme. The enzyme's specificity arises from its active site, an area with a shape corresponding to part of the molecule with which it reacts (the substrate). The shape of the enzyme where the chemical binds only allows the binding of that particular chemical, rather like a specific key only working a specific lock (the lock and key hypothesis). The enzyme and the substrate slot together forming an enzyme–substrate complex that allows the reaction to take place, after which the enzyme falls away unaltered.

The activity and efficiency of enzymes are influenced by various factors, including temperature and acidity (pH). Temperatures above 60°C/140°F damage (denature) the intricate structure of enzymes, inactivating them and causing reactions to stop. Each enzyme operates best – at its maximum rate – within a specific pH range and temperature, and is denatured by excessive acidity or alkalinity or extremes of temperature.

In †digestion, digestive enzymes include amylases (which digest starch), lipases (which digest fats), and proteases (which digest protein). Other enzymes play a part in the conversion of food energy into †ATP, the manufacture of all the molecular components of the body, the replication of †DNA when a cell divides, the production of hormones, and the control of movement of substances into and out of cells.

Enzymes have many uses in medical and industrial †biotechnology, from washing powders to drug production, and as research tools in molecular biology. They are involved in the making of beer, bread, cheese, and yogurt. They can be extracted from bacteria and fungi and †genetic engineering now makes it possible to tailor an enzyme for a specific purpose.

Enzymes http://www.gcsechemistry.com/rc13.htm
Part of a section dealing with rates of reaction, this page considers the role of enzymes in chemical processes. The processes of making beer and bread are used as examples of the enzyme yeast. Key words are highlighted in either red or blue.

enzyme induction stimulation of enzyme formation by the presence of its †substrate or a derivative of the substrate.

Eocene epoch second epoch of the Tertiary period of geological time, roughly 56.5–35.5 million years ago. Originally considered the earliest division of the Tertiary, the name means "early recent," referring to the early forms of mammals evolving at the time, following the extinction of the dinosaurs.

eolian in earth science, describing anything deposited by wind, such as deposits of sand or dust in deserts on the surface of the Earth and on Mars.

eolith naturally shaped or fractured stone found in Lower Pleistocene deposits and once believed by some scholars to be the oldest known artifact type, dating to the pre-Paleolithic era. They are now recognized as not having been made by humans.

eon or **aeon**, in earth science, large amount of geological time consisting of several eras. The term is also used to mean a thousand million years (10^9 y).

eotvos unit symbol E, unit for measuring small changes in the intensity of the Earth's †gravity with horizontal distance.

ephemeral plant plant with a very short life cycle, sometimes as little as six to eight weeks. It may complete several generations in one growing season.

ephemeris plural **ephemerides**, astronomical table that gives the expected positions of the Sun, Moon, planets, comets, and other objects at given times, usually the days in a year.

epicenter point on the Earth's surface immediately above the seismic focus of an †earthquake. Most building damage takes place at an earthquake's epicenter. The term is also sometimes used to refer to a point directly above or below a nuclear explosion ("at ground zero").

epicycle circular orbit of a body round a point that is itself in a circular orbit round a parent body. Such a system was formulated to explain some planetary orbits in the Solar System before they were known to be elliptical.

epicycloid in geometry, a curve resembling a series of arches traced out by a point on the circumference of a circle that rolls around another circle of a different diameter. If the two circles have the same diameter, the curve is a †cardioid.

epicycloid A seven-cusped epicycloid, formed by a point on the circumference of a circle (of diameter d) that rolls around another circle (of diameter 7d/3). Helicon

epidermis outermost layer of †cells on an organism's body. In plants and many invertebrates such as insects, it consists of a single layer of cells. In vertebrates, it consists of several layers of cells.

The epidermis of plants and invertebrates often has an outer noncellular †cuticle that protects the organism from desiccation.

epididymis in male vertebrates, a long coiled tubule in the †testis, in which sperm produced in the seminiferous tube are stored. In men, it is a duct about 6 m/20 ft long, convoluted into a small space.

epigeal describing seed germination in which the †cotyledons (seed leaves) are borne above the soil.

epiglottis small flap located behind the root of the tongue in mammals. It closes off the end of the windpipe during swallowing to prevent food from passing into it and causing choking.

The action of the epiglottis is a highly complex reflex process involving two phases. During the first stage a mouthful of chewed food is lifted by the tongue towards the top and back of the mouth. This is accompanied by the cessation of breathing and by the blocking of the nasal areas from the mouth. The second phase involves the epiglottis moving over the larynx while the food passes down into the esophagus.

Epimetheus in astronomy, small moon of Saturn, discovered in 1980, which orbits just outside the ring system in the same orbit as †Janus. It is irregular in shape, its longest dimension being 140 km/87 mi.

epiphragm in botany, membrane that closes the spore †capsule of different species of moss plant.

epiphragm in zoology, layer of mucus that hardens and seals the opening of the shell of various gastropod mollusks. An epiphragm is usually secreted in dry conditions preventing the animal within the shell from losing excessive amounts of water.

epiphyte any plant that grows on another plant or object above the surface of the ground, and has no roots in the soil. An epiphyte does not parasitize the plant it grows on but merely uses it for support. Its nutrients are obtained from rainwater, organic debris such as leaf litter, or from the air.

The greatest diversity of epiphytes is found in tropical areas and includes many orchids.

epithelium in animals, tissue of closely packed cells that forms a surface or lines a cavity or tube. Epithelial cells line the inside surfaces of fluid or air-filled tubes and spaces within the body. Epithelium may be protective (as in the skin) or secretory (as in the cells lining the wall of the gut). Epithelial cells join each other side to side to make epithelial †tissue. Epithelial cells are specialized cells. They protect cells below them and may also carry out special functions. For example, in the human lung epithelial cells line the alveoli (air spaces). These cells are very thin and form a large surface area in order to facilitate the absorption of oxygen and the loss of carbon dioxide during the process of gas exchange. They also make sure the alveoli remain moist at all times to keep them healthy.

The epithelial cells lining the airways leading to the lung from the mouth and nose are different. Here there are many are specialized ciliated cells. These cells have tiny hairs or cilia which beat backwards and forwards. They sweep fluid up and away from the lungs to the mouth. Any dirt or bacteria that have been inhaled are taken away from the lungs where they may cause damage. Smoking harms these ciliated cells.

epitrochoid or hypotrochoid †locus of a point on a rolling circle moving round the circumference of another circle, that is not on the circumference of the rolling circle. The locus is of importance in the Wankel engine.

epoch subdivision of a geological period in the geological time scale. Epochs are sometimes given their own names (such as the Paleocene, Eocene, Oligocene, Miocene, and Pliocene epochs comprising the Tertiary period), or they are referred to as the late, early, or middle portions of a given period (as the Late Cretaceous or the Middle Triassic epoch).

Geological time is broken up into **geochronological units** of which epoch is just one level of division. The hierarchy of geochronological divisions is eon, †era, period, epoch, age, and chron. Epochs are subdivisions of periods and ages are subdivisions of epochs. Rocks representing an epoch of geological time comprise a **series**.

epoxidation in chemistry, formation of †epoxide compounds by the addition of an oxygen atom across a carbon-carbon double bond. A peroxyacid (an acid containing the —O—O— group) is commonly used to transfer the oxygen to the double bond, for example peracetic acid.

epoxide in chemistry, member of a group of highly reactive, cyclic ethers, for example ethylene oxide C_2H_4O. These compounds all contain a three-member ring of two carbon atoms bonded to an oxygen atom. This configuration is highly strained, resulting in the high reactivity of this class of compounds.

epoxy resin in chemistry, member of a class of thermosetting (do not soften on heating) polymers that are formed by the addition of monomer units via reactions between epoxy groups and compounds. They are used as structural plastics, protective coatings, and adhesives.

EPROM acronym for Erasable Programmable Read-Only Memory, computer memory device in the form of an †integrated circuit (chip) that can record data and retain it indefinitely. The data can be erased by exposure to ultraviolet light, and new data recorded. Other kinds of computer memory chips are †ROM (read-only memory), †PROM (programmable read-only memory), and †RAM (random-access memory).

Epsilon Aurigae †eclipsing binary star in the constellation Auriga. One of the pair is an "ordinary" star, but the other seems to be an enormous distended object whose exact nature remains unknown. The period (time between eclipses) is 27 years, the longest of its kind. The last eclipse was in 1982–84.

Epsilon Eridani small cool star 10.5 light years distant, making it a near neighbor of the Sun. Irregularities in its motion make it possible that it is orbited by a large planet.

Epsom salts $MgSO_4.7H_2O$, hydrated magnesium sulfate, used as a relaxant and laxative and added to baths to soothe the skin. The name is derived from a bitter saline spring at Epsom, Surrey, England, which contains the salt in solution.

Epstein, (Michael) Anthony (1921–) English microbiologist who, in collaboration with his assistant Barr, discovered in 1964 the Epstein–Barr virus (EBV) that causes glandular fever in humans and has been linked to some forms of human cancer.

equation in chemistry, representation of a chemical reaction by symbols and numbers; see †chemical equation. For example, the reaction of sodium hydroxide (NaOH) with hydrochloric acid (HCl) to give sodium chloride and water may be represented by:

$NaOH + HCl \rightarrow NaCl + H_2O$

equation in mathematics, an †expression that represents the equality of two expressions involving constants and/or variables, and thus usually includes an equals (=) sign. For example, the equation $A = pr^2$ equates the area A of a circle of radius r to the product pr^2. This is also known as the †formula for the area of a circle. The algebraic equation $y = mx + c$ is the general one in coordinate geometry for a straight line and is known as a †linear equation. See also †algebra, †quadratic equations, †simultaneous equations, inequations or †inequalities, †graphs, and †substitutions.

solving an equation To solve an equation means to find the value or values of the unknown quantity that satisfy the equation; for example,

$x + 4 = 7$ is true when x is 3

The values of the unknown that make an equation true are called its solutions or roots.

In general, solving an equation depends on transforming it into a simple standard form. This can be achieved by using the following processes:

(i) adding the same quantity to each side of the equation
(ii) subtracting the same quantity from each side of the equation
(iii) multiplying each side of the equation by the same quantity (so long as it is not zero)
These processes can be used to change an equation into a simpler form but they will not alter its solution. For example, to solve the equation $7x - 4 = 3x + 8$:
subtract $3x$ from each side in order to collect the xs on the left-hand side:
$7x - 4 - 3x = 3x + 8 - 3x$ so $4x - 4 = 8$
add 4 to each side in order to collect the numbers on the right-hand side:
$4x - 4 + 4 = 8 + 4$ so $4x = 12$
divide both sides by 4 to obtain the solution:
$4x \div 4 = 12 \div 4$ so $x = 3$
polynomials A type of equation that has been studied particularly intensively is where there is one unknown and the expression involving it is a †polynomial. A polynomial equation has the form:
$f(x) = a_n x^n + a_{n-1} x^{n-1} + \ldots + a_2 x^2 + a_1 x + a_0$
where $a_n, a_{n-1}, \ldots, a_0$ are all constants, n is a positive integer, and $a_n \neq 0$.
The "degree" of a polynomial equation is simply the degree of the polynomial involved. A polynomial of degree one, that is, whose highest power of x is 1, as in $2x + 1$, is called a linear polynomial;
$3x^2 + 2x + 1$ is quadratic;
$4x^3 + 3x^2 + 2x + 1$ is cubic.
indeterminate equations An indeterminate equation is an equation for which there is an infinite set of solutions – for example, $2x = y$. A **diophantine equation** is an indeterminate equation in which both the solution and the terms must be whole numbers (after Diophantus of Alexandria, *c.* AD 250).
identity An equation that is true for all values of the unknown is called an identity, for example $x + x = 2x$. It is denoted by ≡.
Thus $(x + y)^2 \equiv x^2 + 2xy + y^2$ for all real numbers x, y.
Aerobi-graphs http://www.mathsnet.net/aerobics/aerobics.html
Fun look at the graph equation y=mx+c. See what happens to the graph when each variable is manipulated, as demonstrated by human bodies! As the graphs become more complex, so do the actions. However, the good illustrations do make this site quite slow.

equation of motion mathematical equation that gives the position and velocity of a moving object at any time. Given the mass of an object, the forces acting on it, and its initial position and velocity, an equation of motion is used to calculate its position and velocity at any later time. The equation must be based on †Newton's laws of motion or, if speeds near that of light are involved, on the theory of †relativity.
The common equations are:
$v = u + at$
$s = {}^1/_2(u + v)t$
$s = ut + {}^1/_2at^2$
$v^2 = u^2 + 2as$
in which a is the object's constant acceleration, u is its initial velocity, v is its velocity after a time t, and s is the distance traveled by it in that time.

Equator or **terrestrial equator**, †great circle whose plane is perpendicular to the Earth's axis (the line joining the poles). Its length is 40,092 km/24,902 mi, divided into 360 degrees of longitude. The Equator encircles the broadest part of the Earth, and represents 0° latitude. It divides the Earth into two halves, called the northern and the southern hemispheres.
The **celestial equator** is the circle in which the plane of the Earth's Equator intersects the †celestial sphere.

equatorial bulge in astronomy, increase in the diameter of a planet (or the Sun) at its equator, which results from its spinning on its axis. The diameter of the Earth, for example, is 12,756 km/7,926 mi at the equator, whereas its polar diameter is only 12,714 km/7,900 mi.

equatorial coordinates in astronomy, system for measuring the position of astronomical objects on the †celestial sphere with reference to the plane of the Earth's Equator.
†Declination (symbol d), analogous to latitude, is measured in degrees from the Equator to the north (d = 90°) or south (d =–90°) celestial poles. Right †ascension (symbol a), analogous to longitude, is normally measured in hours of time (a = 0 hours to 24 hours) eastward along the Equator from a fixed point known as the first point of †Aries or the vernal †equinox.

equatorial mounting in astronomy, method of mounting a †telescope to simplify the tracking of celestial objects. One axis (the polar axis) is mounted parallel to the rotation axis of the Earth so that the telescope can be turned about it to follow objects across the sky. The †declination axis moves the telescope in declination and is clamped before tracking begins. Another advantage over the simpler †altazimuth mounting is that the orientation of the image is fixed, permitting long-exposure photography.

equilateral describing a geometrical figure, having all sides of equal length.

equilibrium in physics, an unchanging condition in which an undisturbed system can remain indefinitely in a state of balance. In a **static equilibrium**, such as an object resting on the floor, there is no motion. In a **dynamic equilibrium**, in contrast, a steady state is maintained by constant, though opposing, changes. For example, in a sealed bottle half-full of water, the constancy of the water level is a result of molecules evaporating from the surface and condensing on to it at the same rate.

equilibrium constant numerical value that expresses the position of a chemical equilibrium at a given temperature and pressure. It is given by the product of the concentrations of the products divided by the product of the concentrations of the reactants.

equinox time when the Sun is directly overhead at the Earth's †Equator and consequently day and night are of equal length at all latitudes. This happens twice a year: 21 March is the spring, or vernal, equinox and 23 September is the autumn equinox.

The variation in day lengths occurs because the Earth is tilted on its axis with respect to the Sun. However, because the Earth not only rotates on its own axis but also orbits the Sun, at the equinoxes the two bodies are positioned so that the circle of light from the Sun passes through both of the Earth's poles.

equivalent having a different appearance but the same value. $^3/_5$ and $^6/_{10}$ are equivalent fractions. They both have the value 0.6.

equivalent weight mass of a substance that exactly reacts with, or replaces, an arbitrarily fixed mass of another substance in a particular chemical reaction. The combining proportions (by mass) of substances are in the ratio of their equivalent masses (or a multiple of that ratio) and a common standard has been adopted: for elements, the equivalent weight is the quantity that reacts with, or replaces, 1.00797 g/0.035279 oz of hydrogen or 7.9997 g/0.28215 oz of oxygen, or the mass of an element liberated during †electrolysis by the passage of 1 faraday (96,487) coulombs per mole) of electricity. The equivalent weight of an element is given by its gram †atomic mass divided by its †valency. For oxidizing and reducing agents, the equivalent weight is the gram molecular mass divided by the number of electrons gained or lost by each molecule. Some substances have several equivalent weights, depending on the specific reaction in which they are involved.

Equuleus small, faint constellation on the celestial equator (see †celestial sphere), represented as a foal.

era any of the major divisions of geological time that includes several periods but is part of an eon. The eras of the current Phanerozoic in chronological order are the Paleozoic, Mesozoic, and Cenozoic. We are living in the Recent epoch of the Quaternary period of the Cenozoic era.

Geological time is broken up into **geochronological units** of which era is just one level of division. The hierarchy of geochronological divisions is eon, era, period, †epoch, age, and chron. Eras are subdivisions of eons and periods are subdivisions of eras. Rocks representing an era of geological time comprise an **erathem**.

Erasistratus (c. 304–c. 250 BC) Greek physician and anatomist. Regarded as the founder of physiology, he came close to discovering the true function of several important systems of the body, which were not fully understood until nearly 1,000 years later. For example, the principle of blood circulation, though he had it circulating in the wrong direction. Tracing the network of veins, arteries, and nerves, he postulated that the nerves carry the "animal" spirit, the arteries the "vital" spirit, and the veins blood. He did, however, grasp a rudimentary principle of oxygen exchange and condemned bloodletting as a form of treatment.

Eratosthenes (c. 276–c. 194 BC) Greek geographer and mathematician whose map of the ancient world was the first to contain lines of latitude and longitude, and who calculated the Earth's circumference with an error of about 10%. His mathematical achievements include a method for duplicating the cube, and for finding †prime numbers (Eratosthenes' sieve).

Eratosthenes' sieve method for finding †prime numbers. It involves writing in sequence all numbers from 2. Then, starting with 2, cross out every second number (but not 2 itself), thus eliminating numbers that can be divided by 2. Next, starting with 3, cross out every third number (but not 3 itself), and continue the process for 5, 7, 11, 13, and so on. The numbers that remain are primes.

erbium chemical symbol Er, soft, lustrous, grayish, metallic element of the †lanthanide series, atomic number 68, relative atomic mass 167.26. It occurs with the element yttrium or as a minute part of various minerals. It was discovered in 1843 by Carl Mosander (1797–1858), and named after the town of Ytterby, Sweden, near which the lanthanides (rare-earth elements) were first found.

Erbium has been used since 1987 to amplify data pulses in optical fibers, enabling faster transmission. Erbium ions in the fiberglass, charged with infrared light, emit energy by amplifying the data pulse as it moves along the fiber.

erg *c.g.s.* unit of work, replaced in the SI system by the †joule. One erg of work is done by a force of one †dyne moving through one centimeter.

ergodics in dynamics, study of the mathematical principles involved in the kinetic theory of gases.

ergonomics study of the relationship between people and the furniture, tools, and machinery they use at work. The object is to improve work performance by removing sources of muscular stress and general fatigue: for example, by presenting data and control panels in easy-to-view form, making office furniture comfortable, and creating a generally pleasant environment.

Good ergonomic design makes computer systems easier to use and minimizes the health hazards and physical stresses of working with computers for many hours a day: it helps data entry workers to avoid conditions like repetitive strain injury (RSI), eyestrain, and back and muscle aches.

Eridanus sixth-largest constellation, which meanders from the celestial equator (see †celestial sphere) deep into the southern hemisphere of the sky. Eridanus is represented as a river. Its brightest star is †Achernar, a corruption of the Arabic for "the end of the river."

Erikson, Erik (Homburger) (1902–1994) German-born US psychoanalytic theorist who contributed to the understanding of human mental development. Erikson coined the phrase **identity crisis**, and proposed that the ego is not fixed at birth or during childhood, but continues to be molded throughout life by experience and environment. He established his reputation with the influential work *Childhood and Society* (1950), his first major book. His research in developmental psychology led to his theory of eight psychosocial stages in the life cycle.

Erlangen program expression used by German mathematician Felix Klein to denote his unification and classification of geometries Euclidean and non-Euclidean as "members" of one "family," corresponding to the transformations found in each.

Erlanger, Joseph (1874–1965) US physiologist who was awarded a Nobel Prize for Physiology or Medicine in 1944 with Herbert †Gasser for their work on the transmission of impulses by nerve fibers. They found that the smaller nerve fibers were responsible for the conduction of pain, and that the thickness of a nerve fiber dictates the speed at which a nerve can transmit electrical information.

Ernst, Richard Robert (1933–) Swiss physical chemist who was awarded the Nobel Prize for Chemistry in 1991 for his improvements in the technique of nuclear magnetic resonance (NMR) †spectroscopy in the investigation of atomic nuclei by increasing the sensitivity of the NMR instrumentation and interpretation. NMR became the ideal method for examining the structure of proteins and other biological molecules, laying the foundations for †magnetic resonance imaging (MRI).

Eros in astronomy, asteroid discovered in 1898 by German astronomer Gustav Witt. It was the first asteroid to be discovered that has an orbit coming within that of Mars. It passes within 22 million km/14 million mi of the Earth. It is elongated, measures 33 x 13 x 13 km/21 x 8 x 8 mi, rotates around its shortest axis every 5.3 hours, and orbits the Sun every 1.8 years.

NASA's Near-Earth Asteroid Rendezvous spacecraft went into orbit around Eros on 14 February 2000. Pictures showed the asteroid to be heavily cratered, indicating that it is probably older than was previously thought. The surface was discovered to contain aluminum, iron, magnesium, and silicon.

erosion wearing away of the Earth's surface by a moving agent, caused by the breakdown and transport of particles of rock or soil. Agents of erosion include the sea, rivers, glaciers, and wind. By contrast, †weathering does not involve transportation.

The most powerful forms of erosion are water, consisting of sea waves and currents, rivers, and rain; ice, in the form of glaciers; and wind, hurling sand fragments against exposed rocks and moving dunes along. People also contribute to erosion by poor farming practices and the cutting down of forests, which can lead to increased overland water run-off.

There are several processes of river erosion including †hydraulic action, †corrasion, †attrition, and †solution.

Erosion and Deposition http://www.geog.ouc.bc.ca/physgeog/contents/11g.html

Introduction to two important geological processes: erosion and deposition. The hyperlinked text explains the three stages in the process of erosion – detachment, entrainment, and transport – as well as the causes of deposition.

erratic in geology, a displaced rock that has been transported by a glacier or some other natural force to a site of different geological composition.

error in computing, a fault or mistake, either in the software or on the part of the user, that causes a program to stop running (crash) or produce unexpected results. Program errors, or bugs, are largely eliminated in the course of the programmer's initial testing procedure, but some will remain in most programs. All computer operating systems are designed to produce an **error message** (on the display screen, or in an error file or printout) whenever an error is detected, reporting that an error has taken place and, wherever possible, diagnosing its cause.

error in mathematics, incorrect answer to a calculation. Also, the amount by which an incorrect answer differs from the correct one. This is usually denoted by d*x* for a figure whose correct value is *x*.

error detection in computing, the techniques that enable a program to detect incorrect data. A common method is to add a check digit to important codes, such as account numbers and product codes. The digit is chosen so that the code conforms to a rule that the program can verify. Another technique involves calculating the sum (called the hash total) of each instance of a particular item of data, and storing it at the end of the data.

error theory or **theory of errors**, in statistics and probability theory, method of evaluating the effects and the significance of errors, for example when obtaining a mean value from a small sample.

erythroblast in biology, a series of nucleated cells that go through various stages of development in the bone marrow until they form red blood cells (erythrocytes). This process is known as erythropoiesis. Erythroblasts can appear in the blood of people with blood cancers.

erythrocyte another name for †red blood cell.

Esaki, Leo (1925–) born Esaki Reiona, Japanese physicist who in 1957 noticed that electrons could sometimes "tunnel" through the barrier formed at the junctions of certain semiconductors. The effect is now widely used in the electronics industry.

Esaki was awarded the Nobel Prize for Physics in 1973 for this early discovery of the tunneling phenomenon in semiconductors and superconductors. He shared the award with British physicist Brian †Josephson and Norwegian-born US physicist Ivar †Giaever.

escape velocity minimum velocity required for a spacecraft or other object to escape from the gravitational pull of a planetary body. In the case of the Earth, the escape velocity is 11.2 kps/6.9 mps; the Moon, 2.4 kps/1.5 mps; Mars, 5 kps/3.1 mps; and Jupiter, 59.6 kps/37 mps.

escarpment or **cuesta**, large ridge created by the erosion of dipping sedimentary rocks. It has one steep side (scarp) and one gently sloping side (dip). Escarpments are common features of chalk landscapes, such as the Chiltern Hills and the North Downs in England. Certain features are associated with chalk escarpments, including dry valleys (formed on the dip slope), combes (steep-sided valleys on the scarp slope), and springs.

Escherichia coli or **colon bacillus**, rod-shaped Gram-negative bacterium (see †bacteria) that lives, usually harmlessly, in the colon of most warm-blooded animals. It is the commonest cause of urinary tract infections in humans. It is sometimes found in water or meat where fecal contamination has occurred and can cause severe gastric problems.

The mapping of the genome of *E. coli*, consisting of 4,403 genes, was completed in 1997. It is probably the organism about which most molecular genetics is known, and is of pre-eminent importance in recombinant DNA research.

classification *Escherichia coli* is the only species in the bacterial family Enterobacteriaceae.

esker narrow, steep-walled ridge, often meandering, found in formerly glaciated areas. It was originally formed beneath a glacier. It is made of sands and gravels, and represents the course of a river channel beneath the glacier. Eskers vary in height from 3 to 30 m/10 to 100 ft and can be up to 160 km/100 mi or so in length. Eskers are often used for roads as they are areas of high ground above marshy clay lowlands (glacial deposits).

Eskola, Pentti Eelis (1883–1964) Finnish geologist. He was one of the first to apply physicochemical postulates on a far-reaching basis to the study of †metamorphism, thereby laying the foundations of most subsequent studies in metamorphic †petrology.

estrogen any of a group of †hormones produced by the †ovaries of vertebrates; the term is also used for various synthetic hormones that mimic their effects. The principal estrogen in mammals is estradiol. Estrogens control female sexual development, promote the growth of female †secondary sexual characteristics at puberty, stimulate egg (†ovum) production, and, in mammals, prepare the lining of the uterus for pregnancy. In other words, together with another hormone, progesterone, they regulate the growth and functioning of sex organs for †sexual reproduction. Estrogens are also used in female oral contraceptives, to inhibit the production of ova.

Estrogens are used therapeutically for some hormone disorders and to inhibit lactation. US researchers in 1995 observed that estrogen plays a role in the healing of damaged blood vessels. It has also been found that women recover more quickly from strokes if given a low estrogen dose.

ESP abbreviation for †extrasensory perception.

essential amino acid water-soluble organic molecule vital to a healthy diet because it cannot be synthesized from other food molecules; see †amino acid.

essential fatty acid organic compound consisting of a hydrocarbon chain and important in the diet; see †fatty acid.

ester organic compound formed by the reaction between an alcohol and an acid, with the elimination of water. Unlike †salts, esters are covalent compounds.

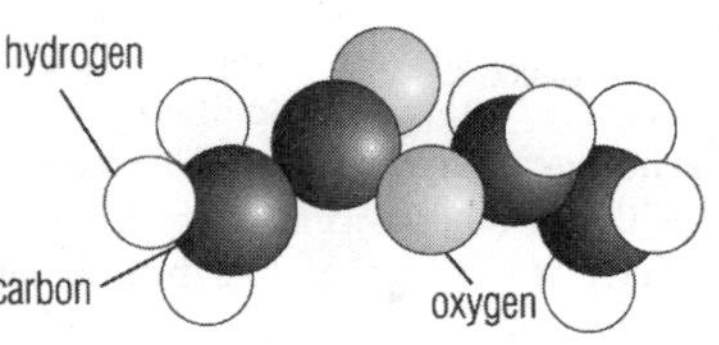

ester Molecular model of the ester ethyl ethanoate (ethyl acetate) $CH_3CO_2CH_2CH_3$. Helicon

esterification in chemistry, a combination reaction between an alcohol and an acid that results in the formation of an †ester.

estradiol alternative spelling of oestradiol, a type of †estrogen (female sex hormone).

esophagus muscular tube through which food travels from the mouth to the stomach. The human esophagus is about 23 cm/9 in long. It extends downwards from the †pharynx, immediately behind the windpipe. It is lined with a mucous membrane made of epithelial cells (see †epithelium), which secretes lubricant fluid to assist the downward movement of food. In its wall is †muscle, which contracts to squeeze the food towards the stomach (†peristalsis).

estrus in mammals, the period during a female's reproductive cycle (also known as the estrus cycle or †menstrual cycle) when mating is most likely to occur. It usually coincides with ovulation.

estuary river mouth widening into the sea, where fresh water mixes with salt water and tidal effects are felt.

Eta Carinae faint luminous blue variable star 100 times more massive than the Sun and approximately 8,000 light years distant in the constellation Carina. It is surrounded by a cool dust cloud that is a powerful infrared source and radiates 5 million times the power of the Sun. The cloud is sometimes called the Homunculus nebula because of its supposed resemblance to a human figure.

It is unstable and some astronomers believe it may explode as a supernova.

etalon instrument that uses †interference phenomena to make possible very high-resolution observations of †spectral lines.

Etesian wind north-northwesterly wind that blows June–September in the eastern Mediterranean and Aegean seas.

ethanal common name **acetaldehyde**; CH_3CHO, one of the chief members of the group of organic compounds known as †aldehydes. It is a colorless flammable liquid boiling at 20.8°C/69.6°F. Ethanal is formed by the oxidation of ethanol or ethene and is used to make many other organic chemical compounds.

ethanal trimer common name **paraldehyde**; $(CH_3CHO)_3$, colorless liquid formed from ethanal (acetaldehyde). It is soluble in water.

ethanamide CH_3CONH_2, solid, crystalline compound produced by distilling ammonium ethanoate in a stream of dry ammonia. As usually prepared it has a strong odor suggestive of mice, but this is due to impurities. It is soluble in water and alcohol, melts at 82°C/179.6°F, and boils at 222°C/431.6°F.

ethane CH_3CH_3, colorless, odorless gas, the second member of the †alkane series of hydrocarbons (paraffins).

ethane-1,2-diol technical name for †glycol.

ethanoate common name **acetate**; $CH_3CO_2^-$, negative ion derived from ethanoic (acetic) acid; any salt containing this ion. In photography, acetate film is a non-flammable film made of cellulose ethanoate. In textiles, it is known as acetate, and is most commonly known in the form of a synthetic fiber which can be woven or knitted to produce a variety of different fabrics including satin, moiré, and taffeta. Fabrics made from acetate absorb moisture, do not shrink, and are cheaper to produce than natural fibers, but they also tend to attract dirt and can be damaged by heat. Acetate filaments are also used to make filters for cigarettes.

To produce acetate, †cellulose is treated with acetic acid in a process called acetylation. This forms a thick solution which is forced through small holes and solidified, and then, through the process of spinning, continuous filaments of acetate are produced.

ethanoic acid common name **acetic acid**; CH_3CO_2H, one of the simplest carboxylic acids (fatty acids). In the pure state it is a colorless liquid with an unpleasant pungent odor; it solidifies to an icelike mass of crystals at 16.7°C/62.4°F, and hence is often called glacial ethanoic acid. In a dilute form, mixed with water, it is the acid found in vinegar. Vinegar contains 5% or more ethanoic acid, produced by fermentation.

Ethanoic acid is produced by the oxidation of †ethanol. It belongs to a homologous series, or family, of organic compounds. Some other members of the series are methanoic acid, propanoic acid, and butanoic acid. They are all weak acids.

Cellulose (derived from wood or other sources) may be treated with ethanoic acid to produce a cellulose ethanoate (acetate) solution, which can be used to make plastic items by injection molding or extruded to form synthetic textile fibers.

ethanioc anhydride or **acetic anhydride**; $(CH_3CO)_2O$, colorless liquid, boiling point 137°C/278.6°F, with a sharp, pungent odor resembling that of ethanoic acid, into which it is slowly decomposed by water. It can be used to substitute the acetyl group (CH_3CO—) into organic compounds, without any complicating side reactions.

ethanol or **ethyl alcohol**; C_2H_5OH, alcohol found in beer, wine, cider, spirits, and other alcoholic drinks. When pure, it is a colorless liquid with a pleasant odor, miscible with water or ether; it burns in air with a pale blue flame. The vapor forms an explosive mixture with air and may be used in high-compression internal combustion engines.

It is produced naturally by the fermentation of carbohydrates by yeast cells. Industrially, it can be made by absorption of †ethene and subsequent reaction with water, or by the reduction of ethanal (acetaldehyde) in the presence of a catalyst, and is widely used as a solvent.

Ethanol is used as a raw material in the manufacture of ethoxyethane (ether), trichloroethanol (chloral), and triiodomethane (iodoform). It can also be added to gasoline, where it improves the performance of the engine, or be used as a fuel in its own right (as in Brazil). In August 2001 it was discovered that when ethanol is added to gasoline it boosts the oxygen content, reducing the emissions of air-polluting gases. Crops such as sugar cane may be grown to provide ethanol (by fermentation) for this purpose.

ethene common name **ethylene**; C_2H_4, colorless, flammable gas, the first member of the †alkene series of hydrocarbons. It is the most widely used synthetic organic chemical and is used to produce the plastics †polythene (polyethylene), polychloroethene, and polyvinyl chloride (PVC). It is obtained from natural gas or coal gas, or by the dehydration of ethanol.

Ethene is produced during plant metabolism and is classified as a plant hormone. It is important in the ripening of fruit and in †abscission. Small amounts of ethene are often added to the air surrounding fruit to artificially promote ripening. Tomato and marigold plants show distorted growth in concentrations as low as 0.01 parts per million. Plants also release ethene when they are under stress. German physicists invented a device in 1997 that measures stress levels in plants by measuring surrounding ethene levels.

ether any of a series of organic chemical compounds having an oxygen atom linking the carbon atoms of two hydrocarbon radical groups (general formula R-O-R¢); also the common name for ethoxyethane $C_2H_5OC_2H_5$ (also called diethyl ether). This is used as an anesthetic and as an external cleansing agent before surgical operations. It is also used as a solvent, and in the extraction of oils, fats, waxes, resins, and alkaloids.

Ethoxyethane is a colorless, volatile, inflammable liquid, slightly soluble in water, and miscible with ethanol. It is prepared by treatment of ethanol with excess concentrated sulfuric acid at 140°C/284°F.

ether (or aether) in the history of science, a hypothetical medium permeating all of space. The concept originated with the Greeks, and has been revived on several occasions to explain the properties and propagation of light. It was supposed that light and other electromagnetic radiation – even in outer space – needed a medium, the ether, in which to travel. The idea was abandoned with the acceptance of †relativity.

ethnobotany study of the relationship between human beings and plants. It combines knowledge of botany, chemistry, and anthropology. Many pharmaceutical companies, universities, and government health agencies have contracted ethnobotanists to conduct research among indigenous peoples, especially in the Amazon, to discover the traditional use of medicinal plants which can lead to the development of new drugs.

Ethnobotanical research is a cheaper way to develop new drugs than random testing of plants. The world's plant species, some of which may contain potential cures for diseases, are rapidly vanishing and there is a rush to learn as much as possible about indigenous plant-use before they disappear. Because the potential for profit exists upon discovery of an active plant compound, and due to the historical appropriation of indigenous resources, there has also been much discussion about the ethics of ethnobotanical research. For many indigenous peoples, the plants are considered sacred and their use is often religious in nature and unrelated to their active properties. Attempts have also been made to protect indigenous knowledge of plants as intellectual property and to compensate the people for revealing their knowledge.

ethology comparative study of animal behavior in its natural setting. Ethology is concerned with the causal mechanisms (both the stimuli that elicit behavior and the physiological mechanisms controlling it), as well as the development of behavior, its function, and its evolutionary history.

Ethology was pioneered during the 1930s by the Austrians Konrad Lorenz and Karl von Frisch who, with the Dutch zoologist Nikolaas Tinbergen, received the Nobel prize in 1973. Ethologists believe that the significance of an animal's behavior can be understood only in its natural context, and emphasize the importance of field studies and an evolutionary perspective. A development within ethology is †sociobiology, the study of the evolutionary function of †social behavior.

ethyl acetate common name **acetic ester**; $CH_3COOC_2H_5$, colorless liquid prepared by adding a mixture of alcohol and ethanoic acid to a mixture of alcohol and strong sulfuric acid, the whole being heated to 140°C/284°F. Ethyl acetate is characterized by a pleasant fruity odor and was formerly used for flavoring sweets and wines, and in perfumes. It is also a useful solvent. It is also known as ethyl ethanoate.

ethyl alcohol common name for †ethanol.

ethylamine colorless organic liquids formed by the †ethylation of ammonia. There are three members of this class of compounds, ethylamine $C_2H_5NH_2$, diethylamine $(C_2H_5)_2NH$, and triethylamine $(C_2H_5)_3N$. Ethylamines are generally formed by passing an ethanol or ethene–ammonia mixture across a heated catalyst. They are strong bases.

Ethylamine is a colorless liquid with a distinctive odor of ammonia; boiling point 19°C/66°F. It can be formed by reacting ethyl chloride and ammonia in an alcoholic solution using heat and pressure. Ethylamine is a strong base and forms salts with chloride compounds, especially metallic chlorides.

ethylation in chemistry, any process that results in the introduction of an ethyl group, $-C_2H_5$, into an organic compound. The †Friedel–Crafts reaction is an example of this type of process.

ethyl benzene $C_6H_5) - C_2H_5$, colorless liquid, prepared by an addition reaction of ethene to benzene; boiling point 136°C/277°F. It is mainly used as a reagent in the manufacture of †styrene.

ethylene common name for †ethene.

ethylene glycol alternative name for †glycol.

ethylene oxide or **oxirane**; C_2H_4O, colorless gas that forms an explosive mixture with air; boiling point approximately 11°C/52°F. It can be produced by oxidation of †ethene at high temperature using a silver catalyst. Ethylene oxide is widely used a reagent in the commercial manufacture of detergents, solvents, and epoxy resins.

ethyne common name **acetylene**; CHCH, colorless inflammable gas produced by mixing calcium carbide and water. It is the simplest member of the †alkyne series of hydrocarbons. It is used in the manufacture of the synthetic rubber neoprene, and in oxyacetylene welding and cutting.

Ethyne was discovered by English chemist Edmund Davy in 1836 and was used in early gas lamps, where it was produced by the reaction between water and calcium carbide. Its combustion provides more heat, relatively, than almost any other fuel known (its calorific value is five times that of hydrogen). This means that the gas gives an intensely hot flame; hence its use in oxyacetylene torches.

etiolation in botany, a form of growth seen in plants receiving insufficient light. It is characterized by long, weak stems, small leaves, and a pale yellowish color (†chlorosis) due to a lack of chlorophyll. The rapid increase in height enables a plant that is surrounded by others to quickly reach a source of light, after which a return to normal growth usually occurs.

Euclid (c. 330–c. 260 BC) Greek mathematician who wrote the *Stoicheia/Elements* in 13 books, nine of which deal with plane and solid geometry and four with number theory. His great achievement lay in the systematic arrangement of previous mathematical discoveries and a methodology based on axioms, definitions, and theorems.

Introduction to the Works of Euclid http://www.obkb.com/dcljr/euclid.html

Thorough introduction to the principles of Euclidean geometry. There is an emphasis on the Elements, but other works associated with, or attributed to, Euclid are discussed. The little that is known of the life of the mathematician is presented. There are a number of sources of further information.

Euclidean geometry see †geometry.

Euclid's fifth postulate states that parallel lines meet only at infinity.

Eudoxus (c. 408–c. 353 BC) also known as **Eudoxus of Cnidus**, Greek mathematician and astronomer. He devised the first system to account for the motions of celestial bodies, believing them to be carried around the Earth on sets of spheres. Work attributed to Eudoxus includes methods to calculate the area of a circle and to derive the volume of a pyramid or a cone.

eukaryote in biology, one of the two major groupings (superkingdoms) into which all organisms are divided. Included are all organisms, except bacteria and cyanobacteria (†blue-green algae), which belong to the †prokaryote grouping.

The cells of eukaryotes possess a clearly defined nucleus, bounded by a membrane, within which DNA is formed into distinct chromosomes. Eukaryotic cells also contain mitochondria, chloroplasts, and other structures (organelles) that, together with a defined nucleus, are lacking in the cells of prokaryotes.

Euler, Leonhard (1707–1783) Swiss mathematician. He developed the theory of differential equations and the calculus of variations, and worked in astronomy and optics. He also enlarged mathematical notation.

Euler-Chelpin, Hans Karl August Simon von (1873–1964) German biochemist who defined the structure of the yeast coenzyme, the nonprotein part of the yeast †enzyme zymase that interlocks with the zymase and affects how it functions. He was awarded the Nobel Prize for Chemistry in 1929 for his work on the fermentation of sugar, and fermentative enzymes.

Euler's number symbol e, the limit of the sequence:

$a_n = 1 + \frac{1}{1!} + \frac{1}{2!} + \frac{1}{3!} + \ldots + \frac{1}{n!}$

An irrational number introduced originally by Swiss mathematician Leonhard Euler, e may be represented to the sixth decimal place as 2.718282; it has useful theoretical properties in differential calculus and serves as a natural base for logarithms (known as "natural logarithms").

Euratom acronym for **European Atomic Energy Community**, forming part of the European Union organization.

Europa in astronomy, fourth-largest moon of the planet Jupiter, diameter 3,140 km/1,950 mi, orbiting 671,000 km/417,000 mi from the planet every 3.55 days. It is covered by ice and was originally thought to be criss-crossed by thousands of thin cracks some 50,000 km/30,000 mi long. These are now known to be low ridges.

NASA's Galileo spacecraft began circling Europa in February 1997. One of the first discoveries from the data it sent back was that the "cracks" covering the surface of the moon are in fact low ridges. These features lend credence to the idea that Europa possesses a hidden subsurface ocean. NASA plans to launch the Europa Orbiter in 2008 to arrive in 2010 and search for water beneath Europa's icy surface, using a radar sounder and altimeter.

Europa http://www.solarviews.com/solar/eng/europa.htm

Comprehensive latest information on Jupiter's moon. There is a full discussion of evidence that Europa may have liquid water under its frozen surface. There are also a number of color images of Europa.

europium chemical symbol Eu, soft, grayish, metallic element of the †lanthanide series, atomic number 63, relative atomic mass 151.96. It is used in lasers and as the red phosphor in color televisions; its compounds are used to make control rods for nuclear reactors. It was named in 1901 by French chemist Eugène Demarçay (1852–1904) after the continent of Europe, where it was first found.

eusociality form of social life found in insects such as honey bees and termites, in which the colony is made up of special castes (for example, workers, drones, and reproductives) whose membership is biologically determined. The worker castes do not usually reproduce. Only one mammal, the naked mole rat, has a social organization of this type. A eusocial shrimp was discovered in 1996 living in the coral reefs of Belize. *Synalpheus regalis* lives in colonies of up to 300 individuals, all the offspring of a single reproductive female. See also †social behavior.

Eustachian tube small air-filled canal connecting the middle †ear with the back of the throat. It is found in all land vertebrates and equalizes the pressure on both sides of the eardrum.

Eustachio, Bartolommeo (1520–1574) Italian anatomist, the discoverer of the Eustachian tube, leading from the middle ear to the pharynx, and of the Eustachian valve in the right auricles of the heart.

eustatic change global rise or fall in sea level caused by a change in the amount of water in the oceans (by contrast, isostatic adjustment involves a rising or sinking of the land, which causes a local change in sea level). During the last ice age, global sea level was lower than today because water became "locked-up" in the form of ice and snow, and less water reached the oceans.

eutectic mixture mixture of chemical compounds, having a specific composition, that melts at a temperature (the **eutectic temperature**) that is lower than the melting points of any of the constituent compounds. The eutectic temperature is well defined and behaves similarly to the melting point expected for a pure compound.

eutectic point position on a †phase diagram that indicates the composition and melting point of a †eutectic mixture.

eutrophication excessive enrichment of rivers, lakes, and shallow sea areas, primarily by municipal sewage, by sewage itself, and by the nitrates and phosphates from †fertilizers used in agriculture. These encourage the growth of algae and bacteria which use up the oxygen in the water, making it uninhabitable for fish and other animal life. In this way eutrophication is responsible for a particular type of †water pollution.

The dissolved fertilizers cause the rapid growth of water plants, especially green algae, which can clog up waterways and prevent light reaching plants below the surface. As the algae die, aerobic bacteria bring about decay, using up oxygen in the water as they do so. Anaerobic bacteria take over and convert part of the dead matter into smelly decay products. Slowly a layer of dead plant material builds up on the bottom of the lake or river. Deprived of oxygen, fish and other life forms die and decay, leaving putrid, poisonous water.

Evans, Alice Catherine (1881–1975) US microbiologist whose research into the bacterial contamination of milk led to the recognition of the danger of unpasteurized milk. As a result of her research the incidence of brucellosis was greatly reduced when the dairy industry accepted that all milk should be pasteurized.

evaporation process in which a liquid turns to a vapor without its temperature reaching boiling point. Evaporation is the †change of state that occurs when a †liquid turns into a †gas. In a liquid the particles are close together, with forces holding them together, yet able to move about. Some particles in a liquid have more energy than others. Even when a liquid is below its boiling point, some particles have enough energy to escape and form a gas. Evaporation is greater when temperatures and wind speeds are high, and the air is dry. It is why puddles dry up in the sun, and clothes dry faster in dry, windy weather.

A liquid left to stand in a saucer eventually evaporates because, at any time, a proportion of its molecules will be fast enough (have enough kinetic energy) to escape from the attractive intermolecular forces at the liquid surface into the atmosphere. The temperature of the liquid tends to fall because the evaporating molecules remove energy from the liquid. The rate of evaporation rises with increased temperature because as the mean kinetic energy of the liquid's molecules rises, so will the number possessing enough energy to escape.

A fall in the temperature of the liquid, known as the **cooling effect**, accompanies evaporation because as the faster molecules escape from the surface, the mean energy of the remaining molecules falls. The effect may be noticed when sweat evaporates from the skin. It plays a part in temperature control of the human body.

evaporite sedimentary deposit precipitated on evaporation of salt water.

With a progressive evaporation of seawater, the most common salts are deposited in a definite sequence: calcite (calcium carbonate), gypsum (hydrous calcium sulfate), halite (sodium chloride), and finally salts of potassium and magnesium.

Calcite precipitates when seawater is reduced to half its original volume, gypsum precipitates when the seawater body is reduced to one-fifth, and halite when the volume is reduced to one-tenth. Thus the natural occurrence of chemically precipitated calcium carbonate is common, of gypsum fairly common, and of halite less common.

Because of the concentrations of different dissolved salts in seawater, halite accounts for about 95% of the chlorides precipitated if evaporation is complete. More unusual evaporite minerals include borates (for example borax, hydrous sodium borate) and sulfates (for example glauberite, a combined sulfate of sodium and calcium).

even number any number divisible by 2, hence the digits 0, 2, 4, 6, 8 are even numbers, as is any number ending in these digits, for example 1736. Any whole number which is not even is odd.

event in statistics, any happening to which a †probability can be attached. Events may be single, such as getting a score of six when one die is thrown, or they may be combined, such as the scores obtained when two dice are thrown. In some cases, such as throwing an unbiased die, each event has the same probability of occurring.

event horizon in astronomy, "edge" of a black hole; the interface between four-dimensional space and a †singularity.

evergreen in botany, a plant such as pine, spruce, or holly, that bears its leaves all year round. Most conifers are evergreen. Plants that shed their leaves in autumn or during a dry season are described as †deciduous.

Evershed, John (1864–1956) English astronomer who made solar observations. In 1909 he discovered the radial movements of gases in sunspots (the **Evershed effect**). He also gave his name to a spectroheliograph, the Evershed spectroscope.

evolution slow gradual process of change from one form to another, as in the evolution of the universe from its formation to its present state, or in the evolution of life on Earth. In biology, it is the process by which life has developed by stages from single-celled organisms into the multiplicity of animal and plant life, extinct and existing, that inhabits the Earth. The development of the concept of evolution is usually associated with the English naturalist Charles †Darwin who attributed the main role in evolutionary change to †natural selection acting on randomly occurring variations. These variations in species are now known to be †adaptations produced by spontaneous changes or †mutations in the genetic material of organisms. In short, evolution is the change in the genetic makeup of a population of organisms from one generation to another. Evidence shows that many species of organisms do not stay the same over generations. The most dramatic evidence of this comes from fossils.

Evolution occurs via the following processes of natural selection: individual organisms within a particular species may show a wide range of †variation because of differences in their †genes; predation, disease, and †competition cause individuals to die; individuals with characteristics most suited to the †environment are more likely to survive and breed successfully; and the genes that have enabled these individuals to survive are then passed on to the next generation, and if the environment is changing, the result is that some genes are more abundant in the next generation and the organism has evolved.

Evolutionary change can be slow, as shown in part of the fossil record. However, it can be quite fast. If a population is reduced to a very small number, evolutionary changes can be seen over a few generations. Because micro-organisms have very short life cycles, evolutionary change in micro-organisms can be rapid. Micro-organisms can evolve resistance to a new antibiotic only a few years after the drug is first used. As a result of evolution from common ancestors, we are able to use †classification of organisms to suggest evolutionary origins.

Coelacanth: the Fish Out of Time http://www.dinofish.com/

Site about the coelacanth, the 400 million-year-old "living fossil" fish thought to have become extinct with the dinosaurs but found alive and well in 1938. It contains a mine of information on coelacanth biology, habitat, and behavior, and reports on current efforts to preserve this unique species. A virtual swimming coelacanth is an additional fun feature.

Evolutionist, The http://www.lse.ac.uk/Depts/cpnss/darwin/evo/

Online magazine devoted to evolutionary ideas which includes features, interviews, and comment. It currently includes an article on the recent political interest in Darwinism and a column on the limits of evolutionary theory.

Introduction to Evolutionary Biology http://www.talkorigins.org/faqs/faq-intro-to-biology.html

Intelligent and readable explanation of the theory of evolution, the cornerstone of modern biology. Author Chris Colby presents the main aspects of evolutionary biology, and robustly confronts the "evolution versus creation" debate. An interesting section dispels common misconceptions about evolution and demonstrates how changes to the environment can affect the way in which a species evolves.

Journey into Phylogenetic Systematics http://www.ucmp.berkeley.edu/clad/clad4.html

Online exhibition about evolutionary theory with a specific emphasis on phylogenetic systematics: the way that biologists reconstruct the pattern of events that has led to the distribution and diversity of life. The site provides an introduction to the philosophy, methodology, and implication of cladistic analysis, with a separate section on the need for cladistics. Many of the scientific terms are included in an illuminating glossary.

Live A Life Page http://alife.fusebox.com/

Fascinating page that brings together interactive programs to simulate ecological and evolutionary processes. Included is an adaptation of Richard Dawkins's Biomorphs program, described in The Blind Watchmaker, which enables the user to select "morphs" for certain qualities and watch as their offspring evolve.

evolutionary toxicology study of the effects of pollution on evolution. A polluted habitat may cause organisms to select for certain traits, as in **industrial melanism** for example, where some insects, such as the peppered moth, evolve to be darker in polluted areas, and therefore better camouflaged against predation.

Pollutants may also trigger mutations, for example, voles living around the Chernobyl exploded nuclear reactor have a very high mutation rate despite appearing healthy and reproducing successfully. Fish in polluted rivers also exhibit mutations.

Ewing, (William) Maurice (1906–1974) US geologist. His studies of the ocean floor provided crucial data for the †plate tectonics revolution in geology in the 1960s. He demonstrated that midocean ridges, with deep central canyons, are common to all oceans.

excitation in physiology, the stimulation of a sense receptor or nerve; in chemistry, the injection of energy into an atom (which may be part of a radical or molecule) which raises it or one of its components to a higher energy level.

excited state, electronic condition of an electron that has absorbed external energy and, as a result, moved from its normal, †ground state energy level to a higher energy level. The excitation energy is the difference in energy between the ground state and the excited state.

excluded middle, law of the law in logic, that a statement is either true or false, leaving no room for any further alternatives. There are nonclassical systems of logic that distinguish between true = proven true and false = proven false, and so allow intermediate values (such as "possibly true"). Boolean-valued logic systems attach probability values to statements that may therefore also have intermediate values other than merely true or false the values form a †Boolean algebra.

exclusion principle in physics, a principle of atomic structure originated by Austrian-born US physicist

Wolfgang †Pauli. It states that no two electrons in a single atom may have the same set of †quantum numbers. Hence, it is impossible to pack together certain elementary particles, such as electrons, beyond a certain critical density, otherwise they would share the same location and quantum number. A white dwarf star, which consists of electrons and other elementary particles, is thus prevented from contracting further by the exclusion principle and never collapses. Elementary particles in the class fermions obey the exclusion principle whilst those in the class bosons do not.

excretion in biology, the removal of the waste products of metabolism from living organisms. In plants and simple animals, waste products are removed by †diffusion. Plants, for example, excrete O_2, a product of photosynthesis. In mammals, waste products are removed by specialized excretory organs, principally the †kidneys, which excrete urea. Water and metabolic wastes are also excreted in the feces and, in humans, through the sweat glands in the skin; carbon dioxide and water are removed via the lungs. The liver excretes bile pigments.

exfoliation in earth science, the successive separation of thin shells of rock from the surface of a massive rock. It is caused by day–night temperature variations or the decrease in compressional forces that accompanies exposure of previous covered rock.

exobiology study of life forms that may possibly exist elsewhere in the universe and of the effects of extraterrestrial environments on Earth organisms. Techniques include space probe experiments designed to detect organic molecules, and the monitoring of radio waves from other star systems.

exocrine gland gland that discharges secretions, usually through a tube or a duct, onto a surface. Examples include sweat glands which release sweat on to the skin, and digestive glands which release digestive juices onto the walls of the intestine. Some animals also have †endocrine glands (ductless glands) that release hormones directly into the bloodstream.

exocytosis in biology, ejection from a cell of undigested remnants of material.

exon in genetics, a sequence of bases in †DNA that codes for a protein. Exons make up only 2% of the body's total DNA. The remainder is made up of †introns. During RNA processing the introns are cut out of the molecule and the exons spliced together.

exoskeleton hardened external skeleton of insects, spiders, crabs, and other arthropods. It provides attachment for muscles and protection for the internal organs, as well as support. To permit growth it is periodically shed in a process called †ecdysis.

exosphere uppermost layer of the †atmosphere. It is an ill-defined zone above the thermosphere, beginning at about 700 km/435 mi and fading off into the vacuum of space. The gases are extremely thin, with hydrogen as the main constituent.

exothermic describing a process or reaction that involves the release of energy (usually in the form of heat). Combustion, for example, is an exothermic reaction. See also †endothermic reaction.

exothermic reaction chemical reaction during which heat is given out (see †energy of reaction).

For example, burning sulfur in air to give sulfur dioxide is an exothermic reaction.

$2S + O_2 \rightarrow SO_2$

expand in algebra, to multiply out. For example, $ac + bc + ad + bd$ is the expanded form of $(a+b)(c+ d)$.

expanded memory in computing, additional memory in an †MS-DOS-based computer, usually installed on an expanded-memory board. Expanded memory requires an expanded-memory manager, which gives access to a limited amount of memory at any one time, and is slower to use than †extended memory. Software is available under both MS-DOS and Windows to simulate expanded memory for those applications that require it.

expansion in physics, the increase in size of a constant mass of substance caused by, for example, increasing its temperature (†thermal expansion) or its internal pressure. The **expansivity**, or coefficient of thermal expansion, of a material is its expansion (per unit volume, area, or length) per degree rise in temperature.

expansion coefficient see †coefficient of expansion.

expectation in statistics, the theoretical †probability of a certain result.

experiment in science, a practical test designed with the intention that its results will be relevant to a particular theory or set of theories. Although some experiments may be used merely for gathering more information about a topic that is already well understood, others may be of crucial importance in confirming a new theory or in undermining long-held beliefs.

The manner in which experiments are performed, and the relation between the design of an experiment and its value, are therefore of central importance. In general, an experiment is of most value when the factors that might affect the results (variables) are carefully controlled; for this reason most experiments take place in a well-managed environment such as a laboratory or clinic.

Exploratorium http://www.exploratorium.edu/index.html

Very well-designed site based on the hands-on science museum in San Francisco. Several of the museum's exhibits are recreated here, including a page on the science of ice hockey and a virtual dissection of a cow's eye. There is practical information about the museum and help for teachers, but most of the pages are interactive: full of images and audio clips to augment the text.

Schools Online Science Cafe http://www.shu.ac.uk/schools/sci/sol/experts/exptin.htm

Part of the Schools Online project, this site has suggestions for science experiments, a facility for posting questions to scientists, and a chance to participate in online experiments by submitting your own data.

experimental petrology field of geology that involves creating or altering rocks and minerals in the laboratory in order to understand rock-forming processes that are impossible to observe, either because they happen out of the physical range of human observation (for example 100 km/62 mi deep), or because they happen too slowly.

experimental psychology application of scientific methods to the study of mental processes and behavior. This covers a wide range of fields of study, including: **human and animal learning**, in which learning theories describe how new behaviors are acquired and modified; **cognition**, the study of a number of functions, such as perception, attention, memory, and language; and **physiological psychology**, which relates the study of cognition to different regions of the brain. **Artificial intelligence** refers to the computer simulation of cognitive processes, such as language and problem-solving.

expert system computer program for giving advice (such as diagnosing an illness or interpreting the law) that incorporates knowledge derived from human expertise. A kind of †knowledge-based system, it contains rules that can be applied to find the solution to a problem. It is a form of †artificial intelligence.

explosive any material capable of a sudden release of energy and the rapid formation of a large volume of gas, leading, when compressed, to the development of a high-pressure wave (blast).

types of explosive Combustion and explosion differ essentially only in rate of reaction, and many explosives (called **low explosives**) are capable of undergoing relatively slow combustion under suitable conditions. **High explosives** produce uncontrollable blasts. The first low explosive was †gunpowder; the first high explosive was †nitroglycerine.

exponent or **index**, in mathematics, a superscript number that indicates the number of times a term is multiplied by itself; for example $x^2 = x \times x$, $4^3 = 4 \times 4 \times 4$.

Exponents obey certain rules. Terms that contain them are multiplied together by adding the exponents; for example, $x^2 \times x^5 = x^7$. Division of such terms is done by subtracting the exponents; for example, $y^5 \div y^3 = y^2$. Any number with the exponent 0 is equal to 1; for example, $x^0 = 1$ and $99^0 = 1$.

See †index notation.

exponential in mathematics, descriptive of a †function in which the variable quantity is an exponent (a number indicating the power to which another number or expression is raised).

Exponential functions and series involve the constant e = 2.71828.... . Scottish mathematician John Napier devised natural †logarithms in 1614 with e as the base.

Exponential functions are basic mathematical functions, written as e^x or exp x. The expression e^x has five definitions, two of which are: (i) e^x is the solution of the differential equation $dx/dt = x$ ($x = 1$ if $t = 0$); (ii) e^x is the limiting sum of the infinite series $1 + x + (x^2/2!) + (x^3/3!) + ... + (x^n/n!)$.

expression in mathematics, a phrase written in symbols. For example, $2x^2 + 3x + 5$ is a quadratic (containing a term or terms raised to the second power but no higher) expression. Equations consist of expressions written around an equals sign.

extend in mathematics, to draw a continuation of a line, usually in order to deduce some property of a figure.

extended memory in computing, memory in an †MS-DOS-based system that exceeds the 1 Mbyte that DOS supports. Extended memory is not accessible to the †operating system and requires an extended memory manager.

extensor a muscle that straightens a limb.

exterior angle one of the four external angles formed when a straight line or transveral cuts through a pair of (usually parallel) lines. Also, an angle formed by extending a side of a polygon.

extinction in biology, the complete disappearance of a species from the planet. Extinctions occur when a species becomes unfit for survival in its natural habitat usually to be replaced by another, better-suited species. An organism becomes ill-suited for survival because its environment is changed or because its relationship to other organisms is altered. For example, a predator's fitness for survival depends upon the availability of its prey.

past extinctions Mass extinctions are episodes during which large numbers of species have become extinct virtually simultaneously in the distant past, the best known being that of the dinosaurs, other large reptiles, and various marine invertebrates about 65 million years ago between the end of the †Cretaceous period and the beginning of the Tertiary period, the latter known as the **K–T extinction**.

There have been several others in the more distant past. There is disagreement about the causes, but one of several major catastrophes have been blamed, including meteorite impact, volcanic eruption, massive lava flows, and significant global warming. Another mass extinction occurred about 10,000 years ago when many giant species of mammal died out. This is known as the "Pleistocene overkill" because their disappearance was probably hastened by the hunting activities of prehistoric humans. The greatest mass extinction occurred about 250 million years ago, marking the Permian–Triassic boundary (see †geological time), when up to 96% of all living species became extinct.

extinction coefficient in chemistry, characteristic quantity of any medium that absorbs light (symbol a). It is defined as the reciprocal of the thickness, in centimeters, of the absorbing substance that light has to pass through until it has only one tenth of its original intensity.

extracellular matrix strong material naturally occurring in animals and plants, made up of protein and long-chain sugars (polysaccharides) in which cells are embedded. It is often called a "biological glue," and forms part of †connective tissues such as bone and skin.

The cell walls of plants and bacteria, and the †exoskeletons of insects and other arthropods, are also formed by types of extracellular matrix.

extraction in chemistry, any process in which a solvent is used to separate a soluble solid from a solid mixture. The term can also be applied to the removal of a component from a liquid mixture by the addition of a solvent in which one liquid phase is soluble and the other is not.

extrasensory perception ESP, any form of perception beyond and distinct from the known sensory processes. The main forms of ESP are clairvoyance (intuitive perception or vision of events and situations without using the senses); precognition (the ability to foresee events); and telepathy or thought transference (communication between people without using any known visible, tangible, or audible medium). Verification by scientific study has yet to be achieved.

extremophile microbe able to thrive in extreme conditions, such as very high temperatures (thermophiles), low temperatures (psychrophiles), high acidity (acidophiles), or high alkalinity (alkaliphiles). Examples of extremophiles are found in both †bacteria and †Archaea.

Thermophiles live at temperatures above 45°C/113°F and some, known as hyperthermophiles, thrive at temperatures greater than 80°C/176°F, and are even found in sea water above its boiling point of 100°C/212°F, such as those living around deep-sea vents. The archaeon *Sulfolobus acidocaldarius* is a hyperthermopile that is also an acidophile. It lives in the hot sulfur springs of Yellowstone Park, Wyoming, USA.

In 2000, new species of psychrophiles were discovered living in Antarctica, where temperatures range between –85°C/–121°F and –13°C/–11°F.

Extremophiles are of commercial interest because they are able to continue producing their enzymes in conditions when other enzymes would have broken down. These **extremozymes** are able to catalyze chemical reactions for the biomedical and other industries at very high temperatures, for example.

extroversion or **extraversion**, personality dimension described by the psychologists Carl †Jung and, later, Hans Eysenck. The typical extrovert is sociable, impulsive, and carefree. The opposite of extroversion is †introversion.

extrusive rock or **volcanic rock**, †igneous rock formed on the surface of the Earth by volcanic activity (as opposed to intrusive, or plutonic, rocks that solidify below the Earth's surface). Magma (molten rock) erupted from volcanoes cools and solidifies quickly on the surface. The crystals that form do not have time to grow very large, so most extrusive rocks are finely grained. The term includes fine-grained crystalline or glassy rocks formed from hot lava quenched at or near Earth's surface, and those made of welded fragments of ash and glass ejected into the air during a volcanic eruption. The formation of extrusive igneous rock is part of the rock cycle.

Large amounts of extrusive rock called †basalt form at the Earth's †ocean ridges from lava that fills the void formed when two tectonic plates spread apart. Explosive volcanoes that deposit pyroclastics generally occur where one tectonic plate descends beneath another. †Andesite is often formed by explosive volcanoes. Magmas that give rise to pyroclastic extrusive rocks are explosive because they are viscous. The island of Montserrat, West Indies, is an example of an explosive volcano that spews pyroclastics of andesite composition. Magmas that produce crystalline or glassy volcanic rocks upon cooling are less viscous. The low viscosity allows the extruding lava to flow easily. Fluidlike lavas that flow from the volcanoes of the Hawaiian Islands have low viscosity and cool to form basalt.

ex vivo describing a technique used in gene therapy, in which genetically defective cells are removed and cultured in vitro. After growth, normal genes are added to the cells to correct the defect. These modified cells are next returned to the donor.

eye organ of vision. In the human eye, the light is focused by the combined action of the curved cornea, the internal fluids, and the lens. The insect eye is compound – made up of many separate facets, known as ommatidia, each of which collects light and directs it separately to a receptor to build up an image. Invertebrates have much simpler eyes, with no lenses. Among mollusks, cephalopods have complex eyes similar to those of vertebrates. The mantis shrimp's eyes contain ten color pigments with which to perceive color; some flies and fish have five, while the human eye has only three.

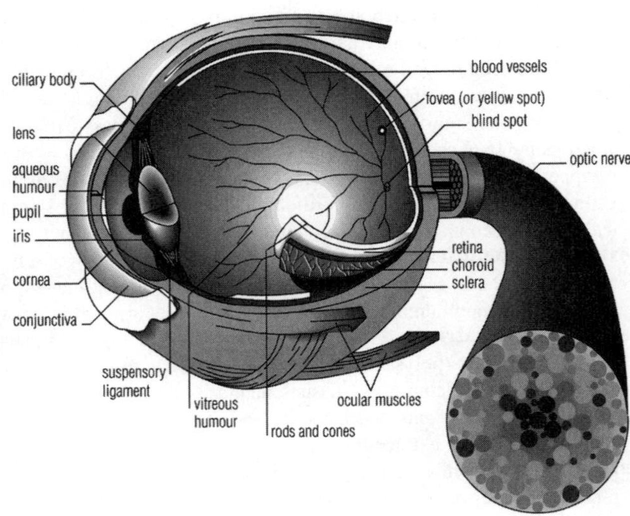

eye The human eye. The retina of the eye contains about 137 million light-sensitive cells in an area of about 650 sq mm/1 sq in. There are 130 million rod cells for black and white vision and 7 million cone cells for color vision. The optic nerve contains about 1 million nerve fibers. The focusing muscles of the eye adjust about 100,000 times a day. To exercise the leg muscles to the same extent would need an 80 km/50 mi walk. Helicon

The human eye is a roughly spherical structure contained in a bony socket. Light enters it through the cornea, a transparent region at the front of the tough outer sclera and passes through the circular opening (pupil) in the iris (the colored part of the eye). The muscular iris controls the size of the pupil and hence the amount of light entering the eye. Light then passes through the lens, which is held in position by suspensory ligaments and ciliary muscles. The ciliary muscles act on the lens (the rounded transparent structure behind the iris) to change its shape, allowing images of objects nearby and at a distance to be focused on the †retina at the back of the eye. The retina is packed with light-sensitive cells (rods and cones). Rods work well in conditions of low light but are unable to sense color. Cones work well in bright light and are responsible for color vision. The part of the retina on which lies the precise point at which the image is focused contains mainly cones. The rods and cones in the retina send impulses to the brain along sensory neurons in the optic nerve.

Eye http://retina.anatomy.upenn.edu/~lance/eye/eye.html
Thorough information on the physical structures of the eye, from the University of Pennsylvania, USA. The page is divided into sections covering the main parts of the eye, such as the retina, cornea, iris, pupil, and lens.

eye, defects of the abnormalities of the eye that impair vision. Glass or plastic lenses, in the form of spectacles or contact lenses, are the usual means of correction. Common optical defects are †short-sightedness or myopia; farsightedness or hypermetropia; lack of †accommodation or presbyopia; and †astigmatism. Other eye defects include color blindness.

Eysenck, Hans Jürgen (1916–1997) German-born British psychologist. His work concentrated on personality theory and testing by developing behavior therapy (treating certain clinical conditions by regarding their symptoms as learned patterns of behavior that therapy can enable the patient to unlearn). He was an outspoken critic of psychoanalysis as a therapeutic method. His theory that intelligence is almost entirely inherited and can be only slightly modified by education aroused controversy, as did his support of the idea that planet positions at birth affect personality, and his belief that smoking does not cause lung cancer.

F

°F symbol for degrees †Fahrenheit.

F in physics, symbol for †farad, the SI unit of capacitance equal to that of a capacitor with a potential difference of 1 volt between plates carrying a charge of 1 coulomb.

Fabricius, Hieronymus (1537–1619) Italian **Geronimo Fabrizio**; also known as **Girolamo Fabrizio**, Italian anatomist and embryologist. From 1574 he made detailed studies of the veins and blood flow and discovered the existence of one-way valves that direct the blood towards the heart. He also studied the development of chick embryos.

Fabry, (Marie Paul Auguste) Charles (1867–1945) French physicist. He specialized in †optics, devising methods for the accurate measurement of †interference effects. He took part in inventing a device known as the **Fabry–Pérot †interferometer**. In 1913, Fabry demonstrated that ozone is plentiful in the upper atmosphere and is responsible for filtering out ultraviolet radiation from the Sun.

face in geometry, a plane surface of a solid enclosed by edges. A cube has six square faces, a cuboid has six rectangular faces, and a tetrahedron has four triangular faces.

facies body of rock strata possessing unifying characteristics usually indicative of the environment in which the rocks were formed. The term is also used to describe the environment of formation itself or unifying features of the rocks that comprise the facies.

Features that define a facies can include collections of fossils, sequences of rock layers, or the occurrence of specific minerals. Sedimentary rocks deposited at the same time, but representing different facies belong to a single **chronostratigraphic unit** (see †stratigraphy). But these same rocks may belong to different **lithostratigraphic units**. For example, beach sand is deposited at the same time that mud is deposited further offshore. The beach sand eventually turns to †sandstone while the mud turns to †shale. The resulting sandstone and shale **strata** comprise two different facies, one representing the beach environment and the other the offshore environment, formed at the same time; the sandstone and shale belong to the same chronostratigraphic unit but distinct lithostratigraphic units.

facsimile transmission full name for †fax or telefax.

factor in algebra, certain kinds of †polynomials (expressions consisting of several or many terms) can be factorized using their common †factors. Brackets are put into an expression, and the common factor is sought. For example, the factors of $2a^2 + 6ab$ are $2a$ and $a + 3b$, since $2a^2 + 6ab = 2a(a + 3b)$. This rearrangement is called **factorization**.

factor number that divides into another †number exactly. It is also known as a †divisor. For example, the factors of 24 are 1, 2, 4, 8, 12, and 24; and the factors of 64 are 1, 2, 4, 8, 16, 32, and 64. The highest factor of both 24 and 64 is 8. This is known as the †highest common factor (HCF) of the two numbers. The largest number to be factorized to date has 167 digits: it was split into its 80- and 87-digit factors in 1997 by US mathematicians, after 100,000 hours of computing.

See also †prime number and †prime factor.

factorial of a positive number, the product of all the whole numbers (integers) inclusive between 1 and the number itself. A factorial is indicated by the symbol "!". Thus $6! = 1 \times 2 \times 3 \times 4 \times 5 \times 6 = 720$. Factorial zero, 0!, is defined as 1.

factorization reduction into constituent factors (which when multiplied together produce the original number or expression).

faculae bright areas on the face of the Sun, commonly in the vicinity of †sunspots. Named by German astronomer Johannes Hevelius, they are thought to be caused by luminous hydrogen clouds close to the †photosphere. They last on average about 15 days.

feces remains of food and other waste material eliminated from the digestive tract of animals by way of the anus. Feces consist of quantities of fibrous material, bacteria and other micro-organisms, rubbed-off lining of the digestive tract, bile fluids, undigested food, minerals, and water.

Fahrenheit, Gabriel Daniel (1686–1736) Polish-born Dutch physicist who invented the first accurate thermometer in 1724 and devised the Fahrenheit temperature scale. Using his thermometer, Fahrenheit was able to determine the boiling points of liquids and found that they vary with atmospheric pressure.

Fahrenheit scale temperature scale invented in 1714 by Gabriel Fahrenheit that was commonly used in English-speaking countries until the 1970s, after which the †Celsius scale was generally adopted, in line with the rest of the world. In the Fahrenheit scale, intervals are measured in degrees (°F); $°F = (°C \times {}^9/_5) + 32$.

Fahrenheit took as the zero point the lowest temperature he could achieve anywhere in the laboratory, and, as the other fixed point, body temperature, which he set at 96°F. On this scale, water freezes at 32°F and boils at 212°F.

Fajans, Kasimir (1887–1975) Polish-born US chemist. He did pioneering work on radioactivity and isotopes, he also formulated rules that help to explain valence and chemical bonding.

Fajans's rules in chemistry, guidelines for estimating if a compound is likely to form †ionic bonds as opposed to †covalent bonds. The rules state that ionic bonds are more probable if (a) the ionic charge is low, (b) the size of the anion is small, (c) the size of the cation is large, and (d) if the cation has the electronic configuration of a noble (rare) gas. The rules were proposed by Polish-born US chemist Kasimir †Fajans.

Fallopian tube or **oviduct**, in mammals, one of two tubes that carry eggs from the ovary to the uterus. An egg is fertilized by sperm in the Fallopian tubes, which are lined with cells whose †cilia move the egg towards the uterus.

fallout harmful radioactive material released into the atmosphere in the debris of a nuclear explosion and descending to the surface of the Earth. Such material can enter the food chain, cause radiation sickness, and last for hundreds of thousands of years (see †half-life).

false of a statement, untrue. Falseness is used in proving propositions by considering the negative of the proposition to be true, then making deductions until a contradiction is reached which proves the negative to be false and the proposition to be true.

false-memory syndrome syndrome occurring in some patients undergoing †psychotherapy or hypnosis, in which the individual "remembers" events that never happened. For example, the patient may recall having been sexually abused as a child. As it is very difficult to establish whether memories are genuine the syndrome is surrounded by controversy.

Proponents of the syndrome believe some therapists may unconsciously implant false memories by their techniques. Psychologists have demonstrated that this is surprisingly easy to do. Painful memories may be blotted out, however, and psychotherapists are concerned that genuine memories may be dismissed as false.

family in biological classification, a group of related genera (see †genus). Family names are not printed in italic (unlike genus and species names), and by convention they all have the ending -idae (animals) or -aceae (plants and fungi). For example, the genera of hummingbirds are grouped in the hummingbird family, Trochilidae. Related families are grouped together in an †order.

farad symbol F, SI unit of electrical capacitance (how much electric charge a †capacitor can store for a given voltage). One farad is a capacitance of one †coulomb per volt. For practical purposes the microfarad (one millionth of a farad, symbol mF) is more commonly used.

faraday unit of electrical charge equal to the charge on one mole of electrons. Its value is 9.648×10^4 coulombs.

Faraday effect in physics, rotation of the plane of polarization of polarized light by passing it through a transparent medium subjected to a transverse magnetic field. It is named after English physicist Michael Faraday.

Faraday's constant symbol F, constant representing the electric charge carried on one mole of electrons. It is found by multiplying Avogadro's constant by the charge carried on a single electron, and is equal to 9.648×10^4 coulombs per mole.

One **faraday** is this constant used as a unit. The constant is used to calculate the electric charge needed to discharge a particular quantity of ions during †electrolysis.

Faraday's laws three laws of electromagnetic induction, and two laws of electrolysis, all proposed originally by English physicist Michael Faraday. The laws of induction are: (1) a changing magnetic field induces an electromagnetic force in a conductor; (2) the electromagnetic force is proportional to the rate of change of the field; and (3) the direction of the induced electromagnetic force depends on the orientation of the field. The laws of electrolysis are: (1) the amount of chemical change during electrolysis is proportional to the charge passing through the liquid; and (2) the amount of chemical change produced in a substance by a given amount of electricity is proportional to the electrochemical equivalent of that substance.

far point farthest point that a person can see clearly. The eye is unable to focus a sharp image on the retina of an object beyond this point. The far point for a normal eye should be at infinity; any eye that has a far point nearer than this is short-sighted (see †short-sightedness).

fast breeder or **breeder reactor**, alternative name for †fast reactor, a type of nuclear reactor.

fast reactor or **fast breeder reactor**, †nuclear reactor that makes use of fast neutrons to bring about fission. Unlike other reactors used by the nuclear-power industry, it has little or no †moderator, to slow down neutrons. The reactor core is surrounded by a "blanket" of uranium carbide. During operation, some of this uranium is converted into plutonium, which can be extracted and later used as fuel.

Fast breeder reactors can extract about 60 times the amount of energy from uranium that thermal reactors do. In the 1950s, when uranium stocks were thought to be dwindling, the fast breeder was considered to be the reactor of the future. Now, however, when new uranium reserves have been found and because of various technical difficulties in their construction, development of the fast breeder has slowed in most parts of the world.

fat in the broadest sense, a mixture of †lipids – chiefly triglycerides (lipids containing three †fatty acid molecules linked to a molecule of glycerol). More specifically, the term refers to a lipid mixture that is solid at room temperature (20°C/68°F); lipid mixtures that are liquid at room temperature are called oils. The higher the proportion of saturated fatty acids in a mixture, the harder the fat. Fats and oils (lipids) are compounds made up of glycerol and fatty acids. Fats are insoluble in †water. Boiling fats in strong alkali forms soaps (saponification).

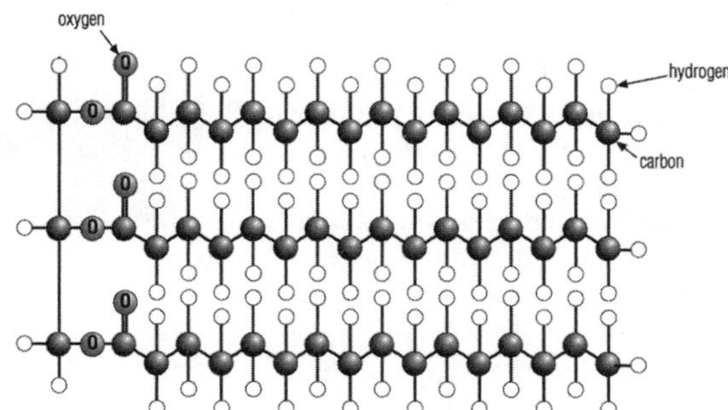

fat The molecular structure of typical fat. The molecule consists of three fatty acid molecules linked to a molecule of glycerol. Helicon

Fats are essential constituents of food for many animals, with a calorific value twice that of carbohydrates. However, eating too much fat, especially fat of animal origin, has been linked with heart disease in humans, where excess fat is deposited in the walls of arteries and may cause heart attacks. In many animals and plants, excess carbohydrates and proteins are converted into fats for storage. Mammals and other vertebrates store fats in specialized connective tissues (†adipose tissues), which not only act as energy reserves but also insulate the body and cushion its organs.

As a nutrient, fat serves five purposes: it is a source of energy (9 kcal/g); makes the diet palatable; provides basic building blocks for cell structure; provides essential fatty acids (linoleic and linolenic); and acts as a carrier for fat-soluble vitamins (A, D, E, and K). Indeed, fatty tissue in the body stores fat-soluble vitamins. Foods rich in fat are butter, lard, and margarine. Products high in monounsaturated or polyunsaturated fats are thought to be less likely to contribute to cardiovascular disease.

fathom (Anglo-Saxon *faethm* "to embrace") in mining, seafaring, and handling timber, a unit of depth measurement (1.83 m/6 ft) used prior to metrication; it approximates to the distance between an adult man's hands when the arms are outstretched.

fatigue in metallurgy, failure of metals under cyclic applications of a stress.

fatigue in muscle, reduced response brought about by the accumulation of lactic acid in muscle tissue due to excessive cellular activity.

fatty acid or **carboxylic acid**, organic compound consisting of a hydrocarbon chain of an even number of carbon atoms, with a carboxyl group (–COOH) at one end. The covalent bonds between the carbon atoms may be single or double; where a double bond occurs the carbon atoms concerned carry one instead of two hydrogen atoms. Chains with only single bonds have all the hydrogen they can carry, so they are said to be saturated with hydrogen. Chains with one or more double bonds are said to be unsaturated (see †polyunsaturate). Fatty acids are produced in the small intestine when fat is digested.

Saturated fatty acids include palmitic and stearic acids; unsaturated fatty acids include oleic (one double bond), linoleic (two double bonds), and linolenic (three double bonds). Linoleic acid accounts for more than one third of some margarines. Supermarket brands that say they are high in polyunsaturates may contain as much as 39%. Fatty acids are generally found combined with glycerol in †lipids such as triglycerides.

fault fracture in the Earth's crust, on either side of which rocks have moved past each other. Faults may occur where rocks are being pushed together (compression) or pulled apart (tension) by †plate tectonics, movements of the †plates of the Earth's crust. When large forces build up quickly in rocks, they become brittle and break; †folds result from a more gradual compression. Faults involve displacements, or offsets, ranging from the microscopic scale to hundreds of kilometers. Large offsets along a fault are the result of the accumulation of many small movements (meters or less) over long periods of time. Large movements cause detectable †earthquakes, such as those experienced along the San Andreas Fault in California, USA.

Faults produce lines of weakness on the Earth's surface (along their strike) that are often exploited by processes of †weathering and †erosion. Coastal caves and geos (narrow inlets) often form along faults and, on a larger scale, rivers may follow the line of a fault. The Great Glen Fault in Scotland is an excellent example of a fault, and Loch Ness on this fault is an example of a fault-line feature.

San Andreas Fault and Bay Area http://sepwww.stanford.edu/oldsep/joe/fault_images/BayAreaSanAndreasFault.html
Detailed tour of the San Andreas Fault and the San Francisco Bay area, with information on the origination of the fault. The site is supported by a full range of area maps.

fault gouge soft, uncemented, pulverized clay-like material found between the walls of a fault. It is created by grinding action (cataclasis) of the fault and subsequent chemical alteration of tiny mineral grains brought about by fluids that flow along the fault.

fax or **facsimile transmission** or **telefax**, transmission of images over a †telecommunications link, usually the telephone network. An original image is scanned by a transmitting device and converted into coded signals, which travel via the telephone lines to the receiving fax machine. The standard transmission takes place at 4,800 or 9,600 bits of information per second.

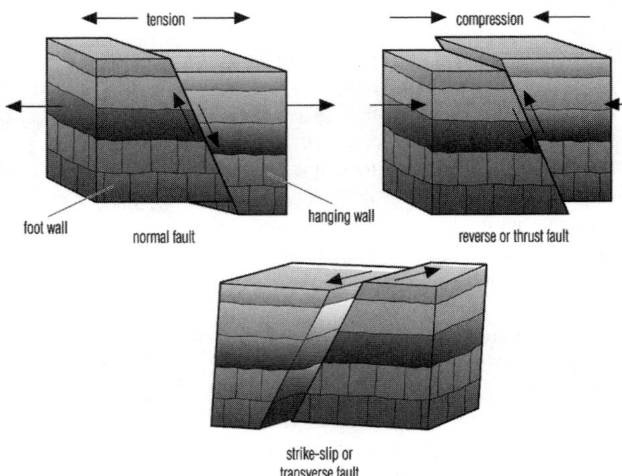

fault Faults are caused by the movement of rock layers, producing such features as block mountains and rift valleys. A normal fault is caused by a tension or stretching force acting in the rock layers. A reverse fault is caused by compression forces. Faults can continue to move for thousands or millions of years. Helicon

The world's first fax machine, the *pantélégraphe*, was invented by Italian physicist Giovanni Caselli in 1866, over a century before the first electronic model came on the market. Standing over 2 m/6.5 ft high, it transmitted by telegraph nearly 5,000 handwritten documents and drawings between Paris and Lyon in its first year.

feather rigid outgrowth of the outer layer of the skin of birds, made of the protein keratin. Feathers provide insulation and facilitate flight. There are several types, including long quill feathers on the wings and tail, fluffy down feathers for retaining body heat, and contour feathers covering the body. The coloring of feathers is often important in camouflage or in courtship and other displays. Feathers are normally replaced at least once a year.

There is an enormous variation between species in the number of feathers; for example a whistling swan has over 25,000 contour feathers, whereas a ruby-throated hummingbird has less than 950.

Feathers generally consist of two main parts, axis and barbs, the former of which is divided into the quill, which is bare and hollow, and the shaft, which bears the barbs. The quill is embedded in the skin, and has at its base a small hole through which the nourishment passes during the growth of the feather. The barbs which constitute the vane are lath-shaped and taper to a point, and each one supports a series of outgrowths known as barbules, so that each barb is like a tiny feather. Adjacent barbs are linked to each other by hooks on the barbules.

Fechner, Gustav Theodor (1801–1887) German psychologist. He became professor of physics at Leipzig in 1834, but in 1839 turned to the study of psychophysics (the relationship between physiology and psychology). He devised **Fechner's law**, a method for the exact measurement of sensation.

fecundity potential rate at which an organism reproduces, as distinct from its ability to reproduce (†fertility). In vertebrates, it is usually measured as the number of offspring produced by a female each year. Specifically, it refers to the quantity of gametes (usually eggs) produced per female over a given time.

feedback in biology, another term for †biofeedback: the influence of the outcome of one process upon the functioning of another process.

feedback general principle whereby the results produced in an ongoing reaction become factors in modifying or changing the reaction; it is the principle used in self-regulating control systems, from a simple †thermostat and steam-engine governor to automatic computer-controlled machine tools. A fully computerized control system, in which there is no operator intervention, is called a **closed-loop feedback** system. A system that also responds to control signals from an operator is called an **open-loop feedback** system.

In self-regulating systems, information about what *is* happening in a system (such as level of temperature, engine speed, or size of workpiece) is fed back to a controlling device, which compares it with what *should* be happening. If the two are different, the device takes suitable action (such as switching on a heater, allowing more steam to the engine, or resetting the tools). The idea that the Earth is a self-regulating system, with feedback operating to keep nature in balance, is a central feature of the †Gaia hypothesis.

Fehling's solution in chemistry, alkaline aqueous solution of copper sulfate and sodium potassium tartrate (Rochelle salt) used as a test (†Fehling's test) for the detection of aldehydes and reducing sugars, such as glucose. These compounds reduce the solution to form a red precipitate of copper oxide.

Fehling's test chemical test to determine whether an organic substance is a reducing agent (substance that donates electrons to other substances in a chemical reaction). It is usually used to detect reducing sugars (monosaccharides, such as glucose, and the disaccharides maltose and lactose) and aldehydes.

If the test substance is heated with a freshly prepared solution containing copper(II) sulfate, sodium hydroxide and sodium potassium tartrate, the production of a brick-red precipitate indicates the presence of a reducing agent.

The test was devised by German chemist Herman von Fehling.

feldspar any of a group of †silicate minerals. Feldspars are the most abundant mineral type in the Earth's crust. They are the chief constituents of †igneous rock and are present in most metamorphic and sedimentary rocks. All feldspars contain silicon, aluminum, and oxygen, linked together to form a framework. Spaces within this framework structure are occupied by sodium, potassium, calcium, or occasionally barium, in various proportions. Feldspars form white, gray, or pink crystals and rank 6 on the †Mohs scale of hardness.

The four extreme compositions of feldspar are represented by the minerals **orthoclase**, $KAlSi_3O_8$; **albite**, $NaAlSi_3O_8$; **anorthite**, $CaAl_2Si_2O_8$; and **celsian**, $BaAl_2Si_2O_8$. **Plagioclase feldspars** contain variable amounts of sodium (as in albite) and calcium (as in anorthite) with a negligible potassium content. **Alkali feldspars** (including orthoclase) have a high potassium content, less sodium, and little calcium.

The type known as †moonstone has a pearl-like effect and is used in jewelry. Approximately 4,000 tonnes of feldspar are used in the ceramics industry annually.

feldspathoid any of a group of silicate minerals resembling feldspars but containing less silica. Examples are nepheline ($NaAlSiO_4$ with a little potassium) and leucite ($KAlSi_2O_6$). Feldspathoids occur in igneous rocks that have relatively high proportions of sodium and potassium. Such rocks may also contain alkali feldspar, but they do not generally contain quartz because any free silica would have combined with the feldspathoid to produce more feldspar instead.

felsic rock a †plutonic rock composed chiefly of light-colored minerals, such as quartz, feldspar, and mica. It is derived from feldspar, lenad (meaning feldspathoid), and silica. The term felsic also applies to light-colored minerals as a group, especially quartz, feldspar, and feldspathoids.

femtochemistry study of chemical reactions using †femtosecond spectroscopy. It was developed by Egyptian-born US physical chemist Ahmed H Zewail.

femtosecond †SI unit of time. It is 10^{-15} seconds (one millionth of a billionth of a second).

femtosecond spectroscopy in chemistry, analytical technique used to study the mechanisms involved in chemical reactions in real time. The technique uses laser flashes that pulse in femtoseconds (10^{-15} s), that is the same time scale as bonds break and form during a chemical reaction.

femur or **thigh-bone**, also the upper bone in the hind limb of a four-limbed vertebrate.

Fermat, Pierre de (1601–1665) French mathematician who, with Blaise †Pascal, founded the theory of †probability and the modern theory of numbers. Fermat also made contributions to analytical geometry. In 1657 Fermat published a series of problems as challenges to other mathematicians, in the form of theorems to be proved.

Fermat, Pierre de http://www-history.mcs.st-and.ac.uk/history/Mathematicians/Fermat.html

Extensive biography of the French mathematical genius Pierre de Fermat. The site contains a description of his contributions to mathematics, including his last theorem and also much of his other works which were of more direct value to mathematics. Also included are the first page of his book Ad Locos Planos et Solidos Isagoge and an extract from his book Maxima et Minima (1679). Several references for further reading are also listed, and the Web site also features several portraits of Fermat.

Fermat's Last Theorem http://www-groups.dcs.st-andrews.ac.uk/history/HistTopics/Fermat's_last_theorem.html

Account of Fermat's last theorem and of the many attempts made to prove it. It is extensively hyperlinked to related mathematicians and also includes a list of 17 references for further reading.

Fermat's last theorem in mathematics, theorem that states that the equation: $x^n + y^n = z^n$ is not solvable in the integers if *n* is greater than 2. Fermat's own proof has never been found and the theorem remained unproved until 1994, when English mathematician Andrew Wiles verified it.

Fermat's principle in physics, the principle that a ray of light, or other radiation, moves between two points along the path that takes the minimum time. The principle is named after French mathematician Pierre de Fermat, who used it to deduce the laws of †reflection and †refraction.

fermentation breakdown of sugars by bacteria and yeasts using a method of respiration without oxygen (†anaerobic). The enzymes in yeast break down glucose to give two products: †ethanol (alcohol) and †carbon dioxide. Fermentation processes have long been utilized in baking bread, making beer and wine, and producing cheese, yogurt, soy sauce, and many other foodstuffs.

In bread baking, the bubbles of carbon dioxide trapped in the dough make the bread rise. In the brewing process, complex sugars, such as sucrose, are first broken down by yeast into simple sugars, such as glucose. Glucose is then further decomposed into ethanol and carbon dioxide. The ethanol produced is the alcohol in alcoholic drinks, such as beer and wine; the carbon dioxide puts bubbles into beers and champagne.

The word equation for fermentation is: glucose ➔ ethanol + carbon dioxide. The symbol equation is: $C_6H_{12}O_6 \rightarrow 2C_2H_5OH + 2CO_2$

Many antibiotics are produced by fermentation; it is one of the processes that can cause food spoilage.

fermenter or **biodigester**, large container in which †fermentation reactions take place. Industrial fermenters may each hold a large volume of nutrient solution (250,000 l/55,000 gal), in which cells are cultured to produce useful substances.

fermi unit of length equal to 10^{-15} m, used in atomic and nuclear physics. The unit is named after Enrico Fermi.

Fermi, Enrico (1901–1954) Italian-born US physicist who was awarded the Nobel Prize for Physics in 1938 for his proof of the existence of new radioactive elements produced by bombardment with neutrons, and his discovery of nuclear reactions produced by low-energy neutrons. This research was the basis for studies leading to the atomic bomb and nuclear energy. Fermi built the first nuclear reactor in 1942 at Chicago University and later took part in the Manhattan Project to construct an atom bomb. His theoretical work included the study of the weak nuclear force, one of the fundamental forces of nature, and beta decay.

Fermi–Dirac statistics mathematical treatment of particles with half-integer †spin in †statistical mechanics, enabling properties of systems made up by such particles to be calculated. They are named after Italian-born US physicist Enrico Fermi and English physicist David Dirac.

Fermilab acronym for FERMI National Accelerator LABoratory, US center for †particle physics at Batavia, Illinois, near Chicago. It is named after Italian-born US physicist Enrico Fermi. Fermilab was opened in 1972, and is the home of the Tevatron, the world's most powerful particle †accelerator. It is capable of boosting protons and antiprotons to speeds near that of light (to energies of 20 TeV).

A new main injector was built in 1998 to enable the Tevatron to function with a higher beam intensity.

fermion in physics, a subatomic particle whose spin can only take values that are half-odd-integers, such as $^1/_2$ or $^3/_2$. Fermions may be classified as leptons, such as the electron, and hadrons, such as the proton, neutron, mesons, and so on. All elementary particles are either fermions or †bosons.

The exclusion principle, formulated by Austrian–US physicist Wolfgang Pauli in 1925, asserts that no two fermions in the same system (such as an atom) can possess the same position, energy state, spin, or other quantized property.

fermium chemical symbol Fm, synthesized, radioactive, metallic element of the †actinide series, atomic number 100, relative atomic mass 257.10. Ten isotopes are known, the longest-lived of which, Fm-257, has a half-life of 80 days. Fermium has been produced only in minute quantities in particle accelerators. It was discovered in 1952 in the debris of the first thermonuclear explosion. The element was named in 1955 in honor of US physicist Enrico †Fermi.

fern any of a group of plants related to horsetails and clubmosses. Ferns are spore-bearing, not flowering, plants and most are perennial, spreading by slow-growing roots. The leaves, known as fronds, vary widely in size and shape. Some taller types, such as tree ferns, grow in the tropics. There are over 7,000 species. (Order Filicales.)

Ferrari, Ludovico (1522–1565) Italian mathematician. His solution, at the age of 15, to a specific quartic equation was published by Girolamo Cardano and credited to Ferrari. He subsequently became professor of mathematics at Bologna.

ferric ion traditional name for the trivalent condition of iron, Fe^{3+}; the modern name is iron(III). Ferric salts are usually reddish or yellow in color and form reddish-yellow solutions. $Fe_2(SO_4)_3$ is iron(III) sulfate (ferric sulfate).

ferrimagnetism form of †magnetism in which adjacent molecular magnets are aligned anti-parallel, but have unequal strength, producing a strong overall magnetization. Ferrimagnetism is found in certain inorganic substances, such as †ferrites.

ferrite ceramic ferrimagnetic material. Ferrites are iron oxides to which small quantities of †transition metal oxides (such as cobalt and nickel oxides) have been added. They are used in transformer cores, radio antennae, and, formerly, in computer memories.

ferro-alloy alloy of iron with a high proportion of elements such as manganese, silicon, chromium, and molybdenum. Ferro-alloys are used in the manufacture of alloy steels. Each alloy is generally named after the added metal – for example, ferrochromium.

ferroelectric material ceramic dielectric material that, like ferromagnetic materials, has a †domain structure that makes it exhibit magnetism and usually the †piezoelectric effect. An example is Rochelle salt (potassium sodium tartrate tetrahydrate, $KNaC_4H_4O_6.4H_2O$).

ferromagnetism form of †magnetism that can be acquired in an external magnetic field and usually retained in its absence, so that ferromagnetic materials are used to make permanent magnets. A ferromagnetic material may therefore be said to have a high magnetic †permeability and †susceptibility (which depends upon temperature). Examples are iron, cobalt, nickel, and their alloys.

Ultimately, ferromagnetism is caused by spinning electrons in the atoms of the material, which act as tiny weak magnets. They align parallel to each other within small regions of the material to form †domains, or areas of stronger magnetism. In an unmagnetized material, the domains are aligned at random so there is no overall magnetic effect. If a magnetic field is applied to that material, the domains align to point in the same direction, producing a strong overall magnetic effect. Permanent magnetism arises if the domains remain aligned after the external field is removed. Ferromagnetic materials exhibit †hysteresis.

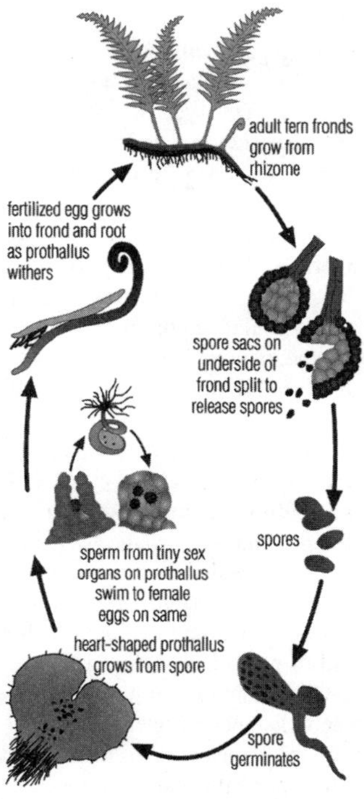

fern The life cycle of a fern. Ferns have two distinct forms that alternate during their life cycle. For the main part of its life, a fern consists of a short stem (or rhizome) from which roots and leaves grow. The other part of its life is spent as a small heart-shaped plant called a prothallus. Helicon

ferrous ion traditional name for the divalent condition of iron, Fe^{2+}; the modern name is iron(II). Ferrous salts are usually green, and form yellow-green solutions. $FeSO_4$ is iron(II) sulfate (ferrous sulfate).

fertility organism's ability to reproduce, as distinct from the rate at which it reproduces (†fecundity). Individuals that can reproduce are fertile. Individuals that cannot reproduce are infertile. Individuals become infertile (unable to reproduce) when they cannot generate gametes (eggs or sperm) or when their gametes cannot yield a viable †embryo after fertilization. Common causes of infertility are: low sperm numbers in the male; blocked oviducts; infrequent ovulation (release of gamete) in the female; consequence of infections in the female reproductive tract; and genetic makeup.

fertilization in †sexual reproduction, the union of two †gametes (sex cells, often called egg or ovum, and sperm) to produce a †zygote, which combines the genetic material contributed by each parent. In self-fertilization the male and female gametes come from the same plant; in cross-fertilization they come from different plants. Self-fertilization rarely occurs in animals; usually even †hermaphrodite animals cross-fertilize each other.

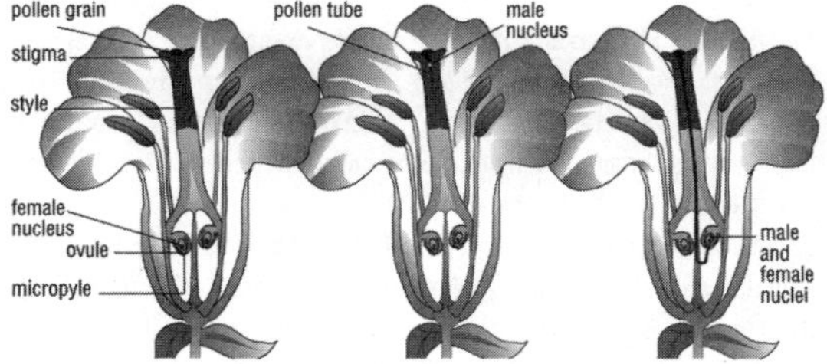

fertilization In a flowering plant pollen grains land on the surface of the stigma, and if conditions are acceptable the pollen grain germinates, forming a pollen tube, through which the male gametes pass, entering the ovule via the micropyle in order to reach the female egg. Helicon

The fusion of gametes combines the genetic material contributed by each parent. To avoid doubling the amount of inherited information every generation, each gamete contains only half the amount of inherited information – it is haploid. This is achieved by halving the number of †chromosomes when gametes are being produced. When the gametes fuse the full amount of information is restored (diploid state). Gametes are therefore produced by a specialized form of cell division, known as meiosis, which is only used for this purpose. This type of cell division has ways of mixing genes before the gametes are produced. This promotes variation (see †natural selection) by producing gametes with different combinations of genes.

fertilizer substance containing some or all of a range of about 20 chemical elements necessary for healthy plant growth, used to compensate for the deficiencies of poor or depleted †soil. Fertilizers may be organic, for example farmyard manure, composts, bonemeal, blood, and fishmeal; or inorganic (synthetic or artificial), in the form of simple compounds, mainly of nitrogen, phosphate, and potash, which have been used on a very much increased scale since 1945. Compounds of nitrogen and phosphorus are of particular importance.

Elements in the soil are taken up through the roots of plants in solution, becoming part of the compounds forming the plant. If plants are allowed to die and decompose, these compounds return to the soil as part of a natural cycle, such as the †nitrogen cycle. However, when crops are harvested the cycle is interrupted, the nutrients are not returned to the soil and are used up. Fertilizers replace these elements, increasing the yield of crops and enabling the soil to be farmed year after year.

Fessenden, Reginald Aubrey (1866–1932) Canadian physicist who worked in the USA. He patented the modulation of radio waves (transmission of a signal using a carrier wave), an essential technique for voice transmission. At the time of his death, he held 500 patents.

fetal alcohol syndrome in medicine, fetal abnormality that may occur if the mother has consumed excessive amounts of alcohol during pregnancy. At birth, the baby is smaller than average, may be mentally retarded, and may suffer heart problems. The baby's facial bones are also incorrectly developed, giving an abnormal look to the face.

fetch distance of open water over which wind can blow to create †waves. The greater the fetch the more potential power waves have when they hit the coast. In the south and west of England the fetch stretches for several thousand kilometers, all the way to South America. This combines with the southwesterly †prevailing winds to cause powerful waves and serious †coastal erosion along south- and west-facing coastlines.

fetch-execute cycle or **processing cycle**, in computing, the two-phase cycle used by the computer's central processing unit to process the instructions in a program. During the fetch phase, the next program instruction is transferred from the computer's immediate-access memory to the instruction register (memory location used to hold the instruction while it is being executed). During the execute phase, the instruction is decoded and obeyed. The process is repeated in a continuous loop.

fetus or **fetus**, stage in mammalian †embryo development after fusion of gametes produces a zygote cell. The human embryo is usually termed a fetus after the eighth week of development, when the limbs and external features of the head are recognizable. The stage ends at birth.

Visible Embryo http://www.visembryo.com/

Learn about developments that take place in the womb during the 40 weeks of pregnancy. For each week of pregnancy the user may view a picture of the embryo and read a description of its development.

Feynman, Richard P(hillips) (1918–1988) US physicist whose work laid the foundations of quantum electrodynamics. He was awarded the Nobel Prize for Physics in 1965 for his work on the theory of radiation. He shared the award with Julian †Schwinger and Sin-Itiro †Tomonaga. He also contributed to many aspects of particle physics, including quark theory and the nature of the weak nuclear force.

Feynman, Richard Phillips http://www-groups.dcs.st-andrews.ac.uk/history/Mathematicians/Feynman.html

Devoted to the life and contributions of physicist Richard Feynman. In addition to biographical information, you will find a list of references about Feynman and links to essays that reference him. The text of this essay includes hypertext links to the essays of those mathematicians and thinkers who influenced Feynman. You will also find an image of him, which you can click on to enlarge, and a map of his birthplace.

Fibonacci, Leonardo (*c.* 1170–*c.* 1250) also known as **Leonardo of Pisa**, Italian mathematician. He published *Liber abaci/The Book of the Calculator* in Pisa in 1202, which was instrumental in the introduction of Arabic notation into Europe. From 1960, interest increased in **Fibonacci numbers**, in their simplest form a sequence in which each number is the sum of its two predecessors (1, 1, 2, 3, 5, 8, 13, ...). They have unusual characteristics with possible applications in botany, psychology, and astronomy (for example, a more exact correspondence than is given by †Bode's law to the distances between the planets and the Sun).

Fibonacci series sequence in which each term is the sum of the preceding two terms immediately preceding it; it begins 1, 1, 2, 3, 5, 8, 13, 21, 34, 55, 89 ... and has a variety of important applications (for example, in search algorithms).

It was first introduced by Italian mathematician Leonardo †Fibonacci (Leonardo of Pisa) in connection with a problem in the growth of a population of rabbits. It occurs in a natural way in many biological problems including the distribution of branches on pine trees and seeds in the head of a sunflower.

fiber, dietary or **roughage**, plant material (outer husks and peel) that cannot be digested by human digestive enzymes. It consists largely of cellulose, a carbohydrate found in plant cell walls. Fiber adds bulk to the gut contents because it can absorb water, assisting the muscular contractions that force food along the intestine. A diet low in fiber causes constipation and is believed to increase the risk of developing diverticulitis, diabetes, gall-bladder disease, and cancer of the large bowel – conditions that are rare in non-industrialized countries, where the diet contains a high proportion of unrefined cereals. It is also known as NSP (non-starch polysaccharide).

Animals are unable to produce the †enzyme necessary to break down cellulose. However, cellulose still forms a necessary part of the human diet as fiber (roughage) – because it is not broken down it acts as a signal that the gut needs to contract to move the contents on. Any undigested plant material acts as fiber.

Soluble fiber consists of indigestible plant carbohydrates (such as pectins, hemicelluloses, and gums) that dissolve in water. A high proportion of the fiber in such foods as oat bran, pulses, and vegetables is of this sort. Its presence in the diet has been found to reduce the amount of cholesterol in blood over the short term, although the mechanism for its effect is disputed.

fiber optics branch of physics dealing with the transmission of light and images through glass or plastic fibers known as †optical fibers. Such fibers are now commonly used in both communications technology and medicine.

Fiber Optic Chronology http://www.sff.net/people/Jeff.Hecht/Chron.html

Time line of the history of fiber optics. Starting with the discovery of glass in around 2500 BC and leading up to the late 1970s, this site details every discovery relevant to the history of the fiber optic cables which, among other things, make the Internet possible.

fibrin insoluble protein involved in blood clotting. When an injury occurs fibrin is deposited around the wound in the form of a mesh, which dries and hardens, so that bleeding stops. Fibrin is developed in the blood from a soluble protein, fibrinogen.

The conversion of fibrinogen to fibrin is the final stage in blood clotting. Platelets, a type of cell found in blood, release the enzyme thrombin when they come into contact with damaged tissue, and the formation of fibrin then occurs. Calcium, vitamin K, and a variety of enzymes called factors are also necessary for efficient blood clotting.

fibula rear lower bone in the hind leg of a vertebrate. It is paired and often fused with a smaller front bone, the tibia.

Fick, Adolf Eugen (1829–1901) German physiologist who worked on the physics of how the human body works, particularly the eyes, heart, and muscles. He devised techniques to quantify body functions, such as muscle contraction, impulse conduction by nerves, and blood pressure in the heart. He also described the physics of how the eye works, especially the blind spot of the retina.

field in physics, region of space in which an object exerts a force on a separate object because of certain properties they both possess. For example, there is a force of attraction between any two objects that have mass when one is within the gravitational field of the other.

Other fields of force include †electric fields (caused by electric charges) and †magnetic fields (caused by circulating electric currents), either of which can involve attractive or repulsive forces.

Field, George Brooks (1929–) US theoretical astrophysicist. His main research has been into the nature and composition of intergalactic matter and the properties of residual radiation in space.

field-effect transistor FET, in physics, solid-state device with three terminals called the source, drain, and gate. An n-type FET, for example, consists of a piece of n-type †semiconductor with the source at one end and the drain at the other. A region of p-type material in the middle forms the gate. A voltage applied to the gate modifies the conducting channel between the source and the drain. Electrons are the only charge carriers and so it is an example of a **unipolar transistor**.

field mapping mapping rocks, rock formations, and geological units while on the ground (as opposed to remote mapping, from airplane or satellite). Field mapping is the basis for understanding the structure of the Earth's crust and its geological history.

field of view angle over which an image may be seen in a mirror or an optical instrument such as a telescope. A wide field of view allows a greater area to be surveyed without moving the instrument, but has the disadvantage that each of the objects seen is smaller. A †convex mirror gives a larger field of view than a plane or flat mirror. The field of view of an eye is called its **field of vision** or visual field.

A telephoto lens has a small field of view, around 14°, and produces a highly magnified image. A "fish-eye" lens has a very wide angle of view, around 180°. Our eyes have a field of vision of around 45°.

field strength see †magnetic field strength.

field studies study of ecology, geography, geology, history, archaeology, and allied subjects, in the natural environment as opposed to the laboratory.

field theory in mathematics, theory involving the mathematical structure known as a field, which displays the operations of addition and multiplication and their inverses (subtraction and division). An elementary example other than the real numbers is constituted by the rational numbers; there are many other types. A ring is much like a field but does not include the inverse operations. A †group is a more restricted concept still.

fifth-generation computer anticipated new type of computer based on emerging microelectronic technologies with high computing speeds and †parallel processing. The development of very large-scale integration (VLSI) technology, which can put many more circuits onto an integrated circuit (chip) than is currently possible, and developments in computer hardware and software design may produce computers far more powerful than those in current use.

It has been predicted that such a computer will be able to communicate in natural spoken language with its user; store vast knowledge databases; search rapidly through these databases, making intelligent inferences and drawing logical conclusions; and process images and "see" objects in the way that humans do.

In 1981 Japan's Ministry of International Trade and Industry launched a ten-year project to build the first fifth-generation computer, the "parallel inference machine," consisting of over a thousand microprocessors operating in parallel with each other. By 1992, however, the project was behind schedule and had only produced 256-processor modules. It has since been suggested that research into other technologies, such as †neural networks, may present more promising approaches to artificial intelligence. Compare earlier †computer generations.

filament in astronomy, dark, winding feature occasionally seen on images of the Sun in hydrogen light. Filaments are clouds of relatively cool gas suspended above the Sun by magnetic fields and seen in silhouette against the hotter †photosphere below. During total †eclipses they can be seen as bright features against the sky at the edge of the Sun, where they are known as †prominences.

file in computing, a collection of data or a program stored in a computer's external memory (for example, on †disk). It might include anything from information on a company's employees to a program for an adventure game. Serial (or sequential) access files hold information as a sequence of characters, so that, to read any particular item of data, the program must read all those that precede it. Random-access (or direct access) files allow the required data to be reached directly. Files are usually located via a †directory.

file server computer on a †network that handles (and usually stores) the data used by other computers on the network. See also †client–server architecture.

filter in chemistry, a porous substance, such as blotting paper, through which a mixture can be passed to separate out its solid constituents.

filter in electronics, a circuit that transmits a signal of some frequencies better than others. A low-pass filter transmits signals of low frequency and also direct current; a high-pass filter transmits high-frequency signals; a band-pass filter transmits signals in a band of frequencies.

filter in optics, a device that absorbs some parts of the visible †spectrum and transmits others. A beam of white light can be made into a beam of colored light by placing a transparent color filter in the path of the beam. For example, a green filter will absorb or block all colors of the spectrum except green, which it allows to pass through. A yellow filter absorbs only light at the blue and violet end of the spectrum, transmitting red, orange, green, and yellow light.

filter press in chemistry, device used for large-scale filtration of liquids, consisting of a series of cloth-covered frames through which a liquid is passed under pressure.

filtrate liquid or solution that has passed through a filter.

filtration technique by which suspended solid particles in a fluid are removed by passing the mixture through a filter, usually porous paper, plastic, or cloth. The particles are retained by the filter to form a residue and the fluid passes through to make up the filtrate. For example, soot may be filtered from air, and suspended solids from water.

fin in aquatic animals, flattened extension from the body that aids balance and propulsion through the water.

In fish they may be paired, such as the pectoral and ventral fins, or singular, such as the caudal and dorsal fins, all being supported by a series of cartilaginous or bony rays.

The fins in cetaceans (whales and dolphins) are simple extensions of the soft tissue and have no bony rays. The flippers of seals are modified five-fingered limbs and contain the same bones as the limbs of other vertebrates.

fine structure splitting of individual †spectral lines seen when viewed under high resolution. This can be produced by external magnetic fields acting on †atoms, for example.

finite having a countable number of elements, the opposite of infinite.

finsen unit **FU**, unit for measuring the intensity of ultraviolet (UV) light; for instance, ultraviolet light of 2 FUs causes sunburn in 15 minutes.

fiord alternative spelling of †fjord.

fireball in astronomy, very bright †meteor, often bright enough to be seen in daylight and occasionally leading to the fall of a †meteorite. Some fireballs are caused by †satellites or other space debris burning up in the Earth's atmosphere.

fire clay a †clay with refractory characteristics (resistant to high temperatures), and hence suitable for lining furnaces (firebrick). Its chemical composition consists of a high percentage of silicon and aluminum oxides, and a low percentage of the oxides of sodium, potassium, iron, and calcium.

firedamp gas that occurs in coal mines and is explosive when mixed with air in certain proportions. It consists chiefly of methane (CH_4, natural gas or marsh gas) but always contains small quantities of other gases, such as nitrogen, carbon dioxide, and hydrogen, and sometimes ethane and carbon monoxide.

fire-danger rating unit index used by the UK Forestry Commission to indicate the probability of a forest fire. 0 means a fire is improbable, 100 shows a serious fire hazard.

fire extinguisher device for putting out a fire. Fire extinguishers work by removing one of the three conditions necessary for fire to continue (heat, oxygen, and fuel), either by cooling the fire or by excluding oxygen.

The simplest fire extinguishers contain water, which when propelled onto the fire cools it down. Water extinguishers cannot be used on electrical fires, as there is a danger of electrocution, or on burning oil, as the oil will float on the water and spread the blaze.

Many domestic extinguishers contain liquid carbon dioxide under pressure. When the handle is pressed, carbon dioxide is released as a gas that blankets the burning material and prevents oxygen from reaching it. Dry extinguishers spray powder, which then releases carbon dioxide gas. Wet extinguishers are often of the soda-acid type; when activated, sulfuric acid mixes with sodium bicarbonate, producing carbon dioxide. The gas pressure forces the solution out of a nozzle, and a foaming agent may be added.

Some extinguishers contain halons (hydrocarbons with one or more hydrogens substituted by a halogen such as chlorine, bromine, or fluorine). These are very effective at smothering fires, but cause damage to the †ozone layer, and their use is now restricted.

firmware computer program held permanently in a computer's †ROM (read-only memory) chips, as opposed to a program that is read in from external memory as it is needed. The BIOS in a personal computer is an example of firmware.

firn or **névé**, snow that has lain on the ground for a full calendar year. Firn is common at the tops of mountains; for example, the Alps in Europe. After many years, compaction turns firn into ice and a †glacier may form.

first filial generation or **F_1 generation**, in genetics, offspring of the parental generation where each parent is †homozygous for the particular characteristic in question. In the F_1 generation, all offspring are †heterozygous for the characteristic.

first order in differential equations, involving only the first †derivative.

Fischer, Emil Hermann (1852–1919) German chemist who produced synthetic sugars and, from these, various enzymes. His descriptions of the chemistry of the carbohydrates and peptides laid the foundations for the science of biochemistry. He was awarded the Nobel Prize for Chemistry in 1902 for his work on the synthesis of sugars and purine compounds.

Fischer, Hans (1881–1945) German chemist awarded the Nobel Prize for Chemistry in 1930 for his work on hemoglobin, the oxygen-carrying, red coloring matter in blood. He determined the molecular structures of three important biological pigments: hemoglobin, chlorophyll, and bilirubin.

Fischer projection in chemistry, two-dimensional representational drawing of the structural arrangement of atoms and groups in a molecule. It was devised by German chemist Emil †Fischer.

Fisher, Ronald Aylmer (1890–1962) English statistician and geneticist. He modernized Charles Darwin's theory of evolution, thus securing the key biological concept of genetic change by natural selection. Fisher developed several new statistical techniques and, applying his methods to genetics, published *The Genetical Theory of Natural Selection* (1930).

fission in physics, the splitting of a heavy atomic nucleus into two or more major fragments. It is accompanied by the emission of two or three neutrons and the release of large amounts of †nuclear energy. Fission occurs spontaneously in nuclei of uranium-235, the main fuel used in nuclear reactors. However, the process can also be induced by bombarding nuclei with neutrons because a nucleus that has absorbed a neutron becomes unstable and soon splits. For example:

$^{235}_{92}U + ^{1}_{0}n \rightarrow ^{236}_{92}U \rightarrow$ 2 nuclei + 2–3 neutrons + energy

The neutrons released spontaneously by the fission of uranium nuclei may therefore be used in turn to induce further fissions, setting up a †chain reaction that must be controlled if it is not to result in a nuclear explosion. In a nuclear power station, heat energy released from the chain reaction is used to boil water to produce steam. The steam is used to drive the turbine of a generator to produce electricity. An atomic bomb uses pure uranium-235 to start an uncontrolled nuclear reaction, producing large amounts of heat energy. The minimum amount of fissile material that can undergo a continuous chain reaction is referred to as the †critical mass.

Fitch, Val Logsdon (1923–) US physicist. He shared the Nobel Prize for Physics in 1980 with James †Cronin for their joint work in particle physics, studying the surprising way certain †mesons change from matter to antimatter.

fitness in genetic theory, a measure of the success with which a genetically determined character can spread in future generations. By convention, the normal character is assigned a fitness of one and variants (determined by other †alleles) are then assigned fitness values relative to this. Those with fitness greater than one will spread more rapidly and will ultimately replace the normal allele; those with fitness less than one will gradually die out.

Fittig, Rudolf (1835–1910) German chemist. Fittig discovered phenanthrene and the lactones, or anhydrides, of oxyacids, and synthesized several hydrocarbons. He re-edited the works of the organic chemist Wöhler, and published his own works, including *Über Aceton/Acetone* (1858) and *Grundriss der anorganischen Chemie/Inorganic Chemistry* (1872).

Fitzgerald, George Francis (1851–1901) Irish physicist known for his work on electromagnetics. He was the first to suggest a method of propagating radio waves, which helped towards the development of wireless telegraphy. In 1892 he explained the anomalous results of the †Michelson–Morley experiment in 1887 by supposing that bodies moving through the ether contracted as their velocity increased. The effect, since known as the **Fitzgerald–Lorentz contraction**, was developed by Albert †Einstein's special theory of †relativity.

Fitzgerald, George Francis http://www-groups.dcs.st-andrews.ac.uk/history/Mathematicians/FitzGerald.html

Part of an archive containing the biographies of the world's greatest mathematicians, this site is devoted to the life and contributions of physicist George Fitzgerald.

fixed point temperature that can be accurately reproduced and used as the basis of a temperature scale. In the Celsius scale, the fixed points are the temperature of melting ice, defined to be 0°C (32°F), and the temperature of boiling water (at standard atmospheric pressure), defined to be 100°C (212°F).

fixed star description sometimes applied to a star, especially in emphasizing its difference from a planet (or "wandering star"). Though most stars are in rapid motion through space, their great distance from Earth ensures that their relative configurations do not change appreciably over a thousand years.

Fizeau, Armand Hippolyte Louis (1819–1896) French physicist. In 1849 he was the first to measure the speed of light on the Earth's surface. He also found that light travels faster in air than in water, which confirmed the wave theory of light, and that the motion of a star affects the position of the lines in its †spectrum.

fjord or **fiord**, narrow sea inlet enclosed by high cliffs. Fjords are found in Norway, New Zealand, Southern Chile, and western parts of Scotland. They are formed when an overdeepened U-shaped glacial valley is drowned by a rise in sea-level. At the mouth of the fjord there is a characteristic lip causing a shallowing of the water. This is due to reduced glacial erosion and the deposition of †moraine at this point.

Fiordland is the deeply indented southwest coast of the South Island, New Zealand; one of the most beautiful inlets is Milford Sound.

flaccidity in botany, the loss of rigidity (turgor) in plant cells, caused by loss of water from the central vacuole so that the cytoplasm no longer pushes against the cellulose cell wall. If this condition occurs throughout the plant then wilting is seen.

Flaccidity can be induced in the laboratory by immersing the plant cell in a strong saline solution. Water leaves the cell by †osmosis causing the vacuole to shrink. In extreme cases the actual cytoplasm pulls away from the cell wall, a phenomenon known as plasmolysis.

flagellate aquatic protozoan of the subphylum †Mastigophora, characterized by possessing one or more †flagella (whiplike organs).

flagellum (plural flagella) small hairlike organ on the surface of certain cells. Flagella are the motile organs of certain protozoa and single-celled algae, and of the sperm cells of higher animals. Unlike †cilia, flagella usually occur singly or in pairs; they are also longer and have a more complex whiplike action.

Each flagellum consists of contractile filaments producing snakelike movements that propel cells through fluids, or fluids past cells. Water movement inside sponges is also produced by flagella.

flame test in chemistry, the use of a flame to identify metal †cations present in a solid.

A nichrome or platinum wire is moistened with acid, dipped in a compound of the element, either powdered or in solution, and then held in a hot non-luminous flame. The color produced in the flame is characteristic of metals present; for example, sodium burns with an orange-yellow flame, and potassium with a lilac one.

Flammarion, Camille (1842–1925) French astronomer and popular writer on astronomy. His *Astronomie populaire*, first published in 1880, has been frequently revised, and was issued in an English translation as the *Flammarion Book of Astronomy* in 1964.

Flamsteed, John (1646–1719) English astronomer. He began systematic observations of the positions of the stars, Moon, and planets at the Royal Observatory he founded at Greenwich, London, in 1676. His observations were published in *Historia Coelestis Britannica* (1725).

Flamsteed, John http://www-history.mcs.st-and.ac.uk/history/Mathematicians/Flamsteed.html

Biography of the English astronomer. Aside from his methodical astronomical observations Flamsteed is most noted for having the Royal Observatory at Greenwich built for him during his period as astronomer royal. Several literature sources for further reading on the life and works of Flamsteed are listed here, and his portrait is also shown at this Web site.

flare star dim red dwarf star that suddenly lights up with great – but brief – luminosity, corresponding to an equally powerful but short-lived burst of radio emission. The cause is thought to be a sudden and intense outburst of radiation on or above the star's surface.

flash flood flood of water in a normally arid area brought on by a sudden downpour of rain. Flash floods are rare and usually occur in mountainous areas. They may travel many kilometers from the site of the rainfall.

Because of the suddenness of flash floods, little warning can be given of their occurrence. In 1972 a flash flood at Rapid City, South Dakota, USA, killed 238 people along Rapid Creek.

flashover small fire that erupts suddenly into a much larger one. It can occur if there is sufficient build up of heat from a small fire to ignite the mixture of gas and smoke produced. This increases temperatures further igniting surroundings. Flashovers can occur extremely rapidly with temperatures of 1,100°C being reached in seconds.

flash point in physics, the lowest temperature at which a liquid or volatile solid heated under standard conditions gives off sufficient vapor to ignite on the application of a small flame.

The **fire point** of a material is the temperature at which full combustion occurs. For safe storage of materials such as fuel or oil, conditions must be well below the flash and fire points to reduce fire risks to a minimum.

flat screen type of display suitable for portable computers such as LCD (†liquid crystal display) or gas plasma screens. Flat-screen, or flat-panel, displays are compact and lightweight compared to traditional cathode-ray tube monitors and TV sets.

It is predicted that eventually all TV screens will be made using this type of technology.

flavone sap-soluble substance that gives the ivory to deep yellow coloring of plants.

Fleming, (John) Ambrose (1849–1945) English electrical physicist and engineer who invented the thermionic valve in 1904 and devised †Fleming's rules.

Fleming, Alexander (1881–1955) Scottish bacteriologist who was awarded a Nobel Prize for Physiology or Medicine in 1945 for his discovery of the bactericidal effect of †penicillin in 1928. In 1922 he had discovered lysozyme, an antibacterial enzyme present in saliva, nasal secretions, and tears. While studying this, he found an unusual mould growing on a culture dish, which he isolated and grew into a pure culture. This led to his discovery of penicillin, which came into use in 1941. He shared the award with Howard W †Florey and Ernst B †Chain, whose research had brought widespread realization of the value of penicillin with its isolation and its development as an antibiotic drug.

Fleming, Alexander http://www.pbs.org/wgbh/aso/databank/entries/bmflem.html

Biography of the Scottish bacteriologist, accompanied by an explanation of his discovery of penicillin.

Fleming, Williamina Paton Stevens (1857–1911) Scottish-born US astronomer. She was an assistant to Edward Pickering, director of the Harvard College Observatory, Cambridge, Massachusetts, with whom she compiled the first general catalogue classifying stellar †spectra, the distinctive pattern produced by each star when its light is passed through a prism.

Fleming's rules memory aids used to recall the relative directions of the magnetic field, current, and motion in an electric generator or motor, using one's fingers. The three directions are represented by the thu*m*b (for *m*otion), *f*orefinger (for *f*ield), and se*c*ond finger (for †conventional *c*urrent), all held at right angles to each other. The right hand is used for generators and the left for motors. The rules were devised by the English physicist John Fleming

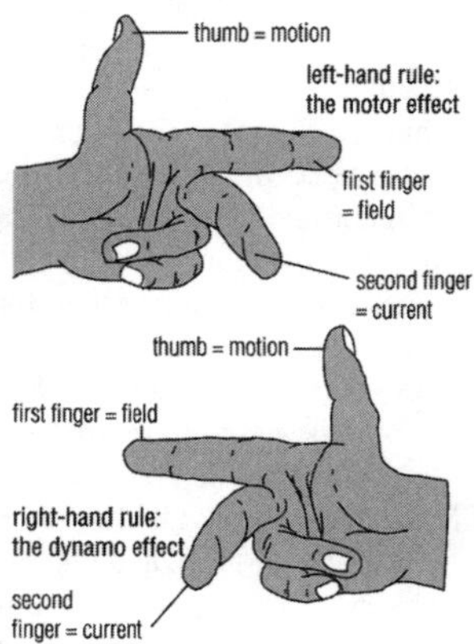

Fleming's rules Fleming's rules give the direction of the magnetic field, motion, and current in electrical machines. The left hand is used for motors, and the right hand for generators and dynamos. Helicon

Flemming, Walther (1843–1905) German biologist who is best known for his work on the way cells divide; a process that he named †mitosis. Using a microscope and aniline dyes Flemming could observe the duplication of chromosomes (the threadlike genetic material in the nucleus) before a cell divides.

flexor any muscle that bends a limb. Flexors usually work in opposition to other muscles, the extensors, an arrangement known as antagonistic.

flight simulator computer-controlled pilot-training device, consisting of an artificial cockpit mounted on hydraulic legs, that simulates the experience of flying a real aircraft. Inside the cockpit, the trainee pilot views a screen showing a computer-controlled projection of the view from a real aircraft, and makes appropriate adjustments to the controls. The computer monitors these adjustments, changes both the alignment of the cockpit on its hydraulic legs, and the projected view seen by the pilot. In this way a trainee pilot can progress to quite an advanced stage of training without leaving the ground.

flint compact, hard, brittle mineral (a variety of chert), brown, black, or gray in color, found as nodules in limestone or shale deposits. It consists of cryptocrystalline (grains too small to be visible even under a light microscope) †silica, SiO_2, principally in the crystalline form of †quartz. Implements fashioned from flint were widely used in prehistory.

The best flint, used for Neolithic tools, is **floorstone**, a shiny black flint that occurs deep within chalk. Because of their hardness (7 on the †Mohs scale), flint splinters are used for abrasive purposes and, when ground into powder, added to clay during pottery manufacture. Flints have been used for making fire by striking the flint against steel, which produces a spark, and for discharging guns. Flints in cigarette lighters are made from cerium alloy.

flip-flop in computing, another name for a †bistable circuit.

floating state of equilibrium in which a body rests on or is suspended in the surface of a fluid (liquid or gas). According to †Archimedes' principle, a body wholly or partly immersed in a fluid will be subjected to an upward force, or upthrust, equal in magnitude to the weight of the fluid it has displaced.

If the †density of the body is greater than that of the fluid, then its weight will be greater than the upthrust and it will sink. However, if the body's density is less than that of the fluid, the upthrust will be the greater and the body will be pushed upwards towards the surface. As the body rises above the surface the amount of fluid that it displaces (and therefore the magnitude of the upthrust) decreases. Eventually the upthrust acting on the submerged part of the body will equal the body's weight, equilibrium will be reached, and the body will float.

floating-point notation system in which numbers are represented by means of a decimal fraction and an exponent. For example, in floating-point notation, 123,000,000,000 would be represented as 0.123×10^{12}, where 0.123 is the fraction, or mantissa, and 12 the exponent. The exponent is the power of 10 by which the fraction must be multiplied in order to obtain the true value of the number.

In computing, floating-point notation enables programs to work with very large and very small numbers using only a few digits; however, it is slower than fixed-point notation and suffers from small rounding errors.

flocculation in soils, the artificially induced coupling together of particles to improve aeration and drainage. Clay soils, which have very tiny particles and are difficult to work, are often treated in this way. The method involves adding more lime to the soil.

flooding the inundation of land that is not normally covered with water. Flooding from rivers commonly takes place after heavy rainfall or in the spring after winter snows have melted. The river's †discharge (volume of water carried in a given period) becomes too great, and water spills over the banks onto the surrounding flood plain. Small floods may happen once a year – these are called **annual floods** and are said to have a one-year return period. Much larger floods may occur on average only once every 50 years, or once every 100 years.

Flooding is least likely to occur in an efficient channel that is semicircular in shape (see †channel efficiency). Flooding can also occur at the coast in stormy conditions (see †storm surge) or when there is an exceptionally high tide. The Thames Flood Barrier was constructed in 1982 to prevent the flooding of London from the sea.

Flood! http://www.pbs.org/wgbh/nova/flood/

Companion to the US Public Broadcasting Service (PBS) television program Nova, this page concerns many aspects of flooding. It takes an historical look at floods and examines the measures that engineers have taken to combat them. Three major rivers are discussed: the Yellow, Nile, and Mississippi. In addition to learning about the negative effects of floods, you can also find out about the benefits that floods bestow on farmland. There are many images dispersed throughout the site, plus an audio clip of a flood in progress.

flood plain area of periodic flooding along the course of a river valley. When river discharge exceeds the capacity of the channel, water rises over the channel banks and floods the surrounding low-lying lands. As water spills out of the channel some †alluvium (silty material) will be deposited on the banks to form †levees (raised river banks). This water will slowly seep onto the flood plain, depositing a new layer of rich fertile alluvium as it does so. Many important flood plains, such as the inner Nile delta in Egypt, are in arid areas where their exceptional productivity is very important to the local economy.

A flood plain is a natural feature, flooded at regular intervals. By plotting floods that have occurred we can speak of the size of flood we would expect once every 10 years, 100 years, 500 years, and so on.

Even the most energetic flood-control plans (such as dams, dredging, and channel modification) sometimes fail, and if towns and villages are built on a flood plain there is always some risk. It is wiser to use flood plains in ways that are compatible with flooding, such as for agriculture or parks. Flood-plain features include †meanders and †oxbow lakes.

floppy disk in computing, a storage device consisting of a light, flexible disk enclosed in a cardboard or plastic jacket. The disk is placed in a disk drive, where it rotates at high speed. Data are recorded magnetically on one or both surfaces.

Floppy disks were invented by IBM in 1971 as a means of loading programs into a computer. They were originally 20 cm/8 in in diameter and typically held about 240 kilobytes of data. Present-day floppy disks, widely used on †microcomputers, are 8.8 cm/3.5 in in diameter, and generally hold up to 2 †megabytes, depending on the disk formatting.

Floppy disks are inexpensive, and light enough to send through the post, but have slower access speeds and are more fragile than hard disks. (See also †disk.)

floral diagram diagram showing the arrangement and number of parts in a flower, drawn in cross section. An ovary is drawn in the center, surrounded by representations of the other floral parts, indicating the position of each at its base. If any parts such as the petals or sepals are fused, this is also indicated. Floral diagrams allow the structure of different flowers to be compared, and are usually shown with the floral formula.

floral formula symbolic representation of the structure of a flower. Each kind of floral part is represented by a letter (K for calyx, C for corolla, P for perianth, A for androecium, G for gynoecium) and a number to indicate the quantity of the part present, for example, C5 for a flower with five petals. The number is in brackets if the parts are fused. If the parts are arranged in distinct whorls within the flower, this is shown by two separate figures, such as A5 + 5, indicating two whorls of five stamens each.

A flower with radial symmetry is known as **actinomorphic**; a flower with bilateral symmetry as **zygomorphic**.

floret small flower, usually making up part of a larger, composite flower head. There are often two different types present on one flower head: disc florets in the central area, and ray florets around the edge which usually have a single petal known as the ligule. In the common daisy, for example, the disc florets are yellow, while the ligules are white.

Florey, Howard Walter (1898–1968) Baron Florey, Australian pathologist who was awarded the Nobel Prize for Physiology or Medicine in 1945 with Ernst †Chain for the isolation of penicillin and its development as an antibiotic drug. His research into lysozyme, an antibacterial enzyme discovered by Alexander †Fleming (who shared the prize), led him to study penicillin (another of Fleming's discoveries), which he and Chain isolated and prepared for widespread use.

Flory, Paul John (1910–1985) US polymer chemist. He was awarded the Nobel Prize for Chemistry in 1974 for his investigations of synthetic and natural macromolecules. With Wallace †Carothers, he developed nylon, the first synthetic polyamide, and the synthetic rubber neoprene.

flotation, law of law stating that a floating object displaces its own weight of the fluid in which it floats. See †Archimedes principle.

flow chart diagram, often used in computing, to show the possible paths that data can take through a system or program.

A **system flow chart**, or **data flow chart**, is used to describe the flow of data through a complete data-processing system. Different graphic symbols represent the clerical operations involved and the different input, storage, and output equipment required. Although the flow chart may indicate the specific programs used, no details are given of how the programs process the data.

A **program flow chart** is used to describe the flow of data through a particular computer program, showing the exact sequence of operations performed by that program in order to process the data. Different graphic symbols are used to represent data input and output, decisions, branches, and †subroutines.

flower reproductive unit of an angiosperm (flowering plant), typically consisting of four whorls of modified leaves: †sepals, †petals, †stamens, and †carpels. These are borne on a central axis or †receptacle. The many variations in size, color, number, and arrangement of parts are closely related to the method of pollination. Flowers adapted for wind pollination typically have reduced or absent petals and sepals and long, feathery †stigmas that hang outside the flower to trap airborne pollen. In contrast, the petals of insect-pollinated flowers are usually conspicuous and brightly colored.

structure The sepals and petals form the **calyx** and **corolla** respectively and together comprise the **perianth** with the function of protecting the reproductive organs and attracting pollinators.

flower Cross section of a typical flower showing its basic components: sepals, petals, stamens (anthers and filaments), and carpel (ovary and stigma). Flowers vary greatly in the size, shape, color, and arrangement of these components. Helicon

The stamens lie within the corolla, each having a slender stalk, or filament, bearing the pollen-containing anther at the top. Collectively they are known as the **androecium** (male organs). The inner whorl of the flower comprises the carpels, each usually consisting of an ovary in which are borne the †ovules, and a stigma borne at the top of a slender stalk, or style. Collectively the carpels are known as the **gynoecium** (female organs).

types of flower In size, flowers range from the tiny blooms of duckweeds scarcely visible to the naked eye to the gigantic flowers of the Malaysian *Rafflesia*, which can reach over 1 m/3 ft across. Flowers may grow either individually or in groups called †inflorescences. The stalk of the whole inflorescence is termed a **peduncle**, and the stalk of an individual flower is termed a **pedicel**. A flower is termed hermaphrodite when it contains both male and female reproductive organs. When male and female organs are carried in separate flowers, they are termed **monoecious**; when male and female flowers are on separate plants, the term **dioecious** is used.

flowering plant term generally used for †angiosperms, which bear flowers with various parts, including sepals, petals, stamens, and carpels.

Sometimes the term is used more broadly, to include both angiosperms and †gymnosperms, in which case the †cones of conifers and cycads are referred to as "flowers." Usually, however, the angiosperms and gymnosperms are referred to collectively as †seed plants, or spermatophytes.

The earliest flowering plant identified so far has been dated as being between 125 and 142 million years old, by Chinese paleontologists, in 1998. The fossil *Archefructus* was found in northeast China. In 1996 UK paleontologists found fossils in southern England of *Bevhalstia pebja*, a wetland herb about 25 cm/10 in high, which has been dated as early Cretaceous, about 130 million years old.

flue-gas desulfurization process of removing harmful sulfur pollution from gases emerging from a boiler. Sulfur compounds such as sulfur dioxide are commonly produced by burning †fossil fuels, especially coal in power stations, and are the main cause of †acid rain.

The process is environmentally beneficial but expensive, adding about 10% to the cost of electricity generation.

fluid any substance, either liquid or gas, in which the molecules are relatively mobile and can "flow." A fluid can be "pushed" by applying a force that is transmitted to different parts of the fluid. The †pressure of a fluid increases with the depth of the fluid. In water, an upward force called the upthrust is the liquid pressure that balances the weight of an object in the water. As liquids can transmit pressure, they are used in hydraulic systems.

fluidization making a mass of solid particles act as a fluid by agitation or by gas passing through. Much earthquake damage is attributed to fluidization of surface soils during the earthquake shock. Another example is ash flow during volcanic eruption.

fluidized bed in chemistry, bed of particles through which a fluid is passed with sufficient velocity that the particles separate and are held in suspension. This technique ensures that the maximum surface area of material is in contact with the fluid. Fluidized beds of catalyst are used in the petrochemical industry to break down crude oil into lighter hydrocarbon compounds.

fluid mechanics study of the behavior of fluids (liquids and gases) in motion. Fluid mechanics is important in the study of the weather, the design of aircraft and road vehicles, and in industries, such as the chemical industry, which deal with flowing liquids or gases.

fluid, supercritical fluid brought by a combination of heat and pressure to the point at which, as a near vapor, it combines the properties of a gas and a liquid. Supercritical fluids are used as solvents in chemical processes, such as the extraction of lubricating oil from refinery residues or the decaffeination of coffee, because they avoid the energy-expensive need for phase changes (from liquid to gas and back again) required in conventional distillation processes.

fluke any of various parasitic flatworms of the classes Monogenea and Digenea, that as adults live in and destroy the livers of sheep, cattle, horses, dogs, and humans. Monogenetic flukes can complete their life cycle in one host; digenetic flukes require two or more hosts, for example a snail and a human being, to complete their life cycle.

An estimated 40 million people worldwide are infected by food-borne flukes, mostly from undercooked or raw fish or shellfish, according to a 1994 WHO report.

fluorescein $C_{20}0H_{12}O_5$, in chemistry, red crystalline solid that decomposes at 314–16°C/597–601°F. It dissolves in alkaline solution to produce an intense green fluorescence that is so strong that it can be detected at dilutions of 1:50,000,000. Fluorescein is used as a water tracer, a chemical indicator, and as a liquid dye.

fluorescence short-lived †luminescence (a glow not caused by high temperature). †Phosphorescence lasts a little longer.

Fluorescence is used in strip and other lighting, and was developed rapidly during World War II because it was a more efficient means of illumination than the incandescent lamp. Recently, small bulb-size fluorescence lamps have reached the market. It is claimed that, if widely used, their greater efficiency could reduce demand for electricity. Other important applications are in fluorescent screens for television and cathode-ray tubes.

fluorescence microscopy technique for examining samples under a †microscope without slicing them into thin sections. Instead, fluorescent dyes are introduced into the tissue and used as a light source for imaging purposes. Fluorescent dyes can also be bonded to monoclonal antibodies and used to highlight areas where particular cell proteins occur.

fluoridation addition of small amounts of fluoride salts to drinking water by certain water authorities to help prevent tooth decay. Experiments in Britain, the USA, and elsewhere have indicated that a concentration of fluoride of 1 part per million in tap water retards the decay of children's teeth by more than 50%.

Much concern has been expressed about the risks of medicating the population at large by the addition of fluoride to the water supply, but the medical evidence demonstrates conclusively that there is no risk to the general health from additions of 1 part per million of fluoride to drinking water.

fluoride F^-, negative ion formed when hydrogen fluoride dissolves in water; compound formed between fluorine and another element in which the fluorine is the more electronegative element (see †electronegativity).

In parts of India, the natural level of fluoride in water is 10 parts per million. This causes fluorosis, or chronic fluoride poisoning, mottling teeth and deforming bones.

fluorine chemical symbol F, pale yellow, gaseous, non-metallic element, atomic number 9, relative atomic mass 19. It is the first member of the halogen group of elements, and is pungent, poisonous, and highly reactive, uniting directly with nearly all the elements. It occurs naturally as the minerals fluorite (CaF_2) and cryolite (Na_3AlF_6). Hydrogen fluoride is used in etching glass, and the freons, which all contain fluorine, are widely used as refrigerants.

Fluorine was discovered by the Swedish chemist Karl Scheele in 1771 and isolated by the French chemist Henri Moissan in 1886. Combined with uranium as UF_6, it is used in the separation of uranium isotopes.

The Infrared Space Observatory detected hydrogen fluoride molecules in an interstellar gas cloud in the constellation Sagittarius in 1997. It was the first time fluorine had been detected in space.

Fluorine http://nobel.scas.bcit.ca/resource/ptable/f.htm

Attractive page describing this highly reactive gas, part of a much larger site containing online resources for teachers. Hypertext links show the importance of fluorine in areas such as the development of non-stick coatings and the prevention of tooth decay.

fluorite or **fluorspar**, glassy, brittle halide mineral, calcium fluoride CaF_2, forming cubes and octahedra; colorless when pure, otherwise violet, blue, yellow, brown, or green.

Fluorite is used as a flux in iron and steel making; colorless fluorite is used in the manufacture of microscope lenses. It is also used for the glaze on pottery, and as a source of fluorine in the manufacture of hydrofluoric acid.

fluorocarbon compound formed by replacing the hydrogen atoms of a hydrocarbon with fluorine. Fluorocarbons are used as inert coatings, refrigerants, synthetic resins, and as propellants in aerosols.

There is concern that the release of fluorocarbons – particularly those containing chlorine (chlorofluorocarbons, CFCs) – depletes the †ozone layer, allowing more ultraviolet light from the Sun to penetrate the Earth's atmosphere, and increasing the incidence of skin cancer in humans.

fluorspar another name for the mineral †fluorite.

fluvial of or pertaining to streams or rivers. A **fluvial deposit** is sedimentary material laid down by a stream or river, such as a sandstone or conglomerate (coarse-grained clastic sedimentary rock composed of rounded pebbles of pre-existing rock cemented in a fine-grained sand or clay matrix).

Streamflow and Fluvial Processes

http://www.geog.ouc.bc.ca/physgeog/contents/11i.html

Clear, hyperlinked explanation of streamflow. This site, which is part of a larger university site on physical geography, uses diagrams and photographs to illustrate the various fluvial processes.

fluvioglacial of a process or landform associated with glacial meltwater. Meltwater, flowing beneath or ahead of a glacier, is capable of transporting rocky material and creating a variety of landscape features, including eskers, kames, and outwash plains.

flux in smelting, a substance that combines with the unwanted components of the ore to produce a fusible slag, which can be separated from the molten metal. For example, the mineral fluorite, CaF_2, is used as a flux in iron smelting; it has a low melting point and will form a fusible mixture with substances of higher melting point such as silicates and oxides.

flux in soldering, a substance that improves the bonding properties of solder by removing contamination from metal surfaces and preventing their oxidation, and by reducing the surface tension of the molten solder alloy. For example, with solder made of lead–tin alloys, the flux may be resin, borax, or zinc chloride.

flux density see †magnetic flux density.

fluxions another name for †calculus. It is no longer used.

FM in physics, abbreviation for †**frequency modulation**.

foam in chemistry, continuous dispersion of a large volume of gas in a much smaller volume of liquid. The gas is held in suspension as bubble pockets surrounded by thin sheets of liquid.

focal length or **focal distance**, distance from the center of a lens or curved mirror to the focal point. For a concave mirror or convex lens, it is the distance at which rays of light parallel to the principal axis of the mirror or lens are brought to a focus (for a mirror, this is half the radius of curvature). For a convex mirror or concave lens, it is the

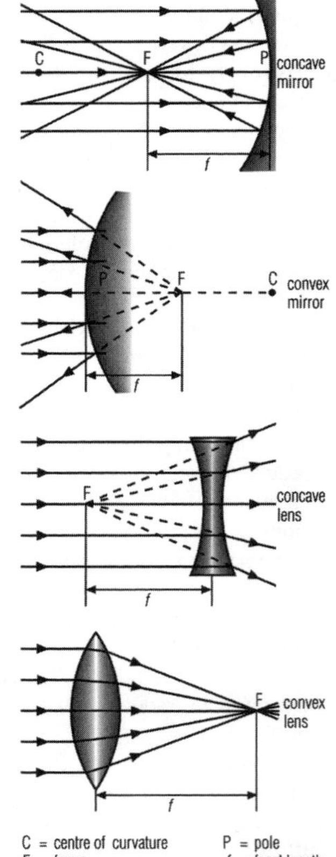

focal length The distance from the pole (P), or optical center, of a lens or spherical mirror to its principal focus (F). The focal length of a spherical mirror is equal to half the radius of curvature ($f = CP/2$). The focal length of a lens is inversely proportional to the power of that lens (the greater the power the shorter the focal length). Helicon distance from the center to the point from which rays of light originally parallel to the principal axis of the mirror or lens diverge after being reflected or refracted.

With lenses, the greater the power (measured in dioptres) of the lens, the shorter its focal length. The human eye has a lens of adjustable focal length to allow the light from objects of varying distance to be focused on the retina.

focus in astronomy, either of two points lying on the major axis of an elliptical †orbit on either side of the center. One focus marks the center of mass of the system and the other is empty. In a circular orbit the two foci coincide at the center of the circle and in a parabolic orbit the second focus lies at infinity. See †Kepler's laws.

focus in earth science, the point within the Earth's crust at which an †earthquake originates. The point on the surface that is immediately above the focus is called the epicenter.

focus or **focal point**, in optics, the point at which light rays converge, or from which they appear to diverge. Other electromagnetic rays, such as microwaves, and sound waves may also be brought together at a focus. Rays parallel to the principal axis of a lens or mirror are converged at, or appear to diverge from, the †principal focus.

fog cloud that collects at the surface of the Earth, composed of water vapor that has condensed on particles of dust in the atmosphere. Cloud and fog are both caused by the air temperature falling below †dew point. The thickness of fog depends on the number of water particles it contains. Officially, fog refers to a condition when visibility is reduced to 1 km/0.6 mi or less, and mist or haze to that giving a visibility of 1–2 km/0.6–1.2 mi.

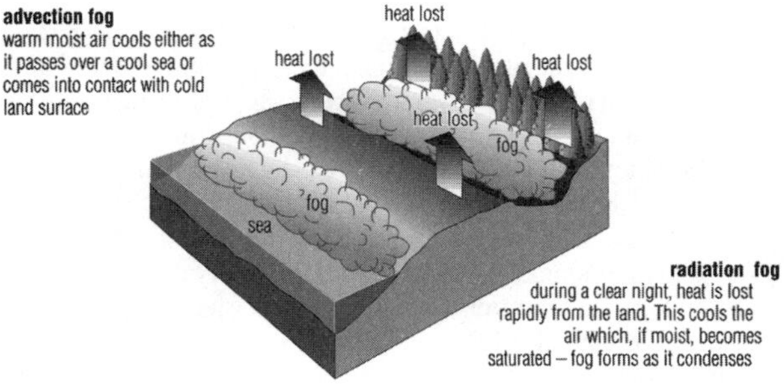

fog Advection fog occurs when two currents of air, one cooler than the other meet, or by warm air flowing over a cold surface. Radiation fog forms through rapid heat loss from the land, causing condensation to take place and a mist to appear. Helicon

There are two types of fog. An **advection fog** is formed by the meeting of two currents of air, one cooler than the other, or by warm air flowing over a cold surface. Sea fogs commonly occur where warm and cold currents meet and the air above them mixes. A **radiation fog** forms on clear, calm nights when the land surface loses heat rapidly (by radiation); the air above is cooled to below its dew point and condensation takes place. A **mist** is produced by condensed water particles, and a haze by smoke or dust.

In some very dry areas, for example Baja California, Canary Islands, Cape Verde Islands, Namib Desert, and parts of Peru and Chile, coastal fogs enable plant and animal life to survive without rain and are a potential source of water for human use (by means of water collectors exploiting the effect of condensation).

Industrial areas uncontrolled by pollution laws have a continual haze of smoke over them, and if the temperature falls suddenly, a dense yellow smog forms.

föhn or **foehn**, warm dry wind that blows down the leeward slopes of mountains.

The air heats up as it descends because of the increase in pressure, and it is dry because all the moisture was dropped on the windward side of the mountain. In the valleys of Switzerland it is regarded as a health hazard, producing migraine and high blood pressure. A similar wind, chinook, is found on the eastern slopes of the Rocky Mountains in North America.

fold in geology, a deformation (bend) in †beds or layers of rock. Folds are caused by pressures within the Earth's crust resulting from †plate-tectonic activity. Rocks are slowly pushed and compressed together, forming folds. Such deformation usually occurs in †sedimentary layers that are softer and more flexible. If the force is more sudden, and the rock more brittle, then a †fault forms instead of a fold.

Folding can result in gentle slopes or mountain chains such as the Rocky Mountains and the Alps. They can be eroded to form escarpments, giving rise to an undulating †topography.

If the bend of the fold is arched up in the middle it is called an **anticline**; if it sags downwards in the middle it is called a **syncline**. The line along which a bed of rock folds is called its axis. The axial plane is the plane joining the axes of successive beds.

fold mountain term no longer used to refer to mountains formed at a convergent margin. See †mountain.

folic acid †vitamin of the B complex. It is found in liver, legumes and green leafy vegetables, and whole grain foods, and is also synthesized by the intestinal bacteria. It is essential for growth, and plays many other roles in the body. Lack of folic acid causes anemia because it is necessary for the synthesis of nucleic acids and the formation of red blood cells.

foliose in botany, covered with leaves or flat and leaflike.

follicle in botany, a dry, usually many-seeded fruit that splits along one side only to release the seeds within. It is derived from a single †carpel. Examples include the fruits of the larkspurs *Delphinium* and columbine *Aquilegia*. It differs from a pod, which always splits open (dehisces) along both sides.

follicle in zoology, a small group of cells that surround and nourish a structure such as a hair (hair follicle) or a cell such as an egg (Graafian follicle; see †menstrual cycle).

follicle-stimulating hormone FSH, †hormone produced by the pituitary gland. It affects the ovaries in women, stimulating the production of an egg cell.

Luteinizing hormone is needed to complete the process. In men, FSH stimulates the testes to produce sperm. It is used to treat some forms of infertility.

Fomalhaut or **Alpha Piscis Austrini**, brightest star in the southern constellation †Piscis Austrinus and the 18th-brightest star in the night sky. It is 25 light years from the Sun, with a true luminosity 13 times that of the Sun.

Fomalhaut is one of a number of stars around which IRAS (the Infra-Red Astronomy Satellite) detected excess infrared radiation, presumed to come from a region of solid particles around the star. This material may be a planetary system in the process of formation.

food anything eaten by human beings and other animals, or absorbed by plants, to sustain life and health. The building blocks of food are nutrients, and humans can utilize the following nutrients: **carbohydrates** as starches found in bread, potatoes, and pasta; as simple sugars in sucrose and honey; and as fibers in cereals, fruit, and vegetables; **proteins** from nuts, fish, meat, eggs, milk, and some vegetables; **fats** as found in most animal products (meat, lard, dairy products, fish), also in margarine, nuts and seeds, olives, and edible oils; **vitamins**, found in a wide variety of foods, except for vitamin B_{12} which is found mainly in foods of animal origin; and **minerals**, found in a wide variety of foods (for example, calcium from milk and broccoli, iodine from seafood, and iron from liver and green vegetables).

From Reactor to Refrigerator http://whyfiles.org/054irradfood/index.html

Part of the Why Files project, published by the National Institute for Science Education (NISE) and funded by the National Science Foundation, this page explains the how, why, and what of food irradiation. The highly readable text, laid out over five pages, includes information about other techniques being offered as means to protect food. Numerous images and diagrams enhance the text throughout. You will also find a comprehensive bibliography of sources for further research.

food chain in ecology, a sequence showing the feeding relationships between organisms in a †habitat or †ecosystem. It shows who eats whom. An organism in one food chain can belong to other food chains. This can be shown in a diagram called a **food web**.

One of the most important aspects of food is that it provides †energy for an organism. So a food chain shows where each organism gets its energy. The arrow in a food chain represents the direction of energy flow. Not all of the energy in all of the organisms at one step of a food chain is available to the organisms later in the chain. In general, fewer organisms are found at each step, or †trophic level, of the chain. A †pyramid of numbers shows this clearly. Some organisms may be small but very numerous, so population size may not be a good measure of how much of an organism there is in a habitat. †Biomass – the total mass of organisms in an area – may be a more useful measure.

food supply availability of food, usually for human consumption. Food supply can be studied at scales ranging from individual households to global patterns. Since the 1940s the industrial and agricultural aspects of food supply have become increasingly globalized. New farming, packaging, and distribution techniques mean that the seasonal aspect of food supply has been reduced in wealthier nations such as the USA and the UK. In some less developed countries there are often problems of food scarcity and distribution caused by climate-related crop failure.

food test any of several types of simple test, easily performed in the laboratory, used to identify the main classes of food.

starch–iodine test Food is ground up in distilled water and iodine solution is added. A dense black color indicates that starch is present.

sugar–Benedict's test Food is ground up in distilled water and placed in a test tube with Benedict's reagent. The tube is then heated in a boiling water bath. If glucose is present the color changes from blue to brick-red.

protein–Biuret test Food is ground up in distilled water and a mixture of copper(II) sulfate and sodium hydroxide is added. If protein is present a mauve color is seen.

food web interconnected and often elaborate mesh of feeding relationships (interconnecting †food chains), within an ecosystem.

foot in geometry, point where a line meets a second line to which it is perpendicular.

foot symbol ft, imperial unit of length, equivalent to 0.3048 m, in use in Britain since Anglo-Saxon times. It originally represented the length of a human foot. One foot contains 12 inches and is one-third of a yard.

foot-and-mouth disease contagious eruptive viral disease of cloven-hoofed mammals, characterized by blisters in the mouth and around the hooves. In cattle it causes deterioration of milk yield and abortions. It is an airborne virus, which makes its eradication extremely difficult.

In the UK, affected herds are destroyed; inoculation is practiced in Europe and the USA. The existing vaccine for the disease leaves vaccinated animals as carriers that can infect those animals that remain unvaccinated, and a more effective vaccine is under development in the USA.

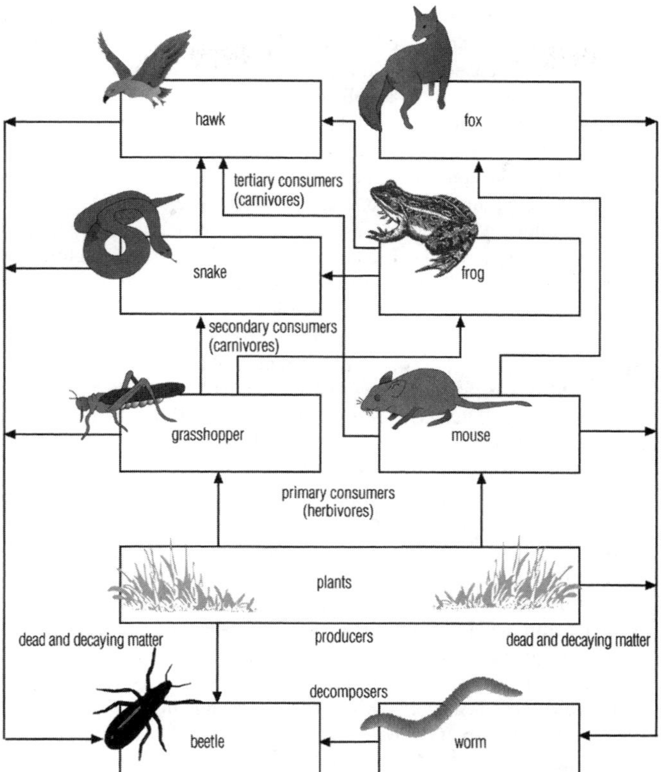

food web The complex interrelationships between animals and plants in a food web. A food web shows how different food chains are linked in an ecosystem. Note that the arrows indicate movement of energy through the web. For example, an arrow shows that energy moves from plants to the grasshopper, which eats the plants. Helicon

Despite health scares, a report in 2001 stated that there had only been 40 confirmed cases of foot-and-mouth disease in humans worldwide.

In February 2001, an outbreak of foot-and-mouth spread rapidly across the UK. By March, cases were discovered in France, Ireland, and the Netherlands. Severe restrictions on the movement of livestock were imposed across Europe; many countries banned imports of British and French meat, and some banned imports from all European Union countries. By the end of June 2001, over 5 million animals in the UK had been slaughtered due to the foot-and-mouth outbreak. The outbreak was the worst in world history.

foot-candle unit of illuminance, now replaced by the †lux. One foot-candle is the illumination received at a distance of one foot from an international candle. It is equal to 10.764 lux.

foot-pound imperial unit of energy (ft-lb), defined as the work done when a force of one pound moves through a distance of one foot. It has been superseded for scientific work by the joule: one foot-pound equals 1.356 joule.

footprint of a satellite or spacecraft, area of the Earth or other body that is being imaged by its instruments, or over which its signals can be received.

Forbes, Edward (1815–1854) British naturalist. He studied mollusks and made significant contributions to oceanography. In palaeobotany, he divided British plants into five groups, and proposed that Britain had once been joined to the continent by a land bridge.

forbidden line in astronomy, emission line seen in the spectra of certain astronomical objects that are not visible under the conditions prevailing in laboratory experiments. They indicate that the hot gas emitting them is at extremely low density. Forbidden lines are seen, for example, in the tenuous gas of the Sun's †corona, in †HII regions (regions of hot, ionized hydrogen), and in the centers of certain active galaxies.

force any influence that tends to change the state of rest of a body its uniform motion in a straight line. The action of an unbalanced or resultant force results in the acceleration of a body in the direction of action of the force, or it may, if the body is unable to move freely, result in its deformation (see †Hooke's law). A force is a push or a pull on an object. A force will cause an object to move if it is stationary, change direction, slow down, or speed up. Force is a vector quantity, possessing both magnitude and direction; its †SI unit is the newton.

Forces that make contact with an object are called contact forces. Examples are normal push/pull force, surface tension, air resistance, and frictional forces. Forces that are able to cause a push/pull without making contact with an object are called non-contact forces. Examples are magnetic force, †gravity, and electric force. A newton is used to measure how much pull the force of gravity has on an object; this corresponds to the †weight of an object. A mass of 1 kg = 9.8 newtons (N). A newton is defined as the amount of force needed to move an object of 1 kg so that it accelerates at 1 meter per second per second.

speed and distance In order to understand movement and what causes it, we need to be able to describe it. †Speed is a measure of how fast something is moving. Speed is measured by dividing the distance traveled by the time taken to travel that distance. Hence speed is distance moved in unit time. Speed is a †scalar quantity in which the direction of travel is not important, only the rate of travel. It is often useful to represent motion using a graph. Plotting distance against time in a †distance–time graph enables one to calculate the total distance traveled. The gradient of the graph represents the speed at a particular point, the instantaneous speed. A straight line on the distance–time graph corresponds to a constant speed. A form of this graph that shows the stages on the journey is called a †travel graph. A †speed–time graph plots speed against time. It shows the instantaneous speed at each point. A horizontal line on the speed–time graph corresponds to a constant speed.

velocity and acceleration †Velocity is the speed of an object in a given direction. Velocity is therefore a †vector quantity, in which both magnitude and direction of movement must be taken into account. †Acceleration is the rate of change of velocity with time. This is also a vector quantity. Acceleration happens when there is a change in speed, or a change in direction, or a change in speed and direction.

forces and motion †Galileo discovered that a body moving on a perfectly smooth horizontal surface would neither speed up nor slow down. All moving bodies continue moving with the same velocity unless a force is applied to cause an acceleration. The reason we appear to have to push something to keep it moving with constant velocity is because of frictional forces acting on all moving objects on Earth. †Friction occurs when two solid surfaces rub on each other; for example, a car tire in contact with the ground. Friction opposes the relative motion of the two objects in contact and acts to slow the velocity of the moving object. A force is required to push the moving object and to cancel out the frictional force. If the forces combine to give a net force of zero, the object will not accelerate but will continue moving at constant velocity. A †resultant force is a single force acting on a particle or body whose effect is equivalent to the combined effects of two or more separate forces. Galileo's work was developed by Isaac †Newton. According to Newton's second law of motion, the magnitude of a resultant force is equal to the rate of change of †momentum of the body on which it acts; the force F producing an acceleration a meters per second per second on a body of mass m kilograms is therefore given by: $F = ma$. Thus Newton's second law states that change of momentum is proportional to the size of the external force and takes place in the direction in which the force acts. Momentum is a function both of the mass of a body and of its velocity. This agrees with our experience, because the idea of force is derived from muscular effort, and we know that we have to exert more strength to stop the motion of a heavy body than a light one, just as we have to exert more strength to stop a rapidly moving body than a slowly moving one. Force, then, is measured by change of momentum, momentum being equal to mass multiplied by velocity. (See also †Newton's laws of motion.) Newton's third law of motion states that if a body A exerts a force on a body B, then body B exerts an equal force on body A but in the opposite direction. This equal and opposite force is called a †reaction force.

free fall and terminal velocity Galileo also established that freely falling bodies, heavy or light, have the same, constant acceleration and that this acceleration is due to †gravity. This acceleration, due to the gravitational force exerted by the Earth, is also known as the acceleration of free fall. It has a value of 9.806 meters per second per second/32.174 feet per second per second. However, air resistance acts when an object falls through the air. This increases greatly as the object's velocity increases, so the object tends to reach a †terminal velocity. It then continues to fall with this same velocity (it has stopped accelerating because its †weight is cancelled out by air resistance) until it reaches the ground. The acceleration due to gravity can be measured using a pendulum.

statics Statics is the branch of mechanics concerned with the behavior of bodies at rest or moving with constant velocity. The forces acting on the body under these circumstances cancel each other out; that is, the forces are in equilibrium.

Air Resistance http://www.curtin.edu.au/curtin/dept/phys-sci/gravity/intro/intro2.htm

Introduction to the topic of air resistance, which presents the basic concepts in a series of stages. This site also features an interactive quiz to test the knowledge of the user.

Animations for Physics http://phys23p.sl.psu.edu/~mrg3/agif/

Simple site with animated GIFs demonstrating different types of acceleration. The user can view one-dimensional positive and negative acceleration, two-dimensional constant acceleration plus uniform circular motion, and more. Each type of motion is demonstrated visually and described.

Balanced and Unbalanced Forces http://www.glenbrook.k12.il.us/gbssci/phys/Class/newtlaws/u2l1d.html

Detailed document explaining exactly what can be defined as a balanced or unbalanced force. It contains numerous diagrams to help illustrate points made in the text. A self-test section is provided, with the facility for checking answers. There are also links to other physics-related topics on the site.

Motion and Force http://library.thinkquest.org/12632/motion/

Clearly explained site covering speed, velocity, acceleration, inertia, momentum, friction, and more. It is a good overview of forces and motion.

force, centrifugal see †centrifugal force.

force, centripetal see †centripetal force.

force, electromagnetic see †electromagnetic force.

force, electromotive see †electromotive force.

force multiplier machine designed to multiply a small effort in order to move a larger load. The ratio by which a machine multiplies the effort is called its †mechanical advantage. Examples of a force multiplier include crowbar, wheelbarrow, nutcrackers, and bottle opener.

force ratio magnification of a force by a machine; see †mechanical advantage.

forces, fundamental in physics, four fundamental interactions currently known to be at work in the physical universe. There are two long-range forces: the **gravitational force**, or **gravity**, which keeps the planets in orbit around the Sun and acts between all particles that have mass; and the **electromagnetic force**, which stops solids from falling apart and acts between all particles with †electric charge. There are two very short-range forces, which operate over distances comparable with the size of the atomic nucleus: the **weak nuclear force**, responsible for the reactions that fuel the Sun and for the emission of †beta particles by some particles; and the **strong nuclear force**, which binds together the protons and neutrons in the nuclei of atoms. The relative strengths of the four forces are: strong, 1; electromagnetic, 10^{-2}; weak, 10^{-6}; gravitational, 10^{-40}.

By 1971, the US physicists Steven Weinberg and Sheldon Glashow, the Pakistani physicist Abdus Salam, and others had developed a theory that suggested that the weak and electromagnetic forces were aspects of a single force called the **electroweak force**; experimental support came from observation at the European particle-physics laboratory †CERN in the 1980s. Physicists are now working on theories to unify all four forces. See †supersymmetry.

foredeep in earth science, an elongated structural basin lying inland from an active mountain system and receiving sediment from the rising mountains. According to plate tectonic theory, a mountain chain forming behind a subduction zone along a continental margin develops a foredeep or gently sloping trough parallel to it on the landward side. Foredeeps form rapidly and are usually so deep initially that the sea floods them through gaps in the mountain range. As the mountain system evolves, sediments choke the foredeep, pushing out marine water. As marine sedimentation stops, only nonmarine deposits from the rapidly eroding mountains are formed. These consist of alluvial fans and also rivers, flood plains, and related environments inland.

Before the advent of plate tectonic theory, such foredeep deposits and changes in sediments had been interpreted as sedimentary troughs, called geosynclines, that were supposed ultimately to build upward into mountains.

forensic entomology branch of †forensic science, involving the study of insects on and around the corpse. Insects rapidly infest a corpse, and do so in an accepted sequence beginning with flies laying eggs. Further insects follow to feed on the decomposing flesh and fly maggots. Forensic entomologists are able to determine time of death by analyzing insect colonization. They can also tell whether or not a corpse has been moved by examining the faunal community in the "seepage area" beneath the body.

Forensic Entomology Pages http://folk.uio.no/mostarke/forens_ent/forensic_entomology.html

Guide to the science of forensic entomology – studying insects found on the corpse to assist in identifying the cause and time of death. This information is supported by case studies.

forensic science use of scientific techniques to solve criminal cases. A multidisciplinary field embracing chemistry, physics, botany, zoology, and medicine, forensic science includes the identification of human bodies or traces. Ballistics (the study of projectiles, such as bullets), another traditional forensic field, makes use of such tools as the comparison microscope and the electron microscope.

Traditional methods such as fingerprinting are still used, assisted by computers; in addition, blood analysis, forensic dentistry, voice and speech spectrograms, and †genetic fingerprinting are increasingly applied. Chemicals, such as poisons and drugs, are analyzed by †chromatography. ESDA (electrostatic document analysis) is a technique used for revealing indentations on paper, which helps determine if documents have been tampered with. †Forensic entomology is also a branch of forensic science.

Forensic Science Web Page http://home.earthlink.net/~thekeither/Forensic/forsone.htm

Aimed at the layperson, this site explains the basics of forensic science. It is broken up into a series of articles, each explaining a different discipline. Topics covered include firearms and toolmark identification, forensic psychiatry, and crime-scene processing. Each article includes a number of diagrams in order to make the science more easily understandable.

forest area where trees have grown naturally for centuries, instead of being logged at maturity (about 150–200 years). A natural, or old-growth, forest has a multistorey canopy and includes young and very old trees (this gives the canopy its range of heights). There are also fallen trees contributing to the very complex ecosystem, which may support more than 150 species of mammals and many thousands of species of insects. Globally forest is estimated to have covered around 68 million sq km/26.25 million sq mi during prehistoric times. By the late 1990s this is believed to have been reduced by half to 34.1 million sq km/13.2 million sq mi.

The Pacific forest of the west coast of North America is one of the few remaining old-growth forests in the temperate zone. It consists mainly of conifers and is threatened by logging – less than 10% of the original forest remains.

Fire Effects Information http://www.fs.fed.us/database/feis/

Developed at the USDA Forest Service Intermountain Research Station's Fire Sciences Laboratory in Missoula, Montana, this site provides up-to-date information on the effects of fire on plants and animals.

forestry science of forest management. Recommended forestry practice aims at multipurpose crops, allowing the preservation of varied plant and animal species as well as human uses (lumbering, recreation). Forestry has often been confined to the planting of a single species, such as a rapid-growing conifer providing softwood for paper pulp and construction timber, for which world demand is greatest. In tropical countries, logging contributes to the destruction of †rainforests, causing global environmental problems. Small unplanned forests are †woodland.

The earliest planned forest dates from 1368 at Nuremberg, Germany; in Britain, planning of forests began in the 16th century. In the UK, Japan, and other countries, forestry practices have been criticized for concentration on softwood conifers to the neglect of native hardwoods.

formaldehyde common name for †methanal.

formalin aqueous solution of formaldehyde (methanal) used to preserve animal specimens.

formatting in computing, short for †disk formatting (laying down a structure for organizing, saving, and retrieving data), or text formatting (changing its appearance). Modern office programs and †desktop publishing packages also allow the formatting of objects such as pictures and frames.

formic acid common name for †methanoic acid.

formula in chemistry, a representation of a molecule, radical, or ion, in which the component chemical elements are represented by their symbols. For example, the formula for carbon dioxide is CO_2, showing that a molecule of carbon dioxide consists of one atom of carbon (C) and two atoms of oxygen (O_2). An **empirical formula** indicates the simplest ratio of the elements in a compound, without indicating how many of them there are or how they are combined. A **molecular formula** gives the number of each type of element present in one molecule. A **structural formula** shows the relative positions of the atoms and the bonds between them. For example, for ethanoic (acetic) acid, the empirical formula is CH_2O, the molecular formula is $C_2H_4O_2$, and the structural formula is CH_3COOH.

formula in mathematics, a set of symbols and numbers that expresses a fact or rule. For example, $A = \pi r^2$ is the formula for calculating the area of a circle. $E = mc^2$ is Einstein's famous formula relating energy and mass. Other common formulae exist for †density, †mass, †volume, and †area.

Fornax inconspicuous constellation of the southern hemisphere, represented as a furnace.

Fortin, Jean Nicolas (1750–1831) French physicist and instrument maker who invented a portable mercury †barometer in 1800. Any barometer in which the mercury level can be adjusted to zero is now known as a Fortin barometer.

FORTRAN or **fortran**; contraction of formula translation, high-level computer-programming language suited to mathematical and scientific computations. Developed by John Backus at IBM in 1956, it is one of the earliest computer languages still in use. †BASIC was strongly influenced by FORTRAN and is similar in many ways.

Fortran FAQ http://www.faqs.org/faqs/fortran-faq/

Comprehensive text-only FORTRAN "Frequently Asked Questions" list. It provides background information on FORTRAN and its evolution, as well as listing FORTRAN books and Internet resources, and answering technical questions.

Fortuna large dark-surfaced asteroid 200 km/124 mi across, the ninth largest in the Solar System.

Fossey, Dian (1932–1985) US zoologist. Almost completely untrained, Fossey was sent by Louis Leakey into the African wild. From 1975, she studied mountain gorillas in Rwanda and discovered that they committed infanticide and that females transferred to nearly established groups but that gorillas led peaceful family lives. Fossey was murdered by poachers whose snares she had cut, after having spent 18 years living in close proximity to the gorillas and fighting for their conservation.

Fossey, Dian http://kroeber.anthro.mankato.msus.edu/information/biography/fghij/fossey_dian.html

Profile of the life and death of the US zoologist. It traces how the young occupational therapist developed an interest in gorillas, taught herself biology, was the first to live intimately with gorillas, went on to establish academic credentials, and was able to highlight the plight of gorillas.

fossil (Latin *fossilis* "dug up") cast, impression, or the actual remains of an animal or plant preserved in rock. Dead animals and plant remains that fell to the bottom of the sea bed or an inland lake were gradually buried under the accumulation of layers of sediment. Over millions of years, the sediment became †sedimentary rock and the remains preserved within the rock became fossilized. Fossils may include footprints, an internal cast, or external impression. A few fossils are preserved intact, as with mammoths fossilized in Siberian ice, or insects trapped in tree resin that is today amber. The study of fossils is called †paleontology. Paleontologists are able to deduce much of the geological history of a region from fossil remains. The existence of fossils is key evidence that organisms have changed with time, that is, evolved (see †evolution).

About 250,000 fossil species have been discovered – a figure that is believed to represent less than 1 in 20,000 of the species that ever lived. **Microfossils** are so small they can only be seen with a microscope. They include the fossils of pollen, bone fragments, bacteria, and the remains of microscopic marine animals and plants, such as foraminifera and diatoms.

Dinosaur Trace Fossils http://www.emory.edu/COLLEGE/ENVS/research/ichnology/dinotraces.html

All about the trace fossils left by dinosaurs. The site is divided into several categories of dinosaur fossils: tracks, eggs and nests, tooth marks, gastroliths, and coprolites, and also includes images and descriptions of each one.

Fossil Collections of the World http://www.geocities.com/CapeCanaveral/Lab/8147/index.html

Comprehensive pages of information about fossils and fossil sites around the world. This is where to find out about geological timescales, the history of fossil collecting, famous fossil collections and collectors, and maps of sites. There are also links to numerous other related sites.

fossil fuel combustible material, such as coal, lignite, oil, †peat, and †natural gas, formed from the fossilized remains of plants that lived hundreds of millions of years ago. Such fuels are †non-renewable resources – once they are burnt, they cannot be replaced.

Fossil fuels are hydrocarbons (they contain atoms of carbon and hydrogen). They generate large quantities of heat when they burn in air, a process known as combustion. In this process carbon and hydrogen combine with oxygen in the air to form carbon dioxide, water vapor, and heat.

Fossil Fuels and their Utilization http://www.ibis.sfu.ca/~rhlogan/fossil.html

Very detailed information on fossil fuels, including information on the evolutionary stages of coal and the disadvantages of using coal. There are links to a "Fossil fuel center" and an "Alternative energy menu."

What Are Fossil Fuels? http://members.aol.com/profchm/fossil.html

Information on what fossil fuels are, and how they are used in society. The site also includes an explanation of how acid rain forms, and the "greenhouse effect."

Foucault, (Jean Bernard) Léon (1819–1868) French physicist who used a pendulum to demonstrate the rotation of the Earth on its axis, and invented the †gyroscope in 1852. In 1862 he made the first accurate determination of the velocity of light.

Foucault pendulum pendulum consisting of a long wire to which is attached a heavy weight, which is then free to swing in any plane. Once it starts swinging in one particular plane, the plane slowly rotates, this being the result of the rotation of the Earth beneath the pendulum. This phenomenon was observed by French physicist Léon Foucault in 1851.

foundations of mathematics or **foundations of arithmetic**, subject of attempts to derive the basic precepts of elementary mathematics from a standpoint of pure logic (thence to derive the more complex principles).

four-color theorem in topology, hypothesis that given any map of regions on a plane or sphere, four colors are sufficient to color the map so that any two regions sharing a boundary are colored differently. It was finally proved in 1976 after being an open question since 1852, when it was first conjectured.

Fourcroy, Antoine Francois, Comte de (1755–1809) French chemist. Fourcroy was one of the earliest converts to Lavoisier's theories, and he prepared *Methode de nomenclature chimique* (1787) with Lavoisier, Berthollet, and Morveau. Fourcry was involved with the introduction of the metric system and he organized the Ecole polytechnique, as well as instituting several medical schools.

Fourier, (Jean Baptiste) Joseph (1768–1830) French applied mathematician whose formulation of heat flow in 1807 contains the proposal that, with certain constraints, any mathematical function can be represented by trigonometrical series. This principle forms the basis of **Fourier analysis**, used today in many different fields of physics. His idea, not immediately well received, gained currency and is embodied in his *Théorie analytique de la chaleur/The Analytical Theory of Heat* (1822).

Fourier, Joseph http://www-history.mcs.st-and.ac.uk/history/Mathematicians/Fourier.html

Extensive biography of the French mathematician Joseph Fourier. The site contains a description of his contributions to mathematics, including his work on the mathematical theory of heat conduction, which led to his discovery of Fourier analysis. Several references for further reading are also listed, and the Web site also features a portrait of Fourier.

Fourier analysis see †Fourier theorem.

Fourier series series in which the terms comprise multiples of the cosine and/or sine of multiple angles. Represented by the formula:

$1/2a_0 + \Sigma(a_n \cos nx + b_n \sin nx$

it is used to analyze periodic functions (that is, functions whose graphs repeat themselves periodically).

Fourier theorem adaptation of the process developed by French mathematician Joseph Fourier as the †Fourier series to the investigation of energy propagated in the form of waves (particularly heat, sound, and light). A further developed version of this method is known as harmonic analysis. Use of the theorem in investigating wave forms is known as Fourier analysis. It was devised by French mathematician Joseph †Fourier.

four-stroke cycle engine-operating cycle of most gasoline and †diesel engines. The "stroke" is an upward or downward movement of a piston in a cylinder. In a gasoline engine the cycle begins with the induction of a fuel mixture as the piston goes down on its first stroke. On the second stroke (up) the piston compresses the mixture in the top of the cylinder. An electric spark then ignites the mixture, and the gases produced force the piston down on its third, power, stroke. On the fourth stroke (up) the piston expels the burned gases from the cylinder into the exhaust.

fourth-generation language in computing, a type of programming language designed for the rapid programming of †applications but often lacking the ability to control the individual parts of the computer. Such a language typically provides easy ways of designing screens and reports, and of using databases. Other "generations" (the term implies a class of language rather than a chronological sequence) are †machine code (first generation); †assembly languages, or low-level languages (second); and conventional high-level languages such as †BASIC and †Pascal (third).

Fowler, William Alfred (1911–1995) US astrophysicist. He shared the Nobel Prize for Physics in 1983 with Indian-born US astrophysicist Subrahmanyan †Chandrasekhar for their work on the life cycle of stars and the nuclear reactions involved in the formation of chemical elements in the universe.

f.p.s. system system of units based on the foot, pound, and second as units of length, mass, and time, respectively. It has now been replaced for scientific work by the †SI system.

fractal (from Latin *fractus* "broken") irregular shape or surface produced by a procedure of repeated subdivision. Generated on a computer screen, fractals are used in creating models of geographical or biological processes (for example, the creation of a coastline by erosion or accretion, or the growth of plants).

Sets of curves with such discordant properties were developed in the 19th century in Germany by Georg †Cantor and Karl †Weierstrass. The name was coined by the French mathematician Benoit Mandelbrot. Fractals are also used for computer art.

fraction in chemistry, a group of similar compounds, the boiling points of which fall within a particular range and which are separated during †fractional distillation (fractionation).

fraction (from Latin *fractus* "broken") in mathematics, a number that indicates one or more equal parts of a whole. Usually, the number of equal parts into which the unit is divided (†denominator) is written below a horizontal or diagonal line, and the number of parts comprising the fraction (†numerator) is written above; for example, $^2/_3$ has numerator 2 and denominator 3. Such fractions are called †vulgar fractions or **simple fractions**. The denominator can never be zero.

proper and improper fractions A †proper fraction is one in which the numerator is less than the denominator. For example, $^2/_5$, $^3/_5$, and $^7/_8$ are all proper fractions.

An †improper (or top heavy) fraction has a numerator that is larger than the denominator. For example:
$^{14}/_5$ means $14 \div 5$
$^{14}/_5 = 2^4/_5$
This is called a †mixed number.

denominator of zero A combination such as $^5/_0$ is not regarded as a fraction (an object cannot be divided into zero equal parts). Zero divided by any number is zero, and any number divided by zero is infinity.

decimal fractions A †decimal fraction has as its denominator a power of 10. Exact †decimals can be changed into fractions using place values. For example:

$$5\overline{)14} \quad 2\text{r}4 \quad \text{or} \quad 2^4/_5$$

0.37 means 3 tenths and 7 hundredths
0.37 is the †equivalent to $^{37}/_{100}$ as a fraction and 37% as a †percentage.

Most fractions can be expressed exactly as decimal fractions ($^1/_3 = 0.333...$). Fractions are also known as †rational numbers; that is, numbers formed by a ratio. †Integers may be expressed as fractions with a denominator of 1, so 6 is $^6/_1$, for example.

addition and subtraction To add or subtract with fractions a †common denominator (a number divisible by both the bottom numbers) needs to be identified. For example:
$^3/_4 + ^5/_6$
First both denominators should be the same. 12 is the lowest number of which both 4 and 6 are factors – it is the lowest common denominator. To change $^3/_4$ into twelfths the denominator is multiplied by 3. The numerator must also be multiplied by 3:
To change $^5/_6$ into twelfths the denominator is multiplied by 2. The numerator must also be multiplied by 2:
$^3/_4 + ^5/_6 = ^9/_{12} + ^{10}/_{12} = ^{19}/_{12} = 1^7/_{12}$
If whole numbers appear in the calculation they can be added/subtracted separately first.

multiplication and division All whole numbers in a division or multiplication calculation must first be converted into improper fractions. For multiplication, the numerators are then multiplied together and the denominators are then multiplied to provide the solution. For example:
$7^2/_3 \times 4^1/_2 = ^{23}/_3 \times ^9/_2 = ^{207}/_6 = 34^1/_2$
In division, the procedure is similar, but the second fraction must be inverted before multiplication occurs. For example,
$5^5/_{12} \div 1^1/_8 = ^{65}/_{12} \div ^9/_8 = ^{65}/_{12} \times ^8/_9 = ^{520}/_{108} = 4^{22}/_{27}$

fraction of an amount For example:
to find $\frac{1}{2}$ of £300 divide by 2 to get £150
to find $\frac{1}{2}$ of 250 m divide by 5 to get 50 m

Fractions http://www.mathleague.com/help/fractions/fractions.htm
Simple but clearly laid out introduction to fractions and decimals. It covers topics such as improper fractions, converting fractions to decimals, and adding and multiplying fractions. The site would be helpful for learning about fractions for the first time, or just for revision.

fractional distillation or **fractionation**, process used to split complex †mixtures (such as †petroleum) into their components, usually by repeated heating, boiling, and condensation; see †distillation. In the laboratory it is carried out using a †fractionating column.

Fractional distillation is used to separate mixtures of miscible liquids, such as ethanol and water. The process depends on the components of the mixture having different boiling points. The liquid is heated so that it turns into a gas. The vapors pass up a fractionating column where they are gradually cooled. As each of the components of the mixture cools to its boiling point, it turns back into a liquid. The different components of the mixture condense at different levels in the fractionating column and thus may be separated.

In industry, fractional distillation is used to separate the compounds in crude oil (unrefined petroleum) into useful fractions, each fraction containing compounds with similar boiling points. Air is also separated by fractional distillation. This is done by cooling air until it condenses and then allowing the temperature of the liquid air to rise. Each gas will distill off at its own boiling point.

fractionating column device in which many separate †distillations can occur so that a liquid mixture can be separated into its components. The technique is known as †fractional distillation, or fractionation. Various designs exist but the primary aim is to allow maximum contact between the hot rising vapors and the cooling descending liquid. As the mixture of vapors ascends the column it becomes progressively enriched in the lower-boiling-point components, so these separate out first.

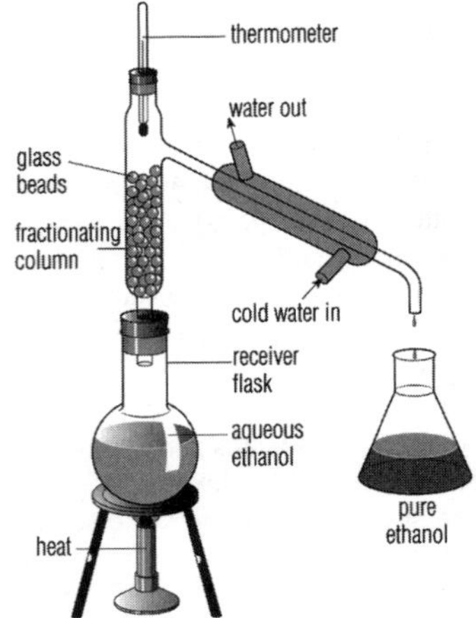

fractionating column Laboratory apparatus for fractional distillation. Fractional distillation is the main means of separating the components of crude oil. Helicon

frame of reference set of axes fixed in such a way as to define uniquely the position of an object in space.

francium chemical symbol Fr, radioactive metallic element, atomic number 87, relative atomic mass 223. It is one of the alkali metals and occurs in nature in small amounts as a decay product of actinium. Its longest-lived isotope has a half-life of only 21 minutes. Francium was discovered and named in 1939 by Marguérite †Perey, to honor her country.

Franck, James (1882–1964) German-born US physicist. He shared the Nobel Prize for Physics in 1925 with his co-worker Gustav †Hertz for their experiments of 1914 on the energy transferred by colliding electrons to mercury atoms, showing that the transfer was governed by the rules of †quantum theory.

Frank, Charles (1911–1998) English physicist who explained the primary mechanism by which crystals grow and who contributed to liquid-crystal and alloy research. He also made contributions to geophysics and polymer physics. He was knighted in 1977.

Frank, Ilya Mikhailoivich (1908–1990) Russian physicist who was known for his work on radiation. In 1934 Pavel†Cherenkov had noted a peculiar blue radiation sometimes emitted as electrons passed through water.

It was left to Frank and his colleague at Moscow University, Igor †Tamm, to realize that this form of radiation was produced by charged particles traveling faster through the medium than the speed of light in the same medium.

Frank shared the Nobel Prize for Physics in 1958 with Cherenkov and Tamm for their discovery and interpretation of Cherenkov radiation.

Frankland, Edward (1825–1899) British chemist. Frankland established the important theory of valency in 1852. Together with Norman Lockyer he discovered the element helium, and concluded that the sun's external layers are gaseous and vaporous. A professor of chemistry at Owens College, Manchester, and the Royal Institute, he received the Royal Society's highest honor – the Copley Medal – in 1897. He was knighted in 1897.

Franklin, Benjamin (1706–1790) US scientist, inventor, statesman, diplomat, writer, printer, and publisher. He proved that lightning is a form of electricity, distinguished between positive and negative electricity, and invented the lightning conductor. He was the first US ambassador to France 1776–85, and negotiated peace with Britain in 1783. As a delegate to the Continental Congress from Pennsylvania 1785–88, he helped to draft the Declaration of Independence and the US Constitution. He was president of the first US abolitionist society in 1775 and was responsible for many improvements in American life, including a modernized postal system, and the first US fire and police departments, hospital, and insurance company.

World of Benjamin Franklin http://sln.fi.edu/franklin/rotten.html

Life and times of Benjamin Franklin. The site includes exhaustive biographical details, a family tree of Franklin and his family, and comprehensive information on his achievements in the many fields in which he was prominent. There is a video clip on the front page and certain other pages have audio clips attached to the articles.

Franklin, Rosalind Elsie (1920–1958) English biophysicist whose research on †X-ray diffraction of †DNA crystals helped Francis †Crick and James D †Watson to deduce the chemical structure of DNA.

Frasch process process used to extract underground deposits of sulfur. Superheated steam is piped into the sulfur deposit and melts it. Compressed air is then pumped down to force the molten sulfur to the surface. The process was developed in the USA in 1891 by German-born Herman Frasch (1851–1914).

Fraunhofer, Joseph von (1787–1826) German physicist who did important work in optics. The dark lines in the solar spectrum (**Fraunhofer lines**), which reveal the chemical composition of the Sun's atmosphere, were accurately mapped by him.

Fraunhofer lines dark lines crossing the solar †spectrum. They are caused by the absorption of light from hot regions of the Sun's surface by gases in the cooler, outer regions. They were first investigated by German physicist Joseph von Fraunhofer.

free association in psychoanalysis, a therapeutic technique developed by Sigmund †Freud in which the patient is encouraged to repeat whatever comes to mind without reservation. Freud claimed that, in the patient's unforced statements, unconscious wishes were unwittingly revealed, allowing analysis to proceed where the resistance or †repression usually manifest in personal disclosure would otherwise interfere.

The process of free association, which is still widely used in psychoanalysis today, is generally facilitated by the nonemotive prompting of the analyst.

free energy sum of †enthalpy and †entropy, that is, the capacity of a system to perform work. The change in free energy accompanying a chemical reaction is a measure of its completeness.

free radical in chemistry, an atom or molecule that has an unpaired electron and is therefore highly reactive. Most free radicals are very short-lived. They are by-products of normal cell chemistry and rapidly oxidize other molecules they encounter. Free radicals are thought to do considerable damage. They are neutralized by protective enzymes.

Free radicals are often produced by high temperatures and are found in flames and explosions.

freeware in computing, free software which may or may not be in the public domain (see †public-domain software). The best-known examples of freeware are the encryption program Pretty Good Privacy (PGP), which is still available free of charge for personal, non-commercial use, and the Web browser Microsoft Internet Explorer.

freeze–thaw form of physical †weathering, common in mountains and glacial environments, caused by the expansion of water as it freezes. Water in a crack freezes and expands in volume by 9% as it turns to ice. This expansion exerts great pressure on the rock, causing the crack to enlarge. After many cycles of freeze–thaw, rock fragments may break off to form †scree slopes.

For freeze–thaw to operate effectively the temperature must fluctuate regularly above and below 0°C/32°F. It is therefore uncommon in areas of extreme and perpetual cold, such as the polar regions.

freezing †change of state from liquid to solid, as when water becomes ice. For a given substance, freezing occurs at a definite temperature, known as the †**freezing point**, that is invariable under similar conditions of pressure, and the temperature remains at this point until all the liquid is frozen; the freezing point and melting point of the substance are the same temperature. By measuring the temperature of a liquid against time as it cools a cooling curve can be plotted; on the cooling curve the temperature levels out at the freezing point.

The amount of heat per unit mass that has to be removed to freeze a substance is a constant for any given substance, and is known as the latent heat of fusion.

freezing point for any given liquid, the temperature at which the liquid changes state from a liquid to a solid. The temperature remains at this point until all the liquid has solidified. It is invariable under similar conditions of pressure – for example, the freezing point of water under standard atmospheric pressure is 0°C/32°F.

For a given liquid under similar conditions, the freezing point and melting point are the same temperature.

freezing point, depression of lowering of a solution's freezing point below that of the pure solvent; it depends on the number of molecules of solute dissolved in it. For a single solvent, such as pure water, all solute substances in the same molar concentration produce the same lowering of freezing point. The depression d produced by the presence of a solute of molar concentration C is given by the equation $d = KC$, where K is a constant (called the cryoscopic constant) for the solvent concerned.

Antifreeze mixtures for car radiators and the use of salt to melt ice on roads are common applications of this principle. Animals in arctic conditions, for example insects or fish, cope with the extreme cold either by manufacturing natural "antifreeze" and staying active, or by allowing themselves to freeze in a controlled fashion, that is, they manufacture proteins to act as nuclei for the formation of ice crystals in areas that will not produce cellular damage, and so enable themselves to thaw out and become active again.

Measurement of freezing-point depression is a useful method of determining the molecular weights of solutes. It is also used to detect the illicit addition of water to milk.

Freon in chemistry, tradename for a group of chlorofluorocarbons (CFCs) that are produced from ethane and methane by partially or fully substituted hydrogen atoms by chlorine and/or fluorine, for example Freon-12, CCl_2F_2, dichlorodifluoromethane. Freons are used mainly as refrigerants, solvents, and aerosol propellants.

frequency in physics, number of periodic oscillations, vibrations, or waves occurring per unit of time. The SI unit of frequency is the hertz (Hz), one hertz being equivalent to one cycle per second. Frequency is related to wavelength and velocity by the equation: $f = v/\lambda$ where f is frequency, v is velocity, and λ is wavelength. Frequency is the reciprocal of the period T: $f = 1/T$

Most alternating current electrical equipment in the UK has an optimum operating frequency of 50 Hz. Frequencies lower than this will cause noticeable voltage variations in electrical appliances. This is because each time the current changes direction, the voltage falls to zero before increasing again as the current moves in the opposite direction. For example, in a television set this will be seen as flickering of the screen. At 50 Hz, the current will switch 100 times a second. This is detected as a continuous image as the human eye cannot react fast enough to detect these changes.

At one end of the electromagnetic spectrum are long radio waves with a frequency in the range 10^4–10^5 Hz and at the other extreme are gamma rays with a frequency in the range 10^{19}–10^{22} Hz.

Human beings can hear sounds from objects vibrating in the range 20–15,000 Hz. Ultrasonic frequencies well above 15,000 Hz can be detected by such mammals as bats. Infrasound (low-frequency sound) can be detected by some mammals and birds. Pigeons can detect sounds as low as 0.1 Hz; elephants communicate using sounds as low as 1 Hz. **Frequency modulation** (FM) is a method of transmitting radio signals in which the frequency of the **carrier wave** is changed and then decoded.

frequency in statistics, the number of times an event occurs. For example, in a survey carried out to find out a group of children's favorite color of the rainbow, the color red is chosen 26 times. This gives the color red a frequency of 26. A table of the †raw data collected, including the frequencies, is called a **frequency distribution**. It is usually presented in a **frequency table** or tally chart. The frequencies can also be shown diagrammatically using a **frequency polygon**.

frequency diagram or **frequency chart**, means of displaying †raw data in a graphical form that is easier to interpret. Commonly used frequency diagrams are †bar charts, †histograms, †pie charts, and †scatter diagrams.

frequency modulation FM, method by which radio waves are altered for the transmission of broadcasting signals. FM varies the frequency of the carrier wave in accordance with the signal being transmitted. Its advantage over AM (†amplitude modulation) is its better signal-to-noise ratio. It was invented by the US engineer Edwin Armstrong.

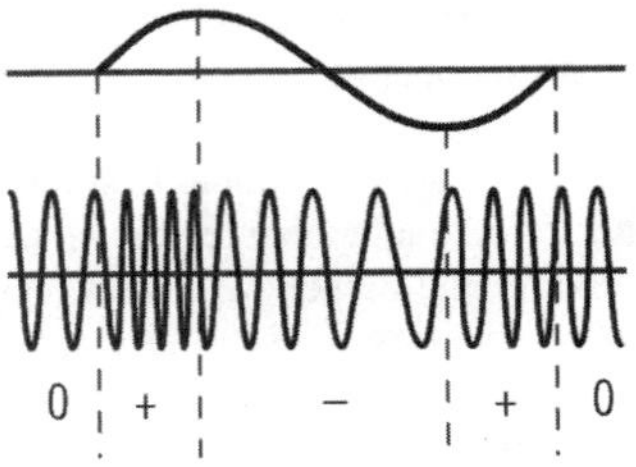

frequency modulation In FM radio transmission, the frequency of the carrier wave is modulated, rather than its amplitude (as in AM broadcasts). The FM system is not affected by the many types of interference which change the amplitude of the carrier wave, and so provides better quality reception than AM broadcasts. Helicon

Fresnel, Augustin Jean (1788–1827) French physicist who refined the theory of †polarized light. Fresnel realized in 1821 that light waves do not vibrate like sound waves longitudinally, in the direction of their motion, but transversely, at right angles to the direction of the propagated wave.

Freud, Anna (1895–1982) Austrian-born founder of child psychoanalysis in the UK. Her work was influenced by the theories of her father, Sigmund Freud. She held that understanding of the stages of psychological development was essential to the treatment of children, and that this knowledge could only be obtained through observation of the child.

Freud, Sigmund (1856–1939) Austrian physician who pioneered the study of the †unconscious mind. He developed the methods of free association and interpretation of dreams that are basic techniques of †psychoanalysis. The influence of unconscious forces on people's thoughts and actions was Freud's discovery, as was his controversial theory of the repression of infantile sexuality as the root of neuroses in the adult. His books include *Die Traumdeutung/The Interpretation of Dreams* (1900), *Jenseits des Lustprinzips/Beyond the Pleasure Principle* (1920), *Das Ich und das Es/The Ego and the Id* (1923), and *Das Unbehagen in der Kultur/Civilization and its Discontents* (1930). His influence has permeated the world to such an extent that it may be discerned today in almost every branch of thought.

Freud Web http://www.victorianweb.org/science/freud/Freud_OV.html

Overview of Sigmund Freud's life and works. The pages include a biography, a chronology, cultural context, and an examination of Freud's theories and techniques. There is also a bibliography and a black-and-white photograph of Freud.

Interpretation of Dreams (3rd edition) by Sigmund Freud

http://www.psychwww.com/books/interp/toc.htm A A Brill's 1911 translation of Freud's classic analysis of dreams.

Freundlich, Erwin Finlay (1885–1964) German astronomer. He collaborated with †Einstein in 1911 in testing predictions of the theory of relativity before its publication. They tested the motion of Mercury's perihelion and were prepared to test the bending of light rays as they passed the Sun during a total solar eclipse in 1914, but were prevented from doing so by the outbreak of war. The test was eventually carried out at the 1919 eclipse by Arthur †Eddington.

friction in physics, the force that opposes the movement of two bodies in contact as they move relative to each other. The **coefficient of friction** is the ratio of the force required to achieve this relative motion to the force pressing the two bodies together.

Two materials with rough surfaces rubbing together will change kinetic energy into heat and sound energy. Friction is greatly reduced by the use of lubricants such as oil, grease, and graphite. A layer of lubricant between two materials reduces the contact, allowing them to slide over each other smoothly. For example, engine oil used in cars reduces friction between metal parts as they move against each other. Air bearings are now used to minimize friction in high-speed rotational machinery. In joints in the human body, such as the knee, synovial fluid plays a key role as a lubricant. In other instances friction is deliberately increased by making the surfaces rough – for example, brake linings, driving belts, soles of shoes, and tires.

Friction is also used to generate static electric charges on different materials by rubbing the materials together.

Friedel, Charles (1832–1899) French organic chemist and mineralogist. Together with US chemist James Mason Crafts (1839–1917) he discovered the **Friedel–Crafts reaction** which uses aluminum chloride as a catalyst to facilitate the addition of an alkyl halide (halogenoalkane) to an aromatic compound. The reaction is now employed in the industrial preparation of triphenylamine dyes. Throughout his career, Friedel successfully combined his interests in chemistry and minerals.

Friedel–Crafts reaction in chemistry, a reaction for the synthesis of alkyl and acyl derivatives of cyclic aromatic compounds catalyzed by anhydrous aluminum chloride. It was developed and named after French chemist Charles †Friedel and US chemist James Mason Crafts.

Friedman, Aleksandr Aleksandrovich (1888–1925) Russian mathematician and cosmologist. He made fundamental contributions to the development of theories regarding the expansion of the universe. His work of the greatest relevance to astronomy was his independent and original approach to the solution of Albert †Einstein's field equation in the general theory of †relativity. Einstein had produced a static solution, which indicated a closed universe. Friedman derived several solutions, all of which suggested that space and time were isotropic (uniform at all points and in every direction), but that the mean density and radius of the universe varied with time – indicating an either expanding or contracting universe.

Friedman, Aleksandr Aleksandrovich http://www-groups.dcs.st-andrews.ac.uk/history/Mathematicians/Friedmann.html
Part of an archive containing the biographies of the world's greatest mathematicians, this site is devoted to the life and contributions of Aleksandr Friedman.

Friedman, Jerome I (1930–) US physicist who, with Henry Way Kendall and Richard E †Taylor, was awarded the Nobel Prize for Physics in 1990 for their pioneering investigations into high-energy electrons colliding with protons and neutrons. These were important in developing the †quark model of particle physics.

fringing reef †coral reef that is attached to the coast without an intervening lagoon.

Frisch, Karl von (1886–1982) Austrian zoologist who was awarded the Nobel Prize for Physiology or Medicine in 1973 with Konrad †Lorenz and Nikolaas †Tinbergen for their work on animal behavior patterns. Karl von Frisch specialized in bees, discovering how they communicate the location of sources of nectar by movements called "dances."

Frisch, Otto Robert (1904–1979) Austrian-born British physicist who first described the fission of uranium nuclei under neutron bombardment, coining the term "fission" to describe the splitting of a nucleus.

Frisch–Peierls memorandum document revealing, for the first time, how small the critical mass (the minimum quantity of substance required for a nuclear chain reaction to begin) of uranium needed to be if the isotope uranium-235 was separated from naturally occurring uranium; the memo thus implied the feasibility of using this isotope to make an atom bomb. It was written by Otto Frisch and Rudolf Peierls at the University of Birmingham in 1940.

Frobenius, (Ferdinand) Georg (1849–1917) German mathematician. He formulated the concept of the abstract group – the first abstract structure of "new" mathematics. His research into the theory of groups and complex number systems would prove useful to the development of †quantum mechanics. He also made contributions to the theory of elliptic functions and to the solution of differential equations.

Frobenius, Ferdinand Georg http://www-groups.dcs.st-andrews.ac.uk/history/Mathematicians/Frobenius.html
Part of an archive containing the biographies of the world's greatest mathematicians, this site is devoted to the life and contribu-

tions of Ferdinand Frobenius. In addition to biographical information, you will find a list of references about Frobenius and links to other essays in the archive that reference him. The text of this essay includes hypertext links to the essays of those mathematicians and thinkers who influenced Frobenius. You will also find an image of him, which you can click on to enlarge, and a map of his birthplace.

Fromm, Erich (1900–1980) German psychoanalyst who moved to the USA in 1933 to escape the Nazis. He believed that human beings experience a separation from nature and from other people which gives them the freedom to decide on the course their lives should take. This gives their lives meaning but also causes anxiety.

frond large leaf or leaflike structure; in ferns it is often pinnately divided. The term is also applied to the leaves of palms and less commonly to the plant bodies of certain seaweeds, liverworts, and lichens.

front in meteorology, the boundary between two air masses of different temperature or humidity. A **cold front** marks the line of advance of a cold air mass from below, as it displaces a warm air mass; a **warm front** marks the advance of a warm air mass as it rises up over a cold one. Frontal systems define the weather of the mid-latitudes, where warm tropical air is continually meeting cold air from the poles.
Warm air, being lighter, tends to rise above the cold; its moisture is carried upwards and usually falls as rain or snow, hence the changeable weather conditions at fronts. Fronts are rarely stable and move with the air mass. An **occluded front** is a composite form, where a cold front catches up with a warm front and merges with it.

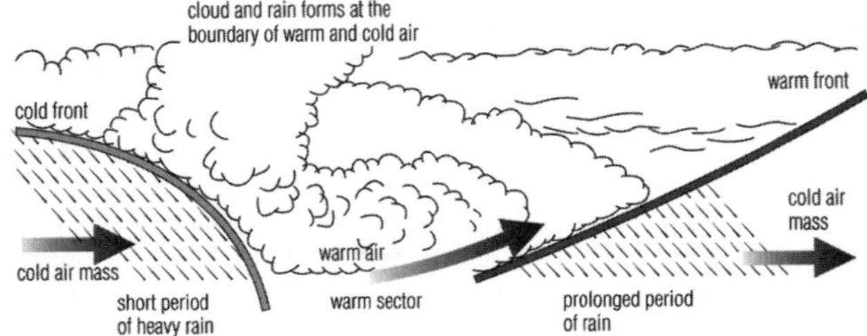

front The boundaries between two air masses of different temperature and humidity. A warm front occurs when warm air displaces cold air; if cold air replaces warm air, it is a cold front. Helicon

frost condition of the weather that occurs when the air temperature is below freezing, 0°C/32°F. Water in the atmosphere is deposited as ice crystals on the ground or on exposed objects. As cold air is heavier than warm air and sinks to the ground, ground frost is more common than hoar (air) frost, which is formed by the condensation of water particles in the air.

frost hollow depression or steep-sided valley in which cold air collects on calm, clear nights. Under clear skies, heat is lost rapidly from ground surfaces, causing the air above to cool and flow downhill (as †katabatic wind) to collect in valley bottoms. Fog may form under these conditions and, in winter, temperatures may be low enough to cause frost.

frost shattering alternative name for †freeze-thaw.

fructose $C_6H_{12}O_6$, sugar that occurs naturally in honey, the nectar of flowers, and many sweet fruits; it is commercially prepared from glucose.
Fructose is a monosaccharide, whereas the more familiar cane or beet sugar is a disaccharide, made up of two monosaccharide units: fructose and glucose. It is sweeter than cane sugar and can be used to sweeten foods for people with diabetes.

fruit (from Latin *frui* "to enjoy") in botany, the ripened ovary in flowering plants that develops from one or more seeds or carpels and encloses one or more seeds. Its function is to protect the seeds during their development and to aid in their dispersal. Fruits are often edible, sweet, juicy, and colorful. When eaten they provide vitamins, minerals, and enzymes, but little protein. Most fruits are borne by perennial plants.
Fruits are divided into three agricultural categories on the basis of the climate in which they grow. **Temperate fruits** require a cold season for satisfactory growth; the principal temperate fruits are apples, pears, plums, peaches, apricots, cherries, and soft fruits, such as strawberries. **Subtropical fruits** require warm conditions but can survive light frosts; they include oranges and other citrus fruits, dates, pomegranates, and avocados. **Tropical fruits** cannot tolerate temperatures that drop close to freezing point; they include bananas, mangoes, pineapples, papayas, and litchis. Fruits can also be divided botanically into **dry** (such as the †capsule, †follicle, †schizocarp, †nut, †caryopsis, pod or legume, †lomentum, and †achene) and those that become **fleshy** (such as the †drupe and the †berry). The fruit structure consists of the †pericarp or fruit wall, which is usually divided into a number of distinct layers. Sometimes parts other than the ovary are incorporated into the fruit structure, resulting in a false fruit or †pseudocarp, such as the apple and strawberry. True fruits include the tomato, orange, melon, and banana. Fruits

may be dehiscent, which open to shed their seeds, or indehiscent, which remain unopened and are dispersed as a single unit. Simple fruits (for example, peaches) are derived from a single ovary, whereas compositae or multiple fruits (for example, blackberries) are formed from the ovaries of a number of flowers. In ordinary usage, "fruit" includes only sweet, fleshy items; it excludes many botanical fruits such as acorns, bean pods, thistledown, and cucumbers.

methods of seed dispersal Efficient seed dispersal is essential to avoid overcrowding and enable plants to colonize new areas; the natural function of a fruit is to aid in the dissemination of the seeds which it contains. A great variety of dispersal mechanisms exist: winged fruits are commonly formed by trees, such as ash and elm, where they are in an ideal position to be carried away by the wind; some wind-dispersed fruits, such as clematis and cotton, have plumes of hairs; others are extremely light, like the poppy, in which the capsule acts like a pepperpot and shakes out the seeds as it is blown about by the wind. Some fruits float on water; the coconut can be dispersed across oceans by means of its buoyant fruit. Geraniums, gorse, and squirting cucumbers have explosive mechanisms, by which seeds are forcibly shot out at dehiscence. Animals often act as dispersal agents either by carrying hooked or sticky fruits (burs) attached to their bodies, or by eating succulent fruits, the seeds passing through the alimentary canal unharmed.

orange
(hesperidium)
exocarp
(rind)
seed
(pip)
pulp
apple
(pome)
receptacle
(flesh)
seed
(pip)
pericarp
(core)
remains of
flower
peach
(drupe)
pulp
seed
(stone)

fruit A fruit contains the seeds of a plant. Its outer wall is the exocarp, or epicarp; its inner layers are the mesocarp and endocarp. The orange is a hesperidium, a berry having a leathery rind and containing many seeds. The peach is a drupe, a fleshy fruit with a hard seed, or "stone," at the center. The apple is a pome, a fruit with a fleshy outer layer and a core containing the seeds. Helicon

Recorded world fruit production in the mid-1980s was approximately 300 million tonnes per year. Technical advances in storage and transport have made tropical fruits available to consumers in temperate areas, and fresh temperate fruits available all year in major markets.

frustum (from Latin "a piece cut off") in geometry, a "slice" taken out of a solid figure by a pair of parallel planes. A conical frustum, for example, resembles a cone with the top cut off. The volume and area of a frustum are calculated by subtracting the volume or area of the "missing" piece from those of the whole figure.

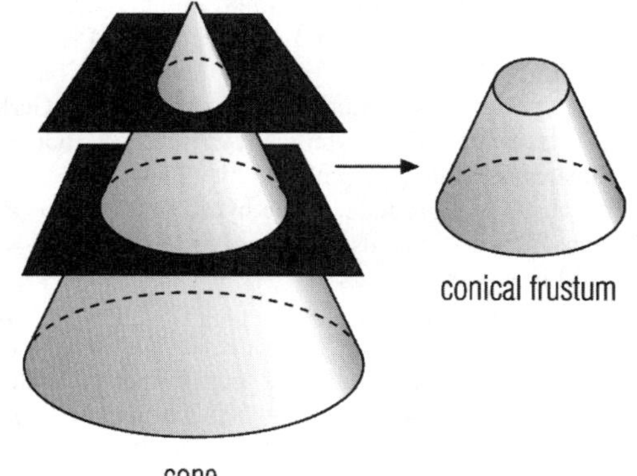

frustum The frustum, a slice taken out of a cone. Helicon

FSH abbreviation for †**follicle-stimulating hormone**.

ft symbol for †**foot**, a measure of distance.

fuel any source of heat or energy, embracing the entire range of materials that burn in air (combustibles). A fuel is a substance that gives out energy when it burns. A **nuclear fuel** is any material that produces energy by nuclear fission in a nuclear reactor. †Fossil fuels are formed from the fossilized remains of plants and animals.

Crude oil (unrefined †petroleum) is purified at an oil refinery by †fractional distillation into fuels such as gasoline and kerosine. The burning of fossil fuels for energy production contributes to environmental problems such as †acid rain and the †greenhouse effect.

fuel cell cell converting chemical energy directly to electrical energy. It works on the same principle as a battery but is continually fed with fuel, usually hydrogen and oxygen. Fuel cells are silent and reliable (no moving parts) but expensive to produce. They are an example of a renewable energy source.

Hydrogen is passed over an †electrode (usually nickel or platinum) containing a †catalyst, which splits the hydrogen into electrons and protons. The electrons pass through an external circuit while the protons pass through a polymer †electrolyte membrane to another electrode, over which oxygen is passed. Water is formed at this electrode (as a by-product) in a chemical reaction involving electrons, protons, and oxygen atoms. A current is generated between the electrodes. If the spare heat also produced is used for hot water and space heating, 80% efficiency in fuel is achieved.

Fuel cells can be used to power cars, replacing the internal combustion engine, and to produce electricity on spacecraft.

fugue in psychology, an abnormal state in which a person under emotional stress suddenly leaves home, apparently forgetting everything about his or her normal life, and assumes a new identity. The state is usually temporary and is probably due to †repression.

Fukui, Kenichi (1918–1998) Japanese industrial chemist who shared the Nobel Prize for Chemistry in 1981 with Roald †Hoffmann for his work on "frontier orbital theory," predicting the change in molecular orbitals (the arrangement of electrons around the nucleus; see †orbital, atomic) during chemical reactions.

fulcrum point of support of a †lever, about which the lever can pivot in lifting and lowering loads applied to it.

Fuller, Solomon Carter (1872–1953) US physician, neurologist, psychiatrist, and pathologist. He worked on degenerative brain diseases including Alzheimer's disease, which he attributed to causes other than arteriosclerosis; this was supported by medical researchers in 1953.

fullerene form of carbon, discovered in 1985, based on closed cages of carbon atoms. The molecules of the most symmetrical of the fullerenes are called †buckminsterfullerenes (or buckyballs). They are perfect spheres made up of 60 carbon atoms linked together in 12 pentagons and 20 hexagons fitted together like those of a spherical football. Other fullerenes with 28, 32, 50, 70, and 76 carbon atoms, have also been identified.

Fullerenes can be made by arcing electricity between carbon rods. They may also occur in candle flames and in clouds of interstellar gas. Fullerene chemistry may turn out to be as important as organic chemistry based on the benzene ring. Already, new molecules based on the buckyball enclosing a metal atom, and "buckytubes" (cylinders of carbon atoms arranged in hexagons), have been made. They were proved to be 200 times tougher than any other known fiber by Israeli and US materials scientists in 1998. Applications envisaged include using the new molecules as lubricants, semiconductors, and superconductors, and as the starting point for making new drugs.

fuller's earth soft, greenish-gray rock resembling clay, but without clay's plasticity. It is formed largely of clay minerals, rich in montmorillonite, but a great deal of silica is also present. Its absorbent properties make it suitable for removing oil and grease, and it was formerly used for cleaning fleeces ("fulling"). It is still used in the textile industry, but its chief application is in the purification of oils. Beds of fuller's earth are found in the southern USA, Germany, Japan, and the UK.

fulminate any salt of fulminic (cyanic) acid (HOCN), the chief ones being silver and mercury. The fulminates detonate (are exploded by a blow).

fulvene C_6H_6, unstable, cyclic hydrocarbon that is an †isomer of benzene. It is the simplest member of a group of compounds of general formula $C_4H_4C = CRR'$, where R and R' can be hydrogen, †alkyl, or †aryl groups.

function in computing, a small part of a program that supplies a specific value – for example, the square root of a specified number, or the current date. Most programming languages incorporate a number of built-in functions; some allow programmers to write their own. A function may have one or more arguments (the values on which the function operates). A **function key** on a keyboard is one that, when pressed, performs a designated task, such as ending a program.

function in mathematics, a function f is a non-empty set of ordered pairs $(x, f(x))$ of which no two can have the same first element. Hence, if $f(x) = x^2$ two ordered pairs are (–2,4) and (2,4). The set of all first elements in a function's ordered pairs is called the **domain**; the set of all second elements is the **range**. Functions are used in all branches of mathematics, physics, and science generally. For example, in the †equation $y = 2x + 1$, y is a function of the symbol x. This can be written as $y = f(x)$.

In another example, the †formula $t = 2\pi\sqrt{(l/g)}$ shows that for a simple pendulum, the time of swing t is a function of its length l and of no other variable quantity (π and g, the acceleration due to gravity, are †constants).

functional group in chemistry, a small number of atoms in an arrangement that determines the chemical properties of the group and of the molecule to which it is attached (for example, the carboxyl group COOH, or the amine group NH_2). Organic compounds can be considered as structural skeletons, with a high carbon content, with functional groups attached.

functions analysis see †function theory.

functions of real variables, theory of involves †functions of which the arguments are †real numbers.

function theory or **theory of functions**, use of †functions, primarily in order to denote mathematical relationships, but also in application to other sciences. Functional analysis, for example, considers problems in which an unknown function is to be found; at this stage †variables may represent functions as opposed to numerical values.

fundamental constant physical quantity that is constant in all circumstances throughout the whole universe. Examples are the electric charge of an electron, the speed of light, Planck's constant, and the gravitational constant.

fundamental forces see †forces, fundamental.

fundamental particle another term for †**elementary particle**.

fundamental theorem of algebra in mathematics, theory that any polynomial (with real or complex coefficients) has a root in the complex plane.

fundamental theorem of analysis in mathematics, result stating the connection between differential and integral †calculus. It states that differentiating an indefinite integral of any function gives the same function again.

fundamental vibration standing wave of the longest wavelength that can be established on a vibrating object such as a stretched string or air column. The sound produced by the fundamental vibration is the lowest-pitched (usually dominant) note heard.

The fundamental vibration of a string has a stationary †node at each end and a single †antinode at the center where the amplitude of vibration is greatest.

fungicide any chemical †pesticide used to prevent fungus diseases in plants and animals. Inorganic and organic compounds containing sulfur are widely used.

fungus plural **fungi**, any of a unique group of organisms that includes moulds, yeasts, rusts, smuts, mildews, mushrooms, and toadstools. There are around 70,000 species of fungi known to science, although there may be as many as 1.5 million actually in existence. They are not considered to be plants for three main reasons: they have no leaves or roots; they contain no chlorophyll (green coloring) and are therefore unable to make their own food by †photosynthesis; and they reproduce by †spores. Some fungi are edible but many are highly poisonous; they often cause damage and sometimes disease to the organic matter on which they live and feed, but some fungi are exploited in the production of food and drink (for example, yeasts in baking and brewing) and in medicine (for example, penicillin).

Fungi are either †parasites, existing on living plants or animals, or †saprotrophs, living on dead matter. Many of the most serious plant diseases are caused by fungi, and several fungi attack humans and animals. Athlete's foot, thrush, and ringworm are fungal diseases. Endophytes are fungi that live inside plants. Almost all plants have endophytes and many have large numbers, for example the grass fescue has 400 species.

Before the classification Fungi came into use, they were included within the division Thallophyta, along with algae and †bacteria. Two familiar fungi are bread mould, which illustrates the typical many-branched body (mycelium) of the organism, made up of threadlike chains of cells called hyphae; and mushrooms, which are the sexually reproductive fruiting bodies of an underground mycelium.

The mycelium of a true fungus is made up of many intertwined hyphae. When the fungus is ready to reproduce, the hyphae become closely packed into a solid mass called the fruiting body, which is usually small and inconspicuous but can be very large; mushrooms, toadstools, and bracket fungi are all examples of large fruiting bodies. These carry and distribute the spores. Most species of fungi reproduce both asexually (on their own) and sexually (involving male and female parents).

Fungi http://www.herb.lsa.umich.edu/kidpage/factindx.htm

University-run network of hyperlinked pages on fungi, from their earliest fossil records to their current ecology and life cycles, from how they are classified systematically to how they are studied. Although this page is not shy of technical terms, there are clear explanations and pictures to help the uninitiated.

Funk, Casimir (1884–1967) Polish-born US biochemist who pioneered research into vitamins. He was the first to isolate niacin (nicotinic acid, one of the vitamins of the B complex).

fur the †hair of certain animals. Fur is an excellent insulating material and so has been used as clothing. This is, however, vociferously criticized by many groups on humane grounds, as the methods of breeding or trapping animals are often cruel. Mink, chinchilla, and sable are among the most valuable, the wild furs being finer than the farmed.

Fur such as mink is made up of a soft, thick, insulating layer called underfur and a top layer of longer, lustrous guard hairs.

Furs have been worn since prehistoric times and have long been associated with status and luxury (ermine traditionally worn by royalty, for example), except by certain ethnic groups like the Inuit. The fur trade had its origin in North America, where in the late 17th century the Hudson's Bay Company was established. The chief centers of the fur trade are New York, London, St Petersburg, and Kastoria in Greece. It is illegal to import furs or skins of endangered species listed by CITES (such as the leopard). Many synthetic fibers are widely used as substitutes.

furan C_4H_2O, colorless, volatile, toxic liquid; boiling point, 32°C/90°F. It is the simplest member of the furan class of heterocyclic aromatic compounds, which are characterized by having a ring structure consisting of four carbon atoms and one oxygen atom. Furans are used as solvents and in the production of plastic resins.

furlong unit of measurement, originating in Anglo-Saxon England, equivalent to 220 yd (201.168 m).

A furlong consists of 40 rods, poles, or perches; 8 furlongs equal one statute †mile. Its literal meaning is "furrow-long," and refers to the length of a furrow in the common field characteristic of medieval farming.

fuse device that prevents the passage of electric †current above a predetermined level in a †circuit, by melting through the consequent temperature rise, and thus breaking the circuit.

fusel oil liquid with a characteristic unpleasant smell, obtained as a by-product of the distillation of the product of any alcoholic fermentation, and used in paints, varnishes, essential oils, and plastics. It is a mixture of fatty acids, alcohols, and esters.

fusion in physics, the fusing of the nuclei of light elements, such as hydrogen, into those of a heavier element, such as helium. The resultant loss in their combined mass is converted into energy. Stars and thermonuclear weapons are powered by nuclear fusion.

Nuclear fusion takes place in the Sun, where hydrogen nuclei fuse at temperatures of about 10 million-°C/18 million-°F producing huge amounts of heat and light energy.

Fusion http://fusedweb.pppl.gov/FAQ/fusion-faq.html

All-text site packed with information about fusion research and its applications. It is quite well organized and includes a glossary of commonly used terms to aid the uninitiated.

Fusion Energy and Plasma Physics Research at PPPL http://www.pppl.gov/

Outlines the work of the Princeton Plasma Physics Laboratory (PPPL). Updated with news from the laboratory, there are announcements of forthcoming meetings and pages detailing the educational programs with which PPPL are involved.

Internet Plasma Physics Experience http://ippex.pppl.gov/

Operate your own virtual Tokamak fusion reactor and get a hands-on feel for this exciting branch of physics. This information-packed site explains the key physics concepts involved in nuclear fusion and plasma physics, such as matter, electricity and magnetism, and energy.

fuzzy logic in mathematics and computing, a form of knowledge representation suitable for notions (such as "hot" or "loud") that cannot be defined precisely but depend on their context. For example, a jug of water may be described as too hot or too cold, depending on whether it is to be used to wash one's face or to make tea.

The central idea of fuzzy logic is **probability of set membership**. For instance, referring to someone 175 cm/5 ft 9 in tall, the statement "this person is tall" (or "this person is a member of the set of tall people") might be about 70% true if that person is a man, and about 85% true if that person is a woman.

G

gabbro mafic (consisting primarily of dark-colored crystals) igneous rock formed deep in the Earth's crust. It contains pyroxene and calcium-rich feldspar, and may contain small amounts of olivine and amphibole. Its coarse crystals of dull minerals give it a speckled appearance.

Gabbro is the plutonic version of basalt (that is, derived from magma that has solidified below the Earth's surface), and forms in large, slow-cooling intrusions.

Gabor, Dennis (1900–1979) Hungarian-born British physicist. He was awarded the Nobel Prize for Physics in 1971 for his invention in 1947 of the holographic method of three-dimensional photography.

Gacrux or **Gamma Crucis**, third-brightest star in the southern-hemisphere constellation of †Crux, marking one of the four corners of the Southern Cross. It is a red-giant star, some 120 light years from the Sun. Together with nearby †Acrux, it points towards the south celestial pole.

gadolinite in chemistry, dark brown to black mineral that was important as an early source of rare-earth metals, notably dysprosium, yttrium, scandium, and †gadolinium. It was discovered by Finnish mineralogist Johan Gadolin.

gadolinium chemical symbol Gd, silvery-white metallic element of the lanthanide series, atomic number 64, relative atomic mass 157.25. It is found in the products of nuclear fission and used in electronic components, alloys, and products needing to withstand high temperatures.

Gaia hypothesis theory that the Earth's living and nonliving systems form an inseparable whole that is regulated and kept adapted for life by living organisms themselves. The planet therefore functions as a single organism, or a giant cell. The hypothesis was elaborated by British scientist James †Lovelock and first published in 1968.

Earthdance: Living Systems in Evolution http://www.ratical.com/LifeWeb/Erthdnce/erthdnce.html

Philosophical work by Elisabet Sahtouris, based on the Gaia hypothesis, published in 1995 with an introduction by James Lovelock. There are chapters on such topics as "Problems for earthlife," "Ecological ethics," and "Survival: the sustainable society."

gain in electronics, the ratio of the amplitude of the output signal produced by an amplifier to that of the input signal. In a †voltage amplifier the voltage gain is the ratio of the output voltage to the input voltage; in an inverting †operational amplifier (op-amp) it is equal to the ratio of the resistance of the feedback resistor to that of the input resistor.

gal symbol for †**gallon**, †**galileo**.

galactic centers phenomena that are now thought to comprise black holes – which would explain why the center of our Galaxy appears strangely obscure, and emits only †infrared radiation.

galactic coordinates in astronomy, system for measuring the position of astronomical objects on the †celestial sphere with reference to the galactic equator (or †great circle).

Galactic latitude (symbol b) is measured in degrees from the galactic equator (b = 0°) to the north (b = 90°) and south (b = – 90°) galactic poles.

Galactic longitude (symbol l) is measured in degrees eastward (l = 0° to 360°) from a fixed point in the constellation of †Sagittarius that approximates to the center of the Galaxy. Galactic coordinates are often used when astronomers are studying the distribution of material in the †Galaxy.

galactic halo in astronomy, outer, sparsely populated region of a galaxy, roughly spheroid in shape and extending far beyond the bulk of the visible stars. In our own Galaxy, the halo contains the globular clusters, and may harbor large quantities of †dark matter.

galactic plane in astronomy, plane passing through the †Sun and the center of our †Galaxy defining the mid-plane of the galactic disc. Viewed from the Earth, the galactic plane is a †great circle (galactic equator) marking the approximate center line of the †Milky Way.

galactic rotation revolving of a galaxy round its central nucleus even as it continues its †proper motion. Such rotation, however, is not uniform but †differential. One revolution of the Sun within our own Galaxy takes about 225 million years, or 1 cosmic year.

galactose $C_6H_{12}O_6$, one of the hexose sugars, an †isomer of †glucose and †fructose.

Galatea small moon of Neptune, discovered in 1989, that orbits just inside the main ring system, where its resonance effects help to keep the ring in position. It orbits 62,000 km/38,525 mi above the planet's surface once every ten hours and has a diameter of 150 km/93 mi.

galaxy grouping of millions or billions of stars, held together by gravity. It is believed that there are billions of galaxies in the †universe. There are different types, including spiral, barred spiral, and elliptical galaxies. Our own Galaxy, the †Milky Way, is about 100,000 light years across (a light year is the distance light travels in a year, about 9.5 billion km/6 billion mi), and contains at least 100 billion stars.

The galaxies are moving away from our own in all directions. The universe is thus expanding in all directions. The evidence for this comes from examining light from the galaxies by splitting the light into a spectrum. A feature known as the red shift appears, where the light is shifted towards the red end of the spectrum due to an increase in wavelength caused by the galaxies' recession.

For the galaxy in the constellation Virgo, which is about 38 million light years away, the red shift indicates the galaxy is moving away from us at 1,200 km/730 mi per second, while the galaxy in the constellation Hydra, which is about 2,000 million light years away, is moving away from us at 56,000 km/35,000 mi per second.

Astro Photos http://www.astro-photos.com/

Enthusiastic site on astrophotography that contains images of star clusters, galaxies, nebulae, and the Moon.

Galaxy Catalogue http://www.astro.princeton.edu/~frei/catalog.htm

Magnificent photographic catalogue of 113 nearby galaxies imaged from the Palomar and Lowell observatories. After selecting a thumbnail, the user is prompted to choose a file format with which to view the full-size galaxy image.

Galen (*c.* 129–*c.* 200) Greek physician and anatomist whose ideas dominated Western medicine for almost 1,500 years. Central to his thinking were the threefold circulation of the blood and the theory of humours (blood, phlegm, choler/yellow bile, and melancholy/black bile) that contributed to mental and physical state. His *On Anatomical Procedures*, a detailed description of animal dissections when work on human corpses was forbidden, became a standard text on anatomy when rediscovered in Western Europe in the 16th century. He remained the highest medical authority until Andreas †Vesalius and William †Harvey exposed the fundamental errors of his system.

On the Natural Faculties http://classics.mit.edu/Galen/natfac.html

Text of this work by the Greek physician Galen.

galena mineral consisting of lead sulfide, PbS, the chief ore of lead. It is lead-gray in color, has a high metallic luster and breaks into cubes because of its perfect cubic cleavage. It may contain up to 1% silver, and so the ore is sometimes mined for both metals. Galena occurs mainly among limestone deposits in Australia, Mexico, Russia, Kazakhstan, the UK, and the USA.

Galilean satellites four-largest moons of Jupiter (†Io, †Europa, †Callisto, and †Ganymede). The moons were seen through a telescope for the first time by the Italian astronomer Galileo Galilei in 1610. The largest Galilean satellite is Ganymede with a diameter (5,262 km/3,270 mi) similar in size to that of the planet Mercury.

Galileo (1564–1642) born Galileo Galilei, Italian mathematician, astronomer, and physicist. He developed the astronomical telescope and was the first to see sunspots, the four main satellites of Jupiter, and the appearance of Venus going through phases, thus proving it was orbiting the Sun. Galileo discovered that freely falling bodies, heavy or light, have the same, constant acceleration and that this acceleration is due to gravity. He also determined that a body moving on a perfectly smooth horizontal surface would neither speed up nor slow down. He invented a thermometer, a hydrostatic balance, and a compass, and discovered that the path of a projectile is a parabola.

IMSS – History of Science Multimedia Catalogue Galileo Galilei

http://galileo.imss.firenze.it/museo/4/index.html

Room IV of the Institute and Museum of the History of Science, which is dedicated to Galileo. This Web site in Florence, Italy, includes a virtual visit (Quick Time format) to this room in the museum along with a multimedia catalogue that includes a descriptive text of the artifacts on display, accompanied by still, video, or animated images. There is an extensive list of objects to choose from, including, rather bizarrely, the middle finger of Galileo's right hand.

galileo symbol gal, unit of acceleration, used in geological surveying. One galileo is 10^{-2} meters per second per second. The Earth's gravitational field often differs by several milligals (thousandths of galileos) in different places, because of the varying densities of the rocks beneath the surface.

gall abnormal outgrowth on a plant that develops as a result of attack by insects or, less commonly, by bacteria, fungi, mites, or nematodes. The attack causes an increase in the number of cells or an enlargement of existing cells in the plant. Gall-forming insects generally pass the early stages of their life inside the gall.

Gall wasps are responsible for the conspicuous bud galls forming on oak trees, 2.5–4 cm/1–1.5 in across, known as "oak apples." The organisms that cause galls are host-specific. Thus, for example, gall wasps tend to parasitize oaks, and sawflies willows.

Gall, Franz Joseph (1758–1828) Austrian anatomist, instigator of the discredited theory of phrenology.

gall bladder small muscular sac, part of the digestive system of most, but not all, vertebrates. In humans, it is situated on the underside of the liver and connected to the small intestine by the bile duct. It stores bile from the liver.

Galle, Johann Gottfried (1812–1910) German astronomer. He located the planet Neptune in 1846, close to the position predicted by French mathematician Urbain †Leverrier and the English astronomer J C †Adams.

gallium chemical symbol Ga, gray metallic element, atomic number 31, relative atomic mass 69.72. It is liquid at room temperature. Gallium arsenide (GaAs) crystals are used for semiconductors in microelectronics, since electrons travel a thousand times faster through them than through silicon. The element was discovered in 1875 by Lecoq de Boisbaudran (1838–1912).

gallium arsenide compound of gallium and arsenic, formula GaAs, used in lasers, photocells, and microwave generators. Its semiconducting properties make it a possible rival to †silicon for use in microprocessors. Chips made from gallium arsenide require less electric power and process data faster than those made from silicon.

Gallo, Robert Charles (1937–) US scientist credited with identifying the virus responsible for †AIDS. Gallo discovered the virus, now known as human immunodeficiency virus (HIV), in 1984; the French scientist Luc †Montagnier of the Pasteur Institute, Paris, discovered the virus, independently, in 1983.

gallon symbol gal, imperial liquid or dry measure, equal to 4.546 liters, and subdivided into four quarts or eight pints. The US gallon is equivalent to 3.785 liters.

Galois, Evariste (1811–1832) French mathematician who originated the theory of groups and greatly extended the understanding of the conditions in which an algebraic equation is solvable.

Galois, Evariste http://www-history.mcs.st-and.ac.uk/history/Mathematicians/Galois.html

Portrait and text-heavy biography of the 19th-century French mathematician. The site, run by St Andrews University, also includes excerpts from original works by Galois and details of his contemporaries.

Galton, Francis (1822–1911) English scientist, inventor, and explorer who studied the inheritance of physical and mental attributes with the aim of improving the human species. He was the first to use twins to try to assess the influence of environment on development, and is considered the founder of eugenics (a term he coined).

Galvani, Luigi (1737–1798) Italian physiologist who discovered galvanic, or voltaic, electricity in 1762, when investigating the contractions produced in the muscles of dead frogs by contact with pairs of different metals. His work led quickly to Alessandro Volta's invention of the electrical †cell, and later to an understanding of how nerves control muscles.

galvanizing process for rendering iron rust-proof, by plunging it into molten zinc (the dipping method), or by electroplating it with zinc.

galvanometer instrument for detecting small electric currents by their magnetic effect.

gamete cell that functions in sexual reproduction by merging with another gamete to form a †zygote. Examples of gametes include sperm and egg cells. In most organisms, the gametes are haploid (they contain half the number of chromosomes of the parent), owing to reduction division or †meiosis.

In higher organisms, gametes are of two distinct types: large immobile ones known as eggs or egg cells (see †ovum) and small ones known as †sperm. They come together at †fertilization. In some lower organisms the gametes are all the same, or they may belong to different mating strains but have no obvious differences in size or appearance.

game theory group of mathematical theories, developed in 1944 by German-born US mathematician Oscar Morgenstern and Hungarian-born US mathematician John †Von Neumann, that seeks to abstract from invented game-playing scenarios and their outcome the essence of situations of conflict and/or cooperation in the real political, business, and social world.

A feature of such games is that the rationality of a decision by one player will depend on what the others do; hence game theory has particular application to the study of oligopoly (a market largely controlled by a few producers).

Strategy and Conflict: An Introductory Sketch of Game Theory

http://william-king.www.drexel.edu/top/eco/game/game.html

Full guide to the history and practice of game theory. This site contains all there is to know about zero-sum games and a variety of dilemmas and conundrums. Game theory is applied to all aspects of human life from marriage vows to companies on the verge of bankruptcy.

gametophyte the †haploid generation in the life cycle of a plant that produces gametes; see †alternation of generations.

gamma globulin one of a group of proteins or †immunoglobulins in the blood that act as †antibodies to specific infections. Gamma globulins extracted from the blood of a patient who has recovered from an infection may be used as †vaccines to stimulate artificial immunity in others.

gamma radiation very high-frequency, high-energy electromagnetic radiation, similar in nature to X-rays but of shorter wavelength, emitted by the nuclei of radioactive substances during decay or by the interactions of high-energy electrons with matter. Cosmic gamma rays have been identified as coming from pulsars, radio galaxies, and quasars, although they cannot penetrate the Earth's atmosphere.

Gamma rays are stopped only by direct collision with an atom and are therefore very penetrating. They can, however, be stopped by about 4 cm/1.5 in of lead or by a very thick concrete shield. They are less ionizing in their effect than alpha and beta particles, but are dangerous nevertheless because they can penetrate deeply into body tissues such as bone marrow. They are not deflected by magnetic or electric fields. Gamma radiation is used to kill bacteria and other micro-organisms, sterilize medical devices, and change the molecular structure of plastics to modify their properties (for example, to improve their resistance to heat and abrasion).

gamma-ray astronomy study of celestial objects that emit gamma rays (energetic photons with very short wavelengths). Much of the radiation detected comes from collisions between hydrogen gas and cosmic rays in our Galaxy. Some sources have been identified, including the Crab Nebula and the Vela pulsar (the most powerful gamma-ray source detected).

Gamma rays are difficult to detect and are generally studied by use of balloon-borne detectors and artificial satellites. The first gamma-ray satellites were SAS II (1972) and COS-B (1975), although gamma-ray detectors were carried on the *Apollo 15* and *Apollo 16* missions, in 1971 and 1972, respectively. SAS II failed after only a few months, but COS-B continued working until 1982, carrying out a complete survey of the galactic disc.

Gamow, George (1904–1968) born Georgi Antonovich Gamow, Russian-born US cosmologist, nuclear physicist, and popularizer of science. His work in astrophysics included a study of the structure and evolution of stars and the creation of the elements. He explained how the collision of nuclei in the solar interior could produce the nuclear reactions that power the Sun. With the "hot †Big Bang" theory, which he co-proposed in 1948, he indicated the origin of the universe.

ganglion plural **ganglia**, solid cluster of nervous tissue containing many cell bodies and †synapses, usually enclosed in a tissue sheath; found in invertebrates and vertebrates.

In many invertebrates, the central nervous system consists mainly of ganglia connected by nerve cords. The ganglia in the head (cerebral ganglia) are usually well developed and are analogous to the brain in vertebrates. In vertebrates, most ganglia occur outside the central nervous system.

gangue part of an ore deposit that is not itself economically valuable; for example, calcite may occur as a gangue mineral with galena.

Ganymede in astronomy, largest moon of the planet Jupiter, orbiting every 7.2 days at a distance of 1.1 million km/700,000 mi. It is the largest moon in the Solar System, 5,260 km/3,270 mi in diameter (larger than the planet Mercury). Its surface is a mixture of extensively cratered and grooved terrain. Molecular oxygen was identified on Ganymede's surface in 1994. It is thought that Ganymede has a water ice crust and possibly a buried water ocean

The space probe Galileo detected a magnetic field around Ganymede in 1996; this suggests it may have a molten core. Galileo photographed Ganymede at a distance of 7,448 km/4,628 mi. The resulting images were 17 times clearer than those taken by Voyager 2 in 1979.

garnet group of †silicate minerals with the formula $X_3Y_3(SiO_4)_3$, where X is calcium, magnesium, iron, or manganese, and Y is usually aluminum or sometimes iron or chromium. Garnets are used as semi-precious gems (usually pink to deep red) and as abrasives. They occur in metamorphic rocks such as gneiss and schist.

Garnets consisting of neodymium, yttrium, and aluminum (referred to as Nd-YAG) produce infrared laser light when sufficiently excited. Nd-YAG lasers are inexpensive and used widely in industry and scientific research.

Garstang, Walter (1868–1949) English zoologist who proposed that the different aquatic larval stages of a variety of invertebrates and chordates are adaptations for dispersal, and not representative of a primitive and ancestral type. It followed that such adaptations might have a profound influence on the subsequent evolution of the species.

gas form of matter, such as air, in which the molecules move randomly in otherwise empty space, filling any size or shape of container into which the gas is put.

A sugar-lump sized cube of air at room temperature contains 30 trillion molecules moving at an average speed of 500 meters per second (1,800 kph/1,200 mph). Gases can be liquefied by cooling, which lowers the speed of the molecules and enables attractive forces between them to bind them together.

gas collection method used to collect a gas in a laboratory preparation. The properties of the gas, and whether it is required dry, dictate the method used. Dry ammonia is collected by downward displacement of air.

gas constant in physics, the constant R that appears in the equation $PV = nRT$, which describes how the pressure P, volume V, and temperature T of an ideal gas are related (n is the amount of gas in moles). This equation combines †Boyle's law and †Charles's law.

R has a value of 8.3145 joules per kelvin per mole.

gas-cooled reactor type of nuclear reactor; see †advanced gas-cooled reactor.

gaseous in chemistry, a term describing the physical state of a substance that is a †gas.

gas exchange movement of gases between an organism and the atmosphere, principally oxygen and carbon dioxide. All aerobic organisms (most animals and plants) take in oxygen in order to burn food and manufacture †ATP (adenosine triphosphate, which provides the energy for all cell reactions). The resultant oxidation reactions release carbon dioxide as a waste product to be passed out into the †environment. Green plants also absorb carbon dioxide during †photosynthesis, and release oxygen as a waste product. It is a characteristic of living organisms that they exchange materials with the environment. As well as the exchange of oxygen and carbon dioxide during †aerobic respiration, nutrients must be taken up from the environment to feed the organism. In humans, these include †carbohydrates, †proteins, minerals (see †mineral salts), and †vitamins. †Water is another important substance that has to be absorbed. However, in plants, only carbon dioxide, water, and minerals are required. The carbon dioxide is used in photosynthesis, which then releases oxygen as a waste product. All oxygen gas in the air on Earth is believed to have been produced by photosynthesis.

Specialized respiratory surfaces have evolved during evolution to make gas exchange more efficient. In humans and other tetrapods (four-limbed vertebrates), gas exchange occurs in the †lungs, aided by the breathing movements of the ribs. Many adult amphibia and terrestrial invertebrates can absorb oxygen directly through the skin. The bodies of insects and some spiders contain a system of air-filled tubes known as †tracheae. Fish have †gills as their main respiratory surface. In plants, gas exchange generally takes place via the †stomata and the air-filled spaces between the cells in the interior of the leaf.

gas giant in astronomy, any of the four large outer planets of the Solar System, †Jupiter, †Saturn, †Uranus, and †Neptune, which consist largely of gas and have no solid surface.

Planetary Rings Node http://ringside.arc.nasa.gov/

Superb planetary science resource focusing on the ring systems of the gas giants. There are reports, scientific papers, and suggestions for further reading. This regularly updated site has plenty of pictures and animations, but it is also possible to search the Rings Node Catalog of Voyager and Hubble Space Telescope images for pictures of planetary rings.

gasification in chemistry, process that converts solid and liquid compounds to gases of lower molecular weight, for example gasification of coal results in the formation of methane, carbon monoxide, and hydrogen.

Gaskell, Walter Holbrook (1847–1914) English physiologist who investigated the anatomy of the nervous system, particularly the †autonomic nervous system, with John †Langley. Gaskell originally worked on the nerve structures in striated muscle and the heart. He then moved on to describe the sympathetic or "voluntary" nervous system that carries signals to the skeletal muscles in response to external stimuli.

gas laws physical laws concerning the behavior of gases. They include †Boyle's law and †Charles's law, which are concerned with the relationships between the pressure (P), temperature (T), and volume (V) of an ideal (hypothetical) gas. These two laws can be combined to give the **general** or **universal gas law**, which may be expressed as: PV/T = constant.

The laws state that gas pressure depends on the temperature and volume of the gas. If the volume of a gas is kept constant, its pressure increases with temperature. When a gas is squeezed into a smaller volume, its pressure increases. If the pressure of a gas is constant, its volume increases with temperature. Such behavior of gases depends on the †kinetic theory of matter, and †particle theory, which states that all matter is composed of particles.

Gases http://chemed.chem.purdue.edu/genchem/topicreview/bp/ch4/1frame.html

Clear explanation of gases and the states of matter, part of an introductory chemistry site aimed at entry-level university students. Click on the sidebar for information on the properties of gases, gas laws, kinetic molecular theory, and deviation from the ideal gas law.

gas mantle illumination device made from impregnating a dome-shaped piece of rayon with compounds of thorium and cerium, which are decomposed by heat.

gasoline mixture of hydrocarbons derived from †petroleum, mainly used as a fuel for internal-combustion engines. It is colorless and highly volatile. Leaded gasoline contains antiknock (a mixture of tetra ethyl lead and dibromoethane), which improves the combustion and the performance of a car engine. The lead from the exhaust fumes enters the atmosphere, mostly as simple lead compounds. There is strong evidence that it can act as a nerve poison on young children and cause mental impairment. This prompted a gradual switch to the use of unleaded gasoline. The British term for gasoline is petrol.

Gaspra (or Gaspar) or **Asteroid 951**, first asteroid observed in close-up by a spacecraft. The Galileo probe made a fly-by in October 1991 on its way to the planet Jupiter. It passed Gaspra at 8 km/5 mi per second, coming within 1,600 km/992 mi to take many images of the S-type asteroid, made up of a mixture of nickel, iron, and silicates.

Gaspra's irregular dimensions are approximately 19x12x 11 km/12x 7x6.8 mi. It is estimated to be between 300 million and 500 million years old. Its orbit averages 205 million km/127 million mi from the Sun. The asteroid was discovered in 1916 by the Russian astronomer Grigoriy Neujmin, who named it after a Black Sea holiday resort.

Gassendi, Pierre (1592–1655) French physicist and philosopher who played a crucial role in the revival of atomism (the theory that the world is made of small, indivisible particles), and the rejection of Aristotelianism so characteristic of the period. He was a propagandist and critic of other views rather than an original thinker.

Gassendi, Pierre http://www-history.mcs.st-and.ac.uk/history/Mathematicians/Gassendi.html

Biography of the French 17th-century philosopher, mathematician, and theologian. This St Andrews University-run site also provides a list of literature references for further study of the life and works of Gassendi.

Gasser, Herbert Spencer (1888–1963) US physiologist who was awarded a Nobel Prize for Physiology or Medicine in 1944 with Joseph †Erlanger for their work on the transmission of impulses by nerve fibers. Gasser was one of the first to demonstrate the chemical transmission of nerve impulses.

gas syringe graduated piece of glass apparatus used to measure accurately the volumes of gases.

gastrolith stone that was once part of the digestive system of a dinosaur or other extinct animal. Rock fragments were swallowed to assist in the grinding process in the dinosaur digestive tract, much as some birds now swallow grit and pebbles to grind food in their crop. Once the animal has decayed, smooth round stones remain – often the only clue to their past use is the fact that they are geologically different from their surrounding strata.

gastrula stage in embryonic development following the †blastula stage in which †gastrulation occurs.

gastrulation cell movements during embryonic development (after cleavage) in which cells move to the positions in which they eventually give rise to the organs of the growing embryo.

gate, logic in electronics, see †logic gate.

gateway in computing, the point of contact between two †wide-area networks.

gauge any scientific measuring instrument – for example, a wire gauge or a pressure gauge. The term is also applied to the width of a railway or tramway track.

gauge boson or **field particle**, any of the particles that carry the four fundamental forces of nature (see †forces, fundamental). Gauge bosons are †elementary particles that cannot be subdivided, and include the photon, the graviton, the gluons, and the W^+, W^-, and Z particles.

gauss symbol Gs, centimeter-gram-second (*c*.g.s.) unit of magnetic induction or magnetic flux density, replaced by the SI unit, the †tesla, but still commonly used. It is equal to one line of magnetic flux per square centimeter. The Earth's magnetic field is about 0.5 Gs, and changes to it over time are measured in gammas (one gamma equals 10^{-5} gauss).

Gauss, Carl Friedrich (1777–1855) German mathematician who worked on the theory of numbers, non-Euclidean geometry, and the mathematical development of electric and magnetic theory. A method of neutralizing a magnetic field, used to protect ships from magnetic mines, is called "degaussing."

Gaussian distribution another name for †normal distribution.

Gay-Lussac, Joseph Louis (1778–1850) French physicist and chemist who investigated the physical properties of gases, and discovered new methods of producing sulfuric and oxalic acids. In 1802 he discovered the approximate rule for the expansion of gases now known as †Charles's law; see also †gas laws.

Gay-Lussac's law in chemistry, law that states when gases are involved in a chemical reaction, at constant temperature and pressure, their volume and that of the product gas is related by a simple ratio. French chemist Joseph-Louis †Gay-Lussac proposed the law in 1808.

gear toothed wheel that transmits the turning movement of one shaft to another shaft. Gear wheels may be used in pairs, or in threes if both shafts are to turn in the same direction. The gear ratio – the ratio of the number of teeth on the two wheels – determines the torque ratio, the turning force on the output shaft compared with the turning force on the input shaft. The ratio of the angular velocities of the shafts is the inverse of the gear ratio.

The common type of gear for parallel shafts is the **spur gear**, with straight teeth parallel to the shaft axis. The **helical gear** has teeth cut along sections of a helix or corkscrew shape; the double form of the helix gear is the most efficient for energy transfer. **Bevel gears**, with tapering teeth set on the base of a cone, are used to connect intersecting shafts.

Gegenschein (German for "counterglow") in astronomy, faint oval patch of light that sometimes appears in the sky opposite the Sun, visible from Earth only at certain times of the year. It is produced when dust in space scatters sunlight back to Earth.

Geiger, Hans (Wilhelm) (1882–1945) German physicist who produced the †Geiger counter. He spent the period 1906–12 in Manchester, England, working with Ernest †Rutherford on †radioactivity. In 1908 they designed an instrument to detect and count alpha particles, positively charged ionizing particles produced by radioactive decay.

Geiger counter any of a number of devices used for detecting nuclear radiation and measuring its intensity by counting the number of ionizing particles produced (see †radioactivity). It detects the momentary current that passes between †electrodes (anode and cathode) in a suitable gas (such as argon) when radiation causes the ionization of the gas. The electrodes are connected to electronic devices that enable the number of particles passing to be measured. The increased frequency of measured particles indicates the intensity of radiation. The device is named after the German physicist Hans †Geiger.

The becquerel (Bq) is the unit used to measure radioactive decay. The number of particles emitted in one second by a radioactive source is known as the activity (or disintegration): 1 Bq = 1 disintegration per second.

The Geiger–Müller, Geiger–Klemperer, and Rutherford–Geiger counters are all devices often referred to loosely as Geiger counters.

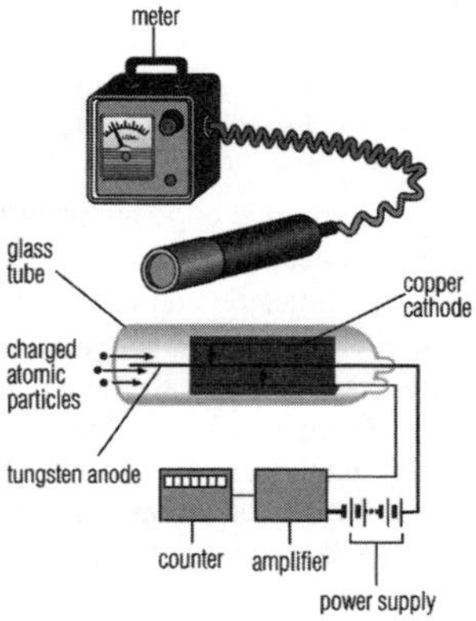

Geiger counter A Geiger–Müller counter detects and measures ionizing radiation (alpha particles, beta particles, and gamma rays) emitted by radioactive materials. Any incoming radiation creates ions (charged particles) within the counter, which are attracted to the anode and cathode to create a measurable electric current. Helicon

Geissler tube high-voltage discharge tube in which traces of gas are ionized and conduct electricity. Since the electrified gas takes on a luminous color characteristic of the gas, the instrument is also used in †spectroscopy. It was developed in 1858 by the German physicist Heinrich Geissler.

gel solid produced by the formation of a three-dimensional cage structure, commonly of linked large-molecular-mass polymers, in which a liquid is trapped. It is a form of †colloid. A gel may be a jellylike mass (pectin, gelatin) or have a more rigid structure (silica gel).

gelatine soluble protein-based substance with the ability to form a jelly on cooling. It is used in photography and the making of glues.

gel filtration in chemistry, a technique for the separation of molecules, according to size, by diffusion through a column of gel. It can be used to determine the molecular weights of molecules. Gel filtration is an important technique for the separation and purification of proteins.

gelifluction type of †solifluction (downhill movement of water-saturated topsoil) associated with frozen ground.

gelignite type of †dynamite.

Gell-Mann, Murray (1929–) US theoretical physicist. In 1964 he formulated the theory of the †quark as one of the fundamental constituents of matter. He was awarded the Nobel Prize for Physics in 1969 for his work on elementary particles and their interaction.

gem mineral valuable by virtue of its durability (hardness), rarity, and beauty, cut and polished for ornamental use, or engraved. Of 120 minerals known to have been used as gemstones, only about 25 are in common use in jewelry today; of these, the diamond, emerald, ruby, and sapphire are classified as precious, and all the others semi-precious; for example, the topaz, amethyst, opal, and aquamarine.

Among the synthetic precious stones to have been successfully produced are rubies, sapphires, emeralds, and diamonds (first produced by General Electric in the USA in 1955). Pearls are not technically gems.

Gemini prominent zodiacal constellation in the northern hemisphere represented as the twins Castor and Pollux. Its brightest star is †Pollux; †Castor is a system of six stars. The Sun passes through Gemini from late June to late July. Each December, the Geminid meteors radiate from Gemini. In astrology, the dates for Gemini are between about 21 May and 21 June (see †precession).

gemma plural **gemmae**, unit of †vegetative reproduction, consisting of a small group of undifferentiated green cells. Gemmae are found in certain mosses and liverworts, forming on the surface of the plant, often in cup-shaped structures, or gemmae cups. Gemmae are dispersed by splashes of rain and can then develop into new plants. In many species, gemmation is more common than reproduction by †spores.

gemmule in early genetic theory, minute particles thought to consist of miniature copies of all parts of the body carried in the blood to the †gametes and from which their larger forms eventually developed. In modern usage, a gemmule is a bud formed on a sponge that may break free and develop into a new animal.

gene basic unit of inherited material, encoded by a strand of †DNA and transcribed by †RNA. In higher organisms, genes are located on the †chromosomes. A gene consistently affects a particular character in an individual – for example, the gene for eye color. Also termed a Mendelian gene, after Austrian biologist Gregor †Mendel, it occurs at a particular point, or †locus, on a particular chromosome and may have several variants, or †alleles, each specifying a particular form of that character – for example, the alleles for blue or brown eyes. Some alleles show †dominance. These mask the effect of other alleles, known as †recessive. Genes can be manipulated using the techniques of †genetic engineering (gene technology).

The inheritance of genes and the way genes work is studied in †genetics. One gene carries the information that describes how one particular †protein is made. This information is stored as a chemical code on a DNA molecule and the genes are found in sequence from one end of the molecule to the other. Each protein that is made helps to determine part of the characteristics of an organism. Between them, all the proteins determine all the inherited characteristics of an organism, though some of these characteristics can be modified by the †environment. The DNA is located in the chromosomes in the †nucleus of a cell. Many thousands of genes are present on each chromosome. The total number of genes in a human, according to estimates published in 2001 by the †Human Genome Project, is thought to be between 27,000 and 40,000, distributed between the 46 chromosomes in each human cell. Occasionally, a gene or a larger part of a chromosome or the number of chromosomes becomes accidentally altered. Such a change is a †mutation. Mutations can cause an individual to have a disease or disorder, such as Huntington's disease, cystic fibrosis, or sickle-cell anemia. Gregor Mendel was the first to understand the mechanism of inheritance by genes, as a result of the study of plant breeding. He did not, however, know about the existence of DNA.

Genetic Technology http://whyfiles.org/075genome/

Wry and thought-provoking look at genetic science. Starting with a dig at the morals of a past US president, who was posthumously exposed by developments in gene mapping, this Web site continues with an intelligent consideration of the potential offered by the human genome project, and the limitations of genetics.

gene amplification technique by which selected DNA from a single cell can be duplicated indefinitely until there is a sufficient amount to analyze by conventional genetic techniques.

Gene amplification uses a procedure called the polymerase chain reaction. The sample of DNA is mixed with a solution of enzymes called polymerases, which enable it to replicate, and with a plentiful supply of nucleotides, the building blocks of DNA. The mixture is repeatedly heated and cooled. At each warming, the double-stranded DNA present separates into two single strands, and with each cooling the polymerase assembles a new paired strand for each single strand. Each cycle takes approximately 30 minutes to complete, so that after 10 hours there is one million times more DNA present than at the start.

The technique can be used to test for genetic defects in a single cell taken from an embryo, before the embryo is reimplanted in †in vitro fertilization.

gene bank collection of seeds or other forms of genetic material, such as tubers, spores, bacterial or yeast cultures, live animals and plants, frozen sperm and eggs, or frozen embryos. These are stored for possible future use in agriculture, plant and animal breeding, or in medicine, genetic engineering, or the restocking of wild habitats where species have become extinct. Gene banks may be increasingly used as the rate of extinction increases, depleting the Earth's genetic variety (biodiversity).

gene imprinting genetic phenomenon whereby a small number of genes function differently depending on whether they were inherited from the father or the mother. If two copies of an imprinted gene are inherited from one parent and none from the other, a genetic abnormality results, whereas no abnormality occurs if, as is normal, a copy is inherited from both parents. Gene imprinting is known to play a part in a number of genetic disorders and childhood diseases, for example, the Prader–Willi syndrome (characterized by mild mental retardation and compulsive eating).

gene pool total sum of †alleles (variants of †genes) possessed by all the members of a given population or species alive at a particular time.

General Catalogue in astronomy, catalogue of several thousand nebulae, compiled by English astronomer John †Herschel in 1864. It was superseded by the *New General Catalogue* in 1888.

generalize in mathematics, to extend a number of results to form a rule. For example, the computations 3 + 5 = 5 + 3 and 1.5 + 2.7 = 2.7 + 1.5 could be generalized to $a + b = b + a$.

generate in mathematics, to produce a sequence of numbers from either the relationship between one number and the next or the relationship between a member of the sequence and its position. For example, $u_{n+1} = 2u_n$ generates the sequence 1, 2, 4, 8, ... ; $a_n = n(n+1)$ generates the sequence of numbers 2, 6, 12, 20, ...

generation in computing, stage of development in computer electronics (see †computer generation) or a class of programming language (see †fourth-generation language).

generator machine that produces †electrical energy from mechanical energy, as opposed to an †electric motor, which does the opposite. A simple generator (known as a dynamo in the UK) consists of a wire-wound coil (†armature) that is rotated between the poles of a permanent magnet. As the coil rotates it cuts across the magnetic field lines and a current is generated. A dynamo on a bicycle is an example of a simple generator.

The current generated can be increased by using a stronger magnet, by rotating the armature faster, or by winding more turns of wire onto a larger armature.

gene-splicing see †genetic engineering.

gene therapy medical technique for curing or alleviating inherited diseases or defects that are due to a gene malfunction; certain infections, and several kinds of cancer in which affected cells from a sufferer are removed from the body, the †DNA repaired in the laboratory (†genetic engineering), and the normal functioning cells reintroduced. In 1990 a genetically engineered gene was used for the first time to treat a patient.

The first human to undergo gene therapy, in 1990, was one of the so-called "bubble babies" – a four-year-old American girl suffering from a rare enzyme (ADA) deficiency that cripples the immune system. Unable to fight off infection, such children are nursed in a germ-free bubble; they usually die in early childhood.

Cystic fibrosis is the most common inherited disorder and the one most keenly targeted by genetic engineers; the treatment has been pioneered in patients in the USA and UK. Gene therapy is not the final answer to inherited disease; it may cure the patient but it cannot prevent him or her from passing on the genetic defect to any children. However, it does hold out the promise of a cure for various other conditions, including some forms of heart disease and some cancers; US researchers have successfully used a gene gun to target specific tumor cells.

By the end of 1990s, although hundreds of people had been treated with gene therapy, nobody had actually been cured. Even in the ADA trials, the most successful to date, the children were still receiving injections of synthetic ADA, possibly the major factor in their improvement. The death of US teenager Jesse Gelsinger during a trial in 1999 presented a serious setback to further trials.

genetically modified organism GMO, organism into which genes from a different species have been transferred. See †genetic engineering.

genetic code way in which instructions for building proteins, the basic structural molecules of living matter, are "written" in the genetic material †DNA. This relationship between the sequence of bases (the subunits in a DNA molecule) and the sequence of †amino acids (the subunits of a protein molecule) is the basis of heredity. The code employs †codons of three bases each; it is the same in almost all organisms, except for a few minor differences recently discovered in some protozoa.

Only 2% of DNA is made up of base sequences, called **exons**, that code for proteins. The remaining DNA is known as "junk" DNA or **introns**.

genetic engineering all-inclusive term that describes the deliberate manipulation of genetic material by biochemical techniques. It is often achieved by the introduction of new †DNA, usually by

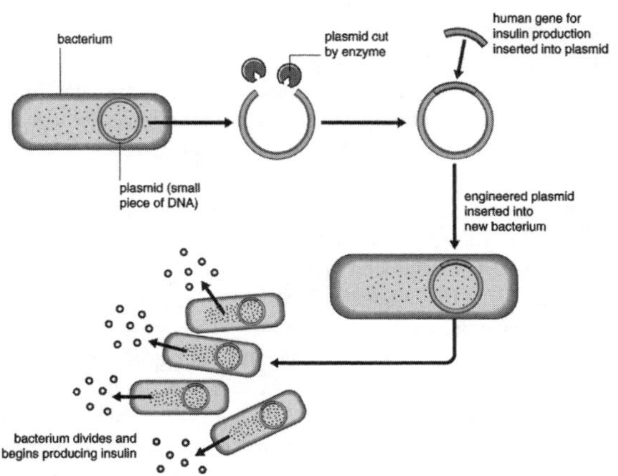

genetic engineering The genetic modification of a bacterium to produce insulin. The human gene for the production of insulin is collected from a donor chromosome and spliced into a vector plasmid (DNA found in bacteria but separate from the bacterial chromosomes). The plasmids and recipient bacteria are mixed together, during which process the bacteria absorb the plasmids. The plasmids replicate as the bacteria divide asexually (producing clones) and begin to produce insulin. Helicon

means of a virus or †plasmid. This can be for pure research, †gene therapy, or to breed functionally specific plants, animals, or bacteria. These organisms with a foreign gene added are said to be transgenic (see †transgenic organism) and the new DNA formed by this process is said to be recombinant. In most current cases the transgenic organism is a micro-organism or a plant, because ethical and safety issues are limiting its use in mammals. At the beginning of 1995 more than 60 plant species had been genetically engineered, and nearly 3,000 transgenic crops had been field-tested.

The process of genetic engineering involves several steps: the formation of DNA fragments, the insertion of DNA fragments into a vector plasmid, cloning of the plasmid, use of the plasmid to introduce the DNA into the organism, and expression of the gene.

One example of genetic engineering that has been very helpful to humans is the production of bacteria that make human insulin – the bacteria were engineered to contain the human gene for insulin. The bacteria are cultivated in fermenters to produce large amounts of insulin, which is then used to treat diabetic patients (see †diabetes). Prior to this, people with diabetes were treated with insulin from other animals. This new procedure removes the need to kill animals for insulin, and in addition the engineered insulin works better.

Biotechnology http://www.foodfuture.org.uk/bio6.htm

Part of the "Food for our future" Web site, which aims to inform consumers about the benefits and concerns over biotechnology. This presentation focuses on the positive side of genetically modified foods.

Question of Genes: Inherited Risks http://www.pbs.org/gene/

Companion to a US Public Broadcasting Service (PBS) television program, this page explores the controversial subject of genetic testing, emphasizing its social, ethical, and emotional consequences. It includes summaries of the seven case studies featured on the television program, including video clips. There is also a section on the basics of genetic testing, which features the transcript of an interview with a genetic counselor. You will also find a list of resources for further research and a forum for discussion.

genetic fingerprinting or **genetic profiling**, technique developed in the UK by Professor Alec Jeffreys (1950–), and now allowed as a means of legal identification. It determines the pattern of certain parts of the genetic material †DNA that is unique to each individual. Like conventional fingerprinting, it can accurately distinguish humans from one another, with the exception of identical siblings from multiple births. It can be applied to as little material as a single cell.

Genetic fingerprinting involves isolating DNA from cells, then comparing and contrasting the sequences of component chemicals between individuals. The DNA pattern can be ascertained from a sample of skin, hair, blood, or semen. Although differences are minimal (only 0.1% between unrelated people), certain regions of DNA, known as **hypervariable regions**, are unique to individuals.

genetics branch of biology concerned with the study of †heredity and variation – inheritance. It attempts to explain how characteristics of living organisms are passed on from one generation to the next. The science of genetics is based on the work of Austrian biologist Gregor †Mendel whose experiments with the cross-breeding (hybridization) of peas showed that the inheritance of characteristics and traits takes place by means of discrete "particles," now known as †genes. These are present in the cells of all organisms, and are now recognized as being the basic units of heredity. All organisms possess †genotypes (sets of variable genes) and †phenotypes (characteristics produced by certain genes). Modern geneticists investigate the structure, function, and transmission of genes.

Before the publication of Mendel's work in 1865, it had been assumed that the characteristics of both parents were blended during inheritance, but Mendel showed that the genes remain intact, although their combinations change. As a result of his experiments with the cultivation of the common garden pea, Mendel introduced the concept of hybridization (see †monohybrid inheritance). Since Mendel, the study of genetics has advanced greatly, first through †breeding experiments and light-microscope observations (classical genetics), later by means of biochemical and electron microscope studies (molecular genetics). In 1944 Canadian-born bacteriologist Oswald Avery, together with his colleagues at the Rockefeller Institute, Colin McLeod and Maclyn McCarty, showed that the genetic material was deoxyribonucleic acid (†DNA), and not protein as was previously thought. A further breakthrough was made in 1953 when James †Watson and Francis †Crick published their molecular model for the structure of DNA, the double helix, based on X-ray diffraction photographs. The following decade saw the cracking of the †genetic code. The genetic code is said to be universal since the same code applies to all organisms from bacteria and viruses to higher plants and animals, including humans. Today the deliberate manipulation of genes by biochemical techniques, or †genetic engineering, takes place.

Natural History of Genetics http://gslc.genetics.utah.edu/

Through a combination of scientific experts and teachers, this site offers an accessible and well-designed introduction to genetics. It includes several guided projects with experiments and explanations aimed initially at young teenage children. However, this site also includes "Intermediate" and "Expert" sections allowing this page to be used by a wide variety of ages and levels of expertise. In addition to the experiments, the site also includes sections on such topics as "Core genetics," "Teacher workshops," and "Fun stuff."

genitalia reproductive organs of sexually reproducing animals, particularly the external/visible organs of mammals: in males, the penis and the scrotum, which contains the testes, and in females, the clitoris and vulva.

genome full complement of †genes carried by a single (haploid) set of †chromosomes. The term may be applied to the genetic information carried by an individual or to the range of genes found in a given species. The human genome is made up of between 27,000 and 40,000 genes, according to a rough draft of the sequenced genome completed by the †Human Genome Project in June 2000.

Genomes have been identified for many other organisms, including the bacteria *Haemofilus influenzae* and *Escherichia coli*, and the mycoplasmas *Mycoplasma genitalium* and *Mycoplasma pneumoniae*. Scientists also completed a genetic blueprint for *Saccharomyces cerevisiae*, common brewer's yeast, which shares a high number of genetic sequences with humans.

genotype particular set of †alleles (variants of genes) possessed by a given organism. The term is usually used in conjunction with †phenotype, which is the product of the genotype and all environmental effects.

genus plural **genera**, group of one or more †species with many characteristics in common. Thus all dog-like species (including dogs, wolves, and jackals) belong to the genus *Canis* (Latin "dog").

Species of the same genus are thought to be descended from a common ancestor species. Related genera are grouped into †families.

geocentric having the Earth at the center.

geochemistry science of chemistry as it applies to geology. It deals with the relative and absolute abundances of the chemical elements and their †isotopes in the Earth, and also with the chemical changes that accompany geologic processes.

geochronology branch of geology that deals with the dating of rocks, minerals, and fossils in order to create an accurate and precise geological history of the Earth. The †geological time scale is a result of these studies. It puts stratigraphic units in chronological order and assigns actual dates, in millions of years, to those units.

geode in geology, a subspherical cavity into which crystals have grown from the outer wall into the center. Geodes often contain very well-formed crystals of quartz (including amethyst), calcite, or other minerals.

geodesic shortest route between two points on any surface.

geodesy science of measuring and mapping Earth's surface for making maps and correlating geological, gravitational, and magnetic measurements. Geodetic surveys, formerly carried out by means of various measuring techniques on the surface, are now commonly made by using radio signals and laser beams from orbiting satellites.

Geoffroy Saint-Hilaire, Etienne (1772–1844) French zoologist who performed some of the earliest embryological experiments on chicks. He tried to halt the development of a chick embryo at the stage of a fish in order to demonstrate that evolutionary stages of development are recapitulated in the embryo.

geography study of the Earth's surface; its topography, climate, and physical conditions, and how these factors affect people and society. It is usually divided into **physical geography**, dealing with landforms and climates, and **human geography**, dealing with the distribution and activities of peoples on Earth.

history Early preclassical geographers concentrated on map-making, surveying, and exploring. In classical Greece theoretical ideas first became a characteristic of geography. Aristotle and †Pythagoras believed the Earth to be a sphere, †Eratosthenes was the first to calculate the circumference of the world, and Herodotus investigated the origin of the Nile floods and the relationship between climate and human behavior.

During the medieval period the study of geography progressed little in Europe, but the Muslim world retained much of the Greek tradition, embellishing the 2nd-century maps of †Ptolemy. During the early Renaissance the role of the geographer as an explorer and surveyor became important once again.

The foundation of modern geography as an academic subject stems from the writings of Friedrich †Humboldt and Johann Ritter, in the late 18th and early 19th centuries, who for the first time defined geography as a major branch of scientific inquiry.

Greatest Places http://www.greatestplaces.org/

Journey to seven of the most geographically dynamic locations on Earth. This site features stylish presentations with extensive pictures and cultural commentary.

The Geography Portal http://www.georesources.co.uk/

Extensive geographical resource, which includes outline maps in GIF format, and geographical photographs, both of which are freely available to download.

geological time time scale embracing the history of the Earth from its physical origin to the present day.

Geological time is traditionally divided into eons (Archaean or Archaeozoic, Proterozoic, and Phanerozoic in ascending chronological order), which in turn are subdivided into eras, periods, epochs, ages, and finally chrons.

The terms eon, era, period, epoch, age and chron are **geochronological units** representing intervals of geological time. Rocks representing an interval of geological time comprise a **chronostratigraphic** (or **time-stratigraphic**) **unit**. Each of the hierarchical geochronological terms has a chronostratigraphic equivalent. Thus, rocks formed during an eon (a geochronological unit) are members of an eonothem (the chronostratigraphic unit equivalent of eon). Rocks of an era belong to an erathem. The chronostratigraphic equivalents of period, epoch, age, and chron are system, series, stage, and chronozone, respectively.

geology science of the Earth, its origin, composition, structure, and history. It is divided into several branches, including **mineralogy** (the minerals of Earth), **petrology** (rocks), **stratigraphy** (the deposition of successive beds of sedimentary rocks), **paleontology** (fossils) and **tectonics** (the deformation and movement of the Earth's crust), **geophysics** and **geochemistry**.

Adventures in the Learning Web http://www.usgs.gov/education/

Huge range of online educational geological resources, including sections on "Finding your way with map and compass," "Monitoring active volcanoes," and "This dynamic earth." The site is run by the US Geological Survey, so some of the material is US-specific. However, a whole host of more general earth science topics are covered on this site in clear, well-illustrated sections. This site is regularly updated with information on any natural disasters as and when they happen.

Ask-A-Geologist http://walrus.wr.usgs.gov/docs/ask-a-ge.html/

Do you have a question about volcanoes, earthquakes, mountains, rocks, maps, ground water, lakes, or rivers? If you do, the US Geological Survey invites you to ask one of their earth scientists a question at this site.

Geologylink http://college.hmco.com/geology/resources/geologylink/index.html

Comprehensive information on geology featuring a daily update on current geologic events, virtual classroom tours, and virtual field trips to locations around the world. The site also provides an image gallery, maps, and a glossary giving clear and concise descriptions.

US Geological Survey http://www.usgs.gov/

US Geological Survey (USGS) site includes information on almost every aspect of earth science you can think of. It is well-organized and contains a useful search engine to make finding things very easy.

geometric curve curve that can be precisely expressed by an equation (unlike a mechanical curve); for example, a circle, parabola, or hyperbola.

geometric mean in mathematics, the *n*th root of the product of *n* positive numbers. The geometric mean *m* of two numbers *p* and *q* is such that $m = \sqrt{(p \times q)}$. For example, the geometric mean of 2 and 8 is $\sqrt{(2 \times 8)} = \sqrt{16} = 4$.

geometric progression or **geometric sequence**, in mathematics, a sequence of terms (progression) in which each term is a constant multiple (called the **common ratio**) of the one preceding it. For example, 3, 12, 48, 192, 768, ... is a geometric progression with a common ratio 4, since each term is equal to the previous term multiplied by 4. Compare †arithmetic progression.

The sum of *n* terms of a **geometric series**

$1 + r + r^2 + r^3 + ... + r^{n-1}$

is given by the formula

$S_n = (1 - r^n)/(1 - r)$ for all $r \neq 1$.

For mod $r < 1$ (see †absolute value), the geometric series can be summed to infinity:

$S_\infty = 1/(1 - r)$.

In nature, many single-celled organisms reproduce by splitting in two so that one cell gives rise to 2, then 4, then 8 cells, and so on, forming a geometric sequence 1, 2, 4, 8, 16, 32, ..., in which the common ratio is 2.

geometric series in mathematics, infinite series of the form $a + ar^2 + ar^3 + ...$, in which each term is a constant multiple *r* of the previous one. Such a series converges if $|r|<1$ and diverges if $|r|>1$.

geometry branch of mathematics concerned with the properties of space, usually in terms of †plane (†two-dimensional, or 2D) and solid (†three-dimensional, or 3D) figures. The subject is usually divided into **pure geometry**, which embraces roughly the plane and solid geometry dealt with in Greek mathematician †Euclid's *Stoicheia/Elements*, and **analytical** or †**coordinate geometry**, in which problems are solved using algebraic methods. A third, quite distinct, type includes the non-Euclidean geometries.

pure geometry This is chiefly concerned with properties of figures that can be measured, such as lengths, †areas, and †angles and is therefore of great practical use. An important idea in Euclidean geometry is the idea of **congruence**. Two figures are said to be congruent if they have the same shape and size (and area). If one figure is imagined as a rigid object that can be picked up, moved and placed on top of the other so that they exactly coincide, then the two figures are congruent. Some simple rules about congruence may be stated: two †line segments are congruent if they are of equal length; two †triangles are congruent if their corresponding sides are equal in length or if two sides and an angle in one is equal to those in the other; two †circles are congruent if they have the same radius; two †polygons are congruent if they can be divided into congruent triangles assembled in the same order.

The idea of picking up a rigid object to test congruence can be expressed more precisely in terms of elementary "movements" of figures: a †translation (or glide) in which all points move the same distance in the same direction (that is, along parallel lines); a †rotation through a defined angle about a fixed point; a †reflection (equivalent to turning the figure over).

Two figures are congruent if one can be transformed into the other by a sequence of these elementary movements. In Euclidean geometry a fourth kind of movement is also studied; this is the †enlargement in which a figure grows or shrinks in all directions by a uniform scale factor. If one figure can be transformed into another by a combination of translation, rotation, reflection, and enlargement then the two are said to be similar. All circles are similar. All squares are similar. Triangles are †similar triangles if corresponding angles are equal.

coordinate geometry A system of geometry in which points, lines, shapes, and surfaces are represented by algebraic expressions. In plane (two-dimensional) coordinate geometry, the plane is usually defined by two axes at right angles to each other, the horizontal *x*-axis and the vertical *y*-axis, meeting at O, the origin. A point on the plane can be represented by a pair of †Cartesian coordinates, which define its position in terms of its distance along the *x*-axis and along the *y*-axis from O. These distances are, respectively, the *x* and *y* †coordinates of the point.

Lines are represented as †equations; for example, $y = 2x + 1$ gives a †straight line, and $y = 3x^2 + 2x$ gives a †parabola (a curve). The †graphs of varying equations can be drawn by plotting the coordinates of points that satisfy their equations, and joining up the points. One of the advantages of coordinate geometry is that geometrical solutions can be obtained without drawing but by manipulating algebraic expressions. For example, the coordinates of the point of intersection of two straight lines can be determined by finding the unique values of *x* and *y* that satisfy both of the equations for the lines, that is, by solving them as a pair of †simultaneous equations. The curves studied in simple coordinate geometry are the †conic sections (circle, ellipse, parabola, and hyperbola), each of which has a characteristic equation.

Geometry probably originated in ancient Egypt, in land measurements necessitated by the periodic inundations of the River Nile, and was soon extended into surveying and navigation. Early geometers were the Greek mathematicians Thales, Pythagoras, and Euclid. Analytical methods were introduced and developed by the French philosopher René †Descartes in the 17th century. From the 19th century, various non-Euclidean geometries were devised by Carl Friedrich Gauss, János Bolyai, and Nikolai Lobachevsky. These were later generalized by Bernhard †Riemann and found to have applications in the theory of relativity.

Areas and Volumes http://www.gcsemaths.fsnet.co.uk/Areas%20and%20Volumes.htm

Introduction to area and volume, with advice on good exam technique. Area is defined first, along with some equations for the areas of common shapes. Volume is then defined, along with the equations for calculating the surface areas and volumes of three-dimensional solids.

Connecting Geometry http://www.k12.hi.us/~csanders/Welcome.html

Geometry lessons online. The site takes an applied look at geometry, covering symmetry, congruent triangles, circles, volume, area, and more! At the end of many of the lessons, there is a suggested project to carry out to extend what has been covered. This is a well-illustrated site with plenty of photographs and diagrams.

Similar Triangles http://www.projectgcse.co.uk/maths/similar.htm

Similar triangles and the ratios between their sides are covered by this concise page. It defines how triangles can be similar, and shows a worked example complete with a diagram. Links at the top of the page lead to other math sections on the project GCSE Web site.

geomorphology branch of geology developed in the late 19th century, dealing with the morphology, or form, of the Earth's surface; nowadays it is also considered to be an integral part of physical geography. Geomorphological studies investigate the nature and origin of surface landforms, such as mountains, valleys, plains, and plateaus, and the processes that influence them. These processes include the effects of tectonic forces, †weathering, running water, waves, glacial ice, and wind, which result in the †erosion, †mass movement (landslides, rockslides, mudslides), transportation, and deposition of †rocks and †soils.

In addition to the natural processes that mould landforms, human activity can produce changes, either directly or indirectly, and cause the erosion, transportation, and deposition of rocks and soils, for example by poor land management practices and techniques in farming and forestry, and in the mining and construction industries.

Geomorphology deals with changes in landforms from the present to the geologic past, and in spatial scales ranging from microscale to mountains. For example, the formation of mountain ranges takes place over millions of years, as the Earth's crust cools and solidifies and the resulting layers, or plates, are folded, uplifted or deformed by the seismic activity of the underlying magma (see †plate tectonics). The gouging out of river valleys by †glacial erosion is a gradual process that takes place over thousands of years. Conversely, volcanic eruptions, by the ejection of rocks and gases and the rapid flow of molten lava down a mountainside, create rapid changes to landforms, as with the volcanic eruptions on the island of Montserrat in the West Indies. Similarly, the eruption of undersea volcanoes can result in the sudden birth of islands, while the consequent and rapidly moving tidal waves (†tsunamis), can produce the unexpected inundation and destruction of low-lying coastal regions in their path.

Landforms and Bodies of Water http://www.enchantedlearning.com/geography/landforms/glossary.shtml

Providing a comprehensive list of different types of landforms, such as capes, islands, and plateaus, this page acts as a starting point for definitions and locations of landforms throughout the world. Each sub-page contains a basic definition, locations where such landforms can be found, pictures, and case study style descriptions. A good resource for revision, or building geography case studies.

Models of Landform Development http://www.geog.ouc.bc.ca/physgeog/contents/11a.html

Discussion of the four types of landform to be found on Earth, and how they develop. Part of a much larger site, Fundamentals of Physical Geography, set up by a Canadian university, the text is supported by a graphical model explaining the relationship between geomorphic processes and landform types.

geophysics branch of earth science using physics (for instance gravity, seismicity, and magnetism) to study the Earth's surface, interior, and atmosphere. Geophysics includes several sub-fields such as seismology, paleomagnetism, and remote sensing.

geostationary orbit or **geostationary Earth orbit**, circular path 35,900 km/22,300 mi above the Earth's Equator on which a †satellite takes 24 hours, moving from west to east, to complete an orbit, thus appearing to hang stationary over one place on the Earth's surface. Geostationary orbits are used particularly for communications satellites and weather satellites.

geothermal energy energy extracted for heating and electricity generation from natural steam, hot water, or hot dry rocks in the Earth's crust. It is a form of †renewable energy. Water is pumped down through an injection well where it passes through joints in the hot rocks. It rises to the surface through a recovery well and may be converted to steam or run through a heat exchanger. Steam may be directed through †turbines to produce †electrical energy. It is an important source of energy in volcanically-active areas such as Iceland and New Zealand.

The heat is produced by volcanic activity and, possibly, by the decay of the Earth's natural radioactive substances.

Gerard of Cremona (1114–1187) Italian translator of the astronomical works of †Ptolemy. His most celebrated work is the translation of Ptolemy's *Almagest*.

germ colloquial term for a micro-organism that causes disease, such as certain †bacteria and †viruses. Formerly, it was also used to mean something capable of developing into a complete organism (such as a fertilized egg, or the †embryo of a seed).

Germain, Sophie (Marie) (1776–1831) French mathematician, born in Paris. Although she was not allowed to study at the newly opened Ecole Polytechnique, she corresponded with the mathematicians Joseph †Lagrange and Karl †Gauss. She is remembered for work she carried out in studying †Fermat's principle.

germanium chemical symbol Ge, brittle, white-white, weakly metallic (†metalloid) element, atomic number 32, relative atomic mass 72.6. It belongs to the silicon group, and has chemical and physical properties between those of silicon and tin. Germanium is a semiconductor material and is used in the manufacture of transistors and integrated circuits. The oxide is transparent to infrared radiation, and is used in military applications. It was discovered in 1886 by German chemist Clemens Winkler (1838–1904).

In parts of Asia, germanium and plants containing it are used to treat a variety of diseases, and it is sold in the West as a food supplement despite fears that it may cause kidney damage.

germination in botany, the initial stages of growth in a seed, spore, or pollen grain. Seeds germinate when they are exposed to favorable external conditions of moisture, light, and temperature, and when any factors causing dormancy have been removed.

The process begins with the uptake of water by the seed. The embryonic root, or radicle, is normally the first organ to emerge, followed by the embryonic shoot, or plumule. Food reserves, either within the †endosperm or from the †cotyledons, are broken down to nourish the rapidly growing seedling.

Germination is considered to have ended with the production of the first true leaves.

germ layer in †embryology, a layer of cells that can be distinguished during the development of a fertilized egg. Most animals have three such layers: the inner, middle, and outer. These differentiate to form the various body tissues.

The inner layer (**endoderm**) gives rise to the gut, the middle one (**mesoderm**) develops into most of the other organs, while the outer one (**ectoderm**) gives rise to the skin and nervous system. Simple animals, such as sponges, lack a mesoderm.

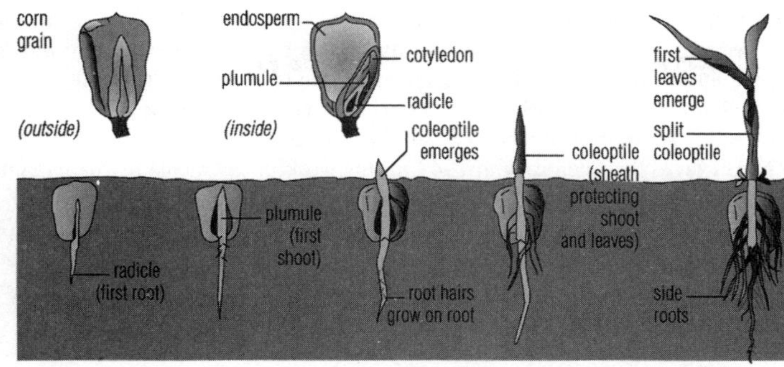

germination The germination of a corn grain. The plumule and radicle emerge from the seed coat and begin to grow into a new plant. The coleoptile protects the emerging bud and the first leaves. Helicon

germ plasm theory in early genetics, the theory that of the two tissue types in multicellular animals (somatoplasm in body cells and germ plasm in reproductive cells) only the integrity of germ plasm is necessary for the inheritance of characteristics.

gestation in all mammals except the †monotremes (platypus and spiny anteaters), the period from the time of implantation of the embryo in the uterus to birth. This period varies among species; in humans it is about 266 days, in elephants 18–22 months, in cats about 60 days, and in some species of marsupial (such as opossum) as short as 12 days.

geyser natural spring that intermittently discharges an explosive column of steam and hot water into the air due to the build-up of steam in underground chambers. One of the most remarkable geysers is Old Faithful, in Yellowstone National Park, Wyoming, USA. Geysers also occur in New Zealand and Iceland.

g-force force experienced by a pilot or astronaut when the craft in which he or she is traveling accelerates or decelerates rapidly. The unit g denotes the acceleration due to gravity, where 1 g is the ordinary pull of gravity.

Early astronauts were subjected to launch and re-entry forces of up to 6 g or more; in the space shuttle, more than 3 g is experienced on lift-off. Pilots and astronauts wear g-suits that prevent their blood pooling too much under severe g-forces, which can lead to unconsciousness.

Giacconi, Riccardo (1931–) Italian-born US physicist. He is the head of a team whose work has been fundamental in the development of †X-ray astronomy. In 1962 a rocket sent up by Giacconi and his group to observe secondary spectral emission (see †spectrum) from the Moon detected strong X-rays from a source evidently located outside the Solar System. X-ray research has since led to the discovery of many types of stellar and interstellar material. Giacconi and his team developed a telescope capable of producing X-ray images. In 1970 they launched a satellite called Uhuru, devoted entirely to the detection of stellar and interstellar X-rays.

Giaever, Ivar (1929–) Norwegian-born US physicist who worked on tunneling (the flow of an electric current through a thin film of insulating material sandwiched between two metal plates when a voltage is applied to the plates) and †superconductivity. He shared the Nobel Prize for Physics in 1973 with Brian †Josephson and Leo †Esaki for their work on the tunneling phenomena in semiconductors and superconductors.

giant molecular structure or **giant covalent structure** or **macromolecular structure**, solid structure made up of many similar atoms joined by covalent bonds in one dimension (long chains, such as †polymers), two dimensions (flat sheets as in graphite, where each layer is a giant molecule), or in three dimensions (such as in diamond and silica).

general properties Giant molecules do not conduct electricity as they have no free electrons as solids and no free ions when molten. They are some of the hardest structures known because of the strong bonds between their atoms. These bonds also give them high melting and boiling points. †Graphite, an allotrope of carbon, is an anomaly here as free electrons between its layers enable it to conduct electricity.

giant planet or **gas giant**, any of the large low-density planets of the outer Solar System: Jupiter, Saturn, Uranus, and Neptune.

giant star member of a class of stars located at the top right pf the †Hertzsprung-Russell diagram, characterized by great size and †luminosity. Giants have exhausted their supply of hydrogen fuel and derive their energy from the fusion of helium and heavier elements. They are roughly 10–300 times bigger than the Sun with 30–1,000 times the luminosity. The cooler giants are known as †red giants.

Giauque, William Francis (1895–1982) Canadian-born US physical chemist who was awarded the Nobel Prize for Chemistry in 1949 for his work in chemical †thermodynamics, in particular the behavior of matter at extremely low temperatures.

gibberellin plant growth substance (see also †auxin) that promotes stem growth and may also affect the breaking of dormancy in certain buds and seeds, and the induction of flowering. Application of gibberellin can stimulate the stems of dwarf plants to additional growth, delay the aging process in leaves, and promote the production of seedless fruit (†parthenocarpy).

gibbous in astronomy, describing the shape of the Moon, Mercury, or Venus when they are between half and full (with more than half illuminated as seen from the Earth).

Gibbs, Josiah Willard (1839–1903) US theoretical physicist and chemist who developed a mathematical approach to thermodynamics and established vector methods in physics. He devised the phase rule and formulated the Gibbs adsorption isotherm.

Gibbs, Josiah http://www-history.mcs.st-and.ac.uk/history/Mathematicians/Gibbs.html

Photograph and biography of the 19th-century US mathematician and chemist. This site, run by St Andrews University, also provides information on Gibbs's constant along with literature references for further study.

Gibbs' function in †thermodynamics, an expression representing part of the energy content of a system that is available to do external work, also known as the free energy G. In an equilibrium system at constant temperature and pressure, $G = H - TS$, where H is the enthalpy (heat content), T the temperature, and S the †entropy (decrease in energy availability). The function was named after US physicist Josiah Willard Gibbs.

Gibbs–Helmholtz equation in chemistry, a thermodynamic formula that relates changes in the free energy and heat of reaction that take place during chemical reactions. It can be used to calculate the heat of reaction if the change in †Gibbs' function and temperature are known.

Gibbs's free energy or **free energy**, in chemistry, a state function of a system in chemical equilibrium. Changes in the value of free energy can be used to determine if a reaction is thermodynamically favorable. See †Gibbs' function.

giga- prefix signifying multiplication by 10^9 (1,000,000,000 or 1 billion), as in **gigahertz**, a unit of frequency equivalent to 1 billion hertz.

gigabyte in computing, a measure of †memory capacity, equal to 1,024 †megabytes. It is also used, less precisely, to mean 1,000 billion †bytes.

GIGO acronym for Garbage In, Garbage Out, expression used in computing to emphasize that inaccurate input data will result in inaccurate output data.

Gilbert, Walter (1932–) US molecular biologist who studied genetic control, seeking the mechanisms that switch genes on and off. By 1966 he had established the existence of the lac repressor, a molecule that suppresses lactose production. He was awarded the Nobel Prize for Chemistry in 1980 for his work on the sequencing of †DNA nucleotides. He shared the award with Frederick †Sanger and Paul †Berg.

Gilbert, William (1540–1603) English scientist who studied magnetism and static electricity, deducing that the Earth's magnetic field behaves as if a bar magnet joined the North and South poles. His book on magnets, published in 1600, is the first printed scientific book based wholly on experimentation and observation.

gill in biology, the main respiratory organ of most fishes and immature amphibians, and of many aquatic invertebrates. In all types, water passes over the gills, and oxygen diffuses across the gill membranes into the circulatory system, while carbon dioxide passes from the system out into the water.

In aquatic insects, these gases diffuse into and out of air-filled canals called tracheae.

gill imperial unit of volume for liquid measure, equal to one-quarter of a pint or five fluid ounces (0.142 liter), traditionally used in selling alcoholic drinks.

In southern England it is also called a noggin, but in northern England the large noggin is used, which is two gills.

Gill, David (1843–1914) Scottish astronomer. He pioneered the use of photography to catalogue stars and also made much use of a heliometer, determining the solar †parallax and measuring the distances of 20 of the brighter and nearer southern stars.

Gilman, Alfred (1941–) US pharmacologist who was awarded a Nobel Prize for Physiology or Medicine in 1994 with Martin Rodbell for their discovery of a family of proteins (G-proteins) that translate messages – in the form of hormones or other chemical signals – into action inside cells.

Gilman, Henry (1893–1986) US organic chemist. He made a comprehensive study of methods of high-yield †synthesis, quantitative and †qualitative analysis, and uses of organometallic compounds, particularly †Grignard reagents.

ginkgo or **maidenhair tree**, tree belonging to the †gymnosperm (or naked-seed-bearing) division of plants. It may reach a height of 30 m/100 ft by the time it is 200 years old. (*Ginkgo biloba.*)

The only living member of its group (Ginkgophyta), widespread in Mesozoic times (245–65 million years ago), it has been cultivated in China and Japan since ancient times, and is planted in many parts of the world. Its leaves are fan-shaped, and it bears fleshy, yellow, foul-smelling fruit enclosing edible kernels.

Ginkgoales order of fossil branching trees dating from the Triassic. They have been found worldwide. Two families are known, the Trichopityaceae and the Ginkgoaceae. See †ginkgo.

Ginzburg, Vitalii Lazarevich (1916–) Russian astrophysicist. His use of †quantum theory in a study of †Cherenkov radiation contributed to the development of nuclear physics. He was one of the first to believe that cosmic background radiation comes from beyond our own Galaxy.

gizzard muscular grinding organ of the digestive tract, below the †crop of birds, earthworms, and some insects, and forming part of the †stomach. The gizzard of birds is lined with a hardened horny layer of the protein keratin, preventing damage to the muscle layer during the grinding process. Most birds swallow sharp grit which aids maceration of food in the gizzard.

glacial deposition laying-down (deposition) of †sediment once carried by a glacier. When ice melts, it deposits the material that it has been carrying. The material deposited by a glacier is called †till, or in Britain **boulder clay**. It comprises angular particles of all sizes from boulders to clay that are unsorted and lacking in stratification (layering).

Unstratified till can be molded by ice to form †drumlins (egg-shaped hills). At the snout of the glacier, till piles up to form a ridge called a terminal †moraine. Glacial deposits occur in many different locations – beneath the ice (subglacial), inside it (englacial), on top of it (supraglacial), at the side of it (marginal), and in front of it (proglacial).

Stratified till that has been deposited by meltwater is termed **fluvioglacial**, because it is essentially deposited by running water. Meltwater flowing away from a glacier will carry some of the till many kilometers away. This sediment will become rounded (by the water) and, when deposited, will form a gently sloping area called an **outwash plain**. Several landforms owe their existence to meltwater (**fluvioglacial landforms**) and include the long ridges called †eskers, which form parallel to the direction of the ice flow. Meltwater may fill depressions eroded by the ice to form **ribbon lakes**. Small depositional landforms may also result from glacial deposition, such as **kames** (small mounds) and **kettle holes** (small depressions, often filled with water).

In †periglacial environments on the margins of an icesheet, freeze–thaw weathering (the alternate freezing and thawing of ice in cracks in the rock) etches the outlines of rock outcrops, exploiting joints and areas of weakness, and results in aprons of †scree.

glacial erosion wearing-down and removal of rocks and soil by a †glacier. Glacial erosion forms impressive landscape features, including †glacial troughs (U-shaped valleys), †arêtes (steep ridges), †corries (enlarged hollows), and †pyramidal peaks (high mountain peaks with three or more arêtes).

Erosional landforms result from †abrasion and †plucking of the underlying bedrock. Abrasion is caused by the rock debris carried by a glacier, wearing away the bedrock. The action is similar to that of sandpaper attached to a block of wood. The results include the polishing and scratching of rock surfaces to form powdered rock flour, and scratches or †striations which indicate the direction of ice movement. Plucking is a form of glacial erosion restricted to the lifting and removal of blocks of bedrock already loosened by †freeze–thaw activity.

The most extensive period of recent glacial erosion was the †Pleistocene epoch (1.6 million to 10,000 years ago) in the †Quaternary period (last 2 million years) when, over a period of 2 to 3 million years, the polar icecaps repeatedly advanced and retreated. More ancient glacial episodes are also preserved in the geological record, the earliest being in the middle †Precambrian era (4.6 billion to 570 million years ago) and the most extensive in Permo-Carboniferous times.

glacial trough or **U-shaped valley**, steep-sided, flat-bottomed valley formed by a glacier. The erosive action of the glacier and of the debris carried by it results in the formation not only of the trough itself but also of a number of associated features, such as †hanging valleys (smaller glacial valleys that enter the trough at a higher level than the trough floor). Features characteristic of glacial deposition, such as †drumlins, are commonly found on the floor of the trough, together with long lakes called ribbon lakes.

U-Shaped Valleys and Truncated Spurs http://www.zephryus.demon.co.uk/geography/resources/glaciers/ushape.html
Information page aimed at GCSE-level geography students, which describes these important physical features and explains the process of glacial erosion that brought them into being.

glacier body of ice, originating in mountains in snowfields above the snowline, that moves slowly downhill and is constantly built up from its source. The geographic features produced by the erosive action of glaciers (†erosion) are characteristic and include †glacial troughs (U-shaped valleys), †corries, and †arêtes. In lowlands, the laying down of debris carried by glaciers (†glacial deposition) produces a variety of landscape features, such as †moraines and †drumlins.

Glaciers form where annual snowfall exceeds annual melting and drainage (see †glacier budget). The area at the top of the glacier is called the zone of **accumulation**. The lower area of the glacier is called the †ablation zone. In the zone of accumulation, the snow compacts to ice under the weight of the layers above and moves downhill under the force of gravity. As the ice moves, it changes its shape and structure. Partial melting of ice at the base of the glacier also produces sliding of the glacier, as the ice travels over the bedrock. In the ablation zone, melting occurs and glacial †till is deposited.

Glacier http://www.glacier.rice.edu/
Find out about the coldest, windiest place on Earth, and the geological processes behind its formation. Full of detail about what it is really like to go on an expedition to Antarctica, this well-designed site has ample information on ice and glaciers, and explains why the ice-sheet that covers all but 2% of Antarctica has expanded and contracted throughout its history.

Glaciers http://www-nsidc.colorado.edu/glaciers/
Comprehensive information about glaciers from the US National Snow and Ice Data Center. There are explanations of why glaciers form, the different kinds of glaciers, and what they may tell us about climate change. There are a number of interesting facts and a bibliography about the compacted tongues of ice that cover 10% of the land surface of our planet.

glacier budget in a glacier, the balance between †accumulation (the addition of snow and ice to the glacier) and †ablation (the loss of snow and ice by melting and evaporation). If accumulation exceeds ablation the glacier will advance; if ablation exceeds accumulation it will probably retreat.

The rate of advance and retreat of a glacier varies from a few centimeters a year to several meters a year.

Glaisher, James (1809–1903) British meteorologist who founded the Royal Meteorological Society, and in the course of his investigations into atmospheric humidity, during 1862–66, he reached a height of over 11 km/6.8 mi in a balloon.

gland specialized organ of the body that manufactures and secretes enzymes, hormones, or other chemicals. In animals, glands vary in size from small (for example, tear glands) to large (for example, the pancreas), but in plants they are always small, and may consist of a single cell. Some glands discharge their products internally, †endocrine glands, and others externally, †exocrine glands. Lymph nodes are sometimes wrongly called glands.

Glaser, Donald Arthur (1926–) US physicist who was awarded the Nobel Prize for Physics in 1960 for his invention in 1952 of the †bubble chamber for detecting high-energy elementary particles.

Glashow, Sheldon Lee (1932–) US particle physicist. In 1964 he proposed the existence of a fourth, "charmed" †quark. He was awarded the Nobel Prize for Physics in 1979 for his work which suggested ways in which the weak nuclear force and the electromagnetic force (two of the fundamental forces of nature) could be unified as a single force now called the electroweak force. He shared the award with Abdus †Salam and Steven †Weinberg.

glass transparent or translucent substance that is physically neither a solid nor a liquid. Although glass is easily shattered, it is one of the strongest substances known. It is made by fusing certain types of sand (silica); this fusion occurs naturally in volcanic glass (see †obsidian).

In the industrial production of common types of glass, the type of sand used, the particular chemicals added to it (for example, lead, potassium, barium), and refinements of technique determine the type of glass produced. Types of glass include: soda glass; flint glass, used in cut-crystal ware; optical glass; stained glass; heat-resistant glass; and glasses that exclude certain ranges of the light spectrum. Blown glass is either blown individually from molten glass (using a tube up to 1.5 m/4.5 ft long), as in the making of expensive crafted glass, or blown automatically into a mould – for example, in the manufacture of light bulbs and bottles; pressed glass is simply pressed into moulds, for jam jars, cheap vases, and light fittings; while sheet glass, for windows, is made by putting the molten glass through rollers to form a "ribbon," or by floating molten glass on molten tin in the "float glass" process; fiberglass is made from fine glass fibers. Metallic glass is produced by treating alloys so that they take on the properties of glass while retaining the malleability and conductivity characteristic of metals.

CMCC – Glass Works http://www.civilization.ca/hist/verre/veint00e.html
Canadian-based site covering all aspects of glass. Out of the ten sections available, the ones of most interest are "What is glass?",

"Natural glass," "Making glass," and "The glass of science and medicine." Each section contains a detailed explanation of the ways in which glass is used, or processed, and photographs of products.

Glauber, Johann Rudolf (1604–1670) German chemist who in about 1625 discovered the salt known variously as **Glauber's salt** and "*sal mirabile*" (sodium sulfate). He made his living selling patent medicines and used the salt to treat almost any complaint.

Glauber's salt crystalline sodium sulfate decahydrate $Na_2SO_4.10H_2O$, produced by the action of sulfuric acid on common salt. It melts at 31°C/87.8°F; the latent heat stored as it solidifies makes it a convenient thermal energy store. It is used in medicine as a laxative.

glaucophane in geology, a blue amphibole, $Na_2(Mg,Fe,Al)_5Si_8O_{22}(OH)_2$. It occurs in glaucophane schists (blue schists), formed from the ocean floor basalt under metamorphic conditions of high pressure and low temperature; these conditions are believed to exist in subduction systems associated with destructive plate boundaries (see †plate tectonics), and so the occurrence of glaucophane schists can indicate the location of such boundaries in geological history.

glia or **neuroglia**, connective tissue of the †central nervous system, composed of cells that "service" the neurons (†nerve cells) with supportive and nutritive activities. Glial cells are far more numerous than neurons.

There are three types of glial cells: astrocytes, oligodendrocytes, and microglia. **Astrocytes** are the largest glial cells. They remove excess neurotransmitter molecules from around nerve cells, preventing stimulatory overload. **Oligodendrocytes** produce the insulating †myelin sheath that surrounds the nerve axon. The smallest glial cells, the **microglia**, play a part in the immunity of the nervous system, removing dead cells and killing microbes.

globalization process by which different parts of the globe become interconnected by economic, social, cultural, and political means. Globalization has become increasingly rapid since the 1970s and 1980s as a result of developments in technology, communications, and trade liberalization. Critics of globalization fear the increasing power of unelected multinational corporations, financial markets, and non-government organizations (NGOs), whose decisions can have direct and rapid effects on ordinary citizens' lives. This has led to growing antiglobalization and anticapitalist protests in the 1990s and early 21st century, which have disrupted international trade talks and meetings of international finance ministers. Supporters of globalization point to the economic benefits of growing international trade and specialization.

NGOs criticized by opponents of globalization include the World Trade Organization (WTO), World Bank, and the International Monetary Fund (IMF). Critics are also concerned about the potential environmental consequences, the risk of eroding distinctive local cultures, and the possible exploitation of workers. Solutions suggested to defuse opposition to globalization include making bodies such as the WTO more politically accountable, and ensuring greater protection of workers' rights.

global warming increase in average global temperature of approximately 0.5°C/0.9°F over the past century. Much of this is thought to be related to human activity. Global temperature has been highly variable in Earth history and many fluctuations in global temperature have occurred in historical times, but this most recent episode of warming coincides with the spread of industrialization, prompting the suggestion that it is the result of an accelerated †greenhouse effect caused by atmospheric pollutants, especially †carbon dioxide gas. The melting and collapse of the Larsen Ice Shelf, Antarctica, since 1995, is a consequence of global warming. Melting of ice is expected to raise the sea level in the coming decades. Natural, perhaps chaotic, climatic variations have not been ruled out as the cause of the current global rise in temperature, and scientists are still assessing the influence of anthropogenic (human-made) pollutants. In 1988 the World Meteorological Organization (WMO) and the United Nations (UN) set up the Intergovernmental Panel on Climate Change (IPCC), a body of more than 2,000 scientists, to investigate the causes of and issue predictions regarding climate change. In June 1996 the IPCC confirmed that global warming was taking place and that human activities were probably to blame.

Assessing the impact of humankind on the global climate is complicated by the natural variability on both geological and human time scales. The present episode of global warming has thus far still left England approximately 1°C/1.8°F cooler than during the peak of the so-called Medieval Warm Period (1000 to 1400 AD). The latter was part of a purely natural climatic fluctuation on a global scale. The interval between this period and the recent rise in temperatures was unusually cold throughout the world, relative to historical temperatures. Scientists predict that a doubling of carbon dioxide concentrations, expected before the end of the 21st century, will increase the average global temperature by 1.4–5.8°C/2.5–10.4°F.

In addition to a rise in average global temperature, global warming has caused seasonal variations to be more pronounced in recent decades. Examples are the most severe winter on record in the eastern USA 1976–77, and the record heat waves in the Netherlands and Denmark the following year. Mountain glaciers have shrunk, late summer Arctic sea-ice has thinned by 40%, and sea levels have risen by 10–20 cm/4–8 in. Scientists have predicted a greater number of extreme weather events and sea levels are expected to rise by 9–88 cm/4–35 in by 2100. 1998 was the warmest year globally of the last millennium, according to US researchers who used tree rings and ice cores to determine temperatures over the past 1,000 years.

A 1995 UN summit in Berlin, Germany, agreed to take action to reduce gas emissions harmful to the environment. Delegates at the summit, from more than 120 countries, approved a two-year negotiating process aimed at setting specific targets and timetables for reducing nations' emissions of carbon dioxide and other greenhouse gases after the year 2000. The Kyoto Protocol of 1997 committed the world's industrialized countries to cutting their annual emissions of harmful gases. However, in June 2001 US president George W Bush announced that the USA would not ratify the protocol.

Global Warming http://pooh.chem.wm.edu/chemWWW/courses/chem105/projects/group1/page1.html

Step-by-step explanation of the chemistry behind global warming. There is information about the causes of global warming, the environmental effects, and the social and economic consequences. The views of those who challenge the assertion that the world is warming up are also presented. The graphics accompanying the site are attractive and easy to follow.

What Hath Man Wrought http://sln.fi.edu/inquirer/warming.html

Extensive account of the causes and consequences of global warming. It offers discussion of issues as diverse as greenhouse gases, ocean currents, flooded coasts, the fluctuations of the Earth's temperature as far back as 15,000 years ago, and the role of human economic activities in the temperature increase. The site has also incorporated links to a series of space institutes, climate centers, and environmental agencies which conduct related research.

globular cluster spherical or near-spherical †star cluster containing from approximately 10,000 stars to many millions. About 120 globular clusters are distributed in a spherical halo around our Galaxy. They consist of old stars, formed early in the Galaxy's history. Globular clusters are also found around other galaxies.

glomerulus in the kidney, the cluster of blood capillaries at the threshold of the renal tubule, or nephron, responsible for filtering out the fluid that passes down the tubules and ultimately becomes urine. In the human kidney there are approximately one million tubules, each possessing its own glomerulus.

The structure of the glomerulus allows a wide range of substances including amino acids and sugar, as well as a large volume of water, to pass out of the blood. As the fluid moves through the tubules, most of the water and all of the sugars are reabsorbed, so that only waste remains, dissolved in a relatively small amount of water. This fluid collects in the bladder as urine.

glucagon in biology, a hormone secreted by the †pancreas, which increases the concentration of glucose in the blood by promoting the breakdown of glycogen in the liver. Secretion occurs in response to a lowering of blood glucose concentrations.

Glucagon injections can be issued to people living with patients with †diabetes who are being treated with insulin. Hypoglycemia may develop in such patients in the event of inadequate control of diabetes. An injection of glucagon can be used to reverse hypoglycemia before serious symptoms, such as unconsciousness, develop.

glucose or **dextrose** or **grape sugar**; $C_6H_{12}O_6$, simple sugar present in the blood and manufactured by green plants during †photosynthesis. It belongs to the group of chemicals known as †carbohydrates. The †respiration reactions inside cells involves the oxidation of glucose to produce †ATP, the "energy molecule" used to drive many of the body's biochemical reactions.

As well as being used in respiration to release useful †energy glucose may also be transported to specific parts of the body for other uses. The liver and muscles can turn it into glycogen, which acts as an energy store. It can also be turned into †fat and stored in fatty tissue. In humans and other vertebrates optimum blood glucose levels are maintained by the †hormone †insulin. An inability to control glucose levels may be caused by the disease †diabetes.

Green plants make glucose during photosynthesis. They can turn it into starch, which acts as an energy store and this starch may form an important part of the human †diet.

Glucose is prepared in syrup form by the hydrolysis of cane sugar or starch, and may be purified to a white crystalline powder. Glucose is a monosaccharide sugar (made up of a single sugar unit), unlike the more familiar sucrose (cane or beet sugar), which is a disaccharide (made up of two sugar units: glucose and fructose).

gluon in physics, a †gauge boson that carries the †strong nuclear force, responsible for binding quarks together to form the strongly interacting subatomic particles known as †hadrons. There are eight kinds of gluon.

Gluons cannot exist in isolation; they are believed to exist in balls ("glueballs") that behave as single particles. Glueballs may have been detected at CERN in 1995 but further research is required to confirm their existence.

glyceride †ester formed between one or more acids and glycerol (propan-1,2,3-triol). A glyceride is termed a mono-, di-, or triglyceride, depending on the number of hydroxyl groups from the glycerol that have reacted with the acids.

Glycerides, chiefly triglycerides, occur naturally as esters of †fatty acids in plant oils and animal fats.

glycerine another name for †glycerol.

glycerol or **glycerine** or **propan-1,2,3-triol**; $HOCH_2CH(OH)CH_2OH$, thick, colorless, odorless, sweetish liquid. It is obtained from vegetable and animal oils and fats (by treatment with acid, alkali, superheated steam, or an enzyme), or by fermentation of glucose, and is used in the manufacture of high explosives, in antifreeze solutions, to maintain moist conditions in fruits and tobacco, and in cosmetics.

glycine or **aminoethanoic acid**; $CH_2(NH_2)COOH$, simplest amino acid, and one of the main components of proteins. When purified, it is a sweet, colorless crystalline compound.

Glycine was found in space in 1994 in the star-forming region Sagittarius B2. The discovery is important because of its bearing on the origins of life on Earth.

glycogen polymer (a polysaccharide) of the sugar †glucose made and retained in the liver as a carbohydrate store, for which reason it is sometimes called animal starch. It is a source of energy when needed by muscles, where it is converted back into glucose by the hormone †insulin and metabolized.

glycol or **ethylene glycol** or **ethane-1,2-diol**; $HOCH_2CH_2OH$, thick, colorless, odorless, sweetish liquid. It is used in antifreeze solutions, in the preparation of ethers and esters (used for explosives), as a solvent, and as a substitute for glycerol.

glycolysis conversion of glucose to †lactic acid. It takes place in the †cytoplasm of cells as part of the process of cellular respiration.

glycoside in biology, compound containing a sugar and a non-sugar unit. Many glycosides occur naturally, for example, digitalis is a preparation of dried and powdered foxglove leaves that contains a mixture of cardiac glycosides. One of its constituents, digoxin, is used in the treatment of congestive heart failure and cardiac arrhythmias.

GMT abbreviation for †**Greenwich Mean Time**.

gneiss coarse-grained †metamorphic rock, formed under conditions of high temperature and pressure, and often occurring in association with schists and granites. It has a foliated, or layered, structure consisting of thin bands of micas and/or amphiboles dark in color alternating with bands of granular quartz and feldspar that are light in color. Gneisses are formed during regional †metamorphism; **paragneisses** are derived from metamorphism of sedimentary rocks and **orthogneisses** from metamorphism of granite or similar igneous rocks.

goblet cell in biology, cup-shaped cell present in the epithelium of the respiratory and gastrointestinal tracts. Goblet cells secrete mucin, the main constituent of mucous, which lubricates the mucous membranes of these tracts.

Goddard, Robert Hutchings (1882–1945) US rocket pioneer. He launched the first liquid-fuelled rocket at Auburn, Massachusetts, in 1926. By 1932 his rockets had gyroscopic control and could carry cameras to record instrument readings. Two years later a Goddard rocket achieved the world altitude record with an ascent of 3 km/1.9 mi.

Gödel, Kurt (1906–1978) Austrian-born US mathematician and philosopher. He proved that a mathematical system always contains statements that can be neither proved nor disproved within the system; in other words, as a science, mathematics can never be totally consistent and totally complete. He worked on relativity, constructing a mathematical model of the universe that made travel back through time theoretically possible.

Goeppert-Mayer, Maria (1906–1972) born Maria Goeppert, German-born US physicist who studied the structure of the atomic nucleus. She was awarded the Nobel Prize for Physics in 1963 for her explanation in 1948 of the stability of particular atoms, which envisaged atomic nuclei as shell-like layers of protons and neutrons, with the most stable atoms having completely filled outermost shells. She shared the award with Eugene †Wigner and Hans †Jensen.

gold chemical symbol Au, heavy, precious, yellow, metallic element, atomic number 79, relative atomic mass 197.0. Its symbol comes from the Latin *aurum* meaning "gold." It occurs in nature frequently as a free metal (see †native metal) and is highly resistant to acids, tarnishing, and corrosion. Pure gold is the most malleable of all metals and is used as gold leaf or powder, where small amounts cover vast surfaces, such as gilded domes and statues.

The elemental form is so soft that it is alloyed for strength with a number of other metals, such as silver, copper, and platinum. Its purity is then measured in †carats on a scale of 24 (24 carats is pure gold). It is used mainly for decorative purposes (jewelry, gilding) but also for coinage, dentistry, and conductivity in electronic devices.

Gold has been known and worked from ancient times, and currency systems were based on it in Western civilization, where mining it became an economic and imperialistic goal. In 1990 the three leading gold-producing countries were South Africa, 605.4 tons; USA, 295 tons; and Russia, 260 tons. In 1989 gold deposits were found in Greenland with an estimated yield of 12 tons per year.

Gold, Thomas (1920–) Austrian-born US astronomer and physicist. In 1948 he formulated, with Fred †Hoyle and Hermann †Bondi, the †steady-state theory regarding the creation of the universe.

golden section or **golden mean**, mathematical relationship between three points, A, B, C, in a straight line, in which the ratio AC:BC equals the ratio BC:AB (about 8:13 or 1:1.618). The area of a rectangle produced by the whole line and one of the segments is equal to the square drawn on the other segment. A **golden rectangle** has sides in the golden mean. Considered a visually satisfying ratio, it was first constructed by the Greek mathematician †Euclid and used in art and architecture, where it was given almost mystical significance by some Renaissance theorists.

The golden section was used extensively by certain painters, above all Piero della Francesca, who had an architectural background. In Vincent van Gogh's picture *Mother and Child* the Madonna's face fits perfectly into a golden rectangle.

In mathematics, the ratio of consecutive †Fibonacci numbers tends to the golden ratio.

Fibonacci Numbers and the Golden Section http://www.mcs.surrey.ac.uk/Personal/R.Knott/Fibonacci/fib.html

Fascinating facts and puzzles involving the Fibonacci series and the golden section. This site also includes biographical information about Fibonacci and some practical applications of his mathematical theories.

Goldstein, Eugen (1850–1930) German physicist. He investigated electrical discharges through gases at low pressures. He discovered canal rays and gave cathode rays their name.

Goldstein, Joseph Leonard (1940–) US geneticist who was awarded a Nobel Prize for Physiology or Medicine in 1985 with Michael Brown for their work on the regulation of †cholesterol metabolism. They discovered that the gene mutated in familial hypercholesterolaemia, a condition in which patients have very high blood cholesterol, is the gene for receptors involved in the removal of cholesterol from the bloodstream.

Golgi, Camillo (1843–1926) Italian cell biologist who was awarded a Nobel Prize for Physiology or Medicine in 1906 with Santiago Ramón y †Cajal for their discovery of the fine structure of the nervous system.

Golgi apparatus or **Golgi body**, stack of flattened membranous sacs found in the cells of †eukaryotes. Many molecules travel through the Golgi apparatus on their way to other organelles or to the endoplasmic reticulum. Some are modified or assembled inside the sacs. The Golgi apparatus is named after the Italian physician Camillo Golgi.

gonad the part of an animal's body that produces the sperm or egg cells (ova) required for sexual reproduction. The sperm-producing gonad is called a †testis, and the egg-producing gonad is called an †ovary.

gonadotrophin any hormone that supports and stimulates the function of the gonads (sex glands); some gonadotrophins are used as fertility drugs.

Gondwanaland or **Gondwana**, southern landmass formed 200 million years ago by the splitting of the single world continent †Pangaea. (The northern landmass was †Laurasia.) It later fragmented into the continents of South America, Africa, Australia, and Antarctica, which then drifted slowly to their present positions. The baobab tree found in both Africa and Australia is a relic of this ancient land mass.

A database of the entire geology of Gondwanaland has been constructed by geologists in South Africa. The database, known as Gondwana Geoscientific Indexing Database (GO-GEOID), displays information as a map of Gondwana 155 million years ago, before the continents drifted apart.

Goodall, Jane (1934–) Baroness van Lawick-Goodall, English primatologist and conservationist who has studied the chimpanzee community on Lake Tanganyika since 1960, and is a world authority on wild chimpanzees.

googol in mathematics, the name given to the number 10^{100} (1 with a hundred zeros). The **googolplex** is (1 with a googol of zeros).

$$10^{10^{100}}$$

gorge narrow steep-sided valley or canyon that may or may not have a river at the bottom. A gorge may be formed as a †waterfall retreats upstream, eroding away the rock at the base of a river valley; or it may be caused by †rejuvenation, when a river begins to cut downwards into its channel for some reason – for example, in response to a fall in sea level. Gorges are common in limestone country (see †karst), where they may be formed by the collapse of the roofs of underground caverns.

Gosset, William Sealey (1876–1937) English industrial research scientist. His work on statistical analysis of the †normal distribution opened the door to developments in the analysis of †variance.

Gould, Benjamin Apthorp (1824–1896) US astronomer. He organized and was the first director of the Argentine National Observatory at Córdoba 1868–85. His *Uranometria Argentina* (1879) gives the estimated magnitudes of 8,000 stars visible from Córdoba, and clearly shows †Gould's Belt, a group of bright stars spread in a broad band inclined at about 20° to the galactic equator.

Gould, Stephen Jay (1941–2002) US paleontologist and writer. In 1972 he proposed the theory of punctuated equilibrium, suggesting that the evolution of species did not occur at a steady rate but could suddenly accelerate, with rapid change occurring over a few hundred thousand years. His books include *Ever Since Darwin* (1977), *The Panda's Thumb* (1980), *The Flamingo's Smile* (1985), *Wonderful Life* (1990), and *The Structure of Evolutionary Theory* (2002).

Gould's Belt band in the sky in which the brighter blue stars appear to concentrate. It is inclined at 20° to the †Milky Way, which it crosses in the constellations of †Vela and †Lacerta, and is especially well marked in the southern hemisphere.

Graaf, Regnier de (1641–1673) Dutch physician and anatomist who discovered the ovarian follicles, which were later named **Graafian follicles**. He named the ovaries and gave exact descriptions of the testicles. He was also the first to isolate and collect the secretions of the pancreas and gall bladder.

Graafian follicle fluid-filled capsule that surrounds and protects the developing egg cell inside the ovary during the †menstrual cycle. After the egg cell has been released, the follicle remains and is known as a corpus luteum.

graben in earth science, long block of rock that has sunk along faults so that it lies lower than the rocks on each side of it. Horsts (long raised blocks) often form the flanking highlands.

graded bedding sedimentary feature in which the sedimentary layer shows a gradual change in particle size, usually coarse at the bottom to fine at the top. It is useful for determining which way was up at the time the bed was deposited.

gradient on a graph, the slope of a straight or curved line. The slope of a curve at any given point is represented by the slope of the †tangent at that point.

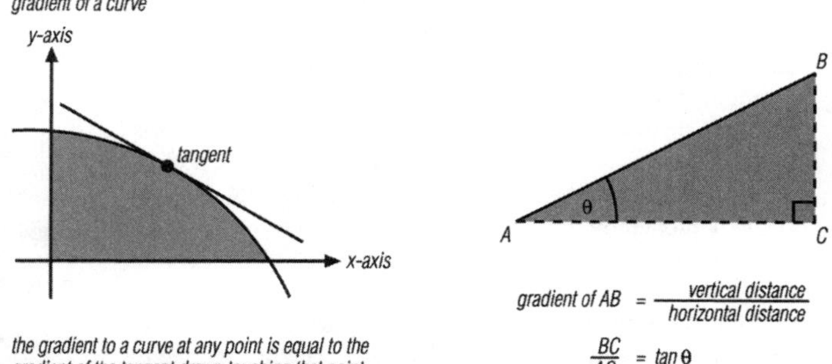

gradient The gradient of a curve keeps changing, so in order to calculate the gradient you have to draw a straight line that touches a point on the curve (the tangent). The gradient for that point on the curve will then be the same as the gradient of the straight line. Helicon

grafting in medicine, the operation by which an organ or other living tissue is removed from one organism and transplanted into the same or a different organism.

In horticulture, it is a technique widely used for propagating plants, especially woody species. A bud or shoot on one plant, termed the **scion**, is inserted into another, the **stock**, so that they continue growing together, the tissues combining at the point of union. In this way some of the advantages of both plants are obtained.

Grafting is usually only successful between species that are closely related and is most commonly practiced on roses and fruit trees. The grafting of nonwoody species is more difficult but it is sometimes used to propagate tomatoes and cacti.

Graham, Thomas (1805–1869) Scottish chemist who laid the foundations of physical chemistry (the branch of chemistry concerned with changes in energy during a chemical transformation) by his work on the diffusion of gases and liquids. **Graham's law** (1829) states that the diffusion rate of a gas is inversely proportional to the square root of its density.

Graham's law in chemistry, law that states that the rate of diffusion of a gas is inversely proportional to the square root of its density. It is named after Thomas †Graham, the Scottish chemist who proposed the law.

grain the smallest unit of mass in the three English systems (avoirdupois, troy, and apothecaries' weights) used in the UK and USA, equal to 0.0648 g. It was reputedly the weight of a grain of wheat. One pound avoirdupois equals 7,000 grains; one pound troy or apothecaries' weight equals 5,760 grains.

gram symbol g, metric unit of mass; one-thousandth of a kilogram.

Gram, Hans Christian Joachim (1853–1938) Danish microbiologist and physician who, in 1884, devised a method of classifying †bacteria by treating them with iodine and staining with aniline-gentian violet. **Gram-positive** bacteria, such as pneumococci, retain the stain whereas **Gram-negative** ones do not.

grand unified theory in physics, sought-for theory that would combine the theory of the strong nuclear force (called †quantum chromodynamics) with the theory of the weak nuclear and electromagnetic forces (see †forces, fundamental). The search for the grand unified theory is part of a larger program seeking a †unified field theory, which would combine all the forces of nature (including gravity) within one framework.

granite coarse-grained intrusive †igneous rock, typically consisting of the minerals quartz, feldspar, and biotite mica. It may be pink or white, depending on the composition of the feldspar. Granites are chiefly used as building materials.

Granite is formed when magma (molten rock) is forced between other rocks in the Earth's crust. It cools and crystallizes deep underground. As it cools slowly large crystals are formed. Granites often form large intrusions in the core of mountain ranges, and they are usually surrounded by zones of †metamorphic rock (rock that has been altered by heat or pressure). Granite areas have characteristic moorland scenery. In exposed areas the bedrock may be weathered along joints and cracks to produce a tor, consisting of rounded blocks that appear to have been stacked upon one another.

graph pictorial representation of numerical †data, such as statistical data, or a method of showing the mathematical relationship between two or more variables by drawing a diagram. There are often two axes, or reference lines, at right angles intersecting at the origin – the zero point, from which values of the variables (for example, distance and time for a moving object) are assigned along the axes. (The axes may not always intersect at the origin; if the data run from, for example, 500 to 1000 on each axis, then the point of intersection could be at (500,500) and this should be clearly labeled.) Pairs of simultaneous values (for example, the distance moved after a particular time) are plotted as points in the area between the axes, and the points then joined by a smooth curve to produce a graph. The horizontal axis is usually referred to as the x-axis, and the vertical axis as the y-axis. The "controllable/flowing" quantity is usually plotted on the x-axis; this is often time.

graph of a straight line

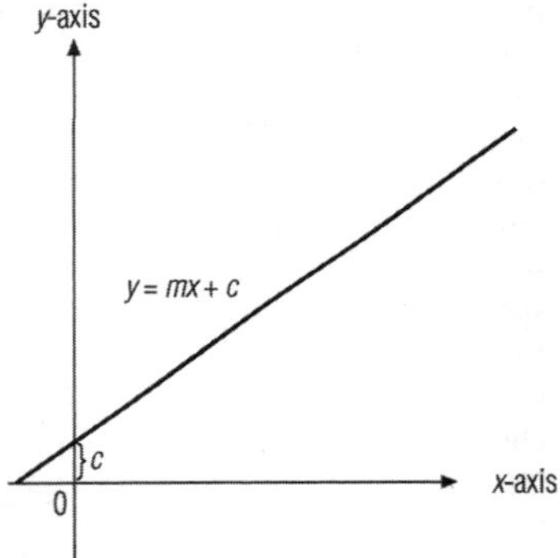

the equation of the straight-line graph takes the form $y = mx + c$, where m is the gradient (slope) of the line, and c is the y-intercept (the value of y where the line cuts the y-axis) for example, a graph of the equation $-y = x + 4$ will have a gradient of –1 and will cut the y-axis at $y = 4$

Cartesian coordinates On a line graph values are plotted using coordinates, components used to define the position of a point by its perpendicular distance from a set of two or more axes, or reference lines. For a two-dimensional area defined by two axes at right angles, the coordinates of a point are given by its perpendicular distances from the y-axis and x-axis, written in the form (x,y). For example, a point P that lies three units from the y-axis and four units from the x-axis has Cartesian coordinates (3,4).

straight-line graph This type of graph is produced by plotting the variables of a †linear equation with the general form

$$y = mx + c$$

where m is the slope of the line represented by the equation and c is the y-intercept, or the value of y where the line crosses the y-axis in the †Cartesian coordinate system.

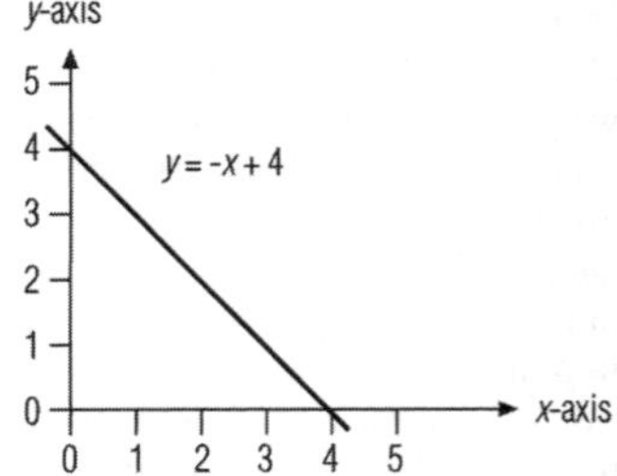

graph A graph is pictorial illustration of numerical data. It is a useful tool for interpreting data and is often used to spot trends or approximate a solution. Helicon

histogram †Histograms are used in statistics, showing frequency of data, in which the horizontal axis details discrete units or class boundaries, and the vertical axis represents the frequency. Blocks are drawn such that their areas (rather than their height as in a †bar chart) are proportional to the frequencies within a class or across several class boundaries. There are no spaces between blocks.

applications Graphs have many practical applications in all disciplines, for example †**distance–time graphs** are used to describe the motion of a body by illustrating the relationship between the distance that it travels and the time taken. Plotting distance (on the vertical axis) against time (on the horizontal axis) produces a graph the gradient of which is the body's speed. If the gradient is constant (the graph is a straight line), the body has uniform or constant speed; if the gradient varies (the graph is curved), then so does the speed and the body may be said to be accelerating or decelerating.

†**speed–time graphs** are used to describe the motion of a body by illustrating how its speed or velocity changes with time. The gradient of the graph gives the object's acceleration: if the gradient is zero (the graph is horizontal) then the body is moving with constant speed or uniform velocity; if the gradient is constant, the body is moving with uniform acceleration. The area under the graph gives the total distance traveled by the body.

Conversion graphs are used for changing values from one unit to another, for example from Celsius to Fahrenheit, with the two axes representing the different units.

graphical methods methods of solving equations and problems using the intersections of curves and lines on graphs.

graphical user interface GUI, in computing, a type of †user interface in which programs and files appear as icons (small pictures), user options are selected from pull-down menus, and data are displayed in windows (rectangular areas), which the operator can manipulate in various ways. The operator uses a pointing device, typically a mouse, to make selections and initiate actions. It is also known as **WIMP** (Windows, Icons, Menus, Pointing device).

The concept of the graphical user interface was developed by the Xerox Corporation in the 1970s, was popularized with the Apple Macintosh computers in the 1980s, and is now available on many types of computer – most notably as Windows, an operating system for IBM PC-compatible microcomputers developed by the software company Microsoft.

graphics tablet or **bit pad**, in computing, an input device in which a stylus or cursor is moved, by hand, over a flat surface. The computer can keep track of the position of the stylus, so enabling the operator to input drawings or diagrams into the computer.

A graphics tablet is often used with a form overlaid for users to mark boxes in positions that relate to specific registers in the computer, although developments in handwriting recognition may increase its future versatility.

graphite blackish-gray, soft, flaky, crystalline form of †carbon. It is used as a lubricant and as the active component of pencil lead.

Graphite, like †diamond and †fullerene, is an allotrope of carbon. The carbon atoms are strongly bonded together in sheets, but the bonds between the sheets are weak, allowing the layers to slide over one another. Graphite has a very high melting point (3,500°C/6,332°F), and is a good conductor of heat and electricity. It absorbs neutrons and is therefore used to moderate the chain reaction in nuclear reactors.

gravel coarse †sediment consisting of pebbles or small fragments of rock, originating in the beds of lakes and streams or on beaches. Gravel is quarried for use in road building, railway ballast, and for an aggregate in concrete. It is obtained from quarries known as gravel pits, where it is often found mixed with sand or clay.

Some gravel deposits also contain †placer deposits of metal ores (chiefly tin) or free metals (such as gold and silver).

gravimetric analysis in chemistry, a technique for determining, by weighing, the amount of a particular substance present in a sample. It usually involves the conversion of the test substance into a compound of known molecular weight that can be easily isolated and purified.

gravimetry measurement of the Earth's gravitational field. Small variations in the gravitational field (gravimetric anomalies) can be caused by varying densities of rocks and structure beneath the surface. Such variations are measured by a device called a gravimeter (or gravity-meter), which consists of a weighted spring that is pulled further downwards where the gravity

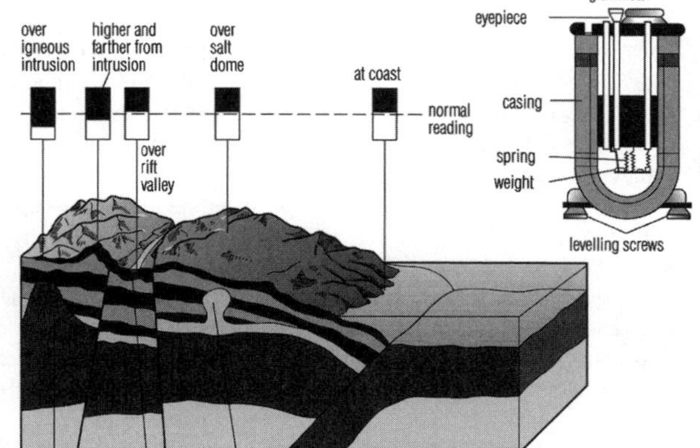

gravimetry The gravimeter is an instrument for measuring the force of gravity at a particular location. Variations in the force of gravity acting on a weight suspended by a spring cause the spring to stretch. The gravimeter is used in aerial surveys. Geological features such as intrusions and salt domes are revealed by the stretching of the spring. Helicon is stronger. Gravimetry is used by geologists to map the subsurface features of the Earth's crust, such as underground masses of dense rock such as iron ore, or light rock such as salt.

gravitational constant symbol *g*, fundamental constant of physics that relates the force of †gravity produced by a body to the masses involved and the separation. According to some physicists, its present value of 6.67 x 10^{-11} $m^3/(kg\ s^2)$ is decreasing at a rate proportional to the age of the universe.

gravitational field region around a body in which other bodies experience a force due to its gravitational attraction. The gravitational field of a massive object such as the Earth is very strong and easily recognized as the force of gravity, whereas that of an object of much smaller mass is very weak and difficult to detect. Gravitational fields produce only attractive forces.

gravitational field strength symbol g, strength of the Earth's gravitational field at a particular point. It is defined as the gravitational force in newtons that acts on a mass of one kilogram. The value of g on the Earth's surface is taken to be 9.806 N kg^{-1}.

The symbol g is also used to represent the acceleration of a freely falling object in the Earth's gravitational field. Near the Earth's surface and in the absence of friction due to the air, all objects fall with an acceleration of 9.8 m s^{-2}/32 ft s^{-2}.

gravitational force or **gravity**, one of the four fundamental †forces of nature, the other three being the electromagnetic force, the weak nuclear force, and the strong nuclear force. The gravitational force is the weakest of the four forces, but acts over great distances. The particle that is postulated as the carrier of the gravitational force is the †graviton.

gravitational lensing bending of light by a gravitational field, predicted by German-born US physicist Albert Einstein's general theory of relativity. The effect was first detected in 1917, when the light from stars was found to bend as it passed the totally eclipsed Sun. More remarkable is the splitting of light from distant quasars into two or more images by intervening galaxies. In 1979 the first double image of a quasar produced by gravitational lensing was discovered and a quadruple image of another quasar was later found.

gravitational potential energy energy possessed by an object when it is placed in a position from which, if it were free to do so, it would fall under the influence of †gravity. If the object is free to fall, then the gravitational potential energy is converted into kinetic (motion) energy. The gravitational potential energy E_p of an object of mass *m* kilograms placed at a height *h* meters above the ground is given by the formula:

$E_p = mgh$

where *g* is the gravitational field strength (in newtons per kilogram) of the Earth at that place. See also †potential energy.

In a †hydroelectric power station, gravitational potential energy of water held in a high-level reservoir is used to drive turbines to produce electricity.

graviton in physics, the †gauge boson that is the postulated carrier of the gravitational force.

gravity force of attraction that arises between objects by virtue of their masses. The larger the mass of an object the more strongly it attracts other objects. On Earth, gravity causes objects to have weight; it accelerates objects (at 9.806 meters per second per second/32.174 ft per second per second) towards the center of the Earth, the ground preventing them falling further.

The Earth's gravity also attracts the †Moon towards the Earth, keeping the Moon in orbit around the Earth. The Moon's gravity is one-sixth that of Earth's, so objects on the Moon weigh less than on Earth. The Sun contains 99.8% of the mass of the Solar System, and the resulting large force of gravity keeps the planets of the Solar System in orbit around the Sun.

A spacecraft launched from Earth must overcome the force of gravity before entering space. This is achieved by using rocket boosters at various stages of the launch. The spacecraft needs an acceleration of three times that of gravity (3 g). The speed required to escape the Earth's gravitational field is 40,000 kph/25,000 mph.

Astro Weight http://www.exploratorium.edu/ronh/weight/

Enter your weight and click on "Calculate," and this site gives your weight on all nine planets of the Solar System plus the largest five moons of Jupiter. It also includes a link to a page where you can calculate your age in all nine planetary years.

Exploring Gravity http://www.curtin.edu.au/curtin/dept/phys-sci/gravity/index2.htm

Interactive tour of all things related to gravity. The site includes quizzes and photos and covers "Introductory," "Intermediate," and "Advanced" stages of study.

From Apples to Orbits: The Gravity Story http://library.thinkquest.org/27585/

Find out all about the force that brought that legendary fruit down on top of Isaac Newton's head, and its importance to the development of space-age science. This nicely packaged site from the impressive ThinkQuest domain considers the history of gravity theory as well as small- and large-scale effects of the force, and includes sample experiments to test them out.

gravity, acceleration due to see †acceleration due to gravity.

gravity, center of see †center of gravity.

gray symbol Gy, SI unit of absorbed radiation dose. It replaces the rad (1 Gy equals 100 rad), and is defined as the dose absorbed when one kilogram of matter absorbs one joule of ionizing radiation. Different types of radiation cause different amounts of damage for the same absorbed dose; the SI unit of **dose equivalent** is the †sievert.

Gray, Henry (*c.* 1827–1861) British anatomist. He compiled a book on his subject, published in 1858 with illustrations by his colleague H Vandyke Carter. What is now known as *Gray's Anatomy* was based on his own dissections. Through its various editions and revisions, it has remained the definitive work on anatomy.

grazing incidence in astronomy, technique that employs the reflection of radiation that strikes a surface at a very small angle. It is employed in telescopes for radiation that cannot be focused by conventional optics, such as X-rays.

Great Artesian Basin largest area of artesian water in the world. It underlies much of Queensland, New South Wales, and South Australia, and in prehistoric times formed a sea. It has an area of 1,750,000 sq km/675,750 sq mi.

Great Bear popular name for the constellation †Ursa Major.

great circle circle drawn on a sphere such that the diameter of the circle is a diameter of the sphere. On the Earth, all meridians of longitude are half great circles; among the parallels of latitude, only the Equator is a great circle.

The shortest route between two points on the Earth's surface is along the arc of a great circle. These are used extensively as air routes although on maps, owing to the distortion brought about by †projection, they do not appear as straight lines.

greatest common divisor or **highest common factor**, in mathematics, for a set of integers, the largest positive integer that divides exactly into each element of the set.

Great Red Spot prominent oval feature, 14,000 km/8,500 mi wide and some 30,000 km/20,000 mi long, in the atmosphere of the planet †Jupiter, south of its equator. It was first observed in 1664. Space probes show it to be an anticlockwise vortex of cold clouds, colored possibly by phosphorus.

Great Wall array of galaxies arranged almost in a perfect plane, consisting of some 2,000 galaxies (about 500 million x 200 million light years across). It was discovered in 1989 by US astronomers at Harvard College Observatory, Cambridge, Massachusetts.

Green, George (1793–1841) English mathematician. He coined the term "potential," now a central concept in electricity, and introduced **Green's theorem**, which is still applied in the solution of partial differential equations; for instance, in the study of †relativity.

Green, George http://www-groups.dcs.st-andrews.ac.uk/history/Mathematicians/Green.html

Part of an archive containing the biographies of the world's greatest mathematicians, this site is devoted to the life and contributions of George Green.

green computing gradual movement by computer companies toward incorporating energy-saving measures in the design of systems and hardware. The increasing use of energy-saving devices, so that a computer partially shuts down during periods of inactivity, but can reactivate at the touch of a key, could play a significant role in †energy conservation.

It is estimated that worldwide electricity consumption by computers amounts to 240 billion kilowatt hours per year, equivalent to the entire annual consumption of Brazil. In the USA, carbon dioxide emissions could be reduced by 20 million tonnes per year – equivalent to the carbon dioxide output of 5 million cars – if all computers incorporated the latest "sleep technology" (which shuts down most of the power-consuming features of a computer if it is unused for any length of time).

Although it was initially predicted that computers would mean "paperless offices," in practice the amount of paper consumed continues to rise. Other environmentally costly features of computers include their rapid obsolescence, health problems associated with monitors and keyboards, and the unfavorable economics of component recycling.

Greengard, Paul (1925–) US neurobiologist who with Swedish pharmacologist Arvid †Carlsson and US neurobiologist Eric R †Kandel shared the Nobel Prize for Physiology or Medicine in 2000 for their discoveries concerning signal transduction in the nervous system. Greengard discovered how dopamine and other transmitters operate in the nervous system.

greenhouse effect phenomenon of the Earth's atmosphere by which solar radiation, trapped by the Earth and re-emitted from the surface as long-wave infrared radiation, is prevented from escaping by various gases (the "greenhouse gases") in the air. These gases trap heat because they readily absorb infrared radiation. As the energy cannot escape, it warms up the Earth, causing an increase in the Earth's temperature (†global warming). The main greenhouse gases are †carbon dioxide, †methane, and †chlorofluorocarbons (CFCs) as well as water vapor. Fossil-fuel consumption and forest fires are the principal causes of carbon dioxide build-up; methane is a by-product of agriculture (rice, cattle, sheep).

The United Nations Environment Program estimates that by 2025, average world temperatures will have risen by 1.5°C/2.7°F with a consequent rise of 20 cm/7.9 in in sea level. Low-lying areas and entire countries would be threatened by flooding and crops would be affected by the change in climate. However, predictions about global warming and its possible climatic effects are tentative and often conflict with each other.

At the 1992 Earth Summit it was agreed that by 2000 countries would stabilize carbon dioxide emissions at 1990 levels, but to halt the acceleration of global warming, emissions would probably need to be cut by 60%. Any increases in carbon dioxide emissions are expected to come from transport. The Berlin Mandate, agreed unanimously at the climate conference in Berlin in 1995, committed industrial nations to the continuing reduction of greenhouse gas emissions after 2000, when the existing pact to stabilize emissions runs out. The stabilization of carbon dioxide emissions at 1990 levels by 2000 would not be achieved by a number of developed countries, including Spain, Australia, and the USA, according to 1997 estimates. Australia is in favor of different targets for different nations, and refused to sign a communiqué at the South Pacific Forum meeting in the Cook Islands in 1997 which insisted on legally-binding reductions in greenhouse gas emissions. The United Nations Framework Convention on Climate Change (UNFCCC) adopted the Kyoto Protocol in 1997, committing the world's industrialized countries to cutting their annual emissions of harmful gases. By July 2001 the Protocol had been signed by 84 parties and ratified by 37; the USA announced its refusal to ratify the Protocol in June 2001.

Dubbed the "greenhouse effect" by Swedish scientist Svante Arrhenius, it was first predicted in 1827 by French mathematician Joseph Fourier.

The Greenhouse Effect: How the Earth Stays Warm

http://globalwarming.enviroweb.org/ishappening/greeneffect/index.html

Explanation of the greenhouse effect, the process by which atmospheric gases trap heat. This page links to a description of how this perfectly normal and essential process is being turned into something harmful by the pollutants humans put into the atmosphere. It also explains the related phenomenon, the hole in the ozone layer.

Greenwich Mean Time GMT, local time on the zero line of longitude (the **Greenwich meridian**), which passes through the Old Royal Observatory at Greenwich, London. It was replaced in 1986 by coordinated universal time (UTC), but continued to be used to measure longitudes and the world's standard time zones.

See also †time.

Gregor, William (1762–1817) English scientist and clergyman. He is chiefly remembered for his discovery of titanium.

Gregorian calendar calendar established with the authority of the Roman Catholic Church by Pope Gregory XIII in 1582. Correcting at a stroke the 10 day accumulated margin of error of the †Julian calendar, the main difference was in fact that century years were discounted as leap years unless they were divisible by 400.

Gregorian telescope reflecting telescope devised – but never constructed – by James Gregory, in which an auxiliary concave mirror reflects the magnified image, the right way up, through a hole in the center of the main objective mirror, that is, through the end of the telescope itself. The †Cassegrain telescope is similar but produces an inverted image.

Gregory, James (1638–1675) Scottish astronomer and mathematician who contributed to the development of †calculus. He anticipated English physicist Isaac Newton by suggesting that an aberration-free astronomical telescope could be made using a curved mirror instead of glass lenses. He also devised a method of finding the Earth–Sun distance by observing transits of Mercury or Venus.

Gregory, Richard Arman (1864–1952) English scientist. His main interest was the popularization of science, and he was the author of several books on general science and school textbooks. He was knighted in 1919 and made a baronet in 1931.

gray matter in biology, those parts of the brain and spinal cord that are made up of interconnected and tightly packed nerve cell nucleuses. The outer layers of the cerebellum contains most of the gray matter in the brain. It is the region of the brain that is responsible for advanced mental functions. Grey matter also constitutes the inner core of the spinal cord. This is in contrast to white matter, which is made of the axons of nerve cells.

grid an †electrode in the form of a mesh that, when placed between the anode and cathode of a thermionic valve, controls the flow of electrons.

grid network of crossing parallel lines. **Rectangular grids** are used for drawing graphs. **Isometric grids**

are used for drawing representations of solids in two dimensions in which lengths in the drawing match the lengths of the object.

grid network by which electricity is generated and distributed over a region or country. It contains many power stations and switching centers and allows, for example, high demand in one area to be met by surplus power generated in another.

The term is also used for any grating system, as in a cattle grid for controlling the movement of livestock across roads, and a conductor in a storage battery or electron gun.

grid reference numbering system used to specify a location on a map. The numbers representing grid lines at the bottom of the map (eastings) are given before those at the side (northings). Successive decimal digits refine the location within the grid system.

Grignard, (François Auguste) Victor (1871–1935) French chemist. He was awarded the Nobel Prize for Chemistry in 1912 for his discovery in 1900 of a series of organic compounds, the **Grignard reagents**, that found applications as some of the most versatile reagents in organic synthesis. Members of the class contain a hydrocarbon radical, magnesium, and a halogen such as chlorine.

Grignard reaction in chemistry, any chemical reaction involving a †Grignard reagent. Their most important use is the conversion of aldehydes and ketones to secondary and tertiary alcohols, respectively.

Grignard reagent in chemistry, a class of versatile organometallic reagents of general formula RMgX, where R is a hydrocarbon radical and X is a halogen atom, for example CH_3MgCl. Grignard reagents are prepared by the reaction of alkyl and aryl halides with magnesium metal in dry ether. Grignard reagents were developed and named after French chemist Victor †Grignard.

Grimaldi, Francesco Maria (1618–1663) Italian physicist. He discovered the diffraction of light and in physiology, he observed muscle action and was the first to note that minute sounds are produced by muscles during contraction.

grooming in biology, the use by an animal of teeth, tongue, feet, or beak to clean fur or feathers. Grooming also helps to spread essential oils for waterproofing. In many social species, notably monkeys and apes, grooming of other individuals is used to reinforce social relationships.

ground state (electronic) the state of an electron in its lowest energy level. When all the electrons orbiting a nucleus are in their lowest energy levels, the atom as a whole has its minimum possible energy and is therefore in its most stable state.

groundwater water present underground in porous rock strata and soils; it emerges at the surface as springs and streams. The groundwater's upper level is called the †water table. Rock strata that are filled with groundwater that can be extracted are called **aquifers**. Aquifers must be both porous (filled with holes) and permeable (full of holes that are interconnected so that the water is able to flow).

Most groundwater near the surface moves slowly through the ground while the water table stays in the same place. The depth of the water table reflects the balance between the rate of infiltration, called recharge, and the rate of discharge at springs or rivers or pumped water wells. The force of gravity makes underground water run "downhill" underground, just as it does above the surface. The greater the slope and the permeability, the greater the speed. Velocities vary from 100 cm/40 in per day to 0.5 cm/0.2 in.

Groundwater Quality and the Use of Lawn and Garden Chemicals by Homeowners

http://www.ext.vt.edu/pubs/envirohort/426-059/426-059.html

Extensive information about the problem of keeping groundwater free from garden chemicals. There is a description of groundwater, as well as detailed information on pesticides and their use. There are also notes on applying lawn and garden chemicals, and what should be done with the leftovers.

group in chemistry, a vertical column of elements in the †periodic table. Elements in a group have similar physical and chemical properties; for example, the group I elements (the †alkali metals: lithium, sodium, potassium, rubidium, cesium, and francium) are all highly reactive metals that form univalent ions. There is a gradation of properties down any group: in group I, melting and boiling points decrease, and density and reactivity increase.

Group 0 consists of the †noble gases (rare gases) and group II consists of the †alkaline-earth metals. Those elements placed between group II and III are the †transition metals and group VII contains the †halogens.

group in mathematics, a finite or infinite set of elements that can be combined by an operation; formally, a group must satisfy certain conditions. For example, the set of all integers (positive or negative whole numbers) forms a group with regard to addition because: (1) addition is associative, that is, the sum of two or more integers is the same regardless of the order in which the integers are added; (2) adding two integers gives another integer; (3) the set includes an identity element 0, which has no effect on any integer to which it is added (for example, $0 + 3 = 3$); and (4) each integer has an inverse (for instance, 7 has the inverse −7), such that the sum of an integer and its inverse is 0. **Group theory** is the study of the properties of groups.

Development of Group Theory www-history.mcs.st-and.ac.uk/history/HistTopics/Development_group_theory.html
History of the development of group theory in mathematics according to St Andrews University, Scotland. Biographical information on the mathematicians responsible and a description of the topic are also included on this site, along with links to many other sources on the history of mathematics.

grouped data in statistics, arrangement of †raw data with a wide range of values into groups. This process makes the †data more manageable. †Graphs and †frequency diagrams can then be drawn showing the †class intervals chosen instead of individual values.

group psychology area of †social psychology dealing with the formation and cohesion of groups, competition and conflict among group members and between groups, the influence of the group on the individual and vice versa, communication within the group and between groups, and many other aspects of social interaction.

group theory investigation and classification of the properties of the mathematical structures known as groups. A group possesses two operations – "multiplication" and "inverting of an element" – and a further designated element called "unity." An example is provided by the set of nonzero real numbers with ordinary multiplication and reciprocation (that is, being multiplied or becoming reciprocals). Another example is provided by the integers for which the operation of multiplication (n x m), is to be addition ($n + m$), with the inverse of n being $-1/n$, and the "unity element" is in fact zero. A third example is provided by a collection of †transformations of the plane where S x T effectively means applying transformation T and then S; such a collection is a group if it includes the identity transformation I (the transformation that leaves all points unchanged) and for each transformation T its inverse, T^{-1} (the transformation of which the effect is to return each point to its original position before it was transformed by T).

groupware in computing, software designed to be used collaboratively by a small group of users, each with his/her own computer and a copy of the software. Examples of groupware are Lotus Notes and Novell GroupWise, both of which provide facilities for sending e-mail and sharing documents.
Standard business applications such as word processors are spoken of as "groupware-enabled" if they provide facilities for a number of users to make revisions and incorporate them all into a final version.

growth in biology, increase in size and mass during the development of an organism over a period of time. Growth is often measured as an increase in †biomass (mass of organic material, excluding water) and is associated with cell division by †mitosis, subsequent increases in cell size, and with the differentiation of cells to perform specific functions, for example red blood cells in mammals and root cells in plants. All organisms grow, although the rate of growth varies over a lifetime. Typically, growth in an organism follows an S-shaped curve, in which growth is at first slow, then fast, then, towards the end of life, non-existent. Growth may even be negative during the period before death, with decay occurring faster than cellular replacement. In humans, there is a short period of rapid growth at puberty. Growth is affected by genetic factors, which dictate the eventual size and appearance of an organism. It is dependent upon an adequate supply of water and †mineral salts (†diet), and, particularly in plants, appropriate conditions of light and temperature, which determine the rate of †photosynthesis.

growth and decay curve graph showing exponential change (growth where the increment itself grows at the same rate) as occurs with compound interest and populations.

growth hormone **GH**; or **somatotrophin**, hormone from the anterior †pituitary gland that promotes growth of long bones and increases protein synthesis.
If it is lacking in childhood, dwarfism results; excess GH causes gigantism.
Levels of GH are high in the fetus and decline throughout early childhood; there is another spurt at adolescence and then production declines throughout adulthood. After the age of about 60 only tiny amounts of GH are produced, and this "deficiency" is responsible for some of the symptoms of aging.
Growth hormone release is controlled by growth hormone releasing hormone (GHRH) and the peptide somatostatin, both produced in the hypothalamus.

growth ring another name for †annual ring.

groyne wooden or concrete barrier built at right angles to a beach in order to block the movement of material along the beach by †longshore drift. Groynes are usually successful in protecting individual beaches, but because they prevent beach material from passing along the coast they can mean that other beaches, starved of sand and shingle, are in danger of being eroded away by the waves. These areas are down-drift of the groyne. This happened, for example, at Barton-on-Sea in Hampshire, England, in the 1970s, following the construction of a large groyne at Bournemouth.

Grus conspicuous constellation of the southern hemisphere, represented as a crane.

gryke enlarged †joint that separates blocks of limestone (clints) in a †limestone pavement.

g-scale scale for measuring force by comparing it with the force due to †gravity (*g*), often called †g-force.

guanine in biochemistry, one of the nucleotide bases. It is part of the genetic code of DNA and RNA, where it pairs with †cytosine. Guanine is a component of molecules of guanosine and is also found in †chromatophores.

guard cell in plants, a specialized cell on the undersurface of leaves for controlling gas exchange and water loss. Guard cells occur in pairs and are shaped so that a pore, or †stomata, exists between them. They can change shape with the result that the pore disappears. During warm weather, when a plant is in danger of losing excessive water, the guard cells close, cutting down evaporation from the interior of the leaf.

Guericke, Otto von (1602–1686) German physicist and politician. He invented the air pump and demonstrated the pressure of the atmosphere. He also constructed the first machine for generating static electricity.

Guérin, Camille (1872–1961) French bacteriologist who, with Albert †Calmette, developed the BCG vaccine for tuberculosis in 1921.

Guillaume, Charles Edouard (1861–1938) Swiss physicist who studied measurement and alloy development. He discovered a nickel-steel alloy, Invar, which showed negligible expansion with rising temperatures. He was awarded the Nobel Prize for Physics in 1920 for his discovery of anomalies in nickel-steel alloys.

Guillemin, Roger Charles Louis (1924–) French-born US endocrinologist who was awarded a Nobel Prize for Physiology or Medicine in 1977 with his co-worker Andrew †Schally for their discovery of hormones produced by the hypothalamus region of the brain. Guillemin isolated and identified various hormones, and also discovered †endorphins. Guillemin and Schally shared the prize with US physicist Rosalyn †Yalow, for her work on radioimmunoassay techniques by which minute quantities of hormone may be detected.

Guldberg, Cato Maximilian (1836–1902) Norwegian chemist who in 1873, with his brother-in-law Peter Waag, first formulated the law of mass action. This states that the rate of a chemical reaction is proportional to the active masses (concentrations) of the reactants. In 1890 Guldberg discovered the law that relates boiling point to a gas's critical temperature (Guldberg's law).

gulf any large sea inlet.

Gulf Stream warm ocean †current that flows north from the warm waters of the Gulf of Mexico along the east coast of America, from which it is separated by a channel of cold water originating in the southerly Labrador current. Off Newfoundland, part of the current is diverted east across the Atlantic, where it is known as the **North Atlantic Drift**, dividing to flow north and south, and warming what would otherwise be a colder climate in the British Isles and northwest Europe.

At its beginning the Gulf Stream is 80–150 km/50–93 mi wide and up to 850 m/2,788 ft deep, and moves with an average velocity of 130 km/80 mi a day. Its temperature is about 26°C. As it flows northwards, the current cools and becomes broader and less rapid.

Gullstrand, Allvar (1862–1930) Swedish physician who was awarded a Nobel Prize for Physiology or Medicine in 1911 for his work on the refraction of light through the different components of the eye, in particular in determining how the lens of the eye is adjusted to focus on both near and far objects. To assist his research Gullstrand invented the slit-lamp which is still used in modern instruments to illuminate the eye.

gully long, narrow, steep-sided valley with a flat floor. Gullies are formed by water erosion and are more common in unconsolidated rock and soils that are easily eroded. They may be formed very rapidly during periods of heavy rainfall and are common in arid areas that have periods of heavy rain. Gully formation in more temperate areas is common where the vegetation cover has been destroyed or reduced, for example as a result of fire or agricultural clearance.

gum in mammals, the soft tissues surrounding the base of the teeth. Gums are liable to inflammation (gingivitis) or to infection by microbes from food deposits (periodontal disease).

gum in botany, complex polysaccharides (carbohydrates) formed by many plants and trees, particularly by those from dry regions. They form four main groups: plant exudates (gum arabic); marine plant extracts (agar); seed extracts; and fruit and vegetable extracts. Some are made synthetically.

Gums are tasteless and odorless, insoluble in alcohol and ether but generally soluble in water. They are used for adhesives, fabric sizing, in confectionery, medicine, and calico printing.

guncotton or **smokeless powder**, common name for the explosive †cellulose nitrate. It is prepared by the nitration of cellulose with nitric or nitrous acid. Guncotton is only formed when the nitrogen content of the material is greater than 13% by weight.

gunpowder or **black powder**, oldest known †explosive, a mixture of 75% potassium nitrate (saltpeter), 15% charcoal, and 10% sulfur. Sulfur ignites at a low temperature, charcoal burns readily, and the potassium nitrate provides oxygen for the explosion. As gunpowder produces lots of smoke and burns quite slowly, it has progressively been replaced since the late 19th century by high explosives, although it is still widely used for quarry blasting, fuses, and fireworks. Gunpowder has high †activation energy; a gun based on gunpowder alone requires igniting by a flint or a match.

It was probably first invented in China, where it was chiefly used for fireworks. It is possible that knowledge of it was transmitted from the Middle East to Europe. The writings of the English monk Roger Bacon show that he was experimenting with gunpowder in 1249. His mixture contained saltpeter, charcoal, and sulfur. The development of effective gunpowder was essential for the growing significance of cannons and handguns in the late medieval period. The Arabs produced the first known working gun, in 1304. Gunpowder was used in warfare from the 14th century but it was not generally adapted to civil purposes until the 17th century, when it began to be used in mining.

Gunter, Edmund (1581–1626) English mathematician who became professor of astronomy at Gresham College, London in 1619. He is reputed to have invented a number of surveying instruments as well as the trigonometrical terms "cosine" and "cotangent."

Gurdon, John Bertrand (1933–) English molecular biologist. He has studied nuclear transplantation and the effects of known protein fractions on gene activity. Transplanting nuclei and nuclear constituents, such as †DNA, into enucleated eggs, mainly frog's eggs, Gurdon showed how the genetic activity changes in the recipient eggs. He concluded that the changes in gene activity induced by nuclear transplantation are indistinguishable from those that occur in normal early development. He also demonstrated how nuclear transplantation and microinjection techniques can be used to elucidate the intracellular movements of proteins.

gust temporary increase in †wind speed, lasting less than two minutes. Gusts are caused by rapidly moving air in higher layers of the atmosphere, mixing with slower air nearer the ground. Gusts are common in urban areas, where winds are funneled between closely-spaced high buildings. Gusting winds do far more damage to buildings and crops than steady winds. The strongest gusts can exceed speeds of 100 m/328 ft per second.

gut or **alimentary canal**, in the †digestive system, the part of an animal responsible for processing food and preparing it for entry into the blood.

The gut consists of a tube divided into segments specialized to perform different functions. The front end (the mouth) is adapted for food intake and for the first stages of digestion. The stomach is a storage area, although digestion of protein by the enzyme pepsin starts here; in many herbivorous mammals this is also the site of cellulose digestion. The small intestine follows the stomach and is specialized for digestion and for absorption. The large intestine, consisting of the colon, cecum, and rectum, has a variety of functions, including cellulose digestion, water absorption, and storage of feces. From the gut nutrients are carried to the liver via the hepatic portal vein, ready for assimilation by the cells.

guttation secretion of water on to the surface of leaves through specialized pores, or †hydathodes. The process occurs most frequently during conditions of high humidity when the rate of transpiration is low. Drops of water found on grass in early morning are often the result of guttation, rather than dew. Sometimes the water contains minerals in solution, such as calcium, which leaves a white crust on the leaf surface as it dries.

guyot flat-topped seamount. Such undersea mountains are found throughout the abyssal plains of major ocean basins, and most of them are covered by an appreciable depth of water, sediment, and ancient coral. They are believed to have started as volcanic cones formed near mid-oceanic ridges or other hot spots, in relatively shallow water, and to have been truncated by wave action as their tops emerged above the surface. As they are transported away from the ridge or other birthplace, the ocean crust ages, cools, and sinks along with the seamounts on top.

Gy symbol for †gray, the SI unit of absorbed ionizing radiation dose, equal to an absorption of 1 joule per kilogram of irradiated matter.

gymnosperm (Greek "naked seed") in botany, any plant whose seeds are exposed, as opposed to the structurally more advanced †angiosperms, where they are inside an ovary. The group includes conifers and related plants such as cycads and ginkgos, whose seeds develop in †cones. Fossil gymnosperms have been found in rocks about 350 million years old.

gynoecium or **gynaecium**, collective term for the female reproductive organs of a flower, consisting of one or more †carpels, either free or fused together.

gypsum common sulfate †mineral, composed of hydrous calcium sulfate, $CaSO_4.2H_2O$. It ranks 2 on the

†Mohs scale of hardness. Gypsum is used for making casts and moulds, and for blackboard chalk.

A fine-grained gypsum, called **alabaster**, is used for ornamental work. Roasted gypsum is known as **plaster of Paris**, because for a long time it was obtained from the gypsum quarries of the Montmartre district of Paris.

gyre circular surface rotation of ocean water in each major sea (a type of †current). Gyres are large and permanent, and occupy the northern and southern halves of the three major oceans. Their movements are dictated by the prevailing winds and the †Coriolis effect. Gyres move clockwise in the northern hemisphere and anticlockwise in the southern hemisphere.

gyroscope mechanical instrument, used as a stabilizing device and consisting, in its simplest form, of a heavy wheel mounted on an axis fixed in a ring that can be rotated about another axis, which is also fixed in a ring capable of rotation about a third axis. Applications of the gyroscope principle include the gyrocompass, the gyropilot for automatic steering, and gyro-directed torpedoes.

The components of the gyroscope are arranged so that the three axes of rotation in any position pass through the wheel's center of gravity. The wheel is thus capable of rotation about three mutually perpendicular axes, and its axis may take up any direction. If the axis of the spinning wheel is displaced, a restoring movement develops, returning it to its initial direction.

H symbol for †henry, the SI unit of inductance, equal to the inductance of a closed circuit with a magnetic flux of 1 weber per ampere of current.

ha symbol for †hectare.

Haber, Fritz (1868–1934) German chemist whose conversion of atmospheric nitrogen to ammonia opened the way for the synthetic fertilizer industry. His study of the combustion of hydrocarbons led to the commercial "cracking" or †fractional distillation of natural oil (petroleum) into its components (for example, diesel fuel, petrol, and paraffin). In electrochemistry, he was the first to demonstrate that oxidation and reduction take place at the electrodes; from this he developed a general electrochemical theory. He was awarded the Nobel Prize for Chemistry in 1918 for his work on the synthesis of ammonia from its elements.

Haber process or **Haber–Bosch process**, industrial process by which †ammonia is manufactured by direct combination of its elements, nitrogen and hydrogen. As the method is a †reversible reaction which reaches †chemical equilibrium, the manufacturing conditions have to be chosen carefully in order to achieve the best yield of ammonia. The reaction is carried out at 400–500°C/752–932°F and at 200 atmospheres pressure. The two gases, in the proportions of 1:3 by volume, are passed over a †catalyst of finely divided iron. Around 10% of the reactants combine, and the unused gases are recycled. The ammonia is separated either by being dissolved in water or by being cooled to liquid form.

$N_2 + 3H_2 \leftrightarrow 2NH_3$

Fritz Haber: Chemist and Patriot http://step.sdsc.edu/projects95/chem.in.history/essays/haber.html

Find out about the early 20th-century German chemist whose discovery of a method of synthesizing ammonia led to the development of artificial fertilizers and explosives. Largely historical in content, this biographical page contains a good, short description of the Haber–Bosch process.

habitat in ecology, the localized †environment in which an organism lives, and which provides for all (or almost all) of its needs. The diversity of habitats found within the Earth's ecosystem is enormous, and they are changing all the time. They may vary through the year or over many years. Many can be considered inorganic or physical; for example, the Arctic icecap, a cave, or a cliff face. Others are more complex; for instance, a woodland, or a forest floor. Some habitats are so precise that they are called **microhabitats**, such as the area under a stone where a particular type of insect lives. Most habitats provide a home for many species, which form a community.

Each species is specially adapted to life in its habitat. For example, an animal is adapted to eat other members of a †food chain or †food web found in the same habitat. Some species may be found in different habitats. They may be found to have different patterns of behavior or structure in these different habitats. For example, a plant such as the blackberry may grow in an open habitat, such as a field, or in a shaded one, such as woodland. Its leaves differ in the two habitats.

hacking unauthorized access to computer systems, either for fun or for malicious or fraudulent purposes. Hackers generally use computers and the internet to obtain access. In computing, the term is used in a wider sense to mean using software for enjoyment or self-education, not necessarily involving unauthorized access. The most destructive form of hacking is the introduction of a computer †virus. In recognition of the potential cost to business that hacking can cause, many jurisdictions have made hacking illegal. One of the most celebrated hacking cases was that of Kevin Mitnick, who spent five years in jail 1995–2000 on 25 counts of computer and wire fraud, and one charge of cracking. Mitnick had hacked into computers at the University of Southern California and tampered with data.

Some hackers call themselves ethical or white hat hackers. Often poachers turned gamekeepers, ethical hackers probe Web sites for holes in the security system, either fixing the holes themselves or alerting the company in question. There is some dispute over the correct use of the term hacker. Some believe hacker should be used to describe someone who develops computer software, with the term cracker being used for a person who breaks into a computer system.

Hacking can be divided into four main areas: †viruses, phreaking, software piracy (stripping away the protective coding that should prevent the software being copied), and accessing operating systems.

A 1996 US survey co-sponsored by the FBI showed 41% of academic, corporate, and government organizations interviewed had had their computer systems hacked into during 1995. White hat hackers are computer experts employed by companies to hack into their networks to expose weaknesses in security.

Hacking Lexicon http://www.ussrback.com/docs/papers/general/hacking-dict.html

A to Z of hacking. This extensive list includes numerous terms "used within the context of information security." Entries are accompanied by "key-point" remarks and examples.

hadal zone deepest level of the ocean, below the abyssal zone, at depths of greater than 6,000 m/19,500 ft. The ocean trenches are in the hadal zone. There is no light in this zone and pressure is over 600 times greater than atmospheric pressure.

Hadamard, Jacques Salomon (1865–1963) French mathematician. He originated functional analysis, one of the most fertile branches of modern mathematics. He also made contributions to †number theory and formulated the concept of a correctly posed problem.

Hadar or **Beta Centauri**, second-brightest star in the constellation of †Centaurus, and the 11th-brightest in the night sky. It is a blue-white giant star of magnitude 0.61, some 320 light years from the Sun. It is a †binary star, comprising two stars of magnitudes 0.7 and 3.9.

Hadley cell in the atmosphere, a vertical circulation of air caused by convection. The typical Hadley cell occurs in the tropics, where hot air over the Equator in the †intertropical convergence zone rises, giving the heavy rain associated with tropical rainforests. In the upper atmosphere this now dry air then spreads north and south and, cooling, descends in the latitudes of the tropics, producing the north and south tropical desert belts. After that, the air is drawn back towards the Equator, forming the northeast and southeast trade winds.

hadron in physics, a subatomic particle that experiences the strong nuclear force. Each is made up of two or three indivisible particles called †quarks. The hadrons are grouped into the †baryons (protons, neutrons, and hyperons), consisting of three quarks, and the †mesons, consisting of two quarks.

Hadron–Electron Ring Accelerator **HERA**, particle †accelerator built under the streets of Hamburg, Germany, occupying a tunnel 6.3 km/3.9 mi in length. It is the world's most powerful collider of protons and electrons, designed to accelerate protons to energies of 820 GeV (billion electron volts), and electrons to 30 GeV. HERA began operating in 1992. HERA can propel electrons into the proton interior, where they interact with the proton's constituent particles: three †quarks and a number of †gluons.

By August 2001, improvements were made that were estimated to increase HERA's proton and electron collision rate by a factor of four.

Haeckel, Ernst Heinrich Philipp August (1834–1919) German zoologist and philosopher. His theory of "recapitulation," expressed as "ontogeny repeats phylogeny" (or that embryonic stages represent past stages in the organism's evolution), has been superseded, but it stimulated research in embryology.

hafnium chemical symbol Hf, (Latin *Hafnia* "Copenhagen") silvery, metallic element, atomic number 72, relative atomic mass 178.49. It occurs in nature in ores of zirconium, the properties of which it resembles. Hafnium absorbs neutrons better than most metals, so it is used in the control rods of nuclear reactors; it is also used for light-bulb filaments.

It was named in 1923 by Dutch physicist Dirk Coster (1889–1950) and Hungarian chemist Georg von Hevesy after the city of Copenhagen, where the element was discovered.

Hagenbeck, Carl (1844–1913) German zoo proprietor. In 1907 he founded Hagenbeck's Zoo, near his native Hamburg. He was a pioneer in the display of animals against a natural setting, rather than in restrictive cages.

Hahn, Otto (1879–1968) German physical chemist who was awarded the Nobel Prize for Chemistry in 1944 for his discovery of nuclear fission (see †nuclear energy). In 1938 with Fritz Strassmann (1902–1980), he discovered that uranium nuclei split when bombarded with neutrons. Hahn did not participate in the resultant development of the atom bomb.

hail precipitation in the form of pellets of ice (hailstones). Water droplets freeze as they are carried upwards. As the circulation continues, layers of ice are deposited around the droplets until they become too heavy to be supported by the air current and they fall as a hailstorm. It is caused by the circulation of moisture in strong convection currents, usually within cumulonimbus †clouds.

hair fine filament growing from mammalian skin. Each hair grows from a pit-shaped follicle embedded in the second layer of the skin, the dermis. It consists of dead cells impregnated with the protein †keratin.

The average number of hairs on a human head varies from 98,000 (red-heads) to 120,000 (blondes). Each grows at the rate of 5–10 mm/0.2–0.4 in per month, lengthening for about three years before being replaced by a new one. A coat of hair helps to insulate land mammals by trapping air next to the body. The thickness of this layer can be varied at will by raising or flattening the coat. In some mammals a really heavy coat may be so effective that it must be shed in summer and a thinner one grown. Hair also aids camouflage, as in the zebra and the white winter coats of Arctic animals; protection, as in the porcupine and hedgehog; bluffing enemies by apparently increasing the size, as in the cat; sexual display, as in humans and the male lion; and its coloring or erection may be used for communication. In 1990 scientists succeeded for the first time in growing human hair in vitro.

Haldane, J(ohn) B(urdon) S(anderson) (1892–1964) Anglo-Indian physiologist, geneticist, and author of popular science books. In 1936 he showed the genetic link between hemophilia and color blindness.

Hale, George Ellery (1868–1938) US astronomer. He made pioneer studies of the Sun and founded three major observatories. In 1889 he invented the spectroheliograph, a device for photographing the Sun at particular wavelengths. In 1917 he established on Mount Wilson, California, a 2.5-m/100-in reflector, the world's largest telescope until superseded in 1948 by the 5-m/200-in reflector on Mount Palomar, which Hale had planned just before he died.

Hale-Bopp C/1995 01, large and exceptionally active comet, which in March 1997 made its closest fly-by to Earth since 2000 BC, coming within 190 million km/118 million mi. It has a nucleus of approximately 40 km/25 mi and an extensive gas coma (when close to the Sun Hale-Bopp released 10 metric tons of gas every second). Unusually, Hale-Bopp has three tails: one consisting of dust particles, one of charged particles, and a third of sodium particles.

Hale-Bopp was discovered independently in July 1995 by two US amateur astronomers, Alan Hale and Thomas Bopp.

Comet Hale-Bopp http://www.jpl.nasa.gov/comet/

Nearly 5,000 images of the comet, which was visible to the naked eye when it came within 190 million km/118 million mi of the Earth in 1997. The site includes information on its discovery and position, and details of current research findings.

Hales, Stephen (1677–1761) English scientist who studied the role of water and air in the maintenance of life. He gave accurate accounts of water movement in plants. He demonstrated that plants absorb air, and that some part of that air is involved in their nutrition. His work laid emphasis on measurement and experimentation.

half-life during †radioactive decay, the time in which the activity of a radioactive source decays to half its original value (the time taken for half the atoms to decay). In theory, the decay process is never complete and there is always some residual radioactivity. For this reason, the half-life of a radioactive isotope is measured, rather than the total decay time. It may vary from millionths of a second to billions of years.

To determine a short half-life a Geiger–Muller tube can be used to count the number of particles emitted by a sample. To determine a longer half-life a mass spectrometer is used. Half-lives can be hours, months, or years. Some examples are: sodium-24, 15 hours; carbon-14, 5,730 years; plutonium-239, 24,000 years; and uranium-238, 4,500 million years.

Rutherford's Discovery of Half-Life http://dbhs.wvusd.k12.ca.us/Chem-History/Rutherford-half-life.html

Transcript of Ernest Rutherford's paper describing his discovery of the half-life of radioactive materials. The paper entitled "A radioactive substance emitted from thorium compounds" first appeared in the Philosophical Magazine in January 1900. The reproduction of it on this Web site is complete with several diagrams and all the relevant equations.

half reaction in chemistry, symbolic representation of the changes that occur to one of the reacting species in a †redox reaction.

halide any compound produced by the combination of a †halogen, such as chlorine or iodine, with a less electronegative element (see †electronegativity). Halides may be formed by †ionic bonds or by †covalent bonds.

In organic chemistry, alkyl halides consist of a halogen and an alkyl group, such as methyl chloride (chloromethane).

halite mineral form of sodium chloride, NaCl. Common †salt is the mineral halite. When pure it is colorless and transparent, but it is often pink, red, or yellow. It is soft and has a low density.

Halite occurs naturally in evaporite deposits that have precipitated on evaporation of bodies of salt water. As rock salt, it forms beds within a sedimentary sequence; it can also migrate upwards through surrounding rocks to form salt domes. It crystallizes in the cubic system.

Hall, Asaph (1829–1907) US astronomer. He discovered the two Martian satellites, Deimos and Phobos, in 1877. He determined the orbits of satellites of other planets and of double stars, the rotation of Saturn, and the mass of Mars.

Hall, Philip (1904–1982) English mathematician. He specialized in the study of group theory. In 1928 Hall began a study of prime power groups. From this work he developed his 1933 theory of regular groups. An investigation of the conditions under which finite groups are soluble led him in 1937 to postulate a general structure theory for finite soluble groups. In 1954, he published an examination of finitely generated soluble groups in which he demonstrated that they could be divided into two classes of unequal size. At the end of the 1950s Hall turned to the subject of simple groups, and later also examined non-strictly-simple groups.

Hall effect production of a voltage across a conductor or semiconductor carrying a current at a right angle to a surrounding magnetic field. It was discovered in 1897 by the US physicist Edwin Hall (1855–1938). It is used in the **Hall probe** for measuring the strengths of magnetic fields and in magnetic switches.

Haller, Albrecht von (1708–1777) Swiss physician and scientist, founder of neurology. He studied the muscles and nerves, and concluded that nerves provide the stimulus that triggers muscle contraction. He also showed that it is the nerves, not muscle or skin, that receive sensation.

Halley, Edmond (1656–1742) English astronomer. He not only identified the comet that was later to be known by his name, but also compiled a star catalogue, detected the †proper motion of stars using historical records, and began a line of research that, after his death, resulted in a reasonably accurate calculation of the astronomical unit.

Halley, Edmond http://es.rice.edu/ES/humsoc/Galileo/Catalog/Files/halley.html

Detailed notes on the life of English astronomer Edmond Halley. The information here includes his life dates, family, and education. There are also descriptions of his religion, scientific disciplines, patronage, and his role in the development of technology.

Halley's Comet comet that orbits the Sun roughly every 75 years, named after English astronomer Edmond Halley, who calculated its orbit. It is the brightest and most conspicuous of the periodic comets, and recorded sightings go back over 2,000 years. The comet travels around the Sun in the opposite direction to the planets. Its orbit is inclined at almost 20° to the main plane of the Solar System and ranges between the orbits of Venus and Neptune. It will next reappear in 2061.

The comet was studied by space probes at its last appearance in 1986. The European probe Giotto showed that the nucleus of Halley's Comet is a tiny and irregularly shaped chunk of ice, measuring some 15 km/10 mi long by 8 km/5 mi wide, coated in a layer of very dark material, thought to be composed of carbon-rich compounds. This surface coating has a very low †albedo, reflecting just 4% of the light it receives from the Sun. Although the comet is one of the darkest objects known, it has a glowing head and tail produced by jets of gas from fissures in the outer dust layer. These vents cover 10% of the total surface area and become active only when exposed to the Sun. The force of these jets affects the speed of the comet's travel in its orbit.

Halley's Comet http://www.solarviews.com/eng/halley.htm

Attractive site devoted to the comet, with facts and statistics, images, and information about the spacecraft that have monitored it.

hallmark official mark stamped on British gold, silver, and (from 1913) platinum, instituted in 1327 (royal charter of London Goldsmiths) in order to prevent fraud. After 1363 personal marks of identification were added. Now tests of metal content are carried out at authorized assay offices in London, Birmingham, Sheffield, and Edinburgh; each assay office has its distinguishing mark, to which is added a maker's mark, date letter, and mark guaranteeing standard.

halo nebulous quality round a celestial body (particularly round a red giant); the galactic halo, however, describes the spherical collection of stars forming a surrounding "shell" for our otherwise compact, discoid Galaxy.

halogen any of a group of five non-metallic elements with similar chemical bonding properties: †fluorine, †chlorine, †bromine, †iodine, and †astatine. They form a linked group (Group 7) in the †periodic table, descending from fluorine, the most reactive, to astatine, the least reactive. They all have colored vapors and are poisonous. Melting points and boiling points increase going down the group. They combine directly with most metals to form salts, such as common salt (NaCl). Each halogen has seven electrons in its valence shell, which accounts for the chemical similarities displayed by the group.

Fluorine and chlorine are gases, bromine is a liquid, and astatine is a solid. The color of the element darkens going down the group from fluorine, which is pale yellow, to astatine, which is black.

Halogens http://www.corrosionsource.com/handbook/periodic/halogens.htm

Definition and properties of halogens: fluorine, chlorine, bromine, iodine, and astatine. The history of their classification is also given and links are provided to related topics, such as the alkali metals.

halogenoalkane any of a group of organic compounds (formerly called alkyl halides) formed by the halogenation of (addition of a halogen to) an alkane. In the presence of ultraviolet light, alkanes react with halides by substitution. In the chlorination of methane, for example, one, two, three or all four of the methane's hydrogens may be substituted by chlorine – depending on how far the reaction is allowed to proceed – with the release of hydrogen chloride at each substitution; the resulting halogenoalkanes are chloromethane (methyl chloride, CH_3Cl), dichloromethane (methylene chloride, CH_2Cl_2), trichloromethane (chloroform, $CHCl_3$) and tetrachloromethane (carbon tetrachloride CCl_4).

halon organic chemical compound containing one or two carbon atoms, together with †bromine and other †halogens. The most commonly used are halon 1211 (bromochlorodifluoromethane) and halon

1301 (bromotrifluoromethane). The halons are gases and are widely used in fire extinguishers. As destroyers of the †ozone layer, they are up to ten times more effective than †chlorofluorocarbons (CFCs), to which they are chemically related.

Levels in the atmosphere are rising by about 25% each year, mainly through the testing of fire-fighting equipment. The use of halons in fire extinguishers was banned in 1994.

halophyte plant adapted to live where there is a high concentration of salt in the soil, for example in salt marshes and mud flats.

haltere in zoology, pair of drumstick-shaped structures that are modifications of the second pair of wings of flies in the order Diptera. They function as gyroscopes, enabling the insect to maintain a correct orientation while flying.

Hamilton, William Rowan (1805–1865) Irish mathematician whose formulation of Isaac †Newton's dynamics proved adaptable to quantum theory, and whose "quarternion" theory was a forerunner of the branch of mathematics known as vector analysis. He was knighted in 1835.

Hamilton, Sir William Rowan http://www-history.mcs.st-and.ac.uk/history/Mathematicians/Hamilton.html

Extensive biography of the Irish mathematician. The site contains a description of his contribution to mathematics, and in particular his discovery of quaternions. The Web site includes a page from his notebook on which there are several examples of the multiplication of quaternions. Several literature references for further reading are also listed, and the Web site also features a portrait of Hamilton.

hand unit used in measuring the height of a horse from front hoof to shoulder (withers). One hand equals 10.2 cm/4 in.

hanging valley minor (tributary) valley that joins a larger †glacial trough at a higher level than the trough floor. During glaciation the ice in the smaller valley was unable to erode as deeply as the ice in the trough, and so the valley was left perched high on the side of the trough when the ice retreated. A river or stream flowing along the hanging valley often forms a †waterfall as it enters the trough. The Bridal Veil Falls in Yosemite National Park (USA) is an excellent example.

Hansteen, Christopher (1784–1873) Norwegian geophysicist. He is best known for his measurements and theories of geomagnetism.

haploid having a single set of †chromosomes in each cell. Most higher organisms are †diploid – that is, they have two sets – but their gametes (sex cells) are haploid. Some plants, such as mosses, liverworts, and many seaweeds, are haploid, and male honey bees are haploid because they develop from eggs that have not been fertilized. See also †meiosis.

hardening of oils transformation of liquid oils to solid products by †hydrogenation.

Vegetable oils contain double covalent carbon-to-carbon bonds and are therefore examples of †unsaturated compounds. When hydrogen is added to these double bonds, the oils become saturated. The more saturated oils are waxlike solids.

hardware mechanical, electrical, and electronic components of a computer system, as opposed to the various programs, which constitute †software.

hard water water that does not lather easily with soap, and produces a deposit or †scale (limescale) in kettles. It is caused by the presence of certain salts of calcium and magnesium.

Temporary hardness is caused by the presence of dissolved hydrogencarbonates (bicarbonates); when the water is boiled, they are converted to insoluble carbonates that precipitate as "scale." **Permanent hardness** is caused by sulfates and silicates, which are not affected by boiling. Water can be softened by †distillation, †ion exchange (the principle underlying commercial water softeners), targeting with low frequency magnetic waves (this alters the crystal structure of calcium salts so that they remain in suspension), addition of sodium carbonate or of large amounts of soap, or boiling (to remove temporary hardness).

Hardy, Godfrey Harold (1877–1947) English mathematician whose research was at a very advanced level in the fields of pure mathematics known as analysis and number theory. His *Course in Pure Mathematics* (1908) revolutionized the teaching of mathematics at senior school and university levels.

Hardy, Godfrey Harold http://www-groups.dcs.st-andrews.ac.uk/history/Mathematicians/Hardy.html

Part of an archive containing the biographies of the world's greatest mathematicians, this site is devoted to the life and contributions of Godfrey Hardy.

Hardy–Weinberg equilibrium in population genetics, the theoretical relative frequency of different †alleles within a given population of a species, when the stable endpoint of evolution in an undisturbed environment is reached.

Hare's apparatus in physics, a specific kind of †hydrometer used to compare the relative densities of two liquids, or to find the density of one if the other is known. It was invented by US chemist Robert

Hare (1781–1858).

It consists of an E-shaped glass tube, with the long limbs dipping into the two liquids and a tap on the short limb. With the tap open, air is pumped out of the apparatus and the liquids are pushed up the long limbs by atmospheric pressure. When the tap is closed, the heights of the liquids are inversely proportional to their relative densities. That is, if a liquid of relative density d_1 rises to a height h_1, and liquid d_2 rises to h_2, then $d_1/d_2 = h_2/h_1$.

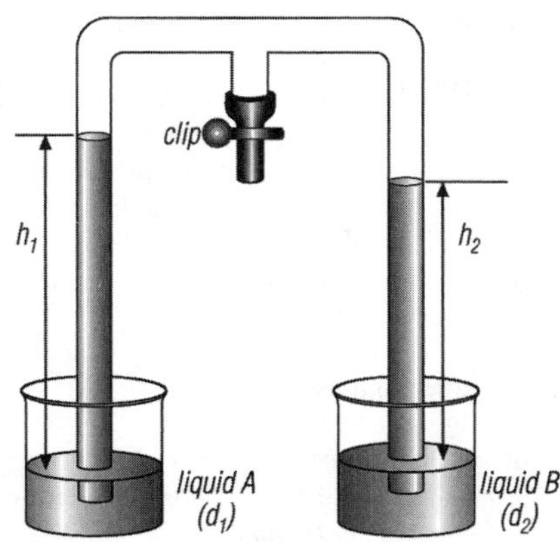

Hare's apparatus Hare's apparatus is used to compare the density of two liquids. When air is removed from the top of the apparatus, the liquids rise in the tubes to heights which are inversely proportional to their densities. Helicon

Harker, Alfred (1859–1939) British petrologist. His main area of research was in the rocks of the Inner Hebrides. Among his many honors were the Murchison Medal of the Geological Society, the Wollaston Medal, and Fellowship of the Royal Society. He was educated at Oxford University, where he taught for 61 years.

harmattan in meteorology, a dry and dusty northeast wind that blows over West Africa.

harmonic analysis see †Fourier theorem.

Harriot (or Hariot), Thomas (*c.* 1560–1621) English mathematician and astronomer. He was tutor to Walter Raleigh, who appointed him to the post of geographer to the second expedition to Virginia in 1585, an account of which voyage was published by Harriot in 1588, and afterwards reprinted in Richard Hakluyt's *Voyages* (1600). He introduced many useful simplifications of algebraic notation. Harriot used telescopes simultaneously with Galileo, and observed sunspots and the comets of 1607 and 1618. Harriot was born in Oxford, England.

Harrison, John (1693–1776) English horologist and instrument maker. He made the first chronometers that were accurate enough to allow the precise determination of longitude at sea, and so permit reliable (and safe) navigation over long distances.

Hartline, Haldan Keffer (1903–1983) US physiologist who was awarded a Nobel Prize for Physiology or Medicine in 1967 with Ragnar Granit and George Wald for their work on the physiology and chemistry of vision. Hartline measured electrical activity in the eye and determined that a decrease in the neurotransmitter (chemical messenger) signals to the brain that the eye has been stimulated by light.

hartzburgite in earth science, coarse-grained ultrabasic igneous rock (a type of †peridotite) containing mainly orthoxypyrene and olivine.

Harvey, Ethel Browne (1885–1965) US embryologist and cell biologist who discovered the mechanisms of cell division, using sea-urchin eggs as her experimental model.

Harvey, William (1578–1657) English physician who discovered the circulation of blood. In 1628 he published his book *De motu cordis/On the Motion of the Heart and the Blood in Animals.* He also explored the development of chick and deer embryos.

Harwell main research establishment of the United Kingdom Atomic Energy Authority, situated near the village of Harwell in Oxfordshire.

Hassel, Odd (1897–1981) Norwegian physical chemist who was awarded the Nobel Prize for Chemistry in 1969 for establishing the technique of **conformational analysis** – the determination of the properties of a molecule by rotating it around a single †bond. Hassel described the conformations of cyclohexane (a saturated hydrocarbon in which the carbonatoms are linked in a ring).

hassium chemical symbol Hs, synthesized, radioactive element of the †transactinide series, atomic number 108, relative atomic mass 265. It was first synthesized in 1984 by the Laboratory for Heavy-Ion Research in Darmstadt, Germany.

Hatch, Marshall Davidson (1932–) Australian biochemist who described the mechanism by which tropical plants trap carbon dioxide in bundle sheath cells, the major sites of photosynthesis. He also published on many other areas of plant metabolism, including how plants break down starch and sugars to form ethanol and carbon dioxide.

Hausdorff, Felix (1868–1942) German mathematician and philosopher who developed the branch of mathematics known as topology, in which he formulated the theory of point sets. He investigated general closure spaces, and formulated Hausdorff's maximal principle in general set theory.

Hausdorff, Felix http://www-groups.dcs.st-andrews.ac.uk/history/Mathematicians/Hausdorff.html

Part of an archive containing the biographies of the world's greatest mathematicians, this site is devoted to the life and contributions of Felix Hausdorff. In addition to biographical information, you will find a list of references about Hausdorff and links to other essays in the archive that reference him. The text of this essay includes hypertext links to the essays of those mathematicians and thinkers who influenced Hausdorff. You will also find an image of him, which you can click on to enlarge, and a map of his birthplace.

hausmannite brownish-black crystalline mineral form of manganese oxide, used as a source of manganese.

haustorium plural **haustoria**, specialized organ produced by a parasitic plant or fungus that penetrates the cells of its host to absorb nutrients. It may be either an outgrowth of hyphae (see †hypha), as in the case of parasitic fungi, or of the stems of flowering parasitic plants, as in dodders (*Cuscuta*). The suckerlike haustoria of a dodder penetrate the vascular tissue of the host plant without killing the cells.

Haüy, René-Just (1743–1822) French mineralogist, the founder of modern crystallography. He regarded crystals as geometrically structured assemblages of units (integrant molecules), and developed a classification system on this basis.

Hawking, Stephen (William) (1942–) English physicist and cosmologist whose work in general †relativity – particularly gravitational field theory – led to a search for a quantum theory of gravity to explain †black holes and the †Big Bang, singularities that classical relativity theory does not adequately explain. His book *A Brief History of Time* (1988) gives a popular account of cosmology and became an international best-seller. He later co-wrote (with Roger †Penrose) *The Nature of Space and Time* (1996), and published *The Universe in a Nutshell* in 2002.

Hawking, Stephen http://www.hawking.org.uk/

Stephen Hawking's own home page, with a brief biography, disability advice, and a selection of his lectures, including "The beginning of time" and a series debating the nature of space and time.

Stephen Hawking's Universe http://www.thirteen.org/hawking/html/home.html

Interesting site with original essays by some of today's leading figures in cosmology. The topics covered include "Strange stuff explained" and "Unsolved mysteries."

Haworth, (Walter) Norman (1883–1950) English organic chemist who was awarded the Nobel Prize for Chemistry in 1937 for his work in synthesizing a vitamin (ascorbic acid, vitamin C) in 1933. He made significant advances in determining the structures of many carbohydrates, particularly sugars. He was knighted in 1947.

Hawthorne effect immediate improvement in industrial performance as a result of a change in conditions. Research at Western Electrical's Chicago plant by Australian-born Elton Mayo between 1927 and 1932 showed that workers would initially interpret change as an indication of managerial concern and so increase productivity.

Hayashi, Chushiro (1920–) Japanese physicist whose research in 1950 exposed a fallacy in the "hot †Big Bang" theory proposed two years earlier by Ralph †Alpher and others. Hayashi has published many papers on the origin of the chemical elements in stellar evolution and on the composition of primordial matter in an expanding universe.

Hayashi track in astronomy, path on the †Hertzsprung-Russell diagram taken by protostars as they emerge from the clouds of dust and gas out of which they were born. A protostar appears on the right (cool) side of the Hertzsprung-Russell diagram and follows a Hayashi track until it arrives on the main sequence where hydrogen burning can start. It is named after the Japanese astrophysicist †Chushiro Hayashi, who studied the theory of protostars in the 1960s.

Hayem, Georges (1841–1920) French physician who pioneered the study of diseases of the blood and was one of the founders of hematology. He was the first to describe in detail the different cell types present in blood. Realizing the importance of blood †platelets and their role in blood clotting, Hayem was the first to perform an accurate blood-platelet count in 1878, and he later published a textbook on the subject.

hazardous waste waste substance, usually generated by industry, that represents a hazard to the environment or to people living or working nearby. Examples include radioactive wastes, acidic resins, arsenic residues, residual hardening salts, lead from car exhausts, mercury, non-ferrous sludges, organic solvents, asbestos, chlorinated solvents, and pesticides. The cumulative effects of toxic waste can take some time to become apparent (anything from a few hours to many years), and pose a serious threat to the ecological stability of the planet; its economic disposal or recycling is the subject of research.

haze factor unit of visibility in mist or fog. It is the ratio of the brightness of the mist compared with that of the object.

headland area of land running out into the sea. Headlands are often high points on the coastline, usually with steep cliffs, and may be made of more resistant rock than adjacent †bays. Erosion is concentrated on the flanks of headlands due to †wave refraction. Good examples include the chalk headland between Handfast Point and Ballard Point near Swanage, England, and the area between Tennyson Down and The

Needles, Isle of Wight, England.

head of water vertical distance between the top of a water stream and the point at which its energy is to be extracted.

headward erosion the backwards erosion of material at the source (start) of a river or stream. Broken rock and soil at the source are carried away by the river, causing the source of the river to move backwards as erosion takes place in the opposite direction to the river's flow. The resulting lowering of the land behind the source may, over time, cause the river to cut backwards into a neighboring valley to "capture" another river (see †river capture).

heart muscular organ that rhythmically contracts to force blood around the body of an animal with a circulatory system. Annelid worms and some other invertebrates have simple hearts consisting of thickened sections of main blood vessels that pulse regularly. An earthworm has ten such hearts. Vertebrates have one heart. A fish heart has two chambers – the thin-walled atrium (once called the auricle) that expands to receive blood, and the thick-walled ventricle that pumps it out. Amphibians and most reptiles have two atria and one ventricle; birds and mammals have two atria and two ventricles. The beating of the heart is controlled by the autonomic nervous system and an internal control center or pacemaker, the sinoatrial node.

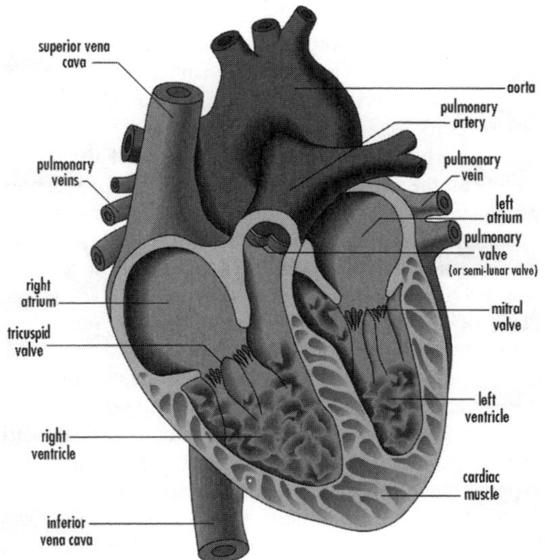

heart The structure of the human heart. During an average lifetime, the human heart beats more than 2,000 million times and pumps 500 million l/110 million gal of blood. The average pulse rate is 70–72 beats per minute at rest for adult males, and 78–82 beats per minute for adult females. Helicon

The human heart is more or less conical in shape and is positioned within the chest, behind the breast bone, above the diaphragm, and between the two lungs. It has flattened back and front surfaces and is, in health, the size of a person's closed fist. However, it varies in size with the person's weight, age, sex, and state of health. Its capacity is about 20 cm^3 in the newborn, reaching 150–160 cm^3 in the midteens. The female heart has a smaller capacity and is lighter than the male. Mammals have a †double circulatory system. In this kind of system the heart is divided into two halves, each half working as a separate pump. One side of the heart pumps blood to the lungs and back to the heart. The other side of the heart pumps blood to all other parts of the body and back to the heart.

Heart http://www.heartpoint.com/theheart.html

Graphic-intense Web site that allows you to follow blood flowing through the heart, using the animated interactive diagram. The names of the various chambers, blood vessels, and valves are all highlighted as the blood progresses. More facts are accessible through the site's links.

Heart: An Online Exploration

http://sln.fi.edu/biosci/heart.html

Explore the heart: discover its complexities, development, and structure; follow the blood on its journey through the blood vessels; learn how to maintain a healthy heart; and look back on the history of cardiology. A picture gallery is also available to preview before the main section of the site.

heartbeat regular contraction and relaxation of the heart, and the accompanying sounds. As blood passes through the heart a double beat is heard. The first is produced by the sudden closure of the valves between the atria and the ventricles. The second, slightly delayed sound, is caused by the closure of the valves found at the entrance to the major arteries leaving the heart. Diseased valves may make unusual sounds, known as heart murmurs.

heat form of energy possessed by a substance by virtue of the vibrational movement (kinetic energy) of its molecules or atoms. Heat energy is transferred by conduction, convection, and radiation. It always flows from a region of higher †temperature (heat intensity) to one of lower temperature. Its effect on a substance may be simply to raise its temperature, or to cause it to expand, melt (if a solid), vaporize (if a liquid), or increase its pressure (if a confined gas).

measurement Quantities of heat are usually measured in units of energy, such as joules (J) or calories (cal). The specific heat of a substance is the amount of heat that is needed to raise the temperature of 1 kg of the substance by 1 K (°C).

conduction, convection, and radiation Conduction is the passing of heat along a medium to neighboring parts with no visible motion accompanying the transfer of heat; for example, when the whole length of a metal rod is heated when one end is held in a fire. Convection is the transmission of heat through a fluid (liquid or gas) in currents; for example, when the air in a room is warmed by a fire or radiator. Radiation is heat transfer by infrared rays. It can pass through a vacuum, travels at the same speed as light, and can be reflected and refracted; for example, heat reaches the Earth from the Sun by radiation. For the transformation of heat, see †thermodynamics.

heat capacity in physics, the quantity of heat required to raise the temperature of an object by one degree. The **specific heat capacity** of a substance is the heat capacity per unit of mass, measured in joules per kilogram per kelvin ($J\ kg^{-1}\ K^{-1}$).

heat death in cosmology, possible fate of the universe in which it continues expanding indefinitely while all the stars burn out and no new ones are formed. See †critical density.

heat island large town or city that is warmer than the surrounding countryside. The difference in temperature is most pronounced during the winter, when the heat given off by the city's houses, offices, factories, and vehicles raises the temperature of the air by a few degrees. The heat island effect is also caused by the presence of surfaces such as black asphalt, that absorb rather than reflect sunlight, and the lack of vegetation, which uses sunlight to photosynthesize rather than radiating it back out as heat energy.

heat of crystallization in chemistry, quantity of heat that is released when a specific amount of †solute crystallizes from a supersaturated solution (see †supersaturation).

heat of formation in chemistry, amount of heat that is absorbed or released when one mole of a substance is formed from its constituent elements in their standard state. It is always measured at one atmosphere pressure and at a specific temperature.

heat of reaction alternative term for †energy of reaction.

heat of solution in chemistry, amount of heat that is absorbed or released when one mole of a substance is dissolved in a sufficient quantity of solvent to allow complete dissolution.

Heaviside, Oliver (1850–1925) English physicist. In 1902 he predicted the existence of an ionized layer of air in the upper atmosphere, which was known as the **Kennelly-Heaviside layer** but is now called the †E-layer of the †ionosphere. Deflection from it makes possible the transmission of radio signals around the world, which would otherwise be lost in outer space.

Heaviside, Oliver http://www-history.mcs.st-and.ac.uk/history/Mathematicians/Heaviside.html

Extensive biography of the English physicist. The site contains a description of his contribution to physics, and in particular his simplification of Maxwell's 20 equations in 20 variables, replacing them by two equations in two variables. Today we call these "Maxwell's equations" forgetting that they are in fact "Heaviside's equations." Several references for further reading are also listed, and the Web site also features a photograph of Heaviside.

heavy hydrogen alternative name for †deuterium, an isotope of hydrogen.

heavy water or **deuterium oxide**; D_2O, water containing the isotope deuterium instead of hydrogen (relative molecular mass 20 as opposed to 18 for ordinary water).

Its chemical properties are identical with those of ordinary water, but its physical properties differ slightly. It occurs in ordinary water in the ratio of about one part by mass of deuterium to 5,000 parts by mass of hydrogen, and can be concentrated by electrolysis, the ordinary water being more readily decomposed by this means than the heavy water. It has been used in the nuclear industry because it can slow down fast neutrons, thereby controlling the chain reaction.

hectare symbol ha, metric unit of area equal to 100 ares or 10,000 square meters (2.47 acres).

Trafalgar Square, London's only metric square, was laid out as one hectare.

hedge or **hedgerow**, row of closely planted shrubs or low trees, generally acting as a land division and windbreak. Hedges also serve as a source of food and as a refuge for wildlife, and provide a †habitat not unlike the understorey of a natural forest.

Heeger, Alan J (1936–) US physicist who with US chemist Alan G †MacDiarmid and Japanese chemist Hideki †Shirakawa shared the Nobel Prize for Chemistry in 2000 for the discovery and development of conductive †polymers. In 1975 Heeger was asked by MacDiarmid and Shirakawa to collaborate with them to study the possibility of an electrically conducting plastic. Conductive plastics, or "synthetic metals" as they are sometimes called, have many potential applications, such as replacing inorganic semiconductor materials to produce more energy-efficient photodiodes and in the antistatic treatment of photographic films.

height of a plane figure or solid, the perpendicular distance from the vertex to the base; see †altitude.

Heisenberg, Werner (Karl) (1901–1976) German physicist who developed †quantum theory and formulated the †uncertainty principle, which places absolute limits on the achievable accuracy of measurement. He was awarded the Nobel Prize for Physics in 1932 for his creation of quantum mechanics, work he carried out when only 24.

Heisenberg, Werner Karl http://www-groups.dcs.st-andrews.ac.uk/history/Mathematicians/Heisenberg.html
Part of an archive containing the biographies of the world's greatest mathematicians, this site is devoted to the life and contributions of physicist Werner Heisenberg.

heliocentric having the Sun at the center.

heliograph device for recording the positions of †sunspots.

heliography former method of signaling, used by armies in the late 19th century, which employed sunlight reflected from a mirror to pass messages in Morse code. On a clear day, a heliograph could send over distances in excess of 50 km/30 mi.

Also, an early photographic process by which a permanent image was formed on a glass plate.

heliometer instrument to measure the apparent diameter of the Sun at different seasons, also used to measure angular distances between stars.

helioseismology study of the Sun's structure by analyzing vibrations and monitoring effects on the Sun's surface. Compare with †seismology.

heliosphere region of space through which the †solar wind flows outwards from the Sun. The **heliopause** is the boundary of this region, believed to lie about 100 astronomical units from the Sun, where the flow of the solar wind merges with the interstellar gas.

helium chemical symbol He, (Greek *helios* "Sun") colorless, odorless, gaseous, non-metallic element, atomic number 2, relative atomic mass 4.0026. It is grouped with the †noble gases (rare gases) in Group 0 of the †periodic table. Helium is nonreactive because of its full outer shell of †electrons and forms no compounds. It is the second most abundant element (after hydrogen) in the universe, and has the lowest boiling (–268.9°C/–452°F) and melting points (–272.2°C/–458°F) of all the elements. It is present in small quantities in the Earth's atmosphere from gases issuing from radioactive elements (from †alpha decay) in the Earth's crust; after hydrogen it is the second-lightest element.

Helium is a component of most stars, including the Sun, where the nuclear-fusion process converts hydrogen into helium with the production of heat and light. It is obtained by the compression and fractionation of naturally occurring gases. It is used for inflating balloons and as a dilutant for oxygen in deep-sea breathing systems. Liquid helium is used extensively in low-temperature physics (cryogenics). Helium was originally discovered in 1868 in the spectrum of the Sun. It was found on Earth in 1895.

helium abundance presence – and dominance – of †helium atoms in the universe. The fact that about 8% of *all* atoms are helium can be traced, through the alpha-beta-gamma theory, to the primordial †Big Bang.

helix in mathematics, a three-dimensional curve resembling a spring, corkscrew, or screw thread. It is generated by a line that encircles a cylinder or cone at a constant angle.

Helmholtz, Hermann Ludwig Ferdinand von (1821–1894) German physiologist, physicist, and inventor of the ophthalmoscope for examining the inside of the eye. He was the first to measure the speed of nerve impulses and the first to explain how the cochlea of the inner ear works. He also specialized in the musical aspects of acoustics. In physics he formulated the law of conservation of energy, and worked in thermodynamics.

helminth in medicine, collective name used to describe parasitic worms. There are several classes of helminth that can cause infections in humans, including ascarids (ascariasis), †tapeworms, and threadworms.

Helmont, Jan Baptist van (1579–1644) Flemish chemist and physician. Helmont was the first to realize that there are gases other than air, and established the present scientific sense of the word "gas" (from Greek *cháos*). He also investigated the chemical properties of the fluids of the human body. His chief work, *Ortus medicinae*, was published by his son in 1648.

hematin pigment that, together with globin, makes up †hemoglobin, which gives red blood cells their color. Hematin has four †ligand groups which form a planar complex with iron in the middle.

hematite or **haematite**, principal ore of iron, consisting mainly of iron(III) oxide, Fe_2O_3. It occurs as **specular hematite** (dark, metallic luster), **kidney ore** (reddish radiating fibers terminating in smooth, rounded surfaces), and a red earthy deposit.

hemisphere half a sphere, produced when a sphere is sliced along a †great circle.

hemoglobin protein used by all vertebrates and some invertebrates for oxygen transport because the two substances combine reversibly. In vertebrates it occurs in red blood cells (erythrocytes), giving them their color.

In the lungs or gills where the concentration of oxygen is high, oxygen attaches to hemoglobin to form **oxyhemoglobin**. This process effectively increases the amount of oxygen that can be carried in the bloodstream. The oxygen is later released in the body tissues where it is at a low concentration, and the deoxygenated blood returned to the lungs or gills. Hemoglobin will combine also with carbon monoxide to form carboxyhemoglobin, which has the effect of reducing the amount of oxygen that can be carried in the blood.

hemolymph circulatory fluid of those mollusks and insects that have an "open" circulatory system. Hemolymph contains water, amino acids, sugars, salts, and white cells like those of blood. Circulated by a pulsating heart, its main functions are to transport digestive and excretory products around the body. In mollusks, it also transports oxygen and carbon dioxide.

hemolysis destruction of red blood cells. Aged cells are constantly being lysed (broken down), but increased wastage of red cells is seen in some infections and blood disorders. It may result in jaundice (through the release of too much hemoglobin) and in anemia.

Hench, Philip Showalter (1896–1965) US physician who was awarded a Nobel Prize for Physiology or Medicine in 1950 with Edward Kendall and Tadeus †Reichstein for their work on the structure and biological effects of hormones of the adrenal cortex. Hench developed a cortisone treatment for rheumatoid arthritis.

Henderson, Thomas (1798–1844) Scottish astronomer. He became the first Astronomer Royal for Scotland in 1834.

Henle, Friedrich Gustav Jakob (1809–1885) German pathologist and anatomist. His *Manual of Rational Pathology* (1846–53) emphasized that pathology depended upon scientific study, marking a fundamentally new approach; it was a precursor of modern microbiology. He published his *Handbook of Systematic Human Anatomy* 1855–71, which contained all that was known of the structure of the human body. His discoveries included the looped portion of the kidney tubule known as Henle's loop. Henle was born at Fürth, Franconia, Germany.

henry symbol H, SI unit of †inductance (the reaction of an electric current against the magnetic field that surrounds it). One henry is the inductance of a circuit that produces an opposing voltage of one volt when the current changes at one ampere per second.

It is named after the US physicist Joseph Henry.

Henry, Joseph (1797–1878) US physicist, inventor of the electromagnetic motor in 1829 and of a telegraphic apparatus. He also discovered the principle of electromagnetic induction, roughly at the same time as Michael Faraday, and the phenomenon of self-induction. The unit of inductance, the henry, is named after him.

Henry, William (1774–1836) English chemist and physician. In 1803 he formulated Henry's law, which states that when a gas is dissolved in a liquid at a given temperature, the mass that dissolves is in direct proportion to the pressure of the gas.

heparin anticoagulant substance produced by cells of the liver, lungs, and intestines. It normally inhibits the clotting of blood by interfering with the production of thrombin, which is necessary for clot formation. Heparin obtained from animals is administered after surgery to limit the risk of thrombosis, or following pulmonary embolism to ensure that no further clots form, and during hemodialysis (using a kidney machine).

heptane C_7H_{16}, saturated hydrocarbon liquid, boiling point 98°C/208°F. Heptane exists in nine isomeric forms; it is flammable and one of the constituent compounds of crude oil and gasoline.

Heraklides of Pontus (388–315 BC) Greek philosopher and astronomer who may have been the first to realize that the Earth turns on its axis, from west to east, once every 24 hours. He also thought that the observed motions of Mercury and Venus suggested that they orbited the Sun rather than the Earth.

herb any plant (usually a flowering plant) tasting sweet, bitter, aromatic, or pungent, used in cooking, medicine, or perfumery; technically, a herb is any plant in which the aerial parts do not remain above ground at the end of the growing season.

herbaceous plant plant with very little or no wood, dying back at the end of every summer. The herbaceous perennials survive winters as underground storage organs such as bulbs and tubers.

herbarium collection of dried, pressed plants used as an aid to identification of unknown plants and by taxonomists in the †classification of plants. The plant specimens are accompanied by information, such as the date and place of collection, by whom collected, details of habitat, flower color, and local names.

Herbaria range from small collections containing plants of a limited region, to the large university and national herbaria (some at botanical gardens) containing millions of specimens from all parts of the world.

herbicide any chemical used to destroy plants or check their growth (see †weedkiller).

Herbig, George Howard (1920–) US astronomer who specialized in spectroscopic research into irregular variable stars, notably those of the T Tauri group. He also worked on binary stars.

Herbig–Haro object or **HH object**, in astronomy, bright region of interstellar dust and gas probably associated with the formation of stars. These nebulae were discovered independently in the 1940s by US astronomer George †Herbig and Mexican astronomer Guillamero Haro.

herbivore animal that feeds on green plants (or photosynthetic single-celled organisms) or their products, including seeds, fruit, and nectar. The most numerous type of herbivore is thought to be the zooplankton, tiny invertebrates in the surface waters of the oceans that feed on small photosynthetic algae. Herbivores are more numerous than other animals because their food is the most abundant. They form a vital link in the food chain between plants and †carnivores.

Hercules fifth-largest constellation, lying in the northern hemisphere. Despite its size it contains no prominent stars. Its most important feature is the best example in the northern hemisphere of a †globular cluster of stars 22,500 light years from the Sun, which lies between Eta and Zeta Herculis.

heredity in biology, the transmission of traits from parent to offspring. See also †genetics.

hermaphrodite organism that has both male and female sex organs. Hermaphroditism is the norm in such species as earthworms and snails, and is common in flowering plants. Cross-fertilization is common among hermaphrodites, with the parents functioning as male and female simultaneously, or as one or the other sex at different stages in their development. Human hermaphrodites are extremely rare.

Hermite, Charles (1822–1901) French mathematician who was a principal contributor to the development of the theory of algebraic forms, the arithmetical theory of quadratic forms, and the theories of elliptic and Abelian functions. Much of his work was highly innovative, especially his solution of the quintic equation through elliptic modular functions, and his proof of the transcendence of e.

Herophilus of Chalcedon (*c.* 330–*c.* 260 BC) Greek physician, active in Alexandria. His handbooks on anatomy make pioneering use of dissection, which, according to several ancient sources, he carried out on live criminals condemned to death.

herpetology the scientific study of †reptiles, including their classification, anatomy, physiology, behavior, and ecology.

Herschel, (Frederick) William (1738–1822) German-born English astronomer. He was a skilled telescope maker, and pioneered the study of binary stars and nebulae. He discovered the planet Uranus in 1781 and infrared solar rays in 1801. He catalogued over 800 double stars, and found over 2,500 nebulae, catalogued by his sister Caroline Herschel; this work was continued by his son John Herschel. By studying the distribution of stars, William established the basic form of our Galaxy, the Milky Way. He was knighted in 1816.

Herschel, Caroline (Lucretia) (1750–1848) German-born English astronomer, sister of William †Herschel, and from 1772 his assistant in Bath, England. She discovered eight comets and worked on her brother's catalogue of star clusters and nebulae.

Herschel, Caroline http://www.agnesscott.edu/lriddle/women/herschel.htm

Profile of the German-born astronomer. It traces her triumph over physical affliction, her interest in music, and her achievements in astronomy. Her role in the discovery of Neptune and comets is also described.

Herschel, John Frederick William (1792–1871) English scientist, astronomer, and photographer who discovered thousands of close double stars, clusters, and nebulae. He coined the terms "photography," "negative," and "positive," discovered sodium thiosulphite as a fixer of silver halides, and invented the cyanotype process; his inventions also include astronomical instruments.

Hershey, Alfred Day (1908–1997) US biochemist who was awarded a Nobel Prize for Physiology or Medicine in 1969 for his work on the replication mechanism and genetic structure of viruses. He used bacteriophages (viruses that infect bacteria) to demonstrate that †DNA, not protein, is the genetic material. His experiments demonstrated that viral DNA is sufficient to transform bacteria.

Hertwig, Wilhelm August Oscar (1849–1922) German zoologist who revealed that fertilization involves the fusion of the nuclei of egg and sperm. He showed for the first time that only one sperm is required to fertilize an egg.

hertz symbol Hz, SI unit of frequency (the number of repetitions of a regular occurrence in one second). A wave source has a frequency of 1 Hz if it produces one wave each second. Human beings have an audible range from approximately 20 Hz to 20,000 Hz. Radio waves are often measured in megahertz (MHz), millions of hertz, and the †clock rate of a computer is usually measured in megahertz. The unit is named after German physicist Heinrich †Hertz.

Hertz, Gustav (1887–1975) German physicist who, with US physicist James †Franck, demonstrated that mercury atoms, when bombarded with electrons, absorb energy in discrete units (or quanta). Following the absorption of energy, the atoms return to their original state by emitting a photon of light. This was the first experimental proof that the †quantum theory of atoms was correct and demonstrated the reality of atomic energy levels. He shared the Nobel Prize for Physics in 1925, with Franck, for the discovery of the laws governing the impact of an electron upon an atom.

Hertz, Heinrich Rudolf (1857–1894) German physicist who studied electromagnetic waves, showing that their behavior resembles that of light and heat waves.

Hertzsprung, Ejnar (1873–1967) Danish astronomer and physicist. He introduced the concept of the absolute magnitude (brightness) of a star, and described the relationship between the absolute magnitude and the temperature of a star, formulating his results in the form of a diagram, known as the †Hertzsprung-Russell diagram, that has become a standard reference.

Hertzsprung-Russell diagram in astronomy, graph on which the surface temperatures of stars are plotted against their luminosities. Most stars, including the Sun, fall into a narrow band called the †main sequence. When a star grows old it moves from the main sequence to the upper right part of the graph, into the area of the giants and supergiants. At the end of its life, as the star shrinks to become a white dwarf, it moves again, to the bottom left area. It is named after the Danish astronomer Ejnar Hertzsprung and the US astronomer Henry Russell, who independently devised it in the years 1911–13.

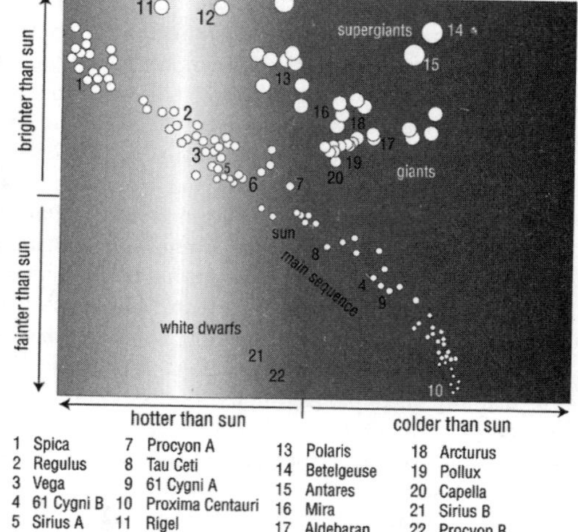

Hertzsprung-Russell diagram The Hertzsprung-Russell diagram relates the brightness (or luminosity) of a star to its temperature. Most stars fall within a narrow diagonal band called the main sequence. A star moves off the main sequence when it grows old. Helicon

Herzberg, Gerhard (1904–1999) German-born Canadian physicist who was awarded the Nobel Prize for Chemistry in 1971 for his research using spectroscopy to determine the electronic structure and geometry of molecules, especially free radicals (atoms or groups of atoms that possess a free, unbonded electron).

Hess, Germain Henri (1802–1850) also known as **German Ivanovich Gess**, Swiss-born Russian chemist, a pioneer in the field of thermochemistry. The law of constant heat summation is named after him.

Hess, Harry Hammond (1906–1969) US geologist who in 1962 proposed the notion of †seafloor spreading. This played a key part in the acceptance of †plate tectonics as an explanation of how the Earth's crust is formed and moves.

Hess, Victor Francis (1883–1964) Austrian physicist who emigrated to the USA shortly after he shared the Nobel Prize for Physics in 1936 for the discovery of cosmic radiation.

heterocyclic in chemistry, describing any organic compound whose molecules contain a closed ring system that is made up from carbon and at least one other type of element. An example is †pyridine.

heterodyne effect superimposition of two waves of different †frequency in a †radio receiver, one of which is being received, the other transmitted within the device itself, producing an intermediate frequency that can be demodulated.

heterogeneous reaction in chemistry, a reaction where there is an interface between the different components or reactants. Examples of heterogeneous reactions are those between a gas and a solid, a gas and a liquid, two immiscible liquids, or two different solids.

heterosis or **hybrid vigor**, improvement in physical capacities that sometimes occurs in the †hybrid produced by mating two genetically different parents.

The parents may be of different strains or varieties within a species, or of different species, as in the mule, which is stronger and has a longer lifespan than either of its parents (donkey and horse). Heterosis is also exploited in hybrid varieties of maize, tomatoes, and other crops.

heterostyly in botany, having †styles of different lengths. Certain flowers, such as primroses (*Primula vulgaris*), have different-sized †anthers and styles to ensure cross-fertilization (through †pollination) by visiting insects.

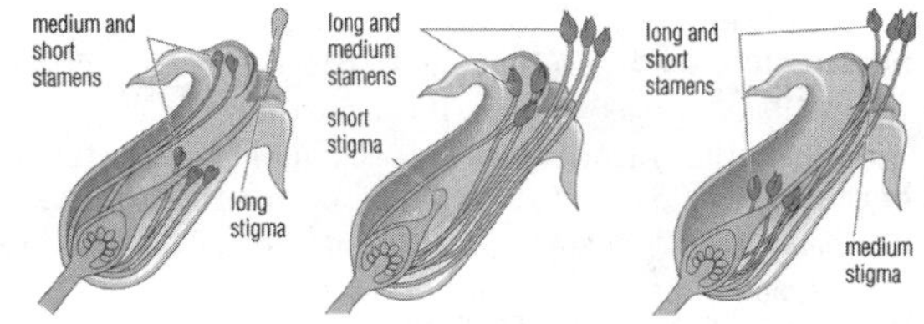

heterostyly Heterostyly, in which lengths of the stamens and stigma differ in flowers of different plants of the same species. This is a device to avoid self-pollination and ensure cross-pollination by visiting insects. Helicon

heterotroph any living organism that obtains its energy from organic substances produced by other organisms. All animals and fungi are heterotrophs, and they include †herbivores, †carnivores, and †saprotrophs (those that feed on dead animal and plant material).

heterozygous in a living organism, having two different †alleles for a given trait. In †homozygous organisms, by contrast, both chromosomes carry the same allele. In an outbreeding population an individual organism will generally be heterozygous for some genes but homozygous for others.

For example, in humans, alleles for both blue- and brown-pigmented eyes exist, but the "blue" allele is †recessive to the dominant "brown" allele.

Only individuals with blue eyes are predictably homozygous for this trait; brown-eyed people can be either homozygous or heterozygous.

heuristics in computing, a process by which a program attempts to improve its performance by learning from its own experience.

Hevelius, Johannes (1611–1687) German **Johann Hewel** or **Howelcke**, German astronomer. He published the first comparatively detailed map of the Moon in his *Selenographia* (1647). He also discovered four comets in 1652, 1661, 1672, and 1677, and suggested that these bodies orbited in parabolic paths about the Sun.

Hevesy, Georg Karl von (1885–1966) Hungarian-born Swedish chemist, discoverer of the element hafnium. He was awarded the Nobel Prize for Chemistry in 1943 for his use of a radioactive isotope (see †radioactivity) to follow the steps of a biological process.

Hewish, Antony (1924–) English radio astronomer who, with Martin †Ryle, was awarded the Nobel Prize for Physics in 1974 for his work on the development of radioastronomy, particularly the aperture-synthesis technique, and the discovery of †pulsars, rapidly rotating neutron stars that emit pulses of energy.

hexachlorobenzene C_6Cl_6, colorless aromatic solid, melting point 227°C/441°F. It can be prepared by fully substituting hydrogen atoms in †benzene with chlorine atoms by a reaction with chlorine gas and iron chloride. It is used in the synthesis of †hexafluorobenzene and as a preservative for wood.

hexachlorophene $(C_6HCl_3OH)_2CH_2$, white odorless bactericide, used in minute quantities in soaps and surgical disinfectants.

Trichlorophenol is used in its preparation, and, without precise temperature control, the highly toxic TCDD (tetrachlorodibenzodioxin; see †dioxin) may form as a by-product.

hexadecimal number system or **hex**, number system to the base 16, used in computing. In hex the decimal numbers 0–15 are represented by the characters 0, 1, 2, 3, 4, 5, 6, 7, 8, 9, A, B, C, D, E, F. Hexadecimal numbers are easy to convert to the computer's internal †binary code and are more compact than binary numbers.

Each place in a number increases in value by a power of 16 going from right to left; for instance, 8F is equal to 15 + (8 x 16) = 143 in decimal. Hexadecimal numbers are often preferred by programmers writing in low-level languages because they are more easily converted to the computer's internal †binary (base-two) code than are decimal numbers, and because they are more compact than binary numbers and therefore more easily keyed, checked, and memorized. (See also †ASCII.)

hexafluorobenzene C_6F_6, colorless aromatic liquid, boiling point 80°C/176°F. It is the simplest of the group of †benzene derivative compounds in which every hydrogen has been replaced by a single type of †halogen atom. It is prepared by reacting †hexachlorobenzene with potassium fluoride using a suitable polar solvent (a †solvent composed of †polar molecules, that can be used to dissolve ionic compounds).

hexagon six-sided †polygon. A regular hexagon has all six sides of equal length, and all six angles of equal size, 120°.

hexane generic name for the five isomeric forms of hexane, C_6H_{14}. Hexane is the simplest member of the group, being an unsaturated aliphatic hydrocarbon (one whose carbon atoms are joined in straight lines or branched chains) whose normal form is that of a colorless liquid. The compound is a constituent of petroleum ether solvents and gasoline.

hexanedioic acid or **adipic acid**; $(CH_2)_4(COOH)_2$, crystalline solid acid, obtained by the oxidation of certain fatty or waxy bodies. It is a dibasic acid, akin to oxalic and succinic acids, with the typical reactions of carboxylic acids, and is used in the manufacture of †nylon 66.

hexanol generic term for the three isomeric forms of the alcohol hexanol, $C_6H_{11}OH$. 4-Methyl-pentan-2-ol and 2-ethyl butan-1-ol are the other two forms. The isomers are variously used as solvents, †plasticizers, and in organic chemical synthesis.

Hey, James Stanley (1909–2000) English physicist whose work in radar led to pioneering research in radioastronomy. He discovered that large sunspots were powerful ultra-shortwave radio transmitters, and pinpointed a radio source in the Milky Way.

Heyrovsky, Jaroslav (1890–1967) Czech chemist who was awarded the Nobel Prize for Chemistry in 1967 for his invention and development of polarography, an electrochemical technique of chemical analysis.

HF in physics, abbreviation for high †frequency. HF radio waves have frequencies in the range 3–30 MHz.

hibernation state of dormancy in which certain animals spend the winter. It is associated with a dramatic reduction in all metabolic processes, including body temperature, breathing, and heart rate.

The body temperature of the Arctic ground squirrel falls to below 0°C/32°F during hibernation. Hibernating bats may breathe only once every 45 minutes, and can go for up to 2 hours without taking a breath.

hide or **hyde**, Anglo-Saxon unit of measurement used to measure the extent of arable land; it varied from about 296 ha/120 acres in the east of England to as little as 99 ha/40 acres in Wessex. One hide was regarded as sufficient to support a peasant and his household; it was the area that could be ploughed in a season by one plough and one team of oxen.

The hide was the basic unit of assessment for taxation and military service; under Norman rule it became the basis for the feudal tax of hidage.

Higgs, Peter Ware (1929–) English theoretical physicist who is best known for his prediction of an as yet undiscovered fundamental particle known as the **Higgs boson**. The Higgs boson, if discovered, will lead to our understanding of the origin of mass.

Higgs boson or **Higgs particle**, postulated †elementary particle whose existence would explain why particles have mass. The current theory of elementary particles, called the †standard model, cannot explain how mass arises. To overcome this difficulty, Peter Higgs of the University of Edinburgh and Thomas Kibble of Imperial College, London, proposed in 1964 a new particle that binds to other particles and gives them their mass.

In September 2000 physicists in Geneva, Switzerland, announced that they had found evidence that they had successfully created a Higgs boson. However, in July 2001 this was revealed to be an error in the data and in December 2001, nuclear physicists at the research organization CERN concluded, after a year analyzing data from the LEP (Large Electron Positron Collider) accelerator, that the Higgs boson probably did not exist. If this is true, it has serious implications for the standard model, the current theory of elementary particles and their interactions.

high-energy particles particles of †electromagnetic radiation that contain high energies, measured in terms of electronvolts. The energy in gamma radiation is of the order of 8×10^7 to 8×10^5 electronvolts and in X-rays of 8×10^3 to 8×10^1 electronvolts.

highest common factor HCF, in a set of numbers, the highest number that will divide every member of the set without leaving a remainder. For example, 6 is the highest common factor of 36, 48, and 72.

high-level language in computing, a programming language designed to suit the requirements of the programmer; it is independent of the internal machine code of any particular computer. High-level languages are used to solve problems and are often described as **problem-oriented languages**; for example, †BASIC was designed to be easily learnt by first-time programmers; †COBOL is used to write programs solving business problems; and †FORTRAN is used for programs solving scientific and mathematical problems. With the increasing popularity of windows-based systems, the next generation of programming languages was designed to facilitate the development of GUI interfaces; for example, †Visual Basic wraps the BASIC language in a graphical programming environment. Support for object-oriented programming has also become more common, for example in †C++ and †Java. In contrast, **low-level languages**, such as †assembly languages, closely reflect the machine codes of specific computers, and are therefore described as **machine-oriented languages**.

Unlike low-level languages, high-level languages are relatively easy to learn because the instructions bear a close resemblance to everyday language, and because the programmer does not require a detailed knowledge of the internal workings of the computer. Each instruction in a high-level language is equivalent to several machine-code instructions. High-level programs are therefore more compact than equivalent low-level programs. However, each high-level instruction must be translated into machine code – by either a †compiler or an interpreter program – before it can be executed by a computer. High-level languages are designed to be **portable** – programs written in a high-level language can be run on any computer that has a compiler or interpreter for that particular language.

high-yield variety crop that has been specially bred or selected to produce more than the natural varieties of the same species. During the 1950s and 1960s, new strains of wheat and corn were developed to reduce the food shortages in poor countries (the Green Revolution). Later, IR8, a new variety of rice that increased yields by up to six times, was developed in the Philippines. Strains of crops resistant to drought and disease were also developed. High-yield varieties require large amounts of expensive artificial fertilizers and sometimes pesticides for best results.

HII region in astronomy, region of extremely hot ionized hydrogen, surrounding one or more hot stars, visible as a bright patch or emission †nebula in the sky. The gas is ionized by the intense ultraviolet radiation from the stars within it. HII regions are often associated with interstellar clouds in which new stars are being born. An example is the †Orion nebula. The HII region takes its name from a spectroscopic notation in which HI represents neutral hydrogen (H) and HII represents ionized hydrogen (H^+).

Hilbert, David (1862–1943) German mathematician, philosopher, and physicist whose work was fundamental to 20th-century mathematics. He founded the formalist school with *Grundlagen der Geometrie/Foundations of Geometry* (1899), which was based on his idea of postulates.

Hilbert, David http://www-history.mcs.st-and.ac.uk/history/Mathematicians/Hilbert.html

Extensive biography of the German physicist, philosopher, and mathematician. The site contains a description of his contributions to physics and mathematics, and in particular describes the fact that he is considered to have had the greatest influence in geometry after Euclid. Several references for further reading are also listed, and the Web site also features a photograph of Hilbert.

Hilda in astronomy, member of a group of asteroids that are separated from the main asteroid belt by the †Kirkwood gaps. Hilda itself was discovered in 1875 and has a diameter of about 180 km/112 mi.

Hildegard of Bingen (1098–1179) German abbess, writer, and composer. Her encyclopedia of natural history, *Liber simplicis medicinae* (1150–60), giving both Latin and German names for the species described as well as their medicinal uses, is the earliest surviving scientific book by a woman.

Hill, Robert (1899–1991) British biochemist who showed that during photosynthesis, oxygen is produced, and that this derived oxygen comes from water. This process is now known as the **Hill reaction**. He also demonstrated the evolution of oxygen in human blood cells by the conversion of hemoglobin to oxyhemoglobin.

hinge joint in vertebrates, a joint where movement occurs in one plane only. Examples are the elbow and knee, which are controlled by pairs of muscles, the †flexors and †extensors.

Hinshelwood, Cyril Norman (1897–1967) English chemist who shared the Nobel Prize for Chemistry in 1956 for his work on chemical chain reactions. He also studied the chemistry of bacterial growth. He was knighted in 1948.

Hipparchus (c. 190–c. 120 BC) Greek astronomer and mathematician. He invented trigonometry and calculated the lengths of the solar year and the lunar month, discovered the precession of the equinoxes, made a catalogue of 850 fixed stars, and advanced Eratosthenes' method of determining the situation of places on the Earth's surface by lines of latitude and longitude.

Hippocrates (c. 460–c. 377 BC) Greek physician, often called the founder of medicine. Important Hippocratic ideas include cleanliness (for patients and physicians), moderation in eating and drinking, letting nature take its course, and living where the air is good. He believed that health was the result of the "humours" of the body being in balance; imbalance caused disease. These ideas were later adopted by †Galen.

Works by Hippocrates http://classics.mit.edu/Browse/browse-Hippocrates.html

Translations of the works of Hippocrates by Francis Adams in downloadable form. Seventeen works by the father of medicine are to be found here, ranging from On Ancient Medicine (this was written in 400 BC!) to On The Surgery. The site also includes the philosopher's best known work The Oath. The text files are very large (with some exceptions, notably The Oath), but the server is very fast so don't be too put off downloading.

His, Wilhelm (1831–1904) Swiss anatomist and physician. He is best known for his research on the embryology of the nervous system, the mode of origin of nerve fibers, and the developing human embryo. He introduced the use of the microtome into microscopical research, which enabled the cutting of serial sections. He was the founder of *Die Zeitschrift fur Anatomie und Entwicklungsgeschichte/*

Journal of Anatomy and Development in 1875.

His, Wilhelm (1863–1934) Swiss physician. He studied the impulse conduction system of the heart, part of which (the nerve fibers connecting atria with ventricles) is known as the **Bundle of His**. He was professor of internal medicine at the University of Berlin in Germany 1907–26.

histamine inflammatory substance normally released in damaged tissues, which also accounts for many of the symptoms of †allergy. It is an amine, $C_5H_9N_3$. Substances that neutralize its activity are known as antihistamines. Histamine was first described in 1911 by British physiologist Henry Dale (1875–1968).

histogram in statistics, a graph showing frequency of data, in which the horizontal axis details discrete units or class boundaries, and the vertical axis represents the frequency. Blocks are drawn such that their areas (rather than their height as in a †bar chart) are proportional to the frequencies within a class or across several class boundaries. There are no spaces between blocks.

If all the columns have the same width, then the element that appears most frequently (the mode) is represented by the tallest column. If the columns are of different widths the areas must be compared. Total frequency can be calculated by adding together the areas of all the columns.

histology study of plant and animal tissue by visual examination, usually with a †microscope.

†Stains are often used to highlight structural characteristics such as the presence of starch or distribution of fats.

HIV abbreviation for human immunodeficiency virus, infectious agent that is believed to cause †AIDS. It was first discovered in 1983 by Luc Montagnier of the Pasteur Institute in Paris, who called it lymphocyte-associated virus (LAV). Independently, US scientist Robert Gallo of the National Cancer Institute in Bethesda, Maryland, claimed its discovery in 1984 and named it human T-lymphocytotrophic virus 3 (HTLV-III).

HLA system abbreviation for human leucocyte antigen system, a series of gene groups (termed A, B, C, and D) that code for proteins on the surface of cells. If two people have the same HLA types, they are histocompatible (having compatible tissue types). HLA types feature in tissue testing before a transplant.

Hoagland, Mahlon Bush (1921–) US biochemist who was the first to isolate transfer †RNA (tRNA), a nucleic acid that plays an essential part in intracellular protein synthesis.

Hodgkin, Alan Lloyd (1914–1998) English physiologist who was awarded a Nobel Prize for Physiology or Medicine in 1963 with Andrew †Huxley for their work on the ionic mechanisms involved in the communication or inhibition of impulses across neuron (nerve cell) membranes. They engaged in research on the mechanism of conduction in peripheral nerves 1945–60, and Hodgkin devised techniques for measuring electric currents flowing across a cell membrane. Hodgkin was knighted in 1972.

Hodgkin, Dorothy Mary Crowfoot (1910–1994) English biochemist who analyzed the structure of penicillin, insulin, and vitamin B_{12}. Hodgkin was the first to use a computer to analyze the molecular structure of complex chemicals, and this enabled her to produce three-dimensional models. She was awarded the Nobel Prize for Chemistry in 1964 for her work in the crystallographic determination of the structures of biochemical compounds, notably penicillin and cyanocobalamin (vitamin B_{12}).

Hodgkin, Thomas (1798–1866) English physician who first recognized **Hodgkin's disease**, a rare form of cancer mainly affecting the lymph nodes and spleen. He pioneered the use of the stethoscope in the UK. He was also the first person to stress the importance of postmortem examinations.

hodometer device that enables the †acceleration of an object moving with known velocity over a path to be determined.

Hoff, Ted (1937–) born Marcian Edward Hoff, Jr, US inventor of the microprocessor. Working at Intel, Hoff conceived the idea of putting a computer on a single integrated circuit or microchip, to simplify the development of a range of pocket calculators for a Japanese company called Busicom.

Hoffmann, Roald (1937–) Polish chemist who worked on **molecular orbital theory** with Robert †Woodward and developed the Woodward–Hoffmann rules for the conservation of orbital symmetry (the arrangement of electrons around the nucleus; see †orbital, atomic), which predict the conditions under which certain organic reactions can occur. He shared the Nobel Prize for Chemistry in 1981 with Kenichi †Fukui for his theories concerning chemical reactions.

Hoffman's voltameter in chemistry, an apparatus for collecting gases produced by the †electrolysis of a liquid.

It consists of a vertical E-shaped glass tube with taps at the upper ends of the outer limbs and a reservoir at the top of the central limb. Platinum electrodes fused into the lower ends of the outer limbs are connected to a source of direct current. At the beginning of an experiment, the outer limbs are completely filled with electrolyte by opening the taps. The taps are then closed and the current switched on. Gases evolved at the electrodes bubble up the outer limbs and collect at the top, where they can be measured.

Hofmeister, Wilhelm Friedrich Benedikt (1824–1877) German botanist. He studied plant development and determined how a plant embryo, lying within a seed, is itself formed out of a single fertilized egg (ovule).

Hofstadter, Robert (1915–1990) US nuclear physicist who was awarded the Nobel Prize for Physics in 1961 for his work in scattering electrons in atomic nuclei, and for his pioneering studies of nuclear structure and the nuclear constituents, the proton and the neutron. He established that the proton and neutron were not pointlike, but had a definite volume and shape.

hogback geological formation consisting of a ridge with a sharp crest and abruptly sloping sides, the outline of which resembles the back of a hog. Hogbacks are the result of differential erosion on steeply dipping rock strata composed of alternating resistant and soft beds. Exposed, almost vertical resistant beds provide the sharp crests.

holdfast organ found at the base of many seaweeds, attaching them to the sea bed. It may be a flattened, suckerlike structure, or dissected and fingerlike, growing into rock crevices and firmly anchoring the plant.

Hollerith, Herman (1860–1929) US inventor of a mechanical tabulating machine, the first device for high-volume data processing. Hollerith's tabulator was widely publicized after being successfully used in the 1890 census. The firm he established, the Tabulating Machine Company, was later one of the founding companies of IBM.

Holley, Robert William (1922– 1993) US biochemist who was awarded a Nobel Prize for Physiology or Medicine in 1968 for his part in the interpretation of genetic code and its function in protein synthesis, and established the existence of transfer †RNA (tRNA) and its function.

Holmes, Arthur (1890–1965) English geologist who helped develop interest in the theory of continental drift. He also pioneered the use of †radioactive decay methods for rock dating, giving the first reliable estimate of the age of the Earth.

holmium chemical symbol Ho, (Latin *Holmia* "Stockholm") silvery, metallic element of the †lanthanide series, atomic number 67, relative atomic mass 164.93. It occurs in combination with other rare-earth metals and in various minerals such as gadolinite. Its compounds are highly magnetic.

The element was discovered in 1878, spectroscopically, by the Swiss chemists Jacques-Louis Soret and Marc Delafontaine, and independently in 1879 by Swedish chemist Per Cleve (1840–1905), who named it after Stockholm, near which it was found.

Holocene epoch period of geological time that began 10,000 years ago, and continues into the present. During this epoch the climate became warmer, the glaciers retreated, and human civilizations developed significantly.

It is the second and current epoch of the Quaternary period.

hologram three-dimensional image produced by †holography. Small, inexpensive holograms appear on credit cards and software licenses to guarantee their authenticity.

Animated Holographer http://www.holoworld.com/holo/demo1.html

Holograms, holography, and lasers are clearly explained in this "slide show" demonstration of how holograms are made. Each slide has an associated RealAudio commentary to help illuminate each stage of this scientific technique.

holography method of producing three-dimensional (3-D) images, called †holograms, by means of †laser light. Holography uses a photographic technique (involving the splitting of a laser beam into two beams) to produce a picture, or hologram, that contains 3-D information about the object photographed. Some holograms show meaningless patterns in ordinary light and produce a 3-D image only when laser light is projected through them, but reflection holograms produce images when ordinary light is reflected from them (as found on credit cards).

Although the possibility of holography was suggested as early as 1947 (by Hungarian-born British physicist Dennis †Gabor), it could not be demonstrated until a pure, coherent light source, the laser, became available in 1963. The first laser-recorded holograms were created by Emmett Leith and Juris Upatnieks at the University of Michigan, USA, and Yuri Denisyuk in the Soviet Union.

The technique of holography is also applicable to sound, and bats may navigate by ultrasonic holography. Holographic techniques also have applications in storing dental records, detecting stresses and strains in construction and in retail goods, detecting forged paintings and documents, and producing three-dimensional body scans. The technique of detecting strains is of widespread application. It involves making two different holograms of an object on one plate, the object being stressed between exposures. If the object has distorted during stressing, the hologram will be greatly changed, and the distortion readily apparent.

Using holography, digital data can be recorded page by page in a crystal. As many as 10,000 pages (100 megabytes) of digital data can be stored in an iron-doped lithium niobate crystal of 1 cubic cm/0.06 cubic in.

Amateur Holography http://members.aol.com/gakall/holopg.html

Features an introduction to the field of holography, as well as a guide to creating your own holograms on a limited budget. There are also numerous links to related Web sites.

homeostasis maintenance of a constant environment around living cells, particularly with regard to pH, salt concentration, temperature, and blood sugar levels. Stable conditions are important for the efficient functioning of the †enzyme reactions within the cells. In humans, homeostasis in the blood (which provides fluid for all tissues) is ensured by several organs. The †kidneys regulate pH, urea, and water concentration. The lungs regulate oxygen and carbon dioxide (see †breathing). Temperature is regulated by the liver and the skin (see †temperature regulation). Glucose levels in the blood are regulated by the †liver and the pancreas.

Body cells need to be in surroundings where the conditions do not change much and never reach extremes that are damaging to them. The surroundings of body cells inside the body can be called an "internal environment." Homeostasis is the keeping of this internal environment stable. Special processes are needed to keep things stable and these may be called homeostatic processes.

Conditions that are regulated in homeostasis include blood glucose level, temperature, water content of the body, and the amount of carbon dioxide and urea being carried by the blood. Body cells are surrounded and bathed in fluid. This fluid – tissue fluid – is made from †blood. This supplies the cells with †mineral salts, such as †glucose. As glucose is needed by cells all the time as a fuel for †respiration, the level of glucose must not be allowed to fall. It is also damaging if the level gets too high. The hormone †insulin is used to help keep blood glucose levels constant, by increasing its uptake into cells and stimulating the conversion of glucose into glycogen in the liver.

Homeostasis http://pespmc1.vub.ac.be/HOMEOSTA.html

Full scientific explanation of homeostasis in animals. This page, part of a larger educational site covering many anatomical topics, includes diagrams and provides links to other sites with further related information. All the organs and body systems are covered here in detail.

homeothermy maintenance of a constant body temperature in endothermic (warm-blooded) animals, by the use of chemical processes to compensate for heat loss or gain when external temperatures change. Such processes include generation of heat by the breakdown of food and the contraction of muscles, and loss of heat by sweating, panting, and other means.

Mammals and birds are homeotherms, whereas invertebrates, fish, amphibians, and reptiles are cold-blooded or poikilotherms. Homeotherms generally have a layer of insulating material to retain heat, such as fur, feathers, or fat (see †blubber). Their metabolism functions more efficiently due to homeothermy, enabling them to remain active under most climatic conditions. See also †temperature regulation.

homocyclic in chemistry, describing a compound whose molecules contain a closed ring system that is composed of only one type of atom, for example cyclohexane.

homogeneous reaction in chemistry, a reaction where there is no interface between the components. The term applies to all reactions where only gases are involved or where all the components are in solution.

homologous in biology, a term describing an organ or structure possessed by members of different taxonomic groups (for example, species, genera, families, orders) that originally derived from the same structure in a common ancestor. The wing of a bat, the arm of a monkey, and the flipper of a seal are homologous because they all derive from the forelimb of an ancestral mammal.

homologous series in chemistry, any of a number of series of organic compounds with similar chemical properties in which members differ by a constant relative molecular mass.

Alkanes (paraffins), alkenes (olefins), and alkynes (acetylenes) form such series in which members differ in mass by 14, 12, and 10 atomic mass units respectively. For example, the alkane homologous series begins with methane (CH_4), ethane (C_2H_6), propane (C_3H_8), butane (C_4H_{10}), and pentane (C_5H_{12}), each member differing from the previous one by a CH_2 group (or 14 atomic mass units).

homology branch of †topology involving the study of †closed curves, closed surfaces, and similar geometric arrangements in two- to *n*-dimensional space, and investigating the ways in which such spatial structures may be dissected. The formulation of the homological theory of dimensionality led to several basic laws of †duality (relating to topological properties of an additional part of space).

homozygous in a living organism, having two identical †alleles for a given trait. Individuals homozygous for a trait always breed true; that is, they produce offspring that resemble them in appearance when bred with a genetically similar individual; inbred varieties or species are homozygous for almost all traits. †Recessive alleles are only expressed in the homozygous condition. †Heterozygous organisms have two different alleles for a given trait.

honeyguide in botany, line or spot on the petals of a flower that indicate to pollinating insects the position of the nectaries (see †nectar) within the flower. The orange dot on the lower lip of the toadflax flower (*Linaria vulgaris*) is an example. Sometimes the markings reflect only ultraviolet light, which can be seen by many insects although it is not visible to the human eye.

Alkane	Alcohol	Aldehyde	Ketone	Carboxylic acid	Alkene
CH_4 methane	CH_3OH methanol	HCHO methanal	—	HCO_2H methanoic acid	—
CH_3CH_3 ethane	CH_3CH_2OH ethanol	CH_3CHO ethanal	—	CH_3CO_2H ethanoic acid	CH_2CH_2 ethene
$CH_3CH_2CH_3$ propane	$CH_3CH_2CH_2OH$ propanol	CH_3CH_2CHO propanal	CH_3COCH_3 propanone	$CH_3CH_2CO_2H$ propanoic acid	CH_2CHCH_3 propene
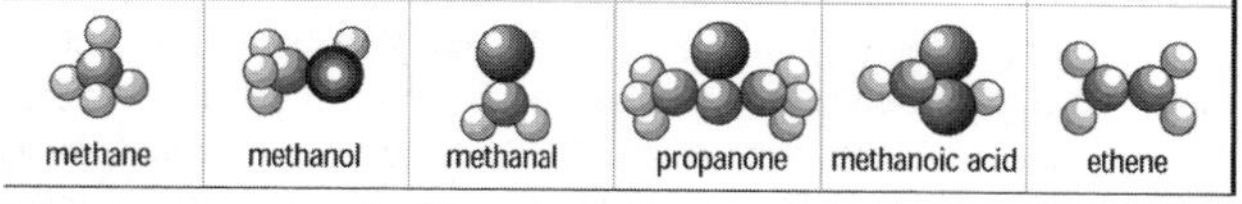 methane	methanol	methanal	propanone	methanoic acid	ethene

homologous series Six different types of homologous series: organic chemicals with similar chemical properties in which members differ by a constant relative atomic mass. For example, all the members of the alkane series differ by a relative atomic mass of 14. Helicon

honey sac in zoology, region of the gut of the worker honeybee, where nectar is converted into honey. The conversion produces glucose, fructose, and other sugars, depending on the source of nectar.

Hooke, Robert (1635–1703) English scientist and inventor, originator of †Hooke's law, and considered the foremost mechanic of his time. His inventions included a telegraph system, the spirit level, marine barometer, and sea gauge. He coined the term "cell" in biology.

Hooker, Joseph Dalton (1817–1911) English botanist who traveled to the Antarctic and India, and made many botanical discoveries. His works include *Flora Antarctica* (1844–47), *Genera plantarum* (1862–83), and *Flora of British India* (1875–97).

Hooke's law law stating that the deformation of a body is proportional to the magnitude of the deforming force, provided that the body's elastic limit (see †elasticity) is not exceeded. If the elastic limit is not reached, the body will return to its original size once the force is removed. The law was discovered by English physicist Robert Hooke in 1676.

A force applied to a spring by adding a weight to it will cause the spring to stretch in proportion to the applied force. If the force is removed and the spring returns to its original size, the material of the spring is said to be elastic. If the spring does not return to its original size, the material is said to be non-elastic. The elastic limit is the point at which a certain amount of force will deform the material so it becomes non-elastic.

For example, if a spring is stretched by 2 cm by a weight of 1 N, it will be stretched by 4 cm by a weight of 2 N, and so on. However, once the load exceeds the elastic limit for the spring, Hooke's law will no longer be obeyed and each successive increase in weight will result in a greater extension until finally the spring breaks.

Hope's apparatus in physics, an apparatus used to demonstrate the temperature at which water has its maximum density. It is named after Thomas Charles Hope (1766–1844).

It consists of a vertical cylindrical vessel fitted with horizontal thermometers through its sides near the top and bottom, and surrounded at the center by a ledge that holds a freezing mixture (ice and salt). When the cylinder is filled with water, this gradually cools, the denser water sinking to the bottom; eventually the upper thermometer records 0°C/32°F (the freezing point of water) and the lower one has a constant reading of 4°C/39°F (the temperature at which water is most dense).

Hopkins, Frederick Gowland (1861–1947) English biochemist who was awarded a Nobel Prize for Physiology or Medicine in 1929 for his discovery of trace substances, now known as vitamins, that stimulate growth. His research into diets revealed the necessity of these vitamins for the maintenance of health. Hopkins shared the prize with Christiaan †Eijkman, who had arrived at similar conclusions. He was knighted in 1925.

Hopkins, Sir Frederick Gowland http://web.calstatela.edu/faculty/nthomas/hopkins.htm

Maintained by Nigel Thomas, PhD, at California State University, Los Angeles, this page is devoted to the life and scientific work of Frederick Hopkins. Biographical information includes text and a time line of important moments in Hopkins' life. You will also find a bibliography of books by and about Hopkins. A special "science section" provides information on essential amino acids and vitamins. There is also a bibliography of texts about vitamins.

Hopper, Grace (1906–1992) born Grace Brewster Murray, US computer pioneer and mathematician who created the first compiler and helped invent the computer language †COBOL. She also coined the term "debug."

horizon in astronomy, the †great circle dividing the visible part of the sky from the part hidden by the Earth.

horizon limit to which one can see across the surface of the sea or a level plain, that is, about 5 km/3 mi at 1.5 m/5 ft above sea level, and about 65 km/40 mi at 300 m/1,000 ft.

horizon in earth sciences, distinct layer in a vertical cross-section of †soil. Most soils consist of layers of particles of different sizes.

horizontal having the same direction as the horizon, flat and level. A marble would not roll off a horizontal table without being pushed. On paper, a horizontal line is shown by a line parallel to the top of the page. In practice, horizontal planes are checked with a spirit level.

hormone in biology, a chemical secretion of the ductless †endocrine glands and specialized nerve cells (see †neurohormones) concerned with control of body functions. Hormones act as chemical messengers and are transported to all parts of the body by the †bloodstream where they affect target organs. The major glands are the thyroid, parathyroid, pituitary, adrenal, pancreas, ovary, and testis. There are also hormone-secreting cells in the kidney, liver, gastrointestinal tract, thymus (in the neck), pineal (in the brain), and placenta. Hormones bring about changes in the functions of various organs according to the body's requirements. The †hypothalamus, which adjoins the pituitary gland at the base of the brain, is a control center for overall coordination of hormone secretion; the thyroid hormones determine the rate of general body chemistry; the adrenal hormones prepare the organism during stress for "fight or flight"; and the sexual hormones such as estrogen and testosterone govern reproductive functions. Plants produce chemicals that affect growth and development. These chemicals can also be called hormones (see †plant hormone).

Hormones often bring about a slower response than messages sent by the †nerve cells, but the response to hormones may be longer lasting.

Examples of hormones include: insulin and glucagon (produced in the pancreas), which are involved in glucose regulation; ADH (produced in the pituitary gland), which regulates the concentration of urine produced by the †kidney; estrogen and progesterone (produced in the ovaries), which regulate the growth and functioning of sex organs for †sexual reproduction; testosterone (produced in the testes), which regulates male sexual development; follicle-stimulating hormone (FSH) and luteinizing hormone (LH) (from the pituitary gland), which regulate the menstrual cycle in females; and adrenaline (released from the adrenal glands), which prepares the body for increased activity.

Hormones: Sources, Targets, and Functions http://wsrv.clas.virginia.edu/~rjh9u/hormlist.html

Useful reference list of hormones showing where they are found, what they target, and what their effects are. This simple text-based page is a great resource for revision.

hornblende green or black rock-forming mineral, one of the †amphiboles. It is a hydrous †silicate composed mainly of calcium, iron, magnesium, and aluminum in addition to the silicon and oxygen that are common to all silicates. Hornblende is found in both igneous and metamorphic rocks and can be recognized by its color and prismatic shape.

hornfels †metamorphic rock formed by rocks heated by contact with a hot igneous body. It is fine-grained, brittle, and lacks foliation (a planar structure).

Hornfels may contain minerals only formed under conditions of great heat, such as andalusite, Al_2SiO_5, and cordierite, $(Mg,Fe)_2Al_4Si_5O_{18}$. This rock, originating from sedimentary rock strata, is found in contact with large igneous †intrusions where it represents the heat-altered equivalent of the surrounding clays. Its hardness makes it suitable for road building and railway ballast.

Horologium inconspicuous constellation of the southern hemisphere, represented as a clock.

Horrocks, Jeremiah (c. 1617–1641) English astronomer. He correctly predicted that there would be a transit of Venus on 4 December 1639, the first time such a phenomenon had ever been seen.

Horsehead nebula in the constellation of †Orion, near the star Zeta Orionis, and about 1,600 light years from the Sun. Part of the nebula consists of a dark cloud resembling a horse's head silhouetted against the glow of an †HII region (a region of hot ionized hydrogen).

horsepower abbreviation hp, imperial unit of power, now replaced by the †watt. It was first used by the engineer James †Watt, who employed it to compare the power of steam engines with that of horses.

horticulture art and science of growing flowers, fruit, and vegetables. Horticulture is practiced in gardens and orchards, along with millions of acres of land devoted to vegetable farming. Some areas, like California, have specialized in horticulture because they have the mild climate and light fertile soil most suited to these crops.

host in biology, an organism that is parasitized by another. In †commensalism, the partner that does not benefit may also be called the host.

hot big bang later, but fundamental, concept within the †Big Bang model, that the primordial explosion occurred in terms of almost unimaginable heat. The concept, formulated by George Gamow, led to considerable study of thermonuclear reactions and the search for †background radiation.

hot spot in earth science, area where a strong current or "plume" of †magma rises upwards below the Earth's crust. The magma spreads horizontally in all directions, and may break through where the crust is thin. Hot spots occur within, rather than on the edges of, lithospheric †plates. However, the magma usually reaches the surface at †plate margins. Examples of hot spots include Hawaii, Iceland, and Yellowstone National Park, Wyoming, USA.

hot spot track volcanic ridge or line of volcanic centers that results when a lithospheric plate moves over a hot spot. Only the volcanoes near the hot spot are active, the others having moved away and become extinct. One major example is the Hawaii-Emperor Seamount chain. Because hot spots are relatively stable with respect to moving plates, hot spot tracks can be used to determine absolute plate motions.

Hounsfield, Godfrey Newbold (1919–) English engineer who was awarded a Nobel Prize for Physiology or Medicine in 1979 for the development of the computed axial †tomography (CAT) scan, the application of computer techniques to X-raying the human body. He was knighted in 1981.

hour period of time comprising 60 minutes; 24 hours make one calendar day.

Houssay, Bernardo Alberto (1887–1971) Argentine physiologist who was awarded a Nobel Prize for Physiology or Medicine in 1947 for determining the function of the pituitary gland in sugar metabolism. He demonstrated that the pituitary gland produces hormones that greatly influence bodily functions, particularly sugar metabolism. One hormone counters the effects of insulin in preventing diabetes and his work gave a new insight into diabetes. He shared the prize with US biochemists Gerty and Carl Cori. Houssay was dismissed from his university post in 1943 for pro-American views. He was reinstated in 1955, after Juan Perón had gone into exile.

Hoyle, Fred(erick) (1915–2001) English astronomer, cosmologist, and writer. His astronomical research dealt mainly with the internal structure and evolution of the stars. In 1948 he developed with Hermann †Bondi and Thomas †Gold the †steady-state theory of the universe. In 1957, with William †Fowler, he showed that chemical elements heavier than hydrogen and helium may be built up by nuclear reactions inside stars. He was knighted in 1972.

hp abbreviation for †horsepower.

HTML abbreviation for hypertext markup language, standard for structuring and describing a document on the †World Wide Web. The HTML standard provides labels for constituent parts of a document (for example headings and paragraphs) and permits the inclusion of images, sounds, and "hyperlinks" to other documents. A †browser program is then used to convert this information into a graphical document on-screen. The specifications for HTML version 4, called †Dynamic HTML, were adopted at the end of 1997.

HTML is a specific example of †SGML (the international standard for text encoding). As such it is not a rigid standard but is constantly being improved to incorporate new features and allow greater freedom of design.

Lightning HTML Editor http://www.lightningsp.com/HTML_Editor/

HTML editor that teaches you about HTML as you use it. This is a well-organized guide to the intricacies of designing Web pages. There is a good list of "Frequently asked questions" and tips and tricks.

Hubble, Edwin (Powell) (1889–1953) US astronomer. He discovered the existence of †galaxies outside our own, and classified them according to their shape. His theory that the universe is expanding is now generally accepted.

Hubble classification in astronomy, scheme for classifying †galaxies according to their shapes, originally devised by US astronomer Edwin †Hubble in the 1920s.

Elliptical galaxies are classed from type E0 to type E7, where the figure denotes the degree of ellipticity. An E0 galaxy appears circular to an observer, while an E7 is highly elliptical (this is based on the apparent shape; the true shape, distorted by foreshortening, may be quite different). **Spiral galaxies** are classed as type Sa, Sb, or Sc: Sa is a tightly wound spiral with a large central bulge, Sc is loosely wound with a small bulge, and Sb is in between. Intermediate types are denoted by Sab or Sbc. **Barred spiral galaxies**, which have a prominent bar across their centers, are similarly classed as type SBa, SBb, or SBc, with intermediates SBab or SBbc. **Lenticular galaxies**, which have no spiral arms, are classed as type S0. **Irregular galaxies**, type Irr, can be subdivided into Irr I, which resemble poorly formed spirals, and Irr II which cannot be classified because of disturbance.

The Hubble classification was once believed to reveal an evolutionary sequence (from ellipticals to spirals) but this is now known not to be the case. Our own †Milky Way is a spiral galaxy, classified as type Sb or Sc, but may have a bar.

Hubble constant in astronomy, measure of the rate at which the universe is expanding, named after US astronomer Edwin Hubble. Observations suggest that galaxies are moving apart at a rate of 50–100 kps/30–60 mps for every million †parsecs of distance. This means that the universe, which began at one point according to the †Big Bang theory, is between 10 billion and 20 billion years old. Observations by the Hubble Space Telescope in 1996 produced a figure for the constant of 73 kps/45 mps.

Hubble expansion apparent recession of galaxies as seen from any point within the universe, the velocity of recession being proportional to the distance of the galaxy from the observer.

Hubble's law law that relates a galaxy's distance from us to its speed of recession as the universe expands, formulated in 1929 by US astronomer Edwin Hubble. He found that galaxies are moving apart at speeds that increase in direct proportion to their distance apart. The rate of expansion is known as the Hubble constant.

Hückel, Erich Armand Arthur Joseph (1896–1980) German physical chemist who, with Peter †Debye, developed in 1923 the modern theory that accounts for the electrochemical behavior of strong electrolytes in solution. Hückel also made discoveries relating to the structures of benzene and similar compounds that exhibit aromaticity (see †aromatic compound).

Huggins, Charles Brenton (1901– 1997) Canadian-born US physician who was awarded a Nobel Prize for Physiology or Medicine in 1966 for his work on the hormonal treatment of prostatic cancer. He shared the prize with Peyton †Rous, who was awarded it for his discovery of tumor-inducing viruses.

Hulse, Russell (1950–) US physicist and radio astronomer who, with Joseph H †Taylor, was awarded the Nobel Prize for Physics in 1993 for the discovery of a new type of †pulsar, called a binary pulsar. The pulsar had an invisible companion orbiting around it and radiated gravitational waves. These waves had been predicted by Einstein's general theory of relativity but this was the first experimental confirmation of their reality.

human body physical structure of the human being. It develops from the single cell of the fertilized ovum, is born at 40 weeks, and usually reaches sexual maturity between 11 and 18 years of age. The bony framework (skeleton) consists of more than 200 bones, over half of which are in the hands and feet. Bones are held together by joints, some of which allow movement. The circulatory system supplies muscles and organs with blood, which provides oxygen and food and removes carbon dioxide and other waste products. Body functions are controlled by the nervous system and hormones. In the upper part of the trunk is the thorax, which contains the lungs and heart. Below this is the abdomen, containing the digestive system (stomach and intestines); the liver, spleen, and pancreas; the urinary system (kidneys, ureters, and bladder); and, in women, the reproductive organs (ovaries, uterus, and vagina). In men, the prostate gland and seminal vesicles only of the reproductive system are situated in the abdomen, the testes being in the scrotum, which, with the penis, is suspended in front of and below the abdomen. The bladder empties through a small channel (urethra); in the female this opens in the upper end of the vulval cleft, which also contains the opening of the vagina, or birth canal; in the male, the urethra is continued into the penis. In both sexes, the lower bowel terminates in the anus, a ring of strong muscle situated between the buttocks.

skeleton The skull is mounted on the spinal column, or spine, a chain of 24 vertebrae. The ribs, 12 on each side, are articulated to the spinal column behind, and the upper seven meet the breastbone (sternum) in front. The lower end of the spine rests on the pelvic girdle, composed of the triangular sacrum, to which are attached the hipbones (ilia), which are fused in front. Below the sacrum is the tailbone (coccyx). The shoulder blades (scapulae) are held in place behind the upper ribs by muscles, and connected in front to the breastbone by the two collarbones (clavicles).

Each shoulder blade carries a cup (glenoid cavity) into which fits the upper end of the armbone (humerus). This articulates below with the two forearm bones (radius and ulna). These are articulated at the wrist (carpals) to the bones of the hand (metacarpals and phalanges). The upper end of each thighbone (femur) fits into a depression (acetabulum) in the hipbone; its lower end is articulated at the knee to the shinbone (tibia) and calf bone (fibula), which are articulated at the ankle (tarsals) to the bones of the foot (metatarsals and phalanges). At a moving joint, the end of each bone is formed of tough, smooth cartilage, lubricated by †synovial fluid. Points of special stress are reinforced by bands of fibrous tissue (ligaments).

Muscles are bundles of fibers wrapped in thin, tough layers of connective tissue (fascia); these are usually prolonged at the ends into strong, white cords (tendons, sinews) or sheets (aponeuroses), which connect the muscles to bones and organs, and by way of which the muscles do their work. Membranes of

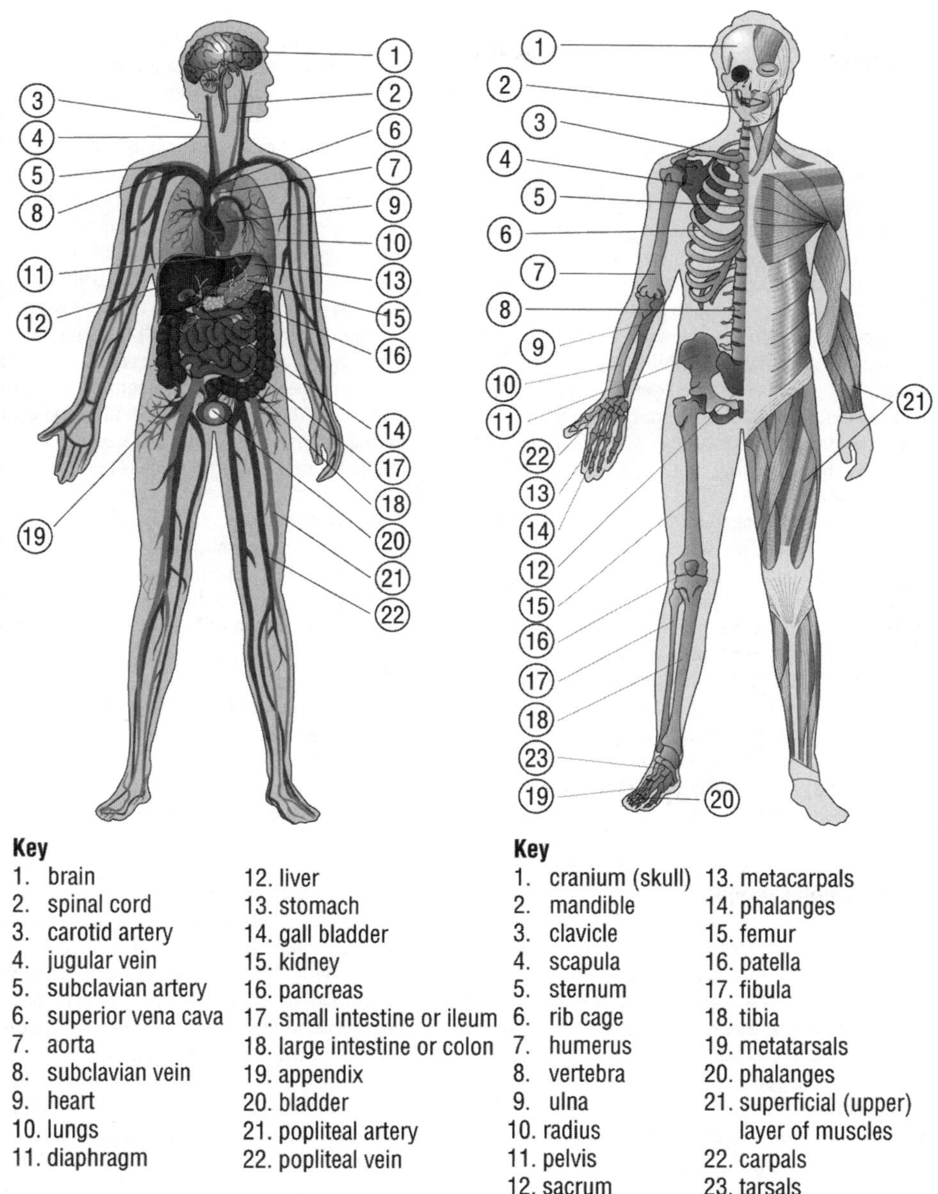

human body The adult human body has approximately 650 muscles, 100 joints, 100,000 km/60,000 mi of blood vessels and 13,000 nerve cells. There are 206 bones in the adult body, nearly half of them in the hands and feet. Helicon

connective tissue also enfold the organs and line the interior cavities of the body. The thorax has a stout muscular floor, the diaphragm, which expands and contracts the lungs in the act of breathing.

The blood vessels of the **circulatory system**, branching into multitudes of very fine tubes (capillaries), supply all parts of the muscles and organs with blood, which carries oxygen and food necessary for life. The food passes out of the blood to the cells in a clear fluid (lymph); this is returned with waste matter through a system of lymphatic vessels that converge into collecting ducts that drain into large veins in the region of the lower neck. Capillaries join together to form veins which return blood, depleted of oxygen, to the heart.

A finely branching **nervous system** regulates the function of the muscles and organs, and makes their needs known to the controlling centers in the central nervous system, which consists of the brain and spinal cord. The inner spaces of the brain and the cord contain cerebrospinal fluid. The body processes are regulated both by the nervous system and by hormones secreted by the endocrine glands. Cavities of the body that open onto the surface are coated with mucous membranes, which secrete a lubricating fluid (mucus).

The exterior surface of the body is covered with **skin**. Within the skin are the sebaceous glands, which secrete sebum, an oily fluid that makes the skin soft and pliable, and the sweat glands, which secrete water and various salts. From the skin grow hairs, chiefly on the head, in the armpits, and around the sexual organs; and nails shielding the tips of the fingers and toes; both hair and nails are modifications of skin tissue. The skin also contains nerve receptors for sensations of touch, pain, heat, and cold.

The human **digestive system** is nonspecialized and can break down a wide variety of foodstuffs. Food is mixed with saliva in the mouth by chewing and is swallowed. It enters the stomach, where it is gently churned for some time and mixed with acidic gastric juice. It then passes into the small intestine. In the first part of this, the duodenum, it is broken down further by the juice of the pancreas and duodenal glands, and mixed with bile from the liver, which splits up the fat. The jejunum and ileum continue the work of digestion and absorb most of the nutritive substances from the food. The large intestine completes the process, reabsorbing water into the body, and ejecting the useless residue as feces.

The body, to be healthy, must maintain water and various salts in the right proportions; the process is called **osmoregulation**. The blood is filtered in the two kidneys, which remove excess water, salts, and metabolic wastes. Together these form urine, which has a yellow pigment derived from bile, and passes down through two fine tubes (ureters) into the bladder, a reservoir from which the urine is emptied at intervals (micturition) through the urethra. Heat is constantly generated by the combustion of food in the muscles and glands, and by the activity of nerve cells and fibers. It is dissipated through the skin by conduction and evaporation of sweat, through the lungs in the expired air, and in other excreted substances. Average body temperature is about 38°C/100°F (37°C/98.4°F in the mouth).

Brainpop – Health http://www.brainpop.com/health/

Excellent site all about the basics of health. There is heavy use of the Flash multimedia plug-in, but it is used very effectively. The user can find out all about human anatomy and how the body changes with age. Each system of the body, such as the immune and digestive systems, is broken down into separate sections, with animations and facts in each.

Human Body http://www.bbc.co.uk/science/humanbody/

Human body Web site from the British Broadcasting Corporation (BBC). It contains articles on current biological issues, with a focus on health. The site is updated daily, so the user should check back regularly for recent issues. Links to other science-based sites, such as Tomorrow's World (the Web site of the BBC's forward-looking science program), are also included.

Human Genome Project HGP, research scheme to map the complete nucleotide (see †nucleic acid) sequence of human †DNA. It was begun in 1990 and a working draft of the genome (a mapping of 97% of the genome, sequencing of 85%, and completion of 24% of the human genome) was achieved in June 2000, with the results being published in February 2001. The publicly-funded Human Genome Organization (HUGO) coordinated the US$300 million project (the largest research project ever undertaken in the life sciences), which took place in over 20 centers around the world. Sequencing was also carried out commercially by US biotechnology company Celera Genomics. The completed detailed mapping of the genome is scheduled for 2003.

Research found the human genome to consist of between 27,000 and 40,000 different †genes – far fewer than expected – and one gene may contain more than 2 million nucleotides. The knowledge gained from mapping all these genes is expected to help prevent or treat many crippling and lethal diseases, but there are potential ethical problems associated with knowledge of an individual's genetic make-up, and fears that it will lead to genetic discrimination.

Human Genome Project Information http://www.ornl.gov/TechResources/Human_Genome/home.html

US-based site devoted to this mammoth project – with news, progress reports, a molecular genetics primer, and links to other relevant sites.

Your Genes, Your Choices: Exploring the Issues Raised by Genetic Research
http://ehrweb.aaas.org/ehr/books/index.html

Learn all about dominant genes and mutation, heredity, and the science behind the Human Genome Project. This site also considers the important ethical, legal, and social issues that are raised by the project.

human reproduction production of offspring by humans. Human reproduction is an example of †sexual reproduction, where the male produces sperm and the female eggs. These gametes contain only half the normal number of chromosomes, 23 instead of 46, so that on fertilization the resulting cell has the correct genetic complement. Fertilization is internal, which increases the chances of conception; unusually for mammals, pregnancy can occur at any time of the year. Human beings are also remarkable for the length of childhood and for the highly complex systems of parental care found in society. The use of contraception and the development of laboratory methods of insemination and fertilization are issues that make human reproduction more than a merely biological phenomenon.

Reproduction http://www.biology.arizona.edu/human_bio/problem_sets/Human_Reproduction/human_reproduction.html
Information designed to help understand the human reproductive system. The site covers such topics as the anatomy of the male and female reproductive systems, the hormones that influence these systems, fertilization, sterilization, and transmission of HIV. Diagrams are also included on the site, accompanying some of the explanations.

human species, origins of evolution of humans from ancestral †primates. The African apes (gorilla and chimpanzee) are shown by anatomical and molecular comparisons to be the closest living relatives of humans. The oldest known hominids (of the human group) had been the australopithecines, found in Africa, dating from 3.5–4.4 million years ago. But in December 2000, scientists unearthed the fossilized remains of a hominid dating back 6 million years. The first hominids to use tools appeared 2 million years ago, and hominids first used fire and moved out of Africa 1.7 million years ago. Modern humans are all believed to descend from one African female of 200,000 years ago, although there is a rival theory that humans evolved in different parts of the world simultaneously.

Miocene apes Genetic studies indicate that the last common ancestor between chimpanzees and humans lived 5 to 10 million years ago. There are only fragmentary remains of ape and hominid fossils from this period. Dispute continues over the hominid status of *Ramapithecus*, the jaws and teeth of which have been found in India and Kenya in late Miocene deposits, dating from between 14 and 10 million years ago. The lower jaw of a fossil ape found in the Otavi Mountains, Namibia, comes from deposits dated between 10 and 15 million years ago, and is similar to finds from East Africa and Turkey. It is thought to be close to the initial divergence of the great apes and humans.

australopithecines *Australopithecus afarensis*, found in Ethiopia and Kenya, date from 3.9 to 4.4 million years ago. These hominids walked upright and they were either direct ancestors or an offshoot of the line that led to modern humans. They may have been the ancestors of *Homo habilis* (considered by some to be a species of *Australopithecus*), who appeared about 2 million years later, had slightly larger bodies and brains, and were probably the first to use stone tools. Also living in Africa at the same time was *A. africanus*, a gracile hominid thought to be a meat-eater, and *A. robustus*, a hominid with robust bones, large teeth, heavy jaws, and thought to be a vegetarian. They are not generally considered to be our ancestors.

A new species of *Australopithecus* was discovered in Ethiopia in 1999. Named *A. garhi*, the fossils date from 2.5 million years ago and also share anatomical features with *Homo* species. The most complete australopithecine skeleton to date was found in South Africa in April 2000. It is about 1.8 million years old and from a female *A. robustus*.

The skull of an unknown hominid species, *Kenyanthropus platyops*, was discovered in Kenya in March 2001. Approximately 3.5 million years old, it is contemporary with the australopithecines, previously the oldest known hominids, leading to the suggestion that humans are descended from *K. platyops*, rather than the australopithecines as has been thought.

Homo erectus Over 1.7 million years ago, *Homo erectus*, believed by some to be descended from *H. habilis*, appeared in Africa. *H. erectus* had prominent brow ridges, a flattened cranium, with the widest part of the skull low down, and jaws with a rounded tooth row, but the chin, characteristic of modern humans, is lacking. They also had much larger brains (900–1,200 cu cm), and were probably the first to use fire and the first to move out of Africa. Their remains are found as far afield as China, West Asia, Spain, and southern Britain. Modern human *H. sapiens sapiens* and the Neanderthals *H. sapiens neanderthalensis* are probably descended from *H. erectus*.

Australian paleontologists announced the discovery of stone tools dated at about 800,000 to 900,000 years old and belonging to *H. erectus* on Flores, an island near Bali, in 1998. The discovery provided strong evidence that *H. erectus* were seafarers and had the language abilities and social structure to organize the movements of large groups to colonize new islands. In 2000 Japanese archaeologists discovered that *H. erectus* were probably building hut-like shelters around 500,000 years ago, the oldest known artificial structures.

Neanderthals Neanderthals were large-brained and heavily built, probably adapted to the cold conditions of the ice ages. They lived in Europe and the Middle East, and disappeared about 40,000 years ago, leaving *H. sapiens sapiens* as the only remaining species of the hominid group. Possible intermediate forms between Neanderthals and *H. sapiens sapiens* have been found at Mount Carmel in Israel and at Broken Hill in Zambia, but it seems that *H. sapiens sapiens* appeared in Europe quite rapidly and either wiped out the Neanderthals or interbred with them.

modern humans There are currently two major views of human evolution: the "out of Africa" model, according to which *H. sapiens* emerged from *H. erectus*, or a descendant species, in Africa and then spread throughout the world; and the multiregional model, according to which selection pressures led to the emergence of similar advanced types of *H. sapiens* from *H. erectus* in different parts of the world at around the same time. Analysis of DNA in recent human populations suggests that *H. sapiens* originated about 200,000 years ago in Africa from a single female ancestor, "Eve." The oldest known fossils of *H. sapiens* also come from Africa, dating from 100,000–150,000 years ago. Separation of human populations occurred later, with separation of Asian, European, and Australian populations taking place between 100,000 and 50,000 years ago.

The human genome consists of between 27,000 to 40,000 genes. Of these only about 1.5% differ between humans and the great apes.

Fossil Hominids FAQ http://www.talkorigins.org/faqs/homs/

Basic information about hominid species, the most important hominid fossils, and creationist arguments, plus links to related sites.

Humboldt, (Friedrich Wilhelm Heinrich) Alexander (1769–1859) Baron von Humboldt, German geophysicist, botanist, geologist, and writer who, with French botanist Aimé Bonpland (1773–1858), explored the regions of the Orinoco and Amazon rivers in South America 1800–04, and gathered 60,000 plant specimens. He was a founder of ecology.

Humboldt Current former name of the †Peru Current.

Hume-Rothery, William (1899–1968) English metallurgist who studied the constitution of alloys. He established that the microstructure of an alloy depends on the different sizes of the component atoms, the valency electron concentration, and electrochemical differences.

humerus upper bone of the forelimb of tetrapods. In humans, the humerus is the bone above the elbow.

humidity quantity of water vapor in a given volume of the atmosphere (absolute humidity), or the ratio of the amount of water vapor in the atmosphere to the saturation value at the same temperature (relative humidity). At †dew point the relative humidity is 100% and the air is said to be saturated. Condensation (the conversion of vapor to liquid) may then occur. Relative humidity is measured by various types of †hygrometer.

humus component of †soil consisting of decomposed or partly decomposed organic matter, dark in color and usually richer towards the surface. It has a higher carbon content than the original material and a lower nitrogen content, and is an important source of minerals in soil fertility.

hundredweight symbol cwt, imperial unit of mass, equal to 112 lb (50.8 kg). It is sometimes called the long hundredweight, to distinguish it from the short hundredweight or **cental**, equal to 100 lb (45.4 kg).

Hunter, John (1728–1793) Scottish surgeon, pathologist, and comparative anatomist who insisted on rigorous scientific method. He was the first to understand the nature of digestion.

hurricane or **tropical cyclone** or **typhoon**, a severe †depression (region of very low atmospheric pressure) in tropical regions, called **typhoon** in the North Pacific. It is a revolving storm originating at latitudes between 5° and 20° north or south of the Equator, when the surface temperature of the ocean is above 27°C/80°F. A central calm area, called the eye, is surrounded by inwardly spiraling winds (anticlockwise in the northern hemisphere and clockwise in the southern hemisphere) of up to 320 kph/200 mph. A hurricane is accompanied by lightning and torrential rain, and can cause extensive damage. In meteorology, a hurricane is a wind of force 12 or more on the †Beaufort scale.

During 1995 the Atlantic Ocean region suffered 19 tropical storms, 11 of them hurricanes. This was the third-worst season since 1871, causing 137 deaths. The most intense hurricane recorded in the Caribbean/Atlantic sector was Hurricane Gilbert in 1988, with sustained winds of 280 kph/175 mph and gusts of over 320 kph/200 mph.

Hurricane & Tropical Storm Tracking http://hurricane.terrapin.com/

Follow the current paths of Pacific and mid-Atlantic hurricanes and tropical storms at this site. Java animations of storms in previous years can also be viewed, and data sets for these storms may be downloaded. Current satellite weather maps can be accessed for the USA and surrounding region.

Tropical Weather and Hurricanes http://www.geog.ouc.bc.ca/physgeog/contents/7u.html

Details how and where tropical storms develop. From a much larger site, Fundamentals of Physical Geography, set up by a Canadian university, the site is well organized and contains images, maps, and diagrams.

Hutton, James (1726–1797) Scottish geologist, known as the "founder of geology," who formulated the concept of †uniformitarianism. In 1785 he developed a theory of the igneous origin of many rocks.

Huxley, Andrew Fielding (1917–) English physiologist who was awarded a Nobel Prize for Physiology or Medicine in 1963 with Alan †Hodgkin for work on the ionic mechanisms involved in the communication or inhibition of impulses across neuron (nerve cell) membranes impulses. He was knighted in 1974.

Huxley, Hugh Esmor (1924–) English physiologist who, using the electron microscope and thin slicing techniques, established the detailed structural basis of muscle contraction.

Huxley, Julian Sorell (1887–1975) English biologist, first director general of UNESCO, and a founder of the World Wildlife Fund (now the World Wide Fund for Nature). He wrote popular science books, including *Essays of a Biologist* (1923). He was knighted in 1958.

Huxley, T(homas) H(enry) (1825–1895) English scientist and humanist. Following the publication of Charles Darwin's *On the Origin of Species* (1859), he became known as "Darwin's bulldog," and for many years was a prominent champion of evolution. In 1869 he coined the word "agnostic" to express his own religious attitude, and is considered the founder of scientific humanism.

Huygens (or Huyghens), Christiaan (1629–1695) Dutch mathematical physicist and astronomer. He proposed the wave theory of light, developed the pendulum clock in 1657, discovered polarization, and observed Saturn's rings. He made important advances in pure mathematics, applied mathematics, and mechanics, which he virtually founded. His work in astronomy was an impressive defense of the Copernican view of the Solar System.

Huygens, Christiaan http://www-history.mcs.st-and.ac.uk/history/Mathematicians/Huygens.html

Extensive biography of the Dutch astronomer, physicist, and mathematician. The site contains a description of his contributions to astronomy, physics, and mathematics. Also included are the title page of his book Horologium Oscillatorium (1673) and the first page of his book De Ratiociniis in Ludo Aleae (1657). Several references for further reading are also listed, and the Web site also features a portrait of Huygens.

Hyades V-shaped cluster of stars that forms the face of the bull in the constellation †Taurus. It is 150 light years away from the Sun, and contains over 200 stars, although only about 12 are visible to the naked eye.

The Hyades is a much older cluster than the Pleiades, for not only have some of the brighter stars evolved into †red giants, some have gone even further and are now †white dwarfs. †Aldebaran, which marks the eye of the bull and which appears to be in the middle of the cluster, is not actually a member of the cluster. It is only 65 light years away from the Sun.

Hyakutake bright, fast-moving comet discovered by Japanese amateur astronomer Yuji Hyakutake in January 1996. The comet became a spectacular sight only two months later. It passed within 15 million km/9 million mi of the Earth on 25 March, one of the closest approaches of a comet in recent history, and within 34 million km/21 million mi of the Sun on 1 May. At its brightest, in mid-April, its broad head reached zero magnitude and its long, narrow tail stretched 100° across the sky. The Ulysses spacecraft passed through the tail of Hyakutake in 1996. Analysis of data from that flight revealed in March 2000 that Hyakutake's tail is 550 million km/342 million mi long, making it the longest comet tail known.

Hyatt, John Wesley (1837–1920) US inventor who in 1869 invented †celluloid, the first artificial plastic, intended as a substitute for ivory. It became popular for making a wide range of products, from shirt collars and combs to toys and babies' rattles, and is still used in the manufacture of table-tennis balls.

hybrid offspring from a cross between individuals of two different species, or two inbred lines within a species. In most cases, hybrids between species are infertile and unable to reproduce sexually. In plants, however, doubling of the chromosomes (see †polyploid) can restore the fertility of such hybrids.

Hybrids between different genera were believed to be extremely rare (an example is the *Cupressocyparis leylandii* cypress which, like some hybrids, shows exceptional vigor, or †heterosis) but research in the late 1990s shows that hybridization is much more common than traditionally represented. One British evolutionary biologist estimated in 1999 that approximately 10% of animal species and 20% of plant species produced fertile offspring through interspecies mating. Blue whales, for example, hybridize with fin whales and different species of birds of paradise also hybridize. In the wild, a "hybrid zone" may occur where the ranges of two related species meet.

hybridization production of a †hybrid.

hydathode specialized pore, or less commonly a hair, through which water is secreted by hydrostatic pressure from the interior of a plant leaf onto the surface. Hydathodes are found on many different plants and are usually situated around the leaf margin at vein endings. Each pore is surrounded by two crescent-shaped cells and resembles an open †stoma, but the size of the opening cannot be varied as in a stoma. The process of water secretion through hydathodes is known as †guttation.

Hydra in astronomy, the largest constellation, winding across more than a quarter of the sky between †Cancer and †Libra in the southern hemisphere. Hydra is named after the multi-headed sea serpent slain by Hercules. Despite its size, it is not prominent; its brightest star is second-magnitude Alphard.

hydrate chemical compound that has discrete water molecules combined with it. The water is known as **water of crystallization** and the number of water molecules associated with one molecule of the compound is denoted in both its name and chemical formula: for example, $CuSO_4.5H_2O$ is copper(II) sulfate pentahydrate.

hydration special type of solvation in which water molecules are attached – either by electrostatic forces or by coordinate (covalent) bonds – to ions or molecules of a solute. Some salts, called hydrates, retain associated water molecules in the solid state (this water is called the water of crystallization); in solid copper(II) sulfate, for example, the hydrated ion is
$[Cu(H_2O)_4]^{2+}$ or $[Cu(H_2O)_6]^{2+}$

hydration in chemistry, the combination of water and another substance to produce a single product. It is the opposite of †dehydration.

hydration in earth science, a form of †chemical weathering caused by the expansion of certain minerals as they absorb water. The expansion weakens the parent rock and may cause it to break up.

hydraulic action in earth science, the erosive force exerted by water (as distinct from erosion by the rock particles that are carried by water). It can wear away the banks of a river, particularly at the outer curve of a †meander (bend in the river), where the current flows most strongly.

Hydraulic action occurs as a river tumbles over a †waterfall to crash onto the rocks below. It will lead to the formation of a plunge pool below the waterfall. The hydraulic action of ocean waves and turbulent currents forces air into rock cracks, and therefore brings about erosion by †cavitation. In coastal areas hydraulic action is often the most important form of †erosion.

hydraulic press device that uses an †incompressible fluid, such as water or oil, to transmit a small downward force applied to a piston of small area to a larger area piston, which then produces a proportionately large upward force. Such a press is a demonstration of †Pascal's law of pressures.

hydraulic radius measure of a river's †channel efficiency (its ability to move water and sediment), used by water engineers to assess the likelihood of flooding. The hydraulic radius of a channel is defined as the ratio of its cross-sectional area to its wetted perimeter (the part of the cross-section – bed and bank – that is in contact with the water).

The greater the hydraulic radius, the greater the efficiency of the channel and the less likely the river is to flood. The highest values occur when channels are deep, narrow, and semi-circular in shape.

hydraulics field of study concerned with utilizing the properties of water and other liquids, in particular the way they flow and transmit pressure, and with the application of these properties in engineering. It applies the principles of †hydrostatics and hydrodynamics. The oldest type of hydraulic machine is the **hydraulic press**, invented by Joseph Bramah in England in 1795. The hydraulic principle of fluid pressure transmitting a small force over a small area in order to produce a larger force over a larger area is commonly used in vehicle braking systems, the forging press, and the hydraulic systems of aircraft and earth-moving machinery.

hydride chemical compound containing hydrogen and one other element, and in which the hydrogen is the more electronegative element (see †electronegativity).

Hydrides of the more reactive metals may be ionic compounds containing a hydride anion (H^-).

hydrocarbon any of a class of chemical compounds containing only hydrogen and carbon (for example, the alkanes and alkenes). Hydrocarbons are obtained industrially principally from petroleum and coal tar. Unsaturated hydrocarbons contain at least one double or triple carbon–carbon †bond, whereas saturated hydrocarbons contain only single bonds.

Hydrocarbon http://www.factmonster.com/ce6/sci/A0824706.html

From the "Fact monster," a page explaining all about hydrocarbons. Starting with an explanation of what exactly a hydrocarbon is, the text then goes on to the differences between alkenes and alkanes. Common hydrocarbons are mentioned, as are common uses for them. This well-designed and neat page also contains links to other educational resources.

hydrochloric acid HCl, highly corrosive solution of hydrogen chloride (a colorless, acidic gas) in water. The concentrated acid is about 35% hydrogen chloride. The acid is a typical strong, monobasic acid forming only one series of salts, the chlorides. It has many industrial uses, including recovery of zinc from galvanized scrap iron and the production of chlorine. It is also produced in the stomachs of animals for the purposes of digestion.

hydrocyanic acid or **prussic acid**, solution of hydrogen cyanide gas (HCN) in water. It is a colorless, highly poisonous, volatile liquid, smelling of bitter almonds.

hydrodynamics branch of physics dealing with fluids (liquids and gases) in motion.

hydroelectric power electricity generated by the motion (†kinetic energy) of water. In a typical scheme, the potential energy of water stored in a reservoir, often created by damming a river, is converted into kinetic energy as it is piped into water †turbines. The turbines are coupled to †generators to produce electricity. Hydroelectric power provides about one-fifth of the world's electricity, supplying more than a billion people. Hydroelectricity is a non-polluting, renewable energy resource, produced from water that can be recycled.

Conventional hydroelectric power stations can provide a constant supply of electricity. They have large dams that can store enough water to last for several years. In pumped storage plants, water flowing through the turbines is recycled. At times of low energy demand, the water is pumped back into an upper reservoir and reused to generate electricity. A tidal power station exploits the rise and fall of the tides.

hydrogen chemical symbol H, (Greek *hydro* + *gen* "water generator") colorless, odorless, gaseous, nonmetallic element, atomic number 1, relative atomic mass 1.00797. It is the lightest of all the elements and occurs on Earth, chiefly in combination with oxygen, as water. Hydrogen is the most abundant element in the universe, where it accounts for 93% of the total number of atoms and 76% of the total mass. It is a component of most stars, including the Sun, whose heat and light are produced through the nuclear-fusion process that converts hydrogen into helium. When subjected to a pressure 500,000 times greater than that of the Earth's atmosphere, hydrogen becomes a solid with metallic properties, as in one of the inner zones of Jupiter. Hydrogen's common and industrial uses include the hardening of oils and fats by hydrogenation, the creation of high-temperature flames for welding, and as rocket fuel. It has been proposed as a fuel for road vehicles.

Its isotopes †deuterium and †tritium (half-life 12.5 years) are used in nuclear weapons, and deuterons (deuterium nuclei) are used in synthesizing elements. The element's name refers to the generation of water by the combustion of hydrogen, and was coined in 1787 by French chemist Louis Guyton de Morveau (1737–1816), after having been discovered in 1766 by English physicist and chemist Henry †Cavendish.

Hydrogen – Coming Soon http://webhome.idirect.com/~famistew/hydrogen/index.html

Interesting, if basic, presentation about the potential for hydrogen to become a significant source of energy in the future. The dangerous effects of fossil fuels on the environment are also discussed.

hydrogenation addition of hydrogen to an unsaturated organic molecule (one that contains †double bonds or †triple bonds). It is widely used in the manufacture of margarine and low-fat spreads by the addition of hydrogen to vegetable oils.

Vegetable oils contain double carbon-to-carbon bonds and are therefore examples of unsaturated compounds. When hydrogen is added to these double bonds, the oils become saturated and more solid in consistency.

hydrogen bond weak electrostatic bond that forms between covalently bonded hydrogen atoms and a strongly †electronegative atom with a lone electron pair (for example, oxygen, nitrogen, and fluorine). Hydrogen bonds (denoted by a dashed line) are of great importance in biochemical processes, particularly the N-H- - -H bond, which enables proteins and nucleic acids to form the three-dimensional structures necessary for their biological activity.

hydrogen burning in astronomy, any of several processes by which hydrogen is converted to †helium by †nuclear fusion in the core of a star. In the Sun, the main process is the proton–proton chain, while in heavier stars the carbon cycle is more important. In both processes, four protons are converted to a helium nucleus with the emission of †positrons, †neutrinos, and gamma †rays. The temperature must exceed several million kelvin for hydrogen burning to start and the least massive stars (†brown dwarfs) never become hot enough.

hydrogencarbonate or **bicarbonate**, compound containing the ion HCO_3^-, an acid salt of carbonic acid (solution of carbon dioxide in water). When heated or treated with dilute acids, it gives off carbon dioxide. The most important compounds are †sodium hydrogencarbonate (bicarbonate of soda), and †calcium hydrogencarbonate.

hydrogen carrier compound that accepts hydrogen ions in biochemical reactions and is therefore important in oxidation–reduction reactions such as the intracellular use of oxygen.

hydrogen cyanide **HCN**, poisonous gas formed by the reaction of sodium cyanide with dilute sulfuric acid; it is used for fumigation.

The salts formed from it are cyanides – for example sodium cyanide, used in hardening steel and extracting gold and silver from their ores. If dissolved in water, hydrogen cyanide gives hydrocyanic acid.

hydrogen ion concentration number of grams of hydrogen ions per liter of solution; denoted by $[H^+]$. It is a measure of the acidity of a solution, in which context it is normally expressed in terms of pH values, given by $pH = \log_{10} (1/[H^+])$.

hydrogen peroxide H_2O_2, colorless syrupy liquid used, in diluted form, as an antiseptic. Oxygen is released when hydrogen peroxide is added to water and the froth helps to discharge dead tissue from wounds and ulcers. It is also used as a mouthwash and as a bleach and as a rocket propellant.

hydrogen sulfate HSO_4^-, compound containing the hydrogen sulfate ion. Hydrogen sulfates are †acid salts.

hydrogen sulfide H_2S, poisonous gas with the smell of rotten eggs. It is found in certain types of crude oil where it is formed by decomposition of sulfur compounds. It is removed from the oil at the refinery and converted to elemental sulfur.

hydrogen trioxide H_2O_3, relatively stable compound of hydrogen and oxygen present in the atmosphere and possibly also in living tissue. It was first synthesized in 1994; previously it had been assumed to be too unstable.

It is produced in a reaction similar to that used for the commercial production of hydrogen peroxide (H_2O_2) but ozone (O_3) is used instead of oxygen. Hydrogen trioxide is stable at low temperatures but begins to decompose slowly at –40°C/–40°F forming the high-energy form of oxygen, singlet oxygen.

hydrograph graph showing how the discharge of a river varies with time (generally over a matter of days). By studying hydrographs, water engineers can predict when flooding is likely and take action to prevent it.

A hydrograph shows the lag time, or delay, between peak rainfall and the resultant peak in discharge, and the length of time taken for that discharge to peak. The shorter the lag time and the higher the peak, the more likely it is that flooding will occur. The peak flow is equal to the **groundwater flow** plus the **storm flow**. Factors likely to give short lag times and high peaks include heavy rainstorms, steep slopes, deforestation, poor soil quality, and the covering of surfaces with impermeable substances such as tarmac and concrete. Actions taken by water engineers to increase lag times and lower peaks includes planting trees in the drainage basin of a river.

hydrography study and charting of Earth's surface waters in seas, lakes, and rivers.

hydrological cycle also known as the †water cycle, by which water is circulated between the Earth's surface and its atmosphere.

hydrology study of the location and movement of inland water, both frozen and liquid, above and below ground. It is applied to major civil engineering projects such as irrigation schemes, dams, and hydroelectric power, and in planning water supply. Hydrologic studies are also undertaken to assess drinking water supplies, to track water underground, and to understand the role of water in geological processes such as fault movement and mineral deposition.

Hydrology Primer http://wwwdutslc.wr.usgs.gov/infores/hydrology.primer.html

Information from the US Geological Survey about all aspects of hydrology. The "clickable" chapters include facts about surface water and ground water, the work of hydrologists, and careers in hydrology. For answers to further questions click on "ask a hydrologist," which provides links to other US national and regional sources.

hydrolysis chemical reaction in which the action of water or its ions breaks down a substance into smaller molecules. Hydrolysis occurs in certain inorganic salts in solution, in nearly all non-metallic chlorides, in esters, and in other organic substances. It is one of the mechanisms for the breakdown of food by the body, as in the conversion of starch to glucose.

hydrolysis in earth science, a form of †chemical weathering caused by the chemical alteration of certain minerals as they react with water. For example, the mineral feldspar in granite reacts with water to form a white clay called †china clay.

hydrometer instrument used to measure the relative density of liquids (the density compared with that of water). A hydrometer consists of a thin glass tube ending in a sphere that leads into a smaller sphere, the latter being weighted so that the hydrometer floats upright, sinking deeper into less dense liquids than into denser liquids. Hydrometers are used in brewing and to test the strength of acid in car batteries.

The hydrometer is based on †Archimedes' principle.

hydrophilic (Greek "water-loving") describing †functional groups with a strong affinity for water, such as the carboxyl group (–COOH).

If a molecule contains both a hydrophilic and a †hydrophobic group (a group that repels water), it may have an affinity for both aqueous and nonaqueous molecules. Such compounds are used to stabilize †emulsions or as †detergents.

hydrophilic colloid †colloid that easily forms an aqueous solution, for example †sols of proteins.

hydrophily type of †pollination where the pollen is transported by water. Water-pollinated plants occur in 31 genera in 11 different families. They are found in habitats as diverse as rainforests and seasonal desert pools. Pollen is either dispersed underwater or on the water's surface.

Pollen may be released directly onto the water's surface, as in the sea grass *Halodule pinifolia*, forming pollen rafts, or as in the freshwater plant *Vallisneria*, the pollen may be released within floating male flowers. In Caribbean turtle grass, *Thalassia testudinum*, pollen is released underwater embedded in strands of mucilage. Denser than water, it is carried by the current.

hydrophobic (Greek "water-hating") describing †functional groups that repel water (the opposite of †hydrophilic).

hydrophobic colloid †colloid that does not readily form an aqueous solution, for example †sols of gold.

hydrophyte plant adapted to live in water, or in waterlogged soil.

Hydrophytes may have leaves with a very reduced or absent †cuticle and no †stomata (since there is no need to conserve water), a reduced root and water-conducting system, and less supporting tissue since water buoys plants up. There are often numerous spaces between the cells in their stems and roots to make †gas exchange with all parts of the plant body possible. Many have highly divided leaves, which lessens resistance to flowing water; an example is spiked water milfoil *Myriophyllum spicatum.*

hydroponics cultivation of plants without soil, using specially prepared solutions of mineral salts. Beginning in the 1930s, large crops were grown by hydroponic methods, at first in California but since then in many other parts of the world.

Julius von Sachs (1832–1897), in 1860, and W Knop, in 1865, developed a system of plant culture in water whereby the relation of mineral salts to plant growth could be determined, but it was not until about 1930 that large crops could be grown. The term was first coined by US scientist W F Gericke.

hydrosphere portion of the Earth made of water, ice, and water vapor, including the oceans, seas, rivers, streams, swamps, lakes, groundwater, and atmospheric water vapor. In some cases its definition is extended to include the water locked up in the Earth's crust and mantle.

hydrostatics branch of †statics dealing with fluids in equilibrium – that is, in a static condition. Practical applications include shipbuilding and dam design.

hydrostatic skeleton in zoology, fluid-filled body space that provides the supportive framework for the bodies of some invertebrates, for example earthworms. It is non-compressible but deformable and as a result, forces generated by the contraction of muscles can be transferred to different parts of the body.

hydrothermal in geology, pertaining to a fluid whose principal component is hot water, or to a mineral deposit believed to be precipitated from such a fluid.

hydrothermal vein crack in rock filled with minerals precipitated through the action of circulating high-temperature fluids. Igneous activity often gives rise to the circulation of heated fluids that migrate outwards and move through the surrounding rock. When such solutions carry metallic ions, ore-mineral deposition occurs in the new surroundings on cooling.

hydrothermal vent or **smoker**, crack in the ocean floor, commonly associated with an †ocean ridge, through which hot, mineral-rich water flows into the cold ocean water, forming thick clouds of suspended material. The clouds may be dark or light, depending on the mineral content, thus producing "white smokers" or "black smokers." In some cases the water is clear.

Sea water percolating through the sediments and crust is heated by the hot rocks and magma below and then dissolves minerals from the hot rocks. The water gets so hot that its increased buoyancy drives it back out into the ocean via a hydrothermal ("hot water") vent. When the water, anywhere from 60°C to over 400°C (kept liquid by the pressure of the ocean above) comes into contact with the frigid sea water, the sudden cooling causes these minerals to precipitate from solution, so forming the suspension. These minerals settle out and crystallize, forming stalagmite-like "chimneys." The chemical-rich water around a smoker gives rise to colonies of primitive bacteria that use the chemicals in the water, rather than the sunlight, for energy. Strange animals that live in such regions include huge tube worms 2 m/6 ft long, giant clams, and species of crab, anemone, and shrimp found nowhere else.

hydroxide any inorganic chemical compound containing one or more hydroxyl (OH) groups and generally combined with a metal. Hydroxides include sodium hydroxide (caustic soda, NaOH), potassium hydroxide (caustic potash, KOH), and calcium hydroxide (slaked lime, $Ca(OH)_2$).

hydroxylamine or **hydroxy-ammonia**; NH_2OH, unstable, deliquescent colorless solid, melting point 33°C/91°F. It can be prepared by reducing nitric oxide or nitric acid with hydrogen and a suitable catalyst. It can explode on heating and is used in the production of nylon.

hydroxylation in chemistry, process that results in the introduction of a †hydroxyl group, –OH, into an organic molecule.

hydroxyl group atom of hydrogen and an atom of oxygen bonded together and covalently bonded to an organic molecule. Common compounds containing hydroxyl groups are alcohols and phenols.

In chemical reactions, the hydroxyl group (—OH) frequently behaves as a single entity.

hydroxypropanoic acid technical name for †lactic acid.

Hydrus constellation of the south polar region, represented as a water snake. It is quite separate from the equatorial constellation †Hydra. Beta Hydri is the nearest bright star to the south pole.

Hygeia large asteroid, which at 300 km/185 mi across is the fourth largest known. It was discovered in 1849 and has an orbital period of 5.6 years.

hygrometer any instrument for measuring the humidity, or water vapor content, of a gas (usually air). A wet and dry bulb hygrometer consists of two vertical thermometers, with one of the bulbs covered in absorbent cloth dipped into water. As the water evaporates, the bulb cools, producing a temperature difference between the two thermometers. The amount of evaporation, and hence cooling of the wet bulb, depends on the relative humidity of the air. Other hygrometers work on the basis of a length of natural fiber, such as hair or a fine strand of gut, changing with variations in humidity. In a dew-point hygrometer, a polished metal mirror gradually cools until a fine mist of water (dew) forms on it. This gives a measure of the †dew point, from which the air's relative humidity can be calculated.

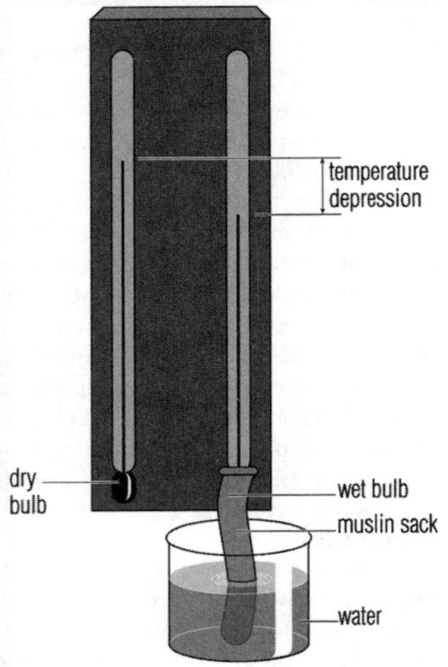

hygrometer The most common hygrometer, or instrument for measuring the humidity of a gas, is the wet and dry bulb hygrometer. The wet bulb records a lower temperature because water evaporates from the muslin, taking heat from the wet bulb. The degree of evaporation and hence cooling depends upon the humidity of the surrounding air or other gas. Helicon

hygroscopic in chemistry, describing solid compounds that readily absorb water from the atmosphere but do not liquefy as a result.

hyperbola in geometry, a curve formed by cutting a right circular cone with a plane so that the angle between the plane and the base is greater than the angle between the base and the side of the cone. All hyperbolae are bounded by two asymptotes (straight lines to which the hyperbola moves closer and closer to but never reaches).

A hyperbola is a member of the family of curves known as †conic sections.

A hyperbola can also be defined as a path traced by a point that moves such that the ratio of its distance from a fixed point (focus) and a fixed straight line (directrix) is a constant and greater than 1; that is, it has an †eccentricity greater than 1.

Hyperbola

http://www-groups.dcs.st-andrews.ac.uk/history/Curves/Hyperbola.html

Introduction to the hyperbola and links to associated curves. If you have a Java-enabled browser you can experiment interactively with this curve and its associates. There are also links to mathematicians who have studied the hyperbola, to the hyperbola's particular attributes, and to related Web sites.

hyperbolic geometry system of non-Euclidean geometry developed by German mathematician Bernhard Riemann, complementary to †elliptic geometry and comprising the geometry of †geodesics in the neighborhood of a point on the (curved) surface at which a tangential plane intersects the surface in a hyperbolic curve.

hypercharge in physics, a property of certain †elementary particles, analogous to electric charge, that accounts for the absence of some expected behavior (such as certain decays).

hypercomplex numbers numbers that expand on †complex numbers; for example, †quaternions.

hyperfine structure very fine splitting of individual lines in a †spectrum, which can be the result of the presence of different †isotopes of an element in the source.

Hyperion or **Saturn VII**, one of the smaller moons of Saturn, discovered in 1848. An irregular-shaped lump measuring 410 x 260 x 220 km/255 x 162 x 137 mi, it orbits Saturn at an average distance of 1,480,000 km/920,000 mi with a period (time taken to circle the planet) of 21.3 days. Hyperion is tumbling chaotically in its orbit, with rotation periods varying from hours to weeks, due to the gravitational influence of Saturn's largest moon, †Titan.

hyperlink in computing, link from one document to another or, within the same document, from one place to another. It can be activated by clicking on the link with a mouse. The link is usually highlighted in some way, for example by the inclusion of a small graphic. Documents linked in this way are described as †hypertext. Examples of programs that use hypertext and hyperlinks are Windows help files, Acrobat, and Mosaic.

hypermetropia or **long-sightedness**, defect of vision in which a person is able to focus on objects in the distance, but not on close objects. It is caused by the failure of the lens to return to its normal rounded shape, or by the eyeball being too short, with the result that the image is focused on a point behind the retina. Hypermetropia is corrected by wearing glasses fitted with †converging lenses, each of which acts like a magnifying glass.

hyperon in physics, any of a group of highly unstable †elementary particles that includes all the †baryons with the exception of protons and neutrons. They are all composed of three quarks. The lambda, xi, sigma, and omega particles are hyperons.

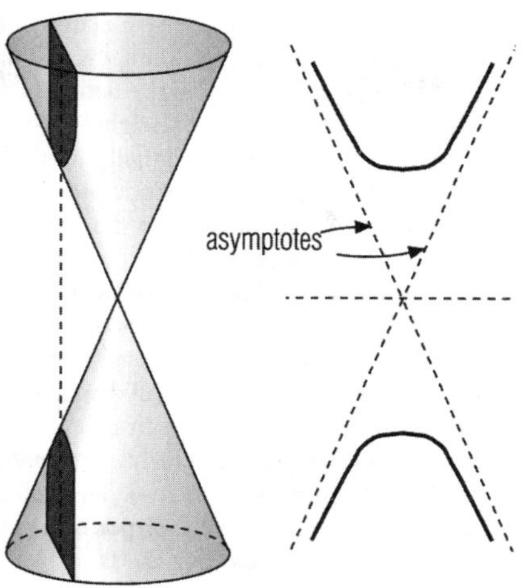

hyperbola The hyperbola is produced when a cone is cut by a plane. It is one of a family of curves called conic sections: the circle, ellipse, and parabola. These curves are produced when the plane cuts the cone at different angles and positions. Helicon

hypertext in computing, a method of forming connections between different files (including office documents, graphics, and Web pages) so that the user can click a "link" with the mouse to jump between them. For example, a software program might display a map of a country; if the user clicks on a particular city the program displays information about that city. The linked files do not need to be on the same computer, or even in the same country, for a hyperlink to be created.

hypha (plural **hyphae**) delicate, usually branching filament, many of which collectively form the mycelium and fruiting bodies of a †fungus. Food molecules and other substances are transported along hyphae by the movement of the cytoplasm, known as "cytoplasmic streaming."

Typically hyphae grow by increasing in length from the tips and by the formation of side branches. Hyphae of the higher fungi (the ascomycetes and basidiomycetes) are divided by cross walls or septa at intervals, whereas those of lower fungi (for example, bread mould) are undivided. However, even the higher fungi are not truly cellular, as each septum is pierced by a central pore, through which cytoplasm, and even nuclei, can flow. The hyphal walls contain †chitin, a polysaccharide.

hypnosis artificially induced state of relaxation or altered attention characterized by heightened suggestibility. There is evidence that, with susceptible persons, the sense of pain may be diminished, memory of past events enhanced, and illusions or hallucinations experienced. Posthypnotic amnesia (forgetting what happened during hypnosis) and posthypnotic suggestion (performing an action after hypnosis that had been suggested during it) have also been demonstrated.

Hypnosis has a number of uses in medicine. Hypnotically induced sleep, for example, may assist the healing process, and hypnotic suggestion (hypnotherapy) may help in dealing with the symptoms of emotional and psychosomatic disorders. The Austrian physician Friedrich Anton Mesmer is said to be the discoverer of hypnosis, but he called it "animal magnetism," believing it to be a physical force or fluid. The term "hypnosis" was coined by James Braid (1795–1860), a British physician and surgeon who was the first to regard it as a psychological phenomenon. The Scottish surgeon James Esdaile (1805–1859), working in India, performed hundreds of operations in which he used hypnosis to induce analgesia (insensitivity to pain) or general anesthesia (total insensitivity).

hypocycloid in geometry, a cusped curve traced by a point on the circumference of a circle that rolls around the inside of another larger circle. (Compare †epicycloid.)

hypogeal describing seed germination in which the †cotyledons remain below ground. It can refer to fruits that develop underground, such as peanuts *Arachis hypogea*.

hypotenuse longest side of a right-angled triangle, opposite the right angle. It is of particular application in Pythagoras' theorem (the square of the hypotenuse equals the sum of the squares of the other two sides), and in trigonometry where the ratios †sine and †cosine are defined as the ratios opposite/hypotenuse and adjacent/hypotenuse respectively.

hypothalamus region of the brain below the †cerebrum which regulates rhythmic activity and physiological stability within the body, including water balance and temperature. It regulates the production of the pituitary gland's hormones and controls that part of the †nervous system governing the involuntary muscles.

hypothesis in science, an idea concerning an event and its possible explanation. The term is one favored by the followers of the philosopher Karl Popper, who argue that the merit of a scientific hypothesis lies in its ability to make testable predictions.

hypsometer (Greek *hypsos* "height") instrument for testing the accuracy of a thermometer at the boiling point of water. It was originally used for determining altitude by comparing changes in the boiling point with changes in atmospheric pressure.

hysteresis phenomenon seen in the elastic and electromagnetic behavior of materials, in which a lag occurs between the application or removal of a force or field and its effect.

If the magnetic field applied to a magnetic material is increased and then decreased back to its original value, the magnetic field inside the material does not return to its original value. The internal field "lags" behind the external field. This behavior results in a loss of energy, called the **hysteresis loss**, when a sample is repeatedly magnetized and demagnetized. Hence the materials used in transformer cores and electromagnets should have a low hysteresis loss. Similar behavior is seen in some materials when varying electric fields are applied (**electric hysteresis**). **Elastic hysteresis** occurs when a varying force repeatedly deforms an elastic material. The deformation produced does not completely disappear when the force is removed, and this results in energy loss on repeated deformations.

Hz in physics, the symbol for †hertz.

I

Iapetus or **Saturn VIII**, one of the larger moons of Saturn, discovered in 1671 by Italian-born French astronomer Giovanni †Cassini. It has a diameter of 1,460 km/900 mi and orbits Saturn at an average distance of 3.6 million km/2.2 million mi with a period (the time it takes to circle the planet) of 79.3 days. As seen from Earth, Iapetus always appears brighter on one side of its orbit than on the other. Voyager probes revealed that the leading hemisphere of Iapetus is very dark, possibly coated with a sooty organic material originating from Saturn's outermost satellite, Phoebe.

Iapetus Ocean or **Proto-Atlantic**, sea that existed in early †Palaeozoic times between the continent that was to become Europe and that which was to become North America. The continents moved together in the late Palaeozoic, obliterating the ocean. When they moved apart once more, they formed the Atlantic.

Ibn an-Nafis (or Ibn al-Nafis) (*c.* 1213–1288) Syrian physician. He practiced medicine at Cairo, Egypt, where he was dean of the Nasiri Hospital. Ibn an-Nafis was the first to explain pulmonary circulation (the circulation of the blood through the lungs for the purpose of oxygenation) correctly. He presented his discovery to the public in 1242, and included it in his 1286 commentary on the anatomy of Avicenna's *Canon Medicinae*. His work, probably unknown to his contemporaries in Europe, was later lost and only rediscovered in 1924.

Icarus in astronomy, †Apollo asteroid 1.5 km/1 mi in diameter, discovered in 1949 by German-born US astronomer Walter Baade. It orbits the Sun every 409 days at a distance of 28–300 million km/18–186 million mi (0.19–2.0 astronomical units). It was the first asteroid known to approach the Sun closer than does the planet Mercury. In 1968 it passed within 6 million km/3.7 million mi of the Earth.

ice solid formed by water when it freezes. It is colorless and its crystals are hexagonal. The water molecules are held together by hydrogen bonds.

The freezing point of ice, used as a standard for measuring temperature, is 0° for the Celsius and Réaumur scales and 32° for the Fahrenheit scale. Ice expands in the act of freezing (hence burst pipes), becoming less dense than water (0.9175 at 5°C/41°F). In 1998 US geologists succeeded in creating ice made up of irregular crystals. The new type of ice was created by exerting an enormous pressure (6,500 atmospheres) on water molecules, until their hydrogen atoms were squeezed into a disorderly state.

Sea Ice http://www.antcrc.utas.edu.au/aspect/seaiceglossary.html

Information about sea ice. Satellite and photographic images are used to help explain the differences between pancake ice, shuga, and ice floes and between sea ice and icebergs.

ice age any period of extensive glaciation (in which icesheets and icecaps expand over the Earth) occurring in the Earth's history, but particularly that in the †Pleistocene epoch (last 2 million years), immediately preceding historic times. On the North American continent, †glaciers reached as far south as the Great Lakes, and an icesheet spread over northern Europe, leaving its remains as far south as Switzerland. In Britain ice reached as far south as a line from Bristol to Banbury to Exeter. There were several glacial advances separated by interglacial (warm) stages, during which the ice melted and temperatures were higher than today. We are currently in an interglacial phase of an ice age.

Other ice ages have occurred throughout geological time: there were four in the Precambrian era, one in the Ordovician, and one at the end of the Carboniferous and beginning of the Permian. The occurrence of an ice age is governed by a combination of factors (the **Milankovitch hypothesis**): (1) the Earth's change of attitude in relation to the Sun – that is, the way it tilts in a 41,000-year cycle and at the same time wobbles on its axis in a 22,000-year cycle, making the time of its closest approach to the Sun come at different seasons; and (2) the 92,000-year cycle of eccentricity in its orbit around the Sun, changing it from an elliptical to a near circular orbit, the severest period of an ice age coinciding with the approach to circularity. There is a possibility that the Pleistocene ice age is not yet over. It may reach another maximum in another 60,000 years.

Cracking the Ice Age http://www.pbs.org/wgbh/nova/ice/

Companion to the US Public Broadcasting Service (PBS) television program Nova, this page provides information about glaciation, the natural changes in climate over the past 60 million years, the greenhouse effect, global warming, and continental movement. There is also a list of related links.

Ice Age, Little period of particularly severe winters that gripped northern Europe between the 13th and 17th centuries. Contemporary writings and paintings show that Alpine glaciers were much more extensive than at present, and rivers such as the Thames, which do not ice over today, were so frozen that festivals could be held on them.

iceberg floating mass of ice, about 80% of which is submerged, rising sometimes to 100 m/300 ft above sea level. Glaciers that reach the coast become extended into a broad foot; as this enters the sea, masses break off and drift towards temperate latitudes, becoming a danger to shipping.

icecap body of ice that is larger than a glacier but smaller than an ice sheet. Such ice masses cover mountain ranges, such as the Alps, or small islands. Glaciers often originate from icecaps.

Iceland spar form of †calcite, $CaCO_3$, originally found in Iceland. In its pure form Iceland spar is transparent and exhibits the peculiar phenomenon of producing two images of anything seen through it (birefringence or †double refraction). It is used in optical instruments. The crystals cleave into perfect rhombohedra.

ice protein gene in biology, gene found in the bacterium *Pseudomonas syringae*. It expresses a protein around which ice crystals form at temperatures between 0°C/32°F and 7°C/45°F. The bacterium is nearly always found on leaves and contributes to frost damage of plants in cold weather.

A strain of *P. syringae* with the ice protein gene absent has been developed ("ice minus strain"). Sprayed over plants it competes with "wild-type" bacteria (with the ice-protein gene), replacing them on plant surfaces. As a result, frost-sensitive plants become less vulnerable to damage.

ice sheet body of ice that covers a large land mass or continent; it is larger than an ice cap. During the last †ice age, ice sheets spread over large parts of Europe and North America. Today there are two ice sheets, covering much of Antarctica and Greenland. About 96% of all present-day ice is in the form of ice sheets. The ice sheet covering western Greenland has increased in thickness by 2 m/6.5 ft 1981–93; this increase is the equivalent of a 10% rise in global sea levels.

ichthyology scientific study of fish, including their classification, general biology, behavior, ecology, and research into commercial fisheries.

icon in computing, a small picture on the computer screen, or †VDU, representing an object or function that the user may manipulate or otherwise use. It is a feature of †graphical user interface (GUI) systems. Icons make computers easier to use by allowing the user to point to and click with a mouse on pictures, rather than type commands.

iconoscope type of television camera where a beam of †electrons scans a special mosaic, which can store an optical image electrically, and converts the image stored to electrical signals for transmission.

icosahedron plural **icosahedra**, regular solid with 20 equilateral (equal-sided) triangular faces. It is one of the five regular †polyhedra, or Platonic solids.

id in Freudian psychology, the mass of motivational and instinctual elements of the human mind, whose activity is largely governed by the arousal of specific needs. It is regarded as the †unconscious element of the human psyche, and is said to be in conflict with the †ego and the †superego.

ideal in projective geometry and algebra, describes a point (one on every line) at infinity in such a way that the point has a coordinate position. In number theory, the term describes a collection of elements in a ring that has specific properties within a universal set, that is, that form a closed system under addition (among themselves) and under scaling by any element from the universal set. For example, the even numbers form an ideal within the universal set of integers in that when added (or multiplied) together, even numbers result.

ideal gas or **perfect gas**, gas obeying the gas laws of Boyle, Charles, and Joule exactly. This would imply that the gas consists of perfectly elastic molecules, each of which has zero volume and no attractive or repulsive interaction with its neighbors.

identity in mathematics, a number or operation that leaves others unchanged when combined with them. Zero is the identity for addition; one is the identity for multiplication. For example:

$7 + 0 = 7$

$7 \times 1 = 7$

identity the distinct and recognizable nature of an individual, which results from a unique combination of characteristics and qualities. In philosophy, identity is the sameness of a person, which may continue in spite of changes in bodily appearance, personality, intellectual abilities, memory, and so on. In psychology, identity refers to one's conception of oneself and sense of continuous being, particularly as an individual distinguishable from, but interacting, with others.

igneous rock rock formed from the cooling and solidification of molten rock called †magma. The acidic nature of this rock type means that areas with underlying igneous rock are particularly susceptible to the effects of acid rain. Igneous rocks that crystallize slowly from magma below the Earth's surface have large crystals. Examples include dolerite and granite.

Types of Rocks: Igneous, Metamorphic, and Sedimentary

http://www.zephryus.demon.co.uk/virtualschool/lessons/Lesson008.htm

Targeted specifically at GCSE geographers, this clear introduction to the three main rock types begins by outlining the history of the "rock cycle" and the relationship between igneous, metamorphic, and sedimentary rocks. Clear explanations of all three types, with photographs to illustrate, are followed by a set of test problems to encourage users to carry out their own research.

ignis fatuus another name for †will-o'-the-wisp.

ignition temperature or **fire point**, minimum temperature to which a substance must be heated before it will spontaneously burn independently of the source of heat; for example, ethanol has an ignition temperature of 425°C/798°F and a †flash point of 12°C/54°F.

ileum part of the small intestine of the †digestive system, between the duodenum and the colon, that absorbs digested food.

Its wall is muscular so that waves of contraction (peristalsis) can mix the food and push it forward. Numerous fingerlike projections, or villi, point inwards from the wall, increasing the surface area available for absorption. The ileum has an excellent blood supply, which receives the food molecules passing through the wall and transports them to the liver via the hepatic portal vein.

illumination or **illuminance**, the brightness or intensity of light falling on a surface. It depends upon the brightness and distance of light sources and the angle at which the light strikes the surface. The SI unit is the †lux.

ilmenite oxide of iron and titanium, iron titanate ($FeTiO_3$); an ore of titanium. The mineral is black, with a metallic luster. It is found as an accessory mineral in mafic igneous rocks and in sands.

image in mathematics, a point or number that is produced as the result of a †transformation or mapping.

image picture or appearance of a real object, formed by light that passes through a lens or is reflected from a mirror. If rays of light actually pass through an image, it is called a **real image**. Real images, such as those produced by a camera or projector lens, can be projected onto a screen. An image that cannot be projected onto a screen, such as that seen in a flat mirror, is known as a **virtual image**.

In a pinhole camera, rays of light traveling from the object through the pinhole of the camera form a real image on the screen of the camera. The image is upside down, or inverted.

image compression in computing, one of a number of methods used to reduce the amount of information required to represent an image, so that it takes up less computer memory and can be transmitted more rapidly and economically via telecommunications systems. It plays a major role in fax transmission and in videophone and multimedia systems.

image intensifier electronic device that brightens a dark image. Image intensifiers are used for seeing at night; for example, in military situations.

The intensifier first forms an image on a photocathode, which emits electrons in proportion to the intensity of the light falling on it. The electron flow is increased by one or more amplifiers. Finally, a fluorescent screen converts the electrons back into visible light, now bright enough to see.

imaginary number term often used to describe the non-real element of a †complex number. For the complex number $(a + ib)$, ib is the imaginary number where $i = \sqrt{(-1)}$, and b any real number.

imago sexually mature stage of an †insect.

imide in chemistry, member of a class of cyclic organic compounds that contain the -CO-NH-CO- group. These compounds are usually prepared by heating †anhydrides with ammonia.

imine in chemistry, member of a class of organic compounds that contain molecules in which an †imino group is attached directly to a single carbon atom. The classification also includes compounds that have the C = N-R group, where R is an †alkyl group.

imino group in chemistry, group consisting of one nitrogen and one hydrogen atom, >NH. Imino compounds have the general formula R-NHR', where R and R' are †alkyl groups. These compounds are secondary †amines and are formed by the †alkylation of ammonia.

immediate access memory in computing, †memory provided in the †central processing unit to store the programs and data in current use.

immigration in ecology, movement of individuals into a population. Immigration contributes to the increase in numbers of a population.

immiscible describing liquids that will not mix with each other, such as oil and water. When two immiscible liquids are shaken together, an emulsion is produced. This normally forms separate layers on being left to stand. Immiscible liquids may be separated using a separating funnel.

immobilized enzyme in biotechnology, enzyme that is bonded to an insoluble matrix. The catalytic properties of the enzyme are unaffected and the enzyme is more easily recovered for re-use and does not contaminate the product. Immobilized enzymes are also effective at temperatures that would denature the non-immobilized forms.

immunity protection that organisms have against foreign micro-organisms, such as bacteria and viruses, and against cancerous cells (see †cancer). The cells that provide this protection are called white blood cells, or leucocytes, and make up the immune system. They include neutrophils and †macrophages, which can engulf invading organisms and other unwanted material, and natural killer cells that destroy cells infected by viruses and cancerous cells. Some of the most important immune cells are the †B cells and †T cells. Immune cells coordinate their activities by means of chemical messengers or †lymphokines, including the antiviral messenger †interferon. The lymph nodes play a major role in organizing the immune response.

Immunology http://www.nutramed.com/immunology/index.htm

Excellent and easy-to-understand introduction to the science of immunology. The clearly written text describes how the immune system works and the function of antibodies. The site also explains what happens when the immune system fails to work, or works too well, and the problems of allergies and hypersensitivity.

immunization process of conferring immunity to infectious disease by artificial methods, in other words making someone not liable to catch a disease. The most widely used technique is vaccination (see †vaccine). Immunization is an important public health measure. If most of the population has been immunized against a particular disease, it is impossible for an epidemic to take hold.

In vaccination people can be immunized against a disease by introducing small quantities of dead or inactive forms of the disease-causing micro-organism into the body. In the vaccine are chemicals which act as antigens. These stimulate the white blood cells to produce antibodies. Antibodies are capable of binding to disease-causing bacteria, resulting in their destruction. If the body is effectively "warned" about the antigen by this means, the body is then able to produce enough of the appropriate antibody whenever the living disease-causing organism enters the body. The micro-organism can then be killed before it harms the body. The person contacted by the disease will probably feel well all the time and will be unaware that he or she has been in contact with the disease. This is called active immunity.

MMR Vaccination can be used to protect against diseases caused by viruses. It is, in fact, the best way to deal with virus diseases, because antibiotics are not effective against them. An example is the MMR vaccine used to protect children against measles, mumps, and rubella. A study of thousands of children in Finland who have had the MMR vaccine has shown that there is only a very low risk of damage being caused to a child as a result of having the vaccine. The risk is much lower than the damage caused by catching one of the diseases.

If vaccination covers a large proportion of the population at risk, a disease can become very rare, or even die out. Smallpox was eliminated in this way. The worries over MMR vaccine have reduced the numbers of children being vaccinated, which increases the risk of measles, mumps, or rubella breaking out in an epidemic.

current research US researchers conducted the first human trials of a vaccine administered by eating genetically engineered potatoes in 1995. In animal trials, mice fed on the potatoes received protection against some common gut infections.

immunocompromised lacking a fully effective immune system. The term is most often used in connection with infections such as †AIDS where the virus interferes with the immune response (see †immunity).

Other factors that can impair the immune response are pregnancy, diabetes, old age, malnutrition, and extreme stress, making someone susceptible to infections by micro-organisms (such as listeria) that do not affect normal, healthy people. Some people are immunodeficient; others could be on †immunosuppressive drugs.

immunodeficient lacking one or more elements of a working immune system. Immune deficiency is the term generally used for patients who are born with such a defect, while those who acquire such a deficiency later in life are referred to as †immunocompromised or immunosuppressed.

A serious impairment of the immune system is sometimes known as SCID, or severe combined immune deficiency. At one time children born with this condition would have died in infancy. They can now be kept alive in a germ-free environment, then treated with a bone-marrow transplant from a relative, to replace the missing immune cells. At present, the success rate for this type of treatment is still fairly low. See also †gene therapy.

immunoglobulin human globulin †protein that can be separated from blood and administered to confer immediate immunity on the recipient. It participates in the immune reaction as the antibody for a specific †antigen (disease-causing agent).

Normal immunoglobulin (gamma globulin) is the fraction of the blood serum that, in general, contains the most antibodies, and is obtained from plasma pooled from about a thousand donors. It is given for short-term (two to three months) protection when a person is at risk, mainly from hepatitis A (infectious hepatitis), or when a pregnant woman, not immunized against German measles, is exposed to the rubella virus.

Specific immunoglobulins are injected when a susceptible (nonimmunized) person is at risk of infection from a potentially fatal disease, such as hepatitis B (serum hepatitis), rabies, or tetanus. These immunoglobulins are prepared from blood pooled from donors convalescing from the disease.

immunological memory in biology, during the immune response, the formation of memory cells (type of †B cell) following the first exposure to an antigen. If there is subsequent exposure to the same antigen, the memory cells rapidly divide forming lymphocytes that quickly destroy the antigen. It is immunological memory ensures that individuals rarely suffer from infectious diseases, such as mumps and chickenpox, more than once in their lives.

immunosuppressive any drug that suppresses the body's normal immune responses to infection or foreign tissue. It is used in the treatment of autoimmune disease (see †autoimmunity); as part of chemotherapy for leukemias, lymphomas, and other cancers; and to help prevent rejection following organ transplantation. Immunosuppressed patients are at greatly increased risk of infection.

impedance symbol Z, total opposition of a circuit to the passage of alternating electric current. It includes the resistance R and the reactance X (caused by †capacitance or †inductance); the impedance can then be found using the equation $Z^2 = R^2 + X^2$.

imperial system traditional system of units developed in the UK, based largely on the foot, pound, and second (f.p.s.) system.

impermeable rock rock that does not allow water to pass through it – for example, clay, shale, and slate. Unlike †permeable rocks, which absorb water, impermeable rocks can support rivers. They therefore experience considerable erosion (unless, like slate, they are very hard) and commonly form lowland areas.

implantation in mammals, the process by which the developing †embryo attaches itself to the wall of the mother's uterus and stimulates the development of the †placenta. In humans it occurs 6–8 days after ovulation.

In some species, such as seals and bats, implantation is delayed for several months, during which time the embryo does not grow; thus the interval between mating and birth may be longer than the †gestation period.

imply in mathematics, to lead logically to. For example, if $2x = 10$, then $x = 5$. The second statement follows from the first.

imprinting in †ethology, the process whereby a young animal learns to recognize both specific individuals (for example, its mother) and its own species.

Imprinting is characteristically an automatic response to specific stimuli at a time when the animal is especially sensitive to those stimuli (the **sensitive period**). Thus, goslings learn to recognize their mother by following the first moving object they see after hatching; as a result, they can easily become imprinted on other species, or even inanimate objects, if these happen to move near them at this time. In chicks, imprinting occurs only between 10 and 20 hours after hatching. In mammals, the mother's attachment to her infant may be a form of imprinting made possible by a sensitive period; this period may be as short as the first hour after giving birth.

improper fraction †fraction whose numerator is larger than its denominator.

impulse in mechanics, the product of a force and the time over which it acts. An impulse applied to a body causes its †momentum to change and is equal to that change in momentum. It is measured in newton seconds (N s).

For example, the impulse J given to a football when it is kicked is given by:

$J = Ft$

where F is the kick force in newtons and t is the time in seconds for which the boot is in contact with the ball.

impulse in biology, a means of carrying information between different parts of the body. Nerve impulses travel at speed along nerve cells (neurones). A sequence of impulses carries information.

A stimulus from the †environment may cause impulses to be sent from the receptor receiving the stimulus, along a sensory neurone, and to the †brain or spinal cord. For example, a touch receptor in the skin of a finger will send impulses that tell the brain an object has been touched. Alternatively, information may be carried from the brain or spinal cord to a muscle or gland along a motor neurone. For example, the information may be carried in a sequence of impulses which causes the contraction of a muscle which causes the finger to be moved away from the object.

A simple example of these processes involving all steps from stimulus to muscle response occurs in any †reflex arc.

in abbreviation for †inch, a measure of distance.

inbreeding in †genetics, the mating of closely related individuals. It is considered undesirable because it increases the risk that offspring will inherit copies of rare deleterious †recessive alleles (genes) from both parents and so suffer from disabilities.

incandescence emission of light from a substance in consequence of its high temperature. The color of the emitted light from liquids or solids depends on their temperature, and for solids generally the higher the temperature the whiter the light. Gases may become incandescent through †ionizing radiation, as in the glowing vacuum discharge tube.

The oxides of cerium and thorium are highly incandescent and for this reason are used in gas mantles. The light from an electric filament lamp is due to the incandescence of the filament, rendered white-hot when a current passes through it.

incenter in mathematics, center of the **incircle** of a triangle (the circle for which each side of the triangle is a tangent). The incenter is the common meeting point of the internal bisectors of the angles of the triangle.

inch imperial unit of linear measure, a twelfth of a foot, equal to 2.54 centimeters.

It was defined in statute by Edward II of England as the length of three barley grains laid end to end.

inch (Gaelic **innis**) Scottish and Irish geographical term denoting a small island or land by a river, or occasionally rising ground on a plain. It is found in place names such as the Inches of Perth on the River Tay, local water meadows, and the islands of Inchcolm, Inch Garvie, and Inchkeith in the Firth of Forth, Scotland.

incised meander in a river, a deep steep-sided †meander (bend) formed by the severe downwards erosion of an existing meander. Such erosion is usually brought about by the †rejuvenation of a river (for example, in response to a fall in sea level).

incisor sharp tooth at the front of the mammalian mouth. Incisors are used for biting or nibbling, as when a rabbit or a sheep eats grass. Rodents, such as rats and squirrels, have large continually-growing incisors, adapted for gnawing. The elephant tusk is a greatly enlarged incisor. In humans, the incisors are the four teeth at the front center of each jaw.

inclination in astronomy, angle between the †ecliptic and the plane of the orbit of a planet, asteroid, or comet. In the case of satellites orbiting a planet, it is the angle between the plane of orbit of the satellite and the equator of the planet.

include in mathematics, to make one set part of another set, possible only if every element in it belongs to the other set. For example, set A includes set B if every element of B also belongs to A; then B is a subset of A.

inclusive fitness in †genetics, the success with which a given variant (or allele) of a †gene is passed on to future generations by a particular individual, after additional copies of the allele in the individual's relatives and their offspring have been taken into account.

The concept was formulated by W D Hamilton as a way of explaining the evolution of †altruism in terms of †natural selection. See also †fitness and †kin selection.

incoherent light light that is not of a single †phase. Daylight is an example.

incompressible fluid fluid that resists changes in density – for example, oil.

indefinite integral in mathematics, for a function *f*, a function *F* that gives *f* after differentiation. It is written $\int f(x)dx$ and is defined only up to the addition of an "arbitrary constant."

independent event in statistics, an unconnected †event whose occurrence has no effect on the †probability of another event happening. For example, if a coin is tossed and a die thrown at the same time, the events of "getting a head on the coin" and "throwing a six on the die" are independent, because they have no connection with each other.

multiplication rule for probability If events *A* and *B* are independent, then $P(A \text{ and } B) = P(A) \times P(B)$. So when a die is thrown and a coin is tossed, $P(\text{H and } 6) = P(\text{H}) \times P(6) = {}^1/_2 \times {}^1/_6 = {}^1/_{12}$.

independent variable the variable that does not depend on another variable for its values. The symbol *x* is usually employed to denote the independent variable, while *y* is used for the dependent variable. Time is always an independent variable.

indeterminacy principle alternative name for †uncertainty principle.

indeterminate problems problems involving one or more unknown or variable quantities. See also †Diophantine equations.

index in statistics, a numerical scale used to summarize a number of changes and by means of which different levels of data can be compared. For example, the cost of living index number compares the cost of a number of present-day ordinary expenses with the same expenses in previous years.

index plural **indices**, (Latin "sign," "indicator") in mathematics, another term for †exponent, the number that indicates the power to which a term should be raised.

index notation plural **indices notation**, the index, or †power of a number indicates how many times the number is to be multiplied by itself. For example, 4^2 (four squared) = 4 x 4 = 16; 2^3 (two cubed) = 2 x 2 x 2 = 8; and 5^4 (five to the power four) = 5 x 5 x 5 x 5 = 625. In these examples, 4^2, 2^3, and 5^4 are all written in index notation.

There are three important rules for the use of indices:

(i) $a^m \times a^n = a^{m+n}$

(ii) $a^m \div a^n = a^{m-n}$

(iii) $(a^m)^n = a^{mn}$

In all these three rules, *m* and *n* are assumed to be positive whole numbers and in rule (ii), *a* is not zero and $m > n$. For example:

$3^3 \times 3^4 = 3^{3+4} = 3^7$; $5^4 \div 5^2 = 5^{4-2} = 5^2$; $(2^3)^4 = 2^{3\times4} = 2^{12}$

When using fractional indices, $x^{1/n} = \sqrt[n]{x}$, so $x^{1/2} = \sqrt[2]{x}$ and $x^{1/4} = \sqrt[4]{x}$; also $x^{m/n} = (\sqrt[n]{x})^m$ or $\sqrt[n]{(x^m)}$. The same rules apply when using fractional indices as for whole number indices.

indicator in chemistry, a compound that changes its structure and color in response to its environment. The commonest chemical indicators detect changes in †pH (for example, †litmus and †universal indicator) or in the oxidation state of a system (redox indicators).

indicator species plant or animal whose presence or absence in an area indicates certain environmental conditions, such as soil type, high levels of pollution, or, in rivers, low levels of dissolved oxygen. Many plants show a preference for either alkaline or acid soil conditions, while certain trees require aluminum, and are found only in soils where it is present. Some lichens are sensitive to sulfur dioxide in the air, and absence of these species indicates atmospheric pollution.

indigo distinctive blue dye that occurs naturally in certain plants, notably the *Indigofera* genus. The dye is also synthetically manufactured on an industrial scale and used to color clothing such as jeans.

indium chemical symbol In, (Latin *indicum* "indigo") soft, ductile, silver-white, metallic element, atomic number 49, relative atomic mass 114.82. It occurs in nature in some zinc ores, is resistant to abrasion, and is used as a coating on metal parts. It was discovered in 1863 by German metallurgists Ferdinand Reich (1799–1882) and Hieronymus Richter (1824–1898), who named it after the two indigo lines of its spectrum.

indole C_6H_7N, heteroaromatic organic crystalline solid with a distinctive unpleasant smell, melting point 52°C/126°F. It is a component of coal tar and feces.

induced current electric current that appears in a closed circuit when there is relative movement of its conductor in a magnetic field. The effect is known as the **dynamo effect**, and is used in all †dynamos and generators to produce electricity. See †electromagnetic induction.

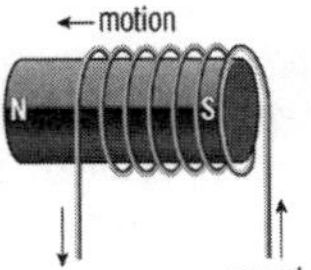

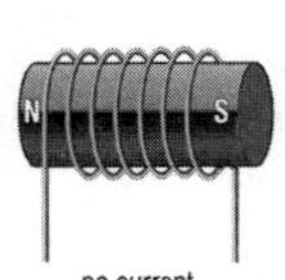

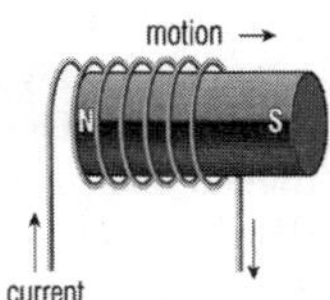

induced current Movement of a magnet in a coil of wire induces a current. Helicon

The magnitude of the induced current depends upon the rate at which the magnetic flux is cut by the conductor, and its direction is given by Fleming's right-hand rule (see †Fleming's rules).

inductance in physics, phenomenon in which a changing current in a circuit builds up a magnetic field which induces an †electromotive force either in the same circuit and opposing the current (self-inductance) or in another circuit (mutual inductance). The SI unit of inductance is the henry (symbol H).

A component designed to introduce inductance into a circuit is called an †inductor (sometimes inductance) and is usually in the form of a coil of wire. The energy stored in the magnetic field of the coil is proportional to its inductance and the current flowing through it. See †electromagnetic induction.

induction in physics, an alteration in the physical properties of a body that is brought about by the influence of a field. See †electromagnetic induction and †magnetic induction.

induction coil type of electrical transformer, similar to an ignition coil, that produces an intermittent high-voltage alternating current from a low-voltage direct current supply.

It has a primary coil consisting of a few turns of thick wire wound around an iron core and subjected to a low voltage (usually from a battery). Wound on top of this is a secondary coil made up of many turns of thin wire. An iron armature and make-and-break mechanism (similar to that in an †electric bell) repeatedly interrupts the current to the primary coil, producing a high-voltage, rapidly alternating current in the secondary circuit.

An **induction motor** employs †alternating current. It comprises a stationary current-carrying coil (stator) surrounding another coil (rotor), which rotates because of the current induced in it by the magnetic field created by the stator; it thus requires no commutator.

induction motor device that produces rotation by induction. An alternating current is fed to a winding of wires which thus induces electrical currents to flow in a second set of windings in a central rotor. Interaction between the two currents and the magnetic flux involved causes rotation.

inductor device included in an electrical circuit because of its inductance.

Indus inconspicuous constellation of the south polar region, represented as an American Indian.

industrial melanism dark or highly pigmented coloring in a "variety" of a species that evolves in a region with high atmospheric pollution.

inequality in mathematics, a statement that one quantity is larger or smaller than another, employing the symbols < and >. Inequalities may be solved by finding sets of numbers that satisfy them. For example, to find the †solution set to the inequality $2x + 3 < 19$, the inequality can be treated like an equation:

$2x + 3 < 19$, so $2x < 19 - 3$, $2x < 16$, and so $x < 8$.

Therefore, the solution set consists of all values of x less than 8 on the †number line. Inequality relationships involving variables are sometimes called **inequations**.

Interactive Algebra http://www.fifeschools.com/math/algebra/Algebra.html

Interactive problems to practice algebra. The questions are generated at random so they are different each time. There are clues given to help with difficult problems, and the answers are displayed online.

inert gas alternative name for †noble gas or rare gas.

inertia in physics, the tendency of an object to remain in a state of rest or uniform motion until an external force is applied, as described by Isaac Newton's first law of motion (see †Newton's laws of motion).

infection invasion of the body by disease-causing organisms (pathogens, or germs) that become established, multiply, and produce symptoms. Bacteria and viruses cause most diseases, but diseases are also caused by other micro-organisms, protozoans, and other parasites.

Most pathogens enter and leave the body through the digestive or respiratory tracts. Polio, dysentery, and typhoid are examples of diseases contracted by ingestion of contaminated foods or fluids. Organisms present in the saliva or nasal mucus are spread by airborne or droplet infection; fine droplets or dried particles are inhaled by others when the affected individual talks, coughs, or sneezes. Diseases such as measles, mumps, and tuberculosis are passed on in this way.

A less common route of entry is through the skin, either by contamination of an open wound (as in tetanus) or by penetration of the intact skin surface, as in a bite from a malaria-carrying mosquito. Relatively few diseases are transmissible by skin-to-skin contact. Glandular fever and herpes simplex (cold sore) may be passed on by kissing, and the group now officially bracketed as sexually transmitted diseases (STDs) are mostly spread by intimate contact.

inferior conjugation in astronomy, closest approach made by Mercury or Venus to the Earth, when the Earth, the planet and the Sun are in a straight line. The planet is nevertheless difficult to observe at this time because it is so close to the Sun in the sky.

inferiority complex in psychology, a †complex or cluster of repressed fears, described by Alfred †Adler, based on physical inferiority. The term is popularly used to describe general feelings of inferiority and the overcompensation that often ensues.

inferior planet planet (Mercury or Venus) whose orbit lies within that of the Earth, best observed when at its greatest elongation from the Sun, either at eastern elongation in the evening (setting after the Sun) or at western elongation in the morning (rising before the Sun).

inferno in astrophysics, a unit for describing the temperature inside a star. One inferno is 1 billion K, or approximately 1 billion °C.

infiltration passage of water into the soil. The rate of absorption of surface water by soil (the infiltration capacity) depends on the intensity of rainfall, the permeability and compactness of the soil, and the extent to which it is already saturated with water. Once in the soil, water may pass into the bedrock to form †groundwater.

infinite series in mathematics, a series of numbers consisting of a denumerably infinite sequence of terms. The sequence $n, n^2, n^3, \ldots$ gives the series $n + n^2 + n^3 + \ldots$. For example, $1 + 2 + 3 + \ldots$ is a divergent infinite arithmetic series, and $8 + 4 + 2 + 1 + \frac{1}{2} + \ldots$ is a convergent infinite geometric series that has a sum to infinity of 16.

infinitesimal number that is not zero but is less than any finite number. Infinitesimal numbers clearly do not exist in the conventional system of real numbers, but modern developments in logic allow the use of an extended system of numbers that includes infinitesimal numbers. Calculations with these adhere to certain restrictions and require a good understanding of †limits.

infinitesimal calculus original name for "calculus," that is, †differential and †integral calculus, and so called because it was thought to rely on "infinitely small" quantities. (It is now seen to be based upon a precise theory of †limits.)

infinity symbol ¥, mathematical quantity that is larger than any fixed assignable quantity. By convention, the result of dividing any number by zero is regarded as infinity.

inflammation defensive reaction of the body tissues to disease or damage, including redness, swelling, and heat. Denoted by the suffix *-itis* (as in appendicitis), it may be acute or chronic, and may be accompanied by the formation of pus. This is an essential part of the healing process.

Inflammation occurs when damaged cells release a substance (†histamine) that causes blood vessels to widen and leak into the surrounding tissues. This phenomenon accounts for the redness, swelling, and heat. Pain is due partly to the pressure of swelling and also to irritation of nerve endings. Defensive white blood cells congregate within an area of inflammation to engulf and remove foreign matter and dead tissue.

inflation in cosmology, a phase of extremely fast expansion thought to have occurred within 10–32 seconds of the †Big Bang and in which almost all the matter and energy in the universe was created. The inflationary model based on this concept accounts for the density of the universe being very close to the †critical density, the smoothness of the †cosmic background radiation, and the homogeneous distribution of matter in the universe. Inflation was proposed by US astrophysicist Alan Guth in the early 1980s.

inflection in mathematics, point on a curve where the curve changes from being convex to concave or vice versa. For differentiable functions the second derivative is zero at such a point.

inflorescence in plants, a branch, or system of branches, bearing two or more individual flowers. Inflorescences can be divided into two main types: cymose (or definite) and racemose (or indefinite). In a **cymose inflorescence**, the tip of the main axis produces a single flower and subsequent flowers arise on lower side branches, as in forget-me-not *Myosotis* and chickweed *Stellaria*; the oldest flowers are, therefore, found at the tip. A **racemose inflorescence** has an active growing region at the tip of its main axis, and bears flowers along its length, as in hyacinth *Hyacinthus*; the oldest flowers are found near the base or, in cases where the inflorescence is flattened, towards the outside. The stalk of the inflorescence is called a peduncle; the stalk of each individual flower is called a pedicel.

Types of racemose inflorescence include the **raceme**, a spike of similar, stalked flowers, as seen in lupin *Lupinus*. A **corymb**, seen in candytuft *Iberis amara*, is rounded or flat-topped because the pedicels of the flowers vary in length, the outer pedicels being longer than the inner ones. A **panicle** is a branched inflorescence made up of a number of racemes; such inflorescences are seen in many grasses, for example, the oat *Avena*. The pedicels of an **umbel**, seen in members of the carrot family (Umbelliferae), all arise from the same point on the main axis, like the spokes of an umbrella. Other types of racemose inflorescence include the †**catkin**, a pendulous inflorescence, made up of many small stalkless flowers; the †**spadix**, in which tiny flowers are borne on a fleshy axis; and the †capitulum, in which the axis is flattened or rounded, bears many small flowers, and is surrounded by large petal-like bracts.

information technology **IT**, collective term for the various technologies involved in processing and transmitting information. They include computing, telecommunications, and microelectronics.

Word processing, databases, and spreadsheets are just some of the computing †software packages that have revolutionized work in the office environment. Not only can work be done more quickly than before, but IT has given decisionmakers the opportunity to consider far more data when making decisions.

Applications & Effects of ICT http://www.dgs.oxon.sch.uk/depts/it/gcse/notes/effects/

Guide to the effects that information and communications technology (ICT) has on society. This is explored through the example of the differences between shops that use ICT and those that do not. It also contains points on the effect of ICT on employment and the consumer. The information is provided in summary form to help with revision.

Do I.T. http://doit.ort.org/

Online Information Technology foundation course for students. A good general site for ICT, the site is split into relevant topics, including computer languages, hardware, software, and networks. Additional resources such as downloadable worksheets, multiple choice test questions, and a glossary are also available.

Free Training with Beginners.com http://www.beginners.co.uk/

Over 200 free online ICT training courses. The site features professionally developed, high-quality pages for a wide range of software packages. They can all be viewed online for free using the latest Adobe Acrobat Reader.

IT Exam Revision http://members.aol.com/EdensorHS/revision/examopen.htm

GCSE revision questions. There are a series of multiple choice, true or false, and short answer questions testing the topics "Hardware" and "Software." The answers are instantly marked giving a good indication of progress.

SamLearning – IT http://www.samlearning.com/examrevision/gcse/it/igtopics.html
Interactive exam revision service for GCSE ICT students. There are multiple choice questions covering key revision points that can be marked online. In addition there are four test exam papers that can be downloaded, completed offline, and then checked against the correct answers provided on the site.

infrared absorption spectrometry technique used to determine the mineral or chemical composition of artifacts and organic substances, particularly amber. A sample is bombarded by infrared radiation, which causes the atoms in it to vibrate at frequencies characteristic of the substance present, and absorb energy at those frequencies from the infrared spectrum, thus forming the basis for identification.

infrared astronomy study of infrared radiation produced by relatively cool gas and dust in space, as in the areas around forming stars. In 1983 the US–Dutch–British Infrared Astronomy Satellite (IRAS) surveyed almost the entire sky at infrared wavelengths. It found five new comets, thousands of galaxies undergoing bursts of star formation, and the possibility of planetary systems forming around several dozen stars.

Planets and gas clouds emit their light in the far- and mid-infrared regions of the spectrum. The Infrared Space Observatory (ISO), launched in 1995, observed a broad wavelength (3–200 micrometers) in these regions. It is 10,000 times more sensitive than IRAS, and searches for †brown dwarfs (cool masses of gas smaller than the Sun).

Darwin: Infrared Space Interferometer http://ast.star.rl.ac.uk/darwin/
All about Darwin, a European infrared interferometer aiming to search for planets outside the Solar System, and for signs of life. The program to launch this spacecraft has been ongoing since 1994, and the ship should be launched around 2015. Its first aim is to characterize the planetary systems orbiting nearby stars by direct imaging and spectroscopy, and then to search any that are Earth-like for signs of life.

Infrared Astronomical Satellite (IRAS) Archive Interface http://space.gsfc.nasa.gov/astro/iras/iras_home.html
Provides information on the wide range of IRAS products, including sky maps, small-field observations of galaxies, and the minor-planet and comet survey. A brief introduction to IRAS is provided, as well as links to IRAS documents and the project Web site at the California Institute of Technology.

infrared radiation **i.r.**, electromagnetic †radiation of wavelength between about 700 nanometers and 1 millimeter – that is, between the limit of the red end of the visible spectrum and the shortest microwaves. All bodies above the †absolute zero of temperature absorb and radiate infrared radiation. Infrared radiation is used in medical photography and treatment, and in industry, astronomy, and criminology.

The human eye cannot detect infrared, but its effects can be demonstrated. For example, an electric hob operates at high temperatures and only the visible light it gives out can be seen. As it cools down, the visible light is no longer seen. However, the heat (infrared radiation) that continues to be given out can be felt. Infrared absorption spectra are used in chemical analysis, particularly for organic compounds. Objects that radiate infrared radiation can be photographed or made visible in the dark on specially sensitized emulsions. This is important for military purposes and in detecting people buried under rubble. The strong absorption by many substances of infrared radiation is a useful method of applying heat.

infrared spectroscopy in chemistry, analytical technique that can be used to obtain information about molecular structure and the atomic groups present in a sample compound. The technique utilizes the fact that interatomic vibrations in molecules absorb characteristic amounts of †infrared radiation.

infrared telescope in astronomy, †telescope designed to receive †electromagnetic waves in the infrared part of the spectrum. Infrared telescopes are always reflectors (glass lenses are opaque to infrared waves) and are normally of the †Cassegrain telescope type.

Since all objects at normal temperatures emit strongly in the infrared, careful design is required to ensure that the weak signals from the sky are not swamped by radiation from the telescope itself. Infrared telescopes are sited at high mountain observatories above the obscuring effects of water vapor in the atmosphere. Modern large telescopes are often designed to work equally well in both visible and infrared light.

Ingenhousz, Jan (1730–1799) Dutch physician and plant physiologist who established in 1779 that in sunlight plants absorb carbon dioxide and give off oxygen. He found that plants, like animals, respire all the time and that respiration occurs in all the parts of plants.

ingestion process of taking food into the mouth. The method of food capture varies but may involve biting, sucking, or filtering. Many single-celled organisms have a region of their cell wall that acts as a mouth. In these cases surrounding tiny hairs (cilia) sweep food particles together, ready for ingestion.

inheritance in biology, the passing of characteristics from parents to offspring. Characteristics that can be passed on in this way are determined by †genes (see also †allele).

The study of inheritance is called genetics and was founded by the Austrian biologist Gregor †Mendel.

He found that crossing two parents having one contrasting characteristic (for example, one tall and the other dwarf) showed that inheritance was the consequence of passing on "particles" or, as we now know them, genes, which determine the characteristics. This kind of cross is called a monohybrid cross.

In †asexual reproduction the offspring inherit the same characteristics as the parent. In †sexual reproduction, the offspring inherit a mix of the characteristics of the two parents.

inherited disease disease resulting from the effect of having defective †genes. These diseases are inherited because genetic material is accurately copied before being passed onto the offspring from a parent. However, they can also arise as a result of a sudden change in the DNA known as a †mutation. Common examples of inherited diseases in humans are cystic fibrosis, sickle-cell anemia, polydactyly, and Huntingdon's disease. Downs syndrome is an example of a disease arising from mutation, but which is not inherited.

Cystic fibrosis, sickle-cell anemia, polydactyly, and Huntingdon's disease are all each the result of a single defective gene. Down's syndrome, however, is usually the result of having an extra chromosome containing thousands of genes. In polydactyly (extra fingers) and Huntingdon's disease (physical and mental degeneration) the diseases show dominance – this means that inheriting a single defective gene will result in the disease. Cystic fibrosis (viscous secretions, especially in the lungs) is recessive, which means that two defective genes have to be inherited for the disease to occur.

In sickle-cell anemia people who inherit a single defective gene are normal most of the time, but can show anemia from time to time. However, if a person inherits two defective genes, they show severe illness. This defective gene exhibits incomplete dominance. The gene for sickle-cell anemia is more common than would be expected considering the anemia it causes. It is more common than expected because inheritance of a single defective gene makes the person partly immune to malaria, so †natural selection favors these individuals where malaria is common.

inhibition, neural in biology, the process in which activity in one †nerve cell suppresses activity in another. Neural inhibition in networks of nerve cells leading from sensory organs, or to muscles, plays an important role in allowing an animal to make fine sensory discriminations and to exercise fine control over movements.

inoculation injection into the body of dead or weakened disease-carrying organisms or their toxins (†vaccine) to produce immunity by inducing a mild form of a disease.

inorganic chemistry branch of chemistry dealing with the chemical properties of the elements and their compounds, excluding the more complex covalent compounds of carbon, which are considered in †organic chemistry.

The origins of inorganic chemistry lay in observing the characteristics and experimenting with the uses of the substances (compounds and elements) that could be extracted from mineral ores. These could be classified according to their chemical properties: elements could be classified as metals or non-metals; compounds as acids or bases, oxidizing or reducing agents, ionic compounds (such as salts), or covalent compounds (such as gases). The arrangement of elements into groups possessing similar properties led to Mendeleyev's †periodic table of the elements, which prompted chemists to predict the properties of undiscovered elements that might occupy gaps in the table. This, in turn, led to the discovery of new elements, including a number of highly radioactive elements that do not occur naturally.

inorganic compound chemical compound that does not contain carbon and is not manufactured by living organisms. Water, sodium chloride, and potassium are inorganic compounds because they are widely found outside living cells. However, carbon dioxide is considered inorganic, contains carbon, and is manufactured by organisms during respiration. Carbonates and carbon monoxide are also regarded as inorganic compounds. See †organic compound.

insanity in medicine and law, any mental disorder in which the patient cannot be held responsible for their actions. The term is no longer used to refer to psychosis.

inscribed circle circle drawn inside a plane figure which touches all the sides of the figure. The center of the inscribed circle of a triangle is the meeting point of the bisectors of the angle of the triangle.

insect any of a vast group of small invertebrate animals with hard, segmented bodies, three pairs of jointed legs, and, usually, two pairs of wings; they belong among the †arthropods and are distributed throughout the world. An insect's body is divided into three segments: head, thorax, and abdomen. On the head is a pair of feelers, or antennae. The legs and wings are attached to the thorax, or middle segment of the body. The abdomen, or end segment of the body, is where food is digested and excreted and where the reproductive organs are located. (Class Insecta.)

Insects vary in size from 0.02 cm/0.007 in to 35 cm/13.5 in in length. The world's smallest insect is believed to be a "fairy fly" wasp in the family Mymaridae, with a wingspan of 0.2 mm/0.008 in.

Many insects hatch out of their eggs as †larvae (an immature stage, usually in the form of a caterpillar, grub, or maggot) and have to pass through further major physical changes (†metamorphosis) before reaching adulthood. An insect about to go through metamorphosis hides itself or makes a cocoon in which to hide, then rests while the changes take place; at this stage the insect is called a †pupa, or a chrysalis if it is a butterfly or moth. When the changes are complete, the adult insect emerges.

The **classification** of insects is largely based upon characteristics of the mouthparts, wings, and metamorphosis. Insects are divided into two subclasses (one with two divisions) and 29 orders. More than 1 million species are known, and several thousand new ones are discovered each year.

The study of insects is called **entomology**.

Introduction to Insect Anatomy http://earthlife.net/insects/six.html

Wealth of colorful facts and illustrations about the anatomy of the insect world. The page is divided into various sections giving detailed information about the basic similarities and differences between various insects. The page begins with an explanation of the main insect body parts – head, thorax, and abdomen – and is illustrated with a useful diagram of a grasshopper.

insecticidal crystal protein **ICP**, in biology, poison produced by the bacterium *Bacillus thuringiensis*, which attacks the lining of the gut of leaf-eating larvae, including caterpillars. Plants have been genetically modified to carry the gene that expresses ICP poison.

The gene is inserted into the *Ti* plasmid of *Agrobacterium tumefasciens*, which infects the plant, causing crown gall disease. Tissue is cut from the gall contain the modified *Ti* plasmid and cultured to form plantlets, which grow into plants that produce ICP poison.

insecticide any chemical pesticide used to kill insects. Among the most effective insecticides are synthetic organic chemicals such as †DDT and dieldrin, which are chlorinated hydrocarbons. These chemicals, however, have proved persistent in the environment and are also poisonous to all animal life, including humans, and are consequently banned in many countries. Other synthetic insecticides include organic phosphorus compounds such as malathion. Insecticides prepared from plants, such as derris and pyrethrum, are safer to use but need to be applied frequently and carefully.

insectivore any animal whose diet is made up largely or exclusively of †insects. In particular, the name is applied to mammals of the order Insectivora, which includes shrews, hedgehogs, moles, and tenrecs. According to the Red List of endangered species published by the World Conservation Union for 1996, 36% of insectivore species are threatened with extinction.

insectivorous plant plant that can capture and digest live prey (normally insects), to obtain nitrogen compounds that are lacking in its usual marshy habitat. Some are passive traps, for example, the pitcher plants *Nepenthes* and *Sarracenia*. One pitcher-plant species has container-traps holding 1.6 l/3.5 pt of the liquid that "digests" its food, mostly insects but occasionally even rodents. Others, for example, sundews *Drosera*, butterworts *Pinguicula*, and Venus flytraps *Dionaea muscipula*, have an active trapping mechanism. Insectivorous plants have adapted to grow in poor soil conditions where the number of micro-organisms recycling nitrogen compounds is very much reduced. In these circumstances other plants cannot gain enough nitrates to grow. See also †leaf.

Near-carnivorous plants are unable to digest insects, but still trap them on their sticky coated leaves. The insects die and decay naturally, with the nutrients eventually becoming washed into the soil where they finally benefit the plant.

inselberg or **kopje**, (German "island mountain") prominent steep-sided hill of resistant solid rock, such as granite, rising out of a plain, usually in a tropical area. Its rounded appearance is caused by so-called onion-skin †weathering (exfoliation), in which the surface is eroded in successive layers.

The Sugar Loaf in Rio de Janeiro harbor in Brazil, and Ayers Rock in Northern Territory, Australia, are well-known examples.

insolation amount of solar radiation (heat energy from the Sun) that reaches the Earth's surface. Insolation varies with season and latitude, being greatest at the Equator and least at the poles. At the Equator the Sun is consistently high in the sky: its rays strike the equatorial region directly and are therefore more intense. At the poles the tilt of the Earth means that the Sun is low in the sky, and so its rays are slanted and spread out. Winds and ocean currents help to balance out the uneven spread of radiation.

insolation weathering type of physical or mechanical weathering that involves the alternate heating and cooling of rocks and minerals. This causes expansion and contraction, particularly of dark minerals (because they absorb more heat), setting up stresses in rocks. As rocks are poor conductors of heat only the surface layer is affected. This may contribute to exfoliation (breaking away of the outer layer of rock).

instinct in †ethology, behavior found in all equivalent members of a given species (for example, all the males, or all the females with young) that is presumed to be genetically determined.

Examples include a male robin's tendency to attack other male robins intruding on its territory and the tendency of many female mammals to care for their offspring. Instincts differ from †reflexes in that they involve very much more complex actions, and learning often plays an important part in their development.

insulation process or material that prevents or reduces the flow of electricity, heat, or sound from one place to another.

Materials that are poor conductors of heat, such as glass, brick, water, or air, are good insulators. They play a vital role, for example, in keeping homes and people warm.

insulator any poor †conductor of heat, sound, or electricity. Most substances lacking free (mobile) †electrons, such as non-metals, are electrical or thermal insulators that resist the flow of electricity or heat through them. Plastics and rubber are good insulators. Usually, devices of glass or porcelain, called insulators, are used for insulating and supporting overhead wires.

The electrons in the outer shells of the atoms of insulators are strongly attracted by the positively-charged nucleus. When external energy is supplied, these electrons are not easily removed from the outer shells of the atoms, which gives these materials their insulating properties.

insulin protein †hormone, produced by specialized cells in the islets of Langerhans in the pancreas, which regulates the metabolism (rate of activity) of glucose, fats, and proteins. In this way it helps to regulate the concentration of †glucose in the †blood of mammals. If the blood glucose concentration is too high, the pancreas releases insulin into the blood. This causes blood glucose levels to fall. This is partly due to increased uptake of glucose into most body cells (except liver and brain) but also because the liver converts glucose into insoluble glycogen and stores it instead of making glucose. In †diabetes a person's blood glucose may rise to such a high concentration that it can kill. This may be because the pancreas does not make enough insulin, or because the body cells respond less to the insulin.

Normally, insulin is secreted in response to rising blood sugar levels (after a meal, for example), stimulating the body's cells to store the excess. Failure of this regulatory mechanism in diabetes mellitus requires treatment with insulin injections or capsules taken by mouth. Types vary from pig and beef insulins to synthetic and bioengineered ones. They may be combined with other substances to make them longer- or shorter-acting. Implanted, battery-powered insulin pumps deliver the hormone at a preset rate, to eliminate the unnatural rises and falls that result from conventional, subcutaneous (under the skin) delivery. Human insulin has now been produced from bacteria by †genetic engineering techniques. In 1990 the Medical College of Ohio developed gelatin capsules and an aspirin-like drug which helps the insulin pass into the bloodstream.

Insulin was discovered by Canadian physician Frederick †Banting and Canadian physiologist Charles †Best, who pioneered its use in treating diabetes.

Discovery of Insulin http://www.discoveryofinsulin.com/

Details the discovery of the hormone insulin. This site includes brief descriptions of the lives of the team that made the discovery. An overview of insulin and diabetes is also included here, describing the hormone and the disease, as well as current developments in the continuing research programs around the world.

Endocrine System http://arbl.cvmbs.colostate.edu/hbooks/pathphys/endocrine/

A good basic introduction to the human endocrine system which is also suitable for more advanced students. Fundamental concepts are followed by a tour of the endocrine system, complete with tables and diagrams showing the location and function of the glands and the hormones that they produce.

integer any whole number. Integers may be positive or negative; 0 is an integer, and is often considered positive. Formally, integers are members of the set

$Z = \{... -3, -2, -1, 0, 1, 2, 3, ... \}$

This is the integer set of the †number line.

†Fractions, such as $^1/_2$ and the †decimal 0.35, are known as non-integral numbers ("not integers").

The †addition (their sum) and †subtraction (their †difference) of two integers will always result in an integer. The †multiplication of two integers (their †product) will always result in an integer; however, the †division of integers may result in a non-integer.

BEATCALC: Beat the Calculator http://mathforum.org/k12/mathtips/beatcalc.html

Handy site containing quick ways to calculate difficult sums, such as dividing six-digit numbers, and squaring. It includes more than 250 BEATCALC methods, plus a page of calculation practice to test mental arithmetic.

Order and Operations of Integers http://www.mathgoodies.com/lessons/toc_vol5.shtm

Excellent site covering the order and operations of integers. Methodical and clear, each of the eight lessons contain problems solved using integers, summaries of the key points, examples, and exercises.

integral calculus branch of mathematics using the process of †integration. It is concerned with finding volumes and areas and summing infinitesimally small quantities.

integral equations equations involving integrals of the unknown function.

integrated circuit **IC**; or **silicon chip**, miniaturized electronic circuit produced on a single crystal, or chip, of a semiconducting material – usually silicon. It may contain many millions of components and yet measure only 5 mm/0.2 in square and 1 mm/0.04 in thick. The IC is encapsulated within a plastic or ceramic case, and linked via gold wires to metal pins with which it is connected to a †printed circuit board and the other components that make up such electronic devices as computers and calculators.

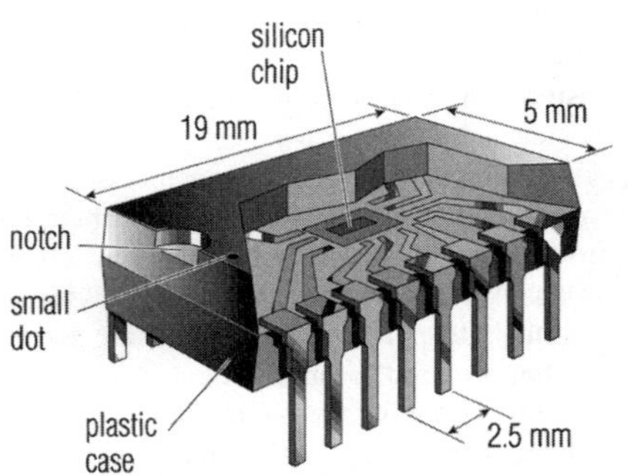

integrated circuit An integrated circuit (IC), or silicon chip. Helicon

integration in mathematics, a method in †calculus of determining the solutions of definite or indefinite integrals. An example of a definite integral can be thought of as finding the area under a curve (as represented by an algebraic expression or function) between particular values of the function's variable. In practice, integral calculus provides scientists with a powerful tool for doing calculations that involve a continually varying quantity (such as determining the position at any given instant of a space rocket that is accelerating away from Earth). Its basic principles were discovered in the late 1660s independently by the German philosopher †Leibniz and the British scientist †Newton.

integument in seed-producing plants, the protective coat surrounding the ovule. In flowering plants there are two, in gymnosperms only one. A small hole at one end, the micropyle, allows a pollen tube to penetrate through to the egg during fertilization.

intellect aspect of the mind concerned with cognitive processes, such as remembering, imagining, conceptualizing, reasoning, understanding, and judging. The term is also used to refer to these rational, or higher, thought processes.

intelligence in psychology, a general concept that summarizes the abilities of an individual in reasoning and problem solving, particularly in novel situations. These consist of a wide range of verbal and nonverbal skills and therefore some psychologists dispute a unitary concept of intelligence.

intelligence quotient **IQ**, intelligence rating ascertained through answers to a test, which are expressed as a score and placed on an index of scores. Formerly, IQ was defined as (mental age/calendar age) x 100.

intelligence test test that attempts to measure innate intellectual ability, rather than acquired ability.

It is now generally believed that a child's ability in an intelligence test can be affected by his or her environment, cultural background, and teaching. There is skepticism about the accuracy of intelligence tests, but they are still widely used as a diagnostic tool when children display learning difficulties. "Sight and sound" intelligence tests, developed by Christopher Brand in 1981, avoid cultural bias and the pitfalls of improvement by practice. Subjects are shown a series of lines being flashed on a screen at increasing speed, and are asked to identify in each case the shorter of a pair; and when two notes are relayed over headphones, they are asked to identify which is the higher. There is a close correlation between these results and other intelligence test scores.

intensity in physics, power (or energy per second) per unit area carried by a form of radiation or wave motion. It is an indication of the concentration of energy present and, if measured at varying distances from the source, of the effect of distance on this. For example, the intensity of light is a measure of its brightness, and may be shown to diminish with distance from its source in accordance with the †inverse square law (its intensity is inversely proportional to the square of the distance).

intensity interferometry in astronomy, the use of two telescopes linked by computer to study the intensity of light received from a star. Analysis of the combined results has enabled measurement of the diameters of stars as apparently small as 2×10^{-4} sec of arc.

interactive computing in computing, a system for processing data in which the operator is in direct communication with the computer, receiving immediate responses to input data. In †batch processing, by contrast, the necessary data and instructions are prepared in advance and processed by the computer with little or no intervention from the operator.

interactive media in computing, new technology such as †CD-ROM and online systems which allow users to interact with other users or to choose their own path through the material.

The newest attempts to create interactive media are books published on the World Wide Web which allow readers to use †hyperlinks to move around the material at will in the order they choose. Other interactive media include plans for films and other projects which allow viewers to choose how to follow the story, which characters to focus on, or which plot threads to follow.

intercept in geometry, point at which a line or curve cuts across a given axis. Also, the segment cut out of a †transversal by the pair of lines it cuts across. If a set of parallel lines make equal intercepts on one transversal, it will make equal intercepts on any other.

intercostal in biology, the nerves, blood vessels, and muscles that lie between the ribs.

interface in computing, the point of contact between two programs or pieces of equipment. The term is most often used for the physical connection between the computer and a peripheral device, which is used to compensate for differences in such operating characteristics as speed, data coding, voltage, and power consumption. For example, a **printer interface** is the cabling and circuitry used to transfer data from a computer to a printer, and to compensate for differences in speed and coding.

Common standard interfaces include the **Centronics interface**, used to connect parallel devices, and the RS-232 interface, used to connect serial devices. For example, in many microcomputer systems, an RS-232 interface is used to connect the microcomputer to a modem, and a Centronics device is used to connect it to a printer.

interference in physics, the phenomenon of two or more wave motions interacting and combining to produce a resultant wave of larger or smaller amplitude (depending on whether the combining waves are in or out of †phase with each other).

Interference of white light (multiwavelength) results in spectral colored fringes; for example, the iridescent colors of oil films seen on water or soap bubbles (demonstrated by †Newton's rings). Interference of sound waves of similar frequency produces the phenomenon of beats, often used by musicians when tuning an instrument. With monochromatic light (of a single wavelength), interference produces patterns of light and dark bands. This is the basis of †holography, for example. Interferometry can also be applied to radio waves, and is a powerful tool in modern astronomy.

interferometer in physics, a device that splits a beam of light into two parts, the parts being recombined after traveling different paths to form an interference pattern of light and dark bands. Interferometers are used in many branches of science and industry where accurate measurements of distances and angles are needed.

In the Michelson interferometer, a light beam is split into two by a semisilvered mirror. The two beams are then reflected off fully silvered mirrors and recombined. The pattern of dark and light bands is sensitive to small alterations in the placing of the mirrors, so the interferometer can detect changes in their position to within one ten-millionth of a meter. Using lasers, compact devices of this kind can be built to measure distances, for example to check the accuracy of machine tools.

In radio astronomy, interferometers consist of separate radio telescopes, each observing the same distant object, such as a galaxy, in the sky. The signal received by each telescope is fed into a computer. Because the telescopes are in different places, the distance traveled by the signal to reach each differs and the overall signal is akin to the interference pattern in the Michelson interferometer. Computer analysis of the overall signal can build up a detailed picture of the source of the radio waves.

In space technology, interferometers are used in radio and radar systems. These include space-vehicle guidance systems, in which the position of the spacecraft is determined by combining the signals received by two precisely spaced antennae mounted on it.

interferometry in astronomy, any of several techniques using interference between multiple beams to obtain high-resolution images or spectra of astronomical objects. The †very long-baseline interferometry system is an example. NASA plans to put radio telescopes, infrared telescopes, and visible-light telescopes in formation in space to produce images of planets around neighboring stars.

interferon or **IFN**, naturally occurring cellular protein that makes up part of mammalian defenses against viral disease. Three types (alpha, beta, and gamma) are produced by infected cells and enter the bloodstream and uninfected cells, making them immune to virus attack.

Interferon was discovered in 1957 by Scottish virologist Alick Isaacs. Interferons are cytokines, small molecules that carry signals from one cell to another. They can be divided into two main types: **type I** (alpha, beta, tau, and omega) interferons are more effective at bolstering cells' ability to resist infection; **type II** (gamma) interferon is more important to the normal functioning of the immune system. Alpha interferon may be used to treat some cancers; interferon beta 1b has been found useful in the treatment of multiple sclerosis.

intergalactic matter hypothetical material within a cluster of galaxies, whose gravitational effect is to maintain the equilibrium of the cluster. Theoretically comprising 10–30 times the mass of the galaxies themselves (in order to have the observed effect), it has yet to be detected in any form – although the most likely form is as †hydrogen.

interhalogen compound chemical compound that is composed of any two different †halogen elements, for example chlorine bromide, ClBr, and iodine heptafluoride, IF_7. Formed by the reaction between the two elements, these compounds are highly reactive.

interior angle one of the four internal angles formed when a transversal cuts two or more (usually parallel) lines. Also, one of the angles inside a †polygon.

interlocking spur one of a series of spurs (ridges of land) jutting out from alternate sides of a river valley. During glaciation its tip may be sheared off by erosion, creating a †truncated spur.

intermediate vector boson member of a group of elementary particles, W^+, W^-, and Z, which mediate the †weak nuclear force. This force is responsible for, among other things, beta decay.

intermolecular force or **van der Waals' force**, force of attraction between molecules. Intermolecular forces are relatively weak; hence simple molecular compounds are gases, liquids, or low-melting-point solids.

internal-combustion engine heat engine in which fuel is burned inside the engine, contrasting with an external-combustion engine (such as the steam engine) in which fuel is burned in a separate unit. The †diesel engine and gasoline engine are both internal-combustion engines. Gas †turbines and †jet and †rocket engines are also considered to be internal-combustion engines because they burn their fuel inside their combustion chambers.

internal resistance or **source resistance**, the resistance inside a power supply, such as a battery of cells, that limits the current that it can supply to a circuit.

international biological standards drugs (such as penicillin and insulin) of which the activity for a specific mass (called the international unit, or IU), prepared and stored under specific conditions, serves as a standard for measuring doses. For penicillin, one IU is the activity of 0.0006 mg of the sodium salt of penicillin, so a dose of a million units would be 0.6 g.

International Date Line **IDL**, imaginary line that approximately follows the 180° line of longitude. The date is put forward a day when crossing the line going west, and back a day when going east. The IDL was chosen at the International Meridian Conference in 1884.

Internet global public computer network that provides the communication infrastructure for applications such as †e-mail, the †World Wide Web, and FTP. The Internet is not one individual network, but an interconnected system of smaller networks using common †protocols to pass packets of information from one computer to another.

Early work on the Internet began in the 1960s at the Advanced Research Projects Agency (ARPA), in the USA. It was based on theories of †packet switching, particularly those of Leonard Kleinrock. In 1967 work by Lawrence Roberts, Vinton Cerf, and Bob Kahn led to the publishing of plans for the ARPANET network. They were put into practice in 1969, when the ARPANET was used to connect four university computers. In 1972 the ARPANET had its first public demonstration at the International Computer Communication Conference. In the same year, Ray Tomlinson designed the first application for this new network: e-mail.

The ARPANET had used a protocol called NCP, but it had compatibility problems. Engineers needed to design protocols that could meet the needs of an open-architecture network. Roberts and Cerf then led the team that created the Transmission Control Protocol/Internet Protocol (†TCP/IP), which swiftly became the standard Internet protocol. In 1983 the ARPANET switched from using NCP to TCP/IP, and the network was separated into military (MILNET) and non-military systems. By 1985 the Internet was well established among scientific researchers and developers, and was beginning to be used for daily computer communications. By the early 1990s, access had become cheap enough for domestic users to have their own links on home personal computers.

The impact of the Internet on the economy has been huge. By June 2000, it was estimated that there were 305 million Internet users worldwide, and that figure was expected to rise to 720 million by the end of 2005.

Basic Guide to the Internet http://library.albany.edu/internet/internet.html

Well-organized introduction for beginners to the Internet. The page is divided into sections dealing with the "Components of the Internet," such as the World Wide Web (WWW), e-mail, telnet, File Transfer Protocol (FTP), and discussion groups. It also contains a number of relevant links to other introductory sites.

Brief History of the Internet http://www.isoc.org/internet-history/brief.html

Detailed and illustrated history of the Internet. It begins, as one would expect, with the "Origins of the Internet" and ends with "History of the future," dealing with numerous other topics on the way, such as "Proving the ideas" and "Commercialization of the technology."

Web Building http://builder.cnet.com/

Guide to HTML, the programming language used to write Web sites. Topics covered here range from the basic – planning your Web site, graphics, and publishing – to more advanced topics, such as JavaScript, commerce over the Internet, and animated graphics. There are also reviews of the other sites, as well as numerous articles relating to Web design.

internode in botany, distance along the stem of a plant between two successive nodes (a node is the point where a leaf stalk or petiole attaches to the stem).

interplanetary matter gas, dust, and charged particles from a variety of sources that occupies the space between the planets. Dust left over from the formation of the Solar System, or from disintegrating comets, orbits the Sun in or near the ecliptic plane and is responsible for scattering sunlight to cause the †zodiacal light (mainly the smaller particles) and for †meteor showers in the Earth's atmosphere (larger particles). The charged particles mostly originate in the †solar wind, but some are cosmic rays from deep space.

interpolation estimate of a value lying between two known values. For example, it is known that 13^2 is 169 and 14^2 is 196. Thus, the square of 13.5 can be interpolated as halfway between 169 and 196, i.e. 182.5 (the exact value is 182.25).

interquartile range in statistics, a measure of the range of data, equaling the difference in value between the upper and lower †quartiles. It provides information on the spread of data in the middle 50% of a distribution and is not affected by freak extreme values.

intersection on a graph, the point where two lines or curves meet. The intersections of graphs provide the graphical solutions of equations.

intersection in set theory, the set of elements that belong to both set A and set B. Intersections in †Venn diagrams are where the circles overlap.

intersex individual that is intermediate between a normal male and a normal female in its appearance (for example, a genetic male that lacks external genitalia and so resembles a female).

interstellar cirrus in astronomy, wispy cloud-like structures discovered in the mid-1980s by the Infrared Astronomy Satellite and believed to be the remains of dust shells blown into space from cool giant or supergiant stars.

interstellar hydrogen hydrogen gas between the stars of a galaxy, thus "filling out" the shape of the galaxy in a way that can be detected by spectral analysis and radio monitoring.

interstellar matter medium of electrons, ions, atoms, molecules, and dust grains that fills the space between stars in our own and other galaxies. Over 100 different types of molecule exist in gas clouds in the Earth's Galaxy. Most have been detected by their radio emissions, but some have been found by the absorption lines they produce in the spectra of starlight. The most complex molecules, many of them based on †carbon, are found in the dense clouds where stars are forming. They may be significant for the origin of life elsewhere in space.

interstellar space space between the stars of a galaxy. It is generally not, however, a void vacuum, and is the subject of considerable spectral research.

interstitial in biology, undifferentiated tissue that is interspersed with the characteristic tissue of an organ. It is often formed of fibrous tissue and supports the organ. Interstitial fluid refers to the fluid present in small amounts in the tissues of an organ.

intertropical convergence zone ITCZ, area of heavy rainfall found in the tropics and formed as the trade winds converge and rise to form cloud and rain. It moves a few degrees northwards during the northern summer and a few degrees southwards during the southern summer, following the apparent movement of the Sun. The ITCZ is responsible for most of the rain that falls in Africa. The †doldrums are also associated with this zone.

interval in statistics, the difference between the smallest and largest measurement in an †class interval.

intestine in vertebrates, the digestive tract from the stomach outlet to the anus. The human **small intestine** is 6 m/20 ft long, 4 cm/1.5 in in diameter, and consists of the duodenum, jejunum, and ileum; the **large intestine** is 1.5 m/5 ft long, 6 cm/2.5 in in diameter, and includes the cecum, colon, and rectum. The contents of the intestine are passed along slowly by †peristalsis (waves of involuntary muscular action). The term intestine is also applied to the lower digestive tract of invertebrates.

intranet in computing, an organization's computer network based on the same †TCP/IP †protocols as the Internet. The intranet resembles an Internet Web site to the user and is effectively an internal Internet. The difference between the Internet and an intranet is that access to an intranet is limited to those with authorization from the organization. This authorization usually extends to all employees. A firewall and other security precautions prevent unwanted external access.

The same technology can be used with an internal network to build an organization-wide web of documents in a format that is familiar, easy to use, and comparatively inexpensive.

intrinsic geometry study of a surface without reference to any point, condition, or space outside it. All measurements and operations carried out on the surface are therefore in terms of its own (intrinsic) form.

intron or **junk DNA**, in genetics, a sequence of bases in †DNA that carries no genetic information. Introns, discovered in 1977, make up approximately 98% of DNA (the rest is made up of †exons). Introns may be present within genes but are removed during translation. Their function is unknown.

about 10% of the human genome is made up of one base sequence, *Alu*, that occurs in about 1 million separate locations. It is made up of 283 nucleotides, has no determinable function (though some do have an effect on nearby genes), and is a †transposon ("jumping gene").

introspection observing or examining the contents of one's own mind or consciousness. For example, "looking" at and describing a "picture" or image in the "mind's eye," or trying to examine what is happening when one performs mental arithmetic.

Its use as an approach to the study of the mind has a history dating back, at least, to Socrates. It was first proposed as an experimental method by Wilhelm †Wundt and employed routinely in his laboratory, established 1879, in accord with his view that psychology is "the science of inward and immediate experience." The method was further developed by Wundt's pupil Edward Bradford Titchener (1867–1927) and by members of the Wurtzberg School around the turn of the century. Wundt eventually became dissatisfied with this method of enquiry and, following severe criticism as to the reliability of introspective data, the method fell into disuse. So grave were the problems with its methodology that, following the advent of †behaviorism, the systematic study of mental processes was largely eschewed by psychologists for half a century, only returning as a course of serious study in the 1960s.

introversion in psychology, preoccupation with the self, generally coupled with a lack of sociability. The opposite of introversion is †extroversion.

The term was introduced by the Swiss psychiatrist Carl Jung in 1924 in his description of schizophrenia, where he noted that "interest does not move towards the object but recedes towards the subject." The term is also used within psychoanalysis to refer to the turning of the instinctual drives towards objects of fantasy rather than the pursuit of real objects. Another term for this sense is fantasy cathexis.

intrusion mass of †igneous rock that has formed by "injection" of molten rock, or magma, into existing cracks beneath the surface of the Earth, as distinct from a volcanic rock mass which has erupted from the surface. Intrusion features include vertical cylindrical structures such as stocks, pipes, and necks; sheet structures such as dykes that cut across the strata and sills that push between them; laccoliths, which are blisters that push up the overlying rock; and batholiths, which represent chambers of solidified magma and contain vast volumes of rock.

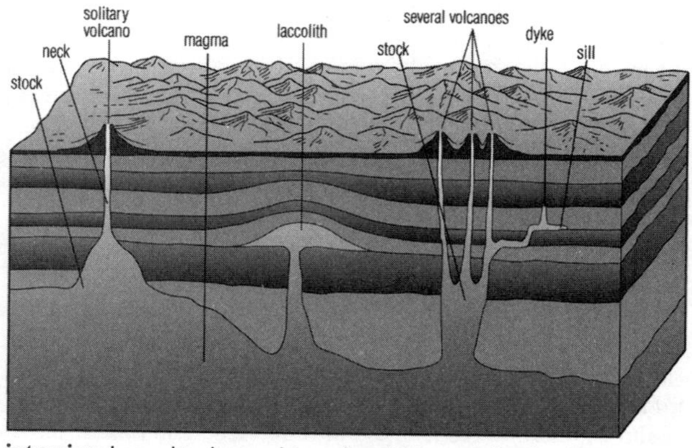

intrusion Igneous intrusions can be a variety of shapes and sizes. Laccoliths are domed circular shapes, and can be many miles across. Sills are intrusions that flow between rock layers. Pipes or necks connect the underlying magma chamber to surface volcanoes. Helicon

intrusive rock or **plutonic rock**, †igneous rock formed beneath the Earth's surface. Magma, or molten rock, cools slowly at these depths to form coarse-grained rocks, such as granite, with large crystals. (†Extrusive rocks, which are formed on the surface, are generally fine-grained.) A mass of intrusive rock is called an intrusion.

intuitional mathematics or **intuitionism**, alternative foundational basis for mathematics that adopts a stricter logic in its approach to proofs concerning the infinite. For example, it dismisses the law of the †excluded middle, and so disregards "proofs" derived by double negatives if the relevant positive statement has not actually been demonstrated to be true. Although this may be considered a sort of philosophical puritanism, the outlook leads to a more refined classification of proof material than merely "true" or "false." The fact that such a critical attitude can itself be formalized in a mathematically sound system is an important achievement. Nevertheless, most practicing mathematicians remain unconcerned by this logical analysis.

invagination formation of an inner pocket within a layer of cells by part of the layer pushing inwards to form a cavity that remains open to the original surface. See also †endocytosis.

Invar †alloy of approximately 64% iron, 36% nickel, and a small amount of carbon, which possesses a very low coefficient of expansion. As a result, it is used in devices such as pendulums whose correct operation depend on maintaining constant length despite temperature changes.

invariant in mathematics, describing a property that is preserved through specified mathematical operations.

inverse function †function that exactly reverses the transformation produced by a function f; it is usually written as f^{-1}. For example, to find the inverse of a function $y = 3x + 2$, x is made the subject of the equation: $3x = y - 2$, so $x = {}^{(y-2)}/_{3}$. So $3x + 2$ and ${}^{(x-2)}/_{3}$ are mutually inverse functions. Multiplication and division are inverse operations (see †reciprocals).

An inverse function is clearly demonstrated on a calculator by entering any number, pressing x^2, then pressing Öx to get the inverse. The functions on a scientific calculator can be inversed in a similar way.

inverse spinel crystal structure of magnetite and other ferrimagnetic materials, so-called because it resembles that of †spinel but with the divalent and half the trivalent ions interchanged.

inverse square law statement that the magnitude of an effect (usually a force) at a point is inversely proportional to the square of the distance between that point and the object exerting the force.

Light, sound, electrostatic force (Coulomb's law), and gravitational force (Newton's law) all obey the inverse square law.

inversion in mineralogy, condition of a mineral that can change from one polymorphous form to another.

invertebrate animal without a backbone. The invertebrates form all of the major divisions of the animal kingdom called phyla, with the exception of vertebrates. Invertebrates include the sponges, coelenterates, flatworms, nematodes, annelids, arthropods, mollusks, and echinoderms.

invert sugar equimolecular mixture of dextrose and laevulose, obtained by reacting cane sugar with dilute acids. It readily ferments and is used in the preparation of sparkling wines.

in vitro fertilization **IVF**, ("fertilization in glass") allowing eggs and sperm to unite in a laboratory to form embryos. The embryos (properly called pre-embryos in their two- to eight-celled state) are stored by cooling to the temperature of liquid air (cryopreservation) until they are implanted into the womb of the otherwise infertile mother (an extension of artificial insemination). The first baby to be produced by this method was Louise Joy Brown, born in 1978 at Oldham General Hospital, Lancashire, UK. In cases where the Fallopian tubes are blocked, fertilization may be carried out by **intra-vaginal culture**, in which egg and sperm are incubated (in a plastic tube) in the mother's vagina, then transferred surgically into the uterus.

in vitro process biological experiment or technique carried out in a laboratory, outside the body of a living organism (literally "in glass," for example in a test tube). By contrast, an in vivo process takes place within the body of an organism.

in vivo process biological experiment or technique carried out within a living organism; by contrast, an in vitro process takes place outside the organism, in an artificial environment such as a laboratory.

involuntary action behavior not under conscious control, for example the contractions of the gut during peristalsis or the secretion of adrenaline by the adrenal glands. Breathing and urination reflexes are involuntary, although both can be controlled voluntarily to some extent. These processes are regulated by the †autonomic nervous system.

involute (Latin "rolled in") in geometry, †spiral that can be thought of as being traced by a point at the end of a taut nonelastic thread being wound onto or unwound from a spool.

Io in astronomy, third-largest moon of the planet Jupiter, 3,630 km/2,260 mi in diameter, orbiting in 1.77 days at a distance of 422,000 km/262,000 mi. It is the most volcanically active body in the Solar System, covered by hundreds of vents that erupt sulfur (rather than lava), giving Io an orange-colored surface. Io and Earth are the only two planetary bodies that are undergoing known high-temperature volcanism.

In July 1995 the Hubble Space Telescope revealed a 320-km/200-mi yellow spot on the surface of Io, located on the volcano Ra Patera. Though clearly volcanic in origin, astronomers are unclear as to the exact cause of the new spot.

Data gathered by the spacecraft Galileo in 1996 indicated that Io has a large metallic core. The Galileo probe also detected a 10-megawatt beam of electrons flowing between Jupiter and Io.

iodate salt of iodic acid, HIO_3. The term can also be applied to any salts that contain iodine oxyanions (any negatively charged species or group of atoms that contains at least one oxygen atom). Iodates are strong oxidizing agents.

iodide compound formed between iodine and another element in which the iodine is the more electronegative element (see †electronegativity, †halide).

iodine chemical symbol I, (Greek *iodes* "violet") black-black non-metallic element, atomic number 53, relative atomic mass 126.9044. It is a member of the †halogen group (Group 7 of the †periodic table). Its crystals sublime, giving off, when heated, a violet vapor with an irritating odor resembling that of chlorine. It occurs only in combination with other elements. Its salts are known as **iodides** and are found in sea water. As a mineral nutrient it is vital to the proper functioning of the thyroid gland, where it occurs in trace amounts as part of the hormone thyroxine. Absence of iodine from the diet leads to goitre. Iodine is used in photography, in medicine as an antiseptic, and in making dyes.

Its radioactive isotope iodine-131 (half-life of eight days) is a dangerous fission product from nuclear explosions and from the nuclear reactors in power plants, since, if ingested, it can be taken up by the thyroid and damage it. It was discovered in 1811 by French chemist B Courtois (1777–1838).

iodoform or **triiodomethane**; CHI_3, antiseptic that crystallizes into yellow hexagonal plates. It is soluble in ether, alcohol, and chloroform, but not in water.

ion atom, or group of atoms, that is either positively charged (†cation) or negatively charged (†anion), as a result of the loss or gain of electrons during chemical reactions or exposure to certain forms of radiation. In solution or in the molten state, ionic compounds such as salts, acids, alkalis, and metal oxides conduct electricity. These compounds are known as †electrolytes.

Ions are produced during †electrolysis, for example the salt zinc chloride ($ZnCl_2$) dissociates into the positively-charged Zn^{2+} and negatively-charged Cl^- when electrolyzed.

ion engine or **ion drive** or **ion thruster**, rocket engine that uses †ions (charged particles) rather than hot gas for propulsion. Ion engines have been successfully tested in space, where they will eventually be used for gradual rather than sudden velocity changes. In an ion engine, atoms of mercury, for example, are ionized (given an electric charge by an electric field) and then accelerated at high speed by a more powerful electric field.

ion exchange process whereby an ion in one compound is replaced by a different ion, of the same charge, from another compound. Ion exchange is used in commercial water softeners to exchange the dissolved ions responsible for the water's hardness with others that do not have this effect. For example, when hard water is passed over an ion-exchange resin, the dissolved calcium and magnesium ions are replaced by either sodium or hydrogen ions, so the hardness is removed. The addition of †washing soda to hard water is also an example of ion exchange:

$Na_2CO_{3(aq)} + CaSO_{4(aq)} \rightarrow CaCO_{3(s)} + Na_2SO_{4(aq)}$

Ion exchange is the basis of a type of †chromatography in which the components of a mixture of ions in solution are separated according to the ease with which they will replace the ions on the polymer matrix through which they flow. The exchange of positively charged ions is called cation exchange; that of negatively charged ions is called anion exchange.

ion half equation equation that describes the reactions occurring at the electrodes of a chemical cell or in electrolysis. It indicates which ion is losing electrons (oxidation) or gaining electrons (reduction).

Examples are given from the electrolysis of dilute hydrochloric acid (HCl):

$2Cl^- - 2e' \rightarrow Cl_2$ (positive electrode)

$2H^+ + 2e' \rightarrow H_2$ (negative electrode)

ionic bond or **electrovalent bond**, bond produced when atoms of one element donate electrons to atoms of another element, forming positively and negatively charged †ions respectively. The attraction between the oppositely charged ions constitutes the bond. Sodium chloride (Na^+Cl^-) is a typical ionic compound.

Each ion has the electronic structure of a †noble gas (rare gas; see †noble gas structure). The maximum number of electrons that can be gained is usually two.

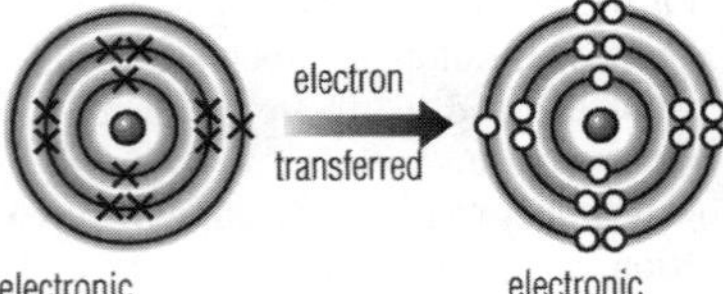

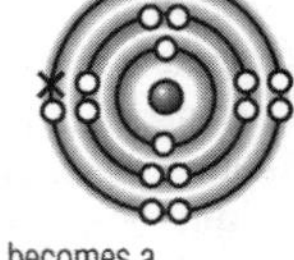

ionic bond The formation of an ionic bond between a sodium atom and a chlorine atom to form a molecule of sodium chloride. The sodium atom transfers an electron from its outer electron shell (becoming the positive ion Na^+) to the chlorine atom (which becomes the negative chloride ion Cl^-). The opposite charges mean that the ions are strongly attracted to each other. The formation of the bond means that each atom becomes more stable, having a full quota of electrons in its outer shell. Helicon

Electric Charge and Ionic Bonds

http://www.southwest.com.au/~jfuller/chemistry/chem4.htm

Introduction to ionic bonds, and the processes involved in creating and destroying them. Animations demonstrate the interaction between atoms when an ionic bond is being formed. Electron valences are also explained, and shown on a simplified periodic table.

ionic compound substance composed of oppositely charged ions. All salts, most bases, and some acids are examples of ionic compounds. They possess the following general properties: they are crystalline solids with a high melting point; are soluble in water and insoluble in organic solvents; and always conduct electricity when molten or in aqueous solution. A typical ionic compound is sodium chloride (Na^+Cl^-).

ionic equation equation showing only those ions in a chemical reaction that actually undergo a change, either by combining together to form an insoluble salt or by combining together to form one or more molecular compounds. Examples are the precipitation of insoluble barium sulfate when barium and sulfate ions are combined in solution, and the production of ammonia and water from ammonium hydroxide.

$Ba^{2+}_{(aq)} + SO_4{}^{2-}_{(aq)} \rightarrow BaSO_{4(s)}$

$NH_4{}^{+}_{(aq)} + OH^{-}_{(aq)} \rightarrow NH_{3(g)} + H_2O_{(l)}$

The other ions in the mixtures do not take part and are called †spectator ions.

ionic radius effective radius of an ion. In positively charged †cations, the ionic radius is less than the †atomic radius (because the electrons are more tightly bound); in †anions the ionic radius is more than the atomic radius. Some elements, such as the transition metals, can have several different ionization states and their ionic radii vary according to the state involved.

ionization process of ion formation. It can be achieved in two ways. The first way is by the loss or gain of electrons by atoms to form positive or negative ions.

$Na - e' \rightarrow Na^+$

$1/2Cl_2 + e' \rightarrow Cl^-$

In the second mechanism, ions are formed when a covalent bond breaks, as when hydrogen chloride gas is dissolved in water. One portion of the molecule retains both electrons, forming a negative ion, and the other portion becomes positively charged. This bond-fission process is sometimes called dissociation.

$HCl_{(g)} + aq \leftrightarrow H^{+}_{(aq)} + Cl^{-}_{(aq)}$

ionization chamber device for measuring †ionizing radiation. The radiation ionizes the gas in the chamber and the ions formed are collected and measured as an electric charge. Ionization chambers are used for determining the intensity of X-rays or the disintegration rate of radioactive materials.

ionization potential measure of the energy required to remove an †electron from an †atom. Elements with a low ionization potential readily lose electrons to form †cations.

ionizing radiation radiation that removes electrons from atoms during its passage, thereby leaving ions in its path. Alpha and beta particles are far more ionizing in their effect than are neutrons or gamma radiation.

ionosphere ionized layer of Earth's outer †atmosphere (60–1,000 km/38–620 mi) that contains sufficient free electrons to modify the way in which radio waves are propagated, for instance by reflecting them back to Earth. The ionosphere is thought to be produced by absorption of the Sun's ultraviolet radiation. The British Antarctic Survey estimates that the ionosphere is decreasing at a rate of 1 km/0.6 mi every five years, based on an analysis of data from 1960 to 1998. Global warming is the probable cause.

ion plating method of applying corrosion-resistant metal coatings. The article is placed in argon gas, together with some coating metal, which vaporizes on heating and becomes ionized (acquires charged atoms) as it diffuses through the gas to form the coating. It has important applications in the aerospace industry.

iridium chemical symbol Ir, (Latin *iridis* "rainbow") hard, brittle, silver-white, metallic element, atomic number 77, relative atomic mass 192.2. It is resistant to tarnish and corrosion. Iridium is one of the so-called platinum group of metals; it occurs in platinum ores and as a free metal (†native metal) with osmium in osmiridium, a natural alloy that includes platinum, ruthenium, and rhodium. It is alloyed with platinum for jewelry and used for watch bearings and in scientific instruments. It was named in 1804 by English chemist Smithson Tennant (1761–1815) for the iridescence of its salts in solution.

iridium anomaly unusually high concentrations of the element iridium found worldwide in sediments that were deposited at the Cretaceous-Tertiary boundary (†K-T boundary) 65 million years ago. Since iridium is more abundant in extraterrestrial material, its presence is thought to be evidence for a large meteorite impact that may have caused the extinction of the dinosaurs and other life at the end of the Cretaceous.

iris in anatomy, the colored muscular diaphragm that controls the size of the pupil in the vertebrate eye. It contains radial muscle that increases the pupil diameter and circular muscle that constricts the pupil diameter. Both types of muscle respond involuntarily to light intensity.

iron chemical symbol Fe, (Germanic *eis* "strong") hard, malleable and ductile, silver-gray, metallic element, atomic number 26, relative atomic mass 55.847. It chemical symbol comes from the Latin *ferrum*. It is the fourth most abundant element in the Earth's crust. Iron occurs in concentrated deposits as the ores hematite (Fe_2O_3), spathic ore ($FeCO_3$), and magnetite (Fe_3O_4). It sometimes occurs as a free metal, occasionally as fragments of iron or iron–nickel meteorites.

Iron is extracted from iron ore in a †blast furnace. The chemical †reactions of iron with oxygen, water, acids, and other substances, can be explained by its position in the middle of the †reactivity series of metals. Iron is the most useful of all metals; it is strongly magnetic and is the basis for †steel (an alloy with carbon and other elements) and †cast iron. Steel is used for buildings, bridges, ships, car bodies, and tools. Stainless steel is used for car parts, kitchen sinks, and cutlery. In electrical equipment iron is used in permanent magnets and electromagnets, and forms the cores of transformers and magnetic amplifiers. The †corrosion of iron is called †rusting and is an example of an †oxidation reaction. In the human body, iron is an essential component of hemoglobin, the molecule in red blood cells that transports oxygen to all parts of the body. A deficiency in the diet causes a form of anemia.

iron ore any mineral from which iron is extracted. The chief iron ores are †magnetite, a black oxide; †**hematite**, or kidney ore, a reddish oxide; †limonite, brown, impure oxyhydroxides of iron; and siderite, a brownish carbonate.

Iron ores are found in a number of different forms, including distinct layers in igneous intrusions, as components of contact metamorphic rocks, and as sedimentary beds. Much of the world's iron is extracted in Russia, Kazakhstan, and the Ukraine. Other important producers are the USA, Australia, France, Brazil, and Canada; over 40 countries produce significant quantities of ore.

iron pyrites or **pyrite**; FeS_2, common iron ore. Brassy yellow, and occurring in cubic crystals, it is often called "fool's gold," since only those who have never seen gold would mistake it.

ironstone type of sedimentary rock that is rich in iron minerals. It forms layers or beds and can be found as nodules.

irradiation subjecting anything to radiation, including cancer tumors (when it is a type of †radiotherapy).

irrational number number that cannot be expressed as an exact †fraction. Irrational numbers include some square roots (for example, $\sqrt{2}$, $\sqrt{3}$, and $\sqrt{5}$ are irrational); numbers such as p (for circles), which is approximately equal to the †decimal 3.14159; and e (the base of †natural logarithms, approximately 2.71828). If an irrational number is expressed as a decimal it would go on for ever without repeating. An irrational number multiplied by itself gives a †rational number.

irregular galaxy class of †galaxy with little structure, which does not conform to any of the standard shapes in the †Hubble classification. The two satellite galaxies of the †Milky Way, the †Magellanic Clouds, are both irregulars. Some galaxies previously classified as irregulars are now known to be normal galaxies distorted by tidal effects or undergoing bursts of star formation (see †starburst galaxy).

irrigation artificial water supply for dry agricultural areas by means of dams and channels. Drawbacks are that it tends to concentrate salts at the surface, ultimately causing soil infertility, and that rich river silt is retained at dams, to the impoverishment of the land and fisheries below them.

Irrigation has been practiced for thousands of years, in Eurasia as well as the Americas. An example is the channeling of the annual Nile flood in Egypt, which has been done from earliest times to its present control by the Aswan High Dam.

Isaacs, Alick (1921–1967) Scottish virologist who, with Swiss colleague Jean Lindemann, in 1957 discovered †interferon, a naturally occurring antiviral substance produced by cells infected with viruses. The full implications of this discovery are still being investigated.

island area of land surrounded entirely by water. Australia is classed as a continent rather than an island, because of its size.

Islands can be formed in many ways. **Continental islands** were once part of the mainland, but became isolated (by tectonic movement, erosion, or a rise in sea level, for example). **Volcanic islands**, such as Japan, were formed by the explosion of underwater volcanoes. **Coral islands** consist mainly of †coral, built up over many years. An **atoll** is a circular coral reef surrounding a lagoon; atolls were formed when a coral reef grew up around a volcanic island that subsequently sank or was submerged by a rise in sea level. **Barrier islands** are found by the shore in shallow water, and are formed by the deposition of sediment eroded from the shoreline.

island arc curved chain of volcanic islands. Island arcs are common in the Pacific where they ring the ocean on both sides; the Aleutian Islands off Alaska are an example. The volcanism that forms island arcs is a result of subduction of an oceanic plate beneath another plate, as evidenced by the presence of ocean trenches on the convex side of the arc, and the Benioff zone of high seismic activity beneath.

Such island arcs are often later incorporated into continental margins during mountain-building episodes.

islets of Langerhans groups of cells within the pancreas responsible for the secretion of the hormone insulin. They are sensitive to blood sugar levels, producing more hormone when glucose levels rise.

ISO alternative name for the International Organization for Standardization, derived from the Greek word *isos* meaning "equal."

isobar line drawn on maps and weather charts linking all places with the same atmospheric pressure (usually measured in millibars). When used in weather forecasting, the distance between the isobars is an indication of the barometric gradient (the rate of change in pressure).

Where the isobars are close together, cyclonic weather is indicated, bringing strong winds and a depression, and where far apart anticyclonic, bringing calmer, settled conditions.

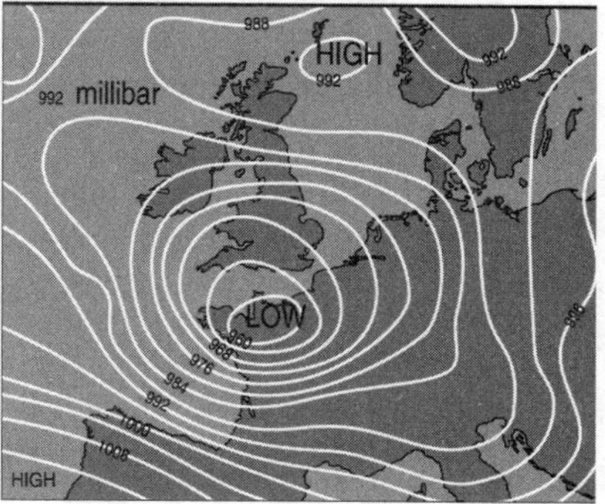

isobar The isobars around a low-pressure area or depression. In the northern hemisphere, winds blow anticlockwise around lows, approximately parallel to the isobars, and clockwise around highs. In the southern hemisphere, the winds blow in the opposite directions. Helicon

isobleth term meaning a line on a †map which connects points of equal value. Specific types of isobleth include: a contour line, connecting points of equal altitude; a isohyet, connecting points of equal rainfall; an isotherm, connecting points of equal temperature at a given time; an isobath, connecting points of equal distance below a water body; and an isobar, connecting points of equal atmospheric pressure.

isochrone on a map, a line that joins places that are equal in terms of the time it takes to reach them.

isohyet on a map, a line joining points of equal rainfall.

isoline on a map, a line that joins places of equal value. Examples are contour lines (joining places of equal height), †isobars (for equal pressure), isotherms (for equal temperature), and isohyets (for equal rainfall). Isolines are most effective when values change gradually and when there is plenty of data.

butane $CH_3(CH_2)_2CH_3$

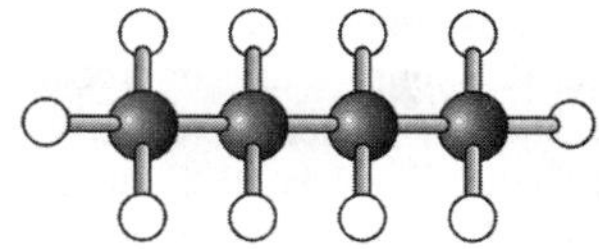

methyl propane $CH_3CH(CH_3)CH_3$

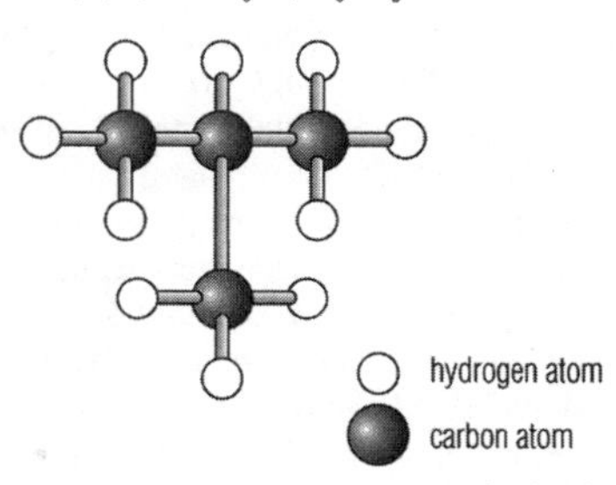

isomer The chemicals butane and methyl propane are isomers. Each has the molecular formula C_4H_{10}, but with different spatial arrangements of atoms in their molecules. Helicon

isomer chemical compound having the same †molecular formula but with different molecular structure. For example, the organic compounds butane ($CH_3(CH_2)_2CH_3$) and methyl propane ($CH_3CH(CH_3)CH_3$) are isomers, each possessing four carbon atoms and ten hydrogen atoms but differing in the way that these are arranged with respect to each other.

Structural isomers have obviously different constructions, but **geometrical** and **optical isomers** must be drawn or modeled in order to appreciate the difference in their three-dimensional arrangement. Geometrical isomers have a plane of symmetry and arise because of the restricted rotation of atoms around a bond; optical isomers are mirror images of each other. For instance, 1,1-dichloroethene ($CH_2 = CCl_2$) and 1,2-dichloroethene ($CHCl = CHCl$) are structural isomers, but there are two possible geometric isomers of the latter (depending on whether the chlorine atoms are on the same side or on opposite sides of the plane of the carbon–carbon double bond).

isometric transformation in mathematics, †transformation in which length is preserved.

isomorphism existence of substances of different chemical composition but with similar crystalline form.

isoperimetry branch of geometry involving the study and measurement of figures with equal perimeters.

isoprene technical name **methylbutadiene**; $CH_2CHC(CH_3)CH_2$, colorless, volatile liquid obtained from petroleum and coal, used to make synthetic rubber.

isoprene polymer compound formed by the †polymerization of †isoprene †monomers. The *cis* structural form of polyisoprene is produced when †Ziegler catalysts are used for polymerization. It is an †elastomer and the synthetic equivalent of natural rubber.

isopropyl ether C_6H_4O, colorless organic liquid widely used as an industrial solvent, boiling point 68°C/154°F. It is usually prepared by reacting iso-propanol with concentrated sulfuric acid.

isosceles triangle †triangle with two sides equal, hence its base angles are also equal. The triangle has an axis of symmetry which is an †altitude of the triangle.

isostasy condition of gravitational equilibrium of all parts of the Earth's †crust. The crust is in isostatic equilibrium if, below a certain depth, the weight and thus pressure of rocks above is the same everywhere. The idea is that the lithosphere floats on the asthenosphere as a piece of wood floats on water. A thick piece of wood floats lower than a thin piece, and a denser piece of wood floats lower than a less dense piece. There are two theories of the mechanism of isostasy, the Airy hypothesis and the Pratt hypothesis, both of which have validity. In the **Airy hypothesis** crustal blocks have the same density but different thicknesses: like ice cubes floating in water, higher mountains have deeper roots. In the **Pratt hypothesis**, crustal blocks have different densities allowing the depth of crustal material to be the same. In practice, both mechanisms are at work.

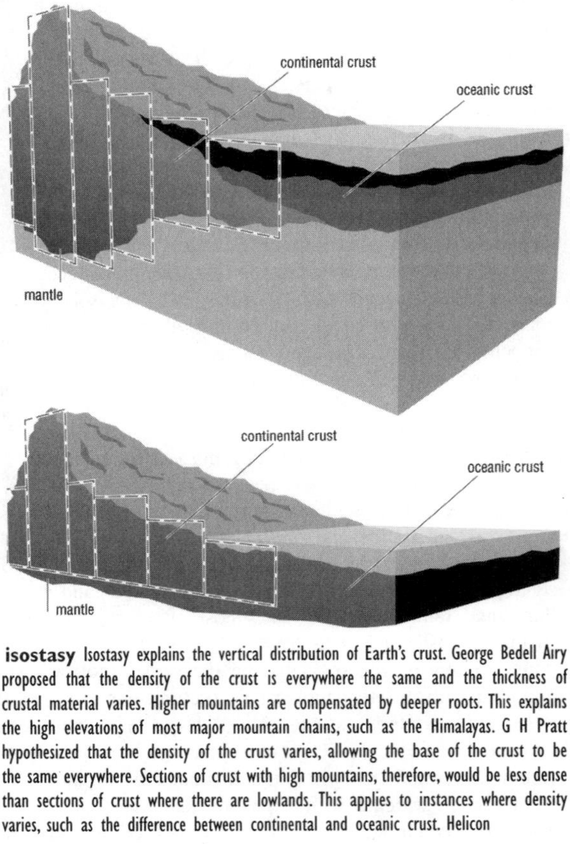

isostasy Isostasy explains the vertical distribution of Earth's crust. George Bedell Airy proposed that the density of the crust is everywhere the same and the thickness of crustal material varies. Higher mountains are compensated by deeper roots. This explains the high elevations of most major mountain chains, such as the Himalayas. G H Pratt hypothesized that the density of the crust varies, allowing the base of the crust to be the same everywhere. Sections of crust with high mountains, therefore, would be less dense than sections of crust where there are lowlands. This applies to instances where density varies, such as the difference between continental and oceanic crust. Helicon

isotactic polymer in chemistry, stereospecific, long-chain hydrocarbon polymer of general formula —$(CH_2CHR)_n$- , where R is an †alkyl group and n is the number of units in the polymer chain. Isotactic polymers are polymerized using certain catalysts, such as †Ziegler catalysts, so that all the alkyl groups are in identical configurations on the same side of the polymer chain.

isotherm line on a map, linking all places having the same temperature at a given time.

isotope one of two or more atoms that have the same atomic number (same number of protons), but which contain a different number of neutrons, thus differing in their †relative atomic mass. They may be stable or radioactive (as a †radioisotope), naturally occurring, or synthesized. For example, hydrogen has the isotopes ^2H (†deuterium) and ^3H (†tritium). The term was coined by English chemist Frederick Soddy, a pioneer researcher in atomic disintegration.

Elements at the lower end of the periodic table have atoms with roughly the same number of protons as neutrons. These elements are called **stable isotopes**. The stable isotopes of oxygen include ^{16}O, ^{17}O, and ^{18}O; those of carbon include ^{12}C and ^{14}C. Elements with high atomic mass numbers have many more neutrons than protons and are therefore less stable. It is these isotopes that are more prone to †radioactive decay. Examples are ^{238}U (uranium-238) and ^{60}Co (cobalt-60).

Fajans on the Concept of Isotopes http://dbhs.wvusd.k12.ca.us/Chem-History/Fajans-Isotope.html

Extract from Kasimir Fajans paper of 1913 titled Radioactive Transformations and the Periodic System of The Elements. The paper describes Fajans's discovery of isotopes and goes on to offer examples of radioactive transformations leading to their production.

Isotopes and Atomic Weights http://dbhs.wvusd.k12.ca.us/Chem-History/Aston-MassSpec.html

Transcript of part of English physicist Francis Aston's paper from 1920 outlining his ideas on atomic weights and isotopes. The paper led directly to the production of the first mass spectrometer, and thus led to the separation of many isotopic elements.

isotropic having equal and uniform properties at all points and in all directions. In astronomy the term describes microwave †background radiation.

isthmus narrow strip of land joining two larger land masses. The Isthmus of Panama joins North and South America.

IT abbreviation for †information technology.

iteration in mathematics, method of solving †equations by a series of approximations which approach the exact solution more and more closely. For example, to find the square root of *N*, start with a guess n_1; calculate

$N/n_1 = x_1$

calculate

$(n_1 + x_1)/2 = n_2$

calculate

$N/n_2 = x_2$

calculate

$(n_2 + x_2)/2 = n_3$

The sequence n_1, n_2, n_3 approaches the exact square root of *N*. Iterative methods are particularly suitable for work with computers and programmable calculators.

J

J in physics, the symbol for †**joule**, the SI unit of energy.

jacinth or **hyacinth**, red or yellowish-red gem, a variety of †zircon, $ZrSiO_4$.

Jackson, John Hughlings (1835–1911) English neurologist and neurophysiologist. As a result of his studies of epilepsy, Jackson demonstrated that specific areas of the cerebral cortex (outer mantle of the brain) control the functioning of particular organs and limbs.

Jacob, François (1920–) French biochemist who was awarded the Nobel Prize in Physiology or Medicine in 1965, with Jacques †Monod and André †Lwoff, for their work on the genetic control of enzyme and virus synthesis. They pioneered research into molecular genetics and showed how the production of proteins from †DNA is controlled.

Jacobi, Carl Gustav Jacob (1804–1851) German mathematician and mathematical physicist, much of whose work was on the theory of elliptical functions, mathematical analysis, number theory, geometry, and mechanics.

Jacobi, Carl Gustav Jacob http://www-groups.dcs.st-andrews.ac.uk/history/Mathematicians/Jacobi.html

Devoted to the life and contributions of Carl Jacobi. In addition to biographical information, you will find a list of references about Jacobi and links to essays that reference him. The text of this essay includes hypertext links to the essays of those mathematicians and thinkers who influenced Jacobi. You will also find an image of him, which you can click on to enlarge, and a map of his birthplace.

Jacquard system use of punched cards to direct the operation of a loom in weaving patterns. It was devised by French engineer Joseph Jacquard, and is similar to a system first used in China in the 3rd century BC, where a loom was controlled by a memory frame with silk threads of varying lengths.

jade semi-precious stone consisting of either jadeite, $NaAlSi_2O_6$ (a pyroxene), or nephrite, $Ca_2(Mg,Fe)_5Si_8O_{22}(OH,F)_2$ (an amphibole), ranging from colorless through shades of green to black according to the iron content. Jade ranks 5.5–6.5 on the Mohs scale of hardness.

The early Chinese civilization discovered jade, bringing it from eastern Turkestan, and carried the art of jade-carving to its peak. The Olmecs, Aztecs, Maya, and the Maori have also used jade for ornaments, ceremony, and utensils.

James, William (1842–1910) US psychologist and philosopher. He was among the first to take an approach emphasizing the ends or purposes of behavior and to advocate a scientific, experimental psychology. His *Varieties of Religious Experience* (1902) is one of the most important works on the psychology of religion.

Classics in the History of Psychology – "Stream of Consciousness"

http://psychclassics.yorku.ca/James/jimmy11.htm

Transcript of James's "Stream of Consciousness" theory as first published in Psychology, Chapter XI, (Cleveland & New York, World). The first sentence of the chapter sums up James's contribution to the science of psychology: "The order of our study must be analytic." James's belief in the value of scientific experimental method changed psychology forever.

Classics in the History of Psychology – The Principles of Psychology

http://psychclassics.yorku.ca/James/Principles/index.htm

Text of William James's classic book.

Varieties of Religious Experience by William James http://www.psychwww.com/psyrelig/james/toc.htm

Online version of James' seminal work on the psychology of religion. A series of well-arranged notes accompany the text and show the ongoing relevance of his analysis.

James–Lange theory in psychology, a theory that sensory impressions resulting from exposure to an emotional situation are caused by physiological reactions. In this way emotions of pleasure, fear, or amusement may be seen as the result of changes of bodily activity; fear is the sensation of sweating and trembling rather than the emotional reaction to a terrifying situation.

It is named after the US psychologist William James and the Danish physician Carl Georg Lange (1834–1900), who arrived at this view independently in the 1880s.

jansky unit of radiation received from outer space, used in radio astronomy. It is equal to 10^{-26} watts per square meter per hertz, and is named after US engineer Karl Jansky.

Jansky, Karl Guthe (1905–1950) US radio engineer who in 1932 discovered that the Milky Way galaxy emanates radio waves; he did not follow up his discovery, but it marked the birth of radioastronomy.

Janssen, (Pierre) Jules César (1824–1907) French astronomer. He studied the solar spectrum. In 1867 he concluded that water vapor was present in the atmosphere of Mars; he developed a spectrohelioscope in 1868.

Janus small low-density moon of Saturn, discovered in 1966, which orbits just outside the ring system. It shares its orbit with †Epimetheus and has an irregular shape (longest dimension 220 km/137 mi).

Japan Current or **Kuroshio**, warm ocean †current flowing from Japan to North America.

jasper hard, compact variety of †chalcedony SiO_2, usually colored red, brown, or yellow. Jasper can be used as a gem.

Java in computing, programming language much like C developed by James Gosling at Sun Microsystems in 1995. Java has been adopted as a multipurpose, cross-platform lingua franca for network computing, including the †World Wide Web. When users connect to a server that uses Java, they download a small program called an †applet onto their computers. The applet then runs on the computer's own processor via a Java Virtual Machine program or JVM. In 2000 Java 2 Micro Edition was released, designed for use in handheld devices, telephones, and pagers.

Shlurrrpp......Java: The First User-Friendly Tutorial On Java http://www.vijaymukhi.com/vmis/java.html
Tutorial to the Internet programming language Java that relates much of the learning process to coffee drinking. Starting with "The first sip" of Java, where you learn the very basics of what Java is and does, the guide goes on to explain the potentials of the language in a detailed but uncomplicated manner. There is also a wealth of advice on writing your own programs, for those who want to try Java firsthand.

JavaScript scripting language commonly used to add interactive elements to Web pages. JavaScript was developed by Netscape Communications as LiveScript (it was not derived from Java) and has been standardized by ECMA (European Computer Manufacturers' Association) as ECMAScript.

jaw one of two bony structures that form the framework of the mouth in all vertebrates except lampreys and hagfishes (the agnathous or jawless vertebrates). They consist of the upper jawbone (maxilla), which is fused to the skull, and the lower jawbone (mandible), which is hinged at each side to the bones of the temple by †ligaments.

Jeans, James Hopwood (1877–1946) English mathematician and scientist. In physics he worked on the kinetic theory of gases, and on forms of energy radiation; in astronomy, his work focused on giant and dwarf stars, the nature of spiral nebulae, and the origin of the cosmos. He did much to popularize astronomy. He was knighted in 1928.

Jeans, Sir James Hopwood http://www-groups.dcs.st-andrews.ac.uk/history/Mathematicians/Jeans.html
Part of an archive containing the biographies of the world's greatest mathematicians, this site is devoted to the life and contributions of James Jeans. In addition to biographical information, you will find a list of references about Jeans and links to other essays in the archive that reference him. The text of this essay includes hypertext links to the essays of those mathematicians and thinkers who influenced Jeans. You will also find an image of him, which you can click on to enlarge, and a map of his birthplace.

Jeans mass in astronomy, mass that a cloud (or part of a cloud) of interstellar gas must have before it can contract under its own weight to form a protostar. The Jeans mass is an expression of the **Jeans criterion**, which says that a cloud will contract when the gravitational force tending to draw material towards its center is greater than the opposing force due to gas pressure. It is named after English mathematician James Hopwood †Jeans, whose work focused on the kinetic theory of gases and the origins of the cosmos.

Jeffreys, Alec John (1950–) English geneticist who discovered the DNA probes necessary for accurate †genetic fingerprinting so that a murderer or rapist could be identified by, for example, traces of blood, tissue, or semen.

jejunum part of the small intestine connecting the †duodenum to the †ileum.

Jenner, Edward (1749–1823) English physician who pioneered vaccination. In Jenner's day, smallpox was a major killer. His discovery in 1796 that inoculation with cowpox gives immunity to smallpox was a great medical breakthrough.

Jensen, (Johannes) Hans (Daniel) (1907–1973) German physicist who, with Maria †Goeppert-Mayer and Eugene †Wigner, was awarded the Nobel Prize for Physics in 1963 for the discovery of the shell-like structure of atomic nuclei.

Jerne, Niels Kaj (1911–1994) British-born Danish microbiologist and immunologist who was awarded a Nobel Prize for Physiology or Medicine in 1984, with Georges †Köhler and César †Milstein, for their work on immunity and their discovery of a technique for producing highly specific, monoclonal antibodies. Jerne profoundly influenced the development of modern immunology by establishing its cellular basis.

jet in astronomy, narrow luminous feature seen protruding from a star or galaxy and representing a rapid outflow of material. See †active galaxy.

jet in earth science, hard, black variety of lignite, a type of coal. It is cut and polished for use in jewelry and ornaments. Articles made of jet have been found in Bronze Age tombs.

jet propulsion method of propulsion in which an object is propelled in one direction by a jet, or stream of gases, moving in the other. This follows from Isaac †Newton's third law of motion: "To every action, there is an equal and opposite reaction." The most widespread application of the jet principle is in the jet (gas turbine) engine, the most common kind of aircraft engine.

jet stream narrow band of very fast wind (velocities of over 150 kph/95 mph) found at altitudes of 10–16 km/6–10 mi in the upper troposphere or lower stratosphere. Jet streams usually occur about the latitudes of the Westerlies (35°–60°).

The jet stream may be used by high flying aircraft to speed their journeys. Their discovery of the existence of the jet stream allowed the Japanese to send gas-filled balloons carrying bombs to the northwestern USA during World War II.

Johanssen, Wilhelm Ludvig (1857–1927) Danish botanist and founder of modern genetics who coined the term **gene** as the unit of heredity (although it was not known at the time to consist of †DNA in cells). He introduced the concept of an organism having a "genotype" (set of variable genes) and "phenotype" (characteristics produced by the presence of certain genes). He also studied the metabolism of †germination (the initial stages of growth in a seed) in plants.

joint in earth science, a vertical crack in a rock, often formed by compression; it is usually several meters in length. A joint differs from a †fault in that no displacement of the rocks on either side has taken place. The weathering of joints in rocks such as limestone and granite is responsible for the formation of features such as †limestone pavements and †tors. Joints in coastal rocks are often exploited by the sea to form erosion features such as caves and geos.

joint point of movement or articulation in any animal with a skeleton. In vertebrates, it is the point where two bones meet. Some joints allow no motion (the sutures between the bones of the skull), others allow a very small motion (the sacroiliac joints in the lower vertebral column), but most allow a relatively free motion. Of these, some allow a gliding motion (one vertebra of the spine on another), some have a hinge action (elbow and knee, see †triceps), and others allow motion in all directions (hip and shoulder joints) by means of a ball-and-socket arrangement.

Movement at a joint is usually produced by †muscles, which are attached to the bones by tendons. The ends of the bones at a moving joint are covered with cartilage for greater elasticity and smoothness (to make movement easy), and enclosed in an envelope (capsule) of tough white fibrous tissue lined with a membrane which secretes a lubricating and cushioning †synovial fluid. The joint is further strengthened by ligaments, which connect bone to bone. These joints are known as synovial joints. In invertebrates with an †exoskeleton, the joints are places where the exoskeleton is replaced by a more flexible outer covering, the arthrodial membrane, which allows the limb (or other body part) to bend at that point.

Major Joints http://www.bonesandjoints.com/joints.htm

This site covers the major joints in the body: ankle, elbow, foot, hand, hip, knee, shoulder, and back. These are listed on the first page, and clicking on a specific joint brings up a description and diagram of it, as well as information on ways to prevent and treat injuries.

Joliot-Curie, Frédéric (1900–1958) born Frédéric Jean Joliot, French physicist. With his wife, Irène †Joliot-Curie, he was awarded the Nobel Prize for Chemistry in 1935 for their discovery of artificial †radioactivity.

Joliot-Curie, Irène (1897–1956) French physicist, daughter of Pierre and Marie †Curie. She and her husband Frédéric †Joliot were jointly awarded the Nobel Prize for Chemistry in 1935 for their discovery of artificial radioactivity. She was professor of physics at the Sorbonne from 1937, and director of the Radium Institute 1946–56.

Jordan, (Marie Ennemond) Camille (1838–1922) French mathematician, the greatest exponent of algebra in his day. He concentrated on research in topology, analysis, and particularly group theory, publishing *Traité des substitutions et des equations algébriques* (1870).

Jordan, Marie Ennemond Camille http://www-groups.dcs.st-andrews.ac.uk/history/Mathematicians/Jordan.html

Part of an archive containing the biographies of the world's greatest mathematicians, this site is devoted to the life and contributions of Camille Jordan. In addition to biographical information, you will find a list of references about Jordan and links to other essays in the archive that reference him. The text of this essay includes hypertext links to the essays of those mathematicians and thinkers who influenced Jordan. You will also find an image of him, which you can click on to enlarge, and a map of his birthplace.

Josephson, Brian David (1940–) Welsh physicist, a leading authority on †superconductivity. He shared the Nobel Prize for Physics in 1973 for his theoretical predictions of the properties of a supercurrent through a tunnel barrier (the **Josephson effect**), which led to the development of the Josephson junction.

Josephson effect in physics, superconducting (see †superconductivity) ring interrupted by a thin layer of insulating material gives rise to an †alternating current (AC) in the barrier when a steady external voltage is applied to it. This AC effect has a direct-current analog, occurring when a steady magnetic field is applied to the insulating material. The effect is named after Welsh physicist Brian Josephson.

joule symbol J, SI unit of work and energy (such as †potential energy, †kinetic energy, or †electrical energy).

Joule, James Prescott (1818–1889) English physicist. His work on the relations between electrical, mechanical, and chemical effects led to the discovery of the first law of †thermodynamics.

Joule–Kelvin effect or **Joule–Thomson effect**, in physics, the fall in temperature of a gas as it expands adiabatically (without loss or gain of heat to the system) through a narrow jet. It can be felt when, for example, compressed air escapes through the valve of an inflated bicycle tire. It is the basic principle of some refrigerators.

Joule's electrical law heat H in joules produced by the passing of a †current of I amperes through a resistance of R ohms for a time t seconds is given by: $H = I^2Rt$

Joule's thermal law internal energy of a gas at constant temperature is independent of its volume, provided the gas is †ideal.

Joy, Alfred Harrison (1882–1973) US astronomer who worked on stellar distances and the radial motions of stars. He observed variable stars and classified them according to their spectra; he also determined the distance and direction of the center of the Galaxy and attempted to calculate its rotation period.

JPEG acronym for Joint Photographic Experts Group, used to describe a compression standard set up by that group and now widely accepted for the storage and transmission of color images. The JPEG compression standard reduces the size of image files considerably, although it is a lossy compression format. Repeating editing and resaving JPEGs will adversely affect their appearance.

jugular vein one of two veins in the necks of vertebrates; they return blood from the head to the superior (or anterior) †vena cava and thence to the heart.

Julian calendar calendar established by Julius Caesar in 46 BC, which overestimated the duration of the †sidereal year by 11 minutes and 14 seconds. It was replaced, from 1582, by the †Gregorian calendar, by which time it was inaccurate by a total of 10 days.

Julian date in astronomy, measure of time used in which days are numbered consecutively from noon †GMT on 1 January 4713 BC. It is useful where astronomers wish to compare observations made over long time intervals. The Julian date (JD) at noon on 1 January 2000 was 2451545.0. The **modified Julian date** (MJD), defined as MJD = JD –2400000.5, is more commonly used since the date starts at midnight GMT and the smaller numbers are more convenient.

junction in mathematics, alternative name for a †node, a point in a network where more than two lines meet.

Jung, Carl Gustav (1875–1961) Swiss psychiatrist. He collaborated with Sigmund †Freud from 1907 until their disagreement in 1914 over the importance of sexuality in causing psychological problems. Jung studied myth, religion, and dream symbolism, saw the unconscious as a source of spiritual insight, and distinguished between introversion and extroversion.

jungle popular name for †rainforest.

junk DNA another name for †intron, a region of DNA that contains no genetic information.

Juno in astronomy, bright asteroid with a diameter of 267 km/166 mi. Its brightness led to its early discovery (in 1804). Juno was the third asteroid to be discovered. It has an orbital period of 4.4 years.

Jupiter fifth planet from the Sun and the largest in the Solar System, with a mass equal to 70% of all the other planets combined and 318 larger than that of the Earth. Its main feature is the Great Red Spot, a cloud of rising gases, 14,000 km/8,500 mi wide and 30,000 km/20,000 mi long, revolving anticlockwise.

mean distance from the Sun: 778 million km/484 million mi

equatorial diameter: 142,800 km/88,700 mi

rotation period: 9 hours 51 minutes

year: 11.86 Earth years

atmosphere: consists of clouds of white ammonia crystals, drawn out into belts by the planet's high speed of rotation (the fastest of any planet). Darker orange and brown clouds at lower levels may contain sulfur, as well as simple organic compounds. Temperatures range from –140°C/–220°F in the upper atmosphere to as much as 24,000°C/43,000°F near the core. This is the result of heat left over from Jupiter's formation, and it is this that drives the turbulent weather patterns of the planet. The Great Red Spot was first observed in 1664. Its top is higher than the surrounding clouds; its color is thought to be due to red phosphorus. The Southern Equatorial Belt in which the Great Red Spot occurs is subject to unexplained fluctuation. In 1989 it sustained a dramatic and sudden fading. Jupiter's strong magnetic field gives rise to a large surrounding magnetic "shell," or magnetosphere, from which bursts of radio waves are detected. Jupiter's faint rings are made up of dust from its moons, particularly the four inner moons

surface: largely composed of hydrogen and helium, which under the high pressure and temperature of the interior behave not as a gas but as a supercritical fluid. Under even more extreme conditions, at a depth of 30,000 km/18,000 mi, hydrogen transforms into a metallic liquid. Jupiter probably has a molten rock core whose mass is 15 to 20 times greater than that of the Earth

In 1995, the Galileo probe revealed Jupiter's atmosphere to consist of 0.2% water, less than previously estimated.

satellites: Jupiter has 28 known moons. The four largest moons, Io, Europa (which is the size of the Moon), Ganymede, and Callisto, are the †Galilean satellites, discovered in 1610 by †Galileo Galilei (Ganymede, which is larger than Mercury, is the largest moon in the Solar System). Three small moons were discovered in 1979 by the US Voyager probes, as was a faint ring of dust around Jupiter's equator 55,000 km/34,000 mi above the cloud tops. One of Jupiter's small inner moons, Almathea (diameter 250 km/155 mi), was shown by pictures from the Galileo probe in April 2000 to have a long, narrow, bright region, as yet unidentified. A new moon was first observed in orbiting Jupiter in October 1999 by US researchers at the Kitt Peak Observatory, Arizona. It was thought to be an asteroid and named S/1999J1, but was confirmed to be a moon in July 2000. The moon is only 5 km/3 mi in diameter and orbits Jupiter once every two years at a distance of 24 million km/15 million mi. Ten previously unobserved moons were discovered orbiting Jupiter in November and December 2000. These moons are all believed to be less than 5 km/3.1 mi in diameter, and were observed by astronomers at the Mauna Kea observatory, Hawaii.

The comet Shoemaker-Levy 9 crashed into Jupiter in July 1994. Impact zones were visible for several months.

Galileo: Journey to Jupiter http://www.jpl.nasa.gov/galileo/

Full details of the ground-breaking mission to the Solar System's largest planet. There are regularly updated reports on instructions being given to the probe by the Jet Propulsion Laboratory, Pasadena, California, and on data being received from Galileo. There are numerous images of Jupiter and its moons and an animation showing fluctuations in the Great Red Spot. If you cannot find an answer to your query in the Frequently Asked Questions, you may e-mail the project.

Jupiter http://www.solarviews.com/eng/jupiter.htm

Full details of the planet and its moons, including a chronology of exploration, various views of the planet and its moons, and links to other planets.

Jurassic period period of geological time 208–146 million years ago; the middle period of the Mesozoic era. Climates worldwide were equable, creating forests of conifers and ferns; dinosaurs were abundant, birds evolved, and limestones and iron ores were deposited.

The name comes from the Jura Mountains in France and Switzerland, where the rocks formed during this period were first studied.

K

K abbreviation for thousand, as in a salary of $30K.

K symbol for **kelvin**, a scale of temperature.

kaleidoscope optical toy invented by the Scottish physicist David Brewster in 1816. It usually consists of a pair of long mirrors at an angle to each other, and arranged inside a triangular tube containing pieces of colored glass, paper, or plastic. An axially symmetrical (hexagonal) pattern is seen by looking along the tube, which can be varied infinitely by rotating or shaking the tube.

Kaleidoscope Heaven http://kaleidoscopeheaven.org/

Central hub of resources for anyone interested in kaleidoscopes. This page has information on the history of the kaleidoscope, where to find materials to make your own, how to make one, a current educational project, and even an essay on the alleged health benefits of the kaleidoscope.

kame geological feature, usually in the form of a mound or ridge, formed by the deposition of rocky material carried by a stream of glacial meltwater. Kames are commonly laid down in front of or at the edge of a glacier (kame terrace), and are associated with the disintegration of glaciers at the end of an ice age.

Kames are made of well-sorted rocky material, usually sands and gravels. The rock particles tend to be rounded (by attrition) because they have been transported by water.

Kamerlingh Onnes, Heike (1853–1926) Dutch physicist who worked mainly in the field of low-temperature physics. He was awarded the Nobel Prize for Physics in 1913 for his discovery in 1911 of the phenomenon of †superconductivity (enhanced electrical conductivity at very low temperatures).

Kammerer, Paul (1880–1926) Austrian zoologist whose work on the genetics of the midwife toad was extremely controversial and his claims that acquired characteristics could be passed to the next generation were eventually disproved.

Kandel, Eric Richard (1929–) Austrian-born US neurobiologist and educator, who with Arvid †Carlsson and Paul †Greengard shared the Nobel Prize in Physiology or Medicine in 2000 for their discoveries concerning signal transduction in the nervous system. Kandel discovered the molecular mechanisms involved in learning and the formation of memories.

Kant, Immanuel (1724–1804) German philosopher. He believed that knowledge is not merely an aggregate of sense impressions but is dependent on the conceptual apparatus of the human understanding, which is itself not derived from experience. In ethics, Kant argued that right action cannot be based on feelings or inclinations but conforms to a law given by reason, the **categorical imperative**.

kaolin or **china clay**, group of clay minerals, such as †kaolinite, $Al_2Si_2O_5(OH)_4$, derived from the alteration of aluminum silicate minerals, such as †feldspars and †mica. It is used in medicine to treat digestive upsets, and in poultices.

Kaolinite is economically important in the ceramic and paper industries. It is mined in the UK, the USA, France, and the Czech Republic.

kaolinite white or grayish †clay mineral, hydrated aluminum silicate, $Al_2Si_2O_5(OH)_4$, formed mainly by the decomposition of feldspar in granite. It is made up of platelike crystals, the atoms of which are bonded together in two-dimensional sheets, between which the bonds are weak, so that they are able to slip over one another, a process made more easy by a layer of water. China clay (kaolin) is derived from it.

Kapitza, Peter Leonidovich (1894–1984) also known as **Pyotr Kapitsa**, Soviet physicist who shared the Nobel Prize for Physics in 1978 for his work on magnetism and low-temperature physics. He worked on the superfluidity of liquid helium and also achieved the first high-intensity magnetic fields.

Kapteyn, Jacobus Cornelius (1851–1922) Dutch astronomer. He analyzed the structure of the universe by studying the distribution of stars using photographic techniques. To achieve more accurate star counts he introduced the technique of statistical astronomy.

Karrer, Paul (1889–1971) Russian-born Swiss organic chemist who shared the Nobel Prize for Chemistry in 1937 for his work in carotenoids, flavins retinal (vitamin A) and riboflavin (vitamin B_2), and determining their structural formulae.

karst landscape characterized by remarkable surface and underground forms, created as a result of the action of water on permeable limestone. The feature takes its name from the Karst (meaning **dry**) region on the Adriatic coast of Slovenia and Croatia, but the name is applied to landscapes throughout the world, the most dramatic of which is found near the city of Guilin in the Guangxi province of China. Karst landscapes are characterized by underground features such as caves, caverns, stalactites, and stalagmites. On the surface, clints, grikes, gorges, and swallow holes are common features.

Limestone is soluble in the weak acid of rainwater. Erosion takes place most swiftly along cracks and joints in the limestone and these open up into gullies called grikes. The rounded blocks left upstanding between them are called clints.

karyotype in biology, the set of †chromosomes characteristic of a given species. It is described as the number, shape, and size of the chromosomes in a single cell of an organism. In humans, for example, the karyotype consists of 46 chromosomes, in mice 40, crayfish 200, and in fruit flies 8.

The diagrammatic representation of a complete chromosome set is called a **karyogram**.

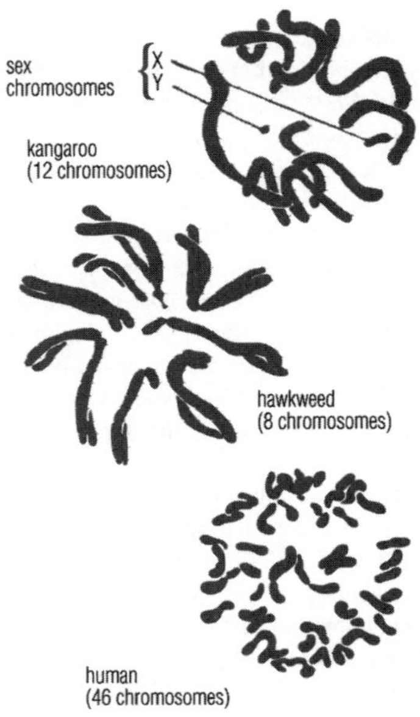

karyotype The characteristics, or karyotype, of the chromosomes vary according to species. The kangaroo has 12 chromosomes, the hawkweed has 8, and a human being has 46. Helicon

Kastler, Alfred (1902–1984) French physicist who worked on double-resonance techniques in †spectroscopy, in which absorption of optical or radio-frequency radiation is used to study the energy levels in atoms. He was awarded the Nobel Prize for Physics in 1966 for his development of the technique known as "optical pumping" in which light is used to excite (energize) an atom to a higher energy level. This work led to the invention of the maser and laser.

katabatic wind cool wind that blows down a valley on calm clear nights. (By contrast, an †anabatic wind is warm and moves up a valley in the early morning.) When the sky is clear, heat escapes rapidly from ground surfaces, and the air above the ground becomes chilled. The cold dense air moves downhill, forming a wind that tends to blow most strongly just before dawn.

Cold air blown by a katabatic wind may collect in a depression or valley bottom to create a †frost hollow. Katabatic winds are most likely to occur in the late spring and autumn because of the greater daily temperature differences.

Katz, Bernard (1911– 2003) German-born British biophysicist who was awarded a Nobel Prize for Physiology or Medicine in 1970 for work on the storage, release, and inactivation of neurotransmitters, research vital in the search for remedies for nervous and mental disorders. He was knighted in 1969.

kayser unit of wave number (number of waves in a unit length), used in spectroscopy. It is expressed as waves per centimeter, and is the reciprocal of the wavelength. A wavelength of 0.1 cm has a wave number of 10 kaysers.

kcal symbol for **kilocalorie** (see †calorie).

Keeler, James Edward (1857–1900) US astrophysicist. He studied the rings of Saturn and the abundance and structure of nebulae. He demonstrated in 1888 that nebulae resembled stars in their pattern of movement.

Kekulé von Stradonitz, Friedrich August (1829–1896) German chemist whose theory in 1858 of molecular structure revolutionized organic chemistry. He proposed two resonant forms of the †benzene ring.

Kelvin, William Thomson (1824–1907) 1st Baron Kelvin, Irish physicist who introduced the **kelvin scale**, the absolute scale of temperature. His work on the conservation of energy in 1851 led to the second law of †thermodynamics. He was knighted in 1866, and made a baron in 1892.

Kelvin, William Thomson http://www-groups.dcs.st-andrews.ac.uk/~history/Mathematicians/Thomson.html

Short biography of Irish physicist William Thomson (Lord Kelvin). The article, which quotes from contemporary sources and links to other important figures, is accompanied by quotes, references, and a list of accolades.

Kelvin scale temperature scale used by scientists. It begins at †absolute zero (–273.15°C) and increases in kelvins, the same degree intervals as the Celsius scale; that is, 0°C is the same as 273.15 K and 100°C is 373.15 K. It is named after the Irish physicist William Thomson, 1st Baron †Kelvin.

Kemeny, John George (1926–1992) Hungarian-born US computer scientist and co-developer of the †BASIC computer language. Kemeny was born in Budapest, Hungary, and emigrated to the USA in 1940 to escape the Nazis. While teaching at Dartmouth College, New Hampshire, Kemeny and Thomas †Kurtz developed the Dartmouth Time-Sharing System and, in 1964, BASIC (Beginners All-purpose Symbolic Instruction Code) to provide computer access to a wide range of undergraduate students. BASIC was based on †Fortran, the language developed by John Backus at IBM.

Kendall, Henry Way (1926–1999) US physicist whose research into the collision of high-energy electrons with protons and neutrons was important in developing the †quark model of particle physics. He was awarded the Nobel Prize for Physics in 1990 for his experiments demonstrating that protons and neutrons are made up of quarks. He shared the award with Jerome I †Friedman and Richard E †Taylor.

Kendrew, John Cowdery (1917–1997) English biochemist who shared the Nobel Prize for Chemistry in 1962 with his colleague Max †Perutz for determining the structure of the muscle protein †myoglobin. He was knighted in 1974.

Kennedy Space Center **KSC**, NASA launch site on Merritt Island, near Cape Canaveral, Florida, used for *Apollo* project and space shuttle launches. The first *Apollo* flight to land on the Moon (1969) and *Skylab*, the first orbiting laboratory (1973), were launched from the site.

The center, named after US president John F Kennedy, is dominated by the Vehicle Assembly Building, 160 m/525 ft tall, used for assembling Saturn rockets and space shuttles. The center is in charge of the checkout, launch, and landing of shuttles.

Kennelly-Heaviside layer former term for the †E-layer of the ionosphere.

Kepler, Johannes (1571–1630) German mathematician and astronomer. He formulated what are now called †Kepler's laws of planetary motion. Kepler's laws are the basis of our understanding of the Solar System, and such scientists as Isaac †Newton built on his ideas.

Kepler, Johannes

http://es.rice.edu/ES/humsoc/Galileo/People/kepler.html

Profile of the life and work of the German astronomer. Kepler's life is traced through his early conversion to Copernican theory, his work with Tycho Brahe, and the support he gave to Galileo. The traumas of his personal life, the accusations of witchcraft leveled against his mother, and the turmoil caused to his career by the Thirty Years' War, are all set out here. This account has pictures of Kepler and a high-resolution image of the model he used to demonstrate the relative distances of the planets from the Sun.

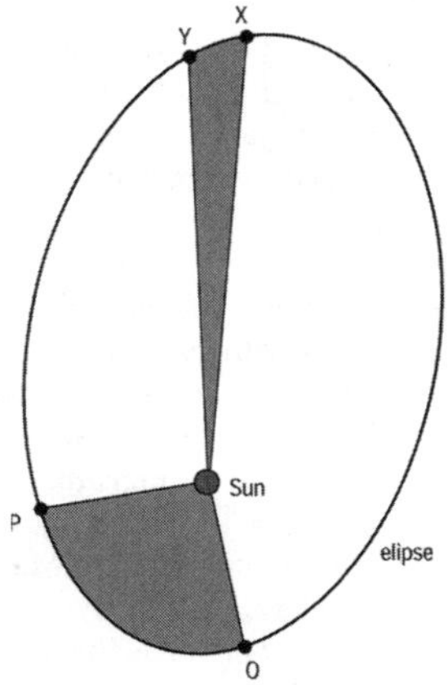

Kepler Kepler's second law states that the two shaded areas are equal if the planet moves from P to O in the same time that it moves from X to Y. The law says, in effect, that a planet moves fastest when it is closest to the Sun. Helicon

Kepler's laws in astronomy, three laws of planetary motion formulated in 1609 and 1619 by German mathematician and astronomer Johannes †Kepler: (1) the orbit of each planet is an ellipse with the Sun at one of the foci; (2) the radius vector of each planet sweeps out equal areas in equal times; (3) the squares of the periods of the planets are proportional to the cubes of their mean distances from the Sun.

Kepler derived the laws after exhaustive analysis of numerous observations of the planets, especially Mars, made by Danish astronomer Tycho †Brahe without telescopic aid. British physicist and mathematician Isaac †Newton later showed that Kepler's laws were a consequence of the theory of universal gravitation.

keratin fibrous protein found in the †skin of vertebrates and also in hair, nails, claws, hooves, feathers, and the outer coating of horns.

If pressure is put on some parts of the skin, more keratin is produced, forming thick calluses that protect the layers of skin beneath.

kernel the inner, softer part of a †nut, or of a seed within a hard shell.

kerosene thin oil obtained from the †fractional distillation of crude oil (unrefined †petroleum; a highly refined form is used in jet aircraft fuel. Kerosene is a mixture of hydrocarbons of the †alkane series, consisting mainly of hydrocarbons with 11 or 12 carbon atoms. Boiling points range from 160°C/320°F to 250°C/480°F. Crude oil contains approximately 10–15% kerosene.

Kerr, John (1824–1907) Scottish physicist who discovered the **Kerr effect**, which produces double refraction in certain media on the application of an electric field.

Kerr cell device making use of the †Kerr effect. The cell consists of a transparent container of a special liquid in which there are two electrodes, placed between two polarizing materials in the container. Only if the planes of polarization of all the various layers in the cell are aligned will light pass, and so the Kerr cell can be used as a shutter device.

Kerr effect elliptical polarization of light as a result of the beam of light being reflected from a pole of an electromagnet. A similar effect also exists for liquids, if a †potential difference is applied to the liquid itself, the angle of polarization depending on the size of the potential difference.

ketene in chemistry, member of a class of organic compounds that contain the >C = C = O group, named after the simplest member of the class, ketene, CH_2CO.

ketone member of the group of organic compounds containing the carbonyl group (C—O) bonded to two atoms of carbon (instead of one carbon and one hydrogen as in †aldehydes). Ketones are liquids or low-melting-point solids, slightly soluble in water. An example is propanone (acetone, CH_3COCH_3), used as a solvent.

ketose in chemistry, †monosaccharide (simple sugar) that contains a keto group, >C—O, in some part of its molecular structure.

kettle hole or **kettle**, pit or depression formed when a block of ice from a receding glacier becomes isolated and buried in glacial debris (till). As the block melts the till collapses to form a hollow, which may become filled with water to form a kettle lake or pond. Kettle holes range from 5 m/15 ft to 13 km/8 mi in diameter, and may exceed 33 m/100 ft in depth.

As time passes, water sometimes fills the kettle holes to form lakes or swamps, features found throughout much of northern North America. Lake Ronkonkoma, the largest lake on Long Island, New York, is an example.

Kettlewell, Henry Bernard David (1907–1979) English geneticist and lepidopterist who carried out important research into the influence of industrial melanism on natural selection in moths, showing why moths are darker in polluted areas.

key small, flat, low-lying island, normally composed of sand, formed on a coral-reef platform. Keys usually reach just a couple of meters above high-tide level. Initially formed from reef debris, they become stabilized by vegetation growth, but always remain susceptible to erosion. Keys are most commonly found in the Gulf of Mexico, particularly off the south coast of Florida, and in the Caribbean (for example, the Abacos islands).

key in biology, means of identifying an unfamiliar organism from a selection of specimens. The easiest type of key to use is a **dichotomous key**. This provides a choice between two statements referring to the specimen in question. Selection of the statement that best describes it leads to further paired statements, each giving a choice that narrows down choice and eventually leads to the identification of the specimen.

kg symbol for †**kilogram**.

khamsin hot southeasterly wind that blows from the Sahara desert over Egypt and parts of the Middle East from late March to May or June. It is called *sharav* in Israel.

Khorana, Har Gobind (1922–) Indian-born US biochemist who was awarded the Nobel Prize for Physiology or Medicine in 1968 for his part in the interpretation of genetic code and its function in protein synthesis. In 1976 he led the team that first synthesized a biologically active †gene. His work provides much of the basis for gene therapy and biotechnology.

Khwarizmi, al-, Muhammad ibn-Musa (*c.* 780–*c.* 850) Persian mathematician. He wrote a book on algebra, from part of whose title (*al-jabr*) comes the word "algebra," and a book in which he introduced to the West the Hindu–Arabic decimal number system. The word "algorithm" is a corruption of his name.

kidney in vertebrates, one of a pair of organs responsible for fluid regulation, excretion of waste products, and maintaining the ionic composition of the blood – in other words the regulation of the concentrations of some chemicals in the blood. The kidneys are situated on the rear wall of the abdomen. Each one consists of a number of long tubules (see †nephron) – the outer parts filter the aqueous components of blood, and the inner parts selectively reabsorb vital salts, leaving waste products in the remaining fluid (urine), which is passed through the ureter to the bladder.

The kidneys tasks can be broken down into a number of stages: filtering of the blood; re-absorbance of all sugar; re-absorbance of the dissolved ions needed by the body; re-absorbance of as much water as the body needs; and release of urea, excess ions, and excess water as urine.

The action of the kidneys is vital and so kidney failure is serious. However, if one kidney fails, the other enlarges to take over its function. A patient with two defective kidneys may still continue near-normal life with the aid of dialysis using a kidney machine or continuous ambulatory peritoneal dialysis (CAPD), or by a kidney transplant. Other diseases of the kidney can include the formation of kidney stones. These hard stones can build up as a result of high levels of blood calcium or high levels of uric acid, and can cause intense pain as they travel down the ureter, as well as causing bleeding in the tissues of the urinary tract.

Kilby, Jack St Clair (1923–) US electrical engineer, co-inventor of the †integrated circuit, popularly known as the microchip, for which he was awarded the 2000 Nobel Prize for Physics (shared with Zhores †Alferov and Herbert †Kroemer). The integrated circuit was developed independently by Kilby at Texas Instruments, Dallas, Texas, in 1958 and by Robert Noyce at Fairchild Semiconductor in California in 1959.

kilo- prefix denoting multiplication by 1,000, as in kilohertz, a unit of frequency equal to 1,000 hertz.

kilogram symbol kg, SI unit of mass equal to 1,000 grams (2.24 lb). It is defined as a mass equal to that of the international prototype, a platinum-iridium cylinder held at the International Bureau of Weights and Measures in Sèvres, France.

kilometer symbol km, unit of length equal to 1,000 meters, equivalent to 3,280.89 ft or 0.6214 (about $^5/_8$) of a mile.

kilowatt symbol kW, unit of power equal to 1,000 watts or about 1.34 horsepower. If an electrical appliance has a power rating of 1 kW, it will change 1000 J of electrical energy into other forms of energy every second; for example, a 1 kW electric heater changes 1000 J of electrical energy into 1000 J of heat energy every second it is turned on.

kilowatt-hour symbol kWh, commercial unit of electrical energy, defined as the work done by a power of 1,000 watts in one hour and equal to 3.6 megajoules. It is used to calculate the cost of electrical energy taken from the domestic supply.

kimberlite igneous rock that is ultramafic (containing very little silica); a type of alkaline †peridotite with a porphyritic texture (larger crystals in a fine-grained matrix), containing mica in addition to olivine and other minerals. Kimberlite represents the world's principal source of diamonds.

Kimberlite is found in carrot-shaped pipelike †intrusions called **diatremes**, where mobile material from very deep in the Earth's crust has forced itself upwards, expanding in its ascent. The material, brought upwards from near the boundary between crust and mantle, often altered and fragmented, includes diamonds. Diatremes are found principally near Kimberley, South Africa, from which the name of the rock is derived, and in the Yakut area of Siberia, Russia.

Kimura, Motoo (1924– 1994) Japanese biologist who, as a result of his work on population genetics and molecular evolution, has developed a theory of neutral evolution that opposes the conventional neo-Darwinistic theory of evolution by natural selection.

kinematic relativity theory proposed by English astrophysicist Edward Milne as a viable alternative to Einstein's general theory of †relativity, and based generally on kinematics (the science of pure motion, without reference to matter or force), from which Milne successfully derived new systems of dynamics and electrodynamics.

kinematics branch of †mechanics that relates †accelerations to the velocities and changes in distance they produce, without considering the forces that generate the accelerations involved. See also †dynamics.

kinesis (plural **kineses**) in biology, a nondirectional movement in response to a stimulus; for example, woodlice move faster in drier surroundings. **Taxis** is a similar pattern of behavior, but there the response is directional.

kinetic energy the energy of a body resulting from motion.

A moving body has kinetic energy. This energy is equal to the work that would have to be done in bringing the body to rest, and is dependent on both the body's mass and speed. The kinetic energy in joules of a mass m kilograms traveling with speed v meters per second is given by the formula:

$KE = 1/2mv^2$

If a moving object collides with another object, then work is done. For example, if a moving car collides with a stationary car, it will cause the stationary car to move. The force from the moving object is used to move the stationary object by a certain distance.

kinetics branch of chemistry that investigates the rates of chemical reactions.

Chemical Kinetics: Terms and Concepts http://www.science.uwaterloo.ca/~cchieh/cact/c123/chmkntcs.html

Introduction to chemical kinetics. This well-organized page defines the subject clearly, and considers topics such as reaction mechanisms and the factors that influence reaction rates.

kinetics branch of †dynamics dealing with the action of forces producing or changing the motion of a body; **kinematics** deals with motion without reference to force or mass.

kinetic theory theory describing the physical properties of matter in terms of the behavior – principally movement – of its component atoms or molecules. It states that all matter is made up of very small particles that are in constant motion, and can be used to explain the properties of solids, liquids, and gases, as well as changes of state. In a solid, the particles are arranged close together in a regular pattern and vibrate on the spot. In a liquid, the particles are still close together but in an irregular arrangement, and the particles are moving a little faster and are able to slide past one another. In a gas, the particles are far apart and moving rapidly, bouncing off the walls of their container. The temperature of a substance is dependent on the velocity of movement of its constituent particles, increased temperature being accompanied by increased movement.

A gas consists of rapidly moving atoms or molecules and, according to kinetic theory, it is their continual impact on the walls of the containing vessel that accounts for the pressure of the gas. The slowing of molecular motion as temperature falls, according to kinetic theory, accounts for the physical properties of liquids and solids, culminating in the concept of no molecular motion at †absolute zero (0 K/–273.15°C).

kingdom primary division in biological †classification. At one time, only two kingdoms were recognized: animals and plants. Today most biologists prefer a five-kingdom system, even though it still involves grouping together organisms that are probably unrelated. One widely accepted scheme is as follows: **Kingdom Animalia** (all multicellular animals); **Kingdom Plantae** (all plants, including seaweeds and other algae, except blue-green); **Kingdom Fungi** (all fungi, including the unicellular yeasts, but not slime moulds); **Kingdom Protista** or **Protoctista** (protozoa, diatoms, dinoflagellates, slime moulds, and various other lower organisms with eukaryotic cells); and **Kingdom Monera** (all prokaryotes – the bacteria and cyanobacteria, or †blue-green algae). The first four of these kingdoms make up the eukaryotes. When only two kingdoms were recognized, any organism with a rigid cell wall was a plant, and so bacteria and fungi were considered plants, despite their many differences. Other organisms, such as the photosynthetic flagellates (euglenoids), were claimed by both kingdoms. The unsatisfactory nature of the two-kingdom system became evident during the 19th century, and the biologist Ernst †Haeckel was among the first to try to reform it. High-power microscopes have revealed more about the structure of cells; it has become clear that there is a fundamental difference between cells without a nucleus (†prokaryotes) and those with a nucleus (†eukaryotes). However, these differences are larger than those between animals and higher plants, and are unsuitable for use as kingdoms. At present there is no agreement on how many kingdoms there are in the natural world.

Although the five-kingdom system is widely favored, some schemes have as many as 20.

kin selection in biology, the idea that †altruism shown to genetic relatives can be worthwhile, because those relatives share some genes with the individual that is behaving altruistically, and may continue to reproduce. See †inclusive fitness.

Alarm-calling in response to predators is an example of a behavior that may have evolved through kin selection: relatives that are warned of danger can escape and continue to breed, even if the alarm caller is caught.

Kipping, Frederick Stanley (1863–1949) English chemist who pioneered the study of the organic compounds of silicon; he invented the term "silicone," which is now applied to the entire class of oxygen-containing polymers.

Kirchhoff, Gustav Robert (1824–1887) German physicist who with R W von †Bunsen developed spectroscopic analysis in the 1850s and showed that all elements, heated to incandescence, have their individual spectra.

Kirchhoff's laws two laws governing electric circuits devised by the German physicist Gustav Kirchhoff. **Kirchhoff's first law** states that the total current entering any junction in a circuit is the same as the total current leaving it. This is an expression of the conservation of electric charge. **Kirchhoff's second law** states that the sum of the potential drops across each resistance in any closed loop in a circuit is equal to the total electromotive force acting in that loop. The laws are equally applicable to DC and AC circuits.

Kirkwood, Daniel (1814–1895) US astronomer. He identified and explained the **Kirkwood gaps**, asteroid-free zones in the Solar System. He used the same theory to explain the nonuniform distribution of particles in the ring system of Saturn.

Kirkwood gaps in astronomy, regions of the asteroid belt, between the planets Mars and Jupiter, where there are relatively few asteroids.

The orbital periods of particles in the gaps correspond to simple fractions, especially $^1/_3$, $^2/_5$, $^3/_7$, and $^1/_2$, of the orbital period of Jupiter, indicating that they are caused by the gravitational influence of the larger planet. The gaps are named after Daniel †Kirkwood, the 19th-century US astronomer who first drew attention to them.

Kitasato, Shibasaburo (1852–1931) Japanese bacteriologist who discovered the plague bacillus while investigating an outbreak of plague in Hong Kong. He was the first to grow the tetanus bacillus in pure culture. He and German bacteriologist Emil von Behring discovered that increasing nonlethal doses of tetanus toxin give immunity to the disease.

kite quadrilateral with two pairs of adjacent equal sides. The geometry of this figure follows from the fact that it has one axis of symmetry.

Klaproth, Martin Heinrich (1743–1817) German chemist who first identified the elements uranium and zirconium, in 1789, and was the second person to isolate titanium, chromium, and cerium. He was a pioneer of analytical chemistry.

Klebs, Edwin (1834–1913) German pathologist and bacteriologist, who helped pioneer the bacterial theory of infection. He was first to see the typhoid bacillus (in 1881) and the diphtheria bacillus (in 1883). He held numerous professorships in pathology, both in Europe and the USA.

Klein, (Christian) Felix (1849–1925) German mathematician and mathematical physicist who unified the various Euclidean and non-Euclidean geometries. He demonstrated that every individual geometry could be constructed purely projectively; such projective models are now called Klein models.

Klein, Felix Christian http://www-groups.dcs.st-andrews.ac.uk/history/Mathematicians/Klein.html

Devoted to the life and contributions of Felix Klein. In addition to biographical information, you will find a list of references about Klein and links to essays that reference him. The text of this essay includes hypertext links to the essays of those mathematicians and thinkers who influenced Klein. You will also find an image of him, which you can click on to enlarge, and a map of his birthplace.

Klein, Melanie (1882–1960) born Melanie Reizes, Austrian child psychoanalyst. She pioneered child psychoanalysis and play studies, and was influenced by Sigmund †Freud's theories. She published *The Psychoanalysis of Children* (1960).

kleptomania (Greek *kleptes* "thief") behavioral disorder characterized by an overpowering desire to possess articles for which one has no need. In kleptomania, as opposed to ordinary theft, there is no obvious need or use for what is stolen and sometimes the sufferer has no memory of the theft.

Klotz, Otto Julius (1852–1923) Canadian astronomer and seismologist. He took up astronomy in 1885 and was appointed director of the Dominion of Canada Observatory in Ottawa in 1917. He wrote numerous papers on terrestrial magnetism and seismology.

Klug, Aaron (1926–) South African molecular biologist who improved the quality of electron micrographs by using laser lighting. He was awarded the Nobel Prize for Chemistry in 1982 for his use of electron microscopy, †X-ray diffraction, and structural modeling to study the structures of different viruses (including the polio virus) that are too small to be visible with a light microscope or to be trapped by filters.

Knopf, Eleanora Frances (1883–1974) born Eleanora Frances Bliss, US geologist who studied metamorphic rocks. She introduced the technique of petrofabrics, a technique for the measurement and interpretation of rock structure, to the USA.

knot in two- and three-dimensional geometry, a †closed curve that loops over or through itself; representations of such structures are commonly presented as congruency problems.

knot in navigation, unit by which a ship's speed is measured, equivalent to one †nautical mile per hour (one knot equals about 1.15 miles per hour). It is also sometimes used in aviation.

knowledge-based system **KBS**, computer program that uses an encoding of human knowledge to help solve problems. It was discovered during research into †artificial intelligence that adding heuristics (rules of thumb) enabled programs to tackle problems that were otherwise difficult to solve by the usual techniques of computer science.

Chess-playing programs have been strengthened by including knowledge of what makes a good position, or of overall strategies, rather than relying solely on the computer's ability to calculate variations.

Koch, (Heinrich Hermann) Robert (1843–1910) German bacteriologist who was awarded a Nobel Prize for Physiology or Medicine in 1905 for his investigations and discoveries in relation to tuberculosis. Koch and his assistants devised the techniques for culturing bacteria outside the body, and formulated the rules for showing whether or not a bacterium is the cause of a disease.

Köhler, Georges Jean Franz (1946–1995) German immunologist who was awarded a Nobel Prize for Physiology or Medicine in 1984 for his work on immunity and the discovery of a technique for producing highly specific †monoclonal antibodies.

Köhler, Wolfgang (1887–1967) Estonian-born German psychologist, cofounder with Max †Wertheimer and Kurt Koffka (1886–1941) of the Gestalt school of psychology. Based on his study of the behavior of apes in a colony on Tenerife, he developed the controversial hypothesis that problem-solving is dependent on a process of insight – a concept central to Gestalt theory – rather than on trial-and-error learning, as was more commonly believed. He published his experiments and observations from this period in *The Mentality of Apes* 1925.

Kohoutek comet discovered by Czech astronomer Lubos Kohoutek at the Hamburg Observatory, Germany, in March 1973. It was expected to become extremely bright towards the end of the year. It passed within 21 million km/13 million mi of the Sun on 28 December 1973 but was much fainter than predicted.

Kolbe, (Adolf Wilhelm) Hermann (1818–1884) German chemist, generally regarded as the founder of modern organic chemistry with his synthesis of acetic acid (ethanoic acid) – an organic compound – from inorganic starting materials. (Previously organic chemistry had been devoted to compounds that occur only in living organisms.)

Kolbe reaction in chemistry, procedure for the preparation of hydrocarbons by †electrolysis. Solutions of the †alkali salts of aliphatic †carboxylic acids are broken down by electrolysis to produce saturated and unsaturated hydrocarbon compounds. It was named after German chemist Hermann †Kolbe.

komatiite oldest volcanic rock with three times as much magnesium as other volcanic rocks. Unlike basaltic lavas, which comprise oceanic crust, komatiites have the chemical composition of †peridotite, the primary constituent rock of the upper †mantle. Komatiites were extruded as a liquid at high temperatures, perhaps more than 1,600°C/2,912°F. They have low titanium and high magnesium, nickel, and chromium content.

Discovered in 1969 by Morris and Richard Viljoen, komatiites are named after the Komati River, Transvaal, South Africa, where they were first identified. The term is also used more loosely to refer to suites of other unusual basaltic rocks with which they are often associated.

Königsberg bridge problem long-standing puzzle that was solved by topology (the geometry of those properties of a figure which remain the same under distortion). In the city of Königsberg (now Kaliningrad in Russia), seven bridges connect the banks of the River Pregol'a and the islands in the river. For many years, people were challenged to cross each of the bridges in a single tour and return to their starting point. In 1736 Swiss mathematician Leonhard Euler converted the puzzle into a topological network, in which the islands and river banks were represented as nodes (junctions), and the connecting bridges as lines. By analyzing this network he was able to show that it is not traversable – that is, it is impossible to cross each of the bridges once only and return to the point at which one started.

Kornberg, Hans Leo (1928–) German-born British biochemist who investigated metabolic pathways and their regulation, especially in micro-organisms. He studied the way in which the cellular economy is balanced to prevent overproduction or waste, and introduced the concept of anaplerotic reactions, whereby metabolic processes are maintained by special enzymes that replenish materials siphoned off for anabolic purposes. He was knighted in 1978.

Korolev, Sergei Pavlovich (1906–1966) Russian designer of the first Soviet intercontinental missile, used to launch the first Sputnik satellite in 1957, and of the *Vostok* spacecraft in which Yuri Gagarin made the world's first space flight in 1961.

Korolev, Sergei P http://www.hq.nasa.gov/office/pao/History/sputnik/korolev.html

Biography of the Soviet rocket designer from NASA, the institution he spent his career attempting to outdo. There is a photograph of Korolev, an account of his achievements, and details of the imprisonment he suffered at the hands of Stalin.

Kossel, Albrecht (1853–1927) German biochemist who was awarded a Nobel Prize for Physiology or Medicine in 1910 for his study of cell proteins and nucleic acids.

Kovalevskaia, Sofya Vasilevna (1850–1891) born Sofya Vasilevna Krukovskya, Russian mathematician and novelist who worked on partial differential equations and Abelian integrals. In 1886 she won the Prix Bordin of the French Academy of Sciences for a paper on the rotation of a rigid body about a point, a problem the 18th-century mathematicians †Euler and †Lagrange had both failed to solve.

Kovalevskaya, Sofia Vasilyevna http://www-history.mcs.st-and.ac.uk/history/Mathematicians/Kovalevskaya.html

Extensive biography of the Russian mathematician. The site contains a description of her contribution to mathematics, and in particular to the theory of differential equations. Several literature references for further reading are also listed, and the Web site also features a photograph of Kovalevskaya.

Krafft-Ebing, Richard, Baron von (1840–1902) German pioneer psychiatrist and neurologist. He published *Psychopathia Sexualis* (1886).

Krebs, Hans Adolf (1900–1981) German-born British biochemist who was awarded a Nobel Prize for Physiology or Medicine in 1953 for his discovery of the citric acid cycle, also known as the †**Krebs cycle**, the final pathway by which food molecules are converted into energy in living tissues. He was knighted in 1958.

Krebs cycle or **citric acid cycle** or **tricarboxylic acid cycle**, final part of the chain of biochemical reactions by which organisms break down food using oxygen to release energy (respiration). It takes place within structures called †mitochondria in the body's cells, and breaks down food molecules in a series of small steps, producing energy-rich molecules of †ATP.

Kremer, Gerhard Flemish mapmaker and mathematician, see †Mercator, Gerardus.

Kroemer, Herbert (1928–) German physicist who with Russian physicist Zhores I †Alferov and US electrical engineer Jack S †Kilby shared the Nobel Prize for Physics in 2000 for their contributions to the field of information and communication technology. Kroemer played a leading role in the development of †semiconductor heterostructures used in high-speed electronics applications and †optoelectronics (the branch of electronics concerned with the development of devices that respond to both electrons and photons).

Krogh, Schack August Steenberg (1874–1949) Danish physiologist who was awarded a Nobel Prize for

Physiology or Medicine in 1920 for his discovery of the mechanism regulating the dilation and constriction of blood capillaries.

Kronecker, Leopold (1823–1891) German mathematician who devised the **Kronecker delta** in linear algebra. He attempted to unify analysis, algebra, and elliptical functions.

Kronecker, Leopold http://www-groups.dcs.st-andrews.ac.uk/history/Mathematicians/Kronecker.html

Part of an archive containing the biographies of the world's greatest mathematicians, this site is devoted to the life and contributions of Leopold Kronecker. In addition to biographical information, you will find a list of references about Kronecker and links to other essays in the archive that reference him. The text of this essay includes hypertext links to the essays of those mathematicians and thinkers who influenced Kronecker. You will also find an image of him, which you can click on to enlarge, and a map of his birthplace.

krypton chemical symbol Kr, (Greek *kryptos* "hidden") colorless, odorless, gaseous, non-metallic element, atomic number 36, relative atomic mass 83.80. It is grouped with the †noble gases (rare gases) in Group 0 of the †periodic table, and was long believed not to enter into reactions, but it is now known to combine with fluorine under certain conditions; it remains inert to all other reagents. As with the other noble gases, krypton's lack of reactivity is due to its full outer shell of electrons. It is present in very small quantities in the air (about 114 parts per million). It is used chiefly in fluorescent lamps, lasers, and gas-filled electronic valves.

Krypton was discovered in 1898 in the residue from liquid air by British chemists William Ramsay and Morris Travers; the name refers to their difficulty in isolating it.

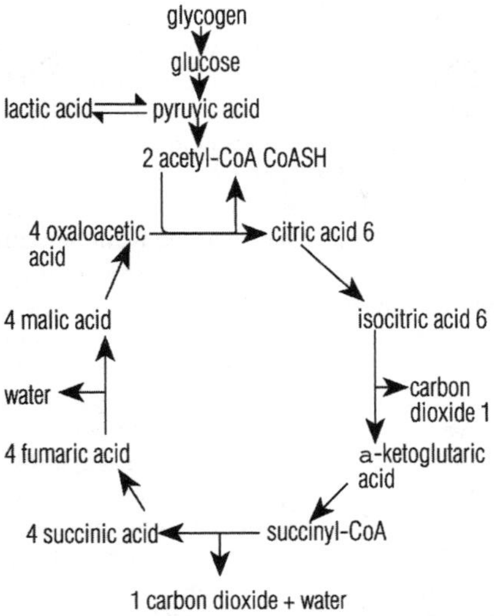

Krebs cycle The purpose of the Krebs (or tricarboxylic acid) cycle is to complete the biochemical breakdown of food to produce energy-rich molecules, which the organism can use to fuel work. Acetyl coenzyme A (acetyl CoA) – produced by the breakdown of sugars, fatty acids, and some amino acids – reacts with oxaloacetic acid to produce citric acid, which is then converted in a series of enzyme-catalyzed steps back to oxaloacetic acid. In the process, molecules of carbon dioxide and water are given off, and the precursors of the energy-rich molecules ATP are formed. (The numbers in the diagram indicate the number of carbon atoms in the principal compounds.) Helicon

K-T boundary geologists' shorthand for the boundary between the rocks of the †Cretaceous and the †Tertiary periods 65 million years ago. It coincides with the end of the extinction of the dinosaurs and in many places is marked by a layer of clay or rock enriched in the element iridium. Extinction of the dinosaurs at the K-T boundary and deposition of the iridium layer are thought to be the result of either impact of an asteroid or comet that crashed into the Yucatán Peninsula (forming the **Chicxulub crater**), perhaps combined with a period of intense volcanism on the continent of India.

Kuhn, Richard (1900–1967) Austrian-born German chemist who determined the structures of vitamins A, B_2, and B_6 in the 1930s, having isolated them from cow's milk. He was awarded the Nobel Prize for Chemistry in 1938 for his research into carotenoids and vitamins, but was unable to receive it until after World War II.

Kuhne, Wilhelme (1837–1900) German physiologist who coined the term †enzyme (a substance produced by cells that is capable of speeding up chemical reactions) and was the first to show the reversible effect of light on the retina of the eye.

Kuiper, Gerard Peter (1905–1973) Dutch-born US astronomer who made extensive studies of the Solar System. His discoveries included the atmosphere of the planet Mars and that of Titan, the largest moon of the planet Saturn. In 1951 he proposed the existence of the †Kuiper belt, a ring of small bodies orbiting the Sun beyond the outermost planet.

Kuiper band one of a number of bands in the spectra of Uranus and Neptune at †wavelengths of 7,500 Å (7.5×10^{-7}m), indicating the presence of †methane. They are named after Dutch-born US astronomer Gerard Kuiper.

Kuiper belt ring of small, icy bodies orbiting the Sun beyond the outermost planet. The Kuiper belt, named after Dutch-born US astronomer Gerard †Kuiper, who proposed its existence in 1951, is thought to be the source of comets that orbit the Sun with periods of less than 200 years. The first member of the Kuiper belt was seen in 1992. In 1995 the first comet-sized objects were discovered; previously the only objects found had diameters of at least 100 km/62 mi (comets generally have diameters of less than 10 km/6.2 mi).

Two new objects were discovered in the Kuiper belt in 1996. The first, 1996 TL66, is 500 km/300 mi in diameter and has an irregular orbit that takes it four to six times further from the Sun than Neptune. The second, 1996 RQ20, is slightly smaller, with an orbit that takes it about three times further from the Sun than Neptune. The orbits of both are at an angle of 20° to the plane of the Solar System.

Kummer, Ernst Eduard (1810–1893) German mathematician who introduced "ideal numbers" in the attempt to prove †Fermat's last theorem. His research into systems of rays led to the discovery of the fourth-order surface known as the **Kummer surface**.

Kundt, August Adolph Eduard Eberhard (1839–1894) German physicist who in 1866 invented **Kundt's tube**, a simple device for measuring the velocity of sound in gases and solids. His later work entailed the demonstration of the dispersion of light in liquids, vapors, and metals.

Kuroshio or **Japan Current**, warm ocean †current flowing from Japan to North America.

Kurtz, Thomas (1928–) Co-developer of the †BASIC computer language. Kurtz joined Dartmouth College, New Hampshire, as an instructor in mathematics in 1956, and with John †Kemeny developed the Dartmouth Time-Sharing System and, in 1964, BASIC (Beginners All-purpose Symbolic Instruction Code) to provide computer access to a wide range of undergraduate students. Kurtz said that if Fortran was the "lingua franca" of computing, BASIC was the "lingua playpen."

Kusch, Polykarp (1911–1933) German-born US physicist who was awarded the Nobel Prize for Physics in 1955 for determining the magnetic moment (or magnetic strength) of the electron.

kyanite aluminum silicate, Al_2SiO_5, a pale-blue mineral occurring as blade-shaped crystals. It is an indicator of high-pressure conditions in metamorphic rocks formed from clay sediments. Andalusite, kyanite, and sillimanite are all polymorphs (see †polymorphism).

kymograph recording produced by an instrument that detects variations such as small muscular contractions or slight changes in arterial blood pressure.

L

labeled compound or **tagged compound**, chemical compound in which a radioactive isotope is substituted for a stable one. The path taken by such a compound through a system can be followed, for example by measuring the radiation emitted.

This powerful and sensitive technique is used in medicine, chemistry, biochemistry, and industry.

labellum lower petal of an orchid flower; it is a different shape from the two lateral petals and gives the orchid its characteristic appearance. The labellum is more elaborate and usually larger than the other petals. It often has distinctive patterning to encourage †pollination by insects; sometimes it is extended backwards to form a hollow spur containing nectar.

laboratory room or building equipped for scientific experiments, research, and teaching.

In Europe, private research laboratories have been used since the 16th century, and by the 18th century laboratories were also used for teaching. The German chemist Justus van Liebig established a laboratory 1826 in Giessen, Germany, where he not only instructed his students in the art of experimental chemistry, but also undertook chemical research with them.

Lacaille, Nicolas Louis de (1713–1762) French astronomer. He determined the positions of nearly 10,000 stars of the southern hemisphere 1750–54. He also performed a number of geodetic investigations; in particular, he made the first measurement of the arc of meridian in the southern hemisphere.

laccolith intruded mass of igneous rock that forces apart two strata and forms a round lens-shaped mass many times wider than thick. The overlying layers are often pushed upward to form a dome. A classic development of laccoliths is illustrated in the Henry, La Sal, and Abajo mountains of southeastern Utah, found on the Colorado Plateau.

Lacerta inconspicuous constellation of the northern hemisphere, represented as a lizard.

Lack, David (1910–1973) English ornithologist who used radar to identify groups of migrating birds. He studied and wrote about the robin, the great tit, the swift, and the finches of the Galapagos Islands.

lactation secretion of milk in mammals, from the mammary glands. In late pregnancy, the cells lining the lobules inside the mammary glands begin extracting substances from the blood to produce milk. The supply of milk starts shortly after birth with the production of colostrum, a clear fluid consisting largely of water, protein, antibodies, and vitamins. The production of milk continues practically as long as the baby continues to suckle.

lacteal small vessel responsible for absorbing fat in the small intestine. Occurring in the fingerlike villi of the †ileum, lacteals have a milky appearance and drain into the lymphatic system.

Before fat can pass into the lacteal, bile from the liver causes its emulsification into droplets small enough for attack by the enzyme lipase. The products of this digestion form into even smaller droplets, which diffuse into the villi. Large droplets re-form before entering the lacteal and this causes the milky appearance.

lactic acid or **2-hydroxypropanoic acid**; $CH_3CHOHCOOH$, organic acid, a colorless, almost odorless liquid, produced by certain bacteria during fermentation and by active muscle cells when they are exercised hard and are experiencing †oxygen debt. An accumulation of lactic acid in the muscles may cause cramp. It occurs in yogurt, buttermilk, sour cream, poor wine, and certain plant extracts, and is used in food preservation and in the preparation of pharmaceuticals.

lactose white sugar, found in solution in milk; it forms 5% of cow's milk. It is commercially prepared from the whey obtained in cheese-making. Like table sugar (sucrose), it is a disaccharide, consisting of two basic sugar units (monosaccharides), in this case, glucose and galactose. Unlike sucrose, it is tasteless.

Laënnec, René Théophile Hyacinthe (1781–1826) French physician, inventor of the stethoscope in 1816. He advanced the diagnostic technique of auscultation (listening to the internal organs) with his book *Traité de l'auscultation médiaté* in 1819, which quickly became a medical classic.

laevulose or **L-fructose** or **fruit sugar**; $CH_2OH(CHOH)_3COCH_2OH$, ketose (simple sugar) contained in most sweet fruits, honey, and starches, together with dextrose.

Laevulose is purified by boiling inulin, a †starch contained in dahlia tubers, with water; or by the breakdown of cane sugar into glucose and laevulose; or by reacting †sucrose with sulfuric acid. It is less easily fermented than glucose, and is a powerful reducing agent. It is sweeter than cane sugar, is more easily absorbed, and can be used instead of cane sugar by people with diabetes.

lagging material used for heat †insulation; it may be wrapped around hot-water tanks to reduce heat loss, or around water pipes to prevent freezing in winter.

lagoon coastal body of shallow salt water, usually with limited access to the sea. The term is normally used to describe the shallow sea area cut off by a †coral reef or barrier islands.

Lagoon bright nebula in the constellation of †Sagittarius. It is about 4,000 light years from the Sun and about 30 light years across. It is an †HII region (a region of extremely hot ionized hydrogen) on the edge of a molecular cloud in which stars are being formed, and contains the open cluster NGC 6530. It is also known as Messier 8 (M8).

Lagrange, Joseph Louis (1736–1813) born Giuseppe Lodovico Lagrange, Italian-born French mathematician. His *Mécanique analytique* (1788) applied mathematical analysis, using principles established by Isaac †Newton, to such problems as the movements of planets when affected by each other's gravitational force. He presided over the commission that introduced the metric system in 1793.

Lagrange, Joseph Louis http://www-groups.dcs.st-andrews.ac.uk/history/Mathematicians/Lagrange.html

Part of an archive containing the biographies of the world's greatest mathematicians, this site is devoted to the life and contributions of Joseph Louis Lagrange.

Lagrangian points five locations in space where the centrifugal and gravitational forces of two bodies neutralize each other; a third, less massive body located at any one of these points will be held in equilibrium with respect to the other two. Three of the points, L1–L3, lie on a line joining the two large bodies. The other two points, L4 and L5, which are the most stable, lie on either side of this line. Their existence was predicted in 1772 by French mathematician Joseph Louis Lagrange.

The **Trojan asteroids** lie at Lagrangian points L4 and L5 in Jupiter's orbit around the Sun. Clouds of dust and debris may lie at the Lagrangian points of the Moon's orbit around the Earth.

lahar mudflow formed of a fluid mixture of water and volcanic ash. During a volcanic eruption, melting ice may combine with ash to form a powerful flow capable of causing great destruction. The lahars created by the eruption of Nevado del Ruiz in Colombia, South America, in 1985 buried 22,000 people in 8 m/26 ft of mud.

Laing, R(onald) D(avid) (1927–1989) Scottish psychoanalyst. He was the originator of the social theory of mental illness; for example, that schizophrenia is promoted by family pressure for its members to conform to standards alien to themselves. His books include *The Divided Self* (1960) and *The Politics of the Family* (1971).

lake body of still water lying in depressed ground without direct communication with the sea. Lakes are common in formerly glaciated regions, along the courses of slow rivers, and in low land near the sea. The main classifications are by origin: **glacial lakes**, formed by glacial scouring; **barrier lakes**, formed by †landslides and glacial †moraines; **crater lakes**, found in †volcanoes; and **tectonic lakes**, occurring in natural fissures.

Crater lakes form in the †calderas of extinct volcanoes, for example Crater Lake, Oregon, USA. Subsidence of the roofs of limestone caves in †karst landscape exposes the subterranean stream network and provides a cavity in which a lake can develop. Tectonic lakes form during tectonic movement, as when a †rift valley is formed. Lake Tanganyika was created in conjunction with the East African Great Rift Valley. Glaciers produce several distinct types of lake, such as the lochs of Scotland and the Great Lakes of North America.

Lakes are mainly freshwater, but salt and bitter lakes are found in areas of low annual rainfall and little surface run-off, so that the rate of evaporation exceeds the rate of inflow, allowing mineral salts to accumulate. The Dead Sea has a salinity of about 250 parts per 1,000 and the Great Salt Lake, Utah, about 220 parts per 1,000. Salinity can also be caused by volcanic gases or fluids, for example Lake Natron, Tanzania.

In the 20th century large artificial lakes have been created in connection with hydroelectric and other works. Some lakes have become polluted as a result of human activity. Sometimes †eutrophication (a state of overnourishment) occurs, when agricultural fertilizers leaching into lakes cause an explosion of aquatic life, which then depletes the lake's oxygen supply until it is no longer able to support life.

Lalande, Joseph Jérome le Français de (1732–1807) French astronomer. He observed the transit of Venus and helped to calculate the size of the Solar System. He compiled a catalogue of 47,000 stars.

Lamarck, Jean Baptiste Pierre Antoine de Monet (1744–1829) Chevalier de Lamarck, French naturalist. His theory of evolution, known as **Lamarckism**, was based on the idea that acquired characteristics (changes acquired in an individual's lifetime) are inherited by the offspring, and that organisms have an intrinsic urge to evolve into better-adapted forms. *Philosophie zoologique/Zoological Philosophy* (1809) outlined his "transformist" (evolutionary) ideas.

Lamarckism theory of evolution, now discredited, advocated during the early 19th century by French naturalist Jean Baptiste Lamarck. Lamarckism is the theory that acquired characteristics, such as the increased body mass of an athlete, were inherited. It differs from the Darwinian theory of evolution.

Lamb, Willis Eugene (1913–) US physicist who revised the quantum theory of Paul †Dirac. The hydrogen atom was thought to exist in either of two distinct states carrying equal energies. More sophisticated measurements by Lamb in 1947 demonstrated that the two energy levels were not equal. He was awarded the Nobel Prize for Physics in 1955 for the discovery of this discrepancy, since known as the **Lamb shift**.

lambert unit of luminance (the light shining from a surface), equal to one †lumen per square centimeter. In scientific work the †candela per square meter is preferred.

Lambert, Johann Heinrich (1728–1777) German physicist, mathematician, and astronomer. Largely self-taught, he traveled widely in Europe before settling in Berlin in 1765 under the patronage of Frederick the Great. He contributed to many branches of science, being the first to suggest that there may be galaxies beyond our own and the first to devise an accurate way of measuring the intensity of light. He also pioneered the sciences of pyrometry and hygrometry. The "lambert," a unit of measurement of illumination equal to 10^4 lx, was named after him.

lamina in flowering plants (†angiosperms), the blade of the †leaf on either side of the midrib. The lamina is generally thin and flattened, and is usually the primary organ of †photosynthesis. It has a network of veins through which water and nutrients are conducted. More generally, a lamina is any thin, flat plant structure, such as the †thallus of many seaweeds.

lamp, electric device designed to convert electrical energy into light energy.

In a **filament lamp** such as a light bulb an electric current causes heating of a long thin coil of fine high-resistance wire enclosed at low pressure inside a glass bulb. In order to give out a good light the wire must glow white-hot and therefore must be made of a metal, such as tungsten, that has a high melting point. The efficiency of filament lamps is low because most of the electrical energy is converted to heat.

A **fluorescent lamp** uses an electrical discharge or spark inside a gas-filled tube to produce light. The inner surface of the tube is coated with a fluorescent material that converts the ultraviolet light generated by the discharge into visible light. Although a high voltage is needed to start the discharge, these lamps are far more efficient than filament lamps at producing light.

Landau, Lev Davidovich (1908–1968) Russian theoretical physicist. He was awarded the Nobel Prize for Physics in 1962 for his theories of condensed matter, especially liquid helium.

Landau, Lev Davidovich http://www-groups.dcs.st-andrews.ac.uk/history/Mathematicians/Landau_Lev.html

Part of an archive containing the biographies of the world's greatest mathematicians, this site is devoted to the life and contributions of theoretical physicist Lev Landau.

land breeze gentle breeze blowing from the land towards the sea and affecting coastal areas. It forms at night in the summer or autumn, and tends to be cool. By contrast, a †sea breeze blows from the sea towards the land.

Landé, Alfred (1888–1975) German-born US quantum physicist. In 1923 he published a formula expressing a factor, now known as the Landé splitting factor, for all multiplicities as a function of the quantum numbers of the stationary state of the atom.

Landé splitting factor calculational device used in †quantum theory that enables the †fine and †hyperfine structure of a †spectral line to be determined by a knowledge of the various †spin and orbital angular momenta of the electrons in the atom.

land reclamation conversion of derelict or otherwise unusable areas into productive land. For example, where industrial or agricultural activities, such as sand and gravel extraction or open-cast mining, have created large areas of derelict or waste ground, the companies involved are usually required to improve the land so that it can be used.

landslide sudden downward movement (see †mass movement) of a mass of soil or rocks from a cliff or steep slope. Landslides happen when a slope becomes unstable, usually because the base has been undercut or because materials within the mass have become wet and slippery.

A **mudflow** happens when soil or loose material is soaked so that it no longer adheres to the slope; it forms a tongue of mud that reaches downhill from a semicircular hollow. A **slump** occurs when the material stays together as a large mass, or several smaller masses, and these may form a tilted steplike structure as they slide. A **landslip** is formed when †beds of rock dipping towards a cliff slide along a lower bed. Earthquakes may precipitate landslides.

Landsteiner, Karl (1868–1943) Austrian-born US immunologist who was awarded a Nobel Prize for Physiology or Medicine in 1930 for his discovery of the ABO †blood group system in the period 1900–02. He also aided in the discovery of the Rhesus blood factors in 1940, and discovered the polio virus.

land use the way in which a given area of land is used. Land is often classified according to its use, for example, for agriculture, industry, residential buildings, and recreation.

Langerhans, Paul (1847–1888) German anatomist who conducted pioneering research into the human pancreas. In his doctoral thesis, written in 1869, he first described the islets in the pancreas that are now named after him.

Langevin, Paul (1872–1946) French physicist who contributed to the studies of magnetism and X-ray emissions, especially paramagnetic (weak attractive) and diamagnetic (weak repulsive) phenomena in gases. During World War I he invented an apparatus for locating enemy submarines, which is the basis of modern echolocation techniques.

Langley, John Newport (1852–1925) English physiologist who investigated the structure and function of the †autonomic nervous system, the involuntary part of the nervous system, that controls the striated and cardiac muscles and the organs of the gastrointestinal, cardiovascular, excretory, and endocrine systems. He went on to divide up the autonomic nervous system into the **sympathetic** and **parasympathetic** branches, with specific functions being apportioned to each.

Langley, Samuel Pierpoint (1834–1906) US astronomer, scientist, and inventor of the †bolometer, an instrument that measures radiation. His steam-driven airplane flew for 90 seconds in 1896 – the first flight by an engine-equipped aircraft.

Langmuir, Irving (1881–1957) US scientist who invented the mercury vapor pump for producing a high vacuum, and the atomic hydrogen welding process; he was also a pioneer of the thermionic valve. He was awarded the Nobel Prize for Chemistry in 1932 for his work on surface chemistry.

Langmuir on the Octet Theory http://dbhs.wvusd.k12.ca.us/Chem-History/Langmuir-1919.html
Transcript of Langmuir's paper in which he outlines the octet theory of electron valency. In the paper Langmuir outlines the way in which electrons arrange themselves in the orbital levels of the atom, and uses their layout to explain many chemical properties of the elements.

Lankester, Edwin Ray (1847–1929) English zoologist who made clear morphological distinctions between the different orders of invertebrates. He distinguished between the hemocoel (blood-containing cavity) in Mollusca and Arthropoda and the coelom (fluid-filled cavity) in worms and vertebrates for the first time, showing that whilst functionally similar they have different origins.

lanthanide any of a series of 15 metallic elements (also known as rare earths) with atomic numbers 57 (lanthanum) to 71 (lutetium). One of its members, promethium, is radioactive. All occur in nature. Lanthanides are grouped because of their chemical similarities (most are trivalent, but some can be divalent or tetravalent), their properties differing only slightly with atomic number.

Lanthanides were called rare earths originally because they were not widespread and were difficult to identify and separate from their ores by their discoverers. The series is set out in a band in the periodic table of the elements, as are the †actinides.

lanthanum chemical symbol La, (Greek *lanthanein* "to be hidden") soft, silvery, ductile and malleable, metallic element, atomic number 57, relative atomic mass 138.91, the first of the lanthanide series. It is used in making alloys. It was named in 1839 by Swedish chemist Carl Mosander (1797–1858).

lapis lazuli rock containing the blue mineral lazurite in a matrix of white calcite with small amounts of other minerals. It occurs in silica-poor igneous rocks and metamorphic limestones found in Afghanistan, Siberia, Iran, and Chile. Lapis lazuli was a valuable pigment of the Middle Ages, also used as a gemstone and in inlaying and ornamental work.

Laplace, Pierre Simon (1749–1827) Marquis de Laplace, French astronomer and mathematician. In 1796 he theorized that the Solar System originated from a cloud of gas (the nebular hypothesis). He studied the motion of the Moon and planets, and published a five-volume survey of †celestial mechanics, *Traité de méchanique céleste* (1799–1825). Among his mathematical achievements was the development of probability theory.

lapse rate in physics, rate of decrease in the value of a quantity with increasing altitude (such as atmospheric pressure).

Large Electron Positron Collider LEP, world's largest particle †accelerator, in operation 1989–2000 at the CERN laboratories near Geneva in Switzerland. It occupies a tunnel 3.8 m/12.5 ft wide and 27 km/16.7 mi long, which is buried 180 m/590 ft underground and forms a ring consisting of eight curved and eight straight sections. In June 1996, LEP resumed operation after a £210 million upgrade. The upgraded machine, known as LEP2, generated collision energy of 161 gigaelectron volts.

Electrons and positrons enter the ring after passing through the Super Proton Synchrotron accelerator. They travel in opposite directions around the ring, guided by 3,328 bending magnets and kept within tight beams by 1,272 focusing magnets. As they pass through the straight sections, the particles are accelerated by a pulse of radio energy. Once sufficient energy is accumulated, the beams are allowed to collide. Four giant detectors are used to study the resulting shower of particles.

ALEPH Experiment http://alephwww.cern.ch/

Home page of one of the four high-energy particle physics experiments, which used the Large Electron Positron Collider at the European Laboratory for Particle Physics. This Web site displays images of the components used in this experiment, including a cutaway schematic diagram of the huge detector assembly. The site also allows the visitor to discover more about the physics the collider and associated instruments are used to observe.

large-number hypothesis theory in †cosmology that tries to understand the basic reason for an apparent coincidence of size in ratios of fundamental quantities in atomic and cosmological theory. If the apparent radius of the universe is divided by the radius of the †electron, the resulting large number (about 10^{43}) is remarkably similar to the ratio of the strengths of the †electrostatic and gravitational force between the †electron and the †proton. The reason for this is not clear at present, but if the relationship is to hold true for all time, it can be shown that on a †Big Bang model of the universe the †gravitational constant *g* must decrease with time, that is, the strength of gravity must decrease.

large numbers, laws of theorems in probability theory that predict that the observed frequencies of events for a large number of repeated trials are more and more likely to approach their theoretical probability as the number of repetitions increases.

Larmor, Joseph (1857–1942) Northern Irish physicist and mathematician, Lucasian professor of mathematics at Cambridge University 1903–32. His most important work was on the interaction of atomically charged matter and the electromagnetic field, though he also wrote extensively on hydrodynamics. He represented Cambridge University in Parliament as a Unionist 1911–22.

larva stage between hatching and adulthood in those species in which the young have a different appearance and way of life from the adults. Examples include tadpoles (frogs) and caterpillars (butterflies and moths). Larvae are typical of the invertebrates, some of which (for example, shrimps) have two or more distinct larval stages. Among vertebrates, it is only the amphibians and some fishes that have a larval stage.

The process whereby the larva changes into another stage, such as a pupa (chrysalis) or adult, is known as †metamorphosis.

larynx in mammals, a cavity at the upper end of the trachea (windpipe) containing the vocal cords. It is stiffened with cartilage and lined with mucous membrane. Amphibians and reptiles have much simpler larynxes, with no vocal cords. Birds have a similar cavity, called the **syrinx**, found lower down the trachea, where it branches to form the bronchi. It is very complex, with well-developed vocal cords.

laser acronym for **light amplification by stimulated emission of radiation**, device for producing a narrow beam of light, capable of traveling over vast distances without dispersion, and of being focused to give enormous power densities (10^8 watts per cm^2 for high-energy lasers). The laser operates on a principle similar to that of the †maser (a high-frequency microwave amplifier or oscillator). The uses of lasers include communications (a laser beam can carry much more information than can radio waves), cutting, drilling, welding, satellite tracking, medical and biological research, and surgery. Sound wave vibrations from the window glass of a room can be picked up by a reflected laser beam. Lasers are also used as entertainment in theatres, concerts, and light shows.

laser material Any substance in which the majority of atoms or molecules can be put into an excited energy state can be used as laser material. Many solid, liquid, and gaseous substances have been used, including synthetic ruby crystal (used for the first extraction of laser light in 1960, and giving a high-power pulsed output) and a helium–neon gas mixture, capable of continuous operation, but at a lower power.

applications Carbon dioxide gas lasers (CO_2 lasers) can produce a beam of 100 watts or more power in the infrared (wavelength 10.6 mm) and this has led to an important commercial application, the cutting of material for suits and dresses in hundreds of thicknesses at a time. Dye lasers, in which complex organic dyes in solution are the lasing material, can be tuned to produce light of any chosen wavelength over a range of a sizeable fraction of the visible spectrum.

latent heat in physics, the heat absorbed or released by a substance as it changes state (for example, from solid to liquid) at constant temperature and pressure.

lateral line system system of sense organs in fishes and larval amphibians (tadpoles) that detects water movement. It usually consists of a row of interconnected pores on either side of the body that divide into a system of canals across the head.

lateral moraine linear ridge of rocky debris deposited near the edge of a †glacier. Much of the debris is material that has fallen from the valley side onto the glacier's edge, having been weathered by †freeze-thaw (the alternate freezing and thawing of ice in cracks); it will, therefore, tend to be angular in nature.

Where two glaciers merge, two lateral moraines may join together to form a **medial moraine** running along the center of the merged glacier.

laterite red residual soil characteristic of tropical rainforests. It is formed by the weathering of basalts, granites, and shales and contains a high percentage of aluminum and iron hydroxides. It may form an impermeable and infertile layer that hinders plant growth.

lateral line system In fishes, the lateral line system detects water movement. Arranged along a line down the length of the body are two water-filled canals, just under the skin. The canals are open to the outside, and water movements cause water to move in the canals. Nerve endings detect the movements. Helicon

latex (Latin "liquid") fluid of some plants (such as the rubber tree and poppy), an emulsion of resins, proteins, and other organic substances. It is used as the basis for making rubber. The name is also applied to a suspension in water of natural or synthetic rubber (or plastic) particles used in rubber goods, paints, and adhesives.

Latin square in mathematics, square array of symbols in which each symbol occurs only once in each row and column. The multiplication table of a group forms such a Latin square but not all Latin squares arise in this way.

latitude and longitude imaginary lines used to locate position on the globe. Lines of latitude are drawn parallel to the Equator, with 0° at the Equator and 90° at the north and south poles. Lines of longitude are drawn at right-angles to these, with 0° (the Prime Meridian) passing through Greenwich, England.

The 0-degree line of latitude is defined by Earth's Equator, a characteristic definable by astronomical observation. It was determined as early as AD 150 by Egyptian astronomer †Ptolemy in his world atlas. The prime meridian, or 0-degree line of longitude, is a matter of convention rather than physics. Prior to the latter half of the 18th century, sailors navigated by referring to their position east or west of any arbitrary meridian. When Nevil Maskelyne (1732–1811), English astronomer and fifth Astronomer Royal, published the *Nautical Almanac* he referred all of his lunar–stellar distance tables to the Greenwich meridian. These tables were relied upon for computing longitudinal position and so the Greenwich meridian became widely accepted.

Chronometers, time-keeping devices with sufficient accuracy for longitude determination, invented by English instrument-maker John Harrison (1693–1776) and perfected in 1759, would gradually replace the lunar distance method for navigation, but reliance on the Greenwich meridian persisted because the *Nautical Almanac* was used by sailors to verify their position. The Greenwich meridian was officially adopted as the Prime Meridian by the International Meridian Conference held in Washington, DC, in 1884.

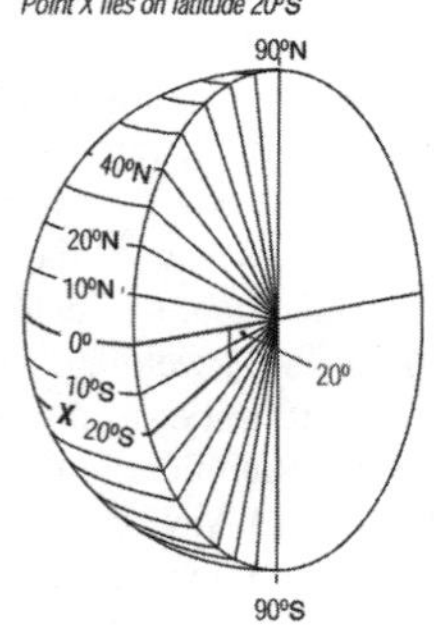

latitude and longitude Locating a point on a globe using latitude and longitude. Longitude is the angle between the terrestrial meridian through a place and the standard meridian 0° passing through Greenwich, England. Latitude is the angular distance of a place from the equator. Helicon

lattice network of straight lines.

lattice points points of intersection of the lines in a lattice.

Laue, Max Theodor Felix von (1879–1960) German physicist who was awarded the Nobel Prize for Physics in 1914 for his pioneering work in measuring the wavelength of X-rays by their diffraction through the closely spaced atoms in a crystal. His work led to the techniques of X-ray spectroscopy, used in nuclear physics, and †X-ray diffraction, used to elucidate the molecular structure of complex biological materials.

Laue pattern in chemistry, symmetrical pattern of spots formed by the diffraction of X-rays passing through a crystalline solid usually recorded on photographic film. Discovered by and named after German physicist Max Theodor Felix von †Laue, it can be used to determine the crystal structure of a substance.

laughing gas popular name for †nitrous oxide, an anesthetic.

Laurasia northern landmass formed 200 million years ago by the splitting of the single world continent †Pangaea. (The southern landmass was †Gondwanaland.) It consisted of what was to become North America, Greenland, Europe, and Asia, and is believed to have broken up about 100 million years ago with the separation of North America from Europe.

lava molten †magma that erupts from a †volcano and cools to form extrusive †igneous rock. Lava types differ in composition, temperature, gas content, and viscosity (resistance to flow).

The three major lava types are basalt (dark, fluid, and relatively low silica content), rhyolite (light, viscous, high silica content), and andesite (an intermediate lava).

Lavoisier, Antoine Laurent (1743–1794) French chemist. He proved that combustion needs only a part of the air, which he called oxygen, thereby destroying the theory of phlogiston (an imaginary "fire element" released during combustion). With astronomer and mathematician Pierre de †Laplace, he showed in 1783 that water is a compound of oxygen and hydrogen. In this way he established the basic rules of chemical combination.

Lavoisier, Antoine Laurent http://www.newadvent.org/cathen/09052a.htm

Account of the life and achievements of the French chemist, philosopher, and economist. It traces his education and growing disillusionment with the theory of phlogiston. There is an account of how in order to pay for his scientific experiments he was forced to enter into government service, a fateful move leading, despite his many successes, to his violent death.

law of universal attraction Isaac Newton's formulation of the law of †gravity.

Lawrence, Ernest O(rlando) (1901–1958) US physicist. He was awarded the Nobel Prize for Physics in 1939 for his invention of the **cyclotron** particle †accelerator which pioneered the production of artificial †radioisotopes, the study of elementary particle interactions, and the synthesis of new transuranic elements.

lawrencium chemical symbol Lr, synthesized, radioactive, metallic element, the last of the actinide series, atomic number 103, relative atomic mass 262. Its only known isotope, Lr-257, has a half-life of 4.3 seconds and was originally synthesized at the University of California at Berkeley in 1961 by bombarding californium with boron nuclei. The original symbol, Lw, was officially changed in 1963.

The element was named after Ernest Lawrence (1901–1958), the US inventor of the cyclotron.

LCD abbreviation for †liquid-crystal display.

LDR abbreviation for †**light-dependent resistor**.

leaching process by which substances are washed through or out of the soil. Fertilizers leached out of the soil drain into rivers, lakes, and ponds and cause †water pollution. In tropical areas, leaching of the soil after the destruction of forests removes scarce nutrients and can lead to a dramatic loss of soil fertility. The leaching of soluble minerals in soils can lead to the formation of distinct soil horizons as different minerals are deposited at successively lower levels.

lead chemical symbol Pb, heavy, soft, malleable, gray, metallic element, atomic number 82, relative atomic mass 207.19. Its chemical symbol comes from the Latin *plumbum*. Usually found as an ore (most often as the sulfide galena), it occasionally occurs as a free metal (†native metal), and is the final stable product of the decay of uranium. Lead is the softest and weakest of the commonly used metals, with a low melting point; it is a poor conductor of electricity and resists acid corrosion. As a cumulative poison, lead enters the body from lead water pipes, lead-based paints, and leaded petrol. (In humans, exposure to lead shortly after birth is associated with impaired mental health between the ages of two and four.) The metal is an effective shield against radiation and is used in batteries, glass, ceramics, and alloys such as pewter and solder.

lead–acid cell type of †accumulator (storage battery).

lead azide $Pb(N_3)_2$, solid used as an initiator for explosives. It can be used to replace †mercury fulminate as a detonator for trinitrotoluene, †TNT.

lead carbonate $PbCO_3$, white solid that decomposes in air around 200°C/392°F to produce carbon dioxide and lead oxide. It is found in nature as the mineral cerussite, and is used as a pigment in paints.

lead chromate or **chrome yellow**; $PbCrO_4$, colored solid that is prepared by precipitation from solutions of lead salts, for example lead nitrate, to which a chromate, for example potassium chromate, has been added. The color of the solid can range from yellow to red depending on the conditions used for precipitation. Lead chromates are important pigments.

leaded petrol petrol that contains †antiknock, a mixture of the chemicals tetraethyl lead and dibromoethane. The lead from the exhaust fumes enters the atmosphere, mostly as simple lead compounds, which are poisonous to the developing nervous systems of children.

lead(II) nitrate $Pb(NO_3)_2$, one of only two common water-soluble compounds of lead (the other is lead ethanoate or acetate). When heated, it decrepitates (see †decrepitation) and decomposes readily into oxygen, brown nitrogen(IV) oxide gas, and the red-yellow solid lead(II) oxide.

$2Pb(NO_3)_2 \rightarrow 2PbO + 4NO_2 + O_2$

lead ore any of several minerals from which lead is extracted. The primary ore is galena or lead sulfite PbS. This is unstable, and on prolonged exposure to the atmosphere it oxidizes into the minerals cerussite $PbCO_3$ and anglesite $PbSO_4$. Lead ores are usually associated with other metals, particularly silver – which can be mined at the same time – and zinc, which can cause problems during smelting.

Most commercial deposits of lead ore are in the form of veins, where hot fluids have leached the ore from cooling †igneous masses and deposited it in cracks in the surrounding country rock, and in thermal †metamorphic zones, where the heat of igneous intrusions has altered the minerals of surrounding rocks. Lead is mined in over 40 countries, but half of the world's output comes from the USA, Canada, Russia, Kazakhstan, Uzbekistan, Canada, and Australia.

lead(II) oxide or **lead monoxide**; PbO, yellow or red solid, an amphoteric oxide (one that reacts with both acids and bases). The other oxides of lead are the brown solid lead(IV) oxide (PbO_2) and red lead (Pb_3O_4).

leaf lateral outgrowth on the stem of a plant, and in most species is the primary organ of †photosynthesis (the process in which the †energy from absorbed sunlight is used to combine †carbon dioxide and †water to form sugars). The chief leaf types are cotyledons (seed leaves), scale leaves (on underground stems), foliage leaves, and bracts (in the axil of which a flower is produced).

Typically, leaves are composed of three parts: the sheath or leaf base, the petiole or stalk, and the lamina or blade. The lamina has a network of veins through which water and nutrients are conducted. A leaf contains several layers of tissue. Structurally the leaf is made up of inner tissues known as †mesophyll cells, surrounded by the epidermis and usually an additional waxy layer, termed the cuticle, which prevents excessive evaporation of water from the leaf tissues by transpiration.

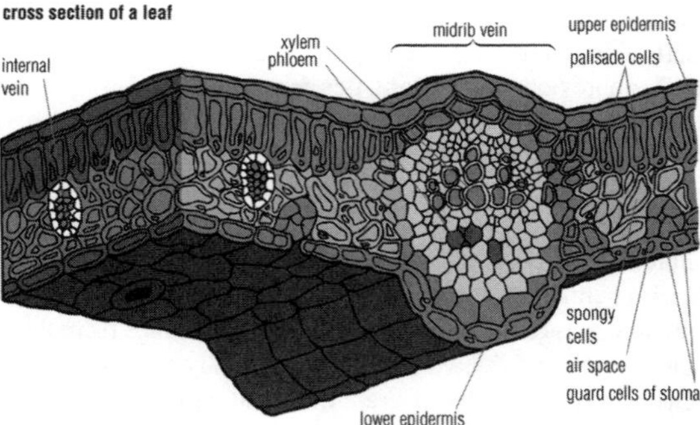

leaf Leaf shapes and arrangements on the stem are many and varied; in cross section, a leaf is a complex arrangement of cells surrounded by the epidermis. This is pierced by the stomata through which gases enter and leave. Helicon

The epidermis is formed of a single layer of cells on the outside of the leaf. It is interrupted by small pores, or stomata, through which gas exchange between the plant and the atmosphere occurs for photosynthesis and respiration. Water is also lost through the stomata by †transpiration. The mesophyll can be divided into two layers. The upper half of the mesophyll has elongated cells densely packed with chloroplasts that carry out photosynthesis – this is the palisade tissue. The lower half has more irregularly arranged cells with fewer chloroplasts – this is called spongy mesophyll. The mesophyll includes many air spaces, which allow carbon dioxide and oxygen to diffuse to and from the leaf cells (see †diffusion). Outside the epidermis is a waxy layer, called the cuticle, which prevents evaporation of water directly from epidermal cells.

Leaf Section http://www.purchon.com/biology/leaf.htm

Part of a larger section on biology, this Web site explains the structure and form of a typical leaf. A detailed diagram is included, together with labels and definitions for the important parts.

leaf litter in ecology, accumulation of leaves and other organic detritus (debris) on a woodland floor following leaf fall. It is a rich source of food for detritovores (animals that feed on detritus) and †decomposers. Their activities eventually lead to the formation of †humus, which improves the texture and structure of soil.

leaf mosaic in botany, arrangement of leaves on a plant such that the ones above do not overshadow those below so that the maximum area of leaf surface is exposed to sunlight.

learning theory in psychology, any theory or body of theories about how behavior in animals and human beings is acquired or modified by experience. Two main theories are classical and operant †conditioning.

least action, principle of in physics, an alternative expression of Newton's laws of motion that states that a particle moving between two points under the influence of a force will follow the path along which its total action is least. Action is a quantity related to the average difference between the kinetic energy and the potential energy of the particle along its path. The principle is only true where no energy is lost from the system; for example an object moving in free fall in a gravitational field. It is closely related to †Fermat's principle of least time which governs the path taken by a ray of light.

least squares, method of method of deriving as exact an average value as possible from a set of approximate or inaccurate values by introducing the errors as unknown variables and requiring the sum of their squares to be minimized. The method was devised by German mathematician Karl Gauss as a precise way of best fitting a straight line through a set of plotted data points that are not collinear.

Leavitt, Henrietta Swan (1868–1921) US astronomer who in 1912 discovered the **period–luminosity law**, which links the brightness of a †Cepheid variable star to its period of variation. This law allows astronomers to use Cepheid variables as "standard candles" for measuring distances in space.

Lebedev, Pyotr Nikolayevich (1866–1912) Russian physicist. He proved by experiment that light exerts a minute pressure upon a physical body, thereby confirming James †Maxwell's theoretical prediction.

Lebesgue, Henri Léon (1875–1941) French mathematician who developed a new theory of integration, now named after him. He also made contributions to set theory, the calculus of variations, and function theory.

Lebesgue, Henri Léon http://www-groups.dcs.st-andrews.ac.uk/history/Mathematicians/Lebesgue.html
Part of an archive containing the biographies of the world's greatest mathematicians, this site is devoted to the life and contributions of Henri Lebesgue.

Leblanc, Nicolas (1742–1806) French chemist who in the 1780s developed a process for making soda ash (sodium carbonate, Na_2CO_3) from common salt (sodium chloride, NaCl). Soda ash was widely used industrially in making glass, paper, soap, and various chemicals.

Le Châtelier, Henri Louis (1850–1936) French physical chemist who formulated the principle now named after him, which states that if any constraint is applied to a system in chemical equilibrium, the system tends to adjust itself to counteract or oppose the constraint.

Le Chatelier's principle or **Le Chatelier-Braun principle**, principle that if a change in conditions is imposed on a system in equilibrium, the system will react to counteract that change and restore the equilibrium.

First stated in 1884 by French chemist Henri le Chatelier (1850–1936), it has been found to apply widely outside the field of chemistry.

Le Chatelier's Principle http://www.chem.uidaho.edu/~honors/lechat.html
Explanation of Le Chatelier's Principle of the reversibility of reactions at equilibrium, both in general terms and then with specific reference to the synthesis of ammonia.

lecithin lipid (fat), containing nitrogen and phosphorus, that forms a vital part of the cell membranes of plant and animal cells. The name is from the Greek *lekithos* "egg yolk," eggs being a major source of lecithin.

Leclanché, Georges (1839–1882) French engineer. In 1866 he invented a primary electrical cell, the **Leclanché cell**, which is still the basis of most dry batteries.

LED abbreviation for †light-emitting diode.

Leda in astronomy, tiny moon of Jupiter discovered in 1974. It is only 16 km/10 mi across and orbits once every 240 days at an average distance of 11 million km/6.8 million mi from the planet.

Lederberg, Joshua (1925–) US geneticist who was awarded the Nobel Prize for Physiology or Medicine in 1958 for work on genetic recombination and the organization of bacterial genetic material. He showed that bacteria can reproduce sexually, combining genetic material so that offspring possess characteristics of both parent organisms. He shared the prize with George †Beadle and Edward †Tatum.

Lederman, Leon (1921–) German-born US physicist who, with Melvin Schwartz and Jack †Steinberger, was awarded the Nobel Prize for Physics in 1988 for the study of elementary particles using †neutrinos, and the discovery of the muon neutrino.

Lee, Yuan Tseh (1936–) Taiwanese chemist who contributed much to the field of chemical reaction dynamics. He put in much of the groundwork for Dudley Herschbach's development of the molecular beam technique and shared the Nobel Prize for Chemistry in 1986 with Herschbach and John †Polanyi for his development of the dynamics of chemical elementary processes.

Lee Tsung-Dao (1926–) Chinese-born US physicist whose research centered on the physics of weak nuclear forces. He was awarded the Nobel Prize for Physics in 1957 for his investigations of weak nuclear forces between elementary particles. He proposed that they might disobey certain key assumptions; for instance, the conservation of parity. He shared the Nobel Prize with his colleague †Yang Chen Ning.

Leeuwenhoek, Anton van (1632–1723) Dutch pioneer of microscopic research. He ground his own lenses, some of which magnified up to 300 times. With these he was able to see individual red blood cells, sperm, and bacteria, achievements not repeated for more than a century.

left-hand rule in physics, a memory aid used to recall the relative directions of motion, magnetic field, and current in an electric motor. It was devised by English physicist John Fleming. (See †Fleming's rules).

Legendre, Adrien-Marie (1752–1833) French mathematician who was particularly interested in number theory, celestial mechanics, and elliptic functions.

Legendre, Adrien-Marie http://www-groups.dcs.st-andrews.ac.uk/history/Mathematicians/Legendre.html

Devoted to the life and contributions of Adrien-Marie Legendre. In addition to biographical information, you will find a list of references about Legendre and links to essays that reference him. The text of this essay includes hypertext links to the essays of those mathematicians and thinkers who influenced Legendre. You will also find an image of him, which you can click on to enlarge, and a map of his birthplace.

Legendre functions functions that satisfy the second-order differential equation:

$(1 - x^2)\, d^2y/dx^2 - 2x\, dy/dx + n(n + 1)y = 0$

legume plant of the family Leguminosae, which has a pod containing dry seeds. The family includes peas, beans, lentils, clover, and alfalfa (lucerne). Legumes are important in agriculture because of their specialized roots, which have nodules containing bacteria capable of fixing nitrogen from the air and increasing the fertility of the soil. The edible seeds of legumes are called **pulses**.

Lehn, Jean-Marie (1939–) French chemist who demonstrated for the first time how metal ions could be made to exist in a non-planar structure, tightly bound into the cavity of a crown ether molecule, explaining a possible mechanism for the transfer of metal ions across biological membranes. He was awarded the Nobel Prize for Chemistry in 1987 with Charles †Pedersen and Donald †Cram for their work on †molecules with highly selective structure-specific interactions.

Leibniz, Gottfried Wilhelm (1646–1716) German mathematician, philosopher, and diplomat. Independently of, but concurrently with, English scientist Isaac †Newton, he developed the branch of mathematics known as †calculus and was one of the founders of symbolic logic. Free from all concepts of space and number, his logic was the prototype of future abstract mathematics.

Leibniz, Gottfried Wilhelm http://mally.stanford.edu/leibniz.html

Maintained by the Metaphysics Research Lab at Stanford University, this page concerns the life and ideas of philosopher, logician, and mathematician Gottfried Leibniz. It features a time line of the major events in Leibniz's life, a list of his principle works, a brief examination of his philosophical contributions, and a list of sources for further reading.

Leishman, William Boog (1865–1926) Scottish army physician who discovered the protozoan parasite that causes the group of diseases now known as leishmaniasis. He was knighted 1909.

lek in biology, a closely spaced set of very small †territories each occupied by a single male during the mating season. Leks are found in the mating systems of several ground-dwelling birds (such as grouse) and a few antelopes, and in some insects.

The lek is a traditional site where both males and females congregate during the breeding season. The males display to passing females in the hope of attracting them to mate. Once mated, the females go elsewhere to lay their eggs or to complete gestation.

Leloir, Luis Federico (1906–1987) Argentine chemist who studied glucose metabolism and discovered an alternative mechanism for †glycogen synthesis, involving a new enzyme. Later, he connected these reactions to glycogen synthesis in the liver and muscles. He showed that a glucose molecule is added by a process in which the reactive intermediate uridine diphosphateglucose (UDPG) transfers glucose to the growing glycogen chain. He found that galactose is broken down to yield glucose in a similar pathway. He was awarded the Nobel Prize for Chemistry in 1970 for his discovery of sugar nucleotides and their role in carbohydrate biosynthesis.

Lemaître, Georges Edouard (1894–1966) Belgian cosmologist. He proposed the †Big Bang theory of the origin of the universe in 1933. US astronomer Edwin †Hubble had shown that the universe was expanding, but it was Lemaître who suggested that the expansion had been started by an initial explosion, the Big Bang, a theory that is now generally accepted.

lemniscate mathematical curve represented by the equation:
$(x^2 + y^2)^2 = a^2(x^2 - y^2)$
where a is constant and x and y are variables.

Lenard, Philipp Eduard Anton von (1862–1947) Hungarian-born German physicist who was awarded the Nobel Prize for Physics in 1905 for his investigations into the †photoelectric effect and cathode rays (the stream of electrons emitted from the cathode in a vacuum tube).

lens in optics, a piece of a transparent material, such as glass, with two polished surfaces – one concave or convex, and the other plane, concave, or convex – that modifies rays of light. A convex lens brings rays of light together; a concave lens makes the rays diverge. Lenses are essential to spectacles, microscopes, telescopes, cameras, and almost all optical instruments.

The image formed by a single lens suffers from several defects or †aberrations, notably **spherical aberration** in which an image becomes blurred, and **chromatic aberration** in which an image in white light tends to have colored edges. Aberrations are corrected by the use of compound lenses, which are built up from two or more lenses of different refractive index.

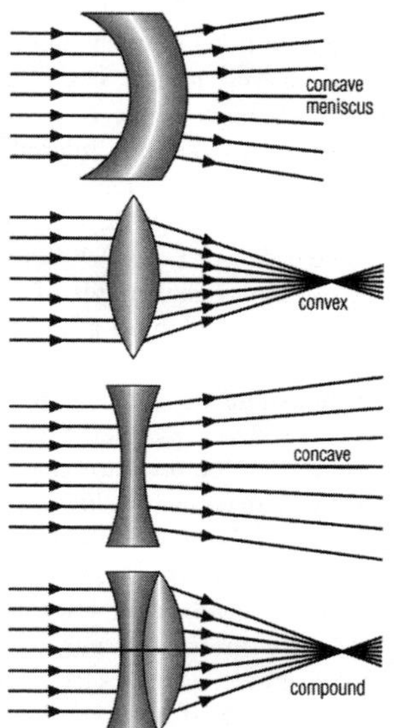

lens The passage of light through lenses. The concave lenses diverges a beam of light from a distant source. The convex and compound lenses focus light from a distant source to a point. The distance between the focus and the lens is called the focal length. The shorter the focus, the more powerful the lens. Helicon

lens, gravitational see †gravitational lensing.

lenticel small pore on the stems of woody plants or the trunks of trees. Lenticels are a means of gas exchange between the stem interior and the atmosphere. They consist of loosely packed cells with many air spaces in between, and are easily seen on smooth-barked trees such as cherries, where they form horizontal lines on the trunk.

lenticular galaxy lens-shaped †galaxy with a large central bulge and flat disc but no discernible spiral arms.

Lenz, Heinrich Friedrich Emil (1804–1865) Russian physicist who in 1833 formulated Lenz's law, a fundamental law of electromagnetism. He also found that the strength of a magnetic field is proportional to the strength of the magnetic induction.

Lenz's law in physics, a law stating that the direction of an electromagnetically induced current (generated by moving a magnet near a wire or a wire in a magnetic field) will be such as to oppose the motion producing it. It is named after the Russian physicist Heinrich Friedrich Lenz, who announced it in 1833.

Leo zodiacal constellation in the northern hemisphere, represented as a lion. The Sun passes through Leo from mid-August to mid-September. Its brightest star is first-magnitude †Regulus at the base of a pattern of stars called the Sickle. In astrology, the dates for Leo are between about 23 July and 22 August (see †precession).

Leo Minor inconspicuous constellation of the northern hemisphere, represented as a lion.

Leonardo da Vinci (1452–1519) Italian painter, sculptor, architect, engineer, and scientist. One of the greatest figures of the Italian Renaissance, he was active in Florence, Milan, and, from 1516, France. As state engineer and court painter to the Duke of Milan, he painted the *Last Supper* mural (*c.* 1495; Sta Maria delle Grazie, Milan), and on his return to Florence painted the *Mona Lisa* (*c.* 1503–05; Louvre, Paris). His notebooks and drawings show an immensely inventive and enquiring mind, studying aspects of the natural and scientific world from anatomy and botany to aerodynamics and hydraulics.

Leonid meteor shower shower of meteors emanating from an apparent point in Leo every 33 years.

lepton any of a class of †elementary particles that are not affected by the strong nuclear force. The leptons comprise the †electron, †muon, and †tau, and their †neutrinos (the electron neutrino, muon neutrino, and tau neutrino), as well as their six †antiparticles.

In July 2000, researchers at Fermilab, in the USA, amassed the first direct evidence for the existence of the tau lepton.

leptoquark in physics, a hypothetical particle made up of a †quark combined with a †lepton, or a new particle created by their interaction.

Lepus small constellation on the celestial equator (see †celestial sphere), represented as a hare.

leucine one of the nine essential †amino acids.

leucite silicate mineral, $KAlSi_2O_6$, occurring frequently in some potassium-rich volcanic rocks. It is dull white to gray, and usually opaque. It is used as a source of potassium for fertilizer.

leucocyte another name for a †white blood cell.

leukotriene in biology, a group of naturally occurring substances that stimulate the activity, for example contraction, of smooth muscles.

levee naturally formed raised bank along the side of a river channel. When a river overflows its banks, the rate of flow is less than that in the channel, and silt is deposited on the banks. With each successive flood the levee increases in height so that eventually the river may be above the surface of the surrounding flood plain. Notable levees are found on the lower reaches of the Mississippi in the USA, and along the Po in Italy. The Hwang Ho in China also has well-developed levees.

lever simple machine consisting of a rigid rod pivoted at a fixed point called the fulcrum, used for shifting or raising a heavy load or applying force. Levers are classified into orders according to where the effort is applied, and the load-moving force developed, in relation to the position of the fulcrum.

A see-saw is an example of a simple lever. An effort force at one end is used to lift a load force at the opposite end, with a turning motion about the pivot. A see-saw can also be used to demonstrate the balancing of the moments of a force. The formula used to work out the force needed and the distance of the force from the pivot to balance the see-saw is: effort force x distance from pivot = load force x distance from pivot.

A **first-order** lever has the load and the effort on opposite sides of the fulcrum – for example, a see-saw or pair of scissors. A **second-order** lever has the load and the effort on the same side of the fulcrum, with the load nearer the fulcrum – for example, nutcrackers or a wheelbarrow. A **third-order** lever has the effort nearer the fulcrum than the load, with both on the same side of it – for example, a pair of tweezers or tongs. The mechanical advantage of a lever is the ratio of load to effort, equal to the perpendicular distance of the effort's line of action from the fulcrum divided by the distance to the load's line of action. Thus tweezers, for instance, have a mechanical advantage of less than one.

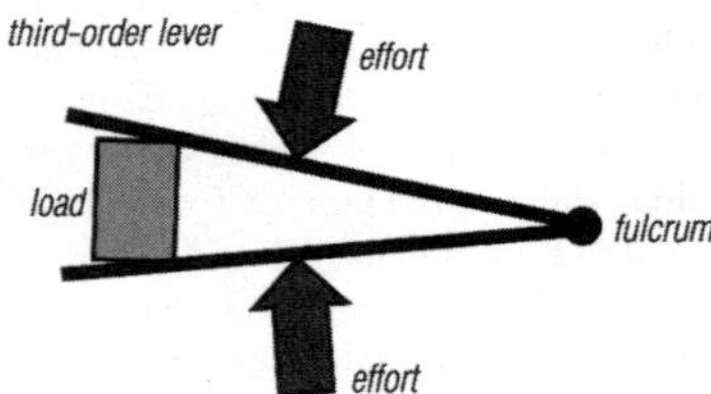

lever Types of lever. Practical applications of the first-order lever include the crowbar, seesaw, and scissors. The wheelbarrow is a second-order lever; tweezers or tongs are third-order levers. Helicon

Leverrier, Urbain Jean Joseph (1811–1877) French astronomer. He predicted the existence and position of the planet †Neptune from its influence on the orbit of the planet †Uranus. It was discovered in 1846.

Leverrier, Urbain Jean Joseph http://www-groups.dcs.st-andrews.ac.uk/history/Mathematicians/Le_Verrier.html

Devoted to the life and contributions of astronomer Urbain Leverrier. In addition to biographical information, you will find a list of references about Leverrier and links to essays that reference him. The text of this essay includes hypertext links to the essays of those mathematicians and thinkers who influenced Leverrier. You will also find an image of him, which you can click on to enlarge, and a map of his birthplace.

Levi-Civita, Tullio (1873–1941) Italian mathematician who developed, in collaboration with Gregorio †Ricci-Curbastro, the absolute differential calculus, published 1900. Levi-Civita also introduced the concept of parallelism in curved space in 1917.

Levi-Civita, Tullio http://www-groups.dcs.st-andrews.ac.uk/history/Mathematicians/Levi-Civita.html

Part of an archive containing the biographies of the world's greatest mathematicians, this site is devoted to the life and contributions of Tullio Levi-Civita.

Levi-Montalcini, Rita (1909–) Italian neurologist who was awarded a Nobel Prize for Physiology or Medicine in 1986 with her co-worker, US biochemist Stanley †Cohen, for their discovery of factors that promote the growth of nerve and epidermal cells. This nerve-growth factor controls how many cells make up the adult nervous system.

Lewis, Edward B (1918–) US geneticist who was awarded a Nobel Prize for Physiology or Medicine in 1995 with Eric F †Wieschaus and Christiane Nusslein-Volhard for their discovery of genes which control the early stages of the body's development.

Lewis, Gilbert Newton (1875–1946) US theoretical chemist who defined a base as a substance that supplies a pair of electrons for a chemical bond, and an acid as a substance that accepts such a pair. He also set out the electronic theory of valency and in thermodynamics listed the free energies of 143 substances.

Lewis acid in chemistry, any compound or ionic species that can accept an electron pair from a donor compound. This definition was proposed by US chemist Gilbert N †Lewis and applies not only to conventional acids, but also to complex species that can accept electron pairs (see also †Lewis base).

Lewis base in chemistry, any compound or ionic species that can donate an electron pair to an acceptor compound. This definition applies not only to conventional bases, but also to complex species that can donate electron pairs (see also †Lewis acid).

lewisite or **dichloro-2chlorovinyl-arsine**; $C_2H_2AsCl_3$, in chemistry, oily, poisonous liquid developed as a chemical weapon during World War I. Contact with the vapor or the liquid causes severe irritation and blistering of the skin. It was developed by and named after US chemist Winford L Lewis.

Leyden jar early form of †capacitor consisting of a glass jar (which acts as the dielectric) with an interior and exterior coating of metal foil (which acts as the plates), used to store static electricity.

LF in physics, abbreviation for low †frequency. LF radio waves have frequencies in the range 30–300 kHz.

LH abbreviation for †**luteinizing hormone**.

L'Hôpital, Guillaume Francois Antoine (1661–1704) Marquis de Sainte-Mesme and Comte d'Autremont, French mathematician. His book, *Analyse des infiniments petits*, (2nd edition, 1715), was the first treatise on calculus. In 1691 he met Johann (Jean) Bernoulli, from whom he learned the principles of the infinitesimal calculus.

Libby, Willard Frank (1908–1980) US chemist who was awarded the Nobel Prize for Chemistry in 1960 for his development in 1947 of †radiocarbon dating as a means of determining the age of organic or fossilized material.

Libby, Willard Frank http://kroeber.anthro.mankato.msus.edu/informati on/biography/klmno/libby_willard.html

Profile of the Nobel prizewinning US chemist. It traces his academic career and official appointments and the process that led to his discovery of the technique of radiocarbon dating.

Libchaber, Albert (1934–) French physicist whose experiment looking at disordered flow within liquid helium revealed one process whereby chaos can arise. Until this experiment chaos had been a mathematical curiosity only. He later demonstrated a second route to chaos.

libido in Freudian psychology, the energy of the sex instinct, which is to be found even in a newborn child. The libido develops through a number of phases, described by Sigmund Freud in his theory of infantile sexuality. The source of the libido is the †id.

The phases of the libido are identified by Freud as the **oral stage**, when a child tests everything by mouth, the **anal stage**, when the child gets satisfaction from control of its body, and the **genital stage**, when the libido becomes concentrated in the sex organs.

Loss of adult libido is seen in some diseases.

Libra faint zodiacal constellation on the celestial equator (see †celestial sphere) adjoining Scorpius, and represented as the scales of justice. The Sun passes through Libra during November. The constellation was once considered to be a part of Scorpius, seen as the scorpion's claws. In astrology, the dates for Libra are between about 23 September and 23 October (see †precession).

libration in astronomy, slight apparent wobble in the rotation of the Moon due to its variable speed of rotation and the tilt of its axis.

Generally, the Moon rotates on its axis in the same time as it takes to complete one orbit, causing it to keep one face turned permanently towards the Earth (see †captured rotation). Its speed in orbit varies, however, because its orbit is not circular but elliptical; hence at times the Moon's axial rotation appears to get either slightly ahead of or slightly behind its orbital motion, so that part of the "dark side" of the Moon is visible around the east and west edges. This is known as **libration in longitude**.

Libration in latitude occurs because the Moon's axis is slightly tilted with respect to its orbital plane, so we can see over the north and south poles. In combination, these effects mean that a total of 59% of the Moon's surface is visible, rather than just 50%, as would be the case if libration did not occur.

lichen any organism of a unique group that consists of associations of a specific †fungus and a specific alga living together in a mutually beneficial relationship. Found as colored patches or spongelike masses on trees, rocks, and other surfaces, lichens flourish in harsh conditions. (Group Lichenes.)

Some lichens are edible, for example, reindeer moss and Iceland moss; others are a source of color dyes, such as litmus, or are used in medicine. They are sensitive to pollution in the air (see †indicator species).

Fun With Lichens http://mgd.nacse.org/hyperSQL/lichenland/

Thorough guide to these unique fungus-alga combinations. Written by a lichen enthusiast at the University of Oregon, this is a good introduction to the subject. The textual content is well supported with photographs.

Lie, (Marius) Sophus (1842–1899) Norwegian mathematician who provided the foundations for the science of topology in transformation groups known as the Lie groups. He was one of the first mathematicians to emphasize the importance of the notion of groups in geometry.

Liebig, Justus (1803–1873) Baron von Liebig, German organic chemist who extended chemical research into other scientific fields, such as agricultural chemistry and biochemistry. He introduced the theory of compound †radicals and discovered chloroform and chloral, and demonstrated the use of fertilizers.

Lie groups or **Lie rings**, collections of mathematical objects in groups or rings that have further (topological) structure under which the collective operations are †continuous, for example, vectors in the plane (where open discs define a topological structure).

life ability to grow, reproduce, and respond to such stimuli as light, heat, and sound. Life on Earth may have begun about 4 billion years ago when a chemical reaction produced the first organic substance. Over time, life has evolved from primitive single-celled organisms to complex multicellular ones. There are now some 10 million different species of plants and animals living on the Earth. The earliest fossil evidence of life is threadlike chains of cells discovered in 1980 in deposits in northwestern Australia; these "stromatolites" have been dated as being 3.5 billion years old.

†Biology is the study of living organisms – their evolution, structure, functioning, classification, and distribution – while †biochemistry is the study of the chemistry of living organisms. Biochemistry is especially concerned with the function of the chemical components of organisms such as proteins, carbohydrates, lipids, and nucleic acids.

Life probably originated in the primitive oceans. The original atmosphere, 4 billion years ago, consisted of carbon dioxide, nitrogen, and water. Laboratory experiments have shown that more complex organic molecules, such as †amino acids and †nucleotides, can be produced from these ingredients by passing electric sparks through a mixture. The climate of the early atmosphere was probably very violent, with lightning a common feature, and these conditions could have resulted in the oceans becoming rich in organic molecules, producing the so-called "primeval soup." These molecules may then have organized themselves into clusters capable of reproducing and eventually developing into simple cells. Soon after life developed, †photosynthesis would have become the primary source of energy for life. By this process, life would have substantially affected the chemistry of the atmosphere and, in turn, that of its own environment. Once the atmosphere had changed to its present composition, life could only be created by the replication of living organisms (a process called †biogenesis).

life cycle in biology, the sequence of developmental stages through which members of a given species pass. Most vertebrates have a simple life cycle consisting of †fertilization of sex cells or †gametes, a period of development as an †embryo, a period of juvenile growth after hatching or birth, an adulthood including †sexual reproduction, and finally death. Invertebrate life cycles are generally more complex and may involve major reconstitution of the individual's appearance (†metamorphosis) and completely different styles of life. Plants have a special type of life cycle with two distinct phases, known as †alternation of generations. Many insects such as cicadas, dragonflies, and mayflies have a long larvae or pupae phase and a short adult phase. Dragonflies live an aquatic life as larvae and an aerial life during the adult phase. In many invertebrates and protozoa there is a sequence of stages in the life cycle, and in parasites different stages often occur in different host organisms.

life-cycle analysis assessment of the environmental impact of a product, taking into account all aspects of production (including resources used), packaging, distribution and ultimate end.

This "cradle-to-grave" approach can expose inconsistencies in many so-called "eco-friendly" labels, applied to products such as soap powders, which may be biodegradable but which are perhaps contained in nonrecyclable containers.

life sciences scientific study of the living world as a whole, a new synthesis of several traditional scientific disciplines including †biology, †zoology, and †botany, and newer, more specialized areas of study such as †biophysics and †sociobiology.

This approach has led to many new ideas and discoveries as well as to an emphasis on †ecology, the study of living organisms in their natural environments.

life table way of summarizing the probability that an individual will give birth or die during successive periods of life. From this, the proportion of individuals who survive from birth to any given age (**survivorship**) and the mean number of offspring produced (**net reproductive rate**) can be determined.

Insurance companies use life tables to estimate risks of death in order to set their premiums and governments use them to determine future needs for education and health services.

ligament strong, flexible connective tissue, made of the protein †collagen, which joins bone to bone at moveable joints and sometimes encloses the joints. Ligaments prevent bone dislocation (under normal circumstances) but allow joint flexion. The ligaments around the joints are composed of white fibrous tissue. Other ligaments are composed of yellow elastic tissue, which is adapted to support a continuous but varying stress, as in the ligament connecting the various cartilages of the †larynx (voice box).

ligand in chemistry, a group that bonds symmetrically to a central atom or ion of a metal; the result is called a **coordination complex**. An example of a neutral ligand is ammonia; the nitrosyl ion NO^+ is a charged ligand. An example of a coordination complex is hexaminocobalt chloride, $[Co(NH_3)_6]Cl_3$, in which the central cobalt ion (Co^{3+}) is surrounded by covalent bonds with six ammonia molecules and ionic bonds with three chloride ions.

Ligands are used in medicine as an antidote to heavy metal poisoning, removing the metal ions by attaching themselves to form a harmless compound.

light †electromagnetic waves (made up of electric and magnetic components) in the visible range, having a wavelength from about 400 nanometers in the extreme violet to about 700 nanometers in the extreme red. Light is considered to exhibit particle and wave properties, and the fundamental particle, or quantum, of light is called the photon. A light wave comprises two transverse waves of electric and magnetic fields traveling at right angles to each other, and as such is a form of electromagnetic radiation. The speed of light (and of all electromagnetic radiation) in a vacuum is approximately 300,000 km/186,000 mi per second, and is a universal constant denoted by *c*.

Light is a form of energy that is mainly visible to the human †eye. Light is radiated by hot objects such as the Sun or an electric light bulb.

The nature of light traveling through a medium can be explained with the help of Young's double-slit experiment. A double slit is placed between a light source and a screen. Light diffracted by the double slit forms a pattern of bright and dark bands (called fringes) on the screen. The fringe pattern is due to the light waves moving through the slits and being diffracted. At a point where light waves are in phase (a crest and crest meeting at the same point) a constructive wave front is formed with a larger amplitude than that of the two waves forming the constructive wave front. This is seen as a bright band on the screen. At a point where the light waves are out of phase (a crest and a trough meeting at the same point) a destructive wave front is formed with zero amplitude (the waves effectively cancel each other out). This is seen as a dark band on the screen.

Light & Optics Readings http://acept.la.asu.edu/PiN/rdg/readings.shtml

Clear illustrated pages introduce concepts in light and optics. The site is part of Arizona State University's Patterns in Nature pages; topics covered include the path of light (and how to make a pinhole camera), color, lenses, lasers, and spectroscopy.

Speed of Light

http://www.thinkquest.org/library/lib/site_sum_outside.html?tname=28160&url=28160/english/speedwave/speed.html

Thought-provoking site on the physics of light. The site provides a short introduction to get the user thinking about the speed of light, then cites historical discoveries to introduce and reinforce the concepts. There are links to pages on color, reflection, refraction, and lasers.

light curve in astronomy, graph showing how the brightness of an astronomical object varies with time. Analysis of the light curves of variable stars, for example, gives information about the physical processes causing the variation.

light-dependent resistor **LDR**, component of electronic circuits whose resistance varies with the level of illumination on its surface. Usually resistance decreases as illumination increases, as the nature of the material is altered by the presence of light. LDRs are used in light-measuring or light-sensing instruments where light intensity is converted to a digital signal (for example, in the exposure-meter circuit of an automatic camera), and in switches (such as those that switch on street lights at dusk).

LDRs are made from †semiconductors, such as cadmium sulfide.

light-emitting diode **LED**, electronic component that converts electrical energy into light or infrared radiation in the range of 550 nm (green light) to 1,300 nm (infrared). They are used for displaying symbols in electronic instruments and devices. An LED is a †diode made of †semiconductor material, such as gallium arsenide phosphide, that glows when electricity is passed through it. The first digital watches and calculators had LED displays, but many later models use †liquid-crystal displays.

In 1993 chemists at the University of Cambridge, England, developed LEDs from the polymer poly(*p*-phenylenevinyl) (PPV) that emit as much light as conventional LEDs and in a variety of colors.

A new generation of LEDs that can produce light in the mid-infrared range (300–1,000 nm) safely and cheaply were developed by British researchers in 1995, using thin alternating layers of indium arsenide and indium arsenide antimonide.

lightning high-voltage electrical discharge between two rainclouds or between a cloud and the Earth, caused by the build-up of electrical charges. Air in the path of lightning ionizes (becomes a conductor), and expands; the accompanying noise is heard as thunder. Currents of 20,000 amperes and temperatures of 30,000°C/54,000°F are common. Lightning causes nitrogen oxides to form in the atmosphere and approximately 25% of atmospheric nitrogen oxides are formed in this way.

According to a 1997 US survey on lightning strength and frequency, using information gathered from satellite images and data from the US Lightning Detection Network, there are 70–100 lightning flashes per second worldwide, with an average peak current of 36 kiloamps.

lightning conductor device that protects a tall building from lightning strike, by providing an easier path for current to flow to earth than through the building. It consists of a thick copper strip of very low resistance connected to the ground below. A good connection to the ground is essential and is made by burying a large metal plate deep in the damp earth. In the event of a direct lightning strike, the current in the conductor may be so great as to melt or even vaporize the metal, but the damage to the building will nevertheless be limited.

light reaction part of the †photosynthesis process in green plants that requires sunlight (as opposed to the †dark reaction, which does not). During the light reaction light energy is used to generate ATP (by the †phosphorylation of ADP), which is necessary for the dark reaction.

light second unit of length, equal to the distance traveled by a beam of light in a vacuum in one second. It is equal to 2.99792458×10^8 m/9.8357103×10^8 ft. See †light year.

light, speed of symbol *c*, fundamental constant of nature, the speed of light is the limiting velocity at which any body can travel. It is equal to 2.997925×10^8 m/186,180 mi per second, and is the same for all observers, no matter how fast they move themselves.

light, theories of the nature of before the advent of modern physics, theories of the basic nature of light fell into two camps: those who viewed light as made up of a stream of particles (corpuscular theory), and those who viewed it as a wave motion (wave theory). Each attempted to explain all phenomena in optics, such as †reflection, †refraction, and so on, on the basis of these two viewpoints. It is now known that light exists in quanta known as photons that exhibit both corpuscular and wavelike behavior in certain circumstances.

light watt unit of radiant power (brightness of light). One light watt is the power required to produce a perceived brightness equal to that of light at a wavelength of 550 nanometers and 680 lumens.

light year distance traveled by a beam of light in a vacuum in one year. It is equal to approximately 9.4605×10^{12} km/5.9128×10^{12} mi.

lignin naturally occurring substance produced by plants to strengthen their tissues. It is difficult for †enzymes to attack lignin, so living organisms cannot digest wood, with the exception of a few specialized fungi and bacteria. Lignin is the essential ingredient of all wood and is, therefore, of great commercial importance.

Chemically, lignin is made up of thousands of rings of carbon atoms joined together in a long chain. The way in which they are linked up varies along the chain.

lignite type of †coal that is brown and fibrous, with a relatively low carbon content. As a fuel it is less efficient because more of it must be burned to produce the same amount of energy generated by bituminous coal. Lignite also has a high sulfur content and is more polluting. It is burned to generate power in Scandinavia and some European countries because it is the only fuel resource available without importing.

lime or **quicklime**; technical name **calcium oxide**; CaO, white powdery substance used in making mortar and cement. It is made commercially by heating calcium carbonate ($CaCO_3$), obtained from limestone or chalk, in a †lime kiln. Quicklime readily absorbs water to become calcium hydroxide $Ca(OH)_2$, known as slaked lime, which is used to reduce soil acidity.

lime kiln oven used to make quicklime (calcium oxide, CaO) by heating limestone (calcium carbonate, $CaCO_3$) in the absence of air. The carbon dioxide is carried away to heat other kilns and to ensure that the reversible reaction proceeds in the right direction: $CaCO_3 \leftrightarrow CaO + CO_2$.

limestone sedimentary rock composed chiefly of calcium carbonate ($CaCO_3$), either derived from the shells of marine organisms or precipitated from solution, mostly in the ocean. Various types of limestone are used as building stone.

†Karst is a type of limestone landscape. Caves commonly occur in limestone. †Marble is metamorphosed limestone. Certain so-called marbles are not in fact marbles but fine-grained fossiliferous limestones that have been polished.

limestone pavement bare rock surface resembling a block of chocolate, found on limestone plateaus. It is formed by the weathering of limestone into individual upstanding blocks, called clints, separated from each other by joints, called grykes. The weathering process is thought to entail a combination of freeze-thaw (the alternate freezing and thawing of ice in cracks) and carbonation (the dissolving of minerals in the limestone by weakly acidic rainwater). Malham Tarn in North Yorkshire is an example of a limestone pavement.

limewater common name for a dilute solution of slaked lime (calcium hydroxide, $Ca(OH)_2$). In chemistry, it is used to detect the presence of carbon dioxide. If a gas containing carbon dioxide is bubbled through limewater, the solution turns milky owing to the formation of calcium carbonate ($CaCO_3$). Continued bubbling of the gas causes the limewater to clear again as the calcium carbonate is converted to the more soluble calcium hydrogencarbonate ($Ca(HCO_3)_2$).

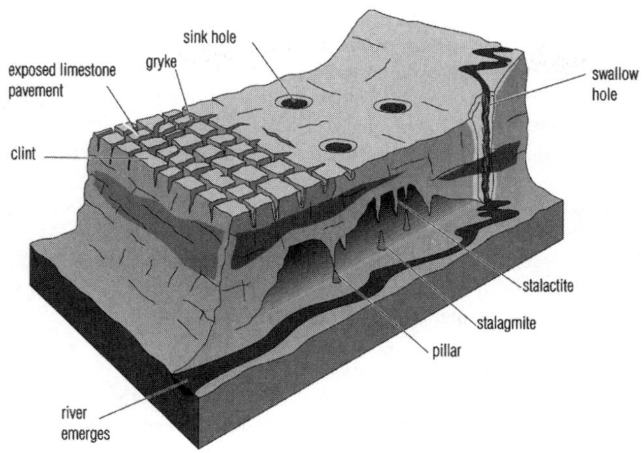

limestone The physical weathering and erosion of a limestone landscape. The freezing and thawing of rain and its mild acidic properties cause cracks and joints to enlarge, forming limestone pavements, potholes, caves, and caverns. Helicon

limit in mathematics, in an infinite sequence, the final value towards which the sequence is tending. For example, the limit of the sequence $^1/_2$, $^3/_4$, $^7/_8$, $^{15}/_{16}$ ... is 1, although no member of the sequence will ever exactly equal 1 no matter how many terms are added together. The limit of the ratios of a †Fibonacci sequence is $^{(\sqrt{5}+1)}/_2$. This number is also the †golden section.

limiting factor in biology, any factor affecting the rate of a metabolic reaction. Levels of light or of carbon dioxide are limiting factors in †photosynthesis because both are necessary for the production of carbohydrates. In experiments, photosynthesis is observed to slow down and eventually stop as the levels of light decrease.

It is believed that the concentrations of carbon dioxide building up in the atmosphere through the burning of fossil fuels will allow faster plant growth.

limnology study of lakes and other bodies of open fresh water, in terms of their plant and animal biology, chemistry, and physical properties.

limonite iron ore, mostly poorly crystalline iron oxyhydroxide, but usually mixed with †hematite and other iron oxides. Also known as brown iron ore, it is often found in bog deposits.

linac contraction of †linear accelerator, a type of particle accelerator in which the particles are accelerated along a straight tube.

Lindblad, Bertil (1895–1965) Swedish astronomer. He demonstrated the rotation of our Galaxy. He went on to stipulate that the speed of rotation of the stars in the Galaxy was a function of their distance from the center (the "differential rotation theory").

Lindemann, (Carl Louis) Ferdinand von (1852–1939) German mathematician whose discussion of the nature of p in 1882 laid to rest the old question of "squaring the circle."

Lindemann, Carl Louis Ferdinand von http://www-groups.dcs.st-andrews.ac.uk/history/Mathematicians/Lindemann.html Devoted to the life and contributions of Ferdinand Lindemann. In addition to biographical information, you will find a list of references about Lindemann and links to essays that reference him. The text of this essay includes hypertext links to the essays of those mathematicians and thinkers who influenced Lindemann. You will also find an image of him, which you can click on to enlarge, and a map of his birthplace.

Lindemann, Frederick Alexander original name of Viscount Cherwell, British physicist.

line in mathematics, a figure produced by joining two or more points in order. The mid-point of any line may be found by †construction of a †perpendicular bisector or by measurement.

linear accelerator or **linac**, in physics, a type of particle †accelerator in which the particles move along a straight tube. Particles pass through a linear accelerator only once – unlike those in a cyclotron or synchrotron (ring-shaped accelerators), which make many revolutions, gaining energy each time.

The world's longest linac is the Stanford Linear Collider, in which electrons and positrons are accelerated along a straight track 3.2 km/2 mi long and then steered into a head-on collision with other particles. The first linear accelerator was built in 1928 by the Norwegian engineer Ralph Wideröe to investigate the behavior of heavy ions (large atoms with one or more electrons removed), but devices capable of accelerating smaller particles such as protons and electrons could not be built until after World War II and the development of high-power radio- and microwave-frequency generators.

linear equation in mathematics, a relationship between two variables that, when plotted on Cartesian axes, produces a straight-line graph; the equation has the general form $y = mx + c$, where m is the slope of the line represented by the equation and c is the y-intercept, or the value of y where the line crosses the y-axis in the †Cartesian coordinate system. Sets of linear equations can be used to describe the behavior of buildings, bridges, trusses, and other static structures.

linear function or **linear transformation**, in its simplest context, a transformation such as $y = mx$ that may be depicted as a line through the origin. More generally, a transformation T defined on a vector space with the property that:

$T(\alpha x + \beta y) = \alpha T(x) + \beta T(y)$

for x, y vectors and α, β scalars.

linear motor type of electric motor, an induction motor in which the fixed stator and moving armature are straight and parallel to each other (rather than being circular and one inside the other as in an ordinary induction motor). Linear motors are used, for example, to power sliding doors. There is a magnetic force between the stator and armature; this force has been used to support a vehicle, as in the experimental maglev linear motor train.

linear programming in mathematics and economics, a set of techniques for finding the maxima or minima of certain variables governed by linear equations or inequalities. These maxima and minima are used to represent "best" solutions in terms of goals such as maximizing profit or minimizing cost.

line of best fit on a †scatter diagram, line drawn as near as possible to the various points so as to best represent the trend being graphed. The sums of the displacements of the points on either side of the line should be equal.

line of force in physics, an imaginary line representing the direction of force at any point in a magnetic, gravitational, or electrical field.

line spectrum †spectrum made up of discrete lines of intensity at certain †wavelengths, characterizing an †atom in a particular state.

line transect in ecology, method used to map plants (or sedentary animals) along a line tied between two trees or other suitable upright supports. The line is marked at 1-m intervals. At each meter mark the distance to the ground is measured and the species recorded.

linkage in genetics, the association between two or more genes that tend to be inherited together because they are on the same chromosome. The closer together they are on the chromosome, the less likely they are to be separated by crossing over (one of the processes of †recombination) and they are then described as being "tightly linked."

Linnaeus, Carolus (1707–1778) (Latinized form of Carl von Linné) Swedish naturalist and physician. His botanical work *Systema naturae* (1735) contained his system for classifying plants into groups depending on shared characteristics (such as the number of stamens in flowers), providing a much-needed framework for identification. He also devised the concise and precise system for naming plants and animals, using one Latin (or Latinized) word to represent the genus and a second to distinguish the species.

linoleic acid $C_{18}H_{32}O_2$, essential polyunsaturated fatty acid present in linseed and certain other vegetable oils as a †glyceride. The compound is one of the most important fatty acids in the human diet.

linolenic acid $C_{18}H_{30}O_2$, an essential †fatty acid found in linseed oil and in most seed fats.

Linux contraction of Linus Unix, in computing, operating system based on an original core program written by Linus Torvalds, a 22-year-old student at the University of Helsinki, Finland, in 1991–92. Linux is a non-proprietary system, made up of freely available ("open source") code created over several years by †Unix enthusiasts all over the world. Each programmer retains the copyright to his or her creation, but makes it freely available on the Internet. Linux retains the flexibility and many of the advanced programming features that make Unix popular for technically minded users, but can run on an ordinary PC instead of an expensive Unix workstation.

Beginners' Guide to Linux http://www.geocities.com/aboutlinux/

Aimed at first-time users of the operating system Linux, this guide explains what Linux is, and what its capabilities are. It also includes information on common commands, firewalls, and Xwindows.

Liouville, Joseph (1809–1882) French mathematician whose main influence was as the founder and first editor 1836–74 of the *Journal de Mathématiques Pures et Appliqués*, which became known as the

Journal de Liouville.

Liouville, Joseph http://www-history.mcs.st-and.ac.uk/history/Mathematicians/Liouville.html

Extensive biography of the French mathematician. The site contains a description of his contribution to mathematics, in particular his work on the existence of a transcendental number in 1844 when he constructed an infinite class of such numbers. Several literature references for further reading are also listed, and the Web site also features a portrait of Liouville.

lipase enzyme responsible for breaking down fats into fatty acids and glycerol. It is produced by the †pancreas and requires a slightly alkaline environment. The products of fat digestion are absorbed by the intestinal wall.

lipid any of a large number of esters of fatty acids, commonly formed by the reaction of a fatty acid with glycerol (see †glyceride). They are soluble in alcohol but not in water. Lipids are the chief constituents of plant and animal waxes, fats, and oils.

Phospholipids are lipids that also contain a phosphate group, usually linked to an organic base; they are major components of biological cell membranes.

lipophilic (Greek "fat-loving") in chemistry, describing †functional groups with an affinity for fats and oils.

lipophobic (Greek "fat-hating") in chemistry, describing †functional groups that tend to repel fats and oils.

liposome in medicine, a minute droplet of oil that is separated from a medium containing water by a †phospholipid layer. Drugs, such as cytotoxic agents, can be incorporated into liposomes and given by injection or by mouth. The liposomes allow the drug to reach the site of action, such as a tumor, without being broken down in the body.

Lippershey, Hans (*c.* 1570–*c.* 1619) Dutch lensmaker, credited with inventing the telescope in 1608.

Lipscomb, William Nunn (1919–) US chemist who studied the relationships between the geometric and electronic structures of molecules and their chemical and physical behavior. He was awarded the Nobel Prize for Chemistry in 1976 for his study of structure and chemical bonding of boranes (compounds of boron and hydrogen).

liquefaction process of converting a gas to a liquid, normally associated with low temperatures and high pressures (see †condensation).

liquefaction in earth science, the conversion of a soft deposit, such as clay, to a jellylike state by severe shaking. During an earthquake buildings and lines of communication built on materials prone to liquefaction will sink and topple. In the Alaskan earthquake of 1964 liquefaction led to the destruction of much of the city of Anchorage.

liquefied petroleum gas LPG, liquid form of butane, propane, or pentane, produced by the distillation of petroleum during oil refining. At room temperature these substances are gases, although they can be easily liquefied and stored under pressure in metal containers. They are used for heating and cooking where other fuels are not available: camping stoves and cigarette lighters, for instance, often use liquefied butane as fuel.

liquid state of matter between a †solid and a †gas. A liquid forms a level surface and assumes the shape of its container. The way that liquids behave can be explained by the †kinetic theory of matter and †particle theory. Its atoms do not occupy fixed positions as in a crystalline solid, nor do they have total freedom of movement as in a gas. Unlike a gas, a liquid is difficult to compress since pressure applied at one point is equally transmitted throughout (Pascal's principle). †Hydraulics makes use of this property.

liquid air air that has been cooled so much that it has liquefied. This happens at temperatures below about –196°C/–321°F. The various constituent gases, including nitrogen, oxygen, argon, and neon, can be separated from liquid air by the technique of †fractional distillation.

Air is liquefied by the **Linde process**, in which air is alternately compressed, cooled, and expanded, the expansion resulting each time in a considerable reduction in temperature. With the lower temperature the molecules move more slowly and occupy less space, so the air changes phase to become liquid.

liquid-crystal display LCD, display of numbers (for example, in a calculator) or pictures (such as on a pocket television screen) produced by molecules of a substance in a semiliquid state with some crystalline properties, so that clusters of molecules align in parallel formations. The display is a blank until the application of an electric field, which "twists" the molecules so that they reflect or transmit light falling on them. There two main types of LCD are **passive matrix** and **active matrix**

liquid–liquid extraction in chemistry, process for the removal of a substance in solution by the addition of a liquid solvent. The substance to be extracted has to be soluble in the added solvent, which has to be immiscible with the original solvent.

liquid phase in chemistry, physical state of a compound or process that is a liquid.

liquidus curve in chemistry, plot of the liquidus point of a †phase diagram as it changes with composition.

liquidus point in chemistry, position on a †phase diagram that indicates the freezing point of a molten substance of mixed composition.

LISP contraction of list processing, high-level computer-programming language designed for manipulating lists of data items. It is used primarily in research into †artificial intelligence (AI).

Developed in the late 1950s, and until recently common only in university laboratories, LISP is used more in the USA than in Europe, where the language PROLOG is often preferred for AI work.

Lissajous, Jules Antoine (1822–1880) French physicist who from 1855 developed **Lissajous figures** as a means of visually demonstrating the vibrations that produce sound waves.

Lissajous, Jules Antoine

http://www-groups.dcs.st-andrews.ac.uk/history/Mathematicians/Lissajous.html

Part of an archive containing the biographies of the world's greatest mathematicians, this site is devoted to the life and contributions of physicist Jules Lissajous.

Lissajous figures path followed by a point subjected to two or more simultaneous simple wave motions, for example, at right angles to one another. They were developed by French physicist Jules Lissajous.

lithification conversion of an unconsolidated (loose) sediment into solid sedimentary rock by **compaction** of mineral grains that make up the sediment, **cementation** by crystallization of new minerals from percolating water solutions, and new growth of the original mineral grains. The term is less commonly used to refer to solidification of †magma to form igneous rock.

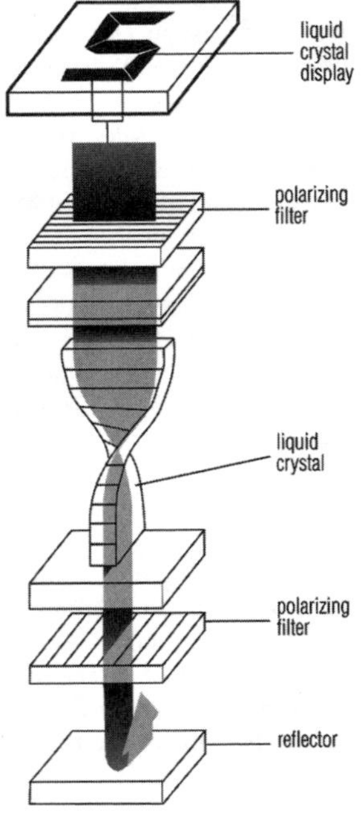

liquid-crystal display A liquid-crystal display consists of a liquid crystal sandwiched between polarizing filters similar to polaroid sunglasses. When a segment of the seven-segment display is electrified, the liquid crystal twists the polarized light from the front filter, allowing the light to bounce off the rear reflector and illuminate the segment. Helicon

lithium chemical symbol Li, (Greek *lithos* "stone") soft, ductile, silver-white, metallic element, atomic number 3, relative atomic mass 6.941. It is one of the †alkali metals (being found at the top of Group 1 of the †periodic table). It has a very low density (far less than most woods), and floats on water (relative density 0.57); it is the lightest of all metals. Although in the †reactivity series it is the least reactive of the alkali metals, it reacts readily with water and the oxygen in air, and so has to be kept under oil. Lithium is used to harden alloys, and in batteries; its compounds are used in medicine to treat manic depression.

Lithium was named in 1818 by Swedish chemist Jöns Berzelius, having been discovered the previous year by his student Johan A Arfwedson. Berzelius named it after "stone" because it is found in most igneous rocks and many mineral springs.

lithium aluminum hydride $LiAlH_4$, important reducing agent used extensively in organic chemistry. It can be used to reduce †carboxylic acids to primary †alcohols.

lithium hydride LiH, reducing agent used as a source of hydrogen. It is produced by strongly heating †lithium metal in the presence of hydrogen gas. It forms †lithium hydroxide and hydrogen when added to water.

lithium hydroxide LiOH, in chemistry, strong base used in powder form as an absorber of carbon dioxide. It is used to purify air in enclosed systems such as submarines.

lithosphere upper rocky layer of the Earth that forms the jigsaw of plates that take part in the movements of †plate tectonics. The lithosphere comprises the †crust and a portion of the upper †mantle. It is regarded as being rigid and brittle and moves about on the more plastic and less brittle †asthenosphere. The lithosphere ranges in thickness from 2–3 km/1–2 mi at mid-ocean ridges to 150 km/93 mi beneath old ocean crust, to 250 km/155 mi under †cratons.

litmus dye obtained from various †lichens and used in chemistry as an indicator to test the acidic or alkaline nature of aqueous solutions; it turns red in the presence of acid, and blue in the presence of alkali.

liter symbol l, metric unit of volume, equal to one cubic decimeter (1.76 imperial pints/2.11 US pints). It was formerly defined as the volume occupied by one kilogram of pure water at 4°C at standard pressure, but this is slightly larger than one cubic decimeter.

Little Dipper another name for the most distinctive part of the constellation †Ursa Minor, the Little Bear.

littoral region in ecology, region of seashore that lies between the high-water mark and low-water mark. Its zones – the upper, middle, and lower shores – each support a community of organisms characterized by their ability to withstand greater or lesser periods of exposure at low tide.

liver in vertebrates, large organ with many regulatory and storage functions. The human liver is situated in the upper abdomen, and weighs about 2 kg/4.5 lb. It is divided into four lobes. The liver receives the products of digestion (food absorbed from the †gut and carried to the liver by the bloodstream), converts †glucose to glycogen (a long-chain carbohydrate used for storage), and then back to glucose when needed. In this way the liver regulates the level of glucose in the blood (see †homeostasis). This is partly controlled by a †hormone, †insulin. The liver removes excess amino acids from the blood, converting them to urea, which is excreted by the kidneys. The liver also synthesizes vitamins, produces bile and blood-clotting factors, and removes damaged red cells and toxins such as alcohol from the blood.

If more †protein is eaten than is needed to make different proteins for the body, the excess is broken down. This breakdown takes place in the liver. One product of this breakdown is urea and this has to be lost from the body in the urine. The liver also stores some vitamins, produces †bile, and breaks down †red blood cells.

Body Systems and Homeostasis http://www.mhhe.com/biosci/genbio/maderbiology/supp/homeo.html

Thorough text-only description of the way in which the systems of the body act to preserve homeostasis (constancy of the body's internal environment irrespective of external factors). The page, a supplement to a textbook written by Sylvia Mader, includes a table of the major endocrine glands and the hormones they produce.

lm in physics, symbol for †lumen, the SI unit of luminous flux, equal to the flux emitted by a 1-candela point source in a 1-steradian solid angle.

ln symbol for †natural logarithm.

load in earth sciences, material transported by a river. It includes material carried on and in the water (suspended load) and material bounced or rolled along the river bed (bedload). A river's load is greatest during a flood, when its discharge is at its highest.

The term "load" can also refer to material transported by a glacier or by the sea.

load generally, the burden inflicted on a system. Thus, in †mechanics, the load is the weight supported by a structure, whereas in electrical engineering it is the output of an electrical device such as a †transformer.

loam type of fertile soil, a mixture of sand, silt, clay, and organic material. It is porous, which allows for good air circulation and retention of moisture.

Lobachevsky, Nikolai Ivanovich (1792–1856) Russian mathematician who founded non-Euclidean geometry, concurrently with, but independently of, Karl †Gauss in Germany and János †Bolyai in Hungary. Lobachevsky published the first account of the subject in 1829, but his work went unrecognized until Georg †Riemann's system was published.

Lobachevsky, Nikolai Ivanovich http://www-groups.dcs.st-andrews.ac.uk/history/Mathematicians/Lobachevsky.html

Part of an archive containing the biographies of the world's greatest mathematicians, this site is devoted to the life and contributions of Nikolai Lobachevsky. In addition to biographical information, you will find a list of references about Lobachevsky and links to other essays in the archive that reference him. The text of this essay includes hypertext links to the essays of those mathematicians and thinkers who influenced Lobachevsky. You will also find an image of him, which you can click on to enlarge, and a map of his birthplace.

local area network LAN, in computing, a †network restricted to a single room or building. Local area networks enable around 500 devices, usually microcomputers acting as workstations, as well as peripheral devices, such as printers, to be connected together.

Central to the LAN is the **file server**, which controls user access to the network. The **print server** processes and controls all print jobs being sent via the network to a connected printer.

Local Group in astronomy, cluster of about 30 galaxies that includes our own, the Milky Way. Like other groups of galaxies, the Local Group is held together by the gravitational attraction between its members, and does not expand with the expanding universe. Its two largest galaxies are the Milky Way and the Andromeda galaxy; most of the others are small and faint.

Lockyer, (Joseph) Norman (1836–1920) English scientist. He studied the spectra of solar prominences and sunspots. Through his pioneering work in spectroscopy, he discovered the existence of helium. He was knighted in 1897.

locomotion ability to move independently from one place to another, occurring in most animals but not in plants. The development of locomotion as a feature of animal life is closely linked to another vital animal feature, that of †nutrition. Animals cannot make their food, as can plants, but must find it first; often the food must be captured and killed, which may require great speed. Locomotion is also important in finding a mate, in avoiding predators, and in migrating to favorable areas.

The mechanism of locomotion in humans depends on the contraction of †muscles stimulated by nerve impulses. Muscles are attached to the †skeleton and create movement at †joints. So, for example, the muscles of the legs contract to move the leg bones at joints such as the knee and hip. This movement is transmitted to the ground by the skeleton and results in locomotion.

locus in biology, the point on a †chromosome where a particular †gene occurs.

locus (Latin "place") in mathematics, traditionally the path traced out by a moving point, but now defined as the set of all points on a curve satisfying given conditions. The locus of points a fixed distance from a fixed point is a circle. The locus of a point equidistant from two fixed points is a straight line that perpendicularly bisects the line joining them. The locus of points a fixed distance from a line is two parallel lines running either side.

lode geological deposit rich in certain minerals, generally consisting of a large vein or set of veins containing ore minerals. A system of veins that can be mined directly forms a lode, for example the mother lode of the California gold rush.

Lodes form because hot hydrothermal liquids and gases from magmas penetrate surrounding rocks, especially when these are limestones; on cooling, veins of ores formed from the magma then extend from the igneous mass into the local rock.

lodestar or **loadstar**, star used in navigation or astronomy, often †Polaris, the Pole Star.

Lodge, Oliver Joseph (1851–1940) British physicist. He developed a system of wireless communication in 1894, and his work was instrumental in the development of radio receivers. He also conducted experiments that suggested that the †ether does not exist, a discovery fundamental to the theory of relativity. After his son was killed in 1915, Lodge became interested in psychic research. He was knighted in 1902.

loess yellow loam, derived from glacial meltwater deposits and accumulated by wind in periglacial regions during the †ice ages. Loess usually attains considerable depths, and the soil derived from it is very fertile. There are large deposits in central Europe (Hungary), China, and North America. It was first described in 1821 in the Rhine area, and takes its name from a village in Alsace.

Loewi, Otto (1873–1961) German physiologist who was awarded a Nobel Prize for Physiology or Medicine in 1936 for his work on the chemical transmission of nerve impulses. He established that a chemical substance is responsible for the stimulation of one nerve cell (neuron) by another.

log in mathematics, abbreviation for †logarithm.

log any apparatus for measuring the speed of a ship; also the daily record of events on board a ship or aircraft.

The log originally consisted of a piece of weighted wood attached to a line with knots at equal intervals that was cast from the rear of a ship. The vessel's speed was estimated by timing the passage of the knots with a sandglass (like an egg timer). Today logs use electromagnetism and sonar.

logarithm or **log**, the †exponent or index of a number to a specified base – usually 10. For example, the logarithm to the base 10 of 1,000 is 3 because $10^3 = 1{,}000$; the logarithm of 2 is 0.3010 because $2 = 10^{0.3010}$. The whole-number part of a logarithm is called the **characteristic**; the fractional part is called the **mantissa**.

Before the advent of cheap electronic calculators, multiplication and division could be simplified by being replaced with the addition and subtraction of logarithms.

For any two numbers x and y (where $x = b^a$ and $y = b^c$), $x \times y = b^a \times b^c = b^{a+c}$; hence one would add the logarithms of x and y, and look up this answer in antilogarithm tables.

Tables of logarithms and antilogarithms are available that show conversions of numbers into logarithms, and vice versa. For example, to multiply 6,560 by 980, one looks up their logarithms (3.8169 and 2.9912), adds them together (6.8081), then looks up the antilogarithm of this to get the answer (6,428,800). **Natural** or **Napierian logarithms** are to the base e, an †irrational number equal to approximately 2.7183.

The principle of logarithms is also the basis of the slide rule. With the general availability of the electronic pocket calculator, the need for logarithms has been reduced. The first log tables (to base e) were published by the Scottish mathematician John Napier in 1614. Base-ten logs were introduced by the Englishman Henry Briggs (1561–1631) and Dutch mathematician Adriaen Vlacq (1600–1667).

logic gate or **logic circuit**, in electronics, one of the basic components used in building †integrated circuits. The five basic types of gate make logical decisions based on the functions NOT, AND, OR, NAND (NOT AND), and NOR (NOT OR). With the exception of the NOT gate, each has two or more inputs.

Information is fed to a gate in the form of binary-coded input signals (logic value 0 stands for "off" or "low-voltage pulse," logic 1 for "on" or "high-voltage"), and each combination of input signals yields a specific output (logic 0 or 1). An **OR** gate will give a logic 1 output if one or more of its inputs receives a logic 1 signal; however, an **AND** gate will yield a logic 1 output only if it receives a logic 1 signal through both its inputs. The output of a **NOT** or **inverter** gate is the opposite of the signal received through its single input, and a **NOR** or **NAND** gate produces an output signal that is the opposite of the signal that would have been produced by an OR or AND gate, respectively. The properties of a logic gate, or of a combination of gates, may be defined and presented in the form of a diagram called a **truth table**, which lists the output that will be triggered by each of the possible combinations of input signals. The process has close parallels in computer programming, where it forms the basis of binary logic.

circuit symbols

truth tables

OR gate

inputs		output
0	0	0
0	1	1
1	0	1
1	1	1

AND gate

inputs		output
0	0	0
0	1	0
1	0	0
1	1	1

NOT gate

inputs	output
0	1
1	0

NOR gate

inputs		output
0	0	1
0	1	0
1	0	0
1	1	0

NAND gate

inputs		output
0	0	1
0	1	1
1	0	1
1	1	0

logic gate The circuit symbols for the five basic types of logic gate: OR, AND, NOT, NOR, and NAND. The truth table displays the output results of each possible combination of input signal. Helicon

logistic curve curve that represents logarithmic functions, from which logarithms of ordinary numbers can be read off.

LOGO (Greek *logos* "word") high-level computer programming language designed to teach mathematical concepts. Developed in about 1970 at the Massachusetts Institute of Technology, it became popular in schools and with home computer users because of its "turtle graphics" feature. This allows the user to write programs that create line drawings on a computer screen, or drive a small mobile robot (a "turtle" or "buggy") around the floor.

LOGO encourages the use of languages in a logical and structured way, leading to "microworlds," in which problems can be solved by using a few standard solutions.

lomentum fruit similar to a pod but constricted between the seeds. When ripe, it splits into one-seeded units, as seen, for example, in the fruit of sainfoin *Onobrychis viciifolia* and radish *Raphanus raphanistrum*. It is a type of †schizocarp.

lone pair in chemistry, a pair of electrons in the outermost shell of an atom that are not used in bonding. In certain circumstances, they will allow the atom to bond with atoms, ions, or molecules (such as boron trifluoride, BF_3) that are deficient in electrons, forming coordinate covalent (dative) bonds in which they provide both of the bonding electrons.

long-chain molecule †aliphatic hydrocarbon in which a large number of atoms are connected to form a single, long molecular unit. Constituent groups added to this chain are referred to as branch groups.

longitude see †latitude and longitude.

longitudinal wave †wave in which the displacement of the medium's particles is in line with or parallel to the direction of travel of the wave motion.

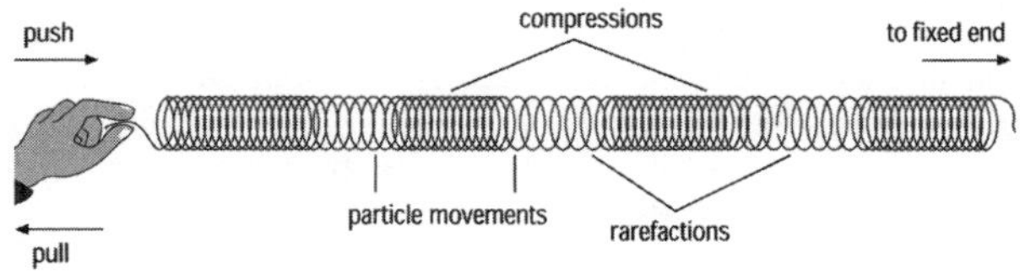

longitudinal wave The diagram illustrates the motion of a longitudinal wave. Sound, for example, travels through air in longitudinal waves: the waves vibrate back and forth in the direction of travel. In the compressions the particles are pushed together, and in the rarefactions they are pulled apart. Helicon

Various methods are used to reproduce waves, such as a ripple tank or a loosely-coiled spring, in order to understand their properties. If a loosely-coiled spring is given a slight "pushing" force, the wave travels in the same direction as the movement of the coils. The waves consist of a series of compressions (where the coils of the spring are pressed together) and rarefactions (where the coils of the spring are more widely spaced). This is known as a longitudinal wave. The moving coils represent waves traveling along the spring. The spring remains as it was after the wave has traveled along the spring; waves carry energy from one place to another but they do not transfer matter. Sound waves are an example of longitudinal waves.

longshore drift movement of material along a †beach. When a wave breaks at an angle to the beach, pebbles are carried up the beach in the direction of the wave (**swash**). The wave returns to the sea at right angles to the beach (**backwash**) because that is the steepest gradient, carrying some pebbles with it. In this way, material moves in a zigzag fashion along a beach. Longshore drift is responsible for the erosion of beaches and the formation of †spits (ridges of sand or shingle projecting into the water). Attempts are often made to halt longshore drift by erecting barriers, or †groynes, at right angles to the shore.

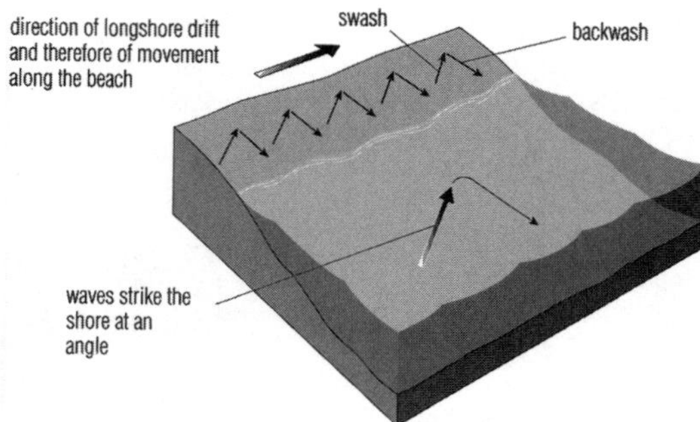

longshore drift Waves sometimes hit the beach at an angle. The incoming waves (swash) carry sand and shingle up onto the shore and the outgoing wave takes some material away with it. Gradually material is carried down the shoreline in the same direction as the longshore current. Helicon

Lonsdale, Kathleen (1903–1971) born Kathleen Yardley, Irish X-ray crystallographer who developed several technologies for the study of crystal structures. She was among the first to determine the structures of organic molecules, and the first to confirm the hexagonal arrangement of carbon in benzene compounds. She derived the structure factor formulas for all space groups. In 1945 she was the first woman to be elected to the Royal Society and she was made DBE in 1956.

loop in computing, short for program loop.

loop in mathematics, part of a curve that encloses a space when the curve crosses itself. In a †flow chart, a path that keeps on returning to the same question.

Lorentz, Hendrik Antoon (1853–1928) Dutch physicist awarded (with his pupil Pieter †Zeeman) the Nobel Prize for Physics in 1902 for his work on the Zeeman effect, in which a magnetic field splits spectral lines.

Lorentz, Hendrik Antoon http://www-history.mcs.st-and.ac.uk/history/Mathematicians/Lorentz.html
Biography of the Dutch physicist. The Web site contains a description of his contribution to physics, and in particular in furthering Maxwell's electromagnetic theory. Several literature references for further reading on the life and works of Lorentz are listed, and his photograph is also shown.

Lorentz transformation means of relating measurements of times and lengths made in one frame of reference to those made in another frame of reference moving at some velocity relative to the first. These formulae, which can be derived directly from Einstein's special theory of †relativity, predict that at velocities approaching that of light, lengths appear to contract (the Lorentz contraction) and time intervals appear to increase (time dilation), as measured in a stationary frame of reference.

Lorenz, Konrad Zacharias (1903–1989) Austrian ethologist who was awarded the Nobel Prize for Physiology or Medicine in 1973 with Nikolaas †Tinbergen and Karl von †Frisch for their work on animal behavior patterns. He studied the relationship between instinct and behavior, particularly in birds, and described the phenomenon of †imprinting in 1935. His books include *King Solomon's Ring* (1952) on animal behavior, and *On Aggression* (1966) on human behavior.

loudness subjective judgment of the level or power of sound reaching the ear. The human ear cannot give an absolute value to the loudness of a single sound, but can only make comparisons between two different sounds. The precise measure of the power of a sound wave at a particular point is called its †intensity. Accurate comparisons of sound levels may be made using sound-level meters, which are calibrated in units called †decibels.

Lovell, (Alfred Charles) Bernard (1913–) English radio astronomer, director 1951–81 of Jodrell Bank Experimental Station (now Nuffield Radio Astronomy Laboratories), Cheshire, England.

Lovelock, James Ephraim (1919–) British scientist who began the study of †chlorofluorocarbons (CFCs) in the atmosphere in the 1960s, and who later elaborated the "Gaia hypothesis" – the concept of the Earth as a single organism, or ecosystem. The Gaia hypothesis, named after an ancient Greek earth goddess, views the planet as a self-regulating system in which all the individual elements coexist in a symbiotic relationship. In developing this theory (first published in 1968), Lovelock realized that the damage effected by humans on many of the Earth's ecosystems was posing a threat to the viability of the planet itself.

Lowell, Percival (1855–1916) US astronomer who predicted the existence of a planet beyond Neptune, starting the search that led to the discovery of Pluto in 1930. In 1894 he founded the Lowell Observatory in Flagstaff, Arizona, where he reported seeing "canals" (now known to be optical effects and natural formations) on the surface of Mars.

lower bound in mathematics, the value that is less than or equal to all of the values of a given set.

lowest common denominator **lcd**, smallest number that is a multiple and thus divides exactly into each of the denominators of a set of fractions.

lowest common multiple **lcm**, smallest number that is a multiple of all of the numbers of a given set. See †lowest common denominator.

low-level language in computing, a programming language designed for a particular computer and reflecting its internal †machine code; low-level languages are therefore often described as **machine-oriented** languages. They cannot easily be converted to run on a computer with a different central processing unit, and they are relatively difficult to learn because a detailed knowledge of the internal working of the computer is required. Since they must be translated into machine code by an †assembler program, low-level languages are also called †assembly languages.

A mnemonic-based low-level language replaces binary machine-code instructions, which are very hard to remember, write down, or correct, with short codes chosen to remind the programmer of the instructions they represent. For example, the binary-code instruction that means "**sto**re the contents of the **a**ccumulator" may be represented with the mnemonic STA.

In contrast, †high-level languages are designed to solve particular problems and are therefore described as **problem-oriented languages**.

LSI abbreviation for large-scale integration, technology that enables whole electrical circuits to be etched into a piece of semiconducting material just a few millimeters square.

By the late 1960s a complete computer processor could be integrated on a single chip, or †integrated circuit, and in 1971 the US electronics company Intel produced the first commercially available †microprocessor. Very large-scale integration (VLSI) results in even smaller chips.

lubricant substance used between moving surfaces to reduce friction. Carbon-based (organic) lubricants, commonly called grease and oil, are recovered from petroleum distillation.

Extensive research has been carried out on chemical additives to lubricants, which can reduce corrosive wear, prevent the accumulation of "cold sludge" (often the result of stop-start driving in city traffic jams), keep pace with the higher working temperatures of aviation gas turbines, or provide radiation-resistant greases for nuclear power plants. Silicon-based spray-on lubricants are also used; they tend to attract dust and dirt less than carbon-based ones.

A solid lubricant is graphite, an allotropic form of carbon, either flaked or emulsified (colloidal) in water or oil.

Lucas, Keith (1879–1916) English neurophysiologist who investigated the transmission of nerve impulses. He demonstrated that the contraction of muscle fibers follows the "all or none" law: a certain amount of stimulus is needed in order to induce a nerve impulse and subsequent muscle contraction. Any stimulus below that threshold has no effect regardless of its duration.

Ludwig, Carl Friedrich Wilhelm (1816–1895) German physiologist who invented graphic methods of recording events within the body. He demonstrated that the circulation of the blood is purely mechanical in nature and involves no occult vital forces.

lumen symbol lm, SI unit of luminous flux (the amount of light passing through an area per second).

The lumen is defined in terms of the light falling on a unit area at a unit distance from a light source of luminous intensity of one †candela. One lumen at a wavelength of 5,550 angstroms equals 0.0014706 watts.

lumen in biology, the space enclosed by an organ, such as the bladder, or a tubular structure, such as the gastrointestinal tract.

luminescence emission of light from a body when its atoms are excited by means other than raising its temperature. Short-lived luminescence is called fluorescence; longer-lived luminescence is called phosphorescence.

When exposed to an external source of energy, the outer electrons in atoms of a luminescent substance absorb energy and "jump" to a higher energy level. When these electrons "jump" back to their former level they emit their excess energy as light. Many different exciting mechanisms are possible: visible light or other forms of electromagnetic radiation (ultraviolet rays or X-rays), electron bombardment, chemical reactions, friction, and †radioactivity. Certain living organisms produce †bioluminescence.

luminosity or **brightness**, in astronomy, amount of light emitted by a star, measured in †magnitudes. The apparent brightness of an object decreases in proportion to the square of its distance from the observer. The luminosity of a star or other body can be expressed in relation to that of the Sun.

luminous paint preparation containing a mixture of pigment, oil, and a phosphorescent sulfide, usually calcium or barium. After exposure to light it appears luminous in the dark. The luminous paint used on watch faces contains radium, is radioactive, and therefore does not require exposure to light.

Lummer, Otto Richard (1860–1925) German physicist who specialized in optics and thermal radiation. His investigations led directly to the radiation formula of Max †Planck, which marked the beginning of †quantum theory.

Lunar Alps conspicuous mountain range on the Moon, northeast of the Sea of Showers (Mare Imbrium), cut by a valley 150 km/93 mi long. The highest peak (about 3,660 m/12,000 ft) is Mont Blanc.

lung in mammals, large cavity of the body, used for †gas exchange. Most tetrapod (four-limbed) vertebrates have a pair of lungs occupying the thorax. The lungs are essentially an infolding of the body surface – a sheet of thin, moist membrane made of a single layer of cells, which is folded so as to occupy less space while having a large surface area for the uptake of †oxygen and loss of †carbon dioxide. The folding creates tiny sacs called †alveoli. Outside the walls of the alveoli there are lots of blood †capillaries for transporting the products of gas exchange. The lung tissue, consisting of multitudes of air sacs and blood vessels, is very light and spongy, and functions by bringing inhaled air into close contact with the blood for efficient gas exchange. The efficiency of lungs is enhanced by †breathing movements, by the thinness and moistness of their surfaces, and by a constant supply of circulating blood.

The lungs inflate and deflate as a result of breathing movements (ventilation). Breathing movements are caused by movements of †muscles between the ribs and the muscles of the diaphragm. Air flows into the mouth and then along ever narrower tubes, trachea, bronchi, and tiny broncheoles. However, the last part of the journey to the alveoli is by diffusion only, as is the exchange with the blood.

Dust in the air is usually trapped by the mucus lining the tubes leading to the lungs. Cells lining the tubes are specialized cells (see †epithelium) and have hair-like structures – cilia – that sweep the trapped dust up to the mouth where it is swallowed. Some dust may reach the lungs where white blood cells may destroy it. However, the lungs can be damaged if dust is not removed. Many miners suffer from lungs damaged by the effects of coal dust, and many other forms of industrial dusts are equally dangerous.

In humans, the principal diseases of the lungs are tuberculosis, pneumonia, bronchitis, emphysema, and cancer. Bronchitis is an irritation of the airways resulting in them becoming narrower than normal so that a person cannot breathe fully. Emphysema is permanent damage to the alveolar walls resulting in too little surface for gas exchange. This too results in difficulties in breathing. The commonest cause of both bronchitis and emphysema is smoking.

Lupus constellation of the southern hemisphere, represented as a wolf. Most of its brighter stars form part of †Gould's Belt.

Luria, Salvador Edward (1912–1991) Italian-born US physician who was awarded the Nobel Prize for Physiology or Medicine in 1969 for his work on the replication mechanism and genetic structure of viruses. Luria was a pacifist and was identified with efforts to keep science humanistic.

luteinizing hormone †hormone produced by the pituitary gland. In males, it stimulates the testes to produce androgens (male sex hormones). In females, it works together with follicle-stimulating hormone to initiate production of egg cells by the ovary. If fertilization occurs, it plays a part in maintaining the pregnancy by controlling the levels of the hormones estrogen and progesterone in the body.

lutetium chemical symbol Lu, (Latin *Lutetia* "Paris") silver-white, metallic element, the last of the †lanthanide series, atomic number 71, relative atomic mass 174.97. It is used in the "cracking," or breakdown, of petroleum and in other chemical processes. It was named by its discoverer, French chemist Georges Urbain, after his native city.

lux symbol lx, SI unit of illuminance or illumination (the light falling on an object). It is equivalent to one †lumen per square meter or to the illuminance of a surface one meter distant from a point source of one †candela.

Lwoff, André Michael (1902–1994) French microbiologist who was awarded a Nobel Prize in Physiology or Medicine in 1965 for his work on genetic control of enzyme and virus synthesis. He proved that enzymes produced by some †genes regulate the functions of other genes. He shared the prize with his fellow researchers Jacques †Monod and François †Jacob.

lx in physics, symbol for †lux, the SI unit of illumination.

lycopene red carotenoid pigment found in tomatoes, rose hips, and many berries. It is the parent substance of all the natural †carotenoids.

Lycopenes have been found to have anticancer properties. In research carried out in 2000 by Israeli researchers, lycopene added to cancer cells taken from mouth tumors killed the cells. There is also a growing body of evidence to suggest that lycopenes in the diet protect against a number of cancers, including breast and prostate. A US study concluded that two or more tomato products eaten per week reduced the risk of prostate cancer by 34%.

Lyell, Charles (1797–1875) Scottish geologist. In his *Principles of Geology* (1830–33), he opposed the French anatomist Georges †Cuvier's theory that the features of the Earth were formed by a series of catastrophes, and expounded the Scottish geologist James †Hutton's view, known as †uniformitarianism, that past events were brought about by the same processes that occur today – a view that influenced Charles †Darwin's theory of †evolution. He was knighted in 1848.

Lyman, Theodore (1874–1954) US physicist whose work was confined to the spectroscopy of the extreme ultraviolet region.

Lyman series series of lines in the †spectrum of hydrogen that lie in the ultraviolet region of the spectrum. It is named after US physicist Theodore Lyman.

lymph fluid found in the lymphatic system of vertebrates.

Lymph is drained from the tissues by lymph capillaries, which empty into larger lymph vessels (lymphatics). These lead to lymph nodes (small, round bodies chiefly situated in the neck, armpit, groin, thorax, and abdomen), which process the †lymphocytes produced by the bone marrow, and filter out harmful substances and bacteria. From the lymph nodes, vessels carry the lymph to the thoracic duct and the right lymphatic duct, which drain into the large veins in the neck. Some vertebrates, such as amphibians, have a lymph heart, which pumps lymph through the lymph vessels.

lymph nodes small masses of lymphatic tissue in the body that occur at various points along the major lymphatic vessels. Tonsils and adenoids are large lymph nodes. As the lymph passes through them it is filtered, and bacteria and other micro-organisms are engulfed by cells known as macrophages.

Lymph nodes are sometimes mistakenly called lymph "glands," and the term "swollen glands" refers to swelling of the lymph nodes caused by infection.

lymphocyte type of white blood cell with a large nucleus, produced in the bone marrow. Most occur in the †lymph and blood, and around sites of infection. **B lymphocytes** or †B cells are responsible for producing †antibodies. **T lymphocytes** or †T cells have several roles in the mechanism of †immunity.

lymphokines chemical messengers produced by lymphocytes that carry messages between the cells of the immune system (see †immunity). Examples include interferon, which initiates defensive reactions to viruses, and the interleukins, which activate specific immune cells.

Lynx inconspicuous constellation of the northern hemisphere. It was named by Johannes †Hevelius, who asserted that it would take the eyes of a lynx to see it.

Lyot, Bernard Ferdinand (1897–1952) French astronomer. He also designed and constructed optical instruments. He concentrated on the study of the solar corona, for which he devised the coronagraph and the photoelectric polarimeter, and he proved that some of the Fraunhofer lines in the solar spectrum represent ionized forms of known metals rather than undiscovered elements.

Lyra small but prominent constellation of the northern hemisphere, represented as the harp of Orpheus. Its brightest star is Vega.

Epsilon Lyrae is a system of four gravitationally linked stars. Beta Lyrae is an eclipsing binary. The Ring nebula, M57, is a †planetary nebula.

Lysenko, Trofim Denisovich (1898–1976) Soviet biologist who believed in the inheritance of †acquired characteristics (changes acquired in an individual's lifetime) and used his position under Joseph Stalin officially to exclude Gregor †Mendel's theory of inheritance. He was removed from office after the fall of Nikita Khrushchev in 1964.

lysis in biology, any process that destroys a cell by rupturing its membrane or cell wall (see †lysosome).

lysosome membrane-enclosed structure, or organelle, inside a †cell, principally found in animal cells. Lysosomes contain enzymes that can break down proteins and other biological substances. They play a part in digestion, and in the white blood cells known as phagocytes the lysosome enzymes attack ingested bacteria.

M

m symbol for †meter.

MacArthur, Robert Helmer (1930–1972) Canadian-born US ecologist who did much to change ecology from a descriptive discipline to a quantitative, predictive science. For example, his index of vegetational complexity (foliage height diversity) of 1961 made it possible to compare habitats and predict the diversity of their species in a definite equation.

McClintock, Barbara (1902–1992) US geneticist who was awarded a Nobel Prize for Physiology or Medicine in 1983 for her discovery of mobile genetic elements, meaning she discovered "jumping" †genes (genes that can change their position on a chromosome from generation to generation). This would explain how originally identical cells take on the specialized functions of skin, muscle, bone, and nerve, and also how evolution could give rise to the multiplicity of species.

McCollum, Elmer Verner (1879–1967) US biochemist and nutritionist who originated the letter system of naming vitamins. He also researched into the role of minerals in the diet. His choice of the albino rat as a laboratory animal was to make it one of the most used animals for research.

MacDiarmid, Alan G (1927–) New Zealand-born US chemist who with US physicist Alan J †Heeger and Japanese chemist Hideki †Shirakawa shared the Nobel Prize for Chemistry in 2000 for the discovery and development of conductive †polymers. The research into conductive polymers could lead to the development of plastic transistors on the molecular scale, which could allow the production of computers both considerably smaller and faster than present models.

Mach, Ernst (1838–1916) Austrian philosopher and physicist. He was an empiricist, believing that science is a record of facts perceived by the senses, and that acceptance of a scientific law depends solely on its standing the practical test of use; he opposed such concepts as Isaac †Newton's "absolute motion." †**Mach numbers** are named after him.

machair low-lying, fertile, sandy coastal plain integral to the crofting agricultural system (a form of subsistence farming) of the Hebrides and parts of the northern Highlands of Scotland.

machine in mechanics, device that allows a small force (the effort) to overcome a larger one (the load). There are three basic machines: the inclined plane (ramp), the lever, and the wheel and axle. All other machines are combinations of these three basic types. Simple machines derived from the inclined plane include the wedge, the gear, and the screw; the spanner is derived from the lever; the pulley from the wheel.

The principal features of a machine are its †mechanical advantage, which is the ratio of load to effort, its velocity ratio, and its †efficiency, which is the work done by the load divided by the work done by the effort; the latter is expressed as a percentage. In a perfect machine, with no friction, the efficiency would be 100%. All practical machines have efficiencies of less than 100%.

machine code in computing, a set of instructions that a computer's central processing unit (CPU) can understand and obey directly, without any translation. Each type of CPU has its own machine code. Because machine-code programs consist entirely of binary digits (bits), most programmers write their programs in an easy-to-use †high-level language. A high-level program must be translated into machine code – by means of a †compiler or interpreter program – before it can be executed by a computer.

Where no suitable high-level language exists or where very efficient machine code is required, programmers may choose to write programs in a low-level, or assembly, language, which is eventually translated into machine code by means of an †assembler program.

Microprocessors (CPUs based on a single integrated circuit) may be classified according to the number of machine-code instructions that they are capable of obeying: CISC (complex instruction set computer) microprocessors support up to 200 instructions, whereas RISC (reduced instruction set computer) microprocessors support far fewer instructions but execute programs more rapidly.

machine-readable of data, readable directly by a computer without the need for retyping. The term is usually applied to files on disk or tape, but can also be applied to typed or printed text that can be scanned for †optical character recognition or †bar codes.

Mach number ratio of the speed of a body to the speed of sound in the medium through which the body travels. In the Earth's atmosphere, Mach 1 is reached when a body (such as an aircraft or spacecraft) "passes the sound barrier," at a velocity of 331 m/1,087 ft per second (1,192 kph/740 mph) at sea level. A space shuttle reaches Mach 15 (about 17,700 kph/11,000 mph an hour) 6.5 minutes after launch.

The Mach number is named after Austrian physicist Ernst †Mach.

MACHO acronym for †massive astrophysical compact halo object.

Maclaurin, Colin (1698–1746) Scottish mathematician who played a leading part in establishing the dominance of Isaac †Newton's calculus in the UK. Maclaurin was the first to present the correct theory for distinguishing between the maximum and minimum values of a function.

Maclaurin, Colin http://www-groups.dcs.st-andrews.ac.uk/history/Mathematicians/Maclaurin.html

Part of an archive containing the biographies of the world's greatest mathematicians, this site is devoted to the life and contributions of Colin Maclaurin. In addition to biographical information, you will find a list of references about Maclaurin and links to other essays in the archive that reference him. The text of this essay includes hypertext links to the essays of those mathematicians and thinkers who influenced Maclaurin. You will also find an image of him, which you can click on to enlarge, and a map of his birthplace.

Macleod, John James Rickard (1876–1935) Scottish physiologist who was awarded a Nobel Prize for Physiology or Medicine in 1923 with Frederick †Banting for their part in the discovery and isolation of the hormone †insulin, the hormone in the pancreas that reduces blood glucose (sugar) levels. Since its discovery, insulin has been used extensively as the main treatment for diabetes.

McMillan, Edwin Mattison (1907–1991) US physicist. In 1940 he discovered neptunium, the first †transuranic element, by bombarding uranium with neutrons. He shared the Nobel Prize for Chemistry in 1951 with Glenn †Seaborg for their discovery and work in the chemistry of transuranic elements.

macro in computer programming, a new command created by combining a number of existing ones. For example, a word-processing macro might create a letterhead or fax cover sheet, inserting words, fonts, and logos with a single keystroke or mouse click. Macros are also useful to automate computer communications – for example, users can write a macro to ask their computer to dial an Internet Service Provider (ISP), retrieve e-mail and Usenet articles, and then disconnect. A **macro key** on the keyboard combines the effects of pressing several individual keys.

macro- prefix meaning on a very large scale, as opposed to micro-.

macromolecule in chemistry, a very large molecule, generally a †polymer.

macrophage type of †white blood cell, or leucocyte, found in all vertebrate animals. Macrophages specialize in the removal of bacteria and other micro-organisms, or of cell debris after injury. Like phagocytes, they engulf foreign matter, but they are larger than phagocytes and have a longer lifespan. They are found throughout the body, but mainly in the lymph and connective tissues, and especially the lungs, where they ingest dust, fibers, and other inhaled particles.

madness state of having a disordered mind; extreme folly or infatuation. Its synonym †insanity is no longer used to mean †psychosis but is used primarily in medical and legal contexts to refer to mental disorders in which patients cannot be held responsible for their actions.

maelstrom whirlpool off the Lofoten Islands, Norway, also known as the Moskenesstraumen, which gave its name to whirlpools in general.

mafic rock plutonic rock composed chiefly of dark-colored minerals such as olivine and pyroxene that contain abundant magnesium and iron. It is derived from **magnesium** and **ferric** (iron). The term **mafic** also applies to dark-colored minerals rich in iron and magnesium as a group. "Mafic rocks" usually refers to dark-colored igneous rocks such as basalt, but can also refer to their metamorphic counterparts.

Magellanic Clouds two galaxies nearest to our own galaxy. They are irregularly shaped, and appear as detached parts of the †Milky Way, in the southern hemisphere.

The Large Magellanic Cloud spreads over the constellations of †Dorado and †Mensa. The Small Magellanic Cloud is in †Tucana. The Large Magellanic Cloud is 169,000 light years from Earth, and about a third the diameter of our Galaxy; the Small Magellanic Cloud, 180,000 light years away, is about a fifth the diameter of our Galaxy. They are named after the early-16th-century Portuguese navigator Ferdinand Magellan, who first described them.

magic numbers in atomic physics certain numbers of †neutrons or †protons (2, 8, 20, 28, 50, 82, 126) in the nuclei of elements of outstanding stability, such as lead and helium.

Such stability is the result of neutrons and protons being arranged in completed "layers" or "shells."

magic square in mathematics, a square array of numbers in which the rows, columns, and diagonals add up to the same total. A simple example employing the numbers 1 to 9, with a total of 15, has a first row of 6, 7, 2, a second row of 1, 5, 9, and a third row of 8, 3, 4.

A **pandiagonal magic square** is one in which all the broken diagonals also add up to the magic constant.

magma molten rock material that originates in the lower part of the Earth's crust, or †mantle, where it reaches temperatures as high as 1,000°C/1,832°F. †Igneous rocks are formed from magma. †Lava is magma that has extruded onto the surface.

magnesia common name for †magnesium oxide.

magnesium chemical symbol Mg, lightweight, very ductile and malleable, silver-white, metallic element, atomic number 12, relative atomic mass 24.305. It is one of the †alkaline-earth metals, and the lightest of the commonly used metals. Magnesium silicate, carbonate, and chloride are widely distributed in nature. The metal is used in alloys, flares, and flash bulbs. It is a necessary trace element in the human diet, and green plants cannot grow without it since it is an essential constituent of the photosynthetic pigment †chlorophyll ($C_{55}H_{72}MgN_4O_5$).

It was named after the ancient Greek city of Magnesia, near where it was first found. It was first recognized as an element by Scottish chemist Joseph †Black in 1755 and discovered in its oxide by English chemist Humphry †Davy 1808. Pure magnesium was isolated in 1828 by French chemist Antoine-Alexandre-Brutus Bussy.

magnesium oxide or **magnesia**; MgO, white powder or colorless crystals, formed when magnesium is burned in air or oxygen; a typical basic oxide. It is used to treat acidity of the stomach, and in some industrial processes; for example, as a lining brick in furnaces, because it is very stable when heated (refractory oxide).

magnet any object that forms a magnetic field (displays †magnetism), either permanently or temporarily through induction, causing it to attract materials such as iron, cobalt, nickel, and alloys of these. It always has two †magnetic poles, called north and south.

The world's most powerful magnet was built in 1997 at the Lawrence Berkeley National Laboratory, California. It produces a field 250,000 times stronger than the Earth's magnetic field (13.5 teslas). The coil magnet is made of an alloy of niobium and tin.

magnetic compass device for determining the direction of the horizontal component of the Earth's magnetic field. It consists of a magnetized needle (a small bar magnet) with its north-seeking pole clearly indicated, pivoted so that it can turn freely in a plane parallel to the surface of the Earth (in a horizontal circle). The needle aligns itself with the lines of force of the Earth's magnetic field, turning so that its north-seeking pole points towards the Earth's magnetic north pole. See also †compass.

Walkers, sailors, and other travelers use a magnetic compass to find their direction. The direction of the geographic, or true, North Pole is, however, slightly different from that of the magnetic north pole, and so the readings obtained from a compass of this sort must be adjusted using tables of magnetic corrections or information marked on local maps.

A magnetic compass can also be used to check the direction of the magnetic field around a wire with a current flowing through it.

magnetic declination see †angle of declination.

magnetic detector device used in early †radio systems, in which high-frequency currents were detected through their demagnetizing effect on a magnetized †iron core surrounded by the wire-carrying currents.

magnetic dip see †dip, magnetic and †angle of dip.

magnetic dipole the pair of north and south magnetic poles, separated by a short distance, that makes up all magnets. Individual magnets are often called "magnetic dipoles." Single magnetic poles, or monopoles, have never been observed despite being searched for. See also magnetic †domain.

magnetic energy property of a magnet described by the multiplication of the †flux density by the †field strength on the demagnetization curve of a permanent magnet.

magnetic field region around a permanent magnet, or around a conductor carrying an electric current, in which a force acts on a moving charge or on a magnet placed in the field. The force cannot be seen; only the effects it produces are visible. The field can be represented by lines of force parallel at each point to the direction of a small compass needle placed on them at that point. These invisible lines of force are called the magnetic field or the flux lines. A magnetic field's magnitude is given by the †magnetic flux density (the number of flux lines per unit area), expressed in †teslas. See also †polar reversal.

Electromagnetism

http://particleadventure.org/particleadventure/frameless/electromagnetism.html

Introduction to the theory of electromagnetism, explaining how and why it occurs, with reference to fundamental particles. There are graphics included to facilitate understanding.

Magnetic Fields and Mars

http://mgs-mager.gsfc.nasa.gov/Kids/magfield.html

In addition to covering magnetic fields and Mars, this site also includes a simple explanation of magnetism and how it occurs. It details different types of

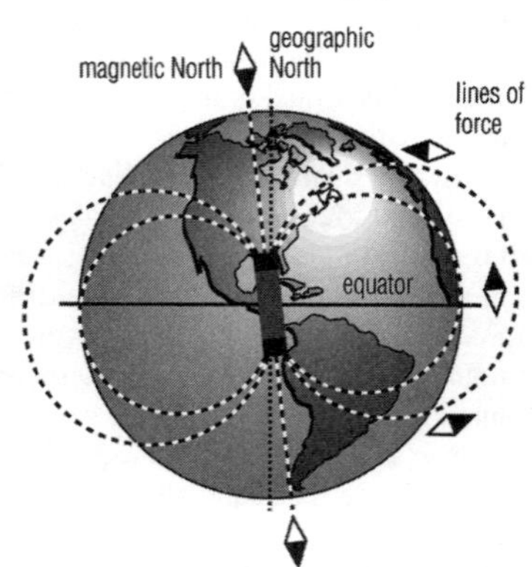

magnetic field The Earth's magnetic field is similar to that of a bar magnet with poles near, but not exactly at, the geographic poles. Compass needles align themselves with the magnetic field, which is horizontal near the equator and vertical at the magnetic poles. Helicon

magnets, such as bar and horseshoe magnets. Comparisons between the Earth's magnetic field and those of other planets are also given, plus links for further exploration of the Earth's magnetosphere.

magnetic field strength property measured to define the strength of a †magnetic field. The †SI unit is the †ampere per meter.

magnetic flux measurement of the strength of the magnetic field around electric currents and magnets. Its SI unit is the †weber; one weber per square meter is equal to one tesla.

The amount of magnetic flux through an area equals the product of the area and the magnetic field strength at a point within that area.

magnetic flux density †magnetic flux passing through one square meter of area of a magnetic field in a direction at right angles to the magnetic force. The †SI unit is the †tesla.

magnetic induction production of magnetic properties in unmagnetized iron or other ferromagnetic material when it is brought close to a magnet. The material is influenced by the magnet's magnetic field. The magnetic lines of force pass through the material causing its particles to align themselves parallel to the lines of the magnetic field. The induced magnetism may be temporary, disappearing as soon as the magnet is removed, or permanent, depending on the nature of the material and the strength of the magnet.

†Electromagnets make use of temporary induced magnetism to lift sheets of steel; the magnetism induced in the steel by the approach of the electromagnet enables it to be picked up and transported. To release the sheet, the current supplying the electromagnet is temporarily switched off and the induced magnetism disappears.

magnetic material one of a number of substances that are strongly attracted by †magnets and can be magnetized. These include iron, nickel, and cobalt, and all those †alloys that contain a proportion of these metals. Nonmagnetic materials are not attracted by magnets. These materials include paper, plastic, and rubber.

Magnetically soft materials can be magnetized very easily, but the magnetism induced in them (see †magnetic induction) is only temporary. They include Stalloy, an alloy of iron with 4% silicon used to make the cores of electromagnets and transformers, and the materials used to make "iron" nails and paper clips. Stroking a magnet over a steel pin from one end to the other will weakly magnetize the steel pin. This is because very large numbers of iron atoms (domains) of the steel become aligned in the same direction.

Magnetically hard materials can be permanently magnetized by a strong magnetic field. Steel and special alloys such as Alcomax, Alnico, and Ticonal, which contain various amounts of aluminum, nickel, cobalt, and copper, are used to make permanent magnets. The strongest permanent magnets are ceramic, made under high pressure and at high temperature from powders of various metal oxides. Iron is an example of a natural hard magnetic material. Its magnetic properties are due to its atomic structure. The electrons in the outer orbit of an iron atom behave as an electric charge and produce a strong magnetic field. In a magnetized piece of iron millions of individual iron atoms, called a domain, are aligned in the same direction. The domains have a north and a south pole.

magnetic moment product of the strength and length of a magnet.

magnetic monopole prediction of one †quantum theory (proposed by Dirac) that involves the existence of individual magnetic poles, analogous to the individual charges found in †electrostatics (that is, †electrons). Such an entity has not yet been definitely observed to exist, but is expected to be rare in any case.

magnetic permeability †magnetic flux density in a body divided by the external magnetic field strength producing it. It can be used as a way of classifying materials into different types of magnetism, such as †ferromagnetic.

magnetic pole region of a magnet in which its magnetic properties are strongest. Every magnet has two poles, called north and south. The north (or north-seeking) pole is so named because a freely-suspended magnet will turn so that this pole points towards the Earth's magnetic north pole. The north pole of one magnet will be attracted to the south pole of another, but will be repelled by its north pole. So unlike poles attract, like poles repel.

Magnetic North Pole http://deeptow.whoi.edu/northpole.html

Pages at the Woods Hole Oceanographic Institute providing a brief introduction to the attributes and behavior of the magnetic poles. There are details of the institute's geomagnetic research and links to lots of other useful Web sites.

magnetic resonance imaging **MRI**, diagnostic scanning system based on the principles of nuclear magnetic resonance (NMR). MRI yields finely detailed three-dimensional images of structures within the body without exposing the patient to harmful radiation. The technique is invaluable for imaging the soft tissues of the body, in particular the brain and the spinal cord.

Claimed as the biggest breakthrough in diagnostic imaging since the discovery of X-rays, MRI is a non-invasive technique based on a magnet that is many thousands of times stronger than the Earth's magnetic field. It causes nuclei within the atoms of the body to align themselves in one direction. When a brief radio pulse is beamed at the body the nuclei spin, emitting weak radio signals as they realign themselves to the magnet. These signals, which are characteristic for each type of tissue, are converted electronically into images on a viewing screen.

Also developed around magnetic technology, **magnetic resonance spectroscopy (MRS)** is a technique for investigating conditions in which there is a disturbance of the body's energy metabolism, including ischaemia and toxic damage due to drugs or other chemicals. MRS is also of value in diagnosing some cancers.

magnetic storm in meteorology, a sudden disturbance affecting the Earth's magnetic field, causing anomalies in radio transmissions and magnetic compasses. It is probably caused by †sunspot activity.

magnetic tape narrow plastic ribbon coated with an easily magnetizable material on which data can be recorded. It is used in sound recording, audio-visual systems (videotape), and computing. For mass storage on commercial mainframe computers, large reel-to-reel tapes are still used, but cartridges are becoming popular. Various types of cartridge are now standard on minicomputers and PCs, while audio cassettes were used with early home computers.

Magnetic tape was first used in **sound recording** in 1947, and made overdubbing possible, unlike the direct-to-disc system it replaced. Two-track tape was introduced in the 1950s and four-track in the early 1960s; today, studios use 16-, 24-, or 32-track tape, from which the tracks are mixed down to a stereo master tape.

In computing, magnetic tape was first used to record data and programs in 1951 as part of the UNIVAC 1 system. It was very popular as a storage medium for external memory in the 1950s and 1960s. Since then it has been largely replaced by magnetic †disks as a working medium, although tape is still used to make back-up copies of important data. Information is recorded on the tape in binary form, with two different strengths of signal representing 1 and 0.

magnetism phenomena associated with †magnetic fields. Magnetic fields are produced by moving charged particles. In electromagnets, electrons flow through a coil of wire connected to a battery; in permanent magnets, spinning electrons within the atoms generate the field.

Substances differ in the extent to which they can be magnetized by an external field (susceptibility). Materials that can be strongly magnetized, such as iron, cobalt, and nickel, are said to be **ferromagnetic**. This is due to the formation of areas called †domains in which atoms, weakly magnetic because of their spinning electrons, align to form areas of strong magnetism. Magnetic materials lose their magnetism if heated to the †Curie temperature. Most other materials are **paramagnetic**, being only weakly pulled towards a strong magnet. This is because their atoms have a low level of magnetism and do not form domains. **Diamagnetic** materials are weakly repelled by a magnet since electrons within their atoms act as electromagnets and oppose the applied magnetic force. **Antiferromagnetic** materials have a very low susceptibility that increases with temperature; a similar phenomenon in materials such as ferrites is called **ferrimagnetism**.

application Apart from its universal application in dynamos, electric motors, and switch gears, magnetism is of considerable importance in advanced technology – for example, in particle †accelerators for nuclear research, memory stores for computers, tape recorders, and †cryogenics.

Materials by Design: Magnetic Materials http://www.mse.cornell.edu/courses/engri111//magnet.htm

Short but effective page explaining why certain materials are magnetic, and how such materials can be used. Complete with complementary graphics, this site also provides links to other relevant topics, such as types of magnetism and the properties of other materials.

magnetite black, strongly magnetic opaque mineral, Fe_3O_4, of the spinel group, an important ore of iron. Widely distributed, magnetite is found in nearly all igneous and metamorphic rocks. Some deposits, called lodestone, are permanently magnetized. Lodestone has been used as a compass since the first millennium BC. Today the orientations of magnetite grains in rocks are used in the study of the Earth's magnetic field (see †paleomagnetism).

magnetohydrodynamics MHD, field of science concerned with the behavior of ionized gases or liquids in a magnetic field. Systems have been developed that use MHD to generate electrical power.

MHD-driven ships have been tested in Japan. In 1991 two cylindrical thrusters with electrodes and niobium–titanium superconducting coils, soaked in liquid helium, were placed under the passenger boat *Yamato 1*. The boat, 30 m/100 ft long, was designed to travel at 8 knots. An electric current passed through the electrodes accelerates water through the thrusters, like air through a jet engine, propelling the boat forward.

magnetometer device for measuring the intensity and orientation of the magnetic field of a particular rock or of a certain area. In geology, magnetometers are used to determine the original orientation of a rock formation (or the orientation when the magnetic signature was locked in), which allows for past plate reconstruction. They are also used to delineate "magnetic striping" on the sea floor in order to make plate reconstruction and to prospect for ore bodies such as iron ore, which can disrupt the local magnetic field.

magnetomotive force or **magnetic potential**, work done in carrying a unit magnetic pole around a magnetic circuit. The concept is analogous to †electromotive force in an electric circuit.

magnetopause in earth science, narrow region that lies between the †magnetosphere and outer space (the interplanetary medium).

magnetosphere volume of space, surrounding a planet, in which the planet's magnetic field has a significant influence. The Earth's magnetosphere extends 64,000 km/40,000 mi towards the Sun, but many times this distance on the side away from the Sun. That of Jupiter is much larger, and, if it were visible, would appear from the Earth to have roughly the same extent as the full Moon. The Russian-led space missions Coronas-I (launched in 1994) and Coronas-F (launched in 2001) were designed to investigate the magnetosphere of the Sun.

The Earth's magnetosphere is not symmetrical. The extension away from the Sun is called the **magnetotail**. The outer edge of the magnetosphere is the **magnetopause**. Beyond this is a turbulent region, the **magnetosheath**, where the †solar wind is deflected around the magnetosphere. Inside the magnetosphere, atomic particles follow the Earth's lines of magnetic force. The magnetosphere contains the †Van Allen radiation belts.

magnetron thermionic †valve (electron tube) for generating very high-frequency oscillations, used in radar and to produce microwaves in a microwave oven. The flow of electrons from the tube's cathode to one or more anodes is controlled by an applied magnetic field.

magnification measure of the enlargement or reduction of an object in an imaging optical system. **Linear magnification** is the ratio of the size (height) of the image to that of the object. **Angular magnification** is the ratio of the angle subtended at the observer's eye by the image to the angle subtended by the object when viewed directly.

magnifying glass simplest optical instrument, a hand-held converging lens used to produce a magnified, erect, and virtual image. The image, being virtual, or an illusion created by the †refraction of light rays in the lens, can only be seen by looking through the magnifying glass.

magnitude in astronomy, measure of the brightness of a star or other celestial object. The larger the number denoting the magnitude, the fainter the object. Zero or first magnitude indicates some of the brightest stars. Still brighter are those of negative magnitude, such as Sirius, whose magnitude is –1.46. **Apparent magnitude** is the brightness of an object as seen from the Earth; **absolute magnitude** is the brightness at a standard distance of 10 parsecs (32.616 light years).

Each magnitude step is equal to a brightness difference of 2.512 times. Thus a star of magnitude 1 is $(2.512)^5$ or 100 times brighter than a sixth-magnitude star just visible to the naked eye. The apparent magnitude of the Sun is –26.8, its absolute magnitude +4.8.

magnitude in mathematics, size irrespective of sign, used especially for †vectors irrespective of direction.

mail merge in computing, a feature offered by some word-processing packages that enables a list of personal details, such as names and addresses, to be combined with a general document outline to produce individualized documents.

For example, a club secretary might create a file containing a mailing list of the names and addresses of the club members. Whenever a letter is to be sent to all club members, a general letter outline is prepared with indications as to where individual names and addresses need to be added. The mail-merge feature then combines the file of names and addresses with the letter outline to produce and print individual letters addressed to each club member.

Maiman, Theodore Harold (1927–) US physicist who in 1960 constructed the first working †laser.

mainframe large computer used for commercial data processing and other large-scale operations. Because of the general increase in computing power, the differences between the mainframe, †supercomputer, †minicomputer, and †microcomputer (personal computer) are becoming less marked.

Mainframe manufacturers include IBM, Amdahl, Fujitsu, and Hitachi. Typical mainframes have from 128 MB to 4 GB of memory and hundreds of gigabytes of disk storage. Their primary use is to conduct very fast transactions with a database and to be split into several independent partitions, each with its own †operating system. A typical mainframe from IBM weighs between 1 and 2 tons.

main sequence in astronomy, part of the †Hertzsprung-Russell diagram that contains most of the stars, including the Sun. It runs diagonally from the top left of the diagram to the bottom right. The most massive (and hence brightest) stars are at the top left, with the least massive (coolest) stars at the bottom right.

major arc larger of the two arcs formed when a circle is divided into two unequal parts by a straight line or chord.

make-and-break circuit circuit that contains a device which, when †current flows in one part of the circuit, causes the device to break that part of the circuit, and allow current to flow in another part, and vice versa.

malachite common †copper ore, basic copper carbonate, $Cu_2CO_3(OH)_2$. It is a source of green pigment and is used as an antifungal agent in fish farming, as well as being polished for use in jewelry, ornaments, and art objects.

malaria infectious parasitic disease of the tropics transmitted by mosquitoes, marked by periodic fever and an enlarged spleen. When a female mosquito of the *Anopheles* genus bites a human who has malaria, it takes in with the human blood one of four malaria protozoa of the genus *Plasmodium*. This matures within the insect and is then transferred when the mosquito bites a new victim. Malaria affects around 300–500 million people each year, in 103 countries, and in 1995 around 2.1 million people died of the disease. In sub-Saharan Africa alone between 1.5 and 2 million children die from malaria and its consequences each year. In November 1998, an agreement was reached to establish a multi-agency program for research and control of the disease. The agencies involved include the World Health Organization (WHO), the World Bank, the United Nations Children's Fund, and the United Nations Development Program. The Roll Back Malaria campaign aims to halve deaths from malaria by 2010.

infection Inside the human body the parasite settles first in the liver, then multiplies to attack the red blood cells. Within the red blood cells the parasites multiply, eventually causing the cells to rupture and other cells to become infected. The cell rupture tends to be synchronized, occurring every 2–3 days, when the symptoms of malaria become evident.

on the increase Global warming is causing a worldwide increase in malaria. For example, in 1998 in Nairobi, where previously malaria cases had been limited to individuals who had traveled to lowland areas of Kenya, doctors were regularly reporting cases in people who had not left the city. In Irian Jaya, New Guinea, thousands of people who have never been exposed to malaria are now affected. According to a WHO report released in September 1999, cases of malaria in Europe rose from 2,882 in 1981 to 12,328 in 1997. African leaders met in Nigeria in April 2000 to discuss ways of fighting the disease, which was spreading across the continent, and, according to a WHO report, which had cost Africa £160 billion/$100 billion in productivity over the past 35 years.

treatment †Quinine, the first drug used against malaria, has now been replaced by synthetics, such as chloroquine, used to prevent or treat the disease. However, chloroquine-resistant strains of the main malaria parasite, *Plasmodium fulciparum*, are spreading rapidly in many parts of the world.

The drug mefloquine (Lariam) is widely prescribed for use in areas where chloroquine-resistant malaria prevails. It is surrounded by controversy, however, as it has been linked to unpleasant side effects, including psychiatric disturbances such as anxiety and hallucinations, epileptic seizures, and memory loss.

Another drug, artemether, derived from the shrub wormwood, was found in 1996 trials to be as effective as quinine in the treatment of cerebral malaria.

The insecticide †DDT remains one of the most effective means of controlling malaria, and consequently is still used despite its persistence in the environment and subsequent danger to wildlife.

vaccine An experimental malaria vaccine SPf66, developed by Colombian scientist Manuel Patarroyo, was trialed in 1994 in rural Tanzania, where villagers are bitten an average of 300 times a year by infected mosquitoes. It reduced the incidence of malaria by one third. However, further trials of SPf66 in the Gambia concluded that the vaccine provided only 8% protection for young children. A further trial in Thailand in 1996 failed to provide any evidence of its effectiveness. A new vaccine was successfully trialed in 1999; The vaccine attacks *Plasmodium falciparum*, the parasite that causes the symptoms of the disease, at each stage of the four main stages of its life cycle.

Malaria http://www.malaria.org/whatismalaria.html

Page from the Malaria Foundation covering frequently asked questions on malaria. The Web site contains answers to a comprehensive list of questions regarding the disease, including thorough descriptions of the virus and its related diseases.

maleic acid or **cis-butenedioc acid**; $C_4H_4O_4$, in chemistry, toxic, irritant, unsaturated dibasic organic acid. It is prepared industrially by the oxidation of benzene in the presence of a catalyst. It is used in the production of †malic acid.

Malfatti's problem medieval European mathematical problem, to inscribe within a triangle three circles tangent to one another.

malic acid or **hydroxysuccinic acid**; $HOOCCH_2CH(OH)COOH$, organic crystalline acid that can be extracted from apples, plums, cherries, grapes, and other fruits, but occurs in all living cells in smaller amounts, being one of the intermediates of the †Krebs cycle.

Malpighi, Marcello (1628–1694) Italian physiologist who made many anatomical discoveries in his pioneering microscope studies of animal and plant tissues. For example, he discovered blood capillaries and identified the sensory receptors (papillae) of the tongue, which he thought could be nerve endings.

Malpighian tubules in zoology, blind-ended tubules found in insects and spiders. They are organs of †excretion (removal of waste products).

There may be two to several hundred tubules depending on species, each growing from and opening into the intestine. The tubules are bathed in †hemolymph (invertebrate circulatory fluid). Sodium, potassium, uric acid (as soluble potassium urate), and other waste substances are secreted from the hemolymph into the tubules, followed by the osmotic movement of water. The excretory wastes are removed from the intestine through the anus, and the water and other useful substances re-absorbed.

maltase enzyme found in plants and animals that breaks down the disaccharide maltose into glucose.

Maltese-cross tube cathode-ray tube used to demonstrate some of the properties of cathode rays. The cathode rays, or electron streams, emitted by the tube's †electron gun are directed towards a fluorescent screen in front of which hangs a metal Maltese cross. Those electrons that hit the screen give up their kinetic energy and cause its phosphor coating to fluoresce. However, the sharply defined cross-shaped shadow cast on the screen shows that electrons are unable to pass through the Maltese cross. Cathode rays are thereby shown to travel in straight lines, and to be unable to pass through metal.

maltose $C_{12}H_{22}O_{11}$, †disaccharide sugar in which both monosaccharide units are glucose.

It is produced by the enzymic hydrolysis of starch and is a major constituent of malt, produced in the early stages of beer and whisky manufacture.

Malus, Etienne Louis (1775–1812) French physicist who discovered the polarization of light by reflection from a surface (see †polarized light). He also found the law of polarization that relates the intensity of the polarized beam to the angle of reflection.

Malus, Etienne Louis http://www-groups.dcs.st-andrews.ac.uk/history/Mathematicians/Malus.html

Part of an archive containing the biographies of the world's greatest mathematicians, this site is devoted to the life and contributions of Etienne Malus. In addition to biographical information, you will find a list of references about Malus and links to other essays in the archive that reference him. The text of this essay includes hypertext links to the essays of those mathematicians and thinkers who influenced Malus. You will also find a map of his birthplace.

Malus's law law giving the intensity of light after having been polarized through a certain angle.

mammal any of a large group of warm-blooded vertebrate animals characterized by having †mammary glands in the female; these are used for suckling the young. Other features of mammals are †hair (very reduced in some species, such as whales); a middle ear formed of three small bones (ossicles); a lower jaw consisting of two bones only; seven vertebrae in the neck; and no nucleus in the red blood cells. (Class Mammalia.)

Mammals are divided into three groups:

placental mammals, where the young develop inside the mother's body, in the †uterus, receiving nourishment from the blood of the mother via the †placenta;

marsupials, where the young are born at an early stage of development and develop further in a pouch on the mother's body where they are attached to and fed from a nipple; and

monotremes, where the young hatch from an egg outside the mother's body and are then nourished with milk.

The monotremes are the least evolved and have been largely displaced by more sophisticated marsupials and placentals, so that there are only a few types surviving (platypus and echidna). Placentals have spread to all parts of the globe, and where placentals have competed with marsupials, the placentals have in general displaced marsupial types. However, marsupials occupy many specialized niches in South America and, especially, Australasia.

According to the Red List of endangered species published by the World Conservation Union for 1996, 25% of mammal species are threatened with extinction.

The theory that marsupials succeed only where they do not compete with placentals was shaken in 1992, when a tooth, 55 million years old and belonging to a placental mammal, was found in Murgon, Australia, indicating that placental animals appeared in Australia at the same time as the marsupials. The marsupials, however, still prevailed.

There are over 4,000 species of mammals, adapted to almost every way of life. The smallest shrew weighs only 2 g/0.07 oz, the largest whale up to 140 tonnes. A 50-million-year-old jaw discovered in 1998 is from the smallest mammal ever to have lived, *Batodonoides*, an insectivore weighing only 1.3 g/0.046 oz.

mammary gland in female mammals, a milk-producing gland derived from epithelial cells underlying the skin, active only after the production of young. In all but monotremes (egg-laying mammals), the mammary glands terminate in teats which aid infant suckling. The number of glands and their position vary between species. In humans there are 2, in cows 4, and in pigs between 10 and 14.

The hatched young of monotremes simply lick milk from a specialized area of skin on the mother's abdomen.

Mandelbrot, Benoit B (1924–) Polish-born French mathematician who coined the term **fractal** to describe geometrical figures in which an identical motif repeats itself on an ever-diminishing scale. The concept is associated with †chaos theory.

Mandelbrot, Benoit http://www-groups.dcs.st-andrews.ac.uk/history/Mathematicians/Mandelbrot.html

Devoted to the life and contributions of Benoit Mandelbrot. In addition to biographical information, you will find a list of references about Mandelbrot and links to other essays that reference him. The text of this essay includes hypertext links to the essays of those mathematicians and thinkers who influenced Mandelbrot.

Mandelbrot set in mathematics, subset of the complex plane consisting of the points c for which repetition of the transformation $z \rightarrow z^2 + c$ applied to the point 0 gives a bounded sequence. It is a set with many interesting fractal properties.

mandible anatomical name for the lower †jaw bone in mammals.

manganese chemical symbol Mn, hard, brittle, gray-white metallic element, atomic number 25, relative atomic mass 54.9380. It resembles iron (and rusts), but it is not magnetic and is softer. It is used chiefly in making steel alloys, also alloys with aluminum and copper. It is used in fertilizers, paints, and industrial chemicals. It is a necessary trace element in human nutrition. The name is old, deriving from the French and Latin forms of Magnesia, a mineral-rich region of Italy.

manganese nodule in earth science, irregularly shaped mass up to 25 cm/10 in across, found mainly on the floor of the Pacific Ocean. The nodules are rich in manganese and iron with some nickel, copper, and sometimes cobalt. They are a potential source of these metals.

manganese ore any mineral from which manganese is produced. The main ores are the oxides, such as **pyrolusite**, MnO_2; **hausmannite**, Mn_3O_4; and **manganite**, $MnO(OH)$.

Manganese ores may accumulate in metamorphic rocks or as sedimentary deposits, frequently forming nodules on the sea floor (since the 1970s many schemes have been put forward to harvest deep-sea manganese nodules). The world's main producers are Georgia, Ukraine, South Africa, Brazil, Gabon, and India.

manganese(IV) oxide or **manganese dioxide**; MnO_2, brown solid that acts as a †depolarizer in dry batteries by oxidizing the hydrogen gas produced to water; without this process, the performance of the battery is impaired.

It acts as a †catalyst in decomposing hydrogen peroxide to obtain oxygen:

$2H_2O_{2\,(aq)} \rightarrow 2H_2O_{(l)} + O_{2\,(g)}$

It oxidizes concentrated hydrochloric acid to produce chlorine:.

$MnO_2 + 4HCl \rightarrow MnCl_2 + Cl_2 + 2H_2O$

manifold in two-dimensional space, a regular surface that locally looks like a flat plane slightly distorted. It can be represented by differentiable functions. There are analogues in spaces of more dimensions.

manometer instrument for measuring the pressure of liquids (including human blood pressure) or gases. In its basic form, it is a U-tube partly filled with colored liquid. Greater pressure on the liquid surface in one arm will force the level of the liquid in the other arm to rise. A difference between the pressures in the two arms is therefore registered as a difference in the heights of the liquid in the arms.

Manson, Patrick (1844–1922) Scottish physician who showed that insects are responsible for the spread of diseases like elephantiasis and malaria. He was knighted in 1903.

Mantell, Gideon Algernon (1790–1852) English geologist. He published many geological works, including *The Wonders of Geology* (1838). He was made a Fellow of the Royal Society in 1825.

mantissa in mathematics, the decimal part of a †logarithm. For example, the logarithm of 347.6 is 2.5411; in this case, the 0.5411 is the mantissa, and the integral (whole number) part of the logarithm, the 2, is the †characteristic.

mantle intermediate zone of the Earth between the †crust and the †core, accounting for 82% of the Earth's volume. The crust, made up of separate tectonic †plates, floats on the mantle which is made of dark semi-liquid rock that is rich in magnesium and silicon. The temperature of the mantle can be as high as 3,700°C/6,692°F. Heat generated in the core causes convection currents in the semi-liquid mantle; rock rises and then slowly sinks again as it cools, causing the movements of the tectonic plates. The boundary (junction) between the mantle and the crust is called the †Mohorovicic discontinuity, which lies at an average depth of 32 km/20 mi. The boundary between the mantle and the core is called the Gutenburg discontinuity, and lies at an average depth of 2,900 km/1,813 mi.

The mantle is subdivided into **upper mantle**, **transition zone**, and **lower mantle**, based upon the different velocities with which seismic waves travel through these regions. The upper mantle includes a zone characterized by low velocities of seismic waves, called the **low-velocity zone**, at 72 km/45 mi to 250 km/155 mi depth. This zone corresponds to the †asthenosphere upon which the Earth's lithospheric plates glide. Seismic velocities in the upper mantle are overall less than those in the transition zone, and those of the transition zone are in turn less than those of the lower mantle. Faster propagation of seismic waves in the lower mantle implies that the lower mantle is more dense than the upper mantle.

The mantle is composed primarily of magnesium, silicon, and oxygen in the form of †silicate minerals. In the upper mantle, the silicon in silicate minerals, such as olivine, is surrounded by four oxygen atoms. Deeper in the transition zone greater pressures promote denser packing of oxygen such that some silicon is surrounded by six oxygen atoms, resulting in magnesium silicates with garnet and pyroxene structures. Deeper still, all silicon is surrounded by six oxygen atoms so that the mineral perovskite $MgSiO_3$ predominates.

mantle keel in earth science, relatively cold slab of mantle material attached to the underside of a continental craton (core of a continent composed of old, highly deformed metamorphic rock), and protruding down into the mantle like the keel of a boat. Their presence suggests that tectonic processes may have been different at the time the cratons were formed.

map diagrammatic representation of an area, for example part of the Earth's surface or the distribution of the stars. Modern maps of the Earth are made using satellites in low orbit to take a series of overlapping stereoscopic photographs from which a three-dimensional image can be prepared. The earliest accurate large-scale maps appeared about 1580.

Conventional aerial photography, laser beams, microwaves, and infrared equipment are also used for land surveying. Many different kinds of †map projection (the means by which a three-dimensional body is shown in two dimensions) are used in map-making. Detailed maps requiring constant updating are kept in digital form on computer so that minor revisions can be made without redrafting.

Graphic Map's World Atlas and Geography Facts http://www.graphicmaps.com/aatlas/world.htm

Essential reference tool for geography students or anyone else interested in learning about the world. Presented as a gazetteer with facts and images, the site contains detailed maps for every country, island, territory, major island, lake, river, and mountain, together with its latitude and longitude, national flag, population, languages, international disputes, environment, and much more. There is a complete list of every country with its capital city, and even links to other sites focusing on the individual countries and cities.

Map Quiz Tutorial: Physical Geography http://www.harper.cc.il.us/mhealy/mapquiz/intro/inphyfr.htm

Interactive site that invites you to locate specific features on a physical map of the world. The answer frame at the bottom of the page tells you if you were correct.

map in mathematics, alternative name for †function.

mapping in mathematics, another name for †transformation.

map projection way of showing the Earth's spherical surface on a flat piece of paper. The most common approach has been to redraw the Earth's surface within a rectangular boundary. The main weakness of this is that countries in high latitudes are shown disproportionately large. The most famous cylindrical projection is the †Mercator projection, which dates from 1569. Although it gives an exaggerated view of the size of northern continents, it is the best map for navigation because a constant bearing appears as a straight line.

In 1973 German historian Arno Peters devised the **Peters projection** in which the countries of the world retain their relative areas. In other projections, lines of longitude and latitude appear distorted, or the Earth's surface is shown as a series of segments joined along the Equator. In 1992 the US physicist Mitchell Feigenbaum devised the **optimal conformal** projection, using a computer program designed to take data about the boundary of a given area and calculate the projection that produces the minimum of inaccuracies.

The theory behind traditional map projection is that, if a light were placed at the center of a transparent Earth, the surface features could be thrown as shadows on a piece of paper close to the surface. This paper may be flat and placed on a pole (azimuthal or zenithal), or may be rolled around the Equator (cylindrical), or may be in the form of a tall cone resting on the Equator (conical). The resulting maps differ from one another, distorting either area or direction, and each is suitable for a particular purpose. For example, projections distorting area the least are used for distribution maps, and those with least distortion of direction are used for navigation charts.

marble rock formed by metamorphosis of sedimentary †limestone. It takes and retains a good polish, and is used in building and sculpture. In its pure form it is white and consists almost entirely of calcite ($CaCO_3$). Mineral impurities give it various colors and patterns. Carrara, Italy, is known for white marble.

Marconi, Guglielmo (1874–1937) Italian electrical engineer and pioneer in the invention and development of radio. In 1895 he achieved radio communication over more than a mile, and in England in 1896 he conducted successful experiments that led to the formation of the company that became Marconi's Wireless Telegraph Company Ltd. He shared the Nobel Prize for Physics in 1909 for the development of wireless telegraphy.

Biography of G Marconi http://www.nobel.se/physics/laureates/1909/marconi-bio.html

Presentation of the life and discoveries of Guglielmo Marconi. The site includes sections on his early years and education, and the many patents that he established in the course of his scientific career. Central emphasis is of course given to his development of wireless telegraphy, from the first attempt to send a signal one and a half miles away, up to his successful wireless transmission across the Atlantic in 1901.

Marcus, Rudolph Arthur (1923–) Canadian chemist who advanced the theory of electron-transfer reactions (involving soluble molecules and/or ions) which drive many biological processes. By focusing on the solvent molecules in the reaction, Marcus realized that electron-transfer reaction rate coefficients could be determined that have an explanatory and predictive value. The theory has also been applied to proton-transfer reactions. He was awarded the Nobel Prize for Chemistry in 1992 for his theoretical discoveries relating to reduction and oxidation reactions.

mare plural **maria**, dark lowland plain on the Moon. The name comes from the Latin word for "sea," because these areas were once wrongly thought to be water.

Mariana Trench lowest region on the Earth's surface; the deepest part of the sea floor. The trench is 2,400 km/1,500 mi long and is situated 300 km/200 mi east of the Mariana Islands, in the northwestern Pacific Ocean. Its deepest part is the gorge known as the Challenger Deep, which extends 11,034 m/36,210 ft below sea level.

Mariotte, Edme (1620–1684) French physicist and priest known for his recognition in 1676 of †Boyle's law about the inverse relationship of volume and pressure in gases, formulated by Irish physicist Robert †Boyle in 1672. He had earlier, in 1660, discovered the eye's blind spot.

Markov, Andrei Andreyevich (1856–1922) Russian mathematician, formulator of the †Markov chain, an example of a stochastic (random) process.

Markov, Andrei Andreyevich http://www-groups.dcs.st-andrews.ac.uk/history/Mathematicians/Markov.html

Devoted to the life and contributions of Andrei Markov. In addition to biographical information, you will find a list of references about Markov and links to essays that reference him. The text of this essay includes hypertext links to the essays of those mathematicians and thinkers who influenced Markov. You will also find a map of his birthplace.

Markov chain in statistics, an ordered sequence of discrete states (random variables) $x_1, x_2, ..., x_i, ..., x_n$ such that the probability of x_i depends only on n and/or the state x_{i-1} which has preceded it. If independent of n, the chain is said to be homogeneous.

marl crumbling sedimentary rock, sometimes called **clayey limestone**, including various types of calcareous †clays and fine-grained †limestones. Marls are often laid down in freshwater lakes and are usually soft, earthy, and of a white, gray, or brownish color. They are used in cement-making and as fertilizer.

Marr, David Courtenay (1945–1980) English psychologist who developed computer-based models of the visual system. Drawing on neurophysiology and the psychology of vision, he applied his models to a number of issues, notably the problem of how objects in the perceptual field are represented within the brain. His findings are summarized in *Vision* (1982), published posthumously.

Mars fourth planet from the Sun. It is much smaller than Venus or Earth, with a mass 0.11 that of Earth. Mars is slightly pear-shaped, with a low, level northern hemisphere, which is comparatively uncratered and geologically "young," and a heavily cratered "ancient" southern hemisphere.

mean distance from the Sun: 227.9 million km/141.6 million mi

equatorial diameter: 6,780 km/4,210 mi

rotation period: 24 hours 37 minutes

year: 687 Earth days

atmosphere: 95% carbon dioxide, 3% nitrogen, 1.5% argon, and 0.15% oxygen. Red atmospheric dust from the surface whipped up by winds of up to 450 kph/280 mph accounts for the light pink sky. The surface pressure is less than 1% of the Earth's atmospheric pressure at sea level

surface: the landscape is a dusty, red, eroded lava plain. Mars has white polar caps (water ice and frozen carbon dioxide) that advance and retreat with the seasons

satellites: two small satellites: †Phobos and Deimos

There are four enormous volcanoes near the equator, of which the largest is Olympus Mons 24 km/15 mi high, with a base 600 km/375 mi across, and a crater 65 km/40 mi wide. To the east of the four volcanoes lies a high plateau cut by a system of valleys, Valles Marineris, some 4,000 km/2,500 mi long, up to 200 km/120 mi wide and 6 km/4 mi deep; these features are apparently caused by faulting and wind erosion. Recorded temperatures vary from –100°C/–148°F to 0°C/32°F.

Mars may approach Earth to within 54.7 million km/34 million mi. The first human-made object to orbit another planet was Mariner 9, launched in 1971. The two Viking probes, which landed in 1976, provided more information.

In December 1996, NASA launched the Mars Pathfinder, which made a successful landing on Mars in July 1997 on a flood plain called Ares Vallis. After initial technical problems, its 0.3-m/1-ft rover, Sojourner, began to explore the Martian landscape and to transmit data back to Earth. Photographs from the Mars Pathfinder indicated that the planet is rusting. NASA announced this in July 1997 and said that a supercorrosive force was eroding rocks on the surface due to iron oxide in the soil.

In May 1997, US scientists announced that Mars is becoming increasingly cold and cloudy. Images from the Hubble Space Telescope showed that dust storms had covered areas of the planet that had been dark features in the early 20th century, including one section as large as California.

The Mars Global Surveyor, launched 7 November 1996, entered Martian orbit in September 1997. Its data revealed that Mars' magnetic field is a mere 800th that of the Earth. In February 1999, the spacecraft established its orbit for mapping the surface of the planet. In 2001 it located two regions on the surface rich in †hematite, providing further evidence for the existence of water at some time in the planet's history.

NASA's Mars Climate Orbiter to monitor weather on Mars was launched from Cape Canaveral, Florida, in December 1998. It was expected to reach its destination in September 1999. However, a measurement error caused the probe to fly too close to Mars and break up. Workers at Lockheed Martin Astronautics in Colorado had given acceleration data in pounds of force, but navigators at NASA's Jet Propulsion Laboratory had assumed the numbers were metric newtons.

In December 2000, a New Zealand scientist revealed that he had managed to grow vegetables in soil taken from Martian meteorites, and that they grew better than vegetables grown in New Zealand farming soil. This supports the contention that human beings can successfully exploit resources on other planets.

The NASA probe Mars Odyssey, launched in April 2001, discovered water in the form of subsurface ice at Mar's south pole in March 2002.

The European Space Agency's Mars Express was launched in June 2003 and discovered methane and water ice during orbits of the planet. It carried a lander for soil collection, Beagle 2, which descended to Mar's surface on 25 December 2003, but all communications were lost.

NASA's Mars Exploration Rover Mission, consisting of two rover-carrying spacecraft, launched in the summer of 2003. The twin rovers, named Spirit and Opportunity, landed on the surface of Mars in January 2004; they found traces of previous water. NASA extended their activity five months beyond their original three-month mission.

Beagle 2 http://beagle2.open.ac.uk/

All about the Beagle 2 project, the UK-led effort to land on the planet Mars as part of the European Space Agency's Mars Express Mission to be launched in June 2003. The site also gives information about Mars itself and about the people involved in the project.

Mars Exploration http://mars.jpl.nasa.gov/

Well-presented NASA site with comprehensive information on current and future missions to Mars. There are fascinating and well-written accounts of Pathfinder and Global Surveyor, and large numbers of images of the planet.

Mars Introduction http://www.solarviews.com/eng/mars.htm

Detailed description of the planet Mars, including statistics and information about its surface, volcanoes, satellites, and clouds, supported by a good selection of images.

Mars Meteorites http://www.jpl.nasa.gov/snc/

Information about and images of the 20 Mars meteorites that have been discovered on Earth, including the latest news on research that debunks NASA's initial claim that a meteorite contained evidence of Martian life. You can research the news archives, and there is information on future missions to Mars.

Virtual Mars http://www.sciencemonster.com/virtualmars

View a three-dimensional image of Mars. Clicking on a land feature centers your view at that spot and allows you to rotate the planet. Click on the button below the image to toggle the names of surface features and landing sites, including the Pathfinder and Viking 2 landing sites. Night and day areas are recalculated every five minutes to adjust to rotation.

Whole Mars Catalog http://www.spaceref.com/mars/

Regularly updated with links to breaking stories of Mars research and events. There are links to other major Mars sites, and to pages that deal with certain key topics such as Mars meteorites and missions to Mars.

marsh low-lying wetland. Freshwater marshes are common wherever groundwater, surface springs, streams, or run-off cause frequent flooding, or more or less permanent shallow water. A marsh is alkaline whereas a †bog is acid. Marshes develop on inorganic silt or clay soils. Rushes are typical marsh plants. Large marshes dominated by papyrus, cattail, and reeds, with standing water throughout the year, are commonly called †swamps. Near the sea, †salt marshes may form.

marsh gas gas consisting mostly of †methane. It is produced in swamps and marshes by the action of bacteria on dead vegetation.

marsupial (Greek *marsupion* "little purse") mammal in which the female has a pouch where young (born tiny and immature) are carried for a considerable time after birth. Marsupials include omnivorous, herbivorous, and carnivorous species, among them the kangaroo, wombat, opossum, phalanger, bandicoot, dasyure, and wallaby.

The Australian marsupial anteater known as the numbat is an exception to the rule in that it has no pouch.

Martin, Archer John Porter (1910–) English biochemist who shared the Nobel Prize for Chemistry in 1952 for work with Richard †Synge on partition †chromatography in 1944.

mascon contraction of **mass concentration**, any of a number of regions on the surface of the Moon where gravity is apparently stronger. The effect is presumed to be due to localized areas of denser rock strata. There have been cases of lunar modules dipping slightly as they pass above mascons.

maser acronym for **microwave amplification by stimulated emission of radiation**, in physics, a high-frequency microwave amplifier or oscillator in which the signal to be amplified is used to stimulate excited atoms into emitting energy at the same frequency. Atoms or molecules are raised to a higher energy level and then allowed to lose this energy by radiation emitted at a precise frequency. The principle has been extended to other parts of the electromagnetic spectrum as, for example, in the †laser.

The two-level ammonia-gas maser was first suggested in 1954 by US physicist Charles Townes at Columbia University, New York, and independently the same year by Nikolai Basov and Aleksandr Prokhorov in Russia. The solid-state three-level maser, the most sensitive amplifier known, was envisaged by Nicolaas Bloembergen at Harvard in 1956. The ammonia maser is used as a frequency standard oscillator, and the three-level maser as a receiver for satellite communications and radio astronomy.

Maskelyne, Nevil (1732–1811) English astronomer. He made observations to investigate the reliability of the lunar distance method for determining longitude at sea. In 1774 he estimated the mass of the Earth by noting the deflection of a plumb line near Mount Schiehallion in Perthshire, Scotland.

Maskelyne, Nevil http://www-groups.dcs.st-andrews.ac.uk/history/Mathematicians/Maskelyne.html

Devoted to the life and contributions of Nevil Maskelyne. In addition to biographical information, you will find a list of references about Maskelyne and links to essays that reference him. The text of this essay includes hypertext links to the essays of those mathematicians and thinkers who influenced Maskelyne. You will also find an image of him, which you can click on to enlarge, and a map of his birthplace.

Maslow, Abraham (Harold) (1908–1970) US psychologist. A professor at Brooklyn College (1937–51) and Brandeis University (1951–61), he is regarded as the founder of humanistic psychology. His seminal *Motivation and Personality* (1954) explicated the new humanistic model and introduced such soon-to-be standard psychological concepts as the hierarchy of needs and self-actualization. Maslow applied his pioneering theory of self-actualization to leadership, management, and organizational development in his publication *Eupsychian Management* (1965).

masochism desire to subject oneself to physical or mental pain, humiliation, or punishment, for erotic pleasure, to alleviate guilt, or out of destructive impulses turned inward. The term is derived from Leopold von Sacher-Masoch.

mass in physics, quantity of matter in a body as measured by its inertia, including all the particles of which the body is made up. Mass determines the acceleration produced in a body by a given force acting on it, the acceleration being inversely proportional to the mass of the body. The mass also determines the force exerted on a body by †gravity on Earth, although this attraction varies slightly from place to place (the mass itself will remain the same). In the SI system, the base unit of mass is the kilogram.

At a given place, equal masses experience equal gravitational forces, which are known as the weights of the bodies. Masses may, therefore, be compared by comparing the weights of bodies at the same place.

The standard unit of mass to which all other masses are compared is a platinum-iridium cylinder of 1 kg, which is kept at the International Bureau of Weights and Measures in Sèvres, France.

mass action, law of in chemistry, a law stating that at a given temperature the rate at which a chemical reaction takes place is proportional to the product of the active masses of the reactants. The active mass is taken to be the molar concentration of each reactant.

mass–energy equation German-born US physicist Albert †Einstein's equation $E = mc^2$, denoting the equivalence of mass and energy. In SI units, E is the energy in joules, m is the mass in kilograms, and c, the speed of light in a vacuum, is in meters per second.

mass extinction event that produces the extinction of many species at about the same time. One notable example is the boundary between the Cretaceous and Tertiary periods (known as the †K-T boundary) that saw the extinction of the dinosaurs and other large reptiles, and many of the marine invertebrates as well. Mass extinctions have taken place frequently during Earth's history.

There have been five major mass extinctions, in which 75% or more of the world's species have been wiped out: End Ordovician period (440 million years ago) in which about 85% of species were destroyed (second most severe); Late Devonian period (365 million years ago) which took place in two waves a million years apart, and was the third most severe, with marine species particularly badly hit; Late Permian period (251 million years ago), the gravest mass Late Triassic (205 million years ago), in which about 76% of species were destroyed, mainly marine; Late Cretaceous period (65 million years ago), in which 75–80 of species became extinct, including dinosaurs.

massive astrophysical compact halo object MACHO, component of the Galaxy's †dark matter. Most MACHOs are believed to be brown dwarfs, tiny failed stars with a mass of about 8% that of the Sun, but they may also include neutron stars left behind after supernova explosions. MACHOs are identifiable when they move in front of stars causing microlensing (magnification) of the star's light. Astronomers first identified MACHOs in 1993 and estimate that they account for 20% of dark matter.

mass movement processes that move weathered material on slopes without the action of running water. Movement is due to gravity, and may be on a large or small scale. Different types of mass movement include rockfalls, †landslides, †soil creep, and †solifluction.

mass number or **nucleon number**; symbol A, sum of the numbers of †protons and †neutrons in the nucleus of an atom. It is used along with the †atomic number (the number of protons) in †nuclear notation: in symbols that represent nuclear †isotopes, such as ${}^{14}_{6}C$, the lower number is the atomic number, and the upper number is the mass number. Since the mass of the †electrons in an atom are negligible, the total number of protons is neutrons (mass number) in an atom determines its mass.

mass spectrometer apparatus for analyzing chemical composition by separating ions by their mass; the ions may be elements, isotopes, or molecular compounds. Positive ions (charged particles) of a substance are separated by an electromagnetic system, designed to focus particles of equal mass to a point where they can be detected. This permits accurate measurement of the relative concentrations of the various ionic masses present. A mass spectrometer can be used to identify compounds or to measure the relative abundance of compounds in a sample.

Introduction to Mass Spectrometry http://www.rmsb.u-bordeaux2.fr/rmsb/ms/IntroMS.html

Good introduction to the mass spectrometer and how it works. Different mass analyzer designs are described, together with sections on ionization and ion detectors.

Mastigophora subphylum of protozoa typified by the possession of one or more †flagella (whiplike organs).

classification Mastigophora is a subphylum in phylum Sarcomastigophora.

The group is composed of two classes, the Phytomastigophora and the Zoomastigophora. The **Phytomastigophora** have chlorophyll and are free-living in the sea, fresh water, and soil. They include *Euglena* and *Volvox*. Possessing chlorophyll, they are able to synthesize organic compounds from inorganic ones using the energy of sunlight; some, however, are colorless, and thus require organic matter to survive. This difference occurs within the members of one genus, *Euglena*, and illustrates the difficulties of deciding whether they are plants or animals.

Members of the **Zoomastigophora** have no chlorophyll and require organic food particles. Some species, such as *Peranema*, obtain these by engulfing large living prey such as *Euglena*. Members of the order Choanoflagellida are filter feeders. They are sessile (fixed) and feed by setting up a current with the flagellum. This current is drawn through the collar, which encloses the base of their flagellum. Here any particles are filtered off and move down through the base of the collar to form food vacuoles. Some Zoomastigophora are parasitic and cause economically important diseases such as leishmaniasis, malaria and trypanosomiasis (sleeping sickness).

mathematical induction formal method of proof in which the proposition $P(n + 1)$ is proved true on the hypothesis that the proposition $P(n)$ is true. The proposition is then shown to be true for a particular value of n, say k, and therefore by induction the proposition must be true for $n = k + 1, k + 2, k + 3, \ldots$. In many cases $k = 1$, so then the proposition is true for all positive integers.

mathematical logic view of mathematics and logic that relates the two disciplines, and to do so uses mathematical or similar notation in the expression of axiomatic statements. †Boolean algebra was the original form, and led to German philosopher Gottlob Frege's symbolic logic which is now in standard use in mathematics.

mathematical structure collection of objects that display (a) one or more relationships, and (b) one or more operations of which the properties may be summarized as a list of †axioms; for example, a group, vector, space or ring.

mathematics science of relationships between numbers, between spatial configurations, and abstract structures. The main divisions of **pure mathematics** include geometry, arithmetic, algebra, calculus, and trigonometry. Mechanics, statistics, numerical analysis, computing, the mathematical theories of astronomy, electricity, optics, thermodynamics, and atomic studies come under the heading of **applied mathematics**.

early history Prehistoric humans probably learned to count at least up to ten on their fingers. The ancient Egyptians (3rd millennium BC), Sumerians (2000–1500 BC), and Chinese (1500 BC) had systems for writing down numbers and could perform calculations using various types of †abacus. They used some fractions. Mathematicians in ancient Egypt could solve simple problems which involved finding a quantity that satisfied a given linear relationship. Sumerian mathematicians knew how to solve problems that involved quadratic equations. The fact that, in a right-angled triangle, the square of the longest side is equal to the sum of the squares of the other two sides (Pythagoras' theorem) was known in various forms in these cultures and also in Vedic India (1500 BC).

The first theoretical mathematician is held to be Thales of Miletus (*c.* 580 BC) who is believed to have proposed the first theorems in plane geometry. His disciple †Pythagoras established geometry as a recognized science among the Greeks. Pythagoras began to insist that mathematical statements must be proved using a logical chain of reasoning starting from acceptable assumptions. Undoubtedly the impetus for this demand for logical proof came from the discovery by this group of the surprising fact that the square root of 2 is a number which cannot be expressed as the ratio of two whole numbers. The use of logical reasoning, the methods of which were summarized by Aristotle, enabled Greek mathematicians to make general statements instead of merely solving individual problems as earlier mathematicians had done.

The spirit of Greek mathematics is typified in one of its most lasting achievements, the *Elements* by Euclid. This is a complete treatise on geometry in which the entire subject is logically deduced from a handful of simple assumptions. The ancient Greeks lacked a simple notation for numbers and nearly always relied on expressing problems geometrically. Although the Greeks were extremely successful with their geometrical methods they never developed a general theory of equations or any algebraic ideas of structure. However, they made considerable advances in techniques for solving particular kinds of equations and these techniques were summarized by Diophantus of Alexandria.

medieval period When the Hellenic civilization declined, Greek mathematics (and the rest of Greek science) was kept alive by the Arabs, especially in the scientific academy at the court of the caliphs of Baghdad. The Arabs also learned of the considerable scientific achievements of the Indians, including the invention of a system of numerals (now called "arabic" numerals) which could be used to write down calculations instead of having to resort to an abacus. One mathematician can be singled out as a bridge between the ancient and medieval worlds: al-†Khwarizmi summarized Greek and Indian methods for solving equations and wrote the first treatise on the Indian numerals and calculating with them. Al-Khwarizmi's books and other Arabic works were translated into Latin and interest in mathematics in Western Europe began to increase in the 12th century. It was the demands of commerce which gave the major impetus to mathematical development and north Italy, the center of trade at the time, produced a succession of important mathematicians beginning with Italian mathematician Leonardo Fibonacci who introduced Arabic numerals. The Italians made considerable advances in elementary arithmetic which was needed for money-changing and for the technique of double-entry bookkeeping invented in Venice. Italian mathematicians began to express equations in symbols instead of words. This algebraic notation made it possible to shift attention from solving individual equations to investigating the relationship between equations and their solutions, and led eventually to the discovery of methods of solving cubic equations (about 1515) and quartic equations. They began to use the square roots of negative numbers (complex numbers) in their solutions to equations.

early modern period In the 17th century the focus of mathematical activity moved to France and Britain though continuing with the major themes of Italian mathematics: improvements in methods of calculation, development of algebraic symbolism, and the development of mathematical methods for use in physics and astronomy. Geometry was revitalized by the invention of coordinate geometry by René Descartes in 1637; Blaise Pascal and Pierre de Fermat developed probability theory; John Napier invented logarithms; and Isaac Newton and Gottfried Leibniz invented calculus, later put on a more rigorous footing by Augustin Cauchy. In Russia, Nikolai Lobachevsky rejected Euclid's parallelism and developed a non-Euclidean geometry; this was subsequently generalized by Bernhard Riemann and later utilized by Einstein in his theory of relativity. In the mid-19th century a new major theme emerged: investigation of the logical foundations of mathematics. George Boole showed how logical arguments could be expressed in algebraic symbolism. Friedrich Frege and Giuseppe Peano considerably developed this symbolic logic.

the present In the 20th century, mathematics became much more diversified. Each specialist subject is being studied in far greater depth and advanced work in some fields may be unintelligible to researchers in other fields. Mathematicians working in universities have had the economic freedom to pursue the subject for its own sake. Nevertheless, new branches of mathematics have been developed which are of great practical importance and which have basic ideas simple enough to be taught in secondary schools. Probably the most important of these is the mathematical theory of statistics in which much pioneering work was done by Karl Pearson. Another new development is operations research, which is concerned with finding optimum courses of action in practical situations, particularly in economics and management. As in the late medieval period, commerce began to emerge again as a major impetus for the development of mathematics.

Higher mathematics has a powerful tool in the high-speed electronic computer, which can create and manipulate mathematical "models" of various systems in science, technology, and commerce.

Modern additions to school syllabuses such as sets, group theory, matrices, and graph theory are sometimes referred to as "new" or "modern" mathematics.

Frequently Asked Questions in Mathematics http://www.cs.unb.ca/~alopez-o/math-faq/mathtext/math-faq.html

Expert answers to frequently asked questions, from "Why is there no Nobel prize in mathematics?" to "What is the current status of Fermat's last theorem?"

History of Mathematics http://www-groups.dcs.st-andrews.ac.uk/history/HistTopics/History_overview.html

Overview of the history of mathematics with numerous links to important mathematicians from ancient times to the present day. The site includes sections on the problems of notation and communication, and on brilliant discoveries. There is also an extensive bibliography.

Mathematics Problem Solving Task Center http://www.mav.vic.edu.au/PSTC/

Australian site, updated monthly, containing not just one, but several problems each month. They are each designed for different levels of understanding. The main problems are divided into lower/middle primary, upper primary/lower secondary, and upper secondary levels. The site also includes a database of previous problems and a "Challenging problem of the month." Good math practice!

Mathilde rocky carbon-rich asteroid. It is dominated by a huge 25-km/15.5-mi crater and its mass is one-millionth the mass of the Moon. According to information from NASA's Near-Earth Asteroid Rendezvous (NEAR), Mathilde's dimensions are 66 x 48 x 46 km/41 x 30 x 29 mi, with a density of 1.3 g per cu cm/0.75 oz per cu in. It has a rotation period of 17.4 days.

matrix in biology, usually refers to the †extracellular matrix.

matrix in mathematics, a square (n x n) or rectangular (m x n) array of elements (numbers or algebraic variables) used to facilitate the study of problems in which the relation between the elements is important. They are a means of condensing information about mathematical systems and can be used for, among other things, solving simultaneous linear equations (see †simultaneous equations and †transformation).

The advantage of matrices is that they can be studied algebraically by assigning a single symbol to a matrix rather than considering each element separately. The symbol used is usually a bold capital letter, but often a matrix is denoted by a symbol like (a_{ij}), meaning "the matrix with element a in row i column j." The size of a matrix is described by stating the number of its rows and then the number of its columns so, for example, a matrix with three rows and two columns is a 3 x 2 matrix. A matrix with equal numbers of rows and columns is called a "square matrix."

Much early matrix theory was developed by the British mathematician Arthur †Cayley, although the term was coined by his contemporary James Sylvester (1814–1897).

Matrices and Determinants http://www-history.mcs.st-and.ac.uk/history/HistTopics/Matrices_and_determinants.html
History of the development and usage of matrices and determinants. This site also provides biographical background on the mathematicians responsible for the advancement of the topic, along with a description of the subject, and some examples.

matrix mechanics mathematical description of subatomic phenomena that views certain characteristics of the particles involved as being matrices, and hence obeying the rules of matrix mathematics that differ in significant ways from the rules obeyed by ordinary arithmetic.

matter in physics, anything that has mass. All matter is made up of †atoms, which in turn are made up of †elementary particles; it ordinarily exists in one of three physical states: solid, liquid, or gas.

states of matter Whether matter exists as a solid, liquid, or gas depends on its temperature and the pressure on it. †Kinetic theory describes how the state of a material depends on the movement and arrangement of its atoms or molecules. In a solid the atoms or molecules vibrate in a fixed position. In a liquid, they do not occupy fixed positions as in a solid, and yet neither do they have the freedom of random movement that occurs within a gas, so the atoms or molecules within a liquid will always follow the shape of their container. The transition between states takes place at definite temperatures, called melting point and boiling point. **conservation of matter** In chemical reactions matter is conserved, so no matter is lost or gained and the sum of the mass of the reactants will always equal the sum of the end products.

Matter and its Changes http://library.thinkquest.org/10429/high/matter/matter.htm
Detailed explanation of what matter is and how it undergoes physical and chemical changes. The site covers atoms and molecules; solids, liquids, and gases; and elements, compounds, and mixtures. It is supported by colorful images and diagrams.

States of Matter http://ull.chemistry.uakron.edu/genobc/Chapter_06/
Summary of the various chemistry concepts associated with the states of matter. Part of a much larger chemistry site from the University of Akron, USA, this page explains kinetic molecular theory and the observed properties of the solid, liquid, and gaseous states. Other features include a set of key equations and an interactive practice exam.

matter, continuous creation of phenomenon invoked in certain cosmological theories, especially the †steady-state model of the universe. In that particular theory, matter is considered to be constantly created, either evenly throughout space, or in localized regions of creation, so as to make up for the diluting effect the †Hubble expansion of the universe has on the average density of matter. By continually creating matter, it is then possible for the universe to maintain a steady-state appearance.

Mauchly, John William (1907–1980) US physicist and engineer who, in 1946, constructed the first general-purpose computer, the ENIAC, in collaboration with John †Eckert. Their company was bought by Remington Rand in 1950, and they built the UNIVAC 1 computer in 1951 for the US census.

Maudsley, Henry (1835–1918) English physician. He specialized in mental disorders, and impressed with the need for research into the heredity and pathology of mental disease, he gave a large sum for the creation of a psychiatric hospital. The Maudsley Hospital, London, was the result, though he did not live to see its completion.

Maunder minimum in astronomy, period 1645–1715 when †sunspots were rarely seen and no †auroras were recorded. The Maunder minimum coincided with a time of unusually low temperature on Earth, known as the Little Ice Age, and is often taken as evidence that changes in solar activity can affect the Earth's climate. The Maunder minimum is named after the English astronomer Edward Walter Maunder, who drew attention to it.

Maupertuis, Pierre Louis Moreau de (1698–1759) French mathematician. He was an ardent Newtonian. In 1736 he led a party seeking to verify Newton's theory that the Earth was flattened at the poles. His analysis of the data was completed in 1739 and the result vindicated Newton's theory (though it was later found to be inaccurate).

Maury, Antonia Caetana de Paiva Pereira (1866–1952) US expert in stellar spectroscopy who specialized in the detection of binary stars. She also formulated a classification system to categorize the appearance of spectral lines.

Maury, Matthew Fontaine (1806–1873) US naval officer, founder of the US Naval Oceanographic Office. His system of recording oceanographic data is still used today.

mauve pale purple †aniline dye originally prepared by English chemist William H †Perkin. This dye has the distinction of being the first dyestuff to be synthetically prepared.

maxilla anatomical name for the upper jaw.

maximum and minimum in †coordinate geometry, points at which the slope of a curve representing a †function changes from positive to negative (maximum), or from negative to positive (minimum). A tangent to the curve at a maximum or minimum has zero gradient.

Maxima and minima can be found by differentiating the function for the curve and setting the differential to zero (the value of the slope at the turning point). For example, differentiating the function for the †parabola $y = 2x^2 - 8x$ gives $dy/dx = 4x - 8$. Setting this equal to zero gives $x = 2$, so that $y = -8$ (found by substituting $x = 2$ into the parabola equation). Thus the function has a minimum at the point (2, –8).

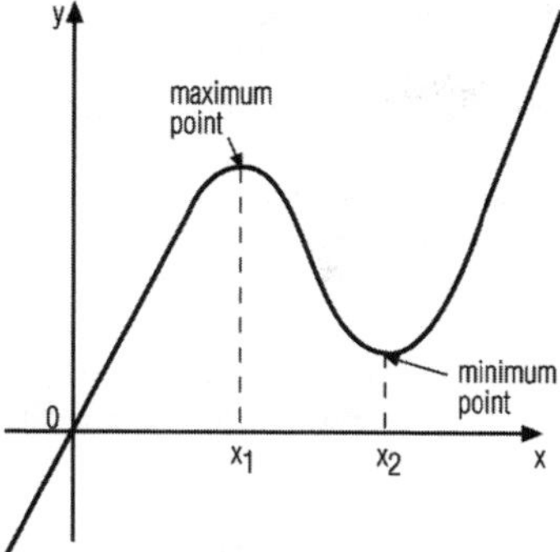

maximum and minimum A maximum point on a curve is higher than the points immediately on either side of it; it is not necessarily the highest point on the curve. Similarly, a minimum point is lower than the points immediately on either side. Helicon

maximum–minimum thermometer thermometer that records the maximum and minimum temperatures occurring over a given period. For example, an alcohol maximum–minimum thermometer has mercury above the alcohol with a steel index at each end of the mercury thread. As the temperature rises, the alcohol pushes up the mercury and the upper index. As the temperature falls, the mercury also falls, pushing down the lower index. In this way the indexes record the highest and lowest temperatures attained.

maxwell symbol Mx, *c.*g.s. unit of magnetic flux (the strength of a †magnetic field in an area multiplied by the area). It is now replaced by the SI unit, the †weber (one maxwell equals 10^{-8} weber).

The maxwell is a very small unit. It is equal to the flux through one square centimeter normal to a magnetic field with an intensity of one gauss.

Maxwell, James Clerk (1831–1879) Scottish physicist. His main achievement was in the understanding of †electromagnetic waves: **Maxwell's equations** bring together electricity, magnetism, and light in one set of relations. He studied gases, optics, and the sensation of color, and his theoretical work in magnetism prepared the way for wireless telegraphy and telephony.

Maxwell, James Clerk http://www-history.mcs.st-and.ac.uk/history/Mathematicians/Maxwell.html

Extensive biography of the Scottish physicist and mathematician. The site contains a description of his contribution to physics, in particular his work on electricity and magnetism. Several literature references for further reading are also listed, and the Web site also features a photograph of Maxwell.

Maxwell–Boltzmann distribution in physics, a statistical equation describing the distribution of velocities among the molecules of a gas. It is named after Scottish physicist James Maxwell and Austrian Physicist Ludwig Boltzmann, who derived the equation, independently of each other, in the 1860s.

One form of the distribution is $n = N\exp(-E/RT)$, where N is the total number of molecules, n is the number of molecules with energy in excess of E, T is the absolute temperature (temperature in kelvin), R is the †gas constant, and exp is the exponential function.

Maxwell's equations set of four †vector equations showing the interdependence of electricity and magnetism. The concepts of charge conservation are built into them, as are the experimental results of Faraday, Gauss, and Ampère.

Maxwell's screw rule in physics, a rule formulated by Scottish physicist James Maxwell that predicts the direction of the magnetic field produced around a wire carrying electric current. It states that if a right-handed screw is turned so that it moves forwards in the same direction as the conventional current, its direction of rotation will give the direction of the magnetic field.

Mayer, Christian (1719–1783) Austrian astronomer, mathematician, and physicist. He was the first to investigate and catalogue double stars, though his equipment was unable to distinguish true binary stars (in orbit round each other) from separate stars seen together only by the coincidence of Earth's viewpoint.

Mayer, Julius Robert von (1814–1878) German physicist who in 1842 anticipated James †Joule in deriving the mechanical equivalent of heat, and Hermann von †Helmholtz in the principle of conservation of energy.

Maynard Smith, John (1920–) English geneticist and evolutionary biologist. He applied †game theory to animal behavior and developed the concept of the evolutionary stable strategy as a mathematical technique for studying the evolution of behavior.

Mayr, Ernst Walter (1904–) German-born US zoologist who was influential in the development of modern evolutionary theories. He led a two-year expedition to New Guinea and the Solomon Islands where he studied the effects of founder populations and speciation on the evolution of the indigenous birds and animals. This research caused him to support †neo-Darwinism, a synthesis of the ideas of †Darwin and †Mendel, being developed at that time.

McDougall, William (1871–1938) English psychologist. He had a great influence on social psychology in the English-speaking world. His main contribution lay in the study of instinct and emotion, which in his classification were interrelated, each instinctive impulse being accompanied by its own quality of emotional experience.

mean in mathematics, a measure of the average of a number of terms or quantities. The simple †**arithmetic mean** is the average value of the quantities, that is, the sum of the quantities divided by their number. The **weighted mean** takes into account the frequency of the terms that are summed; it is calculated by multiplying each term by the number of times it occurs, summing the results and dividing this total by the total number of occurrences. The **geometric mean** of n quantities is the nth root of their product. In statistics, it is a measure of central tendency of a set of data, that is one measure used to express the frequency distribution of a number of recorded events.

meander loop-shaped curve in a †river flowing sinuously across flat country. As a river flows, any curve in its course is accentuated (intensified) by the current. On the outside of the curve the velocity, and therefore the erosion, of the current is greatest. Here the river cuts into the outside bank, producing a **river cliff**. On the inside of the curve the current is slow and so it deposits any transported material, building up a gentle slip-off slope. As each meander migrates in the direction of its outer bank, the river gradually changes its course across the flood plain.

A loop in a river's flow may become so exaggerated that eventually it is cut off from the normal course and forms an †oxbow lake. Meanders are common where the gradient is gentle, the discharge fairly steady (not subject to extremes), and the material that is carried is fine sediment. The word †meander comes from the name of the River Menderes in Turkey.

mean deviation in statistics, a measure of the spread of a population from the †mean.

mean free path in physics, the average distance traveled by a particle, atom, or molecule between successive collisions. It is of importance in the †kinetic theory of gases.

mean life in nuclear physics, the average lifetime of a nucleus of a radioactive isotope; it is equal to 1.44 times the †half-life. See †radioactivity.

measure theory extension of the notion of length, area, or volume (as appropriate) to general sets of points on the line, plane, or in space. Used commonly in analysis (especially integration theory), functional analysis, probability theory, and game theory (in the assessment of the size of coalitions), its definitive form was derived by French mathematician Henri Lebesgue.

mechanical advantage MA, in physics, the ratio by which the load moved by a machine is greater than the effort applied to that machine. In equation terms: MA = load/effort.

The exact value of a working machine's MA is always less than its theoretical value because there will always be some frictional resistance that increases the effort necessary to do the work.

mechanical curve in mathematics, curve that cannot be precisely expressed as an equation (unlike a geometric curve).

mechanical engineering branch of engineering devoted to the study and production of devices, such as tools and vehicles, that are capable of carrying out tasks.

mechanical equivalent of heat in physics, a constant factor relating the calorie (the †*c*.g.s. unit of heat) to the joule (the unit of mechanical energy), equal to 4.1868 joules per calorie. It is redundant in the SI system of units, which measures heat and all forms of energy in †joules.

mechanical weathering in earth science, an alternative name for †physical weathering.

mechanics branch of physics dealing with the motions of bodies and the forces causing these motions, and also with the forces acting on bodies in †equilibrium. It is usually divided into †dynamics and †statics.

Quantum mechanics is the system based on the †quantum theory, which has superseded Newtonian mechanics in the interpretation of physical phenomena on the atomic scale.

Medawar, Peter Brian (1915–1987) Brazilian-born British immunologist who was awarded a Nobel Prize for Physiology or Medicine in 1960 with Macfarlane Burnet for their work on acquired immunological tolerance of transplanted tissues. They discovered that the body's resistance to grafted tissue is undeveloped in the newborn child, and studied the way it is acquired. Medawar was knighted in 1965.

media singular **medium**, in computing, the collective name for materials on which data can be recorded. For example, paper is a medium that can be used to record printed data; a floppy disk is a medium for recording magnetic data.

medial moraine linear ridge of rocky debris running along the center of a glacier. Medial moraines are commonly formed by the joining of two †lateral moraines when two glaciers merge.

median in mathematics and statistics, the middle number of an ordered group of numbers. If there is no middle number (because there is an even number of terms), the median is the †mean (average) of the two middle numbers. For example, the median of the group 2, 3, 7, 11, 12 is 7; that of 3, 4, 7, 9, 11, 13 is 8 (the mean of 7 and 9). The median together with the †mode and †arithmetic mean make up the †average of a set of †data. In addition it is useful to know the †range or spread of the data.

In geometry, the term refers to a line from the vertex of a triangle to the midpoint of the opposite side.

Mediterranean climate climate characterized by hot, dry summers and warm, wet winters. Mediterranean zones are situated in either hemisphere on the western side of continents, between latitudes of 30° and 60°.

During the winter, rain is brought by the †westerlies, which blow from the sea. In summer, Mediterranean climate areas are under the influence of the †trade winds which bring arid conditions with a longer dry season towards the desert margins. Most areas are backed by coastal mountains and so the combined effects of †orographic rainfall and precipitation on the weather †front give high seasonal rainfall totals. The Mediterranean Sea region is noted for its local winds. The †sirocco and †khamsin are two of the hot, dry winds that blow from the Sahara, and can raise temperatures to over 40°C/104°F.

medium in bacteriology, an environment in which micro-organisms can be cultured. Common mediums include agar, broth, and gelatine, often with added salts and trace elements.

medulla central part of an organ. In the mammalian kidney, the medulla lies beneath the outer cortex and is responsible for the reabsorption of water from the filtrate. In plants, it is a region of packing tissue in the center of the stem. In the vertebrate brain, the medulla is the posterior region responsible for the coordination of basic activities, such as breathing and temperature control.

medusa free-swimming phase in the life cycle of a coelenterate, such as a jellyfish or †coral. The other phase is the sedentary **polyp**.

meerschaum aggregate of minerals, usually the soft white clay mineral **sepiolite**, hydrous magnesium silicate. It floats on water and is used for making pipe bowls.

Meer, van der, Simon (1925–) Dutch physicist who was awarded the Nobel Prize for Physics in 1984 for his discovery of the W and Z elementary particles, the particles responsible for propagating the weak nuclear force. The discovery of these particles confirmed a theory developed in the early 1970s by Steven †Weinberg, Sheldon †Glashow, Abdus †Salam, and others that suggested that the †weak nuclear force and the †electromagnetic force were linked. He shared the Nobel Prize shared with Carlo †Rubbia.

mega- prefix denoting multiplication by a million. For example, a megawatt (MW) is equivalent to a million watts.

megabyte MB, in computing, a unit of memory equal to 1,024 kilobytes. It is sometimes used, less precisely, to mean 1 million bytes.

megaton one million (10^6) tons. Used with reference to the explosive power of a nuclear weapon, it is equivalent to the explosive force of one million tons of trinitrotoluene (TNT).

meiosis in biology, a process of cell division in which the number of †chromosomes in the cell is halved. It only occurs in †eukaryotic cells, and is part of a life cycle that involves †sexual reproduction because it allows the genes of two parents to be combined without the total number of chromosomes increasing. Cells in reproductive organs – testes and ovaries in humans – divide to form sex cells (gametes) by meiosis.

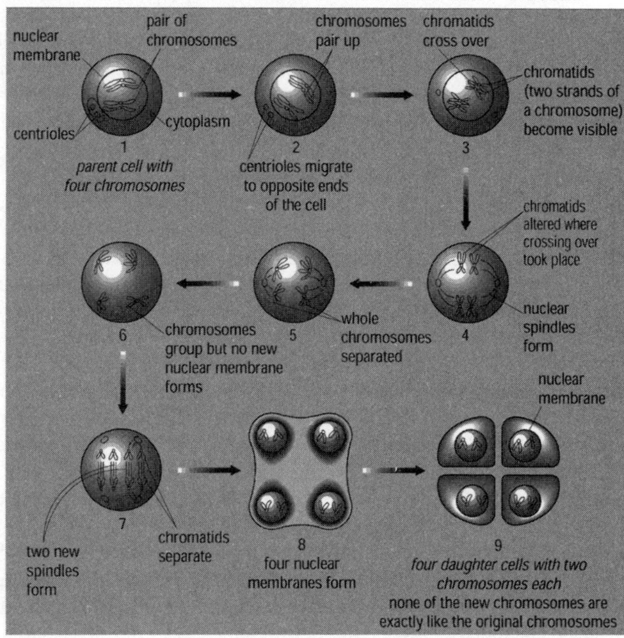

meiosis Meiosis is a type of cell division that produces gametes (sex cells, sperm and egg). This sequence shows an animal cell but only four chromosomes are present in the parent cell (1). There are two stages in the division process. In the first stage (2–6), the chromosomes come together in pairs and exchange genetic material. This is called crossing over. In the second stage (7–9), the cell divides to produce four gamete cells, each with only one copy of each chromosome from the parent cell. Helicon

In meiosis, the nucleus of a cell divides twice. A single cell produces four cells by the end of meiosis. In sexually reproducing †diploid animals (having two sets of chromosomes per cell), meiosis occurs during formation of the †gametes (sex cells – see †sperm and †ovum), so that the gametes are †haploid (having only one set of chromosomes – half the number of chromosomes of the parent cell). When the gametes unite during †fertilization, the diploid condition is restored.

In plants, meiosis occurs just before spore formation. Thus the spores are haploid and in lower plants such as mosses they develop into a haploid plant called a gametophyte which produces the gametes (see †alternation of generations). See also †mitosis.

Meiosis gives rise to variation. This is an important part of sexual reproduction. The variation produced is inherited, which means that evolution can take place as a result of the natural selection of certain variants to suit a changing environment. The way that meiosis gives rise to variation is by recombining genes from chromosomes in new ways. When the number of chromosomes is halved, there is some randomness in the way parts of chromosomes are selected to go into the gametes.

Mitosis and Meiosis http://biology.clc.uc.edu/courses/bio104/meiosis.htm

Clear and detailed description of meiosis, the process of division in reproductive cells. The text is linked to a companion page on mitosis, and there is an animated comparison between the two processes.

Meitner, Lise (1878–1968) Austrian-born Swedish physicist who worked with the German radiochemist Otto †Hahn and was the first to realize that they had inadvertently achieved the fission of uranium. They also discovered †protactinium in 1918. She refused to work on the atom bomb.

meitnerium chemical symbol Mt, synthesized radioactive element of the †transactinide series, atomic number 109, relative atomic mass 266. It was first produced in 1982 at the Laboratory for Heavy-Ion Research in Darmstadt, Germany, by fusing bismuth and iron nuclei; it took a week to obtain a single new, fused nucleus. It was named in 1997 after the Austrian-born Swedish physicist Lise †Meitner.

melamine $C_3H_6N_6$, †thermosetting †polymer based on urea–formaldehyde. It is extremely resistant to heat and is also scratch-resistant. Its uses include synthetic resins.

melanin brown pigment that gives color to the eyes, skin, hair, feathers, and scales of many vertebrates. In humans, melanin helps protect the skin against ultraviolet radiation from sunlight. Both genetic and environmental factors determine the amount of melanin in the skin.

melanism black coloration of animal bodies caused by large amounts of the pigment melanin. Melanin is of significance in insects, because melanic ones warm more rapidly in sunshine than do pale ones, and can be more active in cool weather. A fall in temperature may stimulate such insects to produce more melanin. In industrial areas, dark insects and pigeons match sooty backgrounds and escape predation, but they are at a disadvantage in rural areas where they do not match their backgrounds. This is known as **industrial melanism**.

melanocyte cell that produces †melanin.

melatonin †hormone secreted by the †pineal body that influences the body's †circadian rhythms (metabolic activities centered approximately on a 24-hour cycle). Melatonin secretion is inhibited by bright light and the variations in secretion have been linked to seasonal affective disorder (SAD).

melting point temperature at which a substance melts, or changes from solid to liquid form. A pure substance under standard conditions of pressure (usually one atmosphere) has a definite melting point. If heat is supplied to a solid at its melting point, the temperature does not change until the melting process is complete. The melting point of ice is 0°C or 32°F.

meltwater water produced by the melting of snow and ice, particularly in glaciated areas. Streams of meltwater flowing from glaciers transport rocky materials away from the ice to form †outwash. Features formed by the deposition of debris carried by meltwater or by its erosive action are called **fluvioglacial features**; they include eskers, kames, and outwash plains.

member in mathematics, one of the elements belonging to a set. For example, 25 and 2,500 are both members of the set of square numbers, but 250 is not.

membrane in living things, a continuous layer, made up principally of fat molecules, that encloses a †cell or †organelles within a cell. Small molecules, such as water and sugars, can pass through the cell membrane by †diffusion. Large molecules, such as proteins, are transported across the membrane via special channels, a process often involving energy input. The †Golgi apparatus within the cell is thought to produce certain membranes.

In cell organelles, enzymes may be attached to the membrane at specific positions, often alongside other enzymes involved in the same process, like workers at a conveyor belt. Thus membranes help to make cellular processes more efficient.

membrane potential potential difference that exists across a membrane or cell wall, such as that across the wall of a nerve cell (neuron).

memory in computing, the part of a system used to store data and programs either permanently or temporarily. There are two main types: immediate access memory and backing storage. Memory capacity is measured in †bytes or, more conveniently, in kilobytes (units of 1,024 bytes) or megabytes (units of 1,024 kilobytes).

Immediate access memory, or **internal memory**, describes the memory locations that can be addressed directly and individually by the central processing unit. It is either read-only (stored in ROM, PROM, and EPROM chips) or read/write (stored in RAM chips). Read-only memory stores information that must be constantly available and is unlikely to be changed. It is nonvolatile – that is, it is not lost when the computer is switched off. Read/write memory is volatile – it stores programs and data only while the computer is switched on.

Backing storage, or **external memory**, is nonvolatile memory, located outside the central processing unit, used to store programs and data that are not in current use. Backing storage is provided by such devices as magnetic †disks (floppy and hard disks), †magnetic tape (tape streamers and cassettes), optical disks (such as †CD-ROM), and bubble memory. By rapidly switching blocks of information between the backing storage and the immediate-access memory, the limited size of the immediate-access memory may be increased artificially. When this technique is used to give the appearance of a larger internal memory than physically exists, the additional capacity is referred to as †virtual memory.

memory ability to store and recall observations and sensations. Memory does not seem to be based in any particular part of the brain; it may depend on changes to the pathways followed by nerve impulses as they move through the brain. Memory can be improved by regular use as the connections between †nerve cells (neurons) become "well-worn paths" in the brain. Events stored in **short-term memory** are forgotten quickly, whereas those in **long-term memory** can last for many years, enabling recall of information and recognition of people and places over long periods of time.

Short-term memory is the most likely to be impaired by illness or drugs whereas long-term memory is very resistant to such damage. Memory changes with age and otherwise healthy people may experience a natural decline after the age of about 40. Research is just beginning to uncover the biochemical and electrical bases of the human memory.

Mendel, Gregor Johann (1822–1884) Austrian biologist who founded †genetics. His experiments with successive generations of peas gave the basis for his theory of particulate inheritance rather than blending, involving dominant and recessive characters; see †Mendelism. His results, published 1865–69, remained unrecognized until the early 20th century.

MendelWeb http://www.netspace.org/MendelWeb/

Hefty resource for anyone interested in Gregor Mendel, the origins of classical genetics, and the history and literature of science. View or download Mendel's original paper, browse through glossaries, biographical information, and exercises, or look up the essays, time line, bibliography, and statistical tools.

mendelevium chemical symbol Md, synthesized, radioactive metallic element of the †actinide series, atomic number 101, relative atomic mass 258. It was first produced by bombardment of Es-253 with helium nuclei. Its longest-lived isotope, Md-258, has a half-life of about two months. The element is chemically similar to thulium. It was named by the US physicists at the University of California at Berkeley who first synthesized it in 1955 after the Russian chemist Dmitri Mendeleyev, who in 1869 devised the basis for the periodic table of the elements.

Mendeleyev, Dmitri Ivanovich (1834–1907) Russian chemist who framed the periodic law in chemistry in 1869, which states that the chemical properties of the elements depend on their relative atomic masses. This law is the basis of the †periodic table of the elements, in which the elements are arranged by atomic number and organized by their related groups.

Mendeleyev, Dmitri Ivanovich http://www.chem.msu.su/eng/misc/mendeleev/welcome.html

Detailed biography of Dmitri Mendeleyev, the chemist who framed the periodic table of elements.

Mendelism in genetics, the theory of inheritance originally outlined by Austrian biologist Gregor Mendel. He suggested that, in sexually reproducing species, all characteristics are inherited through indivisible "factors" (now identified with †genes) contributed by each parent to its offspring.

meninges three membranes enclosing the brain and spinal cord. The thickest is the outermost layer, the dura mater; the middle layer is the arachnoid mater; and the innermost layer is the pia mater. Cerebrospinal fluid circulates between the two innermost layers.

meniscus in physics, the curved shape of the surface of a liquid in a thin tube, caused by the cohesive effects of †surface tension (capillary action). When the walls of the container are made wet by the liquid, the meniscus is concave, but with highly viscous liquids (such as mercury) the meniscus is convex.

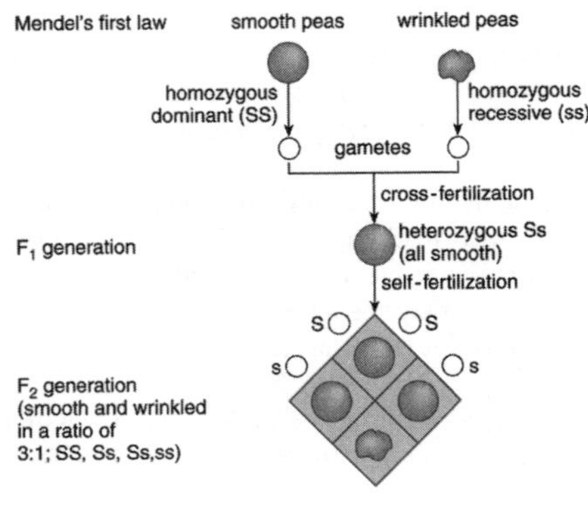

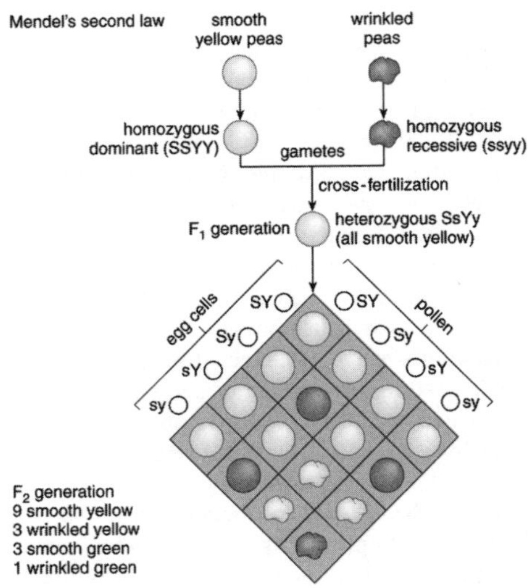

Mendelism Mendel's laws explain the proportion of offspring having various characteristics. When pea plants having smooth yellow peas are crossed with plants with wrinkled green peas, the first-generation offspring all have smooth yellow peas. The second-generation offspring, however, contain smooth yellow, wrinkled green, smooth green, and wrinkled yellow peas. This can be understood by tracing the passage of alleles Y, S, s, y throughout the generations. S and Y are dominant genes. Helicon

Also, a meniscus lens is a concavo-convex or convexo-concave †lens.

menopause in women, the cessation of reproductive ability, characterized by menstruation (see †menstrual cycle) becoming irregular and eventually ceasing. The onset is at about the age of 50, but varies greatly. Menopause is usually uneventful, but some women suffer from complications such as flushing, excessive bleeding, and nervous disorders. Since the 1950s, hormone-replacement therapy (HRT), using †estrogen alone or with progestogen, a synthetic form of †progesterone, has been developed to counteract such effects.

Long-term use of HRT was previously associated with an increased risk of cancer of the uterus, and of clot formation in the blood vessels, but newer formulations using natural estrogens are not associated with these risks. Without HRT there is increased risk of osteoporosis (thinning of the bones) leading to broken bones, which may be indirectly fatal, particularly in the elderly.

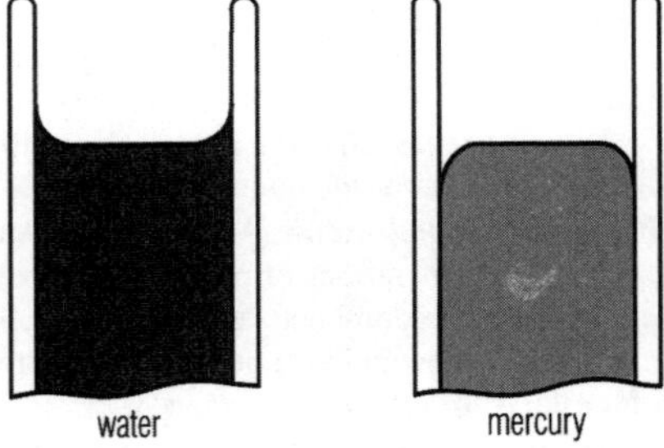

meniscus Comparison of the appearance of the meniscus in water and in mercury. When a liquid is placed in a thin tube, surface tension causes its surface to curve. A liquid, such as water, that wets the walls of the container has a concave meniscus, but a liquid that does not wet the walls, such as mercury, has a convex meniscus. Helicon

The menopause is also known as the "change of life."

Mensa faint constellation of the south polar region.

menstrual cycle biological cycle occurring in female mammals of reproductive age that prepares the body for pregnancy. At the beginning of the cycle, a Graafian (egg) follicle develops in the ovary, and the inner wall of the uterus forms a soft spongy lining. The egg (†ovum) is released from the ovary, and the †uterus lining (endometrium) becomes vascularized (filled with †blood vessels). At this stage fertilization can take place. If fertilization does not occur, the corpus luteum (remains of the Graafian follicle) degenerates, and the uterine lining breaks down, and is shed. This is what causes the loss of blood that marks menstruation. The cycle then begins again. Human menstruation takes place from puberty to menopause, except during pregnancy, occurring about every 28 days.

The monthly release of an egg from a woman's ovaries and the changes in the thickness of the lining of her womb (uterus) are controlled by †hormones produced by the pituitary gland and by the ovaries. If fertilization occurs, the corpus luteum persists and goes on producing progesterone.

mental health well-being and soundness of mind, not only in terms of intellectual abilities, but also in terms of the capability to deal with everyday problems, and the capacity to get on well with other people and to form and sustain relationships.

In †psychoanalysis, much prominence is given to the importance of the relationship between a child and its mother and, since the publication of John Bowlby's *Maternal Care and Mental Health* (1951), a considerable body of research by psychologists and ethologists has confirmed that a warm, intimate, and continuous relationship is essential to a child's intellectual and social development and mental health. Separation from, or loss of, the mother, or surrogate mother, can retard the development of language and intellectual skills as well as severely disrupt a child's emotional life. In puberty and adolescence, the important developmental factor is membership of a peer group. Through peer-group relationships the young person establishes his or her †identity. When a crisis arises, the mentally healthy person is normally able to cope. However, a crisis in the life of a person whose emotional and social development has been inadequate will bring distress and can often result in instability and mental illness.

Mental Health http://www.mentalhealth.com/

Aiming "to improve understanding, diagnosis, and treatment of mental illness throughout the world," this well-designed encyclopedia of mental disorders offers comprehensive information on diagnosing and treating mental health problems, and includes an online magazine.

mental test any of various standardized methods of assessing or measuring mental abilities and personality traits. Where selected for their reliability and validity, mental tests are indispensable aids in assessing educational attainment, in the clinical assessment of mental disorders and their treatment, and in careers guidance and job selection.

Of the various types of mental test, †intelligence tests are among the most widely used, though their reliability and validity have frequently been called into question. In education, tests of attainment have been used for many years to assess language development, reading, elementary mathematics, and other school subjects. There are also tests of aptitude and special abilities, as well as the numerous tests employed by the occupational psychologist whose concern is with personnel selection, performance appraisal, and job analysis. In psychology, personality tests have contributed significantly to the scientific study of †personality, as well as having important applications in clinical and occupational psychology.

menu in computing, a list of options, displayed on screen, from which the user may make a choice – for example, the choice of services offered to the customer by a bank cash dispenser: withdrawal, deposit, balance, or statement. Menus are used extensively in †graphical user interface (GUI) systems, where the menu options are often selected using a pointing device called a mouse.

Mercalli scale qualitative scale describing the **intensity** of an †earthquake. It differs from the †Richter scale, which indicates earthquake **magnitude** and is quantitative. It is named after the Italian seismologist Giuseppe Mercalli (1850–1914).

Intensity is a subjective value, based on observed phenomena, and varies from place to place even when describing the same earthquake.

Modified Mercalli Intensity Scale http://wwwneic.cr.usgs.gov/neis/general/handouts/mercalli.html

Explanation of the Mercalli scale, which is one of the two main scales (with the Richter scale) by which earthquakes are measured.

mercaptan or **thiol**, in chemistry, member of a class of organic compounds that contain molecules in which a carbon atom is directly bonded to an —SH group. The general formula is R—SH, where R is an †alkyl group.

Mercator, Gerardus (1512–1594) Flemish **Gerhard Kremer**, Flemish mapmaker who devised **Mercator's projection** in which the parallels and meridians on maps are drawn uniformly at 90°. The projection continues to be used, in particular for navigational charts, because compass courses can be drawn as straight lines, but the true area of countries is increasingly distorted the further north or south they are from the Equator. For other types, see †map projection.

Mercator, Gerardus

http://www-groups.dcs.st-andrews.ac.uk/history/Mathematicians/Mercator_Gerardus.html

Part of an archive containing the biographies of the world's greatest mathematicians, this site is devoted to the life and contributions of Gerardus Mercator. In addition to biographical information, you will find a list of references about Mercator and links to other essays in the archive that reference him. The text of this essay includes hypertext links to the essays of those mathematicians and thinkers who influenced Mercator. There is also an image of him, which can be enlarged, and a map of his birthplace.

mercuric sulfide HgS, bright red solid that occurs naturally as the mineral cinnabar. It was originally used to make the pigment vermilion red.

Mercury closest planet to the Sun. Its mass is 0.056 that of Earth. On its sunward side the surface temperature reaches over 400°C/752°F, but on the "night" side it falls to –170°C/–274°F.

mean distance from the Sun: 58 million km/36 million mi

equatorial diameter: 4,880 km/3,030 mi

rotation period: 59 Earth days

year: 88 Earth days
atmosphere: Mercury's small mass and high daytime temperature mean that it is impossible for an atmosphere to be retained.
surface: composed of silicate rock often in the form of lava flows. In 1974 the US space probe Mariner 10 showed that Mercury's surface is cratered by meteorite impacts.
satellites: none
NASA's Mariner 10 probe, launched on 3 November 1973, arrived at Mercury on 29 March 1974, and provided the first close-up images of the planet. NASA plans a US$286 million mission to launch the Messenger probe in 2004 to orbit Mercury in 2008, to photograph the planet's surface, analyze its atmospheric composition, and map its magnetic field.
Mercury http://www.solarviews.com/solar/eng/mercury.htm
Detailed description of the planet Mercury. It includes statistics and information about the planet, along with a chronology of its exploration supported by a good selection of images.

mercury chemical symbol Hg; or **quicksilver**, (Latin *mercurius*) heavy, silver-gray, metallic element, atomic number 80, relative atomic mass 200.59. Its symbol comes from the Latin *hydrargyrum*. It is a dense, mobile liquid with a low melting point (–38.87°C/–37.96°F). Its chief source is the mineral cinnabar, HgS, but it sometimes occurs in nature as a free metal.
uses Its alloys with other metals are called amalgams (a silver–mercury amalgam is used in dentistry for filling cavities in teeth). Industrial uses include drugs and chemicals, mercury-vapor lamps, arc rectifiers, power-control switches, barometers, and thermometers.
hazards Mercury is a cumulative poison that can contaminate the food chain, and cause intestinal disturbance, kidney and brain damage, and birth defects in humans. The World Health Organization's "safe" limit for mercury is 0.5 milligrams of mercury per kilogram of muscle tissue. The US Environmental Protection Agency recommended a maximum safe level for mercury of 0.1 mg per kilogram of body weight in January 1998 (a fifth of that recommended by WHO).The discharge into the sea by industry of organic mercury compounds such as dimethylmercury was the chief cause of mercury poisoning in the latter half of the 20th century.
Between 1953 and 1975, 684 people in the Japanese fishing village of Minamata were poisoned (115 fatally) by organic mercury wastes that had been dumped into the bay and had accumulated in the bodies of fish and shellfish.
In a landmark settlement, a British multinational chemical company in April 1997 agreed to pay £1.3 million in compensation to 20 South African workers who were poisoned by mercury. Four of the workers had died and a number of others were suffering severe brain and other neurological damage. The workers had accused Thor Chemical Holdings of adopting working practices in South Africa that would not have been allowed in Britain. The claimants had all worked at Thor's mercury plant at Cato Ridge in Natal. Thor had operated a mercury plant at Margate, in Kent, which during the 1980s was repeatedly criticized by the Health and Safety Executive (HSE) for bad working practices and over-exposure of British workers to mercury. Under pressure from the HSE, Thor closed down its mercury operations in Britain in 1987 and expanded them in South Africa.
history The element was known to the ancient Chinese and Hindus, and is found in Egyptian tombs of about 1500 BC. It was named by the alchemists after the fast-moving god, for its fluidity.

mercury fulminate highly explosive compound used in detonators and percussion caps. It is a gray, sandy powder and extremely poisonous.

meridian half a †great circle drawn on the Earth's surface passing through both poles and thus through all places with the same longitude. Terrestrial longitudes are usually measured from the Greenwich Meridian.
An astronomical meridian is a great circle passing through the celestial pole and the zenith (the point immediately overhead).

meristem region of plant tissue containing cells that are actively dividing to produce new tissues (or have the potential to do so). Meristems found in the tip of roots and stems, the apical meristems, are responsible for the growth in length of these organs.
The †cambium is a lateral meristem that is responsible for increase in girth in perennial plants. Some plants also have intercalary meristems, as in the stems of grasses, for example. These are responsible for their continued growth after cutting or grazing has removed the apical meristems of the shoots.
Meristem culture involves growing meristems taken from shoots on a nutrient-containing medium, and using them to grow new plants.
It is used to propagate infertile plants or hybrids that do not breed true from seed and to generate virus-

free stock, since viruses rarely infect apical meristems.

Merriam, Clinton Hart (1885–1942) US ornithologist and physician who directed the Bureau of Biological Survey and helped to found the National Geographic Society.

Merrifield, R(obert) Bruce (1921–) US biochemist who was awarded the Nobel Prize for Chemistry in 1984 for his development of a method for synthesizing large organic molecules using a solid support or matrix. Merrifield and his co-workers later reached another milestone when they synthesized the enzyme ribonuclease, which contains 124 amino acids.

mesa (Spanish "table") flat-topped, steep-sided plateau, consisting of horizontal weak layers of rock topped by a resistant formation; in particular, those found in the desert areas of the USA and Mexico. A small mesa is called a butte.

Meselson, Matthew Stanley (1930–) US molecular biologist who, with Franklin †Stahl, confirmed that replication of the genetic material †DNA is semiconservative (that is, the daughter cells each receive one strand of DNA from the original parent cell and one newly replicated strand).

mesmerism former term for †hypnosis, after Austrian physician Friedrich Mesmer.

mesoderm central layer of embryonic cells between the †ectoderm and the †endoderm.

mesoglea layer of jellylike noncellular tissue that separates the endoderm and ectoderm in jellyfish and other †coelenterates.

mesomerism or **resonance**, in chemistry, condition of a molecule that cannot be represented as a single structural form because it exists as an intermediate state between a number of nearly equivalent forms; for example, benzene molecules.

meson in physics, a group of unstable subatomic particles made up of a †quark and an antiquark. It is found in cosmic radiation, and is emitted by nuclei under bombardment by very high-energy particles. The mesons form a subclass of the hadrons and include the kaons and pions. Their existence was predicted in 1935 by Japanese physicist Hideki Yukawa.

mesopause layer in the Earth's atmosphere at an altitude of 80–85 km/50–53 mi, at the top of the mesosphere.

mesophyll tissue between the upper and lower epidermis of a leaf blade (†lamina), consisting of parenchyma-like cells containing numerous †chloroplasts.

In many plants, mesophyll is divided into two distinct layers.

The **palisade mesophyll** is usually just below the upper epidermis and is composed of regular layers of elongated cells. Lying below them is the **spongy mesophyll**, composed of loosely arranged cells of irregular shape. This layer contains fewer chloroplasts and has many intercellular spaces for the diffusion of gases (required for †respiration and †photosynthesis), linked to the outside by means of †stomata.

mesophyte in ecology, plant characterized by having only moderate requirements for water.

mesosphere layer in the Earth's †atmosphere above the stratosphere and below the thermosphere. It lies between about 50 km/31 mi and 80 km/50 mi above the ground.

Mesozoic era of geological time 245–65 million years ago, consisting of the Triassic, Jurassic, and Cretaceous periods. At the beginning of the era, the continents were joined together as Pangaea; dinosaurs and other giant reptiles dominated the sea and air; and ferns, horsetails, and cycads thrived in a warm climate worldwide. By the end of the Mesozoic era, the continents had begun to assume their present positions, flowering plants were dominant, and many of the large reptiles and marine fauna were becoming extinct.

messenger RNA mRNA, single-stranded nucleic acid (made up of †nucleotides) found in †ribosomes, †mitochondria, and nucleoli of cells that carries coded information for building chains of †amino acids into polypeptides.

Messier, Charles (1730–1817) French astronomer. He discovered 15 comets and in 1784 published a list of 103 star clusters and nebulae. Objects on this list are given M (for Messier) numbers, which astronomers still use today, such as M1 (the Crab nebula) and M31 (the Andromeda galaxy).

Messier catalogue in astronomy, catalogue of 103 †galaxies, †nebulae, and star clusters (the Messier objects) published in 1784 by French astronomer Charles †Messier. Catalogue entries are denoted by the prefix "M." Well-known examples include M31 (the †Andromeda galaxy), M42 (the †Orion nebula), and M45 (the †Pleiades star cluster).

Messier compiled the catalogue to identify fuzzy objects that could be mistaken for †comets. The list was later extended to 109.

meta- in chemistry, prefix denoting a †benzene compound in which two hydrogen atoms have been substituted by other groups whose positions on the ring are separated by an unsubstituted carbon atom. It is usually denoted in the chemical name as the letter "*m*" separated by a dash, for example *m*-dichlorobenzene.

metabolism chemical processes of living organisms enabling them to grow and to function. It involves a constant alternation of building up complex molecules (**anabolism**) and breaking them down (**catabolism**). For example, green plants build up complex organic substances from water, carbon dioxide, and mineral salts (†photosynthesis); by digestion animals partially break down complex organic substances, ingested as food, and subsequently resynthesize them for use in their own bodies (see †digestive system). Within cells, complex molecules are broken down by the process of †respiration. The waste products of metabolism are removed by †excretion.

metal any of a class of chemical elements with specific physical and chemical characteristics. Metallic elements compose about 75% of the 112 elements in the †periodic table of the elements. Physical properties include a sonorous tone when struck; good conduction of heat and electricity; high melting and boiling points; opacity but good reflection of light; malleability, which enables them to be cold-worked and rolled into sheets; and ductility, which permits them to be drawn into thin wires.

The majority of metals are found in nature in a combined form only, as compounds or mineral ores; about 16 of them also occur in the elemental form, as †native metals. Their chemical properties are largely determined by the extent to which their atoms can lose one or more electrons and form positive ions (cations).

All metals except mercury are solid at ordinary temperatures, and all of them will crystallize under suitable conditions. The chief chemical properties of metals also include their strong affinity for certain non-metallic elements, for example sulfur and chlorine, with which they form sulfides and chlorides. Metals will, when fused, enter into the forming of †alloys.

By comparing the reactions of metals with oxygen, water, acids, and other substances, the metals can be arranged in order of reactivity, known as the †reactivity series of metals.

Refining Metal Ores http://www.moe.edu.sg/schools/rjc/subjects/chemistry/resources/simulations-redox/everyday_refining.html
All about ores and how to extract metals from them. This explanation of extraction processes focuses in particular on the copper-containing ore of chalcocite and the type of electrolysis required. A diagram of the electrolysis is included.

metallic bond force of attraction operating in a †metal that holds the atoms together in a metallic structure. In metallic bonding, metal atoms form a close-packed, regular arrangement. The atoms lose their outer-shell electrons to become positive ions. The outer electrons become a "sea" of mobile electrons surrounding a lattice of positive ions. The lattice is held together by the strong attractive forces between the mobile electrons and the positive ions.

The properties of metals can be explained in terms of metallic bonding. Metals conduct electricity as the electrons are free to move. Conduction of heat occurs by vibration of the positive ions as well as via the mobile electrons. Metals are both ductile and malleable because the bonding is not broken when metals are deformed; instead, the metal ions slide over each other to new lattice positions.

metallic character chemical properties associated with those elements classed as metals. These properties, which arise from the element's ability to lose electrons, are: the displacement of hydrogen from dilute acids; the formation of basic oxides; the formation of ionic chlorides; and their reducing reaction, as in the thermite process (see †reduction).

In the periodic table of the elements, metallic character increases down any group and across a period from right to left.

metallic glass substance produced from metallic materials (non-corrosive alloys rather than simple metals) in a liquid state which, by very rapid cooling, are prevented from reverting to their regular metallic structure. Instead they take on the properties of glass, while retaining the metallic properties of malleability and relatively good electrical conductivity.

metalloid or **semimetal**, any chemical element having some of but not all the properties of metals; metalloids are thus usually electrically semiconducting. They comprise the elements germanium, arsenic, antimony, and tellurium. Metalloids are found in the †periodic table of the elements between metals and non-metals.

metamorphic rock rock altered in structure, composition, and texture by pressure or heat after its original formation. (Rock that actually melts under heat is called †igneous rock upon cooling.) For example, limestone can be metamorphosed by heat into marble, and shale by pressure into slate. The term was coined in 1833 by Scottish geologist Charles †Lyell. Metamorphism is part of the rock cycle, the gradual formation, change, and re-formation of rocks over millions of years.

metamorphism geological term referring to the changes in rocks of the Earth's crust caused by increasing pressure and temperature. The resulting rocks are metamorphic rocks. All metamorphic changes take place in solid rocks. If the rocks melt and then harden, they are considered †igneous rocks.

metamorphosis period during the life cycle of many invertebrates, most amphibians, and some fish, during which the individual's body changes from one form to another through a major reconstitution of its tissues. For example, adult frogs are produced by metamorphosis from tadpoles, and butterflies are produced from caterpillars following metamorphosis within a pupa.

In classical thought and literature, metamorphosis is the transformation of a living being into another shape, either living or inanimate (for example Niobe). The Roman poet Ovid wrote about this theme.

metazoa another name for animals. It reflects a former system of classification, in which there were two main divisions within the animal kingdom, the multicellular animals, or metazoa, and the single-celled "animals" or protozoa. The †protozoa are no longer included in the animal kingdom, so only the metazoa remain.

Metchnikoff, Elie (1845–1916) Russian **Ilya Ilich Mechnikov**, Russian zoologist and immunologist who was a pioneer of cellular immunology and shared the Nobel Prize for Physiology or Medicine in 1908 with Paul †Ehrlich for the discovery of the innate immune response.

meteor flash of light in the sky, popularly known as a **shooting** or **falling star**, caused by a particle of dust, a †meteoroid, entering the atmosphere at speeds up to 70 kps/45 mps and burning up by friction at a height of around 100 km/60 mi. On any clear night, several **sporadic meteors** can be seen each hour. Several times each year, the Earth encounters swarms of dust shed by comets, which give rise to a **meteor shower**. This appears to radiate from one particular point in the sky, after which the shower is named; the **Perseid** meteor shower in August appears in the constellation Perseus. The **Leonids** shoot out from the constellation Leo and are caused by dust from the comet Tempel-Tuttle, which orbits the Sun every 33 years. The Leonid shower reaches its peak when the comet is closest to the Sun.

Meteors, Meteorites, and Impacts http://www.seds.org/billa/tnp/meteorites.html

Informative collection of facts about meteorites: how they are formed, classification, and what happens when one hits Earth. The site includes images of a selection of meteorites.

meteor-burst communications technique for sending messages by bouncing radio waves off the trails of †meteors. High-speed computer-controlled equipment is used to sense the presence of a meteor and to broadcast a signal during the short time that the meteor races across the sky.

The system, first suggested in the late 1920s, remained impracticable until data-compression techniques were developed, enabling messages to be sent in automatic high-speed bursts each time a meteor trail appeared. There are usually enough meteor trails in the sky at any time to permit continuous transmission of a message. The technique offers a communications link that is difficult to jam, undisturbed by storms on the Sun, and would not be affected by nuclear war.

meteorite piece of rock or metal from space that reaches the surface of the Earth, Moon, or other body. Most meteorites are thought to be fragments from asteroids, although some may be pieces from the heads of comets. Most are stony, although some are made of iron and a few have a mixed rock–iron composition.

Stony meteorites can be divided into two kinds: **chondrites** and **achondrites**. Chondrites contain chondrules, small spheres of the silicate minerals olivine and orthopyroxene, and comprise 85% of meteorites. Achondrites do not contain chondrules. Meteorites provide evidence for the nature of the Solar System and may be similar to the Earth's core and mantle, neither of which can be observed directly.

meteoritics branch of geology dealing with the study of †meteorites, their composition, texture, formation, and origin, and applying these observations to a more thorough understanding of the evolution of the Solar System and planets, including Earth.

meteoroid small natural object in interplanetary space. Meteoroids are smaller than †asteroids, ranging from the size of a pebble up, and move through space at high speeds. There is no official distinction between meteoroids and asteroids, although the term "asteroid" is generally reserved for objects larger than 1.6 km/1 mi in diameter.

Meteoroids are believed to result from the fragmentation of asteroids after collisions. Some meteoroids strike the Earth's atmosphere, and their fiery trails are called **meteors**. If they fall to Earth, they are named **meteorites**.

meteorology scientific observation and study of the †atmosphere, so that †weather can be accurately forecast.

Data from meteorological stations and weather satellites are collated by computers at central agencies, and forecast and weather maps based on current readings are issued at regular intervals. Modern analysis, employing some of the most powerful computers, can give useful forecasts for up to six days ahead. At meteorological stations readings are taken of the factors determining weather conditions: atmospheric pressure, temperature, humidity, wind (using the †Beaufort scale), cloud cover (measuring both type of cloud and coverage), and precipitation such as rain, snow, and hail (measured at 12-hour intervals). †Satellites are used either to relay information transmitted from the Earth-based stations, or to send pictures of cloud development, indicating wind patterns, and snow and ice cover.

history Apart from some observations included by Aristotle in his book *Meteorologia*, meteorology did not become a precise science until the end of the 16th century, when Galileo and the Florentine academicians constructed the first thermometer of any importance, and when Evangelista Torricelli in 1643 discovered the principle of the barometer. Robert †Boyle's work on gases, and that of his assistant, Robert †Hooke, on barometers, advanced the physics necessary for the understanding of the weather. Gabriel †Fahrenheit's invention of a superior mercury thermometer provided further means for temperature recording.

weather maps In the early 19th century a chain of meteorological stations was established in France, and weather maps were constructed from the data collected. The first weather map in England, showing the trade winds and monsoons, was made in 1688, and the first telegraphic weather report appeared on 31 August 1848. The first daily telegraphic weather map was prepared at the Great Exhibition in 1851, but the Meteorological Office was not established in London until 1855. The first regular daily collections of weather observations by telegraph and the first British daily weather reports were made in 1860, and the first daily printed maps appeared in 1868.

collecting data Observations can be collected not only from land stations, but also from weather ships, aircraft, and self-recording and automatic transmitting stations, such as the †radiosonde. †Radar may be used to map clouds and storms. Satellites have played an important role in televising pictures of global cloud distribution.

NOAA http://www.noaa.gov/

The US government National Oceanic and Atmospheric Administration (NOAA) site is beautiful and user-friendly, with information and the latest news on weather, climate, oceanography, and remote sensing.

World Meteorological Organization http://www.wmo.ch/

Internet voice of the World Meteorological Organization, a UN division coordinating global scientific activity related to climate and weather. The site offers ample material on the long-term objectives and immediate policies of the organization. It also disseminates important information on WMO's databases, training programs, and satellite activities, as well as its projects related to the protection of the environment.

meteor shower occurrence of †meteors from the same part of the sky (the radiant) at about the same time each year. They are generally named after the constellation where the radiant lies (for example, the Leonids arise from the direction of Leo).

methanal or **formaldehyde**; HCHO, gas at ordinary temperatures, condensing to a liquid at –21°C/–5.8°F. It has a powerful, penetrating smell. Dissolved in water, it is used as a biological preservative. It is used in the manufacture of plastics, dyes, foam (for example, urea-formaldehyde foam, used in insulation), and in medicine.

Formaldehyde http://www.nsc.org/EHC/indoor/formald.htm

Page from the US Environmental Health Center which explains the dangers of formaldehyde, and explains what precautions can be made to decrease exposure to the chemical.

methane CH_4, simplest †hydrocarbon of the †alkane series. Colorless, odorless, and lighter than air, it burns with a bluish flame and explodes when mixed with air or oxygen. As the chief constituent of natural gas, methane's main use is as a fuel. It also occurs in the explosive firedamp of coal mines. Methane emitted by rotting vegetation forms marsh gas, which may ignite by spontaneous combustion to produce the pale flame seen over marshland and known as †will-o'-the-wisp.

Methane causes about 38% of the warming of the globe through the †greenhouse effect; weight for weight it is 60–70 times more potent than carbon dioxide at trapping solar radiation in the atmosphere and so heating the planet. The rate of increase of atmospheric methane is declining and global emissions remained relatively constant over the period 1984–96, so atmospheric levels were predicted, in 1998, to stabilize by the 2020s. An estimated 15% of all methane gas in the atmosphere is produced by cattle and other cud-chewing animals, and 20% is produced by termites that feed on soil.

methanogenic bacteria one of a group of primitive micro-organisms, the †Archaea. They give off methane gas as a by-product of their metabolism, and are common in sewage treatment plants and hot springs, where the temperature is high and oxygen is absent. Archaeons were originally classified as bacteria, but were found to be unique in 1996 following the gene sequencing of the deep-sea vent *Methanococcus jannaschii*.

methanoic acid or **formic acid**; HCOOH, colorless, slightly fuming liquid that freezes at 8°C/46.4°F and boils at 101°C/213.8°F. It occurs in stinging ants, nettles, sweat, and pine needles, and is used in dyeing, tanning, and electroplating.

methanol common name **methyl alcohol**; CH_3OH, simplest of the alcohols. It can be made by the dry distillation of wood (hence it is also known as wood alcohol), but is usually made from coal or natural gas. When pure, it is a colorless, flammable liquid with a pleasant odor, and is highly poisonous.

Methanol is used to produce methanal (formaldehyde, from which resins and plastics can be made), methyl-*tert*-butyl ether (MTB, a replacement for lead as an octane-booster in gasoline), vinyl acetate (largely used in paint manufacture), and gasoline. In 1993 Japanese engineers built an engine, made largely of ceramics, that runs on methanol. The prototype is lighter and has a cleaner exhaust than comparable metal, gasoline -powered engines.

methionine one of the nine essential †amino acids. It is also used as an antidote to paracetamol poisoning.

method of least squares see †least squares, method of.

methyl alcohol common name for †methanol.

methylated spirit alcohol that has been rendered undrinkable, and is used for industrial purposes, as a fuel for spirit burners or a solvent.

It is nevertheless drunk by some individuals, resulting eventually in death. One of the poisonous substances in it is †methanol, or methyl alcohol, and this gives it its name. (The "alcohol" of alcoholic drinks is ethanol.)

methylation in chemistry, any process that results in the introduction of a methyl group, CH_3–, into an organic molecule (see also †Friedel–Crafts reaction).

methyl benzene alternative name for †toluene.

methyl bromide or **bromoethane**; CH_3Br, pesticide gas used to fumigate soil. It is a major †ozone depleter. Industry produces 50,000 tonnes of methyl bromide annually (1995). The European Union (EU) promised a 25% reduction in manufacture by 1998, and the USA intends to ban use by 2001. The EU proposed a total ban on usage by 2001 at a meeting in July 1998. Some countries, however, have failed to significantly reduce their ozone-depleting emissions, such as China, which at the end of 1999 was producing 10% of the world's methyl bromide.

methylcyclohexanol or **hexahydrocresol**; $C_7H_{14}O$, viscous, colorless liquid prepared by methylation of †cyclohexanol. It is used as a solvent for resins, fats, and waxes and as a detergent in soaps.

methylcyclohexanone $C_7H_{12}O$, colorless organic liquid prepared from †methylcyclohexanol by passing the compound in vapor form over a heated catalyst of copper metal. It exists in three isomeric forms and is used as a solvent for resins and in the manufacturing process of lacquers.

methyl methacrylate $C_5H_8O_2$, in chemistry, unsaturated organic compound that is a colorless liquid in its natural state, boiling point 100°C/212°F. It is used in the production of acrylic resin polymers such as †Perspex.

methyl orange $C_{14}H_{14}N_3NaO_3S$, orange-yellow powder used as an acid–base indicator in chemical tests, and as a stain in the preparation of slides of biological material. Its color changes with †pH; below pH 3.1 it is red, above pH 4.4 it is yellow.

Metis small innermost moon of Jupiter, discovered in 1979, that orbits within the planet's main ring. It has an irregular shape, being about 40 km/25 mi across, and orbits once every seven hours. Metis is also the name of a 190-km/118-mi asteroid that was discovered in 1849 (the ninth to be discovered).

Metius, Jacobus (died 1628) Dutch instrument maker. He specialized in grinding lenses and claimed to have invented the telescope. He applied for a patent 1608, a few weeks after Hans †Lippershey.

meter symbol m, SI unit of length, equivalent to 1.093 yards or 39.37 inches. It is defined by scientists as the length of the path traveled by light in a vacuum during a time interval of 1/299,792,458 of a second.

metric system system of weights and measures developed in France in the 18th century and recognized by other countries in the 19th century.

In 1960 an international conference on weights and measures recommended the universal adoption of a revised International System (Système International d'Unités, or SI), with seven prescribed "base units": the meter (m) for length, kilogram (kg) for mass, second (s) for time, ampere (A) for electric current, kelvin (K) for thermodynamic temperature, candela (cd) for luminous intensity, and mole (mol) for quantity of matter.

supplementary units Two supplementary units are included in the SI system – the radian (rad) and steradian (sr) – used to measure plane and solid angles. In addition, there are recognized derived units that can be expressed as simple products or divisions of powers of the basic units, with no other integers appearing in the expression; for example, the watt.

non-SI units Some non-SI units, well established and internationally recognized, remain in use in conjunction with SI units: minute, hour, and day in measuring time; multiples or submultiples of base or derived units that have long-established names, such as ton for mass and liter for volume; and specialist measures such as the metric carat for gemstones.

prefixes Prefixes used with metric units are tera (T), million million times; giga (G), billion (thousand million) times; mega, (M) million times; kilo (k), thousand times; hecto (h), hundred times; deca (da), ten times; deci (d), tenth part; centi (c), hundredth part; milli (m), thousandth part; micro (m), millionth part; nano (n), billionth part; pico (p), trillionth part; femto (f), quadrillionth part; atto (a), quintillionth part.

mho SI unit of electrical conductance, now called the †siemens; equivalent to a reciprocal ohm.

mica any of a group of silicate minerals that split easily into thin flakes along lines of weakness in their crystal structure (perfect basal cleavage). They are glossy, have a pearly luster, and are found in many igneous and metamorphic rocks. Their good thermal and electrical insulation qualities make them valuable in industry.

Their chemical composition is complicated, but they are silicates with silicon–oxygen tetrahedra arranged in continuous sheets, with weak bonding between the layers, resulting in perfect cleavage. A common example of mica is muscovite (white mica), $KAl_2Si_3AlO_{10}(OH,F)_2$.

Michaelis, Leonor (1875–1949) German-born, US biochemist who derived a mathematical model to describe the kinetics of how enzymes catalyze (trigger) reactions. The work of Michaelis and German scientist Maude Menton enabled several subsequent generations of biochemists to correctly assess the nature and efficiency of the key, enzyme-driven steps in cell metabolism.

Michell, John (*c.* 1724–1793) English astronomer and parson who proposed that double stars exist (confirmed by William Herschel in 1803) and predicted the possible existence of black holes. His black hole hypothesis was proposed in 1783 when he suggested that the gravity of a sufficiently massive star would prevent light escaping from it.

Michelson, Albert Abraham (1852–1931) German-born US physicist. With his colleague Edward †Morley, he performed in 1887 the **Michelson–Morley experiment** to detect the motion of the Earth through the postulated ether (a medium believed to be necessary for the propagation of light). The failure of the experiment indicated the nonexistence of the ether, and led Albert †Einstein to his theory of †relativity. Michelson was awarded the Nobel Prize for Physics in 1907 for his measurement of the speed of light through the design and application of precise optical instruments such as the interferometer. He was the first American to be awarded a Nobel prize.

MICR abbreviation for magnetic-ink character recognition.

micro- symbol m, prefix denoting a one-millionth part (10^{-6}). For example, a micrometer, mm, is one-millionth of a meter.

microbe another name for †micro-organism.

microbiology study of micro-organisms, mostly viruses and single-celled organisms such as bacteria, protozoa, and yeasts. The practical applications of microbiology are in medicine (since many micro-organisms cause disease); in brewing, baking, and other food and beverage processes, where the micro-organisms carry out fermentation; and in genetic engineering, which is creating increasing interest in the field of microbiology.

microchip popular name for the silicon chip, or †integrated circuit.

microclimate climate of a small area, such as a woodland, lake, or even a hedgerow. Significant differences can exist between the climates of two neighboring areas – for example, a town is usually warmer than the surrounding countryside (forming a heat island), and a woodland cooler, darker, and less windy than an area of open land.

Microclimates play a significant role in agriculture and horticulture, as different crops require different growing conditions.

microcomputer or **micro or personal computer**, small desktop or portable †computer, typically designed to be used by one person at a time, although individual computers can be linked in a network so that users can share data and programs.

Its central processing unit is a †microprocessor, contained on a single integrated circuit.

Microcomputers are the smallest of the four classes of computer (the others are †supercomputer, †mainframe, and †minicomputer). Since the appearance in 1975 of the first commercially available microcomputer, the Altair 8800, micros have become ubiquitous in commerce, industry, and education.

microglia type of glial cell (see †glia) surrounding the nerve cells in the brain and spinal cord. They have an immune function.

microhabitat in ecology, localized environmental conditions of a population or even an individual. The surface of a pond or the underside of a leaf are examples.

micrometer instrument for measuring minute lengths or angles with great accuracy; different types of micrometer are used in astronomical and engineering work.

The type of micrometer used in astronomy consists of two fine wires, one fixed and the other movable, placed in the focal plane of a telescope; the movable wire is fixed on a sliding plate and can be positioned parallel to the other until the object appears between the wires.

The movement is then indicated by a scale on the adjusting screw.

micrometer symbol mm, one-millionth of a †meter.

micron obsolete name for the micrometer, one-millionth of a meter.

micro-organism or **microbe**, living organism invisible to the naked eye but visible under a microscope. Micro-organisms include †viruses and single-celled organisms such as bacteria and yeasts. Yeasts are fungi, but other fungi are often big enough to see with the naked eye and so are not micro-organisms. The study of micro-organisms is known as microbiology.

Microbe Zoo http://commtechlab.msu.edu/sites/dlc-me/zoo/

Colorful and interactive zoo of some of the microbes that surround us. It includes sections on the "domestic" microbes; the vampire ones that suck the life from other bacteria; the killers that destroy stone buildings; those in aquatic environments; and those that are to be found in foods such as beer, bread, chocolate, and wine.

microphone primary component in a sound-reproducing system, whereby the mechanical energy of sound waves is converted into electrical signals by means of a †transducer. A diaphragm is attached to a coil of wire placed between two poles of a permanent magnet. Sound waves cause the diaphragm to vibrate, which in turn causes the coil of wire to move in the magnetic field of the permanent magnet. An induced electrical current, matching the pattern of the sound waves, flows through the coil and is fed to an amplifier. The amplified signals are either stored or sent to a loudspeaker.

One of the simplest microphones is the telephone receiver mouthpiece, invented by Scottish–US inventor Alexander Graham Bell in 1876; other types of microphone are used with broadcasting and sound-film apparatus.

Telephones have a **carbon microphone**, which reproduces only a narrow range of frequencies. For live music, a **moving-coil microphone** is often used. In it, a diaphragm that vibrates with sound waves moves a coil through a magnetic field, thus generating an electric current. The **ribbon microphone** combines the diaphragm and coil. The **condenser microphone** is most commonly used in recording and works by a †capacitor.

microprocessor complete computer †central processing unit contained on a single †integrated circuit, or chip. The appearance of the first microprocessor in 1971 designed by Intel for a pocket calculator manufacturer heralded the introduction of the microcomputer. The microprocessor has led to a dramatic fall in the size and cost of computers, and dedicated computers can now be found in washing machines, cars, and so on. Examples of microprocessors are the Intel Pentium family and the IBM/Motorola PowerPC, used by Apple Computer.

The earliest chips could process about 60,000 instructions per second. In 1997 Texas Instruments introduced a digital-signal microprocessor chip that could process 1.6 billion instructions a second. This is about 40 times more powerful than a chip now found in today's computer modem. The new chip can reduce the time needed to download a file from the Internet from ten minutes to less than five seconds. Texas Instruments predicted that "system-on-a-chip roadmaps" would achieve 3 trillion instructions per second by 2010. In real terms, this would deliver the functionality of today's laptop computers in a device the size of a wristwatch.

How Microprocessors Work http://www.intel.com/education/mpworks/index.htm

Interactive exploration of how a microprocessor works. It takes you step-by-step through a simple task on the chip to illustrate the function of its different parts. There is also a glossary containing definitions of the component parts of a microprocessor, plus classroom activities, and related topics such as "How transistors work" and "How chips are made."

micropropagation mass production of plants by placing tiny pieces of plant tissue in sterile glass containers along with nutrients. Perfect clones of superplants are produced in sterile cabinets, with filtered air and carefully controlled light, temperature, and humidity. The system is used for the house-plant industry and for forestry – micropropagation gives immediate results, whereas obtaining genetically homogeneous tree seed by traditional means would take over a hundred years.

Micropropagation at Kew http://www.rbgkew.org.uk/ksheets/microprop.html

Report of the work of the Micropropagation Unit at Britain's prestigious Kew Gardens. It contains descriptions of how endangered species are cultured in vitro from seed or vegetative material. There are details of Kew's online publication Micropropagation News.

micropyle in flowering plants, a small hole towards one end of the ovule. At pollination the pollen tube growing down from the †stigma eventually passes through this pore. The male gamete is contained within the tube and is able to travel to the egg in the interior of the ovule. Fertilization can then take place, with subsequent seed formation and dispersal.

microscope instrument for forming magnified images with high resolution for detail. Optical and electron microscopes are the ones chiefly in use; other types include acoustic, †scanning tunneling, and †atomic force microscopes.

The **optical microscope** usually has two sets of glass lenses and an eyepiece. The first true compound microscope was developed in 1609 in the Netherlands by Zacharias Janssen (1580–*c.* 1638). **Fluorescence microscopy** makes use of fluorescent dyes to illuminate samples, or to highlight the presence of particular substances within a sample. Various illumination systems are also used to highlight details.

The †**transmission electron microscope**, developed from 1932, passes a beam of electrons, instead of a beam of light, through a specimen. Since electrons are not visible, the eyepiece is replaced with a fluorescent screen or photographic plate; far higher magnification and resolution are possible than with the optical microscope.

The †**scanning electron microscope** (SEM), developed in the mid-1960s, moves a fine beam of electrons over the surface of a specimen, the reflected electrons being collected to form the image. The specimen has to be in a vacuum chamber.

The **acoustic microscope** passes an ultrasonic (ultrahigh-frequency sound) wave through the specimen, the transmitted sound being used to form an image on a computer screen.

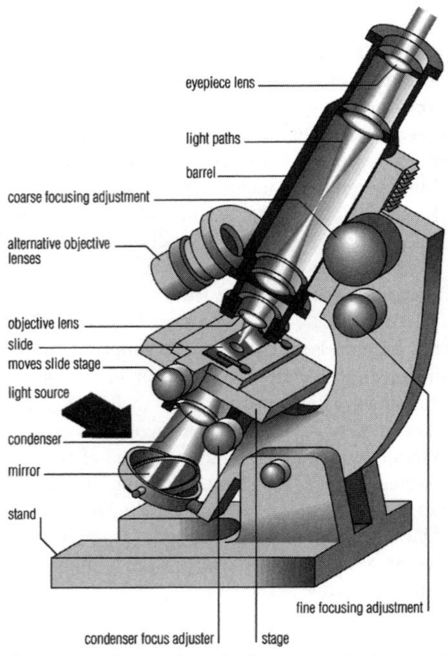

microscope Terms used to describe an optical microscope. In essence, the optical microscope consists of an eyepiece lens and an objective lens, which are used to produce an enlarged image of a small object by focusing light from a light source. Optical microscopes can achieve magnifications of up to 1,500–2,000. Higher magnifications and resolutions are obtained by electron microscopes. Helicon

The **scanned-probe microscope**, developed in the late 1980s, runs a probe, with a tip so fine that it may consist only of a single atom, across the surface of the specimen. In the **scanning tunneling microscope**, an electric current that flows through a probe is used to construct an image of the specimen. In 1988 a scanning tunneling microscope was used to photograph a single protein molecule for the first time. In the **atomic force microscope**, the force felt by a probe is measured and used to form the image. These instruments can magnify a million times and give images of single atoms.

Dennis Kunkel's Microscopy http://www.denniskunkel.com//

This site is a photomicographer's dream – full of pictures taken with both light and electron microscopes. As well as several differing galleries of images, there is also information about microscopy and how the pictures were taken.

Microscopium inconspicuous constellation of the southen hemisphere, represented as a microscope.

microsome minute particle occurring in the cytoplasm of a cell composed of vesicles with attached †ribosomes, which are thought to derive from the †endoplasmic reticulum. Microsomes are also thought to give rise to †mitochondria.

microtome device for cutting extremely thin slices of tissue for microscopic examination. The tissue is embedded in wax or a synthetic resin (or is frozen) for ease of handling.

microtubules tiny tubes found in almost all cells with a nucleus. They help to define the shape of a cell by forming scaffolding for cilia and they also form the fibers of mitotic spindle (see †mitosis).

microwave radiation †electromagnetic wave with a wavelength in the range 0.3 cm to 30 cm/0.1 in to 12 in, or 300–300,000 megahertz (between radio waves and †infrared radiation). Microwaves are used in radar, in radio broadcasting, in satellite communications, and in microwave heating and cooking.

In microwave cooking, microwaves are used to transfer energy to food. The frequency of the microwaves matches the natural frequency at which the water molecules in food vibrate. The vibration energy of the water molecules results in an increase in temperature, which allows the food to be cooked.

Mid-Atlantic Ridge †mid-ocean ridge that runs along the center of the Atlantic Ocean, parallel to its edges, for some 14,000 km/8,800 mi – almost from the Arctic to the Antarctic. Like other ocean ridges, the Mid-Atlantic Ridge is essentially a linear, segmented volcano.

The Mid-Atlantic Ridge runs down the center of the ocean because the ocean crust has continually grown outwards from the ridge at a steady rate during the past 200 million years. Iceland straddles the ridge and was formed by volcanic outpourings.

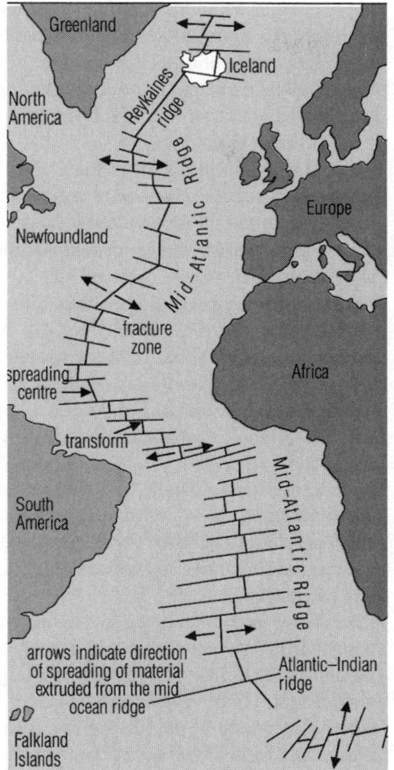

Mid-Atlantic Ridge The Mid-Atlantic Ridge is the boundary between the crustal plates that form America, and Europe and Africa. An oceanic ridge cannot be curved since the material welling up to form the ridge flows at a right angle to the ridge. The ridge takes the shape of small straight sections offset by fractures transverse to the main ridge. Helicon

Midgley, Thomas (1889–1944) US industrial chemist and engineer whose two main discoveries, universally adopted, were later criticized as damaging to the environment. He found in 1921 that tetraethyl lead is an efficient antiknock additive to gasoline (preventing pre-ignition in car engines), and in 1930 introduced Freons (a group of †chlorofluorocarbons) as the working gases in refrigerators, freezers, and air-conditioning units.

midnight sun constant appearance of the Sun (within the Arctic and Antarctic circles) above the †horizon during the summer.

mid-ocean ridge long submarine mountain range that winds along the middle of the ocean floor. The mid-ocean ridge system is essentially a segmented, linear †shield volcano. There are a number of major ridges, including the †Mid-Atlantic Ridge, which runs down the center of the Atlantic; the East Pacific Rise in the southeast Pacific; and the Southeast Indian Ridge. These ridges are now known to be spreading centers, or divergent margins, where two plates of oceanic lithosphere are moving away from one another (see †plate tectonics). Ocean ridges can rise thousands of meters above the surrounding seabed, and extend for up to 60,000 km/37,000 mi in length.

Ocean ridges usually have a †rift valley along their crests, indicating where the flanks are being pulled apart by the growth of the plates of the †lithosphere beneath. The crests are generally free of sediment; increasing depths of sediment are found with increasing distance down the flanks.

migration movement, either seasonal or as part of a single life cycle, of certain animals, chiefly birds and fish, to distant breeding or feeding grounds.

The precise methods by which animals navigate and know where to go are still obscure. Birds have much sharper eyesight and better visual memory of ground clues than humans, but in long-distance flights appear to navigate by the Sun and stars, possibly in combination with a "reading" of the Earth's magnetic field through an inbuilt "magnetic compass," which is a tiny mass of tissue between the eye and brain in birds. Similar cells occur in "homing" honeybees and in certain bacteria that use it to determine which way is "down." Leatherback turtles use the contours of underwater mountains and valleys to navigate by. Most striking, however, is the migration of young birds that have never flown a route before and are unaccompanied by adults. It is postulated that they may inherit as part of their genetic code an overall "sky chart" of their journey that is triggered into use when they become aware of how the local sky pattern, above the place in which they hatch, fits into it. Similar theories have been advanced in the case of fish, such as eels and salmon, with whom vision obviously plays a lesser role, but for whom currents and changes in the composition and temperature of the sea in particular locations may play a part – for example, in enabling salmon to return to the precise river in which they were spawned. Migration also occurs with land animals; for example, lemmings and antelope.

Species migration is the spread of the home range of a species over many years, for example the spread of the collared dove (*Streptopelia decaocto*) from Turkey to Britain over the period 1920–52. Any journey that takes an animal outside of its normal home range is called **individual migration**; when the animal does not return to its home range it is called **removal migration**. An example of **return migration** is the movement of birds that fly south for the winter and return to their home ranges in the spring. Many types of whale also make return migrations. In **remigration**, the return leg of the migration is completed by a subsequent generation, for example locust swarms migrate, but each part of the circuit is completed by a different generation.

Related to migration is the homing ability of pigeons, bees, and other creatures.

Journey North http://www.learner.org/jnorth/

US-based educational site for junior school children on the topic of wildlife migration. As well as the major ongoing project to track the southward migration of Monarch butterflies, there are projects on building garden areas to help migrating animals and the migration stories of a variety of species.

Milankovitch hypothesis combination of factors governing the occurrence of †ice ages proposed in 1930 by the Yugoslav geophysicist M Milankovitch (1879–1958). These include the variation in the angle of the Earth's axis, and the geometry of the Earth's orbit around the Sun.

mile symbol mi, imperial unit of linear measure. A statute mile is equal to 1,760 yards (1.60934 km), and an international nautical mile is equal to 2,026 yards (1,852 m).

milk secretion of the †mammary glands of female mammals, with which they suckle their young (during †lactation). Over 85% is water, the remainder comprising protein, fat, lactose (a sugar), calcium, phosphorus, iron, and vitamins. The milk of cows, goats, and sheep is often consumed by humans, but regular drinking of milk after infancy is principally a Western practice.

milk teeth or **deciduous teeth**, teeth that erupt in childhood between the ages of 6 and 30 months. They are replaced by the permanent teeth, which erupt between the ages of 6 and 21 years. See also †dentition and †tooth.

Milky Way faint band of light crossing the night sky, consisting of stars in the plane of our Galaxy. The name Milky Way is often used for the Galaxy itself. It is a spiral †galaxy, 100,000 light years in diameter and 2,000 light years thick, containing at least 100 billion †stars. The Sun is in one of its spiral arms, about 25,000 light years from the center, not far from its central plane.

The densest parts of the Milky Way, towards the Galaxy's center, lie in the constellation †Sagittarius. In places, the Milky Way is interrupted by lanes of dark dust that obscure light from the stars beyond, such as the Coalsack †nebula in †Crux (the Southern Cross). It is because of these that the Milky Way is irregular in width and appears to be divided into two between Centaurus and Cygnus.

Miller, William (1801–1880) Welsh crystallographer, developer of the **Miller indices**, a coordinate system of mapping the shapes and surfaces of crystals.

milli- symbol m, prefix denoting a one-thousandth part (10^{-3}).

For example, a millimeter, mm, is one thousandth of a meter.

millibar unit of pressure, equal to one-thousandth of a †bar.

Millikan, Robert Andrews (1868–1953) US physicist who was awarded the Nobel Prize for Physics in 1923 for his determination of †Planck's constant (a fundamental unit of quantum theory) in 1916 and the †electric charge on an electron in 1913.

milliliter one-thousandth of a liter (ml), equivalent to one cubic centimeter (cc).

millimeter of mercury symbol mmHg, unit of pressure, used in medicine for measuring blood, pressure defined as the pressure exerted by a column of mercury one millimeter high, under the action of gravity.

million one thousand thousands, 1,000,000 or 10^6.

Mills Cross type of †radio telescope consisting of two rows of aerials at right angles to each other, invented in 1953 by the Australian radio astronomer Bernard Mills. The cross-shape produces a narrow beam useful for pinpointing the positions of radio sources.

Milne, Edward Arthur (1896–1950) English astrophysicist and mathematician. He formulated a theory of †relativity which he called kinematic relativity, parallel to Albert †Einstein's general theory.

Milstein, César (1927–2002) Argentine-born British molecular biologist who was awarded the Nobel Prize for Physiology or Medicine in 1984 for his work on immunity and the discovery of a technique for producing highly specific, †monoclonal antibodies, which give immunity against specific diseases. He shared the prize with Georges †Köhler and Niels †Jerne.

Mimas small, low-density moon of Saturn discovered in 1789 by the German-born English astronomer William †Herschel. It is 400 km/248 mi in diameter and orbits just beyond the ring system, 186,000 km/115,500 mi from the planet, and takes nearly 22 hours for each orbit.

Mimas has a densely cratered surface, including one crater (called Herschel) that is 130 km/81 mi across.

mimicry imitation of one species (or group of species) by another. The most common form is **Batesian mimicry** (named after English naturalist H W †Bates), where the mimic resembles a model that is poisonous or unpleasant to eat, and has aposematic, or warning, coloration; the mimic thus benefits from the fact that predators have learned to avoid the model. Hoverflies that resemble bees or wasps are an example. Appearance is usually the basis for mimicry, but calls, songs, scents, and other signals can also be mimicked.

In **Mullerian mimicry**, two or more equally poisonous or distasteful species have a similar color pattern, thereby reinforcing the warning each gives to predators. In some cases, mimicry is not for protection, but allows the mimic to prey on, or parasitize, the model.

mimicry Batesian mimicry in which a harmless hoverfly is colored like an unpleasant wasp in order to confuse a predator. A predator that has tried to eat a wasp will avoid the hoverfly. Helicon

Mimosa or **Becrux** or **Beta Crucis**, second-brightest star in the southern-hemisphere constellation of †Crux, marking one of the four corners of the Southern Cross, and the 19th-brightest star in the night sky. It is a blue-white giant star of magnitude 0.8 around 460 light years from the Sun.

mineral naturally formed inorganic substance with a particular chemical composition and a regularly repeating internal structure. Either in their perfect crystalline form or otherwise, minerals are the constituents of †rocks. In more general usage, a mineral is any substance economically valuable for mining (including coal and oil, despite their organic origins).

Mineral-forming processes include: melting of pre-existing rock and subsequent crystallization of a mineral to form igneous or volcanic rocks; weathering of rocks exposed at the land surface, with subsequent transport and grading by surface waters, ice, or wind to form sediments; and recrystallization through increasing temperature and pressure with depth to form †metamorphic rocks. The transformation and recycling of the minerals of the Earth's outer layers is known as the rock cycle.

Minerals are usually classified as magmatic, †sedimentary, or †metamorphic. The magmatic minerals, in †igneous rock, include the feldspars, quartz, pyroxenes, amphiboles, micas, and olivines that crystallize from silica-rich rock melts within the crust or from extruded lavas.

The most commonly occurring sedimentary minerals are either pure concentrates or mixtures of sand, clay minerals, and carbonates (chiefly calcite, aragonite, and dolomite).

Minerals typical of metamorphism include andalusite, cordierite, garnet, tremolite, lawsonite, pumpellyite, glaucophane, wollastonite, chlorite, micas, hornblende, staurolite, kyanite, and diopside.

Mineral Gallery http://mineral.galleries.com/

Collection of descriptions and images of minerals, organized by mineral name, class (sulfides, oxides, carbonates, and so on), and grouping (such as gemstones, birth stones, and fluorescent minerals).

mineralogy study of minerals. The classification of minerals is based chiefly on their chemical composition and the kind of chemical bonding that holds their atoms together. The mineralogist also studies their crystallographic and physical characters, occurrence, and mode of formation.

The systematic study of minerals began in the 18th century, with the division of minerals into four classes: earths, metals, salts, and bituminous substances, distinguished by their reactions to heat and water.

Mineralogy Database http://webmineral.com/

Created by a mineral enthusiast, this is an extremely useful Web site if you need to know about a particular mineral. All major mineralogical data of each of the more than 3,700 minerals listed in Dana's Mineralogy, the definitive source on minerals, is in this searchable database.

mineral oil oil obtained from mineral sources, for example coal or petroleum, as distinct from oil obtained from vegetable or animal sources.

mineral salt in nutrition, simple inorganic chemicals that are required, as nutrients, by living organisms. Plants usually obtain their mineral salts from the soil, while animals get theirs from their food. Important mineral salts include iron salts (needed by both plants and animals), magnesium salts (needed mainly by plants, to make †chlorophyll), and calcium salts (needed by animals to make bone or shell). A †trace element is required only in tiny amounts.

Mineral salts are taken up in soluble form. When mineral salts dissolve in †water they separate into particles called ions. Mineral salts do not usually contain the element carbon and are therefore inorganic (organic compounds always contain carbon).

Plant roots absorb individual mineral ions from soil water. Some of the ions travel by †diffusion into the root; others are absorbed by active transport. The minerals required in the greatest amounts are those containing the element nitrogen, for example nitrate ions (or "nitrates"), which are a key component of inorganic fertilizer. A plant uses nitrates in the production of †proteins such as †enzymes, so they are important for plant growth. They are often in short supply in the soil, which is why inorganic fertilizers are required. Plants also require magnesium in order to make chlorophyll, the green chemical that absorbs the †energy of sunlight for †photosynthesis.

Mammals absorb the mineral salts they need from their food. Important mineral salts include iron salts (needed for †hemoglobin) and calcium salts (needed by animals to make bone). Both plants and animals need a range of other minerals in tiny amounts (trace elements).

minicomputer multi-user computer with a size and processing power between those of a †mainframe and a †microcomputer. Nowadays almost all minicomputers are based on †microprocessors.

Minicomputers are often used in medium-sized businesses and in university departments handling †database or other commercial programs and running scientific or graphical applications.

Minkowski, Hermann (1864–1909) Russian-born German mathematician whose introduction of the concept of space-time was essential to the genesis of the general theory of relativity.

Minkowski, Hermann http://www-groups.dcs.st-andrews.ac.uk/history/Mathematicians/Minkowski.html

Part of an archive containing the biographies of the world's greatest mathematicians, this site is devoted to the life and contributions of Hermann Minkowski.

Minkowski, Rudolph Leo (1895–1976) German-born US astrophysicist, responsible for the compilation of a set of photographs now found in every astronomical library, the National Geographic Society Palomar Observatory Sky Survey. A leading authority on novae and †planetary nebulae, he was a pioneer in the science of radio astronomy.

minor in the mathematical theory of †determinants, smaller determinant obtained by deleting one of the rows and one of the columns of a †matrix.

minor arc the smaller of the two arcs formed when a circle is divided into two unequal parts by a straight line or †chord.

minor planet name sometimes given to the larger members of the †asteroid belt.

minute unit of time consisting of 60 seconds; also a unit of angle equal to one sixtieth of a degree.

Miocene ("middle recent") fourth epoch of the Tertiary period of geological time, 23.5–5.2 million years ago. At this time grasslands spread over the interior of continents, and hoofed mammals rapidly evolved.

Mira or **Omicron Ceti**, brightest long-period pulsating †variable star, located in the constellation †Cetus. Mira was the first star discovered to vary periodically in brightness.

In 1596 Dutch astronomer David Fabricus noticed Mira as a third-magnitude object. Because it did not appear on any of the star charts available at the time, he mistook it for a †nova. The German astronomer Johann Bayer included it on his star atlas in 1603 and designated it Omicron Ceti. The star vanished from view again, only to reappear within a year. It was named "Stella Mira," "the wonderful star," by †Hevelius, who observed it 1659–82.

It has a periodic variation between third or fourth magnitude and ninth magnitude over an average period of 331 days. It can sometimes reach second magnitude and once almost attained first magnitude in 1779. At times it is easily visible to the naked eye, being the brightest star in that part of the sky, while at others it cannot be seen without a telescope.

mirage illusion seen in hot weather of water on the horizon, or of distant objects being enlarged. The effect is caused by the †refraction, or bending, of light.

Light rays from the sky bend as they pass through the hot layers of air near the ground, so that they appear to come from the horizon. Because the light is from a blue sky, the horizon appears blue and watery. If, during the night, cold air collects near the ground, light can be bent in the opposite direction, so that objects below the horizon appear to float above it. In the same way, objects such as trees or rocks near the horizon can appear enlarged.

Miranda or **Uranus V**, one of the five major moons of Uranus, discovered by Dutch-born US astronomer Gerard †Kuiper in 1948. It is 480 km/298 mi in diameter and orbits Uranus at an average distance of 129,000 km/80,160 mi with a period (the time it takes to circle the planet) of 1.41 Earth days.

Images from the Voyager 2 probe show that the frozen surface is composed of huge blocks of rock and ice, suggesting that Miranda was once shattered by an asteroid impact and subsequently reformed.

Mississippian US term for the Lower or Early †Carboniferous period of geological time, 363–323 million years ago. It is named after the state of Mississippi.

mist low cloud caused by the condensation of water vapor in the lower part of the †atmosphere. Mist is less thick than †fog, visibility being 1–2 km/0.6–1.2 mi.

mistral cold, dry, northerly wind that occasionally blows during the winter on the Mediterranean coast of France, particularly concentrated along the Rhône valley. It has been known to reach a velocity of 145 kph/90 mph.

Mitchell, Peter Dennis (1920–1992) English chemist. He was awarded the Nobel Prize for Chemistry in 1978 for work on the conservation of energy by plants during respiration and photosynthesis. He showed that the transfer of energy during life processes is not random but directed.

mitochondria singular **mitochondrion**, membrane-enclosed organelles within †eukaryotic cells, containing †enzymes responsible for energy production during †aerobic respiration. They are found in both plant and animal cells. Mitochondria absorb oxygen (O_2) and complete the breakdown of glucose to carbon dioxide (CO_2) and water (H_2O) to produce energy in the form of †ATP, which is used in life processes in the cell. These rodlike or spherical bodies are thought to be derived from free-living bacteria that, at a very early stage in the history of life, invaded larger cells and took up a symbiotic way of life inside them. Each still contains its own small loop of DNA called mitochondrial DNA, and new mitochondria arise by division of existing ones. Mitochondria each have 37 genes.

Mutations in mitochondrial genes are always inherited from the mother. These mutations have been linked to a number of disorders, mainly degenerative, including Alzheimer's disease and diabetes.

mitosis in biology, the process of cell division by which one parent cell produces two genetically identical "daughter" cells. The genetic material of †eukaryotic cells is carried on a number of †chromosomes. During mitosis the DNA is duplicated and the chromosome number doubled – identical copies of the chromosomes are separated into the two daughter cells, which contain the same amount of DNA as the original cell. To control movements of chromosomes during cell division so that both new cells get the correct number, a system of protein tubules, known as the spindle, organizes the chromosomes into position in the middle of the cell before they replicate. The spindle then controls the movement of chromosomes as the cell goes through the stages of division: **interphase**, **prophase**, **metaphase**, **anaphase**, and **telophase**. See also †meiosis.

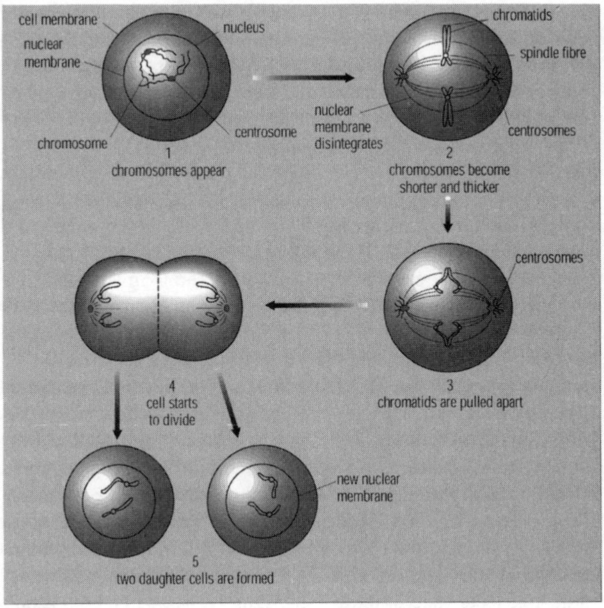

mitosis The stages of mitosis, the process of cell division that takes place when a plant or animal cell divides for growth or repair. The two "daughter" cells each receive the same number of chromosomes as were in the original cell. Helicon

Mitosis is used for growth and for †asexual reproduction. †Growth is the increase in size and weight of an organism over a period of time. In biology growth is often measured as †biomass. Growth results from mitosis followed by the increase in size of the new cells.

Interactive Mitosis Tutorial http://www.sci.sdsu.edu/multimedia/mitosis/

Shockwave-enhanced explanation of mitosis, the process of cell division in non-reproductive eukaryotic cells. All terms are clearly explained. The site also provides animated GIFs for those whose connections are slow.

mixed number number consisting of both whole and fractional parts. For example, $3^1/_2$ and $7^3/_8$ are mixed numbers. When calculating with mixed numbers, it is usual to deal with the whole number parts first, or else to convert them all to decimals or improper fractions.

mixture in chemistry, a substance containing two or more elements or compounds that still retain their separate physical and chemical properties. There is no chemical bonding between them and they can be separated from each other by physical means (compare †compound).

The separation of mixtures can be achieved by a number of methods, such as †filtration and distillation. Examples of mixtures include rocks, air, crude oil (unrefined petroleum), all solutions (such as sea water), and all metal alloys (such as steel).

Separating Mixtures http://www.learn.co.uk/default.asp?WCI=Unit&WCU=9037

This page begins by clarifying terms relevant to the topic and then details the methods of separating mixtures, including filtration, evaporation, distillation, and chromatography. There are excellent diagrams illustrating the processes and the equipment used.

Mizar or **Zeta Ursae Majoris**, second-magnitude star in †Ursa Major, where it marks the middle of the handle of the Plough. It was discovered to be a double star in 1650, and was the first star found to be double by using a telescope. Its brighter component was the first †spectroscopic binary to be discovered, by US astronomer Edward Pickering in 1889. The second component is also a spectroscopic binary.

m.k.s. system system of units in which the base units meter, kilogram, and second replace the centimeter, gram, and second of the †*c*.g.s. system. From it developed the SI system (see †SI units). It simplifies the incorporation of electrical units into the metric system, and was incorporated in SI. For application to electrical and magnetic phenomena, the ampere was added, creating what is called the m.k.s.a. system.

M number designation used in the †Messier catalogue.

mobile ion in chemistry, an ion that is free to move; such ions are only found in the aqueous solutions or melts (molten masses) of an †electrolyte. The mobility of the ions in an electrolyte is what allows it to conduct electricity.

Möbius, August Ferdinand (1790–1868) German mathematician and theoretical astronomer, discoverer of the Möbius strip and considered one of the founders of †topology.

Möbius, August Ferdinand http://www-history.mcs.st-and.ac.uk/history/Mathematicians/Mobius.html

Extensive biography of the mathematician. The site contains a description of his contribution to mathematics, especially of his conception of the Möbius strip, a two-dimensional surface with only one side. Several literature references for further reading are also listed, and the Web site also features a photograph of Möbius.

Möbius strip structure made by giving a half twist to a flat strip of paper and joining the ends together. It has certain remarkable properties, arising from the fact that it has only one edge and one side. If cut down the center of the strip, instead of two new strips of paper, only one long strip is produced. It was invented by the German mathematician August Möbius.

Möbius strip The Möbius strip has only one side and one edge. It consists of a strip of paper connected at its ends with a half-twist in the middle. Helicon

mode in mathematics, the element that appears most frequently in a given set of †data. For example, the mode for the data 0, 0, 9, 9, 9, 12, 87, 87 is 9.

The mode together with the †median and †arithmetic mean make up the †average of a set of data. In addition it is useful to know the †range or spread of the data.

When dealing with †grouped data, the **modal class** represents the most frequent group.

model in computing, set of assumptions and criteria based on actual phenomena, used to conduct a †computer simulation. Models are used to predict the behavior of a system such as the movement of a hurricane or the flow of goods from a store. In industry, they are important tools for testing new products: engineers subject virtual prototypes of aircraft or bridges to various scenarios to find out what adjustments are necessary to the design. However, a model is only as good as the assumptions that underlie it.

Models may run at the same speed of the real situation (**real-time models**) or run at faster or slower speeds.

Models are also the basis for †expert systems, which simulate the knowledge of a human expert.

model simplified version of some aspect of the real world. Models are produced to show the relationships between two or more factors, such as land use and the distance from the center of a town (for example, concentric-ring theory). Because models are idealized, they give only a general guide to what may happen.

modem contraction of modulator/demodulator, device for transmitting computer data over telephone lines. Such a device is used to convert †digital signals produced by computers to †analog signals compatible with the telephone network, and back again.

Modems are used for linking remote terminals to central computers and enable computers to communicate with each other anywhere in the world. In 1997 the fastest standard modems transmitted data at a nominal rate of about 33,600 bps (bits per second), often abbreviated to 33.6K.

56K modems, launched in 1997, achieved higher speeds by using a digital connection to the user's computer, while using a conventional analog connection in the other direction. In theory the downstream link can transfer data at 56 Kbps but in practice, speeds are usually 45–50K or less, depending on such factors as the quality of the telephone line.

Modem HOWTO http://www.linuxdoc.org/HOWTO/Modem-HOWTO.html

Comprehensive "Frequently Asked Question" list on modems. The Web site forms a practical guide and provides technical information on modem set up and operation. The in-depth articles are headed with introductory-level explanations.

moderator in a †nuclear reactor, a material such as graphite or heavy water used to reduce the speed of high-energy neutrons. Neutrons produced by nuclear fission are fast-moving and must be slowed to initiate further fission so that nuclear energy continues to be released at a controlled rate.

Slow neutrons are much more likely to cause †fission in a uranium-235 nucleus than to be captured in a U-238 (nonfissile uranium) nucleus. By using a moderator, a reactor can thus be made to work with fuel containing only a small proportion of U-235.

modulation in radio transmission, the variation of frequency, or amplitude, of a radio carrier wave, in accordance with the audio characteristics of the speaking voice, music, or other signal being transmitted.

modulo in number theory, two numbers are said to be equivalent modulo a fixed number if their difference is divisible by the fixed number.

modulus in mathematics, a number that divides exactly into the difference between two given numbers. Also, the multiplication factor used to convert a logarithm of one base to a logarithm of another base. Also, another name for †absolute value.

Mohole US project for drilling a hole through the Earth's crust, so named from the †Mohorovicic discontinuity that marks the transition from crust to mantle. Initial tests were made in the Pacific in 1961, but the project was subsequently abandoned.

The cores that were brought up illuminated the geological history of the Earth and aided the development of geophysics.

Mohorovicic discontinuity or **Moho** or **M-discontinuity**, seismic discontinuity, marked by a rapid increase in the speed of earthquake waves, that is taken to represent the boundary between the Earth's crust and mantle. It follows the variations in the thickness of the crust and is found approximately 35–40 km/22–25 mi below the continents and about 10 km/6 mi below the oceans. It is named after the Croatian geophysicist Andrija Mohorovicic, who suspected its presence after analyzing seismic waves from the Kulpa Valley earthquake in 1909. The "Moho" is as deep as 70 km/45 mi beneath high mountain ranges.

Mohs, Friedrich (1773–1839) German mineralogist who in 1812 devised the †Mohs scale of minerals, classified in order of relative hardness.

Mohs scale scale of hardness for minerals (in ascending order): 1 talc; 2 gypsum; 3 calcite; 4 fluorite; 5 apatite; 6 orthoclase; 7 quartz; 8 topaz; 9 corundum; 10 diamond.

The scale is useful in mineral identification because any mineral will scratch any other mineral lower on the scale than itself, and similarly it will be scratched by any other mineral higher on the scale.

Moissan, (Ferdinand Frédéric) Henri (1852–1907) French chemist. He was awarded the Nobel Prize for Chemistry in 1906 for his isolation of pure fluorine in 1886, and adoption of the electric furnace. He attempted to create artificial diamonds by rapidly cooling carbon heated to high temperatures. His claims of success were treated with suspicion.

Moivre, Abraham de (1667–1754) French mathematician who pioneered the development of analytical trigonometry, for which he formulated his theorem regarding complex numbers. He also devised a means of research into the theory of probability.

mol symbol for †mole, the SI unit of amount of substance, equal to the amount that contains as many elementary entities (such as atoms, ions, or molecules) as there are atoms in 12 grams of carbon.

molar one of the large teeth found towards the back of the mammalian mouth. The structure of the jaw, and the relation of the muscles, allows a massive force to be applied to molars. In herbivores the molars are flat with sharp ridges of enamel and are used for grinding, an adaptation to a diet of tough plant material. Carnivores have sharp powerful molars called carnassials, which are adapted for cutting meat.

molarity †concentration of a solution expressed as the number of †moles of solute per cubic decimeter of solution.

molar solution solution that contains one †mole of a substance per liter of solvent.

molar volume volume occupied by one †mole (the molecular mass in grams) of any gas at standard temperature and pressure, equal to 2.24136 x 10^{-2} m^3.

mole symbol mol, unit of the amount of a substance. One mole of a substance is the mass that contains the same number of particles (atoms, molecules, ions, or electrons) as there are atoms in 12 grams of the †isotope carbon-12.

One mole of a substance is 6.022045 x 10^{23} atoms, which is †Avogadro's number. It is obtained by weighing out the relative atomic mass (RAM) or relative molecular mass (RMM) in grams (so one mole of carbon weighs 12 g).

molecular biology study of the molecular basis of life, including the biochemistry of molecules such as DNA, RNA, and proteins, and the molecular structure and function of the various parts of living cells.

molecular clock use of rates of †mutation in genetic material to calculate the length of time elapsed since two related species diverged from each other during evolution. The method can be based on comparisons of the DNA or of widely occurring proteins, such as hemoglobin.

Since mutations are thought to occur at a constant rate, the length of time that must have elapsed in order to produce the difference between two species can be estimated. This information can be compared with the evidence obtained from paleontology to reconstruct evolutionary events.

molecular cloud in astronomy, enormous cloud of cool interstellar dust and gas containing hydrogen molecules and more complex molecular species. Giant molecular clouds (GMCs), about a million times as massive as the Sun and up to 300 light years in diameter, are regions in which stars are being born. The †Orion nebula is part of a GMC.

molecular formula in chemistry, formula indicating the actual number of atoms of each element present in a single †molecule of a chemical compound. For example, the molecular formula of carbon dioxide is CO_2, indicating that one molecule of carbon dioxide is made up of one atom of carbon and two atoms of oxygen. This is determined by two pieces of information: the empirical †formula and the †relative molecular mass, which is determined experimentally.

molecular mass or **relative molecular mass**, mass of a molecule, calculated relative to one-twelfth the mass of an atom of carbon-12. It is found by adding the relative atomic masses of the atoms that make up the molecule.

molecular sieve in chemistry, solid with pores and channels of molecular size that can be used to absorb and separate molecules. Separation of specific molecules can be controlled by selection of a molecular sieve of the appropriate pore size. †Zeolites make good materials for molecular sieves.

molecular solid in chemistry, solid composed of molecules that are held together by relatively weak †intermolecular forces. Such solids are low-melting and tend to dissolve in organic solvents. Examples of molecular solids are sulfur, ice, sucrose, and solid carbon dioxide.

molecular structure in chemistry, type, arrangement, and bonding of atoms in a molecule.

molecular weight see †relative molecular mass.

molecule smallest configuration of an element or compound that can exist independently. One molecule is made up of a group of atoms held together by †covalent or †ionic bonds. Several non-metal elements exist as molecules. For example, hydrogen †atoms, at room temperature, do not exist independently. They are bonded in pairs to form hydrogen molecules. A molecule of a compound consists of two or more different atoms bonded together. For example, carbon dioxide is made up of molecules, each containing one carbon and two oxygen atoms bonded together. The †molecular formula is made up of the chemical symbols representing each element in the molecule and numbers showing how many atoms of each element are present. For example, the formula for hydrogen is H_2, and for carbon dioxide is CO_2. Molecules vary in size and complexity from the hydrogen molecule to the large †macromolecules of proteins. In general, elements and compounds with molecular structures have similar properties. They have low melting and boiling points, so that many molecular substances are gases or liquids at room temperatures. They are

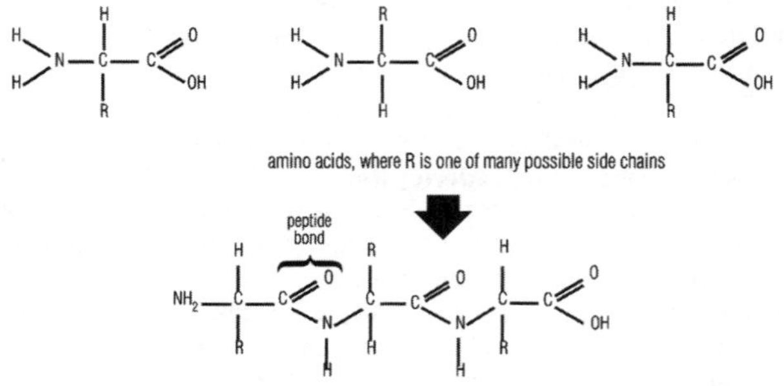

peptide – this is one made of just three amino acid units. Proteins consist of very large numbers of amino acid units in long chains, folded up in specific ways

protein A protein molecule is a long chain of amino acids linked by peptide bonds. The properties of a protein are determined by the order, or sequence, of amino acids in its molecule, and by the three-dimensional structure of the molecular chain. The chain folds and twists, often forming a spiral shape. Helicon usually insoluble in water and do not conduct electricity even when melted.

Molecules may be held together by ionic bonds, in which the atoms gain or lose electrons to form †ions, or by covalent bonds, where electrons from each atom are shared to form the bond.

kinetic theory of matter According to the molecular or †kinetic theory of matter, matter is made up of molecules that are in a state of constant motion, the extent of which depends on their †temperature. Molecules also exert forces on one another. The nature and strength of these forces depends on the temperature and state of the matter (solid, liquid, or gas). The existence of molecules was first inferred from the Italian physicist Amedeo †Avogadro's hypothesis in 1811. He observed that when gases combine, they do so in simple proportions. For example, exactly one volume of oxygen and two volumes of hydrogen combine to produce water. He hypothesized that equal volumes of gases at the same temperature and †pressure contain equal numbers of molecules. Avogadro's hypothesis only became generally accepted in 1860 when proposed by the Italian chemist Stanislao †Cannizzaro. The movement of some molecules can be observed in a microscope. As early as 1827, Robert †Brown observed that very fine pollen grains suspended in water move about in a continuously agitated manner. This continuous, random motion of particles in a fluid medium (gas or liquid) as they are subjected to impact from the molecules of the medium is known as †Brownian movement. The spontaneous and random movement of molecules or particles in a fluid can also be observed as †diffusion occurs from a region in which they are at a high concentration to a region of lower concentration, until a uniform concentration is achieved throughout. No mechanical mixing or stirring is involved. For example, if a drop of ink is added to water, its molecules will diffuse until the color becomes evenly distributed.

kinetic theory of gases The effects of pressure, temperature, and †volume on a gas were investigated during the 17th and 18th centuries. †Boyle's law states that for a fixed mass of gas the volume of the gas is inversely proportional to the pressure at constant temperature. †Charles's law states that for a fixed mass of gas the volume of the gas is proportional to the absolute temperature at constant pressure. The †pressure law states that the pressure of a fixed mass of gas at constant volume is directly proportional to its absolute temperature. These statements together give the †gas laws which can be expressed as: (pressure x volume)/temperature = constant. A plot of the volume of a gas against its temperature gives a straight line, showing that the two are proportional. The line intercepts the temperature axis at –273°C/–459°F. This suggests that, if the gas did not liquefy first, it would occupy zero volume at a temperature of –273°C. This temperature is referred to as †absolute zero, or zero kelvin (0 K) on the †Kelvin scale, and is the lowest temperature theoretically possible. This behavior applies only to ideal gases, which are assumed to occupy negligible volume and contain negligible forces between particles. A real gas often behaves rather differently, and the †van der Waals' law contains a correction to the gas laws to account for the non-ideal behavior of real gases.

change of state As matter is heated its temperature may rise or it may cause a †change of state. As the internal energy of matter increases the energy possessed by each particle increases too. This can be visualized as the †kinetic energy of the molecules increasing, causing them to move more quickly. This movement includes both vibration within the molecule (assuming the substance is made of more than one atom) and rotation. A solid is made of particles that are held together by forces. As a solid is heated, the particles vibrate more vigorously, taking up more space, and causing the material to expand. As the temperature of the solid increases, it reaches its †melting point and turns into a liquid. The particles in a liquid can move around more freely but there are still forces between them. As further energy is added, the particles move faster until they are able to overcome the forces between them. When the †boiling point is reached the liquid boils and becomes a gas. Gas particles move around independently of one another except when they collide. Different objects require different amounts of heat energy to change their temperatures by the same amount. The †heat capacity of an object is the quantity of heat required to raise its temperature by one degree. The †specific heat capacity of a substance is the heat capacity per unit of mass, measured in joules per kilogram per kelvin. As a substance is changing state while being heated, its temperature remains constant, provided that thermal energy is being added. For example, water boils at a constant temperature as it turns to steam. The energy required to cause the change of state is called †latent heat. This energy is used to break down the forces holding the particles together so that the change in state can occur. †Specific latent heat is the thermal energy required to change the state of a certain mass of that substance without any temperature change. †Evaporation causes cooling as a liquid vaporizes. Heat is transferred by the movement of particles (that possess kinetic energy) by conduction, convection, and radiation. †Conduction involves the movement of heat through a solid material by the movement of free electrons. †Convection involves the transfer of energy by the movement of fluid particles. †Convection currents are caused by the expansion of a liquid or gas as its temperature rises.

The expanded material, being less dense, rises above colder and therefore denser material.

the size and shape of molecules The shape of a molecule profoundly affects its chemical, physical, and biological properties. Optical †isomers (molecules that are mirror images of each other) rotate plane †polarized light in opposite directions; isomers of drug molecules may have different biological effects; and †enzyme reactions are crucially dependent on the shape of the enzyme and the substrate on which it acts. A wheel-shaped molecule containing 700 atoms and with a relative molecular mass of about 24,000 was produced by German chemists in 1995. Containing 154 †molybdenum atoms surrounded by oxygen atoms, it belongs to the class of compounds known as metal clusters.

Molecule of the Month http://www.bris.ac.uk/Depts/Chemistry/MOTM/motm.htm

Pages on interesting – and sometimes hypothetical – molecules, contributed by university chemistry departments throughout the world.

Online 3D Visualization of Molecular Molecules http://www.edu.uoc.gr/~baboukas/Java/3Dmol/General.htm

Large collection of 3D chemical structures to explore. The categories of molecules include "Natural," "Inorganics," "Pharmaceuticals," and "Biomolecules."

mole fraction in chemistry, proportion of the number of molecules of one species taking part in a chemical reaction in relation to the sum of the molecules of all the components involved.

Molina, Mario (1943–) Mexican chemist who shared the Nobel Prize for Chemistry in 1995 with Paul †Crutzen and F Sherwood †Rowland for their work in atmospheric chemistry, particularly concerning the formation and decomposition of ozone. They explained the chemical reactions that are destroying the ozone layer.

mollusk any of a group of invertebrate animals, most of which have a body divided into three parts: a head, a central mass containing the main organs, and a foot for movement; the more sophisticated octopuses and related mollusks have arms to capture their prey. The majority of mollusks are marine animals, but some live in fresh water, and a few live on land. They include clams, mussels, and oysters (bivalves), snails and slugs (gastropods), and cuttlefish, squids, and octopuses (cephalopods). The body is soft, without limbs (except for the cephalopods), and cold-blooded. There is no internal skeleton, but many species have a hard shell covering the body. (Phylum Mollusca.)

Mollusks have varying diets, the carnivorous species feeding mainly on other mollusks. Some are vegetarian. Reproduction is by means of eggs and is sexual; many species are hermaphrodite (having both male and female reproductive organs). The shells of mollusks take a variety of forms: single or univalve (like the snail), double or bivalve (like the clam), chambered (like the nautilus), and many other variations. In some cases (for example, cuttlefish and squid), the shell is internal. Every mollusk has a fold of skin, the mantle, which covers either the whole body or only the back, and secretes the chalky substance that forms the shell. The lower ventral surface (belly area) of the body forms the foot, which enables the mollusk to move about.

Mollusca invertebrate animal phylum that contains the †mollusks.

molybdenite molybdenum sulfide, MoS_2, the chief ore mineral of molybdenum. It possesses a hexagonal crystal structure similar to graphite, has a blue metallic luster, and is very soft (1–1.5 on the †Mohs scale).

molybdenum chemical symbol Mo, (Greek *molybdos* "lead") heavy, hard, lustrous, silver-white, metallic element, atomic number 42, relative atomic mass 95.94. The chief ore is the mineral sulfide molybdenite. The element is highly resistant to heat and conducts electricity easily. It is used in alloys, often to harden steels. It is a necessary trace element in human nutrition. It was named in 1781 by Swedish chemist Karl Scheele, after its isolation by another Swedish chemist Peter Jacob Hjelm (1746–1813), for its resemblance to lead ore.

moment in statistics and probability theory, a generalization to a higher power of the mean. The *r*th moment of a random variable is the expected value of the *r*th power of the variable less its mean.

moment of a force in physics, measure of the turning effect, or torque, produced by a force acting on a body. It is equal to the product of the force and the perpendicular distance from its line of action to the point, or pivot, about which the body will turn. The turning force around the pivot is called the moment. Its unit is the newton meter.

The moment of a force can be worked out using the formula: moment = force applied x perpendicular distance from the pivot. If the magnitude of the force is F newtons and the perpendicular distance is d meters then:

moment = Fd

It is easier to undo a bolt using a long spanner than a short spanner. This is because more turning force is produced at the bolt (pivot) with less effort. A long spanner is an example of a force multiplier.

In a simple balanced see-saw, the forces acting on the left- and right-hand sides of the pivot are the same. This is known as balancing moments. Moving the load at one end will cause the see-saw to become unbalanced. To regain balance, the load on the opposite side must either be increased or its position changed. This is known as the principle of moments.

The application of balancing a moment of force (or a turning force) is used by a rope walker. By holding the pole in the middle the rope walker is balancing the turning force on either side of the rope. As the rope walker moves a leg, the turning force on one side is greater and becomes unbalanced. To regain the balance, the pole is shifted to balance the turning forces on both sides of the rope.

See also †couple.

moment of inertia in physics, the sum of all the point masses of a rotating object multiplied by the squares of their respective distances from the axis of rotation.

In linear dynamics, Newton's second law of motion states that the force F on a moving object equals the products of its mass m and acceleration a ($F = ma$); the analogous equation in rotational dynamics is $T = I\alpha$, where T is the torque (the turning effect of a force) that causes an angular acceleration a and I is the moment of inertia. For a given object, I depends on its shape and the position of its axis of rotation.

momentum product of the mass of a body and its velocity. If the mass of a body is m kilograms and its velocity is v m s^{-1}, then its momentum is given by: momentum = mv. Its unit is the kilogram meter-per-second (kg m s^{-1}) or the newton second. The momentum of a body does not change unless a resultant or unbalanced force acts on that body (see †Newton's laws of motion).

According to Newton's second law of motion, the magnitude of a resultant force F equals the rate of change of momentum brought about by its action, or: $F = (mv - mu)/t$ where mu is the initial momentum of the body, mv is its final momentum, and t is the time in seconds over which the force acts. The change in momentum, or †impulse, produced can therefore be expressed as: impulse = $mv - mu = Ft$.

The law of conservation of momentum is one of the fundamental concepts of classical physics. It states that the total momentum of all bodies in a closed system is constant and unaffected by processes occurring within the system. The **angular momentum** of an orbiting or rotating body of mass m traveling at a velocity v in a circular orbit of radius R is expressed as mvR. Angular momentum is conserved, and should any of the values alter (such as the radius of orbit), the other values (such as the velocity) will compensate to preserve the value of angular momentum, and that lost by one component is passed to another.

monadock in earth science, isolated hill or mountain of generally hard rock that rises above a lowland (†peneplain) leveled by water erosion.

monazite mineral, (Ce,La,Th)PO_4, yellow to red, valued as a source of †lanthanides or rare earths, including cerium and europium; generally found in placer deposit (alluvial) sands.

Mond, Ludwig (1839–1909) German-born British chemist who invented a process for recovering sulfur during the manufacture of alkali. He gave his name to a method of extracting nickel from nickel carbonyl, one of its volatile organic compounds.

Monge, Gaspard (1746–1818) French mathematician and chemist, the founder of descriptive geometry. His application of analysis to infinitesimal geometry also paved the way for later developments.

Monge, Gaspard http://www-groups.dcs.st-andrews.ac.uk/history/Mathematicians/Monge.html

Part of an archive containing the biographies of the world's greatest mathematicians, this site is devoted to the life and contributions of Gaspard Monge. In addition to biographical information, you will find a list of references about Monge and links to other essays in the archive that reference him. The text of this essay includes hypertext links to the essays of those mathematicians and thinkers who influenced Monge. There is also an image of him, which can be enlarged, and a map of his birthplace.

monobasic acid acid that has only one replaceable hydrogen atom and can therefore form only one series of salts; for example, HCl.

Monoceros constellation on the celestial equator (see †celestial sphere), represented as a unicorn, which, though it includes no bright stars, lies in the Milky Way and contains many faint stars, clusters, and nebulae.

monochromatic literally, one color, the property of some sources of radiation to emit waves of one †frequency (or wavelength). †Lasers and †masers are examples.

monoclonal describing genetically identical cells produced from one †clone.

monoclonal antibody MAB, antibody produced by fusing an antibody-producing lymphocyte (white blood cell) with a cancerous myeloma (bone-marrow) cell. The resulting fused cell, called a hybridoma, is immortal and can be used to produce large quantities of a single, specific antibody. By choosing antibodies that are directed against antigens found on cancer cells, and combining them with cytotoxic drugs, it is hoped to make so-called magic bullets that will be able to pick out and kill cancers.

It is the antigens on the outer cell walls of germs entering the body that provoke the production of antibodies as a first line of defense against disease. Antibodies "recognize" these foreign antigens, and, in locking on to them, cause the release of chemical signals in the bloodstream to alert the immune system for further action. MABs are copies of these natural antibodies, with the same ability to recognize specific antigens. Introduced into the body, they can be targeted at disease sites.

The full potential of these biological missiles, developed by César †Milstein and others at Cambridge University, England, in 1975, is still under investigation. However, they are already in use in blood-grouping, in pinpointing viruses and other sources of disease, in tracing cancer sites, and in developing vaccines.

monocotyledon angiosperm (flowering plant) having an embryo with a single cotyledon, or seed leaf (as opposed to †dicotyledons, which have two). Monocotyledons usually have narrow leaves with parallel veins and smooth edges, and hollow or soft stems. Their flower parts are arranged in threes. Most are small plants such as orchids, grasses, and lilies, but some are trees such as palms.

monocyte type of white blood cell. They are found in the tissues and the lymphatic and circulatory systems where their purpose is to remove foreign particles, such as bacteria and tissue debris, by ingesting them.

Monod, Jacques Lucien (1910–1976) French biochemist who was awarded the Nobel Prize for Physiology or Medicine in 1965 with his co-workers André †Lwoff and François †Jacob for research into the genetic control of enzyme and virus synthesis.

monoecious having separate male and female flowers on the same plant. Maize (corn) *Zea mays*, for example, has a tassel of male flowers at the top of the stalk and a group of female flowers (on the ear, or cob) lower down. Monoecism is a way of avoiding self-fertilization.

†Dioecious plants have male and female flowers on separate plants.

monohybrid inheritance pattern of inheritance seen in simple †genetics experiments, where the two animals (or two plants) being crossed are genetically identical except for one gene, which is heterozygous. In other words, inheritance is the passing of characteristics from parents to offspring, while monohybrid inheritance is inheritance for a single characteristic.

Characteristics that can be passed on in this way are determined by †genes. Genes may exist in different forms known as †alleles. Since chromosomes in animals and plants are found as matching pairs in the †nucleus of each cell, there will always be two genes for a characteristic in a cell. If the same allele is present twice, the organism is said to be homozygous for this characteristic. If, however, one chromosome contains one allele and the other chromosome a contrasting allele, the organism is said to be heterozygous for that characteristic. In a heterozygous organism the appearance of the organism (phenotype) may be determined by one allele and not the other. The allele that determines the phenotype is said to be dominantly expressed (see †dominance). The expression of the other allele is described as being recessive.

The gene in question may code for some obvious external features such as seed color, with one parent having green seeds and the other having yellow seeds. The offspring are monohybrids, that is, hybrids for one gene only, having received one copy of the gene from each parent. Known as the F1 generation, they are all identical, and usually resemble one parent, whose version of the gene (the dominant allele) masks the effect of the other version (the recessive allele). Although the characteristic coded for by the recessive allele (for example, green seeds) completely disappears in this generation, it can reappear in offspring of the next generation if they have two recessive alleles. On average, this will occur in one out of four offspring from a cross between two of the monohybrids. The next generation (called F2) shows a 3:1 ratio for the characteristic in question, 75% being like the original parent with the recessive allele.

Austrian biologist Gregor †Mendel first carried out experiments of this type (crossing varieties of artificially bred plants, such as peas) and they revealed the principles of genetics. The same basic mechanism underlies all inheritance, but in most plants and animals there are so many genetic differences interacting to produce the external appearance that such simple, clear-cut patterns of inheritance are not evident.

monomer chemical compound composed of simple molecules from which †polymers can be made. Under certain conditions the simple molecules (of the monomer) join together (polymerize) to form a very long chain molecule (macromolecule) called a polymer. For example, the polymerization of †ethene (ethylene) monomers produces the polymer †polythene (polyethylene):

$2nCH_2 = CH_2 \rightarrow (CH_2 - CH_2 - CH_2 - CH_2)_n$

monophyletic in biological classification, term describing a group of species composed of an ancestral species plus all of its descendant species. Examples of such groups are the mammals, birds, and insects. Monophyletic groups, or **clades**, are the basis of the system of classification known as †cladistics.

If a group does not include all the descendants of a common ancestor, it is called †paraphyletic; if it includes some or all of the descendants but not the common ancestor, it is called †polyphyletic.

monosaccharide or **simple sugar**, †carbohydrate that cannot be hydrolyzed (split) into smaller carbohydrate units. Examples are glucose and fructose, both of which have the molecular formula $C_6H_{12}O_6$.

monosodium glutamate **MSG**; $NaC_5H_8NO_4$, white, crystalline powder, the sodium salt of glutamic acid (an †amino acid found in proteins that plays a role in the metabolism of plants and animals). It has no flavor of its own, but enhances the flavor of foods such as meat and fish. It is used to enhance the flavor of many packaged and "fast foods," and in Chinese cooking. Ill effects may arise from its overconsumption, and some people are very sensitive to it, even in small amounts. It is commercially derived from vegetable protein. It occurs naturally in soybeans and seaweed.

monotonic of a mathematical sequence, denoting that its terms are increasing or decreasing along the sequence.

monotreme any of a small group of primitive egg-laying mammals, found in Australasia. They include the echidnas (spiny anteaters) and the platypus. (Order Monotremata.)

In 1995 Australian paleontologists announced a new (extinct) family of monotreme, the Kollikodontidae, following the discovery of a 120-million-year-old jawbone in New South Wales.

Monro, Alexander (1733–1817) also known as **Monro secundus**, Scottish anatomist who determined the role of the lymphatic system, distinguishing it from the circulatory system, and wrote one of the first works of comparative anatomy.

monsoon wind pattern that brings seasonally heavy rain to South Asia; it blows towards the sea in winter and towards the land in summer. The monsoon may cause destructive flooding all over India and Southeast Asia from April to September, leaving thousands of people homeless each year.

The monsoon cycle is believed to have started about 12 million years ago with the uplift of the Himalayas.

Montagnier, Luc (1932–) French molecular biologist who first identified the human immunodeficiency virus (†HIV) in 1983, the single-stranded RNA retrovirus that causes †AIDS. Patients with the disease die from rare infections because their immune systems are impaired by the virus.

Roche-HIV Web Site

http://www.roche.com/home/healthcare/healthcare-therapy/healthcare-therapy-indications/healthcare-therapy-indications-hiv.htm

HIV virus Web site focusing on the properties of the virus and its treatment rather than the illness it causes. Roche say the purpose of the Web site is to "facilitate a better understanding of the virus and treatment issues, specifically resistance and adherence to HIV therapy."

month unit of time based on the motion of the †Moon around the Earth.

The time taken for the Moon to orbit the Earth (from one new or full Moon to the next) is known as a **synodic** or **lunar month** and is 29.53 days. The Western calendar is not based on the lunar month, but rather on the **calendar month**, a human invention devised to fit the calendar year. However, the Islamic calendar is based on the lunar month, and therefore each year a fixed date goes back by about ten days. The time for the Moon to complete one orbit around the Earth relative to the stars (the **sidereal month**) is 27.32 days. The **solar month** equals 30.44 days, and is exactly one-twelfth of the solar or tropical year, the time taken for the Earth to orbit the Sun.

Moog, Bob (1934–) born Robert Moog, US engineer who developed the first synthesizers widely used in popular music. Moog is also known for building theremins, electronic instruments that musicians play by moving their hands between two antennae. In 1996 his company, Big Briar, based in Asheville, North Carolina, announced a MIDI interface for theremins.

Moon natural satellite of Earth, 3,476 km/2,160 mi in diameter, with a mass 0.012 (approximately one-eightieth) that of Earth.

Its surface gravity is only 0.16 (one-sixth) that of Earth. Its average distance from Earth is 384,400 km/238,855 mi, and it orbits in a west-to-east direction every 27.32 days (the **sidereal month**). It spins on its axis with one side permanently turned towards Earth. The Moon has no atmosphere and was thought to have no water until ice was discovered on its surface in 1998.

phases The Moon is illuminated by sunlight, and goes through a cycle of phases of shadow, waxing from **new** (dark) via **first quarter** (half Moon) to **full**, and waning back again to new every 29.53 days (the **synodic month**, also known as a **lunation**). On its sunlit side, temperatures reach 110°C/230°F, but during the two-week lunar night the surface temperature drops to –170°C/–274°F.

origins The origin of the Moon is still open to debate. Scientists suggest the following theories: that it split from the Earth; that it was a separate body captured by Earth's gravity; that it formed in orbit around Earth; or that it was formed from debris thrown off when a body the size of Mars struck Earth.

research 70% of the far side of the Moon was photographed from the Soviet Lunik 3 in October 1959. Much of our information about the Moon has been derived from this and other photographs and measurements taken by US and Soviet Moon probes, from geological samples brought back by US *Apollo* astronauts and by Soviet Luna probes, and from experiments set up by US astronauts 1969–72. The US probe Lunar Prospector, launched in January 1998, examined the composition of the lunar crust, recorded gamma rays, and mapped the lunar magnetic field. It also discovered the ice on the moon in March 1998.

composition The Moon is rocky, with a surface heavily scarred by †meteorite impacts that have formed craters up to 240 km/150 mi across. Seismic observations indicate that the Moon's surface extends downwards for tens of kilometers; below this crust is a solid mantle about 1,100 km/688 mi thick, and below that a silicate core, part of which may be molten. Rocks brought back by astronauts show that the Moon is 4.6 billion years old, the same age as Earth. It is made up of the same chemical elements as Earth, but in different proportions, and differs from Earth in that most of the Moon's surface features were formed within the first billion years of its history when it was hit repeatedly by meteorites.

The youngest craters are surrounded by bright rays of ejected rock. The largest scars have been filled by dark lava to produce the lowland plains called seas, or **maria** (plural of †mare). These dark patches form the so-called "man-in-the-Moon" pattern. Inside some craters that are permanently in shadow is up to 300 million tonnes/330 million tons of ice existing as a thin layer of crystals.

One of the Moon's easiest features to observe is the mare Plato, which is about 100 km/62 mi in diameter and 2,700 m/8,860 ft deep, and at times is visible with the naked eye.

Earth and Moon Viewer http://www.fourmilab.ch/earthview/

Graphical simulations of the Moon and Earth. The Moon (or locations on the Moon) can be viewed from different locations on Earth, from the Sun, or as a map showing day and night across the lunar surface.

Moon http://www.solarviews.com/solar/eng/moon.htm

Detailed description of the Moon. It includes statistics and information about the surface, eclipses, and phases of the Moon, along with details of the Apollo landing missions. The site is supported by a good selection of images.

Phases of the Moon 2001–2005 http://www.stormfax.com/moon.htm

Invaluable site for those interested in observing the phases of the Moon. Presented in calendar format, this site lists extremely precise times and dates for the phases of the Moon day by day up to the year 2005. There is also a list of definitions to explain the terminology used to describe the phases.

moon in astronomy, any natural †satellite that orbits a planet. Mercury and Venus are the only planets in the Solar System that do not have moons.

Moon probe crewless spacecraft used to investigate the Moon. Early probes flew past the Moon or crash-landed on it, but later ones achieved soft landings or went into orbit. Soviet probes included the Luna series. US probes (Ranger space probes, Surveyor, Lunar Orbiter) prepared the way for the *Apollo* project crewed flights.

moonstone translucent, pearly variety of potassium sodium †feldspar, found in Sri Lanka and Myanmar, and distinguished by a blue, silvery, or red opalescent tint. It is valued as a gem.

moor in earth science, a stretch of land, usually at a height, which is characterized by a vegetation of heather, coarse grass, and bracken. A moor may be poorly drained and contain boggy hollows.

Moore, Patrick (Alfred Caldwell) (1923–) English broadcaster, writer, and popularizer of astronomy. He began presenting the BBC television series *The Sky at Night* in 1968.

moraine rocky debris or †till carried along and deposited by a †glacier. Material eroded from the side of a glaciated valley and carried along the glacier's edge is called a **lateral moraine**; that worn from the valley floor and carried along the base of the glacier is called a **ground moraine**. Rubble dropped at the snout (front end) of a melting glacier is called a **terminal moraine**.

When two glaciers converge their lateral moraines unite to form a **medial moraine**. Debris that has fallen down crevasses and becomes embedded in the ice is termed an **englacial moraine**; when this is exposed at the surface due to partial melting it becomes an **ablation moraine**.

mordant in chemistry, compound added to fix a dye to material that would not hold a stable color with the dye alone. Mordants are usually inorganic metal oxides and hydroxides.

Mordell's equation equation named after US-born British mathematician Louis Mordell (1888–1972) that is represented by $y^2 = x^3 + k$.

Morgagni, Giovanni Battista (1682–1771) Italian anatomist who developed the view that disease was not an imbalance of the body's humours but a result of alterations in the organs. His work *De sedibus et causis morborum per anatomen indagatis/On the Seats and Causes of Diseases as Investigated by Anatomy* (1761) formed the basis of †pathology.

Morgan, Augustus De Indian-born British mathematician; see †De Morgan, Augustus.

Morgan, Conwy Lloyd (1852–1936) English psychologist. Of immense influence in the field of comparative psychology, he carried out observational studies of animals in natural settings. In 1894 he was the first to describe **trial-and-error learning** in animals.

Morgan, Thomas Hunt (1866–1945) US geneticist who was awarded the Nobel Prize for Physiology or Medicine in 1933 for his work on the role of chromosomes in heredity. He helped establish that †genes are located on the chromosomes, discovered sex chromosomes, and invented the techniques of genetic mapping. He was the first to work on the fruit fly *Drosophila*, which has since become a major subject of genetic studies.

Morley, Edward Williams (1838–1923) US physicist who collaborated with Albert †Michelson on the **Michelson–Morley experiment** in 1887. In 1895 he established precise and accurate measurements of the densities of oxygen and hydrogen.

morphine narcotic alkaloid, $C_{17}H_{19}NO_3$, derived from opium and prescribed only to alleviate severe pain. Its use produces serious side effects, including nausea, constipation, tolerance, and addiction, but it is highly valued for the relief of the terminally ill.

The risk of addiction arising from the use of morphine for pain relief is much lower than for recreational use (about 1 in 3,000) as the drug is processed differently by the body when pain is present.

morphing metamorphosis of one shape or object into another by computer-generated animation. First used in film-making in 1990, it has transformed cinema special effects. Conventional animation is limited to two dimensions; morphing enables the creation of three-dimensional transformations.

To create such effects, the start and end of the transformation must be specified on screen using a wire-frame model that mathematically defines the object. To make the object three-dimensional, the wire can be extruded from a cross-section or turned as on a lathe to produce an evenly turned surface. This is then rendered, or filled in and shaded. Once the beginning and end objects have been created, the computer can calculate the morphing process.

morphogen in medicine, one of a class of substances believed to be present in the growing embryo, controlling its growth pattern.

It is thought that variations in the concentration of morphogens in different parts of the embryo cause them to grow at different rates.

morphogenesis development of forms and structures in an organism.

morphology in biology, the study of the physical structure and form of organisms, in particular their soft tissues.

Moseley, Henry Gwyn Jeffreys (1887–1915) English physicist. From 1913 to 1914 he established the series of atomic numbers (reflecting the charges of the nuclei of different elements) that led to the revision of Russian chemist Dmitri †Mendeleyev's †periodic table of the elements.

Moseley Articles http://dbhs.wvusd.k12.ca.us/Chem-History/Moseley-article.html

Transcript of Henry Moseley's article The High Frequency Spectra Of The Elements. In the article, Moseley lays the foundations for X-ray spectroscopy, and forms a relationship between atomic number and the frequency of the emitted spectra.

Moseley's law frequency of the characteristic †X-ray line spectra for elements is directly proportional to the square of the †atomic number, Z.

moss small nonflowering plant of the class Musci (10,000 species), forming with the liverworts and the hornworts the order Bryophyta. The stem of each plant bears †rhizoids that anchor it; there are no true roots. Leaves spirally arranged on its lower portion have sexual organs at their tips. Most mosses flourish best in damp conditions where other vegetation is thin. There are 1,000 British species of moss and more than 1,200 North American species.

The peat or bog moss *Sphagnum* was formerly used for surgical dressings. The smallest moss is the Cape pygmy moss *Ephemerum capensi*, only slightly larger than a pin head.

Mössbauer, Rudolf (Ludwig) (1929–) German physicist who was awarded the Nobel Prize for Physics in 1961 for his discovery in 1958 that under certain conditions an atomic nucleus can be stimulated to emit very sharply defined beams of gamma rays – a phenomenon that became known as the †Mössbauer effect.

Mössbauer effect recoil-free emission of gamma rays from atomic nuclei under certain conditions. The effect was discovered in 1958 by German physicist Rudolf Mössbauer, and used in 1960 to provide the first laboratory test of Einstein's general theory of relativity.

The absorption and subsequent re-emission of a gamma ray by an atomic nucleus usually causes it to recoil, so affecting the wavelength of the emitted ray. Mössbauer found that at low temperatures, crystals will absorb gamma rays of a specific wavelength and resonate so that the crystal as a whole recoils while individual nuclei do not. The wavelength of the re-emitted gamma rays is therefore virtually unaltered by recoil and may be measured to a high degree of accuracy. Changes in the wavelength may therefore be studied as evidence of the effect of, say, neighboring electrons or gravity. For example, the effect provided the first verification of the general theory of relativity by showing that gamma-ray wavelengths become longer in a gravitational field, as predicted by Einstein.

motherboard †printed circuit board that contains the main components of a microcomputer. The power, memory capacity, and capability of the microcomputer may be enhanced by adding expansion boards to the motherboard, now more commonly called a mainboard.

motility ability to move spontaneously. The term is often restricted to those cells that are capable of independent locomotion, such as spermatozoa. Many single-celled organisms are motile, for example, the amoeba. Research has shown that cells capable of movement, including vertebrate muscle cells, have certain biochemical features in common. Filaments of the proteins actin and myosin are associated with motility, as are the metabolic processes needed for breaking down the energy-rich compound †ATP (adenosine triphosphate).

motor effect tendency of a wire carrying an electric current in a magnetic field to move. The direction of the movement is given by the left-hand rule (see †Fleming's rules). This effect is used in the †electric motor. It also explains why streams of electrons produced, for instance, in a television tube can be directed by electromagnets.

motor nerve in anatomy, any nerve that transmits impulses from the central nervous system to muscles or organs. Motor nerves cause voluntary and involuntary muscle contractions, and stimulate glands to secrete hormones.

Mott, Nevill Francis (1905–1996) English physicist who researched the electronic properties of metals, semiconductors, and noncrystalline materials. He shared the Nobel Prize for Physics in 1977 for his contributions to understanding the electronic structure of magnetic and disordered systems. He was knighted in 1962.

Mottelson, Ben(jamin Roy) (1926–) US-born Danish physicist who with Aage †Bohr and James †Rainwater shared the Nobel Prize for Physics in 1975 for their discovery of the connection between collective and particle motion in atomic nuclei, and the development of the theory of nuclear structure.

molting periodic shedding of the hair or fur of mammals, feathers of birds, or skin of reptiles. In mammals and birds, molting is usually seasonal and is triggered by changes of day length.

The term is also often applied to the shedding of the †exoskeleton of arthropods, but this is more correctly called †ecdysis.

mountain natural upward projection of the Earth's surface, higher and steeper than a hill. Mountains are at least 330 m/1,000 ft above the surrounding topography. The existing rock below a high mountain may be subjected to high temperatures and pressures, causing metamorphism. Plutonic activity also can accompany mountain building.

mouth cavity forming the entrance to the digestive tract (gut or alimentary canal). In land vertebrates, air from the nostrils enters the mouth cavity to pass down the trachea. The mouth in mammals is enclosed by the jaws, cheeks, and palate. It contains teeth that may have a variety of functions depending on the way of life of the mammal. They may be used to hold things, to kill other organisms, or to cut food.

Teeth are usually also involved in †digestion in that they cut or crush large pieces of food into smaller pieces and mix them with saliva. Saliva is a digestive juice which contains the †enzyme amylase. This enzyme breaks down the large insoluble molecules of †starch into smaller soluble molecules of complex sugar. These sugar molecules are broken into the simpler sugar †glucose later in the alimentary canal.

Starch is rarely in the mouth long enough to be digested completely. However, in the small intestine the pancreas produces a digestive juice that also contains amylase, so it is here that starch digestion is completed.

moving-coil meter instrument used to detect and measure electrical current. A coil of wire pivoted between the

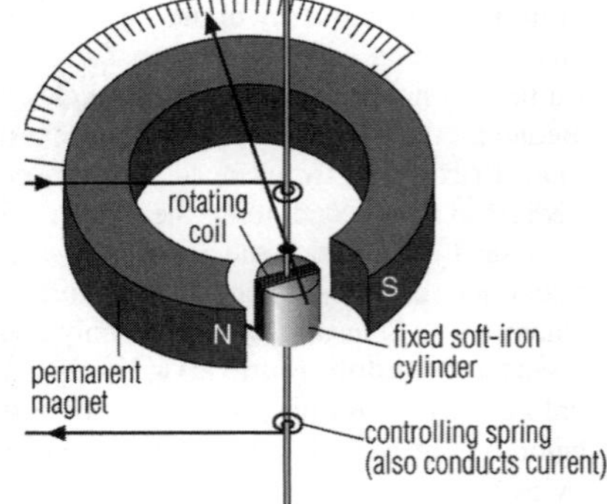

moving-coil meter A simple moving-coil meter. Direct electric current (DC) flowing through the wire coil combined with the presence of a magnetic field causes the coil to rotate; this in turn moves a pointer across a calibrated scale so that the degree of rotation can be related to the magnitude of the current. Helicon

poles of a permanent magnet is turned by the motor effect of an electric current (by which a force acts on a wire carrying a current in a magnetic field). The extent to which the coil turns can then be related to the magnitude of the current.

The sensitivity of the instrument depends directly upon the strength of the permanent magnet used, the number of turns making up the moving coil, and the coil's area. It depends inversely upon the strength of the controlling springs used to restrain the rotation of the coil. By the addition of a suitable large resistor, a moving-coil meter can be adapted to read potential difference in volts.

mp in chemistry, abbreviation for †melting point.

MRI abbreviation for †**magnetic resonance imaging**.

MS-DOS acronym for MicroSoft Disk Operating System, computer †operating system produced by Microsoft Corporation, widely used on †microcomputers with Intel x86 and Pentium family microprocessors. A version called PC-DOS is sold by IBM specifically for its personal computers. MS-DOS and PC-DOS are usually referred to as DOS. MS-DOS first appeared in 1981, and was similar to an earlier system from Digital Research called CP/M. Developments in Windows software have led Microsoft to cease work on any further versions of MS-DOS.

mucin viscid substance found in the secretions of the mucous membranes of the body and in intercellular fluids. One of its main functions is that of a lubricant. It is very important in the feeding mechanisms of many †invertebrates. Mucin is a mucopolysaccharide, which can be split up into a protein and a carbohydrate.

mucous membrane thin skin lining all animal body cavities and canals that come into contact with the air (for example, eyelids, breathing and digestive passages, genital tract). It contains goblet cells that secrete mucus, a moistening, lubricating, and protective fluid. In the air passages mucus captures dust and bacteria. In the gut it helps food slip along, and protects the epithelial cells from being damaged by digestive enzymes. Mucous membranes line the air passages from the †mouth to the †lungs and to the †gut. The layer of cells next to the space in the tubes is an †epithelium. In the air passages many of these cells have hair-like projections called cilia and the epithelium is then called a ciliated epithelium.

mucus lubricating and protective fluid, secreted by mucous membranes in many different parts of the body. In the gut, mucus smooths the passage of food and keeps potentially damaging digestive enzymes away from the gut lining. In the lungs, it traps airborne particles so that they can be expelled.

mud in earth science, mixture of water and particles of clay or silt that may be semi-liquid or soft and plastic. The fine-grained underwater sediment on a continental shelf is also known as mud.

mudflow downhill movement (mass movement) of muddy sediment containing a large proportion of water. Mudflows can be fast and destructive: in 1966 coal waste saturated with water engulfed a school in Aberfan, South Wales, killing 116 children. A †lahar is a form of mudflow associated with volcanic activity.

mudstone fine-grained sedimentary rock made up of clay- to silt-sized particles (up to 0.0625 mm/0.0025 in).

Müller, K Alexander (1927–) Swiss physicist who was awarded the Nobel Prize for Physics in 1987 for his work on high-temperature †superconductivity in ceramic materials. The discovery of these materials was a significant step towards the use of superconductors in computers, magnetic levitation trains, and the more efficient generation and distribution of electricity. He shared the Nobel Prize with Georg †Bednorz.

Müller, Paul Hermann (1899–1965) Swiss chemist who was awarded a Nobel Prize for Physiology or Medicine in 1948 for his discovery of the first synthetic contact insecticide, †DDT, in 1939.

Mulliken, Robert Sanderson (1896–1986) US chemist and physicist who was awarded the Nobel Prize for Chemistry in 1966 for his development of the molecular orbital theory of chemical bonds and structures.

Mullis, Kary Banks (1944–) US molecular biologist who was awarded the Nobel Prize for Chemistry in 1993 for the invention of the †polymerase chain reaction (PCR) technique, which allows specific regions of DNA to be copied many times from a tiny sample, thus amplifying it to a large enough quantity to be analyzed. This technique signaled an end to the laborious method of producing †DNA fragments in vivo. He shared the Nobel Prize with Michael Smith.

multiple any number that is the product of a given number. For example, the sequences 5, 10, 15, 20, 25... are multiples of 5; and 2, 4, 6, 8, 10... are multiples of 2. These sequences form †multiplication tables. Some multiples are common to several multiplication tables. In the sequence 10, 20, 30... all are multiples of 2, 5, and 10. The lowest of this sequence is 10 and this is, therefore, known as the †**lowest common multiple** (lcm) of 2, 5, and 10.

If *N* appears in the multiplication table of *n*, it is a multiple of *n*; it follows from this that *n* is a †factor of *N*. For example, 18 is a multiple of 3 since it appears in the multiplication table of 3, and 3 is a factor of 18 for the same reason.

multiple birth in humans, the production of more than two babies from one pregnancy. Multiple births can be caused by more than two eggs being produced and fertilized (often as the result of hormone therapy to assist pregnancy), or by a single fertilized egg dividing more than once before implantation. See also †twin.

multiple personality disorder **MPD**, psychiatric disorder wherein the patient exhibits two or more personalities. Personalities can be widely differing, with the same patient switching between the personalities of children and adults of different ages, sexes, and temperaments. Each personality is distinct, and often unaware of the others. MPD may occur as a result of child abuse, though not all psychiatrists accept that it is a genuine disorder.

multiple proportions, law of in chemistry, the principle that if two elements combine with each other to form more than one compound, then the ratio of the masses of one of them that combine with a particular mass of the other is a small whole number.

multiplexer in telecommunications, a device that allows a transmission medium to carry a number of separate signals at the same time – enabling, for example, several telephone conversations to be carried by one telephone line, and radio signals to be transmitted in stereo.

In **frequency-division multiplexing**, signals of different frequency, each carrying a different message, are transmitted. Electrical frequency filters separate the message at the receiving station. In **time-division multiplexing**, the messages are broken into sections and the sections of several messages interleaved during transmission. †Pulse-code modulation allows hundreds of messages to be sent simultaneously over a single link.

multiplication one of the four basic operations of arithmetic, usually written in the form $a \times b$ or ab, and involving repeated addition in the sense that a is added to itself b times. Multiplication obeys the associative law (for example, $a \times b \times c = c \times b \times a$), the commutative law (a special case of the associative law where there are only two numbers involved; for example, $ab = ba$), and the distributive law (multiplication over addition; for example, $m(a + b) = ma + mb$), and every number (except 0) has a multiplicative inverse. The number 1 is the identity for multiplication.

multitasking or **multiprogramming**, in computing, a system in which one processor appears to run several different programs (or different parts of the same program) at the same time. All the programs are held in memory together and each is allowed to run for a certain period.

For example, one program may run while other programs are waiting for a peripheral device to work or for input from an operator. The ability to multitask depends on the †operating system rather than the type of computer. Unix is one of the commonest.

muon †elementary particle similar to the electron except for its mass which is 207 times greater than that of the electron. It has a half-life of 2 millionths of a second, decaying into electrons and †neutrinos. The muon was originally thought to be a †meson but is now classified as a †lepton. See also †tau.

mural arc astronomical apparatus, used from the 16th to the 19th century, comprising a carefully oriented wall on which a calibrated device was fixed, by which the altitudes of celestial objects could be measured.

Murchison, Roderick Impey (1792–1871) Scottish geologist responsible for naming the †Silurian period, the subject of his book *The Silurian System* (1839). Expeditions to Russia 1840–45 led him to define another worldwide system, the Permian, named after the strata of the Perm region. He was knighted in 1846, and became a baronet in 1866.

Musca small constellation of the southern hemisphere, represented as a fly.

muscle contractile animal tissue that produces locomotion and power and maintains the movement of body substances. Muscle contains very specialized animal cells – long cells – that can contract to between one-half and one-third of their relaxed length.

Muscle tissue is sometimes found in large amounts, forming muscles, that are organs. Muscle tissue enables movement. It may move the whole body, or part of it, or some material along a tube within it. Muscles can only do this work by contracting. This explains why muscles are usually found in pairs (antagonistic pairs) where the work done in the contraction of one causes the stretching of the other.

Muscle is categorized into three main groups: striped (or striated) muscles are activated by †motor nerves under voluntary control – their ends are usually attached via tendons to bones; involuntary or smooth muscles are controlled by motor nerves of the †autonomic nervous system, and are located in the gut, blood vessels, iris, and various ducts; cardiac muscle occurs only in the †heart, and is also controlled by the autonomic nervous system.

List of Muscles http://www.meddean.luc.edu/lumen/MedEd/GrossAnatomy/dissector/muscles/muscles.html

Resource detailing all the muscles in the body. The page is divided into two frames: the left-hand frame lists all of the muscles alphabetically, and clicking on a specific muscle brings up information and a diagram in the right-hand frame.

Muscular System http://www.innerbody.com/image/musfov.html

Based around annotated diagrams, linked to clear descriptions, this site has lots of useful facts about muscles. Click on a marker to bring up a new window that explains all about that type of muscle.

muscovite white †mica, $KAl_2Si_3AlO_{10}(OH,F)_2$, a common silicate mineral. It is colorless to silvery white with shiny surfaces, and like all micas it splits into thin flakes along its one perfect cleavage. Muscovite is a metamorphic mineral occurring mainly in schists; it is also found in some granites, and appears as shiny flakes on bedding planes of some sandstones.

mutagen any substance that increases the rate of gene †mutation. A mutagen may also act as a †carcinogen.

mutant gene, organism, or population that has undergone a change in character because of †mutation.

mutation in biology, a change in the †genes produced by a change in the †DNA that makes up the hereditary material of all living organisms. It can be a change in a single gene or a change that affects sections of †chromosomes. In the process of DNA replication, which takes place before any cell divides, the two halves of DNA separate and new halves are made. Because of specific base pairing, the inherited information is copied exactly. Despite this, rarely, a mistake occurs and the sequence of bases is altered. This changes the sequence of amino acids in a protein. This is mutation, the raw material of evolution. The consequences of mutation are varied. Due to the redundancy built into genetic code many mutations have no effect upon DNA functions. Genes describe how to make †proteins. As a result of mutation a protein may not be produced, may be produced but act abnormally, or remain fully functional. Only a few mutations improve the organism's performance and are therefore favored by †natural selection. Mutation rates are increased by certain chemicals and by ionizing radiation.

Common mutations include the omission or insertion of a base (one of the chemical subunits of DNA); these are known as **point mutations**. Larger-scale mutations include removal of a whole segment of DNA or its inversion within the DNA strand. Not all mutations affect the organism, because there is a certain amount of redundancy in the genetic information. If a mutation is "translated" from DNA into the protein that makes up the organism's structure, it may be in a non-functional part of the protein and thus have no detectable effect. This is known as a **neutral mutation**, and is of importance in †molecular clock studies because such mutations tend to accumulate gradually as time passes. Some mutations do affect genes that control protein production or functional parts of protein, and most of these are lethal to the organism.

mutual induction in physics, the production of an electromotive force (emf) or voltage in an electric circuit caused by a changing †magnetic flux in a neighboring circuit. The two circuits are often coils of wire, as in a †transformer, and the size of the induced emf depends on the numbers of turns of wire in each of the coils.

mutualism association between two organisms of different species whereby both profit from the relationship; see †symbiosis.

mutually exclusive in probability theory, describing two †events that cannot happen together. For example, it is not possible to throw a 6 and a †prime number on the same die at the same time. The †probability that both events will happen is zero. The probability that either one or the other of two mutually exclusive events will happen is the sum of their separate probabilities. The addition rule is

$P(A \text{ or } B) = P(A) + P(B)$

So when a die is thrown, the probability of throwing a 6 or a 3 is: $P(6 \text{ or } 3) = P(6) + P(3) = {}^1/_6 + {}^1/_6 = {}^2/_6$ or ${}^1/_3$.

If a person buys 5 tickets out of 100 in a lottery with only one winning ticket, the chance of winning the prize is 0.01 + 0.01 + 0.01 + 0.01 + 0.01 = 0.05.

mycelium interwoven mass of threadlike filaments or †hyphae, forming the main body of most fungi. The reproductive structures, or "fruiting bodies," grow from the mycelium.

mycorrhiza mutually beneficial (mutualistic) association occurring between plant roots and a soil fungus. Mycorrhizal roots take up nutrients more efficiently than non-mycorrhizal roots, and the fungus benefits by obtaining carbohydrates from the plant or tree.

An **ectotrophic mycorrhiza** occurs on many tree species, which usually grow much better, most noticeably in the seeding stage, as a result. Typically the roots become repeatedly branched and coral-like, penetrated by hyphae of a surrounding fungal †mycelium. In an **endotrophic mycorrhiza**, the growth of the fungus is mainly inside the root, as in orchids. Such plants do not usually grow properly, and may not even germinate, unless the appropriate fungus is present.

Research by UK ecologists in 1996 showed that mycorrhizal fungi provides protection against †nematode worms that feed on plant roots, as well as pathogenic fungi. Further research in 1998 showed that diversity of mycorrhizal fungi in the soil contributes greatly to plant biodiversity.

myelin fatty substance that forms a sheath around the nerve fibers of vertebrates.

myelin sheath insulating layer that surrounds nerve cells in vertebrate animals. It serves to speed up the passage of nerve impulses.

Myelin is made up of fats and proteins and is formed from up to a hundred layers, laid down by special cells, the **Schwann cells**.

mylonite metamorphic rock composed mainly of feldspar, quartz, and micas and comprising fine-grained highly-deformed layers alternating with layers of less-deformed relict grains often ovoidal in shape. The varied degree of deformation among bands produces a typically streaky appearance. They are found in areas in which rock layers have been deformed by sliding past one another parallel to their boundaries (shear strain).

Once thought to be solely the result of brittle, grinding deformation (cataclasis), it is now recognized that microscopic bending followed by recrystallization has allowed portions of these rocks to deform ductily (see †ductile material). Their unique texture can be used to reconstruct directions of movement that occurred during their formation.

myofibril one of many minute fibrils that together make up a fiber of smooth or striped muscle, running along the length of the muscle.

myoglobin globular protein, closely related to †hemoglobin and located in vertebrate muscle. Oxygen binds to myoglobin and is released only when the hemoglobin can no longer supply adequate oxygen to muscle cells.

myopia or **short-sightedness**, defect of the eye in which a person can see clearly only those objects that are close up. It is caused either by the eyeball being too long or by the cornea and lens system of the eye being too powerful, both of which cause the images of distant objects to be formed in front of the retina instead of on it. Nearby objects are sharply perceived. Myopia can be corrected by suitable glasses or contact lenses.

myosin protein made up of a chain of polypeptides that forms filaments in smooth (or striped) muscle fibrils. During muscle contraction it combines with †actin (another muscle protein, contained in thinner filaments) to form actomyosin; the actin filaments are pulled into the myosin filaments, which shortens the †myofibrils.

myrmecophyte plant that lives in association with a colony of ants and possesses specialized organs in which the ants live. For example, *Myrmecodia*, an epiphytic plant from Malaysia, develops root tubers containing a network of cavities inhabited by ants.

Several species of *Acacia* from tropical America have specialized hollow thorns for the same purpose. This is probably a mutualistic (mutually beneficial) relationship, with the ants helping to protect the plant from other insect pests and in return receiving shelter.

N

n in mathematics, variable used to denote an integer number.

n! factorial *n*; that is, 1 x 2 x 3 x ... *n*.

N abbreviation for **north**; †newton, a unit of force; the chemical symbol for **nitrogen**.

nadir point on the celestial sphere vertically below the observer and hence diametrically opposite the **zenith**.

nail in biology, a hard, flat, flexible outgrowth of the digits of primates (humans, monkeys, and apes). Nails are composed of †keratin.

NAND gate type of †logic gate.

nano- prefix used in †SI units of measurement, equivalent to a one-billionth part (10^{-9}). For example, a nanosecond is one-billionth of a second.

nanotechnology experimental technology using individual atoms or molecules as the components of minute machines, measured by the nanometer, or millionth of a millimeter. Nanotechnology research in the 1990s focused on testing molecular structures and refining ways to manipulate atoms using a scanning tunneling microscope. The ultimate aim is to create very small computers and molecular machines which could perform vital engineering or medical tasks.

The †scanning electron microscope can be used to see and position single atoms and molecules, and to drill holes a nanometer (billionth of a meter) across in a variety of materials. The instrument can be used for ultrafine etching; the entire 28 volumes of the *Encyclopedia Britannica* could be engraved on the head of a pin. In the USA a complete electric motor has been built, which is less than 0.1 mm/0.004 in across with a top speed of 600,000 rpm. It is etched out of silicon, using the ordinary methods of chip manufacturers.

naphtha mixtures of hydrocarbons obtained by destructive distillation of petroleum, coal tar, and shale oil. It is a raw material for the petrochemical and plastics industries. The term was originally applied to naturally occurring liquid hydrocarbons.

naphthalene $C_{10}H_8$, solid, white, shiny, aromatic hydrocarbon obtained from coal tar. The smell of moth balls is due to their naphthalene content. It is used in making indigo and certain azo dyes, as a mild disinfectant, and as a pesticide.

Napier, John (1550–1617) 8th Laird of Merchiston, Scottish mathematician who invented †logarithms in 1614 and "Napier's bones," an early mechanical calculating device for multiplication and division.

Napier, John

http://www-groups.dcs.st-andrews.ac.uk/history/Mathematicians/Napier.html

Devoted to the life and contributions of John Napier. In addition to biographical information, you will find a list of references about Napier and links to other essays in the archive that reference him. The text of this essay includes hypertext links to the essays of those mathematicians and thinkers who influenced Napier. There is also an image of him, which can be enlarged, and a map of his birthplace.

Napier's bones primitive mechanical calculation device created by Scottish mathematician John Napier and consisting of a set of little bone rods with which to multiply or divide.

narcissism in psychology, an exaggeration of normal self-respect and self-involvement which may amount to mental disorder when it precludes relationships with other people.

narcotic pain-relieving and sleep-inducing drug. The term is usually applied to heroin, morphine, and other opium derivatives, but may also be used for other drugs which depress brain activity, including anesthetic agents and hypnotics.

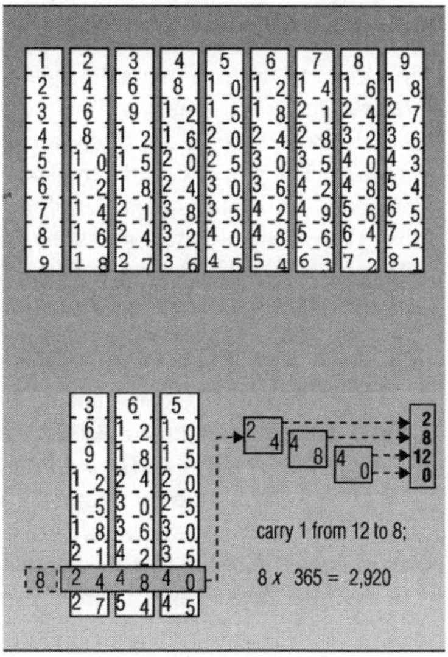

nastic movement plant movement that is caused by an external stimulus, such as light or temperature, but is directionally independent of its source, unlike †tropisms. Nastic movements occur as a result of changes in water pressure within specialized cells or differing rates of growth in parts of the plant.

Examples include the opening and closing of crocus flowers following an increase or decrease in tem-

carry 1 from 12 to 8;

8 x 365 = 2,920

Napier In 1617 Scottish mathematician John Napier published his description of what was arguably the first mechanical calculator – a set of numbered rods, usually made of bone or ivory and therefore known as Napier's bones. Using them, multiplication became merely a process of reading off the appropriate figures and making simple additions. Helicon

perature (**thermonasty**), and the opening and closing of evening primrose *Oenothera* flowers on exposure to dark and light (**photonasty**).

The leaf movements of the Venus flytrap *Dionaea muscipula* following a tactile stimulus, and the rapid collapse of the leaflets of the sensitive plant *Mimosa pudica* are examples of **haptonasty**. Sleep movements, where the leaves or flowers of some plants adopt a different position at night, are described as **nyctinasty**. Other movement types include **hydronasty**, in response to a change in the atmospheric humidity, and **chemonasty**, in response to a chemical stimulus.

Nathans, Daniel (1928–1999) US microbiologist who was awarded a Nobel Prize for Physiology or Medicine in 1978 with his colleague Hamilton †Smith for their discovery of restriction enzymes (enzymes that can cleave genes into fragments) and their application to molecular genetics.

native metal or **free metal**, any of the metallic elements that occur in nature in the chemically uncombined or elemental form (in addition to any combined form). They include bismuth, cobalt, copper, gold, iridium, iron, lead, mercury, nickel, osmium, palladium, platinum, ruthenium, rhodium, tin, and silver. Some are commonly found in the free state, such as gold; others occur almost exclusively in the combined state, but under unusual conditions do occur as native metals, such as mercury. Examples of native non-metals are carbon and sulfur.

Natta, Giulio (1903–1979) Italian chemist who shared the Nobel Prize for Chemistry in 1963 with German chemist Karl †Ziegler for work on the chemistry and technology of high polymers. Natta's early work on heterogeneous catalysts formed the basis for many important industrial syntheses.

natural frequency frequency at which a mechanical system will vibrate freely. A pendulum, for example, always oscillates at the same frequency when set in motion. More complicated systems, such as bridges, also vibrate with a fixed natural frequency. If a varying force with a frequency equal to the natural frequency is applied to such an object the vibrations can become violent, a phenomenon known as †resonance.

natural gas mixture of flammable gases found in the Earth's crust (often in association with petroleum). It is one of the world's three main fossil fuels (with coal and oil).

Natural gas, a †non-renewable resource, is formed from the remains of dead plants and animals. As these plants and animals died they were buried with mud near the sea floor. Over millions of years, heat from the Earth's interior and pressure from overlying rocks slowly changed the dead remains into hydrocarbons (substances containing hydrogen and carbon). The hydrocarbons, being light molecules, moved upwards and became trapped beneath impermeable rocks.

Natural gas is a mixture of †hydrocarbons, chiefly methane (80%), with ethane, butane, and propane. Natural gas reservoirs are often found beneath the seabed and drilling technology is used to locate these supplies. The gas is usually transported from its source by pipeline, although it may be liquefied for transport and storage and is, therefore, often used in remote areas where other fuels are scarce and expensive. Prior to transportation, butane and propane are removed and liquefied to form "bottled gas."

Methane combusts in air to form products with the release of heat energy:

$CH_4 + 2O_2 \rightarrow CO_2 + 2H_2O$ + heat energy

Another source of gas is from the cracking of crude oil into simpler molecules. Methane, propane, and butane are all gases that can be produced from the cracking process. The gases are liquefied and stored under pressure. Propane and butane also undergo combustion when ignited in air. Large amounts of heat energy are released in the reaction. The reaction for the combustion of propane is:

$C_3H_8 + 5O_2 \rightarrow 3CO_2 + 4H_2O$ + heat energy

natural hazard naturally occurring phenomenon capable of causing destruction, injury, disease, or death. Examples include earthquakes, floods, hurricanes, or famine. Natural hazards occur globally and can play an important role in shaping the landscape. The events only become hazards where people are affected. Because of this, natural hazards are usually measured in terms of the damage they cause to persons or property. Human activities can trigger natural hazards, for example skiers crossing the top of a snowpack may cause an avalanche.

natural logarithm in mathematics, the †exponent of a number expressed to base e, where e represents the †irrational number 2.71828... . Natural †logarithms are also called Napierian logarithms, after their inventor, the Scottish mathematician John Napier.

natural number one of the set of numbers used for counting. Natural numbers comprise all the †positive integers, excluding zero.

natural radioactivity radioactivity generated by those radioactive elements that exist in the Earth's crust. All the elements from polonium (atomic number 84) to uranium (atomic number 92) are radioactive. †Radioisotopes of some lighter elements are also found in nature (for example potassium-40). See

†background radiation.

natural selection process by which gene frequencies in a population change through certain individuals producing more descendants than others because they are better able to survive and reproduce in their environment. The accumulated effect of natural selection is to produce †adaptations such as the insulating coat of a polar bear or the spadelike forelimbs of a mole. The process is slow, relying firstly on random variation in the genes of an organism being produced by †mutation and secondly on the genetic †recombination of sexual reproduction. It was recognized by Charles †Darwin and English naturalist Alfred Russel Wallace as the main process driving †evolution.

Natural selection results in evolution in a way that was described by Charles Darwin: individual organisms within a particular species may show a wide range of †variation because of differences in their †genes; predation, disease, and †competition cause individuals to die; individuals with characteristics most suited to the †environment are more likely to survive and breed successfully; and the genes which have enabled these individuals to survive are then passed on to the next generation, and if the environment is changing, the result is that some genes are more abundant in the next generation and the organism has evolved.

nature the living world, including plants, animals, fungi, and all micro-organisms, and naturally formed features of the landscape, such as mountains and rivers.

Nature http://www.nature.com/nature/

Nature's site features recent and archive news articles from this well-established science magazine. Although there is a strong encouragement to subscribe to the magazine, there are several full articles freely available online, as well as a full sample issue of the magazine, and regularly updated "Breaking news" and "Nature science update" sections.

Worldwide Fund for Nature http://www.panda.org/

Online base of the international wildlife charity. As well as giving information about current campaigns and encouraging you to join, there is a wealth of well-presented information here about the natural world, including quizzes, interactive maps, and a regularly updated news section.

nature–nurture controversy or **environment–heredity controversy**, long-standing dispute among philosophers and psychologists over the relative importance of environment, that is, upbringing, experience, and learning ("nurture"), and heredity, that is, genetic inheritance ("nature"), in determining the make-up of an organism, as related to human personality and intelligence.

One area of contention is the reason for differences between individuals; for example, in performing intelligence tests. The environmentalist position assumes that individuals do not differ significantly in their inherited mental abilities and that subsequent differences are due to learning, or to differences in early experiences. Opponents insist that certain differences in the capacities of individuals (and hence their behavior) can be attributed to inherited differences in their genetic make-up.

nautical mile unit of distance used in navigation, an internationally agreed standard (since 1959) equaling the average length of one minute of arc on a great circle of the Earth, or 1,852 m/6,076 ft. The term formerly applied to various units of distance used in navigation.

navel or **umbilicus**, small indentation in the center of the abdomen of mammals, marking the site of attachment of the †umbilical cord, which connects the fetus to the †placenta.

navigation, biological ability of animals or insects to navigate. Although many animals navigate by following established routes or known landmarks, many animals can navigate without such aids; for example, birds can fly several thousand miles back to their nest site, over unknown terrain.

Such feats may be based on compass information derived from the position of the Sun, Moon, or stars, or on the characteristic patterns of Earth's magnetic field.

Biological navigation refers to the ability to navigate both in long-distance †migrations and over shorter distances when foraging (for example, the honey bee finding its way from the hive to a nectar site and back). Where reliant on known landmarks, birds may home on features that can be seen from very great distances (such as the cloud caps that often form above isolated mid-ocean islands). Even smells can act as landmarks. Aquatic species like salmon are believed to learn the characteristic taste of the river where they hatch and return to it, often many years later. Brain cells in some birds have been found to contain †magnetite and may therefore be sensitive to the Earth's magnetic field.

***n*-dimensional** in mathematics, having an unstated but finite number of dimensions. A typical example is the set of all runs of n-numbers (that is, $x_1, ..., x_n$), which is the basis of a coordinate geometry generalizing three-dimensional coordinates (x_1, x_2, x_3).

near point the closest position to the eye to which an object may be brought and still be seen clearly. For a normal human eye the near point is about 25 cm; however, it gradually moves further away with age, particularly after the age of 40.

nebula cloud of gas and dust in space. Nebulae are the birthplaces of stars, but some nebulae are produced by gas thrown off from dying stars (see †planetary nebula; †supernova). Nebulae are classified depending on whether they emit, reflect, or absorb light.

An **emission nebula**, such as the †Orion nebula, glows brightly because its gas is energized by stars that have formed within it. In a **reflection nebula**, starlight reflects off grains of dust in the nebula, such as surround the stars of the †Pleiades cluster. A **dark nebula** is a dense cloud, composed of molecular hydrogen, which partially or completely absorbs light behind it. Examples include the Coalsack nebula in †Crux and the Horsehead nebula in Orion.

Web Nebulae http://www.rog.nmm.ac.uk/astroweb/twn/intro.html

Images of nebulae, plus a short account of the different types of nebulae and a glossary of related terms.

nebular hypothesis hypothesis that the Solar System evolved from a †nebula. 18th-century Swedish mystic and scientist Emanuel Swedenborg and 18th-century German philosopher Immanuel †Kant both put forward explanations of this kind, but 18th–19th-century French astronomer Pierre †Laplace was the first to develop a nebular hypothesis on strictly scientific lines.

Laplace's hypothesis was that rotating nebulae were first formed by the condensation of gaseous matter, and that the Sun was originally a nebula of this type. As the nebula continued to shrink, it rotated faster and faster, until fragments broke away from the main body of the Sun. These fragments condensed under their own gravitational attraction and formed the planets. Laplace's hypothesis is untenable, as it fails to explain the fact that nearly all the angular momentum of the Solar System resides in the planets.

The tidal hypothesis of English mathematician and astrophysicist James †Jeans, that the planets were formed from material torn out of the Sun by the close approach of a passing star, is also untenable, as the ejected material would disperse into space. Modern theories offer no unique solution, but agree that the Sun and planets condensed simultaneously from a rotating cloud of gas and dust.

nebula variable star or **T Tauri variable**, variable star of †spectral classification F, G, or K (a giant above the †main sequence on the †Hertzsprung-Russell diagram) that loses an appreciable proportion of its mass in its (irregular) more luminous periods, and is thus surrounded by volumes of gas and dust.

necessary condition in mathematics, statement A is a necessary condition for a statement B if A is true whenever B is true. It can also be expressed as "B implies A."

neck structure between the head and the trunk in animals. In the back of the neck are the upper seven vertebrae of the spinal column, and there are many powerful muscles that support and move the head. In front, the neck region contains the †pharynx and †trachea, and behind these the esophagus. The large arteries (carotid, temporal, maxillary) and veins (jugular) that supply the brain and head are also located in the neck. The †larynx (voice box) occupies a position where the trachea connects with the pharynx, and one of its cartilages produces the projection known as Adam's apple. The †thyroid gland lies just below the larynx and in front of the upper part of the trachea.

nectar sugary liquid secreted by some plants from a nectary, a specialized gland usually situated near the base of the flower. Nectar often accumulates in special pouches or spurs, not always in the same location as the nectary. Nectar attracts insects, birds, bats, and other animals to the flower for †pollination and is the raw material used by bees in the production of honey.

Néel, Louis Eugène Félix (1904–2000) French physicist who was awarded the Nobel Prize for Physics in 1970 for his work on the magnetic properties of solids and his prediction of the existence of antiferromagnetism, a form of †magnetism.

Néel temperature the temperature at which the †susceptibility of an antiferromagnetic material has a maximum value; see also †magnetism.

negative number number less than †zero. On a †number line, any number to the left of zero is negative. Negative numbers are always written with a minus sign in front, for example, –6 is negative, 6 is positive. Scales often display negative numbers. On a temperature scale, 5°C below freezing is –5°C.

Neher, Erwin (1944–) German cell physiologist who was awarded a Nobel Prize for Physiology or Medicine in 1991 with Bert †Sakmann for their discovery of how gatelike structures (ion channels) regulate the flow of ions into and out of cells, and in particular for their studies on beta-endorphin (a messenger hormone secreted by the pituitary gland).

neighborhood in topology, subsets of points in a topological space defining a "locality" round a specific point, which includes an open set to which the specific point belongs. Axioms describing neighborhoods were first formulated by German mathematician Felix Hausdorff.

nematode any of a group of unsegmented †worms that are pointed at both ends, with a tough, smooth outer skin. They include many free-living species found in soil and water, including the sea, but a large number are parasites, such as the roundworms and pinworms that live in humans, or the eelworms that attack plant roots. They differ from flatworms in that they have two openings to the gut (a mouth and an anus). The group includes *Caenorhabditis elegans* which is a model genetic organism and the first multicellular animal to have its complete genome sequenced. (Phylum Nematoda.)

Most nematode species are found in deep-sea sediment. Around 13,000 species are known, but a 1995 study by the Natural History Museum, London, based on the analysis of sediment from 17 seabed sites worldwide, estimated that nematodes may make up as much as 75% of all species, with there being an estimated 100 million species. Some are anhydrobiotic, which means they can survive becoming dehydrated, entering a state of suspended animation until they are rehydrated.

nemesis theory theory of animal extinction, suggesting that a companion star to the Sun caused the extinction of the dinosaurs and other groups of animals.

The theory holds that the movement of this as yet undiscovered star disrupts the †Oort cloud of comets every 26 million years, resulting in the Earth suffering an increased bombardment from comets at these times. The theory was proposed in 1984 to explain the newly discovered layer of iridium – an element found in comets and meteorites – in rocks dating from the end of dinosaur times. However, many paleontologists deny any evidence for a 26-million-year cycle of extinctions.

neo-Darwinism modern theory of †evolution, built up since the 1930s by integrating the 19th-century English scientist Charles †Darwin's theory of evolution through natural selection with the theory of genetic inheritance founded on the work of the Austrian biologist Gregor †Mendel.

Neo-Darwinism asserts that evolution takes place because the environment is slowly changing, exerting a selection pressure on the individuals within a population. Those with characteristics that happen to adapt to the new environment are more likely to survive and have offspring and hence pass on these favorable characteristics. Over time the genetic make-up of the population changes and ultimately a new species is formed.

neodymium chemical symbol Nd, yellowish metallic element of the †lanthanide series, atomic number 60, relative atomic mass 144.24. Its rose-colored salts are used in coloring glass, and neodymium is used in lasers.

It was named in 1885 by Austrian chemist Carl von Welsbach, who fractionated it away from didymium (originally thought to be an element but actually a mixture of rare-earth metals consisting largely of neodymium, praseodymium, and cerium).

neon chemical symbol Ne, (Greek *neos* "new") colorless, odorless, non-metallic, gaseous element, atomic number 10, relative atomic mass 20.183. It is grouped with the †noble gases (rare gases) in Group 0 of the †periodic table. Neon is nonreactive, and forms no compounds. It occurs in small quantities in the Earth's atmosphere.

Tubes containing neon are used in electric advertising signs, giving off a fiery red glow; it is also used in lasers. Neon was discovered by Scottish chemist William Ramsay and English chemist Morris Travers.

neoprene synthetic rubber, developed in the USA in 1931 from the polymerization of chloroprene monomers, $CH_2 = CCl.CH = CH_2$. The elastic properties of the polymer are enhanced by †vulcanization. It is much more resistant to heat, light, oxidation, and petroleum than is ordinary rubber. It is used for making car tires.

neoteny in biology, the retention of some juvenile characteristics in an animal that seems otherwise mature. An example is provided by the axolotl, a salamander that can reproduce sexually although still in its larval form.

neper unit used in telecommunications to express a ratio of powers and currents. It gives the attenuation of amplitudes as the natural logarithm of the ratio.

nephron microscopic unit in vertebrate kidneys that forms **urine**. A human kidney is composed of over a million nephrons. Each nephron consists of a knot of blood capillaries called a glomerulus, contained in the †Bowman's capsule, and a long narrow tubule enmeshed with yet more capillaries. Waste materials and water pass from the bloodstream into the tubule, and essential minerals and some water are reabsorbed from the tubule back into the bloodstream. The remaining filtrate (urine) is passed out from the body.

Neptune eighth planet in average distance from the Sun. It is a giant gas (hydrogen, helium, methane) planet, with a mass 17.2 times that of Earth. It has the fastest winds in the Solar System.

mean distance from the Sun: 4.4 billion km/2.794 billion mi
equatorial diameter: 48,600 km/30,200 mi
rotation period: 16 hours 7 minutes
year: 164.8 Earth years

atmosphere: methane in its atmosphere absorbs red light and gives the planet a blue coloring. Consists primarily of hydrogen (85%) with helium (13%) and methane (1–2%)
surface: hydrogen, helium, and methane. Its interior is believed to have a central rocky core covered by a layer of ice
satellites: of Neptune's eight moons, two (†Triton and †Nereid) are visible from Earth. Six more were discovered by the Voyager 2 probe in 1989, of which Proteus (diameter 415 km/260 mi) is larger than Nereid (300 km/200 mi)
rings: there are four faint rings: Galle, Le Verrier, Arago, and Adams (in order from Neptune). Galle is the widest at 1,700 km/1,060 mi. Le Verrier and Arago are divided by a wide diffuse particle band called the plateau.

Neptune was located in 1846 by German astronomers Johan †Galle and Heinrich d'Arrest after calculations by English astronomer John Couch Adams and French mathematician Urbain Le Verrier had predicted its existence from disturbances in the movement of Uranus. Voyager 2, which passed Neptune in August 1989, revealed various cloud features, notably an Earth-sized oval storm cloud, the Great Dark Spot, similar to the Great Red Spot on Jupiter, but images taken by the Hubble Space Telescope in 1994 show that the Great Dark Spot has disappeared. A smaller dark spot, DS2, has also gone.

Neptune http://www.solarviews.com/solar/eng/neptune.htm
Detailed description of the planet Neptune. The site includes a chronology of the exploration of the planet, along with statistics and information on its rings, moons, and satellites, supported by a good selection of images.

neptunium chemical symbol Np, silvery, radioactive metallic element of the †actinide series, atomic number 93, relative atomic mass 237.048. It occurs in nature in minute amounts in †pitchblende and other uranium ores, where it is produced from the decay of neutron-bombarded uranium in these ores. The longest-lived isotope, Np-237, has a half-life of 2.2 million years. The element can be produced by bombardment of U-238 with neutrons and is chemically highly reactive.

It was first synthesized in 1940 by US physicists Edwin †McMillan and Philip Abelson (1913–), who named it after the planet Neptune (since it comes after uranium as the planet Neptune comes after Uranus).

Neptunium was the first †transuranic element to be synthesized.

Nereid or **Neptune II**, outermost moon of Neptune, discovered in 1949 by Dutch-born US astronomer Gerard †Kuiper. It is 340 km/210 mi in diameter and orbits Neptune in a highly elliptical orbit ranging from 1.4 million km/0.87 million mi to 9.6 million km/6 million mi from Neptune. It has a period (time taken to circle the planet) of 360 days.

Nernst, (Walther) Hermann (1864–1941) German physical chemist who was awarded the Nobel Prize for Chemistry in 1920 for work on heat changes in chemical reactions. He proposed in 1906 the principle known as the **Nernst heat theorem** or the third law of thermodynamics: chemical changes at the temperature of †absolute zero involve no change of †entropy (disorder).

Nernst effect in physics, phenomenon in which an electric potential develops across a conductor located at right angles to a magnetic field and with heat flowing through it. All three factors – potential, magnetic field, and heat flow – are at right angles to each other.

It is named after German physical chemist Hermann †Nernst.

nerve bundle of nerve cells enclosed in a sheath of connective tissue and transmitting nerve impulses to and from the brain and spinal cord. A single nerve may contain both †motor and sensory nerve cells, but they function independently.

nerve cell or **neuron**, elongated cell that transmits information rapidly between different parts of the body, the basic functional unit of the †nervous system. Each nerve cell has a cell body, containing the nucleus, from which trail processes called dendrites, responsible for receiving incoming signals. The unit of information is the nerve impulse, a traveling wave of chemical and electrical changes involving the membrane of the nerve cell. The cell's longest pro-

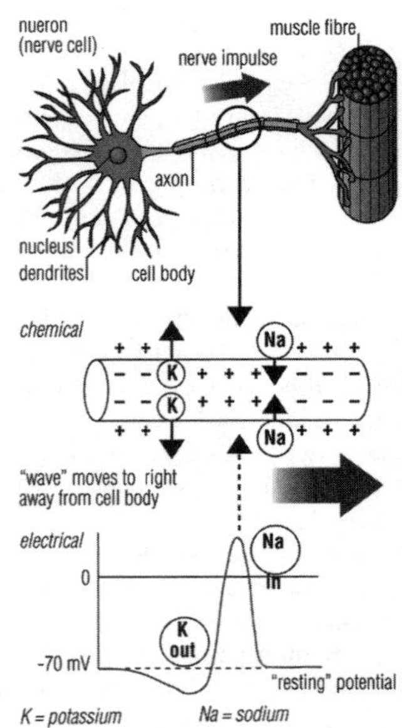

nerve cell The anatomy and action of a nerve cell. The nerve cell or neuron consists of a cell body with the nucleus and projections called dendrites which pick up messages. An extension of the cell, the axon, connects one cell to the dendrites of the next. When a nerve cell is stimulated, waves of sodium (Na^+) and potassium (K^+) ions carry an electrical impulse down the axon. Helicon

cess, the †axon, carries impulses away from the cell body. The †brain contains many nerve cells.

A †reflex arc is a simple example of how nerve cells help to control and coordinate processes in the body. It involves the use of three types of nerve cell. These are the sensory nerve cell, the intermediate (relay) nerve cell and the motor nerve cell. Where each nerve cell connects with the next there is a tiny gap called a †synapse. Impulses can cross this gap.

The impulse involves the passage of sodium and potassium ions across the nerve-cell membrane. Sequential changes in the permeability of the membrane to positive sodium (Na^+) ions and potassium (K^+) ions produce electrical signals called action potentials. Impulses are received by the cell body and passed, as a pulse of electric charge, along the axon. The axon terminates at the synapse, a specialized area closely linked to the next cell (which may be another nerve cell or a specialized effector cell such as a muscle). On reaching the synapse, the impulse releases a chemical †neurotransmitter, which diffuses across to the neighboring cell and there stimulates another impulse or the action of the effector cell.

Nerve impulses travel quickly – in humans, they may reach speeds of 160 m/525 ft per second, although many are much slower.

nervous system system of interconnected †nerve cells of most invertebrates and all vertebrates. It is composed of the †central and †autonomic nervous systems. It may be as simple as the nerve net of coelenterates (for example, jellyfishes) or as complex as the mammalian nervous system, with a central nervous system comprising †brain and †spinal cord and a peripheral nervous system connecting up with sensory organs, muscles, and glands.

In a nervous system, specialized cells called nerve cells or neurones carry messages as nerve impulses quickly from one part of the body to another. These impulses may be carrying information about the outside world (stimuli) which allows the body to respond quickly to them. However, much of the information being carried around is to do with organizing processes inside the body (nervous coordination).

In mammals, some examples of processes regulated by the nervous system are changes in †heart rate, changes in ventilation rate (†breathing rate), the movement of food along the †alimentary canal, and changes in the size of the iris which alter the amount of light entering the †eye.

The nervous system includes the brain, the area in which collected information is used to make decisions and from where responses are initiated. Information is collected from all over the body, including from specialized sense organs such as the eye. The rest of the nervous system includes the spinal cord and nerve cells carrying information to and from the brain. For example, responses to stimuli involve nerve impulses which are carried along nerve cells from the brain, through part of the spinal cord, and then to †muscles or other cells of the body.

Nervous System http://www.innerbody.com/image/nervov.html

Interactive diagram of the nervous system with pop-ups describing the role of the various organs and nerves within the human body.

ness (Old English næse "nose") geographical term signifying promontory. Names with this suffix are common in the Orkney and Shetland Islands, on the coast of Caithness, Scotland, and along the east coast of England as far as Dungeness in Kent. It has been suggested that its use may indicate Scandinavian colonization, *naes* being its Nordic counterpart.

nest place chosen or constructed by a bird or other animal for incubation of eggs, hibernation, and shelter. Nests vary enormously, from saucerlike hollows in the ground, such as the scrapes of hares, to large and elaborate structures, such as the 4-m/13-ft diameter mounds of the megapode birds.

birds' nests Birds that nest in slight depressions in the ground, such as terns and plovers, lay eggs that so closely resemble the ground that they generally escape detection. A slightly more elaborate nest is the burrow, such as that occupied by the sand martin, kingfisher, or puffin. In many cases these underground nests are made in burrows left by rabbits or voles.

Many birds and some animals make their nests in the hollows of trees; the female of the Bornean rhinoceros hornbill is sealed up by the male for many weeks, until her chicks are almost ready to leave the nest. The mud nests of the swallow family and other birds are built mainly with mud; some of these weigh over 2 kg/4.5 lb. Among the crudest nests that are made with collected material, such as sticks, leaves, blades of grass or hair, are those of the wood pigeon, which are so loosely put together that the eggs are visible through them.

Swallows, martins, sparrows, and flycatchers prefer houses as nesting sites; the thrushes, finches, and linnets use evergreen bushes and hedges. Most species construct nests that are highly distinctive.

other nest-builders Among the fish, good examples of nest-builders are the three-spined stickleback *Gasterosteus aculeatus* and the Siamese fighting fish *Betta splendens*. The stickleback's nest is anchored to the river bed and constructed of algae and other aquatic plants stuck together by a secretion from the fish's kidneys. The nest of fighting fish is a floating bubble nest anchored to a piece of plant material into which the eggs are placed by the male after courtship.

Among the insects, nest-builders include the social insects, the ants, bees, and wasps, and their nests or hives, and the huge termite nests, up to 3 m/10 ft tall, found in the tropics.

Of the mammals, good examples are those of the harvest mouse *Micromys minutus* and the red squirrel *Sciurus vulgaris*. The harvest mouse nest is a kind of ball made of interlaced leaves of grass that generally belong to the same plants that support it above the ground. The red squirrel's nest is composed of a layer of twigs with a layer of moss or bark fragments. The whole thing is usually placed at the base of a large branch.

net in mathematics, a plan which can be used to make a model of a solid.

network in computing, a method of connecting computers so that they can share data and †peripheral devices, such as printers. The main types are classified by the pattern of the connections – star or ring network, for example – or by the degree of geographical spread allowed; for example, †local area networks (LANs) for communication within a room or building, and †wide area networks (WANs) for more remote systems. The †Internet is the linking of computer networks of institutions throughout the world: there are around 500 million users.

JANET (joint academic network) is used in Britain. SuperJANET, launched in 1992, is an extension of this that can carry 1,000 million bits (1,000 megabits, or about 100 Mb) of information per second. This is roughly the same speed that the fastest hard drives transfer information inside a PC.

How Ethernet Works http://www.howstuffworks.com/ethernet.htm

Pages from the How Stuff Works Web site explaining how Ethernet works. Easy-to-follow text and clear diagrams explain Ethernet principles and terminology and the pages include discussion of the Ethernet today and alternative network technologies.

network system of †nodes (junctions) and links (transport routes) through which goods, services, people, money, or information flow. Networks are often shown on topological maps.

Neugebauer, Gerry (1932–) born Gerald Neugebauer, German-born US astronomer whose work has been crucial in establishing the study of infrared astronomy. From 1969 he has been closely involved with NASA's interplanetary missions and the design of new infrared telescopes.

neural fold one of two longitudinal (ectodermal) ridges along the dorsal surface of a vertebrate embryo. The ridges fuse to form the †neural tube.

neural network artificial network of processors that attempts to mimic the structure of nerve cells (neurons) in the human brain. Neural networks may be electronic, optical, or simulated by computer software. A basic network has three layers of processors: an input layer, an output layer, and a "hidden" layer in between. Each processor is connected to every other in the network by a system of "synapses"; every processor in the top layer connects to every one in the hidden layer, and each of these connects to every processor in the output layer. This means that each nerve cell in the middle and bottom layers receives input from several different sources; only when the amount of input exceeds a critical level does the cell fire an output signal. The chief characteristic of neural networks is their ability to sum up large amounts of imprecise data and decide whether they match a pattern or not. Networks of this type may be used in developing robot vision, matching fingerprints, and analyzing fluctuations in stock-market prices. However, it is thought unlikely by scientists that such networks will ever be able accurately to imitate the human brain, which is very much more complicated; it contains around 10 billion nerve cells, whereas current artificial networks contain only a few hundred processors.

neural tube embryological structure from which the brain and spinal cord are derived. It forms when two edges of a groove in a plate of primitive nerve tissue fuse. Incomplete fusion results in neural tube defects such as spinal bifida.

neurasthenia obsolete term for nervous exhaustion, covering mild depression and various symptoms of neurosis. Formerly thought to be a bodily malfunction, it is now generally considered to be mental in origin. Dating from the mid-19th century, the term became widely used to describe the symptoms of soldiers returning from the front in World War I.

neurofibril one of the many fibrils in the cytoplasm of a neuron, which extend into its †axon and †dendrites.

neurohormone chemical secreted by nerve cells and carried by the blood to target cells. The function of the neurohormone is to act as a messenger; for example, the neurohormone ADH (antidiuretic hormone) is secreted in the pituitary gland and carried to the kidney, where it promotes water reabsorption in the kidney tubules.

neuron another name for a †nerve cell.

neuropsychology branch of neurology that overlaps with psychiatry and psychology and is mainly concerned with the †cerebral cortex, specifically those disorders of perception, memory, language, and behavior that result from brain injury or disease.

A central concern of neuropsychology has been the question of how such mental functions as perception, memory, and the initiation and control of actions are related to the structure of the brain. Some researchers, for example the French surgeon Paul Broca (1824–1880) and Carl †Wernicke, have favored a holistic approach, believing that the whole brain is involved in any type of mental activity or action. Others, for example the English neurologist John Huglings Jackson (1835–1911) and Kurt Goldstein (1878–1965), have searched for evidence of the localization of specific functions in separate areas. The ultimate aim, however, has always been to understand the mechanisms of cognition and behavior – in short, to discover how the brain works. Studies of the effects of specific brain injuries and the changes in cognition and behavior associated with them – as, for example, in aphasia – together with carefully conducted experimental work with animals, have thrown light on certain aspects of cerebral functioning and also indicated ways of treating brain-damaged patients and helping them to understand their disease.

Neuroptera insect order that includes the lacewings and alder flies.

classification Neuroptera is in subclass Pterygota, class Insecta, phylum Arthropoda.

neurosecretory cell nerve cell that secretes a chemical substance, such as a †hormone.

neuroticism personality dimension described by Hans †Eysenck. People with high neuroticism are worriers, emotional, and moody.

neurotransmitter chemical that diffuses across a †synapse, and thus transmits impulses between †nerve cells, or between nerve cells and effector organs (for example, muscles). Common neurotransmitters are noradrenaline (which also acts as a hormone) and acetylcholine, the latter being most frequent at junctions between nerve and muscle. Nearly 50 different neurotransmitters have been identified.

neutral equilibrium the state of equilibrium possessed by a body that will stay at rest if moved into a new position; it will neither move back to its original position nor move on any further. A sphere placed on a horizontal surface is an example. See †stability.

neutralization in chemistry, a process occurring when the excess acid (or excess base) in a substance is reacted with added base (or added acid) so that the resulting substance is neither acidic nor basic.

In theory neutralization involves adding acid or base as required to achieve †pH 7. When the color of an †indicator is used to test for neutralization, the final pH may differ from pH 7 depending upon the indicator used. It will also differ from pH 7 in reactions between strong acids and weak bases and weak acids and strong bases, because the salt formed will have acidic or basic properties respectively.

neutral oxide oxide that has neither acidic nor basic properties (see †oxide). Neutral oxides are only formed by †non-metals. Examples are carbon monoxide, water, and nitrogen(I) oxide.

neutral solution solution of p H7, in which the concentrations of $H^+_{(aq)}$ and $OH^-_{(aq)}$ ions are equal. It is therefore a solution which is neither acidic nor alkaline. An alkali may be †neutralized by an acid.

neutrino in physics, any of three uncharged †elementary particles (and their antiparticles) of the †lepton class, having a mass that is very small (possibly zero). The most familiar type, the antiparticle of the electron neutrino, is emitted in the beta decay of a nucleus. The other two are the muon and tau neutrinos.

neutron one of the three main subatomic particles, the others being the †proton and the †electron. Neutrons have about the same mass as protons but no electric charge, and occur in the nuclei of all atoms except hydrogen. They contribute to the mass of atoms but do not affect their chemistry.

For instance, the †isotopes of a single element differ only in the number of neutrons in their nuclei but have identical chemical properties. Neutrons and protons have masses approximately 2,000 times those of electrons.

The neutron is a composite particle, being made up of three †quarks, and therefore belongs to the †baryon group of the †hadrons. Outside a nucleus, a free neutron is unstable, decaying with a half-life of 11.6 minutes into a proton, an electron, and an antineutrino. The process by which a neutron changes into a proton is called †beta decay.

The neutron was discovered by the British chemist James Chadwick in 1932.

neutron bomb or **enhanced radiation weapon**, small hydrogen bomb for battlefield use that kills by radiation, with minimal damage to buildings and other structures.

neutron number symbol *N*, number of neutrons possessed by an atomic nucleus. †Isotopes are atoms of the same element possessing different neutron numbers.

neutron star very small, "superdense" star composed mostly of †neutrons. They are thought to form when massive stars explode as †supernovae, during which the protons and electrons of the star's atoms merge, owing to intense gravitational collapse, to make neutrons. A neutron star has a mass two to three times that of the Sun, compressed into a globe only 20 km/12 mi in diameter.

If its mass is any greater, its gravity will be so strong that it will shrink even further to become a †black hole. Being so small, neutron stars can spin very quickly. The rapidly flashing radio stars called †pulsars are believed to be neutron stars. The flashing is caused by a rotating beam of radio energy similar in behavior to a lighthouse beam.

Newcomb, Simon (1835–1909) Canadian-born US mathematician and astronomer. He compiled charts and tables of astronomical data with phenomenal accuracy. His calculations of the motions of the bodies in the Solar System were in use as daily reference all over the world for more than 50 years, and the system of astronomical constants for which he was most responsible is still the standard.

Newlands, John Alexander Reina (1837–1898) English chemist who worked as an industrial chemist; he prepared in 1863 the first †periodic table of the elements arranged in order of relative atomic masses, and pointed out in 1865 the "law of octaves" whereby every eighth element has similar properties. He was ridiculed at the time, but five years later the Russian chemist Dmitri †Mendeleyev published a more developed form of the table, also based on atomic masses, which forms the basis of the one used today (arranged by atomic number).

Newman projection in chemistry, alternative to †Fischer projection as a two-dimensional representational drawing of a molecule. Newman projections are a specialized form of †sawhorse projection and provide a view of the molecule concentrating on the relative orientations of constituent groups on adjacent atoms.

newton symbol N, SI unit of †force. A newton is defined as the amount of force needed to move an object of 1 kg so that it accelerates at 1 meter per second per second. It is also used as a unit of weight. The weight of a medium size (100 g/3 oz) apple is one newton.

Newton, Isaac (1642–1727) English physicist and mathematician who laid the foundations of physics as a modern discipline. During 1665–66, he discovered the †binomial theorem, differential and integral †calculus, and that white light is composed of many colors. He developed the three standard laws of motion (†Newton's laws of motion) and the universal law of gravitation, set out in *Philosophiae naturalis principia mathematica* (1687), usually referred to as the *Principia*. He was knighted for his work in 1705.

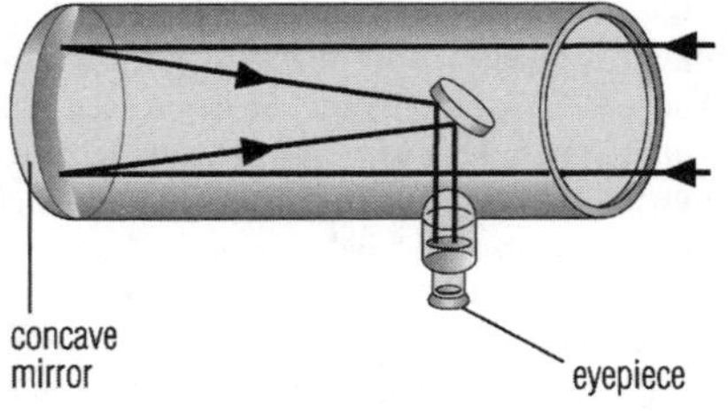

Newton In about 1670 Isaac Newton constructed the first reflecting telescope, which uses a concave mirror to collect and focus incoming light. Helicon

Newton's birthplace: Woolsthorpe Manor, Lincolnshire
http://www.newton.org.uk/
Tour of Isaac Newton's birthplace, with biographical detail and pictures of the interior of the house. There are also plenty of hyperlinks within the text to information about his scientific achievements, and to the rest of his life in sections entitled "Newtonia," "Places," "Bibliography," "Chronology," and "Biographies."

Newtonian physics physics based on the concepts of the English scientist Isaac †Newton, before the formulation of quantum theory or relativity theory.

Newtonian telescope in astronomy, simple reflecting †telescope in which light collected by a parabolic primary mirror is directed to a focus at the side of the tube by a flat secondary mirror placed at 45 degrees to the optical axis. It is named after English physicist and mathematician †Isaac Newton, who constructed such a telescope in 1668.

Newton's law of cooling rate of loss of †heat from a body is directly proportional to the instantaneous temperature difference between the body and the surroundings. This leads to an exponential law of temperature decline.

Newton's laws of motion in physics, three laws that form the basis of Newtonian mechanics, describing the motion of objects. (1) Unless acted upon by an unbalanced force, a body at rest stays at rest, and a moving body continues moving at the same speed in the same straight line. (2) An unbalanced force applied to a body gives it an acceleration proportional to the force (and in the direction of the force) and inversely proportional to the mass of the body. (3) When a body A exerts a force on a body B, B exerts an equal and opposite force on A; that is, to every action there is an equal and opposite reaction.

the first law As an example, if a car is traveling at a certain speed in a certain direction, it will continue to travel at that speed in the same direction unless it is acted upon by an unbalanced force such as †friction in the brake mechanism, which will slow down the car. A person in the car will continue to move forward (in accordance with the first law) unless acted upon by a force; for example, the restraining force of a seat belt.

the second law This can be demonstrated using a ticker timer and a trolley with mass that can be varied. If the mass is kept constant and the amount of force applied in pulling the trolley varies, then the dots on the ticker timer tape become further apart; the trolley is changing velocity and therefore accelerating. The acceleration (*a*) is proportional to the force (*F*) applied. This can be expressed as acceleration/force = a constant, or a/F = a constant.

If the force applied in pulling the trolley is kept the same (constant), and the mass placed on the trolley is varied (from low to high), then the dots on the ticker timer tape become closer together as the mass gets larger; the acceleration decreases as mass increases. This can be expressed as acceleration being inversely proportional to mass (*m*), or $a \propto 1/m$.

The two equations can be combined to give an overall equation, expressing Newton's second law of motion. This is: force = mass x acceleration, or $F = ma$.

the third law As an example, a book placed on a table will remain at rest. The force of gravity acting on the book pulls the book towards the ground. The table opposes the force (weight) exerted on the book by gravity. The table exerts the same amount of force in an upward direction. Hence the book remains at rest on the table.

Free Fall and Air Resistance http://www.glenbrook.k12.il.us/gbssci/phys/Class/newtlaws/u2l3e.html

Detailed explanation of Newton's second law of motion, accompanied by diagrams and animations to help illustrate the concepts.

Newton's rings in optics, an †interference phenomenon seen (using white light) as concentric rings of spectral colors where light passes through a thin film of transparent medium, such as the wedge of air between a large-radius convex lens and a flat glass plate. With monochromatic light (light of a single wavelength), the rings take the form of alternate light and dark bands. They are caused by interference (interaction) between light rays reflected from the plate and those reflected from the curved surface of the lens.

niacin one of the "B group" vitamins; see †nicotinic acid.

niche in ecology, the "place" occupied by a species in its habitat, including all chemical, physical, and biological components, such as what it eats, the time of day at which the species feeds, temperature, moisture, the parts of the habitat that it uses (for example, trees or open grassland), the way it reproduces, and how it behaves.

It is believed that no two species can occupy exactly the same niche, because they would be in direct competition for the same resources at every stage of their life cycle.

Nicholson, William (1753–1815) English scientist who discovered electrolysis in 1800, decomposing water into oxygen and hydrogen by the passage of an electric current using a voltaic cell.

Nichrome trade name for a series of alloys containing mainly nickel and chromium, with small amounts of other substances such as iron, magnesium, silicon, and carbon. Nichrome has a high melting point and is resistant to corrosion. It is therefore used in electrical heating elements and as a substitute for platinum in the †flame test.

nickel chemical symbol Ni, hard, malleable and ductile, silver-white, metallic element, atomic number 28, relative atomic mass 58.71. It occurs in igneous rocks and as a free metal (†native metal), occasionally occurring in fragments of iron–nickel meteorites. It is a component of the Earth's core, which is held to consist principally of iron with some nickel. It has a high melting point, low electrical and thermal conductivity, and can be magnetized. It does not tarnish and therefore is much used for alloys, electroplating, and for coinage.

It was discovered in 1751 by Swedish mineralogist Axel Cronstedt (1722–1765) and the name given is an abbreviated form of *kopparnickel*, Swedish "false copper," since the ore in which it is found resembles copper but yields none.

nickel ore any mineral ore from which nickel is obtained. The main minerals are arsenides such as chloanthite ($NiAs_2$), and the sulfides millerite (NiS) and pentlandite ($(Ni,Fe)_9S_8$), the commonest ore. The chief nickel-producing countries are Canada, Russia, Kazakhstan, Cuba, and Australia.

Nicol, William (1768–1851) Scottish physicist and geologist who invented the first device for obtaining plane-polarized light – the †Nicol prism – in 1828.

Nicol prism device which uses the optical properties of calcite to produce †plane-polarized light by passing rays of light through it. It is named after Scottish physicist William Nicol who invented it in 1828.

nicotinamide adenine dinucleotide **NAD**, naturally occurring compound that acts as a †coenzyme. It consists of two †nucleotides, adenosine mononucleotide and nicotinamide mononucleotide, joined by a phosphate bridge. A similar compound, NAD phosphate, **NADP**, is produced by phosphorylating the ribose part of the adenosine mononucleotide. NAD takes part in oxidation–reduction reactions, being converted to $NADH_2$ by accepting two hydrogen atoms. In this form it is able to enter the †electron transport chain where it is oxidized back to NAD, producing large quantities of †ATP in the process. NADP is an important oxidation–reduction factor in †fatty acid biosynthesis.

nicotine $C_{10}H_{14}N_2$, †alkaloid (nitrogenous compound) obtained from the dried leaves of the tobacco plant *Nicotiana tabacum*. A colorless oil, soluble in water, it turns brown on exposure to the air. Nicotine is found in tobacco smoke. It can be described as a recreational †drug. It stimulates the human body and produces feelings that cause people to carry on smoking. However, nicotine is usually addictive. Regular smokers find that it is very difficult or impossible to give up smoking even though they may try to. This is a problem, in that other chemicals in smoke cause a wide range of diseases and increase the risk of dying early. Nicotine has also been used as an insecticide.

Nicotine in its pure form is one of the most powerful poisons known. It is named after a 16th-century French diplomat, Jacques Nicot, who introduced tobacco to France.

nicotinic acid or **niacin**, water-soluble †vitamin ($C_5H_5N.COOH$) of the B complex, found in meat, fish, and cereals; it can also be formed in small amounts in the body from the essential †amino acid tryptophan. Absence of nicotinic acid from the diet leads to the disease †pellagra.

niobium chemical symbol Nb, soft, gray-white, somewhat ductile and malleable, metallic element, atomic number 41, relative atomic mass 92.906. It occurs in nature with tantalum, which it resembles in chemical properties. It is used in making stainless steel and other alloys for jet engines and rockets and for making superconductor magnets.

Niobium was discovered in 1801 by the English chemist Charles Hatchett (1765–1847), who named it columbium (symbol Cb), a name that is sometimes still used in metallurgy. In 1844 it was renamed after Niobe by the German chemist Heinrich Rose (1795–1864) because of its similarity to tantalum (Niobe is the daughter of Tantalus in Greek mythology).

nitrate salt or ester of nitric acid, containing the NO_3^- ion. Nitrates are used in explosives, in the chemical and pharmaceutical industries, in curing meat (as †niter), and as fertilizers. They are the most water-soluble salts known and play a major part in the nitrogen cycle. Nitrates in the soil, whether naturally occurring or from inorganic or organic fertilizers, can be used by plants to make proteins and nucleic acids. However, run-off from fields can result in †nitrate pollution.

nitrate pollution contamination of water by nitrates. Increased use of artificial fertilizers and land cultivation means that higher levels of nitrates are being washed from the soil into rivers, lakes, and aquifers. There they cause an excessive enrichment of the water (†eutrophication), leading to a rapid growth of algae, which in turn darkens the water and reduces its oxygen content. The water is expensive to purify and many plants and animals die. High levels are now found in drinking water in arable areas. These may be harmful to newborn babies, and it is possible that they contribute to stomach cancer, although the evidence for this is unproven.

niter or **saltpeter**, potassium nitrate, KNO_3, a mineral found on and just under the ground in desert regions; used in explosives. Niter occurs in Bihar, India, Iran, and Cape Province, South Africa. The salt was formerly used for the manufacture of gunpowder, but the supply of niter for explosives is today largely met by making the salt from nitratine (also called Chile saltpeter, $NaNO_3$). Saltpeter is a preservative and is widely used for curing meats.

nitric acid or **aqua fortis**; HNO_3, fuming acid obtained by the oxidation of ammonia or the action of sulfuric acid on potassium nitrate. It is a highly corrosive acid, dissolving most metals, and a strong oxidizing agent. It is used in the nitration and esterification of organic substances, and in the making of sulfuric acid, nitrates, explosives, plastics, and dyes.

nitric oxide or **nitrogen monoxide (NO)**, colorless gas released when metallic copper reacts with nitric acid and when nitrogen and oxygen combine at high temperatures. It is oxidized to nitrogen dioxide on contact with air. Nitric oxide has a wide range of functions in the body. It is involved in the transmission of nerve impulses and the protection of nerve cells against stress. It is released by macrophages in the immune system in response to viral and bacterial infection or to the proliferation of cancer cells. It is also important in the control of blood pressure.

nitrification process that takes place in soil when bacteria oxidize ammonia, turning it into nitrates. Nitrates can be absorbed by the roots of plants, so this is a vital stage in the †nitrogen cycle.

nitrite salt or ester of nitrous acid, containing the nitrite ion (NO_2^-). Nitrites are used as preservatives (for example, to prevent the growth of botulism spores) and as coloring agents in cured meats such as bacon and sausages.

nitrobenzene $C_6H_5NO_2$, yellow, viscous liquid with the distinctive odor of bitter almonds, boiling point 211°C/412°F. It is prepared by the nitration of †benzene using concentrated nitric and sulfuric acids. It is mainly used for the production of †aniline for dye manufacture.

nitrocellulose alternative name for †cellulose nitrate.

nitrogen chemical symbol N, (Greek *nitron* "native soda," sodium or potassium nitrate) colorless, odorless, tasteless, gaseous, non-metallic element, atomic number 7, relative atomic mass 14.0067. It forms almost 80% of the Earth's atmosphere by volume and is a constituent of all plant and animal tissues (in proteins and nucleic acids). Nitrogen is obtained for industrial use by the liquefaction and †fractional distillation of air. Its compounds are used in the manufacture of foods, drugs, fertilizers, dyes, and explosives.

Nitrogen has been recognized as a plant nutrient, found in manures and other organic matter, from early times, long before the complex cycle of †nitrogen fixation was understood. It was isolated in 1772 by the English chemist Daniel Rutherford (1749–1819) and named in 1790 by the French chemist Jean Chaptal (1756–1832).

nitrogen cycle process of nitrogen passing through the ecosystem. Nitrogen, in the form of inorganic compounds (such as nitrates) in the soil, is absorbed by plants and turned into organic compounds (such as proteins) in plant tissue. A proportion of this nitrogen is eaten by †herbivores, with some of this in turn being passed on to the carnivores, which feed on the herbivores. The nitrogen is ultimately returned to the soil as excrement and when organisms die, and is converted back to inorganic forms by †decomposers.

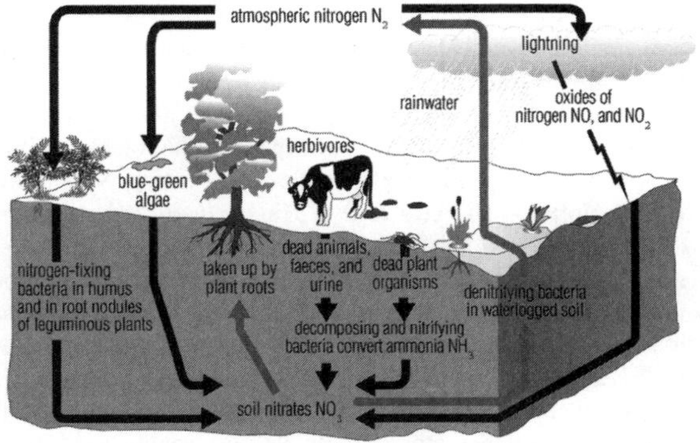

nitrogen cycle The nitrogen cycle is one of a number of cycles during which the chemicals necessary for life are recycled. The carbon, sulfur, and phosphorus cycles are others. Since there is only a limited amount of these chemicals in the Earth and its atmosphere, the chemicals must be continuously recycled if life is to go on. Helicon

Although about 78% of the atmosphere is nitrogen, this cannot be used directly by most organisms. However, certain bacteria and cyanobacteria (†blue-green algae) are capable of nitrogen fixation. Some nitrogen-fixing bacteria live mutually with leguminous plants (peas and beans) or other plants (for example, alder), where they form characteristic nodules on the roots. The presence of such plants increases the nitrate content, and hence the fertility, of the soil.

nitrogen fixation process by which nitrogen in the atmosphere is converted into nitrogenous compounds by the action of micro-organisms, such as cyanobacteria (see †blue-green algae) and bacteria, in conjunction with certain legumes (see †root nodule). Several chemical processes duplicate nitrogen fixation to produce fertilizers; see †nitrogen cycle.

nitrogen oxide any chemical compound that contains only nitrogen and oxygen. All nitrogen oxides are gases. Nitrogen monoxide and nitrogen dioxide contribute to air pollution. See also †nitrous oxide.

nitrogen monoxide NO, or nitric oxide, is a colorless gas released when metallic copper reacts with concentrated †nitric acid. It is also produced when nitrogen and oxygen combine at high temperature. On contact with air it is oxidized to nitrogen dioxide.

Nitrogen monoxide was discovered during the 1980s to act as a chemical messenger in small quantities within the human body, despite being toxic at higher concentrations and its rapid reaction with oxygen. The medical condition of septic shock is linked to overproduction by the body of nitrogen monoxide. Nitrogen monoxide has an unpaired electron, which can be removed to produce the nitrosyl ion, NO^+.

nitrogen dioxide nitrogen(IV) oxide, NO_2, is a brown, acidic, pungent gas that is harmful if inhaled and contributes to the formation of †acid rain, as it dissolves in water to form nitric acid. It is the most common of the nitrogen oxides and is obtained by heating most nitrate salts (for example †lead(II) nitrate, $Pb(NO_3)_2$). If liquefied, it gives a colorless solution (N_2O_4). It has been used in rocket fuels.

nitroglycerine or **glycerol trinitrate**; $C_3H_5(ONO_2)_3$, flammable, explosive oil produced by the action of nitric and sulfuric acids on glycerol. Although poisonous, it is used in cardiac medicine. It explodes with great violence if heated in a confined space and is used in the preparation of dynamite, cordite, and other high explosives.

It was synthesized by the Italian Ascanio Soberro in 1846, and is unusual among explosives in that it is a liquid. Nitroglycerine is an effective explosive because it has low †activation energy, and produces little smoke when burned. However, it was initially so reactive it was virtually unusable. Alfred †Nobel's innovation was to purify nitroglycerine (using water, with which it is immiscible, to dissolve the impurities), and thereby make it more stable, and absorb it into wood flour to make dynamite.

nitrous acid HNO_2, weak acid that, in solution with water, decomposes quickly to form nitric acid and nitrogen dioxide.

nitrous oxide or **dinitrogen oxide**; N_2O, colorless, nonflammable gas that, used in conjunction with oxygen, reduces sensitivity to pain. In higher doses it is an anesthetic. Well tolerated, it is often combined with other anesthetic gases to enable them to be used in lower doses. It may be self-administered; for example, in childbirth. It is a greenhouse gas; about 10% of nitrous oxide released into the atmosphere comes from the manufacture of nylon. It used to be known as "laughing gas."

nivation set of processes, operating beneath or next to snow, believed to be responsible for the development of the hollows in which snow collects. It is also thought to play a role in the early formation of †corries. The processes involved include †freeze–thaw (weathering by the alternate freezing and melting of ice), †mass movement (the downhill movement of substances under gravity), and †erosion by meltwater.

Nobel, Alfred Bernhard (1833–1896) Swedish chemist and engineer. He invented †dynamite in 1867, gelignite in 1875, and ballistite, a smokeless gunpowder, in 1887. Having amassed a large fortune from the manufacture of explosives and the exploitation of the Baku oilfields in Azerbaijan, near the Caspian Sea, he left this in trust for the endowment of five Nobel prizes.

Alfred Nobel – His Life and Work http://www.nobel.se/nobel/alfred-nobel/biographical/life-work/index.html
Presentation of the life and work of Alfred Nobel. The site includes references to Nobel's life in Paris, as well as his frequent travels; his industrial occupations; his scientific discoveries; and especially his work on explosives, which led to the patenting of dynamite. Also covered are his numerous chemical inventions, which included materials such as synthetic leather and artificial silk, his interest in literature and in social and peace-related issues, and of course the Nobel prizes, which came as a natural extension of his lifelong interests.

nobelium chemical symbol No, synthesized, radioactive, metallic element of the †actinide series, atomic number 102, relative atomic mass 259. It is synthesized by bombarding curium with carbon nuclei.
It was named in 1957 after the Nobel Institute in Stockholm, Sweden, where it was claimed to have been first synthesized. Later evaluations determined that this was in fact not so, as the successful 1958 synthesis at the University of California at Berkeley produced a different set of data. The name was not, however, challenged. In 1992 the International Unions for Pure and Applied Chemistry and Physics (IUPAC and IUPAP) gave credit to Russian scientists in Dubna for the discovery of nobelium.

noble gas or **rare gas** or **inert gas**, any of a group of six elements (†helium, †neon, †argon, †krypton, †xenon, and †radon), originally named "inert" because they were thought not to enter into any chemical reactions. This is now known to be incorrect: in 1962, xenon was made to combine with fluorine, and since then, compounds of argon, krypton, and radon with fluorine and/or oxygen have been described.
The extreme unreactivity of the noble gases is due to the stability of their †electronic structure. All the electron shells (†energy levels) of inert gas atoms are full and, except for helium, they all have eight electrons in their outermost (†valency) shell. The apparent stability of this electronic arrangement led to the formulation of the †octet rule to explain the different types of chemical bond found in simple compounds. The noble gases are in Group 0 of the †periodic table of the elements.

noble gas structure configuration of electrons in †noble gases (rare gases): helium, neon, argon, krypton, xenon, and radon.
This is characterized by full electron shells around the nucleus of an atom, which render the element stable. Any ion, produced by the gain or loss of electrons, that achieves an electronic configuration similar to one of the noble gases is said to have a noble gas structure.

noctilucent cloud clouds of ice forming in the upper atmosphere at around 83 km/52 mi. They are visible on summer nights, particularly when sunspot activity is low.

Noddack, Ida Eva (1896–1979) born Ida Tacke, German chemist who in 1925 discovered the element rhenium with German chemist Walter Noddack. She also predicted that nuclear fission could be produced by bombarding uranium with slow neutrons.

node in physics, a position in a †standing wave pattern at which there is no vibration. Points at which there is maximum vibration are called **antinodes**. Stretched strings, for example, can show nodes when they vibrate. Guitarists can produce special effects (harmonics) by touching a sounding string lightly to produce a node.

node in two- and three-dimensional geometry, where a curve intersects itself.

node in botany, point where the petiole (stalk) of a leaf attaches to the stem.

node point where routes meet. It may therefore be the same as a route center. In a topological †network, a node may be the start or crossing point of routes, also called a **vertex**.

node in oceanography, the point on a stationary wave (an oscillating wave with no progressive motion) at which vertical motion is least and horizontal motion is greatest. In earth science, it is the point on a

†fault where the apparent motion has changed direction.

nodule in geology, a lump of mineral or other matter found within rocks or formed on the seabed surface; mining technology is being developed to exploit them.

Noether, Emmy (1882–1935) born Amalie Noether, German mathematician who became one of the leading figures in abstract algebra. Modern work in the field of a general theory of ideals dates from her papers of the early 1920s.

Noether, Emmy http://www-history.mcs.st-and.ac.uk/history/Mathematicians/Noether_Emmy.html

Biography of the German–US mathematician. The site contains a description of her work and her working relationships with her contemporaries, notably Max Noether. Several literature references for further reading on the life and works of Noether are listed, and two photographs of her are also shown.

no hair theorem proposed by John Wheeler, it states that the only properties of matter conserved after entering a black hole are its mass, its angular momentum, and its electrical charge; it thus becomes neither matter nor †antimatter.

noise unwanted sound. Permanent, incurable loss of hearing can be caused by prolonged exposure to high noise levels (above 85 decibels). Over 55 decibels on a daily outdoor basis is regarded as an unacceptable level. In scientific and engineering terms, a noise is any random, unpredictable signal.

noise pollution Noise is a recognized form of pollution, but is difficult to measure because the annoyance or discomfort caused varies between individuals. If the noise is in a narrow frequency band, temporary hearing loss can occur even though the level is below 85 decibels or exposure is only for short periods. Lower levels of noise are an irritant, but seem not to increase fatigue or affect efficiency to any great extent. Loss of hearing is a common complaint of people working on factory production lines or in the construction and road industry. Minor psychiatric disease, stress-related ailments including high blood pressure, and disturbed sleep patterns are regularly linked to noise, although the causal links are in most cases hard to establish. Loud noise is a major pollutant in towns and cities.

electronic noise Electronic noise takes the form of unwanted signals generated in electronic circuits and in recording processes by stray electrical or magnetic fields, or by temperature variations. In electronic recording and communication systems, **white noise** frequently appears in the form of high frequencies, or hiss. The main advantages of digital systems are their relative freedom from such noise and their ability to recover and improve noise-affected signals.

non-aqueous solution in chemistry, any solution in which water is not the solvent, for example ether.

noncommutative in mathematics, describing an operation in which the principle of commutation (see †commutative operation) does not apply.

non-essential amino acid in biology, any †amino acid synthesized in the liver by the transfer of an amino group ($-NH_2$) from an amino acid to part of another molecule, in a process called transamination. Of the 20 naturally occurring amino acids, nine cannot be made by the human body. These are required in food and are the †essential amino acids.

non-Euclidean geometry study of figures and shapes in three-or-more-dimensional (or curved) space, in which Euclid's postulates may not apply fully or at all (see †geometry). There are now many forms of non-Euclidean geometry, probably the best known being those propounded by Bernhard Riemann; the first proponents of such systems, however, were Karl Gauss, Nikolai Lobachevsky, and János Bolyai.

non-metal one of a set of elements (around 20 in total) with certain physical and chemical properties opposite to those of †metal elements. The division between metal and non-metal elements forms the simplest division of the †periodic table of the elements. Common physical properties are that non-metals have low electrical conductivity, are brittle when solid, or are gases or liquids. Exceptions include †graphite, a form of carbon, which is a good electrical conductor.

In structure the non-metals are very diverse. †Hydrogen (H), †oxygen (O), †nitrogen (N), †fluorine (F), †chlorine (Cl), and the noble gases (rare gases), †helium (He), †neon (Ne), †argon (Ar), †krypton (Kr), †xenon (Xe), and radioactive †radon (Rn), are gases. Only †bromine (Br) is a liquid at room temperature, and the rest are solids.

Non-metals are the chemical opposites of metals. Metals form positively charged ions or †cations; non-metals form negatively charged ions or †anions. The exceptions are the chemically unreactive noble gases (rare gases), although xenon does react under certain conditions. Non-metals are electronegative, which means that they are able to gain electrons when bonding with metals. Apart from the noble gases, the non-metal elements have incomplete outer electron shells, and so try to gain enough electrons to fill them. The noble gases do not react because they already have complete outer electron shells. The type of bonding where ions are formed is known as †ionic bonding. Non-metals may also share electrons with other non-metal elements to complete their outer shell; this type of bonding is known as †covalent bonding.

non-renewable resource natural resource, such as coal, oil, or natural gas, that takes millions of years to form naturally and therefore cannot be replaced once it is consumed; it will eventually be used up. The main energy sources used by humans are non-renewable; †renewable resources, such as solar, tidal, wind, and geothermal power, have so far been less exploited.

Fossil fuels like coal, oil, and gas generate a considerable amount of energy when they are burnt (the process of combustion). Non-renewable resources have a high carbon content because their origin lies in the photosynthetic activity of plants millions of years ago. The fuels release this carbon back into the atmosphere as carbon dioxide. The rate at which such fuels are being burnt is thus resulting in a rise in the concentration of carbon dioxide in the atmosphere, a cause of the †greenhouse effect.

non-stoichiometric compound in chemistry, compound in which the ratio of the total number of atoms of each element do not give whole numbers. Many †transition metal oxides are non-stoichiometric.

nonvolatile memory in computing, †memory that does not lose its contents when the power supply to the computer is disconnected.

noradrenaline in the body, a †hormone that acts directly on specific receptors to stimulate the sympathetic nervous system. Released by nerve stimulation or by drugs, it slows the heart rate mainly by constricting arterioles (small arteries) and so raising blood pressure. It is used therapeutically to treat shock.

NOR gate in electronics, a type of †logic gate.

Norma small constellation of the southern hemisphere, formerly known as "Norma et Regula," "the Level and Square."

It was named by 18th-century French scientist and astronomer Nicolas †Lacaille at the same time as the adjacent †Circinus, represented as a pair of compasses.

normal in mathematics, to a curve at a point, a line perpendicular to the tangent at the point.

normal distribution in statistics, a distribution widely used to model variation in a set of data which is symmetrical about its mean value. It can be expressed in the form:

$$f(x) = \frac{1}{\sqrt{2\pi}} \exp\left\{\frac{-(x-\mu)^2}{2\sigma^2}\right\}$$

where *f*(*x*) is the relative frequency of data value *x*, s is the †standard deviation, m is the mean, exp is the exponential function, and p is a mathematical constant. The curve resulting when *f*(*x*) is plotted against *x* is called the **normal distribution curve**.

An example of normal distribution can be found in the intelligence quotients (IQs) found in human populations, distributed around a mean value of 100. The distribution is also known as the Gaussian distribution after German mathematician Karl †Gauss.

normal distribution curve bell-shaped curve that results when a †normal distribution is represented graphically by plotting the distribution *f*(*x*) against *x*. The curve is symmetrical about the mean value.

Norrish, Ronald George Wreyford (1897–1978) English physical chemist who shared the Nobel Prize for Chemistry in 1967 with his co-worker George †Porter for their investigation of fast chemical reactions, particularly those initiated by light. Norrish was largely responsible for the advance of reaction kinetics to a distinct discipline within physical chemistry.

North Atlantic Drift warm †ocean current in the North Atlantic Ocean; an extension of the †Gulf Stream. It flows east across the Atlantic and has a mellowing effect on the climate of northwestern Europe, particularly the British Isles and Scandinavia.

northern lights common name for the †aurora borealis.

North Polar sequence or circumpolar stars, comprises those stars which never set, from the viewpoint of an observer on Earth.

Northrop, John Howard (1891–1987) US chemist. He was awarded the Nobel Prize for Chemistry in 1946 for his work in the 1930s when he crystallized a number of enzymes, including pepsin and trypsin, showing conclusively that they were proteins. He shared the award with Wendell †Stanley and James †Sumner.

nose in humans, the upper entrance of the respiratory tract; the organ of the sense of smell. The external part is divided down the middle by a septum of †cartilage. The nostrils contain plates of cartilage that can be moved by muscles and have a growth of stiff hairs at the margin to prevent foreign objects from entering. The whole nasal cavity is lined with a †mucous membrane that warms and moistens the air as it enters and ejects dirt. In the upper parts of the cavity the membrane contains 50 million olfactory receptor cells (cells sensitive to smell).

nostril in vertebrates, the opening of the nasal cavity, in which cells sensitive to smell are located. (In fish, these cells detect water-borne chemicals, so they are effectively organs of taste.) In vertebrates with lungs, the nostrils also take in air. In humans, and most other mammals, the nostrils are located on the †nose.

NOT gate or **inverter gate**, in electronics, a type of †logic gate.

notochord stiff but flexible rod that lies between the gut and the nerve cord of all embryonic and larval chordates, including the vertebrates. It forms the supporting structure of the adult lancelet, but in vertebrates it is replaced by the vertebral column, or spine.

nova plural **novae**, faint star that suddenly erupts in brightness by 10,000 times or more, remains bright for a few days, and then fades away and is not seen again for very many years, if at all. Novae are believed to occur in close †binary star systems, where gas from one star flows to a companion †white dwarf. The gas ignites and is thrown off in an explosion at speeds of 1,500 kps/930 mps or more. Unlike a †supernova, the star is not completely disrupted by the outburst. After a few weeks or months it subsides to its previous state; it may erupt many more times.

Although the name comes from the Latin "new," photographic records show that such stars are not really new, but faint stars undergoing an outburst of radiation that temporarily gives them an absolute magnitude in the range –6 to –10, at least 100,000 times brighter than the Sun. They fade away, rapidly at first and then more slowly over several years. Two or three such stars are detected in our Galaxy each year, but on average one is sufficiently close to us to become a conspicuous naked-eye object only about once in ten years. Novae very similar to those appearing in our own Galaxy have also been observed in other galaxies.

nuclear energy or **atomic energy**, energy released from the inner core, or †nucleus, of the atom. Energy produced by nuclear †fission (the splitting of certain atomic nuclei) has been harnessed since the 1950s to generate electricity, and research continues into the possible controlled use of †nuclear fusion (the fusing, or combining, of atomic nuclei).

In nuclear power stations, fission of radioactive substances (see †radioactivity) takes place, releasing large amounts of heat energy. The heat is used to produce the steam that drives †turbines and †generators, producing electrical power.

Nuclear energy is produced from the transformation of matter into energy. The amount of energy can be calculated from Einstein's equation, $E = mc^2$, where E is the heat energy given out, m is the mass of radioactive substance, and c is the speed of light.

The Sun is an example of a natural nuclear reactor. Millions of atoms of hydrogen fuse together to form millions of atoms of helium, generating a continuous supply of heat and light energy. This is called a fusion reaction. A hydrogen bomb obtains its energy from a fusion reaction.

International Atomic Energy Agency http://www.iaea.org/worldatom/

Busy Web site brimming with news, reports, and information. The pages look at the role and activities of the International Atomic Energy Agency's departments and focus on key issues such as nuclear waste, depleted uranium, Chernobyl, and arms control.

Nuclear Energy: Frequently Asked Questions http://www-formal.stanford.edu/jmc/progress/nuclear-faq.html

Answers to the most commonly asked questions about nuclear energy, particularly with a view to sustaining human progress. It contains many links to related pages and is a personal opinion that openly asks for comment from visitors.

nuclear fusion process whereby two atomic nuclei are fused, with the release of a large amount of energy. Very high temperatures and pressures are required for the process. Under these conditions the atoms involved are stripped of all their electrons so that the remaining particles, which together make up a **plasma**, can come close together at very high speeds and overcome the mutual repulsion of the positive charges on the atomic nuclei. At very close range the strong nuclear force will come into play, fusing the particles to form a larger nucleus. As fusion is accompanied by the release of large amounts of energy, the process might one day be harnessed to form the basis of commercial energy production. Methods of achieving controlled fusion are therefore the subject of research around the world.

Fusion is the process by which the Sun and the other stars produce their energy.

nuclear isomer atoms of an element of a given mass that differ in their rates of †radioactive decay are known as nuclear isomers.

nuclear notation method used for labeling an atom according to the composition of its nucleus. The atoms or isotopes of a particular element are represented by the symbol ${}^{A}{}_{Z}X$ where A is the mass number of their nuclei, Z is their atomic number, and X is the chemical symbol for that element.

nuclear physics study of the properties of the nucleus of the †atom, including the structure of nuclei; nuclear forces; the interactions between particles and nuclei; and the study of radioactive decay. The study of elementary particles is †particle physics.

Nuclear Physics http://www.scri.fsu.edu/~jac/Nuclear/

"Hyper-textbook" of nuclear physics, with an introduction that includes a graphical description of the size and shape of nuclei and their other properties. The site also includes information about the work of nuclear physicists, and the uses and applications of nuclear physics from medicine, through energy, to smoke detectors.

Nuclear Physics Past Present and Future http://library.thinkquest.org/3471

The site is divided into lessons covering nuclear technology, uses of nuclear energy, general applications and weapons and warfare. A very thorough site with difficult terms explained and colorful illustrations.

nuclear reaction reaction involving the nuclei of atoms. Atomic nuclei can undergo changes either as a result of radioactive decay, as in the decay of radium to radon (with the emission of an alpha particle):

$^{226}{}_{88}Ra \rightarrow {}^{222}{}_{86}Rn + {}^{4}{}_{2}He$

or as a result of particle bombardment in a machine or device, as in the production of cobalt-60 by the bombardment of cobalt-59 with neutrons:

$^{59}{}_{27}Co + {}^{1}{}_{0}n \rightarrow {}^{60}{}_{27}Co + g.$

Nuclear †fission and nuclear †fusion are examples of nuclear reactions. The enormous amounts of energy released arise from the mass–energy relation put forward by Einstein, stating that $E = mc^2$ (where E is energy, m is mass, and c is the velocity of light).

In nuclear reactions the sum of the masses of all the products is less than the sum of the masses of the reacting particles. This lost mass is converted to energy according to Einstein's equation.

nuclear reactor device for producing †nuclear energy in a controlled manner. There are various types of reactor in use, all using nuclear †fission. In a **gas-cooled reactor**, a circulating gas under pressure (such as carbon dioxide) removes heat from the core of the reactor, which usually contains natural uranium. The efficiency of the fission process is increased by slowing neutrons in the core by using a †moderator such as carbon. The reaction is controlled with neutron-absorbing rods made of boron. An **advanced gas-cooled reactor** (AGR) generally has enriched uranium as its fuel. A **water-cooled reactor**, such as the steam-generating heavy water (deuterium oxide) reactor, has water circulating through the hot core. The water is converted to steam, which drives turbo-alternators for generating electricity. The most widely used reactor is the **pressurized-water reactor** (PWR), which contains a sealed system of pressurized water that is heated to form steam in heat exchangers in an external circuit. The **fast reactor** has no moderator and uses fast neutrons to bring about fission. It uses a mixture of plutonium and uranium oxide as fuel. When operating, uranium is converted to plutonium, which can be extracted and used later as fuel. It is also called the fast breeder or breeder reactor because it produces more plutonium than it consumes. Heat is removed from the reactor by a coolant of liquid sodium.

Public concern over the safety of nuclear reactors has been intensified by explosions and accidental release of radioactive materials. The safest system allows for the emergency cooling of a reactor by automatically flooding an overheated core with water. Other concerns about nuclear power center on the difficulties of reprocessing nuclear fuel and disposing safely of nuclear waste, and the cost of maintaining nuclear power stations and of decommissioning them at the end of their lives. The break up of the former USSR raised concerns about the ability of the new nation states to safely manage aging reactors. In 1989 the UK government decided to postpone the construction of new nuclear power stations; in the USA, no new stations have been commissioned in over a decade. Rancho Seco, near Sacramento, California, was the first nuclear power station to be closed, by popular vote, in 1989. Sweden is committed to decommissioning its reactors. Some countries, such as France, are pressing ahead with their nuclear programs. It was reported in 1997 that there were 443 nuclear power plants in 31 countries, representing a net growth of 15 since the 1992 Rio Earth Summit, and a further 36 were under construction. This compares with a total of just over 100 reactors generating electricity in 15 countries in 1972. Nuclear power generates around 17% of the world's electricity.

nuclear safety measures to avoid accidents in the operation of nuclear reactors and in the production and disposal of nuclear weapons and of †nuclear waste. There are no guarantees of the safety of any of the various methods of disposal. Nuclear safety is a controversial subject – some governments do not acknowledge the hazards of †atomic radiation and radiation sickness.

nuclear accidents Windscale (now Sellafield), Cumbria, England. In 1957 fire destroyed the core of a reactor, releasing large quantities of radioactive fumes into the atmosphere. In 1990 a scientific study revealed an increased risk of leukemia in children whose fathers had worked at Sellafield between 1950 and 1985.

Ticonderoga, 130 km/80 mi off the coast of Japan. In 1965 a US Navy Skyhawk jet bomber fell off the deck of this ship, sinking in 4,900 m/16,000 ft of water. It carried a one-megaton hydrogen bomb. The accident was only revealed in 1989.

Three Mile Island, Harrisburg, Pennsylvania, USA. In 1979 a combination of mechanical and electrical failure, as well as operator error, caused a pressurized water reactor to leak radioactive matter.

Church Rock, New Mexico, USA. In July 1979, 380 million litres/100 million gallons of radioactive water containing uranium leaked from a pond into the Rio Purco, causing the water to become over 6,500 times as radioactive as safety standards allow for drinking water.

Chernobyl, Ukraine. In April 1986 there was an explosive leak, caused by overheating, from a non-pressurized boiling-water reactor, one of the largest in Europe. The resulting clouds of radioactive material spread as far as the UK. Thirty-one people were killed in the explosion, and thousands of square kilometers of land were contaminated by fallout. By June 1992, seven times as many children in the Ukraine and Belarus were contracting thyroid cancer as before the accident, and the incidence of leukemia was rising; it was estimated that more than 6,000 people had died as a result of the accident, and that the death toll in the Ukraine alone would eventually reach 40,000.

Tomsk, Siberia, Russia. In April 1993 a tank exploded at a uranium reprocessing plant, sending a cloud of radioactive particles into the air.

Chernobyl http://www.nea.fr/html

This site examines the disaster at the Chernobyl nuclear reactor in August 1986 and what has happened since the accident. In addition to learning about the events leading up to the accident, you can find out about the health impacts of the radiation leak, and learn how the nuclear reactors at Chernobyl were meant to work. This official document, produced by the OECD (Organization of Economic Cooperation and Development), contains the most recent news regarding the accident and its after-effects.

nuclear testing detonation of nuclear devices to verify their reliability, power, and destructive abilities. Although carried out secretly in remote regions of the world, such tests are easily detected by the seismic shock waves produced. The first tests were carried out in the atmosphere during the 1950s by the USA, the USSR, and the UK. The Comprehensive Test Ban Treaty was signed by 149 countries in 1996. These tests produced large quantities of radioactive fallout, still active today. A test ban in 1963 prohibited all tests except those carried out underground. About 2,000 tests have been carried out since World War II, an average of one every nine days. The USA carried out more than 200 secret nuclear tests over the period 1963–90.

France conducted the first of eight planned underground tests in September 1995 at the Mururoa atoll in French Polynesia. However, after carrying out six of these tests, it announced a "definitive end" to its nuclear-testing program in January 1996; the head of the program admitted that radioactive iodine had leaked into the sea around the Mururoa atoll, but in "insignificant amounts."

During the second half of May 1998 India and Pakistan conducted a combined total of 11 nuclear tests (five by India followed by six by Pakistan), provoking widespread international outrage, led by the "Big Five" nuclear powers – Britain, China, France, Russia, and the USA. A campaign by the Group of Eight industrial powers, launched by forceful US sanctions against India and Pakistan, claimed some success with the announcement in early June by both Delhi and Islamabad of test moratoria. On 12 June the G-8 countries announced a freeze on all non-humanitarian loans to the two countries in protest at their recent tests. The freeze was largely symbolic given the halt to all International Monetary Fund and World Bank lending imposed late May.

High-Energy Weapons Archive: A Guide to Nuclear Weapons http://www.fas.org/nuke/hew/

Very comprehensive, frequently updated guide to nuclear weapons. There are in-depth articles, reports, and many photographs. The pages describe and present data detailing the research, testing, and stockpiling activities of nations with, or developing, nuclear weapons.

nuclear waste the radioactive and toxic by-products of the nuclear energy and nuclear weapons industries. Nuclear waste may have an active life of several thousand years. Reactor waste is of three types: **high-level** spent fuel, or the residue when nuclear fuel has been removed from a reactor and reprocessed; **intermediate**, which may be long-or short-lived; and **low-level**, but bulky, waste from reactors, which has only short-lived radioactivity. Disposal, by burial on land or at sea, has raised problems of safety, environmental pollution, and security.

nucleic acid complex organic acid made up of a long chain of †nucleotides, present in the nucleus and sometimes the cytoplasm of the living cell. The two types, known as †DNA (deoxyribonucleic acid) and †RNA (ribonucleic acid), form the basis of heredity. The nucleotides are made up of a sugar (deoxyribose or ribose), a phosphate group, and one of four purine or pyrimidine bases. The order of the bases along the nucleic acid strand contains the genetic code.

nucleolus in biology, an RNA-rich structure found in the nucleus of eukaryotic cells. It produces the RNA that makes up the †ribosomes, from instructions in the DNA.

nucleon in particle physics, either a †proton or a †neutron, when present in the atomic nucleus. **Nucleon**

number is an alternative name for the †mass number of an atom.

nucleon number or **mass number** (symbol *A*), the sum of the numbers of protons and neutrons in the nucleus of an atom. With the †proton number, it is used in nuclear notation – for example, in the symbol $^{14}{}_{6}C$ representing the isotope carbon-14, the lower number is the proton number, and the upper is the nucleon number.

nucleophilic describing an atom, molecule, or ion that seeks a positive center (for example, the atomic nucleus) during a chemical reaction. Nucleophiles react at centers of low electron density because they have electron pairs available for bonding. Common nucleophiles include the hydroxide ion (OH^-), ammonia (NH_3), water (H_2O), and halide anions. See also †electrophilic.

nucleosynthesis cosmic production of all the species of chemical elements by large-scale nuclear reactions, such as those in progress in the Sun or other stars. One element is changed into another by reactions that change the number of protons or neutrons involved.

nucleotide organic compound consisting of a purine (adenine or guanine) or a pyrimidine (thymine, uracil, or cytosine) base linked to a sugar (deoxyribose or ribose) and a phosphate group. †DNA and †RNA are made up of long chains of nucleotides.

nucleus in astronomy, compact central core of a †galaxy, often containing powerful radio, X-ray, and infrared sources. Active galaxies have extremely energetic nuclei.

nucleus in biology, the central, membrane-enclosed part of a eukaryotic cell, containing threads of †DNA. It is found in both plant and animal cells. During cell division the threads of DNA coil up to form †chromosomes. The nucleus controls the function of the cell by determining which proteins are produced within it. It is where inherited information (see †inheritance) is stored as †genes. Because proteins are the chief structural molecules of living matter and, as enzymes, regulate all aspects of metabolism, it may be seen that the genetic code within the nucleus is effectively responsible for building and controlling the whole organism.

The nucleus contains the **nucleolus**, the part of the cell where ribosomes are produced. Movement of molecules into and out of the nucleus occurs through the nuclear pores. An average mammalian nucleus has approximately 3,000 pores.

Nucleus http://www.webref.org/chemistry/n/nucleus.htm

Short, sharp definition of a nucleus. This site is good for revising the very basics of atoms. It contains links to further definitions of structures mentioned, such as electrons, protons, and neutrons.

nucleus plural **nuclei**, in physics, the positively-charged central part of an †atom, which constitutes almost all its mass. Except for hydrogen nuclei, which have only one proton, nuclei are composed of both protons and neutrons. Surrounding the nucleus are electrons, of equal and opposite charge to that of the protons, thus giving the atom a neutral charge. Nuclei that are unstable may undergo †radioactive decay or nuclear †fission. In all stars, including our Sun, small nuclei join together to make more stable, larger nuclei. This process is called nuclear †fusion.

The nucleus was discovered by the New Zealand-born British physicist Ernest Rutherford in 1911 as a result of experiments in firing alpha particles through very thin gold foil.

nuclide in physics, a species of atomic nucleus characterized by the number of protons (*Z*) and the number of neutrons (*N*). Nuclides with identical †proton number but differing †neutron number are called †isotopes.

nuée ardente rapidly flowing, glowing white-hot cloud of ash and gas emitted by a volcano during a violent eruption. The ash and other pyroclastics in the lower part of the cloud behave like an ash flow. In 1902 a *nuée ardente* produced by the eruption of Mount Pelee in Martinique swept down the volcano in a matter of seconds and killed 28,000 people in the nearby town of St Pierre.

nugget piece of gold found as a lump of the †native metal. Nuggets occur in †alluvial deposits where river-borne particles of the metal have adhered to one another.

null hypothesis in probability theory, assumes that events occur on a purely random basis.

null set conventional †set that has no members.

null vector another name for a zero †vector.

number symbol used in counting or measuring. In mathematics, there are various kinds of number. The everyday number system is the decimal ("proceeding by tens") system, using the base ten.

†**Real numbers** include all rational numbers (†integers, or whole numbers, and †fractions) and †irrational numbers (those not expressible as fractions). †**Complex numbers** include the real and imaginary numbers (real-number †multiples of the †square root of –1). The †**binary number system**, used in computers, has two as its base. The †**natural numbers**, 0, 1, 2, 3, 4, 5, 6, 7, 8, and 9, give a counting system that, in the †**decimal number system**, continues 10, 11, 12, 13, and so on. These are whole numbers

(integers), with fractions represented as, for example, $^1/_4$, $^1/_2$, $^3/_4$, or as decimal fractions (0.25, 0.5, 0.75). They are also **rational numbers**.

Irrational numbers cannot be represented in this way and require symbols, such as Ö2, p, and e. They can be expressed numerically only as the (inexact) approximations 1.414, 3.142, and 2.718 (to three places of decimals), respectively. The symbols p and e are also examples of **transcendental numbers**, because they (unlike Ö2) cannot be derived by solving a †polynomial equation (an equation with one †variable quantity) with rational †coefficients (multiplying †factors). Complex numbers, which include the real numbers as well as imaginary numbers, take the general form $a + bi$, where i = Ö–1 (that is, $i^2 = -1$), and a is the real part and bi the imaginary part.

evolution of number systems The ancient Egyptians, Greeks, Romans, and Babylonians all evolved number systems, although none had a †zero, which was introduced from India by way of Arab mathematicians in about the 8th century and allowed a place-value system to be devised on which the decimal system is based. Other number systems have since evolved and have found applications. For example, numbers to base two (binary numbers), using only 0 and 1, are commonly used in digital computers to represent the two-state "on" or "off" pulses of electricity. Binary numbers were first developed by German mathematician Gottfried Leibniz in the late 17th century.

Arabic Numerals http://www.islam.org/Mosque/ihame/Ref6.htm

Account of how numbers developed from their ancient Indian origins to the Modern Arabic numerals that are generally used today. There is a brief description of numbering systems used by the ancient Egyptians, Greeks, and Romans, and some of the problems these systems presented. The page also includes links to aspects of Islamic and Middle Eastern history.

First Use of Zero and Negative Numbers http://www.maxmon.com/300bc.htm

Page that describes how the concept of negative numbers became part of mathematical theory.

number line in mathematics, a way of displaying consecutive numbers diagrammatically. A number line shows †negative numbers to the left of †zero and **positive numbers** to the right of zero, extending in each direction to †infinity. A number line can be used to show †integers, †fractions, and/or †decimal fractions.

number theory in mathematics, the abstract study of the structure of number systems and the properties of positive integers (whole numbers). For example, the theories of factors and prime numbers fall within this area, as does the work of mathematicians Giuseppe Peano (1858–1932), Pierre de Fermat, and Karl Gauss.

numerator number or symbol that appears above the line in a vulgar fraction. For example, the numerator of 5/6 is 5. The numerator represents the fraction's dividend and indicates how many of the equal parts indicated by the denominator (number or symbol below the line) comprise the fraction.

numerical analysis branch of mathematics that is concerned with finding numerical answers to problems and giving estimates for the accuracy of such methods.

nunatak mountain peak protruding through an ice sheet. Such peaks are common in Antarctica.

nut any dry, single-seeded fruit that does not split open to release the seed, such as the chestnut. A nut is formed from more than one carpel, but only one seed becomes fully formed, the remainder aborting. The wall of the fruit, the pericarp, becomes hard and woody, forming the outer shell.

Examples of true nuts are the acorn and hazelnut. The term also describes various hard-shelled fruits and seeds, including almonds and walnuts, which are really the stones of †drupes, and brazil nuts and shelled peanuts, which are seeds. The kernels of most nuts provide a concentrated, nutritious food, containing vitamins, minerals, and enzymes, about 50% fat, and 10–20% protein, although a few, such as chestnuts, are high in carbohydrates and have only a moderate protein content of 5%. Nuts also provide edible and industrial oils. Most nuts are produced by perennial trees and shrubs. Whereas the majority of nuts are obtained from plantations, considerable quantities of pecans and brazil nuts are still collected from the wild. World production in the mid-1980s was about 4 million tonnes per year.

nutation in astronomy, slight "nodding" of the Earth in space, caused by the varying gravitational pulls of the Sun and Moon. Nutation changes the angle of the Earth's axial tilt (average 23.5°) by about 9 seconds of arc to either side of its mean position, a complete cycle taking just over 18.5 years.

nutation in botany, the spiral movement exhibited by the tips of certain stems during growth; it enables a climbing plant to find a suitable support. Nutation sometimes also occurs in tendrils and flower stalks. The direction of the movements, clockwise or anticlockwise, is usually characteristic for particular species.

nutrient member of one of the six classes of substances required for growth and living processes. The classes are †water, †carbohydrates, †lipids, †proteins, †vitamins, and †minerals.

Some nutrients are needed in quite large amounts (**macronutrients**), for example calcium for the formation and maintenance of the human skeleton. **Micronutrients** are needed in much smaller amounts, for example iron as a component of molecules of the blood pigment hemoglobin.

nutrient cycle transfer of nutrients from one part of an †ecosystem to another. Trees, for example, take up nutrients such as calcium and potassium from the soil through their root systems and store them in leaves. When the leaves fall they are decomposed by bacteria and the nutrients are released back into the soil where they become available for root uptake again.

nutrition strategy adopted by an organism to obtain the chemicals it needs to live, grow, and reproduce. These chemicals are nutrients that are absorbed from the environment, such as †mineral salts, or chemicals made inside the body of a plant or animal. Nutrition is a term also used to describe the science of food, and its effect on human and animal life, health, and disease. Nutrition involves the study of the basic nutrients required to sustain life, their bioavailability in foods and overall diet, and the effects upon them of cooking and storage. It is also concerned with dietary deficiency diseases.

Plant nutrition is very different from animal nutrition in several ways. In a plant high-energy food (for example, †carbohydrate) is made inside the plant by †photosynthesis. An animal obtains high-energy food by eating a plant or another animal. Nutrients from the environment are absorbed by different parts of plant and animal bodies. In a plant, nutrients are absorbed from soil water by the roots (see †root hair cell). In an animal, nutrients are absorbed from the gut by the †epithelium cells lining it.

Nyholm, Ronald Sydney (1917–1971) Australian inorganic chemist who worked on the coordination compounds (complexes) of the transition metals. He was knighted in 1967.

nylon synthetic long-chain polymer similar in chemical structure to protein. Nylon was the first fully synthetic fiber. Made from petroleum, natural gas, air, and water, nylon was developed in 1935 by the US chemist W H Carothers and his associates, who worked for Du Pont. It is used in the manufacture of molded articles, textiles, and medical sutures. Nylon fibers are stronger and more elastic than silk and are relatively insensitive to moisture and mildew. Nylon is used for a wide range of different textiles including carpets, and can be used in knitting or weaving. It is also used for simulating other fabrics such as silks and furs.

nymph in entomology, the immature form of insects that do not have a pupal stage; for example, grasshoppers and dragonflies. Nymphs generally resemble the adult (unlike larvae), but do not have fully formed reproductive organs or wings.

O

oasis area of land made fertile by the presence of water near the surface in an otherwise arid region. The occurrence of oases affects the distribution of plants, animals, and people in the desert regions of the world.

Oberon or **Uranus IV**, one of the five major moons of †Uranus, discovered by German-born English astronomer William †Herschel in 1789. It is 1,524 km/947 mi in diameter and orbits Uranus at an average distance of 584,000 km/363,000 mi with a period (time taken to circle the planet) of 13.5 days. Impact craters show fresh, light-colored ice beneath its dark surface.

objective system of lenses in a telescope or microscope nearest to the object being observed.

object program in computing, the †machine code translation of a program written in a †source language.

object relations in psychoanalysis, the emotional relations between subject and object which, through a process of identification, are believed to constitute the developing ego. In this context, the word "object" refers to any person or thing, or representational aspect of them, with which the subject forms an intense emotional relationship.

Object relations were first described by German psychoanalyst Karl Abraham (1877–1925) in an influential paper, published in 1924, in which he developed Freud's ideas on infantile sexuality and the development of the †libido. Object-relations theory has become one of the central themes of post-Freudian psychoanalysis, particularly through the writings of Melanie †Klein, Scottish psychoanalyst Ronald Fairbairn (1889–1964), and Donald †Winnicott, all deeply influenced by Abraham. They have each developed distinct, though complementary, approaches to analysis, evolving theories of personal development based on early parental attachments.

observation in science, the perception of a phenomenon – for example, examining the Moon through a telescope, watching mice to discover their mating habits, or seeing how a plant grows.

Traditionally, observation was seen as entirely separate from theory, free from preconceptions and therefore lending support to the idea of scientific objectivity. However, as the preceding examples show, observations are ordered according to a pre-existing theory; for instance, one cannot observe mating behavior without having decided what mating behavior might look like. In addition, many observations actually affect the behavior of the observed (for instance, of mating mice).

observation sheet in statistics, specially prepared record sheet for experiments, including space for tally marks and for total frequency.

observatory site or facility for observing astronomical or meteorological phenomena. The modern observatory dates from the invention of the telescope. Observatories may be ground-based, carried on aircraft, or sent into orbit as satellites, in space stations, and on the space shuttle.

The erection of observatories was revived in West Asia about AD 1000, and extended to Europe. The observatory built on the island of Hven (now Ven) in Denmark in 1576 for Tycho Brahe (1546–1601) was elaborate, but survived only to 1597. Later, observatories were built in Paris in 1667, Greenwich (the Royal Greenwich Observatory) in 1675, and Kew, England. Most early observatories were near towns, but with the advent of big telescopes, clear skies with little background light, and hence high, remote sites, became essential.

The most powerful optical telescopes covering the sky are at Mauna Kea, Hawaii; Mount Palomar, California; Kitt Peak National Observatory, Arizona; La Palma, Canary Islands; Cerro Tololo Inter-American Observatory, and the European Southern Observatory, Chile; Siding Spring Mountain, Australia; and Zelenchukskaya in the Caucasus.

Radio astronomy observatories include Jodrell Bank, Cheshire, England; the Mullard Radio Astronomy Observatory, Cambridge, England; Arecibo, Puerto Rico; Effelsberg, Germany; and Parkes, Australia. The Hubble Space Telescope was launched into orbit in 1990. The Very Large Telescope, operated by the European Southern Observatory in the mountains of northern Chile, at the Cerro Paranal Space Observatory, is the largest optical telescope in the world.

obsessive-compulsive disorder **OCD**, in psychiatry, anxiety disorder that manifests itself in the need to check constantly that certain acts have been performed "correctly." Sufferers may, for example, feel compelled to wash themselves repeatedly or return home again and again to check that doors have been locked and appliances switched off. They may also hoard certain objects and insist in these being arranged in a precise way, or be troubled by intrusive and unpleasant thoughts. In extreme cases, normal life is disrupted through the hours devoted to compulsive actions. Treatment involves †cognitive therapy and drug therapy with serotonin-blocking drugs such as Prozac.

There are approximately 1 million OCD sufferers in Britain and 5 million in the USA.

Obsessive-Compulsive Disorder http://mentalhelp.net/poc/center_index.php?id=6

Comprehensive resource on this psychiatric disorder. As well as containing a list of symptoms, the site explains in detail the problems an OCD sufferer might face, and discusses the kind of treatments available.

obsidian black or dark-colored glassy volcanic rock, chemically similar to †granite, but formed by cooling rapidly on the Earth's surface at low pressure.

The glassy texture is the result of rapid cooling, which inhibits the growth of crystals. Obsidian was valued by the early civilizations of Mexico for making sharp-edged tools and ceremonial sculptures.

obsidian hydration-rim dating in archaeology, a method of dating artifacts made from the volcanic glass obsidian. Water molecules absorbed by inward diffusion through cut surfaces cause the outer areas of an obsidian article to convert to the mineral perlite. An object may be dated by measuring the thickness of this perlite – the **hydration** (combined with water) rim.

Only a molecule-thick water film is required at the surface to maintain the process, an amount available even in the near-arid zones of Egypt. Temperature, sunlight, and different chemical compositions cause variation in the hydration rate; therefore the method needs to be calibrated against an established chronological sequence for absolute dating. The method has been applied to many periods, including the Aztec age in Mexico, the pre-ceramic era of Japan (about 23,000 BC), and the tribal-war periods of Easter Island before the arrival of traders and missionaries in 1722. Obsidian hydration-rim dating was stimulated by the early research of Irving Friedman of the US Geological Survey from 1955.

obtuse angle angle greater than 90° but less than 180°.

occluded front weather †front formed when a cold front catches up with a warm front. It brings clouds and rain as air is forced to rise upwards along the front, cooling and condensing as it does so.

occultation in astronomy, the temporary obscuring of a star by a body in the Solar System. Occultations are used to provide information about changes in an orbit, and the structure of objects in space, such as radio sources.

The exact shapes and sizes of planets and asteroids can be found when they occult stars. The rings of Uranus were discovered when that planet occulted a star in 1977.

occupational psychology study of human behavior at work. It includes dealing with problems in organizations, advising on management difficulties, and investigating the relationship between humans and machines (as in the design of aircraft controls; see also †ergonomics). Another area is †psychometrics and the use of assessment to assist in selection of personnel.

ocean great mass of salt water. Geographically speaking three oceans exist – the Atlantic, Indian, and Pacific – to which the Arctic is often added. They cover approximately 70% or 363,000,000 sq km/140,000,000 sq mi of the total surface area of the Earth. According to figures released in August 2001, the total volume of the world's oceans is 1,370 million cubic km/329 million cubic mi. Water levels recorded in the world's oceans have shown an increase of 10–15 cm/4–6 in over the past 100 years.

depth (average) 3,660 m/12,000 ft, but shallow ledges (continental shelves) 180 m/600 ft run out from the continents, beyond which the continental slope reaches down to the †abyssal zone, the largest area, ranging from 2,000–6,000 m/6,500–19,500 ft. Only the †deep-sea trenches go deeper, the deepest recorded being 11,034 m/36,201 ft (by the *Vityaz*, USSR) in the Mariana Trench of the western Pacific in 1957

features deep trenches (off eastern and southeastern Asia, and western South America), volcanic belts (in the western Pacific and eastern Indian Ocean), and ocean ridges (in the mid-Atlantic, eastern Pacific, and Indian Ocean).

temperature varies on the surface with latitude (–2°C/28°F to +29°C)/84°F; decreases rapidly to 370 m/1,200 ft, then more slowly to 2,200 m/7,200 ft; and hardly at all beyond that

seawater contains about 3% dissolved salts, the most abundant being †sodium chloride; salts come from the weathering of rocks on land; rainwater flowing over rocks, soils, and organic matter on land dissolves small amounts of substances, which pass into rivers to be carried to the sea. Salt concentration in the oceans remains remarkably constant as water is evaporated by the Sun and fresh water added by rivers. Positive ions present in sea water include sodium, magnesium, potassium, and calcium; negative ions include chloride, sulfate, hydrogencarbonate, and bromide

commercial extraction of minerals includes bromine, magnesium, potassium, salt (sodium chloride); those potentially recoverable include aluminum, calcium, copper, gold, manganese, silver.

pollution Oceans have always been used as a dumping area for human waste, but as the quantity of waste increases, and land areas for dumping diminish, the problem is exacerbated. Today ocean pollutants include airborne emissions from land (33% by weight of total marine pollution); oil from both shipping and land-based sources; toxins from industrial, agricultural, and domestic uses; sewage; sediments from mining, forestry, and farming; plastic litter; and radioactive isotopes. Thermal pollution by cooling water from power plants or other industry is also a problem, killing coral and other temperature-sensitive sedentary species.

ocean current fast-flowing body of seawater forced by the wind or by variations in water density (as a result of temperature or salinity variations) between two areas. Ocean currents are partly responsible for transferring heat from the Equator to the poles and thereby evening out the global heat imbalance.

oceanic ridge mountain range that extends along the center of the bed of an ocean; see †mid-ocean ridge.

oceanography study of the oceans. Its subdivisions deal with each ocean's extent and depth, the water's evolution and composition, its physics and chemistry, the bottom topography, currents and wind, tidal ranges, biology, and the various aspects of human use. Computer simulations are widely used in oceanography to plot the possible movements of the waters, and many studies are carried out by remote sensing.

Oceanography involves the study of water movements – currents, waves, and tides – and the chemical and physical properties of the seawater. It deals with the origin and topography of the ocean floor – ocean trenches and ridges formed by †plate tectonics, and continental shelves from the submerged portions of the continents.

OceanLink http://oceanlink.island.net/index.html

Good site for finding out about many marine animals, with encyclopedic descriptions of selected animals. This site contains sections called "Kidlinks," "Ask a scientist," "Spotlight on...," and "Ocean news."

Ocean Planet http://seawifs.gsfc.nasa.gov/ocean_planet.html

Oceans and the environmental issues affecting their health, based on the Smithsonian Institute's traveling exhibition of the same name. Use the exhibition floor plan to navigate around the different "rooms" – with themes ranging from Ocean Science and Immersion to Heroes and Sea People.

Woods Hole Oceanographic Institution Home Page http://www.whoi.edu/index.html

Site run by a Massachusetts-based oceanographic institute. As well as containing details of their research programs and an overview of the organization, there is a gallery of marine pictures and videos, and contacts for their education programs.

ocean trench submarine valley. Ocean trenches are characterized by the presence of a volcanic arc on the concave side of the trench. Trenches are now known to be related to subduction zones, places where a plate of oceanic †lithosphere dives beneath another plate of either oceanic or continental lithosphere. Ocean trenches are found around the edge of the Pacific Ocean and the northeastern Indian Ocean; minor ones occur in the Caribbean and near the Falkland Islands.

Ocean trenches represent the deepest parts of the ocean floor, the deepest being the †Mariana Trench which has a depth of 11,034 m/36,201 ft. At depths of below 6 km/3.6 mi there is no light and very high pressure; ocean trenches are inhabited by crustaceans, coelenterates (for example, sea anemones), polychaetes (a type of worm), mollusks, and echinoderms.

Ochoa, Severo (1905–1993) Spanish-born US biochemist who was awarded a Nobel Prize for Physiology or Medicine in 1959 for the discovery of enzymes that catalyze the formation of RNA (ribonucleic acid), in 1955, and DNA (deoxyribonucleic acid).

OCR abbreviation for †optical character recognition.

octagon †polygon with eight sides. A regular octagon has all eight sides of equal length, and all eight angles of equal size, 135°.

octahedral in chemistry, describing the most stable spatial arrangement of six identical atoms or groups that are bonded to the same atom. The atoms are arranged such that they are at the corners of a regular octahedron, for example sulfur hexafluoride, SF_6.

octahedron, regular regular solid with eight faces, each of which is an equilateral triangle. It is one of the five regular polyhedra or Platonic solids. The figure made by joining the midpoints of the faces is a perfect cube and the vertices of the octahedron are themselves the midpoints of the faces of a surrounding cube. For this reason, the cube and the octahedron are called dual solids.

octal number system number system to the base eight, used in computing. The highest digit that can appear in the octal system is seven. Whereas normal decimal, or base-ten, numbers may be considered to be written under column headings based on the number ten, octal, or base-eight, numbers can be thought of as written under column headings based on the number eight. The octal number 567 is therefore equivalent to the decimal number 375, since (5 x 64) + (6 x 8) + (7 x 1) = 375.

The octal number system is sometimes used by computer programmers as an alternative to the †hexadecimal number system.

octane in chemistry, any of 18 isomers of the saturated aliphatic hydrocarbon iso-octane, C_8H_{16}. Iso-octane is the most important member of the series and is used as an anti-knocking agent in gasoline and as a standard for the measurement of knocking rating in fuels.

octane rating numerical classification of petroleum fuels indicating their combustion characteristics. The efficient running of an †internal combustion engine depends on the ignition of a gasoline–air mixture at the correct time during the cycle of the engine. Higher-rated gasoline burns faster than lower-rated fuels. The use of the correct grade must be matched to the engine.

Octans faint constellation in the southern hemisphere, represented as an octant. It contains the southern celestial pole. The closest naked-eye star to the south celestial pole is fifth-magnitude Sigma Octantis.

octet rule in chemistry, rule stating that elements combine in a way that gives them the electronic structure of the nearest †noble gas (rare gas). The joining together of elements to form compounds is called bonding. All the noble gases except helium have eight electrons in their outermost shell, hence the term octet.

odd number any number not divisible by 2, thus odd numbers form the infinite sequence 1, 3, 5, 7 ... . Every square number n^2 is the sum of the first n odd numbers. For example, $49 = 7^2 = 1 + 3 + 5 + 7 + 9 + 11 + 13$.

Oedipus complex in psychology, the unconscious antagonism of a son to his father, whom he sees as a rival for his mother's affection. For a girl antagonistic to her mother as a rival for her father's affection, the term is **Electra complex**. The terms were coined by Sigmund †Freud.

Freud saw this as a universal part of childhood development, which in most children is resolved during late childhood. Contemporary theory places less importance on the Oedipus/Electra complex than did Freud and his followers.

oersted symbol Oe, †*c*.g.s. unit of †magnetic field strength, now replaced by the SI unit ampere per meter. The Earth's magnetic field is about 0.5 oersted; the field near the poles of a small bar magnet is several hundred oersteds; and a powerful †electromagnet can have a field strength of 30,000 oersteds.

Oersted, Hans Christian (1777–1851) Danish physicist who founded the science of electromagnetism. In 1820 he discovered the †magnetic field associated with an electric current. In 1825 he was the first to isolate aluminum.

Ogilby, John (1600–1676) English map-maker. He produced the first road map in Britain, *Britannia* (1675), which showed major roads in strip form. These maps were freshly surveyed, and included accurate details of bridges, rivers, and towns. Ogilby concentrated on roads radiating out from London. His atlas was also an innovation in using the statute mile, introduced in 1593 but not in widespread use at that time. His maps remained the standard for almost 100 years.

ogive in statistics, the curve on a graph representing †cumulative frequency.

ohm symbol W, SI unit of electrical †resistance (the property of a conductor that restricts the flow of electrons through it).

It was originally defined with reference to the resistance of a column of mercury, but is now taken as the resistance between two points when a potential difference of one volt between them produces a current of one ampere.

Ohm, Georg Simon (1789–1854) German physicist who studied electricity and discovered the fundamental law that bears his name (see †Ohm's law). The SI unit of electrical resistance, the **ohm**, is named after him, and the unit of conductance (the inverse of resistance) was formerly called the **mho**, which is "ohm" spelled backwards.

Ohm's law law that states that, for many materials over a wide range of conditions, the current flowing in a conductor maintained at constant temperature is directly proportional to the potential difference (voltage) between its ends. The law was discovered by German physicist Georg Ohm in 1827. He found that if the voltage across a conducting material is changed, the current flow through the material is changed proportionally (for example, if the voltage is doubled then the current also doubles).

If a current of I amperes flows between two points in a conductor across which the potential difference is V volts, then V/I is a constant called the †resistance R ohms between those two points. Hence: $V/I = R$ or $V = IR$. Not all conductors obey Ohm's law; those that do are called **ohmic conductors**.

oil flammable substance, usually insoluble in water, and composed chiefly of carbon and hydrogen. Oils may be solids (fats and waxes) or liquids. Various plants produce vegetable oils; mineral oils are based on petroleum.

The crude oil (unrefined †petroleum) found beneath the Earth's surface is formed from the remains of dead plants and animals. As these plants and animals died they were buried with mud near the sea floor. Over millions of years, heat from the Earth's interior and pressure from overlying rocks slowly changed the dead remains into hydrocarbons (substances containing hydrogen and carbon). The hydrocarbons, being light molecules, moved upwards and became trapped beneath impermeable rocks.

Oil reservoirs are often found beneath the seabed and drilling technology is used to locate these supplies of oil. Crude oil extracted from the ground is refined in †fractional distillation columns to produce more useful products such as gasoline, diesel, kerosene, plastics, and chemicals for pharmaceuticals.

What Is Crude Oil? http://www.chevron.com/learning_center/crude/

Beginning with a short history of the uses of crude oil, the text quickly moves on to an explanation of how crude oil is formed over millions of years. An explanation of more contemporary uses of oil, such as in the gasoline industry, is then given. Spaced around the page are pictures of samples of oil from all around the world.

oil spill oil discharged from an ocean-going tanker, pipeline, or oil installation, often as a result of damage. An oil spill kills shore life, clogging the feathers of birds and suffocating other creatures. At sea, toxic chemicals spread into the water below, poisoning sea life. Mixed with dust, the oil forms globules that sink to the seabed, poisoning sea life there as well. Oil spills are broken up by the use of detergents but such chemicals can themselves damage wildlife. The annual spillage of oil is 8 million barrels (280 million gallons) a year. At any given time tankers are carrying 500 million barrels (17.5 billion gallons). The amount of oil entering oceans from shipping operations decreased by 60% 1981–91.

In March 1989 the *Exxon Valdez* (belonging to the Exxon Corporation) ran aground and spilled oil in Alaska's Prince William Sound, covering 12,400 sq km/4,800 sq mi and killing at least 34,400 sea birds, 10,000 sea otters, and up to 16 whales. The incident led to the US Oil Pollution Act of 1990, which requires tankers operating in US waters to have double hulls.

The world's largest oil spill was in the Gulf in 1991 as a direct result of hostilities during the Gulf War. Around 6–8 million barrels (210–280 million gallons) of oil were spilled, polluting 675 km/420 mi of Saudi coastline. In some places, the oil was 30 cm/12 in deep in the sand.

Olah, George Andrew (1927–) Hungarian-born US chemist who was awarded the Nobel Prize for Chemistry in 1994 for his isolation of carbocations, electrically charged fragments of hydrocarbon molecules (molecules containing only hydrogen and carbon). His work launched a new branch of organic chemistry and led to the development of new carbon-based fuels.

Olbers, Heinrich Wilhelm Matthäus (1758–1840) German astronomer. He was a founder member of the †**Celestial Police**, a group of astronomers who attempted to locate a supposed missing planet between Mars and Jupiter. During his search he discovered two †asteroids, Pallas in 1802 and Vesta in 1807. Also credited to Olbers are a number of comet discoveries, a new method of calculating cometary orbits, and the stating of Olbers' paradox.

Olbers's paradox question put forward in 1826 by German doctor, mathematician, and astronomer Heinrich Olbers, who asked: If the universe is infinite in extent and filled with stars, why is the sky dark at night? The answer is that the stars do not live infinitely long, so there is not enough starlight to fill the universe. A wrong answer, frequently given, is that the expansion of the universe weakens the starlight.

olefin common name for †alkene.

Oligocene epoch third epoch of the Tertiary period of geological time, 35.5–3.25 million years ago. The name, from Greek, means "a little recent," referring to the presence of the remains of some modern types of animals existing at that time.

oligodendrocyte type of glial cell (see †glia) surrounding the neurons in the brain and spinal cord. They produce the insulating †myelin sheath surrounding the nerve axon.

oligomer †polymer that is composed of only a few †monomer units.

oligosaccharide †carbohydrate comprising a few †monosaccharide units linked together. It is a general term used to indicate that a carbohydrate is larger than a simple di- or trisaccharide but not as large as a polysaccharide.

oligotrophic in ecology, describing ponds, lakes, and other bodies of still water that lack in nutrients.

Oliphant, Mark (Marcus Laurence Elwin) (1901–2000) Australian nuclear physicist, who from 1927 worked in Cambridge, England. He was involved in the development of radar, which was a device used by British pilots during air-raids of World War II. He was also involved in the Manhattan Project, which raced to develop the atomic bomb during World War II. After the bomb was dropped on Hiroshima, Oliphant worked to promote the peaceful uses of nuclear energy.

olivenite basic copper arsenate, $Cu_2AsO_4(OH)$, occurring as a mineral in olive-green prisms.

olivine greenish mineral, magnesium iron silicate, $(Mg,Fe)_2SiO_4$. It is a rock-forming mineral, present in, for example, peridotite, gabbro, and basalt. Olivine is called **peridot** when pale green and transparent, and used in jewelry.

Olympus Mons volcano in the Tharsis region of †Mars, 600 km/375 mi in diameter and rising 24 km/15 mi above the surrounding plains. The caldera at the summit is 70 km/44 mi across and contains a complex of smaller craters. Olympus Mons is believed to have been extinct for several hundred million years. It is one of several large volcanoes in the Tharsis region, and is the largest known volcano in the Solar System.

omega symbol ω, last letter of the Greek alphabet used in astronomy as a symbol for the mass density of the universe. If w is less than 1.0 the universe will expand forever; if it is more than 1.0 the gravitational pull of its mass will be strong enough to reverse its expansion and cause its eventual collapse. The value of ω is estimated as probably being between 0.1 and 1.0. In physics w stands for angular frequency.

omnivore animal that feeds on both plant and animal material. Omnivores have digestive adaptations intermediate between those of †herbivores and †carnivores, with relatively unspecialized digestive systems and gut micro-organisms that can digest a variety of foodstuffs. Omnivores include humans, the chimpanzee, the cockroach, and the ant.

OMR abbreviation for †optical mark recognition.

oncogene gene carried by a virus that induces a cell to divide abnormally, giving rise to a cancer. Oncogenes arise from mutations in genes (proto-oncogenes) found in all normal cells. They are usually also found in viruses that are capable of transforming normal cells to tumor cells. Such viruses are able to insert their oncogenes into the host cell's DNA, causing it to divide uncontrollably. More than one oncogene may be necessary to transform a cell in this way.

In 1989 US scientists J Michael Bishop and Harold Varmus were jointly awarded the Nobel Prize for Physiology or Medicine for their concept of oncogenes, although credit for the discovery was claimed by a French cancer specialist, Dominique Stehelin.

onco-mouse mouse that has a human †oncogene (gene that can cause certain cancers) implanted into its cells by genetic engineering. Such mice are used to test anticancer treatments and were patented within the USA by Harvard University in 1988, thereby protecting its exclusive rights to produce the animal and profit from its research.

one-one in mathematics, function from one set to another if no two distinct elements are ever mapped to the same element. Equivalently, $f{:}A \rightarrow B$ is one-one if $f(x) = f(y)$ for some x, $y \varepsilon A$ implies $x = y$.

one-one correspondence in mathematics, between two sets, a function that is both one-one and †onto.

online in computing, connected, so that data can be transferred, for example, to a printer or from a network. The opposite of offline.

online system in computing, originally a system that allows the computer to work interactively with its users, responding to each instruction as it is given and prompting users for information when necessary. As almost all the computers used now work this way, "online system" is now used to refer to large database, e-mail, and conferencing systems accessed via a dial-up modem. These often have tens or hundreds of users from different places – sometimes from different countries – "online" at the same time.

Onsager, Lars (1903–1976) Norwegian-born US physical chemist. He worked on the application of the laws of †thermodynamics to systems not in equilibrium, and was awarded the Nobel Prize for Chemistry in 1968 for his discovery of reciprocal relations, fundamental for the thermodynamics of irreversible processes. In 1949 he established a firm statistical basis for the theory of liquid crystals.

onto in mathematics, function from one set to another if every element of the second set has some element of the first set mapped to it.

ontogeny process of development of a living organism, including the part of development that takes place after hatching or birth. The idea that "ontogeny recapitulates phylogeny" (the development of an organism goes through the same stages as its evolutionary history), proposed by the German scientist Ernst Heinrich Haeckel, is now discredited.

Onychophora phylum of invertebrates found in rainforest leaf-litter. See †velvet worm.

onyx semi-precious variety of chalcedonic †silica (SiO_2) in which the crystals are too fine to be detected under a microscope, a state known as cryptocrystalline. It has straight parallel bands of different colors: milk-white, black, and red.

Sardonyx, an onyx variety, has layers of brown or red carnelian alternating with lighter layers of onyx. It can be carved into cameos.

oocyte in medicine, an immature ovum. Only a fraction of the oocytes produced in the ovary survive until puberty and not all of these undergo meiosis to become an ovum that can be fertilized by a sperm.

oolite limestone made up of tiny spherical carbonate particles, called **ooliths**, cemented together. Ooliths have a concentric structure with a diameter up to 2 mm/0.08 in. They were formed by chemical precipitation and accumulation on ancient sea floors.

The surface texture of oolites is rather like that of fish roe. The late Jurassic limestones of the British Isles are mostly oolitic in nature.

Oort, Jan Hendrik (1900–1992) Dutch astronomer. In 1927 he calculated the mass and size of our Galaxy, the †Milky Way, and the Sun's distance from its center, from the observed movements of stars around the Galaxy's center. In 1950 Oort proposed that comets exist in a vast swarm, now called the †Oort cloud, at the edge of the Solar System.

Oort cloud spherical cloud of comets beyond Pluto, extending out to about 100,000 astronomical units (approximately one light year) from the Sun. The gravitational effect of passing stars and the rest of our Galaxy disturbs comets from the cloud so that they fall in towards the Sun on highly elongated orbits, becoming visible from Earth. As many as 10 trillion comets may reside in the Oort cloud, named after Dutch astronomer Jan Oort who postulated its existence in 1950.

oosphere another name for the female gamete, or †ovum, of certain plants such as algae.

ooze sediment of fine texture consisting mainly of organic matter found on the ocean floor at depths greater than 2,000 m/6,600 ft. Several kinds of ooze exist, each named after its constituents.

Siliceous ooze is composed of the †silica shells of tiny marine plants (diatoms) and animals (radiolarians). **Calcareous ooze** is formed from the †calcite shells of microscopic animals (foraminifera) and floating algae (coccoliths).

opal form of hydrous †silica ($SiO_2.nH_2O$), often occurring as stalactites and found in many types of rock. The common opal is translucent, milk-white, yellow, red, blue, or green, and lustrous. Precious opal is opalescent, the characteristic play of colors being caused by close-packed silica spheres diffracting light rays within the stone.

Opal is cryptocrystalline, that is, the crystals are too fine to be detected under an optical microscope. Opals are found in Hungary; New South Wales, Australia (black opals were first discovered there in 1905); and Mexico (red fire opals).

Oparin, Alexandr Ivanovich (1894–1980) Russian biochemist who in the 1920s developed one of the first of the modern theories about the origin of life on Earth, postulating a primeval soup of biomolecules.

opencast mining or **open-pit mining** or **strip mining**, mining at the surface rather than underground. Coal, iron ore, and phosphates are often extracted by opencast mining. Often the mineral deposit is covered by soil, which must first be stripped off, usually by large machines such as walking draglines and bucket-wheel excavators. The ore deposit is then broken up by explosives.

One of the largest excavations in the world has been made by opencast mining at the Bingham Canyon copper mine in Utah, USA, measuring 790 m/2,590 ft deep and 3.7 km/2.3 mi across.

open cluster or **galactic cluster**, loose cluster of young stars. More than 1,200 open clusters have been catalogued, each containing between a dozen and several thousand stars. They are of interest to astronomers because they represent samples of stars that have been formed at the same time from similar material. Examples include the †Pleiades and the †Hyades. See also †globular cluster and †star cluster.

OpenGL abbreviation for open graphics library, 3-D graphics language developed by US-based workstation manufacturer, Silicon Graphics Inc (SGI), and widely used on Unix workstations. Microsoft has developed its own system called Direct3D, which is part of DirectX. However, Microsoft's Windows NT operating system supports OpenGL and Microsoft and SGI are working together to add some OpenGL support to DirectX.

operant conditioning conditioning of an individual's response (to a stimulus) by means of a reward so that the individual eventually behaves so as to be rewarded.

operating system **OS**, in computing, a program that controls the basic operation of a computer. A typical OS controls the peripheral devices such as printers, organizes the filing system, provides a means of communicating with the operator, and runs other programs.

Many operating systems are written to run on specific computers, but some are available from third-party software houses and will run on machines from a variety of manufacturers. Examples include Apple's OS 9, Microsoft's Windows, and Unix.

Unix (developed at AT&T's Bell Laboratories) is the standard on workstations, minicomputers, and supercomputers; it is also used on desktop PCs and mainframes. The relatively recent †Linux is rapidly increasing in popularity.

operation in mathematics, action on numbers, matrices, or vectors that combines them to form others. The basic operations on numbers are addition, subtraction, multiplication, and division. Matrices involve the same operations as numbers (under specific conditions), but two different types of multiplication are used for vectors: vector multiplication and scalar multiplication.

operational amplifier or **op-amp**, processor that amplifies the difference between two incoming electrical signals. It is a current gain device. Operational amplifiers have two inputs, an inverting input (–) and a noninverting input (+). The input signal from a sensor or switch, often as part of a potential divider, is compared with a similar signal at the other input. This then decides the type of output. If a positive signal is applied to the noninverting input, the output will be positive. If the input is negative, the output will be negative. If a positive signal is given to the inverting input, the output will be negative. A negative input will give a positive output. If the signals at the two inputs are the same, the output will be zero. †Feedback loops are attached from the output to the input in order to control the amplification of the output.

Operational amplifiers usually come in an IC (integrated circuit) package. The output from the amplifier is attached to a †transducer driver which in turn activates an †output device.

operculum small cap covering the spore-producing body of many mosses. It is pushed aside when the spores are mature and ready to be ejected.

operon group of genes that are found next to each other on a chromosome, and are turned on and off as an integrated unit. They usually produce enzymes that control different steps in the same biochemical pathway by a single operator gene. Operons were discovered in 1961 (by the French biochemists François Jacob and Jacques Monod) in bacteria.

They are less common in higher organisms where the control of metabolism is a more complex process.

ophiolite type of igneous rock consisting of ancient oceanic crust, now exposed above water, formed when two continental plates collide.

Ophiuchus large constellation along the celestial equator (see †celestial sphere), known as the serpent bearer because the constellation †Serpens is wrapped around it. The Sun passes through Ophiuchus each December, but the constellation is not part of the zodiac. Ophiuchus contains †Barnard's star.

opiate any compound that contains or is derived from opium.

opiate, endogenous naturally produced chemical in the body which has effects similar to morphine and other opiate drugs; a type of neurotransmitter.

Examples include †endorphins and †encephalins.

Öpik, Ernst Julius (1893–1985) Estonian astronomer whose work on the nature of meteors and comets was instrumental in the development of heat-deflective surfaces for spacecraft on their re-entry into the Earth's atmosphere.

Oppenheimer, J(ulius) Robert (1904–1967) US physicist. As director of the Los Alamos Science Laboratory 1943–45, he was in charge of the development of the atom bomb (the Manhattan Project). He objected to the development of the hydrogen bomb, and was alleged to be a security risk in 1953 by the US Atomic Energy Commission (AEC).

opposition in astronomy, moment at which a body in the Solar System lies opposite the Sun in the sky as seen from the Earth and crosses the †meridian at about midnight.

Although the †inferior planets cannot come to opposition, it is the best time for observation of the superior planets as they can then be seen all night.

optical aberration see †aberration, optical.

optical activity in chemistry, the ability of certain crystals, liquids, and solutions to rotate the plane of †polarized light as it passes through them. The phenomenon is related to the three-dimensional arrangement of the atoms making up the molecules concerned. Only substances that lack any form of structural symmetry exhibit optical activity.

optical character recognition **OCR**, in computing, a technique for inputting text to a computer by means of a document reader. First, a scanner produces a digital image of the text; then character-recognition software makes use of stored knowledge about the shapes of individual characters to convert the digital image to a set of internal codes that can be stored and processed by a computer.

OCR originally required specially designed characters but current devices can recognize most standard typefaces and even handwriting. OCR is used, for example, by gas and electricity companies to input data collected on meter-reading cards, and by personal digital assistants to recognize users' handwriting.

optical computer computer in which both light and electrical signals are used in the †central processing unit. The technology is still not fully developed, but such a computer promises to be faster and less vulnerable to outside electrical interference than one that relies solely on electricity.

optical disk in computing, a storage medium in which laser technology is used to record and read large volumes of digital data. Types include †CD-ROM, †WORM, and floptical disk.

optical emission spectrometry another term for emission †spectroscopy.

optical fiber very fine, optically-pure glass fiber through which light can be reflected to transmit images or data from one end to the other. Although expensive to produce and install, optical fibers can carry more data than traditional cables, and are less susceptible to interference. Standard optical fiber transmitters can send up to 10 billion bits of information per second by switching a laser beam on and off.

Optical fibers are increasingly being used to replace metal communications cables, the messages being encoded as digital pulses of light rather than as fluctuating electric current. Research is investigating how optical fibers could replace wiring inside computers.

Bundles of optical fibers are also used in endoscopes to inspect otherwise inaccessible parts of machines or of the living body.

optical instrument instrument that makes use of one or more lenses or mirrors, or of a combination of these, in order to change the path of light rays and produce an image. Optical instruments such as magnifying glasses, †microscopes, and †telescopes are used to provide a clear, magnified image of the very small or the very distant. Others, such as cameras, photographic enlargers, and film projectors may be used to store or reproduce images.

optical isomer or **enantiomorph**, one of a pair of compounds whose chemical composition is similar but whose molecular structures are mirror images of each other. The presence of an asymmetric (usually carbon) atom makes each isomer optically active (that is, its crystals or solutions rotate the plane of polarized light); the direction of rotation is different (left or right, denoted by *d* or *l*) for each isomer.

optical isomerism type of †stereoisomerism in which the isomers differ in their †optical activity because of the different spatial arrangements of their atoms. Enantiomorphs (isomers with asymmetrical structures, each isomer being a mirror image of the other) are optical isomers.

optical mark recognition **OMR**, in computing, a technique that enables marks made in predetermined positions on computer-input forms to be detected optically and input to a computer. An **optical mark reader** shines a light beam onto the input document and is able to detect the marks because less light is reflected back from them than from the paler, unmarked paper.

optic nerve large nerve passing from the eye to the brain, carrying visual information. In mammals, it may contain up to a million nerve fibers, connecting the sensory cells of the retina to the optical centers in the brain. Embryologically, the optic nerve develops as an outgrowth of the brain.

optics branch of physics that deals with the study of †light and vision – for example, shadows and mirror images, lenses, microscopes, telescopes, and cameras. On striking a surface, light rays are reflected or refracted with some absorption of energy, and the study of this is known as geometrical optics.

Online Photonics Resource http://www.optics.org/

News, features, and discussion of developments in optics research and technology. A powerful search facility makes this an invaluable starting point for the finding of optics resources on the Internet.

optoelectronics branch of electronics concerned with the development of devices (based on the †semiconductor gallium arsenide) that respond not only to the †electrons of electronic data transmission, but also to †photons ("particles" of light).

In 1989 scientists at IBM in the USA built a gallium arsenide microprocessor ("chip") containing 8,000 transistors and four photodetectors. The densest optoelectronic chip yet produced, it can detect and process data at a speed of 1 billion bits per second.

orbit path of one body in space around another, such as the orbit of the Earth around the Sun or of the Moon around the Earth. When the two bodies are similar in mass, as in a †binary star, both bodies move around their common center of mass. The movement of objects in orbit follows †Kepler's laws, which apply to artificial satellites as well as to natural bodies.

As stated by the laws, the orbit of one body around another is an ellipse. The ellipse can be highly elongated, as are comet orbits around the Sun, or it may be almost circular, as are those of some planets. The closest point of a planet's orbit to the Sun is called **perihelion**; the most distant point is **aphelion**. (For a body orbiting the Earth, the closest and furthest points of the orbit are called **perigee** and **apogee**.)

orbital, atomic region around the nucleus of an atom (or, in a molecule, around several nuclei) in which an †electron is likely to be found. According to †quantum theory, the position of an electron is uncertain;

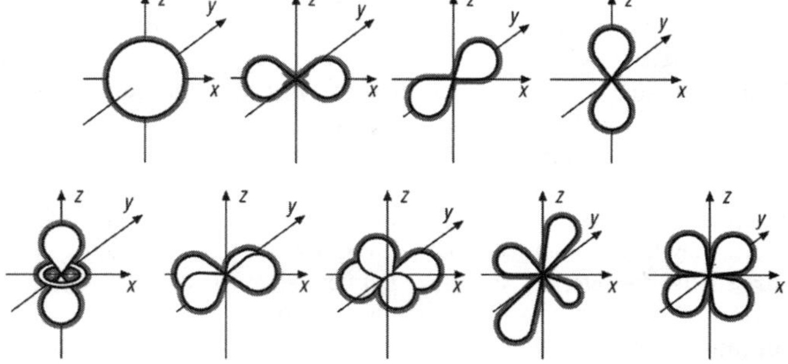

orbital, atomic The shapes of atomic orbitals. An atomic orbital is a picture of the "electron cloud" that surrounds the nucleus of an atom. There are four basic shapes for atomic orbitals: spherical, dumbbell, clover-leaf, and complex (shown at bottom left). Helicon

it may be found at any point. However, it is more likely to be found in some places than in others, and this pattern of probabilities makes up the orbital.

An atom or molecule has numerous orbitals, each of which has a fixed size and shape. An orbital is characterized by three numbers, called †quantum numbers, representing its energy (and hence size), its angular momentum (and hence shape), and its orientation. Each orbital can be occupied by one or (if their spins are aligned in opposite directions) two electrons.

order in biological classification, a group of related †families. For example, the horse, rhinoceros, and tapir families are grouped in the order Perissodactyla, the odd-toed ungulates, because they all have either one or three toes on each foot. The names of orders are not shown in italic (unlike genus and species names) and by convention they have the ending "-formes" in birds and fish; "-a" in mammals, amphibians, reptiles, and other animals; and "-ales" in fungi and plants. Related orders are grouped together in a †class.

order to arrange with regard to size or quantity or other quality, for example alphabetical order. Putting a set of things in order is the same as mapping them on to the set of natural numbers. An infinite set can sometimes be mapped in this way in which case it is said to be countable.

ordered pair in mathematics, any pair of numbers whose order makes a difference to their meaning. Coordinates are an ordered pair because the point (2,3) is not the same as the point (3,2). Vulgar fractions are ordered pairs because the top number gives the quantity of parts while the bottom gives the number of parts into which the unit has been divided.

ordinal number in mathematics, one of the series first, second, third, fourth, ... Ordinal numbers relate to order, whereas †cardinal numbers (1, 2, 3, 4, ...) relate to quantity, or count.

ordinate in †coordinate geometry, the *y* coordinate of a point; that is, the vertical distance of the point from the horizontal or *x*-axis. For example, a point with the coordinates (3,4) has an ordinate of 4. See †abscissa.

Ordovician period period of geological time 510–439 million years ago; the second period of the †Palaeozoic era. Animal life was confined to the sea: reef-building algae and the first jawless fish are characteristic.

The period is named after the Ordovices, an ancient Welsh people, because the system of rocks formed in the Ordovician period was first studied in Wales.

ore body of rock, a vein within it, or a deposit of sediment, worth mining for the economically valuable mineral it contains. The term is usually applied to sources of metals. Occasionally metals are found uncombined (native metals), but more often they occur as compounds such as carbonates, sulfides, or oxides. The ores often contain unwanted impurities that must be removed when the metal is extracted.

Commercially valuable ores include bauxite (aluminum oxide, Al_2O_3) hematite (iron(III) oxide, Fe_2O_3), zinc blende (zinc sulfide, ZnS), and rutile (titanium dioxide, TiO_2).

Hydrothermal ore deposits are formed from fluids such as saline water passing through fissures in the host rock at an elevated temperature. Examples are the "porphyry copper" deposits of Chile and Bolivia, the submarine copper–zinc–iron sulfide deposits recently discovered on the East Pacific Rise, and the limestone lead–zinc deposits that occur in the southern USA and in the Pennines of Britain.

Other ores are concentrated by igneous processes, causing the ore metals to become segregated from a magma – for example, the chromite- and platinum-rich bands within the bushveld, South Africa. Erosion and transportation in rivers of material from an existing rock source can lead to further concentration of heavy minerals in a deposit – for example, Malaysian tin deposits.

Weathering of rocks in situ can result in residual metal-rich soils, such as the nickel-bearing laterites of New Caledonia.

organ in biology, a part of a living body that has a distinctive function or set of functions. Examples include the †esophagus, †liver, or †brain in animals, or a †leaf in plants. An organ is composed of a group of coordinated †tissues. A group of organs working together to perform a function is called an organ system, for example, the †digestive system comprises a number of organs including the esophagus, stomach, the small †intestine, the pancreas, and the liver. The tissues of a leaf include the epidermis, palisade mesophyll, and spongy mesophyll. The tissues of the esophagus include †muscle, and †epithelium.

organelle discrete and specialized structure in a living cell; organelles include mitochondria, chloroplasts, lysosomes, ribosomes, and the nucleus.

organic chemistry branch of chemistry that deals with carbon compounds. Organic compounds form the chemical basis of life and are more abundant than inorganic compounds. In a typical organic compound, each carbon atom forms bonds covalently with each of its neighboring carbon atoms in a chain or ring, and additionally with other atoms, commonly hydrogen, oxygen, nitrogen, or sulfur.

The basis of organic chemistry is the ability of carbon to form long chains of atoms, branching chains, rings, and other complex structures. Compounds containing only carbon and hydrogen are known as **hydrocarbons**.

Introduction to Organic Chemistry: Frequently Asked Questions

http://antoine.frostburg.edu/chem/senese/101/organic/faq.shtml

Simple answers to "Frequently Asked Questions" relating to organic chemistry. The page is part of a larger Web site called "General Chemistry Online!"

Organic Chemistry: Introduction

http://www.netcomuk.co.uk/%7Erpeters1/aufl1oc1.htm

Interesting essay on organic chemistry that includes explanatory diagrams.

organic compound in chemistry, a class of compounds that contain carbon (carbonates, carbon monoxide, and carbon dioxide are excluded). The original distinction between organic and inorganic compounds was based on the belief that the molecules of living systems were unique, and could not be synthesized in the laboratory. Today it is routine to manufacture thousands of organic chemicals both in research and in the drug industry. Certain simple compounds of carbon, such as carbonates, oxides of carbon, carbon disulphide, and carbides are usually treated in inorganic chemistry.

Formula	*Name*	*Atomic bonding*
CH_3	*methyl*	H–C(H)(H)–
CH_2CH_3	*ethyl*	–C(H)(H)–C(H)(H)–H
CC	*double bond*	>C=C<
CHO	*aldehyde*	–C(H)=O
CH_2OH	*alcohol*	–C(H)(H)–OH
CO	*ketone*	>C=O
COOH	*acid*	–C(=O)OH
CH_2NH_2	*amine*	–C(H)(H)–N(H)H
C_6H_6	*benzene ring*	H–C ring of six C, each with H

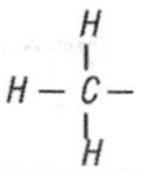

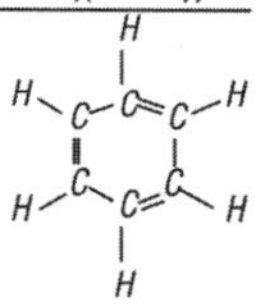

organic chemistry Common organic molecule groupings. Organic chemistry is the study of carbon compounds, which make up over 90% of all chemical compounds. This diversity arises because carbon atoms can combine in many different ways with other atoms, forming a wide variety of rings and chains. Helicon

organizer in embryology, a part of the embryo that causes changes to occur in another part, through induction, thus "organizing" development and †differentiation.

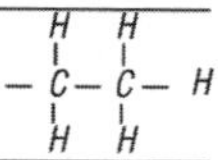

organ of Golgi an elongated structure that occurs at the junction of a muscle and a tendon, which responds to †proprioceptive stimuli.

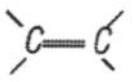

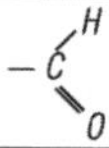

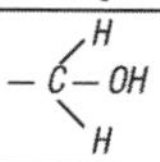

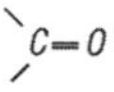

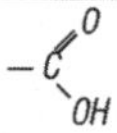

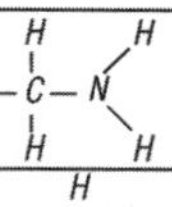

organometallic compound any of a group of substances in which one or more organic †radicals are chemically bonded to a metallic atom – excluding the ionic salts of metals and organic acids. A typical organometallic compound is tetraethyl lead, the "antiknock" substance commonly added to gasoline.

organophosphate insecticide insecticidal compounds that cause the irreversible inhibition of the cholinesterase enzymes that break down acetylcholine. As this mechanism of action is very toxic to humans, the compounds should be used with great care. Malathion and permethrin may be used to control lice in humans and have many applications in veterinary medicine and agriculture. In 1998 organophosphates were the most widely used insecticides, with 40% of the global market.

According to the results of a survey by a UK team of psychiatrists, announced in 1998, one in ten farmers regularly exposed to organophosphates will suffer irreversible physical and mental damage. Sheep dipping is the most common source of exposure. Reported symptoms include tiredness and speech problems. **Organophosphate-induced delayed neuropathy** is a condition of nerve and muscle damage through prolonged contact with organophosphates, which can lead to paralysis.

OR gate in electronics, a type of †logic gate.

origin in mathematics, the point where the *x*–axis meets the *y*–axis. The coordinates of the origin are (0,0).

orimulsion fuel made by mixing †bitumen and water which can be burnt in the same way as heavy oil. It is cheap to make but the smoke produced has a high sulfur content.

Orion very prominent constellation in the equatorial region of the sky (see †celestial sphere), identified with the hunter of Greek mythology.

The bright stars Alpha (†Betelgeuse), Gamma (Bellatrix), Beta (†Rigel), and Kappa Orionis mark the shoulders and legs of Orion. Between them the belt is formed by Delta, Epsilon, and Zeta, three second-magnitude stars equally spaced in a straight line. Beneath the belt is a line of fainter stars marking Orion's sword. One of these, Theta, is not really a star but the brightest part of the †Orion nebula. Nearby is one of the most distinctive dark nebulae, the Horsehead.

Orion nebula luminous cloud of gas and dust 1,500 light years away, in the constellation Orion, from which stars are forming. It is about 15 light years in diameter, and contains enough gas to make a cluster of thousands of stars.

At the nebula's center is a group of hot young stars, called the **Trapezium**, which make the surrounding gas glow. The nebula is visible to the naked eye as a misty patch below the belt of Orion.

ormolu (French *or moulu* "ground gold") alloy of copper, zinc, and sometimes tin, used for furniture decoration.

ornithology study of birds. It covers scientific aspects relating to their structure and classification, and their habits, song, flight, and value to agriculture as destroyers of insect pests. Worldwide scientific banding (or the fitting of coded rings to captured specimens) has resulted in accurate information on bird movements and distribution. There is an International Council for Bird Preservation with its headquarters at the Natural History Museum, London.

ornithophily †pollination of flowers by birds. Ornithophilous flowers are typically brightly colored, often red or orange. They produce large quantities of thin, watery nectar, and are scentless because most birds do not respond well to smell. They are found mostly in tropical areas, with hummingbirds being important pollinators in North and South America, and the sunbirds in Africa and Asia.

orogenesis in its original, literal sense, orogenesis means "mountain building," but today it more specifically refers to the tectonics of mountain building (as opposed to mountain building by erosion). Orogenesis is brought about by the movements of the rigid plates making up the Earth's crust and uppermost mantle (described by †plate tectonics). Where two plates collide at a destructive margin rocks become folded and lifted to form chains of mountains (such as the Himalayas). Processes associated with orogeny are faulting and thrusting (see †fault), folding, metamorphism, and plutonism (see †plutonic rock). However, many topographical features of mountains – cirques, U-shaped valleys – are the result of *non-orogenic* processes, such as weathering, erosion, and glaciation. †Isostasy (uplift due to the buoyancy of the Earth's crust) can also influence mountain physiography.

orographic rainfall rainfall that occurs when an airmass is forced to rise over a mountain range. As the air rises, it cools. The amount of moisture that air can hold decreases with decreasing temperature. So the water vapor in the rising airstream condenses, and rain falls on the windward side of the mountain. The air descending on the leeward side contains less moisture, resulting in a **rainshadow** where there is little or no rain.

orthocenter in mathematics, of a triangle, the common point at which the altitudes (lines from the vertices perpendicular to the opposite side of a triangle) meet.

orthogonal in geometry, having a right angle, right-angled, or perpendicular.

orthohydrogen †molecules of hydrogen in which the †spins of the two atoms are parallel as opposed to parahydrogen in which the spins are opposed.

OS/2 single-user computer †operating system produced jointly by Microsoft Corporation and IBM for use on personal computers, particularly when attached to large IBM computers. Its main features are reliability – useful for running file servers – and a powerful object-oriented graphical user interface, Workplace Shell.

OS/2 was announced in 1987 but sales were dismal. Microsoft abandoned it in 1992 to concentrate on Windows. IBM abandoned attempts to sell it on the mass market in 1997, but supports its use as a corporate and network computer operating system.

oscillating universe in astronomy, theory stating that the gravitational attraction of the mass within the universe will eventually slow down and stop the expansion of the universe. The outward motions of the galaxies will then be reversed, eventually resulting in a "Big Crunch" where all the matter in the universe will be contracted into a small volume of high density. This could undergo a further †Big Bang, thereby creating another expansion phase. The theory suggests that the universe would alternately expand and collapse through alternate Big Bangs and Big Crunches.

oscillation one complete to-and-fro movement of a vibrating object or system. For any particular vibration, the time for one oscillation is called its †period and the number of oscillations in one second is called its †frequency. The maximum displacement of the vibrating object from its rest position is called the †amplitude of the oscillation.

oscillator any device producing a desired oscillation (vibration). There are many types of oscillator for different purposes, involving various arrangements of thermionic †valves or components such as †transistors, †inductors, †capacitors, and †resistors.

An oscillator is an essential part of a radio transmitter, generating the high-frequency carrier signal necessary for radio communication. The †frequency is often controlled by the vibrations set up in a crystal (such as quartz).

oscillograph instrument for displaying or recording rapidly changing oscillations, electrical or mechanical.

oscilloscope another name for †cathode-ray oscilloscope.

osmium chemical symbol Os, (Greek *osme* "odor") hard, heavy, bluish-white, metallic element, atomic number 76, relative atomic mass 190.2. It is the densest of the elements, and is resistant to tarnish and corrosion. It occurs in platinum ores and as a free metal (see †native metal) with iridium in a natural alloy called osmiridium, containing traces of platinum, ruthenium, and rhodium. Its uses include pen points and light-bulb filaments; like platinum, it is a useful catalyst.

It was discovered in 1803 and named in 1804 by English chemist Smithson Tennant (1761–1815) after the irritating smell of one of its oxides.

osmoregulation process whereby the water content of living organisms is maintained at a constant level. If the water balance is disrupted, the concentration of salts will be too high or too low, and vital functions, such as nerve conduction, will be adversely affected.

In mammals, loss of water by evaporation is counteracted by increased intake and by mechanisms in the kidneys that enhance the rate at which water is reabsorbed into the blood before urine production. Both these responses are mediated by hormones, primarily those of the adrenal cortex (see †adrenal gland).

osmosis movement of †water through a partially (selectively) permeable membrane separating solutions of different concentrations. Water passes by †diffusion from a **weak solution** (high water concentration) to a **strong solution** (low water concentration) until the two concentrations are equal. A membrane is partially permeable if it lets water through but not the molecules or ions dissolved in the water (the solute; for example, sugar molecules). Many cell membranes behave in this way, and osmosis is a vital mechanism in the transport of fluids in living organisms. One example is in the transport of water from soil (weak solution) into the roots of plants (stronger solution of cell sap) via the †root hair cells. Another is the uptake of water by the †epithelium lining the †gut in animals. There are also membranes that humans can manufacture that are partially permeable.

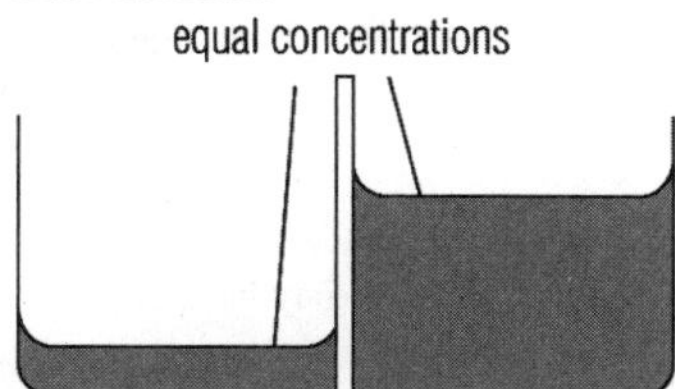

osmosis Apparatus for demostrating osmosis. In 1877 German physicist Wilhelm Pfeffer used a similar apparatus to make the first ever measurement of osmotic pressure and show that osmotic pressure varies according to temperature and the strength of the solute (dissolved substance). Helicon

If solutions have different concentrations of dissolved substance, there will be a difference in the concentration of water as well. Where there is a lot of dissolved substance there is less water. Where there is less dissolved substance there is more water. Water behaves like any other substance in diffusion. It diffuses from an area where there is a high concentration of it to an area where there is a lower concentration of it. Water diffuses down the concentration gradient for water and across the membrane. This means that it will diffuse from the dilute solution into the concentrated solution.

Excessive flow of water into a cell by osmosis can burst the cell. Cells protect against this using processes of osmoregulation. If external pressure is applied to the stronger solution, osmosis is arrested. By this mechanism plant cells can osmoregulate, since the cell wall of a fully turgid cell exerts pressure on the solution within the cell. Animal cells such as the red blood cell cannot osmoregulate in this way since they have no cell wall. Instead, the correct concentration of plasma is maintained by the †kidney. This is important because if the concentration changes significantly, osmosis will take place and a lot of water will either diffuse out of cells into the blood or into cells from the blood, thereby damaging them.

osmotic pressure pressure that must be applied to a solution so that it no longer takes up pure solvent (usually water) across a membrane that is permeable to solvent but not to solute (the dissolved substance). See also †osmosis.

ossification or **osteogenesis**, process whereby bone is formed in vertebrate animals by special cells (**osteoblasts**) that secrete layers of †extracellular matrix on the surface of the existing †cartilage. Conversion to bone occurs through the deposition of calcium phosphate crystals within the matrix.

osteology part of the science of †anatomy, dealing with the structure, function, and development of bones.

Ostwald, (Friedrich) Wilhelm (1853–1932) Latvian-born German chemist who devised the Ostwald process (the oxidation of ammonia over a platinum catalyst to give nitric acid). His work on catalysts laid the foundations of the petrochemical industry. He was awarded the Nobel Prize for Chemistry in 1909 for his work in catalysis, and the principles of equilibria and rates of reaction.

otolith or **ear stone**, tiny calcareous deposit found in the ears of fishes, which are in contact with the sensory cushions supplied by the acoustic nerves. Their function is as an organ of balance.

Otto cycle alternative name for the †four-stroke cycle, introduced by the German engineer Nikolaus Otto (1832–1891) in 1876.

Otto engine an internal-combustion engine that produces power by four strokes: intake, compression, ignition, and expansion, followed by exhaust and further intake. This involves two revolutions of the crankshaft.

Oughtred, William (1575–1660) English mathematician, credited as the inventor of the slide rule in 1622. His major work *Clavis mathematicae/The Key to Mathematics* (1631) was a survey of the entire body of mathematical knowledge of his day. It introduced the "x" symbol for multiplication, as well as the abbreviations "sin" for sine and "cos" for cosine.

Oughtred, William http://www-history.mcs.st-and.ac.uk/history/Mathematicians/Oughtred.html

Biography of the English mathematician. The site contains a description of his works, and includes a page from Clavis mathematicae (1652) showing + and – and the notation for decimal fractions, all of which Oughtred is credited with inventing. Several literature references for further reading on the life and works of Oughtred are listed, and his portrait is also shown.

ounce unit of mass, one-sixteenth of a pound †avoirdupois, equal to 437.5 grains (28.35 g); also one-twelfth of a pound troy, equal to 480 grains.

outcome in †probability theory or statistics, one possible result of an experiment. For example, the possible outcomes from a match between team A and team B are (i) A wins, (ii) B wins, and (iii) a draw.

outlier in statistics, a highly atypical data value that distorts the †mean within a sample.

output device any device for displaying, in a form intelligible to the user, the results of processing carried out by a computer.

Electronic output devices can be mechanical, audio, or visual. Mechanical outputs include motors, solenoids, and other electromagnetic devices. Audio outputs include buzzers, loudspeakers, piezo transducers, and ultrasonic transmitters. Mechanical and audio outputs tend to have a higher power demand than other output devices. Visual outputs include lamps, LEDs (light-emitting diodes), and infrared transmitters.

outwash sands and gravels deposited by streams of meltwater (water produced by the melting of a glacier). Such material may be laid down ahead of the glacier's snout to form a large flat expanse called an **outwash plain**.

Outwash is usually well sorted, the particles being deposited by the meltwater according to their size – the largest are deposited next to the snout while finer particles are deposited further downstream.

ovary in female animals, the organ that generates the †ovum. In humans, the ovaries are two whitish rounded bodies about 25 mm/1 in by 35 mm/1.5 in, located in the lower abdomen on either side of the uterus. Every month, from puberty to the onset of the menopause, an ovum is released from the ovary. This is called ovulation, and forms part of the †menstrual cycle. In botany, an ovary is the expanded basal portion of the †carpel of flowering plants, containing one or more †ovules. It is hollow with a thick wall to protect the ovules. Following fertilization of the ovum, it develops into the fruit wall or pericarp.

The ovaries of female animals secrete the hormones responsible for the secondary sexual characteristics of the female, such as smooth, hairless facial skin and enlarged breasts. An ovary in a half-grown human fetus contains 5 million eggs, and so the unborn baby already possesses the female genetic information for the next generation.

In botany, the relative position of the ovary to the other floral parts is often a distinguishing character in classification; it may be either inferior or superior, depending on whether the petals and sepals are inserted above or below.

over-clocking practice of trying to make a †central processing unit (CPU) run at a faster †clock rate than was originally intended by the manufacturer, and so improve performance with minimal costs.

overfishing fishing at rates that exceed the sustained-yield cropping of fish species, resulting in a net population decline. For example, in the North Atlantic, herring has been fished to the verge of extinction, and the cod and haddock populations are severely depleted. In the developing world, use of huge factory ships, often by fisheries from industrialized countries, has depleted stocks for local people who cannot obtain protein in any other way.

Ecologists have long been concerned at the wider implications of overfishing, in particular the devastation wrought on oceanic †food chains. The United Nations Food and Agriculture Organization estimates that worldwide overfishing has damaged oceanic ecosystems to such an extent that potential catches are on average reduced by 20%. With better management of fishing programs the fishing catch could in principle be increased; it is estimated that, annually, 20 million tonnes of fish are discarded from fishing vessels at sea, because they are not the species sought.

According to an estimate by the Food and Agriculture Organization in 1993, nine of the world's 17 main fishing grounds were suffering a potentially catastrophic decline in some species. In 1994 approximately 17% of fishing waters off the coast of New England, USA, was closed in an attempt to restore dwindling stocks. The area affected covered 17,000 sq km/6,600 sq mi and lay within the Georges Bank region of the Atlantic Ocean.

overland flow another term for †surface run-off of water after rain.

overtone note that has a frequency or pitch that is a multiple of the fundamental frequency, the sounding body's †natural frequency. Each sound source produces a unique set of overtones, which gives the source its quality or timbre.

oviparous method of animal reproduction in which eggs are laid by the female and develop outside the body, in contrast to †ovoviviparous and †viviparous. It is the most common form of reproduction.

ovoviviparous method of animal reproduction in which fertilized eggs develop within the female (unlike †oviparous), and the embryo gains no nutritional substances from the female (unlike †viviparous). It occurs in some invertebrates, fishes, and reptiles.

ovulation in female animals, the process of releasing egg cells (ova) from the †ovary. In mammals it occurs as part of the †menstrual cycle.

ovule structure found in seed plants that develops into a seed after fertilization. It consists of an †embryo sac containing the female gamete (†ovum or egg cell), surrounded by nutritive tissue, the nucellus. Outside this there are one or two coverings that provide protection, developing into the testa, or seed coat, following fertilization.

In †angiosperms (flowering plants) the ovule is within an †ovary, but in †gymnosperms (conifers and their allies) the ovules are borne on the surface of an ovuliferous (ovule-bearing) scale, usually within a †cone, and are not enclosed by an ovary.

ovum plural **ova**, female gamete (sex cell) before fertilization. In mammals it is called an egg, and is produced by a special cell division called †meiosis in the ovaries during the †menstrual cycle. In plants, where it is also known as an egg cell or oosphere, the ovum is produced in an ovule. The ovum does not move by itself. It must be fertilized by a male gamete before it can develop further, except in cases of †parthenogenesis.

Owen, Richard (1804–1892) British anatomist and paleontologist. He attacked the theory of natural selection and in 1860 published an anonymous and damaging review of Charles †Darwin's work. As director of the Natural History Museum, London, he was responsible for the first public exhibition of dinosaurs. He was knighted in 1884.

oxalic acid or **ethamedioic acid**; $(COOH)_2.2H_2O$, white, poisonous solid, soluble in water, alcohol, and ether. Oxalic acid is found in rhubarb leaves, and its salts (oxalates) occur in wood sorrel (genus *Oxalis*, family Oxalidaceae) and other plants. It also occurs naturally in human body cells. It is used in the leather and textile industries, in dyeing and bleaching, ink manufacture, metal polishes, and for removing rust and ink stains.

oxbow lake curved lake found on the flood plain of a river. Oxbows are caused by the loops of †meanders that are cut off at times of flood and the river subsequently adopts a shorter course. In the USA, the term bayou is often used.

oxidation in chemistry, the loss of †electrons, gain of oxygen, or loss of hydrogen by an atom, ion, or molecule during a chemical reaction.

At a simple level, oxidation may be regarded as the reaction of a substance

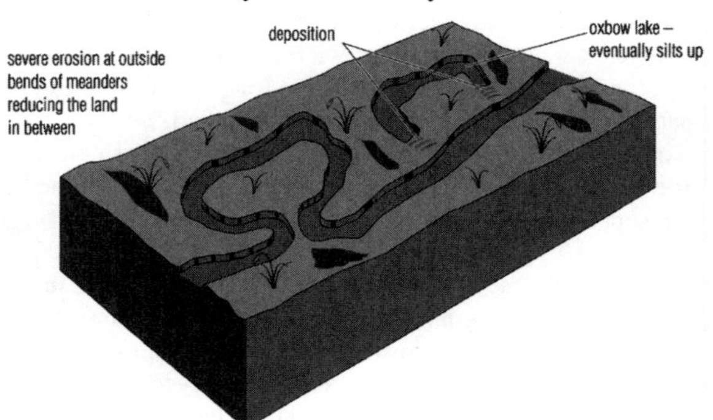

oxbow lake The formation of an oxbow lake. As a river meanders across a flood plain, the outer bends are gradually eroded and the water channel deepens; as the loops widen, the neck of the loop narrows and finally gives way, allowing the water to flow in a more direct route, isolating the old water channel and forming an oxbow lake. Helicon

with oxygen. For example, †rusting, †respiration, and †combustion (burning) are all oxidation reactions. When carbon is burned in air it is oxidized to carbon dioxide:

$C + O_2 \rightarrow CO_2$

Oxidation may be brought about by reaction with another compound (oxidizing agent), which simultaneously undergoes †reduction, or electrically at the anode (positive electrode) of an electrolytic cell.

oxidation in earth science, a form of †chemical weathering caused by the chemical reaction that takes place between certain iron-rich minerals in rock and the oxygen in water. It tends to result in the formation of a red-colored soil or deposit. The inside walls of canal tunnels and bridges often have deposits formed in this way.

oxidation number Roman numeral often seen in a chemical name, indicating the †valency of the element immediately before the number. Examples are lead(II) nitrate, manganese(IV) oxide (manganese dioxide), and potassium manganate(VII) (potassium permanganate).

oxide compound of oxygen and another element, frequently produced by burning the element or a compound of it in air or oxygen.

Oxides of metals are normally †bases and will react with an acid to produce a †salt in which the metal forms the cation (positive ion). Some of them will also react with a strong alkali to produce a salt in which the metal is part of a complex anion (negative ion; see †amphoteric). Most oxides of non-metals are acidic (dissolve in water to form an †acid). Some oxides display no pronounced acidic or basic properties.

oxide film thin film of oxide formed on the surface of some metals as soon as they are exposed to the air. This oxide film makes the metal much more resistant to chemical attack. The considerable lack of reactivity of aluminum to most reagents arises from such a film.

The thickness of the oxide film can be increased by †anodizing the aluminum.

oxidizing agent substance that will oxidize another substance (see †oxidation).

In a redox reaction, the oxidizing agent is the substance that is itself reduced. Common oxidizing agents include oxygen, chlorine, nitric acid, and potassium manganate(VII).

oxyacetylene torch gas torch that burns ethyne (acetylene) in pure oxygen, producing a high-temperature flame (3,000°C/5,400°F). It is widely used in welding to fuse metals. In the cutting torch, a jet of oxygen burns through metal already melted by the flame.

oxyfuel fuel enriched with oxygen to decrease carbon monoxide (CO) emissions. Oxygen is added in the form of chemicals such as methyl tertiary butyl ether (MTBE) and ethanol.

Cars produce CO when there is insufficient oxygen present to convert all the carbon in the gasoline to CO_2. This occurs mostly at low temperatures, such as during the first five minutes of starting the engine and in cold weather. CO emissions are reduced by the addition of oxygen-rich chemicals. The use of oxyfuels in winter is compulsory in 35 US cities. There are fears, however, that MTBE can cause health problems, including nausea, headaches, and skin rashes.

oxygen chemical symbol O, (Greek *oxys* "acid"; *genes* "forming") colorless, odorless, tasteless, †non-metallic, gaseous element, atomic number 8, relative atomic mass 15.9994. It is the most abundant element in the Earth's crust (almost 50% by mass), forms about 21% by volume of the atmosphere, and is present in combined form in water and many other substances. Oxygen is a by-product of †photosynthesis and the basis for †respiration in plants and animals.

Oxygen is very reactive and combines with all other elements except the †noble gases (rare gases) and fluorine. †Combustion (burning) and †rusting are two examples of reactions involving oxygen. It is present in carbon dioxide, silicon dioxide (quartz), iron ore, calcium carbonate (limestone). In nature it exists as a molecule composed of two atoms (O_2); single atoms of oxygen are very short-lived owing to their reactivity. They can be produced in electric sparks and by the Sun's ultraviolet radiation in the upper atmosphere, where they rapidly combine with molecular oxygen to form ozone (O_3), an allotrope of oxygen.

Oxygen is obtained for industrial use by the †fractional distillation of †liquid air, by the †electrolysis of water, or by heating manganese(IV) oxide with potassium chlorate. In the laboratory it is prepared by the action of the †catalyst manganese(IV) oxide on hydrogen peroxide. The simple laboratory test for oxygen is that it relights a glowing spill. Oxygen is essential for †combustion, and is used with ethyne (acetylene) in high-temperature oxyacetylene welding and cutting torches.

The element was first identified by English chemist Joseph Priestley in 1774 and independently in the same year by Swedish chemist Karl Scheele. It was named by French chemist Antoine Lavoisier in 1777.

oxygenation combination of (gaseous or dissolved) oxygen with a substance, such as with the blood in the lungs during respiration.

oxygen debt physiological state produced by vigorous exercise, in which the lungs cannot supply all the oxygen that the muscles need. In other words, the lungs and †bloodstream, pumped by the †heart, cannot supply sufficient †oxygen for †aerobic respiration in the †muscles. In such a situation the muscles can continue to break down †glucose to liberate energy for a short time using anaerobic respiration. This partial breakdown produces †lactic acid, which results in a sensation of fatigue when it reaches certain levels in the muscles and the blood. This explains why it is possible to run faster in a sprint than over longer distances. During the sprint, the muscles can respire anaerobically. Once the vigorous muscle movements cease, the body breaks down the accumulated lactic acid on top of the "normal" breakdown of glucose in aerobic respiration, using up extra oxygen to do so. Panting after exercise is an automatic mechanism to "pay off" the oxygen debt.

oxyhemoglobin oxygenated form of †hemoglobin, the protein found in the red blood cells.

oxytocin hormone that stimulates the uterus in late pregnancy to initiate and sustain labor. It is secreted by the †pituitary gland. After birth, it stimulates the uterine muscles to contract, reducing bleeding at the site where the placenta was attached.

Intravenous injections of oxytocin may be given to induce labor, strengthen contractions, or control hemorrhage after birth. It is also secreted during lactation. Oxytocin sprayed in the nose a few minutes before nursing improves milk production.

ozone gas consisting of three atoms of oxygen (O_3), which is therefore an allotrope of oxygen. It is formed when the molecule of the stable form of oxygen (O_2) is split by ultraviolet radiation or electrical discharge. It forms the †ozone layer in the upper atmosphere, which protects life on Earth from ultraviolet rays, a cause of skin cancer.

Ozone is a highly reactive pale-blue gas with a penetrating odor.

ozone depleter any chemical that destroys the ozone in the stratosphere. Most ozone depleters are chemically stable compounds containing chlorine or bromine, which remain unchanged for long enough to drift up to the upper atmosphere. The best known are †chlorofluorocarbons (CFCs).

Other ozone depleters include halons, used in some fire extinguishers; methyl chloroform and carbon tetrachloride, both solvents; some CFC substitutes; and the pesticide methyl bromide.

CFCs accounted for approximately 75% of ozone depletion in 1995, whereas methyl chloroform (atmo-

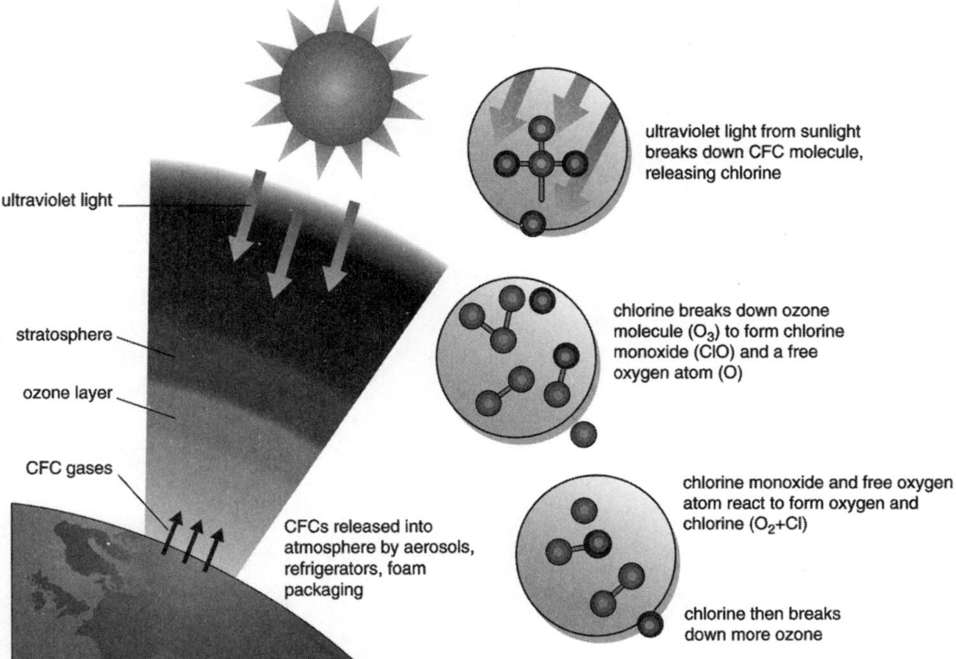

ozone layer The destruction of the ozone layer by chlorofluorocarbons (CFCs). CFCs discharged into the atmosphere break down in sunlight releasing chlorine, which breaks down the ozone to form chlorine monoxide and a free oxygen atom. These products react together to form oxygen and chlorine, leaving the chlorine to break down another ozone molecule, and so on. Helicon

spheric concentrations of which had markedly decreased during 1990–94) accounted for an estimated 12.5%. The ozone depletion rate overall is now decreasing as international agreements to curb the use of ozone-depleting chemicals begin to take effect. In 1996 there was a decrease in ozone depleters in the lower atmosphere. This trend is expected to continue into the stratosphere over the next few years.

ozone layer thin layer of the gas †ozone in the upper atmosphere that shields the Earth from harmful ultraviolet rays. A continent-sized hole has formed over Antarctica as a result of damage to the ozone layer. This has been caused in part by †chlorofluorocarbons (CFCs), but many reactions destroy ozone in the stratosphere: nitric oxide, chlorine, and bromine atoms are implicated.

It is believed that the ozone layer is depleting at a rate of about 5% every 10 years over northern Europe, with depletion extending south to the Mediterranean and southern USA. However, ozone depletion over the polar regions is the most dramatic manifestation of a general global effect. Ozone levels over the Arctic in spring 1997 fell over 10% since 1987, despite the reduction in the concentration of CFCs and other industrial compounds which destroy the ozone when exposed to sunlight. It is thought that this may be because of an expanding vortex of cold air forming in the lower stratosphere above the Arctic, leading to increased ozone loss. It is expected that an Arctic hole as large as that over Antarctica could remain a threat to the northern hemisphere for several decades.

The size of the hole in the ozone layer in October 1998 was three times the size of the USA, larger than it had ever been before. In autumn 2000 the hole in the ozone layer was at its largest ever. Observers had hoped that its 1998 level was due to †El Niño and would not be exceeded.

Chemistry of the Ozone Layer http://pooh.chem.wm.edu/chemWWW/courses/chem105/projects/group2/page1.html

Interesting step-by-step introduction to the ozone layer for those wishing to understand the chemistry of ozone depletion, the role of chlorofluorocarbons, the consequences of increased radiation for life on Earth, and actions to tackle the problem. The information may be readily understood by those with a basic knowledge of chemistry.

ozonide in chemistry, member of a class of explosive organic compounds that are formed by the addition of †ozone across a carbon–carbon double bond. All ozonides have molecules that contain a five-member ring structure consisting of two carbon and three oxygen atoms.

P

Pa in physics, symbol for †pascal, the SI unit of pressure, equal to a pressure of one newton per square meter.

packet switching method of transmitting data between computers connected in a †network. Packet switched networks do not provide a dedicated connection between two locations, as with a circuit switched network. Packet switched networks make more effective use of bandwidth than circuit switched networks, and are more resilient to breaks in network links because there are always multiple routes from source to destination.

It is possible that a large message be broken down into many small packets, each equipped with a sequence number to ensure it can be reassembled properly if some packets take a different route through the network of switches and arrive out of order.

packing fraction in physics, mass defect (the difference between the actual mass of an atomic nucleus and the nearest whole number) divided by the mass number.

paedomorphosis in biology, an alternative term for †neoteny.

pahoehoe or **ropy lava**, type of lava that has a smooth, wavy, or ropy surface.

pair production creation of an †electron and antielectron (†positron) pair by the interaction between a high-energy particle or †photon and the electrostatic field around a nucleus. It can also be used to describe the creation out of the vacuum state of particle–antiparticle pairs as allowed by Heisenberg's †uncertainty principle.

Palade, George (1912–) Romanian cell biologist who was awarded a Nobel Prize for Physiology or Medicine in 1974 with Albert Claude and Christian René de †Duve for work on the structural and functional organization of the cell. Palade determined the parts of the cell involved in protein secretion, and discovered †ribosomes, which he showed to be rich in †RNA, or ribonucleic acid (nucleic acid involved in the process of translating genetic material into proteins).

Paleocene epoch (Greek "old" + "recent") first epoch of the Tertiary period of geological time, 65–56.5 million years ago. Many types of mammals spread rapidly after the disappearance of the great reptiles of the Mesozoic. Flying mammals replaced the flying reptiles, swimming mammals replaced the swimming reptiles, and all the ecological niches vacated by the reptiles were adopted by mammals.

At the end of the Paleocene there was a mass extinction that caused more than half of all bottom-dwelling organisms to disappear worldwide, over a period of around 1,000 years. Surface-dwelling organisms remained unaffected, as did those on land. The cause of this extinction remains unknown, though US paleontologists have found evidence (released in 1998) that it may have been caused by the Earth releasing tonnes of methane into the oceans causing increased water temperatures.

paleomagnetic stratigraphy use of distinctive sequences of magnetic polarity reversals to date rocks. Magnetism retained in rocks at the time of their formation are matched with known dated sequences of †polar reversals or with known patterns of †secular variation.

paleomagnetism study of the magnetic properties of rocks in order to reconstruct the Earth's ancient magnetic field and the former positions of the continents, using traces left by the Earth's magnetic field in igneous rocks before they cool. Paleomagnetism shows that the Earth's magnetic field has reversed itself – the magnetic north pole becoming the magnetic south pole, and vice versa – at approximate half-million-year intervals, with shorter reversal periods in between the major spans.

Starting in the 1960s, this known pattern of magnetic reversals was used to demonstrate seafloor spreading or the formation of new ocean crust on either side of mid-oceanic ridges. As new material hardened on either side of a ridge, it would retain the imprint of the magnetic field, furnishing datable proof that material was spreading steadily outward. Paleomagnetism is also used to demonstrate †continental drift by determining the direction of the magnetic field of dated rocks from different continents.

paleontology the study of ancient life, encompassing the structure of ancient organisms and their environment, evolution, and ecology, as revealed by their †fossils and the rocks in which those fossils are found. The practical aspects of paleontology are based on using the presence of different fossils to date particular rock strata and to identify rocks that were laid down under particular conditions; for instance, giving rise to the formation of oil.

The use of fossils to trace the age of rocks was pioneered in Germany by Johann Friedrich Blumenbach (1752–1830) at Göttingen, followed by Georges †Cuvier and Alexandre †Brongniart in France in 1811.

Museum of Paleontology http://www.ucmp.berkeley.edu/exhibit/exhibits.html

Large amount of detailed information on the subject in a carefully structured and carefully cross-referenced site. You can explore paleontology through the three areas of phylogeny, geology, and evolution.

Strange Science http://www.turnpike.net/~mscott/index.htm
Subtitled "The Rocky Road to Modern Paleontology and Biology," this site examines some of the medieval discoveries that led to the growth of interest in modern-day science. The site is illustrated with images that clearly show how people's perception of the world differed, and how people made up for gaps in their knowledge with a little imagination!

Palaeozoic era era of geological time 570–245 million years ago. It comprises the Cambrian, Ordovician, Silurian, Devonian, Carboniferous, and Permian periods. The Cambrian, Ordovician, and Silurian constitute the Lower or Early Palaeozoic; the Devonian, Carboniferous, and Permian make up the Upper or Late Palaeozoic. The era includes the evolution of hard-shelled multicellular life forms in the sea; the invasion of land by plants and animals; and the evolution of fish, amphibians, and early reptiles.

palate in mammals, the roof of the mouth. The bony front part is the hard palate, the muscular rear part the soft palate. Incomplete fusion of the two lateral halves of the palate (cleft palate) causes interference with speech.

Palencia, Alfonso de (lived 15th century AD) Spanish historical geographer. He studied under George of Trebizond in Italy, and became historiographer to Isabella of Castile. His works include *Chronicle of Henry IV* and *Decades*.

palisade cell cylindrical cell lying immediately beneath the upper epidermis of a leaf. Palisade cells normally exist as one closely packed row and contain many †chloroplasts. During the hours of daylight palisade cells are photosynthetic, using the energy of the sun to create carbohydrates from water and carbon dioxide.

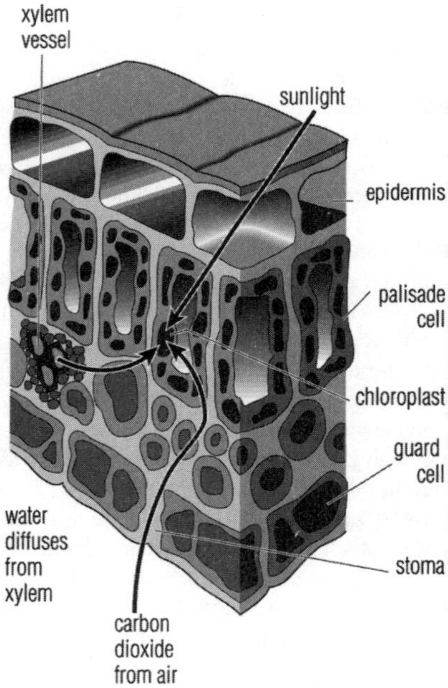

palisade cell Palisade cells are closely packed, columnar cells, lying in the upper surfaces of leaves. They contain many chloroplasts (where photosynthesis takes place) and are well adapted to receive and process the components necessary for photosynthesis – carbon dioxide, water, and sunlight. For instance, their vertical arrangement means that there are few cross-walls to interfere with the passage of sunlight. Helicon

palladium chemical symbol Pd, lightweight, ductile and malleable, silver-white, metallic element, atomic number 46, relative atomic mass 106.4.

It is one of the so-called platinum group of metals, and is resistant to tarnish and corrosion. It often occurs in nature as a free metal (see †native metal) in a natural alloy with platinum. Palladium is used as a catalyst, in alloys of gold (to make white gold) and silver, in electroplating, and in dentistry.

It was discovered in 1803 by English physicist William †Wollaston, and named after the asteroid Pallas (found in 1802).

Pallas second-largest asteroid in the Solar System (after Ceres), discovered by German astronomer Heinrich †Olbers in 1802. It is 522 km/324 mi in diameter. Pallas orbits the Sun at a distance of 414 million km/257 million mi with a period (the time it takes to circle the Sun) of 4.61 years.

palmitic acid $C_{15}H_{31}COOH$, unsaturated fatty acid that occurs naturally as †glycerides in most animal and vegetable fats, notably palm oil. Salts of palmitic acid are used to manufacture soaps.

palp in zoology, jointed sensory structure bearing chemoreceptors that give insects and crustaceans a sense of taste.

pancreas in vertebrates, a gland of the digestive system located close to the duodenum. When stimulated by the hormone secretin, it releases enzymes into the duodenum that digest starches, proteins, and fats. In humans, it is about 18 cm/7 in long, and lies behind and below the stomach. It contains groups of cells called the **islets of Langerhans**, which secrete the hormones insulin and glucagon that regulate the blood sugar level.

Pangaea or **Pangea**, (Greek "all-land") single land mass, made up of all the present continents, believed to have existed between 300 and 200 million years ago; the rest of the Earth was covered by the Panthalassa ocean. Pangaea split into two land masses – †Laurasia in the north and †Gondwanaland in the south – which subsequently broke up into several continents. These then moved slowly to their present positions (see †plate tectonics).

The former existence of a single "supercontinent" was proposed by German meteorologist Alfred Wegener in 1912.

pangenesis erroneous theory that stated that bodies in †gemmules transported by the blood to a parent's reproductive cells represented invisible (but exact) copies of the rest of the organism. After fertilization and combination with the other parent's gemmules these bodies were supposed to develop and grow into the "adult" forms.

Panthalassa ocean that covered the surface of the Earth not occupied by the world continent †Pangaea between 300 and 200 million years ago.

pantothenic acid water-soluble †vitamin ($C_9H_{17}NO_5$) of the B complex, found in a wide variety of foods. Its absence from the diet can lead to dermatitis, and it is known to be involved in the breakdown of fats and carbohydrates.

papilla small growth from the surface of a tissue, such as the papillae on the surface of the tongue.

pappus plural **pappi**, in botany, a modified †calyx comprising a ring of fine, silky hairs, or sometimes scales or small teeth, that persists after fertilization. Pappi are found in members of the daisy family (Compositae) such as the dandelions *Taraxacum*, where they form a parachutelike structure that aids dispersal of the fruit.

Pappus of Alexandria (c. 300–c. 350) Greek mathematician, astronomer, and geographer whose book *Synagogue/Collection* deals with nearly the whole body of Greek geometry, mostly in the form of commentaries on texts the reader is assumed to have to hand.

Pappus of Alexandria http://www-groups.dcs.st-andrews.ac.uk/history/Mathematicians/Pappus.html

Devoted to the life and contributions of Pappus of Alexandria. In addition to biographical information, there is a list of references about Pappus and links to essays that reference him. The text of this essay includes hypertext links to the essays of those mathematicians and thinkers who influenced Pappus. There is also a map of his birthplace.

parabola in mathematics, a curve formed by cutting a right circular cone with a plane parallel to the sloping side of the cone. A parabola is one of the family of curves known as †conic sections. The graph of $y = x^2$ is a parabola.

It can also be defined as a path traced out by a point that moves in such a way that the distance from a fixed point (focus) is equal to its distance from a fixed straight line (directrix); it thus has an †eccentricity of 1.

The trajectories of missiles within the Earth's gravitational field approximate closely to parabolas (ignoring the effect of air resistance). The corresponding solid figure, the paraboloid, is formed by rotating a parabola about its axis. It is a common shape for headlight reflectors, dish-shaped microwave and radar aerials, and radiotelescopes, since a source of radiation placed at the focus of a paraboloidal reflector is propagated as a parallel beam.

Parabola

http://www-groups.dcs.st-andrews.ac.uk/history/Curves/Parabola.html

Introduction to parabolas and links to associated curves. If you have a Java-enabled browser you can experiment interactively with this curve and its associates. There are also links to mathematicians who have studied the parabola, to the parabola's particular attributes, and to related Web sites.

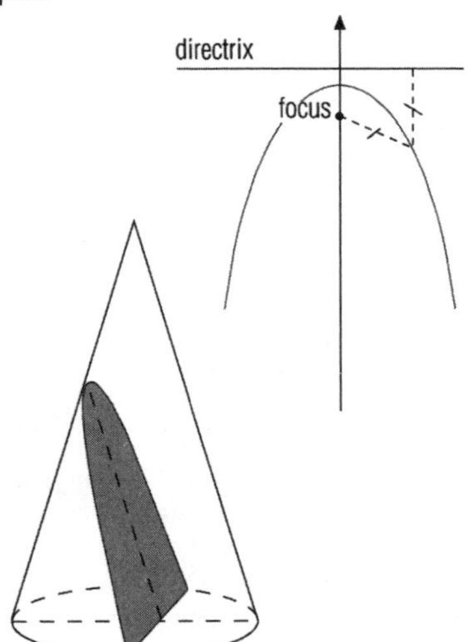

parabola A parabola is a curve produced when a cone is cut by a plane. It is one of the family of curves called conic sections that also includes the circle, ellipse, and hyperbola. These curves are produced when the plane cuts the cone at different angles and positions. Helicon

Paracelsus, Philippus Aureolus (1493–1541) adopted name of Theophrastus Bombastus von Hohenheim, Swiss physician, alchemist, and scientist who developed the idea that minerals and chemicals might have medical uses (iatrochemistry). He introduced the use of laudanum (which he named) for pain-killing purposes.

paradigm all those factors, both scientific and sociological, that influence the research of the scientist. The term, first used by the US historian of science Thomas Kuhn, has subsequently spread to social studies and politics.

paraffin common name for †alkane, any member of the series of hydrocarbons with the general formula C_nH_{2n+2}. The lower members are gases, such as methane (marsh or natural gas). The middle ones (mainly liquid) form the basis of gasoline, kerosene, and lubricating oils, while the higher ones (paraffin waxes) are used in ointment and cosmetic bases.

paraffin wax white, flammable, waxy solid composed of a mixture of saturated hydrocarbons produced by the distillation of petroleum. It is widely used in making candles, polishes, cosmetics, and waterproofing compounds.

parahydrogen †molecules of hydrogen in which the †spins of the two atoms are opposed, in contrast to orthohydrogen in which the spins are parallel.

paraldehyde common name for †ethanal trimer.

parallax change in the apparent position of an object against its background when viewed from two different positions. In astronomy, nearby stars show a shift owing to parallax when viewed from different positions on the Earth's orbit around the Sun. A star's parallax is used to deduce its distance from Earth.

Nearer bodies such as the Moon, Sun, and planets also show a parallax caused by the motion of the Earth.

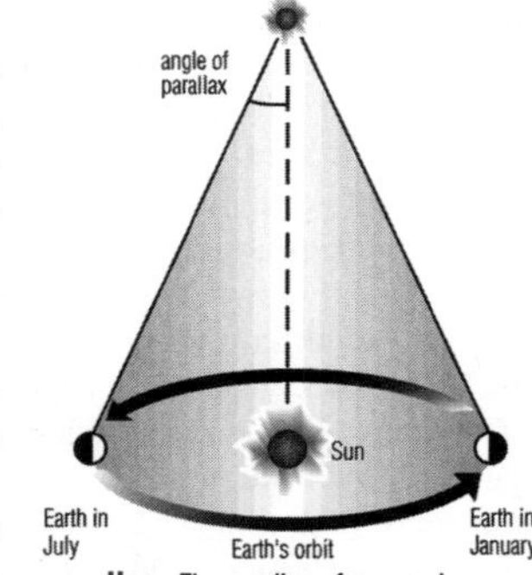

parallax The parallax of a star, the apparent change of its position during the year, can be used to find the star's distance from the Earth. The star appears to change its position because it is viewed at a different angle in July and January. By measuring the angle of parallax, and knowing the diameter of the Earth's orbit, simple geometry can be used to calculate the distance to the star. Helicon

parallel circuit electrical circuit in which the components are connected side by side. The current flowing in the circuit is shared by the components.

voltage in a parallel circuit Components are connected across the same voltage source, and therefore the voltage across each of the components is the same (provided the resistance of each component is also the same). This can be written as $V_T = V_1 = V_2$, where V_T is the total voltage, V_1 is the voltage across component 1, and V_2 is the voltage across component 2. For example, if the voltage across component 1 is 6 volts, then the voltage across component 2 is also 6 volts. The mains supply in the UK is 240 volts. All electrical appliances are designed to operate efficiently at this voltage. Appliances are connected in parallel circuits with the same voltage applied to each appliance.

current in a parallel circuit The current divides to flow through each of the components in the parallel paths. The amount of current that flows through each of the components depends on the resistance of each component. A component with a lower resistance will allow more current to flow through. If I_1 is the current flowing through component 1 and I_2 is the current flowing through component 2, then the total current, I_T, is given by $I_T = I_1 + I_2$.

resistance in a parallel circuit The total resistance R_T of two conductors of resistances R_1 and R_2 connected in parallel is given by:

$1/R_T = 1/R_1 + 1/R_2$

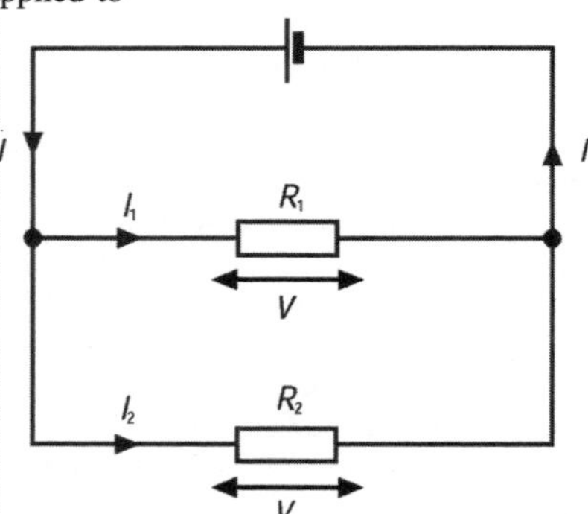

parallel circuit In a parallel circuit, the components are connected side by side, so that the current is split between two or more parallel paths or conductors. Helicon

parallel device in computing, a device that communicates binary data by sending the bits that represent each character simultaneously along a set of separate data lines, unlike a serial device.

parallelepiped six-faced solid in which each pair of opposite faces are equal and parallel †parallelograms.

parallel lines and parallel planes straight lines or planes that always remain a constant distance from one another no matter how far they are extended.

Examples of parallel lines found in everyday life include railway lines, the verticals of goal posts, and shelves in a book-case. Parallel lines are shown in diagrams by using single or double arrows:

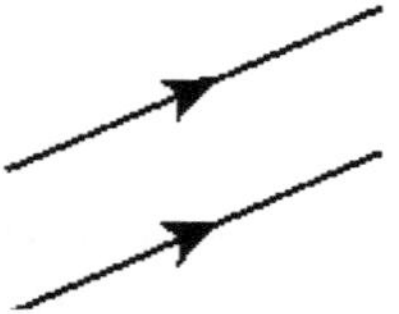

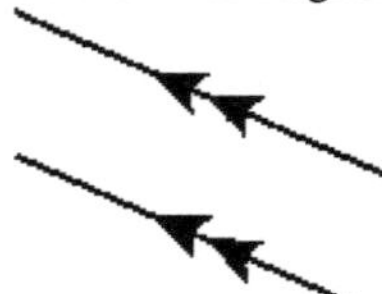

parallelogram a quadrilateral (a shape with four sides) with opposite pairs of sides equal in length and parallel, and opposite angles equal. The diagonals of a parallelogram bisect each other. Its area is the product of the length (l) of one side and the perpendicular distance (height h) between this and the opposite side; the formula is $A = l \times h$. In the special case when all four sides are equal in length, the parallelogram is known as a rhombus, and when the internal angles are right angles, it is a rectangle or square.

For example, to find the area of a parallelogram 12 cm long and 6 cm high:

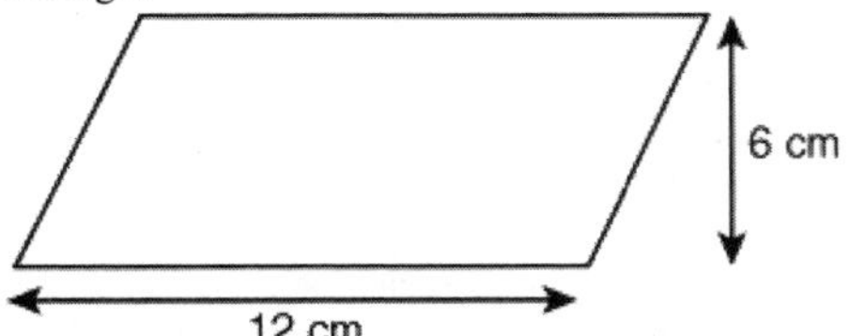

$A = l \times h$, so

$A = 12 \times 6 = 72$ cm^2

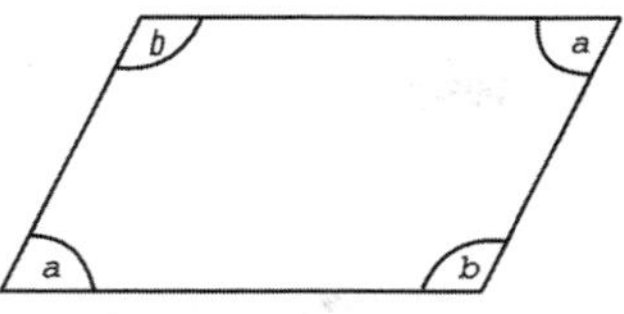

(i) opposite sides and angles are equal

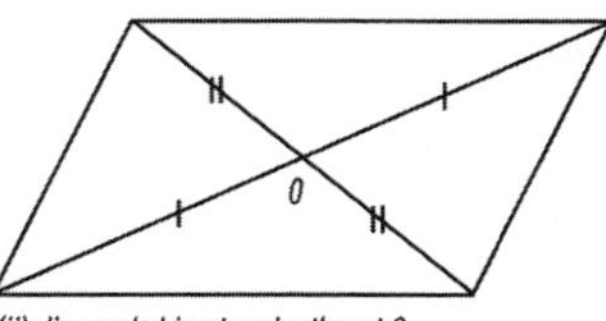

(ii) diagonals bisect each other at 0

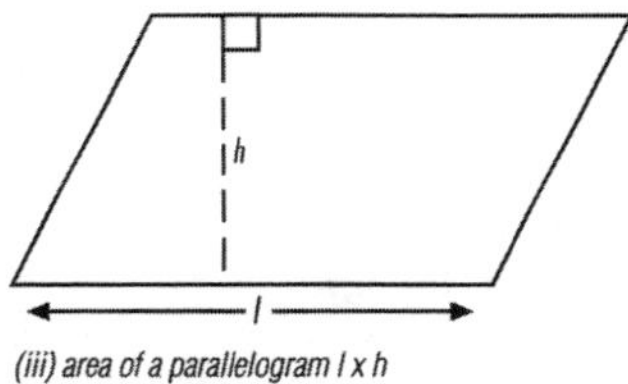

(iii) area of a parallelogram l x h

parallelogram Some properties of a parallelogram. Helicon

parallelogram of forces in physics and applied mathematics, a method of calculating the resultant (combined effect) of two different forces acting together on an object. Because a force has both magnitude and direction it is a †vector quantity and can be represented by a straight line. A second force acting at the same point in a different direction can be represented by another line drawn at an angle to the first. By completing the parallelogram (of which the two lines are sides) a diagonal may be drawn from the original angle to the opposite corner to represent the resultant force vector.

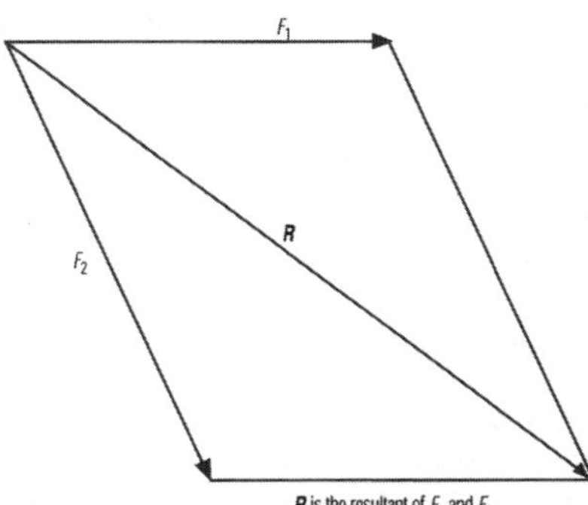

parallelogram of forces The diagram shows how the parallelogram of forces can be used to calculate the resultant (combined effect) of two different forces acting together on an object. The two forces are represented by two lines drawn at an angle to each other. By completing the parallelogram (of which the two lines are sides), a diagonal may be drawn from the original angle to the opposite corner to represent the resultant force vector. Helicon

parallel processing emerging computer technology that allows more than one computation at the same time. Although in the 1990s this technology enabled only a small number of computer processor units to work in parallel, in theory thousands or millions of processors could be used at the same time.

Parallel processing, which involves breaking down computations into small parts and performing thousands of them simultaneously, rather than in a linear sequence, offers the prospect of a vast improvement in working speed for certain repetitive applications.

paramagnetic material material that is weakly pulled towards a strong magnet. The effect is caused by unpaired electrons in the atoms of the material, which cause the atoms to act as weak magnets. A paramagnetic material has a fairly low magnetic †susceptibility that is inversely proportional to temperature. See †magnetism.

paramagnetism property of most elements and some compounds (but excluding the ferromagnetic substances – iron, cobalt, nickel, and their alloys) whereby they are weakly magnetized by relatively strong magnetic fields. See also †ferromagnetism.

parameter variable factor or characteristic. For example, length is one parameter of a rectangle; its height is another. In computing, it is frequently useful to describe a program or object with a set of variable parameters rather than fixed values.

For example, if a programmer writes a routine for drawing a rectangle using general parameters for the length, height, line thickness, and so on, any rectangle can be drawn by this routine by giving different values to the parameters.

Similarly, in a word-processing application that stores parameters for font, page layout, type of justification, and so on, these can be changed by the user.

paranormal not within the range of, or explicable by, established science. Paranormal phenomena include †**extrasensory perception** (ESP) which takes in clairvoyance, precognition, and telepathy;

telekinesis, the movement of objects from one position to another by human mental concentration; and **mediumship**, supposed contact with the spirits of the dead, usually via an intermediate "guide" in the other world. †Parapsychology is the study of such phenomena.

Paranormal phenomena are usually attributed to the action of an unknown factor, †psi.

There have been many reports of sporadic paranormal phenomena, the most remarkable being reports by one person, or occasionally more, of apparitions or hallucinatory experiences associated with another person's death.

Anomalist Site http://www.anomalist.com/

Internet edition of a twice-yearly magazine exploring the mysteries of science, history, and nature. Created by writers, scientists, and investigators intrigued by "that which departs from the common," the online presentation includes contributions on every neglected, unexplained, unexpected, and extraordinary aspect of the world, and invites contributions from visitors.

paraphyletic in biological classification, term describing a group of species that includes the most recent common ancestor of all its members but not all of its descendants. For example, the class Reptilia is a paraphyletic group because it includes the reptiles' sauropsid ancestor but not birds, which are also its descendants. Other examples are invertebrates and fish.

If a group consists of all the descendants of a common ancestor plus the ancestor itself, it is called †monophyletic; if it includes some or all of the descendants but not their common ancestor, it is called †polyphyletic. Distinguishing between such groups is an integral part of the system of classification known as †cladistics, which seeks to classify organisms into a hierarchy of monophyletic groups, or **clades**.

parapsychology (Greek *para* "beyond") study of †paranormal phenomena, which are generally subdivided into two types: †extrasensory perception (ESP), or the paracognitive; and psychokinesis (PK), telekinesis, or the paraphysical – movement of an object without the use of physical force or energy.

Most research into parapsychology has been experimental. The first Society for Psychical Research was established in London in 1882 by scientists, philosophers, classical scholars, and spiritualists. Despite continued skepticism within the scientific establishment, a chair of parapsychology was established in 1984 at Edinburgh University, endowed by the Hungarian author Arthur Koestler.

paraquat technical name **1,1-dimethyl-4,4-dipyridylium**; $CH_3(C_5H_4N)_2CH_3.2CH_3SO_4$ nonselective herbicide (weedkiller). Although quickly degraded by soil micro-organisms, it is deadly to human beings if ingested.

parasite organism that lives on or in another organism (called the host) and depends on it for nutrition, often at the expense of the host's welfare. Parasites that live inside the host, such as liver flukes and tapeworms, are called **endoparasites**; those that live on the exterior, such as fleas and lice, are called **ectoparasites**.

Parasitic wasps, such as ichneumons, are more correctly **parisitoids**, as they ultimately kill their hosts.

parathyroid one of a pair of small †endocrine glands. Most tetrapod vertebrates, including humans, possess two such pairs, located behind the †thyroid gland. They secrete parathyroid hormone, which regulates the amount of calcium in the blood.

Pardee, Arthur Beck (1921–) US biochemist who discovered that enzymes can be inhibited by the final product of the reaction they catalyze (trigger); a regulatory strategy used by cells, called **feedback inhibition**.

parenchyma plant tissue composed of loosely packed, more or less spherical cells, with thin cellulose walls. Although parenchyma often has no specialized function, it is usually present in large amounts, forming a packing or ground tissue. It usually has many intercellular spaces.

parental care in biology, the time and energy spent by a parent in order to rear its offspring to maturity. Among animals, it ranges from the simple provision of a food supply for the hatching young at the time the eggs are laid (for example, many wasps) to feeding and protection of the young after hatching or birth, as in birds and mammals. In the more social species, parental care may include the teaching of skills – for example, female cats teach their kittens to hunt.

parental generation in genetic crosses, the set of individuals at the start of a test, providing the first set of gametes from which subsequent generations (known as the F1 and F2) will arise.

parity concept in quantum theory related to the mirror-symmetry of the functions describing mathematically the behavior of the particles.

parity of a number, the state of being either even or odd. In computing, the term refers to the number of 1s in the binary codes used to represent data. A binary representation has **even parity** if it contains an even number of 1s and **odd parity** if it contains an odd number of 1s.

For example, the binary code 1000001, commonly used to represent the character "A," has even parity because it contains two 1s, and the binary code 1000011, commonly used to represent the character "C," has odd parity because it contains three 1s. A **parity bit** (0 or 1) is sometimes added to each binary representation to give all the same parity so that a validation check can be carried out each time data are transferred from one part of the computer to another. So, for example, the codes 1000001 and 1000011 could have parity bits added and become **0**1000001 and **1**1000011, both with even parity. If any bit in these codes should be altered in the course of processing the parity would change and the error would be quickly detected.

parsec symbol pc, unit used for distances to stars and galaxies. One parsec is equal to 3.2616 †light years, 2.063 x 10^5 †astronomical units, and 3.857 x 10^{13} km.

A parsec is the distance at which a star would have a †parallax (apparent shift in position) of one second of arc when viewed from two points the same distance apart as the Earth's distance from the Sun; or the distance at which one astronomical unit subtends an angle of one second of arc.

Parsons, William (1800–1867) 3rd Earl of Rosse, Irish astronomer, engineer, and politician. He built the largest telescope then in use. He found 15 spiral nebulae and named the †Crab nebula. He was among the first to take photographs of the Moon.

parthenocarpy in botany, the formation of fruits without seeds. This phenomenon, of no obvious benefit to the plant, occurs naturally in some plants, such as bananas. It can also be induced in some fruit crops, either by breeding or by applying certain plant hormones.

parthenogenesis development of an ovum (egg) without any genetic contribution from a male. Parthenogenesis is the normal means of reproduction in a few plants (for example, dandelions) and animals (for example, certain fish). Some sexually reproducing species, such as aphids, show parthenogenesis at some stage in their life cycle to accelerate reproduction to take advantage of good conditions.

partial differentiation form of †differential calculus in which instead of a function $y = f(x)$ – which has one variable – a function of two or more variables is considered; for example, $z = f(x,y)$. Such functions represent a surface in three-dimensional space.

partial eclipse solar †eclipse in which the Sun's disc is not completely covered by the Moon; or a lunar eclipse in which the Earth's shadow does not completely cover the surface of the Moon.

partially permeable membrane or **semipermeable membrane**, in biology, membrane that allows the passage through it of some substances but not others. The passage of substances depends on the size of molecules, diameter of membrane pores, surface area of the membrane, and whether passive or †active transport is involved. In general, substances pass through a membrane if their particles are smaller than the pores of the membrane.

particle detector one of a number of instruments designed to detect subatomic particles and track their paths; they include the †cloud chamber, †bubble chamber, †spark chamber, and multiwire chamber.

The earliest particle detector was the cloud chamber, which contains a super-saturated vapor in which particles leave a trail of droplets, in much the same way that a jet aircraft leaves a trail of vapor in the sky. A bubble chamber contains a superheated liquid in which a particle leaves a trail of bubbles. A spark chamber contains a series of closely-packed parallel metal plates, each at a high voltage. As particles pass through the chamber, they leave a visible spark between the plates. A modern multiwire chamber consists of an array of fine, closely-packed wires, each at a high voltage. As a particle passes through the chamber, it produces an electrical signal in the wires. A computer analyses the signal and reconstructs the path of the particles. Multiwire detectors can be used to detect X-rays and gamma rays, and are used as detectors in †positron emission tomography (PET).

particle physics study of the particles that make up all atoms, and of their interactions. More than 300 subatomic particles have now been identified by physicists, categorized into several classes according to their mass, electric charge, spin, magnetic moment, and interaction. Subatomic particles include the †elementary particles (†quarks, †leptons, and †gauge bosons), which are indivisible, so far as is known, and so may be considered the fundamental units of matter; and the †hadrons (baryons, such as the proton and neutron, and mesons), which are composite particles, made up of two or three quarks. Quarks, protons, electrons, and neutrinos are the only stable particles (the neutron being stable only when in the atomic nucleus). The unstable particles decay rapidly into other particles, and are known from experiments with particle accelerators and cosmic radiation. See †atomic structure.

Pioneering research took place at the Cavendish laboratory, Cambridge, England. In 1897 the English physicist J J Thomson discovered that all atoms contain identical, negatively charged particles (†electrons), which can easily be freed. By 1911 the New Zealand physicist Ernest Rutherford had shown that the electrons surround a very small, positively charged †nucleus. In the case of hydrogen, this was found to consist of a single positively charged particle, a †proton. The nuclei of other elements are made up of protons and uncharged particles called †neutrons. 1932 saw the discovery of a particle (whose existence had been predicted by the British theoretical physicist Paul Dirac in 1928) with the mass of an electron, but an equal and opposite charge – the †positron. This was the first example of †antimatter; it is now believed that all particles have corresponding antiparticles. In 1934 the Italian-born US physicist Enrico Fermi argued that a hitherto unsuspected particle, the †neutrino, must accompany electrons in beta-emission.

particles and fundamental forces By the mid-1930s, four types of fundamental †force interacting between particles had been identified. The †electromagnetic force (1) acts between all particles with electric charge, and is related to the exchange between these particles of †gauge bosons called †photons, packets of electromagnetic radiation. In 1935 the Japanese physicist Hideki Yukawa suggested that the †strong nuclear force (2) (binding protons and neutrons together in the nucleus) was transmitted by the exchange of particles with a mass about one-tenth of that of a proton; these particles, called †pions (originally pi mesons), were found by the British physicist Cecil Powell in 1946. Yukawa's theory was largely superseded from 1973 by the theory of †quantum chromodynamics, which postulates that the strong nuclear force is transmitted by the exchange of gauge bosons called †gluons between the †quarks and antiquarks making up protons and neutrons. Theoretical work on the †weak nuclear force (3) began with Enrico Fermi in the 1930s. The existence of the gauge bosons that carry this force, the W and Z particles, was confirmed in 1983 at CERN, the European nuclear research organization. The fourth fundamental force, †gravity (4), is experienced by all matter; the postulated carrier of this force has been named the †graviton.

leptons The electron, muon, tau, and their neutrinos comprise the †leptons – particles with half-integral spin that "feel" the weak nuclear and electromagnetic force but not the strong force. The muon (found by the US physicist Carl Anderson in cosmic radiation in 1937) produces the muon neutrino when it decays; the tau, a surprise discovery of the 1970s, produces the tau neutrino when it decays.

mesons and baryons The hadrons (particles that "feel" the strong nuclear force) were found in the 1950s and 1960s. They are classified into †mesons, with whole-number or zero spins, and †baryons (which include protons and neutrons), with half-integral spins. It was shown in the early 1960s that if hadrons of the same spin are represented as points on suitable charts, simple patterns are formed. This symmetry enabled a hitherto unknown baryon, the omega-minus, to be predicted from a gap in one of the patterns; it duly turned up in experiments.

quarks In 1964 the US physicists Murray Gell-Mann and George Zweig suggested that all hadrons were built from three "flavors" of a new particle with half-integral spin and a charge of magnitude either $^1/_3$ or $^2/_3$ that of an electron; Gell-Mann named the particle the **quark**. Mesons are quark–antiquark pairs (spins either add to one or cancel to zero), and baryons are quark triplets. To account for new mesons such as the psi (J) particle the number of quark flavors had risen to six by 1985.

Particle Physics http://www.pparc.ac.uk/

Particle Physics and Astronomy Research Council (PPARC) Web site that includes information about its extensive research programs in the fields of astronomy, planetary science, and particle physics, as well as more general information on stars and the origins of the universe.

particle size in chemistry the size of the grains that make up a powder. The grain size has an effect on certain properties of a substance.

Finely divided powders have a greater surface area for contact; they therefore react more quickly, dissolve more readily, and are of increased efficiency as catalysts compared with their larger-sized counterparts.

particle, subatomic a particle that is smaller than an atom; see †particle physics.

particle theory concept that all matter is made up of particles. †Elements are made up of particles called †atoms. Atoms are built up from three types of particle: †protons, †neutrons, and †electrons. Other types of particles in matter are †molecules and †ions. A molecule is a small group of atoms that are held together by †covalent bonds. An ion is a charged particle. It is formed by an atom either gaining or losing electrons. The properties of solids, liquids, and gases can be explained by the †kinetic theory of matter, the theory that all matter is made up of very small particles which are in constant motion.

partition in mathematics, of a set, to divide into parts.

partitioning of an integer in number theory, breaking the integer down into its constituent parts in as many ways as possible; for example, the number 6 can be partitioned in three ways: 5 + 1, 4 + 2, and 3 + 3. Each of these contributory numbers is known as a summand.

Pascal acronym for **Program Appliqué à la Selection et la Compilation Automatique de la Littérature**, high-level computer-programming language. Designed by Niklaus †Wirth in the 1960s as an aid to teaching programming, it is still widely used as such in universities, and as a good general-purpose programming language. Most professional programmers, however, now use †C or †C++. Pascal was named after the 17th-century French mathematician Blaise †Pascal.

Pascal Central http://www.pascal-central.com/

Provides an important gateway to Pascal on the Internet. The easily navigable pages contain articles, news and announcements, and links to dozens of Web sites with Pascal code and tools for programmers.

pascal symbol Pa, SI unit of pressure, equal to one newton per square meter. It replaces †bars and millibars (10^5 Pa equals one bar). It is named after the French mathematician Blaise Pascal.

Pascal, Blaise (1623–1662) French philosopher and mathematician. He contributed to the development of hydraulics, †calculus, and the mathematical theory of †probability.

Pensées http://www.ccel.org/p/pascal/pensees/pensees.htm

Text of Pascal's famous work.

Pascal's pressure law or **Pascal's Principle**, †pressure applied within a fluid is transmitted equally in all directions, the †force per unit area being everywhere the same.

Pascal's triangle triangular array of numbers (with 1 at the apex), in which each number is the sum of the pair of numbers above it. It is named after French mathematician Blaise Pascal, who used it in his study of probability. When plotted at equal distances along a horizontal axis, the numbers in the rows give the binomial probability distribution (with equal probability of success and failure) of an event, such as the result of tossing a coin.

1
1 1
1 2 1
1 3 3 1
1 4 6 4 1
1 5 10 10 5 1
1 6 15 20 15 6 1
1 7 21 35 35 21 7 1

Pascal's triangle In Pascal's triangle, each number is the sum of the two numbers immediately above it, left and right – for example, 2 is the sum of 1 and 1, and 4 is the sum of 3 and 1. Furthermore, the sum of each row equals a power of 2 – for example, the sum of the 3rd row is 4 = 2^2; the sum of the 4th row is 8 = 2^3. Helicon

Pasiphae small outer moon of Jupiter discovered in 1908. It is about 50 km/31 mi across and has a steeply inclined retrograde orbit at an average distance of 23.3 million km/14.4 million mi from the planet. It takes 735 days to complete one orbit.

passive margin in †plate tectonics, a boundary between oceanic and continental crust that is within a single tectonic plate rather than at the boundary between two plates. Very little tectonic activity occurs at passive margins. Examples include the east coast of North America and the entire coast of Australia.

passivity in chemistry, property of certain metals to stop reacting during a chemical reaction due to the formation of a stable oxide layer on the metal surface, for example as can occur with cobalt and nickel. This oxide layer can be formed during the reaction or already be present on the metal surface, as with aluminum.

Pasteur, Louis (1822–1895) French chemist and microbiologist who discovered that fermentation is caused by micro-organisms and developed the germ theory of disease. He also created a vaccine for rabies, which led to the foundation of the Pasteur Institute in Paris in 1888.

patch in computing, modification or update made to a program, consisting of a short segment of additional code. Developers often correct bugs or fine-tune software by releasing a patch which rewrites existing codes and adds new material.

patella or **kneecap**, flat bone embedded in the knee tendon of birds and mammals, which protects the joint from injury.

pathogen (Greek "disease producing") any micro-organism that causes disease. Most pathogens are †parasites, and the diseases they cause are incidental to their search for food or shelter inside the host. Nonparasitic organisms, such as soil bacteria or those living in the human gut and feeding on waste foodstuffs, can also become pathogenic to a person whose immune system or liver is damaged. The larger parasites that can cause disease, such as nematode worms, are not usually described as pathogens.

pathology medical specialty concerned with the study of disease processes and how these provoke structural and functional changes in the body.

Paul, Wolfgang (1913–1993) German nuclear physicist who made fundamental contributions to molecular beam †spectroscopy, mass †spectrometry, and electron acceleration technology. He was awarded the Nobel Prize for Physics in 1989 for his development of the ion trap, or "Paul trap," used to store single atoms long enough to make accurate spectroscopic measurements on them. He shared the award with US scientists Norman †Ramsey and Hans †Dehmelt.

Pauli, Wolfgang (1900–1958) Austrian-born Swiss physicist who was awarded the Nobel Prize for Physics in 1945 for his discovery of the **exclusion principle**: in a given system no two fermions (electrons, protons, neutrons, or other elementary particles of half-integral spin) can be characterized by the same set of †quantum numbers. He also predicted the existence of †neutrinos.

Pauli, Wolfgang http://www-groups.dcs.st-andrews.ac.uk/history/Mathematicians/Pauli.html

Part of an archive containing the biographies of the world's greatest mathematicians, this site is devoted to the life and contributions of physicist Wolfgang Pauli. In addition to biographical information, you will find a list of references about Pauli and links to other essays in the archive that reference him. The text of this essay includes hypertext links to the essays of those mathematicians and thinkers who influenced Pauli.

Pauli exclusion principle in physics, a principle of atomic structure. See †exclusion principle.

Pauling, Linus Carl (1901–1994) US theoretical chemist and biologist. He was awarded the Nobel Prize for Chemistry in 1954 for his study of the nature of chemical bonds, especially in complex substances. His ideas are fundamental to modern theories of molecular structure. He also investigated the properties and uses of vitamin C as related to human health. He was awarded the Nobel Prize for Peace in 1962 for having campaigned for the control of nuclear weapons and nuclear testing.

Dr Linus Pauling Profile http://www.achievement.org/autodoc/page/pau0pro- 1/

Description of the life and works of the double Nobel prizewinner Linus Pauling. Pauling is one of the very few people to have won more than one Nobel prize, having won the prize for chemistry in 1954, followed by the prize for peace in 1962. The Web site contains not only a profile and biographical information, but also holds a lengthy interview with Pauling from 1990 accompanied by a large number of photographs, video sequences, and audio clips.

Pavlov, Ivan Petrovich (1849–1936) Russian physiologist who was awarded a Nobel Prize for Physiology or Medicine in 1904 for his discovery of the physiology of digestion. Pavlov studied conditioned reflexes in animals (see †conditioning). His work had a great impact on behavioral theory (see †behaviorism) and learning theory.

Pavo constellation of the southern hemisphere, represented as a peacock.

Payne-Gaposchkin, Cecilia Helena (1900–1979) born Cecilia Payne, English-born US astronomer who studied stellar evolution and galactic structure. Her investigation of stellar atmospheres during the 1920s gave some of the first indications of the overwhelming abundance of the lightest elements (hydrogen and helium) in the Galaxy.

Peano, Giuseppe (1858–1932) Italian mathematician who was a pioneer in symbolic logic. His concise logical definitions of natural numbers were devised in order to derive a complete system of notation for logic. He also discovered a curve that fills topological space.

Peano, Giuseppe http://www-history.mcs.st-and.ac.uk/history/Mathematicians/Peano.html

Extensive biography of the mathematician. The site contains a description of his contribution to mathematics, and in particular logic, along with an extract from Rivista di Matematica (1895). Several literature references for further reading are also listed, and the Web site also features a photograph of Peano.

Peano axioms axioms that formally introduce the properties of the positive whole numbers (originally devised, despite being named after Italian mathematician Guiseppe Peano, by German mathematician Julius Dedekind).

Pearson, Karl (1857–1936) English statistician who followed Francis †Galton in introducing statistics and probability into genetics and who developed the concept of eugenics (improving the human race by selective breeding). He introduced the term †standard deviation into statistics.

Pearson, Karl http://www-groups.dcs.st-andrews.ac.uk/history/Mathematicians/Pearson.html

Part of an archive containing the biographies of the world's greatest mathematicians, this site is devoted to the life and contributions of statistician Karl Pearson. In addition to biographical information, you will find a list of references about Pearson and links to other essays in the archive that reference him.

peat organic matter found in bogs and formed by the incomplete decomposition of plants such as sphagnum moss. Northern Asia, Canada, Finland, Ireland, and other places have large deposits, which have been dried and used as fuel from ancient times. Peat can also be used as a soil additive. Peat bogs began to be formed when glaciers retreated, about 9,000 years ago. They grow at the rate of only a millimeter a year, and large-scale digging can result in destruction both of the bog and of specialized plants growing there. The destruction of peat bogs is responsible for diminishing fish stocks in coastal waters; the run-off from the peatlands carries high concentrations of iron, which affects the growth of the plankton on which the fish feed.

Approximately 60% of the world's wetlands are peat. In May 1999 the Ramsar Convention on the Conservation of Wetlands approved a peatlands action plan that should have a major impact on the conservation of peat bogs.

peck obsolete unit of dry measure, equaling eight quarts or a quarter bushel (9.002 liters).

pectoral relating to the upper area of the thorax associated with the muscles and bones used in moving the arms or forelimbs, in vertebrates. In birds, the *pectoralis major* is the very large muscle used to produce a powerful downbeat of the wing during flight.

Pedersen, Charles (1904–1990) US organic chemist who shared the Nobel Prize for Chemistry in 1987 with Jean †Lehn and Donald †Cram for the discovery of "crown ether," a cyclic polyether – a molecule with 12 carbon atoms and six oxygen atoms arranged in a crown-like structure. Its discovery opened up the field of guest–host chemistry.

pedicel stalk of an individual flower, which attaches it to the main floral axis, often developing in the axil of a bract.

pediment broad, gently inclined erosion surface formed at the base of a mountain as it erodes and retreats. Pediments consist of bedrock and are often covered with a thin layer of sediments, called alluvium, which have been eroded off the mountain.

pediplain in earth science, plain at the base of a mountain in arid climates, either bare or covered with a thin layer of alluvial deposits.

Pegasus a constellation of the northern hemisphere, near Cygnus, and represented as the winged horse of Greek mythology.

pegmatite extremely coarse-grained †igneous rock of any composition found in veins; pegmatites are usually associated with large granite masses.

Peierls, Rudolf Ernst (1907–1995) German-born British physicist who contributed to the early theory of the neutron–proton system. He helped to develop the atomic bomb 1940–46.

pelagic of or pertaining to the open ocean, as opposed to bottom or shore areas. **Pelagic sediment** is fine-grained fragmental material that has settled from the surface waters, usually the siliceous and calcareous skeletal remains (see †ooze) of marine organisms, such as radiolarians and foraminifera.

pellagra chronic disease mostly seen in subtropical countries in which the staple food is maize (corn). It is caused by deficiency of †nicotinic acid (one of the B vitamins), which is contained in protein foods, beans and peas, and yeast. Symptoms include diarrhea, skin eruptions, and mental disturbances.

Peltier effect in physics, a change in temperature at the junction of two different metals produced when an electric current flows through them. The extent of the change depends on what the conducting metals are, and the nature of change (rise or fall in temperature) depends on the direction of current flow. It is the reverse of the †Seebeck effect. It is named after the French physicist Jean Charles Peltier (1785–1845) who discovered it in 1834.

pelvis in vertebrates, the lower area of the abdomen featuring the bones and muscles used to move the legs or hindlimbs. The **pelvic girdle** is a set of bones that allows movement of the legs in relation to the rest of the body and provides sites for the attachment of relevant muscles.

pendulum weight (called a "bob") swinging at the end of a rod or cord. The regularity of a pendulum's swing was used in making the first really accurate clocks in the 17th century. Pendulums can be used for measuring the acceleration due to gravity (an important constant in physics).

Specialized pendulums are used to measure velocities (ballistic pendulum) and to demonstrate the Earth's rotation (Foucault's pendulum).

Foucault Pendulum http://www.calacademy.org/products/pendulum.html

Information on the Foucault pendulum based at the California Academy of Sciences. The Web site has a fun tutorial on Foucault pendulums.

peneplain in earth science, plain to which the features of a landscape are hypothetically reduced by prolonged erosion.

Penfield, Wilder Graves (1891–1976) Canadian neurosurgeon who, while developing surgical options for the treatment of epilepsy, located several functional areas of the †cerebral cortex by stimulating the patient's brain electrically while he or she was conscious. He demonstrated that sensations and memories, including memories that could not be recalled except by electrical stimulation, had specific locations in the brain that remained constant. He devised the homunculus, a schematic representation showing the site and relative proportions of the cortical areas specialized for motor and sensory functions.

penicillin any of a group of †antibiotic (bacteria killing) compounds obtained from filtrates of moulds of the genus *Penicillium* (especially *P. notatum*) or produced synthetically. Penicillin was the first antibiotic to be discovered (by Alexander †Fleming); it kills a broad spectrum of bacteria, many of which cause disease in humans.

The use of the original type of penicillin is limited by the increasing resistance of †pathogens and by allergic reactions in patients. Since 1941, numerous other antibiotics of the penicillin family have been discovered which are more selective against, or resistant to, specific micro-organisms.

peninsula land surrounded on three sides by water but still attached to a larger landmass. Florida, USA, is an example.

penis male reproductive organ containing the †urethra, the channel through which urine and †semen are voided. It transfers sperm to the female reproductive tract to fertilize the ovum. In mammals, the penis is made erect by vessels that fill with blood, and in most mammals (but not humans) is stiffened by a bone. Snakes and lizards have a paired structure that serves as a penis; other reptiles have a single organ. A few birds, mainly ducks and geese, also have a type of penis, as do snails, barnacles, and some other invertebrates. Many insects have a rigid, nonerectile male organ, usually referred to as an intromittent organ.

Pennsylvanian period US term for the Upper or Late †Carboniferous period of geological time, 323–290 million years ago; it is named after the US state, which contains vast coral deposits.

Penrose, Lionel Sharples (1898–1972) English physician and geneticist who carried out pioneering work on mental retardation and Down's syndrome. He was the first to demonstrate the significance of the mother's age.

Penrose, Roger (1931–) English mathematician who formulated some of the fundamental theorems that describe †black holes, including the singularity theorems, developed jointly with English physicist Stephen †Hawking, which state that once the gravitational collapse of a star has proceeded to a certain degree, singularities (which form the center of black holes) are inevitable. Penrose has also proposed a new model of the universe.

Penston, Michael (1943–1990) British astronomer at the Royal Greenwich Observatory 1965–90. From observations made with the Ultraviolet Explorer Satellite of hot gas circulating around the core of the galaxy NGC 4151, he and his colleagues concluded that a †black hole of immense mass lay at the galaxy's center.

pentadactyl limb typical limb of the mammals, birds, reptiles, and amphibians. These vertebrates (animals with backbone) are all descended from primitive amphibians whose immediate ancestors were fleshy-finned fish. The limb which evolved in those amphibians had three parts: a "hand/foot" with five digits (fingers/toes), a lower limb containing two bones, and an upper limb containing one bone.

This basic pattern has persisted in all the terrestrial vertebrates, and those aquatic vertebrates (such as seals) that are descended from them. Natural selection has modified the pattern to fit different ways of life. In flying animals (birds and bats) it is greatly altered and in some vertebrates, such as whales and snakes, the limbs are greatly reduced or lost. Pentadactyl limbs of different species are an example of †homologous organs.

pentagon five-sided plane figure. A regular pentagon has all five sides of equal length and all five angles of equal size, 108°. It has †golden section proportions between its sides and diagonals. The five-pointed star formed by drawing all the diagonals of a regular pentagon is called a **pentagram**. This star has further golden sections.

pentanol common name **amyl alcohol**; $C_5H_{11}OH$, clear, colorless, oily liquid, usually with a characteristic choking odor. It is obtained by the fermentation of starches and from the distillation of petroleum. There are eight possible isomers.

pentode in physics, thermionic valve (vacuum tube) with five electrodes consisting of anode, cathode, and three grids. A suppressor grid, at the potential of the cathode, is located between the anode and the screen grid where it prevents the loss of secondary electrons emitted by the anode.

penumbra region of partial shade between the totally dark part (umbra) of a †shadow and the fully illuminated region outside. It occurs when a source of light is only partially obscured by a shadow-casting object. The darkness of a penumbra varies gradually from total darkness at one edge to full brightness at the other. In astronomy, a penumbra is a region of the Earth from which only a partial †eclipse of the Sun can be seen.

Penzias, Arno Allan (1933–) German-born US radio engineer who, with the radio astronomer Robert †Wilson, was awarded the Nobel Prize for Physics in 1978 for the discovery in 1964 of †cosmic background radiation. This radiation had been predicted on the basis of the "hot †Big Bang" model of the origin of the universe initially proposed in 1948.

pepsin †enzyme that breaks down proteins during digestion. It is produced by the walls of the †stomach. It requires a strongly acidic environment such as that present in the stomach. It digests large protein molecules into smaller protein molecules (smaller polypeptides) and is therefore a protease – an enzyme that breaks down a protein.

peptide molecule comprising two or more †amino acid molecules (not necessarily different) joined by **peptide bonds**, whereby the acid group of one acid is linked to the amino group of the other (—CO.NH—). The number of amino acid molecules in the peptide is indicated by referring to it as a di-, tri-, or polypeptide (two, three, or many amino acids).

Proteins are built up of interacting polypeptide chains with various types of bonds occurring between the chains. Incomplete hydrolysis (splitting up) of a protein yields a mixture of peptides, examination of which helps to determine the sequence in which the amino acids occur within the protein.

peptide bond bond that joins two peptides together within a protein. The carboxyl (—COOH) group on one †amino acid reacts with the amino (—NH_2) group on another amino acid to form a peptide bond (—CO—NH—) with the elimination of water. Peptide bonds are broken by hydrolytic enzymes called peptidases.

perborate any salt formed by the action of hydrogen peroxide on borates.
Perborates contain the radical BO_3.

percentage way of representing a number as a †fraction of 100. For example, 45 percent (45%) equals $^{45}/_{100}$, and 45% of 20 is $^{45}/_{100}$ x 20 = 9.

percentage increase/decrease In general, if a quantity changes from one value to another then percentage = 100 x $^{\text{(difference in the values)}}/_{\text{old value}}$

For example, in a sale the price of a bicycle is reduced from £120 to £90, that is there is a discount of £30. The percentage decrease is 100 x $^{30}/_{120}$ = 25%.

fractions as percentages To express a fraction as a percentage, its denominator must first be converted to 100. For example, value-added tax (VAT) as a fraction is $^{7}/_{40}$:

$^{7}/_{40} = {}^{17.5}/_{100}$ = 17.5%

The use of percentages often makes it easier to compare fractions that do not have a †common denominator (a number divisible by both the bottom numbers), for instance when comparing rates of inflation or rates of simple interest.

To convert a fraction to a percentage on a calculator, divide the numerator by the denominator and then multiply by 100. The percentage will correspond to the first two figures of the †decimal; for example, $^{7}/_{12}$ = 0.5833333 = 58.3% (correct to 1 decimal place), and $^{7}/_{32}$ = 0.21875 = 21.9% (correct to 1 decimal place).

The percentage sign is thought to have been derived as an economy measure when recording in old counting houses; writing in the numeric symbol for $^{25}/_{100}$ would take two lines of parchment, and hence the "100" denominator was put alongside the 25 and rearranged to "%".

percentile in a cumulative frequency distribution, one of the 99 values of a variable that divide its distribution into 100 parts of equal frequency. In practice, only certain of the percentiles are used. They are the †median (or 50th percentile), the lower and the upper †quartiles (respectively, the 25th and 75th percentiles), the 10th percentile that cuts off the bottom 10% of a frequency distribution, and the 90th that cuts off the top 10%. The 5th and 95th are also sometimes used.

perchloric acid or **chloric acid**; $HClO_4$, colorless, fuming liquid, boiling point 110°C/230°F. It is produced by reacting potassium chlorate with concentrated sulfuric acid. Perchloric acid is a strong oxidizing agent and is explosive in its pure form or in contact with organic compounds.

Peregrinus, Petrus (*c.* 1220–*c.* 1270/90) adopted name of Peregrinus de Maricourt, French scientist and scholar who published *Epistola de magnete* (1269). In it he described a simple compass (a piece of magnetized iron on a wooden disc floating in water) and outlined the laws of magnetic attraction and repulsion.

perennating organ in plants, that part of a †biennial plant or herbaceous perennial that allows it to survive the winter; usually a root, tuber, rhizome, bulb, or corm.

perennial plant plant that lives for more than two years. Herbaceous perennials have aerial stems and leaves that die each autumn. They survive the winter by means of an underground storage (perennating) organ, such as a bulb or rhizome. Trees and shrubs or woody perennials have stems that persist above ground throughout the year, and may be either †deciduous or †evergreen. See also †annual plant, †biennial plant.

Perey, Marguérite (Catherine) (1909–1975) French nuclear chemist who discovered the radioactive element francium in 1939. Her career, which began as an assistant to Marie †Curie in 1929, culminated with her appointment as professor of nuclear chemistry at the University of Strasbourg in 1949 and director of its Center for Nuclear Research in 1958.

perfect number number that is equal to the sum of all its factors (except itself); for example, 6 is a perfect number, being equal to 1 + 2 + 3.

perfusion †in vitro-induced passage of blood or a nutrient fluid through the blood vessels of an organism to keep it supplied with oxygen and nutrients.

perianth in botany, a collective term for the outer whorls of the flower, which protect the reproductive parts during development. In most †dicotyledons the perianth is composed of two distinct whorls, the calyx of †sepals and the corolla of †petals, whereas in many †monocotyledons the sepals and petals are indistinguishable and the segments of the perianth are then known individually as tepals.

periastron in astronomy, point at which an object traveling in an elliptical orbit around a star is at its closest to the star; the point at which it is furthest is known as the †apastron.

pericardium membranous sac surrounding the heart.

pericarp wall of a †fruit. It encloses the seeds and is derived from the †ovary wall. In fruits such as the acorn, the pericarp becomes dry and hard, forming a shell around the seed. In fleshy fruits the pericarp is typically made up of three distinct layers. The **epicarp**, or **exocarp**, forms the tough outer skin of the fruit, while the **mesocarp** is often fleshy and forms the middle layers. The innermost layer or **endocarp**, which surrounds the seeds, may be membranous or thick and hard, as in the †drupe (stone) of cherries, plums, and apricots.

peridot pale-green, transparent gem variety of the mineral †olivine.

peridotite rock consisting largely of the mineral olivine; pyroxene and other minerals may also be present. Peridotite is an ultramafic rock containing less than 45% silica by weight. It is believed to be one of the rock types making up the Earth's upper mantle, and is sometimes brought from the depths to the surface by major movements, or as inclusions in lavas.

perigee in astronomy, the point at which an object, traveling in an elliptical orbit around the Earth, is at its closest to the Earth. The point at which it is furthest from the Earth is the †apogee.

periglacial environment bordering a glacial area but not actually covered by ice all year round, or having similar climatic and environmental characteristics, such as in mountainous areas. Periglacial areas today include parts of Siberia, Greenland, and North America. The rock and soil in these areas is frozen to a depth of several meters (†permafrost) with only the top few centimeters thawing during the brief summer (the active layer). The vegetation is characteristic of †tundra.

During the last ice age all of southern England was periglacial. Weathering by †freeze–thaw (the alternate freezing and thawing of ice in rock cracks) would have been severe, and †solifluction (movement of soil that is saturated by water) would have taken place on a large scale, causing wet topsoil to slip from valley sides.

perihelion in astronomy, the point at which an object, traveling in an elliptical orbit around the Sun, is at its closest to the Sun. The point at which it is furthest from the Sun is the †aphelion.

perimeter or **boundary**, line drawn around the edge of an area or shape. For example, the perimeter of a rectangle is the sum of the lengths of its four sides; the perimeter of a circle is known as its †circumference.

period in chemistry, a horizontal row of elements in the †periodic table. There is a gradation of properties along each period, from metallic (Group 1, the alkali metals) to non-metallic (Group 7, the halogens).

period another name for menstruation; see †menstrual cycle.

period in physics, the time taken for one complete cycle of a repeated sequence of events. For example, the time taken for a pendulum to swing from side to side and back again is the period of the pendulum. The period is the reciprocal of the †frequency.

periodic in mathematics, function that increases its argument by a fixed amount whilst its value remains the same. That is, if for some $a>0$, $f(x) + a = f(x)$ for all x.

periodic law in chemistry, a generalization that there is a recurring pattern in the properties of elements when they are arranged in order of increasing atomic number. The law is most apparent when the elements are arranged in the †periodic table, in which the elements in each vertical column (group) show similar properties.

periodic table of the elements a table in which the †elements are arranged in order of their atomic number. There are eight groups of elements, plus a block of †transition metals in the center. Group 1 contains the †alkali metals; Group 2 the alkaline-earth †metals; Group 7 the †halogens; and Group 0 the †noble gases (rare gases). A zigzag line through the groups separates the metals on the left-hand side of the table from the non-metals on the right. The horizontal rows in the table are called **periods**. The table summarizes the major properties of the elements and enables predictions to be made about their behavior. There are striking similarities in the chemical properties of the elements in each of the †groups (vertical columns), which are numbered 1–7 (I–VII) and 0 to reflect the number of electrons in the outermost unfilled shell and hence the maximum †valency. Reactivity increases down a group. A gradation (trend) of properties may be traced along the periods. Metallic character increases across a period from right to left, and down a group. A large block of elements, between Groups 2 and 3, contains the transition elements, characterized by displaying more than one valence state.

Chemical Elements.com http://www.chemicalelements.com/

Searchable, interactive version of the periodic table. This site provides clear information, pictures of atomic structure, and facts about each element.

Modeling the Periodic Table: An Interactive Simulation http://www.genesismission.org/educate/scimodule/cosmic/ptable.html

Free software to assist students in studying elements and creating their own model of the periodic table of elements. You can create the simulation online (Shockwave plug-in required) or download it.

Periodic Table http://www.schoolchem.com/pt.htm

Virtual periodic table. Each element can be clicked on, which brings up a separate page that includes common reactions, common forms, and properties of the element.

period–luminosity curve graph depicting the variation in †luminosity of a †Cepheid variable star with time. In general, the longer the period, the greater the luminosity. By measuring the period it is possible thus to derive an †absolute magnitude; comparison of this with the star's observed (apparent) †magnitude gives an indication of the distance.

peripheral device in computing, any item connected to a computer's †central processing unit (CPU). Typical peripherals include keyboard, mouse, monitor, and printer. Users who enjoy playing games might add a joystick or a trackball; others might connect a †modem, scanner, or integrated services digital network (ISDN) terminal to their machines.

Worldwide Organization for Mobile Peripherals http://www.pc-card.com/

Official site of the Personal Computer Memory Card International Association, the international trades and standards body for plug-in computer technologies, including PC Card, CardBus, and SmartMedia. Provides a useful guide to existing and upcoming technologies, and contains a bookstore and resource guide for visitors.

periscope optical instrument designed for observation from a concealed position such as from a submerged submarine. In its basic form it consists of a tube with parallel mirrors at each end, inclined at 45° to its axis. The periscope attained prominence in naval and military operations of World War I.

Although most often thought of as a submarine observation device, periscopes were widely used in the trenches during World War I to allow observation without exposing the observer, and special versions were also developed to be attached to rifles.

peristalsis wavelike contractions, produced by the contraction of smooth muscle, that pass along tubular organs, such as the intestines. The same term describes the wavelike motion of earthworms and other invertebrates, in which part of the body contracts as another part elongates.

peritoneum membrane lining the abdominal cavity and digestive organs of vertebrates. **Peritonitis,** inflammation within the peritoneum, can occur due to infection or other irritation. It is sometimes seen following a burst appendix and quickly proves fatal if not treated.

Perkin, William Henry (1838–1907) English chemist. In 1856 he discovered †mauve, the dye that originated the aniline-dye industry and the British synthetic dyestuffs industry generally. He was knighted in 1906.

Perl acronym for Practical Extraction and Report Language, open source programming language developed in 1987 by US linguist Larry Wall for processing text. However, it has proved extremely popular as a way of writing †CGI scripts for Web pages. Hailed as the "duct tape" of the Internet, Perl is used by over half of the sites on the World Wide Web.

Perl, Martin L (1927–) US physicist who was awarded the Nobel Prize for Physics in 1995 for the discovery of the †tau particle, one of the elementary particles which make up all matter. He shared the award with Frederick †Reines.

permafrost condition in which a deep layer of soil does not thaw out during the summer. Permafrost occurs under †periglacial conditions. It is claimed that 26% of the world's land surface is permafrost.

Permafrost gives rise to a poorly drained form of grassland typical of northern Canada, Siberia, and Alaska known as †tundra.

permanent hardness hardness of water that cannot be removed by boiling (see †hard water).

permanent magnet †magnet that retains its †magnetism permanently, and not just when subject to some external †energy source (as is the case with an †electromagnet).

permeability in physics, the degree to which the presence of a substance alters the magnetic field around it. Most substances have a small constant permeability. When the permeability is less than 1, the material is a †diamagnetic material; when it is greater than 1, it is a †paramagnetic material. †Ferrimagnetic materials have very large permeabilities. See also †magnetism.

permeable rock rock which through which water can pass either via a network of spaces between particles or along bedding planes, cracks, and fissures. Permeable rocks can become saturated. Examples of permeable rocks include limestone (which is heavily jointed) and chalk (porous).

Unlike †impermeable rocks, which do not allow water to pass through, permeable rocks rarely support rivers and are therefore subject to less erosion. As a result they commonly form upland areas (such as the chalk downs of southeastern England, and the limestone Pennines of northern England).

Permian period of geological time 290–245 million years ago, the last period of the Palaeozoic era. Its end was marked by a dramatic change in marine life – the greatest mass extinction in geological history – including the extinction of many corals and trilobites. Deserts were widespread, terrestrial amphibians and mammal-like reptiles flourished, and cone-bearing plants (gymnosperms) came to prominence. In the oceans, 49% of families and 72% of genera vanished in the late Permian. On land, 78% of reptile families and 67% of amphibian families disappeared.

permittivity symbol e, in physics, property of a material that affects the electric flux density produced when it is excited by a source of †electromotive force (emf). Absolute permittivity is the electric flux density divided by the electric field strength, expressed in farads per meter. The permittivity of free space is 8.854×10^{-12} Fm^{-1}.

permutation in mathematics, a specified arrangement of a group of objects.

It is the arrangement of *a* distinct objects taken *b* at a time in all possible orders. It is given by $a!/(a - b)!$, where "!" stands for †factorial. For example, the number of permutations of four letters taken from any group of six different letters is $6!/2! = (1 \times 2 \times 3 \times 4 \times 5 \times 6)/(1 \times 2) = 360$. The theoretical number of four-letter "words" that can be made from an alphabet of 26 letters is $26!/22! = 358{,}800$.

permutation group in mathematics, †group consisting of all the transformations (operations) permutating a fixed number of objects among themselves. Such groups were studied first by French mathematician Evariste Galois in connection with permutations of roots (solutions) of a polynominal equation.

perovskite yellow, brown, or grayish-black orthorhombic mineral, $CaTiO_3$, which sometimes contains cerium. Other minerals that have a similar structure are said to have the **perovskite structure**. The term also refers to $MgSiO_3$ with the perovskite structure, the principal mineral that makes up the Earth's lower †mantle.

$CaTiO_3$ perovskite occurs primarily as a minor constituent of some igneous rocks in the Earth's crust and mantle and in some meteorites. $MgSiO_3$ perovskite has the same chemical composition as pyroxene, a principal constituent of the upper mantle. But the extreme pressures in the lower mantle cause the oxygen atoms to be packed more tightly together than in pyroxene, giving rise to the perovskite form of $MgSiO_3$ in the lower mantle.

The perovskite structure of $MgSiO_3$ was discovered in †diamond-anvil cell experiments in which upper mantle materials that are stable at the Earth's surface, such as the mineral pyroxene, are squeezed at high pressures to simulate the Earth's interior. Although this mineral does not occur naturally at the Earth's surface it is thought to be the most abundant mineral in the mantle, and therefore Earth's most abundant mineral.

peroxide in chemistry, member of a class of compounds that contain molecules in which two oxygen atoms are linked together by a hydrogen bond. These compounds are unstable and readily decompose to release oxygen. Inorganic peroxides are used as bleaching agents.

perpendicular at a right angle; also, a †line at right angles to another line or to a plane. Everyday examples include lamp posts, which are perpendicular to the road, and walls, which are perpendicular to the ground.

For a pair of †skew lines (lines in three dimensions that do not meet), there is just one common perpendicular, which is at right angles to both lines; the nearest points on the two lines are the feet of this perpendicular.

perpetual motion idea that a machine can be designed and constructed in such a way that, once started, it will do work indefinitely without requiring any further input of energy (motive power). Such a device would contradict at least one of the two laws of thermodynamics that state that (1) energy can neither be created nor destroyed (the law of conservation of energy) and (2) heat cannot by itself flow from a cooler to a hotter object. As a result, all practical (real) machines require a continuous supply of energy, and no heat engine is able to convert all the heat into useful work.

Perrin, Jean Baptiste (1870–1942) French physicist who produced the crucial evidence that finally established the atomic nature of matter. Assuming the atomic hypothesis, Perrin demonstrated how the phenomenon of †Brownian motion could be used to derive precise values for †Avogadro's number. He was awarded the Nobel Prize for Physics in 1926 for his confirmation of the discontinuous structure of matter.

Perrin On Cathode Rays http://dbhs.wvusd.k12.ca.us/Chem-History/Perrin-1895.html

Transcript of Perrin's paper asserting cathode rays are in fact streams of electrons. The paper was read before the Paris Academy of Sciences on 30 December 1895, and is accompanied on this Web site by Perrin's illustrations.

Perseids strong, long-lived meteor shower whose point of origin (radiant) appears to be in the constellation Perseus. They occur annually in late July to early August.

Perseus a bright constellation of the northern hemisphere, near †Cassiopeia. It is represented as the mythological hero; the head of the decapitated Gorgon, Medusa, is marked by †Algol (Beta Persei), the best known of the eclipsing binary stars.

Perseus lies in the Milky Way and contains the Double Cluster, a twin cluster of stars called h and Chi Persei. They are just visible to the naked eye as two hazy patches of light close to one another.

persistence of vision brief retention of the sensation of light by the brain, once the initial stimulus has been removed. Essential phenomenon for the success of televisual and cinematographic images.

personal computer PC, another name for †microcomputer. The term is also used, more specifically, to mean the IBM Personal Computer and computers compatible with it.

The first IBM PC was introduced in the USA in 1981; it had 64 kilobytes of random access memory (RAM) and one floppy-disk drive. It was followed in 1983 by the XT (with a hard-disk drive) and in 1984 by the AT (based on a more powerful †microprocessor). Many manufacturers have copied the basic design, which is now regarded as a standard for business microcomputers. Computers designed to function like an IBM PC were originally known as **IBM-compatible computers**, but are now more usually called Windows PCs. The power of personal computers has increased year by year, as has their ability to handle sophisticated tasks, such as desktop video editing. The original XT's speed was clocked as 0.25 MIPS (millions of instructions per second), while Pentium-based PCs run in excess of 100 MIPS.

personality individual's characteristic way of behaving across a wide range of situations. Two broad dimensions of personality are †extroversion and †neuroticism. A number of more specific personal traits have also been described, including †psychopathy (antisocial behavior).

Perspex trade name for a clear, lightweight, tough plastic first produced in 1930. It is widely used for watch glasses, advertising signs, domestic baths, motorboat windscreens, aircraft canopies, and protective shields. Its chemical name is polymethylmethacrylate (PMMA). It is manufactured under other names: Plexiglas, Lucite, Acrylite, and Rhoplex (in the USA), and Oroglas (in Europe).

perspiration excretion of water and dissolved substances from the †sweat glands of the skin of mammals. Perspiration has two main functions: body cooling by the evaporation of water from the skin surface, and excretion of waste products such as salts.

In 2001, German researchers discovered an antibiotic protein, dermcidin, that is exuded in sweat. Dermcidin was found to be affective against four common species of bacteria when tested.

perturbation in astronomy, apparent irregularity in an orbit, or occasionally in a star's †proper motion, caused by the gravitational effects of a nearby celestial body.

Peru Current formerly **Humboldt Current**, cold ocean †current flowing north from the Antarctic along the west coast of South America to southern Ecuador, then west. It reduces the coastal temperature, making the western slopes of the Andes arid because winds are already chilled and dry when they meet the coast.

Perutz, Max Ferdinand (1914–2002) Austrian-born British biochemist who shared the Nobel Prize for Chemistry in 1962 with his co-worker John †Kendrew for work on the structure of the hemoglobin molecule.

pervious rock rock that allows water to pass through via bedding planes, cracks, and fissures, but not through pores within the rock (that is, they are non-porous). See †permeable rock.

pest in biology, any insect, fungus, rodent, or other living organism that has a harmful effect on human beings, other than those that directly cause human diseases. Most pests damage crops or livestock, but the term also covers those that damage buildings, destroy food stores, and spread disease.

pesticide any chemical used in farming, gardening, or in the home to combat pests. Pesticides are of three main types: **insecticides** (to kill insects), **fungicides** (to kill fungal diseases), and **herbicides** (to kill plants, mainly those considered weeds). Pesticides cause a number of pollution problems through spray drift onto surrounding areas, direct contamination of users or the public, and as residues on food. The World Health Organization (WHO) estimated in 1999 that 20,000 people die annually worldwide from pesticide poisoning incidents.

The safest pesticides include those made from plants, such as the insecticides pyrethrum and derris. Pyrethrins are safe and insects do not develop resistance to them. Their impact on the environment is very small as the ingredients break down harmlessly.

More potent are synthetic products, such as chlorinated hydrocarbons. These products, including DDT and dieldrin, are highly toxic to wildlife and often to humans, so their use is now restricted by law in some areas and is declining. Safer pesticides such as malathion are based on organic phosphorus compounds, but they still present hazards to health. An international treaty to ban persistent organic pollutants (POPs), including pesticides such as DDT, was signed in Stockholm, Sweden, in May 2001. The

United Nations Food and Agriculture Organization reported in the same month that more than 500,000 tonnes of pesticide waste resides in dumps worldwide.

Biological Control http://www.nysaes.cornell.edu:80/ent/biocontrol/

University-based site on the various methods of biological control used by farmers in the USA. This includes sections on parasitoids, predators, pathogens, and weed feeders. Each section contains images and sections on "Relative effectiveness" and "Pesticide susceptibility."

Molecular Expressions: The Pesticide Collection http://micro.magnet.fsu.edu/pesticides/index.html

Fascinating collection of images showing what pesticides look like when recrystallized and photographed through a microscope. There is also an informative article about pesticides.

petal part of a flower whose function is to attract pollinators such as insects or birds. Petals are frequently large and brightly colored and may also be scented. Some have a nectary at the base and markings on the petal surface, known as honey guides, to direct pollinators to the source of the nectar. In wind-pollinated plants, however, the petals are usually small and insignificant, and sometimes absent altogether. Petals are derived from modified leaves, and are known collectively as a corolla.

Some insect-pollinated plants also have inconspicuous petals, with large colorful †bracts (leaflike structures) or †sepals taking over their role, or strong scents that attract pollinators such as flies.

petiole in botany, the stalk attaching the leaf blade, or †lamina, to the stem. Typically it is continuous with the midrib of the leaf and attached to the base of the lamina, but occasionally it is attached to the lower surface of the lamina, as in the nasturtium (a peltate leaf). Petioles that are flattened and leaflike are termed phyllodes. Leaves that lack a petiole are said to be sessile.

Petit, Alexis(-Thérèse) (1791–1820) French physicist, co-discoverer with Pierre †Dulong of **Dulong and Petit's law**, which states that, for a solid element, the product of †relative atomic mass and †specific heat capacity is approximately constant.

petrochemical chemical derived from the processing of petroleum (crude oil).

Petrochemical industries are those that obtain their raw materials from the processing of petroleum and natural gas. Polymers, detergents, solvents, and nitrogen fertilizers are all major products of the petrochemical industries. Inorganic chemical products include carbon black, sulfur, ammonia, and hydrogen peroxide.

petrol mixture of hydrocarbons derived from †petroleum, mainly used as a fuel for internal-combustion engines. It is colorless and highly volatile.

In the USA, petrol is called †gasoline.

petroleum or **crude oil**, natural mineral oil, a thick greenish-brown flammable liquid found underground in permeable rocks. Petroleum consists of hydrocarbons mixed with oxygen, sulfur, nitrogen, and other elements in varying proportions. It is thought to be derived from ancient organic material that has been converted by, first, bacterial action, then heat, and pressure (but its origin may be chemical also).

From crude petroleum, various products are made by †fractional distillation and other processes; for example, fuel oil, gasoline, kerosene, diesel, and lubricating oil. Petroleum products and chemicals are used in large quantities in the manufacture of detergents, artificial fibers, plastics, insecticides, fertilizers, pharmaceuticals, toiletries, and synthetic rubber.

Petroleum was formed from the remains of marine plant and animal life which existed many millions of years ago (hence it is known as a †fossil fuel). Some of these remains were deposited along with rock-forming sediments under the sea where they were decomposed anaerobically (without oxygen) by bacteria which changed the fats in the sediments into fatty acids which were then changed into an asphaltic material called kerogen. This was then converted over millions of years into petroleum by the combined action of heat and pressure. At an early stage the organic material was squeezed out of its original sedimentary mud into adjacent sandstones. Small globules of oil collected together in the pores of the rock and eventually migrated upwards through layers of porous rock by the action of the oil's own surface tension (capillary action), by the force of water movement within the rock, and by gas pressure. This migration ended either when the petroleum emerged through a fissure as a seepage of gas or oil onto the Earth's surface, or when it was trapped in porous reservoir rocks, such as sandstone or limestone, in anticlines and other traps below impervious rock layers.

The modern oil industry originates in the discovery of oil in western Ontario in 1857 followed by Edwin Drake's discovery in Pennsylvania in 1859. Drake used a steam engine to drive a punching tool to 21 m/68 ft below the surface where he struck oil and started an oil boom. Rapid development followed in other parts of the USA, Canada, Mexico, and then Venezuela where commercial production began in 1878. Oil was found in Romania in 1860, Iran in 1908, Iraq in 1923, Bahrain in 1932, and Saudi Arabia and Kuwait in 1938.

The USA led in production until the 1960s, when the Middle East outproduced other areas, their

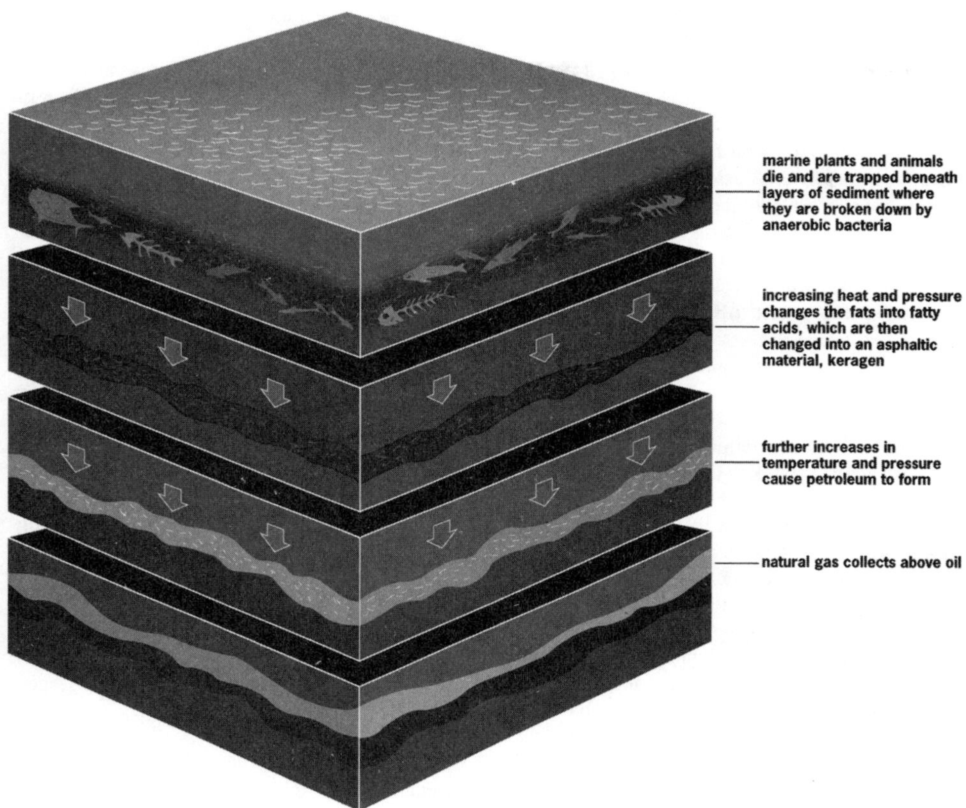

petroleum The formation of oil and natural gas. Oil forms when marine plants and animals die and accumulate in stagnant water lacking in oxygen. They are quickly buried by clay and so do not completely decay but form hydrocarbon-rich muds, broken down by anaerobic bacteria. Increasing heat and pressure transform the hydrocarbons into fatty acids, which are then changed into an asphaltic material, keragen. Further increases in temperature and pressure cause oil to form and natural gas collects above the oil. Helicon

immense reserves leading to a worldwide dependence on cheap oil for transport and industry. In 1961 the Organization of the Petroleum Exporting Countries (OPEC) was established to avoid exploitation of member countries; after OPEC's price rises in 1973, the International Energy Agency (IEA) was established in 1974 to protect the interests of oil-consuming countries. New technologies were introduced to pump oil from offshore and from the Arctic (the Alaska pipeline) in an effort to avoid a monopoly by OPEC. Global consumption of petroleum in 1993 was 23 billion barrels.

As shallow-water oil reserves dwindle, multinational companies have been developing deep-water oilfields at the edge of the continental shelf in the Gulf of Mexico. Shell has developed Mars, a 500-million-barrel oilfield, in 900 m/2,940 ft of water, and the oil companies now have the technology to drill wells of up to 3,075 m/10,000 ft under the sea. It is estimated that the deep waters of Mexico could yield 8–15 million barrels in total; it could overtake the North Sea in importance as an oil source.

In Asia, the oil pipeline from Azerbaijan through Russia to the West, which is the only major pipeline from the Caspian Sea, closed during Russia's conflict with Chechnya but reopened in 1997.

pollution The burning of petroleum fuel is one cause of †air pollution. The transport of oil can lead to catastrophes – for example, the *Torrey Canyon* tanker lost off southwestern England in 1967, which led to an agreement by the international oil companies in 1968 to pay compensation for massive shore pollution. The 1989 oil spill in Alaska from the *Exxon Valdez* damaged the area's fragile environment, despite clean-up efforts. Drilling for oil involves the risks of accidental spillage and drilling-rig accidents. The problems associated with oil have led to the various alternative †energy technologies.

A new kind of bacterium was developed during the 1970s in the USA, capable of "eating" oil as a means of countering oil spills.

petrology branch of geology that deals with the study of rocks, their mineral compositions, their textures, and their origins.

Pfeffer, Wilhelm Friedrich Philipp (1845–1920) German physiological botanist who was the first to measure osmotic pressure, in 1877. He also showed that osmotic pressure varies according to the temperature and concentration of the solute.

pH scale from 0 to 14 for measuring acidity or alkalinity. A pH of 7 indicates neutrality, below 7 is †acid, while above 7 is alkaline. Strong acids, such as those used in car batteries, have a pH of about 2; strong †alkalis such as sodium hydroxide have a pH of about 13.

Acidic fruits such as citrus fruits are about pH 4. Fertile soils have a pH of about 6.5 to 7, while weak alkalis such as soap have a pH of about 9 to 10.

Sören Sörenson and pH http://dbhs.wvusd.k12.ca.us/Chem-History/Sorenson-article.html
Excerpt from a paper on enzymatic processes in which Sörenson defined pH as the relative concentration of hydrogen ions in a solution. This extract from the paper not only explains the origin of the term but also demonstrates its meaning.

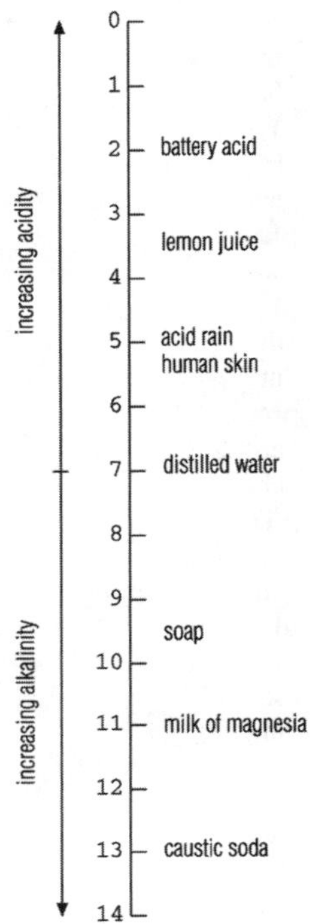

pH The pHs of some common substances. The lower the pH, the more acidic the substance; the higher the pH, the more alkaline the substance. A pH of 7 is neutral. Helicon

Phaeton asteroid discovered in 1983 that is probably the remains of a comet that gave rise to the Geminid meteor shower. A member of the Apollo group of asteroids, it is 6.9 km/4.3 mi in diameter and has a highly eccentric Earth-crossing orbit.

phage another name for a †bacteriophage, a virus that attacks bacteria.

phagocyte type of †white blood cell, or leucocyte, that can engulf a bacterium or other invading micro-organism. Phagocytes are found in blood, lymph, and other body tissues, where they also ingest foreign matter and dead tissue. A †macrophage differs in size and lifespan.

phagocytosis process by which †phagocytes surround foreign particles (by an amoeboid movement), engulf, and digest them. See also †endocytosis.

Phanerozoic eon (Greek *phanero* "visible") eon in Earth history, consisting of the most recent 570 million years. It comprises the Palaeozoic, Mesozoic, and Cenozoic eras. The vast majority of fossils come from this eon, owing to the evolution of hard shells and internal skeletons. The name means "interval of well-displayed life."

pharmaceutical describing any substance that is used in the prevention and treatment of disease or in the treatment and modification of organic functions related to health care.

pharmacognosy branch of pharmacology concerned with the identification of drugs extracted from plant and animal tissues.

pharmacology study of the properties of drugs and their effects on the human body.

pharmacy preparation and dispensing of drugs; as an area of study, pharmacy includes an understanding of the origins, nature, and action of drugs. The place where a pharmacist dispenses drugs is also referred to as a pharmacy.

pharmecogenomics emerging medical science that uses genomic information to diagnose disease and to tailor treatments to suit the individual.

pharynx muscular cavity behind the nose and mouth, extending downwards from the base of the skull. Its walls are made of muscle strengthened with a fibrous layer and lined with mucous membrane. The internal nostrils lead backwards into the pharynx, which continues downwards into the esophagus and (through the epiglottis) into the windpipe. On each side, a Eustachian tube enters the pharynx from the middle ear cavity.

The upper part (nasopharynx) is an airway, but the remainder is a passage for food. Inflammation of the pharynx is named pharyngitis.

phase in astronomy, apparent shape of the Moon or a planet when all or part of its illuminated hemisphere is facing the Earth.

As the Moon orbits the Earth its appearance from Earth changes as different amounts of its surface are illuminated by the Sun. During one orbit of the Earth (29.5 days – a lunar †month) the Moon undergoes a full cycle of phases from new, to first quarter, to full, to last quarter.

The Moon does not reflect sunlight onto the Earth when it is between the Earth and the Sun, and the Moon is not seen. This is the new Moon phase. As the Moon orbits the Earth, part of the Moon reflects sunlight onto the Earth and a crescent is seen, starting from the right-hand side of the Moon. This is the waxing crescent phase. The half Moon phase occurs when half of the Moon's disc is illuminated as seen from Earth. As the Moon continues its orbit, gradually more of the Moon's surface becomes illuminated. At the waxing gibbous phase, three-quarters of the Moon's disc is visible. The full Moon phase occurs when the Earth is between the Moon and the Sun, and the Moon's disc is fully illuminated.

After the full Moon, gradually less of the Moon's surface is illuminated as it continues its orbit. It passes through the waning gibbous phase where three-quarters of its disc is visible, to the third quarter where half is seen, followed by the waning crescent, and finally the new Moon phase again.

Virtual Reality Moon Phase Pictures http://tycho.usno.navy.mil/vphase.html

Image display of the current phase of the Moon, updated every four hours. You can also find the Moon phase for any date and time between AD 1800 and AD 2199.

phase in chemistry, a physical state of matter. For example, ice and liquid water are different phases of water; a mixture of the two is termed a two-phase system.

phase in physics, a stage in an oscillatory motion, such as a wave motion: two waves are in phase when their peaks and their troughs coincide. Otherwise, there is a **phase difference**, which has consequences in †interference phenomena and †alternating current electricity.

phase change in chemistry, any change in the physical state or structure of a material. The term is mostly applied to the gas, liquid, and solid transition of a material, but can also be applied to fundamental changes in the physical structure of certain solids as they pass through a critical temperature.

phase diagram in chemistry, graphical representation of the equilibrium conditions of each phase of a compound or mixture of compounds over a range of different conditions, usually temperature and composition at constant pressure.

phase rule in chemistry, law describing the relationship between the number of variables in a chemical system in thermodynamic equilibrium, developed by US physicist J Willard Gibbs. The law is represented by the formula $P + F = C + 2$, where P is the number of phases or states of the components involved (solid, liquid, or gas), F is the number of degrees of freedom of the system (for example, temperature, pressure), and C is the number of components involved.

phenol member of a group of aromatic chemical compounds with weakly acidic properties, which are characterized by a hydroxyl (OH) group attached directly to an aromatic ring. The simplest of the phenols, derived from benzene, is also known as phenol and has the formula C_6H_5OH. It is sometimes called **carbolic acid** and can be extracted from coal tar.

Pure phenol consists of colorless, needle-shaped crystals, which take up moisture from the atmosphere. It has a strong and characteristic smell and was once used as an antiseptic. It is, however, toxic by absorption through the skin.

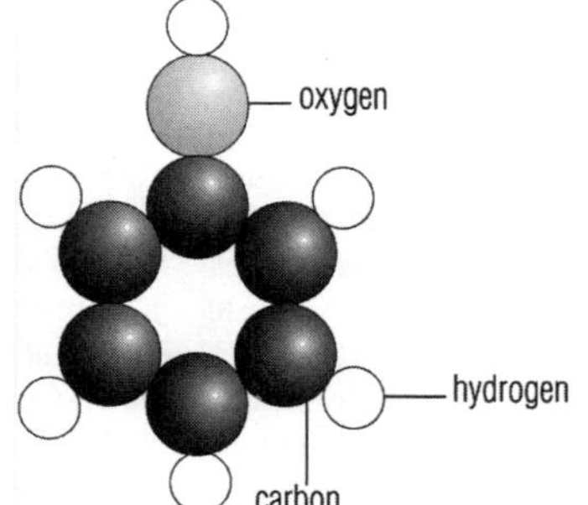

phenol The phenol molecule with its ring of six carbon atoms and a hydroxyl (OH) group attached. Phenol was first extracted from coal tar in 1834. It is used to make phenolic and epoxy resins, explosives, pharmaceuticals, perfumes, and nylon. Helicon

phenolic resin member of a class of synthetic plastics made from †phenol and its derivatives. They are formed by combining a phenolic compound with an †aldehyde in the presence of an acidic or an alkaline catalyst. The first true synthetic plastic, †Bakelite, is a phenolic resin.

phenolphthalein acid–base †indicator that is clear below pH 8 and red above pH 9.6. It is used in titrating (see †titration) weak acids against strong bases.

phenotype in genetics, visible traits, those actually displayed by an organism. The phenotype is not a direct reflection of the †genotype because some alleles are masked by the presence of other, dominant alleles (see †dominance). The phenotype is further modified by the effects of the environment (for example, poor nutrition stunts growth).

phenylanaline one of the nine essential amino acids. Phenylketonuria is a rare genetic disease which results from the inability to metabolize the phenylalanine present in food.

pheromone chemical signal (such as an odor) that is emitted by one animal and affects the behavior of others. Pheromones are used by many animal species to attract mates.

Science of Love http://whyfiles.org/033love/index.html

Part of the Why Files project, published by the National Institute for Science Education (NISE) and funded by the National Science Foundation, this page provides a scientific answer to the age-old query "why do fools fall in love?" The highly readable text, laid out over ten pages, provides "intriguing evidence that humans communicate with unconscious chemical signals," called pheromones. Numerous images and diagrams enhance the text throughout. The more tricky terms are linked to a glossary, and there is a comprehensive bibliography of sources for further research.

phloem tissue found in vascular plants. Its main function is to transport sugars and other food materials such as amino acids (see †protein) from the leaves, where they are produced, to all other parts of the plant. This could be from the leaves to the roots to provide the chemicals needed for growth. However, it could be from a leaf and up to a developing fruit that is rich in sugars. The sugars are made by †photosynthesis, which occurs in green parts of plants, such as leaves (see †leaf). The amino acids are made from sugars and †minerals, such as nitrate absorbed from the soil. Phloem tissue is usually found close to the other transport tissue in plants, †xylem, which transports †water and minerals. In non-woody plants phloem and xylem are found in bundles, such as the veins of a leaf.

Phloem is composed of sieve elements and their associated companion cells, together with some †sclerenchyma and †parenchyma cell types. Sieve elements are long, thin-walled cells joined end to end, forming sieve tubes; large pores in the end walls allow the continuous passage of nutrients. Phloem is usually found in association with †xylem, the water-conducting tissue, but unlike the latter it is a living tissue.

phlogiston hypothetical substance formerly believed to have been produced during combustion. The substance, originally named terra pinguis ("fatty earth") by the German chemist Johann Joachim Becher in 1669, was renamed by Georg †Stahl at the beginning of the 18th century. The phlogiston theory was replaced by the theory of oxygen gain and loss, first enunciated by the French chemist Antoine Lavoisier.

pH meter electronic device used for measuring the †pH value of a solution. It is usually equipped with a hydrogen electrode.

Phobos one of the two moons of Mars, discovered in 1877 by the US astronomer Asaph Hall. It is an irregularly shaped lump of rock, cratered by †meteorite impacts. Phobos is 27 x 22 x 19 km/17 x 13 x 12 mi in size, and orbits Mars every 0.32 days at a distance of 9,400 km/5,840 mi from the planet's center. It is thought to be an asteroid captured by Mars's gravity.

Phoebe Saturn's outermost moon, discovered in 1898. It has a diameter of 230 km/143 mi and a steeply inclined retrograde orbit at an average distance of 13 million km/8 million mi from the planet. It takes 550 days to complete one orbit.

Phoenix constellation of the southern hemisphere.

Pholus or **5145 Pholus**, red centaur star discovered in 1991.

phon unit of loudness, equal to the value in decibels of an equally loud tone with frequency 1,000 Hz. The higher the frequency, the louder a noise sounds for the same decibel value; thus an 80-decibel tone with a frequency of 20 Hz sounds as loud as 20 decibels at 1,000 Hz, and the phon value of both tones is 20. An aircraft engine has a loudness of around 140 phons.

phosphate salt or ester of †phosphoric acid. Incomplete neutralization of phosphoric acid gives rise to acid phosphates (see †acid salt and †buffer). Phosphates are used as fertilizers, and are required for the development of healthy root systems. They are involved in many biochemical processes, often as part of complex molecules, such as †ATP.

phosphate analysis regular sampling and chemical analysis of phosphorus levels in the soil around archaeological sites in order to locate concentrations of human bone and excrement, and hence areas of human activity, settlements, and burial grounds.

phosphate ester chemical compound formed by esterification of an †alkyl or aryl †alcohol with phosphoric acid. Phosphate esters are widely used in solvent extraction processes. Some important biological substances belong to this class of compound, for example †DNA.

phospholipid any †lipid consisting of a glycerol backbone, a phosphate group, and two long chains. Phospholipids are found everywhere in living systems as the basis for biological membranes.

One of the long chains tends to be hydrophobic and the other hydrophilic (that is, they interrelate with water in opposite ways). This means that phospholipids will line up the same way round when in solution.

phosphorescence emission of light by certain substances after they have absorbed energy, whether from visible light, other electromagnetic radiation such as ultraviolet rays or X-rays, or cathode rays (a beam of electrons). When the stimulating energy is removed phosphorescence persists for more than 0.1 nanoseconds before ceasing (unlike †fluorescence, which stops immediately). See also †luminescence.

phosphoric acid acid derived from phosphorus and oxygen. Its commonest form (H_3PO_4) is also known as orthophosphoric acid, and is produced by the action of phosphorus pentoxide (P_2O_5) on water. It is used in rust removers and for rust-proofing iron and steel.

phosphorite or **rock phosphate**, in earth science, type of sedimentary rock that is rich in phosphate minerals (more than 20%). It is a fibrous form of †apatite.

phosphorous pentoxide P_4O_{10}, highly hygroscopic, white crystalline solid prepared by burning phosphorous in air. It is mainly used to manufacture orthophosphoric acid or as a drying and dehydrating agent.

phosphorous ylide organic compound of general formula $R_2C = PR'_3$, where R is an alkyl group and R' is an alkyl or organic group, for example $–C_5H_6$. This compound is formed by reacting a phosphine, the phosphorous analogue of an †amine, with an alkyl halide followed by treatment with a strong base. This produces a compound that is composed of highly polar molecules which readily react with aldehydes and ketones to produce alkenes (see †Wittig reaction).

phosphorus chemical symbol P, (Greek *phosphoros* "bearer of light") highly reactive, non-metallic element, atomic number 15, relative atomic mass 30.9738. It occurs in nature as phosphates (commonly in the form of the mineral †apatite), and is essential to plant and animal life. Compounds of phosphorus are used in fertilizers, various organic chemicals, for matches and fireworks, and in glass and steel.

Phosphorus was first identified in 1674 by German alchemist Hennig Brand (born *c.* 1630), who prepared it from urine. The element has three allotropic forms: a black powder; a white-yellow, waxy solid that ignites spontaneously in air to form the poisonous gas phosphorus pentoxide; and a red-brown powder that neither ignites spontaneously nor is poisonous.

phosphorylation chemical addition of a phosphate group to a (organic) molecule. One of the most important biochemical phosphorylations is the addition of phosphate to ADP to form the energy-rich ATP. Phosphorylation involving light (as in †photosynthesis) is termed photophosphorylation.

photocell or **photoelectric cell**, device for measuring or detecting light or other electromagnetic radiation, since its electrical state is altered by the effect of light. In a **photoemissive** cell, the radiation causes electrons to be emitted and a current to flow (†photoelectric effect); a **photovoltaic** cell causes an †electromotive force to be generated in the presence of light across the boundary of two substances. A **photoconductive** cell, which contains a semiconductor, increases its conductivity when exposed to electromagnetic radiation.

Photocells are used for photographers' exposure meters, burglar and fire alarms, automatic doors, and in solar energy arrays.

photochemical reaction any chemical reaction in which light is produced or light initiates the reaction. Light can initiate reactions by exciting atoms or molecules and making them more reactive: the light energy becomes converted to chemical energy. Many photochemical reactions set up a †chain reaction and produce †free radicals.

This type of reaction is seen in the bleaching of dyes or the yellowing of paper by sunlight. It is harnessed by plants in †photosynthesis and by humans in †photography.

Chemical reactions that produce light are most commonly seen when materials are burned. Light-emitting reactions are used by living organisms in †bioluminescence. One photochemical reaction is the action of sunlight on car exhaust fumes, which results in the production of †ozone. Some large cities, such as Los Angeles, USA, and Santiago, Chile, now suffer serious pollution due to photochemical smog.

photochemistry branch of chemistry that studies chemical reactions that occur as a result of the action of light.

photodiode semiconductor †*p–n* junction diode used to detect light or measure its intensity. The photodiode is encapsulated in a transparent plastic case that allows light to fall onto the junction. When this occurs, the reverse-bias resistance (high resistance in the opposite direction to normal current flow) drops and allows a larger reverse-biased current to flow through the device. The increase in current can then be related to the amount of light falling on the junction.

Photodiodes that can detect small changes in light level are used in alarm systems, camera exposure controls, and optical communication links.

photoelectric cell alternative name for †photocell.

photoelectric effect process by which electromagnetic radiation, including visible light, incident on a material releases an electric charge. It is commonly thought of as the emission of †electrons from a substance (usually a metallic surface) when it is struck by †photons (quanta of electromagnetic radiation), usually those of visible light or ultraviolet radiation.

photoelectric filtering in astronomy, means of measuring the †astronomical color index of a star, involving color filters on photoelectric cells to define the color index between two set wavelengths. The filters correspond to the †UBV photometry system.

photoelectricity phenomenon whereby certain materials, such as †selenium, can produce electricity if exposed to light.

photography process for reproducing permanent images on light-sensitive materials by various forms of radiant energy, including visible light, ultraviolet, infrared, X-rays, atomic radiations, and electron beams.

Photography was developed in the 19th century; among the pioneers were Louis Daguerre in France and William Henry Fox Talbot in the UK. Color photography dates from the early 20th century.

The most familiar photographic process depends upon the fact that certain silver compounds (called †halides) are sensitive to light. A photographic film is coated with these compounds and, in a camera, is exposed to light. An image, or picture, of the scene before the camera is formed on the film because the silver halides become activated (light-altered) where light falls but not where light does not fall. The image is made visible by the process of developing, made permanent by fixing, and, finally, is usually printed on paper. Motion-picture photography uses a camera that exposes a roll of film to a rapid succession of views that, when developed, are projected in equally rapid succession to provide a moving image.

photoluminescence see †luminescence.

photolysis chemical reaction that is driven by light or ultraviolet radiation. For example, the light reaction of †photosynthesis (the process by which green plants manufacture carbohydrates from carbon dioxide and water) is a photolytic reaction.

photometer instrument that measures light, electromagnetic radiation in the visible range, usually by comparing relative intensities from different sources. Bunsen's grease-spot photometer of 1844 compares the intensity of a light source with a known source by each illuminating one half of a translucent area. Modern photometers use †photocells, as in a photographer's exposure meter. A photomultiplier can also be used as a photometer.

photometry measurement of light output, such as the †magnitudes of celestial bodies, originally carried out by expertise of eye alone, but now generally making use of photographic or photoelectric apparatus.

photomicrograph image of a specimen photographed through a microscope. If taken through an electron microscope, it is an electron photomicrograph.

photomultiplier instrument that detects low levels of electromagnetic radiation (usually visible light or †infrared radiation) and amplifies it to produce a detectable signal.

One type resembles a †photocell with an additional series of coated †electrodes (dynodes) between the †cathode and †anode. Radiation striking the cathode releases electrons (primary emission) which hit the first dynode, producing yet more electrons (†secondary emission), which strike the second dynode. Eventually this produces a measurable signal up to 100 million times larger than the original signal by the time it leaves the anode. Similar devices, called image intensifiers, are used in television camera tubes that "see" in the dark.

photon in physics, the †elementary particle or "package" (quantum) of energy in which light and other forms of electromagnetic radiation are emitted. The photon has both particle and wave properties; it has no charge, is considered massless but possesses momentum and energy. It is one of the †gauge bosons, and is the carrier of the †electromagnetic force, one of the fundamental forces of nature.

According to †quantum theory the energy of a photon is given by the formula $E = hf$, where h is Planck's constant and f is the frequency of the radiation emitted.

Beam Me Up: Photons and Teleportation

http://www.sciam.com/article.cfm?articleID=00014CBD-7633-1C76-9B81809EC588EF21&pageNumber=1&catID=4

Part of a larger site maintained by Scientific American, this page reports on the amazing research conducted by physicists at the University of Innsbruck, Austria, who have turned science fiction into reality by teleporting the properties of one photon particle to another. Learn how quantum teleportation was accomplished and find out what the likelihood is that people will soon be teleporting themselves. The text includes hypertext links to further information, and there is also a list of related links at the bottom of the page.

photoperiodism biological mechanism that determines the timing of certain activities by responding to changes in day length. The flowering of many plants is initiated in this way. Photoperiodism in plants is regulated by a light-sensitive pigment, **phytochrome**. The breeding seasons of many temperate-zone animals are also triggered by increasing or declining day length, as part of their †biorhythms.

Autumn-flowering plants (for example, chrysanthemum and soybean) and autumn-breeding mammals (such as goats and deer) require days that are shorter than a critical length; spring-flowering and spring-breeding ones (such as radish and lettuce, and birds) are triggered by longer days.

photosphere visible surface of the Sun, which emits light and heat. About 300 km/200 mi deep, it consists of incandescent gas at a temperature of 5,800 K (5,530°C/9,980°F).

Rising cells of hot gas produce a mottling of the photosphere known as **granulation**, each granule being about 1,000 km/620 mi in diameter. The photosphere is often marked by large, dark patches called †sunspots.

photosynthesis process by which green plants trap light energy from sunlight in order to combine †carbon dioxide and †water to make high-energy chemicals – †glucose and other †carbohydrates. The simple sugar glucose provides the basic food for both plants and animals. For photosynthesis to occur, the plant must possess †chlorophyll (the green chemical that absorbs the energy from the sunlight) and must have a supply of carbon dioxide and water. Photosynthesis takes place inside †chloroplasts, which contain the enzymes and chlorophyll necessary for the process. They are found mainly in the leaf cells of plants. The by-product of photosynthesis, oxygen, is of great importance to all living organisms, and virtually all atmospheric oxygen has originated from photosynthesis.

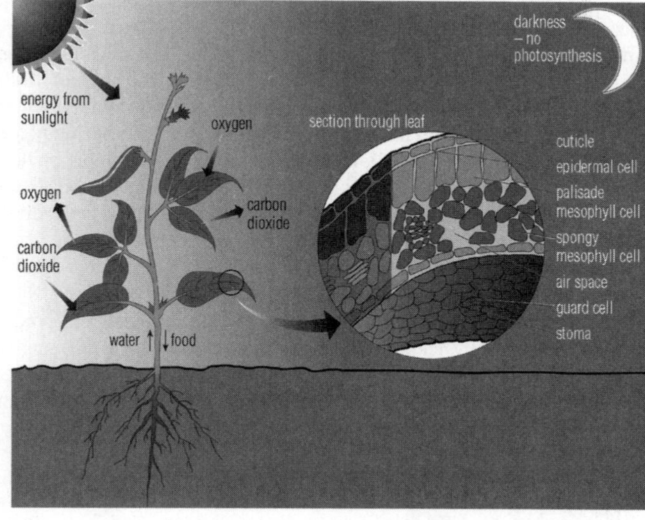

photosynthesis Process by which green plants and some bacteria manufacture carbohydrates from water and atmospheric carbon dioxide, using the energy of sunlight. Photosynthesis depends on the ability of chlorophyll molecules within plant cells to trap the energy of light, in order to split water molecules, giving off oxygen as a by-product. The hydrogen of the water molecules is then used to reduce carbon dioxide to simple carbohydrates. Helicon

Through photosynthesis, plants are able to make some of their own food – this is one way in which the †nutrition of plants differs from that of animals. The glucose they produce may be converted to starch and stored. Starch can then be converted back to glucose to provide energy for the plant at a later stage, for example during the night.

Introduction to Photosynthesis and its Applications http://photoscience.la.asu.edu/photosyn/education/photointro.html

Good general introduction to photosynthesis, including a discussion of the basics, illustrated with photographs and diagrams. This is followed by a more detailed explanation of photosynthetic electron transfer, carbon fixation, and the effects of increasing exposure to carbon dioxide. Helpful pictures aid in visualizing the processes.

Photosynthesis http://www.biology.arizona.edu/biochemistry/problem_sets/photosynthesis_1/photosynthesis_1.html

Increase your understanding of how light energy is converted into different forms of chemical energy during photosynthesis. The user can browse through the tutorials before attempting the tests.

Photosynthesis Directory http://web.mit.edu/esgbio/www/ps/intro.html

Helpful introduction to photosynthesis detailing its "evolution and discovery" and clear explanations of light and dark reactions. Also includes information and a diagram on the structure and function of the chloroplast.

Plant Photobiology Notes http://www.kumc.edu/POL/ASP_Home/what_is.html

Clear, well-designed resource on photobiology, the branch of biology that studies the interaction of plants and light. This evolving site contains helpful pages on the basics of photobiology and the various types of radiation present in the environment.

phototropism movement of part of a plant toward or away from a source of light. Leaves are positively phototropic, detecting the source of light and orientating themselves to receive the maximum amount.

phyllite †metamorphic rock produced under increasing temperature and pressure, in which minute mica crystals are aligned so that the rock splits along their plane of orientation, the resulting break being shiny and smooth. Intermediate between slate and schist, its silky sheen is an identifying characteristic.

phyllotaxis arrangement of leaves on a plant stem. Leaves are nearly always arranged in a regular pattern and in the majority of plants they are inserted singly, either in a **spiral** arrangement up the stem, or on **alternate** sides. Other principal forms are opposite leaves, where two arise from the same node, and whorled, where three or more arise from the same node.

phylogeny historical sequence of changes that occurs in a given species during the course of its evolution. It was once erroneously associated with ontogeny (the process of development of a living organism).

Tree of Life http://tolweb.org/tree/phylogeny.html

Project designed to present information about the phylogenetic relationships and characteristics of organisms, illustrating the diversity and unity of living organisms.

phylum plural **phyla**, major grouping in biological classification. Mammals, birds, reptiles, amphibians, fishes, and tunicates belong to the phylum Chordata; the phylum Mollusca consists of snails, slugs, mussels, clams, squid, and octopuses; the phylum Porifera contains sponges; and the phylum Echinodermata includes starfish, sea urchins, and sea cucumbers. In classifying plants (where the term "division" often takes the place of "phylum"), there are between four and nine phyla depending on the criteria used; all flowering plants belong to a single phylum, Angiospermata, and all conifers to another, Gymnospermata. Related phyla are grouped together in a †kingdom; phyla are subdivided into †classes.

There are 36 different phyla. The most recently identified is the Cycliophora described in 1995. It contains a single known species, *Symbion pandora*, a microscopic organism that lives on the mouths of lobsters.

physical change in chemistry, a type of change that does not produce a new chemical substance, but rather a change physical state (see also †chemical change). Boiling and melting are examples of physical change.

physical chemistry branch of chemistry concerned with examining the relationships between the chemical compositions of substances and the physical properties that they display. Most chemical reactions exhibit some physical phenomenon (change of state, temperature, pressure, or volume, or the use or production of electricity), and the measurement and study of such phenomena has led to many chemical theories and laws.

physical weathering or **mechanical weathering**, form of †weathering responsible for the mechanical breakdown of rocks but involving no chemical change.

Forces acting on rock exposed on the Earth's surface open up any weak points in the rock and cause pieces to be broken off. Piles of jagged rock fragments called †scree are formed. Processes involved include †freeze–thaw (the alternate freezing and melting of ice in rock cracks or pores) and exfoliation (the alternate expansion and contraction of rocks in response to extreme changes in temperature). Physical weathering is also brought about by the growth of plants. A seed falling into a crack in a rock may germinate and the growing plant forces the crack to widen. Similarly, a tree root may grow down into a crack and split the rock. Physical weathering is one of the processes involved in the rock cycle, the formation, change, and re-formation of the Earth's outer layers.

Landforms of Weathering http://www.geog.ouc.bc.ca/physgeog/contents/11c.html

From a much larger site "Fundamentals of Physical Geography" set up by a Canadian university and covering many aspects of the area of study. The site is clear and well organized and contains maps and diagrams.

physics branch of science concerned with the laws that govern the structure of the universe, and the investigation of the properties of matter and energy and their interactions. Before the 19th century, physics was known as **natural philosophy**.

Classical physics consisted of the study of heat, light, electricity, magnetism, optics, acoustics, and mechanics. Since the turn of the 20th century the study of quantum mechanics and relativity has become more important. Today, physics is often divided into branches such as †nuclear physics, †particle physics, †quantum physics, theoretical physics, and †solid-state physics. The study of the planets, stars, and their interactions is known as †astrophysics, the physics of the Earth is called †geophysics, and the study of the physical laws relating to living organisms is called †biophysics.

Bang! Boing! Pop! http://library.thinkquest.org

Interactive learning with "an emphasis on building physical intuition through interaction." Students are encouraged to study this site at their own pace and monitor their progress through short tests.

Eric Weisstein's World of Physics http://scienceworld.wolfram.com/physics/

Searchable database of facts relating to physics compiled by Eric W Weisstein, an "Internet Encyclopedist." You can search by keywords or subject, and the key terms in each definition are linked to their own definition.

How Things Work http://rabi.phys.virginia.edu/HTW/

Well-organized site providing answers to many questions relating to physics. There is a large database of previously asked questions that can be readily accessed. Run by the professor of physics at the University of Virginia, USA, the site is an essential point of reference for anybody with an interest in physics.

Interactive Physics Problem Set
http://www.physics.gatech.edu/academics/tutorial/phys2121/Chapter%202/InteractiveProb.html

Nearly 100 practice problems at various levels of difficulty, with detailed answers. The subjects covered include "Position, velocity, and acceleration," "Forces," and "Energy and work."

NetScience: Physics http://library.thinkquest.org/3616/physics/

Clearly presented site aimed at high school students. The site covers vectors, motion, forces, momentum, work, energy, and power.

Phool Proof Physics http://library.thinkquest.org/15433/

Comprehensive physics site. This is a stylish site containing sections on a variety of physics topics including electromagnetism, waves, and electronic systems. Each section comprises textbook-style pages on numerous subjects and multiple choice quiz questions.

Physics FAQ http://www.weburbia.demon.co.uk/physics/faq.html

Answers to questions on general physics, particle and nuclear physics, and quantum physics.

Wester's Online Guide to Physics http://www.msms.k12.ms.us/teachers/kwester/ap_physics/

Clearly-presented tutorial including lessons, example problems to test your knowledge, and a "Frequently Asked Questions."

physiological psychology aspect of †experimental psychology concerned with physiological and neurological processes as the basis of experience and behavior. It overlaps considerably with such fields as anatomy, physiology, neurology, and biochemistry.

physiology branch of biology that deals with the functioning of living organisms, as opposed to anatomy, which studies their structures.

phytomenadione one form of vitamin K, a fat-soluble chemical found in green vegetables. It is involved in the production of prothrombin, which is essential in blood clotting. It is given to newborns to prevent potentially fatal brain hemorrhages.

pi symbol π, ratio of the circumference of a circle to its diameter. Pi is an irrational number: it cannot be expressed as the ratio of two integers, and its expression as a decimal never terminates and never starts recurring. The value of pi is 3.1415926, correct to seven decimal places. Common approximations to pi are $^{22}/_{7}$ and 3.14, although the value 3 can be used as a rough estimation.

For example, a piece of string tied around the circumference of a mug will measure approximately three times the diameter of the mug.

Pi Through The Ages http://www-history.mcs.st-and.ac.uk/history/HistTopics/Pi_through_the_ages.html

History of the calculation of the number pi. This site gives the figure that mathematicians from the time of Ptolemy to the present day have used for pi, and describes exactly how they calculated their figure. Also included are many historical asides, such as how calculation of pi led to a racial attack on an eminent professor in pre-war Germany. Reference publications for further reading are listed here, and there are also links to other pi-related Web sites.

Piaget, Jean (1896–1980) Swiss psychologist whose studies of the development of thought processes in children have been influential in early-childhood research and on school curricula and teaching methods.

Piazzi, Giuseppe (1746–1826) Italian astronomer. In 1801 he identified the first asteroid, which he named †Ceres.

Picard, Jean (1620–1682) French astronomer. He was professor of astronomy at the Collège Royal de France from 1655, and was a founding member of the Academy of Sciences in 1666. In 1667 he was one of the first astronomers to use cross-wire sights in a telescope, thus establishing a new standard of accuracy in positional measurements. In 1671 he published his measurements of an arc of the meridian of Paris, giving a value of the Earth's radius that was important in verifying †Newton's theory of gravitation.

Picard, Jean http://www-groups.dcs.st-andrews.ac.uk/history/Mathematicians/Picard_Jean.html

Devoted to the life and contributions of astronomer Jean Picard. In addition to biographical information, you will find a list of references about Picard and links to essays that reference him. The text of this essay includes hypertext links to the essays of those mathematicians and thinkers who influenced Picard. There is also find a map of his birthplace.

Piccard, Auguste Antoine (1884–1962) Swiss scientist. In 1931–32 he and his twin brother, **Jean Félix** (1884–1963), made ascents to 17,000 m/55,000 ft in a balloon of his own design, resulting in useful discoveries concerning stratospheric phenomena such as †cosmic radiation. He also built and used, with his son **Jacques Ernest** (1922–), bathyscaphs for research under the sea.

picric acid technical name **2,4,6-trinitrophenol**; $C_6H_2(NO_2)_3OH$, strong acid that is used to dye wool and silks yellow, for the treatment of burns, and in the manufacture of explosives. It is a yellow, crystalline solid.

pictogram pictorial way of presenting statistical data, in which a symbol is used to represent a specific quantity of items.

Pictor constellation of the southern hemisphere, originally named "Equuleus Pictoris," "the Painter's Easel," by 18th-century French astronomer Nicolas †Lacaille. Its name now refers only to the painter.

pie chart method of displaying proportional information by dividing a circle up into different-sized sectors (slices of pie). The angle of each sector is proportional to the size, expressed as a percentage, of the group of data that it represents.

For example, data from a traffic survey could be presented in a pie chart in the following way:
(1) convert each item of data to a percentage figure;
(2) 100% will equal 360 degrees of the circle, therefore each 1% = 360/100 = 3.6 degrees;
(3) calculate the angle of the segment for each item of data by multiplying the percentage by 3.6, and plot this on the circle.
The diagram may be made clearer by adding colors or shadings to each group, together with a key.

piezoelectric effect property of some crystals (for example, quartz) to develop an electromotive force or voltage across opposite faces when subjected to tension or compression, and, conversely, to expand or contract in size when subjected to an electromotive force. Piezoelectric crystal †oscillators are used as frequency standards (for example, replacing balance wheels in watches), and for producing ultrasound. Crystalline quartz is a good example of a piezoelectric material.

pigeon-hole principle in mathematics, another name for †Schubfachprinzip.

pig iron or **cast iron**, iron produced in a †blast furnace. It contains around 4% carbon plus some other impurities.

pigment insoluble colored or white material used to impart color to a suitable medium. A pigment usually takes the form of a powder, which is added to a liquid. Pigments are widely used in the manufacture of paint and to color plastics.

pillow lava type of usually basaltic lava composed of ellipsoidal or spherical structures up to a meter across. It forms when hot molten lava cools rapidly on coming into contact with water, as when a stream of lava flows into the sea.

PIN acronym for personal identification number, in banking, a unique number used as a password to establish the identity of a customer using an automatic cash dispenser. The PIN is normally encoded into the magnetic strip of the customer's bank card and is known only to the customer and to the bank's computer. Before a cash dispenser will issue money or information, the customer must insert the card into a slot in the machine (so that the PIN can be read from the magnetic strip) and enter the PIN correctly at a keyboard. This helps to prevent stolen cards from being used to obtain money from cash dispensers.

pineal body or **pineal gland**, cone-shaped outgrowth of the vertebrate brain. In some lower vertebrates, it develops a rudimentary lens and retina, which show it to be derived from an eye, or pair of eyes, situated on the top of the head in ancestral vertebrates. In fishes that can change color to match their background, the pineal perceives the light level and controls the color change. In birds, the pineal detects changes in daylight and stimulates breeding behavior as spring approaches. Mammals also have a pineal gland, but it is located deeper within the brain. It secretes a hormone, melatonin, thought to influence rhythms of activity. In humans, it is a small piece of tissue attached by a stalk to the rear wall of the third ventricle of the brain.

Pinel, Philippe (1745–1826) French physician who founded †psychiatry and pioneered the humane treatment of the mentally ill.

pingo landscape feature of †tundra terrain consisting of a hemispherical mound about 30 m/100 ft high, covered with soil that is cracked at the top. The core consists of ice, probably formed from the water of a former lake. The lake that forms when such a feature melts after an ice age is also called a pingo.

pinhole camera simplest type of camera, in which a pinhole rather than a lens is used to form an image. Light passes through the pinhole at one end of a box to form an inverted image on the inside surface of the opposite end. A photographic film or plate fitted inside the box will, if exposed for a long time, record the image.

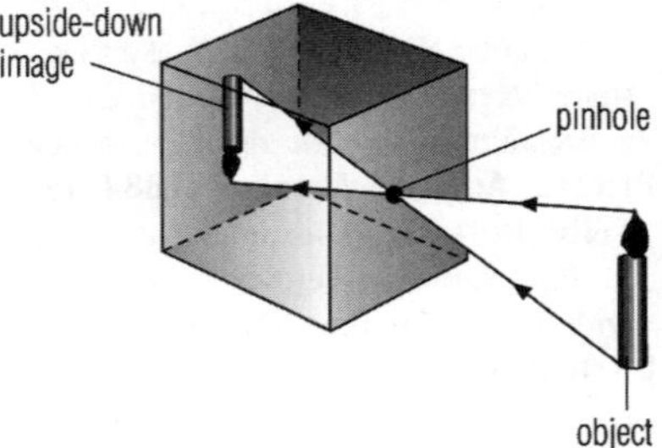

pinhole camera A pinhole camera has no lens but can nevertheless produce a sharp inverted image because only one ray from a particular point can enter the tiny pinhole aperture, and so no blurring takes place. However, the very low amount of light entering the camera also means that the film at the back must be exposed for a long time before a photographic image is produced; the camera is therefore only suitable for photographing stationary objects. Helicon

pinna in botany, the primary division of a †pinnate leaf. In mammals, the pinna is the external part of the ear.

pinnate leaf leaf that is divided up into many small leaflets, arranged in rows along either side of a midrib, as in ash trees (*Fraxinus*). It is a type of compound leaf. Each leaflet is known as a **pinna**, and where the pinnae are themselves divided, the secondary divisions are known as pinnules.

pinocytosis ingestion of the contents of a vesicle by a cell (see †endocytosis).

pint symbol pt, imperial dry or liquid measure of capacity equal to 20 fluid ounces, half a quart, one-eighth of a gallon, or 0.568 liter. In the USA, a liquid pint is equal to 0.473 liter, while a dry pint is equal to 0.550 liter.

pinworm †nematode worm *Enterobius vermicularis*, an intestinal parasite of humans.

pion or **pi meson**, in physics, a subatomic particle with a neutral form (mass 135 MeV) and a charged form (mass 139 MeV). The charged pion decays into muons and neutrinos and the neutral form decays into gamma-ray photons. They belong to the †hadron class of †elementary particles.

The mass of a charged pion is 273 times that of an electron; the mass of a neutral pion is 264 times that of an electron.

pioneer species in ecology, those species that are the first to colonize and thrive in new areas. Coal tips, recently cleared woodland, and new roadsides are areas where pioneer species will quickly appear. As the habitat matures other species take over, a process known as **succession**.

pipeflow movement of water through natural pipes in the soil; it is a form of throughflow and may be very rapid. The pipes used may be of animal or plant origin – for example, worm burrows or gaps created by tree roots.

pipette device for the accurate measurement of a known volume of liquid, usually for transfer from one container to another, used in chemistry and biology laboratories.

A pipette may be a slender glass tube, often with an enlarged bulb, which is calibrated in one or more positions; or it may be a plastic device with an adjustable plunger, fitted with one or more disposable plastic tips.

Pisces inconspicuous zodiac constellation, mainly in the northern hemisphere between †Aries and †Aquarius, near †Pegasus. It is represented as two fish tied together by their tails. The Circlet, a delicate ring of stars, marks the head of the western fish in Pisces. The constellation contains the **vernal equinox**, the point at which the Sun's path around the sky (the **ecliptic**) crosses the celestial equator (see †celestial sphere). The Sun reaches this point around 21 March each year as it passes through Pisces from mid-March to late April. In astrology, the dates for Pisces are between about 19 February and 20 March (see †precession).

Piscis Austrinus or **Southern Fish**, constellation of the southern hemisphere near †Capricornus. Its brightest star is the first-magnitude †Fomalhaut.

pistil general term for the female part of a flower, either referring to one single †carpel or a group of several fused carpels.

pitch in chemistry, a black, sticky substance, hard when cold, but liquid when hot, used for waterproofing, roofing, and paving. It is made by the destructive distillation of wood or coal tar, and has been used since antiquity for caulking wooden ships.

pitch in mechanics, the distance between the adjacent threads of a screw or bolt. When a screw is turned through one full turn it moves a distance equal to the pitch of its thread. A screw thread is a simple type of machine, acting like a rolled-up inclined plane, or ramp (as may be illustrated by rolling a long paper triangle around a pencil). A screw has a †mechanical advantage greater than one.

pitchblende or **uraninite**, brownish-black mineral, the major uranium ore, consisting mainly of uranium oxide (UO_2). It also contains some lead (the final, stable product of uranium decay) and variable amounts of most of the naturally occurring radioactive elements, which are products of either the decay or the fissioning of uranium isotopes. The uranium yield is 50–80%; it is also a source of radium, polonium, and actinium. Pitchblende was first studied by Pierre and Marie †Curie, who found radium and polonium in its residues in 1898.

pitfall trap simple device for trapping small invertebrates. In its simplest form a beaker or jam jar is buried in the ground so that the rim of the jar is flush with the soil. Beetles, millipedes, spiders, and other arthropods tumble into the jar and are unable to escape.

pitot tube instrument that measures fluid (gas or liquid) flow. It is used to measure the speed of aircraft, and works by sensing pressure differences in different directions in the airstream.

It was invented in the 1730s by the French scientist Henri Pitot (1695–1771).

pituitary gland major †endocrine gland of vertebrates, situated in the center of the brain. It is attached to the †hypothalamus by a stalk. The pituitary consists of two lobes. The posterior lobe is an extension of the hypothalamus, and is in effect nervous tissue. It stores two hormones synthesized in the hypothalamus: †ADH and †oxytocin. The anterior lobe secretes six hormones, some of which control the activities of other glands (thyroid, gonads, and adrenal cortex); others are direct-acting hormones affecting milk secretion and controlling growth.

pixel derived from picture element, single dot on a computer screen. All screen images are made up of a collection of pixels, with each pixel being either off (dark) or on (illuminated, possibly in color). The number of pixels available determines the screen's resolution. Typical resolutions of microcomputer screens vary from 320 x 200 pixels to 800 x 600 pixels, but screens with 1,024 x 768 pixels or more are now common for high-quality graphic (pictorial) displays.

Pixii, Hippolyte (1808–1835) French inventor who in 1832 made the first practical electricity generator. It could produce both direct current and alternating current.

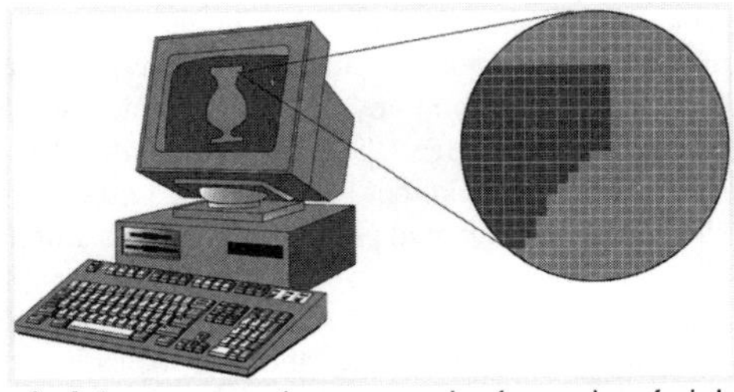

pixel Computer screen images are made of a number of pixels ("dots"). The greater the number of pixels the greater the resolution of the image; most computer screens are set at 640 x 480 pixels, although higher resolutions are available. Helicon

placenta organ that attaches the developing †embryo or †fetus to the †uterus (womb) in placental mammals (mammals other than marsupials, platypuses, and echidnas). Composed of maternal and embryonic tissue, it links the blood supply of the embryo to the blood supply of the mother, allowing the exchange of oxygen, nutrients, and waste products. The two blood systems are not in direct contact, but are separated by thin membranes, with materials diffusing across from one system to the other. The placenta also produces hormones that maintain and regulate pregnancy. It is shed as part of the afterbirth.

The placenta is an organ essential to the development of the fetus. In mammals, the fetus develops inside the body of the mother. The placenta is attached to the fetus by a cord containing blood vessels – the †umbilical cord. The placenta is also firmly attached to the uterus of the mother. The role of the placenta is to bring the †blood of the fetus close to the blood of the mother. This then allows †oxygen and †mineral salts to pass from the mother to the fetus, while †carbon dioxide and other waste materials pass in the opposite direction. A variety of materials, including drugs and viruses, can pass across the placenta and enter the fetus. HIV, the virus that causes †AIDS, can be transmitted in this way. However, the mother provides some protection against infection. There are chemicals that the mother makes to defend against disease that can cross into the fetus, such as some antibodies.

The tissue in plants that joins the ovary to the ovules is also called a placenta.

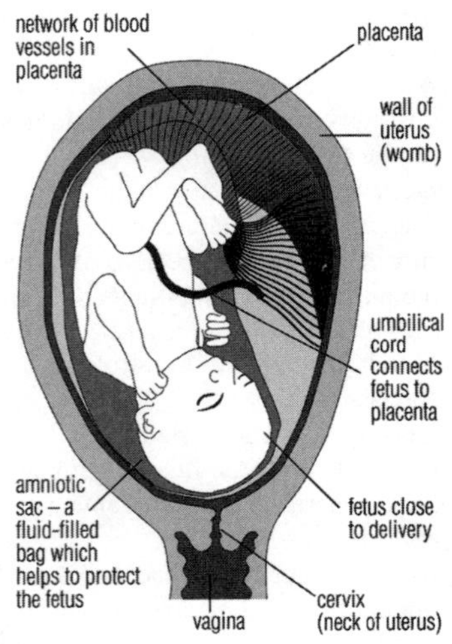

placenta The human placenta is a disc-shaped organ about 25 cm/10 in in diameter and 3 cm/1 in thick. It is connected to the fetus by the umbilical cord. Helicon

placer deposit detrital concentration of an economically important mineral, such as gold, but also other minerals such as cassiterite, chromite, and platinum metals. The mineral grains become concentrated during transport by water or wind because they are more dense than other detrital minerals such as quartz, and (like quartz) they are relatively resistant to chemical breakdown. Examples are the Witwatersrand gold deposits of South Africa, which are gold- and uranium-bearing conglomerates laid down by ancient rivers, and the placer tin deposits of the Malay Peninsula.

place value value given to a digit because of its position within a number. For example, in the decimal number 2,465 the 2 represents two thousands, the 4 represents four hundreds, the 6 represents six tens, and the 5 represents five units.

plage in astronomy, bright patch in the †chromosphere above a group of †sunspots, occasionally seen on images of the Sun in hydrogen light.

plain or **grassland**, land, usually flat, upon which grass predominates. The plains cover large areas of the Earth's surface, especially between the †deserts of the tropics and the †rainforests of the Equator, and have rain in one season only. In such regions the †climate belts move north and south during the year, bringing rainforest conditions at one time and desert conditions at another. Temperate plains include the North European Plain, the High Plains of the USA and Canada, and the Russian Plain (also known as the †steppe).

Planck, Max Karl Ernst Ludwig (1858–1947) German physicist who was awarded the Nobel Prize for Physics in 1918 for his formulation of the †quantum theory in 1900. His research into the manner in which heated bodies radiate energy led him to report that energy is emitted only in indivisible amounts, called "quanta," the magnitudes of which are proportional to the frequency of the radiation. His discovery ran counter to classical physics and is held to have marked the commencement of modern science.

Planck, Max Karl Ernst Ludwig http://www-history.mcs.st-and.ac.uk/history/Mathematicians/Planck.html

Biography of the eminent German physicist Max Planck. Planck is thought of by many to have been more influential than any other in the foundation of quantum physics, and partly as a consequence has a fundamental constant named after him. This Web site contains a description of his working relationships with his contemporaries, and also features a photograph of Planck. Several literature references for further reading on the works of Planck are also listed.

Planck's constant symbol *h*, fundamental constant that relates the energy (*E*) of one quantum of electromagnetic radiation (a "packet" of energy; see †quantum theory) to the frequency (*f*) of its radiation by $E = hf$. Its value is 6.6262 x 10^{-34} joule seconds.

Planck's radiation law energy of †electromagnetic radiation of a certain †frequency is given by the product of the frequency and †Planck's constant, *h*.

plane flat surface. Planes are either parallel or they intersect in a straight line. Vertical planes, for example the join between two walls, intersect in a vertical line. Horizontal planes do not intersect since they are all parallel.

plane figure in geometry, a two-dimensional figure. All †polygons are plane figures.

plane-polarized light light in which the electric and magnetic vibrations of the waves are restricted to a single plane, the plane of the magnetic vibration being at right angles to that of the electric one.

planet (Greek "wanderer") large celestial body in orbit around a star, composed of rock, metal, or gas. There are nine planets in the †Solar System orbiting the †Sun: Mercury, Venus, Earth, Mars, Jupiter, Saturn, Neptune, Uranus, and Pluto. The inner four, called the **terrestrial planets**, are small and rocky, and have few natural †satellites. The outer planets, with the exception of Pluto, are called the **major planets**, and have denser atmospheres consisting mainly of hydrogen and helium gases, and many natural satellites. The largest planet in the Solar System is Jupiter (about 780 million km/490 million mi from the Sun) with a diameter of 140,000 km/87,500 mi, which contains a mass greater than all the other planets combined. The smallest (and furthest from the Sun at about 5,900 million km/3,600 million mi) is Pluto with a diameter of 2,300 km/1,400 mi.

As seen from the Earth, all the historic planets are conspicuous naked-eye objects moving in looped paths against the stellar background. The size of these loops, which are caused by the Earth's own motion around the Sun, are inversely proportional to the planet's distance from the Earth. Planets do not produce light, but reflect the light of their parent star.

recent discoveries In 1995 Italian astronomers believed they had detected a new planet around 51 Pegasi in the constellation Pegasus. It was named 51 Pegasi B and is thought to have a mass comparable to that of Jupiter. The gravitational pull thought to be that of the planet may be caused by pulsation in the parent star. The discoveries of six new planets were announced in 1996. All are outside the Solar System, but two are only about 35 light years from the Earth and orbit stars visible with the naked eye. The first, 70 Vir B, is in the constellation Virgo. The second, 47 UMa B, is in Ursa Major. The third, b Pictoris, is about 50 light years away in the southern constellation Pictor. The fourth is orbiting Rho Cancri in the constellation Cancer. The fifth is orbiting the star Tau Bootis. The sixth has an orbit which is more irregular than that of any other and a mass 1.6 times that of Jupiter, and orbits the star 16 Cygni B. Its distance from the star 16 Cygni B varies from 90 million km/55.9 million mi to 390 million km/242 million mi. In 1998 Australian and New Zealand astronomers discovered a planet 25,000 light years away. Known as 98-351, the planet was discovered at the Mount Stromlo observatory in Australia and the Mount John observatory in New Zealand. By August 2000 astronomers had located 50 planets outside the solar system, nine of which were spotted in 2000.

Space probes to planets include Mariner 2 (launched in 1962 to Venus), Pioneer 10 (launched in 1972 to Jupiter), Mariner 10 (launched in 1973 to Venus and Mercury), Viking 1 and Viking 2 (launched in 1975 to Mars), Voyager 1 and Voyager 2 (launched in 1977) to Jupiter and Saturn, Galileo (launched in 1989 to Jupiter), and Cassini (launched in 1997 to Saturn), and numerous missions to Mars in the 1990s.

Nine Planets http://www.nineplanets.org/

Multimedia tour of the Solar System, with descriptions of each of the planets and major moons. In addition, there are appendices on such topics as astronomical names and how they are assigned, the origin of the Solar System, and hypothetical planets.

Planetary Image Finders http://ic-www.arc.nasa.gov/ic/projects/bayes-group/Atlas/

Page dedicated to listing the best sources of high-resolution, often unprocessed images of each planet or moon in the Solar System.

Planetary Society http://planetary.org/

Home page of the Planetary Society, which was founded in 1980 to encourage the exploration of our Solar System and the search for extraterrestrial life. The site gives the latest news stories on space exploration and encourages visitors to get involved in one of the projects by searching for extraterrestrial life from home.

planetarium optical projection device by means of which the motions of stars and planets are reproduced on a domed ceiling representing the sky.

planetary embryo one of numerous massive bodies thought to have formed from the accretion of †planetesimals during the formation of the Solar System. Embryos in the region of the Earth's orbit would have been about 10^{23} kg in mass, and about 10–100 of them would have coalesced to make the Earth.

planetary geology study of the structure, composition, and geological history of rocky planets, satellites, and asteroids (including, for instance, both the Earth and Moon) and the processes that have shaped them.

planetary nebula shell of gas thrown off by a star at the end of its life. Planetary nebulae have nothing to do with planets. They were named by German-born English astronomer William Herschel, who thought their rounded shape resembled the disc of a planet. After a star such as the Sun has expanded to become a †red giant, its outer layers are ejected into space to form a planetary nebula, leaving the core as a †white dwarf at the center.

planetesimal body of rock in space, smaller than a planet. Scientists study planetesimal images from telescopes in Earth and in space, adhering to the modern theory that planetesimals were the building blocks of planets.

The Sun and the planets are thought to have formed from a rotating dust cloud generated by a †supernova explosion. On condensation, this cloud formed a central sun and a rotating disc, the material of which separated into rings of dust grains that began to stick together. Larger and larger clumps formed in each ring and eventually collected into bodies the size of present-day asteroids, or planetesimals. After numerous collisions, these bodies eventually formed the nucleus of the various planets of our Solar System.

planimeter simple integrating instrument for measuring the area of a regular or irregular plane surface. It consists of two hinged arms: one is kept fixed and the other is traced around the boundary of the area. This actuates a small graduated wheel; the area is calculated from the wheel's change in position.

planisphere in astronomy, graphical device for determining the †aspect of the sky for any date and time in the year. It consists of two discs mounted concentrically so that the upper disc, which has an aperture corresponding to the horizon of the observer, can rotate over the lower disc, which is printed with a map of the sky centered on the north or south celestial pole. In use, the observer aligns the time of day marked around the edge of the upper disc with the date marked around the edge of the lower disc. The aperture then shows which stars are above the horizon.

plankton small, often microscopic, forms of plant and animal life that live in the upper layers of fresh and salt water, and are an important source of food for larger animals. Marine plankton is concentrated in areas where rising currents bring mineral salts to the surface.

plant organism that carries out †photosynthesis, has cellulose cell walls and complex cells, and is immobile. A few parasitic plants have lost the ability to photosynthesize but are still considered to be plants. Plants are †autotrophs, that is, they make carbohydrates from water and carbon dioxide, and are the primary producers in all food chains, so that all animal life is dependent on them. They play a vital part in the carbon cycle, removing carbon dioxide from the atmosphere and generating oxygen. The study of plants is known as †botany.

levels of complexity The simplest plants consist of a body, or thallus, on which the organs of reproduction are borne. Simplest of all are the threadlike algae, for example *Spirogyra*, which consist of a chain of cells.

The seaweeds (algae) and mosses and liverworts (bryophytes) represent a further development, with simple, multicellular bodies that have specially modified areas in which the reproductive organs are carried. Higher in the morphological scale are the ferns, club mosses, and horsetails (pteridophytes). Ferns produce leaflike fronds bearing sporangia on their under surface in which the spores are carried. The spores are freed and germinate to produce small independent bodies carrying the sexual organs; thus the fern, like other pteridophytes and some seaweeds, has two quite separate generations in its life cycle (see †alternation of generations).

The pteridophytes have special supportive water-conducting tissues, which identify them as vascular plants, a group which includes all seed plants, that is the gymnosperms (conifers, yews, cycads, and ginkgos) and the angiosperms (flowering plants).

seed plants The seed plants are the largest group, and structurally the most complex. They are usually divided into three parts: root, stem, and leaves. Stems can grow above or below ground. Their cellular structure is designed to carry water and salts from the roots to the leaves in the †xylem, and sugars from the leaves to the roots in the †phloem. The leaves manufacture the food of the plant by means of photosynthesis, which occurs in the †chloroplasts they contain. Flowers and cones are modified leaves arranged in groups, enclosing the reproductive organs from which the fruits and seeds result.

conservation status The World Conservation Union published a world list of threatened plants in 1998. The list includes 33,798 species of vascular plants (flowering plants, ferns, cycads, and conifers); in other words, one in eight of the estimated 270,000 species of vascular plants are now endangered.

Dr NAD's Mini-Course http://www.geocities.com/CapeCanaveral/5229/n_1.htm

Dr NAD's step-by-step mini-course uses helpful graphics and readable text to guide you through the basics of plant structure and reproduction. It provides a clear explanation of the differences between plant and animal cells.

Plants and their Structure http://www.emc.maricopa.edu/faculty/farabee/BIOBK/BioBookPLANTANAT.html

Informative guide to the structure of plants. The page begins with a hypertext exploration of the general organization of plant species, then moves on to a discussion of plant cells and tissue types. Diagrams and microscopic images accompany the text.

plant cell †cell in a plant, which, like all cells, has a †cell membrane, †cytoplasm, and a †nucleus, but which differs from an animal cell by having a †cell wall outside the cell surface membrane and a large †vacuole. They may also have †chloroplasts in the cytoplasm.

The cell surface membrane keeps the cell together by being strong, although it is still very thin and flexible. This membrane controls what enters and leaves the cell – for example, nutrients must be allowed to enter and waste materials must be allowed to leave. The cell has some cytoplasm, but most of the space inside the cell is generally filled with a large vacuole. The cytoplasm contains the nucleus, which controls the activities of the cell, and often contains many chloroplasts, which capture sunlight and make food for the plant through †photosynthesis. Chloroplasts are green – this explains why most plants look green.

plant classification taxonomy or classification of plants. Originally the plant kingdom included bacteria, diatoms, dinoflagellates, fungi, and slime moulds, but these are not now thought of as plants. The groups that are always classified as plants are the bryophytes (mosses and liverworts), pteridophytes (ferns, horsetails, and club mosses), gymnosperms (conifers, yews, cycads, and ginkgos), and angiosperms (flowering plants). The angiosperms are split into monocotyledons (for example, orchids, grasses, lilies) and dicotyledons (for example, oak, buttercup, geranium, and daisy).

The basis of plant classification was established by the Swedish naturalist Carolus †Linnaeus. Among the angiosperms, it is largely based on the number and arrangement of the flower parts.

The unicellular algae, such as *Chlamydomonas*, are often now put with the protists (single-celled organisms) instead of the plants. Some classification schemes even classify the multicellular algae (seaweeds and freshwater weeds) in a new kingdom, the Protoctista, along with the protists.

plant hormone substance produced by a plant that has a marked effect on its growth, flowering, leaf fall, fruit ripening, or some other process. Examples include †auxin, †gibberellin, †ethene, and †cytokinin.

Unlike animal †hormones, these substances are not produced by a particular area of the plant body, and they may be less specific in their effects. It has therefore been suggested that they should not be described as hormones at all.

Plants lack a nervous system and so, for the most part, are unable to respond quickly to stimuli in the way animals can. The two important stimuli for plants are light and gravity, and plants respond by growing either towards or away from the stimulus (this is termed †tropism). For example, the shoot of a plant grows toward light (phototropism), while root growth is governed by the pull of gravity (geotropism). Tropism plays a very important part in plant development and survival. Shoots need to grow upwards toward the light so that the leaves can photosynthesize, and flowers can be pollinated; whereas roots need to grow downwards into the soil in order to anchor the plant and absorb water and minerals. The response to stimuli is controlled by plant hormones. Plant hormones are organic chemicals, and are usually referred to as "growth substances." These are synthesized by plants, and the five major types are abscisic acid, auxin, cytokinin, ethylene (ethene), and gibberellin. These substances regulate growth and development, and are usually produced in a particular part of the plant, such as the shoot tip, and transported to other parts where they take effect.

plant propagation production of plants. Botanists and horticulturalists can use a wide variety of means for propagating plants. There are the natural techniques of †vegetative reproduction, together with †cuttings, †grafting, and †micropropagation. The range is wide because most plant tissue, unlike animal tissue, can give rise to the complete range of tissue types within a particular species.

Plaskett, John Stanley (1865–1941) Canadian astronomer and engineer. He discovered many new binary stars, including Plaskett's Twins, previously thought to be a single, massive star. He also carried out research into stellar radial velocities.

Plaskett's star most massive †binary star known, consisting of two supergiants of about 40 and 50 solar masses, orbiting each other every 14.4 days. Plaskett's star lies in the constellation Monoceros, named after the unicorn. Plaskett's star is named after Canadian astronomer John S Plaskett (1865–1941), who identified it as a binary star and discovered its massive nature in 1922.

plasma in biology, the liquid component of the †blood. It is a straw-colored fluid, largely composed of water (around 90%), in which a number of substances are dissolved. These include a variety of proteins (around 7%) such as fibrinogen (important in †blood clotting), inorganic mineral salts such as sodium and calcium, waste products such as †urea, traces of †hormones, and †antibodies to defend against infection.

plasma in physics, ionized gas produced at extremely high temperatures, as in the Sun and other stars. It contains positive and negative charges in equal numbers. It is a good electrical conductor. In thermonuclear reactions the plasma produced is confined through the use of magnetic fields.

Perspectives on Plasmas http://www.plasmas.org/

Impressive guide to plasmas. The site has lots of good articles on plasma basics, plasma technology, and the applications of plasma science. The pages contain lots of photographs and there is a gallery of colorful and interesting plasma pictures.

plasmid small, mobile piece of †DNA found in bacteria that, for example, confers antibiotic resistance, used in †genetic engineering. Plasmids are separate from the bacterial chromosome but still multiply during cell growth. Their size ranges from 3% to 20% of the size of the chromosome. Some plasmids carry "fertility genes" that enable them to move from one bacterium to another and transfer genetic information between strains. Plasmid genes determine a wide variety of bacterial properties including resistance to antibiotics and the ability to produce toxins.

plasmolysis separation of the plant cell cytoplasm from the cell wall as a result of water loss. As moisture leaves the vacuole the total volume of the cytoplasm decreases while the cell itself, being rigid, hardly changes. Plasmolysis is induced in the laboratory by immersing a plant cell in a strongly saline or sugary solution, so that water is lost by osmosis. Plasmolysis is unlikely to occur in the wild except in severe conditions.

plaster of Paris form of calcium sulfate $CaCo_3.{}^1/_2H_2O$, obtained from gypsum; it is mixed with water for making casts and moulds.

plastic any of the stable synthetic materials that are fluid at some stage in their manufacture, when they can be shaped, and that later set to rigid or semi-rigid solids. Plastics today are chiefly derived from petroleum. Most are polymers, made up of long chains of identical molecules.

environmental influence Since plastics have afforded an economical replacement for ivory in the manufacture of piano keys and billiard balls, the industrial chemist may well have been responsible for the survival of the elephant.

Most plastics cannot be broken down by micro-organisms, so cannot easily be disposed of. Incineration leads to the release of toxic fumes, unless carried out at very high temperatures.

Plastics http://www.plasticsusa.com/polylist.html

Datasheets on 49 generic plastic materials, showing their comparative chemical and physical properties.

Plastics http://www.sdplastics.com/plastics.html

Good introduction to the nature of plastics from a US plastics distributor. It gives a brief history of the plastics industry, followed by a description of the nature of polymers (the building blocks of plastics) and the properties of different kinds of plastic.

plasticizer in chemistry, additive that is incorporated into a synthetic †polymer to modify its physical properties. The usual effect of a plasticizer is to make the polymer more flexible.

plastid general name for a cell †organelle of plants that is enclosed by a double membrane and contains a series of internal membranes and vesicles. Plastids contain †DNA and are produced by division of existing plastids. They can be classified into two main groups: the **chromoplasts**, which contain pigments such as carotenes and chlorophyll, and the **leucoplasts**, which are colorless; however, the distinction between the two is not always clear-cut.

†Chloroplasts are the major type of chromoplast. They contain chlorophyll, are responsible for the green coloration of most plants, and perform †photosynthesis. Other chromoplasts give flower petals and fruits their distinctive color. Leucoplasts are food-storage bodies and include amyloplasts, found in the roots of many plants, which store large amounts of starch.

plate or **tectonic plate** or **lithospheric plate**, one of several relatively distinct sections of the †lithosphere, approximately 100 km/60 mi thick, which together comprise the outermost layer of the Earth (like the pieces of the cracked shell of a hard-boiled egg).

The plates are made up of two types of crustal material: oceanic crust (sima) and continental crust (sial), both of which are underlain by a solid layer of †mantle. Dense **oceanic crust** lies beneath Earth's oceans and consists largely of †basalt. **Continental crust**, which underlies the continents and the continental shelves, is thicker, less dense, and consists of rocks that are rich in silica and aluminum.

Due to convection in the Earth's mantle (see †plate tectonics) these pieces of lithosphere are in motion, riding on a more plastic layer of the mantle, called the asthenosphere. Mountains, volcanoes, earthquakes, and other geological features and events all come about as a result of interaction between these plates.

plateau elevated area of fairly flat land, or a mountainous region in which the peaks are at the same height. An **intermontane plateau** is one surrounded by mountains. A **piedmont plateau** is one that lies between the mountains and low-lying land. A **continental plateau** rises abruptly from low-lying lands or the sea. Examples are the Tibetan Plateau and the Massif Central in France.

platelet tiny disc-shaped structure found in the blood, which helps it to clot. Platelets are not true cells, but membrane-bound cell fragments without nuclei that bud off from large cells in the bone marrow.

They play a vital role in blood clotting as they release blood clotting factors at the site of a cut. Over twelve clotting factors have been discovered and they produce a complex series of reactions which ultimately leads to fibrinogen, the inactive blood sealant always found in the plasma, being converted into fibrin. Fibrin aggregates into threads which form the fabric of a blood clot.

plate margin or **plate boundary**, the meeting place of one †plate (plates make up the top layer of the Earth's structure) with another plate. There are four types of plate margin – destructive, constructive, collision, and conservative. A †volcano may be found along two of the types of plate margin, and an †earthquake may occur at all four plate margins.

destructive or convergent plate margins At a destructive margin an oceanic plate moves towards (and disappears into the †mantle of) a continental plate or another oceanic plate. This is the **subduction zone**. As it is forced downwards, pressure at the margins increases, and this can result in violent earthquakes. The heat produced by friction turns the crust into †magma (liquid rock). The magma tries to rise to the surface and, if it succeeds, violent volcanic eruptions occur.

constructive or divergent plate margins At a †constructive margin the Earth's crust is forced apart. Magma rises and solidifies to create a new oceanic crust and forms a mid-ocean ridge. This ridge is made from †igneous rock; such ridges usually form below sea level on the sea bed (an exception to this is in Iceland).

collision plate margins A collision margin occurs when two plates moving together are both made from continental crust. Continental crust cannot sink or be destroyed, and as a result the land between them is pushed upwards to form high "fold" mountains like the Himalayas. Earthquakes are common along collision margins but there are no volcanic eruptions.

conservative plate margins At a conservative margin two plates try to slide past each other slowly. Quite often, the two plates stick and pressure builds up; the release of this pressure creates a severe earthquake. There are no volcanic eruptions along conservative plate margins because the crust is neither being created nor destroyed. The San Andreas Fault in California lies above the North American and Pacific plates, and is an example of a conservative plate margin.

plate reconstruction in geology, an illustration of positions of continents and oceans sometime in the geological past.

plate tectonics theory formulated in the 1960s to explain the phenomena of †continental drift and seafloor spreading, and the formation of the major physical features of the Earth's surface. The Earth's outermost layer, the †lithosphere, is seen as a jigsaw puzzle of rigid major and minor plates that move relative to each other, probably under the influence of convection currents in the †mantle beneath. At the margins of the plates, where they collide or move apart or slide past one another, major landforms such as †mountains, †rift valleys, †volcanoes, †ocean trenches, and **mid-ocean ridges** are created. The rate of plate movement is on average 2–3 cm/1 in per year and at most 15 cm/6 in per year.

The concept of plate tectonics brings together under one unifying theory many phenomena observed in the Earth's crust that were previously thought to be unrelated. The size of the crust plates is variable, as they are constantly changing, but six or seven large plates now cover much of the Earth's surface, the remainder being occupied by a number of smaller plates. Each large plate may include both continental and ocean lithosphere. As a result of seismic studies it is known that the lithosphere is a rigid layer extending to depths of about 50–100 km/30–60 mi, overlying the upper part of the mantle (the †asthenosphere), which is composed of rocks very close to melting point. This zone of mechanical weakness allows the movement of the overlying plates. The margins of the plates are defined by major earthquake zones and belts of volcanic and tectonic activity. Almost all earthquake, volcanic, and tectonic activity is confined to the margins of plates, and shows that the plates are in constant motion (see †plate margin).

Paleomap Project http://www.scotese.com/

The goal of the Paleomap Project is to illustrate the plate tectonic development of the ocean basins and continents, as well as the changing distribution of land and sea during the past 1,100 million years.

Plate Tectonics http://volcano.und.nodak.edu/vwdocs/vwlessons/plate_tectonics/introduction.html

US-based site providing a thorough introduction to plate tectonics. Although navigation is rather basic, useful illustrated descriptions of topics such as "The layers of the Earth," "Types of plate motion," "Location of plate boundaries," and "Continental drift" can be found by clicking on the arrow at the bottom of the initial page.

This Dynamic Planet http://pubs.usgs.gov/pdf/planet.html

Online publication about plate tectonics, its history and theory, with a map that shows the Earth's physiographic features, the current movements of its major tectonic plates, and the locations of its volcanoes, earthquakes, and impact craters. The use of color and shaded relief helps the reader to identify significant features of the land surface and the ocean floor. Over 1,500 volcanoes active during the past 10,000 years are plotted on the map in four age categories.

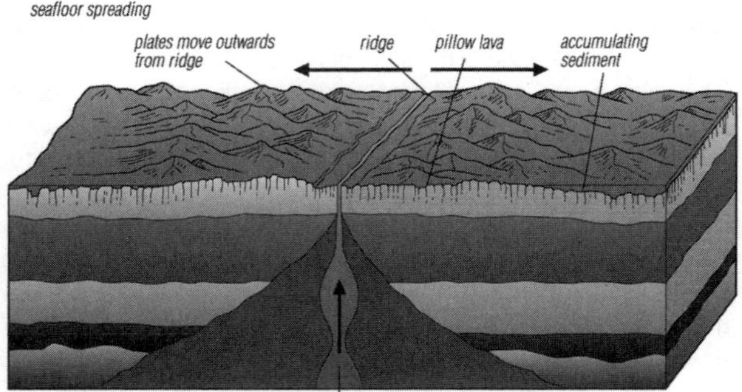

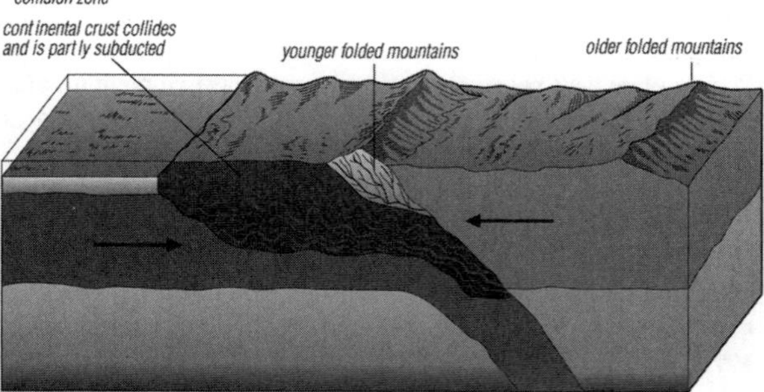

plate tectonics Constructive and destructive action in plate tectonics. (top) Seafloor spreading. The upwelling of magma forces apart the crust plates, producing new crust at the joint. Rapid extrusion of magma produces a domed ridge; more gentle spreading produces a central valley. (middle) The drawing downwards of an oceanic plate beneath a continent produces a range of volcanic fold mountains parallel to the plate edge. (bottom) Collision of continental plates produces immense fold mountains, such as the Himalayas.

Younger mountains are found near the coast with older ranges inland. The plates of the Earth's lithosphere are always changing in size and shape of as material is added at constructive margins and removed at destructive margins. The process is extremely slow, but it means that the tectonic history of the Earth cannot be traced back further than about 200 million years. Helicon

platform in computing, †operating system, together with the †hardware on which it runs.

platinum chemical symbol Pt, (Spanish *platina* "little silver") heavy, soft, silver-white, malleable and ductile, metallic element, atomic number 78, relative atomic mass 195.09. It is the first of a group of six metallic elements (platinum, osmium, iridium, rhodium, ruthenium, and palladium) that possess similar properties, such as resistance to tarnish, corrosion, and attack by acid, and that often occur as free metals (†native metals). They often occur in natural alloys with each other, the commonest of which is osmiridium. Both pure and as an alloy, platinum is used in dentistry, jewelry, and as a catalyst.

platinum resistance thermometer thermometer that uses the change in the electrical †resistance of a platinum coil with temperature to allow measurements over a very wide range (over 1,300 degrees).

Platonic solid in geometry, another name for a regular †polyhedron, one of five possible three-dimensional figures with all its faces the same size and shape.

Platyhelminthes invertebrate phylum consisting of the flatworms.

playa temporary lake in a region of interior drainage (it has no outlet to the sea). Such lakes are common features in arid desert basins that are fed by intermittent streams. The streams bring dissolved salts to the playa, and when it shrinks during dry spells, the salts precipitate as evaporite deposits.

Pleiades in astronomy, an open star cluster about 400 light years away from Earth in the constellation Taurus, represented as the Seven Sisters of Greek mythology. Its brightest stars (highly luminous, blue-white giants only a few million years old) are visible to the naked eye, but there are many fainter ones. It is a young cluster, and the stars of the Pleiades are still surrounded by traces of the reflection †nebula from which they formed, visible on long-exposure photographs.

pleiotropy process whereby a given gene influences several different observed characteristics of an organism. For example, in the fruit fly *Drosophila* the vestigial gene reduces the size of wings, modifies the halteres, changes the number of egg strings in the ovaries, and changes the direction of certain bristles. Many human syndromes are caused by pleiotropic genes, for example Marfan's syndrome where the slender physique, hypermobility of the joints, elongation of the limbs, dislocation of the lens, and susceptibility to heart disease are all caused by one gene.

Pleistocene epoch first part of the Quaternary period of geological time, beginning 1.64 million years ago and ending 10,000 years ago. The polar ice caps were extensive and glaciers were abundant during the ice age of this period, and humans evolved into modern *Homo sapiens sapiens* about 100,000 years ago.

Late Pleistocene Extinctions http://www.museum.state.il.us/exhibits/larson/LP_extinction.html

Exploration of possible causes of the Late Pleistocene extinction of most large mammals in North America. Different theories are discussed. There is also information on the prehistoric inhabitants of North America.

pleura covering of the lungs and of the inner surface of the chest wall; the surfaces are lubricated by fluid to prevent friction from breathing movements.

Pliocene Epoch ("almost recent") fifth and last epoch of the Tertiary period of geological time, 5.2–1.64 million years ago. The earliest hominid, the humanlike ape *Australopithecines*, evolved in Africa.

Plough, the popular name for the most prominent part of the constellation †Ursa Major.

Plücker, Julius (1801–1868) German mathematician and physicist. He made fundamental contributions to the field of analytical geometry and was a pioneer in the investigations of cathode rays that led eventually to the discovery of the electron.

plucking in earth science, a process of glacial erosion. Water beneath a glacier will freeze fragments of loose rock to the base of the ice. When the ice moves, the rock fragment is "plucked" away (literally ripped) from the underlying bedrock. Plucking is thought to be responsible for the formation of steep, jagged slopes such as the backwall of the corrie and the downslope-side of the roche moutonnée.

plumbago alternative name for the mineral †graphite.

plumule part of a seed embryo that develops into the shoot, bearing the first true leaves of a plant. In most seeds, for example the sunflower, the plumule is a small conical structure without any leaf structure. Growth of the plumule does not occur until the †cotyledons have grown above ground. This is †epigeal germination. However, in seeds such as the broad bean, a leaf structure is visible on the plumule in the seed. These seeds develop by the plumule growing up through the soil with the cotyledons remaining below the surface. This is known as †hypogeal germination.

plunge pool deep pool at the bottom of a †waterfall. It is formed by the hydraulic action of the water as it crashes down onto the river bed from a height.

pluteus in echinoderms, an advanced larval stage characterized by bilateral symmetry.

Pluto smallest and, usually, outermost planet of the Solar System. The existence of Pluto was predicted by calculation by US astronomer Percival Lowell and the planet was located by US astronomer Clyde †Tombaugh in 1930. Its highly elliptical orbit occasionally takes it within the orbit of Neptune, as in 1979–99. Pluto has a mass about 0.002 of that of Earth.

mean distance from the Sun: 5.8 billion km/3.6 billion mi
equatorial diameter: 2,300 km/1,438 mi
rotation period: 6.39 Earth days
year: 248.5 Earth years
atmosphere: thin atmosphere with small amounts of methane gas
surface: low density, composed of rock and ice, primarily frozen methane; there is an ice cap at Pluto's north pole
satellites: one moon, Charon

Pluto http://www.solarviews.com/solar/eng/pluto.htm

Detailed description of the planet Pluto. It includes a chronology of the exploration of the planet, along with statistics and information on its rings, moons, and satellites. There is also a good selection of images.

plutonic rock igneous rock derived from magma that has cooled and solidified deep in the crust of the Earth; granites and gabbros are examples of plutonic rocks.

plutonium chemical symbol Pu, silvery-white, radioactive, metallic element of the †actinide series, atomic number 94, relative atomic mass 239.13. It occurs in nature in minute quantities in †pitchblende and other ores, but is produced in quantity only synthetically. It has six allotropic forms (see †allotropy) and is one of three fissile elements (elements capable of splitting into other elements – the others are thorium and uranium). Plutonium dioxide, PuO_2, a yellow crystalline solid, is the compound most widely used in the nuclear industry. It was believed to be inert until US researchers discovered in 1999 that it reacts very slowly with oxygen and water to form a previously unknown green crystalline compound that is soluble in water.

PM10 abbreviation for particulate matter less than 10 micrometers (10 mm) across, clusters of small particles, such as carbon particles, in the air that come mostly from vehicle exhausts. There is a link between increase in PM10 levels and a rise in death rate, increased hospital admissions, and asthma incidence. The elderly and those with chronic heart or lung disease are most at risk.

The smaller the particulates the more dangerous they are: those below 2.5 mm across travel deep into the lungs and lodge there. The particulates themselves may be harmless but they may carry damaging chemicals, such as acids, into the lungs.

A European Commission draft directive released in November 1997 laid down guidelines for acceptable PM10 levels, aiming at maximum concentrations of 50 mg/m^3 of air over each 24-hour period by 2005. The directive stipulated a buffer of 25 days in which this may be exceeded (amended in 1999 to 35 days), to be reduced to seven days by 2010.

pneumatics study and production of devices that rely on air †pressure for their operation.

pneumatophore erect root that rises up above the soil or water and promotes †gas exchange. Pneumatophores, or breathing roots, are formed by certain swamp-dwelling trees, such as mangroves, since there is little oxygen available to the roots in waterlogged conditions. They have numerous pores or †lenticels over their surface, allowing gas exchange.

pneumothorax presence of air in the pleural cavity, between a lung and the chest wall. It may be due to a penetrating injury of the lung or to lung disease, or it may occur without apparent cause (spontaneous pneumothorax) in an otherwise healthy person. Prevented from expanding normally, the lung is liable to collapse.

p-n junction diode in electronics, another name for †semiconductor diode.

pod in botany, a type of fruit that is characteristic of legumes (plants belonging to the Leguminosae family), such as peas and beans. It develops from a single †carpel and splits down both sides when ripe to release the seeds.

In certain species the seeds may be ejected explosively due to uneven drying of the fruit wall, which sets up tensions within the fruit. In agriculture, the name "legume" is used for the crops of the pea and bean family. "Grain legume" refers to those that are grown mainly for their dried seeds, such as lentils, chick peas, and soybeans.

podzol or **podsol**, type of light-colored soil found predominantly under coniferous forests and on moorlands in cool regions where rainfall exceeds evaporation. The constant downward movement of water leaches nutrients from the upper layers, making podzols poor agricultural soils. Podzols are very acidic soils.

The leaching of minerals such as iron and alumina leads to the formation of a bleached zone, which is often also depleted of clay.

These minerals can accumulate lower down the soil profile to form a hard, impermeable layer which restricts the drainage of water through the soil.

poikilothermy condition in which an animal's body temperature is largely dependent on the temperature of the air or water in which it lives. It is characteristic of all animals except birds and mammals, which maintain their body temperatures by †homeothermy (they are "warm-blooded").

Poikilotherms have behavioral means of temperature control; they can warm themselves up by basking in the sun, or shivering, and can cool themselves down by sheltering from the Sun under a rock or by bathing in water.

Poikilotherms are often referred to as "cold-blooded animals," but this is not correct: their internal temperatures, regulated by behavioral means, are often as high as those of birds and mammals during the times they need to be active for feeding and reproductive purposes, and may be higher, for example in very hot climates. The main difference is that their body temperatures fluctuate more than those of homeotherms.

Poincaré, (Jules) Henri (1854–1912) French mathematician who developed the theory of differential equations and was a pioneer in †relativity theory. He suggested that Isaac †Newton's laws for the behavior of the universe could be the exception rather than the rule. However, the calculation was so complex and time-consuming that he never managed to realize its full implication.

Poincaré, Jules Henry http://www-history.mcs.st-and.ac.uk/history/Mathematicians/Poincare.html

Biography of the French mathematician. The site contains a description of his working relationships with his contemporaries, and also features a photograph of Poincaré. Several literature references for further reading on the works of Poincaré are also listed.

point in geometry, a basic element, whose position in the Cartesian system may be determined by its †coordinates.

Mathematicians have had great difficulty in defining the point, as it has no size, and is only the place where two lines meet. According to the Greek mathematician Euclid, (i) a point is that which has no part; (ii) the straight line is the shortest distance between two points.

point sets in geometry or topology, sets comprising some or all of the points of the space under study.

poise symbol P, *c*.g.s. unit of dynamic †viscosity (the property of liquids that determines how readily they flow). It is equal to one dyne-second per square centimeter. For most liquids the centipoise (one hundredth of a poise) is used. Water at 20°C/68°F has a viscosity of 1.002 centipoise.

Poiseuille's formula relationship describing the rate of flow of a fluid through a narrow tube. For a capillary (very narrow) tube of length l and radius r with a pressure difference p between its ends, and a liquid of †viscosity h, the velocity of flow expressed as the volume per second is $\pi pr^4/8l$h. The formula was devised in 1843 by French physicist Jean Louis Poiseuille (1799–1869).

poison or **toxin**, any chemical substance that, when introduced into or applied to the body, is capable of injuring health or destroying life.

The liver removes some poisons from the blood. The majority of poisons may be divided into **corrosives**, such as sulfuric, nitric, and hydrochloric acids; **irritants**, including arsenic and copper sulfate; **narcotics**, such as opium and carbon monoxide; and **narcotico-irritants** from any substances of plant origin including phenol acid and tobacco.

how poisons work Corrosives all burn and destroy the parts of the body with which they come into contact; irritants have an irritating effect on the stomach and bowels; narcotics affect the brainstem and spinal cord, inducing a stupor; and narcotico-irritants can cause intense irritations and finally act as narcotics.

poisoning (of a catalyst) the reduction in effectiveness of a †catalyst as a result of its being contaminated by a reactant or a product of the reaction it catalyses. Although in theory a catalyst is unaffected by the reaction it catalyses, in practice particles ("poisons") accumulate on the surface of the catalyst and reduce its effectiveness.

Poisson, Siméon-Denis (1781–1840) French applied mathematician and physicist. In probability theory he formulated the **Poisson distribution. Poisson's ratio** in elasticity is the ratio of the lateral contraction of a body to its longitudinal extension. The ratio is constant for a given material.

Poisson, Siméon Denis http://www-groups.dcs.st-andrews.ac.uk/history/Mathematicians/Poisson.html

Part of an archive containing the biographies of the world's greatest mathematicians, this site is devoted to the life and contributions of Siméon Poisson. In addition to biographical information, you will find a list of references about Poisson and links to other essays in the archive that reference him. The text of this essay includes hypertext links to the essays of those mathematicians and thinkers who influenced Poisson. There is also an image of him, which can be enlarged, and a map of his birthplace.

Poisson's ratio ratio of the lateral †strain to the longitudinal strain in a wire held under tension.

Polanyi, John Charles (1929–) German physical chemist whose research on infrared light given off during chemical reactions (infrared chemical luminescence) laid the foundations for the development of chemical lasers. He analyzed the chain reaction between hydrogen and chlorine, and showed that two distinct states of hydrochloric acid are formed. He shared the Nobel Prize for Chemistry in 1986 for his development of the dynamics of chemical elementary processes.

polar body in †meiosis, any of three (haploid) egg nuclei that develop from a secondary oocyte (the fourth nucleus is the †ovum); the three polar bodies degenerate.

polar coordinates in mathematics, a way of defining the position of a point in terms of its distance r from a fixed point (the origin) and its angle q to a fixed line or axis. The coordinates of the point are (r,q).

Often the angle is measured in †radians, rather than degrees. The system is useful for defining positions on a plane in programming the operations of, for example, computer-controlled cloth- and metal-cutting machines.

polarimetry in astronomy, any technique for measuring the degree of polarization of radiation from stars, galaxies, and other objects.

Polaris or **Pole Star** or **North Star**, bright star closest to the north celestial pole, and the brightest star in the constellation †Ursa Minor. Its position is indicated by the "pointers" in †Ursa Major. Polaris is a yellow †supergiant about 500 light years away from the Sun. It is also known as **Alpha Ursae Minoris**. It currently lies within 1° of the north celestial pole; †precession (Earth's axial wobble) will bring Polaris closest to the celestial pole (less than 0.5° away) in about AD 2100. Then its distance will start to increase, reaching 1° in 2205 and 47° in 28000. Other bright stars that have been, or will be close to the north celestial pole are Alpha Draconis (2800 BC), Gamma Cephei (AD 4000), Alpha Cephei (AD 7000), and Vega (AD 14000).

polariscope instrument that enables the rotation of the plane of polarization of †polarized light to be determined.

polarization in physics, in an electrolytic cell (battery), the build up of bubbles of hydrogen on the cathode, which leads to a reduction in the cell's output voltage and an increase in its internal resistance. Commercial batteries include a depolarizer (such as manganese oxide) to prevent this effect.

polarization in physics, separation of the charges of the molecules of an insulator in an electric field. One face of the insulator acquires a positive charge while the other face has a negative charge.

polarization in physics, restriction in the directions in which transverse waves can vibrate. Light can be polarized by passing it through certain materials (such as Polaroid) or on being reflected. Radio waves and other types of electromagnetic radiation can also be polarized.

polarized light light in which the electromagnetic vibrations take place in one particular plane. In ordinary (unpolarized) light, the electric fields vibrate in all planes perpendicular to the direction of propagation. After reflection from a polished surface or transmission through certain materials (such as Polaroid), the electric fields are confined to one direction, and the light is said to be **linearly polarized**. In **circularly polarized** and **elliptically polarized** light, the electric fields are confined to one direction, but the direction rotates as the light propagates. Polarized light is used to test the strength of sugar solutions and to measure stresses in transparent materials.

polar molecule in chemistry, covalently bonded molecule in which the electrons are unequally shared by the bonding atoms due to differences in their relative electronegativities. This has the result of making one end of the molecule partially positively charged and the other partially negative, for example H^{d+}–Cl^{d-}.

polarography electrochemical technique for the analysis of oxidizable and reducible compounds in solution. It involves the diffusion of the substance to be analyzed onto the surface of a small electrode, usually a bead of mercury, where oxidation or reduction occurs at an electric potential characteristic of that substance.

polar planimeter device invented by Swiss physicist Jakob Amsler-Laffon to measure area on a curved surface; it could be used to determine Fourier coefficients and was thus particularly valuable to shipbuilders and railway engineers.

polar reversal or **magnetic reversal**, change in polarity of Earth's magnetic field. Like all magnets, Earth's magnetic field has two opposing regions, or poles, positioned approximately near geographical North and South Poles. During a period of normal polarity the region of attraction corresponds with the North Pole. Today, a compass needle, like other magnetic materials, aligns itself parallel to the magnetizing force and points to the North Pole. During a period of reversed polarity, the region of attraction would change to the South Pole and the needle of a compass would point south.

Studies of the magnetism retained in rocks at the time of their formation (like small compasses frozen in time) have shown that the polarity of the magnetic field has reversed repeatedly throughout geological time.

The reason for polar reversals is not known. Although the average time between reversals over the last 10 million years has been 250,000 years, the rate of reversal has changed continuously over geological time. The most recent reversal was 780,000 years ago; scientists have no way of predicting when the next reversal will occur. The reversal process probably takes a few thousand years. Dating rocks using distinctive sequences of magnetic reversals is called †magnetic stratigraphy.

polder area of flat reclaimed land that used to be covered by a river, lake, or the sea. Polders have been artificially drained and protected from flooding by building dykes. They are common in the Netherlands, where the total land area has been increased by nearly one-fifth since AD 1200. Such schemes as the Zuider Zee project have provided some of the best agricultural land in the country.

pole either of the geographic north and south points of the axis about which the Earth rotates. The geographic poles differ from the magnetic poles, which are the points towards which a freely suspended magnetic needle will point.

In 1985 the magnetic north pole was some 350 km/218 mi northwest of Resolute Bay, Northwest Territories, Canada. It moves northwards about 10 km/6 mi each year, although it can vary in a day about 80 km/50 mi from its average position. It is relocated every decade in order to update navigational charts. It is thought that periodic changes in the Earth's core cause a reversal of the magnetic poles (see †polar reversal, †magnetic field). Many animals, including migrating birds and fish, are believed to orient themselves partly using the Earth's magnetic field. A permanent scientific base collects data at the South Pole.

pole, magnetic see †magnetic pole.

Pole Star another name for †Polaris, the northern pole star. There is no bright star near the southern celestial pole.

pollarding type of pruning whereby the young branches of a tree are severely cut back, about 2–4 m/6–12 ft above the ground, to produce a stumplike trunk with a rounded, bushy head of thin new branches. It is similar to coppicing.

pollen grains of †seed plants that contain the male gametes. In †angiosperms (flowering plants) pollen is produced within †anthers; in most †gymnosperms (cone-bearing plants) it is produced in male cones. A pollen grain is typically yellow and, when mature, has a hard outer wall. Pollen of insect-pollinated plants (see †pollination) is often sticky and spiny and larger than the smooth, light grains produced by wind-pollinated species.

The outer wall of pollen grains from both insect-pollinated and wind-pollinated plants is often elaborately sculptured with ridges or spines so distinctive that individual species or genera of plants can be recognized from their pollen. Since pollen is extremely resistant to decay, useful information on the vegetation of earlier times can be gained from the study of fossil pollen. The study of pollen grains is known as palynology.

pollen comb in zoology, row of bristles on each of the front legs of a honeybee. The bristles are used to comb out pollen grains sticking to the bee's body. The pollen is mixed with a little nectar to form balls of pollen paste that are then passed to the pollen sac on each back leg.

pollen tube outgrowth from a pollen grain that grows towards the †ovule, following germination of the grain on the †stigma. In †angiosperms (flowering plants) the pollen tube reaches the ovule by growing down through the †style, carrying the male gametes inside. The gametes are discharged into the ovule and one fertilizes the egg cell.

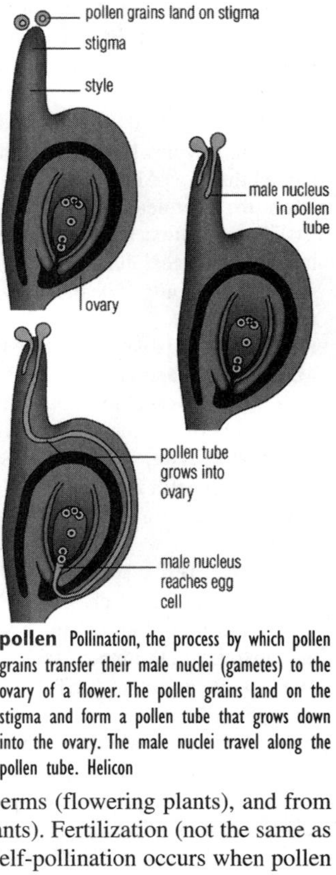

pollen Pollination, the process by which pollen grains transfer their male nuclei (gametes) to the ovary of a flower. The pollen grains land on the stigma and form a pollen tube that grows down into the ovary. The male nuclei travel along the pollen tube. Helicon

pollination process by which pollen is transferred from one plant to another. The male †gametes are contained in pollen grains, which must be transferred from the anther to the stigma in †angiosperms (flowering plants), and from the male cone to the female cone in †gymnosperms (cone-bearing plants). Fertilization (not the same as pollination) occurs after the growth of the pollen tube to the ovary. Self-pollination occurs when pollen is transferred to a stigma of the same flower, or to another flower on the same plant; cross-pollination occurs when pollen is transferred to another plant. This involves external pollen-carrying agents, such as wind (see †anemophily), water (see †hydrophily), insects, birds (see †ornithophily), bats, and other small mammals.

Animal pollinators carry the pollen on their bodies and are attracted to the flower by scent, or by the sight of the petals. Most flowers are adapted for pollination by one particular agent only. Bat-pollinated flowers tend to smell of garlic, rotting vegetation, or fungus. Those that rely on animals generally produce nectar, a sugary liquid, or surplus pollen, or both, on which the pollinator feeds. Thus the relationship between pollinator and plant is an example of mutualism, in which both benefit. However, in some plants the pollinator receives no benefit (as in †pseudocopulation), while in others, nectar may be removed by animals that do not effect pollination.

pollinium group of pollen grains that is transported as a single unit during pollination. Pollinia are common in orchids.

pollution harmful effect on the environment of by-products of human activity, principally industrial and agricultural processes – for example, noise, smoke, car emissions, pesticides, radiation, sewage disposal, household waste, and chemical and radioactive effluents in air, seas, and rivers. †Air pollution contributes to the †greenhouse effect.

Pollution control involves higher production costs for the industries concerned, but failure to implement adequate controls may result in irreversible environmental damage and an increase in the incidence of diseases such as cancer. For example, in agriculture the mismanagement of †fertilizers may result in †eutrophication, an excessive enrichment of lakes and rivers caused by †nitrate pollution; the subsequent rapid growth of algae darkens the water and eventually depletes its oxygen, causing plants and animals to die. Nitrate pollution has also been linked to stomach cancer, although this is unproven. Radioactive pollution results from inadequate †nuclear safety.

Transboundary pollution is when the pollution generated in one country affects another, for example as occurs with †acid rain. Natural disasters may also cause pollution; volcanic eruptions, for example, cause ash to be ejected into the atmosphere and deposited on land surfaces.

Pollux or **Beta Geminorum**, brightest star in the constellation †Gemini and the 17th-brightest star in the night sky. Pollux is a yellow star with a true luminosity 45 times that of the Sun. It is 36 light years away from the Sun.

The first-magnitude Pollux and the second-magnitude †Castor (Alpha Geminorum) mark the heads of the Gemini twins. It is thought that the two stars may have changed their relative brightness since Bayer named them, as Alpha is usually assigned to the brightest star in a constellation.

polonium chemical symbol Po, radioactive, metallic element, atomic number 84, relative atomic mass 210. Polonium occurs in nature in small amounts and was isolated from †pitchblende. It is the element having the largest number of isotopes (27) and is 5,000 times as radioactive as radium, liberating considerable amounts of heat. It was the first element to have its radioactive properties recognized and investigated.

Polonium was isolated in 1898 from the pitchblende residues analyzed by French scientists Pierre and Marie †Curie, and named after Marie Curie's native Poland.

poly- in chemistry, prefix denoting a large number of atoms or groups of the same type being present in a molecule of a compound, usually applied to †polymers, for example polystyrene.

polyacrylonitrile synthetic †elastomer produced by the polymerization of acrylonitrile monomer units. It has the general formula $-(CH_2\text{–}CHCN)_n-$. Polyacrylonitrile is an important synthetic fiber.

polybasic describing an †acid that has more than two atoms of replaceable acidic hydrogen in each of its molecules.

polychlorinated biphenyl PCB, any of a group of chlorinated isomers of biphenyl $(C_6H_5)_2$. They are dangerous industrial chemicals, valuable for their fire-resistant qualities. They constitute an environmental hazard because of their persistent toxicity. Since 1973 their use had been limited by international agreement. In December 2000, 122 nations agreed a treaty to ban the toxic chemicals known as persistent organic polluters (POPs), which include PCBs, although they are unlikely to be totally eliminated until about 2025.

polyembryony reproductive strategy in which one sexually produced embryo is split into a number of genetically identical offspring. Most of the approximately 18 polyembryonic animals are parasitic, including various parasitic wasp species. Two species of armadillo also reproduce by polyembryony.

polyester synthetic resin formed by the †condensation of polyhydric alcohols (alcohols containing more than one hydroxyl group) with dibasic acids (acids containing two replaceable hydrogen atoms). Polyesters are thermosetting †plastics, used for constructional plastics and, with glass fiber added as reinforcement, they are used in car bodies and boat hulls. Polyester is also a major synthetic fiber used for knitting or weaving fabrics which are strong but lightweight, and resist creasing but can be heat-set into pleats. Polyester is often mixed with other fibers and can be found in a wide range of different textiles.

polyethylene alternative term for †polythene.

polygene in genetics, group of genes that determine a particular characteristic. The height of humans is an example of a characteristic determined by a polygene.

polygon in geometry, a plane (two-dimensional) figure with three or more straight-line sides. **Regular polygons** have sides of the same length, and all the †exterior angles are equal. Common polygons have names that define the number of sides (for example, †**triangle** (3), †**quadrilateral** (4), †**pentagon** (5), †**hexagon** (6), **heptagon** (7), †**octagon** (8), and so on). These are all convex polygons, having no †interior angle greater than 180°.

sum of internal angles The sum of the internal angles of a polygon having n sides is given by the formula 180° x (n – 2). For example:

In this case (a hexagon), there are 6 sides. The sum of the interior angles equals:

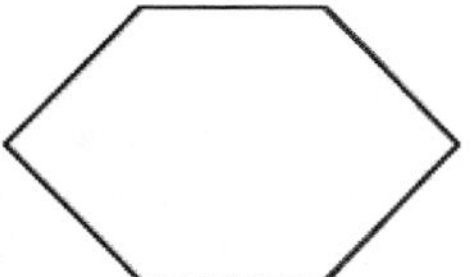

180° x (n – 2) = 180° x (6 – 2) = 180° x 4 = 720°

Therefore, the more sides a polygon has, the larger the sum of its internal angles and, in the case of a convex polygon, the more closely it approximates to a circle.

See also lines of †symmetry.

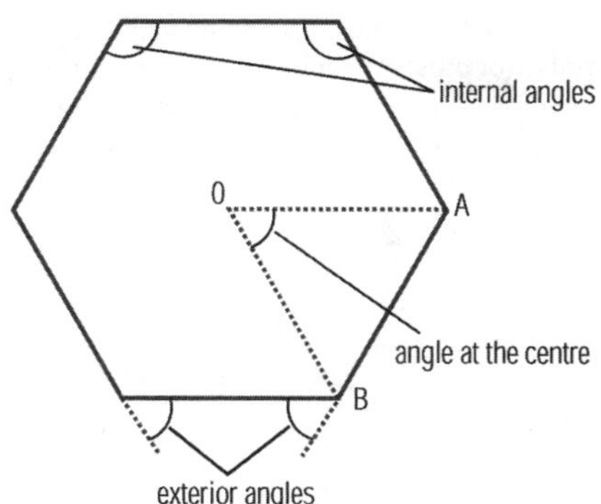

polygon The types of angles in a regular polygon are shown here. These are the angles in a regular hexagon. Helicon

polyhedron in geometry, a solid figure with four or more plane faces. The more faces there are on a polyhedron, the more closely it approximates to a sphere. Knowledge of the properties of polyhedra is needed in crystallography and stereochemistry to determine the shapes of crystals and molecules.

There are only five types of regular polyhedron (with all faces the same size and shape), as was deduced by early Greek mathematicians; they are the tetrahedron (four equilateral triangular faces), cube (six square faces), octahedron (eight equilateral triangles), dodecahedron (12 regular pentagons) and icosahedron (20 equilateral triangles).

Pavilion of Polyhedreality http://www.georgehart.com/pavilion.html

Collection of images and instructions on how to make various polyhedrals, some from quite surprising materials.

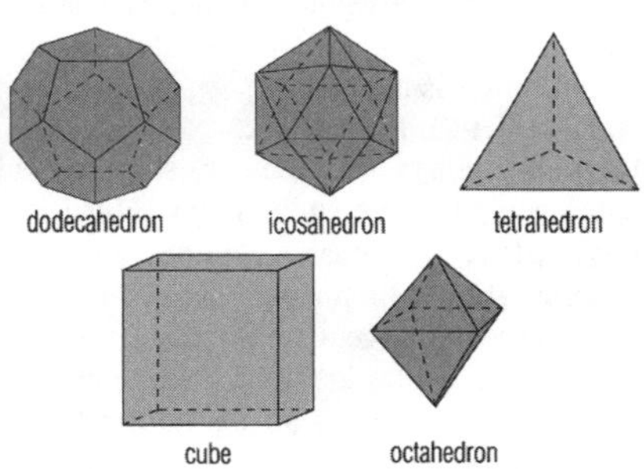

polyhedron The five regular polyhedra or Platonic solids. Helicon

polymer very long-chain molecule made up of many repeated simple units (†**monomers**) linked together by †polymerization. There are many polymers, both natural (cellulose, chitin, lignin, rubber) and synthetic (polyethylene and nylon, types of plastic). Synthetic polymers belong to two groups: thermosoftening and thermosetting (see †plastic).

The size of the polymer matrix is determined by the amount of monomer used; it therefore does not form a molecule of constant molecular size or mass.

Polymers and Liquid Crystals http://plc.cwru.edu/

Online tutorial about two modern physical wonders. The site is divided into a "Virtual textbook" and a "Virtual laboratory," with corresponding explanations and experiments.

polymerase chain reaction PCR, technique developed during the 1980s to clone short strands of DNA from the †genome of an organism. The aim is to produce enough of the DNA to be able to sequence and identify it. It was developed by US biochemist Kary †Mullis in 1983.

polymerization chemical union of two or more (usually small) molecules of the same kind to form a new compound. **Addition polymerization** produces simple multiples of the same compound. **Condensation polymerization** joins molecuies together with the elimination of water or another small molecule.

Addition polymerization uses only a single monomer (basic molecule); condensation polymerization may involve two or more different monomers (**co-polymerization**).

addition of ethane molecules to form polyethene

hydrogen atom
carbon atom
double covalent bond
single covalent bond

polymerization In polymerization, small molecules (monomers) join together to make large molecules (polymers). In the polymerization of ethene to polyethene (polythene), electrons are transferred from the carbon–carbon double bond of the ethene molecule, allowing the molecules to join together as a long chain of carbon–carbon single bonds. Helicon

polymorphism in genetics, the coexistence of several distinctly different types in a †population (groups of animals of one species). Examples include the different blood groups in humans, different color forms in some butterflies, and snail shell size, length, shape, color, and stripiness.

polymorphism in mineralogy, the ability of a substance to adopt different internal structures and external forms, in response to different conditions of temperature and/or pressure. For example, diamond and graphite are both forms of the element carbon, but they have very different properties and appearance. Silica (SiO_2) also has several polymorphs, including quartz, tridymite, cristobalite, and stishovite (the latter a very high-pressure form found in meteoritic impact craters).

polynomial in mathematics, an algebraic expression that has one or more †variables (denoted by letters). A polynomial equation has the form:

$f(x) = a_n x^n + a_{n-1} x^{n-1} + \ldots + a_2 x^2 + a_1 x + a_0$

where a_n, a_{n-1}, ..., a_0 are all constants, n is a positive integer, and a_n x 0. Examples of polynomials are:

$f(x) = 3x^4 + 2x^2 + 1$

or

$f(x) = x^5 - 18x + 71$

or

$f(x) = 2x + 3$

A polynomial of degree one, that is, whose highest †power of x is 1, as in $2x + 1$, is called a linear polynomial;

$3x^2 + 2x + 1$ is quadratic;

$4x^3 + 3x^2 + 2x + 1$ is cubic.

polyp in zoology, the sedentary stage in the life cycle of a coelenterate (such as a †coral or jellyfish), the other being the free-swimming **medusa**.

polypeptide long-chain †peptide.

polyphyletic in biological classification, term describing a group of species that does not include its members' most recent common ancestor. Such groupings are avoided in most modern systems of classification. Examples are the marine mammals, trees, and algae.

If a group consists of all the descendants of a common ancestor plus the ancestor itself, it is called †monophyletic; if it includes its members' common ancestor but not all of its descendants it is called †paraphyletic. Distinguishing between such groups is an important part of the system of classification known as †cladistics, which seeks to classify organisms into monophyletic groups, or **clades**.

polyploid in genetics, possessing three or more sets of chromosomes in cases where the normal complement is two sets (†diploid). Polyploidy arises spontaneously and is common in plants (mainly among flowering plants), but rare in animals. Many crop plants are natural polyploids, including wheat, which has four sets of chromosomes per cell (durum wheat) or six sets (common wheat).

Plant breeders can induce the formation of polyploids by treatment with a chemical, colchicine.

polypropylene or **polypropene**, plastic made by the polymerization, or linking together, of †propene molecules ($CH_2 = CH = CH_3$). Polypropylene is one of the most important plastics and has a large number of different uses, including packaging, house ware, textiles, carpets, artificial turf, and ropes. It is a very light synthetic fiber, which does not absorb moisture, and so stays dry and warm. These properties make it a good fiber for use in clothing designed for cold environments.

polysaccharide long-chain †carbohydrate made up of hundreds or thousands of linked simple sugars (monosaccharides) such as glucose and closely related molecules.

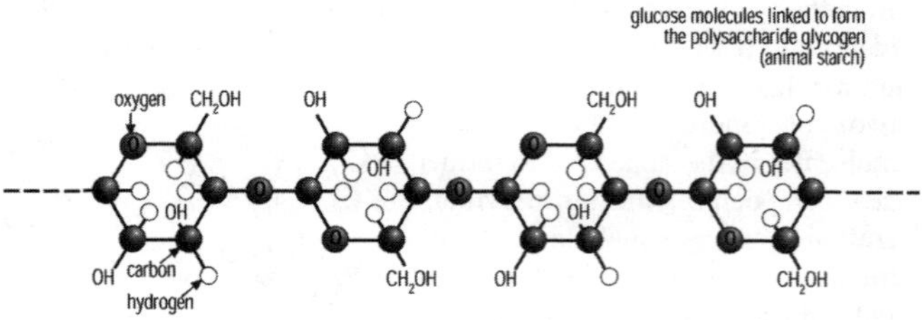

The polysaccharides are natural polymers. They either act as energy-rich food stores in plants (starch) and animals (glycogen), or have structural roles in the plant cell wall (cellulose, pectin) or the tough outer skeleton of insects and similar creatures (chitin). See also †carbohydrate.

polysaccharide A typical polysaccharide molecule, glycogen (animal starch), is formed from linked glucose ($C_6H_{12}O_6$) molecules. A glycogen molecule has 100–1,000 linked glucose units. Helicon

polystyrene type of †plastic used in kitchen utensils or, in an expanded form, in insulation and ceiling tiles. CFCs (†chlorofluorocarbons) are used to produce expanded polystyrene so alternatives are being sought.

polytetrafluoroethene **PTFE**, polymer made from the monomer tetrafluoroethene (CF_2CF_2). It is a thermosetting plastic with a high melting point that is used to produce "nonstick" surfaces on pans and to coat bearings. Its trade name is Teflon.

polythene or **polyethylene** or **polyethene**, †polymer of the gas †ethene (C_2H_4). It is a tough, white, translucent, waxy thermoplastic (which means it can be repeatedly softened by heating). It is used for packaging, bottles, toys, wood preservation, electric cable, pipes, and tubing.

Polythene is produced in two forms: low-density polythene, made by high-pressure †polymerization of ethene gas (the †monomer), and high-density polythene, which is made at lower pressure by using catalysts. This form, first made in 1953 by German chemist Karl Ziegler, is more rigid at low temperatures and softer at higher temperatures than the low-density type. Polythene was first made in the 1930s at very high temperatures by ICI.

polyunsaturate type of †fat or oil containing a high proportion of triglyceride molecules whose †fatty acid chains contain several double bonds. By contrast, the fatty-acid chains of the triglycerides in saturated fats (such as lard) contain only single bonds. Medical evidence suggests that polyunsaturated fats, used widely in margarines and cooking fats, are less likely to contribute to cardiovascular disease than saturated fats, but there is also some evidence that they may have adverse effects on health.

The more double bonds the fatty-acid chains contain, the lower the melting point of the fat. Unsaturated chains with several double bonds produce oils, such as vegetable and fish oils, which are liquids at room temperature. Saturated fats, with no double bonds, are solids at room temperature. The polyunsaturated fats used for margarines are produced by taking a vegetable or fish oil and turning some of the double bonds to single bonds, so that the product is semi-solid at room temperature. This is done by bubbling hydrogen through the oil in the presence of a catalyst, such as platinum. The catalyst is later removed.

Monounsaturated oils, such as olive oil, whose fatty-acid chains contain a single double bond, are probably healthier than either saturated or polyunsaturated fats. Butter contains both saturated and unsaturated fats, together with †cholesterol, which also plays a role in heart disease.

polyurethane polymer made from the monomer urethane. It is a thermoset †plastic, used in liquid form as a paint or varnish, and in foam form for upholstery and in lining materials (where it may be a fire hazard).

polyvinyl chloride **PVC**, type of †plastic used for drainpipes, floor tiles, audio discs, shoes, and handbags. It is derived from vinyl chloride (CH_2—CHCl).

Swedish scientists identified a link between regular exposure to PVC and testicular cancer, in 1998, increasing the demand for a ban on PVC in commercial products.

pome type of †pseudocarp, or false fruit, typical of certain plants belonging to the Rosaceae family. The outer skin and fleshy tissues are developed from the †receptacle (the enlarged end of the flower stalk) after fertilization, and the five †carpels (the true fruit) form the pome's core, which surrounds the seeds. Examples of pomes are apples, pears, and quinces.

pomeron in physics, hypothetical object with energy and momentum, but no color or electrical charge, produced when an electron strikes a proton. Pomerons were first suggested in 1958 by Russian physicist Isaac Pomeranchuk.

Poncelet, Jean-Victor (1788–1867) French mathematician and military engineer who advanced projective geometry. His book *Traité des propriétés projectives des figures* (1822) deals with the properties of plane figures that remain unchanged when projected.

Poncelet, Jean-Victor http://www-groups.dcs.st-andrews.ac.uk/history/Mathematicians/Poncelet.html

Part of an archive containing the biographies of the world's greatest mathematicians, this site is devoted to the life and contributions of Jean-Victor Poncelet. In addition to biographical information, you will find a list of references about Poncelet and links to other essays in the archive that reference him. The text of this essay includes hypertext links to the essays of those mathematicians and thinkers who influenced Poncelet. There is also an image of him, which can be enlarged, and a map of his birthplace.

Pond, John (1767–1836) English astronomer. As Astronomer Royal 1811–35 he reorganized and modernized Greenwich Observatory. Instituting new methods of observation, he went on to produce a catalogue of more than 1,000 stars in 1833.

pools and riffles alternating deeps (pools) and shallows (riffles) along the course of a river. There is evidence to suggest a link between pools and riffles and the occurrence of †meanders (bends in a river), although it is not certain whether they are responsible for meander formation.

pooter small device for collecting invertebrates, consisting of a jar to which two tubes are attached. A sharp suck on one of the tubes, while the other is held just above an insect, will propel the animal into the jar. A filter wrapped around the mouth tube, prevents debris or organisms from being swallowed.

POP3 acronym for Post Office Protocol 3, on the Internet, one of the two most common mail †protocols. Internet Service Providers (ISPs) offer SMTP, POP3, or both. The primary difference to most users is the choice of software available. SMTP is older and more flexible, but POP3 is generally simpler for those accessing the Internet via a dial-up account.

Popov, Alexander Stepanovich (1859–1905) Russian physicist who devised the first †aerial, in advance of †Marconi (although he did not use it for radio communication). He also invented a detector for radio waves.

population in biology and ecology, a group of organisms of one species, living in a certain area. The organisms are able to interbreed. It also refers to the members of a given species in a †community of living things. The area can be small. For example, one can refer to the population of duckweed (a small floating plant found on the surface of ponds) on a pond. Since the pond is a †habitat, one can consider the population of duckweed in a habitat and forming part of the community of plants and animals there. However, it is also possible to use the term population for all the organisms of one species in a large geographical area, for example the elephant population in Africa. It could also be used to describe all the organisms of that species on Earth, for example the world population of humans. Population sizes in habitats change over a period of time. The timescale may be daily, seasonal, or there may be changes over the years.

The success of an organism can be determined by measuring the size of a population or by measuring †biomass. Measuring the size of a population is quite difficult and requires careful sampling of the habitat and careful calculation to estimate population size. It is rarely possible to count directly all the individuals in a population. Typically, sampling techniques require random sampling to get a fair estimate of the population and may use equipment such as quadrats (a way of defining a square sampling area of a certain size). However, these techniques have their limitations – for example, animals move around and may also be difficult to find. Species may also be difficult to identify. There may not be time to take a large enough sample to make the estimate accurate.

Facing the Future: People and the Planet http://www.popinfo.org/

Guide to population issues, aimed at young people. The site tackles issues such as "How many people can the Earth support?" and covers both personal and global solutions.

Population–Environment Dynamics: Ten Case Studies

http://www-personal.umich.edu/%7Ewddrake/545_96.html

US-based site providing case studies to illustrate alternative theoretical constructs useful in studying the interaction between human populations and the environment. Studies on this site include "Urbanization in Egypt," "Population policy and socio-economic development in Vietnam," and "Nine nations, one Nile," which details some of the problems caused by the scarcity and misuse of freshwater.

Urban–Rural Population Distribution

http://www.csudh.edu/global_options/375Students-Sp96/Spain/URB.RURALPOP.HTML

General information on the urban–rural population distribution in the agricultural, industrial, and information ages, with extra insight on Spain. This site could provide a useful resource for a case study.

population in statistics, the universal set from which a sample of data is selected. The chief object of statistics is to find out population characteristics by taking samples.

population cycle in biology, regular fluctuations in the size of a population, as seen in lemmings, for example. Such cycles are often caused by density-dependent mortality: high mortality due to overcrowding causes a sudden decline in the population, which then gradually builds up again. Population cycles may also result from an interaction between a predator and its prey.

population genetics branch of genetics that studies the way in which the frequencies of different †alleles (alternative forms of a gene) in populations of organisms change, as a result of natural selection and other processes.

population II stars older stars, generally formed towards the center of a †galaxy, containing few heavier elements. The brightest of this population are red giants.

population I stars younger stars, generally formed towards the edge of a †galaxy, of the dusty material in the spiral arms, including the heavy elements. The brightest of this population are hot, white stars.

pore in biology, a small opening in the skin that releases sweat and sebum. Sebum acts as a natural lubricant and protects the skin from the effects of moisture or excessive dryness.

porism old-fashioned term for a conclusion or hypothesis following directly upon a statement or proposition.

porosity percentage of empty space existing within a material.

porphyry any †igneous rock containing large crystals in a finer matrix.

Porter, George (1920–2002) Baron Porter of Luddenham, English chemist. He and Ronald †Norrish shared the Nobel Prize for Chemistry in 1967 for the development, from 1947, of a technique by which flashes of high energy are used to bring about extremely fast chemical reactions. He was knighted in 1972 and made a baron in 1990.

position vector vector that defines the position of a point with respect to the origin.

positive integer any whole number from 0 upwards. 0 is included so that the properties of positive integers include those related to the †identity for addition.

positron antiparticle of the electron; an †elementary particle having the same mass as an electron but exhibiting a positive charge. The positron was discovered in 1932 by US physicist Carl Anderson at the California Institute of Technology, USA, its existence having been predicted by the British physicist Paul Dirac in 1928.

positron emission tomography **PET**, imaging technique that enables doctors to observe the metabolic activity of the human body by following the progress of a radioactive chemical that has been inhaled or injected, detecting †gamma radiation given out when †positrons emitted by the chemical are annihilated. The technique has been used to study a wide range of conditions, including schizophrenia, Alzheimer's disease, and Parkinson's disease.

post-traumatic stress disorder **PTSD**, psychiatric disorder caused by the experiencing of a trauma. Diagnosis is based on the presence of a number of symptoms such as recurrent recollections of the trauma, emotional numbness, lack of motivation, and irrational outbursts. Symptoms are intense enough to prevent the sufferer from functioning normally. Treatment is by †psychotherapy. Sufferers include soldiers, rape victims, emergency services staff, and children living in a violent environment.

potash general name for any potassium-containing mineral, most often applied to potassium carbonate (K_2CO_3) or potassium hydroxide (KOH). Potassium carbonate, originally made by roasting plants to ashes in earthenware pots, is commercially produced from the mineral sylvite (potassium chloride, KCl) and is used mainly in making artificial fertilizers, glass, and soap.

The potassium content of soils and fertilizers is also commonly expressed as potash, although in this case it usually refers to potassium oxide (K_2O).

potassium chemical symbol K, (Dutch *potassa* "potash") soft, waxlike, silver-white, metallic element, atomic number 19, relative atomic mass 39.0983. Its chemical symbol comes from the Latin *kalium*. It is one of the †alkali metals (in Group 1 of the †periodic table of the elements), and has a very low density – it floats on water, and is the second lightest metal (after lithium). It is one of the most reactive in the †reactivity series of metals, oxidizing rapidly when exposed to air and reacting violently with water. Of great abundance in the Earth's crust, it is widely distributed. It is found in salt and mineral deposits in the form of potassium aluminum silicates and potassium nitrate (saltpeter). Potassium has to be extracted from its compounds by †electrolysis because of its high reactivity.

Potassium is the main base ion of the fluid in the body's cells. Along with †sodium, it is important to the electrical potential of the nervous system and, therefore, for the efficient functioning of nerve and muscle. Shortage, which may occur with excessive fluid loss (prolonged diarrhea, vomiting), may lead to muscular paralysis; potassium overload may result in cardiac arrest. It is also required by plants for growth. The element was discovered and named in 1807 by English chemist Humphry Davy, who isolated it from potash in the first instance of a metal being isolated by electrolysis.

potassium-argon dating or **K-Ar dating**, isotopic dating method based on the radioactive decay of potassium-40 (^{40}K) to the stable isotope argon-40 (^{40}Ar). Ages are based on the known half-life of ^{40}K, and the ratio of ^{40}K to ^{40}Ar. The method is routinely applied to rock samples about 100,000 to 30 million years old.

The method is used primarily to date volcanic layers in stratigraphic sequences with archaeological deposits, and the paleomagnetic-reversal timescale. Complicating factors, such as sample contamination by argon absorbed from the atmosphere, and argon gas loss by diffusion out of the mineral, limit the application of this technique.

potassium carbonate K_2CO_3, deliquescent white powder used in the preparation of standard alkaline solutions.

potassium dichromate $K_2Cr_2O_7$, orange, crystalline solid, soluble in water, that is a strong †oxidizing agent in the presence of dilute sulfuric acid. As it oxidizes other compounds it is itself reduced to potassium chromate (K_2CrO_4), which is green. Industrially it is used in the manufacture of dyes and glass and in tanning, photography, and ceramics.

potassium hydroxide or **caustic potash**; KOH, caustic, deliquescent, white solid, melting point 306°C/583°F. It can be prepared by the electrolysis of potassium chloride solution. It is used as an electrolyte in batteries and in the production of fertilizers.

potassium manganate(VII) or **potassium permanganate**; $KMnO_4$, dark purple, crystalline solid, soluble in water, that is a strong †oxidizing agent in the presence of dilute sulfuric acid. In the process of oxidizing other compounds it is itself reduced to manganese(II) salts (containing the Mn^{2+} ion), which are colorless.

potassium nitrate or **niter, saltpeter**; KNO_3, white crystalline solid that is a strong oxidizing agent. It is used to make fertilizers and is a component of gunpowder.

potential difference PD, difference in the electrical potential (see †potential, electric) of two points, being equal to the electrical energy converted by a unit electric charge moving from one point to the other. Electrons flow in a conducting material towards the part that is relatively more positive (fewer negative charges). The SI unit of potential difference is the volt (V). The potential difference between two points in a circuit is commonly referred to as voltage (and can be measured with a voltmeter). See also †Ohm's law.

One volt equals one joule of energy used for each coulomb of charge. In equation terms, potential difference V may be defined by: $V = E/Q$, where E is the electrical energy converted in joules and Q is the charge in coulombs. Chemical energy from a battery is converted to electrical energy in a circuit, this energy being given by rearranging the above formula: $E = Q \times V$.

potential divider or **voltage divider**, two resistors connected in series in an electrical circuit in order to obtain a fraction of the potential difference, or voltage, across the battery or electrical source. The potential difference is divided across the two resistors in direct proportion to their resistances.

potential, electric energy required to bring a unit electric charge from infinity to the point at which potential is defined. The SI unit of potential is the volt (V). Positive electric charges will flow "downhill" from a region of high potential to a region of low potential.

A charged †conductor, for example, has a higher potential than the Earth, the potential of which is taken by convention to be zero. An electric cell (battery) has a potential difference between its terminals, which can make current flow in an external circuit. The difference in potential – †potential difference (pd) – is expressed in volts so, for example, a 12 V battery has a pd of 12 volts between its negative and positive terminals.

potential energy PE, †energy possessed by an object by virtue of its relative position or state (for example, as in a compressed spring or a muscle). It can be thought of as "stored" energy. An object that has been raised up has energy stored due to its height. It is described as having †gravitational potential energy.

If a ball is raised to a certain height and released, the ball falls to the ground. The potential energy changes to kinetic energy. As the ball hits the ground some of the kinetic energy is lost as sound and elastic energy. A stretched spring has stored elastic energy; this is known as elastic potential energy. Springs and elastics are designed to store energy and release it either rapidly or slowly. For example, a mechanical toy works by an unwinding spring coil inside the toy. As the spring coil turns, elastic potential energy changes into kinetic and sound energy as the toy operates.

standard potential divider

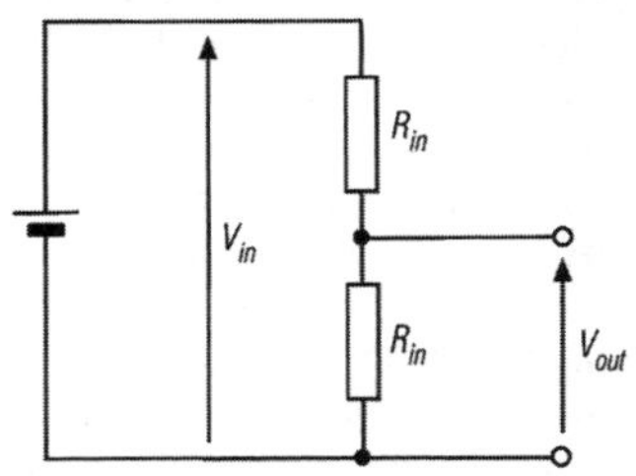

potentiometer used as a potential divider

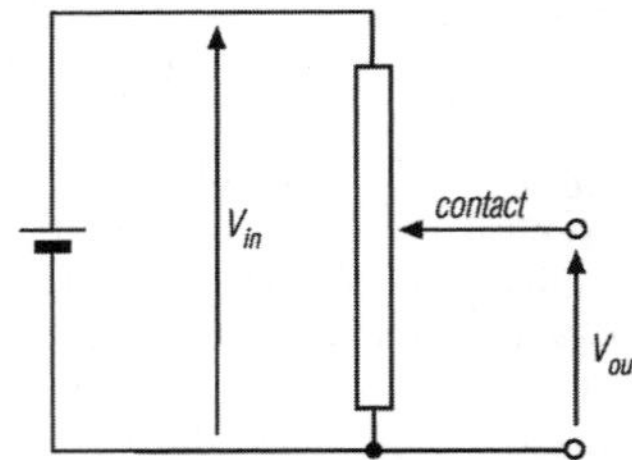

potential divider A potential divider is a resistor or a chain of resistors connected in series in an electrical circuit. It is used to obtain a known fraction of the total voltage across the whole resistor or chain. When a variable resistor, or potentiometer, is used as a potential divider, the output voltage can be varied continuously by sliding a contact along the resistor. Devices like this are used in electronic equipment to vary volume, tone, and brightness control. Helicon

potentiometer electrical †resistor that can be divided so as to compare, measure, or control voltages. In radio circuits, any rotary variable resistance (such as volume control) is referred to as a potentiometer.

A simple type of potentiometer consists of a length of uniform resistance wire (about 1 m/3 ft long) carrying a constant current provided by a battery connected across the ends of the wire. The source of potential difference (voltage) to be measured is connected (to oppose the cell) between one end of the wire, through a †galvanometer (instrument for measuring small currents), to a contact free to slide along the wire. The sliding contact is moved until the galvanometer shows no deflection. The ratio of the length of potentiometer wire in the galvanometer circuit to the total length of wire is then equal to the ratio of the unknown potential difference to that of the battery.

pothole small hollow in the rock bed of a river. Potholes are formed by the erosive action of rocky material carried by the river (corrasion), and are commonly found along the river's upper course, where it tends to flow directly over solid bedrock.

pound abbreviation lb, imperial unit of mass. The commonly used avoirdupois pound, also called the **imperial standard pound** (7,000 grains/0.45 kg), differs from the **pound troy** (5,760 grains/0.37 kg), which is used for weighing precious metals. It derives from the Roman *libra*, which weighed 0.327 kg.

poundal abbreviation pdl, imperial unit of force, now replaced in the SI system by the †newton. One poundal equals 0.1383 newtons.

It is defined as the force necessary to accelerate a mass of one pound by one foot per second per second.

Powell, Cecil Frank (1903–1969) English physicist who was awarded the Nobel Prize for Physics in 1950 for his study of the charged subatomic particles in †cosmic radiation by using photographic emulsions carried in weather balloons, and for his discovery of the pion (p meson) in 1947, a particle whose existence had been predicted by Japanese physicist Hideki †Yukawa in 1935.

power in mathematics, that which is represented by an †exponent or index, denoted by a superior numeral. A number or symbol raised to the power of 2 – that is, multiplied by itself – is said to be squared (for example, 3^2, x^2), and when raised to the power of 3, it is said to be cubed (for example, 2^3, y^3). Any number to the power zero always equals 1.

Powers can be negative. Negative powers produce fractions, with the numerator as one, as a number is divided by itself, rather than being multiplied by itself, so for example $2^{-1} = {}^1/_2$ and $3^{-3} = {}^1/_{27}$.

power in optics, a measure of the amount by which a lens will deviate light rays. A powerful converging lens will converge parallel rays strongly, bringing them to a focus at a short distance from the lens. The unit of power is the **diopter**, which is equal to the reciprocal of focal length in meters. By convention, the power of a converging (or convex) lens is positive and that of a diverging (or concave) lens negative.

power in physics, the rate of doing work or transferring energy from one form to another. It is measured in watts (joules per second) or other units of work per unit time.

power series infinite series of the general form:

$a_0 + a_1x^1 + a_2x^2 + a_3x^3 + ... + a_nx^n + ...$

This is called a power series in x. Examples of power series are the exponential series:

$e^x = 1 + x + x^2/2! + x^3/3! + ...$

the logarithmic series, and the series for the trigonometric functions (sine, cosine, and tangent).

power transmission transfer of electrical power from one location, such as a power station, to another. Electricity is conducted along cables at a high voltage (up to 500 kilovolts (kV) on the super grid) in order to reduce the current in the wires, and hence minimize the amount of energy wasted as heat.

†Transformers are needed to step up the voltages for transmission and to step down these voltages before power can be supplied to consumers. High voltages require special insulators to prevent current from leaking to the ground and these may clearly be seen on the pylons that carry overhead wires.

Poynting, John Henry (1852–1914) English physicist, mathematician, and inventor. He devised an equation by which the rate of flow of electromagnetic energy (now called the **Poynting vector**) can be determined.

In 1891 he made an accurate measurement of Isaac †Newton's gravitational constant.

Poynting–Robertson effect interaction of dust particles in interplanetary space with solar radiation causing a loss in orbital velocity of the dust around the Sun. This causes the dust particles to spiral into the Sun, if the effect is unopposed. However, under certain circumstances, †radiation pressure is large enough to oppose the effect.

pozzolan or **pozzolana**; or **puzzolan**, silica-rich material, such as volcanic tuff or chert, which can be ground up and mixed to form a cement highly resistant to corrosion by salt water. Currently used in Portland cement, it was named after the town of Pozzuoli, Italy, where a nearby volcanic tuff was used as a basis for cement in Roman times.

Praesepe or **the Beehive**, one of the nearer open star clusters, located in the constellation of †Cancer. To the naked eye it appears as a hazy patch, but a small telescope shows it to be a cluster of separate stars. It is very similar to the †Hyades cluster, but is over four times further away from Earth.

prairie central North American plain, formerly grass-covered, extending over most of the region between the Rocky Mountains to the west, and the Great Lakes and Ohio River to the east.

The term was first applied by French explorers to vast, largely level grasslands in central North America, centered on the Mississippi River valley, which extend from the Gulf of Mexico to central Alberta, Canada, and from west of the Appalachian system into the Great Plains. When first seen by explorers, the prairies were characterized by unbroken, waist-high, coarse grasses. Trees were common only along rivers and streams, or in occasional depressions in the land. This prairie is now almost gone, altered by farming to become what is known as the "Corn Belt," much of the "Wheat Belt," and other ploughed lands. Its humus-rich black †loess soils, adequate rainfall, and warm summers foster heavily productive agriculture. In the west – west Kansas, Nebraska, and the Dakotas – is the **short-grass prairie**, occupying large parts of the Great Plains. Higher, drier land here has been used primarily for wheat (aided by deep-well irrigation) and stock raising.

The prairies were formerly the primary habitat of the American bison; other prominent species include prairie dogs, deer and antelope, grasshoppers, and a variety of prairie birds.

Prandtl, Ludwig (1875–1953) German physicist who put fluid mechanics on a sound theoretical basis and originated the boundary-layer theory. His work in aerodynamics resulted in major changes in wing design and streamlining of aircraft.

praseodymium chemical symbol Pr, (Greek *prasios* "leek-green" + *didymos* "twin") silver-white, malleable, metallic element of the †lanthanide series, atomic number 59, relative atomic mass 140.907. It occurs in nature in the minerals monzanite and bastnaesite, and its green salts are used to color glass and ceramics. It was named in 1885 by Austrian chemist Carl von Welsbach.

He fractionated it from dydymium (originally thought to be an element but actually a mixture of rare-earth metals consisting largely of neodymium, praseodymium, and cerium) and named it for its green salts and spectroscopic line.

preadaptation in biology, the fortuitous possession of a character that allows an organism to exploit a new situation. In many cases, the character evolves to solve a particular problem that a species encounters in its preferred habitat, but once evolved may allow the organism to exploit an entirely different situation. The ability to extract oxygen directly from the air evolved in some early fishes, probably in response to life in stagnant, deoxygenated pools; this later made it possible for their descendants to spend time on land, so giving rise eventually to the air-breathing amphibians.

Precambrian era in geology, the time from the formation of the Earth (4.6 billion years ago) up to 570 million years ago. Its boundary with the succeeding Cambrian period marks the time when animals first developed hard outer parts (exoskeletons) and so left abundant fossil remains. It comprises about 85% of geological time and is divided into two eons: the Archaean and the Proterozoic.

precession slow wobble of the Earth on its axis, like that of a spinning top. The gravitational pulls of the Sun and Moon on the Earth's equatorial bulge cause the Earth's axis to trace out a circle on the sky every 25,800 years. The position of the celestial poles (see †celestial sphere) is constantly changing owing to precession, as are the positions of the equinoxes (the points at which the celestial equator intersects the Sun's path around the sky). The **precession of the equinoxes** means that there is a gradual westward drift in the ecliptic – the path that the Sun appears to follow – and in the coordinates of objects on the celestial sphere.

This is why the dates of the astrological signs of the zodiac no longer correspond to the times of year when the Sun actually passes through the constellations. For example, the Sun passes through Leo from mid-August to mid-September, but the astrological dates for Leo are between about 23 July and 22 August.

Precession, first detected in 1891, was finally explained by a US researcher at NASA in 2000. The wobble is caused by changes in ocean pressure, brought about by fluctuations in winds and salinity (salt levels), on the Earth's crust. The pressure change exaggerates the Earth's natural resonance.

Precession also occurs in other planets. Uranus has the Solar System's fastest known precession (264 days) determined in 1995.

precession of the equinoxes apparent movement per year of the two points in the sky representing the †equinoxes; 128 cm/50.26 in per year, also called the constant of precession. Precession is caused mainly by the gravitational effect of the Moon on the Earth's equatorial "bulge."

precipitation in chemistry, the formation of an insoluble solid in a liquid as a result of a reaction within the liquid between two or more soluble substances. For example, if solutions of lead nitrate and potassium iodide are added together, bright yellow, insoluble lead iodide appears as a precipitate in the solution, making it cloudy. The **precipitation reaction** is: lead nitrate + potassium iodide → lead iodide + potassium nitrate. If the newly formed solid settles, it forms a **precipitate**; if the particles of solid are very small, they will remain in suspension, forming a †**colloidal precipitate**.

Precipitation reactions are used in the preparation of †salts; in †qualitative analysis reactions; and in the formation of some †sedimentary rocks.

precipitation in meteorology, water that falls to the Earth from the atmosphere. It is part of the †water (hydrological) cycle. Forms of precipitation include †rain, †snow, †sleet, †hail, †dew, and †frost.

The amount of precipitation in any one area depends on †climate, †weather, and phenomena like trade winds and ocean currents. The cyclical change in the Peruvian Current off the coasts of Ecuador and Peru, known as †El Niño, causes dramatic shifts in the amount of precipitation in South and Central America and throughout the Pacific region.

Precipitation can also be influenced by people. In urban areas dust, smoke, and other particulate pollution that comprise **condensation nuclei**, cause water in the air to condense more readily. †Fog is one example. Precipitation also can react chemically with airborne pollutants to produce †acid rain.

precocial describing animals that are active and mobile from birth. Examples include antelopes, ostriches, and crocodiles.

predation in ecology, taking of prey by a †predator.

predator any animal that hunts, kills, and eats another animal (the prey). Predation is the method of feeding shown by a predator. Examples of predators include lions, which prey on antelope and zebra, owls, which prey on mice and rats, or sparrowhawks, which might prey on sparrows. Predators may affect the population size of their prey, and the population size of the prey may affect that of the predator. In a †food chain predator and prey are placed next to each other to indicate the flow of energy from prey to predator. Predators may compete (see †competition) with each other for prey. If the prey is in short supply the numbers and biomass of predators will fall.

Predators are usually bigger and more powerful than their prey (although a stoat preys on a rabbit, which may be four or five times its size), and their population is smaller – if the predator became more widespread than the prey, the population of the predator would inevitably fall. This is because there is not enough †energy available in the bodies of the prey to sustain more predators. The drop in numbers of organisms along many food chains can be seen using a †pyramid of numbers. Other relationships between species include parasitism, in which the host is not necessarily killed by the †parasite, and †symbiosis, in which both species benefit from the relationship.

predicate calculus in logic, part of the theory of devising models involving deduction and the use of variables and negatives in systems of sentences; the term occurs commonly in †mathematical logic.

preening in biology, bird grooming behavior that keeps feathers in good order. The vein of a †feather consists of numerous barbs attached on each side of the shaft. Each barb carries hundreds of barbules linked to the barbules of the adjacent barb. If the barbules of neighboring barbs separate, splitting the vein, the preening "zips" them together again as the feather is drawn through the beak.

Pregl, Fritz (1869–1930) Austrian chemist who was awarded the Nobel Prize for Chemistry in 1923 for devising, during his research on bile acids, new techniques for microanalysis (the analysis of very small quantities). He scaled down his analytic equipment and designed a new balance that could weigh to an accuracy of 0.001 mg. This breakthrough in organic chemistry paved the way for modern biochemistry.

pregnancy in humans, the process during which a developing embryo grows within the woman's womb. It begins at conception and ends at birth, and the normal length is 40 weeks, or around nine months.

Menstruation usually stops on conception. About one in five pregnancies fails, but most of these failures occur very early on, so the woman may notice only that her period is late. After the second month, the breasts become tender, and the areas round the nipples become darker. Enlargement of the uterus can be felt at about the end of the third month, and after this the abdomen enlarges progressively. Fetal movement can be felt at about 18 weeks; a heartbeat may be heard during the sixth month. Pregnancy in animals is called †gestation.

prehistoric life diverse organisms that inhabited Earth from the origin of life about 3.5 billion years ago to the time when humans began to keep written records, about 3500 BC. During the course of evolution, new forms of life developed and many other forms, such as the dinosaurs, became extinct. Prehistoric life evolved over this vast timespan from simple bacteria-like cells in the oceans to algae and protozoans and complex multicellular forms such as worms, mollusks, crustaceans, fishes, insects, land plants, amphibians, reptiles, birds, and mammals. On a geological timescale humans evolved relatively recently, about 4 million years ago, although the exact dating is a matter of some debate. See also †geological time.

Prelog, Vladimir (1906–1998) Bosnian-born Swiss organic chemist who studied alkaloids and antibiotics. The comprehensive molecular topology that evolved from his work on stereochemistry is gradually replacing classical stereochemistry. He shared the Nobel Prize for Chemistry in 1975 for his work on the stereochemistry of organic molecules and their reactions.

premolar in mammals, one of the large teeth toward the back of the mouth. In herbivores they are adapted for grinding. In carnivores they may be carnassials. Premolars are present in milk †dentition as well as permanent dentition.

pre-operational describing a stage of human development between the ages of two and seven years in which a child learns to imitate and acquires language to describe concrete objects.

presbyopia vision defect, an increasing inability with advancing age to focus on near objects. It is caused by thickening and loss of elasticity in the lens, which is no longer able to relax to the near-spherical shape required for near vision.

pressure in a fluid, the force exerted normally (at right angles) on the surface of a body immersed in the fluid. The SI unit of pressure is the pascal (Pa), equal to a pressure of one newton per square meter. In the atmosphere, the pressure declines with increasing height from about 100 kPa at sea level to zero where the atmosphere fades into space. Pressure is commonly measured with a †barometer, †manometer, or †Bourdon gauge. Other common units of pressure are the bar and the torr.

Pressure can be worked out using the formula: pressure = force/area. For example, farm tractors have wide tires with relatively large areas in contact with the ground. This means that the pressure exerted by these areas is less (the force, or weight, is spread out over a larger area), which helps prevent the tractor wheels from sinking into the ground. Similarly, the head of a safety pin pressed down onto a table will not penetrate into the table; the force exerted by the head is spread out over a relatively large area and the pressure is lower. On the point of the pin, the force is spread over a much smaller area, the pressure is consequently greater, and the point will penetrate into the table.

Absolute pressure is measured from a vacuum; **gauge pressure** is the difference between the absolute pressure and the local †atmospheric pressure. In a liquid, the pressure at a depth *h* is given by r*gh* where r is the density and *g* is the acceleration of free fall.

pressure law law stating that the pressure of a fixed mass of gas at constant volume is directly proportional to its absolute temperature. The law may be expressed as: pressure/temperature = constant or, more usefully, as: $P_1/T_1 = P_2/T_2$ where P_1 and T_1 are the initial pressure and absolute temperature of a gas, and P_2 and T_2 are its final pressure and temperature. See also †gas laws.

pressurized water reactor **PWR**, †nuclear reactor design used in nuclear power stations in many countries, and in nuclear-powered submarines. In the PWR, water under pressure is used in a closed system, as a coolant, passing through the reactor to the generator. Boric acid is added as a †moderator. It circulates through a steam generator, where its heat boils water to provide steam to drive power †turbines.

prevailing wind direction from which the wind most commonly blows in a locality. In northwestern Europe, for example, the prevailing wind is southwesterly, blowing from the Atlantic Ocean in the southwest and bringing moist and warm conditions.

Prévost, Pierre (1751–1839) Swiss physicist who first showed, in 1791, that all bodies radiate heat, no matter how hot or cold they are. In challenging the notion then prevalent that cold was produced by the entry of cold into an object rather than by an outflow of heat, Prévost made a basic advance in our knowledge of energy.

Priestley, Joseph (1733–1804) English chemist and Unitarian minister. He identified oxygen in 1774 and several other gases. Dissolving carbon dioxide under pressure in water, he began a European craze for soda water.

Joseph Priestley on Making Carbonated Water (1772)

http://dbhs.wvusd.k12.ca.us/Chem-History/Priestley-1772/Priestley-1772-Start.html

Transcript of Priestley's paper from 1772 on how to make carbonated water. The paper is split up into individual sub-pages off this main one, and includes the title page, dedication, preface and all illustrations in the total of 26 pages.

Prigogine, Ilya (1917–2003) Viscount Prigogine, Russian-born Belgian chemist who, as a highly original theoretician, has made major contributions to the field of †thermodynamics. Earlier theories had considered systems at or about equilibrium; Prigogine began to study "dissipative" or nonequilibrium structures frequently found in biological and chemical reactions. He was awarded the Nobel Prize for Chemistry in 1977 for his work in the thermodynamics of irreversible and dissipative processes. He became Viscount in 1989.

primary cell voltaic cell (battery) in which the chemical reaction that generates the voltage cannot be reversed (as it can in a secondary cell or accumulator).

primary color in physics, one of three colors of light that mix (in the correct proportions) to give white. There are an infinite number of such sets, though often red, green, and blue are chosen as standard.

primary consumer in ecology, those animals (herbivores) in the †food chain that eat plants.

primary data information that has been collected at first hand. It involves measurement of some sort, whether by taking readings off instruments, sketching, counting, or conducting interviews (using questionnaires).

primary factors most reduced (lowest) form of numbers, which, when multiplied together, produce the principal numbers or expression.

primary sexual characteristic endocrine gland producing maleness and femaleness. In males, the primary sexual characteristic is the †testis; in females it is the †ovary. Both are endocrine glands that produce hormones responsible for secondary sexual characteristics, such as facial hair and a deep voice in males, and breasts in females.

primary tissue in biology, tissue that is the first to be laid down during the developments of a plant. Plants that have only primary tissues are called †herbaceous plants.

primate in zoology, any member of the order of mammals that includes monkeys, apes, and humans (together called **anthropoids**), as well as lemurs, bushbabies, lorises, and tarsiers (together called **prosimians**).

Generally, they have forward-directed eyes, gripping hands and feet, opposable thumbs, and big toes. They tend to have nails rather than claws, with gripping pads on the ends of the digits, all adaptations to the arboreal, climbing mode of life.

In 1996 a new primate genus (probably extinct) was identified by a US anthropologist from a collection of bones believed to belong to a potto. The animal has been named *Pseudopotto martini*.

In the same year, the Red List of endangered species published by the World Conservation Union indicated that 46% of the world's 310 primate species are threatened with extinction. Four years later, this figure had increased to 50% and at least 10% of primate species are likely to become extinct in the wild over the next two decades. The first to be declared extinct, in autumn 2000, was Miss Waldron's red colobus, *Procolobus badius waldroni*.

In June 2000 the number of recognized species of primate was increased from 275 to 310 following a meeting of primatologists, conservationists, and taxonomists in Orlando, Florida. A number of species were reclassified using molecular genetic research, behavioral observations, and anatomical evidence. Brazil has 77 species of primate, far more than any other country. By means of a fossil discovered in China in 2000, it has been established that the smallest primate that ever lived was a prosimian that weighed only 10 g and lived about 45 million years ago.

prime factor any factor of a number that is a †prime number. The fundamental theorem of arithmetic states that every number is either prime or can be expressed as a unique product of primes.

prime number number that can be divided only by 1 and itself, that is, having no other factors. There is an infinite number of primes, the first ten of which are 2, 3, 5, 7, 11, 13, 17, 19, 23, and 29 (by definition, the number 1 is excluded from the set of prime numbers). The number 2 is the only even prime because all other even numbers have 2 as a factor. Numbers other than primes can be expressed as a product of their †prime factors.

Over the centuries mathematicians have sought general methods (algorithms) for calculating primes, from †Eratosthenes' sieve to programs on powerful computers.

Mersenne primes are in the form $2^q - 1$, where q is also a prime. The thirty-eighth Mersenne prime was discovered in 1999. It is 2 million digits long and was discovered as part of the Great Internet Prime Search, where volunteers download software to search for primes while their computers are idle.

Great Internet Mersenne Prime Search http://www.mersenne.org/prime.htm

In January 1996 the "Great Internet Mersenne Prime Search" was launched by George Woltman, a programmer in Florida, to enlist the help of volunteers in tracking down the largest prime number. Software can be downloaded that will search for primes while the computer is idle.

principal focus point at which incident rays parallel to the principal axis of a †lens converge, or appear to diverge, after refraction. The distance from the lens to its principal focus is its †focal length.

Pringsheim, Ernst (1859–1917) German physicist whose experimental work on the nature of thermal radiation led directly to the †quantum theory. In 1881 he developed a spectrometer that made the first accurate measurements of wavelengths in the infrared region.

printed circuit board PCB, electrical circuit created by laying (printing) "tracks" of a conductor such as copper on one or both sides of an insulating board. The PCB was invented in 1936 by Austrian scientist Paul Eisler, and was first used on a large scale in 1948.

Components such as integrated circuits (chips), resistors, and capacitors can be soldered to the surface of the board (surface-mounted) or, more commonly, attached by inserting their connecting pins or wires into holes drilled in the board. PCBs include †motherboards, expansion boards, and adaptors.

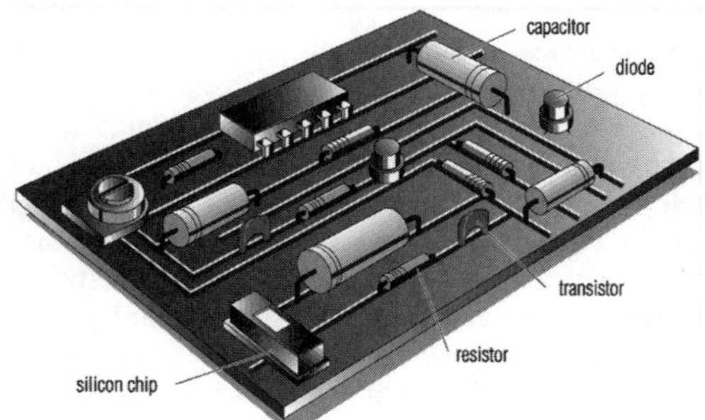

printed circuit board A typical printed circuit board (PCB). The PCB contains sockets for the integrated circuits, or chips, and the connecting tracks. Helicon

printer in computing, an output device for producing printed copies of text or graphics. Types include the daisywheel printer, which produces good-quality text but no graphics; the †dot matrix printer, which produces text and graphics by printing a pattern of small dots; the ink-jet printer, which creates text and graphics by spraying a fine jet of quick-drying ink onto the paper; and the laser printer, which uses electrostatic technology very similar to that used by a photocopier to produce high-quality text and graphics.

Printers may be classified as **impact printers** (such as daisywheel and dot matrix printers), which form characters by striking an inked ribbon against the paper, and **non-impact printers** (such as ink-jet and laser printers), which use a variety of techniques to produce characters without physical impact on the paper.

A further classification is based on the basic unit of printing, and categorizes printers as character printers, line printers, or page printers, according to whether they print one character, one line, or a complete page at a time.

prion acronym, by rearrangement, for PROteinaceous INfectious particle, infectious agent, a hundred times smaller than a virus. Composed of protein, and without any detectable nucleic acid (genetic material), it is strongly linked to a number of fatal degenerative brain diseases in mammals, such as bovine spongiform encephalopathy (BSE) in cattle, scrapie in sheep, and Creutzfeldt–Jakob disease (CJD) and kuru in humans.

The existence of prions was postulated by US neurologist Stanley Prusiner in 1982, when he and his colleagues isolated a single infectious agent for scrapie that consisted only of protein and had no associated nucleic acid (RNA or DNA). His theory remains unproven but has been upheld by subsequent research, which has identified the protein concerned as well as a mechanism for its action.

A US researcher proved that prions are capable of spreading disease, in July 2000, three years after Prusiner was awarded the Nobel Prize for Physiology or Medicine for his prion theory.

prism in mathematics, a solid figure whose cross-section is the same along its length. When a slice is cut through a prism, the size and shape of the cross-section is always the same. A cube, for example, is a rectangular prism with all faces (bases and sides) the same shape and size. A cylinder is a prism with a circular cross section. The name of a prism is often derived from the shape of its prism, such as a triangular prism and a cuboid.

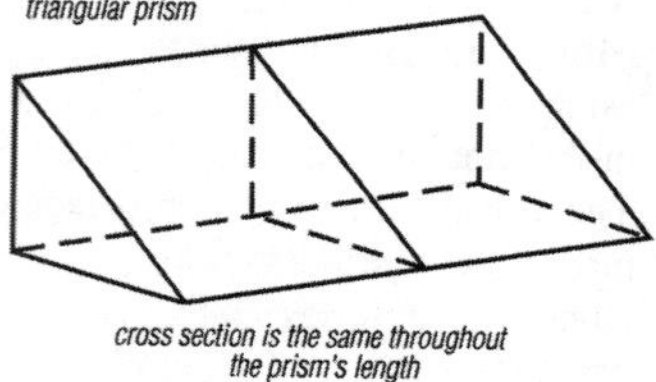

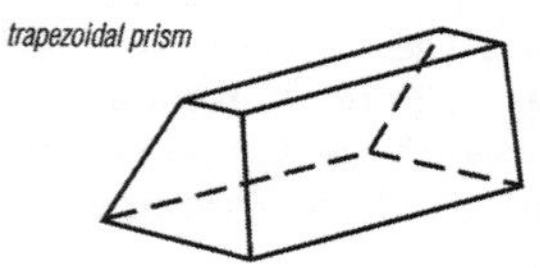

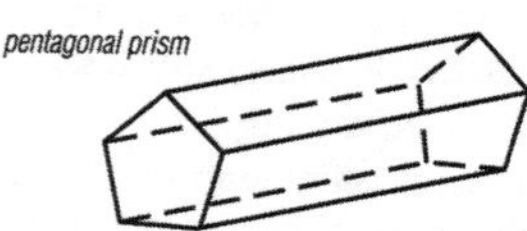

prism The volume of a prism is determined by multiplying the area of the cross section by the length of the prism. Helicon

prism in optics, a triangular block of transparent material (plastic, glass, or silica) commonly used to "bend" a ray of light or split a light beam (for example, white light) into its component colors. Prisms are used as mirrors to define the optical path in binoculars, camera viewfinders, and periscopes. The dispersive property of prisms is used in the †spectroscope.

probability likelihood, or chance, that an event will occur, often expressed as odds, or in mathematics, numerically as a fraction or decimal.

In general, the probability that *n* particular events will happen out of a total of *m* possible events is *n*/*m*. A certainty has a probability of 1; an impossibility has a probability of 0.

Probability = number of successful events/total possible number of events

In tossing a coin, the chance that it will land "heads" is the same as the chance that it will land "tails," that is, 1 to 1 or even; mathematically, this probability is expressed as $^1/_2$ or 0.5. An estimate for probability can be achieved by experiment; this is known as the **relative frequency**. The probability of any chosen number coming up on the roll of a fair die is 5 to 1; the probability is $^1/_6$ or 0.1666. If two dice are rolled there are 6 x 6 = 36 different possible combinations. The probability of a double (two numbers the same) is $^6/_{36}$ or $^1/_6$ since there are six doubles in the 36 events: (1,1), (2,2), (3,3), (4,4), (5,5), and (6,6).

†**Independent events** are those that do not affect each other, for example rolling two dice are independent events, as the rolling of the first die does not affect the outcome of the rolling of the second die. If events are described as †**mutually exclusive** it means that if one happens, then it prevents the other from happening. So tossing a coin is a mutually exclusive event as it can result in a head or a tail but not both. The sum of the probabilities of mutually exclusive events is always equal to 1. For example, if one has a bag containing 3 marbles, each of a different color, the probability of selecting each color would be 1/3.

1/3 + 1/3 + 1/3 = 1

To find out the probability of two or more mutually exclusive events occurring, their individual probabilities are added together. So, in the above example, the probability of selecting either a blue marble or

a red marble is
1/3 + 1/3 = 2/3
The probability of two independent events both occurring is smaller than the probability of one such event occurring. For example, the probability of throwing a 3 when rolling a die is $^1/_6$, but the probability of throwing two 3s when rolling two dice is $^1/_{36}$.
Conditional probability is when the outcome of the first event affects the outcome of the second event. For example, if a ball is chosen at random from a bag of 4 blue balls and 5 red balls, and not replaced, the probability of selecting 2 blue balls is:
$P(b) = 4/9 \times 3/8 = 12/72 = 1/6$
This can be displayed in a †tree diagram.
Probability theory was developed by the French mathematicians Blaise Pascal and Pierre de Fermat in the 17th century, initially in response to a request to calculate the odds of being dealt various hands at cards. Today probability plays a major part in the mathematics of atomic theory and finds application in insurance and statistical studies.
Notes on Tree Diagrams http://www.learn.co.uk/default.asp?WCI=SubUnit&WCU=3622
An explanation of how to use tree diagrams to represent probabilities. There are lots of worked examples given, as well as an explanation of how to interpret tree diagrams to represent probability data.
Stem and Leaf Plots http://dev1.epsb.edmonton.ab.ca/math14_Jim/comphelp/stemleaf.html
Explains what a stem and leaf diagram is, and offers lots of sample data and worked examples. A helpful stem and leaf diagram calculator is included, which enables a diagram to be generated from input data.

processor in computing, another name for the †central processing unit or †microprocessor of a computer.

Procyon or **Alpha Canis Minoris**, brightest star in the constellation †Canis Minor and the eighth-brightest star in the night sky. Procyon is a white star 11.4 light years from the Sun, with a mass 1.7 times that of the Sun. It has a †white dwarf companion that orbits it every 40 years.
The name, derived from Greek, means "before the dog," and reflects the fact that in mid-northern latitudes Procyon rises shortly before †Sirius, the Dog Star. Procyon and Sirius are sometimes called "the Dog Stars." Both are relatively close to us and have white dwarf companions.

produce in mathematics, to extend a straight line indefinitely in one direction or the other.

product in mathematics, result obtained from multiplying two numbers or variables.
For example, *ab* is the product of *a* and *b*.

productivity, biological in an ecosystem, the amount of material in the food chain produced by the primary producers (plants) that is available for consumption by animals. Plants turn carbon dioxide and water into sugars and other complex carbon compounds by means of photosynthesis. Their net productivity is defined as the quantity of carbon compounds formed, less the quantity used up by the respiration of the plant itself.

progesterone †steroid hormone that occurs in vertebrates. In mammals, it regulates the menstrual cycle and pregnancy. Progesterone is secreted by the corpus luteum (the ruptured Graafian follicle of a discharged ovum).

proglottid in zoology, one of many sections that make up the long flat body of a †tapeworm. Each proglottid contains a set of reproductive organs. The mature proglottids detach from the tapeworm and pass out from the host in its feces. In this way the eggs of the tapeworm are dispersed, to be taken up in due course by a new host.

prograde orbit or **direct-motion orbit**, in an orbiting satellite, an orbit in which the satellite travels in the same direction as the planet's rotation. It is the opposite of a †retrograde orbit.
A planet has a prograde orbit or rotation if the sense of rotation is the same as the general sense of rotation of the Solar System. On the †celestial sphere, it refers to motion from west to east against the background of stars.

program in computing, a set of instructions that controls the operation of a computer. There are two main kinds: †applications programs, which carry out tasks for the benefit of the user – for example, word processing; and †systems programs, which control the internal workings of the computer. A †utility program is a systems program that carries out specific tasks for the user. Programs can be written in any of a number of †programming languages but are always translated into machine code before they can be executed by the computer.

program files in computing, files which contain the code used by a computer program.

programmer job classification for computer personnel. Programmers write the software needed for any new computer system or application.

programming writing instructions in a programming language for the control of a computer.

Applications programming is for end-user programs, such as accounts programs or word-processing packages. **Systems programming** is for operating systems and the like, which are concerned more with the internal workings of the computer.

There are several programming styles. **Procedural programming**, in which programs are written as lists of instructions for the computer to obey in sequence, is by far the most popular. It is the "natural" style, closely matching the computer's own sequential operation. **Declarative programming**, as used in the programming language PROLOG, does not describe how to solve a problem, but rather describes the logical structure of the problem. Running such a program is more like proving an assertion than following a procedure. **Functional programming** is a style based largely on the definition of functions. There are very few functional programming languages, HOPE and ML being the most widely used, though many more conventional languages (for example C) make extensive use of functions. **Object-oriented programming**, the most recently developed style, involves viewing a program as a collection of objects that behave in certain ways when they are passed certain "messages." For example, an object might be defined to represent a table of figures, which will be displayed on screen when a "display" message is received.

programming language in computing, a special notation in which instructions for controlling a computer are written. Programming languages are designed to be easy for people to write and read, but must be capable of being mechanically translated (by a †compiler or an interpreter) into the †machine code that the computer can execute. Programming languages may be classified as †high-level languages or †low-level languages. See also †source language.

progression sequence of numbers each occurring in a specific relationship to its predecessor. An **arithmetic progression** has numbers that increase or decrease by a common sum or difference (for example, 2, 4, 6, 8); a **geometric progression** has numbers each bearing a fixed ratio to its predecessor (for example, 3, 6, 12, 24); and a **harmonic progression** has numbers whose †reciprocals are in arithmetical progression, for example 1, $^1/_2$, $^1/_3$, $^1/_4$.

projectile particle that travels with both horizontal and vertical motion in the Earth's gravitational field. If the frictional forces of air resistance are ignored, the two components of its motion can be analyzed separately: its vertical motion will be accelerated due to its weight in the gravitational field; its horizontal motion may be assumed to be at constant velocity. In a uniform gravitational field and in the absence of frictional forces the path of a projectile is a parabola.

projection representation of the surface of the Earth on paper, see †map projection.

projective geometry form of two- and three-dimensional geometry concerned with the geometrical properties that remain constant (invariant) under projection, that is, extended on a single plane, or projected from one plane onto another. Perspective – two-dimensional representation of three-dimensional reality – uses the basic theory of projective geometry.

prokaryote in biology, an organism whose cells lack organelles (specialized segregated structures such as nuclei, mitochondria, and chloroplasts). Prokaryote DNA is not arranged in chromosomes but forms a coiled structure called a **nucleoid**. The prokaryotes comprise only the **bacteria** and **cyanobacteria** (see †blue-green algae); all other organisms are eukaryotes.

Prokhorov, Aleksandr Mikhailovich (1916–2002) Russian physicist whose fundamental work on microwaves with Nikolai †Basov led to the construction of the first practical †maser (the microwave equivalent of the laser). They shared the Nobel Prize for Physics in 1964 with Charles †Townes for their work on quantum electronics leading to the construction of oscillators and amplifiers based on the maser-laser principle.

PROM acronym for Programmable Read-Only Memory, in computing, a memory device in the form of an integrated circuit (chip) that can be programmed after manufacture to hold information permanently. PROM chips are empty of information when manufactured, unlike ROM (read-only memory) chips, which have information built into them. Other memory devices are †EPROM (erasable programmable read-only memory) and †RAM (random-access memory).

promethium chemical symbol Pm, radioactive, metallic element of the †lanthanide series, atomic number 61, relative atomic mass 145. It occurs in nature only in minute amounts, produced as a fission product/by-product of uranium in †pitchblende and other uranium ores; for a long time it was considered not to occur in nature. The longest-lived isotope has a half-life of slightly more than 20 years.

Promethium is synthesized by neutron bombardment of neodymium and is a product of the fission of uranium, thorium, or plutonium; it can be isolated in large amounts from the fission-product debris of uranium fuel in nuclear reactors. It is used in phosphorescent paints and as an X-ray source.

prominence bright cloud of gas projecting from the Sun into space 100,000 km/60,000 mi or more.

Quiescent prominences last for months, and are held in place by magnetic fields in the Sun's corona. **Surge prominences** shoot gas into space at speeds of 1,000 kps/600 mps.

Loop prominences are gases falling back to the Sun's surface after a solar †flare.

proof in mathematics, a set of arguments used to deduce a mathematical theorem from a set of axioms.

proof spirit numerical scale used to indicate the alcohol content of an alcoholic drink. Proof spirit (or 100% proof spirit) acquired its name from a solution of alcohol in water which, when used to moisten gunpowder, contained just enough alcohol to permit it to burn.

propane C_3H_8, gaseous hydrocarbon of the †alkane series, found in petroleum and used as a fuel and as a refrigerant.

propanol or **propyl alcohol**, third member of the homologous series of †alcohols. Propanol is usually a mixture of two isomeric compounds (see †isomer): propan-1-ol ($CH_3CH_2CH_2OH$) and propan-2-ol ($CH_3CHOHCH_3$). Both are colorless liquids that can be mixed with water and are used in perfumery.

propanone common name **acetone**; CH_3COCH_3, colorless flammable liquid used extensively as a solvent, as in nail-varnish remover, and for making acrylic plastics. It boils at 56.5°C/133.7°F, mixes with water in all proportions, and has a characteristic odor.

propenal or **acrolein**; CH_2:CHCHO, colorless liquid formed during the partial combustion of fats. It is an unsaturated aldehyde, boils at 52°C/125.6°F, has an irritating action on the skin, and its vapors cause a copious flow of tears. It was used in gas form as a chemical weapon in World War I by the French under the codename *Papite*.

propene common name **propylene**; $CH_3CH = CH_2$, second member of the alkene series of hydrocarbons. A colorless, flammable gas, it is widely used by industry to make organic chemicals, including polypropylene plastics.

propenoic acid common name **acrylic acid**; $H_2C = CHCOOH$, acid obtained from the aldehyde propenal (acrolein) derived from glycerol or fats. Glasslike thermoplastic resins are made by polymerizing †esters of propenoic acid or methyl propenoic acid and used for transparent components, lenses, and dentures. Other acrylic compounds are used for adhesives, artificial fibers, and artists' acrylic paint.

proper fraction or **simple fraction** or **common fraction**, †fraction whose value is less than 1. The numerator has a lower value than the denominator in a proper fraction.

proper motion gradual change in the position of a star that results from its motion in orbit around our Galaxy, the Milky Way. Proper motions are slight and undetectable to the naked eye, but can be accurately measured on telescopic photographs taken many years apart.

Barnard's Star is the star with the largest proper motion, 10.3 arc seconds per year.

properties in chemistry, the characteristics a substance possesses by virtue of its composition.

Physical properties of a substance can be measured by physical means, for example boiling point, melting point, hardness, elasticity, color, and physical state. **Chemical properties** are the way it reacts with other substances; whether it is acidic or basic, an oxidizing or a reducing agent, a salt, or stable to heat, for example.

proportion relation of a part to the whole (usually expressed as a †fraction or †percentage). In mathematics two variable quantities x and y are proportional if, for all values of x, $y = kx$, where k is a constant. This means that if x increases, y increases in a linear fashion.

direct proportion If A and B are in direct proportion, then as one grows bigger the other also grows bigger by the same proportion (or †ratio). For example, if B doubles then A doubles. This can be written as A μ B or $A = kB$ (where k is a constant multiplier). For example, when costing a number of pens:

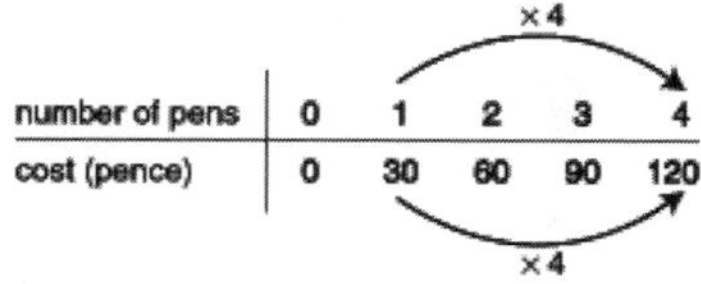

For every extra pen bought, the cost goes up by the same amount (30p). So the cost of the pens will be proportional to the number bought.

If A is proportional to B there are two things that are true:

the multiplier rule – if A is multiplied by a value then B must be multiplied by the same value.

the graph of A against B is always a straight line through (0,0) and the gradient of the graph is the same as the ratio of A:B.

inverse proportion If A is inversely proportional to B, then as B gets bigger A gets smaller by the same factor. For example, if B is increased by a factor of 2 (B doubles) then A decreases by a factor of 2 (A is halved). This can be written as:

$A \propto 1/B$ or $A = k/B$

For example, a group of people want to hire a minibus at a charge of £60 a day. If only 1 person uses the minibus it will cost them £60, for 2 people the cost will be £30 each, and for 3 people the cost will be £20 each.

Many laws of science relate quantities that are proportional (for example, †Boyle's law).

proprioceptor one of the sensory nerve endings that are located in muscles, tendons, and joints. They relay information on the position of the body and the state of muscle contraction.

prop root or **stilt root**, modified root that grows from the lower part of a stem or trunk down to the ground, providing a plant with extra support. Prop roots are common on some woody plants, such as mangroves, and also occur on a few herbaceous plants, such as maize. **Buttress roots** are a type of prop root found at the base of tree trunks, extended and flattened along the upper edge to form massive triangular buttresses; they are common on tropical trees.

propyl alcohol common name for †propanol.

propylene common name for †propene.

prosimian or **primitive primate**, in zoology, any animal belonging to the suborder Strepsirhin of †primates. Prosimians are characterized by a wet nose with slitlike nostrils, the tip of the nose having a prominent vertical groove. Examples are lemurs, pottos, tarsiers, and the aye-aye.

prostaglandin any of a group of complex fatty acids present in the body that act as messenger substances between cells. Effects include stimulating the contraction of smooth muscle (for example, of the womb during birth), regulating the production of stomach acid, and modifying hormonal activity. In excess, prostaglandins may produce inflammatory disorders such as arthritis. Synthetic prostaglandins are used to induce labor in humans and domestic animals.

The analgesic actions of substances such as aspirin are due to inhibition of prostaglandin synthesis.

prostate gland gland surrounding and opening into the †urethra at the base of the †bladder in male mammals.

The prostate gland produces an alkaline fluid that is released during ejaculation; this fluid activates sperm, and prevents their clumping together. Older men may develop **benign prostatic hyperplasia** (BPH), a painful condition in which the prostate becomes enlarged and restricts urine flow. This can cause further problems of the bladder and kidneys. It is treated by prostatectomy.

protactinium chemical symbol Pa, (Latin *protos* "before" + *aktis* "first ray") silver-gray, radioactive, metallic element of the †actinide series, atomic number 91, relative atomic mass 231.036. It occurs in nature in very small quantities in †pitchblende and other uranium ores. It has 14 known isotopes; the longest-lived, Pa-231, has a half-life of 32,480 years.

protandry in a flower, the state where the male reproductive organs reach maturity before those of the female. This is a common method of avoiding self-fertilization. See also †protogyny.

protease general term for a digestive enzyme capable of splitting proteins. Examples include pepsin, found in the stomach, and trypsin, found in the small intestine.

protein large, complex, biologically-important molecules composed of amino acids joined by †peptide bonds. The number of amino acids used can be many hundreds. There are 20 different amino acids and they can be joined in any order. Proteins are essential to all living organisms. As †enzymes they regulate all aspects of metabolism. Structural proteins such as keratin and collagen make up skin, claws, bones, tendons, and ligaments; muscle proteins produce movement; hemoglobin transports oxy-

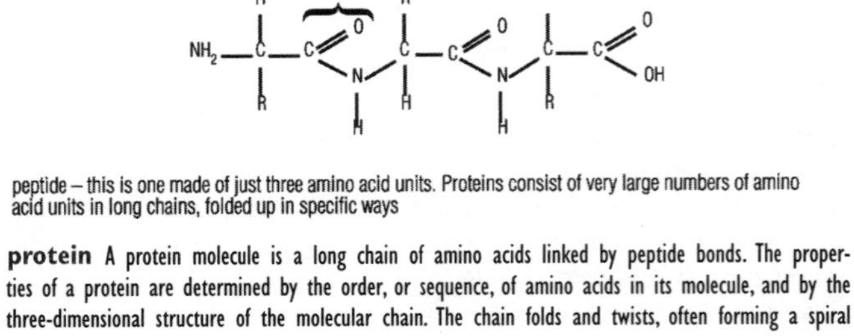

amino acids, where R is one of many possible side chains

peptide bond

peptide – this is one made of just three amino acid units. Proteins consist of very large numbers of amino acid units in long chains, folded up in specific ways

protein A protein molecule is a long chain of amino acids linked by peptide bonds. The properties of a protein are determined by the order, or sequence, of amino acids in its molecule, and by the three-dimensional structure of the molecular chain. The chain folds and twists, often forming a spiral shape. Helicon

gen; and membrane proteins regulate the movement of substances into and out of cells. For humans, protein is an essential part of the diet, and is found in greatest quantity in soy beans and other grain legumes, meat, eggs, and cheese. During digestion protein molecules are broken down into amino acids which are then easily absorbed into the body.

Protein synthesis occurs in cells. The information describing the order in which the different amino acids are joined is found in †DNA in the form of a code. The part of the DNA that carries the code for making one protein is called a †gene. Each protein described in the code has an effect on the appearance and characteristics of the organism.

During digestion in the body, proteases (any enzymes that break down a protein) are produced by the †stomach, the †pancreas, and the small †intestine. These enzymes catalyze the breakdown of protein into amino acids. Pepsin is an enzyme which is produced by the walls of the stomach. It digests large protein molecules into smaller protein molecules (smaller polypeptides). The conditions in the stomach are very acid and pepsin works at its fastest rate in such conditions.

The amino acids produced by the digestion of proteins are absorbed into the blood in the small intestine. They are transported around the body by the †blood for use by cells to make the proteins they need. This protein synthesis needs energy and this energy is provided by †respiration.

Introduction to Proteins http://biotech.icmb.utexas.edu/pages/science/protein_intro.html

Introduction to the world of proteins with discussions of their structures and sequences, as well as descriptions of some major kinds of proteins. The site also includes a useful glossary and offers a variety of photographic shots.

Protein Data Bank WWW Home Page http://www.rcsb.org/pdb/

Expanding scientific site that contains a fully searchable database of molecules. This is useful for students as well as more in-depth researchers.

protein engineering creation of synthetic proteins designed to carry out specific tasks. For example, an enzyme may be designed to remove grease from soiled clothes and remain stable at the high temperatures in a washing machine.

protein synthesis manufacture, within the cytoplasm of the cell, of the proteins an organism needs. The building blocks of proteins are †amino acids, of which there are 20 types. The pattern in which the amino acids are linked decides what kind of protein is produced. In turn it is the genetic code, contained within †DNA, that determines the precise order in which the amino acids are linked up during protein manufacture.

Interestingly, DNA is found only in the nucleus, yet protein synthesis occurs only in the cytoplasm. The information necessary for making the proteins is carried from the nucleus to the cytoplasm by another nucleic acid, †RNA.

proteome full complement of proteins that sustain cell functions within the body. The human proteome consists of millions of proteins and its makeup varies depending on the stage of cell development.

Proterozoic Eon eon of geological time, 3.5 billion to 570 million years ago, the second division of the Precambrian. It is defined as the time of simple life, since many rocks dating from this eon show traces of biological activity, and some contain the fossils of bacteria and algae.

prothallus short-lived †gametophyte of many ferns and other †pteridophytes (such as horsetails or clubmosses). It bears either the male or female sex organs, or both. Typically it is a small, green, flattened structure that is anchored in the soil by several †rhizoids (slender, hairlike structures, acting as roots) and needs damp conditions to survive. The reproductive organs are borne on the lower surface close to the soil. See also †alternation of generations.

prothrombin substance in the plasma from which †thrombin is derived during blood clotting.

protist single-celled organism which has a eukaryotic cell, but which is not a member of the plant, fungal, or animal kingdoms. The main protists are †protozoa.

Single-celled photosynthetic organisms, such as diatoms and dinoflagellates, are classified as protists or algae. Recently the term has also been used for members of the kingdom Protista, which features in certain five-kingdom classifications of the living world (see also †plant classification). This kingdom may include slime moulds, all algae (seaweeds as well as unicellular forms), and protozoa.

protocol in computing, an agreed set of **standards** for the transfer of data between different devices. They cover transmission speed, format of data, and the signals required to synchronize the transfer. See also †interface.

protogyny in a flower, the state where the female reproductive organs reach maturity before those of the male. Like protandry, in which the male organs reach maturity first, this is a method of avoiding self-fertilization, but it is much less common.

proton (Greek "first") in physics, a positively charged subatomic particle, a constituent of the nucleus of all †atoms. It carries a unit positive charge equal to the negative charge of an †electron. Its mass is almost 1,836 times that of an electron, or 1.673 x 10^{-24} g. The number of protons in the atom of an †element is equal to the †atomic number of that element.

A proton belongs to the †baryon group of †hadrons and is composed of two up quarks and one down quark. A proton is extremely long-lived, with a lifespan of at least 10^{32} years.

protonema young †gametophyte of a moss, which develops from a germinating spore (see †alternation of generations). Typically it is a green, branched, threadlike structure that grows over the soil surface bearing several buds that develop into the characteristic adult moss plants.

protonic acid any chemical compound whose molecules dissociate to form hydrogen ions, H^+; for example, hydrochloric acid.

proton number alternative name for †atomic number.

proton–proton cycle process of †nuclear fusion by which relatively cooler stars produce and radiate energy; hotter stars commonly achieve the same result by means of the †carbon–nitrogen cycle.

proto-planet early stage in the formation of a planet according to the theory by which planetary systems evolve through the condensation of gas clouds surrounding a young star. The theory is not, however, generally accepted.

protoplasm contents of a living cell. Strictly speaking it includes all the discrete structures (organelles) in a cell, but it is often used simply to mean the jellylike material in which these float. The contents of a cell outside the nucleus are called †cytoplasm.

protoplast in biology, plant cell with its cell wall removed. Protoplasts can be cultured in a culture medium (solution of all of the substances they need to grow and live). Protoplasts take up substances added to the medium in which they are growing. Different protoplasts can be fused together and grown to produce hybrid plants with new mixtures of characteristics.

protostar in astronomy, early formation of a star that has recently condensed out of an interstellar cloud and which is not yet hot enough for hydrogen burning to start. Protostars derive their energy from gravitational contraction. See †Hayashi track.

prototheria subclass of mammals made up of the egg-laying monotremes. It contains only the echidna (spiny anteater) and the platypus.

prototype experimental version of a system, where the initial design can be tested and improved upon in the light of tests carried out.

protozoa group of single-celled organisms without rigid cell walls. Some, such as amoeba, ingest other cells, but most are †saprotrophs or parasites. The group is polyphyletic (containing organisms which have different evolutionary origins).

protractor instrument used to measure a flat †angle.

Prout, William (1785–1850) British physician and chemist. In 1815 Prout published his hypothesis that the relative atomic mass of every atom is an exact and integral multiple of the mass of the hydrogen atom. The discovery of isotopes (atoms of the same element that have different masses) in the 20th century bore out his idea.

Prout's hypothesis now defunct proposal that all the chemical †elements were made from the combination of different numbers of hydrogen †atoms.

provenance in earth science, area from which sedimentary materials are derived.

provitamin any precursor substance of a vitamin. Provitamins are ingested substances that become converted to active vitamins within the organism. One example is ergosterol (provitamin D_2), which through the action of sunlight is converted to calciferol (vitamin D_2); another example is beta-carotene, which is hydrolyzed in the liver to vitamin A.

Proxima Centauri closest star to the Sun, 4.2 light years away from the Sun. It is a faint †red dwarf, visible only with a telescope, and is a member of the Alpha Centauri triple-star system.

It is called Proxima because it is about 0.1 light years closer to us than its two partners.

prussic acid former name for †hydrocyanic acid.

pseudocarp fruitlike structure that incorporates tissue that is not derived from the ovary wall. The additional tissues may be derived from floral parts such as the †receptacle and †calyx. For example, the colored, fleshy part of a strawberry develops from the receptacle and the true fruits are small †achenes – the "pips" embedded in its outer surface. Rose hips are a type of pseudocarp that consists of a hollow, fleshy receptacle containing a number of achenes within. Different types of pseudocarp include pineapples, figs, apples, and pears.

A **coenocarpium** is a fleshy, multiple pseudocarp derived from an †inflorescence rather than a single flower. The pineapple has a thickened central axis surrounded by fleshy tissues derived from the receptacles and floral parts of many flowers. A fig is a type of pseudocarp called a **syconium**, formed from a hollow receptacle with small flowers attached to the inner wall. After fertilization the ovaries of the female flowers develop into one-seeded achenes.

Apples and pears are †pomes, another type of pseudocarp.

pseudocopulation attempted copulation by a male insect with a flower. It results in †pollination of the flower and is common in the orchid family, where the flowers of many species resemble a particular species of female bee. When a male bee attempts to mate with a flower, the pollinia (groups of pollen grains) stick to its body. They are transferred to the stigma of another flower when the insect attempts copulation again.

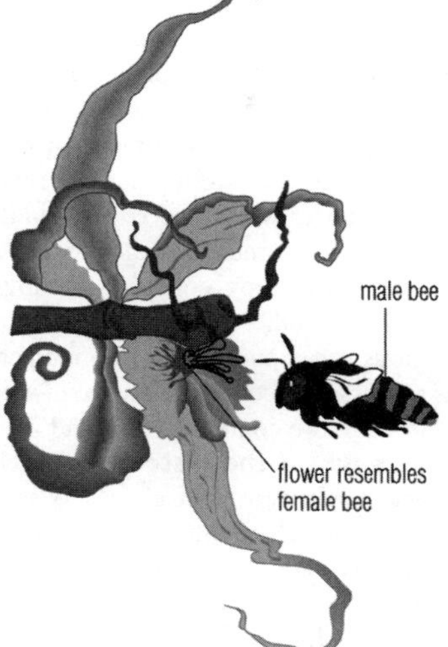

pseudocopulation The male bee, attracted to the orchid because of its resemblance to a female bee, attempts to mate with the flower. The bee's efforts cover its body with pollen, which is carried to the next flower it visits. Helicon

pseudomorph mineral that has replaced another in situ and has retained the external crystal shape of the original mineral.

psi in parapsychology, a hypothetical faculty common to humans and other animals, said to be responsible for †extrasensory perception, telekinesis, and other paranormal phenomena.

psychiatry branch of medicine dealing with the diagnosis and treatment of mental disorder, normally divided into the areas of **neurotic conditions**, including anxiety, depression, and hysteria, and **psychotic disorders**, such as schizophrenia. Psychiatric treatment consists of drugs, analysis, or electroconvulsive therapy.

In practice there is considerable overlap between psychiatry and †clinical psychology, the fundamental difference being that psychiatrists are trained medical doctors (holding an MD degree) and may therefore prescribe drugs, whereas psychologists may hold a doctorate but do not need a medical qualification to practice. See also †psychoanalysis.

psychoanalysis theory and treatment method for neuroses, developed by Sigmund †Freud in the 1890s. Psychoanalysis asserts that the impact of early childhood sexuality and experiences, stored in the †unconscious, can lead to the development of adult emotional problems. The main treatment method involves the free association of ideas, and their interpretation by patient and analyst, in order to discover these long-buried events and to grasp their significance to the patient, linking aspects of the patient's historical past with the present relationship to the analyst. Psychoanalytic treatment aims to free the patient from specific symptoms and from irrational inhibitions and anxieties.

concepts As a theoretical system, psychoanalysis rests on three basic concepts. The central concept is that of the **unconscious**, a reservoir within one's mental state which contains elements and experiences of which one is unaware, but which may to some extent be brought into preconscious and conscious awareness, or inferred from aspects of behavior. The second and related basic concept is that of **resistance**, a process by which unconscious elements are forcibly kept out of the conscious awareness by an active repressive force. Freud came to experience the third basic concept in his work, known as **transference**, with his earliest patients, who transferred to him aspects of their past relationships with others, so that their relationship with him was colored by their previous feelings. The analysis of the transference in all its manifestations has become a vital aspect of current psychoanalytic practice.

id, ego, and superego Freud proposed a model of human psychology based on the concept of the conflicting †id, †ego, and †superego. The id is the mind's instinctual element which demands pleasure and satisfaction; the ego is the conscious mind which deals with the demands of the id and superego; the superego is the ethical element which acts as a conscience and may produce feelings of guilt. The conflicts between these three elements can be used to explain a range of neurotic symptoms.

other schools In the early 1900s a group of psychoanalysts gathered around Freud. Some of these later broke away and formed their own schools, notably Alfred †Adler in 1911 and Carl †Jung in 1913. The significance of early infantile experience has been further elaborated in the field of child analysis, particularly in the work of Melanie †Klein and her students, who pay particular attention to the development of the infant in the first six to eight months of life.

psychology systematic study of human and animal behavior. The first psychology laboratory was founded in 1879 by Wilhelm †Wundt at Leipzig, Germany. The subject includes diverse areas of study and application, among them the roles of instinct, heredity, environment, and culture; the processes of sensation, perception, learning, and memory; the bases of motivation and emotion; and the functioning of thought, intelligence, and language. Significant psychologists have included Gustav Fechner, founder of psychophysics; Wolfgang Köhler, one of the Gestalt or "whole" psychologists; Sigmund Freud and his associates Carl Jung and Alfred Adler; William James, Jean Piaget; Carl Rogers; Hans Eysenck; J B Watson; and B F Skinner.

Experimental psychology emphasizes the application of rigorous and objective scientific methods to the study of a wide range of mental processes and behavior, whereas **social psychology** concerns the study of individuals within their social environment; for example, within groups and organizations. This has led to the development of related fields such as **occupational psychology**, which studies human behavior at work, and **educational psychology**. **Clinical psychology** concerns the understanding and treatment of mental health disorders, such as anxiety, phobias, or depression; treatment may include behavior therapy, cognitive therapy, counseling, psychoanalysis, or some combination of these.

Modern studies have been diverse; for example, the psychological causes of obesity; the nature of religious experience; and the underachievement of women seen as resulting from social pressures. Other related subjects are the nature of sleep and dreams, and the possible extensions of the senses, which leads to the more contentious ground of †parapsychology.

Museum of the History of Psychological Instrumentation

http://www.chss.montclair.edu/psychology/museum/museum.html

Museum of mad scientists' tools. If you expect to see pictures of couches and clipboards on this Web site you are going to be disappointed since a disturbingly high proportion of the tools listed have electrodes attached. The instruments described and pictured on these pages were the tools of a generation of psychology experimenters, and as with all historic science equipment look bizarre to the modern eye. So if you've ever wondered what a "Wundt's tachistoscope" does and what it looks like, this is the place to find out.

psychometrics measurement of mental processes. This includes intelligence and aptitude testing to help in job selection and in the clinical assessment of cognitive deficiencies resulting from brain damage.

psychopathy personality disorder characterized by chronic antisocial behavior (violating the rights of others, often violently) and an absence of feelings of guilt about the behavior.

Because the term "psychopathy" has been misused to refer to any severe mental disorder, many psychologists now prefer the term "antisocial personality disorder," though this also includes cases in which absence or a lesser degree of guilt is not a characteristic feature.

psychosis or **psychotic disorder**, general term for a serious mental disorder in which the individual commonly loses contact with reality and may experience hallucinations (seeing or hearing things that do not exist) or delusions (fixed false beliefs). For example, in a paranoid psychosis, an individual may believe that others are plotting against him or her. A major type of psychosis is schizophrenia.

Schizophrenia http://www.schizophrenia.com/

Facts about schizophrenia – causes and symptoms, the different types of the disease, how it affects sufferers' family members, available treatments, and new developments. The site includes a list of available support resources.

psychotherapy any treatment for psychological problems that involves talking rather than surgery or drugs. Examples include †cognitive therapy and †psychoanalysis.

psychotic disorder another name for †psychosis.

Ptashne, Mark (1940–) US molecular biologist who has contributed much to our understanding of the mechanisms of transcription (formation of messenger RNA from DNA) in prokaryotic cells, such as bacteria or cyanobacteria. He has worked for many years on the mechanisms that switch genes on and off.

pteridophyte simple type of †vascular plant. The pteridophytes comprise four classes: the Psilosida, including the most primitive vascular plants, found mainly in the tropics; the Lycopsida, including the club mosses; the Sphenopsida, including the horsetails; and the Pteropsida, including the ferns. They do not produce seeds.

They are mainly terrestrial, non-flowering plants characterized by the presence of a vascular system; the possession of true stems, roots, and leaves; and by a marked †alternation of generations, with the sporophyte forming the dominant generation in the life cycle. The pteridophytes formed a large and dominant flora during the Carboniferous period, but many are now known only from fossils.

PTFE abbreviation for †**polytetrafluoroethene**.

P3P abbreviation for platform for privacy preferences project, technology that allows the authors of Web sites to express their privacy practices in a standard format that can be retrieved automatically and interpreted easily by a Web †browser. Privacy is one of the major problems facing users of the Web today. Online shopping requires users to submit personal and financial details over the Web, and who has this information, and what they do with it, are issues that often are not appreciated by Web users. The browser can compare the privacy policy of the Web site with that expressed by the user and alert the user if there is a conflict with their pre-defined preferences for the use of their personal information.

Ptolemaic model of the universe †geocentric model in which the Earth remained stationary as the other planets, the Sun, the Moon, and the stars orbited it on their †spheres. It was eventually replaced by the †Copernican model.

Ptolemy (c. AD 100–c. AD 170) born Claudius Ptolemaeus, Egyptian astronomer and geographer. His *Almagest* developed the theory that Earth is the center of the universe, with the Sun, Moon, and stars revolving around it. In 1543 the Polish astronomer †Copernicus proposed an alternative to the **Ptolemaic system**. Ptolemy's *Geography* was a standard source of information until the 16th century.

Ptolemy's Table Of Chords http://hypertextbook.com/eworld/chords.shtml

Demonstration of how trigonometry tables were constructed prior to the invention of infinite series. Ptolemy's methods are deciphered into modern notation, and the standard is such that secondary school students will find it challenging, but not unreadable.

Ptolemy, The Man http://seds.lpl.arizona.edu/billa/psc/theman.html

Profile of the astronomer, mathematician, and geographer. The few known biographical facts are related, together with summaries of his achievements. The text is supported by a number of diagrams and a map. There are links to a number of other Ptolemy, astronomy, and Greek geography sites.

ptomaine any of a group of toxic chemical substances (alkaloids) produced as a result of decomposition by bacterial action on proteins.

puberty stage in human development when the individual becomes sexually mature. It may occur from the age of ten upwards, but each person is individual. The sexual organs take on their adult form and pubic hair grows. In girls, menstruation begins, and the breasts develop; in boys, the voice breaks and becomes deeper, and facial hair develops. Both boys and girls will experience emotional as well as physical changes.

pubes lowest part of the front of the human trunk, the region where the external generative organs are situated. The underlying bony structure, the pubic arch, is formed by the union in the midline of the two pubic bones, which are the front portions of the hip bones. In women this is more prominent than in men, to allow more room for the passage of the child's head at birth, and it carries a pad of fat and connective tissue, the *mons veneris* (mount of Venus), for its protection.

public-domain software any computer program that is not under copyright and can therefore be used freely without charge. Much of this software has been written in US universities, under government contract. Public-domain software should not be confused with shareware, which is under copyright, and may be freely distributed for evaluation purposes, but requires purchasing to use in the longer term.

pulmonary pertaining to the †lungs.

pulsar celestial source that emits pulses of energy at regular intervals, ranging from a few seconds to a few thousandths of a second. Pulsars are thought to be rapidly rotating †neutron stars, which flash at radio and other wavelengths as they spin. They were discovered in 1967 by Jocelyn †Bell Burnell and Antony †Hewish at the Mullard Radio Astronomy Observatory, Cambridge, England. By 1998, 1,000 pulsars had been discovered since the initial identification in 1967.

Pulsars slow down as they get older, and eventually the flashes fade. Of the 500 known radio pulsars, 20 are millisecond pulsars (flashing 1,000 times a second). Such pulsars are thought to be more than a billion years old. Two pulsars, one (estimated to be 1,000 years old) in the Crab nebula and one (estimated to be 11,000 years old) in the constellation Vela, give out flashes of visible light.

pulsating nova or **recurrent nova**, †variable star, probably not a true †nova, in which the change between more and less luminous stages is extreme.

pulsating universe or **oscillating universe**, theory that the universe constantly undergoes a †Big Bang, expands, gradually slows and stops, contracts, and gradually accelerates once more to a Big Bang. Alternative theories include an ever expanding universe and the †steady-state universe.

pulse in biology, impulse transmitted by the heartbeat throughout the arterial systems of vertebrates. When the heart muscle contracts, it forces blood into the †aorta (the chief artery). Because the arteries are elastic, the sudden rise of pressure causes a throb or sudden swelling through them. The actual flow of the blood is about 60 cm/2 ft a second in humans. The average adult pulse rate is generally about 70 per minute. The pulse can be felt where an artery is near the surface, for example in the wrist or the neck.

pulse-code modulation PCM, in physics, a form of digital †modulation in which microwaves or light waves (the carrier waves) are switched on and off in pulses of varying length according to a binary code. It is a relatively simple matter to transmit data that are already in binary code, such as those used by computer, by these means. However, if an analogue audio signal is to be transmitted, it must first be converted to a **pulse-amplitude modulated** signal (PAM) by regular sampling of its amplitude. The value of the amplitude is then converted into a binary code for transmission on the carrier wave.

pumice light volcanic rock produced by the frothing action of expanding gases during the solidification of lava. It has the texture of a hard sponge and is used as an abrasive.

punctuated equilibrium model evolutionary theory developed by Niles Eldredge and US paleontologist Stephen Jay Gould in 1972 to explain discontinuities in the fossil record. It claims that evolution continues through periods of rapid change alternating with periods of relative stability (stasis), and that the appearance of new lineages is a separate process from the gradual evolution of adaptive changes within a species.

The pattern of stasis and more rapid change is now widely accepted, but the second part of the theory remains unsubstantiated.

The **turnover pulse hypothesis** of US biologist Elisabeth Vrba postulates that the periods of rapid evolutionary change are triggered by environmental changes, particularly changes in climate.

Bacterial Evolution: Score one for Punk Eek
http://www.sciam.com/article.cfm?articleID=000DFABC-A1BF-1C76-9B81809EC588EF21&pageNumber=1&catID=4

Part of a larger site maintained by Scientific American, this page reports on research conducted at Michigan State University in which scientists speeded up the evolution of bacteria in order to study the theory of "punctuated equilibrium," which seeks to explain why evolutionary changes sometimes take place with relative suddenness after eons of general stasis. The text includes hypertext links to further information, and there is also a list of links to critical analysis of punctuated equilibrium and related sites on the Web.

PCM microwave

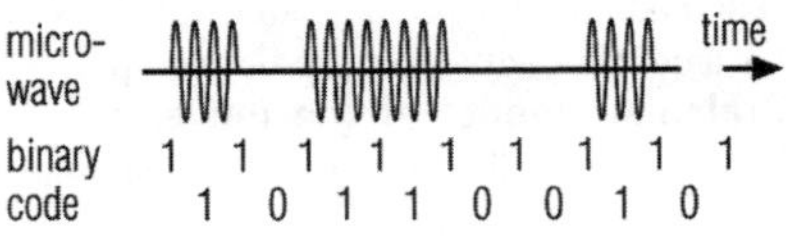

an analogue signal

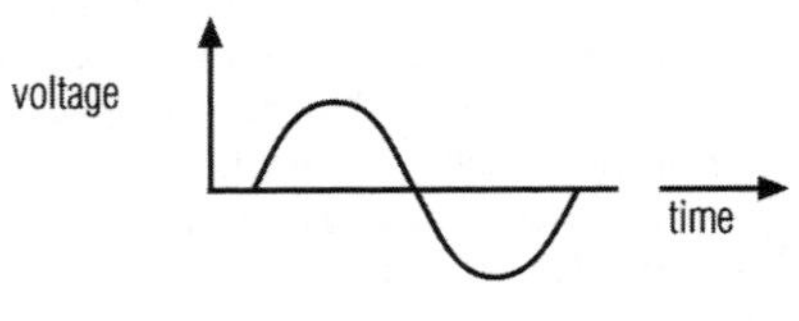

pulse-amplitude-modulated signal (PAM)

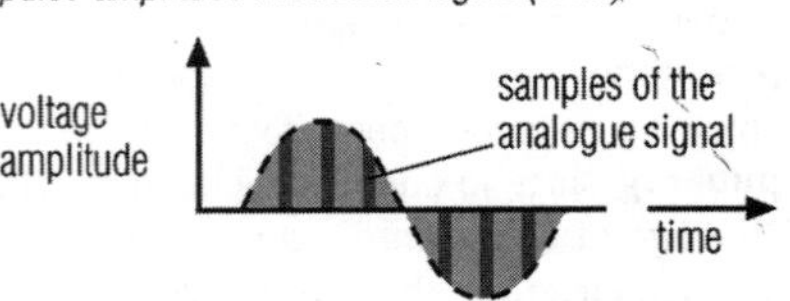

pulse-code modulation The amplitude, duration, and timing of a series of pulses are controlled in pulse-code modulation, which is relatively simple for digital data already in binary code. Analogue signals need to be converted into a recognizable binary code (a pulse-amplitude modulated signal) by regular sampling of its amplitude. Morse code is a very simple example of pulse-code modulation. Helicon

punnett square graphic technique used in genetics for determining the likely outcome, in statistical terms, of a genetic cross. It resembles a game of tic tac toe, in which the genotypes of the parental generation gametes are entered first, so that the subsequent combinations can then be calculated.

pupa nonfeeding, largely immobile stage of some insect life cycles, in which larval tissues are broken down, and adult tissues and structures are formed.

In many insects, the pupa is **exarate**, with the appendages (legs, antennae, wings) visible outside the pupal case; in butterflies and moths, it is called a chrysalis, and is **obtect**, with the appendages developing inside the case.

Insecta http://www.insecta.com/

Beautiful illustrations accompany this site which contains information about insects, drawn from the Spencer Entomological Museum at the University of British Colombia, Canada. There is a "Bug of the Month" feature and a glossary of entomological terms, and the designers hope to cover more areas soon.

pupil circular aperture in the †iris through which light falls onto the lens of the eye.

Puppis bright constellation of the southern hemisphere, representing a ship's poop or stern. It was originally regarded as part of †Argo.

Purcell, Edward Mills (1912–1997) US physicist. He was awarded, with Felix †Bloch, the Nobel Prize for Physics in 1952 for the development of the nuclear magnetic resonance (NMR) method of measuring the magnetism of atomic nuclei in solids and liquids. This method hinges on the fact that any magnetic nuclei will absorb radio-frequency radiation by a resonance effect when in a magnetic field. Measurement of the absorption provides information on the properties of the nuclei and their molecular environment. The technique is now used in chemistry as a powerful analytical tool. Purcell continued applying nuclear magnetic resonance to help develop the radio telescope.

pure breeding in genetics, crossing of organisms that "breed true," their characteristics appearing unchanged generation after generation. To breed "true" the organism must be †homozygous for the characteristic(s) in question.

putrefaction decomposition of organic matter by micro-organisms.

PVC abbreviation for †polyvinyl chloride.

P-wave abbreviation of primary wave, in seismology, a class of †seismic wave that passes through the Earth in the form of longitudinal pressure waves at speeds of 6–7 kps/3.7–4.4 mps in the crust and up to 13 kps/8 mps in deeper layers, the speed depending on the density of the rock. P-waves from an earthquake travel faster than S-waves and are the first to arrive at monitoring stations (hence primary waves). They can travel through both solid rock and the liquid outer core of the Earth.

P-waves are †longitudinal waves; that is, they travel in the same direction as the rock is moving back and forth. P-waves cause one part of the rock to compress together and another section to be pushed apart. This process is known as compression and rarefaction, similar to the propagation of sound waves.

PWR abbreviation for †**pressurized water reactor**, a type of nuclear reactor.

pyramid in geometry, a solid shape with triangular side-faces meeting at a common vertex (point) and with a †polygon as its base. The volume V of a pyramid is given by $V = {}^1/_3Bh$, where B is the area of the base and h is the perpendicular height.

Pyramids are generally classified by their bases. For example, the Egyptian pyramids have square bases, and are therefore called square pyramids. A triangular pyramid is also known as a tetrahedron ("four sides").

pyramidal peak angular mountain peak with three or more arêtes found in glaciated areas; for example, the Matterhorn in Switzerland. It is formed when three or four †corries (steep-sided hollows) are eroded, back-to-back, around the sides of a mountain, leaving an isolated peak in the middle.

pyramid of numbers in ecology, a diagram that shows quantities of plants and animals at different levels (steps) of a †food chain. This may be measured in terms of numbers (how many animals) or biomass (total mass of living matter), though in terms of showing transfer of food, biomass is a more useful measure. Where biomass is measured, the diagram is often termed a pyramid of biomass. There is always far less biomass, or fewer organisms, at the top of the chain than at the bottom, because only about 10% of the food (energy) an animal eats is turned into flesh – the rest is lost through metabolism and excretion. The amount of food flowing through the chain therefore drops with each step up the chain, supporting fewer organisms, hence giving the characteristic "pyramid" shape.

In a pyramid of biomass, the primary producers (usually plants) are represented at the bottom by a broad band, the plant-eaters are shown above by a narrower band, and the animals that prey on them by a narrower band still. At the top of the pyramid are the few "top carnivores" such as lions and sharks.

Pyrenomycetes or **flask fungi**, largest group of fungi (see †fungus) in the class Ascomycetes, containing about 640 genera and 6,000 species. Most are small and inconspicuous, but a few are larger. They are characterized by flask-shaped fruiting bodies that release their spores either by opening at the top or by decaying. Some species are parasitic on plants, others on insect larvae, and some are †saprophytes. *Claviceps purpurea* is the cause of the ergot disease of rye; *Cordyceps* attacks certain caterpillars; *Eurotium aspergillus* is the greenish mould which attacks jam.

pyrethrum popular name for several cultivated chrysanthemums. The ornamental species *C. coccineum*, and hybrids derived from it, are commonly grown in gardens. Pyrethrum powder, made from the dried flower heads of some species, is a powerful pesticide for aphids and mosquitoes. (Genus *Chrysanthemum*, family Compositae.)

pyridine C_5H_5N, heterocyclic compound (see †cyclic compounds). It is a liquid with a sickly smell and occurs in coal tar. It is soluble in water, acts as a strong †base, and is used as a solvent, mainly in the manufacture of plastics.

pyridoxine or **vitamin B$_6$**; $C_8H_{11}NO_3$, water-soluble †vitamin of the B complex. There is no clearly identifiable disease associated with deficiency but its absence from the diet can give rise to malfunction of the central nervous system and general skin disorders. Good sources are liver, meat, milk, and cereal grains. Related compounds may also show vitamin B_6 activity.

pyrite or **fool's gold**, iron sulfide FeS_2. It has a yellow metallic luster and a hardness of 6–6.5 on the Mohs scale. It is used in the production of sulfuric acid.

pyroclastic describing fragments of solidified volcanic magma, ranging in size from fine ash to large boulders, that are extruded during an explosive volcanic eruption; also the rocks that are formed by consolidation of such material. Pyroclastic rocks include tuff (ash deposit) and agglomerate (volcanic breccia).

pyroclastic deposit deposit made up of fragments of rock, ranging in size from fine ash to large boulders, ejected during an explosive volcanic eruption.

pyrogallol technical name **trihydroxybenzene**; $C_6H_3OH_3$, derivative of benzene, prepared from gallic acid. It is used in gas analysis for the measurement of oxygen because its alkaline solution turns black as it rapidly absorbs oxygen. It is also used as a developer in photography.

pyrolysis decomposition of a substance by heating it to a high temperature in the absence of air. The process is used to burn and dispose of old tires, for example, without contaminating the atmosphere.

pyrometer any instrument used for measuring high temperatures by means of the thermal radiation emitted by a hot object. In a **radiation pyrometer** the emitted radiation is detected by a sensor such as a thermocouple. In an **optical pyrometer** the color of an electrically heated filament is matched visually to that of the emitted radiation. Pyrometers are especially useful for measuring the temperature of distant, moving, or inaccessible objects.

pyrophoric finely powdered metals, or mixtures of metals and their oxides, which have a tendency to burst into flame, or oxidize when exposed to air.

pyroxene any one of a group of minerals, silicates of calcium, iron, and magnesium with a general formula X,YSi_2O_6, found in igneous and metamorphic rocks. The internal structure is based on single chains of silicon and oxygen. Diopside (X = Ca, Y = Mg) and augite (X = Ca, Y = Mg,Fe,Al) are common pyroxenes.

Jadeite ($NaAlSi_2O_6$), which is considered the more valuable form of jade, is also a pyroxene.

Pythagoras (*c.* 580–500 BC) Greek mathematician and philosopher who formulated †Pythagoras' theorem.

Pythagoras http://www.utm.edu/research/iep/p/pythagor.htm
Profile of the legendary mathematician, scientist, and philosopher. An attempt is made to disentangle the known facts of his life from the mass of legends about him. His theories of the tripartite soul, transmigration, and the cosmological limit are summarized, alongside his interest in geometry.

Pythagoras' Theorem http://www-groups.dcs.st-andrews.ac.uk/~history/Diagrams/PythagorasTheorem.gif
Simple site giving visual proof of Pythagoras's famous theorem.

Pythagoras' theorem in geometry, a theorem stating that in a right-angled triangle, the square of the †hypotenuse (the longest side) is equal to the sum of the squares of the other two sides. If the hypotenuse is h units long and the lengths of the other sides are a and b, then $h^2 = a^2 + b^2$.

The theorem provides a way of calculating the length of any side of a †right-angled triangle if the lengths of the other two sides are known. For example, to find the length of a bridge constructed over a valley 120 m wide, when the vertical drop of the bridge is 20 m:

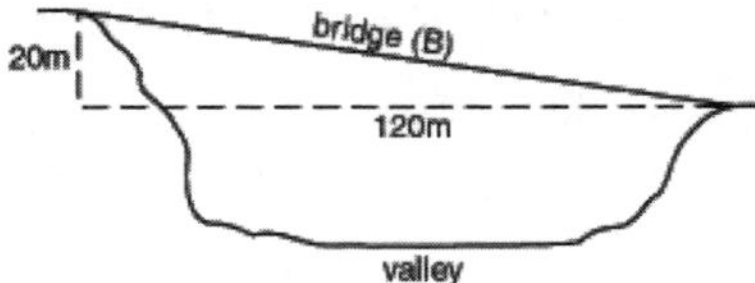

Using Pythagoras' theorem $B^2 = 120^2 + 20^2$, so $B^2 = 14{,}400 + 400 = 14{,}800$, giving $b = \sqrt{14{,}800} = 121.66$ (to two decimal places).

The length of the bridge is 121.66 m.

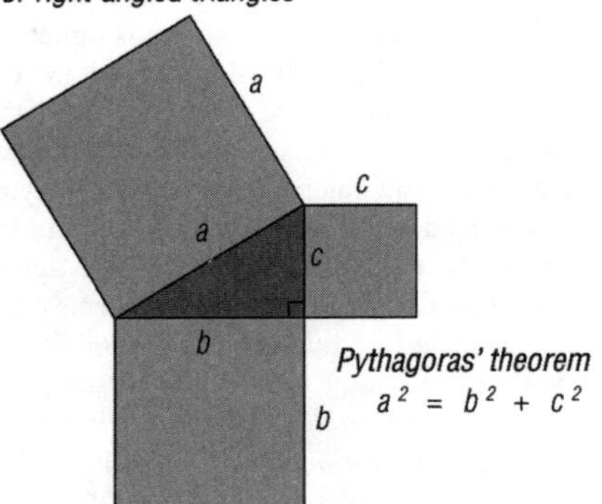

Pythagoras' theorem This states that the area of a square drawn on the longest side of a right-angled triangle (the **hypotenuse**), will be equal to the sum of the areas of the squares drawn on the other two sides. The theorem is likely to have been known long before the time of Pythagoras. It was probably used by the ancient Egyptians to lay out the pyramids. Helicon

Pythagoras' theorem is also used to determine certain †trigonometric identities such as $\sin^2\theta + \cos^2\theta = 1$.

Pyxis small constellation of the southern hemisphere, representing a ship's compass. It was originally regarded as part of †Argo.

Q

quadrant one quarter of a circle. When plotting †coordinates on a †graph, the x- and y-axes may intersect to create four quadrants. The first quadrant lies in the region where both x and y are positive.

The remaining three quadrants are numbered in an anticlockwise direction. Thus in the second quadrant, x is negative and y is positive, in the third quadrant, both are negative and in the fourth quadrant x is positive and y negative.

Quadrantids in astronomy, sharp, short-lived meteor shower whose point of origin (radiant) appears to be in the constellation Boötes, although they are named after the defunct constellation Quadrans Muralis. They occur annually for a few hours on 3 January.

quadrat in environmental studies, a square structure used to study the distribution of plants in a particular place, for instance a field, rocky shore, or mountainside. The size varies, but is usually 0.5 or 1 meter square, small enough to be carried easily. The quadrat is placed on the ground and the abundance of species estimated. By making such measurements a reliable understanding of species distribution is obtained.

quadratic equation in mathematics, a polynomial †equation of second degree (that is, an equation containing as its highest power the square of a variable, such as x^2). The general †formula of such equations is

$ax^2 + bx + c = 0$

in which a, b, and c are real numbers, and only the †coefficient a cannot equal 0.

In †coordinate geometry, a quadratic function represents a †parabola.

Some quadratic equations can be solved by factorization (see †factor (algebra)), or the values of x can be found by using the formula for the general solution

$x = [-b + \sqrt{}(b^2 - 4ac)]/2a$ or
$x = [-b - \sqrt{}(b^2 - 4ac)]/2a$

Depending on the value of the discriminant, $b^2 - 4ac$, a quadratic equation has two real, two equal, or two complex roots (solutions). When $b^2 - 4ac > 0$, there are two distinct real roots. When $b^2 - 4ac = 0$, there are two equal real roots. When $b^2 - 4ac < 0$, there are two distinct complex roots.

Quadratic, Cubic, and Quartic Equations http://www-history.mcs.st-and.ac.uk/history/HistTopics/Quadratic_etc_equations.html

History and usage of quadratic equations in mathematics. The site chronicles the discovery and development of this area of mathematics, and also provides biographical background information on those mathematicians responsible.

quadratic extension in mathematics, extension of a field to a larger field by adjoining the root of a quadratic equation, for example, the set of all p + qÖ2 with both p and q as rational numbers is a quadratic extension of the field of rationals.

quadrature position of the Moon or an outer planet where a line between it and Earth makes a right angle with a line joining Earth to the Sun.

quadrilateral plane (two-dimensional) figure with four straight sides. The sum of all †interior angles is 360°. The following are all quadrilaterals, each with distinguishing properties: †**square** with four equal angles and sides, and four axes of †symmetry; †**rectangle** with four equal angles, opposite sides equal, and two axes of symmetry; †**rhombus** with four equal sides, and two axes of symmetry; †**parallelogram** with two pairs of parallel sides, and rotational symmetry; †**kite** with two pairs of adjacent equal sides, and one 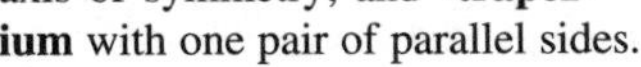
axis of symmetry; and †**trapezium** with one pair of parallel sides.

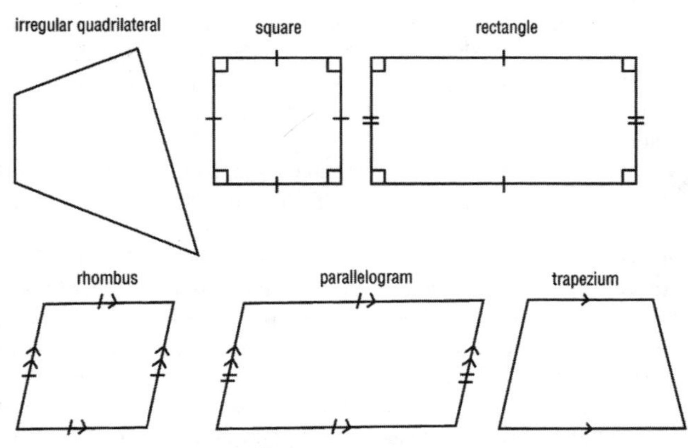

quadrilateral Different types of quadrilateral. Helicon

Types of Quadrilateral http://www.i-t-teacher.freeserve.co.uk/revision/typequadril.html
Simple diagrams and text accompany this explanation of types of quadrilaterals. The types covered are: squares, rectangles, rhombuses, parallelograms, trapeziums, and kites.

qualitative analysis in chemistry, a procedure for determining the identity of the component(s) of a single substance or mixture. A series of simple reactions and tests can be carried out on a compound to determine the elements present.

quantifier in mathematics, one of a pair of symbols ∀} (the universal quantifier) or ∃} (the existential quantifier) used to construct logical statements. The expression (∀}$x \in X$) can be interpreted as "for all x in X," while the expression (∃}$x \in X$) can be translated as "there exists some x in X."

quantitative analysis in chemistry, a procedure for determining the precise amount of a known component present in a single substance or mixture. A known amount of the substance is subjected to particular procedures.

Gravimetric analysis determines the mass of each constituent present; †**volumetric analysis** determines the concentration of a solution by †titration against a solution of known concentration.

quantum chromodynamics QCD, in physics, a theory describing the interactions of †quarks, the †elementary particles that make up all †hadrons (subatomic particles such as protons and neutrons). In quantum chromodynamics, quarks are considered to interact by exchanging particles called gluons, which carry the †strong nuclear force, and whose role is to "glue" quarks together.

The mathematics involved in the theory is complex, and, although a number of successful predictions have been made, the theory does not compare in accuracy with †quantum electrodynamics, upon which it is modeled. See †forces, fundamental.

quantum electrodynamics QED, in physics, a theory describing the interaction of charged subatomic particles within electric and magnetic fields. It combines †quantum theory and †relativity, and considers charged particles to interact by the exchange of photons. QED is remarkable for the accuracy of its predictions; for example, it has been used to calculate the value of some physical quantities to an accuracy of ten decimal places, a feat equivalent to calculating the distance between New York and Los Angeles to within the thickness of a hair. The theory was developed by US physicists Richard Feynman and Julian Schwinger and by Japanese physicist Sin-Itiro Tomonaga in 1948.

quantum mechanics branch of physics dealing with the interaction of †matter and †radiation, the structure of the †atom, the motion of atomic particles, and with related phenomena (see †elementary particle and †quantum theory).

quantum number in physics, one of a set of four numbers that uniquely characterize an †electron and its state in an †atom. The **principal quantum number** n defines the electron's main energy level. The **orbital quantum number** l relates to its angular momentum. The **magnetic quantum number** m describes the energies of electrons in a magnetic field. The **spin quantum number** m_s gives the spin direction of the electron.

The principal quantum number, defining the electron's energy level, corresponds to shells (energy levels) also known by their spectroscopic designations K, L, M, and so on. The orbital quantum number gives rise to a series of subshells designated *s*, *p*, *d*, *f*, and so on, of slightly different energy levels. The magnetic quantum number allows further subdivision of the subshells (making three subdivisions p_x, p_y, and p_z in the p subshell, for example, of the same energy level). No two electrons in an atom can have the same set of quantum numbers (the †Pauli exclusion principle).

quantum physics branch of physics that takes into account the quantum nature of matter, energy, and radiation.

quantum simulation use of †computer simulation to study the quantum behavior of matter. Simulations based on just 50 or 60 atoms are able to mimic the properties of real matter. If a small group of atoms is surrounded by other identical groups every atom behaves as if it were part of a larger material. This makes it is possible, for example, to simulate the behavior of solids and liquids at high temperatures, and to study chemical reactions at surfaces to aid in the design of better catalysts.

quantum theory or **quantum mechanics**, in physics, the theory that †energy does not have a continuous range of values, but is, instead, absorbed or radiated discontinuously, in multiples of definite, indivisible units called quanta. Just as earlier theory showed how light, generally seen as a wave motion, could also in some ways be seen as composed of discrete particles (†photons), quantum theory shows how atomic particles such as electrons may also be seen as having wavelike properties. Quantum theory is the basis of particle physics, modern theoretical chemistry, and the solid-state physics that describes the behavior of the silicon chips used in computers.

The theory began with the work of Max Planck in 1900 on radiated energy, and was extended by Albert

Einstein to electromagnetic radiation generally, including light. Danish physicist Niels Bohr used it to explain the †spectrum of light emitted by excited hydrogen atoms. Later work by Erwin Schrödinger, Werner Heisenberg, Paul Dirac, and others elaborated the theory to what is called quantum mechanics (or wave mechanics).

Quantum Age Begins http://www-history.mcs.st-and.ac.uk/history/HistTopics/The_Quantum_age_begins.html
St Andrews University-run Web site chronicling the discovery of quantum theory. Biographical details of the mathematicians and physicists involved are also provided. The site also includes links to many other history of mathematics-related Web resources.

quark in physics, the †elementary particle that is the fundamental constituent of all †hadrons (subatomic particles that experience the strong nuclear force and divided into baryons, such as neutrons and protons, and mesons). Quarks have electric charges that are fractions of the electronic charge ($+^2/_3$ or $-^1/_3$ of the electronic charge). There are six types, or "flavors": up, down, top, bottom, strange, and charmed, each of which has three varieties, or "colors": red, green, and blue (visual color is not meant, although the analogy is useful in many ways). To each quark there is an antiparticle, called an antiquark. See †quantum chromodynamics (QCD).

Hunting of the Quark http://researchmag.asu.edu/stories/quark.html
Commentary on the research carried out at Arizona State University, USA, into sub-atomic particles. There is also a subsidiary page on the complexities of the subatomic world.

quart imperial liquid or dry measure, equal to two pints or 1.136 liters. In the USA, a liquid quart is equal to 0.946 liter, while a dry quart is equal to 1.101 liters.

quartic to the power of four; it is occasionally replaced by the word "biquadratic."

quartile in statistics, any one of the three values that divide data into four equal parts. They comprise the **lower quartile**, below which lies the lowest 25% of the data; the †**median**, which is the middle 50%, half way through the data; and the **upper quartile**, above which lies the top 25%. The difference of value between the upper and lower quartiles is known as the †interquartile range, which is a useful measure of the dispersion of a statistical distribution because it is not affected by freak extreme values (see †range). These values are usually found using a †cumulative frequency diagram.

quartz crystalline form of †silica SiO_2, one of the most abundant minerals of the Earth's crust (12% by volume). Quartz occurs in many different kinds of rock, including sandstone and granite. It ranks 7 on the Mohs scale of hardness and is resistant to chemical or mechanical breakdown. Quartzes vary according to the size and purity of their crystals. Crystals of pure quartz are coarse, colorless, transparent, show no cleavage, and fracture unevenly; this form is usually called rock crystal. Impure colored varieties, often used as gemstones, include †agate, citrine quartz, and †amethyst. Quartz is also used as a general name for the cryptocrystalline and noncrystalline varieties of silica, such as chalcedony, chert, and opal. Quartz is used in ornamental work and industry, where its reaction to electricity makes it valuable in electronic instruments (see †piezoelectric effect). Quartz can also be made synthetically.

Crystals that would take millions of years to form naturally can now be "grown" in pressure vessels to a standard that allows them to be used in optical and scientific instruments and in electronics, such as quartz wristwatches.

quartzite †metamorphic rock consisting of pure quartz sandstone that has recrystallized under increasing heat and pressure.

quartzite in sedimentology, an unmetamorphosed sandstone composed chiefly of quartz grains held together by silica that was precipitated after the original sand was deposited.

quasar contraction of **quasi-stellar object**; or **QSO**, one of the most distant extragalactic objects known, discovered in 1963. Quasars appear starlike, but each emits more energy than 100 giant galaxies. They are thought to be at the center of galaxies, their brilliance emanating from the stars and gas falling towards an immense †black hole at their nucleus. The Hubble Space Telescope revealed in 1994 that quasars exist in a remarkable variety of galaxies.

Quasar light shows a large †red shift, indicating that the quasars are very distant. The furthest are over 10 billion light years away. A few quasars emit radio waves (see †radio astronomy), which is how they were first identified.

quasi-atom particle assemblage resembling an atom, in which particles not normally found in atoms become bound together for a brief period.

Quasi-atoms are generally unstable structures, either because they are subject to matter–antimatter annihilation (positronium), or because one or more of their constituents is unstable (muonium).

quaternary ammonium ion ion in which the hydrogen atoms of the normal ammonium ion $(NH_4)^+$ have been replaced by organic †alkyl or †aryl radicals; it therefore has the formula $(NR_4)^+$.

Quaternary Period period of geological time from 1.64 million years ago through to the present. It is divided into the †Pleistocene (1.64 million to 10,000 years ago) and †Holocene (last 10,000 years) epochs.

quaternion one of an extended system of †complex numbers, representable in the generalized form:

$a + bi + cj + dk$

where a, b, c and d are real numbers, and in which i, j, and k are additional objects that multiply according to specific rules such that:

$i^2 = j^2 = k^2 = -1$, $ij = -ji = k$, $jk = -kj = i$, and $ki = -ik = j$.

questionnaire question list submitted to a certain number of people whose answers are then used to compile quantitative or qualitative statistical information. Questionnaires are a means of obtaining †primary data and are useful in market research, for example.

Quetelet, (Lambert) Adolphe (Jacques) (1796–1874) Belgian statistician. He developed tests for the validity of statistical information, and gathered and analyzed statistical data of many kinds. From his work on sociological data came the concept of the "average person."

Quetelet, Lambert Adolphe Jacques http://www-groups.dcs.st-andrews.ac.uk/history/Mathematicians/Quetelet.html

Part of an archive containing the biographies of the world's greatest mathematicians, this site is devoted to the life and contributions of statistician Lambert Quetelet. In addition to biographical information, you will find a list of references about Quetelet and links to other essays in the archive that reference him. The text of this essay includes hypertext links to the essays of those mathematicians and thinkers who influenced Quetelet. There is also find an image of him, which can be enlarged, and a map of his birthplace.

queuing theory mathematical tool used for studying and predicting delays in networks. Its applications include modeling human behavior, for example, on motorways or in amusement-park queues, and optimizing the flow of goods through production lines and the flow of input data in telecommunications and computing systems.

Queuing theory analyses situations in which "customers" enter a system and then wait for a particular "service." Formulae bring together variables such as the number of customers in the system, the number in the queue, their arrival rate, and the time spent by customers in the system and in the queue, and can be used to weigh the cost of providing an improved service against the cost of waiting or of losing customers.

quicksilver another name for the element †mercury.

quinine antimalarial drug extracted from the bark of the cinchona tree. Peruvian Indians taught French missionaries how to use the bark in 1630, but quinine was not isolated until 1820. It is a bitter alkaloid, with the formula $C_{20}H_{24}N_2O_2$.

Other antimalarial drugs have since been developed with fewer side effects, but quinine derivatives are still valuable in the treatment of unusually resistant strains.

quintic to the power of five.

quotient result of dividing one number or variable into another.

R

Rabi, Isidor Isaac (1898–1988) Russian-born US physicist who was awarded the Nobel Prize for Physics in 1944 for the development of techniques to measure accurately the strength of the weak magnetic fields generated when charged elementary particles, such as the electron, spin about their axes.

raceme in botany, a type of †inflorescence.

racemic mixture mixture of equal quantities of two enantiomorphs (isomers with mirror-image molecular structures). Because the †optical activity of each component exactly cancels that of the other, the racemic mixture as a whole is optically inactive.

rad unit of absorbed radiation dose, now replaced in the SI system by the †gray (one rad equals 0.01 gray). It is defined as the dose when one kilogram of matter absorbs 0.01 joule of radiation energy.

radar acronym for **radio direction and ranging**, device for locating objects in space, direction finding, and navigation by means of transmitted and reflected high-frequency radio waves.

The direction of an object is ascertained by transmitting a beam of short-wavelength (1–100 cm/0.5–40 in), short-pulse radio waves, and picking up the reflected beam. Distance is determined by timing the journey of the radio waves (traveling at the speed of light) to the object and back again. Radar is also used to detect objects underground, for example service pipes, and in archaeology. Contours of remains of ancient buildings can be detected down to 20 m/66 ft below ground. Radar is essential to navigation in darkness, cloud, and fog, and is widely used in warfare to detect enemy aircraft and missiles. To avoid detection, various devices, such as modified shapes (to reduce their radar cross-section), radar-absorbent paints, and electronic jamming are used. To pinpoint small targets †laser "radar," instead of microwaves, has been developed. Developed independently in Britain, France, Germany, and the USA in the 1930s, it was first put to practical use for aircraft detection by the British, who had a complete coastal chain of radar stations installed by September 1938. It proved invaluable in the Battle of Britain in 1940, when the ability to spot incoming German aircraft did away with the need to fly standing patrols. Chains of ground radar stations are used to warn of enemy attack – for example, North Warning System 1985, consisting of 52 stations across the Canadian Arctic and northern Alaska. Radar is also used in †meteorology and †astronomy.

History of Radar http://www.fi.edu/weather/radar/history.html

A brief history of the development of radar, which gives credit to British researchers for their role in the winning of the Battle of Britain.

radar astronomy bouncing of radio waves off objects in the Solar System, with reception and analysis of the "echoes." Radar contact with the Moon was first made in 1945 and with Venus in 1961. The travel time for radio reflections allows the distances of objects to be determined accurately. Analysis of the reflected beam reveals the rotation period and allows the object's surface to be mapped topographically. The rotation periods of Venus and Mercury were first determined by radar. Radar maps of Venus were obtained first by Earth-based radar and subsequently by orbiting space probes.

radial artery artery that passes down the forearm and supplies blood to the hand and the fingers. The brachial artery, a large artery supplying blood to the arm, divides at the elbow to form the radial and ulnar arteries. The pulsation of blood through the radial artery can be felt at the wrist. This is generally known as the pulse.

radial nerve in biology, the nerve in the upper arm. Nervous impulses to regulate the function of the muscles which extend the arm, the wrist, and some fingers pass along these nerves. They also relay sensation to parts of the arm and hand. The radial nerve arises from the brachial plexus (network of nerves supplying the arm) in the armpit and descends the upper arm before dividing into the superficial radial and interosseous nerves.

radial pulsation periodic expansion and contraction of a star that may be merely an optical effect of recession.

radial symmetry in biology, type of shape and internal organization of an organism's body where there is no front or rear and its parts are arranged evenly "in the round." A section of the body across any diameter results in halves that are mirror images of each other.

radial velocity in astronomy, velocity of an object, such as a star or galaxy, along the line of sight, moving towards or away from an observer. The amount of †Doppler shift (apparent change in wavelength) of the light reveals the object's velocity. If the object is approaching, the Doppler effect causes a †blue shift in its light. That is, the wavelengths of light coming from the object appear to be shorter, tending toward the blue end of the †spectrum. If the object is receding, there is a †red shift, meaning the wavelengths appear to be longer, toward the red end of the spectrum.

radian symbol rad, SI unit of plane angles, an alternative unit to the †degree. It is the angle at the center of a circle when the center is joined to the two ends of an arc (part of the circumference) equal in length to the radius of the circle. There are 2p (approximately 6.284) radians in a full circle (360°).

One radian is approximately 57°, and 1° is p/180 or approximately 0.0175 radians. Radians are commonly used to specify angles in †polar coordinates.

radiant in astronomy, the point in the sky from which a †meteor shower appears to emanate.

radiant heat energy that is radiated by all warm or hot bodies. It belongs to the †infrared part of the †electromagnetic spectrum and causes heating when absorbed. Radiant heat is invisible and should not be confused with the red glow associated with very hot objects, which belongs to the visible part of the spectrum. Infrared radiation can travel through a vacuum and it is in this form that the radiant heat of the Sun travels through space. It is the trapping of this radiation by carbon dioxide and water vapor in the atmosphere that gives rise to the †greenhouse effect.

radiation emission of radiant †energy as particles or waves – for example, heat, light, alpha particles, and beta particles (see †electromagnetic waves and †radioactivity). See also †atomic radiation.

All hot objects radiate heat. Radiated heat does not need a medium through which to travel (it can travel in a vacuum). Most of the energy received on Earth arrives by radiation of †heat energy from the Sun. Of the radiation given off by the Sun, only a tiny fraction of it, called †insolation, reaches the Earth's surface; much of it (for example, radio waves, ultraviolet rays, and X-rays) is absorbed and scattered as it passes through the †atmosphere. Visible light and infrared rays pass through the atmosphere, the infrared rays causing a rise in temperature. The radiation given off by the Earth itself is called **ground radiation**.

How much a surface radiates heat depends on its temperature and the type of surface. Dull black surfaces absorb more heat and therefore radiate more heat that polished shiny surfaces, which reflect heat and are therefore poor radiators of heat. For example, engine-cooling mantles in cars are black so that they radiate heat from the engine. A vacuum flask has a polished, silvery surface so as to keep hot liquids hot and cold liquids cold as no heat is radiated or absorbed.

The human body radiates heat at a rate of 100 joules every second. This is the same energy as radiated by a 100-watt light bulb.

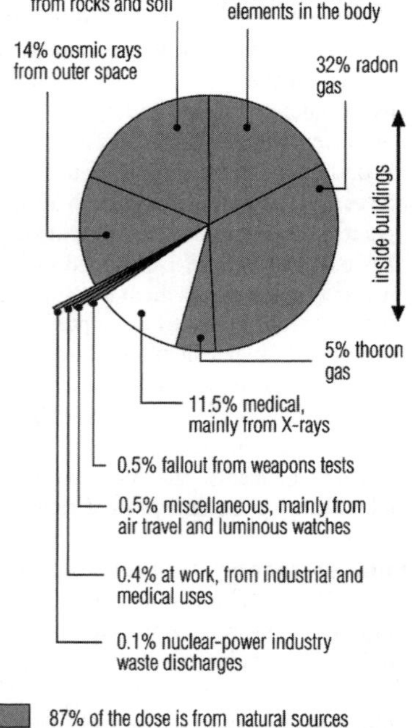

radiation Pie chart showing the various sources of radiation in the environment. Most radiation is from natural sources, such as radioactive minerals, but 13% comes from the by-products of human activities. Helicon

Radiation Reassessed http://whyfiles.org/020radiation/index.html
Part of the Why Files project published by the National Institute for Science Education (NISE) and funded by the National Science Foundation, this page provides insight into the controversy concerning the health effects of ionizing radiation. The highly readable text, laid out over 12 pages, includes information about what radiation studies from Hiroshima, Nagasaki, and Chernobyl have taught us, and what scientists are saying about exposure to small amounts of radiation. Is it always harmful? Numerous images and diagrams enhance the text throughout. The more tricky terms are linked to a glossary, and there is a comprehensive bibliography of sources for further research.

radiation biology study of how living things are affected by radioactive (ionizing) emissions (see †radioactivity) and by electromagnetic (nonionizing) radiation (†electromagnetic waves). Both are potentially harmful and can cause mutations as well as leukemia and other cancers; even low levels of radioactivity are very dangerous. Both, however, are used therapeutically, for example to treat cancer, when the radiation dose is very carefully controlled (†radiotherapy or X-ray therapy).

Radioactive emissions are more harmful. Exposure to high levels produces radiation burns and radiation sickness, plus genetic damage (resulting in birth defects) and cancers in the longer term. Exposure to low-level ionizing radiation can also cause genetic damage and cancers, particularly leukemia.

Electromagnetic radiation is usually harmful only if exposure is to high-energy emissions, for example close to powerful radio transmitters or near radar-wave sources. Such exposure can cause organ damage, cataracts, loss of hearing, leukemia and other cancers, or premature aging. It may also affect the nervous system and brain, distorting their electrical nerve signals and leading to depression, disorientation, headaches, and other symptoms. Individual sensitivity varies and some people are affected by electrical equipment, such as televisions, computers, and refrigerators.

Background radiation is the natural radiation produced by cosmic rays and radioactive rocks such as granite, and this must be taken into account when calculating the effects of nuclear accidents or contamination from power stations.

radiation monitoring system network of monitors to detect any rise in background gamma radiation and to warn of a major nuclear accident within minutes of its occurrence. The accident at Chernobyl in Ukraine in 1986 prompted several Western European countries to begin installation of such systems locally, and in 1994 work began on a pilot system to provide a **gamma curtain**, a dense net of radiation monitors, throughout Eastern and Western Europe.

radiation units units of measurement for radioactivity and radiation doses. In SI units, the activity of a radioactive source is measured in becquerels (symbol Bq), where one becquerel is equal to one nuclear disintegration per second (an older unit is the curie). The exposure is measured in coulombs per kilogram ($C\ kg^{-1}$); the amount of ionizing radiation (X-rays or gamma rays) that produces one coulomb of charge in one kilogram of dry air (replacing the roentgen). The absorbed dose of ionizing radiation is measured in grays (symbol Gy) where one gray is equal to one joule of energy being imparted to one kilogram of matter (the rad is the previously used unit). The dose equivalent, which is a measure of the effects of radiation on living organisms, is the absorbed dose multiplied by a suitable factor that depends upon the type of radiation. It is measured in sieverts (symbol Sv), where one sievert is a dose equivalent of one joule per kilogram (an older unit is the rem).

radiative equilibrium in a star, represents an even process by which energy (heat) is transferred from the core to the outer surface without affecting the overall stability of the star.

radical in mathematics, operation principle in the extraction of roots; that is, square roots, cube roots, and so on.

radical in chemistry, a group of atoms forming part of a molecule, which acts as a unit and takes part in chemical reactions without disintegration, yet often cannot exist alone for any length of time; for example, the methyl radical $—CH_3$, or the carboxyl radical —COOH.

radicle part of a plant embryo that develops into the primary root. Usually it emerges from the seed before the embryonic shoot, or †plumule, its tip protected by a root cap, or calyptra, as it pushes through the soil. The radicle may form the basis of the entire root system, or it may be replaced by adventitious roots (positioned on the stem).

radio transmission and reception of radio waves. In radio transmission a microphone converts †sound waves (pressure variations in the air) into a varying electric current, which is amplified and used to modulate a carrier wave which is transmitted as †electromagnetic waves, which are then picked up by a receiving aerial, amplified, and fed to a loudspeaker, which converts them back into sound waves.

The theory of electromagnetic waves was first developed by Scottish physicist James Clerk †Maxwell in 1864, given practical confirmation in the laboratory in 1888 by German physicist Heinrich †Hertz, and put to practical use by Italian inventor Guglielmo †Marconi, who in 1901 achieved reception of a signal in Newfoundland, Canada, transmitted from Cornwall, England.

To carry the transmitted electrical signal, an †oscillator produces a carrier wave of high frequency; different stations are allocated different transmitting carrier frequencies. A modulator superimposes the audio-frequency signal on the carrier. There are two main ways of doing this: †**amplitude modulation** (AM), used for long- and medium-wave broadcasts, in which the strength of the carrier is made to fluctuate in time with the audio signal; and †**frequency modulation** (FM), as used for VHF broadcasts, in which the frequency of the carrier is made to fluctuate. The transmitting aerial emits the modulated electromagnetic waves, which travel outwards from it.

radioactive decay process of disintegration undergone by the nuclei of radioactive elements, such as radium and various isotopes of uranium and the transuranic elements, in order to produce a more stable nucleus. The three most common forms of radioactive decay are alpha, beta, and gamma decay.

In **alpha decay** (the loss of a helium nucleus – two protons and two neutrons) the atomic number decreases by two and a new nucleus is formed, for example, an atom of uranium isotope of mass 238, on emitting an alpha particle, becomes an atom of thorium, mass 234. In **beta decay** the loss of an electron from an atom is accomplished by the transformation of a neutron into a proton, thus resulting in an increase in the atomic number of one. For example, the decay of the carbon-14 isotope results in the formation of an atom of nitrogen (mass 14, atomic number 7) and the emission of a high-energy electron.

Gamma emission usually occurs as part of alpha or beta emission. In gamma emission high-speed electromagnetic radiation is emitted from the nucleus, making it more stable during the loss of an alpha or beta particle. Certain lighter, artificially-created isotopes also undergo radioactive decay. The associated radiation consists of alpha rays, beta rays, or gamma rays (or a combination of these), and it takes place at a constant rate expressed as a specific half-life, which is the time taken for half of any mass of that particular isotope to decay completely. Less commonly-occurring decay forms include heavy-ion emission, electron capture, and spontaneous fission (in each of these the atomic number decreases). The original nuclide is known as the parent substance, and the product is a daughter nuclide (which may or may not be radioactive). The final product in all modes of decay is a stable element.

radioactive fallout airborne radioactive material resulting from a natural phenomenon or from an artificial occurrence, such as the explosion of a nuclear bomb.

radioactive tracer any of various radioactive †isotopes added to fluids in order to monitor their flow and therefore identify leaks or blockages. See †tracer.

radioactive waste any waste that emits radiation in excess of the background level. See †nuclear waste.

radioactivity spontaneous change of the nuclei of atoms, accompanied by the emission of radiation. Such atoms are called radioactive. It is the property exhibited by the radioactive †isotopes of stable elements and all isotopes of radioactive elements, and can be either natural or induced. See †radioactive decay.

A radioactive material decays by releasing radiation, and transforms into a new substance. The energy is released in the form of †alpha particles and †beta particles or in the form of high-energy electromagnetic waves known as †gamma radiation. Natural radioactive elements are those with an atomic number of 83 and higher. Artificial radioactive elements can also be formed.

Devices such as the Geiger–Muller tube, a photographic plate, or an electroscope can detect radioactivity. An electronic counter on the Geiger–Muller instrument displays a digital reading of the amount of radiation detected.

Decay: Radioactivity http://particleadventure.org/particleadventure/frameless/decay_start.html

Beginning with the history of radioactivity, this page details the development of knowledge about radioactivity, and its impact on the understanding of physics. Further topics relating to decay are available from a menu.

Radiation and Radioactivity http://www.physics.isu.edu/radinf/cover.htm

Lesson on radioactivity. It begins with a definition of radioactivity, and then covers the atom and types of radiation including alpha, beta, and gamma radiation and X-rays, plus the properties of each. A well-presented lesson with clear, helpful diagrams.

Radioactivity in Nature http://www.physics.isu.edu/radinf/natural.htm

Detailed explanation of the different types of radiation found naturally on Earth and in its atmosphere, as well as those produced by humans. It includes tables of the breakdown of nuclides commonly found in soil, the oceans, the air, and even the human body.

Uses of Radiation http://theory.uwinnipeg.ca/mod_tech/node184.html

Short introduction explaining how radiation can be dangerous. Two links are provided detailing medical and non-medical uses of radiation.

radio astronomy study of radio waves emitted naturally by objects in space, by means of a †radio telescope. Radio emission comes from hot gases (**thermal radiation**); electrons spiraling in magnetic fields (**synchrotron radiation**); and specific wavelengths (**lines**) emitted by atoms and molecules in space, such as the 21-cm/8.3-in line emitted by hydrogen gas.

Radio astronomy began in 1932 when US astronomer Karl †Jansky detected radio waves from the center of our Galaxy, but the subject did not develop until after World War II. Radio astronomy has greatly improved our understanding of the evolution of stars, the structure of galaxies, and the origin of the universe. Astronomers have mapped the spiral structure of the Milky Way from the radio waves given out by interstellar gas, and they have detected many individual radio sources within our Galaxy and beyond. Among radio sources in our Galaxy are the remains of †supernova explosions, such as the †Crab nebula and †pulsars. Short-wavelength radio waves have been detected from complex molecules in dense clouds of gas where stars are forming. Searches have been undertaken for signals from other civilizations in the Galaxy, so far without success.

Strong sources of radio waves beyond our Galaxy include †radio galaxies and †quasars. Their existence far off in the universe demonstrates how the universe has evolved with time. Radio astronomers have also detected weak cosmic background radiation, thought to be from the †Big Bang explosion that marked the birth of the universe.

Mullard Radio Astronomy Observatory http://www.mrao.cam.ac.uk/telescopes/index.html

Cambridge University's radio observatory site, which includes a description of current research interests, technical details of the four telescopes, details of the observatory's publications, and links to individual scientists.

radiocarbon cycle production and recycling of the radioisotope carbon-14 (^{14}C). The radioisotope occurs when a neutron flux, caused by cosmic radiation bombarding the upper atmosphere, reacts efficiently with nitrogen present. Carbon-14 intake by living organisms eventually returns to the atmosphere when dead vegetation or animal flesh decomposes, except when it is locked in preserved organic artifacts and remains. Radioactive decay occurs, forming the basis of the †radiocarbon dating method.

Initially the concentration of carbon-14 is nonuniform (levels being higher over polar regions where the Earth's magnetic field is least effective in deflecting cosmic radiation) but air currents at about 10 km/6 mi soon redistribute the newly formed carbon-14 as part of †carbon dioxide gas. Around 7.5 kg/16 lb of carbon-14 is added to the Earth's carbon reservoir each year and distributed throughout the oceans, the biosphere, and the atmosphere, although variations in the magnetic field and sunspot activity can alter the intensity of cosmic radiation, affecting carbon-14 production.

radiocarbon dating or **carbon dating**, method of dating organic materials (for example, bone or wood), used in archaeology. Plants take up carbon dioxide gas from the atmosphere and incorporate it into their tissues, and some of that carbon dioxide contains the radioactive isotope carbon-14 (see †radiocarbon cycle). As this decays at a known rate (half of it decays every 5,730 years), the time elapsed since the plant died can be measured in a laboratory. Animals take carbon-14 into their bodies from eating plant tissues and their remains can be similarly dated. After 120,000 years so little carbon-14 is left that no measure is possible (see †half-life).

Radiocarbon dating was first developed in 1949 by the US chemist Willard †Libby. The method yields reliable ages back to about 50,000 years, but its results require correction since Libby's assumption that the concentration of carbon-14 in the atmosphere was constant through time has subsequently been proved wrong. Discrepancies were noted between carbon-14 dates for Egyptian tomb artifacts and construction dates recorded in early local texts. Radiocarbon dates from tree rings (see †dendrochronology) showed that material before 1000 BC had been exposed to greater concentrations of carbon-14. Now radiocarbon dates are calibrated against calendar dates obtained from tree rings, or, for earlier periods, against uranium/thorium dates obtained from coral. The carbon-14 content is determined by counting beta particles with either a proportional gas or a liquid scintillation counter for a period of time. A new advance, accelerator mass spectrometry, requires only tiny samples and counts the atoms of carbon-14 directly, disregarding their decay.

radiochemistry chemical study of radioactive isotopes and their compounds (whether produced from naturally radioactive or irradiated materials) and their use in the study of other chemical processes.

When such isotopes are used in labeled compounds, they enable the biochemical and physiological functioning of parts of the living body to be observed. They can help in the testing of new drugs, showing where the drug goes in the body and how long it stays there. They are also useful in diagnosis – for example cancer, fetal abnormalities, and heart disease.

radio frequencies and wavelengths see †electromagnetic waves.

radio galaxy galaxy that is a strong source of electromagnetic waves of radio wavelengths. All galaxies, including our own, emit some radio waves, but radio galaxies are up to a million times more powerful. In many cases the strongest radio emission comes not from the visible galaxy but from two clouds, invisible through an optical telescope, that can extend for millions of light years either side of the galaxy. This double structure at radio wavelengths is also shown by some †quasars, suggesting a close relationship between the two types of object. In both cases, the source of energy is thought to be a massive black hole at the center. Some radio galaxies are thought to result from two galaxies in collision or recently merged.

radiography branch of science concerned with the use of radiation (particularly †X-rays) to produce images on photographic film or fluorescent screens. X-rays penetrate matter according to its nature, density, and thickness. In doing so they can cast shadows on photographic film, producing a radiograph. Radiography is widely used in medicine for examining bones and tissues and in industry for examining solid materials; for example, to check welded seams in pipelines.

radio interferometer type of †radio telescope that relies on the use of two or more aerials at a distance from each other to provide a combination of signals from one source which can be analyzed by computer. Such an analysis results in a resolution that is considerably better than that of a parabolic dish aerial by itself because of the greater effective diameter.

radioisotope or **radioactive isotope**, naturally occurring or synthetic radioactive form of an element. Most radioisotopes are made by bombarding a stable element with neutrons in the core of a nuclear reactor (see †fission). The radiations given off by radioisotopes are easy to detect (hence their use as †tracers), can in some instances penetrate substantial thicknesses of materials, and have profound effects (such as genetic †mutation) on living matter.

Most natural isotopes of relative atomic mass below 208 are not radioactive. Those from 210 and up are all radioactive.

uses Radioisotopes have many uses in medicine, for example in †radiotherapy and †radioisotope scanning. The use of radioactive isotopes in the diagnosis, investigation, and treatment of disease is called **nuclear medicine**.

radioisotope decay The nucleus of a radioisotope is unstable and undergoes changes by breaking down into a more stable form. A radioisotope decays over a period of time into a new element. As it decays it emits radiation energy in the form of alpha and beta particles and gamma radiation. The time taken for half of the original atoms to decay is known as the half-life. The product of the radioactive decay is called a "daughter" atom.

radioisotope scanning use of radioactive materials (radioisotopes or radionucleides) to pinpoint disease. It reveals the size and shape of the target organ and whether any part of it is failing to take up radioactive material, usually an indication of disease.

The specialty known as nuclear medicine makes use of the affinity of different chemical elements for certain parts of the body. Iodine, for instance, always makes its way to the thyroid gland. After being made radioactive, these materials can be given by mouth or injected, and then traced on scanners working on the Geiger-counter principle. The diagnostic record gained from radioisotope scanning is known as a **scintigram**.

radiometer device for the detection (and also measurement) of radiant †electromagnetic radiation.

radiometric dating method of dating rock by assessing the amount of †radioactive decay of naturally occurring †isotopes. The dating of rocks may be based on the gradual decay of uranium into lead. The ratio of the amounts of "parent" to "daughter" isotopes in a sample gives a measure of the time it has been decaying, that is, of its age. Different elements and isotopes are used depending on the isotopes present and the age of the rocks to be dated. Once-living matter can often be dated by †radiocarbon dating, employing the half-life of the isotope carbon-14, which is naturally present in organic tissue.

Radiometric methods have been applied to the decay of long-lived isotopes, such as potassium-40, rubidium-87, thorium-232, and uranium-238, which are found in rocks. These isotopes decay very slowly and this has enabled rocks as old as 3,800 million years to be dated accurately. Carbon dating can be used for material between 1,000 and 100,000 years old. **Potassium** dating is used for material more than 100,000 years old, **rubidium** for rocks more than 10 million years old, and **uranium** and **thorium** dating is suitable for rocks older than 20 million years.

radiomicrometer extremely sensitive detector of infrared radiation.

radio scintillation †scintillation in received radio emission; the equivalent of "twinkling" in visible light from the stars.

radiosonde balloon carrying a compact package of meteorological instruments and a radio transmitter, used to "sound," or measure, conditions in the atmosphere. The instruments measure temperature, pressure, and humidity, and the information gathered is transmitted back to observers on the ground. A radar target is often attached, allowing the balloon to be tracked.

radio telescope instrument for detecting radio waves from the universe in †radio astronomy. Radio telescopes usually consist of a metal bowl that collects and focuses radio waves the way a concave mirror collects and focuses light waves. Radio telescopes are much larger than optical telescopes, because the wavelengths they are detecting are much longer than the wavelength of light. The largest single dish is at Arecibo Observatory, Puerto Rico.

A large dish such as that at Jodrell Bank, Cheshire, England, can see the radio sky less clearly than a small optical telescope sees the visible sky. **Interferometry** is a technique in which the output from two dishes is combined to give better resolution of detail than with a single dish. **Very long baseline interferometry** (VBLI) uses radio telescopes spread across the world to resolve minute details of radio sources. The deep-space network (DSN) works in this way to track artificial †satellites.

In **aperture synthesis**, several dishes are linked together to simulate the performance of a very large single dish. This technique was pioneered by English radio astronomer Martin †Ryle at the Mullard Radio Astronomy Observatory, Cambridge, England, site of a radio telescope consisting of eight dishes in a 5-km/3-mi line. The †Very Large Array in New Mexico consists of 27 dishes arranged in a Y-shape, which simulates the performance of a single dish 27 km/17 mi in diameter. Other radio telescopes are shaped like long troughs, and some consist of simple rod-shaped aerials.

National Radio Astronomy Observatory http://www.nrao.edu/

News of the research activities of the National Radio Astronomy Observatory. This large Web site has pages about principles and techniques in radio astronomy and an impressive image gallery.

radiotherapy treatment of disease by †radiation from X-ray machines or radioactive sources. Radiation in the correct dosage can be used to kill cancerous cells and prevent their spreading.

Generally speaking, the rays of a diagnostic X-ray machine are not penetrating enough to be efficient in treatment, so for this purpose more powerful machines are required, operating at from 10,000 volts to over 30 million volts. The lower-voltage machines are similar to conventional X-ray machines. The higher-voltage ones may be of special design; for example, linear accelerators and betatrons. Modern radiotherapy is associated with fewer side effects than formerly, but radiotherapy to the head can cause temporary hair loss, and if the treatment involves the intestine, diarrhea and vomiting may occur. Much radiation now comes from synthesized †radioisotopes. Radioactive cobalt is the most useful, because it produces gamma rays, which are highly penetrating, and it is used instead of very high-energy X-rays.

radio wave electromagnetic wave possessing a long wavelength (ranging from about 10^{-3} to 10^4 m) and a low frequency (from about 10^5 to 10^{11} Hz) that travels at the speed of light. Included in the radio wave part of the spectrum are: †microwaves, used for both communications and for cooking; ultra high- and very high-frequency waves, used for television and FM (†frequency modulation) radio communications; and short, medium, and long waves, used for AM (†amplitude modulation) radio communications. Radio waves that are used for communications have all been modulated (see †modulation) to carry information. Certain astronomical objects emit radio waves, which may be detected and studied using †radio telescopes.

There is a layer in the atmosphere, called the ionosphere, where gas molecules are separated from their electrons by radiation from the Sun. When radio waves reach the ionosphere, they produce movements or oscillations of the electrons. As the electrons oscillate they produce electromagnetic waves that are identical to the radio waves with which the electrons were stimulated. The radio waves are reflected back and can be detected by a receiver. It is important to note that the reflected radio waves come from the oscillating electrons in the ionosphere.

radium chemical symbol Ra, (Latin *radius* "ray") white, radioactive, metallic element, atomic number 88, relative atomic mass 226.02. It is one of the †alkaline-earth metals, found in nature in †pitchblende and other uranium ores. Of the 16 isotopes, the commonest, Ra-226, has a half-life of 1,620 years. The element was discovered and named in 1898 by Pierre and Marie †Curie, who were investigating the residues of pitchblende.

Radium decays in successive steps to produce radon (a gas), polonium, and finally a stable isotope of lead. The isotope Ra-223 decays through the uncommon mode of heavy-ion emission, giving off carbon-14 and transmuting directly to lead. Because radium luminesces, it was formerly used in paints that glowed in the dark; when the hazards of radioactivity became known its use was abandoned, but factory and dump sites remain contaminated and many former workers and neighbors contracted fatal cancers.

radius in biology, one of the two bones in the lower forelimb of tetrapod (four-limbed) vertebrates.

radius straight line from the center of a circle to its circumference, or from the center to the surface of a sphere.

radius of curvature symbol r, in optics, radius of a sphere of which the curved surface of a mirror or lens forms a part. For a mirror, it is twice the focal distance.

radius vector line connecting the center of an orbiting body with the center of the body (or point) that it is orbiting.

radon chemical symbol Rn, colorless, odorless, gaseous, radioactive, non-metallic element, atomic number 86, relative atomic mass 222. It is grouped with the †noble gases (rare gases) and was formerly considered nonreactive, but is now known to form some compounds with fluorine. Of the 20 known isotopes, only three occur in nature; the longest half-life is 3.82 days (Rn-222).

discovery Radon is the densest gas known and occurs in small amounts in spring water, streams, and the air, being formed from the natural radioactive decay of radium. Ernest Rutherford discovered the isotope Rn-220 in 1899, and Friedrich Dorn (1848–1916) in 1900; after several other chemists discovered additional isotopes, William Ramsay and R W Whytlaw-Gray isolated the element, which they named niton in 1908. The name radon was adopted in the 1920s.

radula in zoology, rough-surfaced tongue-like structure in mollusks, used to rasp fragments of food into the mouth. The radula is characteristic of all mollusks except those, such as mussels, that have a body enclosed within two shells (bivalves).

rain form of †precipitation in which separate drops of water fall to the Earth's surface from clouds. The drops are formed by the accumulation of fine droplets that condense from water vapor in the air. The †condensation is usually brought about by rising and subsequent cooling of air.

Rain can form in three main ways: frontal (or cyclonic) rainfall, orographic (or relief) rainfall, and convectional rainfall. **Frontal rainfall** takes place at the boundary, or †front, between a mass of warm air from the tropics and a mass of cold air from the poles. The water vapor in the warm air is chilled and condenses to form clouds and rain. **Orographic rainfall** occurs when an airstream is forced to rise over a mountain range. The air becomes cooled and precipitation takes place. **Convectional rainfall**, associated with hot climates, is brought about by rising and abrupt cooling of air that has been warmed by the extreme heat of the ground surface. The water vapor carried by the air condenses, producing heavy rain. Convectional rainfall is usually accompanied by a thunderstorm, and it can be intensified over urban areas due to higher temperatures there (see †heat island).

rainbow arch in the sky displaying the colors of the †spectrum formed by the refraction and reflection of the Sun's rays through rain or mist. Its cause was discovered by †Theodoric of Freiburg in the 14th century.

About Rainbows http://unidata.ucar.edu/staff/blynds/rnbw.html

Explanation all about rainbows. The site answers such questions as "What is a rainbow?", "Where is the Sun when you see a rainbow?", and "What makes the bow?". It gives a good explanation of the refraction of white light into the seven colors of the visible spectrum.

rainfall gauge instrument used to measure †precipitation, usually rain. It consists of an open-topped cylinder, inside which there is a close-fitting funnel that directs the rain to a collecting bottle inside a second, inner cylinder. The gauge may be partially embedded in soil to prevent spillage. The amount of water that collects in the bottle is measured every day, usually in millimeters.

When the amount of water collected is too little to be measured, **trace rainfall** is said to have taken place. Snow falling into the gauge must be melted before a measurement is taken.

rainforest dense forest usually found on or near the †Equator where the climate is hot and wet. Moist air brought by the converging trade winds rises because of the heat and produces heavy rainfall. More than half the tropical rainforests are in Central and South America, primarily the lower Amazon and the coasts of Ecuador and Columbia. The rest are in Southeast Asia (Malaysia, Indonesia, and New Guinea) and in West Africa and the Congo.

Tropical rainforests once covered 14% of the Earth's land surface, but are now being destroyed at an increasing rate as their valuable timber is harvested and the land cleared for agriculture, causing problems of †deforestation. Although by 1991 over 50% of the world's rainforests had been removed, they still comprise about 50% of all growing wood on the planet, and harbor at least 40% of the Earth's species (plants and animals).

The vegetation in tropical rainforests typically includes an area of dense forest called **selva**; a **canopy** formed by high branches of tall trees providing shade for lower layers; an intermediate layer of shorter trees and tree roots; lianas; and a ground cover of mosses and ferns. The lack of **seasonal rhythm** causes adjacent plants to flower and shed leaves simultaneously. Chemical weathering and leaching take place in the iron-rich soil due to the high temperatures and humidity.

Rainforests comprise some of the most complex and diverse †ecosystems on the planet, deriving their energy from the Sun and photosynthesis. In a hectare (10,000 sq m/107,640 sq ft) of rainforest there are an estimated 200–300 tree species compared with 20–30 species in a hectare of temperate forest. The trees are the main **producers**. †Herbivores such as insects, caterpillars, and monkeys feed on the plants and trees and in turn are eaten by the †carnivores, such as ocelots and puma. Fungi and bacteria, the primary †**decomposers**, break down the dead material from the plants, herbivores, and carnivores with the help of heat and humidity. This decomposed material provides the **nutrients** for the plants and trees.

The rainforest ecosystem helps to regulate global weather patterns – especially by taking up CO_2 (carbon dioxide) from the atmosphere – and stabilizes the soil. Rainforests provide most of the oxygen needed for plant and animal respiration. When deforestation occurs, the microclimate of the mature forest disappears; soil erosion and flooding become major problems since rainforests protect the deep tropical soils. Once an area is cleared it is very difficult for shrubs and bushes to re-establish because soils are poor in nutrients. This causes problems for plans to convert rainforests into agricultural land – after two or three years the crops fail and the land is left bare. Clearing of the rainforests may lead to †global warming of the atmosphere, and contribute to the †greenhouse effect.

Exploring the Tropics http://www.mobot.org/education/tropics/welcome.html

Missouri Botanical Garden guide to tropical rainforests, covering issues such as "Are all tropical forests, rain forests?", "Effects of elevation on climate and vegetation," and "Causes of destruction."

Rainforests of the World http://rain-tree.com/

Resource that explores many aspects of this type of vegetation. The site includes detail on "Climate," "Homeland for Forest Peoples," and "Rainforest Facts."

Rainwater, (Leo) James (1917–1986) US physicist who with the Danes Aage †Bohr and Ben †Mottelson shared the Nobel Prize for Physics in 1975 for their discovery of the connection between collective and particle motion in atomic nuclei, and the development of the theory of nuclear structure.

raised beach beach that has been raised above the present-day shoreline and is therefore no longer washed by the sea. It is an indication of a fall in sea level (eustatic) or of a rise in land level (isostatic).

RAM acronym for **random-access memory**, in computing, a memory device in the form of a collection of integrated circuits (chips), frequently used in microcomputers. Unlike †ROM (read-only memory) chips, RAM chips can be both read from and written to by the computer, but their contents are lost when the power is switched off.

Many modern commercial programs require a great deal of RAM to work efficiently. By 2000 most PCs were supplied with 128 megabytes (MB) of RAM.

Raman, Chandrasekhara Venkata (1888–1970) Indian physicist who was awarded the Nobel Prize for Physics in 1930 for his discovery in 1928 of what became known as the **Raman effect**: the scattering of monochromatic (single-wavelength) light when passed through a transparent substance. The Raman spectra produced are used to obtain information on the structure of molecules. He was knighted in 1929.

Raman effect in spectroscopy, the change in the †wavelength of light scattered by molecules.

Ramanujan, Srinavasa Ayengar (1887–1920) Indian mathematician who did original work especially in function theory and number theory.

Ramanujan, Srinivasa Ayengar http://www-groups.dcs.st-andrews.ac.uk/history/Mathematicians/Ramanujan.html

Part of an archive containing the biographies of the world's greatest mathematicians, this site is devoted to the life and contributions of Srinivasa Ramanujan. In addition to biographical information, you will find a list of references about Ramanujan and links to other essays in the archive that reference him. The text of this essay includes hypertext links to the essays of those mathematicians and thinkers who influenced Ramanujan. There is also an image of him, which can be enlarged, and a map of his birthplace.

Ramón y Cajal, Santiago (1852–1934) Spanish cell biologist and anatomist who was awarded a Nobel Prize for Physiology or Medicine in 1906 for the discovery of the fine structure of the nervous system. Cajal's research revealed that the nervous system is based on units of †nerve cells (neurons).

ramp another name for an inclined plane, a slope used as a simple machine.

Ramsay, William (1852–1916) Scottish chemist who, with Lord †Rayleigh, discovered argon in 1894. In 1895 Ramsay produced helium and in 1898, in cooperation with Morris Travers, identified neon, krypton, and xenon. In 1903, with Frederick †Soddy, he noted the transmutation of radium into helium, which led to the discovery of the density and relative atomic mass of radium. He was awarded the Nobel Prize for Chemistry in 1904 for his discovery of noble gases (rare gases) in air and their locations in the periodic table. He was made a KCB in 1902.

Ramsey, Norman F (1915–) US physicist who invented a method of storing atoms and observing them for long periods of time, and applied this in the hydrogen maser and atomic clock. He was awarded the Nobel Prize for Physics in 1989 for his measurement techniques leading to the discovery of the cesium atomic clock which has been the basis for the definition of the unit of time (the second) since 1967. He shared the award with Hans †Dehmelt and Wolfgang †Paul.

random access in computing, an alternative term for direct access.

random event in statistics, an event that is not affected by either previous or future events. For example, if a new book is opened at random, all page numbers are equally likely.

random number one of a series of numbers having no detectable pattern. Random numbers are used in †computer simulation and computer games. It is impossible for an ordinary computer to generate true random numbers, but various techniques are available for obtaining pseudo-random numbers – close enough to true randomness for most purposes.

Random Number Generator http://www.fourmilab.ch/hotbits/

Internet resource that delivers genuine random numbers, generated by radioactive decay.

range in physical geography, a line of mountains (such as the Alps or Himalayas). In human geography, the distance that people are prepared to travel (often to a central place) to obtain various goods or services. In mathematics, the range of a set of numbers is the difference between the largest and the smallest number; for example, 5, 8, 2, 9, 4 = 9 – 2 = 7; this sense is used in terms like "tidal range" and "temperature range." Range is also a name for an open piece of land where cattle are ranched.

range in statistics, a measure of dispersion in a †frequency distribution, equaling the difference between the largest and smallest values of the variable. The range is sensitive to extreme values in the sense that it will give a distorted picture of the dispersion if one measurement is unusually large or small. The †interquartile range is often preferred.

Rankine cycle cycle used in steam †power plants, where water is introduced under †pressure into a boiler, evaporation taking place followed by expansion without loss of heat to end in condensation and a repeat of the cycle.

Raoult, François Marie (1830–1901) French chemist. In 1882, while working at the University of Grenoble, Raoult formulated one of the basic laws of chemistry. **Raoult's law** enables the relative molecular mass of a substance to be determined by noting how much of it is required to depress the freezing point of a solvent by a certain amount.

rare-earth element alternative name for †lanthanide.

rare gas alternative name for †noble gas.

Chemical Elements – Noble Gases http://www.chemicalelements.com/groups/noblegases.html

All six noble gases are explained in detail. Each page contains basic information on the particular gas, such as atomic number and symbol, and explains the facts behind its discovery, as well as providing related links.

raster graphics computer graphics that are stored in the computer memory by using a map to record data (such as color and intensity) for every †pixel that makes up the image. When transformed (enlarged, rotated, stretched, and so on), raster graphics become ragged and suffer loss of picture resolution, unlike †vector graphics. Raster graphics are typically used for painting applications, which allow the user to create artwork on a computer screen much as if they were painting on paper or canvas.

ratemeter instrument used to measure the count rate of a radioactive source. It gives a reading of the number of particles emitted from the source and captured by a detector in a unit of time (usually a second).

rate of change change of a variable per unit of time. See †distance–time graph.

rate of reaction speed at which a chemical reaction proceeds. It is usually expressed in terms of the concentration (usually in †moles per liter) of a reactant consumed, or product formed, in unit time; so the units would be moles per liter per second (mol l^{-1} s^{-1}). The rate of a reaction may be affected by the concentration of the reactants, the temperature of the reactants (or the amount of light in the case of a photochemical reaction), and the presence of a †catalyst. If the reaction is entirely in the gas state, the rate is affected by pressure, and, where one of the reactants is a solid, it is affected by the particle size.

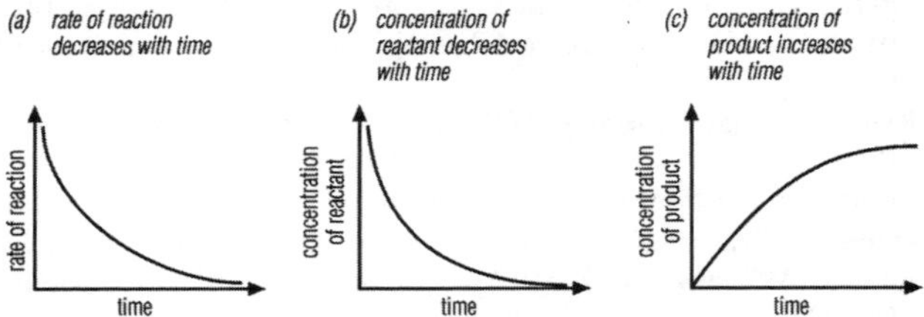

rate of reaction The rate of reaction decreases with time while the concentration of product increases. Helicon

During a reaction at constant temperature the concentration of the reactants decreases and so the rate of reaction gradually slows down. These changes can be represented by drawing graphs.

For an †endothermic reaction (one that absorbs heat) increasing the temperature may produce large increases in the rate of reaction. A 10°C rise can double the rate while a 40°C rise can produce a 50- to 100-fold increase in the rate.

†Collision theory is used to explain these effects. Increasing the concentration or the pressure of a gas means there are more particles per unit volume, therefore there are more collisions and more successful collisions. Increasing the temperature makes the particles move much faster, resulting in more collisions per unit time and more successful collisions; consequently the rate increases.

Rates of Reaction http://www.schoolchem.com/rc.htm

Colorful text-based site that covers all of the topics under the "Rates of reaction" heading, which is necessary for numerous parts of chemistry. Key words and concepts are highlighted in either red or blue.

ratio measure of the relative size of two quantities or of two measurements (in similar units), expressed as a †proportion. For example, the ratio of vowels to consonants in the alphabet is 5:21. As a †fraction $^5/_{26}$ of the letters are vowels. The ratio of 500 m to 2 km is 500:2,000, or in its simplest integer form 1:4 (dividing both sides of the ratio by 500).

Ratios are normally expressed as whole numbers, so 2:3.5 would become 4:7 (the ratio remains the same provided both parts of the ratio are multiplied or divided by the same number).

rationalized units units for which the defining equations conform to the geometry of the system. Equations involving circular symmetry contain the factor 2p; those involving spherical symmetry 4p. †SI units are rationalized, †*c*.g.s. units are not.

rational number in mathematics, any number that can be expressed as an exact fraction (with a denominator not equal to 0), that is, as $a \div b$ where a and b are integers; or an exact decimal. For example, $^2/_1$,

$^1/_4$, $1^5/_4$, $- ^3/_5$ are all rational numbers, whereas p (which represents the constant 3.141592 ...) is not. Numbers such as p are called †irrational numbers.

Raunkiaer system method of classification devised by the Danish ecologist Christen Raunkiaer (1860–1938) whereby plants are divided into groups according to the position of their †perennating (overwintering) buds in relation to the soil surface. For example, plants in cold areas, such as the tundra, generally have their buds protected below ground, whereas in hot, tropical areas they are above ground and freely exposed. This scheme is useful for comparing vegetation types in different parts of the world. The main divisions are **phanerophytes** with buds situated well above the ground; **chamaephytes** with buds borne within 25 cm/10 in of the soil surface; **hemicryptophytes** with buds at or immediately below the soil surface; and **cryptophytes** with their buds either beneath the soil (**geophyte**) or below water (**hydrophyte**).

raw data unprocessed facts and figures collected from surveys, experiments, and other sources, to which a meaning or interpretation has been applied. The †data are usually presented in the form of †frequency diagrams, tables, and †graphs so that they are easier to understand and analyze. For example, meteorological data such as daily temperature or rainfall can be collected over long periods of time and then displayed as line graphs so that comparisons can easily be made.

Ray, John (1627–1705) English naturalist who devised a classification system accounting for some 18,000 plant species. It was the first system to divide flowering plants into †monocotyledons and †dicotyledons, with additional divisions made on the basis of leaf and flower characters and fruit types.

Rayet, George Antoine Pons (1839–1906) French astronomer who, in collaboration with Charles Wolf (1827–1918), detected a new class of peculiar white or yellowish stars whose spectra contain broad hydrogen and helium emission lines. These stars are now called **Wolf–Rayet stars**.

Rayleigh, John William Strutt (1842–1919) 3rd Baron Rayleigh, English physicist who wrote the standard treatise *The Theory of Sound* (1877–78), experimented in optics and microscopy, and, with William †Ramsay, discovered argon. He was awarded the Nobel Prize for Physics in 1904 for his work on the densities of gases and the discovery of argon. He was created a baron in 1873.

Rayleigh–Jeans law formula giving the intensity of †black-body radiation at long †wavelengths for a radiator at a certain temperature. It is thus an approximation to Planck's full formula for the black-body intensity based on quantum concepts.

Rayleigh scattering in physics, phenomenon in which a gas or suspended particles in the air scatter different colors of light differentially. It affects blue light most, which is why the sky is blue. Dust in the air results in red light, as in the thin atmosphere of Mars.

rays original name for radiation of all types, such as †X-rays and gamma rays.

reactance property of an alternating current circuit that together with any †resistance makes up the †impedance (the total opposition of the circuit to the passage of a current).

The reactance to an alternating current of frequency f of an inductance L is $2pfL$ and that of a capacitance C is $1/2pfC$. Reactance is measured in †ohms.

reaction in chemistry, the coming together, or interaction, of two or more atoms, ions, or molecules with the result that a chemical change takes place and a new substance is formed, with a different chemical composition. The nature of the reaction is described by a †chemical equation. For example, in the chemical reaction that occurs when magnesium burns in oxygen, a new substance, magnesium oxide is made:

$2Mg_{(s)} + O_{2(g)} \rightarrow 2MgO_{(s)}$

Chemical equations show the reactants and products of a chemical reaction by using chemical symbols and formulae. State symbols and the energy symbol (DH) can be used to show whether reactants and products are solids (s), liquids (l), gases (g), or are in solution (aq); and whether energy has been released or absorbed during the reaction.

Displacement Reactions of Metals http://www.gcsechemistry.com/xr5.htm

Displacement reactions, in relation to metals, are covered by this page. The main features of this type of reaction are illustrated through examples and there is a table on the subsequent page listing the most and least reactive of metals.

ScienceNet: Chemistry http://www.sci-ctr.edu.sg/ScienceNet/cat_physical/cat_che05887.html

Set up as an answer to a student's question, this site covers what endothermic and exothermic reactions are. Along with helpful explanations are some examples, expressed in equation form, as well as a section at the end, which explains the normal occurrence of exothermic and endothermic reactions in the real world.

reaction force in physics, the equal and opposite force described by Newton's third law of motion that arises whenever one object applies a force (**action force**) to another. For example, if a magnet attracts a piece of iron, then that piece of iron will also attract the magnet with a force that is equal in magnitude but opposite in direction. When any object rests on the ground the downwards contact force applied to the ground always produces an equal, upwards reaction force.

reaction principle principle stated by †Newton as his third law of motion: to every action, there is an equal and opposite reaction.

In other words, a force acting in one direction is always accompanied by an equal force acting in the opposite direction. This explains how rocket and †jet propulsion works and why a gun recoils after firing.

reactivity series chemical series produced by arranging the metals in order of their ease of reaction with reagents such as oxygen, water, and acids. An example of such an arrangement, starting with the most reactive, is: potassium, sodium, calcium, magnesium, aluminum, zinc, iron, tin, lead, copper, silver, gold. This arrangement aids the understanding of the properties of metals, helps to explain differences between them, and enables predictions to be made about a metal's behavior, based on a knowledge of its position or properties. It also allows prediction of the relative stability of the compounds formed by an element: the more reactive the metal, the more stable its compounds are likely to be.

The position of a metal in the series determines the reactions of the metal with various reagents, the displacement of one metal from its compound by another metal, and the method of extraction of a metal from its ore.

real number in mathematics, any of the †rational numbers (which include the integers) or †irrational numbers. Real numbers exclude †imaginary numbers, found in †complex numbers of the general form $a + bi$ where $i = \sqrt{-1}$, although these do include a real component a.

real-time system in computing, a program that responds to events in the world as they happen. For example, an automatic-pilot program in an aircraft must respond instantly in order to correct deviations from its course. Process control, robotics, games, and many military applications are examples of real-time systems.

Reber, Grote (1911–2002) US radio engineer who pioneered radio astronomy. He attempted to map all the extraterrestrial sources of radio emission that could be traced.

receiver, radio component of a radio communication system that receives and processes radio waves. It detects and selects modulated radio waves (see †modulation) by means of an aerial and tuned circuit, and then separates the transmitted information from the carrier wave by a process that involves rectification.

receptacle the enlarged end of a flower stalk to which the floral parts are attached. Normally the receptacle is rounded, but in some plants it is flattened or cup-shaped. The term is also used for the region on that part of some seaweeds which becomes swollen at certain times of the year and bears the reproductive organs.

receptor in biology, receptors are discrete areas of cell membranes or areas within cells with which neurotransmitters, hormones, and drugs interact. Such interactions control the activities of the body. For example, adrenaline transmits nervous impulses to receptors in the sympathetic nervous system, which initiates the characteristic response to excitement and fear in an individual.

Other types of receptors, such as the proprioceptors, are located in muscles, tendons, and joints. They relay information on the position of the body and the state of muscle contraction to the brain.

recessive gene †allele (alternative form of a gene) that will show in the (†phenotype observed characteristics of an organism) only if its partner allele on the paired chromosome is similarly recessive. Such an allele will not show if its partner is dominant, that is if the organism is †heterozygous for a particular characteristic. Alleles for blue eyes in humans and for shortness in pea plants are recessive. Most mutant alleles are recessive and therefore are only rarely expressed.

For every characteristic of a plant or animal that is inherited, there are two genes present in the cells determining this characteristic in all but a few examples. By "characteristic" is meant "height" or "eye color" or "ability to make a particular enzyme." If the two genes are identical (homozygous state) the characteristic seen in the organism is determined by the two genes. However, one gene may be different from the other (heterozygous state). If so, the two genes are alleles – contrasting genes for a characteristic. In this case it is possible that one of them determines the characteristic seen and the other does not. The characteristic seen in this case is said to be dominant. The other allele not expressed in this case will only be expressed when present in the homozygous state. This characteristic is said to be recessive. Sometimes the allele that produces the dominant characteristic is described as being a dominant allele and the one that tends to produce the recessive characteristic as being the recessive allele. This is not really the correct use of the terms dominant and recessive. An allele is one of two or more alternative forms of a gene. This is best explained with examples. A gene which tends to produce blue eyes in a person will have an alternative allele that tends to produce brown eye color. In a plant that may be found in tall and short forms may have an allele that tends to produce tall plants though its alternative allele produces short plants.

reciprocal result of dividing a given quantity into 1. Thus the reciprocal of 2 is $^1/_2$; the reciprocal of $^2/_3$ is $^3/_2$; and the reciprocal of x^2 is $^1/_x{}^2$ or x^{-2}. Reciprocals are used to replace †division by †multiplication, since multiplying by the reciprocal of a number is the same as dividing by that number.

On a calculator, the reciprocals of all numbers except 0 are obtained by using the button marked $^1/_x$.

recombinant DNA in genetic engineering, †DNA formed by splicing together genes from different sources into new combinations.

recombination in genetics, any process that recombines, or "shuffles," the genetic material, thus increasing genetic variation in the offspring. The two main processes of recombination both occur during meiosis (reduction division of cells). One is †**crossing over**, in which chromosome pairs exchange segments; the other is the random reassortment of chromosomes that occurs when each gamete (sperm or egg) receives only one of each chromosome pair.

recording any of a variety of techniques used to capture, store, and reproduce music, speech, and other information carried by sound waves. A microphone first converts the sound waves into an electrical signal that varies in proportion to the loudness of the sound. The signal can be stored in digital or analogue form, or on magnetic tape.

analogue recording In an analogue recording, the pattern of the signal is copied into another form. In a **vinyl** gramophone record, for example, a continuous spiral groove is cut into a plastic disc by a vibrating needle. The recording is replayed by a stylus that follows the track of the groove, and so reproduces the vibrations that are amplified and turned back into sound. In a **magnetic tape** recording, the signal is recorded as a pattern of magnetization on a plastic tape coated with a magnetic powder. When the tape is played back, the magnetic patterns create an electrical signal that, as with the gramophone record, is used to recreate the original sound. All analogue recording techniques suffer from background noise and the quality of reproduction gradually degrades as the format changes and the disc or tape wears out.

digital recording In digital recording, the signals picked up by the microphone are converted into a stream of numbers that can then be stored in several ways. The most well-known of these is the compact disc, in which numbers are coded as a string of tiny pits pressed into a 12 cm/4.7 in plastic disk. When the recording is played back, using a laser, the exact values are retrieved and converted into a varying electrical signal and then back into sound. Digital recording is relatively free from noise and interference and gives a very high quality of reproduction. It is also suitable for storing information to be processed by computers.

recrystallization process used for the purification of crystalline solids. The material is dissolved in a suitable solvent and crystallized a number of times, becoming purer each time the process is repeated.

rectangle quadrilateral (four-sided plane figure) with opposite sides equal and parallel and with each interior angle a right angle (90°). Its area A is the product of the length l and height h; that is, $A = l$ x h. A rectangle with all four sides equal is a †square.

A rectangle is a special case of a †parallelogram. The diagonals of a rectangle are equal and bisect each other.

rectangular axes another name for †Cartesian coordinates.

rectangular hyperbola graph of the †function $y = {}^1/_x$. The x- and y-axes are †asymptotes to this curve. See †hyperbola.

rectangular prism another name for a †cuboid.

rectifier device for obtaining one-directional current (DC) from an alternating source of supply (AC). (The process is necessary because almost all electrical power is generated, transmitted, and supplied as alternating current, but many devices, from television sets to electric motors, require direct current.) Types include plate rectifiers, thermionic †diodes, and †semiconductor diodes.

rectum lowest part of the large intestine of animals, which stores feces prior to elimination (defecation).

recurrence relation in mathematics, method of generating a sequence in which a formula for each member of the sequence is given in terms of the earlier terms. Examples are the recurrence relation $f_{n+1} = f_n + f_{n-1}$, generating the †Fibonacci series.

recurring decimal any fraction that cannot be represented in the decimal system by a finite decimal. All fractions will result in recurring decimals unless the denominator of the fraction is either a power of 2 or 5, or a multiple of these. For example, $^1/_4$ is 0.25 but $^1/_6$ (= $^1/_{2 \times 3}$) is 0.1666... .

recursive in mathematics, a very general description of a function with †natural numbers, as arguments, corresponding to the intuitive notion of computability. The basic recursive functions are $x + y$, xy, $f(x, y, z, ...) = x$, $g(x, y, z, ...) = y$, and a restricted form of the taking of a minimum.

recycling processing of industrial and household waste (such as paper, glass, and some metals and plastics) so that the materials can be reused. This saves expenditure on scarce raw materials, slows down the depletion of †non-renewable resources, and helps to reduce pollution. Aluminum is frequently recycled because of its value and special properties that allow it to be melted down and re-pressed without loss of quality, unlike paper and glass, which deteriorate when recycled.

The USA recycles only around 25% of its waste (1998), compared with around 33% in Japan. However, all US states encourage or require local recycling programs to be set up. It was estimated in 1992 that 4,000 cities collected waste from 71 million people for recycling. Most of these programs were set up 1989–92. Around 33% of newspapers, 22% of office paper, 64% of aluminum cans, 3% of plastic containers, and 20% of all glass bottles and jars were recycled.

Environmental Defense: Recycling http://www.environmentaldefense.org/issues/Recycling.html

Articles on all aspects of recycling. Pages with useful information on how individuals can reduce waste and recycle are presented alongside reports on recycling policy and trends. Recycling is only one of the helpful and interesting areas covered by the Environmental Defense Web site.

Zero Waste America http://www.zerowasteamerica.org/

Site dedicated to the elimination (not reduction) of waste and pollution. The Zero Waste America organization explains its drastic aim of banning garbage. It provides ideas for action by local campaigning groups, and sets out a domestic strategy for composting, recycling and re-using everything the household consumes. This huge resource provides legal reports and a database of political and scientific information relevant to the zero waste ideal.

red blood cell or **erythrocyte**, most common type of blood cell, and responsible for transporting †oxygen around the body. They contain hemoglobin, a red protein, which combines with oxygen from the lungs to form oxyhemoglobin. When transported to the tissues the oxyhemoglobin splits into its original constituents, and the cells are able to release the oxygen. There are about 6 million red cells in every cubic centimeter of blood.

The red cell is a highly specialized cell with a distinctive shape. In mammals they are in the shape of a disc with a depression in the face of the disc and lose their †nucleus before they work as oxygen transporters in order to make maximum space for hemoglobin. In other vertebrates they are oval and nucleated. They are manufactured in the bone marrow. In humans, red cells last for only four months before being destroyed in the †liver. Hemoglobin contains iron, which is why this †mineral must be included as part of a balanced diet. However, the liver can store and re-use iron from old red cells.

red dwarf any star that is cool, faint, and small (about one-tenth the mass and diameter of the Sun). Red dwarfs burn slowly, and have estimated lifetimes of 100 billion years. They may be the most abundant type of star, but are difficult to see because they are so faint. Two of the closest stars to the Sun, †Proxima Centauri and †Barnard's Star, are red dwarfs.

red giant any large bright star with a cool surface. It is thought to represent a late stage in the evolution of a star like the Sun, as it runs out of hydrogen fuel at its center and begins to burn heavier elements, such as helium, carbon, and silicon. Because of more complex nuclear reactions that then occur in the red giant's interior, it eventually becomes gravitationally unstable and begins to collapse and heat up. The result is either explosion of the star as a †supernova, leaving behind a †neutron star, or loss of mass by more gradual means to produce a †white dwarf.

Red giants have diameters between 10 and 100 times that of the Sun. They are very bright because they are so large, although their surface temperature is lower than that of the Sun, about 2,000–3,000 K (1,700–2,700°C/3,000–5,000°F). See also red †supergiants.

Redi, Francesco (1626–1697) Italian physician and poet who disputed the †spontaneous-generation theory, the erroneous belief that animals can be generated spontaneously from decaying matter, and performed one of the first controlled biological experiments.

redox reaction chemical change where one reactant is reduced and the other reactant oxidized. The reaction can only occur if both reactants are present and each changes simultaneously. For example, hydrogen reduces copper(II) oxide to copper while it is itself oxidized to water. The corrosion of iron and the reactions taking place in electric and electrolytic cells are just a few examples of redox reactions.

Oxidation and Reduction http://library.thinkquest.org/10784/chem2.html

Introduction to these key electrochemical concepts from the student-created "ThinkQuest" site. Using tables, equations, and easily understood text, this page explains how oxidation–reduction or redox reactions produce the current that drives batteries. Click forward to access useful pages on other electrochemistry topics such as electrodes and Faraday's constant.

Oxidation/Reduction Chemistry http://www.chem.uidaho.edu/~honors/redox.html

Both oxidation and reduction are included on the page, which covers the basics of redox chemistry. There are helpful methods of remembering the transfer of electrons, by the use of a mnemonic, and example experiments to show the theory in action. Colored scientific diagrams also help visualize the physical chemical changes.

red shift in astronomy, lengthening of the wavelengths of light from an object as a result of the object's motion away from us. It is an example of the †Doppler effect. The red shift in light from galaxies is evidence that the universe is expanding.

Lengthening of wavelengths causes the light to move or shift towards the red end of the †spectrum, hence the name. The amount of red shift can be measured by the displacement of lines in an object's spectrum. By measuring the amount of red shift in light from stars and galaxies, astronomers can tell how quickly these objects are moving away from us. A strong gravitational field can also produce a red shift in light; this is termed **gravitational red shift**.

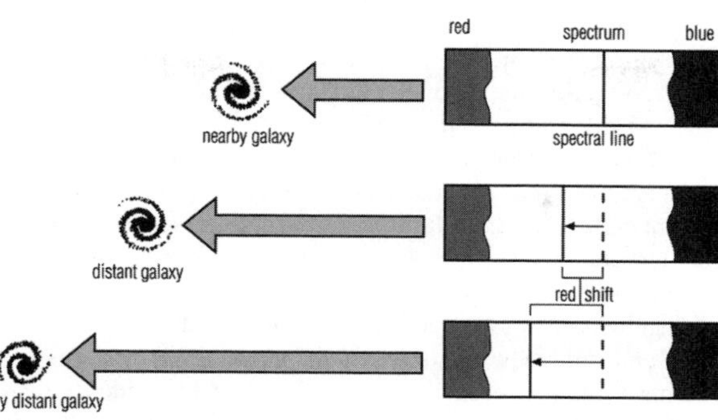

red shift The red shift causes lines in the spectra of galaxies to be shifted towards the red end of the spectrum. More distant galaxies have greater red shifts than closer galaxies. The red shift indicates that distant galaxies are moving apart rapidly, as the universe expands. Helicon

reduce to lowest terms in mathematics, to cancel a fraction until numerator and denominator have no further factors in common.

reduction in chemistry, the gain of electrons, loss of oxygen, or gain of hydrogen by an atom, ion, or molecule during a chemical reaction.

Reduction may be brought about by reaction with another compound, which is simultaneously oxidized (reducing agent), or electrically at the cathode (negative electrode) of an electric cell. Examples include the reduction of iron(III) oxide to iron by carbon monoxide:

$Fe_2O_3 + 3CO \rightarrow 2Fe + 3CO_2$

the hydrogenation of ethene to ethane:

$CH_2{-}CH_2 + H_2 \rightarrow CH_3{-}CH_3$

and the reduction of a sodium ion to sodium:

$Na^+ + e^- \rightarrow Na$

reef underwater ridge or mound composed of the skeletal remains of colonies of †corals and calcareous algae. Types of reefs include atolls, barrier reefs, and fringing reefs.

re-entrant polygon †polygon that is not completely convex. It has at least one interior angle greater than 180°.

reflecting telescope †telescope in which light is collected and brought to a focus by a concave mirror. †Cassegrain and †Newtonian telescopes are examples.

reflection in geometry, a †transformation that reflects every point on a shape to a point that is the same distance on the opposite side of a fixed line – the mirror line, or line of symmetry. Reflections in two perpendicular axes produce a rotation of 180° (a half turn).

For example, in the following diagram, triangle A has been reflected along the *y*-axis (the mirror line) to obtain triangle B:

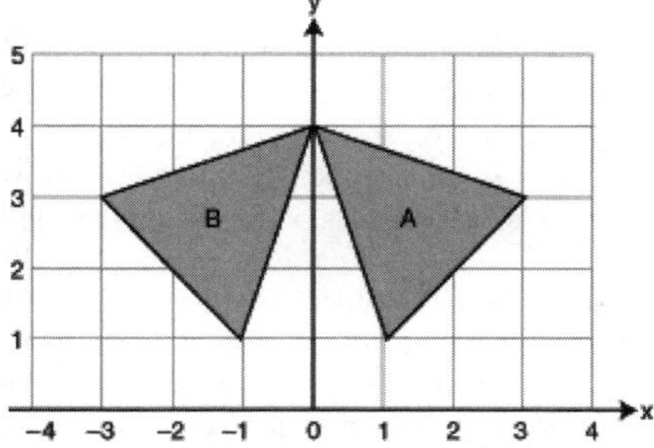

If triangle B is reflected again along the mirror line to triangle A, the movement would be an inverse transformation. Shapes can also be transformed by †translation, †rotation, and †enlargement.

reflection throwing back or deflection of waves, such as †light or †sound waves, when they hit a surface. Reflection occurs whenever light falls on an object. The **law of reflection** states that the angle of incidence (the angle between the ray and a perpendicular line drawn to the surface) is equal to the angle of reflection (the angle between the reflected ray and a perpendicular to the surface).

Looking at an image on the surface of the water in a lake is an example of light rays reflecting towards the observer. Reflection of light takes place from all materials. Some materials absorb a small amount of light and reflect most of it back; for example, a shiny, silvery surface. Other materials absorb most of the light and reflect only a small amount back; for example, a dark, dull surface. Reflected light gives objects their visible texture and color.

Light reflected from a surface can be either regular (plane), where the surface is flat and smooth and light rays are reflected without any scattering; or scattered, where the surface is irregular (in effect, many different surfaces). The color of the sky is due to scattering of sunlight by particles in the atmosphere, such as dust and gas particles, water droplets, or ice crystals. On a clear day the sky appears blue due to the scattering of shorter wavelength light.

When light passes from a dense medium to a less dense medium, such as from water to air, both †refraction and reflection can occur. If the angle of incidence is small, the reflection will be relatively weak compared to the refraction. But as the angle of incidence increases the relative degree of reflection will increase. At some **critical angle of incidence** the angle of refraction is 90°. Since refraction cannot occur above 90°, the light is totally reflected at angles above this critical angle of incidence. This condition is known as **total internal reflection**. Total internal reflection is used in †fiber optics to transmit data over long distances, without the need of amplification.

reflection, laws of incident ray of light, the reflected ray of light, and the normal to the reflecting surface all lie in the same plane. Secondly, the angle between the incident ray and the normal is the same as that between the reflected ray and the normal to the surface.

reflex in animals, a very rapid involuntary response to a particular stimulus. It is controlled by the †nervous system. A reflex involves only a few nerve cells, unlike the slower but more complex responses produced by the many processing nerve cells of the brain.

A **simple reflex** is entirely automatic and involves no learning. Examples of such reflexes include the sudden withdrawal of a hand in response to a painful stimulus, or the jerking of a leg when the kneecap is tapped. Sensory cells (receptors) in the knee send signals to the spinal cord along a sensory nerve cell. Within the spine a **reflex arc** switches the signals straight back to the muscles of the leg (effectors) via an intermediate nerve cell and then a motor nerve cell; contraction of the leg occurs, and the leg kicks upwards. Only three nerve cells are involved, and the brain is only aware of the response after it has taken place. Such reflex arcs are particularly common in lower animals, and have a high survival value, enabling organisms to take rapid action to avoid potential danger. In higher animals (those with a well-developed †central nervous system) the simple reflex can be modified by the involvement of the brain – for instance, humans can override the automatic reflex to withdraw a hand from a source of pain.

A **conditioned reflex** involves the modification of a reflex action in response to experience (learning). A stimulus that produces a simple reflex response becomes linked with another, possibly unrelated, stimulus. For example, a dog may salivate (a reflex action) when it sees its owner remove a can-opener from a drawer because it has learned to associate that stimulus with the stimulus of being fed.

Integration and Control: How Nervous Systems Work http://research.umbc.edu/~farabaug/nerve3.html
Text-based page on the nervous system from the University of Maryland. The page describes concepts such as the reflex arc and explains the main differences between the nervous systems of vertebrate and invertebrate species.

reflex angle angle greater than 180° but less than 360°.

reflex arc in animals, the passage of a nerve †impulse from a receptor to a muscle that results in an automatic response to the stimulus. It is the simplest kind of animal behavior. Examples include the removal of the hand when it touches a hot object, blinking when an object comes close to the eye, and the narrowing of the pupil of the eye in bright light. A reflex arc differs from other responses by being automatic. This is achieved by not allowing the brain to "consider" whether to take action in response to the stimulus or not. In a reflex arc, a particular stimulus virtually always results in the same response.

Reflex arcs involve actions that protect the body. By cutting out having to "consider" whether to take action or not, they do not involve conscious thought. Impulses travel along fewer neurones and so produce a response in the shortest possible time. The slowest steps in the pathway taken by nerve impulses are at the junctions between neurones – the synapses. In many reflex arcs the number of synapses to cross may be as low as three.

reflux in chemistry, process for liquefying vapor liberated by a boiling liquid and returning it to the vessel from which it originally came. This is usually achieved by connecting a vertical condensing column to the reaction vessel.

refracting telescope or **refractor**, †telescope that uses lenses to magnify and focus an image onto an eyepiece.

refraction bending of a wave when it passes from one medium into another. It is the effect of the different speeds of wave propagation in two substances that have different densities. For example, when light passes from air (less dense) into glass (more dense) it slows down (from 300 million to 200 million meters per second) and is refracted. The amount of refraction depends on the densities of the media, the angle at which the wave strikes the surface of the second medium, and the amount of bending and change of velocity corresponding to the wave's frequency (dispersion). Refraction occurs with all types of progressive waves – †electromagnetic waves, sound waves, and water waves – and differs from †reflection, which involves no change in velocity.

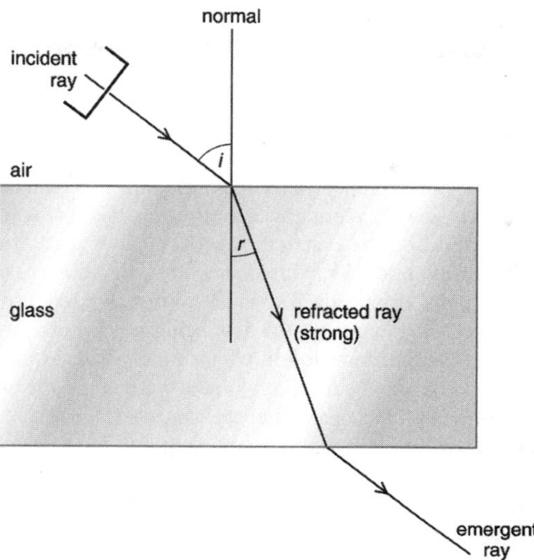

refraction The refraction of light through glass. When the light ray strikes the glass (a denser medium than the air) it is bent towards the normal. When it leaves the glass and re-enters the less dense medium it is bent away from the normal. Helicon

refraction of light The degree of refraction depends in part on the angle at which the light hits the surface of a material. A line perpendicular to that surface is called the **normal**. The angle between the incoming light ray and the normal to the surface is called the **angle of incidence**. The angle between the refracted ray and the normal is called the **angle of refraction**. The angle of refraction cannot exceed 90°. An example of refraction is light hitting a glass pane. When light in air enters the denser medium, it is bent toward the normal. When light passes out of the glass into the air, which is less dense, it is bent away from the normal. The incident light will be parallel to the emerging light because the two faces of the glass are parallel. However, if the two faces are not parallel, as with a prism, the emerging light will not be parallel to the incident light. The angle between the incident ray and the emerging ray is called the **angle of deviation**. The amount of bending and change in velocity of the refracted wave is due to the amount of †dispersion corresponding to the wave's frequency, and the **refractive index** of the material. When light hits the denser material, its †frequency remains constant, but its velocity decreases due to the influence of electrons in the denser medium. Constant frequency means that the same number of light waves must pass by in the same amount of time. If the waves are slowing down, wavelength must also decrease to maintain the constant frequency. The waves become more closely spaced, bending toward the normal as if they are being dragged. The refractive index of a material indicates by how much a wave is bent. It is found by dividing the velocity of the wave in the first medium by the velocity of the wave in the second medium. The **absolute refractive index** of a material is the velocity of light in that material relative to the velocity of light in a vacuum. See also †apparent depth.

refraction of sound Sound waves, unlike light waves, travel faster in denser materials, such as solids and liquids, than they travel in air. When sound waves enter a solid, their velocity and wavelength *increase* and they are bent away from the normal to the surface of the solid.

water waves Water waves are refracted when their velocity decreases. Water waves slow down as water becomes shallower. A good example of this is a wavefront approaching a shore that is shallower in one place and deeper in another. When the wavefront approaches, the part of the wave in shallower water will slow down and its wavelength will decrease, causing it to lag behind the part of the wave in the deeper water.

refraction, laws of incident ray, the refracted ray and the normal to the surface between the two media at the point of incidence lie in the same plane. In addition, †Snel's law states that the ratio of the sines of the angles of incidence and refraction is equal to a constant. See †refractive index.

refractive index measure of the refraction of a ray of light as it passes from one transparent medium to another. If the angle of incidence is *i* and the angle of refraction is *r*, the ratio of the two refractive indices is given by $n_1/n_2 = \sin i/\sin r$. It is also equal to the speed of light in the first medium divided by the speed of light in the second, and it varies with the wavelength of the light.

refractive material material able to stand very high temperature, such as brick and concrete.

refractometer device for determining the †refractive index of a substance.

refractor in astronomy, †telescope in which light is collected and brought to a focus by a convex lens (the object lens or objective).

refrigeration use of technology to transfer heat from cold to warm, against the normal temperature gradient, so that an object can remain substantially colder than its surroundings. Refrigeration equipment is used for the chilling and deep-freezing of food in food technology, and in air conditioners and industrial processes.

Refrigeration is commonly achieved by a vapor-compression cycle, in which a suitable chemical (the refrigerant) travels through a long circuit of tubing, during which it changes from a vapor to a liquid and back again. A compression chamber makes it condense, and thus give out heat. In another part of the circuit, called the evaporator coils, the pressure is much lower, so the refrigerant evaporates, absorbing heat as it does so. The evaporation process takes place near the central part of the refrigerator, which therefore becomes colder, while the compression process takes place near a ventilation grille, transferring the heat to the air outside. The most commonly used refrigerants in modern systems were †chlorofluorocarbons (CFCs), but these are now being replaced by coolants that do not damage the ozone layer.

regelation phenomenon in which water refreezes to ice after it has been melted by pressure at a temperature below the freezing point of water. Pressure makes an ice skate, for example, form a film of water that freezes once again after the skater has passed.

regeneration in biology, regrowth of a new organ or tissue after the loss or removal of the original. It is common in plants, where a new individual can often be produced from a "cutting" of the original. In animals, regeneration of major structures is limited to lower organisms; certain lizards can regrow their tails if these are lost, and new flatworms can grow from a tiny fragment of an old one. In mammals, regeneration is limited to the repair of tissue in wound healing and the regrowth of peripheral nerves following damage.

region space enclosed by arcs in a †network (there is an extra region external to the network if the network is closed). In a closed network the numbers of regions, nodes, and arcs are related by the simple formula

nodes + regions = arcs + 2.

register in computing, a memory location that can be accessed rapidly; it is often built into the computer's central processing unit.

Some registers are reserved for special tasks – for example, an **instruction register** is used to hold the machine-code command that the computer is currently executing, while a **sequence-control register** keeps track of the next command to be executed. Other registers are used for holding frequently used data and for storing intermediate results.

Regnault, Henri Victor (1810–1878) German-born French physical chemist who showed that †Boyle's law applies only to ideal gases. He also invented an air thermometer and a hygrometer, and discovered carbon tetrachloride (tetrachloromethane).

regolith surface layer of loose material that covers most bedrock. It consists of eroded rocky material, volcanic ash, river alluvium, vegetable matter, or a mixture of these known as †soil.

regular describing geometric figures having all angles and sides equal. Also, of solids, having bases comprised of regular †polygons.

Regulus or **Alpha Leonis**, brightest star in the constellation Leo, and the 21st brightest star in the sky. First-magnitude Regulus has a true luminosity 100 times that of the Sun, and is 85 light years from Earth. Regulus was one of the four royal stars of ancient Persia marking the approximate positions of the Sun at the equinoxes and solstices. The other three were †Aldebaran, †Antares, and †Fomalhaut.

Reich, Wilhelm (1897–1957) Austrian physician who emigrated to the USA in 1939. He combined Marxism and †psychoanalysis to advocate the positive effects of directed sexual energies and sexual freedom. His works include *Die Sexuelle Revolution/The Sexual Revolution* (1936–45) and *Die Funktion des Orgasmus/The Function of the Orgasm* (1948).

Reichstein, Tadeus (1897–1996) Swiss biochemist who was awarded a Nobel Prize for Physiology or Medicine in 1950 for work on the structure and biological effects of hormones of the adrenal cortex. By 1946 Reichstein had identified a large number of steroids secreted by the adrenal cortex, some of which would later be used in the treatment of Addison's disease. Reichstein shared the prize with Edward Kendall and Philip †Hench.

Reines, Frederick (1918–1998) US physicist who was awarded the Nobel Prize for Physics in 1995 for experimentally detecting the †neutrino, one of the fundamental particles which make up all matter. He shared the award with Martin †Perl.

rejuvenation in earth science, the renewal of a river's powers of downward erosion. It may be caused by a fall in sea level or a rise in land level, by climate change, by changes in vegetation cover (deforestation), or by the increase in water flow that results when one river captures another (†river capture).

Several river features are formed by rejuvenation. For example, as a river cuts down into its channel it will leave its old flood plain perched up on the valley side to form a †river terrace. Meanders (bends in the river) become deeper and their sides steeper, forming †incised meanders, and †waterfalls and rapids become more common.

relational database †database in which data are viewed as a collection of linked tables. It is the most popular of the three basic database models, the others being **network** and **hierarchical**.

relative atomic mass mass of an atom relative to one-twelfth the mass of an atom of carbon-12. It depends primarily on the number of protons and neutrons in the atom, the electrons having negligible mass. If more than one †isotope of the element is present, the relative atomic mass is calculated by taking an average that takes account of the relative proportions of each isotope, resulting in values that are not whole numbers. The term **atomic weight**, although commonly used, is strictly speaking incorrect.

relative biological effectiveness RBE, relative damage caused to living tissue by different types of radiation. Some radiations do much more damage than others; alpha particles, for example, cause 20 times as much destruction as electrons (beta particles).

The RBE is defined as the ratio of the absorbed dose of a standard amount of radiation to the absorbed dose of 200 kV X-rays that produces the same amount of biological damage.

relative density density (at 20°C/68°F) of a solid or liquid relative to (divided by) the maximum density of water (at 4°C/39.2°F). The relative density of a gas is its density divided by the density of hydrogen (or sometimes dry air) at the same temperature and pressure.

relative humidity concentration of water vapor in the air. It is expressed as the ratio of the partial pressure of the water vapor to its saturated vapor pressure at the same temperature. The higher the temperature, the higher the saturated vapor pressure.

relative molecular mass mass of a molecule, calculated relative to one-twelfth the mass of an atom of carbon-12. It is found by adding the relative atomic masses of the atoms that make up the molecule.

The term **molecular weight**, although commonly used, is strictly speaking incorrect.

relativity in physics, theory of the relative rather than absolute character of mass, time, and space, and their interdependence, as developed by German-born US physicist Albert †Einstein in two phases:

special theory of relativity (1905) Starting with the premises that (1) the laws of nature are the same for all observers in unaccelerated motion and (2) the speed of light is independent of the motion of its source, Einstein arrived at some rather unexpected consequences. Intuitively familiar concepts, like mass, length, and time, had to be modified. For example, an object moving rapidly past the observer will appear to be both shorter and more massive than when it is at rest (that is, at rest relative to the observer), and a clock moving rapidly past the observer will appear to be running slower than when it is at rest. These predictions of relativity theory seem to be foreign to everyday experience merely because the changes are quite negligible at speeds less than about 1,500 kps/930 mps and only become appreciable at speeds approaching the speed of light.

general theory of relativity (1915) The geometrical properties of space-time were to be conceived as modified locally by the presence of a body with mass. A planet's orbit around the Sun (as observed in three-dimensional space) arises from its natural trajectory in modified space-time. Einstein's general theory accounts for a peculiarity in the behavior of the motion of the perihelion of the orbit of the planet Mercury that cannot be explained in Newton's theory. The new theory also said that light rays should bend when they pass by a massive object. The predicted bending of starlight was observed during the eclipse of the Sun in 1919. A third corroboration is found in the shift towards the red in the spectra of the Sun and, in particular, of stars of great density – white dwarfs such as the companion of Sirius.

Einstein showed that, for consistency with the above premises (1) and (2), the principles of dynamics as established by Newton needed modification; the most celebrated new result was the equation $E = mc^2$, which expresses an equivalence between mass (m) and †energy (E), c being the speed of light in a vacuum. In "relativistic mechanics," conservation of mass is replaced by the new concept of conservation of "mass-energy." General relativity is central to modern †astrophysics and †cosmology; it predicts, for example, the possibility of †black holes. General relativity theory was inspired by the simple idea that it is impossible in a small region to distinguish between acceleration and gravitation effects (as in a lift one feels heavier when the lift accelerates upwards), but the mathematical development of the idea is formidable. Such is not the case for the special theory, which a non-expert can follow up to $E = mc^2$ and beyond.

Cambridge Relativity http://www.damtp.cam.ac.uk/user/gr/public/index.html

Discussion of quantum gravity, strings, black holes, inflation, and cosmology, on Cambridge's relativity pages. The site also has information on COSMOS, the national cosmology supercomputer, and links to English physicist Stephen Hawking's pages.

relay in electrical engineering, an electromagnetic switch. A small current passing through a coil of wire wound around an iron core attracts an †armature whose movement closes a pair of sprung contacts to complete a secondary circuit, which may carry a large current or activate other devices. The solid-state equivalent is a thyristor switching device.

relay neuron nerve cell in the spinal cord, connecting motor neurons to sensory neurons. Relay neurons allow information to pass straight through the spinal cord, bypassing the brain. In humans such reflex actions, which are extremely rapid, cause the sudden removal of a limb from a painful stimulus.

rem acronym for **roentgen equivalent man**, unit of radiation dose equivalent.

remainder part left over when one number cannot be exactly divided by another. For example, the remainder of 11 divided by 3 is 2; the remainder may be represented as a fraction or decimal. Decimal remainders are either recurring (0.66666...), cyclic (0.37373737), or terminating (0.125).

remote sensing gathering and recording information from a distance. Aircraft and satellites can observe a planetary surface or atmosphere, and space probes have sent back photographs and data about planets as distant as Neptune. Remote sensing usually refers to gathering data of the electromagnetic spectrum (such as visible light, ultraviolet light, and infrared light). In archaeology, surface survey techniques provide information without disturbing subsurface deposits.

Remote sensing is most commonly taken to refer to the process of photographing the Earth's surface with orbiting satellites. Satellites such as *Landsat* have surveyed all of the Earth's surface from orbit. Computer processing of data obtained by their scanning instruments, and the application of so-called false colors (generated by computer), have made it possible to reveal surface features invisible in ordinary light. This has proved valuable in agriculture, forestry, and urban planning, and has led to the discovery of new deposits of minerals.

REM sleep acronym for **rapid-eye-movement sleep**, phase of sleep that recurs several times nightly in humans and is associated with dreaming. The eyes flicker quickly beneath closed lids.

renal pertaining to the †kidneys.

renewable energy power from any source that can be replenished. Most renewable systems rely on †solar energy directly or through the weather cycle as †wave power, †hydroelectric power, wind power via †wind turbines, or solar energy collected by plants (alcohol fuels, for example). In addition, the gravitational force of the Moon can be harnessed through tidal power stations, and the heat trapped in the center of the Earth is used via †geothermal energy systems. Other examples are energy from †biofuel and †fuel cells. Renewable energy resources have the advantage of being non-polluting. However, some (such as wind energy) can be unreliable and therefore lose their effectiveness in providing a constant supply of energy.

Renewable Energy Trail http://www.dti.gov.uk/renewable/ed_pack/11-16.html

Virtual tour around the subject of renewable energy from the UK Department of Trade and Industry. Select the appropriate age group to access information on how power can be harnessed from Sun, wind, water, wood and crops, and waste.

renewable resource natural resource that is replaced by natural processes in a reasonable amount of time. Soil, water, forests, plants, and animals are all renewable resources as long as they are properly conserved. Solar, wind, wave, and geothermal energies are based on renewable resources.

Hydroelectric Power http://wwwga.usgs.gov/edu/hyhowworks.html

Insight and introduction to how hydroelectric power (HEP) works. The site provides an animated diagram, which allows the path of the water to be traced through a HEP plant, and shows how electricity is generated from it. Further relevant links to resources dealing with other aspects of HEP are provided.

rennin or **chymase**, enzyme found in the gastric juice of young mammals, used in the digestion of milk.

repair in biology, the rectification of damage to a body. If a body is damaged, it needs to be repaired. Bodies are made of †organs, which are themselves made of †tissues. Each tissue is made of many similar †cells. Damage will certainly kill some cells. In most tissues, some surrounding cells will be able to divide to produce new cells identical to the ones that were killed. This cell division repairs the damage. The same kind of cell division is used during †growth.

repellent anything whose smell, taste, or other properties discourages nearby creatures. **Insect repellent** is usually a chemical substance that keeps, for example, mosquitoes at bay; natural substances include citronella, lavender oil, and eucalyptus oils. A device that emits ultrasound waves is also claimed to repel insects and small mammals.

replication in biology, production of copies of the genetic material DNA; it occurs during cell division (†mitosis and †meiosis). Most mutations are caused by mistakes during replication.

During replication the paired strands of DNA separate, exposing the bases. Nucleotides floating in the cell matrix pair with the exposed bases, adenine pairing with thymine, and cytosine with guanine.

repression in psychology, a mental process that ejects and excludes from consciousness ideas, impulses, or memories that would otherwise threaten emotional stability.

In the Austrian psychiatrist Sigmund Freud's early writing, repression is controlled by the censor, a hypothetical mechanism or agency that allows ideas, memories, and so on from the unconscious to emerge into consciousness only if distorted or disguised, as for example in dreams.

repressor substance substance produced by a DNA regulator †gene. When the repressor is inactivated by an inducer, the DNA structural genes are freed for the synthesis of †messenger RNA.

reproduction in biology, the process by which a living organism produces other organisms more or less similar to itself. The ways in which species reproduce differ, but the two main methods are by †asexual reproduction and †sexual reproduction. Asexual reproduction involves only one parent without the formation of †gametes: the parent's cells divide by †mitosis to produce new cells with the same number and kind of †chromosomes as its own. Thus offspring produced asexually are clones of the parent and there is no variation. Sexual reproduction involves two parents, one male and one female. The parents' sex cells divide by †meiosis producing gametes, which contain only half the number of chromosomes of the parent cell. In this way, when two sets of chromosomes combine during †fertilization, a new combination of genes is produced. Hence the new organism will differ from both parents, and variation is introduced. The ability to reproduce is considered one of the fundamental attributes of living things.

sexual reproductive systems The plant organs concerned with sexual reproduction are found in the flowers. These consist of the †stamens (male organ) and †carpels (female organ). In male mammals the reproductive system consists of the †testes, which produce sperm, epididymis, sperm duct, and †penis, and in the females the †ovaries, which produce eggs, †Fallopian tubes, and †uterus.

hermaphrodites These are bisexual organisms, such as earthworms, that have both male and female reproductive organs, or plants whose flowers contain both stamens and carpels. This is the normal arrangement in most plants. Some plant species, such as maize (corn) and birch, which have separate male and female flowers on the same plants are described as **monoeious**; in **dioecious** species, such as willow and holly, the male and female flowers are on separate plants.

reproduction rate or **fecundity**, in ecology, the rate at which a population or species reproduces itself.

reptile any member of a class (Reptilia) of vertebrates. Unlike amphibians, reptiles have hard-shelled, yolk-filled eggs that are laid on land and from which fully formed young are born. Some snakes and lizards retain their eggs and give birth to live young. Reptiles are cold-blooded, and their skin is usually covered with scales. The metabolism is slow, and in some cases (certain large snakes) intervals between meals may be months. Reptiles date back over 300 million years.

Many extinct forms are known, including the orders Pterosauria, Plesiosauria, Ichthyosauria, and Dinosauria. The chief living orders are the Chelonia (tortoises and turtles), Crocodilia (alligators and crocodiles), and Squamata, divided into three suborders: Lacertilia (lizards), Ophidia or Serpentes (snakes), and Amphisbaenia (worm lizards). The order Rhynchocephalia has one surviving species, the lizardlike tuatara of New Zealand.

A four-year study of rainforest in eastern Madagascar revealed 26 new reptile species in 1995.

Reptilia class of vertebrates that comprises the †reptiles.

residue in chemistry, a substance or mixture of substances remaining after the removal of one or more components by a separation process.

The nonvolatile substance left in a container after †evaporation of liquid, the solid left behind after removal of liquid by filtration, and the substances left in a distillation flask after removal of components by †distillation, are all residues.

resin substance exuded from pines, firs, and other trees in gummy drops that harden in air. Varnishes are common products of the hard resins, and ointments come from the soft resins.

Rosin is the solid residue of distilled turpentine, a soft resin. The name "resin" is also given to many synthetic products manufactured by polymerization; they are used in adhesives, plastics, and varnishes.

resistance in physics, that property of a conductor that restricts the flow of electricity through it, associated with the conversion of electrical energy to heat; also the magnitude of this property.

Materials that are good conductors of electricity have electrons held loosely in the outer shells of their atoms. Current can flow easily and these materials have low resistance. In poor conductors of electricity, the electrons in the outer shells of their atoms are more strongly attracted by the positively-charged nucleus. Such materials restrict the flow of electrons and have a high resistance. Resistance (R) is related to current (I) and voltage (V) by the formula: $R = V/I$. The statement that current is proportional to voltage (resistance is constant) at constant temperature is known as †Ohm's law. It is approximately true for many materials that are accordingly described as "ohmic."

Resistance depends on many factors, such as the nature of the material, its temperature, its length and cross-sectional area, and its thermal properties.

resistivity in physics, a measure of the ability of a material to resist the flow of an electric current. It is numerically equal to the †resistance of a sample of unit length and unit cross-sectional area, and its unit is the ohm meter (symbol Wm). A good conductor has a low resistivity (1.7 x 10^{-8} Wm for copper); an insulator has a very high resistivity (10^{15} Wm for polyethane).

resistor in physics, any component in an electrical circuit used to introduce †resistance to a current by restricting the flow of electrons. Resistors are often made from wire-wound coils (higher resistance) or pieces of carbon (lower resistance). †Rheostats and †potentiometers are variable resistors.

When resistors R_1, R_2, R_3,... are connected in a †series circuit, the total resistance of the circuit is $R_1 + R_2 + R_3 + \ldots$. When resistors R_1, R_2, R_3,... are connected in a †parallel circuit, the total resistance of the circuit is R, given by $^1/_R = {}^1/_{R1} + {}^1/_{R2} + {}^1/_{R3} + \ldots$.

resolution of a telescope, the clarity of the final presentation to the observer (in image, radio picture, or X-ray read-out).

resolution of forces in mechanics, the division of a single force into two parts that act at right angles to each other. The two parts of a resolved force, called its **components**, have exactly the same effect when acting together on an object as the single force which they replace.

For example, the weight *W* of an object on a slope, tilted at an angle q, can be resolved into two components: one acting at a right angle to the slope, equal to *W* cos q, and one acting parallel to and down the slope, equal to *W* sin q. The component acting down the slope (minus any friction force that may be acting in the opposite direction) is responsible for the acceleration of the object.

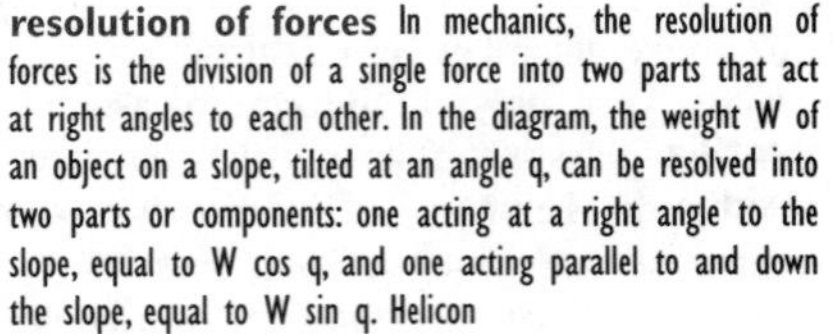

resolution of forces In mechanics, the resolution of forces is the division of a single force into two parts that act at right angles to each other. In the diagram, the weight W of an object on a slope, tilted at an angle q, can be resolved into two parts or components: one acting at a right angle to the slope, equal to W cos q, and one acting parallel to and down the slope, equal to W sin q. Helicon

resolving power ability of an optical device to discern two closely spaced light sources as independent entities.

resonance rapid amplification of a vibration when the vibrating object is subject to a force varying at its †natural frequency. In a trombone, for example, the length of the air column in the instrument is adjusted until it resonates with the note being sounded. Resonance effects are also produced by many electrical circuits. Tuning a radio, for example, is done by adjusting the natural frequency of the receiver circuit until it coincides with the frequency of the radio waves falling on the aerial.

Resonance has many physical applications. Children use it to increase the size of the movement on a swing, by giving a push at the same point during each swing. Soldiers marching across a bridge in step could cause the bridge to vibrate violently if the frequency of their steps coincided with its natural frequency. Resonance caused of the collapse of the Tacoma Narrows Bridge, USA, in 1940, when the frequency of the wind gusts coincided with the natural frequency of the bridge.

resources materials that can be used to satisfy human needs. Because human needs are varied and extend from basic physical requirements, such as food and shelter, to spiritual and emotional needs that are hard to define, resources cover a vast range of items. The intellectual resources of a society – its ideas and technologies – determine which aspects of the environment meet that society's needs, and therefore become resources. For example, in the 19th century uranium was used only in the manufacture of colored glass. Today, with the development of nuclear technology, it is a military and energy resource. Resources are often divided into **human resources**, such as labor, supplies, and skills, and **natural resources**, such as climate, †fossil fuels, and water. Natural resources are divided into †non-renewable resources and †renewable resources.

Non-renewable resources include minerals such as coal, copper ores, and diamonds, which exist in strictly-limited quantities. Once consumed they will not be replenished within the time-span of human history. In contrast, water supplies, timber, food crops, and similar resources can, if managed properly, provide a steady yield virtually forever; they are therefore replenishable or renewable resources. Inappropriate use of renewable resources can lead to their destruction, as for example the cutting down of rainforests, with secondary effects, such as the decrease in oxygen and the increase in carbon dioxide and the resulting †greenhouse effect. Some renewable resources, such as wind or solar energy, are continuous; supply is largely independent of people's actions.

Demands for resources made by rich nations are causing concern that the present and future demands of industrial societies cannot be sustained for more than a century or two, and that this will be at the expense of the developing world and the global environment. Other authorities believe that new technologies will be developed, enabling resources that are now of little importance to replace those being used up.

respiration process that occurs inside cells in which carbohydrate, particularly glucose, is broken down to release †energy that the cell can use. This energy is used for many different processes, but in all of them energy transfer occurs. The processes range from muscle contraction to the manufacture of protein for new cells. Respiration is a key feature of life and is carried out by all living †cells. There are two kinds of respiration in organisms – aerobic and anaerobic respiration. †Aerobic respiration is a complex process of chemical reactions in which oxygen is used to break down glucose into †carbon dioxide and †water. This releases energy in the form of energy-carrying molecules (†ATP). Respiration sometimes occurs without oxygen, and this is called anaerobic respiration. In this case, glucose is only partially broken down, and the end products are energy and either lactic acid or ethanol (alcohol) and carbon dioxide; this process is termed †fermentation.

The word "respiration" should not be used to refer to the air movements in the air passageways. These air movements are called ventilation (†breathing).

The starting and finishing points of aerobic respiration would be the same if glucose was burned. During burning the energy that is released is all in the form of heat. In aerobic respiration, however, the energy is released in a controlled way and less is released as heat. Most of the released energy is used to drive various processes in the cell, such as growth or movement.

In humans, anaerobic respiration can only carry on for a short time. The †muscles producing lactic acid will stop working as it builds up. However, many micro-organisms can respire anaerobically for long periods of time or all the time. Yeast respires aerobically if oxygen is present, but, if there is no oxygen, it respires anaerobically. In anaerobic respiration it produces †alcohol.

Respiration http://www.purchon.com/biology/respire.htm

Outlines the process of respiration, focusing on three main areas of "Tissue Respiration," "Ventilation of the Lung," and "Gaseous Exchange." Key points are highlighted to help revision, and there are further links to a page with tips on remembering the equation for respiration.

Structure of the Human Respiratory System http://www.stemnet.nf.ca/~dpower/resp/main.htm

Short guide to the human respiratory system, including explanations of the functions of the major structures such as the nose and what happens to their effectiveness when they are exposed to tobacco smoke. It is divided into two main sections: "Gas exchange in humans" and "Other factors involved with respiration."

respiratory surface area used by an organism for the exchange of gases, for example the lungs, gills or, in plants, the leaf interior. The gases oxygen and carbon dioxide are both usually involved in respiration and photosynthesis. Although organisms have evolved different types of respiratory surface according to need, there are certain features in common. These include thinness and moistness, so that the gas can dissolve in a membrane and then diffuse into the body of the organism. In many animals the gas is then transported away from the surface and towards interior cells by the blood system.

respiratory tract in animals, the air passageways that lead from the nose and mouth to the lungs, including the air sacs of the lungs. A specialized epithelial tissue lines much of these air passages. This is cilated †epithelium and contains cells with beating hairs called cilia that sweep mucus and trapped particles away from the air sacs in the lungs. Smoking damages this epithelium.

Care should be taken with the term "respiratory tract," because it could be confusing. The word †respiration is now linked to processes that occur inside cells and is no longer used to relate to air movements in the air passageways. The term "air passageway" is probably a less confusing term. Air movements in the air passageways are called ventilation (†breathing).

response any change in an organism occurring as a result of a stimulus. There are many different types of response, some involving the entire organism, others only groups of cells or tissues. Examples include the muscular contractions in an animal, the movement of leaves towards the light, and the onset of hibernation by small mammals at the start of winter.

resting potential potential difference across a nerve or muscle membrane in the absence of a stimulus.

rest mass in physics, the mass of a body when its velocity is zero or considerably below that of light.

According to the theory of †relativity, at very high velocities, there is a relativistic effect that increases the mass of the particle.

restriction enzyme or **endonuclease**, bacterial †enzyme that breaks a chain of †DNA into two pieces at a specific point; used in †genetic engineering. The point along the DNA chain at which the enzyme can work is restricted to places where a specific sequence of base pairs occurs. Different restriction enzymes will break a DNA chain at different points. The overlap between the fragments is used in determining the sequence of base pairs in the DNA chain.

resultant force in mechanics, a single force acting on a particle or body whose effect is equivalent to the combined effects of two or more separate forces. The resultant of two forces acting at one point on an object can be found using the †parallelogram of forces method.

reticulo-endothelial system group of cells that exist in continual contact with the blood and lymph, that is, in the bone marrow, spleen, liver, and lymph nodes. They ingest bacteria, other foreign particles and dead tissue, and aid tissue repair.

Reticulum small constellation of the southern hemisphere, represented as a net or reticule (the grid of lines on the eyepiece of a telescope). It used to be called "Rhombus" or "Reticulum Rhomboidalis."

retina light-sensitive area at the back of the †eye connected to the brain by the optic nerve. It has several layers and in humans contains over a million rods and cones, sensory cells capable of converting light into nervous messages that pass down the optic nerve to the brain.

The **rod cells**, about 120 million in each eye, are distributed throughout the retina. They are sensitive to low levels of light, but do not provide detailed or sharp images, nor can they detect color. The **cone cells**, about 6 million in number, are mostly concentrated in a central region of the retina called the **fovea**, and provide both detailed and color vision. The cones of the human eye contain three visual pigments, each of which responds to a different primary color (red, green, or blue). The brain can interpret the varying signal levels from the three types of cone as any of the different colors of the visible spectrum.

The image actually falling on the retina is highly distorted; research into the eye and the optic centers within the brain has shown that this poor quality image is processed to improve its quality. The retina can become separated from the wall of the eyeball as a result of a trauma, such as a boxing injury. It can be reattached by being "welded" into place by a laser.

retinol or **vitamin A**, fat-soluble chemical derived from b-carotene and found in milk, butter, cheese, egg yolk, and liver. Lack of retinol in the diet leads to the eye disease **xerophthalmia**.

retrograde in astronomy, describing the orbit or rotation of a †planet or †satellite if the sense of rotation is opposite to the general sense of rotation of the Solar System. On the †celestial sphere, it refers to motion from east to west against the background of stars.

retrovirus any of a family of †viruses (Retroviridae) containing the genetic material †RNA rather than the more usual †DNA.

For the virus to express itself and multiply within an infected cell, its RNA must be converted to DNA. It does this by using a built-in enzyme known as reverse transcriptase (since the transfer of genetic information from DNA to RNA is known as †transcription, and retroviruses do the reverse of this). Retroviruses include those causing †AIDS and some forms of leukemia. See †immunity.

Retroviruses are used as vectors in †genetic engineering, but they cannot be used to target specific sites on the chromosome. Instead they incorporate their genes at random sites.

reuse multiple use of a product (often a form of packaging), by returning it to the manufacturer or processor each time. Many such returnable items are sold with a deposit which is reimbursed if the item is returned. Reuse is usually more energy- and resource-efficient than †recycling unless there are large transport or cleaning costs.

reverberation in acoustics, the multiple reflections, or echoes, of sounds inside a building that merge and persist a short time (up to a few seconds) before fading away. At each reflection some of the sound energy is absorbed, causing the amplitude of the sound wave and the intensity of the sound to reduce a little.

Too much reverberation causes sounds to become confused and indistinct, and this is particularly noticeable in empty rooms and halls, and such buildings as churches and cathedrals where the hard, unfurnished surfaces do not absorb sound energy well. Where walls and surfaces absorb sound energy very efficiently, too little reverberation may cause a room or hall to sound dull or "dead." Reverberation is a key factor in the design of theatres and concert halls, and can be controlled by lining ceilings and walls with materials possessing specific sound-absorbing properties.

reverse osmosis movement of solvent (liquid) through a semipermeable membrane from a more concentrated solution to a more dilute solution. The solvent's direction of movement is opposite to that which it would experience during †osmosis, and is achieved by applying an external pressure to the solution on the more concentrated side of the membrane. The technique is used in desalination plants, when water (the solvent) passes from brine (a salt solution) into fresh water via a semipermeable filtering device.

reversible reaction chemical reaction that proceeds in both directions at the same time, as the product decomposes back into reactants as it is being produced. Such reactions do not run to completion, provided that no substance leaves the system. Examples include the manufacture of ammonia from hydrogen and nitrogen, and the oxidation of sulfur dioxide to sulfur trioxide.

reversing layer lower †chromosphere of the Sun, a comparatively cool region in which radiation at certain wavelengths is absorbed from the continuous spectrum emitted from the Sun's †photosphere.

Reynolds, Osborne (1842–1912) Irish physicist and engineer who studied fluid flow and devised the †**Reynolds number**, which gives a numerical criterion for determining whether fluid flow under specified conditions will be smooth or turbulent. His work also led to a radical redesign of boilers, condensers, and turbine pumps.

Reynolds number number used in †fluid mechanics to determine whether fluid flow in a particular situation (through a pipe or around an aircraft body or a fixed structure in the sea) will be turbulent or smooth. The Reynolds number is calculated using the flow velocity, density, and viscosity of the fluid, and the dimensions of the flow channel. It is named after Irish engineer Osborne Reynolds.

The Reynolds number for the wing of a Boeing 747 is 7×10^7 and that for a nuclear submarine is 10^9.

rhe unit of fluidity equal to the reciprocal of the †poise.

Rhea or **Saturn V**, one of the larger moons of †Saturn, discovered in 1672 by Italian-born French astronomer Giovanni †Cassini. It has a diameter of 1,530 km/950 mi and orbits Saturn at an average distance of 527,000 km/327,000 mi with a period (time taken to circle the planet) of 4.52 days.

rhenium chemical symbol Re, (Latin *Rhenus* "Rhine") heavy, silver-white, metallic element, atomic number 75, relative atomic mass 186.2. It has chemical properties similar to those of manganese and a very high melting point (3,180°C/5,756°F), which makes it valuable as an ingredient in alloys.

It was identified and named in 1925 by German chemists Walter Noddack (1893–1960), Ida Tacke, and Otto Berg from the Latin name for the River Rhine.

rheostat variable †resistor, usually consisting of a high-resistance wire-wound coil with a sliding contact. It is used to vary electrical resistance without interrupting the current (for example, when dimming lights). The circular type, which can be used, for example, as the volume control of an amplifier, is also known as a †potentiometer.

rhesus factor group of †antigens on the surface of red blood cells of humans which characterize the rhesus blood group system. Most individuals possess the main rhesus factor (Rh+), but those without this factor (Rh–) produce †antibodies if they come into contact with it. The name comes from rhesus monkeys, in whose blood rhesus factors were first found.

If an Rh– mother carries an Rh+ fetus, she may produce antibodies if fetal blood crosses the †placenta. This is not normally a problem with the first infant because antibodies are only produced slowly. However, the antibodies continue to build up after birth, and a second Rh+ child may be attacked by antibodies passing from mother to fetus, causing the child to contract anemia, heart failure, or brain damage. In such cases, the blood of the infant has to be changed for Rh– blood; a badly affected fetus may be treated in the womb. The problem can be circumvented by giving the mother anti-Rh globulin just after the first pregnancy, preventing the formation of antibodies.

Rhine, Joseph Banks (1895–1980) US parapsychologist who carried out many groundbreaking laboratory experiments in †parapsychology. Some are described in his book *Extra-Sensory Perception* (1934), which made ESP a common term.

rhizoid hairlike outgrowth found on the †gametophyte generation of ferns, mosses, and liverworts. Rhizoids anchor the plant to the substrate and can absorb water and nutrients.

They may be composed of many cells, as in mosses, where they are usually brownish, or may be unicellular, as in liverworts, where they are usually colorless. Rhizoids fulfill the same functions as the †roots of higher plants but are simpler in construction.

rhizome or **rootstock**, horizontal underground plant stem. It is a †perennating organ in some species, where it is generally thick and fleshy, while in other species it is mainly a means of †vegetative reproduction, and is therefore long and slender, with buds all along it that send up new plants. The potato is a rhizome that has two distinct parts, the tuber being the swollen end of a long, cordlike rhizome.

rhm abbreviation of roentgen-hour-meter, the unit of effective strength of a radioactive source that produces gamma rays. It is used for substances for which it is difficult to establish radioactive disintegration rates.

rhodium chemical symbol Rh, (Greek *rhodon* "rose") hard, silver-white, metallic element, atomic number 45, relative atomic mass 102.905. It is one of the so-called platinum group of metals and is resistant to tarnish, corrosion, and acid. It occurs as a free metal in the natural alloy osmiridium and is used in jewelry, electroplating, and thermocouples.

rhombic sulfur allotropic form of sulfur. At room temperature, it is the stable †allotrope, unlike monoclinic sulfur.

rhombus equilateral (all sides equal) †parallelogram. As with a parallelogram, the rhombus has diagonally opposed angles of equal size. Its diagonals bisect each other at right angles, and its area is half the product of the lengths of the two diagonals. The shape is sometimes called a diamond. A rhombus whose internal angles are 90° is called a †square.

rhyolite †igneous rock, the fine-grained volcanic (extrusive) equivalent of granite.

ria long narrow sea inlet, usually branching and surrounded by hills. A ria is deeper and wider towards its mouth, unlike a †fjord. It is formed by the flooding of a river valley due to either a rise in sea level or a lowering of a landmass.

rib long, usually curved bone that extends laterally from the †spine in vertebrates. Most fishes and many reptiles have ribs along most of the spine, but in mammals they are found only in the chest area. In humans, there are 12 pairs of ribs. The ribs protect the lungs and heart, and allow the chest to expand and contract easily.

At the rear, each pair is joined to one of the vertebrae of the spine. The upper seven ("true" or vertebrosternal ribs) are joined by †cartilage directly to the breast bone (sternum). The next three ("false" or vertebro-costal ribs) are joined by cartilage to the end of the rib above. The last two ("floating ribs") are not attached at the front. The diaphragm and muscles between adjacent ribs are responsible for the respiratory movements which fill the lungs with air.

ribbon lake long, narrow lake found on the floor of a †glacial trough. A ribbon lake will often form in an elongated hollow carved out by a glacier, perhaps where it came across a weaker band of rock. Ribbon lakes can also form when water ponds up behind a terminal moraine or a landslide. The English Lake District is named after its many ribbon lakes, such as Lake Windermere and Coniston Water.

riboflavin or **vitamin B_2**, †vitamin of the B complex important in cell respiration. It is obtained from eggs, liver, and milk. A deficiency in the diet causes stunted growth.

ribonucleic acid full name of †RNA.

ribosome protein-making machinery of the cell. Ribosomes are located on the endoplasmic reticulum (ER) of eukaryotic cells, and are made of proteins and a special type of †RNA, ribosomal RNA. They receive messenger RNA (copied from the †DNA) and †amino acids, and "translate" the messenger RNA by using its chemically coded instructions to link amino acids in a specific order, to make a strand of a particular protein.

Ricci-Curbastro, Gregorio (1853–1925) Italian mathematician whose systematization of absolute differential calculus (the **Ricci calculus**) enabled Albert †Einstein to derive the theory of †relativity.

Richards, Theodore William (1868–1928) US chemist who was awarded the Nobel Prize for Chemistry in 1914 for determining as accurately as possible the †relative atomic masses of a large number of elements. He also investigated the physical properties of the elements, such as atomic volumes and the compressibilities of non-metallic solid elements.

Richardson, Owen Willans (1879–1959) British physicist. He studied the emission of electricity from hot bodies, giving the name †thermionics to the subject. At Cambridge University, he worked under J J †Thomson in the Cavendish Laboratory. He was awarded the Nobel Prize for Physics in 1928 for his work on thermionic phenomena and associated law. He was knighted in 1939.

Richter, Burton (1931–) US particle physicist. In the 1960s he designed the Stanford Positron–Electron Accelerating Ring (SPEAR), a machine designed to collide positrons and electrons at high energies. In 1974 Richter and his team used SPEAR to produce a new subatomic particle, the y meson. This was the first example of a particle formed from a charmed quark, the quark whose existence had been postulated by Sheldon †Glashow ten years earlier. Richter was awarded the Nobel Prize for Physics in 1976 for his discovery of the y meson. He shared the award with Samuel †Ting, who had discovered the particle independently. Richter was director of the Stanford Linear Accelerator Center (SLAC) 1984–99.

Richter, Charles Francis (1900–1985) US seismologist, deviser of the †Richter scale used to indicate the strength of the waves from earthquakes.

Richter scale quantitative scale of earthquake magnitude based on the measurement of seismic waves, used to indicate the magnitude of an †earthquake at its epicenter. The Richter scale is logarithmic, so an earthquake of 6.0 is ten times greater than one of 5.0. The magnitude of an earthquake differs from its intensity, measured by the †Mercalli scale, which is qualitative and varies from place to place for the same earthquake. The scale is named after US seismologist Charles Richter.

An earthquake's magnitude is a function of the total amount of energy released, and each point on the Richter scale represents a thirtyfold increase in energy over the previous point. One of the greatest earthquakes ever recorded, in 1920 in Gansu, China, measured 8.6 on the Richter scale.

rickets defective growth of bone in children due to an insufficiency of calcium deposits. The bones, which do not harden adequately, are bent out of shape. It is usually caused by a lack of vitamin D and insufficient exposure to sunlight. Renal rickets, also a condition of malformed bone, is associated with kidney disease.

rickettsia any of a group of rod-shaped micro-organisms similar to bacteria. They are spread to humans by invertebrates such as ticks, fleas, and lice. The diseases they cause include epidemic typhus, and the milder endemic typhus and scrub typhus, Rocky Mountain spotted fever, rickettsial pox, trench fever, and Q fever.

The genome for *Rickettsia* was mapped in 1998. It revealed that *Rickettsia* may well be the closest living relative of the ancestor of †mitochondria, which once lived outside the body.

ridge of high pressure elongated area of high atmospheric pressure extending from an anticyclone. On a synoptic weather chart it is shown as a pattern of lengthened isobars. The weather under a ridge of high pressure is the same as that under an anticyclone.

Riemann, Georg Friedrich Bernhard (1826–1866) German mathematician whose system of non-Euclidean geometry, thought at the time to be a mere mathematical curiosity, was used by Albert †Einstein to develop his general theory of †relativity. Riemann made a breakthrough in conceptual understanding within several other areas of mathematics: the theory of functions, vector analysis, projective and differential geometry, and topology.

Riemann, Georg Friedrich Bernhard http://www-groups.dcs.st-andrews.ac.uk/history/Mathematicians/Riemann.html

Part of an archive containing the biographies of the world's greatest mathematicians, this site is devoted to the life and contributions of Georg Riemann. In addition to biographical information, you will find a list of references about Riemann and links to other essays in the archive that reference him. The text of this essay includes hypertext links to the essays of those mathematicians and thinkers who influenced Riemann. There is also an image of him, which can be enlarged, and a map of his birthplace.

Riemann geometry system of non-Euclidean geometry devised by Bernhard Riemann, developed primarily as †elliptic geometry, but then extended to †hyperbolic geometry.

Riemann hypothesis statement that has as yet not been proved (or disproved), that the †zeta function takes the value zero in the right-half plane of the †Argand diagram only for complex numbers of the form $^1/_2$ + i*a*, where i = Ö–1 and *a* is real.

Riemann space non-Euclidean geometry, using *n*-dimensional coordinates (x_1, ..., x_n) and calculating length according to the formula:

$$ds^2 = \sum g_{ij}\,dx^i dx^j$$

where d*s* is the limiting incremental length along a curve, dx^i is "a limiting increment in the *i* coordinate," and *i*, *j* run through the values 1, 2, 3, …, *n*.

Riemann surface in non-Euclidean geometry, a multi-connected many-sheeted surface that can be dissected by cross-cuts into a singly connected surface. Such a representation of a complex algebraic function is used to study the "behavior" of other complex functions as they are mapped conformally (transformed) onto it. A Riemann surface has been described as topologically equivalent to a box with holes in it.

rift valley valley formed by the subsidence of a block of the Earth's †crust between two or more parallel †faults. Rift valleys are steep-sided and form where the crust is being pulled apart, as at †mid-ocean ridges, or in the Great Rift Valley of East Africa. In cross-section they can appear like a widened gorge with steep sides and a wide floor.

Rigel or **Beta Orionis**, brightest star in the constellation Orion. It is a blue-white supergiant, with an estimated diameter 50 times that of the Sun. It is 910 light years from Earth, and is intrinsically the brightest of the first-magnitude stars, its true luminosity being about 100,000 times that of the Sun. It is the seventh-brightest star in the night sky.

right-angled triangle triangle in which one of the angles is a right angle (90°). It is the basic form of triangle for defining trigonometrical ratios (for example, sine, cosine, and tangent) and for which †Pythagoras' theorem holds true. The longest side of a right-angled triangle is called the hypotenuse; its area is equal to half the product of the lengths of the two shorter sides.

Any triangle constructed with its hypotenuse as the diameter of a circle, with its opposite vertex (corner) on the circumference, is a right-angled triangle. This is a fundamental theorem in geometry, first credited to the Greek mathematician Thales about 580 BC.

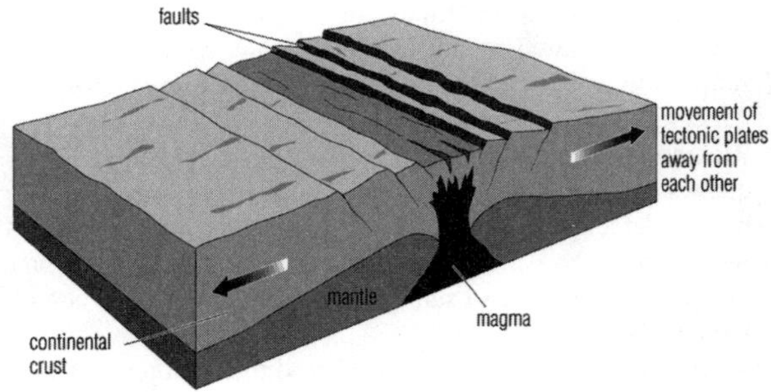

rift valley The subsidence of rock resulting from two or more parallel rocks moving apart is known as a graben. When this happens on a large scale, with tectonic plates moving apart, a rift valley is created. Helicon

right ascension in astronomy, coordinate on the †celestial sphere that corresponds to longitude on the surface of the Earth. It is measured in hours, minutes, and seconds eastwards from the point where the Sun's path, the ecliptic, once a year intersects the celestial equator; this point is called the **vernal equinox**.

right-hand rule in physics, a memory aid used to recall the relative directions of motion, magnetic field, and current in an electric generator. It was devised by English physicist John Fleming. See †Fleming's rules.

Rigil Kent or **Alpha Centauri**, brightest star in the constellation Centaurus and the third-brightest star in the night sky. It is actually a triple star (see †binary star); the two brighter stars orbit each other every 80 years, and the third, Proxima Centauri, is the closest star to the Sun, 4.2 light years away, 0.1 light years closer than the other two.

rille or **rima**, narrow V-shaped valley on the surface of the Moon. Winding, or sinuous, rilles probably formed when the roofs of tunnels in lava flows collapsed when the lava cooled. They may be up to 10 km/6 mi long. Straight or curved rilles correspond to geological fault lines.

Ring †planetary nebula in the constellation of Lyra in the northern hemisphere, appearing as a bright oval-shaped ring with a dark center. The nebula is made up of a shell of hot gas that has been blown away from a faint blue-white dwarf star at the center of the ring. It was discovered by French astronomer Antoine Darquier (1718–1802) in 1779. It is also known as Messier 57 (M57).

ring mathematical structure that constitutes a restricted form of field in which division might be unavailable.

ring circuit household electrical circuit in which appliances are connected in series to form a ring with each end of the ring connected to the power supply. It superseded the radial circuit.

rings, planetary in astronomy, bands of rock particles and dust that orbit the planets Saturn (the most prominent), Jupiter, Uranus, and Neptune. They are very thin (up to 1 km/0.6 mi) compared to their diameter and contain gaps having little or no material, created by the resonance effects of small moons within the rings.

ripple tank in physics, shallow water-filled tray used to demonstrate various properties of waves, such as reflection, refraction, diffraction, and interference.

As light waves are not visible, waves produced in a ripple tank can be used to demonstrate their properties. Water waves are produced using an electric motor to vibrate either a straight bar dipped into the water to produce straight waves or a circular ball to produce circular waves. A stroboscope is used to study the water waves by "freezing" their motion.

If a small piece of polystyrene is placed in a ripple tank while straight water waves are produced, then, viewed from the side, as the ripples move up and down (forming a pattern of crests and troughs) the polystyrene remains in its original position as the water waves travel along. This shows that waves transfer energy from one place to another without transferring matter.

Ritter, Johann Wilhelm (1776–1810) German physicist who carried out early work on electrolytic cells and discovered †ultraviolet radiation.

ritualization in ethology, a stereotype that occurs in certain behavior patterns when these are incorporated into displays. For example, the exaggerated and stylized head toss of the goldeneye drake during courtship is a ritualization of the bathing movement used to wet the feathers; its duration and form have become fixed. Ritualization may make displays clearly recognizable, so ensuring that individuals mate only with members of their own species.

river large body of water that flows down a slope along a channel restricted by adjacent banks and †levees. A river starts at a point called its **source**, and enters a sea or lake at its **mouth**. Along its length it may be joined by smaller rivers called **tributaries**; a river and its tributaries are contained within a †drainage basin. The point at which two rivers join is called the †confluence.

Rivers are formed and molded over time chiefly by the processes of †erosion, and by the **transport** and **deposition** of †sediment. Rivers are able to work on the landscape through erosion, transport, and deposition. The amount of potential energy available to a river is proportional to its initial height above sea level. A river follows the path of least resistance downhill, and deepens, widens and lengthens its channel by erosion. Up to 95% of a river's potential energy is used to overcome friction.

One way of classifying rivers is by their stage of development. An upper course is typified by a narrow V-shaped valley with numerous †waterfalls, †lakes, and rapids. Because of the steep gradient of the topography and the river's height above sea level, the rate of erosion is greater than the rate of deposition, and downcutting occurs by **vertical corrasion** (erosion or abrasion of the bed or bank caused by the load carried by the river).

In the middle course of a river, the topography has been eroded over time and the river's course has a shallow gradient. Such a river is said to be graded. Erosion and deposition are delicately balanced as the river meanders (gently curves back and forth) across the extensive †flood plain. **Horizontal corrasion** is the main process of erosion. The flood plain is an area of periodic †flooding along the course of a river valley where fine silty material called †alluvium is deposited by the flood water. Features of a mature river (or the **lower course** of a river) include extensive †meanders, †oxbow lakes, and braiding.

Many important flood plains, such as the Nile flood plain in Egypt, occur in arid areas where their exceptional fertility is very important to the local economy. However, using flood plains as the site of towns and villages involves a certain risk, and it is safer to use flood plains for other uses, such as agriculture and parks. Water engineers can predict when flooding is likely and take action to prevent it by studying †hydrographs, which show how the †discharge of a river varies with time.

Major rivers of the world include the Ganges, the Mississippi, and the Nile, the world's longest river.

Features Produced by Running Water http://www.zephryus.demon.co.uk/geography/resources/revision/river.html

Revision page aimed at GCSE geography students, explaining essential terms and concepts to do with rivers and running water. The layout is simple and quick to load, and the definitions are clear and concise.

Internal and External Friction http://www.zephryus.demon.co.uk/geography/resources/fieldwork/fluvial/fric.html

Page explaining the concept of friction in the context of rivers. This page, part of a much larger site containing resources for secondary school geography, describes how friction is crucial to the way in which a river is formed and how it acts.

River Systems http://www.riverresource.com/text/river_systems.html

The site begins by defining the major characteristics of a river stream and then classifies an extensive list of concepts including information on drainage patterns, stream flow, and erosion into clearly organized tables and key points to note. A very thorough site, this would make a useful aid to revision.

river capture the diversion (capture) of the headwaters of one river into a neighboring river. River capture occurs when a stream is carrying out rapid †headward erosion (backwards erosion at its source). Eventually the stream will cut into the course of a neighboring river, causing the headwaters of that river to be diverted, or "captured."

The headwaters will then flow down to a lower level (often making a sharp bend, called an elbow of capture) over a steep slope, called a knickpoint. A waterfall will form here. †Rejuvenation then occurs, causing rapid downwards erosion. An excellent example is the capture of the River Burn by the River Lyd, Devon, England.

river cliff or **bluff**, steep slope forming the outer bank of a †meander (bend in a river). It is formed by the undercutting of the river current, which at its fastest when it sweeps around the outside of the meander.

river terrace part of an old †flood plain that has been left perched on the side of a river valley. It results from †rejuvenation, a renewal in the erosive powers of a river. River terraces may be fertile and are often used for farming. They are also commonly chosen as sites for human settlement because they are safe from flooding. Many towns and cities throughout the world have been built on terraces, including London, which is built on the terraces of the River Thames.

Other terraces are formed as a result of glacial outwash. This deposits large amounts of sand and gravel, which are later eroded by rivers. The uneroded parts remain as terraces. In southern England, the Harnborough, Wolvercote, Summertown, and Radley terraces near Oxford are areas of high ground above the flood plains of the River Thames and River Cherwell.

RNA abbreviation for ribonucleic acid, nucleic acid involved in the process of translating the genetic material †DNA into proteins. It is usually single-stranded, unlike the double-stranded DNA, and consists of a large number of nucleotides strung together, each of which comprises the sugar ribose, a phosphate group, and one of four bases (uracil, cytosine, adenine, or guanine). RNA is copied from DNA by the formation of †base pairs, with uracil taking the place of thymine.

RNA occurs in three major forms, each with a different function in the synthesis of protein molecules. **Messenger RNA** (mRNA) acts as the template for protein synthesis. Each †codon (a set of three bases) on the RNA molecule is matched up with the corresponding amino acid, in accordance with the †genetic code. This process (translation) takes place in the ribosomes, which are made up of proteins and **ribosomal RNA** (rRNA). **Transfer RNA** (tRNA) is responsible for combining with specific amino acids, and then matching up a special "anticodon" sequence of its own with a codon on the mRNA.

This is how the genetic code is translated.

Although RNA is normally associated only with the process of protein synthesis, it makes up the hereditary material itself in some viruses, such as †retroviruses.

Roaring Forties nautical expression for regions south of latitude 40°S, in the southern oceans, where strong westerly winds prevail.

Roberts, Richard John (1943–) British molecular biologist who was awarded a Nobel Prize for Physiology or Medicine in 1993 with Phillip †Sharp for the discovery of split genes (genes interrupted by nonsense segments of DNA).

Robinson, Robert (1886–1975) English chemist who was awarded the Nobel Prize for Chemistry in 1947 for his research in organic chemistry on the structure of many natural products, including flower pigments and alkaloids. He formulated the electronic theory now used in organic chemistry. He was knighted in 1939.

Roche limit in astronomy, distance from a planet within which a large moon would be torn apart by the planet's gravitational force, creating a set of rings. The Roche limit lies at approximately 2.5 times the planet's radius (the distance from its center to its surface).

roche moutonnée outcrop of tough bedrock having one smooth side and one jagged side, found on the floor of a †glacial trough (U-shaped valley). It may be up to 40 m/132 ft high. A roche moutonnée is a feature of glacial erosion – as a glacier moved over its surface, ice and debris eroded its upstream side by corrasion, smoothing it and creating long scratches or striations. On the sheltered downstream side fragments of rock were plucked away by the ice, causing it to become steep and jagged.

rock constituent of the Earth's crust composed of †minerals or materials of organic origin that have consolidated into hard masses. There are three basic types of rock: †igneous, †sedimentary, and †metamorphic. Rocks are composed of a combination (or aggregate) of minerals, and the property of a rock will depend on its components. Where deposits of economically valuable minerals occur they are termed †**ores**. As a result of †weathering, rock breaks down into very small particles that combine with organic materials from plants and animals to form †soil. In †geology the term "rock" can also include unconsolidated materials such as †sand, mud, †clay, and †peat.

The change of one type of rock to another is called the rock cycle.

Composition of Rocks http://www.geog.ouc.bc.ca/physgeog/contents/10d.html

From a much larger site "Fundamentals of Physical Geography" set up by a Canadian university and covering many aspects of the area of study. The site is clear and well organized and contains maps and diagrams.

Processes and Products and the Rock Cycle http://www.shoreline.ctc.edu/geology/ProcessesProducts.html

Detailed insight into the rock cycle, which begins with the context of the Earth as a whole, before moving onto specific examples. The stages of the rock cycle are listed, with each stage containing an explanation and definition. A diagram of the cycle is provided, giving a good reference point for the description of the different stages.

Rock Cycle http://www.schoolchem.com/rk1.htm

Graphical depiction of the rock cycle, including the three types of rock: metamorphic, sedimentary, and igneous. Click on the links to find more information on the three types and sub-types that are present. A colorful style highlights key concepts and words in connection with the rock cycle.

rocket projectile driven by the reaction of gases produced by a fast-burning fuel. Unlike jet engines, which are also reaction engines, rockets carry their own oxygen supply to burn their fuel and do not require any surrounding atmosphere. For warfare, rocket heads carry an explosive device.

Rockets have been valued as fireworks since the middle ages, but their intensive development as a means of propulsion to high altitudes, carrying payloads, started only in the interwar years with the state-supported work in Germany (primarily by German-born US rocket engineer Wernher von Braun) and the work of US rocket pioneer Robert Hutchings Goddard. Being the only form of propulsion available that can function in a vacuum, rockets are essential to exploration in outer space. Multistage rockets have to be used, consisting of a number of rockets joined together.

Two main kinds of rocket are used: one burns liquid propellants, the other solid propellants. The fireworks rocket uses gunpowder as a solid propellant. The space shuttle's solid rocket boosters use a mixture of powdered aluminum in a synthetic rubber binder. Most rockets, however, have liquid propellants, which are more powerful and easier to control. Liquid hydrogen and kerosene are common fuels, while liquid oxygen is the most common oxygen provider, or oxidizer. One of the biggest rockets ever built, the Saturn V Moon rocket, was a three-stage design, standing 111 m/365 ft high. It weighed more than 2,700 tonnes on the launch pad, developed a take-off thrust of some 3.4 million kg/7.5 million lb, and could place almost 140 tonnes into low Earth orbit. In the early 1990s, the most powerful rocket system was the Soviet Energiya, capable of placing 190 metric tonnes into low Earth orbit. The US space shuttle can only carry up to 29 metric tonnes of equipment into orbit.

Rocketry Online
http://www.rocketryonline.com/
Searchable site with reference materials, organizations, clubs, upcoming launches, products, and books devoted to rocketry.

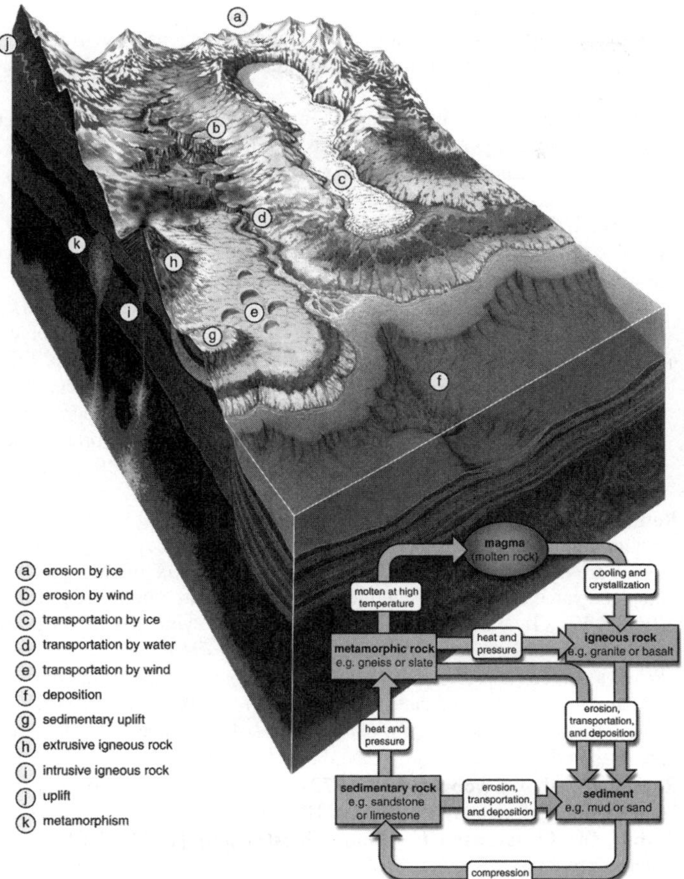

rock The rock cycle. Rocks are not as permanent as they seem but are being constantly destroyed and renewed. When a rock becomes exposed on the Earth's surface, it starts to break down through weathering and erosion. The resulting debris is washed or blown away and deposited, for example in sea or river beds, or in deserts, where it eventually becomes buried by yet more debris. Over time, this debris is compressed and compacted to form sedimentary rock, which may in time become exposed and eroded once more. Alternatively the sedimentary rock may be pushed further towards the Earth's center where it melts and solidifies to form igneous rock or is heated and crushed to such a degree that its mineral content alters and it becomes metamorphic rock. Igneous and metamorphic rock may also become exposed and eroded by the same processes as sedimentary rock, and the cycle continues. Helicon

rod type of light-sensitive cell in the †retina of most vertebrates. Rods are highly sensitive and provide only black and white vision. They are used when lighting conditions are poor and are the only type of visual cell found in animals active at night.

rodent any mammal of the worldwide order Rodentia, making up nearly half of all mammal species. Besides ordinary "cheek teeth," they have a single front pair of incisor teeth in both upper and lower jaw, which continue to grow as they are worn down.

roentgen or **röntgen**; symbol R, unit of radiation exposure, used for X-rays and gamma rays. It is defined in terms of the number of ions produced in one cubic centimeter of air by the radiation. Exposure to 1,000 roentgens gives rise to an absorbed dose of about 870 rads (8.7 grays), which is a dose equivalent of 870 rems (8.7 sieverts).

Rogers, Carl Ransom (1902–1987) US psychologist who developed the client-centered approach to counseling and psychotherapy. This stressed the importance of clients making their own decisions and developing their own potential (self-actualization).

Rohrer, Heinrich (1933–) Swiss physicist who was awarded the Nobel Prize for Physics in 1986 for his involvement in the invention of the scanning tunneling electron microscope (STM), an ultra-powerful microscope capable of imaging individual atoms. He shared the award with Ernst †Ruska and Gerd †Binnig.

ROM acronym for **read-only memory**, in computing, a memory device in the form of a collection of integrated circuits (chips), frequently used in microcomputers. ROM chips are loaded with data and programs during manufacture and, unlike †RAM (random-access memory) chips, can subsequently only be read, not written to, by a computer. However, the contents of the chips are not lost when the power is switched off, as happens in RAM.

ROM is used to form a computer's permanent store of vital information, or of programs that must be readily available but protected from accidental or deliberate change by a user. ROM chips are used in electronic devices, such as calculators and video games. Most personal computers use ROM to store the code that boots (starts up) the computer. The problem with conventional ROM is that small changes are difficult to achieve because they are produced in an assembly line. †PROM (programmable read-only memory) allows a high-voltage electronic pulse to "burn" new sets of instructions, or update old ones. †EPROM (erasable programmable read-only memory) is reprogrammed using ultraviolet light, shone through a window on top. †EEPROM (electronically erasable programmable read-only memory) can be reprogrammed in the machine, a byte at a time. Flash ROM (flash read-only memory) can be erased and rewritten in whole blocks.

Roman numerals ancient European number system using symbols different from Arabic numerals (the ordinary numbers 1, 2, 3, 4, 5, and so on). The seven key symbols in Roman numerals, as represented today, are I (1), V (5), X (10), L (50), C (100), D (500), and M (1,000). There is no zero, and therefore no place-value as is fundamental to the Arabic system. The first ten Roman numerals are I, II, III, IV (or IIII), V, VI, VII, VIII, IX, and X. When a Roman symbol is preceded by a symbol of equal or greater value, the values of the symbols are added (XVI = 16).

When a symbol is preceded by a symbol of less value, the values are subtracted (XL = 40). A horizontal bar over a symbol indicates a multiple of 1,000 ($\overline{x}$ = 10,000). Although addition and subtraction are fairly straightforward using Roman numerals, the absence of a zero makes other arithmetic calculations (such as multiplication) clumsy and difficult.

Romer, Alfred Sherwood (1894–1973) US paleontologist and comparative anatomist who made influential studies of vertebrate evolution. His *The Vertebrate Body* (1949) is still a standard textbook today.

Römer, Ole Christensen (or Olaus Christensen) (1644–1710) Danish astronomer and mathematician. His primary discovery was the finite velocity of light. Observations of †Jupiter's satellites had disagreed with †Cassini's tables, and Roemer demonstrated in 1675 that the error was due to the time taken by light to cross the changing distance between Jupiter and the Earth. He announced that light took 11 minutes to travel the distance between the Earth and the Sun.

röntgen alternative spelling for †roentgen, unit of X- and gamma-ray exposure.

Röntgen (or Roentgen), Wilhelm Konrad (1845–1923) German physicist. He was awarded the Nobel Prize for Physics in 1901 for his discovery of †X-rays in 1895. While investigating the passage of electricity through gases, he noticed the †fluorescence of a barium platinocyanide screen. This radiation passed through some substances opaque to light, and affected photographic plates. Developments from this discovery revolutionized medical diagnosis.

Röntgen, Wilhelm Conrad http://www.nobel.se/physics/laureates/1901/rontgen-bio.html

Biography of Wilhelm Conrad Röntgen, the German physicist who first realized the huge potential of the electromagnetic field of X-rays. The presentation includes sections on Röntgen's early years and education, his academic career and scientific experiments, and the coincidences that led him to his great discovery of X-rays.

root the part of a plant that is usually underground, and whose primary functions are anchorage and the absorption of water and dissolved mineral salts. Roots usually grow downwards and towards water (that is, they are positively geotropic and hydrotropic; see †tropism). Plants such as epiphytic orchids, which grow above ground, produce aerial roots that absorb moisture from the atmosphere. Others, such as ivy, have climbing roots arising from the stems, which serve to attach the plant to trees and walls.

The absorptive area of roots is greatly increased by the numerous slender root hairs formed near the tips. A calyptra, or root cap, protects the tip of the root from abrasion as it grows through the soil.

Symbiotic associations occur between the roots of certain plants, such as clover, and various bacteria that fix nitrogen from the air (see †nitrogen fixation). Other modifications of roots include †contractile roots, †pneumatophores, †taproots, and †prop roots.

evolution The evolution of root systems by land plants was very fast, in evolutionary terms. The earliest plants (410 million years ago) had tiny roots only a few millimeters in length, but within 20 million years roots were as long as 50 cm/20 in.

root of an equation, a value that satisfies the equality. For example, $x = 0$ and $x = 5$ are roots of the equation $x^2 - 5x = 0$.

root in mathematics, a number which when multiplied by itself will equal a given number (the inverse of an †exponent or power). On a calculator, roots may be found by using the buttons marked —or inv x^y. For example, the cubed root of 27 is found by pressing 27, then inv x^y, then 3.

root cap cap at the tip of a growing plant root. It gives protection to the zone of actively dividing cells as the root pushes through the soil.

root directory in computing, the top directory in a tree-and-branch filing system. It contains all the other directories.

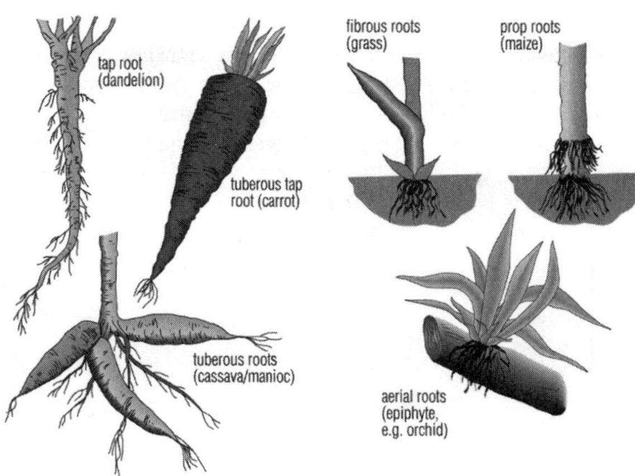

root Types of root. Many flowers (dandelion) and vegetables (carrot) have swollen tap roots with smaller lateral roots. The tuberous roots of the cassava are swollen parts of an underground stem modified to store food. The fibrous roots of the grasses are all of equal size. Prop roots grow out from the stem and then grow down into the ground to support a heavy plant. Aerial roots grow from stems but do not grow into the ground; many absorb moisture from the air. Helicon

root hair tiny, hair-like outgrowth of some surface cells of plant roots that greatly increases the area available for the absorption of water and other materials such as †minerals. It is a delicate structure, which survives for a few days only and does not develop into a root. Root hair cells are found just behind the root tip where they are continually being formed. They grow out into the soil for a few millimeters to produce the root hair. Each root hair is made of a single cell. At the root tip †mitosis cell division is very active making new cells for growth to replace the root hair cells that die.

The majority of land plants have root hairs. The layer of the root's epidermis that produces root hairs is known as the **piliferous layer**.

root mean square rms, in mathematics, value obtained by taking the †square root of the mean of the squares of a set of values; for example the rms value of four quantities a, b, c, and d is $\sqrt{[(a^2 + b^2 + c^2 + d^2)/4]}$.

root nodule clearly visible swelling that develops in the roots of members of the bean family, the Leguminosae. The cells inside this timorous growth have been invaded by the bacteria Rhizobium, a soil microbe capable of converting gaseous nitrogen into nitrate. The nodule is therefore an association between a plant and a bacterium, with both partners benefiting. The plant obtains nitrogen compounds while the bacterium obtains nutrition and shelter.

Nitrogen fixation by bacteria is one of the main ways by which the nitrogen in the atmosphere is cycled back into living things. The economic value of the process is so great that research has been carried out into the possibility of stimulating the formation of root nodules in crops such as wheat, which do not normally form an association with rhizobium.

rootstock another name for †rhizome, an underground plant organ.

Rorschach test in psychology, a method of diagnosis involving the use of inkblot patterns that subjects are asked to interpret, to help indicate personality type, degree of intelligence, and emotional stability. It was invented by the Swiss psychiatrist Hermann Rorschach.

rose in mathematics, symmetrical curve represented by the equation:

$$(x^2 + y^2)^3 = 4a^2x^2y^2$$

where a is constant.

rosette in botany, group of leaves that arise closely together from a short stem. The leaves form a circular cluster near the ground.

Ross, Ronald (1857–1932) Indian-born British physician and bacteriologist who was awarded a Nobel Prize for Physiology or Medicine in 1902 for his work on the role of the *Anopheles* mosquito in transmitting malaria. From 1881 to 1899 he served in the Indian Medical Service, and during 1895–98 identified mosquitoes of the genus *Anopheles* as being responsible for the spread of malaria. He was knighted in 1911.

rotation in geometry, a †transformation in which a figure is turned about a given point, known as the **center of rotation**. A rotation of 180° is known as a half turn.

Three things are needed to describe a rotation: the **angle of rotation**, the **direction of rotation** (clockwise or anticlockwise), and the **center of rotation**.

For example, in the diagram, triangle A is mapped onto triangle B by a rotation of 90° clockwise with center (0,0):

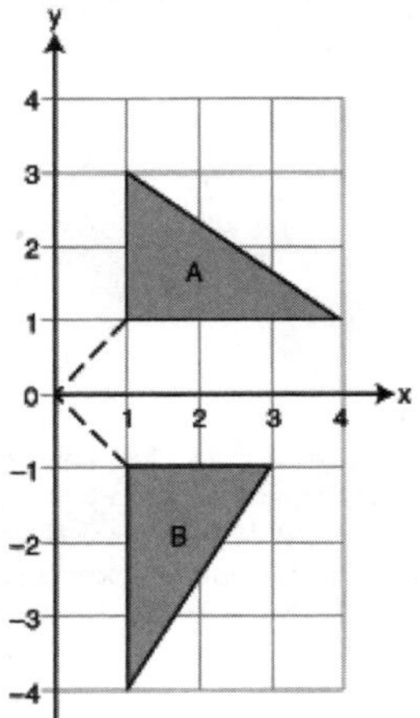

The inverse transformation is the rotation of B back onto A. When a shape is rotated but appears not to have moved it is said to have **rotational symmetry**. †Translation, †reflection, and †enlargement can also transform shapes.

AmbiGrams http://www.ambigram.com/

Unusual site containing ambigrams, which are "Words that can be read in more than one way or from more than a single vantage point." The site demonstrates some interesting principles of rotation, reflection, and symmetry, as well as being fun in its own right.

rotation in astronomy, movement of a planet rotating about its own axis. For the †Earth, one complete rotation takes 23 hours and 56 minutes. The Earth rotates in an anticlockwise direction (as viewed looking along its axis from north to south), leading to the appearance from Earth of the Sun moving from east to west in the daily cycle. The rotation of the Earth produces a surface speed at the Equator of about 1,600 kph/1,000 mph. This speed decreases further north and south of the Equator. Artificial †satellites use the Earth's natural rotation to orbit the Earth.

As the Earth's axis is tilted with respect to the Sun, the lengths of day and night vary across the globe. At the †Equator, day and night are both 12 hours long. During the northern-hemisphere winter, days become shorter with increasing latitude north, until a point where there is continuous night. At the same time, during the southern-hemisphere summer, days become longer with increasing latitude south, until a point where there is continuous daylight. The situation is reversed during the northern-hemisphere summer.

Rotblat, Joseph (1908–) Polish physicist who began working on the atom bomb as part of the Manhattan Project but withdrew in 1944 when he received intelligence that the Germans were not working on a bomb. He was instrumental in the formation of Pugwash, a group of scientists working towards nuclear disarmament. He won the Nobel Peace Prize in 1995 (shared with Pugwash) for campaigning against nuclear weapons.

Rotblat, Joseph http://www.nobel.se/peace/laureates/1995/rotblat-cv.html

Curriculum vitae and photo of the Polish physicist and winner of the Nobel Peace Prize. There is an outline of his professional career and the large number of honors bestowed upon him for his work for nuclear disarmament.

Rothschild, Miriam (1908–) English zoologist and entomologist. She studied fleas and was the first to work out the flea jumping mechanism. She also studied fleas' reproductive cycles and linked this, in rabbits, to the hormonal changes within the host. She has written around 350 papers on entomology, zoology, neurophysiology, and chemistry.

rotifer any of the tiny invertebrates, also called "wheel animalcules," of the phylum Rotifera. Mainly freshwater, some marine, rotifers have a ring of †cilia that carries food to the mouth and also provides propulsion. They are the smallest of multicellular animals – few reach 0.05 cm/0.02 in.

roughage alternative term for dietary †fiber, material of plant origin that cannot be digested by enzymes normally present in the human †gut.

rounding in mathematics, a process by which a number is approximated to the nearest above or below with fewer †decimal places. For example, 34.3583 might be rounded to 3 decimal places – 34.358 (the 3 is below 5), whereas 34.3587 would be rounded to 34.359 (the 7 is above 5). Similarly, 3,587 might be rounded to the nearest thousand (or to 1 †significant figure), giving 4,000. When unwanted decimals are simply left out, the process is known as **truncating**.
Rounding can produce considerable errors, especially when the rounded numbers are multiplied by other rounded numbers.
For example:

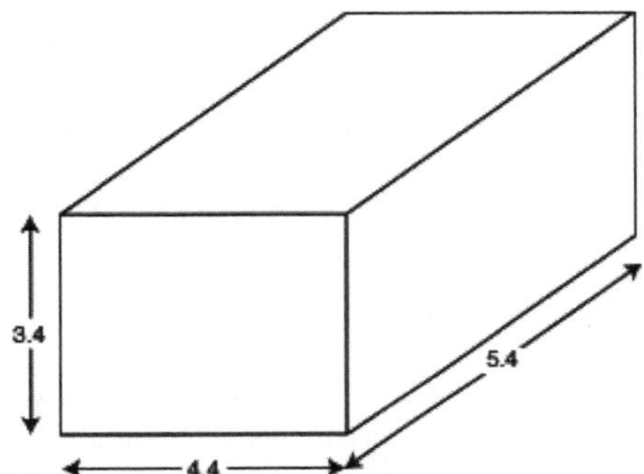

The volume of a cuboid is calculated as 3 x 4 x 5, each dimension being rounded to 0 decimal places, the true dimensions being 3.4 x 4.4 x 5.4. The rounding error is 20.784 out of 80.784, an error of over 25%, which could be an unacceptable degree of †accuracy.

roundworm parasitic †nematode worm, genus *Toxocara*, found in dogs, cats, and other animals.

Rous, (Francis) Peyton (1879–1970) US pathologist who was awarded a Nobel Prize for Physiology or Medicine in 1966 for his discovery of tumor-inducing viruses (from his research on cancer in chickens).

row in mathematics, a †matrix, a horizontal line of numbers. Matrices are made up of rows and columns. A matrix that consists of one row only is called a **row matrix**, while a matrix with one column only is called a **column matrix**.

Rowland, F Sherwood (1927–) US chemist who shared the Nobel Prize for Chemistry in 1995 with Mario †Molina and Paul †Crutzen for their work in atmospheric chemistry, particularly concerning the formation and decomposition of ozone. They explained the chemical reactions that are destroying the ozone layer.

Rowland, Henry Augustus (1848–1901) US physicist who developed the concave diffraction grating in 1882, which made the analysis of spectra much faster and more accurate. He also carried out the precise determination of certain physical constants.

RR Lyrae variable type of short-period †variable star. Spectrally classified as A to F giants, RR Lyrae variables were once called **cluster-Cepheids** (see †Cepheid variable).

rubber coagulated †latex of a variety of plants, mainly from the New World. Most important is Para rubber, which comes from the tree *Hevea brasiliensis*, belonging to the spurge family. It was introduced from Brazil to Southeast Asia, where most of the world supply is now produced, the chief exporters being Peninsular Malaysia, Indonesia, Sri Lanka, Cambodia, Thailand, Sarawak, and Brunei. At about seven years the tree, which may grow to 20 m/60 ft, is ready for tapping. Small cuts are made in the trunk and the latex drips into collecting cups. In pure form, rubber is white and has the formula $(C_5H_8)_n$.
plant rubber Other sources of rubber are the Russian dandelion *Taraxacum koksagyz*, which grows in temperate climates and can yield about 45 kg/100 lb of rubber per ton of roots, and guayule *Parthenium argentatum*, a small shrub that grows in the southwestern USA and Mexico.
treatment Early uses of rubber were limited by its tendency to soften on hot days and harden on colder ones, a tendency that was eliminated by Charles Goodyear's invention of †vulcanization in 1839.
synthetic rubber In the 20th century, world production of rubber increased a hundredfold, and World War II stimulated the production of synthetic rubber to replace the supplies from Malaysian sources overrun by the Japanese. There are a variety of synthetic rubbers adapted to special purposes, but economically foremost is SBR (styrene-butadiene rubber). Cheaper than natural rubber, it is preferable for some purposes, for example in car tires, where its higher abrasion resistance is useful, and it is either blended with natural rubber or used alone for industrial molding and extrusions, shoe soles, hoses, and latex foam.

Rubbia, Carlo (1934–) Italian physicist and director-general of †CERN, the European nuclear research organization 1989–93. Rubbia shared the Nobel Prize for Physics in 1984 with his colleague Simon van der †Meer for contributions to the discovery in 1983 of the W and Z particles (weakons), the agents responsible for transferring the weak nuclear force.

rubidium chemical symbol Rb, (Latin *rubidus* "red") soft, silver-white, metallic element, atomic number 37, relative atomic mass 85.47. It is one of the †alkali metals in Group 1 of the †periodic table. Reactivity of the alkali metals increases down the group and so rubidium is more reactive than lithium, sodium, and potassium. Rubidium ignites spontaneously in air and reacts violently with water. It is used in photocells and vacuum-tube filaments.

Rubidium was discovered spectroscopically by German physicists Robert Bunsen and Gustav Kirchhoff in 1861 and named after the red lines in its spectrum.

ruby red transparent gem variety of the mineral †corundum Al_2O_3, aluminum oxide. Small amounts of chromium oxide, Cr_2O_3, substituting for aluminum oxide, give ruby its color. Natural rubies are found mainly in Myanmar (Burma), but rubies can also be produced artificially and such synthetic stones are used in †lasers.

Rudbeck, Olof (1630–1702) Swedish physiologist who discovered that the thoracic duct is connected to the intestinal lymphatics (the †lymph vessels that carry chyle, the fluid containing all the nutrients derived from digestion).

Ruffini, Paolo (1765–1822) Italian mathematician, philosopher, and physician. He published a theorem stating that it was impossible to give a general solution to equations of greater than the fourth degree using only radicals (such as square roots, cube roots, and so on). This became known as the **Abel–Ruffini theorem** when endorsed by Norwegian mathematician Niels Abel.

Rumford, Benjamin Thompson (1753–1814) Count von Rumford, American-born British physicist and inventor. In 1798, impressed by the seemingly inexhaustible amounts of heat generated in the boring of a cannon, he published his theory that heat is a mode of vibratory motion, not a substance.

Heat is a Form of Motion: An Experiment in Boring Cannon

http://dbhs.wvusd.k12.ca.us/Chem-History/Rumford-1798.html

Transcript taken from Philosophical Transactions (vol. 88), 1798, in which Rumford describes heat as a form of vibrational motion. This paper changed the common scientific perception of heat, and led, in turn, to many new theories for the thermal properties of materials.

ruminant any even-toed hoofed mammal with a rumen, the "first stomach" of its complex digestive system. Plant food is stored and fermented before being brought back to the mouth for chewing (chewing the cud) and then is swallowed to the next stomach. Ruminants include cattle, antelopes, goats, deer, and giraffes, all with a four-chambered stomach. Camels are also ruminants, but they have a three-chambered stomach.

runner or **stolon**, in botany, aerial stem that produces new plants.

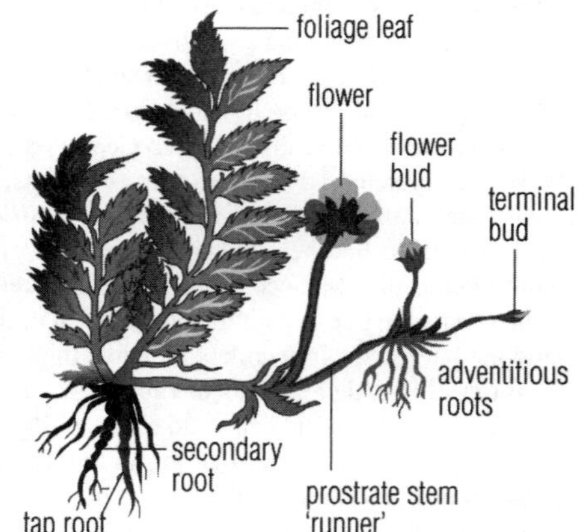

runner A runner, or stolon, grows horizontally near the base of some plants, such as the strawberry. It produces roots along its length and new plants grow at these points. Helicon

run-off water that falls as rain and flows off the land, into streams or rivers.

Ruska, Ernst August Friedrich (1906–1988) German physicist who was awarded the Nobel Prize for Physics in 1986 for his work in opto-electronics and on the construction of the first electron microscope (called the transmission electron microscope). He shared the award with Gerd †Binnig and Heinrich †Rohrer.

Russell, Bertrand Arthur William (1872–1970) 3rd Earl Russell, English philosopher, mathematician, and peace campaigner. He contributed to the development of modern mathematical logic and wrote about social issues. His works include *Principia Mathematica* (1910–13; with A N †Whitehead), in which he attempted to show that mathematics could be reduced to a branch of logic; *The Problems of Philosophy* (1912); and *A History of Western Philosophy* (1946). He was an outspoken liberal pacifist. He was awarded the Nobel Prize for Literature in 1950. He was made an Earl in 1931.

Russell, Henry Norris (1877–1957) US astronomer. He was the first to put in graphic form what became known as the †Hertzsprung-Russell diagram in 1913.

rust reddish-brown oxide of iron formed by the action of moisture and oxygen on the metal. It consists mainly of hydrated iron(III) oxide ($Fe_2O_3.H_2O$) and iron(III) hydroxide ($Fe(OH)_3$). Rusting is the commonest form of †corrosion.

rust prevention There are two main approaches to protect against rusting. **Barrier methods** introduce a barrier between the metal and the air and moisture to minimize the reaction. This is the commonest method of rust prevention and the barrier may consist of a layer of grease, paint, plastic, or an unreactive metal, such as tin, copper, or chromium.

In **sacrificial protection**, the iron is actually covered by a more reactive metal, such a zinc (galvanization), or connected to a more reactive metal, such as magnesium, so that as long as the "sacrificial" metal is present it will corrode first.

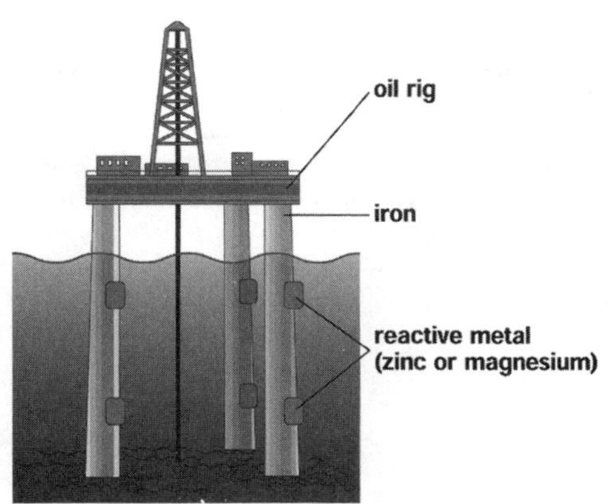

rust Sacrificial metal is used to protect pipes and other exposed metal on oil rigs from corrosion by rust. A more reactive metal than iron, such as zinc or magnesium, is attached in large lumps by conducting wires, at intervals along the pipes. Electrons released during the reaction travel along the conducting wires to the pipes, preventing rust from occurring. It is cheaper to replace the sacrificial metal than it is to replace the pipes. Helicon

ruthenium chemical symbol Ru, hard, brittle, silver-white, metallic element, atomic number 44, relative atomic mass 101.07. It is one of the so-called platinum group of metals; it occurs in platinum ores as a free metal and in the natural alloy osmiridium. It is used as a hardener in alloys and as a catalyst; its compounds are used as coloring agents in glass and ceramics.

Rutherford, Ernest (1871–1937) 1st Baron Rutherford of Nelson, New Zealand-born British physicist. He was a pioneer of modern atomic science. His main research was in the field of †radioactivity, and he discovered alpha, beta, and gamma rays. He was the first to recognize the nuclear nature of the atom in 1911. He was awarded the Nobel Prize for Chemistry in 1908 for his work in atomic disintegration and the chemistry of radioactive substances.

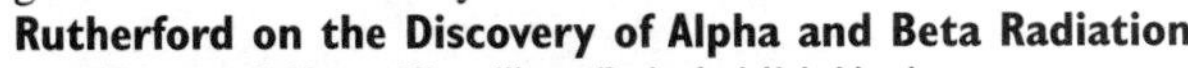

Rutherford on the Discovery of Alpha and Beta Radiation

http://dbhs.wvusd.k12.ca.us/Chem-History/Rutherford-Alpha?.html

Transcript of Rutherford's paper "Uranium radiation and the electrical conduction produced by it," which first appeared in the Philosophical Magazine in January 1899. The full paper with diagrams is reproduced here, and in it Rutherford describes the nature of the two types of radiation he discovered to be emitted from uranium as it decays.

rutherfordium chemical symbol Rf, synthesized, radioactive, metallic element. It is the first of the †transactinide series, atomic number 104, relative atomic mass 262. It is produced by bombarding californium with carbon nuclei and has ten isotopes, the longest-lived of which, Rf-262, has a half-life of 70 seconds.

Two institutions claim to be the first to have synthesized it: the Joint Institute for Nuclear Research in Dubna, Russia, in 1964; and the University of California at Berkeley, USA, in 1969.

Rutherfurd, Lewis Morris (1816–1892) US spectroscopist and astronomical photographer. He produced a classification scheme of stars based on their spectra that turned out to be similar to that of Italian astronomer Angelo †Secchi.

rutile titanium oxide mineral, TiO_2, a naturally occurring ore of titanium. It is usually reddish brown to black, with a very bright (adamantine) surface luster. It crystallizes in the tetragonal system. Rutile is common in a wide range of igneous and metamorphic rocks and also occurs concentrated in sands; the coastal sands of eastern and western Australia are a major source. It is also used as a pigment that gives a brilliant white to paint, paper, and plastics.

Ruzicka, Leopold Stephen (1887–1976) Swiss chemist who began research on natural compounds such as musk and civet secretions. In the 1930s he investigated sex hormones, and in 1934 succeeded in extracting the male hormone androsterone from 31,815 litres/7,000 gallons of urine and synthesizing it. Born in Croatia, Ruzicka settled in Switzerland in 1929. Ruzicka shared the Nobel Prize for Chemistry in 1939 with Adolf †Butenandt for his study of polymethylenes and higher terpenes.

Rydberg, Johannes Robert (1854–1919) Swedish physicist who discovered a mathematical expression that gives the frequencies of spectral lines for elements. It includes a constant named the †**Rydberg constant** after him. Rydberg atoms were also named after him.

Rydberg constant in physics, a constant that relates atomic spectra to the †spectrum of hydrogen. Its value is 1.0977 x 10^7 per meter.

Ryle, Martin (1918–1984) English radio astronomer. At the Mullard Radio Astronomy Observatory, Cambridge, he developed the technique of sky-mapping using "aperture synthesis," combining smaller dish aerials to give the characteristics of one large one. His work on the distribution of radio sources in the universe brought confirmation of the †Big Bang theory. He was awarded with his co-worker, the English radio astronomer Antony †Hewish, the Nobel Prize for Physics in 1974 for his work on the development of radio astronomy, particularly the aperture-synthesis technique, and the discovery of †pulsars, rapidly rotating neutron stars that emit pulses of energy. He was knighted in 1966.

S

S in physics, symbol for †siemens, the SI unit of electrical conductance, equal to a conductance of 1 ohm^{-1}.

Sabatier, Paul (1854–1941) French chemist. He found in 1897 that if a mixture of ethylene and hydrogen was passed over a column of heated nickel, the ethylene changed into ethane. Further work revealed that nickel could be used to catalyze numerous chemical reactions. He was awarded the Nobel Prize for Chemistry in 1912 for finding the method of catalytic hydrogenation of organic compounds.

sabin unit of sound absorption, used in acoustical engineering. One sabin is the absorption of one square foot (0.093 square meter) of a perfectly absorbing surface (such as an open window).

Sabin, Albert Bruce (1906–1993) Russian-born US microbiologist who developed a highly effective, live vaccine against polio. The earlier vaccine, developed by physicist Jonas †Salk, was based on heat-killed viruses. Sabin was convinced that a live form would be longer-lasting and more effective, and in 1957 he succeeded in weakening the virus so that it lost its virulence. The vaccine can be given by mouth.

sabkha flat shoreline zone in arid regions above the high-water mark in which the sediments include large amounts of †evaporites. These occur in the form of nodules, crusts, and crystalline deposits of halite, anhydrite, and gypsum, as well as mineral grains of various sorts. Some of the evaporites form from rapid evaporation of marine waters soaking through from the bordering tidal flats, but some can be derived also from sediment-laden continental waters coming down from adjoining highlands.

saccharide another name for a sugar.

saccharin or **ortho-sulpho benzimide**; $C_7H_5NO_3S$, sweet, white, crystalline solid derived from coal tar and substituted for sugar. Since 1977 it has been regarded as potentially carcinogenic. Its use is not universally permitted and it has been largely replaced by other sweetening agents.

Sachs, Julius von (1832–1897) German botanist and plant physiologist who developed several important experimental techniques and showed that †photosynthesis occurs in the †chloroplasts (the structure in a plant cell containing the green pigment chlorophyll) and produces oxygen. He was especially gifted in his experimental approach; some of his techniques are still in use today, such as the simple iodine test, which he used to show the existence of starch in a whole leaf.

sacrum compound triangular bone at the base of the human spine, formed of five fused vertebrae between the last lumbar vertebra above and the coccyx below. The sacrum forms the back of the †pelvis.

sadism tendency to derive pleasure (usually sexual) from inflicting physical or mental pain on others. The term is derived from the Marquis de Sade.

Saffir–Simpson damage-potential scale scale of potential damage from wind and sea when a hurricane is in progress: 1 is minimal damage, 5 is catastrophic.

Sagan, Carl Edward (1934–1996) US physicist and astronomer who popularized astronomy through writings and broadcasts. His main research was on planetary atmospheres. He wrote or co-wrote 27 books, which include *Broca's Brain: Reflections on the Romance of Science* (1979) and *Cosmos* (1980), based on his television series of that name.

Sagitta small constellation of the northern hemisphere, represented as an arrow. It lies in the †Milky Way.

Sagittarius bright zodiac constellation in the southern hemisphere, represented as an archer aiming a bow and arrow at neighboring Scorpius. The Sun passes through Sagittarius from mid-December to mid-January, including the winter solstice, when it is farthest south of the Equator. The constellation contains many nebulae and †globular clusters, and open †star clusters. Kaus Australis and Nunki are its brightest stars. The center of our Galaxy, the †Milky Way, is marked by the radio source Sagittarius A. In astrology, the dates for Sagittarius are about 22 November–21 December (see †precession).

Saint Elmo's fire bluish, flamelike electrical discharge that sometimes occurs above ships' masts and other pointed objects or about aircraft in stormy weather. Although of high voltage, it is of low current and therefore harmless. St Elmo (or St Erasmus) is the patron saint of sailors.

Sakharov, Andrei Dmitrievich (1921–1989) Soviet physicist. He was an outspoken human-rights campaigner, who with Igor †Tamm developed the hydrogen bomb. He later protested against Soviet nuclear tests and was a founder of the Soviet Human Rights Committee in 1970. In 1975 he was awarded the Nobel Prize for Peace for his advocacy of human rights and disarmament. For criticizing Soviet action in Afghanistan, he was sent into internal exile 1980–86.

Sakharov, Andrei http://www.nobel.se/peace/laureates/1975/sakharov-autobio.html

Brief autobiography written by the physicist and human rights campaigner for the Nobel Committee after he won the 1975 Peace Prize. Sakharov describes how his upbringing inculcated a love of literature, science, and music. He relates his pioneering work in Soviet nuclear physics and the deepening moral doubts this aroused. He movingly describes how it was his conscience that propelled him into an unwilling engagement with politics.

Sakmann, Bert (1942–) German cell physiologist who was awarded a Nobel Prize for Physiology or Medicine in 1991 with Erwin †Neher for their discovery of how gatelike structures (ion channels) regulate the flow of ions into and out of cells. They also determined the role of the neurohormone beta-endorphin.

Salam, Abdus (1926–1996) Pakistani physicist who was awarded the Nobel Prize for Physics in 1979 for his theory linking the electromagnetic and weak nuclear forces. He was the first person from his country to receive a Nobel prize.

sal ammoniac former name for †ammonium chloride.

salicylic acid HOC_6H_4COOH, active chemical constituent of aspirin, an analgesic drug. The acid and its salts (salicylates) occur naturally in many plants; concentrated sources include willow bark and oil of wintergreen.

When purified, salicylic acid is a white solid that crystallizes into prismatic needles at 159°C/318°F. It is used as an antiseptic, in food preparation and dyestuffs, and in the preparation of aspirin.

salinization accumulation of salt in water or soil; it is a factor in †desertification.

saliva in vertebrates, an alkaline secretion from the salivary glands that aids the swallowing and digestion of food in the mouth. In mammals, it contains the enzyme amylase, which converts starch to sugar. The salivary glands of mosquitoes and other blood-sucking insects produce †anticoagulants.

salivary gland or **parotid gland**, in mammals, one of two glands situated near the mouth responsible for the manufacture of saliva and its secretion into the mouth. The salivary glands are stimulated to produce saliva during a meal. Saliva contains an enzyme, ptyalin, and mucous which are essential for the mastication and initial digestion of food.

Salk, Jonas Edward (1914–1995) US physician and microbiologist. In 1954 he developed the original vaccine that led to virtual eradication of paralytic polio in industrialized countries. He was director of the Salk Institute for Biological Studies, University of California, San Diego, 1963–75.

salmonella any of a very varied group of bacteria, genus *Salmonella*, that colonize the intestines of humans and some animals. Some strains cause typhoid and paratyphoid fevers, while others cause salmonella food poisoning, which is characterized by stomach pains, vomiting, diarrhea, and headache. It can be fatal in elderly people, but others usually recover in a few days without antibiotics. Most cases are caused by contaminated animal products, especially poultry meat.

Human carriers of the disease may be well themselves but pass the bacteria on to others through unhygienic preparation of food. Domestic pets can also carry the bacteria while appearing healthy.

Bad Bug Book – Salmonella http://vm.cfsan.fda.gov/~mow/chap1.html

Fact sheet with basic information about salmonella. It offers details on the nature and the diagnostic procedures of the disease, susceptible groups, and the potential complications following the disease, as well as guidelines for cooking eggs, the food mainly associated with salmonella.

Salmonella http://www.salmonella.org/info.html

Information page of a larger salmonella-based Web site. It contains basic information about salmonella, offering details on the nature and the diagnostic procedures of the disease, susceptible groups, and the potential complications following the disease, as well as guidelines on how to prevent salmonella in everyday life.

salt in chemistry, any compound formed from an acid and a base through the replacement of all or part of the hydrogen in the acid by a metal or electropositive radical. **Common †salt** is sodium chloride.

A salt may be produced by a chemical reaction between an acid and a base, or by the displacement of hydrogen from an acid by a metal (see †displacement reaction). As a solid, the ions normally adopt a regular arrangement to form crystals. Some salts form only stable crystals as hydrates (when combined with water). Most inorganic salts readily dissolve in water to give an electrolyte (a solution that conducts electricity).

saltation (Latin *saltare* "to leap") in biology, the idea that an abrupt genetic change can occur in an individual, which then gives rise to a new species. The idea has now been largely discredited, although the appearance of †polyploid individuals can be considered an example.

saltation in earth science, bouncing of rock particles along a river bed. It is the means by which †bedload (material that is too heavy to be carried in suspension) is transported downstream. The term is also used to describe the movement of sand particles bounced along by the wind.

salt, common or **sodium chloride**; NaCl, white crystalline solid, found dissolved in seawater and as rock salt (the mineral halite) in large deposits and salt domes. Common salt is used extensively in the food industry as a preservative and for flavoring, and in the chemical industry in the making of chlorine and sodium.

salt marsh wetland with halophytic vegetation (tolerant to seawater). Salt marshes develop around

†estuaries and on the sheltered side of sand and shingle †spits. They are formed by the deposition of mud around salt-tolerant vegetation. This vegetation must tolerate being covered by seawater as well as being exposed to the air. It also traps mud as the tide comes in and out. This helps build up the salt marsh. Salt marshes usually have a network of creeks and drainage channels by which tidal waters enter and leave the marsh.

saltpeter or **potassium nitrate**; KNO_3, compound used in making gunpowder (from about 1500). It occurs naturally, being deposited during dry periods in places with warm climates, such as India.

Salvarsan proprietary name for **arsphenamine**; technical name **3,3-diamino-4,4-dihydroxyarsenobenzene dichloride**, first specific antibacterial agent, discovered by German bacteriologist Paul Ehrlich in 1909. Because of its destructive effect on *Spirochaeta pallida*, it was used in the treatment of syphilis before the development of antibiotics.

sal volatile another name for †ammonium carbonate.

samara in botany, a winged fruit, a type of †achene.

samarium chemical symbol Sm, hard, brittle, gray-white, metallic element of the †lanthanide series, atomic number 62, relative atomic mass 150.4. It is widely distributed in nature and is obtained commercially from the minerals monzanite and bastnaesite. It is used only occasionally in industry, mainly as a catalyst in organic reactions. Samarium was discovered by spectroscopic analysis of the mineral samarskite and named in 1879 by French chemist Paul Lecoq de Boisbaudran (1838–1912) after its source.

sample in statistics, finite subset of a population. Typically, samples of a population are taken to get estimates for various properties of the population as a whole.

sampling in statistics, taking a small number of cases and assuming that they are representative of the total number. In †survey work in market research, for example, it is usually not practical to survey every customer. It is also important in quality control to avoid the expense of testing every item produced. Instead, a sample is taken. A **random sample** is one where the sample is chosen in a random way.

For example, in a postal survey, one person in every hundred on the electoral register might be chosen. A **stratified random sample** is one where a group is selected and then a random sample is taken from that group. For example, a manufacturer of baby-care products might select pregnant mothers as a group, and then conduct a random-sample survey on that group. A **quota sample survey** is one where the size and composition of the survey matches the total population that is being researched. For example, a psychologist may want to interview 20 boys aged 0–10, 20 girls aged 0–10, 20 males aged 10–20, and so on.

sand loose grains of rock, 0.0625–2.00 mm/0.0025–0.08 in in diameter, consisting most commonly of †quartz, but owing their varying color to mixtures of other minerals. Sand is used in cement-making, as an abrasive, in glass-making, and for other purposes.

Sands are classified into marine, freshwater, glacial, and terrestrial. Some "light" soils contain up to 50% sand. Sands may eventually consolidate into †sandstone.

sandbar ridge of sand built up by the currents across the mouth of a river or bay. A sandbar may be entirely underwater or it may form an elongated island that breaks the surface. A sandbar stretching out from a headland is a **sand spit**.

Coastal bars can extend across estuaries to form **bay bars**.

sandstone †sedimentary rocks formed from the consolidation of sand, with sand-sized grains (0.0625 – 2 mm/0.0025 – 0.08 in) in a matrix or cement. Their principal component is quartz. Sandstones are commonly permeable and porous, and may form freshwater †aquifers. They are mainly used as building materials.

Sandstones are classified according to the matrix or cement material (whether derived from clay or silt).

Sanger, Frederick (1918–) English biochemist. He was awarded the Nobel Prize for Chemistry in 1958 for determining the structure of †insulin, and again in 1980 for work on the chemical structure of †genes. He was the first person to be awarded the chemistry prize twice.

Santa Ana periodic warm Californian †wind.

sap fluids that circulate through †vascular plants, especially woody ones. Sap carries water and food to plant tissues. Sap contains alkaloids, protein, and starch; it can be milky (as in rubber trees), resinous (as in pines), or syrupy (as in maples).

saponification in chemistry, the †hydrolysis (splitting) of an †ester by treatment with a strong alkali, resulting in the liberation of the alcohol from which the ester had been derived and a salt of the constituent fatty acid. The process is used in the manufacture of soap.

sapphire deep-blue, transparent gem variety of the mineral †corundum Al_2O_3, aluminum oxide. Small amounts of iron and titanium give it its color. A corundum gem of any color except red (which is a ruby) can be called a sapphire; for example, yellow sapphire.

saprobiont in biology, fungus or bacterium that obtains its food by secreting digestive enzymes onto dead organic matter and absorbing the products. The process causes the decomposition of the dead material and the release of gases and minerals into the environment. Saprobionts are collectively called †decomposers.

saprophyte in botany, an obsolete term for a †saprotroph, an organism that lives in dead or decaying matter.

saprotroph formerly **saprophyte**, organism that feeds on the excrement or the dead bodies or tissues of others. They include most fungi (the rest being parasites); many bacteria and protozoa; animals such as dung beetles and vultures; and a few unusual plants, including several orchids. Saprotrophs cannot make food for themselves, so they are a type of †heterotroph. They are useful scavengers, and in sewage farms and refuse dumps break down organic matter into nutrients easily assimilable by green plants.

Sarcodina subphylum of protozoa containing the †amoebae.

sarcoma malignant †tumor arising from the fat, muscles, bones, cartilage, or blood and lymph vessels and connective tissues. Sarcomas are much less common than †carcinomas.

sarcomere in a muscle fiber, region of a myofibril (muscle fiber) between two adjacent Z lines (lines that connect thin filaments within a myofibril). Each sarcomere consists of filaments of two types of protein. The thicker filament consists of myosin, the thinner of actin. The filaments of adjacent sarcomeres link at the respective Z lines.

sard yellow or red-brown variety of †chalcedony.

sarin poison gas 20 times more lethal to humans than potassium cyanide. It impairs the central nervous system, blocking the action of an enzyme that removes acetylcholine, the chemical that transmits signals. Sarin was developed in Germany during World War II.

Sarin was used in 1995 in a terrorist attack on the Tokyo underground by a Japanese sect. It is estimated that the USA had a stockpile of 15,000 tonnes of sarin, and more than 1,000 US rockets with sarin warheads were found to be leaking in 1995. There are no known safe disposal methods.

Sarin http://www.geocities.com/CapeCanaveral/Lab/7050/

Information on the nerve agent sarin. The Web site contains sections on the characteristics of sarin, its toxicology, symptoms it causes, cautions and precautions, first aid therapy for victims of the gas, and a list of neutralization and decontamination methods.

saros in astronomy, period in which the apparent relative positions of the Earth, Sun, and Moon repeat, equal to a little more than 18 years (6,595.32 days). It been used since ancient times for predicting eclipses.

satellite any small body that orbits a larger one. **Natural satellites** that orbit planets are called **moons**. The first **artificial satellite**, Sputnik 1, was launched into orbit around the Earth by the USSR in 1957. Artificial satellites can transmit data from one place on Earth to another, or from space to Earth. †Satellite applications include science, communications, weather forecasting, and military use.

An **active satellite** is one that transmits probe signals that return with information. An example was Seasat, which used radar and radio-sensing instruments to monitor the Earth's oceans. A **passive satellite** does not send out probes, but only records information, such as images, about the target object.

Space probes have been sent to natural satellites including the Earth's Moon, Mars's †Deimos, and Jupiter's †Io, †Ganymede, and †Europa.

At any time, there are several thousand artificial satellites orbiting the Earth, including active satellites, satellites that have ended their working lives, and discarded sections of rockets. The brightest artificial satellites can be seen by the naked eye. Artificial satellites eventually re-enter the Earth's atmosphere. Usually they burn up by friction, but sometimes debris falls to the Earth's surface, as with *Skylab* and *Salyut 7*. In 1997 there were 300 active artificial satellites in orbit around Earth, the majority used in communications.

How Satellites Work http://www.howstuffworks.com/satellite5.htm

Entertaining introduction to satellites, with lots of interesting questions answered. There is also information on how to observe satellites.

Satellites of the Outer Planets http://www.lpi.usra.edu/research/outerp/moons.html

Detailed and highly comprehensive page about the satellites of the outer planets. With links to separate pages for every satellite covered, this is more of a resource to dip into than read from start to finish. Images of each satellite are also provided, plus links to a special three-dimensional tour of the Solar System.

satellite applications uses to which artificial satellites are put. These include:

scientific experiments and observation Many astronomical observations are best taken above the disturbing effect of the atmosphere. Satellite observations have been carried out by the Infrared Astronomy Satellite (1983) which made a complete infrared survey of the skies, and Solar Max (1980), which observed solar flares. The Hipparcos satellite, launched in 1989, measured the positions of many stars. The Röntgen Satellite, launched in 1990, examined ultraviolet and X-ray radiation. In 1992, the Cosmic Background Explorer satellite detected details of the Big Bang that mark the first stage in the formation of galaxies. Medical experiments have been carried out aboard crewed satellites, such as the Russian space station *Mir* and the US *Skylab*.

reconnaissance, land resource, and mapping applications Apart from military use and routine map-making, the US Landsat, the French Satellite Pour l'Observation de la Terre, and equivalent Russian satellites have provided much useful information about water sources and drainage, vegetation, land use, geological structures, oil and mineral locations, and snow and ice.

weather monitoring The US National Oceanic and Atmospheric Administration series of satellites, and others launched by the European Space Agency, Japan, and India, provide continuous worldwide observation of the atmosphere.

navigation The US Global Positioning System uses 24 Navstar satellites that enable users (including walkers and motorists) to find their position to within 100 m/328 ft. The US military can make full use of the system, obtaining accuracy to within 1.5 m/4 ft 6 in. The Transit system, launched in the 1960s, with 12 satellites in orbit, locates users to within 100 m/328 ft.

communications A complete worldwide communications network is now provided by satellites such as the US-run Intelsat system.

COBE Home Page http://space.gsfc.nasa.gov/astro/cobe/cobe_home.html

Describes the goals and workings of the Cosmic Background Explorer satellite, with a detailed account of the various instruments the spacecraft carries. Information on cosmology in general is available on the COBE Educational Resource page, which has links to tutorials, documents, and images from COBE.

ESA: Satellite Applications http://www.esa.int/export/esaSA/

Updated with the latest news of European Space Agency satellite technology applied to Earth observation, telecommunications, and navigation. There are also well-written articles, with photographs, maps, and varied spacecraft imagery.

satellite image image of the Earth or any other planet obtained from instruments on a satellite. Satellite images can provide a variety of information, including vegetation patterns, sea surface temperature, weather, and geology.

Landsat 4, launched in 1982, orbits at 705 km/438 mi above the Earth's surface. It completes nearly 15 orbits per day, and can survey the entire globe in 16 days. The instruments on Landsat scan the planet's surface and record the brightness of reflected light. The data is transmitted back to Earth and translated into a satellite image.

saturated compound organic compound, such as propane, that contains only single covalent bonds. †Alkanes are saturated †hydrocarbons that contain only carbon to carbon single bonds. Saturated organic compounds can only undergo further reaction by †substitution reactions, as in the production of chloropropane from propane.

saturated fatty acid †fatty acid in which there are no double bonds in the hydrocarbon chain.

saturated solution in physics and chemistry, a solution obtained when a solvent (liquid) can dissolve no more of a solute (usually a solid) at a particular temperature. Normally, a slight fall in temperature causes some of the solute to crystallize out of solution. If this does not happen the phenomenon is called supercooling, and the solution is said to be **supersaturated**.

saturated vapor pressure †pressure exerted by a †vapor that exists in equilibrium with its liquid.

Saturn sixth planet from the Sun, and the second-largest in the Solar System, encircled by bright and easily visible equatorial rings. Viewed through a telescope it is ochre. Its polar diameter is 12,000 km/7,450 mi smaller than its equatorial diameter, a result of its fast rotation and low density, the lowest of any planet. Its mass is 95 times that of the Earth

Saturn The rings of Saturn, consisting of small pieces of ice and rock, are 62,000 km/38,000 mi wide and just 100 m/300 ft thick. NASA

and its magnetic field 1,000 times stronger.

mean distance from the Sun: 1.427 billion km/0.886 billion mi

equatorial diameter: 120,000 km/75,000 mi

rotational period: 10 hours 14 minutes at equator, 10 hours 40 minutes at higher latitudes

year: 29.46 Earth years

atmosphere: visible surface consists of swirling clouds, probably made of frozen ammonia at a temperature of –170°C/–274°F, although the markings in the clouds are not as prominent as Jupiter's. The Voyager probes, visiting in 1980 and 1981, found winds reaching 1,800 kph/1,100 mph

surface: Saturn is believed to have a small core of rock and iron, encased in ice and topped by a deep layer of liquid hydrogen

satellites: 22 known moons, more than for any other planet. The largest moon, †Titan, has a dense atmosphere. Other satellites include Epimetheus, Janus, Pandora, and Prometheus

rings: The rings visible from Earth begin about 14,000 km/9,000 mi from the planet's cloudtops and extend out to about 76,000 km/47,000 mi. Made of small chunks of ice and rock (averaging 1 m/3.3 ft across), they are 275,000 km/170,000 mi rim to rim, but only 100 m/300 ft thick. The Voyager probes showed that the rings actually consist of thousands of closely spaced ringlets, looking like the grooves in a gramophone record

Apollo Saturn Reference Page http://www.apollosaturn.com/

Apollo-Gemini project resource. There are contemporary press releases, technical reports and diagrams, and photographs. The site features a page with some truly amazing Apollo facts and figures.

Cassini-Huygens Mission to Saturn and Titan http://saturn.jpl.nasa.gov/cassini/english/

Well-organized site documenting the Cassini-Huygens mission to Saturn. Computer simulations show where the probe is currently. There is an impressive gallery featuring the latest pictures returned by the spacecraft, time-lapse movies, and animations. There are also interesting details of what is already known about Saturn's moons.

Saturn http://www.solarviews.com/solar/eng/saturn.htm

How many rings does Saturn have? How many satellites? Find out the answers to these questions and more at this site, which also features a video of a storm in the planet's atmosphere and information on the international Cassini mission to Saturn and Titan.

Saussure, Horace Bénédict de (1740–1799) Swiss geologist who made the earliest detailed and first-hand study of the Alps. He was a physicist at the University of Geneva. The results of his Alpine survey appeared in his classic work *Voyages des Alpes/Travels in the Alps* (1779–86).

savannah or **savanna**, extensive open tropical grasslands, with scattered trees and shrubs. Savannahs cover large areas of Africa, North and South America, and northern Australia. The soil is acidic and sandy and generally considered suitable only as pasture for low-density grazing.

A new strain of rice suitable for savannah conditions was developed in 1992. It not only grew successfully under test conditions in Colombia but also improved pasture quality so that grazing numbers could be increased twentyfold.

Tropical Savannas CRC: Landscape Processes http://savanna.ntu.edu.au/information/information.html

Australian-based site providing information on tropical savannahs including "Savannah explorer," "Savannah map maker," where you can explore relationships between landscape features such as vegetation and soils by overlaying maps, and "Savannah search."

sawhorse projection in chemistry, two-dimensional representational stick diagram of the three-dimensional molecular structure of a substituted aliphatic hydrocarbon molecule. The diagram presents the orientation of the substituted groups with respect to each other (see also †Fischer projection and †Newman projection).

scalar quantity in mathematics and science, a quantity that has magnitude but no direction, as distinct from a †vector quantity, which has a direction as well as a magnitude. Temperature, mass, volume, and †speed are scalar quantities.

scale or **limescale**, in chemistry, †calcium carbonate precipitates that form on the inside of a kettle or boiler as a result of boiling †hard water (water containing concentrations of soluble calcium and magnesium salts). The salts present in hard water also precipitate out by reacting with soap molecules.

Scale may cause damage to water pipes and appliances such as washing machines and water boilers. The build-up of scale causes blockage of pipes, and electrical appliances have to use more electrical energy to operate efficiently.

scale numerical relationship, expressed as a †ratio, between the actual size of an object and the size of an image that represents it on a map, plan, or diagram.

If an object has been enlarged, the amount of increase (scale factor) can be found by dividing a side of the enlarged object by the corresponding side of the original shape. A scale factor of less than one gives a decrease in size. An object is scaled by multiplying it by the scale factor; for example, a model railway may require a scale of 1 in a 100, so all measurements must be multiplied by $^1/_{100}$ (0.01).

scalene in geometry, term used to describe a †triangle that has no two sides equal. Scalene also describes a †cone or †cylinder where the axis is inclined to the base.

scaler instrument that counts the number of radioactive particles passing through a radiation detector such as a Geiger–Müller tube (the scaler and tube together form a †Geiger counter). It gives the total, or cumulative, number of particles counted whereas a †ratemeter gives the number of particles detected in a unit of time.

scalp soft tissue covering the skull (excluding the face).

scandium chemical symbol Sc, silver-white, metallic element of the †lanthanide series, atomic number 21, relative atomic mass 44.956.

Its compounds are found widely distributed in nature, but only in minute amounts. The metal has little industrial importance.

Scandium oxide (scandia) is used as a catalyst, in making crucibles and other ceramic parts, and scandium sulfate (in very dilute aqueous solution) is used in agriculture to improve seed germination. Scandium is relatively more abundant in the Sun and other stars than on Earth.

scanning electron microscope SEM, electron microscope that produces three-dimensional images, magnified 10–200,000 times. A fine beam of electrons, focused by electromagnets, is moved, or scanned, across the specimen. Secondary radiation reflected from the specimen is collected by a detector, giving rise to an electrical signal, which is then used to generate a point of brightness on a television-like screen. As the point moves rapidly over the screen, in phase with the scanning electron beam, an image of the specimen is built up.

The resolving power of an SEM depends on the size of the electron beam – the finer the beam, the greater the resolution. Present-day instruments typically have a resolution of 7–10 nm.

The first scanning electron picture was produced in 1935 by Max Knoll of the German company Telefunken, although the first commercial SEM (produced by the Cambridge Instrument Company in the UK) did not go on sale until 1965.

scanning transmission electron microscope STEM, electron microscope that combines features of the †scanning electron microscope (SEM) and the †transmission electron microscope (TEM). First built in the USA in 1966, the microscope has both the SEM's contrast characteristics and lack of aberrations and the high resolution of the TEM. Magnifications of over 90 million times can be achieved, enough to image single atoms.

A fine beam of electrons, 0.3 nm in diameter, moves across the specimen, as in an SEM. However, because the specimen used is a thin slice, the beam also passes through the specimen (as in a TEM). The reflected electrons and those that penetrated the specimen are collected to form an electric signal, which is interpreted by computer to form an image on a screen.

scanning tunneling microscope STM, microscope that produces a magnified image by moving a tiny tungsten probe across the surface of the specimen. The tip of the probe is so fine that it may consist of a single atom, and it moves so close to the specimen surface that electrons jump (or tunnel) across the gap between the tip and the surface.

The magnitude of the electron flow (current) depends on the distance from the tip to the surface, and so by measuring the current, the contours of the surface can be determined. These can be used to form an image on a computer screen of the surface, with individual atoms resolved. Magnifications up to 100 million times are possible.

The STM was invented in 1981 by Gerd Binning from Germany and Heinrich Rohrer from Switzerland at the IBM Zürich Research Laboratory. With Ernst Ruska, who invented the transmission electron microscope in 1933, they were awarded the Nobel Prize for Physics in 1986.

scapolite group of white or grayish minerals, silicates of sodium, aluminum, and calcium, common in metamorphosed limestones and forming at high temperatures and pressures.

scapula or **shoulder blade**, large, flat, triangular bone which lies over the second to seventh ribs on the back, forming part of the pectoral girdle, and assisting in the articulation of the arm with the chest region. Its flattened shape allows a large region for the attachment of muscles.

scarp and dip in geology, the two slopes that comprise an escarpment. The scarp is the steep slope and the dip is the gentle slope. Such a feature is common when sedimentary rocks are uplifted, folded, or eroded, the scarp slope cuts across the bedding planes of the sedimentary rock while the dip slope follows the direction of the strata. An example is Salisbury Crags in Edinburgh, Scotland.

scatter diagram or **scattergram**, diagram whose purpose is to establish whether or not a relationship or †correlation exists between two variables; for example, between life expectancy and gross national product. Each observation is marked with a dot in a position that shows the value of both variables. The pattern of dots is then examined to see whether they show any underlying trend by means of a †line of best fit (a straight line drawn so that its distance from the various points is as short as possible).

scattering in physics, the random deviation or reflection of a stream of particles or of a beam of radiation such as light, by the particles in the matter through which it passes.

alpha particles Alpha particles scattered by a thin gold foil provided the first convincing evidence that atoms had very small, very dense, positive nuclei. From 1906 to 1908 Ernest Rutherford carried out a series of experiments from which he estimated that the closest approach of an alpha particle to a gold nucleus in a head-on collision was about 10^{-14} m. He concluded that the gold nucleus must be no larger than this. Most of the alpha particles fired at the gold foil passed straight through undeviated; however, a few were scattered in all directions and a very small fraction bounced back towards the source. This result so surprised Rutherford that he is reported to have commented: "It was almost as if you fired a 15-inch shell at a piece of tissue paper and it came back and hit you."

light Light is scattered from a rough surface, such as that of a sheet of paper, by random reflection from the varying angles of each small part of the surface. This is responsible for the dull, flat appearance of such surfaces and their inability to form images (unlike mirrors). Light is also scattered by particles suspended in a gas or liquid. The red and yellow colors associated with sunrises and sunsets are due to the fact that red light is scattered to a lesser extent than is blue light by dust particles in the atmosphere. When the Sun is low in the sky, its light passes through a thicker, more dusty layer of the atmosphere, and the blue light radiated by it is scattered away, leaving the red sunlight to pass through to the eye of the observer.

scavenger in ecology, animal that feeds on plant or animal remains. Crows, woodlice, and vultures are examples.

scent gland gland that opens onto the outer surface of animals, producing odorous compounds that are used for communicating between members of the same species (†pheromones), or for discouraging predators or competitors.

Schally, Andrew V(ictor) (1926–) Polish-born endocrinologist who was awarded a Nobel Prize in Physiology or Medicine in 1977 for the discovery of hormones produced by the hypothalamus region of the brain. A protein chemist and endocrinologist before joining Baylor University in the USA, he spent his career isolating and determining the chemical structures of hypothalamic hormones that affect the thyroid, adrenals, gonads, and somatic growth. He shared the prize with Roger †Guillemin and Rosalyn †Yalow.

Schawlow, Arthur Leonard (1921–1999) US physicist who was awarded the Nobel Prize for Physics in 1981 for the development of laser †spectroscopy and is generally considered, with Charles †Townes, to be co-inventor of the †laser. Schawlow used the laser as a tool to study atomic spectra and their associated energy levels. He derived improved values for atomic constants such as the †Rydberg constant. He shared the Nobel Prize with Nicolaas †Bloembergen and Kai †Siegbahn.

Scheele, Karl Wilhelm (1742–1786) Swedish chemist and pharmacist who isolated many elements and compounds for the first time, including oxygen, about 1772, and chlorine in 1774, although he did not recognize it as an element. He showed that oxygen is involved in the respiration of plants and fish.

Scheiner, Christoph (1573–1650) German astronomer who carried out one of the earliest studies of sunspots and made significant improvements to the helioscope and the telescope. In about 1605 he invented the pantograph, an instrument used for copying plans and drawings to any scale.

Schiaparelli, Giovanni Virginio (1835–1910) Italian astronomer. He drew attention to linear markings on Mars, which gave rise to the popular belief that they were canals. The markings were soon shown by French astronomer Eugène Antoniadi to be optical effects and not real lines. Schiaparelli also gave observational evidence for the theory that all meteor showers are fragments of disintegrating comets, and for this work he was awarded the Gold Medal of the Royal Astronomical Society in 1873.

schist †metamorphic rock containing †mica; its crystals are arranged in parallel layers or bands. Schist may contain additional minerals such as †garnet.

The †sedimentary rocks mudstone and shale become †slate under stress. When slate is subjected to higher temperatures and pressure schist is formed. The temperature causes recrystallization of new minerals, such as mica, and the pressure causes the new minerals to be aligned in bands.

schizocarp dry †fruit that develops from two or more carpels and splits, when mature, to form separate one-seeded units known as mericarps.

The mericarps may be dehiscent, splitting open to release the seed when ripe, as in *Geranium*, or indehiscent, remaining closed once mature, as in mallow *Malva* and plants of the Umbelliferae family, such as the carrot *Daucus carota* and parsnip *Pastinaca sativa.*

Schleiden, Matthias Jakob (1804–1881) German botanist who identified the fundamental units of living organisms when, in 1838, he announced that the various parts of plants consist of cells or derivatives of cells. This was extended to animals by Theodor †Schwann the following year.

Schlesinger, Frank (1871–1943) US astronomer. He developed the first successful technique for observing the trigonometrical parallax of a star photographically. He also developed and applied the technique for obtaining precise stellar positions from photographs taken with wide-angle camera lenses.

Schlieren photography means of photographing turbulences in fast-moving †fluids, through the change of †density and †refractive index that such turbulence produces.

Schmidt, Bernhard Voldemar (1879–1935) Estonian lens- and mirrormaker. He devised a special lens to work in conjunction with a spherical mirror in a reflecting telescope. The effect of this was to nullify "coma," the optical distortion of focus away from the center of the image, and thus to bring the entire image into a single focus.

Schmidt camera telescopic camera incorporating an internal corrective lens or plate that compensates for optical defects and chromatic faults in the main mirror. The system was invented by Estonian lensmaker Bernhard Schmidt.

Schmidt telescope reflecting telescope used for taking wide-angle photographs of the sky. Invented in 1930 by Estonian lensmaker Bernhard Schmidt, it has an added corrector lens to help focus the incoming light. Examples are the 1.2-m/48-in Schmidt telescope on Mount Palomar and the UK Schmidt telescope, of the same size, at the Anglo-Australian Observatory, Siding Spring Mountain, New South Wales, Australia.

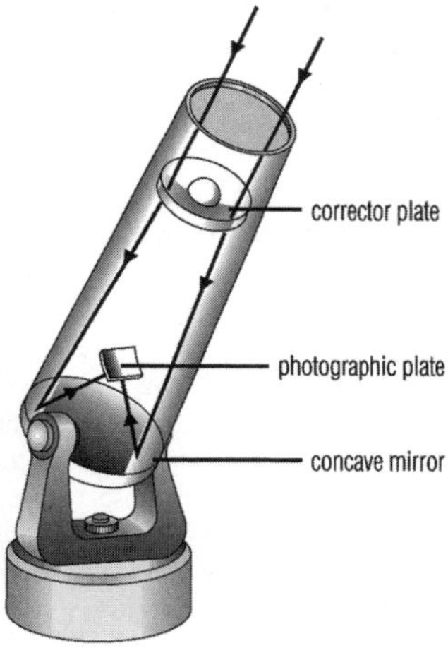

Schmidt telescope The Schmidt telescope uses a corrective lens (corrector plate) to achieve a wide field of view. It is one of the most widely used tools of astronomy. Helicon

Schonbein, Christian Friedrich (1799–1868) German chemist. He became professor at Basel University in 1828. In 1839 he discovered ozone, which he so named because of its peculiar smell (Greek *ozein*, "to smell"). In 1845 he invented gun-cotton, from which he obtained collodion.

Schrieffer, John Robert (1931–) US physicist who, with John †Bardeen and Leon †Cooper, was awarded the Nobel Prize for Physics in 1972 for developing the first satisfactory theory of †superconductivity (the resistance-free flow of electrical current which occurs in many metals and metallic compounds at very low temperatures). He has also worked on ferromagnetism, surface physics, and dilute alloys.

Schrödinger, Erwin (1887–1961) Austrian physicist. He advanced the study of wave mechanics to describe the behavior of electrons in atoms. In 1926 he produced a solid mathematical explanation of the †quantum theory and the structure of the atom. He shared the Nobel Prize for Physics in 1933 for his work in the development of quantum mechanics.

Schrödinger, Erwin http://www-groups.dcs.st-andrews.ac.uk/history/Mathematicians/Schrodinger.html

Part of an archive containing the biographies of the world's greatest mathematicians, this site is devoted to the life and contributions of physicist Erwin Schrödinger. In addition to biographical information, you will find a list of references about Schrödinger and links to other essays in the archive that reference him. The text of this essay includes hypertext links to the essays of those mathematicians and thinkers who influenced Schrödinger. There is also an image of him, which can be enlarged, and a map of his birthplace.

Schrödinger's equation equation that considers the †electron in terms of a wave of probability, and which enables the behavior of the electron in atoms and electric potentials to be calculated, and also the spectra of atoms to be predicted. It is the basis of †wave mechanics.

Schubfachprinzip or **pigeon-hole principle**, principle that states that if in *n* boxes one distributes more than *n* objects, at least one box must contain more than one object. The principle is used in the logic of number theory.

Schultze, Max Johann Sigismund (1825–1874) German zoologist who adopted the term †protoplasm to refer to the contents of cells. His work emphasized its role in both plant and animal cells. He also worked on the retina, correctly arguing that the cones were responsible for color vision and the rods for the perception of light.

Schwabe, Samuel Heinrich (1789–1875) German astronomer. He was the first to measure the periodicity of the sunspot cycle. This may be considered as marking the beginning of solar physics.

Schwann, Theodor (1810–1882) German physiologist who, with Matthias †Schleiden, is credited with formulating the cell theory, one of the most fundamental concepts in biology. Schwann also did important work on digestion, fermentation, and the study of tissues.

Schwann cell in vertebrates, the neurilemma cell of myelinated peripheral nerve fibers, important in the manufacture of †myelin. On myelin-coated fibers a Schwann cell occurs between each pair of adjacent nodes.

Schwartz, Melvin (1932–) US physicist who, with Jack †Steinberger and Leon †Lederman, was awarded the Nobel Prize for Physics in 1988 for the development of the use of †neutrinos to study †elementary particles and for the discovery of the muon neutrino.

Schwarzschild, Karl (1873–1916) German astronomer and theoretical physicist. He was the first to substitute a photographic plate at the telescope in place of the eye and then measure densities with a photometer. He designed and constructed some of his own instruments.

Schwarzschild, Karl http://www-groups.dcs.st-andrews.ac.uk/history/Mathematicians/Schwarzschild.html

Devoted to the life and contributions of Karl Schwarzschild.

Schwarzschild radius in †astrophysics, the radius of the event horizon surrounding a †black hole within which light cannot escape its gravitational pull.

For a black hole of mass m, the Schwarzschild radius Rs is given by Rs = $2gm/c^2$, where g is the gravitational constant and c is the speed of light. The Schwarzschild radius for a black hole of solar mass is about 3 km/1.9 mi. It is named after Karl †Schwarzschild, the German astronomer who deduced the possibility of black holes from †Einstein's general theory of relativity in 1916.

Schwinger, Julian Seymour (1918–1994) US quantum physicist. His research concerned the behavior of charged particles in electrical fields. This work, expressed entirely through mathematics, combines elements from †quantum theory and †relativity theory into a new theory called **quantum electrodynamics**, the most accurate physical theory of all time. Schwinger was awarded the Nobel Prize for Physics in 1965 for his development of the basic principles of quantum electrodynamics. He shared the award with Richard †Feynman and Sin-Itiro †Tomonaga.

science (Latin *scientia* "knowledge") any systematic field of study or body of knowledge that aims, through experiment, observation, and deduction, to produce reliable explanations of phenomena, with reference to the material and physical world.

history Activities such as healing, star-watching, and engineering have been practiced in many societies since ancient times. Pure science, especially physics (formerly called natural philosophy), had traditionally been the main area of study for philosophers. The European scientific revolution between about 1650 and 1800 replaced speculative philosophy with a new combination of observation, experimentation, and rationality.

philosophy of science Today, scientific research involves an interaction between tradition, experiment and observation, and deduction. The subject area called philosophy of science investigates the nature of this complex interaction, and the extent of its ability to gain access to the truth about the material world. It has long been recognized that induction from observation cannot give explanations based on logic. In the 20th century Karl Popper described †scientific method as a rigorous experimental testing of a scientist's ideas or hypotheses (see †hypothesis). The origin and role of these ideas, and their interdependence with observation, have been examined, for example, by the US thinker Thomas S Kuhn, who places them in a historical and sociological setting.

sociology of science The sociology of science investigates how scientific theories and laws are produced, and questions the possibility of objectivity in any scientific endeavor. One controversial point of view is the replacement of scientific realism with scientific relativism, as proposed by Paul K Feyerabend. Questions concerning the proper use of science and the role of science education are also restructuring this field of study.

Andy Darvill's Science Site http://www.darvill.clara.net/index.htm

Constructed by an ex-teacher, this site contains a variety of resources. In the "Online materials" section there are some excellent physics Web pages created by the teacher. There are online quizzes with hints available, links to relevant downloadable science resources, recommendations of other science software, plus a useful section on revision tips.

Elementary Science http://www.lme.mankato.msus.edu/ci/elem.sci.html

Brief biographies of famous scientists, some experiments to do in the home, a featured animal of the month, information on the night sky, and a "Tell me why" feature. It seems to be an irregularly updated site, but there are plenty of back issues still available.

Mad Scientist Network http://www.madsci.org/

Indispensable resource for those in need of answers to scientific questions. The Mad Scientist Network is a group of scientists prepared to field queries on a vast range of subjects. There is an online archive of previous questions and answers, and helpful guides to accessing other general interest science Web sites.

Museum of the History of Science http://www.mhs.ox.ac.uk/

Based in Oxford, UK. This site offers exhibitions – both from the galleries and especially for the Web – together with news of museum activities, publications, events, and courses.

Science Club http://www.eskimo.com/~billb/

General science site, with loads of science projects to build, from a "Quick and simple telescope" to an electrostatic loudspeaker. This site also includes a "Science fair ideas exchange" section and pages of links in an attempt to answer any science questions you might have.

Why Files http://whyfiles.org/index.html

Well-designed and topical science site tackling a couple of recent issues in each bimonthly update. As well as having a lively presentation, it also includes a lot of useful background information on a wide range of topics. This site also retains many of its previous entries, which are fully searchable.

scientific law principles that are taken to be universally applicable.

Laws (for instance, †Boyle's law and †Newton's laws of motion) form the basic theoretical structure of the physical sciences, so that the rejection of a law by the scientific community is an almost inconceivable event. On occasion a law may be modified, as was the case when Einstein showed that Newton's laws of motion do not apply to objects traveling at speeds close to that of light.

scientific method belief that experimentation and observation, properly understood and applied, can avoid the influence of cultural and social values and so build up a picture of a reality independent of the observer.

Techniques and mechanical devices that improve the reliability of measurements may seem to support this theory; but the realization that observations of subatomic particles influence their behavior has undermined the view that objectivity is possible in science (see †uncertainty principle).

scientific notation alternative term for †standard form.

scintillation in radio astronomy, a rapid oscillation in the detected intensity of radiation emitted by stellar radio sources, caused by disturbances in ionized gas at some point between the source and the Earth's surface (usually in the Earth's own upper atmosphere).

scintillation counter instrument for measuring very low levels of radiation. The radiation strikes a scintillator (a device that emits a unit of light when a charged elementary particle collides with it), whose light output is "amplified" by a †photomultiplier; the current pulses of its output are in turn counted or added by a scaler to give a numerical reading.

sclerenchyma plant tissue whose function is to strengthen and support, composed of thick-walled cells that are heavily lignified (toughened). On maturity the cell inside dies, and only the cell walls remain.

Sclerenchyma may be made up of one or two types of cells: **sclereids**, occurring singly or in small clusters, are often found in the hard shells of fruits and in seed coats, bark, and the stem cortex; **fibers**, frequently grouped in bundles, are elongated cells, often with pointed ends, associated with the vascular tissue (†xylem and †phloem) of the plant.

Some fibers provide useful materials, such as flax from *Linum usitatissimum* and hemp from *Cannabis sativa.*

scolex in zoology, anterior (front) end of a †tapeworm. It bears hooks and suckers that fasten the tapeworm to the inner lining of the host's intestine. The attachment prevents the tapeworm from being dislodged by the intestine's peristaltic movements.

Scorpius bright zodiacal constellation in the southern hemisphere between †Libra and †Sagittarius, represented as a scorpion. The Sun passes briefly through Scorpius in the last week of November. The heart of the scorpion is marked by the bright red supergiant star †Antares. Scorpius contains rich †Milky Way star fields, plus the strongest †X-ray source in the sky, Scorpius X-1. The whole area is rich in clusters and nebulae. In astrology, the dates for Scorpius are about 24 October–21 November (see †precession).

Scott, Dukinfield Henry (1854–1934) English botanist who studied the anatomy of plants and, with William Crawford Williamson, described the evolutionary links between ferns and cycads, research that led to the development of phylogenetic theories of plants. His best known studies were in the field of paleobotany, including an excellent account of the fruiting bodies of fossil plants in 1904.

scree pile of rubble and sediment that collects at the foot of a mountain range or cliff. The rock fragments that form scree are usually broken off by the action of frost (†freeze–thaw weathering).

With time, the rock-waste builds up into a heap or sheet of rubble that may eventually bury even the upper cliffs, and the growth of the scree then stops. Usually, however, erosional forces remove the rock waste so that the scree stays restricted to lower slopes.

screw-worm fly in biology, parasitic fly *Chrysomya bezziana* that is a serious pest of cattle. A wound (a small scratch, for example) attracts the female fly to lay her eggs around the edges of the damaged area. After a few days the larvae hatch from the eggs and eat into the tissue increasing the size of the wound. Unless treated quickly with insecticide to kill the larvae, the victim loses condition and eventually dies. The range of the fly includes Africa, India, and southeast Asia.

scrotum paired sac in male mammals enclosing the testes and the epididymes (coiled tubes in which the sperm mature).

SCSI acronym for Small Computer System Interface; pronounced "scuzzy," in computing, one of the methods for connecting peripheral devices (such as printers, scanners, and CD-ROM drives) to a computer. A group of peripherals linked in series to a single SCSI port is called a **daisy-chain**.

SCSIs are common in file servers and high-performance computers. One of the main advantages of a SCSI is that devices attached to it can communicate with each other independently of the CPU. Another important feature is the ability to connect devices that are external to the main computer unit.

Sculptor inconspicuous constellation of the southern hemisphere, containing the south pole of the Galaxy. It was originally named by French astronomer Nicolas †Lacaille as "Apparatus Sculptoris," "the Sculptor's Tools."

scurvy disease caused by deficiency of vitamin C (ascorbic acid), which is contained in fresh vegetables and fruit. The signs are weakness and aching joints and muscles, progressing to bleeding of the gums and other spontaneous hemorrhage, and drying-up of the skin and hair. It is reversed by giving the vitamin.

Scutum small constellation of the southern hemisphere, represented by a shield. It lies in the †Milky Way, which in this region consists of uniform clouds of faint stars.

SDRAM abbreviation for **synchronous dynamic random-access memory**, in computing, the latest generation of DRAM memory modules, rapidly replacing EDO-RAM technology. SDRAM †memory chips are able to transfer data in a single clock cycle, because their operations are synchronized with the system clock.

Seaborg, Glenn Theodore (1912–1999) US nuclear chemist. He shared the Nobel Prize for Chemistry in 1951 with his co-worker Edwin †McMillan for the discovery of plutonium and research on the †transuranic elements.

seaborgium chemical symbol Sg, synthesized radioactive element of the †transactinide series, atomic number 106, relative atomic mass 263. It was first synthesized in 1974 in the USA and given the temporary name unnilhexium. The discovery was not confirmed until 1993. It was officially named in 1997 after US nuclear chemist Glenn †Seaborg.

The University of California, Berkeley, bombarded californium with oxygen nuclei to get isotope 263; the Joint Institute for Nuclear Research, Dubna, Russia, bombarded lead with chromium nuclei to obtain isotopes 259 and 260.

sea breeze gentle coastal wind blowing off the sea towards the land. It is most noticeable in summer when the warm land surface heats the air above it and causes it to rise. Cooler air from the sea is drawn in to replace the rising air, so causing an onshore breeze. At night and in winter, air may move in the opposite direction, forming a †land breeze.

seafloor spreading growth of the ocean †crust outwards (sideways) from ocean ridges. The concept of seafloor spreading has been combined with that of continental drift and incorporated into †plate tectonics. Seafloor spreading was proposed in 1960 by US geologist Harry Hess (1906–1969), based on his observations of ocean ridges and the relative youth of all ocean beds. In 1963 British geophysicists Fred Vine and Drummond Matthews observed that the floor of the Atlantic Ocean was made up of rocks that could be arranged in strips, each strip being magnetized either normally or reversely (due to changes in the Earth's polarity when the North Pole becomes the South Pole and vice versa, termed †polar reversal). These strips were parallel and formed identical patterns on both sides of the ocean ridge. The implication was that each strip was formed at some stage in geological time when the magnetic field was polarized in a certain way. The seafloor magnetic-reversal patterns could be matched to dated magnetic reversals found in terrestrial rock. It could then be shown that new rock forms continuously and spreads away from the ocean ridges, with the oldest rock located farthest away from the midline. The observation was made independently in 1963 by Canadian geologist Lawrence Morley, studying an ocean ridge in the Pacific near Vancouver Island.

sea level average height of the surface of the oceans and seas measured throughout the tidal cycle at hourly intervals and computed over a 19-year period. It is used as a datum plane from which elevations and depths are measured.

Factors affecting sea level include temperature of seawater (warm water is less dense and therefore takes up a greater volume than cool water) and the topography of the ocean floor.

seamount in earth science, isolated volcanic mount on the sea bed at least 700 m/2,300 ft tall. Seamounts vary from smallish conical peaks to large flat-topped masses (†guyots) and may occur singly or in chains or groups.

search engine in computing, remotely accessible program to help users find information on the Internet. Commercial search engines such as Google and Lycos comprise databases of documents, †URLs, Usenet articles, and more, which can be searched by keying in a key word or phrase. The databases are compiled by a mixture of automated agents (spiders) and webmasters registering their sites.

US researchers estimate that search engines in 1999 are indexing only approximately 16% of the 800 million publicly available Web pages when used individually; when the 11 main search engines are combined to conduct a metasearch they still fail to index more than half of the Web pages. This is partly because many Web sites generate pages on request, filling in standard Web page templates with information from databases.

searching in computing, extracting a specific item from a large body of data, such as a file or table. The method used depends on how the data are organized. For example, a binary search, which requires the data to be in sequence, involves first deciding which half of the data contains the required item, then which quarter, then which eighth, and so on until the item is found.

season period of the year having a characteristic climate. The change in seasons is mainly due to the Earth's axis being tilted in relation to the Sun, and hence the position of the Sun in the sky at a particular place changes as the Earth orbits the Sun.

When the northern hemisphere is tilted away from the Sun (winter) the Sun's rays have further to travel through the atmosphere (they strike the Earth at a shallower angle) and so have less heating effect, resulting in colder weather. The days are shorter and the nights are longer. At the same time, the southern hemisphere is tilted towards the Sun (summer) and experiences warmer weather, with longer days and shorter nights. The opposite occurs when the northern hemisphere is tilted towards the Sun and the southern hemisphere away from the Sun.

Earth's Seasons: Equinoxes, Solstices, Perihelion, and Aphelion

http://aa.usno.navy.mil/data/docs/EarthSeasons.html

Part of a larger site on astronomical data maintained by the US Naval Observatory, this site gives the dates and hours (in Universal Time) of the changing of the seasons from 1992 through to 2005. It also includes sections of "Frequently Asked Questions" and research information.

seawater water of the seas and oceans, covering about 70% of the Earth's surface and comprising about 97% of the world's water (only about 3% is freshwater). Seawater contains large numbers of electrically charged particles, or †ions. These may be positively or negatively charged. The most common positive ions are sodium, potassium, magnesium, and calcium. These positive ions are balanced by negative ions, for example chloride, iodide, fluoride, bromide, sulfates, carbonates, bicarbonates, phosphates, nitrates, and others. As a result the physical chemistry of seawater is extremely complex.

As seawater evaporates, the positive and negative ions are attracted to one another to form crystals of various types, the most common of these being sodium chloride (common salt). Seawater also contains a large amount of dissolved carbon dioxide, and thus the oceans act as carbon "sinks" that may help to reduce the greenhouse effect.

sebaceous gland gland in the skin producing waxlike †sebum. It opens into a hair follicle. Distribution of sebaceous glands varies in different parts of the body.

sebum oily secretion from the sebaceous glands that acts as a skin lubricant. Acne is caused by inflammation of the sebaceous glands and over-secretion of sebum.

sec or **s**, abbreviation for **second**, a unit of time.

SECAM acronym for *Système Electronique Couleur Avec Mémoire*, "electronic color system with memory"; or *SEquentiel Couleur A Mémoire*, "sequential color to memory," television and video standard used in France, some states in Eastern Europe, and a few other countries. It is broadly similar to the PAL system used in most of Europe.

secant in trigonometry, the function of a given angle in a right-angled triangle, obtained by dividing the length of the hypotenuse (the longest side) by the length of the side adjacent to the angle. It is the †reciprocal of the †cosine (sec = 1/cos).

Secchi, Pietro Angelo (1818–1878) Italian astronomer and astrophysicist. He classified stellar spectra into four classes based on their color and spectral characteristics. He was the first to classify solar †prominences, huge jets of gas projecting from the Sun's surface.

second symbol sec or s, basic †SI unit of time, one-sixtieth of a minute. It is defined as the duration of

9,192,631,770 cycles of regulation (periods of the radiation corresponding to the transition between two hyperfine levels of the ground state) of the cesium-133 isotope. In mathematics, the second is a unit (symbol ") of angular measurement, equaling one-sixtieth of a minute, which in turn is one-sixtieth of a degree.

secondary color color of light that is created by mixing two primary colors. The primaries red and blue combine to form the secondary magenta; blue and green form cyan; and red and green produce yellow.

secondary consumer in ecology, animals (primary carnivores) that eat herbivores.

secondary data information that has been collected by another agency. Examples of secondary data include government reports and statistics, company reports and accounts, and weather reports in newspapers.

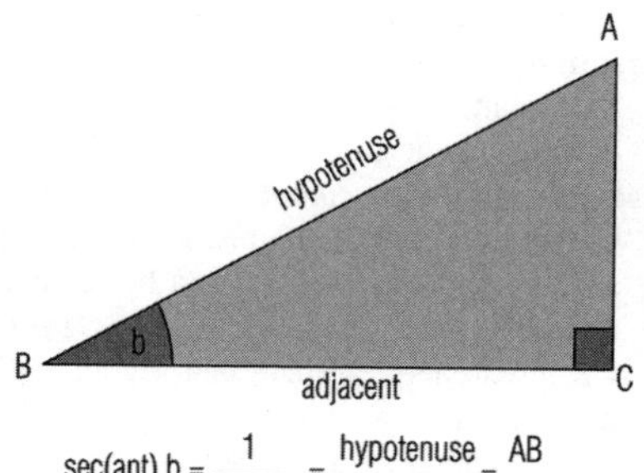

secant The secant of an angle is a trigonometrical function. If the secant of angle B is known, then the hypotenuse can be found given the length of the adjacent side, or the adjacent side can be found from the hypotenuse. Helicon

secondary emission emission of electrons from the surface of certain substances when they are struck by high-speed electrons or other particles from an external source. It can be detected with a †photomultiplier.

secondary growth or **secondary thickening**, increase in diameter of the roots and stems of certain plants (notably shrubs and trees) that results from the production of new cells by the †cambium. It provides the plant with additional mechanical support and new conducting cells, the secondary †xylem and †phloem. Secondary growth is generally confined to †gymnosperms and, among the †angiosperms, to the dicotyledons. With just a few exceptions, the monocotyledons (grasses, lilies) exhibit only primary growth, resulting from cell division at the apical †meristems.

secondary sexual characteristic external feature of an organism that is indicative of its gender (male or female), but not the reproductive organs themselves. They include facial hair in men and breasts in women, combs in cockerels, brightly colored plumage in many male birds, and manes in male lions. In many cases, they are involved in displays and contests for mates and have evolved by †sexual selection. Their development is stimulated by sex †hormones – in humans the change in concentrations of these hormones at puberty results not only in the development of the physical secondary sexual characteristics, but also emotional changes.

In humans the development of secondary sexual characteristics is stimulated at †puberty by the hormone †estrogen in women and †testosterone in men. Testosterone is produced in the testes of males, while estrogen and progesterone are produced in the ovaries of females. Women develop breasts and wider hips; men develop a deeper voice, body and facial hair, and stronger body odor; and both develop hair in underarm and genital regions. However, at puberty the reproductive organs also mature as a consequence of growth and changes in circulating hormones, ready for sexual reproduction.

second filial generation or **F_2 generation**, in genetics, offspring of the †first filial generation (F_1) resulting from a typical Mendelian cross. The F_2 generation exhibit the true breeding character of the parents in a ration of 3 to 1.

second order in mathematics, describing a differential equation, involving only first and second †derivatives. The term is occasionally used in algebraic contexts to mean "of the second degree," that is, involving expressions raised to at most the power of two (squares).

secretin †hormone produced by the small intestine of vertebrates that stimulates the production of digestive secretions by the pancreas and liver.

secretion in biology, any substance (normally a fluid) produced by a cell or specialized gland, for example, sweat, saliva, enzymes, and hormones. The process whereby the substance is discharged from the cell is also known as secretion.

sector in geometry, part of a circle enclosed by two radii and the arc that joins them. A **minor sector** has an angle at the center of the circle of less than 180°. A **major sector** has an angle at the center of the circle of more than 180°.

The area of a sector is found by dividing the angle at the center by 360 and multiplying by pr^2, where r is the radius of the circle.

secular in astronomy, describing something that is gradual, taking eons to accomplish.

secular acceleration of the Moon, of the Sun apparent acceleration of the Moon and Sun across the sky, caused by extremely gradual reduction in speed of the Earth's rotation (one 50-millionth of a second per day).

secular variation changes in the position of Earth's magnetic poles measured with respect to geographical positions, such as the North Pole, throughout geological time.

sedative any drug that has a calming effect, reducing anxiety and tension.

Sedatives will induce sleep in larger doses. Examples are barbiturates, †narcotics, and benzodiazepines.

Sedgwick, Adam (1785–1873) English geologist who contributed greatly to understanding the †stratigraphy of the British Isles, using fossils as an index of relative time. Together with Scottish geologist Roderick †Murchison, he identified the Devonian system in southwest England.

sediment any loose material that has "settled" after deposition from suspension in water, ice, or air, generally as the water current or wind speed decreases. Typical sediments are, in order of increasing coarseness: clay, mud, silt, sand, gravel, pebbles, cobbles, and boulders.

Sediments differ from †sedimentary rocks, in which deposits are fused together in a solid mass of rock by a process called †lithification (solidification). Pebbles are cemented into †conglomerates; sands become sandstones; muds become mudstones or shales; peat is transformed into coal.

sedimentary rock rock formed by the accumulation and cementation of deposits that have been laid down by water, wind, ice, or gravity. Sedimentary rocks cover more than two-thirds of the Earth's surface and comprise three major categories: clastic, chemically precipitated, and organic (or biogenic). Clastic sediments are the largest group and are composed of fragments of pre-existing rocks; they include clays, sands, and gravels.

Chemical precipitates include some limestones and evaporated deposits such as gypsum and halite (rock salt). Coal, oil shale, and limestone made of fossil material are examples of organic sedimentary rocks.

Most sedimentary rocks show distinct layering (stratification), because they are originally deposited as more or less horizontal layers.

sedimentology branch of geology dealing with the structure, composition, and evolution of †sedimentary rocks, and what they can tell us about the history of the Earth.

Seebeck, Thomas Johann (1770–1831) Estonian-born German physicist who discovered the thermoelectric effect named after him. In 1822 he discovered that a current flows in a circuit made from two dissimilar metals when each junction is kept at a different temperature. Seebeck also studied luminescence, the heating effect of infrared radiation, and magnetism associated with a flowing electric current.

Seebeck effect generation of a voltage in a circuit containing two different metals, or semiconductors, by keeping the junctions between them at different temperatures. Discovered by the German physicist Thomas Seebeck (1770–1831), it is also called the thermoelectric effect, and is the basis of the †thermocouple. It is the opposite of the †Peltier effect (in which current flow causes a temperature difference between the junctions of different metals).

seed reproductive structure of higher plants (†angiosperms and †gymnosperms). It develops from a fertilized ovule and consists of an embryo and a food store, surrounded and protected by an outer seed coat, called the testa. The food store is contained either in a specialized nutritive tissue, the †endosperm, or in the †cotyledons of the embryo itself. In angiosperms the seed is enclosed within a †fruit, whereas in gymnosperms it is usually naked and unprotected, once shed from the female cone.

Following †germination the seed develops into a new plant.

Seeds may be dispersed from the parent plant in a number of different ways. Agents of dispersal include animals, as with †burs and fleshy edible fruits, and wind, where the seed or fruit may be winged or plumed. Water can disperse seeds or fruits that float, and various mechanical devices may eject seeds from the fruit, as in the pods of some leguminous plants (see †legume).

There may be a delay in the germination of some seeds to ensure that growth occurs under favorable conditions (see †after-ripening, †dormancy). Most seeds remain viable for at least 15 years if dried to about 5% water and kept at –20°C/–4°F, although 20% of them will not survive this process.

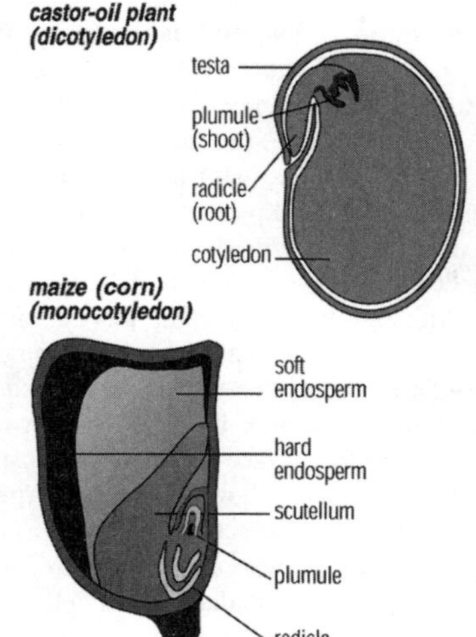

seed The structure of seeds. The castor is a dicotyledon, a plant in which the developing plant has two leaves, developed from the cotyledon. In maize, a monocotyledon, there is a single leaf developed from the scutellum. Helicon

seed bank repository for seeds. The seeds of crop plants and wild species of plants threatened with extinction are kept in cold storage, forming a reference collection that is an invaluable source of genetic material for breeding and conservation programs. The world's largest seed bank is connected to Kew Gardens, the popular name for the Royal Botanic Gardens at Kew, Surrey, in England.

seedling in botany, structure produced during the early stages of the germination of a seed. It develops from the growth of the embryo within the seed and consists of the growing radicle (root), plumule (shoot), and the cotyledon(s). When the first foliage leaves appear on the shoot and photosynthesis begins, the seedling becomes an established plant.

seed plant any seed-bearing plant; also known as a **spermatophyte**.

The seed plants are subdivided into two classes: the †angiosperms, or flowering plants, and the †gymnosperms, principally the cycads and conifers.

Together, they comprise the major types of vegetation found on land.

Angiosperms are the largest, most advanced, and most successful group of plants at the present time, occupying a highly diverse range of habitats. There are estimated to be about 250,000 different species. Gymnosperms differ from angiosperms in their ovules which are borne unprotected (not within an †ovary) on the scales of their cones.

The arrangement of the reproductive organs, and their more simplified internal tissue structure, also distinguishes them from the flowering plants. In contrast to the gymnosperms, the ovules of angiosperms are enclosed within an ovary and many species have developed highly specialized reproductive structures associated with †pollination by insects, birds, or bats.

segment in geometry, part of a circle cut off by a straight line or †chord, running from one point on the circumference to another. All angles in the same segment are equal.

segmentation in biology, division of the animal body into portions each containing a set of body structures. The †annelid worms (phylum Annelida) demonstrate segmentation the most clearly.

The adult body is divided into segments marked externally by grooves called annuli, which locate internally the septa dividing the fluid-filled cavity (coelom) into compartments. Internally the tissues and organs are repeated in most segments (metamerism) but the fusion of the structures occurs in some (the reproductive organs between segments 9 and 15 of earthworms for example).

Segrè, Emilio Gino (1905–1989) Italian-born US physicist who was awarded the Nobel Prize for Physics in 1959 for his discovery in 1955 of the antiproton, a new form of †antimatter. He shared the award with his co-worker Owen †Chamberlain. Segrè discovered the first synthetic element, technetium (atomic number 43), in 1937.

segregation in chemistry, settling over time of solid particles into layers according to their size that occurs in dry solids and liquid suspensions.

seiche pendulous movement seen in large areas of water resembling a †tide. It was originally observed on Lake Geneva and is created either by the wind, earth tremors or other atmospheric phenomena.

seif dune or **longitudinal dune**, in earth science, long narrow ridge of sand with sinuous crests, aligned for up to 10 km/6 mi along the direction of the prevailing wind.

seismic gap theory theory that along faults that are known to be seismically active, or in regions of high seismic activity, the locations that are more likely to experience an †earthquake in the relatively near future are those that have not shown seismic activity for some time. When records of past earthquakes are studied and plotted onto a map, it becomes possible to identify **seismic gaps** along a fault or plate margin. According to the theory, an area that has not had an earthquake for some time will have a great deal of stress building up, which must eventually be released, causing an earthquake.

Although the seismic gap theory can suggest areas that are likely to experience an earthquake, it does not enable scientists to predict when that earthquake will occur.

seismic wave energy wave generated by an †earthquake or an artificial explosion. There are two types of seismic waves: **body waves** that travel through the Earth's interior; and **surface waves** that travel through the surface layers of the crust and can be felt as the shaking of the ground, as in an earthquake. Seismic waves show similar properties of reflection and refraction as light and sound waves. Seismic waves change direction and speed as they travel through different densities of the Earth's rocks.

body waves There are two types of body waves: P-waves and S-waves, so-named because they are the primary and secondary waves detected by a seismograph. **P-waves**, or compressional waves, are longitudinal waves (wave motion in the direction the wave is traveling), whose compressions and rarefactions resemble those of a sound wave. **S-waves** are transverse waves or shear waves, involving a back-and-forth shearing motion at right angles to the direction the wave is traveling (see †wave).

Because liquids have no resistance to shear and cannot sustain a shear wave, S-waves cannot travel through liquid material. The Earth's outer core is believed to be liquid because S-waves disappear at the mantle-core boundary, while P-waves do not.

surface waves Surface waves travel in the surface and subsurface layers of the crust. **Rayleigh waves** travel along the free surface (the uppermost layer) of a solid material. The motion of particles is elliptical, like a water wave, creating the rolling motion often felt during an earthquake. **Love waves** are transverse waves trapped in a subsurface layer due to different densities in the rock layers above and below. They have a horizontal side-to-side shaking motion transverse (at right angles) to the direction the wave is traveling.

seismogram or **seismic record**, trace, or graph, of ground motion over time, recorded by a seismograph. It is used to determine the magnitude and duration of an earthquake.

seismograph instrument used to record ground motion. A heavy inert weight is suspended by a spring and attached to this is a pen that is in contact with paper on a rotating drum. During an earthquake the instrument frame and drum move, causing the pen to record a zigzag line on the paper; held steady by inertia, the pen does not move.

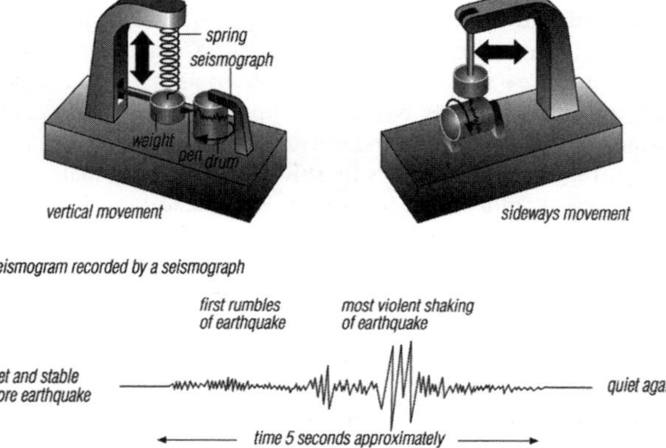

seismograph A seismogram, or recording made by a seismograph. Such recordings are used to study earthquakes and in prospecting. Helicon

seismology study of †earthquakes, the seismic waves they produce, the processes that cause them, and the effects they have. By examining the global pattern of waves produced by an earthquake, seismologists can deduce the nature of the materials through which they have passed. This leads to an understanding of the Earth's internal structure.

On a smaller scale, artificial earthquake waves, generated by explosions or mechanical vibrators, can be used to search for subsurface features in, for example, oil or mineral exploration. Earthquake waves from underground nuclear explosions can be distinguished from natural waves by their shorter wavelength and higher frequency.

Seki Kowa (or Seki Takakazu) (*c.* 1642–1708) Japanese mathematician who created a new mathematical notation system and used it to discover many of the theorems and theories that were being – or were shortly to be – discovered in the West.

selection pressure in biology, degree to which †natural selection acts on a population. A stable environment reduces selection pressure and therefore evolutionary change. A rapidly changing environment increases selection pressure and therefore evolutionary change as species adapt.

selenium chemical symbol Se, (Greek *Selene* "Moon") gray, non-metallic element, atomic number 34, relative atomic mass 78.96. It belongs to the sulfur group and occurs in several allotropic forms that differ in their physical and chemical properties. It is an essential trace element in human nutrition.

Obtained from many sulfide ores and selenides, it is used as a red coloring for glass and enamel.

Because its electrical conductivity varies with the intensity of light, selenium is used extensively in photoelectric devices. It was discovered in 1817 by Swedish chemist Jöns Berzelius and named after the Moon because its properties follow those of tellurium, whose name derives from Latin *Tellus* "Earth."

self individual as an experiencing being, the subject of contemplation, the object of †introspection, and the agent of thought and action. †Personality and †ego are commonly used synonyms, though they do not have exactly the same meaning. The personality is more outwardly observable (by others, that is) and the ego, as a psychoanalytical term at least, contains unconscious elements that the self does not recognize.

self-inductance or **self-induction**, in physics, the creation of an electromotive force opposing the current. See †inductance.

self-pollination pollination of a plant by itself, whether intra- or interfloral.

selva equatorial rainforest, such as that in the Amazon basin in South America.

semelparity in biology, the occurrence of a single act of reproduction during an organism's lifetime. Most semelparous species produce very large numbers of offspring when they do reproduce, and normally die soon afterwards. Examples include the Pacific salmon and the pine looper moth. Many plants are semelparous, or monocarpic. Repeated reproduction is called iteroparity.

semen fluid containing †sperm from the testes and secretions from various sex glands (such as the prostate gland) that is ejaculated by male animals during copulation. The secretions serve to nourish and activate the sperm cells, and prevent them clumping together.

Semenov, Nikolai Nikolaevich (1896–1986) Russian physical chemist who was the first Russian to be awarded the Nobel Prize for Chemistry in 1956 for his study of chemical chain reactions, particularly branched-chain reactions, which can accelerate with explosive velocity.

semicircular canal one of three looped tubes that form part of the labyrinth in the inner †ear. They are filled with fluid and detect changes in the position of the head, contributing to the sense of balance.

semiconductor material with electrical conductivity intermediate between metals and insulators and used in a wide range of electronic devices. Certain crystalline materials, most notably silicon and germanium, have a small number of free electrons that have escaped from the bonds between the atoms. The atoms from which they have escaped possess vacancies, called holes, which are similarly able to move from atom to atom and can be regarded as positive charges. Current can be carried by both electrons (negative carriers) and holes (positive carriers). Such materials are known as **intrinsic semiconductors**. Conductivity of a semiconductor can be enhanced by doping the material with small numbers of impurity atoms which either release free electrons (making an **n-type semiconductor** with more electrons than holes) or capture them (a **p-type semiconductor** with more holes than electrons). When p-type and n-type materials are brought together to form a p–n junction, an electrical barrier is formed that conducts current more readily in one direction than the other. This is the basis of the †semiconductor diode, used for rectification, and numerous other devices including †transistors, rectifiers, and †integrated circuits (silicon chips).

semiconductor diode or **p–n junction diode**, in electronics, a two-terminal semiconductor device that allows electric current to flow in only one direction, the **forward-bias** direction. A very high resistance prevents current flow in the opposite, or **reverse-bias**, direction. It is used as a †rectifier, converting alternating current (AC) to direct current (DC).

seminiferous tubule in male vertebrates, one of a number of tightly packed, highly coiled tubes in the testis. The tubes are lined with germinal epithelium cells, from which †sperm are produced by cell division (†meiosis) and in the midst of which the sperm mature, nourished and protected in the folds of large Sertoli cells. After maturation, the sperm are stored in a long tube called the epididymis.

semipermeable membrane membrane that allows certain substances in solution, such as †crystalloids, to pass through it but is impervious to others, such as †colloids. Semipermeable membranes are used in dialysis.

Senebier, Jean (1742–1809) Swiss botanist, plant physiologist, and pastor, whose research on †photosynthesis (the process by which green plants use light energy to make carbohydrates) showed that "fixed air" (now known to be carbon dioxide) was converted to "pure air" (oxygen) in a light-dependent process.

senescence in biology, the deterioration in physical and (sometimes) mental capacities that occurs with aging.

sense in mathematics, the orientation of a vector. Each vector has an equivalent vector of the opposite sense. The combined effect of two vectors of opposite sense is a zero vector.

sense organ any organ that an animal uses to gain information about its surroundings. All sense organs have specialized receptors (such as light receptors in the eye) and some means of translating their response into a nerve impulse that travels to the brain. The main human sense organs are the eye, which detects light and color (different wavelengths of light); the ear, which detects sound (vibrations of the air) and gravity; the nose, which detects some of the chemical molecules in the air; and the tongue, which detects some of the chemicals in food, giving a sense of taste. There are also many small sense organs in the skin, including pain, temperature, and pressure sensors, contributing to our sense of touch.

animal senses Research suggests that our noses may also be sensitive to magnetic forces, giving us an innate sense of direction. This sense is well developed in other animals, as are a variety of senses that we do not share. Some animals can detect small electrical discharges, underwater vibrations, minute vibrations of the ground, or sounds that are below (infrasound) or above (ultrasound) our range of hearing. Sensitivity to light varies greatly. Most mammals cannot distinguish different colors, whereas some birds can detect the polarization of light. Many insects can see light in the ultraviolet range, which is beyond our spectrum, while snakes can form images of infrared radiation (radiant heat). In many animals, light is also detected by another organ, the †pineal body, which "sees" light filtering through the skull, and measures the length of the day to keep track of the seasons.

It's All in the Brain http://www.hhmi.org/senses/a/a110.htm

As part of a much larger site called "Seeing, Hearing, and Smelling the World," here is a set of pages introducing the way in which we perceive the world through our senses. It is divided into five sections called "Illusions reveal the brain's assumptions," "Sensing change in the environment," "Vision, hearing, and smell: the best-known senses," "A language the brain can understand," and "More than the sum of its parts."

New Imaging Techniques Show Brain at Work http://www.hhmi.org/senses/e/e210.htm

As part of a much larger site called "Seeing, Hearing, and Smelling the World," here is a page examining the advances in brain scanning techniques that have helped scientists understand the difference between the senses. It is divided into three sections called "The PET scan difference between seeing and hearing words," "Giant magnet reveals the brain's activity," and "What is this person hearing – music or just meaningless clicks?" This site makes good use of images and animations to help with the explanations, so it is best viewed with an up-to-date browser.

Seeing, Hearing, and Smelling the World http://www.hhmi.org/senses/

Produced by the Howard Hughes Medical Institute, this site provides useful articles on three of the human senses. You can learn how humans see, hear, and smell, and also visit pages on sense experts and current research.

sensitivity in biology, the ability of an organism, or part of an organism, to detect changes in the environment. All living things are capable of some sensitivity, and any change detected by an organism is called a †stimulus. Plant response to stimuli (for example, light, heat, moisture) is by directional growth (†tropism). In animals, the body cells that detect the stimuli are called †receptors, and these are often contained within a †sense organ. For example, the eye is a sense organ, within which the retina contains rod and cone cells which are receptors. The part of the body that responds to a stimulus, such as a muscle, is called an effector, and the communication of stimuli from receptors to effectors is termed "coordination"; messages are passed from receptors to effectors either via the †nerves or by means of chemicals called †hormones. Rapid communication and response to stimuli, such as light, sound, and scent, can be essential to an animal's well-being and survival, and evolution has led to the development of highly complex mechanisms for this purpose.

nervous systems Most animals have a nervous system that coordinates communication between stimulus and response. Nervous systems consist of special cells called neurones (see †nerve cell) which are fundamentally the same as other body cells in that each contains a nucleus, cytoplasm, and cell membrane. In addition, in order to receive and pass messages, they also have long thin fibers of cytoplasm extending out from the cell body termed "nerve fibers." The longest of these, which can be more than a meter long, are called **axons**. The shorter fibers are called **dendrites**.

nerve nets Small animals, such as jellyfish, which do not need to coordinate complex messages between stimuli and response mechanisms, have simple nervous systems, termed "nerve nets." In a nerve net, each neurone is connected by fibers to adjacent neurones, so that a message received in any one part of the nervous system is relayed from neurone to neurone throughout the whole of the organism's body.

central nervous systems The evolution of larger and more complex animals, such as humans, has necessitated the development of far more elaborate nervous systems, and most animals have a †central nervous system (CNS). The main difference between a simple nerve net and a central nervous system is the addition of a brain and spinal cord to coordinate and relay messages between receptors and the appropriate effectors, without involving the whole body. Thus rapid responses to specific stimuli are triggered.

sensor in computing, a device designed to detect a physical state or measure a physical quantity, and produce an input signal for a computer. For example, a sensor may detect the fact that a printer has run out of paper or may measure the temperature in a kiln.

In electronics, a sensor is triggered by changes outside an electronic system. There are many different types of sensors including heat sensors (thermistors, thermocouples), light sensors (LDR, infrared), sound sensors (microphones, ultrasound), moisture sensors, and many more. Most sensors change their internal resistance as conditions change. This allows them to let more or less electricity to pass through, depending on the condition. Although it is possible to fit some sensors directly into a circuit, they are usually arranged with a resistor as part of a potential divider subsystem. This increases their sensitivity. The subsystem can then be used to activate a †transistor, logic gate, or an †operational amplifier.

sensorimotor phase first stage in human mental development from birth to about two years of age in which reflex actions lead to an awareness of the permanence of objects.

sensory nerve afferent nerve of the peripheral nervous system, made up of sensory neurons, which carries impulses to the central nervous system.

sepal part of a flower, usually green, that surrounds and protects the flower in bud. The sepals are derived from modified leaves, and are collectively known as the †calyx.

In some plants, such as the marsh marigold *Caltha palustris*, where true †petals are absent, the sepals are brightly colored and petal-like, taking over the role of attracting insect pollinators to the flower.

sequence list of †numbers that are generated by a rule. Each number in the sequence is a term. For example, 3, 7, 11, 15, 19 is a sequence of five terms generated by adding 4 to the previous term. To generate a sequence, it is necessary to know the first term, the number of terms in the sequence, and the rule. Examples of special sequences include:

†**odd numbers** 1, 3, 5, 7, 9... the rule is add 2

even numbers 2, 4, 6, 8, 10... the rule is add 2

†**square numbers** 1, 4, 9, 16, 25... the rule is 1^2, 2^2, 3^2, and so on

triangle numbers 1, 3, 6, 10, 15... the rule is 1, 1 + 2, 1 + 2 + 3, and so on

sequencing in biochemistry, determining the sequence of chemical subunits within a large molecule. Techniques for sequencing amino acids in proteins were established in the 1950s, insulin being the first for which the sequence was completed. The †Human Genome Project is attempting to determine the sequence of the 3 billion base pairs within human †DNA. A rough draft of the whole genome had been sequenced by June 2000.

In July 2001 the Global Musa Genomics Consortium, which is made up of publicly funded institutes, announced that the banana was to become the first fruit to have its genome sequenced. The process was expected to take five years and all sequences were to be made available on the Internet.

sere plant †succession developing in a particular habitat. A **lithosere** is a succession starting on the surface of bare rock. A **hydrosere** is a succession in shallow freshwater, beginning with planktonic vegetation and the growth of pondweeds and other aquatic plants, and ending with the development of swamp. A **plagiosere** is the sequence of communities that follows the clearing of the existing vegetation.

series sum of the terms of a sequence. Series may be convergent (the limit of the sum is a finite number) or divergent (the limit is infinite). For example, $1 + {}^1/_2 + {}^1/_4 + {}^1/_8 + ...$ is a convergent series because the limit of the sum is 2; 1 + 2 + 3 + 4 + ... is a divergent series.

series circuit electrical circuit in which the components are connected end to end, so that the current flows through them all one after the other.

current in a series circuit The current flowing through each of the components is the same. This can be written as $I_T = I_1 = I_2$, where I_T is the total current flow, I_1 is the current flowing through component 1, and I_2 is the current flowing through component 2. To measure the current flow an ammeter is connected to the components in series. For example, if the current at component 1 is 2 amps and at component 2 is 2 amps, then the total current is 2 amps.

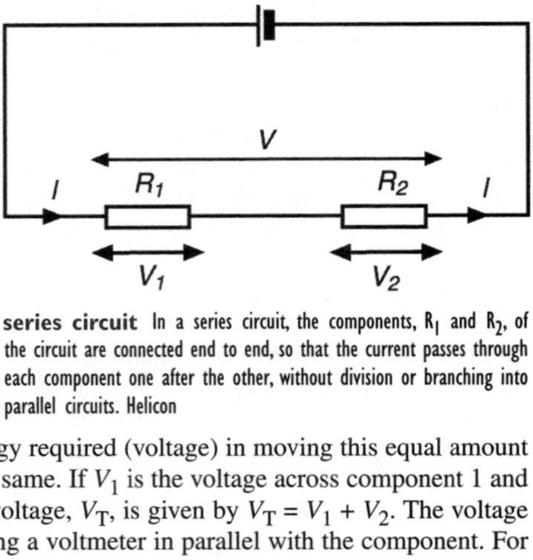

series circuit In a series circuit, the components, R_1 and R_2, of the circuit are connected end to end, so that the current passes through each component one after the other, without division or branching into parallel circuits. Helicon

voltage in a series circuit As the current flowing through each of the components is the same, the energy required (voltage) in moving this equal amount of current through each of the components is also the same. If V_1 is the voltage across component 1 and V_2 is the voltage across component 2, then the total voltage, V_T, is given by $V_T = V_1 + V_2$. The voltage can be measured across each component by connecting a voltmeter in parallel with the component. For example, if the voltage across component 1 is 3 volts and across component 2 is 3 volts, then the total voltage is 6 volts.

serotonin or **5-hydroxytryptamine (5-HT)**, chemical widely distributed in the body tissues – in the blood platelets, the wall of the intestine, and the †central nervous system. It is believed to be implicated in the inflammatory process and, in the nervous system, it acts as a †neurotransmitter, controlling sleep. Serotonin is derived from the amino acid tryptophan.

serous membranes layers of tissue lining the closed cavities of the body, and characterized by the secretion of a serous fluid; they allow the organ to slide within its cavity.

Serpens constellation on the celestial equator (see †celestial sphere), represented as a serpent coiled around the body of †Ophiuchus. It is the only constellation divided into two halves: **Serpens Caput**, the head (on one side of Ophiuchus), and **Serpens Cauda**, the tail (on the other side). Its main feature is the Eagle nebula.

serpentine member of a group of minerals, hydrous magnesium silicate, $Mg_3Si_2O_5(OH)_4$, occurring in soft †metamorphic rocks and usually dark green. The fibrous form **chrysotile** is a source of †asbestos; other forms are **antigorite** and **lizardite**. Serpentine minerals are formed by hydration of ultramafic rocks during metamorphism. Rare snake-patterned forms are used in ornamental carving.

serum clear fluid that separates out from clotted blood. It is blood plasma with the anticoagulant proteins removed, and contains †antibodies and other proteins, as well as the fats and sugars of the blood. It can be produced synthetically, and is used to protect against disease.

Servetus, Michael (1511–1553) Spanish **Miguel Serveto**, Spanish Anabaptist theologian and physician. He was a pioneer in the study of the circulation of the blood and found that it circulates to the lungs from the right chamber of the heart. He was, however, as noted for his death as for his life: passing through Calvin's Geneva, he was arrested for his unorthodox views on the Trinity (expressed in his treatise *On the Errors of the Trinity* (1531)). After Servetus had been found guilty of heresy, Calvin had him burnt.

servo mechanism mechanism that uses relatively low †power to control the behavior of a much larger output device in a proportionate way.

sesamoid bones small, rounded bones which are developed in tendons subjected to much pressure where they glide over bony prominences. The largest example in the human body is the †patella or kneecap.

sessile in botany, describing a leaf, flower, or fruit that lacks a stalk and sits directly on the stem, as with the sessile acorns of certain oaks. In zoology, it is an animal that normally stays in the same place, such as a barnacle or mussel. The term is also applied to the eyes of †crustaceans when these lack stalks and sit directly on the head.

set or **class**, in mathematics, any collection of defined things (elements), provided the elements are distinct and that there is a rule to decide whether an element is a member of a set. It is usually denoted by a capital letter and indicated by curly brackets { }.

For example, *L* may represent the set that consists of all the letters of the alphabet. The symbol Î stands for "is a member of"; thus *p* Î *L* means that *p* belongs to the set consisting of all letters, and 4 Ï *L* means that 4 does not belong to the set consisting of all letters.

There are various types of sets. A **finite set** has a limited number of members, such as the letters of the alphabet; an **infinite set** has an unlimited number of members, such as all whole numbers; an **empty** or **null set** has no members, such as the number of people who have swum across the Atlantic Ocean, written as { } or ø; a **single-element set** has only one member, such as days of the week beginning with M, written as {Monday}. **Equal sets** have the same members; for example, if *W* = {days of the week} and *S* = {Sunday, Monday, Tuesday, Wednesday, Thursday, Friday, Saturday}, it can be said that *W* = *S*. Sets with the same number of members are **equivalent sets**. Sets with some members in common are **intersecting sets**; for example, if *R* = {red playing cards} and *F* = {face cards}, then *R* and *F* share the members that are red face cards. Sets with no members in common are **disjoint sets**. Sets contained within others are **subsets**; for example, *V* = {vowels} is a subset of *L* = {letters of the alphabet}.

Sets and their interrelationships are often illustrated by a †Venn diagram.

Beginnings of Set Theory http://www-history.mcs.st-and.ac.uk/history/HistTopics/Beginnings_of_set_theory.html

Run by St Andrews University, Scotland, this site details the beginnings of set theory in mathematics. It includes biographical details of the mathematicians involved and gives a brief description of the topic with a few examples.

sex chromosome †chromosome that differs between the sexes and serves to determine the sex of the individual. In humans, whether a person is male or female is determined by the combination of the two sex chromosomes in the body cells. In females both chromosomes are the same – two X chromosomes (XX). In males the two chromosomes are different – one X chromosome and one Y chromosome (XY). The Y chromosome is shorter than the X. †Genes on these chromosomes determine sex.

As a result of †meiosis gametes from a female each contain one X chromosome. However, gametes from a male are of two kinds. Half of the gametes contain an X chromosome and half contain a Y chromosome. If an X-carrying gamete from a male fertilizes a female gamete the result will be a female. If a Y-carrying gamete from a male fertilizes a female gamete, the result will be a male.

sex determination process by which the sex of an organism is determined. In many species, the sex of an individual is dictated by the two sex chromosomes (X and Y) it receives from its parents. In mammals, some plants, and a few insects, males are XY, and females XX; in birds, reptiles, some amphibians, and butterflies the reverse is the case. In bees and wasps, males are produced from unfertilized eggs, females from fertilized eggs. In 1991 it was shown that maleness is caused by a single gene, 14 base pairs long, on the Y chromosome.

Environmental factors can affect some fish and reptiles, such as turtles, where sex is influenced by the temperature at which the eggs develop.

sex hormone steroid hormone produced and secreted by the gonads (testes and ovaries). Sex hormones control development and reproductive functions and influence sexual and other behavior.

sex linkage in genetics, the tendency for certain characteristics to occur exclusively, or predominantly, in one sex only. Human examples include red-green color blindness and hemophilia, both found predominantly in males. In both cases, these characteristics are †recessive and are determined by genes on the †X chromosome.

Since females possess two X chromosomes, any such recessive †allele on one of them is likely to be masked by the corresponding allele on the other. In males (who have only one X chromosome paired with a largely inert †Y chromosome) any gene on the X chromosome will automatically be expressed. Color blindness and hemophilia can appear in females, but only if they are †homozygous for these traits, due to inbreeding, for example.

Sextans faint constellation on the celestial equator (see †celestial sphere), represented as a sextant. Johannes †Hevelius originally named it "Sextans Uraniae," after his sextant was destroyed in a fire 1679.

sextic to the power of six.

sexuality attribute or characteristic of being male or female, usually taken to involve more than the ability or disposition to play the appropriate role in sexual reproduction. Today, as much emphasis is placed on an individual's awareness of and response to culturally and socially derived gender differences as on biological factors in the development of sexuality.

Sexuality has been an important topic in †developmental psychology. In psychoanalysis, in particular, Sigmund †Freud's ideas on infantile sexuality have been immensely influential. Recent years have seen a growth of interest in sexuality and gender differences in such varied fields of study as sociology, social historical and cultural studies, and politics.

males have two different sex chromosomes X and Y
females have two similar sex chromosomes X and X
MEIOSIS
X sperm cells Y
X ova X
FERTILIZATION
boy XY
girl XX

sex determination In humans, sex is determined by the male. Sperm cells contain an X or a Y chromosome but egg cells contain only X chromosomes. If a sperm cell carrying an X chromosome fertilizes the egg the resulting baby will be female; if the sperm cell is carrying a Y chromosome then the baby will be male. Helicon

sexual reproduction reproductive process in organisms that requires the union, or †fertilization, of gametes (such as eggs and sperm). These are usually produced by two different individuals, although self-fertilization occurs in a few †hermaphrodites such as tapeworms. Most organisms other than bacteria and cyanobacteria (†blue-green algae) show some sort of sexual process. Except in some lower organisms, the gametes are of two distinct types called the egg (†ovum) and the †sperm. The organisms producing the eggs are called females, and those producing the sperm, males. The fusion of a male and female gamete produces a **zygote**, from which a new individual develops. See †reproduction. Gametes are produced by †meiosis. If the male and female gametes are produced by two different individuals, sexual reproduction combines inherited information from the two parents. In mammals, the male gametes, sperm, are made in the testes, while the female gametes, eggs (ova), are made in the ovaries. In the human female, gametes called ova are released regularly in the ovaries after puberty. The Fallopian tubes carry the ova to the uterus or womb, in which the fetus will develop. In the male, sperm is produced inside the testes after puberty; about 10 million sperm cells are produced each day, enough to populate the world in six months. The sperm duct or vas deferens, a continuation of the epididymis, carries sperm to the urethra during ejaculation.

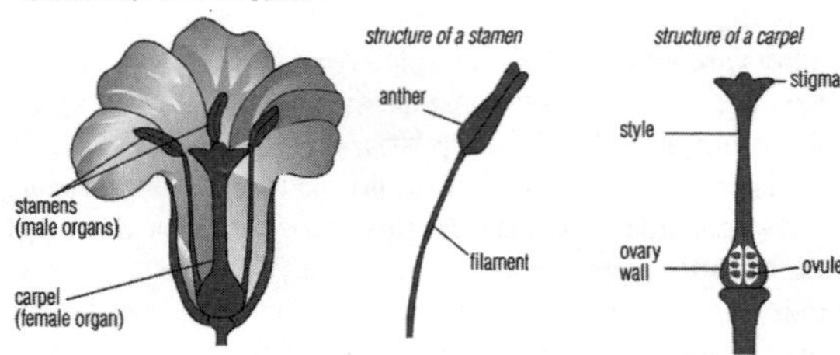

sexual reproduction Reproductive organs in flowering plants. The stamens are the male parts of the plant. Each consists of a stalklike filament topped by an anther. The anther contains four pollen sacs which burst to release tiny grains of pollen, the male sex cells. The carpels are the female reproductive parts. Each carpel has a stigma which catches the pollen grain. The style connects the stigma to the ovary. The ovary contains one or more ovules, the female sex cells. Helicon

Most animals and plants reproduce sexually, though quite a few plants also reproduce asexually. The male and female sex organs of a plant are usually found in a flower. Many flowers contain both male and female organs. The male gametes of a plant are inside the pollen grains, and the female gametes are inside the ovules.

In micro-organisms, †asexual reproduction is very common. Yeasts and a few bacteria may reproduce sexually, although most reproduce asexually

Sexual Reproduction in Animals http://www.learn.co.uk/default.asp?WCI=Unit&WCU=19648

Clearly organized discussion of sexual reproduction. This page compares sexual reproduction in humans with animals and discusses puberty, listing primary and secondary sexual characteristics. There are links to articles in the left-hand margin which cover each stage of the reproductive process.

sexual selection process similar to †natural selection but relating exclusively to success in finding a mate for the purpose of sexual reproduction and producing offspring. Sexual selection occurs when one sex (usually but not always the female) invests more effort in producing young than the other. Members of the other sex compete for access to this limited resource (usually males competing for the chance to mate with females).

Sexual selection often favors features that increase a male's attractiveness to females (such as the pheasant's tail) or enable males to fight with one another (such as a deer's antlers). More subtly, it can produce hormonal effects by which the male makes the female unreceptive to other males, causes the abortion of fetuses already conceived, or removes the sperm of males who have already mated with a female.

Seyfert, Carl Keenan (1911–1960) US astronomer and astrophysicist who studied the spectra of stars and galaxies, and identified and classified the type of galaxy that now bears his name.

Seyfert galaxy galaxy in which a small, bright center is caused by hot gas moving at high speed around a massive central object, possibly a †black hole. Almost all Seyferts are spiral galaxies. They seem to be closely related to †quasars, but are about 100 times fainter. They are named after their discoverer, US astronomer Carl Seyfert.

SGML abbreviation for standard generalized markup language, International Standards Organization standard describing how the structure (features such as headers, columns, margins, and tables) of a text can be identified so that it can be used, probably via filters, in applications such as †desktop publishing and †electronic publishing. †HTML and †VRML are both types of SGML.

shadow area of darkness behind an opaque object that cannot be reached by some or all of the light coming from a light source in front. Its presence may be explained in terms of light rays traveling in straight lines and being unable to bend around obstacles. The light in front of the object is blocked. A point source of light produces an †umbra, a completely black shadow with sharp edges. An extended source of light produces both a central umbra and a †penumbra, a region of semidarkness with blurred edges where darkness gives way to light.

†Eclipses are caused by the Earth passing into the Moon's shadow or the Moon passing into the Earth's shadow. During a solar eclipse the Moon forms a shadow on the Earth as it lines up between the Sun and the Earth. A total eclipse occurs when the Moon completely blocks out the light from the Sun. This is called the umbra. A partial eclipse occurs when some of the light from the Sun reaches the Earth around the edges of the Moon. This is called a penumbra.

shale fine-grained and finely layered †sedimentary rock composed of silt and clay. It is a weak rock, splitting easily along bedding planes to form thin, even slabs (by contrast, mudstone splits into irregular flakes). Oil shale contains kerogen, a solid bituminous material that yields †petroleum when heated.

Shannon, Claude Elwood (1916–2001) US mathematician who founded the science of information theory. He argued that †entropy is equivalent to a shortage of information content (a degree of uncertainty), and obtained a quantitative measure of the amount of information in a given message.

shape form of an object, defined by its outline. **Plane shapes** are two-dimensional; **solid shapes** are three-dimensional.

Angle and Parallel Lines http://www.ies.co.jp/math/java/geo/angle.html

Collection of interactive applets illustrating some principles about angles. There are two angle problems to try, and a "Polygon creator" to experiment with drawing different shapes.

Shapley, Harlow (1885–1972) US astronomer. He established that our Galaxy was much larger than previously thought. His work on the distribution of †globular clusters showed that the Sun was not at the center of the Galaxy as then assumed, but two-thirds of the way out to the rim; globular clusters were arranged in a halo around the Galaxy.

Shapley-Curtis Debate in 1920 http://antwrp.gsfc.nasa.gov/diamond_jubilee/debate20.html

Description of the debate that led to the scale of our Galaxy being greatly modified. This Web site describes the history of the argument, and also covers Cepheid variables – the phenomenon that led to the discovery of the universe outside our spiral galaxy. To spoil the plot it turns out that both astronomers were actually wrong, but Shapley more so than Curtis.

Sharp, Phillip Allen (1944–) US molecular biologist who was awarded a Nobel Prize for Physiology or Medicine in 1993 with Richard †Roberts for their discovery of split genes (genes interrupted by nonsense segments of DNA). Using his technique for measuring nucleic acid fragments on genes, he found that genes consist of regions of †DNA (genetic material) separated by regions that do not contain genetic information, called †introns.

Sharpey-Schafer, Edward (1850–1935) born Edward Albert Schäfer, English physiologist, one of the founders of endocrinology. He made important discoveries relating to the hormone †adrenaline, and to the †pituitary and other †endocrine, or ductless, glands. He was knighted in 1913.

Shaula or **Lambda Scorpii**, second-brightest star in the constellation of †Scorpius. It is a blue-white star of magnitude 1.6, some 330 light years from the Sun. Spectroscopic observations reveal it to be a close †binary, the two stars orbiting about each other with a period (the time they take to circle each other) of 5.6 days.

shelf sea relatively shallow sea, usually no deeper than 200 m/650 ft, overlying the continental shelf around the coastlines. Most fishing and marine mineral exploitations are carried out in shelf seas.

shell in zoology, hard outer covering of a wide variety of invertebrates. The covering is usually mineralized, normally with large amounts of calcium. The shell of birds' eggs is also largely made of calcium.

shell burning in astronomy, burning of hydrogen that takes place outside the core of an old star, which has contracted because all of its hydrogen has been used up (and converted to helium). The outer layers swell and cool to create a red giant or supergiant star.

Sherrington, Charles Scott (1857–1952) English neurophysiologist who was awarded a Nobel Prize for Physiology or Medicine in 1932 for the discovery of the function of neurons (nerve cells). His book *The Integrative Action of The Nervous System* (1906) formulated the principles of reflex action.

shield in geology, alternative name for †craton, the ancient core of a continent.

shield volcano broad, flat †volcano formed at a †constructive margin between tectonic plates or over a †hot spot. The †magma (molten rock) associated with shield volcanoes is usually basalt – thin and free-flowing. An example is Mauna Loa on the Pacific island of Hawaii. A †composite volcano, on the other hand, is formed at a destructive margin.

Shield volcanoes are found along the rift valleys and ocean ridges of constructive plate margins. The lava flows for some distance over the surface before it sets, so forming low, broad volcanoes. The lava of a shield volcano is not ejected violently, but simply flows over the crater rim.

Shimbel index mathematical measurement of the accessibility of a transportation network. It is the total of the number of links or edges that form the shortest path between each †node (junctions or places) and all other nodes in the network.

Shirakawa, Hideki (1936–) Japanese chemist who with US physicist Alan J †Heeger and US chemist Alan G †MacDiarmid shared the Nobel Prize for Chemistry in 2000 for the discovery and development of conductive polymers. Plastics that can conduct electricity have many potential commercial applications, such as shields against electromagnetic radiation from computer screens, mobile phone displays, and in the development of light-emitting diodes.

Shockley, William Bradford (1910–1989) US physicist and computer scientist who was awarded the Nobel Prize for Physics in 1956 for his study of semiconductors and the discovery of the †transistor. He shared the award with his co-workers John †Bardeen and Walter †Brattain.

shock wave narrow band or region of high pressure and temperature in which the flow of a fluid changes from subsonic to supersonic.

Shock waves are produced when an object moves through a fluid at a supersonic speed. See †sonic boom.

Shoemaker-Levy 9 comet that crashed into Jupiter in July 1994. The fragments crashed at 60 kps/37 mps over the period 16–22 July 1994. The impacts occurred on the far side of Jupiter, but the impact sites came into view of Earth about 25 minutes later. Analysis of the impacts shows that most of the pieces were solid bodies about 1 km/0.6 mi in diameter, but that at least three of them were clusters of smaller objects.

When first sighted on 24 March 1993 by US astronomers Carolyn and Eugene Shoemaker and David Levy, it was found to consist of at least 21 fragments in an unstable orbit around Jupiter. It is believed to have been captured by Jupiter in about 1930, and fragmented by tidal forces on passing within 21,000 km/ 13,050 mi of the planet in July 1992.

Comet Shoemaker-Levy Collision with Jupiter http://www.jpl.nasa.gov/sl9/

Description of the comet's collision with Jupiter in 1994, the first collision of two Solar System bodies ever to be observed. The site includes background information, the latest theories about the effects of the collision, and even some animations of Jupiter and the impact.

shoot in botany, the parts of a †vascular plant growing above ground, comprising a stem bearing leaves, buds, and flowers. The shoot develops from the †plumule of the embryo.

shooting star another name for a †meteor.

short circuit unintended direct connection between two points in an electrical circuit. †Resistance is proportional to the length of wire through which current flows. By bypassing the rest of the circuit, the short circuit has low resistance and a large current flows through it. This may cause the circuit to overheat dangerously.

short-sightedness nontechnical term for †myopia.

shrub perennial woody plant that typically produces several separate stems, at or near ground level, rather than the single trunk of most trees. A shrub is usually smaller than a tree, but there is no clear distinction between large shrubs and small trees.

Shull, Clifford G (1915–2001) US physicist who, with Bertram †Brockhouse, was awarded the Nobel Prize for Physics in 1994 for the development of neutron diffraction techniques used for studying the structure and properties of matter and which led to advances in semiconductor technology. Shull used neutron scattering techniques to answer questions that the similar technique of X-ray diffraction had failed to answer, such as where the atoms of the light element hydrogen are located in an ice crystal. He also showed how neutrons can reveal the magnetic properties of metals and alloys.

shunt in electrical engineering, a conductor of very low resistance that is connected in parallel to an †ammeter in order to enable it to measure larger electric currents. Its low resistance enables it to act like a bypass, diverting most of the current through itself and away from the ammeter.

SI abbreviation for **Système International d'Unités** (French "International System of Metric Units"); see †SI units.

sial in geochemistry and geophysics, the substance of the Earth's continental †crust, as distinct from the †sima of the ocean crust. The name, now used rarely, is derived from **si**lica and **al**umina, its two main chemical constituents. Sial is often rich in granite.

sickle-cell disease or **sickle-cell anemia**, hereditary chronic blood disorder common among people of black African descent; also found in the eastern Mediterranean, parts of the Gulf, and in northeastern India. It is characterized by distortion and fragility of the red blood cells, which are lost too rapidly from the circulation. This often results in anemia.

People with this disease have abnormal red blood cells (sickle cells), containing defective †hemoglobin. The presence of sickle cells in the blood is called **sicklemia**. The disease is caused by a recessive allele. Those with two copies of the allele suffer debilitating anemia; those with a single copy paired with the normal allele, suffer with only mild anemia and have a degree of protection against †malaria because fewer normal red blood cells are available to the parasites for infection.

In the USA there were approximately 65,000 African-Americans suffering from sickle-cell disease in 1996; there were about 5,500 British sufferers. Worldwide, 100,000 babies are born with the disease annually. Those born in developing countries are unlikely to survive for long.

Bone marrow transplantation can provide a cure, but the risks (a fatality rate of 10% and a complications rate of 20%) are so great that it is only an option for the severely ill. US researchers announced in April 1995 that patients treated with a drug called hydroxyurea showed a reduction in the number of sickle cells. The drug works by reducing the amount of defective hemoglobin produced, and reviving the production of fetal hemoglobin. Fetal hemoglobin is not affected by sickling. Partial (40%) bone marrow transplants used during early 2001 in sickle-cell treatment trials in the USA also proved successful in combating the disease.

side reaction most chemical reactions are a combination of processes and do not exclusively produce the predicted final products. Those reactions that do not produce the expected products are called side reactions.

sidereal period orbital period of a planet around the Sun, or a moon around a planet, with reference to a background star. The sidereal period of a planet is in effect a "year" for that planet. A †synodic period is a full circle as seen from Earth.

sidereal time in astronomy, time measured by the rotation of the Earth with respect to the stars. A sidereal day is the time taken by the Earth to turn once with respect to the stars, namely 23 hours 56 minutes 4 seconds. It is divided into sidereal hours, minutes, and seconds, each of which is proportionally shorter than the corresponding SI unit.

Siegbahn, Kai (1918–) Swedish physicist who was awarded the Nobel Prize for Physics in 1981 for the development of the use of electron spectroscopy for chemical analysis. He studied the electrons emitted from substances irradiated by X-rays by focusing a narrow beam of X-rays of a single wavelength onto a specimen and measuring the energy spectrum of the ejected electrons. The spectrum showed peaks and troughs which were characteristic of the atoms in the specimen. He shared the Nobel Prize with Nicolaas †Bloembergen and Arthur †Schawlow.

Siegbahn, Karl Manne Georg (1886–1978) Swedish physicist who was awarded the Nobel Prize for Physics in 1924 for his work in the field of X-ray spectroscopy – the study of the X-rays produced by bombardment of a sample with electrons or photons. Each element emits X-rays with characteristic wavelengths and these can be used to identify the sample. The X unit or Siegbahn unit, used for expressing wavelengths of X-rays and gamma rays, is named after him. One Siegbahn unit equals 1.00202 x 10^{-13} m.

siemens symbol S, SI unit of electrical conductance, the reciprocal of the †resistance of an electrical circuit. One siemens equals one ampere per volt. It was formerly called the mho or reciprocal ohm.

sievert symbol Sv, SI unit of radiation dose equivalent. It replaces the rem (1 Sv equals 100 rem). Some types of radiation do more damage than others for the same absorbed dose – for example, an absorbed dose of alpha radiation causes 20 times as much biological damage as the same dose of beta radiation. The equivalent dose in sieverts is equal to the absorbed dose of radiation in grays multiplied by the relative biological effectiveness. Humans can absorb up to 0.25 Sv without immediate ill effects; 1 Sv may produce radiation sickness; and more than 8 Sv causes death.

sight the detection of light by an †eye, which can form images of the outside world.

How We See Things That Move http://www.hhmi.org/senses/b/b210.htm

As part of a much larger site called "Seeing, Hearing, and Smelling the World," here is a set of pages examining the way we are able to see and understand movement. It is divided into three sections called "The strange symptoms of blindness to motion," "A hot spot in the brain's motion pathway," and "Integrating information about movement." This site makes good use of images and animations to help with the explanations, so it is best viewed with an up-to-date browser.

Visual Pathway http://www.hhmi.org/senses/b/b150.htm

As part of a much larger site called "Seeing, Hearing, and Smelling the World," here is a page explaining the way light travels from the eye to the brain and how it is converted into images our brain can understand. This site makes good use of images and animations to help with the explanations, so it is best viewed with an up-to-date browser.

sigma symbol S, †summation symbol.

sigma bond type of chemical bond in which an electron pair (regarded as being shared by the two atoms involved in the bond) occupies a molecular orbital situated between the two atoms; the orbital is located along a hypothetical line linking the atoms' nuclei. See also †hybridization.

Sigma Octantis star closest to the south celestial pole (see †celestial sphere), in effect the southern equivalent of †Polaris, although far less conspicuous. Situated just less than 1° from the south celestial pole in the constellation Octans, Sigma Octantis is 120 light years away from the Sun.

sign symbol that indicates whether a number is positive or negative (+ or –), the operation that is to be carried out on a number of set of numbers (for example, +, x, ÷, and √), or the relationship that exists between two numbers or sets of numbers (for example, =, ≠, <, and >).

significant figures figures in a number that, by virtue of their position, express the magnitude of that number to a specified degree of accuracy. In a number, the first significant figure is the first figure that is not a 0. The final significant figure is rounded up if the following digit is greater than 5. For example, 5,463,254 to three significant figures is 5,460,000; 3.462891 to four significant figures is 3.463; 0.00347 to two significant figures is 0.0035.

silica silicon dioxide, SiO_2, the composition of the most common mineral group, of which the most familiar form is quartz. Other silica forms are †chalcedony, chert, opal, tridymite, and cristobalite. Common sand consists largely of silica in the form of quartz.

silica gel SiO_2, amorphous, hygroscopic solid form of hydrated silica, usually taking the form of hard granules. It readily absorbs water from the atmosphere and when saturated can be regenerated by heating. It is widely used as a desiccant.

silicate one of a group of minerals containing silicon and oxygen in tetrahedral units of SiO_4, bound together in various ways to form specific structural types. Silicates are the chief rock-forming minerals. Most rocks are composed, wholly or in part, of silicates (the main exception being limestones). Glass is a manufactured complex polysilicate material in which other elements (boron in borosilicate glass) have been incorporated.

Generally, additional cations are present in the structure, especially Al^{3+}, Fe^{2+}, Mg^{2+}, Ca^{2+}, Na^{+}, K^{+}, but quartz and other polymorphs of SiO_2 are also considered to be silicates; stishovite (a high-pressure form of SiO_2) is a rare exception to the usual tetrahedral coordination of silica and oxygen.

In **orthosilicates**, the oxygens are all ionically bonded to cations such as Mg^{2+} or Fe^{2+} (as olivines), and are not shared between tetrahedra. All other silicate structures involve some degree of oxygen sharing between adjacent tetrahedra. For example, beryl is a **ring silicate** based on tetrahedra linked by sharing oxygens to form a circle. Pyroxenes are single **chain silicates**, with chains of linked tetrahedra extending in one direction through the structure; amphiboles are similar but have double chains of tetrahedra. In micas, which are **sheet silicates**, the tetrahedra are joined to form continuous sheets that are stacked upon one another. **Framework silicates**, such as feldspars and quartz, are based on three-dimensional frameworks of tetrahedra in which all oxygens are shared.

silicon chemical symbol Si, (Latin *silex* "flint") brittle, non-metallic element, atomic number 14, relative atomic mass 28.086. It is the second-most abundant element (after oxygen) in the Earth's crust and occurs in amorphous and crystalline forms. In nature it is found only in combination with other elements, chiefly with oxygen in silica (silicon dioxide, SiO_2) and the silicates. These form the mineral †quartz, which makes up most sands, gravels, and beaches.

Pottery glazes and glassmaking are based on the use of silica sands and date from prehistory. Today the crystalline form of silicon is used as a deoxidizing and hardening agent in steel, and has become the basis of the electronics industry because of its †semiconductor properties, being used to make "silicon chips" for microprocessors.

The element was isolated by Swedish chemist Jöns Berzelius in 1823, having been named in 1817 by Scottish chemist Thomas Thomson by analogy with boron and carbon because of its chemical resemblance to these elements.

silicon chip †integrated circuit with microscopically small electrical components on a piece of silicon crystal only a few millimeters square.

One chip may contain more than a million components. A chip is mounted in a rectangular plastic package and linked via gold wires to metal pins, so that it can be connected to a printed circuit board for use in electronic devices, such as computers, calculators, television sets, car dashboards, and domestic appliances.

silicon chip http://courses.nus.edu.sg/course/phyweets/Projects98/Masking/newpage11.htm

Illustrated guide to the manufacture of silicon chips. This well-designed Web site describes a range of processes using clear diagrams and many excellent photographs.

silicone member of a class of synthetic polymers whose molecular structure is based on long chains of alternating silicon and oxygen atoms, $-(Si-O-Si-O)_n$, usually with connecting organic groups attached at silicon atoms at various points on the chain. Silicones form synthetic rubbers and lubricants.

sill sheet of igneous rock created by the intrusion of magma (molten rock) between layers of pre-existing rock. (A †dyke, by contrast, is formed when magma cuts *across* layers of rock.) An example of a sill in the UK is the Great Whin Sill, which forms the ridge along which Hadrian's Wall was built.

A sill is usually formed of **dolerite**, a rock that is extremely resistant to erosion and weathering, and often forms ridges in the landscape or cuts across rivers to create †waterfalls.

sillimanite aluminum silicate, Al_2SiO_5, a mineral that occurs either as white to brownish prismatic crystals or as minute white fibers. It is an indicator of high-temperature conditions in metamorphic rocks formed from clay sediments. Andalusite, kyanite, and sillimanite are all polymorphs of Al_2SiO_5.

silt sediment intermediate in coarseness between clay and sand; its grains have a diameter of 0.002–0.02 mm/0.00008–0.0008 in. Silt is usually deposited in rivers, and so the term is often used generically to mean a river deposit, as in the silting-up of a channel.

Silurian Period period of geological time 439–409 million years ago, the third period of the Paleozoic era. Silurian sediments are mostly marine and consist of shales and limestone. Luxuriant reefs were built by coral-like organisms. The first land plants began to evolve during this period, and there were many ostracoderms (armored jawless fishes). The first jawed fishes (called acanthodians) also appeared.

silver chemical symbol Ag, white, lustrous, extremely malleable and ductile, metallic element, atomic number 47, relative atomic mass 107.868. Its chemical symbol comes from the Latin *argentum*. It occurs in nature in ores and as a free metal; the chief ores are sulfides, from which the metal is extracted by smelting with lead. It is the best metallic conductor of both heat and electricity; its most useful compounds are the chloride and bromide, which darken on exposure to light and are the basis of photographic emulsions.

Silver is used ornamentally, for jewelry and tableware, for coinage, in electroplating, electrical contacts, and dentistry, and as a solder. It has been mined since prehistory; its name is an ancient non-Indo-European one, *silubr*, borrowed by the Germanic branch as *silber*.

silver nitrate $AgNO_3$, white, crystalline substance used in chemical analysis and inks.

silver plate silverware made by depositing a layer of silver on another metal, usually copper, by the process of †electroplating.

sima in geochemistry and geophysics, the substance of the Earth's oceanic †crust, as distinct from the †sial of the continental crust. The name, now used rarely, is derived from **si**lica and **ma**gnesia, its two main chemical constituents.

similar of plane or solid figures, having the same shape but a different size or orientation. Two similar figures will have the same corresponding angles, and the ratio of the lengths of their corresponding sides or edges will be constant. They are, therefore, identical shapes but different sizes. Similar figures may be produced by †enlargement.

For example, †triangles ABC and DEF are **similar triangles**:

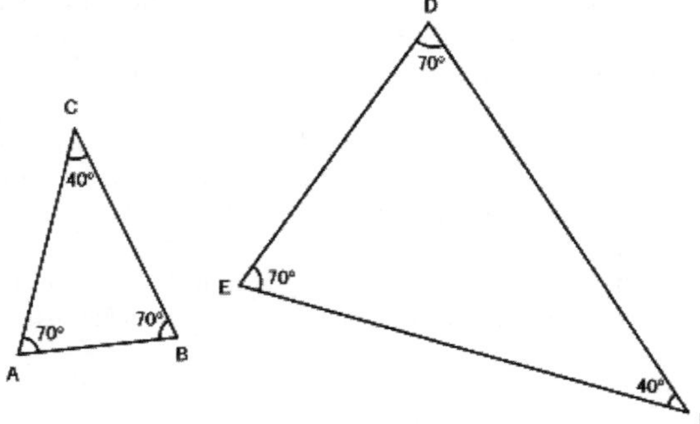

The ratio of the areas of similar plane shapes is equal to the square of the ratio of their corresponding sides. For example, a rectangle of length 8 cm and height 6 cm and a larger rectangle of length 12 cm and height 9 cm have equal ratios of length (8:12 or 1:1.5) and equal ratios of height (6:9 or 1:1.5), while the ratio of the areas of the two rectangles is 48:108, which can be reduced to 1:2.25 or $1:(1.5)^2$.

Similarity can apply to any †two-dimensional (2D) shapes if their corresponding angles are the same.

Simon, Franz Eugen (1893–1956) German-born British physicist who developed methods of achieving extremely low temperatures (nearly as low as one millionth of a degree above absolute zero). He experimentally established the validity of the third law of thermodynamics. He was knighted in 1954.

simple harmonic motion SHM, oscillatory or vibrational motion in which an object (or point) moves so that its acceleration towards a central point is proportional to its distance from it. A simple example is a pendulum, which also demonstrates another feature of SHM, that the maximum deflection is the same on each side of the central point.

A graph of the varying distance with respect to time is a sine curve, a characteristic of the oscillating current or voltage of an alternating current (AC), which is another example of SHM.

simplex method in linear computer programming, an algorithm designed to find the optimum solution in a finite number of steps.

simplify of a fraction, to reduce to lowest terms by dividing both numerator and denominator by any number that is a factor of both, until there are no common factors between the numerator and denominator. Also, in algebra, to condense an algebraic expression by grouping similar terms and reducing constants to their lowest terms. For example, the expression $a + 2b + b + 2a - 2(a + b)$ can be simplified to $a + b$.

Simpson, Thomas (1710–1761) English mathematician and writer who devised **Simpson's rule**, which simplifies the calculation of areas under graphic curves. He also worked out a formula that can be used to find the volume of any solid bounded by a ruled surface and two parallel planes.

Simpson, Thomas http://www-groups.dcs.st-andrews.ac.uk/history/Mathematicians/Simpson.html

Part of an archive containing the biographies of the world's greatest mathematicians, this site is devoted to the life and contributions of Thomas Simpson. In addition to biographical information, you will find a list of references about Simpson and links to other essays in the archive that reference him. The text of this essay includes hypertext links to the essays of those mathematicians and thinkers who influenced Simpson. There is also an image of him, which can be enlarged, and a map of his birthplace.

simultaneous equations two or more algebraic equations that contain two or more unknown quantities that may have a unique solution. For example, in the case of two linear equations with two unknown variables, such as:

(i) $x + 3y = 6$

and

(ii) $3y - 2x = 4$

the solution will be those unique values of x and y that are valid for both equations. Linear simultaneous equations can be solved by using algebraic manipulation to eliminate one of the variables, †coordinate geometry, or matrices (see †matrix).

For example, by using algebra, both sides of equation (i) could be multiplied by 2, which gives $2x + 6y = 12$. This can be added to equation (ii) to get $9y = 16$, which is easily solved: $y = {}^{16}/_{9}$. The variable x can now be found by inserting the known y value into either original equation and solving for x. Another method is by plotting the equations on a graph, because the two equations represent straight lines in coordinate geometry and the coordinates of their point of intersection are the values of x and y that are true for both of them. A third method of solving linear simultaneous equations involves manipulating matrices. If the equations represent either two parallel lines or the same line, then there will be no solutions or an infinity of solutions, respectively.

sine sin, in trigonometry, a function of an angle in a right-angled †triangle that is defined as the ratio of the length of the side opposite the angle to the length of the hypotenuse (the longest side). This function can be used to find either angles or sides in a right-angled triangle.

sine (left) The sine of an angle; (right) constructing a sine wave. The sine of an angle is a function used in the mathematical study of the triangle. If the sine of angle b is known, then the hypotenuse can be found given the length of the opposite side, or the opposite side can be found from the hypotenuse. Within a circle of unit radius (left), the height P_IA_I equals the sine of angle P_IOA_I. This fact and the equalities below the circle allow a sine curve to be drawn, as on the right. Helicon

sine rule in trigonometry, a rule that relates the sides and angles of a triangle, stating that the ratio of the length of each side and the sine of the angle opposite is constant (twice the radius of the circumscribing circle). If the sides of a triangle are a, b, and c, and the angles opposite are A, B, and C, respectively, then the sine rule may be expressed as

$a/\sin A = b/\sin B = c \sin C$

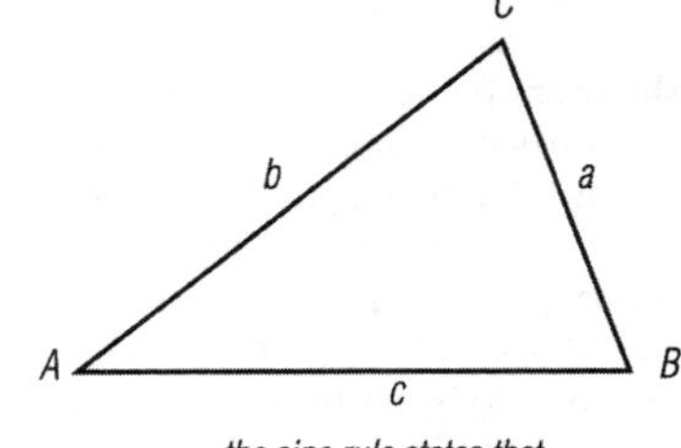

$$\frac{a}{\sin A} = \frac{b}{\sin B} = \frac{c}{\sin C}$$

or

$$\frac{\sin A}{a} = \frac{\sin B}{b} = \frac{\sin C}{c}$$

sine rule The sine rule relates the sides and angles of a triangle, stating that the ratio of the length of each side and the sine of the angle opposite is constant. Helicon

sine wave or **sine curve**, in physics, graph demonstrating properties that vary sinusoidally. It is obtained by plotting values of angles against the values of their †sines. Examples include †simple harmonic motion, such as the way alternating current (AC) electricity varies with time.

single phase describing electrical †power transmission involving a single sinusoidally varying †potential difference.

singularity in astrophysics, point in †space-time at which the known laws of physics break down. Singularity is predicted to exist at the center of a black hole, where infinite gravitational forces compress the infalling mass of a collapsing star to infinite density. It is also thought, according to the Big Bang model of the origin of the universe, to be the point from which the expansion of the universe began.

sink hole funnel-shaped hollow in an area of limestone. A sink hole is usually formed by the enlargement of a joint, or crack, by †carbonation (the dissolving effect of water). It should not be confused with a †swallow hole, or swallet, which is the opening through which a stream disappears underground when it passes onto limestone.

sinoatrial node heart's natural pacemaker.

Sinope tiny outermost moon of Jupiter discovered in 1914. It has a diameter of about 35 km/22 mi and an eccentric, highly inclined retrograde orbit at an average distance of 24 million km/15 million mi from the planet. One complete orbit takes 758 days.

siphon tube in the form of an inverted U with unequal arms. When it is filled with liquid and the shorter arm is placed in a tank or reservoir, liquid flows out of the longer arm provided that its exit is below the level of the surface of the liquid in the tank.

The liquid flows through the siphon because low pressure develops at the apex as liquid falls freely down the long arm. The difference between the pressure at the tank surface (atmospheric pressure) and the pressure at the apex causes liquid to rise in the short arm to replace that falling from the long arm.

Sirius or **the Dog Star** or **Alpha Canis Majoris**, brightest star in the night sky, 8.6 light years from the Sun in the constellation †Canis Major. Sirius is a double star: Sirius A is a white star with a mass 2.3 times that of the Sun, a diameter 1.8 times that of the Sun, and a true luminosity of 23 Suns. It is orbited every 50 years by a †white dwarf, Sirius B, also known as the Pup.

Sirius B is an eighth-magnitude companion which is sometimes known as "the Dark Companion" as it was first detected in the 19th century by German astronomer Friedrich †Bessel from its gravitational effect on the proper motion of Sirius A. "Dog Star" is the alternative name for Sirius, dating back to ancient Egypt. The unpleasantness of the hot summer season known as the "dog days" was attributed to the influence of Sirius being in conjunction with the Sun. Sirius B was seen for the first time in 1862 but it was only in the 1920s that it was recognized as the first known example of a white dwarf.

sirocco hot, normally dry and dust-laden wind that blows from the deserts of North Africa across the Mediterranean into southern Europe. It occurs mainly in the spring. The name "sirocco" is also applied to any hot oppressive wind.

Sitter, Willem de (1872–1934) Dutch astronomer, mathematician, and physicist. He contributed to the birth of modern cosmology and was influential in English-speaking countries in bringing the relevance of the general theory of †relativity to the attention of astronomers.

Sitter, Willem de http://www-groups.dcs.st-andrews.ac.uk/history/Mathematicians/Sitter.html

Part of an archive containing the biographies of the world's greatest mathematicians, this site is devoted to the life and contributions of astronomer Willem de Sitter. In addition to biographical information, you will find a list of references about de Sitter and links to other essays in the archive that reference him. The text of this essay includes hypertext links to the essays of those mathematicians and thinkers who influenced de Sitter. There is also an image of him, which can be enlarged, and a map of his birthplace.

SI units French **Système International d'Unités**, standard system of scientific units used by scientists worldwide. Originally proposed in 1960, it replaces the †m.k.s., †c.g.s., and †f.p.s. systems. It is based on seven basic units: the meter (m) for length, kilogram (kg) for mass, second (s) for time, ampere (A) for electrical current, kelvin (K) for temperature, mole (mol) for amount of substance, and candela (cd) for luminosity.

skeleton framework of bones that supports and gives form to the body, protects its internal organs, and provides anchorage points for its †muscles. It is composed of about 200 bones. Each bone is made of a †mineral, calcium phosphate, and †protein. Bones of the skeleton are joined to each other by ligaments. In the human body, walking, running, arm and leg movements, hand actions, and even just standing, are all achieved by the operation of muscles attached to bones of the skeleton. Movement of the body is brought about by the moveable †joints of the body. The elbow joint is a good example. Muscles are attached to bones by tendons, and contractions of the muscle bring about movement.

The skeleton may be composed of bone and cartilage (vertebrates), chitin (arthropods), calcium carbonate (mollusks and other invertebrates), or silica (many protists). The human skeleton is composed of 206 bones, with the †vertebral column (spine) forming the central supporting structure.

A skeleton may be internal, forming an †endoskeleton, or external, forming an †exoskeleton, as in the shells of insects or crabs. Another type of skeleton, found in invertebrates such as earthworms, is the hydrostatic skeleton. This gains partial rigidity from fluid enclosed within a body cavity. Because the fluid cannot be compressed, contraction of one part of the body results in extension of another part, giving peristaltic motion.

skew distribution in statistics, a distribution in which frequencies are not balanced about the mean. For example, low wages are earned by a great number of people, while high wages are earned by very few. However, because the high wages can be very high they pull the average up the scale, making the average wage look unrepresentatively high.

skew lines straight lines that are not parallel and yet do not meet since they lie in a different plane. Every pair of skew lines has a minimum distance between them, which is the length of their common perpendicular.

skin covering of the body of a vertebrate. In mammals, the outer layer (epidermis) is dead and its cells are constantly being rubbed away and replaced from below; it helps to protect the body from infection and to prevent dehydration. The lower layer (dermis) contains blood vessels, †nerves, hair roots, and sweat and sebaceous glands (producing oil), and is supported by a network of fibrous and elastic cells. The medical specialty concerned with skin diseases is called dermatology.

The skin helps to protect the body from drying out. It is waterproof and covered with dry, dead cells, so little water is lost from skin cells. However, water is lost from a human body when the body sweats. The skin helps to regulate body temperature. Body temperature is monitored and controlled by the thermoregulatory center in the †brain. This center has special cells sensitive to the temperature of blood flowing through the brain (receptors). Temperature receptors in the skin also send nerve impulses to this center giving information about skin temperature.

If the body temperature is too high blood vessels supplying the skin capillaries expand (dilate) so that more blood flows through the capillaries and more heat is lost. To further assist heat loss sweat glands release sweat, which cools the body as it evaporates. If the body temperature is too low blood vessels supplying the skin capillaries constrict to reduce the flow of blood through the capillaries.

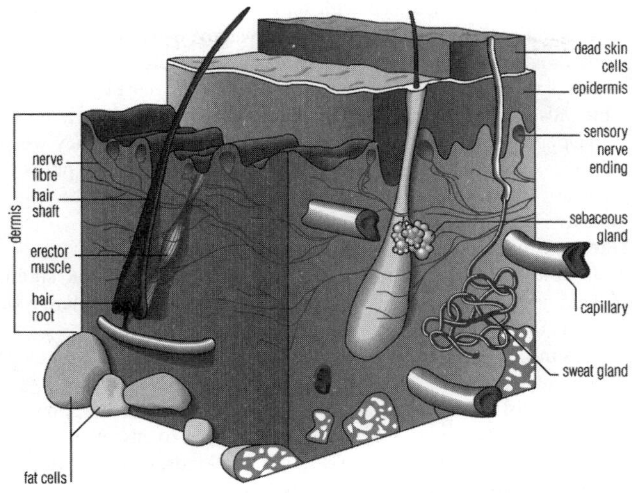

skin The skin of an adult man covers about 1.9 sq m/20 sq ft; a woman's skin covers about 1.6 sq m/17 sq ft. During our lifetime, we shed about 18 kg/40 lb of skin. Helicon

Other than measuring the temperature, the skin senses the environment in several other ways. Some receptors in the skin are sensitive to touch and pressure. If they are stimulated, nerve impulses are sent to the brain carrying information. It also protects the body from disease-causing organisms, which find it hard to penetrate the skin.

Skinner, B(urrhus) F(rederic) (1904–1990) US psychologist. He was a radical behaviorist who rejected mental concepts, seeing the organism as a "black box" where internal processes are not significant in predicting behavior. He studied operant conditioning (influencing behavior patterns by reward or punishment) and held that behavior is shaped and maintained by its consequences.

skull in vertebrates, the collection of flat and irregularly shaped bones (or cartilage) that enclose the brain and the organs of sight, hearing, and smell, and provide support for the jaws. In most mammals, the skull consists of 22 bones joined by fibrous immobile joints called sutures. The floor of the skull is pierced by a large hole (*foramen magnum*) for the spinal cord and a number of smaller apertures through which other nerves and blood vessels pass.

The skull comprises the cranium (brain case) and the bones of the face, which include the upper jaw, enclosing the sinuses, and form the framework for the nose, eyes, and the roof of the mouth cavity. The lower jaw is hinged to the middle of the skull at its lower edge. The opening to the middle ear is located near the jaw hinge. The plate at the back of the head is jointed at its lower edge with the upper section of the spine. Inside, the skull has various shallow cavities into which fit different parts of the brain.

slag in chemistry, the molten mass of impurities that is produced in the smelting or refining of metals. The slag produced in the manufacture of iron in a †blast furnace floats on the surface above the molten iron. It contains mostly silicates, phosphates, and sulfates of calcium. When cooled, the solid is broken up and used as a core material in the foundations of roads and buildings.

slaked lime common name for †calcium hydroxide.

slate fine-grained, usually gray metamorphic rock that splits readily into thin slabs along its †cleavage planes. It is the metamorphic equivalent of †shale.

Slate is highly resistant to atmospheric conditions and can be used for writing on with chalk (actually gypsum). Quarrying slate takes such skill and time that it is now seldom used for roof and sill material except in restoring historic buildings.

sleep state of natural unconsciousness and activity that occurs at regular intervals in most mammals and birds, though there is considerable variation in the amount of time spent sleeping. Sleep differs from hibernation in that it occurs daily rather than seasonally, and involves less drastic reductions in metabolism. The function of sleep is unclear. People deprived of sleep become irritable, uncoordinated, forgetful, hallucinatory, and even psychotic.

In humans, sleep is linked with hormone levels and specific brain electrical activity, including delta waves, quite different from the brain's waking activity. REM (rapid eye movement) phases, associated with dreams, occur at regular intervals during sleep, when the eyes move rapidly beneath closed lids.

Species that do not have distinct periods of sleep (most invertebrates and fishes, for example) have short intervals of reduced activity throughout a given 24-hour period. Extensive periods of sleep may have developed to make animals inconspicuous at times when they would be vulnerable to predators.

US researchers into REM sleep showed in 1998 that platypuses experience more REM (dream) sleep than any other animal. It had previously been thought that dreaming had evolved only in placental mammals.

sleet precipitation consisting of a mixture of water and ice.

slide rule mathematical instrument with pairs of logarithmic sliding scales, used for rapid calculations, including multiplication, division, and the extraction of square roots. It has been largely superseded by the electronic calculator.

It was invented in 1622 by the English mathematician William Oughtred. A later version was devised by the French army officer Amédée Mannheim (1831–1906).

slime mould or **myxomycete**, extraordinary organism that shows some features of †fungus and some of †protozoa. Slime moulds are not closely related to any other group, although they are often classed, for convenience, with the fungi. There are two kinds, cellular slime moulds and plasmodial slime moulds, differing in their complex life cycles.

Cellular slime moulds go through a phase of living as single cells, looking like amoebae, and feed by engulfing the bacteria found in rotting wood, dung, or damp soil. When a food supply is exhausted, up to 100,000 of these amoebae form into a colony resembling a single sluglike animal and migrate to a fresh source of bacteria. The colony then takes on the aspect of a fungus, and forms long-stalked fruiting bodies which release spores. These germinate to release amoebae, which repeat the life cycle.

Plasmodial slime moulds have a more complex life cycle involving sexual reproduction. They form a slimy mass of protoplasm with no internal cell walls, which slowly spreads over the bark or branches of trees.

Myxo Web http://www.wvonline.com/myxo/

All there is to know about myxomycetes. For the general reader there is a good description of these extraordinary organisms and how to identify, collect, and cultivate fungi. There are large numbers of dramatic photographs of fruiting specimens. For professional myxomycologists this site also includes links to international myxo sites.

Slipher, Vesto Melvin (1875–1969) US astronomer who established that spiral nebulae lie beyond our Galaxy. He also discovered the existence of particles of matter in interstellar space. His work in †spectroscopy increased our knowledge of planetary and nebular rotation, planetary and stellar atmospheres, and diffuse and spiral nebulae.

slip-off slope gentle slope forming the inner bank of a †meander (bend in a river). It is formed by the deposition of fine silt, or alluvium, by slow-flowing water.

As water passes round a meander the fastest current sweeps past the outer bank, eroding it to form a steep river cliff. Water flows more slowly past the inner bank, and as it reduces speed the material it carries is deposited around the bank to form a slip-off slope.

slope another name for †gradient.

slug obsolete unit of mass, equal to 14.6 kg/32.17 lb. It is the mass that will have an acceleration of one foot per second when under a force of one pound weight.

Smalley, Richard E(rrett) (1943–) US chemist who, with colleagues Robert Curl and Harold Kroto, was awarded the Nobel Prize for Chemistry in 1996 for their discovery of †buckminsterfullerene (carbon 60) in 1985. Smalley also pioneered the technique used to discover buckminsterfullerene, supersonic jet laser-beam †spectroscopy.

smart fluid or **electrorheological fluid**, liquid suspension that solidifies to form a jellylike solid when a high-voltage electric field is applied across it and that returns to the liquid state when the field is removed. Most smart fluids are †zeolites or metals coated with polymers or oxides.

smell sense that responds to chemical molecules in the air. It works by having receptors for particular chemical groups, into which the airborne chemicals must fit to trigger a message to the brain.

A sense of smell is used to detect food and to communicate with other animals (see †pheromone and †scent gland). Humans can distinguish between about 10,000 different smells. Aquatic animals can sense chemicals in water, but whether this sense should be described as "smell" or "taste" is debatable. See also †nose.

Mystery of Smell http://www.hhmi.org/senses/d/d110.htm

As part of a much larger site called "Seeing, Hearing, and Smelling the World," here is a page examining the way our sense of smell works. It is divided into four sections called "The vivid world of odors," "Finding the odorant receptors," "How rats and mice – and probably humans – recognize odors," and "The memory of smells." This site makes good use of images and animations to help with the explanations, so it is best viewed with an up-to-date browser.

Secret Sense in the Human Nose? http://www.hhmi.org/senses/d/d210.htm

As part of a much larger site called "Seeing, Hearing, and Smelling the World," here is a page examining the way our sense of smell detects odors of which we are not aware. It is divided into three sections called "Sniffing out social and sexual signals," "Triggers of innate behavior," and "Pheromones and mammals." This site makes good use of images and animations to help with the explanations, so it is best viewed with an up-to-date browser.

smelling salts mixture of ammonium carbonate, bicarbonate, and carbamate together with other strong-smelling substances, formerly used as a restorative for dizziness or fainting.

smelting processing a metallic ore in a furnace to produce the metal. Oxide ores such as iron ore are smelted with coke (carbon), which reduces the ore into metal and also provides fuel for the process.

A substance such as limestone is often added during smelting to facilitate the melting process and to form a slag, which dissolves many of the impurities present.

Smith, Francis Graham (1923–) English radio astronomer who with his colleague Martin †Ryle mapped the radio sources in the sky in the 1950s. Smith discovered the strongly polarized nature of radiation from pulsars in 1968, and estimated the strength of the magnetic field in interstellar space. He was Astronomer Royal 1982–90 and was knighted in 1986.

Smith, Hamilton Othanel (1931–) US microbiologist who was awarded a Nobel Prize for Physiology or Medicine in 1978 with his colleague Daniel †Nathans for their discovery of restriction enzymes and their application to molecular genetics. Restriction enzymes are enzymes that can cleave genes into fragments.

Smith, John Maynard English biologist, see †Maynard Smith.

Smith, William (1769–1839) English geologist. He produced the first geological maps of England and Wales, setting the pattern for geological cartography. Often called the founder of stratigraphical geology, he determined the succession of English strata across the whole country, from the Carboniferous up to the Cretaceous. He also established their fossil specimens.

smog natural fog containing impurities, mainly nitrogen oxides (NO_x) and volatile organic compounds (VOCs) from domestic fires, industrial furnaces, certain power stations, and internal-combustion engines (gasoline or diesel). It can cause substantial illness and loss of life, particularly among chronic bronchitics, and damage to wildlife.

photochemical smog is mainly prevalent in the summer as it is caused by chemical reaction between strong sunlight and vehicle exhaust fumes. Such smogs create a build-up of ozone and nitrogen oxides which cause adverse symptoms, including coughing and eye irritation, and in extreme cases can kill.

The London smog of 1952 lasted for five days and killed more than 4,000 people from heart and lung diseases. The use of smokeless fuels, the treatment of effluent, and penalties for excessive smoke from poorly maintained and operated vehicles can be effective in reducing smog but it still occurs in many cities throughout the world.

smoke suspension of a solid in a gas; the solid is in the form of extremely small particles and the smoke may be a †colloid.

smokeless fuel fuel that does not give off any smoke when burned, because all the carbon is fully oxidized to carbon dioxide (CO_2). Natural gas, oil, and coke are smokeless fuels.

smoker hot fissure in the ocean floor, known as a †hydrothermal vent.

Smoot, George (1945–) US astrophysicist. He specializes in †cosmic background radiation, and was the project leader for the differential microwave radiometers, one of the three types of instrument aboard the Cosmic Background Explorer satellite that detected "ripples" in the background radiation in 1992, thought to trace back to the formation of galaxies.

smoothing capacitor large electronic †capacitor connected across the output of a rectifier circuit that has the effect of smoothing out the voltage variations to give a nearly steady DC voltage supply.

The voltage and current output from a rectifier circuit fitted with a smoothing capacitor is similar to that provided by a battery.

smooth muscle involuntary muscle capable of slow contraction over a period of time. It is present in hollow organs, such as the intestines, stomach, bladder, and blood vessels. Its presence in the wall of the alimentary canal allows slow rhythmic movements known as †peristalsis, which cause food to be mixed and forced along the gut. Smooth muscle has a microscopic structure distinct from other forms.

Snel, Willebrord van Roijen (1581–1626) Dutch mathematician and physicist who devised the basic law of refraction, known as **Snel's law**, in 1621. He also founded the method of determining distances by triangulation.

Snel, Willebrord van Roijen http://www-history.mcs.st-and.ac.uk/history/Mathematicians/Snell.html

Part of an archive containing the biographies of the world's greatest mathematicians, this site is devoted to the life and contributions of Willebrord Snel. In addition to biographical information, you will find a list of references about Snel and links to other essays in the archive that reference him. The text of this essay includes hypertext links to the essays of those mathematicians and thinkers who influenced Snel. There is also an image of him, which can be enlarged, and a map of his birthplace.

Snell, George Davies (1903–1996) US geneticist who was awarded a Nobel Prize for Physiology or Medicine in 1980 with Jean †Dausset and Baruj †Benacerraf for their work on genetically determined structures on the cell surface that regulate immunological reactions, and in particular their identification of histocompatibility genes, genes that control the acceptance or rejection of tissue and organ transplants, in mice.

snellen unit expressing the visual power of the eye.

Snel's law of refraction in optics, the rule that when a ray of light passes from one medium to another, the sine of the angle of incidence divided by the sine of the angle of refraction is equal to the ratio of the indices of refraction in the two media. For a ray passing from medium 1 to medium 2: $n_2/n_1 = \sin i/\sin r$ where n_1 and n_2 are the refractive indices of the two media. The law was devised by the Dutch physicist, Willebrord Snel.

snout in earth science, the front end of a †glacier, representing the furthest advance of the ice at any one time. Deep cracks, or †crevasses, and ice falls are common.

Because the snout is the lowest point of a glacier it tends to be affected by the warmest weather. Considerable melting takes place, and so it is here that much of the rocky material carried by the glacier becomes deposited. Material dumped just ahead of the snout may form a †terminal moraine.

The advance or retreat of the snout depends upon the glacier budget – the balance between †accumulation (the addition of snow and ice to the glacier) and †ablation (their loss by melting and evaporation).

snow precipitation in the form of soft, white, crystalline flakes caused by the condensation in air of excess water vapor below freezing point. Light reflecting in the crystals, which have a basic hexagonal (six-sided) geometry, gives snow its white appearance.

Questions and Answers About Snow http://nsidc.org/snow/index.html

Comprehensive information about snow from the US National Snow and Ice Data Center. Among the interesting subjects discussed are why snow is white, why snow flakes can be up to two inches across, what makes some snow fluffy, why sound travels farther across snowy ground, and why snow is a good insulator.

Snyder, Solomon Halbert (1938–) US pharmacologist and neuroscientist who has studied the chemistry of the brain, and co-discovered the receptor mechanism for the body's own opiates, the †encephalins.

soap mixture of the sodium salts of various †fatty acids: palmitic, stearic, and oleic acid. It is made by the action of sodium hydroxide (caustic soda) or potassium hydroxide (caustic potash) on fats of animal or vegetable origin. Soap makes grease and dirt disperse in water in a similar manner to a †detergent.

Soap was mentioned by Galen in the 2nd century for washing the body, although the Romans seem to have washed with a mixture of sand and oil. Soap was manufactured in Britain from the 14th century, but better-quality soap was imported from Castile or Venice. The Soapmakers' Company, London, was incorporated in 1638. Soap was taxed in England from the time of Cromwell in the 17th century to 1853.

soapstone compact, massive form of impure †talc.

social behavior in zoology, behavior concerned with altering the behavior of other individuals of the same species. Social behavior allows animals to live harmoniously in groups by establishing hierarchies of dominance to discourage disabling fighting. It may be aggressive or submissive (for example, cowering and other signals of appeasement), or designed to establish bonds (such as social grooming or preening).

The social behavior of mammals and birds is generally more complex than that of lower organisms, and involves relationships with individually recognized animals. Thus, courtship displays allow individuals to choose appropriate mates and form the bonds necessary for successful reproduction. In the social systems of bees, wasps, ants, and termites, an individual's status and relationships with others are largely determined by its biological form, as a member of a caste of workers, soldiers, or reproductives; see †eusociality.

social psychology branch of †psychology concerned with the behavior of individuals in groups and the ways in which they relate to one another and to the societies of which they are a part.

Different kinds of social act are studied, as are the attitudes and assumptions that lie behind them and make them meaningful. There are two main approaches. One considers individuals as viewers, or perceivers, of the social environment, learning about it and forming judgments and attitudes, and so links up with general experimental psychology and other biological sciences. The other views individuals as participants in the social environment, influencing it and being influenced by it, and thus overlaps with sociology and other social sciences.

sociobiology study of the biological basis of all social behavior, including the application of †population genetics to the evolution of behavior. It builds on the concept of †inclusive fitness, contained in the notion of the "selfish gene." Contrary to some popular interpretations, it does not assume that all behavior is genetically determined.

soda ash former name for †sodium carbonate.

soda lime powdery mixture of calcium hydroxide and sodium hydroxide or potassium hydroxide, used in medicine and as a drying agent.

Soddy, Frederick (1877–1956) English physical chemist who pioneered research into atomic disintegration and coined the term †isotope. He was awarded the Nobel Prize for Chemistry in 1921 for investigating the origin and nature of isotopes.

sodium chemical symbol Na, soft, waxlike, silver-white, metallic element, atomic number 11, relative atomic mass 22.989. Its chemical symbol comes from the Latin *natrium*. It is one of the †alkali metals (in Group 1 of the †periodic table) and has a very low density, being light enough to float on water. It is the sixth-most abundant element (the fourth-most abundant metal) in the Earth's crust. Sodium is highly reactive, oxidizing rapidly when exposed to air and reacting violently with water. It is one of the most reactive metals in the †reactivity series of metals. Its most familiar compound is sodium chloride (common salt), which occurs naturally in the oceans and in salt deposits left by dried-up ancient seas.

sodium carbonate or **soda ash**; Na_2CO_3, anhydrous white solid. The hydrated, crystalline form ($Na_2CO_3.10H_2O$) is also known as washing soda.

It is made by the †Solvay process and used as a mild alkali, as it is hydrolyzed in water:

$$CO_3{}^{2-}{}_{(aq)} + H_2O_{(l)} \rightarrow HCO_3{}^{-}{}_{(aq)} + OH^{-}{}_{(aq)}$$

It is used to neutralize acids, in glass manufacture, and in water softening.

sodium chloride or **common salt** or **table salt**; NaCl, white, crystalline compound found widely in nature. The crystals are cubic in shape. It is a typical ionic solid with a high melting point (801°C/1,474°F); it is soluble in water, insoluble in organic solvents, and is a strong electrolyte when molten or in aqueous solution. Found in concentrated deposits as the mineral halite, it is widely used in the food industry as a flavoring and preservative, and in the chemical industry in the manufacture of sodium, chlorine, and sodium carbonate.

Chlorine and sodium hydroxide are produced by the †electrolysis of brine, a solution of sodium chloride in water. Salt for use in the food industry includes added iodine to reduce the incidence of the iodine deficiency diseases such as goiter and cretinism.

sodium hydrogencarbonate chemical name for †bicarbonate of soda.

sodium hydroxide or **caustic soda**; NaOH, commonest alkali. The solid and the solution are corrosive. It is used to neutralize acids, in the manufacture of soap, and in drain and oven cleaners. It is prepared industrially from sodium chloride by the †electrolysis of concentrated brine.

sodium hyposulfite see †sodium thiosulfate.

sodium metasilicate Na_2SiO_4, amorphous solid made by the fusion of sodium carbonate and silicon dioxide at high temperature. It is one of a class of compounds called sodium silicates. These compounds are used in the production of detergents, adhesives, and †silica gel.

sodium nitrate $NaNO_3$, deliquescent, colorless, crystalline solid found naturally as the mineral Chile saltpeter. It is used in the production of explosives and fertilizers.

sodium pump in biology, hypothetical mechanism that maintains the asymmetry of the ionic (concentration) balance across a nerve cell membrane, reflected in the cell's †resting potential.

sodium sulfate Na_2SO_4, chemical name for †Glauber's salt.

sodium thiosulfate $Na_2S_2O_3$, correct chemical name for sodium hyposulfite (hypo), the substance used to fix photographic images after developing.

soft iron form of iron with a low carbon content that does not retain †magnetism once the current in a †coil surrounding it has been removed.

software in computing, a collection of programs and procedures for making a computer perform a specific task, as opposed to †hardware, the physical components of a computer system. Software is created by programmers and is either distributed on a suitable medium, such as the †floppy disk, or built into the computer in the form of †firmware. Examples of software include †operating systems, †compilers, and applications programs such as payrolls or word processors. No computer can function without some form of software.

To function, computers need two types of software: application software and systems software. **Application software**, such as a payroll system or a †word processor, is designed for the benefit of the end user. **Systems software** performs tasks related to the operation and performance of the computer system itself. For example, a systems program might control the operation of the display screen, or control and organize backing storage.

soft water water that contains very few dissolved metal ions such as calcium (Ca^{2+}) or magnesium (Mg^{2+}). It lathers easily with soap, and no †scale is formed inside kettles or boilers.

soil loose covering of broken rocky material and decaying organic matter overlying the bedrock of the Earth's surface. It is composed of minerals (formed from †physical weathering and †chemical weathering of rocks), organic matter (called †humus) derived from decomposed plants and organisms, living organisms, air, and water. Soils differ according to climate, parent material, rainfall, relief of the bedrock, and the proportion of organic material. The study of soils is **pedology**.

Soils influence the type of agriculture employed in a particular region – light well-drained soils favor arable farming, whereas heavy clay soils give rise to lush pasture land. Plant roots take in nutrients (in the form of †ions) dissolved in the water in soil. The main elements that plants need to absorb through their roots are †nitrogen, †phosphorus, and †potassium.

Soil pH – What It Means http://www.esf.edu/pubprog/brochure/soilph/soilph.htm

Explanation of soil pH. The Web site also describes how to measure the pH of soil using simple experimental equipment, and goes on to describe methods that may be used to modify the acidity of alkalinity of soil.

soil creep gradual movement of soil down a slope in response to gravity. This eventually results in a mass downward movement of soil on the slope.

Evidence of soil creep includes the formation of terracettes (steplike ridges along the hillside), leaning walls and telegraph poles, and trees that grow in a curve to counteract progressive leaning.

soil depletion decrease in soil quality over time. Causes include loss of nutrients caused by overfarming, erosion by wind, and chemical imbalances caused by acid rain.

soil erosion wearing away and redistribution of the Earth's soil layer. It is caused by the action of water, wind, and ice, and also by improper methods of †agriculture. If unchecked, soil erosion results in the formation of deserts (†desertification). It has been estimated that 20% of the world's cultivated topsoil was lost between 1950 and 1990.

If the rate of erosion exceeds the rate of soil formation (from rock and decomposing organic matter), then the land will become infertile. The removal of forests (†deforestation) or other vegetation often leads to serious soil erosion, because plant roots bind soil, and without them the soil is free to wash or blow away, as in the American †dust bowl. The effect is worse on hillsides, and there has been devastating loss of soil where forests have been cleared from mountainsides, as in Madagascar.

Improved agricultural practices such as contour ploughing are needed to combat soil erosion. Windbreaks, such as hedges or strips planted with coarse grass, are valuable, and organic farming can reduce soil erosion by as much as 75%.

Soil degradation and erosion are becoming as serious as the loss of the rainforest. It is estimated that more than 10% of the world's soil lost a large amount of its natural fertility during the latter half of the 20th century. Some of the worst losses are in Europe, where 17% of the soil is damaged by human activity such as mechanized farming and fallout from acid rain. In Mexico and Central America, 24% of soil is highly degraded, mostly as a result of deforestation.

Soil Erosion – Causes and Effects http://www.gov.on.ca/OMAFRA/english/engineer/facts/87-040.htm

Detailed page from the Canadian Ontario Ministry of Agriculture on the causes and effects of erosion. The site provides a good summary of the various factors affecting the rate and magnitude of soil erosion including rainfall intensity, slope gradient and length, and vegetation.

soil mechanics branch of engineering that studies the responses of soils with different water and air contents, to loads. Soil is investigated during construction work to ensure that it has the mechanical properties necessary to support the foundations of dams, bridges, and roads.

sol †colloid of very small solid particles dispersed in a liquid that retains the physical properties of a liquid.

solar constant mean radiation received from the Sun at the top level of Earth's atmosphere: 1.95 calories per square centimeter per minute.

solar cycle variation of activity on the †Sun over an 11-year period indicated primarily by the number of †sunspots visible on its surface. The next period of maximum activity is expected round 2011.

solar energy energy derived from the light and heat from the Sun. The amount of energy falling on just 1 sq km/0.4 sq mi is about 4,000 megawatts, enough to heat and light a small town. In one second the Sun gives off 13 million times more energy than all the electricity used in the USA in one year. **Solar heaters** usually consist of concave mirrors or reflective parabolic surfaces that concentrate the Sun's rays onto a black (heat-absorbing) panel containing pipes through which air or water is circulated, either by thermal †convection or by a pump. The heat energy of the air or water is converted into †electrical energy via a †turbine and a †generator. Hot water for industrial and domestic use can be produced by circulating water through panels, the water absorbing heat from the Sun as it passes through the panels.

Solar energy may also be harnessed indirectly using **solar cells** (photovoltaic cells) made of panels of †semiconductor material (usually silicon), which generate electricity when illuminated by sunlight. Although it is difficult to generate a high output from solar energy compared to sources such as nuclear or fossil fuels, it is a major nonpolluting and renewable energy source used as far north as Scandinavia as well as in the southwestern USA and in Mediterranean countries.

solar flare brilliant eruption on the Sun above a †sunspot, thought to be caused by release of magnetic energy. Flares reach maximum brightness within a few minutes, then fade away over about an hour. They eject a burst of atomic particles into space at up to 1,000 kps/600 mps. When these particles reach Earth they can cause radio blackouts, disruptions of the Earth's magnetic field, and †auroras.

solar parallax †parallax of the Sun, now measured as 22.34 cm/8.794 in.

solar pond natural or artificial "pond," such as the Dead Sea, in which salt becomes more soluble in the Sun's heat. Water at the bottom becomes saltier and hotter, and is insulated by the less salty water layer at the top. Temperatures at the bottom reach about 100°C/212°F and can be used to generate electricity.

solar prominence mass of hot hydrogen rising from the Sun's †chromosphere, best observed during a total †eclipse. Eruptive prominences are violent in force and may reach heights of 2,000,000 km/1,243,000 mi; quiescent prominences are relatively pacific but may last for months.

solar radiation radiation given off by the Sun, consisting mainly of visible light, †ultraviolet radiation, and †infrared radiation, although the whole spectrum of †electromagnetic waves is present, from radio waves to X-rays. High-energy charged particles, such as electrons, are also emitted, especially from solar †flares. When these reach the Earth, they cause magnetic storms (disruptions of the Earth's magnetic field), which interfere with radio communications.

Solar Energy: Basic Facts http://www.brookes.ac.uk/other/uk-ises/facts.htm

Quick guide to this renewable energy source. Taken from a series of fact sheets published by the Solar Energy Society, this page explains concepts such as solar water heating, photovoltaics, and passive solar architecture. Find out why solar power works better in Scotland than in the south of France.

solar rotation rotation of the Sun about its axis. It is differential, the equatorial rotation taking less time than the polar by up to 9.4 Earth-days.

Solar System †Sun (a star) and all the bodies orbiting it: the nine †planets (Mercury, Venus, Earth, Mars, Jupiter, Saturn, Uranus, Neptune, and Pluto), their moons, and smaller objects such as †asteroids and †comets. The Sun contains 99.86% of the mass of the Solar System. The planets orbit the Sun in elliptical paths, and in the same direction as the Sun itself rotates. The planets nearer the Sun have shorter orbital times than those further away since the distance they travel in each orbit is less. The inner planets (Mercury, Venus, Earth, and Mars) have solid, rocky surfaces; relatively slow periods of rotation (Mercury takes 59 days to complete one rotation, Venus 243 days, Earth nearly 24 hours, and Mars 24.5 hours); very few natural †satellites; and diameters less than 13,000 km/8,000 mi. Venus can be seen with the unaided eye, appearing in the evening as the brightest "star" in the sky. In contrast, the outer planets (Jupiter, Saturn, Uranus, and Neptune) have denser, gaseous atmospheres composed mainly of hydrogen and helium; fast periods of rotation (Jupiter takes 10 hours for one rotation, Saturn nearly 10.5 hours, and Uranus 11 hours); and many natural satellites (Jupiter and Saturn have more than 30 between them and Uranus has 15). Uranus, Neptune, and Pluto were discovered after the development of the telescope.

Earth View http://www.fourmilab.ch/cgi-bin/uncgi/Earth?opt=-p&img=learth.evif

Real-time picture of the Earth, showing night and day regions. This site provides a variety of links to other Earth- and space-related sites, including a Moon view and views from any satellite orbiting the Earth. Zooming-in and panning controls are also provided, offering close-up detail of different countries.

Lunar and Planetary Institute http://www.lpi.usra.edu/

Wonderful, extremely well-organized site, which includes information about planetary exploration and transcripts from astronauts' lunar field trips. The site also includes lots of materials for teachers.

Solar System Live http://www.fourmilab.ch/solar/solar.html

Take a look at the entire Solar System as it might be seen at different times and dates or from different viewpoints.

Views of the Solar System http://www.solarviews.com/eng/homepage.htm
Educational tour of the Solar System. It contains information and statistics about the planets, moons, asteroids, comets, and meteorites found within the Solar System, and is supported by images.

solar time time of day as determined by the position of the †Sun in the sky.

Apparent solar time, the time given by a sundial, is not uniform because of the varying speed of the Earth in its elliptical orbit. **Mean solar time** is a uniform time that coincides with apparent solar time at four instants through the year. The difference between them is known as the equation of time, and is greatest in early November when the Sun is more than 16 minutes fast on mean solar time. Mean solar time on the Greenwich meridian is known as †Greenwich Mean Time and is the basis of civil timekeeping.

solar wind stream of atomic particles, mostly protons and electrons, from the Sun's corona, flowing outwards at speeds of between 300 kps/200 mps and 1,000 kps/600 mps.

The fastest streams come from "holes" in the Sun's corona that lie over areas where no surface activity occurs. The solar wind pushes the gas of comets' tails away from the Sun, and "gusts" in the solar wind cause geomagnetic disturbances and auroras on Earth. Other planets are also affected: the European Space Agency's probe Mars Express (to be launched in 2003) will investigate how the solar wind affects that planet's atmosphere.

solenoid coil of wire, usually cylindrical, in which a magnetic field is created by passing an electric current through it (see †electromagnet). This field can be used to temporarily magnetize, and so move, an iron rod placed on its axis. Mechanical valves attached to the rod can be operated by switching the current on or off, so converting electrical energy into mechanical energy. Solenoids are used to relay energy from the battery of a car to the starter motor by means of the ignition switch.

The solenoid has a magnetic field, one end being the north pole and the other end the south pole. A solenoid behaves as any other magnet; by changing the direction of the current the position of the north and south poles can be switched. If a larger current is passed through the coil it becomes a stronger magnet.

solid in physics, a state of matter that holds its own shape (as opposed to a liquid, which takes up the shape of its container, or a gas, which totally fills its container). According to the †kinetic theory of matter, the atoms or molecules in a solid are packed closely together in a regular arrangement, and are not free to move but merely vibrate about fixed positions, such as those in crystal lattices.

solidification change of state from liquid (or vapor) to solid that occurs at the †freezing point of a substance.

solid of revolution in mathematics, solid generated by revolving a plane figure about an axis. Its surface is called a **surface of revolution**.

solid solution in chemistry, solid in which two or more elements or compounds share the same structural lattice, usually applied to metal †alloys. Where the solute atoms replace those of the solvent, a substitutional alloy is formed, for example bronze.

solid-state circuit electronic circuit in which all the components (resistors, capacitors, transistors, and diodes) and interconnections are made at the same time, and by the same processes, in or on one piece of single-crystal silicon. The small size of this construction accounts for its use in electronics for space vehicles and aircraft.

solid-state physics study of materials in the solid state, investigating the magnetic, thermal, and electrical properties, for example.

solid-state reaction in chemistry, reaction in which all the components are solids. The term is also applied to systems in which there is only one component, such as the decomposition of a solid.

solifluction movement (flow) of topsoil that has become saturated with water. It is an important process of transportation in †periglacial environments where water released by the spring thaw cannot percolate downwards because the ground below is permanently frozen (permafrost). The saturated topsoil may then flow slowly downhill to form a **solifluction lobe**.

soliton solitary wave that maintains its shape and velocity, and does not widen and disperse in the normal way. The mathematical equations that sum up the behavior of solitons are used to further research in nuclear fusion and superconductivity.

solstice either of the days on which the Sun is farthest north or south of the celestial equator each year. In the northern hemisphere, the **summer solstice**, when the Sun is farthest north, occurs around 21 June and the **winter solstice** around 22 December.

solubility measure of the amount of solute (usually a solid or gas) that will dissolve in a given amount of solvent (usually a liquid) at a particular temperature. Solubility may be expressed as grams of solute per 100 grams of solvent or, for a gas, in parts per million (ppm) of solvent.

solubility curve graph that indicates how the solubility of a substance varies with temperature. Most salts increase their solubility with an increase in temperature, as they dissolve endothermically. These curves can be used to predict which salt, and how much of it, will crystallize out of a mixture of salts.

soluble group notion introduced to extend theorems concerning commutative groups to a wider class of groups that, in an intuitive sense, can be constructed out of commutative "pieces." Given a group G the commutator subgroup C is introduced (generated by elements of the form $xyx^{-1}y^{-1}$); G can then be "collapsed" to an Abelian group by a process that reduces C to the identity element; C is then collapsed in an identical fashion, and the process continually repeated. If the process, through a finite number of steps, eventually leads to a last group that is already commutative, G is said to be soluble.

solute substance that is dissolved in another substance (see †solution).

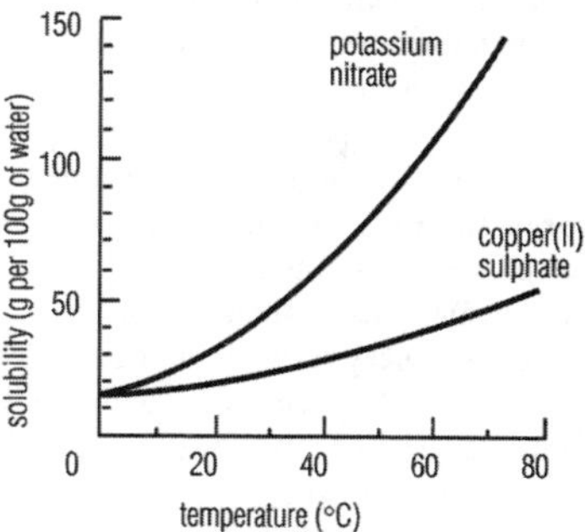

solubility curve Comparative aqueous solubility curves for copper(II) sulfate and potassium nitrate. Helicon

solution two or more substances mixed to form a single, homogenous phase. One of the substances is the **solvent** and the others (**solutes**) are said to be dissolved in it.

The constituents of a solution may be solid, liquid, or gaseous. The solvent is normally the substance that is present in greatest quantity; however, if one of the constituents is a liquid this is considered to be the solvent even if it is not the major substance.

solution or **dissolution**, in earth science, process by which the minerals in a rock are dissolved in water. Solution is one of the processes of †erosion as well as †weathering (in which the breakdown of rock occurs without transport of the dissolved material). An example of this is when weakly acidic rainfall dissolves calcite.

Solution commonly affects †limestone and †chalk, which are both formed of calcium carbonate. It can occur in coastal environments along with corrasion and hydraulic action, producing features like the white cliffs of Dover, as well as fluvial (river) environments, like the one that formed Cheddar Gorge. Solution is also responsible for the weathering of many buildings, monuments and other structures.

solution in algebra, the value of a variable that satisfies a given equation; see †root.

solution set in mathematics, set of values that satisfies an †inequality relationship, or equation(s).

Solvay, Ernest (1838–1922) Belgian industrial chemist who in the 1860s invented the ammonia-soda process, also known as the **Solvay process**, for making the alkali sodium carbonate.

Solvay process industrial process for the manufacture of sodium carbonate.

It is a multistage process in which carbon dioxide is generated from limestone and passed through †brine saturated with ammonia. Sodium hydrogencarbonate is isolated and heated to yield sodium carbonate. All intermediate by-products are recycled so that the only ultimate by-product is calcium chloride.

solvent substance, usually a liquid, that will dissolve another substance (see †solution). Although the commonest solvent is water, in popular use the term refers to low-boiling-point organic liquids, which are harmful if vapor from them is breathed in as a result of their use in a confined space, or from skin contact with the liquid. They can give rise to respiratory problems, liver damage, and neurological complaints.

Typical organic solvents are petroleum distillates (in glues), xylol (in paints), alcohols (for synthetic and natural resins such as shellac), esters (in lacquers, including nail varnish), ketones (in cellulose lacquers and resins), and chlorinated hydrocarbons (in paint stripper and dry-cleaning fluids). The fumes of some solvents, when inhaled (glue-sniffing), affect mood and perception. In addition to damaging the brain and lungs, repeated inhalation of solvent from a plastic bag can cause death by asphyxiation.

somatic cell any cell in an organism, excluding the reproductive cells.

Sommerfeld, Arnold Johannes Wilhelm (1868–1951) German physicist who demonstrated that difficulties with Niels †Bohr's model of the atom, in which electrons move around a central nucleus in circular orbits, could be overcome by supposing that electrons adopt elliptical orbits.

Sommerfeld, Arnold Johannes Wilhelm http://www-groups.dcs.st-andrews.ac.uk/history/Mathematicians/Sommerfeld.html
Part of an archive containing the biographies of the world's greatest mathematicians, this site is devoted to the life and contributions of Arnold Sommerfeld. In addition to biographical information, you will find a list of references about Sommerfeld and links to other essays in the archive that reference him.

sonar acronym for SOund Navigation And Ranging, method of locating underwater objects by the reflection of ultrasonic waves. The time taken for an acoustic beam to travel to the object and back to the source enables the distance to be found since the velocity of sound in water is known. Sonar devices, or **echo sounders**, were developed in 1920, and are the commonest means of underwater navigation.

sone unit of subjective loudness. A tone of 40 decibels above the threshold of hearing with a frequency of 1,000 hertz is defined as one sone; any sound that seems twice as loud as this has a value of two sones, and so on. A loudness of one sone corresponds to 40 †phons.

sonic boom noise like a thunderclap that occurs when an aircraft passes through the sound barrier, or begins to travel faster than the speed of sound. It happens when the cone-shaped shock wave caused by the plane touches the ground.

sonochemistry branch of chemistry based on the properties of chemicals when subjected to high-intensity ultrasound. The ultrasound causes bubbles to form, grow, and implode. The implosion generates intense local heat and extreme pressure, but for less than a millionth of a second. Light is also emitted (†sonoluminescence).

Applications include the production of amorphous metals, molten metals cooled so fast they freeze before crystals have a chance to form properly. This gives them unique electromagnetic properties and resistance to corrosion.

sonoluminescence emission of light by a liquid that is subjected to high-frequency sound waves. The rapid changes of pressure induced by the sound cause minute bubbles to form in the liquid, which then collapse. Light is emitted at the final stage of the collapse, probably because it squeezes and heats gas inside the bubbles.

sorbic acid $CH_3CH = CHCH = CHCOOH$, tasteless acid found in the fruit of the mountain ash (genus *Sorbus*) and prepared synthetically. It is widely used in the preservation of food – for example, cider, wine, soft drinks, animal feeds, bread, and cheese.

Sorby, Henry Clifton (1826–1908) English geologist whose discovery in 1863 of the crystalline nature of steel led to the study of metallography. Thin-slicing of hard minerals enabled him to study the constituent minerals microscopically in transmitted light. He later employed the same techniques in the study of iron and steel under stress.

Sørensen, Søren Peter Lauritz (1868–1939) Danish chemist who in 1909 introduced the concept of using the †pH scale as a measure of the acidity of a solution. On Sørensen's scale, still used today, a pH of 7 is neutral; higher numbers represent alkalinity, and lower numbers acidity.

sorus in ferns, a group of sporangia, the reproductive structures that produce †spores. They occur on the lower surface of fern fronds.

Sosigenes (lived 1st century BC) Alexandrian astronomer and mathematician. He was employed by Julius Caesar to reform the calendar, and it appears from Pliny that he taught that Mercury moves around the Sun. All that remains of his work are extracts from his *Revolving Spheres*.

sound physiological sensation received by the ear, originating in a vibration causing sound waves. The sound waves are pressure variations in the air and travel in every direction, spreading out as an expanding sphere. Sound energy cannot travel in a vacuum.

All sound waves in air travel with a speed dependent on the temperature; under ordinary conditions, this is about 330 m/1,080 ft per second. The pitch of the sound depends on the number of vibrations imposed on the air per second (†frequency), but the speed is unaffected. The loudness of a sound is dependent primarily on the amplitude of the vibration of the air.

How We Perceive Sound: The Ear http://library.thinkquest.org/19537/Ear2.html

Details of the workings of the ear. The user can click on a part of the diagram of the ear to find out what it is and how it works. This is a very clearly laid out site with easy-to-understand definitions.

Physics of Sound http://www.glenbrook.k12.il.us/gbssci/phys/Class/sound/u11l2c.html

Exploration of the speed of sound. This site deals with the different speeds that occur in solids, liquids, and gases. The equation for calculating speed and the calculation of differences between different materials is defined. A "Check your understanding" section tests the areas covered in the text, and answers are available from the pull-down menus under each question.

sound absorption in acoustics, the conversion of sound energy to heat energy when sound waves strike a surface. The process reduces the amplitude of each reflected sound wave (echo) and thus the degree to which †reverberation takes place. Materials with good sound-absorbing properties are often fitted on walls and ceilings in buildings such as offices, factories, and concert halls in order to reduce or control sound levels.

sound-level indicator instrument used to measure the intensity or loudness of sound.

Readings are given on a †decibel (dB) scale that compares the sound level with the threshold of human hearing (standardized as an intensity of 1.0 x 10^{-12} watts per square meter). A specialized scale called the dBA scale gives a weighted reading that takes into account the ear's sensitivity to different frequencies.

sound wave longitudinal wave motion with which sound energy travels through a medium. It carries energy away from the source of the sound without carrying the material itself with it. Sound waves are mechanical; unlike electromagnetic waves, they require vibration of their medium's molecules or particles (manifested in air as compressions and rarefactions of the air), and this is why sound cannot travel through a vacuum.

In air, the pressure variations as an object vibrates travel at a speed of 330 m/1,080 ft per second, are detected by the ear, and interpreted by the brain as sound. A person with normal hearing can detect sounds with frequencies in the range of 20 to 20,000 hertz.

The energy of the air vibrating travels along the wave without transferring matter. A loosely-coiled spring can be used to demonstrate how sound waves travel through air. The disturbance produced by the vibrating object causes compressions and rarefactions of air particles to move in the same direction as the waves; they are called longitudinal waves. The pattern of longitudinal waves is produced when the spring is given a forward "push." Sound can be produced by the vibrations of objects such as the stretched strings of a violin or air particles in wind instruments; the sounds of the voice are produced by air causing the "strings" of the vocal cords to vibrate.

Sound can also travel through solids and liquids. For example, voices can be heard through a wall between one room and another. This effect is used when listening with a stethoscope to hear the sounds of heart and lungs through the walls of the chest. Porpoises use ultrasonic sound in water as an echo-guiding system.

Sound waves can be reflected to produce echoes, and diffracted to produce interference patterns of louder and softer sound.

source language in computing, the language in which a program is written, as opposed to †machine code, which is the form in which the program's instructions are carried out by the computer. Source languages are classified as either †high-level languages or †low-level languages, according to whether each notation in the source language stands for many or only one instruction in machine code.

Programs in high-level languages are translated into machine code by either a †compiler or an interpreter program. Low-level programs are translated into machine code by means of an †assembler program. The program, before translation, is called the **source program**; after translation into machine code it is called the **object program**.

source resistance alternative term for †internal resistance, the resistance inside an electric power supply.

souring change that occurs to wine on prolonged exposure to air. The ethanol in the wine is oxidized by the air (oxygen) to ethanoic acid. It is the presence of the ethanoic (acetic) acid that produces the sour taste:

$CH_3CH_2OH_{(aq)} + O_{2(g)} \rightarrow CH_3COOH_{(aq)} + H_2O_{(l)}$

Milk turns sour with the production of lactic acid, brought about by bacterial action.

Southern Cross popular name for the constellation †Crux.

southern lights common name for the †aurora australis, colored light in southern skies.

space or **outer space**, void that exists beyond Earth's atmosphere. Above 120 km/75 mi, very little atmosphere remains, so objects can continue to move quickly without extra energy. The space between the planets is not entirely empty, but filled with the tenuous gas of the †solar wind as well as dust.

Ask the Space Scientist http://image.gsfc.nasa.gov/poetry//ask/askmag.html

Well-organized NASA site with answers to questions about astronomy and space. The site endeavors to send a reply to any question about space and astronomy within three days. There are useful links to question archives, general files, and other space-related sites.

space collection of mathematical objects (referred to as points) with an associated structure resembling (or analogous to) the properties of the space of everyday experience.

space constant in mathematics, characterizing the line in non-Euclidean geometry that relates the †angle of parallelism of two lines – a concept formulated by Hungarian mathematician János Bolyai.

space group in chemistry, designation classifying the symmetry of the molecular structure of a crystalline material. There are 230 different space groups, representing all possible crystal structures.

space probe any instrumented object sent beyond Earth to collect data from other parts of the Solar System and from deep space. The first probe was the Soviet Lunik 1, which flew past the Moon in 1959. The first successful planetary probe was the US Mariner 2, which flew past Venus in 1962, using a

†transfer orbit. The first space probe to leave the Solar System was Pioneer 10 in 1983. Space probes include Galileo, Giotto, Magellan, Mars Observer, Ulysses, the †Moon probes, and the Mariner, Pioneer, Viking, and Voyager series.

space-time in physics, combination of space and time used in the theory of †relativity. When developing relativity, Albert Einstein showed that time was in many respects like an extra dimension (or direction) to space. Space and time can thus be considered as entwined into a single entity, rather than two separate things.

Space-time is considered to have four dimensions: three of space and one of time. In relativity theory, events are described as occurring at points in space-time. The **general theory of relativity** describes how space-time is distorted by the presence of material bodies, an effect that we observe as gravity.

Light Cone: An Illuminating Introduction to Relativity
http://www.phy.syr.edu/courses/modules/LIGHTCONE/index.html
Informal and entertaining site introducing space-time. Easy-to-digest tutorials, illustrated with simple diagrams and animations, explain the concepts and how they have evolved.

Spacetime Wrinkles http://archive.ncsa.uiuc.edu/Cyberia/NumRel/NumRelHome.html
Masses of space-time information from the National Center for Supercomputing Applications at the University of Illinois. There are also illustrated easy-to-follow tutorials organized by theme, and lots of video clips and animations.

spadix in botany, an †inflorescence consisting of a long, fleshy axis bearing many small, stalkless flowers. It is partially enclosed by a large bract or †spathe. A spadix is characteristic of plants belonging to the family Araceae, including the arum lily *Zantedeschia aethiopica*.

Spallanzani, Lazzaro (1729–1799) Italian biologist. He disproved the theory that microbes spontaneously generate out of rotten food by showing that they would not grow in flasks of broth that had been boiled for 30 minutes and then sealed.

spark chamber electronic device for recording tracks of charged subatomic particles, decay products, and rays. In combination with a stack of condenser plates, a spark chamber enables the point where an interaction has taken place to be located, to within a cubic centimeter. At its simplest, it consists of two smooth threadlike †electrodes that are positioned 1–2 cm/0.5–1 in apart, the space between being filled with a mixture of neon and helium gas. Sparks jump through the gas along the ionized path created by the radiation. See †particle detector.

spathe in flowers, the single large bract surrounding the type of inflorescence known as a †spadix. It is sometimes brightly colored and petal-like, as in the brilliant scarlet spathe of the flamingo plant *Anthurium andreanum* from South America; this serves to attract insects.

spawn extruded egg mass of such egg-laying animals as fish, amphibians, and mollusks.

Spawn is produced in very variable quantities (for example, the ling lays about 150,000,000 eggs, and the American oyster 60,000,000 eggs) and is much preyed upon, even, as in the case of the stickleback, by the female herself. A variety of means have been devised for its protection.

In fish, spawning is the reproductive phase, variously involving aggregation of the fish, courtship, extrusion of eggs and sperm into the water, and fertilization of the eggs. Large migrations may precede spawning, as in salmon and herring. In some groups the spawn is not released, fertilization being internal.

The name is also sometimes given to the mycelium of mushrooms and other fungi, seen as white threads in decaying matter.

Spearman's rank in statistics, an index of the strength of the relationship between two sets of information; for example, altitude and temperature. The value of the index varies between +1 (a perfect positive correlation) and –1 (a perfect negative correlation); 0 = no correlation.

The formula is:

$R = 1 - (6 \times \sum d^2)/(n^3 - n)$

where n is the number of pairs of data and d is the difference in rank.

As a first step to investigate correlation, a †scatter diagram can be plotted, to see how near to a straight line the points lie.

speciation emergence of a new species during evolutionary history. One cause of speciation is the geographical separation of populations of the parent species, followed by reproductive isolation and selection for different environments so that they no longer produce viable offspring when they interbreed. Other causes are †assortative mating and the establishment of a †polyploid population.

species in biology, a distinguishable group of organisms that resemble each other or consist of a few distinctive types (as in †polymorphism), and that can all interbreed to produce fertile offspring. Species are the lowest level in the system of biological classification.

Related species are grouped together in a genus. Within a species there are usually two or more separate

†populations, which may in time become distinctive enough to be designated subspecies or varieties, and could eventually give rise to new species through †speciation. Around 1.4 million species have been identified so far, of which 750,000 are insects, 250,000 are plants, and 41,000 are vertebrates. In tropical regions there are roughly two species for each temperate-zone species. It is estimated that one species becomes extinct every day through habitat destruction.

species in chemistry, collective term for a specific type of atom, molecule, particle, or ion; for example Na^+ ions are a species of metallic ion, protein molecules are a species of colloidal particle.

specific gravity alternative term for †relative density.

specific heat capacity quantity of heat required to raise unit mass (1 kg) of a substance by one †kelvin (1 K). The unit of specific heat capacity in the SI system is the †joule per kilogram per kelvin (J kg^{-1} K^{-1}).

specific latent heat heat that changes the physical state of a unit mass (one kilogram) of a substance without causing any temperature change.

The **specific latent heat of fusion** of a solid substance is the heat required to change one kilogram of it from solid to liquid without any temperature change. The **specific latent heat of vaporization** of a liquid substance is the heat required to change one kilogram of it from liquid to vapor without any temperature change.

speckle interferometry technique whereby large telescopes can achieve high resolution of astronomical objects despite the adverse effects of the atmosphere through which light from the object under study must pass. It involves the taking of large numbers of images, each under high magnification and with short exposure times. The pictures are then combined to form the final picture. The technique was introduced by the French astronomer Antoine Labeyrie in 1970.

spectator ion in a chemical reaction that takes place in solution, an ion that remains in solution without taking part in the chemical change. For example, in the precipitation of barium sulfate from barium chloride and sodium sulfate, the sodium and chloride ions are spectator ions.

$BaCl_{2\,(aq)} + Na_2SO_{4\,(aq)} \rightarrow BaSO_{4\,(s)} + 2NaCl_{(aq)}$

spectral classification in astronomy, classification of stars according to their surface temperature and †luminosity, as determined from their spectra. Stars are assigned a spectral type (or class) denoted by the letters O, B, A, F, G, K, and M, where O stars (about 40,000 K/39,700°C/71,500°F) are the hottest and M stars (about 3,000 K/2,700°C/5,000°F) are the coolest.

Each letter may be further divided into ten subtypes, B0, B1, B2, and so on. Stars are also assigned a luminosity class denoted by a Roman numeral attached to the spectral type: I (†supergiants), II (bright giants), III (giants), IV (subgiants), V (main sequence), VI (subdwarfs), or VII (†white dwarfs). The Sun is classified as type G2V. See also †Hertzsprung-Russell diagram.

spectral line dark line visible in an absorption †spectrum, or one of the bright lines that make up an emission spectrum. Spectral lines are caused by the transference of an †electron in an atom from one energy level to another; strong lines are produced at levels at which such transference occurs easily, weak where it occurs with difficulty. †Ionization of certain elements can affect such transferences and cause problems in spectral analysis.

spectroheliograph device with which spectra of the various regions of the Sun are obtained and photographed.

spectrometer in physics and astronomy, instrument used to study the composition of light emitted by a source. The range, or †spectrum, of wavelengths emitted by a source depends upon its constituent elements, and may be used to determine its chemical composition.

The simpler forms of spectrometer analyze only visible light. A **collimator** receives the incoming rays and produces a parallel beam, which is then split into a spectrum by either a †diffraction grating or a prism mounted on a turntable. As the turntable is rotated each of the constituent colors of the beam may be seen through a **telescope**, and the angle at which each has been deviated may be measured on a circular scale. From this information the wavelengths of the colors of light can be calculated. Spectrometers are used in astronomy to study the electromagnetic radiation emitted by stars and other celestial bodies. The spectral information gained may be used to determine their chemical composition, or to measure the †red shift of light associated with the expansion of the universe and thereby calculate the speed with which distant stars are moving away from the Earth.

spectrometry in analytical chemistry, a technique involving the measurement of the spectrum of energies (not necessarily electromagnetic radiation) emitted or absorbed by a substance.

spectrophotometer analytical instrument that measures the relationship between the amount of electromagnetic radiation absorbed by a sample and the frequency or wavelength of that radiation.

Instruments usually cover a specific area of the electromagnetic spectrum, for example infrared and ultraviolet.

spectroscopic binary †binary star in which two stars are so close together that they cannot be seen separately, but their separate light spectra can be distinguished by a spectroscope.

The first spectroscopic binary to be discovered, in 1889 by US astronomer Edward Pickering, was the brighter component of †Mizar. Many hundreds are now known, with orbital periods ranging from 82 minutes to 15 years. The most common period is a few days.

spectroscopy study of spectra (see †spectrum) associated with atoms or molecules in the solid, liquid, or gaseous phase. Spectroscopy can be used to identify unknown compounds and is an invaluable tool in science, medicine, and industry (for example, in checking the purity of drugs).

Emission spectroscopy is the study of the characteristic series of sharp lines in the spectrum produced when an †element is heated. Thus an unknown mixture can be analyzed for its component elements. Related is **absorption spectroscopy**, dealing with atoms and molecules as they absorb energy in a characteristic way. Again, dark lines can be used for analysis. More detailed structural information can be obtained using **infrared spectroscopy** (concerned with molecular vibrations) or **nuclear magnetic resonance (NMR) spectroscopy** (concerned with interactions between adjacent atomic nuclei). **Supersonic jet laser beam spectroscopy** enables the isolation and study of clusters in the gas phase. A laser vaporizes a small sample, which is cooled in helium, and ejected into an evacuated chamber. The jet of clusters expands supersonically, cooling the clusters to near absolute zero, and stabilizing them for study in a †mass spectrometer.

Spectroscopy On Line http://www.spectroscopymag.com/spectroscopy/

Spectroscopy pages, the on-line counterpart to Spectroscopy magazine. This busy Web site has news, articles, tutorials, glossary, and links to spectroscopy-related materials on the Internet.

spectrum plural **spectra**, in physics, the pattern of frequencies or wavelengths obtained when electromagnetic radiations are separated into their constituent parts. Visible light is part of the †electromagnetic spectrum and most sources emit waves over a range of wavelengths that can be broken up or "dispersed"; white light can be separated (for example, using a triangular prism) into red, orange, yellow, green, blue, indigo, and violet. The visible spectrum was first studied by English physicist Isaac †Newton, who showed in 1666 how white light could be broken up into different colors.

speculum alloy of mainly copper and tin formerly used to make the mirrors of reflecting astronomical telescopes. It is still used to make diffraction gratings.

speech recognition or **voice input**, in computing, any technique by which a computer can understand ordinary speech. Spoken words are divided into "frames," each lasting about one-thirtieth of a second, which are converted to a wave form. These are then compared with a series of stored frames to determine the most likely word. Research into speech recognition started in 1938, but the technology did not become sufficiently developed for commercial applications until the late 1980s.

There are three types: **separate word recognition** for distinguishing up to several hundred separately spoken words; **connected speech recognition** for speech in which there is a short pause between words; and **continuous speech recognition** for normal but carefully articulated speech.

Speech Recognition http://www.tifaq.com/speech.html

List of speech recognition materials on the Internet. There are links to articles, speech recognition in the news, user groups, and mailing lists and information on dictation software for Windows, Apple, and Unix systems.

speed rate at which an object moves, or how fast an object moves. The average speed v of an object may be calculated by dividing the distance s it has traveled by the time t taken to do so, and may be expressed as:

$$v = s/t$$

The usual units of speed are meters per second or kilometers per hour.

Speed is a †scalar quantity in which direction of motion is unimportant (unlike the vector quantity †velocity, in which both magnitude and direction must be taken into consideration). Movement can be described by using motion graphs. Plotting distance against time in a †distance–time graph allows the total distance covered to be worked out. See also †speed–time graph.

speed of light speed at which light and other †electromagnetic waves travel in a vacuum. Its value is 299,792,458 m per second/186,282 mi per second but for most calculations 3×10^8 m s^{-1} (300 million meters per second) suffices. In glass the speed of light is two-thirds of its speed in air, about 200 million meters per second. The speed of light is the highest speed possible, according to the theory of †relativity, and its value is independent of the motion of its source and of the observer. It is impossible to accelerate any material body to this speed because it would require an infinite amount of energy.

speed of reaction alternative term for †rate of reaction.

speed of sound speed at which sound travels through a medium, such as air or water. In air at a temperature of 0°C/32°F, the speed of sound is 331 m/1,087 ft per second. At higher temperatures, the speed of sound is greater; at 18°C/64°F it is 342 m/1,123 ft per second. It is also affected by the humidity of the air. It is greater in liquids and solids; for example, in water it is about 1,440 m/4,724 ft per second, depending on the temperature.

speed–time graph graph used to describe the motion of a body by illustrating how its speed or velocity changes with time. The gradient of the graph gives the object's †acceleration. If the gradient is zero (the graph is horizontal) then the body is moving with constant speed or uniform velocity (the acceleration is zero); if the gradient is constant, the body is moving with uniform acceleration. If the gradient is positive then the body is accelerating; if the gradient is negative then the body is decelerating. The area under the graph gives the total distance traveled by the body.

speleology scientific study of caves, their origin, development, physical structure, flora, fauna, folklore, exploration, mapping, photography, cave-diving, and rescue work. **Potholing**, which involves following the course of underground rivers or streams, has become a popular sport.

Speleology first developed in France in the late 19th century, where the Société de Spéléologie was founded in 1895.

Spemann, Hans (1869–1941) German embryologist who was awarded a Nobel Prize for Physiology or Medicine in 1935 for his work on the organizer effect in embryonic development. He discovered the phenomenon of embryonic induction – the influence exerted by various regions of an embryo that controls the development of cells into specific organs and tissues.

Spencer Jones, Harold (1890–1960) English astronomer who made a determination of solar parallax, using observations of the asteroid Eros. He also studied the speed of rotation of the Earth, and the motions of the Sun, Moon, and planets. He was knighted in 1943.

sperm or **spermatozoon**, in biology, the male †gamete of animals before fertilization in †sexual reproduction. Each sperm cell has a head capsule containing a nucleus, a middle portion containing †mitochondria (which provide energy), and a long tail (flagellum). In mammals sperm cells are produced in the testes of a male. They are produced by a special kind of cell division called †meiosis, which halves the number of †chromosomes present.

Sperm cells are highly specialized. When they are introduced into the vagina of the female by the penis of the male, they swim up the †uterus and oviducts. The swimming is a result of the beating of the tail of the sperm cell. At the other end is the †nucleus, but there is little else in the cell, making it fairly small and light.

The human sperm is 0.005 mm/0.0002 in long and can survive inside the female for 2–9 days. Sperm counts have fallen by 50% worldwide since 1940, according to a Danish study in 1990. This reduction may be due to increases in pollution – a number of pollutants, including some petroleum by-products and PCBs (†polychlorinated biphenyls), appear to have similar effects to the female hormone estrogen.

spermatophore small capsule containing †sperm and other nutrients produced in invertebrates, newts, and cephalopods.

spermatophyte another name for a †seed plant.

sperm competition competition between the sperm of rival males within the body of a female. This may involve actual removal of a rival's sperm before mating, for example the penis of a male dragonfly is shaped to scoop out existing sperm; or internal competition between rival sperm cells.

In the bug *Xylocoris maculipennis*, sperm competition actually takes place within the male's body. Males inject sperm directly into the females body cavity by piercing her body wall, but males also mate with other males, injecting sperm into the body cavity so that it will migrate to the testes, where it remains until injected into a female.

In 1999 English biologists discovered "mating by proxy," a previously unknown form of sperm competition (in flour beetles *Trilobium castaneum*). When a male flour beetle mates with a female that has been previously mated, he uses his spiny genitals to extract any sperm already there, but the sperm can still compete with his own by remaining viable so that when he mates next he introduces the rival sperm as well as his own. Female flour beetles can therefore be fertilized by males they have never met.

sphalerite mineral composed of zinc sulfide with a small proportion of iron, formula (Zn,Fe)S. It is the chief ore of zinc. Sphalerite is brown with a non-metallic luster unless an appreciable amount of iron is present (up to 26% by weight). Sphalerite usually occurs in ore veins in limestones, where it is often associated with galena. It crystallizes in the cubic system but does not normally form perfect cubes.

sphenoid bone wedge-shaped †bone lying across the base of the skull near the middle and taking part in the formation of the cavity of the †cranium, the orbits (eye sockets), and the posterior nares.

sphere perfectly round object with all points on its surface the same distance from the center. This distance is the radius of the sphere. For a sphere of radius r, the volume $V = {}^4/_3\pi r^3$ and the surface area $A = 4\pi r^2$.

spherical aberration optical error occurring when a lens or curved mirror does not bring all the incident rays of light to a sharp focus. See also †aberration.

spherical collapse initial stage in the collapse of a star, followed by gravitational collapse and finally †singularity.

spherical geometry system of non-Euclidean geometry devised by German mathematician Bernhard Riemann as an extension of †elliptic geometry and comprising two-dimensional geometry as effected on the outer surface of a sphere.

sphincter ring of muscle, such as is found at various points in the †alimentary canal, that contracts and relaxes to open and close the canal and control the movement of food. The **pyloric sphincter**, at the base of the stomach, controls the release of the gastric contents into the †duodenum. After release the sphincter contracts, closing off the stomach. The **external anal sphincter** closes the †anus; the **internal anal sphincter** constricts the rectum; the **sphincter vesicae** controls the urethral orifice of the bladder. In the eye the **sphincter pupillae** contracts the pupil in response to bright light.

Spica or **Alpha Virginis**, brightest star in the constellation Virgo and the 16th-brightest star in the night sky. Spica has a true luminosity of over 1,500 times that of the Sun and is 260 light years from the Sun. It is a †spectroscopic binary star, the components of which orbit each other every four days.

spicules, solar in astronomy, short-lived jets of hot gas in the upper †chromosphere of the Sun. Spiky in appearance, they move at high velocities along lines of magnetic force to which they owe their shape, and last for a few minutes each. Spicules appear to disperse material into the †corona.

spikelet in botany, one of the units of a grass †inflorescence. It comprises a slender axis on which one or more flowers are borne.

Each individual flower or floret has a pair of scalelike bracts, the glumes, and is enclosed by a membranous lemma and a thin, narrow palea, which may be extended into a long, slender bristle, or **awn**.

spin in physics, the intrinsic †angular momentum of a subatomic particle, nucleus, atom, or molecule, which continues to exist even when the particle comes to rest. A particle in a specific energy state has a particular spin, just as it has a particular electric charge and mass. According to †quantum theory, this is restricted to discrete and indivisible values, specified by a spin †quantum number. Because of its spin, a charged particle acts as a small magnet and is affected by magnetic fields.

spinal cord major component of the †central nervous system in vertebrates. It consists of bundles of nerves enveloped in three layers of membrane (the meninges) and is bathed in cerebrospinal fluid. The spinal cord is encased and protected by the vertebral column, lying within the vertebral canal formed by the posterior arches of successive vertebrae.

In humans, the spinal cord is about 45 cm/18 in long, extending from the bottom of the skull, where it is continuous with the medulla oblongata, only to about waist level. It consists of nerve cell bodies (gray matter) and their myelinated processes or nerve fibers (white matter). In cross-section, the gray matter is arranged in an H-shape around the central canal of the spinal cord, and it is surrounded in turn by the white matter.

Paired spinal nerves arise from the cord at each vertebra. Each is a mixed nerve, consisting of both sensory and motor nerve fibers. The sensory fibers enter the spinal cord at a dorsal root and the motor fibers enter at a ventral root. This arrangement enables the spinal cord to relay impulses coming in and out at the same level, to relay impulses going up and down the cord to other levels, and relay impulses to and from the brain. The first of these involves a **reflex arc**, by which a sensory impulse can create a very rapid, involuntary response to a particular stimulus.

spine backbone of vertebrates. In most mammals, it contains 26 small bones called **vertebrae**, which enclose and protect the **spinal cord** (which links the peripheral nervous system to the brain). The spine articulates with the skull, ribs, and hip bones, and provides attachment for the back muscles.

In humans it is made up of individual vertebrae, separated by intervertebral discs. In the adult there are seven cervical vertebrae in the neck; twelve thoracic in the upper trunk; five lumbar in the lower back; the sacrum (consisting of five rudimentary vertebrae fused together, joined to the hipbones); and the coccyx (four vertebrae, fused into a tailbone). The human spine has four curves (front to rear), which allow for the increased size of the chest and pelvic cavities, and for a degree of spring so as to minimize jolting of the internal organs.

spinel any of a group of "mixed oxide" minerals consisting mainly of the oxides of magnesium and aluminum, $MgAl_2O_4$ and $FeAl_2O_4$. Spinels crystallize in the cubic system, forming octahedral crystals. They are found in high-temperature igneous and metamorphic rocks. The aluminum oxide spinel contains gem varieties, such as the ruby spinels of Sri Lanka and Myanmar (Burma).

spin-orbit interaction interaction between †electrons orbiting a †nucleus in an †atom that arises from the magnetic field produced by the nucleus interacting with the spinning electron is to split the individual †spectral lines in a †spectrum into a number of components.

spiracle in insects, the opening of a †trachea, through which oxygen enters the body and carbon dioxide is expelled. In cartilaginous fishes (sharks and rays), the same name is given to a circular opening that marks the remains of the first gill slit.

In tetrapod vertebrates, the spiracle of early fishes has evolved into the Eustachian tube, which connects the middle ear cavity with the pharynx.

spiral plane curve formed by a point winding round a fixed point from which it distances itself at regular intervals, for example the spiral traced by a flat coil of rope. Various kinds of spirals can be generated mathematically – for example, an equiangular or logarithmic spiral (in which a tangent at any point on the curve always makes the same angle with it) and an †involute. Spirals also occur in nature as a normal consequence of accelerating growth, such as the spiral shape of the shells of snails and some other mollusks.

spiral galaxy one of the main classes of †galaxy in the Hubble classification comprising up to 30% of known galaxies. Spiral galaxies are characterized by a central bulge surrounded by a flattened disc containing (normally) two spiral arms composed of hot young stars and clouds of dust and gas. In about half of spiral galaxies (barred spirals) the arms originate at the ends of a bar across the central bulge. The bar is not a rigid object but consists of stars in motion about the center of the galaxy.

spiral nebula spiral †galaxy – not really a †nebula at all (although many do appear nebulous).

spirillium bacterium with a spiral shape (see †bacteria).

spirits of salts former name for †hydrochloric acid.

spirochete spiral-shaped bacterium. Some spirochetes are free-living in water, others inhabit the intestines and genital areas of animals. The sexually transmitted disease syphilis is caused by a spirochete.

spit ridge of sand or shingle projecting from the land into a body of water. It is formed by a combination of †longshore drift, tides, river currents, and/or a bend in the coastline. The decrease in wave energy causes more material to be deposited than is transported down the coast, building up a finger of sand that points in the direction of the longshore drift. Deposition in the brackish water behind a spit may result in the formation of a †salt marsh.

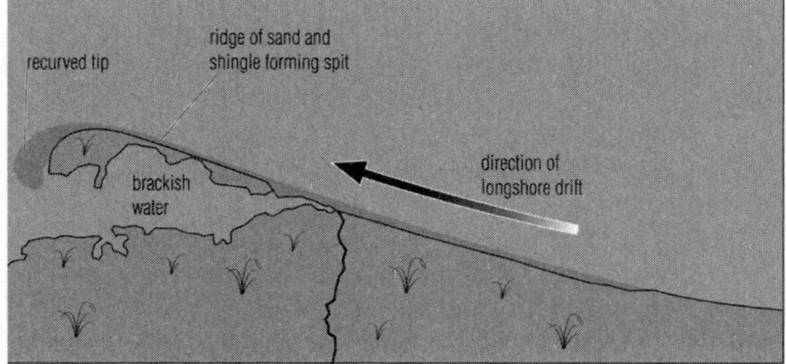

spit Longshore drift carries sand and shingle up coastlines. Deposited material gradually builds up over time at headlands forming a new stretch of land called a spit. A spit that extends across a bay is known as a bar. Helicon

Spitzer, Lyman (1914–1997) US astrophysicist whose contributions included work on the interstellar medium, which concerns the gas and dust between stars from which new stars form. Early on, Spitzer suggested that the brightest stars in spiral galaxies have formed recently from the gas and dust there. He also noted the presence and importance of interstellar magnetic fields, the likelihood of a multiphase medium – with hot, warm and cool components – and the significance of dust grains.

spleen organ in vertebrates, part of the reticuloendothelial system, which helps to process †lymphocytes. It also regulates the number of red blood cells in circulation by destroying old cells, and stores iron. It is situated on the left side of the body, behind the stomach.

sponge any saclike simple invertebrate of the phylum Porifera, usually marine. A sponge has a hollow body, its cavity lined by cells bearing flagella, whose whiplike movements keep water circulating, bringing in a stream of food particles. The body walls are strengthened with protein (as in the bath sponge) or small spikes of silica, or a framework of calcium carbonate.

Fossil sponges only 1 mm/0.04 in in length were found in China and described in 1998. They are estimated to be 580 million years old, the oldest discovered to date.

spontaneous combustion burning that is not initiated by the direct application of an external source of heat. A number of materials and chemicals, such as hay and sodium chlorate, can react with their surroundings, usually by oxidation, to produce so much internal heat that combustion results.

Special precautions must be taken for the storage and handling of substances that react violently with moisture or air. For example, phosphorus ignites spontaneously in the presence of air and must therefore be stored under water; sodium and potassium are stored under kerosene in order to prevent their being exposed to moisture.

spontaneous generation or **abiogenesis**, erroneous belief that living organisms can arise spontaneously from nonliving matter. This survived until the mid-19th century, when the French chemist Louis Pasteur demonstrated that a nutrient broth would not generate micro-organisms if it was adequately sterilized. The theory of †biogenesis holds that spontaneous generation cannot now occur; it is thought, however, to have played an essential role in the origin of †life on this planet 4 billion years ago.

spooling in computing, the process in which information to be printed is stored temporarily in a file, the printing being carried out later. It is used to prevent a relatively slow printer from holding up the system at critical times, and to enable several computers or programs to share one printer.

sporangium structure in which †spores are produced.

spore small reproductive or resting body, usually consisting of just one cell. Unlike a †gamete, it does not need to fuse with another cell in order to develop into a new organism. Spores are produced by the lower plants, most fungi, some bacteria, and certain protozoa. They are generally light and easily dispersed by wind movements.

Plant spores are haploid and are produced by the sporophyte, following †meiosis; see †alternation of generations.

sporophyte diploid spore-producing generation in the life cycle of a plant that undergoes †alternation of generations.

sporozoite protozoon of the class Sporozoa, such as the malarial parasite Plasmodium.

spreadsheet in computing, a program that mimics a sheet of ruled paper, divided into columns down the page, and rows across. The user enters values into cells within the worksheet, then instructs the program to perform some operation on them, such as totaling a column or finding the average of a series of numbers. Calculations are made by using a formula. Highly complex numerical analyses may be built up from these simple steps.

Columns and rows in a spreadsheet are labeled; although different programs use different methods, columns are often labeled with alpha characters, and rows with numbers. When a cell containing a formula is copied and pasted within a spreadsheet, the formula is said to be **relative**, meaning the cell references from which it takes its values are relative to its new position. An **absolute** reference does not change.

The pages of a spreadsheet can be formatted to make them easier to read; the height of rows, the width of columns, and the typeface of the text may all be changed. Number formats may also be changed to display, for example, fractions as decimals or numbers as integers. Spreadsheets are widely used in business for forecasting and financial control. The first spreadsheet program, Software Arts' VisiCalc, appeared in 1979. The best known include Lotus 1-2-3 and Microsoft Excel.

Spreadsheets http://mathforum.org/sum95/math_and/spreadsheets/basics.html

Outlines how to input data into a spreadsheet in easy steps. There is also information on how to make charts from spreadsheet data.

spring in geology, a natural flow of water from the ground, formed at the point of intersection of the water table and the ground's surface. The source of water is rain that has percolated through the overlying rocks. During its underground passage, the water may have dissolved mineral substances that may then be precipitated at the spring (hence, a mineral spring).

A spring may be continuous or intermittent, and depends on the position of the water table and the topography (surface features).

spring balance instrument for measuring weight that relates the weight of an object to the extent to which it stretches or compresses a vertical spring. According to †Hooke's law, the extension or compression will be directly proportional to the weight, providing that the spring is not overstretched. A pointer attached to the spring indicates the weight on a scale, which may be calibrated in newtons (the SI unit of force) for physics experiments, or in grams, kilograms, or pounds (units of mass) for everyday use.

spring line geological feature where water springs up in several places along the edge of a permeable rock escarpment. Springline settlements may become established around this.

sprite in computing, a graphics object made up of a pattern of †pixels (picture elements) defined by a computer programmer. Some †high-level languages and †applications programs contain routines that allow a user to define the shape, colors, and other characteristics of individual graphics objects. These objects can then be manipulated and combined to produce animated games or graphic screen displays.

sprite in earth science, rare thunderstorm-related luminous flash. Sprites occur in the mesosphere, at altitudes of 50–90 km/30–55 mi. They are electrical, like lightning, and arise when the electrical field that occurs between the thunder cloud top and the ionosphere (ionized layer of the Earth's atmosphere) draws electrons upwards from the cloud. If the air is thin and this field is strong the electrons accelerate rapidly, transferring kinetic energy to molecules they collide with. The excited molecules then discharge this energy as a light flash.

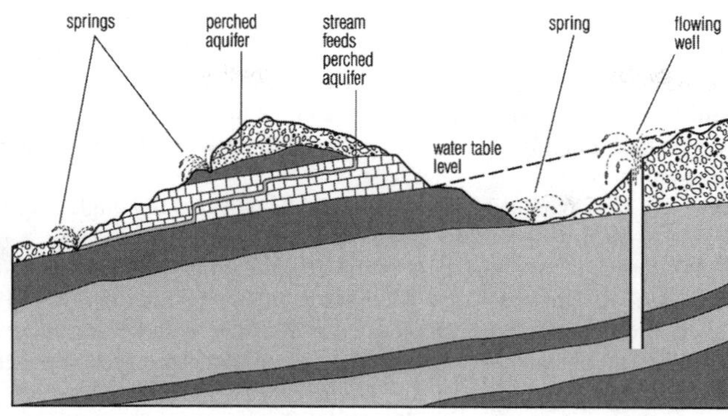

spring Springs occur where water-laden rock layers (aquifers) reach the surface. Water will flow from a well whose head is below the water table. Helicon

spur ridge of rock jutting out into a valley or plain. In mountainous areas rivers often flow around †interlocking spurs because they are not powerful enough to erode through the spurs. Spurs may be eroded away by large and powerful glaciers to form †truncated spurs.

sq abbreviation for **square** (measure).

SQL abbreviation for structured query language, high-level computer language designed for use with †relational databases. Although it can be used by programmers in the same way as other languages, it is often used as a means for programs to communicate with each other. Typically, one program (called the "client") uses SQL to request data from a database "server."

Although originally developed by IBM, SQL is now widely used on many types of computer.

square in geometry, a quadrilateral (four-sided) plane figure with all sides equal and each angle a right angle. Its diagonals bisect each other at right angles. The area A of a square is the length l of one side multiplied by itself ($A = l \times l$).

Also, any quantity multiplied by itself is termed a square, represented by an †exponent of power 2; for example, $4 \times 4 = 4^2 = 16$ and $6.8 \times 6.8 = 6.8^2 = 46.24$.

An algebraic term is squared by doubling its exponent and squaring its coefficient if it has one; for example, $(x^2)^2 = x^4$ and $(6y^3)^2 = 36y^6$. A number that has a whole number as its †square root is known as a **perfect square**; for example, 25, 144, and 54,756 are perfect squares (with roots of 5, 12, and 234, respectively).

spur Interlocking spurs (above) and truncated spurs (below). Interlocking spurs are formed by meandering rivers; truncated spurs are interlocking spurs eroded by glaciation. Helicon

square root number that when it is †squared (multiplied by itself) equals a given number. For example, the square root of 25 (written as $\sqrt{25}$) is +5 or –5. This is because $+5 \times +5 = 25$, and $-5 \times -5 = 25$. A square root can be written as $^1/_2$. For example, $16^{1/2} = +4$ or $16^{1/2} = -4$. This is called an †exponent.

Negative numbers (less than 0) do not have square roots that are †real numbers. Their roots are represented by †complex numbers, in which the square root of –1 is given the symbol i (that is, $\pm i^2 = -1$). All square roots generate two answers and complex square roots occur in **conjugate pairs**, that is, $\sqrt{(a + ib)} = c \pm id$. Thus the square root of –4 is $\sqrt{[(-1) \times 4]} = \sqrt{-1} \times \sqrt{4} = \pm 2i$.

squaring the circle ancient Greek problem in geometrical construction, to describe a square of exactly the same area as a given circle, using ruler and compass only. German mathematician Ferdinand von Lindemann established that p was a †transcendental number; unable thus to be the root of an equation, it cannot be constructed by ruler and compass – and the problem is therefore not solvable.

squid acronym for **s**uperconducting **qu**antum **i**nterference **d**evice that makes use of the †Josephson effect to produce a highly sensitive magnetic field detection system, or an ultra-fast switching device for use in computers.

sr in physics, symbol for †steradian, the SI unit of solid angle. For example, a sphere subtends a solid angle of 4π steradians at its center.

SRAM acronym for **static random-access memory**, computer memory device in the form of a silicon chip used to provide †immediate access memory. SRAM is faster but more expensive than DRAM (dynamic random-access memory), and does not require such frequent refreshing.

stability measure of how difficult it is to move an object from a position of balance or †equilibrium with respect to gravity.

An object displaced from equilibrium does not remain in its new position if its weight, acting vertically downwards through its †center of mass, no longer passes through the line of action of the †contact force (the force exerted by the surface on which the object is resting), acting vertically upwards through the object's new base. If the lines of action of these two opposite but equal forces do not coincide they will form a couple and create a moment (see †moment of a force) that will cause the object either to return to its original rest position or to topple over into another position.

An object in **stable equilibrium** returns to its rest position after being displaced slightly. This form of equilibrium is found in objects that are difficult to topple over; these usually possess a relatively wide base and a low center of mass – for example, a cone resting on its flat base on a horizontal surface. When such an object is tilted slightly its center of mass is raised and the line of action of its weight no longer coincides with that of the contact force exerted by its new, smaller base area. The moment created will tend to lower the center of mass and so the cone will fall back to its original position.

An object in **unstable equilibrium** does not remain at rest if displaced, but falls into a new position; it does not return to its original rest position. Objects possessing this form of equilibrium are easily toppled and usually have a relatively small base and a high center of mass – for example, a cone balancing on its point, or apex, on a horizontal surface. When an object such as this is given the slightest push its center of mass is lowered and the displacement of the line of action of its weight creates a moment. The moment will tend to lower the center of mass still further and so the object will fall on to another position.

An object in **neutral equilibrium** stays at rest if it is moved into a new position – neither moving back to its original position nor on any further. This form of equilibrium is found in objects that are able to roll, such as a cone resting on its curved side placed on a horizontal surface. When such an object is rolled its center of mass remains in the same position, neither rising nor falling, and the line of action of its weight continues to coincide with the contact force; no moment is created and so its equilibrium is maintained.

stable equilibrium state of equilibrium possessed by a body that will return to its original rest position if displaced slightly. See †stability.

stack in computing, a method of storing data in which the most recent item stored will be the first to be retrieved. The technique is commonly called "last in, first out."

Stacks are used to solve problems involving nested structures; for example, to analyze an arithmetical expression containing subexpressions in parentheses, or to work out a route between two points when there are many different paths.

stack isolated pillar of rock that has become separated from a headland by †coastal erosion. It is usually formed by the collapse of an †arch. Further erosion will reduce it to a stump, which is exposed only at low tide.

Stahl, Franklin William (1929–) US molecular biologist who, with Matthew †Meselson, confirmed that replication of the genetic material †DNA is semiconservative (that is, the daughter cells each receive one strand of DNA from the original parent cell and one newly replicated strand).

Stahl, Georg Ernst (1660–1734) German chemist who developed the theory that objects burn because they contain a combustible substance, †phlogiston. Substances rich in phlogiston, such as wood, burn almost completely away. Metals, which are low in phlogiston, burn less well. Chemists spent much of the 18th century evaluating Stahl's theories before these were finally proved false by Antoine †Lavoisier.

stain colored compound that will bind to other substances. Stains are used extensively in microbiology to color micro-organisms and in histochemistry to detect the presence and whereabouts in plant and animal tissue of substances such as fats, cellulose, and proteins.

stalactite and stalagmite cave structures formed by the deposition of calcite dissolved in ground water. **Stalactites** grow downwards from the roofs or walls and can be icicle-shaped, straw-shaped, curtain-shaped, or formed as terraces. **Stalagmites** grow upwards from the cave floor and can be conical, fir-cone-shaped, or resemble a stack of saucers. Growing stalactites and stalagmites may meet to form a continuous column from floor to ceiling.

Stalactites are formed when ground water, hanging as a drip, loses a proportion of its carbon dioxide into the air of the cave. This reduces the amount of calcite that can be held in solution, and a small trace of calcite is deposited. Successive drips build up the stalactite over many years. In stalagmite formation the calcite comes out of the solution because of agitation – the shock of a drop of water hitting the floor is sufficient to remove some calcite from the drop. The different shapes result from the splashing of the falling water.

stamen male reproductive organ of a flower. The stamens are collectively referred to as the †androecium. A typical stamen consists of a stalk, or filament, with an anther, the pollen-bearing organ, at its apex, but in some primitive plants, such as *Magnolia*, the stamen may not be markedly differentiated.

The number and position of the stamens are significant in the classification of flowering plants. Generally the more advanced plant families have fewer stamens, but they are often positioned more effectively so that the likelihood of successful pollination is not reduced.

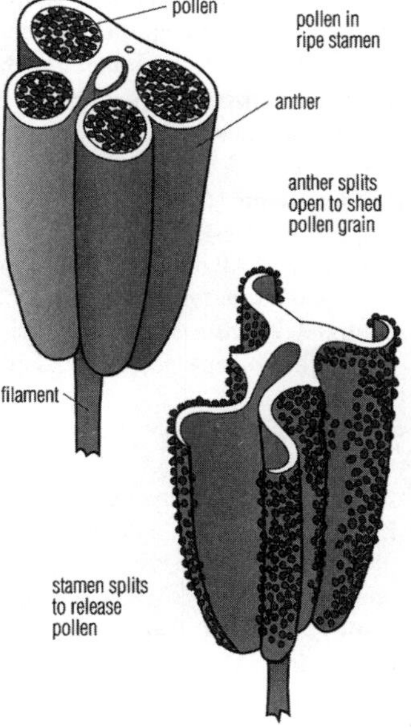

stamen The stamen is the male reproductive organ of a flower. It has a thin stalk called a filament with an anther at the tip. The anther contains pollen sacs, which split to release tiny grains of pollen. Helicon

standard measure against which others are compared. For example, until 1960 the standard for the meter was the distance between two lines engraved on a platinum–iridium bar kept at the International Bureau of Weights and Measures in Sèvres, France.

standard atmosphere alternative term for †atmosphere, a unit of pressure.

standard cell in physics, voltaic cell whose emf (†electromotive force) is always the same and can be used as a standard voltage. Examples include the Clark cell and the Weston cadmium cell.

standard deviation in statistics, a measure (symbol s or *s*) of the spread of data. The deviation (difference) of each of the data items from the mean is found, and their values squared. The mean value of these squares is then calculated. The standard deviation is the square root of this mean.

If *n* is the number of items of data, *x* is the value of each item, and m is the mean value, the standard deviation s may be given by the formula

$$\sigma = \sqrt{\frac{\sum (x-\mu)^2}{n}}$$

where S indicates that the differences between the value of each item of data and the mean should be summed.

To simplify the calculations, the formula may be rearranged to

$$\sigma = \sqrt{\frac{\sum x^2}{n} - \mu^2}$$

As a result, it becomes necessary only to calculate Sx and Sx^2.

For example, if the ages of a set of children were 4, 4.5, 5, 5.5, 6, 7, 9, and 11, Sx would be 52, m would be $52/n = 52/8 = 6.5$, and Sx^2 would be 378.5 ($= 4^2 + 4.5^2 + 5^2 + 5.5^2 + 6^2 + 7^2 + 9^2 + 11^2$). Therefore, the standard deviation s would be

$$\sqrt{\frac{378.5}{8} - (6.5)^2} = \sqrt{5.0625} = 2.25$$

standard form or **scientific notation**, method of writing numbers often used by scientists, particularly for very large or very small numbers. The numbers are written with one digit before the decimal point and multiplied by a power of 10. The number of digits given after the decimal point depends on the accuracy required. For example, the †speed of light is 2.9979×10^8 m/1.8628×10^5 mi per second.

standard gravity acceleration due to gravity, generally taken as 9.81274 m/32.38204 ft per second per second. See also †*g* scale.

standard illuminant any of three standard light intensities, A, B, and C, used for illumination when phenomena involving color are measured. A is the light from a filament at 2,848 K (2,575°C/4,667°F), B is noon sunlight, and C is normal daylight. B and C are defined with respect to A. Standardization is necessary because colors appear different when viewed in different lights.

standard model in physics, modern theory of †elementary particles and their interactions. According to the standard model, elementary particles are classified as leptons (light particles, such as electrons), †hadrons (particles, such as neutrons and protons, that are formed from quarks), and gauge bosons. Leptons and hadrons interact by exchanging †gauge bosons, each of which is responsible for a different fundamental force: photons mediate the electromagnetic force, which affects all charged particles; gluons mediate the strong nuclear force, which affects quarks; gravitons mediate the force of gravity; and the intermediate vector bosons mediate the weak nuclear force. See also †forces, fundamental, †quantum electrodynamics, and †quantum chromodynamics.

standard temperature and pressure STP, in chemistry, a standard set of conditions for experimental measurements, to enable comparisons to be made between sets of results. Standard temperature is 0°C/32°F (273 K) and standard pressure 1 atmosphere (101,325 Pa).

standard time time in any of the 24 time zones, each an hour apart, into which the Earth is divided. The respective times depend on their distances, east or west of Greenwich, England. In North America the eight zones (Atlantic, Eastern, Central, Mountain, Pacific, Alaska, Hawaii-Aleutian, and Samoa) use the mean solar times of meridians 15° apart, starting with 60° longitude. (See also †time.)

standard volume in physics, the volume occupied by one kilogram molecule (the molecular mass in kilograms) of any gas at standard temperature and pressure. Its value is approximately 22.414 cubic meters.

standing wave in physics, a wave in which the positions of †nodes (positions of zero vibration) and antinodes (positions of maximum vibration) do not move. Standing waves result when two similar waves travel in opposite directions through the same space.

For example, when a sound wave is reflected back along its own path, as when a stretched string is plucked, a standing wave is formed. In this case the antinode remains fixed at the center and the nodes are at the two ends. Water and †electromagnetic waves can form standing waves in the same way.

Stanley, Wendell Meredith (1904–1971) US biochemist who crystallized the tobacco mosaic virus (TMV) in 1935. He demonstrated that, despite its crystalline state, TMV remained infectious. Together with John †Northrop and James †Sumner, Stanley was awarded the Nobel Prize for Chemistry in 1946 for the preparation of pure enzymes and virus proteins.

staphylococcus spherical bacterium that occurs in clusters. It is found on the skin and mucous membranes of humans and other animals. It can cause abscesses and systemic infections that may prove fatal. *Staphylococcus aureus* is a very common bacterium, present in the nose in 30% of people. Normally it gives no trouble, but, largely due to over-prescribing of antibiotics, strains have arisen that are resistant to the drugs used to treat them, principally methicillin, a semisynthetic form of penicillin. Methicillin-resistant *S. aureus* (MRSA) strains represent a serious hazard to the critically ill or immunosuppressed. MRSA normally responds to two antibiotics which are considered too toxic for use in any but life-threatening infections (vancomycin and teicoplanin) but it still causes fatalities.

star luminous globe of gas, mainly hydrogen and helium, which produces its own heat and light by nuclear reactions. Although stars shine for a very long time – many billions of years – they change in appearance at different stages in their lives (they are said to have a "life cycle"). Stars seen at night belong to our †galaxy, the Milky Way. The Sun is the nearest star to Earth; other stars in the Milky Way are large distances away (to get to the nearest would take about 4 years traveling at the speed of light).

The smallest mass possible for a star is about 8% that of the Sun (80 times that of †Jupiter), otherwise nuclear reactions do not occur. Objects with less than this critical mass shine only dimly, and are termed **brown dwarfs**.

Stars http://www.hotliquidmagma.com/space/html/star.html

This mainly text-based page offers concise yet useful information about those objects that are only visible at night: stars! The types of stars covered include red and white dwarfs, and red and super giants.

starburst galaxy in astronomy, spiral galaxy that appears unusually bright in the infrared part of the spectrum due to a recent burst of star formation, possibly triggered by the gravitational influence of a nearby companion galaxy.

starch widely distributed, high-molecular-mass †carbohydrate, produced by plants as a food store; main dietary sources are cereals, legumes, and tubers, including potatoes. It consists of varying proportions of

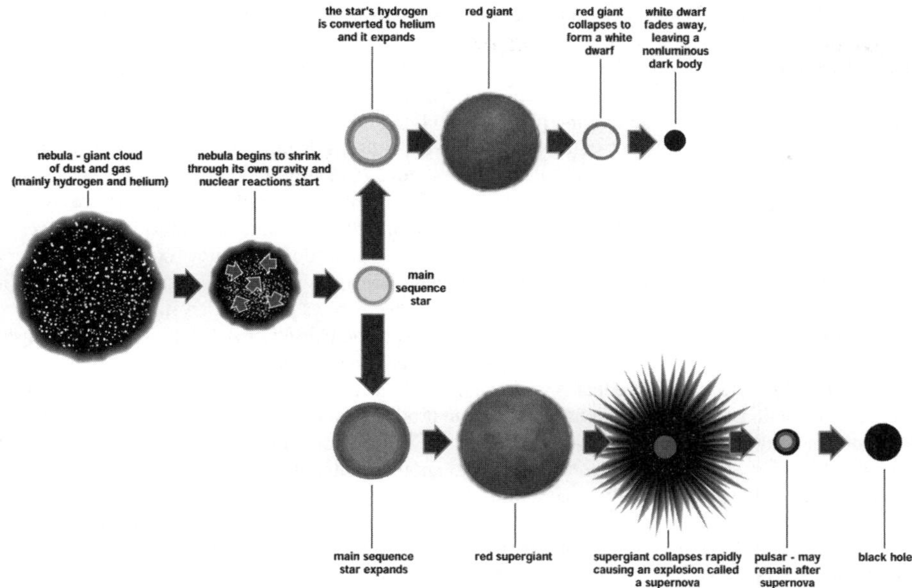

star The life cycle of a star. New stars are being formed all the time when nebulae (giant clouds of dust and gas) contract due to the action of gravity. As the star contracts and heats up eventually nuclear reactions begin and the star becomes a main sequence star. If the star is less than 1.2 times the mass of the Sun, it eventually forms a white dwarf that finally fades to a dark body. If it is a massive star, then the main sequence star expands to become a red supergiant that eventually explodes as a supernova. It leaves part of the core as a neutron star (pulsar), or as a black hole if the mass of the collapsing supernova core is three times greater than the Sun. Helicon

two †glucose polymers (†polysaccharides): straight-chain (amylose) and branched (amylopectin) molecules.

Purified starch is a white powder used to stiffen textiles and paper and as a raw material for making various chemicals. It is used in the food industry as a thickening agent. Chemical treatment of starch gives rise to a range of "modified starches" with varying properties. Hydrolysis (splitting) of starch by acid or enzymes generates a variety of "glucose syrups" or "liquid glucose" for use in the food industry. Complete hydrolysis of starch with acid generates the †monosaccharide glucose only. Incomplete hydrolysis or enzymic hydrolysis yields a mixture of glucose, maltose, and nonhydrolyzed fractions called dextrins.

The chemical test for starch consists of adding a drop of iodine solution to the substance. It will turn bright blue if starch is present.

star cluster group of related stars, usually held together by gravity. Members of a star cluster are thought to form together from one large cloud of gas in space. **Open clusters** such as the †Pleiades contain from a dozen to many hundreds of young stars, loosely scattered over several light years. †Globular clusters are larger and much more densely packed, containing perhaps 10,000–1,000,000 stars.

The more conspicuous clusters were originally catalogued with the nebulae, and are usually known by their Messier or New General Catalogue (NGC) numbers. A few clusters like the Pleiades, †Hyades, and †Praesepe are also known by their traditional names.

Stark, Johannes (1874–1957) German physicist. He was awarded the Nobel Prize for Physics in 1919 for his discovery in 1902 that high-velocity rays of positive ions (canal rays) would demonstrate the †Doppler effect, and his demonstration in 1913 that a strong electric field can alter the wavelength of light emitted by atoms (the **Stark effect**).

Stark effect splitting of †spectral lines into a number of components by a strong electric field.

starlight energy (seen as light) produced by a star through †nuclear fusion.

Starling, Ernest Henry (1866–1927) English physiologist who, with William †Bayliss, discovered †secretin and in 1905 coined the word "hormone." He formulated **Starling's law**, which states that the force of the heart's contraction is a function of the length of the muscle fibers. He is considered one of the founders of endocrinology.

startup screen in computing, screen displayed by a PC while it loads its †operating system and other resident software. It is also possible to create a custom start-up screen, perhaps incorporating a favorite image or corporate logo.

states of matter forms (solid, liquid, or gas) in which material can exist. Whether a material is solid, liquid, or gaseous depends on its temperature and pressure. The transition between states takes place at definite temperatures, called the melting point and boiling point.

†Kinetic theory describes how the state of a material depends on the movement and arrangement of its atoms or molecules. The atoms or molecules of **gases** move randomly in otherwise empty space, filling any size or shape of container. Gases can be liquefied by cooling as this lowers the speed of the molecules and enables attractive forces between them to bind them together. A **liquid** forms a level surface and assumes the shape of its container; its atoms or molecules do not occupy fixed positions, nor do they have total freedom of movement. **Solids** hold their own shape as their atoms or molecules are not free to move about but merely vibrate about fixed positions, such as those in †crystal lattices.

state symbol symbol used in chemical equations to indicate the physical state of the substances present. The symbols are: (s) for solid, (l) for liquid, (g) for gas, and (aq) for aqueous.

static electricity †electric charge that is stationary, usually acquired by a body by means of electrostatic induction or friction. Rubbing different materials can produce static electricity, as seen in the sparks produced on combing one's hair or removing a nylon shirt. The frictional force causes electrons to move out of their orbits. The electrons are then transferred to another material. The material that gains electrons becomes negatively charged and the material that loses electrons becomes positively charged. In some processes static electricity is useful, as in paint spraying where the parts to be sprayed are charged with electricity of opposite polarity to that on the paint droplets, and in xerography.

Static Electricity Page http://www.amasci.com/emotor/statelec.html

Simple Web pages with tutorials and information on static electricity. There are also pages detailing science projects and a good list of electronics and electrostatic resources on the Internet.

statics branch of mechanics concerned with the behavior of bodies at rest and forces in equilibrium, and distinguished from †dynamics.

statistical mechanics branch of physics in which the properties of large collections of particles are predicted by considering the motions of the constituent particles. It is closely related to †thermodynamics.

statistics branch of mathematics concerned with the collection and interpretation of data. For example, to determine the †mean age of the children in a school, a statistically acceptable answer might be obtained by calculating an average based on the ages of a representative sample, consisting, for example, of a random tenth of the pupils from each class. †Probability is the branch of statistics dealing with predictions of events.

mean, median, and mode The mean, median, and mode are different ways of finding a "typical" or "central" value of a set of data. The †mean is obtained by adding up all the observed values and dividing by the number of values; it is the number that is commonly used as an average value. The †median is the middle value, that is, the value that is exceeded by half the items in the sample. The †mode is the value that occurs with greatest frequency, the most common value. The mean is the most useful measure for the purposes of statistical theory. The idea of the median may be extended and a distribution can be divided into four quartiles. The first quartile is the value that is exceeded by three-quarters of the items; the second quartile is the same as the median; the third quartile is the value that is exceeded by one-quarter of the items.

standard deviation and other measures of dispersion The mean is a very incomplete summary of a group of observations; it is useful to know also how closely the individual members of a group approach the mean, and this is indicated by various measures of dispersion. The **range** is the difference between the maximum and minimum values of the group; it is not very satisfactory as a measure of dispersion. The **mean deviation** is the arithmetic mean of the differences between the mean and the individual values, the differences all being taken as positive. However, the mean deviation also does not convey much useful information about a group of observations. The most useful measure of dispersion is the **variance**, which is the arithmetic mean of the squares of the deviations from the mean. The positive square root of the variance is called the †standard deviation, a measure (symbol s or s) of the spread of data. The deviation (difference) of each of the data items from the mean is found, and their values squared. The mean value of these squares is then calculated. The standard deviation is the square root of this mean.

It is usual to standardize the measurements by working in units of the standard deviation measured from the mean of the distributions, enabling statistical theories to be generalized. A standardized distribution has a mean of zero and a standard deviation of unity. Another useful measure of dispersion is the semi-interquartile range, which is one-half of the distance between the first and third quartiles, and can be considered as the average distance of the quartiles from the median. In many typical distributions the semi-interquartile range is about two-thirds of the standard deviation and the mean deviation is about four-fifths of the standard deviation.

applications One of the most important uses of statistical theory is in testing whether experimental data support hypotheses or not. For example, an agricultural researcher arranges for different groups of cows to be fed different diets and records the milk yields. The milk-yield data are analyzed and the means and standard deviations of yields for different groups vary. The researcher can use statistical tests to assess whether the variation is of an amount that should be expected because of the natural variation in cows or whether it is larger than normal and therefore likely to be influenced by the difference in diet.

correlation Correlation measures the degree to which two quantities are associated, in the sense that a variation in one quantity is accompanied by a predictable variation in the other. For example, if the pressure on a quantity of gas is increased then its volume decreases. If observations of pressure and volume are taken then statistical correlation analysis can be used to determine whether the volume of a gas can be completely predicted from a knowledge of the pressure on it.

Staudinger, Hermann (1881–1965) German organic chemist who was awarded the Nobel Prize for Chemistry in 1953 for his discoveries in macromolecular chemistry, of which he was the founder. He carried out pioneering research into the structure of albumen and cellulose.

staurolite silicate mineral, $(Fe,Mg)_2(Al,Fe)_9Si_4O_{20}(OH)_2$. It forms brown crystals that may be twinned in the form of a cross. It is a useful indicator of medium grade (moderate temperature and pressure) metamorphism in metamorphic rocks formed from clay sediments.

steady-state theory in astronomy, rival theory to that of the †Big Bang, which claims that the universe has no origin but is expanding because new matter is being created continuously throughout the universe. The theory was proposed in 1948 by Austrian-born British cosmologist Hermann †Bondi, Austrian-born US astronomer Thomas Gold, and English astronomer, cosmologist, and writer Fred †Hoyle, but it was dealt a severe blow in 1965 by the discovery of †cosmic background radiation (radiation left over from the Big Bang and the formation of the universe) and is now largely rejected.

steam dry, invisible gas formed by vaporizing water. The visible cloud that normally forms in the air when water is vaporized is due to minute suspended water particles. Steam is widely used in chemical and other industrial processes and for the generation of power.

steam engine engine that uses the power of steam to produce useful work. The first successful steam engine was built in 1712 by English inventor Thomas Newcomen at Dudley, West Midlands; it was developed further by Scottish instrument maker James †Watt from 1769 and by English mining engineer Richard Trevithick, whose high-pressure steam engine of 1802 led to the development of the steam locomotive.

In Newcomen's engine, steam was admitted to a cylinder as a piston moved up, and was then condensed by a spray of water, allowing air pressure to force the piston downwards. James Watt improved Newcomen's engine in 1769 by condensing the steam outside the cylinder (thus saving energy formerly used to reheat the cylinder) and by using steam to move the piston. Watt also introduced the **double-acting engine**, in which steam is alternately sent to each side of the piston forcing it up and down. The **compound engine** (1781) uses the exhaust from one cylinder to drive the piston of another. A later development was the steam †turbine, still used today to power ships and generators in power stations. In other contexts, the steam engine was superseded by the †internal-combustion engine or the †electric motor.

stearic acid $CH_3(CH_2)_{16}COOH$, saturated long-chain †fatty acid, soluble in alcohol and ether but not in water. It is found in many fats and oils, and is used to make soap and candles and as a lubricant. The salts of stearic acid are called stearates.

stearin mixture of stearic and palmitic acids, used to make soap.

steel alloy or mixture of iron and up to 1.7% carbon, sometimes with other elements, such as manganese, phosphorus, sulfur, and silicon. The USA, Russia, Ukraine, and Japan are the main steel producers. Steel has innumerable uses, including ship and car manufacture, skyscraper frames, and machinery of all kinds.

Steels with only small amounts of other metals are called **carbon steels**. These steels are far stronger than pure iron, with properties varying with the composition. **Alloy steels** contain greater amounts of other metals. Low-alloy steels have less than 5% of the alloying material; high-alloy steels have more. Low-alloy steels containing up to 5% silicon with relatively little carbon have a high electrical resistance and are used in power transformers and motor or generator cores, for example. **Stainless steel** is a high-alloy steel containing at least 11% chromium. Steels with up to 20% tungsten are very hard and are used in high-speed cutting tools. About 50% of the world's steel is now made from scrap.

Steel is produced by increasing the carbon content of wrought iron, or decreasing that of cast iron, produced by a †blast furnace. The main industrial process is the **basic–oxygen process**, in which molten pig iron and scrap steel is placed in a container lined with heat-resistant, alkaline (basic) bricks. A pipe or lance is lowered near to the surface of the molten metal and pure oxygen blown through it at high pressure. The surface of the metal is disturbed by the blast and the impurities are oxidized (burned out). The **open-hearth process** is an older steelmaking method in which molten iron and limestone are placed in a shallow bowl or hearth (such as an open-hearth furnace). Burning oil or gas is blown over the surface of the metal, and the impurities are oxidized. High-quality steel is made in an **electric furnace**. A large electric current flows through electrodes in the furnace, melting a charge of scrap steel and iron. The quality of the steel produced can be controlled precisely because the temperature of the furnace can be maintained exactly and there are no combustion by-products to contaminate the steel. Electric furnaces are also used to refine steel, producing the extra-pure steels used, for example, in the petrochemical industry.

The steel produced is cast into ingots, which can be worked when hot by hammering (forging) or pressing between rollers to produce sheet steel. Alternatively, the **continuous-cast process**, in which the molten metal is fed into an open-ended mould cooled by water, produces an unbroken slab of steel.

Steele, Edward John (1948–) Australian immunologist whose research into the inheritance of immunity has lent a certain amount of support to the Lamarckian theory of the inheritance of acquired characteristics, thus challenging modern theories of †heredity and †evolution.

Stefan, Josef (1835–1893) Austrian physicist who established one of the basic laws of heat radiation in 1879, since known as the **Stefan–Boltzmann law**. This states that the heat radiated by a hot body is proportional to the fourth power of its absolute temperature.

Stefan–Boltzmann constant in physics, a constant relating the energy emitted by a black body (a hypothetical body that absorbs or emits all the energy falling on it) to its temperature. Its value is 5.6697 x 10^{-8} W m^{-2} K^{-4}.

Stefan–Boltzmann law in physics, a law that relates the energy, E, radiated away from a perfect emitter (a †black body), to the temperature, T, of that body. It has the form $E = \sigma T^4$, where E is the energy radiated per unit area per second, T is the temperature, and s is the **Stefan–Boltzmann constant**. Its value is 5.6697 x 10^{-8} W m^{-2} K^{-4}. The law was derived by the Austrian physicists Josef Stefan and Ludwig Boltzmann.

Steinberger, Jack (1921–) US physicist who, with Melvin Schwartz and Leon †Lederman, was awarded the Nobel Prize for Physics in 1988 for the development of the use of neutrinos to study †elementary particles and for the discovery of the muon †neutrino.

Steiner, Jakob (1796–1863) Swiss mathematician, the founder of modern synthetic, or projective, geometry. He discovered the **Steiner surface** (also called the Roman surface), which has a double infinity of conic sections on it, and the Steiner theorem.

Steiner, Jakob http://www-groups.dcs.st-andrews.ac.uk/history/Mathematicians/Steiner.html

Part of an archive containing the biographies of the world's greatest mathematicians, this site is devoted to the life and contributions of Jakob Steiner. In addition to biographical information, you will find a list of references about Steiner and links to other essays in the archive that reference him. The text of this essay includes hypertext links to the essays of those mathematicians and thinkers who influenced Steiner. There is also an image of him, which can be enlarged, and a map of his birthplace.

stellar of a star, of the stars.

stellar population in astronomy, classification of stars according to their chemical composition as determined by †spectroscopy.

Population I stars have a relatively high abundance of elements heavier than hydrogen and helium, and are confined to the spiral arms and disc of the Galaxy. They are believed to be young stars formed from material that has already been enriched with elements created by †nuclear fusion in earlier generations of stars. Examples include open clusters and †supergiants. **Population II** stars have a low abundance of heavy elements and are found throughout the Galaxy but especially in the central bulge and outer halo. They are among the oldest objects in the Galaxy and include †globular clusters. The Sun is a Population II star.

stem main supporting axis of a plant that bears the leaves, buds, and reproductive structures; it may be simple or branched. The plant stem usually grows above ground, although some grow underground, including †rhizomes, †corms, †rootstocks, and †tubers. Stems contain a continuous vascular system that conducts water and food to and from all parts of the plant.

The point on a stem from which a leaf or leaves arise is called a node, and the space between two successive nodes is the internode. In some plants, the stem is highly modified; for example, it may form a leaf-like †cladode or it may be twining (as in many climbing plants), or fleshy and swollen to store water (as in cacti and other succulents). In plants exhibiting †secondary growth, the stem may become woody, forming a main trunk, as in trees, or a number of branches from ground level, as in shrubs.

stem and leaf display in mathematics, way of displaying two-digit data graphically. The data is grouped according to the first digit and so forms a histogram representing the distribution of the data. It was invented by the US mathematician John Wilder Tukey in 1977.

stem cell embryonic cell that can develop into different tissues. Stem cells are used in medical research but their use is ethically controversial as the donor †embryo is destroyed.

There are three basic kinds of stem cell, which are known as totipotent, pluripotent, and multipotent. **Totipotent** stem cells form when a fertilized egg first divides, and can form a complete organism by forming all the necessary tissues, such as bone, muscle, or nerve tissue, and also form the placenta. **Pluripotent** stem cells form as part of the †blastocyst and can form most kinds of tissue, but are not able to generate the whole organism. **Multipotent** cells are generated by an embryo and give rise to only specific kinds of cells.

In Britain in August 2000 proposals were put forward to allow researchers to harvest stem cells from the embryo. In the USA medical guidelines are more restrictive.

Steno, Nicolaus (1638–1686) Danish **Niels Steensen**, Danish anatomist and naturalist, one of the founders of stratigraphy. To illustrate his ideas, Steno sketched what are probably the earliest geological sections.

steppe temperate grasslands of Europe and Asia. The term is sometimes used to refer to other temperate grasslands and semi-arid desert edges.

steradian symbol sr, SI unit of measure of solid (three-dimensional) angles, the three-dimensional equivalent of the †radian. One steradian is the angle at the center of a sphere when an area on the surface of the sphere equal to the square of the sphere's radius is joined to the center.

stereochemistry branch of chemistry that is concerned with a study of the shapes of molecules.

stereocilium sensory hair cell found on the †cochlea in the inner †ear. When the cochlear fluid vibrates in response to sound the hairs move together pushing on protein molecules on their surface. On each stereocilium the protein molecule triggers the opening of an ion gate that enables the flow of potassium ions through the hair. The ion flow causes an electrical signal to be transmitted to the brain as a nerve impulse, where it is interpreted as sound.

stereoisomer chemical compound that has the same elemental composition and atomic weight as another compound but differs as a result of the spatial arrangement of its molecules. Stereoisomers can be characterized into two classes of compound, optical and geometrical isomers. Optical isomers are compounds that have one or more †chiral centers and exist in two types, enantiomers and diastereoisomers. Enantiomers have a molecular structure that can be superimposed on its mirror image, whereas diastereoisomers have a molecular structure that cannot. Geometrical stereoisomers, or cis-trans stereoisomers, are compounds in which there are differences in the relative positions of dissimilar groups attached to two adjacent atoms, usually across a carbon-to-carbon double bond.

stereoisomerism type of isomerism in which two or more substances differ only in the way that the atoms of their molecules are oriented in space.

stereoscope device that can produce the effect of three-dimensional images using only two-dimensional images, using the human brain to carry out the merging necessary.

stereospecific reaction in chemistry, reaction that produces only one type of †stereoisomer. Reactions of this type usually involve the breaking and formation of a bond at a single asymmetric atom.

steric hindrance in chemistry, influence of neighboring groups, due to their size, on reactions between organic molecules. The spatial arrangement of large neighboring groups can prevent reacting molecules coming close enough to interact.

sterilization killing or removal of living organisms such as bacteria and fungi. A sterile environment is necessary in medicine, food processing, and some scientific experiments. Methods include heat treatment (such as boiling), the use of chemicals (such as disinfectants), irradiation with gamma rays, and filtration.

sterilization in medicine, any surgical operation to terminate the possibility of reproduction. In women, this is normally achieved by sealing or tying off the †Fallopian tubes (tubal ligation) so that fertilization can no longer take place. In men, the transmission of sperm is blocked by vasectomy.

According to the results of a long-term US study released in 1996, the failure rate for female sterilization is 1 in 50, higher than previously believed, with some pregnancies occurring as long as 14 years after the operation.

Sterilization may be encouraged by governments to limit population growth or as part of a selective-breeding policy, for example approximately 600,000 women are sterilized every year in India (1998).

Stern, Otto (1888–1969) German physicist who demonstrated by means of the **Stern–Gerlach apparatus** that elementary particles have wavelike properties as well as the properties of matter that had been demonstrated. He was awarded the Nobel Prize for Physics in 1943 for his development of the molecular-ray method of investigating elementary particles, and for the discovery of the magnetic moment of the proton.

sternum or **breastbone**, large flat bone, 15–20 cm/5.9–7.8 in long in the adult, at the front of the chest, joined to the ribs. It gives protection to the heart and lungs. During open-heart surgery the sternum must be split to give access to the thorax.

steroid any of a group of cyclic, unsaturated alcohols (lipids without fatty acid components), which, like sterols, have a complex molecular structure consisting of four carbon rings. Steroids include the sex hormones, such as †testosterone, the corticosteroid hormones produced by the †adrenal gland, bile acids, and †cholesterol.

The term is commonly used to refer to †anabolic steroid. In medicine, synthetic steroids are used to treat a wide range of conditions.

Steroids are also found in plants. The most widespread are the **brassinosteroids**, necessary for normal plant growth.

sterol any of a group of solid, cyclic, unsaturated alcohols, with a complex structure that includes four carbon rings; cholesterol is an example. Steroids are derived from sterols.

Stevens, Nettie Maria (1861–1912) US biologist whose experiments with a species of beetle showed that sex was determined by a specific chromosome. This was the first direct evidence that the units of heredity postulated by Austrian biologist Gregor †Mendel were associated with chromosomes.

Stevenson screen box designed to house weather-measuring instruments such as thermometers. It is kept off the ground by legs, has louvered sides to encourage the free passage of air, and is painted white to reflect heat radiation, since what is measured is the temperature of the air, not of the sunshine.

Stevin, Simon (1548–1620) Flemish engineer and mathematician best known for his 1585 book on arithmetic, which introduced decimal notation. He is also considered to be the founder of †hydrostatics, after showing that the pressure exerted by a liquid upon a given surface depends on the height of the liquid and the area of the surface.

stibnite shiny gray mineral form of antimony sulfide, used as an ore of antimony. It often occurs associated with ores of lead, mercury, and silver.

stigma in a flower, the surface at the tip of a †carpel that receives the †pollen. It often has short outgrowths, flaps, or hairs to trap pollen and may produce a sticky secretion to which the grains adhere.

stimulus plural **stimuli**, change in the †environment of an organism that the organism can detect and respond to. These changes may be noise, light, heat, or pressure. The changes are detected by specialized cells called receptors. The †skin contains many receptors for heat, touch, pressure, and pain. Sometimes the receptors are found in an organ which is so specialized for this purpose it is called a sense organ – for example the †eye. See †sensitivity.

stipule outgrowth arising from the base of a leaf or leaf stalk in certain plants. Stipules usually occur in pairs or fused into a single semicircular structure.

Stirling, James (1692–1770) Scottish mathematician and engineer, best known for his formula approximating n! and for the so-called **Stirling numbers** that occur in many combinatorial problems.

stishovite highest density, highest pressure polymorph (substance with the same chemical composition but different crystal structure) of silica, SiO_2, in which the crystal structure is the same as that of the mineral †rutile, titanium dioxide. Rare in nature it is found primarily near meteor impact craters. It is thought to be a rare constituent of the lower †mantle.

Stock, Alfred (1876–1946) German inorganic chemist who prepared many of the hydrides of boron (called boranes). He introduced sensitive tests for mercury and devised improved laboratory techniques for dealing with the metal to minimize the risk of poisoning.

stoichiometric in chemistry, describing compounds in which the ratio of the total number of atoms of each element to each other is a ratio of small whole numbers. Most compounds are stoichiometric, for example the stoichiometic ratio of elements in potassium carbonate, K_2CO_3, is 2K:1C:3O.

stoichiometry in chemistry, relative ratios of compounds involved in a chemical reaction.

stokes symbol St, *c*.g.s. unit of kinematic viscosity (a liquid's resistance to flow).

Liquids with higher kinematic viscosity have higher turbulence than those with low kinematic viscosity. It is found by dividing the dynamic viscosity in †poise by the density of the liquid.

Stokes, George Gabriel (1819–1903) Irish physicist who studied the †viscosity (resistance to relative motion) of fluids. This culminated in **Stokes' law**, $F = 6phrv$, which applies to a force acting on a sphere falling through a liquid, where h is the liquid's viscosity and r and v are the radius and velocity of the sphere. He became a baronet in1889.

Stokes, George Gabriel http://www-groups.dcs.st-andrews.ac.uk/history/Mathematicians/Stokes.html

Part of an archive containing the biographies of the world's greatest mathematicians, this site is devoted to the life and contributions of physicist George Stokes.

Stokes' law in physics, law that quantifies the force that acts to oppose the motion of an object falling through a viscous liquid. If the object is a sphere of radius r falling at a speed v in a fluid of viscosity h, the force equals $6phrv$,

stolon in botany, a type of †runner.

stoma plural **stomata**, in botany, a pore (tiny hole) in the epidermis (outer layer of tissue) of a plant. There are lots of these holes, usually in the lower surface of the †leaf. A leaf contains several layers of †tissue. The outer layer is the epidermis and is only one cell thick. Stomata occur in the lower epidermis. Each stoma is surrounded by a pair of guard cells that are crescent-shaped when the stoma is open but can collapse to an oval shape, thus closing off the opening between them. Stomata allow the exchange of carbon dioxide and oxygen (needed for †photosynthesis and †respiration) between the internal tissues of the plant and the outside atmosphere. They are also the main route by which water is lost from the plant, and they can be closed to conserve water, the movements being controlled by changes in turgidity of the guard cells.

At night the stomata may allow oxygen to diffuse in and carbon dioxide out because only respiration is taking place. However in many plants the stomata close at night, preventing this gas exchange.

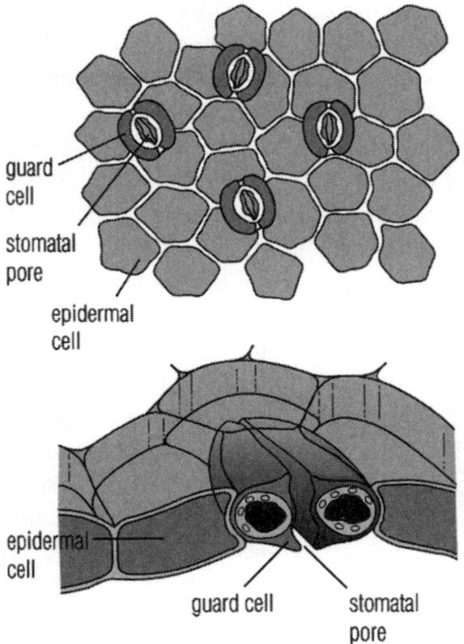

stoma The stomata, tiny openings in the epidermis of a plant, are surrounded by pairs of crescent-shaped cells, called guard cells. The guard cells open and close the stoma by changing shape. Helicon

stomach organ that forms the first cavity in the digestive system of animals. In mammals it is a bag of †muscle situated just below the diaphragm. Food enters it from the †esophagus, is digested by the acid and †enzymes secreted by the stomach lining. The wall of the stomach contracts to mix the food with the digestive juice to help digestion of †protein. After a while, partly digested food is then passed into the small †intestine (duodenum). Some plant-eating mammals have multi-chambered stomachs that harbor bacteria in one of the chambers to assist in the digestion of †cellulose.

The gizzard is part of the stomach in birds.

stomach acid gastric juices secreted by specialized cells lining pits in the wall of the stomach. Gastric juice contains an enzyme called pepsin that digests protein. It also contains hydrochloric acid and water. The hydrochloric acid is very strong and is needed to help kill bacteria that may cause illness and provides acid conditions in which pepsin works quickly.

On a scale of acidity called the pH scale, the acid secretions from the stomach wall are between pH 0.8 and pH 1.0. This makes the secretions so strong that, if bottled, they would be too dangerous to have on an open shelf in a school laboratory and would have to be labeled "corrosive." This is strong enough normally to destroy human tissues on contact. The stomach is protected against damage from the acid by a lining of mucus, unless a person suffers from an ulcer.

stone plural **stone**; abbreviation st, imperial unit of mass. One stone is 14 pounds (6.35 kg).

storm surge abnormally high tide brought about by a combination of a deep atmospheric †depression (very low pressure) over a shallow sea area, high spring tides, and winds blowing from the appropriate direction. It can be intensified by snowmelt and/or river flooding. A storm surge can cause severe flooding of lowland coastal regions and river estuaries.

Bangladesh is particularly prone to surges, because it is sited on a low-lying †delta where the Indian Ocean funnels into the Bay of Bengal. In May 1991, 125,000 people were killed there in such a disaster. In February 1953 more than 2,000 people died when a North Sea surge struck the Dutch and English coasts.

Strabo (c. 63 BC – c. AD 24) Greek geographer and historian who traveled widely to collect first-hand material for his *Geography*.

straight line line that does not bend or curve. The graph of a linear relationship is a straight line and is often presented in the form $y = mx + c$, where m is the slope, or gradient, of the line and c is the y-intercept (the point at which the line cuts the y-axis).

strain in the science of materials, the extent to which a body is distorted when a deforming force (stress) is applied to it. It is a ratio of the extension or compression of that body (its length, area, or volume) to its original dimensions (see †Hooke's law). For example, linear strain is the ratio of the change in length of a body to its original length.

strange attractor in mathematics, a point that moves chaotically within a given region at all times. Such movement describes, for example, the motion of turbulent fluids.

strangeness †quantum number assigned to certain unstable †elementary particles that decay much more slowly than was originally expected. Stable particles, such as †protons, have a strangeness quantum number of zero. Others, such as the †hyperons, have nonzero quantum numbers.

Strasburger, Eduard Adolf (1844–1912) German botanist who discovered that the nucleus of plant cells divides during cell division and clarified the role that chromosomes play in heredity. It had previously been thought that the nucleus disappeared during cell division until Strasburger saw the nucleus of a dividing cell divide.

strata singular **stratum**, layers or †beds of sedimentary rock.

stratigraphy branch of geology that deals with †sedimentary rock layers (strata) and their sequence of formation. Its basis was developed by English geologist William †Smith. The basic principle of superposition establishes that upper layers or deposits accumulated later in time than the lower ones.

Stratigraphy involves both the investigation of sedimentary structures to determine past environments represented by rocks, and the study of fossils for identifying and dating particular beds of rock. A body of rock strata with a set of unifying characteristics indicative of an environment is called a †facies.

Stratigraphic units can be grouped in terms of time or lithology (rock type). Strata that were deposited at the same time belong to a single **chronostratigraphic unit** but need not be the same lithology. Strata of a specific lithology can be grouped into a **lithostratigraphic unit** but are not necessarily the same age.

Stratigraphy in the interpretation of archaeological excavations provides a relative chronology for the levels and the artifacts within rock beds. It is the principal means by which the context of archaeological deposits is evaluated.

stratopause layer in the Earth's atmosphere at an altitude of 50–55 km/31–34 mi, at the top of the stratosphere.

stratosphere that part of the atmosphere 10–40 km/6–25 mi from the Earth's surface, where the temperature slowly rises from a low of –55°C/–67°F to around 0°C/32°F. The air is rarefied and at around 25 km/15 mi much †ozone is concentrated.

stratovolcano another term for †composite volcano, a type of volcano made up of alternate ash and lava layers.

streaming in computing, sending data, for example video frames or radio broadcasts, in a steady flow over the Internet. Streaming requires data to pass through a special channel or dedicated connection; conventional packets, which travel by a multiplicity of routes, may arrive in the wrong order or be duplicated on the way. Streaming is used in multimedia presentations, such as RealAudio, RealVideo, RealText, and Flash animations; it enables presentations to be commenced before files are finished downloading. The data is placed in a temporary memory area until it is downloaded.

streamline line in a fluid such that the †tangent to it at every point gives the direction of flow, and its †speed, at any instant.

Street, J(abez) C(urry) (1906–1989) US physicist who, with Edward C Stevenson, discovered the muon (an †elementary particle) in 1937.

strength of acids and bases in chemistry, the ability of †acids and †bases to dissociate in solution with water, and hence to produce a low or high †pH, respectively.

A strong acid is fully dissociated in aqueous solution, whereas a weak acid is only partly dissociated. Since the †dissociation of acids generates hydrogen ions, a solution of a strong acid will have a high concentration of hydrogen ions and therefore a low pH. A strong base will have a high pH, whereas a weaker base will not dissociate completely and will have a pH of nearer 7.

stress and strain in the science of materials, measures of the deforming force applied to a body (stress) and of the resulting change in its shape (†strain). For a perfectly elastic material, stress is proportional to strain (†Hooke's law).

stress protein one of a group of proteins produced by a cell when it is under stress. Stress proteins are created in response to a wide variety of stresses including illness, high temperatures, alcohol, and poisons. They act as a buffer for the cell and repair damage.

striation scratch formed by the movement of a glacier over a rock surface. Striations are caused by the scraping of rocky debris embedded in the base of the glacier (†corrasion), and provide an useful indicator of the direction of ice flow in past †ice ages.

They are common features of †roche moutonnées.

stridulatory organs in insects, organs that produce sound when rubbed together. Crickets rub their wings together, but grasshoppers rub a hind leg against a wing. Stridulation is thought to be used for attracting mates, but may also serve to mark territory.

strike compass direction of a horizontal line on a planar structural surface, such as a fault plane, bedding plane, or the trend of a structural feature, such as the axis of a fold. Strike is 90° from †dip.

strike-slip fault common name for a lateral fault in which the motion is sideways in the direction of the †strike of the fault.

string in computing, a group of characters manipulated as a single object by the computer. In its simplest form a string may consist of a single letter or word – for example, the single word SMITH might be established as a string for processing by a computer. A string can also consist of a combination of words, spaces, and numbers – for example, 33 HIGH STREET ANYTOWN ANYSTATE could be established as a single string.

Most high-level languages have a variety of string-handling †functions. For example, functions may be provided to read a character from any given position in a string or to count automatically the number of characters in a string.

string theory mathematical theory developed in the 1980s to explain the behavior of †elementary particles; see †superstring theory.

string, vibrations of standing waves set up in a stretched string or wire when it is plucked, or stroked with a bow. They are formed by the reflection of progressive waves at the fixed ends of the string. Waves of many different †frequencies can be established on a string at the same time. Those that match the natural frequencies of the string will, by a process called †resonance, produce large-amplitude vibrations. The vibration of lowest frequency is called the †fundamental vibration; vibrations of frequencies that are multiples of the fundamental frequency are called harmonics.

strip mining another term for †opencast mining.

strobilus in botany, a reproductive structure found in most †gymnosperms and some †pteridophytes, notably the club mosses. In conifers the strobilus is commonly known as a †cone.

stroboscope instrument for studying continuous periodic motion by using light flashing at the same frequency as that of the motion; for example, rotating machinery can be optically "stopped" by illuminating it with a stroboscope flashing at the exact rate of rotation.

stromatolite mound produced in shallow water by mats of algae that trap mud particles. Another mat grows on the trapped mud layer and this traps another layer of mud and so on. The stromatolite grows to heights of a meter or so. They are uncommon today but their fossils are among the earliest evidence for living things – over 2,000 million years old.

Strömgren, Bengt Georg Daniel (1908–1987) Swedish-born Danish astronomer whose 1940 hypothesis about the so-called **Strömgren spheres** – zones of ionized hydrogen gas surrounding hot stars embedded in gas clouds – proved fundamental to our understanding of the structure of interstellar material.

Strömgren sphere or **H II zone**, zone of ionized hydrogen gas that surrounds hot stars embedded in interstellar gas clouds.

strong nuclear force one of the four fundamental †forces of nature, the other three being the gravitational force or gravity, the electromagnetic force, and the weak nuclear force. The strong nuclear force was first described by the Japanese physicist Hideki Yukawa in 1935. It is the strongest of all the forces, acts only over very small distances within the nucleus of the atom (10^{-13} cm), and is responsible for binding together †quarks to form †hadrons, and for binding together protons and neutrons in the atomic nucleus. The particle that is the carrier of the strong nuclear force is the †gluon, of which there are eight kinds, each with zero mass and zero charge.

strontium chemical symbol Sr, soft, ductile, pale-yellow, metallic element, atomic number 38, relative atomic mass 87.62. It is one of the †alkaline-earth metals, widely distributed in small quantities only as a sulfate or carbonate. Strontium salts burn with a red flame and are used in fireworks and signal flares. The radioactive isotopes Sr-89 and Sr-90 (half-life 25 years) are some of the most dangerous products of the nuclear industry; they are fission products in nuclear explosions and in the reactors of nuclear power plants. Strontium is chemically similar to calcium and deposits in bones and other tissues, where the radioactivity is damaging. The element was named in 1808 by English chemist Humphry Davy, who isolated it by electrolysis, after Strontian, a mining location in Scotland where it was first found.

structural geology branch of geology dealing with the physical shape of rock formations, from large-scale features such as mountain belts and rift valleys, to microscopic features such as deformed mineral grains. Structural geologists deal with the brittle and ductile deformation of planetary surfaces and the processes that cause deformation.

Strutt, John William English physicist; see †Rayleigh.

Struve, (Gustav Wilhelm) Ludwig (Ottovich) von (1858–1920) Russian astronomer. He was an expert on the occultation of stars during a total lunar eclipse, and on stellar motion.

Struve, (Karl) Hermann (Ottovich) von (1854–1920) Russian astronomer. He was an expert on Saturn. His other work was largely concerned with features of the Solar System, although he shared the family interest in stellar astronomy. He was the brother of Ludwig von Struve.

Struve, F(riedrich) G(eorg) W(ilhelm) von (1793–1864) German-born Russian astronomer. He was a pioneer in the observation of double stars and one of the first to measure stellar parallax in 1830. He had 18 children and was the founder of a family of astronomers that spanned four generations.

Struve, Otto von (1897–1963) Russian-born US astronomer. He developed a nebular spectrograph to study interstellar gas clouds. In 1938 he showed that ionized hydrogen is present in interstellar matter. He also determined that the interstellar hydrogen is concentrated in the galactic plane. He was the son of Ludwig von Struve.

Struve, Otto Wilhelm von (1819–1905) Russian astronomer. He made an accurate determination of the constant of precession. He discovered about 500 double stars. He was the son of F G W von Struve and father of Hermann and Ludwig von Struve.

strychnine $C_{21}H_{22}O_2N_2$, bitter-tasting, poisonous alkaloid. It is a poison that causes violent muscular spasms, and is usually obtained by powdering the seeds of plants of the genus *Strychnos* (for example *S. nux vomica*). Curare is a related drug.

stump low outcrop of rock formed by the erosion of a coastal †stack. Unlike a stack, which is exposed at all times, a stump is exposed only at low tide. Eventually it is worn away completely.

Sturm, Jacques Charles François (1803–1855) Swiss mathematician. In 1829 he discovered the theorem which is named after him. This theorem completes the resolution of equations of any degree by an intermediate determination of the number of real roots lying between two given limits. He measured the velocity of sound in water by using a bell submerged in Lake Geneva. He became professor at the Ecole polytechnique in 1838. His *Cours d'analyse/Course on analysis* (1857–59) and *Cours de mecanique/Course on mechanics* (1861) were published posthumously and were widely used as textbooks.

Sturtevant, Alfred Henry (1891–1970) US geneticist who was the first, in 1911, to map the position of genes on a chromosome. He worked with US biologist Thomas †Morgan.

style in flowers, the part of the †carpel bearing the †stigma at its tip. In some flowers it is very short or completely lacking, while in others it may be long and slender, positioning the stigma in the most effective place to receive the pollen.

styrene C_8H_8, colorless, unsaturated aromatic liquid mainly used as a monomer in the production of the †polystyrene class of synthetic polymers.

subarachnoid describing the fluid-filled region between the arachnoid membrane and the pia mater membrane which surround the brain and spinal cord.

subatomic particle in physics, a particle that is smaller than an atom. Such particles may be indivisible †elementary particles, such as the †electron and †quark, or they may be composites, such as the †proton, †neutron, and †alpha particle. See also †particle physics.

subconscious state of being partially conscious. The mind can perceive impressions or events that may be forgotten at the time but can continue to influence the conscious mind or may enter the conscious mind a long time after the event has taken place.

Psychoanalysis places much importance on the influence of painful and unpleasant experiences that, although forgotten, continue to have a detrimental influence on the mind.

subduction zone in †plate tectonics, a region where two plates of the Earth's rigid †lithosphere collide, and one plate descends below the other into the weaker †asthenosphere. Subduction results in the formation of ocean trenches, most of which encircle the Pacific Ocean.

Ocean trenches are usually associated with volcanic †island arcs and deep-focus earthquakes (more than 300 km/185 mi below the surface), both the result of disturbances caused by the plate subduction.

suberin in botany, waxy substance that impregnates the walls of plant cells. It is a waterproof material found in the cells of the outer layers of older roots and stems. It is also found in the endodermis, forming a band (the †casparation strip) that encircles the central core of vascular material, separating it from the cells of the cortex. Suberin prevents water and solutes from entering the vascular tissue, thereby controlling the passage of water into the xylem.

subglacial beneath a glacier. Subglacial rivers are those that flow under a glacier; subglacial material is debris that has been deposited beneath glacier ice. Features formed subglacially include †drumlins and †eskers.

sublimation conversion of a solid to vapor without passing through the liquid phase. It is one of the †changes of state of matter.

Sublimation depends on the fact that the boiling point of the solid substance is lower than its melting point at atmospheric pressure. Thus by increasing pressure, a substance that sublimes can be made to go through a liquid stage before passing into the vapor state.

Some substances that do not sublime at atmospheric pressure can be made to do so at low pressures. This is the principle of freeze-drying, during which ice sublimes at low pressure.

subliminal message any message delivered beneath the human conscious threshold of perception. It may be visual (words or images flashed between the frames of a cinema or TV film), or aural (a radio message broadcast constantly at very low volume).

subroutine in computing, a small section of a program that is executed ("called") from another part of the program. Subroutines provide a method of performing the same task at more than one point in the program, and also of separating the details of a program from its main logic. In some computer languages, subroutines are similar to †functions or procedures.

subset in mathematics, set that is part of a larger set. For example, the girls in a class make up a subset if the class contains both boys and girls.

subsidence downward movement of a block of rock. Subsidence is usually due to the removal of material from below the surface, and can be caused by faults, †erosion, or by human activities such as mining.

substitute in mathematics, to put values in the place of variables in an algebraic expression or formula. An algebraic expansion or simplification can be checked by substituting simple values for each variable. For example, to check that

$x^3 + y^3 = (x + y)(x^2 + y^2 - xy)$

the value $x = 1$ and $y = 2$ might be substituted in both sides of the expression, giving

left-hand side: $1^3 + 2^3 = 1 + 8 = 9$

right-hand side: $(1 + 2)(1^2 + 2^2 - 2) = 3 \times 3 = 9$.

The two sides are the same, so the expansion of $x^3 + y^3$ is correct.

substitution reaction in chemistry, the replacement of one atom or †functional group in an organic molecule by another.

substrate in biochemistry, a compound or mixture of compounds acted on by an enzyme. The term also refers to a substance such as †agar that provides the nutrients for the metabolism of micro-organisms. Since the enzyme systems of micro-organisms regulate their metabolism, the essential meaning is the same.

substratum or **substrate**, in biology, medium or surface in which or on which an organism lives. The substratum may supple the organism with its nourishment.

subtraction taking one number or quantity away from another, or finding the difference between two quantities; it is one of the four basic operations of arithmetic. Subtraction is neither commutative:

$a - b \neq b - a$

nor associative:

$a - (b - c) \neq (a - b) - c$.

For example,

$8 - 5 \neq 5 - 8$

$7 - (4 - 3) \neq (7 - 4) - 3$.

succession in ecology, a series of changes that occur in the structure and composition of the vegetation in a given area from the time it is first colonized by plants (**primary succession**), or after it has been disturbed by fire, flood, or clearing (**secondary succession**).

If allowed to proceed undisturbed, succession leads naturally to a stable †climax community (for example, oak and hickory forest or savannah grassland) that is determined by the climate and soil characteristics of the area.

succinic acid or **butanedioic acid**; $C_4H_6O_4$, colorless crystalline solid, melting point 182°C/360°F. Succinic acid is an intermediate compound in the metabolism of glucose in living organisms and is present in nearly all plant and animal tissues.

succulent plant thick, fleshy plant that stores water in its tissues; for example, cacti and stonecrops *Sedum*. Succulents live either in areas where water is very scarce, such as deserts, or in places where it is not easily obtainable because of the high concentrations of salts in the soil, as in salt marshes. Many desert plants are †xerophytes.

suckering in plants, reproduction by new shoots (suckers) arising from an existing root system rather than from seed. Plants that produce suckers include elm, dandelion, and members of the rose family.

sucrase enzyme capable of digesting sucrose into its constituent molecules of glucose and fructose.

In mammals this action takes place within the wall of the intestine, the products of the reaction being liberated into the lumen. This is an example of intracellular digestion.

sucrose or **cane sugar** or **beet sugar**; $C_{12}H_{22}O_{11}$, sugar found in the pith of sugar cane and in sugar beet. It is popularly known as sugar.

Sucrose is a disaccharide sugar, each of its molecules being made up of two simple sugar (monosaccharide) units: glucose and fructose.

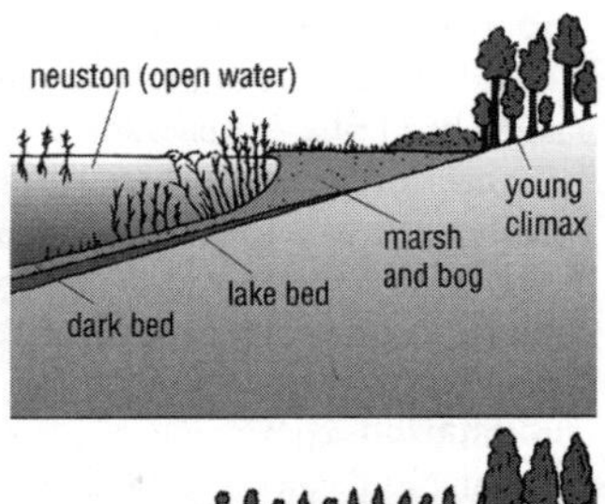

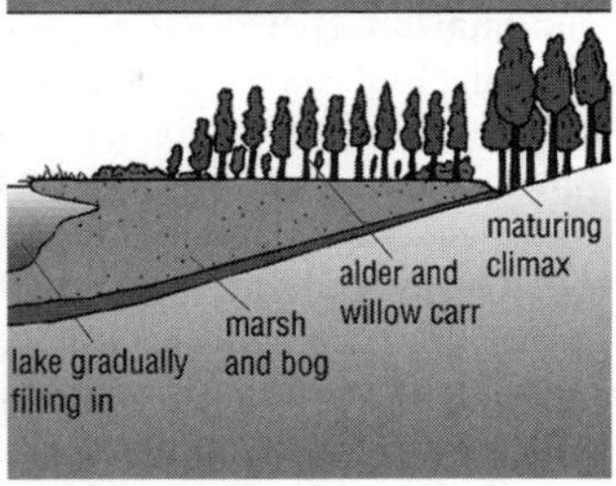

succession The succession of plant types along a lake. As the lake gradually fills in, a mature climax community of trees forms inland from the shore. Extending out from the shore, a series of plant communities can be discerned with small, rapidly growing species closest to the shore. Helicon

Suess, Eduard (1831–1914) Austrian geologist who helped pave the way for the theories of †continental drift. He suggested that there had once been a great supercontinent, made up of the present southern continents; this he named Gondwanaland, after a region of India.

sufficient condition in mathematics, statement *A* is a sufficient condition for a statement *B* if *B* is true whenever *A* is true. It can also be expressed as *A* implies *B*.

Sufi, al- (903–986) Persian astronomer whose importance lies in his compilation of a valuable catalogue of 1,018 stars with their approximate positions, magnitudes, and colors.

sulfate SO_4^{2-}, salt or ester derived from sulfuric acid. Most sulfates are water soluble (the chief exceptions are lead, calcium, strontium, and barium sulfates) and require a very high temperature to decompose them.

The commonest sulfates seen in the laboratory are copper(II) sulfate ($CuSO_4$), iron(II) sulfate ($FeSO_4$), and aluminum sulfate ($Al_2(SO_4)_3$). The ion is detected in solution by using barium chloride or barium nitrate to precipitate the insoluble barium sulfate.

sulfide compound of sulfur and another element in which sulfur is the more electronegative element (see †electronegativity). Sulfides occur in many minerals. Some of the more volatile sulfides have extremely unpleasant odors (hydrogen sulfide smells of bad eggs).

sulfite SO_3^{2-}, salt or ester derived from sulfurous acid.

sulfonamide any of a group of compounds containing the chemical group sulfonamide (SO_2NH_2) or its derivatives, which were, and still are in some cases, used to treat bacterial diseases. Sulfadiazine ($C_{10}H_{10}N_4O_2S$) is an example.

Sulfonamide was the first commercially available antibacterial drug, the forerunner of a range of similar drugs. Toxicity and increasing resistance have limited their use chiefly to the treatment of urinary-tract infection.

sulfonation in chemistry, any process that results in the introduction of a sulfonic group, —SO_3H, into an organic molecule.

sulfonic acid in chemistry, member of a class of organic compounds whose molecules contain a sulfonic group, —SO_3H. These compounds are usually produced by reactions with concentrated sulfuric acid.

sulfur chemical symbol S, brittle, pale-yellow, non-metallic element, atomic number 16, relative atomic mass 32.064. It occurs in three allotropic forms: two crystalline (called rhombic and monoclinic, following the arrangements of the atoms within the crystals) and one amorphous. It burns in air with a blue flame and a stifling odor. Insoluble in water but soluble in carbon disulphide, it is a good electrical insulator. Sulfur is widely used in the manufacture of sulfuric acid (used to treat phosphate rock to make fertilizers) and in making paper, matches, gunpowder and fireworks, in vulcanizing rubber, and in medicines and insecticides.

sulfur dioxide SO_2, pungent gas produced by burning sulfur or sulfide ores in air or oxygen. It is widely used for making sulfuric acid and for disinfecting food vessels and equipment for bleaching paper, and as a preservative in some food products. It occurs in industrial flue gases and is a major cause of †acid rain.

sulfuric acid or **oil of vitriol**; H_2SO_4, dense, viscous, colorless liquid that is extremely corrosive. It gives out heat when added to water and can cause severe burns. Sulfuric acid is used extensively in the chemical industry, in the refining of gasoline, and in the manufacture of fertilizers, detergents, explosives, and dyes. It forms the acid component of car batteries.

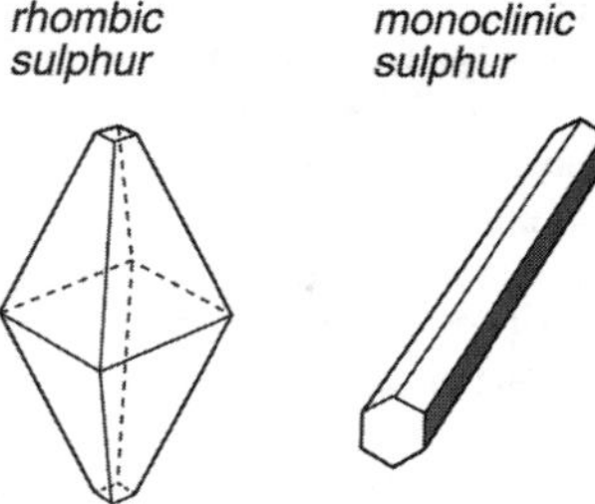

sulfur Two common allotropes of sulfur: rhombic and monoclinic crystals. A reactive element, sulfur combines with most other elements and has a wide range of industrial uses. It often occurs around hot springs and in volcanic regions, and there are large deposits in the USA (Texas and Louisiana), Japan, Sicily, and Mexico. Helicon

sulfurous acid H_2SO_3, solution of sulfur dioxide (SO_2) in water. It is a weak acid.

sulfur trioxide SO_3, colorless solid prepared by reacting sulfur dioxide and oxygen in the presence of a vanadium(V) oxide catalyst in the †contact process. It reacts violently with water to give sulfuric acid. The violence of its reaction with water makes it extremely dangerous. In the contact process, it is dissolved in concentrated sulfuric acid to give oleum ($H_2S_2O_7$).

summation symbol sign (Σ) representing the sum taken over all instances that accompany the sign, indicated above and below it. The symbol is the capital form of the Greek letter sigma.

Sumner, James (Batcheller) (1887–1955) US biochemist. Sumner shared the Nobel Prize for Chemistry in 1946 with John †Northrop and Wendell †Stanley for his work in 1926 when he succeeded in crystallizing the enzyme urease and demonstrating its protein nature.

Sun †star at the center of our Solar System. It is about 5 billion years old, with a predicted lifetime of 10 billion years; its diameter is 1.4 million km/865,000 mi; its temperature at the surface (the †photosphere) is about 5,800 K/5,530°C/9,986°F, and at the center 15 million K/about 15 million°C/about 27 million°F. It is composed of about 70% hydrogen and 30% helium, with other elements making up less than 1%. The Sun's energy is generated by nuclear fusion reactions that turn hydrogen into helium, producing large amounts of light and heat that sustain life on Earth.

Space probes to the Sun have included NASA's series of Orbiting Solar Observatory satellites, launched between 1963 and 1975, the Ulysses space probe, launched in 1990, and Genesis, launched in 2001.

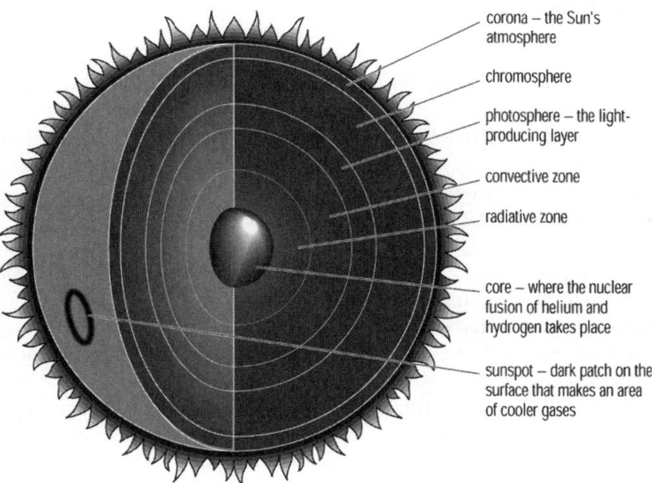

Sun The structure of the Sun. Nuclear reactions at the core release vast amounts of energy in the form of light and heat that radiate out to the photosphere and corona. Surges of glowing gas rise as prominences from the surface of the Sun and cooler areas, known as sunspots, appear as dark patches on the star's surface. Helicon

Sun http://www.solarviews.com/eng/sun.htm

All you ever wanted to know about our closest star, including cross-sections, photographs, a history of exploration, animations of eclipses, and much more. The user can also take a multimedia tour of the Sun or find out what the weather is like on the Sun.

Sun http://www.hotliquidmagma.com/space/html/sun.html

Short history of the Sun, and an insight into its context within the contemporary universe. This site is mainly text-based but includes the odd dramatic picture, making this a useful site.

Sun http://seds.lpl.arizona.edu/nineplanets/nineplanets/sol.html

Clear and concise information about the Sun. There are key facts and a summary of what we still have to learn about the Sun. The site features a very good resource list, with links to many other excellent Sun-related sites.

sungrazer in astronomy, type of bright comet that makes a very close approach to the Sun, passing through the Sun's outer layers (corona) and possibly destroying itself when it collides with the Sun. Examples include the long-period Kreutz group of comets.

sunshine recorder device for recording the hours of sunlight during a day. The **Campbell-Stokes sunshine recorder** consists of a glass sphere that focuses the Sun's rays on a graduated paper strip. A track is burned along the strip corresponding to the time that the Sun is shining.

sunspot dark patch on the surface of the Sun, actually an area of cooler gas, thought to be caused by strong magnetic fields that block the outward flow of heat to the Sun's surface. Sunspots consist of a dark central **umbra**, about 4,000 K (3,700°C/6,700°F), and a lighter surrounding **penumbra**, about 5,500 K (5,200°C/9,400°F). They last from several days to over a month, ranging in size from 2,000 km/1,250 mi to groups stretching for over 100,000 km/62,000 mi.

Sunspots are more common during active periods in the Sun's magnetic cycle, when they are sometimes accompanied by nearby †flares. The number of sunspots visible at a given time varies from none to over 100, in a cycle averaging 11 years. There was a lull in sunspot activity, known as the Maunder minimum, 1645–1715, that coincided with a cold spell in Europe.

Sunspots http://www.exploratorium.edu/sunspots/

Well-written and easily understandable explanation of sunspots. There is a high-resolution image of a group of sunspots and a link for further sunspot pictures. There is information on how to observe sunspots and the Sun's cycle of magnetic activity. For more detailed information, there are links to solar observatories.

Sunyaev-Zel'dovich effect slight dip in the †cosmic background radiation when observed through the gas that surrounds clusters of galaxies.

First mapped in 1993 by British astronomers around the Abell 2218 galaxy cluster. The size of the Sunyaev-Zel'dovich effect is a measure of the thickness of the gas through which the cosmic background radiation passes.

superactinide any of a theoretical series of superheavy, radioactive elements, starting with atomic number 113, that extend beyond the †transactinide series in the periodic table. They do not occur in nature and none has yet been synthesized.

It is postulated that this series has a group of elements that have half-lives longer than those of the transactinide series.

This group, centered on element 114, is referred to as the "island of stability," based on the nucleon arrangement. The longer half-lives will, it is hoped, allow enough time for their chemical and physical properties to be studied when they have been synthesized.

supercharging heating of a liquid above its boiling point.

supercluster in astronomy, grouping of several clusters of galaxies to form a structure about 100–300 million light years across. Our own Galaxy and its neighbors lie on the edge of the local supercluster of which the †Virgo cluster is the dominant member.

supercomputer fastest, most powerful type of computer, capable of performing its basic operations in picoseconds (trillionths of a second), rather than nanoseconds (billionths of a second), like most other computers.

To achieve these extraordinary speeds, supercomputers use many processors working together and techniques such as cooling processors down to nearly †absolute zero temperature, so that their components conduct electricity many times faster than normal. Supercomputers are used in weather forecasting, fluid dynamics, and aerodynamics. Manufacturers include Cray Research, Fujitsu, and NEC.

TOP 500 Supercomputer Sites http://www.netlib.org/benchmark/top500.html

Regularly updated list of the top 500 supercomputers around the world. The computer type, manufacturer, country, and performance are given, as well as the broad area of use and a link to the organization using it.

superconductivity increase in electrical conductivity at low temperatures. The resistance of some metals and metallic compounds decreases uniformly with decreasing temperature until at a critical temperature (the superconducting point), within a few degrees of absolute zero (0 K/–273.15°C/–459.67°F), the resistance suddenly falls to zero. The phenomenon was discovered by Dutch scientist Heike Kamerlingh Onnes in 1911.

Some metals, such as platinum and copper, do not become superconductive; as the temperature decreases, their resistance decreases to a certain point but then rises again. Superconductivity can be nullified by the application of a large magnetic field. In the superconducting state, an electric current will continue indefinitely once started, provided that the material remains below the superconducting point. In 1986 IBM researchers achieved superconductivity with some ceramics at –243°C/–405°F, opening up the possibility of **"high-temperature" superconductivity**; a year later Paul Chu at the University of

Houston, Texas, achieved superconductivity at –179°C/–290°F, a temperature that can be sustained using liquid nitrogen. Researchers are trying to find a material that will be superconducting at room temperature; in 2001 the material magnesium diboride (MgB_2) was found to become a superconductor when cooled to only 39 K/–234.15°C/–389.47°F.

supercooling cooling of a liquid below its freezing point without freezing taking place; or the cooling of a †saturated solution without crystallization taking place, to form a supersaturated solution. In both cases supercooling is possible because of the lack of solid particles around which crystals can form. Crystallization rapidly follows the introduction of a small crystal (seed) or agitation of the supercooled solution.

supercritical fluid fluid that combines the properties of a gas and a liquid, see †fluid, supercritical.

superego in Freudian psychology, the element of the human mind concerned with the ideal, responsible for ethics and self-imposed standards of behavior. It is characterized as a form of conscience, restraining the †ego, and responsible for feelings of guilt when the moral code is broken.

superfluid fluid that flows without viscosity or friction and has a very high thermal conductivity. Liquid helium at temperatures below 2 K (–271°C/–456°F) is a superfluid: it shows unexpected behavior. For instance, it flows uphill in apparent defiance of gravity and, if placed in a container, will flow up the sides and escape.

German physicists discovered in 1998 that as few as 60 atoms will exhibit superfluidity.

supergiant largest and most luminous type of star known, with a diameter of up to 1,000 times that of the Sun and an apparent magnitude of between 0.4 and 1.3. Supergiants are likely to become †supernovae.

superheterodyne or **supersonic heterodyne**, means of receiving †radio transmissions involving the changing of the †frequency of the carrier wave to an intermediate frequency above the limit of audible sound by a heterodyne process. In this, the received wave is combined with a slightly different frequency wave produced within the receiver. Once the intermediate frequency has been formed, the combined waves are amplified and the signal taken off by a demodulator.

superior planet planet that is farther away from the Sun than the Earth is: that is, Mars, Jupiter, Saturn, Uranus, Neptune, and Pluto.

supernova explosive death of a star, which temporarily attains a brightness of 100 million Suns or more, so that it can shine as brilliantly as a small galaxy for a few days or weeks. Very approximately, it is thought that a supernova explodes in a large galaxy about once every 100 years. Many supernovae – astronomers estimate some 50% – remain undetected because of obscuring by interstellar dust.

The name "supernova" was coined in 1934 by Swiss Astronomer Fritz Zwicky and German-born US astronomer Walter Baade. Zwicky was also responsible for the division into types I and II. **Type I** supernovae are thought to occur in †binary star systems, in which gas from one star falls on to a †white dwarf, causing it to explode. **Type II** supernovae occur in stars ten or more times as massive as the Sun, which suffer runaway internal nuclear reactions at the ends of their lives, leading to explosions. These are thought to leave behind †neutron stars and †black holes. Gas ejected by such an explosion causes an expanding radio source, such as the †Crab nebula. Supernovae are thought to be the main source of elements heavier than hydrogen and helium.

supernova remnant **SNR**, in astronomy, glowing remains of a star that has been destroyed in a †supernova explosion. The brightest and most famous example is the †Crab nebula.

superphosphate phosphate fertilizer made by treating apatite (calcium phosphate mineral) with sulfuric or phosphoric acid. The commercial mixture contains largely monocalcium phosphate. Single-superphosphate obtained from apatite and sulfuric acid contains 16–20% available phosphorus, as P_2O_5; triple-superphosphate, which contains 45–50% phosphorus, is made by treating apatite with phosphoric acid.

supersaturation state of a solution that has a higher concentration of †solute than would normally be obtained in a †saturated solution.

Many solutes have a higher †solubility at high temperatures. If a hot saturated solution is cooled slowly, sometimes the excess solute does not come out of solution. This is an unstable situation and the introduction of a small solid particle will encourage the release of excess solute.

supersonic speed speed greater than that at which sound travels, measured in †Mach numbers. In dry air at 0°C/32°F, sound travels at about 1,170 kph/727 mph, but decreases its speed with altitude until, at 12,000 m/39,000 ft, it is only 1,060 kph/658 mph.

When an aircraft passes the sound barrier, shock waves are built up that give rise to †sonic boom, often heard at ground level. US pilot Captain Charles Yeager was the first to achieve supersonic flight, in a Bell

VS-1 rocket plane on 14 October 1947.

superstring theory in physics and cosmology, a mathematical theory developed in the 1980s to explain the properties of †elementary particles and the forces between them (in particular, gravity and the nuclear forces) in a way that combines †relativity and †quantum theory. In string theory, the fundamental objects in the universe are not pointlike particles but extremely small stringlike objects. These objects exist in a universe of ten dimensions, but since the earliest moments of the Big Bang six of these have been compacted or "rolled up," so that now, only three space dimensions and one dimension of time are discernible.

There are many unresolved difficulties with superstring theory, but some physicists think it may be the ultimate "theory of everything" that explains all aspects of the universe within one framework.

Superstrings! String Theory Home Page http://www.physics.ucsb.edu/~jpierre/strings/

Excellent introduction to string theory. This simple, uncluttered Web site has string theory tutorials, lists reference materials and links, and includes a useful glossary of technical terms.

supersymmetry in physics, a theory that relates the two classes of elementary particle, the †fermions and the †bosons. According to supersymmetry, each fermion particle has a boson partner particle, and vice versa. It has not been possible to marry up all the known fermions with the known bosons, and so the theory postulates the existence of other, as yet undiscovered fermions, such as the photinos (partners of the photons), gluinos (partners of the gluons), and gravitinos (partners of the gravitons). Using these ideas, it has become possible to develop a theory of gravity – called **supergravity** – that extends Einstein's work and considers the gravitational, nuclear, and electromagnetic forces to be manifestations of an underlying superforce. Supersymmetry has been incorporated into the †superstring theory, and appears to be a crucial ingredient in the "theory of everything" sought by scientists.

supersynthesis †radio interferometer system in which two †synthesis aerials are used; one is static and utilizes the rotation of the Earth to provide a field of scan, the other is mobile.

suprachiasmatic nucleus in anatomy, one of two small regions in the †hypothalamus that play a part in controlling †circadian rhythm.

supraglacial on top of a glacier. A supraglacial stream flows over the surface of the glacier; supraglacial material collected on top of a glacier may be deposited to form †lateral and †medial moraines.

suprarenal gland alternative name for the †adrenal gland.

surd in mathematics, expression containing the root of an †irrational number that can never be exactly expressed – for example, $\sqrt{3} = 1.732050808...$.

$\sqrt{3}$ can be expressed in †index notation as $3^{1/2}$ or 3 to the †power half.

simplifying surds (roots) Expressions involving surds can be factorized numerically to simplify them. For example, to simplify $\sqrt{20} + \sqrt{5}$:

$\sqrt{20} = \sqrt{(4 \times 5)} = \sqrt{4} \times \sqrt{5} = 2\sqrt{5}$, so

$\sqrt{20} + \sqrt{5} = 2\sqrt{5} + \sqrt{5} = 3\sqrt{5}$

surface activity in chemistry, ability of a substance to influence the surface tension of a liquid. An example is the reduction in surface tension in water by the addition of a detergent.

surface area area of the outside surface of a solid.

surface-area-to-volume ratio in biology, the ratio of an animal's surface area (the area covered by its skin) to its total volume. This is high for small animals, but low for large animals such as elephants.

The ratio is important for endothermic (warm-blooded) animals because the amount of heat lost by the body is proportional to its surface area, whereas the amount generated is proportional to its volume. Very small birds and mammals, such as hummingbirds and shrews, lose a lot of heat and need a high intake of food to maintain their body temperature. Elephants, on the other hand, are in danger of overheating, which is why they have no fur.

surface chemistry branch of chemistry concerned with chemical reactions that occur at the interface between two phases, such the reaction of a gas at the surface of a liquid.

surface compound chemical compound that is formed only on the surface of a material. Such compounds usually have chemical properties that differ from the material they are covering. The oxide layers that cover metals such as aluminum and iron are examples of surface compounds.

surface pressure difference in the surface tension of a liquid and that of a †surfactant adsorbed on it. In general, this usually results in the surfactant spreading across the surface of the host liquid, for example as occurs with detergents and water.

surface run-off overland transfer of water after a rainfall. It is the most rapid way in which water reaches a river. The amount of surface run-off increases given (1) heavy and prolonged rainfall, (2) steep gradients, (3) lack of vegetation cover, (4) saturated or frozen soil. A †hydrograph can indicate the time the run-off takes to reach the river. Throughflow is another way water reaches a river.

surface tension property that causes the surface of a liquid to behave as if it were covered with a weak elastic skin; this is why a needle can float on water. It is caused by the exposed surface's tendency to contract to the smallest possible area because of cohesive forces between †molecules at the surface. Allied phenomena include the formation of droplets, the concave profile of a meniscus, and the †capillary action by which water soaks into a sponge.

surfactant contraction of surface-active agent, substance added to a liquid in order to increase its wetting or spreading properties. Detergents are surfactants.

surge abnormally high tide; see †storm surge.

surgical spirit †ethanol to which has been added a small amount of methanol to render it unfit to drink. It is used to sterilize surfaces and to cleanse skin abrasions and sores.

surging glacier glacier that has periods, generally lasting one to four years, of very rapid flow (up to several meters per hour compared to normal glaciers which move at a rate of several meters per year) followed by periods of stagnation lasting up to ten years. Surging glaciers are heavily crevassed.

surface area of common three-dimensional shapes

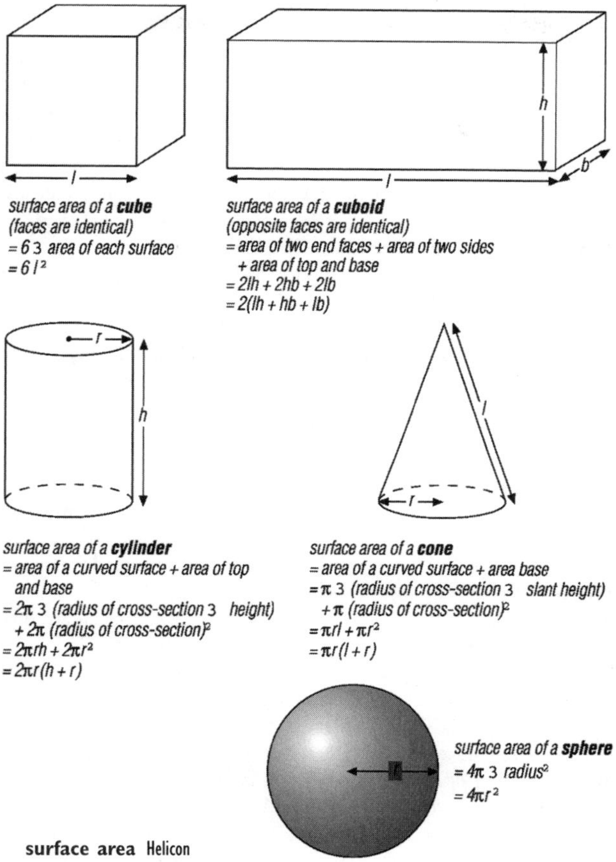

surface area Helicon

survey in statistics, a method of collecting data in which people are asked to answer a number of questions (usually in the form of a questionnaire). An opinion poll is a survey. The reliability of a survey's results depends on whether the sample from which the information has been collected is free from bias and sufficiently large.

surveying accurate measuring of the Earth's crust, or of land features or buildings. It is used to establish boundaries, and to evaluate the topography for engineering work. The measurements used are both linear and angular, and geometry and trigonometry are applied in the calculations.

susceptibility in physics, ratio of the intensity of magnetization produced in a material to the intensity of the magnetic field to which the material is exposed. It measures the extent to which a material is magnetized by an applied magnetic field. †Diamagnetic materials have small negative susceptibilities; †paramagnetic materials have small positive susceptibilities; †ferromagnetic materials have large positive susceptibilities. See also †magnetism.

suspension mixture consisting of small solid particles dispersed in a liquid or gas, which will settle on standing. An example is milk of magnesia, which is a suspension of magnesium hydroxide in water.

suspensory ligament in the †eye, a ring of fiber supporting the lens. The ligaments attach to the ciliary muscles, the circle of muscle mainly responsible for changing the shape of the lens during †accommodation. If the ligaments are put under tension, the lens becomes flatter, and therefore able to focus on objects in the far distance.

sustainable capable of being continued indefinitely. For example, the sustainable yield of a forest is equivalent to the amount that grows back. Environmentalists made the term a catchword, in advocating the sustainable use of resources.

sustainable development in ecology, development that could continue, in theory, for ever without seriously damaging the †environment. Environmental damage may involve, for example, pollution, loss of †habitats, extinction of †species, and reduction in the diversity of species (†biodiversity). In 1992 the greatest summit of leaders the world had ever seen took place in Rio de Janeiro – the Rio Summit. The intention was to set a framework to ensure that humans did not destroy the world through overexploitation and pollution. A key part of the conclusions of the summit was that development should be sustainable.

It is not easy to determine whether a proposed development is sustainable or not. Developments may be so pressing that they have to take place even though they damage the environment. However, the importance of the term "sustainable development" is that it should now be known and used worldwide because most world leaders were signatories at the Rio Summit.

SUSY in physics, an abbreviation for †supersymmetry.

Sutherland, Earl Wilbur, Jr (1915–1974) US physiologist who was awarded a Nobel Prize for Physiology or Medicine in 1971 for his discovery of cyclic AMP, a chemical messenger that plays a role in the action of many hormones.

Sutherland, Gordon Brims Black McIvor (1907–1980) Scottish physicist who used infrared †spectroscopy to study molecular structure. He elucidated the structure of a wide range of substances, from proteins to diamonds. He was knighted in 1960.

Sv in physics, symbol for †sievert, the SI unit of dose equivalent of ionizing radiation, equal to the dose in grays multiplied by a factor that depends mainly on the type of radiation and its effects.

Svedberg, Theodor (1884–1971) Swedish chemist. In 1923 he constructed the first ultracentrifuge, a machine that allowed the rapid separation of particles by mass. This can reveal the presence of contaminants in a sample of a new protein, or distinguish between various long-chain polymers. He was awarded the Nobel Prize for Chemistry in 1926 for his investigation of dispersed systems.

swallet alternative name for a †swallow hole.

swallow hole or **swallet**, hole, often found in limestone areas, through which a surface stream disappears underground. It will usually lead to an underground network of caves.

Swammerdam, Jan (1637–1680) Dutch naturalist who is considered a founder of both comparative anatomy and entomology. Based on their metamorphic development, he classified insects into four main groups, three of which are still used in a modified form in insect classification. He was also probably the first to discover red blood cells when he observed oval particles in frog's blood in 1658.

swamp region of low-lying land that is permanently saturated with water and usually overgrown with vegetation; for example, the everglades of Florida, USA. A swamp often occurs where a lake has filled up with sediment and plant material. The flat surface so formed means that run-off is slow, and the water table is always close to the surface. The high humus content of swamp soil means that good agricultural soil can be obtained by draining.

swash advance of water and sediment up a beach as a †wave breaks. Swash plays a significant role in the movement of beach material by †longshore drift, and is responsible for throwing shingle and pebbles up a beach to create ridges called †berms.

S-wave abbreviation of secondary wave, in seismology, a class of †seismic wave that passes through the Earth in the form of transverse shear waves. S-waves from an earthquake travel at roughly half the speed of P-waves (about 3 kps/1.9 mps), the speed depending on the density of the rock, and arrive later at monitoring stations (hence secondary waves) though with greater amplitude. They can travel through solid rock but not through the liquid outer core of the Earth. Rocks move perpendicular (at right angles) to the direction of travel of the wave, and as such an S-wave is an example of a †transverse wave.

sweat gland †gland within the skin of mammals that produces surface perspiration. In primates, sweat glands are distributed over the whole body, but in most other mammals they are more localized; for example, in cats and dogs they are restricted to the feet and around the face.

Swift-Tuttle, comet large, bright short-period comet discovered in 1862 then lost and rediscovered in 1992. It is thought to be the source of the Perseid meteor shower.

swim bladder thin-walled, air-filled sac found between the gut and the spine in bony fishes. Air enters the bladder from the gut or from surrounding capillaries (see †capillary), and changes of air pressure within the bladder maintain buoyancy whatever the water depth.

In evolutionary terms, the swim bladder of higher fishes is a derivative of the lungs present in all primitive fishes (not just lungfishes).

syenite gray, crystalline, plutonic (intrusive) †igneous rock, consisting of feldspar and hornblende; other minerals may also be present, including small amounts of quartz.

Sylow, Ludwig Mejdell (1832–1918) Norwegian mathematician who in 1872 published his fundamental theorem on groups and the type of subgroups now named after him.

Sylvester, James Joseph **(1814–1897)** English mathematician who was one of the discoverers of the theory of algebraic invariants. He coined the term "matrix" in 1850 to describe a rectangular array of numbers out of which determinants can be formed.

Sylvester, James Joseph

http://www-history.mcs.st-and.ac.uk/history/Mathematicians/Sylvester.html

Extensive biography of the mathematician. The site contains a description of his working relationships with his contemporaries, and also includes a page from his book A new Class of Theorems. Several literature references for further reading on the mathematician are also listed, and the Web site also features a photograph of Sylvester.

symbiosis any close relationship between two organisms of different species, and one where both partners benefit from the association. A well-known example is the pollination relationship between insects and flowers, where the insects feed on nectar and carry pollen from one flower to another. This is sometimes known as †mutualism.

symbol, chemical letter or letters used to represent a chemical element, usually derived from the beginning of its English or Latin name. Symbols derived from English include B, boron; C, carbon; Ba, barium; and Ca, calcium. Those derived from Latin include Na, sodium (Latin *natrium*); Pb, lead (Latin *plumbum*); and Au, gold (Latin *aurum*).

symmetry exact likeness in shape about a given line (axis), point, or plane. A figure has symmetry if one half can be rotated and/or reflected onto the other. (Symmetry preserves length, angle, but not necessarily orientation.) In a wider sense, symmetry exists if a change in the system leaves the essential features of the system unchanged; for example, reversing the sign of electric charges does not change the electrical behavior of an arrangement of charges.

line symmetry In the diagram, the letter A has one **line of symmetry**, or mirror line, shown by the dotted line. This is the line of reflection:

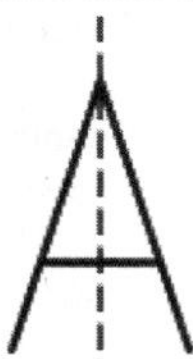

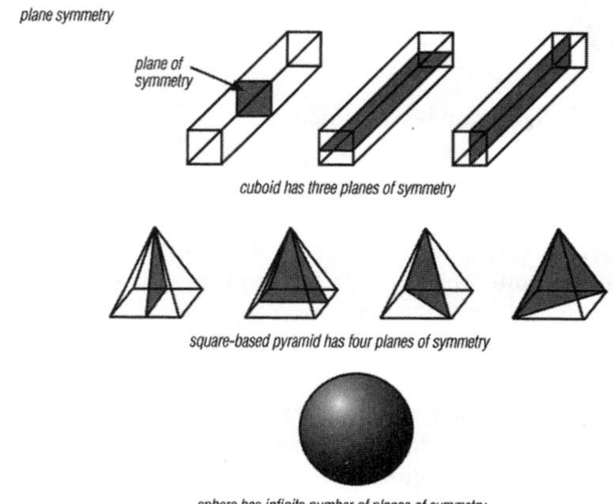

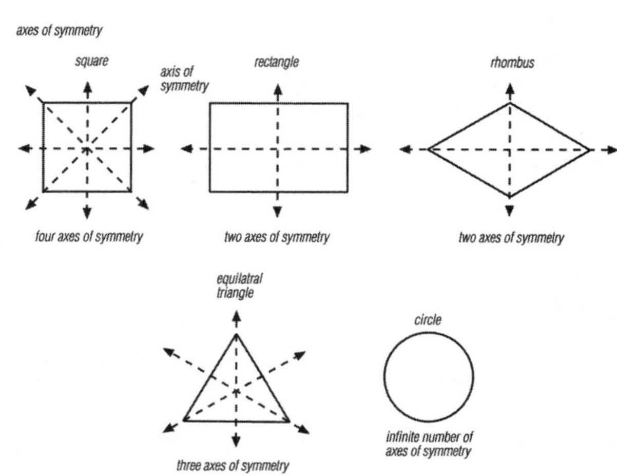

symmetry Helicon

rotational symmetry The diagram shows that the shape may be rotated about O into three identical positions. It has rotational symmetry of order three:

†Transformation of shapes can also take place by †translation, †rotation, †reflection, and †enlargement.

sympatric in biology, formation of new species that takes place when various factors (isolating mechanisms) prevent the members of populations of the same species in close proximity from interbreeding.

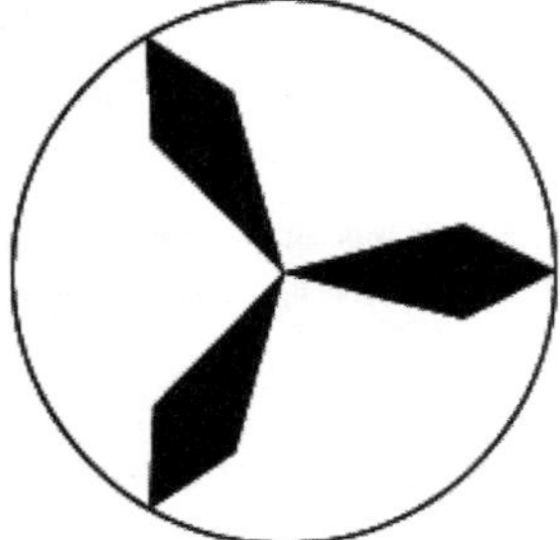

Differences in courtship behavior and sperm and egg incompatibility are examples.

synapse junction between two †nerve cells, or between a nerve cell and a muscle (a neuromuscular junction), across which a nerve impulse is transmitted. The two cells are separated by a narrow gap called the **synaptic cleft**. The gap is bridged by a chemical †neurotransmitter, released by the nerve impulse. The threadlike extension, or †axon, of the transmitting nerve cell has a slightly swollen terminal point, the **synaptic knob**. This forms one half of the synaptic junction and houses membrane-bound vesicles, which contain a chemical neurotransmitter. When nerve impulses reach the knob, the vesicles release the transmitter and this flows across the gap and binds itself to special receptors on the receiving cell's membrane. If the receiving cell is a nerve cell, the other half of the synaptic junction will be one or more extensions called †dendrites; these will be stimulated by the neurotransmitter to set up an impulse, which will then be conducted along the length of the nerve cell and on to its own axons. If the receiving cell is a muscle cell, it will be stimulated by the neurotransmitter to contract.

Synapsida group of mammal-like reptiles living 315–195 million years ago, whose fossil record is largely complete, and who were for a while the dominant land animals, before being replaced by the dinosaurs. The true mammals are their descendants.

synchronous orbit another term for †geostationary orbit.

synchronous rotation in astronomy, another name for †captured rotation.

synchrotron particle †accelerator in which particles move, at increasing speed, around a hollow ring. The particles are guided around the ring by electromagnets, and accelerated by electric fields at points around the ring. Synchrotrons come in a wide range of sizes, the smallest being about 1 m/3.3 ft across while the largest is 27 km/17 mi across. The Tevatron synchrotron at †Fermilab is some 6 km/4 mi in circumference and accelerates protons and antiprotons to 1 TeV.

The European Synchrotron Radiation Facility (ESRF) opened in Grenoble, France, in September 1994, funded by £400 million from 12 European countries.

synchrotron radiation polarized form of †radiation produced by high-speed †electrons in a magnetic field; it is this type of radiation that is emitted by the Crab nebula (in Taurus).

syncline geological term for a fold in the rocks of the Earth's crust in which the layers or †beds dip inwards, thus forming a trough-like structure with a sag in the middle. The opposite structure, with the beds arching upwards, is an †anticline.

Synge, Richard Laurence Millington (1914–1994) British biochemist who improved paper †chromatography (a means of separating mixtures) to the point where individual amino acids could be identified. He shared the Nobel Prize for Chemistry in 1952 with his colleague Archer †Martin for the development in 1944 of the technique known as partition chromatography.

synodic period time taken for a planet or moon to return to the same position in its orbit as seen from the Earth; that is, from one †opposition to the next. It differs from the †sidereal period because the Earth is moving in orbit around the Sun.

synoptic chart weather chart in which symbols are used to represent the weather conditions experienced over an area at a particular time. Synoptic charts appear on television and newspaper forecasts, although the symbols used may differ.

synovial fluid viscous colorless fluid that bathes movable joints between the bones of vertebrates. It nourishes and lubricates the †cartilage at the end of each bone.

Synovial fluid is secreted by a membrane, the synovium, that links movably jointed bones. The same kind of fluid is found in bursae, the membranous sacs that buffer some joints, such as in the shoulder and hip region.

synovial membrane membrane covering the articular extremities of †bones and the inner surface of †ligaments that make up a †joint. It secretes a clear lubricating fluid with an alkaline reaction, the †synovial fluid. Inflammation of the synovial membrane is **synovitis**.

syntactic polymer stereoregular synthetic †polymer polymerized using certain catalysts, such as †Ziegler catalysts, so that the substituent groups lie in alternating configurations on opposite sides of the main carbon chain.

synthesis in chemistry, the formation of a substance or compound from more elementary compounds. The synthesis of a drug can involve several stages from the initial material to the final product; the complexity of these stages is a major factor in the cost of production.

synthesis aerial †radio interferometer system utilizing a number of small aerials to achieve the effect of an impossibly large single one.

synthetic any material made from chemicals. Since the 1900s, more and more of the materials used in everyday life are synthetics, including plastics (polythene, polystyrene), synthetic fibers (nylon, acrylics, polyesters), synthetic resins, and synthetic rubber. Most naturally occurring organic substances can now be made synthetically, especially pharmaceuticals.

synthetic rubber any compound synthesized artificially that has the characteristics of natural rubber, for example †neoprene.

syrinx voice-producing organ of a bird. It is situated where the trachea divides in two and consists of vibrating membranes, a reverberating capsule, and numerous controlling muscles.

systematics science of naming and identifying species, and determining their degree of relatedness. It plays an important role in preserving †biodiversity; only a small fraction of existing species have been named and described. See also †classification.

Système International d'Unités official French name for †SI units.

systems analysis in computing, the investigation of a business activity or clerical procedure, with a view to deciding if and how it can be computerized. The analyst discusses the existing procedures with the people involved, observes the flow of data through the business, and draws up an outline specification of the required computer system. The next step is †systems design. A recent system is Unified Modeling Language (UML), which is specifically designed for the analysis and design of object-oriented programming systems.

Systems analysis and design methodologies currently in use include Yourdon, SSADM (Structured Systems Analysis and Design Methodology), and Soft Systems Methodology.

Systems Analysis Interesting Web Sites List http://www.umsl.edu/~sauter/analysis/ba215start.html

Systems analysis at the University of Missouri, St Louis. This collection of readings in systems analysis and software design was assembled to assist students but the notes are well presented and generally accessible and so of use to anyone with an interest in the field.

systems design in computing, the detailed design of an applications package. The designer breaks the system down into component programs, and designs the required input forms, screen layouts, and printouts. Systems design forms a link between systems analysis and †programming.

Systems are a series of blocks fitted together in a logical sequence. All systems have an input where information is put into a system, a process where the information is altered, and an output which communicates with the outside world. There are many different types of systems, including computer systems, mechanical systems, and electronic systems. Some systems have subsystems within the main system. Some have †feedback loops. Feedback loops pass information from one part of the system back to an earlier part. Positive feedback senses a signal and sends information back so that the signal is amplified. Negative feedback senses the signal and sends information back so that the signal is reduced. Feedback loops are a way of controlling a system. Systems can also be controlled by using sensors and switches. Control of a system can be external, using inputs such as switches, or it can be internal using devices such as feedback loops.

systems program in computing, a program that performs a task related to the operation and performance of the computer system itself. For example, a systems program might control the operation of the display screen, or control and organize backing storage. In contrast, an †applications program is designed to carry out tasks for the benefit of the computer user.

systole in biology, the contraction of the heart. It alternates with diastole, the resting phase of the heart beat.

syzygy in astronomy, alignment of three celestial bodies, usually the †Sun, †Earth, and †Moon or the Sun, Earth, and another planet. A syzygy involving the Sun, Earth, and Moon usually occurs during solar and lunar †eclipses.

The term also refers to the Moon or another planet when it is in †conjunction or †opposition.

Szasz, Thomas (1920–) Hungarian-born US psychiatrist. In his book *The Myth of Mental Illness* (1961), he argued that the concept of mental illness was false. Believing that individuals should take responsibility for their own actions, he supported the policy of moving mental patients out of hospital and into the community.

Szent-Györgyi, Albert von Nagyrapolt (1893–1986) Hungarian-born US biochemist who was awarded the Nobel Prize for Physiology or Medicine in 1937 for his investigation of biological oxidation processes and of the action of ascorbic acid (vitamin C).

Szilard, Leo (1898–1964) Hungarian-born US physicist who, in 1934, was one of the first scientists to realize that nuclear fission, or atom splitting, could lead to a chain reaction releasing enormous amounts of instantaneous energy. He emigrated to the USA in 1938 and there influenced Albert †Einstein to advise President Roosevelt to begin the nuclear-arms program. After World War II he turned his attention to the newly emerging field of molecular biology.

Leo Szilard Home Page http://www.dannen.com/szilard.html

Home page for the Hungarian-born physicist who was one of the first to realize the significance of splitting the atom. This site includes an illustrated biography, audio clips of people who worked with him, and documentation relating to the USA's decision to develop and then use the A-bomb.

T

T in physics, symbol for †tesla, the SI unit of magnetic flux density, equal to a flux density of one weber of flux per square meter.

table collection of information, usually numbers, arranged so that information is easy to find. **Two-way tables** are designed to be read from the top and from the side, for example a multiplication table, which gives all the products of numbers from one to ten.

tachyon in physics, hypothetical elementary particle that can travel faster than the speed of light, a condition that is not admitted by the special theory of relativity.

taiga or **boreal forest**, Russian name for the forest zone south of the †tundra, found across the northern hemisphere. Here, dense forests of conifers (spruces and hemlocks), birches, and poplars occupy glaciated regions punctuated with cold lakes, streams, bogs, and marshes. Winters are prolonged and very cold, but the summer is warm enough to promote dense growth.

The varied fauna and flora are in delicate balance because the conditions of life are so precarious. This ecology is threatened by mining, forestry, and pipeline construction.

talc $Mg_3Si_4O_{10}(OH)_2$, mineral, hydrous magnesium silicate. It occurs in tabular crystals, but the massive impure form, known as **steatite** or **soapstone**, is more common. It is formed by the alteration of magnesium compounds and is usually found in metamorphic rocks. Talc is very soft, ranked 1 on the Mohs scale of hardness. It is used in powdered form in cosmetics, lubricants, and as an additive in paper manufacture.

French chalk and potstone are varieties of talc. Soapstone has a greasy feel to it, and is used for carvings such as Inuit sculptures.

Tamm, Igor Yevgenyevich (1895–1971) Russian theoretical physicist who, with Paul †Cherenkov and Ilya †Frank, was awarded the Nobel Prize for Physics in 1958 for their discovery and interpretation of the blue light (†Cherenkov radiation) emitted from water exposed to radioactivity from radium. They showed that the radiation arises when a charged particle travels through a medium (liquid or solid) at greater than the speed of light in the medium. They were able to predict the direction and polarization of the radiation.

tangent in graphs, a straight line that touches a curve and gives the †gradient of the curve at the point of contact. At a maximum, minimum, or point of inflection, the tangent to a curve has zero gradient.

To find the gradient of the tangent of the curve at point P, extend the tangent AC and construct the right-angled triangle ABC:

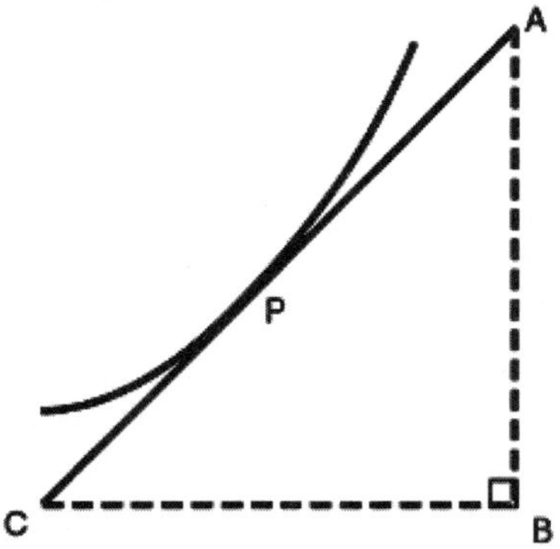

The gradient of the tangent at the point P is $^{AB}/_{BC}$. This is only an approximation. To make the answer as accurate as possible, the length of the tangent AC should be large.

For example, to find the gradient of the curve $y = x^2 - 2x$ at the point P(2,0):

P = AB/BC = 5/2.5 = 2

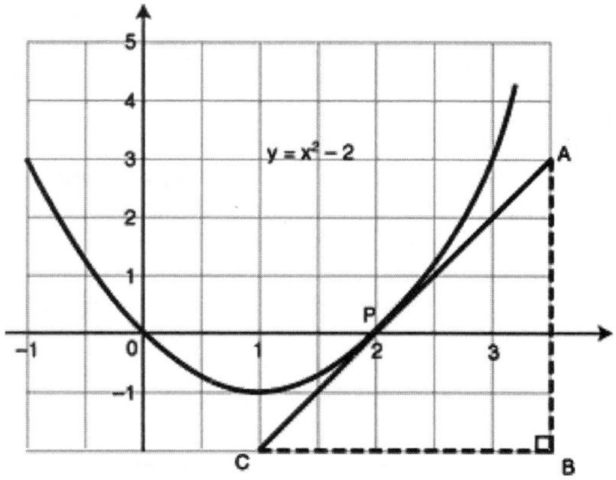

tangent **tan**, in trigonometry, a function of an acute angle in a right-angled †triangle, defined as the ratio of the length of the side opposite the angle to the length of the side adjacent to it; a way of expressing the gradient of a line. This function can be used to find either sides or angles in a right-angled triangle. For example, to work out the height (h) of a 5-m ramp set at an angle of 8°:

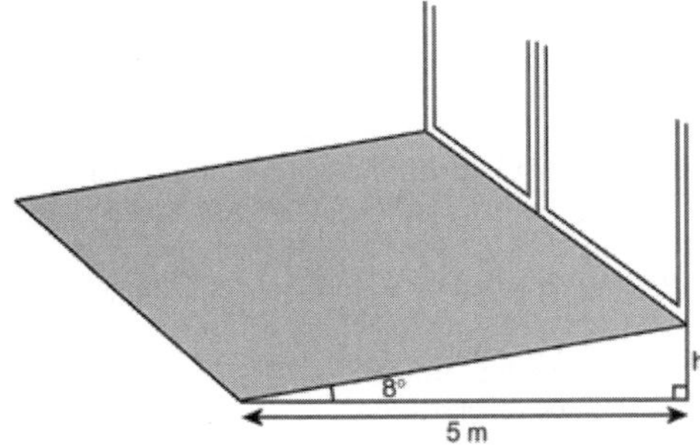

tan 8° = opposite/adjacent = $h/5$
$h = 5 \times \tan 8° = 0.70$ m (to the nearest cm)
The height of the ramp is 70 cm.
See also †sine, †cosine, and †tangent (graph).

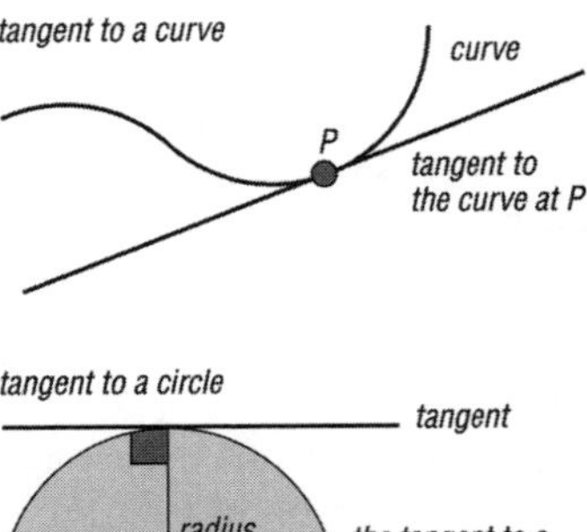

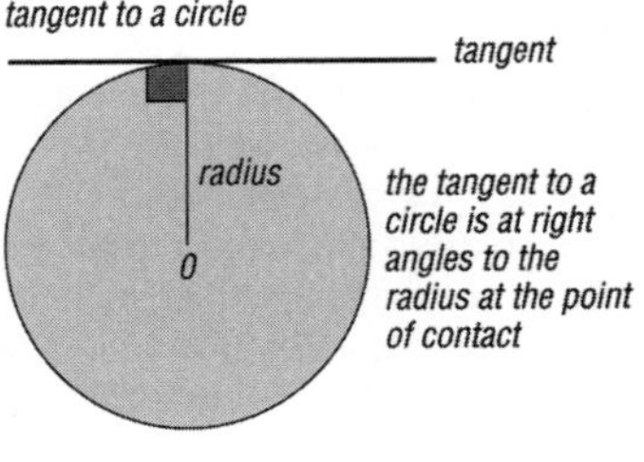

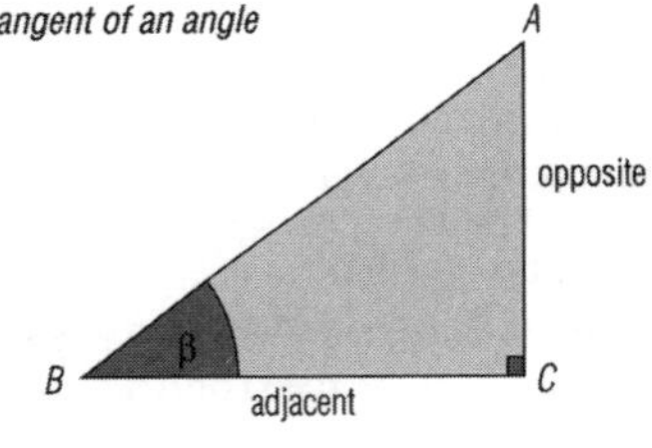

$$\text{tangent } \beta = \frac{\sin \beta}{\cos \beta} = \frac{\text{opposite}}{\text{adjacent}} = \frac{AC}{BC}$$

tangent The tangent of an angle is a mathematical function used in the study of right-angled triangles. If the tangent of an angle b is known, then the length of the opposite side can be found given the length of the adjacent side, or vice versa. Helicon

tangram puzzle made by cutting up a square into seven pieces.

tannic acid or **tannin**; $C_{14}H_{10}O_9$, yellow astringent substance, composed of several †phenol rings, occurring in the bark, wood, roots, fruits, and galls (growths) of certain trees, such as the oak. It precipitates gelatin to give an insoluble compound used in the manufacture of leather from hides (tanning).

Tansley, Arthur George (1871–1955) English botanist, a pioneer in the science of plant ecology. He coordinated a large project to map the vegetation of the British Isles; the results were published in *Types of British Vegetation* (1911). He was also instrumental in the formation of organizations devoted to the study of ecology and the protection of wildlife. He was knighted in 1950.

tantalum symbol Ta, hard, ductile, lustrous, gray-white, metallic element, atomic number 73, relative atomic mass 180.948. It occurs with niobium in tantalite and other minerals. It can be drawn into wire with a very high melting point and great tenacity, useful for lamp filaments subject to vibration. It is also used in alloys, for corrosion-resistant laboratory apparatus and chemical equipment, as a catalyst in manufacturing synthetic rubber, in tools and instruments, and in rectifiers and capacitors.

Tantalum was discovered and named in 1802 by Swedish chemist Anders Ekeberg (1767–1813) after the mythological Greek character Tantalus. It is mined as tantalite ore and most of the world's supply comes from Australia, with significant deposits in Brazil, Canada, and Nigeria. Tantalum is used in mobile phone capacitors, and as a consequence demand for the element has risen dramatically.

tapeworm any of various parasitic flatworms of the class Cestoda. They lack digestive and sense organs, can reach 15 m/50 ft in length, and attach themselves to the host's intestines by means of hooks and suckers. Tapeworms are made up of hundreds of individual segments, each of which develops into a functional hermaphroditic reproductive unit capable of producing numerous eggs. The larvae of tapeworms usually reach humans in imperfectly cooked meat or fish, causing anemia and intestinal disorders.

taproot in botany, a single, robust, main †root that is derived from the embryonic root, or †radicle, and grows vertically downwards, often to considerable depth. Taproots are often modified for food storage and are common in biennial plants such as the carrot *Daucus carota*, where they act as †perennating organs.

tar dark brown or black viscous liquid obtained by the destructive distillation of coal, shale, and wood. Tars consist of a mixture of hydrocarbons, acids, and bases. †Creosote and †paraffin oil are produced from wood tar. See also †coal tar.

Tarantula nebula extensive luminous nebula in the Large Magellanic Cloud. It is 1,000 light years across and includes a star of up to 200 solar masses, one of the most massive ever observed.

target tissue in biology, tissue on which a hormone (or group of hormones) acts. The cells of the target tissue have cell surface receptors (or receptors within the cytoplasm) to which a particular hormone binds. The receptor-hormone complexes initiate the series of reactions that form the cells' response to the hormone.

tartaric acid $HOOC(CHOH)_2COOH$, organic acid present in vegetable tissues and fruit juices in the form of salts of potassium, calcium, and magnesium. It is used in carbonated drinks and baking powders.

tartrazine E102, yellow food coloring produced synthetically from petroleum. Many people are allergic to foods containing it. Typical effects are skin disorders and respiratory problems. It has been shown to have an adverse effect on hyperactive children.

taste sense that detects some of the chemical constituents of food. The human †tongue can distinguish four basic tastes (sweet, sour, bitter, and salty) but it is supplemented by the sense of smell. What we refer to as taste is really a composite sense made up of both taste and smell. In 2000 US researchers confirmed the existence of the fifth taste, "umami," which was first proposed in the early 19th century by Japanese researcher Kikunae Ikeda. Umami is now called L-glutamate and a specific molecule receptor for it has been identified in taste buds.

Tatum, Edward Lawrie (1909–1975) US microbiologist who was awarded the Nobel Prize for Physiology or Medicine in 1958 for the discovery that genes regulate precise chemical effects. He shared the prize with his co-workers George †Beadle and Joshua †Lederberg.

tau †elementary particle with the same electric charge as the electron but a mass nearly double that of a proton. It has a lifetime of around 3×10^{-13} seconds and belongs to the †lepton family of particles – those which interact via the electromagnetic, weak nuclear, and gravitational forces, but not the strong nuclear force.

Taube, Henry (1915–) US chemist who established the basis of inorganic chemistry through his study of the loss or gain of electrons by atoms during chemical reactions. He was awarded the Nobel Prize for Chemistry in 1983 for his work on electron transference between molecules in chemical reactions.

Tau Ceti one of the nearest stars visible to the naked eye, 11.9 light years from Earth in the constellation Cetus. It has a diameter slightly less than that of the Sun, and an actual luminosity of about 45% of the Sun's. Its similarity to the Sun is sufficient to suggest that Tau Ceti may possess a planetary system, although observations have yet to reveal evidence of this.

Taurids minor meteor shower whose point of origin (radiant) appears to be in the constellation Taurus. They occur annually in October and early November.

Taurus conspicuous zodiacal constellation in the northern hemisphere near †Orion, represented as a bull. The Sun passes through Taurus from mid-May to late June. In astrology, the dates for Taurus are between about 20 April and 20 May (see †precession).

The V-shaped †Hyades open †star cluster forms the bull's head, with †Aldebaran as the red eye. The †Pleiades open cluster is in the shoulder. Taurus also contains the †Crab nebula, the remnants of the supernova of AD 1054, which is a strong radio and X-ray source and the location of one of the first †pulsars to be discovered.

tautomerism form of isomerism in which two interconvertible †isomers are in equilibrium. It is often specifically applied to an equilibrium between the keto ($-CH_2-C=O$) and enol ($-CH=C-OH$) forms of carbonyl compounds.

taxis plural **taxes**; or **tactic movement**, in botany, the movement of a single cell, such as a bacterium, protozoan, single-celled alga, or gamete, in response to an external stimulus. A movement directed towards the stimulus is described as positive taxis, and away from it as negative taxis. The alga *Chlamydomonas*, for example, demonstrates positive **phototaxis** by swimming towards a light source to increase the rate of photosynthesis.

Chemotaxis is a response to a chemical stimulus, as seen in many bacteria that move towards higher concentrations of nutrients.

taxonomy another name for the †classification of living organisms.

Taylor, Brook (1685–1731) English mathematician and secretary of the Royal Society.

Taylor, Joseph H (1941–) US radio astronomer who, with Russell †Hulse, was awarded the Nobel Prize for Physics in 1993 for their discovery of a new type of †pulsar, a binary pulsar. The pulsar has an invisible companion orbiting around it and radiates gravitational waves. This was the first experimental confirmation of the reality of waves predicted by Einstein's general theory of relativity.

Taylor, Richard E (1929–) Canadian physicist who worked with Jerome I †Friedman and Henry Way Kendall conducting pioneering research into the collision of high-energy electrons with protons and neutrons, which were important in developing the †quark model of particle physics. He shared the Nobel Prize for Physics in 1990 for experiments demonstrating that protons and neutrons are made up of quarks.

Taylor's theorem or **Brook Taylor's theorem**, expands a function of x as an infinite power series in powers of x.

TBT abbreviation for tributyl tin, chemical used in antifouling paints that has become an environmental pollutant.

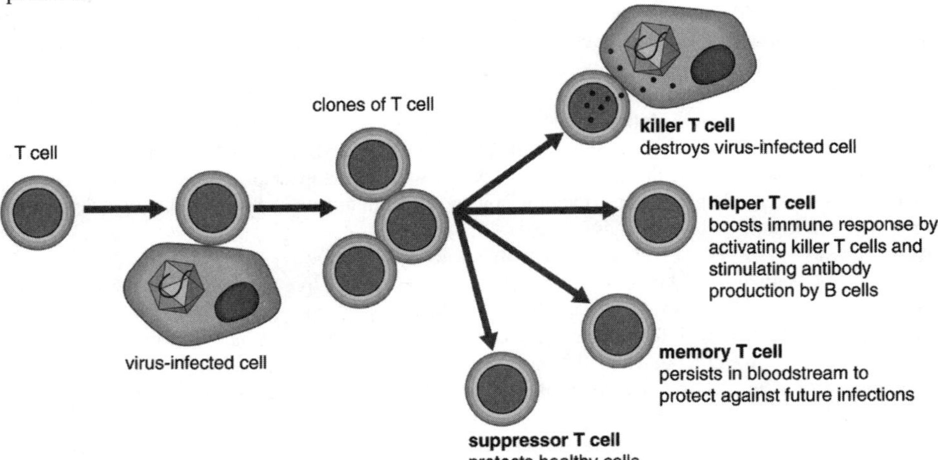

T cell T cells, a type of lymphocyte (white blood cell), have an important role in the body's immune system. When a T cell encounters an invading virus it begins to divide, forming four different types of T cell, each with a different function. Killer T cells destroy cells that have become infected with the virus by lysis. Helper T cells activate more killer T cells and also stimulate B cells to begin antibody production (unlike B cells, T cells do not produce antibodies to destroy invading pathogens). Suppressor T cells protect healthy cells from viral attack and memory T cells persist in the bloodstream to guard against re-infection. Helicon

T cell or **T lymphocyte**, immune cell (see †immunity and †lymphocyte) that plays several roles in the body's defenses. T cells are so called because they mature in the †thymus.

There are three main types of T cells: T helper cells (Th cells), which allow other immune cells to go into action; T suppressor cells (Ts cells), which stop specific immune reactions from occurring; and T cytotoxic cells (Tc cells), which kill cells that are cancerous or infected with viruses. Like †B cells, to which they are related, T cells have surface receptors that make them specific for particular antigens.

TCP/IP abbreviation for transmission or transfer control protocol/Internet protocol, set of network protocols, developed principally by the US Department of Defense. TCP/IP is the protocol used by the Internet, and is the technology that underpins Internet services like the †World Wide Web, Internet Relay Chat, and †e-mail. TCP/IP has always been the principal networking protocol used by †Unix, and is now supported by almost all types of operating system.

tears salty fluid exuded by lacrimal glands in the eyes. The fluid contains proteins that are antibacterial, and also absorbs oils and mucus. Apart from cleaning and disinfecting the surface of the eye, the fluid supplies nutrients to the cornea, which does not have a blood supply.

If insufficient fluid is produced, as sometimes happens in older people, the painful condition of "dry-eye" results and the eye may be damaged.

technetium chemical symbol Tc, (Greek *technetos* "artificial") silver-gray, radioactive, metallic element, atomic number 43, relative atomic mass 98.906. It occurs in nature only in extremely minute amounts, produced as a fission product from uranium in †pitchblende and other uranium ores. Its longest-lived isotope, Tc-99, has a half-life of 216,000 years. It is a superconductor and is used as a hardener in steel alloys and as a medical tracer.

It was synthesized in 1937 (named in 1947) by Italian physicists Carlo Perrier and Emilio Segrè, who bombarded molybdenum with deuterons, looking to fill a gap in the †periodic table of the elements (at that time it was considered not to occur in nature). It was later isolated in larger amounts from the fission product debris of uranium fuel in nuclear reactors.

tectonics in geology, the study of the movements of rocks on the Earth's surface. On a small scale tectonics involves the formation of †folds and †faults, but on a large scale †plate tectonics deals with the movement of the Earth's surface as a whole.

Science Odyssey: You Try It: Plate Tectonics http://www.pbs.org/wgbh/aso/tryit/tectonics/intro.html

Well-presented site to "the theory that Earth's outer layer is made up of plates." This site, labeled "Mountain Maker, Earth Shaker" has science-related interactive activities that use the Shockwave plug-in.

Teflon trade name for polytetrafluoroethene (PTFE), a tough, waxlike, heat-resistant plastic used for coating nonstick cookware and in gaskets and bearings.

In 2001 Canadian scientists discovered that Teflon degraded when heated to form trifluoroacetic acid (TFA), a pollutant that persists in the environment for many years.

tektite (from Greek *tektos* "molten") small, rounded glassy stone, found in certain regions of the Earth, such as Australasia. Tektites are probably the scattered drops of molten rock thrown out by the impact of a large †meteorite.

telecommunications communications over a distance, generally by electronic means. Long-distance voice communication was pioneered in 1876 by Scottish scientist Alexander Graham Bell when he invented the telephone. The telegraph, radio, and television followed. Today it is possible to communicate internationally by telephone cable or by satellite or microwave link, with over 100,000 simultaneous conversations and several television channels being carried by the latest satellites.

telescope optical instrument that magnifies images of faint and distant objects; any device for collecting and focusing light and other forms of electromagnetic radiation. A telescope with a large aperture, or opening, can distinguish finer detail and fainter objects than one with a small aperture. The **refracting telescope** uses lenses, and the **reflecting telescope** uses mirrors. A third type, the **catadioptric telescope**, is a combination of lenses and mirrors. See also †radio telescope.

refractor In a refractor, light is collected by a †lens called the **object glass** or **objective**, which focuses light down a tube, forming an image magnified by an **eyepiece**. Invention of the refractor is attributed to a Dutch optician, Hans †Lippershey, in 1608. Hearing of the invention in 1609, †Galileo quickly constructed one for himself and went on to produce a succession of such instruments which he used from 1610 onwards for astronomical observations. The largest refracting telescope in the world, at Yerkes Observatory, Willimas Bay, Wisconsin, has an aperture of 102 cm/40 in.

reflector In a reflector, light is collected and focused by a concave mirror. The first reflector was built about 1670 by Isaac †Newton. Large mirrors are cheaper to make and easier to mount than large lenses, so all the largest telescopes are reflectors. The largest reflector with a single mirror, 6 m/236 in, is at Zelenchukskaya, Russia. Telescopes with larger apertures composed of numerous smaller segments have been built, such as the Keck Telescopes on Mauna Kea, Hawaii. A **multiple-mirror telescope** was installed on Mount Hopkins, Arizona, USA, in 1979. It originally consisted of six mirrors of 1.8 m/72 in aperture, which performed like a single 4.5-m/176-in mirror. The six mirrors were replaced in 1996 by a single 6.5-m/21.3-ft mirror. †**Schmidt telescopes** are used for taking wide-field photographs of the sky. They have a main mirror plus a thin lens at the front of the tube to increase the field of view.

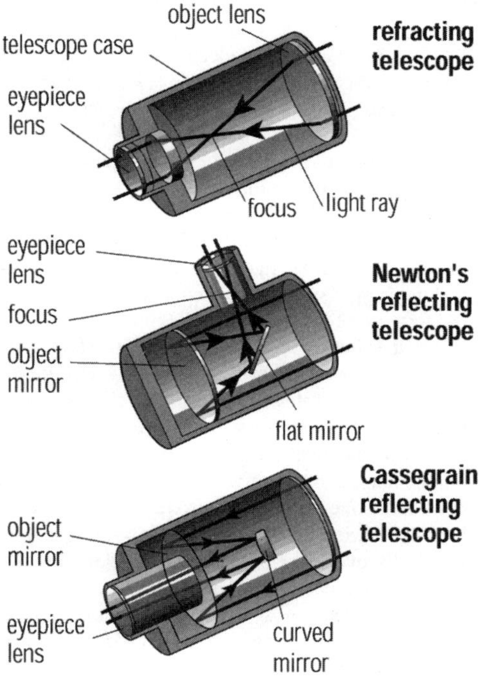

telescope Three kinds of telescope. The refracting telescope uses a large objective lens to gather light and form an image which the smaller eyepiece lens magnifies. A reflecting telescope uses a mirror to gather light. The Schmidt telescope uses a corrective lens to achieve a wide field of view. It is one of the most widely used tools of astronomy. Helicon

In 1995 NASA completed a 3-m/9.8-ft **liquid-mirror telescope** at its Orbital Debris Observatory in New Mexico. The liquid-mirror telescope is a reflecting telescope constructed with a rotating mercury mirror.

telescopes in space Large telescopes can now be placed in orbit above the distorting effects of the Earth's atmosphere. Telescopes in space have been used to study infrared, ultraviolet, and X-ray radiation that does not penetrate the atmosphere but carries much information about the births, lives, and deaths of stars and galaxies. The 2.4-m/94-in Hubble Space Telescope, launched in 1990, can see the sky more clearly than can any telescope on Earth.

An X-ray telescope for satellites, developed by UK, US, and Australian astronomers, is scheduled to be launched in 2005. It is based on the structure of a lobster's eye, which has thousands of square tubes reflecting light onto the retina. The $6.4 million Lobster Eye Telescope will contain millions of tubes 10–20 micrometers across.

NASA launched the Far Ultraviolet Spectroscopic Explorer (FUSE), an Earth-orbiting telescope, in June 1999.

Telescope http://es.rice.edu/ES/humsoc/Galileo/Things/telescope.html

Early history of the telescope, documenting its first astronomical use by Italian astronomer Galileo and English astronomer Thomas Harriot. There are photographs of Galileo's and English physicist Isaac Newton's telescopes and interesting drawings and engravings from old manuscripts.

Telescopium inconspicuous constellation of the southern hemisphere, represented as a telescope.

Telesto tiny moon of Saturn discovered in 1980. It has an irregular shape, maximum dimension 34 km/21 mi, and shares its orbit with Calypso and Tethys, 175,000 km/108,000 mi above the surface of the planet.

television TV, reproduction of visual images at a distance using radio waves. For transmission, a television camera converts the pattern of light it takes in into a pattern of electrical charges. This is scanned line by line by a beam of electrons from an electron gun, resulting in variable electrical signals that represent the picture. These signals are combined with a radio carrier wave and broadcast as electromagnetic waves. The TV aerial picks up the wave and feeds it to the receiver (TV set). This separates out the vision signals, which pass to a cathode-ray tube where a beam of electrons is made to scan across the screen line by line, mirroring the action of the electron gun in the TV camera. The result is a recreation of the pattern of light that entered the camera.

Teller, Edward (1908–2003) Hungarian-born US physicist known as the father of the hydrogen bomb (H-bomb). He worked on the Manhattan Project developing the fission bomb – the first atomic bomb 1942–46, and on the fusion, or hydrogen, bomb 1946–52. Vigorous in his promotion of nuclear weapons and in his opposition to communism, he was, in the 1980s, one of the leading advocates of the Star Wars program (the Strategic Defense Initiative).

Edward Teller Profile http://www.achievement.org/autodoc/page/tel0pro-1

Description of the life and work of the "father of the hydrogen bomb," physicist Edward Teller. The Web site contains not only a profile and biographical information, but also holds a lengthy interview with Teller from 1990 accompanied by a large number of photographs, video sequences, and audio clips.

tellurium chemical symbol Te, (Latin *Tellus* "Earth") silver-white, semi-metallic (†metalloid) element, atomic number 52, relative atomic mass 127.60. Chemically it is similar to sulfur and selenium, and it is considered one of the sulfur group. It occurs naturally in telluride minerals, and is used in coloring glass blue-brown, in the electrolytic refining of zinc, in electronics, and as a catalyst in refining petroleum.

It was discovered in 1782 by the Austrian mineralogist Franz Müller (1740–1825), and named in 1798 by the German chemist Martin Klaproth.

telomere (Greek *telos* "end"; *meros* "part") chromosome tip. Telomeres prevent chromosomes sticking together. Like DNA they are made up of nucleotides, usually rich in thymine and guanine. Every time a cell divides, the telomeres shorten. They trigger the cell's senescence (inability to reproduce) when they reach a threshold length. This process is prevented from happening during the replication of cancer cells by the presence of the enzyme **telomerase**. Telomerase replaces the segments of the telomeres, maintaining their length so that cell replication is no longer controlled.

Temin, Howard Martin (1934–1994) US virologist who was awarded a Nobel Prize for Physiology or Medicine in 1975 for work on the interactions between tumor-inducing viruses and the genetic material of the cell. He shared the prize with David Baltimore and Renato †Dulbecco.

temperature measure of how hot an object is. It is temperature difference that determines whether heat transfer will take place between two objects and in which direction it will flow, that is from warmer object to cooler object. The temperature of an object is a measure of the average kinetic energy possessed by the atoms or molecules of which it is composed. The SI unit of temperature is the kelvin (symbol K) used with the Kelvin scale. Other measures of temperature in common use are the Celsius scale and the Fahrenheit scale. The Kelvin scale starts at absolute zero (0 K = –273°C). The Celsius scale starts at the freezing point of water (0°C = 273 K). 1 K is the same temperature interval as 1°C.

The normal temperature of the human body is about 36.9°C/98.4°F. Variation by more than a degree or so indicates ill health, a rise signifying excessive activity (usually due to infection), and a decrease signifying deficient heat production (usually due to lessened vitality).

Explanation of Temperature Related Theories http://www.unidata.ucar.edu/staff/blynds/tmp.html

Detailed explanatory site on the laws and theories of temperature. It explains what temperature actually is, what a thermometer is, and the development of both, complete with illustrations and links to pioneers in the field. There is a temperature conversion facility and explanations of associated topics such as kinetic theory and thermal radiation.

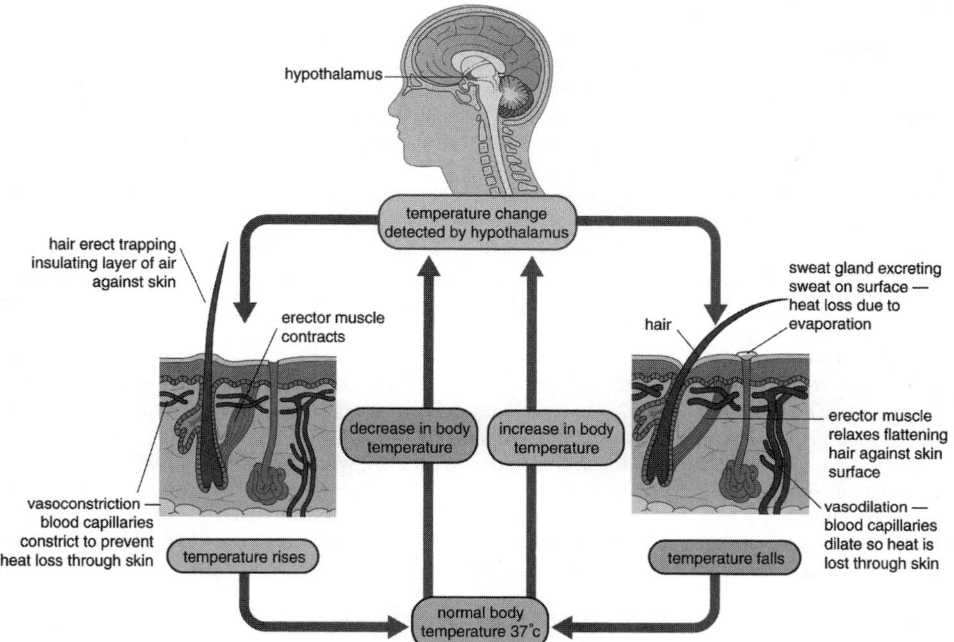

temperature regulation The ability of warm-blooded animals to control their body temperature by physiological means is an example of homeostasis (maintenance of a constant environment around living cells). As temperature increases, the change is detected by the hypothalamus and the skin responds by sweating and dilating capillaries near the skin surface so heat is lost by evaporation and convection. If the temperature falls, blood capillaries constrict and sweating ceases. Shivering causes heat generation within the muscles. Helicon

temperature regulation ability of an organism to control its internal body temperature. Animals that rely on their environment for their body temperature (and therefore have a variable temperature) are known as **ectotherms** or "cold-blooded" animals (for example, lizards). Animals with a constant body temperature irrespective of their environment are known as **endotherms** or "warm-blooded" animals (for example, birds and mammals). Their temperature is regulated by the hypothalamus in the brain.

Ectotherms have behavioral means of temperature control; they can warm themselves up by basking in the sun, or shivering, and can cool themselves down by sheltering from the sun under a rock or by bathing in water. In cold weather their metabolism slows and they become less active.

Endotherms too can regulate temperature by behavioral means (for example, seeking sun or shade), but also make use of physical and metabolic processes of regulation. The skin plays an important role – heat can be lost by sweating, whereby the water secreted evaporates from the skin and causes cooling. Heat can also be lost through capillaries near the surface of the skin, which flush with blood and lose heat by radiation. In the same way, heat can be retained by diverting blood from the surface capillaries, which gives the skin a paler color in the cold. Heat can also be retained by hairs which become erect and trap an insulating layer of air next to the skin (producing "goose bumps"). In some animals, there is a more specialized insulating layer such as fur, feathers, or †blubber. Endotherms can also gain heat metabolically through shivering, whereby involuntary muscle contractions release heat, and by an increased metabolism in the liver.

template specially cut piece of card, metal, or Perspex used to facilitate the drawing of shapes.

temporal bone bone forming part of the base and side walls of the †skull.

temporal niche diversification in ecology, sharing of a niche by different species separated in time, so that only one species occupies the niche for a particular period. In this way the resources of a habitat are fully exploited but inter-specific competition is reduced.

temporary hardness hardness of water that is removed by boiling (see †hard water).

tendon or **sinew**, in vertebrates, a cord of very strong, fibrous connective tissue that joins muscle to bone. Tendons are largely composed of bundles of fibers made of the protein collagen, and because of their inelasticity are very efficient at transforming muscle power into movement.

tendril in botany, a slender, threadlike structure that supports a climbing plant by coiling around suitable supports, such as the stems and branches of other plants. It may be a modified stem, leaf, leaflet, flower, leaf stalk, or stipule (a small appendage on either side of the leaf stalk), and may be simple or branched. The tendrils of Virginia creeper *Parthenocissus quinquefolia* are modified flower heads with suckerlike pads at the end that stick to walls, while those of the grapevine *Vitis* grow away from the light and thus enter dark crevices where they expand to anchor the plant firmly.

tension in physics, the stress (force) set up in a stretched material. In a stretched string or wire it exerts a pull that is equal in magnitude but opposite in direction to the stress being applied at the string ends. Tension is measured in newtons.

tensor calculus way of working with complicated mathematical objects. Position in ordinary space usually requires specification of three coordinates, singly indexed, for example, x_1, x_2, x_3. In describing mathematical objects more complicated than position, a generalized type of coordinate system may be used: for example, X_{ijk} where i, j, k can each take values 1, 2, or 3. Tensor calculus is a systematized use of such awkward objects.

terbium chemical symbol Tb, soft, silver-gray, metallic element of the †lanthanide series, atomic number 65, relative atomic mass 158.925. It occurs in gadolinite and other ores, with yttrium and ytterbium, and is used in lasers, semiconductors, and television tubes. It was named in 1843 by Swedish chemist Carl Mosander (1797–1858) after the town of Ytterby, Sweden, where it was first found.

terminal in computing, a device consisting of a keyboard and display screen (†VDU) to enable the operator to communicate with the computer. The terminal may be physically attached to the computer or linked to it by a telephone line (remote terminal). A dumb terminal has no processor of its own, whereas an intelligent terminal has its own processor and takes some of the processing load away from the main computer.

terminal moraine linear, slightly curved ridge of rocky debris deposited at the front end, or snout, of a glacier. It represents the furthest point of advance of a glacier, being formed when deposited material (till), which was pushed ahead of the snout as it advanced, became left behind as the glacier retreated.

terminal velocity or **terminal speed**, the maximum velocity that can be reached by a given object moving through a fluid (gas or liquid) under the action of an applied force. As the speed of the object increases so does the total magnitude of the forces resisting its motion. Terminal velocity is reached when the resistive forces exactly balance the applied force that has caused the object to accelerate; because there is now no resultant force, there can be no further acceleration.

For example, an object falling through air experiences air resistance. It will reach a terminal velocity and cease to accelerate under the influence of gravity when the air resistance equals the force of gravity (the object's weight).

Parachutes are designed to increase air resistance so that the acceleration of a falling person or package ceases more rapidly, thereby limiting terminal velocity to a safe level.

terminal voltage potential difference (pd) or voltage across the terminals of a power supply, such as a battery of cells. When the supply is not connected in circuit its terminal voltage is the same as its †electromotive force (emf); however, as soon as it begins to supply current to a circuit its terminal voltage falls because some electric potential energy is lost in driving current against the supply's own †internal resistance. As the current flowing in the circuit is increased the terminal voltage of the supply falls.

terminating decimal decimal fraction with a finite number of digits (a †recurring decimal, by contrast, has an infinite number of digits). Only those fractions with denominators that are powers of 2 and 5 can be converted into terminating decimals.

ternary compound in chemistry, compound composed of three elements, for example H_2SO_4.

tetrode thermionic valve (vacuum tube) with four electrodes consisting of anode, cathode, and two grids. A screen grid, at a fixed positive potential, is located between the anode and the control grid where it acts to improve the high-frequency performance of the valve.

terpene any of a class of organic compounds, originally derived from plant oils, that contain only carbon and hydrogen and are empirically regarded as derivatives of isoprene (C_5H_8). They are classified according to the number of isoprene units in the molecule – for example, monoterpenes contain two isoprene units and have the formula $C_{10}H_{16}$, sesquiterpenes contain three units ($C_{15}H_{24}$), and diterpenes contain four units ($C_{20}H_{32}$). Turpentine consists of a mixture of several monoterpenes. Rubber is a polyterpene with between 1,000 and 5,000 isoprene units.

terrane tract of land with a distinct geological character. The term **exotic terrane** is commonly used to describe a rock mass that has a very different history from others near by. The exotic terranes of the Western †Cordillera of North America represent old island chains that have been brought to the North

American continent by the movements of plate tectonics, and welded to its edge.

terrestrial planet any of the four small, rocky inner †planets of the Solar System: †Mercury, †Venus, †Earth, and †Mars. The †Moon is sometimes also included, although it is a satellite of the Earth and not strictly a planet.

terrigenous derived from or pertaining to the land. River sediment composed of weathered rock material and deposited near the mouth of the river on the ocean's continental shelf (the shallow ledge extending out from the continent) is called **terrigenous sediment**.

territorial behavior in biology, any behavior that serves to exclude other members of the same species from a fixed area or †territory. It may involve aggressively driving out intruders, marking the boundary (with dung piles or secretions from special scent glands), conspicuous visual displays, characteristic songs, or loud calls.

territory in animal behavior, a fixed area from which an animal or group of animals excludes other members of the same species. Animals may hold territories for many different reasons; for example, to provide a constant food supply, to monopolize potential mates, or to ensure access to refuges or nest sites.

The size of a territory depends in part on its function: some nesting and mating territories may be only a few square meters, whereas feeding territories may be as large as hundreds of square kilometers.

Tertiary period period of geological time 65 to 1.64 million years ago, divided into five epochs: Paleocene, Eocene, Oligocene, Miocene, and Pliocene. During the Tertiary period, mammals took over all the ecological niches left vacant by the extinction of the dinosaurs, and became the prevalent land animals. The continents took on their present positions, and climatic and vegetation zones as we know them became established. Within the geological time column the Tertiary follows the Cretaceous period and is succeeded by the Quaternary period.

Terylene trade name for a synthetic polyester fiber produced by the chemicals company ICI. It is made by polymerizing ethylene glycol and terephthalic acid. Cloth made from Terylene keeps its shape after washing and is hard-wearing.

Terylene was the first wholly synthetic fiber invented in Britain. It was created by the chemist J R Whinfield of Accrington in 1941. In 1942 the rights were sold to ICI (DuPont in the USA, where it is called Dacron) and bulk production began in 1955. Since 1970 it has been the most widely produced synthetic fiber, often under the generic name polyester. In 1989 8.4 million tonnes were produced, constituting over 50% of world synthetic fiber output.

tesla symbol T, SI unit of †magnetic flux density. One tesla represents a flux density of one †weber per square meter, or 10^4 †gauss. It is named after the Croatian-born US physicist Nikola Tesla.

Tesla, Nikola (1856–1943) Croatian-born US physicist and electrical engineer who invented fluorescent lighting, the Tesla induction motor (1882–87), and the Tesla coil, and developed the †alternating current (AC) electrical supply system.

Tesla, Nikola http://www.neuronet.pitt.edu/~bogdan/tesla/

Short biography of the electrical inventor plus quotations by and about Tesla, anecdotes, a photo gallery, and links to other sites of interest.

tessellation any tiling pattern that covers space without leaving any gaps.

testa outer coat of a seed, formed after fertilization of the ovule. It has a protective function and is usually hard and dry. In some cases the coat is adapted to aid dispersal, for example by being hairy. Humans have found uses for many types of testa, including the fiber of the cotton seed.

test cross in genetics, a breeding experiment used to discover the genotype of an individual organism. By crossing with a double recessive of the same species, the offspring will indicate whether the test individual is homozygous or heterozygous for the characteristic in question. In peas, a tall plant under investigation would be crossed with a double recessive short plant with known genotype tt. The results of the cross will be all tall plants if the test plant is TT. If the test plant is in fact Tt then there will be some short plants (genotype tt) among the offspring.

testis plural **testes**, organ that produces †sperm in male (and hermaphrodite) animals. In vertebrates it is one of a pair of oval structures that are usually internal, but in mammals (other than elephants and marine mammals), the paired testes (or testicles) descend from the body cavity during development, to hang outside the abdomen in a scrotal sac. The testes also secrete the male sex hormone †androgen.

testosterone †hormone secreted chiefly by the testes, but also by the ovaries and the cortex of the adrenal glands. It promotes the development of †secondary sexual characteristics in males at puberty. It is also needed for the development of the male sex organs and for male fertility. The hormone is partly responsible for the difference in behavior that may be seen between males and females. In animals with a breeding season, the onset of breeding behavior is accompanied by a rise in the level of testosterone in the blood.

Synthetic or animal testosterone is used to treat inadequate development of male characteristics or (illegally) to aid athletes' muscular development. Like other sex hormones, testosterone is a †steroid.

Tethys or **Saturn III**, one of the larger moons of †Saturn, discovered in 1684 by the Italian-born French astronomer Giovanni †Cassini. It has a diameter of 1,060 km/659 mi and orbits Saturn at an average distance of 295,000 km/183,000 mi with a period (time taken to circle the planet) of 1.89 days. Tethys is heavily cratered, with a notable 400-km/249-mi crater called Odysseus. A 2,000-km/1,243-mi complex of valleys called Ithaca Chasma stretches three-quarters of the way around the moon and may be a crack caused by the same impact that produced Odysseus. Tethys is in the same orbit as two smaller moons, Telesto and Calypso, each about 20–30 km/12–19 mi in size.

Tethys Sea sea that in the Mesozoic era separated †Laurasia from †Gondwanaland. The formation of the Alpine fold mountains caused the sea to separate into the Mediterranean, the Black, the Caspian, and the Aral seas.

tetrachloromethane or **carbon tetrachloride**; CCl_4, chlorinated organic compound that is a very efficient solvent for fats and greases, and was at one time the main constituent of household dry-cleaning fluids and of fire extinguishers used with electrical and gasoline fires. Its use became restricted after it was discovered to be carcinogenic and it has now been largely removed from educational and industrial laboratories.

tetraethyl lead $Pb(C_2H_5)_4$, compound added to leaded gasoline as a component of †antiknock to increase the efficiency of combustion in car engines. It is a colorless liquid that is insoluble in water but soluble in organic solvents such as benzene, ethanol, and gasoline.

tetragonal coordination in chemistry, most stable spatial arrangement of four identical atoms or groups that are bonded to the same atom. The atoms are arranged such that they are at the corners of a tetrahedron, as in methane, CH_4.

tetrahedron plural **tetrahedra**, in geometry, a solid figure (†polyhedron) with four triangular faces; that is, a †pyramid on a triangular base. A regular tetrahedron has equilateral triangles as its faces.

The volume (V) of a tetrahedron is given by

$V = 1/3 - Bh$

where B is the area of the base and h is the perpendicular height.

In chemistry and crystallography, tetrahedra describe the shapes of some molecules and crystals; for example, the carbon atoms in a crystal of diamond are arranged in space as a set of interconnected regular tetrahedra.

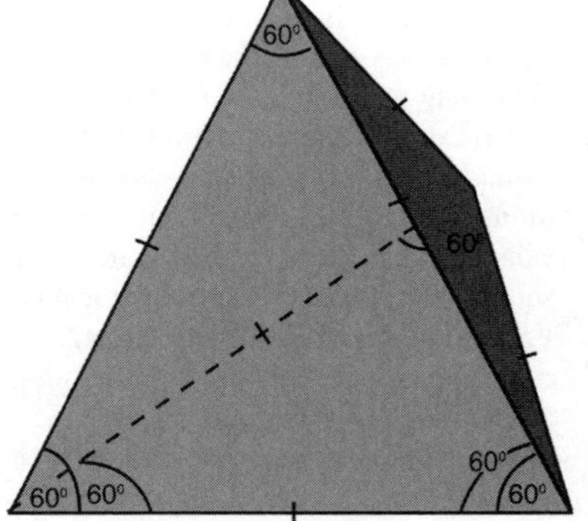

tetrahedron A regular tetrahedron is a pyramid on a triangular base with all its sides equal in length. Helicon

tetrahydrofuran or **tetramethylene oxide**; C_4H_8O, colorless organic liquid. It is a member of the †furan class of †heterocyclic compounds and is used industrially as a solvent for plastics and resins.

tetrahydrofuran polymer member of a class of synthetic polymers based on the polymerization of †tetrahydrofuran. It is mainly used as a †plasticizer with †polyurethane to produce a range of †elastomers and †thermoplastics.

tetrapod (Greek "four-legged") type of †vertebrate. The group includes mammals, birds, reptiles, and amphibians. Birds are included because they evolved from four-legged ancestors, the forelimbs having become modified to form wings. Even snakes are tetrapods, because they are descended from four-legged reptiles.

text editor in computing, a program that allows the user to edit text on the screen and to store it in a file. Text editors are similar to †word processors, except that they lack the ability to format text into paragraphs and pages and to apply different typefaces and styles.

thalamus one of two oval masses of gray matter lying in the cerebral hemispheres. It is an integrating center for sensory information on its way to the †cerebral cortex and is also important in motor control.

Thales (*c.* 624–*c.* 547 BC) also known as **Thales of Miletus**, Greek philosopher and scientist. He made advances in geometry, predicted an eclipse of the Sun in 585 BC, and, as a philosophical materialist, theorized that water was the first principle of all things. He speculated that the Earth floated on water, and so proposed an explanation for earthquakes. He lived in Miletus in Asia Minor.

thallium chemical symbol Tl, (Greek *thallos* "young green shoot") soft, bluish-white, malleable, metallic element, atomic number 81, relative atomic mass 204.38. It is a poor conductor of electricity. Its compounds are poisonous and are used as insecticides and rodent poisons; some are used in the optical-glass and infrared-glass industries and in photocells.

Discovered spectroscopically by its green line, thallium was isolated and named by William Crookes in 1861.

thallus any plant body that is not divided into true leaves, stems, and roots. It is often thin and flattened, as in the body of a seaweed, lichen, or liverwort, and the gametophyte generation (†prothallus) of a fern. Some flowering plants (†angiosperms) that are adapted to an aquatic way of life may have a very simple plant body which is described as a thallus (for example, duckweed *Lemna*).

thaumatin naturally occurring, non-carbohydrate sweetener derived from the bacterium *Thaumatococcus danielli*. Its sweetness is not sensed as quickly as that of other sweeteners, and it is not as widely used in the food industry.

thebaine $C_{19}H_{21}NO_3$, highly poisonous extract of opium.

Thebe tiny inner moon of Jupiter discovered in 1979. It is ellipsoidal in shape (longest dimension 110 km/68 mi) and it orbits 79,000 km/49,000 mi above the surface of the planet.

theodolite instrument for the measurement of horizontal and vertical angles, used in surveying. It consists of a small telescope mounted so as to move on two graduated circles, one horizontal and the other vertical, while its axes pass through the center of the circles.

Theodoric of Freiburg (*c.* 1250–1310) German scientist and monk. He studied in Paris 1275–77. In his work *De iride/On the Rainbow* he describes how he used a water-filled sphere to simulate a raindrop, and determined that colors are formed in the raindrops and that light is reflected within the drop and can be reflected again, which explains secondary rainbows.

Theon of Smyrna (lived *c.* AD 130) Greek astronomer and mathematician. In his celestial mechanics, the planets, Sun, Moon, and the sphere of fixed stars were all set at intervals congruent with an octave. His only surviving work, *Expositio rerum mathematicarum ad legendum Platoneum utilium*, is in two manuscripts, one on mathematics and one on astronomy and astrology.

theorem mathematical proposition that can be deduced by logic from a set of axioms (basic facts that are taken to be true without proof). Advanced mathematics consists almost entirely of theorems and proofs, but even at a simple level theorems are important.

theory in science, a set of ideas, concepts, principles, or methods used to explain a wide set of observed facts. Among the major theories of science are †relativity, †quantum theory, †evolution, and †plate tectonics.

theory of everything ToE, another name for †grand unified theory.

therm unit of energy defined as 10^5 British thermal units; equivalent to 1.055×10^8 J. It is no longer in scientific use.

thermal column of warm air, which is of a lower †density than its surroundings, and contains rising currents of air.

thermal analysis in chemistry, analytical technique that uses the direct action of heat to identify and obtain information about compounds. Differential thermal analysis, DTA, and differential scanning calorimetry, DSC, are examples of thermal analytical techniques.

thermal capacity another name for †heat capacity

thermal conductivity ability of a substance to conduct heat. Good thermal conductors, like good electrical conductors, are generally materials with many free electrons, such as †metals. A poor conductor, called an †insulator, has low conductivity.

Thermal conductivity is expressed in units of joules per second per meter per kelvin ($J\ s^{-1}\ m^{-1}\ K^{-1}$). For a block of material of cross-sectional area a and length l, with temperatures T_1 and T_2 at its end faces, the thermal conductivity λ equals $Hl/at(T_2 - T_1)$, where H is the amount of heat transferred in time t.

thermal dissociation reversible breakdown of a chemical compound into simpler substances by heating it (see †dissociation). The splitting of ammonium chloride into ammonia and hydrogen chloride is an example. On cooling, they recombine to form the salt.

thermal expansion expansion that is due to a rise in temperature. It can be expressed in terms of linear, area, or volume expansion.

The coefficient of linear expansion a is the fractional increase in length per degree temperature rise; area, or superficial, expansion b is the fractional increase in area per degree; and volume, or cubic, expansion g is the fractional increase in volume per degree. To a close approximation, b = 2a and g = 3a.

thermal reactor nuclear reactor in which the neutrons released by fission of uranium-235 nuclei are slowed down in order to increase their chances of being captured by other uranium-235 nuclei, and so induce further fission. The material (commonly graphite or heavy water) responsible for doing so is called a **moderator**. When the fast newly emitted neutrons collide with the nuclei of the moderator's atoms, some of their kinetic energy is lost and their speed is reduced. Those that have been slowed down to a speed that matches the thermal (heat) energy of the surrounding material are called **thermal neutrons**, and it is these that are most likely to induce fission and ensure the continuation of the chain reaction. See †nuclear reactor and †nuclear energy.

thermionics branch of electronics dealing with the emission of electrons from matter under the influence of heat.

thermionic tube or **thermionic valve**, evacuated metal or glass container enclosing a system of electrodes. The †cathode emits †electrons when heated, and these are attracted to a positively charged †anode. Perforated grid electrodes within the tube can be used to control the electron †current.

thermistor resistor whose †resistance changes significantly when its temperature changes. The resistance of a †semiconductor thermistor decreases with increase of temperature. As temperature rises, the resistance of the material decreases so that the current passing through a thermistor increases rapidly. They are used to electronically detect changes in temperature (electrical thermometers), as in fire alarms and thermostats. They also are used in lamp filaments and electric motors to stop large currents flowing through them when they are initially turned on.

When a lamp is switched on, the thermistor is cold initially and conducts poorly and has a high resistance. As the bulb heats up, the **intrinsic conduction** of the thermistor improves because the increasing thermal energy of vibrating atoms liberates electrons. These electrons flow more freely as current as the resistance of the lamp's thermistor decreases. The resistance of metals increases with temperature. When a metal is heated, the atoms vibrate more, thus impeding the flow of electrons.

thermochemistry branch of chemistry that deals with the heat changes that accompany chemical reactions.

thermocouple electric temperature measuring device consisting of a circuit having two wires made of different metals welded together at their ends. A current flows in the circuit when the two junctions are maintained at different temperatures (†Seebeck effect). The electromotive force generated – measured by a millivoltmeter – is proportional to the temperature difference.

thermodynamic equilibrium condition of a system where no heat flows between its component parts. See also †equilibrium.

thermodynamics branch of physics dealing with the transformation of heat into and from other forms of energy. It is the basis of the study of the efficient working of engines, such as the steam and internal combustion engines. The three laws of thermodynamics are: (1) energy can be neither created nor destroyed, heat and mechanical work being mutually convertible; (2) it is impossible for an unaided self-acting machine to convey heat from one body to another at a higher temperature; and (3) it is impossible by any procedure, no matter how idealized, to reduce any system to the †absolute zero of temperature (0 K/–273.15°C/–459.67°F) in a finite number of operations. Put into mathematical form, these laws have widespread applications in physics and chemistry.

thermography photographic recording of heat patterns. It is used medically as an imaging technique to identify "hot spots" in the body – for example, tumors, where cells are more active than usual. Thermography was developed in the 1970s and 1980s by the military to assist night vision by detecting the body heat of an enemy or the hot engine of a tank. It uses detectors sensitive to infrared (heat) radiation.

thermoluminescence **TL**, release, in the form of a light pulse, of stored nuclear energy in a mineral substance when heated to perhaps 500°C/930° F. The energy originates from the radioactive decay of uranium and thorium, and is absorbed by crystalline inclusions within the mineral matrix, such as quartz and feldspar. The release of TL from these crystalline substances is used in archaeology to date pottery, and by geologists in studying terrestrial rocks and meteorites.

thermoluminescent dating Crystalline substances found their way into the clay fabric of ancient pottery as additives designed to strengthen the material and allow it to breathe during kiln-firing at 600°C and above. Firing erased the huge level of TL energy accrued in geological times and sets a "time-zero" for fresh energy accumulation over archaeological times, the TL intensity measured today being proportional to the pottery's age. TL can date inorganic materials, including stone tools left as burnt flint, older than about 50,000–80,000 years, although it is regarded as less precise in its accuracy than radiocarbon dating.

thermometer instrument for measuring temperature. There are many types, designed to measure dif-

Monomer	Polymer	Polymer name	Uses
CH_2=CH_2 ethene	$[CH_2–CH_2]_n$	poly(ethene), polythene	bottles, packaging, insulation, pipes
CH_2=CH–CH_3 propene	$[CH_2–CH]_n$ (with CH_3 side group)	poly(propene), polypropylene	mouldings, film, fibres
CH_2=CH–CL chloroethene (vinyl chloride)	$[CH_2–CH]_n$ (with Cl side group)	polyvinylchloride (PVC), poly(chloroethene)	insulation, flooring, household fabric
CH_2=CH–C_6H_5 phenylethene (styrene)	$[CH_2–CH]_n$ (with C_6H_5 side group)	polystyrene, poly(phenylethene)	insulation, packaging
CF_2=CF_2 tetrafluoroethene	$[CF_2–CF_2]_n$ (n = 1000+)	poly(tetrafluoroethene) (PTFE)	high resistance to chemical and electrical reaction, low-friction applications

thermoplastic Examples of thermosoftening plastics, their basic monomer origins, polymer names, and everyday uses. Helicon

Monomer I	Monomer II	Polymer name	Uses
formaldehyde (methanal)	phenol	PF resins (Bakelites)	electrical fittings, radio cabinets
formaldehyde	urea	UF resins	electrical fittings, insulation, adhesives
formaldehyde	melamine	melamines	laminates for furniture

thermoset Unlike thermoplastics, thermosets remain rigid when set and do not soften when heated. Helicon

ferent temperature ranges to varying degrees of accuracy. Each makes use of a different physical effect of temperature. Expansion of a liquid is employed in common **liquid-in-glass thermometers**, such as those containing mercury or alcohol. The more accurate **gas thermometer** uses the effect of temperature on the pressure of a gas held at constant volume. A **resistance thermometer** takes advantage of the change in resistance of a conductor (such as a platinum wire) with variation in temperature. Another electrical thermometer is the †thermocouple. Mechanically, temperature change can be indicated by the change in curvature of a **bimetallic strip** (as commonly used in a †thermostat).

thermopile instrument for measuring radiant heat, consisting of a number of †thermocouples connected in series with alternate junctions exposed to the radiation. The current generated (measured by an †ammeter) is proportional to the radiation falling on the device.

thermoplastic or **thermosoftening plastic**, type of †plastic that always softens on repeated heating. Thermoplastics include polyethene (polyethylene), polystyrene, nylon, and polyester.

thermoset type of †plastic that remains rigid when set, and does not soften with heating. Thermosets have this property because the long-chain polymer molecules cross-link with each other to give a rigid structure. Examples include Bakelite, resins, melamine, and urea–formaldehyde resins.

thermosphere layer in the Earth's †atmosphere above the mesosphere and below the exosphere. Its lower level is about 80 km/50 mi above the ground, but its upper level is undefined. The ionosphere is located in the thermosphere. In the thermosphere the temperature rises with increasing height to several thousand degrees Celsius. However, because of the thinness of the air, very little heat is actually present.

thermostat temperature-controlling device that makes use of feedback. It employs a temperature sensor (often a bimetallic strip) to operate a switch or valve to control electricity or fuel supply. Thermostats are used in central heating, ovens, and engine cooling systems.

At the required preset temperature (for example of a room or gas oven), the movement of the sensor switches off the supply of electricity to the room heater or gas to the oven. As the room or oven cools down, the sensor turns back on the supply of electricity or gas.

theta functions in mathematics, four types of †elliptic function devised by German mathematician Carl Jacobi. Each function is defined as a Fourier series, and written q_1, q_2, q_3, and q_4; any q can be converted by translation of the argument into another q multiplied by a simple factor. (The quotient of any two q is then periodic twice.)

thiamine or **vitamin B$_1$**, a water-soluble vitamin of the B complex. It is found in seeds and grain. Its absence from the diet causes the disease †beriberi.

thiazole C_3H_3NS, volatile, colorless organic liquid composed of heterocyclic molecules, boiling point 117°C/243°F. It is mainly used in the production of dyes.

thio- in chemistry, prefix that indicates the presence of sulfur in a compound. It is usually used to denote that an oxygen atom has been substituted by sulfur.

thiol group in chemistry, sulfur analogue of the hydroxyl group, having the formula —SH. Compounds that contain this group are known as thiols (see †mercaptan).

thixotropy property of a substance that enables it to form a jellylike †colloid that reverts to a liquid on mechanical agitation. Nondrip paints are common thixotropic materials.

Thompson, Benjamin American-born British physicist; see Count von †Rumford.

Thompson, D'Arcy Wentworth (1860–1948) Scottish biologist and classical scholar who interpreted the structure and growth of organisms in terms of the physical forces to which every individual is subjected throughout its life. He also hypothesized, in his book *On Growth and Form* (1917), that the evolution of one species into another results mainly from transformations involving the entire organism.

Thompson, D'Arcy Wentworth http://www-groups.dcs.st-andrews.ac.uk/history/Mathematicians/Thompson_D'Arcy.html
Part of an archive containing the biographies of the world's greatest mathematicians, this site is devoted to the life and contributions of biologist D'Arcy Thompson.

Thomson, George Paget (1892–1975) English physicist who shared the Nobel Prize for Physics in 1937 for his work on †interference phenomena in the scattering of electrons by crystals which helped to confirm the wavelike nature of particles. He was knighted in 1943.

Thomson, J(oseph) J(ohn) (1856–1940) English physicist. He discovered the †electron in 1897. His work inaugurated the electrical theory of the atom, and his elucidation of positive rays and their application to an analysis of neon led to the discovery of †isotopes. He was awarded the Nobel Prize for Physics in 1906 for his theoretical and experimental work on the conduction of electricity by gases. He was knighted in 1908.

Thomson on the Number of Corpuscles in an Atom
http://dbhs.wvusd.k12.ca.us/Chem-History/Thomson-1906/Thomson-1906.html
J J Thomson's paper on the number of particles, or as he called them corpuscles, in an atom. The Web site is a reproduction of Thomson's publication in Philosophical Magazine vol. 11, in June 1906. The paper is reproduced with all the relevant equations and diagrams, but is not quite complete as the last two pages are missing. This does not detract from the work since all the deductions and the majority of the mathematics are complete before that point.

Thomson, William Irish physicist, see Lord †Kelvin.

thorax part of the body in four-limbed vertebrates containing the †heart and †lungs, and protected by the ribcage. It is separated from the abdomen by the diaphragm. During †breathing (ventilation) the volume inside the thorax is changed. This then causes air to move in or out of the air passages that lead to the lungs. The volume of the thorax is altered by the contraction of †muscles in the diaphragm and the contraction of muscles between the ribs – the intercostal muscles.

In arthropods, the thorax is the middle part of the body, between the head and abdomen. In insects the thorax bears the legs and wings. The thorax of spiders and crustaceans, such as lobsters, is fused with the head, to form the cephalothorax.

thorium chemical symbol Th, dark-gray, radioactive, metallic element of the †actinide series, atomic number 90, relative atomic mass 232.038. It occurs throughout the world in small quantities in minerals such as thorite and is widely distributed in monazite beach sands. It is one of three fissile elements (the others are uranium and plutonium) and its longest-lived isotope has a half-life of 1.39×10^{10} years. Thorium is used to strengthen alloys. It was discovered by Jöns Berzelius in 1828 and was named by him after the Norse god Thor.

Thorndike, Edward Lee (1874–1949) US educational psychologist whose experiments in behavior of cats and dogs in a "puzzle box" brought him to the conclusion that learning was improved when it achieved a satisfactory result. He extended this theory to human learning and found that students were encouraged by good results, but that being wrong did not teach them to correct their errors.

three-body problem mathematical problem in astronomy, to describe the gravitational effects of three interacting celestial bodies on each other, and the shape of their orbits round each other.

3-D graphics in computing, graphics defined by width, height, and depth (in mathematical terms, x, y, and z axes). In **business applications** such as spreadsheets, 3-D graphics allow users to display complex relationships between several different types of data. In **computer animation**, animators can make films composed entirely of computer-generated 3-D graphics, such as *A Bug's Life* (1999) and *Toy Story 2* (2000).

three-dimensional **3D**, describing a solid shape having three measurements: length, width, and depth. Examples of three-dimensional shapes are a †cube, †cuboid, and †prism.

throat in human anatomy, the passage that leads from the back of the nose and mouth to the †trachea and †esophagus. It includes the †pharynx and the †larynx, the latter being at the top of the trachea. The word "throat" is also used to mean the front part of the neck, both in humans and other vertebrates; for example, in describing the plumage of birds. In engineering, it is any narrowing entry, such as the throat of a carburetor.

thrombin substance that converts the soluble protein fibrinogen to the insoluble fibrin in the final stage of blood clotting. Its precursor in the bloodstream is †prothrombin.

thulium chemical symbol Tm, soft, silver-white, malleable and ductile, metallic element of the †lanthanide series, atomic number 69, relative atomic mass 168.94. It is the least abundant of the rare earth metals, and was first found in gadolinite and various other minerals. It is used in arc lighting.

The X-ray-emitting isotope Tm-170 is used in portable X-ray units. Thulium was named by Swedish chemist Per Cleve in 1879 after the northland, Thule.

thunderstorm severe storm of very heavy rain, thunder, and lightning. Thunderstorms are usually caused by the intense heating of the ground surface during summer. The warm air rises rapidly to form tall cumulonimbus clouds with a characteristic anvil-shaped top. Electrical charges accumulate in the clouds and are discharged to the ground as flashes of lightning. Air in the path of lightning becomes heated and expands rapidly, creating shock waves that are heard as a crash or rumble of thunder.

The rough distance between an observer and a lightning flash can be calculated by timing the number of seconds between the flash and the thunder. A gap of three seconds represents about a kilometer; five seconds represents about a mile.

Thunderstorms and Tornadoes http://www.geog.ouc.bc.ca/physgeog/contents/7t.html

Striking photographs, one of a 21-cm/8-in hailstone, another of the anvil-shaped cloud typical of a severe thunderstorm, illustrate this hyperlinked description of some dramatic weather events. Part of a much larger physical geography site set up by a Canadian university, the page gives a clear explanation of the various types of thunderstorm and the conditions that bring them about.

thymine in biochemistry, a colorless crystalline solid, also known as 5-methyl-2,6-dioxytetrahydropyrimidine. Thymine is one of four molecules that form the †base pairs in the DNA molecule, where it is always paired with the purine †adenine. Its base pairing with adenine occurs through hydrogen bonding.

thymus organ in vertebrates, situated in the upper chest cavity in humans. The thymus processes †lymphocyte cells to produce T-lymphocytes (T denotes "thymus-derived"), which are responsible for binding to specific invading organisms and killing them or rendering them harmless.

The thymus reaches full size at puberty, and shrinks thereafter; the stock of T-lymphocytes is built up early in life, so this function diminishes in adults, but the thymus continues to function as an †endocrine gland, producing the hormone thymosin, which stimulates the activity of the T-lymphocytes.

thyristor type of †rectifier, an electronic device that conducts electricity in one direction only. The thyristor is composed of layers of †semiconductor material sandwiched between two electrodes called the anode and cathode. The current can be switched on by using a third electrode called the gate.

Thyristors are used to control mains-driven motors and in lighting dimmer controls.

thyroid †endocrine gland of vertebrates, situated in the neck in front of the trachea. It secretes several hormones, principally thyroxine, an iodine-containing hormone that stimulates growth, metabolism, and other functions of the body. The thyroid gland may be thought of as the regulator gland of the body's metabolic rate. If it is overactive, as in hyperthyroidism, the sufferer feels hot and sweaty, has an increased heart rate, diarrhea, and weight loss. Conversely, an underactive thyroid leads to myxoedema, a condition characterized by sensitivity to the cold, constipation, and weight gain. In infants, an underactive thyroid leads to cretinism, a form of mental retardation.

thyroxine hormone containing iodine that is produced by the thyroid gland. It is used to treat conditions that are due to deficiencies in thyroid function, such as myxoedema.

tibia anterior of the pair of bones in the leg between the ankle and the knee. In humans, the tibia is the shinbone. It articulates with the †femur above to form the knee joint, the †fibula externally at its upper and lower ends, and with the talus below, forming the ankle joint.

tidal energy energy derived from the tides. The tides mainly gain their potential energy from the gravitational forces acting between the Earth and the Moon. If water is trapped at a high level during high tide, perhaps by means of a barrage across an estuary, it may then be gradually released and its associated †gravitational potential energy exploited to drive †turbines and generate electricity. Several schemes have been proposed for the Bristol Channel, in southwestern England, but environmental concerns as well as construction costs have so far prevented any decision from being taken.

tidal heating in astrophysics, process in which one body is heated internally by tidal stresses set up by the gravitational pull of another body. Tidal heating is common among the moons of the giant planets, and is the heat source for volcanic activity on †Io, one of the moons of †Jupiter.

tidal wave common name for a †tsunami.

tide rhythmic rise and fall of the †sea level in the Earth's oceans and their inlets and estuaries due to the gravitational attraction of the Moon and, to a lesser extent, the Sun, affecting regions of the Earth unequally as it rotates. Water on the side of the Earth nearest to the Moon feels the Moon's pull and accumulates directly below the Moon, producing a high tide.

High tide occurs at intervals of 12 hr 24 min 30 sec. The maximum high tides, or spring tides, occur at or near new and full Moon when the Moon and Sun are in line and exert the greatest combined gravitational pull. Lower high tides, or neap tides, occur when the Moon is in its first or third quarter and the Moon and Sun are at right angles to each other.

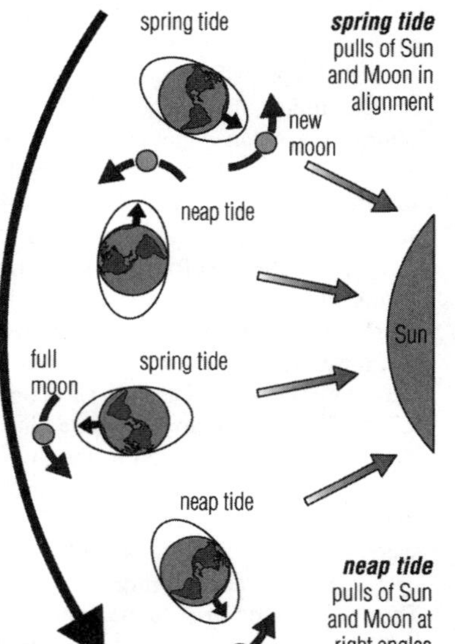

tide The gravitational pull of the Moon is the main cause of the tides. Water on the side of the Earth nearest the Moon feels the Moon's pull and accumulates directly under the Moon. When the Sun and the Moon are in line, at new and full Moon, the gravitational pull of Sun and Moon are in line and produce a high spring tide. When the Sun and Moon are at right angles, lower neap tides occur. Helicon

till or **boulder clay**, deposit of clay, mud, gravel, and boulders left by a †glacier. It is unsorted, with all sizes of fragments mixed up together, and shows no stratification; that is, it does not form clear layers or †beds.

time continuous passage of existence, recorded by division into hours, minutes, and seconds. Formerly the measurement of time was based on the Earth's rotation on its axis, but this was found to be irregular. Therefore the second, the standard †SI unit of time, was redefined in 1956 in terms of the Earth's annual orbit of the Sun, and in 1967 in terms of a radiation pattern of the element cesium.

Walk Through Time: The Evolution of Time Measurement http://physics.nist.gov/GenInt/Time/time.html

Designed by the National Institute of Standards and Technology (NIST) Physics Laboratory to provide a historic understanding of the evolution of time measurement. There are sections on "Ancient calendars," "Early clocks," "World time scales," the "Atomic age," and "The revolution in timekeeping." The site also explains Greenwich Mean Time and world time zones. Visitors can synchronize their clocks to NIST time.

time-sharing in computing, a way of enabling several users to access the same computer at the same time. The computer rapidly switches between user †terminals and programs, allowing each user to work as if he or she had sole use of the system.

Time-sharing was common in the 1960s and 1970s before the spread of cheaper computers, and is now making something of a comeback, with the advent of **application service providers**.

time zone longitudinal strip of the Earth's surface, stretching from pole to pole and sharing the same time of day or night. In a 24-hour period the Earth makes one complete rotation on its axis; thus the direct rays of the Sun pass through one degree of longitude every 4 minutes. To allow for time changes on an hourly basis, each time zone covers 15 degrees of longitude in width. In practice, however, zone boundary lines are adjusted to accommodate political units. (See also †international date line).

Worldtime http://www.worldtime.com/

Interactive world atlas featuring information on local time and sunrise and sunsets in hundreds of cities, and a database of public holidays around the world. You can rotate the globe to view areas of daylight and night, zoom in on areas of interest, or display national borders.

tin chemical symbol Sn, soft, silver-white, malleable and somewhat ductile, metallic element, symbol Sn, atomic number 50, relative atomic mass 118.69. Its chemical symbol comes from the Latin *stannum*.

Tin exhibits †allotropy, having three forms: the familiar lustrous metallic form above 13.2°C/55.8°F, a brittle form above 161°C/321.8°F, and a gray powder form below 13.2°C/55.8°F (commonly called tin pest or tin disease). The metal is quite soft (slightly harder than lead) and can be rolled, pressed, or hammered into extremely thin sheets; it has a low melting point. In nature it occurs rarely as a free metal. It resists corrosion and is therefore used for coating and plating other metals.

Tin and copper smelted together form the oldest desired alloy, bronze; since the Bronze Age (3500 BC) that alloy has been the basis of both useful and decorative materials. Tin is also alloyed with metals other than copper to make solder and pewter. It was recognized as an element by Antoine Lavoisier, but the name is very old and comes from the Germanic form *zinn*. The mines of Cornwall were the principal Western source of tin until the 19th century, when rich deposits were found in South America, Africa, South-East Asia, and Australia. Tin production is concentrated in Malaysia, Indonesia, Brazil, and Bolivia.

Tinbergen, Niko(laas) (1907–1988) Dutch-born British zoologist who was awarded a Nobel Prize for Physiology or Medicine in 1973 for his work in animal behavior patterns. He specialized in the study of instinctive behavior in animals, and was one of the founders of †ethology, the scientific study of animal behavior in natural surroundings. He shared the prize with Konrad †Lorenz (with whom he worked on several projects) and Karl von †Frisch.

Ting, Samuel Chao Chung (1936–) US physicist. In 1974 he and his team at the Brookhaven National Laboratory, New York, detected a new subatomic particle, which he named the J particle. It was found to be identical to the y particle discovered in the same year by Burton †Richter and his team at the Stanford Linear Accelerator Center, California. Ting and Richter shared the Nobel Prize for Physics in 1976 for their discovery of the y meson.

tin ore mineral from which tin is extracted, principally cassiterite, SnO_2. The world's chief producers are Malaysia, Thailand, and Bolivia.

Tiselius, Arne Wilhelm Kaurin (1902–1971) Swedish chemist who developed a powerful method of chemical analysis known as electrophoresis. He applied his new techniques to the analysis of animal proteins. He was awarded the Nobel Prize for Chemistry in 1948 for his researches in electrophoresis and adsorption analysis, and discoveries concerning serum proteins.

tissue in biology, any kind of cellular fabric that occurs in an organism's body. It is a group of similar †cells that are carrying out a function in a plant or animal. Several kinds of tissue can usually be distinguished, each consisting of cells of a particular kind bound together by cell walls (in plants) or extracellular matrix (in animals). Thus, nerve and muscle are different kinds of tissue in animals, as are †parenchyma and †sclerenchyma in plants. Tissues of different kinds may be found in a distinct structure, which is then called an †organ. The leaf of a plant or the heart of a mammal is an organ.

One example of tissue is the †epithelium. Epithelial tissue is made up of epithelial cells of several kinds, such as ciliated epithelial cells and cells producing mucus. Epithelial cells line the air passages and the gut. In the air passages the epithelial tissue protects the walls from dust and micro-organisms that are trapped in the mucus and swept up to the mouth by the beating, hair-like cilia. In the gut, the epithelium protects the wall by producing mucus, but there are no ciliated cells. The mucus also helps food move along. In addition, digestive enzymes which digest food are produced by the epithelium.

tissue culture process by which cells from a plant or animal are removed from the organism and grown under controlled conditions in a sterile medium containing all the necessary nutrients. Tissue culture can provide information on cell growth and differentiation, and is also used in plant propagation and drug production.

tissue fluid in biology, fluid that forms from the blood plasma, which is forced through the walls of the capillary blood vessels by the high hydrostatic pressure at the arteriole end of a capillary network. The fluid bathes the surrounding tissues, providing cells not in direct contact with a capillary with oxygen, nutrients, and other materials.

tissue plasminogen activator tPA, naturally occurring substance in the body tissues that activates the enzyme plasmin that is able to dissolve blood clots. Human tPA, produced in bacteria by genetic engineering, has, like streptokinase, been used to dissolve blood clots in the coronary arteries of heart-attack victims. It has been shown to be more effective than streptokinase when used in conjunction with heparin, but it is much more expensive.

tissue rejection in medicine, series of events in the immune response that lead to the failure of transplanted tissue to survive within the body of a recipient. The risk of rejection is reduced if the antigens on the surface of the cells of the tissues of donor and recipient closely matched.

To reduce the risk of tissue rejection, a match between the human lymphocyte antigens (HLA) of donor and recipient is particularly important. Following the transplant, †immunosuppressive drugs are taken to further reduce the risks.

tissue typing in medicine, procedures that aim to identify the degree of match between the human lymphocyte antigens (HLA) of the donor and those of the recipient of a transplant. The closer the match, the less likely it is that the transplanted tissue will provoke an immune response in the recipient, leading to rejection of the donated tissue.

Titan in astronomy, largest moon of the planet Saturn, with a diameter of 5,150 km/3,200 mi and a mean distance from Saturn of 1,222,000 km/759,000 mi. It was discovered in 1655 by Dutch mathematician and astronomer Christiaan †Huygens, and is the second-largest moon in the Solar System (only Ganymede, of Jupiter, is larger).

Titan is the only moon in the Solar System with a substantial atmosphere (mostly nitrogen), topped with smoggy orange clouds that obscure the surface, which may be covered with liquid ethane lakes. Its surface atmospheric pressure is greater than Earth's. Radar signals suggest that Titan has dry land as well as oceans (among the planets, only Earth has both in the Solar System).

Titania or **Uranus III**, largest of the five major moons of †Uranus, discovered by German-born English astronomer William †Herschel in 1789. It is 1,580 km/982 mi in diameter, and orbits Uranus at an average distance of 436,000 km/271,000 mi with a period (the time it takes to circle the planet) of 8.71 days. Impact craters have exposed fresh, light-colored ice beneath its dark surface.

titanium chemical symbol Ti, strong, lightweight, silver-gray, metallic element, atomic number 22, relative atomic mass 47.90. The ninth most abundant element in the Earth's crust, its compounds occur in practically all igneous rocks and their sedimentary deposits. It is very strong and resistant to corrosion, so it is used in building high-speed aircraft and spacecraft; it is also widely used in making alloys, as it unites with almost every metal except copper and aluminum. Titanium oxide is used in high-grade white pigments.

Titanium bonds with bone in a process called **osseointegration**. As the body does not react to the titanium it is valuable for permanent implants such as prostheses.

The element was discovered in 1791 by English mineralogist William Gregor (1761–1817) and was named by German chemist Martin Klaproth in 1796 after the Titans, the giants of Greek mythology. It was not obtained in pure form until 1925.

titanium dioxide TiO_2, highly stable, white solid that occurs naturally as the mineral rutile. It is used in powder form as a white pigment for paints and plastics.

titanium ore any mineral from which titanium is extracted, principally ilmenite ($FeTiO_3$) and rutile (TiO_2). Brazil, India, and Canada are major producers. Both these ore minerals are found either in rock formations or concentrated in heavy mineral sands.

titration in analytical chemistry, a technique to find the concentration of one compound in a solution by determining how much of it will react with a known amount of another compound in solution.

One of the solutions is measured by †pipette into the reaction vessel. The other is added a little at a time from a †burette. The end-point of the reaction is determined with an †indicator or an electrochemical device.

TNT abbreviation for **trinitrotoluene**, $CH_3C_6H_2(NO_2)_3$, a powerful high explosive. It is a yellow solid, prepared in several isomeric forms from †toluene by using sulfuric and nitric acids.

tocopherol or **vitamin E**, fat-soluble chemical found in vegetable oils. Deficiency of tocopherol leads to multiple adverse effects on health. In rats, vitamin E deficiency has been shown to cause sterility.

Todd, Alexander Robertus (1907–1997) Baron Todd, Scottish organic chemist who was awarded the Nobel Prize for Chemistry in 1957 for his work on the role of nucleic acids in genetics. He also synthesized vitamins B_1, B_{12}, and E. He was knighted in 1954 and became baron in 1962.

tog unit of measure of thermal insulation used in the textile trade; a light summer suit provides 1.0 tog.

The tog value of an object is equal to ten times the temperature difference (in °C) between its two surfaces when the flow of heat is equal to one watt per square meter; one tog equals 0.645 †clo.

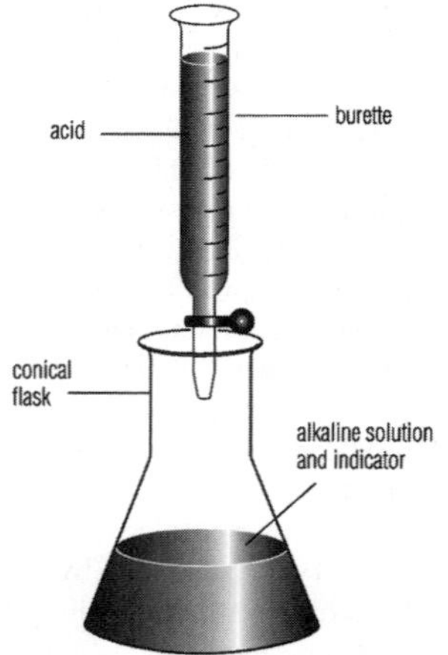

titration A method used to find the concentration of an acid or an alkali. Typically a burette is filled with an acid of unknown concentration which is slowly (drop by drop) added to a known volume of an alkali of a known concentration, mixed with an indicator (such as phenolphthalein). The volume of acid need to neutralize the alkali in the flask can be used to calculate the concentration of the acid. Helicon

toggle in computing, to switch between two settings. In software a toggle is usually triggered by the same code, so it is important that this code only has two meanings. An example is the use of the same character in a text file to indicate both opening and closing quotation marks; if the same character is also used to mean an apostrophe, then conversion, via a toggle switch, for a †desktop publishing system that uses different opening and closing quotation marks, will not be carried out correctly.

Tolansky, Samuel (1907–1973) English physicist who analyzed spectra to investigate nuclear spin and magnetic and quadrupole moments. He used multiple-beam interferometry to explore the fine details of surface structure.

tolerance in measurement, range in the physical dimensions of an object within which the true dimensions lie. Often expressed in the form of, for example, ± 3 mm for a length.

toluene or **methyl benzene**; $C_6H_5CH_3$, colorless, inflammable liquid, insoluble in water, derived from petroleum. It is used as a solvent, in aircraft fuels, in preparing phenol (carbolic acid, used in making resins for adhesives, pharmaceuticals, and as a disinfectant), and the powerful high explosive †TNT.

Tombaugh, Clyde William (1906–1997) US astronomer who discovered the planet †Pluto, the most distant planet in the Solar System, in 1930.

Dr Clyde Tombaugh Profile http://www.achievement.org/autodoc/page/tom0pro-1

Description of the life and work of the discoverer of Pluto, astronomer Dr Clyde Tombaugh. The Web site contains not only a profile and biographical information, but also holds a lengthy interview with Tombaugh from 1991 accompanied by a large number of photographs, video sequences, and audio clips.

tomography technique of using X-rays or ultrasound waves to procure images of structures deep within the body for diagnostic purposes. In modern medical imaging there are several techniques, such as the †CAT scan (computerized axial tomography).

Tomonaga, Sin-Itiro (or Shin'ichiro) (1906–1979) Japanese theoretical physicist who was awarded the Nobel Prize for Physics in 1965 for developing the theory of †quantum electrodynamics (QED). Tomonaga, Richard †Feynman, and Julian †Schwinger independently developed methods for calculating the interaction between electrons, positrons, and photons. The three approaches were essentially the same and QED remains one of the most accurate physical theories known.

ton symbol t, imperial unit of mass. The **long ton**, used in the UK, is 1,016 kg/2,240 lb; the **short ton**, used in the USA, is 907 kg/2,000 lb. The **metric ton** or **tonne** is 1,000 kg/2,205 lb.

ton in shipping, unit of volume equal to 2.83 cubic meters/100 cubic feet. **Gross tonnage** is the total internal volume of a ship in tons; **net register tonnage** is the volume used for carrying cargo or passengers. **Displacement tonnage** is the weight of the vessel, in terms of the number of imperial tons of seawater displaced when the ship is loaded to its load line; it is used to describe warships.

Tonegawa, Susumu (1939–) Japanese molecular biologist who was awarded a Nobel Prize for Physiology or Medicine in 1987 for his work on the process by which genes alter to produce a range of different antibodies. He discovered that around 1,000 genes are used by the human body to generate billions of different antibodies. Antibodies are molecules that defend the body against infection. When a foreign substance, called an antigen, enters the body, it is attacked and destroyed by antibodies.

tongue in tetrapod vertebrates, a muscular organ usually attached to the floor of the mouth. It has a thick root attached to a U-shaped bone (hyoid), and is covered with a †mucous membrane containing nerves and taste buds. It is the main organ of taste. The tongue directs food to the teeth and into the throat for chewing and swallowing. In humans, it is crucial for speech; in other animals, for lapping up water and for grooming, among other functions. In some animals, such as frogs, it can be flipped forwards to catch insects; in others, such as anteaters, it serves to reach for food found in deep holes.

tonne symbol t, metric ton of 1,000 kg/2,204.6 lb; equivalent to 0.9842 of an imperial †ton.

tonsils in higher vertebrates, masses of lymphoid tissue situated at the back of the mouth and throat (palatine tonsils), and on the rear surface of the tongue (lingual tonsils). The tonsils contain many †lymphocytes and are part of the body's defense system against infection.

The †adenoids are sometimes called pharyngeal tonsils.

toolbar in computing, area usually at the top or side of a screen with push-buttons and other features to perform frequently used tasks. For example, the toolbar of a paint program might offer quick access to different brushes, spray cans, erasers, and other useful tools. "Floating" and "smart" toolbars have become more common, where a toolbar only appears when a relevant object is selected and can be repositioned at will.

tooth in vertebrates, one of a set of hard, bonelike structures in the mouth, used for biting and chewing food, and in defense and aggression. In humans, the first set (20 milk teeth) appear from age six months to two and a half years. The permanent †dentition replaces these from the sixth year onwards, the wisdom teeth (third molars) sometimes not appearing until the age of 25 or 30. Adults have 32 teeth: two incisors, one canine (eye tooth), two premolars, and three molars on each side of each jaw. Each tooth consists of an enamel coat (hardened calcium deposits), dentine (a thick, bonelike layer), and an inner pulp cavity, housing nerves and blood vessels. Mammalian teeth have roots surrounded by cementum, which fuses them into their sockets in the jawbones.

The neck of the tooth is covered by the †gum, while the enamel-covered crown protrudes above the gum line. The chief diseases of teeth are misplacements resulting from defect or disturbance of the tooth-germs before birth, eruption out of their proper places, and caries (decay).

topaz mineral, aluminum fluorosilicate, $Al_2(F_2SiO_4)$. It is usually yellow, but pink if it has been heated, and is used as a gemstone when transparent. It ranks 8 on the Mohs scale of hardness.

topography surface shape and composition of the landscape, comprising both natural and artificial features, and its study. Topographical features include the relief and contours of the land; the distribution of mountains, valleys, and human settlements; and the patterns of rivers, roads, and railways.

topology branch of geometry that deals with those properties of a figure that remain unchanged even when the figure is transformed (bent, stretched) – for example, when a square painted on a rubber sheet is deformed by distorting the sheet.

Topology has scientific applications, as in the study of turbulence in flowing fluids.

The topological theory, proposed in 1880, that only four colors are required in order to produce a map in which no two adjoining countries have the same color, inspired extensive research, and was proved in 1972 by Kenneth Appel and Wolfgang Haken.

Topology in Mathematics http://www-history.mcs.st-and.ac.uk/~history/HistTopics/Topology_in_mathematics.html

History of topology in mathematical situations. Starting from the earliest application of the four color theorem in mathematics, this Web site chronicles both the developments and the developers through to the formulation of the complete 1912 topology theory of Brouwer. Also included are several references for further reading and links to the biographies of those involved in the subject.

topsoil upper, cultivated layer of soil, which may vary in depth from 8 to 45 cm/3 to 18 in. It contains organic matter – the decayed remains of vegetation, which plants need for active growth – along with a variety of soil organisms, including earthworms.

tor isolated mass of rock, often granite, left upstanding on a hilltop after the surrounding rock has been broken down. Weathering takes place along the joints in the rock, reducing the outcrop into a mass of rounded blocks.

tornado extremely violent revolving storm with swirling, funnel-shaped clouds, caused by a rising column of warm air propelled by strong wind. A tornado can rise to a great height, but with a diameter of only a few hundred meters or less. Tornadoes move with wind speeds of 160–480 kph/100–300 mph, destroying everything in their path. They are common in the central USA and Australia.

Tornado Project Online http://www.tornadoproject.com/

All about tornadoes – including myths and oddities, personal experiences, tornado chasing, tornado safety, and tornadoes past and present.

torque turning effect of force on an object. A turbine produces a torque that turns an electricity generator in a power station. Torque is measured by multiplying the force by its perpendicular distance from the turning point.

torr unit of pressure equal to 1/760 of an †atmosphere, used mainly in high-vacuum technology. One torr is equivalent to 133.322 pascals, and for practical purposes is the same as the millimeter of mercury. It is named after the Italian physicist Evangelista †Torricelli.

Torricelli, Evangelista (1608–1647) Italian physicist who established the existence of atmospheric pressure and devised the mercury †barometer in 1644.

Torricelli, Evangelista http://www-groups.dcs.st-andrews.ac.uk/history/Mathematicians/Torricelli.html

Part of an archive containing the biographies of the world's greatest mathematicians, this site is devoted to the life and contributions of Evangelista Torricelli. In addition to biographical information, you will find a list of references about Torricelli and links to other essays in the archive that reference him. The text of this essay includes hypertext links to the essays of those mathematicians and thinkers who influenced Torricelli. There is also an image of him, which can be enlarged, and a map of his birthplace.

torsion in physics, the state of strain set up in a twisted material; for example, when a thread, wire, or rod is twisted, the torsion set up in the material tends to return the material to its original state. The **torsion balance**, a sensitive device for measuring small gravitational or magnetic forces, or electric charges, balances these against the restoring force set up by them in a torsion suspension.

torus in nuclear physics, ring-shaped chamber used to contain †plasma in nuclear fusion reactors such as the Joint European Torus (JET) reactor.

total eclipse solar eclipse in which the Sun's disc is completely covered by the Moon, or a lunar eclipse in which the Moon's disc is completely covered by the shadow of the Earth.

total internal reflection complete reflection of a beam of light that occurs from the surface of an optically "less dense" material. For example, a beam from an underwater light source can be reflected from the surface of the water, rather than escaping through the surface. Total internal reflection can only happen if a light beam hits a surface at an angle greater than the †critical angle for that particular pair of materials.

Total internal reflection is used as a means of reflecting light inside †prisms and †optical fibers. Light is contained inside an optical fiber not by the cladding around it, but by the ability of the internal surface of the glass-fiber core to reflect 100% of the light, thereby keeping it trapped inside the fiber.

touch sensation produced by specialized nerve endings in the skin. Some respond to light pressure, others to heavy pressure. Temperature detection may also contribute to the overall sensation of touch. Many animals, such as nocturnal ones, rely on touch more than humans do. Some have specialized organs of touch that project from the body, such as whiskers or antennae.

touch screen in computing, an input device allowing the user to communicate with the computer by touching a display screen with a finger. In this way, the user can point to a required †menu option or item of data. Touch screens are used less widely than other pointing devices such as the mouse or joystick. A typical application is in restaurants, where sales staff simply need to touch the items sold on the screen; a total is displayed for the customer, while the computing system connected to the screen calculates remaining stock levels.

Typically, the screen is able to detect the touch either because the finger presses against a sensitive membrane or because it interrupts a grid of light beams crossing the screen surface.

touch sensor in a computer-controlled robot, a device used to give the robot a sense of touch, allowing it to manipulate delicate objects or move automatically about a room. Touch sensors provide the feedback necessary for the robot to adjust the force of its movements and the pressure of its grip. The main types include the strain gauge and the microswitch.

tourmaline hard, brittle mineral, a complex silicate of various metals, but mainly sodium aluminum borosilicate.

Small tourmalines are found in granites and gneisses. The common varieties range from black (schorl) to pink, and the transparent gemstones may be colorless (achromatic), rose pink (rubellite), green (Brazilian emerald), blue (indicolite, verdelite, Brazilian sapphire), or brown (dravite).

Townes, Charles Hard (1915–) US physicist who shared the Nobel Prize for Physics in 1964 for his work in 1953 to design and construct the first †maser.

town gas fuel gas mixture of methane, hydrogen, and carbon monoxide produced by the †carbonization of coal.

Townsend, John Sealy Edward (1868–1957) Irish mathematical physicist who studied the kinetics of electrons and ions in gases. He was the first to obtain a value for the charge on the electron, in 1898, and to explain how electric discharges pass through gases. He was knighted in 1941.

toxic describing anything that is poisonous to life.

toxicity tests tests carried out on new drugs, cosmetics, food additives, pesticides, and other synthetic chemicals to see whether they are safe for humans to use. They aim to identify potential toxins, carcinogens, teratogens, and mutagens.

Traditionally such tests use live animals such as rats, rabbits, and mice. Animal tests have become a target for criticism by antivivisection groups, and alternatives have been sought. These include tests on human cells cultured in a test tube and on bacteria.

In Europe in 1990, 4,365 animals were used for testing cosmetic products, out of a total of 276,674 animals used to test products other than pharmaceuticals. Including pharmaceuticals, a total of around 3.2 million animals were used in experimentation. The US Office of Technology Assessment estimates that around 1.6 million animals are used annually in government research laboratories, of which 90% are rats or mice.

In October 1999 the British government announced that it would not issue licenses for animal testing until it was satisfied that alternative tests were unsatisfactory to test toxins. The license, known as LD_{50}, involves determining the chemical dose of a toxin that will kill half of the laboratory animals tested. Despite widespread belief among industrialized nations that the test is unnecessary and cruel, the test remains standard in the USA and Japan.

toxic waste †hazardous waste, especially when it has been dumped.

toxin any poison produced by another living organism (usually a bacterium) that can damage the living body. In vertebrates, toxins are broken down by †enzyme action, mainly in the liver.

trace in mathematics, sum of the diagonal elements of a †matrix. Like the †determinant it is preserved by many of the important transformations applied to matrices.

trace element chemical element necessary in minute quantities for the health of a plant or animal. For example, magnesium, which occurs in chlorophyll, is essential to photosynthesis, and iodine is needed by the thyroid gland of mammals for making hormones that control growth and body chemistry.

tracer in science, a small quantity of a radioactive †isotope (form of an element) used to follow the path of a substance. Certain chemical reactions or physical or biological processes can then be monitored. The location (and possibly concentration) of the tracer is usually detected by using a Geiger–Muller counter. For example, the activity of the thyroid gland can be monitored by giving the patient an injection containing a small dose of a radioactive isotope of iodine, which is selectively absorbed from the bloodstream by the gland.

trachea tube that forms an airway in air-breathing animals. In land-living †vertebrates, including humans, it is also known as the **windpipe** and runs from the larynx to the upper part of the chest. Its diameter is about 1.5 cm/0.6 in and its length 10 cm/4 in. It is strong and flexible, and reinforced by rings of †cartilage. In the upper chest, the trachea branches into two tubes: the left and right bronchi, which enter the lungs. Insects have a branching network of tubes called tracheae, which conduct air from holes (†spiracles) in the body surface to all the body tissues. The finest branches of the tracheae are called tracheoles.

Some spiders also have tracheae but, unlike insects, they possess gill-like lungs (book lungs) and rely on their circulatory system to transport gases throughout the body.

tracheid cell found in the water-conducting tissue (†xylem) of many plants, including gymnosperms (conifers) and pteridophytes (ferns). It is long and thin with pointed ends.

The cell walls are thickened by †lignin, except for numerous small rounded areas, or pits, through which water and dissolved minerals pass from one cell to another. Once mature, the cell itself dies and only its walls remain.

tracking agent chemical compound (most usually NPPD, nitrophenyl pentadiene) used in espionage. It is applied to people, clothes, and cars to trace their movements and contacts.

Tradescant, John (1570–*c.* 1638) English gardener and botanist who traveled widely in Europe and is thought to have introduced the cos lettuce to England from the Greek island of that name. He was appointed gardener to Charles I and was succeeded by his son, **John Tradescant the Younger** (1608–1662). The younger Tradescant undertook three plant-collecting trips to Virginia in North America.

trade wind prevailing wind that blows towards the Equator from the northeast and southeast. Trade winds are caused by hot air rising at the Equator and the consequent movement of air from north and south to take its place. The winds are deflected towards the west because of the Earth's west-to-east rotation.

The unpredictable calms known as the †doldrums lie at their convergence.

The trade-wind belts move north and south about 5° with the seasons. The name is derived from the obsolete expression "blow trade" meaning to blow regularly, which indicates the trade winds' importance to navigation in the days of cargo-carrying sailing ships.

trajectory path of a projectile moving under gravity. More generally, the term is used for the solution curve of any differential equation or dynamical system.

trance mental state in which the subject loses the ordinary perceptions of time and space, and even of his or her own body.

In this highly aroused state, often induced by rhythmic music, "speaking in tongues" (glossolalia) may occur; this usually consists of the rhythmic repetition of apparently meaningless syllables, with a euphoric return to normal consciousness. It is also practiced by American Indian and Australian Aboriginal healers, Afro-Brazilian spirit mediums, and in shamanism.

trans- prefix used in †stereochemistry to indicate that two groups or substituents lie on opposite sides of the main axis or plane of a molecule (as opposed to **cis**-, which indicates that they are on the same side).

transactinide element any of a series of eight radioactive, metallic elements with atomic numbers that extend beyond the †actinide series, those from 104 (rutherfordium) upwards. They are grouped because of their expected chemical similarities (they are all bivalent), the properties differing only slightly with atomic number. All have †half-lives of less than two minutes.

transamination in biochemistry, transfer of an amino group ($—NH_2$) to a molecule of another substance, forming an †amino acid. In mammals, transanimation occurs in the tissues of the liver (see †nonessential amino acid).

transcendental curve curve for which there is no representative algebraic equation; examples are logarithmic curves and trigonometric curves.

transcendental number †real number that is not an †algebraic number, and can therefore not be expressed as a root (solution) of an algebraic equation with integral coefficients; an example is †Euler's number (e).

transcription in living cells, the process by which the information for the synthesis of a protein is transferred from the †DNA strand on which it is carried to the messenger †RNA strand involved in the actual synthesis.

It occurs by the formation of †base pairs when a single strand of unwound DNA serves as a template for assembling the complementary nucleotides that make up the new RNA strand.

transducer device that converts one form of energy into another. For example, a thermistor is a transducer that converts heat into an electrical voltage, and an electric motor is a transducer that converts an electrical voltage into mechanical energy. Transducers are important components in many types of †sensor, converting the physical quantity to be measured into a proportional voltage signal.

transduction transfer of genetic material between cells by an infectious mobile genetic element such as a virus. Transduction is used in †genetic engineering to produce new varieties of bacteria.

transference in psychoanalysis, the patient's transfer of feelings and wishes experienced in earlier relationships into the relationship with the analyst.

First described by Sigmund Freud in 1895, the transference relationship is often viewed as taking a positive or negative form. In **positive transference** the patient is compliant or unrealistically overvalues the analyst. In **negative transference** the patient is defiant or dislikes the analyst. When positive, transference can be used as a means of overcoming resistance to the recall of unpleasant material but, when negative, often endangers the continuum of treatment and so is generally isolated. Freud regarded the transference relationship as an essential tool in analysis, but some subsequent schools regard it as a side effect to be countered early in treatment.

transfermium element or **superheavy element**, chemical element with an atomic number greater than 100 (the atomic number of fermium). Transfermium elements are all highly unstable and have existed only briefly, and because of this there is a continuing controversy over who should be credited with their discoveries. See also †transuranic element.

transfer orbit elliptical path followed by a spacecraft moving from one orbit to another, designed to save fuel although at the expense of a longer journey time. Space probes travel to the planets on transfer orbits. A probe aimed at Venus has to be "slowed down" relative to the Earth, so that it enters an elliptical transfer orbit with its perigee (point of closest approach to the Sun) at the same distance as the orbit of Venus; towards Mars, the vehicle has to be "speeded up" relative to the Earth, so that it reaches its apogee (furthest point from the Sun) at the same distance as the orbit of Mars. **Geostationary transfer orbit** is the highly elliptical path followed by satellites to be placed in †geostationary orbit around the Earth (an orbit coincident with Earth's rotation). A small rocket is fired at the transfer orbit's apogee to place the satellite in geostationary orbit.

transfer orbit The transfer orbit used by a spacecraft when traveling from Earth to Mars. The orbit is chosen to minimize the fuel needed by the spacecraft; the craft is in free fall for most of the journey. Helicon

transfer RNA tRNA, relatively small molecule of ribonucleic acid, the function of which is to carry †amino acids to †ribosomes where †protein synthesis occurs. Each amino acid is borne by a different tRNA molecule. tRNA is complementary to †messenger RNA (mRNA).

transfinite see †set theory.

transformation in mathematics, a mapping or †function, especially one which causes a change of shape or position in a geometric figure. †Reflection, †rotation, †enlargement, and †translation are the main geometrical transformations.

Two or more transformations performed on one shape are called **combined transformations**. The transformation of a shape from A to B can be reversed by an **inverse transformation** from B back onto A.

transformation of function rearrangement of a mathematical †formula or formulae to give a different subject. For example, in the formula $A = bh$, A is the subject of the formula. It can be rearranged as $b = A \div h$, where b is the subject, or as $h = A \div b$, where h is the subject.

The variable that becomes the subject of the formula should be written on its own on the left-hand side of the formula.

use of inverses To transform a formula †inverse functions can be used. Multiplication and division are inverses, as are addition and subtraction.

flowcharts A flowchart is a good method that uses inverses to transform a formula. For example, to make r the subject of the formula $D = 2r$:

Start with r → [× 2] → 2D

$\frac{D}{2}$ ← [÷ 2] ← Start with D

So $r = {}^{D}/_{2}$.

In a more complicated example, the formula used to convert temperatures from degrees Celsius to degrees Fahrenheit is

$F = 9/5\ C + 32$

To make C the subject of the formula:

Start with C → [× 9] → 9C → [÷ 5] → $\frac{9C}{5}$ → [+ 32] → $\frac{9C}{5} + 32$

$\frac{5(F-32)}{9}$ ← [÷ 9] ← 5(F − 32) ← [× 5] ← F − 32 ← [− 32] ← Start with F

Therefore, $C = 5(F - 32)/9$.

transformer device in which, by electromagnetic induction, an alternating current (AC) of one voltage is transformed to another voltage, without change of †frequency. Transformers are widely used in electrical apparatus of all kinds, and in particular in power transmission where high voltages and low currents are utilized.

A transformer has two coils, a primary for the input and a secondary for the output, wound on a common iron core. The ratio of the primary to the secondary voltages is directly proportional to the number of turns in the primary and secondary coils; the ratio of the currents is inversely proportional.

step-up transformer This type of transformer has more turnings on the secondary coil than on the primary coil and therefore the voltage induced in the secondary coil is larger than the primary coil voltage. If the number of turns on the primary coil is N_P and the number of turns on the secondary coil is N_S, and the respective voltages are V_P and V_S, then $N_S/N_P = V_S/V_P$. For example, if there are 200 turnings on the primary coil and 2,000 turnings on the secondary coil, then the voltage induced in the secondary coil is ten times larger than the primary coil voltage.

step-down transformer This type of transformer has more turnings on the primary coil than on the secondary coil and therefore the voltage induced in the secondary coil is smaller than the primary coil voltage. The same formula applies as for a step-up transformer, so if there are 2,000 turnings on the primary coil and 200 turnings on the secondary coil, then the voltage induced in the secondary coil is ten times smaller than the primary coil voltage.

transform margin or **transform boundary**, in †plate tectonics, a boundary between two lithospheric plates in which the plates move past each other rather than towards or away from one another. Transform boundaries are characterized by large †strike-slip faults called **transform faults**, and powerful and destructive earthquakes.

Transform margins are named as such because they act as a connection, or "transform," between two other types of margins. The boundary between two plates differs from place to place: some segments are †convergent margins, some †divergent margins, and some transform. Transforms link convergent with divergent segments, and similar segments with each other. The most famous example is the San Andreas fault, which marks the boundary between the North American and Pacific plate.

transfusion intravenous delivery of blood or blood products (plasma, red cells) into a patient's circulation to make up for deficiencies due to disease, injury, or surgical intervention.

Cross-matching is carried out to ensure the patient receives the right blood group. Because of worries about blood-borne disease, there is a growing interest in autologous transfusion with units of the patient's own blood "donated" over the weeks before an operation.

Blood is rarely transfused whole. Blood cells and platelets are separated and resuspended in solution. Plasma can be frozen and is used to treat clotting deficiencies.

Blood transfusion, first successfully pioneered in humans in 1818, remained a high risk until the discovery of blood groups, by Austrian-born immunologist Karl †Landsteiner in 1900, which indicated the need for compatibility of donated blood.

transgene in biology, a gene that is "foreign" to the cell into which it is transferred. A †transgenic organism carries a transgene that is foreign to its normal genotype.

transgenic organism plant, animal, bacterium, or other living organism that has had a foreign gene added to it by means of †genetic engineering.

transistor solid-state electronic component, made of †semiconductor material, with three or more electrical contacts that can regulate a current passing through it. A transistor can act as an amplifier, †oscillator, †photocell, or switch, and (unlike earlier thermionic valves) usually operates on a very small amount of power. Transistors commonly consist of a tiny sandwich of †germanium or †silicon, alternate layers having different electrical properties because they are impregnated with minute amounts of different impurities.

A crystal of pure germanium or silicon would act as an insulator (nonconductor). By introducing impurities in the form of atoms of other materials (for example, boron, arsenic, or indium) in minute amounts, the layers may be made either **n-type**, having an excess of electrons, or **p-type**, having a deficiency of electrons. This enables electrons to flow from one layer to another in one direction only. Transistors have had a great impact on the electronics industry, and thousands of millions are now made each year. They perform many of the functions of the thermionic valve, but have the advantages of greater reliability, long life, compactness, and instantaneous action – no warming-up period being necessary. They are widely used in most electronic equipment, including portable radios and televisions, computers, and satellites, and are the basis of the †integrated circuit (silicon chip). They were invented at Bell Telephone Laboratories in the USA in 1948 by John †Bardeen and Walter †Brattain, developing the work of William †Shockley.

Transistorized! The History of the Invention of the Transistor http://www.pbs.org/transistor/index.html

Superb, in-depth history of the invention of the transistor. This attractive Web site has lots of interesting, well-researched articles, biographies of key people, an event time line, audio and video clips, and more.

transistor–transistor logic **TTL**, in computing, the type of integrated circuit most commonly used in building electronic products. In TTL chips the bipolar transistors are directly connected (usually collector to base). In mass-produced items, large numbers of TTL chips are commonly replaced by a small number of uncommitted logic arrays (ULAs), or logic gate arrays.

transit in astronomy, passage of a smaller object across the visible disc of a larger one. Transits of the inferior planets occur when they pass directly between the Earth and the Sun, and are seen as tiny dark spots against the Sun's disc.

Other forms of transit include the passage of a satellite or its shadow across the disc of Jupiter and the passage of planetary surface features across the central †meridian of that planet as seen from Earth. The passage of an object in the sky across the observer's meridian is also known as a transit.

transition metal any of a group of metallic †elements that have incomplete inner electron shells and exhibit variable valency – for example, cobalt, copper, iron, and molybdenum. They form a long block in the middle of the †periodic table of the elements, between groups 2 and 3. They are excellent conductors of electricity, and generally form highly colored compounds.

They include most of the hard, dense, and less reactive metals used as building materials, such as iron, lead, and copper. The precious metals, gold, platinum, and silver are also transition metals. They are rare, beautiful and costly, and so unreactive that they can be used as jewelry. Transition metals and their compounds are often used in industry as catalysts. For example, iron is used as a catalyst in the †Haber process to produce ammonia. They are also used to make †alloys.

Transition Metals http://chemed.chem.purdue.edu/genchem/topicreview/bp/ch12/trans.html

Transition metals, and their place within the periodic table, are explored on this page. The method of classification of the transition metals is explained, with increasing detail as the page progresses.

transition temperature in chemistry, temperature at which an allotropic element or compound changes from one allotropic form to another at a given pressure, for example trigonal a-quartz becomes hexagonal b-quartz at a transition temperature of 573°C/1,063°F (see †allotropy).

translation in living cells, the process by which proteins are synthesized. During translation, the information coded as a sequence of nucleotides in messenger †RNA is transformed into a sequence of amino acids in a peptide chain. The process involves the "translation" of the †genetic code. See also †transcription.

translation in mathematics, the movement of a point, line, or shape along a straight line. Translations may be combined with †rotations and †reflections.

For example:

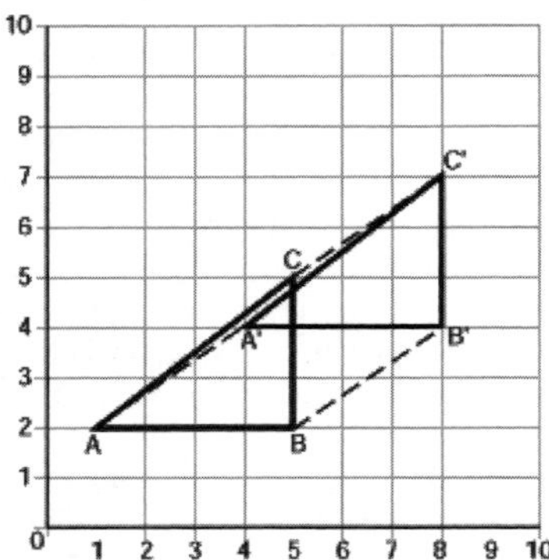

The triangle A'B'C' is a translation of triangle ABC. All the points on triangle ABC have been moved the same distance.

The amount and direction of a translation can be described using vector notation. In the example, the vector for the translation would be:

$\binom{3}{2}$

that is, a translation of 3 squares right and 2 squares up.

The translation of A'B'C' to ABC would be described as:

$\binom{-3}{-2}$

that is, 3 squares left and 2 squares down.

translation program in computing, a program that translates another program written in a high-level language or assembly language into the machine-code instructions that a computer can obey. See †assembler and †compiler.

translocation in genetics, the exchange of genetic material between chromosomes. It is responsible for congenital abnormalities, such as Down's syndrome.

transmission electron microscope TEM, most powerful type of †electron microscope, with a resolving power ten times better than that of a †scanning electron microscope and a thousand times better than that of an optical microscope. A fine electron beam passes through the specimen, which must therefore be sliced extremely thinly – typically to about one-thousandth of the thickness of a sheet of paper (100 nanometers). The TEM can resolve objects 0.001 micrometers (0.04 millionth of an inch) apart, a gap that is 100,000 times smaller than the unaided eye can see.

A TEM consists of a tall, evacuated column at the top of which is a heated filament that emits electrons. The electrons are moved at uniform high velocity down the column by a high voltage (around 100,000 volts) and pass through the slice of specimen at a point roughly half-way down. Because the density of the specimen varies, the "shadow" of the beam falls on a fluorescent screen near the bottom of the column and forms an image. A camera is mounted beneath the screen to record the image.

The electron beam is controlled by magnetic fields produced by electric coils, called electron lenses. One electron lens, called the condenser, controls the beam size and brightness before it strikes the specimen. Another electron lens, called the objective, focuses the beam on the specimen and magnifies the image about 50 times. Other electron lenses below the speci-

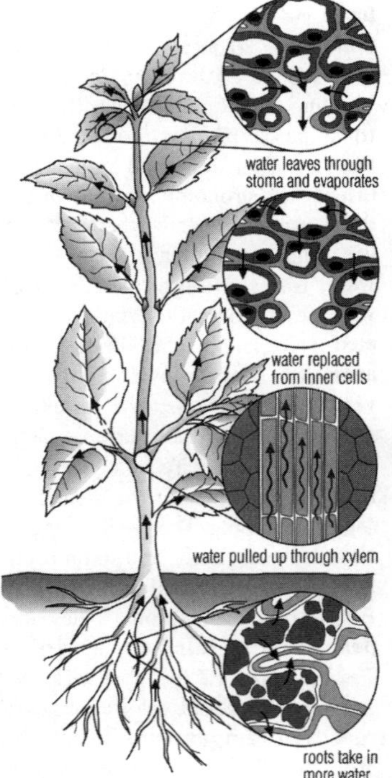

transpiration The loss of water from a plant by evaporation is known as transpiration. Most of the water is lost through the surface openings, or stomata, on the leaves. The evaporation produces what is known as the transpiration stream, a tension that draws water up from the roots through the xylem, water-carrying vessels in the stem. Helicon

men then further magnify the image.

The **high voltage transmission electron microscope** (HVEM) uses voltages of up to 3 million volts to accelerate the electron beam. The largest of these instruments is as tall as a three-story building.

The first experimental TEM was built in 1931 by German scientists Max Knoll and Ernest Ruska of the Technische Hochschule, Berlin, Germany. They produced a picture of a platinum grid magnified 117 times. The first commercial electron microscope was built in England in 1936.

transpiration loss of water from a plant by evaporation. Most water is lost by †diffusion of water vapor from the leaves through pores known as †stomata to the outside air. The primary function of stomata is to allow †gas exchange between the plant's internal tissues and the atmosphere. Transpiration from the leaf surfaces causes a continuous upward flow of water from the roots via the †xylem, which is known as the transpiration stream. This replaces the water that is lost, and allows minerals absorbed from the soil to be transported through the xylem to the leaves. This is important because many plant cells need the minerals as nutrients.

transposon or **jumping gene**, segment of DNA able to move within or between chromosomes. Transposons trigger changes in gene expression by shutting off genes or causing insertion †mutations. The origins of transposons are obscure, but geneticists believe some may be the remnants of viruses that have permanently integrated their genes with those of their hosts. They were first identified by US geneticist Barbara †McClintock in 1947.

transuranic element or **transuranium element**, chemical element with an atomic number of 93 or more – that is, with a greater number of protons in the nucleus than has uranium. All transuranic elements are radioactive. Neptunium and plutonium are found in nature; the others are synthesized in nuclear reactions.

transversal line cutting two or more (usually parallel) lines in the same plane.

transverse wave †wave in which the displacement of the medium's particles, or in electromagnetic waves the direction of the electric and magnetic fields, is at right angles to the direction of travel of the wave motion.

Various methods are used to reproduce waves, such as a ripple tank or a rope, in order to understand their properties. If one end of a rope is moved in an up and down motion (the other end being fixed), a wave travels along the rope. The particles in the rope oscillate at right angles to the direction of the wave, moving up and down as the wave travels along the rope. This is known as a transverse wave. The rope remains as it was after the wave has traveled along the rope; waves carry energy from one place to another but they do not transfer matter. Examples of transverse waves are water waves and electromagnetic waves.

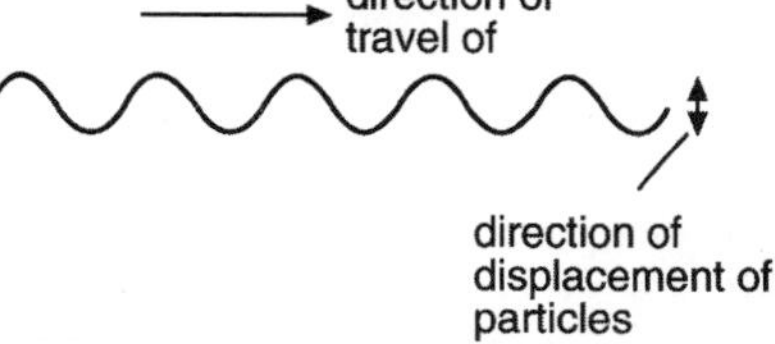

transverse wave The diagram illustrates the motion of a transverse wave. Light waves are examples of transverse waves: they undulate at right angles to the direction of travel and are characterized by alternating crests and troughs. Simple water waves, such as the ripples produced when a stone is dropped into a pond, are also examples of transverse waves. Helicon

trapezius triangular muscle running from high up in the back of the neck, across the back of the shoulder and down the upper spine. It is involved in movements of the †scapula (shoulder blade) and also moves the head from side to side.

trapezoid in geometry, a quadrilateral (a shape with four sides) with two of its sides parallel. If the parallel sides have lengths *a* and *b* and the perpendicular distance between them is *h* (the height of the trapezoid), its area $A = {}^1/_2h(a + b)$.

An isosceles trapezoid has its sloping sides equal, is symmetrical about a line drawn through the midpoints of its parallel sides, and has equal base angles:

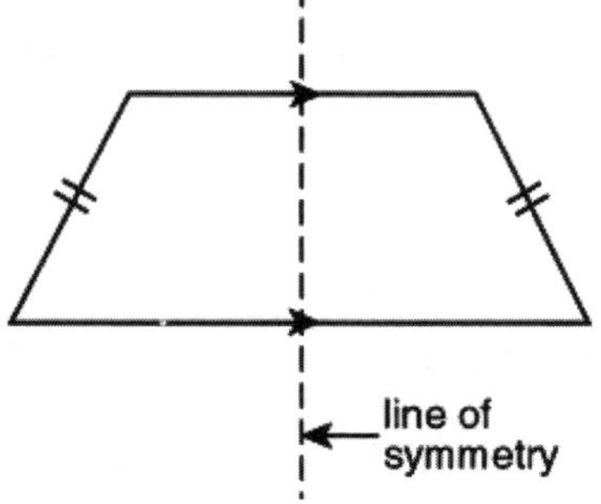

travel graph type of †distance–time graph which shows the stages of a journey. Stops on the journey are shown as horizontal lines. Travel graphs can be used to find out where vehicles will overtake each other or meet if traveling in opposite directions.

tree perennial plant with a woody stem, usually a single stem (trunk), made up of †wood and protected by an outer layer of †bark. It absorbs water through a †root system. There is no clear dividing line between shrubs and trees, but sometimes a minimum achievable height of 6 m/20 ft is used to define a tree.

angiosperms A treelike form has evolved independently many times in different groups of plants. Among the †angiosperms, or flowering plants, most trees are †dicotyledons. This group includes trees such as oak, beech, ash, chestnut, lime, and maple, and they are often referred to as broad-leaved trees because their leaves are broader than those of conifers, such as pine and spruce. In temperate regions angiosperm trees are mostly †deciduous (that is, they lose their leaves in winter), but in the tropics most angiosperm trees are evergreen. There are fewer trees among the †monocotyledons, but the palms and bamboos (some of which are treelike) belong to this group.

gymnosperms The †gymnosperms include many trees and they are classified into four orders: Cycadales (including cycads and sago palms), Coniferales (the conifers), Ginkgoales (including only one living species, the ginkgo, or maidenhair tree), and Taxales (including yews). Apart from the ginkgo and the larches (conifers), most gymnosperm trees are evergreen.

tree ferns There are also a few living trees in the †pteridophyte group, known as tree ferns. In the swamp forests of the Carboniferous era, 300 million years ago, there were giant treelike horsetails and club mosses in addition to the tree ferns.

oldest trees The world's oldest living trees are found in the Pacific forest of North America, some more than 2,000 years old.

conservation According to a 1998 report by the World Conservation Monitoring Center (WCMC) and the World Conservation Union, approximately 8,750 tree species (around 10% of all known tree species) are in danger of extinction. In the 20th century 77 species became extinct and 7 more species were reduced to fewer than a dozen specimens.

British Trees Home Page http://www.british-trees.com/

Introductory guide to native British trees. It can be searched by common name or Latin name, and there is a photographic collection of the mature trees, as well as seeds and flowers in various stages of their development.

Wonderful World of Trees http://www.domtar.com/arbre/english/start.htm

Well-structured site leading you down interesting pathways where you can learn about a year in the life of a tree, uses for trees, and how to identify different species.

tree diagram in probability theory, a branching diagram consisting only of arcs and nodes (but not loops curving back on themselves), which is used to establish †probabilities. Each branch represents a possible outcome of one event. These possible outcomes are known as †mutually exclusive events. The final probability depends on the path taken through the tree.

For example, when two cakes are baked the probability that one cake will not be burnt is 0.95. The tree diagram shows all the possible outcomes:

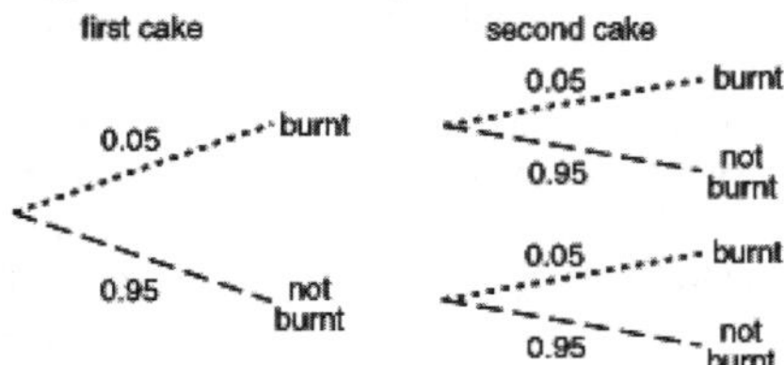

The outcome of each cake can be described as an †independent event as they have no effect on each other. Each path through the tree gives a different outcome. In this case there are four outcomes.

The dotted path gives the outcome of both cakes burning. The probabilities are multiplied to obtain this outcome:

Probability of both cakes burning = 0.05 x 0.05 = 0.0025

The dashed path gives the outcome of both cakes not burning:

Probability of both cakes not burning = 0.95 x 0.95 = 0.9025

The probability of one cake burning has two outcomes: (i) the first burnt and second not burnt, and (ii) the first not burnt and the second burnt. These probabilities are:

(i) 0.05 x 0.95 = 0.0475 and (ii) 0.95 x 0.05 = 0.0475

The two probabilities are added together to obtain the probability of one cake burning = 0.0475 + 0.0475 = 0.095.

The four probabilities should add up to 1.

Probabilities of independent events are always multiplied, characterized by *and*, for example, first cake burnt and second cake burnt. Probabilities of mutually exclusive events are always added, characterized by *or*, for example, first cake burnt or not burnt.

treeline highest point on a mountain where trees will grow, and above which it is too cold for them to survive.

tree rings rings visible in the wood of a cut tree; see †annual rings.

trematode parasitic flatworm with an oval non-segmented body, of the class Trematoda, including the †fluke.

tremor minor †earthquake.

trial in a probability experiment, a term denoting each experiment or observation.

trial and improvement method in mathematics, method of problem solving that uses a series of estimations, getting closer and closer to the actual answer. For example, when calculating $\sqrt{10}$, the answer must lie between 3 and 4, because $3^2 = 9$ and $4^2 = 16$. So a first guess might be 3.5:

$3.5^2 = 12.25$ – too large, so improve the guess; $3.2^2 = 10.24$ – too large; $3.1^2 = 9.61$ – too small; $3.15^2 = 9.9225$ – too small; $3.17^2 = 10.0489$ – too large; $3.16^2 = 9.9856$ – too small; $3.165^2 = 10.017225$ – too large; $3.164^2 = 10.010896$

The final answer, 3.164, although still too large, is probably accurate enough for most purposes. The problem solver does not expect to solve the problem outright but only to get progressively nearer to a solution.

triangle in geometry, a three-sided plane figure, the sum of whose †interior angles is 180°. Triangles can be classified by the relative lengths of their sides. A †**scalene triangle** has three sides of unequal length. An †**isosceles triangle** has at least two equal sides; it has one line of †symmetry. An †**equilateral triangle** has three equal sides (and three equal angles of 60°); it has three lines of symmetry.

triangle of forces method of calculating the force (the resultant) produced by two other forces. It is based on the fact that if three forces acting at a point can be represented by the sides of a triangle, the forces are in equilibrium. See †parallelogram of forces.

triangular number any number that will form an equilateral triangle when arranged in dots; for example, 3, 6, 10 ....

Triangulum small constellation of the northern hemisphere, represented as a triangle.

Triangulum Australe or **Southern Triangle**, small bright constellation of the southern hemisphere, more conspicuous than its northern counterpart, Triangulum.

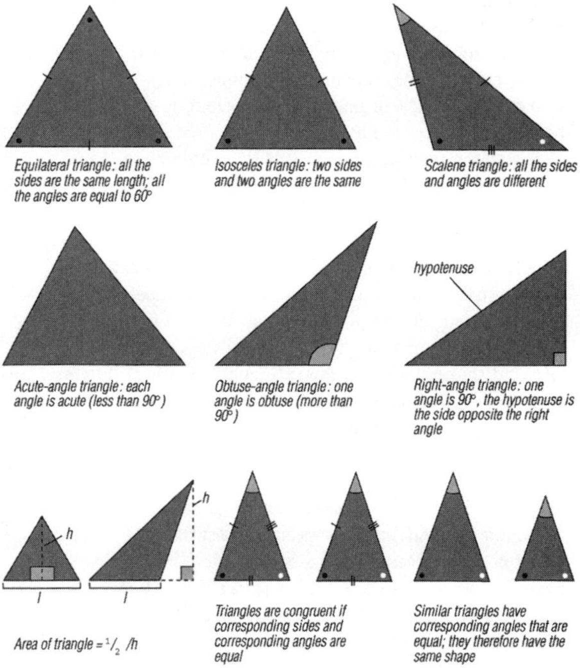

Equilateral triangle: all the sides are the same length; all the angles are equal to 60°

Isosceles triangle: two sides and two angles are the same

Scalene triangle: all the sides and angles are different

Acute-angle triangle: each angle is acute (less than 90°)

Obtuse-angle triangle: one angle is obtuse (more than 90°)

Right-angle triangle: one angle is 90°, the hypotenuse is the side opposite the right angle

Area of triangle = $\frac{1}{2}lh$

Triangles are congruent if corresponding sides and corresponding angles are equal

Similar triangles have corresponding angles that are equal; they therefore have the same shape

angles in triangles The exterior angle of a triangle is equal to the sum of the opposite interior angles. Helicon

triangle Types of triangle. Helicon

Triassic Period period of geological time 245–208 million years ago, the first period of the Mesozoic era. The present continents were fused together in the form of the world continent †Pangaea. Triassic sediments contain remains of early dinosaurs and other animals now extinct. By late Triassic times, the first mammals had evolved.

There was a mass extinction of 95% of plants at the end of the Triassic possibly caused by rising temperatures.

tribology science of friction, lubrication, and lubricants. It studies the origin of frictional forces, the wearing of interacting surfaces, and the problems of efficient lubrication.

tributyl tin TBT, chemical used in antifouling paints on ships' hulls and other submarine structures to deter the growth of barnacles. The tin dissolves in sea water and enters the food chain. It can cause reproductive abnormalities such as exposed female whelks developing penises; the use of TBT has therefore been banned in many countries, including the UK.

In October 2001, the United Nations (UN) body the International Maritime Organization announced plans to phase out the use of TBT. All shipping companies were ordered to abandon the use of TBT by January 2003 and to resurface all TBT-covered hulls by 2008.

triceps †muscle in the upper arm that causes the arm to straighten at the elbow (extend). To bend the arm its †antagonistic muscle – the biceps – has to contract. The bending of the arm in this movement occurs at the elbow †joint.

trichloromethane technical name for †chloroform.

tricuspid valve flap of tissue situated on the right side of the †heart between the atrium and the ventricle. It prevents blood flowing backwards when the ventricle contracts.

As in all valves, its movements are caused by pressure changes during the beat rather than by any intrinsic muscular activity. As the valve snaps shut, a vibration passes through the chest cavity and is detectable as the first sound of the heartbeat.

Trifid nebula in the constellation of †Sagittarius, about 3,000 light years from the Sun and about 12 light years across. It is also known as Messier 20 (M20).

triglyceride chemical name for †fat comprising three fatty acids reacted with a glycerol.

trigonometry branch of mathematics that concerns finding lengths and angles in †triangles. In a right-angled triangle the sides and angles are related by three trigonometric ratios: †sine, †cosine, and †tangent. Trigonometry is used frequently in navigation, surveying, and simple harmonic motion in physics.

Using trigonometry, it is possible to calculate the lengths of the sides and the sizes of the angles of a right-angled triangle as long as one angle and the length of one side are known, or the lengths of two sides. The longest side, which is always opposite to the right angle, is called the **hypotenuse**. The other sides are named depending on their position relating to the angle that is to be found or used: the side opposite this angle is always termed **opposite** and that adjacent is the **adjacent**. So the following trigonometric ratios are used:

sine = opposite/hypotenuse
cosine = adjacent/hypotenuse
tangent = opposite/adjacent

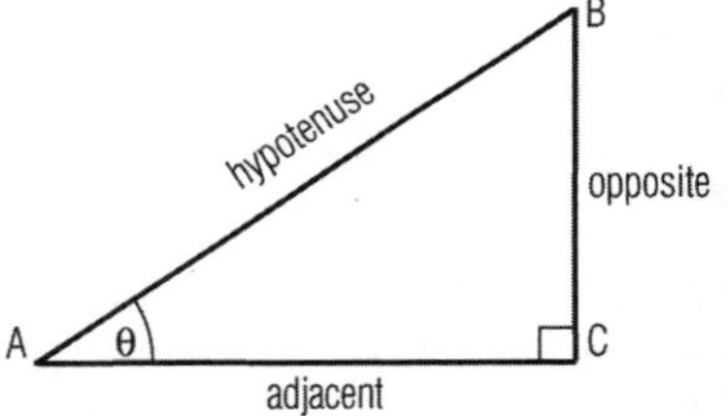

for any right-angled triangle with angle θ as shown the trigonometrical ratios are

$$\sin(e)\ \theta = \frac{BC}{AB} = \frac{\text{opposite}}{\text{hypotenuse}}$$

trigonometry At its simplest level, trigonometry deals with the relationships between the sides and angles of triangles. Unknown angles or lengths are calculated by using trigonometrical ratios such as sine, cosine, and tangent. The earliest applications of trigonometry were in the fields of navigation, surveying, and astronomy, and usually involved working out an inaccessible distance such as the distance of the Earth from the Moon. Helicon

Trigonometric Functions http://www-history.mcs.st-and.ac.uk/history/HistTopics/Trigonometric_functions.html

History of the development of the trigonometric functions in mathematics. The site also explains the basic principles of the use of these functions and details the role they have played in the advancement of other branches of mathematics. Biographical details of the mathematicians chiefly responsible for the discovery of trigonometry are also available here.

Trigonometry http://www.krysstal.com/trigonometry.html

Essay on trigonometry. It introduces Pythagoras' theorem and trigonometric functions, and then explains the more complex sine and cosine rules for finding the angles in non-right-angled triangles.

triiodomethane technical name for †iodoform.

trilobite any of a large class (Trilobita) of extinct, marine, invertebrate arthropods of the Paleozoic era, with a flattened, oval body, 1–65 cm/0.4–26 in long. The hard-shelled body was divided by two deep furrows into three lobes. Some were burrowers, others were swimming and floating forms. Their worldwide distribution, many species, and the immense quantities of their remains make them useful in geological dating.

There were more than 1,500 genera of trilobites, which existed for about 300 million years. They disappeared 250 million years ago, probably due to the evolution of new predators. According to a 1997 study of fossils by a US paleontologist, the trilobite molted its body segments individually; this molting may have taken several days, increasing vulnerability. The largest trilobite found was 70 cm/28 in long, and lived 445 million years ago, in what is now Hudson Bay, Canada. It was discovered by Canadian paleontologists in 2000.

trim in †aerodynamics, the slight actions on the controls needed to achieve stability in a particular mode of flight.

triode three-electrode thermionic †valve containing an anode and a cathode (as does a †diode) with an additional negatively-biased control grid. Small variations in voltage on the grid bias result in large variations in the output current. The triode was commonly used in amplifiers but has now been almost entirely superseded by the †transistor.

tripeptide sequence of three †amino acids, often occurring in the biochemical synthesis or breakdown of †proteins.

triphenylmethane dye member of a class of dyestuffs that are derived from the organic compound triphenylmethane, $(C_6H_5)_3CH$. These dyes produce brilliant colors ranging from green to violet.

triple bond three covalent bonds between adjacent atoms in a chemical compound, as in the †alkynes $(-C \equiv C-)$.

triploblastic in biology, having a body wall composed of three layers. The outer layer is the **ectoderm**, the middle layer the **mesoderm**, and the inner layer the **endoderm**. This pattern of development is shown by most multicellular animals (including humans).

trisect in mathematics, to divide into three equal parts. The angle trisection problem was one of the three problems (together with squaring the circle and duplicating the cube) much studied by ancient Greek mathematicians.

triticale cereal crop of recent origin that is a cross between wheat *Triticum* and rye *Secale*. It can produce heavy yields of high-protein grain, principally for use as animal feed.

tritium radioactive isotope of hydrogen, three times as heavy as ordinary hydrogen, consisting of one proton and two neutrons. It has a half-life of 12.5 years.

Triton in astronomy, largest of Neptune's moons. It has a diameter of 2,700 km/1,680 mi, and orbits Neptune every 5.88 days in a retrograde (east to west) direction at a distance of 354,000 km/220,000 mi. It takes the same time to rotate about its own axis as it does to make one revolution of Neptune.

It is slightly larger than the planet Pluto, which it is thought to resemble in composition and appearance. Probably Triton was formerly a separate body like Pluto but was captured by Neptune. Triton was discovered in 1846 by English astronomer William Lassell (1799–1880) only weeks after the discovery of Neptune. Triton's surface, as revealed by the Voyager 2 space probe, has a temperature of 38 K (–235°C/–391°F), making it the coldest known place in the Solar System. It is covered with frozen nitrogen and methane, some of which evaporates to form a tenuous atmosphere with a pressure only 0.00001 that of the Earth at sea level. Triton has a pink south polar cap, probably colored by the effects of solar radiation on methane ice. Dark streaks on Triton are thought to be formed by geysers of liquid nitrogen. The surface has few impact craters (the largest is the Mazomba, with a diameter of 27 km/17 mi), indicating that many of the earlier craters have been erased by the erupting and freezing of water (cryovulcanism).

troilite FeS, iron sulfide, mineral abundant in meteorites and probably present in the Earth's core.

Trojan either of two groups of more than 470 asteroids that share Jupiter's orbit, located 60° ahead of or behind the planet (at the two Lagrangian points). The largest Trojan is the asteroid Hektor, discovered in 1907, which is a greatly elongated object that measures 300 km/185 mi by 100 km/60 mi.

trophallaxis in social insects (ants, bees, wasps, and termites), the transfer of regurgitated food between colony members. This behavior plays a role in transfer of nutrients and pheromones within the colony but it also has a social role in averting aggression by appeasing dominant individuals and in recognition of colony members.

trophic level in ecology, the position occupied by a species (or group of species) in a †food chain. The main levels are **primary producers** (photosynthetic plants), **primary consumers** (herbivores), **secondary consumers** (carnivores), and †**decomposers** (bacteria and fungi).

tropical cyclone another term for †hurricane.

tropics area between the tropics of Cancer and Capricorn, defined by the parallels of latitude approximately 23°30¢ north and south of the Equator. They are the limits of the area of Earth's surface in which the Sun can be directly overhead. The mean monthly temperature is over 20°C/68°F.

Climates within the tropics lie in parallel bands. Along the Equator is the †intertropical convergence zone, characterized by high temperatures and year-round heavy rainfall. Tropical rainforests are found here. Along the tropics themselves lie the tropical high-pressure zones, characterized by descending dry air and desert conditions. Between these, the conditions vary seasonally between wet and dry, producing the tropical grasslands.

tropine $C_8H_{15}NO$, poisonous crystalline solid formed by the hydrolysis of the †alkaloid atropine.

tropism or **tropic movement**, directional growth of a plant, or part of a plant, in response to an external stimulus such as gravity or light. If the movement is directed towards the stimulus it is described as positive; if away from it, it is negative. **Geotropism** for example, the response of plants to gravity, causes the root (positively geotropic) to grow downwards, and the stem (negatively geotropic) to grow upwards.

Phototropism occurs in response to light, **hydrotropism** to water, **chemotropism** to a chemical stimulus, and **thigmotropism**, or **haptotropism**, to physical contact, as in the tendrils of climbing plants when they touch a support and then grow around it.

Tropic movements are the result of greater rate of growth on one side of the plant organ than the other. Tropism differs from a †nastic movement in being influenced by the direction of the stimulus.

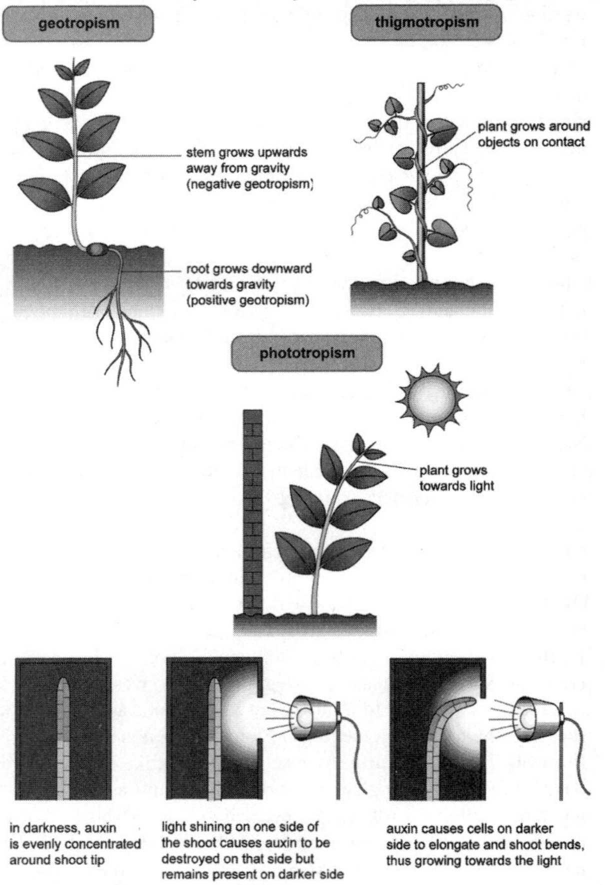

tropism Tropisms are plant movements in response to external stimuli. Geotropism is the movement in response to gravity, either towards it in most root systems (positive geotropism), or away as in most stems (negative geotropism). Most plants exhibit positive phototropism and grow towards light. Thigmotropism is found in most climbing plants. Helicon

tropopause thin layer in the Earth's atmosphere whose altitude varies from about 7 km/4.3 mi at the poles to 28 km/17 mi at the equator. It is located at the top of the troposphere.

troposphere lower part of the Earth's †atmosphere extending about 10.5 km/6.5 mi from the Earth's surface, in which temperature decreases with height to about –60°C/–76°F except in local layers of temperature inversion. The **tropopause** is the upper boundary of the troposphere, above which the temperature increases slowly with height within the atmosphere. All of the Earth's weather takes place within the troposphere.

troy system system of units used for precious metals and gems. The pound troy (0.37 kg) consists of 12 ounces (each of 120 carats) or 5,760 grains (each equal to 65 mg).

Trumpler, Robert Julius (1886–1956) Swiss-born US astronomer who studied and classified star clusters found in our Galaxy. He also took part in observational tests of the general theory of †relativity in 1922.

truncated spur blunt-ended ridge of rock jutting from the side of a †glacial trough, or valley. As a glacier moves down a former river valley it is unable to flow around the †interlocking spurs that project from either side, and so it erodes straight through them, shearing away their tips and forming truncated spurs.

truth table in electronics, a diagram showing the effect of a particular †logic gate on every combination of inputs.

Every possible combination of inputs and outputs for a particular gate or combination of gates is described, thereby defining their action in full. When logic value 1 is written in the table, it indicates a "high" (or "yes") input of perhaps 5 volts; logic value 0 indicates a "low" (or "no") input of 0 volts.

trypanosome any parasitic flagellate protozoan of the genus *Trypanosoma* that lives in the blood of vertebrates, including humans. They often cause serious diseases, called trypanosomiases.

trypsin enzyme in the vertebrate gut responsible for the digestion of protein molecules. It is secreted by the pancreas but in an inactive form known as trypsinogen. Activation into working trypsin occurs only in the small intestine, owing to the action of another enzyme, enterokinase, secreted by the wall of the duodenum. Unlike the digestive enzyme pepsin, found in the stomach, trypsin does not require an acid environment.

tryptophan $C_{11}H_{12}N_2O_2$, amino acid essential in the diet of animals. It is used in biological systems to produce the vitamin †niacin.

tsunami (Japanese "harbor wave") ocean wave generated by vertical movements of the sea floor resulting from †earthquakes or volcanic activity or large submarine landslides. Unlike waves generated by surface winds, the entire depth of water is involved in the wave motion of a tsunami. In the open ocean the tsunami takes the form of several successive waves, rarely in excess of 1 m/3 ft in height but traveling at speeds of 650–800 kph/400–500 mph. In the coastal shallows tsunamis slow down and build up producing huge swells over 15 m/45 ft high in some cases and over 30 m/90 ft in rare instances. The waves sweep inland causing great loss of life and damage to property.

Before each wave there may be a sudden withdrawal of water from the beach. Used synonymously with tsunami, the popular term "tidal wave" is misleading: tsunamis are not caused by the gravitational forces that affect †tides.

Double Whammy http://www.sciam.com/article.cfm?articleID=000A231F-EBFC-1CDF- B4A8809EC588EEDF&catID=4

Part of a larger site maintained by Scientific American, this page explores the catastrophic impact that an asteroid's crashing into the sea would have on civilization and the environment. Animated simulations show the effects of impact and the effect that a tsunami would have on the eastern seaboard of the USA if an asteroid struck the Atlantic Ocean. Learn about the tsunami that struck Prince William Sound, Alaska, in 1964 after an underwater earthquake. The text includes hypertext links to further information and a list of related links.

Tsunami! http://www.geophys.washington.edu/tsunami/intro.html

Description of many aspects of tsunamis. Included are details on how a tsunami is generated and how it propagates, how they have affected humans, and how people in coastal areas are warned about them. The site also discusses if and how you may protect yourself from a tsunami and provides "near real-time" tsunami information bulletins.

Tswett, Mikhail Semyonovich (1872–1919) Italian-born Russian botanist who made an extensive study of plant pigments and developed the technique of †chromatography to separate them, in 1906.

T Tauri phase in astronomy, phase named after a variable star in the constellation of †Taurus, believed to be a protostar that has condensed out of a cloud of dust and gas and has not yet started hydrogen burning. It is the prototype for a class of similar stars known as T Tauri stars that are characterized by irregular variability and the emission of substantial amounts of gas and dust at speeds of 100–200 kps/62–124 mps. T Tauri stars have masses in the range of 0.2–2.0 times the mass of the Sun and can lose up to half the mass of the Sun before settling down to hydrogen burning. It is thought that most stars, including the Sun, have passed through a T Tauri phase.

T Tauri variable another name for †nebular variable star.

TTL abbreviation for †transistor–transistor logic, a family of integrated circuits.

tuber swollen region of an underground stem or root, usually modified for storing food. The potato is a **stem tuber**, as shown by the presence of terminal and lateral buds, the "eyes" of the potato. **Root tubers**, for example dahlias, developed from †adventitious roots (growing from the stem, not from other roots) lack these. Both types of tuber can give rise to new individuals and so provide a means of †vegetative reproduction.

Unlike a bulb, a tuber persists for one season only; new tubers developing on a plant in the following year are formed in different places. See also †rhizome.

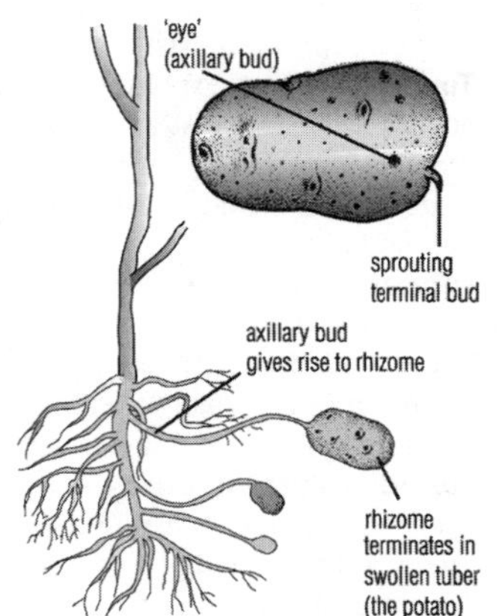

tuber Tubers are produced underground from stems, as in the potato, or from roots, as in the dahlia. Tubers can grow into new plants. Helicon

tube, thermionic see †thermionic tube.

Tucana constellation of the southern hemisphere, represented as a toucan. It contains the second most prominent †globular cluster in the sky, 47 Tucanae, and the Small †Magellanic Cloud.

Tucana is one of the 11 constellations named by Johann Bayer early in the 17th century to complement the 65 constellations delineated in ancient times.

tufa or **travertine**, soft, porous, †limestone rock, white in color, deposited from solution from carbonate-saturated ground water around hot springs and in caves.

Undersea tufa columns, such as those in the Ikka Fjord in southwestern Greenland, that form over alkaline springs, can reach 20 m/65 ft in height and grow at about 50 cm/20 in per year. They provide a habitat for a wide variety of marine life.

Tullgren funnel in biology, device used to extract mites, springtails, fly larvae, and other small invertebrates from a sample of soil.

Soil resting on a net inside a funnel is illuminated from above, so that the top layers of the soil warm and begin to dry out. The invertebrates, trying to escape the desiccating effect of the heat burrow downwards and in the process drop out of the soil and through the net, to be collected in a beaker resting below the funnel.

tumor overproduction of cells in a specific area of the body, often leading to a swelling or lump. Tumors are classified as **benign** or **malignant** (see †cancer). Benign tumors grow more slowly, do not invade surrounding tissues, do not spread to other parts of the body, and do not usually recur after removal. However, benign tumors can be dangerous in areas such as the brain. The most familiar types of benign tumor are warts on the skin. In some cases, there is no sharp dividing line between benign and malignant tumors.

tundra region in high latitudes with almost no trees – they cannot grow because the ground is permanently frozen (†permafrost). The vegetation consists mostly of grasses, sedges, heather, mosses, and lichens. Tundra stretches in a continuous belt across northern North America and Eurasia. Tundra is also used to describe similar conditions at high altitudes.

tung oil oil used in paints and varnishes, obtained from tung trees native to China. (Genus *Aleurites*, family Euphorbiaceae.)

tungstate any salt that contains an anion that is composed of both oxygen and †tungsten atoms. Tungstates are usually produced by cationic exchange in solutions of sodium hydroxide and †tungsten oxide. Some salts of this class are used to manufacture tungsten-based catalysts.

tungsten chemical symbol W, (Swedish *tung sten* "heavy stone") hard, heavy, gray-white, metallic element, atomic number 74, relative atomic mass 183.85. Its chemical symbol comes from the German *Wolfram*. It occurs in the minerals wolframite, scheelite, and hubertite. It has the highest melting point of any metal (3,410°C/6,170°F) and is added to steel to make it harder, stronger, and more elastic; its other uses include high-speed cutting tools, electrical elements, and thermionic couplings. Its salts are used in the paint and tanning industries.

Tungsten was first recognized in 1781 by Swedish chemist Karl Scheele in the ore scheelite. It was isolated in 1783 by Spanish chemists Fausto D'Elhuyar (1755–1833) and his brother Juan José (1754–1796).

tungsten ore either of the two main minerals, wolframite $(FeMn)WO_4$ and scheelite, $CaWO_4$, from which tungsten is extracted. Most of the world's tungsten reserves are in China, but the main suppliers are Bolivia, Australia, Canada, and the USA.

tungsten oxide WO_3, yellow solid produced by heating †tungsten in an atmosphere containing excess oxygen. It is used in the ceramics industry to produce a yellow glaze.

Tunguska Event explosion at Tunguska, central Siberia, Russia, in June 1908, which devastated around 6,500 sq km/2,500 sq mi of forest. It is thought to have been caused by either a cometary nucleus or a fragment of †Encke's Comet about 200 m/660 ft across, or possibly an asteroid. The magnitude of the explosion was equivalent to an atom bomb (10–20 megatons) and produced a colossal shock wave; a bright falling object was seen 600 km/375 mi away and was heard up to 1,000 km/625 mi away.

tunicate any marine †chordate of the subphylum Tunicata (Urochordata), for example the sea squirt. Tunicates have transparent or translucent tunics made of cellulose. They vary in size from a few millimeters to 30 cm/1 ft in length, and are cylindrical, circular, or irregular in shape. There are more than 1,000 species.

Pyrosomes are small tunicates, only a few millimeters in length, which live in free-floating tube-shaped colonies up to 10 m/33 ft in length. The colony is bioluminescent (light-producing), and emits a greenish glow.

turbidity current gravity-driven current in air, water, or other fluid resulting from accumulation of suspended material, such as silt, mud, or volcanic ash, and imparting a density greater than the surrounding fluid. Marine turbidity currents originate from tectonic movement, storm waves, tsunamis (tidal waves) or earthquakes and move rapidly downward, like underwater avalanches, leaving distinctive deposits called **turbidites**. They are thought to be one of the mechanisms by which submarine canyons are formed.

turbine engine in which steam, water, gas, or air is made to spin a rotating shaft by pushing on angled blades, like a fan. There are two sets of blades, the stator (does not rotate) and the rotor (does rotate).

The rotating turbine shaft can be connected to an electricity generator. Turbines are among the most powerful machines.

Steam turbines are used to drive generators in power stations and ships' propellers; water turbines spin the generators in hydroelectric power plants; and gas turbines (as jet engines; see †jet propulsion) power most aircraft and drive machines in industry.

turbofan jet engine of the type used by most airliners, so called because of its huge front fan. The fan sends air not only into the engine for combustion but also around the engine for additional thrust. This results in a faster and more fuel-efficient propulsive jet (see †jet propulsion).

turbulence irregular fluid (gas or liquid) flow, in which vortices and unpredictable fluctuations and motions occur. Streamlining reduces the turbulence of flow around an object, such as an aircraft, and reduces drag. Turbulent flow of a fluid occurs when the †Reynolds number is high.

turgor rigid condition of a plant caused by the fluid contents of a plant †cell exerting a mechanical pressure against the cell wall. Turgor supports plants that do not have woody stems. Plants lacking in turgor visibly wilt. The process of †osmosis plays an important part in maintaining the turgidity of plant cells.

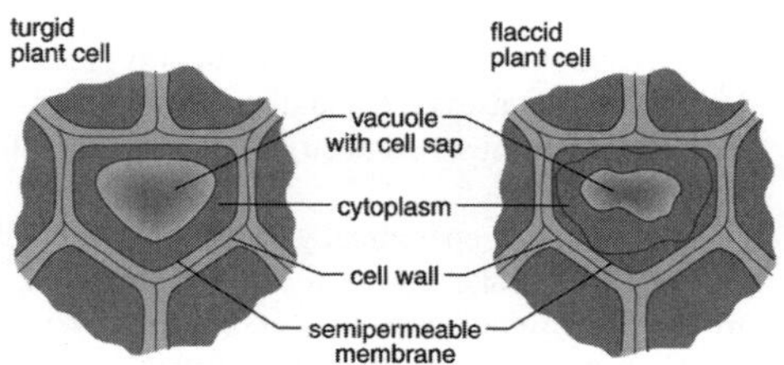

turgor A turgid plant cell (left) and a flaccid plant cell. Water leaves and enters the cell by osmosis. If too much water leaves the cell, for example during drought or saline conditions, then turgor is lost and the cell becomes flaccid. As turgor gives the plant rigidity, loss of turgidity results in the plant wilting. Helicon

When water moves into plant cells by osmosis it makes the cell expand. However, the †cell walls are very strong and cannot expand much. So the pressure rises inside the cell – it becomes tightly filled, like a balloon. This makes the cells rigid and this helps to support parts of plants like leaves. Plant cells that are rigid like this are said to be turgid.

If a plant loses too much water, or runs short of water in the soil, it will wilt (leaves and stems droop). If this continues parts of the plant will die because of water loss.

Turing, Alan Mathison (1912–1954) English mathematician and logician. In 1936 he described a "universal computing machine" that could theoretically be programmed to solve any problem capable of solution by a specially designed machine. This concept, now called the †Turing machine, foreshadowed the digital computer.

Alan Turing Home Page http://www.turing.org.uk/turing/

Authoritative illustrated biography of the computer pioneer, plus links to related sites. This site contains information on his origins and his code-breaking work during World War II, as well as several works written by Turing himself.

Turing machine abstract model of an automatic problem-solving machine, formulated by Alan Turing in 1936. It provides the theoretical basis of modern digital computing.

turpentine solution of resins distilled from the sap of conifers, used in varnish and as a paint solvent but now largely replaced by †white spirit.

turquoise mineral, hydrous basic copper aluminum phosphate, $CuAl_6(PO_4)_4(OH)_8 5H_2 O$. Blue-green, blue, or green, it is a gemstone. Turquoise is found in Australia, Egypt, Ethiopia, France, Germany, Iran, Turkestan, Mexico, and southwestern USA. It was originally introduced into Europe through Turkey, from which its name is derived.

twilight period of faint light that precedes sunrise and follows sunset, caused by the reflection of light from the upper layers of the atmosphere. The limit of twilight is usually regarded as being when the Sun is 18° below the horizon. The length of twilight depends on the latitude – at the tropics, it only lasts a few minutes; near the poles, it may last all night.

Sun and Moon Data for One Day http://aa.usno.navy.mil/data/docs/RS_OneDay.html

Part of a larger site on astronomical data maintained by the US Naval Observatory, this site includes the times of sunrise, sunset, moonrise, moonset, transits of the Sun and Moon, and the beginning and end of civil twilight, along with information on the Moon's phase. Information can be accessed by filling out one of two forms, depending on whether you live in or outside of the USA. For US cities and towns, there are links to US Census Bureau maps of the area for which astronomical data has been given. There are also sections of "Frequently Asked Questions" and research information.

twin one of two young produced from a single pregnancy. Human twins may be genetically identical (monozygotic), having been formed from a single fertilized egg that splits into two cells, both of which became implanted. Nonidentical (fraternal or dizygotic) twins are formed when two eggs are fertilized at the same time.

twinning in chemistry, denoting the presence of two or more †phases in a crystalline material that have identical molecular structures and properties but have different orientations with respect to each other.

twin primes in mathematics, pair of primes whose difference is 2. For example, the primes 101 and 103 are twin primes. The (unproved) **twin prime conjecture** is that there are infinitely many such pairs.

twistor theory model of the universe proposed by English mathematician Roger Penrose, based on the application of complex numbers (involving the square root of –1) used in calculations in the microscopic world of atoms and quantum theory to the macroscopic ordinary world of physical laws and relativity. The result is an eight-dimensional concept of reality.

2-D graphics in computing, graphics defined only by width and height (in mathematical terms, x and y axes). Examples of 2-D graphics are those generated by spreadsheets or animated characters with no shading.

two-dimensional **2D**, describing a flat shape having two measurements, or dimensions: length and width. Examples of 2D shapes are †square, †rectangle, †hexagon, and †triangle. Accurate drawings of 2D shapes can be constructed and 2D representations of †three-dimensional (3D) shapes, consisting of plan, elevations (side views), and cross-sections, serve to show the full structure of the shape. These views are known as projections.

two-stroke cycle operating cycle for internal combustion piston engines. The engine cycle is completed after just two strokes (up or down) of the piston, which distinguishes it from the more common †four-stroke cycle. Some power mowers and lightweight motorcycles use two-stroke gasoline engines, which are cheaper and simpler than four-strokes.

Most marine diesel engines are also two-stroke. In a typical two-stroke motorcycle engine, fuel mixture is drawn into the crankcase as the piston moves up on its first stroke to compress the mixture above it. Then the compressed mixture is ignited, and hot gases are produced, which drive the piston down on its second stroke. As it moves down, it uncovers an opening (port) that allows the fresh fuel mixture in the crankcase to flow into the combustion space above the piston. At the same time, the exhaust gases leave through another port.

tympanic membrane another name for the eardrum.

Tyndall, John (1820–1893) Irish physicist who 1869 studied the scattering of light by invisibly small suspended particles in †colloids. Known as the **Tyndall effect**, it was first observed with colloidal solutions, in which a beam of light is made visible when it is scattered by minute colloidal particles (whereas a pure solvent does not scatter light). Similar scattering of blue wavelengths of sunlight by particles in the atmosphere makes the sky look blue (beyond the atmosphere, the sky is black).

Tyndall effect scattering of light to produce a visible beam.

typhoon violent revolving storm, a †hurricane in the western Pacific Ocean.

Tyrian purple natural purple dye extracted from the shellfish *Murex brandaris*.

U

UBV photometry measurement of the †astronomical color index of a star, utilizing the ultraviolet, blue, and yellow visual images over two pre-set wavelengths obtained by †photoelectric filtering. Other standardized filter wavebands are also used.

ulna one of the two bones found in the lower limb of the tetrapod (four-limbed) vertebrate. It articulates with the shorter †radius and †humerus (upper arm bone) at one end and with the radius and wrist bones at the other.

ultrabasic in geology, an igneous rock with a lower silica content than basic rocks (less than 45% silica). Part of a system of classification based on the erroneous concept of silica acidity and basicity. Once used widely it has now been largely replaced by the term **ultramafic**.

ultrafiltration process by which substances in solution are separated on the basis of their molecular size. A solution is forced through a membrane with pores large enough to permit the passage of small solute molecules but not large ones.

Ultrafiltration is a vital mechanism in the vertebrate kidney: the cell membranes lining the Bowman's capsule act as semipermeable membranes, allowing water and substances of low molecular weight such as urea and salts to pass through into the urinary tubules but preventing the larger proteins from being lost from the blood.

ultramafic in earth science, describing a usually dark-colored igneous rock that contains more than 90% mafic minerals (that is, those rich in magnesium and iron).

ultramicroscope modified microscope arrangement used to study colloidal particles normally too small to be observed by conventional microscope-based techniques. It was developed by Austrian-born German chemist Richard †Zsigmondy.

ultrasonics branch of physics dealing with the theory and application of ultrasound: sound waves occurring at frequencies too high to be heard by the human ear (that is, above about 20 kHz).

The earliest practical application of ultrasonics was the detection of submarines during World War I by reflecting pulses of sound from them (see †sonar). Similar principles are now used in industry for non-destructive testing of materials and in medicine to produce images of internal organs and developing fetuses (ultrasound scanning). High-power ultrasound can be used for cleaning, welding plastics, and destroying kidney stones without surgery.

ultrasound pressure waves, known as ultrasonic waves, similar in nature to sound waves but occurring at frequencies above 20,000 Hz (cycles per second), the approximate upper limit of human hearing (15–16 Hz is the lower limit).

Ultrasonics is concerned with the study and practical application of these phenomena.

Some animals, such as dogs and bats, and insects are able to detect ultrasonic waves. Bats, as well as detecting ultrasonic waves, also emit them. Sound waves reflected from objects allow the bat to estimate the distance of objects, from the time taken for the echo to return.

Ultrasound can also be used for cleaning. A material is immersed in water through which ultrasonic waves are passed. The high-frequency vibrations cause dirt particles to be removed. Another use of ultrasound is in developing emulsion-based products in the cosmetics and healthcare industries. High-frequency sounds used in a mixture of oil and water form cavitations (formation of large bubbles). This helps the oil to mix with water to form an emulsion.

The earliest practical application was to detect submarines during World War I, but recently the field of ultrasonics has greatly expanded. Frequencies above 80,000 Hz have been used to produce echoes as a means of measuring the depth of the sea or to detect flaws in metal, and in medicine, high-frequency pressure waves are used to investigate various body organs. High-power ultrasound has been used with focusing arrangements to destroy deep-lying tissue in the body, and extremely high frequencies of 1,000 MHz (megahertz) or more are used in ultrasonic microscopes.

How Ultrasound Works http://www.howstuffworks.com/ultrasound1.htm

This site gives an insight into how ultrasound works, and where and why it is used. It contains an example of an ultrasound image of a fetus. There are further links provided to other sections of the site, explaining the medical uses of ultrasound.

ultraviolet astronomy study of cosmic ultraviolet emissions using artificial satellites. The USA launched a series of satellites for this purpose, receiving the first useful data in 1968. Only a tiny percentage of solar ultraviolet radiation penetrates the atmosphere, this being the less dangerous longer-wavelength ultraviolet radiation. The dangerous shorter-wavelength radiation is absorbed by gases in the ozone layer high in the Earth's upper atmosphere.

The US Orbiting Astronomical Observatory (OAO) satellites provided scientists with a great deal of information regarding cosmic ultraviolet emissions. OAO-1, launched in 1966, failed after only three days, although OAO-2, put into orbit in 1968, operated for four years instead of the intended one year, and carried out the first ultraviolet observations of a supernova and also of Uranus. OAO-3 (Copernicus), launched in 1972, continued transmissions into the 1980s and discovered many new ultraviolet sources. The International Ultraviolet Explorer, which was launched in January 1978 and ceased operation in September 1996, observed all the main objects in the Solar System (including Halley's comet), stars, galaxies, and the interstellar medium.

ultraviolet excess screening technique devised by English astronomer Martin Ryle and US astronomer Allan Sandage to measure the spectral red shift of suspected †quasars. It was this process that resulted in the discovery of quasi-stellar objects.

ultraviolet radiation electromagnetic radiation of wavelengths from about 400 to 10 nanometers (where the †X-ray range begins). Physiologically, ultraviolet radiation is extremely powerful, producing sunburn and causing the formation of vitamin D in the skin. Ultraviolet radiation is invisible to the human eye, but its effects can be demonstrated.

Levels of ultraviolet radiation have risen an average of 6.8% a decade in the northern hemisphere and 9.9% in the southern hemisphere 1972–96, according to data gathered by the Total Ozone Mapping Spectrometer on the Nimbus 7 satellite. Ultraviolet rays are strongly germicidal and may be produced artificially by mercury vapor and arc lamps for therapeutic use. The radiation may be detected with ordinary photographic plates or films. It can also be studied by its fluorescent effect on certain materials. The desert iguana *Disposaurus dorsalis* uses it to locate the boundaries of its territory and to find food.

Effects of Ultraviolet Radiation on Plants and Marine Organisms

http://pooh.chem.wm.edu/chemWWW/courses/chem105/projects/group2/page7.html

Useful outline of the risks to life on Earth posed by the build-up of radiation reaching the surface of the Earth. The information can be understood by those without a specialized knowledge of chemistry.

umbilical cord connection between the †embryo (or fetus) and the †placenta inside the †uterus of placental mammals. It has one vein and two arteries, transporting †oxygen and nutrients to the developing young, and removing waste products. †Blood is carried from the fetus along the umbilical cord and into the placenta. Here it is brought close to the mother's blood. Oxygen, nutrients, and antibodies from the mother diffuse (see †diffusion) into the fetal blood. Waste materials from the fetus pass into the mother's blood. The fetal blood, which has been enriched with nutrients, oxygenated, and cleaned of waste, is carried back to the fetus by another blood vessel in the umbilical cord. At birth, the connection between the young and the placenta is no longer necessary. The umbilical cord drops off or is severed, leaving a scar called the navel.

umbra central region of a †shadow that is totally dark because no light reaches it, and from which no part of the light source can be seen (compare †penumbra). In astronomy, it is a region of the Earth from which a complete †eclipse of the Sun can be seen.

Umbriel or **Uranus II**, one of the five major moons of †Uranus, discovered by English astronomer William Lassell in 1851. It is 1,172 km/728 mi in diameter and orbits Uranus at an average distance of 266,000 km/165,000 mi with a period (time taken to circle the planet) of 4.14 days. One of the darkest moons in the Solar System, Umbriel is thought to be covered in a thin sooty coating that has been undisturbed by impacts.

uncertainty principle or **indeterminacy principle**, in quantum mechanics, the principle that it is impossible to know with unlimited accuracy the position and momentum of a particle. The principle arises because in order to locate a particle exactly, an observer must bounce light (in the form of a †photon) off the particle, which must alter its position in an unpredictable way.

It was established by German physicist Werner †Heisenberg, and gave a theoretical limit to the precision with which a particle's momentum and position can be measured simultaneously: the more accurately the one is determined, the more uncertainty there is in the other.

unconditioned reflex behavioral reflex or response that is natural and not acquired by training or conditioning.

unconditioned stimulus natural stimulus unassociated with behavioral training; it evokes a natural or unconditioned reflex or response (see †conditioned stimulus).

unconformity surface of erosion or nondeposition eventually overlain by younger †sedimentary rock strata and preserved in the geologic record. A surface where the †beds above and below lie at different angles is called an **angular unconformity**. The boundary between older igneous or metamorphic rocks that are truncated by erosion and later covered by younger sedimentary rocks is called a **nonconformity**.

unconscious in psychoanalysis, a part of the personality of which the individual is unaware, and which contains impulses or urges that are held back, or repressed, from conscious awareness.

unicellular organism animal or plant consisting of a single cell. Most are invisible without a microscope but a few, such as the giant †amoeba, may be visible to the naked eye. The main groups of unicellular organisms are bacteria, protozoa, unicellular algae, and unicellular fungi or yeasts. Some become disease-causing agents (†pathogens).

Unicode 16-bit character encoding system, intended to cover all characters in all languages (including Chinese and similar languages) and to be backwards compatible with †ASCII.

Unlike ASCII, which is 8-bit and can therefore represent only 256 characters – insufficient for many diacritics outside the English language – Unicode can represent 65,536 characters, big enough to handle almost all written languages, including Japanese, Tibetan, and the International Phonetic Alphabet (IPA). It was created in the 1980s by Apple and Xerox in the USA.

unified field theory in physics, the theory that attempts to explain the four fundamental forces (strong nuclear, weak nuclear, electromagnetic, and gravity) in terms of a single unified force (see †particle physics).

Research was begun by Albert Einstein, and by 1971 a theory developed by US physicists Steven Weinberg and Sheldon Glashow, Pakistani physicist Abdus Salam, and others, had demonstrated the link between the weak and electromagnetic forces. The next stage is to develop a theory (called the †grand unified theory) that combines the strong nuclear force with the electroweak force. The final stage will be to incorporate gravity into the scheme. Work on the †superstring theory indicates that this may be the ultimate "theory of everything."

uniformitarianism in geology, the principle that processes that can be seen to occur on the Earth's surface today are the same as those that have occurred throughout geological time. For example, desert sandstones containing sand-dune structures must have been formed under conditions similar to those present in deserts today. The principle was formulated by Scottish geologists James †Hutton and expounded by Charles †Lyell.

unimolecular describing a chemical reaction that involves only one reacting molecule or species.

unimolecular film in chemistry, film of a substance that is only one molecule thick.

union in mathematics, a set which is formed by joining up two or more other sets.

unit standard quantity in relation to which other quantities are measured. There have been many systems of units. Some ancient units, such as the day, the foot, and the pound, are still in use. †SI units, the latest version of the metric system, are widely used in science.

Measurement Unit Converter http://www.people.virginia.edu/~rmf8a/convert.html

Simple, but very effective site that allows conversion from values in any of an astonishingly wide variety of units to any other relevant unit. All you have to do is select the category, the units you want to convert from and into, and type in the value you want to convert to get a fairly immediate answer. A Java-enabled browser is required for this site to work.

unit cell in chemistry, smallest unit that contains the characteristic configuration of molecules that define a crystal structure. The structure of a unit cell can be repeated in three dimensions to produce the molecular structure of the whole crystal material.

unit matrix †matrix which is the †identity for multiplication in a family of matrices.

$$\begin{pmatrix} 1 & 0 \\ 0 & 1 \end{pmatrix}$$

is the unit matrix for 2 by 2 matrices.

unit theory or **theory of units**, in a field or ring, involves an element that possesses an inverse. In a field, every nonzero element is a unit; in a ring, 1 and –1 represent a unit.

universal indicator in chemistry, a mixture of †pH indicators, used to gauge the acidity or alkalinity of a solution. Each component changes color at a different pH value, and so the indicator is capable of displaying a range of colors, according to the pH of the test solution, from red (at pH 1, strong acid) through green (neutral) to purple (at pH 13, strong alkali).

Universal Serial Bus **USB**, in computing, a royalty-free connector intended to replace the out-of-date COM and parallel printer ports that have been used in PCs since 1981. The USB allows up to 127 peripherals – including joysticks, scanners, printers, and keyboards – to be daisy-chained from a single socket, offering higher speeds and improved plug-and-play facilities.

The USB includes hardware and software specifications controlled by the multivendor USB-IF (Universal Serial Bus Implementors Forum) formed in March 1995. This original specification is now known as USB1 and it operates at between 1 and 12 megabits per second. USB2 has been developed as a more comparable alternative to IEEE 1394 (or FireWire), with speeds of up to 400 megabits per second. USB1 peripherals can be linked to USB2 sockets, avoiding the need to change the peripherals that the switch to IEEE 1394 would require.

USB: Universal Serial Bus http://www.usb.org/faq/

Good introduction to the features and applications of the Universal Serial Bus (USB) computer hardware connector, a standardized connector which allows up to 127 peripherals to be connected to a single computer at one time. The site includes general product information, installation advice, and trouble-shooting tips.

universal set in mathematics, the set from which subsets are taken, the set of all the objects under consideration in a problem. The odd numbers 1, 3, 5, 7 ... and the square numbers 1, 4, 9, 16 ... are taken from the universal set of natural numbers.

universal set in set theory, with regard to a †mathematical structure, a set of objects in the structure. More generally, it represents the "universe of discourse" appropriate to the discourse – for example, the set of vowels and the set of consonants are disjoint sets within the universal set comprising the alphabet.

Universal Time **UT**, another name for †Greenwich Mean Time. It is based on the rotation of the Earth, which is not quite constant. Since 1972, UT has been replaced by Universal Time Coordinated (UTC), which is based on uniform atomic time. See †time.

universe all of space and its contents, the study of which is called †cosmology. The universe is thought to be between 10 billion and 20 billion years old, and is mostly empty space, dotted with †galaxies for as far as telescopes can see. These galaxies are moving further apart as the universe expands. Several theories attempt to explain how the universe came into being and evolved; for example, the †Big Bang theory of an expanding universe originating in a single explosive event (creating hydrogen and helium gases), and the contradictory †steady-state theory.

Unix multiuser †operating system designed for minicomputers but becoming increasingly popular on microcomputers, workstations, mainframes, and supercomputers.

Unix was developed by AT&T's Bell Laboratories in the USA during the late 1960s, using the programming language †C. It could therefore run on any machine with a C compiler, so ensuring its wide portability. Its wide range of functions and flexibility, together with the fact that it was available free from 1976 to 1983, have made it widely used by universities and in commercial software.

In the 1990s, AT&T's Unix System Laboratories was taken over by Novell, which later sold it to the Santa Cruz Operation.

unleaded gasoline gasoline manufactured without the addition of †antiknock. It has a slightly lower octane rating than leaded gasoline, but has the advantage of not polluting the atmosphere with lead compounds. Many cars can be converted to run on unleaded gasoline by altering the timing of the engine, and most new cars are designed to do so. Cars fitted with a †catalytic converter must use unleaded fuel. Aromatic hydrocarbons and alkenes are added to unleaded gasoline instead of lead compounds to increase the octane rating. After combustion the hydrocarbons produce volatile organic compounds. These have been linked to cancer, and are involved in the formation of phytochemical smog. A low-lead fuel is less toxic than unleaded gasoline for use in cars that are not fitted with a catalytic converter.

The use of unleaded gasoline has been standard in the USA for some years.

unsaturated compound chemical compound in which two adjacent atoms are linked by a double or triple covalent bond. †Alkenes, such as †ethene, are unsaturated †hydrocarbons that contain a carbon to carbon double bond.

Other examples are †ketones, where the unsaturation exists between atoms of different elements (carbon and oxygen). The laboratory test for unsaturated compounds is the addition of bromine water; if the test substance is unsaturated, the bromine water will be decolorized.

unsaturated solution solution that is capable of dissolving more solute than it already contains at the same temperature.

unstable equilibrium state of equilibrium possessed by a body that will not remain at rest if displaced slightly, but will topple over into a new position; it will not return to its original rest position. See †stability.

ununnilium symbol Uun, temporary name for a synthesized radioactive element of the †transactinide series, atomic number 110, relative atomic mass 269. It was discovered in October 1994, detected for a millisecond, at the GSI heavy-ion cyclotron, Darmstadt, Germany, while lead atoms were bombarded with nickel atoms.

unununium symbol Uuu, temporary name for a synthesized radioactive element of the †transactinide series, atomic number 111, relative atomic mass 272. It was detected at the GSI heavy-ion cyclotron, Darmstadt, Germany, in December 1994, when bismuth-209 was bombarded with nickel.

upgrade in computing, improved version of an existing software program. Upgrades are sometimes available free or at low cost to registered owners of previous versions.

upper bound in mathematics, a value that is greater than or equal to all the values of a given set. The upper bound of a measurement is taken as the top extreme of the possible values. For example, for a length given as 3.4 cm correct to one decimal place, the upper bound is 3.45 correct to two decimal places. The limiting value of an infinite set is the **least upper bound**. See also †approximation.

upper respiratory tract in biology, region of the respiratory tract that consists of the trachea (windpipe) and the length of each bronchus lying outside of the lungs. It is an air-conditioning system, which warms and filters inhaled air.

upthrust upward force experienced by all objects that are totally or partially immersed in a fluid (liquid or gas). It acts against the weight of the object, and, according to Archimedes' principle, is always equal to the weight of the fluid displaced by that object. An object will float when the upthrust from the fluid is equal to its weight (see †floating).

uracil in biochemistry, pyrimidine base that is part of the genetic code of ribonucleic acid (RNA). It pairs with †adenine during the process of †transcription, in which a strand of messenger RNA (mRNA) is produced against a sense strand of deoxyribonucleic acid (DNA) acting as a template. Uracil is replaced by thymine in DNA.

uraninite uranium oxide, UO_2, an ore mineral of uranium, also known as **pitchblende** when occurring in massive form. It is black or brownish-black, very dense, and radioactive. It occurs in veins and as massive crusts, usually associated with granite rocks.

uranium chemical symbol U, hard, lustrous, silver-white, malleable and ductile, radioactive, metallic element of the †actinide series, atomic number 92, relative atomic mass 238.029. It is the most abundant radioactive element in the Earth's crust, its decay giving rise to essentially all radioactive elements in nature; its final decay product is the stable element lead. Uranium combines readily with most elements to form compounds that are extremely poisonous. The chief ore is †pitchblende, in which the element was discovered by German chemist Martin Klaproth in 1789; he named it after the planet Uranus, which had been discovered in 1781.

Small amounts of certain compounds containing uranium have been used in the ceramics industry to make orange-yellow glazes and as mordants in dyeing; however, this practice was discontinued when the dangerous effects of radiation became known.

Uranium is one of three fissile elements (the others are thorium and plutonium). It was long considered to be the element with the highest atomic number to occur naturally on Earth. The isotopes U-238 and U-235 have been used to help determine the age of the Earth.

Uranium-238, which comprises about 99% of all naturally occurring uranium, has a half-life of 4.51×10^9 years. Because of its abundance, it is the isotope from which fissile plutonium is produced in breeder †nuclear reactors. The fissile isotope U-235 has a half-life of 7.13×10^8 years and comprises about 0.7% of naturally occurring uranium; it is used directly as a fuel for nuclear reactors and in the manufacture of nuclear weapons.

What is Uranium? http://www.uic.com.au/uran.htm

Comprehensive and informative page on uranium, its properties and uses, mainly in nuclear reactors and weapons, provided by the Uranium Information Council.

uranium halide chemical compound that contains only uranium and a halogen atom. Uranium forms halides in a number of valence states, IV–VI, for example UCl_4, UBr_5, UF_6. Uranium hexafluoride is used in the process for the separation of the fissile isotope U-235 from non-fissile U-238.

uranium ore material from which uranium is extracted, often a complex mixture of minerals. The main ore is uraninite (or pitchblende), UO_2, which is commonly found with sulfide minerals. The USA, Canada, and South Africa are the main producers in the West.

uranium oxide chemical compound that contains only uranium and oxygen atoms. Uranium forms oxides in a number of valence states, IV–VI. The more stable forms are UO_2, U_3O_8, and U_4O_9. Derivatives of uranium oxides are used as coloring agents in the ceramics industry.

Uranus seventh planet from the Sun, discovered by German-born British astronomer William †Herschel in 1781. It is twice as far out as the sixth planet, Saturn. Uranus has a mass 14.5 times that of Earth. The spin axis of Uranus is tilted at 98°, so that one pole points towards the Sun, giving extreme seasons.

mean distance from the Sun: 2.9 billion km/1.8 billion mi

equatorial diameter: 50,800 km/31,600 mi

rotation period: 17 hours 12 minutes

year: 84 Earth years

atmosphere: deep atmosphere composed mainly of hydrogen and helium

surface: composed primarily of rock and various ices with only about 15% hydrogen and helium, but may also contain heavier elements, which might account for Uranus's mean density being higher than that of Saturn

satellites: 18 moons (two discovered in 1997, one in 1999)

rings: 11 rings, composed of rock and dust, around the planet's equator, were detected by the US space probe Voyager 2 in 1977. The rings are charcoal black and may be debris of former "moonlets" that have broken up. The ring furthest from the planet center (51,000 km/31,800 mi), Epsilon, is 100 km/62 mi at its widest point. In 1995, US astronomers determined that the ring particles contained long-chain hydrocarbons. Looking at the brightest region of Epsilon, they were also able to calculate the precession of Uranus as 264 days, the fastest known precession in the Solar System

Uranus has a peculiar magnetic field, in that it is tilted at 60° to the axis of spin, and is displaced about a third of the way from the planet's center to its surface. Uranus spins from east to west, the opposite of the other planets, with the exception of Venus and possibly Pluto. The rotation rate of the atmosphere varies with latitude, from about 16 hours in mid-southern latitudes to longer than 17 hours at the equator.

Uranus http://www.solarviews.com/solar/eng/uranus.htm

Did you know that Uranus is tipped on its side? Find out more about Uranus, its rings, and its moons at this site. Also included are a table of statistics about the planet, photographs, and animations of its rotation.

urban ecology study of the †ecosystems, animal and plant communities, soils and microclimates found within an urban landscape.

Parks are important for many organisms, such as song birds, while birds of prey (the kestrel being a notable example) find ample food in the wasteland around estates and offices. Mammals, including foxes and badgers, are regular visitors, especially if there is an undisturbed corridor penetrating into the town, such as a disused railway.

Urban Ecology http://www.urbanecology.org/

Outlines the urban planning and regeneration programs of the US society Urban Ecology. The Web site has articles from their journal Urban Ecology and a very good selection of urban ecology links.

urea $CO(NH_2)_2$, waste product formed in the mammalian liver when nitrogen compounds are broken down. It is filtered from the blood by the kidneys, and stored in the bladder as urine prior to release. When purified, it is a white, crystalline solid. In industry it is used to make urea-formaldehyde plastics (or resins), pharmaceuticals, and fertilizers.

urea cycle biochemical process discovered by German-born British biochemist Hans †Krebs and German physician Kurt Henseleit in 1932, by which nitrogenous waste is converted into urea, which is easily excreted. When †proteins and †amino acids break down, †ammonia, which is highly toxic, is formed. Most of the ammonia is converted into glutamate, and becomes usable for the synthesis of more amino acids and proteins. Any excess is converted into the water-soluble compound †urea, which can be excreted as †urine. Free ammonia, carbon dioxide, and †ATP react to form carbamyl phosphate. This compound then forms †citrulline by reacting with the amino acid ornithine, which can then accept another amino group giving arginine. Arginine can then break down, giving urea, for excretion, and ornithine, which can take part in the cycle again.

urease enzyme in plants that catalyses the hydrolysis of †urea to form carbon dioxide and ammonia.

ureter tube connecting the kidney to the bladder. Its wall contains fibers of smooth muscle whose contractions aid the movement of urine out of the kidney.

urethra in mammals, a tube connecting the bladder to the exterior. It carries urine and, in males, semen.

Urey, Harold Clayton (1893–1981) US chemist. In 1932 he isolated †heavy water and was awarded the Nobel Prize for Chemistry in 1934 for his discovery of †deuterium (heavy hydrogen).

uric acid $C_5H_4N_4O_3$, nitrogen-containing waste substance, formed from the breakdown of food and body protein.

It is only slightly soluble in water. Uric acid is the normal means by which most land animals that develop in a shell (birds, reptiles, insects, and land gastropods) deposit their waste products. The young are unable to get rid of their excretory products while in the shell and therefore store them in this insoluble form.

Humans and other primates produce some uric acid as well as urea, the normal nitrogenous waste product of mammals, adult amphibians, and many marine fishes. If formed in excess and not excreted, uric acid may be deposited in sharp crystals in the joints and other tissues, causing gout; or it may form stones (calculi) in the kidneys or bladder.

urinary system system of organs that removes nitrogenous waste products and excess water from the bodies of animals. In vertebrates, it consists of a pair of kidneys, which produce urine; ureters, which drain the kidneys; and (in bony fishes, amphibians, some reptiles, and mammals) a bladder that stores the urine before its discharge. In mammals, the urine is expelled through the urethra; in other vertebrates, the urine drains into a common excretory chamber called a †cloaca, and the urine is not discharged separately.

urine amber-colored fluid filtered out by the kidneys from the blood. It contains excess water, salts, proteins, waste products in the form of urea, a pigment, and some acid.

The kidneys pass it through two fine tubes (ureters) to the bladder, which may act as a reservoir for up to 0.7 l/1.5 pt at a time. In mammals, it then passes into the urethra, which opens to the outside by a sphincter (constricting muscle) under voluntary control. In reptiles and birds, nitrogenous wastes are discharged as an almost solid substance made mostly of †uric acid, rather than urea.

URL abbreviation for uniform resource locator, series of letters and/or numbers specifying the location of a document on the †World Wide Web. Every URL consists of a domain name, a description of the document's location within the host computer, and the name of the document itself, separated by full stops and backslashes. Thus *The Times* Web site can be found at http://www.the-times.co.uk/news/pages/home.html, and a tribute to Elvis Presley is at http://sunsite.unc.edu/elvis/ elvishom.html. The complexity of URLs explains why bookmarks and links, which save the user from the chore of typing them in, are so popular.

Beginners' Guide to URLs http://www.lancs.ac.uk/help/url_primer.htm

Guide to uniform resource locators (URLs) describing what they do. The site differentiates between tie different types of URL, covering "file," "partial," "gopher," "HTTP URLSs," and "news" URLs. The bottom of the page contains a link to more detailed information about URLs.

uronic acid acid formed from a sugar that has had its primary alcohol group oxidized. Uronic acids are named after their original sugar form; for example, glucuronic acid is formed by the selective oxidation of glucose.

Ursa Major (Latin "Great Bear") third-largest constellation in the sky, in the north polar region. Its seven brightest stars make up the familiar shape or †asterism of the **Big Dipper** or **Plough**. The second star of the handle of the dipper, called †Mizar, has a companion star, Alcor.

Two stars forming the far side of the dipper bowl act as pointers to the north pole star, †Polaris. Dubhe, one of them, is the constellation's brightest star.

Ursa Minor (Latin "Little Bear") small constellation of the northern hemisphere, popularly known as the Little Dipper. It is shaped like a dipper, with the bright north pole star †Polaris at the end of the handle. Two other bright stars in this group, Beta and Gamma Ursae Minoris, are called "the Guards" or "the Guardians of the Pole." The constellation also contains the orange subgiant Kochab, about 95 light years from Earth.

USB in computing, abbreviation for †Universal Serial Bus.

user interface in computing, the procedures and methods through which the user operates a program. These might include †menus, input forms, error messages, and keyboard procedures. A †graphical user interface (GUI or WIMP) is one that makes use of icons (small pictures) and allows the user to make menu selections with a mouse.

A **command line interface** is a character-based interface in which a prompt is displayed on the screen at which the user types a command, followed by carriage return, at which point the command, if valid, is executed. An example of a command line interface is the **DOS prompt**.

A **menu-driven interface** presents various options to the user in the form of a list, from which commands may be selected. Types of menu include the **menu bar**, which displays the top-level options available to the user as a single line across the top of the screen; selecting one of these options displays a **pull-down** menu. Programs such as Microsoft Word use menus in this way.

In a **graphical user interface** programs and files appear as icons (small pictures), user options are selected from pull-down menus, and data is displayed in windows (rectangular areas), which the operator can manipulate in various ways. The operator uses a pointing device, typically a mouse, to make selections and initiate actions.

The study of the ways in which people interact with computers is a subbranch of ergonomics. It aims to make it easier for people to use computers effectively and comfortably, and has become a focus of research for many national and international programs.

U-shaped valley another term for a †glacial trough, a valley formed by a glacier.

uterus hollow muscular organ of female mammals, located between the bladder and rectum, and connected to the Fallopian tubes above and the vagina below. The embryo develops within the uterus, and in placental mammals is attached to it after implantation via the †placenta and umbilical cord. The lining of the uterus changes during the †menstrual cycle to prepare it for pregnancy. In humans and other higher primates, it is a single structure, but in other mammals it is paired.

In †fertilization a sperm fuses with an ovum and produces a zygote. The zygote sinks into the lining of the uterus. It then grows a placenta which brings the fetus's †blood close to the mother's blood so that the fetus can obtain †oxygen and nutrients. The outer wall of the uterus is composed of smooth †muscle, capable of powerful contractions (induced by hormones) during childbirth.

utility program in computing, a systems program designed to perform a specific task related to the operation of the computer when requested to do so by the computer user. For example, a utility program might be used to complete a screen dump, format a disk, or convert the format of a data file so that it can be accessed by a different applications program.

U-tube U-shaped tube that may be partly filled with liquid and used as a †manometer.

U-value measure of a material's heat-conducting properties. It is used in the building industry to compare the efficiency of insulating products, a good insulator having a low U-value. The U-value of a material is defined as the rate at which heat is conducted through it per unit surface area per unit temperature difference between its two sides; it is measured in watts per square meter per kelvin ($W\ m^{-2}\ K^{-1}$). In mathematical terms, it may be calculated as the rate of loss of heat/surface area x temperature difference.

uvula muscular structure descending from the soft palate. Its length is variable (5–20 mm/0.2–0.8 in in humans). It is raised during swallowing and is sometimes used in the production of speech sounds.

V

V in physics, symbol for †volt, the SI unit of emf or potential difference, equal to the potential difference between two points in an electrical circuit when a current of 1 ampere flowing between them dissipates a power of 1 watt.

vaccine any preparation of modified pathogens (viruses or bacteria) that is introduced into the body, usually either orally or by a hypodermic syringe, to induce the specific †antibody reaction that produces †immunity against a particular disease.

In 1796 Edward †Jenner was the first to inoculate a child successfully with cowpox virus to produce immunity to smallpox. His method, the application of an infective agent to an abraded skin surface, is still used in smallpox inoculation.

vacuole in biology, a fluid-filled, membrane-bound cavity inside a cell. It may be a reservoir for fluids that the cell will secrete to the outside, or may be filled with excretory products or essential nutrients that the cell needs to store.

Plant cells usually have a large central vacuole containing call sap (sugar and salts in solution) which serves both as a store of food and as a key factor in storing water and in maintaining †turgor. Absorbing more water to make a bigger vacuole adds bulk to the plant. This expansion of cells is very important in plant †growth. In amoebae (single-celled animals), vacuoles are the sites of digestion of engulfed food particles. Animal cells may only have small vacuoles, which are usually called vesicles.

vacuum in general, a region completely empty of matter; in physics, any enclosure in which the gas pressure is considerably less than atmospheric pressure (101,325 pascals).

vacuum flask or **Dewar flask** or **Thermos flask**, container for keeping things either hot or cold. It has two silvered glass walls with a vacuum between them, in a metal or plastic outer case. This design reduces the three forms of heat transfer: radiation (prevented by the silvering), conduction, and convection (both prevented by the vacuum). A vacuum flask is therefore equally efficient at keeping cold liquids cold or hot liquids hot. It was invented by the British scientist James Dewar in about 1872, to store liquefied gases.

vagina lower part of the reproductive tract in female mammals, linking the uterus to the exterior. It admits the penis during sexual intercourse, and is the birth canal down which the baby passes during delivery.

valence in chemistry, the measure of an element's ability to combine with other elements, expressed as the number of atoms of hydrogen (or any other standard univalent element) capable of uniting with (or replacing) its atoms. The number of electrons in the outermost shell of the atom dictates the combining ability of an element.

The elements are described as uni-, di-, tri-, and tetravalent when they unite with one, two, three, and four univalent atoms respectively. Some elements have **variable valence**: for example, nitrogen and phosphorus have a valence of both three and five. The valency of oxygen is two: hence the formula for water, H_2O (hydrogen being univalent).

valence electron electron in the outermost shell of an †atom. It is the valence electrons that are involved in the formation of ionic and covalent bonds (see †molecule). The number of electrons in this outermost shell represents the maximum possible valence for many elements and matches the number of the group that the element occupies in the †periodic table of the elements.

valence shell outermost shell of electrons in an †atom. It contains the †valence electrons. Elements with four or more electrons in their outermost shell can show variable valence. Chlorine can show valences of 1, 3, 5, and 7 in different compounds.

valine one of the nine essential †amino acids .

Valles Marineris complex system of rift valleys and canyons on the surface of †Mars. It was first delineated by the Mariner 9 spacecraft that surveyed the planet from its orbit in 1971. It is 4,500 km/2,800 mi long and up to 8 km/5 mi deep. Part of the system is visible from Earth-bound telescopes as a dark streak known as Coprates.

value in mathematics, a number or other fixed quantity applied to a †variable. The value of an expression will depend on the numbers which are substituted for the variables.

For example, $x^2 + y^2$ has the value 25 when $x = 3$ and $y = 4$.

valve in animals, a structure for controlling the direction of the blood flow. In humans and other vertebrates, the contractions of the beating heart cause the correct blood flow into the arteries because a series of valves prevents back flow. Diseased valves, detected as "heart murmurs," have decreased efficiency. The tendency for low-pressure venous blood to collect at the base of limbs under the influence of gravity is counteracted by a series of small valves within the veins. It was the existence of these valves that prompted the 17th-century physician William Harvey to suggest that the blood circulated around the body.

valve or **electron tube**, in electronics, a glass or metal tube containing gas at low pressure, which is used to control the flow of electricity in a circuit. The electron tube valve was developed by US radio engineer Lee de Forest (1873–1961) and is used to modify electrical signals. Three or more metal electrodes are inset into the tube. By varying the voltage on one of them, called the **grid electrode**, the current through the valve can be controlled, and the valve can act as an amplifier. It is defined as a valve because it allows a undirectional flow of electrons.

Valves have been replaced for most applications by †transistors. However, they are still used in high-power transmitters and amplifiers, and in some hi-fi systems.

valve in technology, a device that controls the flow of a fluid. Inside a valve, a plug moves to widen or close the opening through which the fluid passes.

Common valves include the cone or needle valve, globe valve, and butterfly valve, all named after the shape of the plug. Specialized valves include the one-way valve, which permits fluid flow in one direction only, and the safety valve, which cuts off flow under certain conditions.

vanadate any salt that contains an anion that is composed of both vanadium and oxygen atoms.

vanadium chemical symbol V, silver-white, malleable and ductile, metallic element, atomic number 23, relative atomic mass 50.942. It occurs in certain iron, lead, and uranium ores and is widely distributed in small quantities in igneous and sedimentary rocks. It is used to make steel alloys, to which it adds tensile strength.

Spanish mineralogist Andrés del Rio (1764–1849) and Swedish chemist Nils Sefström (1787–1845) discovered vanadium independently, the former in 1801 and the latter in 1831. Del Rio named it "erythronium," but was persuaded by other chemists that he had not in fact discovered a new element; Sefström gave it its present name, after the Norse goddess of love and beauty, Vanadis (or Freya).

vanadium(V) oxide or **vanadium pentoxide**; V_2O_5, crystalline compound used as a catalyst in the †contact process for the manufacture of sulfuric acid.

Van Allen, James Alfred (1914–) US physicist whose instruments aboard the first US satellite Explorer 1 in 1958 led to the discovery of the Van Allen belts, two zones of intense radiation around the Earth. He pioneered high-altitude research with rockets after World War II.

Van Allen radiation belts two zones of charged particles around the Earth's magnetosphere, discovered in 1958 by US physicist James Van Allen. The atomic particles come from the Earth's upper atmosphere and the †solar wind, and are trapped by the Earth's magnetic field. The inner belt lies 1,000–5,000 km/620–3,100 mi above the Equator, and contains †protons and †electrons. The outer belt lies 15,000–25,000 km/9,300–15,500 mi above the Equator, but is lower around the magnetic poles. It contains mostly electrons from the solar wind.

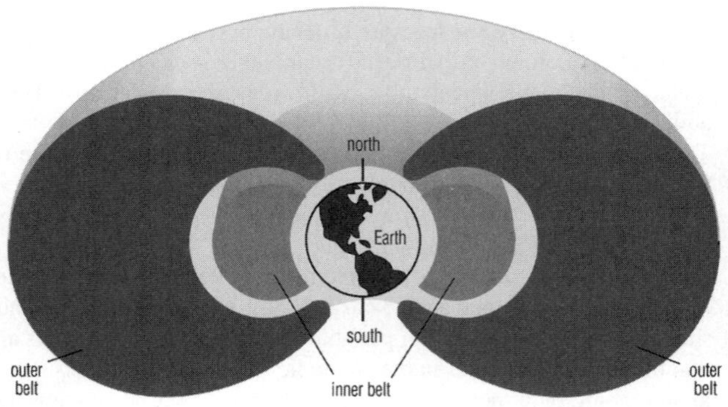

Van Allen radiation belts The Van Allen belts of trapped charged particles are a hazard to spacecraft, affecting on-board electronics and computer systems. Similar belts have been discovered around the planets Mercury, Jupiter, Saturn, Uranus, and Neptune. Helicon

van de Graaff generator electrostatic generator capable of producing a voltage of over a million volts. It consists of a continuous vertical conveyor belt that carries electrostatic charges (resulting from friction) up to a large hollow sphere supported on an insulated stand. The lower end of the belt is earthed, so that charge accumulates on the sphere. The size of the voltage built up in air depends on the radius of the sphere, but can be increased by enclosing the generator in an inert atmosphere, such as nitrogen.

Vandermonde, Alexandre-Théophile (1735–1796) French musician and musical theorist who wrote original and influential papers on algebraic equations and determinants.

van der Waals' forces in chemistry, short-range, weak attractive forces between molecules. They are thought to arise from the formation of temporary †dipoles caused by variations in the electronic configuration of the molecules.

van der Waals' law modified form of the †gas laws that includes corrections for the non-ideal behavior of real gases (the molecules of ideal gases occupy no space and exert no forces on each other). It is named after Dutch physicist J D van der Waals (1837–1923).

The equation derived from the law states that:

$(P + a/V^2)(V - b) = RT$

where P, V, and T are the pressure, volume, and temperature (in kelvin) of the gas, respectively; R is the †gas constant; and a and b are constants for that particular gas.

vanillin or **4-hydroxy-3methoxybenzaldehyde**; $C_8H_8O_3$, in chemistry, white crystalline solid, melting point 82°C/180°F. It has the distinctive smell and taste of vanilla and occurs naturally in a number of plants, notably the vanilla pod. It is used as a flavoring agent and in cosmetics.

van't Hoff, Jacobus Henricus (1852–1911) Dutch physical chemist. He explained the "asymmetric" carbon atom occurring in optically active compounds. His greatest work – the concept of chemical affinity as the maximum work obtainable from a reaction – was shown with measurements of osmotic and gas pressures, and reversible electrical cells. He was the first person to be awarded the Nobel Prize for Chemistry in 1901 for his study of the laws of chemical dynamics and osmotic pressure.

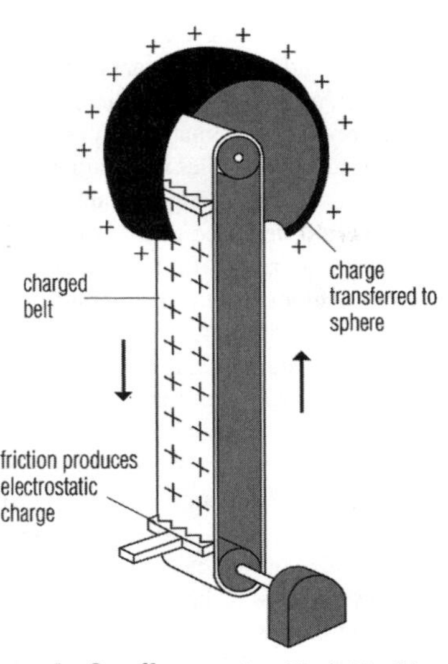

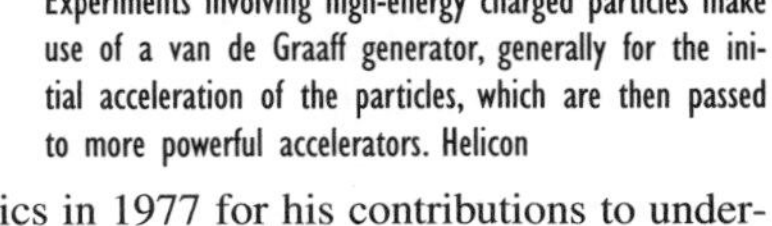
van de Graaff generator US physicist Robert Jemison van de Graaff developed this high-powered generator, which can produce more than a million volts. Experiments involving high-energy charged particles make use of a van de Graaff generator, generally for the initial acceleration of the particles, which are then passed to more powerful accelerators. Helicon

Van Vleck, John Hasbrouck (1899–1980) US physicist, considered one of the founders of modern magnetic theory (see †magnetism). He was awarded the Nobel Prize for Physics in 1977 for his contributions to understanding the electronic structure of magnetic and disordered systems. He shared the award with his student Philip †Anderson.

vapor one of the three states of matter (see also †solid and †liquid). The molecules in a vapor move randomly and are far apart, the distance between them, and therefore the volume of the vapor, being limited only by the walls of any vessel in which they might be contained. A vapor differs from a †gas only in that a vapor can be liquefied by increased pressure, whereas a gas cannot unless its temperature is lowered below its †critical temperature; it then becomes a vapor and may be liquefied.

vapor density density of a gas, expressed as the †mass of a given volume of the gas divided by the mass of an equal volume of a reference gas (such as hydrogen or air) at the same temperature and pressure. If the reference gas is hydrogen, it is equal to half the relative molecular weight (mass) of the gas.

vapor phase or **gaseous phase**, physical state of a substance or process that is a gas.

vapor pressure pressure of a vapor given off by (evaporated from) a liquid or solid, caused by atoms or molecules continuously escaping from its surface. In an enclosed space, a maximum value is reached when the number of particles leaving the surface is in equilibrium with those returning to it; this is known as the **saturated vapor pressure** or **equilibrium vapor pressure**.

variable in mathematics, a changing quantity (one that can take various values), as opposed to a †constant. For example, in the algebraic expression $y = 4x^3 + 2$, the variables are x and y, whereas 4 and 2 are constants. A variable may be dependent or independent. Thus if y is a †function of x, written $y = f(x)$, such that $y = 4x^3 + 2$, the domain of the function includes all values of the **independent variable** x while the range (or co-domain) of the function is defined by the values of the **dependent variable** y.

variable star star whose brightness changes, either regularly or irregularly, over a period ranging from a few hours to months or years. The †Cepheid variables regularly expand and contract in size every few days or weeks.

Stars that change in size and brightness at less precise intervals include **long-period variables**, such as the red giant †Mira in the constellation †Cetus (period about 331 days), and **irregular variables**, such as some red supergiants. **Eruptive variables** emit sudden outbursts of light. Some suffer flares on their surfaces, while others, such as a †nova, result from transfer of gas between a close pair of stars. A †supernova is the explosive death of a star. In an †**eclipsing binary**, the variation is due not to any change in the star itself, but to the periodic eclipse of a star by a close companion. The different types of variability are closely related to different stages of stellar evolution.

variance in statistics, the square of the †standard deviation, the measure of spread of data. Population variance and sample variance are denoted by s^2 and s^2, respectively. Variance provides a measure of the dispersion of a set of statistical results about the mean or average value.

variation in mathematics, a practical relationship between two †variables. **Direct variation** (when the values of the variables maintain a constant ratio) corresponds to $y = kx$ on a straight line graph, for example distance traveled at a steady speed. **Inverse variation** (when an increase in the value of one variable results in a decrease in that of the other) corresponds to $y = {}^{k}/_{x}$ on a rectangular hyperbolic graph, for example the price of an article and the quantity that can be bought for a fixed sum of money. In problems of direct and inverse variation, the first step is to find the value of k.

variation in biology, a difference between individuals of the same species, found in any sexually reproducing population. Variations may be almost unnoticeable in some cases, obvious in others, and can concern many aspects of the organism. Typically, variations in size, behavior, biochemistry, or coloring may be found. The cause of the variation is genetic (that is, inherited), environmental, or more usually a combination of the two. Some variation is the result of the environment modifying inherited characteristics. The origins of variation can be traced to the recombination of the genetic material during the formation of the gametes, and, more rarely, to mutation.

An example of inherited variation is blood groups in humans. An extreme example of variation produced by the environment is the loss of a part of the body from injury. An example of variation that involves the environment modifying inherited characteristics is height. The characteristic of being tall or short is inherited, but the availability of food in the environment determines the actual height of the individual.

variegation description of plant leaves or stems that exhibit patches of different colors. The term is usually applied to plants that show white, cream, or yellow on their leaves, caused by areas of tissue that lack the green pigment †chlorophyll. Variegated plants are bred for their decorative value, but they are often considerably weaker than the normal, uniformly green plant. Many will not breed true and require †vegetative reproduction.

The term is sometimes applied to abnormal patchy coloring of petals, as in the variegated petals of certain tulips, caused by a virus infection. A mineral deficiency in the soil may also be the cause of variegation.

variety in biology, a stable group of organisms within a single species, clearly different from the rest of the species. Such a group would generally be called a variety for plants and a breed for animals. The differences lie in their genetic make-up and could have arisen naturally – through †natural selection or as a result of selective †breeding by humans. Most varieties have been produced by selective breeding, for example "Cox's," "Golden Delicious," and "Bramley" apple varieties.

In most species there are often genetic differences between individuals that cause them to be distinct. These are not varieties – a variety is a group of individuals. However, not every genetically distinct group can be a variety. There must be a greater degree of variation between varieties than within the varieties for the term to be appropriate. This can be shown by collecting data on the mass, length, and diameter of varieties of tomato, potato, or runner bean, for example.

Selective breeding results in the production of new varieties if enough useful characteristics are brought together in one group. The "Bramley" apple variety had the characteristics of vigorous growth, disease resistance, large apples, good storage, and soft texture on cooking. This has made it a popular cooking apple in some countries.

Varmus, Harold Elliot (1939–) US molecular biologist who was awarded a Nobel Prize for Physiology or Medicine in 1989 with John Michael †Bishop for the discovery of oncogenes, genes carried by viruses that can trigger cancerous growth in normal cells. Varmus's work showed that a single viral gene is sufficient to transform a cell into a cancerlike cell, and that this gene is also present in normal mammalian cells. Since the discovery of the first oncogene, many more have been found and used in cancer research.

varve in geology, a pair of thin sedimentary beds, one coarse and one fine, representing a cycle of thaw followed by an interval of freezing, in lakes of glacial regions.

Each couplet thus constitutes the sedimentary record of a year, and by counting varves in glacial lakes a record of absolute time elapsed can be determined. Summer and winter layers often are distinguished also by color, with lighter layers representing summer deposition, and darker layers being the result of dark clay settling from water while the lake is frozen.

varve analysis method of archaeological dating using annual varve (glacial deposit) thickness patterns formed in lakes near the edge of a glacier's retreat. A hot summer results in a thick varve, owing to a greater meltwater rate and discharge of gravel and clay, whereas a cold summer results in a thin one. By analysis of thickness patterns and cross-linking to adjacent regions, a chronology has been developed back to the end of the last Ice Age.

The method was pioneered by the Swedish geologist Gerard de Geer, who first noted the regular patterns of glacial deposits in lakes in 1878. He tracked and dated the complete 3,600-year glacial retreat in southern Scandinavia. This sequence was then tied to river estuary varves (caused by rhythmic annual flooding) spanning about 7,500 years up to AD 700.

vascular bundle in botany, strand of primary conducting tissue (a "vein") in vascular plants, consisting mainly of water-conducting tissues, metaxylem and protoxylem, which together make up the primary †xylem, and nutrient-conducting tissue, †phloem. It extends from the roots to the stems and leaves. Typically the phloem is situated nearest to the epidermis and the xylem towards the center of the bundle. In plants exhibiting †secondary growth, the xylem and phloem are separated by a thin layer of vascular †cambium, which gives rise to new conducting tissues.

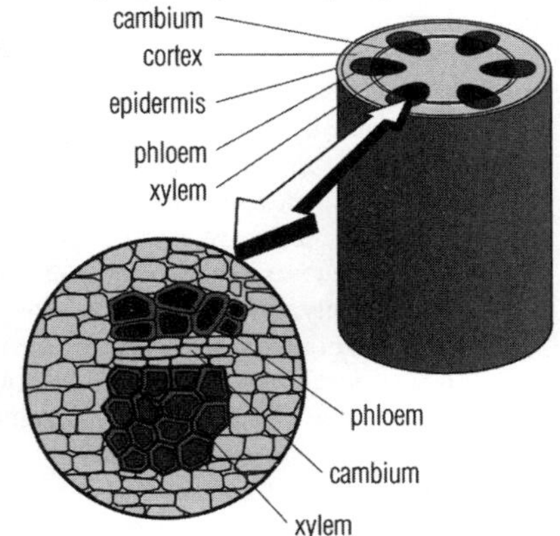

vascular bundle The fluid-carrying tissue of most plants is normally arranged in units called vascular bundles. The vascular tissue is of two types: xylem and phloem. The xylem carries water up through the plant; the phloem distributes food made in the leaves to all parts of the plant. Helicon

vascular plant plant containing vascular bundles. †Pteridophytes (ferns, horsetails, and club mosses), †gymnosperms (conifers and cycads), and †angiosperms (flowering plants) are all vascular plants.

vas deferens in male vertebrates, a tube conducting sperm from the testis to the urethra. The sperm is carried in a fluid secreted by various glands, and can be transported very rapidly when the smooth muscle in the wall of the vas deferens undergoes rhythmic contraction, as in sexual intercourse.

vasoconstriction narrowing of blood vessels. It is mediated by the vasomotor center in the †brainstem, causing reflex changes in the heart rate with an adjustment of blood pressure. Some drugs act as vasoconstrictors.

vasodilation widening of the blood vessels; the opposite of †vasoconstriction. Some drugs act as vasodilators.

vasopressin another name for the antidiuretic †hormone †ADH.

Vaucouleurs, Gerard Henri De French-born US astronomer; see †De Vaucouleurs.

Vauquelin, Louis Nicolas (1763–1829) French chemist who discovered chromium (in 1797 in the mineral crocoite) and beryllium (in 1798 in the minerals beryl and emerald), although beryllium was not isolated as the pure metal until 1828 (by Friedrich Wöhler). Vauquelin also isolated the first amino acid asparagine (from asparagus).

VDU abbreviation for †visual display unit.

vector in medicine, disease-carrying agent. The term is usually applied to insects, ticks, and the like, that transmit parasites from one person to another or from animals to human beings. The *Anopheles* mosquito is a vector, transmitting the †malaria parasite in its salivary glands.

vector graphics computer graphics that are stored in the computer memory by using geometric formulae. Vector graphics can be transformed (enlarged, rotated, stretched, and so on) without loss of picture resolution. It is also possible to select and transform any of the components of a vector-graphics display because each is separately defined in the computer memory. In these respects vector graphics are superior to †raster graphics. Vector graphics are typically used for drawing applications, allowing the user to create and modify technical diagrams such as designs for houses or cars.

Vector and Bitmap Images http://graphicssoft.about.com/library/weekly/aa000327a.htm?once=true&

Learn about the differences between the two major types of 2D graphics: scalable vector graphics and pixel-based bitmap images. Further links cover other useful information, including which format to use when, and the differences between various graphics tools.

vector quantity any physical quantity that has both magnitude (size) and direction, such as velocity, acceleration, or force, as distinct from a †scalar quantity such as speed, density, or mass, which has magnitude but no direction. A vector is represented either geometrically by an arrow whose length corresponds to its †magnitude and points in an appropriate direction, or by two or three numbers representing the magnitude of its components. Vectors can be added graphically by constructing a parallelogram of vectors (such as the †parallelogram of forces commonly employed in physics and engineering). This will give a **resultant vector**.

The †position vector of a point A(*x*,*y*) represents the move from the origin to A, that is a †translation. A free vector has magnitude and direction but no fixed position in space.

If two forces *p* and *q* are acting on a body at A, then the parallelogram of forces is drawn to determine the resultant force and direction *r*. *p*, *q*, and *r* are vectors. In technical writing, a vector is denoted by **bold** type, underlined, or overlined.

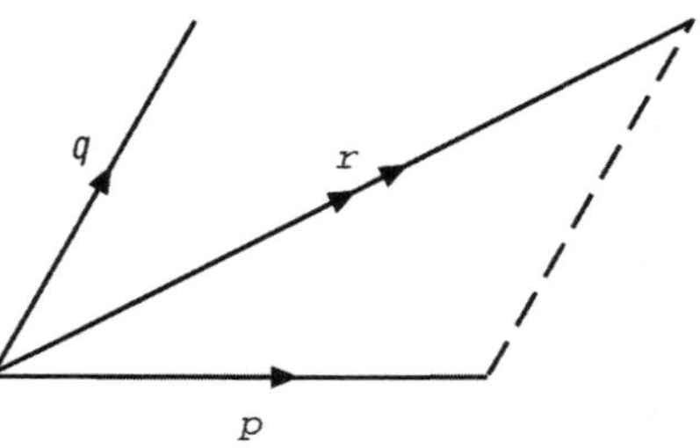

vector quantity A parallelogram of vectors. Vectors can be added graphically using the parallelogram rule. According to the rule, the sum of vectors p and q is the vector r which is the diagonal of the parallelogram with sides p and q. Helicon

vector space or **linear space**, mathematical structure comprising two types of objects: vectors and scalars. Vectors can be added by themselves; scalars lengthen or shorten them (that is, scale the vector length up or down) and are commonly either real or complex numbers. In general, scalars of a vector space comprise a field. The prime example of a vector space is the collection of elementary vectors (see †vector) in the two- or three-dimensional space of everyday experience. (The operation of addition follows the parallelogram law: to add two vectors, complete the parallelogram defined by the two lines – the sum is given by the diagonal through the common origin.) Vector spaces provide a framework for the study of linear transformations – which can be represented by †matrices in finite-dimensional spaces – and are important in mathematical modeling of complicated systems (engineering, biological, and so on) where general transformations are approximated by linear ones, with recourse to the apparatus of †differential calculus.

vegetation plant life of a particular area, such as shrubs, trees, and grasses. The type of vegetation depends on the soil, which may become impoverished by deforestation; climate, especially the amount of moisture available and the possibility of frost, since these factors influence the types of vegetation that can survive; and human activity, such as overgrazing or the destruction of grasslands.

vegetative reproduction type of †asexual reproduction in plants that relies not on spores, but on multicellular structures formed by the parent plant. Some of the main types are †stolons and runners, †gemmae, †bulbils, sucker shoots produced from roots (such as in the creeping thistle *Cirsium arvense*), †tubers, †bulbs, †corms, and †rhizomes. Vegetative reproduction has long been exploited in horticulture and agriculture, with various methods employed to multiply stocks of plants. See also †plant propagation.

vein vessel that carries †blood from the body to the †heart in animals with a circulatory system. Veins contain valves that prevent the blood from running back when moving against gravity. They carry blood at low pressure, so their walls are thinner than those of arteries. They always carry deoxygenated blood, with the exception of the pulmonary vein, leading from the lungs to the heart in birds and mammals, which carries newly oxygenated blood.

The term is also used more loosely for any system of channels that strengthens living tissues and supplies them with nutrients – for example, leaf veins (see †vascular bundle), and the veins in insects' wings. In leaves they make up the network that can normally be seen, especially from the underside of the leaf. These veins are made up of the two transport tissues of a plant, †xylem and †phloem.

Vela bright constellation of the southern hemisphere near Carina, represented as the sails of a ship. It contains large wisps of gas – called the Gum nebula after its discoverer, the Australian astronomer Colin Gum (1924–1960) – believed to be the remains of one or more †supernovae. Vela also contains the second optical †pulsar (a pulsar that flashes at a visible wavelength) to be discovered.

Vela was originally regarded as part of †Argo. Its four brightest stars are second-magnitude, one of them being Suhail, about 490 light years from Earth.

veldt subtropical grassland in South Africa, equivalent to the Pampas of South America.

velocity symbol *v*, speed of an object in a given direction, or how fast an object changes its position in a given direction. Velocity is a †vector quantity, since its direction is important as well as its magnitude (or speed). For example, a car could have a speed of 48 kph/30 mph and a velocity of 48 kph/30 mph northwards. Velocity = change in position/time taken.

The velocity at any instant of a particle traveling in a curved path is in the direction of the tangent to the path at the instant considered. The velocity v of an object traveling in a fixed direction may be calculated by dividing the distance s it has traveled by the time t taken to do so, and may be expressed as: $v = s/t$

velvet worm or **peripatus** invertebrate with soft body (15–150 mm/0.5–5.9 in in length), velvety skin and paired unsegmented legs. They are found in leaf-litter in rainforests in Africa, Australia, Indonesia,

Malaysia, and South America, and feed on small invertebrates. They are unlike any other invertebrates and so the 200 species occupy their own phylum, Onychophora. Velvet worms date back 400 million years and may be a "missing link" between the arthropods (insects and crustaceans) and annelids (soft-bodied segmented worms, including earthworms). The 20–40 million-year-old fossils of two new kinds of velvet worm were discovered by US paleontologists in 1996.

vena cava either of the two great veins of the trunk, returning deoxygenated blood to the right atrium of the †heart. The **superior vena cava**, beginning where the arches of the two innominate veins join high in the chest, receives blood from the head, neck, chest, and arms; the **inferior vena cava**, arising from the junction of the right and left common iliac veins, receives blood from all parts of the body below the diaphragm.

Vening Meinesz, Felix Andries (1887–1966) Dutch geophysicist who originated the method of making very precise gravity measurements in the stable environment of a submarine. The results he obtained were important in the fields of geophysics and geodesy. He was able to discount the model of the Earth's shape that proposed a flattening at the Equator.

Venn, John (1834–1923) English logician whose diagram, known as the †Venn diagram, is much used in the teaching of elementary mathematics.

Venn, John http://www-groups.dcs.st-andrews.ac.uk/history/Mathematicians/Venn.html
Devoted to the life and contributions of logician John Venn.

Venn diagram in mathematics, a diagram representing a †set or sets and the logical relationships between them. The sets are drawn as circles. An area of overlap between two circles (sets) contains elements that are common to both sets, and thus represents a third set. Circles that do not overlap represent sets with no elements in common (disjoint sets). The method is named after the English logician John †Venn.

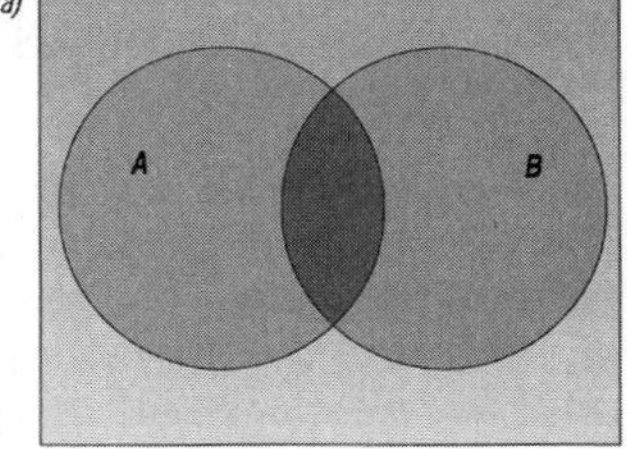

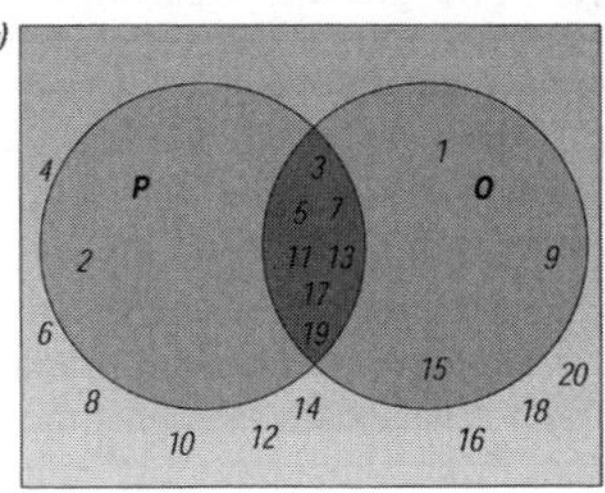

Venn diagram Sets and their relationships are often represented by Venn diagrams. The sets are drawn as circles – the area of overlap between the circles shows elements that are common to each set, and thus represent a third set. (Top) A Venn diagram of two intersecting sets and (bottom) a Venn diagram showing the set of whole numbers from 1 to 20 and the subsets P and O of prime and odd numbers, respectively. The intersection of P and O contains all the prime numbers that are also odd. Helicon

ventral surface front of an animal. In vertebrates, the side furthest from the backbone; in invertebrates, the side closest to the ground. The positioning of the main nerve pathways on the ventral side is a characteristic of invertebrates.

ventricle in zoology, either of the two lower chambers of the heart that force blood to circulate by contraction of their muscular walls. The term also refers to any of four cavities within the brain in which cerebrospinal fluid is produced.

Venturi tube device for measuring the rate of fluid flow through a pipe. It consists of a tube with a constriction (narrowing) in the middle of its length. The constriction causes a drop in pressure in the fluid flowing in the pipe. A pressure gauge attached to the constriction measures the pressure drop and this is used to find the rate of fluid flow. Venturi tubes are also used in the carburetor of a motor car to draw gasoline into the engine.

venule in anatomy, narrow diameter blood vessels that branch from veins and further branch to form capillaries of even narrower diameter.

Venus second planet from the Sun. It can approach Earth to within 38 million km/24 million mi, closer than any other planet. Its mass is 0.82 that of Earth. Venus rotates on its axis more slowly than any other planet, from east to west, the opposite direction to the other planets (except Uranus and possibly Pluto).

mean distance from the Sun: 108.2 million km/67.2 million mi

equatorial diameter: 12,100 km/7,500 mi

rotation period: 243 Earth days

year: 225 Earth days

atmosphere: Venus is shrouded by clouds of sulfuric acid droplets that sweep across the planet from east to west every four days. The atmosphere is almost entirely carbon dioxide, which traps the Sun's heat by the †greenhouse effect and raises the planet's surface temperature to 480°C/900°F, with an atmospheric pressure of 90 times that at the surface of the Earth

surface: consists mainly of silicate rock and may have an interior structure similar to that of Earth: an iron-nickel core, a †mantle composed of more mafic rocks (rocks made of one or more ferromagnesian, dark-colored minerals), and a thin siliceous outer †crust. The surface is dotted with deep impact craters. Some of Venus's volcanoes may still be active

satellites: no moons

The first artificial object to hit another planet was the Soviet probe Venera 3, which crashed on Venus on 1 March 1966. Later Venera probes parachuted down through the atmosphere and landed successfully on its surface, analyzing surface material and sending back information and pictures. In December 1978, NASA's Pioneer Venus probe went into orbit around the planet and mapped most of its surface by radar, which penetrates clouds. In 1992 the US space probe Magellan mapped 99% of the planet's surface to a resolution of 100 m/330 ft.

The largest highland area is Aphrodite Terra near the equator, half the size of Africa. The highest mountains are on the northern highland region of Ishtar Terra, where the massif of Maxwell Montes rises to 10,600 m/35,000 ft above the average surface level. The highland areas on Venus were formed by volcanoes.

Venus has an ion-packed tail 45 million km/28 million mi in length that stretches away from the Sun and is caused by the bombardment of the ions in Venus's upper atmosphere by the solar wind. It was first discovered in the late 1970s but it was not until 1997 that the Solar Heliospheric Observatory revealed its immense length.

Magellan Images http://www.jpl.nasa.gov/magellan/images.html

Library of images of Venus from NASA's Magellan Project home page. Each image has a full explanation of the subject matter.

Venus Introduction http://www.solarviews.com/solar/eng/venus.htm

All you ever wanted to know about the planet Venus can be found at this site, which includes a table of statistics, photographs of the planet, animations, information about its volcanic features and impact craters, plus a chronology of exploration.

verdigris green-blue coating of copper ethanoate that forms naturally on copper, bronze, and brass. It is an irritating, poisonous compound made by treating copper with ethanoic acid, and was formerly used in wood preservatives, antifouling compositions, and green paints.

vermilion HgS, red form of mercuric sulfide; a scarlet that occurs naturally as the crystalline mineral †cinnabar.

vernal equinox see †equinox.

vernalization stimulation of flowering by exposure to cold. Certain plants will not flower unless subjected to low temperatures during their development. For example, winter wheat will flower in summer only if planted in the previous autumn. However, by placing partially germinated seeds in low temperatures for several days, the cold requirement can be supplied artificially, allowing the wheat to be sown in the spring.

vernier device for taking readings on a graduated scale to a fraction of a division. It consists of a short divided scale that carries an index or pointer and is slid along a main scale. It was invented by French engineer Pierre Vernier.

vertebra in anatomy, an irregularly shaped bone that forms part of the †vertebral column. Children have 33 vertebrae, 5 of which fuse in adults to form the sacrum and 4 to form the coccyx. There are 7 cervical vertebrae in the neck, 12 thoracic vertebrae in the thorax with the ribs attached, and 5 lumbar vertebrae in the lower back.

vertebral column backbone, giving support to an animal and protecting its spinal cord. It is made up of a series of bones or vertebrae running from the skull to the tail, with a central canal containing the nerve fibers of the spinal cord.

In tetrapods the vertebrae show some specialization with the shape of the bones varying according to position. In the chest region the upper or thoracic vertebrae are shaped to form connections to the ribs. The backbone is only slightly flexible to give adequate rigidity to the animal structure.

vertebrate any animal with a backbone. The 41,000 species of vertebrates include mammals, birds, reptiles, amphibians, and fishes. They include most of the larger animals, but in terms of numbers of species are only a tiny proportion of the world's animals. The zoological taxonomic group Vertebrata is a subgroup of the †phylum Chordata.

A giant fossil conodont (an eel-like organism from the Cambrian period) was discovered in South Africa in 1995, and is believed to be one of the first vertebrates. Conodonts evolved 520 million years ago, predating the earliest fish by about 50 million years.

vertex plural **vertices**, in geometry, a point shared by three or more sides of a solid figure; the point farthest from a figure's base; or the point of intersection of two sides of a plane figure or the two rays of an angle.

vertical at right angles to the horizontal plane.

vertically opposite angles pair of angles that lie vertically opposite each other on the same side of a transveral (a line that intersects two or more lines).

Very Large Array **VLA**, largest and most complex single-site †radio telescope in the world. It is located on the Plains of San Augustine, 80 km/50 mi west of Socorro, New Mexico. It consists of 27 dish antennae, each 25 m/82 ft in diameter, arranged along three equally spaced arms forming a Y-shaped array. Two of the arms are 21 km/13 mi long, and the third, to the north, is 19 km/11.8 mi long. The dishes are mounted on railway tracks enabling the configuration and size of the array to be altered as required.

Pairs of dishes can also be used as separate interferometers (see †interferometry), each dish having its own individual receivers that are remotely controlled, enabling many different frequencies to be studied. There are four standard configurations of antennae ranging from A (the most extended) through B and C to D. In the A configuration the antennae are spread out along the full extent of the arms and the VLA can map small, intense radio sources with high resolution. The smallest configuration, D, uses arms that are just 0.6 km/0.4 mi long for mapping larger sources. Here the resolution is lower, although there is greater sensitivity to fainter, extended fields of radio emission.

very long-baseline array **VLBA**, in astronomy, group of ten 25-m/82-ft †radio telescopes spread across North America and Hawaii that operate as a single instrument using the technique of †very long-baseline interferometry to obtain high-resolution images. The longest baseline (distance between pairs of telescopes) is about 8,000 km/4,970 mi. The first observation using all ten sites occurred 29 May 1993.The total construction cost was over US$85 million.

very long-baseline interferometry **VLBI**, in radio astronomy, a method of obtaining high-resolution images of astronomical objects by combining simultaneous observations made by two or more radio telescopes thousands of kilometers apart.

Vesalius, Andreas (1514–1564) Belgian physician who revolutionized anatomy by performing postmortem dissections and making use of illustrations to teach anatomy. Vesalius upset the authority of †Galen, and his book – the first real textbook of anatomy – marked the beginning of biology as a science.

vesiculation in earth science, process by which the expansion of gases in molten extrusive igneous rocks forms small spherical cavities (vesicles), which may later become filled with various minerals.

Vesta third-largest asteroid in the Solar System, 500 km/310 mi in diameter, discovered in 1807 by German astronomer Heinrich †Olbers. Vesta orbits the Sun at a distance of 353 million km/219 million mi, with a period (time taken to circle the Sun) of 3.63 years. It is the only asteroid that ever becomes bright enough to be seen without a telescope.

In July 1999, the NASA space probe Deep Space 1 photographed the small asteroid Braille. The spectrum of light given off by Braille leads US astronomers to conclude that Braille has the same composition as Vesta and so was probably chipped off during a past impact. Vesta has a shallow crater at its South Pole indicating that a large chunk of rock has been dislodged at some point.

vestigial organ in biology, an organ that remains in diminished form after it has ceased to have any significant function in the adult organism. In humans, the appendix is vestigial, having once had a digestive function in our ancestors.

vetch any of a group of trailing or climbing plants belonging to the pea family, usually having seed pods and purple, yellow, or white flowers, and including the fodder crop alfalfa (*Medicago sativa*). (Several genera, especially *Vicia*, family Leguminosae.)

VHF abbreviation for very high frequency, referring to radio waves that have very short wavelengths (10 m–1 m). They are used for interference-free FM transmissions (see †frequency modulation). VHF transmitters have a relatively short range, producing frequencies in the 30–300 MHz band.

Viète, François (1540–1603) French mathematician who developed algebra and its notation. He was the first mathematician to use letters of the alphabet to denote both known and unknown quantities, and is credited with introducing the term "coefficient" into algebra.

Viète, François http://www-groups.dcs.st-andrews.ac.uk/history/Mathematicians/Viete.html

Part of an archive containing the biographies of the world's greatest mathematicians, this site is devoted to the life and contributions of François Viète.

villus plural **villi**, small fingerlike projection extending into the interior of the small †intestine and increasing the absorptive area of the intestinal wall. Digested nutrients, including sugars and amino acids, pass into the villi and are carried away by the circulating blood.

vinyl acetate or **vinyl ethanoate**; $CH_2COOCH = CH_2$, in chemistry, colorless liquid, boiling point 73°C/163°F. It is produced by a reaction between †ethene, †ethyne, and †ethanoic acid in the presence of an appropriate catalyst. It is used in the production of polyvinyl acetate polymers.

vinyl chloride or **chloroethene**; $CH_2 = CHCl$, in chemistry, carcinogenic colorless gas. It can be produced by reacting †ethyne with hydrochloric acid in the presence of a catalyst. It is mainly used for the production of †vinyl chloride polymers and co-polymers.

vinyl chloride polymer member of a class of synthetic †thermoplastics based on the polymerization of †vinyl chloride, general formula $-(CH_2 - CHCl)_n-$. †Polyvinyl chloride (PVC) is a member of this polymer group. These polymers are frequently modified by the addition of †plasticizers. They are used as construction materials, clothing, packaging, and furnishings.

vinyl ester polymer member of a class of synthetic polymers produced by the polymerization of vinyl ethers. The most important is polyvinyl acetate, produced by free-radical initiated polymerization of †vinyl acetate. These polymers are used in plasticizers, solvents, adhesives, and in the production of polyvinyl alcohol.

vinyl group or **ethenyl group**, unsaturated organic group of general formula $- CH_2 = CH -$.

Virchow, Rudolf Ludwig Carl (1821–1902) German pathologist, the founder of cellular pathology. Virchow was the first to describe leukemia (cancer of the blood). In his book *Die cellulare Pathologie/Cellular Pathology* (1858), he proposed that disease is not due to sudden invasions or changes, but to slow processes in which normal cells give rise to abnormal ones.

Virgo zodiacal constellation of the northern hemisphere, the second-largest in the sky. It is represented as a maiden holding an ear of wheat, marked by first-magnitude †Spica, Virgo's brightest star. The Sun passes through Virgo from late September to the end of October. In astrology, the dates for Virgo are between about 23 August and 22 September (see †precession).

Virgo contains the nearest large cluster of galaxies to us, 50 million light years away, consisting of about 3,000 galaxies centered on the giant elliptical galaxy M87. Also in Virgo is the nearest †quasar, 3C 273, an estimated 3 billion light years away.

Virgo Cluster cluster of more than 2,500 galaxies in the direction of the †Virgo constellation. The cluster is about 9 million light years across and 60 million light years away from the Sun. It contains three giant elliptical galaxies, each 20 times the size of our Galaxy. One is a powerful radio galaxy, a strong source of electromagnetic waves known as M87. Virgo is the nearest rich cluster to our own Galaxy.

virion the smallest unit of a mature †virus.

virtual memory in computing, a technique whereby a portion of the computer backing storage, or external, †memory is used as an extension of its immediate-access, or internal, memory. The contents of an area of the immediate-access memory are stored on, say, a hard disk while they are not needed, and brought back into main memory when required.

The process, called **paging** or **segmentation**, is controlled by the computer †operating system and is hidden from the programmer, to whom the computer's internal memory appears larger than it really is. The technique can be successfully implemented only if very fast backing store is available, so that "pages" of memory can be rapidly switched into and out of the immediate-access memory.

virtual particle theory theory devised by English physicist Stephen Hawking to account for apparent thermal radiation from a black hole (from which not even light can escape). It supposes that space is full of "virtual particles" in a particle–antiparticle relationship, being created out of "nothing" and instantly destroying each other. At an †event horizon, however, one particle may be gravitationally drawn into the †singularity, and the other appear to radiate as heat.

virtual reality advanced form of computer simulation, in which a participant has the illusion of being part of an artificial environment. The participant views the environment through two tiny television screens (one for each eye) built into a visor. Sensors detect movements of the participant's head or body, causing the apparent viewing position to change. Gloves (datagloves) fitted with sensors may be worn, which allow the participant seemingly to pick up and move objects in the environment.

The technology is still under development but is expected to have widespread applications; for example, in military and surgical training, architecture, and home entertainment.

virus in computing, a piece of †software that can replicate and transfer itself from one computer to another, without the user being aware of it. Some viruses are relatively harmless, but others can damage or destroy data.

Viruses are written by anonymous programmers, often maliciously, and are spread on †floppy disks, CD-ROMs, and via networks and e-mail (†worm). Most viruses hide in the boot sectors of floppy or hard disks or infect program files, but there has been a huge growth in macro viruses that infect Microsoft

Word or Excel, for example the virus Melissa that infected over 100,000 computers via the Internet in April 1999. Melissa is a macro virus, a virus that exists as a small program attached to a word-processing document. It arrives by e-mail hidden in a Word file. If the recipient opens the file Melissa tampers with Word's security settings and uses Microsoft's Outlook to send copies of itself to the first 50 contacts in the recipients' address book. The author of the Melissa virus was caught and charged shortly after its release. On 4 May 2000 the "I Love You" worm virus affected an estimated 45 million computers worldwide, causing around $7 billion/£4.4 billion worth of damage. The virus, in a similar way to Melissa, exploited a feature in the Microsoft Outlook e-mail program to spread itself around the world faster than had ever been seen before. It was written by Onel de Guzman, a 22-year-old computer school dropout from Manila, Philippines.

Antivirus software can be used to detect and destroy well-known viruses, but new viruses continually appear and these may bypass existing antivirus programs.

Computer viruses may be programmed to operate on a particular date, such as Dark Avenger's Michelangelo Virus, which was triggered on 6 March 1992 (the anniversary of the birth of Italian artist Michelangelo) and erased hard disks. An estimated 5,000–10,000 PCs were affected.

CERT Coordination Center http://www.cert.org/

Site tracking and summarizing problems with the Internet, such as virus activity, scans, probes, and attacks. Based at the Carnegie Mellon University in the USA, the CERT Coordination Center studies vulnerabilities in Internet systems, publishes security alerts, and offers advice and training for those looking to improve security on their Web sites. Find out which viruses are active now, and take steps to prevent such viruses from damaging a site.

Dr Solomon's All About Viruses http://www.humboldtlaw.net/computers/aboutvirus.html

Informative guide to viruses. The pages take the user through what a virus is and what to do if a computer gets one. It differentiates viruses from other programs, such as trojans, bugs, and jokes, and then goes on to detail the different types of virus that exist. It is all presented in clear and straightforward language.

Vmyths.com http://www.vmyths.com/

Avoid falling for computer virus hoaxes by checking suspects at this hysteria-free site. Written and maintained by a former computer security expert, this site provides detailed and up-to-date information about current hoax virus alerts as well as genuine viruses.

virus infectious particle consisting of a core of nucleic acid (†DNA or RNA) enclosed in a protein shell. They are extremely small and cause disease. They differ from all other forms of life in that they are not cells – they are acellular. They are able to function and reproduce only if they can invade a living cell to use the cell's system to replicate themselves. In the process they may disrupt or alter the host cell's own DNA. They use the cell they invade to make more virus particles that are then released. This usually kills the cell. The healthy human body reacts by producing an antiviral protein, †interferon, which prevents the infection spreading to adjacent cells. There are around 5,000 species of virus known to science (1998), although there may be as many as half a million actually in existence.

Examples of diseases in humans caused by viruses are the common cold, chickenpox, influenza, AIDS, herpes, mumps, measles, and rubella. Recent evidence implicates viruses in the development of some forms of cancer (see †oncogenes). Antibiotics do not work against viruses. The best protection against diseases caused by viruses is †immunization.

Viruses can change by mutation. When they do so, a human body is sometimes unable to fight the new virus very well. This happens regularly with the influenza virus. A small change results in a small influenza epidemic, but a big change results in a pandemic that can kill millions of people worldwide. Many viruses mutate continuously so that the host's body has little chance of developing permanent resistance; others transfer between species, with the new host similarly unable to develop resistance. The viruses that cause †AIDS and Lassa fever are both thought to have "jumped" to humans from other mammalian hosts.

Hidden Killers http://library.thinkquest.org/23054/gather/index.shtml

Viruses are exposed in all their "villainy" in this sophisticated and dramatic Web site. An excellent and detailed guide to the basic facts on viruses is presented alongside further pages explaining the lethal potential of these micro-organisms and the dangers of their use as weapons.

viscera general term for the organs contained in the chest and abdominal cavities.

viscose yellowish, syrupy solution made by treating cellulose with sodium hydroxide and carbon disulphide or with cuprammonium liquor and sodium hydroxide. The solution is then regenerated as continuous filament (rayon) by forcing it through very small holes and solidifying it (in a process known as spinning). Forcing viscose through a slit instead of small holes produces cellophane.

viscosity the resistance of a fluid to flow, caused by its internal friction, which makes it resist flowing past a solid surface or other layers of the fluid. Treacle and other thick, sticky liquids are highly viscous liquids. Water and gasoline are runny liquids and have low viscosity.

The term viscosity applies to the motion of an object moving through a fluid as well as the motion of a fluid passing by an object. For the purposes of calculation, many fluids in physics are considered to be perfect, or nonviscous.

viscous force drag that occurs on an object placed in a viscous medium. For example, air causes a viscous force on an aircraft traveling through it. For two parallel layers of fluid close to each other, the viscous force is proportional to the difference in velocity between the two layers.

vision defect any abnormality of the eye that causes less-than-perfect sight. Common defects are short-sightedness or †myopia; long-sightedness or †hypermetropia; lack of †accommodation or presbyopia; and †astigmatism. Other eye defects include color blindness.

vision, persistence of see †persistence of vision.

Visual Basic Microsoft computer language based on †BASIC. Visual Basic provides a rich set of visual programming tools that allow programmers to copy and paste common Windows components on to forms, in order to build Windows applications. BASIC is the language that is used to combine all the graphical components, and to code subroutines.

Gary Beene's Visual Basic Information Center http://www.garybeene.com/vb/

Features a number of tutorials for those just starting to program in Visual Basic, as well as information for the more advanced programmer. There are links to related sites and files.

visual display unit **VDU**, computer terminal consisting of a keyboard for input data and a screen for displaying output. The screen and its housing are now more usually termed a monitor.

vitalism idea that living organisms derive their characteristic properties from a universal life force. In the 20th century, this view is associated with the French philosopher Henri Bergson.

vitamin any of various chemically-unrelated organic (carbon-containing) compounds that are necessary in small quantities for the normal functioning of the mammalian body. Many act as coenzymes, small molecules that enable †enzymes to function effectively. Vitamins must be supplied by the diet because the body generally cannot make them. Deficiency of a vitamin may lead to a metabolic disorder ("deficiency disease"), which can be remedied by sufficient intake of the vitamin. Vitamins are generally classified as **water-soluble** (B and C) or **fat-soluble** (A, D, E, and K).

Vitamin A (retinol) is found in milk, cheese, butter, eggs, liver, kidney, oily fish (such as herring), and cod liver oil. However, the chemical carotene (found in most plants) can be converted to vitamin A in the human body. This vitamin is needed to make a chemical used by rod cells in the retina of the eye that respond in dim light. Shortage of this vitamin results in poor vision, especially at night.

Vitamin D is found in the greatest amounts in fish, but is present in tiny amounts in milk, cheese, butter, eggs, and liver. However, it can be made in the human skin when it is exposed to sunlight and this is sufficient for most people. It is needed to maintain bone in the body. Without the vitamin, bones are weak and can bend (rickets). Young children, pregnant women, and women breastfeeding a baby need much more than most other people. People who are housebound and do not get into the sun may become short of this vitamin.

Most people have balanced diets and so a real need to obtain extra vitamins from tablets is not common. Some vitamins are poisonous in high doses and there have been cases of people dying as a result of taking excessive amounts of vitamin pills.

Molecular Expressions: The Vitamin Collection http://micro.magnet.fsu.edu/vitamins/index.html

Fascinating collection of images showing what all the known vitamins look like when recrystallized and photographed through a microscope. There is also a brief article about vitamins.

Reference Guide for Vitamins http://www.realtime.net/anr/vitamins.html

Quick reference guide to vitamins. After an overview of vitamins in general, the page describes the functions of common and obscure vitamins, and explains what happens when the body does not get enough of them.

Vitamins and Dietary Supplements http://www.vitamins.net/

US-based health site that offers extensive information about the benefits of different vitamins and supplements to your diet.

vitamin A another name for †retinol.

vitamin B complex in chemistry, vitamin originally thought to be only one compound, but later shown to be a complex of vitamins, including B_1, B_6, and B_{12}.

vitamin B_1 another name for †thiamine.

vitamin B_6 another name for †pyridoxine.

vitamin B_{12} another name for †cyanocobalamin.

vitamin B_2 another name for †riboflavin.

vitamin C another name for †ascorbic acid.

vitamin D another name for †cholecalciferol.

vitamin E another name for †tocopherol.

vitamin H another name for †biotin.

vitamin K another name for †phytomenadione.

vitreous describing a substance that has the physical properties and appearance of a glass.

vitreous humor transparent jellylike substance behind the lens of the vertebrate †eye. It gives rigidity to the spherical form of the eye and allows light to pass through to the retina.

vitriol former name for any of a number of sulfate salts. Blue, green, and white vitriols are copper, ferrous, and zinc sulfate, respectively.

Oil of vitriol is a former name for sulfuric acid.

viviparous in animals, a method of reproduction in which the embryo develops inside the body of the female from which it gains nourishment (in contrast to †oviparous and †ovoviviparous). Vivipary is best developed in placental mammals, but also occurs in some arthropods, fishes, amphibians, and reptiles that have placentalike structures. In plants, it is the formation of young plantlets or bulbils instead of flowers. The term also describes seeds that germinate prematurely, before falling from the parent plant. Premature germination is common in mangrove trees, where the seedlings develop sizable spearlike roots before dropping into the swamp below; this prevents their being washed away by the tide.

vivisection literally, cutting into a living animal. Used originally to mean experimental surgery or dissection practiced on a live subject, the term is often used by antivivisection campaigners to include any experiment on animals, surgical or otherwise.

VLBA abbreviation for †very long-baseline array.

vocal cords paired folds, ridges, or cords of tissue within a mammal's larynx, and a bird's syrinx. Air expelled from the lungs passes between these folds or membranes and makes them vibrate, producing sounds. Muscles in the larynx change the pitch of the sounds produced, by adjusting the tension of the vocal cords.

Vogel, Hermann Carl (1842–1907) German astronomer. He discovered spectroscopic binary stars. By measuring the †Doppler effect in the spectral lines of stars to ascertain their velocity, he ended the controversy over the value of Christian Doppler's theory for investigating motion in the universe.

Vogt, Marthe Louise (1903–2003) Neuropharmacist who was involved in proving that †acetylcholine is a neurotransmitter (chemical that transmits impulses between nerve cells) involved in the stimulation of muscles by nerves. Acetylcholine is now known to be one of the principal neurotransmitters that communicates nerve impulses across synapses, the junctions between individual nerves.

voice sound produced through the mouth and by the passage of air between the †vocal cords. In humans the sound is much amplified by the hollow sinuses of the face, and is modified by the movements of the lips, tongue, and cheeks.

Volans small constellation of the southern hemisphere, represented as a flying fish. It was formerly known as "Piscis Volans."

volatile describing a substance that readily passes from the liquid to the vapor phase. Volatile substances have a high †vapor pressure.

volatile memory in computing, †memory that loses its contents when the power supply to the computer is disconnected.

volcanic rock another name for †extrusive rock, igneous rock formed on the Earth's surface.

volcano crack in the Earth's crust through which hot †magma (molten rock) and gases well up. The magma is termed **lava** when it reaches the surface. A volcanic moun-

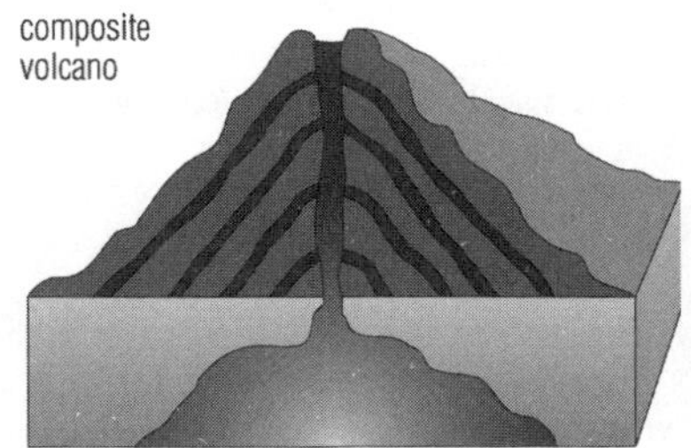

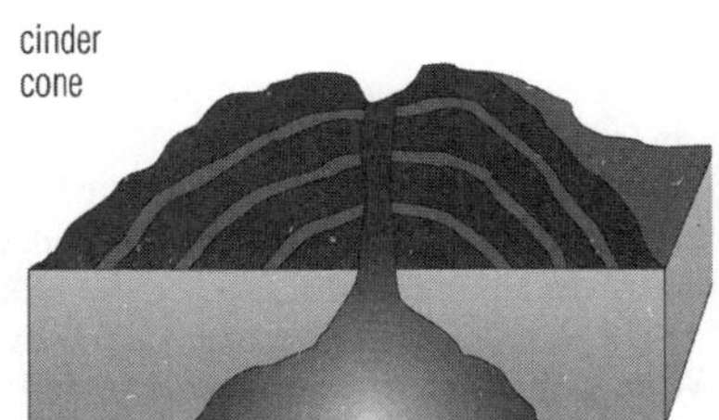

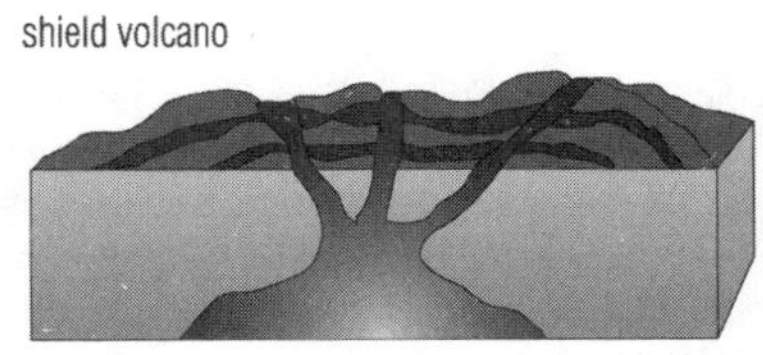

volcano There are two main types of volcano, but three distinctive cone shapes. Composite volcanoes emit a stiff, rapidly solidifying lava which forms high, steep-sided cones. Volcanoes that regularly throw out ash build up flatter domes known as cinder cones. The lava from a shield volcano is not ejected violently, flowing over the crater rim forming a broad low profile. Helicon

tain, usually cone-shaped with a crater on top, is formed around the opening, or vent, by the build-up of solidified lava and ash (rock fragments). Most volcanoes occur on plate margins (see †plate tectonics), where the movements of plates generate magma or allow it to rise from the mantle beneath. However, a number are found far from plate-margin activity, on †hot spots where the Earth's crust is thin, for example in Hawaii. There are two main types of volcano: †composite volcanoes and †shield volcanoes.

The type of volcanic activity also depends on the age of the volcano. The first stages of an eruption are usually vigorous as the magma forces its way to the surface. As the pressure drops and the vents become established, the main phase of activity begins. Composite volcanoes emit †pyroclastic debris, while shield volcanoes produce lava flows. When the pressure from below ceases, due to exhaustion of the magma chamber, activity wanes and is confined to the emission of gases, and in time this also ceases. The volcano then enters a period of quiescence, after which activity may resume after a period of days, years, or even thousands of years. Only when the root zones of a volcano have been exposed by erosion can a volcano be said to be truly extinct.

Many volcanoes are submarine and occur along †mid-ocean ridges. The main volcanic regions are around the Pacific rim (Cape Horn to Alaska); the central Andes of Chile (with the world's highest active volcano, Guallatiri, 6,063 m/19,892 ft); North Island, New Zealand; Hawaii; Japan; and Antarctica. There are more than 1,300 potentially active volcanoes on Earth. Volcanism has also helped shape other members of the Solar System, including the Moon, Mars, Venus, and Jupiter's moon Io.

Katmai National Park and Preserve http://www.alaskanet.com/Tourism/Parks/Katmai/

Guide to the National Park, which contains the famous Valley of Ten Thousand Smokes, where the 20th century's most dramatic volcanic episode took place.

MTU Volcanoes Page http://www.geo.mtu.edu/volcanoes/

Provided by Michigan Technological University, this site includes a world map of volcanic activity with information on recent eruptions, the latest research in remote sensing of volcanoes, and many spectacular photographs.

Vesuvius, Italy http://volcano.und.nodak.edu/vwdocs/volc_images/img_vesuvius.html

Site examining the complex geology of Vesuvius and its famous eruption of AD 79. There are images of the volcano and historical drawings. There is also a link to a local site campaigning for an improved civil defense plan as the volcano prepares once more to explode.

VolcanoWorld http://volcano.und.edu/

Comprehensive site on volcanoes, with details of the most recent eruptions, currently active volcanoes, a glossary, images and video clips, and even a list of extraterrestrial volcanoes. If you cannot find out what you want to know, you can "Ask a volcanologist."

volcanology study of †volcanoes, the lava, rocks, and gases that erupt from them, and the geological phenomena that cause them.

volt symbol V, SI unit of electromotive force or electric potential (see †potential, electric). A small battery has a potential of 1.5 volts, while a high-tension transmission line may carry up to 765,000 volts. The domestic electricity supply in the UK is 230 volts (lowered from 240 volts in 1995); it is 110 volts in the USA.

The absolute volt is defined as the †potential difference (PD) necessary to produce a current of one ampere through an electric circuit with a resistance of one ohm. It can also be defined as the potential difference that requires one joule of work to move a positive charge of one coulomb from the lower to the higher potential. It is named after the Italian scientist Alessandro Volta.

Volta, Alessandro Giuseppe Antonio Anastasio (1745–1827) Count Volta, Italian physicist who invented the first electric cell (the voltaic pile, in 1800), the electrophorus (an early electrostatic generator, in 1775), and an †electroscope.

Electricity Excited by The Mere Contact of Conducting Substances of Different Kinds
http://dbhs.wvusd.k12.ca.us/Chem-History/Volta-1800.html

Transcript reconstructed from Philosophical Magazine (September 1800), in which Volta describes the electrical phenomena that led to the discovery of the battery as a source of electricity. The Web master makes the following note for readers, "This article has been reconstructed from extensive quotes of it in a book on the history of the Galvani-Volta controversy. The author also quoted bits of the article within several of his own paragraphs. I have enclosed those portions in square brackets."

voltage commonly used term for †potential difference (PD).

voltage amplifier electronic device that increases an input signal in the form of a voltage or †potential difference, delivering an output signal that is larger than the input by a specified ratio.

voltaic cell another name for an electric cell.

Volterra, Vito (1860–1940) Italian mathematician whose chief work was in the fields of function theory and differential equations. His chief method, hit upon as a young boy, was based on dividing a problem into a small interval of time and assuming one of the variables to be constant during each time period.

Volterra, Vito http://www-groups.dcs.st-andrews.ac.uk/history/Mathematicians/Volterra.html

Part of an archive containing the biographies of the world's greatest mathematicians, this site is devoted to the life and contributions of Vito Volterra. In addition to biographical information, you will find a list of references about Volterra and links to other essays in the archive that reference him. The text of this essay includes hypertext links to the essays of those mathematicians and thinkers who influenced Volterra. There is also an image of him, which can be enlarged, and a map of his birthplace.

voltmeter instrument for measuring the †potential difference (voltage) between two points in a circuit. It should not be confused with an †ammeter, which measures current. A voltmeter has a high internal resistance (so that it passes only a small current), and is connected in parallel with the component across which potential difference is to be measured – that is, the current divides and passes through both the voltmeter and the component at the same time.

A common type of voltmeter is constructed from a sensitive current-detecting moving-coil †galvanometer placed in series with a high-value resistor (multiplier). To measure an AC (†alternating current) voltage, the circuit must usually include a rectifier; however, a moving-iron instrument can be used to measure alternating voltages without the need for such a device.

volume in geometry, the space occupied by a †three-dimensional (3D) solid object. A †prism such as a †cube, †cuboid, or a †cylinder has a volume equal to the †area of the base multiplied by the height. For a †pyramid or †cone, the volume is equal to one-third of the area of the base multiplied by the perpendicular height. The volume of a †sphere is equal to $^4/_3$ x pr^3, where r is the radius. Volumes of irregular solids may be calculated by the technique of †integration.

volumetric analysis chemical procedure used for determining the concentration of a solution. A known volume of a solution of unknown concentration is reacted with a known volume of a solution of known concentration (standard). The standard solution is delivered from a †burette so the volume added is known. This technique is known as †titration. Often an indicator is used to show when the correct proportions have reacted. This procedure is used for acid–base, †redox, and certain other reactions involving solutions.

voluntary reaction in biology, an individual's response to a stimulus under the control of the brain. Voluntary reactions are the basis of modified behavior in the light of experience. Involuntary reactions are †reflexes.

von Klitzing, Klaus (1943–) German physicist who was awarded the Nobel Prize for Physics in 1985 for his discovery of the quantized Hall effect. This effect involves the behavior of electric currents in thin films of material held in a strong magnetic field at a low temperature. If the magnetic field is varied, the current changes in a step-like fashion, rather than in a smooth regular way. This steplike behavior indicates that quantum effects are taking place. The quantized Hall effect is one of the few situations where quantum effects can be studied using ordinary macroscopic measurements. The effect is used to make very accurate measurements of electrical voltages.

Von Neumann, John (or Johann) (1903–1957) Hungarian-born US scientist and mathematician, a pioneer of computer design. He invented his "rings of operators" (called **Von Neumann algebras**) in the late 1930s, and also contributed to set theory, game theory, quantum mechanics, cybernetics (with his theory of self-reproducing automata, called **Von Neumann machines**), and the development of the atomic and hydrogen bombs.

Vries, Hugo (Marie) de (1848–1935) Dutch botanist who conducted important research on osmosis in plant cells and was a pioneer in the study of plant evolution. His work led to the rediscovery of Austrian biologist Gregor †Mendel's laws and the discovery of spontaneously occurring †mutations.

VRML abbreviation for virtual reality modeling language, in computing, method of displaying three-dimensional images on a †Web page. VRML, which functions as a counterpart to †HTML, is a platform-independent language that creates a †virtual reality scene which users can "walk" through and follow links much like a conventional Web page.

In some contexts, VRML can replace conventional computer interfaces with their icons, menus, files, and folders. It is possible to use VRML to create, for example, a virtual museum with all the elements of a real museum, including corridors, display cases, and multimedia demonstrations. Other possibilities include a Web market containing stalls with goods that can be "handled" using a mouse, or a virtual library of "books" that can be taken off "shelves."

V-shaped valley river valley with a V-shaped cross-section. Such valleys are usually found near the source of a river, where the steeper gradient means that there is a great deal of †corrasion (grinding away by rock particles) along the stream bed, and erosion cuts downwards more than it does sideways. However, a V-shaped valley may also be formed in the lower course of a river when its powers of down-

volume of common three-dimensional shapes

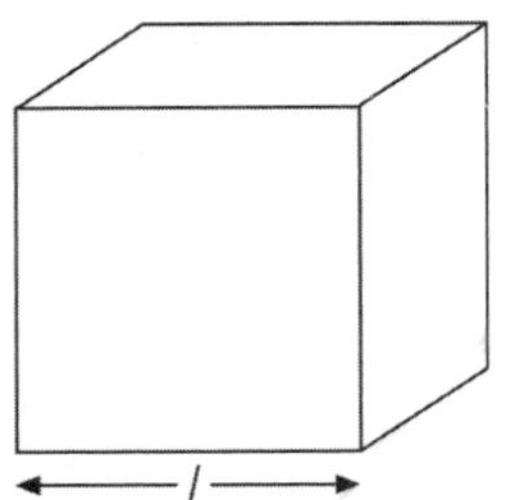

volume of a ***cube***
= length^3
= l^3

volume of a ***cuboid***
= length X breadth X height
= $l \times b \times h$

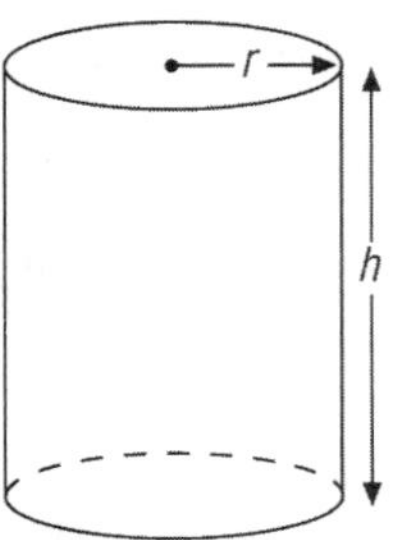

volume of a ***cylinder***
= π X (radius of cross section)^2 X height
= $\pi r^2 h$

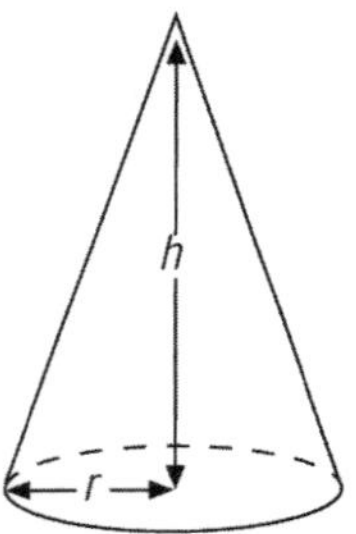

volume of a ***cone***
= $\frac{1}{3}\pi$ X (radius of cross section)^2 X height
= $\frac{1}{3}\pi r^2 h$

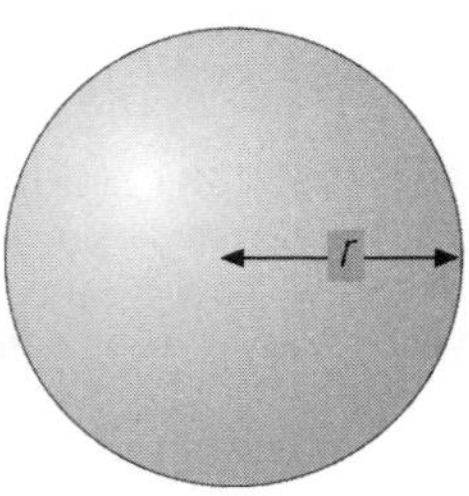

volume of a ***sphere***
= $\frac{4}{3}\pi$ radius^3
= $\frac{4}{3}\pi r^3$

volume Helicon

ward erosion become renewed by a fall in sea level, a rise in land level, or the capture of another river (see †rejuvenation).

The angle of the V-shaped cross-section depends on the rate of uplift of rock, the type of rock, the erosive ability of the river, the type of climate, and the stage of the river.

vulcanization technique for hardening rubber by heating and chemically combining it with sulfur. The process also makes the rubber stronger and more elastic. If the sulfur content is increased to as much as 30%, the product is the inelastic solid known as ebonite. More expensive alternatives to sulfur, such as selenium and tellurium, are used to vulcanize rubber for specialized products such as vehicle tires. The process was discovered by US inventor Charles Goodyear in 1839 and patented in 1844.

Accelerators can be added to speed the vulcanization process, which takes from a few minutes for small objects to an hour or more for vehicle tires. Molded objects are often shaped and vulcanized simultaneously in heated moulds; other objects may be vulcanized in hot water, hot air, or steam.

vulgar fraction or **common fraction**; or **simple fraction**, fraction comprising natural or whole numbers and written as a ratio, for example $^a/_b$, such that *a* is less than *b*.

Vulpecula small constellation in the northern hemisphere just south of †Cygnus, represented as a fox. It contains a major planetary †nebula, the Dumbbell, and the first †pulsar (pulsating radio source) to be discovered.

vulva female external genitalia. Two folds of fleshy tissue, the labia majora and labia minora, extend downwards from the clitoris to surround the entrances to the †urethra and the vagina.

Vygotsky, Lev Semionovich (1896–1934) Soviet psychologist whose work on language and linguistic development is based on his supposition that higher cognitive processes are a product of social development. From early research into the rules and development of tool-use and sign-use behavior, Vygotsky turned to symbolic processes in language, focusing on the semantic structure of words and the way in which meanings of words change from emotive to concrete and then become more abstract.

W in physics, symbol for †watt, the SI unit of power, equal to a power output of 1 joule per second. Multiple units include the kilowatt (kW, 1,000 watts) and megawatt (MW, 1,000,000 watts).

Waddington, Conrad Hal (1905–1975) English geneticist and biologist who examined the evolutionary significance of the interrelations between the genetic constitution, characteristics, and environment of an organism. Waddington performed experiments that demonstrated the theory that natural selection is the cause of hereditary change.

wadi in arid regions of the Middle East, a steep-sided valley containing an intermittent stream that flows in the wet season.

Waldsterben (German "forest death") tree decline related to air pollution, common throughout the industrialized world. It appears to be caused by a mixture of pollutants; the precise chemical mix varies between locations, but it includes acid rain, ozone, sulfur dioxide, and nitrogen oxides.

Waldsterben was first noticed in the Black Forest of Germany during the late 1970s, and is spreading to many countries in the developing world, such as China.

Wallace, Alfred Russel (1823–1913) Welsh naturalist who collected animal and plant specimens in South America and Southeast Asia, and independently arrived at a theory of evolution by natural selection similar to that proposed by Charles †Darwin.

Wallace line imaginary line running down the Lombok Strait in Southeast Asia, between the island of Bali and the islands of Lombok and Sulawesi. It was identified by English naturalist Alfred Russel Wallace as separating the South Asian (Oriental) and Australian biogeographical regions, each of which has its own distinctive animals.

Subsequently, others have placed the boundary between these two regions at different points in the Malay archipelago, owing to overlapping migration patterns.

Wallach, Otto (1847–1931) German analytic chemist who isolated a new class of compounds, called **terpenes**, from essential oils (oils extracted from plants and used in medicine, aromatherapy, and perfume). Terpenes are made up of isoprene units containing five carbon atoms, with the general formula $(C_5H_8)_n$. He went on to determine the formula of the most important terpene, limonene, $C_{10}H_{16}$. He was awarded the Nobel Prize for Chemistry in 1910 for his work in alicyclic compounds.

Wallis, John (1616–1703) English mathematician and cleric who made important contributions to the development of algebra and analytical geometry. He was one of the founders of the Royal Society.

Wallis, John http://www-history.mcs.st-and.ac.uk/history/Mathematicians/Wallis.html

Extensive biography of the mathematician John Wallis. The site contains a clear description of his working relationships with his contemporaries, and also includes pages from his books Arithmetica Infinitorum and De Postulato Quinto: et Definitione Quinta. Several literature references for further reading on the mathematician are also listed, and the Web site also features a portrait of Wallis.

wall pressure in plants, the mechanical pressure exerted by the cell contents against the cell wall. The rigidity (turgor) of a plant often depends on the level of wall pressure found in the cells of the stem. Wall pressure falls if the plant cell loses water.

Walton, Ernest Thomas Sinton (1903–1995) Irish physicist who collaborated with John †Cockcroft on investigating the structure of the atom. They shared the Nobel Prize for Physics in 1951 for their success in 1932 in splitting the nucleus of an atom for the first time.

WAP acronym for Wireless Application Protocol, initiative started in the 1990s by Unwired Planet and mobile phone manufacturers Motorola, Nokia, and Ericsson to develop a standard for delivering Web-like applications on mobile phones and other wireless devices. In theory WAP phones can be used for e-mail and messaging, reading Web pages, shopping, booking tickets, and making other financial transactions, as well as for phone calls. WAP as been superseded to some extent by 3G (third generation) mobile phone technology.

The WAP protocol has many similarities to Internet technologies. For instance, the Wireless Markup Language (WML) used to create WAP pages is very similar to †HTML, which is used to create Web pages. Similarly, the WMLScript is based on JavaScript. Both WML and WMLScript are adapted and designed for a wireless environment, to address issues such as limited bandwidth and limited processing power in the mobile phone.

WAP Forum http://www.wapforum.org/

Official home of the Wireless Application Protocol (WAP) Forum, an industry association comprising 90% of all organizations involved in WAP technology development, and a leading source of information about WAP and telephony. If you want to know more about WAP technology, there is a useful "Frequently Asked Questions" section, details of the Forum's work, and a news and events area.

Warburg, Otto Heinrich (1883–1970) German biochemist who was awarded a Nobel Prize for

Physiology or Medicine in 1931 for the discovery of respiratory enzymes that enable cells to process oxygen. In 1923 he devised a manometer (pressure gauge) sensitive enough to measure oxygen uptake of respiring tissue. By measuring the rate at which cells absorb oxygen under differing conditions, he was able to show that enzymes called †cytochromes enable cells to process oxygen.

Warming, Johannes Eugenius Bülow (1841–1924) Danish botanist whose pioneering studies of the relationships between plants and their natural environments established plant ecology as a new discipline within botany.

warning coloration in biology, an alternative term for †aposematic coloration.

washing soda chemical name **sodium carbonate decahydrate**; $Na_2CO_3.10H_2O$, substance added to washing water to "soften" †hard water. The calcium and magnesium ions in hard water react with the carbonate ions from the sodium carbonate and a precipitate of calcium carbonate results.

$Na_2CO_{3(aq)} + CaSO_{4(aq)} \rightarrow CaCO_{3(s)} + Na_2SO_{4(aq)}$

waste materials that are no longer needed and are discarded. Examples are household waste, industrial waste (which often contains toxic chemicals), medical waste (which may contain organisms that cause disease), and †nuclear waste (which is radioactive). By †recycling, some materials in waste can be reclaimed for further use. In 1990 the industrialized nations generated 2 billion tonnes of waste. In the USA, 40 tons of solid waste are generated annually per person, roughly twice as much as in Europe or Japan.

There has been a tendency to increase the amount of waste generated per person in industrialized countries, particularly through the growth in packaging and disposable products, creating a "throwaway society."

Environment: EU Focus on Waste Management http://europa.eu.int/comm/environment/waste/facts_en.htm

Details of European Union policy on waste management made available in a downloadable Adobe Acrobat document. There is information on waste-related EU events, strategy documents, scientific reports, and access to other areas of the EU Environment Commission.

waste disposal depositing of waste. Methods of waste disposal vary according to the materials in the waste and include incineration, burial at designated sites, and dumping at sea. Organic waste can be treated and reused as fertilizer. †Nuclear waste and †toxic waste are usually buried or dumped at sea, although this does not negate the danger.

environmental pollution Waste disposal is an increasing problem. Environmental groups, such as Greenpeace and Friends of the Earth, are campaigning for more recycling, a change in lifestyle so that less waste (from packaging and containers, to nuclear materials) is produced, and safer methods of disposal.

Although incineration cuts down on landfill and can produce heat as a useful by-product it is still a wasteful method of disposal in comparison with recycling. For example, recycling a plastic bottle saves twice as much energy as is obtained by burning it.

waste disposal, USA The USA burns very little of its rubbish as compared with other industrialized countries. Most of its waste, 80%, goes into landfills.

European Commission: Environment http://europa.eu.int/comm/environment/index_en.htm

Huge collection of materials documenting the activities and concerns of the EU Environment Commission. A plain but well-organized Web site, it has its own search engine to help locate items of interest.

water chemical compound of hydrogen and oxygen elements – H_2O. It can exist as a solid (ice), liquid (water), or gas (water vapor). Water is the most common compound on Earth and vital to all living organisms. It covers 70% of the Earth's surface, and provides a habitat for large numbers of aquatic organisms. It is the largest constituent of all living organisms – the human body consists of about 65% water. It is found in all cells and many chemicals involved in processes such as respiration and photosynthesis need to be in solution in water in order to react. Pure water is a colorless, odorless, tasteless liquid which freezes at 0°C/32°F, and boils at 100°C/212°F. Natural water in the environment is never pure and always contains a variety of dissolved substances. Some 97% of the Earth's water is in the oceans; a further 2% is in the form of snow or ice, leaving only 1% available as freshwater for plants and animals. The recycling and circulation of water through the †biosphere is termed the **water cycle**, or "hydrological cycle"; regulation of the water balance in organisms is termed †osmoregulation. Water becomes more dense when it cools but reaches maximum density at 4°C/39°F. When cooled below this temperature the cooler water floats on the surface, as does ice formed from it. Animals and plants can survive under the ice.

The recycling and circulation of water on Earth is called the water cycle. Water occurs on the Earth's surface as standing water in oceans and lakes, as running water in rivers and streams, as rain, and as water vapor in the atmosphere. Together these sources comprise the †hydrosphere which is in a constant state of flux – water vapor condenses as it cools to form clouds, droplets of water in the clouds merge to form raindrops that fall to earth (precipitation), and after flowing through rivers and streams into lakes and oceans water is returned to the atmosphere by evaporation; and so the cycle continues. Since the hydrological cycle is a closed system, the amount of water in the Earth's hydrosphere is constant. The cycle is powered by solar radiation which provides the energy to maintain the flow through the processes of evaporation, transpiration, precipitation, and run-off.

Physical Properties of Water and Ice http://www.nyu.edu/pages/mathmol/modules/water/water_student.html
Well-designed guide to the molecular properties of water in its various states. Click on the introduction for a clear outline of the concepts, then take a look at the various molecular models and select which is water and which is ice. This graphics-rich site is part of a much larger resource for science students.

Water http://www.bris.ac.uk/Depts/Chemistry/MOTM/water/water.htm
"The aqueous environment influences all of the chemistry that takes place in it," according to this Molecule of the Month page from Bristol University, UK, which takes a challenging look at the chemistry of water. One of the most exciting aspects considered is the discovery of the "hot life" found in thermal vents on the open floor.

Water Science for Schools: Earth's Water http://ga.water.usgs.gov/edu/mearth.html
Information on the Earth's water. Part of a larger site run by the US Geological Survey, this section includes clear and well-illustrated information on topics such as the water cycle, water on the Earth's surface, and glaciers and icecaps. The main site is regularly updated with information on any natural disasters as and when they happen.

water cycle or **hydrological cycle**, natural circulation of water through the upper part of the Earth. It is a complex system involving a number of physical and chemical processes (such as †evaporation, †precipitation, and †infiltration) and stores (such as rivers, oceans, and soil).

Water is lost from the Earth's surface to the atmosphere by evaporation caused by the Sun's heat on the surface of lakes, rivers, and oceans, and through the †transpiration of plants. This atmospheric water is carried by the air moving across the Earth, and **condenses** as the air cools to form clouds, which in turn deposit moisture on the land and sea as †precipitation. The water that collects on land flows to the ocean overland – as streams, rivers, and glaciers – or through the soil (†infiltration) and rock (†groundwater). The boundary that marks the upper limit of groundwater is called the †water table.

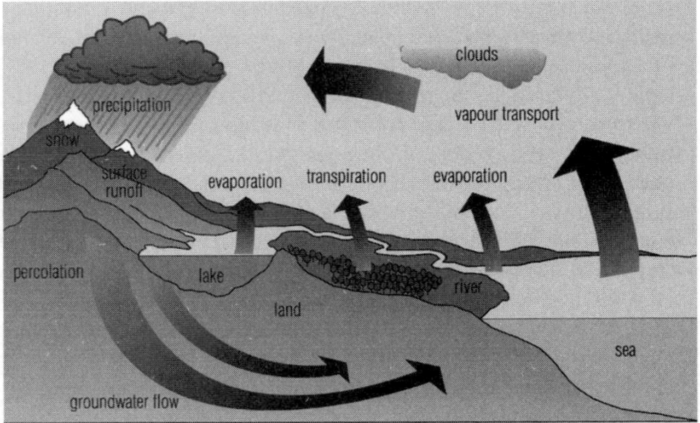

water cycle About one-third of the solar energy reaching the Earth is used in evaporating water. About 380,000 cubic km/95,000 cubic mi is evaporated each year. The entire contents of the oceans would take about one million years to pass through the water cycle. Helicon

The oceans, which cover around 70% of the Earth's surface, are the source of most of the moisture in the atmosphere.

Hydrologic Cycle http://ww2010.atmos.uiuc.edu/(Gh)/guides/mtr/hyd/home.rxml
Easy-to-understand, well-illustrated guide to the water cycle. The site is part of an online meteorology guide provided by the University of Illinois and includes useful diagrams to aid understanding.

waterfall cascade of water in a river or stream. It occurs when a river flows over a bed of rock that resists erosion; weaker rocks downstream are worn away, creating a steep, vertical drop and a plunge pool into which the water falls. Over time, continuing erosion causes the waterfall to retreat upstream forming a deep valley, or †gorge. Good examples of waterfalls include Victoria Falls (Zimbabwe/Zambia), Niagara Falls (USA/Canada), and Angel Falls (Venezuela).

water gas fuel gas consisting of a mixture of carbon monoxide and hydrogen, made by passing steam over red-hot coke. The gas was once the chief source of hydrogen for chemical syntheses such as the Haber process for making ammonia, but has been largely superseded in this and other reactions by hydrogen obtained from natural gas.

water glass common name for sodium metasilicate (Na_2SiO_3). It is a colorless, jellylike substance that dissolves readily in water to give a solution used for preserving eggs and fireproofing porous materials such as cloth, paper, and wood. It is also used as an adhesive for paper and cardboard and in the manufacture of soap and silica gel, a substance that absorbs moisture.

water of crystallization water chemically bonded to a salt in its crystalline state. For example, in copper(II) sulfate, there are five moles of water per mole of copper sulfate: hence its formula is $CuSO_4.5H_2O$.

This water is responsible for the color and shape of the crystalline form. When the crystals are heated gently, the water is driven off as steam and a white powder of the anhydrous salt is formed.

$CuSO_4.5H_2O_{(s)} \rightarrow CuSO_{4\,(s)} + 5H_2O_{(g)}$

water pollution any addition to fresh or seawater that interferes with biological processes or causes a health hazard. Common pollutants include nitrates, pesticides, and sewage (resulting from poor sewage disposal methods), although a huge range of industrial contaminants, such as chemical by-products and residues created in the manufacture of various goods, also enter water – legally, accidentally, and through illegal dumping.

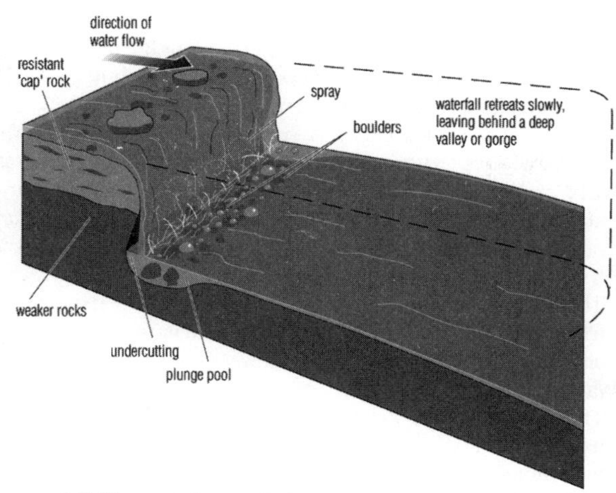

waterfall When water flows over hard rock and soft rock, the soft rocks erode creating waterfalls. As the erosion processes continue, the falls move backwards, in the opposite direction of the water Helicon

In 1980 the UN launched the "Drinking Water Decade," aiming for cleaner water for all by 1990. However, in 1994 it was estimated that roughly half of all people in the developing world did not have safe drinking water. A 1995 World Bank report estimated that some 10 million deaths in developing countries were caused annually by contaminated water.

water potential in biology, measure of the kinetic energy of an aqueous system and hence the tendency of the system to lose water. The water potential of a system containing pure water is zero.

The kinetic energy of the system is high and there will be a net loss of water from the system if it is separated by a permeable membrane from an adjacent aqueous system of lower water potential. The addition of solute lowers water potential, which is measured as a negative value. The more negative the water potential the less is the loss of water from the system to an adjacent aqueous system of higher water potential.

water softener any substance or unit that removes the hardness from water. Hardness is caused by the presence of calcium and magnesium ions, which combine with soap to form an insoluble scum, prevent lathering, and cause deposits to build up in pipes and cookware (kettle fur). A water softener replaces these ions with sodium ions, which are fully soluble and cause no scum.

waterspout funnel-shaped column of water and cloud that is drawn from the surface of the sea or a lake by a †tornado.

Waterston, John James (1811–1883) Scottish physicist who first formulated the essential features of the kinetic theory of gases 1843–45. He also estimated the temperature of the Sun in 1857.

water table upper level of †groundwater (water collected underground in porous rocks). Water that is above the water table will drain downwards; a spring forms where the water table meets the surface of the ground. The water table rises and falls in response to rainfall and the rate at which water is extracted, for example for irrigation and industry.

In many irrigated areas the water table is falling due to the extraction of water. Below northern China, for example, the water table is sinking at a rate of 1 m/3 ft a year. Regions with a high water table and dense industrialization have problems with †pollution of the water table. In the USA, New Jersey, Florida, and Louisiana have water tables that are contaminated by both industrial †wastes and saline seepage from the ocean.

Watson, James Dewey (1928–) US biologist who was awarded a Nobel Prize for Physiology or Medicine in 1962 for the discovery of the double-helical structure of DNA and determining the significance of this structure in the replication and transfer of genetic information. He shared the prize with his co-worker Francis †Crick.

Dr James D Watson Profile http://www.achievement.org/autodoc/page/wat0pro-1

Description of the life and works of the discoverer of the structure of the DNA molecule, James Watson. Watson won the Nobel prize for his most famous discovery (made with Francis Crick) in 1962. The Web site contains not only a profile and biographical information, but also holds a lengthy interview with Watson from 1991 accompanied by a large number of photographs, video sequences, and audio clips.

Watson, John Broadus (1878–1958) US psychologist, founder of behaviorism. He rejected introspection (observation by an individual of his or her own mental processes) and regarded psychology as the study of observable behavior, within the scientific tradition.

Classics In The History Of Psychology – Psychology As The Behaviorist Views It

http://psychclassics.yorku.ca/Watson/glossary.htm

Full transcript of John B Watson's 1913 paper Psychology As The Behaviorist Views It. This is Watson's most famous publication, and probably did more than anything else to turn psychology into a mainstream science and away from its introspective origins.

watt symbol W, SI unit of power (the rate of expenditure or transformation of energy from one form to another) defined as one joule per second. A light bulb, for example, may use 40, 60, 100, or 150 watts of power; an electric heater will use several kilowatts (thousands of watts). The watt is named after the Scottish engineer James †Watt.

The absolute watt is defined as the power used when one joule of work is done in one second. In electrical terms, the flow of one ampere of current through a conductor whose ends are at a potential difference of one volt uses one watt of power (watts = volts x amperes).

Watt, James (1736–1819) Scottish engineer who developed the †steam engine in the 1760s, making Thomas Newcomen's engine vastly more efficient by cooling steam in a condenser separate from the main cylinder. He eventually made a double-acting machine that supplied power with both directions of the piston and developed rotary motion. He also invented devices associated with the steam engine, artistic instruments and a copying process, and devised the horsepower as a description of an engine's rate of working. The modern unit of power, the **watt**, is named after him.

wave in the oceans, a ridge or swell formed by wind or other causes. The power of a wave is determined by the strength of the wind and the distance of open water over which the wind blows (the fetch). Waves are the main agents of †coastal erosion and deposition: sweeping away or building up beaches, creating †spits and †berms, and wearing down cliffs by their †hydraulic action and by the †corrosion of the sand and shingle that they carry. A †tsunami (misleadingly called a "tidal wave") is formed after a submarine earthquake.

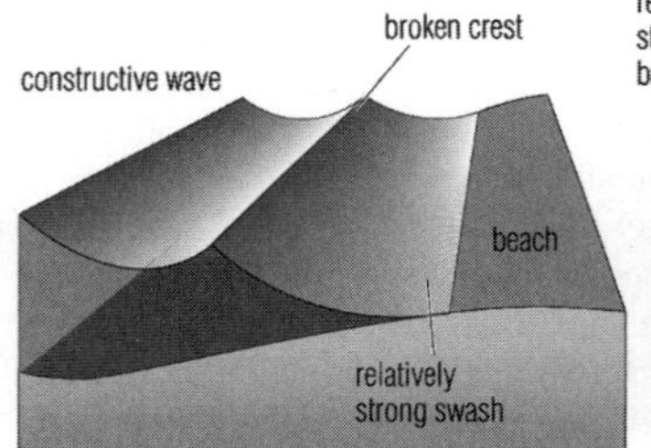

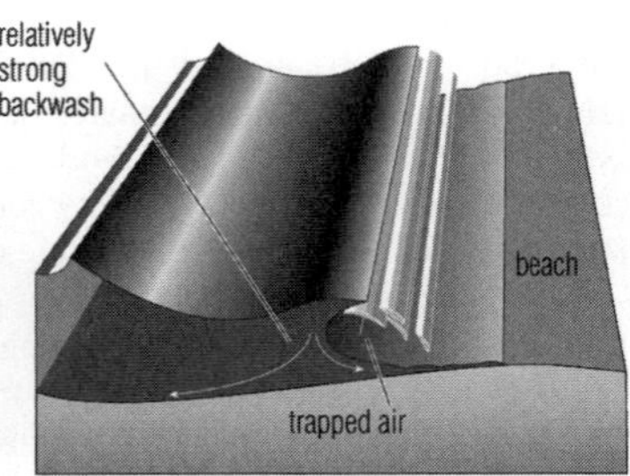

wave The low, gentle crests of a constructive wave, with the energy of the wave flowing up the beach in a strong swash and depositing material, contrasts with the high, steep-crested, more forceful motions of destructive waves which crash in at an angle to the beach directing all their energy into plunging waves which tear up the sand and shingle and carry it out with the strong backwash. Helicon

As a wave approaches the shore it is forced to break as a result of friction with the seabed. When it breaks on a beach, water and sediment are carried up the beach as **swash**; the water then drains back as **backwash**.

A **constructive wave** causes a net deposition of material on the shore because its swash is stronger than its backwash. Such waves tend be low and have crests that spill over gradually as they break. The backwash of a **destructive wave** is stronger than its swash, and therefore causes a net removal of material from the shore. Destructive waves are usually tall and have peaked crests that plunge downwards as they break, trapping air as they do so.

If waves strike a beach at an angle the beach material is gradually moved along the shore (†longshore drift), causing deposition of material in some areas and erosion in others.

Atmospheric instability caused by the †greenhouse effect and †global warming appears to be increasing the severity of Atlantic storms and the heights of the ocean waves. Waves in the South Atlantic are shrinking – they are on average half a meter smaller than in the mid-1980s – and those in the northeast Atlantic have doubled in size over the last 40 years. As the height of waves affects the supply of marine food, this could affect fish stocks, and there are also implications for shipping and oil and gas rigs in the North Atlantic, which will need to be strengthened if they are to avoid damage.

Freak or "episodic" waves form under particular weather conditions at certain times of the year, traveling long distances across the Atlantic, Indian, and Pacific oceans. They are considered responsible for the sudden unexplained disappearance of many ships.

Freak waves become extremely dangerous when they reach the shallow waters of the continental shelves at 100 fathoms (180 m/600 ft), especially when they meet currents: for example, the Agulhas Current to the east of South Africa, and the Gulf Stream in the North Atlantic. A wave height of 34 m/112 ft has been recorded.

wave in physics, oscillation that is propagated from a source. Mechanical waves require a medium through which to travel. Electromagnetic waves do not; they can travel through a vacuum. Waves carry energy but they do not transfer matter. The medium (for example the Earth, for seismic waves) is not permanently displaced by the passage of a wave. The model of waves as a pattern is used to help understand the properties of light and sound. Experiments conducted in a ripple tank with water waves can explain how waves slow down as water becomes shallower, how waves change direction when traveling through another medium, and how waves are reflected from different surfaces. See also †standing wave.

types of wave There are various ways of classifying wave types. One of these is based on the way the wave travels. In a †transverse wave, the displacement of the medium is perpendicular to the direction in which the wave travels. An example of this type of wave is a mechanical wave projected along a tight string. The string moves at right angles to the wave motion. †Electromagnetic waves are another example of transverse waves. The directions of the electric and magnetic fields are perpendicular to the wave motion. In a †longitudinal wave the disturbance takes place parallel to the wave motion. A longitudinal wave consists of a series of compressions and rarefactions (states of maximum and minimum density and pressure, respectively). Such waves are always mechanical in nature and thus require a medium through which to travel. †Sound waves are an example of longitudinal waves. Waves that result from a stone being dropped into water appear as a series of circles. These are called circular waves and can be generated in a †ripple tank for study. Waves on water that appear as a series of parallel lines are called plane waves.

characteristics of waves All waves have a †wavelength. This is measured as the distance between successive crests (or successive troughs) of the wave. It is given the Greek symbol l. The †frequency of a wave is the number of vibrations per second. The reciprocal of this is the wave period. This is the time taken for one complete cycle of the wave oscillation. The speed of the wave is measured by multiplying wave frequency by the wavelength.

properties of waves When a wave moves from one medium to another (for example a light wave moving from air to glass) it moves with a different speed in the second medium. This change in speed causes it to change direction. This property is called †refraction. The †angle of refraction depends on whether the wave is speeding up or slowing down as it changes medium. †Reflection occurs whenever a wave hits a barrier. The wave is sent back, or reflected, into the medium. The †angle of incidence (the angle between the ray and a perpendicular line drawn to the surface) is equal to the †angle of reflection (the angle between the reflected ray and a perpendicular to the surface). See also †total internal reflection. An echo is the repetition of a sound wave by reflection from a surface. All waves spread slightly as they travel. This is called †diffraction and it occurs chiefly when a wave interacts with a solid object. The degree of diffraction depends on the relationship between the wavelength and the size of the object (or gap through which the wave travels). If the two are similar in size, diffraction occurs and the wave can be seen to spread out. Large objects cast †shadows because the difference between their size and the wavelength is so large that light waves are not diffracted around the object. A dark shadow results. When two or more waves meet at a point, they interact and combine to produce a resultant wave of larger or smaller amplitude (depending on whether the combining waves are in or out of phase with each other). This is called †interference. Transverse waves can exhibit polarization. If the oscillations of the wave take place in many different directions (all at right angles to the directions of the wave) the wave is unpolarized. If the oscillations occur in one plane only, the wave is polarized. Light, which consists of transverse waves, can be polarized.

Making Waves http://www.smgaels.org/physics/home.htm

Easy-to-follow site created by students at a US high school. The topics covered include "Sound," "Microwave," "Light and lasers," and "Gamma rays."

Wavelength http://www.glenbrook.k12.il.us/gbssci/phys/Class/waves/wavestoc.html

General explanation of the basics behind waves. The information is offered in a series of lessons, which cover the nature, properties, and behavior of waves, and also looks at standing waves.

Waves http://members.aol.com/nicholashl/waves/waves.htm

Animations demonstrating different types of wave behavior. There are a series of Java applets demonstrating moving, stationary, and circular waves, plus reflection, refraction, and diffraction. This is a very simple but effective site.

wave, carrier see †carrier wave.

wave-cut platform gently sloping rock surface found at the foot of a coastal cliff. Covered by water at high tide but exposed at low tide, it represents the last remnant of an eroded headland (see †coastal erosion). Wave-cut platforms have a gentle gradient (usually less than 1°) and can be up to about 1 km/0.6 mi in length. Over time, wave-cut platforms are lengthened and protect the cliffs behind them.

wave, electromagnetic see †electromagnetic wave.

wavefront line of points in a wave motion that are all of equal †phase.

waveguide hollow metallic tube, either empty or containing a †dielectric, used to guide a high-frequency electromagnetic wave (microwave) traveling within it. The wave is reflected from the internal surfaces of the guide. Waveguides are extensively used in radar systems.

wavelength distance between successive crests or troughs of a †wave. The wavelength of a light wave determines its color; red light has a wavelength of about 700 nanometers, for example. The complete range of wavelengths of electromagnetic waves is called the electromagnetic †spectrum.

wave, longitudinal see †longitudinal wave.

wave mechanics branch of quantum theory that derives the various properties of †atoms on the basis of every particle having an associated wave existing in a multidimensional space, representing probabilities of certain properties of the particles involved. †Schrödinger's equation is the basis of wave mechanics, which has been shown to be equivalent to †matrix mechanics.

wave power power obtained by harnessing the energy of water waves. Various schemes have been advanced since 1973 when oil prices rose dramatically and an energy shortage threatened. In 1974 the British engineer Stephen Salter developed the "duck" – a floating boom, the segments of which nod up and down with the waves. The nodding motion can be used to drive pumps and spin generators. Another device, developed in Japan, uses an oscillating water column to harness wave power. A major technological breakthrough will be required if wave power is ever to contribute significantly to the world's energy needs, although several ideas have reached prototype stage.

wave refraction distortion of waves as they reach the coast, due to variations in the depth of the water and shape of the coastline. It is particularly evident where there are headlands and bays.

The bending of a wave crest as it approaches a headland concentrates the energy of the wave in the direction of that headland, and increases its power of erosion. By contrast, the bending that a wave crest experiences when it moves into a bay causes its energy to be dissipated away from the direction of the shore. As a result the wave loses its erosive power and becomes more likely to deposit sediment on the shore.

wave, transverse see †transverse wave.

wax solid fatty substance of animal, vegetable, or mineral origin.

Waxes are composed variously of †esters, †fatty acids, free †alcohols, and solid hydrocarbons.

Mineral waxes are obtained from petroleum and vary in hardness from the soft petroleum jelly (or petrolatum) used in ointments to the hard paraffin wax employed for making candles and waxed paper for drinks cartons.

Animal waxes include beeswax, the wool wax lanolin, and spermaceti from sperm-whale oil; they are used mainly in cosmetics, ointments, and polishes. Another animal wax is tallow, a form of suet obtained from cattle and sheep's fat, once widely used to make candles and soap. Sealing wax is made from lac or shellac, a resinous substance obtained from secretions of scale insects.

Vegetable waxes, which usually occur as a waterproof coating on plants that grow in hot, arid regions, include carnauba wax (from the leaves of the carnauba palm) and candelilla wax, both of which are components of hard polishes such as car waxes.

Wb in physics, symbol for †weber, the SI unit of magnetic flux, equal to the flux that (linking a circuit of one turn) produces an emf (†electromotive force) of 1 volt when the flux is reduced at a constant rate to zero in 1 second.

weak acid acid that only partially ionizes in aqueous solution (see †dissociation). Weak acids include ethanoic acid and carbonic acid. The pH of such acids lies between pH 3 and pH 6.

weak base base that only partially ionizes in aqueous solution (see †dissociation); for example, ammonia. The pH of such bases lies between pH 8 and pH 10.

weak interacting massive particle **WIMP**, hypothetical subatomic particle found in the Galaxy's †dark matter. These particles could constitute the 80% of dark matter unaccounted for by †massive astrophysical compact halo objects.

weak nuclear force or **weak interaction**, one of the four fundamental †forces of nature, the other three being the gravitational force or gravity, the electromagnetic force, and the strong nuclear force. It causes radioactive beta decay and other subatomic reactions. The particles that carry the weak force are called †weakons (or intermediate vector bosons) and comprise the positively and negatively charged W particles and the neutral Z particle.

weakon or **intermediate vector boson**, member of a group of elementary particles, see † intermediate vector boson.

weather variation of atmospheric conditions at any one place over a short period of time. Such conditions include humidity, precipitation, temperature, cloud cover, visibility, and wind. Weather differs from †climate in that the latter is a composite of the average weather conditions of a locality or region over a long period of time (at least 30 years). †Meteorology is the study of short-term weather patterns and data within a particular area; climatology is the study of weather over longer timescales on a zonal or global basis. Weather forecasts are based on current meteorological data, and predict likely weather for a particular area.

Department of Atmospheric Sciences http://www.atmos.uiuc.edu/

Masses of data and other information on climate around the planet. Among the site's more attractive features is the use of multimedia instructional modules, customized weather maps, and real-time weather data.

Weather: What Forces Affect Our Weather? http://www.learner.org/exhibits/weather/

User-friendly, US-based site exploring aspects of climate including "The water cycle," "Powerful storms," and "Our changing climate." The site includes "hands-on" activities such as "Try your hand at tornado chasing."

weather area any of the divisions of the sea around the British Isles for the purpose of weather forecasting for shipping. The areas are used to indicate where strong or gale-force winds are expected.

weathering process by which exposed rocks are broken down on the spot (in situ) by the action of rain, frost, wind, and other elements of the weather. It differs from †erosion in that no movement or transportion of the broken-down material takes place. Two types of weathering are recognized: †physical (or mechanical) weathering and †chemical weathering. They usually occur together, and are an important part of the development of landforms.

Physical Weathering http://library.thinkquest.org/20035/physical.htm

After a quick definition of weathering and erosion the main body of this page deals with physical weathering. The types of physical weathering covered are "Pressure release," "Exfoliation," "Frost shattering," "Crystallization," and "Biological weathering." The locations in which the different types of weathering take place are covered, and links are provided to other types of weathering.

Weathering and Erosion http://www.hbcumi.cau.edu/tqp/301/301-11/301-11.html

Simple summary of "Physical or mechanical weathering," "Chemical weathering," and "Rates of weathering."

Web browser in computing, client software that allows access to the World Wide Web. See †browser.

weber symbol Wb, SI unit of †magnetic flux (the magnetic field strength multiplied by the area through which the field passes). It is named after German chemist Wilhelm Weber. One weber equals 10^8 †maxwells.

A change of flux at a uniform rate of one weber per second in an electrical coil with one turn produces an electromotive force of one volt in the coil.

Weber, Ernst Heinrich (1795–1878) German anatomist and physiologist. He applied hydrodynamics to study blood circulation, and formulated **Weber's law**, relating response to stimulus.

Weber, Heinrich (1842–1913) German mathematician whose chief work was in the fields of algebra and number theory. He demonstrated Norwegian mathematician Niels †Abel's theorem in its most general form.

Web page in computing, a †hypertext document on the †World Wide Web.

weedkiller or **herbicide**, chemical that kills some or all plants. Selective herbicides are effective with cereal crops because they kill all broad-leaved plants without affecting grasslike leaves. Those that kill all plants include sodium chlorate and †paraquat. The widespread use of weedkillers in agriculture has led to an increase in crop yield but also to pollution of soil and water supplies and killing of birds and small animals, as well as creating a health hazard for humans.

Wegener, Alfred Lothar (1880–1930) German meteorologist and geophysicist whose theory of †continental drift, expounded in *Origin of Continents and Oceans* in 1915, was originally known as "Wegener's hypothesis." His ideas can now be explained in terms of †plate tectonics, the idea that the Earth's crust consists of a number of plates, all moving with respect to one another.

Weierstrass, Karl Theodor Wilhelm (1815–1897) German mathematician who deepened and broadened the understanding of functions. He demonstrated in 1871 that there exist continuous functions in an interval which have derivatives nowhere in the interval.

Weierstrass, Karl Theodor Wilhelm http://www-groups.dcs.st-andrews.ac.uk/history/Mathematicians/Weierstrass.html

Part of an archive containing the biographies of the world's greatest mathematicians, this site is devoted to the life and contributions of Karl Weierstrass.

weight force exerted on an object by †gravity. The weight of an object depends on its mass – the amount of material in it – and the strength of the Earth's gravitational pull (the acceleration due to gravity), which decreases with height. Consequently, an object weighs less at the top of a mountain than at sea level. On the surface of the Moon, an object has only one-sixth of its weight on Earth (although its mass is unchanged), because the Moon's surface gravity is one-sixth that of the Earth's.

weightlessness apparent loss in weight of a body in free fall. Astronauts in an orbiting spacecraft do not feel any weight because they are falling freely in the Earth's gravitational field (not because they are beyond the influence of Earth's gravity). The same phenomenon can be experienced in a falling lift or in an aircraft imitating the path of a freely falling object.

On Station http://esapub.esrin.esa.it/onstation/onstation.htm

Current and archived pages of the online magazine On Station. Journal articles are clearly written and accompanied by photographs. They contain announcements of events and meetings, as well as updates on both flight- and ground-based microgravity research. The site also includes links to archive issues of Microgravity News.

weights and measures see under †*c*.g.s. system, †f.p.s. system, †m.k.s. system, †SI units.

Weinberg, Steven (1933–) US physicist who, together with Abdus †Salam, was awarded the Nobel Prize for Physics in 1979 for demonstrating in 1967 that the weak nuclear force and the electromagnetic force (two of the fundamental forces of nature) are variations of a single underlying force, now called the electroweak force, and for the prediction of the existence of the weak neutral current.

Weismann, August Friedrich Leopold (1834–1914) German biologist, one of the founders of †genetics. He postulated that every living organism contains a special hereditary substance, the "germ plasm," and in 1892 he proposed that changes to the body do not in turn cause an alteration of the genetic material.

Weizsäcker, Carl Friedrich von (1912–) German theoretical physicist who investigated the way in which energy is generated in the cores of stars. He also developed a theory that planetary systems were formed as a natural by-product of stellar evolution, condensing from vortices of gas.

Werner, Abraham Gottlob (1749–1817) German geologist, one of the first to classify minerals systematically. He also developed the later discarded theory of neptunism – that the Earth was initially covered by water, with every mineral in suspension; as the water receded, layers of rocks "crystallized."

Werner, Alfred (1866–1919) French-born Swiss chemist. He was awarded the Nobel Prize for Chemistry in 1913 for his work on valency theory, which gave rise to the concept of coordinate bonds and coordination compounds.

Wernicke, Carl (1848–1905) German neurologist and psychiatrist. He is known for his study of aphasia. In *The Aphasic Syndrome* (1874), he described what later became known as sensory aphasia (that is, defects in, or loss of, speech and expression) as distinct from motor aphasia, first described by French surgeon Paul Pierre Broca (1824–1880).

Wertheimer, Max (1880–1943) Czech-born psychologist and founder, with Koffka and Kohler, of gestalt psychology. While traveling on a train in 1910 he saw that a light flashing rapidly from two different positions seemed to be one light in motion. This type of perception became the basis for his gestalt concept.

Classics In The History Of Psychology – Laws of Organization in Perceptual Forms
http://psychclassics.yorku.ca/Wertheimer/Forms/forms.htm

Transcript of Max Wertheimer's 1923 paper Laws of Organization in Perceptual Forms. This paper outlined the principles of Gestalt psychology, and therefore formed the basis for much of the development in the science in the early part of this century.

West bright comet discovered by Danish astronomer Richard West at the European Southern Observatory in 1975. It became a prominent sight in the spring of 1976.

West passed within 30 million km/19 million mi of the Sun on 25 February 1976, and within 120 million km/75 million mi of Earth on 29 February. At its brightest in early March it reached zero magnitude and had a broad tail 30° long. A few days later it broke up into four pieces.

westerlies prevailing winds from the west that occur in both hemispheres between latitudes of about 35° and 60°. Unlike the †trade winds, they are very variable and produce stormy weather.

The westerlies blow mainly from the southwest in the northern hemisphere and the northwest in the southern hemisphere, bringing moist weather to the west coast of the landmasses in these latitudes.

West Nile virus viral disease carried by birds that spreads to humans via mosquitoes (which suck the blood of both birds and humans). The virus is endemic in Africa, Asia, and Europe and spreads only to humans when infection levels in birds (where the virus is largely harmless in many species) are high enough for levels of infected mosquitoes to also become high. In humans the virus causes encephalitis, with a mortality rate of around 10%.

In June 2001 US researchers developed a virus that successfully protected horses against the virus.

Weston cadmium cell standard cell (battery) with a constant emf (†electromotive force) at 20°C of 1.019 volts. It has a cadmium-mercury amalgam as cathode and mercury as anode, located in the legs of an H-shaped glass vessel. The electrolyte occupying the rest of the vessel is a saturated solution of cadmium sulfate.

wetland permanently wet land area or habitat. Wetlands include areas of †marsh, fen, †bog, flood plain, and shallow coastal areas. Wetlands are extremely fertile. They provide warm, sheltered waters for fisheries, lush vegetation for grazing livestock, and an abundance of wildlife. Estuaries and seaweed beds are more than 16 times as productive as the open ocean.

The term is often more specifically applied to a naturally flooding area that is managed for agriculture or wildlife. A water meadow, where a river is expected to flood grazing land at least once a year thereby replenishing the soil, is a traditional example.

wetted perimeter length of that part of a river's cross-section that is in contact with the water. The wetted perimeter is used to calculate a river's hydraulic radius, a measure of its †channel efficiency.

wetting agent surface active agent that decreases the surface tension of water to allow it to wet a surface. Detergents are wetting agents.

Weyl, Hermann (1885–1955) German mathematician and mathematical physicist who studied mainly topological space and the geometry of Bernhard †Riemann. He also studied quantum mechanics and number theory.

Weyl, Hermann Klaus Hugo http://www-groups.dcs.st-andrews.ac.uk/history/Mathematicians/Weyl.html

Part of an archive containing the biographies of the world's greatest mathematicians, this site is devoted to the life and contributions of Hermann Weyl.

Wheatstone, Charles (1802–1875) English physicist and inventor. With William Cooke, he patented a railway telegraph in 1837, and, developing an idea of Samuel Christie (1784–1865), devised the **Wheatstone bridge**, an electrical network for measuring resistance. He also invented the concertina.

Wheatstone bridge electric circuit that is divided into sections enabling the relative †resistances or devices placed in the sections to be deduced.

Whipple, Fred Lawrence (1906–) US astronomer whose hypothesis in 1949 that the nucleus of a comet is like a dirty snowball was confirmed in 1986 by space-probe studies of †Halley's Comet.

Whirlpool Galaxy spiral galaxy in the constellation of †Canes Venatici. It has a smaller galaxy at the end of one arm. It is also known as Messier 51 (M51).

whirlwind rapidly rotating column of air, often synonymous with a †tornado. On a smaller scale it produces the dust-devils seen in deserts.

white arsenic alternative name for †arsenic oxide. It is prepared by burning arsenic in air.

white blood cell or **leucocyte**, one of a number of different cells that play a part in the body's defenses and give immunity against disease. Some (neutrophils and †macrophages) engulf invading micro-organisms, others kill infected cells, while †lymphocytes produce more specific immune responses. White blood cells are colorless, with clear or granulated cytoplasm, and are capable of independent amoeboid movement. They occur in the blood, †lymph, and elsewhere in the body's tissues. Unlike mature red blood cells they contain a nucleus.

Human blood contains about 11,000 leucocytes to the cubic millimeter – about one to every 500 red blood cells. White blood cell numbers may be reduced (leucopenia) by starvation, pernicious anemia, and certain infections, such as typhoid and malaria. An increase in their numbers (leucocytosis) is a reaction to normal events such as digestion, exertion, and pregnancy, and to abnormal ones such as loss of blood, cancer, and most infections.

Some white cells can produce antibodies. Antibodies are specific chemicals that can attach to chemicals that do not belong to the body, such as toxins or chemicals on disease-causing bacteria. Chemicals that do not belong to the body are said to be "foreign." If these chemicals cause the body to produce antibodies, they are called antigens. Once these foreign chemicals have contacted the blood, some white blood cells can retain the memory of the particular antibody that is needed for defense. The next time the antibody is needed it is produced quickly and in large amounts to prevent the body being harmed – the body shows immunity to the disease. This is also the basis of †immunization by a vaccine.

white dwarf small, hot †star, the last stage in the life of a star such as the Sun. White dwarfs make up 10% of the stars in the Galaxy; most have a mass 60% of that of the Sun, but only 1% of the Sun's diameter, similar in size to the Earth. Most have surface temperatures of 8,000°C/14,400°F or more, hotter than the Sun. However, being so small, their overall luminosities may be less than 1% of that of the Sun. The Milky Way contains an estimated 50 billion white dwarfs.

Whitehead, Alfred North (1861–1947) English philosopher and mathematician. In his "theory of organism," he attempted a synthesis of metaphysics and science. His works include *Principia Mathematica* (1910–13) (with Bertrand †Russell), *The Concept of Nature* (1920), and *Adventures of*

Ideas (1933).

Whitehead, Alfred North http://plato.stanford.edu/entries/whitehead/

Good summary of the life and legacy of the English mathematician and logician. His influence in mathematics, metaphysics, and the philosophy of science is examined. There is a full list of his publications and a good bibliography.

Whitehead, John Henry Constantine (1904–1960) British mathematician who studied the more abstract areas of differential geometry, and of algebraic and geometrical topology.

whiteout "fog" of grains of dry snow caused by strong winds in temperatures of between –18°C/0°F and –1°C/30°F. The uniform whiteness of the ground and air causes disorientation in humans.

white spirit colorless liquid derived from gasoline; it is used as a solvent and in paints and varnishes.

Whittaker, Edmund Taylor (1873–1956) British mathematician and physicist, professor of mathematics at Edinburgh University 1912–46. He worked principally on differential equations, but was also interested in relativity theory and analytical dynamics. In his principal study, *A History of the Theories of Aether and Electricity: the Modern Theories* (1910, revised 1931, 1953), he gave a comprehensive review of theories of atomism and electromagnetism.

wide area network **WAN**, in computing, a †network that connects computers distributed over a wide geographical area. "Dumb" terminals and microcomputers act as workstations, which connect to remote systems via a local host computer.

Wieland, Heinrich Otto (1877–1957) German organic chemist who was awarded the Nobel Prize for Chemistry in 1927 for determining the structures of †steroids and related compounds. He also studied other natural compounds, such as alkaloids and pterins, and contributed to the investigation of biological oxidation.

Wien, Wilhelm Carl Werner Otto Fritz Franz (1864–1928) German physicist who studied radiation and established the principle, since known as **Wien's law**, that the wavelength at which the radiation from an idealized radiating body is most intense is inversely proportional to the body's absolute temperature. (That is, the hotter the body, the shorter the wavelength.) He was awarded the Nobel Prize for Physics in 1911 for his study of the laws governing the radiation of heat.

Wien, Wilhelm Carl Werner Otto Fritz Franz http://www-groups.dcs.st-andrews.ac.uk/history/Mathematicians/Wien.html

Part of an archive containing the biographies of the world's greatest mathematicians, this site is devoted to the life and contributions of physicist Wilhelm Wien.

Wiener, Norbert (1894–1964) US mathematician, credited with the establishment of the science of cybernetics in his book *Cybernetics* (1948). In mathematics, he laid the foundation of the study of stochastic processes (those dependent on random events), particularly †Brownian motion.

Wiener, Norbert http://www-groups.dcs.st-andrews.ac.uk/history/Mathematicians/Wiener_Norbert.html

Part of an archive containing the biographies of the world's greatest mathematicians, this site is devoted to the life and contributions of Norbert Wiener.

Wien's displacement law in physics, a law of radiation stating that the wavelength carrying the maximum energy is inversely proportional to the absolute temperature of a †black body: the hotter a body is, the shorter the wavelength. It has the form $\lambda_{max}T$ = constant, where λ_{max} is the wavelength of maximum intensity and T is the temperature. The law is named after German physicist Wilhelm Wien.

Wieschaus, Eric F (1947–) US geneticist who was awarded a Nobel Prize for Physiology or Medicine in 1995 with Edward †Lewis and Christiane Nüsslein-Volhard for their discoveries concerning the genes that control the early stages of the body's development.

Wigglesworth, Vincent Brian (1899–1994) English entomologist whose research covered many areas of insect physiology, especially the role of hormones in growth and metamorphosis. He was knighted in 1964.

Wigner, Eugene P(aul) (1902–1995) Hungarian-born US physicist who shared the Nobel Prize for Physics in 1963 for his work which introduced the notion of parity, or symmetry theory, into nuclear physics, showing that all nuclear processes should be indistinguishable from their mirror images.

Wigner, Eugene Paul http://www-groups.dcs.st-andrews.ac.uk/history/Mathematicians/Wigner.html

Part of an archive containing the biographies of the world's greatest mathematicians, this site is devoted to the life and contributions of physicist Eugene Wigner.

wilderness area of uninhabited land that has never been disturbed by humans, usually located some distance from towns and cities. According to estimates by US group Conservation International, 52% (90 million sq km/35 million sq mi) of the Earth's total land area was still undisturbed in 1994.

Wilderness Society: Protecting America's Wilderness http://www.wilderness.org/

Frequently updated site with news of the Wilderness Society's events and campaigns. Attractive pages present the work of the Wilderness Society's regional offices and include detailed information on wildfires and forest management.

wildlife corridor passage between habitats. See †corridor, wildlife.

wild type in genetics, the naturally occurring gene for a particular character that is typical of most individuals of a given species, as distinct from new genes that arise by mutation.

Wiles, Andrew John (1952–) English mathematician famous for finally completing the proof of Fermat's last theorem in 1994, more than 460 years after †Fermat claimed to have proved it.

Wilkins, Maurice Hugh Frederick (1916–) New Zealand-born British molecular biologist who was awarded the Nobel Prize for Physiology or Medicine in 1962 with Francis †Crick and James †Watson for the discovery of the double-helical structure of DNA and of the significance of this structure in the replication and transfer of genetic information.

Williamson, Alexander William (1824–1904) English organic chemist who made significant discoveries concerning alcohols and ethers, catalysis, and reversible reactions. He was the first to explain the action of a †catalyst in terms of the formation of an intermediate compound.

will-o'-the-wisp light sometimes seen over marshy ground, believed to be burning gas containing methane from decaying organic matter.

An Australian scientist put forward the hypothesis in 1995 that the phenomenon may be caused by barn owls, which sometimes develop a ghostly glow due to a light-emitting honey fungus that they pick up from rotting trees.

willy-willy Australian Aboriginal term for a cyclonic whirlwind.

Wilson, Charles Thomson Rees (1869–1959) Scottish physicist who shared the Nobel Prize for Physics in 1927 for his invention in 1911 of the Wilson †cloud chamber, an apparatus for studying subatomic particles.

Wilson, Edward Osborne (1929–) US zoologist whose books have stimulated interest in biogeography, the study of the distribution of species, and sociobiology, the evolution of behavior. He is a world authority on ants.

Wilson, John Tuzo (1908–1993) Canadian geologist and geophysicist who established and brought about a general understanding of the concept of †plate tectonics.

Wilson, Robert Woodrow (1936–) US radio astronomer who in 1964, with the German-born US radio engineer Arno †Penzias, detected †cosmic background radiation, which is thought to represent a residue of the primordial †Big Bang. He and Penzias shared the Nobel Prize for Physics in 1978 for their work on cosmic background radiation.

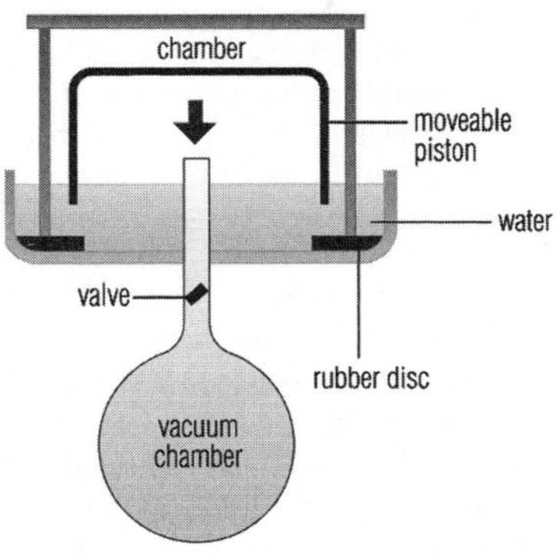

Wilson The cloud chamber devised by C T R Wilson was the first instrument to detect the tracks of atomic particles. It consisted originally of a cylindrical glass chamber fitted with a hollow piston, which was connected, via a valve, to a large evacuated flask. The piston falls rapidly when the valve is opened, reducing the pressure in the chamber, and water vapor condenses along the tracks of any particles in the chamber. Helicon

wilting loss of rigidity (†turgor) in plants, caused by a decreasing wall pressure within the cells making up the supportive tissues. Wilting is most obvious in plants that have little or no wood.

WIMP acronym for †weak interacting massive particle.

WIMP acronym for **Windows, icons, menus, pointing device**, in computing, another name for †graphical user interface (GUI).

wind lateral movement of the Earth's atmosphere from high-pressure areas (anticyclones) to low-pressure areas (depressions). Its speed is measured using an †anemometer or by studying its effects on, for example, trees by using the †Beaufort scale. Although modified by features such as land and water, there is a basic worldwide system of †trade winds, †westerlies, and polar easterlies.

A belt of low pressure (the †doldrums) lies along the Equator. The trade winds blow towards this from the horse latitudes (areas of high pressure at about 30° north and south of the Equator), blowing from the northeast in the northern hemisphere, and from the southeast in the southern hemisphere. The westerlies (also from the horse latitudes) blow north of the Equator from the southwest, and south of the Equator from the northwest.

Cold winds blow outwards from high-pressure areas at the poles. More local effects result from landmasses heating and cooling faster than the adjacent sea, producing onshore winds in the daytime and offshore winds at night.

The †monsoon is a seasonal wind of southern Asia, blowing from the southwest in summer and bringing the rain on which crops depend. It blows from the northeast in winter.

Famous warm winds include the **chinook** of the eastern Rocky Mountains, North America; the **föhn** of Europe's Alpine valleys; the **sirocco** (Italy)/**khamsin** (Egypt)/**sharav** (Israel) – spring winds that bring warm air from the Sahara and Arabian deserts across the Mediterranean; and the **Santa Ana**, a periodic warm wind from the inland deserts that strikes the California coast.

The dry northerly **bise** (Switzerland) and the **mistral** (which strikes the Mediterranean area of France) are unpleasantly cold winds.

The fastest wind speed ever measured on Earth, 512 kph/318 mph, occurred on 3 May 1999 in a †tornado that struck the suburbs of Oklahoma City, Oklahoma, USA.

Global Wind Patterns http://kids.earth.nasa.gov/archive/nino/global.html

Definition and explanation of the different types of wind patterns present around the world. Basic terms are defined in bold, such as trade winds, doldrums, prevailing westerlies, and polar easterlies. An activity is also included, which enables the user to see visually the world's wind patterns.

Windaus, Adolf Otto Reinhold (1876–1959) German chemist who was awarded the Nobel Prize for Chemistry in 1928 for his research on the structure of †cholesterol, its relationship to vitamin D, and his discovery that †steroids are precursors of vitamins. He also linked the roles of sunlight and vitamin D in the prevention of †rickets.

window in computing, a rectangular area on the screen of a †graphical user interface. A window is used to display data and can be manipulated in various ways by the computer user.

wind power power produced from the harnessing of wind energy. The wind has long been used as a source of energy: sailing ships and windmills are ancient inventions. After the energy crisis of the 1970s †wind turbines began to be used to produce electricity on a large scale.

wind tunnel enclosure that contains a large †turbine capable of sending air streams over any object (such as a model aircraft) whose †aerodynamic performance is to be assessed. Lift, drag, and airflow patterns are observed by the use of special cameras and sensitive instruments. Wind-tunnel testing is used to assess aerodynamic design, prior to full-scale construction.

wind turbine windmill of advanced aerodynamic design connected to a generator producing †electrical energy and used in wind-power installations. Wind is a form of †renewable energy that is used to turn the †turbine blades of the windmill. Wind turbines can be either large propeller-type rotors mounted on a tall tower, or flexible metal strips fixed to a vertical axle at top and bottom.

wing in biology, the modified forelimb of birds and bats, or the membranous outgrowths of the †exoskeleton of insects, which give the power of flight. Birds and bats have two wings. Bird wings have feathers attached to the fused digits ("fingers") and forearm bones, while bat wings consist of skin stretched between the digits. Most insects have four wings, which are strengthened by wing veins.

The wings of butterflies and moths are covered with scales. The hind pair of a fly's wings are modified to form two knoblike balancing organs (halteres).

Winnicott, Donald Woods (1896–1971) English psychoanalyst and child psychiatrist. In his early work, Winnicott studied the relationship between mother and child, developing the view that, for the infant, the mother mediates development of the self. His theories were developed in three volumes summarizing his clinical experience, entitled *Collected Papers: Through Pediatrics to Psychoanalysis* (1958), *The Maturational Process and the Facilitating Environment* (1965), and *Playing and Reality* (1971).

Wirth, Niklaus (1934–) Swiss-born developer of computer languages such as Euler, Pascal, Modula-2 and OBERON. His programming languages have been designed to encourage good programming practices leading to reliability and the re-usability of code. This restrictive approach has found favor in the academic world, where Pascal is widely taught. So far, however, most commercial programmers working on personal computer software have largely ignored his languages and his advice.

Wittig, Georg (1897–1987) German chemist who shared the Nobel Prize for Chemistry in 1979 for his method of synthesizing olefins (alkenes) from carbonyl compounds. This reaction is often termed the **Wittig synthesis**.

Wittig reaction in chemistry, reaction between a †phosphorous ylide and an †aldehyde or †ketone to produce an †alkene. Phosphorous ylides that take part in this reaction have the general formula RR'C = PR", where R and R' are †alkyl groups, and R" is either an alkyl group or an organic group, for example C_5H_6. It was developed by and named after German chemist Georg †Wittig.

Wöhler, Friedrich (1800–1882) German chemist who in 1828 became the first person to synthesize an organic compound (†urea) from an inorganic compound (ammonium cyanate). He also devised a method in 1827 that isolated the metals aluminum, beryllium, yttrium, and titanium from their ores.

On The Artificial Production of Urea http://dbhs.wvusd.k12.ca.us/Chem-History/Wohler-article.html

Excerpts from Friedrich Wöhler's paper in the Annalen der Physik und Chemie, 88, Leipzig, 1828, describing his synthesis of urea from inorganic materials. The production of urea in this manner was the first synthesis of a so-called animal or organic product from a non-organic source.

wold (Old English *wald* "forest") open, hilly country. The term refers specifically to certain areas in England, notably the Yorkshire and Lincolnshire Wolds and the Cotswold Hills.

Wolf, Maximilian Franz Joseph Cornelius (1863–1932) German astronomer. He developed new photographic methods for observational astronomy. He discovered several new nebulae, both within the Milky Way and outside our Galaxy; more than 200 asteroids; and in 1883 a comet, which now bears his name.

Wolf, Rudolph (1816–1893) Swiss astronomer. He calculated that the mean period of the sunspot cycle was 11.1 years.

Wolff, Kaspar Friedrich (1733–1794) German surgeon and physiologist who is regarded as the founder of †embryology. He introduced the idea that initially unspecialized cells later differentiate to produce the separate organs and systems of the plant or animal body.

wolfram alternative name for †tungsten.

wolframite iron manganese tungstate, $(Fe,Mn)WO_4$, an ore mineral of tungsten. It is dark gray with a submetallic surface luster, and often occurs in hydrothermal veins in association with ores of tin.

Wollaston, William Hyde (1766–1828) English chemist and physicist who discovered in 1804 how to make malleable platinum. He went on to discover the new elements palladium in 1804 and rhodium in 1805. He also contributed to optics through the invention of a number of ingenious and still useful measuring instruments.

womb common name for the †uterus.

wood hard tissue beneath the bark of many perennial plants; it is composed of water-conducting cells, or secondary †xylem, and gains its hardness and strength from deposits of †lignin. **Hardwoods**, such as oak, and **softwoods**, such as pine, have commercial value as structural material and for furniture.

The central wood in a branch or stem is known as **heartwood** and is generally darker and harder than the outer wood; it consists only of dead cells. As well as providing structural support, it often contains gums, tannins, or pigments, which may impart a characteristic color and increased durability. The surrounding **sapwood** is the functional part of the xylem that conducts water.

The **secondary xylem** is laid down by the vascular †cambium which forms a new layer of wood annually, on the outside of the existing wood and visible as an †annual ring when the tree is felled; see †dendrochronology.

Commercial wood can be divided into two main types: hardwood, containing xylem vessels and obtained from angiosperms (for example, oak); and softwood, containing only †tracheids, obtained from gymnosperms (for example, pine). Although in general softwoods are softer than hardwoods, this is not always the case: balsa, the softest wood known, is a hardwood, while pitch pine, very dense and hard, is a softwood. A superhard wood is produced in wood–plastic combinations (WPC), in which wood is impregnated with liquid plastic (monomer) and the whole is then bombarded with gamma rays to polymerize the plastic.

woodland area in which trees grow more or less thickly; generally smaller than a †forest. Temperate climates, with four distinct seasons a year, tend to support a mixed woodland habitat, with some conifers but mostly broad-leaved and deciduous trees, shedding their leaves in autumn and regrowing them in spring. In the Mediterranean region and parts of the southern hemisphere, the trees are mostly evergreen. Temperate woodlands grow in the zone between the cold coniferous forests and the tropical forests of the hotter climates near the Equator. They develop in areas where the closeness of the sea keeps the climate mild and moist.

Old woodland can rival tropical rainforest in the number of species it supports, but most of the species are hidden in the soil. A study in Oregon, USA, in 1991 found that the soil in a single woodland location contained 8,000 arthropod species (such as insects, mites, centipedes, and millipedes), compared with only 143 species of reptile, bird, and mammal in the woodland above.

wood pitch by-product of charcoal manufacture, made from **wood tar**, the condensed liquid produced from burning charcoal gases. The wood tar is boiled to produce the correct consistency. It has been used since ancient times for caulking wooden ships (filling in the spaces between the hull planks to make them watertight).

Woodward, Arthur Smith (1864–1944) English geologist who published important research into the fossils of fish, amphibia, birds, mammals, and reptiles. He was the author of an influential textbook, *Vertebrate Paleontology* (1898).

Woodward, Robert Burns (1917–1979) US chemist who worked on synthesizing a large number of complex molecules. These included quinine in 1944, cholesterol in 1951, chlorophyll in 1960, and vitamin B_{12} in 1971. He was awarded the Nobel Prize for Chemistry in 1965 for his work in organic synthesis.

Woodworth, Robert Sessions (1869–1962) US psychologist. He collaborated with Edward †Thorndike in examining individual learning differences and later attempted to derive objective tests of emotional stability. Although his contributions to research were limited, he published a number of general texts on psychology including *Experimental Psychology* (1938), *Contemporary Schools of Psychology* (1931), and *Dynamics of Behavior* (1958).

Woolley, Richard van der Riet (1906–1986) English astronomer whose work included observational and theoretical astrophysics, stellar dynamics, and the dynamics of the Galaxy. He was knighted in 1963.

word processing input, amendment, manipulation, storage, and retrieval of text. A computer system that runs such software is known as a **word processor**. Since word-processing programs became available to microcomputers, the method has largely replaced the typewriter for producing letters or other text. Typical facilities include insert, delete, cut and paste, reformat, search and replace, copy, print, mail merge, and spelling check.

The leading word-processing programs include Microsoft Word, the market leader, Lotus WordPro, and Corel WordPerfect.

work in physics, a measure of the result of transferring energy from one system to another to cause an object to move. Work should not be confused with †energy (the capacity to do work, which is also measured in joules) or with †power (the rate of doing work, measured in joules per second).

Work is equal to the product of the force used and the distance moved by the object in the direction of that force. If the force is F newtons and the distance moved is d meters, then the work W is given by: $W = Fd$. For example, the work done when a force of 10 newtons moves an object 5 m against resistance is 50 joules (50 newton-meters).

working fluid †fluid that is used in such a way that its internal †energy is converted into external energy. One example is water in †hydroelectricity generation.

workstation high-performance desktop computer with strong graphics capabilities, traditionally used for engineering (†CAD and †CAM), scientific research, and desktop publishing. From 1985 to 1995 workstations were frequently based on fast RISC (reduced instruction-set computer) chips running the Unix operating system. However, the market is under attack from "Wintel" PCs with Intel Pentium processors running Microsoft Windows NT, which are cheaper and run PC software as well as workstation programs. By 1997, four of the five leading workstation manufacturers – DEC, Hewlett-Packard, IBM, and Silicon Graphics Inc., but not Sun Microsystems – had committed themselves to supporting NT.

world conservation strategy in ecology, plan prepared in 1980 by the United Nations Environmental Program (UNEP), the then World Wildlife Fund (now the World Wide Fund for Nature, WWF), and the International Union for the Conservation of Nature and Natural Resources (IUCN), which established an agenda that protected the environment while allowing the exploitation of its resources for human use. The plan has been updated and aims to influence politicians whose decisions affect the environment and its natural resources.

World Wide Web **WWW**, †hypertext system for publishing information on the †Internet. World Wide Web documents ("Web pages") are text files coded using †HTML to include text and graphics, and are stored on a Web server connected to the Internet. Web pages may also contain dynamic objects and †Java applets for enhanced animation, video, sound, and interactivity. In 2000 it was estimated that there were over one billion pages on the Web.

The Web server can be any computer, providing Web server software is available. Every Web page has a †URL (Uniform Resource Locator) – a unique address (usually starting with http://www) which tells a †browser program (such as Netscape Navigator or Microsoft Internet Explorer) where to find it. An important feature of the World Wide Web is the facility to link between documents. This enables readers to follow whatever aspects of a subject interest them most. These links may connect to different computers all over the world. Interlinked or nested Web pages belonging to a single organization are known as a Web site.

About the WWW http://www.w3.org/WWW/

Definition of the World Wide Web and lots of information on what it is all about. This is a good starting point for exploring the Internet as it includes "talks" by members of the WWW Consortium team, and explains some different aspects of using the Web, such as using newsgroups and how to put up a site, as well as containing a brief history of the World Wide Web.

History of the Web http://dbhs.wvusd.k12.ca.us/Chem-History/Hist-of-Web.html

Transcript of Birthplace of the Web by Eric Berger, Office of Public Affairs at FermiLab. The text covers the origins of the Web as a means of communication between scientists at CERN and at FermiLab, and describes how one person's idea in 1991 has brought

about a social and cultural revolution in just a few years.

worm any of various elongated, limbless invertebrates belonging to several phyla. Worms include the flatworms, such as †flukes and †tapeworms; the roundworms or †nematodes, such as the eelworm and the hookworm; the marine ribbon worms or nemerteans; and the segmented worms or †annelids.

WORM acronym for **write once read many times**, in computing, a storage device, similar to a †CD-ROM. The computer can write to the disk directly, but cannot later erase or overwrite the same area. WORMs are mainly used for archiving and back-up copies.

worm in computing, †virus designed to spread from computer to computer across a network. Worms replicate themselves while "hiding" in a computer's memory, causing systems to "crash" or slow down, but do not infect other programs or destroy data directly.

The most celebrated worm was the "Internet worm" of November 1988. Released onto the Internet by Robert Morris, Jr, a graduate student at Cornell University, it infected some 6,000 systems via a loophole in †Unix e-mail and finger procedures. Morris claimed that a programming bug had caused the worm to replicate far more virulently than he had intended, and took swift measures to publish an "antidote" on the network – but by then, many machines had already been disconnected from it. Morris was later convicted, fined, and sentenced to 400 hours community service.

In June 1999 the worm virus ExploreZip overwrote files on computer systems worldwide, causing thousands of companies to close down their e-mail systems. ExploreZip reproduces by mailing itself as replies to any mails waiting in the inbox of the affected computer.

W particle †elementary particle, one of the † intermediate vector bosons responsible for transmitting the †weak nuclear force. The W particle exists as both W^+ and W^-

Wright, Sewall (1889–1988) US geneticist and statistician. During the 1920s he helped modernize Charles †Darwin's theory of evolution, using statistics to model the behavior of populations of †genes.

Wright, Thomas (1711–1786) English astronomer and teacher. He was the first to propose that the Milky Way is a layer of stars, including the Sun, in his *An Original Theory or New Hypothesis of the Universe* (1750).

write protection device on computer disks and tapes that provides data security by allowing data to be read but not deleted, altered, or overwritten.

wrought iron fairly pure iron containing some beads of slag, widely used for construction work before the days of cheap steel. It is strong, tough, and easy to machine. It is made in a puddling furnace, invented by Henry Colt in England in 1784. Pig iron is remelted and heated strongly in air with iron ore, burning out the carbon in the metal, leaving relatively pure iron and a slag containing impurities. The resulting pasty metal is then hammered to remove as much of the remaining slag as possible. It is still used in fences and gratings.

Wundt, Wilhelm Max (1832–1920) German physiologist who regarded psychology as the study of internal experience or consciousness. His main psychological method was introspection; he also studied sensation, perception of space and time, and reaction times.

Wurtz, Charles Adolphe (1817–1884) French organic chemist who discovered ethylamine, the first organic derivative of ammonia, in 1849, and ethylene glycol (1,2-ethanediol) in 1856.

wurtzite naturally occurring brown to black mineral usually found in the form of pyramidal crystals. It is mainly composed of zinc sulfide, ZnS, and an important structural type is named after this mineral, due to its distinctive crystalline structure.

Wurtz synthesis in chemistry, reduction reaction of alkyl halides, in ether solution, by the action of sodium metal. The reaction produces †hydrocarbons corresponding to the alkyl halide used. The reaction was developed by and named after the French chemist Charles Adolphe †Wurtz.

Wynne-Edwards, Vero Copner (1906–1997) English zoologist who argued that animal behavior is often altruistic and that animals will behave for the good of the group, even if this entails individual sacrifice. His study *Animal Dispersion in Relation to Social Behavior* was published in 1962.

WYSIWYG acronym for **what you see is what you get**, in computing, a program that attempts to display on the screen a faithful representation of the final printed output is said to use a WYSIWYG display. For example, a WYSIWYG †word processor would show actual page layout – line widths, page breaks, and the sizes and styles of type.

x in mathematics, an unknown quantity.

xanthophyll yellow pigment in plants that, like †chlorophyll, is responsible for the production of carbohydrates by photosynthesis.

X chromosome larger of the two †sex chromosomes, the smaller being the †Y chromosome. These two chromosomes are involved in sex determination. In humans, whether a person is male or female is determined by the particular combination of the two sex chromosomes in the body cells. In females both the sex chromosomes are the same – two X chromosomes (XX). In males the two are different – one X chromosome and one Y chromosome (XY). The Y chromosome is shorter than the X. †Genes on these chromosomes determine a person's sex. Genes carried on the X chromosome produce the phenomenon of †sex linkage.

As a result of †meiosis gametes from a female each contain one X chromosome. However, gametes from a male are of two kinds. Half of the gametes contain an X chromosome and half contain a Y chromosome. If an X carrying gamete from a male fertilizes a female gamete the result will be a female. If a Y carrying gamete from a male fertilizes a female gamete, the result will be a male.

Early in the development of a female embryo, one of the X chromosomes becomes condensed so that most of its genes are inactivated. If this inactivation is incomplete, skeletal defects and mental retardation result.

xenon chemical symbol Xe, (Greek *xenos* "stranger") colorless, odorless, gaseous, non-metallic element, atomic number 54, relative atomic mass 131.30. It is grouped with the †noble gases (rare gases) and was long believed not to enter into reactions, but is now known to form some compounds, mostly with fluorine. It is a heavy gas present in very small quantities in the air (about one part in 20 million).

Xenon is used in bubble chambers, light bulbs, vacuum tubes, and lasers. It was discovered in 1898 in a residue from liquid air by Scottish chemists William Ramsay and Morris Travers.

xenotime yellow-brown mineral found in certain types of igneous rock. It is mostly composed of yttrium phosphate, YPO_4, but also contains small quantities of other rare-earth metals such as cerium and erbium.

xerophyte plant adapted to live in dry conditions. Common adaptations to reduce the rate of †transpiration include a reduction of leaf size, sometimes to spines or scales; a dense covering of hairs over the leaf to trap a layer of moist air (as in edelweiss); water storage cells; sunken stomata; and permanently rolled leaves or leaves that roll up in dry weather (as in marram grass). Many desert cacti are xerophytes.

XHTML abbreviation for **extensible hypertext markup language**, latest version of HTML to be released by the W3 Consortium (World Wide Web Consortium). Most of the markup tags that were available in HTML 4.0 are still available in XHTML. The major development is that the entire language has been rewritten in †XML (eXtensible Markup Language). The design is modular, with different modules for different classes of markup tags; for example, there are separate modules for form tags, and for frame tags. This modularity means that devices that do not need all the functions of HTML can be designed to handle only the core modules. This makes their programming easier and more compact, and suits the new breed of mobile Web browsers, and devices like TV set-top boxes.

Because XML has been used to design XHTML, Web authors have to be more careful with HTML syntax and presentation; XHTML is case sensitive for example.

XML abbreviation for **eXtensible markup language**, in computing, a simplified subset of †SGML for defining languages for specific purposes or specific industries for use on the World Wide Web. XML is more powerful than †HTML, because the formatting tags are user-defined, but less cumbersome than SGML. XML has been developed through the W3 Consortium, who published XML 1.0 in December 1997.

XML provides the last piece of technology required to develop truly portable applications. The Internet provides the global network, †Java provides portable programs, and XML enables programmers to create portable data.

XML Tutorial http://www.w3schools.com/xml/

Informative guide to basic and advanced XML, packed with information on syntax, elements, browsers, and much more. The topics are presented in a clear format even if they are slightly disjointed to read through due to the site's overuse of subheadings in some of the chapters. There are links to similar guides to HTML and Web scripting languages, such as JavaScript, of which visitors to the XML site are advised to have some basic understanding.

X-ray band of electromagnetic radiation in the wavelength range 10^{-12} to 10^{-8} m (between gamma rays and ultraviolet radiation; see †electromagnetic waves). Applications of X-rays make use of their short wavelength (as in †X-ray diffraction) or their penetrating power (as in medical X-rays of internal body tissues). X-rays are dangerous and can cause cancer.

X-rays with short wavelengths pass through most body tissues, although dense areas such as bone prevent their passage, showing up as white areas on X-ray photographs. The X-rays used in †radiotherapy have very short wavelengths that penetrate tissues deeply and destroy them.

Because of their short wavelength, X-rays can be diffracted by the atoms in crystalline substances. An arrangement of a pattern of dots is formed on a photographic plate that provides information about the structure of the crystal.

X-rays were discovered by German experimental physicist Wilhelm Röntgen in 1895 and formerly called röntgen rays. They are produced when high-energy electrons from a heated filament cathode strike the surface of a target (usually made of tungsten) on the face of a massive heat-conducting anode, between which a high alternating voltage (about 100 kV) is applied.

X-ray astronomy detection of X-rays from intensely hot gas in the universe. Such X-rays are prevented from reaching the Earth's surface by the atmosphere, so detectors must be placed in rockets and satellites. The first celestial X-ray source, Scorpius X-1, was discovered by a rocket flight in 1962.

Since 1970, special satellites have been put into orbit to study X-rays from the Sun, stars, and galaxies. These include the Chandra X-Ray Observatory and the Röntgen Satellite. Many X-ray sources are believed to be gas falling in to †neutron stars and †black holes.

ROSAT Guest Observer Facility http://astroe.gsfc.nasa.gov/docs/rosat/rosgof.html

Profiles the nine-year Röntgen Satellite (ROSAT) mission and describes the spacecraft and its science payload. There is extensive technical documentation, including the ROSAT users' handbook, access to data from the mission, and a very good image gallery.

X-Ray Astronomy http://www.xray.mpe.mpg.de/

Impressive X-ray resource at the Max Planck-Institut für Extraterrestrische Physik. There are details of X-ray missions, institute projects, and publications, and a well-organized image gallery.

X-ray diffraction method of studying the atomic and molecular structure of crystalline substances by using †X-rays. X-rays directed at such substances spread out as they pass through the crystals owing to †diffraction (the slight spreading of waves around the edge of an opaque object) of the rays around the atoms. By using measurements of the position and intensity of the diffracted waves, it is possible to calculate the shape and size of the atoms in the crystal. The method has been used to study substances such as †DNA that are found in living material.

X-ray diffraction analysis use of X-rays to study the atomic and molecular structure of crystalline substances such as ceramics, stone, sediments, and weathering products on metals. The sample, as a single crystal or ground to powder, is exposed to X-rays at various angles; the diffraction patterns produced are then compared with reference standards for identification.

X-ray diffractometer instrument used to determine the crystalline structure (or the atomic structure) of a material such as a mineral. X-ray diffractometers are used to identify finely crystalline minerals such as clays, and to determine the crystal structure of newly discovered minerals.

X-ray fluorescence spectrometry technique used to determine the major and trace elements in the chemical composition of such materials as ceramics, obsidian, and glass. A sample is bombarded with X-rays, and the wavelengths of the released energy, or fluorescent X-rays, are detected and measured. Different elements have unique wavelengths, and their concentrations can be estimated from the intensity of the released X-rays. This analysis may, for example, help an archaeologist in identifying the source of a material.

xylem transport tissue found in †vascular plants, whose main function is to conduct water and dissolved mineral nutrients from the roots to other parts of the plant. The water is ultimately lost by †transpiration from the leaves (see †leaf). Xylem is composed of a number of different types of cell, and may include long, thin, usually dead cells known as †tracheids; fibers (schlerenchyma); thin-walled †parenchyma cells; and conducting vessels.

Transpiration is the loss of water, in the form of water vapor, from leaves of a plant to the air outside the leaf. The loss of water from the leaves drives the movement of water and minerals in the xylem. This is important because many plant cells need the minerals as nutrients. Xylem tissue is usually found close to the other transport tissue in plants, †phloem, which transports sugars and amino acids (see †protein). In non-woody plants phloem and xylem are found in bundles (called veins if they are in a leaf).

In most †angiosperms (flowering plants) water is moved through these vessels. Most †gymnosperms and †pteridophytes lack vessels and depend on tracheids for water conduction. Non-woody plants contain only primary xylem, derived from the procambium, whereas in trees and shrubs this is replaced for the most part by secondary xylem, formed by †secondary growth from the actively dividing vascular †cambium. The cell walls of the secondary xylem are thickened by a deposit of †lignin, providing mechanical support to the plant; see †wood.

xylene or **dimethylbenzene**; $C_6H_4(CH_3)_2$, volatile organic liquid that exists as three isomers, 1,2-xylene, 1,3-xylene, and 1,4-xylene. It is used in the production of polyester synthetic fibers, for example Terylene.

Y

Yalow, Rosalyn Sussman (1921–) US physicist who was awarded a Nobel Prize for Physiology or Medicine in 1977 for the development of radioimmunoassay (RIA) techniques by which minute quantities of hormone may be detected. It can be used to discover a range of hormones produced in the hypothalamic region of the brain.

Yang Chen Ning (1922–) Chinese-born US physicist who, with †Lee Tsung-Dao, overthrew a longstanding principle in theoretical physics called the conservation of parity. This principle states that physical laws are the same in all mirror-image systems. The work of Yang and Lee showed that the †weak nuclear force caused particles called K-mesons to decay in a way which violated parity conservation. They shared the Nobel Prize for Physics in 1957 for their investigations of weak interactions between elementary particles.

Yanofsky, Charles (1925–) US geneticist who demonstrated a linear correspondence between the sequence of bases along the †DNA of a gene and the amino acid sequence of the gene's protein product.

yard symbol yd, unit of length, equivalent to 3 feet (0.9144 m).

In the USA, it is sometimes used to denote a cubic yard (0.7646 cubic meters), as of topsoil.

yardang ridge formed by wind erosion from a dried-up riverbed or similar feature, as in Chad, China, Peru, and North America. On the planet Mars yardangs occur on a massive scale.

Y chromosome smaller of the two †sex chromosomes. In male mammals it occurs paired with the other type of sex chromosome (X), which carries far more genes. The Y chromosome is the smallest of all the mammalian chromosomes and is considered to be largely inert (that is, without direct effect on the physical body), apart from containing the genes that control the development of the testes. There are only 20 genes discovered so far on the human Y chromosome, far fewer than on all other human chromosomes.

In humans, whether a person is male or female is determined by the particular combination of the two sex chromosomes in the body cells. In females both the sex chromosomes are the same – two X chromosomes (XX). In males the two are different – one X chromosome and one Y chromosome (XY). The Y chromosome is shorter than the X. †Genes on these chromosomes determine a person's sex (†sex determination).

As a result of †meiosis gametes from a female each contain one X chromosome. However, gametes from a male are of two kinds. Half of the gametes contain an X chromosome and half contain a Y chromosome. If an X carrying gamete from a male fertilizes a female gamete the result will be a female. If a Y carrying gamete from a male fertilizes a female gamete, the result will be a male.

In humans, about one in 300 males inherits two Y chromosomes at conception, making him an XYY triploid. Few if any differences from normal XY males exist in these individuals, although at one time they were thought to be emotionally unstable and abnormally aggressive. In 1989 the gene determining that a human is male was found to occur on the X as well as on the Y chromosome; however, it is not activated in the female.

year unit of time measurement, based on the orbital period of the Earth around the Sun. The **tropical year** (also called equinoctial and solar year), from one spring †equinox to the next, lasts 365.2422 days (nearly $365^1/_4$ days). It governs the occurrence of the seasons, and is the period on which the calendar year is based. Every four years is a leap year, when the four quarters of a day are added as one extra day. A year on Mercury is only 88 days; a year on Mars is 23 months.

The **sidereal year** is the time taken for the Earth to complete one orbit relative to the fixed stars, and lasts 365.26 days (about 20 minutes longer than a tropical year). The difference is due to the effect of †precession, which slowly moves the position of the equinoxes. The **anomalistic year** is the time taken by any planet in making one complete revolution from perihelion to perihelion; for the Earth this period is about five minutes longer than the sidereal year due to the gravitational pull of the other planets. The **calendar year** consists of 365 days, with an extra day added at the end of February each leap year. **Leap years** occur in every year that is divisible by four, except that a century year is not a leap year unless it is divisible by 400. Hence 1900 was not a leap year, but 2000 was.

yeast one of various single-celled fungi (see †fungus) that form masses of tiny round or oval cells by budding. When placed in a sugar solution the cells multiply and convert the sugar into alcohol and carbon dioxide. Yeasts are used as fermenting agents in baking, brewing, and the making of wine and spirits. Brewer's yeast (*S. cerevisiae*) is a rich source of vitamin B. (Especially genus *Saccharomyces*; also other related genera.)

yeast artificial chromosome YAC, fragment of †DNA from the human genome inserted into a yeast cell. The yeast replicates the fragment along with its own DNA. In this way the fragments are copied to be preserved in a gene library. YACs are characteristically between 250,000 and 1 million base pairs in length. A †cosmid works in the same way.

yield point or **elastic limit**, stress beyond which a material deforms by a relatively large amount for a small increase in stretching force. Beyond this stress, the material no longer obeys †Hooke's law.

Yixian formation in paleontology, a Chinese geological formation in the rural province of Liaoning that is yielding a wealth of extraordinarily well-preserved fossils. The fossils date from 150 to 120 million years ago (late Jurassic or early Cretaceous) and include those of hundreds of early birds, such as *Confuciusornis*, with feathers, lizards with full skin, and mammals with hair.

ylem hypothetical primordial state of matter – neutrons and their decay products (protons and electrons) – that might have existed before the Big Bang. The term was taken from Aristotle and forms part of the alpha-beta-gamma theory.

yolk store of food, mostly in the form of fats and proteins, found in the †eggs of many animals. It provides nourishment for the growing embryo.

yolk sac sac containing the yolk in the egg of most vertebrates. The term is also used for the membranous sac formed below the developing mammalian embryo and connected with the umbilical cord.

Young, Charles Augustus (1834–1908) US astronomer who made some of the first spectroscopic investigations of the Sun. He was the first person to observe the spectrum of the solar †corona.

Young, J(ohn) Z(achary) (1907–1997) English zoologist who discovered and studied the giant nerve fibers in squids, contributing greatly to knowledge of nerve structure and function. He also did research on the central nervous system of octopuses, demonstrating that memory stores are located in the brain.

Young, Thomas (1773–1829) English physicist, physician, and Egyptologist who revived the wave theory of light and identified the phenomenon of †interference in 1801. He also established many important concepts in mechanics.

ytterbium chemical symbol Yb, soft, lustrous, silvery, malleable, and ductile metallic element of the †lanthanide series, symbol Yb, atomic number 70, relative atomic mass 173.04. It occurs with (and resembles) yttrium in gadolinite and other minerals, and is used in making steel and other alloys.

In 1878 Swiss chemist Jean-Charles de Marignac gave the name ytterbium (after the Swedish town of Ytterby, near where it was found) to what he believed to be a new element. French chemist Georges Urbain (1872–1938) discovered in 1907 that this was in fact a mixture of two elements: ytterbium and lutetium.

yttrium chemical symbol Y, silver-gray, metallic element, atomic number 39, relative atomic mass 88.905. It is associated with and resembles the rare-earth elements (†lanthanides), occurring in gadolinite, xenotime, and other minerals. It is used in color-television tubes and to reduce steel corrosion.

The name derives from the Swedish town of Ytterby, near where it was first discovered in 1788. Swedish chemist Carl Mosander (1797–1858) isolated the element in 1843.

Yukawa, Hideki (1907–1981) Japanese physicist. He was awarded the Nobel Prize for Physics in 1949 for his discovery in 1935 of the †strong nuclear force that binds protons and neutrons together in the atomic nucleus, and his prediction of the existence of the subatomic particle called the †meson.

Z

Zeeman, Pieter (1865–1943) Dutch physicist who discovered in 1896 that when light from certain elements, such as sodium or lithium (when heated), is passed through a spectroscope in the presence of a strong magnetic field, the †spectrum splits into a number of distinct lines. He was awarded the Nobel Prize for Physics in 1902 for his discovery, known as the **Zeeman effect**, of the influence of magnetism on radiation phenomena. He shared the award with Hendrik †Lorentz.

Effect of Magnetization on the Nature of Light Emitted by a Substance

http://dbhs.wvusd.k12.ca.us/Chem-History/Zeeman-effect.html

Excerpt from Pieter Zeeman's paper published in Nature, vol. 55 on 11 February 1897. The effect strong magnetic fields have on light is described in the paper, and it is accompanied here by a photograph Zeeman himself took which illustrates the effect. Also shown is a photograph of Zeeman.

Zeeman effect splitting of †spectral lines of a substance placed in an intense magnetic field.

zeitgeber (German "time giver") in biology, signal that synchronizes †circadian rhythms ensuring that internal and external time are linked. In most species the most important zeitgeber is light and dark changes during the cycle of night and day.

Zener diode solid-state device consisting of a semiconductor diode that is highly doped each side of the junction. Under reverse current it breaks down at a defined potential, making it useful as a voltage regulator.

zenith uppermost point of the celestial horizon, immediately above the observer; the †nadir is below, diametrically opposite. See †celestial sphere.

zeolite any of the hydrous aluminum silicates, also containing sodium, calcium, barium, strontium, or potassium, chiefly found in igneous rocks and characterized by a ready loss or gain of water. Zeolites are used as "molecular sieves" to separate mixtures because they are capable of selective absorption. They have a high ion-exchange capacity and can be used to separate gasoline, benzene, and toluene from low-grade raw materials, such as coal and methanol.

Zermelo, Ernst Friedrich Ferdinand (1871–1953) German mathematician who made important contributions to the development of †set theory, particularly in developing the axiomatic set theory that now bears his name.

Zernike, Frits (1888–1966) Dutch physicist who was awarded the Nobel Prize for Physics in 1953 for his invention of the phase-contrast microscope in 1935. Earlier microscopes allowed many specimens to be examined only after they had been transformed by heavy staining and other treatment. The phase-contrast microscope allowed living cells to be directly observed by making use of the difference in refractive indices between specimens and medium.

zero number (written 0) that when added to any number leaves that number unchanged. It results when any number is subtracted from itself, or when any number is added to its negative. The product of any number with zero is itself zero.

zero matrix in mathematics, †matrix whose entries are all zero.

zeta function function that may be represented as the value of the infinite series:

$\xi(s) = 1 + 1/2^S + 1/3^S + 1/4^S + \ldots$

where s is a complex number. The function was significant in Swiss mathematician Leonhard Euler's study of prime numbers. See also the †Riemann hypothesis.

Z_{He} symbol for the total mass density (in a star) of elements heavier than helium; values of Z_{He} are small for †population II stars, large for †population I stars.

Zhubov scale scale for measuring ice coverage, developed in the former USSR. The unit is the **ball**; one ball is 10% coverage, two balls 20%, and so on.

Ziegler, Karl (1898–1973) German organic chemist. He shared the Nobel Prize for Chemistry in 1963 with Giulio †Natta of Italy for work on the chemistry and technology of large polymers. He combined simple molecules of the gas ethylene (ethene) into the long-chain plastic †polyethene (polyethylene).

Ziegler catalyst catalyst used to produce †stereospecific polymerization of polymers. Ziegler catalysts are prepared by a reaction between an organometallic derivative and a derivative of a transition metal. The catalysts were developed by German chemist Karl †Ziegler.

zinc chemical symbol Zn, (Germanic *Zinke* "point") hard, brittle, bluish-white, metallic element, atomic number 30, relative atomic mass 65.37. The principal ore is sphalerite or zinc blende (zinc sulfide, ZnS). Zinc is hardly affected by air or moisture at ordinary temperatures; its chief uses are in alloys such as brass, in coating metals (for example, galvanized iron), and in making batteries. Its compounds include zinc oxide, used in ointments (as an astringent) and cosmetics, paints, glass, and printing ink.

Zinc is an essential trace element in most animals; adult humans have 2–3 g/0.07–0.1 oz of zinc in their bodies. There are more than 300 known enzymes that contain zinc.

Zinc has been used as a component of brass since the Bronze Age, but it was not recognized in Europe as a separate metal until the 16th century. It was isolated in 1746 by German chemist Andreas Sigismund Marggraf (1709–1782). The name derives from the shape of the crystals on smelting.

zinc ore mineral from which zinc is extracted, principally sphalerite (Zn,Fe)S, but also zincite, ZnO_2, and smithsonite, $ZnCO_3$, all of which occur in mineralized veins. Ores of lead and zinc often occur together, and are common worldwide; Canada, the USA, and Australia are major producers.

zinc oxide ZnO, white powder, yellow when hot, that occurs in nature as the mineral zincite. It is used in paints and as an antiseptic in zinc ointment; it is the main ingredient of calamine lotion.

zinc sulfide ZnS, yellow-white solid that occurs in nature as the mineral sphalerite (also called zinc blende). It is the principal ore of zinc, and is used in the manufacture of fluorescent paints.

zircon zirconium silicate, $ZrSiO_4$, a mineral that occurs in small quantities in a wide range of igneous, sedimentary, and metamorphic rocks. It is very durable and is resistant to erosion and weathering. It is usually colored brown, but can be other colors, and when transparent may be used as a gemstone.

Zircons contain abundant radioactive isotopes of uranium and so are useful for uranium–lead dating to determine the ages of rocks.

zirconium chemical symbol Zr, (Germanic *zircon*, from Persian *zargun* "golden") lustrous, grayish-white, strong, ductile, metallic element, atomic number 40, relative atomic mass 91.22. It occurs in nature as the mineral zircon (zirconium silicate), from which it is obtained commercially. It is used in some ceramics, alloys for wire and filaments, steel manufacture, and nuclear reactors, where its low neutron absorption is advantageous.

It was isolated in 1824 by Swedish chemist Jöns Berzelius. The name was proposed by English chemist Humphry Davy in 1808.

zodiac zone of the heavens containing the paths of the Sun, Moon, and planets. When this was devised by the ancient Greeks, only five planets were known, making the zodiac about 16° wide. In astrology, the zodiac is divided into 12 signs, each 30° in extent: Aries, Taurus, Gemini, Cancer, Leo, Virgo, Libra, Scorpio, Sagittarius, Capricorn, Aquarius, and Pisces. These do not cover the same areas of sky as the astronomical constellations.

The 12 astronomical constellations are uneven in size and do not between them cover the whole zodiac, or even the line of the †ecliptic, much of which lies in the constellation of †Ophiuchus.

zodiacal light cone-shaped light sometimes seen extending from the Sun along the †ecliptic (that is, the path that the Sun appears to follow each year as it is orbited by Earth), visible after sunset or before sunrise. It is due to thinly spread dust particles in the central plane of the Solar System. It is very faint, and requires a dark, clear sky to be seen.

The dust is thought to be produced mainly by colliding asteroids and active comets, though a small amount probably originates from outside the Solar System. Astronomers estimate that some 10,000 tonnes/11,023 tons of this dust hits the Earth each year.

zoidogamy type of plant reproduction in which male gametes (antherozoids) swim in a film of water to the female gametes. Zoidogamy is found in algae, bryophytes, pteridophytes, and some gymnosperms (others use siphonogamy).

zone of avoidance apparent lack of distant †galaxies in the plane of our own Galaxy, now explained as being caused by optical interference of dust and interstellar debris on the rim of the Galaxy.

zoology branch of biology concerned with the study of animals. It includes any aspect of the study of animal form and function – description of present-day animals, the study of evolution of animal forms, †anatomy, †physiology, †embryology, behavior, and geographical distribution.

Cryptozoology http://www.ncf.carleton.ca/~bz050/HomePage.cryptoz.html

Fascinating page, with an introduction to cryptozoology, sections on personalities in the field, and the latest news. This site also contains a list of many cryptozoological publications.

Z particle in physics, an †elementary particle, one of the †intermediate vector bosons responsible for carrying the †weak nuclear force. The Z particle is neutral.

Zsigmondy, Richard Adolf (1865–1929) Austrian-born German chemist who devised and built an ultramicroscope in 1903. The microscope's illumination was placed at right angles to the axis. (In a conventional microscope the light source is placed parallel to the instrument's axis.) Zsigmondy's arrangement made it possible to observe particles with a diameter of one 10-millionth of a millimeter. He was awarded the Nobel Prize for Chemistry in 1925 for the elucidation of heterogeneity of colloids.

Zuckerman, Solly (1904–1993) Baron Zuckerman of Burnham Thorpe, South African-born British zoologist, educationalist, and establishment figure. He did extensive research on primates, publishing a number of books that became classics in their field, including *The Social Life of Monkeys and Apes* (1932) and *Functional Affinities of Man, Monkeys and Apes* (1933). He was chief scientific adviser to the British government 1964–71

Zuse, Konrad (1910–1995) German inventor of the world's first binary digital computer. Zuse worked on building computers in Berlin, Germany, from 1935 onwards, starting in his parents' living room. He built the first binary digital computer, the Z1, in 1938, and in 1941 completed the world's first program-controlled electromechanical digital computer, the Z3.

Zweig, George (1937–) Russian-born US theoretical physicist who, along with US physicist Murray Gell-Mann, developed the quark theory. They developed a theory based on three quarks as the fundamental building blocks of the hadrons.

Zwicky, Fritz (1898–1974) Swiss astronomer. He predicted the existence of †neutron stars in 1934. He discovered 18 supernovae and determined that cosmic rays originate in them.

zwitterion ion that has both a positive and a negative charge, such as an †amino acid in neutral solution. For example, glycine contains both a basic amino group (NH_2) and an acidic carboxyl group (COOH); when these are both ionized in aqueous solution, the acid group loses a proton to the amino group, and the molecule is positively charged at one end and negatively charged at the other.

zygomatic arch arch of bone on each side of the face, just below the eyes.

zygote †ovum (egg) after †fertilization but before it undergoes cleavage to begin embryonic development.

APPENDIX

Contents

MAJOR AIR POLLUTANTS

Air pollution is contamination of the atmosphere caused by the discharge, accidental or deliberate, of a wide range of toxic airborne substances. Often the amount of the released substance is relatively high in a certain locality, so the harmful effects become more noticeable. The cost of preventing any discharge of pollutants into the air is prohibitive, so attempts are more usually made to reduce the amount of discharge gradually and to disperse it as quickly as possible by using a very tall chimney, or by intermittent release.

Pollutant	Sources	Effects
Sulphur dioxide (SO2)	oil, coal combustion in power stations	acid rain formed, which damages plants, trees, buildings, and lakes
Oxides of nitrogen (NO, NO2)	high-temperature combustion in cars, and to some extent power stations	acid rain formed
Lead compounds	from leaded petrol used by cars	nerve poison
Carbon dioxide (CO2)	oil, coal, petrol, diesel combustion	greenhouse effect
Carbon monoxide (CO)	incomplete combustion of oil, coal, petrol, diesel fuels	poisonous, leads to photochemical smog in some areas
Nuclear waste	nuclear power plants, nuclear weapon testing, war	radioactivity, contamination of locality, cancers, mutations, death
Ozone (O3)	formed by the reaction between sunlight and NOx compounds	greenhouse effect, acid rain by-product
Particulates (PM10, 'smog')	vehicle exhausts, power generation, combustion processes	respiratory effects, acid rain
Volatile organic compounds (VOC)	industrial processes, vehicle emissions, petrol evaporation	greenhouse effect, ozone production

COMMON ALLOYS

Name	Approximate composition	Uses
brass	35–10% zinc, 65–90% copper	decorative metalwork, plumbing fittings, industrial tubing
bronze – common	2% zinc, 6% tin, 92% copper	machinery, decorative work
bronze – aluminium	10% aluminium, 90% copper	machinery castings
bronze – coinage	1% zinc, 4% tin, 95% copper	coins
cast iron	2–4% carbon, 96–98% iron	decorative metalwork, engine blocks, industrial machinery
dentist's amalgam	30% copper, 70% mercury	dental fillings
duralumin	0.5% magnesium, 0.5% manganese, 5% copper, 94% aluminium	framework of aircraft
gold – coinage	10% copper, 90% gold	coins
gold – dental	14–28% silver, 14–28% copper, 58% gold	dental fillings
lead battery plate	6% antimony, 94% lead	car batteries
manganin	1.5% nickel, 16% manganese, 82.5% copper	resistance wire
nichrome	20% chromium, 80% nickel	heating elements
pewter	20% lead, 80% tin	utensils
silver – coinage	10% copper, 90% silver	coins
solder	50% tin, 50% lead	joining metals
steel – stainless	8–20% nickel, 10–20% chromium, 60–80% iron	kitchen utensils
steel – armour	1–4% nickel, 0.5–2% chromium, 95–98% iron	armour plating
steel – tool	2–4% chromium, 6–7% molybdenum, 90–95% iron	tools

ASCII CODES

Character	Binary (base 2)	Decimal (base 10)	Hexadecimal (base 16)
Space	00100000	32	20
!	00100001	33	21
"	00100010	34	22
#	00100011	35	23
$	00100100	36	24
%	00100101	37	25
&	00100110	38	26
`	00100111	39	27
(	00101000	40	28
)	00101001	41	29
*	00101010	42	2A
+	00101011	43	2B
,	00101100	44	2C
-	00101101	45	2D
.	00101110	46	2E
/	00101111	47	2F
0	00110000	48	30
1	00110001	49	31
2	00110010	50	32
3	00110011	51	33
4	00110100	52	34
5	00110101	53	35
6	00110110	54	36
7	00110111	55	37
8	00111000	56	38
9	00111001	57	39
:	00111010	58	3A
;	00111011	59	3B
<	00111100	60	3C
=	00111101	61	3D
>	00111110	62	3E
?	00111111	63	3F
@	01000000	64	40
A	01000001	65	41
B	01000010	66	42
C	01000011	67	43
D	01000100	68	44
E	01000101	69	45
F	01000110	70	46
G	01000111	71	47
H	01001000	72	48
I	01001001	73	49
J	01001010	74	4A
K	01001011	75	4B
L	01001100	76	4C
M	01001101	77	4D
N	01001110	78	4E
O	01001111	79	4F
P	01010000	80	50
Q	01010001	81	51
R	01010010	82	52
S	01010011	83	53
T	01010100	84	54
U	01010101	85	55
V	01010110	86	56
W	01010111	87	57
X	01011000	88	58
Y	01011001	89	59
Z	01011010	90	5A
[	01011011	91	5B
\	01011100	92	5C
]	01011101	93	5D
^	01011110	94	5E
_	01011111	95	5F
'	01100000	96	60
a	01100001	97	61
b	01100010	98	62
c	01100011	99	63
d	01100100	100	64
e	01100101	101	65
f	01100110	102	66
g	01100111	103	67
h	01101000	104	68
i	01101001	105	69
j	01101010	106	6A
k	01101011	107	6B
l	01101100	108	6C
m	01101101	109	6D
n	01101110	110	6E
o	01101111	111	6F
p	01110000	112	70
q	01110001	113	71
r	01110010	114	72
s	01110011	115	73
t	01110100	116	74
u	01110101	117	75
v	01110110	118	76
w	01110111	119	77
x	01111000	120	78
y	01111001	121	79
z	01111010	122	7A
{	01111011	123	7B
|	01111100	124	7C
}	01111101	125	7D
~	01111110	126	7E
Delete	01111111	127	7F

LARGEST ASTEROIDS

An asteroid is a small body, composed of rock and iron, that orbits the Sun. Most lie in a belt between the orbits of Mars and Jupiter, and are thought to be fragments left over from the formation of the Solar System.

Name	Diameter km	Diameter mi	Average distance from Sun (Earth = 1)	Orbital period (years)
Ceres	940	584	2.77	4.6
Pallas	588	365	2.77	4.6
Vesta	576	358	2.36	3.6
Hygeia	430	267	3.13	5.5
Interamnia	338	210	3.06	5.4
Davida	324	201	3.18	5.7

THE BEAUFORT SCALE

The Beaufort scale is a system of recording wind velocity (speed) devised in 1806 by Francis Beaufort (1774–1857). It is a numerical scale ranging from 0 to 12, calm being indicated by 0 and a hurricane by 12. The scale received international recognition in 1874. Measurements are made at 10 metres/33 feet above ground level.

Force	Description	Features	Air speed kph	Air speed mph
0	calm	smoke rises vertically; water smooth	0–2	0–1
1	light air	smoke shows wind direction; water ruffled	2–5	1–3
2	light breeze	leaves rustle; wind felt on face	6–11	4–7
3	gentle breeze	loose paper blows around	12–19	8–12
4	moderate breeze	branches sway	20–29	13–18
5	fresh breeze	small trees sway, leaves blown off	30–39	19–24
6	strong breeze	whistling in telephone wires; sea spray from waves	40–50	25–31
7	near gale	large trees sway	51–61	32–38
8	gale	twigs break from trees	62–74	39–46
9	strong gale	branches break from trees	75–87	47–54
10	storm	trees uprooted; weak buildings collapse	88–101	55–63
11	violent storm	widespread damage	102–117	64–73
12	hurricane	widespread structural damage	above 118	above 74

BIOLOGY: KEY DATES

c. 500 BC	First studies of the structure and behaviour of animals, by the Alcmaeon of Croton.
c. 450	Hippocrates of Kos undertakes the first detailed studies of human anatomy.
c. 350	Aristotle lays down the basic philosophy of the biological sciences and outlines a theory of evolution.
c. 300	Theophrastus carries out the first detailed studies of plants.
c. AD 175	Galen establishes the basic principles of anatomy and physiology.
c. 1500	Leonardo da Vinci studies human anatomy to improve his drawing ability and produces detailed anatomical drawings.
1628	William Harvey describes the circulation of the blood and the function of the heart as a pump.
1665	Robert Hooke uses a microscope to describe the cellular structure of plants. This is the first description of cells.
1672	Marcello Malpighi undertakes the first studies in embryology by describing the development of a chicken egg.
1677	Anton van Leeuwenhoek greatly improves the microscope and uses it to describe spermatozoa as well as many micro-organisms.
1736	Carolus Linnaeus (Carl von Linné) publishes his systematic classification of plants, so establishing taxonomy.
1768–79	James Cook's voyages of discovery in the Pacific reveal a great diversity of living species, prompting the development of theories to explain their origin.
1796	Edward Jenner establishes the practice of vaccination against smallpox, laying the foundations for theories of antibodies and immune reactions.
1809	Jean-Baptiste Lamarck advocates a theory of evolution through inheritance of acquired characteristics.
1812	Mary Anning discovers the fossilized skeleton of an ichthyosaurus, one of the first finds proving that huge reptiles once dominated the Earth. The discovery paves the way for a new understanding of the Earth's remote past, challenging the Biblical account of creation.
1822	Gideon and Anne Mantell discover the fossil of an iguanadon.
1839	Theodor Schwann proposes that all living matter is made up of cells.
1842	Richard Owen coins the word 'dinosaur' (from the Greek for terrible lizard) to describe the animals belonging to what we now know as two groups of reptiles that dominated the Earth for 175 million years during the Triassic, Jurassic and Cretaceous periods.
1857	Louis Pasteur establishes that micro-organisms are responsible for fermentation, creating the discipline of microbiology.
1859	Charles Darwin publishes 'On the Origin of Species', expounding his theory of the evolution of species by natural selection.
1865	Gregor Mendel pioneers the study of inheritance with his experiments on peas, but achieves little recognition.
1869	Alfred Russell Wallace, who arrived at a theory of the origin of species independently of Darwin, describes the differences between the flora and fauna in the western and eastern parts of the Malay archpelago. Those in the west are related to plants and animals in Asia, those of the east are related to species found in Australia. An imaginary line, called Wallace's Line, divides the two populations.
1883	August Weismann proposes his theory of the continuity of the germ plasm.
1900	Mendel's work is rediscovered and the science of genetics founded.
1910	T H Morgan discovers that genes are physically located on chromosomes within cell nuclei.
1935	Konrad Lorenz publishes the first of many major studies of animal behaviour, creating the discipline of ethology.
1944	Oswald Avery and co-workers discover that genes consist of DNA.
1953	James Watson and Francis Crick describe the molecular structure of DNA.

Biology: Key Dates continued

1964 William Hamilton recognizes the importance of inclusive fitness, so paving the way for the development of sociobiology.

1969 Jonathan Beckwith and co-workers isolate a single gene for the first time.

1973 Stanley H Cohen and Herbert W Boyer demonstrate the use of restrictive enzymes for slicing up DNA.

1975 Discovery of endogenous opiates (the brain's own painkillers) opens up a new phase in the study of brain chemistry.

1976 Har Gobind Khorana and his colleagues construct the first artificial gene to function naturally when inserted into a bacterial cell, a major step in genetic engineering.

1982 Gene databases are established at Heidelberg, Germany, for the European Molecular Biology Laboratory, and at Los Alamos, USA, for the US National Laboratories.

1985 The first human cancer gene, retinoblastoma, is isolated by researchers at the Massachusetts Eye and Ear Infirmary and the Whitehead Institute, Massachusetts, USA.

1988 The Human Genome Organization (HUGO) is established in Washington, DC, USA, with the aim of mapping the complete sequence of human DNA.

1991 Biosphere 2, an experiment that attempts to reproduce the world's biosphere in miniature within a sealed glass dome, is launched in Arizona, USA.

1992 Researchers at the University of California, USA, stimulate the multiplication of isolated brain cells of mice, overturning the axiom that mammalian brains cannot produce replacement cells once birth has taken place. The world's largest organism, a honey fungus with underground hyphae (filaments) spreading across 600 hectares/1,480 acres, is discovered in Washington State, USA.

1994 Scientists from Pakistan and the USA unearth a 50-million-year-old fossil whale with hind legs that would have enabled it to walk on land.

1996 The sequencing of the genome of brewer's yeast Saccharomyces cerevisiae is completed, the first time this has been achieved for an organism more complex than a bacterium. The 12 million base pairs of genes took 300 scientists six years to map. A new muscle is discovered by two US dentists. It is 3 cm/1 in long, and runs from the jaw to behind the eye socket.

1997 The first mammal to be cloned from a nonreproductive cell is born. The lamb (named Dolly) has been cloned from an udder cell from a six-year-old ewe.

1999 Canadian researchers engineer an artificial chromosome that can be inserted into mammal cells and then transmitted from one generation to the next. The development has important implications towards germ-line therapy where a defect is corrected in the gametes and the change passed on to future generations.

2000 Scientists complete a rough working draft of the map of the entire human genome.

BOILING POINTS OF SELECTED LIQUIDS

Liquid	Boiling point °C	°F	Liquid	Boiling point °C	°F
Methanol	64.7	148.5	Octane	125.6	258.1
Ethanol	78.2	172.8	Turpentine	156.0	312.8
Benzene	80.1	176.2	Glycerol	290.0	554.0
Water	100.0	212.0	Mercury	356.9	674.4

BONES OF THE HUMAN BODY

Bone	Number
Cranium (skull)	
Occipital	1
Parietal: 1 pair	2
Sphenoid	1
Ethmoid	1
Inferior nasal conchae	2
Frontal: 1 pair, fused	1
Nasal: 1 pair	2
Lacrimal: 1 pair	2
Temporal: 1 pair	2
Maxilla: 1 pair, fused	1
Zygomatic: 1 pair	2
Vomer	1
Palatine: 1 pair	2
Mandible (jawbone): 1 pair, fused	1
Total	**21**
Ear	
Malleus (hammer)	1
Incus (anvil)	1
Stapes (stirrups)	1
Total (2 _ 3)	**6**
Vertebral column (spine)	
Cervical vertebrae	7
Thoracic vertebrae	12
Lumbar vertebrae	5
Sacral vertebrae: 5, fused to form the sacrum	1
Coccygeal vertebrae: between 3 and 5, fused to form the coccyx	1
Total	**26**
Ribs	
Ribs, 'true': 7 pairs	14
Ribs, 'false': 5 pairs, of which 2 pairs are floating	10
Total	**24**

Bone		Number
Sternum (breastbone)		
Manubrium, sternebrae ('body'), and xiphisternum		3
Total		**3**
Throat		
Hyoid		1
Total		**1**
Pectoral girdle		
Clavicle: 1 pair (collarbone)		2
Scapula (including coracoid): 1 pair (shoulder-blade)		2
Total		**4**
Upper extremity (each arm)		
Forearm	humerus	1
	radius	1
	ulna	1
Carpus (wrist)	scaphoid	1
	lunate	1
	triquetral	1
	pisiform	1
	trapezium	1
	trapezoid	1
	capitate	1
	hamate	1
	metacarpals	5
Phalanges (fingers)	first digit	2
	second digit	3
	third digit	3
	fourth digit	3
	fifth digit	3
Total (2 _ 30)		**60**

Bone		Number
Pelvic girdle		
Ilium, ischium, and pubis (combined): 1 pair of hip bones, innominate		2
Total		**2**
Lower extremity (each leg)		
Leg	femur (thighbone)	1
	tibia (shinbone)	1
	fibula	1
	patella (kneecap)	1
Tarsus (ankle)	talus	1
	calcaneus	1
	navicular	1
	cuneiform, medial	1
	cuneiform, intermediate	1
	cuneiform, lateral	1
	cuboid	1
	metatarsals (foot bones)	5
Phalanges (toes)(toes)	first digit	2
	second digit	3
	third digit	3
	fourth digit	3
	fifth digit	3
Total (2 _ 30)		**60**
TOTAL		**207**

CHEMISTRY: KEY EVENTS

c. 3000 BC	Egyptians begin producing bronze – an alloy of copper and tin.
c. 450 BC	Empedocles proposes that all substances are made up of a combination of four elements – earth, air, fire, and water – an idea that is developed by Plato and Aristotle and persists for over 2,000 years.
c. 400 BC	Democritus theorizes that matter consists ultimately of tiny, indivisible particles, atomoi. The Greek word atomos means uncuttable.
AD 1	Gold, silver, copper, lead, iron, tin, and mercury are known.
200	The techniques of solution, filtration, and distillation are known.
7th–17th centuries	Chemistry is dominated by alchemy, the attempt to transform nonprecious metals such as lead and copper into gold. Though misguided, it leads to the discovery of many new chemicals and techniques.
12th century	Alcohol is first distilled in Europe.
1242	Gunpowder is introduced to Europe from the Far East.
1620	The scientific method of reasoning is expounded by Francis Bacon in his Novum Organum.
1650	Leyden University in the Netherlands sets up the first chemistry laboratory.
1661	Robert Boyle defines an element as any substance that cannot be broken down into still simpler substances and asserts that matter is composed of 'corpuscles' (atoms) of various sorts and sizes, capable of arranging themselves into groups, each of which constitutes a chemical substance.
1662	Boyle describes the inverse relationship between the volume and pressure of a fixed mass of gas (Boyle's law).
1697	Georg Stahl proposes the erroneous theory that combustible materials are rich in a substance called phlogiston, which is released when they burn.
1755	Joseph Black discovers carbon dioxide.
1774	Joseph Priestley discovers oxygen, which he calls 'dephlogisticated air'. Antoine Lavoisier demonstrates his law of conservation of mass.
1777	Lavoisier shows air to be made up of a mixture of gases, and shows that one of these – oxygen – is the substance necessary for combustion (burning) and rusting to take place.
1781	Henry Cavendish shows water to be a compound.
1792	Alessandro Volta demonstrates the electrochemical series.
1807	Humphry Davy passes an electric current through molten compounds (the process of electrolysis) in order to isolate elements, such as potassium, that have never been separated by chemical means. Jöns Berzelius proposes that chemicals produced by living creatures should be termed 'organic'.
1808	John Dalton publishes his atomic theory. It states that every element consists of similar indivisible particles – called atoms – which differ from the atoms of other elements in their mass. Dalton also draws up a list of relative atomic masses. Joseph Gay-Lussac announces that the volumes of gases that combine chemically with one another are in simple ratios.
1811	Amedeo Avogadro's hypothesis on the relation between the volume and number of molecules of a gas, and its temperature and pressure, is published.
1813–14	Berzelius devises the chemical symbols and formulae still used to represent elements and compounds.
1828	Franz Wöhler converts ammonium cyanate into urea – the first synthesis of an organic compound from an inorganic substance.
1832–33	Michael Faraday expounds the laws of electrolysis, and adopts the term 'ion' for the particles believed to be responsible for carrying current.
1846	Thomas Graham expounds his law of diffusion.
1853	Robert Bunsen invents the Bunsen burner.
1858	Stanislao Cannizzaro differentiates between atomic and molecular weights (masses).
1861	Organic chemistry is defined by German chemist Friedrich Kekulé as the chemistry of carbon compounds.
1864	John Newlands devises the first periodic table of the elements.
1869	Dmitri Mendeleyev expounds his periodic table of the elements (based on atomic mass), leaving gaps for elements that are predicted but have not yet been discovered.

Chemistry: Key Events continued

1874 Jacobus van't Hoff suggests that the four bonds of carbon are arranged tetrahedrally, and that carbon compounds can therefore be three-dimensional and asymmetric.

1884 Svante Arrhenius suggests that electrolytes (solutions or molten compounds that conduct electricity) dissociate into ions, atoms, or groups of atoms that carry a positive or negative charge.

1894 William Ramsey and Lord Rayleigh discover the first inert gas, argon.

1897 The electron is discovered by J J Thomson.

1901 Mikhail Tsvet invents paper chromatography as a means of separating pigments.

1909 Sören Sörensen devises the pH scale of acidity-alkalinity.

1912 Max von Laue shows crystals to be composed of regular, repeating arrays of atoms by studying the patterns in which they diffract X-rays.

1913–14 Henry Moseley equates the atomic number of an element with the positive charge on its nuclei, and draws up the periodic table, based on atomic number, that is used today.

1916 Gilbert Newton Lewis explains covalent bonding between atoms as a sharing of electrons.

1927 Nevil Sidgwick publishes his theory of valency, based on the numbers of electrons in the outer shells of the reacting atoms.

1930 Electrophoresis, which separates particles in suspension in an electric field, is invented by Arne Tiselius.

1932 Deuterium (heavy hydrogen), an isotope of hydrogen, is discovered by Harold Urey.

1940 Edwin McMillan and Philip Abelson show that new elements with a higher atomic number than uranium can be formed by bombarding uranium with neutrons, and synthesize the first transuranic element, neptunium.

1942 Plutonium is first synthesized by Glenn T Seaborg and Edwin McMillan.

1950 Derek Barton deduces that some properties of organic compounds are affected by the orientation of their functional groups (the study of which becomes known as conformational analysis).

1954 Einsteinium and fermium are synthesized.

1955 Ilya Prigogine describes the thermodynamics of irreversible processes (the transformations of energy that take place in, for example, many reactions within living cells).

1962 Neil Bartlett prepares the first compound of an inert gas, xenon hexafluoroplatinate; it was previously believed that inert gases could not take part in a chemical reaction.

1965 Robert B Woodward synthesizes complex organic compounds.

1981 Quantum mechanics is applied to predict the course of chemical reactions by US chemist Roald Hoffmann and Kenichi Fukui of Japan.

1985 Fullerenes, a new class of carbon solids made up of closed cages of carbon atoms, are discovered by Harold Kroto and David Walton at the University of Sussex, England.

1987 Donald Cram, Charles Pederson, and Jean-Marie Lehn create artificial molecules that mimic the vital chemical reactions of life processes.

1990 Jean-Marie Lehn, Ulrich Koert, and Margaret Harding report the synthesis of a new class of compounds, called nucleohelicates, that mimic the double helical structure of DNA, turned inside out.

1993 Chemists at the University of California and the Scripps Institute synthesize rapamycin, one of a group of complex, naturally occurring antibiotics and immunosuppressants that are being tested as anticancer agents.

1995 German chemists build the largest ever wheel molecule, made up of 154 molybdenum atoms surrounded by oxygen atoms.It has a relative molecular mass of 24,000 and is soluble in water.

1997 The International Union of Pure and Applied Chemistry (IUPAC) officially names elements 104–109. Element 104 (discovered in Russia in 1964 and the USA in 1969) is named rutherfordium, element 105 (discovered in Russia in 1967 and the USA in 1970) is named dubnium, element 106 (discovered in 1973) is named seaborgium, element 107 (discovered in 1976) is named bohrium, element 108 (discovered in 1984) is named hassium, and element 109 (discovered in 1982) is named meitnerium.

1999 Russian scientists at the Institute of Nuclear Research at Dubna create element 114 by colliding isotopes calcium 48 and plutonium 44. Shortly afterwards, US physicists create element 118, which decays into another new element, 116, by bombarding lead with krypton.

SOME MAJOR COMETS

A comet is a small, icy body orbiting the Sun, usually on a highly elliptical path. Comets consist of a central nucleus a few kilometres across, and are often likened to dirty snowballs because they consist mostly of ice mixed with dust. As a comet approaches the Sun, it warms up, releasing gas and dust to form a coma (halo). Many comets have tails – these always point away from the Sun and are made up of gas and dust particles.
(– = not known or not applicable.)

Name	First recorded sighting	Orbital period (yrs)	Interesting facts
Comet Halley	240 BC	76	parent of Eta Aquarid and Orionid meteors
Comet Tempel-Tuttle	1366	33	parent of Leonid meteors
Biela's comet	1772	6.6	broke in half in 1846; not seen since 1852
Encke's comet	1786	3.3	parent of Taurid meteors
Comet Swift-Tuttle	1862	130	parent of Perseid meteors; reappeared 1992
Comet Ikeya-Seki	1965	880	so-called 'Sun-grazing' comet, passed 500,000 km/300,000 mi above surface of the Sun on 21 October 1965
Comet Kohoutek	1973	–	observed from space by Skylab astronauts
Comet West	1975	500,000	nucleus broke into four parts
Comet Bowell	1980	–	ejected from Solar System after close encounter with Jupiter
Comet IRAS-Araki-Alcock	1983	–	passed only 4.5 million km/2.8 million mi from the Earth on 11 May 1983
Comet Austin	1989	–	passed 32 million km/20 million mi from the Earth in 1990
Comet Shoemaker-Levy 9	1993	–	made up of 21 fragments; crashed into Jupiter in July 1994
Comet Hale-Bopp	1995	3,500	produces a coma, a surrounding hazy cloud of gas and dust, of greater volume than the Sun; the bright coma is due to an outgassing of carbon monoxide; clearly visible with the naked eye in March 1997
Comet Hyakutake	1996	–	passed 15 million km/9.3 million mi from the Earth in 1996; has the longest-known comet tail (550 million km/342 million mi)

CONSTELLATIONS

A constellation is one of the 88 areas into which the sky is divided for the purposes of identifying and naming celestial objects. The first constellations were simple, arbitrary patterns of stars in which early civilizations visualized gods, sacred beasts, and mythical heroes. (– = not applicable.)

Constellation	Abbreviation	Popular name
Andromeda	And	–
Antlia	Ant	Airpump
Apus	Aps	Bird of Paradise
Aquarius	Aqr	Water-bearer
Aquila	Aqi	Eagle
Ara	Ara	Altar
Aries	Ari	Ram
Auriga	Aur	Charioteer
Boötes	Boo	Herdsman
Caelum	Cae	Chisel
Camelopardalis	Cam	Giraffe
Cancer	Cnc	Crab
Canes Venatici	CVn	Hunting Dogs
Canis Major	CMa	Great Dog
Canis Minor	CMi	Little Dog
Capricornus	Cap	Goat
Carina	Car	Keel
Cassiopeia	Cas	–
Centaurus	Cen	Centaur
Cepheus	Cep	–
Cetus	Cet	Sea Monster, Whale
Chamaeleon	Cha	Chameleon
Circinus	Cir	Compasses
Columba	Col	Dove
Coma Berenices	Com	Berenice's Hair
Corona Australis	CrA	Southern Crown
Corona Borealis	CrB	Northern Crown
Corvus	Crv	Crow
Crater	Crt	Cup
Crux	Cru	Southern Cross
Cygnus	Cyn	Swan
Delphinus	Del	Dolphin
Dorado	Dor	Goldfish
Draco	Dra	Dragon
Equuleus	Equ	Foal
Eridanus	Eri	River
Fornax	For	Furnace
Gemini	Gem	Twins
Grus	Gru	Crane
Hercules	Her	–
Horologium	Hor	Clock
Hydra	Hya	Watersnake
Hydrus	Hyi	Little Snake
Indus	Ind	American Indian

Constellation	Abbreviation	Popular name
Lacerta	Lac	Lizard
Leo	Leo	Lion
Leo Minor	LMi	Little Lion
Lepus	Lep	Hare
Libra	Lib	Balance
Lupus	Lup	Wolf
Lynx	Lyn	–
Lyra	Lyr	Harp
Mensa	Men	Table
Microscopium	Mic	Microscope
Monoceros	Mon	Unicorn
Musca	Mus	Fly
Norma	Nor	Level, Square
Octans	Oct	Octant
Ophiuchus	Oph	Serpent-bearer
Orion	Ori	–
Pavo	Pav	Peacock
Pegasus	Peg	Flying Horse
Perseus	Per	–
Phoenix	Phe	Phoenix
Pictor	Pic	Painter
Pisces	Psc	Fishes
Piscis Austrinus	PsA	Southern Fish
Puppis	Pup	Poop, Stern
Pyxis	Pyx	Compass
Reticulum	Ret	Net
Sagitta	Sge	Arrow
Sagittarius	Sgr	Archer
Scorpius	Sco	Scorpion
Sculptor	Scl	Sculptor's Tools
Scutum	Sct	Shield
Serpens	Ser	Serpent
Sextans	Sex	Sextant
Taurus	Tau	Bull
Telescopium	Tel	Telescope
Triangulum	Tri	Triangle
Triangulum Australe	TrA	Southern Triangle
Tucana	Tuc	Toucan
Ursa Major	UMa	Great Bear
Ursa Minor	UMi	Little Bear
Vela	Vel	Sails
Virgo	Vir	Virgin
Volans	Vol	Flying Fish
Vulpecula	Vul	Fox

METRIC CONVERSION FACTORS

Standard to metric			Metric to standard	
To convert from	**Multiply by**		**To convert from**	**Multiply by**
		Length		
inches to millimetres	25.4		millimetres to inches	0.0393701
feet to metres	0.3048		metres to feet	3.28084
yards to metres	0.9144		metres to yards	1.09361
furlongs to kilometres	0.201168		kilometres to furlongs	4.97097
miles to kilometres	1.609344		kilometres to miles	0.621371
		Area		
square inches to square centimetres	6.4516		square centimetres to square inches	0.1550
square feet to square metres	0.092903		square metres to square feet	10.7639
square yards to square metres	0.836127		square metres to square yards	1.19599
square miles to square kilometres	2.589988		square kilometres to square miles	0.386102
acres to square metres	4,046.856422		square metres to acres	0.000247
acres to hectares	0.404685		hectares to acres	2.471054

Imperial and Metric Conversion Factors continued

To convert from	Multiply by		To convert from	Multiply by
		Volume/capacity		
cubic inches to cubic centimetres	16.387064		cubic centimetres to cubic inches	0.061024
cubic feet to cubic metres	0.028317		cubic metres to cubic feet	35.3147
cubic yards to cubic metres	0.764555		cubic metres to cubic yards	1.30795
cubic miles to cubic kilometres	4.1682		cubic kilometres to cubic miles	0.239912
fluid ounces (imperial) to millilitres	28.413063		millilitres to fluid ounces (imperial)	0.035195
fluid ounces (US) to millilitres	29.5735		millilitres to fluid ounces (US)	0.033814
pints (imperial) to litres	0.568261		litres to pints (imperial)	1.759754
pints (US) to litres	0.473176		litres to pints (US)	2.113377
quarts (imperial) to litres	1.136523		litres to quarts (imperial)	0.879877
quarts (US) to litres	0.946353		litres to quarts (US)	1.056688
gallons (imperial) to litres	4.54609		litres to gallons (imperial)	0.219969
gallons (US) to litres	3.785412		litres to gallons (US)	0.364172
		Mass/weight		
ounces to grams	28.349523		grams to ounces	0.035274
pounds to kilograms	0.453592		kilograms to pounds	2.20462
stones (14 lb) to kilograms	6.350293		kilograms to stones (14 lb)	0.157473
tons (imperial) to kilograms	1,016.046909		kilograms to tons (imperial)	0.000984
tons (US) to kilograms	907.18474		kilograms to tons (US)	0.001102
tons (imperial) to metric tonnes	1.016047		metric tonnes to tons (imperial)	0.984207
tons (US) to metric tonnes	0.907185		metric tonnes to tons (US)	1.10231
		Speed		
miles per hour to kilometres per hour	1.609344		kilometres per hour to miles per hour	0.621371
feet per second to metres per second	0.3048		metres per second to feet per second	3.28084
		Force		
pounds-force to newtons	4.44822		newtons to pounds-force	0.224809
kilograms-force to newtons	9.80665		newtons to kilograms-force	0.101972

Metric Conversion Factors continued

Standard to metric To convert from	Multiply by	Metric to standard To convert from	Multiply by
Pressure			
pounds-force per square inch to kilopascals	6.89476	kilopascals to pounds-force per square inch	0.145038
tons-force per square inch (imperial) to megapascals	15.4443	megapascals to tons-force per square inch (imperial)	0.064779
atmospheres to newtons per square centimetre	10.1325	newtons per square centimetre to atmospheres	0.098692
atmospheres to pounds-force per square inch	14.695942	pounds-force per square inch to atmospheres	0.068948
Energy			
calories to joules	4.1868	joules to calories	0.238846
watt hours to joules	3,600	joules to watt hours	0.000278
Power			
horsepower to kilowatts	0.7457	kilowatts to horsepower	1.34102
Fuel consumption			
miles per gallon (imperial) to kilometres per litre	0.3540	kilometres per litre to miles per gallon (imperial)	2.824859
miles per gallon (US) to kilometres per litre	0.4251	kilometres per litre to miles per gallon (US)	2.3521
gallons per mile (imperial) to litres per kilometre	2.824859	litres per kilometre to gallons per mile (imperial)	0.3540
gallons per mile (US) to litres per kilometre	2.3521	litres per kilometre to gallons per mile (US)	0.4251

THE DECIBEL SCALE

The decibel (dB) scale is used primarily to measure relative sound intensities. It is used for audibility measurements because one decibel, representing an increase of about 25%, is about the smallest change the human ear can detect. An increase of 10 decibels is equivalent to a ten-fold increase in sound intensity or power. Decibels are also used to compare electric voltages or currents.

Decibels	Typical sound	Decibels	Typical sound
0	threshold of human hearing	65–90	train
10	quiet whisper	75–80	factory (light/medium work)
20	average whisper	90	heavy traffic
20–50	quiet conversation	90–100	thunder
40–45	light traffic	110–140	jet aircraft at take-off
50–65	loud conversation	130	threshold of pain for humans
65–70	traffic on busy street	140–190	space rocket at take-off

DENSITIES OF SOME COMMON SUBSTANCES

Substance	Density in kg per cu m	Substance	Density in kg per cu m	Substance	Density in kg per cu m
Solids		**Liquids**		**Gases**	
balsa wood	200	water	1,000	(at standard temperature and pressure of 0°C and 1 atm)	
oak	700	petrol, kerosene	800	air	1.30
butter	900	olive oil	900	hydrogen	0.09
ice	920	milk	1,030	helium	0.18
ebony	1,100	sea water	1,030	methane	0.72
sand (dry)	1,600	glycerine	1,260	nitrogen	1.25
concrete	2,400	Dead Sea brine	1,800	oxygen	1.43
aluminium	2,700			carbon dioxide	1.98
steel	7,800			propane	2.02
copper	8,900			butane (iso)	2.60
lead	11,300				
uranium	19,000				

LARGEST DESERTS IN THE WORLD

Desert	Location	Area[1] sq km	sq mi
Sahara	northern Africa	9,065,000	3,500,000
Gobi	Mongolia/northeastern China	1,295,000	500,000
Patagonian	Argentina	673,000	260,000
Rub al-Khali	southern Arabian peninsula	647,500	250,000
Kalahari	southwestern Africa	582,800	225,000
Chihuahuan	Mexico/southwestern USA	362,600	140,000
Taklimakan	northern China	362,600	140,000
Great Sandy	northwestern Australia	338,500	130,000
Great Victoria	southwestern Australia	338,500	130,000
Kyzyl Kum	Uzbekistan/Kazakhstan	259,000	100,000
Thar	India/Pakistan	219,000	84,556
Sonoran	Mexico/southwestern USA	181,300	70,000
Simpson	Australia	103,600	40,000
Mojave	southwestern USA	65,000	25,000

[1] Desert areas are very approximate because clear physical boundaries may not occur.

SCIENTIFIC DISCOVERIES

Discovery	Date	Discoverer	Nationality
Absolute zero, concept	1851	William Thomson, 1st Baron Kelvin	Irish
Adrenalin, isolation	1901	Jokichi Takamine	Japanese
Alizarin, synthesized	1869	William Perkin	English
Allotropy (in carbon)	1841	Jöns Jakob Berzelius	Swedish
Alpha particles	1899	Ernest Rutherford	New Zealand-born British
Alternation of generations (ferns and mosses)	1851	Wilhelm Hofmeister	German
Aluminium, extraction by electrolysis of aluminium oxide	1886	Charles Hall, Paul Héroult	US, French
Aluminium, improved isolation	1827	Friedrich Wöhler	German
Anaesthetic, first use (ether)	1842	Crawford Long	US
Anthrax vaccine	1881	Louis Pasteur	French
Antibacterial agent, first specific (Salvarsan for treatment of syphilis)	1910	Paul Ehrlich	German
Antiseptic surgery (using phenol)	1865	Joseph Lister	English
Argon	1892	William Ramsay	Scottish
Asteroid, first (Ceres)	1801	Giuseppe Piazzi	Italian
Atomic theory	1803	John Dalton	English
Australopithecus	1925	Raymond Dart	Australian-born South African
Avogadro's hypothesis	1811	Amedeo Avogadro	Italian
Bacteria, first observation	1683	Anton van Leeuwenhoek	Dutch
Bacteriophages	1916	Felix D'Herelle	Canadian
Bee dance	1919	Karl von Frisch	Austrian
Benzene, isolation	1825	Michael Faraday	English
Benzene, ring structure	1865	Friedrich Kekulé	German
Beta rays	1899	Ernest Rutherford	New Zealand-born British
Big Bang theory	1948	Ralph Alpher, George Gamow	US
Binary arithmetic	1679	Gottfried Leibniz	German
Binary stars	1802	William Herschel	German-born English
Binomial theorem	1665	Isaac Newton	English
Blood, circulation	1619	William Harvey	English
Blood groups, ABO system	1900	Karl Landsteiner	Austrian-born US
Bode's law	1772	Johann Bode, Johann Titius	German
Bohr atomic model	1913	Niels Bohr	Danish
Boolean algebra	1854	George Boole	English
Boyle's law	1662	Robert Boyle	Irish
Brewster's law	1812	David Brewster	Scottish
Brownian motion	1827	Robert Brown	Scottish
Cadmium	1817	Friedrich Strohmeyer	German
Caesium	1861	Robert Bunsen	German
Carbon dioxide	1755	Joseph Black	Scottish
Charles' law	1787	Jacques Charles	French
Chlorine	1774	Karl Scheele	Swedish
Complex numbers, theory	1746	Jean d'Alembert	French
Conditioning	1902	Ivan Pavlov	Russian
Continental drift	1912	Alfred Wegener	German
Coriolis effect	1834	Gustave-Gaspard Coriolis	French
Cosmic radiation	1911	Victor Hess	Austrian
Decimal fractions	1576	François Viète	French
Dinosaur fossil, first recognized	1822	Mary Ann Mantell	English

Scientific Discoveries continued

Discovery	Date	Discoverer	Nationality
Diphtheria bacillus, isolation	1883	Edwin Krebs	US
DNA	1869	Johann Frederick Miescher	Swiss
DNA and RNA	1909	Phoebus Levene	Russian-born US
DNA, double-helix structure	1953	Francis Crick, James Watson	English, US
Doppler effect	1842	Christian Doppler	Austrian
Earth's magnetic pole	1546	Gerardus Mercator	Flemish
Earth's molten core	1916	Albert Michelson	German-born US
Earth's molten core, proof	1906	Richard Oldham	Welsh
Earth's rotation, demonstration	1851	Léon Foucault	French
Eclipse, prediction	585 BC	Thales of Miletus	Greek
Electrolysis, laws	1833	Michael Faraday	English
Electromagnetic induction	1831	Michael Faraday	English
Electromagnetism	1819	Hans Christian Oersted	Danish
Electron	1897	J J Thomson	English
Electroweak unification theory	1967	Sheldon Lee Glashow, Abdus Salam, Steven Weinberg	US, Pakistani, US
Endorphins	1975	John Hughes	US
Enzyme, first animal (pepsin)	1836	Theodor Schwann	German
Enzyme, first (diastase from barley)	1833	Anselme Payen	French
Enzymes, 'lock and key' hypothesis	1899	Emil Fischer	German
Ether, first anaesthetic use	1842	Crawford Long	US
Eustachian tube	1552	Bartolomeo Eustachio	Italian
Evolution by natural selection	1858	Charles Darwin	English
Exclusion principle	1925	Wolfgang Pauli	Austrian-born Swiss
Fallopian tubes	1561	Gabriello Fallopius	Italian
Fluorine, preparation	1886	Henri Moissan	French
Fullerines	1985	Harold Kroto, David Walton	English
Gay-Lussac's law	1808	Joseph-Louis Gay-Lussac	French
Geometry, Euclidean	300 BC	Euclid	Greek
Germanium	1886	Clemens Winkler	German
Germ theory	1861	Louis Pasteur	French
Global temperature and link with atmospheric carbon dioxide	1896	Svante Arrhenius	Swedish
Gravity, laws	1687	Isaac Newton	English
Groups, theory	1829	Evariste Galois	French
Gutenberg discontinuity	1914	Beno Gutenberg	German-born US
Helium, production	1896	William Ramsay	Scottish
Homo erectus	1894	Marie Dubois	Dutch
Homo habilis	1961	Louis Leakey, Mary Leakey	Kenyan, English
Hormones	1902	William Bayliss, Ernest Starling	English
Hubble's law	1929	Edwin Hubble	US
Hydraulics, principles	1642	Blaise Pascal	French
Hydrogen	1766	Henry Cavendish	English
Iapetus	1671	Giovanni Cassini	Italian-born French
Infrared solar rays	1801	William Herschel	German-born English
Insulin, isolation	1921	Frederick Banting, Charles Best	Canadian
Insulin, structure	1969	Dorothy Hodgkin	English
Interference of light	1801	Thomas Young	English
Irrational numbers	450 BC	Hipparcos	Greek
Jupiter's satellites	1610	Galileo	Italian

Scientific Discoveries continued

Discovery	Date	Discoverer	Nationality
Kinetic theory of gases	1850	Rudolf Clausius	German
Krypton	1898	William Ramsay, Morris Travers	Scottish, English
Lanthanum	1839	Carl Mosander	Swedish
Lenses, how they work	1039	Ibn al-Haytham Alhazen	Arabic
Light, finite velocity	1675	Ole Römer	Danish
Light, polarization	1678	Christiaan Huygens	Dutch
Linnaean classification system	1735	Linnaeus	Swedish
'Lucy', hominid	1974	Donald Johanson	US
Magnetic dip	1576	Robert Norman	English
Malarial parasite in Anopheles mosquito	1897	Ronald Ross	British
Malarial parasite observed	1880	Alphonse Laveran	French
Mars, moons	1877	Asaph Hall	US
Mendelian laws of inheritance	1866	Gregor Mendel	Austrian
Messenger RNA	1960	Sydney Brenner, François Jacob	South African, French
Micro-organisms as cause of fermentation	1856	Louis Pasteur	French
Monoclonal antibodies	1975	César Milstein, Georges Köhler	Argentine-born British, German
Motion, laws	1687	Isaac Newton	English
Natural selection	1859	Charles Darwin	English
Neon	1898	William Ramsay, Morris Travers	Scottish, English
Neptune	1846	Johann Galle	German
Neptunium	1940	Edwin McMillan, Philip Abelson	US
Nerve impulses, electric nature	1771	Luigi Galvani	Italian
Neutron	1932	James Chadwick	English
Nitrogen	1772	Daniel Rutherford	Scottish
Normal distribution curve	1733	Abraham De Moivre	French
Nuclear atom, concept	1911	Ernest Rutherford	New Zealand-born British
Nuclear fission	1938	Otto Hahn, Fritz Strassman	German
Nucleus, plant cell	1831	Robert Brown	Scottish
Ohm's law	1827	Georg Ohm	German
Organic substance, first synthesis (urea)	1828	Friedrich Wöhler	German
Oxygen	1774	Joseph Priestley	English
Oxygen, liquefaction	1894	James Dewar	Scottish
Ozone layer	1913	Charles Fabry	French
Palladium	1803	William Hyde Wollaston	English
Pallas (asteroid)	1802	Heinrich Olbers	German
Pendulum, principle	1581	Galileo	Italian
Penicillin	1928	Alexander Fleming	Scottish
Penicillin, widespread preparation	1940	Ernst Chain, Howard Florey	German, Australian
Pepsin	1836	Theodor Schwann	German
Periodic law for elements	1869	Dmitri Mendeleyev	Russian
Period–luminosity law	1912	Henrietta Swan	US
Phosphorus	1669	Hennig Brand	German
Piezoelectric effect	1880	Pierre Curie	French
Pi meson (particle)	1947	Cecil Powell, Giuseppe Occhialini	English, Italian
Pistils, function	1676	Nehemiah Grew	English
Planetary nebulae	1790	William Herschel	German-born English
Planets, orbiting Sun	1543	Copernicus	Polish

Scientific Discoveries continued

Discovery	Date	Discoverer	Nationality
Pluto	1930	Clyde Tombaugh	US
Polarization of light by reflection	1808	Etienne Malus	French
Polio vaccine	1952	Jonas Salk	US
Polonium	1898	Marie and Pierre Curie	French
Positron	1932	Carl Anderson	US
Potassium	1806	Humphry Davy	English
Probability theory	1654	Blaise Pascal, Pierre de Fermat	French
Probability theory, expansion	1812	Pierre Laplace	French
Proton	1914	Ernest Rutherford	New Zealand-born British
Protoplasm	1846	Hugo von Mohl	German
Pulsar	1967	Jocelyn Bell Burnell	Irish
Pythagoras' theorem	550 BC	Pythagoras	Greek
Quantum chromodynamics	1972	Murray Gell-Mann	US
Quantum electrodynamics	1948	Richard Feynman, Seymour Schwinger, Shin'chiro Tomonaga	US, US, Japanese
Quark, first suggested existence	1963	Murray Gell-Mann, George Zweig	US
Quasar	1963	Maarten Schmidt	Dutch-born US
Rabies vaccine	1885	Louis Pasteur	French
Radioactivity	1896	Henri Becquerel	French
Radio emissions, from Milky Way	1931	Karl Jansky	US
Radio waves, production	1887	Heinrich Hertz	German
Radium	1898	Marie and Pierre Curie	French
Radon	1900	Friedrich Dorn	German
Refraction, laws	1621	Willibrord Snell	Dutch
Relativity, general theory	1915	Albert Einstein	German-born US
Relativity, special theory	1905	Albert Einstein	German-born US
Rhesus factor	1940	Karl Landsteiner, Alexander Wiener	Austrian, US
Rubidium	1861	Robert Bunsen	German
Sap circulation	1846	Giovanni Battista Amici	Italian
Sap flow in plants	1733	Stephen Hales	English
Saturn, 18th moon	1990	Mark Showalter	US
Saturn's satellites	1656	Christiaan Huygens	Dutch
Smallpox inoculation	1796	Edward Jenner	English
Sodium	1806	Humphry Davy	English
Stamens, function	1676	Nehemiah Grew	English
Stars, luminosity sequence	1905	Ejnar Hertzsprung	Danish
Stereochemistry, foundation	1848	Louis Pasteur	French
Stratosphere	1902	Léon Teisserenc	French
Sunspots	1611	Galileo, Christoph Scheiner	Italian, German
Superconductivity	1911	Heike Kamerlingh-Onnes	Dutch
Superconductivity, theory	1957	John Bardeen, Leon Cooper, John Schrieffer	US
Thermodynamics, second law	1834	Benoit-Pierre Clapeyron	French
Thermodynamics, third law	1906	Hermann Nernst	German
Thermoelectricity	1821	Thomas Seebeck	German
Thorium-X	1902	Ernest Rutherford, Frederick Soddy	New Zealand-born British, English
Titius–Bode law	1772	Johan Bode, Johann Titius	German
Tranquillizer, first (reserpine)	1956	Robert Woodward	US
Transformer	1831	Michael Faraday	English
Troposphere	1902	Léon Teisserenc	French

Scientific Discoveries continued

Discovery	Date	Discoverer	Nationality
Tuberculosis bacillus, isolation	1883	Robert Koch	German
Tuberculosis vaccine	1923	Albert Calmette, Camille Guérin	French
Uranus	1781	William Herschel	German-born English
Urea cycle	1932	Hans Krebs	German
Urease, isolation	1926	James Sumner	US
Urea, synthesis	1828	Friedrich Wöhler	German
Valves, in veins	1603	Geronimo Fabricius	Italian
Van Allen radiation belts	1958	James Van Allen	US
Virus, first identified (tobacco mosaic disease, in tobacco plants)	1898	Martinus Beijerinck	Dutch
Vitamin A, isolation	1913	Elmer McCollum	US
Vitamin A, structure	1931	Paul Karrer	Russian-born Swiss
Vitamin B, composition	1955	Dorothy Hodgkin	English
Vitamin B, isolation	1925	Joseph Goldberger	Austrian-born US
Vitamin C	1928	Charles Glen King, Albert Szent-Györgi	US, Hungarian-born US
Vitamin C, isolation	1932	Charles Glen King	US
Vitamin C, synthesis	1933	Tadeus Reichstein	Polish-born Swiss
Wave mechanics	1926	Erwin Schrödinger	Austrian
Xenon	1898	William Ramsay, Morris Travers	Scottish, English
X-ray crystallography	1912	Max von Laue	German
X-rays	1895	Wilhelm Röntgen	German

COMMON DRUGS DERIVED FROM PLANTS

Many of these plants are poisonous and if swallowed can cause serious illness or unconsciousness. They should only be used if administered by a medically trained professional.

Plant	Drug	Use
Amazonian liana	curare	muscle relaxant
Annual mugwort	artemisinin	antimalarial
Autumn crocus	colchicine	antitumour agent
Camphor tree	camphor	rubefacient; used in insect repellents and medicinal inhalants and liniments
Coca	cocaine	local anaesthetic
Common thyme	thymol	antifungal
Deadly nightshade (belladonna)	atropine	anticholinergic
Dog button (nux-vomica)	strychnine	central nervous system stimulant
Ergot fungus	ergotamine	analgesic
Foxglove	digitoxin, digitalis	cardiotonic
Indian snakeroot	reserpine	antihypertensive
Meadowsweet	salicylate	analgesic
Mexican yam	diosgenin	birth control pill
Mint, peppermint	menthol	rubefacient
Opium poppy	codeine, morphine	analgesic (codeine is also antitussive)
Pacific yew	taxol	antitumour agent
Recured thornapple	scopolamine	sedative
Rosy periwinkle	vincristine, vinblastine	antileukaemia
Tea, coffee, and kola nuts	caffeine	central nervous system stimulant
Velvet bean	L-dopa	antiparkinsonian
White willow	salicylic acid	topical analgesic
Yellow cinchona	quinine	antimalarial, antipyretic

MOST COMMON ELEMENTS IN THE EARTH'S CRUST

Element	Average percentage by volume	Element	Average percentage by volume
oxygen	46.6	sodium	2.8
silicon	27.7	potassium	2.6
aluminium	8.1	magnesium	2.1
iron	5.0	titanium	0.4
calcium	3.6	hydrogen	0.1

SOLAR AND LUNAR ECLIPSES

This table does not include partial or penumbral eclipses of the Moon.

Month	Day	Type of eclipse	Duration of totality	Region for observation
2003				
May	16	lunar total	0 hr 53 min	the Americas, eastern and western Africa
May	31	solar annular	3 min 37 sec	Iceland, Greenland
November	9	lunar total	0 hr 24 min	the Americas, Africa, Europe
November	23	solar total	1 min 57 sec	Antarctica
2004				
April	19	solar partial	–	Antarctica, southern Africa
May	4	lunar total	1 hr 16 min	South America, Africa, Europe, Asia, Australia
October	14	solar partial	–	northeast Asia, north Pacific as far south as Hawaii, Alaska
October	28	lunar total	1 hr 21 min	the Americas, Europe, Africa, central Asia
2005				
April	8	solar annular/total	0 min 42 sec	New Zealand, part of Antarctica, southern USA, Central America, western South America
October	3	solar annular	4 min 32 sec	Africa, Europe, western Asia (incl. western India)

THE CHEMICAL ELEMENTS

An element is a substance that cannot be split chemically into simpler substances. The atoms of a particular element all have the same number of protons in their nuclei (their atomic, or proton, number).
(– = not applicable.)

Name	Symbol	Atomic number	Atomic mass (amu)[1]	Relative density[2]	Melting or fusing point (°C)
Actinium	Ac	89	227[3]	–	–
Aluminium	Al	13	26.9815	2.58	658
Americium	Am	95	243[3]	–	–
Antimony	Sb	51	121.75	6.62	629
Argon	Ar	18	39.948	gas	–188
Arsenic	As	33	74.9216	5.73	volatile, 450
Astatine	At	85	210[3]	–	–
Barium	Ba	56	137.34	3.75	850
Berkelium	Bk	97	249[3]	–	–
Beryllium	Be	4	9.0122	1.93	1,281
Bismuth	Bi	83	208.9806	9.80	268
Bohrium	Bh	107	262[3]	–	–
Boron	B	5	10.81	2.5	2,300
Bromine	Br	35	79.904	3.19	–7.3
Cadmium	Cd	48	112.40	8.64	320
Caesium	Cs	55	132.9055	1.88	26
Calcium	Ca	20	40.08	1.58	851
Californium	Cf	98	251[3]	–	–
Carbon	C	6	12.011	3.52	infusible
Cerium	Ce	58	140.12	6.68	623
Chlorine	Cl	17	35.453	gas	–102
Chromium	Cr	24	51.996	6.5	1,510
Cobalt	Co	27	58.9332	8.6	1,490
Copper	Cu	29	63.546	8.9	1,083
Curium	Cm	96	247[3]	–	–
Dubnium	Db	105	262[3]	–	–
Dysprosium	Dy	66	162.50	–	–
Einsteinium	Es	99	254[3]	–	–
Erbium	Er	68	167.26	4.8	–
Europium	Eu	63	151.96	–	–
Fermium	Fm	100	253[3]	–	–
Fluorine	F	9	18.9984	gas	–223
Francium	Fr	87	2233	–	–
Gadolinium	Gd	64	157.25	–	–
Gallium	Ga	31	69.72	5.95	30
Germanium	Ge	32	72.59	5.47	958
Gold	Au	79	196.9665	19.3	1,062
Hafnium	Hf	72	178.49	12.1	2,500
Hassium	Hs	108	265[3]	–	–
Helium	He	2	4.0026	gas	–272

The Chemical Elements continued

Name	Symbol	Atomic number	Atomic mass (amu)[1]	Relative density[2]	Melting or fusing point (°C)
Holmium	Ho	67	164.9303	–	–
Hydrogen	H	1	1.0080	gas	–258
Indium	In	49	114.82	7.4	155
Iodine	I	53	126.9045	4.95	114
Iridium	Ir	77	192.22	22.4	2,375
Iron	Fe	26	55.847	7.86	1,525
Krypton	Kr	36	83.80	gas	–169
Lanthanum	La	57	138.9055	6.1	810
Lawrencium	Lr	103	260[3]	–	–
Lead	Pb	82	207.2	11.37	327
Lithium	Li	3	6.941	0.585	186
Lutetium	Lu	71	174.97	–	–
Magnesium	Mg	12	24.305	1.74	651
Manganese	Mn	25	54.9380	7.39	1,220
Meitnerium	Mt	109	266[3]	–	–
Mendelevium	Md	101	256[3]	–	–
Mercury	Hg	80	200.59	13.596	–38.9
Molybdenum	Mo	42	95.94	10.2	2,500
Neodymium	Nd	60	144.24	6.96	840
Neon	Ne	10	20.179	gas	–248.6
Neptunium	Np	93	237[3]	–	–
Nickel	Ni	28	58.71	8.9	1,452
Niobium	Nb	41	92.9064	8.4	1,950
Nitrogen	N	7	14.0067	gas	–211
Nobelium	No	102	254[3]	–	–
Osmium	Os	76	190.2	22.48	2,700
Oxygen	O	8	15.9994	gas	–227
Palladium	Pd	46	106.4	11.4	1,549
Phosphorus	P	15	30.9738	1.8–2.3	44
Platinum	Pt	78	195.09	21.5	1,755
Plutonium	Pu	94	242[3]	–	–
Polonium	Po	84	210[3]	–	–
Potassium	K	19	39.102	0.87	63
Praseodymium	Pr	59	140.9077	6.48	940
Promethium	Pm	61	145[3]	–	–
Protactinium	Pa	91	231.0359	–	–
Radium	Ra	88	226.0254	6.0	700
Radon	Rn	86	222[3]	gas	–150
Rhenium	Re	75	186.2	21	3,000
Rhodium	Rh	45	102.9055	12.1	1,950
Rubidium	Rb	37	85.4678	1.52	39
Ruthenium	Ru	44	101.07	12.26	2,400

The Chemical Elements continued

Name	Symbol	Atomic number	Atomic mass (amu)[1]	Relative density[2]	Melting or fusing point (°C)
Rutherfordium	Rf	104	262[3]	–	–
Samarium	Sm	62	150.4	7.7	1,350
Scandium	Sc	21	44.9559	–	–
Seaborgium	Sg	106	263[3]	–	–
Selenium	Se	34	78.96	4.5	170–220
Silicon	Si	14	28.086	2.0–2.4	1,370
Silver	Ag	47	107.868	10.5	960
Sodium	Na	11	22.9898	0.978	97
Strontium	Sr	38	87.62	2.54	800
Sulphur	S	16	32.06	2.07	115–119
Tantalum	Ta	73	180.9479	16.6	2,900
Technetium	Tc	43	99[3]	–	–
Tellurium	Te	52	127.60	6.0	446
Terbium	Tb	65	158.9254	–	–
Thallium	Tl	81	204.37	11.85	302
Thorium	Th	90	232.0381	11.00	1,750
Thulium	Tm	69	168.9342	–	–
Tin	Sn	50	118.69	7.3	232
Titanium	Ti	22	47.90	4.54	1,850
Tungsten	W	74	183.85	19.1	2,900–3,000
Ununbium	Uub[4]	112	277[3]	–	–
Ununhexium	Uuh[4]	116	289[3]	–	–
Ununnilium	Uun[4]	110	269[3]	–	–
Ununoctium	Uno[4]	118	293[3]	–	–
Ununquadium	Unq[4]	114	289[3]	–	–
Unununium	Uuu[4]	111	272[3]	–	–
Uranium	U	92	238.029	18.7	–
Vanadium	V	23	50.9414	5.5	1,710
Xenon	Xe	54	131.30	gas	–140
Ytterbium	Yb	70	173.04	–	–
Yttrium	Y	39	88.9059	3.8	–
Zinc	Zn	30	65.37	7.12	418
Zirconium	Zr	40	91.22	4.15	2,130

1 Atomic mass units.

2 Also known as specific gravity. Relative density is the density (at 20°C/68°F) of a solid or liquid relative to the maximum density of water (at 4°C/39.2°F).

3 The number given is that for the most stable isotope of the element.

4 Elements as yet unnamed; temporary identification assigned until a name is approved by the International Union for Pure and Applied Chemistry.

ENDANGERED SPECIES

Species	Observation
plants	a quarter of the world's plants are threatened with extinction by the year 2010
amphibians	worldwide decline in numbers; half of New Zealand's frog species are now extinct; 25% of species threatened with extinction (1996); 38% of US amphibians are endangered (1997)
birds	three-quarters of all bird species are declining; 11% are threatened with extinction (1996)
carnivores	almost all species of cats and bears are declining in numbers
fish	one-third of North American freshwater fish are rare or endangered; half the fish species in Lake Victoria, Africa's largest lake, are close to extinction due to predation by the introduced Nile perch; 33% of species are threatened with extinction (1996)
invertebrates	about 100 species are lost each day due to deforestation; half the freshwater snails in the southeastern USA are now extinct or threatened; 50% of crayfish and 56% of mussel species are endangered in the USA; a quarter of German invertebrates are threatened
mammals	half of Australia's mammals are threatened; 40% of mammals in France, the Netherlands, Germany, and Portugal are threatened; 25% of species are threatened with extinction (1996)
primates	two-thirds of primate species are threatened
reptiles	over 40% of reptile species are threatened; 20% with extinction (1996)

ENRICO FERMI AWARD

Named in honour of Enrico Fermi, the atomic pioneer, the US$100,000 award is given in recognition of outstanding scientific and technical achievement in the field of nuclear energy. The award is announced by the White House.

Year	Winner(s)
1954	Enrico Fermi
1955	no award
1956	John von Neumann
1957	Ernest O Lawrence
1958	Eugene P Wigner
1959	Glenn T Seaborg
1960	no award
1961	Hans A Bethe
1962	Edward Teller
1963	J Robert Oppenheimer
1964	Hyman G Rickover
1965	no award
1966	Otto Hahn, Lise Meitner, and Fritz Strassman
1967	no award
1968	John A Wheeler
1969	Walter H Zinn
1970	Norris E Bradbury
1971	Shields Warren and Stafford L Warren
1972	Manson Benedict
1973	no award
1974	no award
1975	no award
1976	William L Russell
1977	no award
1978	Harold M Agnew and Wolfgang K H Panofsky
1979	no award
1980	Alvin M Weinberg and Rudolf E Peiris
1981	W Bennett Lewis
1982	Herbert Anderson and Seth Neddermeyer
1983	Alexander Hollaender and John Lawrence
1984	Robert R Wilson and Georges Vendryès
1985	Norman C Rasmussen and Marshall N Rosenblath
1986	Ernest D Courant and M Stanley Livingston
1987	Luis W Alvarez and Gerald F Tape
1988	Richard B Setlow and Victor F Weisskopf
1989	no award
1990	George A Cowan and Robley D Evans
1991	no award
1992	Leon M Lederman, Harold Brown, and John S Foster Jr
1993	Freeman J Dyson and Liane B Russell
1994	no award
1995	Ugo Fano and Martin Kamen
1996	Richard Garwin, Mortimer Elkind, and H Rodney Withers
1997	no award
1998	Maurice Goldhaber and Michael E Phelps
1999	Michael E Phelps
2000	Sheldon Datz, Sidney Drell, and Herbert York
2001	no award
2002	no award
2003	John Bahcall, Raymond Davis, Jr., Seymour Sack

FIELDS MEDAL

This international prize for achievement in the field of mathematics is awarded every four years by the International Mathematical Union.

Year	**Winner(s)**
1936	Lars Ahlfors (Finland); Jesse Douglas (USA)
1950	Atle Selberg (USA); Laurent Schwartz (France)
1954	Kunihiko Kodaira (USA); Jean-Pierre Serre (France)
1958	Klaus Roth (UK); René Thom (France)
1962	Lars Hörmander (Sweden); John Milnor (USA)
1966	Michael Atiyah (UK); Paul J Cohen (USA); Alexander Grothendieck (France); Stephen Smale (USA)
1970	Alan Baker (UK); Heisuke Hironaka (USA); Sergei Novikov (USSR); John G Thompson (USA)
1974	Enrico Bombieri (Italy); David Mumford (USA)
1978	Pierre Deligne (Belgium); Charles Fefferman (USA); G A Margulis (USSR); Daniel Quillen (USA)
1982	Alain Connes (France); William Thurston (USA); S T Yau (USA)
1986	Simon Donaldson (UK); Gerd Faltings (West Germany); Michael Freedman (USA)
1990	Vladimir Drinfeld (USSR); Vaughan F R Jones (USA); Shigefumi Mori (Japan); Edward Witten (USA)
1994	L J Bourgain (USA/France); P-L Lions (France); J-C Yoccoz (France); E I Zelmanov (USA)
1998	Richard E Borcherds (UK); W Timothy Gowers (UK); Maxim Kontsevich (Russia); Curtis T McMullen (USA)
2002	Laurent Lafforgue (France); Vladimir Voevododsky (Russia)

FRACTIONS AS DECIMALS

Fraction	Decimal	Fraction	Decimal	Fraction	Decimal
1/2	0.5000	8/9	0.8889	15/16	0.9375
1/3	0.3333	1/10	0.1000	1/20	0.0500
2/3	0.6667	3/10	0.3000	3/20	0.1500
1/4	0.2500	7/10	0.7000	7/20	0.3500
3/4	0.7500	9/10	0.9000	9/20	0.4500
1/5	0.2000	1/11	0.0909	11/20	0.5500
2/5	0.4000	2/11	0.1818	13/20	0.6500
3/5	0.6000	3/11	0.2727	17/20	0.8500
4/5	0.8000	4/11	0.3636	19/20	0.9500
1/6	0.1667	5/11	0.4545	1/32	0.0312
5/6	0.8333	6/11	0.5455	3/32	0.9038
1/7	0.1429	7/11	0.6364	5/32	0.1562
2/7	0.2857	8/11	0.7273	7/32	0.2188
3/7	0.4286	9/11	0.8182	9/32	0.2812
4/7	0.5714	10/11	0.9091	11/32	0.3438
5/7	0.7143	1/12	0.0833	13/32	0.4062
6/7	0.8571	5/12	0.4167	15/32	0.4688
1/8	0.1250	7/12	0.5833	17/32	0.5312
3/8	0.3750	11/12	0.9167	19/32	0.5938
5/8	0.6250	1/16	0.0625	21/32	0.6562
7/8	0.8750	3/16	0.1875	23/32	0.7188
1/9	0.1111	5/16	0.3125	25/32	0.7812
2/9	0.2222	7/16	0.4375	27/32	0.8438
4/9	0.4444	9/16	0.5625	29/32	0.9062
5/9	0.5556	11/16	0.6875	31/32	0.9688
7/9	0.7778	13/16	0.8125		

NEAREST GALAXIES

Name	Constellation	Type	Approximate dista nce (light-years)
Sagittarius Dwarf	Sagittarius	spheroidal	80,000
Large Magellanic Cloud (LMC)	Dorado/Mensa	irregular	160,000
Small Magellanic Cloud (SMC)	Tucana	irregular	190,000
Ursa Minor Dwarf	Ursa Minor	spheroidal	205,000
Draco Dwarf	Draco	spheroidal	247,000
Sculptor Dwarf	Sculptor	spheroidal	254,000
Sextans Dwarf	Sextans	spheroidal	258,000
Carina Dwarf	Carina	spheroidal	284,000
Fornax Dwarf	Fornax	spheroidal	427,000

GENETICS: KEY EVENTS

1856 Austrian monk and botanist Gregor Mendel begins experiments breeding peas, research that will lead him to discover the laws of heredity.

1865 Gregor Mendel publishes a paper in the Proceedings of the Natural Science Society of Brünn that outlines the fundamental laws of heredity.

1869 Swiss biochemist Johann Miescher discovers a nitrogen and phosphorous material in cell nuclei that he calls nuclein butwhich is now known as the genetic material DNA.

1888 Dutch geneticist Hugo Marie de Vries uses the term 'mutation' to describe varieties that arise spontaneously in cultivated primroses.

1902 US geneticist Walter Sutton and German zoologist Theodor Boveri find the chromosomal theory of inheritance when they show that cell division is connected with heredity.

1906 English biologist William Bateson introduces the term 'genetics'.

1910 US geneticist Thomas Hunt Morgan discovers that certain inherited characteristics of the fruit fly Drosophila melanogaster are sex linked. He later argues that because all sex-related characteristics are inherited together they are linearly arranged on the X chromosome.

1934 Norwegian biochemist Asbjrn Fölling discovers the genetic metabolic defect phenylketonuria, which can cause retardation; his discovery stimulates research in biochemical genetics and the development of screening tests for carriers of deleterious genes.

1944 The role of deoxyribonucleic acid (DNA) in genetic inheritance is first demonstrated by US bacteriologist Oswald Avery, US biologist Colin MacLeod, and US biologist Maclyn McCarthy; it opens the door to the elucidation of the genetic code.

1945 Working in Japan, US geneticist Samuel G Salmon discovers Norin 10, a semidwarf wheat variety that grows quickly, responds well to fertilizer, does not fall over from the weight of the grains, and, when crossed with disease-resistant strains in the USA, results in a wheat strain that increases wheat harvests by more than 60% in India and Pakistan.

25 April 1953 English molecular biologist Francis Crick and US biologist James Watson announce the discovery of the double helix structure of DNA, the basic material of heredity. They also theorize that if the strands are separated then each can form the template for the synthesis of an identical DNA molecule. It is one of the most important discoveries in biology.

1954 Russian-born US cosmologist George Gamow suggests that the genetic code consists of the order of nucleotide triplets in the DNA molecule.

1958 US geneticists George Beadle, Edward Tatum, and Joshua Lederberg share the Nobel Prize for Physiology or Medicine: Beadle and Tatum for their discovery that genes act by regulating definite chemical events; and Lederberg for his discoveries concerning genetic recombination.

1961 French biochemists François Jacob and Jacques Monod discover messenger ribonucleic acid (mRNA), which transfers genetic information to the ribosomes, where proteins are synthesized.

1967 US scientist Charles Caskey and associates demonstrate that identical forms of messenger RNA produce the same amino acids in a variety of living beings, showing that the genetic code is common to all life forms.

1967 US biochemist Marshall Nirenberg establishes that mammals, amphibians, and bacteria all share a common genetic code.

Genetics: Key Events continued

October 1968 US geneticists Mark Ptashne and Walter Gilbert separately identify the first repressor genes.

1969 US geneticist Jonathan Beckwith and associates at Harvard Medical School isolate a single gene for the first time.

1969 The Nobel Prize for Physiology or Medicine is awarded jointly to US physiologists Max Delbrück, Alfred Hershey, and Salvador Luria for their discoveries concerning the replication mechanism and genetic structure of viruses.

1970 US geneticist Hamilton Smith discovers type II restriction enzyme that breaks the DNA strand at predictable places, making it an invaluable tool in recombinant DNA technology.

1970 US biochemists Howard Temin and David Baltimore separately discover the enzyme reverse transcriptase, which allows some cancer viruses to transfer their RNA to the DNA of their hosts turning them cancerous – a reversal of the common pattern in which genetic information always passes from DNA to RNA.

1972 US microbiologist Daniel Nathans uses a restriction enzyme that splits DNA molecules to produce a genetic map of the monkey virus (SV40), the simplest virus known to produce cancer; it is the first application of these enzymes to an understanding of the molecular basis of cancer.

1972 Venezuelan-born US immunologist Baruj Benacerraf and Hugh O'Neill McDevitt show immune response to be genetically determined.

1973 US biochemists Stanley Cohen and Herbert Boyer develop the technique of recombinant DNA. Strands of DNA are cut by restriction enzymes from one species and then inserted into the DNA of another; this marks the beginning of genetic engineering.

1975 The gel-transfer hybridization technique for the detection of specific DNA sequences is developed; it is a key development in genetic engineering.

1976 US biochemist Herbert Boyer and US venture capitalist Robert Swanson found Genentech in San Francisco, California, the world's first genetic engineering company.

28 August 1976 Indian-born US biochemist Har Gobind Khorana and his colleagues announce the construction of the first artificial gene to function naturally when inserted into a bacterial cell. This is a major breakthrough in genetic engineering.

1977 US biochemist Herbert Boyer, at Genentech, fuses a segment of human DNA into the bacterium Escherichia coli, which begins to produce the human protein somatostatin; this is the first commercially produced genetically engineered product.

1980 A new vaccine for the prevention of hepatitis B is tested in the USA. It is the first genetically engineered vaccine and has a success rate of 92%. It wins Federal Drug Administration approval in 1986.

16 June 1980 The US Supreme Court rules that a microbe created by genetic engineering can be patented.

1981 The US Food and Drug Administration grants permission to Eli Lilley and Co to market insulin produced by bacteria, the first genetically engineered product to go on sale.

1981 The genetic code for the hepatitis B surface antigen is discovered, creating the possibility of a bioengineered vaccine.

1981 US geneticists Robert Weinberg, Geoffrey Cooper, and Michael Wigler discover that oncogenes (genes that cause cancer) are integrated into the genome of normal cells.

1982 Using genetically engineered bacteria, the Swedish firm Kabivitrum manufactures human growth hormone.

1983 Geneticist James Gusella identifies the gene for Huntington's disease.

1984 British geneticist Alec Jeffreys discovers that a core sequence of DNA is almost unique to each person; this examination of DNA, known as 'genetic fingerprinting', can be used in criminal investigations and to establish family relationships.

1986 The US Department of Agriculture permits the Biological Corporation of Omaha to market a virus produced by genetic engineering; it is the first living genetically altered organism to be sold. The virus is used against a form of swine herpes.

1986 The US Department of Agriculture permits the first outdoor test of genetically altered high-yield plants (tobacco plants).

1987 German-born British geneticist Walter Bodmer and associates announce the discovery of a marker for a gene that causes cancer of the colon.

1987 The first genetically altered bacteria are released into the environment in the USA; they protect crops against frost.

1987 Foxes in Belgium are immunized against rabies by using bait containing a genetically engineered vaccine, dropped from helicopters. The success of the experiment leads to a large-scale vaccination programme.

Genetics: Key Events continued

April 1987	The US Patent and Trademark Office announces its intention to allow the patenting of animals produced by genetic engineering.
10 October 1987	The New York Times announces US geneticist Helen Donis-Keller's mapping of all 23 pairs of human chromosomes, allowing the location of specific genes for the prevention and treatment of genetic disorders.
April 1988	The US Patent and Trademark Office grants Harvard University a patent for a mouse developed by genetic engineering.
1989	Scientists in the UK introduce genetically engineered white blood cells into cancer patients, to attack tumours.
1990	The Human Genome Project to map the complete nucleotide sequence of human DNA is formally launched.
1991	British geneticists Peter Goodfellow and Robin Lovell-Badge discover the gene on the Y chromosome that determines sex.
1992	US biologist Philip Leder receives a patent for the first genetically engineered animal, the oncomouse, which is sensitive to carcinogens.
1993	US geneticist Dean Hammer and colleagues at the US National Cancer Institute publish the approximate location of a gene that could predispose human males to homosexuality.
1994	Trials using transfusions of artificial blood begin in the USA. The blood contains genetically engineered haemoglobin.
February 1994	The US Food and Drug Administration approves the use of genetically engineered bovine somatotropin (BST), which increases a cow's milk yield by 10–40%. It is banned in Europe.
May 1994	The first genetically engineered food goes on sale in the USA in California and Chicago, Illinois. The 'Flavr Savr' tomato is produced by the US biotechnology company Calgene.
1995	A genetically engineered potato is developed that contains the gene for Bt toxin, a natural pesticide produced by a soil bacterium. The potato plant produces Bt within its leaves.
1995	US embryologists Edward Lewis and Eric Wieschaus and German embryologist Christiane Nüsslein-Volhard are jointly awarded the Nobel Prize for Physiology or Medicine for their discoveries concerning the genetic control of early embryonic development.
1995	Australian geneticists produce a genetically engineered variety of cotton that contains a gene from a soil bacteria that kills the cotton bollworm and native budworm.
1995	Trials begin in the USA to treat breast cancer by gene therapy. The patients are injected with a virus genetically engineered to destroy their tumours.
April 1995	US surgeons report the successful transplant of genetically altered hearts of pigs into baboons, a notable advance in trans-species operations.
July 1995	The US government approves experimentation of genetically altered animal organs in humans.
August 1995	The US Environmental Protection Agency approves the sale of genetically modified maize, which contains a gene from a soil bacterium that produces a toxin fatal to the European corn borer, a pest that causes approximately US$1 billion's-worth of damages annually.
January 1996	The first genetically engineered salmon are hatched, at Loch Fyne in Scotland. The salmon contain genes from the ocean pout as well as a salmon growth hormone gene that causes them to grow five times as fast as other salmon.
9 May 1996	Scientists at the National Institute of Allergy and Infectious Disease discover a protein, fusin, that allows the HIV virus to fuse with a human immune system cell's outer membrane and inject genetic material. Its presence is necessary for the AIDS virus to enter the cell.
August 1996	US geneticists clone two rhesus monkeys from embryo cells.
27 February 1997	British researcher Ian Wilmut of the Roslin Institute in Edinburgh, Scotland, announces that British geneticists have cloned an adult sheep. A cell was taken from the udder of the mother sheep and its DNA combined with an unfertilized egg that had had its DNA removed. The fused cells were grown in the laboratory and then implanted into the uterus of a surrogate mother sheep. The resulting lamb, Dolly, came from an animal that was six years old. This is the first time cloning has been achieved using cells other than reproductive cells. The news is met with international calls to prevent the cloning of humans.
February 1997	US genetic scientist Don Wolf announces the production of monkeys cloned from embryos. It is a step closer to cloning humans and intensifies ethical debate.
16 May 1997	US geneticists identify a gene 'clock' in chromosome 5 in mice that regulates the circadian rhythm (the metabolic rhythm found in most organisms).

Genetics: Key Events continued

3 June 1997 US geneticist Huntington F Wilard constructs the first artificial human chromosome. He inserts telomeres (DNA and protein found on the tips of chromosomes) and centromeres (specialized regions of DNA within a chromosome) removed from white blood cells into human cancer cells. The cells are then assembled into chromosomes, which are about one-tenth the size of normal chromosomes. The artificial chromosome is successfully passed on to all daughter cells.

11 June 1997 English behavioural scientist David Skuse claims that boys and girls differ genetically in the way they acquire social skills. Girls acquire social skills intuitively and are 'pre-programmed', while boys have to be taught. This has important implications for education.

August 1997 US geneticist Craig Venter and colleagues publish the genome of the bacterium Helicobacter pylori, a bacterium that infects half the world's population and is the leading cause of stomach ulcers. It is the sixth bacterium to have its genome published, but is the most clinically important. It has 1,603 putative genes, encoded in a single circular chromosome that is 1,667,867 nucleotide base-pairs of DNA long. Complete genomes are increasingly being published as gene-sequencing techniques improve.

18 September 1997 US geneticist Bert Vogelstein and colleagues demonstrate that the p53 gene, which is activated by the presence of carcinogens, induces cells to commit suicide by stimulating them to produce large quantities of poisonous chemicals, called 'reactive oxygen species' (ROS). The cells literally poison themselves. It is perhaps the human body's most effective way of combating cancer. Many cancers consist of cells with a malfunctioning p53 gene.

November 1997 The US Food and Drug Administration approves Rituxan, the first anticancer monoclonal antibody made from genetically engineered mouse antibodies. The antibody binds itself to non-Hodgkin's lymphoma (a cancer of the lymph system) cells and triggers the immune system to kill the cells.

October 1998 US scientist French Anderson announces a technique that could cure inherited diseases by inserting a healthy gene to replace a damaged one. He calls for a full debate on the issue of gene therapy, which brings with it the dilemma of whether it is ethical to enable the choice of physical attributes such as eye colour and height.

8 December 1998 The Human Fertilization and Embryology Authority and the Human Genetics Advisory Commission publish a joint report in the UK on cloning. While they oppose cloning for reproductive purposes, they leave the door open for cloning for curing intractable diseases.

10 December 1998 In a joint effort by scientists around the world, the first genetic blueprint for a whole multicellular animal is completed. The 97 million-letter code, which is published on the Internet, is for a tiny worm called Caenorhabditis elegans. The study began 15 years ago and cost £30 million.

24 January 1999 US scientist Craig Venter of the Institute for Genomic Research in Maryland announces the possibility of creating a living, replicating organism from an artificial set of genes, at a meeting of the American Association for the Advancement of Science, in Anaheim, California. The experiment is put on hold until the moral question is discussed by religious leaders and ethicists at the University of Pennsylvania.

April 1999 European Union legislation is implemented in the UK, requiring that some foods containing genetically modified (GM) protein or DNA be labelled in restaurants and food shops. In September, the UK passes its own legislation to make labelling compulsory at all food outlets.

18 May 1999 A group of scientists at a specially convened Royal Society meeting finds that the experiments of Hungarian-born doctor Arpád Pusztai on GM foods were 'fundamentally flawed'. In August 1998, Pusztai claimed that his experiments demonstrated that genetically modified potatoes stunted the growth of laboratory rats, strengthening public opinion in the UK against genetically modified foods.

September 1999 A volunteer dies while participating in a gene therapy trial at the University of Pennsylvania, Philadelphia, USA. The Food and Drug Administration suspends all such trials at the university pending an investigation.

December 1999 Chromosome 22 is the first human chromosome to have its DNA sequence decoded, by scientists from the Sanger Institute.

May 2000 A team of German and Japanese scientists reveal they have sequenced chromosome 21, the second human chromosome to be decoded.

26 June 2000 A major scientific landmark is reached as the first draft of the complete human genome is completed. This is announced jointly by the publicly funded Human Genome Project and the commercial Celera Genomics, which have competed to produce the human DNA sequence first.

September 2000 Taco shells across the USA are found to contain GM StarLink corn, which is authorized for use as animal feed but not for human consumption. This prompts widespread product recalls.

14 December 2000 The entire DNA sequence of thale cress Arabidopsis thaliana is published. This completes five years of work by an international team, and is the first complete genome of a plant to be mapped.

Genetics: Key Events continued

11 January 2001	A team of US scientists led by Gerald Schatten announces the birth of the world's first genetically modified primate, a rhesus monkey dubbed Andi. The monkey's DNA includes a gene taken from a jellyfish included to act as a marker. This raises the possibility of using primates as models for the study of human genetic diseases. A fierce debate about the ethical implications of genetically modifying primates follows.
24 January 2001	The UK becomes the first country to legalize the cloning of human embryos for research purposes.
26 January 2001	Supermarket chains Tesco and Asda, which hold 42% of the UK grocery market between them, declare that all their meat and dairy products are to be sourced from animals fed a non-GM diet. This mirrors a general consumer mistrust of GM foods.
26 January 2001	Two corporations, Syngenta of Switzerland and Myriad Genetics of the USA, announce the publication of the complete genome of the rice plant. This is the first crop plant to be sequenced.
February 2001	Celera Genomics and the Human Genome Project publish their data on the human genetic code in different scientific publications. The most striking conclusion is that the genome consists of around 30,000 genes – much fewer than anticipated, and around the same number as a mouse.
August 2001	Researchers at the University of Abertay, Dundee, Scotland, announce that they have grown the world's first genetically modified elm trees, designed to be resistant to Dutch elm disease.

GEOLOGICAL ERAS, PERIODS, AND EPOCHS

Eon	Era	Period	Epoch	Began	Stage of life
Phanerozoic	Cenozoic	Quaternary	Holocene	10,000 years ago	end of last ice age; civilization develops
			Pleistocene	1.64 million years ago	Homo sapiens evolves
		Tertiary	Pliocene	5.2 million years ago	earliest hominids evolve
			Miocene	23.5 million years ago	grasslands spread; hoofed mammals evolve
			Oligocene	35.5 million years ago	first anthropoid apes evolve
			Eocene	56.5 million years ago	predecessors of most modern mammals evolve
			Palaeocene	65 million years ago	mammals flourish
	Mesozoic	Cretaceous		146 million years ago	seed-bearing plants evolve; mass extinction (including dinosaurs) at end of period
		Jurassic		208 million years ago	birds evolve
		Triassic		245 million years ago	early dinosaurs; first mammals; mass plant extinction at end of period
	Palaeozoic	Permian		290 million years ago	mass extinction
		Carboniferous		360 million years ago	reptiles evolve
		Devonian		409 million years ago	amphibians and insects evolve
		Silurian		439 million years ago	land plants and jawed fishes evolve
		Ordovician		510 million years ago	reef-building algae and jawless fish evolve
		Cambrian		570 million years ago	invertebrate animals evolve
Precambrian	Proterozoic			2.5 billion years ago	first biological activity; bacteria and algae
	Archaean			4.6 billion years ago	none

HUMAN BODY: COMPOSITION

Chemical element or substance		Body weight (%)	
Pure elements		**Water and solid matter**	
Oxygen	65	Water	60–80
Carbon	18	Total solid material	20–40
Hydrogen	10	**Organic molecules**	
Nitrogen	3	Protein	15–20
Calcium	2	Lipid	3–20
Phosphorus	1.1	Carbohydrate	1–15
Potassium	0.35	Other	0–1
Sulphur	0.25		
Sodium	0.15		
Chlorine	0.15		
Magnesium, iron, manganese, copper, iodine, cobalt, zinc	traces		

MAJOR ICE AGES

name	date (years ago)
Pleistocene	1.64 million–10,000
Permo-Carboniferous	330–250 million
Ordovician	440–430 million
Verangian	615–570 million
Sturtian	820–770 million
Gnejso	940–880 million
Huronian	2,700–1,800 million

UNITS IN THE IMPERIAL/U.S. STANDARD SYSTEM

Length

1 foot	= 12 inches
1 yard	= 3 feet
1 rod, pole, perch	= 51/2 yards (= 161/2 feet)
1 chain	= 4 rods (= 22 yards)
1 furlong	= 10 chains (= 220 yards)
1 mile	= 5,280 feet (=1,760 yards/8 furlongs)

Nautical

1 fathom	= 6 feet
1 cable length	= 100 fathoms (UK)/120 fathoms (USA)
1 nautical mile (international)	= 6,076 feet

Area

1 square foot	= 144 square inches
1 square yard	= 9 square feet
1 square rod	= 301/4 square yards
1 rood	= 1,210 square yards
1 acre	= 4,840 square yards (= 4 roods)
1 square mile	= 640 acres

Volume

1 cubic foot	= 1,728 cubic inches
1 cubic yard	= 27 cubic feet

Shipping

1 register ton	= 100 cubic feet

Capacity

1 fluid ounce	= 8 fluid drachms
1 gill	= 5 fluid ounces
1 pint	= 4 gills (= 20 fluid ounces)
1 quart	= 2 pints
1 gallon	= 4 quarts (= 8 pints)
1 peck	= 2 gallons
1 bushel	= 4 pecks (= 8 gallons)
1 quarter	= 8 bushels (= 64 gallons)

Weight (avoirdupois)

1 ounce	= 4371/2 grains
1 ounce	= 16 drams
1 pound	= 16 ounces
1 stone	= 14 pounds
1 quarter	= 2 stones (= 28 pounds)
1 hundredweight	= 4 quarters (= 112 pounds (UK)/100 pounds (USA))
1 ton (UK long ton)	= 20 hundredweights (= 2,240 pounds)
1 ton (US short ton)	= 20 hundredweights (= 2,000 pounds)

LARGEST LAKES IN THE WORLD

Lake	Location	Area sq km	sq mi
Caspian Sea	Azerbaijan/Russian Federation/Kazakhstan/Turkmenistan/Iran	370,990	143,239
Superior	USA/Canada	82,071	31,688
Victoria	Tanzania/Kenya/Uganda	69,463	26,820
Aral Sea	Kazakhstan/Uzbekistan	64,500	24,903
Huron	USA/Canada	59,547	22,991
Michigan	USA	57,735	22,291
Tanganyika	Tanzania/Democratic Republic of Congo/Zambia/Burundi	32,880	12,695
Baikal	Russian Federation	31,499	12,162
Great Bear	Canada	31,316	12,091
Malawi (or Nyasa)	Malawi/Tanzania/Mozambique	28,867	11,146
Great Slave	Canada	28,560	11,027
Erie	USA/Canada	25,657	9,906
Winnipeg	Canada	25,380	9,799
Ontario	USA/Canada	19,010	7,340
Balkhash	Kazakhstan	18,421	7,112
Ladoga	Russian Federation	17,695	6,832
Chad	Chad/Cameroon/Nigeria	16,310	6,297
Maracaibo	Venezuela	13,507	5,215

MATHEMATICS: KEY EVENTS

BC

c. 2500 The people of Mesopotamia (now Iraq) develop a positional numbering (place-value) system, in which the value of a digit depends on its position in a number.

c. 2000 Mesopotamian mathematicians solve quadratic equations (algebraic equations in which the highest power of a variable is 2).

876 A symbol for zero is used for the first time, in India.

c. 550 Greek mathematician Pythagoras formulates a theorem relating the lengths of the sides of a right-angled triangle. The theorem is already known by earlier mathematicians in China, Mesopotamia, and Egypt.

c. 450 Hipparcos of Metapontum discovers that some numbers are irrational (cannot be expressed as the ratio of two integers).

300 Euclid lays out the laws of geometry in his book Elements, which is to remain a standard text for 2,000 years.

c. 230 Eratosthenes develops a method for finding all prime numbers.

c. 100 Chinese mathematicians begin using negative numbers.

c. 190 Chinese mathematicians use powers of 10 to express magnitudes.

AD

c. 210 Diophantus of Alexandria writes the first book on algebra.

c. 600 A decimal number system is developed in India.

829 Persian mathematician Muhammad ibn-Musa al-Khwarizmi publishes a work on algebra that makes use of the decimal number system.

1202 Italian mathematician Leonardo Fibonacci studies the sequence of numbers (1, 1, 2, 3, 5, 8, 13, 21, ...) in which each number is the sum of the two preceding ones.

1550 In Germany, Rheticus publishes trigonometrical tables that simplified calculations involving triangles.

1614 Scottish mathematician John Napier invents logarithms, which enable lengthy calculations involving multiplication and division to be carried out by addition and subtraction.

1623 Wilhelm Schickard invents the mechanical calculating machine.

1637 French mathematician and philosopher René Descartes introduces coordinate geometry.

1654 In France, Blaise Pascal and Pierre de Fermat develop probability theory.

1666 Isaac Newton develops differential calculus, a method of calculating rates of change.

1675 German mathematician Gottfried Wilhelm Leibniz introduces the modern notation for integral calculus, a method of calculating volumes.

1679 Leibniz introduces binary arithmetic, in which only two symbols are used to represent all numbers.

1684 Leibniz publishes the first account of differential calculus.

1718 Jakob Bernoulli in Switzerland publishes his work on the calculus of variations (the study of functions that are close to their minimum or maximum values).

1742 German mathematician Christian Goldbach conjectures that every even number greater than two can be written as the sum of two prime numbers. Goldbach's conjecture has still not been proven.

1746 In France, Jean le Rond d'Alembert develops the theory of complex numbers.

1747 D'Alembert uses partial differential equations in mathematical physics.

1798 Norwegian mathematician Caspar Wessel introduces the vector representation of complex numbers.

1799 Karl Friedrich Gauss of Germany proves the fundamental theorem of algebra: the number of solutions of an algebraic equation is the same as the exponent of the highest term.

1810 In France, Jean Baptiste Joseph Fourier publishes his method of representing functions by a series of trigonometric functions.

1812 French mathematician Pierre Simon Laplace publishes the first complete account of probability theory.

1822 In the UK, Charles Babbage begins construction of the first mechanical computer, the difference machine, a device for calculating logarithms and trigonometric functions.

1827 Gauss introduces differential geometry, in which small features of curves are described by analytical methods.

Mathematics: Key Events continued

1829 In Russia, Nikolai Ivanonvich Lobachevsky develops hyperbolic geometry, in which a plane is regarded as part of a hyperbolic surface, shaped like a saddle. In France, Evariste Galois introduces the theory of groups (collections whose members obey certain simple rules of addition and multiplication).

1844 French mathematician Joseph Liouville finds the first transcendental number, which cannot be expressed as an algebraic equation with rational coefficients. In Germany, Hermann Grassmann studies vectors with more than three dimensions.

1854 George Boole in the UK publishes his system of symbolic logic, now called Boolean algebra.

1858 English mathematician Arthur Cayley develops calculations using ordered tables called matrices.

1865 August Ferdinand Möbius in Germany describes how a strip of paper can have only one side and one edge.

1892 German mathematician Georg Cantor shows that there are different kinds of infinity and studies transfinite numbers.

1895 Jules Henri Poincaré publishes the first paper on topology, often called 'the geometry of rubber sheets'.

1931 In the USA, Austrian-born mathematician Kurt Gödel proves that any formal system strong enough to include the laws of arithmetic is either incomplete or inconsistent.

1937 English mathematician Alan Turing publishes the mathematical theory of computing.

1944 John Von Neumann and Oscar Morgenstern develop game theory in the USA.

1945 The first general purpose, fully electronic digital computer, ENIAC (electronic numerator, integrator, analyser, and computer), is built at the University of Pennsylvania, USA.

1961 Meteorologist Edward Lorenz at the Massachusetts Institute of Technology, USA, discovers a mathematical system with chaotic behaviour, leading to a new branch of mathematics – chaos theory.

1962 Benoit Mandelbrot in the USA invents fractal images, using a computer that repeats the same mathematical pattern over and over again.

1975 US mathematician Mitchell Feigenbaum discovers a new fundamental constant (approximately 4.669201609103), which plays an important role in chaos theory.

1980 Mathematicians worldwide complete the classification of all finite and simple groups, a task that has taken over a hundred mathematicians more than 35 years to complete and whose results take up more than 14,000 pages in mathematical journals.

1989 A team of US computer mathematicians at Amdahl Corporation, California, discovers the highest known prime number (it contains 65,087 digits).

1993 British mathematician Andrew Wiles publishes a 1,000-page proof of Fermat's last theorem, one of the most baffling challenges in pure mathematics.

1996 Wiles's proof is accepted after revision.

1997 The largest number to be fractorized to date has 167 digits: (3349–1)/2 is split into its 80- and 87-digit factors by a team of US mathematicians after 100,000 hours of computing.

MATHEMATICAL SIGNS

Symbol	Meaning
$a \rightarrow b$	a implies b
∞	infinity
lim	limiting value
$a \equiv b$	a approximately equal to b
$a=b$	a equal to b
$a>b$	a greater than b
$a<b$	a smaller than b
$a \neq b$	a not equal to b
$b<a<c$	a greater than b and smaller than c, that is a lies between b and c, but cannot equal either
$a \geq b$	a equal to or greater than b, that is a at least as great as b
$a \leq b$	a equal to or less than b, that is, a at most as great as b
$b \leq a \leq c$	a lies between the values of b and c and could take the values of b and c
$\lvert a \rvert$	absolute value of a; this is always positive, for example $\lvert -5 \rvert = 5$
$+$	addition sign, positive
$-$	subtraction sign, negative
x	multiplication sign, times
: or $\div$ or /	division sign, divided by
$a+b=c$	a+b, read as 'a+b', denotes the addition of a and b. The result of the addition, c, is also known as the sum.
$\int$	indefinite integral
$_a\int^b f(x)dx$	definite integral, or integral between x=a and x=b
$a-b=c$	a–b, read as 'a minus b', denotes subtraction of b from a. a–b, or c, is the difference. Subtraction is the opposite of addition.
$a \times b=c$ $ab=c$ $a.b=c$	a x b, read as 'a multiplied by b', denotes multiplication of a by b. a x b, or c, is the product; a and b are factors of c.
$a:b=c$ $a \div b=c$ $a/b=c$	a:b, read as 'a divided by b', denotes division. a is the dividend, b is the divisor. a:b, or c, is the quotient. One aspect of division – repeated subtraction, is the opposite of multiplication – repeated addition.
a^b	ab, read as 'a to the power b'. a is the base, b the exponent
$b\sqrt{a}=c$	$b\sqrt{a}$ is the bth root of a, b being known as the root exponent. In the special case of $2\sqrt{a}=c$, $2\sqrt{a}$ or c is known as the square root of a, and the root exponent is usually omitted, that is $2\sqrt{a}=\sqrt{a}$.
e	exponential constant and is the base of natural (napierian) logarithms = 2.7182818284...
π	ratio of the circumference of a circle to its diameter = 3.1415925535...

MERCALLI SCALE

The Mercalli scale is a measure of the intensity of an earthquake. It differs from the Richter scale, which measures magnitude. It is named after the Italian seismologist Giuseppe Mercalli (1850–1914). The scale shown here is the Modified Mercalli Intensity Scale, developed in 1931 by US seismologists Harry Wood and Frank Neumann.

Intensity value	Description
I	not felt, except by a very few under especially favourable conditions
II	felt only by a few persons at rest, especially on upper floors of buildings
III	felt quite noticeably by persons indoors, especially on upper floors of buildings; many people do not recognize it as an earthquake; standing motor cars may rock slightly
IV	felt indoors by many, outdoors by a few persons during the day; at night, some awakened; dishes, windows, doors disturbed; walls make cracking sound; standing motor cars rock noticeably
V	felt by nearly everyone; many awakened; some dishes, windows broken; unstable objects overturned; pendulum clocks may stop
VI	felt by all; some heavy furniture moved; a few instances of fallen plaster; damage slight
VII	damage negligible in buildings of good design and construction; slight to moderate in well-built ordinary structures; considerable damage in poorly built or badly designed structures; some chimneys broken
VIII	damage slight in specially designed structures; considerable damage in ordinary substantial buildings with partial collapse; damage great in poorly built structures; fall of chimneys, factory stacks, columns, monuments, walls; heavy furniture overturned
IX	damage considerable in specially designed structures; damage great in substantial buildings, with partial collapse; buildings shifted off foundations
X	some well built wooden structures destroyed; most masonry and frame structures with foundations destroyed; rails bent
XI	few, if any (masonry) structures remain standing; bridges destroyed; rails bent greatly
XII	damage total; lines of sight and level are distorted; objects thrown into the air

METEORITES: WORLD'S LARGEST

Location	Country	Approximate weight (tonnes)
Hoba West, Grootfontein	Namibia	60
Tent (Abnighito), Cape York	Greenland	30
Bacuberito	Mexico	27
Mbosi	Tanzania	26
Armanty	Mongolia	20
Williamette (1902), Oregon	USA	14
Chupaero, Chihuahua	Mexico	14
Campo de Cielo	Argentina	13
Morito, Chihuahua	Mexico	11

UNITS IN THE METRIC SYSTEM

Length

1 centimetre	= 10 millimetres	
1 decimetre	= 10 centimetres	= 100 millimetres
1 metre	= 10 decimetres	= 1,000 millimetres
1 decametre	= 10 metres	
1 hectometre	= 10 decametres	= 100 metres
1 kilometre	= 10 hectometres	= 1,000 metres

Area

1 square centimetre	= 100 square millimetres	
1 square metre	= 10,000 square centimetres	= 1,000,000 square millimetres
1 are	= 100 square metres	
1 hectare	= 100 ares	= 10,000 square metres
1 square kilometre	= 100 hectares	= 1,000,000 square metres

Mass (avoirdupois)

1 centigram	= 10 milligrams	
1 decigram	= 10 centigrams	= 100 milligrams
1 gram	= 10 decigrams	= 1,000 milligrams
1 decagram	= 10 grams	
1 hectogram	= 10 decagrams	= 100 grams
1 kilogram	= 10 hectograms	= 1,000 grams
1 metric ton	= 1,000 kilograms	

Volume

1 cubic centimetre	= 1,000 cubic millimetres	
1 cubic decimetre	= 1,000 cubic centimetres	= 1,000,000 cubic millimetres
1 cubic metre	= 1,000 cubic decimetres	= 1,000,000,000 cubic millimetres

Capacity

1 centilitre	= 10 millilitres	
1 decilitre	= 10 centilitres	= 100 millilitres
1 litre	= 10 decilitres	= 1,000 millilitres
1 decalitre	= 10 litres	
1 hectolitre	= 10 decalitres	= 100 litres
1 kilolitre	= 10 hectolitres	= 1,000 litres

GEMS AND SEMI-PRECIOUS STONES

Stone	Colour	Hardness	Composition
amethyst	purple or blue-violet	7.0	silica (quartz)
aquamarine	pale blue-green	7.5	beryl
carnelian	red or red-brown	7.0	silica (quartz)
citrine	yellow	7.0	silica (quartz)
diamond	colourless, yellow, green, or black	10.0	carbon
emerald	green	7.5	beryl
lapis lazuli	blue	5.5	lazurite
malachite	green	4.0	copper carbonate
moonstone	silver/blue opalescent	6.0–6.5	alkali feldspar
morganite	pink	7.5	beryl
opal	multicoloured	6.0	silica (amorphous)
pearl	white or black	3.5	calcium carbonate
peridot	olive-green	6.0–7.0	olivine
ruby	red	9.0	corundum
sapphire	blue, green, or yellow	9.0	corundum
topaz	green, blue, or yellow	8.0	complex silicate
tourmaline	dark green or pink	7.5	complex borosilicate
turquoise	blue	6.0	aluminium/copper phosphate
zircon	colourless or pink	6.5–7.5	zirconium silicate

MULTIPLICATION TABLE

	2	3	4	5	6	7	8	9	10	11	12	13	14	15	16	17	18	19	20	21	22	23	24	25
2	4	6	8	10	12	14	16	18	20	22	24	26	28	30	32	34	36	38	40	42	44	46	48	50
3	6	9	12	15	18	21	24	27	30	33	36	39	42	45	48	51	54	57	60	63	66	69	72	75
4	8	12	16	20	24	28	32	36	40	44	48	52	56	60	64	68	72	76	80	84	88	92	96	100
5	10	15	20	25	30	35	40	45	50	55	60	65	70	75	80	85	90	95	100	105	110	115	120	125
6	12	18	24	30	36	42	48	54	60	66	72	78	84	90	96	102	108	114	120	126	132	138	144	150
7	14	21	28	35	42	49	56	63	70	77	84	91	98	105	112	119	126	133	140	147	154	161	168	175
8	16	24	32	40	48	56	64	72	80	88	96	104	112	120	128	136	144	152	160	168	176	184	192	200
9	18	27	36	45	54	63	72	81	90	99	108	117	126	135	144	153	162	171	180	189	198	207	216	225
10	20	30	40	50	60	70	80	90	100	110	120	130	140	150	160	170	180	190	200	210	220	230	240	250
11	22	33	44	55	66	77	88	99	110	121	132	143	154	165	176	187	198	209	220	231	241	253	264	275
12	24	36	48	60	72	84	96	108	120	132	144	156	168	180	192	204	216	228	240	252	264	276	288	300
13	26	39	52	65	78	91	104	117	130	143	156	169	182	195	208	221	234	247	260	273	286	299	312	325
14	28	42	56	70	84	98	112	126	140	154	168	182	196	210	224	238	252	266	280	294	308	322	336	350
15	30	45	60	75	90	105	120	135	150	165	180	195	210	225	240	255	270	285	300	315	330	345	360	375
16	32	48	64	80	96	112	128	144	160	176	192	208	224	240	256	272	288	304	320	336	352	368	384	400
17	34	51	68	85	102	119	136	153	170	187	204	221	238	255	272	289	306	323	340	357	374	391	408	425
18	36	54	72	90	108	126	144	162	180	198	216	234	252	270	288	306	324	342	360	378	396	414	432	450
19	38	57	76	95	114	133	152	171	190	209	228	247	266	285	304	323	342	361	380	399	418	437	456	475
20	40	60	80	100	120	140	160	180	200	220	240	260	280	300	320	340	360	380	400	420	440	460	480	500
21	42	63	84	105	126	147	168	189	210	231	252	273	294	315	336	357	378	399	420	441	462	483	504	525
22	44	66	88	110	132	154	176	198	220	242	264	286	308	330	352	374	396	418	440	462	484	506	528	550
23	46	69	92	115	138	161	184	207	230	253	276	299	322	345	368	391	414	437	460	483	506	529	552	575
24	48	72	96	120	144	168	192	216	240	264	288	312	336	360	384	408	432	456	480	504	528	552	576	600
25	50	75	100	125	150	175	200	225	250	275	300	325	350	375	400	425	450	475	500	525	550	575	600	625

NOBEL PRIZE FOR CHEMISTRY

Year	Winner(s)I	Awarded for
1901	Jacobus van't Hoff (Netherlands)	laws of chemical dynamics and osmotic pressure
1902	Emil Fischer (Germany)	sugar and purine syntheses
1903	Svante Arrhenius (Sweden)	theory of electrolytic dissociation
1904	William Ramsay (UK)	discovery of rare gases in air and their locations in the periodic table
1905	Adolf von Baeyer (Germany)	work in organic dyes and hydroaromatic compounds
1906	Henri Moissan (France)	isolation of fluorine and adoption of electric furnace
1907	Eduard Buchner (Germany)	biochemical research and discovery of cell-free fermentation
1908	Ernest Rutherford (UK)	work in atomic disintegration and the chemistry of radioactive substances
1909	Wilhelm Ostwald (Germany)	work in catalysis and principles of equilibria and rates of reaction
1910	Otto Wallach (Germany)	work in alicyclic compounds
1911	Marie Curie (France)	discovery of radium and polonium, and the isolation and study of radium
1912	Victor Grignard (France)	discovery of Grignard reagents
	Paul Sabatier (France)	finding method of catalytic hydrogenation of organic compounds
1913	Alfred Werner (Switzerland)	work in bonding of atoms within inorganic molecules
1914	Theodore Richards (USA)	accurate determination of the atomic masses of many elements
1915	Richard Willstätter (Germany)	research into plant pigments, especially chlorophyll
1916	no award	
1917	no award	
1918	Fritz Haber (Germany)	synthesis of ammonia from its elements
1919	no award	
1920	Walther Nernst (Germany)	work in thermochemistry
1921	Frederick Soddy (UK)	work in radioactive substances, especially isotopes
1922	Francis Aston (UK)	work in mass spectrometry of isotopes of radioactive elements, and enunciation of the whole-number rule
1923	Fritz Pregl (Austria)	method of microanalysis of organic substances
1924	no award	
1925	Richard Zsigmondy (Austria)	elucidation of heterogeneity of colloids
1926	Theodor Svedberg (Sweden)	investigation of dispersed systems
1927	Heinrich Wieland (Germany)	research on constitution of bile acids and related substances
1928	Adolf Windaus (Germany)	research on constitution of sterols and related vitamins
1929	Arthur Harden (UK) and Hans von Euler-Chelpin (Sweden)	work on fermentation of sugar and fermentative enzymes
1930	Hans Fischer (Germany)	analysis of haem (the iron-bearing group in haemoglobin) and chlorophyll and the synthesis of haemin (a compound of haem)
1931	Carl Bosch (Germany) and Friedrich Bergius (Germany)	invention and development of chemical high-pressure methods
1932	Irving Langmuir (USA)	discoveries and investigations in surface chemistry
1933	no award	
1934	Harold Urey (USA)	discovery of deuterium (heavy hydrogen)
1935	Irène and Frédéric Joliot-Curie (France)	synthesis of new radioactive elements
1936	Peter Debye (Netherlands)	work in molecular structures by investigation of dipole moments and the diffraction of X-rays and electrons in gases
1937	Norman Haworth (UK)	work in carbohydrates and ascorbic acid (vitamin C)
	Paul Karrer (Switzerland)	work in carotenoids, flavins, retinol (vitamin A) and riboflavin (vitamin B2)
1938	Richard Kuhn (Germany) (declined)	carotenoids and vitamins research
1939	Adolf Butenandt (Germany) (declined)	work in sex hormones
	Leopold Ruzicka (Switzerland)	polymethylenes and higher terpenes
1940	no award	
1941	no award	
1942	no award	

Nobel Prize for Chemistry continued

Year	Winner(s) I	Awarded for
1943	Georg von Hevesy (Hungary)	use of isotopes as tracers in chemical processes
1944	Otto Hahn (Germany)	discovery of nuclear fission
1945	Artturi Virtanen (Finland)	work in agriculture and nutrition, especially fodder preservation
1946	James Sumner (USA)	discovery of crystallization of enzymes
	John Northrop (USA) and Wendell Stanley (USA)	preparation of pure enzymes and virus proteins
1947	Robert Robinson (UK)	investigation of biologically important plant products, especially alkaloids
1948	Arne Tiselius (Sweden)	researches in electrophoresis and adsorption analysis, and discoveries concerning serum proteins
1949	William Giauque (USA)	work in chemical thermodynamics, especially at very low temperatures
1950	Otto Diels (West Germany) and Kurt Alder (West Germany)	discovery and development of diene synthesis
1951	Edwin McMillan (USA) and Glenn Seaborg (USA)	discovery and work in chemistry of transuranic elements
1952	Archer Martin (UK) and Richard Synge (UK)	development of partition chromatography
1953	Hermann Staudinger (West Germany)	discoveries in macromolecular chemistry
1954	Linus Pauling (USA)	study of nature of chemical bonds, especially in complex substances
1955	Vincent Du Vigneaud (USA)	investigations into biochemically important sulphur compounds, and the first synthesis of a polypeptide hormone
1956	Cyril Hinshelwood (UK) and Nikolai Semenov (USSR)	work in mechanism of chemical reactions
1957	Alexander Todd (UK)	work in nucleotides and nucleotide coenzymes
1958	Frederick Sanger (UK)	determination of the structure of proteins, especially insulin
1959	Jaroslav Heyrovsk_ (Czechoslovakia)	discovery and development of polarographic methods of chemical analysis
1960	Willard Libby (USA)	development of radiocarbon dating in archaeology, geology, and geography
1961	Melvin Calvin (USA)	study of assimilation of carbon dioxide by plants
1962	Max Perutz (UK) and John Kendrew (UK)	determination of structures of globular proteins
1963	Karl Ziegler (West Germany) and Giulio Natta (Italy)	chemistry and technology of producing high polymers
1964	Dorothy Crowfoot Hodgkin (UK)	crystallographic determination of the structures of biochemical compounds, notably penicillin and cyanocobalamin (vitamin B12)
1965	Robert Woodward (USA)	organic synthesis
1966	Robert Mulliken (USA)	molecular orbital theory of chemical bonds and structures
1967	Manfred Eigen (West Germany), Ronald Norrish (UK), and George Porter (UK)	investigation of rapid chemical reactions by means of very short pulses of light energy
1968	Lars Onsager (USA)	discovery of reciprocal relations, fundamental for the thermodynamics of irreversible processes
1969	Derek Barton (UK) and Odd Hassel (Norway)	concept and applications of conformation in chemistry
1970	Luis Federico Leloir (Argentina)	discovery of sugar nucleotides and their role in carbohydrate biosynthesis
1971	Gerhard Herzberg (Canada)	research on electronic structure and geometry of molecules, particularly free radicals
1972	Christian Anfinsen (USA), Stanford Moore (USA), and William Stein (USA)	work in amino-acid structure and biological activity of the enzyme ribonuclease
1973	Ernst Fischer (West Germany) and Geoffrey Wilkinson (UK)	work in chemistry of organometallic sandwich compounds
1974	Paul Flory (USA)	studies of physical chemistry of macromolecules
1975	John Cornforth (UK)	work in stereochemistry of enzyme-catalysed reactions
	Vladimir Prelog (Switzerland)	work in stereochemistry of organic molecules and their reactions
1976	William Lipscomb (USA)	study of structure and chemical bonding of boranes (compounds of boron and hydrogen)

Nobel Prize for Chemistry continued

Year	Winner(s)[1]	Awarded for
1977	Ilya Prigogine (Belgium)	work in thermodynamics of irreversible and dissipative processes
1978	Peter Mitchell (UK)	formulation of a theory of biological energy transfer and chemiosmotic theory
1979	Herbert Brown (USA) and Georg Wittig (West Germany)	use of boron and phosphorus compounds, respectively, in organic syntheses
1980	Paul Berg (USA)	biochemistry of nucleic acids, especially recombinant DNA
	Walter Gilbert (USA) and Frederick Sanger (UK)	base sequences in nucleic acids
1981	Kenichi Fukui (Japan) and Roald Hoffmann (USA)	theories concerning chemical reactions
1982	Aaron Klug (UK)	determination of crystallographic electron microscopy: structure of biologically important nucleic-acid–protein complexes
1983	Henry Taube (USA)	study of electron-transfer reactions in inorganic chemical reactions
1984	Bruce Merrifield (USA)	development of chemical syntheses on a solid matrix
1985	Herbert Hauptman (USA) and Jerome Karle (USA)	development of methods of determining crystal structures
1986	Dudley Herschbach (USA), Yuan Lee (USA), and John Polanyi (Canada)	development of dynamics of chemical elementary processes
1987	Donald Cram (USA), Jean-Marie Lehn (France), and Charles Pedersen (USA)	development of molecules with highly selective structure-specific interactions
1988	Johann Deisenhofer (West Germany), Robert Huber (West Germany), and Hartmut Michel (West Germany)	discovery of three-dimensional structure of the reaction centre of photosynthesis
1989	Sidney Altman (USA) and Thomas Cech (USA)	discovery of catalytic function of RNA
1990	Elias James Corey (USA)	new methods of synthesizing chemical compounds
1991	Richard Ernst (Switzerland)	improvements in the technology of nuclear magnetic resonance (NMR) imaging
1992	Rudolph Marcus (USA)	theoretical discoveries relating to reduction and oxidation reactions
1993	Kary Mullis (USA)	invention of the polymerase chain reaction technique for amplifying DNA
	Michael Smith (Canada)	invention of techniques for splicing foreign genetic segments into an organism's DNA in order to modify the proteins produced
1994	George Olah (USA)	development of technique for examining hydrocarbon molecules
1995	F Sherwood Rowland (USA), Mario Molina (USA), and Paul Crutzen (Netherlands)	explaining the chemical process of the ozone layer
1996	Robert Curl Jr (USA), Harold Kroto (UK), and Richard Smalley (USA)	discovery of fullerenes
1997	John Walker (UK), Paul Boyer (USA), and Jens Skou (Denmark)	study of the enzymes involved in the production of adenosine triphosphate (ATP), which acts as a store of energy in bodies called mitochondria inside cells
1998	Walter Kohn (USA), John Pople (USA)	research into quantum chemistry
1999	Ahmed Zewail (USA) spectroscopy	studies of the transition states of chemical reactions using femtosecond
2000	Alan J Heeger (USA), Alan G MacDiarmid (New Zealand), and Hideki Shirakawa (Japan)	their roles in the development of electrically conductive polymers
2001	William S Knowles (USA) and Ryoji Noyori (Japan)	work on chirally catalysed hydrogenation reactions
	K Barry Sharpless (USA)	work on chirally catalysed oxidation reactions
2002	John B Fenn (USA) and Koichi Tanaka (Japan)	work on analyses of biological macromolecules
	Kurt Wüthrich (Switzerland)	development of nuclear magnetic resonance spectroscopy for the study of biological macromolecules
2003	Peter Agre (USA)	discovery of water channels in cell membranes
	Roderick Mackinnon (USA)	discovery of structural and mechanistic srudies of ion channels

[1] Nationality given is the citizenship of recipient at the time award was made.

NOBEL PRIZE FOR PHYSICS

Year	Winner(s) I	Awarded for
1901	Wilhelm Röntgen (Germany)	discovery of X-rays
1902	Hendrik Lorentz (Netherlands) and Pieter Zeeman (Netherlands)	influence of magnetism on radiation phenomena
1903	Henri Becquerel (France)	discovery of spontaneous radioactivity
	Pierre Curie (France) and Marie Curie (France)	research on radiation phenomena
1904	John Strutt (Lord Rayleigh, UK)	densities of gases and discovery of argon
1905	Philipp von Lenard (Germany)	work on cathode rays
1906	Joseph J Thomson (UK)	theoretical and experimental work on the conduction of electricity by gases
1907	Albert Michelson (USA)	measurement of the speed of light through the design and application of precise optical instruments such as the interferometer
1908	Gabriel Lippmann (France)	photographic reproduction of colours by interference
1909	Guglielmo Marconi (Italy) and Karl Ferdinand Braun (Germany)	development of wireless telegraphy
1910	Johannes van der Waals (Netherlands)	equation describing the physical behaviour of gases and liquids
1911	Wilhelm Wien (Germany)	laws governing radiation of heat
1912	Nils Dalén (Sweden)	invention of light-controlled valves, which allow lighthouses and buoys to operate automatically
1913	Heike Kamerlingh Onnes (Netherlands)	studies of properties of matter at low temperatures
1914	Max von Laue (Germany)	discovery of diffraction of X-rays by crystals
1915	William Bragg (UK) and Lawrence Bragg (UK)	X-ray analysis of crystal structures
1916	no award	
1917	Charles Barkla (UK)	discovery of characteristic X-ray emission of the elements
1918	Max Planck (Germany)	formulation of quantum theory
1919	Johannes Stark (Germany)	discovery of Doppler effect in rays of positive ions, and splitting of spectral lines in electric fields
1920	Charles Guillaume (Switzerland)	discovery of anomalies in nickel–steel alloys
1921	Albert Einstein (Switzerland)	theoretical physics, especially law of photoelectric effect
1922	Niels Bohr (Denmark)	discovery of the structure of atoms and radiation emanating from them
1923	Robert Millikan (USA)	discovery of the electric charge of an electron, and study of the photoelectric effect
1924	Karl Siegbahn (Sweden)	X-ray spectroscopy
1925	James Franck (Germany) and Gustav Hertz (Germany)	discovery of laws governing the impact of an electron upon an atom
1926	Jean Perrin (France)	confirmation of the discontinuous structure of matter
1927	Arthur Compton (USA)	transfer of energy from electromagnetic radiation to a particle
	Charles Wilson (UK)	invention of the Wilson cloud chamber, by which the movement of electrically charged particles may be tracked
1928	Owen Richardson (UK)	work on thermionic phenomena and associated law
1929	Louis Victor de Broglie (France)	discovery of the wavelike nature of electrons
1930	Chandrasekhara Raman (India)	discovery of the scattering of single-wavelength light when it is passed through a transparent substance
1931	no award	
1932	Werner Heisenberg (Germany)	creation of quantum mechanics
1933	Erwin Schrödinger (Austria) and Paul Dirac (UK)	development of quantum mechanics
1934	no award	
1935	James Chadwick (UK)	discovery of the neutron

Nobel Prize for Physics continued

Year	Winner(s)I	Awarded for
1936	Victor Hess (Austria)	discovery of cosmic rays
	Carl Anderson (USA)	discovery of the positron
1937	Clinton Davisson (USA) and George Thomson (UK)	diffraction of electrons by crystals
1938	Enrico Fermi (Italy)	use of neutron irradiation to produce new elements, and discovery of nuclear reactions induced by slow neutrons
1939	Ernest Lawrence (USA)	invention and development of the cyclotron, and production of artificial radioactive elements
1940	no award	
1941	no award	
1942	no award	
1943	Otto Stern (USA)	molecular-ray method of investigating elementary particles, and discovery of magnetic moment of proton
1944	Isidor Isaac Rabi (USA)	resonance method of recording the magnetic properties of atomic nuclei
1945	Wolfgang Pauli (Austria)	discovery of the exclusion principle
1946	Percy Bridgman (USA)	development of high-pressure physics
1947	Edward Appleton (UK)	physics of the upper atmosphere
1948	Patrick Blackett (UK)	application of the Wilson cloud chamber to nuclear physics and cosmic radiation
1949	Hideki Yukawa (Japan)	theoretical work predicting existence of mesons
1950	Cecil Powell (UK)	use of photographic emulsion to study nuclear processes, and discovery of pions (pi mesons)
1951	John Cockcroft (UK) and Ernest Walton (Ireland)	transmutation of atomic nuclei by means of accelerated subatomic particles
1952	Felix Bloch (USA) and Edward Purcell (USA)	precise nuclear magnetic measurements
1953	Frits Zernike (Netherlands)	invention of phase-contrast microscope
1954	Max Born (UK)	statistical interpretation of wave function in quantum mechanics
	Walther Bothe (West Germany)	coincidence method of detecting the emission of electrons
1955	Willis Lamb (USA)	structure of hydrogen spectrum
	Polykarp Kusch (USA)	determination of magnetic moment of the electron
1956	William Shockley (USA), John Bardeen (USA), and Walter Houser Brattain (USA)	study of semiconductors, and discovery of the transistor effect
1957	Tsung-Dao Lee (China) and Chen Ning Yang (China)	investigations of weak interactions between elementary particles
1958	Pavel Cherenkov (USSR), Ilya Frank (USSR), and Igor Tamm (USSR)	discovery and interpretation of Cherenkov radiation
1959	Emilio Segrè (USA) and Owen Chamberlain (USA)	discovery of the antiproton
1960	Donald Glaser (USA)	invention of the bubble chamber
1961	Robert Hofstadter (USA)	scattering of electrons in atomic nuclei, and structure of protons and neutrons
	Rudolf Mössbauer (West Germany)	resonance absorption of gamma radiation
1962	Lev Landau (USSR)	theories of condensed matter, especially liquid helium
1963	Eugene Wigner (USA)	discovery and application of symmetry principles in atomic physics
	Maria Goeppert-Mayer (USA) and Hans Jensen (Germany)	discovery of the shell-like structure of atomic nuclei
1964	Charles Townes (USA), Nikolai Basov (USSR), and Aleksandr Prokhorov (USSR)	work on quantum electronics leading to construction of oscillators and amplifiers based on maser–laser principle
1965	Sin-Itiro Tomonaga (Japan), Julian Schwinger (USA), and Richard Feynman (USA)	basic principles of quantum electrodynamics

Nobel Prize for Physics continued

Year	Winner(s)1	Awarded for
1966	Alfred Kastler (France)	development of optical pumping, whereby atoms are raised to higher energy levels by illumination
1967	Hans Bethe (USA)	theory of nuclear reactions, and discoveries concerning production of energy in stars
1968	Luis Alvarez (USA)	elementary-particle physics, and discovery of resonance states, using hydrogen bubble chamber and data analysis
1969	Murray Gell-Mann (USA)	classification of elementary particles, and study of their interactions
1970	Hannes Alfvén (Sweden)	work in magnetohydrodynamics and its applications in plasma physics
	Louis Néel (France)	work in antiferromagnetism and ferromagnetism in solid-state physics
1971	Dennis Gabor (UK)	invention and development of holography
1972	John Bardeen (USA), Leon Cooper (USA), and John Robert Schrieffer (USA)	theory of superconductivity
1973	Leo Esaki (Japan) and Ivar Giaever (USA)	tunnelling phenomena in semiconductors and superconductors
	Brian Josephson (UK)	theoretical predictions of the properties of a supercurrent through a tunnel barrier
1974	Martin Ryle (UK) and Antony Hewish (UK)	development of radioastronomy, particularly the aperture-synthesis technique, and the discovery of pulsars
1975	Aage Bohr (Denmark), Ben Mottelson (Denmark), and James Rainwater (USA)	discovery of connection between collective motion and particle motion in atomic nuclei, and development of theory of nuclear structure
1976	Burton Richter (USA) and Samuel Ting (USA)	discovery of the psi meson
1977	Philip Anderson (USA), Nevill Mott (UK), and John Van Vleck (USA)	contributions to understanding electronic structure of magnetic and disordered systems
1978	Peter Kapitza (USSR)	invention and application of low-temperature physics
	Arno Penzias (USA) and Robert Wilson (USA)	discovery of cosmic background radiation
1979	Sheldon Glashow (USA), Abdus Salam (Pakistan), and Steven Weinberg (USA)	unified theory of weak and electromagnetic fundamental forces, and prediction of the existence of the weak neutral current
1980	James W Cronin (USA) and Val Fitch (USA)	violations of fundamental symmetry principles in the decay of neutral kaon mesons
1981	Nicolaas Bloembergen (USA) and Arthur Schawlow (USA)	development of laser spectroscopy
	Kai Siegbahn (Sweden)	high-resolution electron spectroscopy
1982	Kenneth Wilson (USA)	theory for critical phenomena in connection with phase transitions
1983	Subrahmanyan Chandrasekhar (USA)	theoretical studies of physical processes in connection with structure and evolution of stars
	William Fowler (USA)	nuclear reactions involved in the formation of chemical elements in the universe
1984	Carlo Rubbia (Italy) and Simon van der Meer (Netherlands)	contributions to the discovery of the W and Z particles (weakons)
1985	Klaus von Klitzing (West Germany)	discovery of the quantized Hall effect
1986	Ernst Ruska (West Germany)	electron optics, and design of the first electron microscope
	Gerd Binnig (West Germany) and Heinrich Rohrer (Switzerland)	design of scanning tunnelling microscope
1987	Georg Bednorz (West Germany) and Alex Müller (Switzerland)	superconductivity in ceramic materials
1988	Leon M Lederman (USA), Melvin Schwartz (USA), and Jack Steinberger (USA)	neutrino-beam method, and demonstration of the doublet structure of leptons through discovery of muon neutrino
1989	Norman Ramsey (USA)	measurement techniques leading to development of caesium atomic clock
	Hans Dehmelt (USA) and Wolfgang Paul (Germany)	ion-trap method for isolating single atoms

Nobel Prize for Physics continued

Year	Winner(s)1	Awarded for
1990	Jerome Friedman (USA), Henry Kendall (USA), and Richard Taylor (Canada)	experiments demonstrating that protons and neutrons are made up of quarks
1991	Pierre-Gilles de Gennes (France)	work on disordered systems including polymers and liquid crystals; development of mathematical methods for studying the behaviour of molecules in a liquid on the verge of solidifying
1992	Georges Charpak (France)	invention and development of detectors used in high-energy physics
1993	Joseph Taylor (USA) and Russell Hulse (USA)	discovery of first binary pulsar (confirming the existence of gravitational waves)
1994	Clifford Shull (USA) and Bertram Brockhouse (Canada)	development of technique known as 'neutron scattering' which led to advances in semiconductor technology
1995	Frederick Reines (USA)	discovery of the neutrino
	Martin Perl (USA)	discovery of the tau lepton
1996	David Lee (USA), Douglas Osheroff (USA), and Robert Richardson (USA)	discovery of superfluidity in helium-3
1997	Claude Cohen-Tannoudji (France), William Phillips (USA), and Steven Chu (USA)	discovery of a way to slow down individual atoms using lasers for study in a near-vacuum
1998	Robert B Laughlin (USA), Horst L Störmer (USA), and Daniel C Tsui (USA)	discovery of a new form of quantum fluid with fractionally charged excitations
1999	Gerardus 't Hooft (Netherlands) and Martinus Veltman (Netherlands)	elucidating the quantum structure of electroweak interactions in physics
2000	Zhores I Alferov (Russia) and Herbert Kroemer (Germany)	development of semiconductor heterostructures, which lead to faster transistors and more efficient laser diodes
	Jack St Clair Kilby (USA)	co-invention of the integrated circuit
2001	Eric A Cornell (USA), Wolfgang Ketterle (Germany), and Carl E Wieman (USA)	the achievement of Bose-Einstein condensation in dilute gases of alkali atoms, and early fundamental studies of the condensates
2002	Raymond Davis, Jr (USA) and Masatoshi Koshiba (Japan)	pioneering contributions to astrophysics, particularly the detection of cosmic neutrinos
	Riccardo Giacconi (USA)	pioneering contributions to astrophysics which have led to the discovery of cosmic X-ray sources
2003	Alexei A. Abrikosov (USA), Vitaly L. Ginsburg (Russia), and Anthony J. Leggett (USA)	pioneering contributions to the theory of superconductors and superfluids

1 Nationality given is the citizenship of recipient at the time award was made.

NUMBER SYSTEMS

Binary (Base 2)	Octal (Base 8)	Decimal (Base 10)	Hexadecimal (Base 16)
0	0	0	0
1	1	1	1
10	2	2	2
11	3	3	3
100	4	4	4
101	5	5	5
110	6	6	6
111	7	7	7
1000	10	8	8
1001	11	9	9
1010	12	10	A
1011	13	11	B
1100	14	12	C
1101	15	13	D
1110	16	14	E
1111	17	15	F
10000	20	16	10
11111111	377	255	FF
11111010001	3721	2001	7D1

MAJOR OCEANS AND SEAS OF THE WORLD

Ocean/sea	Area[1]		Average depth	
	sq km	sq mi	m	ft
Pacific Ocean	166,242,000	64,186,000	3,939	12,925
Atlantic Ocean	86,557,000	33,420,000	3,575	11,730
Indian Ocean	73,429,000	28,351,000	3,840	12,598
Arctic Ocean	13,224,000	5,106,000	1,038	3,407
South China Sea	2,975,000	1,149,000	1,464	4,802
Caribbean Sea	2,754,000	1,063,000	2,575	8,448
Mediterranean Sea	2,510,000	969,000	1,501	4,926
Bering Sea	2,261,000	873,000	1,491	4,893
Sea of Okhotsk	1,580,000	610,000	973	3,192
Gulf of Mexico	1,544,000	596,000	1,614	5,297
Sea of Japan	1,013,000	391,000	1,667	5,468
Hudson Bay	730,000	282,000	93	305
East China Sea	665,000	257,000	189	620
Andaman Sea	565,000	218,000	1,118	3,667
Black Sea	461,000	178,000	1,190	3,906
Red Sea	453,000	175,000	538	1,764
North Sea	427,000	165,000	94	308
Baltic Sea	422,000	163,000	55	180
Yellow Sea	294,000	114,000	37	121
Gulf	230,000	89,000	100	328
Gulf of California	153,000	59,000	724	2,375
English Channel	90,000	35,000	54	177
Irish Sea	89,000	34,000	60	197

1 All figures are approximate as boundaries of oceans and seas cannot be exactly determined.

PERCENTAGES AS FRACTIONS AND DECIMALS

%	Decimal	Fraction	%	Decimal	Fraction	%	Decimal	Fraction
1	0.01	1/100	20	0.20	1/5	41	0.41	41/100
2	0.02	1/50	21	0.21	21/100	42	0.42	21/50
3	0.03	3/100	22	0.22	11/50	43	0.43	43/100
4	0.04	1/25	23	0.23	23/100	44	0.44	11/25
5	0.05	1/20	24	0.24	6/25	45	0.45	9/20
6	0.06	3/50	25	0.25	1/4	46	0.46	23/50
7	0.07	7/100	26	0.26	13/50	47	0.47	47/100
8	0.08	2/25	27	0.27	27/100	48	0.48	12/25
81/3	0.083	1/12	28	0.28	7/25	49	0.49	49/100
9	0.09	9/100	29	0.29	29/100	50	0.50	1/2
10	0.10	1/10	30	0.30	3/10	55	0.55	11/20
11	0.11	11/100	31	0.31	31/100	60	0.60	3/5
12	0.12	3/25	32	0.32	8/25	65	0.65	13/20
121/2	0.125	1/8	33	0.33	33/100	662/3	0.667	2/3
13	0.13	13/100	331/3	0.333	1/3	70	0.70	7/10
14	0.14	7/50	34	0.34	17/50	75	0.75	3/4
15	0.15	3/20	35	0.35	7/20	80	0.80	4/5
16	0.16	4/25	36	0.36	9/25	85	0.85	17/20
162/3	0.167	1/6	37	0.37	37/100	90	0.90	9/10
17	0.17	17/100	38	0.38	19/50	95	0.95	19/20
18	0.18	9/50	39	0.39	39/100	100	1.00	1
19	0.19	19/100	40	0.40	2/5			

PHYSICAL CONSTANTS

Physical constants, or fundamental constants, are physical quantities that are constant in all circumstances throughout the whole universe.

Constant	Symbol	Value in SI units
acceleration of free fall	g	9.80665 m s^{-2}
Avogadro's constant	N_A	6.0221367 x 1023 mol–1
Boltzmann's constant	k	1.380658×10^{-23} J K^{-1}
elementary charge	e	$1.60217733 \times 10^{-19}$ C
electronic rest mass	m_e	$9.1093897 \times 10^{-31}$ kg
Faraday's constant	F	9.6485309×10^{4} C mol^{-1}
gas constant	R	8.314510 J K^{-1} mol^{-1}
gravitational constant	G	6.672×10^{-11} N m2 kg^{-2}
Loschmidt's number	N_L	2.686763×10^{25} m^{-3}
neutron rest mass	m_n	$1.6749286 \times 10^{-27}$ kg
Planck's constant	h	$6.6260755 \times 10^{-34}$ J s
proton rest mass	m_p	$1.6726231 \times 10^{-27}$ kg
speed of light in a vacuum	c	2.99792458×10^{8} m s^{-1}
standard atmosphere	atm	1.01325×10^{5} Pa
Stefan–Boltzmann constant	σ	$5.67051\ \sigma\ 10^{-8}$ W m^{-2} K^{-4}

PHYSICS: KEY DATES

c. 400 BC The first 'atomic' theory is put forward by the Greek philosopher Democritus.

c. 340 Arisotle adapts and elaborates an earlier theory of his earlier Greek compatriot Empedocles that all matter consists of four elements – earth, air, fire, and water – with a fifth element, ether, making up the heavens. He also develops ideas about motion.

c. 250 Archimedes' principle of buoyancy is established. Archimedes also discovers the principle of the lever.

c. 60 The Roman poet Lucretius writes De rerum natura / On the nature of the Universe, which includes an exposition of the atomic theory of Democritus as interpreted by Epicurus, a Greek philosopher of the late 4th and early 3rd centuries.

AD 1600 Magnetism is described by William Gilbert of England.

1608 Hans Lippershey, a Dutch lens-maker, patents the refracting telescope.

c. 1610 The principle of falling bodies descending to earth at the same speed is established by the Italian astronomer and physicist Galileo Galilei.

1642 The principles of hydraulics are put forward by French physicist and mathematician Blaise Pascal.

1643 The mercury barometer is invented by the Italian physicist Evangelista Torricelli.

1656 The pendulum clock is invented by Christiaan Huygens of the Netherlands.

1662 Boyle's law concerning the behaviour of gases is established by Irish physicist and chemist Robert Boyle.

c. 1665 The English physicist and mathematician Isaac Newton works out that the Earth exerts a constant force on falling bodies. This force is called gravitation or gravity.

1685 Newton publishes his Principia mathematica philosophiae naturalis / Principles of natural philosophy, containing an outline of his system of mechanics and including his law of universal gravitation and his three laws of motion.

1690 The wave theory of light is propounded by Christiaan Huygens.

1704 The corpuscular theory of light is put forward by Newton.

1714 The mercury thermometer is invented by a Polish-born Dutch physicist, Daniel Fahrenheit. He also invents a temperature scale that is later refined to register the freezing point of water at 32° and its boiling point at 212°.

1724 Fahrenheit describes the phenomenon of supercooled water.

1729 English physicist Stephen Gray discovers that static electric charges exist on the surfaces of objects, not within them, and that electricity can pass from one object to another and over distances by means of conductors.

1733 French physicist Charles François de Cisternay du Fay discovers that there are two kinds of static electric charge and that like charges repel, while unlike charges attract.

1742 Anders Celsius, a Swedish astronomer, invents a centigrade temperature scale, fixing 0° as the boiling point of water and 100° as the freezing point. In 1743, these points are swapped to give the centigrade scale still in use.

1751 The American polymath Benjamin Franklin describes electricity as a fluid, differentiates between positive and negative electricity, and shows that electricity can magnetize and demagnetize iron. In 1752 he performs his famous experiment, flying a kite into a thunder cloud and proving that lightning is a form of electricity.

1764 Specific and latent heats are described by the Scottish chemist Joseph Black.

1771 The link between nerve action and electricity is discovered by the Italian physicist Luigi Galvani.

c. 1787 Charles's law relating the pressure, volume, and temperature of a gas is established by French physicist and chemist Jacques Charles.

1795 The metric system is adopted in France.

1798 The link between heat and friction is discovered by the American-born physicist Benjamin Thompson, Count Rumford.

1800 Italian physicist Alessandro Volta invents the Voltaic cell.

1801 Interference of light is discovered by the English physicist Thomas Young.

1808 The 'modern' atomic theory is propounded by the English chemist John Dalton.

1811 Avogadro's hypothesis relating volumes and numbers of molecules of gases is proposed by Italian chemist Amedeo Avogadro.

1814 Dark lines in the solar spectrum are mapped by the German physicist and optician Joseph von Fraunhofer, from whom they receive their name.

1815 Refraction of light is explained by French physicist Augustin Fresnel.

1820 The discovery of electromagnetism is made by Danish physicist Hans Oersted.

1821 The dynamo principle is described by the English physicist Michael Faraday.

1822 The laws of electrodynamics are established by the French physicist André Ampère.

1824 Thermodynamics as a branch of physics is foreshadowed in an essay on heat by the French physicist Sadi Carnot.

Physics: Key Dates continued

1826 French physiologist Henri Dutrochet discovers and explains osmosis.

1827 Ohm's law of electrical resistance is established by the German physicist Georg Ohm; Brownian movement (rapid movement resulting from molecular vibrations) is observed by the Scottish botanist Robert Brown in pollen grains suspended in a liquid.

1829 The law of gaseous diffusion is established by Scottish chemist and physicist Thomas Graham.

1831 Electromagnetic induction is discovered independently by Michael Faraday and US physicist Joseph Henry.

1834 Faraday discovers self-induction.

1836 English chemist John Frederic Daniell invents the Daniell cell, a device supplying a steady, long-lasting current through the chemical interaction between zinc and copper electrodes.

1842 The principle of conservation of energy is observed by German physicist Julius von Mayer.

c. 1847 The mechanical equivalent of heat is described by the English physicist James Joule.

1849 A measurement of the velocity of light is put forward by French physicist Armand Fizeau. Fizeau discovers in 1851 that light travels slower in water when the direction of the light beam is against the current of water than when it is following the current.

Scottish mathematician and physicist William Thomson, later Lord Kelvin, invents the term 'thermodynamics' in an article on the theory of heat developed by Sadi Carnot.

1850 Rudolf Clausius publishes the first generalized statement of the second law of thermodynamics, later restated by him as 'In a closed system entropy tends to increase'.

1851 The rotation of the Earth is demonstrated by French physicist Jean Foucault using a pendulum.

William Thomson, later Lord Kelvin, proposes the concept of absolute zero, the temperature at which all molecular movement stops. He determines this temperature to be –273°C/–459.4°F, which is later designated the zero point on a temperature scale named in his honour.

1858 Thomson invents the mirror galvanometer, an instrument for measuring small electric currents.

1859 Spectrographic analysis carried out by Robert Bunsen and Gustav Kirchhoff of Germany reveals that each chemical element is uniquely associated with a specific spectral line or set of lines.

1864 In his book A Dynamic Theory of the Electromagnetic Field, Scottish physicist James Clerk Maxwell presents the first mathematical treatment of electricity and magnetism using Faraday's idea of a field as a set of lines of force.

1873 Maxwell conceives light as electromagnetic radiation.

1877 A theory of sound as vibrations in an elastic medium is propounded by the English physicist John Strutt, 3rd Baron Rayleigh.

1880 Piezoelectricity is discovered by the French physicist Pierre Curie.

1887 The existence of radio waves is predicted by the German physicist Heinrich Hertz.

The US physicists Albert Michelson and Edward Morley measure the velocity of light in two different directions in an attempt to detect the Earth's motion through the ether, the supposed medium allowing the propagation of light. They find no difference in the two measurements, proving that the speed of light is constant and independent of the motion of the observer and that the ether does not exist.

1895 X-rays are discovered by the German physicist Wilhelm Röntgen.

1896 The French physicist Antoine Becquerel discovers that uranium gives off rays. This is the first observation of natural radioactivity.

1897 The English physicist Joseph Thomson discovers the electron.

1898 French physicists Marie and Pierre Curie discover that the element thorium gives off rays like uranium. Marie Curie names the phenomenon radioactivity.

1899 New Zealand-born physicist Ernest Rutherford discovers alpha and beta rays. Becquerel discovers in 1900 that beta rays in fact consist of electrons.

1900 Quantum theory is propounded by Max Planck of Germany; the discovery of gamma rays is made by French physicist Paul-Ulrich Villard.

1902 The English physicist and electrical engineer Oliver Heaviside discovers the ionosphere.

1904 The theory of radioactivity is put forward by Rutherford and English chemist Frederick Soddy.

1905 The German-born physicist Albert Einstein propounds his special theory of relativity.

1908 The Geiger counter is invented by German physicist Hans Geiger and Rutherford.

1911 Rutherford discovers the atomic nucleus.

Physics: Key Dates continued

1911 The Dutch physicist Heike Kamerlingh Onnes observes that mercury, when cooled to just above absolute zero, loses its electrical resistance. The phenomenon becomes known as superconductivity.

1913 The orbiting electron atomic theory is propounded by Danish physicist Niels Bohr.

1915 X-ray crystallography is discovered by the English physicists William and Lawrence Bragg.

1916 Einstein puts forward his general theory of relativity; mass spectrography is discovered by the Englishman William Aston.

1926 Wave mechanics is introduced by the Austrian physicist Erwin Schrödinger.

1927 The uncertainty principle of quantum physics is established by the German physicist Werner Heisenberg.

1931 The cyclotron is developed by US physicist Ernest Lawrence.

1932 James Chadwick of England discovers the neutron; the electron microscope is developed by the Russian-born US physicist Vladimir Zworykin.

1933 The positron, the antiparticle of the electron, is discovered by US physicist Carl Anderson.

1934 Artificial radioactivity is developed by the French physicists Frédéric and Irène Joliot-Curie.

1939 The discovery of nuclear fission is made by the German physicists Otto Hahn and Fritz Strassmann.

1942 The first controlled nuclear chain reaction is achieved by the Italian-born US physicist Enrico Fermi.

1956 The neutrino, an elementary particle, is discovered by Clyde Cowan and Fred Reines of the USA.

1960 The Mössbauer effect of atom emissions is discovered by German physicist Rudolf Mössbauer; the first laser and the first maser are developed by US physicist Theodore Maiman.

1964 Murray Gell-Mann and George Zweig of the USA discover and name the quark.

1967 English astronomers Jocelyn Bell (now Jocelyn Bell Burnell) and Antony Hewish discover pulsars (rapidly rotating neutron stars that emit pulses of energy).

1979 The discovery of the asymmetry of elementary particles is made by US physicists James W Cronin and Val L Fitch.

1982 The discovery of processes involved in the evolution of stars is made by the astrophysicists Subrahmanyan Chandrasekhar of India and William Fowler of USA.

1983 Evidence of the existence of weakons (W and Z particles) is confirmed at CERN, the European centre for nuclear research, validating the link between the weak nuclear force and the electromagnetic force.

1986 The first high-temperature superconductor is discovered, able to conduct electricity without resistance at a temperature of –238°C/–396°F.

1989 CERN's Large Electron Positron Collider (LEP), a particle accelerator with a circumference of 27 km/16.8 mi, comes into operation.

1991 LEP experiments demonstrate the existence of three generations of elementary particles, each with two quarks and two leptons.

1995 Top quark is discovered at Fermilab, the US particle-physics laboratory, near Chicago. US researchers announce the discovery of a material which is superconducting at the temperature of liquid nitrogen – a much higher temperature than previously achieved.

1996 CERN physicists create the first atoms of antimatter (nine atoms of antihydrogen). The Lawrence Livermore National Laboratory, California, USA, produces a laser of 1.3 petawatts (130 trillion watts).

1997 A new subatomic particle, an exotic meson, is possibly discovered at Brookhaven National Laboratory, Upton, New York, USA. The exotic meson is made up of either a quark, an antiquark and a gluon, or two quarks and two antiquarks. US physicists display the first atomic laser. It emits atoms that act like lightwaves.

1999 Scientists succeed in slowing down the speed of light from its normal speed of 299,792 km/186,282 mi per second to 61 km/38 mi per hour, opening up potential for the development of high-precision computer and telecommunications technologies, as well as for the advanced study of quantum mechanics.

2000 CERN scientists announce a new state of matter produced when lead atoms are made to collide in a particle accelerator. It is 20 times as dense as nuclear matter and is believed to have existed about 10 microseconds after the Big Bang. In this new state of matter, quarks are not bound together but roam freely.

PERIODIC TABLE OF THE ELEMENTS

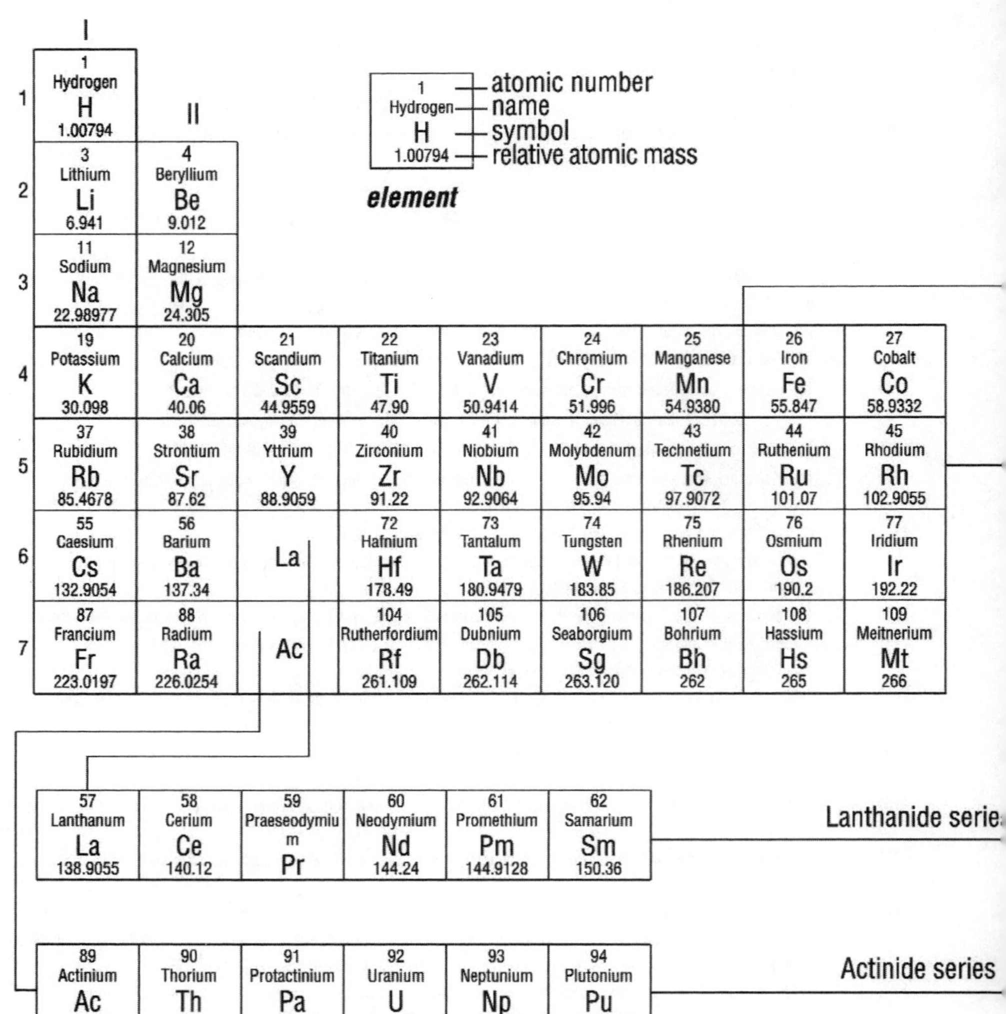

	I	II							
1	1 Hydrogen H 1.00794								
2	3 Lithium Li 6.941	4 Beryllium Be 9.012							
3	11 Sodium Na 22.98977	12 Magnesium Mg 24.305							
4	19 Potassium K 30.098	20 Calcium Ca 40.06	21 Scandium Sc 44.9559	22 Titanium Ti 47.90	23 Vanadium V 50.9414	24 Chromium Cr 51.996	25 Manganese Mn 54.9380	26 Iron Fe 55.847	27 Cobalt Co 58.9332
5	37 Rubidium Rb 85.4678	38 Strontium Sr 87.62	39 Yttrium Y 88.9059	40 Zirconium Zr 91.22	41 Niobium Nb 92.9064	42 Molybdenum Mo 95.94	43 Technetium Tc 97.9072	44 Ruthenium Ru 101.07	45 Rhodium Rh 102.9055
6	55 Caesium Cs 132.9054	56 Barium Ba 137.34	La	72 Hafnium Hf 178.49	73 Tantalum Ta 180.9479	74 Tungsten W 183.85	75 Rhenium Re 186.207	76 Osmium Os 190.2	77 Iridium Ir 192.22
7	87 Francium Fr 223.0197	88 Radium Ra 226.0254	Ac	104 Rutherfordium Rf 261.109	105 Dubnium Db 262.114	106 Seaborgium Sg 263.120	107 Bohrium Bh 262	108 Hassium Hs 265	109 Meitnerium Mt 266

57 Lanthanum La 138.9055	58 Cerium Ce 140.12	59 Praseodymiu m Pr	60 Neodymium Nd 144.24	61 Promethium Pm 144.9128	62 Samarium Sm 150.36	Lanthanide series
89 Actinium Ac 227.0278	90 Thorium Th 232.0381	91 Protactinium Pa 231.0359	92 Uranium U 238.029	93 Neptunium Np 237.0482	94 Plutonium Pu 244.0642	Actinide series

			III	IV	V	VI	VII	0
								2 Helium He 4002.60
			5 Boron B 10.81	6 Carbon C 12.011	7 Nitrogen N 14.0067	8 Oxygen O 15.9994	9 Fluorine F 18.99840	10 Neon Ne 20.179
			13 Aluminium Al 26.98154	14 Silicon Si 28.066	15 Phosphorus P 30.9738	16 Sulphur S 32.06	17 Chlorine Cl 35.453	18 Argon Ar 39.948
28 Nickel Ni 58.70	29 Copper Cu 63.546	30 Zinc Zn 65.38	31 Gallium Ga 69.72	32 Germanium Ge 72.59	33 Arsenic As 74.9216	34 Selenium Se 78.96	35 Bromine Br 79.904	36 Krypton Kr 83.80
46 Palladium Pd 106.4	47 Silver Ag 107.868	48 Cadmium Cd 112.40	49 Indium In 114.82	50 Tin Sn 118.69	51 Antimony Sb 121.75	52 Tellurium Te 127.75	53 Iodine I 126.9045	54 Xenon Xe 131.30
78 Platinum Pt 195.09	79 Gold Au 196.9665	80 Mercury Hg 200.59	81 Thallium Tl 204.37	82 Lead Pb 207.37	83 Bismuth Bi 207.2	84 Polonium Po 210	85 Astatine At 211	86 Radon Rn 222.0176
110 Ununnilium Uun 269	111 Unununium Uuu 272	112 Ununbium Uub 277						

63 Europium Eu 151.96	64 Gadolinium Gd 157.25	65 Terbium Tb 158.9254	66 Dysprosium Dy 162.50	67 Holmium Ho 164.9304	68 Erbium Er 167.26	69 Thulium Tm 168.9342	70 Ytterbium Yb 173.04	71 Lutetium Lu 174.97

95 Americium Am 243.0614	96 Curium Cm 247.0703	97 Berkelium Bk 247	98 Californium Cf 251.0786	99 Einsteinium Es 252.0828	100 Fermium Fm 257.0951	101 Mendelevium Md 258.0986	102 Nobelium No 259.1009	103 Lawrencium Lr 260.1054

THE PLANETS

(– = not applicable.)

Planet	Main constituents	Atmosphere	Average distance from the Sun (in millions)		Orbital period (Earth yrs)	Day length	Diameter (in thousands)		Density[1]
			km	mi			km	mi	
Mercury	rock, ferrous	–	58	36	0.241	58d 15h 30m	4.88	3.03	5.4
Venus	rock, ferrous	carbon dioxide	108	67	0.615	243d 0h 27m	12.10	7.51	5.2
Earth	rock, ferrous	nitrogen, oxygen	150	93	1.000	23h 56m	12.76	7.92	5.5
Mars	rock	carbon dioxide	228	141	1.880	24h 37m	6.78	4.21	3.9
Jupiter	liquid hydrogen, helium	–	778	483	11.860	9h 51m	142.80	88.73	1.3
Saturn	hydrogen, helium	–	1,427	886	29.460	10h 14m	120.00	74.56	0.7
Uranus	ice, hydrogen, helium	hydrogen, helium	2,870	1,783	84.000	17h 14m	50.80	31.56	1.3
Neptune	ice, hydrogen, helium	hydrogen, helium	4,497	2,794	164.800	16h 7m	48.60	30.20	1.6
Pluto	ice, rock	methane	5,900	3,666	248.50	6d 9h 17m	2.27	1.41	~2.0

1 Relative density: 1 = density of water.

PRIME NUMBERS

A prime number is a number that can be divided only by 1 and itself, that is, having no other factors. There is an infinite number of primes. Below are all the prime numbers between 1 and 1,000.

All the prime numbers between 1 and 1,000

2	3	5	7	11	13	17	19	23	29
31	37	41	43	47	53	59	61	67	71
73	79	83	89	97	101	103	107	109	113
127	131	137	139	149	151	157	163	167	173
179	181	191	193	197	199	211	223	227	229
233	239	241	251	257	263	269	271	277	281
283	293	307	311	313	317	331	337	347	349
353	359	367	373	379	383	389	397	401	409
419	421	431	433	439	443	449	457	461	463
467	479	487	491	499	503	509	521	523	541
547	557	563	569	571	577	587	593	599	601
607	613	617	619	631	641	643	647	653	659
661	673	677	683	691	701	709	719	727	733
739	743	751	757	761	769	773	787	797	809
811	821	823	827	829	839	853	857	859	863
877	881	883	887	907	911	919	929	937	941
947	953	967	971	977	983	991	997		

PROGRAMMING LANGUAGES

Language	Main uses	Description
Ada	defence applications	high-level
assembler languages	jobs needing detailed control of the hardware, fast execution, and small program size	fast and efficient but require considerable effort and skill
ALGOL (algorithmic language)	mathematical work	high-level with an algebraic style; no longer in current use, but has influenced languages such as Ada and Pascal
BASIC (beginners' all-purpose symbolic instruction code)	mainly in education, business, and the home, and among non-professional programmers, such as engineers	easy to learn; early versions lacked the features of other languages
C	systems and general programming	fast and efficient; widely used as a general-purpose language; especially popular among professional programmers
C++	systems and general programming; commercial software development	developed from C, adding the advantages of object-oriented programming
COBOL (common business-oriented language)	business programming	strongly oriented towards commercial work; easy to learn but very verbose; widely used on mainframes
FORTH	control applications	reverse Polish notation language
FORTRAN (formula translation)	scientific and computational work	based on mathematical formulae; popular among engineers, scientists, and mathematicians
Java	developed for consumer electronics; used for many interactive Web sites	multipurpose, cross-platform, object-oriented language with similar features to C and C++ but simpler
LISP (list processing)	artificial intelligence	symbolic language with a reputation for being hard to learn; popular in the academic and research communities
LOGO	teaching of mathematical concepts	high-level; popular with schools and home computer users
Modula-2	systems and real-time programming; general programming highly-structured	intended to replace Pascal for 'real-world' applications
OBERON	general programming	small, compact language incorporating many of the features of Pascal and Modula-2
Pascal (program appliqué à la sélection et la compilation automatique de la littérature)	general-purpose language	highly-structured; widely used for teaching programming in universities
Perl (practical extraction and report language)	systems programming and Web development	easy manipulation of text, files, and processes, especially in Unix environment
PROLOG (programming in logic)	artificial intelligence	symbolic-logic programming system, originally intended for theorem solving but now used more generally in artificial intelligence

THE RICHTER SCALE

The Richter scale is based on measurement of seismic waves, used to determine the magnitude of an earthquake at its epicentre. The magnitude of an earthquake differs from its intensity, measured by the Mercalli scale, which is subjective and varies from place to place for the same earthquake. The Richter scale was named after US seismologist Charles Richter (1900–1985). The relative amount of energy released indicates the ratio of energy between earthquakes of different magnitude.

Magnitude	Relative amount of energy released	Examples	Year
1	1		
2	31		
3	960		
4	30,000	Carlisle, England (4.7)	1979
5	920,000	Wrexham, Wales (5.1)	1990
6	29,000,000	San Fernando (CA), USA (6.5)	1971
		northern Armenia (6.8)	1988
7	890,000,000	Loma Prieta (CA), USA (7.1)	1989
		Kobe, Japan (7.2)	1995
		Izmit, Turkey (7.4)	1999
		Taichung and Nantou counties, Taiwan (7.6)	1999
		Rasht, Iran (7.7)	1990
		San Francisco (CA), USA (7.7–7.9)1	1906
8	28,000,000,000	Tangshan, China (8.0)	1976
		Gansu, China (8.6)	1920
		Lisbon, Portugal (8.7)	1755
9	850,000,000,000	Prince William Sound (AK), USA (9.2)	1964

1 Richter's original estimate of a magnitude of 8.3 was revised by two studies carried out by the California Institute of Technology and the US Geological Survey.

LONGEST RIVERS IN THE WORLD

River	Location	Approximate length	
		km	mi
Nile	Africa	6,695	4,160
Amazon	South America	6,570	4,083
Chang Jiang (Yangtze)	China	6,300	3,915
Mississippi–Missouri–Red Rock	USA	6,020	3,741
Huang He (Yellow River)	China	5,464	3,395
Ob–Irtysh	China/Kazakhstan/Russian Federation	5,410	3,362
Amur–Shilka	Asia	4,416	2,744
Lena	Russian Federation	4,400	2,734
Congo	Africa	4,374	2,718
Mackenzie–Peace–Finlay	Canada	4,241	2,635
Mekong	Asia	4,180	2,597
Niger	Africa	4,100	2,548
Yenisei	Russian Federation	4,100	2,548
Paraná	Brazil	3,943	2,450
Mississippi	USA	3,779	2,348
Murray–Darling	Australia	3,751	2,331
Missouri	USA	3,726	2,315
Volga	Russian Federation	3,685	2,290
Madeira	Brazil	3,241	2,014
Purus	Brazil	3,211	1,995
São Francisco	Brazil	3,199	1,988
Yukon	USA/Canada	3,185	1,979
Rio Grande	USA/Mexico	3,058	1,900
Indus	Tibet/Pakistan	2,897	1,800
Danube	central and eastern Europe	2,858	1,776
Japura	Brazil	2,816	1,750
Salween	Myanmar/China	2,800	1,740
Brahmaputra	Asia	2,736	1,700
Euphrates	Iraq	2,736	1,700
Tocantins	Brazil	2,699	1,677
Zambezi	Africa	2,650	1,647
Orinoco	Venezuela	2,559	1,590
Paraguay	Paraguay	2,549	1,584
Amu Darya	Tajikistan/Turkmenistan/Uzbekistan	2,540	1,578
Ural	Russian Federation/Kazakhstan	2,535	1,575
Kolyma	Russian Federation	2,513	1,562
Ganges	India/Bangladesh	2,510	1,560
Arkansas	USA	2,344	1,459
Colorado	USA	2,333	1,450
Dnieper	Russian Federation/Belarus/Ukraine	2,285	1,420
Syr Darya	Asia	2,205	1,370
Irrawaddy	Myanmar	2,152	1,337
Orange	South Africa	2,092	1,300

LARGEST NATURAL PLANETARY SATELLITES

Planet	Satellite	Diameter		Mean distance from centre of primary planet		Orbital period (Earth days)
		km	mi	km	mi	
Jupiter	Ganymede	5,262	3,300	1,070,000	664,898	7.16
Saturn	Titan	5,150	3,200	1,221,800	759,226	15.95
Jupiter	Callisto	4,800	3,000	1,883,000	1,170,096	16.69
Jupiter	Io	3,630	2,240	421,600	261,982	1.77
Earth	Moon	3,476	2,160	384,400	238,866	27.32
Jupiter	Europa	3,138	1,900	670,900	416,897	3.55
Neptune	Triton	2,700	1,690	354,300	220,162	5.88

SI PREFIXES

To avoid the use of zeros in large or small numbers, prefixes are added to each of the SI units to indicate multiples and submultiples of ten.

Multiple	Prefix	Symbol	Example
1,000,000,000,000,000,000 (10^{18})	exa-	E	Eg (exagram)
1,000,000,000,000,000 (10^{15})	peta-	P	PJ (petajoule)
1,000,000,000,000 (10^{12})	tera-	T	TV (teravolt)
1,000,000,000 (10^{9})	giga-	G	GW (gigawatt)
1,000,000 (10^{6})	mega-	M	MHz (megahertz)
1,000 (10^{3})	kilo-	k	kg (kilogram)
100 (10^{2})	hecto-	h	hm (hectometre)
10 (10^{1})	deca-	da	daN (decanewton)
1/10 (10^{-1})	deci-	d	dC (decicoulomb)
1/100 (10^{-2})	centi-	c	cm (centimetre)
1/1,000 (10^{-3})	milli-	m	mm (millimetre)
1/1,000,000 (10^{-6})	micro-	µ	µF (microfarad)
1/1,000,000,000 (10^{-9})	nano-	n	nm (nanometre)
1/1,000,000,000,000 (10^{-12})	pico-	p	ps (picosecond)
1/1,000,000,000,000,000 (10^{-15})	femto-	f	frad (femtoradian)
1/1,000,000,000,000,000,000 (10^{-18})	atto-	a	aT (attotesla)

SI UNITS

Quantity	SI unit	Symbol
Base units		
length	metre	m
mass	kilogram	kg
time	second	s
electric current	ampere	A
temperature, thermodynamic	kelvin	K
amount of substance	mole	mol
luminous intensity	candela	cd
Derived units		
absorbed radiation dose	gray	Gy
electric capacitance	farad	F
electric charge	coulomb	C
electric conductance	siemens	S
energy or work	joule	J
force	newton	N
frequency	hertz	Hz

Quantity	SI unit	Symbol
Derived units continued		
illuminance	lux	lx
inductance	henry	H
luminous flux	lumen	lm
magnetic flux	weber	Wb
magnetic flux density	tesla	T
plane angle	radian	rad
potential difference	volt	V
power	watt	W
pressure	pascal	Pa
radiation dose equivalent	sievert	Sv
radiation exposure	roentgen	R
radioactivity	becquerel	Bq
resistance	ohm	Ω
solid angle	steradian	sr
sound intensity	decibel	dB
temperature	degree Celsius	°C

SPACE PROBES: KEY DATES

13 September 1959	Luna 2 hits the Moon, the first craft to do so
10 October 1959	Luna 3 photographs the far side of the Moon
14 December 1962	Mariner 2 flies past Venus; launch date 26 August 1962
31 July 1964	Ranger 7 hits the Moon, having sent back 4,316 pictures before impact
14 July 1965	Mariner 4 flies past Mars; launch date 28 November 1964
3 February 1966	Luna 9 achieves the first soft landing on the Moon, having transmitted 27 close-up panoramic photographs; launch date 31 January 1966
2 June 1966	Surveyor 1 lands softly on the Moon and returns 11,150 pictures; launch date 30 May 1965
13 November 1971	Mariner 9 enters orbit of Mars; launch date 30 May 1971
3 December 1973	Pioneer 10 flies past Jupiter; launch date 3 March 1972
29 March 1974	Mariner 10 flies past Mercury; launch date 3 November 1973
22 October 1975	Venera 9 lands softly on Venus and returns its first pictures; launch date 8 June 1975
20 July 1976	Viking 1 first lands on Mars; launch date 20 August 1975
3 September 1976	Viking 2 transmits data from the surface of Mars
20 August 1977	Voyager 2 is launched
5 September 1977	Voyager 1 is launched. With a faster trajectory, it will reach Jupiter sooner than Voyager 2
4 December 1978	Pioneer-Venus 1 orbits Venus; launch date 20 May 1978
5 March 1979	Voyager 1 encounters Jupiter
9 July 1979	Voyager 2 encounters Jupiter
12 November 1980	Voyager 1 reaches Saturn
25 August 1981	Voyager 2 flies past Saturn
1 March 1982	Venera 13 transmits its first colour pictures of the surface of Venus; launch date 30 October 1981
10 October 1983	Venera 15 maps the surface of Venus from orbit; launch date 2 June 1983
2 July 1985	The European Space Agency (ESA) probe Giotto is launched to Halley's comet
24 January 1986	Voyager 2 encounters Uranus
13–14 March 1986	Giotto meets Halley's comet, closest approach 596 km/370 mi, at a speed 50 times faster than that of a bullet
4 May 1989	Magellan launches from space shuttle Atlantis on a 15-month cruise to Venus across 15 million km/9 million mi of space
25 August 1989	Voyager 2 reaches Neptune (4,400 million km/2,700 million mi from Earth), approaching it to within 4,850 km/3,010 mi
18 October 1989	Galileo is launched from space shuttle Atlantis for a six-year journey to Jupiter
10 August 1990	Magellan arrives at Venus and transmits its first pictures on 16 August 1990
6 October 1990	The ESA probe Ulysses is launched from space shuttle Discovery, to study the Sun
29 October 1991	Galileo makes the closest-ever approach to an asteroid, Gaspra, flying within 1,600 km/990 mi
8 February 1992	Ulysses flies past Jupiter at a distance of 380,000 km/236,000 mi from the surface, just inside the orbit of Io and closer than 11 of Jupiter's 16 moons
10 July 1992	Giotto flies at a speed of 14 kms/8.5 mps to within 200 km/124 mi of comet Grigg-Skjellerup, 12 light years (240 million km/150 mi) away from Earth
25 September 1992	Mars Observer is launched from Cape Canaveral, the first US mission to Mars for 17 years
10 October 1992	Pioneer Venus 1 burns up in the atmosphere of Venus
21 August 1993	Contact is lost with Mars Observer three days before it is due to drop into orbit around Mars
28 August 1993	Galileo flies past the asteroid Ida
December 1995	Galileo's probe enters the atmosphere of Jupiter. It radios information back to the orbiter for 57 minutes before it is destroyed by atmospheric pressure

Space Probes: Key Dates continued

17 February 1996	NASA's Near-Earth Asteroid Rendezvous (NEAR) is launched to study Eros
1997	Galileo begins orbiting Jupiter's moons. It takes photographs of Europa for a potential future landing site, and detects molecules containing carbon and nitrogen on Callisto, suggesting that life once existed there
17 June 1997	The US NEAR spacecraft flies within 1,200 km/746 mi of the asteroid Mathilde, taking high-resolution photographs and revealing a 25-km/15.5-mi crater covering the 53-km/33-mi asteroid
4 July 1997	The NASA Mars Pathfinder becomes the third spacecraft to soft land on Mars. It deploys a micro-roving vehicle called Sojourner. The Mars Pathfinder images are posted on Internet Web sites and accessed by millions of people all over the world
11 September 1997	NASA's Mars Global Surveyor enters orbit around the Red Planet to begin a highly successful mission which continues into 2000, taking high-resolution images of almost the entire Martian surface, as well as gathering vital climate data
15 October 1997	The NASA–ESA Cassini-Huygens mission gets underway with the successful launch of a Titan IV Centaur booster from Cape Canaveral. Cassini will enter orbit around Saturn in July 2004 and will deploy the ESA Huygens spacecraft to land on Titan, the largest moon of the ringed planet
1998	Analysis of high-resolution images from the Galileo spacecraft suggests that the icy crust of Europa, Jupiter's fourth-largest moon, may hide a vast ocean warm enough to support life
6 January 1998	NASA's Lunar Prospector is launched by an Athena 2 rocket from Cape Canaveral and enters orbit around the Moon to conduct a comprehensive survey of the Moon's chemical composition, including a discovery, announced in March 1998, that there could be frozen water beneath the surface of the polar regions. Lunar Prospector later impacts the south pole of the Moon on 31 July 1999 but no trace of water is found in the resulting impact plume
April 1998	Daniel Goldin, the administrator of NASA, gives the go-ahead for the development of the fourth and last Great Observatory series spacecraft, the Space Infrared Telescope Facility, to be launched in 2001. The Great Observatory series includes the Hubble Space Telescope which was launched in 1990
4 July 1998	Japan launches its first probe to explore Mars using an M5 booster from Kagoshima. Called Planet B or Nozomi, the spacecraft was scheduled to reach the Red Planet via fly-bys of the Moon in October and December 1998 and a fly-by of the Earth in October 1999 but will not reach Mars until 2003 due to flight-path deviations
24 October 1998	The first spacecraft in NASA's New Millennium programme, called Deep Space 1, is launched aboard a Delta II booster from Cape Canaveral, to demonstrate a range of new technologies needed for future Solar System exploration, including ion propulsion and 'intelligent spacecraft' systems during a flight that featured a fly-by of an asteroid in 1999
11 December 1998	The Mars Climate Orbiter is launched by a Delta II from Cape Canaveral as part of NASA's long-term exploration of Mars. However, due to a metric–imperial value data error – made during the manufacture of the spacecraft – the craft makes a wrong approach to Mars and crashes into the Martian surface during the planned orbital insertion burn on 23 September 1999
3 January 1999	The second Mars '99 programme spacecraft, the Mars Polar Lander, is launched by another Delta II from Cape Canaveral but is also lost (on 6 December 1999) as it attempts to make a soft landing in a region close to the Red Planet's south pole. The dual loss of the Mars missions is a major blow to NASA and to the future exploration of Mars
7 April 1999	ESA's Mars Express mission is given formal approval with a contract to build the Mars orbiting spacecraft. Mars Express will be launched in 2003, entering orbit in December of that year. A UK 'piggy-back' spacecraft, Beagle 2, will land on the Martian surface
June 1999	NASA gives formal approval to the development of the Next-Generation Space Telescope (NGST) to succeed the Hubble Space Telescope and to be launched in about 2008. NASA also says that Hubble will remain operational until 2010 and may be brought back to Earth aboard the space shuttle
23 July 1999	The space shuttle Columbia STS-93 is launched and deploys the Chandra X-Ray Observatory, NASA's third Great Observatory series spacecraft, into orbit
July 1999	NASA gives approval for two new Discovery programme spacecraft: Messenger, which will make two close fly-bys of the planet Mercury in 2008, the first such mission since Mariner 10 in 1974; and Deep Impact, a mission to penetrate the comet P/Tempel 1 in July 2005
10 December 1999	ESA's X-Ray Multi-Mirror telescope is launched by an Arianespace Ariane 5 booster from Kourou, French Guiana

Space Probes: Key Dates continued

7 February 2000	NASA's Stardust spacecraft, en route to the comet Wild 2 in 2004, collects its first samples of interstellar dust, which are due to be returned to Earth in 2006, along with samples of comet dust from Wild 2
14 February 2000	NASA's NEAR spacecraft becomes the first to orbit an asteroid – Eros. NEAR takes thousands of high-resolution images and makes detailed scientific observations during a year-long mission
March 2000	NASA scientists announce that they may have to crash the Galileo probe in order to prevent it from contaminating Europa, one of Jupiter's moons, with microbes from Earth. They believe their best chance of locating extraterrestrial life within our Solar System is on Europa
4 June 2000	NASA's Gamma-Ray Observatory is deliberately de-orbited by the space agency as a precaution after gyroscope control problems
22 June 2000	Images from the Mars Global Surveyor revealing 'water gullies' on the planet are released by NASA
July 2000	NASA announces that twin Mars Exploration Rover missions will be launched in 2003 to land at separate locations on the Red Planet in 2004
September 2000	ESA approves the BepiColombo mission to explore the planet Mercury, to be launched in 2009
December 2000	As NASA's Cassini spacecraft flies past Jupiter en route to Saturn, making observations of Jupiter, the first dual-space exploration of a planet is conducted, with the Galileo Jupiter orbiter
January 2001	Contact with the first deep-space interplanetary explorer, Pioneer 10, is lost. The spacecraft, which explored Jupiter in 1973, was over 11 billion km/7 billion mi from Earth
12 February 2001	NASA's NEAR makes a spectacular finale to its mission at the asteroid Eros by making a carefully thruster-controlled soft landing on the surface. The originally unplanned manoeuvre makes the craft the first to land on an asteroid
March 2001	NASA budget cuts threaten to cancel the only planned mission to explore the final planet in the solar system, Pluto. The Pluto Kuiper Express, to fly past Pluto and its moon, Charon, and to explore the region beyond Pluto in which other objects orbit the Sun, was to have reached Pluto by 2020
7 April 2001	NASA's Mars Odyssey is successfully launched by a Delta rocket from Cape Canaveral. The craft is due to reach Mars in October 2001, and is expected to attain its working orbit around the planet in January 2002. The probe carries instruments designed to examine from orbit the chemical make-up of the planet's surface, and it is hoped that the information gathered may help to determine whether life has ever existed on Mars. The success of this mission is important to NASA, after the loss of both the Mars Climate Orbiter and the Mars Polar Lander in 1999
June 2003	NASA launches two Mars Exploration Rovers, 'Spirit' and 'Opportunity'. The European Space Agency launches its 'Mars Express' Orbiter carrying the British built lander 'Beagle 2'. Japan launches its Mars space probe 'Nozomi'.
Jan 2004	NASA's two rovers begin transmitting information back to Earth; they find traces of previous water.

SQUARES, CUBES, AND ROOTS

Number	Square	Cube	Square root	Cube root
1	1	1	1.000	1.000
2	4	8	1.414	1.260
3	9	27	1.732	1.442
4	16	64	2.000	1.587
5	25	125	2.236	1.710
6	36	216	2.449	1.817
7	49	343	2.646	1.913
8	64	512	2.828	2.000
9	81	729	3.000	2.080
10	100	1,000	3.162	2.154
11	121	1,331	3.317	2.224
12	144	1,728	3.464	2.289
13	169	2,197	3.606	2.351
14	196	2,744	3.742	2.410
15	225	3,375	3.873	2.466
16	256	4,096	4.000	2.520
17	289	4,913	4.123	2.571
18	324	5,832	4.243	2.621
19	361	6,859	4.359	2.668
20	400	8,000	4.472	2.714
25	625	15,625	5.000	2.924
30	900	27,000	5.477	3.107
40	1,600	64,000	6.325	3.420
50	2,500	125,000	7.071	3.684

BRIGHTEST STARS

Scientific name	Common name	Distance from Sun (light years)	Apparent magnitude	Absolute magnitude
Alpha Canis Majoris	Sirius	9	–1.46	1.4
Alpha Carinae	Canopus	1,170	–0.72	–2.5
Alpha Centauri	Rigil Kent	4	–0.27	4.4
Alpha Boötis	Arcturus	36	–0.04	0.2
Alpha Lyrae	Vega	26	0.03	0.6
Alpha Aurigae	Capella	42	0.08	0.4
Beta Orionis	Rigel	910	0.12	–8.1
Alpha Canis Minoris	Procyon	11	0.38	2.6
Alpha Eridani	Achernar	85	0.46	–1.3
Alpha Orionis	Betelgeuse	310	0.50[1]	–7.2
Beta Centauri	Hadar	460	0.61[1]	–4.4
Alpha Aquilae	Altair	17	0.77	2.3
Alpha Crucis	Acrux	360	0.83	–4.6
Alpha Tauri	Aldebaran	68	0.85[1]	–0.3
Alpha Scorpii	Antares	330	0.96[1]	–5.2
Alpha Virginis	Spica	260	0.98[1]	–3.2
Beta Geminorum	Pollux	36	1.14	0.7
Alpha Piscis Austrini	Fomalhaut	22	1.16	2.0
Beta Crucis	Mimosa	490	1.25[1]	–4.7
Alpha Cygni	Deneb	1,830	1.25	–7.2

1 Variable star.

TABLE OF EQUIVALENT TEMPERATURES

°C	°F	°C	°F	°C	°F	°C	°F
100	212	70	158	40	104	10	50
99	210.2	69	156.2	39	102.2	9	48.2
98	208.4	68	154.4	38	100.4	8	46.4
97	206.6	67	152.6	37	98.6	7	44.6
96	204.8	66	150.8	36	96.8	6	42.8
95	203	65	149	35	95	5	41
94	201.2	64	147.2	34	93.2	4	39.2
93	199.4	63	145.4	33	91.4	3	37.4
92	197.6	62	143.6	32	89.6	2	35.6
91	195.8	61	141.8	31	87.8	1	33.8
90	194	60	140	30	86	0	32
89	192.2	59	138.2	29	84.2	–1	30.2
88	190.4	58	136.4	28	82.4	–2	28.4
87	188.6	57	134.6	27	80.6	–3	26.6
86	186.8	56	132.8	26	78.8	–4	24.8
85	185	55	131	25	77	–5	23
84	183.2	54	129.2	24	75.2	–6	21.2
83	181.4	53	127.4	23	73.4	–7	19.4
82	179.6	52	125.6	22	71.6	–8	17.6
81	177.8	51	123.8	21	69.8	–9	15.8
80	176	50	122	20	68	–10	14
79	174.2	49	120.2	19	66.2	–11	12.2
78	172.4	48	118.4	18	64.4	–12	10.4
77	170.6	47	116.6	17	62.6	–13	8.6
76	168.8	46	114.8	16	60.8	–14	6.8
75	167	45	113	15	59	–15	5
74	165.2	44	111.2	14	57.2	–16	3.2
73	163.4	43	109.4	13	55.4	–17	1.4
72	161.6	42	107.6	12	53.6	–18	–0.4
71	159.8	41	105.8	11	51.8	–19	–2.2

MISCELLANEOUS UNITS OF MEASURE

Unit	Definition
acoustic ohm	cgs unit of acoustic impedance (the ratio of sound pressure on a surface to sound flux through the surface)
acre	traditional English land measure; 1 acre = 4,480 sq yd (4,047 sq m or 0.4047 ha)
acre-foot	unit sometimes used to measure large volumes of water such as reservoirs; 1 acre-foot = 1,233.5 cu m/43,560 cu ft
astronomical unit	unit (symbol AU) equal to the mean distance of the Earth from the Sun: 149,597,870 km/92,955,808 mi
atmosphere	unit of pressure (abbreviation atm); 1 standard atmosphere = 101,325 Pa
barn	unit of area, especially the cross-sectional area of an atomic nucleus; 1 barn = 10–28 sq m
barrel	unit of liquid capacity; the volume of a barrel depends on the liquid being measured and the country and state laws. In the USA, 1 barrel of oil = 42 gal (159 l/34.97 imperial gal), but for federal taxing of fermented liquor (such as beer), 1 barrel = 31 gal (117.35 l/25.81 imperial gal). Many states fix a 36-gallon barrel for cistern measurement and federal law uses a 40-gallon barrel to measure 'proof spirits'. 1 barrel of beer in the UK = 163.66 l (43.23 US gal/36 imperial gal)
base box	imperial unit of area used in metal plating; 1 base box = 20.232 sq m/31,360 sq in
baud	unit of electrical signalling speed equal to 1 pulse per second
brewster	unit (symbol B) for measuring reaction of optical materials to stress
British thermal unit	imperial unit of heat (symbol Btu); 1 Btu = approximately 1,055 J
bushel	measure of dry and (in the UK) liquid volume. 1 bushel (struck measure) = 8 dry US gallons (64 dry US pt/35.239 l/2,150.42 cu in). 1 heaped US bushel = 1,278 bushels, struck measure (81.78 dry pt/45.027 l/2,747.715 cu in), often referred to a 11/4 bushels, struck measure. In the UK, 1 bushel = 8 imperial gallons (64 imperial pt); 1 UK bushel = 1.03 US bushels
cable	unit of length used on ships, taken as 1/10 of a nautical mile (185.2 m/607.6 ft)
calorie	cgs unit of heat, now replaced by the joule; 1 calorie = 4.1868 J
carat	unit for measuring mass of precious stones; 1 carat = 0.2 g/0.00705 oz
carat	unit of purity in gold; pure gold is 24-carat
carcel	obsolete unit of luminous intensity
cental	name for the short hundredweight; 1 cental = 45.36 kg/100 lb
chaldron	obsolete unit measuring capacity; 1 chaldron = 1.309 cu m/46.237 cu ft
clausius	in engineering, a unit of entropy; defined as the ratio of energy to temperature above absolute zero
cleanliness unit	unit for measuring air pollution; equal to the number of particles greater than 0.5 mm in diameter per cu ft of air
clo	unit of thermal insulation of clothing; standard clothes have insulation of about 1 clo, the warmest have about 4 clo per 2.5 cm/1 in of thickness
clusec	unit for measuring the power of a vacuum pump
condensation number	in physics, the ratio of the number of molecules condensing on a surface to the number of molecules touching that surface
cord	unit for measuring the volume of wood cut for fuel; 1 cord = 3.62 cu m/128 cu ft, or a stack 2.4 m/8 ft long, 1.2 m/4 ft wide and 1.2 m/4 ft high
crith	unit of mass for weighing gases; 1 crith = the mass of 1 litre of hydrogen gas at standard temperature and pressure
cubit	earliest known unit of length; 1 cubit = approximately 45.7 cm/18 in, the length of the human forearm from the tip of the middle finger to the elbow
curie	former unit of radioactivity (symbol Ci); 1 curie = 3.7 _ 1010 becquerels
dalton	international atomic mass unit, equivalent to 1/12 of the mass of a neutral carbon-12 atom
darcy	cgs unit (symbol D) of permeability, used mainly in geology to describe the permeability of rock
darwin	unit of measurement of evolutionary rate of change
decontamination factor	unit measuring the effectiveness of radiological decontamination; the ratio of original contamination to the radiation remaining
demal	unit measuring concentration; 1 demal = 1 gram-equivalent of solute in 1 cu dm of solvent
denier	unit used to measure the fineness of yarns; 9,000 m of 15 denier nylon weighs 15 g/0.5 oz
dioptre	optical unit measuring the power of a lens; the reciprocal of the focal length in metres
dram	unit of apothecaries' measure; 1 dram = 60 grains/3.888 g

Miscellaneous Units of Measure continued

Unit	Definition
dyne	cgs unit of force; 105 dynes = 1 N
einstein unit	unit for measuring photoenergy in atomic physics
eotvos unit	unit (symbol E) for measuring small changes in the intensity of the Earth's gravity with horizontal distance
erg	cgs unit of work; equal to the work done by a force of 1 dyne moving through 1 cm
erlang	unit for measuring telephone traffic intensity; for example, 90 minutes of carried traffic measured over 60 minutes = 1.5 erlangs ('carried traffic' refers to the total duration of completed calls made within a specified period)
fathom	unit of depth measurement in mining and seafaring; 1 fathom = 1.83 m/6 ft
finsen unit	unit (symbol FU) for measuring intensity of ultraviolet light
fluid ounce	measure of capacity; equivalent in the USA to 1/16 of a pint (1/20 of a pint in the UK and Canada)
foot	imperial unit of length (symbol ft), equivalent to 0.3048 m
foot-candle	unit of illuminance, replaced by the lux; 1 foot-candle = 10.76391 lux
foot-pound	imperial unit of energy (symbol ft-lb); 1 ft-lb = 1.356 joule
frigorie	unit (symbol fg) used in refrigeration engineering to measure heat energy, equal to a rate of heat extraction of 1 kilocalorie per hour
furlong	unit of measurement, originating in Anglo-Saxon England, equivalent to 201.168 m/220 yd
galileo	unit (symbol Gal) of acceleration; 1 galileo = 10–2 m s–2
gallon	imperial liquid or dry measure subdivided into 4 quarts or 8 pints; 1 US gal = 3.785 l; 1 imperial gal = 4.546 l
gauss	cgs unit (symbol Gs) of magnetic flux density, replaced by the tesla; 1 gauss = 1 _ 10–4 tesla
gill	imperial unit of volume for liquid measure; equal to 1/4 of a pint (in the USA, 4 fl oz/0.118 l; in the UK, 5 fl oz/0.142 l)
grain	smallest unit of mass in the three English systems of measurement (avoirdupois, troy, apothecaries' weights) used in the UK and USA; 1 grain = 0.0648 g
hand	unit used in measuring the height of a horse from front hoof to shoulder (withers); 1 hand = 10.2 cm/4 in
hardness number	unit measuring hardness of materials. There are many different hardness scales: Brinell, Rockwell, and Vickers scales measure the degree of indentation or impression of materials; Mohs's scale measures resistance to scratching against a standard set of minerals
hartree	atomic unit of energy, equivalent to atomic unit of charge divided by atomic unit of length; 1 hartree = 4.850 _ 10–18 J
haze factor	unit of visibility in mist or fog; the ratio of brightness of mist compared with that of the object
Hehner number	unit measuring concentration of fatty acids in oils; a Hehner number of 1 = 1 kg of fatty acid in 100 kg of oil or fat
hide	unit of measurement used in the 12th century to measure land; 1 hide = 60–120 acres/25–50 ha
horsepower	imperial unit (abbreviation hp) of power; 1 horsepower = 746 W
hundredweight	imperial unit (abbreviation cwt) of mass; 1 cwt = 45.36 kg/100 lb in the USA and 50.80 kg/112 lb in the UK
inch	imperial unit (abbreviation in) of linear measure, 1/12 of a ft; 1 in = 2.54 cm
inferno	unit used in astrophysics for describing the temperature inside a star; 1 inferno = 1 billion K (kelvin)
iodine number	unit measuring the percentage of iodine absorbed in a substance, expressed as grams of iodine absorbed by 100 grams of material
jansky	unit used in radio astronomy to measure radio emissions or flux densities from space; 1 jansky = 10–26 W m–2 Hz–1. Flux density is the energy in a beam of radiation which passes through an area normal to the beam in a single unit of time. A jansky is a measurement of the energy received from a cosmic radio source per unit area of detector in a single time unit
kayser	unit used in spectroscopy to measure wave number (number of waves in a unit length); a wavelength of 1 cm has a wave number of 1 kayser
knot	unit used in navigation to measure a ship's speed; 1 knot = 1 nautical mile per hour, or about 1.15 miles per hour
league	obsolete imperial unit of length; 1 league = 3 nautical mi/5.56 km or 3 statute mi/4.83 km
light year	unit used in astronomy to measure distance; the distance travelled by light in one year, approximately 9.46 _ 1012 km/5.88 _ 1012 mi

Miscellaneous Units of Measure continued

Unit	Definition
mache	obsolete unit of radioactive concentration; 1 mache = 3.7 _ 10–7 curies of radioactive material per cu m of a medium
maxwell	cgs unit (symbol Mx) of magnetic flux, the strength of a magnetic field in an area multiplied by the area; 1 maxwell = 108 weber
megaton	measurement of the explosive power of a nuclear weapon; 1 megaton = 1 million tons of trinitrotoluene (TNT)
mil	(a) one-thousandth of a litre; contraction of the word millilitre; (b) imperial measure of length, equal to one-thousandth of an inch; also known as the thou
mile	imperial unit of linear measure; 1 statute mile = 1.60934 km/5,280 ft; 1 international nautical mile = 1.852 km/6,076 ft
millimetre of mercury	unit of pressure (symbol mmHg) used in medicine for measuring blood pressure
morgan	arbitrary unit used in genetics; 1 morgan is the distance along the chromosome in a gene that gives a recombination frequency of 1%
nautical mile	unit of distance used in navigation, equal to the average length of 1 minute of arc on a great circle of the Earth; 1 international nautical mile = 1.852 km/6,076 ft
neper	unit used in telecommunications; gives the attenuation of amplitudes of currents or powers as the natural logarithm of the ratio of the voltage between two points or the current between two points
oersted	cgs unit (symbol Oe) of magnetic field strength, now replaced by amperes per metre (1 Oe = 79.58 amp per m)
ounce	unit of mass, 1/16 of a pound avoirdupois, equal to 437.5 grains/28.35 g; or 14.6 pound troy, equal to 480 grains/31.10 g
parsec	unit (symbol pc) used in astronomy for distances to stars and galaxies; 1 pc = 3.262 light years, 2.063 _ 105 astronomical units, or 3.086 _ 1013 km
peck	obsolete unit of dry measure, equal to 8 imperial quarts or 1 quarter bushel (8.1 l in the USA or 9.1 l in the UK)
pennyweight	imperial unit of mass; 1 pennyweight = 24 grains = 1.555 _ 10–3 kg
perch	obsolete imperial unit of length; 1 perch = 51/2 yards = 5.029 m, also called the rod or pole
pint	imperial unit of liquid or dry measure; in the USA, 1 liquid pint = 16 fl oz/0.473 l, while 1 dry pint = 0.551 l; in the UK, 1 pt = 20 fl oz, 1/2 quart, 1/8 gal, or 0.568 l
point	metric unit of mass used in relation to gemstones; 1 point = 0.01 metric carat = 2 _ 10–3 g
poise	cgs unit of dynamic viscosity; 1 poise = 1 dyne-second per sq cm
pound	imperial unit (abbreviation lb) of mass; the avoirdupois pound or imperial standard pound = 0.45 kg/7,000 grains, while the pound troy (used for weighing precious metals) = 0.37 kg/5,760 grains
poundal	imperial unit (abbreviation pdl) of force; 1 poundal = 0.1383 newton
quart	imperial liquid or dry measure; in the USA, 1 liquid quart = 0.946 l, while 1 dry quart = 1.101 l; in the UK, 1 quart = 2 pt/1.137 l
rad	unit of absorbed radiation dose, replaced in the SI system by the gray; 1 rad = 0.01 joule of radiation absorbed by 1 kg of matter
relative biological effectiveness	relative damage caused to living tissue by different types of radiation
roentgen	unit (symbol R) of radiation exposure, used for X- and gamma rays
rood	imperial unit of area; 1 rood = 1/4 acre = 1,011.7 sq m
rydberg	atomic unit of energy; 1 rydberg = 2.425 _ 10–18 J
sabin	unit of sound absorption, used in acoustical engineering; 1 sabin = absorption of 1 sq ft (0.093 sq m) of a perfectly absorbing surface
scruple	imperial unit of apothecaries' measure; 1 scruple = 20 grains = 1.3 _ 10–3 kg
shackle	unit of length used at sea for measuring cable or chain; 1 shackle = 15 fathoms (27 m/90 ft)
slug	obsolete imperial unit of mass; 1 slug = 14.59 kg/32.17 lb
snellen	unit expressing the visual power of the eye
sone	unit of subjective loudness
standard volume	in physics, the volume occupied by 1 kilogram molecule (molecular mass in kilograms) of any gas at standard temperature and pressure; approximately 22.414 cu m
stokes	cgs unit (symbol St) of kinematic viscosity; 1 stokes = 10–4 m2 s–1

Miscellaneous Units of Measure continued

Unit	Definition
stone	imperial unit (abbreviation st) of mass; 1 stone = 6.35 kg/14 lb
strontium unit	measures concentration of strontium-90 in an organic medium relative to the concentration of calcium
tex	metric unit of line density; 1 tex is the line density of a thread with a mass of 1 gram and a length of 1 kilometre
tog	measure of thermal insulation of a fabric, garment, or quilt; the tog value is equivalent to 10 times the temperature difference (in °C) between the two faces of the article, when the flow of heat across it is equal to 1 W per sq m
ton	unit of mass; the long ton (UK) = 1,016 kg/2,240 lb; 1 short ton (USA) = 907 kg/2,000 lb; 1 metric tonne = 1000 kg/2205 lb
yard	imperial unit (symbol yd) of length, equivalent to 0.9144 m/3 ft

HIGHEST WATERFALLS IN THE WORLD

Waterfall	Location	Total drop m	ft
Angel Falls	Venezuela	979	3,212
Yosemite Falls	USA	739	2,425
Mardalsfossen–South	Norway	655	2,149
Tugela Falls	South Africa	614	2,014
Cuquenan	Venezuela	610	2,000
Sutherland	New Zealand	580	1,903
Ribbon Fall, Yosemite	USA	491	1,612
Great Karamang River Falls	Guyana	488	1,600
Mardalsfossen–North	Norway	468	1,535
Della Falls	Canada	440	1,443
Gavarnie Falls	France	422	1,385
Skjeggedal	Norway	420	1,378
Glass Falls	Brazil	404	1,325
Krimml	Austria	400	1,312
Trummelbach Falls	Switzerland	400	1,312
Takkakaw Falls	Canada	366	1,200
Silver Strand Falls, Yosemite	USA	357	1,170
Wallaman Falls	Australia	346	1,137
Wollomombi	Australia	335	1,100
Cusiana River Falls	Colombia	300	984
Giessbach	Switzerland	300	984
Skykkjedalsfossen	Norway	300	984
Staubbach	Switzerland	300	984

TROY WEIGHTS

The troy system, used for weighing gold, silver, and gemstones, is thought to have derived its name from the town of Troyes, France, where the system was probably first used at the medieval fairs held in the town. The pound troy has been legally obsolete since 1878, but for transactions in precious metals the troy ounce is authorized for use without time limit.

Troy weight	Metric equivalent	
1 pennyweight	= 24 grains 1.555 grams	
1 ounce troy	= 20 pennyweights (= 480 grains)	31.103 grams
1 pound troy	= 12 ounces troy (= 5,760 grains)	373.242 grams